Handbook of Sociology of Aging

Handbooks of Sociology and Social Research

Series Editor:
Howard Kaplan, *Texas A & M University, College Station, Texas*

HANDBOOK OF SOCIOLOGY OF AGING
Edited by Richard A. Settersten, Jr. and Jacqueline L. Angel

HANDBOOK OF COMMUNITY MOVEMENTS AND LOCAL ORGANIZATIONS
Edited by Ram. A Cnaan and Carl Milofsky

HANDBOOK OF DISASTER RESEARCH
Theory, Science, and Prevention
Edited by Zili Sloboda and William J. Bukoski

HANDBOOK OF THE LIFE COURSE
Edited by Jeylan T. Mortimer and Michael J. Shanahan

HANDBOOK OF POPULATION
Edited by Dudley L. Poston and Michael Micklin

HANDBOOK OF RELIGION AND SOCIAL INSTITUTIONS
Edited by Helen Rose Ebaugh

HANDBOOK OF SOCIAL PSYCHOLOGY
Edited by John Delamater

HANDBOOK OF SOCIOLOGICAL THEORY
Edited by Jonathan H. Turner

HANDBOOK OF THE SOCIOLOGY OF EDUCATION
Edited by Jan E. Stets and Jonathan H. Turner

HANDBOOK OF THE SOCIOLOGY OF GENDER
Edited by Janet Saltzman Chafetz

HANDBOOK OF THE SOCIOLOGY OF MENTAL HEALTH
Edited by Carol S. Aneshensel and Jo C. Phelan

HANDBOOK OF THE SOCIOLOGY OF THE MILITARY
Edited by Giuseppe Caforio

For other titles published in this series, go to
www.springer.com/series/6055

Richard A. Settersten, Jr. • Jacqueline L. Angel
Editors

Handbook of Sociology of Aging

Springer

Editors
Richard A. Settersten, Jr.
Department of Human Development and Family Sciences
Oregon State University
Corvallis, OR
USA
richard.settersten@oregonstate.edu

Jacqueline L. Angel
Department of Sociology
The University of Texas at Austin
Austin, TX
USA
jangel@austin.utexas.edu

ISBN 978-1-4419-7373-3 (hardcover)
ISBN 978-1-4614-4095-6 (softcover)
ISBN 978-1-4419-7374-0 (eBook)
DOI 10.1007/978-1-4419-7374-0
Springer New York Dordrecht Heidelberg London

Library of Congress Control Number: 2011924339

Printed on acid-free paper

Springer is part of Springer Science+Business Media (www.springer.com)

Scholarly Foreword

Sometime in the relatively near future, the human species will reach a watershed moment in which people over the age of 60 outnumber children. This is a world in which there are both low birth rates and declining death rates at all ages, and in which people live much longer than had ever been true in the past. Technological advances and scientific discoveries over the past few centuries have resulted in a near doubling of life expectancy in developed regions of the world – increases estimated to be roughly 3 months per year since 1840 in most industrial societies (Oeppen and Vaupel 2002). In areas where low levels of mortality are combined with low levels of fertility, population aging is a demographic fact of life – and an unprecedented social achievement.

In the modern context, these profound trends demand new knowledge on the *social* forces and factors that shape aging patterns and processes and on the *social* consequences of aging patterns and processes. And this first *Handbook of Sociology of Aging*, co-edited by Richard Settersten and Jacqueline Angel, provides that knowledge. It provides a comprehensive and in-depth view into a wide range of social phenomena and reveals just how vibrant the sociology of aging has become. Its chapters also show that the contributions of sociologists are crucial for fostering interdisciplinary research on aging.

At present, the study of older age and older people is a burgeoning industry in all contemporary industrialized countries, due in part to increases in population aging. This has not always been the case. This is not only because of lack of good demographic data, but also because of the phenomenon itself. Historical demographers can tell us little about the lives and social roles of the elderly population over the past several centuries, and yet the consideration of age and aging is not a new phenomenon, even if the study of older people (i.e., gerontology) may not have existed. Indeed, social theorists from the eighteenth and nineteenth centuries pointed to the significance of understanding the social meanings and consequences of age and the succession of generations. This handbook, centuries later, is a testament to these early theorists of age and generation.

This handbook is also a testament to the legacy of twentieth-century pioneers of the field of aging, such as Bernice Neugarten, Matilda White Riley, Anne Foner, Norman Ryder, Glen Elder, and others. These scholars articulated the importance of understanding not only the age structuring of society, but also the processes of individual-level changes associated with biological maturation and change, along with the events and transitions associated with the life course. The range of topics included in the present volume eclipses what early sociological researchers on age and aging probably imagined, including scores of new topics and contributions by both stalwart members of the field and many younger scholars who are reshaping its contours.

Health, aging, and mortality present some of the most profound puzzles in modern human science, and sociologists of aging have crucial roles to play in helping to solve those puzzles. The demographic realities of population aging will substantially increase the number of elderly persons with disability and in need of long-term care over the next several decades. We will have unprecedented numbers of people living longer, but more often living with chronic disease and disability.

We do not necessarily expect future generations of the "aged" to experience aging (e.g., physical, psychological, cognitive, social) and later life development as their parents did. Hence, we might expect a decreasing prevalence of dementia and other types of low or impaired cognitive functioning in the worlds of the future. However, keeping people alive for longer through improvements in medical technology may create some unanticipated consequences, what some have called the *failure of success* (Vaupel and Yasin 1985). Health advances allow frailer, more vulnerable individuals to live longer. We do not know the consequences of this for levels of cognitive performance and physical well-being among the aged in future worlds. Nor are we doing enough to anticipate their consequences for families, work, education, policies, or the welfare of a society. Many of the chapters in this handbook help us begin to undertake these challenges.

These possibilities may raise the specter of massive limitations in daily and social roles in the vast numbers of postage 65 citizens who will inhabit our future world. But it may also raise an entirely different set of possibilities, once we consider cohort differences in how the lives of the future and past aged have played out. How much can our current knowledge of the post-World War II birth cohorts (e.g., their health, nutrition, activities, work lives) be extended to future cohorts? The research of sociologists of aging and the life course is crucial in making historical changes such as these visible, and in tracing the legacies of these historical changes on individuals, families, and social institutions. Similar questions can and should be asked about many other domains of aging.

Despite our collective interest in the consequences of population aging, not enough is known about why some individuals live to middle age and others to extreme old age; nor is enough known about why some people develop dementia in old age and others do not. One area sociologists have been studying for decades, for example, involves the study of disability, which is an interdisciplinary study of loss of physical and cognitive function. The study of disability is, thus, an increasing focus of studies of aging, and new formulations, such as the concept of "disability-free life expectancy," have been introduced to distinguish among statuses along a quality of life dimension. Another new development involves the study of components of "successful aging."

The chapters of this handbook repeatedly show that aging involves a complex interplay of multiple layers of potential and experience that are often anchored in the social world. Its publication is also a sign of the need to explicitly treat the social aspects of aging in interdisciplinary research and to incorporate it alongside biological, neurological, and psychological factors that influence processes involved in differential aging. The chapters of this handbook also show how the social processes that affect aging interact with the life course – events, transitions, and trajectories of roles and relationships. They bring to life the essential proposition that in order to understand the nature of human aging, it is imperative to take into account the many layers of social context in which it occurs, from interpersonal relationships all the way out to historical and demographic parameters.

Indeed, the monotonic and irreversible trends associated with the first demographic transition led to the inevitable second-order growth in the proportion of elderly people in modern day populations. That is, it is the dramatic declines in mortality, morbidity, and fertility that have *permitted* a long and generally healthy old age. Increasingly, individuals at the early end of "old age" are considered to have entered a new "third age" in their lives. This term came from France in the 1970s and 1980s, where "Universities of the Third Age" were designed for the enrichment of the lives of pensioners, making reference to a new life stage viewed "as distinct from both the second age [adulthood] and the fourth age [old age]." The emergence of this new life stage – between the traditional age of retirement and what many consider to be "old age" – challenges many of our preconceptions about the "expected" life course (Moen and Spencer 2006). This is a new area for research, inasmuch as the traditional blueprints for how persons pattern their lives are diverging from the ways in which future cohorts will develop their later life choices. Many of the chapters in this handbook deal directly with these issues.

As the proportion of the world's population in the older age ranges continues to increase, there is a need for new and improved information about the factors that contribute to health and longevity – a need to reconcile current bodies of knowledge with the changing experiences of an aging population. This handbook, with each chapter providing a 30-year window on the topic, is not only exemplary of present-day knowledge, but also the necessary foundation for moving our field forward.

Age continues to be the greatest risk factor for most major chronic diseases, and there are new policy challenges as well, although a focus on social policy is not part of my mission here. Suffice it to say that in order to preempt what some see as a major "global aging crisis," it may be necessary to avert the negative consequences of population aging, by intervening in the lives of aging humans to "retard, arrest and even reverse aging damage" (Rae et al. 2010). With their focus on the social explanations and consequences of human aging, the chapters of this handbook contain important insights for translational research and for possible interventions meant to improve human aging. In bringing "the social" back, and in putting it front and center, the *Handbook of Sociology of Aging* will be an invaluable resource for research on aging for many years to come.

Duane F. Alwin

References

Moen, Phyllis, and Donna Spencer. 2006. "Converging Divergences in Age, Gender, Health, and Well-being: Strategic Selection in the Third Age." Pp. 129–40 in *Handbook of Aging and the Social Sciences*, edited by R.H. Binstock & L.K. George. New York: Academic Press.

Oeppen, Jim and James W. Vaupel. 2002. "Broken Limits to Life Expectancy." *Science* 296:1029–31.

Rae, Michael, J., Robert N. Butler, Judith Campisi, Aubrey de Grey, Caleb E. Finch, Michael Gough, George M. Martin, Jan Vijg, Kevin M. Perrott, and Barbara Logan. 2010. "The Demographic and Biomedical Case for Late-Life Interventions in Aging." *Science of Translational Medicine* 2:40cm21.

Vaupel, James W., and Anatoli K.Yashin. 1985. "Heterogeneity's Ruses: Some Surprising Effects of Selection on Population Dynamics." *The American Statistician* 39:176–85.

Policy Foreword

The field of sociology has certainly gone through significant changes over the past half century. As an undergraduate sociology major in the 1960s, I gained great insights about the challenges facing individuals and groups, and I was touched by the conditions of vulnerable populations. While I was deeply affected by my course work, I found myself continually asking: what are we going to *do* to deal with these challenges? And how have our social policies affected these individuals? Too often, my courses stopped short of assessing policy options, or the implications of policy actions, to improve the lives of real people.

During my graduate school training in public policy, I often relied on the base of my undergraduate coursework in sociology to provide "real world" content to compliment my analytical policy skills. I remember thinking during my graduate education that a closer interconnection between sociology and policy-making would make both better.

The *Handbook of Sociology of Aging* shows how far we have come over the past half century in connecting the field of sociology to policy. To mark the thirtieth anniversary of the American Sociological Association's Section on Aging and the Life course, the editors have mobilized nearly 80 scholars to highlight scholarship on over forty topics. One thing that makes this handbook so special is the recurrent focus on policy matters. The contributors have been asked to emphasize how each of their topics is affected by social phenomena, the life course, and social policy. These three prongs bring much-needed attention to the social forces and factors that shape aging patterns. For me, it is this third prong – social policy – that makes this handbook so distinct and important, as contributors make explicit how the phenomena being considered are affected by and have implications for policy and practice. Because the chapters rest on cutting-edge research, the work of the authors will help guide policy-makers in the formulation and implementation of policy. As such, this handbook provides a rich and clear roadmap for future policy-makers, especially in the United States.

I have spent the better part of the past three decades focusing on social policy, with a large share of that time directed at aging policy. Over all those years, I carried with me both my sociology background as well as my skills in policy analysis. Without both, my ability to shape and influence policy would be lessened. Indeed, I find it hard how policy can be effectively shaped without the voices of sociologists.

Examples of the linkage between sociology and policy abound. What are the benefits of establishing a solid system of home- and community-based care for the elderly, as well as respite care services for their caregivers? How do we provide a secure foundation of economic support for vulnerable populations? What are the implications of establishing a strong hospice benefit for Medicare beneficiaries? What are the individual, family, and societal implications of requiring more work later in life? How should changes in race and ethnicity characteristics as well as family composition shape future policy? And how have social policies related to older Americans shaped our definition of who we are as a people, and how we define our collective responsibilities to one another?

Over my career working on Capitol Hill or at the U.S. Department of Health and Human Services and the Office of Management and Budget, or during my service as the Commissioner of the Social

Security Administration, I used sociological research and analysis to help shape policy options. And it should be clear from this volume that I was certainly not alone: sociological research is now deeply integrated into the policy-making process. How could one "do" good policy without it?

Given my own background, I am especially pleased to see sizable attention in the handbook devoted to health and retirement security, which I see as two of the most pressing domains for public policies related to aging. The United States faces major challenges in financing the retirement and healthcare needs of the elderly while also financing defense, infrastructure improvement, and the education of future generations, and in stimulating economic growth more generally. The challenges of meeting the retirement and healthcare needs of the elderly will only grow in the future. Let me highlight just a few issues.

Pension Security: Balancing Adequacy and Affordability

As America ages, we face twin public pension challenges that need to be addressed, and fortunately, the work of sociologists is being used to address both challenges. The first challenge is to finance our public pension systems. There are currently about three workers for every person receiving Social Security, and within a generation, a retiree will depend on the contributions of about two workers. Much has been written on the dimensions of the long-term fiscal challenge of Social Security, and the work of sociologists is used by Social Security actuaries in their projections because understanding societal trends is a critical part of those models.

The second Social Security challenge is one that receives considerably less attention but is equally compelling: to ensure the adequacy of income supports, particularly for vulnerable populations. The work of sociologist is even more important in addressing this policy challenge. As just one example, how should we think about the progressivity in our social insurance system? We know that people are living longer lives, but more substantial gains are taking place at upper income levels. As the share of benefits increases over time to white collar and higher income beneficiaries, our social insurance system becomes increasingly less progressive. Sociologist can clearly help policy-makers think through the myriad of underlying issues, which require an understanding of the combined effects of minority group status, occupational disadvantage, and demographic transitions on the labor force's carrying capacity, as well as the sources of support for elderly Americans.

Other issues abound. How can we assess the needs of the oldest old, whose sources of non-Social Security income erode over time? How will family composition change the outlook for widows and single elderly women? What are the future needs of persons with low lifetime earnings, given the near absence of substantial sources of private retirement income? Since low income workers experience greater risk of becoming disabled prior to retirement, what policies are needed to assist these individuals if the retirement age is increased? And how should our social insurance system deal with the reality that our working-age population is becoming increasingly minority, while the older retired population remains predominantly non-Hispanic and White? All of these issues have profound consequences for individuals, families, society as a whole, and the social contract between generations. Sociologists can and do influence the policy development process in these areas.

Health Care: Balancing Access to Care and Controls on Cost

Most immediately, aging populations raise issues of critical importance to health policy-makers. Sociologists can play critical roles in framing and providing practical information to inform the social, ethical, and legal implications of healthcare issues. For example, with continuing advances in medical care, and the extraordinary price tag with which they come, the potential cost of health care could bankrupt the nation. Within this context, the need to control healthcare budgets grows – and, with it, proposals for the rationing of care emerge. In an economic and political environment in which some form of rationing is deemed necessary, weak political or social groups are vulnerable and certain to lose. These concerns seem especially likely to emerge in times of great economic retrenchment, like now. It is critical to understand the intended and unintended consequences of healthcare cost-control decisions.

Nations also face big challenges in developing resources and infrastructure to promote intergenerational health equity in communities. The volume addresses several topics that are pertinent to this issue, including how to improve health policies and practice for the most vulnerable elderly groups and curb inequalities in health care and health insurance.

Caregiving and Long-Term Care

Responses to the challenges of long-term care also require the contributions of sociologists. Aging brings pervasive caregiving demands for spouses and middle-aged children. These responsibilities can jeopardize financial resources, make it impossible to maintain paid work, strain psychological well-being, and have ripple effects on younger generations. Sociological research emphasizes the need to develop "family friendly" policies to offset some of this burden and make it more possible for family members to manage their caregiving responsibilities alongside other work and family responsibilities. Policies related to paid leave, sick time and vacation time, and flexible hours are just a few of many examples.

Aging also brings significant challenges for a range of long-term care settings – from continuing care retirement facilities, assisted living facilities, nursing homes, hospice, and others. Policies that govern the organization and finance of these institutions have a direct impact on the quality of care received in these settings. The volume addresses several topics that are pertinent to this issue, in particular, averting the "old-age welfare crisis," the hidden costs of caregiving, politics of aging and disability, and the role of civil society and nongovernmental actors in eldercare. Quality of care is also, of course, directly affected by the care workforce – and the adequacy and quality of that workforce are, in turn, shaped by policies related to the compensation and training of nurses, paraprofessionals, and other staff in long-term care facilities.

Social Policy in a Globalized World

Approaches to aging policies in the United States must also be understood within an international context. Sociological research raises our awareness of the interdependence of policies in a global world and of the lessons we can learn from other nations, and vice versa. We often assess comparative policies primarily with an eye to European experience. But lessons can be learned by casting our lenses even wider, and again the role of sociological research becomes essential in understanding how:

- Aging processes in Latin America influence social policies in this important part of the world. For example, Chile recently adopted new social assistance programs for poor elderly to make up for some of the glaring inadequacies of their privatized pension system. There are lessons from Chile's recent experience to help guide U.S. policies in an effort to provide stronger protections for poor elderly.
- Immigration poses multiple challenges to aging. Given high levels of immigration to the United States, many Latin Americans will age in the United States and draw upon social services here. The aging of older immigrants in the United States will have a major influence on social delivery

here, and the movement of elderly persons both ways across national boundaries needs careful policy examination. As our labor markets become more globalized – with more workers moving across borders for parts of their careers – we also need to consider how to better integrate pension and health systems with other countries, such as Mexico. This is a natural extension of efforts now underway to integrate the trade of goods and services across borders.

- In Asia, with rapid declines in fertility and a history of early retirement, many countries will in the near future face massive problems related to the financing and care of the elderly. An oft-repeated quote in Asian social policy circles is that "Asia is becoming old before it becomes rich." Systems are not in place to deal with looming aging challenges. Countries such as China have yet to establish a foundation of economic support for the elderly, and the small share of workers who are pension-eligible in China have overly generous pension systems as well as a very early retirement age. Understanding family and work changes and immigration patterns within China is an essential precondition in assessing the policy challenges. In addition, the experiences of other countries can help countries such as China develop appropriate policy solutions. For example, there are important lessons that China can learn about the design and implementation of the universal Social Security system in the United States, as well as our efforts to encourage more work later in life.

Policy in an Increasingly Diverse World

The new diversity of the U.S. population poses direct challenges to aging policies now and in the future. This new diversity not only relates to race and ethnicity, but also to assumptions reinforced in our policies about gender roles, the stability of marriage, and the definition and functions of "family." As the editors also note, these old assumptions need to be revisited and brought into alignment with the realities of contemporary life, and we must also be more future-oriented as we make policies today. We can do this by keeping a more careful watch on younger generations, whose lives do not and will not fit the molds of generations past. They will surely arrive at and move through old age in ways that are different from what we have known until now. Policy-makers need to understand these trends, and policy-makers need sociologists to help understand these trends. Several of the chapters shed light on the implications of the ethnic age grading of the U.S. population. The work force of tomorrow will become disproportionately minority and more specifically Hispanic, and this has the potential for ethnically based intergenerational conflict. The spiraling cost of the old-age welfare state will inevitably affect the healthcare entitlements for future generations.

A Wide Range of Other Issues for Policy and Practice

Besides the issues I have already noted, policy and practice audiences will find a remarkable range of pressing concerns in the pages of this handbook. These include how work organizations, educational institutions, and communities can be redesigned to better respond to the challenges of aging and to better nurture productivity, learning, and social experiences for older people. Other chapters highlight crucial social problems to be solved, including elder mistreatment, how elders are affected by or involved in crime, how elders manage natural disasters such as hurricane Katrina, the social isolation of those with fragile or fractured family relationships, or the significant physical or mental health needs of veterans, of those who are obese, have mental illnesses, or live with or are at risk of contracting HIV/AIDS. There are so many ways in which policies and practices can be developed to address these problems.

Final Thoughts

The chapters in this handbook, written by top sociologists who are acute observers of the world around us, repeatedly return to this simple truth: that the social world is rapidly changing and our policies must keep pace with those realities. It is our task as policy-makers to not only refresh our policies to better match current realities, but also to keep one eye to the future so that we can better prepare for the world tomorrow. This handbook leaves us well positioned to make our way. If it had been published a half century ago, when I was an undergraduate sociology major, it would have had a much smaller link to policy and practice. But as this volume makes clear, sociologists are now firmly connected to the policy world. One can only imagine the contents of the handbook that will be published in the future, say a half century from now. Predictions of the future are never easy, but my hunch is that sociology and policy will be even more deeply connected than they are today. And our world will be the better for it.

Kenneth S. Apfel

Preface

The idea for this handbook was sparked in anticipation of the 30th anniversary of the American Sociological Association's Section on Aging and the Life Course (SALC) in August of 2009. We received an invitation from Howard Kaplan at Texas A&M University to edit a *Handbook of Sociology of Aging* as part of Springer's series on *Handbooks of Sociology and Social Research*. We both have long histories with SALC and, at the time, we were both in leadership roles (with Jacqui as Immediate Past Chair and Rick as a member of Council and, today, Chair-Elect). The two of us had always hoped we might one day have the chance to collaborate – our respective areas of scholarship and skills are complementary and our professional conversations often turned to the need to ensure a vibrant future for our field, especially in nurturing leadership and finding ways to bridge generations of senior and junior scholars. The invitation to edit a handbook presented that opportunity.

Far more important, though, was the fact that the 30-year anniversary of SALC offered a perfect moment to reflect on the history of the field, trace the evolution of ideas, synthesize knowledge, and offer compelling new directions for its future. Indeed, one of the greatest strengths of the sociology of aging and the life course is the wide range of topics and methods that characterize it. Yet the breadth and depth of the field had not been considered, let alone well represented, in a single volume.

To this end, we mobilized top scholars in the field – both long-standing and more junior ones, even doing some collaborative match-making along the way – to turn much-needed attention to the *social* forces and factors that shape aging patterns and processes and to the *social* consequences of aging patterns and processes.

In our opening chapter as well as in our final chapter, we take a "birds-eye" view of the field, with the former describing important trends in scholarship on the sociology of aging over the past three decades, and the latter highlighting some of the most exciting new prospects for the decade ahead. In between, readers will find chapters on many specific topics allocated across eight sections – Theories and Methods in the Sociology of Aging; Social Diversity and Inequalities of Aging; Social Relationships and Aging; Social Institutions and Aging; Economies, Government, and Aging; Social Vulnerabilities and Aging; Public Health and Aging; and Care Arrangements and Aging.

We asked the authors of each chapter to keep in mind the 30-year metric and consider the following questions as they treated their topics:

- Where has the field been and where is it now? What specific ideas have generated excitement and transformed thinking? What important ideas have been lost along the way and might now be reclaimed in light of contemporary conditions?
- What new intellectual frontiers should be pursued to ensure a vibrant decade ahead? What opportunities exist or might be made to propel the field in these directions? What barriers might be anticipated and how might they be overcome?

- What are the central social sources and consequences of the topic under study? How might greater attention to "the social" revolutionize theories and empirical research on this topic?
- What are the implications for social policy?

The section "Sociological Lives" contains intimate, and often playful, essays written by a group of senior statespeople who share their personal experiences in the field and their hopes for its future.

As we were getting the handbook launched, Janet Wilmoth and then-Chair Eliza Pavalko also began to organize a special session for the 2009 annual meeting to commemorate the 30th anniversary of SALC. We worked with Janet and Eliza to feature in that session a handful of papers that were being also planned for the handbook. We are grateful for their willingness to collaborate in this way, and we are pleased that they both contributed chapters here in their areas of expertise.

We chose Duane Alwin to write the scholarly foreword, as he was Chair of SALC when we designed the handbook and began commissioning chapters. Duane has quick wit, shrewd observation, and a long and broad view of both the substance and methods of our field. We asked Ken Apfel to write the policy foreword, as he had served as commissioner of the Social Security Administration during the Clinton years and was a colleague of Jacqui's at the Lyndon B. Johnson School of Public Affairs at the University of Texas. Ken has unwavering commitment to interdisciplinary scholarship and respect for the important and unique contributions that sociological perspectives bring to policy-making related to aging. We are honored to have Duane and Ken's insights open the volume.

We would like to express our gratitude to the nearly 80 authors who contributed to the project. Each one has inspired us. Each was also patient as we rigorously reviewed their chapters and asked for revisions. We are especially indebted to our senior colleagues who, in the final section of the book, permit us a glimpse into their personal and intellectual lives as they look back on their many years in the field. What a treat it is to read their essays.

At Springer Science, we also wish to thank our editor, Teresa Krauss, and assistant editor, Katie Chabalko, for their steadfast support of the project from the inception of the idea to its completion. Teresa is a champion of the sociology of aging in the publishing world, and she has the special ability to see the momentum in a field and seize it. Katie did the yeomen's work in house to ensure that the project stayed on course. We also wish to thank Morgan Ryan for her editorial assistance in the final phase.

One person deserves a very special acknowledgment: Liz Bayler Levaro, who served as project manager and is a doctoral student in human development and family sciences at Oregon State University. Words cannot express how much we appreciated Liz, especially as she was called to assume caregiving responsibilities alongside the handbook and her doctoral dissertation. We would not have completed the project were it not for Liz's generous spirit, contagious enthusiasm, editorial acumen, technical wizardry, and unfailing patience.

It is our hope that the *Handbook of Sociology of Aging* will be of great use to researchers, emergent scholars, and policy-makers whose work focuses on aging and the life course. We also hope it will be of use to those who study other periods of life (e.g., infancy, childhood, adolescence, early adulthood, or midlife) – especially in being more sensitized to how experiences earlier in the life course might be affected by older people or an aging society or have a legacy for individuals or families decades later. The subject matter will, of course, speak directly to sociologists working on the many topics covered here, but it will also be of interest to members of other disciplines who also work on issues related to aging.

Most of all, we hope that the chapters of the handbook not only showcase the extraordinary advances in our field and how vibrant it now is, but also how much excitement we have as we imagine the decades ahead. Thanks to all who have contributed to our rich history, and to those who will make its future, in ways big and small.

Corvallis, OR
Austin, TX

Richard A. Settersten, Jr.
Jacqueline L. Angel

Contents

About The Editors and Contributors

Editors

Richard A. Settersten, Jr., is Professor of Human Development and Family Sciences at Oregon State University, Corvallis, OR, USA

Jacqueline L. Angel is Professor of Sociology and Public Affairs at the University of Texas at Austin, Austin, TX, USA

Contributors

Duane F. Alwin is the McCourtney Professorship in Sociology and Director of the Center for Life Course and Longitudinal Studies at Pennsylvania State University

Ronald J. Angel is Professor of Sociology at the University of Texas at Austin

Kenneth S. Apfel is Professor at the University of Maryland School of Public Policy and the former Commissioner of the Social Security Administration

Sara Arber is Professor of Sociology and Co-Director of the Centre for Research on Ageing and Gender (CRAG) at the University of Surrey, UK

Vern L. Bengtson is AARP/University Professor of Gerontology and Sociology Emeritus at the University of Southern California , Los Angeles, CA, USA

Susan L. Brown is Professor of Sociology and National Center for Family and Marriage Research, Bowling Green State University

Amy M. Burdette is Assistant Professor of Sociology at Florida State University

Jeffrey A. Burr is Professor of Gerontology at the University of Massachusetts, Boston

Deborah Carr is Professor of Sociology and the Institute for Health, Health Care Policy and Aging Research at Rutgers University, New Brunswick, NJ, USA

Stephen J. Cutler is Professor of Sociology and Emeritus and Emeritus Bishop Robert F. Joyce Distinguished University Professor of Gerontology at the University of Vermont

Dale Dannefer is Selah Chamberlain Professor of Sociology Chair, Department of Sociology, Case Western Reserve University, Cleveland, OH, USA

Kate Davidson is Visiting Senior Fellow and Co-Director of the Centre for Research on Ageing and Gender (CRAG) at the University of Surrey, UK

Sarah Desai is a doctoral candidate at the State University of New York at Buffalo

Cheryl Elman is Professor of Sociology at the University of Akron, Akron, OH, USA

Valerie Episcopo is Assistant Professor in the School of Behavioral and Social Sciences at St. Edward's University, TX

Carroll L. Estes is Board Chair of the National Committee to Preserve Social Security and Medicare and Professor and Founding Director of the Institute for Health & Aging at the University of California, San Francisco

Jean-Simon Farrah is a Master's candidate in Political Science at McGill University

Kenneth F. Ferraro is Distinguished Professor of Sociology and Director of the Center on Aging and the Life Course at Purdue University, West Lafayette, IN, USA

Anne Foner is Professor Emerita of Sociology at Rutgers University, New Brunswick, NJ, USA

Linda K. George is Professor of Sociology and Associate Director of the Center for the Study of Aging at Duke University, Durham, NC, USA

Kerstin Gerst is Assistant Professor of Public Health and the Institute of Gerontology at the University of Georgia, Athens

Roseann Giarrusso is Associate Professor of Sociology at California State University, Los Angeles

Chris Gilleard is Honorary Research Fellow in the Division of Research Strategy at the University College London Medical School, London, UK

Megan Gilligan is a doctoral candidate in Sociology and an NIH predoctoral trainee in the Center on Aging and the Life Course at Purdue University

Gunhild O. Hagestad is Senior Researcher/Professor at NOVA/Norwegian Social Research, Oslo, Norway

Melissa Hardy is Distinguished Professor of Sociology, Demography and Family Studies at The Pennsylvania State University, University Park, PA, USA

Paul Higgs is Professor of the Sociology of Ageing in the Division of Research Strategy, University College London Medical School, UK

Masa Higo is Research Associate at Sloan Center on Aging and Work at Boston College, Chestnut Hill, MA, USA

Terrence D. Hill is Assistant Professor of Sociology at Florida State University, Tallahassee, FL, USA

Christine L. Himes is Maxwell Professor of Sociology at Syracuse University, Austin, TX, USA

Ellen L. Idler is Professor of Sociology and the Institute for Health, Health Care Policy and Aging Research at Rutgers University

Tania Jenkins is a doctoral student in Sociology at Brown University

Emily Jovic is a Postdoctoral Researcher in Sociology at The University of Western Ontario, London, Ontario, Canada

Boaz Kahana is Professor of Psychology at Cleveland State University, Cleveland, OH, USA

Eva Kahana is Robson Professor of Sociology, Humanities, Applied Social Science, Medicine and Nursing at Case Western Reserve University

Ben Lennox Kail is a doctoral student in Sociology at Florida State University

Jessica A. Kelley-Moore is Associate Professor of Sociology at Case Western Reserve University, Cleveland, OH, USA

James N. Laditka is Associate Professor of Public Health Sciences at the University of North Carolina, Charlotte, NC, USA

Sarah B. Laditka is Associate Professor of Public Health Sciences at the University of North Carolina, Charlotte

Allen J. LeBlanc is Professor of Sociology and the Health Equity Institute for Research, Practice, and Policy at San Francisco State University, San Francisco, CA, USA

Jielu Lin is doctoral student in Sociology at Case Western Reserve University

Andrew S. London is Chair and Professor of Sociology, Co-Director of the Lesbian, Gay, Bisexual, and Transgender (LGBT) Studies Program, and Senior Research Associate in the Center for Policy Research at Syracuse University

Loren Lovegreen is Assistant Professor of Gerontology at Simon Fraser University

Wendy D. Manning is Professor of Sociology & National Center for Family and Marriage Research at Bowling Green State University, Bowling Green, OH, USA

Christopher Steven Marcum is a doctoral candidate in Sociology at the University of California, Irvine

Kyriakos S. Markides is Annie & John Gnitzinger Professor of Aging Studies, University of Texas Medical Branch, Galveston, TX, USA

Victor W. Marshall is Professor of Sociology and Senior Research Scientist in the Institute on Aging at the University of North Carolina at Chapel Hill, NC, USA

Duane Matcha is Professor of Sociology at Siena College, Loudonville, NY

Donna D. McAlpine is Associate Professor in the Division of Health Policy & Management at the University of Minnesota School Of Public Health

Julie McMullin is Professor of Sociology at The University of Western Ontario, London, Ontario, Canada

David Mechanic is René Dubos Professor of Behavioral Sciences and Director of the Institute for Health, Health Care Policy and Aging Research at Rutgers University, New Brunswick, NJ, USA

Madonna Harrington Meyer is Laura J. and L. Douglas Meredith Professor in Sociology and Senior Research Associate at the Center for Policy Research at Syracuse University, Syracuse, NY, USA

Phyllis Moen is McKnight Presidential Chair and Professor of Sociology at the University of Minnesota, South Minneapolis, MN, USA

Sara M. Moorman is Assistant Professor of Sociology and the Institute on Aging at Boston College

Nancy Morrow-Howell is Ralph and Muriel Pumphrey Professor of Social Work at Washington University

Jan E. Mutchler is Professor of Gerontology at University of Massachusetts, Boston, MA, USA

Greg O'Neill is Director of the National Academy on an Aging Society, Washington, DC, USA

Wendy M. Parker is Assistant Professor of Sociology at the Albany College of Pharmacy and Health Sciences

Eliza K. Pavalko is Allen D. and Polly S. Grimshaw Professor of Sociology at Indiana University, Bloomington, IN, USA

Chris Phillipson is Professor of Applied Social Studies and Social Gerontology at Keele University, Keele, UK

Karl Pillemer is Hazel E. Reed Professor in the Department of Human Development at Cornell University

Jill Quadagno is Claude Pepper Eminent Scholar Chair in Social Gerontology and Professor of Sociology at Florida State University, Tallahassee, FL, USA

Amélie Quesnel-Vallée is Assistant Professor of Sociology, Epidemiology, Biostatistics and Occupational Health at McGill University, Montreal, Canada

R. Corey Remle is Visiting Assistant Professor in the Department of Sociology at Wake Forest University, Winston-Salem, NC, USA

Sonia Salari is Associate Professor of Human Development and Family Studies at the University of Utah, Salt Lake City, UT, USA

Jori Sechrist is Postdoctoral Research Associate in Sociology at Purdue University

K. Russell Shekha is a doctoral student in Sociology at Florida State University

Dena Shenk is Professor of Anthropology and Director of the Gerontology Program at the University of North Carolina Charlotte

Merril Silverstein is Professor of Gerontology and Sociology at the University of Southern California, 3715 Mc Clintock Ave., Los Angeles, CA 90089-0191

Debra Street is Associate Professor of Sociology at the State University of New York at Buffalo, NY, USA

J. Jill Suitor is Professor of Sociology, Center on Aging and the Life Course at Purdue University, West Lafayette, IN, USA

Judith Treas is Professor of Sociology and Director of the Center for Demographic and Social Analysis at the University of California, Irvine, CA, USA

Susan Venn is Researcher in Sociology and Co-Director of the Centre for Research on Ageing and Gender (CRAG) at the University of Surrey, UK

Janice I. Wassel is Director of the Gerontology Program and Co-Director of the Dual Degree MS in Gerontology-MBA Program at the University of North Carolina, Greensboro, NY, USA

John B. Williamson is Professor of Sociology at Boston College

Janet M. Wilmoth is Professor of Sociology and Director of the All-University Gerontology Center, and Senior Research Associate in the Center for Policy Research at Syracuse University, Syracuse, NY, USA

Sarah F. Wilson is Program Manager at the National Academy on an Aging Society

Part I
Historical Trends in the Sociology of Aging

Chapter 1
Trends in the Sociology of Aging: Thirty Year Observations

Richard A. Settersten, Jr. and Jacqueline L. Angel

This chapter takes a broad view of scholarship on the sociology of aging, highlighting how the contours and content of the field have changed over the last 30 years. The chapters of the *Handbook of Sociology of Aging* provide fertile grounds for these observations. Each chapter traces the evolution of important ideas, synthesizes knowledge, and offers compelling new directions for future inquiry on specific topics. This handbook illustrates the fact that one of the greatest strengths of the sociology of aging is the wide range of topics and methods that characterizes the field. To generate additional observations on the field, we examined a wide range of books and articles on behavioral and social aspects of aging, including three decades of the *Journal of Gerontology: Social Sciences, The Gerontologist,* and *Research on Aging*.

In taking 30 years as the lens for our observations here and in the handbook – 1979–2009 – we became quickly aware of the extraordinary growth that has occurred in our field during this time. Our most senior colleagues will, of course, immediately recognize this. But many readers may not appreciate just how significant these advances have been. To begin, we therefore ask readers to simply imagine a field:

- Where scholarship on aging is relegated a marginal status in most disciplines, including sociology, and struggles for its legitimacy as an area of inquiry
- Without *Research on Aging, Journal of Aging Studies, Ageing and Society* and other aging-related journals that emerged at the start of this period or well into it
- Without all but the first of seven editions of the *Handbook of Aging and the Social Sciences* (e.g., Binstock and Shanas 1976), and without the two editions of the *Handbook of Theories of Aging* (Bengtson and Schaie 1999; Bengtson et al. 2009) and their predecessor, *Emergent Theories of Aging* (Birren and Bengtson 1988)
- Without major secondary datasets, including longitudinal ones, we now take for granted – the Health and Retirement Study, the National Survey of Families and Households, the Longitudinal Studies of Aging, Midlife in the United States, and many waves of the Panel Study of Income Dynamics, Current Population Surveys, Integrated Public Use Microdata Series, and other data sets and supplements sponsored by federal agencies
- Without so many of the advanced research methods, statistical techniques, and widely used and well validated measures we know today, including multi-level modeling, structural equation modeling, event history analysis, and advanced methods for measuring and analyzing change

R.A. Settersten, Jr. (✉)
Department of Human Development and Family Sciences,
Oregon State University, Corvallis, OR, USA
e-mail: richard.settersten@oregonstate.edu

R.A. Settersten, Jr. and J.L. Angel (eds.), *Handbook of Sociology of Aging*, Handbooks of Sociology and Social Research, DOI 10.1007/978-1-4419-7374-0_1, © Springer Science+Business Media, LLC 2011

- Without the infrastructure of the National Institute on Aging (NIA) and other public and private agencies and foundations, including initiatives of the MacArthur Foundation and the Retirement Research Foundation, which have provided extensive funding for behavioral and social research on aging
- Without many of the training programs and research centers that have educated generations of graduate students, both within sociology proper or with core contributions from sociologists – including NIA-sponsored predoctoral and postdoctoral training programs, Centers on the Demography and Economics of Aging, Centers on Minority Aging Research, and the Roybal Centers for Translation Research in the Behavioral and Social Sciences of Aging
- Without so many other ground-breaking articles, books and special issues of journals, especially edited projects or commissioned papers, too many to name, in which authors were playful with ideas that nourished the field. Indeed, many of the readings we consider to be "classic" today were written not long before the start of this period and even into it
- Without many textbooks for teaching undergraduate students about the sociology of aging – or, for that matter, undergraduate courses on the sociology of aging
- Without the later works of so many of the pioneers of our field, some of whom are still with us, and without the contributions of so many scholars since
- Without the controversy over the name change of the American Sociological Association's Section on the Sociology of Aging to the Sociology of Aging and the Life Course, which was ultimately resolved after seven years in 1997. One proposal, to add "and the life course" to the name, was adopted, while another proposal, to change "aging" to "age," ultimately was not

Space will not permit us to discuss the many important works that have punctuated these three very rich decades of scholarship. However, the chapters of this handbook will do so, as each takes an in-depth view into the history and future of sociological theories and research on particular topics. We begin by describing larger historical trends in theories, methods, and topics.

Historical Trends in Theories, Methods, and Topics

Broad Trends in Theories

As sociologists arrived at the study of aging, they naturally relied on their toolbox of classical and contemporary theories. Besides the classical writings of Emile Durkheim, Karl Marx, and Max Weber, who were acknowledged as the "holy trinity" of founding fathers, sociologists of aging also drew upon theories of structural functionalism (e.g., Talcott Parsons, Robert Merton), social conflict (e.g., Gerhard Lenski, Ralph Dahrendorf), social interaction and exchange (e.g., Georg Simmel, George Homans, Peter Blau), symbolic interactionism (e.g., George Herbert Mead, Charles Cooley, Irving Goffman), and phenomenology and social constructivism (e.g., Alfred Schütz, Peter Berger and Thomas Luckmann).

Central starting points for developing theories were also found in social gerontology, especially disengagement and modernization theories of aging in the 1960s and 1970s (e.g., Cumming and Henry 1961, Cowgill and Holmes 1972, respectively). The former postulated that physical decline in old age, and the social withdrawal of old people, is inevitable and functional for both individuals and society. In generating spirited debates ever since, disengagement theory would become crucial in the evolution of alternative views (e.g., activity theory, Havighurst 1963; continuity theory, Atchley 1971; functional equilibrium theory, Rosow 1963). These theories focused on individual behavior, but with an eye to social expectations and the greater social good. In contrast, modernization theory was crucial in taking a purer societal view on aging. It examined the status of the aged across cultures and identified how the social changes associated with modernization contribute to the disadvantaged position of older people (e.g., changes in the economy, medicine, technology, education, and urbanization).

Within sociology, the age stratification framework was also conceived not long thereafter and would become a dominant heuristic device for advancing inquiry into the social, and especially structural, aspects of aging. Although the age stratification framework is associated with Riley et al. (1972), the often overlooked work of Leonard Cain (1964) was an important precursor of both the age stratification framework and the life course perspective, as were much earlier works by sociologists and anthropologists of age status, including Shmuel Eisenstadt (1956), Ralph Linton (1942), Arnold van Gennep (1908/1960), and Anselm Strauss (1959).

Each in their own way, these theorists emphasized that aging and the life course are social processes and that age is a structural feature of societies, with both people and roles allocated on the basis of it. These theorists also pointed to the dynamic aspects of aging at both individual and societal levels. That is, new cohorts of people are born, grow up and older together, and move through the age structure of the population. These ideas also heightened the awareness of the need to not only understand the unique characteristics of particular cohorts as they aged, but also to understand the differential effects of social change on adjacent cohorts. This brought new thinking about how to better conceptualize cohorts and measure cohort effects (Rosow 1978; Riley 1973; Ryder 1965), building especially on Karl Mannheim's (1928/1952) classic essay on "The Problem of Generations." The emphasis on understanding the legacy of historical events and social change on individual and collective life histories would also become central to Glen Elder's *Children of the Great Depression* at about this time (1974) and, later, a hallmark of the life course perspective.

As the life course perspective evolved, it explicitly built on two paradigms – the *personological* and *institutional* paradigms (see Dannefer and Settersten 2010). The personological paradigm, of which Elder's body of work is a good example (and that of John Clausen too, e.g., 1972), attempts to use key features of early life experience to predict and account for outcomes later in life. Although this paradigm is often focused on individuals, it can address the life experiences of whole cohorts or populations (recent theories of cumulative advantage and disadvantage are good examples; Dannefer 2003). The *institutional* paradigm, in contrast, needs not focus on individuals at all, but instead analyzes the life course as a social and political construct, often consisting of more or less explicitly defined age-graded stages that are created or reinforced in institutions and social policy (e.g., Kohli and Meyer 1986; Mayer and Müller 1986). It therefore refers to a part of the social and cultural definition of reality that broadly organizes both people's lives and their "knowledge" about age and aging. In the field of aging, the institutional paradigm has strong roots in the writings on formal and informal age norms and the rise of age consciousness in modern societies (e.g., Neugarten et al. 1965; Kohli 1986).

Over these years, other major theoretical traditions in the sociology of aging would also emerge, especially theories informed by the political economy of aging (e.g., Kail et al. 2009), theories of cumulative advantage and disadvantage (e.g., Ferraro et al. 2009), feminist theories (e.g., Calasanti 2009), and critical theories of gerontology (e.g., Baars et al. 2006).

Social theories of aging can be characterized as having experienced a pendulum swing away from (1) at one extreme, "grand" theories in an early era of research that was "theory rich but data poor," to use James Birren and Vern Bengtson's (1988) phrase, to (2) a subsequent era that was, at the other extreme, "data rich but theory poor" – what C. Wright Mills (1959) might also have called "abstracted empiricism," in which too much attention is given to data over theory, to (3) the era of research today, which has perhaps swung back toward theories of the "middle range," to use Robert Merton's (1968) term, built around narrow topics and a good dose of data. The presence of middle-range theories in the field is evident across the chapters of this handbook. A larger window into theories of aging, including *social* theories of aging, can be found in the recent *Handbook of Theories of Aging* (Bengtson et al. 2009).

Social theories of aging can also be characterized as having experienced a pendulum swing moved through eras with differential attention to micro issues, macro issues, or the connections between them (micro-macro linkages): from (1) an early era of theories that were focused on individual issues, especially activity and life satisfaction, to (2) a subsequent era that was focused on larger contextual issues, especially how structural conditions determine the parameters of aging and the life situations of older people, to (3) an era of theorizing today that has attempted to synthesize micro and macro perspectives, especially in explaining differences in aging experiences by larger social forces or through intermediate contexts (see also Bengtson et al. 1997).

Broad Trends in Methods

Over these decades, several important methodological shifts have occurred in the field:

- From studies that emphasize basic description, to explanation, and to causality
- From studies that emphasize qualitative methods or relatively simple quantitative ones, to advanced quantitative methods, and to multi-method studies (at least in principle, though this ideal, at least in doing right by *all* of the methods brought to bear in a single study, has proven to be difficult to accomplish in practice)
- From a reliance on cross-sectional studies to the eventual building of panel and longitudinal data sets
- From studies based on small, original data sets to a preponderance of publications based on the analyses of large secondary data sets\designed for the scientific community

The field has also shifted from a reliance on crude measures to the development of measures that are more refined, more reliable, and better validated – even though there is much distance to go in creating truly meaningful measures that capture the complexities and realities of social aspects of aging. In fact, the most frequently read and cited articles across these decades pertain to scales meant to measure aspects of health or quality of life – specifically, life satisfaction, psychological well being, caregiver strain, caregiver burden, and cognitive performance, self-reported physical health, the use of services, and loneliness (see also Ferraro and Schafer 2008).

The investments in gathering longitudinal data are particularly important to note, as these investments have been made alongside the growing interest in the life course and recognition of the need to understand aging as a long-term process. Although longitudinal data permit new kinds of analyses, they also demand new kinds of methods, and this time period has brought much attention to methods for doing so.

Consistent with shifts in theories, the field has also moved beyond the use of age as a causal variable and instead toward understanding age in ways that are mediating or contextual. With this, the field has also turned greater attention to the possible *processes* and *mechanisms* that drive change, rather than simply demonstrating the empirical connection across the variables that are arrayed in a model.

However, some of these shifts have also fractured whole people and phenomena in favor of a narrow "peephole" perspective on small sets of variables, though the growing recognition and respect for interpretive and qualitative approaches has helped to counter these limitations. Sociological research has become more specialized over these decades, which also results in what seem like small windows into narrow phenomena of interest – though it has simultaneously become more interdisciplinary, which has similarly helped to counter these limitations.

Trends in Topics and Associated Terms

In conducting our review, we were immediately struck by how much language structures and reflects our realities. The language of our field becomes powerful in signaling our intellectual preoccupations and some of the assumptions that we make about our subject matter. Some of the topics and terminology that were present in the earlier years have vanished, some have persisted, but most have emerged and taken shape in the last three decades.

First, we no longer use some of the terms that were a prevalent part of our lexicon in the beginning of this period – terms we immediately recognize as outdated today. Many of these terms shock the contemporary inclination in the academy to deny or defy age: terms such as "the aged," "the elderly," "old people," and "old age" were very present in early articles and have now faded away, though they continue to be used by policymakers and in the media. Their disappearance is also surely linked to the fact that statements about "*the* old" or "*the* elderly" homogenize large groups of people who may be more different from one another than they are similar – a theme that gerontologists have sounded strongly in recent years, along with the reminder that chronological age is itself a poor proxy for the biological, psychological, or social statuses of individuals. The increasingly taboo nature of these terms in scholarship today serves as a powerful reminder that the meanings of age, and in this case *old* age and the things associated with it, are *socially* constructed.

Some topics that were very salient at the start of our review period have now largely vanished as major points of inquiry – including "interstate migration," "snowbirds," "NORCs" (naturally occurring retirement communities), "SROs" (single room occupancy hotels), "elder abuse," "elderly drivers," "senior suicide," "homes for the aged," "senior centers," "institutionalized aged," "fear of victimization," and "aging group consciousness." Several of these terms reflect the strong early emphasis of sociologists on the social problems of aging, problems that now receive much less attention in the face of a strong countertrend toward positive aging.

Second, many topics have been persistent or grown in significance, especially those that relate to health, family, and retirement. The strongest specific example is the explosion of interest in successful aging over this time period. However, it will surprise some readers to know that the term "successful aging" was, to our knowledge, first used by Robert Havighurst in a 1963 article that preceded our period, and again at the start of our period by Erdman Palmore in a 1979 article. These articles appeared well before Jack Rowe and Robert Kahn's first landmark article (1987) and later works (1997, 1998). Since Havighurst's and Palmore's times, many other variants of successful aging have emerged, even before or around the same time that Rowe and Kahn came onto the scene with their oft-cited articles and book. These include references to "optimal," "productive," "vital," "proactive," "robust," and "healthful" aging, among others.

Third, there are terms that are very much a part of scholarship now, but were largely absent in the early years of our review period. This language signals significant trends in the intellectual preoccupations and commitments of our field in the past 30 years. These include the following:

Anti-aging speak: Terms such as "old people," "old age," and "elderly" have been replaced by neutral language intentionally meant to avoid the sense of "old." In the face of concerns about ageism, the field of gerontology has, ironically, become rather ageless. We now speak of "older people" and "later life." This reflects a growing sense that age is something that can be defied or transcended, and the accompanying emphasis among gerontologists and in our society on successful aging and positive images of aging. Yet in lauding the potentials of age and aging, the field has deemphasized the social problems of aging and old age that captured the attention of the field in the early years, problems that we, as sociologists, are uniquely positioned to help solve.

These trends also make apparent another striking fact: aging and anti-aging industries have become big business. Aging (and anxiety about aging) has demanded that new institutions be designed (e.g., residential settings to meet the full spectrum of needed care; educational settings to meet the need for "lifelong learning") and that services and products be brought to market (e.g., aesthetic services, hormone treatments, vitamins and supplements, legal services). The marketing and consumption of these institutions, services, and products has also brought a wave of new legal and regulatory concerns.

Care speak: With the long era of caregiving research from the late 1980s through the 1990s, an extensive language of "social support" grew with it, especially to reflect the more negative aspects of giving care – including "caregiver strain" and "caregiver burden."

Generational relations speak: Longstanding interests in intergenerational relationships, especially between older people and middle aged children, and between older people and grandchildren, grew alongside the long era of caregiving research. With it grew attention to the dimensions to characterize family relationships – for example, many types of "solidarity" and "conflict" and, more recently, "ambivalence."

Health and disability speak: Health also becomes a dominant point of focus over these years. Here, we see the emergence of notions of "health span," "healthful aging," and "healthy life expectancy"; attention to "activities of daily living," "functional status," and the "disability cascade"; concern about a wide variety of specific disability or illness conditions, some of which also mark the times (including HIV/AIDS, Alzheimer's disease, cancer, osteoporosis, arthritis, and obesity); and language related to health care and institutions (including "independent living," "assisted living," "long-term care," "home health care," "rationing," and "person-environment fit"). Alongside major attention to health grew significant attention to the connections between health, religiosity, and spirituality (and their measurement).

Technology speak: With advances in technology came new hopes for the "built environment" and the use of new technologies to help people "age in place" and monitor their health, as well as terms associated with advances in computing, the internet, and digital social networking.

Life course speak: With the growing significance of the life course perspective came much attention to "pathways," "trajectories," "antecedents," "consequences," "event histories," "linked lives," "timing," "dynamics," and "human agency." Attention to human agency is also captured in a variety of related social-psychological concepts that also became salient over these decades – such as "self-efficacy," "self-determination," "locus of control," "effort," "mindfulness," "resourcefulness," "mastery," and "autonomy."

Methods speak: With major methodological advances came new language to reflect the strategies of the day that would become endorsed in journals: "multi-level modeling," "structural equation modeling," "longitudinal" methods, "growth curves," "latent classes," "life history" and "event history" analyses, and the like. The emergence of the life course perspective, apart from aging, also reinforced the need for advanced temporal and contextual methods, including strategies for better disentangling "age," "period," and "cohort" effects.

Diversity speak: As diversity became part of our scientific lenses in the late 1980s and through the 1990s, many new terms appeared to reflect those commitments – "aged heterogeneity," and regular references to Blacks, Hispanics, and other special populations (e.g., the "differently abled," "rural elderly," and "gays and lesbians"). With this came sensitivity as well to the combined risks of being in multiple vulnerable statuses or positions – for example, the "double jeopardy" or "triple jeopardy" hypotheses.

Gender speak: With greater attention to diversity also came a stronger focus on women – the "feminization" of later life, the "sandwich generation" of "women in the middle" who are caregivers to both parents and children, and women as "kin-keepers" of family relationships and traditions.

Inequality speak: Explanations of the differences among older people prompted a closer look at social inequalities. These were signaled especially by attention to the "Matthew effect" and theories of "cumulative advantage and disadvantage" over the life course, and to the "poor," "near poor," and "underserved" aged; widespread concern about "health disparities" across different groups of older people; and heated controversies about "generational equity" between the young and the old.

Taken together, the commitments to diversity, gender, and inequality in the intervening years have brought significant breakthroughs in understanding inter-individual variability in aging. In having made great strides to include women, minorities, and, to a lesser degree, children in our research, the field has gone far beyond the white, middle class, male perspective that was both the norm and the source of much complaint in our early science.

The Institutionalization of the Sociology of Aging

This handbook takes as its starting point the birth date of the Section on Aging and the Life Course (SALC) of the American Sociological Association (ASA) in 1979. The establishment of the Section marks the formal recognition and institutionalization of the sociology of aging in our discipline's primary professional organization. But the forces that led to its installation were underway before then, and its history since has been punctuated by some tensions that mirror waves of intellectual tension in the field. (For further information on the history of the Section, see Ferraro et al. 2005.)

The organization of sociologists with interests in aging grew out of activities of Harold (Hal) Orbach and Leonard Breen in the fall of 1961. They both explored the idea of launching a section on aging in the ASA. But it was not until nearly 20 years later that a petition, signed by eighty charter members, many of whom are past and present luminaries, to create a Section on the Sociology of Aging was officially approved by the ASA.

Yet as George Maddox, one of the field's pioneers, observed: "In the early days of the Section, [sociologists of aging] sometimes had concerns about their identity. Were they mainly sociologists and incidentally gerontologists? Could one be both? How one answered the identity query was usually based on who was asking. An accommodation was achieved in which sociologists and gerontologists largely ignored each other. In "Sociology of Later Life," Maddox (1979) documented that articles on aging rarely appeared in the principal journals of sociology and, when they did, citations to gerontological journals were rare. Unfortunately, gerontological journals and authors largely returned the disregard" (as cited in Ferraro et al. 2005:13–14).

Today, of course, sociologists of aging can find homes in vibrant sections of professional organizations that allow us to nurture both the "sociology" and the "aging" facets of our identity – in the SALC of the ASA, and in the Behavioral and Social Sciences Section of the Gerontological Society of America. Relative to the time of Maddox's observations, citations to gerontology journals today have also seen extraordinary leaps as the significance of gerontology and the quality and range of journals has expanded. But it is still the case that sociologists of aging do not have a journal to formally institutionalize the intersection between sociology and aging. Instead, we have the option of publishing either in sociology journals, with our papers emphasizing aging and life course issues, or in gerontology journals, with our papers emphasizing social issues. Two of a handful of gerontology journals have traditionally had strong sociological content – the *Journal of Gerontology: Social Sciences,* which in 1995 officially became its own section in "Series B" (though the *Journal of Gerontology* was launched five decades earlier in 1946), and *Research on Aging,* which began in 1979, and even for a while carried the subtitle "A Quarterly Journal of Aging and the Life Course."

As the field grew and diversified, many scholars felt that aging, and especially the term "gerontology," signaled interests that were too cohort-centric. That is, "aging" is a long-term process and need not – and, indeed, should not – be restricted to the study of the population of older people at any given time. The emergence of the life course perspective, as we will later discuss, both heightened that concern and strengthened that intellectual thread in the sociology of age and aging. Thus, Matilda White Riley (who had fashioned the "age stratification" framework and founded the Behavioral and Social Research Program at the National Institute on Aging), along with other leaders and members of the Section, began in 1990 to advocate a name change to what was then the Section on Aging.

Their proposal was to change "Aging" to "Age" and add "and the Life Course." The reason for the change from "aging" to "age" was at least twofold. One reason was that the inclusion of the "life course" would repeat the attention to processes already captured by the term "aging." But a second and more important reason was that "age" would keep attention focused on age as a property of social life and social systems – and therefore topics of inquiry that had been central to the field and its evolution (e.g., how social roles and activities are allocated on the basis of age, how age underlies the organization of social institutions, how age determines legal rights and responsibilities or eligibility for social programs, how age is used to determine expectations eligibility for social programs of the self and others).

Seven years later, after much controversy, the "life course" part was eventually adopted, but the "age" part ultimately rejected. However, the emphases on the sociology of age were nonetheless very present in the revised mission of the Section, which holds to this day:

> Sociology of Aging and the Life Course provides an analytical framework for understanding the interplay between human lives and changing social structures. Its mission is to examine the interdependence between (a) aging over the life course as a social process and (b) societies and groups as stratified by age, with succession of cohorts as the link connecting the two. This special field of age draws on sociology as a whole and contributes to it through reformulation of traditional emphases on process and change, on the multiple interdependent levels of the system, and on the multidimensionality of sociological concerns as they touch on related aspects of other disciplines. The field is concerned with both basic sociological research on age and its implications for public policy and professional practice.

The Life Course as Both Friend and Foe to the Sociology of Aging

The renaming of the Section to include "the life course" foreshadowed what would perhaps become the most significant development in our field in the years that followed to today. And it is a development around which we sense some new tensions growing.

The life course has gained tremendous momentum in our field. This is evidenced in a search we conducted using Sociological Abstracts from 1975 through 2008. The number of articles indexed with the term "life course" was essentially nonexistent in the 1970s; grew very slowly in the 1980s, reaching an annual peak of 88 publications by the late 1980s; and stayed fairly steady over the early 1990s, growing to no more than 148 annual articles by the late 1990s. It was in the 2000s that the presence of the life course surged: with annual numbers between 200 and 300 in the early 2000s, the numbers nearly tripled between 2003 and 2004 (to almost 800) and have remained relatively steady since, with a shift upward in 2008 and no doubt jumping further today.

From our perspective, the life course is both a friend and a foe to the sociology of aging. On the one hand, the life course perspective, at its most basic level, reminds us that the biological, social, and psychological aspects of aging outcomes are often not determined by chronological age itself but by the constellation of social factors that accumulate over a lifetime. It has therefore been central to the task of reclaiming the "social" in social gerontology, particularly in emphasizing how experiences in late life are shaped by those in earlier periods, and how aging experiences are shaped by a range of social institutions and forces (for further discussion, see Dannefer and Settersten 2010).

It has emphasized the great degree and types of diversity among older people, and that dynamics of aging are a collective process characterized by the accumulation of inequality over the life course. It has also emphasized that although age is important in every society, societies vary dramatically in how they use age and the meanings they attach to it. These are all very crucial lessons for sociologists of aging.

On the other hand, so much attention to the life course could also threaten the sociology of aging. As attention to the life course seems certain to grow exponentially in the years ahead, it is our hope that this attention will not ultimately compromise the scope and clarity of the sociology of *aging*. There are questions about aging that do not entail the life course, and many more questions about the life course that do not entail aging. They are not one and the same. Both are naturally treated in our work, but how we put the two together has tremendous implications for future scholarship. The life course perspective has offered valuable insights and transformed scholarship in our field. It is important for researchers to consider what is gained and what is lost or put at risk in our understanding of the sociology of aging if too great an emphasis is placed on the life course. In pursuing an understanding of "aging and the life course," the field seems to become ever bigger and broader, and therefore difficult to draw boundaries around. Are we becoming a field of every possible age, and every possible transition, in every possible domain of life? As the life course perspective gains momentum in other areas of sociology (e.g., family, education, work, health, and criminology), the overlap between these fields and ours also grows. What is it, then, that leaves our own subject matter distinct?

To explore the status of publications on "aging," we conducted a parallel search of Sociological Abstracts for the same period. It tracks in exactly the same way. But more important, the number of annual publications on "aging" are surprisingly higher by hundreds more articles (as many as 500 more at the peak). The fact that sociological articles on "aging" far outnumber those on the life course is, in some ways, reassuring. Nonetheless, we sense growing concern that the life course might also compromise the sociology of aging. And it is our prediction that the tensions that sociologists of aging feel around these difficult and critical questions will become increasingly strong in the near future.

The Social Organization of the Field

Finally, in reviewing 30 years of scholarship on the sociology of aging, we were often aware of the social organization of our science and the power of social relationships in producing knowledge. We were aware of the strong roles that leaders in the sociology of aging expectably play in shaping the intellectual agenda of the field and the theories and methods used to advance it. These dynamics were especially apparent in the early years of the field, before there was an infrastructure to support it. Keep in mind the conditions we noted at the beginning of this chapter. The actions of individuals and small groups in education, government, or practice were instrumental in forming the infrastructure we take for granted today. The history of the Section on the Sociology of Aging and the Life Course of the American Sociological Association is a good example of that.

We could trace generational lineages in the sociology of aging and see how the intellectual preoccupations of particular times are rooted in those relationships and exert their influence over time. We could see the influence of mentors at work, and how their capital and visibility played roles not only in promoting their success but also ultimately in forwarding the field. A "sociometrics" of our field becomes visible – the emergence of stars, networks of in-groups and out-groups, old guards and new ones; the control and circulation of the elite; and the statuses of the universities where members train or work in shaping their mobility or the allocation of resources.

We wondered about the infamous file drawer problem: the sociological research on aging that we will never know because it does not make it into the published scientific record, especially if it does not reinforce or if it critically challenges the ideas and methods that are in fashion at any given moment. We were also aware of the bias to publish findings of difference rather than findings of no-difference, though the latter may be just as important as the former to developing theory and our knowledge base on the sociology of aging.

We wondered about the things that are also not revealed in the record of published papers: how nonlinear the research process is; the complex and nuanced decisions that are made on the spot as research is being conducted; the leaps of faith in interpreting data; findings that are downplayed or dismissed, even unconsciously, because they run counter to theoretical lenses, prior findings, or personal convictions; admissions of what went wrong or was not done well.

We wondered, too, about the ambivalence so many scholars seem to have about writing, and about writing in ways that often seem artificial and distanced – as if we are in the world but not of it, as if we are bystanders to our work, wholly detached from the people and topics we study. At the same time, sociologists of aging are not fully insiders to the people and topics we are seeking to understand: That is, while we are all aging, most of us are not yet old. And yet we play roles in creating expectations, conducting research, making policies, and designing practice related to both aging and old age – and we carry assumptions and values with us as we do.

We have much to gain in more often turning a critical sociological lens on the social forces that promote or inhibit the advancement of knowledge, careers, and professional organizations. With greater consciousness of the social organization of our field comes the possibility to change that which is questionable or problematic.

Concluding Comment

In this chapter, we have outlined some of the broad trends in the sociology of aging over the past 30 years. In the final chapter, we will highlight some of the most provocative and pressing prospects for the decade ahead. In between, readers will find 36 chapters devoted to 30 years of inquiry on specific topics, and seven additional reflections from senior statespeople about their personal experiences in the field. The future holds great promise for the field of sociology, but there is much work to be done. With the new generations of scholars making their way into the field, we know that the next 30 years of the sociology of aging will be as exciting as the last thirty.

References

Atchley, Robert C. 1971. "Retirement and Leisure Participation: Continuity or Crisis?" *The Gerontologist* 11:13–17.

Baars, Jan, Dale Dannefer, Chris Phillipson, and Alan Walker. 2006. "Introduction: Critical Perspectives in Social Gerontology." Pp. 1–16 in *Aging, Globalization and Inequality: The New Critical Gerontology*, edited by J. Baars, D. Dannefer, C. Phillipson, and A. Walker. Amityville, NY: Baywood.

Bengtson, Vern L. and K. Warner Schaie, eds. 1999. *Handbook of Theories of Aging*. New York: Springer.

Bengtson, Vern L., Elisabeth O. Burgess, and Tonya M. Parrott. 1997. "Theory, Explanation, and a Third Generation of Theoretical Development in Social Gerontology." *Journal of Gerontology: Social Sciences* 52(2):S72–88.

Bengtson, Vern L., Daphna Gans, Norella M. Putney, and Merril Silverstein. 2009. *Handbook of Theories of Aging*, 2nd edition. New York: Springer.

Binstock, Robert H. and Ethel Shanas. 1976. *Handbook of Aging and the Social Sciences*, 1st edition. New York: Van Nostrand Reinhold.

Birren, James E. and Vern L. Bengtson, eds. 1988. *Emergent Theories of Aging*. New York: Springer.

Cain, Leonard D., Jr. 1964. "Life Course and Social Structure." Pp. 272–309 in *Handbook of Modern Sociology*, edited by R. E. L. Faris. Chicago: Rand McNally.

Calasanti, Toni. 2009. "Theorizing Feminist Gerontology, Sexuality, and Beyond: An Intersectional Approach." Pp. 471–86 in *Handbook of Theories of Aging,* 2nd edition, edited by V. L. Bengtson, D. Gans, N. M. Putney, and M. Silverstein. New York: Springer.

Clausen, John. 1972. "The Life Course of Individuals." Pp. 457–514 in *Aging and Society: A Sociology of Age Stratification, Volume III,* edited by M. W. Riley, M. Johnson, and A. Foner. New York, NY: Russell Sage Foundation.

Cowgill, Donald O. and Lowell H. Holmes. 1972. *Aging and Modernization.* New York: Appleton-Century-Crofts.

Cumming, Elaine and William Henry. 1961. *Growing Old: The Process of Disengagement.* New York: Basic Books.

Dannefer, Dale. 2003. "Cumulative Advantage/Disadvantage and the Life Course: Cross-fertilizing Age and the Social Science Theory." *Journal of Gerontology: Social Sciences* 58:S327–37.

Dannefer, Dale and Richard A. Settersten, Jr. 2010. "The Study of the Life Course: Implications for Social Gerontology." Pp. 3–19 in *The SAGE Handbook of Social Gerontology,* edited by D. Dannefer and Chris Phillipson. Los Angeles: Sage.

Elder, Glen H., Jr. 1974. *Children of the Great Depression,* 2nd edition. Chicago: The University of Chicago Press.

Eisenstadt, Shmuel N. 1956. *From Generation to Generation: Age Groups and Social Structure.* Glencoe, IL: Free Press.

Ferraro, Kenneth F. and Markus H. Schafer. 2008. "Gerontology's Greatest Hits." *Journal of Gerontology: Social Sciences* 63(1):S3–6.

Ferraro, Kenneth F., Tetyana Pylypiv, and Shalon M. Irving. 2005. *Commemorating 26 Years of the Section on Aging and the Life Course.* Washington, D.C.: American Sociological Association.

Ferraro, Kenneth F., Tetyana Pylypiv Shippee, and Markus Schafer. 2009. "Cumulative Inequality Theory for Research on Aging and the Life Course." Pp. 413–34 in *Handbook of Theories of Aging,* 2nd edition, edited by V. L. Bengtson, D. Gans, N. M. Putney, and M. Silverstein. New York: Springer.

Havighurst, Robert. 1963. "Successful Aging." Pp. 299–320 in *Processes of Aging,* edited by R. Williams, C. Tibbitts, and W. Donahue. New York: Atherton.

Kail, Ben Lennox, Jill Quadagno, and Jennifer Reid Keene. 2009. "The Political Economy Perspective of Aging." Pp. 555–72 in *Handbook of Theories of Aging,* 2nd edition, edited by V. L. Bengtson, D. Gans, N. M. Putney, and M. Silverstein. New York: Springer.

Kohli, Martin. 1986. "The World We Forgot: A Historical Review of the Life Course." Pp. 271–303 in *Later Life: The Social Psychology of Aging*, edited by V.W. Marshall. Beverly Hills: Sage.

Kohli, Martin and John Meyer. 1986. "Social Structure and the Social Construction of Life Stages." *Human Development* 29:145–9.

Linton, Ralph. 1942. "Age and Sex Categories." *American Sociological Review* 7:589–603.

Maddox, George L. 1979. "Sociology of Later Life." *Annual Review of Sociology* 5:113–35.

Mannheim, Karl. 1928 [1952]. "The Problem of Generations." Pp. 276–320 in *Essays on the Sociology of Knowledge.* London: Routledge and Kegan Paul.

Mayer, Karl Ulrich and Walter Müller. 1986. "The State and the Structure of the Life Course." Pp. 217–45 in *Human Development and the Life Course: Multidisciplinary Perspectives*, edited by A. Sørensen, F. Weinert and L. Sherrod. Hillsdale, NJ: Lawrence Erlbaum.

Merton, Robert K. 1968. *Social Theory and Social Structure,* 3rd edition. New York: Free Press.

Mills, C. Wright. 1959. *The Sociological Imagination.* London: Oxford University Press.

Neugarten, Bernie L., Joan W. Moore, and John C. Lowe. 1965. "Age Norms, Age Constraints, and Adult Socialization." *American Journal of Sociology* 70(6):710–17.

Palmore, Erdman E. 1979. "Predictors of Successful Aging." *The Gerontologist* 19(5):427–31.

Riley, Matilda White. 1973. "Aging and Cohort Succession: Interpretations and Misinterpretations." *Public Opinion Quarterly* 37:35–49.

Riley, Matilda White, Marilyn Johnson, and Anne Foner. 1972. *Aging and Society, Volume III: A Sociology of Age Stratification.* New York: Russell Sage Foundation.

Rosow, Irving. 1963. "Adjustment of the Normal Aged." Pp. 195–233 in *Processes of Aging: Psychological and Social Perspectives,* Volume 2, edited by R. H. Williams, C. Tibbitts, W. Donohue. New York: Atherton.

———. 1978. "What is a Cohort and Why?" *Human Development* 21:65–75.

Rowe, John W. and Robert L. Kahn. 1987. "Human Aging: Usual and Successful." *Science* 237(4811):143–49.

———. 1997. "Successful Aging." *The Gerontologist* 37(4):433–40.

———. 1998. *Successful Aging.* New York: Pantheon Books.

Ryder, Norman B. 1965. "The Cohort as a Concept in the Study of Social Change." *American Sociological Review* 30(6):843–61.

Strauss, Anselm. 1959. *Mirrors and Masks: The Search for Identity.* New York: Free Press.

van Gennep, Arnold. 1908 [1960]. *The Rites of Passage.* Chicago: University of Chicago Press.

Part II
Theories and Methods in the Sociology of Aging

Chapter 2
Theoretical Perspectives on the Sociology of Aging

Victor W. Marshall and Vern L. Bengtson

Throughout the relatively short history of the sociology of aging, many scholars have lamented the lack of theory and common conceptualization of issues (Orbach 1974; Maddox and Wiley 1976; Passuth and Bengtson 1988; Marshall and Mueller 2003; Bengtson et al. 1997). This reflects the growing opportunity for theorizing about age and age structure, but at the same time points to the continuing lack of agreement in theoretical paradigms or perspective – a problem that has persisted throughout the years. Our approach in this chapter is to provide a brief history of theoretical perspectives in the sociology of aging, and to indicate some areas where promising theoretical developments are emerging. First, we review theory development up to the late 1970s, when the field began to turn to the life course perspective. Second, we explore some of the more significant developments in theory over the past 30 years. Lastly, we preview a few challenges in theorizing for the future.

This chapter is about theoretical *perspectives*, which we define as systematic "ways of looking" at complex and interrelated processes, such as social aspects of aging. Theories are *explanations* or systematic statements that provide generalizations about causal relations. Perspectives frame, and consequently guide, theorizing; and theorizing, developing an understanding of the *how* and the *why* behind the social manifestations of aging we observe, is the ultimate goal. We cannot possibly do full justice to this topic in one chapter, so we recommend several more detailed overviews and discussions of the development of theory in aging. For general overviews see Marshall (1999); for the life course perspective, see Marshall and Mueller (2003), Macmillan (2005), Mayer (2009), and Bengtson et al. (2005). For demography and the life course see Hogan and Goldscheider (2003); for families and aging see Bengtson and Allen (1993); for reviews of the latest theory in nine substantive areas in the sociology of aging, see chapters in Bengtson et al. (2009). For a description of 12 classic theories in social gerontology see Bengtson et al. (1997), and for an examination of trends in the use of theories in our field since 1990 see Alley et al. (2010).

Early Developments in the Sociology of Aging

Theoretical perspectives in the sociology of aging can be categorized as macro-level (structural) or micro-level (social psychological), and further classified as to whether the individual-society assumptions are normative (individual behavior is seen as following norms) or interpretive (individuals construct norms but may not necessarily adhere to them) (Marshall 1996). The 1960s were a dazzling period of theoretical developments in the sociology of aging at both the macro and micro

V.W. Marshall (✉)
Department of Sociology and Institute on Aging,
University of North Carolina at Chapel Hill, Chapel Hill, NC, USA
e-mail: victor_marshall@unc.edu

R.A. Settersten, Jr. and J.L. Angel (eds.), *Handbook of Sociology of Aging*, Handbooks of Sociology and Social Research, DOI 10.1007/978-1-4419-7374-0_2, © Springer Science+Business Media, LLC 2011

levels, largely within the normative tradition. Burgess (1960) described an historical shift to a "role-less role" for the aged since the nineteenth century, and laid the foundation for the theory of aging and modernization, later developed more systematically by Cowgill and Holmes (1972). Cumming and Henry (1961) advanced the first truly multidisciplinary theory of aging in proposing that disengagement of older individuals from society was a normal part of the aging process. The theory posited that biological, psychological, and social withdrawals were universal, inevitable, and adaptive in preparing both the individual and the society for eventual death. But disengagement theory clashed with "activity theory," which implicitly treated aging as a decrement to be "adjusted to" by attempting to maintain as much as possible the activity levels of mid-life. This "keep active!" theory of successful aging was much more consistent with American values of individualism and autonomy than the disengagement perspective and was in keeping with concrete applications, such as the development of senior citizens centers and Leisure World housing developments.

The two competing theories stirred up debate in the previously uncontroversial sociology of aging; and this in turn led to healthy progress in the explicit development and testing of concepts and theories over the next two decades (Maddox 1965; Lemon et al. 1972; Longino and Cart 1982; Palmore 1968). Activity theory is still visible in current advice about "successful aging." In the best-seller of that name, Rowe and Kahn (1998) argue that the key to aging successfully lies in "keeping active" in four ways: active social engagement, active exercise, proactive diet, and avoiding disease. Critics note three difficulties with their prescription: (1) It is based on the old activity theory model (unfortunately never acknowledged or referenced), which had received little empirical support over the previous three decades of research; (2) It assumes that successful aging is under the control of an individual's own efforts and initiative; (3) It ignores social constraints and inequalities – race, poverty, gender, isolation – that could make it structurally impossible to follow these prescriptions (Schmeeckle and Bengtson 1999).

In the early 1960s, Bernice Neugarten and her students began conducting systematic research on age norms, concluding that in America there exists a prescriptive timetable for the ordering of major life events, such as the appropriate age to finish education, get married, have children, and retire (Neugarten et al. 1965). Moreover, the age expectations and the actual age at which these events occurred were highly related. They concluded that age norms operate as accelerators and brakes on individuals' behavior, sometimes hastening an event and sometimes delaying it. Regarding old age, however, Neugarten (1974) subsequently argued that normative roles for later life were changing, and that "old age" needed to be differentiated into "young old" and "old old" categories, based largely on a good health-frail health transition that, she suggested, typically happened around the age of 75. Neugarten later expressed regret at having set age 75 or any particular age for the young old-old old division, since this could lead to stereotyping, but the terminology was widely adopted, creating two reified groupings that failed to capture diversity within and across these groups.

A few years later, Suzman and Riley (1985) proposed that a third group be recognized, the "oldest old" at age 85. This led Leonard Cain Jr., the pioneering sociologist of aging, to playfully dub the three categories "the frisky, the frail, and the fragile." Perhaps having second thoughts about their tripartite division of the aged, Suzman and Riley (1985:180) tried to have it both ways: "Also widely unrecognized is the pronounced diversity of the population aged 85 and over. At this age many people still function effectively, while others have outlived their social and financial supports and have become dependent upon society for their daily living."

Rosow (1974) presented the most nuanced theorizing to date in this normative tradition in social psychology of aging. He argued that for various reasons socialization for old age was not highly successful. Riley introduced the age stratification perspective in 1971 (elaborated in Riley et al. 1972). Her model included both structural and social-psychological levels of analysis, with links between the society and the individual through role allocation processes (structural) and socialization. The main contribution of this perspective was to conceptualize age stratification as a component of the social structure, a succession of age statuses through which cohorts or individuals flowed as they aged. The model depicted two sources of change over time – the aging of individuals and the differentiation of cohorts. The way this conceptualized social structure as groupings of role positions followed the predominant perspective of general sociology at that time. Parsons's (1951) structural–functional "general theory of action" downplayed more materialistic aspects of social structure that constrain or create life chances for individuals or for cohorts (Marshall and Clarke 2010; Sewell 1992) and presented the aging individual as a somewhat passive receptor of socialization (Dowd 1987, Marshall and Tindale 1978–1979; Settersten and Gannon 2005).

Through the 1960s and 1970s, most research in social gerontology focused on "the aged" and a narrow range of such later-life transitions as retirement, and widowhood was considered at the social psychological or social structural levels. The unit of analysis was primarily the individual, and the processes investigated usually focused on the short term. The dependent variable most frequently investigated was "life satisfaction" (including "morale," "happiness," and "adjustment"), examined principally in relation to short-term causal mechanisms, such as current economic well-being, health status, and social integration (Marshall and Tindale 1978–1979). The temporal scope of investigation largely was confined to single events in these domains – retirement, health decrements, and widowhood – leading to role loss that affected life satisfaction.

At the same time, at the structural level of social institutions, aging and modernization theorists saw macro-level social forces acting on individuals through highly deterministic processes. In this theory (Cowgill 1974), forces of modernization (the rise of scientific technology, urbanization, literacy and mass education, and health technology) created population-level changes in life expectancy, the emergence of new occupations, geographical and occupational mobility, and changing marital status. These, in turn, fostered residential and social distancing between the generations, including the shift to a primarily nuclear family form, leading to cultural changes and to the emergence of the social institution of retirement. Finally, the theory maintained, development of the institution of retirement and accompanying cultural changes led to a lower status for the aged. Burgess (1960) had laid systematic foundations for this view in coining the phrase, the "role-less role of the aged." The processes postulated by Cowgill involved long periods of historical and family time. However, theories are made for testing, and the modernization theory did not stand up well to subsequent empirical investigation.

At the micro or social psychological level, *several* investigators began research in the interpretive theoretical tradition. Matthews (1979) drew on symbolic interactionism to understand *The Social World of Older Women*, and Kuypers and Bengtson (1973) used labeling theory principles to develop their "Social Breakdown Theory of Aging." Gubrium (1975) drew on phenomenological sociology for his study of life in a nursing home. Explicit recommendations for the application of Mead's perspective to the study of aging were given by Chappel and Orbach (1986) and Spence (1986). Marshall (1978) analyzed socialization for impending death in a retirement community from the perspective of Schutz and the symbolic interactionists. Atchley (1971) proposed a continuity theory of aging to explain why older adults were able to maintain a strong and consistent sense of self, emphasizing the active role individuals take in developing personal constructs.

Developments from the 1980s

Reviewing the "sociology of later life" as the decade of the 1970s drew to a close, Maddox (1979:113) suggested it was "…at the edges of mainstream sociology." However, the foundation for change had been laid, and the decades to come would see significant changes to the definition of the field and decreasing marginalization. Much of this progress would be shaped by three developments: increased adherence of sociologists to the political economy perspective, theoretical formalization of the life course perspective, and the maturation of interpretive sociology applied to aging.

Establishment of the Political Economy Perspective

As the 1970s ended, the framework of political economy became widely accepted in sociology and scholars in the sociology of aging soon found it a useful corrective to previous macro-social perspectives in the field. Townsend (1981:5) suggested that the dependency of the elderly in the twentieth century was "socially manufactured," calling it "structured dependency." Walker (1981:74) noted that poverty had wrongly been accepted as an inevitable consequence of old age and therefore something to which elderly people must simply "adjust." Estes (1979) linked the social construction of reality with a political economy perspective that focuses on the function of social policies and programs in bolstering existing power arrangements. Gerontological theories, she felt, support the status quo by locating the "problem" within the older individual rather than in the socio-political production of the problem. Guillemard (1983) brought together social scientists from Europe, America, and Canada to examine old age and the welfare state, stimulating much new work: Guillemard and Rein (1993) on the welfare state and retirement; Kohli et al. (1991) on state policies and the retirement transition; Myles (1984) on the political economy of pension policies; Myles and Quadagno (1991) on aging and labor markets; and Leisering and Leibfried (1999) on welfare state policies and their impact across the life course in a comparative context (see Mayer 2009).

Quadagno (1982) suggested that the aging and modernization theory as developed by Burgess (1960) and Cowgill and Holmes (1972) was ahistorical and idealized the past. In a historical analysis of nineteenth-century England, she showed that the aged were not generally venerated; rather, their status was highly conditioned by structured power relations. She also provided evidence that negated the timing of some events in the causal chain hypothesized by Cowgill.

Aboderin (2007) applied a political economy perspective to refine the earlier theorizing of aging and modernization. She suggested two central mechanisms leading to a decline in the status of the aged with modernization: increasing individualism, which weakens traditional norms of filial obligation by emphasizing the value of independence in old age, and "modernization," which erodes conformity with filial obligation norms by eroding older people's powers to wield sanctions that traditionally enforced the norms, and the resources they can offer in exchange for support. She criticized the aging and modernization theory for ignoring family exchanges other than filial support from children to parents, and for ignoring the impact of constraints such as poverty or economic crisis on the provision of support.

A variant of the political economy perspective, the "moral economy" of aging, suggests that social movements, political and economic processes are shaped not only by material interests in a Marxian sense but also by cultural conceptions of legitimacy, equity, citizens' rights, and moral contracts across generations (Hendricks 2005; Marshall and Clarke 2007; Minkler and Estes 1999). In its many variants, the political economy perspective has had a great impact on aging research since 1980. Moreover, because it addresses the ways in which individual lives are affected by changing social structures, the perspective is frequently merged with the life course perspective.

Development and Formalization of the Life Course Perspective

The first formal statement of the life course perspective was provided by Cain (1964), who drew on other sociologists (Strauss for a dynamic view of negotiating age structure and Eisenstadt for the importance of generations), anthropologists (van Gennep for the concept of structured age statuses), and other disciplines (Marshall and Mueller 2003). A decade later, Abeles and Riley (1977) outlined three central premises that still form the core of the life course perspective: (1) developmental change and aging represent a continuous process; (2) change occurs in interrelated social, psychological, and biological domains, and (3) life-course development is multidetermined.

They urged a cross-disciplinary research agenda covering domains such as the sense of self and self-esteem; health, physical functioning and stress; social networks and relationships; and work and retirement. Riley's approach to the life course built directly on her prior work on age stratification but with greater emphasis on social processes of individuals and cohorts passing through the age structure (Riley 1987).

Elder (1985) described the life course perspective as emanating from Chicago-school sociologists who made prominent use of case studies of individual lives and advocated a longitudinal approach to research methods. Incorporating aspects of Bronfenbrenner's ecological perspective, Elder (1995) pointed to five principles of the life course perspective; expanding those originally suggested by Abeles and Riley to emphasize the importance of historical time and place, how linked lives influence behavior, and human agency. Placing the individual and structural life course in historical time is perhaps Elder's major contribution to theorizing the life course, as exemplified by his classic, *Children of the Great Depression* (Elder 1974).

Although the concept of "linked lives" refers to any social linkage, which creates contingencies in life decisions and actions, the principle has most frequently been employed in the intersection of life course sociology and family sociology, with issues such as family caregiving or grandparent–grandchild relationships. Building on earlier work by Shanas et al. (1968), Bengtson developed, refined, and measured a typology of dimensions of intergenerational solidarity and conflict (Mangan et al. 1988) for use in the Longitudinal Study of Generations, which began in 1971. The typology was widely used by others (notably Rossi and Rossi 1990). Although Shanas's work was in effect a test of the modernization theory of aging as applied to the family domain, Bengtson linked his work to social change through his conceptualization of generational succession, employing the concept of generation in both its kinship and cohort usages to examine stability and change in kinship generation relations as successive cohorts age.

The term "generation" is much debated because it can refer to both kinship generations and generational cohorts (Bengtson et al. 1985; Alwin 1995). As Uhlenberg and Miner (1995) observe, "cohort" has been a critical concept in the life course perspective. Ryder's (1965) influential paper, The Cohort as a Concept in The Study of Social Change, recommends confining the term generation to the kinship realm, a recommendation seconded by Kertzer (1983). However, European social scientists, paying attention to Mannheim's conceptualization of generations, mostly disagree. Maddox and Wiley (1976) and Marshall (1983) recommend restricting "cohort" to a methodological device to array data by year of birth or some other date. Bengtson et al. (1985) suggest the term "generational cohort" to distinguish a Mannheimian usage from the methodological or kinship usages. Being able to distinguish between cohorts and generational cohorts has profound implications for aging and life course theorizing because it allows generations to be treated structurally.

Interpretive Sociology and Social Construction of Aging and the Life Course

In a third major contribution shaping aging research since 1980, symbolic interactionist, phenomenological, and related "interpretive" sociologies come together in what is often now called the social constructionist perspective; these emphasize that individuals exercise agency over their life courses.

Handel (2000) recently revived the "Chicago School" tradition of life narratives with his study of a working class man in New York, situating him in time and place and recognizing that the life course is a social institution and that the individual exercises choice while navigating it. Handel states life course principles in the language of symbolic interactionism (using "career" rather than "trajectory") and situates the "story" in historical time and place, illuminating the extensive life-story data that provide the empirical base for the book. Newman's (2003) life-course study of minority women's journeys from adulthood into old age, struggling for economic security and family stability in poverty areas of New York, used a set of narrative case studies.

In Europe, the University of Bremen's "Status Passages and Risks in the Life Course" Research Center has promoted this approach for almost 15 years, theorizing and pursuing life course studies that integrated individual lives with social structure in the context of the risk society, distinguishing its approach from the early life course theorizing in North America (Heinz et al. 2009; Marshall and Mueller 2003).

Phenomenological sociology represents an important variant of the interpretive perspective in aging and life course studies (Holstein and Gubrium 2007). Common to various approaches in phenomenological sociology is the notion of meaning-making through what Berger and Luckmann (1966) had called the "social construction of reality," drawing heavily on the phenomenological sociology of Alfred Schutz (1967). In its various guises, the interpretive perspective has influenced aging and life course theorists to emphasize the active part played by individuals as they struggle to make their lives.

Current Theoretical Development in the Sociology of Aging

We feel that a number of perspectives are currently emerging that appear promising for significant theory development in the sociology of aging. These will provide an agenda for theory-relevant research in the next decade, including – and this will be particularly important – efforts to make links among these topics. We confine ourselves here to a selected number of these developments.

Stress Theory, Aging, and the Life Course

The largest and most rapidly growing substantive area in aging research recently has been in aspects of social health and aging, often theorized in terms of social models of stress. Pearlin (Pearlin and Skaff 1996) contributed to theoretical development in this area by his own research as well as by providing a publication vehicle as editor for many years of the *Journal of Health and Social Behavior*. Departing from previous perspectives in stress research, especially in the critical life events tradition (Holmes and Rahe 1967), he emphasized that life changes are not necessarily negative, that it is a normal part of aging to experience them, and that life course transitions are not necessarily associated with decreased psychological well-being (Pearlin 1982). Instead, he theorized that life course transitions do lead to changes in social and economic conditions, and these conditions can pose "life strains." Furthermore, he issued a sociologist's challenge to the prevailing wisdom at the time: "Given the variety of ways in which different people respond to being at the same stage (of life), any research that is worth its salt has to attend to the parts played by social and economic factors in creating and channeling the effects of life-cycle transitions" (Pearlin 1980:359).

As Wheaton and Clarke (2003) note, recognizing that the impact of stressors can spin out over long periods of time led to a "natural alliance" of the life course and stress process perspectives. Structurally, even early-life neighborhood contexts continue to be operative throughout transitions over many stages of the life course, with the strongest effects in relation to gains and losses of status and roles (Clarke and Wheaton 2005). A major way of theorizing how earlier life course conditions and events can have outcomes later in life is through the concept of "cumulative advantage" or "cumulative inequality," a concept applied to stress but to many other outcomes in later life.

Cumulative Inequality

Possibly the most active theorizing within the sociology of aging recently has been in the areas of cumulative advantage, disadvantage, and inequality (Dannefer 2003; DiPrete and Eirich 2006; Ferraro et al. 1999; O'Rand 2003). Theory development in this area was initiated by Merton's (1968) description of the "Matthew Effect" in shaping scientific careers ("To he who hath much, much will be given") and applied to the problem of accounting for differential aging trajectories by Dannefer (1987), O'Rand (1996), and O'Rand and Henretta (1999). Building on this work, Ferraro and his colleagues have engaged in formal theory development, formulating specific hypotheses linking life course principles, through cumulative inequality processes, to later-life outcomes such as variability in self-efficacy or mental and physical health (Ferraro et al. 2009). O'Rand and Hamil-Luker (2005) traced a "sequentially contingent" process of cumulative disadvantage in terms of heart attack risk, while Hayward and Gorman (2004) examined the influence of early-life social conditions on men's mortality. Willson and Shuey (2007) muster cautious support for both path dependent and duration dependent processes of cumulative disadvantage affecting self-rated health in later life.

Cumulative inequality outcomes are seen in later life, making this an important theoretical development for the sociology of aging. However Mayer (2009:417) notes that few studies demonstrate linkages between early initial conditions to late life outcomes. Thus, in their enthusiasm for the plausibility of the cumulative advantage hypothesis – and it must still be considered a hypothesis, given the absence of data demonstrating long-term causal linkages – theorists in aging should not forget the importance of history, including the effects of changing social institutions and policies.

The Standardized Life Course

Cain (1964) had suggested that the life course experienced by an individual included a preparation for work stage, a "breadwinner" stage, and a retirement stage. Decades later, Kohli described the life course as "periods of preparation," "activity," and "retirement" (Kohli 1986:72). O'Rand (2003:693) observed that "The life course – when defined as interdependent sequences of age-related social roles across life domains (family, education, work, health, leisure) – is a product of the linkages among state (welfare), market and familial (gender) institutions, and demographic behaviors across the life span." She notes that "… when these linkages are tightly coupled and universally salient in a population, their coherence and normative strength lead to a more highly institutionalized, age-graded life course. Alternatively, when these linkages are loosely coupled, variability (de-institutionalization) in the life course increases: The relationship of age to role transitions weakens, and the synchronization of roles across life domains becomes less standardized."

Since 1978, empirical and theoretical studies have questioned the extent to which the social institution of the life course is now or was ever highly standardized (Bruckner and Mayer 2005; Hogan 1978; Rindfuss et al. 1987). Wilensky (1960) debated the consequences of not following a standardized life course for individuals and suggested that a focus on the time or the order of entry to states is insufficient to understand standardization of life course issues and their impact on later life (see Settersten et al. 2005).

Some industrial or occupational sectors offer stable and predictable opportunities for a career more than others. In labor market segmentation theory, the primary distinction has been that between firms with internal labor markets and firms that employ unstructured, open labor markets (Sorensen 1986). Blossfeld and Mayer (1988) showed that even a more complex typology of types of labor market structures is not highly effective to characterize the occupational life course.

There have been major changes in the past century in the age patterning of labor force participation and in the timing of retirement (Cahill et al. 2006; Carr 1966; Hardy 2006). Age-related labor force participation and retirement timing have been much in flux since 1995 in response to changing economic conditions and public and corporate policy responses (Marshall 2009a), but these have not been well theorized.

A similar three-part view of the life course is assumed in welfare state provisions for income security, education, and family supports (Marshall 2009a; Moen et al. 2005, Myles and Street 1995), yet the life course can be destandardized through the reinforcing effects of industrial change, weakening of the welfare state, and globalization, which devolve risk from social institutions to individuals (O'Rand 2006).

Risk Society

The concept of the risk society has been seldom applied in theorizing of aging (Phillipson and Powell 2004), despite important areas of cross-fertilization between aging and life course and risk society theorizing. Giddens (1991) argued the need for greater life course reflexivity in late modern risk societies. The risk society is one in which social institutions provide less "insurance" against the vicissitudes of life, such as job loss or loss of one's health, and individuals are expected to assume responsibility to navigate these risks. Risks such as income loss with retirement cannot be anticipated with certainty (e.g., unexpected "early retirement" because of job loss or the need to provide family care). DiPrete addresses "unanticipated events" in the context of stratification and life course mobility under different welfare regimes, which provide various degrees of protection for such instability. He argues that highly educated people orient their lifestyle to the long-term expected living conditions corresponding to the wealth associated with their human capital. But to do so requires long-range estimates, while the future cannot be anticipated accurately. As DiPrete argues, "unpredictable …. changes that have large and potentially durable effects offer a challenge to the behavioral theory implied in the concept of permanent income" (DiPrete 2002:272).

Chance Events and the Life Course

The concept of a structured and predictable (institutionalized) life course has been an important element in the life course perspective and assumed by social gerontologists to be the key to a good old age. However, Pearlin (1982:57) noted the distinction between the "relatively durable strains" of daily life; scheduled and transitional life events, and the less-expected, unscheduled events. Notable are "…the more eruptive and less predictable events of life, which stand in sharp contrast to the normative events that can be forecast far in advance of their actual occurrence." To understand both anticipated and unplanned life course events, we need to appreciate them as events organized through time and space. We need to consider not only the stimulus of the event, but also the ways in which people experience, adapt, and find social support to cope with them (Pearlin 1982).

However, Pearlin's call for attention to unanticipated life course events has been infrequently heeded. Shanahan and Porfeli (2007) observe that "chance" is a tricky concept that, when applied to the course of human lives, goes well beyond probability or statistical odds. Our understanding of how people experience aging and cope with it changes will be enhanced if we can better theorize the role that unanticipated events and chance play. This is particularly true in the context of a risk society, which provides few institutional supports.

Structure and Agency in the Life Course

Understanding the relationship of agency (individual choice) and social structure (societal constraints) has been the goal of much sociological theorizing. This has recently surfaced as an important focus for life course and aging researchers (Dannefer 1989, Settersten and Gannon 2005). In the normative perspective that characterized much early sociology, sets of roles organized into social institutions constituted the social structure, defined in terms of norms, or expectations for behavior. The individual appeared mostly as a role player. The metaphor was that of the theatre, in which actors learned their scripts through what were presumed to be highly efficient socialization mechanisms and then acted them out. Those employing the interpretive perspective took as metaphor free theatre or "theatre improv," in which individuals worked jointly and collectively to produce a play, simultaneously writing and enacting the scripts (Marshall 1986). The human agency principle was Elder's conceptual device to argue against extremes of normative determination in the shaping of individual's biographies.

How to define, measure, and use a concept such as "agency" (or "social action," as it has been otherwise termed) has been a matter of considerable debate in the social sciences and the aging and the life course literature (Marshall 2005, Marshall and Clarke 2010). We will not be able to cover the complexity of this issue; see Dannefer (1989), Marshall and Clarke (2010) for applications of agency to aging and the life course and Sewell (1992) for a general overview of the problem. Settersten and Gannon (2005:36) call for models of "agency within structure, which explicitly seek to understand how individuals set goals, take action, and create meanings within – and often despite – the parameters of social settings, and even how individuals may change those parameters through their own actions."

The Family and the Life Course

Some of the clearest theoretical applications of life course concepts can be seen in research on families (Bengtson and Allen 1993). Although many aspects of the life course perspective are useful in theorizing family processes, the concept of "linked lives" has proven the most useful tool in understanding the actions and interactions of family members over time. For example, research on grandparents and their influence has expanded the concept of linked lives beyond the boundaries of the nuclear family (Bengtson 2001; Roberto et al. 2001). Grandparents are increasingly important for the survival and well-being of growing numbers of children worldwide. As grandparents go on to live longer and in better health, they are called upon for longer years of both caregiving and mentoring (Mueller et al. 2002) and they serve as core conveyers of values and moral teachings, as well as personal and family identity (King 2003). Religion is an area where grandparents appear to be particularly influential (Bengtson et al. 2008). We suggest that grandparenthood be theorized much as parent–child socialization has been, that is, with a focus on processes or mechanisms reflected in socialization theories: role modeling, status inheritance, reflected appraisal such as attachment or solidarity (Bengtson et al. 2002a).

A second example concerns caregiving for dependent elders, also a growing concern in nations both East and West. Despite variations between nations in state welfare provisions, the primary responsibility for providing care and support remains with the family (Bengtson and Lowenstein 2006). Elder-caregiving can be theorized as a career. It epitomizes the long-term linkage between lives, with consequences that continue up to dying and beyond death to widowhood and bereavement. Its demands begin slowly but may progress to the highest levels of stress and burden experienced during the life course, resulting in exhaustion, depression, and declining physical health

(Pinquart and Sorensen 2006). Agency also can be seen in "caregiver growth." Positive consequences of caregiving mentioned by many respondents include increased self-understanding and a sense of pride (Roberto and Jarrott 2008).

The linked lives in caregiving careers often extend well beyond the marital dyad, and then inequities can surface and conflict may result. Siblings will negotiate who will provide care, what kind of care, and with what exchanges or rewards – negotiations that are often marked with perceptions of inequalities (Connidis and Kemp 2008). There are marked gender inequalities in caregiving. Its stresses are more evident in women than in men, a reflection of disadvantage in balancing demands of work and family (Heinz 2003). Thus we suggest caregiving be theorized as a career in itself, a story of linked lives involving earlier family antecedents, cyclical engagement, negotiations of roles, short- and long-term impacts, structural constraints and agency.

Beyond Theory? Critical Gerontology and the Critical Feminist Perspective

Achenbaum describes "critical gerontology" as having "entered the gerontological mainstream without losing its edge" (Achenbaum 2009:35). This probably exaggerates the extent to which critical gerontology has been accepted by sociologists of aging today. It is not adhered to by the vast majority of researchers producing empirical studies that characterize almost all of the aging research published in major sociology or social gerontology journals, whose work, incidentally, would be criticized by the critical theorists.

Marshall (2009b) has asked, "What is new about critical gerontology?" Except for the self-labeling as a critique of the status quo in society and their claim to be struggling for social justice and a better world, an observer may have difficulty identifying perspectives or concepts that are unique to critical theory. All self-labeled critical gerontology incorporates earlier traditions, especially the political economy tradition in the sociology of aging, and employs narrative approaches that relate closely to symbolic interactionism and related interpretive sociologies. The same may be said of the varieties of "feminist theory" (Marshall and Clarke 2007). An emphasis on social justice and efforts to usefully apply research-based knowledge is common to both of these. However, many other sociologists in aging could argue that their scholarship is supporting social justice and human betterment. In the tradition of Dewey's pragmatism, many interpretive sociologists see practice as the test of knowledge. It is not always necessary to proclaim humanitarian goals as a unique contribution of one's own perspective.

Estes (2006) links critical theory and feminist theory in articulating four premises of the "critical feminist perspective": (1) the experiences and situations of women across the life span are socially constructed; (2) the lived experiences and problems of older women are not solely, or even largely, the product of individual behavior and decisions; (3) the disadvantages of women are cumulative across the life course; and (4) the feminization of poverty is inextricably linked to the interlocking oppressions of race, ethnicity, class, sexuality and nation that produce the marginalization of women. What apparently makes all this different from the existing political economy of aging perspective is a "critical feminist epistemology" that requires that the world must be understood from the standpoint of women (and not men). The reader can judge whether men could ever be able to take a similar standpoint, or whether men and women could possibly be able to see things the same way. Moreover, one need not be a feminist to recognize that gender and the other interlocking sources of inequality mentioned by Estes, as well as age stratification, are the proper subject matter of all good sociology (McMullin 1995).

Notably absent from our discussion here is postmodernism, which might be considered a perspective in the sense that it represents something like a "way of looking." But as postmodernism eschews development of explicit theory, in fact is anti-theory in many formulations, it is not considered here. See Baars (2006) for a stimulating discussion from the perspective of critical theory, and Marshall and Clarke (2007) for a less sympathetic discussion.

Globalization, Aging, and the Life Course

Attention to globalization in aging research can enrich perspectives guiding theory by directing attention to neglected phenomena. Attending to globalization requires going beyond a comparative perspective to address how social institutions and the movement of people, products, and ideas across nations influence the life course and aging (Marshall 2011; Phillipson and Vincent 2007).

A political economy of aging and the life course in this era must extend beyond national boundaries or comparative analysis to develop theory that addresses complex linkages of people and social institutions globally (Walker 2005; Walker and Deacon 2003; Vincent 2006). The concept of globalization and its effects have been addressed over the working life course in an international study of information technology firms (McMullin and Marshall 2010). Estes (2006) describes the emergence of "global families," with generations separated and moving back and forth between the developed and less developed nations, accompanied by the economic marginalization of women and the emergence of international caregiving networks.

Where Does Theory Need to Go?

In the past decade, there has been considerable progress in the development of theory to help understand social processes related to aging. This is reflected in the increased use of theory in journal publications recently. For example, in the *Journal of Gerontology: Social Sciences*, the rate of theory-related articles grew from 20% of the papers published between 1990 and 1994 to 47% published from 2000 to 2004 (Alley et al. 2010). In light of the theoretical progress summarized in this chapter, we suggest five issues worth addressing in future theory development.

First, we need more explanatory theory. We have much research that is descriptive and much less that is confirmative or theory-supportive. Theories are explanations, descriptions of general causal relationships among phenomena of interest to us. The rigor and maturity of a scholarly or scientific field can be judged by the strength of its theories. Mayer (2009:423) suggests that "In contrast to other disciplines dealing with human lives, life course sociology lacks a coherent body of theory." This has been a consistent criticism of our field since the 1960s. By contrast, Mayer notes the fields of economics, life span psychology, and biology integrate coherent theory with the life course and aging perspectives (e.g., the life cycle theory of consumption and savings; Baltes's Selection, Optimization and Compensation psychological theory of aging; biological theories of cellular maturation and senescence). Mayer (2009:423) states the dilemma: "Indeed, because there is not just one mechanism underlying the social structuring of human lives but rather manifold mechanisms, one might contend that a simple, unified sociological theory of the life course is not possible at all. Nonetheless, I claim that the lack of explanatory theory(ies) accounts for much of the mostly descriptive or illustrative rather than confirmative nature of life course research."

Second, researchers should pay more attention to previous theorizing lest they repeat the mistakes of the past or attempt to reinvent the wheel. As sociologists, we and our students are trained to be highly critical, but we would benefit from not rejecting out of hand insights from earlier theoretical approaches, as it often seems fashionable to do (structural–functionalism being an example).

A genuinely dialectical approach will distill the best and most productive insights from both the thesis and the antithesis into a synthesis. To remain ignorant of past theorizing may waste precious scientific resources. In Rowe and Kahn's (1998) earlier-mentioned widely-cited prescription for "successful aging," – that keeping active is what propels individuals into a trajectory of a satisfying old age – the authors seemed unaware of the research tradition three decades earlier questioning the linkage between such activity and a "good old age," or of evidence supporting the competing theory, disengagement, as one of several "styles" of aging, nor did they benefit from insights of political economy theorists concerning the social constraints that block some elders from engaging in the activities they prescribed. Talent and energy would better be spent refining and redeveloping, rather than ignoring or rejecting, theoretical perspectives that can help us understand the current social and technological climate of aging.

Third, we will need to foster more interdisciplinary research and theory. The conclusion of the recent decade review of theory development in the biology, psychology, sociology, social policy, and practice of aging is that the most striking theoretical trend since the publication of the previous edition of the *Handbook of Theories of Aging* is the development of interdisciplinary theories of aging: " Despite the difficulties in bridging traditional disciplinary boundaries and despite the challenges of working with different research paradigms, there have been significant breakthroughs in explanations of aging that take approaches from several disciplinary perspectives and blend them together into a unified theory" (Bengtson et al. 2009:6). Unfortunately, too much current theorizing in the sociology of aging is still insular, with sociologists reluctant to venture beyond their comfort zone. Disciplinary boundaries have increasingly become barriers to aging and life course theorizing. This is ironic, since from its earliest beginnings both gerontologists and life course scholars drew not only on sociological concepts but also on anthropological constructs of age grading, psychological perspectives on developmental stages, and biological theories of epigenetic development throughout the life cycle. Aging and life course scholars are well positioned to cross over disciplinary boundaries and profit from the proliferating knowledge about aging mechanisms from other disciplines.

Fourth, as theorizing in the sociology of aging progresses, we should take care that the predominance of the life course perspective does not obscure our vision, preventing us seeing the value of different perspectives – for example, continuing research on the social institutions that create opportunities and constraints for people's lives. These social institutions need to be understood in terms of their own dynamics, and trying to squeeze theorizing into the life course perspective might actually limit our vision (Marshall 1995).

Fifth, we should continue awareness of diversity in social aspects of aging. Early contributions toward this goal were to highlight diversity on the basis of age, cohort, and generation, which in turn emphasized the diversity of history and of place, and then to recognize diversity across age strata. Feminist theorists prodded researchers to take not only gender differences but also gender relations seriously, and this mission has expanded to a recognition that we need to simultaneously consider class, age, gender, race and ethnicity, place, and other social factors as they shape individual life courses (McMullin 1995; 2000).

Finally, we should keep ourselves and our students conscious of the fact that there *has* been important theory development in the relatively few years since the beginning of the sociology of aging; moreover, the pace of theory development has increased in the past decade. Sociological perspectives and theories of aging – many of them insightful, many of them generative, and a few of them not – have considerably helped our understanding of social processes related to age and aging at a time when aging societies across the world are in great need of such knowledge. We need more efforts directed at understanding the how and the why underlying social aspects of aging.

In this chapter, we have noted some promising first steps in developing theory concerning a number of emerging issues: the accumulation of advantage and disadvantage; theorizing the life course in terms of risk and chance events; the social psychological balancing act between social constraints and individual initiative; globalization; generational and family negotiations with aging and the succession of cohorts. This is a rich agenda for future research; but the collection of information – data – about these important issues will not add much to our deeper knowledge unless we pay attention to the development of *theory* about social processes and the course of life to help us understand the data we accumulate.

References

Abeles, Ronald P. and Matilda White Riley. 1977. "A Life-Course Perspective on the Later Years of Life: Some Implications for Research." *Social Science Research Council Annual Report* 1976–77. (Reprinted pamphlet in author's files).

Aboderin, Isabella. 2007. "Modernization and Economic Strain: The Impact of Social Change on Material Family Support for Older People in Ghana." Pp. 284–302 in *Global Aging and Challenges to Families*, edited by V. L. Bengtson and A. Lowenstein. New York: Aldine De Gruyter.

Achenbaum, W. Andrew. 2009. "A Metahistorical Perspective on Theories of Aging." Pp. 25–38 in *Handbook of Theories of Aging* 2nd ed., edited by V. L. Bengtson, D. Gans, N. M. Putney, and M. Silverstein. New York: Springer.

Alley, Dawn H., Putney, Norella, Rice, Melissa & Vern L. Bengtson. 2010. "The Increasing Use of Theory in Social Gerontology: 1990–2004." *Journals of Gerontology: Social Sciences* 65:583–90.

Alwin, Duane F. 1995. Taking Time Seriously: Studying Social Change, Social Structure, and Human Lives. Pp. 211–62 in *Examining Lives in Context. Perspectives on the Ecology of Human Development*, edited by P. Moen, G. H. Elder, Jr., & K. Luescher. Washington D.C.: American Psychological Association.

Atchley, Robert C. 1971. "Aging and Leisure Participation: Continuity or Crisis?" *The Gerontologist* 11(1):13–17.

Baars, Jan. 2006. "Beyond Neomodernism, Antimodernism, and Postmodernism: Basic Categories of Critical Gerontology." Pp. 94–121 in *Aging, Globalization, and Inequality: The New Critical Gerontology*, edited by J. Baars, D. Dannefer, C. Phillipson, and A. Walker. Amityville, NY: Baywood.

Bengtson, V. L. 2001. "Beyond the Nuclear Family: The Increasing Importance of Multigenerational Relationships in American Society." *Journal of Marriage and Family* 63(1):1–16.

Bengtson, Vern L., and Katherine R. Allen. 1993. "The Life Course Perspective Applied to Families over Time." Pp. 469–98 in *Sourcebook of Family Theories and Methods: A Contextual Approach*, edited by P. Boss, W. J. Doherty, R. LaRossa, W. R. Schumm, and S. K. Steinmetz. New York: Plenum.

Bengtson, Vern L., Timothy J. Biblarz, And Robert E. Lee. Roberts. 2002. *How Families Still Matter: A Longitudinal Study Of Youth In Two Generations*. New York: Cambridge University Press.

Bengtson, Vern L., Elizabeth O. Burgess, and Tonya M. Parrott. 1997. "Theory, Explanation, and a Third Generation of Theoretical Development in Social Gerontology." *Journal of Gerontology: Social Sciences* 52B:S72–88.

Bengtson, Vern L., Casey Copen, Norella Putney and Merril Silverstein. 2008. "The Religious Influence of Grandparents on Grandchildren." *International Journal of Sociology* 27:136–49.

Bengtson, Vern L., Neal E. Cutler, David J. Mangen, and Victor W. Marshall. 1985. "Generations, Cohorts, and Relations between Age Groups." Pp. 304–38 in *Handbook of Aging and the Social Sciences*. 2nd ed., edited by R. H. Binstock and E. Shanas. New York: Van Nostrand Reinhold.

Bengtson, Vern L., Glen H. Elder Jr., and Norella M. Putney. 2005. "The Life Course Perspective on Aging: Linked Lives, Timing, and History." Pp. 329–47 in *The Cambridge Handbook of Age and Ageing*, edited by M. Johnson, V. L. Bengtson, P. Coleman, and T. Kirkwood. Cambridge: Cambridge University Press.

Bengtson, Vern L., Daphna Gans, Norella Putney, and Merril Silverstein, eds. 2009. *Handbook of Theories of Aging*. New York: Springer.

Bengtson, Vern L., and Ariela Lowenstein. 2006. *Global Aging and the Challenge to Families*. New York: Aldine de Gruyter.

Berger, Peter L., and Thomas Luckmann. 1966. *The Social Construction of Reality*. New York: Doubleday.

Blossfeld, Hans-Peter and Karl Ulrich Mayer. 1988. "Labor Market Segmentation in the Federal Republic of Germany: An Empirical Study of Segmentation Theories from a Life Course Perspective." *European Sociological Review* 4(2):123–40.

Bruckner, Hannah and Karl Ulrich Mayer. 2005. "De-standardization of the Life Course: What it Might Mean? And if it Means Anything, Whether it Actually Took Place?" *Advances in Life Course Research* 9:27–53.

Burgess, Ernest W. 1960. "Aging in Western Culture." Pp. 3–38 in *Aging in Western Societies*, edited by E. W. Burgess. Chicago and London: The University of Chicago Press.

Cahill, Kevin E., Michael D. Giandrea, and Joseph F. Quinn. 2006. "Retirement Patterns from Career Employment." *The Gerontologist* 46:514–23.

Cain, Leonard D., Jr. 1964. "Life Course and Social Structure." Pp. 272–309 in *Handbook of Modern Sociology*, edited by R. E. L. Faris. Chicago: Rand McNally.

Carr, Debora. 1966. "Two Paths to Self-Employment." *Work and Occupations* 23(1):26–53.
Chappell, Neena L., and Harold L. Orbach. 1986. "Socialization in Old Age: A Median Perspective." Pp. 75–106 in *Later Life: The Social Psychology of Aging*, edited by V. W. Marshall. Beverly Hills: Sage.
Clarke, Philippa J., and Blaire Wheaton. 2005. "Mapping Social Context on Mental Health Trajectories through Adulthood." *Advances in Life Course Research* 9:269–301.
Connidis, Ingrid A. and Candace Kemp. 2008. "Negotiating Actual and Anticipated Parental Support: Multiple Sibling Voices in Three-generation Families." *Journal of Aging Studies* 22(3):229–38.
Cowgill, Donald O. 1974. "Aging and modernization: A Revision of the Theory." Pp. 123–45 in *Late Life: Communities and Environmental Policy*, edited by J. F. Gubrium. Springfield, IL: Charles C. Thomas.
Cowgill, Donald O. and Lowell H. Holmes. 1972. *Aging and Modernization*. New York: Appleton-Century-Crofts.
Cumming, Elaine and William Henry. 1961. *Growing Old: The Process of Disengagement*. New York: Basic Books.
Dannefer, Dale. 1987. "Aging as Intracohort Differentiation: Accentuation, the Matthew Effect, and the Life Course." *Sociological Forum* 2:211–36.
———. 1989. "Human Action and its Place in Theories of Aging." *Journal of Aging Studies* 3(1):1–20.
———. 2003. "Cumulative Advantage/Disadvantage and the Life Course: Cross-fertilizing Age and the Social Science Theory." *Journal of Gerontology: Social Sciences* 58:S327–37.
DiPrete, Thomas A. 2002. "Life Course Risks, Mobility Regimes, and Mobility Consequences: A Comparison of Sweden, Germany, and the United States." *American Journal of Sociology* 108(2):267–309.
DiPrete, Thomas A. and Gregory M. Eirich. 2006. "Cumulative Advantage as a Mechanism for Inequality: A Review of Theoretical and Empirical Developments." *Annual Review of Sociology* 32:271–97.
Dowd, James J. 1987. "The Reification of Age: Age Stratification Theory and the Passing of the Autonomous Subject." *Journal of Aging Studies* 1(4):317–35.
Elder, Glen H., Jr. 1974. *Children of the Great Depression*, 2nd ed. Chicago: The University of Chicago Press.
———. 1985. "Perspectives on the Life Course." Pp. 147–69 in *Life Course Dynamics: Trajectories and Transitions, 1968–1980*, edited by G. H. Elder, Jr. Ithaca: Cornell University Press.
———. 1995. The life course paradigm: Social change and individual development. Pp. 101–36 in *Examining Lives in Context: Perspectives on the Ecology of Human Development*, edited by P. Moen, G.H. Elder, Jr., and K. Luscher. Washington DC: American Psychological Association.
Estes, Carroll L. 1979. *The Aging Enterprise*. San Francisco: Jossey-Bass.
———. 2006. "Critical Feminist Perspectives, Aging, and Social Policy." Pp. 81–101 in *Aging, Globalization, and Inequality: The New Critical Gerontology* edited by J. Baars, D. Dannefer, C. Phillipson, and A. Walker. Amityville, NY: Baywood.
Ferraro, Kenneth F., Tetyana Pylypiv Shippee, and Markus H. Shafer. 2009. "Cumulative Inequality Theory for Research in Aging and the Life Course." Pp. 413–33 in *Handbook of Theories of Aging*, 2nd ed., edited by V. L. Bengtson, D. Gans, N. M. Putney, and M. Silverstein. New York: Springer.
Ferraro, Kenneth F., & Tetyana Pylypiv Shippee. 2009. Aging and cumulative inequality: How does inequality get under the skin? *The Gerontologist 49* (3): 333–343.
Giddens, Anthony. 1991. *Modernity and Self-Identity: Self and Society in the Late Modern Age*. Stanford, CA: Stanford University Press.
Gubrium, Jaber F. (1975). *Living and Dying at Murray Manor*. Charlottesville: University Press of Virginia.
Guillemard, Anne-Marie. 1983. *Old Age and the Welfare State*. Beverly Hills & London: Sage.
Guillemard, Anne-Marie and Martin Rein. 1993. "Comparative Patterns of Retirement: Recent Trends in Developed Societies." *Annual Review of Sociology* 19:469–503.
Handel, Gerald. 2000. *Making a Life in Yorkville: Experience and Meaning in the Life-Course Narrative of an Urban Working-Class Man*. Westport, CT: Greenwood Press.
Hardy, Melissa. 2006. "Older workers." Pp. 201–18 in *Handbook of Aging and the Social Sciences*,6th ed., edited by R. H. Binstock, Linda K. George, and Associates. Amsterdam: Elsevier Academic.
Hayward, Mark D. and Bridget K. Gorman. 2004. "The Long Arm of Childhood: The Influence of Early-life Social Conditions on Men's Mortality." *Demography* 41:87–107.
Heinz, Walter R., Johannes Huinink, and Ansgar Weymann. 2009. *The Life Course Reader: Individuals and Societies across Time*. Frankfurt and New York: Campus.
Heinz, Walter R. and Victor W. Marshall. 2003. Social Dynamics of the Life Course: Transitions, Institutions, and Interrelations. New York: Aldine de Gruyter.
Hendricks, Jon. 2005. "Moral Economy and Ageing." Pp. 510–17 in *The Cambridge Handbook of Age and Ageing*, edited by M. L. Johnson in association with V. L. Bengtson, P. G. Coleman, and T. B. L. Kirkwood. Cambridge: Cambridge University Press.
Hogan, Dennis P. 1978. "The Variable Order of Events in the Life Course." *American Sociological Review* 43(August):573–86.
Hogan, Dennis P. and Frances K. Goldscheider. 2003. "Success and Challenge in Demographic Studies of the Life Course." Pp. 681–91 in *Handbook of the Life Course*, edited by J. T. Mortimer and M. J. Shanahan. New York: Kluwer Academic/Plenum.

Holmes, Thomas H., and Richard H. Rahe. 1967. "Social Readjustment Rating Scale." *Journal of Psychosomatic Research* 11:213–18.

Holstein, James A. and Jaber F. Gubrium. 2007. "Constructionist Perspectives on the Life Course." *Sociology Compass* 1(1):335–52.

Kertzer, David I. 1983. "Generation as a Sociological Problem." *Annual Review of Sociology* 9:125–49.

King, Valerie. 2003. "The Legacy of a Grandparent's Divorce: Consequences for Ties between Grandparents and Grandchildren." *Journal of Marriage and Family* 65:170–83.

Kohli, Martin. 1986. "The World We Forgot: A Historical Review of the Life Course." Pp. 271–303 in *Later Life: The Social Psychology of Aging*, edited by V.W. Marshall. Beverly Hills: Sage.

Kohli, Martin, Martin Rein, Anne-Marie Guillemard, and Herman van Gunsteren. 1991. *Time for Retirement: Comparative Studies of Early Exit from the Labor Force.* Cambridge: Cambridge University Press.

Kuypers, Joseph A., and Vern L. Bengtson. 1973. Social breakdown and competence: A model of normal aging. *Human Development 16* (3): 181–201.

Leisering, Lutz and S. Leibfried. 1999. *Time and Poverty in Western Welfare States*. Cambridge: Cambridge University Press.

Lemon, Bruce W., Vern L. Bengtson, and James A. Peterson. 1972. "An Exploration of the Activity Theory of Aging: Activity Types and Life Satisfaction among In-movers to a Retirement Community." *Journal of Gerontology* 27:511–23.

Longino, Charles H. Jr. and Cary S. Cart. 1982. "Explicating Activity Theory: A Formal Replication." *Journal of Gerontology* 37:713–22.

Macmillan, Ross. 2005. "The Structure of the Life Course: Classic Issues and Current Controversies." *Advances in Life Course Research* 9:3–24.

Maddox, George L. 1965. "Fact and Artifact: Evidence Bearing on Disengagement Theory." *Human Development* 8:117–30.

———. 1979. "Sociology of later life." *Annual Review of Sociology* 5:113–35.

Maddox, George L. And James Wiley. 1976. "Scope, Concepts And Methods In The Study Of Aging." Pp. 3–34 In *Handbook Of Aging And The Social Sciences*, Edited By R. H. Binstock And E. Shanas. New York: Van Nostrand Reinhold.

Mangan, David J., Vern L. Bengtson, and Pierre H. Landry, Jr., eds. 1988. *The Measurement of Intergenerational Relations.* Beverly Hills, CA: Sage.

Marshall, Victor W. 1978–9. "No Exit: A Symbolic Interactionist Perspective on Aging." *International Journal of Aging and Human Development* 9(4):345–58.

———. 1983. "Generations, Age Groups and Cohorts: Conceptual Distinctions." *Canadian Journal on Aging* 2(2):51–61.

———. 1986. "Dominant and Emerging Paradigms in the Social Psychology of Aging." Pp. 9–31 in *Later Life: The Social Psychology of Aging*, edited by V. W. Marshall. Beverly Hills: Sage.

———. 1995. "Social Models of Aging." *Canadian Journal on Aging* 14(1):12–34.

———. 1996 "The State of Theory in Aging and the Social Sciences." Pp. 12–30 in *Handbook of Aging and the Social Sciences*, 4th ed., edited by R. H. Binstock and G. L. Maddox. San Diego, CA: Academic.

———. 1999. "Analyzing Social Theories of Aging." Pp. 434–55 in *Handbook of Theories of Aging*, edited by V. L. Bengtson and K. W. Schaie. New York: Springer.

———. 2005. "Agency, Events, and Structure at the End of the Life Course." Pp. 57–91 in *Advances in Life Course Research*, edited by P. Ghisletta, J.-M. LeGoff, R. Levy, D. Spini, and E. Widmer. New York: Elsevier.

———. 2009a. "Theory Informing Public Policy: The Life Course Perspective as a Policy Tool." Pp. 573–92 in *Handbook of Theories of Aging*, 2nd ed., edited by V. L. Bengtson, D. Gans, N. M. Putney, and M. Silverstein. New York: Springer.

———. 2009b. "What is *New* about Critical Gerontology?" *Ageing and Society* 29(4):651–2.

———. 2011. "Global Aging and Families: Some Policy Concerns about the Global Aging Perspective." *From Generation to Generation: Continuity and Discontinuity in Aging Families*, edited by M. Silverstein. Baltimore: Johns Hopkins University Press.

Marshall, Victor W. and Philippa J. Clarke. 2007. "Theories of Aging: Social." Pp. 621–30 in *Encyclopedia of Gerontology*, edited by J. E. Birren. Amsterdam: Academic

———. 2010. "Agency and social structure in aging and life course research." In *International Handbook of Social Gerontology*, edited by D. Dannefer and C. Phillipson. London: Sage.

Marshall, Victor W. and Margaret M. Mueller. 2003. "Theoretical Roots of the Life-Course Perspective." Pp. 3–32 in *Social Dynamics of the Life Course*, edited by W. R. Heinz and V. W. Marshall. New York: Aldine De Gruyter.

Marshall, Victor W. and Joseph A. Tindale. 1978–1979. "Notes for a Radical Gerontology." *International Journal of Aging and Human Development* 9(2):163–75.

Matthews, Sarah H. 1979. *The Social World of Old Women: Management of Self-Identity.* Beverly Hills and London: Sage.

Mayer, Karl Ulrich. 2009. "New Directions in Life Course Research." *Annual Review of Sociology* 35:413–33.

McMullin, Julie A. 1995. "Theorizing Age and Gender Relations." Pp.30–41 in *Connecting Gender and Ageing. A Sociological Approach*, edited by S. Arber and J. Ginn. Buckinham, UK and Philadelphia, USA: Open University Press.

———. 2000. "Diversity and the State of Gerontological Ageing Theory." *The Gerontologist* 40(5):517–30.

McMullin, Julie. A, and Victor W. Marshall. 2010. *Aging and Working in the New Economy: Careers and Changing Structures in Small and Medium Size Information Technology Firms*. Camberly: Edward Elgar.

Merton, Robert K. 1968. The Matthew effect in science: The reward and communication systems of science are considered. *Science* 159: 56–63.

Minkler, Meredith and Carroll L. Estes. 1999. *Critical Gerontology: Perspectives from Political and Moral Economy*. Amityville, NY: Baywood.

Moen, Phyllis, Stephen Sweet, and Raymond Swisher. 2005. "Embedded Career Clocks: The Case of Retirement Planning." *Advances in Life Course Research* 9:237–65.

Mueller, Mary, Bernard Wilhelm, and Glen H. Elder Jr. 2002. "Variations in Grandparenting." *Research on Aging* 24:360–88.

Myles, John F. 1984. *Old Age in the Welfare State: The Political Economy of Public Pensions*. Boston: Little, Brown.

Myles, John F. and Jill Quadagno. 1991. *States, Labor Markets, and the Future of Old-Age Policy*. Philadelphia: Temple University Press.

Myles, John and Debra Street. 1995. "Should the Economic Life Course be Redesigned? Old Age Security in a Time of Transition." *Canadian Journal on Aging* 14(2):335–59.

Neugarten, Bernice L. 1974. "Age Groups in American Society and the Rise of the Young-old." *Annals of the American Academy of Political and Social Science* 415(September):187–98.

Neugarten, Bernie L., Joan W. Moore, and John C. Lowe. 1965. "Age Norms, Age Constraints, and Adult Socialization." *American Journal of Sociology* 70(6):710–17.

Newman, Katherine S. 2003. *A Different Shade of Gray: Midlife and Beyond in the Inner City*. New York: The New Press.

O'Rand, Angela M. 1996. "The Precious and the Precocious: Understanding Cumulative Disadvantage and Cumulative Advantage over the Life Course." *The Gerontologist* 36:230–38.

———. 2003. The future of the life course. Late modernity and life course risks. Ch. 33, pp. 693–701 in Jeylan T. Mortimer & Michael J. Shanahan (Eds.), *Handbook of the Life Course*. New York: Kluwer Academic/Plenum Publishers.

———. 2003. "Cumulate Advantage Theory in Life Course Research." *Annual Review of Gerontology and Geriatrics* 22:14–30.

———. 2006. "Stratification and the Life Course: Life Course Capital, Life Course Risks, and Social Inequality." Pp. 146–62 in *Handbook of Aging and the Social Sciences*, 6th ed., edited by R. H. Binstock and L. K. George, Boston: Academic.

O'Rand, Angela M., and Jennifer Hamil-Luker. 2005. "Processes of Cumulative Adversity Linking Childhood Disadvantage to Increased Risk of Health Attack across the Life Course." *Special issue of Journal of Gerontology* B 60:117–24.

O'Rand, Angela M., and John C. Henretta. 1999. *Age and Inequality: Diverse Pathways Through Later Life*. Boulder, CO: Westview Press.

Orbach, Harold. L. 1974. "The Disengagement Theory of Aging, 1960–1970: A Case Study of a Scientific Controversy." Unpublished doctoral dissertation, University of Michigan.

Palmore, Erdman B. 1968. "The Effects of Aging on Activities and Attitudes." *The Gerontologist* 8:259–63.

Parsons, Talcott. 1951. *The Social System*. New York: The Free Press of Glencoe.

Passuth, Patricia M. and Vern. L. Bengtson (1988). "Sociological Theories of Aging: Current Perspectives and Future Directions." Pp. 333–55 in *Emergent Theories of Aging*, edited by J. E. Birren and V. L. Bengtson. New York: Springer.

Pearlin, Leonard I. 1982. Discontinuities in the study of aging. Pp. 55–74 in Tamara K.Hareven & Kathleen J. Adams (Eds.), *Aging and Life Course Transitions*. New York: The Guilford Press.

———. 1980. The life cycle and life strains. Ch. 28, pp. 349–360 in Hubert M. Blalock (Ed.) Sociological Theory and Research: *A Critical Approach*. New York: The Free Press.

Pearlin, Leonard, I. and Marilyn M. Skaff. 1996. "Stress and the Life Course: A Paradigmatic Alliance." *Gerontologist* 36:239–47.

Phillipson, Chris and Jason L. Powell. 2004. "Risk, Social Welfare and Old Age." In *Old Age and Human Agency*, edited by E. Tulle. Hauppauge NY: Nova Science.

Phillipson, Chris and John Vincent. 2007. "Globalization and Aging." Pp. 630–35 in *Encyclopedia of Gerontology*, 2nd ed., edited by J. E. Birren. Amsterdam: Elsevier.

Pinquart, Peter and Stephen Sorensen. 2003. "Associations of Stressors and Uplifts with Caregiver Burden and Depressive Mood: A Meta-Analysis." *Journals of Gerontology: Psychological Sciences* 58B(2):P112–28.

Quadagno, Jill. 1982. *Aging in Early Industrial Society: Work, Family and Social Policy in Nineteenth-Century England.* New York: Academic.

Riley, Matilda White. 1987. "On the Significance of Age in Sociology." *American Sociological Review* 52(February):1–14.

Riley, Matilda White, Marilyn Johnson, and Anne Foner. 1972. *Aging and Society, Vol. 3: A Sociology of Age Stratification.* New York: Russell Sage Foundation.

Rindfuss, Ronald R., C. Gray Swicegood, and Rachel A. Rosenfeld. 1987. "Disorder in the Life Course: How Common and Does it Matter?" *American Sociological Review* 52(6):785–801.

Roberto, James and William Jarrott. 2008. "Caregiver Burdens and Uplifts." *Journal of Gerontology: Psychological Sciences* 63P:127–69.

Roberto, Karen A., Allen, Katherine R. and Rosemary Blieszner. 2001. "Grandfathers' Perceptions and Expectations of Relationships with Their Adult Children." *Journal of Family Issues* 57:100–11.

Rosow, Irving. 1974. *Socialization to Old Age.* Berkeley: The University of California Press.

Rossi, Alice S. and Peter H. Rossi. 1990. *Of Human Bonding: Parent-child Relationships across the Life Course.* Hawthorne, New York: Aldine De Gruyter.

Rowe, John W. and Robert L. Kahn. 1998. *Successful Aging.* New York: Pantheon Books.

Ryder, Norman B. 1965. "The Cohort as a Concept in the Study of Social Change." *American Sociological Review* 30(6):843–61.

Schmeeckle, Maria and Vern L. Bengtson. 1999. "Successful Aging? Review of a prescription by Rowe and Kahn." *Contemporary Gerontology* 5(3):86–90.

Schutz, Alfred. 1967 [originally published in German, 1932]. *The Phenomenology of the Social World.* Translated by George Walsh & Frederick Lehnert. Chicago IL: Northwestern University Press.

Settersten, Richard A., Jr., Frank F. Furstenberg Jr., and Ruben G. Rumbaut. 2005. *On the Frontier of Adulthood: Theory, Research, and Public Policy.* Chicago: The University of Chicago Press.

Settersten, Richard A., Jr. and Lynn Gannon. 2005. "Structure, Agency, and the Space Between: On the Challenges and Contradictions of a Blended View of the Life Course." *Advances in Life Course Research* 10:35–55.

Sewell, William A. 1992. "A Theory of Structure: Duality, Agency, and Transformation." *American Journal of Sociology* 98(1):1–29.

Shanahan, Michael J. and Erik J. Porfeli. 2007. "Chance Events in the Life Course." *Advances in Life Course Research* 11:97–109.

Shanas, Ethel, Peter Townsend, D. Wedderburn, H. Friis, P. Milhoj, and J. Stehouwer. 1968. *Old People in Three Industrial Societies.* New York: Atherton.

Sorensen, Aage B. 1986. "Social Structure and Mechanisms of Life-Course Processes." Pp. 177–197 in *Human Development and the Life Course: Multidisciplinary Perspectives*, edited by A. B. Sorensen, F. E. Weinert, and L. R. Sherrod. Hillsdale, NJ: Lawrence Erlbaum.

Spence, Donald L. 1986. "Some contributions of symbolic interaction to the study of growing old." Pp. 107–123 in *Later Life: The Social Psychology of Aging*, edited by V. W. Marshall. Beverly Hills: Sage.

Suzman, Richard and Matilda White Riley. 1985. "Introducing the 'Oldest Old.'" *Milbank Memorial Fund Quarterly* 63(2):177–86.

Townsend, Peter. 1981. "The Structured Dependency of the Elderly: Creation of Social Policy in the Twentieth Century." *Ageing and Society* 1(1):5–28.

Uhlenberg, Peter and Sonia Miner. 1995. "Life Course and Aging: A Cohort Perspective." Pp. 208–223 in *Handbook of Aging and the Social Science*, 4th ed., edited by R. H. Binstock and L. K. George. San Diego: Academic.

Vincent, John A. 2006. "Globalization and Critical Theory: Political Economy of World Population Issues." Pp. 245–276 in *Aging, Globalization and Inequality*, edited by J. Baars, D. Dannefer, C. Phillipson, and A. Walker. Amityville, NY: Baywood.

Walker, Alan. 1981. "Towards a Political Economy of Old Age." *Ageing and Society* 1:73–94.

———. 2005. "Towards an International Political Economy of Ageing." *Ageing and Society* 25:815–96.

Walker, Alan and B. Deacon. 2003. "Economic Globalization and Policies on Ageing." *Journal of Societal and Public Policy* 2(2):1–18.

Wheaton, Blair and Philippa Clarke. 2003. "Space Meets Time: Integrating Temporal and Contextual Influences on Mental Health in Early Adulthood." *American Sociological Review* 68:680–706.

Wilensky, Harold L. 1960. "Work, Careers, and Social Integration." *International Social Science Journal* 12(4):543–60.

Willson, Andrea E. and Kim M. Shuey. 2007. "Cumulative Advantage Processes as Mechanisms of Inequality in Life Course Health." *American Journal of Sociology* 112(6):1886–924.

Chapter 3
Aging Individuals, Families, and Societies: Micro–Meso–Macro Linkages in the Life Course

Merril Silverstein and Roseann Giarrusso

As a conceptual lens with which to view and understand changes over the human life span, the life course perspective has achieved a dominant, some might say near hegemonic, status in social gerontology over the last several decades. It is difficult to overstate the importance of life course concepts to the sociology of aging and the science of human development. In simple terms, the life course framework considers the process of human growth and senescence within historical context, producing unique life experiences and trajectories for different birth cohorts (Elder 1994; Riley and Riley 1994). For its sweeping scope and focus on biographical and historical dynamics, the life course perspective is a powerful tool in the social scientific investigation of aging, but it is not without conceptual and empirical challenges.

The purpose of this chapter is to bring institutional contexts more fully into the purview of the life course framework by developing a dynamic *biographical–institutional–societal* model of the life course that hierarchically links micro, meso, and macro levels of analyses. In our conceptualization, these three temporal metrics simultaneously and interdependently exert a dynamic force on aging individuals. By presenting an empirical example of this model, we hope to increase the likelihood that researchers will consider human development as guided simultaneously by individual, institutional, and societal forces.

We begin the chapter with an overview of the life course perspective in its twin guises as the interplay of individual development with changing historical conditions and institutional structures. Next, we present a model with three levels of analysis that formally incorporates social institutions as pivotal links between individual development and historical conditions. Finally, we discuss a unique methodological design that allows us to test the three-level model by treating the *aging family* as a set of micro-level experiences, a meso-level social institution, and a macro-level societal construct, each fully dynamic with respect to its respective time metric.

Dynamic and Structured Aspects of the Life Course

Underlying the life course perspective is the notion that individuals advance through time in two ways: (1) ontologically through chronological aging and the structured dynamics of development and decline and (2) historically by passing through epochal periods that shape value orientations and give rise to unique opportunities and constraints. When aging and historical time meet in particular configurations – that is, when particular historical periods are experienced

M. Silverstein (✉)
Davis School of Gerontology, University of Southern California,
3715 Mc Clintock Ave., Los Angeles, CA 90089-0191
e-mail: merrils@usc.edu

R.A. Settersten, Jr. and J.L. Angel (eds.), *Handbook of Sociology of Aging*, Handbooks of Sociology and Social Research, DOI 10.1007/978-1-4419-7374-0_3,

at key junctions in life – individuals may bear the indelible imprint of the conditions of those periods in their dispositions and value orientations as well as in their stock of social, health, and economic resources. A fundamental principle of the life course perspective is that exposure of birth cohorts to unique events and social conditions variously constrains or expands opportunities that structure life circumstances and pathways. Ryder's (1965) treatise on inter-cohort change as the engine of societal transformation formalized in earlier discussions by Mannheim (1952) about how "fresh contact" with the world by newly minted cohorts brought about social change through questioning the implicit arrangements and taken-for-granted conventions forged by their predecessors.

Treatment of the role of history in shaping human lives has several traditions in life course scholarship that we divide into three general approaches: those focusing on *great-events*, *cultural Zeitgeists,* and *institutional change*. We discuss each approach in turn below.

Great Events

The life course perspective has gathered some of its most compelling support through evidence showing how human lives are abruptly influenced by great events such as wars and economic depressions. Consideration of cataclysmic historical events of World War II, the Great Depression, and the farm crisis of the 1980s has shed light on how life choices and life chances are structured by far-reaching events (Elder and Conger 2002; Clausen 1993; Conger and Elder 1994). Elder's epic research into the children of the Great Depression (Elder 1974) is a transformative piece in life course scholarship for demonstrating the ripple effect of an economic catastrophe on the life paths of a whole generation in terms of subsequent timing of marriage, family formation, schooling, and careers. The macro-policy environment has been an object of study as well in life course research, such as in investigations of how military service and the periodicity of benefits affect career development and long-term quality of life among veterans (London and Wilmoth 2006; Settersten 2006).

Political transformations provide the opportunity to conduct natural experiments that test how sudden macro-societal change alters the social institutions that govern the metabolism of human lives. Notable in this regard is the fall of the Berlin Wall in 1989 and the reunification of Germany, a seminal historical event that obliterated the social order of the former East Germany, shifting schedules of educational achievement, career advancement, family formation, and fertility in subpopulations at risk (Diewald et al. 2006). As a result of reunification, retired East Germans also saw increases to their state pensions that in conjunction with a strong sense of family duty resulted in increased economic transfers from older to younger generations (Kohli 2005).

Cultural Zeitgeist

A second stream of life course scholarship that we label the *cultural Zeitgeist* approach most often focuses on the social climate to which birth cohorts are exposed at critical junctions in their lives. For instance, shifting gender norms and lifestyle preferences may be incorporated by individuals differently depending on when in the lifespan those shifts were encountered. Formally, a cohort-based model focuses on how cohort turnover – the process by which new birth cohorts replace older ones – changes attitudes and behaviors in the population. Key to this logic is that birth cohorts are indelibly shaped in their world views by the historical conditions into which they emerge as newly formed adults. Such a mechanism is suggested in the *impressionable youth hypothesis* positing that fundamental values are absorbed from the wider culture during the early adult years (when world views and preferences are beginning to coalesce) and then remain fairly stable over the remainder of the life span (Alwin et al. 1991). The concept of cohort turnover has become a staple in the study of how social change results from cohort differentiation, and has been related to such diverse outcomes as socio-political orientations (Alwin and Krosnick 1991), gender role attitudes (Brooks and Bolzendahl 2004), crime rates (O'Brien et al. 1999), and perceived age discrimination (Gee et al. 2007).

Another application of the *Zeitgeist* approach traces evolutionary change in society and its influence on the entire population (a "period" effect). In this version, the force of history and aging may be seen as simultaneously and independently influencing individuals as they age across the lifespan. One multi-generational longitudinal study of competing values of individualism and collectivism from the 1970s through the 1990s found that both values strengthened over time. Collectivism strengthened due to chronological aging (family and social ties gaining salience in middle and old age) and individualism strengthened due to advancing historical time (presumably a post-1970s shift in the socio-political climate toward favoring self-interest over altruism) (Roberts and Bengtson 1999). What is clear from the overall evidence is that several time clocks operating at personal and historical levels of analysis guide the pace and junctions of human development.

Institutional Change

An *institutional* approach to the life course stresses how the major institutions of society – social organizations such as the workplace, the educational system, and the family – guide orderly transitions in role and status positions across the lifespan by fostering collective expectations for an orderly life path, and by providing incentives and disincentives for particular actions. When institutional arrangements change or their rules are in flux, new social conventions may emerge as older ones are discarded. For instance, institutional age markers for appropriate role transitions in families are malleable based on the changing normative basis of family life guiding if and when to marry and have children (Bengtson and Allen 1993). Much contemporary commentary has focused on the *deinstitutionalization* of the life course, by which it is generally meant that the coercive power of social institutions to standardize life transitions by age has ebbed, in favor of greater individual discretion and flexibility in the timing of important life transition decisions of marriage, fertility, schooling, work, and retirement (Settersten 2003).

Age constraints on role transitions have weakened leading to a variegated set of lifestyle options available to individuals (Bruckner and Mayer 2005; MacMillan and Copher 2005). A study of family and non-family role configurations in two cohorts of British young adults separated by 12 years found a distinct move away from traditional social arrangements, related to marriage and children, to a diversified set of arrangements that were more focused on career and less on family (Ross et al. 2009). Some social theorists suggest that de-standardization of the life course is the result of economic turbulence and globalization which rendered the normal life course and its institutionalized pathways less certain. In late modern society, labor markets are increasingly fluid and contingent, while individuals have greater freedom but also less certainty (Phillipson 2003). Because of greater difficulty in building a work career with continuity and security, family life and human capital acquisition are also less certain. The timing of individual life course transitions has been affected, as reflected in the increasingly later ages of marriage and childbearing, delayed completion of education, and later ages of retirement (Heinz 2003). That the de-evolution of a single dominant cultural script is predicated on interdependent structural social changes – such as increased economic inequality, entry of women into the labor force, diversity in family structure – lies at the heart of the idea that life courses are institutionally embedded and multi-level in form.

While institutional approaches include cohort-turnover and historical change in their formulations, they tend to focus on social institutions as mediating bridges between macro-historical conditions and individual outcomes, providing explicit middle-range explanations for *how* social change occurs. Institutional approaches are also consistent with ecological orientations toward the life course that conceptualize human lives as hierarchically embedded within families, communities, political regimes, and, the most distal of all, global systems (Bronfenbrenner 1995; Dannefer 2003).

The life course perspective has roots in the institutionally focused conceptual framework developed by Riley and Riley (1994) known as the age and society paradigm. Their model specifies that lifespan development could be understood as the process by which human lives are shaped by the social institutions – family, education, work, and leisure – through which they passed on their developmental journey (Riley et al. 1972). This approach provided a sharp analytic lens to describe how key life transitions are organized by the expectations and opportunities structured into the institutions of society at the particular historical time they are encountered. A key tenet of this perspective is that social institutions, while not static, change at a considerably slower pace than emergent needs and preferences rising out of cohort replacement. In other words, lifestyles transform faster than the institutions meant to accommodate them, a phenomenon known as structural lag (Riley et al. 1994). Such different rates of change between cohort flows and institutional transformation may lead to dilemmas such as workplace polices incapable of accommodating older workers who prefer to remain in the labor force, and family support systems unable to serve elders in divorced and remarried families.

A Dynamic Biographical–Institutional–Societal Model of the Life Course

Using the life course perspective as a sensitizing concept, we propose a *biographical–institutional–societal* model of life span development that hierarchically links three levels of analysis: micro, meso, and macro. Our model resembles Bronfenbrenner's ecological theory (1995) positing that human development unfolds within multiple social spaces and embedded contextual systems that spiral outward from micro to macro-realms. In our conceptualization, three temporal metrics simultaneously and interdependently exert dynamic force on aging individuals through time (see Fig. 3.1).

At the micro-biographical level, we note that physical, mental, social, and financial forms of well-being evolve over biographical time as individuals grow up and grow old through the life-span. At the meso-institutional level, family, education, work, and leisure institutions guide expectations and provide incentive (and disincentive) structures for making (or delaying) life transitions and taking (or not taking) certain pathways. Finally at the macro-societal level, the wider social environment exerts influence over the population by virtue of society-wide norms, economic conditions, public policies, and demographic structures (cataclysmic events such as depressions and wars fit into this category). Three important contributions of the model are that (1) social institutions are considered to be dynamic, evolving with individuals and societies through time and most likely at different rates, (2) each of the three levels is considered mutually interdependent with the other two (for example, changes at the micro-biographical level when aggregated can produce institutional and societal change), and (3) substantive areas within each level interact to produce unique outcomes (for example, at the meso-institutional level, incompatibility between work and family institutions may result in individual manifestations of stress).

As with many conceptual schemes with a multiplicity of inputs, the three-level model presented in Fig. 3.1 requires quite elaborate data to empirically articulate. Few, if any, studies are

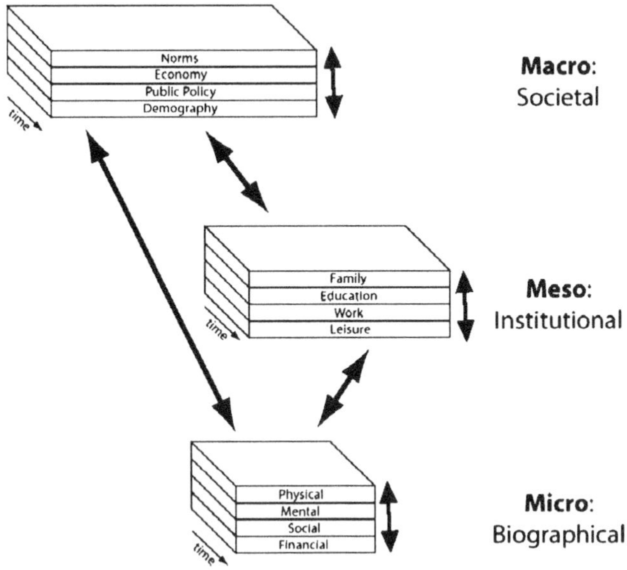

Fig. 3.1 Dynamic biographical–institutional–societal model of the life course

capable of providing a test of such a model; most studies only examine two levels of the model. Therefore, before describing a unique research design and data set capable of investigating this three-level model, we briefly review an example of recent empirical work that assesses one of the two levels of the model: macro–micro, macro–meso, and meso–micro.

Multi-Level Life Course Approaches

Macro-Level Effects

The widest distance across levels is that between national state policies and the manner in which intimate social relationships are maintained. The case has been made that cultural styles of interaction, national in character, map onto the social welfare gradient (Turner 1988). Using a more structural orientation, Mayer (2009) writes that "progress may be achieved by…moving away from developmental origins towards a specific understanding of life courses as institutionally embedded purposive action," which, he adds in the European tradition, has focused on how welfare state policies have shaped the organization of social life.

At the macro-level the demands placed on families are structured by the generosity of welfare state provisions variably releasing or obligating filial duties of adult children toward older parents and affecting the emotional tenor of their relationships. Micro-interactions in the family may be shaped by the political economies and cultures within which those interactions are embedded, specifically the way in which welfare production is allocated among state, market, and family. A commonly used gradient of nation-states puts *social democratic* nations at one end of the spectrum (citizens are incorporated under a single universal benefits system) and *residualist* nations at the other end (benefits provided by the state only when personal resources are exhausted), with the middle ground occupied by *liberal-market* nations (benefits are means-tested and modest social insurance plans exist) (Esping-Anderson 1999). Welfare production influences helping behavior, interaction, and coresidence between adult generations with generally weaker filial behavior in the social democratic states of northern Europe than in the residualist states of southern Europe (e.g., Brandt et al. 2009; Broese van Groenou et al. 2006; Hank 2007).

Policy changes within a welfare regime may motivate behavioral change at the micro-level. For instance, a study in Sweden found that an increase in the proportion of elders supported by family members followed the restriction of eligibility rules for home-help services (Sundstrom et al. 2002). In another example – but from a less-developed country – Yi (1996) found that fertility rates declined in an area of rural China where an experimental pension program was initiated, concluding that pension availability weakened the belief that having children was necessary to ensure old age security.

In a study of the influence of national membership on intimacy in close personal relationships (arguably representing the largest gap between macro- and micro-environments), Silverstein et al. (2010) examined how state structures influence the emotional quality of older parent–child relations in England, Germany, Israel, Norway, Spain, and the United States. The authors found four relational styles across the six nations: (1) *detached* – low affection and low conflict, (2) *amicable* – high affection and low conflict as, (3) *disharmonious* – low affection and high conflict, and (4) *ambivalent* – high affection and high conflict. Further, the styles were distributed differently across the six nations studied. What macro-level characteristics might explain these national differences? Relationships measured in the United States were more than twice as likely as those in the other national samples to be characterized as *disharmonious* and *detached* – the two relational styles with higher levels of conflict. Compared to its national counterparts, the United States embraces a stronger individualistic ideology with respect to kinship ties and a relatively weak public sector that may serve to obligate children to a degree not seen in the other countries (Hollinger and Haller 1990). The fact that relationships in Israel were more likely than average-to-be *ambivalent* reflects a culture that emphasizes interpersonal engagement and legitimates mild forms of conflict in intimate relationships. The greater prevalence of emotional *detachment* among parent–child relationships in Germany possibly reflects an intergenerational schism rooted in the association of older parents with the National Socialist regime, a hypothesis suggested by Szydlik (1996). The finding that intergenerational relationships in England are more likely than those in other nations to be emotionally close and free of conflict could result from a cultural tendency to inhibit the expression of strong negative emotion (Peabody 1985).

Meso-Level Effects

The family represents an institution that has been in flux over the last several decades with profound implications for how individuals live their daily lives and the resources that are available to them. A large amount of attention has been devoted to tracing the micro-level consequences of divorce and remarriage, and non-marital childbearing, forcing a reconsideration of how families and kinship relations are to be defined. No longer is there a single culturally dominant family pattern, but rather a multiplicity of family and household arrangements whose forms are fluid. Indeed, the very definition of "family" has become ambiguous (Stacey 1991). Schmeeckle et al. (2006), for instance, found substantial cohort variation in the extent to which adult children perceived their parents, step-parents, and former step-parents as full family members. These findings have implications for the caregiving of aging stepparents by adult stepchildren in cohorts where family structures have increasingly diversified and become more complex.

As we discussed earlier, institutional domains are rarely independent and it is often unrealistic to partition an institution as if it were isolated from the others in society. The institutions of family and work, for instance, mutually interact but also compete with each other for the time and energy of their constituent members. Within a multi-level perspective the protections and benefits afforded workers (e.g., equal opportunity laws, family-leave policies), and macro-economic conditions can be seen to influence interfaces between the institutions of work and family life.

One of the greatest social changes of the last half-century has been the increase of women and mothers in the paid labor force (Sayer et al. 2005). While this trend is undeniably favorable for women and the financial well-being of families, it also means that there are fewer family members available for care of family dependents such as the very young and the very old. Changes to the economy – particularly stagnation in real wages since the 1970s and relative growth of the service economy – have increased economic uncertainty in many families, consequently increasing the amount of time devoted to labor market activities. These changes have made it increasingly more difficult for families to simultaneously meet the demands of work and the obligations of family life (Moen 2003).

Empirical Test of a Biographical–Institutional–Societal Model of the Life Course

Our multi-level approach to human development presented in Fig. 3.1 takes its cue from versions of the life course that speak to historical events and epochs, as well as social institutions in structuring individual outcomes. In this section, we use intergenerational family relations as our social realm of interest and present an empirical example to explore the linkage between meso- and macro-realms on emotional cohesion between older parents and their adult children. We examine long-term change in the strength of emotional cohesion between generations as a function of changing family structures at the institutional level, and weakening strength of filial obligation at the societal level.

We rely on the dialectic between individual and contextual characteristics to examine how intergenerational relationships are shaped by the historical epochs within which they are embedded. In our empirical example we focus on two fundamental changes in families over the last few decades: (1) change at the meso-level in family structure as a result of increases in divorce, remarriage, and step-family formation and (2) change at the macro-level in norms of filial responsibility that reflect a societal drift from values of collectivism to those of individualism. We review these trends briefly below.

Change in Family Structures (Meso-Level)

Increases in marital disruption and remarriage since the 1970s have introduced uncertainties about the quality of intergenerational relations and have cast doubt on whether adult children will continue to serve as resources to their aging parents. The structure of the family within which a child is raised defines a host of social, economic, and psychological factors that set a standard for how intergenerational relations are maintained through the life span. Non-traditional family structures may adversely affect the quality of intergenerational relationships by engendering interpersonal strains with divorced parents and step parents (Amato 1994; Cherlin 1978) that may put children at risk of having distant relations with them into adulthood (Aquilino 1994; Furstenberg and Cherlin 1991). Evidence is accumulating that the legacy of divorce and remarriage extends to later life, suppressing the economic and instrumental support exchanged between adult children and their divorced parents, stepparents, and remarried biological parents (Aquilino 2005; Ganong and Coleman 1999; Pezzin and Schone 1999). Yet cross-generational studies capable of examining the historical effects of the divorce-revolution are rare, as almost all studies to date have relied on inferences about divorce and remarriage that have occurred in a single generation.

Weakening Norms of Filial Responsibility (Macro-Level)

Filial responsibility as a social norm reflects the generalized expectation that family members should be central in each other's lives and perform particular functions with respect to each other. Much has been written about the declining importance of the family in American society over recent decades. Social scientists, historians, and cultural critics have noted that values of collectivism – including those of familism – have been overtaken by values of individualism (Bellah et al. 1985; Popenoe 1993; Putnam 1995). Gans and Silverstein (2006), for instance, found that the strength of parent-care norms weakened between the 1980s and 1990s, a trend that is not disconnected from marital instability. Some research suggests that children of divorcees feel less filial obligation to their aging parents than those who have not experienced a marital disruption (Ganong and Coleman 1999).

The Generational–Sequential Design

The superficial simplicity of the model in Fig. 3.1 belies the complexity of its underlying assumption that individuals may age differently depending on *the historical time and personal age when the institutions are experienced.* Timing of exposure to history is a key feature of our rendering of the life course paradigm, but cross-generational comparisons are difficult because generations pass asynchronously through historical time such that earlier and later generations experience the same historical periods at different life stages, or age through the same life stages in different historical periods. To remedy this difficulty, we advance a *generational–sequential design* that examines parents and their adult children at the same chronological age but in different historical periods, essentially standardizing age of exposure to different macro-institutional conditions. An example of this design is presented in Fig. 3.2 where parents and their adult children are represented as arrows being

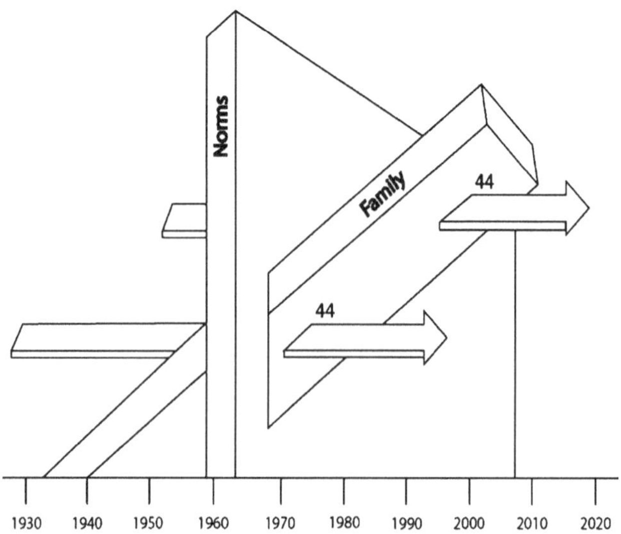

Fig. 3.2 Generation-sequential design: examines parents and adult children at the same chronological age (44 years) but in different historical periods (1970 vs. 1997)

exposed to different macro-institutional conditions (the intersection of changing norms of filial responsibility and changing family structures) at the same chronological age (44 years) but in different historical periods (1970 and 1997, respectively).

Data Set

To examine cross-cohort variation in intergenerational relations we use data from the Longitudinal Study of Generations (LSOG) (see Bengtson 1975). The LSOG began in 1971 with a survey of 2044 respondents from 328 three-generation families living in southern California and including grandparents (G1s), parents (G2), and adult children (G3). Beginning in 1985, follow-up surveys were administered every 3–5 years until 2005.

The LSOG is unique in that it follows individuals over biographical time, includes family members in linked generations, and follows them over a period of history when many of the social changes affecting families occurred. Combining these special design features leads to an important feature of the LSOG: the capacity to compare human development of adjacent generations within the same family lineages and at the same ages but under different socio-historical conditions. With data collected over three decades we provide an empirical example that examines age-standardized historical differences between generations in the strength of emotional cohesion with aging parents, as well as potential explanations for observed differences based on several social trends over the period of study.

The LSOG due to its unique multi-generational design and long time horizon allows such cross-generational comparisons. Unlike typical cohort models, the generational–sequential design is capable of accounting for the fact that individuals are intimately tied to members of other cohorts through common family membership. Approaches that use repeated cross-sectional data are highly useful for separating aging and cohort effects in *aggregates* of individuals, but are not fully able to investigate human development as an intra-familial process. A core principle of the life course paradigm is that human lives are interdependent with each other particularly as they are linked by family bonds.

Methodological Considerations

There are not insignificant practical difficulties when life course propositions are tested empirically. Attributing individual phenomena to the character of historical periods is a well-known challenge in contextual analyses given the myriad number of ways that contexts differ from one another and the often large amount of heterogeneity within them; data demands may be quite substantial.

In approaching our topic, we rely on the principle of *methodological individualism* – the idea that contexts may be represented by the aggregation of their constituent elements (Weber 1968). Extending (or reversing) this logic, we argue that if the size of contextual variation or change in a social phenomenon is reduced after statistically controlling for individual characteristics, then those characteristics hold the promise of being explanatory factors or causal agents at the contextual level (even though they are measured at the unit level). This logic is implicit in Schooler's (1996) argument that the influence of national context on a seemingly individualized psychological process is identifiable if variables representing alternative theoretical explanations are controlled *at the individual level.* Such an approach also can be used to make cross-cohort comparisons where historical location can be considered a context.

Our empirical model includes 554 G2 parents and their G3 children measured between 1971 and 1997 when the large majority of children reached the age their parents were at baseline. The mean age of generational convergence is 44 years. The outcome variable of interest is emotional closeness with mothers and fathers. To test for historical effects, we match each generational pair so they are aligned at a common age. In this approach, "generation" becomes a dummy variable to assess G3–G2 differences in the strength of intergenerational ties when the generations are approximately at the same life stage: a negative value indicates weaker intergenerational cohesion of G3 children compared to their G2 parents. Estimates are generated using Hierarchical Linear Models that consider age, historical time, and family membership.

Results

We describe the most important results of our analysis in Fig. 3.3. The first set of bars (for relations with mothers and fathers) shows a statistically significant historical difference, with G3s feeling less close to their parents than G2s felt toward *their* parents several decades earlier. In other words, using age-matched controls we found that intergenerational cohesion has weakened over historical time. We then assessed whether this generational difference was mediated by marital disruption in either of the two generations as a manifestation of institutional change in the family. The second set of bars in Fig. 3.3 shows that with marital disruption controlled there is a discernable and significant reduction in the historical loss of intergenerational cohesion, though the decline is still statistically significant. Next we examine the contribution of weakened norms of familism as an element of societal change in the salience of family life measured as the importance of discussing important life decisions with family members, and degree to which personal life-styles choices should conform to family wishes. The third set of bars shows that when norms of familism are controlled, the historical differences between generations shrinks toward zero and loses their statistical significance. Thus, we attribute historical weakening in the strength of emotional bonds between generations primarily to a secular decline in the centrality of family life. Beyond the particulars of our analysis, we demonstrate the potential of a generational–sequential design to isolate historical effects in micro-level data by using age-standardized differences and then attribute these effects to specific institutional and societal forces to which adjacent generations have differential exposure. In doing so, we also intend to alleviate anxiety over the sometimes daunting complexity and theoretical abstractness of the life course perspective – in terms of representing multiple dynamic and hierarchically embedded systems – by connecting it to the more practical realm of operational definitions, empirical modeling, and statistical method.

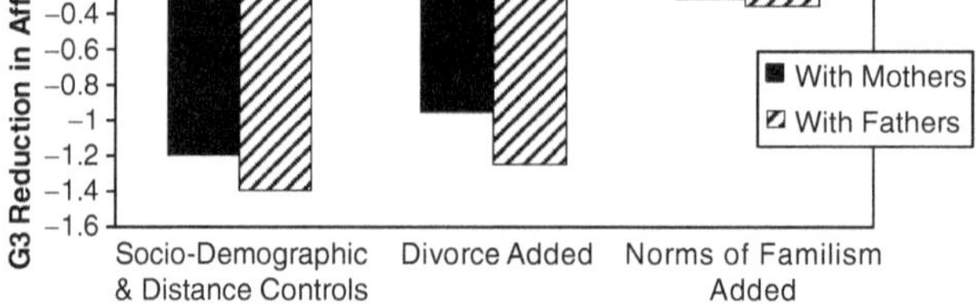

Fig. 3.3 Average G3–G2 difference in affection for parents when generations are age-matched (M=44 years)

Conclusion: The Life Course as a Sensitizing Concept in Multi-Level Studies

In this chapter we considered the life course paradigm in aging research as a system of conditions that includes micro-biographical, meso-institutional, and macro-societal dimensions all moving dynamically through time and subject to change. The study of human development has deep roots in the biographical–historical framework, relying on age-period-cohort models to study the extent to which cohort replacement is responsible for social change and cohort constancy is responsible for social stability. Less considered are the institutional forces that directly influence micro-conditions and that mediate or explain the influence of macro-conditions on human lives. In synthesizing these perspectives – what might be summarized crudely as dynamic and structural perspectives – we developed a fully dynamic model of nested social processes based on the aging of individuals, the reconfiguring of social institutions, and the transformation of societies.

To bridge the division between the theoretical and empirical, we presented an example of research that showed how historically comparative research can gain explanatory leverage by selectively including theoretically relevant variables into predictive models. We employed a generational–sequential model showing that contemporary adult children maintain lower-quality intergenerational relationships when compared to those maintained by *their* parents up to one-quarter century earlier. Mapping marital disruption and norms of familism onto our multi-level model of the life course suggests the importance of institutional-level and societal-level factors as potent historical explanations for the decline in the emotional tone of intergenerational relationships within families.

The term *life course* has been applied in numerous ways to important questions in sociology, social psychology, and gerontology. While many life-course studies employ longitudinal data, a longitudinal design – even if long-term – is not sufficient, but neither is it necessary for research to gain the life course appellation. For the *life course* to have utility as an operational concept – beyond being a general heuristic to describe long-term change – it needs to be distinguished from the many other terms used to describe dynamic systems in the social and behavioral sciences (the most common being life-span development). Here the multi-level distinction may be of use. In our opinion, any formulation of a research question in the life course tradition must at minimum consider how a social context (macro and/or meso) exerts a force on more micro-level outcomes, and at its best treat social contexts as evolving through time.

We also note that the method of data collection is of lesser concern as studies invoking the life course need not be longitudinal. Retrospectively reported data may invoke the life course paradigm when accounting for the influence of divergent social contexts experienced in youth on contemporary adult behaviors (see Axinn and Yabiku 2001 for an excellent example using retrospective reports to examine how dynamic community contexts affect fertility in Nepal; see also Mayer 2007 for an application using retrospective data). A comparison of groups that experienced the same historical conditions but with different types of exposure – such as the disadvantages experienced by aging WW II combat veterans (Elder et al. 1997) – provides purchase on the specific aspects of historical contexts that matter for life trajectories.

In reviewing the literature and looking ahead, we make three general suggestions by which life course research will benefit from considering aging as a multi-level, hierarchical phenomenon and move the “field” of life course studies closer to achieving its promise.

Multi-National Studies

Our impression from reviewing the empirical literature in life course studies is that publications making use of multi-level designs have proliferated, offering insights into how variation in state (or lower level) structures influence life choices and life chances throughout the life span. Multi-national studies, in particular, are essential for documenting the importance of the macro-level environments for micro-level processes and are to be celebrated. However, as a general rule, the larger and more heterogeneous the contexts of interest, the more difficult it is to make valid causal inferences about their influence on life choices and life chances over the life-span. Contextual level variables such as gross domestic product or public spending on dependent populations are certainly useful in documenting how nations differ on dimensions of theoretical interest, but remain essentially proxies for what remains a complex set of pathways leading from welfare regimes to individual outcomes. In addition, the attribution of causal agency to macro-level characteristics is best avoided. Consider the following counter-factual thought experiment: were the United States to instantly possess a Scandinavian-type welfare system, would outcomes (defined in terms of less inequality, improved life expectancy, and increased women's labor participation) be the same, or of the same magnitude, as that achieved by Sweden, Norway, and Denmark over the second half of the 20th century? Our guess is probably not, given the isomorphism between politics, economics, culture, and historical timing that determine how a national program of this type comes into being and is maintained over time. Nevertheless, increasing diversity in represented contexts and better targeting macro-measures to suit the object of study will better insure that multi-level studies will converge on fundamental truths about how social environments influence individuals – whether those environments are nations, regions, or birth cohorts.

Institutional Change

Consideration of dynamic institutions represents one of the most promising areas in hierarchical representations of life course studies. As we discussed, naturally occurring historical change presents opportunities to explore natural experiments at the population level. Yet few societal level factors fluctuate sufficiently to have demonstrable impacts over time (with the possible exception of the impact of German reunification and more generally the breakup of the Soviet empire) and those that do produce a myriad of changes, the effects of which are difficult to isolate. Specific changes in public policy present opportunities to more precisely link the macro to the meso and then to the micro. For instance, demographic change in the aging of populations may engender policy changes such as reduced and/or delayed retirement benefits (or in the case of less-developed Asian countries the initiation of such benefits) that ultimately alter work and family decisions. Japan's universal long-term care insurance, instituted in part because of dramatic population aging and alarming fertility reductions, may have the unintended consequence of further reducing fertility and shifting the responsibility of caregiving farther away from family members (see Takagi and Silverstein 2006 for a similar discussion with regard to older parent–child coresidence in Japan).

Cross-Generational Correspondence

Finally we note that much of the literature on social change and societal transformation is rooted in the idea that each birth cohort that arrives into adulthood (re)invents its social milieu. Although many studies have shown cohort replacement to be the engine of social change in a given population over time, they tend to ignore the main societal institution that promotes continuity from one generation to another – namely, the multi-generational family. The life course principle of *linked lives* provides a useful metaphor for describing how family members are interdependent with each other, particularly as they are linked across generations by bonds of kinship. The family as the primary meso-social institution charged with maintaining societal stability serves as the conduit through which values and resources are reproduced in subsequent generations through time. The paradox that continuity is maintained in society even during times of great transformation has yet to be theoretically or empirically integrated into life course scholarship on social change and represents a promising direction for future development.

At its core, the life course perspective is a conceptualization of life span development as guided by the twin axes of biographical and historical time. In this chapter we strived to bring social institutions more fully under the life course umbrella and present an empirical strategy that teases out institutional explanations underlying descriptive differences in life pathways between social aggregates – in our example generations rooted in historical time. We hope that by laying out the issues in this way researchers will be more likely to consider human development as guided by micro-individual, meso-institutional, and macro-societal forces, and, where the data exist, to render as complete a depiction of the embedded life span as possible.

References

Alwin, Duane F., Ronald L. Cohen, and Theodore M. Newcomb. 1991. *Political Attitudes over the Life Span: The Bennington Women After Fifty Years*. Madison, WI: University of Wisconsin Press.

Alwin, Duane F. and Jon A. Krosnick. 1991. "Aging, Cohorts, and the Stability of Sociopolitical Orientations over the Life Span." *American Journal of Sociology* 97:169–95.

Amato, Paul R. 1994. "Father-Child Relations, Mother-Child Relations, and Offspring Psychological Well-Being in Early Adulthood." *Journal of Marriage and the Family* 56:1031–42.

Aquilino, William S. 1994. "Impact of Childhood Family Disruption on Young Adults Relationships with Parents." *Journal of Marriage and the Family* 56(2):295–313.

———. 2005. "Impact of Family Structure on Parental Attitudes Toward the Economic Support of Adult Children over the Transition to Adulthood." *Journal of Family Issues* 26:143–67.

Axinn, William G. and Scott T. Yabiku. 2001. "Social Change, the Social Organization of Families, and Fertility Limitation." *American Journal of Sociology* 106:1219–61.

Bellah, Robert N., Richard Madsen, William M. Sullivan, Ann Swidler and Steven M. Tipton. 1985. *Habits of the Heart: Individualism and Commitment in American Life*. Berkeley, CA: University of California Press.

Bengtson, Vern L. 1975. "Generation and Family Effects in Value Socialization." *American Sociological Review* 40:358–71.

Bengtson, Vern L. and Katherine R. Allen. 1993. The Life Course Perspective Applied to Families over Time. Pp. 469–98 in *Sourcebook of Family Theories and Methods: A Contextual Approach*, edited by P. Boss, W. Doherty, R. LaRossa, W. Schumm, and S. Steinmetz. New York: Plenum.

Brandt, Martina, Klaus Haberkern, and Marc Szydlik. 2009. "Intergenerational Help and Care in Europe." *European Sociological Review* 25(5):585–601.

Broese van Groenou, Marjolein, Karen Glaser, Cecilia Tomassini, and Thérèse Jacob. 2006. "Socio-Economic Status Differences in Older People's Use of Informal and Formal Help: A Comparison of Four European Countries." *Ageing and Society* 26:745–66.

Bronfenbrenner, Urie. 1995. "Developmental Ecology Through Space and Time: A Future Perspective." Pp. 619–47 in *Examining Lives in Context: Perspectives on the Ecology of Human Development*, edited by P. Moen, G.H. Elder, Jr., and K. Luscher. Washington, DC: APA Books.

Brooks, Clem and Catherine Bolzendahl. 2004. "The Transformation of US Gender Role Attitudes: Cohort Replacement, Social-Structural Change, and Ideological Learning." *Social Science Research* 33:106–33.

Bruckner, Hannah and Karl Ulrich Mayer. 2005. "De-Standardization Of The Life Course: What Might It Mean? And If It Means Anything, Whether It Actually Took Place." *Advances in Life Course Research* 9:27–53.

Cherlin, Andrew J. 1978. "Remarriage as an Incomplete Institution." *American Journal of Sociology* 84:634–60.

Clausen, John A. 1993. *American Lives: Looking Back at the Children of the Great Depression*. New York: Free Press.

Conger, Rand D. and Glen H. Elder, Jr. 1994. *Families in Troubled Times: Adapting to Change in Rural America*. Hawthorne, NY: Aldine DeGruyter.

Dannefer, Dale. 2003. "Whose Life Course Is It, Anyway? Linked Lives in Global Perspective." In *Invitation to the Life Course*, edited by R. A. Settersten, Jr. Amityville, NY: Baywood.

Diewald, Martin, Anne Goedicke, and Karl U. Mayer, eds. 2006. *After the Fall of the Wall: Life Courses in the Transformation of East Germany.* Palo Alto: Stanford University Press.

Elder, Glen H., Jr. 1974. *Children of the Great Depression: Social Change in Life Experience*. Chicago, IL: University of Chicago Press.

———. 1994. "Time, Human Aging, and Social Change: Perspectives on the Life Course." *Social Psychology Quarterly* 57:4–15.

Elder, Glen H., Jr. and Rand D. Conger. 2002. *Children of the Land: Adversity and Success at Century's End.* Chicago, IL: University of Chicago Press.

Elder, Glen H., Jr., Michael J. Shanahan and Elizabeth C. Clipp. 1997. "Linking Combat and Physical Health: The Legacy of World War II in Men's Lives." *American Journal of Psychiatry* 154(3):330–6.

Esping-Anderson, Gøsta. 1999. *Social Foundations of Postindustrial Economies.* London: Oxford University Press.

Furstenberg, Frank and Andrew Cherlin. 1991. *Divided Families: What Happens to Children When Parents Part.* Cambridge, MA: Harvard University Press.

Ganong, Lawrence H. and Marilyn Coleman, M. 1999. *Changing Families, Changing Responsibilities: Family Obligations Following Divorce and Remarriage*. Mahwah, NJ: Lawrence Erlbaum.

Gans, Daphna and Merril Silverstein. 2006. "Norms of Filial Responsibility for Aging Parents across Time and Generations." *Journal of Marriage and Family* 68:961–76.

Gee, Gilbert C., Eliza K. Pavalko, and J. Scott Long. 2007. "Age, Cohort, and Perceived Age Discrimination: Using the Life Course to Assess Self-Reported Age Discrimination." *Social Forces* 86:265–90.

Hank, Karsten. 2007. "Proximity and Contacts Between Older Parents and Their Children: A European Comparison." *Journal of Marriage and Family* 69:157–73.

Heinz, Walter R. 2003. "From Work Trajectories to Negotiated Careers: The Contingent Work Life Course." Pp. 185–204 in *Handbook of the Life Course*, edited by J. T. Mortimer and M. J. Shanahan. New York: Kluwer Academic/Plenum.

Hollinger, Franz and Max Haller. 1990. "Kinship and Social Networks in Modern Societies: A Cross-Cultural Comparison among Seven Nations." *European Sociological Review* 6:103–24.

Kohli, Martin. 2005. "Intergenerational Transfers and Inheritance: A Comparative View." Pp. 266–89 in *Intergenerational Relations Across Time and Place, Annual Review of Gerontology and Geriatrics,* Vol. 24, edited by M. Silverstein. New York: Springer.

London, Andrew and Janet Wilmoth. 2006. "Military Service and (Dis)Continuity in the Life Course: Evidence on Disadvantage and Mortality from the HRS and AHEAD." *Research on Aging* 28:35–59.

MacMillan, Ross and Rhonda Copher. 2005. "Families in the Life Course: Interdependency of Roles, Role Configurations, and Pathways." *Journal of Marriage and Family* 67:858–79.

Mannheim, Karl. 1952. "The Problem of Generations." Pp. 276–320 in *Essays on the Sociology of Knowledge.* London: Routledge and Kegan Paul.

Mayer Karl U. 2004. "Whose Lives? How History, Societies, and Institutions Define and Shape Life Courses." *Research in Human Development* 1:161–87.

———. 2006. "After the Fall of the Wall: Living Through the Post-Socialist Transformation in East Germany." Pp. 1–28 in *After the Fall of the Wall: Life Courses in the Transformation of East Germany*, edited by M. Diewald, A. Goedicke, and K.U. Mayer. Palo Alto: Stanford University Press.

———. 2007. "Retrospective Longitudinal Research: The German Life History Study," Pp. 85–106 in *Handbook of Longitudinal Research: Design, Measurement, and Analysis*, edited by S. Menard. San Diego: Elsevier.

———. 2009. "New Directions in Life Course Research." *Annual Review of Sociology* 35(1):413–33.

Moen, Phyllis. (2003). *It's About Time: Career Strains, Strategies, and Successes.* Ithaca, NY: Cornell University Press.

O'Brien, Richard M., Jeanne Stockard, and Lynne Isaacson. 1999. "The Enduring Effects of Cohort Characteristics on Age-Specific Homicide Rates, 1960–1995." *American Journal of Sociology* 104(4):1061–95.

Peabody, Dean. 1985. *National Characteristics.* New York: Cambridge University Press.

Pezzin, Liliana E. and Barbara Schone. 1999. "Parental Marital Disruption and Intergenerational Transfers: An Analysis of Lone Elderly Parents and Their Children." *Demography* 36:287–97.

Phillipson, Chris. 2003. "Globalization and the Reconstruction of Old Age: New Challenges for Critical Gerontology." Pp. 163–79 in *The Need for Theory: Critical Approaches to Social Gerontology*, edited by S. Biggs, A. Lowenstein, and J. Hendricks. New York: Baywood.

Popenoe, David. 1993. "American Family Decline, 1960–1990: A Review and Appraisal." *Journal of Marriage and the Family* 55:527–55.

Putnam, Robert D. 1995. "Bowling Alone: America's Declining Social Capital." *Journal of Democracy* 6(1):65–71.

Riley, Matilda W., Marilyn Johnson, and Anne Foner. 1972. *Aging and Society, Vol. 3. A Sociology of Age Stratification.* New York: Russell Sage.

Riley, Matilda W., Robert L. Kahn, and Anne Foner. 1994. *Age and Structural Lag*. New York: Wiley.

Riley, Matilda W., and John W. Riley. 1994. "Structural Lag: Past and Future." Pp. 15–36 in *Age and Structural Lag*, edited by M. W. Riley, R. L. Kahn, and A. Foner. New York: Wiley.

Roberts, Robert E. L. and Vern L. Bengtson. 1996. "Affective Ties to Parents in Early Adulthood and Self-Esteem across 20 Years." *Social Psychology Quarterly* 59:96–106.

———. 1999. The Social Psychology of Values: Effects of Individual Development, Social Change, and Family Transmission over the Life Span. Pp. 453–82 in *The Self and Social Processes in Aging,* edited by C. Ryff and V. W. Marshall. New York: Springer.

Ross, Andy, Ingrid Schoon, Peter Martin, and Amanda Sacker. 2009. "Family and Nonfamily Role Configurations in Two British Cohorts." *Journal of Marriage and Family* 71:1–14.

Ryder, Norman B. 1965. "The Cohort as a Concept in the Study of Social Change." *American Sociological Review* 30:843–61.

Sayer, Liana, Philip Cohen, and Lynne Casper. 2005. "Women, Men, and Work." Pp. 76–106 in *The American People*, edited by R. Farley and J. Haaga. New York: Russell Sage.

Schmeeckle, Maria, Roseann Giarrusso, Du Feng, and Vern L. Bengtson. 2006. "What Makes Someone Family? Adult Children's Perceptions of Current and Former Stepparents." *Journal of Marriage and Family* 68:595–610.

Schooler, Carmi. 1996. "Cultural and Social-Structural Explanations of Cross-National Psychological Differences." *Annual Review of Sociology* 22:323–49.

Settersten, Richard A., Jr., ed. 2003. *Invitation to the Life Course: Toward New Understandings of Later Life.* Amityville, NY: Baywood.

———. 2006. "When Nations Call: How Wartime Military Service Matters for the Life Course and Aging." *Research on Aging* 28(1):12–36.

Silverstein, Merril, Daphna Gans, Ariela Lowenstein, Roseann Giarrusso, and Vern L. Bengtson. 2010. "Older Parent-Child Relationships in Six Developed Nations: Comparisons at the Intersection of Affection and Conflict." *Journal of Marriage and Family* X:XX–XX.

Stacey, Judith. 1991. *Brave New Families: Stories of Domestic Upheaval in Late Twentieth Century America.* New York: Basic Books.

Sundstrom, Gerdt, Lennarth Johansson, and Linda B. Hassing. 2002. "The Shifting Balance of Long-Term Care in Sweden." *The Gerontologist* 42:350–55.

Szydlik, Marc. 1996. "Parent-Child Relations in East and West Germany Shortly after the Fall of the Wall." *The International Journal of Sociology and Social Policy* 16:63–88.

Takagi, Emiko, and Merril Silverstein. 2006. "Intergenerational Coresidence of the Japanese Elderly: Are Cultural Norms Proactive or Reactive?" *Research on Aging* 28:473–92.

Turner, Ralph H. 1988. "Personality in Society: Social Psychology's Contribution to Sociology." *Social Psychology Quarterly* 51:1–10.

Weber, Max. 1968. *Economy and Society*, edited by G. Roth and C. Wittich, Berkeley: University of California Press.

Yi, Zeng. 1996. "Is Fertility in China in 1991–92 Far Below Replacement Level?" *Population Studies* 50:27–54.

Chapter 4
Widening the View: Capturing "Unobserved" Heterogeneity in Studies of Age and the Life Course

Jessica A. Kelley-Moore and Jielu Lin

In the first volume of their three-volume work entitled *Aging and Society* (1968), Matilda White Riley and Anne Foner produced the first comprehensive inventory of empirical studies on aging. Demonstrating the growing interest in age and aging processes, the scope of this work spanned many disciplines and included a wide array of topics including adjustment to retirement (Thompson et al. 1960), response to institutional care (Beyer and Woods 1963), and age composition of the labor force (Bancroft 1958). The varied approaches to questions of the nature and process of aging had resulted in a significant body of empirical information but, to date, as Riley and Foner accounted, as yet there was "…no unified body of knowledge, no general theory of aging, that can be transmitted to students, applied in professional practice, or tested and amplified through further research" (Riley and Foner 1968:1).

In the 40 years since, great advances have been made in research on age and the life course, largely due to significant improvement in research designs, particularly the explosion of high-quality longitudinal studies – panel, rotating panel, and repeated cross-sections – that have allowed investigators to separate the effects of age, chronological time, and historical placement of cohorts on a range of observable outcomes. The composition of samples have become more representative of the population, because of concerted efforts to draw more diverse samples of older adults (and oversamples of under-represented groups), and to include multiple birth cohorts in the same study design. Further, administrative data sources (e.g., Social Security records; Medicare claims), geographic indicators (e.g., Census tracts), and vital statistics data have been put to use in myriad new ways to address questions related to the age distribution of society and the social distribution of resources based on age. International collaborations have sought to standardize study designs and measurement, such as the Health and Retirement Survey (HRS), Survey of Health, Ageing, and Retirement in Europe (SHARE), English Longitudinal Study of Aging (ELSA), and the China Health and Retirement Longitudinal Survey (CHARLS), opening a new vista of cross-national research to further understand the social experience of aging.

Over the past several decades, sociologists have been instrumental in shifting the logic of inquiry away from chronological age as *explananda* and toward the socially-constructed meaning and consequences of age and the life course in which lives are embedded. The significance of age as a critically important organizing tool for society, justifying allocation of resources, rights, and responsibilities – has been recognized. At the same time, chronological age itself is a weak predictor of human development and behavior. By identifying elements of social context with increasing precision (e.g., the social distribution of opportunity structures), and then locating individuals and

J.A. Kelley-Moore (✉)
Department of Sociology, Case Western Reserve University, Cleveland, OH, USA
e-mail: jak119@case.edu

R.A. Settersten, Jr. and J.L. Angel (eds.), *Handbook of Sociology of Aging*, Handbooks of Sociology and Social Research, DOI 10.1007/978-1-4419-7374-0_4, © Springer Science+Business Media, LLC 2011

groups within that structure based on their age, birth cohort, and stage in the life course, sociologists have systematically reduced the scope of phenomena that can be attributed to organismic aging. The increasing sophistication in study designs and an ever-growing bank of social indicators have allowed researchers to identify processes that may correlate with chronological age, the onset of chronic disease for example, but are not in fact caused by age.

Through this advancement in methodological tools, social research has made a significant contribution toward the "unified body of knowledge and general theory of aging" that Riley and Foner noted was lacking nearly a half a century ago. Indeed, critical inquiry about the socially-constructed meaning of age and the life course – and the decreasing utility of chronological age as an explanatory factor – has begun to pervade other disciplines, including developmental biology (Gluckman and Hanson 2006), economics (Heckman 2006), and health services research (Halfon and Hochstein 2002). Sociologists are in a unique position to document, characterize, and explain the patterns associated with age in a population and continue to make substantial contributions across the many subfields and lines of inquiry.

Across a range of domains and specific outcomes related to human lives, we observe a robust association with chronological age. We observe substantial age patterning in the distribution of social roles, socioeconomic resources, disease risk, and movement into and out of social institutions. This occurs at the individual level, such as age at labor force exit (Warner et al. 2010) and at higher levels of aggregation, such as within-cohort rates of disablement by age (Hoffman 2009). The significant improvement in study designs, measurement, and statistical modeling that have advanced the study of age and the life course in the past few decades means that social scientists have access to information in a wide array of domains that are sensitive enough to decompose causal factors that are sociogenic or organismic in origin. Yet, this focus on normative, or "average," trends in age-associated phenomena should be coupled with empirical attention to patterns of *variability* that may also be associated with chronological age.

Sociologists have for some time now emphasized the importance of characterizing and acknowledging variability in the aging experience (Maddox 1987; Riley 1983), exemplified in discussions of heterogeneity (Nelson and Dannefer 1992), diversity (Bass et al. 1990), differential gerontology (Dannefer 1988) or inter-individual differences in intra-individual change (Baltes and Nesselroade 1979). It is important to note that these discussions cut across different levels of aggregation, making it critically important in empirical studies to specify which type of variability pattern one is testing and at what level (individual, cohort, etc.). At the individual-level, for example, one may examine an average rate of change in a phenomenon with age, but such an analysis could be enhanced by further asking: do the between-person differences in within-person variability diminish or increase with advancing age? At the level of a birth cohort, age-based patterns of variability can include increasing differentiation with age (e.g., health or income), decreasing differentiation with age (e.g., cognitive functioning due to selective mortality), and constant variability with age (e.g., personality).

In this chapter, we refer to "unobserved" heterogeneity as the systematic patterns of variability in observational studies that has been largely under-recognized or under-specified in social science, and our special concern is with unobserved heterogeneity within cohorts or age categories. At least in the context of age, this neglect has been largely due to limitations in (1) standard statistical practices, (2) current prospective study designs, and (3) single-methodological approaches. This is not to say that we should focus on age-based heterogeneity to the exclusion of "average" trends in aging. Rather, a rigorous accounting of phenomena that is arrayed across the life span must necessarily examine the "main effect" of age, as well as potential age-based patterns in variability. The precise characterization of patterns of age-based heterogeneity, within or between individuals or cohorts, opens the range of explanatory frameworks for phenomena arrayed and organized over the life course to the "sociological imagination" ranging from the social context in utero to continued development and growth in adulthood (Dannefer 2011).

Drawing our focus to age-based variability in empirical research will allow for finer identification of social and individual characteristics or patterns that may influence well being over the entire life course. Refocusing social inquiry to include sociogenic origins of age-based heterogeneity can yield significant benefits for the field of aging. First, characterizing and understanding how social context may lead to differential outcomes for individuals within and across cohorts over the life course reduces the phenomena that may be mistakenly attributed to organismically-based age-related change or other individual-level factors (e.g., genetics; Freese 2008). Second, systematic attention in empirical studies to variability in life paths reduces the tendency to over-emphasize "normal" aging explanations. For the remainder of the chapter, we explain how three current streams of methodological inquiry (standard statistical practice, current prospective study designs, single methodological approaches) have limited our ability to examine age-based patterns of variability, then we present potential solutions to these limitations and note exemplary research that has already "widened the view" of social inquiry to include unobserved heterogeneity.

Unobserved Heterogeneity Due to Standard Statistical Practice

With the recent explosion of high-quality longitudinal data sources that follow both individuals and cohorts over time, social scientists now have significant potential to identify and characterize patterns of variability that we observe as a population ages. Despite the availability of both the data and statistical tools for documenting patterns of age-based heterogeneity, standard statistical practice continues to rely heavily on measures of central tendency to the neglect of the distribution around these estimates. In the current statistical paradigm, indicators of range (e.g., confidence intervals) have been used solely to document statistical significance, rather than precision of the estimate (Cohen 1994; Ziliak and McCloskey 2008). Interpretations of the findings in observational studies that tend to focus exclusively on the central tendency neglect the opportunity to integrate these indicators of variability into the causal interpretations regarding the phenomenon under study.

Although this exclusion of range and variability is a general problem when interpreting statistics in scientific inquiry across disciplines, it is particularly consequential for sociology of age and the life course. The degree of heterogeneity in the population, particularly patterns that may be associated with age, has received scant attention in empirical studies rather emphasizing comparison of "average" levels across age groups, estimation of a global age-based trajectory without accounting for birth cohort, or estimation of average rate of change over time. If there is a high degree of variability in the population, one risks a weak characterization of that particular phenomenon when using a central tendency estimate or age-based trajectory. Further, emphasis on the "average" in observational studies (without acknowledging variability) implies the existence of a "normal" aging pathway that may be driven by physiologic or organismic processes. Characterizing the extent of variability in a population in conjunction with measures of central tendency opens the opportunity to examine social processes arrayed over the life course that may lead to both the global pattern observed with age and age-based patterns of variability.

Although sociologists are generally cautious about attributing observed individual or collective patterns of change solely to organismic aging processes, there is a tendency to relax this caution where the subject matter has an inherently organismic dimension such as age-related physiological decline (e.g., frailty; Hoffman 2009). Standard statistical practices that focus on increasingly more precise central tendency estimates – and neglect age-associated patterns of variability – falsely imply that there is, indeed, a normative "aging" effect. In recent years, significant advancement in statistical modeling, specifically mixed models, multi-level models, and growth curves, have allowed investigators to estimate unexplained variance in observational studies. Yet these robust patterns of variability around the average trajectory, many of which have some age patterning, continue to be relegated statistically and conceptually to *deviations* from supposed the global estimated trajectory by age.

Recently, sociologists have begun to address questions of heterogeneity and to make creative use of the statistical modeling techniques available to do so. Indeed, there is a growing practice to estimate age-based trajectories by social subgroup using either a group-based trajectory modeling technique (Sampson and Laub 2003) or stratified modeling (Kim and Durden 2007; Willson 2003). There are statistical techniques that explicitly model heteroskedasticity and complex patterns of variation, accounting for differing levels of variance across subgroups (Browne et al. 2002; Goldstein 2005). Some notable work has decomposed the unexplained variance around a global age trajectory by gender, race, and socioeconomic status to help account for the social patterning of variability (Yang and Lee 2009). This work is at the frontier of understanding social patterns of differentiation and variability, but the statistical tools exist to push this line of inquiry further. A largely-presumed but unchecked assumption is that the degree of variability in inter-individual differences and intra-individual change does not vary across social groups such as gender or race. Systematic examination of subgroup differences in age-based heterogeneity could contribute to our understanding of the complex social patterning over the life course that influences individual level and cohort level patterns with age.

Among the numerous types of heterogeneity in the population that are associated with time-based and age-based processes, we illustrate three specific types in this chapter that could be characterized using existing longitudinal studies but have to date been neglected due to standard statistical practice that emphasizes central tendencies or "average" trends. The first source is *intra-individual variability over time*, which refers to the patterns of change (or stability) within a particular individual over time. In panel data with only two observation points, analytic strategies generally center on whether change occurred for the individual and the amount of change. Yet, with an increasing number of observation points, analytic techniques typically *smooth* an individual's trajectory of change, estimating the average rate of change over the period. This estimated trajectory of change for an individual obscures the variability within individuals over time, and this dynamic pattern of change is relegated to individual-level residual error or the disturbance term in statistical modeling.

As an empirical example, Fig. 4.1 illustrates the high degree of variability that may exist in repeated measurements of the CES-D scale in a 12-year period. Drawn from the Health and Retirement Study (HRS), the figure shows the raw scores for a single respondent at each observation point, overlaid with the estimated trajectory over time. Although the trajectory for this individual is

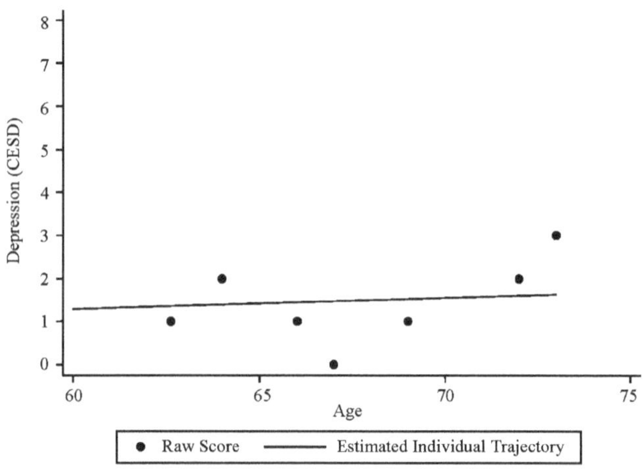

Fig. 4.1 Intra-individual variability over time

nearly flat, it is clear that she/he did, indeed, fluctuate over time in response to the same questions. In standard statistical practice, this intra-individual variability is relegated to the residual error term, and given that total fluctuation was between 0 and 2, this can seem inconsequential. However, when considered in aggregation across many respondents with varying levels of fluctuation over time, it amounts to a significant amount of unaccounted variability in individual life paths. Some portion of the observed within-person variability may reflect measurement error. Yet, if there is, indeed, an age-based pattern to the intra-individual variability, such as greater within-person variability at older ages relative to younger ages, systematic attention to the individual-level error term could yield important information about age-related processes.

Intra-individual variability over time is an area of inquiry that has long been considered (see Baltes and Nesselroade 1979 for a discussion) but with the improvement of panel designs, namely expansion of observational windows and increase in number of observation points, it is worthy to reemphasize the importance of analyzing within-person variability. Understanding what may fluctuate, remain steady, or decline for individuals over a period of time – and if there are associated age patterns – could provide a richer understanding of well being over the entire life course. As Hooker (1991) has demonstrated, there is significant theoretical and empirical value in understanding variability within a single individual over time. A number of statistical models have been refined for the exact purpose of characterizing variability in individual life paths (Halaby 2003; Macmillan and Eliason 2003).

The second source of heterogeneity is *age-based patterns of intra-cohort variability*, which are patterns of heterogeneity over the collective life course of a single birth cohort that are correlated with chronological age. There are a number of age-based patterns of heterogeneity depending on what phenomenon is under study. This could include increasing heterogeneity or other systematic fluctuations in variability over the collective life course of a cohort (Dannefer and Sell 1988). For example, social opportunity structures may cause a cohort to increase in heterogeneity with age (e.g., income and wealth in Wolff 2002) or selective mortality may cause the heterogeneity among survivors to decrease (e.g., chronic conditions in Dupre 2007). Further, age-graded policy interventions may restrict or otherwise alter patterns of heterogeneity as well (e.g., Medicare age eligibility). This is especially important because presumptions that such age-based patterns do not exist, or failing to characterize them, risks the over-reliance on normative aging as the primary explanatory framework.

To illustrate an age-based pattern in variability, Fig. 4.2 shows an estimated age-based trajectory of physical function for adults ages 65 and older, adjusting for selection bias due to mortality. Data for this figure were drawn from four waves of the North Carolina *Established Populations of*

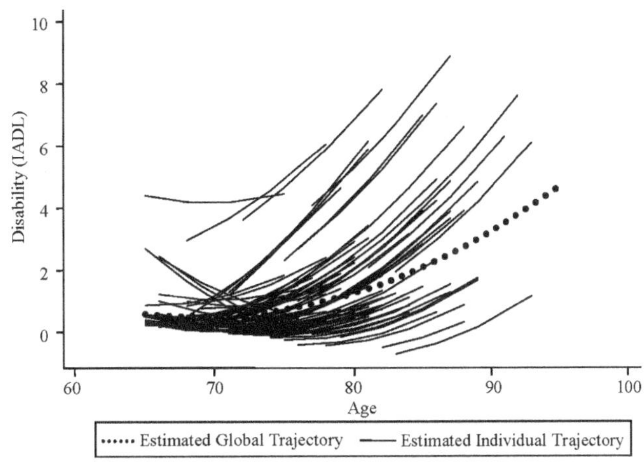

Fig. 4.2 Age-based pattern of intra-cohort variability

Epidemiologic Study of the Elderly, a community-based sample of more than 4,000 black and white older adults. For illustrative purposes, arrayed around the global age-based trajectory of IADL disability is a random sample of individual trajectories. In this example, the estimated variance of total residuals increases over time, as evident by individuals becoming more heterogeneous at upper ages. In very old age, the mean estimation poorly characterizes the "average" or "typical" individual's physical function. This example demonstrates the danger of relying on a global or average age effect because of its decreasing utility for characterizing physical function. On the contrary, it imposes a "created" trajectory associated with age and largely neglects the existing heterogeneity caused by other social processes, thereby reifying age as an ontogentic process.

In sum, assuming homoskedastistic variance around age-based trajectories (constant variance across age) results in significant loss in both empirical and theoretical contributions to our understanding of aging processes, particularly associated with age-based patterns of intra-cohort differentiation. Much of this empirical neglect of variability in growth curve or trajectory models may stem from the early application of this technique, which focused on early-life phenomena such as verbal acquisition, where a developmental or "age" effect is presumed. It is tenuous to presume commensurate "normative aging" effects for outcomes in later points in the life span, such as disability, income, and depression.

The third source is *inter-cohort differences in patterns of intra-cohort variability*. As Dannefer and Kelley-Moore (2009) noted, a rarely-addressed question in research on age and the life course is the extent to which patterns of change or stability with age that are observed in one birth cohort can be generalized across cohorts. The historical placement of a birth cohort, which represents a specific set of social influences on opportunity structures over the life course, may produce an age-based pattern of heterogeneity unique only to that cohort or to a narrow band of adjacent ones. For example, did the unprecedented governmental assistance in education and home-buying for World War II veterans *narrow* or *delay* age-based patterns of heterogeneity of health or wealth among Baby Boomers, relative to subsequent or previous birth cohorts?

Considerable conceptual attention has been paid to the problem of inter-cohort differences in intra-cohort variability (Alwin et al. 2008; Riley 1973; Riley et al. 1972) but, until recently, it has not been matched with empirical attention, largely due to the lack of panel studies designed sufficiently to examine potential inter-cohort differences in age-based patterns of intra-cohort variability. Such an examination requires data that measures (1) *distinct* cohorts, (2) on *multiple* occasions (3), at the *same* ages. It is the only way to determine whether there is systematic change in variability with age and whether the patterns of that systematic change differ by cohort. These data requirements are extensive and to date, there have been few available sources for study. There are now several high-quality panel studies with which this can be accomplished.

The HRS is perhaps the most recognized choice for research questions about inter-cohort differences in intra-cohort patterns of variability. The HRS is the largest on-going longitudinal study of older persons' health and labor force activity in the United States (HRS 2008). With its transition in 1998 to a steady-state cohort design, there is sufficient data to study inter-cohort differences in age-based patterns of change for persons in their 50s and 60s from at least three different cohorts with at least five (for the Early Baby Boom cohort) and up to ten (for the original HRS cohort) measurement occasions. The potential for studying inter-cohort differences with the HRS continues to increase with biennial interviews. Moreover, the steady-state panel design means that new birth cohorts are added to the panel every 6 years, with Early Baby Boomers (b. 1948–1953) sampled in 2004 and the Mid Baby Boomers (b. 1954–1959) sampled in 2010 (HRS 2008).

A second data source, much less utilized by sociologists studying aging but that meets the standards for inter-cohort comparisons of intra-cohort variability, is the *National Longitudinal Surveys of Youth* (NLSY 2009). The original NLSY cohort came under observation in 1979 and has followed the experiences of the Late Baby Boomers (b. 1957–1964; ages 14–22 in 1979). A new NLSY cohort of persons born between 1980 and 1984 (roughly corresponding to the Echo Boom Cohort) came under observation in 1997 when they were in the ages 12–16. Both NLSY cohorts have been observed for more than eleven waves of data collection, which allows for cohort comparison of patterns of change or stability *plus* the potential age-based patterns in variability in emerging adulthood, i.e., persons in their mid-teens to late 20s. As with the HRS, data collection for these NLSY cohorts is continuing and the potential usefulness of these data sources for understanding inter-cohort differences in intra-cohort variability has yet to be fully tapped.

Despite the availability of *data sources* that can be used to make cohort comparisons in age-based patterns of variability, *standard statistical practice* continues to utilize an accelerated longitudinal design, also called a synthetic cohort design. In these models, investigators estimate a global age-based trajectory of a particular phenomenon that *assumes* that there are no differences in these age-based patterns of variability across cohorts, yet Alwin and colleagues have noted that birth cohort itself is a potential source of heterogeneity (Alwin et al. 2006). Consistent with this, many studies now test whether the global age-based trajectory does, indeed vary by birth cohort (Yang 2007). The general conclusion is that if the trajectories (average rate of change for the same age range) are not significantly different, then birth cohort is not consequential. However, few studies press further and examine whether the *patterns of age-based heterogeneity* vary by birth cohort.

Figure 4.3 illustrates the differences in age-based variability across cohorts; specifically, this is the estimated trajectory of number of functional limitations for the birth cohorts HRS and CODA in the HRS for the age range of 65–74. It is important to note that the "average" trajectory is not greatly dissimilar across cohorts, indeed the estimated slope is not significantly different across cohorts. Yet there are notable differences in the age-based patterns of variability, compelling the question about the differential social experiences and opportunities across cohorts that led to these patterns of variability. Our illustration uses adjacent birth cohorts, which may not be that dissimilar given some shared social experiences. However, for one to consider birth cohorts that were further apart, it would be tenuous to presume similar patterns of age-based variability. The cost of such assumptions is a prime example of the neglect of unobserved heterogeneity due to standard statistical practice.

Careful examination of the patterns of variability around global trajectories has a number of implications for the study of stability and change in phenomena with age and over time. First, a dominant assumption, both statistically and conceptually, is that population variation around age-based or time-based trajectories is normally distributed but this is rarely confirmed (Lumley et al. 2002). The presence of a few outlier respondents with extremely high or low values on an outcome can skew the estimate of the "average" level. Indeed, returning to Fig. 4.2, a few respondents had

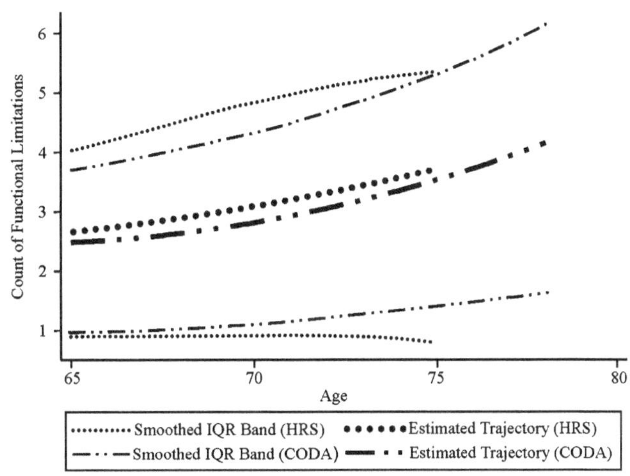

Fig. 4.3 Inter-cohort difference in age-based pattern of intra-cohort variability

extremely high levels of functional limitations, biasing the global estimate upward relative to the actual population. Neglect of the distribution of observations around measures of central tendency can directly affect findings of a range of empirical questions, including comparison of average differences between age groups, average rates of change over time, and average age-based trajectories within a cohort. Researchers risk over- or under-estimating the phenomenon under study when there is no systematic consideration of patterns of variability around the central tendency estimate.

Second, an empirically well-established pattern of an age-based trajectory of a certain characteristic perhaps even now considered axiomatic (e.g., frailty with age) may need revision when the patterns of variability around that trajectory are taken into account. However, in a study reporting the raw data along with the predicted value of verbal ability with age (Alwin and McCammon 2001), the great amount of variability, particularly in old age, suggests that cognitive functioning may actually decrease at a much slower rate than what we have assumed when the focus was on "average" rates of decline with age, which may be biased by a small number of persons with faster-than-average cognitive decline. Documenting variability in what is largely-presumed to be an "age effect" represents an opportunity to think critically about the role of the social environment and life course experiences that lead to differentiation, rather than focusing solely on the organismic aging processes that lead to homogeneity.

The foregoing discussion has been based on the premise that all observed variability in studies is "true" heterogeneity in the population. We want to caution that statistical models can capture the amount of unexplained variance remaining but cannot distinguish between actual heterogeneity in the population and variability due to measurement error (Alwin 2007), study design (Kelley-Moore 2006), or imputation procedures (Allison 2000). For the purpose of our arguments here, this means that a sociologist needs both theoretical frameworks and a solid understanding of the fallibility of data and measurement to aptly consider potential sources of variability in an observational sample. Our major task as social scientists is not to "correct for" or "explain away" the heterogeneity that exists in our studies. Rather, researchers need to consider critically the founding assumptions of the standard statistical practices – which rely heavily on measures of central tendency – and incorporate explicit tests for patterns of variability, whether within individuals over time, within a cohort with age, or across cohorts with age. Continued emphasis on just the "middle" or "average" effects, to the neglect of potential age-based patterns of variability, limits the contribution that social scientists can make in understanding phenomena arrayed over the life course.

Unobserved Heterogeneity Due to Study Design

There is a strong theoretical literature articulating the influence of life course processes on late-life outcomes (Ferraro and Shippee 2009; George 2005), particularly the potential ways that social structure and individual characteristics may interact in early and mid-life, yet there has not been much empirical research accounting for these social selection patterns in studies of older adults. This type of unobserved heterogeneity, which is the variability in individual life course paths that we do not observe due to limits in study design, can have substantial impact on phenomena under study in samples of older adults. One of the key contributors to this gap in our knowledge is the left-truncation or censoring of sampling frames, where studies of aging start at ages 50, 65, or even older. When an accelerated longitudinal design is employed, many respondents first come under observation at even higher ages amounting to further loss of information. To further exacerbate the problem, some study designs impose additional selectivity on their samples by restricting eligibility to those who, for example, have no functional limitations at baseline (e.g., Women's Health and Aging Study I).

The consequence of these limits in study design is that we know little about cohort-specific patterns in social selection over the collective life course. A bevy of empirical work has articulated complex social selection mechanisms in domains such as mortality (Hoffman 2009), disablement (Verbrugge and Jette 1994), and labor-force exits (Brown and Warner 2008). Although this work has had untold value in the study of aging, attention to selection has been almost exclusively limited to the prospective observation window of the panel study. This focus on "selection moving forward" virtually ignores the mechanisms of selection that occurred *prior* to the observation window.

The one social selection mechanism that has received systematic attention is premature mortality, the process whereby individuals are selected out by high mortality risk prior to old age, leaving a robust cohort with less overall variability and low subsequent mortality risk (Markides and Machalek 1984). Mortal attrition bias in studies of older adults is not only a problem of who may remain in the panel study, but also who would appear at the time of study origin. Men, persons of lower socioeconomic status, and race/ethnic minorities tend to have higher rates of premature mortality, so the survivors in these groups likely appear substantially more robust in older adulthood compared with their socially-advantaged counterparts (Markides and Machalek 1984; Willson et al. 2007). Additionally, the social distribution of premature mortality shapes not only *who* is available for observation at the oldest ages but also *when* these individuals may experience particular phenomena (e.g., end of life processes). This problem becomes more acute as the age-eligibility threshold for the sampling frame is set at higher ages. These patterns of mortality selection have received considerable conceptual attention (Geronimus 1992) and the empirical patterns of premature mortality are well-known (Hayward et al. 1997; Lantz et al. 1998; Mansfield et al. 1999). Given the substantial attention selective mortality has received, researchers who utilize prospective studies of older adults are generally quite informed about the social distribution of mortality risk over the entire life course and the differential effect on social subgroups within a particular cohort. Indeed, many studies acknowledge that the observed *homogeneity* in late-life outcomes may obscure the *heterogeneity* that existed much earlier in the life course.

Beyond selective mortality, however, there are myriad other social selection mechanisms that influence the well being and composition of a cohort, many of which remain largely unobserved in studies of older adults. These include health processes such as early onset of disability (Doblhammer 2003), socioeconomic circumstances such as lack of employer-based health insurance (Baker et al. 2001) or unsteady labor force participation (Moen and Chermack 2005); and social processes such as assortative mating (Lillard et al. 1995; Mare 1991) or residential moves (Sampson and Sharkey 2008). Some notable work has begun examining selection processes over the life course, demonstrating the complex social patterning of structural opportunity, individual circumstances, and social constructions of age. Wilmoth and Koso (2002) reconstructed individual marital histories and identified the pathways to late-life financial well being across life course marital status typologies. Elman and O'Rand (2002) demonstrated that mid-life educational attainment has a substantial impact on income trajectories. There has even been some consideration of inter-generational contributions to social and health inequality over the life course (Gluckman and Hanson 2006; Oliver and Shapiro 1995; Uhlenberg 2009). Characterizing patterns of social selection over the life course is both an empirical and conceptual frontier in need of systematic attention.

With the rising recognition of the "long reach" of early-life circumstances, a number of studies have incorporated survey questions or other data sources on childhood, including abuse (Irving and Ferraro 2006), neighborhood stability (Bures 2003), family socioeconomic circumstances (Hayward and Gorman 2004), and birth weight (Barker 2001). Although these studies that examine the potential influence of early- or mid-life social circumstances have been extremely valuable in identifying life-course selection pathways and their impact on later-life outcomes, the specification and measurement of these circumstances typically remains limited to major life circumstances that are easy for respondents to recall (e.g., date of marriage) or can be measured through administrative records (e.g., hospitalization; birth). Unfortunately, due to study design limitations, these earlier-life indicators are often treated as "turning point" events comparing those who did and did not experience a selected early-life exposure. Such an approach does not allow for the consideration of the *cascade* of events or circumstances that follow, which likely shapes the outcomes seen in later life.

Figure 4.4 presents a conceptual representation of a selection of socially-constructed life course processes that can influence not only the composition of the cohort at a given age, but also within that cohort, when individuals may be at risk of adverse outcomes such as compromised health. Study designs that do not capture patterns of social selection across the life course may lead to three types of mis-specification in analyses and interpretations. First, there may be an over-emphasis of current social, health, and economic circumstances and perhaps even over-estimation of their impact, since these indicators may "absorb" the effect of these earlier-life selection processes. A prime example is the current explosion of research on the influence of neighborhood context on older adults' well being. Although current residential environment may influence physical and mental health among older adults through mechanisms such as frequency of recreational walking (Li et al. 2005), fear of crime (Chiricos et al. 1997) or distress (Hill et al. 2005), a fundamental cause of these environment-health correlates is chronic resource deprivation in disadvantaged neighborhoods over the entire life span, which suggests intersecting influences from race, poverty, and family structure of the residents. Over-emphasizing the current residential environment may obscure the interim social processes such as selective residential patterning and/or housing discrimination.

The second type of potential mis-specification is because temporally-proximate indicators of social disadvantage may show only marginal impact on late-life outcomes largely due to the neglect of their influence in early and mid-life, which shaped the composition of the cohort at advanced ages. For example, Hoffman (2009) demonstrated that mid-life socioeconomic status had a greater impact on patterns of old-age mortality than later-life socioeconomic circumstances because social disadvantage shapes the composition of the birth cohort at all ages, not just in older adulthood. Third, we may risk the "Time 1 Encapsulation" problem identified by Dannefer and colleagues (Dannefer and Kelley-Moore 2009; Dannefer and Uhlenberg 1999) whereby late-life outcomes are over-attributed to early-life social circumstances to the exclusion of interim social context. Neglect of the ongoing and interactive nature of social structure and individual characteristics that occur *throughout the life course* over-emphasizes starting berth (in tournament framework, c.f., Rosenbaum 1978) or early-life circumstances.

Despite the growing recognition that social structural circumstances provide and limit opportunities for individuals over the entire life course, scarce attention has been paid to the *mechanisms of selection*, particularly in mid-life, that influence later-life outcomes (George 2005). To date, solutions to the unobserved heterogeneity "problem" due to age-truncated (left-censored) study designs have been statistical in nature, even using simulated data to adjust estimates in an observed sample (Gutierrez 2002). Although study designs have greatly improved in recent decades, there are still many gaps in our knowledge about early and mid-life circumstances. Beyond selective mortality, there are myriad social selection processes that occur at all ages and influence what we are able to observe in older adulthood. Investigators should consider carefully appropriate study designs that may allow the identification of potential selection processes over the entire life course, articulate the mechanisms at work, and characterize the impact of this "left selection" on later-life outcomes.

Unobserved Heterogeneity Due to Single Methodologies

The growing availability of large-scale panel studies in the past decades has changed the scope of inquiry and the specificity of measurement available to scholars interested in age and the life course. Although this type of quantitative, survey-based methodology has been our primary focus up to this point in the chapter, it is important to note that the exclusive use of any single methodology has the

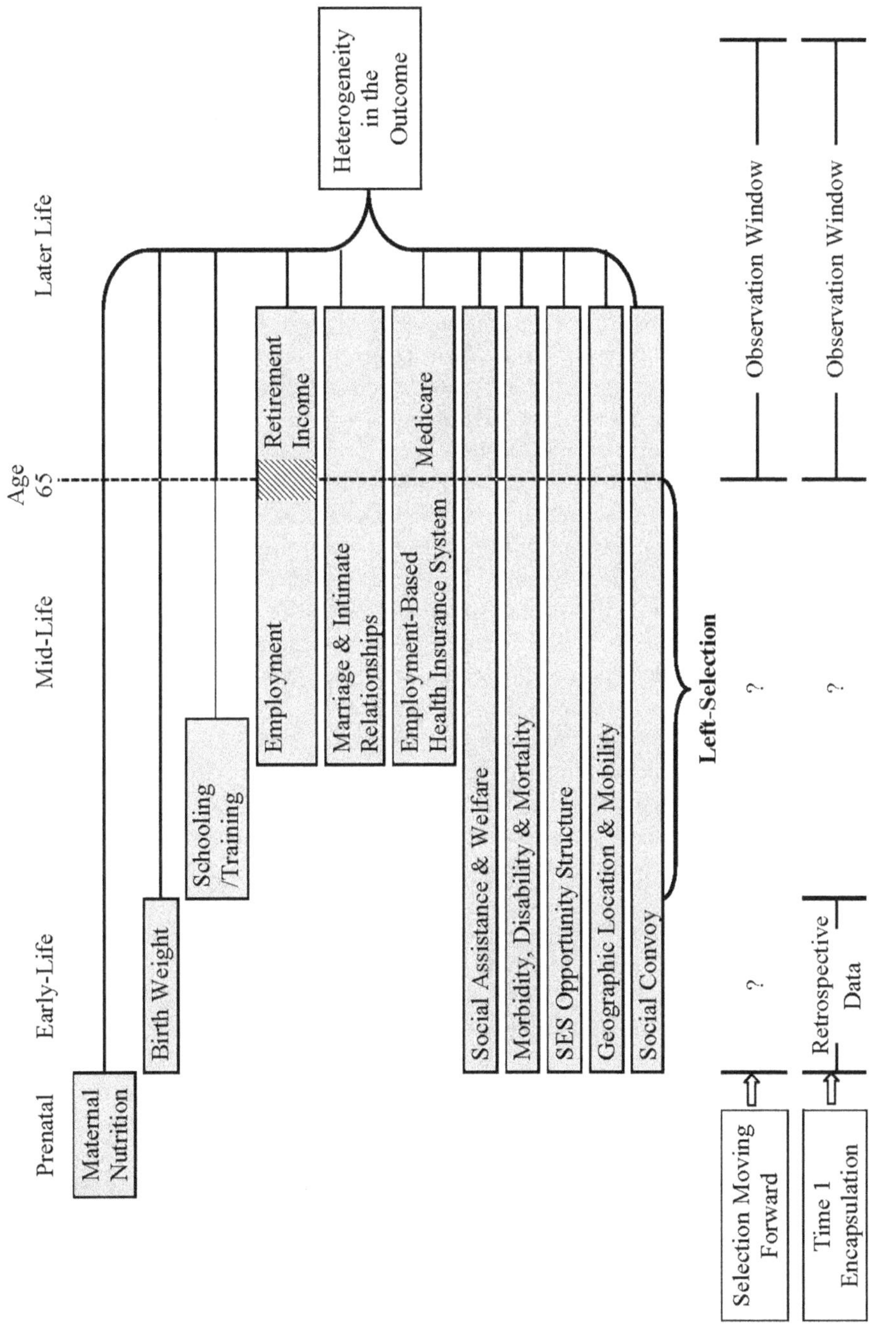

Fig. 4.4 Life course selection processes

risk of obscuring underlying social processes that may contribute to differentiation in life paths. The third type of "unobserved" heterogeneity discussed here is the variability in lived experiences that cannot be explained or confirmed due to reliance on a single methodological approach to the question under study. This type of unobserved heterogeneity, aptly addressed, represents a rich source of new theorizing about the influence of the social world on individual lives over the life course.

Scholars of age and the life course widely acknowledge that there are limitations in existing studies that do not allow for identification of underlying processes or finer characterizations of phenomena (Holstein and Gubrium 2000; Riley 1983; Sørensen 1986). This critique is commonly leveled on quantitative studies such as surveys and administrative databases since many important concepts are limited to only a handful of indicators (Dannefer 2011). Yet the utilization of any single methodology, including qualitative approaches, can leave sources and patterns of population variability over the life course largely unstudied – even undetected. The type and pattern of variability that remains obscured varies by the phenomenon under study and the methodology put to use. In this section, we discuss the scope of the problem stemming from the utilization of a single methodology in studies of age and the life course and then provide examples how a mixed-methods approach could either *explain observed variability* or *confirm hypothesized variability* in the population, opening the door for new and better-articulated social explanations.

Just as we noted in Section "Unobserved Heterogeneity Due to Standard Statistical Practice" about standard statistical practices, the challenges of relying on a single methodology are not unique to sociology, but do have significant consequences for the types of inquiry we make and the conclusions that we draw. Because of the complex social world in which individual lives are embedded, age-based patterns of variability may not be fully detected or explained with a single methodology. A multi-dimensional study could open the vista for inquiry into the sociogenic influences that lead to differentiation (or lack thereof) in the population at any point in the life course, allowing for a better articulation of social patterns or the discovery of new social paths that had been previously obscured due to the primary methodological lens applied to the problem.

Singer and Ryff (2001:44) were among the first to advocate for the blending of qualitative and quantitative methodologies in the study of age and the life course, specifically noting the "... dearth of analytic strategies for producing representations of whole lives and for aggregating them into meaningful taxonomies that facilitate the understanding of how given outcomes come about." More than simply an ad hoc approach where a supplemental qualitative component is added to a largely quantitative study (or vice versa), mixed methodological approaches are emerging as a paradigm of inquiry that triangulates evidence regarding the phenomenon under study. Mixed-methodology is accomplished through interacting elements of qualitative and quantitative approaches to inquiry and integrating methodologies that utilize a specific type of reasoning, including inductive, deductive, or abductive (Creswell 2009). Although there are a number of other important contributions that mixed-methods can make to scientific inquiry, we focus here on the potential gains in studies of age and the life course. Below, we present two examples where a mixed methodological design would help explicate the underlying processes and mechanisms that lead to age-related variability in the population – one that utilizes an explanatory approach and one that utilizes a confirmatory approach.

First, a mixed-methods design can be useful when a quantitative study documents substantial variability in a particular outcome and may even detect subgroup (e.g., gender, race) differences but cannot capture sufficiently the underlying forces that drive the pattern of variability. A secondary methodology can be used to *explain* the patterns of variability observed in the primary study. For example, surveys and administrative databases (e.g., MDS) that capture segments of the older adult population typically use the standardized assessment scales of Activities of Daily Living (Katz et al. 1963) and Instrumental Activities of Daily Living (Lawton and Brody 1969) to assess dependency or need for assistance on a limited set of tasks. These measures have been critiqued because they poorly identify those with mild to moderate limitations (Guccione et al. 1990), those whose functional limitations fluctuate over time (Gill and Kurland 2003), or those who have developed successful adaptation strategies (Atchley 1998; Gignac et al. 2000). Empirical studies continually note that *despite equivalent levels of ADL or IADL dependency*, we observe significant variability in social engagement (Mendes de Leon et al. 2003), mental health (Albrecht and Devlieger 1999), and self-rated health (Cott et al. 1999). Integrating a qualitative component into the quantitative study could capture the narratives individuals construct about their functional limitations, which may *explain* the observed variability in well being and social engagement despite similar levels of functional dependency.

Second, a mixed-method approach can be used to *confirm hypothesized* patterns of age-based variability. This is particularly critical in cases where there is a strong conceptual foundation for hypothesizing variability in a particular outcome but the primary methodology cannot detect it. For example, both qualitative and quantitative studies have demonstrated that when one partner of a married couple becomes critically ill or develops severe functional limitations, there tends to be a decrease in mental health for both partners (Northhouse et al. 2007) and an increase in caregiver stress for the other partner (Sherwood et al. 2005). However, subsequent changes in marital satisfaction after a partner becomes ill or disabled yields mixed results (Suitor and Pillemer 1994; Wolff and Agree 2004; Wright and Aquilino 1995). Drawing from conceptual work on the socially-constructed life course, one could reasonably hypothesize that there is an *age-based pattern* in the degree of change in marital satisfaction when one partner becomes ill. Specifically, couples who consider disability or illness to be an "off time" event, perhaps even biographical disruption (Bury 1982), may experience greater marital strain or take longer to make adjustments in the dyad that are satisfactory to both. Degenerating health conditions and functional limitations are negotiated within a marital dyad daily, weekly, or monthly such that annual follow-up surveys may be too far apart temporally to capture variability in marital adjustment, obscuring an age-based pattern that unfolded between observation points. A complementary methodology that examines marital dynamics repeatedly in a shorter time window could *confirm* whether there are truly age differences in subsequent marital satisfaction and adaptation to an ill or limited partner.

This type of unobserved heterogeneity, which cannot be explained or confirmed without a complementary methodology, is a greater challenge to the field of age and the life course than the previous two types discussed. This is because it requires careful consideration of conceptual and theoretical frameworks in which we work to even develop a mixed-methodological approach to the empirical question. Mixed-methods research attempts to incorporate insights offered by both qualitative and quantitative methods into a single study design to provide the most accurate answer to the research question under study. These selected methodologies are part of a broader integrated research design, where the methodological components may be sequential, parallel, simultaneous, or even unbalanced (Creswell 2009; Creswell et al. 2003). With such powerful tools and flexibility in research design, social scientists are now able to identify, document, and address heterogeneity in the population, including age-based patterns of variability.

Conclusion

Social inquiry into phenomena associated with age has grown in the past decades, fueled by increased sophistication in statistical methodology and study designs. As a result, sociologists have had more voice in the broader inquiry about what it means to age in society by documenting the myriad ways that social structure influences the individual and collective life course. The success in generating a broader recognition of social influences in human lives is evident by the range of disciplines that now acknowledge the social environment as a critically important causal factor for human development and later-life outcomes (Guo and Stearns 2001; Shanahan and Hofer 2005). Yet, as this chapter has set out to demonstrate, we are only at the frontier of recognizing and identifying the range of social influences that frame individual and collective experience over the life course. To date, the dominant paradigm of inquiry for sociologists of age and the life course has focused on the "average" experience using prospective data. As a discipline, we should work to integrate empirical and conceptual consideration of age-based patterns of variability around these presumably normative phenomena.

Our description of "unobserved" heterogeneity in standard statistical practice, study designs, and single-methodological approaches was designed to demonstrate the range and scope of social inquiry that has yet to be addressed systematically but has the potential to radically shift not just our understanding of individual or collective lives over the life course but the complete paradigm of inquiry about what it means to age in society. This age-based heterogeneity, which is undocumented and even undetected in some cases, is within the grasp of social inquiry through systematic empirical investigation. Yet it requires commensurate conceptual attention to the influence of the social environment on individual lives and potential age-based patterning in the variability that may or may not be directly captured in studies.

In section "Unobserved Heterogeneity Due to Standard Statistical Practice," we present unobserved heterogeneity due to the standard statistical practice of relying on measures of central tendency to the neglect of age-based patterns of variability around those central measures. In this case, we argue that the variability is already captured in observational studies, but has been neglected in empirical tests due to over-reliance on middle observations leading to a lack of conceptual attention to the social constructs within which individual lives are embedded. This type of unobserved heterogeneity can be addressed most readily because much of the variability is already captured in existing data sources and the statistical tools exist to conduct such systematic observation.

In section "Unobserved Heterogeneity Due to Study Design," we discuss unobserved heterogeneity due to limitations of study designs, specifically age-truncation of samples and few, if any, indicators of early-life, mid-life, or ongoing processes that may lead to age-based patterns in variability within a cohort. This type of heterogeneity is unobserved because it is not captured in current study designs but is ripe for conceptual and empirical attention – particularly to ongoing life course processes rather than encapsulating the effects into life stages or over-emphasizing a single early life event. In section "Unobserved Heterogeneity Due to Single Methodologies," we introduce a third type of unobserved heterogeneity, which is due to the utilization of a single methodology. Variability in process and outcomes may remain unexplained or unconfirmed in studies that utilize a single mode of inquiry. Creating an integrated design that would triangulate information around the phenomenon under study can strengthen our understanding about range of social influences over the life course – even those that are subtle or short-lasting. Although there is great value in large-scale omnibus panel surveys, sociologists must recognize that these cannot detect, confirm, or explain all patterns of age-based variability that exist.

In sum, the future challenges for sociology of age and the life course include the following: sufficiently characterizing the variability in phenomena that are already captured in observational studies; addressing potential life course (collective and individual) pathways and social selection processes that produce heterogeneity or homogeneity in late-life outcomes, but may be predominantly unobserved in studies currently; and designing multi-methodological studies to identify and characterize the myriad life course pathways. By widening the view of social inquiry to include these as-yet minvestigated sources and patterns of age-based heterogeneity, studies of age and the life course can reclaim the discourse about what it means to age "normally."

Acknowledgments Special thanks to Dale Dannefer, David Warner, and Duane Alwin who provided comments on a previous draft.

References

Albrecht, Gary L. and Patrick J. Devlieger. 1999. "The Disability Paradox: High Quality of Life Against All Odds." *Social Science & Medicine* 48:977–88.

Allison, Paul D. 2000. "Multiple Imputation for Missing Data: A Cautionary Tale." *Sociological Methods and Research* 28:301–09.

Alwin, Duane F. 2007. *Margins of Error: A Study of Reliability in Survey Measurement.* Hoboken, NJ: Wiley.

Alwin, Duane F. and Ryan J. McCammon. 2001. "Aging, Cohorts and Verbal Ability." *Journal of Gerontology: Social Sciences* 56B:S151–61.

Alwin, Duane F., Ryan J. McCammon, Linda A. Wray and Willard L. Rodgers. 2008. "Population Processes and Cognitive Aging." Pp. 69–89 in *Handbook of Cognitive Aging: Interdisciplinary Perspectives*, edited by S. M. Hofer and D. F. Alwin. Thousand Oaks, CA: Sage.

Alwin, Duane F., Scott M. Hofer, and Ryan J. McCammon. 2006. "Modeling the Effects of Time: Integrating Developmental and Demographic Perspectives." Pp. 20–38 in *Handbook of Aging and the Social Sciences*, 6th ed., edited by R. H. Binstock and L. K. George. San Diego, CA: Academic.

Atchley, Robert C. 1998. "Activity Adaptations to the Development of Functional Limitations and Results for Subjective Well-Being in Later Adulthood: A Qualitative Analysis of Longitudinal Panel Data Over a 16-year Period." *Journal of Aging Studies* 12:19–38.

Baker, David W., Joseph J. Sudano, Jeffrey M. Albert, Elaine A. Borawski, and Avi Dor. 2001. "Lack of Health Insurance and Decline in Overall Health in Late Middle Age." *The New England Journal of Medicine* 345:1106–12.

Baltes, Paul B. and John R. Nesselroade. 1979. History and Rationale of Longitudinal Research. Pp. 1–39 in Longitudinal Research in the Study of Behavior and Development, edited by J. R. Nesselroade and P. B. Baltes. New York: Academic.

Bancroft, Gertrude. 1958. *The American Labor Force: Its Growth and Changing Composition.* New York: Wiley.

Barker, D. J. P. 2001. "Fetal and Infant Origins of Adult Disease." *Monatsschrift Kinderheilkunde* 1:S2–6.

Bass, Scott A., Elizabeth Ann Kutza and Fernando M. Torres-Gil. 1990. *Diversity in Aging.* Glenview, IL: Scott, Foresman.

Beyer, Glenn H. and Margaret Woods. 1963. *Living and Activity Patterns of the Aged. Research Report No. 6.* Ithaca, NY: Center for Housing and Developmental Studies, Cornell University.

Brown, Tyson H. and David F. Warner. 2008. "Divergent Pathways? Racial/Ethnic Differences in Older Women's Labor Force Withdrawal." *Journal of Gerontology: Social Sciences* 63B:S122–34

Browne, William J., David Draper, Harvey Goldstein, and Jon Rasbash. 2002. "Bayesian and Likelihood Methods for Fitting Multilevel Models with Complex Level-1 Variation." *Computational Statistics and Data Analysis* 39:203–25.

Bures, Regina M. 2003. "Childhood Residential Stability and Health at Midlife." *American Journal of Public Health* 193:1144–48.

Bury, Michael. 1982. "Chronic Illness as Biographical Disruption." *Sociology of Health and Illness* 4:167–82.

Chiricos, Ted, Michael Hogan and Marc Gertz. 1997. "Racial Composition of Neighborhood and Fear of Crime." *Criminology* 35:107–31.

Cohen, Jacob. 1994. "The Earth is Round (p < .05)." *American Psychologist* 49:997–1003.

Cott, Cheryl, Monique A. M. Gignac and Elizabeth M. Badley. 1999. "Determinants of Self Rated Health for Canadians with Chronic Disease and Disability." *Journal of Epidemiology and Community Health* 53:731–36.

Creswell, John W. 2009. *Research Design: Qualitative, Quantitative, and Mixed Methods Approaches.* New York: Sage.

Creswell, John W., Vicki L. Plano Clark, Michelle L. Gutmann and William E. Hanson. 2003. "Advanced Mixed Methods Research Design." Pp. 209–40 in *Handbook of Mixed Methods in Social & Behavioral Research*, edited by A. Tashakkori and C. Teddlie. New York: Sage.

———. 1988. "Differential Aging and the Stratified Life Course." Pp. 3–36 in *Annual Review of Gerontology and Geriatrics: Varieties of Aging*, Vol. 8, edited by M. P. Lawton and G. Maddox. New York: Springer.

———. 2011. "Aging, the Life Course and the Sociological Imagination: Prospects for Theory." Forthcoming in *Handbook of Aging and Social Sciences*, 7th ed., edited by R. H. Binstock and L. K. George. San Diego, CA: Academic.

Dannefer, Dale and Jessica A. Kelley-Moore. 2009. "Theorizing the Life Course: New Twists in the Paths." Pp. 389–412 in *Handbook of Theories of Aging*, 2nd ed., edited by V. L. Bengtson, D. Gans, N. M. Putney and M. Silverstein. New York: Springer.

Dannefer, Dale and Ralph R. Sell. 1988. "Age Structure, the Life Course and 'Aged Heterogeneity': Prospects for Research and Theory." *Comprehensive Gerontology* 2:1–10.

Dannefer, Dale and Peter Uhlenberg. 1999. "Paths of the Life Course: A Typology." Pp. 306–26 in *Handbook of Theories of Aging*, 1st ed., edited by V. L. Bengtson and K. W. Schaie. New York: Springer.

Doblhammer, Gabriele. 2003. *The Late Life Legacy of Very Early Life.* New York: Springer.

Dupre, Matthew E. 1997. "Educational Differences in Age-Related Patterns of Disease: Reconsidering the Cumulative Disadvantage and Age-As-Leveler Hypotheses."*Journal of Health and Social Behavior* 48:1–15.
Elman, Cheryl and Angela M. O'Rand. 2002. "Perceived Job Insecurity and Entry into Work-Related Education and Training among Adult Workers." *Social Science Research* 31:49–76.
Ferraro, Kenneth F. and Tetyana Pylypiv Shippee. 2009. "Aging and Cumulative Inequality: How Does Inequality Get Under the Skin?" *The Gerontologist* 49:333–43.
Freese, Jeremy. 2008. "Genetics and the Social Science Explanation of Individual Outcomes." *American Journal of Sociology* 114:S1–35.
George, Linda K. 2005. "Socioeconomic Status and Health across the Life Course: Progress and Prospects." *Journal of Gerontology: Social Sciences* 60B:S135–39.
Geronimus, Arline T. 1992. "The Weathering Hypothesis and the Health of African-American Women and Infants: Evidence and Speculations." *Ethnicity and Disease* 2:207–21.
Gignac, Monique A. M., Cheryl Cott and Elizabeth M. Badley. 2000. "Adaptation to Chronic Illness and Disability and Its Relationship to Perceptions of Independence and Dependence." *Journal of Gerontology: Psychological Sciences* 55B:P362–72.
Gill, Thomas M. and Brenda Kurland. 2003. "The Burden and Patterns of Disability in Activities of Daily Living Among Community-living Older Persons." *Journal of Gerontology: Medical Sciences* 58A:M70–75.
Gluckman, Peter D. and Mark A. Hanson. 2006. *Mismatch: Why Our World No Longer Fits Our Bodies*. New York: Oxford University Press.
Goldstein, Harvey. 2005. "Heteroscedasticity and Complex Variation." *Encyclopedia of Statistics in Behavioral Science* 2:790–95.
Guccione, Andrew A., David T. Felson and Jennifer J. Anderson. 1990. "Defining Arthritis and Measuring Functional Status in Elders: Methodological Issues in the Study of Disease and Physical Disability." *American Journal of Public Health* 80:945–49.
Guo, Guang and Elizabeth Stearns. 2002. "The Social Influences on the Realization of Genetic Potential for Intellectual Development." *Social Forces* 80:881–910.
Gutierrez, Roberto G. 2002. "Parametric Frailty and Shared Frailty Survival Models." *The Stata Journal* 2:22–44.
Halaby, Charles. 2003. "Panel Models for the Analysis of Change and Growth in Life Course Studies." Pp. 503–27 in *Handbook of the Life Course*, edited by Jeylan Mortimer and M. Shanahan. New York: Plenum.
Halfon, Neal and Miles Hochstein. 2002. "Life Course Health Development: An Integrated Framework for Developing Health, Policy, and Research." *The Milbank Quarterly* 80:433–79.
Hayward, Mark D., and Bridget K. Gorman. 2004. "The Long Arm of Childhood: The Influence of Early-Life Social Conditions on Men's Mortality." *Demography* 41:87–107.
Hayward, Mark D., Amy Pienta, and Diane K. McLaughlin. 1997. "Inequality in Men's Mortality: The Socioeconomic Status Gradient and Geographic Context." *Journal of Health and Social Behavior* 38:213–34.
Heckman, James J. 2006. "Skill Formation and the Economics of Investing in Disadvantaged Children." *Science* 312:1900–02.
Hill, Terrence, Catherine Ross, and Ronald Angel. 2005. "Neighborhood Disorder, Psychophysiological Distress, and Health." *Journal of Health and Social Behavior* 46:170–86.
Hoffman, Rasmus. 2009. "Socioeconomic Differences in Old Age Mortality in Denmark and the USA – with Special Emphasis on the Impact of Unobserved Heterogeneity on the Change of Mortality Differences over Age." PhD dissertation, University of Rostock, Germany.
Holstein, James A. and Jaber F. Gubrium. 2000. *Constructing the Life Course*, 2nd ed. Dix Hills, NY: General Hall.
Hooker, Karen. 1991. "Change and Stability in Self During Transition to Retirement: An Intraindividual Study Using P-Technique Factor Analysis." *International Journal of Behavior Development* 14:209–33.
Health and Retirement Study (HRS). 2008. "Design History." Retrieved May 18, 2010 (http://hrsonline.isr.umich.edu/sitedocs/DesignHistory.pdf).
Irving, Shalon M. and Kenneth F. Ferraro. 2006. "Reports of Abusive Experiences during Childhood and Adult Health Ratings: Personal Control as a Pathway?" *Journal of Health and Social Behavior* 18:458–85.
Katz, Sidney, Amassa B. Ford, Roland W. Moskowitz, Beverly A. Jackson, and Marjorie W. Jaffee. 1963. "Studies of Illness in the Aged: The Index of ADL, a Standardized Measure of Biological and Psychological Function." *JAMA*185:914–19.
Kelley-Moore, Jessica A. 2006. "Assessing Racial Health Inequality in Older Adulthood: Comparisons from Mixed-Mode Panel Interviews." *Journal of Gerontology: Social Sciences* 61B:S212–20.
Kim, Jinyoung and Emily Durden. 2007. "Socioeconomic Status and Age Trajectories of Health." *Social Science & Medicine* 65:2489–502.
Lantz, Paula, James S. House, James M. Lepkowski, David R. Williams, Richard P. Mero and Jieming Chen. 1998. "Socioeconomic Factors, Health Behaviors, and Mortality: Results from a Nationally Representative Prospective Study of U.S. Adults" *JAMA* 279:1703–08.

Lawton, M. P. and E. M. Brody. 1969. "Assessment of Older People: Self-Maintaining and Instrumental Activities of Daily Living." *The Gerontologist* 9:179–86.
Li, Fuzhong, K. John Fisher, Ross C. Brownson, Mark Bosworth. 2005. "Multilevel Modeling of Built Environmental Characteristics in Relation to Neighborhood Walking Activity in Older Adults." *Journal of Epidemiology and Community Health* 59:558–64.
Lillard, Lee A. Michael J. Brien and Linda J. Waite. 1995. "Premarital Cohabitation and Subsequent Marital Dissolution: A Matter of Self-Selection?" *Demography* 32:437–57.
Lumley, Thomas, Paula Diehr, Scott Emerson, and Lu Chen. 2002. "The Importance of the Normality Assumptions in Large Public Health Data Sets." *Annual Review of Public Health* 23:151–69.
MacMillan, Ross and Scott R. Eliason. 2003. "Characterizing the Life Course as Role Configurations and Pathways: A Latent Structure Approach." Pp. 529–54 in *Handbook of the Life Course*, edited by J. Mortimer and M. Shanahan. New York: Plenum.
Maddox, George L. 1987. "Aging Differently." *The Gerontologist* 27:557–64.
Mansfield, Christopher J., James L. Wilson, Edward J. Kobrinski and Jim Mitchell. 1999. "Premature Mortality in the United States: The Roles of Geographic Area, Socioeconomic Status, Household Type, and Availability of Medical Care." *American Journal of Public Health* 89:893–98.
Mare, Robert D. 1991. "Five Decades of Educational Assortative Mating." *American Sociological Review* 56:15–32.
Markides, Kyriakos S. and Richard Machalek.1984. "Selective Survival, Aging and Society." *Archives of Gerontology and Geriatrics* 3:207–22.
Mendes de Leon, Carlos F., Thomas A. Glass and Lisa Berkman. 2003. "Social Engagement and Disability in a Community Population of Older Adults: The New Haven EPESE." *American Journal of Epidemiology* 157:633–42.
Moen, Phyllis and K. Chermack. 2005. "Gender Disparities in Health: Strategic Selection, Careers and Cycles of Control." *Journal of Gerontology: Social Sciences* 60B:S99–108.
Nelson, E. Anne and Dale Dannefer. 1992. "Aged Heterogeneity: Fact of Fiction? The Fate of Diversity in Gerontological Research." *The Gerontologist* 32:17–23.
NLSY. 2009. "NLSY79 User's Guide." Retrieved from http://www.nlsinfo.org/nlsy79/docs/79html/79text/front.htm on 18 May 2010.
Northhouse, Laurel L., Darlene W. Mood, James E. Montie, Howard M. Sandler, Jeffrey D. Forman, Maha Hussain, Kenneth J. Pienta, David C. Smith, Martin G. Sanda and Trace Kershaw. 2007. "Living With Prostate Cancer: Patients' and Spouses' Psychosocial Status and Quality of Life." *Journal of Clinical Oncology* 25:4171–77.
Oliver, Melvin L. and Thomas M. Shapiro. 1995. *Black Wealth/White Wealth: A New Perspective on Racial Inequality.* New York: Routledge.
Riley, Matilda White. 1973. "Aging and Cohort Succession: Interpretations and Misinterpretations." *Public Opinion Quarterly* 37:35–49.
———. 1983. "On the Significance of Age in Sociology." *American Sociological Review* 52:1–14
Riley, Matilda White and Anne Foner. 1968. *Aging and Society, Vol. 1: An Inventory of Research Findings.* New York: Russell Sage.
Riley, Matilda White, Marilyn Johnson and Anne Foner, eds. 1972. *Aging and Society, Vol. 3: A Sociology of Age Stratification.* New York: Russell Sage.
Rosenbaum, James E. 1978. "The Structure of Opportunity in School." *Social Forces* 57:236–56.
Sampson, Robert J. and John H. Laub. 2003. "Life-Course Desisters? Trajectories of Crime among Delinquent Boys followed to Age 70." *Criminology* 41:301–39.
Sampson, Robert J. and Patrick Sharkley. 2008. "Neighborhood Selection and the Social Reproduction of Concentrated Racial Inequality." *Demography* 45:1–29.
Shanahan, Michael J. and Scott M. Hofer. 2005. "Social Context in Gene-Environment interactions: Retrospect and Prospect." *Journal of Gerontology: Social Sciences* 60B:65–76.
Sherwood, Paula R., Charles W. Given, Barbara A. Given, and Alexander von Eye. 2005. "Caregiver Burden and Depressive Symptoms: Analysis of Common Outcomes in Caregivers of Elderly Patients." *Journal of Aging and Health* 17:125–47.
Singer, B., and Ryff, Carol D. 2001. "Understanding Aging via Person-Centered Methods and the Integration of Numbers and Narratives." Pp. 44–65 in *Handbook of Aging and Social Sciences*, 5th ed., edited by R. H. Binstock and L. K. George. San Diego, CA: Academic.
Sørensen, Aage Bøttger. 1986. "Social Structure and Mechanisms of Life Course Processes." Pp. 177–97 in *Human Development and the Life Course: Multidisciplinary Perspectives*, edited by Aage Bøttger Sørensen, Franz E. Weinert and Lonnie R. Sherrod. Hillsdale, NJ: Erlbaum.
Suitor, J. Jill and Karl Pillemer. 1994. "Family Caregiving and Marital Satisfaction: Findings from a 1-Year Panel Study of Women Caring for Parents with Dementia." *Journal of Marriage and the Family* 56:681–90.
Thompson, Wayne E., Gordon F. Streib and John Kosa. 1960. "The Effect of Retirement on Personal Adjustment: A Panel Analysis." *Journal of Gerontology* 15:165–69.

Uhlenberg, Peter. 2009. "Children in an Aging Society." *Journal of Gerontology: Social Sciences* 64B:S489–96.
Verbrugge, Lois M. and Alan M. Jette. 1994. "The Disablement Process." *Social Science & Medicine* 38:1–14.
Warner, David F., Mark Hayward and Melissa Hardy. 2010. "The Retirement Life Course in America at the Dawn of the Twenty-First Century." *Population Research and Policy Review* 29:893–919.
Willson, Andrea E. 2003. "Race and Women's Income Trajectories: Employment, Marriage, and Income Security over the Life Course." *Social Problems* 50:87–110.
Willson, Andrea E., Kim M. Shuey, and Glen H. Elder, Jr. 2007. "Cumulative Advantage Processes as Mechanisms of Inequality in Life Course Health." *American Journal of Sociology* 112:1886–924.
Wilmoth, Janet and Gregor Koso. 2002. "Does Marital History Matter? Marital Status and Wealth Outcomes among Preretirement Adults." *Journal of Marriage and Family* 64:254–68.
Wolff, Edward N. 2002. "Income, Wealth and Late-Life Inequality in the United States." Pp. 31–58 in *Annual Review of Gerontology and Geriatrics: Varieties of Aging*, Vol. 8, edited by M. P. Lawton and G. Maddox. New York: Springer.
Wolff, Jennifer L. and Emily M. Agree, 2004. "Depression among Recipients of Informal Care: The Effects of Reciprocity, Respect, and Adequacy of Support." *Journal of Gerontology: Social Sciences* 59B:S173–80.
Wright, Debra L. William S. Aquilino. 1995. "Influence of Emotional Support Exchange in Marriage on Caregiving Wives' Burden and Marital Satisfaction." *Family Relations* 47:195–204.
Yang, Yang. 2007. "Is Old Age Depressing? Growth Trajectories and Cohort Variations in Late-Life Depression." *Journal of Health and Social Behavior* 48:16–32.
Yang, Yang and Linda C. Lee. 2009. "Sex and Race Disparities in Health: Cohort Variations in Life Course Patterns." *Social Forces* 87:2093–124.
Ziliak, Stephen T. and Deirdre N. McCloskey. 2008. *The Cult of Statistical Significance: How the Standard Error Costs Us Jobs, Justice, and Lives.* Ann Arbor, MI: The University of Michigan Press.

Part III
Social Diversity and Inequalities of Aging

Chapter 5
Gender and Aging

Susan Venn, Kate Davidson, and Sara Arber

Over the last 30 years, the field of aging has been the site of an exceptional growth of research interest, yet it is only really within more recent social gerontology that the many varied experiences of older people are being acknowledged and explored through the intersection of race, ethnicity, sexuality, class, and, as many have argued, most importantly, gender. This chapter will show the importance of recognizing the gender dimension within the study of aging by exploring how using a "gendered eye," or adopting a "gender lens" (Calasanti and Slevin 2001) not only reveals neglected issues for older people, but is also fundamentally important in thinking about the study of old age at a time when the growth in the aging population is unprecedented.

The chapter commences by stepping back 30 years to reflect on theoretical approaches to the study of aging and approaches that initially neglected gender. We then move on to demonstrate how the contribution of feminist scholarship raised awareness of the need to identify and address the neglected area of older women's issues, and at the same time neglecting older men's issues (Thompson 1994). Next, we consider how this approach ultimately evolved to take account of the interlinkages between age and gender relations in recognition of the need to examine men's and women's experiences relationally.

Emerging issues within this new era of age and gender relations are introduced in the form of the changing nature of roles, relationships, and older people's identity. The following section focuses on the aging, gendered body through discussion of the tension between aging identities and narratives of bodily decline as experienced differently by men and women (Gullette 1997). Finally, the chapter concludes by highlighting challenges for the future of aging for men and women, such as the growing numbers of the oldest old, who are predominantly widowed women, implications for unmarried or childless older men and women, and the policy implications of such challenges.

Introducing the Gendered Dimension

Reflecting the dominant cultural views of the time, earlier theoretical approaches to aging focused on retirement as a marker for the commencement of old age (Midwinter 1997; Phillipson 1998), rather than as a physiological process (Townsend 1981; Gilleard and Higgs 1998). With old age thus categorized by retirement, and therefore socially constructed, approaches to the field of aging were

S. Venn (✉)
Centre for Research on Ageing and Gender (CRAG) Department of Sociology
the University of Surrey, Surrey, UK
e-mail: s.venn@surrey.ac.uk

R.A. Settersten, Jr. and J.L. Angel (eds.), *Handbook of Sociology of Aging*, Handbooks of Sociology and Social Research, DOI 10.1007/978-1-4419-7374-0_5, © Springer Science+Business Media, LLC 2011

primarily policy oriented. Subsequently, "old age," and the issues that accompany aging such as poverty and health problems became perceived as a "problem" or "burden" to society.

Having conceptualized that old age is "problematic," several different theoretical approaches to studying the "problem" of aging were posited, and disputed over the last half century. Disengagement Theory (Cumming and Henry 1961), for example, although conceived some 50 years ago, continues to resonate with lay perceptions of how and why older people come to terms with withdrawal from the labor force, society, and ultimately, life. Criticized widely as an outdated positivist approach, the theory was challenged by those proponents of the Activity (Havighurst 1963) and Continuity (Atchley 1971, 1987) theories. The Structured Dependency theory of the 1980s saw old age as socially rather than biologically or psychologically constructed, predicated on mandatory retirement with the ensuing loss of income, social status, and self-identity.

However, these theories, ground-breaking in their conceptualization of later life, were largely gender blind. Cumming and Henry's research was carried out in mid-western conservative Kansas in the 1950s, and the vast majority of their participants were (white, physically able) men who had experienced long-term employment. Women in mid-century post-World War II America were less likely to have been long-term employed, and thus did not experience the disengagement stimulated by retirement. Similarly, Havighurst (1963) and Atchley (1971) concentrated on older people who were mainly physically fit and could afford to continue active life after retirement, without recourse to the constraints that impact men and women differently in later life. There was little recognition of those people with disabilities and/or in poverty who are, in the main, women.

In the 1980s, writers such as Walker (1981), Phillipson (1982), and Estes (1979) moved away from traditional gerontological approaches toward a political economy perspective, which called attention to how an individual's position in the labor market is a determining factor for subsequent inequalities in life post-retirement. For example, Calasanti (1986) was among early feminists, who challenged the male paradigm of production in a formal economy by highlighting the vital, unpaid work carried out particularly by older women. This was later echoed by Estes (1991), who pointed out how the gendered nature of the division of labor is an extra causal factor in the post-retirement position of women. In essence, as Bury (1995) quoting Walker (1981), explains, "the effects of a weak position in the labor market, prior to retirement, meant restricted access to a 'wide range of resources' and resulted in 'the imposition of depressed social status in old age'" (Bury 1995:18; Walker 1981:88). However, Estes (1991) recognized that this perspective is rooted in the predominantly male experience of employment and economic security and does not adequately address issues of women and members of minority ethnic groups. Their different work pattern means that women and other minority groups are often excluded from the labor market long before the official age of retirement. Thus, the political economy perspective expanded in the 1990s to take into account other dimensions of inequality.

Earlier social structural approaches failed to address the varied and diverse nature of the aging experience for men and women, including diversity *within* the gender categories such as according to race, ethnicity, class, and sexual orientation. The political economy approach to aging expanded in the later 1980s and 1990s to take into account other dimensions of inequality by including race and ethnic inequalities as well as gender.

During the 1990s, feminist insights sought to further raise awareness of the diversity of the aging experience by examining these many cross-cutting influences on the aging experience for women (Arber and Ginn 1995). Although remaining within a broadly political economy focus, they demonstrated the extent of structured inequalities experienced by women in later life by highlighting how women do not merely "do gender," they also "do" race, ethnicity, class, and sexual orientation (Krekula 2007).

The numerical predominance of women in later life, and the particular "problems" faced by older women became identified as "the feminization" of later life. Such feminist emphasis was placed largely on the disadvantaged position of older women in relation to their pensions, health status, and access to care (Arber and Ginn 1991; Estes 1991). Thus, a feminist political economy approach sought to redress the earlier neglect of issues that are predominantly related to women in later life, as, they observed, women are more likely to face greater constraints, and have access to fewer resources (Arber and Ginn 1991; Bernard and Phillips 1998; Bury 1995).

Yet, though feminists challenged the theoretical perspectives that omitted aging women, they have also themselves been subject to criticism for not only neglecting to recognize the diversity of the aging experience, but also for ignoring men (Katz 2005). As an attempt to counter these criticisms and to integrate age, with gender, and other dimensions of inequality, such as race and ethnicity, a fresh approach was adopted, whereby men's and women's experiences of aging are seen in relation to each other.

Emergence of Gender Relations Through a "Gender Lens"

The adoption of a "gender lens" arose out of a move to counter critics of a feminist approach to aging, where only women's issues were seen to be addressed, albeit in recognition of other, intersecting, social identities such as race, class, and ethnicity. This approach to gender and aging was introduced in recognition of how men's and women's experiences of aging are shaped in relation to one another, rather than as a separate and unique experience for each gender individually (Calasanti and Slevin 2001). Thus utilizing a "gender lens" leads to a more all-inclusive and wide-ranging understanding of old age and offers a framework within which to understand men and women's varied experiences of aging.

McMullin (1995) emphasized that it is important not just to "add" women into research where men are the reference group, with women seen as an "add-on" after the fact, and their experiences are only viewed in light of their difference from men's experiences. Calasanti and Slevin (2001:34) explain:

> Research conducted in this vein not only constructs women as "other," but also obscures the similarities between women and men. It also mystifies gender relations. For example, research on employed women often considers the impact of their family roles, but this is often neglected in research on employed men. Yet family plays an important role in the careers of both.

Emphasizing the need to understand gender relationally, Calasanti and Slevin (2001) also argued that gender should not be the remit of women solely, such as when asking only women about gendered roles within the family, as this would, in essence, imply that such matters do not influence men.

An example of research that adopted a "gender" lens is a recent study on older people's experiences of poor sleep (SomnIA, Sleep in Aging), where equal numbers of older men and women who were aged 65–95 years were interviewed in their own homes to gain an understanding of how their diverse lifecourse experiences shaped and influenced their experiences and management of poor sleep (Venn and Arber 2011). Additionally, varying health status, class, and age groups were incorporated in the study design, to ensure that other social dimensions were included. Caregiving for a partner was often given as a reason for disturbed sleep, and, echoing Calasanti's (2006) study on spousal caregiving (see later), both women and men were engaged in physical and emotional support for their partners during the night, as well as during the day.

Disadvantages and Advantages Facing Older Men and Women

Adopting the use of a "gender lens" by feminists has allowed for an exploration of both the disadvantages *and* advantages faced by men and women in later life. For example, examining employment and retirement has not only drawn attention to gender inequalities in pension provision, but also raised awareness of the neglected issue of unpaid work, largely undertaken by women. Access to pension provision has previously been male focused, which is largely designed with men in mind, assuming long term, continuous employment leading to access to an income in retirement. Women, on the contrary, were expected to receive income and benefits indirectly through their marital relationship (Ginn 2003).

Changes to existing pension schemes have attempted to address gender inequalities, but inherent disadvantages faced by women pre-retirement still lead to them tending to have lower retirement income than men. Several factors contribute to the disadvantaged status of women with regard to pensions:

- The influence of life events such as marriage, divorce, and childbirth affect women's participation in the labor market and therefore access to pension provisions (Ginn 2003; DWP 2005). Married women are more likely to be dependent on their husband or the state for their pension provision than men, and women are more likely to work part-time or withdraw from the labor market when they have children (Ginn 2003; Price and Ginn 2003).
- Women are also more likely to undertake unpaid labor, such as when caring for sick or elderly relatives and therefore are more reliant on state pensions (Ginn 2003).
- Women's pay does not increase over time at the same rate as men's, resulting in a gendered pay gap, which continues after retirement (McMullin 1995).
- Women live longer than men so that their pension will have to pay out for longer (DWP 2005).
- The majority of older women are widowed, and alongside divorced older women, widows are particularly financially disadvantaged in later life because of their reduced employment participation (Ginn 2003; Arber 2004).

However, not all pension provision disadvantages women. For example, in the UK, where employees have the right to swap some pension income for a lump sum on retirement, men face disadvantages, as the lump sum they may be entitled to withdraw could be less than women's, because of the expectation that women will live longer, based on actuarial calculations by insurance companies (Goldsmith 1993).

Furthermore, Arber et al. (2003) highlighted other *advantages* older women often face, in relation to some older men, such as in better social relationships with friends and family through their role as "kin-keepers" (Finch and Mason 1993). Older women living alone are also more likely to draw on their social networks and neighbors to sustain connections within their local communities (Walker and Hiller 2007). Following widowhood, on the one hand, many older women also report experiencing a greater sense of freedom and autonomy, and would choose not to give this up in order to re-marry. On the other hand, older men do not always have the same opportunities to access social networks as women, and may not experience the same positive perceptions of widowhood (Davidson et al. 2003).

Redressing Imbalances: Changing Partnership Status and Older Men

Building on their previous work in the 1990s of integrating gender into the analysis of aging, Arber et al. (2003) sought to redress two imbalances within earlier feminist scholarship. First, they acknowledged the importance of the changing nature of relationships in later life since the latter years of the twentieth century. Second, they highlighted the long-term neglect of the meanings of masculinity and identity to older men. Men and masculinities have been the subject of much sociological research in recent years, such as Kimmel et al. (2004), yet very little of this research has addressed older men (Meadows and Davidson 2006).

Partnership Status

Those entering later life in the early years of the twenty-first century differ from earlier cohorts, not only just in economic terms, but also in the nature of their relationships (Arber et al. 2003). Changes to family structure are increasing, such as a rise in the number of couples who divorce, more couples cohabiting and subsequently separating, and more people in Living Apart Together relationships, resulting in an increase in multiple partnership experiences over the lifecourse (Price and Ginn 2003; Borell and Karlsson 2003). In Sweden those aged between 65 and 69 are more likely to be divorced than widowed. These changes in partnership and family status have many implications in later life, especially in terms of gender disparities (Borell and Karlsson 2003).

For example, in terms of pensions, as mentioned earlier, women over the age of 60 living on their own in the UK are at greater risk of poverty compared with other older people, especially if they are divorced, largely because of lower pension levels. Additionally, divorced older women do not have access to any proportion of their husbands' pension, unlike widows (Ginn 2003; Bardasi and Jenkins 2004). In the US, pension provision similarly reflects a gender bias with substantially fewer women receiving a pension than men, and unmarried women expecting a pension of about half that of unmarried men (Calasanti and Slevin 2001). However, being divorced or separated may also have a negative effect on men's pensions, as this group are more likely to have experienced unemployment and loss of wealth or capital to a former spouse (Price and Ginn 2003), whereas the income of widowers differs little from that of married men of equivalent age (Arber 2004).

Changes in later life relationships and living arrangements not only have financial implications. Davidson (2002) found that older widowed women often did not choose or desire to be married again. Rather they expressed a preference to maintain autonomy over their lives and not to return to looking after a man in their later years; whereas most men who had been widowed wanted to marry again out of a desire for companionship, although opportunities to remarry may decrease with age and frailty. The wish for autonomy in later life is also exemplified by "Living Apart Together" relationships, where couples can experience intimacy and companionship, yet have the freedom to undertake their own daily activities through maintaining their own living space. This emerging type of relationship is particularly favored by older women (Karlsson and Borell 2002; Borell and Karlsson 2003).

Finally, a growing body of research is attempting to redress the neglect of older gay and lesbian relationships in the field of gerontology. Although there has been an increasing recognition of the need to understand diversity and heterogeneity in the study of aging such as including race, ethnicity, and gender, it is only more recently that sexuality in later life has been discussed, and within discussions of sexuality, the experiences of older gay and lesbian adults have largely been ignored (Cronin 2004; Calasanti and Slevin 2001). Cronin (2004) and Heaphy (2007) have shown that although many of the experiences of heterosexual older people are the same for lesbian and gay older people, there are additional issues that pertain specifically to the latter. For example, lesbian or gay older adults are more likely to live on their own, are less likely to receive social support from family members, but are more likely to have stronger friendship networks that provide support (Cronin 2004).

Challenges for Men in Later Life

Although there has been a growth in research into issues that pertain specifically to men and masculinities in recent years, the focus has largely been on younger men (Kimmel and Messner 2001; Whitehead 2002), and challenges to masculinity in later life has, with a few exceptions, been ignored (see Thompson 1994; Meadows and Davidson 2006). It has been contended that aging for men is less problematic than for women, both in terms of financial implications and fewer expectations of men to maintain a youthful appearance in later life (Sontag 1978). More recently, however, there has been a suggestion that older men may have to face their own particular challenges arising from changes that occur through the loss of normative roles and relationship status, which usually emphasizes masculine autonomy and power.

Particular challenges face divorced older men who experience greater loneliness because of reduced contact with children and other family members following marital breakdown (Dykstra and de Jong Gierveld 2004). Trends indicate that the percentage of divorced men aged 65–74 will increase from 9 to 17% between 2007 and 2031, and the proportion of divorced men over 75 will increase from 4% in 2007 to 12% in 2031, whereas the proportion of widowed women aged 65–74 will decrease from 21 to 10% and those aged 75+ will fall from 61 to 40% (ONS 2009). Therefore, more recent scholarship has emphasized the importance of recognizing those older men who are divorced or widowed and the implications for their social and physical health and wellbeing (Arber et al. 2003).

For example, Davidson et al. (2003) addressed the issue of the social participation of older divorced and never married men, who are more likely to be socially isolated and to have poorer health than those who remain married in later life. They found that men who lived on their own were reluctant to join local community groups such as day centers, largely because they were perceived as being aimed at older people, and at older working-class women in particular. Policy changes, which adapt such community groups to make them more attractive to older men, as well as older women, could potentially make a difference, as they suggest:

> In order to attract older men, attention should be paid by local authority and voluntary organizations to offering appropriate facilities and activities for older men so that they may be supported in leading socially integrated and independent lives within the community (Davidson et al. 2003:88).

Attention has also been drawn to the challenges that men face in later life as they struggle to remain within the dominant hegemonic ideals of masculine space, as exemplified by younger, fitter, and employed men (Connell 2000; Meadows and Davidson 2006). They may even come to occupy, to some extent, feminine space, as when undertaking domestic activities such as cooking (Calasanti and Slevin 2001). Both of these challenges may cause men to renegotiate their status in terms of "manliness," as they attempt to come to terms with decreasing physical strength and position within society, and the loss of power that accompanies such altered status (Meadows and Davidson 2006).

The potential challenge to notions of "manliness" and masculine identity is particularly salient when examining caregiving experiences by older men, an area that has largely been overlooked in gerontology because of its traditional dominance by women (Calasanti 2003, 2006). Comparisons have traditionally been made between women caregivers who not only provide physical care, but also emotional support, and men caregivers, who are more likely to provide practical instrumental support. However, Calasanti (2003) and Thompson (1997) have pointed out that it is important not only to study and judge men's caregiving solely in relation to women's caregiving, as this may lead to an overemphasis on what women do, to the detriment of what men do. Indeed, men are caregivers, but little has been done to understand what caregiving *means* to them and their masculine identity (Thompson 1994). Suggestions that men are less effective at caring for their spouse or partner because of the rational nature of their caregiving has been challenged by Davidson et al. (2000) and Russell (2001), who argue that men do not solely adopt a professional or managerial approach to caring, but rather they see caring as "logical" and "natural" (Russell 2001:306), in recognition of the bond of a long-term marital relationship, which engenders a sense of responsibility to care sensitively and appropriately.

Feminist scholarship has also contributed to raising awareness of the gendered dimension of the aging body. The next section discusses changes in our understanding of the differential ways in which aging bodies are experienced by men and women.

The Aging (Gendered) Body

Understanding the body has increasingly become an arena for study within contemporary societies, such as Elias's (1978) discussion of civilizing bodies, as in etiquette and manners, and Foucault's (1979) disciplined or "docile" bodies in institutions. However, it is only in recent years that the *aging* body has received the same attention, which is surprising given that the physical changes to the body are some of the more overt markers of aging (Twigg 2000; Calasanti and Slevin 2001). Ageism and sexism are embedded within contemporary societies, with the media portraying the young as beautiful, and the old as ugly, with a strong emphasis on women maintaining youthfulness (Calasanti and Slevin 2001).

Building on Susan Sontag (1978), Arber and Ginn (1991) highlighted the "double standard" of aging, where the definition of old age shifts according to gender, that is, women are seen to physically age sooner than men. With the discourse negatively focusing on (aging) women's bodies, it is no surprise that women express greater dissatisfaction with their aging bodies than do men. Feminist analysis has argued that the pressure on women to look younger in the (literal) face of the negative aspects of aging means they are more likely to adopt various forms of anti-aging strategies, such as cosmetic surgery, hair transplants, and hair dyeing (Gilleard and Higgs 2000; Ballard et al. 2005). Notwithstanding the importance of the media in defining the presentation of appropriate body images, Hurd Clarke et al. (2007) have suggested that older women also learn how to manage the presentation of their bodies and the proper way to "do gender" through their mothers. Paradoxically, women also face additional pressure from feminists, with on the one hand, women being urged to reject age-resisting activities and to "grow old gracefully," and on the other, to take hold of the increasing opportunities available to them and reject an aging identity (Ballard et al. 2005).

But female bodies are not the only subjects of negative discourses in both academic literature and the media. Although women are more likely to be dissatisfied with their body weight and attractiveness, older men also experience bodily dissatisfaction through reduced physical strength and sexual potency, all of which are related to their masculine identity.

As mentioned previously, the body has, until relatively recently, been ignored within gerontology, possibly, as Twigg (2004) argues that there can be negative consequences from focusing on (aging) bodies: "From this perspective, attempting to emphasize the bodily can seem a retrogressive step, one that takes us back into the territory of biological determinism and the narrative of decline" (2004:59). But Twigg emphasizes the importance of including the body in the study of old age, especially for women, who are more subject to ageist attitudes in terms of the aging body.

Challenges for the Future

It is undoubtedly an achievement that more people are living longer and healthier lives, in virtually all parts of the world. However, the questions now to be raised are what are the likely changes and challenges to be faced over the next 30 years for an aging population, and how can gender relations be integrated into effective policy planning to address these issues. The following are three areas that present challenges for the future.

Gender Imbalances

Global projections for an aging population indicate an overall rise in men and women over the age of 60 (Gist and Velkoff 1997). For example, in 1997, 10% of the world's population was over 60; this will most likely increase to 15% in 2025. In nearly all countries of the world, women comprise the majority of the older population, accounting for over half of the 60–69 age group and 65% of those aged 80 and above (Gist and Velkoff 1997). However, the gender balance is set to change in different ways in developed and developing countries. At one time in developed countries, the life expectancy of women increased at a greater rate than that of men, whereas the trend currently is that the life expectancy of men is increasing at a faster rate than for women. For example in the UK, the proportions of women at older ages will decrease. But within developing countries, the current trend is expected to echo the earlier trend of developed countries, so that there will be a widening gap between male and female life expectancies (Gist and Velkoff 1997). This means that developing countries are set to experience the "feminization of later life" and will undoubtedly face the same issues, such as increased poverty and poorer health of older women. This is compounded by a decrease in multigenerational households, smaller families, and rural-urban migration, leading to a reduction in the availability of caregivers for frail older people, the majority of whom will be women.

Within this globally projected gender imbalance, further complicating anomalies exist. For example, China is set to have a proportionately larger percentage of older people by 2040 than, for example, the United States. Further to this, the one-child policy in China has led to an "unnatural" gender imbalance, with a ratio of 120 boys to 100 girls being born in 2000 (Li et al. 2007). As these young men age and have expectations of marriage and partnership, this raises important concerns about the impact on the stability of the country. The implications are that there may not be enough younger men and women to support the increasing aging population.

The Oldest Old

The fastest growing age group in the majority of countries is the oldest old, that is those who are aged over 80 (Gist and Velkoff 1997), and this trend is likely to continue. In the UK, the proportion of those aged 85 and above has increased from the 0.1% in the early 1900s to 2% of the total population by the start of the twenty-first century (Dini and Goldring 2008). Globally, the number of those aged 80 years and above is expected to grow from 69 million in 2000 to 379 million in 2050 (UN Population Division 2001). The implications of this rapid increase in the "oldest old" are numerous and are likely to have a significant impact on policy planning, such as the need for increased pension provision and other retirement incomes to cover longer lives. The percentage of women over 80 is smaller in developing countries, but it is anticipated that their numbers will increase more rapidly than those in developed countries (Gist and Velkoff 1997). As mentioned previously, women are more likely to experience disability and poorer health in later life, so an increase in the numbers of women over the age of 80 will inevitably lead to growth in the need for access and utilization of health care services.

Unmarried and Childless Older People

Declining fertility rates, increased childlessness, and changing social norms surrounding divorce and marital separation are leading to an increasing number of women and men without some sort of family support in later life (Kneale and Joshi 2008). The number of childless men and women is rising, particularly in Europe and North America, but also in Latin America and South East Asia (Li et al. 2007). This has significant implications, particularly in terms of costs of health, social, and long-term care. If the numbers needing long-term care, both in the community and in assisted living or residential settings increase, there will also be a need to increase the numbers of paid caregivers, who if current trends continue, are likely to be female, migrant workers from lower income countries (Li et al. 2007). This will ultimately have follow-on effects for long-term care of older people in lower income countries.

There are gender differences in the implications of childlessness. Childless never-married men are more likely to be disadvantaged than comparable with older women because they are more likely to have smaller social support networks than women, and more likely to experience isolation. However, childless never-married women are more liable to have been better educated, have stable employment, and therefore have greater access to income and pensions in old age than women who have had children. Childless married couples, though, are more likely to be financially better off in retirement because of dual earnings, and a larger income and pensions (Koropeckyj-Cox and Call 2007; Plotnick 2009).

Conclusions

This chapter has examined how the incorporation of gender as a social dimension within the field of gerontology has grown steadily over the last three decades. Feminist perspectives sought to develop early theoretical frameworks on aging to focus on women's issues, but ultimately received criticism for neglecting men (Katz 2005; Arber et al. 2003). Later feminist approaches raised awareness of the need to not only include men but also incorporate gender relations alongside other dimensions that intersect with age, such as ethnicity, race, class, and sexual orientation. The adoption of a "gender lens" to address later life issues has highlighted emerging areas that have previously been neglected, such as the changing nature of social relationships in later life, and the lack of research on older men and masculine identities. Extending existing sociological studies of the (younger) body into aging bodies opens up further ways to explore how gender and aging intersect in later life within such arenas as body control and embellishment, body maintenance, and how the "social" influences the body, as in responses to ageist behaviors (Tulle 2003; Hurd Clarke et al. 2007).

Although there are clearly challenges for the future study of aging, it is important not to over-emphasize the problematic aspects of an increasingly larger global aging population. It would undoubtedly be a retrograde step in aging research, if the growing aging population was conceptualized as a "burden," especially as the vast majority of the "oldest old" are women. One way of avoiding this is to maintain the current focus on all dimensions of aging, which means keeping gender, along with ethnicity, race, sexuality and class, firmly in the spotlight when addressing issues that surround aging and later life. The pertinence of this approach is demonstrated in recent policy frameworks that emphasize "Active Aging" (World Health Organization 2002). It is argued that men and women should be targeted differently, as when addressing lung cancer, for example. Smoking was formerly predominant in men in developing countries, but is increasing amongst women as they are particularly targeted by cigarette advertisers following women's increasing status and emancipation. Deaths from lung cancer in men are declining in some developed countries, whereas those of women are increasing (WHO 2002).

The primacy of gender for aging issues is illustrated by the WHO's approach to the increasing global aging population, "A gender analysis should be applied no matter the issue being considered, for example, abuse, emergencies, health promotion, primary health care, supportive environments and income security" (WHO 2002). Therefore, policies aimed at addressing the issues that impact on older people should be informed by an understanding that men and women, both separately and relationally, may experience old age very differently.

References

Arber, Sara. 2004. "Gender and Marital Status in Later Life." *Journal of Aging Studies* 18:91–108.

Arber, Sara and Jay Ginn. 1991. *Gender and Later Life; A Sociological Analysis of Resources and Constraints.* Newbury Park: Sage.

———. 1995. *Connecting Gender and Aging: A Sociological Approach.* Buckingham: Open University Press.

Arber, Sara, Kate Davidson, and Jay Ginn. 2003. *Gender and Aging: Changing Roles and Relationships*. Maidenhead: Open University Press.

Atchley, Robert. 1971. "Retirement and leisure participation: continuity or crises?" *The Gerontologist* 11(1):13–17.

———. 1987. *Aging: Continuity and Change*. Belmont, CA: Wadsworth.

Ballard, Karen, Mary Ann Elston, and Jonathan Gabe. 2005. "Beyond the Mask: Women's Experiences of Public and Private Aging during Midlife and Their Use of Age-Resisting Activities." *Health* 9:169–87.

Bardasi, Elena and Stephen Jenkins. 2004. "The Gender Gap in Private Pensions." *ISER Working Paper* 29. Colchester: University of Essex.

Bernard, Miriam and Judith Phillips. 1998. *The Social Policy of Old Age: Moving into the 21st Century*. London :Centre for Policy on Aging.

Borell, Klas and Sofie Ghazanfareeon Karlsson. 2003. "Reconceptualizing Intimacy and Aging: Living Apart Together." Pp. 47–62 in *Gender and Aging: Changing Roles and Relationships* edited by S. Arber, K. Davidson and J. Ginn. Maidenhead: Open University Press.

Bury, Mike. 1995. "Aging, Gender and Sociological Theory." Pp.15–29 in *Connecting Gender and Aging: A Sociological Approach*, edited by S. Arber and J. Ginn. Buckingham: Open University Press.

Calasanti, Toni. 2006. "Gender and Old Age: Lessons from Spousal Care Work." Pp. 269–94 in *Age Matters*, edited by T. Calasanti and K. Slevin. New York: Routledge.

———. 2003. "Masculinities and Care Work in Old Age." Pp. 15–30 in *Gender and Aging: Changing Roles and Relationships*, edited by S. Arber, K. Davidson and J. Ginn. Maidenhead: Open University Press.

———. 1986. "The Social Creation of Dependence, Dependency Ratios, and the Elderly in the United States: A critical analysis." *Social Science and Medicine* 23(12):1229–36.

Calasanti, Toni and Kathleen Slevin. 2001. *Gender, Social Inequalities and Aging*. Walnut Creek, CA: AltaMira.

Connell, Raewyn. 2000. *The Men and the Boys*. Berkeley: University of California Press.

Cronin, Ann. 2004. "Sexuality in Gerontology: a Heteronormative Presence, a Queer Absence." Pp. 107–22 in *Aging and Diversity: Multiple Pathways and Cultural Migrations*, edited by S. O. Daatland and S. Biggs. Bristol: The Policy Press.

Cumming, Elaine and W. E. Henry. 1961. *Growing Old: The Process of Disengagement*. New York: Basic Books.

Davidson, Kate. 2002. "Gender differences in new partnership choices and constraints of older widows and widowers." *Aging International* 27(4):43–60.

Davidson, Kate, Sara Arber, and Jay Ginn. 2000. "Gendered Meanings of Care Work within Late Life Marital Relationships." *Canadian Journal on Aging* 19:536–53.

Davidson, Kate, Tom Daly and Sara Arber. 2003. "Older Men, Social Integration, and Organisational Activities." *Social Policy and Society* 2(2):81–89.

Department for Work and Pensions. 2005. *Women and Pensions: The Evidence*. London: DWP, http://www.dwp.gov.uk

Dini, Ercilia and Shayla Goldring. 2008. "Estimating the changing population of the 'oldest old.'" *Office for National Statistics*. Retrieved (http://www.statistics.gov.uk/articles/population_trends/PT132OldestOldArticle.pdf).

Dykstra, Pearl Annette and Jenny de Jong Gierveld. 2004. "Gender and Marital-History Differences in Emotional and Social Loneliness among Dutch Older Adults," *Canadian Journal on Aging* 23(2):141–55.

Elias, Norbert. 1978. *The Civilizing Process: Sociogenetic and Psychogenetic Investigations*. Oxford: Blackwell.

Estes, Carroll. 1979. *The Aging Enterprise*. San Francisco, CA: Jossey Bass.

———. 1991. "The New Political Economy of Aging: Introduction and Critique." Pp. 19–36 in *Critical Perspectives on Aging: the Political and Moral Economy of Growing Old*, edited by M. Minkler and C. Estes. Amityville, NY: Baywood.

Finch, Janet and Jennifer Mason. 1993. *Negotiating Family Responsibilities*. London: Routledge.

Foucault, Michel. 1979. *Discipline and Punish: The Birth of the Prison*. New York: Random House.

Gilleard, Chris and Paul Higgs. 1998. "Aging and the Limiting Conditions of the Body." *Sociological Research Online* 3(4). Retrieved http://www.socresonline.org.uk/3/4/4.html.

———. 2000. *Cultures of Aging: Self, Citizen and the Body*. London: Prentice Hall.

Ginn, Jay. 2003. *Gender, Pensions and the Lifecourse*. Bristol: Policy Press.

Gist, Yvonne and Victoria Velkoff. 1997. *Gender and Aging: Demographic Dimensions*. Retrieved (http://purl.access.gpo.gov/GPO/LPS27838e).

Goldsmith, V. 1993. "Police Lead Fight for Pensions Sex Quality: Men are Claiming Discrimination over Lump-Sum Rights." *The Independent Online*. Retrieved February 22, 2010.

Gullette, Margaret. 1997. *Declining to Decline: Cultural Combat and the Politics of the Midlife*. Charlottesville, VA: University of Virginia.

Havighurst, Robert. 1963. "Successful Aging." Pp. 299–320 in *Processes of Aging* edited by R. Williams, C. Tibbitts, and W. Donahue, New York: Atherton.

Heaphy, Brian. 2007. "Sexualities, Gender and Aging: Resources and Social Change," *Current Sociology* 55(2): 193–210.

Hurd Clarke, Laura and Meredith Griffin. 2007. "Becoming and Being Gendered through the Body: Older Women, Their Mothers and Body Image." *Aging and Society* 27:701–18.

Karlsson, Sofie G. and Klas Borell. 2002. "Intimacy and Autonomy, Gender and Aging: Living Apart Together." *Aging International* 24(4):11–26.

Katz, Stephen. 2005. *Cultural Aging: Life Course, Lifestyle, and Senior Worlds.* Ontario: Broadview.

Kimmel, Michael, Jeff Hearn, and Raewyn Connell. 2004. *Handbook of Studies on Men and Masculinities.* Thousand Oaks, CA: Sage.

Kimmel, Michael and Michael Messner. 2007. *Men's Lives.* 6th ed. New York: Allyn and Bacon

Kneale, Dylan and Heather Joshi. 2008. "Postponement and Childlessness: Evidence from Two British Cohorts." *Demographic Research* 19:1935–68.

Koropeckyj-Cox, Tanya and Vaughn R. A. Call, 2007. "Characteristics of Older Childless Persons and Parents: Cross-National Comparisons." *Journal of Family Issues* 28(10):1362–415.

Krekula, C. 2007. "The Intersection of Age and Gender: Reworking Gender Theory and Social Gerontology." *Current Sociology* 55(2):155–72.

Li, Rose Maria, Amy Idarola, and Christine Maisano. 2007. "Why Population Aging Matters: A Global Perspective." National Institutes on Aging and National Institutes of Health. Retrieved (http://www.nia.nih.gov/NR/rdonlyres/9E91407E-CFE8-4903-9875-D5AA75BD1D50/0/WPAM.pdf).

Meadows, Robert and Kate Davidson. 2006. "Maintaining Manliness in Later Life: Hegemonic Masculinities and Emphasized Feminities." Pp. 295–312 in *Age Matters*, edited by T. Calasanti and K. Slevin. New York: Routledge.

Midwinter, Eric. (1997). *Pensioned Off Retirement and Income Examined.* Buckingham: Open University Press.

McMullin, Julie Ann. 1995. "Theorising Age and Gender Relations". Pp. 30–41 in *Connecting Gender & Aging: A Sociological Approach* edited by S. Arber and J. Ginn. Buckingham, UK: Open University Press.

Office for National Statistics. 2009. "2006-Based Marital Status Projections for England and Wales" Retrieved (http://www.statistics.gov.uk/downloads/theme_population/MaritalStatusProjection06.pdf).

Phillipson, Chris. 1982. *Capitalism and the Construction of Old Age.* London: Macmillan.

———. 1998. *Reconstructing Old Age: New Agendas in Social Theory and Practice.* London: Sage.

Plotnick, Robert. 2009. "Childlessness and the Economic Well-Being of Older Americans." *The Journals of Gerontology* 64B(6):767–76.

Price, Debora and Jay Ginn. 2003. "Sharing the Crust: Gender, Partnership Status and Inequalities in Pension Accumulation." Pp. 127–47 in *Gender and Aging: Changing Roles and Relationships*, edited by S. Arber, K. Davidson, and J. Ginn. Maidenhead: Open University Press.

Russell, Richard. 2001. "In Sickness and in Health: A Qualitative Study of Elderly Men Who Care for Wives with Dementia." *Journal of Aging Studies* 15(4):351–67.

Sontag, Susan. 1978. "The Double Standard of Aging." Pp. 72–80 in *An Aging Population*, edited by V. Carver and P. Liddiard. Milton Keynes. Open University Press.

Thompson, Edward, ed. 1994. *Older Men's Lives.* Thousand Oaks, CA: Sage.

———. 1997. "The Gendered Caregiving of Husbands and Sons and the Social Construction of Men Caregivers as Deviants." Presented at the annual meeting of the American Sociological Association, Toronto.

Townsend, Peter. 1981. "The Structured Dependency of the Elderly: A Creation of Social Policy in the Twentieth Century." *Aging and Society* 1(1):5–28.

Tulle, Emmanuelle. 2003. "Sense and Structure: Toward a Sociology of Old Bodies." Pp. 91-104 in *The Need for Theory: Critical Approaches to Social Gerontology*, edited by S. Biggs, A. Lowenstein, and J. Hendricks. Amityville, NY: Baywood.

Twigg, Julia. 2000. *Bathing: The Body and Community Care.* London: Routledge.

———. 2004. "The Body, Gender, and Age: Feminist Insights in Social Gerontology." *Journal of Aging Studies* 18:59–73.

United Nations Population Division. 2001. *World Population Prospects: The 2000 Revision.* Retrieved (http://www.un.org/spanish/esa/population/wpp2000h.pdf).

Venn, Susan and Sara Arber. 2011. "Daytime Sleep and Active Aging in Later Life." *Ageing and Society* 31(2):197–216.

Walker, Alan. 1981. "Towards a Political Economy of Old Age." *Aging and Society* 1(1):73–94.

Walker, Ruth and Janet Hiller. 2007. "Places and health: A Qualitative Study to Explore How Older Women Living Alone Perceive the Social and Physical Dimensions of Their Neighbourhoods." *Social Science & Medicine* 65(6):1154–65.

Whitehead, Stephen. 2002. *Men and Masculinities: Key Themes and New Directions.* Oxford: Blackwell.

World Health Organization. 2002. "Gender and Ageing." http://www.who.int/ageing/gender/en/index.html.

Chapter 6
Race, Ethnicity, and Aging

Jan E. Mutchler and Jeffrey A. Burr

Sociologists recognize age and race as key dimensions of social reality, especially in the American context, with its heterogeneous population, turbulent race relations history, and rapidly changing age structure. As so-called "master variables," age and race (along with sex) are regarded as statuses shaping and penetrating nearly every aspect of social life, including but not limited to economic opportunity and overall well-being (for reviews, see Leicht 2008; Mayer 2009; Riley and Riley 2000; Winant 2000). Sociologists who focus their research at the intersection of age and race sometimes struggle to define the "social" boundaries of these constructs, as independent from perspectives found in the biological, anthropological, and psychological sciences. Understanding how race and ethnic group status, along with the related issue of culture, shape the experience of aging itself remains an elusive but worthwhile goal.

A series of literature reviews dating from the 1980s laments the theoretical and research limitations in the field of race and ethnic aging (Angel and Angel 2006b; Markides and Black 1996; Markides et al. 1990; Williams and Wilson 2001). Among the earliest of these commentators, Jackson defines the scope of the field as "the study of the causes, processes, and consequences of race, national origin, and culture on individual and population aging" (Jackson 1985:265). Jackson notes that a substantial share of the scientific literature at the time focused on *describing differences* among socially defined groups rather than developing theories and garnering evidence to help explain diverse outcomes in the aging context. Not much has changed in this regard. In the mid-1980s as well as currently, a substantial share of the published studies explores *dimensions of disadvantage* associated with race or ethnic group membership, with far less attention directed toward identifying commonalities across groups or areas of strength derived from association with and membership in specific race and ethnic groups (intra-ethnic sources of social capital being one prominent example). As a result, we know quite a lot from descriptive studies about the disadvantages faced by older members of some racial and ethnic groups in later life, especially in terms of the social factors that correlate with health, economic security, and family life. In contrast, the potential benefits associated with ethnic group membership are poorly documented. Moreover, far less knowledge has been accumulated relating to the causal pathways linking race or ethnic group membership to these outcomes as people age.

The field of sociology offers uniquely valuable insights in two areas critical to understanding race and ethnic status as markers of diversity in aging. First, the foundational insights relating to the role of social context and social forces in shaping individual experience and behavior are central (e.g., Durkheim 1951; Mannheim 1952). We outline below a number of ways in which this is the

J.E. Mutchler (✉)
University of Massachusetts, Department of Gerontology, Boston, MA, USA
e-mail: jan.mutchler@umb.edu

R.A. Settersten, Jr. and J.L. Angel (eds.), *Handbook of Sociology of Aging*, Handbooks of Sociology and Social Research, DOI 10.1007/978-1-4419-7374-0_6, © Springer Science+Business Media, LLC 2011

case, and we emphasize the opportunities posed by a life course framework for expanding this understanding. Second, although we acknowledge the potential drawbacks associated with an *over*-emphasis on dimensions of disadvantage, we note that understanding the forces linking racial and ethnic group inequality and aging is, in fact, useful in further developing the sociology of aging, race, and ethnicity (Markides and Black 1996). Significant and growing inequality based on social class characterizes American society, with the result being that many individuals, regardless of race and ethnic status, reach old age with insufficient economic and health resources (Crystal and Shea 1990; Gerst and Mutchler 2009). Sociological insights can be effective in helping to understand the processes that yield disadvantage for racial and ethnic minorities, as well as in drawing our attention to strengths associated with positive outcomes in later life.

The purpose of this chapter is to review briefly the limited scholarly literature in the sociology of aging that identifies social processes linking race and ethnic status to the aging experience. Illustrations are drawn from the empirical literature, focusing primarily on income security and health-related outcomes. Due to space limitations, many other dimensions of race and ethnic group status and aging are not covered here (e.g., impact of cultural beliefs and spirituality on coping and health outcomes). Following a description of the demographic context, we introduce the reader to several conceptual frameworks: double jeopardy, age as leveler, and cumulative (dis)advantage. We also briefly outline life course sociology and recommend that this approach has the potential to emerge as a viable framework for organizing research in this area. Cultural distinctiveness and assimilation are posed as potentially useful intervening and modifying constructs in developing a life course perspective on the sociology of aging, race, and ethnicity. Opportunities posed by recent methodological developments are discussed as avenues for enriching our understanding of race and ethnic diversity in aging. We also offer some potential linkages with policy consistent with this approach.

The Demographic Context

For much of the twentieth century, sociological studies of race focused primarily on comparing the experiences of African Americans to whites (e.g., Park 1950), while investigations into ethnicity were based largely on the "old ethnic" populations such as Italians or the Irish (e.g., Alba 1990; Lieberson 1980; Thomas and Znaniecki 1974). The foci of these earlier explorations were motivated in part by the demographic characteristics of the U.S. at the time. Most of the population was non-Hispanic and white or African American, and as late as 1980 only 9% of the population of all ages reported Asian, Native American, or another race, or Hispanic ethnicity (Gibson and Jung 2002). During the last 30 years, the share of the U.S. population that is both white and non-Hispanic has dropped precipitously and now represents less than two-thirds of the population for all ages combined. These changes are largely due to patterns in fertility and recent immigration, including large migration flows from Latin America and Asia and relatively small flows from Europe and Africa. [The topic of immigration as it relates to the older population is taken up in a subsequent chapter in this volume (Markides), and readers are advised to consult this work for a more extensive discussion.]

The spatial settlement patterns of both U.S.-born and immigrant racial and ethnic populations have resulted in geographic clustering of ethnic groups, residential segregation by race and ethnicity, and the development of ethnic enclaves in many areas of the country. Although race and ethnic minority populations are becoming increasingly dispersed geographically (Frey 2006), distinctive spatial patterns continue to characterize most groups and this geographic patterning contributes to differential experiences and outcomes among ethnic groups.

The Hispanic population now exceeds the African American population in size; Hispanics represent 16% of the population, whereas African Americans represent 12% of the total U.S. population.

Looking ahead, it is expected that population growth in the coming 30 years will be even more heavily Hispanic as well as Asian. More than half of total population growth will be among Hispanics, a group that is expected to represent 30% of the U.S. population by 2050 according to projections from the U.S. Census Bureau (U.S. Bureau of the Census 2008). The Asian population is expected to double in size from 4% in 2000 to 8% in 2050, while the share of the population that is African American is projected to remain steady at about 12%. By the middle of this century, the share of the population that is white and non-Hispanic is expected to drop below 50%.

The growing diversity of the older population parallels that of the overall population, although for some groups the expansion is advancing at a somewhat slower pace. For the elderly population (age 65 years and over), the largest rate of growth is among Hispanics, a group that is expected to increase from its current level of 5% of the elderly population to almost 20% by 2050 (U.S. Bureau of the Census 2008). The Asian population is expected to quadruple in relative size, increasing from 2 to 8% by 2050. The older African American population will grow from 8 to 11%; and although the older Native American population will grow numerically, it will remain a relatively small share – about 1% – of the older population. Only relatively recently have sociologists of aging begun to consider the minority aging population in broader terms (e.g., Angel and Hogan 2004). Yet these compositional changes have implications for a diverse aging society and the manner in which sociologists frame their research.

As noted above, immigration plays a sizable role in the growth of both the older and the total populations. Within the Hispanic population, 40% of the year-to-year change in size is due to immigration (U.S. Bureau of the Census 2008). Higher still, 70% of the growth of the Asian population is directly due to immigration. Immigrants from Latin America or Asia who arrived as youths or young adults are aging in place within their new U.S. communities. Other immigrants arrive in the U.S. after they are already older, often following younger family members, and thus may be described as seeking a place to age. In 2008, an estimated 96,000 persons aged 60 years and over obtained legal permanent resident status; this age group accounted for nearly 9% of the admissions that year. Most of these individuals (82%) were admitted as immediate relatives of U.S. citizens, with another 7% admitted under more broadly defined family-sponsored preferences (Department of Homeland Security 2008). The bulk of the projected U.S. population growth depends on immigration, which will simultaneously yield steady increases in the size of the Hispanic and Asian populations. The growing race and ethnic diversity in the aging population is, therefore, inextricably linked with immigration processes; and many of the factors shaping incorporation of younger immigrants also shape the well-being of older immigrants, the vast majority of whom are non-white or Hispanic (Angel and Torres-Gil 2010; Markides and Mindel 1987; Treas and Mazumdar 2002).

Some Conceptual Issues

The development of sociological theory specifically addressing the intersection of race, ethnicity, *and* aging is nascent, at best. However, scholarship informing this topic draws on a rich set of general sociological perspectives and concepts that have the potential to lend insight to this combination of social statuses. Many of these ideas focus on socially structured sources of disadvantage in later life (e.g., cohort location; discrimination; policies that work against the maintenance of families), and some also provide insight into sources of cultural variability that have implications for aging (e.g., health beliefs; linguistic acculturation). The sociological concepts introduced below share a focus on several central themes, including a longitudinal view of the issues, the importance of social status to life outcomes, and implications of being embedded in a social structure (Lynch 2008).

Early Sociological Perspectives

Modernization Theory, outlined initially by Burgess (1960) and developed more extensively by Cowgill (1972), highlights structural sources of distinctiveness in the significance and implications of age prevalent among different societies. Most relevant for this discussion, this perspective proposes that in developing regions of the world, where economic well-being depends on mutual support, cultural values arise and are reinforced that emphasize familial solidarity and filial piety. Given that sizable segments of the fastest growing race and ethnic groups in the U.S. are composed of immigrants coming from less industrialized countries in which elderly family members are respected and held in high esteem, it is likely that these values are retained in one form or another, even after some level of incorporation into the majority culture occurs. The maintenance of family norms from country of origin may act as a buffer for the difficulties that arise during the incorporation process, including offsetting some of the resource limitations.

Social Demographic Contributions

Themes drawn from the social demographic research literature also offer conceptual guidance with respect to race, ethnicity, and aging. The cohort concept (especially birth cohorts; Ryder 1965), for example, highlights the experiences and circumstances to which people who share the time period of their birth are exposed; people who subsequently move through stages of life together, including transitioning into later life. Through the process of cohort succession – by which birth cohorts with unique life course experiences are, over time, replaced by subsequent birth cohorts having a different set of life course experiences – the understanding of what it means to be "old" shifts over time. A familiar example of cohort succession compares persons who grew up during the "Great Depression" to those who came of age during the Vietnam War era. In a similar way, we anticipate that increases in ethnic and race heterogeneity among the older population will potentially change norms and expectations about aging. Evidence shows that different minority groups face different levels of discrimination and blocked opportunities, and that these experiences have evolved over time and across cohorts – yielding different levels of well-being in later life, different expectations about retirement, and different expectations about family responsibilities, including care receipt and delivery.

Another important insight offered in the social demographic literature relates to the understanding that any comparison of groups in later life is necessarily a comparison of survivors. A well-developed picture of this process occurs in the research literature on mortality differences by race, as depicted by the so-called "crossover effect." Crossover effects have been documented most extensively for African American populations compared to non-Hispanic white populations, but have been identified for Native American and other groups as well (e.g., Hummer et al. 2004). Social demographers have described extensively the risks to mortality posed by social and economic disadvantage, disadvantages that frequently are more concentrated in minority populations. As a result of these risks, for example, African Americans are more likely than whites to die at virtually every age. Some evidence suggests that survivorship rates may reverse or "crossover" in later life, at which point the risk of dying for African Americans (and, in some studies, for other groups with a history of disadvantage) drops below the mortality risk for more privileged groups (Eberstein et al. 2008). The extent to which this "crossover" is the result of selective survival, or alternatively, reflective of poor data quality, has not been resolved entirely. Because of this selectivity effect, comparisons of some minority groups to the majority group may be biased because only the healthiest minority group members survive. This is one reason why data are needed that cover long periods of the life course.

The principle of selectivity is reflected also in populations that are heavily shaped by immigration, such as among the Hispanic and Asian populations. Unlike the African American population, immigrant populations appear to have lower morbidity (for some conditions) and mortality risk than their non-Hispanic and white counterparts across the life course. To some extent, characteristics of the immigrant segments of these populations are affected by positive selection, a process by which the more advantaged segments of a population are more likely to migrate to the U.S. – particularly, those with the most to gain by moving. Inasmuch as good health and other forms of highly valued capital may be a prerequisite to initiating and benefitting from an international move, Hispanics and Asians who migrate to the U.S. may be more healthy and less likely to die than their U.S.-born, white counterparts (the "healthy immigrant" effect; see discussion in Mutchler et al. 2007b). Moreover, immigrants who experience declines in health or the onset of disabilities may choose to engage in reverse migration (the "salmon effect"); leaving behind in the U.S. their counterparts who are in better health, and who have a lower risk of dying (see discussion in Markides and Eschbach 2005 and Turra and Elo 2008). The research literature indicates that in some cases, healthy lifestyles among immigrant populations that are influenced by cultural practices (e.g., healthy nutrition behaviors) may contribute to relatively high levels of health and survivorship; however, these lifestyle advantages may erode with time (e.g., Frisbie et al. 2001).

A central insight offered by social demographers is that when segments of the older population are compared with respect to their well-being only in later life, what is observed is behavior that may be related to differential survivorship rates across race and ethnic groups (Lynch 2003; Shuey and Willson 2008). Because so many older minority group members are also immigrants, the well-being of race and ethnic populations is conditioned by processes of mortality *and* immigration, both of which are selective with respect to who survives to a given age.

Double Jeopardy Perspective

Among the sociological perspectives focusing explicitly on the intersection of age and race, ideas relating to double jeopardy are perhaps the most often referenced in the research literature. The general concept of double jeopardy has been widely used throughout the social, behavioral, and medical sciences to describe how two factors interact to influence a wide range of outcomes. In the sociological treatment of age and race, double jeopardy suggests that these characteristics combine to create a "double disadvantage" for aging members of minority groups (Dowd and Bengtson 1978). Among the individual characteristics examined with respect to the double jeopardy thesis are income, social interaction and engagement, health status, life satisfaction, and mortality (e.g., Dilworth-Anderson et al. 2002; Lynch 2008; Markides and Mindel 1987). The hypothesis holds that the disadvantage associated with being a member of a race or ethnic minority group increases with age; as a result, longitudinal data are required to adequately test this hypothesis (Ferraro and Farmer 1996). The double jeopardy concept offers a convenient, but largely descriptive way of summarizing expectations for widening gaps between minority and majority groups in later life. Racism and ageism are implied as the causal mechanisms for these expanding inequalities. The possibility of a triple jeopardy effect, based on gender and undergirded by sexism, has also been forwarded (Ferraro and Farmer 1996).

The double jeopardy perspective, as applied by sociologists interested in aging, often does not acknowledge the considerable heterogeneity within specific race and ethnic groups. For example, the African American aging experience is different from the experience of black persons from the Caribbean (e.g., Taylor et al. 2007) and the experience of aging is different when comparing Korean and Chinese elders (the Chinese group is further differentiated by persons whose background may be from the People's Republic of China or from Taiwan; Mui and Shibusawa 2008). The predictive power of the double jeopardy hypothesis, as well as any other perspective that does not account for this heterogeneity, is thus limited.

An alternative to the double jeopardy perspective is the age-as-leveler perspective (Dowd and Bengtson 1978). Mixed support in the research literature has been generated for this hypothesis, which proposes that race and ethnic group gaps in critical dimensions of well-being decline with age (e.g., Herd 2006; Kim and Miech 2009; Willson et al. 2007). Strong evidence has been offered in support of a "persistent inequality" thesis, at least with respect to health (Ferraro and Farmer 1996). However, it is likely that race and ethnic group differences in survival rates and income security would be even larger if not for selectivity in the mortality experiences of members of minority groups, return migration of less healthy immigrants, and the social insurance policies of the federal, state, and local governments aimed at older Americans (e.g., Social Security, Medicare, Supplemental Security Income, Medicaid).

Cumulative Advantage and Disadvantage

The underlying premises of the double jeopardy and age-as-leveler hypotheses are expanded within the more fully developed cumulative advantage–disadvantage perspective. This concept has also been applied to a wide variety of issues within sociology, as well as economics, psychology, epidemiology, and criminology. A central hypothesis derived from this framework holds that over the life course, *initial inequalities* – in financial resources, in health, in social status, in educational opportunity, and in other dimensions of well-being – are heightened. Disadvantages experienced in childhood or young adulthood are accentuated, while advantages experienced by members of privileged groups multiply. Scientific evidence suggests that accumulated advantages and disadvantages are carried into old age, resulting in a persistently high level of economic inequality among those 65 years old and over (e.g., Crystal et al. 1992; Dannefer 2003; Henretta and Campbell 1976; O'Rand 1996; Walesmann et al. 2008). The original idea for this perspective flows from Robert Merton's observations surrounding the trajectory of scientists' careers, also known as the "Matthew Effect" (DiPrete and Eirich 2006; Merton 1968). Researchers more often emphasize the intersection of aging and inequality as defined by economic class rather than by race or ethnic status (Dannefer 1987); yet this approach has recently been usefully applied to dimensions of structured inequality in later life as defined by race and ethnic group membership, especially in the area of health (e.g., see Lynch 2008 for an introduction to a special issue of *Research on Aging* on this topic).

On average, members of some race and ethnic groups (most notably, African Americans, Native Americans, and Hispanics) are more likely to be born into families with fewer advantages than are their white counterparts, and they accumulate fewer resources as they age due in part to blocked opportunities for the development of human, financial, and cultural capital, yielding less prosperous outcomes in early adulthood (see Duncan and Brooks-Dunn 1997; McClanahan and Percheski 2008). We know from early seminal research in sociology that the intergenerational transfer of inequality among Americans is jointly determined by class and race (Blau and Duncan 1967). Unequal access to education, employment options, high status and high paying occupations, health care, and enriching life experiences are based at least partly on overt and institutional discrimination and partially on unequal returns to human capital (Leicht 2008). Gaps in the accumulation of wealth and in homeownership levels are also evident across race and ethnic groups (Burr et al. 2010; Sykes 2003), having substantial impacts on income security in later life and placing limits on the intergenerational transfer of wealth.

One advantage of the cumulative advantage/disadvantage thesis, as compared to the double jeopardy hypothesis and the age-as-leveler hypothesis, may be that by explicitly directing attention to broader stratification processes associated with accumulation of human and other forms of capital (Nee and Sanders 2001), work and career trajectories, and the intergenerational transmission of poverty, it offers opportunities to specify within-group differentiation, while still acknowledging differences between race and ethnic groups as related to important trajectories and outcomes (Leicht 2008). For example, race and ethnic group membership shapes the values attached to (or benefit received from) important cultural characteristics such as family ties and other culturally defined conditions such as gender roles (O'Rand 1996). Further, a systematic devaluation or discounting of the human capital held by some race and ethnic groups (including immigrants), coupled with inflated valuation of human capital held by dominant groups, may over time be reflected in diverging patterns of well-being in later life consistent with the cumulative advantage/disadvantage hypothesis. Notably, processes of cumulative advantage/disadvantage are expected to result in divergence within racial and ethnic groups as well. African Americans, Latinos, or others who achieve early accomplishments through obtaining access to higher education, building a business within an ethnic enclave or elsewhere, or embarking on a promising career ladder, would be expected to reap higher levels of asset accumulation throughout their lifetimes, with enhanced well-being in later life an expected result.

Our reading of the empirical literature identifies a common issue in findings from research on late-life social group diversity on a variety of dimensions (e.g., living arrangements, income, and health). It is often the case that these differences remain statistically significant even after controlling for the factors believed to underlie these differences (e.g., Shuey and Willson 2008). Researchers typically identify methodological limitations in their research designs as being responsible for this outcome, including poor measurement, small samples and limited time frames of observation. Another limitation that is both related to research design and conceptual development in the field is the impact of unobserved variables (e.g., cultural and community variables).

Cultural Distinctiveness and Assimilation

One way to address the issue posed above is to expand our theoretical models by including explicitly in research projects the concepts of cultural distinctiveness and assimilation. A large scientific literature dating back many decades highlights the *cultural distinctiveness* of ethnic and race groups (e.g., Cox 1948). Members of social groups share a cultural heritage, often reflected in religious practices and beliefs, language, value systems, and norms; these differences distinguish them from the majority group and should shape aging experiences and behavior in distinctive ways.

The assimilation perspective (Gordon 1964) holds that over time, and with increased involvement in the broader society and culture, this distinctiveness will disappear. Whether in fact cultural distinctiveness disappears or evolves in form and significance are topics of considerable debate (e.g., Alba and Nee 1999; Portes and Zhou 1993). Nonetheless, for older members of many ethnic groups, culturally based norms and values may be related to distinct behaviors and expectations that impact social support and social engagement, income security, health and disability, and life satisfaction. For example, researchers find that older African Americans and Hispanics report stronger attitudes about family values than their non-Hispanic white counterparts, and that group variability in intergenerational co-residence is accounted for by these attitudinal characteristics (Burr and Mutchler 1999). Evidence relating to the role of assimilative processes has also been reported showing that the influence of cultural markers (i.e., English language proficiency and duration of residence in the U.S.) on the likelihood of living alone among older Hispanic women diminishes with higher economic status, but does not completely disappear (e.g., Burr and Mutchler 1993).

Life Course Sociology

The life course perspective as applied by sociologists is a central thesis for understanding social aspects of the aging process (e.g., Mortimer and Shanahan 2003). Although falling short of a formalized body of theory (e.g., Mayer 2009; Settersten 2003, 2006), the rich concepts and logic of the life course perspective provide a well-developed foundation for a sociological understanding of the dynamics of the aging process. The life course perspective provides a way to conceptualize the "twists and turns" of an individual's life through a focus on the transitions and trajectories that mark taking on, playing out, and relinquishing roles and statuses (Hagestad 1990).

Changes in roles and statuses are a hallmark of moving into the later years of an individual's life brought on by the relinquishment of work and family roles, the adoption of roles of family patriarch and matriarch, and the eventual acceptance of the retiree status (Weiss 2005). Although each individual's path is somewhat unique and is modified by individual free will and choice (*human agency*), the life course perspective highlights the social forces that influence and lend significance to those individual choices. Each individual's life course is also shaped by those of family members and significant others with whom he or she is linked both inter- and intra-generationally (*interconnected lives*). Features of the broader social structure relating to educational and work organizations, labor markets, and public policies attach significance to the age at which transitions occur (*timing*), the occurrence of transitions relative to one another (*sequence*), and the length of time spent in given statuses (*duration*) (Elder 1985).

Settersten (2006:4) writes that the life course perspective invites attention to "differentiation in aging-related experiences across cohort, sex, *race*, and social class groups, generations within families, and nations" (emphasis added). Lifetime experiences generate life course capital that is shaped by membership in race or ethnic groups. This life course capital is carried into later life, where its significance influences resource, health-related, or social support-related vulnerabilities (O'Rand 2006). Yet attention to differentiation defined by race and ethnic group status has occurred in the late-life life course research literature in a piecemeal fashion rather than as an integrated approach. In fact, in a recent, otherwise comprehensive review of life course sociology, very little attention is paid to race and ethnic group issues (Mayer 2009). One example where these insights are usefully applied relates to the differential accumulation of human capital over the life course. Barriers to obtaining schooling or training associated with race or ethnic group membership filter into the diversity of life-long accumulation of occupational and economic benefits, which have implications for later-life resources in the form of wealth, pensions, Social Security credits, and health (see Walsemann et al. 2008). Thus, the life course approach is linked to some of the central tenets of the cumulative advantage/disadvantage thesis, which in turn is linked to the double jeopardy and age-as-leveler hypotheses.

A more refined appreciation of how race and ethnic group membership among older persons shapes other aspects of the life course is under-developed. For example, the life course perspective attaches significance to the timing and sequencing of events, suggesting that events that occur "off-time" or out of sequence relative to other life course events result in negative consequences (e.g., having children before completing one's education; or retiring from the labor force at an early age due to health or employment discrimination). Yet for some group, distinctive norms may support an alternative sequencing or timing of events. Gibson's (1987) introduction of the "unretired-retired" concept, describing a scenario in which middle-aged African Americans find neither the "retired" nor the "worker" statuses fully available, is an example. Several questions need to be addressed. Are the consequences of off-time transitions consistently negative for older members of specific groups? And, if so, to what extent are these negative consequences enforced by social policies or social institutions that may increasingly be out of step with the way that people live their lives? Continued research combined with conceptual development and theorizing regarding the life course as structured by race and ethnic group status is required, along with a fuller evaluation of the implications of those structures in later life.

Opportunities Ahead

Several emerging frontiers for increasing our understanding of race and ethnic social issues as they relate to aging are identified here. Some of these opportunities involve building on the strengths offered by unique characteristics of the sociological imagination, relating to the importance of social context and specifying dimensions of inequality that are identified with race and ethnic group status. Other opportunities are defined by the ways in which sociologists may advance their contributions to interdisciplinary investigations of aging and diverse populations, investigations that increasingly characterize the study of aging. Methodological issues that represent opportunities for propelling the field forward are also identified, as well as barriers that challenge these efforts.

New Intellectual Frontiers

One active area of sociological study in recent years surrounds the immigrant experience, including the processes and degrees of incorporation of the so-called "new immigrants" into the broader society. Included here are the debates surrounding whether over periods of time the new immigrants will experience levels of assimilation similar to earlier European immigrants or whether they will achieve a kind of segmented assimilation versus various forms of entrenched stratification, highlighted by little progress in residential segregation, income inequality, and so forth. Scholarship on immigrants and immigration offers promise for the sociology of aging, race and ethnicity, representing a potentially rich area of theoretical development that has not yet been systematically incorporated into the discussion. For example, we know much about the residential segregation of younger immigrants but virtually nothing about the implications of this form of spatial segregation for older immigrants. A well-developed literature exists on the implications for working-age immigrants of living in ethnic enclaves (e.g., Logan et al. 2003; Xie and Gough 2009), but little is known about the implications for older immigrants. We have a significant body of research on homeownership, residential crowding, and home values for younger immigrants (e.g., Friedman and Rosenbaum 2004), but little for older immigrants. We also know very little about how the variable nature of reception into the structure of American culture impacts quality of life among immigrant elders.

Sharpening our understanding of how the advantages and disadvantages associated with race and ethnic group membership play out across the life course may yield better insights regarding the differences that we observe in later life, and help us identify appropriate targets of intervention. One step toward this goal may be the continued development of the cumulative advantage and cumulative disadvantage perspective as represented in the cumulative inequality theory offered by Ferraro et al. (2009). Ferraro et al. (2009) conceptualize the implications of linked stratification processes occurring over a lifetime. This approach invites a focus on childhood effects on late-life well-being (e.g., Crosnoe and Elder 2004; Palloni 2006; Warner and Hayward 2006), acknowledges the importance of life-long inequalities, and directs our attention to how the impact of discrimination in its many forms helps to accelerate the aging process among some minority groups; this is sometimes referred to as the "weathering hypothesis" (Geronimus 2001; Taylor 2008).

Useful linkages may also be forged between a more comprehensive theory of life-long inequality over the life course and stress theory (Pearlin 2010). Specifically with respect to race or ethnicity, experiencing chronic discrimination and subtler forms of bias may be a source of stress that yields a cascading trajectory of negative mental and physical health outcomes that last a lifetime (see also Williams 2004). However, some research suggests that African Americans and perhaps some other ethnic groups may be more successful in coping with stress. For example, African American women may more effectively cope with the stress of caregiving – for grandchildren as well as for family members in need of long-term care (Musil and Ahmad 2002; Roff et al. 2004; Sands and

Goldberg-Glen 1998). The way in which racial and ethnic group membership shapes the matrix of exposures to potentially stressful circumstances, as well as the coping behaviors and social capital that may mediate stress response, is a topic for further consideration.

Another promising area is the expansion of theory relating to the importance of community or neighborhood characteristics, coupled with improved methods for analyzing these relationships. Understanding the importance of the connection between individual and social context, a hallmark of life course sociology, may be particularly important when studying race and ethnic differences in health in later life. Thus, increased attention is also usefully directed toward spatial and environmental features that extend beyond one's intimate social network, into the neighborhood, the community, and the service environment. The extent to which differences in behavior or experiences across older members of different race and ethnic groups – such as use of medical services and income supports, or co-residence with younger relatives – are conditioned by features of the communities in which they live (e.g., cost of living, availability of services, transportation options, crime) has not been fully explored (however, see Mutchler and Burr 2003). The availability of methodological tools that permit more robust examinations of these multilevel effects promises to support the expansion of sociological insights relating to macro-level and meso-level influences on micro-level behavior. Recent research using such techniques demonstrates, for example, that both individual and neighborhood SES contribute to differences in self-rated health among older African Americans and whites (Yao and Robert 2008). Similarly, among both Hispanics and Chinese Americans, immigrant ethnic composition of the neighborhood has been shown to be related to health behaviors (specifically, healthier food choices, but less physical activity) (Osypuk et al. 2009).

One area in which sociologists are making substantial contributions to multidisciplinary investigations of aging and diverse populations is in the area of health disparities (Williams 2004). Members of some groups, most notably African Americans and Native Americans, are less likely to survive to old age, and more likely to enter old age in a disabled state than are older whites and Asian Americans (e.g., Goins et al. 2007; Hayward and Heron 1999). The Institute of Medicine concludes that disparities exist in the quality of formal care received by members of ethnic minorities, and suggests that eliminating these disparities will require a multi-dimensional approach that considers providers, patients, and service environments (Smedley et al. 2003). Yet much of the research focus has been on documenting differences in health care and health outcomes, rather than explaining them (LaVeist 2004). Over time, it has become clear that no single discipline can provide a comprehensive understanding of the origin of health disparities relating to race and ethnic group status, nor adequate remedies or interventions.

The biopsychosocial approach to health research highlights the combination of biological, psychological, and social structural influences on well-being in later life; this approach lends itself to the inclusion of race and ethnic group status, immigration, and cultural context (e.g., Bengtson et al. 2009; Berkman et al. 2000). Sociologists have made and will continue to make important contributions to understanding and correcting the modifiable bases of health disparities by highlighting the role that race and ethnic group membership may play in influencing health outcomes, through health behaviors such as diet and exercise, as well as along pathways relating to bias and inequality, such as unequal access to health care and poor interactions with health care providers. We need additional research on how psychological characteristics (e.g., anger, hostility, self-efficacy) and health behaviors and health life styles (e.g., obesity, smoking, alcohol consumption, physical activity) mediate and modify the relationships between race and ethnic group status and a large range of indicators of well-being. Finally, we need to explore and explain how being a member of a minority group is related to so-called "under the skin" biological factors (e.g., immune system and sympathetic nervous system) that impact health and mortality differences. Fortunately, several new data sources are available to help us begin this research journey. An example of a question that needs continued evaluation is whether long-term exposure to racial discrimination impacts the psychological characteristics, health behaviors, and biological functions that lead to disparities in health.

On balance, the literature on race and ethnic issues in aging emphasizes the disadvantages experienced by members of racial and ethnic minority groups, vis-à-vis their non-Hispanic white peers. A more limited literature suggests sources of strength within minority populations, lending new insights to build on in future investigations. Cultural attributes – such as strong religious or spiritual beliefs or healthy behaviors relating to diet or exercise – along with meaningful culturally defined roles may yield measurable advantages to older members of ethnic groups. Resilience, or the capacity to generate well-being despite adversity (Ryff and Singer 2009), may be associated with many cultural beliefs or practices in ways that are not yet understood. For example, some evidence suggests that African Americans may have better mental health than their white counterparts, despite circumstances that would be expected to yield more negative outcomes (Keyes 2009). Specific religious beliefs may not only be more common among some race and ethnic groups, but also have a more positive impact on emotional well-being (Krause 2005).

Cultural meanings associated with being an older person, or playing age-graded roles, may yield benefit to older members of some ethnic groups. The familistic values expressed by members of some ethnic groups, and the resulting support received by many older family members, are frequently highlighted as defining cultural features (Dilworth-Anderson and Burton 1999). Some ethnic groups, such as Native Americans, value grandparents and other older family members as carriers of traditional practices and values, and older individuals may draw great satisfaction from participating in the intergenerational transmission of cultural beliefs (Schweitzer 1999). Yet culturally based behaviors may not always prove beneficial. Although it is well established that older Asians and Latinos are less likely than their non-Hispanic white counterparts to become institutionalized, and more likely to live in intergenerational households (Angel and Hogan 2004; Burr and Mutchler 1993, 2003), some of these individuals may be housed in environments that are inadequately supportive, or in homes in which they are socially isolated or economically dependent (Treas 2008–2009). Although the expectation of intergenerational family support may not always benefit older individuals, familistic norms may prove to be advantageous if policies shift to place more emphasis on "private" sources of support for economic security or for long-term care in later life. Should policy shifts in this direction occur, groups with stronger norms for family-based support may experience unexpected benefits.

Methodological Barriers and Opportunities

As noted above, the development of specific theories focusing on race and ethnic group status and aging processes and outcomes is only beginning. Inductive theory, a common form of theory building in the social sciences, is based on the accumulation of sound information from rigorous exploratory and descriptive studies; to date this approach has yielded some verifiable hypotheses in this area. However, a clearer, comparative portrait of the landscape of aging among ethnic and race groups is limited in part by the shortage of data adequate to test those hypotheses that have emerged. Our strongest conceptual material focuses on processes relating to aging that may take different forms across subgroups, and about processes that occur across decades and even lifetimes. Questions that arise from these frames require longitudinal data spanning many decades (lifetimes would be ideal, but generally are not available) with adequate sample sizes. Key aspects of the aging process are put in motion before old age begins – in the womb, infancy, childhood, early adulthood, or middle age – making clear that long spans of high quality, comprehensive longitudinal data are necessary for methodological as well as theoretical reasons (Lynch 2008). Relatively few longitudinal data sources are available focusing on targeted race or ethnic groups that permit in-depth examination of life course features and cultural characteristics as they relate to life course outcomes and trajectories (the Hispanic Established Populations for the Epidemiologic Studies of the Elderly is one excellent example, although it does not offer comparative data for other ethnic groups; for a review of the longitudinal data available for life course sociology see Mayer 2009).

One of the best examples of a data source with adequate sample size capturing the U.S. experience is the Health and Retirement Study (HRS; another example is the Panel Study of Income Dynamics). Although the sample size of the HRS is sufficiently large to permit comparisons of non-Hispanic whites and African Americans with Hispanics, for confidentiality reasons the data do not allow for examination of other major racial groups (e.g., Asians; Native Americans) or national-origin groups within Hispanic ethnic categories (e.g., Puerto Ricans, Cubans). The HRS also begins observing its sample members in mid-life, making it more difficult to explore fully the impact of earlier life experiences, although some retrospective questions on childhood experiences are available. Moreover, multi-purpose data sets such as the HRS often do not include sufficient measures of cultural traits, such as reliance on a language other than English; measures of culturally based health beliefs; or indicators of the extent to which social support networks are composed of members of the same ethnic group. Themes highly relevant to immigrant and minority populations, such as how social capital impacts social support in later life, are thus difficult to examine (Angel and Angel 2006a).

Several methodological innovations offer optimism with respect to advancing the field. Of special interest is the broader application of growth curve models that allow for the examination of change over time within and between individuals, and hierarchical models that offer the ability to properly estimate the independent and joint effects of individual level and contextual level factors. Strategies for examining the effects of mortality selection bias are also being implemented. While continuing to look at the main effects of race and ethnic status, it is also necessary to examine more thoroughly the interaction effects of race and ethnic status with other characteristics, such as gender, immigration status, kinship characteristics, economic status, and so forth. The implementation of interaction models is employed with increasing frequency to examine race and ethnic differences in the developmental trajectory of health and disability-related outcomes (Kelley-Moore and Ferraro 2004; Shuey and Willson 2008; Yang and Lee 2010).

Social Origins and Consequences

The intersection of race, ethnicity, and aging provides an especially fertile area for observing the implications of social processes. Two areas of particular strength in the sociological approach – attention to the importance of structural or contextual influences on individual behavior and experiences, and a focus on the forces that produce inequality across social groups – are especially relevant. At this intersection, social processes related to group membership, intergroup relations, environmental settings, and access to opportunities and resources converge.

Settersten (2006) has issued a call for bringing "the social" back into the study of the sociology of aging and social gerontology. We agree that a greater focus on the life course provides a vehicle for achieving this goal. For this to be beneficial in the area of race, ethnicity, and aging, the conceptualization of the life course must be broadened and must be flexible. We need to identify places in the life course where race, ethnicity, and culture matter the most, and to explore the social forces at work in those areas that result in different consequences for members of different groups. Moreover, we need to understand the points at which forces relating to inequality and discrimination intersect with the life course, and in fact shape the life course, having life-long implications. It is likely that remedies for many of the problems experienced in later life will be found in interventions occurring much earlier in the life course.

Several challenges must be overcome to ensure the accomplishment of this goal. One of the challenges relates to the sometimes competing goals of theoretical breadth and the acknowledgement of meaningful differences among cultural groups. The descriptive literature on diversity in aging is replete with studies that target particular race or ethnic groups quite narrowly, often in specific, local contexts. Even among theoretically motivated studies relating to culture and process, difficulties in obtaining data and lack of consensus surrounding the measurement of key constructs relating to cultural beliefs and ethnic identity prevent the investigation of large-scale questions that span subgroups. To be both relevant and theoretically informative, our thinking about race and ethnic dimensions of aging needs to reflect aspects of the "particular" (e.g., of culturally based beliefs associated with specific race or ethnic groups) while also informing conceptual themes that span subgroups, such as those relating to discrimination, linguistic barriers, or other commonly shared experiences.

An emerging issue that makes this all the more challenging, yet timely, relates to the theme of multiple race and ethnic group identities. A number of familiar data collections now include opportunities for respondents to report more than one race (e.g., in the U.S. Census and other federal data). This provides researchers with both challenges and opportunities. Although multiple race group identification has been invited in the Census Bureau data collections since the 2000 Census, a relatively small share of the population actually reports multiple races (U.S. Bureau of the Census 2009). In the 2008 American Community Survey, just over 2% of the population reported two or more races, but the percentage doing so differs substantially by age. Whereas 4.6% of children under the age of 18 are reported as having two or more races, less than 1% of adults 65 and over report multi-racial identities. Although the source of this age difference remains unclear, it is likely that it reflects increasing levels of inter-racial coupling as well as a higher acceptance of multiple racial identities among younger cohorts. The multiple race issue is far more important for some racial groups than for others. For example, the size of the American Indian and Alaska Native (AIAN) population for all ages combined reporting a single race is 2.4 million, but this increases by almost double to 4.7 million when those reporting AIAN in combination with another race are also included (U.S. Bureau of the Census 2009). Because the behavior of individuals reporting AIAN only may differ in meaningful ways from those reporting AIAN in combination with another race (e.g., Mutchler et al. 2007a), the evolving measurement of race and ethnic group membership promises to challenge continued efforts to strengthen our understanding of diversity in later life. Both the shrinking share of the older population that is non-Hispanic white and increases in the population reporting multiple racial and ethnic identities suggest that the significance of ethnic identity for social and economic experiences throughout the life course is likely to evolve in coming decades.

Policy Implications

The expanding diversity of the older population will have implications for virtually every aspect of social policy (Angel and Torres-Gill 2010). Differences among older members of racial and ethnic groups are well described in the literature: in terms of their financial resources (such as Social Security credits, pensions, and wealth), their family configurations (including partner status, coresident household members, and number and geographic proximity of children), and their levels of health and disability. Sociological insights yielding better understandings of the sources and scope of these differences are valuable in building better and more comprehensive social policy. As well, a better understanding of the needs, preferences and goals of the full complement of the expanding older population, rather than just segments of it, will provide important bases for evaluating the success of policies in development.

The major public programs targeting the older population (Social Security, Medicare, and Medicaid) will be heavily stressed in the coming decades as the absolute numbers of individuals entering those systems increases dramatically and as federal and state governments wrestle with mounting debt and competing priorities. As the financing difficulties associated with these programs become even more pressing, there is a strong likelihood that continuation of benefits in their current form will be challenged. Because minority populations – especially African Americans and Hispanics – are reliant on these programs more than the white majority, they are likely to be most strongly affected by program modifications (Burr et al. 2010). Higher risk of employment instability and disability among middle-aged and older minority workers means that these individuals may be most inclined to leave the labor force early; but with the increasing age at which full eligibility for Social Security occurs – people born at the peak of the baby boom in 1957 cannot retire with full Social Security benefits until age 66.5 years, and those born in 1960 or later are not eligible for full benefits until age 67 (Social Security Administration 2009) – this may impose financial disadvantages on members of minority groups. For older workers with chronic disease and disability, an extended work life is unlikely to be a realistic solution to late-life economic shortfalls. Anticipated demographic shifts make it likely that the absolute size of the groups who could most benefit from needs-based transfer programs such as SSI and Medicaid will increase just at a time when policy makers and some members of the general public will be inclined to reduce these programs or shift the burden to families or the private sector (Wilmoth and Longino 2006). These changes may challenge the pact between generations that formed the basis for developing these programs (Angel and Angel 2006b).

Summary

Although scholars have been interested in race and ethnic issues as they relate to the older population for several decades, the development of concepts and especially theory that would help guide research in these important areas is limited. We believe that useful inroads may be made by blending existing conceptual frameworks such as cumulative advantage and the life course, as well as by reflecting on how cultural features of ethnic groups intersect with broad theoretical ideas such as the stress process, resilience, and adaptation. Longitudinal examinations drawing out the elements of well-being that are structured by race and ethnicity not only will inform our understanding of late-life disparities in health, social support, and economic security, but will also yield insight to the ways in which macro-level social forces shape outcomes over the life course more generally. A great deal can be learned about how social processes and policies have differential effects, and how advantages and disadvantages impact the life course and well-being outcomes, by examining race and ethnic issues relating to aging.

Yet it is likely that historical advantages, and disadvantages, attached to racial and ethnic group membership will shift over time. "Majority–minority" communities are becoming increasingly common (Frey 2006), and attracting racially diverse and immigrant populations is increasingly viewed as essential to the continued vitality of some areas (Myers 2007). The advantages historically experienced by non-Hispanic whites may erode when that group is no longer numerically in the majority. However, should the significance of racial or ethnic group membership for later-life well-being become more limited in the future, it is likely to take multiple generations to be realized. Sizable disparities in education, earnings, household income, and asset accumulation, especially among African Americans and Hispanics, are evident among Baby Boomers and are likely to persist in later life (Mutchler and Burr 2008). Gains for some segments of the minority population – in the form of rising economic well-being and movement into the middle class – may be countered by losses in other areas, such as engaging in poorer health behaviors or increased alienation from valued roles and identities. Whether the gradual improvements in life circumstances experienced by segments of the racial and ethnic minority population will ultimately result in a lessening of minority group-based inequality, or just expanded heterogeneity within ethnic groups, remains to be seen (Angel and Torres-Gil 2010).

We anticipate that the sociological literature on race, ethnicity, and aging will look quite different in the coming decades. Shifting demographics will yield a U.S. population that is older and that includes more persons of color; but for now, these processes are occurring at opposite ends of the age distribution. For some time, the non-Hispanic white population will be far older than the African American, Latino, and Asian populations, on average, giving rise to potentially challenging political debates about public funding for education (which will disproportionately benefit populations of color, who are younger) as well as for Social Security and other old-age programs (which disproportionately benefit white populations, who are older) (Angel and Torres-Gil 2010; Myers 2007). Within a few decades, however, the racial and ethnic diversity that currently characterizes the younger population will be realized in the older population as well. The significance of this demographic shift for reshaping what we know about aging and aging policy will ultimately depend on the extent to which the interests are shaped more heavily by age, by racial and ethnic group membership, or by some unique combination of these two aspects of identity. As non-Hispanic whites become a proportionally smaller share of the total older population, will their "majority group" identity become more solidified, or more diffused? Will older Hispanics perceive more affinity among younger Latinos, or among their older, non-Latino peers when choosing among programs to support? When choosing between public schools and senior centers in public funding decisions, how will more diverse older populations weight the value attached to each option? If in the future the older population is to successfully secure public support for programs and services that it values, common ground must be identified that transcends race and ethnicity, and that respects the "intergenerational social contract" (Myers 2007) linking well-being across cohorts and ethnic groups.

References

Alba, Richard D. 1990. *Ethnic Identity: The Transformation of White America.* New Haven: Yale University Press.

Alba, Richard D. and Victor Nee. 1999. "Rethinking Assimilation Theory for a New Era of Immigration." Pp. 137–60 in *The Handbook of International Migration: The American Experience*, edited by C. Hirschman, P. Kasinitz and J. DeWind. New York: Russell Sage.

Angel, Jacqueline L. and Ronald J. Angel. 2006a. "Minority Group Status and Healthful Aging: Social Structure Still Matters." *American Journal of Public Health* 96:1152–59.

Angel, Ronald J. and Jacqueline L. Angel. 2006b. "Diversity and Aging in the United States." Pp. 94–110 in *Handbook of Aging and the Social Sciences*, 6th ed., edited by R. H. Binstock and L. K. George. New York: Springer.

Angel, Jacqueline L. and Dennis P. Hogan. 2004. "Population Aging and Diversity in a New Era." Pp. 1–12 in *Closing the Gap: Improving the Health of Minority Elders in the New Millennium*, edited by K. E. Whitfield. Washington, DC: The Gerontological Society of America.

Angel, Jacqueline L. and Fernando Torres-Gil. 2010. "Hispanic Aging and Social Policy." Pp. 1–19 in *Aging in America, Vol. 3, Societal Issues*, edited by J. C. Cavanaugh and C. Cavanaugh. Santa Barbara, CA: Praeger.

Bengtson, Vern L., Daphna Gans, Norella M. Putney, and Merril Silverstein. 2009. "Theories about Age and Aging." Pp. 3–24 in *Handbook of Theories of Aging*, edited by V. L. Bengtson, D. Gans, N. M. Putney, and M. Silverstein. New York: Springer.

Berkman, Lisa F., Thomas Glass, Ian Brissettte, and Teresa E. Seeman. 2000. "From Social Integration to Health: Durkheim in the New Millennium." *Social Science and Medicine* 51:843–58.

Blau, Peter M., and Otis D. Duncan. 1967. *The American Occupational Structure*. New York: Wiley.

Burgess, Ernest W., ed. 1960. *Ageing in Western Societies*. Chicago: University of Chicago Press.

Burr, Jeffrey A., Kerstin Gerst, and Jan E. Mutchler. 2010. "Public Policy and Populations of Color," in *The New Politics of Old Age Policy*, edited by R. Hudson. Baltimore, MD: Johns Hopkins University Press.

Burr, Jeffrey A. and Jan E. Mutchler. 1993. "Ethnic Living Arrangements: Cultural Convergence or Cultural Manifestation?" *Social Forces* 72:169–80.

———. 1999. "Race and Ethnic Variation in Norms of Filial Responsibility among Older Persons." *Journal of Marriage and the Family* 61:674–87.

———. 2003. "English Language Skills, Ethnic Concentration, and Household Composition: Older Mexican Immigrants." *Journal of Gerontology: Social Sciences* 58B:S83–S92.

Cowgill, Donald O. 1972. "A Theory of Aging in Cross-Cultural Perspective." Pp. 1–14 in *Ageing and Modernization*, edited by D. O. Cowgill and L. D. Holmes. New York: Appleton-Century-Crofts.

Cox, Oliver. 1948. *Race, Caste and Class*. New York: Monthly Review Press.

Crosnoe, Robert and Glen H. Elder, Jr. 2004. "From Childhood to the Later Years: Pathways of Human Development." *Research on Aging* 26:623–54.

Crystal, Stephen and Dennis Shea. 1990. "Cumulative Advantage, Cumulative Disadvantage, and Inequality among Elderly People." *The Gerontologist* 30:437–43.

Crystal, Stephen, Dennis Shea, and Shreeram Krishnaswami. 1992. "Educational Attainment, Occupational History, and Stratification: Determinants of Later-Life Economic Outcomes." *Journal of Gerontology: Social Sciences* 47:S213–21.

Dannefer, Dale. 1987. "Aging as Intracohort Differentiation: Accentuation, the Matthew Effect, and the Life Course." *Sociological Forum* 2:211–36.

———. 2003. "Cumulative Advantage/Disadvantage and the Life Course: Cross-Fertilizing Age and Social Science Theory." *Journal of Gerontology: Social Sciences* 58:S327–38.

Department of Homeland Security. 2008. *Yearbook of Immigration Statistics: 2008*. Retrieved on 12/21/09 (http://www.dhs.gov/files/statistics/publications/LPR08.shtm).

Dilworth-Anderson, Peggye and Linda Burton. 1999. "Critical Issues in Understanding Family Support and Older Minorities." Pp. 93–105 in *Full-Color Aging: Facts, Goals, and Recommendations for America's Diverse Elders*, edited by Toni P. Miles. Washington, DC: Gerontological Society of America.

Dilworth-Anderson, Peggye, Ishan C. Williams, and Brent E. Gibson. 2002. "Issues of Race, Ethnicity, and Culture in Caregiving Research: A 20-year Review (1980–2000)." *The Gerontologist* 42:237–72.

DiPrete, Thomas and Gregory Eirich. 2006. "Cumulative Advantage as a Mechanism for Inequality: A Review of Theoretical and Empirical Developments." *Annual Review of Sociology* 32:271–97.

Dowd, James and Vern Bengtson. 1978. "Aging In Minority Populations: An Examination of the Double Jeopardy Hypothesis." *Journal of Gerontology* 33:427–36.

Duncan, Greg J. and Jeanne Brooks-Gunn, eds. 1997. *Consequences of Growing Up Poor*. New York: Russell Sage.

Durkheim, Emile. 1951. *Suicide: A Study in Sociology*. Glencoe, IL: Free Press.

Eberstein, Isaac W., Charles B. Nam, and Kathleen M. Heyman. 2008. "Causes of Death and Mortality Crossovers by Race." *Biodemography and Social Biology* 54:214–28.

Elder, Glen H., Jr. 1985. "Perspectives on the Life Course." Pp. 23–49 in *Life Course Dynamics: Trajectories and Transitions, 1968-1980*, edited by G. H. Elder, Jr. Ithaca, NY: Cornell University Press.

Ferraro, Kenneth F. and Melissa M. Farmer. 1996. "Double Jeopardy, Aging as Leveler or Persistent Health Inequality? A Longitudinal Analysis of White and Black Americans." *Journal of Gerontology: Social Sciences* 51B:S319–28.

Ferraro, Kenneth F., Tetyana Pylypiv Shippee, and Markus H. Schafer. 2009. "Cumulative Inequality Theory for Research on Aging and the Life Course." Pp. 413–34 in *Handbook of Theories of Aging*, edited by V. L. Bengtson, D. Gans, N. M. Putney, and M. Silverstein. New York: Springer.

Frey, William H. 2006. "Diversity Spreads Out: Metropolitan Shifts in Hispanic, Asian, and Black Populations since 2000." The Brookings Institution, Living Cities Census Series. Retrieved April 2, 2010 (http://www.frey-demographer.org/reports/R-2006-1_DiversitySpreadsOut.pdf).

Friedman, Samantha and Emily Rosenbaum. 2004. "Nativity Status and Racial/Ethnic Differences in Access to Quality Housing: Does Homeownership Bring Greater Parity?" *Housing Policy Debate* 15:865–901.

Frisbie, W. Parker, Youngtae Cho, and Robert Hummer. 2001. "Immigration and the Health of Asian and Pacific Islander Adults in the United States." *American Journal of Epidemiology* 153:372–80.

Geronimus, Arline T. 2001. "Understanding and Eliminating Racial Inequalities in Women's Health in the United States: The Role of the Weathering Conceptual Framework." *Journal of the American Medical Women's Association* 56:133–36.

Gerst, Kerstin and Jan E. Mutchler. 2009. "Poverty, Later Life." Pp. xx–xx in *Encyclopedia of the Life Course and Human Development*, Vol. 3, edited by D. Carr. New York: Macmillan Reference USA.

Gibson, Rose C. 1987. "Reconceptualizing Retirement for Black Americans." *The Gerontologist* 27:691–98.

Gibson, Campbell and Kay Jung. 2002. *Historical Census Statistics on Population Totals by Race, 1790 to 1990, and by Hispanic Origin, 1970 to 1990, for the United States, Regions, Divisions, and States*. Population Division Working Paper No. 56. U.S. Census Bureau.

Goins, R. Turner, Margaret P. Moss, Dedra Buchwald, and Jack Guralnik. 2007. "Disability among Older American Indians and Alaska Natives: An Analysis of the 2000 Census Public Use Microdata Sample." *The Gerontologist* 47:690–6.

Gordon, Milton. 1964. *Assimilation in American Life: The Role of Race, Religion and National Origins*. Oxford: Oxford University Press.

Hagestad, Gunhild O. 1990. "Social Perspectives on the Life Course." Pp. 151–168 in *Handbook of Aging and the Social Sciences*, 3rd ed., edited by R. H. Binstock and L. K. George. San Diego: Academic.

Hayward, Mark D. and Melonie Heron. 1999. "Racial Inequality in Active Life among Adult Americans." *Demography* 36:77–91

Henretta, John C. and Richard T. Campbell. 1976. "Status Attainment and Status Maintenance: A Study of Stratification in Old Age." *American Sociological Review* 41:981–92.

Herd, Pamela. 2006. "Do Functional Health Inequalities Decrease in Old Age? Educational Status and Functional Decline among the 1931–1941 Birth Cohort." *Research on Aging* 28:375–92.

Hummer, Robert, Maureen Benjamins, and Richard Rogers. 2004. "Racial and Ethnic Disparities in Health and Mortality among the U.S. Elderly Population." Pp. 53–94 in *Critical Perspectives on Racial and Ethnic Differences in Health in Later Life,* edited by Norman Anderson, Rodolfo Bulatao, and Barney Cohen. Washington, DC: The National Academies Press.

Jackson, James J. 1985. "Race, National Origin, Ethnicity, and Aging." Pp. 264–303 in *Handbook of Aging and the Social Sciences,* 2nd ed., edited by R. H. Binstock and E. Shanas. New York: Van Nostrand Reinhold.

Kelley-Moore, Jessica A. and Kenneth F. Ferraro. 2004. "The Black/White Disability Gap: Persistent Inequality in Later Life?" *Journal of Gerontology: Social Sciences* 59B:S34–43.

Keyes, Corey L. M. 2009. "The Black-White Paradox in Health: Flourishing in the Face of Social Inequality and Discrimination." *Journal of Personality* 77:1677–705.

Kim, Jinyoung and Richard Miech. 2009. "The Black–White Differences in Age Trajectories of Functional Health over the Life Course." *Social Science and Medicine* 68:717–25.

Krause, Neal. 2005. "God-Mediated Control and Psychological Well-Being in Late Life." *Research on Aging* 27:136–64.

LaVeist, Thomas A. 2004. "Conceptualizing Racial and Ethnic Disparities in Access, Utilization, and Quality of Care." Pp. 87–93 in *Closing the Gap: Improving the Health of Minority Elders in the New Millennium*, edited by K. E. Whitfield. Washington, DC: The Gerontological Society of America.

Leicht, Kevin T. 2008. "Broken Down by Race and Gender? Sociological Explanations of New Sources of Earnings Inequality." *Annual Review of Sociology* 34:236–56.

Lieberson, Stanley. 1980. *A Piece of the Pie: Blacks and White Immigrants since 1880*. Berkeley: University of California Press.

Logan, John R., Richard D. Alba, and Brian J. Stults. 2003. "Enclaves and Entrepreneurs: Assessing the Payoff for Immigrants and Minorities." *International Migration Review* 37:344–88.

Lynch, Scott. 2003. "Cohort and Life Course Patterns in the Relationship between Education and Health: A Hierarchical Approach." *Demography* 40:309–31.

———. 2008. "Race, Socioeconomic Status and Health in Life Course Perspective: Introduction to the Special Issue." *Research on Aging* 30:127–36.

Mannheim, Karl. 1952. "The Sociological Problem of Generations." Pp. 276–322 in *Essays on the Sociology of Knowledge*, edited by P. Kecskemeti. New York: Oxford University Press.

Markides, Kyriakos S. and Sandra A. Black. 1996. "Race, Ethnicity, and Aging: The Impact of Inequality." Pp. 153–170 in *Handbook of Aging and the Social Sciences*, 4th ed., edited by R. H. Binstock & L. K. George. San Diego: Academic.

Markides, Kyriakos S. and Karl Eschbach. 2005. "Aging, Migration, and Mortality: Current Status of Research on the Hispanic Paradox." *Journal of Gerontology Series B*: Special Issue II 60B:68–75.

Markides, Kyriakos S., Jersey Liang, and James S. Jackson. 1990. "Race, Ethnicity, and Aging: Conceptual and Methodological Issues." Pp. 112–29 in *Handbook of Aging and the Social Sciences*, 3rd ed., edited by R. H. Binstock and L. K. George. New York: Academic.

Markides, Kyriakos S. and Charles Mindel. 1987. *Aging and Ethnicity*. Newbury Park, CA: Sage.

Mayer, Karl U. 2009. "New Directions in Life Course Research." *Annual Review of* Sociology 35:413–33.

McClanahan, Sara and Christine Percheski. 2008. "Family Structure and the Reproduction of Inequalities." *Annual Review of Sociology* 34:257–76.

Merton, Robert K. 1968. "The Matthew Effect in Science: Theoretical and Empirical Investigations." *Science* 199:55–63.

Mortimer, Jeylan T. and Michael J. Shanahan, eds. 2003. *Handbook of the Life Course*. New York: Kluwer Academic.

Mui, Ada C. and Tazuko Shibusawa. 2008. *Asian American Elders in the Twenty-first Century: Key Indicators of Well-Being*. New York: Columbia University Press.

Musil, Carol M. and Muayyad Ahmad. 2002. "Health of Grandmothers: A Comparison by Caregiver Status." *Journal of Aging and Health* 14:96–121.

Mutchler, Jan E., Lindsey A. Baker, and SeungAh Lee. 2007a. "Grandparents Responsible for Grandchildren in Native American Families." *Social Science Quarterly* 88:990–1009.

Mutchler, Jan E. and Jeffrey A. Burr. 2003. "Living Arrangements among Older Persons: A Multilevel Analysis of Housing Market Effects." *Research on Aging* 25:531–58.

———. 2008. "Varieties of Wellbeing: Race, Class, Gender, and Age." Pp. 23–46 in *Boomer Bust? Economic and Political Dynamics of the Graying Society*, edited by R. B. Hudson. New York: Praeger.

Mutchler, Jan E., Archana Prakash, and Jeffrey A. Burr. 2007b. "The Demography of Disability and the Effects of Immigration History: Older Asians in the United States." *Demography* 44:251–63.

Myers, Dowell. 2007. *Immigrants and Boomers: Forging a New Social Contract for the Future of America.* New York: Russell Sage.

Nee, Victor and Jimy Sanders. 2001. "Understanding the Diversity of Immigrant Incorporation: A Forms-of-Capital Model." *Ethnic and Racial Studies* 24:386–411.

O'Rand, Angela M. 1996. "The Precious and the Precocious: Understanding Cumulative Disadvantage and Cumulative Advantage over the Life Course." *The Gerontologist* 36:230–8.

———. 2006. "Stratification and the Life Course: Life Course Capital, Life Course Risks, and Social Inequality." Pp. 145–62 in *Handbook of Aging and the Social Sciences*, edited by R. H. Binstock and L. K. George. San Diego: Elsevier.

Osypuk, Theresa L., Ana V. Diez Roux, Craig Hadley, and Namratha Kandula. 2009. "Are Immigrant Enclaves Healthy Places to Live? The Multi-Ethnic Study of Atherosclerosis." *Social Science and Medicine* 69:110–20.

Palloni, Alberto. 2006. "Reproducing Inequalities: Luck, Wallets, and the Enduring Effects of Childhood Health." *Demography* 43:587–615.

Park, Robert. 1950. *Race and Culture*. Glencoe, IL: Free Press.

Pearlin, Leonard I. 2010. "The Life Course and the Stress Process: Some Conceptual Comparisons." *Journal of Gerontology: Social Sciences* 65B:207–15.

Portes, Alejandro and Min Zhou. 1993. "The New Second Generation: Segmented Assimilation and Its Variants." *Annals of the American Academy of Political and Social Science* 530:74–96.

Riley, Matilda White and John W. Riley, Jr. 2000. "Age Integration: Conceptual and Historical Background." *The Gerontologist* 40:266–70.

Roff, Lucinda L., Lewis D. Burgio, Laura Gitlin, Linda Nichols, William Chaplin, and J. Michael Hardin. 2004. "Positive Aspects of Alzheimer's Caregiving: The Role of Race." *Journal of Gerontology: Psychological Sciences* 59B, P185–90.

Ryder, Norman. 1965. "The Cohort as a Concept in the Study of Social Change." *American Sociological Review* 30:843–61.

Ryff, Carol D. and Burton Singer. 2009. "Understanding Healthy Aging: Key Components and Their Integration." Pp. 117–44 in *Handbook of Theories of Aging*, edited by V. L. Bengtson, D. Gans, N. M. Putney, and M. Silverstein. New York: Springer.

Sands, Roberta G. and Robin S. Goldberg-Glen. 1998. "The Impact of Employment and Serious Illness on Grandmothers Who Are Raising Their Grandchildren." *Journal of Women and Aging* 10(3):41–58.

Schweitzer, Marjorie M. 1999. "Oteo-Missouria Grandmothers: Linking Past, Present, and Future." Pp. 159–80 in *American Indian Grandmothers: Traditions and Transitions,* edited by M. Schweitzer. Albuquerque, NM: University of New Mexico Press.

Settersten, Richard A., Jr., ed. 2003. *Invitation to the Life Course: Toward New Understandings of Later Life.* Amityville, NY: Baywood.

———. 2006. "Aging and the Life Course." Pp. 3–19 in *Handbook of Aging and the Social Sciences*, 6th ed., edited by R. H. Binstock and L. K. George. New York: Academic.

Shuey, Kim M. and Andrea E. Willson. 2008. "Cumulative Disadvantage and Black-White Disparities in Life-Course Health Trajectories." *Research on Aging* 30:200–25.

Smedley, Brian D. Adrienne Stith, and Alan R. Nelson, eds. 2003. *Unequal Treatment: Confronting Racial and Ethnic Disparities in Health Care*. Washington, DC: The National Academies Press.

Social Security Administration. 2009. *Retirement Benefits by Year of Birth.* Retrieved on December 20, 2009 (http://www.socialsecurity.gov/retire2/agereduction.htm).

Sykes, Lori L. 2003. "Income Rich and Asset Poor: A Multilevel Analysis of Racial and Ethnic Differences in Housing Values among Baby Boomers." *Population Research and Policy Review* 22:1–20.

Taylor, Miles G. 2008. "Timing, Accumulation, and the Black/White Disability Gap in Later Life: A Test of Weathering." *Research on Aging* 30:226–50.

Taylor, Robert J., Linda M. Chatters, and James S. Jackson. 2007. "Religious Participation among Older Black Caribbeans in the United States." *Journal of Gerontology: Social Sciences* 62B:S251–6.

Thomas, William I. and Florian Znaniecki. 1974 [1918–1920]. *The Polish Peasant in Europe and America.* New York: Octagon Books.

Treas, Judith. 2008–2009. "Four Myths about Older Adults in America's Immigrant Families." *Generations* 32(4):40–5.

Treas, Judith and Shampa Mazumdar. 2002. "Older People in America's Immigrant Families: Dilemmas of Dependence, Integration, and Isolation." *Journal of Aging Studies*16:243–58.

Turra, Cassio and Irma Elo. 2008. "The Impact of Salmon Bias on the Hispanic Mortality Advantage: New Evidence from Social Security Data." *Population Research and Policy Review* 7:515–30.

U.S. Bureau of the Census. 2008. U.S. Population Projections. Retrieved December 11, 2009 (http://www.census.gov/population/www/projections/downloadablefiles.html).

———. 2009. American Community Survey, 2008. Retrieved December 13, 2009 (http://www.census.gov/acs/www/Products/index.html).

Walesmann, Katrina M., Arline T. Geronimus, and Gilbert C. Gee. 2008. "Accumulating Disadvantage over the Life Course: Evidence from a Longitudinal Study Investigating the Relationship between Educational Advantage in Youth and Health in Middle Age." *Research on Aging* 30:169–99.

Warner, David F. and Mark D. Hayward. 2006. "Early-life Origins of the Race Gap in Men's Mortality." *Journal of Health Social Behavior* 47:209–26.

Weiss, Robert. 2005. *The Experience of Retirement*. Ithaca, NY: ILR Press.

Williams, David R. 2004. "Racism and Health." Pp. 69–80 in *Closing the Gap: Improving the Health of Minority Elders in the New Millennium*, edited by K. Whitfield. Washington, DC: The Gerontological Society of America.

Williams, David R. and Colwick M. Wilson. 2001. "Race, Ethnicity, and Aging." Pp. 160–78 in *Handbook of Aging and the Social Sciences,* 5th ed., edited by R. H. Binstock and L. K. George. New York: Academic.

Willson, Andrea E., Kim M. Shuey, and Glen H. Elder, Jr. 2007. "Cumulative Advantage Processes as Mechanisms of Inequality in Life Course Health." *American Journal of Sociology* 112:1886–924.

Wilmoth, Janet M. and Charles F. Longino. 2006. "Demographic Trends that Will Shape U.S. Policy in the 21st Century." *Research on Aging* 28:269–88.

Winant, Howard. 2000. "Race and Racism." *Annual Review of Sociology* 26:169–85.

Xie, Yu and Margaret Gough. 2009. "Ethnic Enclaves and the Earnings of Immigrants." Population Studies Center Research Report #09-685, University of Michigan. Retrieved April 10, 2010 (http://www.psc.isr.umich.edu/pubs/pdf/rr09–685.pdf).

Yang, Yang and Linda C. Lee. 2010. "Dynamics and Heterogeneity in the Process of Human Frailty and Aging: Evidence from the U.S. Older Adult Population." *Journal of Gerontology: Social Sciences* 65B:S246–55.

Yao, Li and Stephanie A. Robert. 2008. "The Contributions of Race, Individual Socioeconomic Status and Neighborhood Socioeconomic Context on the Self-Rated Health Trajectories and Mortality of Older Adults." *Research on Aging* 30:251–73.

Chapter 7
Immigration, Aging, and Health in the United States

Kyriakos S. Markides and Kerstin Gerst

Given the rising numbers of immigrants to the United States, in recent decades there has been increasing interest in better understanding the health status and the health care needs of immigrants and how they impact the host societies' health and social service systems. Recent evidence has suggested the existence of a health advantage among immigrants, especially those from non-western origins, which has challenged our previous and often stereotypical notions regarding immigrants from poor countries. Early research in North America was almost exclusively focused on the negative impact of immigration on mental health (Malzberg 1967). The negative aspects of immigration were also the dominant theme of early studies in Europe (Friis et al. 1998). It is now commonly assumed that early research was often culturally biased, methodologically weak, and based on small numbers of immigrants (Friis et al. 1998; Markides 2001).

Recent studies, which include adequate numbers of immigrants and employ better study designs, provide compelling evidence that most immigrants to the United States and other western societies enjoy a health advantage. Below we focus on the health of immigrants to the United States, how their health changes over time, and we examine evidence regarding the association between aging and the health of immigrants. Since the majority of immigrants to the United States are from Mexico and Latin America, we give them particular attention. There has also been currently increasing interest in the health of immigrants of Asian origins, but studies of the influence of aging on health in such populations are lacking. Because of problems inherent in early research, we are unable to identify significant cohort differences in the impact of immigration on health. Although we may be unable to identify cohort differences in the health of immigrants, the volume of immigration is so much higher than it was just a few decades ago so that the impact of immigration on the overall health status of Americans as well as its impact on the health care system is considerably higher now.

Explanations of the health advantages of immigrants have focused on several major factors, including migration selection, return migration of less healthy people or "salmon bias," strong families and social networks, and better health behaviors. Some time ago Friis et al. (1998) proposed that the association between migration and health can be approached using the "stress-illness" model where immigration is a major life event conceptualized as a source of stress. At the same time, the process of acculturation into the host society can be a stressful experience that can impact physical and mental health. As immigrants become more acculturated and assimilated into the larger society, the level of stress they experience and its impact on health are reduced. It has also been suggested that immigrants arriving later in life often become linguistically and culturally isolated as

K.S. Markides (✉)
University of Texas Medical Branch, Dept. of Preventive Medicine and Community Health, Galveston, TX, USA
e-mail: kmarkide@utmb.edu

R.A. Settersten, Jr. and J.L. Angel (eds.), *Handbook of Sociology of Aging*, Handbooks of Sociology and Social Research, DOI 10.1007/978-1-4419-7374-0_7, © Springer Science+Business Media, LLC 2011

their children become more acculturated into the large society, often leading to intergenerational frictions (Markides and Black 1995).

While immigrants arrive with better health status than the native-born, they appear to experience a convergence to native-born health status levels with time in the United States, and these health advantages disappear by the next generation. Reasons for convergence have focused on worsening health behaviors resulting from the stresses of acculturation (Friis et al. 1998), or "unhealthy assimilation," or the adoption of unhealthy behaviors (Antecol and Bedard 2006). With respect to the worsening of health status into old age, there has been some focus on the influence of physical labor over a life time accompanied by substandard medical care (Markides et al. 2008–2009), as well as on the impact of cumulative social and economic disadvantages over the life course (Wakabayashi 2009).

Below we review recent evidence regarding the health status of immigrants to the United States and evaluate the viability of explanations offered in the literature. We focus on major health status indicators such as mortality and chronic diseases and disabilities. We also examine how the health of immigrants changes with age and time in the United States. Special focus is given to the Hispanic population especially Mexican Americans on whom adequate data are available. We also refer to research in Canada and Australia which corroborates research findings in the United States.

Overview of Immigration Trends

Current international migration levels are at an all-time high, with 175 million persons (about 3% of the world's population) residing in a country that they were not born in (United Nations 2002; Jackson and Howe 2008). Most advanced western societies are receiving increasing numbers of immigrants, many of whom are arriving from nonwestern countries (Markides et al. 2008–2009). While immigrants tend to be relatively young, their large numbers will assure rising numbers of foreign-born older people who may have special health concerns. The United States, Canada, and Australia have been the three major immigrant destinations where most immigrants have gone and stayed, unlike some other European nations where return migration to the country of origin is common. In 2006, the United States received 1.2 million immigrants, Canada 250,000, and Australia 125,000 (Martin and Zürcher 2008). Although Canada and the United States make up only approximately 5% of the worlds' population, these two countries receive over half of the world's immigrants (Martin and Zürcher 2008).

In 2008, the United States had nearly 38 million persons that were foreign born (12.5% of the total populations), of whom approximately 12.3% were aged 65 years and older (Migration Policy Institute 2010). The immigration patterns to the United States have shifted dramatically in recent decades. Whereas immigrants used to be predominantly from Europe, increasing numbers are arriving from Latin America and Asia. In 1960, 75% of all immigrants to the United States were from Europe, less than 10% were from Latin America, and only 5% came from Asia. This is in stark contrast to 2008, where only 13% of immigrants were from Europe, over half (53.1%) were from Latin America, and 27.3% came from Asia (Migration Policy Institute 2010).

Mortality and Life Expectancy

A great deal of attention has been given recently to mortality and life expectancy among ethnic and minority populations in the United States. This literature has consistently shown higher mortality rates among African Americans at every age except at very advanced ages, typically 85 or 90 years and older when African Americans appear to have lower mortality rates than non-Hispanic whites.

The black/white mortality crossover has been controversial primarily because of questions about the quality of data for African Americans at advanced ages. Others feel that the crossover is real and can be explained in terms of greater selective survival among African Americans (Markides and Black 1995; Manton and Stallard 1997). That is, the few African Americans that survive to very advanced ages are "hardier" and experience lower mortality rates than non-Hispanic whites of the same age. A relatively new phenomenon has been the attention to the recent increase of foreign-born African Americans who appear to have quite favorable mortality rates. Favorable rates have been attributed primarily to immigrant selection forces including good health and good healthy behaviors (Hummer et al. 1999; Singh and Hiatt 2006).

The overall mortality disadvantage of African Americans has been attributed to socioeconomic forces including low education, low financial resources, as well as the stresses associated with racial prejudice and discrimination. In contrast, the negative effects of such socioeconomic forces have not translated into poorer health and mortality among most Hispanic/Latino populations who now together constitute the largest minority population in the United States at around 48 million. Some 25 years ago Markides and Coreil (1986) proposed an "epidemiological paradox" with respect to the health and mortality situation of Southwestern Hispanics who were overwhelmingly of Mexican origin. Data from around 1980 suggested that mortality rates for Southwestern Hispanics were similar to those of the non-Hispanic white population and were considerably lower than mortality rates of African Americans with whom they shared similar socioeconomic circumstances. This seemed paradoxical at that time because of the high rates of poverty among Mexican Americans and most other Hispanics given the established association between poverty and socioeconomic status with health and mortality. There were other risk factors that made similar mortality rates paradoxical including high rates of obesity, diabetes, and sedentary life styles among Mexican Americans especially at older ages. Mexican Americans had lower mortality rates from cardiovascular diseases and major cancers, especially among men. Explanations suggested included strong family supports, certain cultural practices, and health selective immigration. Hispanics were disadvantaged on certain health factors including diabetes and infectious and parasitic diseases (see also Hays-Bautista 1992; Vega and Amaro 1994), but overall they were in relatively good health.

By the 1990s, studies began showing a mortality advantage among Mexican Americans and other Hispanic groups. Puerto Ricans living on the mainland were an exception. The literature commonly referred to the epidemiological paradox as the "Hispanic Paradox" or "Latino Paradox." Franzini et al. (2001) performed a review of the literature over a 20-year period and concluded that the mortality advantage was most pronounced among infants and among older people. They suggested that the paradox might be explained by problems with vital statistics data, a healthy immigrant effect, and a "salmon bias," or return migration to Mexico by less healthy older people. They concluded that these three factors may explain part but not all of the mortality advantage. Abraido-Lanza et al. (1999) employed data from the National Longitudinal Mortality Study and found evidence of salmon bias. However it was too small to account for the mortality advantage. One study using the National Health Interview Survey – Multiple Cause of Death (NHIS-MCD) data set found evidence of selective outmigration of unhealthy Mexican Americans and argued that indeed such salmon bias explains the Mexican American mortality advantage (Palloni and Arias 2004). However, questions about limitations of the NHIS sample cautioned against such definitive conclusions (Markides and Eschbach 2005, 2011).

Hummer et al. (2007) examined a potential salmon bias in infancy. They employed data from the United States birth and infant death cohort files from 1995 to 2000 and found that death rates for infants born to Mexican immigrant women were approximately 10% lower than for infants born to non-Hispanic white U.S.-born women or to U.S.-born Mexican origin women. The large sample yielded stable rates for the first hour of life, first day, and first week. The authors concluded that favorable rates so early in life were unlikely to result from outmigration of immigrant women and their infants.

Perhaps the most definitive test of the salmon bias in old age was performed by Turra and Elo (2008) using data from the Master Beneficiary Record and the NUDIMENT data files of the Social Security Administration. They found higher mortality rates among foreign-born Hispanic beneficiaries living abroad than foreign-born beneficiaries living in the United States. At the same time, a significant number of older Hispanics living outside the United States appear to return to the United States when their health worsens and thus have high mortality rates. The authors note that the influence of salmon bias on Hispanic death rates in the United States is partially offset by the high mortality rates of Hispanic emigrants who return to the United States. They conclude that indeed a salmon bias exists but it is too small to account for the Hispanic mortality advantage (see, also, Abraido-Lanza et al. 1999).

Earlier we noted limitations of the NHIS-MCD data used by Palloni and Arias (2004). An enlarged public use file that included more years of survey from 1986 through 2000, with mortality follow-up through 2002 has recently become available. Borrell and Crawford (2009) used the data and found evidence corroborating earlier findings of low mortality among Hispanics. As in previous reports much of the advantage occurs in old age with findings being inconsistent at ages 25–44 regardless of nativity. As in previous studies Puerto Ricans had higher death rates than other Hispanic groups.

Eschbach et al. (2007) used vital registration data from Texas and California for 1999–2001 linked to 2000 Census population data to examine mortality at younger ages, namely at ages 15–44. Data presented by 5-year age groups showed a mortality advantage among foreign-born Hispanics of both genders relative to non-Hispanic whites. Among men the majority of the immigrant advantage was due to lower suicide rates and lower substance abuse (other than alcohol) than among non-Hispanic whites. At the same time, U.S.-born Hispanic men had higher death rates than non-Hispanic white men from both social and behavioral causes and chronic diseases. Social and behavioral causes were related to HIV and other sexually transmitted diseases, substance abuse, alcohol, and homicide. Female Hispanic immigrant advantages over non-Hispanic white women were attributable to social and behavioral causes but also to mortality from circulatory disease and major cancers. U.S.-born Hispanic women had higher death rates from homicide, HIV, and infectious and parasitic diseases than non-Hispanic white women and lower death rates from suicide, substance abuse, and unintentional accidents. No differences were observed with respect to chronic disease mortality. The authors concluded that the Hispanic mortality paradox at younger ages is primarily an immigrant phenomenon.

The inconsistency between these findings and those based on the NDI linkage discussed above (Borrell and Crawford 2009) may be related to the accumulation of mortality over many years in the latter and thus the inability of such studies to be responsive to period effects. For example, recent rates of homicide mortality and motor vehicle accidents showed greater declines among Hispanics than among non-Hispanic whites in Texas. Now younger Hispanics regardless of nativity have lower mortality from motor vehicle accidents a reversal from earlier patterns (Markides and Eschbach 2011).

The immigrant mortality advantage in the United States is also present in other immigrant populations (see Jasso et al. 2004; Singh and Siahpush 2002; Singh and Hiatt 2006; Cunningham et al. 2008). There is an overall significant immigrant advantage which may have increased in recent years from 2.3 years in life expectancy in 1979–1981 (76.2 vs. 73.9 years) to 3.4 years in 1991–2001 (80.9 vs. 76.6 years) (Singh and Hiatt). In 1999–2001, immigrants had significantly lower mortality from lung and esophageal cancer, COPD, HIV/AIDS, and suicide, but higher mortality from stomach and liver cancer. Among women estimated life expectancy at birth was the highest among U.S.-born Asian/Pacific Islanders (86.0 years), followed by immigrant Asian/Pacific Islanders (85.0 years), and Hispanic immigrant women (84.1 years). Among men Asian/Pacific Islanders had the highest life expectancy (80.7 years), followed by Hispanic immigrants (79.0), followed by U.S.-born Asian/Pacific Islanders (78.9), and by non-Hispanic white and black immigrants (both at 75.6). For each ethnic origin, there was an immigrant advantage (except for Asian/Pacific Islander women) which likely reflects compositional differences. This is especially the case at older ages wherein immigrants are increasingly Filipino and Vietnamese while a substantial percent of the native born are of Japanese origin (see, also, Markides et al. 2007). Singh and Hiatt (2006) note that the largest nativity advantages were among blacks and Hispanics for both men and women. They attributed the immigrant advantage to health selection and better health behaviors including smoking and obesity, as well as lower chronic disease prevalence.

The large mortality advantages of Asian/Pacific Islanders led Lauderdale and Kestenbaum (2002) to suggest that very low mortality rates may very well be the result of healthy immigrant selection and relatively high socioeconomic status. At the same time, they raised the possibility that the rates may be understated because of underreporting of Asian/Pacific Islander race on death certificates. They employed the Master Beneficiary Record of the Social Security Administration as well as the NUDIMENT data files discussed earlier which avoid problems of misclassification of ethnicity. Elderly Asian Americans from six ethnic origins – Chinese, Filipinos, Indian, Japanese, Korean, and Vietnamese – were estimated to have lower death rates than those computed for elderly non-Hispanic whites. They raised the question of whether a "healthy immigrant" effect might account for the Asian mortality advantages but found inconsistent results. One issue is that older people of some Asian origins (Vietnamese, Filipino, Korean, and Indian) are overwhelmingly foreign born, so that meaningful comparisons with their native-born counterparts are not possible. As suggested earlier, comparing native-born and foreign-born mortality rates for all Asian origin groups lumped together is not advisable because of ethnic compositional differences between elderly foreign-born and native-born Asian Americans.

Physical Health and Disability

Much of the literature in the United States shows that most immigrants arrive with better health than native-born populations and experience lower mortality rates. Their health status appears to converge to native levels over time with any health advantages disappearing in the next generation. It has been suggested that such convergence might reflect improved access to health care, which can lead to increased diagnosis of preexisting conditions (Antecol and Bedard 2006; Jasso et al. 2004) thus leading to reduced reported health status. Data from Canada do not support this hypothesis. McDonald and Kennedy (2004) suggested that an observed convergence in Canada reflects actual convergence in physical health rather than a convergence in screening and diagnosis of existing health problems. They base this conclusion on data showing that immigrants' use of health care services converges with native-born levels faster than health outcomes do. An assimilation hypothesis suggests that since with time income and employment rates increase and at some point converge to native levels, one could predict that the health status of immigrants would improve (Antecol and Bedard 2006; Jasso et al. 2004; Sorlie et al. 1993). However, the evidence is that the opposite takes place with immigrants becoming less healthy with time in the country (Antecol and Bedard 2006; Cunningham et al. 2008; Cho et al. 2004; Stephen et al. 1994).

Explanations of such convergence have focused primarily on changes in health behaviors associated with increasing acculturation into the host society, including higher rates of smoking, changes in diet, and increasing rates of obesity (Cunningham et al. 2008; Markides 2001; Stephen et al. 1994). In fact obesity might be the central mechanism through which length of time since immigration leads to worsening health. Antecol and Bedard (2006) used data from National Health Interview Survey (NHIS) for 1989–1996 and found that male and female immigrants enter the United States with lower body mass index (BMI) levels than native-born men and women. They estimate that the BMI's of foreign-born women almost completely converge to native-born levels within 10 years while men close about one-third of the gap within 15 years. However, the convergence among females appears to be largely driven by Hispanics. There is also a convergence among Hispanic males, but only with respect to overweight and not with respect to obesity (BMI 30+). Interestingly, black immigrants do not appear to converge to native BMI levels. These patterns appear to mirror those for health conditions, self-reported health, and activity limitations. Again the conclusion is that most of the convergence in BMI and overall health appears to take place among Hispanic immigrants. Importantly, most immigrants entering the United States in recent decades have come from Latin America, especially Mexico.

Huh et al. (2008) examined nativity differences in chronic conditions and other health indicators among Asian and Hispanic populations in the United States using NHIS data from 2000 to 2001. As expected foreign-born persons reported fewer chronic conditions (hypertension, asthma, heart disease, cancer, and diabetes) than U.S.-born non-Hispanic whites. They also found evidence that the initial health advantage of immigrants diminishes over time suggesting possible adverse effects of acculturation into American society. Frisbie et al. (2001) reached similar conclusions using data from the 1992–1995 NHIS. They found that immigrant health advantages were greatest for the first 5 years since immigration had declined consistently with time in the United States. The evidence was not uniformly positive in that Vietnamese immigrants had less than average health. It is possible that since Vietnamese-origin immigrants to the United States were mostly refugees, they were not subject to the same barriers as other immigrants.

Huh et al. (2008) caution that nativity is a crude indicator and may not fully capture the complexity of the acculturation process. While both Hispanic and Asian immigrants reported better physical health, they nevertheless were more likely to rate their health as being poorer than did non-Hispanic whites. This was especially so among Asian immigrants. While such a finding may suggest greater negative consequences of poor health among certain ethnic groups (Markides and Black 1995), it also cautions against relying only on self-rated health to study health disparities, which is often the case in the literature. Self-ratings of health have been found to be less valid among Hispanic immigrants (Finch et al. 2009) as well as among older Hispanics and African Americans who tend to be health pessimistic (Markides and Black 1995; Markides et al. 1997).

In recent years, there has been interest in understanding health disparities in old age using a cumulative advantage/disadvantage perspective (Crystal and Shea 2002; Dannefer 2003; O'Rand and Hamil-Luker 2005). Health advantages/disadvantages in old age are thought to result from accumulation of stressors over the life course. It has been found, for example, that the experience of economic strains early in life is associated with higher levels of disability, prevalence of serious medical conditions, depression, and poorer general health in later life (Kahn and Pearlin 2006; Lynch et al. 1997; Wakabayashi 2009). Conversely, it has also been found that people of high socioeconomic status are less likely to experience disability and morbidity as they age than persons of lower socioeconomic status (House et al. 1990, 1994). Thus, differences in socioeconomic status and economic and other stressors over the life course may explain poor health outcomes in old age in relatively long-living immigrant and ethnic populations such as Hispanics (Markides et al. 2008–2009).

Wakabayashi (2009) examined the viability of cumulative disadvantage/advantage theory to understand the health trajectories of immigrants using data from the Health and Retirement Study (HRS) for 1996, 1998, 2000, 2002, 2004, and 2006. She found that women immigrating after the age of 34 were more likely to have poor health trajectories with respect to limitations in activities of daily living possibly because they had less time to accumulate financial resources which would be protective of health in late life. More specifically it was Hispanic men and women regardless of age at immigration who experienced the most disadvantageous health trajectories with respect to reporting poor health because of limited opportunities to accumulate economic and other resources that would benefit health in late life. Thus, while Hispanic immigrants arrive with a health advantage, they exhibit worsening health profiles relative to non-Hispanic whites with time in the United States, a pattern observed with several large data bases, as we saw earlier. Similar data on health trajectories from middle into old age are not currently available on Asian-origin immigrants.

Trends in the Health of Older Mexican Americans

Of the major immigrant populations there has been more interest in the health of older Mexican Americans. The Hispanic Established Population for the Epidemiological Study of the Elderly (Hispanic EPESE) was launched in 1993–1994 when data were collected on a representative sample of 3,050 Mexican Americans aged 65 and over from the Southwestern United States (Texas, New Mexico, Colorado, Arizona, and California). Subjects have been followed every 2–3 years. By Wave 5 in 2004–2005, there were 1,167 surviving subjects from the original cohort who were then aged 75 and over. A new cohort of 902 Mexican Americans aged 75 years and over was drawn from the same region using similar procedures giving us the opportunity to examine trends in the health of very old Mexican Americans over an 11-year period.

It has been well-documented that the health of older Americans as well as the health of older people in other western societies began showing improvements in the mid-1980s (Crimmins et al. 1997; Manton 2008; Manton and Gu 2001; Freedman et al. 2002; Waidman and Liu 2000; Zunzunequi et al. 2006). This was a reversal from the 1970s and early 1980s, a period when increases in life expectancy were accompanied by increases in morbidity and disability. There is some recent evidence that the declines in old age disability and poor health in the United States may have stopped or may have reversed (Seeman et al. 2010). We have suggested elsewhere that the Mexican American population, much like Latin American and other developing countries' populations, are at a similar point in the epidemiological transition that the more advantaged western populations were in during the 1970s and early 1980s, a period of rising life expectancy accompanied by increases in the prevalence of chronic conditions and disabilities (Markides et al. 2011).

In one analysis using data from the Hispanic EPESE (Beard et al. 2009), we observed a significant increase in the prevalence of self-reported diabetes among Mexican Americans aged 75 years and over. Such an increase was attributed to better management of diabetes in older Mexican Americans and thus increased survival with the disease. Using the same data, we also found slight increases in hypertension prevalence over the 11-year period (Al Ghatrif et al. 2011). There was also a significant increase in awareness and control of hypertension. Our results suggest better management of chronic conditions among older Mexican Americans in recent years and increased survival to advanced ages. Below we examine whether there was also a corresponding increase in disability among Mexican Americans aged 75 years and over during the same time period.

Table 7.1 presents data on trends in disability and other health indicators using data from two cohorts of Mexican Americans aged 75 years and over from the Hispanic EPESE from 1993–1994 to 2004–2005. Bivariate comparisons were made for reporting any ADL disability (toileting, walking across a room, eating, transferring from bed to chair, bathing, and dressing). The table shows that the percent with any ADL disability increased from 20.2 to 29.7% among men and from 21.5 to 41.2% among women. Also shown are significant increases in prevalence of diabetes, hypertension, obesity, and cognitive impairment. The increase in disability can be attributed partly to increases in the prevalence of diabetes and cognitive impairment. No doubt they can also be attributed to increases in life expectancy and in frailty among very old Mexican Americans (Palloni 2007).

Further analysis by nativity status (not shown) showed that immigrant women were significantly less likely to self-report diabetes than were native-born women aged 75 years and over at Wave 5 in 2004–2005. However, immigrant women were more likely to be cognitively impaired using the Mini Mental State Examination (MMSE<21). No such differences were observed among men. In both genders, both the foreign-born and native-born older Mexican Americans experienced significant increases from 1993–1994 to 2004–2005 in the prevalence of diabetes, obesity, cognitive impairment, and ADL disability. Thus, our data support the hypothesis of increases in morbidity and disability in a relatively disadvantaged population experiencing significant increases in life expectancy.

Table 7.1 Sociodemographic characteristics and prevalence of medical conditions in older Mexican American men and women in 1993–1994 and 2004–2005

	Men		Women	
	1993–1994 (n=470)	2004–2005 (n=371)	1993–1994 (n=662)	2004–2005 (n=531)
Age (mean ± SD)	80.9 ± 5.2	81.3 ± 4.7	81.0 ± 5.0	81.5 ± 5.4
Married	67.50	66.00	25.10	29.80
Chronic diseases				
Hypertension	57.00[a]	66.00	66.80[a]	72.60
Self-reported diabetes	21.30[b]	32.70	21.50[b]	38.10
Self-reported heart attack	14.60	13.40	13.40	10.10
Self-reported stroke	9.60	9.80	10.00	8.30
Self-reported cancer	6.60	8.40	6.80	7.30
Self-reported hip fracture	3.60	4.30	7.30	7.20
Any ADL limitation	19.90[b]	28.80	26.60[b]	42.30
Years of education (mean ± SD)	4.4 ± 3.9	4.9 ± 4.4	4.2 ± 3.7	5.1 ± 4.2
Obesity (BMI ≥ 30 kg/m²)	17.90	22.40	26.70	28.90

[a] $p<0.05$
[b] $p<0.01$

Census Disability Rates for Older Hispanics by Nativity

While the Hispanic EPESE data are interesting and suggest important trends, they are limited only to the Mexican-origin population. A useful resource with data on large samples of older people from all major ethnic groups can be found in the 2000 United States Census. In a recent analysis, we computed disability rates for people 65 and by race/ethnicity and by type of Hispanic origin using the 2000 Census 5% public use micro data sample (PUMS) file, which includes data on approximately 1.8 million Americans aged 65 years and over (Markides et al. 2007). Rates were directly standardized to the total United States population of people aged 65 years and over in 2000 for ages 65–69, 70–74, 75–79, 80–84, 85–89, and 90+.

The 2000 Census obtained data on each person aged 5 years and over in the sampled households on six disabilities items (Stern 2004). Items were (1) *Sensory disability*, defined as blindness, deafness, or a severe vision or hearing impairment; (2) *Physical disability*, defined as a long-lasting condition which substantially limits one or more basic activities; (3) *Mental disability* referred to difficulty in learning, remembering, or concentrating; (4) *Self-Care disability* referred to difficulty in dressing and bathing; (5) *Going outside the home disability* asked about difficulty going outside the home alone to shop or to visit a doctor; (6) *Employment disability* asked about difficulty working at a job or business (persons aged 16 years or older). We used the first five items only and dropped the employment disability item because it has little relevance to most older people.

Our analysis showed that older Native Americans of both genders were the most disabled, followed by African Americans, Hispanics, Asian Americans, and Non-Hispanic whites. Among Hispanic groups, Puerto Ricans were the most disabled followed by those of Dominican origin, those of Mexican origin, and by Cuban, Central, and South American origin groups, the latter three having rather similar rates. As expected, the most favorable rates were for those whose origin was Spain (Markides et al. 2007).

Below we used the PUMS data to compute "any disability" rates (any of the five items) by nativity for major Hispanic groups at ages 65 and over. We had previously hypothesized (Markides et al. 2007, 2008–2009; Markides and Eschbach 2005) that there are likely to be gender differences in migration selection at least in the older generation. The rationale was that men immigrated for occupational reasons while women more often immigrated to be with their spouses and their families. Thus, one would expect possible health advantages among men but not among women. Figures 7.1 and 7.2 present any disability rates by nativity for men and women, respectively. Rates were computed for Non-Hispanic whites, all Hispanics, and people of Mexican origin. Other Hispanic groups excluded because they were overwhelmingly immigrant in old age were Cubans, Central Americans, South Americans, and Dominicans. Puerto Ricans were also excluded because there are no immigrants among them since all are U.S. citizens. Thus migrating to the United States mainland from the island is not subject to any barriers and thus migration would not be as selective. Figure 7.1 indeed shows a small but noteworthy health advantage of foreign-born men among all Hispanics,

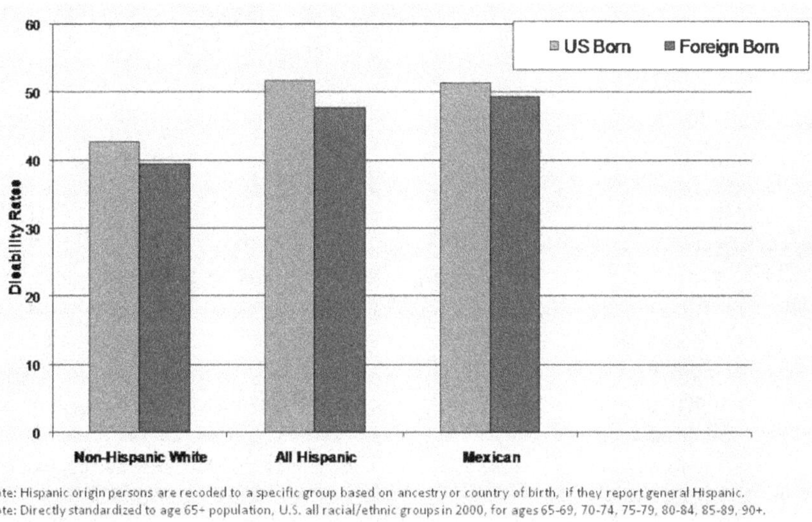

Fig. 7.1 Age-standardized census disability rates (percents) for U.S. born and foreign born males aged 65 years and over by type of Hispanic origin: United States Census, 2000

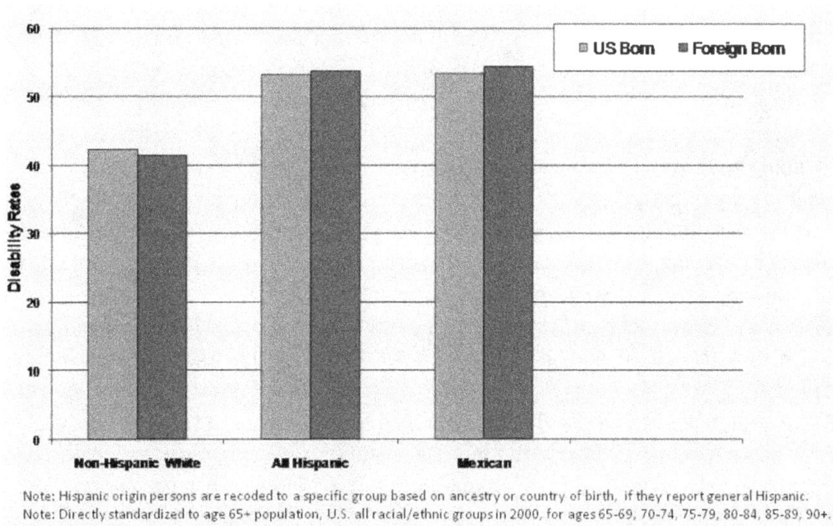

Fig. 7.2 Age-standardized census disability rates (percents) for U.S. born and foreign born females aged 65 years and over by type of Hispanic origin: United States Census, 2000

the Mexican origin, as well as among non-Hispanic whites. Figure 7.2 shows no nativity differences for the three groups giving support to the theory of gender differentials in migration selection in the older generations. Such differences are not thought to exist among younger immigrants in more recent years (Markides et al. 2007).

Conclusion

To summarize, it appears that while immigrants to the United States arrive with relatively good health and health behaviors, their health status appears to converge to native levels, a process that is more rapid among women than among men. A key mechanism of the convergence appears to be obesity. Such convergence is driven primarily by rising obesity rates among Hispanics. By old age Hispanics (and to a lesser extent other immigrants) appear to generally have more health problems than the general population even though they may experience favorable mortality rates. Poorer health can be attributed to a life-time of socioeconomic disadvantages as well as substandard medical care. The data lend support to cumulative disadvantage theory. Analysis of trends in the health of Mexican Americans aged 75 years and over from 1993–1994 to 2004–2005 using data from the Hispanic EPESE shows increases in the prevalence of obesity, diabetes, cognitive impairment, and ADL disability. The data suggest that Mexican Americans, like Latin American and other developing country populations, might be at a similar stage of the epidemiologic transition as were western nations during the 1970s and early 1980s when rising life expectancy was accompanied by increasing health problems and higher rates of disability. Finally, analysis of 2000 Census disability data on older Hispanics lend support to the hypothesis that immigrant health selection favors men over women at least at this generation of older people.

We began by noting the increased interest in the health of immigrants to the United States. The vast and rapidly increasing numbers of immigrants are having a major impact on American society. Because of their youth, they are changing the ethnic complexion of our schools, cities, as well as the occupational structure of both urban and rural areas. They are coming primarily from Latin America, especially from Mexico. Asian immigrants are moving primarily to the west coast where their presence and impact have been growing steadily. Immigration has been the topic of considerable political attention primarily because large numbers of immigrants are undocumented. There has also been considerable opposition to providing medical coverage to undocumented immigrants through public funds. The argument often is based, at least partially, on the notion that immigrants bring with them significant health problems and would constitute a burden on the health care system. As we have seen such stereotypical notions are not supported by the evidence. In fact, some of the healthiest people in the United States are immigrants originating in Mexico as well as Central and South America (Vega et al. 2009).

We also noted that much of the earlier literature in North America through the middle of the twentieth century gave great focus to the negative effects of immigration on mental health. This literature has been challenged more recently on the basis of cultural and scientific bias of much of the early research. In more recent decades, immigration has been viewed from a social stress-illness perspective wherein immigration is a major stressor (Friis et al. 1998). From such a social stress perspective, one would expect that immigrants, especially early on, would experience health problems resulting from the stresses of adjusting to the host society. There is no evidence that this is so because new immigrants, those in the United States less than 5 years, enjoy the greatest health advantage relative to the native born. Data from Canada and Australia show similar results. It could be argued that since most immigrants are young, it is unlikely that acculturative stresses will significantly influence their physical health in such a short time. However, an argument could be made that acculturative stresses are likely to have an impact on immigrant mental health.

This perspective has been dominant in the field. With respect to Mexican Americans, Escobar et al. (2000) examined the psychiatric and mental health literature and found no evidence in support of the notion of a negative effect of acculturation on the mental health of Mexican Americans. In fact, the shorter the time since immigration and the lower the level of acculturation, the lower the prevalence of psychiatric disorders (see, also, Vega et al. 2009). It has also been found that the mental health advantages of foreign-born Mexican Americans may extend to foreign-born non-Hispanic whites (Grant et al. 2004). As with physical health, most major immigrant populations appear to enjoy mental health advantages over the native born. Again, healthy immigrant selection is likely a major factor in such advantages as is the protective influence of traditional Mexican culture retention (Escobar et al. 2000).

The good health of immigrants from poor countries has also challenged our traditional notions regarding the association of social class and socioeconomic status with health. It could be hypothesized that a significant number of immigrants in an ethnic population is likely to depress the usual inverse SES gradient in ethnic populations. In fact, this was the case in the mortality study using national data by Turra and Goldman (2007) who found a much lower SES gradient among Hispanics than among Non-Hispanic whites. Similar findings were obtained with respect to other health indicators and health behaviors in the Mexican-origin population using data from three studies. Goldman et al. (2006) found an absence of significant educational differentials for several health-related variables among Mexican-origin adults as well as among adolescents and infants. The absence of such differentials was especially present among immigrants. Both studies conclude that the Hispanic health advantage pertains primarily to lower SES people.

Again, immigrant health advantages disappear by the next generation, driven primarily by psychosocial factors related to health behaviors, substance abuse, HIV and other sexually transmitted diseases, as well as homicide (Eschbach et al. 2007). As also shown in an examination of differences in biological health profiles, the Hispanic Epidemiologic Paradox is primarily an immigrant phenomenon (Crimmins et al. 2007).

We also saw that immigrant health advantages appear to converge to native levels with time. This finding was also replicated in Canada (Chen et al. 1996; McDonald and Kennedy 2004; Gee et al. 2004) and Australia (Biddle et al. 2007). Because of a number of factors we outlined earlier, by the time immigrants reach old age, they do not exhibit any health advantages and at least most Hispanic populations appear to be more disabled than older non-Hispanic whites.

At present, older immigrants are having a relatively small influence on the larger society. However, their numbers are projected to increase dramatically between now and the middle of the century because of continuing high rates of immigration and relatively high life expectancies. If their morbidity and disability rates remain at the current high levels, the impact of older immigrants on the health care system, as well as on their families, is likely to be substantial. Increased attention by scholars and policy makers to the rapidly growing and rapidly aging immigrant population is paramount. An unresolved issue in the literature in the Hispanic population is the presence of high morbidity and disability among Hispanics in old age when most of the evidence suggests the existence of favorable mortality in old age. It would be highly important to monitor trends in life expectancy as well as in morbidity and disability at all ages and in all ethnic and immigrant populations as their presence and impact on the larger society are likely to grow steadily between now and the middle of the century. It is also possible that the Hispanic mortality and health advantage, driven by immigrants from Mexico and Central and South America, is a cohort phenomenon which may well disappear in the future. For example, obesity rates in Mexico are only slightly lower than obesity rates in the United States which raises the possibility that future immigrants may be less health selected in the years to come. Monitoring trends in the health of the Mexican population as well as in the health of other immigrant sending countries would be important to health and social policy in the United States.

Acknowledgements Supported by grant R01 AG10939 from the National Institute of Aging and by the UTMB Claude D. Pepper Older Americans Independence Center, National Institute of Aging grant P30 AG24832.

References

Albraido-Lanza, Ana F., Bruce P. Dohrenwend, Daisy S. Ng-Mak, and J. Blake Turner. 1999. "The Latino Mortality Paradox: A Test of the Salmon Bias and Health Migrant Hypotheses." *American Journal of Public Health* 89:1534–48.

Al Ghatrif, Majd, Yong-Fang Kuo, Soham Al Snih, Mukaila A. Raji, Laura A. Ray, and Kyriakos S. Markides. 2011. "Trends in Hypertension Prevalence, Awareness, Treatment and Control in Older Mexican Americans 1993–2005." *Annals of Epidemiology* 21(1):15–25.

Antecol, Heather, and Kelly Bedard. 2006. "Unhealthy Assimilation: Why Do Immigrants Converge to American Health Status Levels?" *Demography* 3(2):337–60.

Beard, Holly A., Majd Al Ghatrif., Rafael A. Samper-Ternent, Kerstin Gerst, and Kyriakos S. Markides. 2009. "Trends in Diabetes Prevalence and Diabetes-Related Complications in Older Mexican Americans from 1993/1994 to 2004/2005." *Diabetes Care* 32(12):2112–17.

Biddle, Nicholas, Steven Kennedy, and James T. McDonald. 2007. "Health Assimilation Patterns amongst Australian Immigrants." *The Economic Record* 83(260):16–30.

Borrell, Luisa N. and Natalie D. Crawford. 2009. "All-Cause Mortality among Hispanics in the United States: Exploring Heterogeneity by Nativity Status, Country of Origin, and Race in the National Health Interview Survey-Linked Mortality Files." *Annals of Epidemiology* 19:336–43.

Chen, Jiajian, Edward Ng, and Russell Wilkins. 1996. "The Health Expectancy by Immigrant Status." *Health Reports*, 8(3):29–37.

Cho, Youngtae, W. Parker Frisbie, and Richard G. Rogers. 2004. "Nativity, Duration of Residence, and the Health of Hispanic Adults in the United States." *International Migration Review* 38:184–211.

Crimmins, Eileen M., Yasuhiko Saito, and Sandra I. Reynolds. 1997. "Further Evidence on Recent Trends in the Prevalence and Incidence of Disability among Older Americans from Two Sources: The LSOA and the NHIS." *Journals of Gerontology, Series B: Psychological Sciences* 52(2):X59–71.

Crimmins, Eileen M., Jung K. Kim, Dawn W. Alley, Arun Karlamangla, and Teresa Seeman. 2007. "Hispanic Paradox in Biological Profiles." *American Journal of Public Health* 97:1305–10.

Crystal, Stephen and Dennis G. Shea 2002. "Introduction: Cumulative Advantage, Public Policy, and Late-life Inequality." *Annual Review of Gerontology with Geriatrics* 22:1–13.

Cunningham, Solveig. A., Julia D. Ruben, and K. M. Vengat Narayan. 2008. "Health of the Foreign-Born People in the United States: A Review." *Health and Place* 14:623–35.

Dannefer, Dale. 2003. "Cumulative Advantage/Disadvantage and the Life Course: Cross-Fertilizing Age and Social Science Theory." *Journal of Gerontology: Social Sciences* 58B(6):5327–37.

Eschbach, Karl, Jim P. Stimpson, Yong-Fang Kuo, and James S. Goodwin. 2007. "Mortality of Foreign-born and U.S.-Born Hispanic Adults at Younger Ages: A Re-Examination of Recent Patterns." *American Journal of Public Health* 97:1297–1304.

Escobar, Javier, Constanza Hoyos Nervi, and Michael A. Gara. 2000. "Immigration and Mental Health: Mexican Americans in the United States." *Harvard Review of Psychiatry* 8(2):64–72.

Finch, Brian K., Diem P. Do, Roanne Frankj, and Teresa Seeman. T. 2009. "Could 'Acculturation' Effects Be Explained by Latent Health Disadvantages among Mexican Immigrants?" *International Migration Review* 43(3):471–95.

Franzini, Luisa, John C. Ribble, and Arlene M. Keddie, 2001. "Understanding the Hispanic Paradox." *Ethnicity and Disease* 11:496–518.

Freedman, Vicki A., Linda G. Martin, and Robert F. Schoeni. 2002. "Recent Trends in Disability and Functioning among Older Adults in the United States: A Systematic Review." *Journal of American Medical Association* 288(24):3137–46.

Friis, Robert, Agneta Yngue, and Viveka Perssan. 1998. "Review of Social Epidemiologic Research on Migrants' Health: Findings, Methodological Cautions, and Theoretical Perspectives." *Scandinavian Journal of Social Medicine* 26:173–80.

Frisbie, W. Parker, Youngtae Cho, and Robert A. Hummer. 2001. "Immigration and the Health of Asian and Pacific Islander Adults in the United States." *American Journal of Epidemiology* 153(4):372–80.

Gee, Ellen M., Karen M. Kobayashi, and Steven G. Prus. 2004. "Examining the Healthy Immigrant Effect in Mid to Later Life: Findings from the Canadian Community Health Survey" *Canadian Journal on Aging / La Revue canadienne du vieillissement,* 23 Supplement/Supplément: S5563.

Goldman, Noreen, Rachel T. Kimbro, Cassio M. Turra, and Anne R. Pebley. 2006. "Another Hispanic Paradox: Differences in Socioeconomic Gradients in Health between White and Mexican-Origin Populations." *American Journal of Public Health* 96(12):2186–93.

Grant, Bridget F., Frederick S. Stinson, Deborah S. Hasin, Deborah A. Dawson, Patricia Chou, and Karyn Anderson. 2004. "Immigration and Lifetime Prevalence of DSM-IV Psychiatric Disorders among Mexican Americans and Non-Hispanic Whites in the United States." *Archives of General Psychiatry* 61(Dec.):1226–33.

Hayes-Bautista, David. 1992. "Latino Health Indicators and the Underclass Model: From Paradox to New Policy Models." Pp. 32–47 in *Health Policy and the Hispanic*, edited by A Furino. Boulder, Colorado: Westview Press.
House, James S., Ronald C. Kessler, and A. Regula Herzog. 1990. "Age Socioeconomic Status, and Health." *Milbank Quarterly 68*:383–411.
House, James S., James M. Lepkowski, Ann M. Kinney, Richard P. Mero, Ronald C. Kessler, and A. Regula Herzog. 1994. "The Social Stratification of Aging and Health." *Journal of Health and Social Behavior* 35:213–34.
Huh, Jimi, Jo Ann Prause, and C. David Dooley. 2008. "The Impact of Nativity on Chronic Diseases, Self-Rated Health, and Co-morbidity Status of Asian and Hispanic Immigrants." *Journal of Immigrant and Minority Health* 10:103–18.
Hummer, Robert A., Daniel A. Powers, Starling G. Pullum, Ginger L. Grossman, and W. Parker Frisbie, 2007. "Paradox Found Again: Infant Mortality among the Mexican-Origin Population in the United States." *Demography* 44:441–58.
Hummer, Robert A., Richard G. Rogers and Charles B. Nam. 1999. "Race/Ethnicity, Nativity, and U.S. Adult Mortality." *Social Science Quarterly* 80(1):136–53.
Jackson, Richard and Neil Howe. 2008. *The Graying of the Great Powers: Demography and Geopolitics in the 21st Century*. Washington, DC: Center for Strategic & International Studies.
Jasso, Guillermina, Douglas Massey, Mark Rosenzweig, and James Smith. 2004. "Immigrant Health-selectivity and Acculturation." Pp. 227–266 in *Critical Perspectives on Racial and Ethnic Differences in Health in Late Life,* edited by N.B. Anderson, R.A. Bulatao, and B. Cohen. Washington, DC: National Academy Press.
Kahn Joan R., and Leonard I. Pearlin. 2006. "Financial Strain over the Life Course and Health among Older Adults." *Journal of Health and Social Behavior* 47:17–31.
Lauderdale, Diane S., and Bert Kestenbaum, 2002. "Mortality Rates in Elderly Asian American Populations Based on Medicare and Social Security Data." *Demography* 39(3):529–40.
Lynch, John W., George A. Kaplan, and Sarah J. Shema. 1997. "Cumulative Impact of Sustained Economic Hardship on Physical, Cognitive, Psychological, and Social Functioning." *New England Journal of Medicine* 337:1889–95.
Malzberg, Benjamin. 1967. "Internal Migration and Mental Disease among the White Population of New York State, 1960–1961". *International Journal of Social Psychiatry* 13:184–191.
Manton, Kenneth G. and Eric Stallard. 1997. "Health and Disability Differences among Racial and Ethnic Groups." Pp. 48–104 in *Racial and Ethnic Differences in the Health of Older Americans,* edited by L. G. Martin and B.J. Soldo. Washington, D.C.: National Academy Press.
Manton, Kenneth G. and Xi Liang Gu. 2001. "Changes in the Prevalence of Chronic Disability in the United States Black and Non-Black Population above Age 65 from 1982 to 1999." *Proceeding of the National Academy of Science USA* 98(11):6354–59.
Manton, Kenneth G. 2008. "Recent Declines in Chronic Disability in the Elderly U.S. Population: Risk Factors and Future Dynamics." *Annual Review of Public Health* 29:91–113.
Markides, Kyriakos S. 2001 "Migration and Health" in *International Encyclopedia of the Social and Behavioral Sciences,* edited by M. Paul Baltes and Neil J. Smelser. New York and Amsterdam: Elsevier.
Markides, Kyriakos S., and Sandra A. Black. 1995. "Race, Ethnicity, and Aging: The Impact of Inequality." Pp. 137–73 in *Handbook of Aging and the Social Sciences,* edited by R. H. Binstock and L. K. George. San Diego: Academic Press.
Markides, Kyriakos S. and Jeannine Coreil. 1986. "The Health of Southwestern Hispanics: An Epidemiologic Paradox." *Public Health Reports* 101:253–65.
Markides, Kyriakos S., and Karl Eschbach. 2005. "Aging, Migration, and Mortality: Current Status of Research on the Hispanic Paradox." *Journals of Gerontology*: *Psychological and Social Sciences* 60B:68–72.
———. 2011. "Hispanic Paradox in Adult Mortality in the United States." In *International Handbook of Adult Mortality*, edited by R. Rogers and E. Crimmins. New York: Springer Publications
Markides, Kyriakos S., Karl Eschbach, Laura A. Ray, and M. Kristen Peek. 2007. "Census Disability Rates among Older People by Race/Ethnicity and Type of Hispanic Origin." In *The Health of Aging Hispanics: The Mexican Origin Population*, edited by J. L. Angel and K. W. Whitfield. New York: Springer Publications.
Markides, Kyriakos S., Laura Rudkin, Ronald J. Angel, and David V. Espino. 1997. "Health Status of Hispanic Elderly." Pp. 285–300 in *Racial and Ethnic Differences in the Health of Older Americans*, edited by L. G. Martin and B. J. Soldo. Washington, D.C.: National Academy Press.
Markides, Kyriakos S., Jennifer Salinas, and Kristin Sheffield. 2008–2009. "The Health of Older Immigrants." *Generations* 32(4):46–52.
Markides, Kyriakos S., Jennifer Salinas and Rebeca Wong. 2010. "Ageing and Health among Hispanics/Latinos in the Americas." In *Handbook of Aging and the Life Course,* edited by D. Dannefer and C. Phillipson. London: Sage Publications.
Martin, Philip and Gottfried Zürcher. 2008. "Managing Migration: The Global Challenge" *Population Bulletin* 63(1). Washington, D.C.: Population Reference Bureau.
McDonald, James T. and Steven Kennedy. 2004. "Insights into the 'Healthy Immigrant Effect': Health Status and Health Service Use of Immigrants to Canada." *Social Science and Medicine* 59:1613–27.

Migration Policy Institute 2010. "Foreign-Born Population by Region of Birth as a Percentage of the Total Foreign-Born Population, for the United States: 1960 To 2008" Retrieved (http://www.migrationinformation.org/datahub/historicaltrends.cfm#source).

O'Rand, Angela M., and Jenifer Hamil-Luker. 2005. "Process of Cumulative Adversity: Childhood Disadvantage and Increased Risk of Heart Attack across the Life Course." *Journal of Gerontology:* Series B, 60B (Special Issue II):117–24.

Palloni, Alberto and Elizabeth Arias. 2004. "Paradox Lost: Explaining the Hispanic Adult Mortality Advantage." *Demography* 41:385–415.

Palloni, Alberto. 2007. "Health Status of Elderly Hispanics in the United States." Pp. 17–25 in *The Health of Aging Hispanics: The Mexican Origin Population*, edited by J. L. Angel and K.W. Whitfield. New York: Springer Publications.

Seeman, Teresa E., Sharon S. Merkin, Eileen M. Crimmins, and Arun S. Karlamangla. 2010. "Disability Trends among Older Americans: National Health and Nutrition Examination Surveys, 1988–1994 and 1998–2004." *American Journal of Public Health* 100(1):100–7.

Singh, Gopel K. and Robert A. Hiatt. 2006. "Trends and Disparities in Socioeconomic and Behavioral Characteristics, Life Expectancy, and Cause-Specific Mortality of Native-Born and Foreign-Born Populations in the United States, 1979–2003." *International Journal of Epidemiology* 35:903–19.

Singh, Gopal K. and Mohammad Siahpush, 2002. "Ethnic-Immigrant Differentials in Health Behaviors, Morbidity, and Cause-Specific Mortality in the United States: An Analysis of Two National Data Bases." *Human Biology* 74:83–109.

Stephen, Elizabeth. H., Karen Foote, Gerry E. Hendershot, and Charlotte A. Schoenborn. 1994. Health of the Foreign-born Population: United States, 1987–90. *Advanced Data from Vital and Health Statistics*, No. 241. Hyattsville, MD: National Center for Health Statistics.

Stern, Sharon M. 2004. *Counting People with Disabilities: How Survey Methodology Influence Estimates In Census 2000 and the Census 2000 Supplementary Survey*. Washington, D.C.: U.S. Bureau of the Census.

Sorlie, Paul D., Eric Backlund, Norman J. Johnson, and Eugene Rogot. 1993. "Mortality by Hispanic Status in the United States." *Journal of the American Medical Association* 270(2):2464–68.

Turra, Cassio M. and Irma T. Elo. 2008. "The Impact of Salmon Bias on the Hispanic Mortality Advantage. New Evidence from Social Security Data." *Population Research and Policy Review* 27:515–30.

Turra, Cassio M. and Noreen Goldman. 2007. "Socioeconomic Differences in Mortality among U.S. Adults: Insights into the Hispanic Paradox." *Journal of Gerontology: Social Sciences* 62B:S184–92.

United Nations Report. 2002. *Number of World's Migrants Reaches 175 Million Mark*. Retrieved (http://www.un.org/esa/population/publications/ittmig2002/press-releaseeng.htm).

Vega, William A. and Hortensia Amaro 1994. "Latino Outlook: Good health, Uncertain Prognosis." *Annual Review of Public Health* 15:39–67.

Vega, William A., Michael R. Rodriquez, and Elisabeth Gruskin. 2009. "Health Disparities in Latino Population." *Epidemiologic Reviews* 31:99–112.

Waidmann, Timothy A. and Korbin Liu. 2000. "Disability Trends among Elderly Persons and Implications for the Future." The *Journals of Gerontology, Series B: Psychological Sciences* 55(5): S298–307.

Wakabayashi, Chizuko. 2009. "Effects of Immigration and Age on Health of Older People in the United States." *Journal of Applied Gerontology* 20:1–23.

Zunzunegui, Maria V., Olivier Nunez, María Durbán, M. Jesús Yébenes, and Ángel Otero. 2006. "Decreasing Prevalence of Disability in Activities of Daily Living Functional Limitations and Poor Self-Rated Health: A 6 Year Follow-Up Study in Spain." *Aging Clinical and Experimental Research* 18(5):352–358.

Chapter 8
Global Aging

Masa Higo and John B. Williamson

In the sociology of aging and life course, global aging is relatively a new issue; but it has become an increasingly important subject since the early 1990s (Estes and Phillipson 2002). Many factors are contributing to population aging. The single most important factor is the decline of fertility that is taking place in many countries around the globe (Lloyd-Sherlock 2010). Another major factor is decline in mortality rates at all age levels, a trend which translates into increases in life expectancy (Uhlenberg 2009). Because people have come to live longer, including more disability-free years, population aging can and should be viewed as a positive achievement associated with improved healthcare, more government spending on social security, and other age-related social welfare programs, as well as changes in life style, such as less smoking and more exercise (National Research Council 2001).

However, the unprecedented increase in the relative size of the older population has also created major challenges for many countries around the world. It strains existing systems of social and financial support for older people as the burdens of intergenerational dependency increase. In addition to these demographic changes, the trend toward ever-increasing economic globalization is in many countries transforming existing social institutions including the social welfare policies that support older people (Phillipson 2005). Population aging and economic globalization are two major social forces that will be shaping social change around the world throughout the twenty-first century. Research on global aging must give considerable attention to the ways in which these two major social forces affect the lives of older people around the world (Uhlenberg 2009).

As the processes of population aging and economic globalization vary among countries and regions of the world, the consequences of changes linked to these social forces can be expected to vary as well (Phillipson 2005). In general, the economically developed parts of the world – Europe and North America in particular – have been experiencing gradual changes in the age composition of their populations for decades. In contrast, many developing (as well as many transitional) countries are currently experiencing demographic changes at a much faster pace than are the developed countries of the world (United Nations 2009a). In the early twenty-first century, the developing countries are being referred to by some as the *new-old world*, whereas the developed countries are being viewed as the *old-old world* (AARP 2003; Perkins et al. 2004).

Global aging is going to create new challenges and risks, particularly with respect to the allocation of resources to older people for informal care-giving, formal healthcare provision, financial security, and long-term care (LTC). To date, much of the research on global aging has focused on assessing how the implications of population aging and economic globalization for the developed countries differ from those for the developing countries. In the decades ahead, the risks and challenges attributed to global aging are likely to be much greater for developing countries than for

M. Higo (✉)
Boston College Department of Sociology, Chestnut Hill, MA, USA
e-mail: higo@bc.edu

R.A. Settersten, Jr. and J.L. Angel (eds.), *Handbook of Sociology of Aging*, Handbooks of Sociology and Social Research, DOI 10.1007/978-1-4419-7374-0_8, © Springer Science+Business Media, LLC 2011

developed countries (United Nations 2009a). Global aging may create or increase risks for vulnerable populations, such as unmarried older women and widows, particularly in developing countries (Browne and Braun 2008).

The purposes of this chapter are twofold: First, this chapter reviews the main areas of research to date on global aging, including issues such as (1) family structure and living arrangements; (2) epidemiological transition and healthcare burdens; (3) retirement and old-age financial security; and (4) LTC and healthcare worker migration. In each of these areas, the research suggests that in the decades ahead developing countries are likely to be confronted with even greater challenges than the developed countries linked to population aging in an ever more competitive global economy. Second, this chapter suggests three directions for future research on global aging.

Global Population Aging and Economic Globalization

The twenty-first century will witness a rate of population aging without parallel in past history. Considering the global population, young children (aged 5 and younger) have always outnumbered older people (aged 65 and older). However, between 2020 and 2025, for the first time in history, older people will outnumber young children. In 2000, the numbers of young children and older people worldwide were about 627 and 473 million, respectively; by 2025, the figures are projected to increase to about 650 and 714 million. By 2050, the figures are projected to increase to 592 million and 1.5 billion (United Nations 2009a).

Between 1975 and 2000, the median age of the world population increased from 22.4 to 26.4 years. It is projected to rise to 32.8 years by 2025 and to 38.4 years by 2050 (United Nations 2009a). The old-age dependency ratio is another important measure of population aging. Calculated by dividing the number of people aged 65 years and older by that of those of working age (aged 15–64 years), the old-age dependency ratio is an indicator of both the formal and informal burdens of providing for an older population's economic security and well-being (Schulz and Binstock 2006). Globally, the ratio increased from about 8.5% in 1950 to about 10.9% in 2000, and it is projected to steadily increase to about 15.8% by 2025 and to 25.3% by 2050 (Organization for Economic Co-operations and Development 2009a).

Two major determinants of global population aging are increasing life expectancy and decreasing fertility (United Nations 2009b). Globally, the average life expectancy at birth increased from 46.6 years for the 1950–1955 birth cohort to 66.4 for the 2000–2005 cohort. The figure is projected to further increase to 72.1 years for the 2025–2030 cohort and to 75.5 for the 2050–2055 cohort. The world has also experienced an overall decrease in fertility rates since the mid-twentieth century. The world's total fertility rate (average number of children born to each woman over the course of her lifetime) was 4.92 between 1950 and 1955 and the figure decreased to 2.67 between 2000 and 2005. It is currently projected to fall to 2.21 between the years 2025 and 2030 and to 2.02 between the years of 2050 and 2055 (United Nations 2009a).

Generally, the process of population aging differs between the economically developed and developing parts of the world (Lloyd-Sherlock 2010). In most developed countries, European countries in particular, population aging began slowly during the late nineteenth century as birth rates entered a phase of sustained decline and life expectancies began to gradually increase. As of 2008, the percentages of older population (age 65 and older) are highest for Japan (21.6%), Italy (20.0%), Germany (20.0%), Greece (19.1%), and Sweden (18.3%) (United Nations 2009a).

The number and proportion of older people have also been growing in developing regions of the world. By 2015, there will be more people aged 65 and over living in China alone (132 million) than in all of Europe (128 million) (United Nations 2009a). As of 2008, some 62% of the world's population

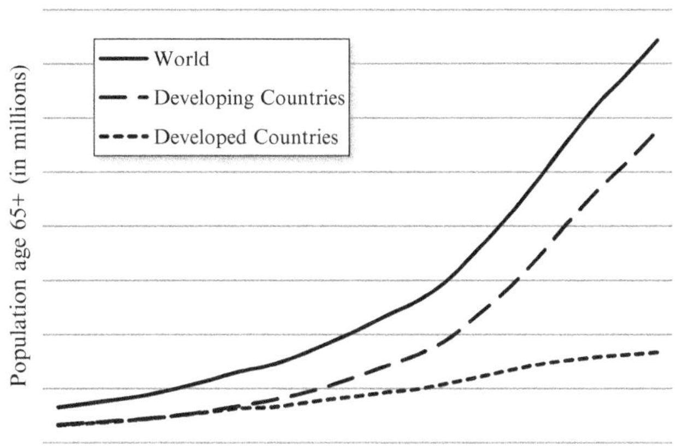

Fig. 8.1 Population age 65 and older, 1950–2050, World, Developed Countries and Developing Countries (United Nations 2009a). *Note*: In United Nations' (2009a) definition, the category of developed countries includes all regions of Europe, Northern America, Australia, New Zealand, and Japan, and that of developing countries comprises all regions of Africa, Asia (excluding Japan), Latin America and the Caribbean plus Melanesia, Micronesia and Polynesia

aged 65 and over lived in developing countries – an estimated 313 million people. This figure is projected to increase to 71% (690 million) by 2030. By 2050, almost 1.2 billion of the expected 1.5 billion people aged 65 or older are projected to live in countries that today are classified as developing nations, a designation which very likely will change for some of these countries by 2050 (see Fig. 8.1). Old-age dependency ratios are projected to at least double between 2000 and 2030 in many East and Southeast Asian and Latin American countries, and to triple in South Korea (United Nations 2009a).

The demographic changes that many developing countries are now experiencing have happened at a much faster rate than experienced by most countries currently classified as developed (Uhlenberg 2009). Most of today's developed countries had many decades to adjust to the gradual graying of their age structures. For example, it took 115 years (from 1865 to 1980) for France's population aged 65 and over to increase from 7 to 14% of the total population. To make this same shift it took Sweden 85 years (1890–1975), Canada 65 years (1944–2009), and current projections suggest that it will take Australia 73 years (1938–2011) and the United States 69 years (1944–2013). In contrast, many developing countries are projected to experience similar increases in the proportion that are old much more rapidly, with much less time to make the adjustments needed to deal with the policy implications of these changes. The following illustrate how much more rapid the shift is, according to current projections, going to be in many developing nations: Colombia (2017–2036) 19 years, Brazil (2011–2032) 21 years, Thailand (2002–2024) 22 years, Sri Lanka (2002–2026) 24 years, and China (2000–2026) 26 years (United Nations 2009a).

In addition to these demographic changes, economic globalization has become a social force with major implications for social institutions and public policies that impact the well-being of older people (Higo and Yamada 2009; Phillipson 2005). Globalization takes many forms, both cultural and economic. Of particular relevance for this chapter are the consequences of economic globalization including the process by which local, regional, or national markets become integrated into the global economy and the consequences of this process (Tonkiss 2006). By some accounts, economic globalization has been taking place since the fourteenth century (since the end of restrictive merchant economies). The rate of this process has increased quite dramatically over the past 20–30 years. Modern economic globalization is characterized by integration of developing countries into the global market economy, under terms which has been dominated largely by the developed countries and their multinational corporations (Estes and Phillipson 2002).

This economic globalization involves increases in international trade, foreign direct investment (FDI), and migration. *World trade* as a percentage of *gross world product rose from 8.5% in 1970 to 16.1% in 2001* (World Bank 2008). FDI for the United States, the world's largest economy, has been on the rise around the world since the 1970s. Both inflows and outflows of FDI for the United States have grown from an annual average of $45.3 billion in the 1970s to an average of $117.5 billion in the first half of the 1990s. Between 2000 and 2006, inward FDI stock in Brazil, Russia, India, and China grew from 8 to 14% of global FDI stock. Today, China is by far the leading emerging destination of FDI; in 2008, China ranked second behind the United States in the inflow of FDI, receiving $67.3 billion in FDI inflow (World Bank 2008). In tandem with the flow of trade and investment, economic globalization in the form of the flow of people among countries has also been increasing. Between 2000 and 2005, some 16 million people emigrated form developing to developed nations. This figure is nearly triple the number for the period between 1970 and 1975 (United Nations 2009b).

Family Structure and Living Arrangements

The global trend of decline of birth rates has implications for family size and structure. Most people in the world today eventually have children, grandchildren, and siblings. However, as the proportion of women and men who have only one or two children has been steadily increasing, in future generations it is likely that many who are old will have few, if any, siblings (Uhlenberg 1996). Changes in family structure will have implications for living arrangements. These changes in turn will affect the availability of economic resources and potential sources of informal care (care and support provided by family members) for older people. It is generally known that it will be a problem for the elderly in the developed nations. What is less well known is that this is going to become a big issue in developing countries as well (Agree and Glaser 2009).

In developed countries, over roughly the past three decades, multigenerational living arrangements have been steadily declining. Reflecting rising rates of divorce, delayed marriage, and increasing percentages of never-married and childless adults, an increasing proportion of older people are living alone. In Sweden, Finland, Denmark, and the United Kingdom, well over one-third of older people (aged 65 and older) live alone (Agree and Glaser 2009). In Greece, the proportion of unmarried older people living with a married child dropped from 23% in 1974 to less than 9% in1999. The share of unmarried older people living alone almost doubled (10–19%) during the same 25-year period (Karagiannaki 2005). The growth in the number of households consisting of one older person in developed countries has increased, particularly for women; as of 2006, of all the older people (aged 65 and older) living alone in the United States, women account for about 77% (U.S. Census Bureau 2009). In developed countries, where social support systems outside the immediate family have been well developed, this trend has been fueled by a combination of factors: greater longevity and increases in pension benefits as well as greater emphasis on care and support in the community (Lloyd-Sherlock 2010).

In contrast, older people in less developed countries rely heavily on their family members for well-being and survival. The majority of older people reside with their adult children and/or grandchildren. In Bangladesh, Indonesia, Singapore, and Taiwan, more than 80% of older parents live with one or more children (Ghuman and Ofstedal 2004). While many older people in developed countries report that they prefer living alone, those in developing countries prefer to live with, or at least close to, their children and to see them on a daily basis (Bongaarts and Zimmer 2002). In addition to these cultural preferences and the availability of kin, the lesser development of old-age financial security or social welfare programs is another reason for the prevalence of multigenerational co-residence in many developing countries (Frankenberg et al. 2002).

Such changes in the family size and structure are global phenomena. Over the next few decades, however, population aging in many developing countries is going to create problems particularly for the elderly poor due to the relative lack of familial resources for informal care and support in later life. Due to the decline of family size and traditional living arrangements, some older people, particularly unmarried older women or widows without any children and with fewer social resources, can be left with little support and nowhere to live if extended family members are not available to provide care for them (Lloyd-Sherlock 2010).

Epidemiological Transition and Healthcare Burdens

The world is experiencing an epidemiological transition due to changes in leading causes of death (Wilkinson 2004). Globally, while populations have become less vulnerable to acute and infectious diseases and are living longer, the number of deaths due to chronic and non-communicable disease has been rapidly increasing. Chronic and non-communicable diseases, including ischemic heart disease, cerebral-vascular disease, chronic obstructive pulmonary disease, and lower respiratory infections, are characteristic of old age (World Health Organization 2008). The number of deaths from cardiovascular diseases and cancers in 2004 worldwide was about 19 million; that number is projected to increase to more than 40 million by 2030. Overall, chronic and non-communicable diseases are projected to account for about three quarters of causes of all deaths worldwide in 2030 (WHO 2008).

The epidemiological transition takes on major importance in connection with research on global aging. Changes to leading diseases and causes of death posit challenges for allocating resources for formal care and social intervention such as public expenditures on healthcare (Crystal and Siegel 2009). Worldwide, healthcare costs are rapidly rising – in many countries, the growth rates of expenditures on public healthcare are projected to exceed the national economic growth rates in the decade ahead (WHO 2008).

Although most countries around the world have undergone such an epidemiological transition, developed and developing parts of the world have taken different paths for this transition (Wilkinson 2004). In developed countries, Europe and the United States in particular, the transition began in the late eighteenth century, when birth rates began gradually falling in some countries. In most developing countries, the transition did not start until well into the twentieth century, but since the mid-twentieth century it has been taking place at a much faster pace than in developed countries (Crystal and Siegel 2009). In developing countries, while infectious diseases accounted for about 40% of all deaths in 1990, by 2020 non-communicable diseases are projected to account for more than three quarters of all deaths (WHO 2008).

Research on global aging often refers to *burden of disease* as a measure of the region-specific economic burden caused by diseases in a baseline year. Burden of disease is calculated by combining years of life lost due to premature mortality and years of life lost due to time lived in less than full health (WHO 2008). During the next several decades, the burden of non-communicable disease (as opposed to the burden of infectious disease), already a major problem in the developed nations, is projected to increase rapidly in developing countries as well. The World Health Organization (2008) estimates that in 2004 the share of the total burden of disease attributed to non-communicable diseases was about 85% for developed countries and about 44% for developing countries. By 2030, those figures are projected to increase to 89 and 54%, respectively.

The challenges of the burden of diseases associated with the epidemiological transition are likely to be greater for developing countries than for developed countries (Crystal and Siegel 2009). Chronic and non-communicable diseases are imposing a growing burden on developing countries, many of which have limited resources for informal healthcare relative to developed countries. Simultaneously, in many developing countries, infectious diseases such as malaria, tuberculosis, and HIV/AIDS are likely to remain a devastating health issue; by 2030, while the share of the burden attributed to infectious diseases in developed countries is estimated to be about 3%, the figure for developing countries is estimated to remain quite high at 32% (WHO 2008). While their health systems are already being stressed due to limited funding, many developing countries are likely to face a double burden of disease – high rates of infectious diseases including HIV/AIDS will co-exist with increasing rates of non-communicable diseases (Wilkinson 2004). Thus, developing countries need to mobilize and allocate resources to address the non-communicable diseases that are characteristically the leading cause of death in aging societies while they continue to struggle with the high prevalence rates for infectious diseases.

Retirement and Old-Age Financial Security

Changes in the patterns of work and retirement among older people have implications for their financial security in later life and for the younger, working-age people who support them (Kohli and Rein 1991). Population aging around the world has contributed to what many analysts refer to as a global pension crisis. At the core of this discussion is the debate over the sustainability of existing public pension programs (Blackburn 2006). The fear of possible public pension fund insolvency is common among policy makers in many countries around the world today. Research suggests that older workers in developing countries are particularly vulnerable as they are more likely than those in developed countries to be exposed to high levels of individual financial risk (Walker 2006).

Public pension schemes were first institutionalized in what are today considered the developed countries. Later, such schemes became widespread throughout most parts of the world including many currently referred to as developing countries. While the origin of public pension schemes can be traced back to Germany at the end of the nineteenth century, such schemes have come to play an increasingly important role in providing financial security to retired workers in developed countries, particularly since the 1950s (Macnicol 2006). By 2000, public pension programs covered more than 90% of the workforce in OECD countries (OECD 2009b). Most public old-age pension schemes around the world are based on the pay-as-you-go defined benefit (PAYG-DB) model. With a DB scheme, the pension benefit is based primarily on some measure of average or final wage and the number of years the worker has contributed; it is not dependent on fluctuations in financial markets. Under the PAYG-DB model, pension benefits are for the most part not pre-funded, although in some countries such as the United States today, a modest amount of pre-funding is introduced by building up substantial reserves in a trust fund, often with a plan to draw down those assets to deal with an anticipated demographic bubble such as the retirement of the baby-boom generation (Williamson, forthcoming). With a PAYG scheme, revenues from the current working population's payroll taxes are used to finance the benefits of current retirees, with a very modest trust fund used to assure that there will be sufficient funds on hand to pay pensions during dips in the revenues collected due to short-term fluctuations in unemployment rates (Macnicol 2006).

From the 1950s through the end of the 1980s, retirement was institutionalized in most developed countries due largely to availability of relatively generous public pension benefits for many retirees (Gruber and Wise 1998). Old age became synonymous with retirement as a phase of individuals' life courses in which workers were encouraged to leave the labor force (Phillipson 2005). During this period, paid labor force activities of older men declined sharply in most develop countries. In the United Kingdom, the labor force participation rate for men aged 65 years and older in 1921 was 60%, but by 1951 the rate had dropped to 32%. The figure continued to fall to 23% by 1971 and then to only 11% in 1981 (Macnicol 2006). With the increasing paid labor force participation of women, they too have increasingly been incorporated into the institutionalization of retirement in later life (Phillipson 2005).

Today, the number of pensioners (retirees) relative to contributors (workers) is increasing and this has raised serious questions about the sustainability of the traditional, PAYG defined-benefit pension schemes in many countries. Many are calling into question the viability of the implicit intergenerational contract that such schemes are based on. In 2003, total public pension expenditures in 25 European Union (EU) countries consumed one-eighth of the EU's total gross domestic product. In Italy in 2004, old-age pension expenditures absorbed 15% of the nation's gross domestic product (OECD 2007). In recent years, pension reform has become a highly charged issue in most developed countries; more than half of the OECD countries have made major changes in their public pension schemes during the past 15 years (OECD 2009b).

In recent years, one pension reform option that has received a great deal of attention has been to shift from a traditional PAYG-DB to a multi-pillar scheme that includes a defined contribution (DC) pillar (Williamson 2004). With a DC pillar what is promised is that a specified amount will be contributed each month, but no promise is made with respect to the size of that actual pension that will be paid based on those contributions. An individual or personal account is created for each covered worker with the funding based on contributions from that worker (often supplemented by contributions from the employer) via payroll taxes and the earnings (or loses) on those assets over the years when those assets are invested by private sector money management organizations in financial markets. This is typically done as part of a set of reforms designed to reduce the government's pension obligations and shift much of the risks associated with paying those pensions from the government to the individual worker.

Although public pension schemes in a majority of OECD countries are still based on the PAYG-DB model, an increasing number are introducing mandatory DC schemes (Cushing-Daniels and Johnson 2008). This trend has been particularly true in some developing parts of the world such as Latin America and Eastern Europe (Bockman and Eyal 2002; Brooks 2004). Because a DC scheme links pension income to the contributions to the account, this alternative provides workers with incentives to delay retirement and remain economically active longer than is the case with the traditional PAYG-DB alternative. DC schemes also reduce demographic risks, such as growth of the older population relative to the younger one, since pension benefits under a defined-contribution scheme do not rely on the earnings of younger generations (Williamson 2004).

Researchers on global aging are starting to pay increasing attention to a new pension model which some countries such as Sweden, Italy, Poland, and Russia have introduced; it is a new type of DC pillar. These countries have implemented notional defined-contribution (NDC) schemes which are based on PAYG financing, but like the DC schemes they also more closely link pension benefits to the individual worker's lifetime contributions than is typically the case with DB schemes. Workers have individual accounts, which record how much has been paid into the pension system via the payroll tax (sometimes including matching contributions from the employer). Under the NDC scheme, the individual account is notional (unfunded) and the analog of interest is the annual credit added based on the size of the account and trends in wage rates, not trends in financial markets (Williamson 2004).

By 2004, public old-age social insurance programs had come to be established in 167 countries, including some of the economically poorest countries (United Nations 2009b). However, in developing countries, public pension programs typically cover a much smaller fraction of workers than in most developed countries. Coverage rates of under 10% are not uncommon in very poor countries. For example, in Malaysia and Thailand, public pension coverage is restricted to certain categories of workers such as public sector employees, civil servants, and military personnel. Nearly one-third of countries in Africa that offer public pension benefits have a life expectancy less than the statutory pensionable age both for men and women. While the projected future costs of public pension schemes are a major concern, many developing countries are even more concerned about how they are going to finance their strategic development plans, particularly those linked to infrastructure, security, education, and health (United Nations 2009b).

Many countries around the world are facing less than enough government resources to cover all workers with pension benefits or in many cases even to pay the promised benefits to the few who are covered (Williamson, forthcoming). Globally, many are looking for ways to reduce the government obligation via reforms that at least partially privatize their existing PAYG-DB schemes. Many of these governments, including those of both developed and developing countries, are attracted to market-oriented partial privatization schemes that: promise less pension support from the government, shift much of the risks to workers, and creative incentives for works both to save more and remain in the labor force longer (Madrid 2003). More than 20 developing countries, including Chile, and many other countries in Latin America and Eastern Europe have introduced funded individual accounts DC pillars thereby partially privatizing their national pension schemes (Phillipson 2005).

The increasing prevalence of pension privatization worldwide, that in developing countries in particular, is of concern to some sociologists and economists who study global trends in pension policy (Schulz and Borrowski 2006). With such schemes financial security in old age depends on a number of factors that involve different forms of risk, of particular note in this context are: what worker is able to save, how those savings are invested, the fees assessed for managing these assets, and fluctuations in financial markets. The combined impacts of population aging, increasing global competition, and the potentially dramatic corrections in financial markets just before retirement, represent a huge shift of risk in many countries from the government to individual workers, many of whom can expect to end up with less than an adequate pension in retirement (Williamson, forthcoming).

In the midst of economic globalization, an increasing number of developing countries such as China are experiencing economic expansion and urbanization. As we know based on the past experience of many of the now developed countries, such social structural transformations tend to weaken informal, familial support for the elderly (Tirrito 2003). Without the protections provided by the traditional DB or the new NDC schemes, the trend toward greater pension privatization is putting millions of workers in both developed and developing countries at financial risk. These risks are problematic for most workers but particularly problematic for low-wage workers, older workers, and single widows in today's developing countries (Walker 2006).

Long-Term Care and Healthcare Worker Migrations

Over the last few decades, the demand for healthcare workers for older people has been increasing in many countries around the world (Kingma 2006). There will be a demand for home and community-based LTC workers to service the frail and disabled elderly. In 2004, there were 60 million healthcare workers worldwide, the majority of whom were LTC workers (WHO 2006).

Since the mid-1980s, many developed countries including Canada, the United Kingdom, and the United States, have promoted aging-in-place for the care of the elderly. In many of these countries, an effort is being made to shift the major site for elderly healthcare from institutions (e.g., hospitals and nursing home) to home and community-based locations (Center for Health Workforce Studies 2006). This promotion of aging-in-place has increased the demand for home and community-based LTC services designed to provide older people with social, medical, and health support in their residential settings and local environments (Connell 2008). In addition, the ever-increasing cost of LTC in an institutional context has also contributed to the demand for home and community-based LTC services. Compared to traditional, institutional care, home and community LTC is much less costly. In the United States, whereas the LTC average cost is $70,900 per year for nursing home care, 4 h/day of care from a home health aide only averages $36,500 a year (Feder et al. 2007).

Many countries around the world have been formulating a global market of LTC workforce. Due partly to declining birth rates, increasing divorce rates, and increasing female employment, developed countries are increasingly facing shortages of LTC workers. In the United States, over $207 billion was spent on LTC in 2005, and by 2010 the demand for LTC workers is projected to increase by about 64% (Center for Health Workforce Studies 2006). This growing demand has made the developed countries major employers and importers of LTC workers from the global LTC workforce. Over the last three decades, the migration of healthcare workers has increased significantly. In the United States, where foreign-born workers accounted for only 5% of the total LTC workforce in 1980, the figure increased to 17% by 2003 (Clearfield and Batalova 2007).

In the midst of global concern over the LTC workforce shortage, developing countries are being gradually integrated into the global market for LTC workers. Many are becoming major suppliers of LTC workers for the developed countries (Browne and Braun 2008). Over the last two decades, the governments of many developing countries have encouraged their younger workers, women in particular, to provide LTC services for the elderly in developed countries. The Philippines is the world's leading LTC workforce exporter. Since the mid-1990s, the Philippine government has supported the education and exportation of many Filipinas as LTC workers, mainly as nurses and nurse aides. The reason is the financial benefits to the country from money remitted by these workers back to their families in the Philippines (Ball 2008).

Many developing countries currently benefit economically from their role as exporters of LTC workers (United Nations 2008). However, there are also potential long-term costs to these same countries do to labor shortages in their own countries (Hussein and Manthorpe 2006). By some projections, the need for elderly healthcare in developing countries will increase by as much as 400% over the next 20 years (WHO 2006). According to the World Health Organization (2006), health worker density (the ratio of health workers to the total population) needs to be at least 2.5 workers per 1,000 people. Among 186 countries worldwide, 75 countries do not meet this condition. About 45 of these 75 countries are in sub-Saharan Africa area (WHO 2008). While countries in Sub-Saharan Africa today have about 11% of the world's population and 24% of the global burden of disease, these countries have only 3% of the world's healthcare workers (WHO 2006). As developing countries are further integrated into the global labor market for LTC providers, more of their younger workers will be immigrating to developed countries in search of economic opportunities in the LTC area. This trend is likely to make it more difficult from many of these countries to adequately meet their own needs for long-term elder care (Browne and Braun 2008).

Directions for Future Research

We have presented evidence suggesting that, in the decade ahead, population aging, economic globalization, and the intersection of these two global trends are likely to be powerful determinants of the quality of life for the older population around the world. As mentioned in the introduction of this chapter, global aging is partly a result of improved healthcare and development of age-related social welfare programs around the world. It can be viewed as an indicator of a society's ability and commitment to providing well for its older population. Many of these older people continue to make important contributions to society, some as paid workers, some as family caregivers, and some through their unpaid service to the local community (Harper 2006; National Research Council 2001).

However, the global aging and economic globalization are also creating major challenges in both the developed and the developing countries around the world. Elders living in the developing countries will generally be at greater risk than those in the developed countries. Today's developing countries are relative newcomers to the challenges of population aging. Many developing countries are confronting the burden of adequately providing for the well-being of their elderly while at the same time contending with both the fiscal problems associated with rapid population aging and the economic challenges with respect to their integration into the ever more competitive global economy.

The burdens of providing for their elders are likely to be particularly difficult in the areas of living arrangements, healthcare policies, financial security in later life, and LTC for the elderly. These burdens will be very challenging when dealing with the needs of some high-risk groups such as the disabled and various categories of economically vulnerable and socially marginalized elderly women, such as widows, the never married, those without adult children, and, in some regions, women responsible for the care of their grandchildren due to the death of their adult children often linked to diseases such as AIDS. In many developing countries around the world, during the decades ahead, population aging in combination with economic globalization is going to pose a number of major risks for the older population (Brooks 2004; Estes and Phillipson 2002).

While there has already been a substantial amount of research and much has been said about the potential global impact of population aging on the well-being of the older population around the world, there has been very little research that has attempted to deal with this set of issue using life course analysis. Further research is needed to better understand and help policy makers prepare for the impact of population aging and the risks associated with population aging in the context of economic globalization at different life course stages. This is particularly true for the developing countries. For those interested in conducting research designed to help meet this need, we suggest the following three directions for future research on global aging.

First, it would be very useful to have data, particularly for developing countries, on both objective and subjective quality of life measures suitable for life course analysis, including data from both men and women. The relative lack of such data makes it difficult to compare various forms of inequality across the life course in developing countries and between those in developed and developing countries at comparable life course stages (Phillipson 2005). More high quality data would also enable researchers who deal with the issue of global aging to come up with more useful and well-informed policy suggestions.

Second, research on global aging would benefit from paying greater attention to the roles of transnational financial organizations in shaping the future of old-age policy and the related experience of being old, particularly in developing countries. The International Monetary Fund, World Bank, and World Trade Organization, to name a few, have played key roles in facilitating economic globalization and in promoting market-oriented old-age social policies (Estes and Phillipson 2002). Since World War II, these transnational organizations have actively lowered barriers to international trade between countries and have been attempting to limit spending on social security programs by introducing market-oriented alternatives (Deacon et al. 1997). In the area of public pension reform, for instance, these organizations have argued for reducing state PAYG-DB schemes to the minimal or residual role of providing a very modest old-age pension and promoting an expanded role for individualized and capitalized private pensions (Brooks 2004; Estes and Phillipson 2002). The World Trade Organization has also placed enormous pressure on its member countries to further open-up health and social welfare programs to competition from global corporate providers (Campbell and Pedersen 2001).

As global social forces, population aging and economic globalization are best analyzed in tandem, rather than independently. Transnational financial organizations have taken active roles linked to both these social forces (Phillipson 2005). In order to gain a better understanding of the unequal distribution of the risks associated with aging and old age, particularly the likely differences between old-old and new-old parts of the world for the decade ahead, research on global aging would benefit from a close observation and detailed critical examinations of the roles played by these organizations. The impacts of these organizations are not limited to the elderly; it would also be useful to apply life course analysis to the assessment of these impacts, particularly for those living in the developing world.

Finally, theoretical work is needed to more adequately theorize the links between global aging, economic globalization, and social policies designed to deal with the well-being of workers across the life course up to and including old age. The social forces associated with population aging and economic globalization have implications for the forms that inequality takes, the consequences of those inequalities, and the risks that workers and their families are subjected to at all stages of the life course.

What is needed is more effort to modify or extend existing theories to better account for the trends that are emerging. To dates the links are under theorized. Hitherto, research on global aging has largely failed to theoretically address the ways in which inequalities between developed and developing parts of the world and vulnerable populations in developing countries are being produced (Estes and Phillipson 2002; Phillipson 2005). In their theorizing about globalization, Bauman (1998) and Beck (2001), characterize our contemporary societies as operating in an age of individualization, privatization, and of each worker being responsible for his or her own risks. In this era of rapid global aging, there has been a ideological shift to the right in thinking about social welfare issues, a shift from less focus on looking to the government programs for answers and more focus on individuals being responsible for coping with risk during old age and over the life course (Beattie and McGillivray 1995). At the same time that demographic pressures associated with social welfare programs are increasing, transnational financial organizations have been fostering a shift to greater individual responsibility for securing resources for healthcare, financial security, and LTC in later life, particularly for those living in developing countries (Deacon et al. 1997; Walker 2006). From Bauman's (1998) and Beck's (2001) theoretical view, these transnational financial organizations can be understood as one of the main agents that have been promoting an ideology of individual or local solutions for globally generated problems.

In light of a twenty-first century characterized by unprecedented global population aging and an ever intensifying global economy, research on global aging may need to reclaim the sociological imagination that Mills (1959) called for. Future research on global aging is needed to construct theoretical accounts that trace the links between individual risks in later life and the broader social structural forces including global population aging and economic globalization. Despite the growing ideology of individual solutions for global problems, future research on global aging framed by critical sociological imagination is needed to better inform social policies calling for greater attention to social, collective, solidaristic, and communitarian solutions for the globally generated problems associated with aging and old age around the world.

References

AARP. 2003. "The New Old World: Challenges and Opportunities of Aging Populations Speech April 2003." Retrieved September 7, 2009 (www.aarp.org/research).

Agree, Emily M. and Karen Glaser. 2009. "Demography of Informal Caregiving." Pp. 647–70 in *International Handbook of Population Aging*, edited by P. Uhlenberg. New York: Springer.

Ball, Rochelle E. 2008. "Globalized Labor Markets and the Trade of Filipino Nurses: Implications for International Regulatory Governance." Pp. 30–46 in *The International Migration of Health Workers*, edited by J. Connell. New York: Routledge.

Bauman, Zygmunt. 1998. *Globalization: The Human Consequences*. Oxford, UK: Polity Press.

Beattie, Roger and Warren McGillivray. 1995. "A Risky Strategy: Reflections on the World Bank Report Averting the Old Age Crisis." *International Social Security Review* 48(3):5–22.

Beck, Ulrich. 2001. *What is Globalization?* Oxford, UK: Polity Press.

Blackburn, Robin. 2006. "The Global Pension Crisis: From Gray Capitalism to Responsible Accumulation." *Politics & Society* 34(2):135–86.

Bockman, Johanna and Gil Eyal. 2002. "Eastern Europe as a Laboratory for Economic Knowledge: The Transnational Roots of Neoliberalism." *American Journal of Sociology* 108(2):310–52.

Bongaarts, John and Zachary Zimmer. 2002. "Living Arrangements of the Elderly in the Developing World: An Analysis of DHS Household Surveys." *Journal of Gerontology Series B: Psychological Sciences and Social Sciences* 57(3):145–57.

Brooks, Sarah M. 2004. "International Financial Institutions and the Diffusion of Foreign Models for Social Security Reform in Latin America." Pp. 53–80 in *Learning from Foreign Models in Latin American Policy Reform*, edited by K. Weyland. Baltimore, MD: Johns Hopkins University Press.

Browne, Colette V. and Kathryn L. Braun. 2008. "Globalization, Women's Migration, and the Long-Term-Care Workforce." *The Gerontologist* 48(1):16–24.

Campbell, John L. and Ove Kaj Pedersen. 2001. "The Rise of Neoliberalism and Institutional Analysis." Pp. 1–14 in *The Rise of Neoliberalism and Institutional Analysis*, edited by J. L. Campbell and O. K. Pedersen. Princeton, NJ: Princeton University Press.

Center for Health Workforce Studies. 2006. *The Impact of the Aging Population on the Health Workforce in the United States: Summary of Key Findings March 2006*. Rensselaer, NY: Center for Health Workforce Studies.

Clearfield, Esha and Jeanne Batalova. 2007. *Foreign-Born Health-Care Workers in the United States*. Washington, DC: Migration Policy Institute.

Connell, John, 2008. "Towards a Global Health Care System?" Pp. 1–29 in *The International Migration of Health Workers*, edited by J. Connell. New York: Routledge.

Crystal, Stephen and Michele J. Siegel. 2009. "Health Care Policy and the Demography of Aging in Cross-National Perspective." Pp. 607–30 in *International Handbook of Population Aging*, edited by P. Uhlenberg. New York: Springer.

Cushing-Daniels, Brendan, and Richard W. Johnson. 2008. "Employer-Sponsored Pensions: A Primer." Washington, DC: The Urban Institute. Retrieved September 17, 2009 (http://www.urban.org/url.cfm?ID=901144).

Deacon, Bob, Michelle Hulse, and Paul Stubbs, 1997. *Global Social Policy: International Organizations and the Future of Welfare*. London, UK: Sage.

Estes, Carroll and Chris Phillipson. 2002. "The Globalization of Capital: The Welfare State and Old Age Policy." *International Journal of Health Services* 32(2):279–97.

Feder, Judith, Harriet L. Komisar, and Robert B. Friedland. 2007. "Long-Term Care Financing: Policy Options for the Future." Georgetown University Long-Term Care Financing Project, June 2007. Retrieved July 21, 2009 (http://ltc.georgetown.edu/pdfs/finalsummary.pdf).

Frankenberg, Elizabeth, Angelique Chan, and Mary Beth Ofstedal. 2002. "Stability and Change in Living Arrangements: Evidence from Indonesia, Singapore and Taiwan." *Population Studies: A Journal of Demography* 56(2):201–13.

Ghuman, Sharon and Mary Beth Ofstedal. 2004. "Gender and Family Support for Older Adults in Bangladesh." PSC Research Report, Report No. 04563. Retrieved August 12, 2009 (http://www.psc.isr.umich.edu/pubs/pdf/rr04-563.pdf).

Gruber, Jonathan and David Wise. 1998. "Social Security and Retirement: An International Comparison." *The American Economic Review* 88(2):158–63.

Harper, Sarah. 2006. *Ageing Societies: Myths, Challenges and Opportunities*. London, UK: Hodder Arnold.

Higo, Masa and Atsuhiro Yamada. 2009. "Japan: Public Policy." Sloan Center on Aging & Work at Boston College, Global Policy Brief No 2. July 2009. Retrieved December 17, 2009 (http://agingandwork.bc.edu/documents/GPB02_Japan_2009-07-02.pdf).

Hussein, Shereen and Jill Manthorpe. 2006. "An International Review of the Long-Term Care Workforce Policies and Shortages." *Journal of Aging & Social Policy* 17(4):75–94.

Karagiannaki, Eleni. 2005. "Changes in the Living Arrangements of Elderly People in Greece: 1974-1999." Centre for Analysis of Social Exclusion, CASE/104 November 2005. Retrieved September 2, 2009 (http://sticerd.lse.ac.uk/dps/case/cp/CASEpaper104.pdf).

Kingma, Mireille. 2006. *Nurses on the Move: Migration and the Global Health Care*. Ithaca, NY: Cornell University Press.

Kohli, Martin and Martin Rein. 1991. "The Changing Balance of Work and Retirement." Pp. 1–35 in *Time for Retirement: Comparative Studies of Early Exit from the Labor Force*, edited by M. Kohli, M., Rein, A. M. Guillemard, and H. V. Gunstern. Cambridge, UK: Cambridge University Press.

Lloyd-Sherlock, Peter. 2010. *Population Ageing and International Development: From Generalization to Evidence*. Bristol, UK: Polity Press.

Macnicol, John. 2006. *Age Discrimination: An Historical and Contemporary Analysis*. Cambridge, UK: Cambridge University Press.

Madrid, Paul L. 2003. *Retiring the State: The Politics of Pension Privatization in Latin America and Beyond*. Princeton, CA: Stanford University Press.

Mills, C. Wright. 1959. *Sociological Imagination*. New York: Oxford University Press.

National Research Council. 2001. *Preparing for an Aging World: The Case for Cross-National Research*. Washington, DC: National Academy Press.

Organization for Economic Co-operations and Development. 2007. *Pensions at a Glance 2007*. Paris: OECD Publications.

———. 2009a. *OECD in Figures 2009*. Paris: OECD Publications.

———. 2009b. *Pensions at a Glance 2009: Retirement-Income Systems in OECD Countries*. Paris: OECD Publications.

Perkins, Bradford, J. David Hoglund, Douglas King, and Eric Cohen. 2004. *Building Type Basics for Senior Living*. Hoboken, NJ: Wiley.
Phillipson, Chris. 2005. "The Dynamic Nature of Societal Aging in a Global Perspective." Pp. 131–158 in *Enduring Questions in Gerontology*, edited by D. J. Sheets, D. B. Bradley, and J. Hendricks. New York: Springer.
Schulz, James H. and Robert H. Binstock. 2006. *Aging Nation: The Economics and Politics of Growing Older in America*. Baltimore, MD: John Hopkins University Press.
Schulz, James H. and Allan Borrowski. 2006. "Economic Security in Retirement: Reshaping the Public-Private Pension Mix." Pp. 360–79 in *Handbook of Aging and the Social Sciences, 6th edition*, edited by R. H. Binstock and L.K. George. San Diego, CA: Academic.
Tirrito, Terry. 2003. *Aging in the New Millennium: A Global View*. Columbia, SC: University of South Carolina Press.
Tonkiss, Fran. 2006. *Contemporary Economic Sociology: Globalization, Work and Inequality*. New York: Routledge.
Uhlenberg, Peter. 1996. "Mortality Decline in the Twentieth Century and Supply of Kin over the Life Course." *The Gerontologist* 36(5):681–85.
———. 2009. "Introduction." Pp. 1–4 in *International Handbook of Population Aging*, edited by P. Uhlenberg. New York: Springer.
United Nations. 2008. *Assessing the Costs and Impacts of Migration Policy: An International Comparison*. New York: United Nations, Department of Economic and Social Affairs.
———. 2009a. *World Population 2008*. New York: United Nations, Department of Economic and Social Affairs.
———. 2009b. *World Economic and Social Survey 2009: Promoting Development, Saving the Planet*. New York: United Nations, Department of Economic and Social Affairs.
U.S. Census Bureau. 2009. "United States Census 2009." Retrieved October 19, 2009 (http://www.census.gov/).
Walker, Alan. 2006. "Reexamining the Political Economy of Aging: Understanding the Structure/Agency Tension." Pp. 59–80 in *Aging, Globalization and Inequality: The New Critical Gerontology*, edited by J. Baars, C. Phillipson, A. Walker, and D. Dannefer. Amityville, New York: Baywood.
Wilkinson, Richard G. 2004. "The Epidemiological Transition: From Material Scarcity to Social Disadvantage?" Pp. 112–21 in *The Sociology of Health and Illness: A Reader*, edited by M. Bury and J. Gabe. London, UK: Routledge.
Williamson, John B. 2004. "Assessing the Pension Reform Potential of a Notional Defined Contribution Pillar." *International Social Security Review* 57(1):47–64.
———. Forthcoming. "The Future of Retirement Security." In *Handbook of Aging and the Social Sciences*, 7th ed., edited by R. H. Binstock and L. K. George. San Diego, CA: Academic.
World Bank. 2008. *Globalization and Technology Absorption in Europe and Central Asia: The Role of Trade, FDI, and Cross-border Knowledge Flows*. Washington, DC: The World Bank.
World Health Organization. 2006. "The Global Shortage of Health Workers and Its Impact." Retrieved September 12, 2009 (http://www.who.int/mediacentre/factsheets/fs302/en/print.html).
———. 2008. "Global Burden of Disease: The 2004 Update." Retrieved August 12, 2009 (http://www.searo.who.int/LinkFiles/Reports_GBD_report_2004update_full.pdf).

Chapter 9
Diversity and Family Relations in an Aging Society

Judith Treas and Christopher Steven Marcum

To appreciate the changes that have occurred in American families over the last three decades, consider the occupants of 1600 Pennsylvania Avenue. In the 1980s, Nancy and Ronald Reagan called the White House home. Members of the "Greatest Generation" shaped by World War II, the Reagans were prototypical empty nesters with grown children pursuing their own, sometimes awkward, way in the world. Famously devoted, the President and First Lady seemed to have a storybook marriage, albeit one with a twist. He had been married before, creating what sociologists call a blended family of divorce and remarriage.

Today, three decades later, the First Family is a prototypical once-married, heterosexual couple with two young children. This family type is now a minority in U.S. households due to fewer and later marriages, more divorces, fewer remarriages, more cohabitation, and more non-marital births. While they represent one family ideal, the First Family's diversity is illuminating. Besides being the first African-Americans to call the White House home, other aspects of the Obamas address the growing multiplicity of American family life. The President was raised by a single, working mother. With children born in her mid-thirties, the First Lady has been a working mother and speaks from personal experience about the difficulties of balancing work and parenting. Her own mother moved into the White House to help out with the grandchildren. Also telling is the diversity of the President's kin. With a Kenyan father, Barack Obama is closer to his immigrant heritage than Ronald Reagan was to his Irish-Scots-English roots. Thirty years ago, a family with ties to three continents was almost unimaginable, but transnational families are familiar to the many Americans today whose lives have been touched by the great immigration of the past half century.

The changing context and growing diversity of American family life have fascinated sociologists of aging and the life course. In this chapter, we explore four broad changes in American society which have impacted family relationships as we grow older. First, we consider generational change, particularly the Baby Boom which is arriving at old age with different family experiences than generations that went before. Second, we address growing family diversity. New family forms, whether stemming from racial-ethnic immigrant diversity or increased visibility of sexual minorities, point to less uniformity for families. Third, we examine the changing terrain of gender. Men's lives and women's lives have converged with notable consequences for family relationships. Fourth, we consider the demographic and technological developments that impact the intergenerational relationships that sustain us through childhood, old age, and adversity. We conclude by commenting on how developments from life course research will continue to inform scholarship on these changes in the future.

J. Treas (✉)
Department of Sociology, and Center for Demographic and Social Analysis,
University of California, Irvine, CA, USA
e-mail: jktreas@uci.edu

R.A. Settersten, Jr. and J.L. Angel (eds.), *Handbook of Sociology of Aging*, Handbooks of Sociology and Social Research, DOI 10.1007/978-1-4419-7374-0_9, © Springer Science+Business Media, LLC 2011

Aging Baby Boomers

The Baby Boomers are a defining feature of American society and a touchstone for research on aging and the life course. Boomers have embodied many trends, from the sexual revolution to rising women's workforce participation to changes in intergenerational relationships. While unique for its sheer size, the cohort is also unique for its trajectory. The Baby Boomers came of age during a period of great technological and social innovation. With important implications for family life, their aging is a testing ground for theories of cohort and generation, and it underscores the importance of research on the older population and later life.

To review facts well known to sociologists, the Baby Boomers were born between 1946 and 1964. Their defining feature is the cohort's size. Much to the surprise of demographers accustomed to the low fertility of the 1930s, the U.S. birth rate remained high for nearly 20 years before it fell to the low levels we know today. The Baby Boom has been hypothesized to stem from post-war prosperity, which encouraged young adults who grew up during the Great Depression to marry earlier and raise larger families (Easterlin 1961). Whatever the causes, the Baby Boom would set the context for family politics in the twentieth century.

In childhood, the Baby Boomers taxed the educational system. New schools, colleges, and universities were built to educate them. With liberal and technical educations, more Boomers obtained white-collar and service sector jobs. While hardly homogeneous in attitudes and values, they fueled social reform by protesting the Vietnam War and fighting for gender and racial equality. Despite general parent–child value congruence (Glass et al. 1986), a vanguard of Baby Boomers questioned the model of American society dominated by white males. A smaller cohort might not have had as big an impact on social change.

One distinguishing feature of educated Baby Boomers was an orientation towards self-actualization (Hughes and Waite 2007). Their values were more individualistic than their parents' and more concerned with personal gratification than with adherence to institutional demands (Bengtson 1975). Rather than just sharing common birth years, their values set them apart as a self-aware "generation" in the tradition of Mannheim (1953). Abetted by economic constraints, individualism was a force for later marriage and smaller families. Higher education and women's work force participation, along with liberal family and divorce laws, transformed family life (Caldwell 1980). As *U.S. Monthly Vital Statistics* reports, marriage rates between 1964 and 2004 plummeted while mean age at first marriage rose from 21 to about 26. The divorce rate rose to a record high of 5.3 per thousand people in 1981, before declining to 3.5 per thousand in 2008. Total fertility stabilized near replacement level (i.e., 2.1 children per woman); fertility rates for unmarried women more than doubled. Today, nearly four in ten births are non-marital.

Looking forward, family will remain important, but aging Baby Boomers will have different support systems than older people today. There will be: (1) more childless and more never married or divorced seniors (Himes 2002); (2) smaller sibships but more siblings surviving together into old age; and (3) a deteriorating economic support ratio as relatively few workers, taxpayers, and kin support growing numbers of older adults (Knickman and Snell 2002).

The Baby Boomers' retirement brings new challenges. Foremost is funding living costs and healthcare for a very large older population. Besides a smaller tax base, fewer children mean fewer family resources for aging parents. However, the demands of retiring Baby Boomers may be no greater than the strain they placed on schools in the 1960s (Knickman and Snell 2002). The trend towards working longer, due to changes in policies and expectations (Pienta and Hayward 2002), may offset some of the immediate costs of an imbalanced age-pyramid. Whether focusing on late-life employment, volunteerism, civic engagement, or family contributions, "productive aging" offers a conceptual counterweight to the stress on old age dependency and incapacity (Uhlenberg 1992). At least until the recent economic crisis, demographic adaptations by the Baby Boomers (e.g., fewer children, women's higher labor force participation) compensated for any competitive disadvantage of large cohort size (Easterlin et al. 1990). Baby Boomers accumulated more wealth than their parents; as of 2001, they were in good shape for retirement (Keister and Deeb-Sossa 2001). Their wealth stemmed from the prosperity of whites, however (Lusardi and Mitchell 2007). Blacks and Hispanics had median net worth only 1/8th and 1/4 whites', respectively. It remains to be seen whether non-white and Hispanic racial and ethnic groups will be able to afford a retirement characteristic of the "second youth" (to borrow the name of the AARP's Spanish language magazine for seniors) that white Baby Boomers appear economically prepared for.

These findings on Baby Boomer wealth need to be updated in light of the global economic collapse. Their wealth is concentrated in home ownership and retirement savings accounts that dropped in value between 2007 and 2009 while home values plummeted 30–40% (Rosnick and Baker 2009). With high unemployment, reduced employee benefits, and lower pay for people of all ages, the crisis led families to look to each other for support. At the end of the recession, Baby Boomers may wind up less wealthy, albeit it perhaps still better off than the young whom they may be called on to support.

Immigration and Diversity

Changes in U.S. immigration law in 1965 led to a remarkable increase in Americans who are foreign born – from about 6 to 15% between 1980 and 2008 (Migration Policy Institute 2009). This new wave of immigration contributed to racial and ethnic diversity. Looking ahead raises questions about how immigration will impact older adults. Are we investing enough in the schooling of the coming generation of Hispanic taxpayers? Can the intergenerational contract of the young supporting the old be sustained if taxpayers are increasingly minority and older adults are largely non-Hispanic whites? Meanwhile, the very definition of "minority" includes more and more racial and ethnic groups.

The growth in the population of Asian and Hispanic origins has blurred the racial color line (Lee and Bean 2007). Inter-racial relations no longer mean just black and white. Growing numbers of Americans define themselves as multi-racial. Race still plays an important role in pairing off (Feliciano et al. 2009), but racial intermarriage has increased in recent decades, as has marriage between immigrants and non-immigrants within racial-ethnic groups (Qian and Lichter 2007). By birth, marriage, or adoption, more Americans point to family members of a different race. Many bring together different heritages within their own homes. Families negotiate different cultural expectations even as diversity offers new possibilities for parent–child support, marital relationships, and elder care.

Since *The Polish Peasant* (Thomas and Znaniecki 1996), an immigration research tradition has addressed adaptation, intergenerational socialization, family relations, and other life course topics. Immigration research has focused on the adaptation of working age immigrants and their children, because most people immigrate at young ages and their incorporation determines their well-being and their contributions to the broader society. This research is rich in life course insights, including how age-related variation in the pace of acculturation within immigrant families contributes to divergent expectations between generations (Pyke 2004; Treas and Mazumdar 2002). Incorporation depends not only on time in the host country but also on age at arrival. Children who immigrate at older ages lag behind younger siblings. Therefore, those who immigrate before adolescence are labeled the 1.5 generation, whose acculturation falls between first (foreign-born) and second (U.S.-born) immigrant generations.

Older immigrants receive little attention outside their ethnic communities. As Treas (2009:40) observes, "They never win spelling bees. They do not join criminal gangs. Nobody worries about

Americans losing jobs to Korean grandmothers." Until recently, older immigrants were mostly long-time U.S. residents with socio-demographic profiles much like U.S.-born seniors. Because of the recent arrival of Asians and Latinos, immigrant aging means that older Americans are growing more diverse. Immigration law, which puts no numerical limits on aging parents sponsored by U.S. citizen children, adds about 80,000 older newcomers annually. Compared to U.S.-born counterparts, Hispanic and Asian immigrants, 65 years and older, are more likely to be poor and to receive public assistance, despite 1996 welfare reform measures curtailing non-citizen eligibility (Burr et al. 2009).

If immigration research has neglected older immigrants, so have studies of aging and the life course in which sociologists' interests focused on older minority populations, not immigrant status. The proposed 2007 immigration policy reforms died without mention of aging parents whose numbers would have been cut in half by the law (Treas 2008b). Yet, older newcomers are a textbook case of the intersection of age, life course events, and history in human lives. Without the influence of U.S. schools and workplaces, late-life immigrants seldom master the English language or embrace American customs as thoroughly as younger kin. They do not catch up to long-term immigrants who arrived at young ages and grew old here. We have dubbed these older newcomers the ".5 (point five) generation" in recognition of their limited acculturation.

Older newcomers challenge assumptions about intergenerational relations. Unless poor health makes them too frail to contribute to their families, they are often important resources, not just family dependents. Grown children sponsor aging immigrants, in part, because they keep house and look after grandchildren for busy, two-earner couples (Treas and Mazumdar 2004). As hands-on caregivers, they offer a model of intergenerational ties contrasting with U.S.-born grandparents' largely fun-loving, or sometimes distant, relations with grandchildren (Cherlin and Furstenberg 1986). Older newcomers rely on families for companionship, financial support, and help navigating social institutions, because they have few opportunities to make friends of their own age, are unfamiliar with American society, cannot get paid jobs, and are mostly ineligible for government benefits and services.

Being more likely than other older Americans to live with younger kin, elderly newcomers enjoy the benefits and protections of close family ties. Their experiences, however, point out the limitations of family support. Despite high rates of multigenerational residence, older immigrants – especially recent ones – are more vulnerable to depression than younger immigrants or other older Americans. Being more likely to be widowed, they report being lonely and bored (Treas and Mazumdar 2002). Family members are often too busy going to school and earning a living to provide the desired companionship or to help to navigate an unfamiliar culture. Valued for transmitting cultural traditions and sustaining transnational ties, older relations are not household authority figures and must often defer to others' needs. Older immigrants present an opportunity for aging and life course studies that tests our understanding of how sweeping historical events such as immigration reform impact the course of human lives, how age intersects with social structures such as labor markets and pension systems, and how our intimate relations with others channel our development and aging.

New Family Forms

Although racial and ethnic diversification calls for new understandings of family relationships in an aging society, family life has also diversified along other dimensions. Cohabitation, single-mother families, and non-marital births point out the new family forms and norms that have emerged. These trends underscore the significance of starting points and demonstrate the growing heterogeneity of the life course. The earliest childhood experiences have life-long consequences for health, socioeconomic well-being, and intimate relationships (Shapiro and Cooney 2007). Furthermore, new family forms show that normative pressures and social institutions (e.g., welfare rules, inheritance laws) that once standardized family behavior are less compelling today. As Americans marry late, if ever, and often only after cohabitation and parenthood, sociologists point to a deinstitutionalization of marriage (Cherlin 2004). Life course sociologists stress the "destandardization" of the family life course (Brückner and Mayer 2005), the growing disarray in the orderly attainment of marriage and parenthood that once certified adult status.

Life course destandardization has far-reaching implications. Once confined to the young, cohabitation is now favored by middle-aged divorced persons, older adults, and even parents raising children. More children will spend some time growing up without two parents, sometimes creating lifetime disadvantages. Serial partnering complicates family relationships, fusing blended families and creating ghost relationships, ties to meaningful kin (e.g., grandparents) that are ruptured when romantic unions end. With the rise in childlessness, a larger share of successive generations will lack intergenerational support in old age (Himes 2002). Increased divorce has negative consequences for older fathers' relations with grown children and for children's own marriages (Shapiro and Cooney 2007). The rise in age at first marriage, divorce, and cohabitations (with their higher rates of dissolution) points to new groups to study (singles and non-cohabiting romantic partners "living apart together"). More needs to be learned about the potential of new family forms to bring meaning and support to our lives. New welfare initiatives, for example, have identified distant kin willing to adopt previously unknown relations, teenagers who have languished for years in foster care (Eckholm 2010).

The increased acceptance of homosexuality highlights the interplay between institutionalization and deinstitutionalization in new family forms. Non-heterosexuals have challenged traditional institutions by insisting on new definitions of family that recognize voluntary networks of reciprocity and affection, rather simply ascribed relationships based on blood, marriage, or adoption. But, same-sex couples have also asserted their rights to marry and enjoy all the benefits traditionally accorded by marriage. Of the half million same-sex couples living together in 2008, a quarter reported their partner as their husband or wife (although the meaning of the term is in flux due to a patchwork of state and federal marriage and domestic partnership laws). Research is quicker to acknowledge the family relations of non-heterosexuals today. In the past, gays were stereotyped as family-less "confirmed bachelors" or adults whose sexual orientation distanced them from disapproving kin. In a domestication of homosexuality, discourse about gays and lesbians no longer focuses on sexual lifestyles but rather on familiar family and relationship concerns (adoption, parenting, blended families, getting health coverage for dependents), issues with which many heterosexuals can identify. Aging and life course principles are valuable tools in understanding these changes. At the macro-social level, separating cohort turnover from intra-cohort change shows that acceptance of same-sex relationships increased less because liberal cohorts replaced conservative ones and more because cohorts changed their views over time (Treas 2002). At the micro-social level, post-Stonewall cohorts of gays and lesbians – consistent with Mannheim's (1953) concept of self-aware generations – pride themselves on living open lives in contrast to older generations for whom successful identity management meant passing as heterosexual (Rosenfeld 1999). Thus, the life course perspective informs and is informed by the growing diversity in family relationships.

The Changing Gender Terrain

A tenet of life course sociology recognizes that men's and women's lives traverse different paths. This is still the case, but their life courses have converged and even crossed in recent decades (Brückner and Mayer 2005). Women once left school earlier than men, getting less education. Today, more women than men enroll in higher education in developed nations. Gender convergence is seen in women's greater workforce involvement and men's greater involvement in the home. Most young American women will work, typically full-time, over much of their lives. Poorly educated women with poor employment prospects were once more likely to marry, but today well-educated women have the best marriage prospects. As women have spent more time in paid work, they have cut back on the time they spend in housekeeping (Treas and Drobnic 2010). Their husbands have taken on more of the household chores. Couples have fewer children, but fathers are more active in their care. Indeed, single-fathers are not as rare a family type as just 30 years ago. Given social expectations that middle-class parents spend time with offspring, mother–child relationships continue to be intensive ones despite the competing demands of employment.

Inequality and difference remain part of the gendered life course, but the lives of men and women are more alike than in the past. Their skill sets are less specialized. Their experiences share more in common. Where gender specialization once called on partners to complement one another's activities (he worked for pay, she kept house), they now substitute (she works more hours if he loses his job) (Treas 2008a). More women have parity with their partners, and many even earn more than their husbands. With male and female life expectancies converging, widowhood will be a less gendered experience.

Rethinking Kinship

With demographic changes, kinship may be a more important resource for young and old. Improved infant mortality and healthier lifestyles raised life expectancy from 65 to 79 years over the last century. The number of centenarians and super-centenarians grew by a factor of 7 between 1950 and 1990 (Krach and Velkoff 1999). The lives of children, parents, and grandparents overlap more often. Higher divorce and remarriage rates among Baby Boomers and their grown children bring step-parents and step-grandparents into children's networks. Some say that kinship relations will become more multigenerational (Bengtson 2001). As Settersten (2007) cautions, however, episodic and contingent relations may not translate into meaningful support. Not all ties will be positive ones, nor will the circumstances under which multigenerational households are formed.

Grandparents and grandchildren have become more likely to share lives and households (Fuller-Thomson et al. 1997). By 1999, over 10% of grandparents were responsible for the full-time care of at least one grandchild (Minkler 1999). More than 4.5 million children under 18 were living with grandparents, whether in their grandparent's or parent's home (Simmons and Lawler 2000; Mader 2009). Life course sociologists stress the benefits, as well as burdens, of multigenerational households. While Baby Boomers may be "sandwiched" between filial and parental responsibilities, they can benefit from an extra pair of hands and eyes around the house. Children growing up in extended family households are sometimes seen to be healthier than children in single generation homes (Kanaìaupuni et al. 2005). Grandparents benefit from social support in the face of loss and from learning about developments in popular culture from grandchildren.

Social networks decline in size and become more kin oriented with age. To explain how aging affects individual network size, membership, and frequency of interaction, social psychologists point to changes in goal-attainment strategies at the end-of-life (Carstensen 1991), risk minimization (Rohr and Lang 2009), and maintenance of companionship in the face of loss (Rook and Schuster 1996). Many explanations imply that age-related changes in social interaction reflect older adults' preferences, rather than structural changes that shape opportunities to make and keep social ties. For instance, the age-structure constrains interaction based on the availability of cohort and non-cohort members. Group size can explain a lot (Mayhew 1973). Numerically, younger and older cohorts should have a higher likelihood of interacting with Baby Boomers, because there are so many Boomers. Assuming random mixing, small cohorts have fewer opportunities to maintain relationships with each other (all things being equal, the young are likely to have elderly neighbors to befriend). Of course, because these generalizations are challenged everyday by retirement communities, universities, and workplaces that integrate or segregate age groups, life course sociologists need to test the limits of the assumption that young and old mix freely.

Why do older people's social networks have a concentration of kin? Older adults' networks are trimmed of age-peers, because they are less likely to replace a friend lost to declining health or death and to be sought out by age-peers for the same reasons. Continuing ageism, differing physical capacities, and generational differences in tastes mean that older adults are less likely to be sought out by younger people outside the family. This results in kin-dominated networks. Current old-age cohorts are smaller, had more children, are less mobile, and probably have more health problems than the Baby Boomers will have in old age. Reluctant to identify as old – with fewer children but more surviving siblings – Baby Boomers may have more within-group and between-group diversity in their social networks than older people in the past, a hypothesis to challenge sociologists chronicling the trajectory of the cohort.

Staying Connected

Technology has changed the way families interact. In 1960, 22% of households had no telephones. By 1980, only 7% lacked phones. Today, nearly 20% of U.S. households are "wireless" with no landline at all; only 2% have no telephones (Blumberg and Luke 2009). Cellular phones highlight how remote communication for families has shifted, very quickly, from household-centric to egocentric. Seniors account for the largest share of households with landlines, and wireless-only households are most prevalent among 18-to-35-year-olds. The Internet allows family members to live further apart but remain in touch. At home, families use baby monitors and webcams to keep watch over one another. YouTube™ videos share the baby's first steps or the nephew's wedding with kin across the globe.

Despite stereotypes of tech-savvy teenagers, older people now use technology to stay connected and informed. Many Baby Boomers have had jobs that required computers, exposing them to information technology long before they reached retirement age. Earlier cohorts pick up technology in an ad hoc way (e.g., being taught by grandchildren). According to the Pew Internet American Life Project, Baby Boomers use on-line banking, travel booking, and emails to kin and coworkers more than younger and older generations. Younger people do homework, socialize via chat and social networking sites, play games, and use email. Older Internet users, ages 73 and up, limit their use to email and information searches, particularly for health information. Very old people are also less likely to use cellular phones.

Information technology has implications for how family members stay in touch. Although the geographical distance between adult children and their parents has increased over time, a study of seven Western nations shows 1988–2001 increases in the frequency of "other" (not face to face) contact between adults and their mothers, a trend tracking the diffusion of telephones and email (Treas and Gubernskaya 2009). Distance has always mattered for maintaining relationships, but technology decreases its significance for many forms of interaction (Mok et al. 2007). Even face-to-face interaction is possible at a distance with videochat services, but the need for in-person contact may diminish as technology changes the content of interactions. For instance, the 1990s' increase in older adults' use of assistive technology (grab bars, walkers, etc.) coincided with declines in the personal care (e.g., help bathing) often provided in person by kin (Freedman et al. 2006).

The risks of not having localized kin may be greater for older people than for younger ones, however. Preferring phone and in-person visits over newer technologies, very old adults risk social isolation – a major predictor of bad health and mortality. Geriatric care technology, however, has evolved from assisted devices, such as hearing-aides, to health and environmental solutions incorporating the Internet and monitoring devices embedded in "Smart Houses." Technologies allow doctors and kin to monitor the health, safety, and well-being of older adults when local contact is impossible. As elder outreach programs teach seniors how to use computers, the Internet promotes successful aging by allowing people to connect with family, health, work, and consumer relations – all while aging in place, at home.

In summary, new technologies compensate for the growing geographic distance between young and old. In the home and on the go, they bridge disconnected kin and become a key feature of successful and productive aging. While the oldest-old have been slower to adopt new communication tools, the Baby Boomers will narrow the generation gap in technology.

Looking Back, Looking Forward

Diversity and change may challenge our knowledge base, but they also validate the theoretical perspectives that guide research on aging and the life course. Social change calls into question taken-for-granted assumptions. Thirty years ago, it seemed that contact between aging parents and grown children might become a casualty of declining residential proximity between the generations. Few anticipated that immigrant families would buoy the numbers of multigenerational households, that grandparents would step in to care for grandchildren when parents could not, and that new communication technology would sustain daily contact across vast distances. Although new family forms were recognized, few in the early 1980s anticipated how quickly same-sex relations would win acceptance. Nor did they guess that family issues would come to dominate narratives about gays and lesbians. The current financial crisis has shaken what we know about our largest cohort, but it has reaffirmed the importance of intergenerational support in adversity. Foreclosures, unemployment, and shrinking 401Ks demand that we revisit the conclusion that most Baby Boomers (despite or because of their unique family life course) will enter old age on solid financial footing.

Rapid social change is the friend of life course research, which illuminates the change in individual lives. From social changes of the past three decades have come new conceptual tools. The "destandardization" of the life course brought balance to historical accounts of "standardization," the growing regimentation of individual trajectories brought about by the state and other social institutions. "Productive aging," the social and economic contributions of older adults, unified studies on caregivers, community volunteers, and older workers. At the same time, social changes validated studies of aging and the life course. As the Baby Boomers grow older, they are driving home the practical importance of knowledge about family relations, work, health, and satisfaction in later life. Theoretically and methodologically, the concepts of cohorts and generations permit sociologists to gain a purchase on the rise of self-actualizing value orientations, the family life course adaptations of the Baby Boomers, and the growing acceptance of same-sex couples. Pursuing parent–child relationships beyond childhood, research confirms that social changes affecting early life have lasting consequences.

These social changes create new opportunities for research on aging and the life course. The twin processes of life course standardization and destandardization point to the need to understand how phenomena such as mass incarceration and international migration emerge as normative transitions in certain populations, how they color the lives of those who experience these events, and how they impact parents, children, and other linked lives. The convergent life courses of men and women invite a second generation of research on how these changes play out across different social classes and different countries. Increasing rates of labor force participation in later life, coupled with the financial crisis and the demographic challenges to social insurance systems, call for new attention to the socioeconomic life course.

The coming years offer an opportunity to influence the direction of change with research that informs public opinion and social policy. Take immigration law. Research on the dependence of an aging society on foreign-trained healthcare workers points up the need for immigration policies based on employment considerations, but studies of the interdependent fortunes of immigrant family members highlight the value in current laws reuniting kin. Research reveals thorny issues of justice, such as the lack of legal protections for young undocumented immigrants who have lived virtually their entire lives in the U.S.

As the contest over same-sex marriage makes clear, most family law is made by states, not the federal government. Given this patchwork tradition, the support for family forms – from covenant marriages to single-parent adoptions to custodial grandparenting – depends on where one lives. These variations make for natural experiments that can inform debates on how policies and contexts promote the well-being of families over the life course. New family constellations are often victims of cultural lags, whereby existing institutional arrangements are out of step with the realities of life. Sociologists can sensitize others to the challenges which the life course poses for foster families whose children age out of the system, elderly immigrants who long to go home, or same-sex couples confronting end-of-life issues. By bringing our theories and methods to the task, the sociologists of the life course can most assuredly contribute to society and sociology in coming decades.

References

Bengtson, Vern L. 1975. "Generation and Value Effects in Socialization." *American Sociological Review* 40(3):358–71.

———. 2001. "The Burgess Award Lecture: Beyond the Nuclear Family: The Increasing Importance of Multigenerational Bonds." *Journal of Marriage and Family* 63(1):1–16.

Blumberg, Stephen J. and Julian V. Luke. 2009. "Wireless Substitution: Early Release of Estimates from the National Health Interview Survey, July–December 2008." *Technical Report*, Center for Disease Control, National Center for Health Statistics.

Brückner, Hannah and Karl Ulrich Mayer. 2005. "The De-Standardization of the Life Course: What It Might Mean and If It Means Anything Whether It Actually Took Place." *Advances in Life Course Research* 9:27–54.

Burr, Jeffery A., Kirsten K. Gerst, Ngai Kwan, and Jan E. Mutchler. 2009. "The Economic Well-Being and Welfare Program Participation of Older Immigrants." *Generations* 32(4):53–60.

Caldwell, John C. 1980. "Mass Education as a Determinant of the Timing of Fertility Decline." *Population and Development Review* 6(2):225–55.

Carstensen, Laura L. 1991. "Socioemotional Selectivity Theory: Social Activity in Life-Span Context." Pp. 195–217 in *Annual Review of Gerontology and Geriatrics*, volume 11, edited by K. W. Schaie and M. P. Lawton. New York: Springer.

Cherlin, Andrew J. 2004. "The Deinstitutionalization of American Marriage." *Journal of Marriage and the Family* 66:848–61.

Cherlin, Andrew and Frank F. Furstenberg Jr. 1986. *The New American Grandparent: A Place in the Family, a Life Apart*. New York: Basic Books.

Easterlin, Richard A. 1961. "The American Baby Boom in Historical Perspective." *The American Economic Review* 51(5):869–911.

Easterlin, Richard A., Christine Macdonald, and Diane J. Macunovich. 1990. "Retirement Prospects of the Baby Boom Generation: A Different Perspective." *The Gerontologist* 30(6):776–83.

Eckholm, Erik. 2010. "A Determined Quest to Bring Adoptive Ties to Foster Teenagers." *The New York Times*. Retrieved January 30, 2010. (http://www.nytimes.com/2010/01/31/us/31adopt.html?scp=2&sq=foster%20care&st=cse).

Feliciano, Cynthia, Belinda Robnet, and Golnaz Komaie. 2009. "Gendered Racial Exclusion among White Internet Daters." *Social Science Research* 38(1):39–54.

Fuller-Thomson, Esme, Meredith Minkler, and Diane Driver. 1997. "A Profile of Grandparents Raising Grandchildren in the United States." *The Gerontologist* 37(3):406–11.

Freedman, Vicki A., Emily M. Agree, Linda G. Martin, and Jennifer C. Cornman. 2006. "Trends in the Use of Assistive Technology and Personal Care for Late-Life Disability, 1992–2001." *The Gerontologist* 46:124–27.

Glass, Jennifer L., Vern L. Bengtson, and Charlotte Chorn Dunham. 1986. "Attitude Similarity in Three-Generation Families: Socialization, Status Inheritance, or Reciprocal Influence?" *American Sociological Review* 51(5):685–98.
Himes, Christine L. 2002. "Elderly Americans." *Population Bulletin* 4 (June). Washington, DC: U.S. Bureau of the Census.
Hughes, Mary Elizabeth and Linda J. Waite. 2007. "The Aging of the Second Demographic Transition." Pp. 179–211 in *Social Structures: Demographic Changes and the Well-Being of Older Persons*, edited by K. W. Schaie and P. Uhlenberg. New York: Springer.
Kanaìaupuni, Shawn Malia, Katharine M. Donato, Theresa Thompson-Col6n, and Melissa Stainback. 2005. "Counting on Kin: Social Networks, Social Support, and Child Health Status." *Social Forces* 83(3):1137–64.
Keister, Lisa A. and Natalia Deeb-Sossa. 2001. "Are Baby Boomers Richer Than Their Parents? Intergenerational Patterns of Wealth Ownership in the United States." *Journal of Marriage and Family* 53(2):569–79.
Knickman, James R. and Emily K. Snell. 2002. "The 2030 Problem: Caring for Aging Baby Boomers." *Health Services Research* 37(4):849–84.
Krach, Constance A., Victoria A. Velkoff. 1999. "Centenarians in the United States." *Current Population Reports Series P23-199RV*. Washington, DC: U.S. Bureau of the Census.
Lee, Jennifer and Frank D. Bean. 2007. "Reinventing the Color Line: Immigration and America's New Racial/Ethnic Divide." *Social Forces* 86(2):561–86.
Lusardi, Annamaria and Olivia S. Mitchell. 2007. "Baby Boomer Retirement Security: The Roles of Planning, Financial Literacy, and Housing Wealth." *Journal of Monetary Economics* 54:205–24.
Mader, Sharon L. 2009. "Grandparents Raising Their Grandchildren." *Fact Sheet FLM-FS-25-01-R09, Family and Consumer Sciences*, Ohio State University.
Mannheim, Karl. 1927[1953]. "The Problem of Generations." Pp. 275–322 in *Essays on the Sociology of Knowledge*, edited by P. Kecskemeti. London: Routledge.
Mayhew, Bruce H. 1973. "System Size and Ruling Elites." *American Sociological Review* 38(3):468–75.
Migration Policy Institute. 2009. *Immigration and America's Future: A New Chapter*. Washington, DC: Migration Policy Institute.
Minkler, Meredith. 1999. "Intergenerational Households Headed by Grandparents: Contexts, Reality, and Implications for Policy." *Journal of Aging Studies* 13(2):199–218.
Mok, Diana, Barry Wellman, and Ranu Basu. 2007. "Did Distance Matter before the Internet? Interpersonal Contact and Support in the 1970s." *Social Networks* 29(3):430–61.
Pienta, Amy Mehraban and Mark D. Hayward. 2002. "Who Expects to Continue Working after Age 62? The Retirement Plans of Older Couples." *Journal of Gerontology: Social Sciences* 57:S199–208.
Pyke, Karen. 2004. "Immigrant Families in the U.S." Pp. 253–69 in *The Blackwell Companion to the Sociology of the Family*, edited by J. L. Scott, J. Treas, and M. P. M. Richards. New York: Blackwell.
Qian, Zhenchao and Daniel T. Lichter. 2007. "Social Boundary and Marital Assimilation: Evaluating Trends in Racial and Ethnic Intermarriage." *American Sociological Review* 72:68–94.
Rohr, Margund K. and Frieder R. Lang. 2009. "Aging Well Together: A Mini-Review." *Gerontology* 55(3):333–43.
Rook, Karen S. and Tonya L. Schuster. 1996. "Compensatory Processes in the Social Networks of Older Adults." Pp. 219–48 in *Handbook of Social Support and the Family*, edited by G. R. Pierce, B. R. Sarason, and I. G. Sarason. New York: Plenum.
Rosenfeld, Dana. 1999. "Identity Work among Lesbian and Gay Elderly." *Journal of Aging Studies* 13(2):121–44.
Rosnick, David and Dean Baker. 2009. "The Wealth of the Baby Boom Cohorts after the Collapse of the Housing Bubble." *Technical Report, Center for Economic and Policy Research*, Washington, D.C. Available at: http://www.flhsa.org/Sage/baby
Settersten, Richard A., Jr. 2007. "Social Relationships in the New Demographic Regime: Potentials and Risks, Reconsidered." *Advances in Life Course Research* 12:3–28.
Shapiro, Adam and Teresa M Cooney. 2007. "Divorce and Intergenerational Relations across the Life Course." *Advances in Life Course Research* 12:191–220.
Simmons, Tavia and Jane Lawler. 2002. "Grandparents Living with Their Grandchildren." *Technical Report, Census 2000 Brief C2KBR-31*, US Bureau of the Census, October.
Thomas, William and Florian Znaniecki. 1918[1996]. *The Polish Peasant in Europe and America*. Urbana, IL: University of Illinois Press.
Treas, Judith. 2002. "How Cohorts, Education, and Ideology Shaped a New Sexual Revolution on American Attitudes toward Nonmarital Sex." *Sociological Perspectives* 43(3):267–83.
———. 2008a. "The Dilemma of Gender Specialization: Substituting and Augmenting Wives' Household Work." *Rationality and Society* 20:259–82.
———. 2008b. "Transnational Older Adults and Their Families." *Family Relations* 57:468–78.
———. 2009. "Four Myths about Older Adults in America's Immigrant Families." *Generations: Immigration in an Aging Society* 32(4):40–45.

Treas, Judith and Sonja Drobnic, eds. 2010. *Dividing the Domestic: Women, Men, and Housework in Cross-National Perspective*. Stanford, CA: Stanford University Press.

Treas, Judith and Zoya Gubernskaya. 2009. "Has Maternal Contact Declined? Evidence for Seven Countries, 1986–2001." XXVI IUSSP International Population Conference. Marrakech.

Treas, Judith and Shampa Mazumdar. 2002. "Older People in America's Immigrant Families: Dilemmas of Dependence, Integration, and Isolation." *Journal of Aging Studies* 16:243–58.

———. 2004. "Kinkeeping and Caregiving: Contributions of Older People in Immigrant Families." *Journal of Comparative Family Studies* 35(1):105–22.

Uhlenberg, Peter. 1992. "Population Aging and Social Policy." *Annual Review of Sociology* 18(1):449–74.

Part IV
Social Relationships and Aging

Chapter 10
Social Relations and Aging

Deborah Carr and Sara M. Moorman

Studies dating back to Emile Durkheim's (1897) *Suicide* demonstrate that social relationships provide emotional, social, and economic supports that enhance physical and emotional well-being throughout the life course (House et al. 1988). Over the past three decades, however, researchers have discovered that social relationships are not universally protective for late-life well-being; rather, the protective effects of social ties vary based on the structure, nature, and quality of the relationship. Methodological advances also have enabled researchers to ascertain whether the association between relationships and health reflects social causation (i.e., direct benefits of social relationships), or selection (i.e., characteristics of those people who form and maintain relationships over the life course). Social gerontologists no longer ask, "*Do* social relationships affect the well-being of older adults?" Rather, they now ask, "*Why, how, when, and for whom* do social relations affect the health of older adults?"

In this chapter, we first discuss recent innovations in the conceptualization and measurement of older adults' relationships, including their objective and subjective properties, structures, and functions. We then provide an overview of the characteristics and health implications of four types of relationships: marriage and romantic partnerships (and the dissolution thereof through widowhood and divorce); intergenerational relationships (i.e., parent–child and grandparent–grandchild); friendships; and lack of satisfying relationships (i.e., loneliness). We then discuss the consequences of late-life relationships for health and well-being., We conclude by highlighting the research topics and methods that hold great promise for future exploration, and the implications of recent research for social and health policy.

Conceptualizing and Measuring Social Relations in Late Life

General Properties of Social Relationships

Social relationships are a multifaceted concept, encompassing a variety of structures, functions, and qualities. They may be based on legal ties, blood relations, coresidential status, or simply a fondness for and a desire to affiliate with one another. Relationships also vary with respect to their voluntariness, permanence, and duration. Voluntariness refers to whether one chooses to enter a relationship; permanence indicates whether one is able to terminate a relationship; and duration describes the amount of time that the relationship has existed.

D. Carr (✉)
Department of Sociology and Institute for Health, Health Care Policy and Aging Research,
Rutgers University, New Brunswick, NJ, USA
e-mail: carrds@rci.rutgers.edu

R.A. Settersten, Jr. and J.L. Angel (eds.), *Handbook of Sociology of Aging*, Handbooks of Sociology and Social Research, DOI 10.1007/978-1-4419-7374-0_10, © Springer Science+Business Media, LLC 2011

Family ties, particularly relationships with blood relatives such as parents, siblings, and children, are involuntary and are based on a powerful sense of obligation, thus they cannot be abandoned easily (Litwak 1985). By contrast, friendships and romantic relationships are formed voluntarily in contemporary western societies; most individuals choose to befriend or marry persons with whom they are compatible. Social ties with friends and even romantic partners today may be terminated without violating important norms regarding obligation and commitment, although adults may incur some emotional or financial costs from severing these ties (Litwak 1985). For older adults, sibling relationships generally have the longest duration of any relationship; they are formed at the birth of the younger sibling and typically persist until one's death. Marriages, parent–child relations, and friendships, by contrast, typically begin later in the life course – when one marries, gives birth, or forms a new friendship.

Sociologists and psychologists have developed a range of measures to further capture important quantitative and qualitative aspects of social relationships. Early studies typically focused on simple quantitative characteristics such as the number of persons in one's social network, to capture one's social embeddedness and integration (Antonucci 1990). In the past decade, social networks researchers have developed innovative new measures to further capture quantitative aspects of one's relationships, including their density (i.e., number of people in an individual's network who know one another) and multiplexity (i.e., number of different types of interactions exchanged within a single dyadic relationship) (Smith and Christakis 2008).

One of the most influential research advances, however, has been the widespread recognition that qualitative (or subjective) aspects of relationships are more important to one's well-being than are simple counts of significant others. For instance, measures of perceived support are stronger predictors of well-being than the number of persons in one's network available to provide help. Similarly, subjective appraisals of positive and negative relationship characteristics are powerful predictors of older adults' well-being (Rook 1998). Positive aspects include feeling loved, cared for, and understood, whereas negative aspects include demands, criticism, and conflict. A recent innovation is the conceptualization and operationalization of "ambivalence," which refers to having both positive and negative sentiments toward a single person or relationship (Lüscher and Pillemer 1998). For example, a spouse's well-intended desire to provide advice may be appreciated, yet also perceived as critical. Ambivalence is a common feature of late-life relationships, especially among spouses and parents and their adult children.

Functions of Social Relationships in Later Life

Social relationships also vary widely with respect to the functions they serve in older adults' lives. Older adults often rely on members of their social network for emotional and instrumental (i.e., practical or financial) support, yet they may expect and desire different types of support based on the nature of a particular relationship. Individuals usually expect both instrumental and emotional support from family members, but emotional support only from friends. Two influential models have been developed over the past three decades to characterize older adults' support preferences. The hierarchical compensatory model proposes that older people have a rank-ordered preference for receiving support from others (Cantor 1979). Most will turn first to family members and turn to nonfamily only when kin are unavailable. Cantor (1979) further specifies that people prefer to receive support from their spouse, followed by children, other relatives, friends, and professionals or formal organizations. Empirical studies generally support the hierarchical model; older adults are more likely to both prefer and receive assistance from a spouse, followed by children, other relatives, and friends.

The functional specificity model counters that supportive relationships are best matched according to each partner's needs and resources (Litwak 1985). For example, a married woman may turn to her best friend for emotional support or household help, if she feels that her husband is ill equipped to do so. Studies consistently show that emotional support from friends is more strongly associated with emotional well-being than comparable support from family, suggesting that support is most effective when provided by persons who are most qualified to do so.

Both the hierarchical compensatory and functional specificity models were developed to characterize relationships at one point in time, however. The "convoy model," by contrast, emphasizes that individuals maintain convoys of significant others who provide both instrumental and emotional support over the life course (Kahn and Antonucci 1980). The desirability, value, and types of support needed from a member of the convoy may change over the life course. The number of persons included in one's convoy, as well as one's level of closeness to convoy members, shifts over the life course. The model also emphasizes reciprocity and "support banks," where deposits are made early in the life course in anticipation of future needs or "withdrawals." This is one mechanism through which support to older family members becomes obligatory; adult children may feel obliged to "give back" to the parents and grandparents who supported them earlier in the life course.

All three models imply that older adults desire support from significant others. However, several theorists acknowledge that individual-level and cultural factors may affect the comfort with which one solicits and accepts help. Cohler (1983) has argued that self-reliance and autonomy are core values of capitalist societies, and they inhibit expression of dependency among older adults in the United States. For example, studies repeatedly show that older adults – even those with serious physical health limitations – do not want to reside with their adult children but prefer to remain independent and reside in their own homes: that is, they want "intimacy at a distance." Feeling dependent on significant others, especially for health-related caregiving, may be highly distressing to frail older adults and even hasten their desire for death. In sum, older adults' social relationships are not a monolithic entity; rather, they differ with respect to their number, structure, subjective qualities, and functions. As we shall see next, each of these aspects of relationships carries important implications for the well-being of older adults.

Marriage and Other Romantic Relationships

Concepts and Patterns

Heterosexual marriage is the most common type of romantic relationship among older Americans. In 2008, 73.2% of men and 42.9% of women aged 65 and older were married. A larger proportion of women (41.9%) than men (13.8%) are currently widowed, reflecting men's higher mortality risk and greater tendency to remarry. Only 10.3% of older men and women are currently divorced. Divorce was relatively rare and stigmatized during the young adult years of current cohorts of older adults, and most of those who did divorce ultimately remarried. Only 4.1% have never married (U.S. Bureau of the Census 2008).

As of February 2011, *same-sex marriage* is legally performed in five U.S. states and is recognized, though not legally performed, in three states. Exact numbers of legally married gay and lesbian couples are difficult to determine, because many couples report being "married" despite having no U.S. marriage license and because the legal status of unions may shift as new state legislation is implemented. Of those gay men and women who say they are married, 21.0 and 16.5% are 65 or older, respectively (O'Connell and Lofquist 2009). By contrast, 13% of all Americans are aged 65 and older, thus older adults are overrepresented among gay and lesbian partners who self-identify as married.

Many older adults, like younger persons, are in committed nonmarital romantic relationships. In 2000, *heterosexual cohabitation* was the relationship of choice for 4% of unmarried persons over aged 50 in the United States (Brown et al. 2006). According to estimates from the U.S. Census, *same-sex cohabitation* is an uncommon status for older Americans. Only 5.5% of gay male and 4.5% of lesbian cohabiting couples are aged 65 or older, although this may reflect the way that gay couples are measured in the Census. Reports are based on one's own gender, one's report of the cohabitant's gender, and the householder's report of the relationship between the two (O'Connell and Lofquist 2009). By contrast, studies based on nonrandom samples indicate that as many as one-quarter to two-thirds of coupled gay and lesbian persons aged 50 or older live with their partner (Grossman et al. 2000).

Some older couples in committed romantic relationships do not coreside – an arrangement called *living apart together*. In the United States, 7% of women and 6% of men report that they "live apart together" (Strohm et al. 2008). Older adults who choose this arrangement tend to own their own homes, and neither partner wants to move and combine their possessions nor create inheritance complications for their children. Older women also may not want to take on the homemaking and caregiving responsibilities that often accompany coresidence.

Researchers know relatively little about *dating* in later life. Although the American Association of Retired Persons commissioned a national study of dating in 2003, their sample of more than 3,000 persons focused on those aged 40–69 only. Carr (2004) estimated that 18 months after becoming widowed, only one-fifth of men and one-tenth of women in their 60s or 70s residing in the Detroit area were dating – although nearly 40% of men and 15% of women said that they were interested in dating. In the next decade, we suspect that much more will be learned about dating. Aging Baby Boomers, many of whom divorced and re-entered the dating market in later life, may feel more comfortable and less stigmatized in discussing their nonmarital, perhaps even casual, sexual, and romantic relationships.

Implications for Health and Well-Being

Empirical studies consistently show that married persons are healthier than their unmarried counterparts; strong effects are found for a range of outcomes including all-cause mortality (Johnson et al. 2000) and psychological distress (Johnson and Wu 2002). *Social selection* and *causation* are the dominant explanations proposed for the so-called "marriage benefit." The social selection perspective holds that healthy and financially secure people are more likely to marry and remain married over the life course, thus accounting for the association between marital status and health. The social causation perspective counters that marriage provides social control, economic, and psychosocial benefits that directly enhance health.

Marriage is a key source of social control; married people are less likely than unmarried persons to smoke, drink excessively, and engage in risky behaviors like not wearing seat belts. Spouses, especially wives, may help their partner with health-enhancing behaviors, such as regular visits to the doctor, exercise, healthy eating, and complying with medication regimens (Schone and Weinick 1998). Partnered adults also have higher household incomes than single individuals, and economic well-being predicts good health. In general, persons in a committed relationship are more likely than the unpartnered to report that they have a confidante, and that they feel loved and supported. Although representative studies of long-term gay and lesbian relationships are rare, mounting research suggests that they provide at least some of the same benefits for older adults' well-being as do heterosexual marriages (Grossman et al. 2000).

One of the most provocative discoveries in recent decades is the recognition that the benefits of a romantic relationship vary considerably based on legal and structural aspects of the relationships. First marriages are more protective than remarriages, legal marriage is more protective than cohabitation or dating, and heterosexual relationships are somewhat more protective than gay and lesbian relationships. For example, the "benefits" of remarriage for mental health and self-rated physical health are more modest than for first marriages (Barrett 2000). These protective effects also are short-lived, appearing only in the early stages of the remarriage transition (Blekesaune 2008). Similarly, cohabitors fare better than unpartnered persons yet worse than married persons in terms of depressive symptoms (Brown 2000), all-cause mortality (Koskinen et al. 2007), and self-rated physical health (Wu et al. 2003). A recent analysis of merged data from the General Social Survey, National Health and Social Life Survey, and the Chicago Health and Social Life Survey found that partnered gays and lesbians were similar to married persons and straight unmarried cohabitors in terms of self-rated health but fared poorer on measures of happiness (Wienke and Hill 2009).

Most scholars concur that these patterns reflect social selection more than social causation, however. For example, older heterosexual cohabitors have a greater mortality risk than their married peers, due largely to the cohabitors' relatively lower socioeconomic status. However, some studies suggest that cohabiting relationships are qualitatively different from marriages. Marcussen (2005) found that older cohabiting men receive fewer caregiving benefits than their married peers, perhaps because older cohabiting women felt less obligation than married women to provide this time-intensive care. Moreover, gay partnerships and remarriages face distinctive stressors that persons in heterosexual first marriages are spared of, including homophobia and strains of negotiating relationships with ex-spouses (and stepchildren), respectively. Future studies should explore the distinctive stressors experienced by a range of romantic relationships, including cohabitation, same-sex unions, and higher order marriages, to help pinpoint precisely why their health benefits are not comparable to those reaped in marriage.

Across all types of romantic relationships, however, health benefits are contingent upon relationship quality. Survey-based studies show that negative marital interactions increase one's risk of poor physical and emotional health (Liu and Umberson 2008). A pathbreaking development over the past decade has been the identification of the physiological pathways through which relationships "get under our skin," with particular attention to cardiovascular, endocrine, immune, metabolic, and sympathetic nervous systems (Ryff and Singer 2001).

Experimental studies typically induce either conflict or closeness among couples in laboratory settings, and then gauge physiological responses. Relationship conflict can impair immune response, slow wound healing, heighten susceptibility to infectious agents, and increase cardiovascular reactivity – all factors that may increase mortality risk among older adults (Robles and Kiecolt-Glaser 2003). Experiments focusing on positive interactions show that inducing physical contact and closeness under a stressful condition may decrease blood pressure and heart rate, and increase oxytocin, a hormone that weakens the impact of stress (Grewen et al. 2005). Negative interactions typically have a more powerful impact on health than positive interactions, and effects are stronger for women than men (Robles and Kiecolt-Glaser 2003). In sum, while long-term committed relationships provide emotional and health-enhancing supports, these benefits vary widely based on the legal status and quality of the relationship, and the personal characteristics of those who enter into (and remain in) such unions.

The Loss of Romantic Relationships: Divorce and Widowhood

Later life is marked by the loss of important social relationships, including the deaths of friends, siblings, parents, and spouses. Relationship loss via divorce is very rare among current cohorts of older adults; the vast majority of divorced older adults dissolved their marriages in young or middle adulthood, and many subsequently remarried. Studies using cross-sectional and administrative data show that currently divorced older adults have an elevated risk of all-cause, cardiovascular disease, cancer, and suicide mortality relative to married persons (Johnson et al. 2000). However, studies based on multiwave data showed that much of this gap is due to social selection, particularly the disadvantageous health and personality traits of those who divorce and do not ultimately remarry (Sbarra and Nietert 2009). As later life divorce becomes more normative among members of the Baby Boom and subsequent cohorts, however, researchers will need to delve more fully into the distinctive ways that late-life marital dissolution affects health and well-being.

Widowhood, by contrast, is a distressing and health-compromising transition faced overwhelmingly by older adults; two-thirds of the 2 million deaths occurring in the United States each year befall persons aged 65 and older (Federal Interagency Forum on Aging-Related Statistics 2009). Older bereaved spouses lose their primary source of emotional, instrumental, and financial support along with the disruption of daily routines and practices. While early stress theories suggested that widowhood was universally and intensely distressing, emerging evidence shows that the physical and emotional consequences of widowhood vary widely, based on one's gender and the nature of one's late relationship.

Men are more likely than women to experience physical health declines, increased disability, and heightened risk of mortality after their wives die. While popular lore claims that widowers may "die of a broken heart," research shows that it is the loss of a helpmate and caretaker that is really the culprit. Wives typically monitor their husbands' diets, remind them to take their daily medications, and urge them to give up vices like smoking and drinking (Umberson et al. 1992). Widowers are more likely than married men to die of accidents, alcohol-related deaths, lung cancer, and chronic ischemic heart disease during the first 6 months after their loss, but not from other causes that are less closely linked to health behaviors (Martikainen and Valkonen 1996). Even worse for older men is that their wives often are their primary (or only) source of social support and integration; when a man loses his wife, he also loses an important connection to his social networks.

By contrast, women's more emotionally intimate social relations over the life course are an important resource as they adjust to spousal loss. Older widows typically receive more instrumental and emotional support from their children than do widowers, given mothers' closer relationships with their children throughout the life course. Women also are more likely to have larger and more varied friendship networks than men, and these friendships provide an important source of support to women as they cope with their loss (Ha 2008).

The well-being of older widow(er)s also is linked to the emotional climate of the late marriage. Early writings, based on the psychoanalytic tradition, proposed that bereaved persons with the most troubled marriages would suffer heightened and pathological grief (Parkes and Weiss 1983). This perspective held that persons who had conflicted marriages would find it hard to let go of their spouses, yet also feel angry at the deceased for abandoning them. However, longitudinal studies that track married persons over time through the widowhood transition find that the loss of high-quality marriages is most distressing. Wheaton (1990) found that the emotional consequences of role loss are contingent upon one's "role history"; widowhood is distressing when the marital relationship had been satisfying. In sum, while classic bereavement theories proposed that a romantic partner's death is universally distressing, more recent work suggests that reactions to loss are contingent upon precisely what was lost: the loss of a beloved helpmate has a more profound impact than the death of a difficult or distant spouse.

Life-Long Singlehood

Never married persons have been nearly absent from social relations research. This absence reflects the fact that only 4% of persons aged 65 and older in the United States have never married (U.S. Census Bureau 2008). As such, researchers using sample surveys often do not have enough cases to study the health and well-being of the never married. Mortality is one of the few outcomes studied, because mortality and marital status data are available on vital registries and large administrative data sets such as the National Longitudinal Mortality Study (Johnson et al. 2000). However, these sources include only basic demographic measures, so researchers cannot explore the psychosocial pathways through which singlehood affects health.

The few studies exploring other indicators of single older adults' well-being reveal quite a positive picture. Older never married women enjoy mental health (Pudrovska et al. 2006) and physical health (Cwikel et al. 2006) equal to their married peers, and superior to their formerly married peers. These patterns partly reflect selection, where older cohorts of never married women are better educated than their married and formerly married peers, and have richer economic resources than their divorced or widowed peers. Moreover, never married women have adjusted to their status over time; they have chosen relationships that offer socioemotional support (Pudrovska et al. 2006), and they rely on formal services such as meal preparation services to help manage their instrumental needs (Cwikel et al. 2006). In sum, while research on the health of never married persons is sparse, the evidence generally reveals that singlehood may carry health advantages for women, yet these advantages typically reflect social selection factors including greater educational and economic resources of never married women in current cohorts of older adults.

Parent/Child Relationships and Grandparent/Grandchild Relationships

Thus far, we have focused solely on relationships between romantic partners; however, intergenerational relations also are a critical source of support (and strain). We now discuss characteristics and health consequences of parent–child and grandparent–grandchild relations.

Patterns and Concepts

Approximately 85% of American adults aged 65 and older are parents. Childless older adults are a heterogeneous group, including those who are child-free by choice, and those who desired children yet did not ultimately become parents. In 2010, 80% of women aged 60–64 will have at least one grandchild, but they will have fewer grandchildren than same-age women in prior cohorts. Due to increases in longevity, children will have more living grandparents than in the past: In 2020, nearly half of 10-year-olds will have four living grandparents, and four-fifths of 30-year-olds will have at least one living grandparent (Uhlenberg 2005).

Implications for Health and Well-Being

Nearly two-thirds of parents aged 50 and older rate their relationships with all of their adult children as excellent, and relationships tend to become even better as parents and children age (Birditt et al. 2009). However, even the best filial relationships can experience tensions, which may threaten older adults' well-being. Older adults who rate even one of their relationships with an adult child to be anything less than excellent report less happiness and more depressive symptoms (Ward 2008). Older parents may experience ambivalent feelings in their relationships with adult children when children face major problems, such as an illness, divorce, a legal problem, or substance abuse.

Such major problems in the life of an adult child also may affect older adults' health and well-being indirectly – via taking on the role of caregiving or custodial grandparent. Grandparents who are raising their grandchildren are a topic of heightened concern in the United States today. More than 1.5 million children (2.0% of all children) now live with at least one grandparent and no parent, and this figure underestimates the number of children whose primary noncoresidential caregiver is a grandparent (U.S. Census Bureau 2008). Grandparents step into a parenting role, or are asked to step in by a family member or the state, when their children (i.e., parents of the grandchildren) experience serious problems such as physical or mental illness, substance abuse, incarceration, or homelessness. Love and obligation may motivate grandparents to keep their grandchildren in the family and out of the foster care system.

Serving as primary caregiver to a grandchild is highly stressful: Grandparents worry about their children's problems and the financial demands of raising grandchildren – many of whom have emotional or developmental difficulties – all while navigating their own aging process. The early stages of caregiving are especially difficult. New custodial grandparents may forgo their own preventive behaviors, and experience increased depressive symptoms and poor physical health (Hughes et al. 2007). Despite the strains imposed by coresidence with their grandchildren, grandparents provide an indispensable safety net. Moreover, for many older adults, grandchildren provide an important source of happiness and purpose.

Friendships in Later Life

Older adults are more than romantic partners or caregivers to the younger generations; most maintain vibrant friendships with peers. Research on late-life friendship has flourished over the past three decades. Friendship is distinct from family relationships in that it is voluntary, nonobligatory, and typically based on the exchange of emotional rather than instrumental support (Blieszner and Roberto 2004). Friends are usually of similar ages, whereas family relationships often are cross-generational and thus may be marked by an imbalance of power. Because of their voluntary nature, friendships require more agency and motivation than familial relationships; they must be sought out, cultivated, and maintained – or in the case of unsatisfying relationships – dissolved. Healthy friendships also tend to have an equitable "give and take." Reciprocity, or being able to give back support equivalent to what was received, is more important to the quality and sustenance of friendships than kin relations (Rook 1987). Because friendships are less institutionalized than family relationships, however, norms for affiliation may be unclear, creating potential for discord and misunderstanding (Blieszner and Roberto 2004).

Friendships carry more rewards than costs, including companionship, shared leisure, emotional support, social integration, and informational assistance. Friendship is a more powerful predictor of older adults' psychological well-being and life satisfaction than family relationships. Persons with rich friendship networks tend to have better physical health than those with more tenuous ties, yet this relationship is mutually influential; those with good health also are best equipped to maintain friendships (Rawlins 2004).

Like family relationships, friendships differ in the levels of instrumental and expressive support they provide. Rawlins (2004) has observed that friendships have "two general modes": confidantes and companions. The former are based on intimate conversation, caring, strong emotional attachment, commitment, and loyalty. The latter, by contrast, involve socializing, social interactions with groups of individuals, limited emotional attachment, and reciprocity.

Late-life friendships vary by gender and social class. Older women's friendships involve more intimate self-disclosure, whereas men's friendships often are based on shared activities. As a result, women tend to have closer same-sex friendships than men do, and this is particularly true for married individuals. Most married women have close confidants beyond their spouse, but men typically do not (Antonucci 1990). Middle class older adults tend to have more friends than their working class peers (Phillipson 1997), perhaps due to their greater financial resources and greater likelihood of living geographically far from their kin.

An important discovery over the past three decades is that friendships change in quantity, quality, and importance over the life course, particularly as one faces late-life transitions including widowhood and the onset of health problems, and as one's needs change (Litwak 1985). Laura Carstensen's influential socioemotional selectivity theory proposes that older adults selectively choose to maintain relatively fewer, but higher quality, relationships as they age and experience declines in health. Casual, less rewarding ties may lapse while only the most meaningful relationships are maintained (Carstensen et al. 1999). Empirical evidence confirms that as adults age they evidence a reduction in the number of friends reported (Phillipson 1997). This decrease in the number of one's friendships may not entirely reflect preferences, however. As older adults retire, their ties to workplace friends weaken, while physical health declines, caregiving demands, and functional limitations may impede one's ability to maintain social ties. Some scholars speculate that online relationships may become increasingly important to older adults, especially those with serious mobility limitations, as they enable social interaction with acquaintances and "weak tie" relationships (Wright 2000).

An important avenue that researchers are only starting to explore is the identification of specific tasks for which friends are "substitutable" when family ties are not available. For example, a recent study of older adults' choices for health care proxy designations revealed that nearly all (85–90%) married persons and parents named a spouse or child to make end-of-life decisions for them. Among unmarried childless persons, by contrast, nearly one-quarter appointed a friend to play this important role (Carr and Khodyakov 2007). Other studies reveal that widows and widowers are less likely to seek out new romantic partnerships if their social and emotional needs are met by supportive friends (Carr 2004). Among future cohorts of older adults, for whom divorce rates are higher and fertility rates lower than for current cohorts, friendships may be an essential source of instrumental and emotional support.

Absent and Unfulfilling Social Relationships: Loneliness and Social Isolation

Old age historically has been considered a time of social isolation. Disengagement theory proposed that it was beneficial for both the older adult and society if the elder were to gradually withdraw from his or her social roles and relationships (Cumming and Henry 1961). Similarly, classic role theories held that the loss of the work role for men (via retirement) and loss of the wife role for women (via widowhood) would leave older adults socially isolated and despondent (see Biddle 1986, for review). More recent research counters, however, that while loneliness and social isolation are problematic, they are neither inevitable nor universal features of aging.

Over the past three decades, researchers have recognized that loneliness is not triggered by a quantitative lack of relationships, but by a lack of satisfaction with the number or quality of one's relationships. Contemporary researchers have identified two statistically and conceptually distinct subtypes of loneliness: emotional loneliness refers to the absence of an intimate confidante, while social loneliness refers to the absence of a broader social network. The two types are highly correlated; widowed persons, those living alone, or those living far away from their friends and families consistently report higher levels of both types of loneliness than persons who are more socially integrated (de Jong Gierveld and Havens 2004).

The mere presence of proximate relationships does not ward off loneliness, however. An estimated 25% of older married persons report emotional and social loneliness; this pattern is particularly common among persons whose spouses are ill, who have a dissatisfying (or nonexistent) sexual relationship, and who have infrequent or conflicted conversations (De Jong et al. 2009).

As de Jong Gierveld and Havens (2004) noted, loneliness depends on one's "standards as to what constitutes an optimal network of relationships."

Despite its subjective nature, loneliness is a serious problem for many older adults; it is linked to sleep problems, poor cardiovascular health, and elevated blood pressure, each of which carries long-term consequences for mortality risk (Cacioppo et al. 2002). Loneliness also may be a particularly acute social problem for older adults in future cohorts. Smaller families and an increased prevalence of divorce and childlessness among future cohorts of older adults may create a context where one maintains objectively fewer relationships (see Chap. 13), thus triggering social loneliness. More importantly, however, some have argued that current cohorts of midlife adults have unrealistically high expectations for what their social relationships should provide (e.g., one's partner should be their "soulmate"); if these lofty expectations go unfulfilled, then older adults may report higher levels of emotional loneliness, as well. In sum, it is not how many relationships one maintains that matters for one's late-life well-being, but the extent to which those relationships are deemed personally satisfying and fulfilling.

Looking Forward to the Next Three Decades

Over the past three decades, scholars have made major advances in conceptualizing and measuring the multifaceted nature of older adults' social relationships, and documenting the implications of these relationships for health and well-being. We anticipate that five topics are ripe for further exploration in the coming decades: sexuality; population-based studies of elder abuse; racial and ethnic differences in late-life social relationships; social relations of Baby Boomers; and the use of research methods that incorporate multiple reporters and multiple sources of health and relationship data.

Sexuality and Romantic Relationships in Later Life

Even at the turn of the 21st century, most studies of older adults' romantic relationships focus on the protective effects of instrumental and emotional support, and fail to consider another critically important component: sexuality. In the future, we expect that scholars will explore more fully the role of sexuality in older adults' relationships – regardless of whether the sexual relationship occurs within the context of a legal or coresidential relationship. Of particular interest is how sexual aspects of older adults' relationships are protective for health and well-being, and how aging-related changes in physical functioning may affect the quality and nature of sexual relations among long-term partners. Waite et al. (2009) suggest that older adults with high-quality sexual and intimate relations will have better trajectories of physical and mental health than those whose relationships function less well (or who lack such relationships). The recently collected National Social Life, Health and Aging Project (NSHAP), a nationally representative sample of community-dwelling individuals aged 57–85, provides in-depth measures of sexual behavior, practices, and health and will be an invaluable resource as social gerontologists further investigate both heterosexual and homosexual older adults' sexual relationships.

The Dark Side of Relationships: Elder Abuse

In the past three decades, researchers have documented that even high-quality relationships may be strained or ambivalent. Yet relatively little population-based research focuses on extreme negativity in older adults' relationship. Elder abuse and neglect comprise intentional physical, emotional, financial, or sexual abuse, as well as failure by a caregiver to meet an older adult's basic needs (Bonnie and Wallace 2005). Elder abuse is a serious though understudied social problem; data historically have come from small, nonrepresentative samples; the criminal justice system; or agency or caregiver reports.

In 2004, however, the NSHAP obtained reports of respondents' recent experiences of mistreatment, and for those who reported mistreatment, their relationship to the perpetrator. NSHAP also obtained detailed demographic, health, cognitive functioning, and social relationships data (Laumann et al. 2008). These data reveal that family members are the most common perpetrators of elder abuse; verbal and physical abuse are most often committed by romantic partners and children, whereas financial abuse is most frequently perpetuated by children and siblings. In the next decade, sociologists will have the resources to fully explore the risk factors for and consequences of multiple types of elder mistreatment, with particular attention to the ways that family relations may elevate or buffer against the risk of multiple types of late-life abuse.

Racial and Ethnic Differences in Late-Life Social Relationships

Most research on late-life social relationships focuses on white Americans. This pattern partly reflects the fact that older blacks and Latinos are underrepresented in large-scale sample surveys, given their elevated rates of mortality and morbidity. However, documenting the nature and consequences of social relationships for ethnic minorities is an important inquiry. Some scholars have proposed that blacks' low rates of marriage relative to whites contribute to blacks' elevated risk of mortality and morbidity (Kaplan and Kronick 2006). However, this argument rests on the assumption that marriage benefits blacks' and whites' health similarly.

This assumption requires interrogation. Studies reveal that blacks report poorer marital quality and more marital conflict than whites, and that the economic gains to marriage are less for blacks than whites given black men's disadvantaged economic prospects (Broman 1993). Given these patterns, marriage may be less protective to blacks compared to whites. Further, given the very high levels of intergenerational integration and support in Latino and Asian families, it is plausible that the benefits of marriage *vis a vis* other social relationships are weaker than they are for whites. Identifying the distinctive contributions of marriage, parenthood, friendships, and sibling relationships to the health of ethnic minorities will be a valuable line of inquiry in the future decades. In 2006, whites, blacks, Latinos, and Asians accounted for 81, 9, 6, and 3% of the U.S. population, respectively. By 2050, these proportions will be 61, 12, 18, and 8%, respectively (Federal Interagency Forum on Aging-Related Statistics 2009).

As the demographics of older Americans shift, social gerontologists should identify the distinctive correlates of health and well-being for all ethnic and racial groups. Small-scale qualitative studies may be particularly useful in revealing the ways that the distinctive cultural views and practices of ethnic families affect late-life health and well-being. For example, Confucian values including filial piety affect the ways adult children monitor the health of aging parents among Chinese Americans (Park and Chesla 2007). Gendered cultural views such as *machismo* (i.e., men's adherence to traditionally masculine, high-risk behaviors) and *marianismo* (i.e., women's self-sacrifice for spouse and children) in Latino families affect both family relations and health practices – which may have important implications for later-life health (Cianelli et al. 2008). We are optimistic that future research blending qualitative and quantitative research may better illuminate the ways that cultural context shapes the relationship between family and health among ethnic and racial subgroups.

Social Relationships of Baby Boomers

Much of what scholars know about older adults' social relationships is based on the experiences of current cohorts of older adults, those men and women born in the early 20th century who were socialized into rigid gender-typed social roles during their formative years and who often maintained a traditional division of labor in their marriages, where men were primary breadwinners, and women were primary caretakers for their husbands and children. It is not surprising, then, that the studies reported in this chapter reveal that women have closer ties to friends and children than their husbands, that women are an important source of health control to their husbands, and that men typically fare worse than women when their spouse dies.

For members of the Baby Boom cohort, the 75 million persons born between 1946 and 1964, late-life relationships may be reinvented. Future cohorts of older women are more likely than their predecessors to have received college degrees, held professional occupations, and shared childrearing tasks with their husbands. Future cohorts of men, by contrast, may have more emotionally intimate friendships than prior cohorts of men who were socialized to be self-sufficient and independent. Thus, we might expect future cohorts to fare better in the face of spousal loss – given that women may have the economic resources and men the interpersonal resources to cope with bereavement. Further, given that Baby Boomers faced fewer social obstacles to divorce during their young and midlife years, we might expect that their late-life marriages will be of higher quality than current cohorts; older adults today may have faced social or economic pressure to remain in dissatisfying marriages. Given major shifts in family structure and gender role socialization over the past 30 years, we expect that scholars' understanding of older adults' social relationships may be transformed in coming decades.

Methodological Innovations

Sociologists overwhelmingly rely on large-scale sample survey data to document the nature and consequences of older adults' relationships. However, this approach has a number of limitations. First, survey-based studies of social relations typically focus on a single individual's report, rather than information from both members of a romantic dyad, or multiple members of a family or social network. Ironically, research purportedly exploring spousal dynamics or parent–child relationships typically relies on a single person's appraisal of the relationship.

Further, research reveals that a parent may have a very different relationship with each of his or her children, yet most standard survey items rely on an overall appraisal of one's relationship with "children" (Davey et al. 2009). It is not clear whether these measures tap one's assessment of the child with whom one is closest, most emotionally distant, or an average across children. New analytic techniques, such as dyadic data analysis, allow researchers to use information from multiple reporters to estimate how much each person's outcome is associated with both own and partner characteristics (Kenny et al. 2006). This approach enables researchers to explore questions such as: how do both spouses' reports of marital conflict affect each spouse's health and health behaviors, and to what extent does one partner's health behaviors and practices affect the other partner's outcomes?

Second, survey-based studies of relationships and health historically have relied on self-reports of symptoms and conditions, rather than physiological indicators that may capture short-term responses to relationship characteristics and strains. As such, scholars still do not fully understand how older adults' social relationships "get under the skin" to affect physical and emotional health. Laboratory research, conducted primarily by psychologists, has made important advances, by measuring the physiological responses of older couples placed in either stressful or supportive settings (Robles and Kiecolt-Glaser 2003). In the past decade, a number of large representative sample surveys of older adults including the Health and Retirement Study, Midlife Development in the United States, National Health and Nutrition Examination Survey, NSHAP, and Wisconsin Longitudinal Study have supplemented their self-reported health data with extensive genetic and biological indicators, such as immune response measures. We are optimistic about the scientific discoveries that may develop in the next three decades, as interdisciplinary teams of researchers continue to investigate the complex ways that demographic, socioeconomic, biological, psychosocial, and genetic factors link social relationships to health and well-being among older adults.

Implications for Policy and Practice

Over the past 30 years, researchers have demonstrated that older adults maintain diverse social networks including but not limited to spouses, former spouses, children, grandchildren, siblings, and friends. Although socially integrated persons enjoy better health than those who are more isolated, the protective effects of social relations vary widely based on the quality of those relationships. These findings have important implications for social policy.

Current policies tend to privilege legal and biological ties over all other relationships. For instance, if an incapacitated older adult has not appointed a health care proxy, many states have policies that give priority to specific family members as substitute decision makers; spouses and children are typically at the top of the hierarchy. Likewise, the majority of U.S. states do not grant gay and lesbian partners the legal right to formalize their union. Public policies, and especially health care policy, should be based on an expansive definition of "family," and allow older adults to include as their decision makers, advocates, and beneficiaries whomever they consider their closest and most meaningful ties.

Although the health-enhancing effects of high-quality social relationships cannot be understated, it is essential to recognize that at least part of these benefits reflect social selection characteristics; that is, the preexisting traits of persons who enter in and out of particular relationships. For example, researchers have found that the deleterious health effects of divorce and grandparent–grandchild coresidence and the relatively weak health protection provided by cohabitation reflect the fact that those who divorce, serve as custodial grandparents, and cohabiters tend to have fewer economic resources than their counterparts. As such, a financial safety net that provides at least minimal quality housing, food, health care, and economic security may be the most effective policy for promoting physical and emotional well-being among older adults.

References

Antonucci, Toni C. 1990. "Social Supports and Social Relationships." Pp. 205–26 in *The Handbook of Aging and the Social Sciences,* 3rd ed., edited by R. H. Binstock and L. K. George. San Diego, CA: Academic.

Barrett, Anne. 2000. "Marital Trajectories and Mental Health." *Journal of Health and Social Behavior* 41:451–64.

Biddle, Bruce J. 1986. "Recent Developments in Role Theory." *Annual Review of Sociology* 1267–92.

Birditt, Kira S., Lisa M. H. Jackey, and Toni C. Antonucci. 2009. "Longitudinal Patterns of Negative Relationship Quality across Adulthood." *Journals of Gerontology: Psychological Sciences* 64B:P55–64.

Blekesaune, Morten. 2008. "Partnership Transitions and Mental Distress: Investigating Temporal Order." *Journal of Marriage and Family* 70:879–90.

Blieszner, Rosemary and Karen A. Roberto 2004. "Friendship across the Life Span: Reciprocity in Individual and Relationship Development." Pp. 159–82 in *Growing Together: Personal Relationships across the Lifespan*, edited by F. R. Lang and K. L. Fingerman. Cambridge: Cambridge University Press.

Bonnie, Richard J., and Robert B. Wallace, eds. 2005. *Elder Mistreatment: Abuse, Neglect and Exploitation in an Aging America.* Washington, D.C.: National Research Council.

Broman, Clifford L. 1993. "Race Differences in Marital Well-Being." *Journal of Marriage and the Family* 55:724–32.

Brown, Susan L. 2000. "The Effect of Union Type on Psychological Well-being: Depression among Cohabitors versus Marrieds." *Journal of Health and Social Behavior* 41:241–55.

Brown, Susan L., Gary R. Lee, and Jennifer R. Bulanda. 2006. "Cohabitation among Older Adults: A National Portrait." *Journals of Gerontology: Social Sciences* 61B:S71–9.

Cacioppo, J. T., Hawkley, L. C., Crawford, L. E., Ernst, J. M., Burleson, M. H., Kowalewski, R. B., Malarkey, W. B., Van Cauter, E., and Berntson, G. G. 2002. "Loneliness and Health: Potential Mechanisms." *Psychosomatic Medicine* 64:407–17.

Cantor, Marjorie H. 1979. "Neighbors and Friends: An Overlooked Resource in the Informal Support Systems." *Research on Aging* 1:434–63.

Carr, Deborah. 2004. "The Desire to Date and Remarry Among Older Widows and Widowers." *Journal of Marriage and Family* 66:1051–68.

Carr, Deborah and Dmitry Khodyakov. 2007. "Health Care Proxies in Later Life: Whom Do We Choose and Why?" *Journal of Health and Social Behavior* 48:180–94.

Carstensen, Laura L., Derek Isaacowitz, and Susan Turk Charles. 1999. "Taking Time Seriously: A Theory of Socioemotional Selectivity." *American Psychologist* 54:165–81.

Cianelli, Rosina, Lilian Ferrer, Beverly J. McElmurry. 2008. "HIV Prevention and Low-Income Chilean Women: Machismo, Marianismo, and HIV Misconceptions." *Culture, Health & Sexuality: An International Journal for Research, Intervention, and Care* 10:297–306.

Cohler, Bertram J. 1983. "Autonomy and Interdependence in the Family of Adulthood." *The Gerontologist* 3:33–39.

Cumming, Elaine and William H. Henry. 1961. *Growing Old: The Process of Disengagement.* New York: Basic Books.

Cwikel, Julie, Helen Gramotnev and Christina Lee. 2006. "Never-married Childless Women: Health and Social Circumstances in Older Age." *Social Science & Medicine* 62:1991–2001.

Davey, Adam, Corrine Jenkins Tucker, Karen L. Fingerman and Jyoti Savla. 2009. "Within-Family Variability in Representations of Past Relationships with Parents." *Journals of Gerontology: Social Sciences* 64B: S125–36.

de Jong Gierveld, Jenny and Marjolein Broese van Groenou, Adriaan W. Hoogendoorn, and Johannes H. Smit. 2009. "Quality of Marriages in Later Life and Emotional and Social Loneliness." *Journals of Gerontology: Social Sciences* 64B:S497–506.

de Jong Gierveld, Jenny and Betty Havens. 2004. "Cross-National Comparisons of Social Isolation and Loneliness: Introduction and Overview." *Canadian Journal on Aging* 23:109–13.

Durkheim, Emile 1897 [1951]. *Suicide: A Study in Sociology.* J. Spaulding & G. Simpson (Trans.). New York: Free Press.

Federal Interagency Forum on Aging-Related Statistics. 2009. *Older Americans 2009: Key Indicators of Well-Being.* Washington, D.C.: U.S. Government Printing Office.

Grewen, Karen M., Susan S. Girdler, Janet Amico, and Kathleen C. Light. 2005. "Effects of Partner Support on Resting Oxytocin, Cortisol, Norepinephrine, and Blood Pressure Before and After Warm Partner Contact." *Psychosomatic Medicine* 67:531–8.

Grossman, Arnold H., Anthony R. D'Augelli, and Scott L. Hershberger. 2000. "Social Support Networks of Lesbian, Gay, and Bisexual Adults 60 Years of Age and Older." *Journals of Gerontology: Psychological Sciences* 55B:P171–9.

Ha, Jung-Hwa. 2008. "Changes in Support from Confidantes, Children, and Friends Following Widowhood." *Journal of Marriage and Family* 70:306–18.

House, James S., Karl R. Landis, and Debra Umberson. 1988. "Social Relationships and Health." *Science* 241:540–5.

Hughes, Mary E., Linda J. Waite, Tracey A. LaPierre, and Ye Luo. 2007. "All in the Family: The Impact of Caring for Grandchildren on Grandparents' Health." *Journals of Gerontology: Social Sciences* 62B:S108–19.

Johnson, Norman J., Eric Backlund, Paul D. Sorlie, and Catherine A. Loveless. 2000. "Marital Status and Mortality: The National Longitudinal Mortality Study." *Annals of Epidemiology* 10:224–38.

Johnson, David R. and J. Wu. 2002. "An Empirical Test of Crisis, Social Selection, and Role Explanations of the Relationship between Marital Disruption and Psychological Distress: A Pooled Time-Series Analysis of Four-Wave Panel Data." *Journal of Marriage and Family* 64:211–24.

Kahn Robert L., and Toni Antonucci. 1980. "Convoys over the Life Course: Attachment, Roles and Social Support." Pp. 253–86 in *Life-span Development and Behavior*, Vol. 3, edited by P. B. Baltes, and O. G. Brim. New York: Academic.

Kaplan, Robert M., and Richard G. Kronick, R.G. 2006. "Marital Status and Longevity in the United States Population." *Journal of Epidemiology and Community Health* 60:760–5.

Kenny, David A., Deborah A., Kashy, and William L. Cook. 2006. *Dyadic Data Analysis*. New York: Guilford.

Koskinen, Seppo, Kaisla Joutsenniemi, Tuija Martelin, and Pekka Martikainen, P. 2007. "Mortality Differences According to Living Arrangements." *International Journal of Epidemiology* 36:1255–64.

Laumann, Edward O., Sara A. Leitsch, and Linda J. Waite. 2008. "Elder Mistreatment in the United States: Prevalence Estimates from a Nationally Representative Study." *Journal of Gerontology: Social Sciences* 63:S248–54.

Litwak, Eugene. 1985. *Helping the Elderly: The Complementary Roles of Informal Networks and Formal Systems*. New York: Guilford.

Liu, Hui and Debra J. Umberson. 2008. "The Times They are a Changin': Marital Status and Health Differentials from 1972 to 2003." *Journal of Health & Social Behavior* 49:239–53.

Lüscher, Kurt and Karl Pillemer. 1998. "Intergenerational Ambivalence. A New Approach to the Study of Parent-Child Relations in Later Life." *Journal of Marriage and the Family* 60:413–25.

Marcussen, Kristen. 2005. "Explaining Differences in Mental Health between Married and Cohabitating Individuals." *Social Psychology Quarterly* 68:239–57.

Martikainen, Pekka and Tapani Valkonen. 1996. "Mortality after the Death of a Spouse: Rates and Causes of Death in a Large Finnish Cohort." *American Journal of Public Health* 86:1087–93.

O'Connell, Martin and Daphne Lofquist. 2009. *Counting Same-Sex Couples: Official Estimates and Unofficial Guesses*. Washington, D.C.: U.S. Bureau of the Census.

Park, Mijung and Catherine Chesla. 2007. "Revisiting Confucianism as a Conceptual Framework for Asian Family Study." *Journal of Family Nursing* 13:293–311.

Parkes, C. Murray and Robert S. Weiss. 1983. *Recovery from Bereavement*. New York: Basic.

Phillipson, Chris. 1997. "Social Relationships in Later Life: A Review of the Research Literature." *International Journal of Geriatric Psychiatry* 12:505–12.

Pudrovska, Tetyana, Scott Schieman, and Deborah Carr. 2006. "The Strains of Singlehood in Later Life: Do Race and Gender Matter?" *Journal of Gerontology: Social Sciences* 61:P315–22.

Robles, Theodore F. and Janice Kiecolt-Glaser. 2003. "The Physiology of Marriage: Pathways to Health." *Physiology and Behavior* 79:409–16.

Rawlins, William K. 2004. "Friendships in Later Life." Pp. 273–99 in *Handbook of Communication and Aging Research*, 2nd ed., edited by J. F. Nussbaum and J. Coupland. Mahwah, NJ: Lawrence Erlbaum.

Rook, Karen S. 1987. "Reciprocity of Social Exchange and Social Satisfaction among Older Women." *Journal of Personality and Social Psychology* 52:145–54.

———. 1998. "Investigating the Positive and Negative Sides of Personal Relationships: Through a Lens Darkly?" Pp. 369–93 in *The Dark Side of Close Relationships*, edited by B. H. Spitzberg and W. R. Cupach. Mahwah, NJ: Erlbaum

Ryff, Carol D. and Burton H. Singer. 2001. *Emotion, Social Relationships, and Health*. New York: Oxford University Press.

Sbarra, David A. and Paul J. Nietert. 2009. "Divorce and Death: Forty Years of the Charleston Heart Study." *Psychological Science* 20:107–13.

Schone, Barbara S. and Robin M. Weinick. 1998. "Health-related Behaviors and the Benefits of Marriage for Elderly Persons." *The Gerontologist* 38:618–27.

Smith, Kirsten P. and Nicholas A. Christakis. 2008. "Social Networks and Health." *Annual Review of Sociology* 34:405–29.

Strohm, Charles, Judith A. Seltzer, Susan D. Cochran, and Vickie Mays. 2008. "Living Apart Together Relationships in the United States." Working Paper No. 042-08, California Center for Population Research, University of California, Los Angeles.

U.S. Bureau of the Census. 2008. *Marital Status of People 15 Years and Over, by Age, Sex, Personal Earnings, Race, and Hispanic Origin, 2008*. Washington, D.C.: U.S. Bureau of the Census.

Uhlenberg, Peter. 2005. "Historical Forces Shaping Grandparent-Grandchild Relationships: Demography and Beyond." *Annual Review of Gerontology and Geriatrics* 24:77–97.

Umberson, Debra, Camille B. Wortman, and Ronald C. Kessler. 1992. "Widowhood and Depression: Explaining Long-Term Gender Differences in Vulnerability." *Journal of Health and Social Behavior* 33:10–24.

Waite, Linda, Edward O. Laumann, Aniruddha Das, and L. Philip Schumm. 2009. "Sexuality: Measures of Partnerships, Practices, Attitudes, and Problems in the National Social Life, Health, and Aging Study." *Journal of Gerontology: Social Sciences* 61B:S71–9.

Ward, Russell A. 2008. "Multiple Parent-Adult Child Relations and Well-Being in Middle and Later Life." *Journals of Gerontology: Social Sciences* 63B:S239–47.

Wheaton, Blair. 1990. "Life Transitions, Role Histories, and Mental Health." *American Sociological Review* 55:209–23.

Wienke, Chris and Gretchen J. Hill. 2009. "Does the 'Marriage Benefit' Extend to Partners in Gay and Lesbian Relationships? Evidence from a Random Sample of Sexually Active Adults." *Journal of Family Issues* 30:259–89.

Wright, Kevin B. 2000. "Computer-Mediated Social Support, Older Adults and Coping." *Journal of Communication* 50:100–18.

Wu, Zheng, Margaret J. Penning, Michael S. Pollard, and Randy Hart. 2003. "In Sickness and In Health: Does Cohabitation Count?" *Journal of Family Issues* 24:811–38.

Chapter 11
Intergenerational Relations in Later-Life Families

J. Jill Suitor, Jori Sechrist, Megan Gilligan, and Karl Pillemer

Relations between the generations have been a central feature in literature and popular culture throughout recorded history. The dramatic increase in life expectancy across the last century, combined with more recent changes in divorce, child-bearing, and women's employment, has challenged old assumptions and created new inquiries into intergenerational relations in later life. Thus, it is not surprising that the study of these relations in the later years has grown exponentially across the last three decades as scholars have rushed to identify and explain these new patterns of relations and their consequences on family members.

In reviewing the study of families in later life across the past 30 years, this chapter has three major goals. First, we highlight the major theoretical developments of this period. We then review the empirical research on intergenerational relationships in the family, including relations between parents and adult children and between grandparents and grandchildren. Because the literature on these topics is so extensive, we will focus our review in the following ways. First, we will concentrate primarily on interpersonal relationships among family members, referring when appropriate to related chapters in this volume that address other dimensions of intergenerational relations (e.g., caregiving, coresidence, diversity in later-life families, and demographic changes). Second, we will devote the most space to parent–adult child relations; this choice reflects the importance of this topic in the literature on later-life families. Third, we focus on later-life families in the United States; given the large body of work on cross-national differences in families, placing them in an appropriate historical and cultural context is beyond the scope of this review. Finally, consistent with the focus of the volume, we emphasize advances in the sociological study of later-life families across the past three decades, although research from other disciplines is noted where relevant.

Theoretical Roots and Conceptual Advances

The 1980s saw the emergence of an important body of theoretical work that influenced the study of relations between parents and their adult children, including Bengtson and colleagues' extension of the family solidarity model (Bengtson and Schrader 1982); Riley's (1987) essays on the significance of age in sociology; and Hagestad's (1986) and Nydegger's (1986) discussions of off-time transitions, optimum timing preferences, and filial maturity. This scholarship played a key

J.J. Suitor (✉)
Department of Sociology, Center on Aging and the Life Course, Purdue University,
West Lafayette, IN, USA
Karl Pillemer's details are: Department of Human Development Cornell University (Ithaca, New Yourk)
e-mail: jsuitor@purdue.edu

R.A. Settersten, Jr. and J.L. Angel (eds.), *Handbook of Sociology of Aging*, Handbooks of Sociology and Social Research, DOI 10.1007/978-1-4419-7374-0_11, © Springer Science+Business Media, LLC 2011

role in expanding the range of approaches to studying parent–adult child relations, including new theoretical frameworks that emphasized dissensus rather than consensus and empirical studies that focused on life course transitions and within-family variations in parent–adult child relations.

Intergenerational Solidarity

Bengtson's intergenerational solidarity model was introduced just prior to the period on which we are focusing in this review, quickly becoming the dominant theoretical framework in the study of the intergenerational relations. Not only has Bengtson and colleagues' own body of work had a major impact on the field (cf. Bengtson et al. 2002), their framework has been incorporated in almost all of the major programs of research on intergenerational relations across this period (Rossi and Rossi 1990; Silverstein and Giarrusso in press).

Bengtson and colleagues posited that family solidarity is comprised of several interrelated components: (1) contact, (2) exchange of support, (3) norms of obligation, (4) value similarity, (5) relationship quality, and (6) opportunity structure. Empirical research framed by this model has been developed in three increasing complex directions. The first of these directions focused on single concepts, such as affect and support, emphasizing how they develop within the family (Rossi and Rossi 1990) and influence outcomes for parents and children (Silverstein, Chen, and Heller 1996; Giarrusso et al. 2001). In the second direction, individual concepts have been combined into a broader single measure of family solidarity. In these studies, solidarity is often the outcome, rather than the predictor (Silverstein and Bengtson 1997). In the third, most complex direction, scholars have explored the interdependence among concepts (Bengtson and Roberts 1991; Rossi and Rossi 1990), in the theoretical tradition of Homans (1950).

Most of the research drawing upon the solidarity model has explored parent–adult child relationships. However, there is evidence of the utility of this model for other intergenerational relations, particularly relations between grandparents and grandchildren (Giarrusso et al. 2001). In the early 1990s, Bengtson and colleagues (Bengtson et al. 2002) also expanded the model to incorporate negative affect. This development may have helped to fuel new theoretical developments on aging families across the past decade which have, paradoxically, challenged Bengtson's assumption that solidarity is the central factor shaping intergenerational relations.

The Life Course Perspective in Later-Life Family Relationships

Across the past three decades, the life course perspective has been one of the most influential approaches in the social sciences. Marshall and Bengtson's chapter in this volume provides a comprehensive discussion of the life course perspective; therefore, we will focus specifically on the way in which life course theories have shaped the study of intergenerational relations in recent decades.

As discussed in greater detail in Marshall and Bengtson's chapter, the life course perspective draws from both sociological theories of social change and psychological theories of individual and family development. This perspective highlights the importance of historical and social contexts and individual time and development on family relationships (Settersten 2003); further, it addresses individual change within the family context as well as how these changes are linked to other family members (Elder 1994). This perspective is complementary to Bengtson's solidarity model in that they both emphasize the importance of time and generation in explaining the relationships between members of all dyads within the family at any point in the life course.

Empirical studies of intergenerational relations across the past three decades have typically drawn upon the life course perspective to address two issues. First, studies have examined continuity in family relations across the life course, reporting that closer and more harmonious relationships between parents and children in early life were associated with higher relationship quality and exchange of support across the life course (Rossi and Rossi 1990).

The second line of inquiry framed by the life course perspective addresses the notion of "linked lives," focusing on the ways in which life events experienced by family members shape their intergenerational relations. In some cases, the effects of transitions appear to be consistent, such as in the case of parental divorce, which almost variably lowers the quality of relations between fathers and their adult children (Connidis 2003; Wethington and Dush 2007). However, in many cases, the direction and extent of the effects are conditional. For example, adult children's divorce generally does not affect relations between the generations (Kaufman and Uhlenberg 1998), but it does when the event leads offspring to return to their parents' homes needing of high levels of support (Aquilino and Supple 1991). Further, consistent with Hagestad's classic work on timing (Hagestad 1986), the same transition may improve parent–child relations when the timing fits with normative life course expectations, but may have deleterious effects when it does not. For example, children's completion of higher education may improve relations with parents (Aquilino 1997), but not necessarily when an adult daughter returns to school while raising her own family (Suitor 1987).

In sum, the life course perspective has become an important conceptual tool in understanding relations between and within generations in later-life families. Thus far, much of the research that has drawn upon this framework has been cross-sectional, despite the obvious longitudinal character of its basic tenets; its influence is likely to become even stronger when more studies of intergenerational relations follow families across longer periods.

Ambivalence in Later-Life Family Relationships

A more recent theoretical development in the study of later-life families is intergenerational ambivalence. The concept of ambivalence has roots in classic theories in both sociology and psychology beginning in the early 1900s (Freud 1913; Merton and Barber 1963; Coser 1966); however, only since the late 1990s (Luescher and Pillemer 1998) has this concept come to play a central role in the study of intergenerational relations. Although this framework has been applied predominantly in the area of parent–adult child relations, it has also been used to shed light on other relationships in middle and later life such as those between grandparents and grandchildren.

In contrast to theoretical perspectives that focus heavily on positive aspects of intergenerational relationships, the ambivalence framework is based on the assumption that family relationships are characterized by both positive and negative feelings or attitudes. This perspective posits that family roles are often contradictory, thus producing ambivalent feelings (Pillemer and Suitor 2005, 2008). The majority of empirical work on intergenerational ambivalence has focused on parent–adult child relations, exploring the prevalence of ambivalence and identifying characteristics of parents, children, and dyads which predict this dimension of relationship quality (Pillemer et al. 2007; Wilson et al. 2006). A variety of direct and indirect measures of ambivalence have been used across these studies. Recent research comparing the most commonly used direct and indirect measures has shown that although they are moderately strongly associated, the association is not sufficiently strong to demonstrate that they capture the same underlying construct; further, the findings suggest that direct and indirect measures have different meanings for particular subgroups of parents and adult children (Suitor, Gilligan, and Pillemer 2009).

Taken together, this developing line of scholarship suggests that ambivalence is an important line of inquiry for understanding later-life families, primarily because it captures many complexities and nuances in later-life family relationships that previous research has not. Nevertheless, it is important to note that it complements both the solidarity and life course perspectives in its focus on understanding how current ambivalence is shaped by the complexity in role relationships across the life course.

Substantive Advances

Parent–Adult Child Relations

The early 1980s were a pivotal time in the study of parent–adult child relations. For much of the three previous decades, research on intergenerational relations was concerned with exploring patterns of contact and the provision of support to parents in need of care (Albrecht 1953; Winch 1970). This line of work was fueled by concern regarding the broad social changes occurring across that period, including the civil rights movement, the anti-war movement, and the women's movement. These societal upheavals led to skyrocketing increases in both women's employment and divorce, both were viewed to be particularly threatening to later-life families. However, by the early 1980s, research had demonstrated that parents and children continued to stay in regular contact and children, particularly daughters, continued to provide care to older parents in need. Thus, concerns about the demise of traditional intergenerational relations declined (Bengtson and DeTerre 1980), allowing scholars to turn to new questions regarding exchanges and other dimensions of parent–adult child relations.

Exchanges Between the Generations

From the 1980s through the first decade of the century, scholars of later-life families continued to monitor children's care to parents and also turned from asking merely *whether* children provide support to their parents to what factors lead particular children to provide support to particular parents and the consequences of parent care on adult children's physical, psychological, and social well-being. Research also turned to exploring the flow of support between the generations, as opposed to only from children to parents. In this section, we will focus primarily on instrumental exchanges with parents, including our discussion of expressive support in the section on relationship quality between parents and adult children.

Support from Parents to Adult Children. One of the patterns shown most consistently in this line of work is that, despite the concern that adult children will become overburdened with elder care, parents typically give more support than they receive until their 70s or 80s (Cooney and Uhlenberg 1992; Rossi and Rossi 1990; Umberson 2006). In fact, the flow of support generally does not change until parents' health begins to decline (Eggebeen and Hogan 1990; Rossi and Rossi 1990). The type of assistance that parents provide to their adult children varies as a function of the children's point in the life course, moving from assistance with childcare and routine housekeeping and maintenance tasks when children are raising young families to financial assistance as adult children move into later middle age (Cooney and Uhlenberg, 1992; Swartz 2009). However, it is important to remember that the support provided to adult children in the form of childcare and household tasks can also be considered to be financial assistance, in that they reduce the financial resources that children would otherwise have to direct toward these tasks. This point is best made by Silverstein and colleagues' calculation that the childcare provided by grandparents saves parents between 17 and 29 billion dollars (Gans and Silverstein 2006).

It might appear that parents provide more assistance to married adult children raising families; however, children who are not married are often in greater need, and therefore receive more support (Spitze et al. 1994; Suitor, Pillemer, and Sechrist 2006). Other life events and conditions that increase adult children's needs also lead to more parental support. For example, adult children who experience serious physical and mental health problems also receive greater support from parents (Seltzer et al. 2008; Suitor et al. 2006a, b), as do children who are divorced (Spitze et al. 1994), unemployed (Suitor, Sechrist, and Pillemer 2007), or who have engaged in deviant behaviors as adults

(Suitor et al. 2006a, b). It is also important to note that the support given to children in need is, in a sense, more costly to parents because the children's needs often render them unable to reciprocate in either the short-term or long-term (Pillemer and Suitor 1991; Greenfield and Marks 2006), which becomes more problematic as parents age and their needs increase while their ability to provide support to needy children declines (Seltzer et al. 2008).

Another way in which older parents may provide support for adult children is housing. Several studies during the 1980s highlighted the continued coresidence or return of adult children to their parents' homes not related to care for the older parent, but to the financial or emotional status of the adult child (Mancini and Blieszner 1985). Although coresidence may be beneficial for both generations, it is most often fueled by children's rather than the parents' needs and circumstances (Choi 2003; Ward, Logan, and Spitze 1992).

Some scholars argued that such coresidence would have detrimental effects on parents' marital quality and well-being (Clemens and Axelson 1985); however, findings from most large-scale surveys reported that such negative effects appeared primarily in the presence of high levels of conflict between parents and coresident children (Suitor and Pillemer 1988) and when children's coresidence resulted from problems in their own lives (Aquilino and Supple 1991; Pudrovska 2009). However, these findings suggest that the current economic crisis, involving the highest level of unemployment and home loss in nearly three decades and leading to increased coresidence (Fleck 2009), may be setting the stage for increased intergenerational coresidence and accompanying interpersonal and psychological stress resulting from the stressful circumstances that led to this arrangement.

In summary, research from the past three decades has shown that parents continue to assist their adult children, providing both routine support and support in times of particular need from the point when children enter adulthood until the parents' health or financial resources require that the flow of support begins to reverse toward the older generation. It is also worth noting that under normal circumstances, the provision of support to their children has few negative effects on parents' well-being, unless that support is needed due to children's serious problems, as discussed above, or when parents are experiencing their own stressful transitions, such as widowhood or retirement (Davey and Eggebeen 1998).

Support from Children to Parents. As noted earlier, one of the issues of greatest concern to sociologists studying older families across the past several decades has been whether adult children provide adequate support to their parents, and more recently, the effects of providing that support on children's physical, psychological, and social well-being. Both scholarly and popular interests in this subject have been so great that family caregiving has become one of the most rapidly growing bodies of literature in the social sciences since the early 1980s.

One reason for the rapidly growing interest in support to parents is the increasing number of adult children in their 40s, 50s, and 60s who have living parents (US Bureau of the Census 2010), relative to earlier decades. However, that does not mean that all of these parents are in need of care. As discussed earlier, the flow of support is disproportionately from parents to children typically until parents are in their 70s, at which point it begins to reverse. However, recent data on the health and activity of adults can be used to suggest that studies may soon find either that the flow continues toward children for a longer period than in earlier decades, or that there is an interlude in which there is little flow in either direction.

The phenomenon of "better aging" has found its way into popular culture as well as scholarly research, as illustrated by a recent cartoon in *The New Yorker* (2009) magazine which shows a wife saying to her husband with great enthusiasm, "70 is the new 50." Although that may be an oversimplification, the trends are certainly moving in that direction. First, data on chronic conditions and disability indicate that individuals are aging with fewer limitations, and thus are less likely to need care from their adult children. In fact, by 2007, only about 3% of persons ages 65–74 and slightly more than 10% of those 75 and older in the United States had limitations in their Activities of Daily Living (ADLs); only about 6% of those 65–74 and less than 20% of those 75 or older had limitations in their Instrumental Activities of Daily Living (IADLS) (US Bureau of the Census 2010). Second, due to better health, individuals who are 65 and over have become increasingly likely to remain active and independent. For example, in the past three decades, there has been a substantial increase in the percent of individuals in this age group in the labor force, and the projections are that this trend will continue, particularly among women. Between 1980 and 2008, men 65 and over had a 13% increase in labor force participation, compared to a 64% increase for women in that age group (US Bureau of the Census 2010). By 2016, it is projected that more than one quarter of men in this age group will remain in the labor force, as will nearly one in five women.

These trends call into question whether the picture of care to aging parents is in transition. Despite recent interest in the "sandwich generation" of women caring simultaneously for parents and minor children (Spitze and Logan 1990), it is more likely that over the next few decades, parental caregivers will be well beyond raising minor children; in fact, many of these "children" will themselves be in their late 50s or early 60s when they begin parent care. Further, it may be important that adult children do not begin zealously providing support to parents who are not yet in need. In fact, several studies using data collected since the late 1980s, when the health of individuals over 65 had improved from earlier decades, have found that high levels of support from adult children had negative effects on psychological well-being (Silverstein et al. 1996), even when controlling on current health and previous depression. Perhaps, this is because many of those parents were not in need, yet received high levels of helping; such an interpretation would be consistent with Davey and Eggebeen's (1998) finding that adult children's support had positive effects on parents' well-being only when the support was warranted by the parents' circumstances.

The patterns we have just discussed raise an interesting question: When should children's support to parents be characterized as "caregiving" as opposed to part of an exchange relationship? When do parents' become sufficiently "old" to render the support they receive "caregiving" or "elder support?" The difficulty that scholars face when drawing these distinctions can best be illustrated by the age distributions of subsamples used in some of the investigations of intergenerational exchange and caregiving. In many studies, the "adult children" range well into their 60s (e.g., Cooney and Uhlenberg 1992); in fact, sometimes up to age 70 (Wakabayashi and Donato 2006). In other studies, however, the range of care *recipients* begins well below the upper limit in studies of caregivers. In fact, Davey and Eggebeen used a subsample from the National Survey of Families and Households (NSFH) with an age range beginning at 50 to study the effects of exchanges on "older adults' psychological well-being" (1998:92). Thus, there continues to be confusion about when parents become "care recipients" as opposed to exchange partners. This issue is likely to become more common as the trend toward healthy aging continues and the "sandwich generation" becomes the cohort of women 55–75 who are providing care *to* their older parents while beginning to receive care from their adult children.

Despite these changes in older parents' health and activity levels, the focus of studies of support to parents has changed remarkably little across the past three decades, continuing throughout the period to emphasize describing and explaining the flow of exchanges between parents and their adult children. This body of work has revealed several consistent patterns that we believe are worthy of note.

First, gender of both parents and children continues to play the greatest role in the study of support to the older generation. Mothers receive more support from their adult children than do fathers (Silverstein et al. 2002), and daughters are more likely than sons to be the source of that support (Spitze and Logan 1990; Chesley and Poppie 2009). On one hand, this pattern is not surprising, based on classic feminist arguments regarding women's greater investment in relationships and sensitivity to others' needs (Gilligan 1982; Chodorow 1978). On the other hand, it might be expected that this pattern would become less pronounced as a consequence of changes in gender-role attitudes among Americans across the past three decades (Powers et al. 2004). However, this does not appear to be the case. Recent studies are equally as likely as older studies to report that both mothers and daughters are the most likely to provide and receive support than are fathers and sons (Suitor et al. 2006a, b; Chesley and Poppie 2009). Further, not only do daughters provide more support than do sons, but daughters are also typically both mothers' and fathers' preferred source of emotional support and help during illness (Suitor and Pillemer 2006; in press).

Because of daughters' prominent role in providing support to parents, one area of concern has been whether women's increasing labor force participation would reduce their ability to provide care. However, studies across the past three decades have shown that women's employment has fewer effects than was feared (Pavalko and Artis 1997; Pavalko, Chap. 37), despite the fact that women have become increasingly committed to the labor force.

Another demographic change that has been feared would shape patterns of support to parents is divorce. Most of this concern has centered on whether divorced daughters would continue to provide support to their parents; however, evidence from throughout the past three decades has shown that this is not the case (Cicirelli 1986; Spitze et al. 1994). In contrast, parents' marital instability does affect patterns of support. In earlier generations, almost all marriages ended with the death of one partner; however, for the first time in American history, the skyrocketing divorce rate of the 1970s created a notable population of older divorced parents. The consequences of parental divorce fall far more heavily on fathers than mothers. Most studies find little difference in support to older mothers (Lye 1996); however divorced fathers are far less likely to receive support than are their counterparts who remain married to their children's mothers (Lye 1996). Even if these divorced fathers remarry, support is less likely to be provided by their children (Lin 2008), although this is less pronounced when children have a long and close relationship with the stepparent (Ganong et al. 2009). Although no studies to date have followed divorced families from childhood through the later years, it is likely that the lower support to fathers follows a pattern of less closeness and contact in the early years following parents' marital disruption (Aquilino 2006; Scott et al. 2007).

The other demographic trend across the past three decades that may have greater consequences for support to parents in future decades is the increasing level of educational attainment, particularly among women. In 1970, only 8% of women had completed college, a figure that increased to only 13% by 1980; however by 2008, that figure had increased to 29% (US Bureau of the Census 2010). Although adult children who are better educated likely have greater resources to provide support to parents, college graduates are substantially more likely to be geographically mobile, thus reducing residential proximity to parents. This pattern is particularly consequential because, following gender, proximity is typically found to be the best predictor of intergenerational support, despite the greater ease of travel and communication in recent decades. It is likely to be another decade or two before we can assess the effects of women's increasing educational attainment on support to parents through changes in proximity.

Costs of Caring on Adult Children's Well-Being. Beginning in the 1980s, Pearlin and colleagues' groundbreaking theoretical work on caregiving provided a basis for viewing parental caregiving as a life course transition with much in common with other status transitions and life events studied by family scholars. One component of Pearlin's conceptualization of the stress processes that is particularly relevant to adult children's caregiving is that life events often intensify preexisting strains while also bringing older problems to the forefront (Pearlin 1989). One way in which to conceptualize the transition to caregiving is as a process that involves adhering to the norms of providing physical and emotional support to the care recipient (George 1986; Suitor and Pillemer 1990). Further, it has been recognized that successful adoption of this role includes some role renegotiation with the parent (Brody 2004). However, an aspect of the transition that is seldom considered involves the changes in role relationships with siblings, spouses, other kin, and even friends.

New caregivers must negotiate the expectations of their new role with all of these role partners, typically leading to changes in their relationships with the core of their preexisting social support networks (Litvin et al. 1995; Suitor and Pillemer 1987), a process that often rekindles conflict and ambivalence from earlier points in the relationship.

For example, the parent may serve as a source of expressive support early in the caregiving career (Walker, Pratt, Oppy 1992); however, as the parent's physical or cognitive health declines and the balance of exchange changes, the role relationship with the parent may well become a source of stress (Kramer 1997; Aquilino 1998), particularly in the case of dementia. Relationships with other role partners who were previously sources of support may also become sources of stress. In the case of siblings, caregiving often ignites tension in the relationship often stemming from earlier points in their relationship (Suitor and Pillemer 1987; Ingersoll-Dayton et al. 2003; Merrill 1996); in the case of spouses, the responsibilities of caregiving often interfere with the performance of marital and parenting roles (Suitor and Pillemer 1994; Spitze and Logan 1990). Whether preexisting role partners serve as a source of support or stress is shaped to a great extent by whether they have had direct experience themselves in the role of caregiver; consistent with studies of other status transitions, experientially similar role partners are much more likely to have a positive impact on caregivers' lives (Suitor, Pillemer, and Keeton 1995a; Umberson 2006).

Quality of Relations Between Parents and Adult Children

Have Parents and Children Remained Close? One question that has been investigated throughout the past three decades is whether parents and adult children have continued to have the high levels of closeness that was found in earlier studies (Adams 1968; Troll 1971). The answer to this question has been remarkably consistent: Dozens of studies across this period have found that most members of both generations report the relationship as very close (cf. Silverstein and Bengtson 1997; Kaufman and Uhlenberg 1998; Rossi and Rossi 1990; Sechrist et al. 2007) and relatively free of conflict (Szydik 2008; Umberson 1992). The consistency of this finding is of practical as well as scholarly significance, given that close and harmonious relations between the generations have salutary effects on both parents and children, whereas conflict and the absence of closeness have deleterious effects (Koropeckj-Cox 2002).

The Generational Stake in The Twenty-First Century. A second question enduring across the past three decades involved the generational stake proposed by Bengtson and Kuypers in the early 1970s (Bengtson and Kuypers 1971). Family scholars in the 1960s and 1970s (cf. Neugarten 1970; Hill 1970) identified differences between the behaviors and attitudes of parents and their adult children – differences that many were concerned might erode affect and support between the generations. Bengtson and Kuypers proposed that parents' and children's perceptions of one another might be as important, if not more important, in predicting relationship quality than actual differences in attitudes and behaviors. Using data from a study of students and their parents, they found that parents perceived their relations with their offspring as closer and more harmonious than did the students, and parents viewed greater similarity between the generations than did their young adult children. Bengtson and Kuypers argued that the explanation for these patterns lay in parents' stake in continuity and stability in their relations with their children, as opposed to their children's stake in individuation and change (Bengtson and Kuypers 1971:258).

Their argument, albeit compelling, was developed in a historical context in which such divisions between the generations were evident to all Americans, regardless of their political stance. Not only did the 1960s and early 1970s see divisions between parents and adult children – they also saw the divisions that were a part of the Civil Rights Movement, the Women's Movement, and the Anti-War Movement. Perhaps neither the theory nor the data would apply to later cohorts who were not struggling so intensely with these issues. With this in mind, many studies, including several by Bengtson and his colleagues, have reinvestigated the generational stake, or as it was later renamed, the "intergenerational stake," across the past three decades. Each of these studies, regardless of whether they have relied upon data from the USC Longitudinal Study of Generations (Giarrusso et al. 1995), the NFSH (Shapiro 2004), or other sets (Rossi and Rossi 1990) have revealed the same general pattern, suggesting that the intergenerational stake is as relevant a concept as it was when first introduced nearly 40 years ago.

Explaining Affectional Closeness in Parent–Child Relations. A third question that has continued to be asked across the past three decades is how to best predict which parents and adult children have high levels of positive sentiment toward one another. The factor found to predict the quality of affective relations between parents and children most consistently is gender. Closest ties have almost always been found between mothers and adult children, beginning with the earliest studies of parent–child affect in the later years (cf. Adams 1968; Suitor et al. 1995b). Further, the preponderance of studies have reported stronger affectional ties and greater confiding between mothers and daughters than mothers and sons (for example, Rossi and Rossi 1990; Spitze et al. 1994; Suitor and Pillemer 2006).

Research on other structural characteristics of children has yielded much less consistent results. Classic theories of similarity and interpersonal relationships (Homans 1950; Lazarsfeld and Merton 1954) would lead to the expectation that parents and children would be closer when they share structural characteristics, such as education, parental status, marital status, and religion. Further, sharing parents' sociodemographic characteristics often indicates that adult children have achieved normative benchmarks in development that are highly valued by parents. However, the literature across the past three decades does not support this hypothesis. Whereas some studies have found greater closeness and harmony when adult children become parents themselves (Fischer 1986; Spitze, et al. 1994; Umberson, 1992), other studies found either no positive effects of parenthood (Suitor and Pillemer 2006) or effects specific only to particular parent–child combinations (Kaufman and Uhlenberg 1998; Rossi and Rossi 1990), some of which were negative (Kaufman and Uhlenberg 1998). Findings regarding the effects of other dimensions of status similarity, such as marital status, occupational status, and educational attainment, provide an equally inconsistent picture.

We believe that there are two reasons why similarity of normative adult social statuses is less predictive of relationship quality for parents and adult children than expected based on homophily theory. First, achieving some of these social statuses that increase their similarity to parents, such as marriage and parenthood, produces responsibilities that make them less available to their parents and may sometimes create tension, for example, conflict over child-rearing, or between parents and children-in-law (Fischer 1983; Merrill 2007).

Second, as Suitor and colleagues' (Suitor, Pillemer, and Keeton 1995a) research on support and conflict with kin and nonkin has shown, the reason that status similarity leads to better relationship quality is because such similarity increases the likelihood that associates will have similar experiences, leading to shared values and perspectives. Such similarity of values and experiences has been shown to be associated with greater closeness and less conflict and ambivalence (Rossi and Rossi 1990; Pillemer et al. 2007; Suitor and Pillemer 2006); further, it is substantially more important than is similarity of social statuses (Suitor and Pillemer 2006). Not surprisingly, studies of parent–child relations that take into consideration only structural similarity do not produce the consistent findings that might be expected based on theories of homophily.

Ambivalence in Parent–Adult Child Relations. As noted earlier in this chapter, over the past decade there has been increasing interest in examining the prevalence and predictors of ambivalence in parent–adult child relations. Because this line of research has developed only since the publication of Luescher and Pillemer's essay on intergenerational ambivalence in 1998, there have been relatively few studies to date. This set of studies has shown that ambivalence is very common in later-life families, regardless of whether the reports are provided by mothers or their offspring (Pillemer et al. 2007; Wilson et al. 2006). Further, consistent with findings discussed above regarding the intergenerational stake, mothers report less ambivalence than do their adult children (Wilson et al. 2006).

In some cases, the findings from this line of research mirror those of studies of other dimensions of parent–adult child relationships quality. For example, higher levels of ambivalence have been found when either adult children experience problems that increase dependence or concerns about possible dependence, such as when they engage in deviant behaviors (Pillemer et al. 2007) or require financial assistance (Pillemer and Suitor 2002; Wilson et al. 2006). However, in some cases, the findings diverge from the broader literature; for example, parents have been found to express less ambivalence regarding relationships with adult children who are married, whereas marital status is an inconsistent predictor of closeness or conflict. In other cases, the findings from the extant set of ambivalence studies mirrors the broader literature in the degree of inconsistency across studies with similar methodologies. For example, Willson and colleagues (2006) found greater ambivalence between dyads of women, whereas gender did not predict ambivalence in either of Pillemer and colleagues' studies (Pillemer and Suitor 2002; Pillemer et al. 2007). There are also inconsistencies in the findings for both mothers' and children's health and several demographic characteristics of parents and children.

In sum, the study of ambivalence is greatly expanding our understanding of the complexity of parent–adult child relations, a topic that was greatly ignored in research prior to the late 1990s. However, the inconsistency found across studies suggests that we must wait for further study before we can draw many firm conclusions about how to explain this intriguing aspect of later-life family relations.

Diversity and Parent–Child Relations

The study of race differences in family relations has been of great interest among scholars for several decades. Given this level of attention, it is perplexing that there are so few consistent patterns by race. Although it has become standard to include race as a predictor in studies of parent–adult child relations, there has been almost no attempt to examine whether the same set of factors explain intergenerational relations across racial and ethnic subgroups. Instead, most studies have focused on whether there are differences in affect, support, and coresidence among these groups.

Many early studies showed strong ties and large kin support networks among minority families (Stack 1974; Taylor 1986), implying that minority families had stronger ties and support systems than White families (Hofferth 1984; Mutran 1985). Most studies making direct *comparisons* in parent–adult child relations among subgroups did not appear until the 1990s. This research has provided compelling evidence that race differences in intergenerational support are fueled primarily by structural differences between Blacks and Whites (Sarkisian and Gerstel 2004). For instance, White parents may provide higher levels of financial assistance to their adult children compared to minority groups (Berry 2006), yet multigenerational households (Choi 2003) and childcare provided by grandparents (Berry 2006) are more common among minority families. Thus, structural differences such as socioeconomic status and family structure influence the type of support exchanged, but exchanges of support are still common in most parent–adult child relationships regardless of race or ethnicity.

In contrast to findings regarding differences in intergenerational support among subgroups, studies of race differences in affect have found greater closeness between mothers and adult children in Black than White families even after controlling on structural characteristics (Aquilino 1997; Kaufman and Uhlenberg 1998; Umberson 1992). However, there do not appear to be consistent race variations in mothers' feelings of ambivalence toward their adult children (Pillemer et al. 2007), and Black and White mothers are equally likely to differentiate among children in terms of positive affect (Suitor et al. 2007; Ward, Spitze, and Deane 2009).

Grandparent–Grandchild Relations

Research on grandparenting has experienced increasing attention in recent decades, primarily as the result of three sociodemographic trends that shaped the experience of grandparenting, and thus research on this topic. First and most important, increasing life expectancy meant that by the 1980s, most adults would occupy the role of grandparent for nearly one third of their lives. Second, high rates of divorce beginning in the 1970s and continuing through the first decade of this century affected ties between grandparents and grandchildren as well as parents and children. Third, sky-rocketing rates of birth to single mothers across the past 30 years, particularly to African–American women, led many grandmothers to return to the role of primary caretaker in their middle and later years. In response to these sociodemographic patterns, research on grandparenting in the 1980s began by documenting and explaining patterns of contact, closeness, and support, followed by studies of the effects of marital instability on these patterns, and later turning to grandparents raising grandchildren. Interestingly, scholars studying grandparenting gave greater attention to race and ethnicity than was typical in the broader literature on American families across this period, thus shedding important and unique light on diversity in family relations.

The first large-scale sociological survey of grandparenting was conducted by Cherlin and Furstenberg in the early 1980s and published in 1986. This work provided a comprehensive picture of variations in contemporary grandparent–grandchild relationships and the ways in which these patterns were shaped by grandchildren's age, proximity, race, and the parents' marital status. More recent studies have corroborated these findings (Hodgson 1992; Silverstein and Marenco 2001), suggesting that these patterns have continued across the intervening years. Further, consistent with the pattern that has predominated throughout this chapter, gender played an important role in grandparent–grandchild relations, with almost uniformly greater contact and positive affect from grandchildren toward grandmothers (Eisenberg 1988; Hodgson 1992), particularly maternal grandparents in families in which parents divorced (Cherlin and Furstenberg 1986; Gladstone 1988; Matthews and Sprey 1985).

Grandparents have traditionally participated in the informal care of grandchildren, particularly when mothers were employed full-time (Cherlin and Furstenberg 1986; Jendrek 1994; US Bureau of the Census 2010); however, in recent years there has been a sharp increase in the practice of grandparents taking primary responsibility for raising grandchildren. Data from the US Bureau of the Census show instances of grandchildren living in a grandparent-maintained household have increased from 3.6% in 1980 to about 6% in 2008 (US Bureau of the Census 2010). This trend coincides with the percent of births to unmarried mothers, which has increased steadily from 18% in 1980 to 52% in 2007 (Ventura 2009). Although coresidence does not necessarily indicate that grandparents take on sole or even shared responsibility for the raising of grandchildren, of those children living in a grandparent's home in 2008, 35% did not have any parents present (US Bureau of the Census 2010) Taking on sole responsibility of grandchildren, with or without the presence of the parent, continues to occur in large part due to parents' emotional problems, substance abuse, or the need for greater support because of the absence of a partner (Jendrek 1994; Pruchno 1999; Sands and Goldberg-Glen 2000). Grandparents who provide primary care differ from those who do not by gender, economic status, and race. Most of the grandparent caregivers are women, Black or Native

American, and have low incomes and educational attainment (Fuller-Thompson, Minkler, and Driver 1997; Minkler and Fuller-Thompson 2000).

Although many grandparents report positive outcomes from raising grandchildren (Hayslip et al. 1998), this role often has negative consequences on caregivers' physical, social, and psychological well-being (Giarrusso et al. 2001), particularly when grandchildren have behavioral or emotional problems (Hayslip et al. 1998; Sands and Goldberg-Glen 2000). Further, many minority grandparent caregivers are at an even greater disadvantage because they are more at risk for being single, living below the poverty-line, and having health limitations (Burnette 1999; Pruchno 1999). Paradoxically, some studies have found that minority caregiving grandparents report less burden than do White caregiving grandparents (Pruchno 1999; Pruchno and McKenney 2002); this pattern can be accounted for by the fact that minority grandparents are more likely to have stronger family networks, family histories of grandparent coresidence, and peers within their social networks who also experience this role (Burnette 1999; Pruchno 1999).

Next Steps: Capturing the Complexity of Intergenerational Relations

The past three decades have seen cohorts of researchers turning their attention to relations between the generations in midlife and beyond. As noted at the outset of this chapter, genuinely new demographic realities have spurred this extraordinary growth in scientific interest. Because of the increased life span, in contemporary society we experience both the benefits and challenges of the lengthened shared lifetimes of generations. Indeed, many of us can look forward to continued relationships with our parents until well into late middle age, a historically unprecedented situation. Looking to the future, we anticipate that attention to the nature and dynamics of intergenerational relations will continue to expand in sociology, as well as in related disciplines such as psychology and economics.

What direction should the field take to build on the solid foundations created over the past three decades? More than a half century ago, in a classic article Weaver argued that all scientific fields engage in a predictable progression from simpler models to more complex ones. In the early stages of a scientific discipline, concern is with questions of categorization, description, and relatively simple hypotheses. As the field progresses, however, the "organized complexity" of systems is acknowledged and investigations increasingly take such complexity into account (Weaver 1948).

It is clear that such a movement is underway in the scientific study of intergenerational relations among adults (Pillemer and Suitor 2008). Scholars from a variety of disciplines are looking beyond simple models of older parent–adult child relationships to orientations and approaches that recognize the complex and sometimes contradictory world of the family in later life. Thus, over time, scholarship on this topic has moved from concerns about the weakening of intergenerational ties commonly expressed in the 1960s to an emphasis on the continued importance and influence of these linkages. Similarly, research has progressed from describing the amount and type of contact, interaction, and exchanges between the generations to more complex conceptual and empirical approaches.

What kind of research is needed over the coming years to capture the complexity of intergenerational relations? We offer a few suggestions here, but we note that our goal is to be provocative rather than definitive. One pressing need is for studies that recognize and exploit powerful within-family designs in studying intergenerational relations. Until very recently, most research in intergenerational relations involved between-family studies involving one parent and one target child or children in the aggregate. A growing body of research suggests that it is fundamentally necessary to collect data from both generations and from multiple members of each generation to fully understand intergenerational relations in later life (Davey et al. 2009; Suitor et al. 2006a, b; Ward et al. 2009).

With the knowledge that there is a great deal of variation within the family concerning most aspects of intergenerational relations, it is important to revisit past investigations with an eye to refining our understanding of these processes in light of within-family differences. Along these lines, researchers should revisit life course analyses concerning status transitions and parent–child relations; for example, by identifying how status transitions of adult children can influence changes in parental differentiation and favoritism across time. Further, current theoretical frameworks concerning dyadic processes across generations should be revisited to incorporate the complex influences of individual relationships on other relationships within the family; for example, how parental differentiation among adult children impacts grandparent–grandchild relations. Inquiries such as these can build a greater understanding of individuals' and dyads' roles within the family as well as the family as a whole.

A second tendency of past research has been to conceptualize and measure parent–adult child relations in a unidimensional manner focusing on either closeness *or* conflict. This work has been very valuable and has provided the foundation for much of our understanding of intergenerational relations. However, given what is now known about the interplay of positive and negative feelings, attitudes, and behaviors that are embodied in intergenerational relationships, future research must take such complexity into account. To be sure, longitudinal research designs that include multiple family members and assessments of positive and negative components of multiple relationships are likely to be resource-intensive. However, to advance scientific knowledge about intergenerational relations in twenty-first century families, such designs are likely to be required.

Further, just as research across the past thirty years responded to the major demographic, social, and economic changes of that time, future research must pay particular attention to current shifting trends. For instance, the characteristics of the older cohorts on which previous research is based differ considerably from those of the baby boomer cohort which is currently in late middle age and entering old age. Unlike their predecessors, the baby boom cohort has fewer children and is more likely to have experienced divorce, giving them fewer resources for support in later life. Given the projected increase in the older population overall, a shortage of caregiving resources in both the formal and the informal sectors is likely to result. Further, the baby boom cohort is experiencing their 60s and 70s in far better health than did earlier cohorts, raising new issues about the timing and meaning of caregiving, as well as about intergenerational relations more generally. The outcomes of these demographic changes as well as other social and economic changes are likely to present fertile grounds for scholarship over the coming three decades.

Finally, researchers in the field of intergenerational relations work in an area that is of considerable interest to policy makers and to the general public and is likely to increase given the growth in the older population. Throughout the social and behavioral sciences, there is an emerging movement to link progress in basic research to advances in application, described by the term "translational research." Translational research models emphasize the systematic translation of basic social science findings into rigorously tested interventions, the results of which in turn inform basic science (Pillemer, Suitor, and Wethington 2003). We suggest that translational research models are particularly appropriate for the field of intergenerational relations, and that such models can enhance both fundamental science and the development of interventions to improve family relationships in later life. Although it is typical for sociological articles on intergenerational relations to touch on policy or practice implications, there is in fact a substantial gap between research and application; in particular, interventions to improve family relations in later life are often not based on research evidence. We would argue that translational research models could be applied very productively to this field, bridging the gap between research themes discussed in this chapter and evidence-based intervention.

Acknowledgment This project was supported by grants from the National Institute on Aging (RO1 AG18869-01; 2RO1 AG18869-04), J. Jill Suitor and Karl Pillemer, Co-Principal Investigators. Karl Pillemer also acknowledges support from an Edward R. Roybal Center grant from the National Institute on Aging (1 P50 AG11711-01). Megan Gilligan and Jill Suitor also wish to acknowledge support from an NIH Pre-Doctoral Training Grant awarded to the Center on Aging and the Life Course at Purdue University (Kenneth Ferraro, Principal Investigator, T32AG025671). We would like to thank Abigail Howard, Seoyoun Kim, and Michael Steinhour for their assistance in the preparation of this chapter.

References

Adams, Bert. 1968. *Kinship in an Urban Setting.* Chicago, IL: Markham.

Albrecht, Ruth. 1953. "Relationships of Older People with Their Own Parents." *Marriage and Family Living* 4:296–98.

Aquilino, William S. 1997. "From Adolescent to Young Adult: A Prospective Study of Parent-Child Relations during the Transition to Adulthood." *Journal of Marriage and the Family* 59:670–86.

———. 1998. "Effects of Interview Mode on Measuring Depression in Younger Adults." *Journal of Official Statistics* 14:15–29.

———. 2006. "Family Relationships and Support Systems in Emerging Adulthood." Pp.193–217 in *Emerging Adults in America: Coming of Age in the 21st Century,* edited by J. J. Arnett and J. L. Tanner (Eds.) Washington, D.C.: American Psychological Association.

Aquilino, Williams and K. Supple. 1991. "Parent-Child Relations and Parental Satisfaction with Living Arrangements when Adult Children Live at Home." *Journal of Marriage and the Family* 53:13–27.

Bengtson, Vern L. and Edythe DeTerre. 1980. "Aging and Family Relations." *Marriage and Family Review* 3:51–76.

Bengtson, Vern L. and Joseph A Kuypers. 1971. "Generational Difference and the 'Developmental Stake.'" *Aging and Human Development* 2:249–60.

Bengtson, Vern L. and Robert E. L. Roberts. 1991. "Intergenerational Solidarity in Aging Families: An Example of Formal Theory Construction." *Journal of Marriage and the Family* 53:856–70.

Bengtson, Vern L. and Sandi S. Schrader. 1982. "Parent-Child Relations." Pp. 115–85 in *Handbook of Research Instruments in Social Gerontology, Vol. 2,* edited by, D. Mangen and W. Peterson. Minneapolis: University of Minnesota Press.

Bengtson, Vern L., Roseann Giarrusso, Beth J. Mabry, and Merril Silverstein. 2002. "Solidarity, Conflict, and Ambivalence: Complementary or Competing Perspectives on Intergenerational Relationships?" *Journal of Marriage and Family* 64:568–76.

Berry, Brent. 2006. "What Accounts for Race and Ethnic Differences in Parental Financial Transfers to Adult Children in the United States?" *Journal of Family Issues* 27:1583–1604.

Brody, Elaine. 2004. *Women in the Middle: Their Parent Care Years*, 2nd ed. New York: Springer.

Burnette, Denise. 1999. "Social Relationships of Latino Grandparent Caregivers: A Role Theory Perspective." *The Gerontologist* 39(1):49–58.

Chesley, Noelle and Kyle Poppie. 2009. "Assisting Parents and In-Laws: Gender, Type of Assistance, and Couples' Employment." *Journal of Marriage and Family* 71:247–62.

Cherlin, Andrew and Frank F. Furstenberg. 1986. *The New American Grandparent: A Place in the Family, A Life Apart.* Boston, MA: Harvard University Press.

Chodorow, Nancy J. 1978. *The Reproduction of Mothering*. Berkeley and Los Angeles, CA: University of California Press.

Choi, Namkee. 2003. "Coresidence between Unmarried Aging Parents and Their Adult Children: Who Moved in with Whom and Why." *Research on Aging* 25:384–404.

Cicirelli. Victor G. 1986. "The Relationship of Divorced Adult Children with Their Elderly Parents." *Journal of Divorce* 9(4):39–54.

Clemens, Audra and Leland J. Axelson. 1985. "The Not-So-Empty-Nest: The Return of the Fledging Adult." *Family Relations* 34(2):259–64.

Connidis, Ingrid A. 2003. "Divorce and Union Dissolution: Reverberations over Three Generations." *Canadian Journal of Aging* 22(4):353–68.

Cooney, Teresa. M and Peter Uhlenberg. 1992. "Support from Parents over the Life Course: The Adult Child's Perspective." *Social Forces* 71:63–84.

Coser, Rose L. 1966. "Role Distance, Sociological Ambivalence, and Transitional Status Systems." *The American Journal of Sociology* 72(2):173–87.

Davey, Adam and David Eggebeen 1998. "Patterns of Intergenerational Exchange and Mental Health." *Journal of Gerontology* 53B(2):86–95.

Davey, Adam, Corinna Jenkins Tucker, Karen Fingerman, and Jyoti Savla. 2009. "Within-Family Variability in Representations of Past Relationships with Parents." *Journal of Gerontology: Social Sciences* 64B:125–36.

Eggebeen, David J. and Dennis P. Hogan. 1990. "Giving Between Generations in American Families." *Human Nature* 1:211–22.

Eisenberg, Ann. 1988. "Grandchildren's Perspectives on Relationships with Grandparents: The Influence of Gender across Generations." *Sex Roles* 19(3):205–17.

Elder, Glen H. 1994. "Time, Human Agency, and Social Change: Perspectives on the Life Course." *Social Psychology Quarterly* 57(1), 4–15.

Fischer, Lucy Rose. 1983. "Mothers and Mothers-in-Law." *Journal of Marriage and Family* 45:187–92.

———. 1986. *Linked Lives: Adult Daughters and Their Mothers*. New York: Harper & Row.

Fleck, Carole. 2009. "All Under One Roof" *AARP Bulletin*, 50:24–25.

Freud, Sigmund. 1913. *Totem and Taboo*, edited by T. A. Brill. New York: New Republic.

Fuller-Thompson, Esme, Meredith Minkler, and Diane Driver. 1997. "Profile of Grandparents Raising Grandchildren in the United States." *The Gerontologist* 37(3):406–11.

Ganong, Lawrence, Marilyn Coleman, Annette K. McDaniel, and Tim Killian. 2009. "Attitudes Regarding Obligations to Assist an Older Parent or Stepparent Following Later-Life Remarriage." *Journal of Marriage and Family* 60:595–610.

Gans, Daphna and Merril Silverstein. 2006. "Norms of Filial Responsibility for Aging Parents across Time and Generations." *Journal of Marriage and Family* 68:961–76.

Giarrusso, Roseanne, Du Feng, Merril Silverstein, and Vern Bengtson. 2001. "Grandparent-Adult Grandchild Affection and Consensus." *Journal of Family Issues* 22(4):456–77.

Giarrusso, Roseanne, Michael Stallings, and Vern L. Bengtson. 1995. "The Intergenerational Stake Hypothesis Revisited: Parent-Child Differences in Perceptions of Relationships 20 Years Later." Pp. 227–96 in *Adult Intergenerational Relations: Effects of Societal Change*, edited by V. L. Bengtson, K. W. Schaie, and L. M. Burton. New York: Springer.

Gilligan, Carol. 1982. *In a Different Voice: Psychological Theory and Women's Development*. Cambridge, MA: Harvard University Press.

Gladstone, James W. 1988. "Perceived Changes in Grandmother-Grandchild Relations Following a Child's Separation or Divorce." *The Gerontologist* 28(1):66–72.

George, Linda. 1986. "Caregiver Burden: Conflict between Norms of Reciprocity and Solidarity." Pp. 67–92 in *Elder Abuse: Conflict in the Family*, edited by K. Pillemer and R. Wolf. New York: Auburn House.

Greenfield, Emily A. and Nadine F. Marks. 2006. "Linked Lives: Adult Children's Problems and Their Parents' Psychological and Relational Well-Being." *Journal of Marriage and the Family* 68:442–54.

Hagestad, Gunhild O. 1986. "Dimensions of Time and the Family." *American Behavioral Scientist* 29:679–94.

Hayslip, Bert, Jerald Shore, Craig E. Henderson, and Paul Lambert. 1998. "Custodial Grandparenting and the Impact of Grandchildren with Problems on Role Satisfaction and Role Meaning." *Journal of Gerontology* 53B:164–73.

Hill, Ruben. 1970. *Family Development in Three Generations: A Longitudinal Study of Changing Family Patterns of Planning and Achievement*. Cambridge, MA: Schenkman.

Hodgson, Lynne G. 1992. "Adult Grandchildren and Their Grandparents: The Enduring Bond." *International Journal of Aging and Human Development* 34(3):209–25.

Hofferth, Sandra L. 1984. "Kin Networks, Race, and Family Structure." *Journal of Marriage and the Family* 46:791–806.

Homans, George C. 1950. *The Human Group*. England: Harcourt, Brace.

Ingersoll-Dayton, Berit, Margaret B. Neal, Jung-Hwa Ha and Leslie B. Hammer. 2003. "Redressing Inequality in Parent Care among Siblings." *Journal of Marriage and Family* 65(1):201–12.

Jendrek, Margaret P. 1994. "Grandparents Who Parent Their Grandchildren: Circumstances and Decisions." *The Gerontologist* 34(2):206–16.

Kaufman, Gayle and Peter Uhlenberg. 1998. "Effects of Life Course Transitions on the Quality of Relationships between Adult Children and Their Parents." *Journal of Marriage and the Family* 60:924–38.

Kramer, Betty. 1997. "Gain in the Caregiving Experience: Where Are We? What Next?" *The Gerontologist* 37:218–32.

Koropeckj-Cox, Tanya. 2002. "Beyond Parental Status: Psychological Well-Being in Middle and Old Age." *Journal of Marriage and Family* 64:957–71.

Lazarsfeld, Paul F. and Robert K. Merton. 1954. "Friendship as Social Process: A Substantive and Methodological Analysis." Pp. 18–66 in *Freedom and Control in Modern Society*, edited by M. Berger, T. Abel, and C. H. Page. New York: Litton.

Lin, I-Fen. 2008. "Consequences of Parental Divorce for Adult Children's Support of Their Frail Parents." *Journal of Marriage and Family* 70(1):113–28.

Litvin, Sandra J. Steven M Albert, Elaine Brody M, and Christine Hoffman.1995. "Marital Status, Competing Demands, and Role Priorities of Parent-Caring Daughters." *Journal of Applied Gerontology* 14:372–390.

Luescher, Kurt and Karl Pillemer. 1998. "Intergenerational Ambivalence: A New Approach to the Study of Parent-Child Relations in Later-Life." *Journal of Marriage and the Family* 60:413–25.
Lye, Diane N. 1996. "Adult Child-Parent Relationships." *Annual Review of Sociology* 22:79–102
Mancini, Jay A., and Rosemary Blieszner. 1985. "Return of Middle Aged Children to the Parental Home." *Medical Aspects of Human Sexuality* 19:192–4.
Matthews, Sarah H. and Jetse Sprey. 1985. "Adolescents Relationships with Grandparents: An Empirical Contribution to Conceptual Clarification." *Journals of Gerontology* 40:621–6.
Merton, Robert K. and Elinor Barber. 1963. "Sociological Ambivalence." in *Sociological Theory: Values and Sociocultural Change*, edited by E. A. Tiryakian. New York: Free Press.
Merrill, Deborah M. 1996. "Conflict and Cooperation among Adult Siblings during the Transition to the Role of Filial Caregiver." *Journal of Social and Personal Relationships* 13:399–413.
———. 2007. *Mothers-in-law and Daughters-in-law: Understanding the Relationship and What makes them Friends or Foe*. Westport, CT: Praeg.
Minkler, Meredith and Esme Fuller-Thomson. 2000. "Second Time Around Parenting: Factors Predictive of Grandparents Becoming Caregivers for Their Grandchildren." *International Journal of Aging and Human Development* 50:185–200.
Mutran, Elizabeth. 1985. "Intergenerational Family Support among Blacks and Whites: Response to Culture or to Socioeconomic Differences." *Journals of Gerontology* 40:382–9.
Neugarten, Bernice 1970. "The Old and Young in Modern Societies." *American Behavioral Scientist* 14:13–24.
Nydegger, Corrine N. 1986. "Timetables and Implicit Theory." *American Behavioral Scientist* 29:710–29.
Pavalko, Eliza K. and Julie E. Artis. 1997. "Women's Caregiving and Paid Work: Causal Relationships in Late Midlife. *Journal of Gerontology: Social Sciences* 52B:S170–9.
Pearlin, Leonard I. 1989. "The Sociological Study of Stress." *Journal of Health and Social Behavior* 30:241–56.
Pillemer, Karl and J. Jill Suitor. 1991. "Will I Ever Escape My Child's Problems?' Effects of Adolt Childern's Problems on Elderly Parente. *Journal of Marriage and the Family* 53:585–594.
———. 2002. "Explaining Mothers' Ambivalence toward Their Adult Children." *Journal of Marriage and Family* 64:602–13.
———. 2005. "Ambivalence in Intergenerational Relations over the Life-Course." Pp. 3–28 in *Annual Review of Gerontology and Geriatrics, Vol. 2: Intergenerational Relations across Time and Place*, edited by M. Silverstein. New York: Springer.
———. 2008. "Collective Ambivalence: Considering New Approaches to the Complexity of Intergenerational Relations." *Journals of Gerontology: Social Sciences* 63:394–6.
Pillemer, Karl, J. Jill Suitor, Steven Mock, Myra Sabir, and Jori Sechrist. 2007. "Capturing the Complexity of Intergenerational Relations Exploring Ambivalence within Later-Life Families." *Journal of Social Issues* 63:775–91.
Pillemer, Karl, J. Jill Suitor, and Elaine Wethington. 2003. "Integrating Theory, Basic Research, and Intervention: Two Case Studies from Caregiving Research." *The Gerontologist* 43:19–28.
Powers, Rebecca, Jill. J. Suitor, Susana. Guerra, Monisa Shackelford, Dorothy Mecom, and Kim Gusman. 2004. "Regional Differences in Gender Role Attitudes: Variations by Gender and Race." *Gender Issues* 21:40–54.
Pruchno, Rachel. 1999. "Raising Grandchildren: The Experiences of Black and White Grandmothers." *The Gerontologist* 39(2):209–21.
Pruchno, Rachel A. and Dorothy McKenney. 2002. "Psychological Well-Being of Black and White Grandmothers Raising Grandchildren: Examination of a Two-Factor Model." *Journal of Gerontology* 57:444–52.
Pudrovska, Tetyana. 2009. "Parenthood, Stress, and Mental Health in Late Midlife and Early Old Age." *The International Journal of Aging and Human Development* 68:127–47.
Riley, Matilda white. 1987. "On the Significance of Age in Sociolgy: American Sociological Association, 1986 Presidential Address." *American Sociological Review* 52:1–14.
Rossi, Alice S. and Peter H. Rossi. 1990. *Of Human Bonding: Parent-Child Relations across the Life Course*. Hawthorne, NY: Aldine de Gruyter.
Sands, Roberta G. and Robin Golberg–Glen. 2000. "Factors Associated with Stress among Grandparents Raising Their Grandchildren." *Family Relations* 49:97–105.
Sarkisian, Natalia, and Naomi Gerstel. 2004. "Kin Support among Blacks and Whites: Race and Family Organization." *American Sociological Review* 69:812–37.
Scott, Mindy E., Alan Booth, Valerie King, and David R. Johnson. 2007. "Postdivorce Father-Adolescent Closeness." *Journal of Marriage and the Family* 69:1194–1209.
Sechrist, Jori, J. Jill Suitor, Angela Henderson, Krista Cline, and Michael Steinhour. 2007. "Regional Differences in Mother–Adult-Child Relations: A Brief Report." *Journal of Gerontology: Social Sciences* 62(6):S388–91.
Seltzer, Marsha M., Jan S. Greenberg, Gael I. Orsmond, Julie. Lounds, and Matthew J. Smith. 2008. "Unanticipated lives: Inter- And Intra-Generational Relationships in Families with Children with Disabilities. Pp. xx-xx in *Intergenerational Caregiving*, edited by A. Booth, A. Crouter, S. Bianchi, and J.A. Seltzer. Washington, D.C.: Urban Institute Press.

Shapiro, Adam. 2004. "Revisiting the Generation Gap: Exploring the Relationships of Parent/Adult-Child Dyads." *The International Journal of Aging and Human Development* 58:127–46.
Silverstein, Merril and Vern L. Bengtson. 1997. "Intergenerational Solidarity and the Structure of Adult Child-Parent Relationships in American Families." *American Journal of Sociology* 103:429–60.
Silverstein, Merril, Xuan Chen., and Kenneth Heller. 1996. "Too Much of a Good Thing? Intergenerational Social Support and the Psychological Well-Being of Aging Parents." *Journal of Marriage and the Family* 58:970–82.
Silverstein, Merril, Stephen J. Conroy, Haitao Wang, Roseann Giarrusso, and Vern Bengtson. 2002. "Reciprocity in Parent-Child Relations over the Adult Life Course." *Journal of Gerontology: Social Sciences* 57B:S3–13.
Silverstein, Merril and Roseanne Giarrusso, eds. In Press. *From Generation to Generation: Continuity and Discontinuity in Aging Families*. Baltimore, MD: Johns Hopkins University Press.
Silverstein, Merril and Anne Marenco. 2001. "How Americans Enact the Grandparent Role." *Journal of Family Issues* 22: 493–522.
Settersten, Richard A. 2003. "Age Structuring and the Rhythm of the Life Course." Pp. 81-102 in *Handbook of the Life Course*, edited by J. Mortimer and M. Shanahan. New York: Kluwer Academic/Plenum.
Spitze, Glenna and John R. Logan. 1990. "More Evidence on Women (and Men) in the Middle." *Research on Aging* 12:182–98.
Spitze, Glenna, John. R. Logan, Glenn Deane, and Suzanne Zerger. 1994. "Adult Children's Divorce and Intergenerational Relationships." *Journal of Marriage and the Family* 56:279–93.
Stack, Carol. 1974. *All Our Kin: Strategies for Survival in Black Communities*. New York: Harper & Row.
Suitor, J. Jill. 1987. "Mother-Daughter Relations When Married Daughters Return to School: Effects of Status Similarity." *Journal of Marriage and the Family* 49:355–44.
Suitor, J. Jill, Megan Gilligan, Karl Pillemer. 2009. "Are You Sure I'm Ambivalent?' Conceptualizing and Measuring Intergenerational Ambivalence in Later-Life Families." Presented at the annual meeting of the Gerontological Society of America, Atlanta, GA.
Suitor, J. Jill, and Karl Pillemer. 1987. "The Presence of Adult Children: A Source of Stress for Elderly Couples' Marriages?" *Journal of Marriage and the Family* 49:717–25.
———. 1988. "Explaining Intergenerational Conflict when Adult Children and Elderly Parents Live Together." *Journal of Marriage and the Family* 50:1037–47.
———. 1990. "Transition to the Status of Family Caregiver: A New Framework for Studying Social Support and Well-Being." Pp. 310–20 in *The Legacy of Longevity: Health, Illness, and Long-Term Care in Later Life*, edited by S. M. Stahl. Beverly Hills: Sage.
———. 1994. "Family Caregiving and Marital Satisfaction: Findings from a One-Year Panel Study of Women Caring for Parents with Dementia." *Journal of Marriage and the Family* 56:681–90.
———. 2006. "Choosing Daughters: Exploring Why Mothers Favor Adult Daughters over Sons" *Sociological Perspectives* 49:139–60.
———. In press. "Differences in Mothers' and Fathers' Parental Favoritism in Later-Life: A Within-Family Analysis." In *From Generation to Generation: Continuity and Discontinuity in Aging Families*, edited by M. Silverstein and R. Giarrusso. Baltimore, MD: Johns Hopkins University Press.
Suitor, J. Jill, Karl Pillemer, and Shirley Keeton. 1995. "When Experience Counts: The Effects of Experiential and Structural Similarity on Patterns of Support and Interpersonal Stress." *Social Forces* 73:1573–88.
Suitor, J. Jill, Karl Pillemer, Shirley Keeton, and Julie Robison. 1995. *Aged Parents and Aging Children: Determinants of Relationship Quality*. Westport, CT: Greenwood.
Suitor, J. Jill, Karl Pillemer, and Jori Sechrist. 2006. "Within-Family Differences in Mothers' Support to Adult Children." *Journal of Gerontology: Social Science* 16B:S10–17.
Suitor, J. Jill, Jori Sechrist, and Karl Pillemer. 2007. "When Mothers Have Favorites: Conditions under Which Mothers Differentiate among Their Adult Children." *Canadian Journal on Aging* 26:85–99.
Suitor, J. Jill, Jori Sechrist, Mike Steinhour, and Karl Pillemer. 2006. "'I'm Sure She Chose Me!' Consistency in Intergenerational Reports of Mothers' Favoritism in Later-Life Families." *Family Relations 55:*526–38.
Swartz, Teresa T. 2009. "Intergenerational Family Relations in Adulthood: Patterns, Variations, and Implications in the Contemporary United States." *Annual Review of Sociology* 35:191–212.
Szydik, Marc. 2008. "Intergenerational Solidarity and Conflict." *Journal of Comparative Family Studies* 1:97–114.
Szinovacz, Maximiliane E. Stanley DeViney, and Maxine P. Atkinson 1999. "Effects of Surrogate Parenting on Grandparents' Well-Being." *Journal of Gerontology: Social Sciences* 54B:S376–88
Taylor, Robert. 1986. "Receipt of Support from Family among Black Americans: Demographic and Familial Differences." *Journal of Marriage and the Family* 48:67–77.
Troll, Lillian E. 1971. "The Family of Later Life: A Decade Review." *Journal of Marriage and Family* 33:263–90.
Umberson, Debra. 1992. "Relationships between Adult Children and Their Parents: Psychological Consequences for Both Generations." *Journal of Marriage and the Family* 54:664–74.
———. 2006. "Parents, Adult Children, and Mortality." *Context* 5:48–53.
US Bureau of the Census. 2010. *2010 Census*. Washington, D.C.: U.S. Government Printing Office.

Ventura, Stephanie J. 2009. *Changing Patterns of Nonmarital Childbearing in the United States*. NCHS data brief, no 18. Hyattsville, MD: National Center for Health Statistics.
Wakabayashi, Chizuko and Katherine M. Donato. 2006. "Does Caregiving Increase Poverty Among Women in Later Life? Evidence from the Health and Retirement Survey." *Journal of Health and Social Behavior* 47:258–74.
Walker, Alexis J. Clara C Pratt, and Nancy Chun Oppy. 1992. "Perceived Reciprocity in Family Caregiving." *Family Relations* 41:82–5.
Ward, Russell, John Logan, and Glenna Spitze. 1992. "The Influence of Parent and Child Needs on Coresidence in Middle and Later Life." *Journal of Marriage and the Family* 54:209–21.
Ward, Russell, Glenna Spitze, and Glenn Deane. 2009. "The More the Merrier? Multiple Parent-Adult Child Relations" *Journal of Marriage and Family* 71:161–73.
Weaver, W. 1948. "Science and Complexity." *American Scientist* 36:536.
Wethington, Elaine and Claire M, Kamp Dush. 2007. "Assessments of Parenting Quality and Experiences across the Life Course." *Advances in Life Course Research* 12:123–52.
Wilson, Andrea E., Kim M.Shuey, Glen H. Elder. Jr., and K. A. S. Wickrama. 2006. "Ambivalence in Parent-Adult Child Relations: A Dyadic Analysis." *Social Psychology Quarterly* 69:235–52.
Winch, Robert F. 1970. "Permanence and Change in the History of the American Family and Some Speculations as to the Its Future." *Journal of Marriage* 32:6–15.

Chapter 12
The Midlife Financial Squeeze: Intergenerational Transfers of Financial Resources Within Aging Families

R. Corey Remle

Due to demographic changes such as increased longevity and decreased family size, midlife adults can now expect to spend more years relating to both ascendant and descendent generations sharing co-biographical experiences (Soldo 1996; Uhlenberg 1996 Bengtson et al. 1990). Intergenerational family relations are becoming increasingly important to Americans in response to these demographic shifts (Bengtson 2001). We must seek to understand "families as context" – that is, as dynamic units with members who age concurrently and share co-biographical experiences – rather than examining the isolated life course patterns of individuals (Davey et al. 2005; Hagestad 2003). Studying "families as context" suggests that intergenerational financial exchanges deserve closer attention as concerns rise about job security, postsecondary education costs, retirement savings, and public support programs such as Social Security and Medicare.

Life course researchers must consider how multigenerational families respond collectively to economic shocks and transitions into and out of financial self-sufficiency that are experienced by one or more family members. The type, timing, and extent of support exchanges vary widely, and they are moderated by structurally ambivalent expectations and norms that influence family decision-making regarding the distribution of private resources across generations (Eggebeen and Hogan 1990; Rossi and Rossi 1990). It is unclear if, when, and how midlife adults should provide assistance to children during the transition to adulthood or to aging parents with reduced financial resources, but multiple public policies implicitly rely on the middle generation to support ascendant and descendent family members in need (Schoeni and Ross 2005; Hooyman and Gonyea 1995). Additionally, the economic costs to donors are rarely considered in sociological studies of intergenerational support.

Financial self-sufficiency is the ability to meet one's financial obligations (e.g., mortgages, credit card debts, and daily living expenses) without the need for monetary assistance from others. This can be a precarious balancing act for individuals who may have a spillover effect for others in one's kin network. In particular, the middle generation (broadly defined as aged 40–65) must adapt to a financial "midlife squeeze" that involves the conflicting responsibilities of (1) preparing for its future financial self-sufficiency via retirement savings; (2) providing support for ascendant generation members who experience declines in their financial self-sufficiency; and/or (3) assisting descendent generation members who have not yet gained full financial self-sufficiency themselves (Settersten 2007; Soldo 1996; Cheal 1983). Oftentimes, the sociological literature fails to acknowledge how financial support to family members can hinder necessary retirement preparations for those in the middle generation. To be sure, other types of intergenerational support (e.g., emotional support, advice, and practical assistance) are significant research topics in their own right that have received much attention in the literature on kin relationships (for examples, see Sarkisian and Gerstel 2004; Logan and

R.C. Remle (✉)
Department of Sociology, Wake Forest University, Winston-Salem, NC, USA
e-mail: remlerc@wfu.edu

R.A. Settersten, Jr. and J.L. Angel (eds.), *Handbook of Sociology of Aging*, Handbooks of Sociology and Social Research, DOI 10.1007/978-1-4419-7374-0_12, © Springer Science+Business Media, LLC 2011

Spitze 1996; Cooney and Uhlenberg 1992; Rossi and Rossi 1990). However, as these types may be given while individuals remain financially self-sufficient or in conjunction with material assistance (and can admittedly influence the flow of monetary exchanges), the focus on transitions into and out of financial independence requires that monetary transfers deserve separate attention.

This chapter has five sections. First, I describe intergenerational financial transfers – focusing specifically on inter vivos gifts or loans. Second, I discuss how various theoretical approaches are used to explain factors that influence the motives and transfer behaviors of middle generation donors in their family roles as parents *and* children. Third, I identify several life course transitions where ambiguity creates challenges for determining donors' responsibilities to assist across generational lines – the transition to adulthood, the transition out of financial self-sufficiency for elders, and marital status transitions. Fourth, I explain the significant potential costs that public policies in the United States force upon the middle generation to provide financial assistance to other generations in their social roles as both parents and children within an aging population. Finally, I suggest two areas for future research to expand our understanding of the life course effects of inter vivos transfers.

Intergenerational Financial Transfers

The most common explanation for intergenerational financial transfers is that they offer assurance for family members' financial self-sufficiency and overall well-being via a private safety net (Kohli and Künnemund 2003; Eggebeen and Davey 1998; Eggebeen and Hogan 1990). Financial status transitions are never smooth or easy. Instead, the process may involve difficult periods where an individual moves into and out of financial self-sufficiency multiple times and may receive financial support from others specifically as a means to return to a stable financial position. Material resources can flow in two directions across generations either "upward" to older family members or "downward" to descendents (Soldo and Hill 1993). Money transfers come in two forms – inter vivos transfers and bequests. Inter vivos transfers are intermittent and routine monetary gifts or loans. (See Angel (2008) for an excellent review of the literature about families, bequests, and related social policies.)

Assistance to adult children is the most common purpose for inter vivos transfers. The support may be used for educational costs, living expenses, sudden crises, and/or "upfront costs" to attain financial self-sufficiency (Schoeni and Ross 2005; Wong et al. 1999; Eggebeen 1992). A longitudinal examination of downward transfers found that 63% of all households in the Health and Retirement Study (HRS) gave money at least once between 1992 and 1998 (Shapiro and Remle 2011). From the viewpoint of recipients, Remle and O'Rand (forthcoming) demonstrated that 36.2% of adult children received one or more transfers between 1992 and 1998. This is in general agreement with Schoeni and Ross (2005) who found 34% of adult children received cash assistance, using data from the 1988 Panel Study of Income Dynamics (PSID). As adult children age, the likelihood they will receive assistance declines rapidly. For example, over 65% of adult children in their early 20s received money for schooling and/or general living expenses while 40% of those in their late 20s and early 30s, and less than one-tenth above age 35 received assistance from midlife parents (Schoeni and Ross 2005).

By comparison, upward financial transfers are rare. Estimates indicate that only 2–7% of elderly parents receive assistance from their children (Lin 2008; Furstenberg et al. 1995; Eggebeen 1992). Instead, a much greater percentage of upward intergenerational support is provided via time transfers such as caregiving for parents in declining health and practical assistance with various activities of daily living (Grundy and Henretta 2006; Sarkisian and Gerstel 2004; Furstenberg et al. 1995). Oftentimes, monetary assistance to parents varies based on the marital status and gender of the recipient. Lin (2008) found a prevalence rate of 3.9–5.1% for transfers to divorced mothers while the range for divorced fathers was 2.4–3.6%. Furstenberg et al. (1995) also showed that transfers were unevenly distributed as more widowed and divorced mothers received financial assistance (5.3 and 3.0%, respectively) than married elderly couples, widowed fathers, and divorced fathers. When they included coresident parents who received monetary support as well, McGarry and Schoeni (1995) reported that 7.1% of elderly parents received financial assistance from midlife adults. Many more coresident parents received assistance than nonresident parents.

Multiple studies have demonstrated general consistency in the amounts transferred to other generations when education costs for adult children are excluded. For the HRS sample, Cao (2006) estimated that the average transfer to adult children in 1996 was $1,585, and in 2000, it was $1,519. For the 1988 PSID sample, Furstenberg et al. (1995) estimated the average transfer values to be $1,550 for adult children and $1,194 for elderly parents. In one of the earliest studies known to estimate transfer values, Cox and Raines (1985) found an average intergenerational transfer in 1979 to be $2,100. It is important to recognize that while cohort comparisons indicate similar amounts, inflation significantly alters the financial value of transfers across time. While transfers estimated by Cao (2006) equate to approximately $2,200 in 2010 constant dollars, the financial value of average transfers in 1979 would equal nearly $7,000 today. If middle-aged donors had instead made sound investments for retirement planning, their future financial status would have significantly better prospects.

Once they complete the transition to adulthood, many individuals may never experience a transition out of financial self-sufficiency that requires assistance from family members. However, job loss, divorce, or health-related traumas may create economic shocks that can lead individuals to reach out to family members (Rossi and Rossi 1990). Gradual transitions (e.g., postsecondary education, job market entry, retirement, disability, or entry into institutional care) can also elicit financial assistance from family members (Swartz 2009; Fingerman et al. 2011). In response to these economic shocks and transitions for older or younger generations, it is becoming increasingly common for the middle generation to provide some level of support for *either* older parents *or* adult children at any one time (and, in rare cases, both generations) (Kronebusch and Schlesinger 1994; Silverstein 2006). Research about precautionary saving among middle-aged adults has demonstrated insufficient knowledge and preparedness for retirement-associated expenses (Lusardi et al. 2009; Sullivan et al. 2000). As shown above, this does not stop some from providing financial assistance to other generations despite future repercussions, and sociologists of aging can play an important role in understanding why.

Theoretical Explanations for Inter Vivos Transfers

In this section, I review several theories that have been used by sociologists to explain family transfer behaviors. The heterogeneity of within-family contexts combined with the variety of transfer types requires a broad consideration of how the theoretical approaches complement each other in relation to intergenerational financial transfers. There are several limitations in the existing literature that should be acknowledged first. One challenge to understanding financial transfers is that many studies have examined multiple types of support simultaneously or have combined types of support to examine "intergenerational support" as a broader category (e.g., Davey et al. 2005; Sarkisian and Gerstel 2004). The high level of interdependence between support types affects how families decide to distribute time, space, and financial resources most effectively, particularly for families low on resources in one of the three domains (Silverstein 2006). However, these methods limit what we can know about financial transfers specifically. Another challenge is determining how to examine exchanges across the life course compared to measuring transfers at one point in time. One response is to use multilevel modeling techniques with longitudinal data (Remle and O'Rand forthcoming; Fingerman et al. 2011). Additional studies have examined reciprocity between parents and children by determining how receipt of financial transfers in early adulthood influenced subsequent upward transfers of caregiving or social support by midlife adults (e.g., Henretta et al. 1997; Silverstein et al. 2002). Other researchers have studied parents' *expectations* of future monetary, caregiving, or social support in return for financial assistance in the present (Eggebeen and Davey 1998; Ganong and Coleman 2006). Even so, recipient characteristics, family structure, family-oriented transitions (e.g., parental divorce), and normative expectations have been the most common contextual factors examined by researchers in this area to date (Davey et al. 2005, Shapiro and Remle 2011). However, this has diverted attention away from donors' decision-making processes and the effects on preparations for their own financial futures.

Contingency Theory

Researchers have defined contingency theory as sharing resources across generational lines based upon the recipients' resource needs and the availability of donors' resources (Fingerman et al. 2009; Eggebeen 1992; Rossi and Rossi 1990). The theory is not always explicitly stated by researchers because it is often implied in the underlying purpose – identifying under what conditions financial *assistance* is provided. A contingency-based support system is often assumed as part of a familistic social contract – both in public policies and in privately held beliefs within families (Hooyman and Gonyea 1995; Riley and Riley 1993). Because of their higher likelihood for socioeconomic and labor force stability as well as their position in the extended kin network, the middle generation is much more likely to assist others than to receive assistance themselves (Davey et al. 2005). As we might expect, parents with higher incomes and higher educational attainment are more likely to assist adult children financially (Remle and O'Rand forthcoming; Fingerman et al. 2011; Eggebeen and Davey 1998; Eggebeen 1992). However, donors' wealth resources, risk assessments of their own life expectancy, assessments of their existing or planned retirement resources or their financial literacy have never been considered as antecedents of financial transfers, and these contingencies should be considered relevant to family decision-making processes. Recipient needs and the availability of donor resources are not sufficient comprehensive explanations, so theorists find it important to understand also why transfers are made.

Within the contingency approach, two competing explanations have been given for donors' motives: the altruism hypothesis and the exchange hypothesis. Higher levels of emotional closeness and attachment between family members lead some researchers to hypothesize that inter vivos transfers are altruistically motivated (Becker 1991; MacDonald and Kuo 2003). The second explanation proposes that donors give financial assistance while expecting reciprocation with a support transfer of similar value in the future through contact, caregiving, money, or social support (Altonji et al. 1997; Cox and Rank 1992). In many of these studies, researchers have indirectly surmised parental motives as altruistic or self-interested exchange based on the static contexts immediately surrounding transfer behaviors.

Tests of each hypothesis have provided mixed results. Some have shown unmarried adult children and those with lower incomes are more likely to receive assistance, suggesting an altruistic response to greater financial need (Altonji et al. 1997; Furstenberg et al. 1995; McGarry and Schoeni 1995). For upward transfers, Hogan et al. (1993) found that adult children gave more financial support to parents who were in poor health or widowed compared to children with healthy or married parents, also suggesting altruistic motives. MacDonald and Koh (2003) argued that all upward transfers to parents are altruistic because there is a lower likelihood of reciprocal exchange from dependent elderly family members.

On the other hand, a long-term view of the family life course reveals that reciprocal exchanges have been severely underestimated by gerontologists, and this erroneously enhances the view of the elderly as dependent on other generations when many elderly individuals actually remain independent and may reliably provide assistance of other types to younger generations (Spitze and Logan 1992). Goldscheider et al. (2001) found that unmarried mothers planning to assist children financially favored daughters over sons because they believed daughters were more likely to provide care or practical assistance if the mothers became ill or disabled. Ganong and Coleman (2006) showed that expectations to assist elderly parents financially were higher among middle-aged adults who viewed inter vivos transfers as reciprocity for past exchanges. Earlier financial transfers from parents have been shown to increase the likelihood of caregiving and social support from adult children later in their respective lives (Henretta et al. 1997; Silverstein et al. 2002).

The life course approach and longitudinal research designs allow altruism and exchange motives to coexist. Both motives operate at various stages of the family life course because the timing, frequency, and types of family transfers vary so widely (Silverstein 2006). While there may be altruistic motives for one transfer, subsequent transfers or transfers of another currency that may be considered equitable to financial assistance (e.g., time, space, or social support) may include reciprocity expectations for "insurance" in later life (Silverstein et al. 2002). For example, intermittent financial assistance to adult children who experience short-term economic shocks through unemployment or a divorce may be altruistically motivated, but parents may find the assistance to be reciprocated by adult children later (Ganong and Coleman 2006; Henretta et al. 1997). Altruistic motives guiding the redistribution of family resources may reflect the intensity of intergenerational ties but redistribution planning by family members may also affect how such ties develop over time if financial giving is accompanied by reciprocity expectations (Kronebusch and Schlesinger 1994; Silverstein et al. 2002). Thus, theoretical developments regarding family life course dynamics suggest that the dichotomization of motives may be unnecessary.

Intergenerational Solidarity Theory

The term "intergenerational solidarity" represents a multidimensional model regarding the interdependence and intensity of ties between family members. Solidarity reflects the underlying potential of members to share instrumental, emotional, and other resources that can be activated primarily when someone within the network is in need (Silverstein et al. 1997). This theoretical model has been the primary approach applied to within-family support across extended kin over the past 25 years. It has enhanced family and aging research profoundly because its core principles acknowledge the complexity of intergenerational relationships as well as their likelihood to endure changes in individuals' lives across the life course (Bengtson et al. 2002). Much of the solidarity-based literature describes the social cohesiveness and emotional attachment between donors and recipients, and/or family structural factors that enhance or impede the likelihood of inter vivos transfers. Researchers have also examined influential demographic characteristics such as race, ethnicity, age, and socioeconomic status. (Due to space constraints, these factors cannot be addressed here. For excellent reviews, see Davey et al. (2005) and Swartz (2009).) The broad applicability of the theory across many types of support exchanges allows researchers to examine multiple contexts and family structures, and the depth of its development over time makes intergenerational solidarity theory compatible with contingency theory and both motive hypotheses.

Three of the six dimensions of the solidarity model represent what Silverstein et al. (1997) labeled "latent solidarity" and are more related to affinity between individuals. These dimensions are emotional closeness (affective), shared opinions (consensual), and the dialectic between familistic and individualistic orientations (normative). The other dimensions of "manifest solidarity" primarily represent behavioral processes. The level of contact family members have with each other (association) and co-residence or living in geographic proximity (structural) represent opportunity structures.

Finally, functional solidarity – the dimension that includes inter vivos transfers – represents instrumental assistance between generations. A latent class analysis of intergenerational solidarity in families yielded five types (Silverstein et al. 1997). Tight-knit families were those who shared higher affinity, more opportunity structures, and were more likely to exchange assistance across generational lines. The sociable and "intimate but distant" types were less likely to engage in functional assistance, but the authors point out potential recipients may not need assistance (i.e., low contingency), which may be one key reason that these family types could be differentiated from tight-knit families. The detached family type was least likely to have assistance-oriented relationships. Finally, though they were significantly less likely to exchange time or money than to do so at all, the frequency of exchanges was still higher for obligatory families than for sociable and "intimate but distant" families, suggesting intimate bonds may be strong without being positive in nature.

Ambivalence Theory

As a counterpoint to the emphasis on the strength of family bonds inherent to solidarity theory, Lüscher and Pillemer (1998) theorized that conflicts and ambivalent feelings are common in multigenerational relationships and that they influence support exchanges. Because of its life course emphasis, this theory is also compatible with contingency theory, altruism, and the exchange hypothesis. Ambivalence theorists address the lack of mutual understanding between family members, the mix of positive and negative feelings individuals may have for other family members, and the emotional and practical trade-offs that affect decision-making about sharing resources (Connidis and McMullin 2002). For example, donors' favoritism toward one child over others indicates parental ambivalence (Fingerman et al. 2009; Suitor et al. 2006; Goldscheider et al. 2001). For individuals who experience the transition out of financial self-sufficiency, the loss of autonomy also presents emotional and practical strains that are likely to become more common for many families as the population ages and older adults' retirement incomes suffer significant losses as part of the recent global recession.

Beyond the psychosocial component, Connidis and McMullin (2002) proposed ambivalence theory should also include "sociological ambivalence" – the conflicts, contradictions, and confusion that are embedded in ambiguous social structures about how families *should* behave. The influences of public policies and socially determined moral beliefs about assistance to needy family members become manifested in their intimate social and emotional ties. Multigenerational families must negotiate their relationships and uncertain fiscal responsibilities in the midst of costly life course transitions such as an older relative's entry into institutional care or a young adult's advancement to college. Ambiguity about familial responsibility in such situations forces many middle-generation kin into difficult decisions regarding the availability of private resources and their willingness to relinquish some financial security as a trade-off for helping other family members (Kohli and Künemund 2003). Donors must decide how to combine potential public assistance (e.g., Medicaid, college loan programs) with family-based financial support and usually must do so with incomplete information. For families today, preparations for others' financial status transitions and the redistribution of family resources must be flexible, are often context-dependent, and may be structured, at least in part, by sociological ambivalence.

Theoretical Synthesis

Next, we must bring together ideas from each theoretical explanation to best explain family transfer behaviors comprehensively. First, it is generally safe to assume that support exchanges are made contingent upon recipients' needs and the availability of donors' resources. Notably, a small percentage of transfers occur in wealthy families that are not contingent upon need. However, contingency theory is limited by the relativity of defining "need" for recipients and the challenges of conflicting needs within families that require multiple contingencies be balanced. Second, lifetime emotional bonds and normative solidarity may be strengthened through inter vivos transfers, but functional solidarity reflects a dialectical juxtaposition between dependence and autonomy (Bengtson et al. 2002). This may be difficult to differentiate from the emotional and practical strains within families that can be associated with psychological ambivalence. Structural changes surrounding families such as marital instability, extended transitions into adulthood, and the risks associated with potential transitions out of financial self-sufficiency challenge the reach of intergenerational solidarity theory to offer comprehensive explanations for monetary support (Remle and O'Rand, forthcoming; Soldo 2006 1996; Wong et al., 1999). Finally, conflict and ambivalence are characteristic of family interactions and ambiguous social responsibilities (Connidis and McMullin 2002). In other words, intergenerational financial assistance is (1) most commonly contingent on needs and resource availability, (2) may be a function of solidarity within the family but (3) may also create conflict and may be affected by social structural factors that obscure the normative responsibilities of intergenerational support.

However, even this synthesis of theoretical approaches does not allow researchers to determine accurately the eventual financial costs for donors. Conceptually, we do not recognize or give sufficient attention to tradeoffs made by one generation in order to support another generation. The current global economic recession has weakened the long-term financial stability of lower-class and middle-class families. Costs have increased substantially for postsecondary education, insurance coverage, and long-term health care needs. Subsequently, for many families, one generation is more likely to be "squeezed" financially and must make tradeoffs such as reductions in overall wealth, lowered retirement savings, or limited preparation for uncertain future financial needs. This midlife financial squeeze will be more likely to occur as the population ages.

Ambiguous Transitions and Ambiguous Responsibilities

The Transition to Financial Self-Sufficiency

Establishing financial self-sufficiency and independence from one's parents is the pivotal experience identified as indicative of "adulthood" (Arnett 2004). However, social structural factors such as an increasing need for postsecondary education and difficulties in labor market entry (particularly during the recent economic recession and slow recovery) can significantly alter the timing of the transition to self-sufficiency. Additionally, the sociological ambivalence regarding when and how the transition to financial self-sufficiency is complete originates from structural contradictions as to how long parents are expected to support adult children financially, affecting parents' own financial statuses in the process. Schoeni and Ross (2005) demonstrated that parents spend approximately one-third again as much in monetary assistance to adult children aged 18–34 as they had spent raising their children.

Parents may view their material support to children as a means to provide a quicker route to financial self-sufficiency by supplementing incomes or paying "up-front" costs of housing, transportation, and insurance (Swartz 2009). Shapiro (2004) describes these as "transformative assets" that enhance children's abilities to gain human and financial capital for further gains than those whose parents cannot provide resources. Little has been said, however, of what midlife parents may be giving up as a result of providing support to others in relation to their future financial needs as aging adults (i.e., retirement, living expenses, health care, and insurance costs). When midlife parents provide financial assistance to adult children, the future value for their own financial status is rarely considered despite the opportunity costs involved.

The Transition out of Financial Self-Sufficiency

Fingerman et al. (2011) found that middle-aged adults were more likely to provide financial assistance to elderly parents who experienced crises (e.g., financial problems, crime) or to parents with functional disabilities rather than to help them meet everyday needs, suggesting contingency-based support. As described above, monetary transfers to address the transition out of financial self-sufficiency are rare in the United States. There are several possible reasons for this. First, many adults remain relatively healthy into old age and are able to provide sufficient care for themselves (and/or their spouses). Though they may experience increased disability over time and develop greater caregiving needs, health declines may be slow enough not to disrupt financial independence. Second, in our age-graded social system, elderly adults become eligible to receive funds from private pensions or annuities as well as public Social Security pensions. Combined with Medicare coverage for many formal healthcare needs, this mix of resources allows most elderly adults to maintain financial self-sufficiency beyond 75 years of age. Third, the caregiving literature has dominated research on intergenerational support to elderly adults as many middle-aged adults who help parents may be using time and coresidence as their initial "currencies of choice" and providing financial transfers only in extreme cases. Finally, financial transfers to older generations are less likely to have been examined than other types of transfers and especially less likely to have been examined separately from time and space transfers. The necessary data may simply not yet be available for researchers to examine upward transfers more fully but with an aging population, such research is more likely to occur in the near future.

The Influence of Marital Status Transitions

The widening diversity of intergenerational kinship structures in recent decades also encourages sociological ambivalence for financial status transitions. An increased prevalence of remarriages and blended families accompanied the increased divorce rate (Cherlin and Furstenberg 1994). The lack of social norms regarding how much responsibility the middle generation has toward adult children from previous marriages or their divorced or widowed elderly parents has led to ambivalence about financial and social obligations in exchanges within multigenerational kinship structures (Lin 2008; Killian and Ferrell 2005; Furstenberg et al. 1995). Parental divorce when children are young results in significant declines in intergenerational solidarity between parents and children, and especially between noncustodial parents and their children (Silverstein et al. 1997; White 1994). Other studies have found that parental divorce results in lower levels of affection and reduced feelings of obligation (Ganong et al. 1995; Cooney and Uhlenberg 1992). Obligations to assist biological children are felt more strongly than those to assist stepchildren – though not to the exclusion of stepchildren (Aquilino 2005; Killian 2004). Rossi and Rossi (1990) attributed this to what they called "differentiated norms of family obligations."

How Public Policies Influence Private Family Transfers

Current public policies may have an extensive impact on private family financial resources over the long term that puts current midlife and retirement-age individuals at risk to remain financially independent into their later years. Despite the common definition of college-age children (18–23 years old) as adults, various events associated with the transition to adulthood (e.g., postsecondary education, leaving the parental home) often require intergenerational financial support. Additionally, families formed through remarriage face mounting pressures from socially ambiguous policies to support biological children from previous marriages, stepchildren, and in-laws (Hans 2008; Hooyman and Gonyea 1995). Such policies are based on familistic assumptions about kinship obligations that do not accommodate complex family structures where nonresident parents, step-kin, and in-laws are ambivalent about their financial obligations to others. Policymakers in the United States have attempted to ease cost concerns through legislation to expand the number of Americans with health insurance, to increase the availability of federally funded grants and loans for post-secondary education, and to institute employment guarantees for caregivers who must take time off from work (i.e., the Family Medical Leave Act of 1993). However, these watered-down solutions have exposed a pervasive reliance on private family resources stretching across generational lines to support various financial needs. The eventual effects may also unduly influence families' fiscal responsibilities over time such that the "midlife squeeze" is transmitted to future generations (Künemund 2006). Though taxation and the long-term viability of Social Security and Medicare are among the policy concerns of the American public, these are socially broader than three policies that have a more direct impact on multigenerational families' financial transfers.

First, nearly half of all uninsured Americans are young adults aged 18–30 years (Levy 2007). The 2010 Patient Protection and Affordable Care Act signed into law by President Obama reduces barriers for young adults to have health insurance. Previous eligibility of adult children was generally discontinued upon graduation from high school or college but the eligibility age of adult children has now been raised to 26 years for coverage through their parents' employer-provided health insurance, regardless of student status (Jost 2010). However, we should recognize that the requirement that all individuals be insured – a central aspect of the reform legislation – inherently relies on parents to finance adult children's insurance costs if the adult children are unable to do so for themselves.

Second, the value of parents' overall wealth and their ability to cosign loans are often included in formulas used to determine how much financial aid will be made available to college and university students. The Federal Application for Student Aid (FAFSA) collects information about parents' incomes and assets from unmarried, childless applicants under age 23 based on assumptions of parental responsibility, but as Hans (2008) points out, only residential parents' incomes and assets are considered relevant and while some state policies require nonresidential parents to maintain child support for children in college, many states have no policies to address this issue, and two states (New Hampshire and Pennsylvania) have rejected the practice of obligating nonresidential parents to support children in college. Rising tuition costs and increased expectations that job applicants have a postsecondary education force parents to make difficult choices while balancing the trade-offs of investing in children's future earning potentials and their personal future financial needs. In response to these policies, many parents are subsidizing their adult children's incomes to cover education, insurance, and general living expenses as well as acting as a safety net for unexpected emergencies. From another vantage point, public policies may encourage a "failure to launch" such that many adult children maintain prolonged dependence on midlife parents for financial assistance through at least the college-age years.

Third, the Family Medical Leave Act of 1993 (FMLA) includes only provisions for unpaid leave to care for family members (e.g., parents, young children, spouses). Thus, the financial and time-related costs associated with caregiving to the elderly are not compensated by publically funded resources. Recent attempts by some state policymakers to expand requirements for paid leave to family caregivers have succeeded by limiting paid leave requirements to accommodate the birth or adoption of a young child and ignoring compensation for those caring for spouses or elderly parents (Wisensale 2006). Either through financial support or various caregiving activities, many middle-aged children offer support to their aging parents with healthcare needs at the expense of their own future financial statuses in relation to lost wages, lost advancement opportunities, and lost pension incomes. Meanwhile, federal policymakers have paid attention to potential fraud that could result from intergenerational transfers in late life. Regulations for entry into Medicaid require a "look-back period" where officials examine within-family financial transactions over the previous 5 years to verify that recent transfers have not occurred so that the applicant may appear erroneously impoverished for eligibility (Gearon 2006).

Each of these examples demonstrates that policy changes are necessary to offer relief and support to the many couples and individuals who privately support other generations as they move into and out of financial self-sufficiency. Otherwise, the combination of family conflicts, ambiguous policies, and diverse structures may overwhelm families, and the pattern of a financial "mid-life squeeze" may be perpetuated from one generation to the next.

Future Research Topics

Skipped-Generation Transfers

Nearly all of the research on financial transfers has focused on resource sharing between parents and adult children. However, as kin relationships change over time and grandchildren experience the transition to adulthood while grandparents remain healthy and active, they may receive financial assistance directly from grandparents when previous support was indirect (Schoeni and Ross 2005; Silverstein and Marenco 2001). Past research has been able to examine time, space, and financial transfers from grandparents to grandchildren indirectly by analyzing time spent caring for young grandchildren (e.g., babysitting), coresidence in multigenerational households, and financial assistance provided to adult children who are raising young children themselves. However, there has been little research done to date that explores direct financial transfers between grandparents and grandchildren. Hoff (2007) demonstrated that the likelihood of skipped-generation transfers increased within German families – primarily as one-time gifts from grandparents to grandchildren without a reciprocated support exchange. Even-Zohar and Sharlin (2009) examined which factors perceived by Israeli grandparents and grandchildren led grandchildren to provide upward assistance. They found that grandparents focused on the more immediate exchange of giving material assistance in return for emotional or practical assistance while grandchildren focused on reciprocating grandparents' past emotional support. While some adult grandchildren may provide physical care or emotional support in response, the general assumption has been that grandparents offer altruistic financial assistance (Silverstein 2006). Research about skipped-generation transfers in American families would greatly enhance our understanding of multigenerational kin relationships.

Providing Cumulative Advantages via Transfers

Sociologists of the life course are uniquely able to examine how wealth stratification is perpetuated from one generation to future generations. Wealth transfers are more commonly associated with bequests than inter vivos transfers and suggest cumulative advantages are accrued via inheritance (Angel 2008). The common understanding of "old money" reflects the sharing of family resources across generations that is available to a small percentage of the population based on high wealth levels experienced by the oldest generation. However, few researchers have considered how bequest and transfer patterns have changed in recent years among the country's wealthiest citizens to avoid tax penalties and maintain wealth within multigenerational families (Wolff 2002).

Studies of financial transfers that consider parental resources may include income but have not yet included wealth as a measure of the family's ability to support its own members. Rarely have retirement assets, accumulated or anticipated pensions, or other elements such as employee benefits and health insurance coverage been included in calculations of overall wealth (Spilerman 2000). For many Americans, wealth may not be accessible in the short term to provide financial assistance to a family member. In particular, housing wealth, which often accounts for the major portion of the wealth of retired individuals, is not easily convertible into cash even though it has a high monetary value. Instead, liquid wealth that can be easily accessed for multiple purposes on short notice may be a better determinant of the likelihood to provide financial assistance to kin. Net financial assets may be a valuable wealth measure, particularly in cases where the household has no occupational income due to retirement or disability but has access to other resources (Oliver and Shapiro 1995).

Conclusion

Intergenerational relationships will become increasingly important to American families as they respond to the demographic pressures imposed by an aging population and decrease in family size. To fully comprehend the dynamics of "families as context," it is essential to understand how multigenerational families respond collectively to members' transitions into and out of financial self-sufficiency. In this chapter, I have outlined the relevance of inter vivos transfers to life course research and their practical significance for aging families. The financial transfer literature to date has leaned heavily toward identifying recipient characteristics, family structure characteristics, and affinity for or commitments to other family members that elicit financial gifts and loans. Valuable and insightful observations have resulted from such studies, but we have paid insufficient attention to the present and future tradeoffs incurred by donors responding to others' needs and expectations. Donors experience a "midlife financial squeeze" by sharing finite resources and potentially losing out on the greater accumulation of valuable resources for their own financial statuses in late life. The sociological ambivalence (i.e., conflicts, contradictions, and confusion) engendered by ambiguous life course transitions and ambiguous public policies challenges multigenerational families to work together in new ways in order to respond best to economic uncertainties for all involved.

References

Altonji, Joseph G., Fumio Hayashi, and Laurence J. Kotlikoff. 1997. "Parental Altruism and Inter Vivos Transfers." *Journal of Political Economy* 105:1121–66.

Angel, Jacqueline L. 2008. *Inheritance in Contemporary America: The Social Dimensions of Giving Across Generations.* Baltimore, MD: The Johns Hopkins University Press.

Aquilino, William S. 2005. "Impact of Family Structure on Parental Attitudes toward the Economic Support of Adult Children over the Transition to Adulthood." *Journal of Family Issues* 26:143–67.

Arnett, James J. 2004. *Emerging Adulthood: The Winding Road from the Late Teens through the Twenties.* New York: Oxford University Press.

Becker, Gary S. 1991. *A Treatise on the Family.* Cambridge, MA: Harvard University.

Bengtson, Vern L. 2001. "Beyond the Nuclear Family: The Increasing Importance of Multigenerational Bonds." *Journal of Marriage and Family* 63:1–16.

Bengtson, Vern L., Roseann Giarrusso, J. Beth Mabry, and Merril Silverstein. 2002. "Solidarity, Conflict, and Ambivalence: Complimentary or Competing Perspectives on Intergenerational Relationships?" *Journal of Marriage and Family* 64:568–76.

Bengtson, Vern L., Carolyn J. Rosenthal, and Linda M. Burton. 1990. "Families and Aging: Diversity and Heterogeneity." Pp. 263–87 in *Handbook of Aging and the Social Sciences,* 3rd ed., edited by R. H. Binstock and L. K. George. New York: Academic.

Cao, Honggao. 2006. "Time and Financial Transfers Within and Beyond the Family." *Journal of Family and Economic Issues* 27:375–400.

Cheal, David J. 1983. "Intergenerational Family Transfers." *Journal of Marriage and the Family* 45:805–13.

Cherlin, Andrew J. and Frank F. Furstenberg, Jr. 1994. "Stepfamilies in the United States: A Reconsideration." *Annual Review of Sociology* 20:359–81.

Connidis, Ingrid A. and Julie McMullin. 2002. "Sociological Ambivalence and Family Ties: A Critical Perspective." *Journal of Marriage and Family* 64:558–67.

Cooney, Teresa M. and Peter Uhlenberg. 1992. "Support from Parents over the Life Course: The Adult Child's Perspective." *Social Forces* 71:63–84.

Cox, Donald and Frederic Raines. 1985. "Interfamily Transfers and Income Redistribution." Pp. 393–426 in *Horizontal Equity, Uncertainty, and Economic Well-Being* edited by M. David and T. Smeeding. Retrieved January 12, 2010 (http://www.nber.org/chapters/c6157).

Cox, Donald and Mark R. Rank. 1992. "Inter-Vivos Transfers and Intergenerational Exchange." *Review of Economics and Statistics* 74:305–14.

Davey, Adam, Megan Janke, and Jyoti Savla. 2005. "Antecedents of Intergenerational Support: Families in Context and Families as Context." Pp. 29–54 in *Annual Review of Gerontology and Geriatrics, Vol. 24, Intergenerational Relations across Time and Place,* edited by M. Silverstein. New York: Springer.

Eggebeen, David J. 1992. "Family Structure and Intergenerational Exchanges." *Research on Aging* 14:427–47.

Eggebeen, David J. and Adam Davey. 1998. "Do Safety Nets Work? The Role of Anticipated Help in Times of Need." *Journal of Marriage and the Family* 60:939–50.

Eggebeen, David J. and Dennis P. Hogan. 1990. "Giving Between Generations in American Families." *Human Nature* 1:211–32.

Even-Zohar, Ahuva and Sholmo Sharlin. 2009. Grandchildhood: Adult Grandchildren's Perception of Their Role towards Their Grandparents from an Intergenerational Perspective. *Journal of Comparative Family Studies* 40:167–85.

Fingerman, Karen, Laura Miller, Kira Birditt, and Steven Zarit. 2009. "Giving to the Good and the Needy: Parental Support of Grown Children." *Journal of Marriage and Family* 71:1220–33.

Fingerman, Karen, Lindsay M. Pitzer, Wai Chan, Kira Birditt, Melissa M. Franks, and Steven Zarit. 2011. "Who Gets What and Why? Help Middle-Aged Adults Provide to Parents and Grown Children." *Journal of Gerontology: Social Sciences* 66:87–98.

Furstenberg, Frank F. Jr., Saul D. Hoffman, and Laura Shrestha. 1995. "The Effect of Divorce on Intergenerational Transfers: New Evidence." *Demography* 32:319–33.

Ganong, Lawrence H. and Marilyn Coleman. 2006. "Patterns of Exchange and Intergenerational Responsibilities after Divorce and Remarriage." *Journal of Aging Studies* 20:265–78.

Ganong, Lawrence H., Marilyn Coleman, and D. Mistina. 1995. "Normative Beliefs about Parents' and Stepparents' Financial Obligations to Children Following Divorce and Remarriage." *Family Relations* 44:306–35.

Gearon, Christopher J. 2006. "Getting Stuck with the Tab: Tighter Asset Spend-Down Rules Will Force More Families to Cover Nursing Homes Costs Alone." *The Washington Post*, February 21. Retrieved April 6, 2009 (http://www.washingtonpost.com/wp-dyn/content/article/2006/02/20/ARTICLE/gettingstuck/).

Goldscheider, Frances K., Arland Thornton, and Li-Shou Yang. 2001. "Helping Out the Kids: Expectations about Parental Support in Young Adulthood." *Journal of Marriage and Family* 63:727–40.

Grundy, Emily and John C. Henretta. 2006. "Between Elderly Parents and Adult Children: A New Look at the Intergenerational Care Provided by the 'Sandwich Generation'." *Ageing & Society* 26:707–22.

Hagestad, Gunhild O. 2003. "Interdependent Lives and Relationships in Changing Times: A Life-Course View of Families and Aging." Pp. 135–59 in *Invitation to the Life Course: Toward New Understandings of Later Life,* edited by R. A. Settersten, Jr. Amityville, NY: Baywood.

Hans, Jason D. 2008. "Should Child Support Obligations Continue for Children Attending College?" *Journal of Divorce and Remarriage* 48:109–25.

Henretta, John C., Martha S. Hill, Wei Li, Beth J. Soldo, and David A. Wolff. 1997. "The Selection of Children to Provide Care: The Effect of Earlier Parental Transfers." *Journal of Gerontology: Social Sciences* 52B:110–19.

Hoff, Andreas. 2007. "Patterns of Intergenerational Support in Grandparent-Grandchild and Parent-Child Relationships in Germany." *Ageing & Society* 27:643–55.

Hogan, Dennis P., David J. Eggebeen, and Clifford C. Clogg. 1993. "The Structure of Intergenerational Exchanges in American Families." *American Journal of Sociology* 98:1428–58.

Hooyman, Nancy R. and Judith Gonyea. 1995. *Feminist Perspectives on Family Care.* Thousand Oaks, CA: Sage.

Jost, Tomothy. 2010. "Implementing Health Reform: Young Adult Coverage." The Health Affairs Blog, May 11, 2010. Retrieved May 15, 2010 (http://healthaffairs.org/blog/2010/05/11/implementing-health-reform-young-adult-coverage/).

Killian, Timothy S. 2004. "Intergenerational Money Transfers to Adult Children and Stepchildren: A Household-Level Analysis." *Journal of Divorce and Remarriage* 42:105–30.

Killian, Timothy S. and Jana Ferrell. 2005. "Perceived Obligations of Remarried Households to Provide Assistance to Younger Family Members." *Journal of Intergenerational Relationships* 3:23–43.
Kohli, Martin and Harald Künemund. 2003. "Intergenerational Transfers in the Family: What Motivates Giving?" Pp. 123–42 in *Global Aging and Challenges to Families,* edited by V. L. Bengtson and A. Lowenstein. New York: Aldine de Gruyter.
Kronebusch, Karl and Mark Schlesinger. 1994. "Intergenerational Transfers." Pp. 112–151 in *Intergenerational Linkages: Hidden Connections in American Society,* edited by V. L. Bengtson and R. A. Harootyan. New York: Springer.
Künemund, Harald. 2006. "Changing Welfare States and the 'Sandwich Generation': Increasing Burden for the Next Generation?" *International Journal of Ageing and Later Life* 1:11–29.
Levy, Helen. 2007. "Health Insurance and the Transition to Adulthood." Pp. 84–106 in *The Price of Independence,* edited by S. Danzinger and C. E. Rouse. New York: Russell Sage Foundation.
Lin, I-Fen. 2008. "Consequences of Parental Divorce for Adult Children's Support of Their Frail Parents." *Journal of Marriage and Family* 70:113–28.
Logan, John R. and Glenna D. Spitze. 1996. *Family Ties: Enduring Relations between Parents and Their Grown Children.* Philadelphia, PA: Temple University Press.
Lusardi, Annamaria, Olivia S. Mitchell, and Vilsa Curto. 2009. "Financial Literacy and Financial Sophistication among Older Americans." Working Paper No. 15469, National Bureau of Economic Research, Cambridge, MA. Retrieved November 10, 2009 (http://www.nber.org/papers/w15469).
Lüscher, Kurt and Karl Pillemer. 1998. "Intergenerational Ambivalence: A New Approach to the Study of Parent-Child Relations in Later Life." *Journal of Marriage and the Family* 60:413–25.
MacDonald, Maurice and Sun-Kang Koh. 2003. "Consistent Motives for Inter-Family Transfers: Simple Altruism." *Journal of Family and Economic Issues* 24:73–97.
McGarry, Kathleen and Robert F. Schoeni. 1995. "Transfer Behavior: Measurement and the Redistribution of Resources within the Family." *The Journal of Human Resources* 30 (special issue):S184–226.
Oliver, Melvin L. and Thomas M. Shapiro. 1995. *Black Wealth/White Wealth: A New Perspective on Racial Inequality.* New York: Routledge.
Remle, R. Corey and Angela M. O'Rand. Forthcoming. "Intergenerational Solidarity in Blended Families: The Inequality of Financial Transfers to Adult Children and Stepchildren" in *Continuity and Discontinuity between Generations in an Aging Society,* edited by M. Silverstein and R. Giarrusso. Baltimore MD: Johns Hopkins University Press.
Riley, Matilda W. and John W. Riley Jr. 1993. "Connections: Kin and Cohort." Pp. 169–89 in *The Changing Contract across Generations* edited by V. L. Bengtson and W. A. Achenbaum. New York: Aldine de Gruyter.
Rossi, Alice S. and Peter H. Rossi. 1990. *Of Human Bonding: Parent-Child Relations Across the Life Course.* New York: Aldine de Gruyter.
Sarkisian, Natalia and Naomi Gerstel. 2004. "Kin Support Among Blacks and Whites: Race and Family Organization." *American Sociological Review* 69:812–37.
Schoeni, Robert and Karen Ross. 2005. "Material Assistance Received from Families During the Transition to Adulthood." Pp. 396–416 in *On the Frontier of Adulthood: Theory, Research, and Public Policy,* edited by R. A. Settersten, Jr., F. F. Furstenberg, Jr., and R. G. Rumbaut. Chicago, IL: University of Chicago Press.
Settersten, Richard A. Jr. 2007. "Social Relationships in the New Demographic Regime." Pp. 3–28 in *Advances in Life Course Research: Interpersonal Relations across the Life Course,* edited by T. Owens and J. J. Suitor. Oxford, UK: Elsevier.
Shapiro, Adam and R. Corey Remle. 2011. "Generational Jeopardy? Parental Marital Transitions and the Provision of Financial Transfers to Adult Children." *Journal of Gerontology: Social Sciences,* 66:99–108.
Shapiro, Thomas M. 2004. *The Hidden Cost of Being African American: How Wealth Perpetuates Inequality.* New York: Oxford University Press.
Silverstein, Merril. 2006. "Family Structure and Transfers." Pp. 140–62 in *Handbook of Aging and the Social Sciences,* 6th ed., edited by R. H. Binstock and L. K. George. San Diego, CA: Academic.
Silverstein, Merril, Vern L. Bengtson, and Leora Lawton. 1997. "Intergenerational Solidarity and the Structure of Adult Child-Parent Relationships in American Families." *American Journal of Sociology* 103:429–60.
Silverstein, Merril, Stephen J. Conroy, Haitao Wang, Roseann Giarrusso, and Vern L. Bengtson. 2002. "Reciprocity in Parent-Child Relations over the Adult Life Course." *Journal of Gerontology: Social Sciences* 57B:S3–13.
Silverstein, Merril and Anne Marenco. 2001. "How Americans Enact the Grandparent Role across the Family Life Course." *Journal of Family Issues* 22:493–522.
Soldo, Beth J. 1996. "Cross Pressures on Middle-aged Adults: A Broader View." *Journal of Gerontology: Social Sciences* 51:S271–3.
Soldo, Beth J. and Martha S. Hill. 1993. "Intergenerational Transfers: Economic, Demographic, and Social Perspectives." Pp. 187–206 in *Annual Review of Gerontology and Geriatrics, Vol. 13, Focus on Kinship, Aging, and Social* Change, edited by G. Maddox and M. P. Lawton. New York: Springer.

Spilerman, Seymour. 2000. "Wealth and Stratification Processes." *Annual Review of Sociology* 26:497–24.
Spitze, Glenna D. and John R. Logan. 1992. "Helping as a Component of Parent-Adult Child Relations." *Research on Aging* 14:291–312.
Suitor, J. Jill, Karl Pillemer and Jori Sechrist. 2006. "Within-Family Differences in Mothers' Support to Adult Children." *Journals of Gerontology: Social Sciences* 61:S10–17.
Sullivan, Teresa A., Elizabeth Warren, and J. Lawrence Westbrook. 2000. *The Fragile Middle Class: Americans in Debt*. New Haven, CT: Yale University Press.
Swartz, Teresa Toguchi. 2009. "Intergenerational Family Relations in Adulthood: Patterns, Variations, and Implications in the Contemporary United States." *Annual Review of Sociology* 35:191–212.
Uhlenberg, Peter. 1996. "Mortality Decline in the Twentieth Century and Supply of Kin over the Life Course." *The Gerontologist* 36:681–5.
White, Lynn. 1994. "Growing Up with Single Parents and Stepparents: Long-term Effects on Solidarity." *Journal of Marriage and the Family* 56:935–48.
Wisensale, Steven K. 2006. "What Role for the Family and Medical Leave Act in Long-Term Care Policy?" *Journal of Aging and Social Policy* 18:79–93.
Wolff, Edward N. 2002. *Top Heavy: The Increasing Inequality of Wealth in America and What Can Be Done About It.* New York: The New Press.
Wong, Rebeca, Chiara Capoferro, and Beth J. Soldo. 1999. "Financial Assistance from Middle-Aged Couples to Parents and Children: Racial-Ethnic Differences." *The Journals of Gerontology: Social Sciences* 54B:S145–53.

Chapter 13
The Demography of Unions Among Older Americans, 1980–Present: A Family Change Approach

Wendy D. Manning and Susan L. Brown

The current generation of older Americans faces more complex family and marital histories than any prior generation. Moreover, baby boomers, the first cohort to experience high levels of divorce, single parenthood, and remarriage, are now moving into older adulthood. This movement will likely exacerbate the trend away from marriage among older adults. Researchers are uncovering greater heterogeneity and complexity in the family life of older Americans, which in turn portends a shift in the benefits and rewards offered by certain family circumstances (Allen et al. 2000; Cooney and Dunne 2001). The growing diversity of living arrangements characterizing older adulthood is likely to have important consequences for individual health and well-being as well as policy ramifications for the changing types of institutional support older adults require (Wilmoth and Longino 2006).

In this chapter, we document changes in the marital status and household living arrangements of older Americans over the past 4 decades, integrating explanations for and consequences of these changing patterns. The increasingly varied family life course trajectories experienced in early and middle adulthood have enduring consequences. Namely, older adults are much less likely to be married now than were previous cohorts. This trend is expected to accelerate with a declining share of the older adult population being married in coming decades (e.g., Allen et al. 2000; Cooney and Dunne 2001). A decade ago, gerontologists predicted that older men more likely will be never married, and older women increasingly will be divorced rather than widowed (Cooney and Dunne 2001).

The purpose of this chapter is threefold: describe how family change approaches can elucidate recent demographic shifts in the union behaviors of older adults; provide empirical evidence that describes the recent trends in marital status and living arrangements for older adults; and discuss the implications of these new marriage and family patterns for individuals, families, and society. In addition to examining marriage and widowhood, which have been the primary foci of earlier work on the family status of older adults, we also consider union experiences which fall outside these typical marital status categories. Specifically, we investigate cohabitation, living apart together (LAT), and same-sex unions, all of which appear to be increasing among the older adult population but have received limited attention from researchers (Bennett and Gates 2004; Brown et al. 2005, 2006; de Jong Gierveld 2005; Gates 2003; Huyck 1995).

W.D. Manning (✉)
Department of Sociology & National Center for Family and Marriage Research,
Bowling Green State University, Bowling Green, OH, USA
e-mail: wmannin@bgsu.edu

R.A. Settersten, Jr. and J.L. Angel (eds.), *Handbook of Sociology of Aging*, Handbooks of Sociology and Social Research, DOI 10.1007/978-1-4419-7374-0_13,

Family Change Approaches

There is wide recognition of rapid changes in marriage and family, but there is not a consensus about the mechanisms underlying these changes. The most commonly cited reasons for changes in marriage and family behavior include economic, normative, and institutional explanations. While the recent changes in family and marriage have been widely noted, it is important to recognize that social scientists have long been interested in the causes of family change and have invoked these theories of change across a broad range of time periods (Smock 2004).

The life course perspective is a popular approach used to study marriage and family behavior. Among older Americans, this perspective is important because it is recognizes that family behavior is age-graded and integrates social, historical, cultural contexts. Life course theory posits that one's life is a sequence of events, social roles, and changes that begin in infancy and continue until death (Caspi et al. 1989). The lives of family members are "linked" in the decisions of one member who have implications for the other family members (Elder 1985). These linked lives imply that generations are connected (e.g., parents and children) as well as family members of the same generation (e.g., siblings, spouses). Further, an individual's actions that occur earlier in the life course both directly and indirectly influence behavior in later life stages (Elder 1985). This perspective highlights interactions between individuals and their social environments. Age, timing, and sequencing are key concepts within a life course framework. The timing of transitions (e.g., parenthood or marriage) can be considered early, on-time or late which subsequently has implications in later life. A fundamental feature of the life course perspective is the individual-level progression or trajectories through life and the recognition that trajectories may shift across time or context. While the life course approach showcases how individual family experience progress over their life, it does not provide a direct explanation for the mechanisms or reasons for family change.

Most research examining family change focuses on earlier points in the life course (e.g., the twenties) with an emphasis on how young adults form and sustain relationships. This attention on young adults is based on concerns about the well-being of children, i.e., providing stable home environments for children. As many changes in families have occurred among young adults, they have reverberations among older adults. For example, divorce in her early thirties may have implications for the accumulation of wealth for a woman in her sixties. Yet studies of marriage are often based on realities of young adults, ignoring the distinct patterns of union formation and stability among older adults. The generations that have experienced the most family change have not yet reached their older years, and we have much to learn about their experiences. As a result, our theories of union formation may be useful for explaining current behavior; however, they may not adequately predict or explain future trends (Seltzer et al. 2005).

The reasons for changes in marital behavior are interrelated and include the following: shifts in economic gains to marriage, ideational changes, and institutional recognition of a broad array of families (e.g., Cherlin 2004; Seltzer et al. 2005; Waite et al. 2000). First, the gendered basis to marriage has changed. Women are increasingly valued for their economic contributions to families. Women with higher incomes and education are actually more likely to marry than their counterparts with fewer economic resources (Sweeney 2002). This shift in the traditional marital bargain calls for new approaches to studying the criteria for marriage as well as the complicated balance required to sustain marriage. These new economic realities challenge a traditional home economics perspective (Becker 1974), which focuses on a gender-based traditional division of labor with the decision-making power assigned to the male head of the household (Bianchi et al. 2008). The shift in the gendered economics surrounding marriage and the value of women's economic position are broad and have implications not only for younger but also for older Americans today.

A second explanation for family change is based on normative or cultural change brought about in part by the greater secularization or individualization of American society (Lesthaeghe and Neidert 2006). These types of normative changes are often measured in terms of attitudes, such as beliefs about gender equity. However, it remains a challenge to measure the accepted rules of behavior and discern subgroup differences (Rossi and Rossi 1990). The change in attitudes about cohabitation, single parenthood, and the fragility of marriage have been well documented (Axinn and Thornton 2000; Cherlin 2004). A related normative shift is the fundamental reason for or meaning underlying marriage. The focus of contemporary marriage is on emotional fulfillment and love leading to shifts in the purpose and stability of marriages (Cherlin 2004). Thus, the nature of couple interactions may be altered in part because of this new normative climate. As the demands of marriage increase, there may be further shifts in the timing and stability of marriage.

A third reason for change in families is that adults now may enjoy many of the instrumental and social benefits of marriage without actually marrying. For example, social welfare benefits, property rights, and custody arrangements are not based solely on marital status. As we observe more legal arrangements, social policies, and institutions that do not discriminate based on marital status, the benefits of marriage become less clear. Further, marriage may require a level of interdependence or enmeshment (financial and emotional) that is not desirable which in turn may encourage new family forms. Certainly, marriage offers a legal commitment and bond that secures some stability and permanence. Cherlin (2004:855) refers to this as "enforceable trust" and claims that it could be "eroding." As the traditional supports for marriage are shifting, the uncertainty or ambiguity surrounding the meaning and definition of marriage may have implications for marital and family decisions and contribute to the rise in unmarried family forms (Settersten 2009).

The ramifications of these economic, normative, and institutional changes for the family behaviors of older adults remain largely unexplored both theoretically and empirically, but family changes are identified by gerontologists and family scholars as integral to our understanding of the demographic shifts taking place in an aging society (Allen et al. 2000; Cooney and Dunne 2001). This chapter aims to interpret the latest demographic patterns through the lens of the family change approach to provide a richer portrait of aging families in the contemporary U.S. context.

Measures and Data

Our empirical analyses focus on older Americans defined as men and women aged 65 and older. We compared older Americans in 1980 (65-year-olds were born in 1915) to 2008 (65-year-olds were born in 1943). The early cohort became adults during the Great Depression, and the later cohort became adults during the economic boom years (1960s). This time span includes those adults who produced the peak of the baby boom.

The United States is an aging society with older Americans representing an increasingly greater share of the total population. In 1980, persons over age 65 comprised 11% of the American population, while in 2008 those over age 65 were 13% of the population. Over the next 50 years, the population of adults aged 65 and older is predicted to double. By 2050, the U.S. older adult population will reach 88 million or one-fifth of the population (U.S. Census Bureau 2008).

Life expectancy is increasing for women and men alike, but the two groups continue to have different life expectancies. Currently, life expectancy for men is about 75 years, whereas for women is 80 years. Consequently, the majority of older adults are women (Heron et al. 2009). In 2007, there were 137 women over age 65 for every 100 men over age 65 in the United States. This ratio increases among the oldest old. By the time older adults reached their mid-eighties, there were 210 women to every 100 men (Administration on Aging 2009). The lengthening life span of Americans presents the possibility of more relationships in older adulthood including the potential for expanded intergenerational family ties and at the same time potentially more opportunities for family instability. As Settersten (2007) argues, divorce rather than death is the primary cause of family disruption.

Apart from gender, life expectancy also varies by race-ethnicity. Although the racial gap in life expectancy has declined over time as minorities are living longer, nonetheless, whites enjoy greater life expectancies than do either blacks or Hispanics in the United States. For instance, current life expectancy among whites is about 78 years, whereas for blacks it is 73 years. Today, among 65-year-olds, life expectancy is roughly 2 years greater for white than black men and 1 year greater for white than black women (Heron et al. 2009). Consequently, the older adult population is now more racially and ethnically diverse than in the past. In 1980, 8% of the 65 and older population was black, and less than 3% was Hispanic. Today, nearly 9% of older adults are black, and more than 6% are Hispanic (National Center for Health Statistics 2009). Population projections estimate that by 2050, about 58% of the older population will be non-Hispanic white, 20% of the older population will be Hispanic, and 11% black (Administration on Aging 2009). Changes in immigration in the United States portend an increasing share of older Americans will be foreign-born, and nativity status is likely to play a role in the lives of older Americans.

Gender and race-ethnicity structure the experience of aging. Therefore, we provide gender-specific comparisons of marital status and living arrangements across age groups. We focus on similarities and differences in the experiences of older men and women. Where possible, results are shown for specific racial and ethnic groups.

This chapter draws on several data sources. A primary source of data is the United States decennial Census, which offers a view of the trends in marital status and living arrangements of older Americans. The recent American Community Survey provides specific measures about marriage that are not available in the Census data. We also present findings from published results from surveys, such as the Current Population Survey March Supplement and the Health and Retirement Study. In the concluding section, we discuss future data needs for research on the family demography of an aging population.

Marital Status

Marriage

Overall, there has been a decline between 1980 and 2008 in the proportion of older men who are married, as shown in Fig. 13.1. In 1980, more than three quarters (76%) of older men were married. By 2008, there was a modest decrease, and 71% of older men were married. In 1980, the proportion of men married was negatively associated with age. In 2008, the proportions married were less sensitive to age than in 1980, with those ranging in age from 65 to 79 almost equally likely to be married. The trend in proportion married between 1980 and 2008 varied by age, such that among the young old (65–74), smaller proportions of men were married in 2008 than in 1980, whereas among the old old (75–84) and oldest old (85 and older), married men were a bit more prevalent in 2008 than in 1980. For instance, 83% of 65–69-year-old men were married in 1980, but just 75% of men in this age group were married in 2008. In 1980, 48% of men 85 and older were married. By 2008, 54% of oldest old men were married.

Among older women, the proportion married appears to have changed little over the past 4 decades Fig. 13.2. In 1980, 37% of older women were married compared to 40% in 2008. This overall pattern characterizes young old women, too, but not old old and oldest old women, who now compose larger shares of the population. Among women 65–69, there was essentially no change between 1980 and 2008 in the proportion married (55 and 56%, respectively). All other age groups witnessed increases in the proportion married over time. Among 75–79-year-olds, the proportion married grew from 29% in 1980 to 41% in 2008. Similarly, among the oldest old, the proportion married rose by 50% from 8% in 1980 to 12% in 2008.

Comparing the proportions of men vs. women who were married reveals that more men were married than women, regardless of age or time period. Gender distinctions in marital status were

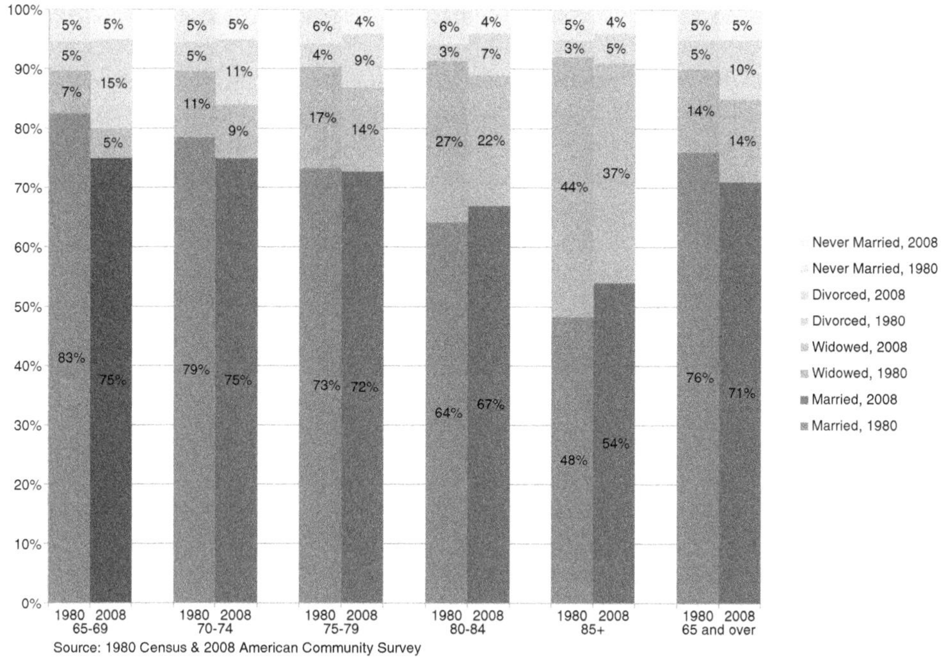

Fig. 13.1 Marital status by age for males, 1980 and 2008

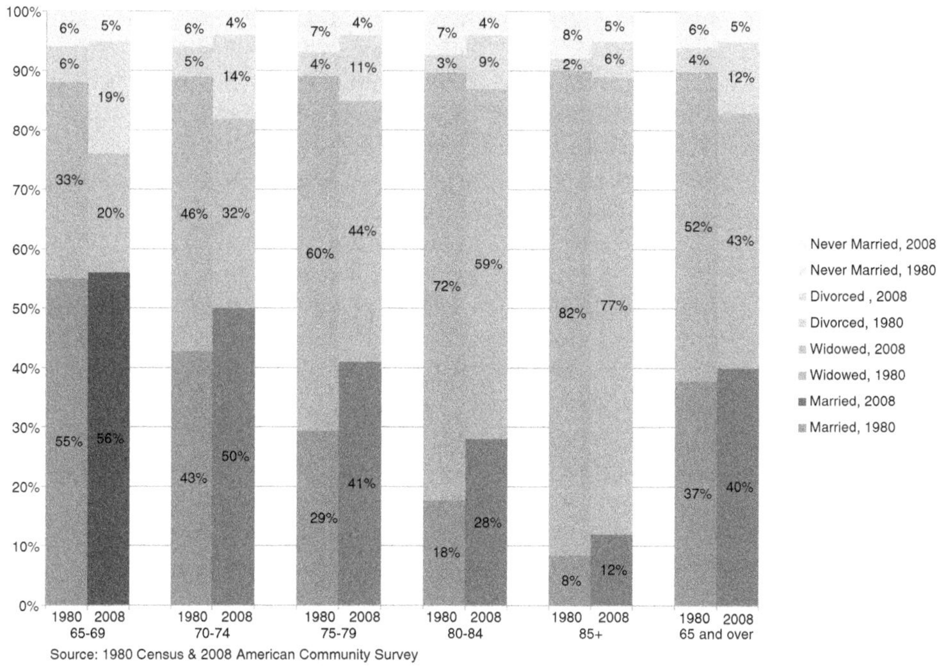

Fig. 13.2 Marital status by age for females, 1980 and 2008

due in part to the substantial gender differentials in mortality and life expectancy and the tendency for women to marry men a few years older than themselves. Indeed, at all ages, older men were more likely to be married than older women, and this differential became more pronounced with age. Nonetheless, the gender differential was smaller in 2008 than it was in 1980. Six times as many men as women aged 85 and older were married in 1980 (48 and 8%, respectively) vs. 4.5 times in 2008 (54 and 12%, respectively).

These trends reveal the distinct gendered patterns characterizing marriage in later life. Married is the modal marital status for men, but not for women. It is notable that the proportions married have declined modestly among older men at the same time they have increased slightly among older women. This convergence is consistent with the family change approach, which suggests increasing gender similarity in the mate selection process.

Widowhood

Widowhood remained consistent among older men over the past 4 decades: 14% in 1980 and 2008 (Fig. 13.1). Despite this overall stability, the pattern varies by age group, with the growth in widowers concentrated among the oldest old men. In 1980, less than one-fifth of men aged 65–79 were widowers, about one-quarter of 80–84-year-old men were widowers, and nearly half of men over age 85 were widowers. In 2008, less than 15% of men aged 65–79 were widowers, 22% of men 80–84, and 37% of men over age 85 were widowers. Stated differently, widowhood has declined most precipitously among the oldest men.

By contrast, there have been declines in widowhood among all older women, from 52% in 1980 to 43% in 2008 (Fig. 13.2). In 1980, the proportion of women widowed ranged from one-third among 65–69-year-olds to 82% among women aged 85 and older. The proportion of women who were married outpaced those who were widowed only at ages 65–69. At every other age, the proportion of women married was less than the proportion widowed. In 2008, one-fifth of the youngest older American women were widowed, and 77% of women aged 85 and older were widowed, but the proportion married was higher than the proportion widowed not only for 65–69-year-olds but also for 70–74-year-olds. These results suggest that we may be observing a slight delay in widowhood among women, which is consistent with lengthening life expectancy.

As expected, widowhood increased with age, but the age gradient was steeper among women than men. A greater proportion of women were widowed than men at every age. Among the oldest old, widowhood was 50% higher among women than men (77 vs. 37%, respectively, in 2008). The modal marital status category among women is widowed, although the proportion of older women who are widowed has declined since 1980 and a corresponding increase over this time period occurred for the proportion divorced. This pattern is in line with predictions by Cooney and Dunne (2001). Widowhood, while common, brings a host of stress in terms of emotional loss, social network shifts, financial concerns, and instrumental support. Men and women respond differently to widowhood with men much more likely to remarry (see Chap. 10 for further discussion). Family change approaches do not provide much insight into the changing patterns of widowhood, but they reflect both lengthening life expectancy and arguably shifts in preferences for remarriage formation. Future research should attend to whether the propensity to remarry following widowhood has changed in recent decades as well as the extent to which these changes are gendered.

Divorce

Sustained high levels of divorce over the past few decades in the U.S. population have contributed to the declines in the proportions of older adults who are married. We combine separated and divorced into one category. In fact, the proportion of older men who were divorced (or separated) doubled between 1980 and 2008 (Fig. 13.1). Only 5% of older men were divorced in 1980, whereas the figure rose to 10% in 2008. This growth is concentrated among the young old. In 1980, 5% of men ages 65–69 and 6% of men ages 70–74 were divorced. In 2008, the corresponding figures were 15 and 11%, respectively. At older ages, the proportions of divorced men did not change much over time. Among the oldest old, 3% of men were divorced in 1980, and 5% were divorced in 2008.

Among older women, there is a similar pattern. The proportion of older women who were divorced climbed from 4% in 1980 to 12% in 2008, nearly identical to the proportions documented earlier for older men (Fig. 13.2). The rise occurred among all older women. Among 65–69-year-old women, 6% were divorced in 1980 vs. 19% in 2008. For women ages 80–84, 3% were divorced in 1980, and 6% were divorced in 2008. The proportion of the oldest old that is divorced increased from 2% in 1980 to 6% in 2008.

Comparing the trends for men and women, the proportions divorced in each 5-year age interval were essentially the same in 1980. By 2008, older women were somewhat more likely to be divorced than were older men, at least at younger ages. There were no gender differences in divorce among the oldest old. With sustained high levels of divorce in the U.S. population coupled with the weakening propensity to remarry following divorce (the divorced increasingly favor postmarital cohabitation), it is likely that the proportions of older men and women divorced will rise in the coming years. The implications of divorce most likely depend on its timing (Shapiro and Cooney 2007). Early divorce in the life course has been found to influence later relationships between adult children and older fathers (Shapiro and Cooney 2007), and divorce in older adulthood may translate into declines in social and economic support. Throughout the life course, the costs of divorce appear to differ according to gender. From a family change approach, the normative and institutional constraints supporting marriage continue to erode, opening up the possibility of other forms of family outside of marriage (e.g., cohabitation) and increasing the acceptability of living alone.

Never Married

The percentage of the older population never married has not shifted over the past 4 decades, nor does it differ for men and women (Figs. 13.1 and 13.2). Between 5 and 6% of the older adult population was never married in 1980 and 2008. In 1980, only 2 or 3% of old old and oldest old men and women were never married. By 2008, between 4 and 5% of both older women and men in all age groups were never married. Even though men and women are equally likely to remain never-married, there is a gender gap in the well-being of never-married men and women. Never-married women appear to fare as well as married women in old age while never-married men appear to fare worse than their married counterparts (see Chap. 10). The proportion of older adults who are never married is expected to rise in the future, especially among men (Cooney and Dunne 2001), reflecting more inclusive normative expectations about families and loosening institutional constraints.

Racial and Ethnic Variation

Most research on the marriage and living arrangement patterns of older Americans focuses on gender and age distinctions. However, given the striking racial differences in marriage and divorce rates, it is important to consider race and ethnicity when studying marriage and other close relationships among older Americans.

In 2008, the majority of older white, black, and Hispanic men were married. Nearly three-quarters of white men, two-thirds of Hispanic men, and about half of black men were married. Widowhood levels were similar across race and ethnic groups, with slightly higher levels among black men (18%) than Hispanic (14%) or white (13%) men. The proportion of older men who were divorced was twice as high among black as white men. A substantial minority (9%) of black men had never married in contrast to 5% of Hispanic and 4% of white men.

Older women were less likely to be married than men, and less than half of each racial and ethnic group was married. White older women were more often married (42%) than Hispanic (36%) or black (23%) women. Nearly half of black older women were widowed, and about two-fifths of white and Hispanic women were widowed. The levels of divorce were higher among black (20%) and Hispanic (17%) women than that of white (12%) women. The vast majority of older American women was ever married; however, black (9%) and Hispanic (7%) older women were more often never married than white (4%) older women. Racial and ethnic variation in family formation behaviors earlier in the life course is likely to have persistent effects as people age, yet few researchers have considered racial and ethnic differences in family behaviors among older adults (although see Calasanti and Kiecolt 2007; Coward et al. 1996). Indeed, from a family change perspective, the economic and normative factors may vary considerably by race and ethnicity for older men and women, but this remains largely unexplored.

Marriage and Divorce in 2008

We rely on the American Community Survey to specifically examine marriage and divorce among older Americans. While we typically think of brides and grooms as young men and women in their twenties, according to the American Community Study there were about 91,000 marriages among older Americans, including 31,500 brides and 59,300 grooms, over the age of 65 in 2008. In 1985, nearly 25 years ago, 71,000 persons over age 65 married (Meyers and Wilson 1988). The increase was not necessarily due to an increase in marriage rates among older Americans but a shift in the age structure of the population. The American Community Survey data indicate that most older Americans who married in the last year were not first time brides or grooms; only 10% were first marriages. About half (54%) of marriages in the last year to older Americans were second marriages, and one-third (35%) were third marriages. The patterns and levels are similar for men and women. In 1985, one-quarter of grooms over age 65 were divorced, and three-fourths of the brides were widowed (Meyers and Wilson 1988). In 2008, about three-quarters of married older Americans were still in their first marriage, which is similar to 76% among the population over age 15 (National Center for Family and Marriage Research 2009).

In the American Community Survey 2008 data, there were approximately 119,700 divorces among older Americans. The ratio of marriages to divorces among older Americans in 2008 is opposite of what it is among the total population. Among older Americans, there were only 0.8 marriages for every one divorce, indicating divorce is more common than marriage. Among the total population over age 15, the ratio is two marriages for every one divorce. Given the stabilization of high divorce rates among the total population (Raley and Bumpass 2003), we expect divorce rates among older Americans to follow the broader population trends. What is notable for older Americans is that they are relatively unlikely to marry again following divorce. Whether they form an unmarried cohabiting union or remain single is unclear, but both pathways are consistent with the family change approach.

The most common marital status change in older adulthood is widowhood. As indicated earlier, widowhood is prevalent among older Americans, and over one million older Americans were widowed in 2008. Women were 2.3 times more likely to experience widowhood than men.

Projections provide a glimpse into the relationship patterns of older Americans in the next few decades and indicate that by 2040, 42% of older women and 69% of older men will be married (Wade 1989). Wade argues that women's decline in marriage will be the result of declines in widowhood and growth in divorce, while men's decline will be due to older men's lack of marriage (never married status). These predictions mirror those of Cooney and Dunne (2001). Updated projections suggest that the proportion of older adults never married will increase, the proportion married will decline, and the proportion divorced will stabilize in recent years (Tamborini 2007).

Living Arrangements

The living arrangements of older Americans are tied to their marital status but living arrangements offer a unique lens on the potential sources of support available to older adults within the household. For instance, marital status per se does not reveal whether an individual lives alone or with other family members. The rapid rise in older adults living alone is of considerable policy interest to the extent that it portends growth in the share of elderly without the supports required to delay or avoid institutional care (Mutchler 1992). Transitions in living arrangements among older adults sometimes follow a different pattern than that documented at a single point in time (Wilmoth 1998), but they are beyond the scope of this chapter.

We examine changes in living arrangements among older adults over time, distinguishing among four categories: living alone, married and living with spouse, living with family members, and living with nonfamily members. Older Americans also may live in group quarters but are not included in our Census estimates below. The term "group quarters" may include a variety of circumstances. The Administration on Aging (2009) reports that, in 2007, 1.57 million or 4.4% of older Americans lived in institutional settings, and 2–5% lived in senior housing with support services.

Living Alone

Figure 13.3 shows that the proportion of older men living alone has increased slightly from 15% in 1980 to 19% in 2008. As men age, a greater proportion were living on their own; the oldest old were about twice as likely to live alone as the youngest old. In 1980, 11% of men ages 65–69 lived alone, and 26% of the oldest old men lived alone. Similarly, in 2008, 15% of men 65–69 years old lived alone, 32% of men aged 85 or older also did so.

There was a consistent pattern in the proportion of older women who lived alone in 1980 and 2008: 40% in 1980 and 37% in 2008 (Fig. 13.4). The proportion of older women who lived alone increased sharply according to age. In 1980, 30% of women 65–69 lived alone, and 46% of the oldest old lived alone. In 2008, there was a steeper age gradient, 26% of women ages 65–69 lived alone, while 56% of women over age 85 lived alone.

Women more often live alone than men, and this is true for every age group of older Americans. In fact, the gender gap in living alone increases with age. In 2008, among the oldest old, about one-third (32%) of men and over half (56%) of women lived alone. Solo living in older adulthood is related to fewer economic resources in addition to lower levels of social support. As the population ages and life expectancy increases, living alone will be more prevalent among older adults and its consequences for society more salient.

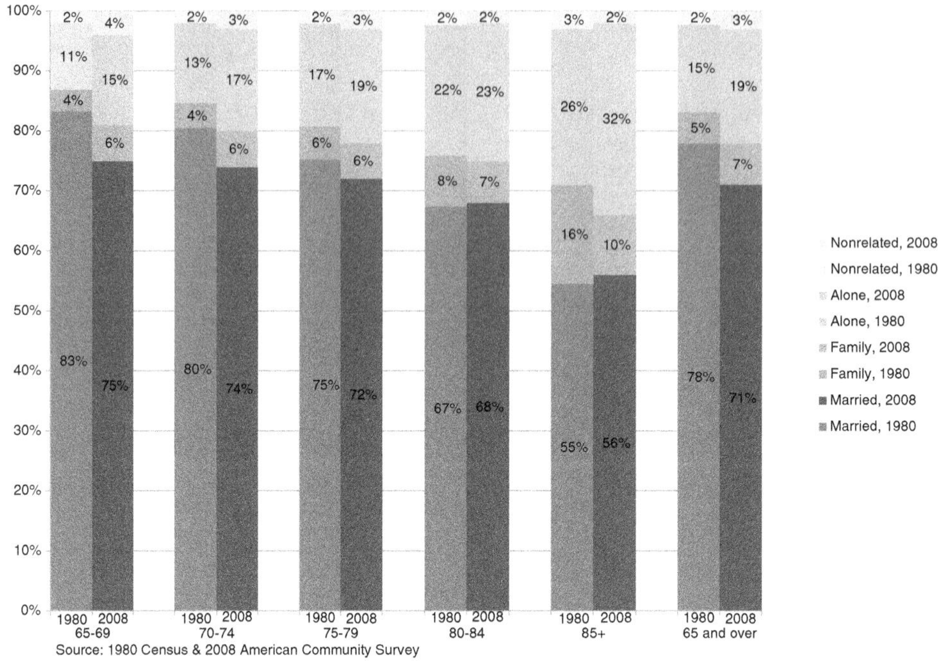

Fig. 13.3 Living arrangements by age for males, 1980 and 2008

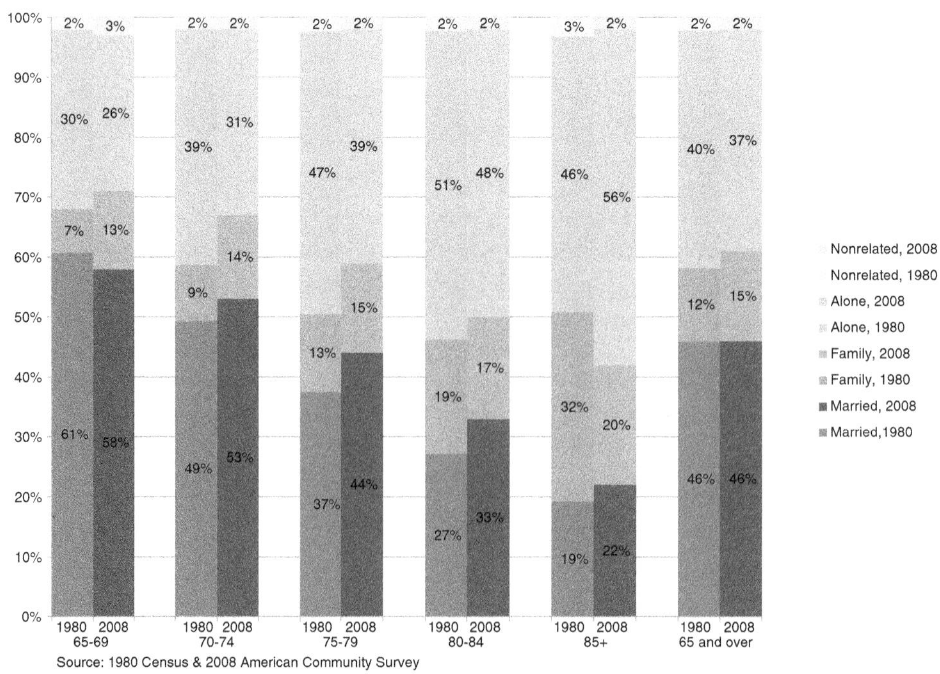

Fig. 13.4 Living arrangements by age for females, 1980 and 2008

Married with Spouse

These estimates differ slightly from the levels discussed earlier because we exclude in the denominator older Americans living in group quarters. Accordingly, these estimates are limited to older Americans who are not living in group quarters. Over the last 40 years, there has been a small decline in the proportion of older American men who were married, from 78% in 1980 to 71% in 2008 (Fig. 13.3). In 1980, young old men experienced the highest levels of living with a spouse (80–83%), and this declined to 55% among the oldest old. A similar age pattern existed in 2008.

There has been no change in the proportion of older women who live with a spouse (46%) in 1980 and in 2008 (Fig. 13.4). About three-fifths (61% in 1980 and 58% in 2008) of women 65–69 lived with their spouse. In both 1980 and 2008, oldest old women experienced a sharp decline in coresidence with their spouse; only 19% in 1980 and 22% in 2008 lived with their husband.

As discussed earlier, a greater proportion of older men than women lived with their spouse. In 1980, 46% of women and 77% of men were living with their spouse. The gender gradient in marriage increased with age, and the levels remained consistent in 1980 and 2008. In 2008, men 65–69 years old were 1.3 times more likely to be living with a spouse than women, while men 85 years old and older were 2.5 times more likely to be living with their spouse than women. Most women not residing with a spouse live alone, meaning they are at risk of lower levels of economic, socioemotional, and physical well-being.

Living with Family or Nonfamily Members

In 1980, it was relatively rare for older men to live with family members (5%) or nonfamily members (2%) (Fig. 13.3). Coresiding with family (7%) or nonfamily (3%) members remained uncommon in 2008. The proportion of men living with nonfamily members remained nearly constant across age groups. By contrast, in 1980, there was fourfold gap in the proportion of men living with family between the youngest and oldest age groups. About 4% men ages 65–59 and 16% of oldest old men lived with family members. In 2008, 6% of men ages 65–69 lived with family, and only 10% of oldest old men lived with family. Thus, there was a decline in the proportion of the oldest old men who were living with family members.

In 1980 and 2008, a similar proportion (2%) of older women lived with nonfamily members (Fig. 13.4). The low level of nonfamily coresidence persisted across age groups. In 1980, 12% of women lived with family members, and in 2008, about 15% lived with family. In 1980, 7% of 65–69-year-old women lived with family, and 32% of 85+ year-old women lived with family. By 2008, there appears to be an increase in family living among young older women (65–69) and a decline among older (85+) women. Consequently, in 2008, 13% of 65–69-year-old women lived with family members, and 20% of women aged 85 or older lived with family members.

There is no gender gap in the proportion of older Americans living with nonfamily members. However, in 2008, older women more often lived with family members (12%) than men (7%). At each age, women were more likely to live with family than men, and the proportion living with family increased with age. The gender gap is consistent across age groups at both time periods; women experience about 2–3 times higher levels of family coresidence than men. These patterns illustrate the closer family ties enjoyed by women relative to men. Among those residing without a spouse, both men and women are most likely to live with family members, but this is more pronounced among women and accounts for a larger share of all women. As family instability weakens men's ties to their children and other family members, we can anticipate that older men's coresidence with family members will proceed at a slower pace than that for older women.

Racial and Ethnic Variation

As discussed earlier, there were racial and ethnic distinctions in the proportion of older Americans who were living with a spouse. White men and women more often lived with a spouse than did black men or women. Very few (5%) white older men lived alone, while 15% of black and 16% of Hispanic men lived alone. About one-quarter of black men lived with family members in contrast to only 18% of white and 16% of Hispanic older men. Living with nonfamily members is relatively rare and quite similar across race and ethnic groups.

The majority of older women did not live with a spouse; two-fifths (42%) of white women, 36% of Hispanic women, and only about one-quarter of black women lived with their spouse. One-third of older black women lived alone, while 28% of older Hispanic women and 12% of older white women lived alone. Nearly two-fifths (38%) of black women lived with family members, representing the most common living arrangement among older black women. Similar proportion of black and white older women lived with family, and 28% of Hispanic older women lived with family. Only 2% of older women lived with nonfamily members, and this was similar for each race and ethnic group considered here.

New Union and Couple Forms

There are several union and couple experiences that are not captured with the traditional indicators of marital status and living arrangements, including cohabitation, LAT, and same sex unions. Although these less traditional couple relationships have received comparatively little attention in the gerontological literature, aging scholars have begun to call attention to the importance of examining these emerging union types (Cooney and Dunne 2001; Huyck 1995). As fewer older adults are married, a larger share is available to form nonmarital relationships. Whether these relationships offer benefits akin to marriage is largely unknown although some research has compared either the relationship quality (Brown and Kawamura forthcoming) or the psychological well-being of older cohabitors and marrieds (Brown et al. 2005) and shows that the union type gap in well-being is relatively modest. Here, we document the demographic trends in cohabitation, LAT, and same sex relationships.

Cohabitation

In recent decades, cohabitation, or the sharing of a household by an unmarried opposite-sex couple, has increased dramatically in the United States. This growth is not limited to younger adults but rather extends through older adulthood (Brown et al. 2006; Chevan 1996). The reasons for cohabiting later in life are distinctive. For instance, older adults may prefer to cohabit to retain financial autonomy and protect their wealth for eventual transfer to their heirs (Brown et al. 2005; Chevan 1996; King and Scott 2005). Among older adults, cohabitors tend to be younger than marrieds. Cohabitors are disproportionately black. Most older cohabitors are divorced (71%) as opposed to widowed (18%) or never married (11%) (Brown et al. 2006). Chevan's (1996) estimates using indirect measures of cohabitation revealed sustained growth in cohabitation between 1960 and 1990 for adults aged 60 and older. Direct measures indicate that by 2000, more than 400,000 persons in this age group were cohabiting (Brown et al. 2006). In 2008, the number of older Americans who were cohabiting increased, 4% of cohabiting couples had a member aged 65 or older (U.S. Census

Bureau 2009). Among older Americans living in an opposite sex relationships, about 2% were cohabiting, and 98% were married (U.S. Census Bureau 2009).

Living Apart Together

The concept of LAT relationships evolved from research in Europe (Levin 2004; Haskey and Lewis 2006) and has extended to the United States (Strohm et al. 2009). These relationships are committed, long-term intimate unions in which couples do not share a home but rather maintain separate residences. Most of the United States and European studies are limited to younger age groups, and only a few examine patterns among older adults. De Jong Gierveld (2004) focused on Dutch respondents aged 50 and older who experienced widowhood or divorce and found that one-third of those who repartnered after age 50 were in LAT relationships. She concluded that older adults are more likely to live apart together than their younger counterparts (De Jong Gierveld 2004). Research in Sweden indicates that this arrangement is increasingly common among older couples as a strategy to engage in emotional support with some level of autonomy (Karlsson and Borell 2002). This seems to be a relationship that may be especially well suited to older Americans who can afford to live separately and want to maintain some autonomy while experiencing the benefits of a close intimate bond. Unfortunately, estimates of LAT relationships among older adults in the United States are not available.

Gays and Lesbians

Research on gay and lesbian family experiences is typically restricted to young ages, excluding older Americans. Of course, older Americans are part of same sex couples, too. Recent data permit measurement of the prevalence of same sex households in the United States. In the 2007 American Community Survey, just over 10% (11.3%) of same sex couple households had one member over the age of 65. A greater proportion of these were male (60%) than female (40%) couples. Seventy-one percent of same-sex older couple households had coresided for 5 years or more (Bennett and Gates 2004). According to the 2000 census, almost every county (97%) had an older same-sex American (Gates 2003).

Given new legislation about marriage among same-sex couples and the development of domestic partnership agreements, there are more options available for older Americans. The legal definitions of these relationships are significant when one member of the couple dies (Bennett and Gates 2004). The social security benefits are lower, and taxes on inheritance and retirement plans are larger among gays and lesbians than among married couples (Bennett and Gates 2004). Gay and lesbian older Americans also face constraints due to Medicaid and property laws that may make it difficult for partners to remain in their home following their partner's move to a nursing home or death (Bennett and Gates 2004).

Conclusion

The aging of the U.S. population coupled with rapid changes in family formation and dissolution earlier in the life course has contributed to shifts in the marital status and living arrangements of older men and women over the last 30 years (Chevan 1996; Cooney and Dunne 2001). This chapter documents changes in the marital status and living arrangement distributions of women and men between 1980 and 2008 using data from the decennial census and the American Community Survey. We also describe racial and ethnic variation in these patterns. These changes are discussed in terms of the family change approach, which emphasizes how shifting economic, normative, and institutional constraints have contributed to family change. These patterns are informative for researchers and policymakers alike, as an increasingly diverse older population likely will require an array of both informal and formal supports as they age. Specifically, with fewer older adults married and more living alone, increased institutional supports will be needed. New family forms, such as cohabitation, LAT, and same sex relationships also merit careful study to determine the costs and benefits of these unions relative to other living arrangements. These new living arrangements appear to be here to stay and represent the future of families. Moreover, much can be learned by considering the full range of families which move beyond just legal heterosexual marriages. For example, family change means that traditionally taken for granted intergenerational family ties are more complex and disrupted. Among younger samples, there is lower social support for gay and lesbian couples (Kurdek 2004) and cohabiting couples (Eggebeen 2005), which is suggestive of what may happen among older Americans who do not engage in traditional marital relationships.

Similar to the overall U.S. population, marriage among older adults has declined since 1980. This decline has been more pronounced among men than women. For women, the proportion married has changed very little. Consistent with the projections made in prior research (Wade 1989), widowhood has fallen, whereas divorce has risen among women. Men are also much more likely to be divorced today than they were in 1980. The share of men who are never married remains essentially unchanged, even though it was projected to double between 1990 and 2040 (Wade 1989). More recent projections indicate that the proportion of older adults married will continue to decline with a corresponding increase in the never married. The proportion divorced should stabilize in the coming years (Tamborini 2007). These levels of divorce will most likely have negative implications. Lin (2010) reports that divorced parents expect less support from children, and divorced men are most likely to experience a retraction of their children's support in old age.

Living alone is more common today than in 1980, and a larger share of women than men reside solo. Family and nonfamily living arrangements remain rare. The vast majority of older adults are residing either with a spouse or alone, and this is especially true among whites. Blacks are relatively less likely to be married and more likely to reside with family, reflecting the racial and ethnic variation characterizing the living arrangements of older adults (Himes et al. 1996).

This examination of changes in marital status and living arrangements does not capture all family and relationship types that exist among older Americans. For instance, our measure of marriage does not distinguish between first and higher order marriages. Remarriage among older Americans in 1980 was largely a function of widowhood rather than divorce. In 1980, most widowed older Americans did not remarry; just 1% of women and 25% of men remarried (Moss and Moss 1981). More recent analysis of remarriage is warranted, as remarriage and stepfamilies are linked to variation in support in older adulthood (Curran et al. 2003; Lin 2010).

Our descriptive analyses rely on cross-sectional data to provide a snapshot of older adults at two points in time. This approach is useful because it illustrates the distribution of older adults across marital status or living arrangement categories, but it obscures the broader trajectories of relationship histories that unfolded earlier in the life course. Increasingly diverse family and living arrangement experiences during young and middle adulthood have enduring consequences for later life (Cooney and Dunne 2001). At the same time, older adults experience living arrangement transitions and, from a life course perspective, these transitions are contingent upon earlier events and experiences (Wilmoth 1998). Voluntary (e.g., marriage or divorce) vs. involuntary (e.g., widowhood) transitions may have differential effects on older adult well-being.

For these reasons and to better test union formation and stability theories, future work on the demography of unions in later life would benefit from a longitudinal approach that incorporates family and living arrangement transitions both prior to and during older adulthood to provide a more nuanced portrait of the family and relationship experiences of older adults. In the last 30 years, new and on-going data collections have offered longitudinal (e.g. NLSY, NSFH, WLS) and periodic cross-sectional (e.g., NSFG, CPS, Census) assessments. To best understand the experiences of family life among older Americans in the next 30 years requires investment in longitudinal data collection of young and middle age adults today. In addition, men and women experience relationships differently, but few large-scale studies include information from both men and women or couples. Indeed, intimate relationships outside of marriage have risen in recent years. Growing attention has been paid to unmarried intimate relationships among older adults, including cohabitation, LAT, and same sex relationships (Cooney and Dunne 2001). Still, little is known about the prevalence of these union types or the determinants of entry and exit from these unions (Brown et al. 2006). As they become more common, especially with the movement of the baby boom cohort into older adulthood, it will be more feasible to incorporate these emerging relationship types in large, national surveys to generate new scholarship on the demography of these unions. This line of inquiry will be enriched by theoretical developments concerning the meaning and significance of these new partnerships for the health and well-being of older adults.

We focus on *coresidential* relationships, including marriage, cohabitation, and family household membership and therefore do not examine romantic or sexual relationships. Qualitative evidence suggests that dating among older adults following widowhood or divorce is fairly common (Bulcroft and Bulcroft 1991; Cooney and Dunne 2001), although women are less likely to desire remarriage than men (Talbott 1998). There are few empirical studies of dating, with Carr's (2004) recent work a notable exception. She finds that widowers' interest in dating is greatest when they lack other forms of social support, whereas widows' interest in dating does not depend on levels of support from friends.

In addition to dating relationships, new research is emerging on sexual activity among older adults. Waite (2009) maintains that sexual functioning is integral to overall well-being in later life. Even though research often assumes that older Americans are not sexually active, Lindau et al. (2007) report that in a sample of partnered older adults, 67% of men and 40% of women aged 65–74 report being sexually active in the last 12 months, while 39% of men and 17% of women aged 75–84 did. These figures are from the recently fielded National Social Life, Health, and Aging Project (NSHAP), which provides detailed data on the relationship and sexual behaviors of Americans ages 57–84. The declining share of married older adults translates into rising proportions at risk of forming new sexual partnerships, which not only may have ramifications for access to social support but also for physical and mental health.

Older adults are a growing share of the U.S. population. In fact, the oldest old are the fastest growing age group today (National Center for Health Statistics 2009). The older adult population, which is diverse in terms of racial and ethnic composition, has witnessed significant changes in marriage and living arrangements since 1980. The decline in marriage coupled with increase in divorce and living alone have consequences for the well-being, care, and support of older adults. Shifts to more complex family living arrangements, including the emergence of unmarried partnerships, have led to more varied types of support networks. Mounting evidence suggests that men will pay a higher cost for less stable family life by having limited support networks in old age (Lin 2010). Theoretical perspectives on family change must pay attention to the shifting demographic foundations of families and households across the life course.

Family change has not been experienced in the same way by all Americans. We have shown differentials according to gender and race/ethnicity, but there are also likely socioeconomic differences in living arrangements and relationships among older Americans. Consistent with both the family change approach and a life course perspective, these social inequalities in family formation and stability have ripple effects and continue to exist among older Americans.

The demography of unions in older adulthood has important social policy implications. Institutional supports will be needed to serve a growing, heterogeneous population that will have fewer forms of informal support to draw from. Lengthening life expectancies will only exacerbate these trends. Policymakers must be aware of the wider array of family situations characterizing older adulthood to ensure responsive policies are developed and maintained that reach across the range of relationship and living circumstances experienced by older adults. Close relationships play an integral role in the health and well-being of older adults, and formal institutional supports, particularly large government-sponsored initiatives, have the potential to greatly enhance the lives of the aged.

Acknowledgement This research was supported in part by the Center for Family and Demographic Research, Bowling Green State University, which has core funding from the Eunice Kennedy Shriver National Institute of Child Health and Human Development (R24HD050959-01).

References

Administration on Aging 2009. Aging Statistics. Retrieved July 30, 2010 (http://www.aoa.gov/AoARoot/Aging_Statistics/index.aspx).

Allen, Katherine R., Rosemary Blieszner, and Karen A. Roberto. 2000. "Families in the Middle and Later Years: A Review and Critique of Research in the 1990s." *Journal of Marriage and the Family* 62:911–26.

Axinn, William and Arland Thornton. 2000. "The Transformation in the Meaning of Marriage." Pp. 147–65 in *The Ties that Bind: Perspectives on Marriage and Cohabitation*, edited by L. Waite, C. Bachrach, M. Hindin, E. Thomson and A. Thornton. New York: Walter de Gruyter, Inc.

Becker, Gary S. 1974. Marriage, Family Human Capital and Fertility. *The Journal of Political Economy* 82:S11–26.

Bennett, Lisa and Gary J. Gates 2004. "The Cost of Marriage Inequality to Gay, Lesbian, and Bisexual Seniors." Washington, D.C.: Human Rights Campaign.

Bianchi, Suzanne M., V. Joseph Hotz, Kathleen McGarry, and Judith A. Seltzer 2008. *Intergenerational Ties: Alternative Theories, Empirical Findings and Trends, and Remaining Challenges*. Pp. xx–xx in *Intergenerational Caregiving*, edited by A. Booth, N. Crouter, S. Bianchi, and J. Seltzer. Washington DC: Urban Institute Press.

Brown, Susan L. and Sayaka Kawamura. Forthcoming. "Relationship Quality among Cohabitors and Marrieds in Older Adulthood." Social Science Research 39:777–86.

Brown, Susan L., Jennifer Roebuck Bulanda, and Gary Lee 2005. "The Significance of Nonmarital Cohabitation: Marital Status and Mental Health Benefits among Middle-Aged and Older Adults." *Journal of Gerontology: Social Sciences* 60B:S21–9.

Brown, Susan L., Gary R. Lee, and Jennifer Roebuck Bulanda. 2006. "Cohabitation among Older Adults: A National Portrait." *Journal of Gerontology: Social Sciences* 61B:S71–9.

Bulcroft, Richard A. and Kris A. Bulcroft. 1991. "The Nature and Functions of Dating in Later Life." *Research on Aging* 13:244–60.

Carr, Deborah 2004. "The Desire to Date and Remarry among Older Widows and Widowers." *Journal of Marriage and Family* 66:1051–68.

Calasanti, Toni and K. Jill Kiecolt. 2007. "Diversity among Late-Life Couples." *Generations*, 31, 10–17.

Caspi, Avshalom, Daryl J. Bem, and Glen H. Elder. 1989. "Continuities and Consequences of Interactional Styles across the Life Course." *Journal of Personality* 57:375–406.

Cherlin, Andrew. 2004. "The Deinstitutionalization of American Marriage" *Journal of Marriage and Family* 66:848–61.

Chevan, Albert. 1996. "As Cheaply as One: Cohabitation in the Older Population." *Journal of Marriage and the Family* 58:656–67.

Cooney, Teresa M. and Kathleen Dunne. 2001. "Intimate Relationships in Later Life: Current Realities, Future Prospects." *Journal of Family Issues* 22:838–58.

Coward, Raymond T., Gary R. Lee, Julie K. Netzer, Stephen J. Cutler, and Nicholas L. Danigelis 1996. "Racial Differences in the Household Composition of Elders by Age, Gender, and Area of Residence." *International Journal of Aging and Human Development* 42:205–27.

Curran, Sara R., Sara McLanahan, and Jean Knab. 2003. "Does Remarriage Expand Perceptions of Kinship Support Among the Elderly?" *Social Science Research* 32:171–90.

De Jong Gierveld, Jenny. 2004. "Remarriage, Unmarried Cohabitation, Living Apart Together: Partner Relationships Following Bereavement or Divorce." *Journal of Marriage and Family* 66:236–43.

Elder, Glen H., Jr. 1985. "Perspectives on the Life Course." Pp. 23–49 in *Life Course Dynamics: Trajectories and Transitions, 1968–1980*, edited by G. H. Elder, Jr. Ithaca, NY: Cornell University Press.

Eggebeen, David J. 2005. "Cohabitation and Exchanges of Support." *Social Forces* 83:3:1097–1110.

Gates, Gary. 2003. "Gay and Lesbian Families in the Census: Gay and Lesbian Seniors." Washington, D.C.: Urban Institute.

Haskey, John and Jane Lewis. 2006. "Living-Apart-Together in Britain: Context and Meaning." *International Journal of Law in Context* 2(1):37–48.

Heron, Melanie P., Donna Hoyert, Sherry Murphy, Jiaquan Xu, Kenneth Kochanek, and Betzaida Tejada-Vera. 2009. *Deaths: Final Data for 2006*. National Vital Statistics Reports, 57(14). Hyattsville, MD: National Center for Health Statistics.

Himes, Christine L., Dennis P. Hogan, and David J. Eggebeen. 1996. "Living Arrangements of Minority Elders." *Journal of Gerontology: Social Sciences* 51B(1):S42–8.

Huyck, Margaret Hellie. 1995. "Marriage and Close Relationships of the Marital Kind." Pp. 181–200 in *Handbook of Aging and the Family*, edited by R. Blieszner and V. H. Bedford. Westport, CT: Greenwood Press.

Karlsson, Sofie Ghazanfareeon and Klas Borell. 2002. "Intimacy and Autonomy, Gender and Ageing: Living Apart Together." *Ageing International* 27:11–26.

King, V., & Scott, M.E. 2005. A Comparison of Cohabiting Relationships Among Older and Younger Adults. *Journal of Marriage and Family* 67:271–85.

Kurdek, Lawrence A. 2004. "Are Gay and Lesbian Cohabiting Couples Really Different from Heterosexual Married Couples?" *Journal of Marriage and Family* 66:4:880–900.

Lesthaeghe, Ron and Lisa Neidert 2006. "The Second Demographic Transition in the United States: Exception or Textbook Example?" *Population and Development Review* 32:4:669–98.

Levin, Irene. 2004. "Living Apart Together: A New Family Form." *Current Sociology* 52:223–40.

Lin, I-Fen. 2010. "Gender, Divorce, and Expectations of Support from Adult Children." Center for Family and Demographic Research Working Paper 2010–04. Bowling Green State University Bowling Green, Ohio.

Lindau, Stacy Tessler, L. Philip Schumm, Edward O. Laumann, Wendy Levinson, Colm A. O'Muircheartaigh, and Linda J. Waite. 2007. "A Study of Sexuality and Health among Older Adults in the United States." *The New England Journal of Medicine* 357:762–74.

Meyers, George C. and Barbara Foley Wilson. 1988. "Marriage and Divorce Trends and Patterns among Older Americans." Presented at the annual Meeting of the Population Association of America, New Orleans, LA, April.

Moss, Miriam S., and Sidney Z. Moss. 1981. "The Image of the Deceased Spouse in Remarriage of Elderly Widow(ers)." *Journal of Gerontological Social Work* 3:59–70.

Mutchler, Jan E. 1992. "Living Arrangements and Household Transitions among the Unmarried in Later Life." *Social Science Quarterly* 73:565–80.

National Center for Family and Marriage Research. 2009. "Marital Status in the U.S., 2008." Family Profile 09-04.v. Retrieved (http://ncfmr.bgsu.edu/family_percent20marriage_lit/Family percent20Profiles/family_profiles.html).

National Center for Health Statistics. 2009. Health, United States, 2008 with Chartbook Hyattsville, MD.

Raley, R. Kelly and Larry Bumpass. 2003. "The Topography of the Divorce Plateau: Levels and Trends in Union Stability in the United States after 1980." *Demographic Research* 8:245–60.

Rossi, Alice S. and Peter H. Rossi. 1990. *Of Human Bonding: Parent-Child Relations across the Life Course.* New York: Walter de Gruyter, Inc.

Seltzer, Judith, Christine Bachrach, Suzanne M. Bianchi, Caroline H. Bledsoe, Lynne M. Casper, P. Lindsay Chase-Lansdale, Thomas A. DiPrete, V. Joseph Hotz, S. Philip Morgan, Seth G. Sanders and Duncan Thomas. 2005. "Explaining Family Change and Variation: Challenges for Family Demographers." *Journal of Marriage and Family* 67:4:908–25.

Settersten, Richard 2007. "Social Relationships in the New Demographic Regime: Potentials and Risks, Reconsidered." *Advances in Life Course Research* 12:3–28.

Settersten, Richard A., Jr. 2009. "It Takes Two to Tango: The (Un)Easy Dance between Life-Course Sociology and Life-Span Psychology." *Advances in Life Course Research* 14(1–2):74–81.

Shapiro, Adam and Teresa M. Cooney. 2007. "Divorce and Intergenerational Relations across the Life Course." *Advances in Life Course Research* 12:191–219.

Smock, Pamela. 2004. "The Wax and Wane of Marriage: Prospects for Marriage in the 21st Century." *Journal of Marriage and Family* 66(4):966–73.

Strohm, Charles Q., Judith A. Seltzer, Susan D. Cochran, and Vickie M. Mays. 2009. "Living Apart Together: Relationships in the United States." *Demographic Research* 21:177–214.

Sweeney, Megan M. 2002. "Two Decades of Family Change: The Shifting Economic Foundations of Marriage." *American Sociological Review* 67:132–47.

Talbott, M.M. 1998. Older Widows' Attitudes Towards Men and Remarriage. *Journal of Aging Studies* 12:429–49.

Tamborini, Christopher R. 2007. "The Never-Married in Old Age: Projections and Concerns for the Near Future." *Social Security Bulletin* 67(2):25–40.

U.S. Bureau of the Census. 2009. "Same-Sex Couples, Data from the American Community Survey: 2008." Retrieved July 30, 2010 (http://www.census.gov/population/www/socdemo/hh-fam.html).

Wade, A. 1989. Social Security Area Population Projections. Actuarial Study No. 105, Washington, D.C.: Office of the Actuary, Social Security Administration.

Waite, Linda. 2009. *Marriage & Health, Especially at Older Ages.* Families and Health Conference. Washington, D.C., June. Retrieved July 30, 2010 (http://ncfmr.bgsu.edu/pdf/families_health/file78452.pdf).

Waite, Linda J., Christine Bachrach, Michelle Hindin, Elizabeth Thomson, and Arland Thornton. 2000. *The Ties that Bind: Perspectives on Marriage and Cohabitation.* New York: Walter de Gruyter, Inc.

Wilmoth, Janet M. 1998. "Living Arrangement Transitions among America's Older Adults." *The Gerontologist* 38:434–44.

Wilmoth, Janet M. and Charles F. Longino, Jr. 2006. "Demographic Trends That Will Shape U.S. Policy in the Twenty-First Century." *Research on Aging* 28:269–88.

Part V
Social Institutions and Aging

Chapter 14
Rethinking Retirement

Melissa Hardy

Because retirement is continuously being shaped by social policy, labor markets, economic systems, family dynamics, and personal predilections, retirement research has been the study of a changing social institution. Macro and micro level social processes have transformed what once was a status unavailable to the majority of older workers into a highly anticipated and often eagerly awaited stage of the life course. As the means of production shifted away from agriculture and the scale of enterprises grew, bureaucratic standardization of job entries and exits became a key component of labor force management. Defining the composition of that labor force relative to skill, speed, and compliance was linked with demographic traits such as age, gender, and education. Although the specifics have changed, retirement continues to be a key component of the social organization of work and the distribution of income. As categories of nonemployment, retirement, unemployment, disability, home production, and leisure are responsive to broader labor market conditions, prevailing economic policies, and social norms.

As a relatively new field, studies of retirement relied on prevailing theoretical perspectives adapted from the broader landscape of social science disciplines. As retirement research developed, interest in the personal and social consequences of allocating a growing proportion of the adult life course to "years spent in retirement" shifted from a primary focus on adjustment and adaptation to an emphasis on the social and monetary costs of assigning an expanding number of years to a stage of the life course that primarily involved consumption rather than production or accumulation. The political discourse has emphasized issues of distribution and equity relative to generational membership, social class, and demographic subgroups as points of contention, questioning the fairness and, in some cases, the basic morality of a redistributive regime that is both deliberate, hierarchical, and self-consciously productive of social relations in contrast to one that is market driven and indifferent to consequential inequalities.

During the last century, government retirement policies, industrial retirement programs, and personal retirement expectations were adapting to an age-structured approach to employment that moved retirement to younger ages, gained many twentieth century advocates, and now has many twenty-first century critics (Ekerdt and DeViney 1990). One issue at the core of retirement research has been the tension between human agency and social structure that is often framed in the terminology of voluntary decisions, personal preference, and individual behavior vs. unequally distributed opportunities and constraints, and the complex social relationships in which they are embedded. A basic division in the literature, therefore, is among studies that address what actors did, when the action occurred, and the circumstances and characteristics that influenced the likelihood and timing

M. Hardy (✉)
Department of Sociology and Population Research Institute,
Pennsylvania State University, University Park, PA, USA
e-mail: Melissa.Hardy@psu.edu

R.A. Settersten, Jr. and J.L. Angel (eds.), *Handbook of Sociology of Aging*, Handbooks of Sociology and Social Research, DOI 10.1007/978-1-4419-7374-0_14, © Springer Science+Business Media, LLC 2011

of these actions; studies that attempt to incorporate the voices of the actors, their explanations for their actions, how they viewed their options, how they evaluate different life domains, and how they see themselves; and studies that take a broader often comparative view of retirement (and retirement policy) as a symptom of labor market reorganization and industrial restructuring, reflective of culturally based perceptions of age, collective responsibility, governments, and markets.

Regardless of whether one thinks of retirement as a status, a transition, or a stage of the life course, classifying people as retired (or in the process of retiring) is a messy process. The use of labor force status has been a favored strategy, since it allows a consistent basis for historical and international comparisons of aggregate trends, but does not necessarily capture the complexities of individual transition behavior. Comparative researchers suggested employment status as a better indicator than labor force status, since a number of European countries (e.g., France, Germany, The Netherlands) had muddied the retirement transition by using unemployment or disability programs to replace wages for those too young for "early" retirement benefits, making the distinction programmatic rather than behavioral (Guillemard and Rein 1993). Add self-reports of transitions, sources of income, work history and identity to the mix, and possible operational definitions multiply (Szinovaca and DeViney 1999).

As retirement became more established as a distinct stage of the life course and the size of the retired population grew, retirement research was dominated by four sorts of questions. The first set of questions focused on the behavioral patterns: When do people retire? Why do they retire when they do? What do they do after retirement? The second set of questions tackled issues of inequality: Who were able to retire when they wanted to, and who worked longer or retired earlier than planned? How was the financial security of retirement affected by one's place in the occupational structure? By one's race, gender, marital status, health, or work history? To what extent did inequality increase, decrease, or remain the same after retirement? The third set of questions dealt with issues of adjustment: Were retirees happy with their lives? Did health (both mental and physical) improve or deteriorate because of retirement? How did retirement affect marriages and marital satis faction? A fourth set of questions addressed how retirement and retirement transitions differed across countries: How did the state respond to the challenges of industrial reorganization? What role did firms or unions play in orchestrating these transitions? How were different retirement regimes reflected in the security vs. vulnerability of the older population? In the remainder of this chapter, I will discuss our questions, our answers, and our approaches using an organizational heuristic – historical decades – to align for the reader the ongoing research process with changes in policy, politics, the economy, social attitudes, and societal expectations. Neither the prominence of the issues nor the appearance of the research can be neatly packaged into this framework. Changes in political leadership, presidential terms, economic conditions, and demographic shifts do not coincide with 10-year time spans. Even so, this approach invokes a broader context in which the reader can loosely situate the retirement research enterprise and appreciate the major themes of inquiry.

The 1970s: Building a Financially Secure Retirement

The 1970s ushered in a new era of retirement – perhaps what could be considered the modern era of retirement – as the early wave of baby boomers shifted from school to work and older workers moved into retirement with somewhat higher Social Security benefits and Medicare coverage. Retirement research was tracking these changes from the perspective of workers who had been born in the first part of the twentieth century and had fueled the postwar economic growth and the rising living standards the booming economy made possible. The research literature became more interdisciplinary, as many of the social sciences as well as some schools of business became more interested in the phenomenon. Although earlier studies had focused on mandatory retirement policies, passage of the American Discrimination in Employment Act in 1967 began the process of largely eliminating mandatory ages. A branch of retirement research began to identify the "push" and "pull" factors that could move older workers into retirement, while a second branch of research noted insufficient institutional support for the new status. Researchers began to demonstrate how retirement behavior was responding to changes in policies, with ages of pension entitlement mirrored in how retirement was timed. As the circumstances of retirement improved, the public image of retirement was changing to that of an eagerly anticipated escape from the routine of work to the discretion of leisure.

Research studies were spurred by the development of several large longitudinal data collection projects, including the National Longitudinal Surveys (NLS) and the Retirement History Survey (RHS). This research concentrated largely on the retirement behavior of white men and in the RHS, unmarried women. Based on a prospective longitudinal design, these data sets allowed an observational window on "before" and "after" retirement characteristics, activities, and events. They included large national samples so that population estimates of this transition behavior could be developed, and their over-samples of African Americans allowed comparisons by race, although the concept of retirement did not fit minority experiences as well as those of white men.

A few studies focused on specific firms (Barfield and Morgan 1969; Rogers and Friedman 1980) to study retirement when modes of production were changing in the workplace. Transitions to new technologies that required new skills inspired a firm-based calculus that balanced the costs of retaining and retraining older workers against the costs of sweetening pensions and hiring younger replacement workers. Employers often assumed that older workers would be unable to make the transition to a new production technology. That many older workers lamented the demise of a labor process that involved skilled craftsmanship contributed to prevailing stereotypes of older workers – that they resisted change and were too rigid in their approaches to their work.

The general trend in research was toward developing reliable population estimates, which meant that the retirement research enterprise would increasingly shift to these large, secondary data sets and necessarily be limited by the information that was available in them. This period also marked disagreement over how to define retirement. Early reliance on self-reports was criticized as being too subjective. Whether these self-reports were used to sort people into workers vs. retirees or distinguish those with or without health impairments, or whether they described why the retirement occurred, such measures were viewed as inferior to behavior-based measures, such as labor force participation or labor supply, or functional measures, such as diagnosed disease, activity limitations, or formal disability status.

The 1980s: Understanding the Democratization of Retirement

In the early 1980s, we experienced the first modern crisis in social security; on the heels of legislation that improved the predictability and level of benefits, stabilized replacement rates, and secured the real value of benefits, the high levels of unemployment and inflation in the late 1970s produced a revenue shortfall, which required the temporary transfer of general funds. Putting Social Security back on sound fiscal footing led to another fundamental shift in funding strategy – preemptive tax increases in FICA, the reemergence of the trust fund, and scheduled delays in the full retirement age (FRA) to be staged in for the baby boom cohorts (Hardy and Hazelrigg 2007). The economic impact of retirement – on nations, firms, and households – was receiving negative attention and although ambivalence about financial support for early retirement was rising, workers still expected to retire before age 65.

As Sociologists tried to understand why some people retired earlier and others retired later, the general expectation was that virtually all workers would retire at some point. Some sociologists still emphasized social structure – opportunities available to older workers, the apparent lack of fit between those older workers who wanted to stay employed and the older workers whose employers wanted to hire or retain, and the impact of an ever lengthening retirement period on couples, families, and communities; others focused more on person-level processes of anticipation, adjustment, and adaptation. Studies that addressed the relationship between health and retirement searched for better measures of health (compared to self-reported work limitations) as predictors of retirement, particularly early retirement, and noted that the health declines experienced in retirement could reflect the more negative health trajectories of those moving into retirement.

Viewing the attributes of the retirement status and the character and timing of the transition as an outcome of a career pattern also fit within the life course framework. Workers' location in the occupational structure correlated with a number of retirement-relevant characteristics. Job security, access to promotion ladders, coverage by employer-sponsored pensions and health insurance, total compensation packages, and the nature of job tasks were allocated through jobs rather than assigned to specific workers. Workers with secure jobs who engaged in substantively complex work tasks, exercised discretion over their work day, and engaged in positive social interactions tended to retire later than workers whose jobs were physically demanding and whose work days were highly structured (Hayward et al. 1989).

Researchers began to unfold retirement as an increasingly complicated and heterogeneous set of transitions and adjustments. Once viewed as a one-time, complete and permanent exit from paid employment, various retirement patterns started to emerge. Workers might end a career job through a "retirement" exit and then, within a year or two, reenter the labor force, sometime in an entirely different sort of job (Beck 1985; Hayward, Hardy, and Liu 1994). Older workers who were displaced from their jobs, spent some time looking for work, and then move into retirement; older workers who retired or were displaced from jobs moved into "bridge" jobs until they retired "completely," often working part-time schedules and at lower pay. Workers (primarily white-collar) claiming age discrimination – in hiring or retention decisions – went to court (Quadagno and Hardy 1991), and older workers caught up in a spiral of plant closings and downsizing took advantage of early retirement packages and either remained retired or shifted to different jobs. The extent to which older workers were "pushed" or "pulled" into retirement remained a point of debate, and whether differences in the nature of retirement transitions were linked to more fundamental inequalities was also explored, since different transitional processes were linked to different levels of satisfaction and financial security in retirement. "Discouraged" workers as a different category of labor force status resurfaced briefly (Hardy 1991), but "evidence" of this status relied on self-reported reasons for not actively searching for employment. Given that unemployment rates among older workers were quite low in relative and absolute terms, researchers continued to emphasize the supply of older workers rather than demand factors: what sorts of incentives could entice an older worker into early retirement or a retiree with desirable skills back into the labor force? The language of retirement also shifted from something that happened to people ("I was retired" or "it was time" for me to retire) to something they "took," a linguistic shift that suggested a change in agency and a focus on income replacement (taking the pension in lieu of wages or taking the retirement "package").

Recognition that women – married and unmarried – were also moving into retirement prompted researchers to question whether the transition, the process, and the status may differ for women and men (Hayward, Grady, and McLaughlin 1988). As the rising employment rates of women pushed the study of women and work (and gender differences in work) to the forefront, gendered responsibilities for housework and family care also received more attention. Whereas work was seen as central to the life course of men, family responsibilities occupied a central position for women. This assumption framed research on both the retirement transition and on retirement adjustment for women. Researchers started asking whether women's experiences should be viewed through a different lens (Calasanti 1993), or at least an expanded lens, that incorporated work and family roles as well as other life course dimensions on which men's and women's experiences and expectations differed.

Through cohort replacement, growing proportions of women had longer employment spells, so research began to focus on the impact of the gender gap in wages and the gendered composition of occupations and industries for women's financial security in retirement. Because of women's greater involvement in parenting, women's work histories were less continuous then men's; therefore, the study of women's retirement had to take family responsibilities and the joint trajectories of women's work and family roles into account (O'Rand and Henretta 1982; Henretta and O'Rand 1983). Compared to men, women generally had shorter work histories, lower annual and lifetime earnings, and lower rates of participation in employer-provided pensions (DeViney and O'Rand 1988). The discontinuity in women's work histories explained some of the gender gap in retirement income, but the underlying processes that sorted women into jobs that offered no pension plans reflected the gendered nature of occupational structure. Therefore, in retirement women were more reliant on spousal and survivor's benefits and particularly vulnerable to the policies that governed such benefits. That women faced higher risks of poverty in old age and that African American women faced much higher poverty risks reflected these life course patterns of gender and race inequalities.

The terrain of employer-sponsored pensions was also changing, as companies began to shift from defined benefit to defined contribution plans and 401(k)s. Put into effect on January 1, 1980, the amendment to the Internal Revenue Code provided tax exemptions to employees on income they chose to receive as deferred compensation rather than direct compensation. By 1983, almost half of large firms already offered or were considering a 401(k) plan (EBRI 2005), a retirement plan easier to administer and cheaper for employers, who paid only plan administration and support costs (in the absence of matching contributions) and were able to pass on much of that cost to plan participants.

But by the end of the 1980s, pension plan coverage had peaked and long-term liabilities in defined benefit pension plans were rising. The shift to defined contribution plans was, in part, attributed to employers' reactions to increases in reporting requirements, funding regulations, and attendant rising costs of administration, but it also signaled changes in institutional structure, the perceived advantages of long-term contracts, and employers' desire for additional "flexibility" in retention decisions. Loss of market share in the manufacturing industry was increasing the rate of plant closures, just-in-time production, and outsourcing. The vertically integrated enterprise offering long-term contracts for worker loyalty and the development of firm-specific skills was being replaced by more limited inventories of products and workers, shorter term contracts, and limited payroll liabilities – both current and deferred. In retrospect, we would see that the mid-1980s as the low point in labor force participation rates for older men and women. "Black Monday" (October 19, 1987), which marked the largest drop in the U.S. stock market up to that time, temporarily shook the confidence of U.S. investors, who had been experiencing long-term growth in equities; more and larger drops were to follow, but these events helped make the economy a major concern as we moved into the last decade of the millennium.

The 1990s: The Accelerating Process of Individuation

By the 1990s, the retirement research process had become increasingly politicized. Conservative think tanks emphasized intergenerational equity issues and the need to privatize Social Security. Counter arguments emphasized how fragile retirement was for many U.S. households, the importance of social insurance in combination with private investment, and the benefits of widely pooled risk. Frequent stories in the public media privileged the sound bite over more nuanced explanations, and researchers learned to parse their words carefully when talking about the topic.

Policy changes in both public and private spheres accelerated the process of greater individuation, manifested in greater within-cohort heterogeneity in when and how major life course transitions occurred. Although the timing of retirement was still influenced by ages of Social Security entitlement, the early retirement incentives offered through private industry had shifted many retirement exits to younger ages. Pathways into retirement had also become more varied and more elaborate (Han and Moen 1999). Growing evidence of this heterogeneity challenged the concept of the "institutionalized life course" (Meyer 1986). Pathways to early retirement in seven industrialized nations illustrated the mix of public and private institutional arrangements that shaped the retirement process and revealed a wide range of age intersections (Kohli et al. 1991). Retirement as the entry point into "old age" was also being contested, as growing numbers of younger, healthier retirees rejected the label of "senior citizen," and an "anti-aging" industry offered a dizzying array of products to help buttress their claim.

Growth of defined contribution pensions further reduced the age-standardization of pension entitlement. Positive features of these plans included their portability, flexibility in when benefits could be claimed, and the possibility of "borrowing" funds if necessary. But these advantages were counterbalanced by the increased risk that was placed on workers and households. Different investment strategies, different economic conditions, different patterns of contributions could all produce different rates of accrual. When firms converted their DB plans to DC plans, the expected retirement age for older workers often changed substantially, since the rate of accumulation in DC plans could not produce the benefits promised by DB plans for workers nearing retirement age. What had been a clear connection between age, service, and benefits now depended on contributions, markets, returns, and annuities, an unfamiliar environment for many U.S. households.

Why people retired when they did was an important policy question, since reversing the pattern required an ability to manipulate the factors that prompted the behavior. Early in the decade, a new national survey was launched to provide longitudinal data to researchers interested in studying these issues. The Health and Retirement Study (HRS) fielded in 1992 became the primary U.S. data resource for studying retirement. Consistent with the view that retirement primarily involved the substitution of one income stream (earnings) for another (pension and savings income), HRS collected detailed information on income and assets as well as employment. Measures of "pension wealth," which were based on anticipated lifetime benefits, economic assumptions about inflation and discount rates, and demographic assumptions about longevity, convinced many households that early retirement was achievable even as studies demonstrated that median nonpension asset levels in U.S. households were primarily bound up in housing equity. Housing prices were rising, too; therefore accumulating higher levels of nonhousing, nonpension, and more liquid forms of savings seemed less important, especially when credit was so readily available. Although some threshold of retirement income was a prerequisite for voluntary retirement, understanding why people retired when they did also involved family considerations, social expectations, normative behaviors, and the organizational culture of the workplace. To the extent that firms aligned their pension plans with desired terms of labor force management, corporate downsizing could be handled through retirement rather than layoff, but the long-term costs of that strategy had been mounting. Employers routinely complained that ERISA regulations requiring them to offer ERIs to entire classes of workers eliminated their effectiveness in targeting reductions, with too many of the workers they hoped to "shed" or replace remaining on the job (Hardy, Hazelrigg, and Quadagno 1996). More skilled workers had better reemployment prospects and could retire and move to different jobs while simultaneously collecting their pension benefits. Accepting an ERI was therefore a reflection of household and employment trajectories that were more complicated than timing the generosity of the benefit package.

Older workers were not only retiring at earlier ages from career jobs. They faced growing risks of displacement, termination, and long-term unemployment. Previous studies of displaced workers had indicated that older workers faced lower risks of displacement than younger workers, but once out of work, they experienced longer spells of unemployment and larger wage reductions when they did find jobs. As a lengthening list of Supreme Court decisions narrowed the interpretation of the ADEA, involuntary job separation, which was more common among women and minorities

(Flippen and Tienda 2000), coupled with older workers' slightly higher rates of displacement compared to younger workers increased the volatility in late career employment behavior. Those factors highly correlated with age could be used to terminate employees made it much less risky for employers to focus layoffs and terminations on older workers. Using salary level or internal reorganization as criteria, employers could easily target more highly paid older workers for termination, inflicting disparate impact while avoiding disparate treatment.

Higher job exit rates, reductions in older men's job tenure, occupancy of short-term bridge jobs, and returns to work after exiting the labor force blurred work-to-retirement transitions. Beginning in the mid-1980s, the labor force participation rates for older men had stabilized at a historically low level, but in the 1990s rates for men had begun to rebound, even for men aged 65 and older. The labor force participation rates for older women also showed a small upward trend, as older workers continued in full-time or adopted part-time schedules and enacted transitional sequences reflecting opportunities and choices.

Disparate pathways to retirement had been more common in European countries, where public programs of retirement, disability, and unemployment provided wage replacement income to older workers who were considered "redundant." In countries such as Germany (Kohli 1987) and France (Guillemard and Rein 1993), the public sector financed early retirement as a method of easing unemployment. Although some researchers increasingly equated "early" retirement with voluntary retirement and viewed virtually all retirement exits as the behavioral manifestation of workers' preferences, sociologists continued to parse the difference between exits that were initiated through changing institutional demands, those prompted by poor health, and those that seemed to be proactively and prospectively planned by workers.

Within that framework, some retirements appeared to signal acceptance of one option among a choice set limited by protracted spells of illness or disability, by repeated failure to secure subsequent employment, because of ageism or skill obsolescence, or by increased family responsibilities (DeViney and O'Rand 1988). Were workers acting in accordance with their own timetables? Or were they adapting to changing circumstances, less secure jobs, or managers' signaling behavior? Earlier studies of how planned retirement ages matched (though perhaps not exactly) subsequent behavior had indicated a fairly high level of accuracy in such predictions, but that accuracy was arguably a function of the institutional predictability of retirement income and the security it offered. Even then, one third of the sample experienced either unfulfilled plans or unanticipated retirements (Ekerdt, Vinick, and Bosse 1989). Women's expected age of retirement reflected uncertainty as well. Initially responsive to when husbands would retire, women's expectations changed across time, becoming more or less certain, and more often shifting to older rather than younger ages (Wong and Hardy 2009).

Here again, the language mattered. Surveys differed in whether they asked about "expected" retirement age, "planned" retirement age, or when workers would "like" to retire; other questions were framed in terms of expectations of employment behavior, whether workers preferred full-time or part-time employment now of "after" they retired, and what kind of work they would do or would "like" to do. As U.S. researchers tried to grasp whether, when, and in what sorts of jobs people in this age group would continue to work for pay, European researchers were demonstrating the effectiveness of "partial pensions" and formal provisions for combining earnings and public pension income for keeping older workers in the labor force. Studies documented that earnings had become an increasingly important source of retirement income, a combination that to some was a contradiction. Pension benefits as wage replacement had been the conventional rationale; pension benefits as earnings supplements were something different. Even so, legislation modifying the earnings test in Social Security and employer attempts to circumvent regulations restricting in-service distributions supported phased retirement options.

As concerns about financial security in retirement increased, older workers' expectations about their retirement ages seemed to shift upward, as well. Whereas prior changes to pension benefits made early retirement (before age 62) affordable, more recent changes encouraged retirement delays. Concerns about continued access to health insurance, uncertainty about the time horizon for benefits generated by defined contribution plans, reduction of work disincentives embedded in the benefit structure of Social Security, and ADEA amendments removing the upper age limit for protection coupled with relatively low unemployment rates during the 1990s correlated with rising rates of labor force participation that were, in part, due to a retirement process characterized by jobs shifts interspersed with multiple exits and reentries for a significant minority of workers. These blurred vs. crisp exits (Elder and Pavalko 1993; Mutchler et al. 1997), characterized by multiple jobs and switches in labor force status (Hirsch, MacPherson, and Hardy 2000), made the retirement process more difficult to define (Ekerdt and DeViney 1990). What was the difference between a "bridge" job and a routine job change? How was "partial" retirement different from part-time work? Although most workers still retired in one step, the ambiguities of characterizing late career employment behavior made post-hoc or age based-definitions more common and receipt of pension benefits more determinative. Framing it as a "bridge" process to retirement gave it a more positive spin, but whether older workers were easing themselves out of the labor force or trying to hang on was often unclear.

As cohorts of married women who had experienced higher labor force participation rates moved into older ages, studying how couples timed their retirements gained additional attention (Szinovacz, Ekerdt, and Vinick 1992). Sometimes approached as a problem of joint utility maximization, sociologists looked at how spouses linked their transitions, the quality of the marital relationship, the past family and career patterns of wives (Pienta 1999), and subsequent satisfaction with the experience of retirement. That husbands were often older than their wives, earned more than their wives, and were more likely to participate in an employment-based pension plan factored into the timing of retirement behavior. Retiring at the same time, despite differences in age, was a common pattern and the older the spouse, the more common this behavior. But a lack of pension coverage and the lower the wage of one spouse, the more likely couples retired at different times, with the lower wage, uncovered spouse working longer. Further study emphasized the importance of the early "family organizational economy" for the sequential synchronization of retirement. Although proximal work-related characteristics such as health, pension benefits, and work features constrain the timing of retirement for couples, these factors act in conjunction with longer-term patterns in the division of work and family roles (Henretta, O'Rand, and Chan 1993).

Whether couples act with equal influence or whether the timing of retirement displays an asymmetry in marital power continues to be explored as cohorts aging into retirement demonstrate different work and marital histories and different understandings of gender roles (Smith and Moen 1998). The inequity in responsibility for domestic labor also affects women's retirement. Women's disproportionate responsibility in this domain not only constrains their opportunities to build retirement income, but women are more likely than men to retire in response to caregiving obligations (Brubaker and Brubaker 1992; Dorfman 1992).

Not only should women's retirement be studied on its own merits, but the retirement process for race/ethnic minorities also arguably diverges from the model for white men. Among men, African Americans have lower labor force participation rates, less continuity in attachment, and higher unemployment (Gibson 1992). Rates of disability are also higher for African American men, suggesting that their lower levels of labor force participation are more a function of health limitations than early retirement. The retirement experience is therefore different for these men, since they spend a greater portion of their lives both working and disabled than white men, a difference that also reflects their higher mortality rates and shorter life expectancy (Hayward, Friedman, and Chen 1996).

That retirement as a transition and as a life course stage reflects substantial inequality also has been an important focus of sociological research. Acknowledgement that public programs such as Social Security redistributed income from high wage earners to low wage earners suggested that perhaps income inequality after retirement would narrow. But retirement income sources were diversifying through the same sorts of market mechanisms that were increasing inequality in other areas of compensation. Stratified along several key dimensions that tended to be mutually reinforcing, the foundation for increasing inequality after retirement was being laid. The first layer of inequality involved access to employer provided pensions, health insurance, and relatively high wages that allowed substantial personal savings in pre-tax dollars, which was aligned with other favorable traits of "good" jobs. Among those with access, the level of benefits was further stratified along similar lines. Finally, those who expected or received relatively high benefits enjoyed different levels of reliability in the continuation of those benefits. For those not yet retired, plans for a particular retirement date could evaporate when employers froze existing defined benefit plans and replaced them with defined contribution plans based on benefits payable at that point in time rather than benefits promised by the typical back-loading of pension accrual. For those already retired, a bankruptcy could send the plan to the Pension Benefit Guaranty Corporation for administration. As dual-earner couples became the norm and the gender gap in pension participation narrowed, the retirement assets accumulated by couples rapidly outpaced those of the nonmarried.

Although defined benefit plans differed in benefit levels and eligibility rules, defined contribution pensions differed in both institutional and individual terms. Whether employers contributed, how much they contributed, and whether they matched employee contributions combined with whether employees chose to participate, whether and how much they contributed, how funds were invested, and whether pension assets were rolled over or cashed out when changing jobs. Women were increasing their participation, but they tended to invest more conservatively regardless of whether the market was rising or falling, and the frequency of job shifts made taking the relatively small balances in lump sum payments quite common (Hardy and Shuey 2000; O'Rand and Henretta 1999).

2000: Risk, Uncertainty, and Changing Expectations

The year 2000 started with a budget surplus that made the financial imbalance of Social Security seem easily solvable. But the surplus soon turned into steadily rising budget deficits under the priorities of a new administration. A changing economic and political environment fueled concerns about the future, and the discourse of privatizing Social Security was renewed. By the end of the decade, financial scandals, steep (though geographically uneven) declines in the housing market, rising unemployment, and a plunging stock market schooled a nation on the ephemeral nature of paper profits and job security. As the United States struggled to recover from what was likely its longest post-war recession (federal reserve), the elimination of old jobs and the failure to add new ones frayed public confidence. Pension wealth and housing equity plummeted, and though consumers were asked to buy to help reboot the economy, the mounting toll of earlier excesses was ushering in a new reality. Now, even for the average American, retirement had become a question of budget spreadsheets and disappointing quarterly reports, bound up with global financial markets, the strength of the dollar, and international banking. Expectations for early retirements were changing to delayed or partial retirements in an indefinite future. As the first wave of members of the baby boom cohorts turned 60, an event foreshadowed by apocalyptic predictions of bankrupt social programs and generational warfare, what had been a dire prediction – that wave after wave of baby boomers would flood into early retirement to "live off their children and grandchildren" – became a question: how will baby boomers change retirement?

Comparative studies of retirement indicated how European countries were struggling to shift from earlier to later retirement (and to manage the financial consequences of that trend within prevailing economic environments) by changing their public pension systems. Whereas employment rates among older workers had started moving upwards in the mid-1980s in the US, many European nations had continued to see erosion of labor force participation in older age groups. Some of the early retirement trend could be attributed to the maturation of pension programs, but European governments had also provided generous support to those taking early retirement during times of high unemployment. Concern over the sustainability of social welfare programs had resulted in the European Union's setting ambitious goals for delaying retirement, calling for a near doubling (to 50%) of labor force participation rates among those aged 55–64 within a 10-year window (Taylor 2006). Although Finland, Sweden, and the UK reached this target in 2004, Austria, Belgium, and Italy still hovered at 30%.

An early developer of a universal public pension system, Sweden passed pension reform in the mid-1990s, replacing their defined benefit system with a combination pay-as-you-go notional defined contribution plan, an advance funded individual account system, and a guaranteed supplemental pension payable at age 65 for those with low lifetime earnings. Contribution rates were set at 18.5%, with earnings-replacement benefits (received during periods of illness, unemployment, disability, or work injury) and child-care credits factored into pension rights. An advantage of the Swedish system is that it easily accommodates partial retirement, allowing earnings and partial-pensions to be combined (Palmer 2000). On the other hand, the system is so closely tied to economic growth, that existing pensioners can see their pensions reduced during economic downturns, introducing unpredictability to retirees' future benefits (Simpson 2009).

Workplace "flexibility" has been promoted as one strategy to extend the working lives of older workers. Flexible work options can benefit workers of all ages, but the options for older workers differ because of their access to earnings replacement income, retirement health care, and asset income. For younger workers, particularly those raising children, health insurance coverage and accumulating vs. spending savings remain important, particularly where child care credits are not available. Questions about the advisability of pushing leisure to older ages rather than distributing it more gradually throughout adulthood is part of an "age integrated" life course that would considerably reduce the age limitations associated with adult behavior (Riley and Riley 1994). If we assume a given proportion of the average lifespan will be spent as "leisure," is it rational to allocate so little leisure time to midlife? The affordability of reallocating leisure across the lifespan poses a difficult societal dilemma. Earnings replacement for "family leave," childcare, or elder care was a European approach; in the US, these "leisure" activities had to be self-financed. Employers' reactions to increased workplace flexibility echoed their early responses to pensions and early retirement incentives. They preferred to have the discretion to offer these options sometimes but not always, to some but not all workers.

The future of phased retirement remains uncertain. Implemented as reduced hours per day, fewer days per week, fewer weeks per year, extended leave, job sharing, or the rehiring of retired workers, workers in government, and higher education are more likely to have these opportunities. Employers can exercise discretion when shifting an older worker into a "phased" schedule by being selective in the offer and setting a termination date. One of the major obstacles for employees, however, is maintaining access to employer-provided benefits, since cutting back on hours may compromise their eligibility for inclusion. In addition, the current law is silent on phased retirement, but provisions in ERISA and the tax code have limited the ability of workers (and employers) to use pension benefits as earnings supplements. To the extent that workplace flexibility is not available to all workers, but only to a subset of workers to whom employers choose to extend the option, it becomes another source of workplace inequality (Hardy 2008).

Whether delaying retirement through retention in career jobs, mobility to new jobs, hours reductions through phased retirement, or other aspects of workplace flexibility, much depends on whether employers want to rely on older workers. Industries can have different age profiles, depending on when they expand or contract. While government workers tend to be older, those in biotechnology are younger. The insurance, manufacturing, mining, public administration, real estate, transportation, wholesale trade, and utilities industries are highly dependent on baby boom workers, with education and health care services being particularly vulnerable. Within these industries, a high percentage of baby boomers are in managerial or office and administrative support positions. The different age profiles of occupations and industries point to the types of positions that will open up when baby boomers retire from their career jobs. Assuming that productivity continues to increase, a significant number of retirements occur in declining or contracting industries, and sufficiently skilled workers from younger cohorts are available for promotion or hire, baby boom retirements should not lead to any labor shortage (Levine 2008).

At this point, however, it remains to be seen how and when baby boom workers will retire. Employers' decisions to eliminate health benefit coverage for retirees, the legislated increases in actuarial penalties for early Social Security benefits, and changes in the earnings test all encourage delayed retirement. Baby boomers' plans to postpone retirement beyond age 62 have been empirically linked to these factors as well as to higher levels of educational attainment and reduced entitlement to defined benefit pension plans compared to cohorts born earlier (Mermin, Johnson, and Murphy 2007). By the end of the decade, three-quarters of 51–61-year-old men, more than half 62–64-year-old men, and more than one third 65–69-year-old men were in the labor force, with rates of increased participation steepest at older ages (Johnson 2007). But a tight labor market and high unemployment for men and women aged 55 and older also produced a surge in early retirements, with a 20% jump in newly entitled Social Security beneficiaries in 2009 compared to 2008. Contributing to this increase in benefit claims was the leading edge of the baby boom cohorts, who celebrated their 62nd birthdays in 2009 and chose early benefits rather than wait for full benefits at age 66. Whether their earlier plans to work longer will motivate a return to employment once unemployment rates go down remains to be seen, but in tough economic times, even those who had planned later retirements may turn to retirement benefits to replace lost income (Johnson and Mommaerts 2010).

Future Prospects

Retirement is shaped less by aging than by institutional mechanisms that provide incentives and systematic procedures for workers' exits from the labor force. Historically linked to industrialization, by 1980, retirement was supported by a substantial infrastructure. Although the features of that infrastructure varied across nations and changed over time as social, demographic, and economic circumstances evolved, high income countries had been relying on retirement as a labor management tool. By reallocating long-term unemployment to older workers, renaming it retirement, and providing incentives through earnings replacement, a new life course stage was created. Decades of improved health, extended longevity, and declines in the median age of retirement had increased the duration of this stage. As the twentieth century drew to a close, however, early retirement was seen as unsustainable through federal programs. In the U.S., rates of employment among older workers had reversed course, showing significant increases. Policy revisions encouraged older workers to delay retirement, and the self-financing of retirement became a more prominent component of late life financial security. But by 2010, an extended global recession and high unemployment rates made even the sharper reductions in early Social Security benefits an attractive alternative to joblessness.

The high rates of early benefit entitlement have prompted some policy analysts to call for an increase in the entitlement age for receiving reduced benefits. Making such changes to entitlement age may or may not reduce program expenditures. Because OASI benefits are actuarially adjusted, the benefit formula as well as the entitlement age would have to be changed; otherwise, somewhat higher benefits would be paid out over fewer years. The portrait of those who claim actuarially reduced benefits at age 62 also suggests the picture is more complicated. Less well educated, in poorer health, more likely to be limited in the kind of work they can do, having stopped working before age 62, and more likely to retire from physically demanding jobs, they may be candidates for disability benefits if denied access to early retirement benefits (Li, Hurd, and Loughran 2008).

Whether in private or government sector industries, the expansion of early retirement in the United States was funded primarily through provisions in employer-sponsored defined benefit plans, with some accommodating revisions in Social Security. Within the manufacturing industry, for example, many blue-collar workers were able to afford early retirement because of company plan incentives that provided higher benefits until age 62 when Social Security benefits could be claimed. Firms' abilities to meet their long-term commitments depend on their profitability and on the government's ability to enforce regulations designed to ensure the sound investment of pension funds. Complaints that ERISA required overly conservative investment strategies may be less compelling in light of recent investment scandals and market performance. But the factors that produced disappointing statements for 401(k) plan holders also produced sharp declines in DB pension funds, which means that a number of DB pensions are currently underfunded (i.e., hold assets insufficient to pay promised benefits in the long run, though able to maintain benefits in the short term). ERISA regulations prevent employers from breaking pension promises, and the Pension Benefit Guaranty Corporation, which is funded by premiums paid by firms offering DB plans, steps in when plans go bankrupt; but as more DB pension plans fail, the PBGC absorbs higher and higher long-term benefit commitments. Whether existing DB plans will be able to meet their obligations is unclear, but one prediction for an accelerated rate of plan conversions (from DB to DC) would have important negative financial consequences for baby boom retirees (Butrica et al. 2009).

Questions about how to distribute resources not only across people but also across generations – those currently living and those not yet born – should be an important focus of societal debates. Although the retirement of the baby boom cohorts has fueled talk of generational consequences, the baby boom cohorts may be a watershed cohort for retirement in another sense. While previous birth cohorts lived and worked much of their lives with one understanding of the "rules" of retirement, 1970s formulas for calculating Social Security benefits changed the rules in their favor, allowing them to collect higher benefits than they had expected. Baby boomers began working as these changes were being debated, passed, and implemented; but then the rules changed again, and they may change still more. After following the company "rules," their pension wealth may be redefined and responsibility for generating the benefits they expected placed in their hands.

Whereas the costs of DB pensions are necessarily pushed into the future, the value of DC plans is also pushed into the future: DB funding depends on the continuing ability to generate a revenue stream to pay benefits; DC accounts depend on an ongoing ability to build the asset and a future ability to exchange investment holdings for an expected benefit flow. Both involve risk and uncertainty, but while a DB approach can distribute risk widely across people and time, DC plans are individual and much more limited in timeframe. Balancing the two is an important element of societal variation, as is the extent of economic inequality, the impact of redistribution policies, the role of government, and the value of different sorts of activities.

The societal view of retirees as an "untapped resource" has introduced a new expectation for retirees. Although the "leisure lifestyle" of retirees has received considerable media play, for most retirees the reality has been more modest and less adventurous. Retirees are now called to "civic engagement," volunteerism, and other forms of unpaid labor. Older volunteers can reinvigorate our communities, staff our underfunded social programs, volunteer as mentors, and serve as the remedy for any number of nagging social problems. All societies rely on their citizens for various forms of productive behaviors. Where societies differ is in the sort of work (and workers) worth paying, and what sorts of work (and workers) are kept unpaid or underpaid. Much of the work in the service of home and family has been unpaid work, and as unpaid work, the responsibility has not been shared.

Labor force statuses are social constructs, designating salient distinctions among people vis-à-vis paid employment, as distinct from productive activity. Including "retired" as one of these categories recognized a distinction among those not employed that was primarily a function of age and receipt of pension benefits. Statuses of "non-employment" have been the focus of legislative battles and court cases for the better part of a century, as women, people of color, older people, gays and lesbians, and those with disabilities fought for the right to work on equal terms. How to financially maintain subpopulations of adults with inadequate access to earnings, on the one hand, and dispensing accumulated claims to deferred income, on the other, raises fundamental issues of distribution. Negotiating the terms by which people become eligible for earnings replacement and claims to future income raises additional societal issues. How societies distribute the responsibilities of family, caregiving, programs for the poor, mentoring, tutoring, and the like speaks to value systems, patterns of exploitation and dependency, and the distribution of power. Before we can determine what sort of society we want to be, we must understand the interests served by the status quo and by the vectors of change being proposed and opposed.

What we know about retirement transitions is largely from the retrospective information and the prospective study of people as they approach and move into retirement. The perspective of employers is limited to a few firms, federally mandated reports, or interviews with human resource managers. Employer cooperation in these surveys is notoriously poor, and it is difficult to know what sort of work climate older employees face based on these data. However, organizational studies can produce important insights, and they provide a necessary balance to the person-based studies utilizing national samples of persons of households.

Survey instruments are central to the development of population estimates. If sociologists want to argue for a different understanding of retirement as a status or a process, they must move more aggressively on measurement issues. The concept of social context has been picked up by any number of disciplines, but the analytic treatment of this construct is often viewed as an exogenous factor that influences individual behavior. For sociologists to argue for endogeneity or system "embeddedness," we must develop testable theories with measures and techniques powerful enough to generate convincing empirical proof. Deciding how to define retirement likely will get more complicated. Rather than think in terms of categories, we should emphasize the joint patterns of employment, income sources, and health. If age is becoming a less reliable predictor of when people enter this stage of the life course, then we should move away from using it as our societal indicator of population subgroups. Does it matter what proportion of the population is aged 65 or older, or are we more concerned with the proportion of the population unable to be self-supporting and live independently?

Questions of timeframes, risk, redistribution, and equity should all be addressed by both macro and micro level research perspectives. Behaviors are an important focus of study, but so are meaning structures, social norms, distributional regimes, and cultural values. Much of retirement research has also become individualized. Responses to different systems of incentives, choices among alternatives, the correspondence between plans and actions, the use of information, and the decision-making process have received increased amounts of attention, especially by economists working from basic utility theory. Social psychological studies of satisfaction, adjustment, roles and relationships have also focused on individual behaviors and responses. This approach will continue to yield interesting and useful results, but a sociological perspective also has much to add to our understanding of retirement as a social institution.

Future decades of retirement research will need to track how changes in major life transitions intersect with shifts in demographic structure, cultural values, global markets, health, illness, and longevity. "Exceptionalism" in public policy on the part of any nation will have different implications in this new millennium. The mobility of labor and capital has already changed the terms of economic competition, and whether and how older workers will fit into the employment picture remains unclear. Changes in the family – the frequency of divorce, the growing numbers of gay and lesbian couples, and the extended period of adolescence and young adulthood – already demonstrate important inequities in our policy orientation. The interest of funding agencies for the quantification of behavioral responses to policy features or potential policy changes has become a dominant component of retirement research. Research that implicitly offers a cultural critique of growing inequality, late life vulnerability, exploitation or abuse tells a different sort of story, and depending on the political climate, may have difficulty finding a receptive audience. Addressing social problems, in whatever guise, is as much about making society as describing it.

How changes in later life work patterns will shape the experience of aging depends not only on the pathways on which people have already embarked but also on how the terrain changes as they move on with their lives. Differences in education, first jobs, job changes, job losses, family structure, health, and social networks are just one layer of stratifications that will be reinforced, redirected, overcome, or underachieved. If education is the prelude to the career, then retirement is the conclusion. The conditions that determine how retirement will be experienced can change gradually or suddenly and, if the latter, allow little time for adjustment. Much of retirement behavior is therefore premised on trust, on confidence, on hope – trust that contracts, social or employment based or familial, will be honored; confidence that nest eggs will not shift overnight from goose to sparrow-sized, that the national economy will be strong, standards of living maintained, and services available; hope that good days are still ahead, that a good quality of life will be enjoyed, and that those we care about will enjoy it with us.

References

Barfield, Richard and James Morgan. 1969. *Early Retirement, The Decision and the Experience.* Ann Arbor: University of Michigan Press.

Beck, Scott. 1985. "Determinants of Labor Force Activity among Retired Men." *Research on Aging* 7:251–80.

Brubaker, Ellie and Timothy Brubaker. 1992. "The Context of Retired women as Caregivers." Pp. 222–235 in *Families and Retirement*, edited by M. Szinovacz, D. J. Ekerdt, and B. Vinick. Newbury Park: Sage Publications.

Butrica, Barbara, Howard Iams, Karen Smith, and Eric Toder. 2009. "The Disappearing Defined Benefit Pension and Its Potential Impact On The Retirement Incomes of Baby Boomers." *Social Security Bulletin* 69(3):1–27.

Calasanti, Toni. 1993. "Bringing in Diversity: Toward an Inclusive Theory of Retirement." *Journal of Aging Studies* 7(2):133–50.

DeViney, Stanley and Angela O'Rand. 1988. "Gender-Cohort Succession and Retirement among Older Men and Women, 1951–1984." *Sociological Quarterly* 29:525–40.

Dorfman, Lorraine. 1992. "Couples in Retirement: Division of Household Work." Pp. 159–173 in *Families and Retirement,* edited by M. Szinovacz, D. J. Ekerdt, and B. Vinick. Newbory Park: Sage Publications.

EBRI. (2005). *History of 401(k) Plans: An Update.* Washington, DC: EBRI.

Ekerdt, David, Barbara Vinick, and Raymond Bossé. 1989. "Orderly Endings: Do Men Know When They Will Retire?" *Journal of Gerontology: Social Sceinces* 44:S28–35.

Ekerdt, David and Stanley DeViney. 1990. "On Defining Persons as Retired." *Journal of Aging Studies* 4(3):211–29.

Elder, Glen and Eliza Pavalko. 1993. "Work Careers in Men's Later Years: Transitions, Trajectories,and Historical Change." *Journal of Gerontology: Social Sciences 48:*S180–91.

Flippen, Chenoa and Marta Tienda. 2000. "Pathways to Retirement: Patterns of Labor Force Participation and Labor Market Exit among The Pre-Retirement Poplation by Race, Hispanic Origin, And Sex." *Journal of Gerontology: Social Sciences* 55B:S14–27.

Gibson, Rose. 1992. "Reconceptualizing Retirement for Black Americans." *The Gerontologist* 27:691–98.

Guillemard, Anne-Marie and Martin Rein. 1993. "Comparative Patterns of Retirement: Recent Trends in Developed Societies." *Annual Review of Sociology* 19:469–503.

Han, Shin-Kap and Phyllis Moen. 1999. "Clocking Out: Temporal Patterning of Retirement." *American Journal of Sociology* 105(1):191–236.

Hardy, Melissa. 2008. *Making Work More Flexible: Opportunities and Evidence.* Washington, DC: Public Policy Institute, AARP.

Hardy, Melissa and Lawrence Hazelrigg. 2007. *Pension Puzzles: Social Security and the Great Debate.* New York, NY: Russell Sage Foundation.

Hardy, Melissa and Kim Shuey. 2000. "Pensions Decisions in a Changing Economy: Gender, Structure and Choice." *Journal of Gerontology: Social Sciences* 55:S271–77.

Hardy, Melissa. 1991. "Employment After Retirement: Who Gets Back In?" *Research on Aging* 13(3):267–88.
Hardy, Melissa, Lawrence Hazelrigg and Jill Quadagno. 1996. *Ending a Career in the Auto Industry: "30 and Out."* New York: Plenum.
Hayward, Mark, William Grady and Steven MCLaughlin. 1988. "The Retirement Process among Older Women in the United States: Changes in the 1970s." *Research on Aging* 10(3):358–82.
Hayward, Mark, William Grady, Melissa Hardy and David Sommers. 1989. "Occupational Influences on Retirement, Disability, and Death." *Demography,* 26(3):393–409.
Hayward, Mark, Melissa Hardy and Mei-Chen Liu. 1994. "Work after Retirement: The Experiences of Older Men in the U.S." *Social Science Research* 23:82–107.
Hayward, Mark, Samantha Friedman and Hsinmu Chen. 1996. "Race Inequities in Men's Retirement." *Journal of Gerontology: Social Sciences* 51B:S1–10.
Henretta, John and Angela O'Rand. 1983. "Joint Retirement in the Dual Worker Family." *Social Forces* 62(2):504–20.
Henretta, John, Angela O'Rand and Christopher Chan. 1993. "Joint Role Investments and Synchronization of Retirement: A Sequential Approach to Couples' Retirement Timing." *Social Forces* 71(4):981–1000.
Hirsch, Barry, David MacPherson and Melissa Hardy. 2000. "Occupational Age Structure and Access for Older Workers: The Role of Compensation, Skills, and Working Conditions." *Industrial and Labor Relations Review 53*:401–18.
Johnson, Richard and Corina Mommaerts. 2010. *Fact Sheet on Retirement Policy.* Retrieved February 12, 2010, from Urban Institute: http://www.urban.org/UploadedPDF/412010_retirement_benefit_awards.pdf
Kohli, Martin. 1987. "Retirement and the Moral Economy: An Historical Interpretation of the German Case." *Journal of Aging Studies* 1:125–44.
Kohli, Martin and Martin Rein, Anne-Marie Guillemard, and Herman van Gunstere, eds. 1991. *Time for Retirement: Comparative Studies of Early Exit from The Labor Force.* New York, NY: Cambridge University Press.
Levine, Linda. 2008. *Retiring Baby-Boomers = a Labor Shortage?* Congressional Research Service, Domestic Social Policy Division. Washington, D.C.: Government Printing Office.
Li, Xiaoyan, Michael Hurd and David Loughran. 2008. *The Characteristics of Social Security Beneficiaries Who Claim Benefits at the Early Entitlement Age.* Washington, D.C.: AARP.
Mermin, Gordon, Richard Johnson and Dan Murphy. 2007. "Why Do Boomers Plan to Work Longer?" *The Journals of Gerontology: Social Sciences* 62B(5):S286–94.
Mutchler, Jan, Jeffrey Burr, Amy Pienta and Michael Massagli. 1997. "Pathways to Labor Force Exit: Work Transitions and Work Instability." *Journal of Gerontology: Social Sciences* 52B:S4–12.
O'Rand, Angela and John Henretta. 1999. *Age and Inequality: Diverse Pathways Through Later Life.* Boulder: Westview Press.
O'Rand, Angela and John Henretta. 1982. "Delayed Career Entry, Industrial Pension Structure, and Early Retirement in a Cohort of Unmarried Women." *American Sociological Review* 47:365–73.
Palmer, Edward. 2000. *The Swedish Pension Reform Model Framework and Issues.* Washington, D.C.: The World Bank.
Pienta, Amy. 1999. "Early Childbearing Pattern and Women's Labor Force Behavior in Later Life." *Journal of Women and Aging* 11:69–83.
Quadagno, Jill and Melissa Hardy. 1991. "Regulating Retirement through the Age Discrimination in Employment Act." *Research on Aging* 13(4):470–75.
Riley, Matilda White and John Riley. 1994. "Age Integration and the Lives of Older People." *The Gerontologist* 34:110–15.
Rogers, Teresa and Natalie Friedman. 1980. *Printers Face Automation: The Impact of Technology on Work and Retirement among Skilled Craftsmen.* Lexington, MA: Lexington Books.
Simpson, Peter. 2009. *Swedish Pension Fund Warns of Payment Cuts.* Retrieved February 10, 2010 (http://www.thelocal.se/17012/20090119).
Smith, Deborah and Phillis Moen. (1998). "Spousal Influence on Reitrement: His, Her, and Their Perceptions." *Journal of Marriage and Family* 60(3):734–44.
Szinovacz, Maximilliane and Stanely DeViney. 1999. "The Retiree Identity: Gender and Race Differences." *Journal of Gerontology: Social Sciences* 54B:S207–18.
Szinovacz, Maximilliane, David Ekerdt and Barbara Vinick. 1992. *Families and Retirement.* Newbury Park: Sage Publications.
Taylor, Phillip. 2006. *Development in Employment Initiatives for an Ageing Workforce: Cnsoloidated Analytical Report.* Luxembourg: Office for Official Publications of the European Communities.
Wong, Jen and Melissa Hardy. 2009. "Women's Retirement Expectations: How Stable Are They?" *Journal of Gerontology: Social Sciences* 64B(1):77–86.

Chapter 15
Learning and Aging

Emily Jovic and Julie McMullin

When we think about learning we often think about it in relation to formal education. Yet, education and learning are not necessarily the same things; one can go to school and not learn anything, while others learn without participating in formal schooling. Nonetheless, in sociology, the study of learning has largely centered on the study of education and its institutions and structural features, with an attendant focus on inequality and the individuals and groups most typically served by the system, namely, younger people. Although the adult educational complex has certainly expanded in recent years, 248% from the 1970s to 1990s in the United States (Hamil-Luker and Uhlenberg 2002:S324), there remains a considerable degree of age-based compartmentalization when it comes to learning, with an emphasis on training young people for the labor market. Hence, there continues to be an overall lack of scholarly attention to learning later in life.

A long tradition of research and theorizing affirm that industrialized societies and individual lives are largely organized around school and work relationships and that education is a significant determinant of the social and economic success of people and of nations (Comeau and McMullin 2010; Canada Council on Learning/CCL 2009a; Kohli 1986). Disparities in educational and occupational attainment based on class, gender, race, and ethnicity are a primary focus in research and policy; however, age is often treated as a constant or neglected altogether (McMullin 2010). The prevalence and duration of lifelong career jobs in many industrial sectors are being reduced, which has implications for patterns and experiences of learning and working over the life course. Yet, little is known about the extension of educational careers beyond adolescence and early adulthood. These matters are increasingly salient in the context of a changing economy, given the centrality of information and as people switch jobs or careers more frequently throughout their lives.

Lifelong learning (LLL) represents a recent and popular policy initiative that conceives of learning as an ongoing and voluntary pursuit having either personal or professional motivations. Ongoing learning is largely perceived as beneficial and can positively influence income, job satisfaction, political participation, and health and well-being as well as a nation's economic productivity and competitiveness (CCL 2008). Importantly, LLL recognizes learning over the entire life course and in many situations, not just early in life or in the classroom. The basic goals are appealing, to widen participation in learning activity and reduce inequalities (Morgan-Klein and Osborne 2007).

Although LLL is an attractive concept in theory, significant barriers and concerns emerge in its interpretation and application (e.g., Jackson 2003; Martin 2003; Schultz 2000). In some policy circles, it takes on the character of a buzzword and has been contested and criticized for its conflation with adult education (Morgan-Klein and Osborne 2007) and for the way in which it discursively masks certain social, economic, and political shifts (Martin 2003). Schultz (2000:72) points to an abundance of "lip service" but an overall lack of serious commitment to providing authentic learning

E. Jovic (✉)
Department of Sociology, University of Western Ontario, London, ON, Canada
e-mail: ejovic@gmail.com

R.A. Settersten, Jr. and J.L. Angel (eds.), *Handbook of Sociology of Aging*, Handbooks of Sociology and Social Research, DOI 10.1007/978-1-4419-7374-0_15, © Springer Science+Business Media, LLC 2011

opportunities over the life course. Similarly, Jackson (2003:367) argues that "a learning society that is 'grafted on' can only continue and replicate the structural inequalities of gender, class and other differences, where only certain types of knowledge, skills and work are valued."

While education and learning comprise a wide base, including, for example, personal growth, community development, social justice, and esthetic endeavors, LLL policy tends to be tied closely to skill development associated with the economy and labor market. In a new economy context in particular, there is a heightened emphasis on LLL insofar as it presumably enables workers to keep pace with technological and workplace changes and gain or maintain employability. Returning to school is one way people may "catch up" and better their economic and life circumstances. This is not always a choice, though; restructuring trends and the nature of work in knowledge-intensive fields often compel ongoing learning. To remain economically active, many workers are advised, and in some cases, obligated, to upgrade their skills and education (CCL 2008; Elman 1998).

In this chapter, we examine what it means to learn, work, and age under changing social and economic conditions. The inquiry begins with the new economy concept and the current context for learning over the life course. This is followed by an overview of sociological perspectives on learning, largely through the study of formal education, which is institutionally structured for the early part of the life course. In the third section, we address some ideas that generate promise for learning throughout life, including advancements in brain science and a better understanding of the social correlates and conditions for excellence in learning. The chapter closes with a consideration of policy implications and new directions for research and thinking.

Changing Contexts for Learning, Working, and Aging

Much of the learning that occurs beyond initial and compulsory schooling is job-related. Thus, work, employment, and larger economic forces are salient in considering later life learning. In the sociology of work, learning is often taken up in the context of an emergent "new" economy and a focus on the skills development of individuals and workforces. The new economy concept taps the idea that old ways of doing business are waning, attributable to technological advancement, the commodification of knowledge and the need to be globally competitive (McMullin and Marshall 2010; Ranson 2003). Under such circumstances, individuals are confronted with shifting labor markets and changing skill requirements, which can have ramifications that extend across the life course.

Technology is critical to the new social and economic context. Information and Communication technologies (ICTs) have essentially transformed life in many industrialized societies. They are now an integral part of everyday life, penetrating deeply into learning, work, and communication activities as well as culture and entertainment. One of the typical canons of the rate of technological change in recent years is its propensity to create new jobs (e.g., 20 years ago there were no web developers) and growth in certain employment sectors (e.g., information technology [IT]), fundamentally altering and even destroying old ones (e.g., manufacturing). The adoption of new technologies can thus render the knowledge and skills of some workers obsolete, making ongoing learning and the ability to reskill more important today than in the past.

The case of IT serves as an example. In this sector, knowledge and innovation are key sources of economic activity and jobs require increasingly higher order skills and a wider knowledge base. Workers need to be "self-programmable" in Castells' terms – flexible, adaptable, and quick to retrain (Adams and Demaiter 2010; Castells 2000). The rapid pace of technological change in this industry not only transforms types of jobs, work content, and skill requirements; it also drives the need for workers to stay current, especially if they wish to maintain or improve their labor market prospects. Thus, compulsory schooling and initial forays into postsecondary education, generally concentrated in the earlier stages of life, no longer sustain an individual throughout his or her life (McMullin 2010). Returning to school and engaging in concentrated periods of learning in order to gain new or more relevant knowledge and skills are increasingly a requirement for many IT workers. As work continues to be transformed by technology, the likelihood is that this trend will emerge in other industries as well.

Another feature associated with a new economy is the reorganization of employment relationships (e.g., McMullin and Marshall 2010; Smith 2001). Among certain segments of the population, education and work careers have been quite stable and predictable, shaped by the bureaucratic tendencies of large organizations and a fairly linear understanding of time and achievement: "The sheer passage of time necessary to accumulate skill [gave] a person standing and rights – value in a material sense" (Sennett 1998:96). Within the current context, this seems to be somewhat unraveling and sustaining lifelong employment in one firm is increasingly rare (McMullin and Marshall 2010). Traditional industrial sectors like manufacturing are in decline and large corporations are restructuring, displacing many workers in the process. As a result, educational and employment trajectories may take on different, often more disjointed forms, even among formerly stable groups.

Transformation and trends in technology and the global economy have coincided with particular social and demographic changes. For example, there have been shifts in the incidence and frequency of marital and parental events over the life course, which influence educational and work trajectories (McMullin 2010). Related demographic patterns (e.g., smaller families) and increased life expectancies have led to population and workforce aging, which are occurring in industrialized nations at unprecedented rates. An aging population is not a crisis per se, but it can introduce potential challenges, including perceived threats to prosperity and the possibility of labor and skill shortages in certain industrial sectors (McMullin and Marshall 2010; Statistics Canada 2008a). The primary source of new skills and workers in the economy used to entail the integration of "new" people into the labor force, usually young adults, and sometimes immigrants; however, population aging is reducing the supply of workers. Retention and training of those already in the labor force emerges as a strategy for addressing workforce aging (McMullin and Marshall 2010). Thus, in order to remain employed and adapt to changing labor markets, workers of all ages may need to retrain or upgrade. Learning may no longer be a luxury (Eisen 2005) and learning throughout the life course is required of more and more people.

These shifting economic and social contexts generate new uncertainties and demands, which bring us to another significant trend, and the last one we address here: individualization and the resultant need for people to become more active in managing to safe-guard their own lives. Some thinkers argue that there is a loosening of traditional institutions and a decline in biographical continuity, which necessitates more life planning and self-monitoring (e.g., O'Rand 2003; Beck and Beck-Gernsheim 2002). The increasing diversity of the social milieu calls for openness to the ideas and beliefs of others, the ability to question institutional assumptions, and a greater capacity for self-reflection (Taylor and Lamoreaux 2008). People are expected to be flexible, to tolerate uncertainty, and to handle greater diversity, new ideas and new situations (Butler 2005:67).

Under such conditions, individuals are able to pursue a wider variety of life trajectories; however, they also face a large number and variety of choices, a great deal of instability, and few guidelines or roadmaps to help (Macmillan 2005; Beck and Beck-Gernsheim 2002). There is pressure to take *personal* control of one's life and *personal* responsibility for successes and, especially, failures. Individuals are therefore required to bear more of the risks in navigating the life course (Shuey and O'Rand 2004; O'Rand 2003; Smith 2001). This is particularly so given that social institutions are providing less protection. It is generally left up to individuals to alleviate personal risk via accessing, monitoring, and interpreting information and choosing among alternatives (O'Rand 2003:695). One way to mitigate risk may be to engage in ongoing learning but as the next section suggests, there can be significant challenges associated with doing so.

Education, Learning, and Life Course Structure

In this section, we draw on select insights from the sociology of education, which has primarily focused on formal schooling and the early stages of the life course. Indeed, McMullin and Marshall (2010:19) note that work and aging are interrelated, dynamic processes that influence personal development and well-being; yet, aging has been generally ignored in studies of paid work. This observation can be extended to learning, training, and education. Hence, after considering educational sociology, we turn to a discussion of learning and life course structure, followed by an overview of studies in the field of adult education.

The Sociology of Education

Learning is highly institutionalized and educational systems, through their age-graded and standardizing functions, contribute to the construction of normative life course trajectories. Schools are quintessential social institutions and systems of education are primary sites for youth-centric learning activity. They support processes of socialization and selection via mandated and hidden curricula and contribute to the organization and legitimation of knowledge (Comeau and McMullin 2010; Lehmann 2007; Davies and Guppy 2006). Theorizing about education draws from sociology's foundational thinkers, including Durkheim, Marx, and Weber, and ranges from the functional to the interpretive to the critical. While some lines of thought have fallen out of favor more generally (i.e., functionalism), all of these strands are present in contemporary theories and research.

Human capital theory, originating in the field of economics, is highly influential in education policy and holds currency in the popular imagination as well. It offers a perspective on why people invest in education and why education may provide differential returns. Human capital amounts to a stock of resources – skills, knowledge, abilities, etc. – embodied in people (Becker 1993). This approach is premised on a direct link between education, skills acquisition, and income. Education and training are thus seen as investments that will yield returns collectively (e.g., competitive advantage and economic growth) and personally (e.g., higher earnings and better labor market prospects). In this view, education primarily imparts skills and knowledge associated with productivity and employability and the role of learning is largely economic (Davies and Guppy 2006; Jackson 2003). Though important, job-related skills and instrumental value are not the sole motivators for learning (Pallas 2003). Nonetheless, most educational participation by adults tends to involve work-related training (e.g., Statistics Canada 2008a; Hamil-Luker and Uhlenberg 2002).

That workers almost invariably wish to optimize their labor market potential is a key assumption. Training and education are believed to equip people with the skills and knowledge required to do this. Yet, skill development is limited to the supply side, which places the onus on individuals to pursue appropriate training in order to realize personal returns. Employers and economies also stand to gain, even if they only minimally support participation in education. Human capital also helps justify and legitimate the allocation of people to various occupations and social standings, largely on the basis of their individual efforts and educational investments. This point ties into perceptions of accessibility and the presumption that everybody has a roughly equal chance of acquiring and leveraging education; however, critical perspectives point to differential experiences among social groups.

Reproduction theories of education, derived from Marxian lines of thinking, contend that dominant groups shape educational systems in self-interested ways (e.g., Bowles and Gintis 1976). Taking a conflictive stance rooted in Weber's work on power, monopoly, and social closure, educational attainment also conveys and maintains status (Davies and Guppy 2006). Education and credentials may indeed serve as a rough measure of skill or ability (human capital); however, they are also found to tap attitude and potential (signaling) and status (credentialing) (Adams and Demaiter 2010; Arkes 1999; Collins 1979). Thus, educational systems are also places in which societal inequalities are reproduced and where already-privileged groups solidify and maintain their advantages (Comeau and McMullin 2010; Lehmann 2007).

A large body of literature points to connections between socioeconomic status (SES) and education. Put simply, those whose families have higher levels of education and greater financial security are more likely to participate in higher education (e.g., Comeau and McMullin 2010; Lehmann 2007; Davies and Guppy 2006; Jackson 2003). People with more education are also likely to enjoy better health and live longer, and those who are healthy are more likely to be in the workforce, further opening up opportunities to learn and achieve (McMullin 2010; Leighton 2007). SES is fundamental in understanding variation in educational attainment; however, critical perspectives have adopted a wider focus on inequality and structured social relations, including patterns and experiences that vary along gender, racial, and ethnic lines as well as according to time, place, and citizenship (e.g., McMullin 2010; Jackson 2003; Martin 2003; Elman and O'Rand 1998).

More recently, *intersections* among structured social relations and the application of other forms of capital have been leveraged to better understand disparities in educational achievement (e.g., McMullin 2010; Davies and Guppy 2006). There is a complex interplay between personal agency and structural factors; social background simultaneously structures the choices individuals make and shapes the structures in which they can exercise choice (McMullin 2010; Pallas 2003; Settersten and Lovegreen 1998). In addition to financial aspects, social, and family backgrounds also mediate social and cultural capital, such as the form and degree to which education is valued and supported in a household and exposure to people and experiences which enhance learning (Davies and Guppy 2006).

Age and historical period also shape one's chances in higher education, although these factors are less often the focus in educational research and theorizing (Comeau and McMullin 2010; Jovic 2009). Given current economic, social, and demographic changes and enduring patterns of inequality over time, issues of age, time, accumulation, and social change are particularly salient (McMullin 2010). Using logic akin to "the rich get richer while the poor get poorer," invoked sociologically by Robert Merton in 1968, Dannefer (2003) extends the "Matthew effect" to aging and the life course, noting the interaction of age and social structure in producing heterogeneity within cohorts. Referred to in life course studies as cumulative advantage/disadvantage (CAD), this perspective proposes a systemic tendency for initial inequalities in a given characteristic (e.g., education, health, money, status) to be magnified over time (McMullin 2010; Dannefer 2003; Elman and O'Rand 1998).

Cumulative (dis)advantage is highly relevant for educational participation. In fact, many later life outcomes can be traced back to early stratification processes in elementary and secondary schools, which have been amplified over the life course (Comeau and McMullin 2010; Davies and Guppy 2006; Elman and O'Rand 1998). Empirical studies are consistent with the presence of CAD in multiple life domains, including education. There are systemic and institutional barriers to later life learning and disproportionate access to higher education by more advantaged groups, including middle and upper classes and professionals (e.g., Comeau and McMullin 2010; Pallas 2003; Elman 1998; Elman and O'Rand 1998; Settersten and Lovegreen 1998; Quinnan 1997).

Although learning occurs throughout the life course, the timing, sequencing, and outcomes of educational participation, especially later in life, have been examined less frequently (Jovic 2009; Hamil-Luker and Uhlenberg 2002; Elman and O'Rand 1998; Jacobs and Stoner-Eby 1998). We know little about processes of human capital acquisition and the experience of education and learning beyond early adulthood (Elman and O'Rand 2007). And, as the next section shows, it seems age-graded structures in education may have increasingly less bearing on contemporary lives.

Learning, Earning, and Returning: The Institutionalized Life Course

In most societies, there is a rough outline of how lives are expected to unfold over time across interrelated life domains (e.g., education, family, work, health, etc.). Many life events, transitions and domains of activity are linked to socially constructed meanings of age, which can be quite strongly entrenched in institutions, organizations, and culture. Thus, the life course tends to be structured into stages based, at least loosely, on chronological age. Learning activity is highly institutionalized through systems of education and within the life course. Age has traditionally been a reliable determinant of educational participation and this holds to some degree. However, in the North American context in particular, there is a greater tolerance for educational "late blooming" (Levin and Levin 1991). Moreover, broader social and economic changes and associated shifts in institutional provisions and social policies are leading to declines in the uniformity and predictability of learning and work trajectories (O'Rand 2003). There is an increasing array of possible life course configurations, which has ramifications for learning over the life course.

State and economic institutions impose a considerable degree of order on lives, framing and indexing aging and life course patterns. In one influential view, the life course, at least in many industrialized countries, is divided into three periods of activity: education/preparation, work/activity, retirement/leisure (e.g., Kohli 1986; Cain 1964), or more colloquially, learning, earning, and returning (or serving). This tripartite or "three-box" model of life links age and life stage, with activity domains roughly corresponding with childhood/adolescence (and increasingly, young adulthood – Macmillan 2005), adulthood and old age, respectively. The three boxes are linked via their relationship to paid work and there is a presumed temporal lockstep between them (Kohli 1986), with learning largely confined to the early part of the life course. Certain achievements as well as cultural exemplars of lives are predicated on this model and the assumption that it is normatively, if not universally, experienced.

The tripartite view captures major stages and transitions in life as well as some of the institutionally-mediated relationships to the rhythms of chronological aging, presenting them as a familiar and relatively common life pattern. Yet, it also implies considerable homogeneity, a gendered demarcation of stages and a one-way linearity that does not correspond with many people's lives, especially in light of recent changes. Although there is no longer (and may never have been) a rigid order, pace, or sequence regarding participation in education and work, lives continue to be divided into a time and place to acquire knowledge and skills (youth, school) and a time and place to apply them (adulthood, the workplace). This model also fails to adequately capture the complexity of experiences of education and learning. Nonetheless, it continues to anchor social policy and institutions (Marshall and Mueller 2002; Settersten and Lovegreen 1998) and represents an influential timetable for how lives "should" unfold over time – neatly, steadily, and predictably.

In the three-box life course, education precedes work activity, particularly a person's first "real" job. Following high school, the typical options are to enter the labor force or to matriculate to postsecondary training. Increasingly, people are opting for higher education and there has been tremendous growth in college and university attendance (Comeau and McMullin 2010; Davies and Guppy 2006). Compulsory, early educational pathways are institutionalized, ensuring that young people between the ages of 18 and 24 comprise the bulk of the traditional postsecondary population (Comeau and McMullin 2010). There is a corresponding, intense concentration on education and learning processes in the first 20 or so years of life, and research and policy initiatives overwhelmingly address patterns, access, and outcomes among young people, usually not past age 30. This focus is not entirely misplaced; younger people indeed make up a substantial proportion of student populations and significant numbers of youth face difficulties with labor market integration and un- and underemployment (e.g., McMullin 2010). However, there has been a steady increase in the number and proportion of older students enrolled in postsecondary institutions (e.g., Hamil-Luker and Uhlenberg 2002; Elman 1998; Jacobs and Stoner-Eby 1998).

Clearly, age is not always a reliable indicator of involvement in learning and education (e.g., Levin and Levin 1991). It is increasingly important and necessary to consider multiple sequences of learning and educational participation and their consequences across the life course. Many adults follow varied and disjointed educational paths, returning to school to increase job security, improve current prospects, pursue new career possibilities, or prepare for future job changes (Jovic 2009; Elman 1998; Settersten and Lovegreen 1998). Sociological research has tended to focus on structural factors and patterns of participation (Jacobs and Stoner-Eby 1998), and in many cases, the experiences of women (e.g., Bradburn et al. 1995). Less is known about the personal and career-related experiences and outcomes of educational reentry and transitions back to the workforce later in life.

Corresponding shifts in patterns of educational consumption are often acknowledged; however, the tripartite view is the implicit life course structure. Learning efforts throughout life and the experiences of learners outside the youth demographic tend to be examined less frequently in research and surface less often in educational and workplace policy (Jovic 2009; Elman and O'Rand 2007; Settersten and Lovegreen 1998). That said, there has been some telling recent work in adult education, a topic to which we now turn.

Adult Education

More adults are engaging in postcompulsory education and learning endeavors, in part because of the imperatives of a shifting economy and labor market. In fact, educators are well aware that large numbers of people turn to education during times of change, as they seek opportunities to enhance skill sets and improve employment prospects (Butler 2005). The most prominent form of adult learning is workplace or job-related training (e.g., CCL 2009a; Hamil-Luker and Uhlenberg 2002). There is a rich tradition of research and theorizing in adult education (e.g., Morgan-Klein and Osborne 2007; Selwyn et al. 2006; Jarvis 2004), that is associated and sometimes conflated with aspects of LLL. Yet, in much educational research, adult learning is often treated as a separate chapter, a footnote or a specialist niche. Schultz (2000:80) reminds us that learning comes in various forms "with access never to be limited by age." Thus, there is thus a need to balance the unique needs and barriers faced by these learners (e.g., Settersten and Lovegreen 1998; Quinnan 1997) within a more integrated and inclusive framework for learning over the life course. In order to get a sense of the field of adult education, we distinguish three forms of learning and consider some of the challenges and barriers adult learners may confront.

Table 15.1 presents three categories of settings in which purposeful learning takes place (Statistics Canada 2008b; Hamil-Luker and Uhlenberg 2002). This distinction is salient for adult learners as they often do not have the same compulsory obligation and institutional attachment to education as youth and young adults.

Table 15.1 Types of learning

Type	Location	Credential[a]	Structured[b]	Intentional[c]
Formal	Educational/training institutions, schools	Yes	Yes	Yes
Nonformal	Workplaces, community settings, civic organizations, etc.	No	Yes	Yes
Informal	Daily life activities in work, family, leisure, etc.	No	No	Yes/no

[a]Leads to certification in the form of a degree or credential
[b]Structured in terms of learning objectives, learning time, or learning support
[c]Intentional from the learner's perspective
Source: Derived from Statistics Canada (2008b)

Formal learning covers the official, age-graded education system from elementary through postsecondary and comprises general academic study as well as specialized technical or professional training. Nonformal education has a much broader base, occurring in places like libraries and religious institutions or within community groups (Hamil-Luker and Uhlenberg 2002). Although informal modes of learning may impart relevant skills and knowledge, it is generally not recognized by governments and employers (CCL 2008).

About half of the adult population in the United States and Canada were enrolled in organized forms of adult learning and training in 2002 (Statistics Canada 2008a). As they age, people tend to engage in learning less frequently and become even less involved with career-related learning, speaking to the association of adult education with work-related training (Statistics Canada 2008b). Younger people are thus most likely to take advantage of opportunities for learning and training (CCL 2009a; Statistics Canada 2008b; Hamil-Luker and Uhlenberg 2002); in fact, in Canada, younger and single workers are more likely than older and married or divorced workers to participate in training and to obtain a credential (Zhang and Palameta 2006).

Vulnerable groups, including those with low literacy levels, low SES, older adults and less educated adults, are least likely to engage in formal education (Comeau and McMullin 2010; Rubenson et al. 2007). There is evidence of cumulative advantage; in short, learning begets more learning. Later educational participation reflects stratification processes that start early in life and initial levels of attainment, literacy, and skill are thus significant determinants of participation in adult education (Comeau and McMullin 2010; CCL 2009a; Statistics Canada 2008b). So, contrary to its purpose, adult education may actually amplify rather than attenuate inequalities in education and labor market outcomes (Rubenson et al. 2007:38; Morgan-Klein and Osborne 2007).

Marshall and Mueller (2002:11) suggest an effective model of lifelong education which entails a vaccination approach, "with repeated exposures over time needed to maintain an optimal level of functioning." However, unlike inoculation programs that promote public health more broadly, the cost, insecurity, and responsibility for postcompulsory learning are placed largely on individuals. A large proportion of adult learners thus pursue training on their own, without the financial assistance of employers (Statistics Canada 2008a). Firm size is a strong determinant of participation in employer-sponsored training, with workers in small firms less likely to be offered such opportunities (McMullin and Marshall 2010; CCL 2009a; Statistics Canada 2008a; Gorard and Selwyn 2005). Similarly, employer support for training tends to favor already high-skill workers in jobs with high skill requirements, keeping with the cumulative advantage hypothesis.

Assumptions and stereotypes about age and optimal learning and the perceived value of making later educational investments may present additional barriers to participation. Older learners tend to face more challenges in returning to school, personally and structurally (Comeau and McMullin 2010; Settersten and Lovegreen 1998; Quinnan 1997). This is in part due to time and financial costs and the fact that most adult learners take on learning activity on top of existing work and family responsibilities. Additional barriers include fiscal practices, such as rules relating to pensions, scholarships, and student financial aid (Settersten and Lovegreen 1998; Quinnan 1997). Institutional mechanisms are similarly implicated, as age biases and administrative imperatives in schools, such as class scheduling or admissions procedures, can discourage would-be learners (Jovic 2009; Elman and O'Rand 2007). Moreover, there are further challenges with "fit" and socialization with one's learning cohort as well as anxieties regarding academic performance, fear of failure, concerns about rusty academic skills, and a general sense that one may be "too old" to learn well or efficiently (Comeau and McMullin 2010; Jovic 2009; Leighton 2007; Butler 2005).

Although these latter concerns may weigh heavily on the minds of individual learners and run rampant in educational and employment systems, recent research suggests that they are unfounded or at least necessitate some qualification. In the next section, we consider emergent fields of inquiry and address how their development and application is expanding our understanding of learning and aging.

Learning and the Aging Brain

Thus far, this discussion has centered on some of the social structural factors that shape patterns and experiences of learning over the life course. We have established that adult participation in education is mostly work-related; however, learning certainly goes beyond the transmission of labor market skills and bodies of knowledge. It also encompasses the cultivation of identities, citizenship, and culture. As the previous section indicates, chronological age does not necessarily tell us much about an individual's capabilities or potential (McMullin 2010; Butler 2005). Although talents, abilities, and attitudes are distributed unevenly and can change over time, age-related declines are not inevitable (Corna 2009). Everybody can become a better learner (Leighton 2007). In this section, we address developments in brain science and metacognitive knowledge that can be integrated with sociological perspectives in order to better understand and potentially enhance learning throughout the life course.

Learning physically alters the brain, which lends credence to the edict of "use it or lose it" (Zull 2006). Advances in imaging technology and rapid growth in cognitive neuroscience are enabling researchers to observe the detailed architecture of the brain and its functions. For example, they can identify which areas of the brain are activated during different learning tasks (Zull 2002). This is useful for confirming and enhancing optimal strategies for learning. We are not going to delve further into physiological, cognitive, or neuroscientific detail here; suffice it to say that the brain changes as we age *and* as we learn.

Metacognition involves knowledge and awareness about learning, thinking, and problem-solving, generally and personally (Pintrich 2002). Educators and psychologists have made strides in understanding the practices and characteristics that promote excellence in learning. Many of the ideas and strategies target younger people, but they can be extended or adapted for all learners. As with educational attainment, SES is significant in promoting or thwarting conditions for optimal learning. For children, SES is thought to represent a rough measure of environmental stability and the degree to which basic human needs, such as food, shelter and love, are being met (Leighton 2007). Though this may appear self-evident, fundamental needs are sometimes easy for older learners to overlook – being tired or hungry or dehydrated, for example. Generally, less is known about this mechanism of SES in the case of adults; however, the situation is presumably complicated as many older learners participate in education precisely to improve their stability and security via income and labor market prospects.

Leighton (2007) offers a succinct overview of three elements that promote excellence in learning: deliberate practice, mentorship, and the nurturing of personal dispositions. Although these factors are articulated with young and exemplary learners in mind, they capture insights that, with some tweaking and recognition of barriers unique to older students, may apply to learning throughout life. As people become aware of metacognition and the conditions for optimal learning and enact them, they will tend to learn better (Leighton 2007; Pintrich 2002).

Deliberate practice refers to the fact that learning entails hard work as well as the motivation to engage in hard work (Leighton 2007). This is reflected in what is popularly known as the "10,000-hour rule" (e.g., Gladwell 2008). The most accomplished individuals in a given field have spent tremendous amounts of time – thousands of hours – engaging in intense and focused practice, whether in music, sport or computer programming. Obviously, the typical adult does not have that many free hours to sink into dedicated learning, nor do they necessarily need to in order to learn the new office accounting system. In most cases, they are not writing the software program. However, the principle of learning as labor applies here and it is important for older learners to recognize this, since they are often layering learning on top of other responsibilities, such as working and caring.

Mentorship is another condition for optimal learning. Good mentors inspire, motivate, and support, helping learners understand what they need to do and provide detailed feedback to direct and enhance learning (Leighton 2007). For adults, mentorship can be complicated by a tendency to engage in individualized learning (e.g., self-teaching) without mentors as well as what might be termed a paradox of age and competence. There is an inherent assumption that mentorship typically flows one way, from older expert teacher to younger student. Moreover, learning has an emotional foundation and feelings of uncertainty or fear can impact thinking, decision-making, and ability to retain information (Zull 2006). In a culture that favors competence, dislikes failure, and often dismisses the value of experience, particularly in some new economy fields, older learners who find themselves as neophytes in a given subject may experience some discomfort (Butler 2005; Sennett 1998). Thus, it is useful for learners to seek out mentor relationships and for both mentors and learners to think through age-related presumptions and biases.

Personal dispositions also promote optimal learning. There are four characteristics at the core of metacognition (Leighton 2007): reflecting (thinking about learning goals and strategies for achieving them); leveraging (taking stock of abilities, identifying and using strengths and improving weaknesses); framing (thinking constructively about performance, learning from failure and mistakes, taking risks); and control (taking charge of one's learning). Settersten and Lovegreen (1998:526–528) identified a series of similar personal attributes, including planful competence, motivation, coping, and risk-taking as well as contextual ones related to resources and responsibilities. These orientations and habits permit learners to grasp the importance of deliberate practice and the benefit of constructive feedback and mentorship. They also provide strategies for approaching learning tasks and improving performance, and thus underpin transferable skills and critical "learning to learn" qualities (Leighton 2007). Like many habits, these dispositions are perhaps more easily instilled earlier in life, but can be acquired at any time.

Those who are aware of conditions and strategies that support optimal learning will be more likely to use them and Pintrich (2002:223) identifies a need to explicitly relate this knowledge to learners. Metacognition is particularly salient for adult learners, who may have been estranged from formal learning for some time (Leighton 2007). Older people may approach learning with strategies that are less efficient, outdated, or inappropriate. Further, internalized age stereotypes about trainability and attitudes toward change can also impact one's perceptions of one's abilities. Adult learners require current and accurate knowledge about different learning strategies (e.g., memorizing, extracting meaning from text) and cognitive tasks (e.g., solving general vs. specific problems), as well as knowledge about themselves and their dispositions (Pintrich 2002). Metacognitive knowledge tends to be directed at individual learners; however, mechanisms of cultural and social capital (e.g., Comeau and McMullin 2010) presumably shape tendencies in both individuals and groups. Sociological perspectives can thus provide additional insight regarding the social correlates of optimal learning. Policy-makers and providers of educational services can help maximize learning at all ages by recognizing barriers and taking them into account in activity design and program delivery.

As social and economic imperatives push and pull more people toward ongoing learning, concern about age-related declines emerge. Workers wonder about their capacity and desire to learn and ability to keep skills and knowledge current, especially when it comes to new technology (McMullin and Comeau 2011). For example, a reader poses the following question to a newspaper advice column: "Given the fact that memory and concentration skills tend to decline with age, how difficult would it be for a person in their 50s to work on a master's degree?" (Troyer 2009). The simple answer is not too difficult, given some basic metacognitive knowledge and appropriate compensatory strategies.

Cognitive ability can be understood as a combination of two types of intelligence: "*Crystallized intelligence*, or acculturated knowledge, is believed to accumulate over time through educational, occupational and cultural experiences and pursuits. It is reflected in tests of vocabulary, information accumulation and specific knowledge-based activities. *Fluid intelligence*, on the other hand, encompasses information processing, reasoning, short-term memory, abstract thinking, reaction time and regulatory processes" (Corna 2009:3). People can experience certain cognitive declines as they age, but this is not an inevitable aspect of aging (Corna 2009; Zull 2006). In fact, research is revealing that the human brain is "almost infinitely malleable" – even in adults (Carr 2008:2).

Windows for learning in select areas, such as visual development or language, may begin to close off in adulthood; however, research strongly indicates that "there is no age of finality for any learning" (Zull 2006:7). Moreover, there is presumably enough variation in baseline abilities related to learning, such as memory and recall, that difficulties are not necessarily linked to age. When age-related declines are observed, they tend to be associated with fluid intelligence – processing speed, ability to concentrate, and short-term memory (Corna 2009; Troyer 2009; Zull 2006). However, losses in these areas tend to occur slowly, allowing for gradual compensation (Corna 2009; Eisen 2005). Many people are thus able to maintain cognitive function as they age. Further, stereotype-driven self-doubts and the fast pace of a lot of educational programming are believed to be larger deterrents to later life learning than any changes in intellectual capacity (Eisen 2005).

While certain cognitive abilities may decline with age for some people, other capacities actually improve (Troyer 2009; Corna 2009; Zull 2006). Knowledge and experience help build neural networks, essential for thinking and decision-making, which become denser and more elaborate with age (Taylor and Lamoreaux 2008; Zull 2002). Learning entails continuous modification of what is already known and when faced with new information, the brain seeks patterned connections to existing information (Taylor and Lamoreaux 2008; Zull 2002, 2006). Over time, it becomes increasingly efficient at "cutting to the chase" when analyzing complex connections. With age, then, people have the capacity to gain wisdom via the ability to delineate the critical elements of complex knowledge, identifying what is important and discarding what is not (Zull 2006).

Keeping with the metacognitive perspective, adult learners can assess their strengths and weaknesses, including potential changes in cognitive ability, and adjust their learning practices accordingly. There are a range of techniques and strategies that will help people proactively compensate for abilities that do diminish as well as those they might lack in the first place (Troyer 2009; Leighton 2007; Zull 2006). Although there is evidence supporting the effectiveness of cognitive training, more research is needed to understand how these practices relate to learning (Corna 2009). Returning to would-be master's candidate, the expert respondent, a clinical neuroscientist, identifies three potential challenges for older learners: slower intake and processing of new information, increased vulnerability to distraction, and difficulty with short-term memory. The advice she provides is simple and highly practical: stay on top of the workload, allow extra time in new learning situations, focus on one task at a time, study in a quiet place, and employ repetition strategies to retain new information (Troyer 2009:L3). Insights from cognitive psychology and brain science indicate that learning may be enhanced if educators and learners align practice with how the brain functions (Zull 2002), particularly if social structural factors are also taken into account (Leighton 2007).

Discussion

Sociologists have often turned to the economy and the domain of paid work to understand patterns and experiences in education and learning. Links between educational systems and labor markets receive a lot of attention, in part because of their connection to social stratification and inequality. Put simply, education is a major conduit to occupational attainment, financial reward, and social advancement, as well as personal and social well-being (e.g., Comeau and McMullin 2010). People tend to look to education for solutions to many social problems and higher learning is perceived as one of the more accessible avenues for upward social mobility. Yet, the egalitarian and socially inclusive aspects of education and LLL directives often take a backseat to economic imperatives (e.g., Martin 2003).

Information is increasingly salient across life domains, and more and more people are expected to pilot complex, knowledge-intensive terrains, including employment, health, well-being, finance, civic participation, entertainment, and culture. Informed and engaged citizens are better equipped to confront the critical issues of the day, such as poverty, conflict, and ecological responsibility. Education thus has important and lasting effects on future life chances and also on the shape and experience of social and individual lives. Economic opportunity and social engagement are increasingly linked to a person's ability to direct his or her own life (Beck and Beck-Gernsheim 2002). Learning may serve as a protective factor in this regard because it can enhance one's sense of control in managing one's life, creating opportunities, solving problems, and mitigating risks (Leighton 2007). Thus, the focus in education and learning cannot be confined to youth and employment preparation.

People of all ages are being advised to engage in ongoing learning in order to remain socially and economically active. Further, chronological age reveals increasingly less about one's capabilities, state of health, level of active engagement, or future potential (Butler 2005:63). It is troubling, then, that the sociological study of learning is so robustly associated with educational systems, labor market preparation, and the early part of the life course. There have been important economic and demographic shifts producing tighter links between learning and work in many employment sectors as well as unprecedented numbers of later life learners (Eisen 2005; Elman 1998; Jacobs and Stoner-Eby 1998; Levin and Levin 1991). The retention and retraining of workers over the life course is a significant issue in many industrialized countries (McMullin and Marshall 2010). This entails a complex set of challenges that require policies and programs which better attend to the changing learning needs and capacities of people over time and throughout their lives. In this regard, Settersten and Lovegreen (1998) suggest that despite a myriad of barriers to adult participation in education, there are opportunities for greater flexibility in the life course more generally.

Government policy tends to approach education and learning largely economically, with an emphasis on building a skilled, productive and flexible workforce, and a focus on the economically active and their competitiveness (CCL 2009a; Rubenson et al. 2007). Perhaps as a result, education is often analytically treated as fixed at the "highest" level of schooling. Although this approach is convenient and measurement-friendly, it does not adequately capture informal learning, upgrading, and retraining throughout life, learning pursued for nonwork-related goals, or circumstances in which general education (e.g., a bachelor's degree in sociology) may be sought at one type of institution, followed by occupationally-specific training at another (e.g., a certificate in database administration). Thus, we need more systematic examinations of understudied aspects related to learning, retraining, and requalification over the life course.

There are indications that returning to school and efforts to provide retraining may not be meeting the needs of many older learners (e.g., Jovic 2009; Settersten and Lovegreen 1998; Quinnan 1997). We need to work toward identifying and creating more favorable conditions for ongoing participation in learning, to allow people to better harness their capacities in order to meet everyday circumstance as well as life's challenges. In this regard, attention to the quality of programs, instruction, and learning, especially that which targeting adult students, is significant. Rising enrolment and higher levels of education inflate the value of credentials and additional educational investment does not guarantee improved or even similar prospects. There is a need to better understand how credentials and learning opportunities are evaluated and applied by employers and workers. In knowledge-intensive fields like IT, skills acquisition does not necessarily happen in formal settings; rather, alternative, leisure-based methods (e.g., hobby programming) can be more salient (Adams and Demaiter 2010; Jovic 2009).

Career management and development, including the acquisition and maintenance of relevant skills, are encouraged by governments and employers, yet responsibility is largely offloaded onto individuals. There are tax subsidies and programs to encourage retirement savings and investment for children's future educational needs, but fewer incentives for people to seek out the training they need over the life course (Schultz 2000:79) – to navigate changing labor markets as well as the increasingly complicated and information-heavy domains of health, finance, and culture. To help people engage in the learning necessary to function in social life, more inclusive approaches to responsibility for learning are needed, ones that bridge employers, educational institutions and governments, as well as individuals and families. Optimal learning is predicated on social relationships and personal dispositions and we need to foster more favorable conditions for learning over the life course, including a solid foundation via the cultivation of basic skills and attitudes as well as guidance and support regarding how to direct one's learning efforts.

There remains a need to rethink so-called "normative" educational and career pathways (e.g., Marshall and Mueller 2002; Settersten and Lovegreen 1998) and provide more and better opportunities for learning and (re)training across the life course, "when people are young, before workers lose their jobs, when they lose their jobs" (Schultz 2000:79). This requires greater flexibility and responsiveness to needs that change over time, including the impact of events like parenthood, marital breakdown, and unemployment as well as provisions for smoother transitions between ages and stages. If this is not the case, then learning has the potential to become yet another demand stacked upon already-overburdened individuals. Research shows that employer support for training tends to favor those already at the higher end of the spectrum in terms of skills and skill requirements. There is a need to better balance the investments of government, employers, and individuals in learning and training.

At the same time, it is important to broaden the focus of learning beyond economic and labor market applications. Many people feel they do not have a pressing economic need for or interest in ongoing learning (Gorard and Selwyn 2005) and this is a valid position; however, given the ubiquity of information and technology, and especially, the internet, certain skill sets like literacy, numeracy, and critical thinking, are needed not only in the workplace but in most life domains. The increased flow of data tips heavily into social and cultural environments and people are left with little choice but to deal with it. This is especially the case in areas that become increasingly salient as we age, such as health, health care, and financial planning. People need to be able to take in, process, and evaluate large amounts of highly specialized information, or at least know where to seek out simplified yet reliable sources.

This leads to an emergent and highly interesting perspective on some profound changes in how we learn: "In becoming information-rich, we have become attention-poor." Technology commentator Peter Nicholson (2009:A21) cuts to the quick of a key contemporary challenge for learning – the tradeoff between the depth of what we know and how fast we can retrieve it. He employs an apt river-vs.-lake metaphor to capture the difference between a flow and a stock of knowledge. Similarly, Nicholas Carr (2008) wonders if *Google* is making us stupid, proposing that the internet and ICTs may be "reprogramming" us, shaping our brains to process information rather than understand or even remember it. Both commentators expound on the social and biological consequences of a glut of information, including scattered attention, diffused concentration, and potential impacts on deep thinking and the cultivation and application of expertise. The volume and flow of information is rapid, leaving less time to think and reflect. Thus, there may be an eroding of the "deep, integrative mode of knowing" associated with the 10,000-hour rule (Nicholson 2009:A21). Fostering the skills required to navigate information flows, particularly the ability to read, cut through, and access what is needed, will be an important challenge for learners of all ages and exploration of these themes is a significant future direction for researchers in science and social science.

The intersections of learning, work and technology, particularly ICTs, are drawing much academic and political interest (e.g., CCL 2009b; Selwyn et al. 2006). ICTs are believed to offer solutions to many barriers to learning, especially with respect to access and the provision of flexible, convenient and ongoing learning opportunities for people of all ages. However, as we have seen, determinants of participation in education and learning are deeply rooted in structured social relations and family background as well as matters of time and place (e.g., Comeau and McMullin 2010). Research indicates that these enduring patterns seem to hold, whether learning is aided by technology, leading Gorard and Selwyn (2005:85) to observe that, so far, the "e-learning" society seems to be "remarkably similar to its nontechnological predecessor."

Aging and learning are biological, psychological, and social processes. As the knowledge base grows in each area and their various dimensions and intersections are fleshed out, we uncover new problems and questions. It is as necessary as ever to recognize and target enduring barriers to learning in the form of structured social relations and to take into account processes of time and cumulative (dis)advantage. Experiences with and orientations toward learning early in life shape later life learning. Similarly important is the attenuation and possibly reversal of individualizing trends with respect to learning, both in terms of cultural values (such as the personalization of success and achievement) and structural factors (educational institutions, workplaces, policies); otherwise, we will miss, in Gladwell's (2008:32) words, "opportunities to lift others up onto to the top rungs." Promoting optimal learning over the life course is not an end in itself; rather it is a means to an end, which is a better life (Leighton 2007) – a goal we ideally have for everybody.

References

Adams, Tracey and Erin I. Demaiter. 2010. "Knowledge Workers in The New Economy: Skill, Flexibility and Credentials." Pp. 119–42 in *Ageing and Working in the New Economy*, edited by J.A. McMullin and V.W. Marshall. Northampton: Edward Elgar.

Arkes, Jeremy. 1999. "What Do Educational Credentials Signal and Why Do Employers Value Credentials?" *Economics of Education Review* 18(1):133–41.

Beck, Ulrich and Elisabeth Beck-Gernsheim. 2002. *Individualization*. Thousand Oaks, CA: Sage.

Becker, Gary S. 1993. *Human Capital: A Theoretical and Empirical Analysis with Special Reference to Education*. Chicago: University of Chicago Press.

Bowles, Samuel and Herbert Gintis. 1976. *Schooling in Capitalist America*. New York: Basic Books.

Bradburn, Ellen M., Phyllis Moen and Donna Dempster-McClain. 1995. "Women's Return to School Following the Transition to Motherhood." *Social Forces*, 73(4):1517–51.

Butler, Christina Baroody. 2005. "Age-Related Paradigms." *New Directions for Adult and Continuing Education*, 108:61–68.

Cain, Leonard D., Jr. 1964. "Life Course and Social Structure." In R. E. L. Faris, ed. *Handbook of Modern Sociology*, 272–309. Chicago: Rand McNally.

Carr, Nicholas. 2008. "Is Google Making Us Stupid? What The Internet Is Doing to Our Brains." *The Atlantic Online*, Jul/Aug:1–5. Retrieved September 2009 (www.theatlantic.com/doc/print/200807/).

Castells, Manuel. 2000. "Materials for an Exploratory Theory of the Network Society." *British Journal of Sociology* 51(1):5–24.

CCL/Canadian Council on Learning. 2009a. *Securing Prosperity through Canada's Human Infrastructure*. Ottawa, ON: Canadian Council on Learning.

———. 2009b. *The State of e-Learning in Canada*. Ottawa ON: Canadian Council on Learning.

———. 2008. *State of Learning in Canada: Toward a Learning Future*. Ottawa, ON: Canadian Council on Learning.

Collins, Randall. 1979. *The Credential Society: An Historical Sociology of Education and Stratification*. New York: Academic Press.

Comeau, Tammy Duerden and Julie McMullin. 2010. "CAGE(s) and education." Pp. 209–50 in *Understanding Social Inequality, second edition*, edited by J.A. McMullin and Don Mills, Oxford: Oxford University Press.

Corna, Laurie M. with David Haardt. 2009. "Cognitive Function, Aging And Paid Work." Population Change and Life Course Strategic Knowledge Cluster, Policy Brief no.1 (November 2009). London: University of Western Ontario. Retrieved November, 2009 (http://www.ssc.uwo.ca/sociology/cluster/en/documents/Research%20Briefs/PolicyBrief1.pdf).

Dannefer, Dale. 2003. "Cumulative Advantage/Disadvantage and The Life Course: Cross-Fertilizing Age and Social Science Theory." *Journal of Gerontology*, 58B(6):S327–37.

Davies, Scott and Neil Guppy. 2006. *The Schooled Society: An Introduction to the Sociology of Education*. Don Mills, ON: Oxford University Press.

Eisen, Mary-Jane. 2005. "Shifts in the Landscape Of Learning: New Challenges, New Opportunities." *New Directions for Adult and Continuing Education* 108(winter):15–26.

Elman, Cheryl. 1998. "Adult Education: Bringing in a Sociological Perspective." *Research on Aging* 20(4):379–90.
Elman, Cheryl and Angela M. O'Rand. 2007. "The Effects of Social Origins, Life Events, and Institutional Sorting on Adults' School Transitions." *Social Science Research*, 36:1276–99.
———. 1998. "Midlife Work Pathways and Educational Entry." *Research on Aging* 20(4):475–505.
Gorard, Stephen and Neil Selwyn. 2005. "Towards a learning society? The Impact of Technology on Patterns of Participation in Lifelong Learning." *British Journal of Sociology of Education* 26(1):71–89.
Gladwell, Malcolm. 2008. *Outliers*. New York: Little, Brown and Company.
Hamil-Luker, Jenifer and Peter Uhlenberg. 2002. "Later Life Education in the 1990s: Increasing Involvement and Continuing Disparity." *Journal of Gerontology: Social Sciences* 57B(6):S324–31.
Jackson, Sue. 2003. "Lifelong Earning: Working-Class Women and Lifelong Learning." *Gender and Education* 15(4):365–76.
Jacobs, Jerry A. and Scott Stoner-Eby. 1998. "Adult Enrollment and Educational Attainment." *The Annals of the American Academy of Political and Social Science* 559:91–108.
Jarvis, Peter. 2004. *Adult Education and Lifelong Learning*. New York: Routledge Falmer.
Jovic, Emily. 2009. "On Time and Off Time Career Trajectories in the New Economy: The Case of Information Technology Work." PhD dissertation, University of Western Ontario, London, ON.
Kohli, Martin. 1986. "The World We Forgot: A Historical Review of The Life Course." Pp. 271–303 in *Later Life: The Social Psychology of Aging*, edited by V. W. Marshall. Beverly Hills CA: Sage Publications.
Lehmann, Wolfgang. 2007. *Choosing to Labour? School: Work Transitions and Social Class*. Montreal, QC: McGill-Queen's University Press.
Leighton, Jacqueline P. 2007. "Cultivating a Flexible Mind: Can You Learn How to Learn For Life?" Retrieved August, 2009 (www.ccl-ca.ca/CCL/Events/Minerva/PastLectures/PastLecturesLeighton.htm.0)
Levin, Jack and William C. Levin. 1991. "Sociology of Educational Late Blooming." *Sociological Forum* 6(4):661–79.
Macmillan, Ross, ed. 2005. *The Structure of the Life Course: Standardized? Individualized? Differentiated? Advances in Life Course Research*, volume 9. New York: Elsevier.
Marshall, Victor W. and Margaret M. Mueller. 2002. "Rethinking Social Policy for an Aging Workforce and Society: Insights from the Life Course Perspective." *CPRN Discussion Paper*, W/18. Ottawa: Canadian Policy Research Networks.
Martin, Ian. 2003. "Adult Education, Lifelong Learning and Citizenship: Some Ifs and Buts." *International Journal of Lifelong Education* 22(6):566–79.
McMullin, Julie. 2010. *Understanding Social Inequality: Intersections of Class, Age, Gender, Ethnicity, and Race in Canada*, 2nd ed. Don Mills: Oxford University Press.
McMullin, Julie Ann and Victor W. Marshall, eds. 2010. *Ageing and Working in the New Economy*. Northampton: Edward Elgar.
McMullin, Julie Ann and Tammy Duerden Comeau. 2011. "Ageism and Age Discrimination." In *Age Gender, and Work Information Technology Firms in the New Economy*, edited by J.A. McMullin. Kelowna: UBC Press.
Morgan-Klein, Brenda and Michael Osborne. 2007. *The Concepts and Practices of Lifelong Learning*. New York: Routledge.
Nicholson, Peter. 2009. "Information-Rich and Attention-Poor." *The Globe and Mail* 12 September 2009:A21.
O'Rand, Angela M. 2003. "The Future of the Life Course: Late Modernity and Life Course Risks." Pp. 693–701 in *Handbook of the Life Course*, edited by J.T. Mortimer and M.J. Shanahan. New York: Springer.
Pallas, Aaron M. 2003. "Educational Transitions, Trajectories and Pathways." Pp. 165–84 in *Handbook of the Life Course*, edited by J.T. Mortimer and M.J. Shanahan. New York: Springer.
Pintrich, Paul R. 2002. "The Role of Metacognitive Knowledge in Learning, Teaching and Assessing." *Theory into Practice* 41(4):219–25.
Ranson, Gillian. 2003. "Understanding the 'New Economy': a Conceptual Journey." *WANE Working Paper*, No.2. London, ON: Workforce Aging in the New Economy.
Rubenson, Kjell, Richard Desjardins and Ee-Seul Yoon. 2007. *Adult Learning in Canada: A Comparative Perspective*. Catalogue no. 89-552-MIE #17. Ottawa ON: Statistics Canada.
Quinnan, Timothy William. 1997. *Adult Students "At-Risk": Culture Bias in Higher Education*. Westport CT: Bergin and Garvey.
Schultz, James H. 2000. "'The Full Monty' and Life-Long Learning in the 21st Century." *Journal of Aging and Social Policy* 11(2/3):71–82.
Selwyn, Neil, Stephen Gorard and John Furlong. 2006. *Adult Learning in the Digital Age: Information Technology and the Learning Society*. New York: Routledge.
Sennett, Richard. 1998. *The Corrosion of Character: The Personal Consequences of Work in the New Capitalism*. New York: W.W. Norton and Company.
Settersten, Richard A. and Loren D. Lovegreen. 1998. "Educational Experiences throughout Adult Life: New Hopes or No Hope for Life Course Flexibility?" *Research on Aging*, 20(4):506–38.

Shuey, Kim M. and Angela M. O'Rand. 2004. "New Risks for Workers: Pensions, Labor Markets, and Gender." *Annual Review of Sociology* 30:453–77.

Smith, Vicki. 2001. *Crossing the Great Divide: Worker Risk and Opportunity in the New Economy*. Ithaca NY: ILR Press.

Statistics Canada. 2008a. "Adult Learning and the World of Work." *Education Matters* 5(2):1–4.

———. 2008b. "Adult Learning in Canada: Characteristics of Learners." *Education Matters* 5(1):1–4.

Taylor, Kathleen and Annalee Lamoreaux. 2008. "Teaching with the Brain in Mind." *New Directions for Adult and Continuing Education* 119(fall):49–59.

Troyer, Angela. 2009. "Can aging Brains Cut it in the classroom?" *The Globe and Mail*, 17 November 2009:L3.

Zhang, Xuelin and Boris Palameta. 2006. "Participation in Adult Schooling and Its Earning Impacts in Canada." *Analytical Studies Branch Research Paper Series*, catalogue no. 11F0019MIE No.276. Ottawa: Statistics Canada.

Zull, James E. 2006. "Key Aspects of How the Brain Learns." *New Directions for Adult and Continuing Education* 110(summer):3–9.

———. 2002. *The Art of Changing the Brain: Enriching the Practice of Teaching by Exploring the Biology of Learning*. Sterling, VA: Stylus.

Chapter 16
The Midlife Years: Human Capital and Job Mobility

Cheryl Elman

What does midlife hold for today's U.S. working adults? Until recently, the modern life course was arranged around three distinct life stages: schooling in youth, work/family responsibilities over the middle adult years, and retirement in the later adult years. This tripartite division of the life course (Kohli 1986) was underpinned by normative expectations about role transitions due to age-grading (Mayer and Schoepflin 1989; Settersten 1998) and cohort socialization (Easterlin 1987). The timing and sequencing of role transitions within each life stage – although often gender-segregated into midlife work careers for men and family careers for women – then constituted, until recently, fairly predictable life course pathways, interwoven as the substance of individual lives (Elder 1975). However, the tripartite division, also rooted in state policies and industrial production methods of the mid-twentieth century, is becoming more blurred, associated with recent economic restructuring (Heinz 2003; Mayer and Schoepflin 1989; Myles 1990). A shift in work regimes, from mass production to flexible production, is reshaping midlife work arrangements and perhaps the conditions of employability itself (Myles 1990; Vallas 1999).

Although this transformation affects workers of all ages, it uniquely impacts middle-aged workers, about ages 40–59 who, well into their work careers, may find themselves newly "free" agents in a flexible economy (Heinz 2003). Midlife work careers no longer unfold under the institutional and contextual conditions from which they emerged and so adults may feel unprepared for new job searches or the need to cross educational or labor market boundaries for employment (Elman and O'Rand 2007; Johnson et al. 2009). But, as will be described in this chapter, these changes also carry promise. They may open up new pathways of flexible employment, human capital gains, increased work-engagement and phased retirement at the same time that they impose burdens, such as the reintroduction to roles and activities once thought to be relegated to one's personal past.

This chapter examines midlife human capital, work arrangements, and job mobility within and across two twentieth century periods marked by different methods of economic production: a mid-twentieth century period of mass production or Fordism and a post-1970s period of flexible production. It is important to study midlife work arrangements in the context of both the old period and the new period because macroeconomic transformations are slow and uneven. Some middle-aged workers remain employed under the rules of the old, even as many more traverse the divide to the new. The first two sections of this chapter are overviews of the older industrial and newer flexible production regimes, respectively. Each section describes key theories that sociologists and other social scientists use to understand adult human capital, job mobility, and work arrangements. A third section details current trends in midlife job displacement, reemployment, and work-to-retirement pathways, in the light of theoretical perspectives. A fourth section examines work-related human capital as adult education: a growing midlife response to today's transforming markets.

C. Elman (✉)
Department of Sociology, University of Akron, Akron, OH, USA
e-mail: cheryl2@uakron.edu

R.A. Settersten, Jr. and J.L. Angel (eds.), *Handbook of Sociology of Aging*, Handbooks of Sociology and Social Research, DOI 10.1007/978-1-4419-7374-0_16, © Springer Science+Business Media, LLC 2011

Institutionalizing Employment: Mass Production Under Fordism

Fordist mass production in the United States emerged as a structured system of labor markets, product markets, and state policies by the mid-twentieth century (Meyer 1995; Mayer and Schoepflin 1989; Myles 1990). It is instructive to place Fordism in long-term perspective. By the early twentieth century nonagricultural workers were already employed in mid-to-large size – and growing – firms and state agencies (e.g., food service, department stores, railroads, financial institutions, government service) (Graebner 1980; Whitten 1983). Although large firm characteristics differed, early twentieth century employers faced similar problems in coordinating increasingly complex mass production and service delivery systems (Nelson 1975). They adopted Fordist techniques which restructured production processes and reallocated workers into newly redesigned jobs (Meyer 1995; Myles 1990). These included a core of managerial/administrative jobs to coordinate decision-making and a large base of "assembly-line" production or "shop-counter" service jobs with limited, repetitive job tasks or skills, subject to reorganization under managerial control (Gordon et al. 1982; Jacoby 1985). Additional firm-level jobs emerged over time, in human resources, marketing, clerical services, transportation, maintenance, and other areas of support.

Of course, many factors led to what became a mid-century U.S. Fordist accord, or social contract between employers and employees (Kalleberg 2000). For example, the early twentieth century reorganization of production might have tilted toward market or hierarchical form, but, perhaps because firm size was already large, Fordist expansion was hierarchical (Kalleberg 2000). Career ladders and labor contracts were adopted to regulate and rationalize hiring, promotion, and job turnover (Althauser 1989; Graebner 1980; Petersen et al. 1989); work rewards were adopted, based on wage formulas, job tenure, and age rules (O'Rand and Henretta 1999; Wise 1997). Fordist system growth also reflected industrial relations and the unionization of workers and mid-century geopolitical, often war-related, pressures that required national employer-employee cooperation, if not social cohesion (Myles 1990).

For workers, Fordism represented the potential of *lifetime work trajectories* of full-time employment until retirement, under a given employer's supervision and place of business, as long as performance expectations were met (Kalleberg 2000). As Davis (1988) noted, this was the Western "Standard Path of Retirement." Most adult workers – especially the leading edge of the baby boom cohort that entered labor markets at Fordism's height by the late 1960s – expected that these work arrangements would hold over their work lives. Indeed, women and ethnic minorities who were largely excluded from the Standard Path prior to the 1970s pushed for entry into these very arrangements (O'Rand and Henretta 1999).

Theoretical Approaches and Empirical Studies

This subsection highlights theoretical approaches used by sociologists and other social scientists to study human capital and work under the Fordist regime and summarizes key findings. As background, the United States has an open "market model" of job matching relative to the "institutional models" of other nations where school, apprenticeships, and work careers are more coordinated or tightly linked (Heinz 2003). It is in this open market model context that *human capital theories* arose by the 1960s (Becker 1964). Its tenets implicitly upheld the tripartite life course by differentiating lifetime streams of human capital, as education and work experience, across three life stages. Formal schooling was viewed as the occupation of youth, pursued to the point where it would be rewarded by employers. On-the-job training in firm-specific skills occurred over the work life stage (Becker 1993) and retirement, a reward for past productivity, occurred in old age.

Human capital theories apportioned the phases of work careers over the work life stage as well. School exit was followed by several years of job changes, testing "job matches." With a good job match early-career workers self-subsidized training costs for another period of time (Becker 1964). They earned significantly lower wages and benefits than did experienced workers but expected to reap relatively larger wage returns to experience over mid-to-late career (Kotlikoff 1988). Employers provided firm-specific training for their early-career workers, then "locked-in" or vested these trained employees with contracted benefits such as pensions to reduce labor turnover costs (Hall and Lazear 1984; Petersen et al. 1989, for the organizational perspective). Employers also placed age and tenure clocks into contracts so relatively higher-paid older workers, as a group, would not overstay their welcome on-the-job (Graebner 1980). Workers' retirement decisions then reflected the relatively higher cost of continuing to work vs. retiring, also informed by tastes for leisure and cyclical conditions (Hall and Lazear 1984; Wise 1997).

To summarize, human capital models arose under the Fordist production regime and highlighted rational decision-making of employers and employees, expecting job changes to be more likely for the young, as market entrants, and less likely in later life, due to work experience with wage penalties for job exits (Lazear 1976). A classic study illustrating this distinguished job separations for men ages 45–59 as voluntary quits vs. involuntary layoffs (Bartel and Borjas 1977; NLS 1966–1971). Longer job tenures reduced layoffs, indicating that employers valued experienced employees. Yet about 22% of older men changed jobs over the period, half due to quits and half due to involuntary displacements. The quits predominated in good economic times, especially for high human capital men, but a 3 year job match pattern of higher turnover was still present for most of the older men who changed jobs. Whether because the older men were chronic movers or had selected into insecure jobs, a recent layoff or quit increased the likelihood of a subsequent layoff or quit. The time to reemployment, as now, was longer for older vs. younger men (also Rones 1983).

Sociological research around this time also found support for – or had theoretical parallels to – human capital hypotheses, including higher rates of early- vs. late-career job mobility (Tuma 1976), positive wage returns to job tenure (Hachen 1992), employer-limited turnover due to cost (Petersen et al. 1989) and retirement contracts with incentives to exit work (Hayward 1986; O'Rand and MacLean 1986). But sociologists also identified processes less consistent with human capital or Fordist principles, more generally. For example, studies found that hiring, on-the-job training, and promotions often reflected bureaucratic mechanisms or labor market divisions more than human capital theory would predict (Hachen 1992; Petersen et al. 1989; Rosenfeld 1992). Also, even though Fordist centralization of decision-making with the reorganization of other jobs into routine tasks was promoted on economic efficiency grounds, studies found that putative organizational efficiencies often were sacrificed for even greater managerial control (Gordon et al. 1982; Jacoby 1985; Myles 1990).

Market segmentation research studies soon focused on labor market divisions, finding that the social-demographic composition of work forces and structured work arrangements such as job entry opportunities, retention, and work rewards diverged across market sectors (e.g., firms, occupations, industries and/or cross-cutting domains) (Althauser 1989; Hachen 1992; Hodson 1984; Petersen et al. 1989). *Dual labor market* studies differentiated primary sectors from secondary sectors, finding that primary sector workers earned higher wages and had promotion opportunities – via career ladders with benefits – that most supported Fordist lifetime work trajectories (Althauser 1989). Secondary sector workers more often were women, ethnic minorities, very young or old workers and had lower-skilled, less-secure, lower-paid jobs with fewer benefits. Moreover, employers appeared to reward education and skill differently across sectors, even rationing access to primary sector jobs by setting up "job queues" or barriers to sector entry for secondary sector workers (Thurow 1975).

Occupational segmentation came to be of theoretic concern, with segments denoted by the occupational skills workers performed (Spenner 1990), the different rewards to skills (Althauser 1989; Rosenfeld 1992), the degree of occupational expansion or contraction (Hayward and Grady 1986) and occupational group control over work (O'Rand 1986). Studies found that occupational sector and skill effects explained much about mid-to-late-life job separation and retirement patterns, net of human capital hypotheses. For example, men in primary sector occupations or in self-employment were less likely to retire if their job tasks engaged them, *net of* wealth or expected postretirement income (Elder and Pavalko 1993; Hayward and Hardy 1985). Middle-aged and older men in occupations marked by expansion, task complexity, and control over work also were less likely to experience involuntary job loss, net of education, and other human capital effects (Hayward and Grady 1986; Hayward et al. 1988a).

Also, where human capital theories proposed retirement patterns to mirror "benefit promises" in firm retirement packages and Social Security, this more likely occurred for primary segment employees (O'Rand and McLean 1986; O'Rand and Henretta 1999). For example, a relatively even proportion of 41% of women ages 55–64 worked over the 1965–1980 period (Purcell 2000, Table 2), will most ineligible for job-related pensions because pension-eligibility was unevenly allocated across industrial and occupational sectors and firms (O'Rand 1986; O'Rand and McLean 1986; Slevin and Wingrove 1995 for review of women and retirement). Women tended to work in low-wage service sector jobs or in firms with poor pension benefit coverage (O'Rand and McLean 1986). Therefore, women and minority elders – who also had less access to pension benefits because of sector of employment – often could not retire due to low income and so worked part-time into old age (Morrison 1983).

Job mobility and *life course* theories, which I treat here due to space concerns as interlocking perspectives, also fueled research on middle-aged and older workers by the 1980s. With regard to job mobility, the units of study include jobs with a given employer and/or the movement from one job to another (Hachen 1992; Petersen et al. 1989; Rosenfeld 1992). Studies also explore how adjacent jobs might differ, why job transitions occur, rates of change in job moves, and how jobs combine to form careers, in terms of directionality or coherence (Rosenfeld 1992). Life course research focuses on changes in state as well, but over the human life course more broadly (Elder 1975). Both perspectives are multilevel and dynamic and assume that decision-making is contingent on historical time, place, personal resources, and interpersonal linkages in life course studies (Elder 1975, 1994) or human capital and the opportunity structure that firms and labor markets present in mobility studies (Hachen 1992; Rosenfeld 1992).

Life course and job mobility studies uncovered gender segregation in occupations and work careers; the existence of wage penalties, net of human capital and productivity, due to family roles; and "glass ceilings" that limited hiring and promotions across firms, occupational and industrial sectors (O'Rand and Henretta 1999; Rosenfeld 1992). Life course studies also uncovered variability in job mobility and retirement timing due to nonwork influences, such as family or spousal circumstances (Henretta and O'Rand 1983; O'Rand and Henretta 1999; Slevin and Wingrove 1995) and cumulative life events (Szinovacz and Washo 1992). Human capital theorists Bartel and Borjas (1977) and later Ruhm (1990) noted that mature workers who were not securely attached to labor markets moved in and out of employment like young workers in a "job matching" process. Hayward et al. (1988b) also noted that older women in secondary sector jobs had high rates of *post-retirement* job mobility, while the patterns of older women retiring from primary sector jobs resembled men's Standard Path to retirement exits. Older African-American men under Fordism also had high rates of job mobility; life table analysis of work transitions found that they spent more time working and less time retired than European-American men, in both absolute and relative terms, due to higher rates of disability and mortality (Hayward et al. 1996). Older African-American workers also held different retirement expectations because they tended to be nonpension eligible. As Gibson (1986) noted, there was a third life stage of "unretired-retired" for workers who could not participate in the social institution of retirement.

Fordist Breakdown and the "Flexible" Turn

Economic restructuring captured sociological attention by the late 1970s (Bluestone and Harrison 1982; Piore and Sabel 1984). Economic stagnation (rising unemployment) with inflation was a concern during this period, initially attributed to exogenous market shocks (i.e., nonstructural changes) such as increased energy prices, a labor market supply glut of baby boomers and/or rising numbers of career-oriented women (Thurow 1975). But a different employment pattern appeared by the late 1980s and 1990s, marked by the new types of workers at risk of unemployment and the types of jobs involved (DiPrete 1993; Farber 2003; Johnson et al. 2009). This second section discusses economic restructuring that has occurred in the last few decades with the transformation from Fordist to flexible production methods. The next (third) section of this chapter details recent trends in job mobility and work/nonwork transitions.

Recent labor market transformations reflect many factors such as global wage pressures with falling product pricing (Bills 1995; Smith 1997), declining union strength (Mayer and Schoepflin 1989; Myles 1990), corporate financial rewards for merging and downsizing (Morris and Western 1999; Hardy and Quadagno 1995), and new information systems that geographically decouple production from a managerial core (Prechel 1994). Also important is rising demand for niche and specialized products and services (Bills 1995; Vallas 1999), associated with technology shifts: in computerized hardware and software, the on-line production or delivery of goods and services and the like. Employers find it less necessary or feasible to maintain large, standing hierarchies of narrowly-trained Fordist production workers as a long-term labor force. They seek greater *numerical flexibility* in staffing (Kalleberg et al. 2003; Smith 1997) and so limit Fordist-style long-term employment contracts. They use retention and job separations as tools to gain numerical flexibility (Hachen 1992) with job separations occurring as layoffs or displacements. They also externalize employment obligations by outsourcing or subcontracting similar or new jobs to new workers or firms, or limit the duration of work contracts by hiring contingent or temp agency employees. Yet employers also seek greater *functional flexibility* in production and adopt new technologies, tools, reorganize shop-floor activity (Smith 1997), and/or reshape organizational networks to rapidly produce or modify specialized products for markets (Kalleberg et al. 2003; Smith 1997).

What this means for middle-aged workers is that jobs and work careers are being recast, from the *lifetime work trajectories* that kept job turnover low and anchored a tripartite life course division, to *flexible career trajectories* that accommodate short term or "just-in-time" staffing (Polivka 1996) and may override this division (Heinz 2003; Myles 1990). Even though many middle-aged and older workers continue to work under the old wage and benefit contracts of the Fordist regime, many more take new jobs at midlife, "flexible" jobs which increasingly proliferate under myriad "spot-market", or shorter-term work contracts (DiPrete et al. 2002; Kalleberg 2000; Smith 1997). In line with the change to flexible work arrangements, portable retirement contracts have emerged, such as 401Ks or defined contribution plans, less linked to specific employers or the age-scheduling of retirement (O'Rand and Henretta 1999).

Theoretical Approaches

This subsection describes new and evolving perspectives, introduced earlier, that are used to study work arrangements under flexible production. New *market flexibility perspectives* view the changing conditions of work as employer or market-rational responses to the macroeconomic challenges described above. Flexible hiring – rather than hiring workers into hierarchical internal labor markets with long training/vesting periods and career ladders – may better allocate worker productivity and human capital in the new economy. Also, the breakdown of rigidly structured Fordist production may foster more fluid organizational forms and career lines; this, in turn, may improve work conditions by supporting the creation of jobs that are less routine, more creative, and provide workers with more autonomy and decision-making power (DiPrete et al. 2002; Kalleberg et al. 2003; Smith 1997). Workers enter jobs: if jobs are more fulfilling then job satisfaction might offset the job insecurity that is inherent in flexible work (Smith 1997).

What is flexible work? Flexible work arrangements by one definition, take place outside of a workplace or at a time other than the standard workday (Rau 2003). This type of work can include off-shift work, flextime, telecommuting or home-based work, part-time or contingent work (also see Fenwick 2009). In another definition of nonstandard work arrangements, the Bureau of Labor Statistics (BLS) categories include: independent contract work, on-call work, temp agency work, and contract company work (Kalleberg 2000 for overview). About 9.3% of U.S. workers in the late 1990s (DiNatale 2001; Polivka 1996) and about 10.7% in 2005 (BLS 2005) fell into one of these four categories. Also, in overlapping categories, the BLS reports contingent jobs (i.e., jobs not expected to last a year). About 4% of employees were contingent in 2005 but this group was diverse; members were more likely to be nonhigh school graduates but they were also more likely to have B.A. degrees than noncontingent workers (BLS 2005). Few "flexible work" studies as yet focus on older workers, but flexible work more generally can provide positive work conditions (DiNatale 2001; Johnson et al. 2009; Smith 1997), increase market rewards (DiPrete et al. 2002) and enhance skill sets (Spenner 1995). As described in a previous section, mature workers who are engaged in their jobs tend to delay retirement. Older workers also desire flexible work schedules and so might prefer nonstandard work as they near retirement (DiNatale 2001; Purcell 2000; Rau 2003). This suggests that recent restructuring, despite uncertainties felt by current older workers, might increase working life expectancy in the *long* run.

However, many adult workers in BLS nonstandard arrangements, with the exception perhaps of independent contract workers, prefer the longer, more durable employer–employee contracts that predominated earlier in their careers (BLS 2005; DiNatale 2001; Polivka 1996). The Fordist regime promoted a significant measure of job security, culminating in the Standard Path of retirement, especially for primary sector workers. As noted by McCall (2000) many employees define the new, flexible employment as insecure employment. *Market accumulation perspectives* view changed conditions, under the "flexible turn," as a change in the relative Fordist balance of employer vs. worker power. Fordism was anchored in durable employment agreements over the mid-twentieth century, and if it conferred job ladders and retirement benefits for workers it benefitted employers as well, reducing turnover and training costs. In this view, deinstitutionalization of Fordist arrangements may represent a continued shift of relative power back to employers' structured advantage. As Fordist arrangements fade, even advantaged sector workers may lose previously gained wage rewards, workplace power or even skills – if employers redesign jobs or allocate mass job losses relative to their particular skill sets (Bluestone and Harrison 1982; Prechel 1994; Spenner 1995).

In this critical vein, recent *market segmentation* studies find that some market divisions have grown in scope with the decades-long shedding of Fordist-style work contracts, especially among midlevel employees and union workers (Hudson 2007). *Dual market* studies now highlight a polarized "good-jobs" vs. "bad-jobs" divide (Hudson 2007; Kalleberg 2000) with significant differences in job security, wages, and control over work across this divide. Of concern is that stratification resulting from flexible staffing can develop, including structures of "insider and outsider" employees (Kalleberg et al. 2003) where insiders populating the organizational core are better-trained with relative autonomy in work, higher wages, and job benefits. In contrast, workers at the organizational periphery, as "outsiders," tend to have less control over work, be contingent, and do less skilled work with weaker occupational ties even if highly skilled (Brand 2006; Kalleberg et al. 2003). The share of middle-aged and older workers as "outsiders" in nonstandard work arrangements has grown (Hudson 2007; Kalleberg et al. 2003).

Critical studies also highlight forms of *flexible work trajectories* - vs. organizational careers that have emerged. A "*knowledge worker*" perspective identifies polarization in flexible careers which also may be associated with a growing educational divide in wages and employability (Morris and Western 1999). Middle-aged workers with some limits to education or skills – although many have considerable work experience or postsecondary degrees – follow disadvantaged flexible career pathways marked by high rates of job mobility without advancement in low-wage market sectors. Many of these workers are women (Han and Moen 1999) or men displaced from blue-collar and midlevel jobs (Farber 2003). In contrast, middle-age workers with higher levels – or different mixes – of education or skills form more advantaged protean careers, such as "portfolio workers" who build careers across work settings and jobs (Kalleberg 2000; also see DiPrete et al. 2002; Han and Moen 1999; Kleinman and Vallas 2001; Uchitelle 2006). Protean work trajectories can encompass job-hopping across managerial, employee, and self-employment roles and across education, government, and other industry segments. Employers loosen managerial control due to the creative nature of their production and seek to retain them (DiPrete et al. 2002). In turn, protean careers allow knowledge workers to develop skills, improve task-based skill performance, and align themselves to projects in their areas of expertise rather than to organizational boundaries.

Empirical Studies of Mid- to Later-Life Job Mobility

How do middle-aged workers fare given such changes? A rapidly expanding social science literature finds that, with the "flexible turn," job tenure or career job length has declined (Munnell et al. 2006) and job mobility rates and work/nonwork transitions have markedly increased and grown more blurred (Elder and Pavalko 1993; Han and Moen 1999; Hayward et al. 1988a, b; Giandrea et al. 2009; Sterns and Chang 2009). Given the growing heterogeneity in work patterns at midlife, this section and the next section on adult education provide overviews of fields of study in flux. I first define types of job mobility and work-retirement status changes that are addressed in this review, including the limitations of these definitions.

Excluding firing for cause, nonretired older workers might exit jobs due to quits or involuntary separations. The reasons for quits at midlife are diverse and understudied (Bartel and Borjas 1977; Hachen 1992). Involuntary separations can be defined as job displacements or exits from jobs due to the elimination of positions or shifts, plant closings, employers going out of business, layoffs, and/or reductions in force due to slack work for workers older than age 20. Job loss is an event that results in a state of unemployment. This state traditionally penalizes older more than younger workers, due to longer times to reemployment. As will be shown, involuntary job loss is not uncommon, yet excellent databases still underestimate its incidence and prevalence. This is because the line between quits and involuntary separations is often invisible. Employers can keep wages low, reduce hours, limit benefits, and make work conditions difficult in order to induce – or at least not discourage – quits (Farber 2003).

It is as difficult to estimate final retirements. According to Purcell (2000) retirement reflects two characteristics, nonparticipation in paid labor and income receipt from Social Security or another form of pension plan. Final retirement takes place when nonparticipation and pension receipt occur together as a stable state. But middle-aged workers seek or reenter employment while collecting Social Security or pensions (Johnson et al. 2009). They also are economically inactive in the absence of pension or other income (Burr et al. 1996). Also important, then, is the perception, planning and/or the intention to retire (Ekerdt et al. 1996; Han and Moen 1999; Szinovacz and Davey 2005) or stages of retirement, such as from self-reported partial to complete work exits (Ruhm 1990, 1991). Yet, beyond this, studies distinguish voluntary vs. involuntary retirements because even preretirement age workers (well under age 62) retire in response to economic inducements, poor health or disability, threat of layoff, or actual job displacement (Han and Moen 1999; Hardy and Quadagno 1995; Szinovacz and Davey 2005). A career job is a worker's longest spell of employment with a single firm (Ruhm 1990).

Because middle-aged workers enter postcareer jobs with pensions or after spells of inactivity, the jobs taken between a career job exit and a final retirement are often termed "bridge jobs" (Ruhm 1990). "Re-careering jobs" involving change in occupation and employer are also studied (Johnson et al. 2009). Unfortunately few, if any, studies of job displacement or bridge jobs incorporate the BLS definitions of nonstandard work so cross-classification of the prevalence of "flexible" postcareer jobs is unknown. But postcareer activities constitute (often blurred) pathways to a final retirement. The next two subsections describe job displacement and work-retirement status changes, respectively. I focus as much as possible on midlife. My separation of these processes is not entirely by choice, but, beyond definitional differences, treatment of each process markedly differs across the social sciences.

Job Displacement

Recent studies largely come from labor economics and use panel data such as the Displaced Worker Supplements (DWS); the Wisconsin Longitudinal Survey (WLS); or the prospective longitudinal Health and Retirement Survey (HRS). Most study samples are adults with 3–5 years of job tenure or in a baseline career job, a point beyond human capital theory's critical "job-match" period. How common is displacement? Couch (1998, HRS) found 2-year displacement rates of 4.7 for men aged 51–60, similar to BLS data of the period (Gardner 1995). Chan and Stevens (2001, HRS) found that 17.2% of workers aged 50 and over were displaced at least once due to layoffs or firm closings between 1984 and 1996; about 14% of the WLS cohort experienced a job displacement between 1972 and 1992 (Brand 2006). In a sample of retirees aged 51 and over, 13% of men and 15% of women were displaced prior to retirement (Szinovacz and Davey 2005, HRS). Finally, 25% of full-time workers, ages 51–55 in 1992, had been displaced by 2006 (Johnson et al. 2009, HRS).

Of importance is whether displacement reflects cyclical economic conditions (recessions), more deeply-rooted or structural change, or both. Also important is whether displacement risk differs by age. Studies show that displacement [incidence] rates rose in the early 1980s and quickly returned to baseline, for workers of all ages, when recession-related conditions improved. In contrast, displacement rates for all workers jumped over the early 1990s recession but then remained elevated for older workers to about middecade (Farber 2003, ten DWS waves; Gardner 1995), especially men and women aged 45–59 in blue-collar, and less educated men aged 45–59 in white-collar jobs (Siegal et al. 2000, DWS). Siegal et al.'s (2000) examination of relative displacement rates of younger (age 25–39) vs. older (age 45–59) workers found that older men had lower rates than younger men until the early 1990s when rates then converged (also Gardner 1995). Farber (2003) concludes that, beyond the cyclical job losses due to layoffs and plant closings over the early 1980s, early 1990s and early 2000s, older workers, and especially educated older workers became subject to secular increases in job displacement over the mid-1990s and beyond, largely due to "shifts/positions abolished."

To be sure, the high unemployment rates of youth and declining labor force participation of young adults – under age 40 – are of great concern (Gendell 2008; Uchitelle 2006). But, as Munnell et al. (2006, DSW and HRS) note, a positive and significant relationship between age and the incidence of job displacement, has emerged, holding job tenure constant. If job tenure continues to decline at midlife, job displacement will become more common (DiPrete et al. 2002; Munnell et al. 2006).

Consistent with "good jobs, bad jobs" labor market models, part-time workers – who are more likely to be in poor quality jobs – have relatively high displacement rates, as do semiskilled blue-collar workers (Chan and Stevens 2001; Couch 1998). Yet men have a higher risk of displacement than women (Brand 2006; Munnell et al. 2006). And, although relatively high 1980s displacement rates in blue-collar jobs continued into the 1990s (Siegal et al. 2000), job loss by the 1990s strongly impacted white-collar, primary sector jobs (DiPrete 1993). Human capital effects on displacement also became less protective over time: college degrees reduced the probability of job displacement by 5% points in the 1984 DSW wave but were not significant by the 2002 wave (Munnell et al. 2006). These latter findings, especially with regard to the weakening protective effects of seniority, male gender, human capital, and primary sector location on involuntary job loss are consistent with waning of the Fordist labor regime, and market accumulation processes of work-reward retrenchment for formerly advantaged employees.

Displacement results in unemployment; unemployment at midlife is associated with relatively longer times to reemployment and entry into lower-quality jobs (Couch 1998). In a simulation with HRS data, Chan and Stevens (2001) estimate that, if displaced at aged 55, only 50% of men and 45% of women find work 1 year out (also Johnson et al. 2009). Munnell et al. (2006, HRS) find that non-Whites are 13% less likely than Whites and women 7% less likely than men to become reemployed or else may enter a series of low-wage jobs (also Couch 1998; Mutchler et al. 1997). Postdisplacement jobs then tend to be of short duration (Chan and Stevens 2001) and pay significantly less than lost jobs, with wages reduced for 6 years out or more (Chan and Stevens 2002). However, consistent with the market flexibility perspective, many older workers enter flexible work arrangements by choice; if independent contractors, they often are self-employed or employers of others (DiNatale 2001). Job quality can be high with control over the tempo and organization of work (Kalleberg 2000; Polivka 1996); job satisfaction can be high in new flexible jobs, even if older workers earn less than they did in a prior job (Johnson et al. 2009; Sterns and Chang 2009). Finally, although displacement might lead to final retirement (Szinovacz and Davey 2005) most displaced workers do become reemployed, about 75%, with one-third of these new jobs being career-changes (Johnson et al. 2009). And, with reemployment, displaced workers enter more blurred pathways to final retirement.

Pathways to Final Retirement

Relatively few workers aged 50 and over move directly from career jobs into final retirement, however modeled. More than 50% move from full-time career jobs into self-reported partial retirement (e.g., at least part-time work) (Ruhm 1990, RHS) and more than 60% from full-time career into bridge jobs (Cahill et al. 2006, HRS). Also, about 44% of workers (age 51–55) who had retired from full-time career jobs in 1992 took new jobs by 2006 (Johnson et al. 2009, HRS). The variability in retirement timing has increased over time, measured across cohorts (Elder and Pavalko 1993; Han and Moen 1999; Giandrea et al. 2009). And, although critical human capital influences such as health, wealth, and pension availability remain important predictors of work vs. retirement (Cahill et al. 2006; Wise 1997), complex mixes of social characteristics, past job mobility, market segmentation, and life course influences also explain pathways to retirement.

As under Fordism, social heterogeneity in worker characteristics allocates workers across divergent market segments that then shape work-to-retirement pathways (Couch 1998; Flippen and Tienda 2000; Han and Moen 1999; Mutchler et al. 1997; Quinn and Kozy 1996). For example, careers – marked by employment continuity, direction (e.g., upward mobility) and work rewards – shape postcareer job pathways, although gender and ethnicity play a role in these work trajectories (Flippen and Tienda 2000; Han and Moen 1999; Quinn and Kozy 1996). For example, men in "high geared" careers tend to end their careers in "crisp" final retirements of the Standard Path while women in "intermittent" careers tend to enter postretirement work (Han and Moen 1999). More generally, relatively advantaged workers, with higher education, greater past earnings and pension benefits and generally European American men, tend to still retire in a "crisp" pattern of the Standard Path (Johnson et al. 2009; Mutchler et al. 1997); they are less likely to change careers or change employers if they do become reemployed (Johnson et al. 2009). In contrast, women, ethnic minorities, workers with less wealth or pension access or workers with more midlevel (12–16 years) education are more likely to "re-career" or change employers and occupations after career job exits, although these jobs often have lower status and wages (Johnson et al. 2009). The least advantaged group, often women with delayed work experience, former blue-collar males, secondary sector or part-time workers, (Han and Moen 1999; Mutchler et al. 1997; Quinn and Kozy 1996) and a large subset of ethnic minorities has the least secure market attachment (Burr et al. 1996; Flippen and Tienda 2000). Then, as poorly attached older workers, depending on economic need and health, they are either more likely to withdraw from activity in midlife as economic nonparticipation becomes an absorbing state (Burr et al. 1996; Hayward et al. 1996) or have sporadic work spells over later life (Han and Moen 1999; Mutchler et al. 1997). Work-to-retirement patterns remain influenced by spousal and family events and relations (Cahill et al. 2006; Han and Moen 1999).

Retirement is examined well elsewhere in the handbook, so I highlight two changing midlife patterns, across displacement and retirement pathways studies, relevant to the "flexible turn." A first pattern is a changing of directions in the traditional correlations between human capital and work/retirement. As noted, job tenure, or firm-based work experience, once valued by employers (Bartel and Borjas 1977) is becoming less so (Munnell et al. 2006), as also tracked by displacement rate trends. Also, age until recently was negatively correlated with job displacement and positively correlated with final retirement. But, at some point in the 1990s, the relative risk [incidence] of displacement changed for older vs. younger workers (Farber 2003; Siegal et al. 2000) while, at the aggregate level, older workers (age 60+) began to delay final retirement or reattach to markets postretirement (Gendell 2008) as employment, especially among men ages 45–59, remained stagnant or declined (Gendell 2008; Uchitelle 2006). Although rising old age participation reflects growing economic need with recent (post 2000s) market volatility, changing pension types/rules and perhaps better health, middle-aged workers have similar, if not greater, economic need to work and greater exposure to defined contribution pensions. Finally, human capital as education and work experience, once associated with stable and lengthy work careers, also appears to be a changing indicator as it is often a nonsignificant predictor of job retention vs. displacement (Couch 1998; Munnell et al. 2006) or of retirement or nonparticipation (Burr et al. 1996; Flippen and Tienda 2000; Han and Moen 1999), especially for men (Giandrea et al. 2009). Known correlations have weakened if not reversed.

A second pattern is one of growing inequality in work outcomes of employability and wages. Middle-aged men have long had more orderly, upwardly mobile careers than middle-aged women (Flippen and Tienda 2000; Han and Moen 1999), including institutional supports for retirement. But, well before the recent economic downturn in 2007, adult men had growing problems with labor force attachment and employability, especially if under age 60 (Uchitelle 2006). In contrast, women's labor force attachment has increased more than men's, especially at older ages (Gendell 2008) but women continue to earn lower wages and benefits than men (O'Rand and Henretta 1999). In addition, within – gender inequalities may be growing, in terms of complex race/ethnicity and labor market structure effects. Men of color have shorter career job durations and higher rates of persistent nonparticipation than European American men (Burr et al. 1996; Mutchler et al. 1997; Flippen and Tienda 2000). Among women, although middle-aged and older Hispanic women work less than other women, (Flippen and Tienda 2000), marked differentials in wages, more than employability, are present. And, wage differences reflect differences in quality of nonstandard work arrangements more than in level of education or skill (McCall 2000).

Both the patterns of changing human capital effects and rising inequality and what they portend needs study; policy recommendations for middle-aged workers will differ, depending on findings. With regard to changing human capital effects, or the loss of employability for workers of once-advantaged status, employers do not likely disvalue work experience, education, or skill – indeed, recent growing wage inequality reflects growing education and skills-gaps (Elman and O'Rand 2004; Morris and Western 1999). But educated, skilled and, especially, experienced older workers have much *firm-specific training* and employers may value firm-specific skills less. Market flexibility perspectives suggest that employers seek to add innovation and value in production; if so, they may gravitate toward workers with a wider range of cross-sector or transferable skill, for functional flexibility ends (DiPrete et al. 2002; Kalleberg et al. 2003; Smith 1997). This is what Fordist-trained middle-aged workers may lack and so adult education and general skills training (discussed in the next section) may allow better job matches (Sterns and Chang 2009). Beyond retraining, midlife workers may also has to reconceptualize "work": as employers seek numerical flexibility and employability in the new century, "work" may refer less to new career jobs and more to gaining entry into any job, including nonstandard arrangements.

But, to the degree that market accumulation forces are dominant, employers may reap much of the rewards from flexible employment, limiting employment opportunities, as middle-aged, often skilled employees become subject to displacement, lower wages in reemployment, and less engaging, "outsider" work. If so, then mature workers, especially women, minorities, and those with few resources might work sporadically into old age because of economic need and an inability to retire (Burr et al. 1996; Hayward et al. 1988b; Gibson 1986; Johnson et al. 2009). But, in the meantime, workers with more education and skill – if able to exit economic activity via the "Standard Path" – will do so rather than find reemployment, given the option of taking *low quality* flexible jobs (Cahill et al. 2006; Johnson et al. 2009). Notwithstanding the societal, aggregate losses of human capital and contractions in working life expectancy (Davis 1988; Wise 1997), this outcome is not ideal for middle-aged workers because they potentially constitute a leading edge of employees that might benefit from high-quality nonstandard jobs.

In summary, a mix of policies is needed that can benefit heterogeneous mixes of midlife workers. In general and especially for those weakly attached to markets, job creation, and targeted (re)training is important; for workers who seek a broader educational foundation or new skills for recareering, formal education programs might facilitate employment. To encourage extended working life expectancy for the more advantaged, skilled workers whose relatively early retirements tend to be final, policies that support work-engagement, job-sharing, phased retirement or other high-quality flexible job redesigns might be as or more important than retraining. This said, many middle aged workers, especially women, seek retraining to increase employability. I turn to this below.

Adult Education New Human Capital Streams in the Flexible Economy

The U.S. educational system grew rapidly over the twentieth century. The early twentieth century "nontraditional" students, children of immigrant families, used education to enter the semiskilled blue-collar and white-collar Fordist jobs that were emerging (Nasaw 1979). And, by mid-century the educational system had become a "licensing agency for Americans who wanted to enter professions" (Lazerson 1998: 64), mostly European-American men. Then, between mid-century and the late 1970s a different type of educational expansion occurred as the number of baccalaureate, community college, and vocational institutions of higher education doubled in the United States (Lazerson 1998). This post-World War 2 school expansion or "educational massification" (Zemsky 1998) represented a U.S. public commitment to schooling and contributed to the education-to-work life stage division that arose over the mid-century.

But, another group of students appeared in the 1970s: young to middle-aged, largely female, often of ethnic minority background, as the new "nontraditional" students. Members of the baby boom cohort (Jacobs and Stoner-Aby 1998), they sought professions and other good jobs, too. Postsecondary education would help them enter the economic mainstream, including the white-collar or service, albeit less-secure, non-Fordist jobs emerging by century's end. Although both men and women entered 2- and 4-year colleges at higher rates after the 1970s, women's enrollment rates, then degree completion rates, rapidly accelerated across all educational settings, surpassing men's rates by the late twentieth century (Buchmann et al. 2008).

By the end of the twentieth century, the U.S. educational system had changed from being a licensing system for professional jobs into a gateway for many jobs – or sometimes any job (Bills 2003). Moreover, adult demand for education continues to grow even as demassification occurs: an erosion of state financial support for higher education amidst rising tuitions (Zemsky 1998).

Adult students absorb most of the cost of schooling and so adult educational participation needs to be explored in the context of future employability, job availability, job quality, and wage rewards. Under the mid-twentieth century Fordist accord, educated and less educated workers could be matched to jobs, even if not always "good jobs." But, with a growing share of post-Fordist jobs as flexible or nonstandard low-wage and low-skilled jobs, facilitating employment for newly skilled adults will become critically important.

Theoretical Perspectives and Empirical Studies

As noted in the previous section, human capital and employment relationships have changed, singly and in the context of each other, but these new relationships have yet to be fully understood or theoretically integrated. Adult education is not well addressed in human capital theory, other than in variants such as *screening* or *signaling* theories. These theories propose that employers notice degrees [credentials] and who holds them more than educational content or learning, when assessing the productivity of potential hires (Bills 2003). At issue is whether an older age of graduation sends neutral or negative signals to employers that affect hiring and wages (Monks 1997). In contrast, *cumulative dis/advantage theory* makes variability in the life course timing of education the central problem, in broader social, historical, and institutional context (Pallas 1993). It highlights educational stratification processes by exploring institutionalized tracks of advantage and how life course timing differences in the access to such structural resources produces growing intragroup inequalities over time (Dannefer 1987).

Consistent with a process of cumulative disadvantage/advantage, most adults who graduate with baccalaureate and higher degrees "on time" – in a trajectory of postsecondary persistence – had, as youth, advantaged social backgrounds (Bozick and DeLuca 2005; Elman and O'Rand 2004; O'Rand et al. 2009). Adult schooling, if education occurs, is then likely to be episodic with entries and exits across different semesters, years and/or schools (Goldrick-Rab 2006; Zemsky 1998). While most adult postsecondary school entrants do not complete degrees (Jacobs and King 2002), about 6% of B.A. degrees and 20% of Associate degrees are obtained after age 40 (O'Rand et al. 2009). Adult postsecondary completion rates are low because most adults go to school part-time (Jacobs and King 2002) and have high intensity co-occurring economic activities (Jacobs and King 2002; Kane and Rouse 1999; NCES 2002). Most adult community college students report that work is their primary activity (Kane and Rouse 1999).

Program-of-entry patterns resemble those found in *market segmentation* studies: differential selections by race/ethnicity, class, and/or gender into postsecondary 2- vs. 4-year degree programs (Elman and O'Rand 2007; Karen 2002; Kane and Rouse 1999). Adult women and minorities tend to enter non-baccalaureate programs, partly because they appear to cost less. However, when the opportunity costs of paid work and time are added to total costs, tuition is a relatively small part of the total cost (Kane and Rouse 1999). As *life course* theory would predict, the timing of education and the type of program entered is constrained by family roles (Astone et al. 2000; Bozick and DeLuca 2005; Jacobs and King 2002). And, as signaling theory suggests, adults may earn less for equivalent degrees earned at older ages (Elman and O'Rand 2004; Monks 1997).

Yet adult educational participation has rapidly increased over the last few decades. Cross-sectional National Household Education Surveys (NHES) highlight the range of adult educational activities, including basic skills training, apprenticeships, work-related courses, and part-time college or university degree programs. About 37% of adults ages 45–54 and 27% ages 55–64 took a *work-related course* in 2005 (Levesque et al. 2008). Considering only economically active adults, about 40–45% of adults ages 45–49, 50–54, 55–59, and 60–64 took a work-related course in 2005. Comparable percentages by age group were lower in 1995 than in 2005, but comparable percentages in *both* 1995 and 2005 were lower than in 2003, when participation jumped near the 2001 recession (Snyder et al. 2009). Also, NHES adults entered school to do more than "brush up" on a single skill: working adults aged 45–64 were more likely to take two to four courses than working younger adults (Levesque et al. 2008).

In summary, adult educational participation is growing as adult career job length declines, as flexible replacement jobs proliferate and as the greatest wage gains and employability accrue to workers with baccalaureate and higher degrees (McCall 2000; Morris and Western 1999). But adult students find educational completion to be a difficult accomplishment; for completion rates to increase, research and programmatic change will be needed to address ways in which school settings and practices are not tailored to adult student needs. Also, financing is costly, whether in community or baccalaureate settings (Kane and Rouse 1999). Given the cost, research is needed to understand the type(s) of education that enhance employability and earnings for middle-aged workers – and for what subgroups or under what circumstances – and whether working life expectancy is extended (Davis 1988).

Concluding Remarks

Transformative change in the social organization of work has occurred. This has uniquely impacted middle-aged adults, who bridge changing production regimes. Some subgroups will do well post-career: in self-employment, reemployment in similar economic sectors, or in flexible work arrangements, including in the nonstandard arrangements that can provide job satisfaction (DiNatale 2001; Johnson et al. 2009). Other subgroups will experience midlife downward mobility, into lower-wage or less engaging jobs or will experience reduced employability with inactivity for long spells and early economic withdrawal (Burr et al. 1996; Flippen and Tienda 2000; Uchitelle 2006). What middle-aged adults now need is information about "what works" for job matching at midlife. Many will and can retrain (Sterns and Chang 2009), partly motivated by the perceived job insecurity associated with flexible or contingent jobs (Elman and O'Rand 2002). But what should the content of adult human capital be under flexible production regimes? Should adults pursue general education and cross-training? Or renew or update the narrower firm-or skill-specific types of training that, under Fordist production, was mostly provided by employers?

Overall, middle-aged adults have higher levels of human capital, work experience, and computer skills than past midlife cohorts, even perhaps comparable levels to current young market entrants (Munnell et al. 2006). They want to work longer than they actually do (Hardy and Quadagno 1995; Han and Moen 1999), to at least age 65 (Munnell et al. 2006), but face barriers to employment. How can midlife employment, job satisfaction and working life expectancy best be facilitated? Overall, we need to explore jobs, job matching, and employability in the context of cyclical *and* transformational changes. This will require refinement of human capital, labor market, and life course theories to fit new conditions. This will also require the implementation of policies, including the development of mechanisms – given our open "market system" vs. an "institutional system" of employment – that facilitate job matches in the hiring process for older workers who (re)enter labor markets. This will also require longitudinal databases, to track midlife education and work trajectories in the context of each other, and to map these long-term trajectories to later-life wealth, health, and work-related outcomes.

References

Althauser, Robert P. 1989. "Internal Labor Markets." *Annual Review of Sociology* 15:143–61.

Astone, Nan, Robert Schoen, Margaret Ensminger, and Kendra Rothbert. 2000. "School Reentry in Early Adulthood: the Case of Inner-City African Americans." *Sociology of Education* 73:133–154.

Bartel, Ann P. and George J. Borjas. 1977. "Middle-Age Job Mobility: Its Determinants and Consequences." Working Paper #161. NBER Working Paper series.

Becker, Gary S. 1964. *Human Capital: A Theoretical and Empirical Analysis, with Special reference to Education*. New York: Columbia University Press.

———. 1993. *Human Capital: A Theoretical and Empirical Analysis with Special Reference to Education*. New York: Columbia Press.

Bills, David. 1995. *The New Modern Times: Factors Reshaping the World of Work*. Albany: State University of New York Press.

———. 2003. "Credentials, Signals and Screens: Explaining the Relationship between Schooling and Job Assignment." *Review of Educational Research* 73:441–69.

Bureau of Labor Statistics. 2005. *Contingent and Alternative Employment Arrangements*, February 2005. Release USDL 05-1433. Washington, D.C.: Bureau of Labor Statistics.

Bluestone, Barry and Bennett Harrison. 1982. *The Deindustrialization of America: Plant Closings, Community Abandonment and the Dismantling of Basic Industry*. New York: Basic Books.

Bozick, Robert and Stefanie DeLuca. 2005. "Better Late Than Never? Delayed Enrollment in the High School to College Transition." *Social Forces* 84:528–50.

Brand, Jennie E. 2006. "The Effects of Job Displacement on Job Quality: Findings from the Wisconsin Longitudinal Study." *Research in Social Stratification and Mobility* 24:275–98.

Buchmann, Claudia, Thomas A. DiPrete and Anne McDaniel. 2008. "Gender Inequalities in Education." *Annual Review of Sociology* 34:319–37.

Burr Jeffrey A., Michael P. Massagli, Jan E. Mutchler, Amy M. Pienta. 1996. "Labor Force Transitions among Older African American and White Men." *Social Forces* 74: 963–82.

Cahill, Kevin E., Michael D. Giandrea and Joseph F. Quinn. 2006. "Retirement Patterns from Career Employment." *The Gerontologist* 46:514–23.

Chan, Sewin and Ann Huff Stevens. 2001. "Job Loss and Employment Patterns of Older Workers." *Journal of Labor Economics* 19:484–521.

Chan, Sewin and Ann Huff Stevens. 2002. *How Does Job Loss Affect the Timing of Retirement?* Working Paper #8780. National Bureau of Economic Research.

Couch, Kenneth A. 1998. "Late life job displacement." *The Gerontologist* 38:7–17.

Dannefer, Dale. 1987. "Aging as Intracohort Differentiation: Accentuation, the Matthew Effect, and the Life Course." *Sociological Forum* 2:211–36.

Davis, Kingsley. 1988. "Retirement as a Dubious Paradise." Pp. 191–203 in *Issues in Contemporary Retirement*, edited by R. Ricardo-Campbell and E. P. Lazear. Stanford, CA: Hoover Institute Press.

DiNatale, Marisa. 2001. "Characteristics of and preference for alternative work arrangements, 1999." *Monthly Labor Review* 124: March 28–49.

DiPrete, Thomas A., Dominique Goux and Eric Maurin. 2002. "Internal Labor Markets and Earnings Trajectories in the Post-Fordist Economy: An Analysis of Recent Trends." *Social Science Research* 31:175–96.

DiPrete, Thomas. 1993. "Industrial Restructuring and the Mobility Response." *American Sociological Review* 58:74–96.

Easterlin, Richard A. 1987. *Birth and Fortune: The Impact of Numbers on Personal Welfare*. Chicago: University of Chicago Press.

Ekerdt, David J., Stanley DeViney and Karl Kosloski. 1996. "Profiling Plans for Retirement." *Journal of Gerontology* 5I:SI40–49.

Elder, Glen H. Jr. 1975. "Age Differentiation and the Life Course." *Annual Review Sociology* 1:165–190.

———. 1994. "Time, Human Agency, and Social Change: Perspectives on the Life Course." *Social Psychology Quarterly* 57:4–15.

Elder, Glen H. Jr. and Eliza Pavalko. 1993. "Work Careers in Men's Later Years: Transitions, Trajectories." *Journal of Gerontology* 48: S180–91.

Elman, Cheryl and Angela O'Rand 2002. "Perceived Job Insecurity and Entry into Work-Related Education and Training among Adult Workers." *Social Science Research* 31:49–76.

———. 2004. "The Race is to the Swift: Socioeconomic Origins, Adult Education, and Wage Attainment." *American Journal of Sociology* 110:123–60.

———. 2007. "The Effects of Social Origins, Life Events and Institutional Sorting on Adults' School Transitions." *Social Science Research* 36:1276–99

Farber, Henry S. 2003. *Job Loss in the United States 1981–2001*. Working Paper 9707. NBER Working Paper Series.

Fenwick, Rudy. 2009. "Flexible Work Arrangements." In *Encyclopedia of the Life Course and Human Development*, edited by Deborah Carr. New York: Gale Press.

Flippen, Chenoa and Marta Tienda. 2000. "Pathways to Retirement: Patterns of Labor Force Participation and Labor Market Exit." *Journal of Gerontology* 55:S14–S27.

Gardner, J. M. 1995. "Worker Displacement: A Decade of Change." *Monthly Labor Review* 118:14–23.

Gendell, Murray. 2008. "Older Workers: Increasing Their Labor Force Participation and Hours of Work." *Monthly Labor Review* 131(January):41–54.

Giandrea, Michael D., Kevin E. Cahill and Joseph F. Quinn. 2009. "Bridge Jobs: A Comparison Across Cohorts." *Research on Aging* 31:549–76.

Gibson, Rose C. 1986. "Blacks in an Aging Society." *Daedalus* 115:349–71.

Goldrick-Rab, Sara. 2006. "Following Their Every Move: An Investigation of Social-Class Differences in College Pathways." *Sociology of Education* 79:61–79.

Gordon, David M., Richard Edwards and Michael Reich. 1982. *Segmented Work, Divided Workers: the Historical Transformation of Labor in the United States*. New York: Cambridge University Press.

Graebner, William. 1980. *A History of Retirement*. New Haven: Yale Univ. Press.

Hachen, David S. 1992. "Industrial Characteristics and Job Mobility Rates." *American Sociological Review* 57:39–55.

Hall, Robert E. and Edward P Lazear. 1984. "The Excess Sensitivity of Layoffs and Quits to Demand." *Journal of Labor Economics* 2:233–57.

Han, Shin-Kap and Phyllis Moen. 1999. "Clocking Out: Temporal Patterning of Retirement." *American Journal of Sociology* 105:191–236.

Hardy, Melissa A and Jill Quadagno. 1995. "Satisfaction with Early Retirement: Making Choices in the Auto Industry." *Journal of Gerontology* 50:S217–28.

Hayward, Mark D. 1986. "The Influence of Occupational Characteristics on Men's Early Retirement." *Social Forces* 64:1032–45.

Hayward, Mark D. and Melissa A. Hardy. 1985. "Early Retirement Processes among Older Men: Occupational Differences." *Research on Aging* 7:491–515.

Hayward, Mark D., Samantha Friedman and Hsinmu Chen. 1996. "Race Inequities in Men's Retirement." *Journal of Gerontology* 51:S1–S1O.

Hayward, Mark D., William R. Grady and Steven D. McLaughlin. 1988a. "Changes in the Retirement Process Among Older Men in the United States: 1972–1980." *Demography* 25:371–86.

———. 1988b. "The Retirement Process Among Older Women in the United States." *Research on Aging* 10:358–82.

Hayward, Mark D. and William R. Grady. 1986. "The Occupational Retention and Recruitment of Older Men: The Influence of Structural Characteristics of Work." *Social Forces* 64: 644–66.

Heinz, Walter R. 2003. "From Work Trajectories to Negotiated Careers: The Contingent Work Life Course." Pp. 185–204 in *Handbook of the Life Course*, edited by J.T. Mortimer and M.J. Shanahan. New York: Kluwer Academic/Plenum Publishers.

Henretta, John C. and Angela M. O'Rand. 1983. "Delayed Career Entry, Industrial Pension Structure, and Early Retirement in a Cohort of Unmarried Women." *American Sociological Review* 47:365–73.

Hodson, Randy. 1984. "Companies, Industries, and the Measurement of Economic Segmentation." *American Sociological Review* 49:335–48.

Hudson, Kenneth. 2007. "The new labor market segmentation: Labor Market Dualism in the New Economy." *Social Science Research* 36:286–312.

Jacobs, Jerry A. and Rosalind B. King. 2002. "Age and College Completion: A Life-History Analysis of Women Aged 15-44." *Sociology of Education* 75:211–30.

Jacobs, Jerry A. and Scott Stoner-Aby. 1998. "Adult Enrollment and Educational Attainment." *Annals of the American Academy of Political and Social Science* 559:91–108.

Jacoby, Sanford M. 1985. *Employing Bureaucracy: Managers, Unions, and the Transformation of Work in American Industry 1900–1945*. New York: Columbia University Press.

Johnson, Richard W., Janette Kawachi and Eric K. Lewis. 2009. "Older Workers on the Move: Recareering in Later Life." Washington, D.C.: AARP Public Policy Institute.

Kalleberg, Arne L. 2000. "Nonstandard Employment Relations: Part-Time, Temporary and Contract Work." *Annual Review of Sociology* 26:341–65.

Kalleberg, Arne L., Jeremy Reynolds and Peter V. Marsden. 2003. "Externalizing employment: Flexible Staffing Arrangements in U.S. Organizations." *Social Science Research* 32:525–52.

Kane, Thomas J. and Cecelia E. Rouse. 1999. "The Community College: Educating Students at the Margin between College and Work." *Journal of Economic Perspectives* 13:63–84.

Karen, David. 2002. "Changes in Access to Higher Education in the United States: 1980–1992." *Sociology of Education* 75:191–210.

Kleinman, Daniel L. and Steven P. Vallas. 2001. "Science, Capitalism and the Rise of the Knowledge Worker: The Changing Structure of Knowledge Production in the United States." *Theory and Society* 30:451–92.

Kohli, Martin. 1986. "The World We Forgot: A Historical View of the Life Course." Pp. 271–303 in *Later Life: The Social Psychology of Aging*, edited by V. Marshall. Beverly Hills: Sage.

Kotlikoff, Lawrence. 1988. "The Relationship of Productivity to Age." Pp. 100–129 in *Issues in Contemporary Retirement*, edited by R. Ricardo-Campbell and E.P. Lazear. Stanford: Hoover Institute Press.

Lazear, Edward P. 1976. "Age, Experience, and Wage Growth." *American Economic Review* 66: 548–58.

Lazerson, Marvin. 1998. "The Disappointments of Success: Higher Education after World War II." *Annals of the American Academy of Political and Social Science* 559:64–76.

Levesque, K. et al. 2008. *Career and Technical Education in the United States: 1990 to 2005*. NCES 2008-035. U.S. Dept. of Education. Washington, D.C.

Mayer, Karl Ulrich and Urs Schoepflin.1989. "The State and the Life Course." *Annual Review of Sociology* 15:187–209.

McCall, Leslie. 2000. "Gender and the New Inequality: Explaining the College/Non-College Wage Gap." *American Sociological Review* 65:234–55.

Meyer, Stephen. 1995. "Technology and Work: A Social and Historical Overview of U.S. Industrialization." Pp. 29–50 in *The New Modern Times: Factors Reshaping the World of Work*, edited by David B. Bills. Albany, NY: State University of New York Press.

Monks, James. 1997. "The Impact of College Timing on Earnings." *Economics of Education Review* 16:419–23.

Morrison, Malcolm. 1983. "The Aging of the U.S. Population: Human Resource Implications." *Monthly Labor Review* 106: May 13–19.

Morris, Martina and Bruce Western. 1999. "Inequality in Earnings at the Close of the Twentieth Century." *Annual Review of Sociology* 25:623–57.

Munnell, Alice H., Steven Sass, Mauricio Soto and Natalia Zhivan. 2006. *Has the Displacement of Older Workers Increased?* Working paper 2006-17. Center for Retirement Research.

Mutchler, Jan E., Jeffrey A. Burr, Amy M. Pienta, and Michael P. Massagli. 1997. "Pathways to Labor Force Exit: Work Transitions and Work Instability." *Journal of Gerontology* 52:S4–S12.

Myles, John. 1990. "States, Labor Markets and Life Cycles." Pp. 271–97 in *Beyond the Marketplace: Rethinking Economy and Society*, edited by R. Friedland and A. Robertson. New York: Aldine de Gruyter.

Nasaw, David. 1979. *Schooled to Order: A Social History of Schooling in the United States*. Oxford: Oxford University Press.

NCES. 2002. *Stats in Brief: The Persistence of Employees Who Pursue Postsecondary Study*. Washington, DC: National Center for Education Statistics.

Nelson, Daniel. 1975. *Managers and Workers*. Madison: University of Wisconsin Press.

O'Rand, Angela M., Jenifer Hamil-Luker and Cheryl Elman. 2009. "Childhood Adversity, Educational Trajectories and Self-Reported Health in Later Life among U.S. Women and Men at the Turn of the Century." *Zeitschrift für Erziehungswissenschaft* 12:357–84.

O'Rand, Angela M. and John C. Henretta. 1999. *Age and Inequality: Diverse Pathways through Later Life*. Boulder: Westview Press.

O'Rand, Angela M. and Vicky M. MacLean. 1986. "Labor Market, Pension Rule, Structure and Retirement Benefit Promise for Long-Term Employees." *Social Forces* 65:224–40.

O'Rand, Angela M. 1986. "The Hidden Payroll: Employee Benefits and the Structure of Workplace Inequality." *Sociological Forum* 1:657–83.

Pallas, Aaron M. 1993. "Schooling in the Course of Human Lives: the Social Context of Education and the Transition to Adulthood in Industrial Society." *Review of Educational Research* 63:409–47.

Petersen, Trond, Seymour Spilerman, and Svenn-Age Dahl. 1989. "The Structure of Employment Terminations among Clerical Employees in a Large Bureaucracy." *Acta Sociologica* 32:319–38.

Piore, Michael J. and Charles F. Sabel. 1984. *The Second Industrial Divide: Possibilities for Prosperity*. New York: Basic Books.

Polivka, Anne E. 1996. "Contingent and Alternative Work Arrangements, Defined and Profile of Contingent Workers." *Monthly Labor Review* 119:3–21.

Prechel, Harland. 1994. "Economic Crisis and the Centralization of Control over the Managerial Process: Corporate Restructuring and Neo-Fordist Decision-Making." *American Sociological Review* 59:723–45.

Purcell, Patrick J. 2000. "Older workers: employment and retirement trends." *Monthly Labor Review* 123: October 19–30.

Quinn, Joseph F. and Michael Kozy. 1996. "The Role of Bridge Jobs in the Retirement Transition: Gender, Race and Ethnicity." *The Gerontologist* 36:363–72.

Rau, Barbara. 2003. "Flexible Work Arrangements." In *Work and Family Encyclopedia*, edited by E. Kossek and M. Pitt-Catsouphes. Chestnut Hill, MA: Sloan Work and Family Research Network. Retrieved November 2, 2009 (http://wfnetwork.bc.edu).

Rones, Philip L.1983. "The Labor Market Problems of Older Workers." *Monthly Labor Review* 106:3–12.

Rosenfeld, Rachel A. 1992. "Job Mobility and Career Processes." *Annual Review of Sociology* 18:39–61.

Ruhm, Christopher J. 1990. "Bridge Jobs and Partial Retirement." *Journal of Labor Economics* 8:482–501.

———. 1991. "Career Employment and Job Stopping." *Industrial Relations* 30:193–208.

Settersten, Richard A. 1998. "Time, Age and the Transition to Retirement: New Evidence on Life-Course Flexibility?" *International Journal of Aging and Human Development* 47:177–203.

Siegal, Michele, Charlotte Muller and Marjorie Honig. 2000. *The Incidence of Job Loss: The Shift from Younger to Older Workers, 1981–1996*. NBER Working Paper. #279095.

Slevin, Kathleen and C. Ray Wingrove. 1995. "Women in Retirement: A Review and Critique of Empirical Research since 1976." *Sociological Inquiry* 65:1–21.

Smith, Vicki. 1997. "New Forms of Work Organization." *Annual Review of Sociology* 23:315–39.

Snyder, T. D., S. A. Dillow, and C. Hoffman. 2009. *Digest of Education Statistics 2008*. National Center for Educational Statistics. U.S. Department of Education. Washington, DC.

Spenner, Kenneth I. 1990. "Skills." *Work and Occupations* 17:422–48.

———. 1995. "Technological Change, Skill Requirements and Education: The Case for Uncertainty." Pp. 81–138 in *The New Modern Times: Factors Reshaping the World of Work,* edited by D. Bills. Albany: State University of New York Press.

Sterns, Harvey and Boin Chang. 2009. "Workforce Issues and Retirement." Working Paper. The University of Akron.

Szinovacz, Maximiliane E. and Christine Washo. 1992. "Gender Differences in Exposure to Life Events and Adaptation to Retirement." *Journal of Gerontology* 47:S191–96.

Szinovacz, Maximiliane E. and Adam Davey. 2005. "Predictors of Perceptions of Involuntary Retirement." *The Gerontologist* 45:36–47.

Thurow, Lester C. 1975. *Generating Inequality: Mechanisms of Distribution in the U.S. Economy*. New York: Basic Books.

Tuma, Nancy B. 1976. "Rewards, Resources and the Rate of Mobility: A Nonstationary Multivariate Stochastic Model." *American Sociological Review* 41:338–60.

Uchitelle, Louis. 2006. *The Disposable American*: Layoffs and Their Consequences. New York: Vintage Books.

Vallas, Steven P. 1999. "Rethinking Post-Fordism: The Meaning of Workplace Flexibility." *Sociological Theory* 17:68–101.

Whitten, David O. 1983. *The Emergence of Giant Enterprise, 1860–1914*. Westport, Connecticut: Greenwood Press.

Wise, David. 1997. "Retirement against the Demographic Trend: More Older People Living Longer, Working Less, and Saving Less." *Demography* 34:83–95.

Zemsky, Robert. 1998. "Labor, Markets, and Educational Restructuring." *Annals of the American Academy of Political and Social Science* 559:77–90.

Chapter 17
The Changing Worlds of Family and Work

Madonna Harrington Meyer and Wendy M. Parker

In the United States, the worlds of family and work are often incongruent. Much has changed since the *Leave It to Beaver* era of the 1950s, when men worked primarily outside the home and women worked primarily inside the home. In the past several decades, aging and life course scholars have documented and analyzed unprecedented increases in single mothers raising children; widespread entry of women, particularly mothers of young children, into the labor force; and modest increases in men's contributions to the unpaid house work and child care at home (Glass 2000). In some respects, these trends have narrowed gender gaps, but in other respects these trends have collided, leaving many women, and a growing share of men, feeling mounting pressures at home and at the office. Moreover, the efforts to juggle work and family responsibilities with few supports from the welfare state or employers leaves many ill prepared for the financial challenges of their own old age. This chapter explores these changing trends and their impact on the ways in which families balance paid and unpaid work over the life course. We pay special attention to differences between women and men and to differences among women by race, class, and marital status. We focus on how the presence of certain US policies, and the absence of others, shapes both the ways that families balance work and family obligations and the implications of that balance on economic security in later life.

Theoretical Approaches

Aging and life course scholars tend to highlight the extent to which social structures and institutions constrain individual actions across the life course, generating inequality in old age. The *life course perspective*, which emphasizes the long-term implications of factors such as historical events, cohort differences, timing and sequencing of personal decisions, and structural opportunities, is widely deployed among aging and life course scholars (Elder 2006). An emphasis on life course ramifications highlights the ongoing impact of various choices, opportunities, policies, and programs at different stages of life (Settersten 2003; Moen and Spencer 2006). For example, when women take time out from work to raise children, the immediate consequence is loss of salary, but the long-term consequence is reduced access to public pensions, private pensions, and savings. People traverse different pathways throughout their lives. The *cumulative (dis)advantage perspective* points out that along those pathways, they may accumulate advantages and disadvantages. Multiple layers of accumulating advantage and disadvantage across the life course lead to inequality

M. Harrington Meyer (✉)
Center for Policy Research, Syracuse University, Syracuse, NY, USA
e-mail: mhm@maxwell.syr.edu

R.A. Settersten, Jr. and J.L. Angel (eds.), *Handbook of Sociology of Aging*, Handbooks of Sociology and Social Research, DOI 10.1007/978-1-4419-7374-0_17, © Springer Science+Business Media, LLC 2011

in old age and those trajectories are not randomly distributed. Persistent organizing features of social life in the U.S., notably gender, race, class, and marital status, generate significant income and health inequality across the life course (Acker 2006; Lorber 2005; Padavic and Reskin 2002; Estes 2001; Harrington Meyer and Pavalko 1996; Settersten 2003; Baca Zinn and Thornton Dill 2005).

Aging and life course scholars analyze how best to use social policies and programs to reduce inequality in old age. *Feminist scholars*, who are united in their attention to gender inequality in old age, are divided in their views of how best to minimize that inequality. *Liberal, or gender-equity, feminists* focus on the continued gender gap in wages and support equal opportunity and comparable worth in the labor market. This focus has helped to shrink the wage gap and launch more women into positions of authority in a wider array of fields (Acker 1989). Because their aim is for women to compete fairly with men on the current playing field, liberal feminists have typically not supported policies like paid maternity or family leaves, flexible scheduling, or universal child care (Jacobs and Gerson 1998). They argue that the development of benefits specifically for women or those that women are much more likely than men to take undermines efforts toward gender equity by further entrenching the gendered division of labor at home and by concentrating women in part time and lower paying traditionally female occupations (Mandel and Semyonov 2005, 2006; Morgan and Zippel 2003; Hewlett 1986; Padavic and Reskin 2002). By contrast, *socialist, or gender-accommodative feminists,* support welfare policies that help women and men balance work and family across the life course, arguing that policies such as paid parental leave, flexible scheduling, and universal child care are needed to offset the unequal amounts of unpaid work that most women, and some men, continue to perform (Hewlett 1986; Pascall 1986; Herd and Harrington Meyer 2002). Moreover, they suggest that these programs are needed for the growing legions of single mothers raising children on their own. Their aim is to rewrite the rules at work and at home to reward women and men for both the paid and unpaid work they are doing (O'Rand and Henretta 1999; Lorber 2005; Mandel and Semyonov 2006).

Evidence from Norway suggests that neither of these approaches may be sufficient to entice the majority of men to take a leave from work to stay home with their newly born children. In the 1990s, Norway introduced a gender neutral policy that offers cash-for-care, and working parents may choose which of them will stay home to care for the children (Brandth and Kvande 2009). This policy has not increased fathers' participation in child care. Norway also introduced a policy known as the fathers' quota policy, which provides up to 6 weeks of paid leave only to fathers (Brandth and Kvande 2009). The gendered quota policy has increased fathers' participation in child care and helped offset the otherwise highly gendered practice of child care. Only the use-it-or-lose-it nature of the paternity leave provided sufficient impetus to encourage a substantial portion of men to take time at home with their children.

The *political economic* perspective, which emphasizes the structural influences that shape peoples' lives, points out that both the old age polices we do have, and those we do not have, are often shaped more by the business interests of those who lobby effectively to shape the legislation rather than the day-to-day interests of older people and their families. Indeed, supporters of *market-friendly policies* suggest that the government should bow out of the retirement business altogether and allow consumer needs for income security, health care, and long-term care to be addressed via the market in ways that maximize business interests (Harrington Meyer and Herd 2007; Hacker 2002; Pauly et al. 1991;Yergin and Stanislaw 1998; Gilbert 2002). Supporters of *family-friendly policies* favor a full slate of policies that help families juggle work and family at all ages and enable people to enter old age with sufficient economic resources (Korpi and Palme 1998; Korpi 2000). They caution that the political and economic power of business interests are evident both in the unequal distribution of employer-based benefits and in the quick dismissal of efforts to develop policies for paid family leave, universal day care, or universal long-term care programs, with little regard for those in need of such assistance, who are disproportionately women (Katz Olson 2003).

Changing Social and Demographic Trends

In recent decades, aging and life course scholars have written extensively about sweeping changes in our patterns of marriage, parenting, paid work, and unpaid care work. In many instances, gender gaps have grown smaller, but economic insecurity has remained persistent for some, particularly single mothers and Blacks and Hispanics.

Changing Family Patterns

Families have changed remarkably during the past 30 years with implications at all stages of the life course (Goldstein and Kenney 2001). Most notably, the U.S. has experienced a retreat from marriage coupled with an increase in cohabitation and a dramatic increase in single parent families. These trends vary significantly by race and ethnicity. Between 1960 and 2008, the percent of women married dropped from 66 to 53%, the percent of women divorced rose from 3 to 11% (U.S. Census Bureau 2008a, b, c). These trends are more pronounced among Blacks. The percent of married White women dropped from 67 in 1960 to 55 in 2008, while the percent of married Black women dropped from 60 to 34% (U.S. Census Bureau 2008a, b, c). The retreat from marriage is linked to an increase in single parenting. As Fig. 17.1 shows, between 1960 and 2008, among families with children under age 18, the percent of married couple families dropped from 91 to70%, while the percent of single mother families rose from 8 to 23%.

As Fig. 17.2 shows, in 2005, fully 39% of all U.S. births were to unmarried women. The rate varied dramatically by race. Asians had the lowest rate of births to single mothers, 17%, compared to Whites with 28%, Latinos with 51%, and Blacks with 72% of births to single mothers (Hamilton et al. 2009; U.S. Census Bureau 2008a, b, c).

The rise in single parenting is particularly problematic for women, with implications across the life course. In 2005, 80% of single parents were women (U.S. Census Bureau 2008a, b, c). Single parenting is linked to poverty, especially for women, because it often impedes advanced education or working the number of hours needed for many career advancements. Moreover, many single mothers raise children on a single income when noncustodial fathers provide little or no child support

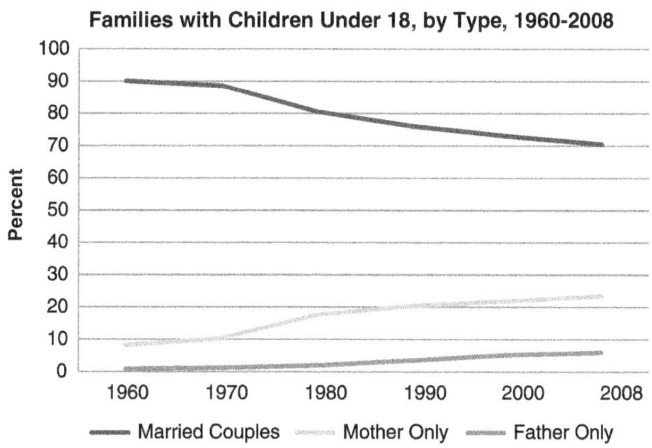

Fig. 17.1 Families with children under 18, by Type, 1960–2008. U.S. Census Bureau 2008. Table FM-1. Families by Presence of Own Children Under 18: 1950 to Present

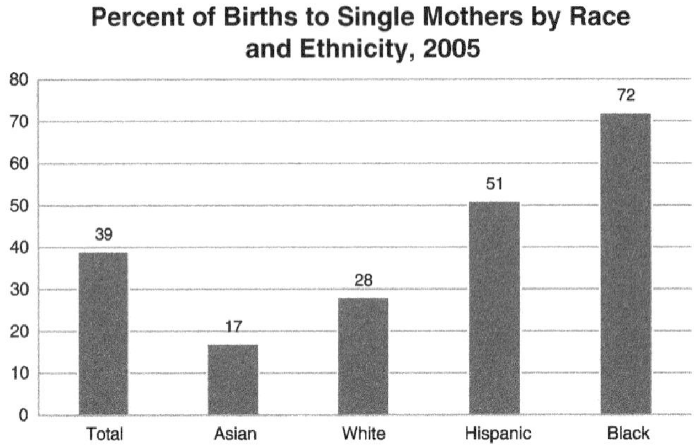

Fig. 17.2 Percent of birth to single mothers by race, 2005. Source: U.S. Census Bureau 2008

(Sorensen and Zibman 2000; Carlson 2006). About one-quarter of single mothers are not receiving child support payments to which they are legally entitled, and another one-third are not receiving the full award (Sorensen and Zibman 2000; Sorensen and Hill 2004). Among families with a child under 18, 7% of all married couples, compared to 36% of female-headed households, are poor (U.S. Census Bureau 2006). Single parenting is especially difficult for Black and Hispanic mothers; 45% of Hispanic and 42% of Black single mothers live in poverty (U.S. Census Bureau 2006). Many of these single mothers continue to be poor throughout the life course and as they reach old age.

At all ages, women are more likely to be single and the gender gap grows with age, often with adverse effects on economic security. Among those aged 65 and older, 72% of the men and only 42% of the women are married (U.S. Census Bureau 2008a, b, c). Among those 85 and older, 54% of the men and only 14% of the women are married. There is a strong link between marriage and poverty in old age. Older people who live alone do not enjoy the economies of scale afforded to those who live together, such as sharing the rent, heating, and grocery bills. Figure 17.3 shows that among those aged 65 and older, married couples have poverty rates of 5% while 40% of single Black and Hispanic women are poor (He et al. 2005).

Several aspects of the retreat from marriage have led to more complex family trees. Divorce rates rose throughout the 1960s and 1970s and then leveled off at a fairly high rate. One half of all those aged 65 and older have been divorced, and many of them, particularly men, have remarried (Bengtson et al. 1990; Cherlin 1996). The impact of divorce and remarriage on various generations is somewhat mixed. While some blended families fare quite well, some research shows that blended families report lower well-being and weaker family ties than nonblended families (Goldscheider and Goldscheider 1993). Following divorce, children are more likely to move between parents, have no significant father role model, or live in blended families where nonfamily or extended family play a sizeable role in their upbringing (McLanahan and Sandefur 1994; Rainwater and Smeeding 2004). Following a divorce or remarriage, some grandparents find their roles severed, while others play a central role in helping grandchildren adjust (Cooney and Smith 1996). During particularly difficult times, including parental drug abuse or incarceration, some grandparents become legal guardians and raise their grandchildren. The percent of children being raised by grandparents rose from 3% in 1970 to 6% in 2004 (He et al. 2005). Blacks are more likely than other groups to be custodial grandparents – they make up 12% of the elderly but account for 29% of custodial grandparents (He et al. 2005). The job can be rewarding, leading to very close relationships and a better quality of life for the grandchildren. But it can be daunting as well, particularly if the children have special needs or are struggling in school. Many grandparents report that while they are glad they are

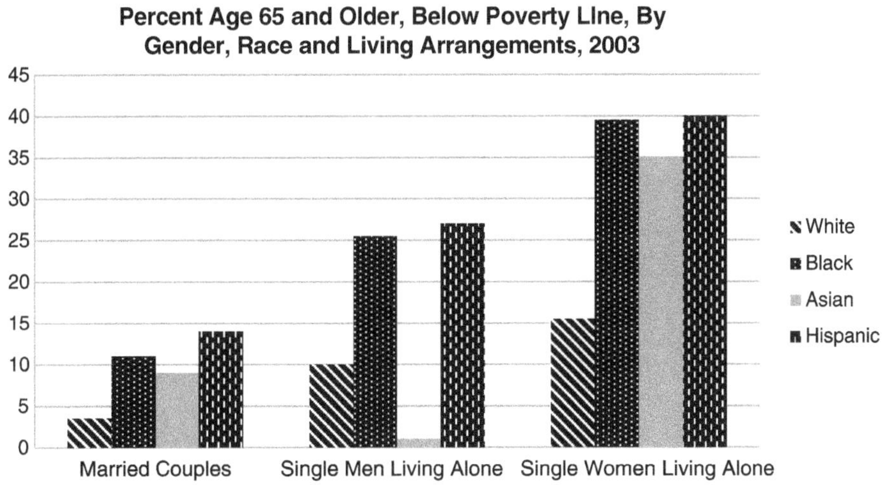

Fig. 17.3 Percent of age 65 and older, below poverty line, by gender, race, and living arrangements, 2003. Source: U.S. Census Bureau 2005. Civilian non-institutionalized persons

contributing, they experienced declining physical or emotional well-being, and rising economic concerns, after taking on the responsibility (Burton and deVries 1993).

Changing Work Patterns

Over the past several decades the gender gap in employment and in wages has narrowed. Nonetheless women continue to work fewer hours, in part because of their disproportionate responsibility for care work, and for lower pay, in part because they continue to be concentrated in traditionally female and lower paying jobs (Glass 2000). Despite a sizable increase in women's labor force participation, women remain more likely than men to work part time, less likely to work year-round, and more likely to have entire years out of the labor force (Rose and Hartmann 2004). Nearly 60% of all adult women work in the U.S., but only 75% of them work full time and the remainder work part time (U.S. Department of Labor 2008). By comparison, 90% of employed men work full time. While men's labor force participation has dropped from 76% in 1970 to 69% in 2004, women's overall participation has risen from 40 to 56%. Figure 17.4 tracks the changes in labor force participation over the decades for women with children. Women with children ages 6–17 are the most likely to work, with the percent employed rising from 55% in 1975 to 78% in 2004. Note that the percent of women with children under age 3 working rose from 34 to 57% between 1975 and 2004.

As family and work patterns changed, there has been a decline in the traditional breadwinner family. Historically, marriage was an event that led many women to leave the labor force. Over time, marriage lost that impact, but childbirth prompted many women to exit. Today, neither transition has much impact on employment. Among married couples, the percent with only the husband working has dropped from 29% in 1986 to 22% in 2008 (U.S. Census Bureau 2008a, b, c). Among married couples with children under age 18, 60% of the wives were working in 2008, and among married couples with children under age 6, 62% of the wives were working in 2008. With the exception of

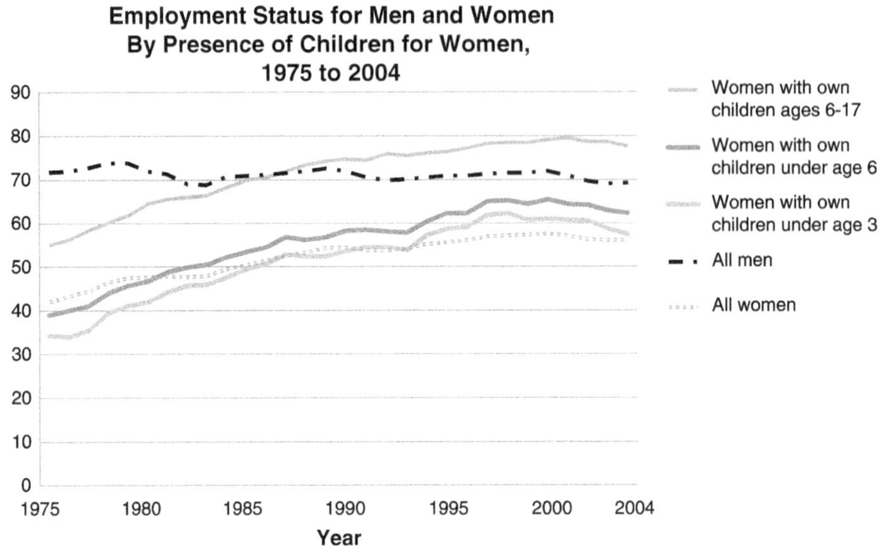

Fig. 17.4 Employment status for men and women by presence of children for women, 1975–2004. Source: U.S. Bureau of Labor Statistics. (2005). "Women in the Labor Force: A Databook. Report 985." Table 7. Employment status of women by presence and age of youngest child, 1975–2004."

widowed women, married and unmarried women are now equally likely to be employed, though those with younger children work fewer hours (U.S. Bureau of Labor Statistics 2006a, b).

Historic race differences in work patterns have also narrowed. In the early 1900s, Black women were nearly three times more likely than White women to be employed; currently 62% of Black and 59% of White women work (Goldin 1992; U.S. Department of Labor 2008; U.S. Bureau of Labor Statistics 2005a, b). Black women remain, however, significantly more likely than Whites to work full time, to be unemployed, and to receive smaller wages. Race differences in wages are pronounced. In 2004, median weekly earnings for Black women were 14% less, and for Latino women were 28% less, than for White women (U.S. Bureau of Labor Statistics 2005a, b).

The gender gap in wages among full time year round workers has shrunk from 62% in 1979 to nearly 80% in 2008, marking important gains in some women's economic security (U.S. Department of Labor 2008; Institute for Women's Policy Research 2006). But the cumulative wage gap may be more telling of the life course consequences of juggling work and family responsibilities for women. Hartmann et al. (2006) traced the cumulative wage gap, which captures both fewer hours of employment and lower wages. They analyzed PSID data from 1983 to 1998 and found that for all women of ages 26–59 with at least 1 year of earnings, women earn just 38% of what men earn. They then traced a cohort of college educated women and men who were of ages 25–29 in 1984. By 2004, the women were $440,000 behind the men in cumulative wages. The gender gap in wages increased over these 20 years. When they were in their 20s, women's average annual full-time year-round earnings were about 75% of men's; by their late 40s, 62%. These data, which exclude women who are not working full time, year-round, suggest that as they try to juggle work and family responsibilities, women fall further behind men. Lower wages during the younger years translate into smaller savings and investments, smaller private pensions, and smaller public pensions in old age.

The gender gap in wages continues into old age. Relatively few older people are employed; among those aged 65 and older in 2007, just 20% of men and 12% of women were employed (U.S. Department of Labor 2008). Because most have retired, earnings account for only 22% of all old age income. Employment is highest among those who are younger. Among people aged 65–69, 33% of men and 23% of women are employed (He et al. 2005). Though the gap has narrowed in recent decades, older women are still only 57% as likely as older men to be employed. When employed, they work fewer hours and accrue smaller earnings (He et al. 2005; Wu 2005; SSA 2009). Generally, single older people plan their retirement based on individual income and health concerns, but married older couples often try to plan to retire jointly, rather than sequentially, if they can afford to do so (Pienta and Hayward 2002; Elder and Pavalko 1993; Henretta et al. 1993). Because women tend to marry men 2–3 years older, they tend to retire at younger ages so that they can spend their retirement years with their husbands and this tendency toward earlier retirement undermines their financial security when they are older. When the economy and the stock markets are in a downturn, however, many older persons may strive to continue working. They may find it hard to obtain or retain employment, which is particularly troubling for those who are aware that their nest egg will not be big enough (Gendell 2008).

These changing demographic trends increase the need for high-quality and readily available child care, parental leave, flexible work hours, and early childhood education and other programs that will help families juggle work and care work responsibilities. Thus far, however, many of the responses to these growing demands, both through employment-based and welfare state programs, have been aimed at either the very poor or the upper middle class leaving the majority in the middle with little support. For example, because the US Family and Medical Leave Act (FMLA) offers only unpaid leave, few in the middle and lower income brackets are able to take advantage of the benefit (Rudd 2004). The overall impact of the rise in single mothers, and in middle-aged and older single women, is increased financial insecurity across the life course for many women, particularly those who are Black and Hispanic.

Changing Care Work Patterns

No matter how demanding the work schedule, every family has to find ways to take care of the day-to-day unpaid care work of dishes and laundry, raising children, and caring for disabled or frail older adults. At every point in the life course, women are more likely than men to provide that care for family members. Indeed, though the gender gap in care work has shrunk, women still perform the majority of unpaid work within families for children, people with disabilities, and the elderly. And the work itself remains devalued (Hochschild 1989; Hooyman and Gonyea 1995; Cancian and Oliker 2000). The gendered division of labor among younger families is not as pronounced as it was in the 1960s and 1970s, men are spending significantly more time with children than their fathers did, but women continue to spend about twice as much time as men caring for children and performing household chores like cooking and cleaning (Krants-Kent 2009; Bianchi et al. 2006). Moreover, the nature of parenting has changed considerably, especially for middle and upper class families. In contrast to her mother 30 years earlier, a good mother now is expected to provide a very intensive type of mothering that involves hands on help with activities such as homework, sports, music, languages, and summer camps (Stone 2007). Childcare and homemaking responsibilities lead many women to reduce or eliminate employment and education, with lifelong implications for financial security. Even though there has been some closing of the care work gap, Stone (2007) shows just how firmly held traditional expectations about women's responsibilities for raising children and running the household continue to be for many. She describes how women with very high status and high paying jobs tend to cut back or eliminate paid work following the birth of children in part because their husbands increase paid work and stop performing most house and child care work altogether. Single mothers face a different set of obstacles: there is no division of labor for the growing numbers of women who are heading households on their own.

The gender gap in providing care to frail older relatives has also eased, but women continue to perform about twice as much care work and face greater consequences for their own well-being (Chesley and Moen 2006). Overall, families are providing close to 80% of all long-term care and many of caregivers are on call 7 days a week for at least 3 hours a day (Brody 2004; National Alliance for Caregiving and AARP 2004; Navaie-Waliser et al. 2002). Care work may include cooking, cleaning, helping with finances, running errands, providing medical care, and assisting with eating, bathing, dressing, and mobility. Women are more likely to provide care both because of long held perceptions that they are natural caregivers, and because they tend to have more tenuous and lower paying relationships to the labor force (Stoller 1994; Stone et al. 1987; Moen et al. 1994). Nearly 70% of spousal caregivers are wives and 60–80% of children who care for their older parents are daughters. Compared to sons, daughters do an additional 10–18 hours a month of care work (National Alliance for Caregiving and AARP 2004; Navaie-Waliser et al. 2002). Compared to White women, Black and Hispanic women perform carework more often, for longer periods, and at more medically intense levels, to relatives that are in worse health. Given that they have lower incomes and worse health, these additional duties add up to an even greater burden (Wolf et al. 1997).

The nature and scope of carework for the frail elderly changed dramatically in 1984 when Medicare and later Medicaid altered the reimbursement structures in ways that caused hospitals to shorten stays and export unprofitable care out of the hospital (Estes 2001). Hospitals discharged patients quicker and sicker and with little training, families, mainly wives and daughters, were expected to take on highly technical work including chemotherapy, apnea monitoring, phototherapy, oxygen tents, tubal feedings, dressing changes, and more. Estes (1989) calculated that in the first 5 years of the prospective payment system, more than 21 million days of care work had been transferred from hospitals to families. In recent decades, Medicare and Medicaid coverage of community based long-term care has shrunk considerably. Since 1997, the percent of the Medicare budget devoted to home care has dropped by 50% and the number of recipients and the number of home visits have dropped significantly (Harrington Meyer and Herd 2007). The net effect is that more, and more intensive, care work is pushed on to families. Stone (2000) estimated that there are 40 unpaid informal care workers for every paid formal care worker. The total value of annual care work is now estimated to be between $50 and $103 billion (Holtz-Eakin 2005; National Alliance for Caregiving and AARP 2004; Stone 2000).

While carework for frail older relatives can be rewarding, it can also lead to caregiver burden and stress (Pearlin et al. 1995). Caregivers report adverse effects on physical and mental health, marriages and other relationships, and employment and earnings (Cantor 1989; Brody 1990, 2004; Hooyman and Gonyea 1995; Stone 2000). Caregivers, particularly those who report doing more than their share, have higher rates of physical and psychological distress, anxiety, depression, loneliness, family tension, sleeplessness, exhaustion, inadequate exercise, increases in chronic conditions, and drug misuse (Bird 1999; National Alliance for Caregiving and AARP 2004; Pavalko and Woodbury 2000). Despite poorer health, caregivers are twice as likely to lack the time, energy, or resources to obtain care or fill a prescription for themselves (Commonwealth Fund 2005). Prolonged carework may lead to a weakened immune system, increased psychological distress, and accelerated aging at the cellular level (Pavalko and Woodbury 2000; Epel et al. 2004). Chesley and Moen (2006) found that carework is associated with diminished mental health for dual earner women but improved mental health for dual earner men. That gender difference appears to be due, at least in part, to gender differences in the number of hours, and years, spent providing care. The impact of carework on women's work and wages is mixed. Some women add unpaid carework to their paid responsibilities without taking any cut in pay, while others tend to reduce or eliminate paid work (Pienta et al. 1994; Scharlach 1994). Pavalko and Henderson (2006), found that women whose jobs have less flexible hours, unpaid family leave, and paid sick leave are more likely to reduce or end employment, which reduces not only current income but old age income from public and private pensions and savings. Employers can do a lot to help employees balance work and family commitments, but for the most part such benefits are available only to higher paid and better educated workers.

Aging and the life course scholars note that to the extent that it impedes paid work, unpaid care work across the life course contributes to many women's, and some men's, economic hardship in old age. Moreover, given that it remains devalued, and that single motherhood continues to be on the rise, carework is likely to continue to be disproportionately the responsibility of women.

Changing Policies at Work and Home

Aging and the life course scholars have explored various policy proposals that might help younger families by increasing their ability to balance work and family with minimal economic penalties and decreasing the likelihood that they will arrive at old age with meager economic resources. These policies, including paid leave, flexible scheduling, and universal day care, are particularly important for women because even when they are not actually providing the care, they are often responsible for arranging that care. And their importance for single mothers can hardly be overstated because the range of affordable alternatives available to them is slim indeed.

Paid Leave

Under the provisions of the FMLA male and female employees in the US are eligible to take up to 12 weeks of unpaid leave each year for either the birth or care of their newborn or adopted child – or to care for immediate family with serious health problems – if they work in a company with more than 50 employees, have been employed for at least 12 months with that employer, and average at least 24 hours of work per week (U.S. Department of Labor 2000; Dorman 2001). Though the policy was developed to help families balance the needs of work and family and to encourage economic stability and security for families, most studies have shown relatively little impact for women or men (Waldfogel 2001). The act covers just about 60% of private sector workers, and just fewer than 50% are both covered and eligible. Among those who were eligible but did not take the leave, most reported that they could not afford time off without pay (Baum 2006; Armenia and Gerstel 2006; U.S. Department of Labor 2000). Though FMLA policies in the US are gender neutral in design, women are significantly more likely than men to take the unpaid leave (Rudd 2004). Additionally, those with lower incomes and who are single parents, who are least likely to be able to afford time off without pay, are least likely to take leave (Rudd 2004; Levine Sheriff 2007; Han et al. 2009).

Critics suggest that the unpaid family leave does not provide meaningful choices for single mothers and those at the lower end of the socioeconomic spectrum (Baum 2006; Harrington Meyer and Herd 2007). Several states now offer 6 weeks of partial wage replacement through their disability insurance programs and studies show the take-up rates are relatively high (Levine Sheriff 2007). California offers 55% wage replacement, up to a maximum of $728 per week, for up to 6 weeks. The benefit is paid for by a 0.08% tax on the first $70,000 in earnings, with a maximum cost of $55 per worker per year (Appelbaum and Milkman 2004).

In the absence of federal or state mandates, many workers rely on employment-based leave policies. But workers with higher levels of education and salaries are more likely than those of lower socioeconomic status to be offered such benefits. Reliance on employment-based paid leaves creates greater inequality among workers because those with lower incomes are both less likely to be offered paid leave through their jobs, and less likely to be able to take unpaid leave through the FMLA. Proposals to expand unemployment insurance to cover paid parental leave estimate that costs would range from $11 to $28 per worker per year (Institute for Women's Policy Research 2000).

Most European countries have long offered paid time off for the birth or adoption of a child. Most EU countries offer 3–12 months at full or partial pay. Norway offers 42 weeks at full pay or 52 weeks at 80% pay (Gornick and Meyers 2003). In addition to more generous paid leaves,

European nations also establish a more regulated and structured work schedule that allows more time for family (Bosch et al. 1994). Europeans average more than a month of paid vacation time per year and enjoy fairly broad coverage of sick days. Lower income and less educated workers in the US often have just 2 weeks of paid vacation and no coverage whatsoever of sick days. For example, Bond and Galinsky (2006) found that 79% of middle and higher waged women had some paid sick time, compared to just 39% of lower earners.

Many aging and life course scholars have concluded that workplaces in the US are often incompatible with family life and called for paid time off that would permit women and men to juggle, rather than chose between, work and family. Neither US employers nor welfare state policies have made much progress in providing meaningful family leave supports for those who are unable to take time off without pay (Chesley and Moen 2006). In the years ahead, aging and life course scholars will likely focus more sharply on gender, race, class, and marital status differences in taking paid and unpaid leaves in both the short run and in the long run, as well as models for providing more supports for parents with the greatest needs for those supports.

Flexible Scheduling

During the past few decades, some employers have introduced family-friendly forms of flexible scheduling into the workplace to help families accommodate work and family lives. As methods of communication have multiplied, including e-mail, texting, and cell phone calls, so have methods of conducting business. Flexible scheduling includes compressed work weeks such as working four 10 hours days or working from home on some days (Weber and Curlew 2009). Employees often have to pursue such flexibility through their own initiative, and some find that it hampers career advancement, but those with children or frail older relatives in need of care may find the flexibility worth the trouble (Stone 2007). Flexible schedules are not available to all; those in manufacturing and service sectors are often excluded. Employees who are Black or Hispanic, have less education, or earn lower wages, are less likely to be in positions that are offered or allowed to arrange flexible scheduling. Indeed, middle and higher earners are significantly more likely than lower earners to be able to set their start and end times through traditional or daily flex time benefits (Weber and Curlew 2009; Bond and Galinsky 2006). But those who can get them appreciate the flexible schedules. The National Study of the Changing Workforce indicates that employees who have more control over their schedule or access to flexible schedules typically report greater involvement in their workplace, commitment to their jobs, and well-being (Galinsky et al. 2008).

Another flexible option that permits families to juggle work and family involves working part time or job sharing. But these options tend to be combined with lower pay and fewer fringe benefits. They also tend to be combined with less meaningful work, slower advancements, and less recognition (Stone 2007). Moreover, in the current throes of economic downturn, conversion to part time work is being used as an alternative to lay-offs, often with negative short and long-term economic consequences, which include less pay now and less Social Security and private pensions later (Hewlett 2009 interview; Weber and Curlew 2009; Goldin 2006).

Child Care

More women with young children are in the labor market than at any other time in history, hence there is an increasing demand for flexible, affordable, and high-quality child care (U.S. Census Bureau 2006; Presser 2003). In 2005, about one-fourth of children of employed mothers were in organized child care facilities, and one-fifth was cared for regularly by a grandparent (Federal Interagency Forum on Child and Family Statistics 2006). Child care is often difficult to find, expensive, and of variable quality. Those with irregular work patterns, requiring child care during late afternoons, evenings, or on a part time or rotational basis often find that there are few organized child care options for them (Presser 2003). Moreover, when children are sick, they are not permitted at day care and working parents have to make other arrangements or stay home from work. The costs for child care can add up, eroding family earnings. According to an Urban Institute report (2000), families with children under the age of 13 averaged nearly $300 a month for child care accounting for the second largest household expensive, after the house mortgage (Giannarelli and Barsimantov 2000). Concerns about the quality of child care are persistent. Studies show that care is variable and often only fair to poor in quality (Vandell and Wolfe 2000). Aging and life course scholars have shown that when child care is difficult to arrange, costly, and poor in quality, women are more likely than are men to reduce hours or leave the labor force altogether (Crittenden 2001; Presser 2003).

The state provides some supports for families. Middle-income families may be eligible for tax subsidies through the dependent care tax credit. Low-income families may be eligible for tax credits and subsidized child care, but finding openings in quality settings especially for young children may be problematic. Programs including Head Start and universal pre-K have garnered attention in recent years as a way to bridge families, and particularly moms, back into the labor force and to offer a better start for children from low-income households. Head Start has demonstrated success, but budgetary constraints that have reduced its reach and waiting lists are often long. Universal pre-K programs have also demonstrated cost effectiveness and academic benefits for children in some states, but there has been little movement toward a national pre-K program to date. Most of these programs are targeted to nearly school age children and families living below the poverty line, leaving a number of critical child development years and families behind (Zigler and Muenchow 1992).

Despite evidence that universal day care throughout Western Europe fosters early childhood development, and despite evidence that readily available good child care prompts mothers and fathers to remain employed, there has been little movement by the US toward universal or expanded subsidized child care (Crittenden 2001; Gornick et al. 1997). Over the decades, aging and life course scholars have concluded that providing accessible and affordable child care is one of the most important strategies for improving income security, particularly for women, both in the short run as they are raising their children and in the long run as they reach old age and rely on their Social Security, private pensions, and private savings.

Discussion

As aging and life course scholars continue to analyze changing trends, and somewhat unresponsive social policies, they will work to place gender, race, class, and marital status inequality at the center of those analyses. While some gender differences have narrowed, they have not disappeared and show no signs of doing so. Moreover, many differences due to other factors such as race and marital status are increasing. Our social policies have not kept pace with key demographic changes, particularly the increase in single mothers and in employment among mothers. Though policy solutions are discussed, few are acted upon, in part because there is debate about the best way to minimize inequality in old age and in part because business interests play a central role in the development of contemporary policies (Crittenden 2001; Moon with Herd 2002; Harrington Meyer and Herd 2007). Americans tend to avoid wholesale policy shifts in favor of more incremental changes. Expanded social policies for younger family, particularly paid family leave, increased flexibility options, and flexible, affordable, and high-quality day care would go a long way toward assuring that vulnerable people were less likely to arrive at old age with relatively few resources. But such proposals are rarely considered seriously and are often preemptively silenced by powerful business interests, those who favor the traditionally gendered division of labor, and those who prefer gender neutral welfare state policies. If house work and care work for the young and old alike were sufficiently revalued, encouraging men and women to do their share, it would help married and cohabitating women. But even then, many families, and nearly all single women, who continue to bear the economic burden of providing care work, would benefit from stronger social supports across the life course.

References

Acker, Joan. 2006. *Class Questions: Feminist Answers*. Lanham, MD: Rowman & Littlefield.

———. 1989. *Doing Comparable Worth: Gender, Class and Pay Equity*. Philadelphia: Temple University Press.

Appelbaum, Eileen and Ruth Milkman. 2004. *Paid Family Leave in California: New Research Findings*. University of California Institute for Labor and Employment. The State of California Labor, 2004 Paper 02 (November 1). Retrieved (http://repositories.cdlib.org/ile/scl2004/02/).

Armenia, Amy and Gerstel, Naomi. 2006. "Family Leaves, the FMLA and Gender Neutrality: The Intersection of Race and Gender." *Social Science Research*, 35:871–891.

Baca Zinn, Maxine and Bonnie Thornton Dill. 2005. "What Is Multiracial Feminism?" Pp. 202–207 in *Gender Inequality: Feminist Theories and Politics*, 3rd ed., edited by Judith Lorber. Los Angeles, CA: Roxbury.

Baum, Charles L. 2006. "The Effects of Government-Mandated Family Leave on Employer Family Leave Policies." *Contemporary Economic Policy* 24(3):432–445.

Bengtson, Vern, Carolyn Rosenthal, and Linda Burton. 1990. "Families and Aging: Diversity and Heterogeneity." Pp. 263–281 in *Handbook of Aging and the Social Sciences*, edited by R. Binstock and L. George. San Diego CA: Academic.

Bird, Chloe E. 1999. "Gender, Household Labor, and Psychological Distress: The Impact of the Amount and Division of Housework." *Journal of Health and Social Behavior* 40(1):32–45.

Bianchi, Suzanne M., John P. Robinson, and Melissa A. Milkie. 2006. *Changing Rhythms of American Family Life*. New York: Russell Sage Foundation.

Bond, James T. and Ellen Galinsky. 2006. "What Workplace Flexibility is Available to Entry-Level, Hourly Employees?" Families and Work Institute, Research Brief No. 3, New York.

Bosch, G., Dawkins, P., and Michon, F. 1994. "Times are Changing: Working Time in 14 Industrialised Countries." Geneva: International Institute for Labour Studies.

Brody, Elaine M. 1990. *Women in the Middle: Their Parent-Care Years*. New York: Springer.

———. 2004. *Women in the Middle: Their Parent Care Years*, 2nd ed. New York: Springer.

Brandth, Berit and Elin Kvande. 2009. "Gendered or Gender-Neutral Care Politics for Fathers?" *The Annals of the American Academy of Political and Social Science* 624(1):177–189.

Burton, Linda, Cynthia de Vries. 1993. "Challenges and Rewards: African American Grandparents as Surrogate Parents." Pp. 101–8 in *Families and Aging*, edited by L. Burton San Francisco: Baywood.

Cantor, Marjorie H. 1989. "Social Care: Family and Community Support Systems." *The Annals of the American Academy of Political and Social Science* 503(May):99–112.

Cancian, Francesca M. and Stacey J. Oliker. 2000. *Caring and Gender*. Thousand Oaks, CA: Pine Forge Press.

Carlson, Marcia J. 2006. "Family Structure, Father Involvement, and Adolescent Behavioral Outcomes." *Journal of Marriage and Family* 68(1):137–154.

Cherlin, Andrew. 1996. *Public and Private Families*. New York: McGraw Hill.

Chesley, Noelle and Phyllis Moen. 2006. "When Workers Care: Dual-Earner Couples' Caregiving Strategies, Benefit Use, and Psychological Well-Being." *American Behavioral Scientist* 49(9):1248–1269.

Commonwealth Fund. 2005. *The Long-Term Budget Outlook*. Washington: Congressional Budget Office.

Cooney, Teresa M. and Lori Ann Smith. 1996. Young Adults' Relations with Grandparents Following Recent Parental Divorce. *Journal of Gerontology: Social Sciences* 51B:S91–S95.

Crittenden, Ann. 2001. *The Price of Motherhood: Why the Most Important Job in the World Is Still the Least Valued*. New York: Metropolitan Books.

Dorman, Pat 2001. "Maternity and Family Leave Policies: A Comparative Analysis." *The Social Science Journal*, 38:189–201.

Elder, Glen. 2006. "Life Course." Pp. 109–131 in *The Blackwell Encyclopedia of Sociology*, edited by George Ritzer. Massachusetts: Blackwell.

Elder, Glen and Eliza Pavalko. 1993. "Work Careers in Men's Later Years: Transition, Trajectories, and Historical Change." *Journal of Gerontology* 48:80–191

Epel, Elissa S., Elizabeth H. Blackburn, Jue Lin, Firdaus S. Dhabhar, Nancy E. Adler, Jason D. Morrow, and Richard M. Cawthon. 2004. "Accelerated Telomere Shortening in Response to Life Stress." *Proceedings of the National Academy of Sciences of the United States of America* 101(49):17312–17315.

Estes, Carroll. 1989. "Aging, Health and Social Policy: Crisis and Crossroads." *Journal of Aging and Social Policy* 1:17–32.

Estes, Carroll and Associates. 2001. *Social Policy and Aging: A Critical Perspective*. Thousand Oaks, CA: Sage.

Federal Interagency Forum on Child and Family Statistics 2006. *America's Children: Key National Indicators of Well-Being, 2006*. Washington, DC: U.S. Government Printing Office. (http://childstats.gov/americaschildren/tables.asp)

Galinsky, Ellen, James T. Bond, and Kelly Sakai. 2008. *2008 National Study of Employers*. New York: Families and Work Institute.

Gendell, Murray. 2008. "Older Workers Increasing Their Labor Force Participation and Hours of Work." *Monthly Labor Review* 131(1): 41–54.

Gilbert, Neil. 2002. *Transformation of the Welfare State: The Silent Surrender of Public Responsibility*. New York: Oxford University Press.

Giannarelli, Linda and James Barsirmantov. 2000. *Child Care Expenses of America's Families*. Washington, DC: Urban Institute. (http://www.urban.org/UploadedPDF/310028_occa40.pdf)

Glass, Jennifer 2000. "Envisioning the Integration of Family and Work: Toward a Kinder, Gentler Workplace" *Contemporary Sociology* 29(1):129–143

Goldin, Claudia D. 1992. *Understanding the Gender Gap: An Economic History of American Women*. New York: Oxford University Press.

———. 2006. "The Rising (and Then Declining) Significance of Gender." Pp. 67–101 in *The Declining Significance of Gender?*, edited by F. D. Blau, M. C. Brinton, and D. B. Grusky. New York: Russell Sage Foundation.

Goldscheider, Frances Kobrin and Calvin Goldscheider. 1993. *Leaving Home Before Marriage*. Madison, WI: University of Wisconsin Press.

Goldstein Joshua R. and Catherine Kenney. 2001. "Marriage Delayed or Marriage Forgone? New Cohort Forecasts of First Marriage for U.S. Women." *American Sociological Review* 66(4):509–519.

Gornick, Janet C. and Marcia Meyers. 2003. *Families That Work: Policies for Reconciling Parenthood and Employment*. New York: Russell Sage Foundation.

Gornick, Janet C., Marcia Meyers, and Katherine E. Ross. 1997. "Supporting the Employment of Mothers: Policy Variation across Fourteen Welfare States." *Journal of European Social Policy* 7(1):45–70.

Hacker, Jacob S. 2002. *The Divided Welfare State: The Battle over Public and Private Social Benefits in the United States*. New York: Cambridge University Press.

Hamilton, Brady E. Joyce A. Martin, and Stephanie J. Ventura. 2009. *Births: Preliminary data for 2007*. National vital statistics reports, Web release; vol 57 no 12. Hyattsville, MD: National Center for Health Statistics. (http://www.cdc.gov/nchs/data/nvsr/nvsr57/nvsr57_12.pdf)

Han, Wen-Jui, Christopher Ruhm, and Jane Waldfogel. 2009. *Journal of Policy Analysis and Management* 28(1):29–54.

Harrington Meyer, Madonna and Pamela Herd. 2007. *Market Friendly or Family Friendly? The State and Gender Inequality in Old Age*. NY: Russell Sage.

Harrington Meyer, Madonna and Eliza Pavalko. 1996. "Family, Work, and Access to Health Insurance among Mature Women." *Journal of Health and Social Behavior*, 37(4):311–325.

Hartmann, Heidi, Stephen J. Rose, and Vicky Lovell. 2006. "How Much Progress in Closing the Long-Term Earnings Gap?" In *The Declining Significance of Gender?*, edited by F. D. Blau, M. C. Brinton, and D. B. Grusky. New York: Russell Sage Foundation.

He, Wan, Manisha Sangupta, Victoria A. Velkoff, and Kimberly A. Debaros. 2005. *65+ in the United States*. Current Population Reports, Special Studies Series P23, no. 209. Washington, D.C.: U.S. Census Bureau.

Henretta, John, Angela O'Rand and Christopher Chan. 1993. "Joint Role Investments and Synchronization of Retirement: A Sequential Approach to Couples' Retirement Timing." *Social Forces* 71:981–1000.

Herd, Pamela and Madonna Harrington Meyer. 2002. "Carework: Invisible Civic Engagement." *Gender and Society* 16(5):665–688.

Hewlett, Sylvia Ann. 1986. *A Lesser Life: The Myth of Women's Liberation in America*. New York: W. Morrow.

———. 2009. "Working Moms Concerned About Looming Budget Cuts." Interviewed by Michel Martin on *Tell Me More* on National Public Radio, March 25, 2009.

Hochschild, Arlie Russell (with Anne Machung). 1989. *The Second Shift: Working Parents and the Revolution at Home*. New York: Viking Penguin.

Holtz-Eakin, Douglas. 2005. "The Cost and Financing of Long-Term Care Services." Statement of Douglas Holtz-Eakin, Director, before the Subcommittee on Health Committee on Ways and Means U.S. House of Representatives, April 27. Washington: Congressional Budget Office.

Hooyman, Nancy R. and Judith Gonyea. 1995. *Feminist Perspectives on Family Care: Policies for Gender Justice.* Thousand Oaks, CA: Sage.

Institute for Women's Policy Research. 2000. "Paid Family and Medical Leave: Essential Support for Working Women and Men." IWPR Publication #A124. IWPR. (http://www.iwpr.org/pdf/famlve2.pdf)

Institute for Women's Policy Research. 2006. "Memo to John Roberts: The Gender Wage Gap is Real." IWPR #C362. (http://www.iwpr.org/pdf/C362.pdf)

Jacobs, Jerry A. and Kathleen Gerson. 1998. "Integrating Family and Work in the 21st Century." In *Gender Inequality: Feminist Theories and Politics*, edited by J. Lorber. Los Angeles, CA: Roxbury, 35–39.

Katz Olson, Laura. 2003. *The Not-So-Golden Years: Caregiving, the Frail Elderly, and the Long-Term Care Establishment.* Lanham, MD: Rowman and Littlefield.

Krants-Kent, Rachel. 2009. "Measuring Time Spent in Unpaid Household Work: Results from the American Time Use Survey." *Monthly Labor Review* July:46–59.

Korpi, Walter. 2000. "Faces of Inequality: Gender, Class and Patterns of Inequalities in Different Types of Welfare States." Luxembourg Income Study Working Paper No. 224. Syracuse, NY: Syracuse University.

Korpi, Walter and Joakim Palme. 1998. "The Paradox of Redistribution and Strategies of Equality: Welfare State Institutions, Inequality, and Poverty in the Western Countries." *American Sociological Review* 63:661–687.

Levine Sheriff, Rebecca. 2007. *Balancing Work and Family*. California Senate Office of Research Publication 1396-S. Sacramento, CA.

Lorber, Judith. 2005. *Gender Inequality: Feminist Theories and Politics.* 3rd edition. Los Angeles, CA: Roxbury.

Mandel, Hadas and Moshe Semyonov. 2006. "A Welfare State Paradox: State Interventions and Women's Employment Opportunities in 22 Countries." *American Journal of Sociology* 111(6):1910–1949.

Mandel, Hadas and Moshe Semyonov. 2005. "Family Policies, Wage Structures, and Gender Gaps: Sources of Earnings Inequality in 20 Countries." *American Sociological Review* 70(6):949–967.

McLanahan, Sara and Gary Sandefur. 1994. *Growing Up with a Single Parent: What Hurts, What Helps*. Cambridge: Harvard University Press.

Moen, Phyllis, Julie Robison, and Vivian Fields. 1994. "Women's Work and Caregiving Roles: A Life Course Approach." *Journals of Gerontology* 49(4):S176–S186.

Moen, Phyllis and Donna Spencer. 2006. "Converging Divergences in Age, Gender, Health, and Well-Being: Strategic Selection in the Third Age." In *Handbook of Aging and the Social Sciences*. Oxford: Elsevier Science.

Moon, Marilyn with Pamela Herd. 2002. *A Place at the Table: Women's Needs and Medicare Reform*. New York: The Century Foundation.

Morgan, Kimberly J. and Kathrin Zippel. 2003. "Paid to Care: The Origins and Effects of Care Leave Policies in Western Europe." *Social Politics* 10(1):49–85.

National Alliance for Caregiving and AARP. 2004. *Caregiving in the U.S.* Washington DC: National Alliance for Caregiving and AARP.

Navaie-Waliser, Maryam, Aubrey Spriggs, and Penny H. Feldman. 2002. "Informal Caregiving – Differential Experiences by Gender." *Medical Care* 40(12):1249–1259.

O'Rand, Angela M. and John C. Henretta. 1999. *Age and Inequality: Diverse Pathways through Later Life.* Boulder, CO: Westview Press.

Padavic, Irene and Barbara F. Reskin. 2002. *Women and Men at Work.* 2nd ed. Newbury Park: Pine Forge Press.

Pauly, Mark, Patricia Danzon, Paul Feldstein, and John Hoff. 1991. "A Plan for 'Responsible National Health Insurance'." *Health Affairs* 10(1):5–25.

Pascall, Gillian. 1986. *Social Policy: A Feminist Analysis*. London: Tavistock.

Pavalko, Eliza and Kathryn Henderson. 2006. "Combining Care Work and Paid Work: Do Workplace Policies Make a Difference?" *Research on Aging* 28(3):359–374.

Pavalko, Eliza K. and Shari Woodbury. 2000. "Social Roles as Process: Caregiving Careers and Women's Health." *Journal of Health and Social Behavior* 41(1): 91–105.

Pearlin, Leonard I., Aneshensel, Carol A., Mullan, Joseph T., and Whitlatch, Carol J. 1995. "Social support and the stress process: The case of caregiving." In *Handbook of Aging and the Social Sciences*, 4th ed., edited by R. H. Binstock and L. K. George. Orlando, FL: Academic.

Pienta, Amy, Jeffrey Burr, and Jan Mutchler 1994. "Women's Labor Force Participation in Later Life: The Effects of Early Work and Family Experiences." *Journal of Gerontology* 49:S231–S239.

Pienta, Amy and Mark Hayward. 2002. "Who Expects to Continue Working after Age 62? The Retirement Plans of Couples." *Journal of Gerontology* 57B(4):ES199–208.

Presser, Harriett B. 2003. *Working in a 24/7 Economy: Challenges for American Families*. New York: Russell Sage Foundation.

Rainwater, Lee and Timothy M. Smeeding. 2004. "Lone Parent Poverty, Inequality, and the Welfare State." Pp. 96–115 in *The Future of the* Family, edited by D. P. Moynihan, T. M. Smeeding, and L. Rainwater. New York, NY: Russell Sage Foundation.

Rose, Stephen and Heidi Hartmann. 2004. *Still A Man's Labor Market: The Long Term Earning's Gap*. Washington DC: The Institute for Women's Policy Research.

Rudd, Elizabeth. 2004. "Family Leave: A Policy Concept Made in America." In *Work-Family Encyclopedia*, edited by M. Pitt-Catsouphes and E. Kossek. (http://www.bc.edu/bc_org/avp/wfnetwork/rft/wfpedia/index.html).

Scharlach, Andrew E. 1994. "Caregiving and Employment: Competing or Complementary Roles?" *The Gerontologist* 34(3):378–385.

Settersten, Richard A., Jr. 2003. "Propositions and Controversies in Life Course Scholarship." Pp. 15–48 in *Invitation to the Life Course: Toward New Understandings of Later Life*, edited by R. A. Settersten. Amityville, NY: Baywood.

Social Security Administration (SSA). 2009. "Annual Statistical Supplement 2008." *Social Security Bulletin*. Washington: Department of Health and Human Services.

Sorensen, Elaine and Ariel Hill. 2004. "Single Mothers and Their Child-Support Receipt: How Well Is Child-Support Enforcement Doing?" *Journal of Human Resources* 39(1):135–154.

Sorensen, Elaine and Chava Zibman. 2000. "Child Support Offers Some Protection against Poverty." New Federalism: National Survey of America's Families. B-10. Washington, DC: Urban Institute. March 15.

Stoller, Eleanor Palo. 1994. "Why Women Care: Gender and the Organization of Lay Care." Pp. 187–93 in *Worlds of Difference: Inequality in the Aging Experience*, edited by E. P. Stoller and R. C. Gibson. Thousand Oaks, CA: Pine Forge Press.

Stone, Pamela. 2007. *Opting Out: Why Women Really Quit Careers and Head Home.* Berkeley, CA: University of California Press.

Stone, Robyn I. 2000. *Long-Term Care for the Elderly with Disabilities: Current Policy, Emerging Trends, and Implications for the Twenty-First Century*. Washington: Milbank Memorial Fund.

Stone, Robyn, Gail Lee Cafferata, and Judith Sangl. 1987. "Caregivers of the Frail Elderly: A National Profile." *The Gerontologist* 27(5):616–626.

U.S. Bureau of Labor Statistics. 2005a. Women in the Labor Force: A Databook. Report 985. Table 2. *Employment Status of the Civilian Noninstitutional Population 16 Years and Over by Sex, 1970–2004 Annual Averages*. Washington, DC. (http://www.bls.gov/cps/wlf-databook2005.htm)

———. 2005b. Women in the Labor Force: A Databook. Report 985. Table 7. *Employment status of women by presence and age of youngest child, 1975–2004.* Washington, DC. (http://www.bls.gov/cps/wlf-databook2005.htm)

U.S. Census Bureau. 2006. Poverty status by type of family, presence of related children, race and Hispanic origin; Historical poverty tables. Retrieved December 10, 2009, from (http://www.census.gov/hhes/www/poverty/histpov/famindex.html)

———. 2008a. Table MS-1. *Marital Status of the Population 15 Years Old and Over, by Sex and Race: 1950 to Present*. (http://www.census.gov/population/www/socdemo/hh-fam.html#ht)

———. 2008b. Table FM-1. *Families by Presence of Own Children Under 18: 1950 to Present*. (http://www.census.gov/population/www/socdemo/hh-fam.html#ht)

———. 2008c. Table MC1 *Married Couples by Labor Force Status of Spouses 1986 to Present.* (http://www.census.gov/population/www/socdemo/hh-fam.html#ht)

U.S. Department of Labor. 2000. *The Family and Medical Leave Act of 1993*. (http://www.dol.gov/whd/regs/statutes/fmla.htm)

U.S. Department of Labor, U.S. Bureau of Labor Statistics. 2008. *Highlights of Women's Earnings in 2008*. Report 1017. (http://stats.bls.gov/cps/cpswom2008.pdf)

Vandell, Deborah Lowe and Barbara Wolfe. 2000. *Child Care Quality: Does it Matter and Does it Need to be Improved?* Report to the Assistant Secretary for Planning and Evaluation, US Department of Health and Human Services. (http://aspe.hhs.gov/hsp/ccquality00/index.htm)

Waldfogel, Jane. 2001. "Family and Medical Leave: Evidence from the 2000 Surveys." *Monthly Labor Review* 124:17–23.

Weber, Julie and Mary Curlew. 2009. "The Impact of the Recession on Work and Family." Sloan Work and Family Research Network Policy Brief, Issue 19.

Wolf, Douglas A., Vicki Freedman, and Beth J. Soldo. 1997. "Division of Family Labor: Care for Elderly Parents." *Journals of Gerontology: Psychological and Social Sciences* 52B(Special Issue):102–109

Wu, Ke Bin. 2005. *African Americans Age 65 and Older: Their Sources of Income.* Fact Sheet. FS 100. Washington: AARP Public Policy Institute.

Yergin, Daniel and Joseph Stanislaw. 1998. *The Commanding Heights: The Battle between Government and the Marketplace that is Remaking the Modern World.* New York: Simon & Schuster.

Zigler, Edward and Susan Muenchow. 1992. *Head Start: The Inside Story of American's Most Successful Educational Experiment*. New York: Basic Books.

Chapter 18
Developing Age-Friendly Communities: New Approaches to Growing Old in Urban Environments

Chris Phillipson

Developing what has been termed "age-friendly" communities has become a significant issue for social policy, embracing questions ranging across urban as well as rural environments. The reasons for such attention are not hard to discern and will be assessed in some detail in this chapter. In brief, however, they include: first, the complexity of demographic change, with the emergence of a wide spectrum of housing and community needs among those in the 50 plus age group. Second, the pressures affecting different types of localities, with the impact of accelerated urbanization for some and deindustrialization for others. Third, is acceptance of the importance of the physical and social environment as a factor influential in maintaining the quality of life of older people (Wahl and Oswald 2010). Fourth, is the policy debate about what constitutes "good" or "optimal" places to age, as reflected in the work of the World Health Organization (WHO) around "age-friendly" cities, these defined as encouraging: "...active ageing by optimizing opportunities for health, participation and security in order to enhance the quality of life as people age" (WHO 2007:1).

Despite the importance attached to building age-friendly communities the approach itself requires better understanding and elaboration at conceptual and operational levels. Some important and critical questions that might be raised include:

- How does the age-friendly theme link to research covering environmental issues and aging?
- What are the origins of the age-friendly approach within social and public policy?
- How viable is the idea given the extent of change operating within communities?
- What needs to be done to make age-friendly communities a realistic option for older people?
- What are some of the barriers that might be encountered in attempting to implement the policy?

The above questions will be examined in the context of the process of urbanization affecting communities across the world. Population aging and urbanization have in their different ways become the dominant social trends of the twenty-first century, with their interaction raising issues for all types of communities – from the most isolated to the most densely populated. By 2030, two-thirds of the world's population will be residing in cities; by that time the major urban areas of the developed world will have 25% or more of their population people aged 60 and over. Cities and the metropolitan regions of which they are a part are themselves changing. Soja and Kanai (2008:58) use the term "global city region" to refer to a: "...new metropolitan form characterized by sprawling polycentric networks of urban centres clustered around one or more 'historic' urban cores". Such networks vary in size from one million at the lower level to ten million and beyond at the upper, covering well over a billion of the world's residents.

C. Phillipson (✉)
Professor of Applied Social Studies and Social Gerontology, Centre for Social Gerontology, Keele University, Keele, UK
e-mail: spa05@keele.ac.uk

R.A. Settersten, Jr. and J.L. Angel (eds.), *Handbook of Sociology of Aging*, Handbooks of Sociology and Social Research, DOI 10.1007/978-1-4419-7374-0_18, © Springer Science+Business Media, LLC 2011

Accelerated urbanization brings advantages and opportunities of different kinds. Soja and Kanai (2008:68) note the extent to which: "Dense and heterogeneous cities and city regions have become the driving forces of the global economy, generating enormous wealth as well as technological innovation and cultural creativity". Cities are regarded as central to economic development, attracting waves of migrants and supporting new knowledge-based industries (Savage et al. 2003). The rebuilding of many cities – notwithstanding economic recession – provides opportunities for innovations in housing and services suitable for a range of age and income groups (Rogers and Power 2000). However, the extent to which the "new urban age" will produce "age-friendly" communities remains uncertain. Cities produce advantages for older people in respect of easy access to medical services, provision of cultural and leisure facilities, shopping and general necessities for daily living (Katz et al. 2008). However, they are also seen as threatening environments, often creating insecurity and feelings of vulnerability arising from changes to neighborhoods and communities. Assessing the challenge of creating age-friendly communities within urban environments is the central aim of this chapter. The discussion is organized under the following headings:

- What have been the main approaches used by sociologists to understand the impact of physical and social environments on older people?
- What have been the main factors driving the debate on creating age-friendly cities?
- How age-friendly is the urban environment?
- What are the options for improving urban environments for older people?
- What are the restrictions and barriers likely to be encountered implementing an age-friendly policy?

The chapter will conclude with a brief summary of the arguments and provide an assessment of the next stage in the research and policy debate.

Environmental Issues and Aging

The environmental context of aging – and issues relating to place and location in particular – has resurfaced as a major theme within the sociology of aging. Drawing on classic studies from Rosow (1967), Rowles (1978), Lawton (1980), and others, environmental gerontology is presented as developing a systematic approach to understanding the physical and spatial contexts influencing people through the life course (Wahl and Weisman 2003). The theoretical questions examined in this body of work include: the influence of the physical environment on the aging self (Wahl and Lang 2006); the impact of neighborhood characteristics on health and well-being (Krause 2004; Smith 2009); the relationship between place and identity (Rowles 1978; Laws 1997); and the impact of urban change on social networks in later life (Deeg and Thomése 2005; Phillipson et al. 2000). These and related questions have emerged as influential themes within social gerontology, and have stimulated diverse empirical projects with a comparative as well as single-country focus (Phillips et al. 2005; Rodwin and Gusmano 2006). Interest in the effects of the physical and spatial environment has been a long-standing feature of research into aging with studies covering:

- "Integration" vs. "segregation" debates in the field of housing.
- Theoretical models examining person–environment interactions.
- Research on attachment to place and the construction of identity.
- The impact of community and neighborhood change.

The field of housing provided the initial focus for research, with studies in the 1950s and 1960s using this topic to explore arguments around "age integrated" vs. "age segregated" communities (Kleemier 1956; Webber and Osterbind 1960). Such debates reflected competing theories in social gerontology during this period – notably activity and disengagement theory (see Marshall and Bengtson this volume) – which presented contrasting hypotheses about the degree to which older people wished to remain part of mainstream society (Friis et al. 1968). An additional factor influencing this work, however, was the rapid development of retirement communities – from the 1950s onwards – in Florida, California, and beyond. Such communities raised issues about the advantages of "age dense environments" for older people, especially regarding security, recreational facilities, and access to specialist support (Streib et al. 1986). This work was subsequently extended with researchers, such as (Rosow 1967) and Teaff et al. (1973), examining the benefits for friendship formation and reduced isolation arising from different types of age-mix within public and private housing.

By the 1970s researchers had moved on from focusing on the advantages and advantages of specific housing arrangements or types, to modeling the way in which older people might be influenced by their physical environment. In particular, the Lawton and Nahemow (1973) "press-competence" (PC) model was widely adopted as an approach to understanding person–environment relationships (Schiedt and Norris-Baker 2004; Oswald et al. 2005). The approach taken by the model is that individual behavior and satisfaction are contingent upon the dynamic balance between the demand character of the environment (press) and the individual's ability to deal with that demand (competence). With this approach, those with low competence encountering strong environmental press are more likely to have maladaptive behavior compared with those having high competence encountering weak environmental press where behavior is likely to be positive (Lawton 1980). The PC model, along with its variants (see, for example, Carp and Carp 1984), has been especially influential in examining the extent to which a mismatch between environments and basic needs can undermine emotional well-being and mental health (Kahana 1982). Peace et al. (2007) highlight the extent to which this work has been dominated by studies of age segregated settings (such as in residential homes), with a particular focus on the impact of lack of privacy and control in creating stress-inducing environments.

The person–environment tradition became the dominant paradigm in environmental gerontology in the 1970s and 1980s, and has retained its influence (Wahl and Oswald 2010). Through the 1980s and 1990s, however, further approaches emerged, drawing on perspectives from cultural and human geography and urban sociology (Rowles 1983; Laws 1997; Peace et al. 2007). These raised issues both about the meanings individuals attach to their environment as well as the structural forces affecting communities and neighborhoods. Rowles (e.g. 1978, 1993) has been especially influential on the first of these, applying phenomenological perspectives to an understanding of the impact of environment on self-identity. His work has drawn attention to the way in which attachment to particular places – such as home and neighborhood – accumulate over time, creating what he views as an inherent "body awareness" of the physical configuration of the environment. According to Rowles (1978:163), this has important implications for aging in place as the "lessons learned through repeatedly traversing familiar space facilitate… continued participation in environments which might otherwise have been almost impossible to negotiate". Rowles (1978, 1983) developed the term "insideness" to refer to the different types of attachment that people maintain to their environment. These include the "*physical insideness*" already referred to but also *social insideness* arising from everyday social exchange over long periods and *autobiographical insideness* reflecting the way in which place can become "a landscape of memories" which can assist in the maintenance of identity in old age (Rowles 1983:114).

At the same time, attachment to the environment may itself be compromised by structural changes affecting communities, these undermining the strategies identified by Rowles for maintaining identity in old age. This may be especially the case in urban environments marked by a rapid turnover of people and buildings, and in unpopular urban neighborhoods characterized by low housing demand and abandonment by all but the poorest and least mobile residents (Wacquant 2008). Here, urban sociology drawing on the Chicago School has provided an important intellectual framework with a range of studies examining problems of aging within inner-city contexts (e.g. Townsend 1957; Birren 1969; Cantor 1975; Stephens 1978; Phillipson et al. 2000; Smith 2009). This work has been extended over recent years, with studies examining the impact of economic recession on inner-city areas (Ogg 2005; Scharf et al. 2002). Such research has been further linked to issues of inequality and racial discrimination affecting minority groups in metropolitan districts (Smith 2009).

The various strands of research identified above have been especially important in highlighting the influence of environmental change at meso- and microlevels. However, the narrow focus of research within most of the approaches raises problems given the larger-scale forces affecting urban communities. Much of the "integration/segregation" debate was itself applied to a limited range of housing-related issues, drawing on theoretical models now largely discredited within the field. The person–environment perspective remains important but its testing has been confined to institutional or domestic settings and, as with phenomenological approaches, this approach is poorly equipped to deal with change at a macrosociological level. Even research examining problems of older people in urban areas such as inner-cities, while exploring issues about the impact of global change, has yet to incorporate findings into a coherent theoretical model.

At the same time, the need for an approach linking different levels of environmental change is especially urgent given the pressures affecting many of the communities in which older people live. In this chapter, the idea of "age friendly" communities is used as way of exploring macro-, meso- and microlevels of change affecting older people. The next section examines some of the main factors behind the development of this approach.

Developing Age-Friendly Communities

Debates about securing optimum community environments for aging populations emerged from a number of organizations during the 1990s (Evans 2009). The theme of age-friendly communities arose from policy initiatives launched by the WHO. A precursor was the notion of "active ageing" developed during the United Nations' Year of Older People in 1999 and elaborated by the European Union (1999) and the WHO (2002). The idea of maintaining "active ageing" referred to the notion of older people's "continuing participation in social, economic, cultural, spiritual and civic affairs, not just the ability to be physically active or to participate in the labour market" (WHO 2002:12). Achieving this was seen as requiring interventions at a number of levels, including maintaining effective supports within the physical and built environment. Here, the WHO (2002) acknowledged that:

Physical environments that are age-friendly can make the difference between independence and dependence for all individuals but are of particular importance for those growing older. For example, older people who live in an unsafe environment or areas with multiple physical barriers are less likely to get out and therefore more prone to isolation, depression, reduced fitness, and increased mobility problems (P. 27).

The theme of age-friendly environments was subsequently applied to urban contexts, with work beginning in 2005 around the theme of "Global Age-friendly Cities". Subsequent work by the WHO, based upon focus groups with older people, caregivers, and service providers, produced a guide and checklist of action points focused on producing an "ideal" city relevant to all age groups. This work concluded that: "It should be normal in an age-friendly city for the natural and built environment to anticipate users with different capacities instead of designing for the mythical 'average' (i.e. young) person. An age-friendly city emphasizes enablement rather than disablement; it is friendly for all ages and not just 'elder-friendly' (WHO 2007:72)."

Such ideas were consistent with new perspectives influencing urban development over the course of the 1990s and early 2000s, notably ideas around "sustainable" (Satterthwaite 1999) and "harmonious cities" (UN-Habitat 2008). The former raised questions about managing urban growth in a manner able to meet the needs of future as well as current generations. The idea of "harmonious" development emphasized values such as "tolerance, fairness, social justice and good governance" (UN-Habitat 2008), these regarded as essential in achieving sustainable development in urban planning. These themes were also influential in the elaboration of ideas associated with "lifetime homes" and "lifetime neighbourhoods" (Department of Community and Local Government 2008; Atlanta Regional Commission 2009), which emerged alongside recognition of the need for more systematic interventions to support population aging at a community level. An additional influence was recognition of the development in many localities of what has been termed "naturally occurring retirement communities" (NORCS), i.e. neighborhoods that, with the migration of younger people, have effectively evolved into communities of older people. The key issue behind the "lifetime" concept was an understanding that effective support for older people within neighborhoods would require a range of interventions linking different parts of the urban system – from housing and the design of streets to transportation and improved accessibility to shops and services (see further below).

Finally, the extent to which environments are "age-friendly" or "hostile" to groups such as older people has been raised in theoretical debates around globalization and its influence on the routines and structures influencing everyday life (Phillipson 2007). One view of the impact of globalization, identified in the research literature, is that it has fragmented and distorted the experience of community and place for older people and other social groups. Rather than the mutual solidarities presented as typical of locality ties in the 1950s and 1960s, community life is now regarded, in the words of Beck (2000), as "unsettled [and] friable". In Bauman's view (2001:47), the present period is best characterized as "times of disengagement": "Gone are most of the steady and solidly dug-in orientation points which suggested a social setting that was more durable, more secure and more reliable than the time-span of an individual life". More generally, some commentators have presented such changes as reflecting the decline of "social capital" or the "disembeddness" of individuals and families from a stable community existence (Putnam 2000; Blokland 2003; Charles et al. 2008).

The link between globalization and themes associated with rootlessness, mobility, and impermanence have become familiar and might be judged as a factor undermining the development of age-friendly communities. An alternative view, however, would be that many of the changes associated with globalization, and its consequences for urban environments in particular, reinforce the need to rethink the way communities provide support through the life course. This is especially important given evidence about the significance of the physical environment in the lives of older people, as well as evidence from research about its influence on the quality of life in old age. A review of this aspect follows below, together with a survey of research findings on the impact of urbanism on older people.

How Age-Friendly is the Urban Environment?

Pressures and Constraints of Urban Living

Physical environments are significant influences for all age groups, but may be especially important for the old and young with the former becoming increasingly reliant on their immediate community. Wight et al. (2009) suggest that there are good reasons to expect that with increased age there is increased vulnerability to the "press" (see above) of the neighborhood environment. Glass and

Balfour highlight four mechanisms that may be at work: "longer duration of exposure; increased biological, psychological, and cognitive vulnerability; changing patterns of spatial use; and reliance on community sources of integration" (cited in Wight et al. 2009:247). Older people may in fact be especially sensitive to changes to the physical and built environment, given its significance for the maintenance of identity, and because of the length of time spent at home – 80% of the time of those 70 and above according to one study (Horgas et al. 1998).

Older people may experience urban areas as "unfriendly" for a variety of reasons – many of these shared with other age groups but in some cases experienced in a more intense form because of the vulnerabilities associated with age. Brown et al. (2009:235) highlight a number of research findings demonstrating that older adults residing "in physically deteriorated neighborhoods [are more likely] to perceive that social support is less available to them…[in comparison with] elders who reside in better-maintained neighbourhoods". The authors also note the extent to which problems associated with the built environment (such as poor maintenance of buildings, limited access to shops and facilities within the neighborhood) have been shown to increase levels of psychological distress even after controlling for variables such as age, gender, and financial strain. Neighborhoods characterized by high levels of economic and social deprivation may present a particular challenge to older people. This can in part be attributed to the relatively high rates of both personal and property crime that characterize such neighborhoods (Smith 2009). The experience of crime and the fear of being a victim of crime can act as direct barriers to the maintenance of a "normal" daily life for many older people. As a consequence, older people may be less likely – especially in inner-city areas – to leave their homes after dark (Phillipson et al. 2000; Scharf et al. 2002). There may even be some parts of neighborhoods, such as parks and cemeteries, which are effectively out of bounds to older people even during daylight hours (Smith 2009).

Older people may be especially susceptible to problems arising from poor-quality housing (Hunt and McKenna 1992; Department for Communities and Local Government 2008; Brown et al. 2009). In England, 33% of homes occupied by older people fail the official "decent homes standard" (Evans 2009). Particular types of urban housing have been linked to mental health problems in later life. Research in the USA found that elderly residents living in high-rise dwellings were more depressed, had higher rates of psychiatric disorder, and were more socially isolated than those living in detached homes in the community (Husaini et al. 1991). Evandrou (2003) highlights housing problems faced by older people living in Inner London where high density terraced housing damaged during the Second World War was replaced with high-rise flats:

Many of these were erected during (the 1950s) and the 1960s and their tenants have aged with them. Flats that were suitable for people earlier in their life course are no longer suitable for them in later life. Poor communal infrastructure, with lifts that frequently break down, entry phones subject to vandalism and graffiti-ridden communal landings all serve to heighten older people's sense of isolation and exclusion (P. 3).

Studies of older people in inner-city areas have identified the ways in which they can be affected by the side effects of population turnover and pressures on public space (Hannan Foundation 2001; Scharf et al. 2002; Newman 2003; Smith 2009). At the same time, we might also consider the extent to which contemporary urbanization creates extreme pressures on everyday living for certain groups. Klinenberg (2002) examined the 1995 heat wave in Chicago that over 1 month alone killed around 600 people, of whom three-quarters were aged 65 and over. As well as the acute factors that caused such high rates of mortality among the old, the author noted chronic attributes of the urban environment that reduced the quality of life of elderly residents:

In recent years, a number of studies have shown that older people living in violent and deteriorated urban areas tend to be more isolated and afraid of crime than those in more robust regions. Among the mechanisms producing this concentrated fear and isolation in ecologically depleted and politically underserved places are the lack of local commercial venues and service providers to draw people into the streets; barriers to physical mobility, such as broken stairs, crumbling sidewalks, and poor lighting;

the psychological impact of living among signs of disorder; indifferent government agencies who neglect the local infrastructure; and the decrease of trusting and reciprocal relationships in areas with high levels of crime.

Restrictions on daily living, as detailed in the research, were reinforced by degraded public space, a product of abandoned buildings, poor-quality infrastructure, and the loss of local businesses – "institutional disengagement" in the words of Herbert Gans (1972). Such conditions became especially perilous for older people when matched by declines in health and social support networks. Newman (2003) illustrated this in her research on older people in inner-city New York, where she argues that, unlike younger people and families who move in and out of neighborhoods, for older people "the home place sets the tone for their daily lives" (Newman 2003:198).

The structure of the modern city also raises issues for groups such as older people. Mike Davis (2002) highlights the development of a "post-urban metropolis" where traditional central-place functions (culture and sports, government, shopping and administration) are radically dispersed among different locations. Along with this, he points to the emergence of "dead cities" stripped of the functions and activities that contribute to the maintenance of lively and diverse public spaces, or what Jane Jacobs (1961) referred to as the "daily ballet" of city sidewalks safe from "barbarism and fear". In Davis's world, by contrast, pedestrian expeditions become an ordeal for young and old alike in cities dominated by cars. Geographical disparities within urban areas may lead to age-and class-segregated neighborhoods, with significant tensions between the needs of groups at different stages of the life course (Rogers and Power 2000).

Sassen (2001) has identified the way in which large cities concentrate both the leading sectors of global capital, along with a growing share of disadvantaged populations (see also Rodwin and Gusmano 2006). Cities, she argues, have become a strategic terrain for a series of conflicts and contradictions – among which the management and support of vulnerable populations is one of the most acute. In like vein, Wacquant (2008:257) refers to what he terms the appearance of "advanced marginality" within urban areas, with the multiplication of "unstable social positions [with] vulnerable populations at an increasing remove from the middle and upper tiers of the structure of classes and places".

An important question concerns the extent to which the construction of the modern (or late-modern) city as the "site for the new consumerism" (Savage et al. 2003:149) results in social exclusion for groups such as older people. Rodwin, Gusmano, and Butler (2006:7) make the point that while world cities offer extensive cultural and entertainment opportunities, they are expensive places in which to live. They illustrate this point by citing a study of New York City that found that only 1-in-20 older households had sufficient money to take full advantage of the quality of life offered by the city. Comparable data is unavailable for British cities, although a relevant finding from the English Longitudinal Study of Ageing (ELSA) for the Social Exclusion Unit was that a larger percentage of older people living in London than in the rest of the country experienced multiple types of poverty and deprivation (ODPM 2006). This is consistent with an analysis of global cities that emphasizes the increasing divergence of the lifestyles and opportunities of wealthy and poor residents, itself a manifestation of growing inequalities linked to social class, ethnicity and, in some respects, age (Gordon and Townsend 2000).

The Benefits of Urban Environments

The pressures associated with urban environments appear to be well documented, and would seem to pose a considerable challenge for implementing an age-friendly approach. On the other hand, the advantages – both existing and potential – of urban areas for supporting aging populations must also be highlighted: First, the resources characteristic of urban economies suggests major opportunities for developing policies of relevance to a more diverse population of older people. Mike Davis

(2002), despite emphasizing the crisis affecting many urban environments, reminds us nonetheless of the facilities they provide, notably in respect of libraries, museums, parks, and communal spaces. He contrasts the "public affluence" of cities with the limits of "privatized consumerism", viewing the former as able to challenge some of the worst excesses associated with the latter. In similar fashion, Smith (2009) points to the cultural and social diversity associated with urban living, with cities able to develop spaces and places associated with different lifestyle choices. Cities can thus present a supportive environment to groups with specialist needs and interests, an aspect that may be especially valued by groups in later life.

Second, even in areas of high economic and social deprivation (such as those in the study by Scharf et al. 2002), older people may report a strong sense of identification with their community. In the Scharf et al. (2002) study of three districts in major cities of the United Kingdom (Liverpool, London and Manchester), three out of four respondents identified positive features about their neighborhood, with most of these commenting on the presence of good neighbors, friends, and family. More than three-quarters of respondents indicated that they had at least one friend in their neighborhood. Of those people with local friends, almost half had a chat or did something with a friend every day.

Third, rather than providing limited social support, urban environments may, as the above would indicate, offer assistance from a wider range of networks as compared with rural areas. Friendship networks, for example, appear to be especially robust in urban communities and may offer an important support mechanism for those who are single or widowed (Phillipson et al. 2000). Urban settings are of particular importance to migrant groups, especially in respect of access to special forms of cultural, social, religious and economic support.

Fourth, although much of the discussion thus far has emphasized the problems associated with urban living, this environment may also represent advantages for those moving into retirement. Ogg and Bonvalet (2007) make this point when examining residential preferences among the baby-boom generation (those born between 1945 and 1955) living in inner-city London and Paris. They highlight in particular the experience of those who chose to move to the city when starting their careers. Among this group are the "pioneers" of gentrification (i.e. house improvements) within inner-cities who benefited from the rise in property values associated with urban regeneration. Some in this group may well have "second homes" in the county, with dividing their time between the "the city" and "the country" becoming a preferred strategy to leaving the urban environment altogether. The city may thus retain its appeal for those who moved earlier in their lives from provincial towns or rural areas. Conversely, "local inhabitants", who were born in and who have spent most of their life living in and around a particular urban area, may be more divided about the changes affecting their locality. Nonetheless, even among this group, despite concern with changes in the population and the apparent loss of social cohesion, attachment to life in an urban environment appears strong (Ogg and Bonvalet 2007).

Constructing Age-Friendly Communities

Creating a better "fit" between urban environments on the one side and aging populations on the other is assuming some urgency within social policy. The WHO (2007:4) develops the point that "making cities more age-friendly is a necessary and logical response to promote the well-being and contributions of older urban residents and keep cities thriving". Equally, measures to support the inclusion of elderly people within cities must be viewed as a key part of the agenda for creating sustainable and harmonious urban environments. Implementing this agenda will, however, demand radical interventions across urban areas. A number of themes can be identified here:

- First, developing new forms of "urban citizenship" which recognize and support changing social needs across the life course.
- Second, understanding the dilemmas of the urban environment in terms of the changing needs of the body.
- Third, developing an age-friendly approach within the context of lifelong/lifetime communities.
- Fourth, ensuring the engagement of older people in the replanning and regeneration of neighborhoods.

The first argument concerns the need to link the discussion about age-friendly cities to ideas about urban citizenship and rights to the benefits which living in a city brings. Painter (2005), for example, cites the work of Henri Lefebvre, who explored issues relating to citizenship and rights in an urban context (see also Holston 1999; Soja 2010). Lefebvre stressed:

The use-value of the city over its exchange value, emphasizing that citizens have a right to make use of the city, and that it is not just a collection of resources to enable economic activity. The uses of the city by citizens should be seen as valid ends in themselves, not merely as a means to produce economic growth. The right to the city is the right to live a fully urban life, with all the liberating benefits it brings. [Lefebvre] believed the majority of city residents are denied this right because their lives are subordinated to economic pressures – despite being *in* the city, they are not fully *of* the city (P. 9).

This last point applies especially well to older people, who may find that despite having contributed to an urban world in which they have spent most of their life, it may present major obstacles to achieving a fulfilling existence in old age. On the one hand, cities are increasingly viewed as key drivers of a nation's economic and cultural success. On the other hand, the reconstruction of cities is often to the detriment of those outside the labor market, especially those on low incomes (see further below). Achieving recognition of the needs of different generations within cities, and exploiting the potential of the city for groups of whatever age, will be central to implementing an age-friendly approach.

Second, linking "age-friendliness" with "urban citizenship" also draws attention to changing needs associated with age, and in particular recognition of the frailty of the human body. Geographical perspectives on cities assert that "all urban dwellers have to negotiate the city *practically* [author's emphasis], and work through the dilemmas, problems, and possibilities of 'getting by' in the city" (Hubbard 2006:113). And feminist perspectives on the use of urban space highlight the way in which "women's spatial practice is constrained by geographies of violence and fear" (Tonkiss 2005:94–95). Such ideas, and the extent to which older people are to some degree "prisoners of space" (Rowles 1978), are especially relevant to aging in the city – both in global cities undergoing accelerated change and those cities facing deindustrialization. Both contexts raise dilemmas in the context of mental and physical vulnerabilities. Elderly men and women may experience difficulties "creating" space within cities. Global cities raise tensions between a "hypermobile" minority and those aging in place; deindustrializing cities (with shrinking populations) create problems arising from the withdrawal of an economic base which can maintain sustainable networks for different social groups. The challenge here then is creating an urban environment that supports the autonomy of the aging body and the equal rights of older people with others to a "share" of urban space. This issue will be especially important to implement at a local level, with a particular focus on improving the quality of urban design and promoting safety and inclusion as key features of urban living.

Third, some of the issues associated with the above are being developed through the ideas associated with lifetime communities and neighborhoods. In planning for lifetime neighborhoods Harding (2007:8) suggests the need to consider:

- Accessibility of the built environment.
- Appropriateness of available housing.

- Fostering social capital.
- Location and accessibility of services.
- Creating esthetically pleasing public spaces which promote a sense of place and social cohesion.
- Cross-sectoral integration and planning of services.
- Building intergenerational relationships by shared site usage.
- Better use of information technology.

Work by the Atlanta Regional Commission (ARC) (2009:7) made the point that although mobility begins inside the individual unit or house, it must carry on throughout the entire built environment "... from inside the dwelling, down the street, and into the restaurant, theatre or store ... continuously across the entire urban environment". This argument applies equally to all types of communities – suburban as well as inner and outer city. In relation to suburbs, for example, these have frequently been designed with families and commuters in mind, rather than the specific needs of older people and/or smaller households. There is scope here to explore the urban design implications of a different population mix together with neighborhoods that explicitly have a longer lifespan. Strategic guidance on urban design might be developed further to indicate how "lifelong" adaptability for an aging population can be built into communities from the start (Brook Lyndhurst 2004).

Fourth, a critical issue for an "age-friendly" approach concerns ensuring the involvement of older people in urban regeneration policies. A study in the United Kingdom by Riseborough and Sribjilanin (2000) found that older people were often "invisible" in regeneration policies. The problem here was less the absence of older people in consultations around policies, more an underlying "ageism" which viewed them only as "victims" of neighborhood change. Again, this goes back to the argument from Rowles (1978:216) of the "need to break free from prevailing social attitudes, which have served to alienate the elderly and to install within us a view of their lives as one of inevitable spatial withdrawal". Riseborough and Sribjilanin (2000) make the point that regeneration practice could benefit from the experience of older people, their attachment to their neighborhoods, and their involvement in community organizations. At the same time, there is also a need to develop urban regeneration strategies targeted at different groups within the older population, with awareness, for example, of contrasting issues faced by different ethnic groups, people with particular physical/mental health needs, and those living in areas with poor housing alongside high population turnover.

Obstacles to an Age-Friendly Perspective

Despite the benefits of applying an age-friendly approach, some critical questions also need to be faced to ensure effective implementation of such a policy. At the present time, discussions around age-friendliness have been largely disconnected from the pressures on urban environments in the Global North, where private developers remain the dominant influence on urban planning (Solnit and Schwartzenburg 2003; Minton 2009). The result, according to Harvey (2008:31), is that the "quality of urban life has become a commodity, as has the city itself, in a world where consumerism, tourism, cultural and knowledge-based industries have become major aspects of the urban political economy". Blokland and Savage (2008:38) argue that such processes are leading to a different type of urbanism, one that is "confirming rather than challenging inequalities within cities". The tension here is between the social needs of older people, as an increasingly important constituent of urban populations, and the pressures on public space arising from private ownership. This may lead to a distortion in provision in terms of meeting the needs of competing groups within the urban system (Tonkiss 2005; Minton 2009).

A second important issue concerns applying "age-friendliness" in a way that recognizes the complexity of the urban environment. The techniques for ensuring an age-friendly approach will vary considerably depending on the characteristics of urban change and development. While the trend towards urban living is world-wide, the pattern of urban growth demonstrates huge variation: shrinking city populations in the developed world (Europe especially); and accelerating urbanization in Africa and Asia, with both continents demonstrating a mix of rapidly expanding cities in some cases, declining ones in others (UN-Habitat 2008). "Age-friendliness" will also need to reflect the size of a city: The approach might, for example, be different in Europe where small cities with fewer than 500,000 residents are the norm, as compared with the USA where large urban agglomerations (with populations of between two and five million) are much more common. Securing "age-friendliness" in the context of the rise of "mega-cities" and "hyper-cities" (the latter with populations of 20 million or more) provide another variation. At the same time, processes for developing "age-friendliness" will need radical adaptation given the "slum cities" prevalent in Southern Asia and sub-Saharan Africa (UN-Habitat 2008). The bulk of population growth in these continents has taken place largely through the rise of slums, many of these located on the periphery of capital cities (Davis 2006). The problem of reaching older people and migrants who are "ageing in place" and housed in temporary accommodation bereft of basic facilities, underlines the need for new models of intervention which can respond to the highly unequal contexts experienced by the urban elderly across the world.

A third issue concerns linking the debate around environmental change with that concerned with urbanization and population aging. The vulnerability of older people in periods when urban environments are challenged through extremes of temperature has already been highlighted. But other problems may also be cited. Air pollution is a major hazard, especially in newly industrializing countries in the Global South. WHO estimates that more than one billion people in Asia alone are exposed to outdoor air pollutant levels that exceed WHO guidelines, leading to the premature death of half a million people annually. Older people, especially those with chronic health conditions – these exacerbated or caused by environmental degradation – will be among the worst affected in terms of their quality of life. Looking ahead, rising sea levels associated with climate change may bring particular risks for older people living in cities. Fourteen of the world's 19 largest cities are port cities located along a coastline or river delta. In Asia, the dominance of port cities is even greater; 17 of the region's 20 largest cities are coastal or on a river bank or in a delta. Many of the world's most prominent global cities (e.g. Mumbai, Shanghai, New York City) will be among the most exposed to surge-induced flooding in the event of sea level rise (UN-HABITAT 2008:141). The consequences for their substantial populations of elderly and very elderly people are immense, and will require detailed planning and assessment. Older people are especially vulnerable in periods of environmental crisis, with the potential for displacement from their home, from relatives, and services and support (Rodwin and Gusmano 2006). These aspects were clearly demonstrated in crises such as Hurricane Katrina in the USA (Bytheway 2006) and the 2003 heat wave in France (Ogg 2005). In both cases, elderly people were disproportionately affected compared with other age groups, but failed to receive the special help and assistance required. In these instances, environmental disaster undermined the ability of urban areas to respond quickly and effectively to the needs of their elderly habitants.

Conclusion

Securing age-friendly cities remains an important goal for economic and social policy. The future of communities across the world will in large part be determined by the response made to achieving a higher quality of life for their older citizens. A crucial part of this response must lie in creating supportive environments providing access to a range of facilities and services. However, the research and policy agenda will need to change in at least three ways if this is to be realized: First, the issues raised by developing age-friendly communities within complex urban environments will require a more coherent link between research and policy than has thus far been achieved. Research on environmental aspects of aging has an impressive literature to its name, yet it remains detached from analyzing the impact of powerful global and economic forces transforming the physical and social context of cities. As remedy, this will require closer integration with developments in disciplines such as urban sociology, urban economics, and human geography. Understanding optimum environments for aging must be seen as an interdisciplinary enterprise requiring understanding of the impact on older people of developments such as the changing dynamics of urban poverty; the impact of urban renewal and regeneration; the influence of transnational networks; and changing relations between different class, gender, ethnic, and age-based groups.

Second, in keeping with the approach taken in this chapter, given the rapid changes affecting many urban areas, new approaches to understanding older people's relationship to urban change – and city development in particular – is urgently required. In particular, there is a strong case for more research in "urban ethnography" to capture the disparate experiences of those living in cities that are now experiencing intense global change and that are strongly influenced by complex patterns of migration on the one side and population aging on the other. Sassen (2000:146) pointed to the need for detailed fieldwork as a "necessary step in capturing many aspects of the urban condition" (see also Wacquant 2008); such work will be especially important for understanding the impact of urban growth on groups such as migrants aging in place, single people, and those on low incomes. Urban sociology was founded (through the work of the Chicago School from the 1920s) upon detailed studies of experiences of urban life, particular of disadvantaged and insecure people from different migrant populations. Ethnographies would bring to the surface the attitudes, motivations, and experiences of older people who are "ageing in place" and will deepen our understanding about the way in which cities are changing, and about the positive and negative contributions that the changes have on the quality of daily life in old age.

Thirdly, it might be argued that the benefit of thinking about age-friendliness lies more in its challenge to reassessing the values (and ideals) that might be nurtured within urban communities. From the 1960s onwards, writers such as Jane Jacobs (1961) and Richard Sennett (1970) argued the case for celebrating the diversity of city life. Giradet (1999:424) put forward his vision of the city "as a place of culture and creativity, of conviviality and above all else of sedentary living". In the United Kingdom, Richard Rogers and Anne Power (2000) developed a new approach to urban planning, one calling for a sharing of spaces for the collective good and for a reversal of the drift towards suburbanization. All of these – and similar ideas – are relevant to developing age-friendly cities and arguably need closer integration to the sociology of aging. Thus, despite the many obstacles to implementing this approach, its potential for reminding us of the values to be nurtured for harmonious city living are important and certainly relevant for building communities fit for populations of older people.

References

Atlanta Regional Commission. 2009. *Lifelong Communities: A Regional Guide to Growth and Longevity.* Atlanta: ARC.
Bauman, Zigmunt. 2001. *Community: Seeking Safety in an Insecure World.* Cambridge: Polity.
Beck, Ulrich. 2000. *What is Globalization?* Cambridge: Polity Press.
Birren, James. 1969. "The Aged in Cities." *The Gerontologist* 9:163–9.
Blokland, Talja. 2003. *Urban Bonds.* Cambridge: Cambridge University Press.
Blokland, Talja and Michael Savage. 2008. *Networked Urbanism.* Aldershot: Ashgate.

Brook Lyndhurst Ltd. 2004. Sustainable Cities and the Ageing Society: The Role of Older People in an Urban Renaissance. Retrieved May 11, 2009 (http://www.communities.gov.uk/documents/corporate/pdf/142412.pdf).

Brown, Scott C., Craig A. Mason, Joanna L. Lombard, Frank Martinez, Elizabeth Plater-Zyberk, Arnold R. Spokane, Frederick L. Newman, Hilda Pantin, and José Szapocznik. 2009. "The Relationship of Built Environment to Perceived Social Support and Psychological Distress in Hispanic Elders." *Journal of Gerontology: Psychological Sciences* 64B:234–46.

Bytheway, Bill. 2006. "The Evacuation of Older People: The Case of Hurricane Katrina." Presented at the annual conference of the Royal Geographical Society and the Institute of Geography, August 31, London.

Cantor, Margaret. 1975. "Lifespace and The Support System of the Inner-City Elderly." *The Gerontologist* 15:23–7.

Carp, Francis and Abraham Carp. 1984. "A Complementary/Congruence Model of Well-Being or Mental Health for the Community Elderly." Pp. 279–336 in *Human Behaviour and Environment*, edited by I. Altman and M. P. Lawton. New York: Plenum Press.

Charles, Nickie, Charlotte Davies, and Chris Harris. 2008 *Families in Transition: Social Change, Family Formation and Kin Relationships*. Bristol: Policy Press.

Davis, Mike. 2002. *Dead Cities*. New York: The New Press.

Davis, Mike. 2006. *Planet of Slums*. London: Verso.

Deeg, Dorly and G.C. Fleur Thomése. 2005. "Discrepancies between Personal Income and Neighbourhood Status: Effects on Physical and Mental Health". *European Journal of Ageing* 2:98–107.

Department of Communities and Local Government. 2008. "Lifetime Homes, Lifetime Neighbourhoods: A National Strategy for Housing in an Ageing Society." London: DCLG.

European Union. 1999. "Active Ageing: Pivot of Policies for Older People in the New Millennium." Retrieved August 8, 009 (http://europa.eu.int/comm/dgs/employment-social/speeches/991116ad.doc).

Evandrou, Maria. 2003. "Growing Old in London: Socio-Economic Inequalities". Sage Discussion Paper No. 15, ESRC SAGE Research Group, London: London School of Economics.

Evans, Simon. 2009. *Community and Ageing*. Bristol: Policy Press.

Friis, Henning, Peter Townsend, and Ethel Shanas. 1968. "Old People in Three Industrial Societies: An Introduction." Pp. 2–18 in *Old People in Three Industrial Societies*, edited by E. Shanas, P. Townsend, D. Wedderburn, H. Friis, Poul Milhøj, and J. Stenhower. London: Routledge.

Gans, Herbert. 1972. *People and Plans: Essays on Urban Problems and Solutions*. London: Routledge.

Giradet, Herbert. 1999. "Sustainable Cities: A Contradiction in Terms?" Pp. 403–25 in *Sustainable Cities*, edited by D. Satterthwaite. London: Earthscan.

Gordon, David and Peter Townsend. 2000. *Breadline Europe: The Measurement of Poverty*. Bristol: Policy Press.

Hannan Foundation. 2001. "The Hannan Study of Older Adults in Detroit's Central City." Final Report to the Luella Hannan Memorial Foundation Board of Directors, Center for Health Care Effectiveness. Detroit, Michigan: Wayne State University.

Harding, Ed. 2007. "Towards Lifetime Neighbourhoods: Designing Sustainable Communities For All." London: International Longevity Centre/Department for Communities and Local Government.

Harvey, David. 2008. "The Capitalist City." *New Left Review*. 53:23–42.

Holston, James, ed. 1999. *Cities and Citizenship*. Durham: Duke University Press.

Horgas, Ann L., Hans-Ulrich Wilms, and Margaret. M. Baltes. 1998. "Daily life in Very Old Age: Everyday Activities as Expressions of Daily Life". *The Gerontologist* 38:556–68.

Hubbard, Phil. 2006. *City*. London: Routledge.

Hunt, Sonja and Stephen McKenna. 1992. "The Impact of Housing Quality on Mental and Physical Health." Housing Review 41:47–9.

Husaini, Bagar, Stephen Moore, and Robert Castor. 1991. "Social and Psychological Well-being of Black Elderly Living in High Rise for the Elderly." *Journal of Gerontological Social Work* 16:57–78.

Jacobs, Jane. 1961. *The Death and Life of Great American Cities*. New York: Vintage Books.

Kahana, Eva. 1982. "A congruence model of person-environment interaction." Pp. 97–121 in *Aging and the Environment: Theoretical Approaches*, edited by M. Powell Lawton, Paul. G. Windley and Thomas. O. Byerts. New York: Springer.

Katz, B., Altman, A and Wagner, J. 2008. "An Agenda for the Urban Age: Urban Age Glossary." Pp. 474-484 in The Endless City, edited by Burdett, R. and Sudjic, D. London: Phaidon.

Kleemier, R.W. 1956. "Environmental Settings and the Ageing Process." Pp. 105–16 in *Psychological Aspects of Aging*, edited by J. E. Anderson. Washington D.C.: American Psychological Association.

Klinenberg, Eric. 2002. *Heatwave: A Social Autopsy of Disaster in Chicago*. Chicago: University of Chicago Press.

Krause, Neil. 2004. "Neighbourhoods, Health, and Well-Being in Later Life." Pp. 223–49 in *Aging in Context: Socio-Physical Environments*, edited by H.-W. Wahl, R. Scheidt, and P. Windley. New York: Springer.

Laws, G. 1997. "Spatiality and Age Relations." Pp. 90–100 in *Critical Approaches to Ageing and Later Life*, edited by A. Jamieson, S. Harper, and C. Victor. Buckingham: Open University Press

Lawton, M. Powell and Lucille Nahemow. 1973. "Ecology of the aging process." Pp. 619–24 in *Psychology of Adult Development and Aging*, edited by C. Eisdorfer and M.P. Lawton. Washington, DC: American Psychological Association.
Lawton, M. Powell. *1980 Environment and Aging*. Pacific Grove, CA: Brooks/Cole.
Minton, Anna. 2009. *Ground Control: Fear and Happiness in the Twenty-first Century City*. London: Penguin Books.
Newman, Kathleen. 2003. *A Different Shade of Gray*. New York: New Press.
Office of the Deputy Prime Minister. 2006. *A Sure Start to Later Life: Ending Inequalities for Older People*. London: Social Exclusion Unit.
Ogg, Jim. 2005. *Heat Wave*. London: The Young Foundation.
Ogg, Jim and Catherine Bonvalet. 2007. "Ageing in Inner Cities: The Residential Dilemmas of the Baby Boomer Generation." *International Journal of Ageing and Later Life* 2:61–90.
Oswald, Frank, Annette Hieber, Hans-Werner Wahl, and Heidrun Mollenkopf. 2005. "Ageing and Person Environment Fit in Different Urban Neighbourhoods." *European Journal of Ageing* 2:88–97.
Painter, Joe. 2005. Urban Citizenship and Rights to the City. Durham: Durham University International Centre for Regional Regeneration and Development Studies.
Peace, Sheila, Hans-Werner Wahl, Heidrun Mollenkopf, and Frank Oswald. 2007. "Environment and Ageing." Pp. 209–234 in *Ageing in Society*, edited by J. Bond, S. Peace, F. Dittman-Kohli, and G. Westerhof. London: Sage.
Phillips, David. R., Oi-Ling Siu, Anthony Yeh, and Kevin Cheng. 2005. "Ageing and the Urban Environment." Pp. 147–63 in *Ageing and Place: Perspectives, Policy, Practice*, edited by G. J Andrews and D. R. Phillips. London: Routledge.
Phillipson, Chris. 2007. "The 'Elected' and the 'Excluded': Sociological Perspectives on the Experience of Place and Community in Old Age." *Ageing and Society*, 27:321–42.
Phillipson, Chris, Miriam Bernard, Judith Phillips, and Jim Ogg. 2000. *The Family and Community Life of Older People*. Routledge, London.
Rogers, Richard and Anne Power. 2000. *Cities for a Small Country*. London: Faber and Faber.
Putnam, Robert. 2000. *Bowling Alone*. New York: Simon and Schuster.
Riseborough, Moyra and Alan Sribjilanin. 2000. *Overlooked and Excluded? Older People and Regeneration: A Review of Policy and Practice*. London: Age Concern England.
Rodwin, Victor and Michael Gusmano. 2006. *Growing Older in World Cities*. Nashville, TN: Vanderbilt University Press.
Rodwin, Victor, Michael Gusmano, and Robert Butler. 2006. "Growing Older in World Cities: Implications for Health and Long-Term Policy." Pp. 1–16 in *Growing Older in World Cities*, edited by V. Rodwin and M. Gusmano. Nashville, Tennessee: Vanderbilt University Press.
Rosow, Irving. 1967. *Social Integration of the Aged*. New York: Free Press.
Rowles, Graham. 1978. *Prisoners of Space? Exploring the Geographical Experience of Older People*. Boulder, Colorado: Westview.
———. 1983. "Place and Personal Identity in Old Age: Observations from Appalachia." *Journal of Environmental Psychology* 3:299–313.
———. 1993. "Evolving Images of Place in Aging and 'Aging in Place'." *Generations* 17:65–70.
Sassen, S. 2000. Cities in a World Economy, Thousand Oaks, Calif: Pine Forge Press.
Sassen, Saskia. 2001. *The Global City*. Princeton, NJ: Princeton University Press.
Satterthwaite, David, ed. 1999. *Sustainable Cities*. London: Earthscan.
Soja, Edward. 2010. *Seeking Social Justice*. Minneapolis MN: University of Minnesota Press.
Soja, Edward and Miguel Kanai. 2008. "The Urbanization of the World." Pp. 54–69 in *The Endless City*, edited by R. Burdett and D Sudjic. London: Phaidon Books.
Stephens, Joyce. 1978. *Loners, Losers, and Lovers*. Seattle WA: University of Washington Press.
Streib, Gordon, Anthony J. LaGreca, and Williams Folts. 1986. "Retirement Communities: People, Planning and Prospects." Pp. 94–103 in *Housing and Aging Society*, edited by R. Newcomer, M. P. Lawton, and T. Byerts. New York: Van Nostrand Rheinhold.
Savage Michael, Alan Warde, and Kevin Ward. 2003. *Urban Sociology, Capitalism, and Modernity*. 2nd ed. London: Macmillan.
Scharf, Thomas, Chris Phillipson, Allison Smith, and Paul Kingston. 2002. *Growing Older in Socially Deprived Areas*. London: Help the Aged.
Schiedt, Rick, J. and Carolyn Norris-Baker. 2004. "The General Ecological Model Re-Visited: Evolution, Current Status and Continuing Challenges." Pp. 34–58, in *Aging in Context: Socio-Physical Environments*, edited by H.-W. Wahl, R. J. Schiedt, and P. G. Windley. New York: Springer.
Sennett, Richard. 1970. *The Uses of Disorder: Personal Identity and City Life*. New Haven, CT: Yale University Press.
Smith, Allison. 2009. *Ageing in Urban Neighbourhoods*. Bristol: Policy Press.

Solnit, Rebecca and S. Schwartenberg. 2000. *Hollow City*. London: Verso.

Teaff, Joseph, M. Powell Lawton, Lucille Nahemow, and Diane Carlson. 1973. "Impact of Age Segregation on the Well-Being of Elderly Tenants in Public Housing". *Journal of Gerontology*, 33:130–33.

Tonkiss, Fran. 2005. *Space, the City and Social Theory*. Cambridge: Polity Press.

Townsend, Peter. 1957. *The Family Life of Old People*. London: Routledge.

UN-HABITAT. 2008. *State of the World's Cities 2008/2009*. London: Earthscan.

Wacquant, Loic. 2008. *Urban Outcasts*. Cambridge: Cambridge University Press.

Wahl, Hans-Werner and Gerald Weisman. 2003. "Environmental Gerontology at the Beginning of the New Millennium: Reflections on Its Historical, Empirical, and Theoretical Development." *The Gerontologist* 43:616–27.

Wahl, Hans-Werner and Freider Langer. 2006. "Psychological Aging: A Contextual View." Pp. 881–95 in *Handbook of Human Aging*, edited by P. Michael Conn. Amsterdam: Elsevier.

Wahl, Hans-Werner and Frank Oswald. 2010 "Environmental Perspectives on Ageing". Pp. 111–23 in *The Sage Handbook of Social Gerontology*, edited by D. Dannefer and C. Phillipson. London: Sage.

Webber, Irving L. and Carter C. Osterbind. 1960. "Types of Retirement Communities." Pp. 157–71 in *Housing the Elderly in Retirement Communities*, edited by E. W. Burgess. Ann Arbor, MI: University of Michigan, Division of Gerontology.

World Health Organization. 2002. *Active Aging: A Policy Framework*. Geneva: WHO.

———. 2007. *Global Age-Friendly Cities*: A Guide. Geneva: WHO.

Wight, Richard G, Janet Cummings, Arun. Karlamangla, and Carol Aneshensel. 2009. "Urban Context and Change in Depressive Symptoms in Later Life. *Journal of Gerontology: Psychological Sciences* 64B:247–51.

Part VI
Economies, Government, and Aging

Chapter 19
Crises and Old Age Policy

Carroll L. Estes

The theme of crisis is a central motif resonating throughout the contemporary U.S. welfare state that is particularly salient for old age politics and policy. To understand these issues, analysis must attend to crisis construction and crisis management by the state and the role of other interests and actors in the economic, political, and civil sectors of society.

During the past three decades, much has been written about crises related to old age and aging. Perspectives on crises surrounding aging span the topics of individual human development, the family, demographic and population aging, health, health disparities, health costs, social inequalities, and national and global economics, among others.

The perspective in this chapter draws from the framework of critical theory as incorporated and reflected in critical gerontology and the political economy of aging. Critical theory is "designed with a practical intent to criticize and subvert domination in all its forms" (Bottomore 1983:183). For critical gerontology, "the overall project ... has been to provide alternative theoretical frameworks and emancipatory knowledge, address[ed] toward concerns of social inequalities and social justice" (Estes and Phillipson 2007:330–1). The political economy of aging is employed to understand the stakes of the state, capital, and democracy (citizens and the public) in the constructions of old age policy and politics in crisis.

Working within variants of this critical approach, the processes of aging and societal aging may be analyzed as a crisis, a problem, a triumph, or some version of all three. The study of beliefs, values, and ideology is central to understanding competing constructions of reality regarding old age and societal and global aging. These constructions signal whether and how old age and aging are defined as issues, problems, and crises (or not) as well as whether or which lines of public or private action are considered appropriate or even feasible.

Critical gerontology draws from a *conflict perspective* that posits that the social order is held together by the dominance of certain interests and groups over others (Weber 1946; Estes, Swan, and Gerard 1982; Collins 1988). Power struggles and their outcomes shape how society is organized, its key institutions, institutional functions, and resource allocations. The social order (the status quo) is held together by constraint rather than by consensus. The most powerful impose their ideas, material interests, and actions on the less powerful. The state and its key agents in their multiple institutional forms (e.g., legislative, executive, judicial, military, and educational) also actively participate in these power struggles (Skocpol 1992). Key actors and interests that comprise the economic system of capitalism and its multiple sectors (e.g., the medical industrial complex, the military industrial complex, and the prison industrial complex), political parties and organizations, and the public (including communities of color, gender, and disability) are major forces in combat.

C.L. Estes (✉)
Institute for Health & Aging, University of California, San Francisco, CA, USA
e-mail: carroll.estes@gmail.com

R.A. Settersten, Jr. and J.L. Angel (eds.), *Handbook of Sociology of Aging*, Handbooks of Sociology and Social Research, DOI 10.1007/978-1-4419-7374-0_19, © Springer Science+Business Media, LLC 2011

Crisis Defined

As "turning points" in which preexisting relationships and meanings can no longer be assumed, crises are "moments of truth" that signify "the restructuring of social relationships that occurs when new power centers confront existing structures of domination; the outcome is generally unknown and existing institutions and social practices can no longer be taken for granted" (O'Connor 1987:54–5).

Critical theorists weave together individual and system level crisis theories in ways that implicate the political, economic, and socio-cultural realms (Habermas 1975; Offe and Keane 1984; O'Connor 1987). O'Connor posits "that modern economic, social, political, and cultural crises interpenetrate one another in ways which transform them into different dimensions of the same historical process … [and] the modern crisis becomes one 'general crisis'" (pp. 11, 54). Habermas explores crisis tendencies specific to advanced capitalism that originate in the political and socio-cultural as well as the economic system. Offe and Keane posit the centrality of state political and administrative power for understanding crisis tendencies in light of the state's coordination and "steering" problems (see below).

Ideology and Crisis

Power struggles over beliefs and ideology reflect, reproduce, challenge, and/or alter the dominant social relations that frame the nation's treatment of old age and aging via its societal institutions, especially as advanced through state policy. As belief systems, ideologies are competing worldviews that reflect the social position and structural advantages of their adherents. All political, cultural, and economic regimes use ideology as the discourse with which to communicate and impose their reflection of the dominant social relations. The contest for *ideological hegemony* (Gramsci 1971) is about achieving acquiescence and maintaining power through the means of the production and inculcation of ideas in the broader public rather than through overt violent coercion or revolution. When a dominant group achieves widespread support for its priorities, the dominant group attains hegemony (Sassoon 2001). For example, "When older people are perceived by other, dominant groups, to be irrelevant to social stability, or a threat to economic viability, they may be coerced into certain roles and institutions" (Estes, Biggs, and Phillipson 2003:64).

Ideology is integral to three processes by which dominant views of social policy and aging are produced and sustained (Estes 1991a, b): (1) the successful creation of cultural images by policymakers, experts, and the media, for example, that the elderly are "greedy geezers" who steal our children's futures and unfairly enrich themselves through Social Security; (2) the appeal to the necessities of the economic system, for example, claims that the elderly are responsible for the nation's economic problems and "busting the budget"; and (3) the implementation of policy and the application of expertise in ways that transform conflicts over goals and means into systems of rational problem-solving. The processes of "rational problem resolution by technical experts" (e.g., commissions) obfuscate the substance of class, gender, racial/ethnic, religious, and other conflicts (George and Wilding as cited in Manning 1985) that are codified and reified through crisis definitions and solutions.

Ideologies shape and delimit public understanding of what is, what is good, and what is possible (Therborn 1980). A crucial dimension of ideology formation is how it shapes what is seen as conceivable in terms of the root problem and the solutions deemed as appropriate. Two challenges arise: (1) to locate the systems of beliefs and values within specific social formations (such as old age policy) and examine how these systems articulate with the economic system of capitalism, the state, and social class, gender, race/ethnic, and other struggles; and (2) to investigate ideological communities that exist within specific, organized collectivities (e.g., nation states, economic and cultural sectors) and how these communities relate to and mask past social and class struggles and contradictory norms and values (Thompson 1986:14).

Several ideologies are particularly relevant to old age crisis politics and policy in the United States policy. First in relevance is the hegemonic *market ideology*, which some call *neo-liberalism* or *market fundamentalism* (Soros 1998; Stiglitz 2010). Neoliberalism, attributed to Adam Smith and neoclassical economics, is hostile to anything that may impede the (assumed) natural superiority of the market (Levitas 1986). O'Connor (1984) observes that *American individualism* is:

> capital's most powerful weapon of ideological domination of labor [with its]...myths of the 'self,' private property and natural rights [that] 'mediate back' [i.e., obfuscate] ethnicity, race, and sex with a fateful and unconscious power which subverts ... ethnic and sexual identities. 'Free market laissez faire' and 'individual responsibility' reinforce the two great pillars of social domination, 'blame the victim' and 'scapegoating.' (P. 14)

Both the laissez faire free market and individual responsibility stand in direct opposition to the alternative, inherently social and collective framework that values and acknowledges the commons, interdependence, social solidarity, and the intergenerational chains of exchange and care over many lifetimes that make possible individual opportunity and success at any one time and place.

Gender ideology supports the separate public and private spheres in which women's work in the home is seen as separate, hence private (off limits to state policy but subject to traditional patriarchal and religious family values). In contrast, men's work is seen in the public sphere, an issue that Gita Sen (1980:77) affirms in observing the existence of "patriarchal ideology."

Classical neoliberal economists fail "to accurately analyze (or even recognize) the role of the market in creating intractable inequalities in power relations within the family" (Bergmann as cited in Ciscel and Heath 2001:408). Folbre (2001) further emphasizes the unrecognized imbalance of familial forces by describing an *invisible heart of care*:

> The invisible hand of the market depends upon the invisible heart of care. Markets cannot function effectively outside the framework of families and communities built on values of love, obligation, and reciprocity ... The invisible hand [of Adam Smith] is about achievement. The invisible heart is about care for others. The hand and the heart are interdependent, but they are also in conflict. The only way to balance them successfully is to find fair ways of rewarding those who care for other people. This is not a problem that economists or business people take seriously. (P. 4)

Neoconservative ideology complements gender ideology in advancing the religious right's cause of recapturing traditional patriarchal family structure and norms. The resurgence of efforts to reinstate women's subjugation is manifested not only in conflicts over reproductive choice, but also in increased demands for more unpaid reproductive labor of caregiving from women of all generations. The *family responsibility* (i.e., women's caring work and sacrifice attendant thereto) is deeply inscribed in the U.S. nation state's long-term care policy (Estes, Swan, and Associates 1993). Well-documented consequences are unequal, sex-based work opportunities; wage discrimination; grinding poverty; and negative health effects, with the most deleterious disadvantages accumulated by women of color, unmarried women, and women not living in the Ozzie and Harriet type nuclear family.

Racial ideology, according to Feagin (2006), is a white racial frame in which "the dominant discourse denies racism" (p. 259), as it fosters and rationalizes "racial injustice and social institutions that privilege one racial group over another" (p. 310). From neocolonial days of the United States, this ideology defended "white privilege as meritorious and accenting" (Feagin 2006:230) the alleged superiority of whites and the inferiority of those who are racially oppressed. This "elite color-coded ideology ... permeated the legitimating discourse of the whites who were dominant in major social institutions including the economy, law, politics, education, and religion" (Feagin 2006:28). Economic domination by white males is imposed via processes of discrimination, exclusion, and marginalization. The *racial state* (Omi and Winant 1986, 1994) is a structural manifestation of the ideological hegemony of the white racial frame that has enormous consequences for the meaning and experience of race in the daily lives of people of color and ethnic minorities across the life course as well as profound inequalities. *Racial formation* is "the process by which social, economic, and political forces determine the content and importance of racial categories, and by which they are … shaped by racial meanings" (Omi and Winant 1986:61).

Ageism as Ideology

In 1975, Butler coined the term, *ageism*. Both he and Maggie Kuhn recognized ageism and its commonality with racism and sexism as a highly negative attribution that provide the basis for discrimination. Ageism is manifested both structurally and ideologically. Structurally, ageism influences the agenda setting, formation, and implementation processes surrounding public policy. Ideologically, ageism bolsters images of elders as undeserving persons who do not merit the benefits (e.g., Social Security) that they have earned and paid into for decades (Estes et al. 2009b). Maggie Kuhn campaigned against ageism through the Gray Panthers (Estes and Portacolone 2009), showing that "[a] geism afflicts … institutions serving older people with a deep, insidious paternalism … Ageism infects us when we reject ourselves and despair our own powerlessness, wrinkled skin and physical limitations" (Kuhn 1974).

In its contemporary form and at the collective structural level, ageism is the denial of basic and civil rights of elders (Anti-Ageism Taskforce 2006). As evidenced by more than three decades of attacks on American elders as greedy geezers, the slander against older persons has garnered such media acceptance and promotion that it qualifies as an ideology.

Crisis Construction and Impact

Subjective and Objective Crisis Construction

The subjective and objective dimensions of crisis formation have both a symbolic and material impact. *Crises* are socially constructed in relation to the capacity and power of strategically located agents and interests to define particular issues as problems and to press their views into public consciousness and law (Estes 1979). The manufacture of crises (objective and subjective) and crisis recognition occurs at all levels, the individual (micro), the organizational and institutional (meso), and the systemic (macro).

Subjectively, crises are socially constructed through social perception and definition. Crisis ideation is influenced by dominant ways of conceptually ordering the world, especially and by ideologies. In a sense, a crisis may be said to exist *if it is perceived to exist*. Conversely, a crisis does not exist if the condition or situation is not perceived as a crisis and if people do not act as though a crisis exists. This is not to assert that the elderly face no problems independent of those that are perceived as real. Demographic aging and structural conditions (e.g., stock market crash and global warming) may be said to be objectively real, regardless of how they are perceived. Social action, however, is indivisible from the socially constructed ideas that define and provide images of (or deny) these phenomena (Estes 1979:14).

Objectively, crisis and crisis tendencies are an integral part of the workings of the economy according to both classical and critical economists. Crisis tendencies occur with the slowdown or breakdown in profit-making, bad economic times, massive unemployment, and threats to financial institutions, as with the recession-depression unfolding from 2007 to the present. Most critical theorists argue that there is a "'logic of crisis' in crisis ridden growth and unstable accumulation"

(Bottomore 1983:102). Significantly, crises are potential avenues leading to societal transformation insofar as they "undermin[e] core organizational principles of society" (Bottomore 1983:102). This is the case, regardless of how crises are generated.

Symbolic and Material Impact of Crisis Construction

Crises are flashpoints for conflicts over the allocation and control of symbolic and material resources. *The symbolic impact* of the labeling of crises and other social problems is in the designation of cause and effect. Crisis construction with a negative symbolic and material impact is exemplified by portrayals of old people as unproductive drains on society. Linking the symbol of crisis to the perception of old people as *the problem* is a central dimension of age-related versions of crisis. As research on labeling theory illustrates, negative labels, when applied to any group, have consequences. The empirical case for lasting individual and psychological damage of racial and sexual stereotyping is well demonstrated. For older persons to be stereotyped as weak, sexless, toothless, worthless, and senile is similarly damaging (Estes 1979:12–14). In each case, personal self-esteem, efficacy, and individual performance are likely to be impaired. *Learned helplessness* (Seligman 1975) and the *looking-glass self* (Thomas [1923] 1969) describe the fact that negative perceptions, labels, and attributes ascribed to individuals in certain population groupings (and to the group at a systemic institutional level) not only provide for discrimination that is structured into our way of life, but also may teach the labeled parties to think and act as they are labeled (Mead 1934; Blumer 1969). A large body of literature on ageism and low self-efficacy attests to the destructive effects of such labeling.

The material impact of socially manufactured crises is lodged in the disparate societal and human consequences as crises: (1) generate a climate of uncertainty that legitimates the rejection of old and familiar assumptions, opening the way for previously unaccepted (even radical) solutions; (2) provide a sense of urgency to *do something* and prepare the public for the idea that sacrifices have to be made; and (3) generate public anxiety, thus expanding the authority of public officials to act or enact laws or regulations previously thought to be extreme (Edelman 1964). The disparity between those who are required to make sacrifices and those who are not is determined by those who exert the most powerful political muscle (Weber 1946; Estes 1979). Ultimately, the accumulation and disaccumulation of economic and other material advantages from crises are most likely to be unevenly distributed (Edelman 1964).

The State

The study of the state is central to the understanding of old age politics and policy and especially for the life chances of older people. The state has the power to allocate and distribute scarce resources to ensure the survival and growth of the economy, mediate between the different needs and demands across different social groupings (according to gender, race, ethnicity, age, class, [dis]ability, and sexuality), and ameliorate social conditions that could threaten the existing order (Estes 1991a).

Four theoretical frameworks are relevant to critical gerontology in studying the state, crisis, and old age policy: (1) critical theory of capitalism and the state (class theories), (2) feminist theories of the state, (3) theories of systemic racism (Feagin 2006) and racial formation (Omi and Winant 1994), and (4) theories on the political economy of aging and ageism (Estes 1979; Estes et al. 2009b). These are not mutually exclusive but rather highlight interlocking oppressions (Collins 1991) that are highly consequential for old age policy design.

Capitalism and the State

The capitalist state has three major functions which are contradictory according to a number of critical theorists (O'Connor 1973; Alford and Friedland 1985) and those writing on the political economy of aging (Estes 1991a, 1999; Estes and Associates 2001; Phillipson 2006; Walker 2006) and the welfare state and aging (Myles 1984; Pampel 1998).

The first and major function of the state is to ensure conditions favorable to economic growth and private profit. The ability of the U.S. state to ensure favorable conditions for capital accumulation within its borders is challenged by the globalization of capital. Contemporary issues concern the extent and effects of the decline in state power and control over key policy levers designed to ensure continuing U.S. corporate hegemony over the global economy.

The second state function is to ensure continuing social harmony across the nation and the undisturbed legitimacy of the existing social order (the status quo). The state does this by alleviating or mediating the worst effects of problems generated by the free enterprise system (e.g., unemployment, hunger, and destitution of those who are left behind by, or expelled from, the market). As Piven and Cloward (1971) argue, the state intervenes in social problems and wrests social control in order to stem potential social unrest.

A contradiction arises between these first two-state functions, each of which requires the expenditure of increasingly large public resources that must be raised through taxation. Jointly (as in the Great Recession of 2008), the cost of maintaining these two functions may spend the state into fiscal crisis (O'Connor 1973). Paradoxically, when federal bailouts are needed most, there is less access to tax revenue, as free market ideologues rage against taxation.

The third function of the state is to ensure continuation of democracy and protection of the democratic process (Alford and Friedland 1985). Responsibility for meeting the needs of the human community and the assurance of opportunities for human development are associated principles. Many critical scholars cite the inherent tensions between the state's democratic function and the state's function of ensuring conditions favorable to the advancement of capital accumulation (Myles 1984; Alford and Friedland 1985). This tension is specifically illuminated in issues surrounding the economic and health security of the aged. Pampel's observation that "national differences in pension policy programs (and the welfare state more generally) reflect two basic and fundamentally contradictory principles" (1998:114–15) is supported by Myles (1984). One principle, according to Pampel (1998), is based on:

> the growth of capitalist economies [and] reflects the logic of the free market [where] pensions represent wages set aside for old age in place of current payment and should reflect wage differences during the work life of pensioners … The other principle stems from the presence of political democracy in capitalist economies. Democracy emphasizes equality and citizenship rights rather than inequality in wages and salaries … Democratic pressures for equality … tend to favor universal programs for pensions and other social needs rather than programs based on economic contributions. (Pp. 114–5)

Feminist Theories of the State

Acker (1988) argues that theories of the state and social class that do not explicitly address the subordination of women and the *privileging of men* fail as comprehensive frameworks for understanding social phenomena. Social class is produced through gendered processes that are structured by production and distribution. Distribution, in particular, is vitally affected by the dominance of market relations as the basis of distribution; however, distribution is also affected by the indifference of the economic system to the reproduction of the working class and to the demands of working class daily life (Acker 1988), the demands of which fall most heavily on women.

Quadagno (1994) faults class and state theory for its inattention to the role of state policy in mediating race relations and for its blindness to "a defining feature of social provision: its organization around gender" (p. 14). Connell (1987) observes that the state not only regulates institutions and relations like marriage and motherhood but also manages them. The state actually constitutes "the social categories of the gender order" as "patriarchy is both constructed and contested" through the state (Connell 1987:130).

Women's dependency has shifted from the man to the state (Brown 1995) and from the family to the state (Dickinson and Russell 1986; Estes 1991a, b; Orloff 1993). "Instead of private patriarchy dependent on a husband, women are subject to public patriarchy of a paternalistic state" (Lorber 1998:44). This situation renders women of all ages, particularly older women and younger, single mothers, extremely vulnerable to state policies that are subject to politically charged partisan conflicts and uncertain policy changes and swings. The outcomes of these struggles subject those who are most dependent upon the state (ethnic minorities, women, low-income individuals, the disabled, and elders) to erratic and sometimes radical and regressive policies. For the large majority of older women dependent upon the state, there is the triple threat of *cumulative disadvantage* which is brought on by a lifetime of sex-gender and racial-ethnic discrimination that is further exacerbated by widowhood, divorce, living alone, ill health, and the *spend down* of financial resources in old-old age (Crystal and Shea 2003; Dannefer 2003; O'Rand 2003; Harrington Meyer and Herd 2008). Current, threatened *entitlement reforms* imperil recipients of Social Security, Medicare, and Medicaid who are the most significantly affected (e.g., women, ethnic and racial minorities, and the oldest old).

The Racial State

In their work on racial formation theory, Omi and Winant (1994) contend, "the state is inherently racial. Far from intervening in racial conflicts, the state is itself increasingly the preeminent site of racial conflict" (p. 82). The racial state consists of institutions, policies, conditions, and rules backed by "legitimating ideologies" enforced by a "repressive apparatus" (Omi and Winant 1994:84–5). Racial politics and racially based social movements engage with the state over time to produce racial formation, through patterns of conflict and accommodation. These struggles and outcomes are linked to hegemony, "the way in which society is organized and ruled" (Omi and Winant 1994:56). Macrolevel political processes are linked to the microlevel, everyday experience; "[t]o recognize the racial dimension in social structure is to interpret the meaning of race" (Omi and Winant 1994:57). Socio-historically, the meaning of race is fluid and a source of contest.

At its outset, Social Security excluded domestic workers and migrant labor from coverage and consigned African-Americans to punitive and highly variable state level welfare policies and programs. Poole (2006) identifies the shared interest of white policy makers in "preserving white race privilege," illustrating the "existing marginality of the country's 12 million African Americans … [since] Social Security rewarded those who were already privileged" (p. 174).

Contradictions and Crisis Tendencies in the State

Over the past 30 years, critical scholars (Offe 1973; Habermas 1975; O'Connor 1973, 1984, 1987) have concurred that the contradictions between the different functions of the state (support for the processes of capital accumulation, social integration, and democracy) produce crisis tendencies that are an integral part of late capitalism. A major conundrum is that resources to pay for the state infrastructure and functions (taxes and other revenues) are dependent upon the success of the capitalist economy and the willingness of its economic dominants to be taxed (O'Connor 1973; Offe and Ronge 1982). The Bush tax cuts of 2000 exemplify such difficulties: Revenues lost from the tax cuts of 2000 are a substantial cause of the U.S. deficit in 2010, which in turn provides fiscal hawks with a rationale for cutting Social Security.

Legitimacy and Legitimation Crises

The fate and ultimate viability of all institutions and organizations are contingent, to a major degree, upon their legitimacy (Weber 1946). To understand crisis and old age policy and politics, attention must be given to the legitimacy of the state, to the nation's economic and democratic institutions, and to civil society. Without legitimacy, institutions may be defined as being in crisis.

Legitimation refers to "the process by which 'power is not only institutionalized but more importantly is given moral grounding'" (Marshall 1998:363). It is "the authority of an institution, person, or practice to command obedience" (Silby as cited in Turner 2006:332). For Habermas (1975, 1984), legitimacy means hegemony, although once challenged, true claims of legitimacy are undermined. *Political legitimacy* is "the quality of 'oughtness' that is perceived by the public to inhere in a political regime. That government is legitimate which is viewed as morally proper for a society" (Merelman 1966:548).

A feature of modern capitalist societies is that there are multiple sources of legitimacy for its major institutional sectors that include the state, capital, and democracy (including civil society). Public activity is legitimated almost universally for functions of the police and national security, while profit-making entities, historically, have been deemed inappropriate for these purposes. An increasingly contested sector is the nonprofit sector (NPS). To the extent that the NPS represents pluralism (voluntary and participatory action), the state's support has been a significant source of legitimation. However, since the early 1980s, state support of the NPS has eroded as pressures to fund for-profits versus nonprofits have accelerated, and the state has bolstered this trend under Medicare, the Older Americans Act, and other state policies (Estes et al. 1984; Estes, Alford, and Egan 2001).

The State as Site of Economic, Political, and Social Struggle

Systemic crises arise not only from the sphere of production but also from the inability of the political system to prevent or compensate for economic and social crisis tendencies that inhere in (and/or are exacerbated by) capitalism. Critical scholars generally concur that under modern capitalism the struggles between workers and capital (traditional class struggle) have been transported into the state. Concurrently, other major social struggles also have increasingly been transported into the state for resolution, heightening challenges to the legitimacy of the state on multiple fronts (O'Connor 1973; Offe 1973; Habermas 1984). It is not surprising that in the present socio-historical moment social struggles involving the economy and race, ethnicity, gender, and age are at a fever pitch, embroiling the state, congress, the presidency, political parties, and the public in intense ideological conflicts. With both economic and social crises together displaced into and within state administration, additional political crises may be produced, as the state is required to:

> organize the dysfunctional social consequences of private production, [while] state policy is not supposed to infringe on the primacy of private production. [The contradiction is that] if state policy is to be adequate, however, it is forced to rely on means which either violate the dominant capital relation or undermine [the state's] functional requirements—the legitimacy and administrative competence of state regulation itself. (Offe 1977)

Critical Sociology, Crisis, and Old Age Policy

Estes (1979, 1982) and others (Estes and Associates 2001; Walker 2006; Svihula and Estes 2009; Phillipson (2010)) have identified the import of the social construction and reproduction of crises surrounding the aging society and old age policy, including Social Security. Gramsci's construct of hegemony underscores the crucial roles of discourse, rhetoric, and speech that frame policy, its processing, and its management. Social problems related to age, gender, race, home, family, women's reproduction rights, and access to work are increasingly being adjudicated through the state (Estes and Associates 2001:92, 97; Harrington Meyer and Herd 2008).

Recent socio-cultural theorizing includes postcolonial and postmodern theories that underscore the significance of identity politics of race, ethnicity, sexuality and gender, (dis)ability, and age and generation through new social movements. Critical gerontological scholarship on identity, alienation, and meaning has proliferated in the last two decades with exploration of the stability and fluidity of aging identities and the relationship between biological and social aspects of aging (Biggs 2001, 2003), as has work on the concept of the *colonization of the life world* (Habermas 1986:392).

Baars (2006), Dannefer (2006), Walker (2009), and others offer trenchant critiques of postmodernist work in aging that promotes the *individualization of the social* (Ferge 1997), emphasizing "presumably autonomous preferences and choices" to the neglect of profound "problems of social inequality ... material reproduction ... in favor of conflicts that arise in the domain of cultural reproduction" (Baars 2006:26–7).

Crises and Old Age Policy

Old age politics and policy have both provided a series of institutional buffers and served as a battleground for framing, defining, and resolving different crisis tendencies in past and present socio-historical moments. Construction of the aging population in crisis terms serves two ideological purposes. First, the "demographic imperative" has been a rallying point for those who argue that the elderly are living too long, consuming too many societal resources, and robbing the young, thus justifying rollbacks of state benefits for the aged. Second, ideological attacks on health care as an earned right for the elderly use projections of a pandemic of chronic illness – yet another version of crisis.

The social contract embodied in the founding of Social Security is based on the commitment that we each take responsibility and contribute to universal earned benefits through our wages, in order to insure ourselves and our families against the unpredictable events of death of a parent or spouse or of being orphaned, disabled, or retired. Kingson, Cornman, and Torre-Norton (2009) emphasize the intergenerational ties implicit in programs like Social Security, in that "[t]he social insurance approach rests on an understanding that generations depend on each other; they are interdependent. It gives concrete expression to values supporting mutual aid and the obligation to care for our families, our neighbors, and ourselves." (p. 107); furthermore, "[t]he debate is about how much added risk [that] society, through public policies and private practices, wants to place on families now and into the future, and the place of our country's current social insurance programs like Social Security and Medicare in that balance" (p. 95).

Contested ideas about what one generation owes another generation and pointed questions about fairness between different age cohorts penetrate and shape political discourse. The specter of *intergenerational war* emerged as an important product of the Reagan years (Estes 1991b:75). A reenergized *gender war*, Estes (2004) asserts, also "laid the affective base for increased pressures for family responsibility," that is, women's free labor over a lifetime of never-ending caregiving for multiple generations. At the center of these conflicts are what may be characterized as a fierce attack on the nation's bedrock entitlement programs of Social Security and Medicare.

Social Security and Medicare are the U.S. welfare state, according to Myles (1984), Quadagno (1999), and Walker (1999). Hence, legitimacy problems that attend the state would be expected to extend to programs of the state. Social Security and Medicare are now the focal point of intense struggles of class, gender, race, and generation, involving both the corporate sector and the state. Deeply implicated in these struggles are the cultural institutions of media, religion, and education (intellectuals' think tanks).

Corporate demands that the state redirect its resources away from safety net programs and earned rights, such as Social Security, and instead press for deficit reduction and additional state subsidies for the financial and other sectors of capital are a response to economic crisis tendencies as well as to the crises of administration, legitimation, and rationality described by Habermas (1975) and others (O'Connor 1973; Ronge 1974; Offe 1977; Offe and Keane 1984). Since these crises are interrelated (although each may each have its own trajectory), the state must respond in some measure to the appearance of each.

Market Fundamentalism, Crisis, and Aging

Myles and Boyd (1982) underscore the economic significance of a *graying America*, in that it "increases the size of the public economy and reduces the share of the national income directly subject to market forces. Thus, while population aging is unlikely to break the "national bank," it will alter the bank's structure of ownership and control" (p. 19). This may explain the relentless attacks on Social Security. Market ideologists contend that Social Security reduces the public's reliance on the market, increases the individual's dependency on government, and diminishes incentives for private sector saving and investing that are a major source of the capital deemed essential for economic growth (Rahn and Simonson 1980). The fact that the retirement wage in the form of Social Security is administered by the state (Myles 1991:304) means that market actors do not retain control over the deferred wage for distribution during retirement. Myles (1991) notes the fact that:

> public pensions are now the major source of income for the retired in all capitalist democracies … [is] an event of enormous significance in the … distributive practices of these nations … [and tantamount to] … an ever-growing and increasingly important portion of the national wage bill … removed from the market and subject to a democratic political process, one in which workers, in their capacity as citizens are able to claim a share of the social product … independent of … their capacity as wage earners. (Pp. 304–5)

In effect, this form of the retirement wage is actually "a citizen's wage, [with] an income entitlement partially independent of the commodity value of the worker's labor power" (Myles 1991:305). This fact mediates the raw discipline that is imposed by the market on wage workers.

Legitimacy Crises of the State, Capitalism, and Democracy

Old age and ageism are deeply implicated in the legitimacy crises of the state, capitalism, and democracy. Ageism has been virulent and violent, yet unrelenting for three decades, although rarely acknowledged. Ageism is intertwined with sexism, racism, class struggle, and abelism as ideological and structural forces. The socially constructed debt crisis is an identifiable vehicle by which a divisive war among opposing forces is underway.

> A giant window of potential social change has opened. And with it ... the "mother" of all class struggles, gender struggles, and race and ethnic struggles. The issues ... starkly underscore the crucial importance of social insurance and the peril of the choices that will be made concerning the nation's economic and health security. (Estes et al. 2009a:436)

Legitimacy Crisis of the State

Beginning with the vigorous claims of President Reagan (Estes 2004) that "government is the problem," legitimacy problems of the state have intensified with the incessant rhetoric by neoliberal politicians and think tanks that government is incompetent and/or inappropriate to deal with most (if not all) problems of the society. In the face of the pervasive market ideology, the state is confronted by real and profound problems in its attempt (rather than its inability) to effectively perform its steering functions, in view of the state's multiple and contradictory functions outlined above. Habermas (1975) defines the results of such a predicament as a *rationalization crisis* that contributes to legitimacy problems of the state, including charges that the state is incompetent.

Legitimacy Crisis of Capitalism

The "meltdown" of Wall Street and the Great Recession of 2008 are burned into the memory and psyche of millions of people around the globe as more than 14 trillion dollars in the wealth of individual Americans was destroyed. Bail outs for *too big to fail* banking, insurance, and Wall Street corporations cost trillions more. The recession itself and double-digit unemployment are an additional high state cost including millions of workers forced into early retirement and millions of more workers jobless and not paying into Social Security or Medicare. Uncertainty is particularly threatening to middle and working class Americans in view of the success of Market Fundamentalism in instantiating "individual autonomy as official state policy" (Dean 2003).

A core element of the legitimacy crisis of capitalism is the objective fiscal crisis of capitalism produced by the combination of deregulation, speculation, derivatives, corporate greed, and a host of perverse incentives, instruments, and contradictions initiated in the United States spread globally. Stiglitz (2010) describes this *freefall* as *the sinking of the world economy* (see also O'Connor 1987; Krugman [2007] 2009; Tabb 2008).

Legitimacy Crisis of Democracy

A major contributor to tensions surrounding the legitimacy of democracy in the United States is the inherent contradiction between democracy and capitalism (Myles 1984; Alford and Friedland 1985; Dean 2003).

To comprehend the relationships between crises and old age policy and politics, attention must be given to legitimacy issues confronting the nation's democratic institutions. In a democracy, all individuals existing within the geo-political jurisdiction of the democracy have equal and inalienable rights, regardless of property. Democracy accords rights of citizenship to everyone who is a member of the society. Recently, scholars note that the current sway of market fundamentalism is being used to alter the concept of citizenship rights to mean property rights. (Note the Supreme Court's historic and recent rulings on corporations with the rights of individuals.) From the neoliberal perspective, citizens are seen primarily as taxpayers (as are corporations) who have the right to decide what the state does with its resources, while the state is there to protect and advance the defense of the nation and the property rights therein (in contrast to the individual person and the democratic community).

Democracy is predicated on the principle that every individual and every vote counts: one person, one vote. Democracy assumes the will of the people and government by and for the people, all people. Basic equality across all members of the democratic society is a core value, belief, and principle. Capitalism is predicated upon economic competition to accumulate private wealth, producing inevitable inequality in the distribution of resources and the stratification of society. Survival of the fittest is an essential element of capitalism. In contrast, democracy is about the greatest good for the greatest number. Capitalism thrives on individual entrepreneurship and private shareholder *bottom line*, theoretically, in free markets, unencumbered by responsibility to larger social collective and communal responsibility embedded in the notion of *the common good*. In short, democracy relies on inclusion of all members of society, while capitalism relies on exclusionary privileges bestowed according to property and wealth and the power to dictate "choices" flowing wherefrom.

More direct challenges to democracy reside in mounting charges of voter fraud and disenfranchisement arising with the contested 2000 Bush presidential election, Diebold voting machine controversies, and accusations of vote suppression designed to intimidate people of color and ethnicity. Examples, accusations, and deep distrust abound: Florida's massive purge of eligible voters, Ohio's eight hour waits for voting, Georgia's photo IDs, and other mandates, such as Arizona's "Papers Please" law. These tactics exploit racial and other societal divisions (Krugman [2007] 2009:197). The Tea Party and screaming Town Hall meetings, reeking with disgust for members of Congress, is a symptom of legitimacy problems in democracy. Attacks on, and disrespect for, democracy and democratic institutions are both a cause and an effect of widespread public cynicism.

These signs of legitimacy crisis tendencies in the United States have serious implications for old age politics and policy. Economic and health security across the life course are key elements of citizenship (Grossman et al. 2009:115). This is consistent with the concept of the social rights of citizenship grounded in the notion of *life course interdependence* (Twine 1994). Indeed, the "quality of life in old age strongly reflects the different costs of life course interdependencies, of childrearing and employment [and long-term caregiving in old age], and these vary by social class and gender" (1994:34).

In opposition, the neoliberal ideology of individualism posits a different view of citizenship rights, supporting the concept that we are independent, *as if* independent of society. This is "a form of *freedom from society* or more specifically, *freedom from the state*" (Twine 1994:29). The reluctance of neoliberals to embrace the principle of *interdependence* stems from the fact that to do so "would present us with moral obligations to compensate those who bear the costs of our progress" (Twine 1994:29).

A critical approach to social policy and aging is consistent with Twine's assumption that "the material and human reproduction of society involves relations and processes of interdependence" (1994:29). It is appropriate for social policy on aging to reflect (and even more importantly, to account for – and where inequitable burdens arise – to compensate for) these life course and society-wide interdependencies. "The burdens of not doing so are profoundly unequally borne" (Estes et al. 2003:143).

Somers (2008) observes that "access to public services and social insurance are institutionalized expressions of rights internal to the *inclusionary ethic of citizenship* and the common good" (p. 33). An "inclusive citizenship regime" (Somers, 2008:38) rejects the narrow concept of citizenship advanced by anti-immigration forces, which is both exclusive and xenophobic. Somers argues that the concept of *privatized citizenship* is tantamount to a market morality where individual responsibility is substituted for social morality – responsibility to the community and society and to the national and global commons.

Somers (2008) argues against the current threat to *the right to have rights* through *the contractualization of citizenship,* by which "the poor and low-income members ... experience *statelessness* in a particularly virulent way ... They ... become today's socially excluded. As such, they no longer have meaningful membership in civil society – that which confers recognizable human identity" (p. 134). This represents a "conquest by market powers of a state's institutions of inclusion, protections, and rights" (Somers 2008:135). Likewise, Dean (2003) critiques the *individual responsibility project* under capitalist social relations. Describing the deleterious effects upon both citizenship and democracy, she states that " ... self-reliance is demanded under threat of future impoverishment. The structurally irrelevant [workers/individuals] are just that: irrelevant. Their fate depends on the readiness of governments and their electorates to dole out ever more meager and grudgingly offered sustenance" (2003:162).

Debt Crisis Framing and Old Age Politics

A frame of *apocalyptic deficits* (author's term) is the new mantra applied by opponents of government spending on health care, the social safety net, jobs, and Social Security. The deficit *rant* of fiscal conservatives is compatible with the *generational equity* argument (Williamson, Watts-Roy, and Kingson 1999) that is being advanced by the Social Security opponents who argue that elders are getting more than their fair share of benefits from the program. As these Social Security opponents frame arguments within terms of generational equity and the economic unsustainability of this bedrock social insurance program, little or no attention is given to income adequacy during old age or to how retirees can sustain a habitable living standard. It is noteworthy that 20% of all retirees and 40% of retired African-Americans and Hispanic-Americans count on Social Security for 100% of their annual income, while two-thirds of all Americans depend on Social Security for more than 60% of their retirement income (U.S. Social Security Administration 2009).

The successful deployment of deficit crises, structurally produced and politically constructed by the state and capital from the 1980s to the present, underscore the inherent tendencies to the legitimacy problems of the state, capitalism, and democracy. Such crises, objective and subjective, are an effective mechanism that calls into question the legitimacy of the state "and even of democratic actions," particularly at the federal level. Utilizing Habermas's (1975) framework, the deficit situation foregrounds the steering and rationalization crisis tendencies of the state in a capitalist society. In the case of the United States, legitimation crisis tendencies inhere in the failed state oversight and regulatory powers regarding finance capital of gigantic proportions and offshore oil drilling. The confluence of multiple crisis tendencies operating at several levels of society and the state may produce a near-toxic extreme crisis.

Presently, such crisis tendencies (however produced) have provided the rationale for sustained ideological attacks and deficit crisis rhetoric aimed at retrenchment in welfare state commitments to the American public. This "debt crisis" is often blamed on older U.S. citizens who had nothing to do with it. The primary solution proposed is to cut the only program that has not contributed to the deficit, Social Security (currently with a surplus of $2.6 trillion). The entrenchment of the *deficit crisis frame* on both sides of the aisle (in Congress, the White House, corporate America, and the media) reflects the strength of adherents of market fundamentalism who unalterably opposed to safety net domestic spending.

With the debt crisis framing and the legislative and administrative wrangling over "entitlement reform," it appears that the state's legitimation problems are being "hived off" from the state writ-large. With the support of market ideologists and fearful deficit hawk democrats, the primary alternative solutions are being guided away from the White House and pressure for tax increases. Instead, the preferred solutions are cuts to the specific entitlement programs of Social Security,

Medicare, and Medicaid. Significantly, as well, federal entitlement programs are defined as budgetary problems to be solved by budgetary means only (Quadagno 1999) rather than as programs designed to provide economic security and a collective community of pooled social risk at times of grave economic vulnerability for the majority of the nation's middle class during the demonstrated failure of the market.

The attention of policymakers, economists, and the media to the conflicts and challenges facing Social Security has shaken the confidence of many Americans that Social Security will be there for future generations. Nevertheless, there remains strong consensus that Social Security is crucial, especially in the U.S. market economy during precarious economic times. Rockeymoore and Maitin-Shepart (2010) confirm a long line of prior research, showing that:

> Americans want to make sure that Social Security is strengthened for current and future generations … [More than 65 percent of Americans] … support strengthening Social Security over cutting its benefits … [while] 88 percent of African Americans, 84 percent of Hispanics, and 74 percent of whites agree that preserving Social Security for future generations is critical, even if it means increasing Social Security taxes on workers. (P. 1)

In summary, the legitimacy problems of the state, capitalism, and democracy have been transported into all aspects of social policy for the aging, and particularly targeted against the bedrock programs of Social Security and Medicare. The state's responses to its legitimacy problems are numerous and the source of conflicts among contending parties. Responses include: the imposition of deficit reduction ideology and law; tax cuts for the wealthy and the call for more of them; funding cuts in politically weak programs; policies to stimulate the market; erosion of government entitlements; devolution of federal fiscal responsibility; deficit reduction to constrain social spending; and a health policy and health reform agenda of market stimulation, privatization, for-profit managed care, and individual responsibility (Estes 1991b). A capstone illustration of the attacks on the legitimacy of the state came from the ranking minority member of the Senate Finance Committee, Senator Charles Grassley, as he declared during the Health Care Reform debate that "[g]overnment is not a competitor for health insurance. With a public option, government is a predator" (PBS News Hour Online 2009).

Thirty Years of Critical Thought: The Future

Future scholarship in the sociology of aging, crisis, and the attendant policy and politics of old age is paramount. Despite advances in work on diversity and risk, including the cultural turn of postmodernism, unfolding global and national events and conditions underscore the urgent need for frameworks to guide research on the construction and uses of crises in the welfare state politics of societal aging and their distributive consequences. The pressures are to "responsibilize a new senior citizenry to care for itself" (Katz 2003:26) and to engage in "productive aging" by "aging well." Failure to meet such expectations threatens to provoke negative moral sanctions (Estes, Mahakian, and Weitz 2001; Holstein 1993; Estes et al. 2003).

A weakness in the sociology of aging over the past three decades is its inadequate attention to macrolevel societal forces that are transforming the experience of growing old and the responses of societal institutions to these transitions. In the developed West, normative aging patterns involving work, family, and leisure are seriously challenged by economic and political events (and struggles) that produce subjectively and objectively rendered crises.

New challenges to both "usual aging" and "successful aging" (Rowe and Kahn 1997) of older persons stem from: forced early retirements caused by high unemployment; the need to work well into old age as private pensions and home equity crash; low interest rates to savings; augmented resource transfers from old to young due to the bleak economic situation of young and middle age adult children; and increased, unceasing caregiving work and costs of long-term care and grandparent caregiving. These shifts are occurring in the historical moment of power struggles over the social position, rights, responsibilities, and the decline in the perceived deservingness of older people. Research in needed on:

- Theorizing the aging society, ageism, and crisis
- Globalization, state theory, and social rights
- Social inequality and social policy, age and intergenerational relations
- Social movements and old age politics
- Praxis and public sociology in old age politics and policy

Theorizing the Aging Society, Ageism, and Crisis

The idea of society as a bounded, self-sufficient entity (a theme most associated with functionalist theory and Talcott Parsons (1951)) is generally assumed, within mainstream sociology of aging, as being built around questions of how social integration in old age is possible given the shocks of late life (e.g., retirement, bereavement, and chronic illness). Such formulations assume a coherent and bounded society into which integration is attainable, an idea rendered problematic by globalization. Minimally, it is essential to recognize the diverse and unequal networks which interact throughout the life course to produce differences in national, global, and subcultural aging (and for some, the wish or the probability of not aging at all).

A cornerstone of the political economy of public policy and aging is the unpacking of social issues that are framed as crises and remedies to deal with the dominant and marginalized subgroups within the aging population. A second cornerstone is concerned with the processes by which – and for whom – these frames, definitions, and ideologies serve as political, economic, and cultural determinants of the policy agenda, policy formation, and policy implementation affecting the aged.

A task for sociological theory is to examine the national and global power dynamics that perpetuate the dominant, competing, and repressed (Alford 1975) frames of old age, aging, and ageism. Priority areas include studies of discursive constructions of aging, intergenerational relations, and associated crises; the politics of representation; and the emerging (and dissembling) forms of human and political agency, especially the effects on social policy and societal aging.

Frameworks and research must advance a creative amalgam of two streams of the sociology of aging: *agency and structure*, as linked to crises and social policy. Without a valid account of social structure, a decidedly microfoundational approach to aging is isolated from the effects of period, cohort, and history (Hagestad and Dannefer 2001). At the same time, macro approaches that ignore the everyday lives of people also offer limited insight into human development, self-reflection, and real, existing human agency and emancipation.

The intersection of agency and social structure is acknowledged in Marshall's (2009) perspective on *the life course as a policy tool*, particularly in reference to cumulative advantage and disadvantage. Scholarship is needed to explicate the successes and limits of agency, particularly since the realization of agency is key to social change. "Situational events and interactional opportunities" for agency occur within the context of "structural constraints that limit the range of possible interactions … reinforcing certain lines of action while barring others" (Estes 1981:400).

In addition, research on ageism and its relationship to old age politics and policy is essential. There has been little detailed empirical work on the topic either at the meso- or macrolevel. Measurement of ageism and its empirical and theoretical specification are necessary at the level of particular social institutions and at the broader social system level. Attitudinal and public opinion research on the aging society, policy, and crisis is similarly lacking and necessary. Media studies are needed to investigate the framing, rhetoric, and ideology used in political discourse via congressional testimony, government reports, think tanks, and the debates of political, economic, and cultural elites (Svihula 2005; Svihula and Estes 2007).

Globalization, State Theory, and Social Rights

The phenomenon of globalization renders many problematic concepts employed by sociologists, such as the collective and individual meanings of the state, citizen, gender, social class, ethnicity, and generation. Each of these concepts is substantially challenged at the level of global actors and institutions (Bauman 1998). With some exceptions, older persons have been marginalized or made invisible in research on globalization (Estes and Phillipson 2002; Vincent 2006; Phillipson 2009, 2010; Walker 2006, 2009) except for the perceived threat of demographic aging.

The social production of crisis as a vehicle for promoting political and economic objectives has taken center stage in old age politics as global shifts portend a world in which growing old is becoming an increasing risk, threatened by detachment from the protective welfare states that once comprised the core of Western society (Phillipson 2009). The paradox for older, as well as younger, generations is that the macrolevel has become *more* rather than less important as a factor influencing daily life.

Major areas for investigation must concern the effects on elders and other vulnerable populations of: the changing and contested forms of the nation state and its welfare provisioning; threats to democracy posed by nationalism and issues surrounding citizenship; the enhanced and largely opaque roles of globalization and supranational bodies which remain unencumbered by formal democratic processes; the concentration and power of financial capital and multinational corporate interests; and the persistence and emergence of new forms of class-, gender-, race-, ethnic-, (dis) ability-, generation-, and age-based inequalities and divisions. The new social construction (and contradiction) of old age and aging societies is to socially reconstruct old age as a personal rather than collective responsibility, on the one hand, and to wring our hands about global aging, on the other. Nevertheless, the risks associated with old age and aging are relatively unchanged and increasing only in intensity and severity in terms of the threat of poverty, the need for long-term care, and the likelihood of serious illness. What has *changed* is that the duty and the necessity to cope with these risks is being transferred to individual elders and individual families. These shifts in risk apply particularly to individual elders in attempting to sustain finances during old age and to women as caregivers (Bauman 2000; Hacker [2006] 2008; Estes 2008; Harrington Meyer and Herd 2008; Settersten and Trauten 2009).

Aging as a global phenomenon is transforming developed as well as developing nations (Estes and Wallace 2010). It is imperative to identify the ways in which global institutions and global governance (chaotic or not) are employed to promote (or not) the needs and rights of older citizens. Studies are needed not only in single societies but also comparatively in the global community of the first, second, and third worlds.

Social Inequality and Social Policy, Age and Intergenerational Relations

Calasanti (2009) observes that age relations encompass "the ways …age serves a social organizing principle such that different age groups gain identities in relation to one another" and that "being old, in and of itself, confers a loss of power of all those designated as 'old' regardless of their possible advantages on other social hierarchies" (p. 475), for example, labor market discrimination, ageism, and gender-based lower job opportunities for women. This is a reminder that work on social inequalities and social policy must incorporate, conceptualize, and explicate frameworks for age relations and for intergenerational relations. Theoretical developments need to incorporate the excellent developments already made on intersectionality and interlocking oppressions (Collins 1991) that are essential in the study of inequality and difference.

A critical perspective theorizes and interrogates power relations and the social, economic, and political processes and structures that are institutionalized and insinuated into everyday encounters and relationships between older adults and other generations and between older adults and the professionals who interact with them. Links must be made between macro, meso, and microlevels of analysis so that the sources and processes of the production and reproduction of social inequalities may be fully delineated. Only then will it be possible to identify the potential malleability of persistent inequalities in old age to policy or other social interventions.

Sociological work on inequality and aging, as developed by scholars of cumulative advantage and disadvantage theory (Crystal and Shea 1990; Crystal and Waehrer 1996; Dannefer 2003; O'Rand 2003, 2006), has been advanced by the *Cumulative Inequality Theory* (CIT) of Ferraro et al. (2009). The axioms, propositions, and methods of CIT are promising in light of the authors' insistence that social systems are a vital part of the production and reproduction of inequality across the life course. Also, the theory attends to life course trajectories, population aging, and "cohorts [that] provide the context for development, structuring of risks, and opportunities" (Ferraro et al. 2009:419).

There remains a need for understanding how social inequalities in old age may be exacerbated or ameliorated in society and addressed through social policy (J. Angel and R. Angel 2006; R. Angel and J. Angel 2007). Scholarly research will be enriched by attention to two macro areas: (1) the roles of dominant political, economic, and culturally elite organizations and institutions as they manufacture, reproduce, perpetuate, change, and/or exploit social divisions in the society; and (2) the role of the state and of global politics and economics in these processes.

A cautionary note is needed. Current social and policy sciences studies of inequalities in aging (generally) take as "given" the existing systems of medicine under capitalism, as scholars work largely within "definitions of the situation" framed by classical economic paradigms and functionalist sociological assumptions, models of cost effectiveness, and individual level outcomes (Clarke et al. 2003). The end result is that such investigations consider only a limited array of potentially viable policy options; thus ensuring, almost exclusively, the serious consideration of primarily incremental (status quo) changes that will do little to alter underlying root problems that confront older people or the societies in which they age.

Social Movements and Old Age Politics

Social movements, as a broad social alliance of persons who seek to effect or block an element of social change within a society, have been instrumental in U.S. politics and policy related to aging from the nation's earliest days. Much of such social movement literature focuses on the Townsend Movement of 1933, the Social Security Act of 1935, and the 1965 passage of Medicare, Medicaid, and the Older Americans Act. Binstock (1972), Marmor (1973), Estes (1979), Quadagno (1988), Campbell (2003), and Amenta (2006), Schultz and Binstock (2006) have produced many of the best treatises on policy and aging. Campbell (2003) argues that senior organizations were not influential in creating Social Security but rather were an outgrowth of the legislation itself. The result was the development of a constituency of interest group entrepreneurs with increased income, time in retirement, and political interests and opportunities. The founding of AARP in 1958 is seen as a secondary effect of policy. Amenta (2006) contends that the Townsend Movement was formative in producing Social Security. His major work focuses on identifying the short- and long-term consequences of social movements.

The growth in the number and scope of providers and organizations in "the aging enterprise" (Estes 1979) after 1965 coincided not only with Medicare and Medicaid but also with the Older Americans Act. As the numbers of senior organizations expanded throughout the 1980s and 1990s, they became synonymous with the *gray lobby* (Pratt 1993) and are presently loosely organized under the national umbrella of the Washington, DC-based Leadership Council of Aging Organizations (LCAO). Social movement theories have embraced multiple potential explanations for their trajectories, including struggles over meaning and interpretations, identity politics, resource mobilization, and political opportunity structures, among others (McAdam, McCarthy, and Zald 1996). As Hudson (2009) observes, the "aged's new policy and political presence will represent an impressive counterweight against those who would pare away existing benefits" (p. 550). Highlighting the import of continuing research in this area are the deficit crisis and swirling attacks on Social Security and entitlements for the aged.

Praxis and Public Sociology in Old Age Politics and Policy

Praxis is a key element of critical and public sociology in aging. Critical gerontology supplies evidence and perspectives designed to assist groups and individuals in assessing their situations to make the links between personal experience and structural inequities and to engage stakeholders in collective change. Praxis is "practical action on the world, or … the practical transformation of the world in accord with a desired end or finality" (Sartre 1960 as cited by Macey 2000:311). Praxis is a vital source of agency, calling attention to "the socially constructed nature of economic and social institutions and the possibility of changing these – humanity's capacity for freedom, which cannot be achieved entirely at the individual level" (Jary and Jary 2005:485).

Burawoy (2005) contrasts four types of sociological knowledge: professional (classical-traditional) sociology, policy sociology (e.g., applied and health services research), critical sociology, and public sociology. In mapping the division of sociological labor, Burawoy identifies critical sociology and public sociology as reflexive knowledge that "interrogates the value premises of society as well as our profession" (Burawoy 2005:11). Critical sociology is "a dialogue about ends within the academic community about the foundations of its research" (Burawoy 2005:11). Public sociology is a conversation "between academics and various publics about the direction of society" (Burawoy 2005:11). The "sociological field of power" is comprised of "a ruling coalition of professional and policy sociology and a subaltern mutuality of critical and public sociology," which Burawoy (2005) defines as "a pattern of domination … [reflecting that] in our society power and money speak louder than values and influence … and [has] a public sphere [that] is not only weak but overrun by armies of experts and a plethora of media" (p. 18). Being *subaltern* means that critical sociology and public sociology are subordinate, marginalized, and lower in status. Burawoy (2007) continues by defining critical sociology as "the conscience of professional sociology" and public sociology as "the conscience of policy sociology" (p. 33). Scholars of all sociological types have contributed, and will continue to contribute, to much that is understood about the relationship between the construction and consequences of the insertion of crisis discourses that drive old age politics and policy.

A major foundation for critical and public sociology in aging is directly attributable to Maggie Kuhn as a sociologist and cofounder of the Gray Panthers. Maggie's speeches, writings, and activism contributed to the identification and early study of identity politics, intersectionality, cultural and media studies, the sociology of knowledge of gerontology, the political economy of aging, and global and world imperialism (Estes and Portacolone 2009). Kuhn and Tish Sommers (co-founder of the Older Women's League) demonstrated the real meaning of agency and public sociology in aging. For Maggie and Tish, praxis and possibility framed the adventure that was daily life for each of them (Estes 2008). They worked through living as public sociologists, as Burawoy describes, "in close connection with visible, thick, active, local and often counter-publics" (2005:7).

Serious attention in the sociology of aging must be given to the promise and limits of agency, because the realization of agency is key to public sociology and public gerontology (Estes 2008). Discourse, social constructions of reality, ideologies, and framing are integral to the relationship between meaning and action. Framing is all about power relations that eventually become embedded in the mind and the structure of our worlds (Estes 2008). "Framing itself is an action," (Lakoff and the Rockridge 2006:25), while "reframing *is* social change" (Lakoff 2004).

Organic intellectuals (Gramsci 1971) are integral to mobilizing for social change and to praxis in its pursuit. Intellectuals may serve roles in the production of both (or either) hegemony and resistance to its domination. They may work in relation to dominant or oppositional groups or classes that seek *a new hegemony* designed to generate widespread public consent for political and social reform. The promise of emancipatory knowledges and critical pedagogy in the sociology of aging, crisis, politics, and policy is possible through critical and public sociology. Critical gerontologists are already engaged in larger discourses, emancipatory knowledges, and critical pedagogy (Moody 1993; Katz 2001; Ray 2003; Baars, Dannefer, Phillipson and Walker 2006; Powell 2006; Estes and Phillipson 2007). Examples of emancipatory knowledges appear in the literature on regimes of truth that comprise critical feminist, race, ethnic, class, and generation discourses. Critical pedagogy engages education as the practice of freedom, possibility, and social transformation, much of it building off Paulo Freire's *Pedagogy of the Oppressed* (1972) and bell hooks's (1994) invitation to *teach to transgress*.

It would be remiss to conclude this chapter without acknowledging the deep epistemological and ontological tensions in sociology and other social sciences and humanities that are raised by some of the work considered here. Dorothy Smith (1990) warns of the danger of a social science that is captured by and envisioned through the "relations of ruling" and the limitations of gendered and raced knowledge and epistemologies. Work on feminist standpoint theory and the epistemic advantage (Harding 1996), queer theory, heteronormativity, and (dis)ability signal additional important developments for inclusion in the sociology of aging, politics, and policy. A blooming literature on the sociology of science (sociology of knowledges) includes work in grounded theory (Clarke 2005) and symbolic interactionism (Clarke and Star 2003) that attests to the vitality and significance of such concerns for the sociology of aging.

References

Acker, Joan. 1988. "Class, Gender, and the Relations of Distribution." *Signs* 13(3):473–93.

Alford, Robert R. 1975. *Health Care Politics: Ideological and Interest Group Barriers to Reform*. Chicago, IL: University of Chicago Press.

Alford, Robert R. and Roger Friedland. 1985. *Powers of Theory: Capitalism, the State, and Democracy*. New York: Cambridge University Press.

Amenta, Edwin. 2006. *When Movements Matter: The Townsend Plan and the Rise of Social Security*. Princeton, NJ: Princeton University Press.

Angel, Ronald J. and Jacqueline L. Angel. 2007. "The Health Care Safety Net for Hispanics." Pp. 263–279 in *The Health of Aging Hispanics: The Mexican-Origin Population*, edited by J. L. Angel and K. E. Whitfield. New York: Springer.

Angel, Jacqueline L. and Ronald J. Angel. 2006. "Minority Group Status and Healthful Aging: Social Structure Still Matters." *American Journal of Public Health* 96(7):1152–9.

Anti-Ageism Taskforce. (2006). *Ageism in America*. New York: The International Longevity Center-USA. Retrieved June 24, 2010 (http://www.ilcusa.org/media/pdfs/Ageism%20in%20America%20-%20The%20ILC%20Report.pdf).

Baars, Jan. 2006. "Beyond Neomodernism, Antimodernism, and Postmodernism: Basic Categories for Contemporary Critical Gerontology." Pp. 17–42 in *Aging, Globalization and Inequality: The New Critical Aging*, edited by J. Baars, D. Dannefer, C. Phillipson, and A. Walker. Amityville, NY: Baywood.

Baars, Jan, Dale Dannefer, Chris Phillipson, and Alan Walker. 2006. "Introduction: Critical Perspectives in Social Gerontology." Pp. 1–16 in *Aging, Globalization and Inequality: The New Critical Gerontology*, edited by J. Baars, D. Dannefer, C. Phillipson, and A. Walker. Amityville, NY: Baywood.

Bauman, Zygmunt. 1998. *Globalization: The Human Consequences*. Cambridge, UK: Polity Press.

———. 2000. *Liquid Modernity*. Cambridge, UK: Polity Press.

Biggs, Simon. 2001. "Toward Critical Narrativity: Stories of Aging in Contemporary Social Policy." *Journal of Aging Studies* 15(4):303–16.

———. 2003. "Negotiating Aging Identity: Surface, Depth, and Masquerade." Pp. 145–59 in *The Need for Theory: Critical Approaches to Social Gerontology*, edited by S. Biggs, A. Lowenstein, and Jon Hendricks. Amityville, NY: Baywood.

Binstock, Robert H. 1972. "Interest Group Liberalism and the Politics of Aging." *The Gerontologist* 12(3):265–80.

Blumer, Herbert. 1969. *Symbolic Interactionism: Perspective and Method.* Berkeley, CA: University of California Press.

Bottomore, Tom, ed. 1983. *A Dictionary of Marxist Thought.* Cambridge, MA: Harvard University Press.

Brown, Wendy. 1995. *States of Injury: Power and Freedom in Late Modernity.* Princeton, NJ: Princeton University Press.

Burawoy, Michael. 2005. "2004 American Sociological Association Presidential Address: For Public Sociology." *American Sociological Review* 70(1):4–28.

———. 2007. "For Public Sociology." Pp. 23–64 in *Public Sociology: Fifteen Eminent Sociologists Debate Politics and the Profession in the Twenty-First Century*, edited by D. Clawson, R. Zussman, J. Misra, N. Gerstel, and R. Stokes. Berkeley, CA: University of California Press.

Butler, Robert N. 1975. *Why Survive? Being Old in America.* New York: Harper & Row.

Calasanti, Toni. 2009. "Theorizing Feminist Gerontology, Sexuality, and Beyond: An Intersectional Approach." Pp. 471–485 in *Handbook of Theories of Aging*, 2nd ed., edited by V. L. Bengtson, D. Gans, N. M. Putney, and M. Silverstein. New York: Springer.

Campbell, Andrea Louise. 2003. *How Policies Make Citizens: Senior Citizen Activism and the American Welfare State*. Princeton, NJ: Princeton University Press.

Ciscel, David H. and Julia A. Heath. 2001. "To Market, To Market: Imperial Capitalism's Destruction of Social Capital and the Family." *Review of Radical Political Economics* 33(4):401–14.

Clarke, Adele E. 2005. *Situational Analysis: Grounded Theory After the Postmodern Turn*. Thousand Oaks, CA: Sage Publications.

Clarke, Adele E. and Susan Leigh Star. 2003. "Science, Technology, and Medicine Studies." Pp. 539–74 in *Handbook of Symbolic Interactionism*, edited by L. T. Reynolds and N. J. Herman-Kinney. Walnut Creek, CA: AltaMira Press.

Clarke, Adele E., Janet K. Shim, Laura Mamo, Jennifer Ruth Fosket, and Jennifer R. Fishman. 2003. "Biomedicalization: Technoscientific Transformations of Health, Illness, and U.S. Biomedicine." *American Sociological Review* 68(2):161–94.

Collins, Patricia Hill. 1991. *Black Feminist Thought: Knowledge, Consciousness, and the Politics of Empowerment.* New York: Routledge.

Collins, Randall. 1988. *Theoretical Sociology.* San Diego, CA: Harcourt Brace Jovanovich.

Connell, Raewyn W. 1987. *Gender and Power: Society, the Person, and Sexual Politics.* Stanford, CA: Stanford University Press.

Crystal, Stephen and Dennis G. Shea. 1990. "Cumulative Advantage, Cumulative Disadvantage, and Inequality among Elderly People." *The Gerontologist* 30(4):437–43.

———. 2003. "Introduction: Cumulative Advantage, Public Policy, and Inequality." Pp. 1–13 in *Annual Review of Gerontology and Geriatrics*, Vol. 22, edited by S. Crystal and D. Shea. New York: Springer.

Crystal, Stephen and Keith Waehrer. 1996. "Later-life Economic Inequality in Longitudinal Perspective." *Journals of Gerontology* 51B(6):S307–18.

Dannefer, Dale. 2003. "Cumulative Advantage/Disadvantage and the Life Course: Cross-Fertilizing Age and Social Science." *Journal of Gerontology: Social Sciences* 58B(6):S327–37.

———. 2006. "Reciprocal Co-Optation: The Relationship of Critical Theory and Social Gerontology." Pp. 103–20 in *Aging, Globalization and Inequality: The New Critical Gerontology,* edited by J. Baars, D. Dannefer, C. Phillipson, and A. Walker. Amityville, NY: Baywood.

Dean, Kathryn. 2003. *Capitalism and Citizenship: The Impossible Partnership.* London, UK: Routledge.

Dickinson, James and Bob Russell, eds. 1986. *Family, Economy & State: The Social Reproduction Process under Capitalism.* New York: St. Martin's Press.

Edelman, Murray J. 1964. *The Symbolic Uses of Politics.* Urbana, IL: University of Illinois Press.

Estes, Carroll. 2008. "A First Generation Critic Comes of Age: Reflections of a Critical Gerontologist." *Journal of Aging Studies* 22(2):120–31.

Estes, Carroll L. 1979. *The Aging Enterprise: A Critical Examination of Social Policies and Services for the Aged.* With contributions by P. R. Lee, L. Gerard, and M. Noble. San Francisco, CA: Jossey-Bass.

———. 1981. "The Social Construction of Reality: A Framework for Inquiry." Pp. 395–402 in *The Nation's Health*, edited by P. R. Lee, N. B. Ramsay, and I. Red. San Francisco, CA: Boyd & Fraser.

———. 1982. "Austerity and Aging in the United States: 1980 and Beyond." *International Journal of Health Services* 12(4):573–84.

———. 1991a. "The New Political Economy of Aging: Introduction and Critique." Pp. 19–36 in *Critical Perspectives on Aging: The Political and Moral Economy of Growing Old,* edited by M. Minkler and C. L. Estes. Amityville, NY: Baywood.

———. 1991b. "The Reagan Legacy: Privatization, the Welfare State, and Aging in the 1990's." Pp. 59–83 in *States, Labour Markets and the Future of Old Age Policy,* edited by J. Myles and J. Quadagno. Philadelphia, PA: Temple University Press.

———. 1999. "Critical Gerontology and the New Political Economy of Aging." Pp. 17–35 in *Critical Gerontology: Perspectives from Political and Moral Economy*, edited by M. Minkler and C. L. Estes. Amityville, NY: Baywood.

———. 2004. "Social Security Privatization and Older Women: A Feminist Political Economy Perspective." *Journal of Aging Studies* 18(1):9–26.

Estes, Carroll L. and Associates. 2001. *Social Policy & Aging: A Critical Perspective.* Thousand Oaks, CA: Sage.

Estes, Carroll L. and Chris Phillipson. 2002. "The Globalization of Capital, the Welfare State, and Old Age Policy." *International Journal of Health Services* 32(2):279–97.

———. 2007. "Critical Gerontology." Pp. 330–36 in *Encyclopedia of Gerontology: Age, Aging, and the Aged*, 2nd ed., Vol. A–K, edited by J. E. Birren. San Diego, CA: Academic.

Estes, Carroll L. and Elena Portacolone. 2009. "Maggie Kuhn: Social Theorist of Radical Gerontology." *International Journal of Sociology and Social Policy* 29(1/2):15–26.

Estes, Carroll L. and Steven P. Wallace. 2010. "Globalization, Social Policy, and Ageing: A North American Perspective." Pp. 513–24 in *The SAGE Handbook of Social Gerontology*, edited by D. Dannefer and C. Phillipson. London, UK: Sage.

Estes, Carroll L., James H. Swan, and Associates. 1993. *The Long Term Care Crisis: Elders Trapped in the No-Care Zone*. Newbury Park, CA: Sage.

Estes, Carroll L., James H. Swan, and Lenore E. Gerard. 1982. "Dominant and Competing Paradigms: Towards a Political Economy of Aging." *Ageing and Society* 2(2):151–64.

Estes, Carroll L., Jane L. Mahakian, and Tracy A. Wietz. 2001. "A Political Economy Critique of 'Productive Aging.'" Pp. 187–199 in *Social Policy & Aging: A Critical Perspective*. Thousand Oaks, CA: Sage.

Estes, Carroll L., Leah Rogne, Brian R. Grossman, Brooke A. Hollister, and Erica Solway. 2009a. "Epilogue: From the Audacity of Hope to the Audacity of Action." Pp. 435–39 in *Social Insurance and Social Justice: Social Security, Medicare and the Campaign Against Entitlements*, edited by L. Rogne, C. L. Estes, B. R. Grossman, B. A. Hollister, and E. Solway. New York: Springer.

Estes, Carroll L., Leah Rogne, Brian R. Grossman, Erica Solway, and Brooke A. Hollister. 2009b. "Introduction: We're All in This Together: Social Insurance, Social Justice, and Social Change." Pp. xxv–xxxiii in *Social Insurance and Social Justice: Social Security, Medicare and the Campaign Against Entitlements,* edited by L. Rogne, C. L. Estes, B. R. Grossman, B. A. Hollister, and E. Solway. New York: Springer.

Estes, Carroll L., Lenore E. Gerard, Jane S. Zones, and James H. Swan. 1984. *Political Economy, Health, and Aging*. Boston, MA: Little Brown.

Estes, Carroll L., Robert R. Alford, and Anne Hays Egan. 2001. "The Transformation of the Nonprofit Sector: Systemic Crisis and the Political Economy of Aging Services." Pp. 61–94 in *Social Policy & Aging: A Critical Perspective.* Thousand Oaks, CA: Sage.

Estes, Carroll L., Simon Biggs, and Chris Phillipson. 2003. *Social Theory, Social Policy, and Ageing: A Critical Introduction.* Maidenhead, Berkshire, UK: Open University Press.

Feagin, Joe R. (2006). *Systemic Racism: A Theory of Oppression*. New York: Routledge.

Ferge, Zsuzsa. 1997. "The Changed Welfare Paradigm – The Individualization of the Social." *Social Policy and Administration* 31(1):20–44.

Ferraro, Kenneth F., Tetyana Pylypiv Shippee, and Markus H. Schafer. 2009. "Cumulative Inequality Theory for Research on Aging and the Life Course." Pp. 413–33 in *Handbook of Theories of Aging*, 2nd ed., edited by V. L. Bengtson, D. Gans, N. M. Putney, and M. Silverstein. New York: Springer.

Freire, Paulo. 1972. *Pedagogy of the Oppressed*. New York: Penguin Books.

Folbre, Nancy. 2001. *The Invisible Heart: Economics and Family Values*. New York: New Press.

Gramsci, Antonio. 1971. *Selections from the Prison Notebooks.* Edited and translated by Q. Hoare and G. Nowell-Smith. London, UK: Lawrence and Wishart.

Grossman, Brian R., Erica Solway, Brooke A. Hollister, Carroll L. Estes, and Leah Rogne. 2009. "One Nation, Interdependent: Exploring the Boundaries of Citizenship in the History of Social Security and Medicare." Pp. 115–47 in *Social Insurance and Social Justice: Social Security, Medicare and the Campaign Against Entitlements*, edited by L. Rogne, C. L. Estes, B. R. Grossman, B. A. Hollister, and E. Solway. New York: Springer.

Habermas, Jürgen. 1975. *Legitimation Crisis*. Translated by Thomas McCarthy. Boston, MA: Beacon.

———. 1984. "What Does a Legitimation Crisis Mean Today? Legitimation Problems in Late Capitalism." Pp. 134–155 in *Legitimacy and the State,* edited by W. E. Connolly. New York: New York University Press.

———. 1986. *Autonomy and Solidarity: Interviews with Jürgen Habermas.* Edited and introduced by P. Dews. London, UK: Verso Press.

Hacker, Jacob S. [2006] 2008. *The Great Risk Shift: The New Economic Insecurity and the Decline of the American Dream*. New York: Oxford University Press.

Hagestad, Gunhild O. and Dale Dannefer. 2001. "Concepts and Theories of Aging: Beyond Microfication in Social Science Approaches." Pp. 3–21 in *Handbook of Aging and the Social Sciences*, 5th ed., edited by R. H. Binstock and L. K. George. San Diego, CA: Academic.

Harding, Sandra. 1996. "Standpoint Epistemology (a Feminist Version): How Social Disadvantage Creates Epistemic Advantage." Pp. 146–60 in *Social Theory and Sociology: The Classics and Beyond*, edited by S. P. Turner. Cambridge, MA: Basil Blackwell.

Harrington Meyer, Madonna and Pamela Herd. 2008. "Is Poverty A Disappearing Problem for Older Women?" Op-ed, Huffington Post (Online), February 20. Retrieved June 24, 2010 (http://www.huffingtonpost.com/madonna-harrington-meyer-and-pamela-herd/is-poverty-a-disappearing_b_87684.html)

Holstein, Martha. 1993. "Productive Aging: A Feminist Critique." *Journal of Aging & Social Policy* 4(3/4):17–34.

hooks, bell. 1994. *Teaching to Transgress: Education as the Practice of Freedom*. New York: Routledge.

Hudson, Robert B. 2009. "From Industrialization to Institutionalism: Theoretical Accounts of Aging Policy Development in the United States." Pp. 531–53 in *Handbook of Theories of Aging*, edited by V. L. Bengtson, D. Gans, N. M. Putney, and M. Silverstein. 2nd ed. New York: Springer.

Jary, David and Julia Jary, eds. 2005. *Collins Web-Linked Dictionary of Sociology*. 3rd ed. Glasgow, UK: Collins.

Katz, Stephen. 2001. "Growing Older Without Aging? Positive Aging, Anti-Ageism, and Anti-Aging." *Generations* 25(4):27–32.

———. 2003. "Critical Gerontological Theory: Intellectual Fieldwork and the Nomadic Life of Ideas." Pp. 15–31 in *The Need for Theory: Critical Approaches to Social Gerontology*, edited by S. Biggs, A. Lowenstein, and J. Hendricks. Amityville, NY: Baywood.

Kingson, Eric R., John M. Cornman, and Amanda L. Torre-Norton. 2009. "The Future of Social Insurance: Values and Generational Interdependence." Pp. 95–108 in *Social Insurance and Social Justice: Social Security, Medicare and the Campaign Against Entitlements*, edited by L. Rogne, C. L. Estes, B. R. Grossman, B. A. Hollister, and E. Solway. New York: Springer.

Krugman, Paul. 2009 [2007]. *The Conscience of a Liberal*. New York: W. W. Norton.

Kuhn, Maggie and Gray Panthers. 1974. *New Life for the Elderly*. Philadelphia, PA: Gray Panthers.

Lakoff, George. 2004. *Don't Think of An Elephant! Know Your Values and Frame the Debate: The Essential Guide for Progressives*. White River Junction, VT: Chelsea Green.

Lakoff, George and the Rockridge Institute. 2006. *Thinking Points: Communicating our American Values and Vision: A Progressive's Handbook*. New York: Farrar, Straus, and Giroux.

Levitas, Ruth. 1986. "Competition and Compliance: The Utopias of the New Right." Pp. 80–106 in *The Ideology of the New Right*, edited by R. Levitas. Cambridge, UK: Polity Press.

Lorber, Judith. 1998. *Gender Inequality: Feminist Theories and Politics*. Los Angeles, CA: Roxbury.

Macey, David. 2000. *The Penguin Dictionary of Critical Theory*. London: Penguin Books.

Manning, Nick P., ed. 1985. *Social Problems and Welfare Ideology*. Brookfield, VT: Gower.

Marmor, Theodore R. 1973. *The politics of Medicare*. With the assistance of Jan S. Marmor. Chicago, IL: Aldine.

Marshall, Gordon, ed. 1998. *A Dictionary of Sociology*, 2nd ed. New York: Oxford University Press.

Marshall, Victor W. 2009. "Theory Informing Public Policy: The Life Course Perspective as a Policy Tool." Pp. 573–93 in *Handbook of Theories of Aging*, 2nd ed., edited by V. L. Bengtson, D. Gans, N. M. Putney, and M. Silverstein. New York: Springer.

McAdam, Doug, John D. McCarthy, and Mayer N. Zald, eds. 1996. *Comparative Perspectives on Social Movements: Political Opportunities, Mobilizing Structures, and Cultural Framings*. New York: Cambridge University Press.

Mead, George Herbert. 1934. *Mind, Self, and Society from the Standpoint of a Social Behaviorist*. Edited by C.W. Morris. Chicago, IL: University of Chicago Press.

Merelman, Richard M. 1966. "Learning and Legitimacy." *American Political Science Review* 60(3):548–61.

Moody, Harry R. 1993. "Overview: What is Critical Gerontology and Why is It Important?" Pp. xv–xli in *Voices and Visions of Aging: Toward a Critical Gerontology*, edited by T. R. Cole., W. A. Achenbaum, P. L. Jakobi, and R. Kastenbaum. New York: Springer.

Myles, John. 1984. *Old Age in the Welfare State: The Political Economy of Public Pensions*. Boston, MA: Little, Brown.

———. 1991. "Postwar Capitalism and the Extension of Social Security into a Retirement Wage." Pp. 293–309 in *Critical Perspectives on Aging: The Political and Moral Economy of Growing Old*, edited by M. Minkler and C. L. Estes. Amityville, NY: Baywood.

Myles, John F. and Monica Boyd. 1982. "Population Aging and the Elderly." Pp. 258–85 in *Social Issues: Sociological Views of Canada*, edited by D. Forcese and S. Richer. Scarborough, ON: Prentice-Hall.

O'Connor, James. 1973. *The Fiscal Crisis of the State*. New York: St. Martin's Press.

———. 1984. *Accumulation Crisis*. New York: Basil Blackwell.

———. 1987. *The Meaning of Crisis: A Theoretical Introduction*. New York: Basil Blackwell.

Offe, Claus. 1973. "The Abolition of Market Control and the Problems of Legitimacy." Working Papers on the Kapitalstate, I and II, Max Planck Institute, Saarbrucken, Germany.

———. 1977. *Industry and Inequality: The Achievement Principle in Work and Social Status*. Translated by James Wickham. New York: St. Martin's Press.

Offe, Claus and John Keane. 1984. *Contradictions of the Welfare State*. Cambridge, MA: MIT Press.

Offe, Claus and Volker Ronge. 1982. "Theses on the Theory of the State." Pp. 249–56 in *Classes, Power, and Conflict: Classical and Contemporary Debates*, edited by A. Giddens and D. Held. Berkeley, CA: University of California Press.

Omi, Michael and Howard Winant. 1986. *Racial Formation in the United States: From the 1960s to the 1980s*. New York: Routledge and Kegan Paul.

———. 1994. *Racial Formation in the United States: From the 1960s to the 1990s*, 2nd ed. New York: Routledge.

O'Rand, Angela M. 2003. "Cumulative Advantage Theory in Aging Research." Pp. 14–30 in *Annual Review of Gerontology and Geriatrics*, Vol. 22, edited by S. Crystal and D. Shea. New York: Springer.

———. 2006. "Stratification and the Life Course: Life Course Capital, Life Course Risks, and Social Inequality. Pp. 145–62 in *Handbook of Aging and the Social Sciences*, 6th ed., edited by R. H. Binstock and L. K. George. San Diego, CA: Academic.

Orloff, Ann Shola. 1993. "Gender and the Social Rights of Citizenship: The Comparative Analysis of Gender Relations and Welfare States." *American Sociological Review* 58(3):303–28.

Pampel, Fred C. 1998. *Aging, Social Inequality, and Public Policy*. Thousand Oaks, CA: Pine Forge Press.

Parsons, Talcott. 1951. *The Social System*. New York: Free Press of Glencoe.

PBS News Hour Online. 2009. "Grassley Outlines Obstacles Facing Health Care in Senate." Transcript originally aired on August 5, 2009. Retrieved June 24, 2010 (http://www.pbs.org/newshour/bb/health/july-dec09/health_08-05.html)

Phillipson, Chris. 2006. "Aging and Globalization: Issues for Critical Gerontology and Political Economy." Pp. 43–58 in *Aging, Globalization and Inequality: The New Critical Gerontology*, edited by J. Baars, D. Dannefer, C. Phillipson, and A. Walker. Amityville, NY: Baywood.

———. 2009. "Reconstructing Theories of Aging: The Impact of Globalization on Critical Gerontology." Pp. 615–28 in *Handbook of Theories of Aging*, 2nd ed., edited by V. L. Bengtson, D. Gans, N. M. Putney, and M. Silverstein. New York: Springer.

———. 2010. "Ageing and Urban Society: Growing Old in the 'Century of the City.'" Pp. 597–606 in *The SAGE Handbook of Social Gerontology*, edited by D. Dannefer and C. Phillipson. London, UK: Sage.

Piven, Francis Fox and Richard A. Cloward. 1971. *Regulating the Poor: The Functions of Public Welfare*. New York: Pantheon Books.

Poole, Mary. 2006. *The Segregated Origins of Social Security: African Americans and the Welfare State*. Chapel Hill, NC: University of North Carolina Press.

Powell, Jason L. 2006. *Social Theory and Aging*. With forward by Charles Lemert. Oxford, UK: Rowman & Littlefield.

Pratt, Henry J. 1993. *Gray Agendas: Interest Groups and Public Pensions in Canada, Britain, and the United States*. Ann Arbor, MI: University of Michigan Press.

Quadagno, Jill. 1999. "1998 American Sociological Association Presidential Address: Creating a Capital Investment Welfare State: The New American Exceptionalism." *American Sociological Review* 64(1):1–11.

Quadagno, Jill S. 1988. *The Transformation of Old Age Security: Class and Politics in the American Welfare State*. Chicago, IL: University of Chicago Press.

———. 1994. *The Color of Welfare: How Racism Undermined the War on Poverty.* New York: Oxford University Press.

Rahn, Richard W. and Kenneth. D. Simonson. *1980. Tax Policy for Retirement Programs*. In *Retirement Income: Who Gets How Much and Who Pays? (National Journal Issues Book)*. Washington, DC: Government Research Corporation.

Ray, Ruth E. 2003. "The Perils and Possibilities of Theory." Pp. 33–44 in *The Need for Theory: Critical Approaches to Social Gerontology*, edited by S. Biggs, A. Lowenstein, and J. Hendricks. Amityville, NY: Baywood.

Rockeymoore, Maya and Melissa Maitin-Shepart. 2010. *Tough Times Require Strong Social Security Benefits: Views on Social Security Among African Americans, Hispanic Americans, and White Americans*. Washington, DC: National Academy of Social Insurance. Retrieved June 24, 2010 (http://www.nasi.org/sites/default/files/research/SS_Brief_032.pdf).

Ronge, Volker. 1974. "The Politicization of Administration in Advanced Capitalist Societies." *Political Studies* 22(1):86–93.

Rowe, John W. and Robert L. Kahn. 1997. "Successful Aging." *The Gerontologist* 37(4):433–40.

Sassoon, Anne Showstack. 2001. "Globalization, Hegemony, and Passive Revolution." *New Political Economy* 6(1):5–17.

Schulz, James H. and Robert H. Binstock. 2006. *Aging Nation: The Economics and Politics of Growing Older in America*. Westport, CT: Praeger.

Seligman, Martin E. P. 1975. *Helplessness: On Depression, Development, and Death*. San Francisco, CA: W. H. Freeman.

Sen, Gita. 1980. "The Sexual Division of Labor and the Working Class Family: Towards a Conceptual Synthesis of Class Relations and the Subordination of Women." *Review of Radical Political Economics* 12(2):76–86.
Settersten, Richard A., Jr. and Molly E. Trauten. 2009. "The New Terrain of Old Age: Hallmarks, Freedoms, and Risks." Pp. 455–469 in *Handbook of Theories of Aging*, 2nd ed., edited by V. L. Bengtson, D. Gans, N. M. Putney, and M. Silverstein. New York: Springer.
Skocpol, Theda. (1992). *Protecting Soldiers and Mothers: The Political Origins of Social Policy in the United States*. Cambridge, MA: Belknap Press of Harvard University Press.
Smith, Dorothy. 1990. *The Conceptual Practices of Power: A Feminist Sociology of Knowledge*. Boston, MA: Northeastern University Press.
Somers, Margaret R. 2008. *Genealogies of Citizenship: Markets, Statelessness, and the Right to Have Rights*. New York: Cambridge University Press.
Soros, George. 1998. *The Crisis of Global Capitalism: Open Society Endangered*. New York: Public Affairs.
Stiglitz, Joseph E. 2010. *Freefall: America, Free Markets, and the Sinking of the World Economy*. New York: W. W. Norton.
Svihula, Judie and Carroll L. Estes. 2007. "Social Security Politics: Ideology and Reform." *Journals of Gerontology: Psychological Sciences and Social Sciences* 62B(2):S79–89.
———. 2009. "Social Security Privatization: The Institutionalization of an Ideological Movement." Pp. 217–31in *Social Insurance and Social Justice: Social Security, Medicare, and the Campaign Against Entitlements,* edited by L. Rogne, C. L. Estes, B. R. Grossman, B. A. Hollister, and E. Solway. New York: Springer.
Svihula, Judie Lynn. 2005. "Policy and Politics of Reform: Social Security in the United States." Ph.D. dissertation, School of Social Welfare, University of California, Berkeley, CA.
Tabb, William K. 2008. "Four Crises of the Contemporary World Capitalist System." *Monthly Review* 60(5):43–59.
Therborn, Göran. 1980. *The Ideology of Power and the Power of Ideology*. London: New Left Books.
Thomas, William Isaac. 1969[1923]. *The Unadjusted Girl: With Cases and Standpoint for Behavior Analysis*. Foreword by Mrs. W. F. Dummer. Montclair, N.J: Patterson Smith.
Thompson, Kenneth. 1986. *Beliefs and Ideology*. New York: Tavistock.
Turner, Bryan S., ed. 2006. *Cambridge Dictionary of Sociology.* New York: Cambridge University Press.
Twine, Fred. 1994. *Citizenship and Social Rights: The Interdependence of Self and Society*. London, UK: Sage.
U.S. Social Security Administration. (2009). *Income of the population 55 or older, 2006*. Social Security Administration Publication No. 13-11871. Washington, DC: U.S. Social Security Administration.
Vincent, John A. 2006. "Globalization and Critical Theory: Political Economy of World Population Issues." Pp. 245–71 in *Aging, Globalization and Inequality: The New Critical Gerontology*, edited by J. Baars, D. Dannefer, C. Phillipson, and A. Walker. Amityville, NY: Baywood.
Walker, Alan. 1999. "Public Policy and Theories of Aging: Constructing and Reconstructing Old Age." Pp. 361–78 in *Handbook of Theories of Aging*, edited by V. L. Bengtson and K. W. Schaie. 1st ed. New York: Springer.
———. 2006. "Reexamining the Political Economy of Aging: Understanding the Structure/Agency Tension." Pp. 59–79 in *Aging, Globalization and Inequality: The New Critical Gerontology*, edited by J. Baars, D. Dannefer, C. Phillipson, and A. Walker. Amityville, NY: Baywood.
———. 2009. "Aging and Social Policy: Theorizing the Social." Pp. 595–613 in *Aging, Globalization and Inequality: The New Critical Gerontology*, edited by J. Baars, D. Dannefer, C. Phillipson, and A. Walker. Amityville, NY: Baywood.
Weber, Max. 1946. *From Max Weber: Essays in Sociology*. Translated and edited by H. H. Gerth and C. W. Mills, with an introduction by H. H. Gerth and C. Wright Mills. New York: Oxford University Press.
Williamson, John B., Diane M. Watts-Roy, and Eric R. Kingson, eds. 1999. *The Generational Equity Debate.* New York: Columbia University Press.

Chapter 20
Welfare States: Protecting or Risking Old Age

Jill Quadagno, Ben Lennox Kail, and K. Russell Shekha

The modern welfare state was created in the long economic expansion of the post-World War II era, as rising productivity and economic growth made it possible for governments to fund an array of benefits that provided a modicum of income security across the life course. Although welfare states everywhere included benefits for unemployment, health and disability, the bulk of expenditures went to public pensions. In all western capitalist democracies, expenditures for the aged represented the single largest item in national budgets. As Myles (1989:2) states, "the contemporary welfare state in the capitalist democracies is largely a welfare state for the elderly." Public pensions made it possible for older people to retire before any actual physiological decline occurred. More importantly, they exempted the wages of the old from the control of the market with its laws of supply and demand and instead subjected them to a process of political decision-making that led to ever more generous pensions as elected officials competed to claim credit for benefit hikes.

The public pensions created in this period represented a break with the poor law tradition of means-tested benefits for the needy. Rather, they were designed around the concept of social insurance; individuals who contributed during their working years would automatically be eligible for benefits in retirement. These benefits would not merely allow for a bare subsistence but instead would provide income replacement at levels that allowed recipients to maintain the same standard of living in retirement that they had had during their working years. Social insurance systems were distinctive also in that they coupled the private insurance principle of shared risk with the power of the state to require participation (Hacker 2006).

Beginning in the 1970s, the era of welfare state expansion slowed. The immediate cause was a stagnant economy and rampant inflation, but long-term social trends associated with population aging and expanding public budgets also put strains on the fiscal capacity of states. Many European nations and the United States enacted modest measures to reduce public pension expenditures, for the most part without abandoning the underlying principles of social insurance. In Latin America, by contrast, the World Bank provided a model and incentives for countries to privatize social insurance programs in hopes of promoting economic stability (World Bank 1994).

Rising public budgets were not the only challenge faced by all western nations. Rather, welfare states designed for an industrial economy and the male breadwinner family type that accompanied it needed to be retooled for service-oriented economies where dual-earner and female-headed households were becoming the norm (Esping-Andersen 2009). In response to new needs associated with changing family structure and the rising labor force participation of women, some nations began also

J. Quadagno (✉)
Department of Sociology, Florida State University, Tallahassee, FL, USA
e-mail: jquadagno@fsu.edu

R.A. Settersten, Jr. and J.L. Angel (eds.), *Handbook of Sociology of Aging*, Handbooks of Sociology and Social Research, DOI 10.1007/978-1-4419-7374-0_20, © Springer Science+Business Media, LLC 2011

restructuring public expenditures. These reforms not only targeted benefit formulas for retirees but also involved measures to improve support systems for working women across the life course.

In this chapter, we first discuss theories of the origins of social insurance systems. We then examine patterns of welfare state restructuring in Europe and the United States, where the commitment to social insurance has endured, and in Latin America where most pension reforms included privatization. In the process, we ask whether gender inequality can be alleviated by adjustments to public pensions or whether more comprehensive changes are needed that take into account the needs of working women.

The Origins of Welfare States

The Logic of Industrialism

When social scientists first sought to explain the phenomenal post-World War II growth of the welfare state, manifested foremost in public pension programs, they drew upon modernization theory, a broader theory of economic development. Modernization theory presumed that nations could be arrayed on a continuum from most to least modern and that all would converge along a similar trend line.

Elements of modernization theory were then modified to explain the development of the welfare state. According to the "logic of industrialism" theory, in preindustrial economies the social character of old age was structured by the family, community and local labor market. Industrialization disrupted these traditional relationships by eroding kinship ties and the patrimonial traditions of agrarian societies and increasing dependence on wage labor. As declining mortality resulted in rising life expectancy and population aging, local institutions lost the capacity to care for growing numbers of the old and the sick and disabled. Industrialization also spurred rapid economic growth, creating surplus funds that made it possible for governments to provide the support that families and local communities could no longer provide. Every industrial nation created public pensions that seemed to support the thesis of convergence.

Thus, some theorists concluded that population aging was the main motor of welfare state development: "If there is one source of welfare spending that is most powerful – a single proximate cause, it is the proportion of old people in the population" (Wilensky 1975:40).

The logic of industrialization theory had one fatal flaw. It could not explain variations around the trend line. Yet numerous studies showed that public pensions varied significantly in generosity and structure, even in nations with comparable levels of economic development (Myles and Quadagno 2002). By the end of the 1970s, social scientists began looking for an alternative explanation that could take into account human agency. The central premise of this new approach was that political forces, not level of development, were responsible for welfare state diversity.

Explaining Variation in Welfare States

Social democratic theory represented an attempt to explain why nations vary in the structure and quality of benefits. Its central claim is that, in an industrial economy, the programs that protect the elderly against income loss and allow for a dignified retirement are determined by politics, not demography. Public pensions do not arise as just an automatic response to need but rather are produced by political groups and economic institutions that establish the parameters of the distributional choices available in the policymaking process (Myles 1989).

What determines the organization of benefits and the degree of generosity? One argument, termed power resource theory, contends that variations across nations in welfare state structure and spending are due to the relative success of left parties aligned with strong trade unions in shaping the democratic class struggle. The most generous welfare states occur in nations where trade unions are able to form a political party and also create cross-class coalitions with other groups. They then use their numerical superiority to enact programs that protect the working class against the vagaries of the market and provide a secure income in old age (Korpi and Palme 1998). Thus, according to power resource theory, variations across nations in public pensions can be explained by the strength of trade unions and degree of working class solidarity.

Welfare States and Social Stratification

Welfare states are not only produced in the struggle between classes but also become mechanisms of social stratification. Once enacted, social programs can reduce inequality produced by markets and even out income over the life course across class, gender, and racial lines (O'Connor et al. 1999). The extent to which welfare states redistribute resources depends on the allocation of responsibility for social needs between the state, the market, and the family. These distributional choices have made it possible to devise typologies that group together nations that adopt similar strategies. One widely-cited typology (Esping-Andersen 1990) identifies three regime types characteristic of Europe and North America: liberal, conservative, and social democratic. Liberal welfare states are characterized by a reverence for the market, a preference for government subsidies to encourage private benefits, and extensive reliance on means-testing in public benefits. Conservative regimes are less concerned with market efficiency and rather emphasize supporting the traditional family and maintaining a hierarchical social order inherited from the past. Finally, social democratic regimes provide universal welfare benefits granted to all as a right of citizenship with private benefits having a marginal role.

Analyses of the distributional consequences of regime types show they can have a significant impact on the well-being of older people. Tai and Treas (2009) found that older adults are significantly less likely to be poor in social democratic and conservative welfare regimes than in liberal regimes. However, in a country classified as liberal, such as Taiwan, multigenerational living arrangements and the associated family support compensates for the lack of public support. Thus, regime type alone is not the only factor in determining how older people fare. Familial traditions also matter.

Determinants of Welfare State Restructuring

The era of welfare state expansion ended in the 1970s, as governments everywhere faced rising deficits and growing awareness of the fiscal strains of population aging on pension and healthcare budgets. Although expenditures in the western, capitalist democracies continued to rise in the 1980s, they rose at a slower rate than in earlier decades and then leveled off in the 1990s. In recent years rollbacks have been commonplace, although they have not occurred without a political struggle (Myles and Quadagno 2000). As many elected officials have learned, it is easier to give benefits than to take them away. This is especially true of benefits that not only protect the poor against destitution but provide income security for the middle class.

As nations have sought to rein in public expenditures, a key question is whether pensions have retained the social insurance principle or rather shifted toward increased private responsibility for income security in old age (Hacker 2006). Privatization refers to a reduction in public responsibility and an increase in private activity in the financing and provision of social benefits (Gilbert 2002). As a general trend, the private share of spending for retirement income has increased everywhere, even in countries traditionally associated with a collectivist ideology, redrawing the boundaries between the state and the market. Although privatization has occurred to some extent in many European countries and the United States, it is most pervasive in Latin America, a region not included in most typologies.

Budget reduction is not the only item on the political agenda for welfare states. Gender relations and family forms have become central to welfare state restructuring in the contemporary period in ways that they were not during the age of expansion. The social insurance systems created in the postwar period were designed to protect workers against the loss of income in old age and presumed that the typical household consisted of a male breadwinner and a dependent spouse. In the male breadwinner model, married women were subordinated to their husbands in terms of social insurance eligibility and were expected to care for children and other family members with no public support (Lewis 1998). Policies developed around this model took the form of protection for widows and spouses who were not employed outside the home.

Widow and spouse benefits did improve women's income security in old age, but have proved insufficient for an era marked by a dramatic rise in women's education and labor force participation. Working women not only require protection against the loss of a spouse; they also need social benefits that support their multiple familial responsibilities and resultant intermittent employment patterns (Esping-Andersen 2009). Rising divorce rates and increasing numbers of out-of-wedlock births and single parent households also have generated new needs (Castles et al. 2010). Some nations have consciously abandoned the male breadwinner model and enacted policies designed to do more than even out income inequality in old age but also support working women across the life course.

Welfare State Restructuring in Europe

Since the 1980s, many nations have sought to reduce expenditures on public pensions by encouraging greater private responsibility for income security in old age. The transition toward private responsibility comes through an increase in individual savings vehicles and through shifts in the structure of employer-based schemes, which have been stimulated by tax incentives and legal mandates (Gilbert 2002).

Trends in Privatization

Since the 1950s, Sweden, the prototype of the social democratic welfare state, has had a universal, guaranteed basic pension coupled with earnings-related pensions funded through payroll taxes, which also pay for disability benefits and coverage for widows and orphans. Although pension reform moved to the top of the political agenda in the mid-1970s, it was not until 1994 that a new framework was established. In that year Sweden revised the existing social insurance scheme by coupling the basic pension with mandatory private accounts. Individuals and employers "contribute" to these accounts during their working years and accrue pension credits with benefits based on lifetime earnings (Anderson et al. 2008). While this may appear to represent a move toward privatization, the important point about pension reform in Sweden is that no contributions are actually deposited in individual accounts. Rather, retirement benefits are paid based on pay-as-you-go financing. Further, although Sweden's pension reform seemingly shifted responsibility for retirement income from the public sector toward the individual, balances in individual accounts are protected by the government (Schulz and Binstock 2006). Thus, Sweden retains a substantial amount of government responsibility for the well-being of the aged.

Pension reforms have also taken place in nations classified as conservative welfare state regimes. Italy has enacted reforms that are similar to Sweden's (Schulz and Binstock 2006). In 2000, Germany restructured its pension system, reducing benefits from 70% of average wages to 67%, while offering substantial tax breaks and subsidies to encourage workers to deposit 4% of their earnings into a private pension fund (Jochem 2008). In 1997, France passed a law authorizing private firms to establish employee pension funds. The public pension system remains protected, however, because contributions to private pensions are not deducted from the payroll taxes that fund it (Gilbert 2002).

Although some European countries have introduced or expanded private options, the structure of social insurance remains intact. What is happening is a modest trend toward greater reliance on private funds in some countries. However, this is no guarantee that more radical changes will not take place in the future.

Welfare State Restructuring for Gender Equity

Many countries in Europe have sought to reduce gender inequity in old age by making it easier for women to obtain pension credits during adulthood. These measures reflect the recognition that the male breadwinner model no longer accurately describes the typical family. Throughout most of Europe, female educational attainment now exceeds that of males, and, in several countries, female labor force participation, even among mothers with young children, has soared. For example, in Sweden only 38% of mothers with small children were employed in the early 1960s compared to 82% in the 1980s, a level that has remained stable since then. In Great Britain the change has been more gradual and stabilized at 60%, while Italy and Spain lag behind.

In some European countries, high rates of female labor force participation have been fostered by an activist state and the adoption of family policies that encourage work. Much of the increase has been in part-time work, but there is a significant difference in support for part-time work across countries. Beginning in the late 1960s, Sweden consciously moved toward a dual-breadwinner nation by introducing separate taxation for married couples, generous parental leaves, universal, state-run day care, and policies to encourage greater involvement of fathers in childrearing. Sweden also expanded the definition of work to allow pension credits to be earned for childrearing as well as for military service, spells of illness, unemployment, and disability (Anderson et al. 2008). Most Swedish women working part-time have full benefits and the right to work three-quarter time while their children are young. Thus, Swedish women remain full participants in the public pension system. In the Nordic countries as a whole, the full-time housewife has basically disappeared with part-time work serving as a bridge between childbirth and full-time employment (Esping-Andersen 2009). France also has an array of services and subsidies to support dual-earner families.

This trend is not universal, however, for the Netherlands upholds the male breadwinner model and encourages mothers to stay home with young children (Morgan 2006). Similarly, Great Britain created an allowance for unpaid care work for infirm dependents in the 1970s, but married women were ineligible for this benefit, because caregiving was considered part of their normal duties (Lewis 1998). In Great Britain, part-time work involves short hours and few benefits (Lewis 1998). British women are largely excluded from the public pension system during periods of caregiving.

Variation in approaches to gender equity have inspired alternative typologies based on the extent to which welfare policies are used to encourage a male breadwinner model or promote female labor force participation. Although some nations have allowed women to earn pension credits for care work, these measures are insufficient to guarantee gender equity. Rather family policies that support working mothers and encourage men to take a more active familial role reduce the penalty women pay for caregiving responsibilities. Such policies recognize that the ability to have secure retirement income for most women depends on interventions earlier in the life course.

Welfare State Restructuring in the United States

In all models that classify nations into regime types, the United States is the archetype of a liberal welfare state, because of its preference for private sector solutions to public problems. In the past quarter century, the U.S. has adopted policies that have gradually eroded the value of Social Security and shifted private pensions from a guaranteed benefit to tax-subsidized savings vehicles (Schulz and Binstock 2006). Thus, privatization in the U.S. involves both a trend toward less public and greater private responsibility for income in old age and a shift in risk from employers to individuals and families (Hacker 2006).

The Trend Toward Privatization

When Social Security was enacted in 1935, it included a strict earnings test. Any beneficiary between the ages of 65 and 69 who earned more than $15 a month lost all benefits. Beneficiaries were expected to retire. In 1956 an early retirement option was extended to women who could retire at 62 with a 20% benefit reduction; in 1961 this privilege was extended to men. The early retirement option decreased the incentive to work and led a long term decline in the average age of retirement (Schulz and Binstock 2006).

Since the 1970s, policies designed to restore the fiscal solvency of the Social Security trust fund have begun to reverse the decline in labor force participation and encourage people to work longer. The 1983 amendments to the Social Security Act, enacted in response to a shortfall in revenues, raised the age of eligibility for full benefits from 65 to 67, and increased the early retirement penalty from 20 to 30%. These changes represented a significant benefit cut over the long term but the pain was not felt immediately because they were phased in gradually. Then in 1994 Social Security benefits for higher income retirees were taxed for the first time. The net effect has been to reduce the value of Social Security benefits relative to wages (Gilbert 2002). The elimination of the earnings test in 2000 for people 65–69 provided a further incentive to encourage work. Thus, incremental changes in Social Security benefits over two decades increased the value of work over retirement and made individuals more responsible for income security in later life.

In 2004 President George W. Bush proposed diverting a portion of Social Security payroll tax revenues into private accounts. His plan created a public uproar, as polls showed that most Americans feared that privatization would destroy the safety net for the elderly poor and reduce income security for the middle class. The proposal was subsequently withdrawn (Beland 2007). The Bush administration's failure to partly privatize Social Security illustrates the principle of middle class universalism (Esping-Andersen 1990). Welfare programs that provide income security for the middle class create a constituency of beneficiaries who mobilize against any threat to their benefits (Pierson 1994).

How, then, did previous efforts to reduce Social Security succeed? There are two explanations. First, the 1983 benefit reductions were enacted during a trust fund crisis, had bi-partisan support and were phased in gradually in the distant future, while the 1994 taxation of benefits affected only a small percentage of beneficiaries at the time. Second, unlike Sweden, the U.S. already has a private pension system and numerous tax-subsidized options for saving for retirement. That made it difficult to make the case that an additional savings vehicle was needed.

Changes in the nature of private pensions have also increased work incentives. Since the 1940s, private pensions have been an important source of retirement income for many workers (Quadagno 1988). These took the form of defined benefit plans (DB), which were subsidized by the tax system and provided a fixed income for the duration of a retiree's life (McDonnell 2008). A new provision in the tax code, Section 401(k) of the Revenue Act of 1978, shifted employers' incentives away from DB plans to defined contribution (DC) plans (Even and Macpherson 2007). Under DC plans employers make contributions to a tax-protected retirement savings account in the employee's name. Employees are allowed to match a portion of these funds. DC plans encourage work, because the more money in the account, the higher the retirement income. Yet DC plans also increase risk, because income in the funds is not guaranteed for any period of time and can be spent down early. Further, since the funds may be invested in the stock market, retirees could lose a significant portion of their savings in an economic downturn, as occurred in 2008–2009 (Hardy and Shuey 2000; O'Rand and Shuey 2007). Thus, the growth of DC plans has increased private responsibility for income in old age.

Distributional Issues

From the 1930s to the 1970s, Social Security was based on a vision of a male breadwinner and a dependent spouse, and the main policy objective was to provide better protection for wives and widows (Harrington Meyer et al. 2005; Harrington Meyer and Herd 2007). Initially, only workers were eligible for Social Security, but in 1939 a benefit for spouses and widows was added. The benefit paid spouses 50% of the worker benefit while widows received 75% of the combined worker/spouse benefit. In 1961, the widow's benefit was increased to 82% of the deceased workers benefit and then in 1972 to 100%. In response to rising divorce rates, in 1965 spouse benefits were granted to divorced wives, if the marriage had lasted at least 20 years; in 1977 this was reduced to 10 years (Harrington Meyer et al. 2006).

While spouse benefits are available *de jure* to both men and women, *de facto* it is almost always women who receive them. Recently, however, more women have drawn Social Security benefits based on their own earnings, but they receive lower benefits than men, because they earn less and have more breaks in employment (Harrington Meyer 1990; Herd 2005). Social Security has maintained a breadwinner model that has not responded to rising numbers of single parent families and increasing female labor force participation.

The Family and Medical Leave Act of 1993 did allow employees to take job-protected leave for a serious illness or to care for a family member but the leave is unpaid. Further, the FMLA does not include businesses with fewer than 50 employees, part-time workers, or workers who need time off to recover from a common illness like a cold, or to care for a family member with a short-term illness. More importantly, workers who do take leave receive no child care or elder care credits in Social Security to compensate for breaks in employment. In terms of child care, public funds are available only for the very poorest families, mainly those who are recipients of Temporary Assistance for Needy Families (TANF) (Morgan 2006).

Social Security evens out income inequality by providing higher replacement rates for low-wage workers but disparities in benefit levels remain, especially for racial and ethnic minorities. Compared to whites, African-Americans receive less in Social Security income over the life course, because they have lower wages and higher mortality rates (Kail et al. 2008). This disadvantage is exacerbated among African-American women who are less likely than white women to be married, more likely to be divorced and thus less likely to be eligible for spousal and widow benefits (Harrington Meyer et al. 2006; Herd 2005). Moreover, because African-American women are usually married to African-American men (Jacobs and Labov 2002; Kalmijn 1998), they receive smaller spousal benefits when eligible at all (Harrington Meyer 1996).

Welfare State Restructuring in Latin America

The western, capitalist democracies have introduced some policy initiatives to reduce public pension expenditures and respond to women's rising labor force participation. Yet these initiatives have often been fiercely resisted and in some cases modified or even rescinded. As noted above, in mature welfare states, benefits are difficult to cut because they generate constituencies with a vested interest in the status quo. Compared to Europe, Latin America has moved more rapidly toward privatization, for several reasons: existing social insurance systems had not reached maturity or were failing entirely, public funds were poorly managed, inflation had reduced the real value of benefits, and rates of contribution to the old system had declined. Thus, some changes were necessary, even inevitable, as aging populations, rising life expectancy, and the weakening of the traditional family increased the need for secure income in old age (Barrientos 1998a). The question was what form restructuring would take.

In 1994 the World Bank issued a report, *Averting the Old Age Crisis: Policies to Protect the Old and Promote Growth*, which criticized the failing social insurance schemes and promoted instead a multitier pension system. As an incentive to adopt the new model, developing countries were promised large loans for restructuring purposes (Brooks 2009). Many Latin American countries quickly adopted the multitier system, though to differing degrees (Holzman and Hinz 2005). The consequence was a new regime type.

The Latin American Welfare State Regime

The World Bank program was based on a plan first adopted by Chile in 1981, when the military dictator, General Pinochet, transformed Chile's public pension from a social insurance scheme to a multipillar system (Arenas de Mesa et al. 2008). The first pillar consisted of a means-tested benefit to prevent destitution in old age. The second pillar involved mandatory contributions of 10% of earnings to privately-managed individual accounts, and another 2–3% for administration and survivor and disability benefits. Those who contributed for at least 20 years were guaranteed a minimum pension (MPG), with a government subsidy for low wage workers. The third pillar was a voluntary scheme that allowed workers to contribute additional tax-subsidized savings to privately managed funds (Arenas de Mesa et al. 2008; Williamson 2001).

Many nations followed suit. The pension reforms adopted in Latin American consist of three types. In the first instance, the new system completely replaces the existing program (Bolivia and Mexico in 1997, El Salvador in 1998, and the Dominican Republic in 2003–2006). A second type involves parallel schemes where the private and public systems compete (Peru in 1993 and Colombia in 1994). Finally, in the mixed scheme (Argentina in 1994, Uruguay in 1996 and Costa Rica in 2001), everyone is eligible for a flat pension and those who contribute to private accounts receive an extra, supplementary benefit (Arenas de Mesa and Mesa-Lago 2006; Arza 2008).

Overall, Latin American pension reforms are similar across the region, although specific rules regarding coverage, contributions, and regulations vary from country to country. For example, in Argentina the first pillar is a flat-rate Universal Basic Pension (UBP). After 30 years of work, men can retire at 65 and women at 60 and receive a benefit that replaces 27.5% of average wages. One percent is added to the pension for each additional year of work up to a maximum of 31.62% (Arenas de Mesa and Bertranou 1997). At age 70, workers with at least 10 years of contributions and widows receive 70% of the UBP (Arenas de Mesa and Mesa-Lago 2006; Arza 2008; James et al. 2008).

Argentina's second pillar allows employees to choose between two systems. The first is a defined-contribution scheme, similar to the Chilean system, where workers contribute 11% of wages to privately managed individual accounts (James et al. 2008). Alternatively, workers may instead invest in another publicly-funded system for additional retirement income (AP). The UBP combined with the AP pays the flat-rate and then an additional percentage for each extra year of service.

Mexico uses a multitier system that includes a Social Quota, which is a prefunded benefit to induce low-income workers to participate in the formal economy, and a second pillar of individually funded retirement accounts (James et al. 2008). Workers contribute 11.5% of wages of which 6.5% goes to an individual retirement account, and 5% to a fund to purchase housing. If the funds are not used for a home purchase, they are refunded to the individual's retirement account (James et al. 2008). While Chile relies solely on individual contributions to fund accounts, Mexico also requires employers and the government to contribute. Workers who contribute for 24 years, and retire (age 60 for women and 65 for men) may arrange for scheduled withdrawals or purchase annuities (Dion 2006).

Distributional Consequences of Privatization in Latin America

One of the World Bank's (1994:20) major criticisms of social insurance systems was that they were unfair to women and low-wage workers:

> Since access to earnings-related benefits is based on employment history, these plans are of limited help to women and others who have had little or low-paid labor market experience. In OECD countries with earnings-related benefits, old women living alone have a much higher poverty rate than other groups.

A stated goal of privatization was to increase coverage rates among women and low-wage workers. Under the former social insurance schemes, coverage rates were low for some groups, participation for the self-employed was voluntary, and women had lower employment rates and wages due to familial responsibilities and employment discrimination. In addition, a flourishing underground economy allowed low wage workers to escape the payroll taxes levied on workers in the formal sector (Barrientos 1998a, b; James et al. 2008; Arza 2008). Yet claims that the multipillar scheme would reduce distributional inequities have not been realized.

In non-Latin American country did privatization lead to the large increase in coverage that policy makers had expected. Rather, in Latin American countries, privatization has had a negative effect on coverage and benefit levels for low-income and women workers. For example, coverage rates in Chile before reform were low and declining. In 1970, the social insurance program covered 75.6% of the labor force compared to just 61.2% in 1980 (Mesa-Lago 1994). Following the 1981 reform, coverage rates increased temporarily to 79.2% in 1985 (Mesa-Lago 1994) but then decreased to 63.6% by 2000 (Arenas de Mesa et al. 2008). Coverage rates for the self-employed, always low, dropped from 12.4% in 1986 to 5.4% in 2004, while the gap in coverage between the richest and the poorest increased from 14.7 to 23.4% from 1992 to 2003 (Arenas de Mesa and Mesa-Lago 2006).

Privatization failed to improve benefits for women. For example, a 2002 survey in Chile found that, on average, women receive pensions that are 50% lower than the average for men, and that

45% of covered women receive benefits that are lower than the minimum pension (Arenas de Mesa and Mesa-Lago 2006). Other countries that pursued the Chilean model have had similar outcomes. There are several reasons why gender inequity remains a problem. In Latin America, as in most other countries, women receive lower pay than men, have higher unemployment rates, and participate more in the informal sector where their work effort is invisible. All these factors result in lower coverage, contribution, and accumulation rates that can't be solved by pension policy alone. For example, nearly all the countries that adopted the World Bank model allowed women to retirement earlier, but lower retirement ages give women have fewer years to qualify for minimum pension guarantees or accumulate sufficient funds to receive high benefits (Arenas de Mesa and Montecinos 1999). The rules regarding eligibility and payout for the contributory schemes exacerbated these problems. Further, many countries use gender-specific tables for benefit calculation. Because women live longer than men on average, the new formulas ensure that they receive lower monthly benefits (Arenas de Mesa and Bertranou 1997; Bertranou 2001; Dion 2006). Guatemala was one of the few Latin American countries that did not adopt the World Bank model but retained a contributory social insurance scheme. Interestingly, although overall pension benefits are lower in Guatemala than in other Latin American countries, female-headed households receive higher benefits compared to Mexico, Peru, and Uruguay (Pederson and Sheka 2010). This suggests that gender inequality in old age cannot be addressed simply through restructuring pension schemes but rather must involve equalizing women's employment opportunities across the life course.

Conclusion

In most western capitalist democracies, the welfare states created in the post-World War II era were social insurance systems designed to provide a guaranteed income in old age at replacement rates that would allow workers to maintain a preretirement standard of living. As welfare states reached maturity, they faced fiscal strains associated with population aging, resulting in fewer workers supporting more retirees. This conundrum was accompanied by fundamental challenges to the core assumptions around which pension programs were structured – that the primary beneficiaries were male workers with a dependent spouse. Thus, nations were faced with dual objectives, to reduce expenditures for future generations and to reconfigure benefits to more adequately reflect the needs of dual-earner families and single parent households.

Some European nations have restructured public pensions in an attempt to reduce gender inequality in benefit levels. These efforts have taken the form of pension credits for years out of the labor force for caregiving responsibilities or reductions in the number of years required for full benefits. Yet these measures are insufficient from a life course perspective. Gender inequity is not a problem that can be resolved solely through pension restructuring. Rather what is needed are policy changes that support women's employment during their adult years and thus reduce the penalty women pay for family caregiving. Higher earned income during adulthood leads to gains in income in old age in both the public and private sphere.

In response to fiscal pressures, many countries have modified the public pensions that were organized around the principles of social insurance and experimented with a variety of options that have increased the importance of private income sources in retirement. The most dramatic changes have occurred in Latin America. Facing ever increasing budget deficits, mismanaged public funds, corrupt government agencies and the fiscal insolvency of social insurance programs, nations from this region created a new regime type based on the World Bank multipillar model. While women and low-income workers were part of the focus, the main goal of privatization was to reduce budget deficits, remove the control of funds from the hands of government, and increase financial stability.

The result did not solve the problems but instead increased old age income inequality for women and the poor.

The Latin American experience provides lessons for nations seeking to cut public expenditures and achieve greater distributional equity. Given that the promise of reducing gender inequality and improving income security for low-wage workers has not materialized, western democracies involved in restructuring welfare states need to be wary. The solution to reducing costs and evening out disparities need not come by destroying social insurance and privatizing a program that spreads the risk across generations and social classes.

References

Anderson, Karen, Paula Blomqvist, and Ellen Immergut. 2008. "Sweden: Markets within Politics." Pp. 169–89 in *Public and Private Social Policy*, edited by D. Beland and B. Gran. Palgrave. New York: Macmillan.

Arenas de Mesa, A., David Bravo, Jere R. Behrman, Olivia S. Mitchell, and Petra E. Todd. 2008. "The Chilean Pension Reform Turns 25: Lessons from the Social Protection Survey." Pp. 23–58 in *Lessons from Pension Reform in the Americas,* edited by S. J. Kay and T. Sinha. Oxford: Oxford University Press.

Arenas de Mesa, A., and C. Mesa-Lago. 2006. "The Structural Pension Reform in Chile: Effects, Comparisons with Other Latin American Reforms, and Lessons." *Oxford Review of Economic Policy* 22 (1):149–167.

Arenas de Mesa, A., and V. Montecinos. 1999. "The Privatization of Social Security and Women's Welfare: Gender Effects of the Chilean Reform." *Latin American Research Review* 34 (3):7–37.

Arenas de Mesa, A., and F. Bertranou. 1997. "Learning from Social Security Reforms: Two Different Cases, Chile and Argentina." *World Development* 25:329–348.

Arza, Camila. 2008. "The Limits of Pension Privatization: Lessons from Argentine Experience." *World Development* 36(12):2696–2712.

Barrientos, A. 1998a. *Pension Reform in Latin America.* Brookfield: Ashgate.

———. 1998b. "Pension Reform, Personal Pensions and Gender Differences in Pension Coverage." *World Development* 26(1):125–137.

Beland, Daniel. 2007. *Social Security: History and Politics from the New Deal to the Privatization Debate.* Lawrence, KS: University Press of Kansas.

Bertranou, F. M. 2001. "Pension Reform and Gender Gaps in Latin America: What are the Policy Options?" *World Development* 29(5):911–923.

Brooks, Sarah M. 2009. *Social Protection and the Market in Latin America.* Cambridge: Cambridge University Press.

Castles, Frank, Stephan Leibfried, Jane Lewis, Herbert Obinger, and Christopher Pierson. 2010. *The Oxford Handbook of the Welfare State.* Oxford: Oxford University Press.

Dion, Michelle. 2006. "Women's Welfare and Social Security Privatization in Mexico." *Social Politics* 13(3):400–426.

———. 2008. "Pension Reform and Gender Inequality." Pp. 134–63 in *Lessons from Pension Reform in the Americas*, edited by S. J. Kay and T. Sinha. Oxford: Oxford University Press.

Esping-Andersen, Gosta. 1990. *The Three Worlds of Welfare Capitalism.* Cambridge: Polity Press.

———. 2009. *The Incomplete Revolution: Adapting to Women's New Roles.* Cambridge: Polity Press.

Even, William and David Macpherson. 2007. "Defined Contribution Plans and the Distribution of Pension Wealth." *Industrial Relations* 46:551–581.

Gilbert, Neil. 2002. *Transformation of the Welfare State.* New York: Oxford University Press.

Hacker, Jacob. 2006. *The Great Risk Shift.* New York: Oxford University Press.

Hardy, Melissa A. and Kim Shuey. 2000. "Pension Decisions in a Changing Economy Gender, Structure, and Choice." *Journals of Gerontology Series B: Psychological Sciences and Social Sciences* 55:271–277.

Harrington Meyer, M. 1990. "Family Status and Poverty among Older Women: The Gendered Distribution of Retirement Income in the United States." *Social Problems* 37:551–563.

Harrington Meyer, Madonna and Pamela Herd. 2007. *Market Friendly or Family Friendly*? New York: Russell Sage.

Harrington Meyer, Madonna, Doug A. Wolf, and Christine L. Himes. 2005. "Linking Benefits to Marital Status: Race and Social Security in the US." *Feminist Economics* 11:145–62.

———. 2006. "Declining Eligibility for Social Security Spouse and Widow Benefits in the United States?" *Research on Aging* 28:240.

Herd, Pamela. 2005. "Reforming a Breadwinner Welfare State." *Social Forces* 83:1365–1393.
Holzmann, Robert, and Richard Hinz. 2005. *Old Age Income Support in the 21st Century*. Washington, D.C.: The World Bank.
Jacobs, J. A. and T. G. Labov. 2002. "Gender Differentials in Intermarriage among Sixteen Race and Ethnic Groups." *Sociological Forum* 17:621–646.
James, Estelle, Alejandra Cox Edwards, and Rebeca Wong. 2008. "Reflections on Pension Reform in the Americas: From 'Averting the Old-Age Crisis' to 'Keeping the Promise of Old-Age Security' and Beyond." Pp. 164–84 in *Lessons from Pension Reform in the Americas*, edited by S. J. Kay and T. Sinha. Oxford: Oxford University Press.
Jochem, Sven. 2008. "Germany: The Public-Private Dichotomy in the Bismarckian Welfare Regime." Pp. 190–206 in *Public and Private Social Policy*, edited by D. Beland and B. Gran. Basingstoke: Palgrave.
Kail, Ben Lennox, Jill Quadagno, and Jennifer Reid Keene. 2008. "The Political Economy Perspective in Aging." in *Handbook of Theories of Aging*, edited by V. Bengtson, M. Silverstein, and N. Putney. New York: Springer.
Kalmijn, Matthijs. 1998. "Intermarriage and Homogamy: Causes, Patterns, Trends." *Annual Review of Sociology* 24:395–421.
Korpi, Walter and Joakim Palme. 1998. "The Paradox of Redistribution and Strategies of Equality: Welfare State Institutions, Inequality, and Poverty in the Western Countries." *American Sociological Review* 63(5):661–687.
Lewis, Jane. 1998. "Gender and the Development of Welfare Regimes." Pp. 229–49 in *Power Resource Theory and the Welfare State*, edited by J. O'Connor and G. Olsen. Toronto: University of Toronto Press.
Mesa-Lago, Carmelo. 1994. *Changing Social Security in Latin America: Towards Alleviating the Social Costs of Economic Reform*. Colorado: Lynne Rienner Publishers, Inc.
McDonnell, K. J. 2008. "Retirement Annuity and Employment-Based Pension Income, Among Individuals Age 50 and Over: 2006." *EBRI Notes* 29.
Morgan, Kimberly. 2006. *Working Mothers and the Welfare State*. Palo Alto: Stanford University Press.
Myles, John. 1989. *Old Age and the Welfare State: The Political Economy of Public Pensions*. Lawrence, KS: University Press of Kansas.
Myles, John and Jill Quadagno. 2000. "Envisioning a Third Way: The Welfare State in the 21st Century." *Contemporary Sociology* 29(1):156–167
———. 2002. "Political Theories of the Welfare State." *Social Service Review* (March):1–22.
O'Connor, Julia, Ann Orloff and Sheila Shaver. 1999. *States, Markets, Families: Gender, Liberalism and Social Policy in Australia, Canada, Great Britain and the United States*. Cambridge, England: Cambridge University Press.
O'Rand, Angela M. and Kim M. Shuey. 2007. "Gender and the Devolution of Pension Risks in the US." *Current Sociology* 55:287.
Pederson, JoEllen and K. Russell Sheka. 2010. "Pension Policy in Latin America: The Effect of Privatization on Gender Inequality in Old Age." Florida State University, Tallahassee, FL. Unpublished manuscript.
Pierson, Paul. 1994. *Dismantling the Welfare State*. Cambridge, England: Cambridge University Press.
Quadagno, Jill. 1988. *The Transformation of Old Age Security: Class and Politics in the American Welfare State*. Chicago: University of Chicago Press.
———. 2005. *One Nation, Uninsured: Why the US Has No National Health Insurance*. New York: Oxford University Press.
Schulz, James and Robert Binstock. 2006. *Aging Nation*. Westport, CT: Praeger.
Social Security Administration. 2007. *Fast Facts & Figures about Social Security, 2007*. Retrieved (http://www.ssa.gov/policy/docs/chartbooks/fast_facts/2007/fast_facts07.html).
Tai, Tsui-O and Judith Treas. 2009. "Does Household Composition Explain Welfare Regime Poverty Risks for Older Adults and Other Household Members?" *Journal of Gerontology* 64B(6):777–787.
Wilensky, Harold. 1975. *The Welfare State and Equality*. Berkeley: University of California Press.
Williamson, John B. 2001. "Privatizing Public Pension Systems – Lessons from Latin America." *Journal of Aging Studies* 15(3):285–302.

Chapter 21
Volunteering in Later Life: From Disengagement to Civic Engagement

Greg O'Neill, Nancy Morrow-Howell, and Sarah F. Wilson

The topic of volunteering in later life has a long tradition in gerontology, perhaps because historically volunteering was one of the few formal roles available to older adults after leaving the workforce. Volunteer activity fit well with the "busy ethic" that shaped modern retirement (Ekerdt 1986) and was in line with the involvement promoted by activity theory (Havighurst 1963). Yet volunteering was considered a leisure activity (Musick and Wilson 2008), a discretionary role that might fill in for roles losses in employment and parenting. Recently, the discussion has changed, as an upsurge of academic and political interest in volunteering in later life begins to dominate the discourse on civic engagement.

There are several explanations for this change in perspective. First, dramatic gains in health, education, and longevity over the past half-century have increased individuals' capacity and desire for civic roles in later life. Second, the nonprofit sector and many government agencies are struggling to meet increasing demands with stagnant or decreasing resources. In this environment, volunteer labor has become crucial, and maximizing the deployment of older adults depends on knowledge about volunteers that have grown old and volunteers who come forward for the first time in later years. Finally, Robert Putnam's (2000) argument that social capital and civic participation have declined in the later part of the twentieth century has sparked considerable academic and political interest in civic engagement.

Civic engagement is a broad concept with many definitions, but the term usually associated with membership in voluntary associations, volunteering, and political participation. Volunteering through formal organizations has dominated discussions of civic engagement and drawn the most attention from researchers (Martinson and Minkler 2006). In this paper, we focus on formal volunteering, defined as an activity undertaken by an individual that is uncoerced, unpaid (or paid minimal compensation to offset costs), structured by an organization, and directed toward a community concern (Cnaan, Handy, and Wadsworth 1996). We discuss definitional struggles later, but our working definition excludes informal helping and caregiving, two productive activities of great importance to families and communities. This exclusion is in line with the view that volunteering is an altruistic behavior that is aimed at others to whom the volunteer owes no contractual, familial, or friendship obligation (Musick and Wilson 2008). It is, in part, that characteristic that has attracted scholarly attention to volunteering among older adults as a way simultaneously to address societal needs and promote the health of the aging population.

Research on volunteering among older adults has evolved with the emergence of new paradigms about old age and aging, the development of gerontological theories, the production of more and better social science on volunteering, and the creation of new public policies. This

G. O'Neill (✉)
National Academy on an Aging Society, Washington, DC, USA
e-mail: goneill@agingsociety.org

R.A. Settersten, Jr. and J.L. Angel (eds.), *Handbook of Sociology of Aging*, Handbooks of Sociology and Social Research, DOI 10.1007/978-1-4419-7374-0_21, © Springer Science+Business Media, LLC 2011

chapter begins with a brief overview of concepts and theories about volunteering that have emerged over the last several decades. We then summarize the current knowledge base on older volunteers. We end with a review of challenges and opportunities for older adult civic engagement research and policy.

Central Concepts and Developments in Volunteering

Gerontologists long have argued that active engagement in society is related to well-being in later life, and accordingly have promoted various forms of activity and engagement – social, productive, religious, educational, and intergenerational. Today, as the demographic, economic, and social realities of our world shift, another type of active engagement is gaining attention: civic engagement. The confluence of increasingly complex social problems and the burgeoning numbers of people who can expect to live 20 or 30 years past retirement calls us to consider the possibilities of a civically engaged older population. Having older adults vitally involved in volunteer activities aimed at the social good is a win–win scenario that has captured the imagination of many gerontologists, policymakers, and government and nonprofit leaders.

That said, our national efforts to invigorate volunteering have focused largely on youth. Volunteer opportunities for young people have seen impressive growth and improved incentive structures to attract participants. They include volunteer work as a graduation requirement, service learning curricula in schools and universities, and selective young–adult services corps like Teach for America and City Year. Older adults participate in our national service programs, but they are underrepresented in the largest and most widely known programs. For example, volunteers over age 50 account for only 6% of the Peace Corps volunteers (Peace Corps 2009), and fewer than 9% of AmeriCorps volunteers are over the age of 50 (O'Neill 2006–2007). The notable exception is the federal government's network of Senior Corps programs: RSVP, Foster Grandparents, and Senior Companions. Strict income and age eligibility requirements for the latter two programs, however, limit the involvement of large numbers of older adults.

Recently, the focus of civic engagement initiatives has shifted to include older adults. Service programs have emerged that specifically recruit older adults and that hold older age criteria for inclusion (see Morrow-Howell, Carden, and Sherraden 2005). To counter the prevailing view of aging as loss and decline and to combat the focus of public discussions on the significant costs of population aging, gerontologists have begun to use new terms like the new gerontology (Rowe 1997), active aging (World Health Organization 2002), and positive aging (Katz 2001–2002). Scholars now emphasize the health and vitality of the older population, the growing human capital and experience reserves within the older population, and the potential for ongoing and new involvements in later life. As Hudson (2008) summarizes, today's seniors are classified by production and contribution, not just consumption and need. In this changing context, older adults have been called the "new trustees of civic life" (Freedman 1999:19).

Volunteering fits squarely with emerging paradigms of successful and productive aging. Rowe and Kahn (1998) proposed three characteristics of successful aging: physical and functional health; high cognitive function; and active involvement in society. Societal involvement includes both close personal relationships with family and friends and continued involvement in productive activities. In the productive aging perspective (Morrow-Howell, Hinterlong, and Sherraden 2001), volunteering takes center stage – along with working and caregiving – as an activity that should be promoted because it produces economic and social benefits to communities. In fact, the term "unpaid work" gained popularity as a way to recognize the value of volunteering. In these paradigms, volunteering gained new respect, and now is viewed not just as a leisure activity but as a vital function in society and as an indicator of successful aging and a key to wellness in later life.

Volunteering as a health-producing activity is particularly compelling, given the rising costs of health care and the aging of the population. It is a win–win: volunteering not only improves civil society, it also improves the lives of older adults. Advocates call for expanding volunteer involvement of older adults – for the sake of communities as well as older adults themselves – and researchers have supplied the evidence that volunteering has positive effects on older adults, perhaps more so than on younger people (Van Willigen 2000). Of all the aspects of volunteering that social scientists study, the effects of volunteering on older adults' physical health, mental health, and life satisfaction have received the most attention (see Grimm, Spring, and Dietz 2007 for a review). This literature, which we review later, has accumulated to the point that volunteering is viewed as a public health strategy for an aging society (Fried et al. 2004; Tang 2009).

Theoretical Roots Within the Study of Volunteering

One of the original formal theories in social gerontology focused on engagement in later life and proposed that disengagement of older adults from social roles benefited both older adults and society as a whole because it facilitated an orderly transfer of power from one generation to the next (Cumming and Henry 1961). Yet disengagement theory stood in contrast to activity theory with its central proposition that active engagement was a key to a satisfying later life (Havighurst 1963). The positive outcomes associated with volunteering in later life have been explained by activity theory; and this theory is the conceptual base of programs aimed at recruiting and sustaining older adults in volunteer roles. Challenging both disengagement and activity theories, scholars put forth empirical evidence suggesting that late life satisfaction was related to continuing patterns of engagement and that the maintenance of previous activity patterns over time was related to positive outcomes for the individual (Neugarten, Havighurst, and Tobin 1968). Continuity theory (Atchley 1971) continues to explain patterns of volunteering; recent data clearly show that volunteering is a stable activity from midlife on, with past volunteering being the most powerful predictor of current volunteering (Chambré and Einolf 2008). A life course perspective also has been instructive in understanding how motivations, opportunities, and social roles affect the dynamics of volunteering over a lifetime. In the following sections, we review trends, social contexts, and outcomes of volunteering, presenting empirical and conceptual work that has advanced since these theoretical beginnings.

Volunteering in Later Life

Trends and Life Course Patterns

The number of people volunteering in the United States has fluctuated over the past 30 years, causing some observers to express concern about the decline of civic life in America. The volunteer rate for adults 65 years and older, however, has shown a continuous, upward trajectory throughout the last three decades. About 23.9% of adults aged 65 and older who participated in the current population survey (CPS) reported volunteering for an organization in 2009, which was 67% higher than in 1974 (Bureau of Labor Statistics (BLS) 2010; Grimm, Dietz, and Foster-Bey 2006). No other age group experienced such a large rise in its volunteer rate over the same time period.

Research consistently shows a distinct life course pattern of volunteering. Among the U.S. population, the rate of volunteering is shaped like an inverted U, peaking in midlife. On the other hand, the number of hours per volunteer rises linearly with age (Musick and Wilson 2008). September 2009

CPS data show that 26.8% of the U.S. adult population (age 16 and over) volunteered with an organization at least once over the previous year. When delineated by age group, adults aged 35–44 years are most likely to volunteer (31.5%), while adults aged 20–24 years are least likely (18.8%). Older adults (age 65 and over) fall in between, with 23.9% having volunteered over the past year (BLS 2010). The data, however, shows a different trend when one considers the annual number of hours each volunteer devotes to service activities. Older volunteers invest the most hours of any age group, boasting an average of 90 h per year; all other age groups fail to match this, with averages between 36 (ages 25 to 34) and 60 (ages 55 to 64) hours per year. In addition, those aged 65 and over are more likely than any other age group to volunteer more than 500 h per year (BLS).

When asked which type of organization volunteers considered their primary organization, 44.8% of CPS respondents aged 65 and over selected religious organizations. This was followed by social or community service organizations (18% of respondents) and hospital or other health-related organizations (10.1% of respondents). Volunteers aged 65 and older identified public safety organizations or environmental organizations as their primary organizations least frequently (1 and 1.9%, respectively) (BLS).

Although almost one quarter of older adults devote some time to formal volunteer activities, their engagement varies widely by individual characteristics (BLS 2010; Zedlewski and Schaner 2005). Older adults who are in the workforce – especially those in part-time positions – are more likely than their nonworking counterparts to volunteer; older adults with comparatively high levels of education and managerial experience are more likely to begin volunteering than older adults with low levels of education and without managerial experience; older adults who are in excellent or good health are more likely to begin volunteering than their peers with fair or poor health; married older adults are more likely to volunteer than are their peers who are unmarried; and retirees who consider religion important and who are active in their religious communities are more likely to volunteer than their nonreligious counterparts (BLS; Musick and Wilson 2008; Wilson 2000; Zedlewski 2007; Zedlewski and Schaner 2006). The tendency for these individual characteristics to be associated with greater likelihood of volunteering can be explained both by applying a life course continuity theory that asserts that in old age, individuals maintain previously held levels of activity (Atchley 1999; Chambré and Einolf 2008), and by exploring the social context in which an individual resides (Musick and Wilson 2008; Tang 2006).

To test continuity theory, Zedlewski (2007) employed both cross-sectional and longitudinal methods to explore characteristics of older adult volunteers. Her data, taken from the 1996 through 2004 waves of the health and retirement study (HRS), suggest that 81% of the older adults who volunteered before retirement (one-third of adult workers aged 55–64 years) continued to volunteer after retirement, while only 27% of the older adults who did not volunteer before retirement took on volunteer positions after retirement. Similarly, Chambré and Einolf (2008) analysis of data from the 1995 and 2005 waves of the midlife in the United States (MIDUS) survey reveals that the strongest single predictor of whether or not a participant volunteered in 2005 was the participant's volunteer status in 1995; those who volunteered in 2005 likely also volunteered in 1995.

Butrica, Johnson, and Zedlewski's (2009) analysis of HRS data collected over the four consecutive 2-year periods between 1996 and 2004 also reveals stability of volunteer participation over time. Their results show that almost 40% of participants who volunteered in 1996 also volunteered in all three subsequent periods. In contrast, almost every participant who did not volunteer in 1996 also did not volunteer in any subsequent period. Butrica et al.'s data also reveal that the rate at which older volunteers dropout starts relatively high, but declines over time – meaning that the longer an older person volunteers, the less likely he or she is to quit volunteering. Similarly, the likelihood that an older nonvolunteer will begin volunteering starts out low and declines over time, meaning that a nonvolunteer is unlikely to begin volunteering, and the longer an older nonvolunteer refrains from volunteering, the less likely he or she is to start volunteering. In looking closely at the characteristics of the participants who volunteered throughout all four periods, Butrica et al. characterize the most consistent volunteers as those with the highest education, greatest religiosity, and marriage to a spouse who also volunteers.

Social Context

When considering social context, researchers theorize that the traits associated with higher rates of volunteering – continued workforce involvement, high socioeconomic status and educational attainment, good health, marriage, and high religiosity – leading to both greater awareness of and access to volunteer opportunities (Zedlewski 2007) and an increased likelihood of being asked to volunteer (Mutchler, Burr, and Caro 2003; Musick and Wilson 2008; Rotolo and Wilson 2004; Tang 2006). Older adults who remain in the workforce, are married, or are active in their religious communities, have larger social networks and experience greater social connectedness than older adults who are retired, unmarried, or nonreligious. This connectedness increases the likelihood that an older adult both will become aware of volunteer opportunities and will be asked to volunteer (BLS 2010; Gonyea and Googins 2006–2007; Musick and Wilson; Wilson 2000; Zedlewski 2007; Zedlewski and Schaner 2006). Further, older adults who work in managerial or salaried positions tend to have more flexible work schedules and have likely acquired more civic organizational and leadership skills than their lower-status counterparts, making them more desirable recruits, and therefore more likely to be asked to volunteer (Chambré and Einolf 2008; Mutchler et al. 2003; Rotolo and Wilson; Wilson 2004).

Being asked personally to volunteer is more effective than an impersonal appeal (Wilson 2000). According to September 2009 CPS data, 43.5% of volunteers over age 65 became involved with their main organization after being asked to volunteer, often by someone already involved in the organization (BLS 2010). Further, the Committee on an Aging Society reports that when not asked, only one of five people volunteer, but when asked, four of five people volunteer (Independent Sector 2000).

Older adults' volunteering practices and motivations are rooted in their social networks and goals. Tang (2006) found that the importance of one's social network in inducing volunteering is positively related to age. Studies have shown that older volunteers are much more likely than non-volunteers to have friends who volunteer (Wymer 1999). In comparison to younger adults, older adults rely more heavily on their social networks to become involved in volunteering. Plus, compared to younger adults, older adults are more motivated to volunteer by religious beliefs, a desire to interact with friends, and a need to be valued by others, whereas younger adults are more likely to adopt volunteer roles to explore their strengths and interests and to gain advancement in their careers (Chambré and Einolf 2008; Morrow-Howell 2006–2007; Musick and Wilson 2008; Okun and Schultz 2003). These age differences in types of volunteer organizations and activities may be due to different avenues of recruitment as well as different motivations for volunteering.

Older adults adjust their volunteer obligations based on each obligation's relevance for and effect on their well-being. Erikson's "generativity versus stagnation" stage of human development is rooted in an older adult's desire to leave behind a legacy and to pass wisdom and knowledge to younger generations (Erikson, Erikson, and Kivnick 1986). Chambré and Einolf's (2008) analysis of the MIDUS survey showed that older adults' sense of generativity – measured by six statements that discerned a participant's level of concern for the next generation – was highly correlated with the likelihood of volunteering. Cheng (2009: 46) hypothesizes that "because generativity is often manifested in a social context (e.g., assistance to others in need), the judgment of impact [of a volunteer activity], therefore, largely depends on others' feedback." Cheng's results show that when older adults volunteer to benefit younger generations, older adults' belief that they have achieved their generative goals depends on their perception of respect from younger people. If older people do not feel valued or fail to receive adequate recognition for their volunteer efforts, they likely will discontinue their service (Wilson 2000).

Carstensen's socioemotional selectivity theory explains that later in life, when adults recognize that their time and energy are limited, they pursue only the most meaningful social interactions and avoid those that are least meaningful (Carstensen 1992). When applied to volunteerism, socioemotional selectivity predicts that older people will have fewer overall volunteer commitments than younger adults, but their commitments will be more concentrated. CPS data fit this theory, showing that with age, a volunteer's total number of service hours increases, but these hours are concentrated among fewer organizations than for younger volunteers (BLS 2010; Hendricks and Cutler 2004).

Societal and Personal benefits of Volunteering

Societal Benefits

Older adults' volunteering is tremendously valuable to communities, measured both in economic terms and by the benefits reaped by the recipients of older adults' services (Zedlewski and Butrica 2007). Using data from the 2002 HRS and moderate-cost assumptions for estimating the value of work, Johnson and Schaner (2005) approximate the value of older adults' formal volunteer activities (defined as volunteering through or for an organization) at $44.3 billion per year.

Further, in modeling the Experience Corps Baltimore program's short- and long-term economic contributions to society, Frick and colleagues (2004) calculated that over a 2-year time period, each volunteer showed an average medical expenditure savings of $273 (when compared to a control group), which offset a portion of the program costs. Their models also showed that, in the long-term, if the program could cause even a mere 0.4% of students who would not have graduated otherwise to graduate eventually, the long-term benefits to the economy would far offset the program costs. These few additional graduates could expect increased lifetime earnings, and could expect to reap the health and sociocultural benefits associated with higher education.

Although these economic estimates are striking, they do not capture fully the personal benefits reaped by the recipients of older adults' services. Wheeler, Gorey, and Greenblatt (1998) conducted a meta-analysis of 37 studies published in the prior 25 years, nine of which measured the benefits to recipients of older adult volunteerism. Out of these nine studies, care recipients in four were nursing home residents, recipients in two were community-dwelling older adults, recipients in two were disabled children, and recipients in one were elderly caregivers. Studies measured various outcomes, but all included one of the following: life satisfaction and happiness, depression and isolation, client-assessed helpfulness, and goal attainment. When controlling for sample size, selection of participants, country, major design typology, and type or validity of dependent measures, Wheeler et al. found that 85% of people served by older adult volunteers showed improvement in the target area.

Additional research has found that older adult volunteers in preschool classrooms greatly help struggling kids adjust to school, and promote manners, appropriate behavior, and language development under stricter guidelines than are set by the professional educators (Larkin and Newman 2001). Older adult volunteers in primary and secondary schools significantly help improve attendance, academic achievement, and well-being, and significantly help decrease behavioral problems and substance use and abuse (Blieszner and Artale 2001; Rebok et al. 2004; Rogers and Taylor 1997). Older volunteers help increase graduation and retention rates of high school children and college freshmen, respectively (Coleman and DeRosa 2006; Muir 2006); and older volunteers' services help frail older adults remain independent and delay institutionalization (Barker 2002).

Personal Benefits

In addition to its value to the community, volunteering significantly benefits older adults. In fact, Van Willigen (2000) found that older volunteers reaped greater benefits than their younger counterparts. Volunteering in later life has been associated with improved physical health and decreased functional dependency (Fried et al. 2004; Luoh and Herzog 2002; Morrow-Howell et al. 2003; Tan et al. 2009; Tang 2009; Thoits and Hewitt 2001; Van Willigen), decreased morbidity (Fengler 1984; Luoh and Herzog 2002; Musick, Herzog, and House 1999), increased psychological well-being and fewer depressive symptoms (Greenfield and Marks 2004; Hao 2008; Jirovec and Hyduk 1998; Morrow-Howell et al. 2003; Morrow-Howell, Hong, and Tang 2009; Newman, Vasudev, and Onawola 1985; Sugihara et al. 2008; Thoits and Hewitt 2001; Windsor, Anstey, and Rodgers 2008), increased life satisfaction and quality of life (Van Willigen; Wheeler et al. 1998), improved cognitive ability and slower age-related cognitive decline (Carlson et al. 2008; Fried et al. 2004; Hao 2008; Park et al. 2007; Stine-Morrow et al. 2007), increased self efficacy (Li 2007), increased resilience and protection against role loss and negative life events (Greenfield and Marks 2004; Li 2007; Sugihara et al. 2008), and decreased mortality (Glass et al. 1999; Musick et al. 1999.). What is more, if an older adult is not civically and socially engaged, they risk experiencing decreased purpose in life (Greenfield and Marks 2004), health decline due to stagnant physical activity levels (Tan et al. 2009), and increased mortality, especially in the case of an emergency (Cannuscio, Block, and Kawachi 2003). In an extreme example, Semenza and colleagues (1996) studied deaths related to the July 1995 Chicago heat wave – three-fourths of which occurred among adults aged 65 and older – and found that socially isolated older adults (i.e., those who didn't participate in community groups) faced the highest risk of death; social isolation prevented many older residents from seeking shelter in community cooling centers and from being drawn out of their homes by concerned neighbors, friends, and colleagues.

Although countless studies identify the benefits of volunteering for older adults, research that explores the optimal amount of volunteering is inconsistent. Many researchers' data support a linear pattern, in which increased volunteering is associated with increasing benefits (Hinterlong, Morrow-Howell, and Rozario 2007), whereas others' data show that the positive effects of engagement persist to an upper limit, and any engagement beyond the limit has no further effect (Butrica and Schaner 2005). In contrast, Musick and colleagues' (1999) and Windsor and colleagues' (2008) data reveals a curvilinear pattern, in which increased volunteering is associated with increased benefits until a moderate number of hours (Musick et al. assign this upper limit at 40 annual hours and Windsor et al. at 800 annual hours), at which point increased volunteering hours are associated with decreasing benefits.

Explanations

Several mechanisms have been suggested as being responsible for the benefits of volunteering. For example, volunteering has been associated with greater social support and a stronger sense of responsibility and purpose that accompanies social and productive engagement. According to role theory, ebbing social roles (i.e., student, parent, spouse, employee) associated with later life can lead to decreased well-being. Therefore, adopting a formal volunteer role may serve to moderate the negative effect of role loss, especially for older adults who have experienced the most significant role losses (Adelmann 1994; Greenfield and Marks 2004; Sugihara et al. 2008; Van Willigen 2000). In fact, Musick and colleagues' (1999) data show the strongest protective effect of volunteering for participants who previously engaged in the least informal social interaction; specifically, lower-income, lower-educated, single, and minority older adults reaped more benefits than their higher-income, higher-educated, married, and nonminority counterparts (Carlson et al. 2008; Fengler 1984; Morrow-Howell et al. 2009). This supports widely-accepted notions that social interaction and the accompanying affection, behavioral confirmation, and support attained through engagement actually is responsible for volunteerism's beneficial effects (Fried et al. 2004; Glass et al. 2006; Grundy et al. 2007; Li 2007; Luoh & Herzog 2002; Park et al. 2007; Rowe and Kahn 1997; Steverink and Lindenberg 2006; Stine-Morrow et al. 2007; Sun and Liu 2006; Tang 2009; Thoits and Hewett 2001).

An alternative explanation for the benefits of engagement is the sense of responsibility, purpose, productivity, and usefulness one feels when engaged in volunteering (Greenfield and Marks 2004; Langer and Rodin 1976; Luoh and Herzog 2002; Mannell 1993; Rowe and Kahn 1997). To demonstrate the power of a sense of responsibility and purpose, Wilson et al. (2007) analyzed and reported 12 years of data from The Religious Orders Study; participants included 997 older Catholic nuns, priests, and brothers recruited from more than 40 groups across the country. Upon enrollment, each participant underwent a uniform clinical evaluation that included a medical history, a complete neurological examination, and a set of 20 cognitive evaluations. This set of evaluations was repeated annually for up to 12 years. In the case of death, a participant's brain underwent a full neuropathologic evaluation.

Results show that high conscientiousness, defined as "an individual's tendency to control impulses and be goal directed . . . be self-disciplined, scrupulous, and purposeful" (Wilson et al. 2007:1204–1205), is associated with reduced risk for Alzheimer's disease, even when controlling for all covariates individually and collectively: age, sex, education, the effect of other personality traits (neuroticism, extraversion, openness, and agreeableness), physical activity, frequency of cognitive activity, size of social network, depressive symptoms, and mild cognitive impairment. Further, when examining participants' brains during autopsies, many highly conscientious participants had lesions and tangles that would meet criteria for Alzheimer's, but those patients did not exhibit any symptoms of dementia before death. Although Wilson et al.'s sample was relatively homogenous and does not represent the general population, their data strongly suggest that reduced risk of Alzheimer's disease is associated with the sense of purpose and discipline that accompanies holding volunteer roles in later life.

Challenges and Directions for Future Research

Institutional Lag

Aging advocates long have pointed to the asynchrony between the human capital resources of the aging population and the capacity of organizational structures to engage older adults in socially valued and productive roles (Riley and Riley 1994). A current concern is that organizations are not ready to take advantage of the growing numbers of older volunteers (Casner-Lotto 2007; Freedman 1999). In a large survey of nonprofits conducted by the Urban Institute (Hager and Brudney 2004), researchers noted that organizational adoption of recommended volunteer management practices was not widespread. About 30% of older adults drop out of volunteering after 1 year of service (Foster-Bey, Grimm, and Dietz 2007), a loss of volunteer labor that is estimated at about $38 billion a year (Eisner et al. 2009). Plus, turnover tends to be associated with the nature of the volunteer work. For example, volunteer tenure was the shortest for those providing general labor or supplying transportation and longest for those providing professional or management services (Foster-Bey et al.).

To advance understanding of these issues, new models of volunteer engagement have focused on organizational structures rather than individual volunteers' characteristics. Invoking Riley and Riley's (1994) concept of structural lag, researchers have begun systematically to define and measure aspects of institutional capacity that affect the recruitment, retention, and effective use of older volunteers (Sherraden et al. 2001; Hong et al. 2009). For example, in a study of 374 volunteers across 51 service programs, Tang, Morrow-Howell, and Hong (2008) document the importance of flexibility in terms of time commitment, schedule, and type of volunteer activity to older volunteers in general, and to low-income and non-White volunteers in particular. In another study using the institutional perspective, McBride et al. (2009) demonstrate the role of stipends in increasing diversity among older volunteers and ensuring retention. Other work suggests that volunteer turnover can be reduced by ensuring that volunteers gain a sense of accomplishment and are recognized for their contributions by effectively monitoring and supervising their activities and by providing various cash or in-kind compensation to meet expenses (Cnaan and Cascio 1998; Finkelstein, Penner, and Brannick 2005; Wilson and Musick 1999; Netting et al. 2004). In addition, characteristics of volunteer programs like adequate training, supervision, recognition, flexibility, and stipends, are associated with more positive outcomes in terms of meeting volunteer expectations, intensity and duration of volunteer service, and perceived benefits of participation (McBride et al. 2009).

As Musick and Wilson (2008) point out, individual characteristics, such as capacities, values, and attitudes, are necessary but not sufficient to move people into volunteering. Given the low likelihood of modifying *individual* capacity, especially in older populations, improving *institutional* capacity may be the most promising direction for future research aimed at maximizing volunteer engagement and outcomes.

Critical Perspectives

Evidence to date indicates that civic engagement benefits both participants and the broader community. That fact notwithstanding, scholars must heed concerns from critical gerontology (Holstein 2006; Hudson 2008; Martinson and Minkler 2006). Martinson and Minkler have argued that the "politics of [welfare-state] retrenchment and devolution" over the past three decades have served as a "powerful motivator for the current promotion of volunteering in later life" (p. 321); there is concern that older adults are being called on as substitutes for critical public funding support to nonprofit and public organizations that meet community needs. Also, researchers express concern that civic engagement among older adults is dominated by high-resource individuals and fails to include the ethnic and socioeconomic diversity of the older population. Because the term civic engagement commonly favors formal volunteer activity, it devalues informal volunteering or helping outside of an official volunteer structure. Neighborhood- and family-based helping, however, is common among minority groups, including African-Americans and Hispanics, who often do not define helping as "volunteering," and therefore are unlikely to report it as such on a survey. Nor do most immigrant groups, which have rich traditions of helping others, self-describe their roles in terms of "volunteering" (Yoshida, Gordon, and Henkin 2008). Studies have noted that "volunteering" and "service" are cultural constructs of specific segments of American society that may be understood differently across different groups. For example, in low-income populations, the term "community service" may have negative connotations, conjuring images of court-ordered community service. Studies also have found that the term "volunteer" does not resonate well among African-American and Hispanics who traditionally have been viewed as recipients of services rather than as contributors engaged in community problem solving (Center for Health Communication, Harvard School of Public Health 2004). In presenting volunteering as a normative retirement role, older adults who do not engage risk judgment (Kaskie et al. 2008). Holstein (2006) and Ekerdt (1986) express concern that coercive social expectations, comparable to the notion of a work ethic or a "busy ethic" around older adult civic engagement, may denigrate those who are too ill, too poor, or otherwise unable to join the movement. The challenge for policymakers, program leaders, and researchers is to expand older adults' opportunities for civic contribution while addressing their economic, social, cultural, and physical diversity.

Bowling Alone or Blogging Together?

Putnam's claim that participation in virtually every traditional civic, social, and fraternal arena – including politics, churches, labor unions, parent–teacher associations, and even organized bowling leagues – has declined since the 1960s has generated vigorous discussion and debate among academics and policymakers alike. Critics claim that his argument relies too heavily on statistics showing declines in membership in fraternal organizations like the Boy Scouts, 4-H Clubs, Shriners, and Elks, but overlooks newer organizations that have risen to take their places, such as Habitat for Humanity and Meals on Wheels (Rich 1999; Schudson 2006). The United States, Putnam's detractors maintain, simply is experiencing a transformation in its forms of civic engagement rather than a decline. More traditional civic associations, such as the PTA., the League of Women Voters, or the American Legion, are giving way to new and alternative forms of political and civic engagement, particularly those facilitated by the Internet. Volunteering, in particular, is expanding through the Internet. Indeed, some researchers have faith that the Internet will launch and nurture forms of civic engagement that will be as important to communities as were the Rotary, Kiwanis, and Lions Clubs in past decades (Ester and Vinken 2003).

The Internet is creating new ways for people to connect and engage with one another. According to data from VolunteerMatch.com, an online database of volunteer jobs, interest in "virtual" or remote volunteering is rising dramatically. In 2005, volunteers matched themselves to nearly 40,000 virtual positions on VolunteerMatch.org, compared with 15,000 in 2000. The increase of websites like onlinevolunteering.org, an initiative by the United Nations Volunteers program, helps nonprofits around the world find online volunteers for jobs like editing or translating documents, writing newsletters, mentoring, and creating databases. Nonprofits and charities find that recruiting online volunteers not only helps them expand their searches beyond their immediate geographic areas but also helps them appeal to people with busy schedules and attract a more diverse and skilled volunteer corps than they otherwise might. Also, online volunteering offers those with physical limitations, home-based obligations, or transportation barriers opportunities to engage. People who volunteer through the Internet find that because they work around their own schedules – and avoid commuting – they are able to donate more time and energy to causes that really matter to them (Vail 2008).

Although Putnam worries about the Internet's tendency to separate people from each other rather than bring them together, others see the Internet as an opportunity to reestablish connections – through email lists, affinity groups, chat rooms, and blogs – that have been lost as fewer people join civic organizations (Uslaner 2004). Scholars like Ester and Vinken (2003:667) argue that the evidence about the Internet, taken together, makes "a strong case that the potential for advancing civil society is more likely to be rising than declining." Moreover, a study by Best and Krueger (2006) that uses nationally representative data and more refined measures of online social interactions than previously available finds that online interactions do foster connections critical to expanding social networks and producing residuals – jointly known as "social capital" – such as generalized trust, integrity, and reciprocity. Finally, although Gilleard, Hyde, and Higgs (2007) found a decreased attachment to the physical community with increased technology use, their data assert that communication technology use does not decrease the sense of trust and friendliness of a neighborhood.

Clearly, there is a need for research and theory on civic engagement to consider how the Internet is developing as an alternative to traditional civic engagement structures. As Ester and Vinken (2003) note, however, standard political and civic engagement participation scales found in mainstream survey research do not cover such new forms of political and civic involvement. The authors argue that scholars must develop ways to measure these contemporary forms of civic engagement before conclusions can be drawn regarding the decline of civic life in the twenty-first century.

A "Civic Core"?

Despite the popularity of Putnam's work, recent research finds little support for his claim that levels of civic engagement are declining in the general population (Foster-Bey et al. 2007) or that baby boomers will volunteer in smaller numbers than previous elderly cohorts (Einolf 2009; Rotolo and Wilson 2004). Putnam's observation that some individuals are considerably more engaged in civic activities than others, however, has stimulated new research that raises questions about the challenges of citizen engagement. Using data from the 2000 National Survey of Giving, Volunteering, and Participating conducted by Statistics Canada, Reed and Selbee (2001) identify the principal contributors in each of three modes of civic engagement: volunteering, charitable giving, and participation in community organizations. Their data reveal a small cadre – a civic core – of individuals who are involved heavily in giving, volunteering, and participating. Indeed, the authors find a remarkably high degree of concentration: one quarter of adults account for about three quarters of all time and money contributed to communal and charitable activities. Furthermore, the disparity in engagement levels likely is not unique to Canada. Several U.S. studies suggest that a small proportion of older adult volunteers account for the majority of volunteer time (Fischer, Mueller, and Cooper 1991; Kaskie et al. 2008). Burr, Mutchler, and Caro's (2007) finding that a small group of "super helpers" are engaged heavily in a cluster of productive activities in later life lends support to the thesis that most civic activity is done by a small proportion of the population. Although the finding may not seem surprising – and conforms to popular notions of "super volunteers" – its implications for strengthening civic engagement among older adults have not yet been addressed by researchers or policymakers. Clearly, further research is required to determine accurately the disparity in levels of civic activity among older Americans.

The Search for a Consensual Definition of Civic Engagement

Before researchers truly can measure civic engagement, they must agree on a common definition. To date, there is no prevailing definition, although many have been offered (National Academy on an Aging Society 2009a). The term currently encompasses a range of activities, "[f]rom volunteering to voting, from community organizing to political advocacy." Central to most characterizations is the notion that civic engagement is *action* that contributes "to the improvement of one's community, neighborhood, and nation" (Philanthropy for Active Civic Engagement 2009, ¶ 7). The types of activities identified as civic engagement fall into two broad categories: social (including most behaviors classified as volunteering and public service) and political (including participation in all levels of the political process; McBride 2006–2007). The multifaceted character of civic engagement likely is responsible for a lack of consensus regarding its definition. Not only does the concept of civic engagement cover a number of different activities, but it encompasses volunteering done on behalf of formal organizations, including schools, churches, hospitals, and non-profit organizations, and on behalf of informal entities, such as friends, neighbors, and relatives (Putnam 2000).

Further complicating matters is that some observers include both paid and unpaid contributions under the same rubric. The advent of incentive programs, such as educational vouchers, stipends, and other forms of compensation has blurred the line between paid and unpaid participation. Moreover, some scholars argue that unpaid caregiving constitutes a vital form of civic engagement (Herd and Meyer 2002; Martinson and Minkler 2006). Others suggest that civic engagement includes the acquisition of knowledge and skills required to perform various civic actions (Fisher, McInerney, and Petersen 2005).

Recently, Toppe and Galaskiewicz (2006) proposed that civic engagement be conceptualized and measured along four dimensions: (1) helping, including volunteering and neighboring; (2) giving to people, organizations, and causes; (3) influencing, including political and cause-related activities; and (4) participating, as in group memberships or bowling leagues. Their detailed typology has been applied in recent work (e.g., Yoshida et al. 2008) and shows promise as a starting point for a systematic approach to the conceptualization and measurement of civic engagement. Given widespread efforts to promote civic engagement, a more precise definition will help policymakers and program administrators target the activities that receive public and private support. Further, as Kaskie et al. (2008: 370) note, "being able to define and differentiate individuals who occupy this role may help researchers conduct a more rigorous evaluation of the relationships among civic engagement, social capital, and individual health outcomes."

Policy Implications

The retirement of the baby boom cohort over the next 20 years offers an unprecedented opportunity to tap into a large base of potential volunteers. But policy interventions are needed to ensure that all older adults have opportunities to engage in volunteer activities that can benefit the community as well as the volunteers themselves (Kahana and Force 2008). Key findings we have highlighted can guide policy initiatives to maximize volunteer recruitment, retention, and benefits. Evidence suggests that the most effective way to recruit a volunteer is to ask them (Independent Sector 2000). In light of the fact that the American Time Use Survey finds that adults aged 65 and over spend much more time watching television (4 h) than do younger adults (2.5 h), it may prove effective to disseminate such information through public awareness campaigns that would educate older adults about volunteer opportunities and the positive health outcomes of engagement (Kaskie et al. 2008). Further, media campaigns targeting those of lower socioeconomic status might yield the largest payoffs because these individuals report the lowest levels of volunteer activity. Another recruitment strategy could come from the finding that past volunteering is the leading predictor of future volunteering. Zedlewski (2007), for example, showed that the share of working persons who start volunteering after retirement is much smaller than those who continue to volunteer after retirement. With little evidence that retirement inspires volunteering, recruiting initiatives may experience greater payoffs by targeting young and middle-aged adults while they are still in the workforce. The strong link between work and volunteering spotlights the critical role of the business sector in promoting volunteering among the aging population. Work-based volunteer programs in particular have the potential to promote a commitment to volunteering among workers that will extend into retirement (Gonyea and Googins 2006–2007). Recognizing the importance of volunteering continuity, policymakers might offer subsidies, tax credits, and other incentives to encourage employers to expand community service opportunities for their employees as they approach retirement (O'Neill 2006–2007).

From an institutional perspective, research reviewed here has demonstrated that voluntary organizations can facilitate the inclusion of older adults from diverse backgrounds by offering stipends, increasing role flexibility in the volunteer position, and providing public recognition for the volunteers' contributions. Furthermore, expanding home-based volunteer opportunities (via telephone or computer) could increase access for older adults with disabilities, transportation difficulties, or mobility challenges. Policymakers also can use rewards and incentives to encourage activities that benefit the public. Studies suggest that small incentives – such as education credits, access to group health insurance, or a modest monthly stipend – might reap large benefits by attracting more adults into community service (Bridgeland, Putnam, and Wofford 2008). Indeed, the utilization of stipended volunteers long has been seen as a route to maximizing the participation of low-income older adults. For example, older adults at or below 125% of the poverty line can serve in the voluntary programs of the national Senior Corps for 15–20 h per week in exchange for a modest stipend of $2.65 an hour. Also, under Experience Corps, a school-based program, older adults work one-on-one with young children, create and run before- and after-school programs, and receive modest stipends for their services. Not only does this program reach out to an underutilized group of volunteers (generally older, low-income African-Americans), but it also helps to address their economic needs. Morrow-Howell and colleagues (2003, 2009) go so far as to recommend that because of the benefits reaped, it might make sense to select low-income and minority volunteers because of the greater likelihood of maximizing returns to the individuals.

Although advocates for these programs agree that they do a great deal to facilitate the involvement of low-income older adults, the age, income, and commitment requirements often are viewed as exclusive of a broad population of older adults – especially among baby boomers (Freedman 1999). In April 2009, however, President Obama signed the Edward M. Kennedy Serve America Act; it calls for the increased engagement of older adults in volunteer roles, allotting 10% of AmeriCorps' total FY 2010 budget of $372.5 million for organizations enrolling adults aged 55 and older, and funding $1,000 higher education awards – transferable to children and grandchildren – to older volunteers who contribute 350 h of service annually (Edward M. Kennedy Serve America Act 2009, § 2142). The new law also expands service options for older Americans by lowering the age requirement for the Foster Grandparent and Senior Companion programs from 60–55, and increasing hourly stipend eligibility for these programs from 125 to 200% of the federal poverty level.

The passage of this landmark legislation – the largest expansion of national service since the Civilian Conservation Corps during the Great Depression – highlights the need to increase the amount of science-based research aimed at improving recruitment and retention of older volunteers.

Conclusion

Civic engagement among older adults is gaining attention in both popular and academic press. As the health and education of aging Americans continue to increase, so does the opportunity to engage this population in civic activities aimed at improving communities. Furthermore, older adults' volunteer activities help governments and nonprofit organizations meet the growing demand for social services. As we have highlighted, increased volunteerism has a palpable effect on the economy, and potentially reduces government costs, as the resulting healthier older adults require less health care (Zedlewski and Butrica 2007).

This win–win scenario has captured the attention of social scientists, policymakers, and nonprofit leaders who promote civic engagement and has spawned a host of related activities. The three major professional associations focused on aging in this country – the American Society on Aging, The Gerontological Society of America, and the National Council on Aging – have made civic engagement a programmatic priority. Research on volunteering has benefitted from keen interest by private foundations. The Atlantic Philanthropies and the MetLife Foundation, in particular, have invested significant resources in program and policy developments. These developments have created opportunities for applied research and community-academic partnerships. Research has played an important role in building a civic engagement movement and expanding the Civic Enterprise (National Academy on an Aging Society 2009b), the network of interrelated initiatives focused on civic engagement among older adults. For example, the accumulation of more and better evidence on the health benefits of volunteering is responsible for the push to expand volunteer opportunities as a public health strategy. In other ways, however, action is outstripping knowledge development, especially in regards to effective strategies to recruit and retain older volunteers. Applied social science knowledge about engaging older volunteers to guide program and policy initiatives has never been more important.

References

Adelmann, Pamela K. 1994. "Multiple Roles and Psychological Well-Being in a National Sample of Older Adults." *Journal of Gerontology: Social Sciences* 49:S277–85.

Atchley, Robert C. 1999. *Continuity and Adaptation in Aging: Creating Positive Experiences*. Baltimore, MD: Johns Hopkins University Press.

Atchley, Robert C. 1971. "Retirement and Leisure Participation: Continuity or Crisis?" *The Gerontologist* 11(1):13–17.

Barker, Judith C. 2002. "Neighbors, Friends, and Other Nonkin Caregivers of Community-Living Dependent Elders." *Journal of Gerontology: Social Sciences* 57:S158–67.

Best, Samuel J. and Brian S. Krueger. 2006. "Online Interactions and Social Capital: Distinguishing between New and Existing Ties." *Social Science Computer Review* 24(4):395–410.

Blieszner, Rosemary, and Lisa M. Artale. 2001. "Benefits of Intergenerational Service-Learning to Human Services Majors." *Educational Gerontology* 27:71–87.

Bridgeland, John M., Robert D. Putnam, and Harris L. Wofford. 2008. *More to Give: Tapping the Talents of the Baby Boomer, Silent and Greatest Generations*. Washington, D.C.: AARP.

Bureau of Labor Statistics. 2010. *Volunteering in the United States, 2009*. Washington, D.C.: Bureau of Labor Statistics, United States Department of Labor.

Burr, Jeffrey A., Jan E. Mutchler, and Francis G. Caro. 2007. "Productive Activity Clusters among Middle-Aged and Older Adults: Intersecting Forms and Time Commitments." *Journal of Gerontology: Social Sciences* 62B:267–75.

Butrica, Barbara A., Richard W. Johnson, and Sheila R. Zedlewski. 2009. "Volunteer Dynamics of Older Americans." *Journal of Gerontology: Social Sciences* 64B:644–55.

Butrica, Barbara A. and Simone G. Schaner. 2005. *Satisfaction and Engagement in Retirement*. Washington, D.C.: The Urban Institute.

Cannuscio, Carolyn, Jason Block, and Ichiro Kawachi. 2003. "Social Capital and Successful Aging: The Role of Senior Housing." *Annals of Internal Medicine* 139(5, Part 2):395–400.

Carlson, Michelle C., Jane S. Saczynski, George W. Rebok, Teresa Seeman, Thomas A. Glass, Sylvia McGill, James Tielsch, Kevin D. Frick, Joel Hill, and Linda P. Fried. 2008. "Exploring the Effects of an 'Everyday' Activity Program on Executive Function and Memory in Older Adults: Experience Corps®." *The Gerontologist* 48:793–801.

Carstensen, Laura L. 1992. "Social and Emotional Patterns in Adulthood: Support for Socioemotional Selectivity Theory." *Psychology and Aging* 7(3):331–338.

Casner-Lotto, Jill. 2007. *Boomers are ready for nonprofits, but are nonprofits ready for them?* Washington, D.C.: The Conference Board. Retrieved January 2, 2010 (http://www.conference-board.org/publications/describe.cfm?id=1319).

Center for Health Communication, Harvard School of Public Health. 2004. *Reinventing Aging: Baby Boomers and Civic Engagement*. Boston: Harvard School of Public Health.

Chambré, Susan M. and Christopher J. Einolf. 2008. *Is Volunteering Work, Prosocial Behavior, or Leisure?* (Working Paper Series). New York: Center for Nonprofit Strategy and Management, School of Public Affairs, City University of New York.

Cheng, Sheung-Tak. 2009. "Generativity in Later Life: Perceived Respect from Younger Generations as a Determinant of Goal Disengagement and Psychological Well-Being." *Journal of Gerontology: Psychological Sciences* 64B:45–54.

Cnaan, Ram A., Femida Handy, and Margaret Wadsworth. 1996. "Defining Who is a Volunteer: Conceptual and Empirical Considerations." *Nonprofit and Voluntary Sector Quarterly* 25(3):364–83.

Cnaan, Ram A. and Toni A. Cascio. 1998. "Performance and Commitment: Issues in Management of Volunteers in Human Service Organizations." *Journal of Social Service Research* 24:1–37.

Coleman, Patricia and Susan DeRosa. 2006. "High School Students Take Classes in a Nursing Home." *The American Journal of Nursing* 106(12):49–50.

Cumming, Elaine and William E. Henry, eds. 1961. *Growing Old: The Process of Disengagement*. New York: Basic Books.

Edward M. Kennedy Serve America Act. Pub. L. No. 111–13, § 2142 2009.

Einolf, Christopher J. 2009. "Will the Boomers Volunteer during Retirement? Comparing the Baby Boom, Silent, and Long Civic Cohorts." *Nonprofit and Voluntary Sector Quarterly 38*:181–99.

Eisner, David, Robert T. Grimm, Jr., Shannon Maynard, and Susannah Washburn. 2009. "The New Volunteer Workforce." *Stanford Innovation Review* Winter.

Ekerdt, David. 1986. "The Busy Ethic: Moral Continuity between Work and Retirement." *The Gerontologist* 26:239–44.

Erikson, Erik H., Joan M. Erikson, and Helen Q. Kivnick. 1986. *Vital Involvement in Old Age.* New York: Norton.
Ester, Peter and Henk Vinken. 2003. "On the Fear For Civic Decline and Hope for the Internet Alternative." *International Sociology* 18(4):659–80.
Fengler, Alfred P. 1984. "Life Satisfaction of Subpopulations of Elderly: The Comparative Effects of Volunteerism, Employment, and Meal Site Participation." *Research on Aging* 6:189–212.
Finkelstein, Marcia A., Louis A. Penner, and Michael T. Brannick. 2005. "Motive, Role Identity, and Prosocial Personality as Predictors of Volunteer Activity." *Social Behavior and Personality* 33(4):403–18.
Fischer, Lucy Rose, Daniel P. Mueller, and Philip W. Cooper. 1991. "Older Volunteers: A Discussion of the Minnesota Senior Study." *The Gerontologist* 31:183–94.
Fisher, Dana, Paul-Brian McInerney, and Kevin Petersen. 2005, August. *Reconceptualizing the Notion of Civic Engagement.* Paper presented at the annual meeting of the American Sociological Association, Philadelphia, PA. Retrieved January 2, 2010 (http://www.allacademic.com/meta/p22852_index.html).
Foster-Bey, John, Robert Grimm, Jr., and Nathan Dietz. 2007. *Keeping Baby Boomers Volunteering: A Research Brief on Volunteer Retention and Turnover.* Retrieved January 2, 2010 from the Corporation for National and Community Service, Office of Research and Policy Development Web site (http://www.nationalservice.gov/pdf/07_0307_boomer_report.pdf).
Freedman, Marc. 1999. *Prime time: How Baby Boomers Will Revolutionize Retirement and Transform America.* New York: Public Affairs.
Frick, Kevin D., Michelle C. Carlson, Thomas A. Glass, Sylvia McGill, George W. Rebok, Crystal Simpson, and Linda P. Fried. 2004. "Modeled Cost-Effectiveness of the Experience Corps Baltimore Based on a Pilot Randomized Trial." *Journal of Urban Health: Bulletin of the New York Academy of Medicine* 81:106–17.
Fried, Linda P., Michelle C. Carlson, Marc Freedman, Kevin D. Frick, Thomas A. Glass, Joel Hill, Sylvia McGill, George W. Rebok, Teresa Seeman, James Tielsch, Barbara A. Wasik, and Scott Zeger. 2004. "A Social Model for Health Promotion for an Aging Population: Initial Evidence on the Experience Corps Model." *Journal of Urban Health: Bulletin of the New York Academy of Medicine* 81:64–78.
Gilleard, Chris, Martin Hyde, and Paul Higgs. 2007. "Community and Communication in the Third Age: The Impact of Internet and Cell Phone Use on Attachment to Place in Later Life in England." *Journal of Gerontology: Social Sciences* 62B:276–83.
Glass, Thomas A., Carlos F. Mendes de Leon, Shari S. Bassuk, and Lisa F. Berkman. 2006. "Social Engagement and Depressive Symptoms in Late Life: Longitudinal Findings." *Journal of Aging and Health* 18:604–28.
Glass, Thomas A., Carlos Mendes de Leon, Richard A. Marottoli, and Lisa F. Berkman. 1999. "Population Based Study of Social and Productive Activities as Predictors of Survival among elderly Americans." *British Medical Journal* 319:478–83.
Gonyea, Judith G. and Bradley K. Googins. 2006–2007. "Expanding the Boundaries of Corporate Volunteerism: Tapping the Skills, Talent, and Energy of Retirees." *Generations* 30(4):78–84.
Greenfield, Emily A. and Nadine F. Marks. 2004. "Formal Volunteering as a Protective Factor for Older Adults' Psychological Well-Being." *Journal of Gerontology: Social Sciences* 59B:S258–64.
Grimm, Robert Jr., Nathan Dietz, and John Foster-Bey. 2006. *Volunteer Growth in America: A Review of Trends since 1974.* Washington, D.C.: Corporation for National and Community Service.
Grimm, Robert Jr., Kimberly Spring, and Nathan Dietz. 2007. *The Health Benefits of Volunteering.* Washington, D.C.: Corporation for National and Community Service.
Grundy, Emily, Astrid Fletcher, Sarah Smith, and Donna Lamping. 2007. *Successful Ageing and Social Interaction: A Policy Brief.* London: The International Longevity Centre, United Kingdom.
Hager, Mark A. and Jeffrey L. Brudney. 2004. *Volunteer Management Capacity in America's Charities and Congregations: A Briefing Report.* Washington, D.C.: Urban Institute.
Havighurst, Robert J. 1963. "Successful Aging." Pp. 299–320 in *Processes in Aging*, Vol. 1, edited by R. H. Williams, C. Tibbetts, & W. Donohue. New York: Atherton.
Hao, Yanni. 2008. "Productive Activities and Psychological Well-Being among Older Adults." *Journal of Gerontology: Social Sciences* 63B:S64–72.
Hendricks, Jon and Stephen J. Cutler. 2004. "Volunteerism and Socioemotional Selectivity in Later Life." *Journal of Gerontology: Social Sciences* 59B:251–57.
Herd, Pamela and Madonna Harrington Meyer. 2002. "Care work: Invisible Civic Engagement." *Gender & Society* 16(5):665–88.
Hinterlong, James E., Nancy Morrow-Howell, and Philip A. Rozario. 2007. "Productive Engagement and Late Life Physical and Mental Health: Findings from a Nationally Representative Panel Study." *Research on Aging* 29:348–70.
Holstein, Martha. 2006. "A Critical Reflection on Civic Engagement." *Public Policy & Aging Report* 16(4):21–6.
Hong, Song-Iee, Nancy Morrow-Howell, Fengyan Tang, and James Hinterlong. 2009. "Engaging Older Adults in Volunteering: Conceptualizing and Measuring Institutional Capacity." *Nonprofit and Voluntary Sector Quarterly* 38(2):200–19.
Hudson, Robert B. 2008. "Engaging Boomers: Hope or Hype?" *The Gerontologist* 48(1):124–26.

Independent Sector. 2000. *America's Senior Volunteer.* Washington, D.C.: Independent Sector.
Jirovec, Ronald L. and Christine A. Hyduk. 1998. "Type of Volunteer Experience and Health among Older Adult Volunteers." *Journal of Gerontological Social Work* 30(3):29–42.
Johnson, Richard W. and Simone G. Schaner. 2005. *Value of Unpaid Activities by Older Americans Tops $160 Billion per Year.* Washington, D.C.: The Urban Institute.
Kahana, Jeffrey and Lawrence T. Force. 2008. "Toward Inclusion: A Public-Centered Approach to Promote Civic Engagement by The Elderly." *Public Policy & Aging Report* 18(3):30–35.
Kaskie, Brian, Sara Imhof, Joseph Cavanaugh, and Kennith Culp. 2008. "Civic Engagement as a Retirement Role for Aging Americans." *The Gerontologist* 48(3):368–77.
Katz, Stephen. 2001–2002. "Growing Older Without Aging? Positive Aging, Anti-Ageism, and Anti-Aging." *Generations* 25(4):27–32.
Langer, Ellen J. and Judith Rodin. 1976. "The Effects of Choice and Enhanced Personal Responsibility for The Aged: A Field Experiment in an Institutional Setting." *Journal of Abnormal & Social Psychology* 34:191–98.
Larkin, Elizabeth and Sally Newman. 2001. "Benefits of Intergenerational Staffing in Preschools." *Educational Gerontology* 27:373–85.
Li, Yunqing. 2007. "Recovering From Spousal Bereavement in Later Life: Does Volunteer Participation Play a Role?" *Journals of Gerontology: Social Sciences 62B*:S257–66.
Luoh, Ming-Ching and A. Regula Herzog. 2002. "Individual Consequences of Volunteer and Paid Work in Old Age: Health and Mortality." *Journal of Health and Social Behavior* 43:490–509.
Mannell, Roger C. 1993. "High-Investment Activity and Life Satisfaction among Older Adults: Committed, Serious Leisure, and Flow Activities." Pp. 123–45 in *Activity and Aging; Staying Involved in Later Life*, edited by J. R. Kelly. Thousand Oaks, CA: Sage Publications.
Martinson, Marty and Meredith Minkler. 2006. "Civic Engagement and Older Adults: A Critical Perspective." *The Gerontologist* 46:318–24.
McBride, Amanda Moore 2006–2007. "Civic Engagement, Older Adults and Inclusion." *Generations* 30(4):66–71.
McBride, Amanda Moore, Ernest Gonzales, Nancy Morrow-Howell, and Stacey McCrary. 2009. *A Case for Stipends.* Working paper. St. Louis: Center for Social Development, Washington University.
Morrow-Howell, Nancy. 2006–2007. "Civic Service across the Life Course." *Generations* 30(4):37–42.
Morrow-Howell, Nancy, James Hinterlong, and Michael Sherraden. 2001. *Productive Aging: Concepts and Challenges*. Baltimore, MD: The Johns Hopkins University Press.
Morrow-Howell, Nancy, Melinda Carden, and Michael Sherraden. 2005. "Volunteerism, philanthropy, and service." in *Perspectives on productive aging*, edited by L. Kaye. Washington, D.C.: NASW.
Morrow-Howell, Nancy, James Hinterlong, Philip A. Rozario, and Fengyan Tang. 2003. "Effects of Volunteering on the Well-Being of Older Adults." *Journals of Gerontology: Social Sciences* 53B:S137–45.
Morrow-Howell, Nancy, Song-Iee Hong, and Fengyan Tang. 2009. "Who Benefits from Volunteering? Variations in Perceived Benefits." *The Gerontologist* 49:91–102.
Muir, Kenneth B. 2006. "Measuring the Benefits of Mentoring for Foster Grandparents: A Research Note." *Educational Gerontology* 32:379–387.
Musick, Marc A., A. Regula Herzog, and James S. House. 1999. "Volunteering and Mortality among Older Adults: Findings from a National Sample." *Journals of Gerontology: Social Sciences* 54B:S173–80.
Musick, Marc A. and John Wilson. 2008. *Volunteers: A Social Profile.* Bloomington, IN: Indiana University Press.
Mutchler, Jan E., Jeffrey A. Burr, and Francis G. Caro. 2003. "From Paid Worker to Volunteer: Leaving the Paid Workforce and Volunteering in Later Life." *Social Forces* 81:1267–93.
National Academy on an Aging Society. 2009a. *Definitions of civic engagement.* Retrieved December 22, 2009 Retrieved (http://www.civicengagement.org/agingsociety/links/CEdefinitions.pdf).
———. 2009b. *The Civic Enterprise.* Retrieved August 18, 2009 (http://www.agingsociety.org/agingsociety/civic-engagement-diagram.html).
Netting, F. Ellen, H. Wayne Nelson, Jr., Kevin Borders, and Ruth Huber. 2004. "Volunteer and Paid Staff Relationships: Implications for Social Work Administration." *Administration in Social Work* 28(3):69–89.
Neugarten, Bernice, Robert Havighurst, and Sheldon Tobin. 1968. "Personality and Patterns of Aging." In *Middle Age and Aging*, edited by B. L. Neugarten. Chicago: University of Chicago Press.
Newman, Sally, Jyotsna Vasudev, and Roland Onawola. 1985. "Older Volunteers' Perceptions of Impacts of Volunteering on Their Psychological Well-Being." *Journal of Applied Gerontology* 4:123–127.
Okun, Morris A. and Amy Schultz. 2003. "Age And Motives for Volunteering: Testing Hypotheses Derived from Socioemotional Selectivity Theory." *Psychology and Aging* 18:231–239.
O'Neill, Greg. 2006–2007. "Civic Engagement on the Agenda at the 2005 White House Conference on Aging." *Generations* 30(4):101–8.
Park, Denise C., Angela H. Gutchess, Michelle L. Meade, and Elizabeth A. L. Stine-Morrow. 2007. "Improving Cognitive Function in Older Adults: Nontraditional Approaches." *Journals of Gerontology: Social Sciences* 62B:Special Issue 45–52.

Peace Corps. 2009. *Fast Facts*. Retrieved January 28, 2010 (http://www.peacecorps.gov/index.cfm?shell=learn.whatispc.fastfacts).

Philanthropy for Active Civic Engagement. 2009. *Welcome to PACE - Philanthropy for Active Civic Engagement*. Retrieved December 22, 2009 (http://www.pacefunders.org/index.html).

Putnam, Robert D. 2000. *Bowling Alone*. New York: Simon and Schuster.

Rebok, George W., Michelle C. Carlson, Thomas A. Glass, Sylvia McGill, Joel Hill, Barbara A. Wasik, Nicholas Ialongo, Kevin D. Frick, Linda P. Fried, and Meghan D. Rasmussen. 2004. "Short-Term Impact of Experience Corps Participation on Children and Schools: Results from a Pilot Randomized Trial." *Journal of Urban Health: Bulletin of the New York Academy of Medicine* 81:79–93.

Reed, Paul B. and L. Kevin Selbee. 2001. "The Civic Core in Canada: Disproportionality in Charitable Giving, Volunteering, and Civic Participation." *Nonprofit and Voluntary Sector Quarterly* 30:761–80.

Rich, Paul. 1999. "American Voluntarism, Social Capital, and Political Culture." *The Annals of the American Academy of Political and Social Science* 565:15–34.

Riley, Matilda White and John W. Riley, Jr. 1994. "Age Integration and the Lives of Older People." *The Gerontologist* 34:110–15.

Rogers, Anita M. and Andrea S. Taylor. 1997. "Intergenerational Mentoring: A Viable Strategy for Meeting the Needs of Vulnerable Youth." *Journal of Gerontological Social Work* 28:125–140.

Rotolo, Thomas and John Wilson. 2004. "What Happened to the 'Long Civic Generation'? Explaining Cohort Differences in Volunteerism." *Social Forces* 82:1091–21.

Rowe, John W. 1997. "The New Gerontology." *Science* 278(5337):367.

Rowe, John W. and Robert L. Kahn. 1997. "Successful aging." *The Gerontologist* 37:433–40.

———. 1998. *Successful Aging*. New York: Pantheon.

Schudson, Michael. 2006. "The Varieties of Civic Experience." *Citizenship Studies* 10(5):591–606.

Semenza, Jan C., Carol H. Rubin, Kenneth H. Falter, Joel D. Selanikio, W. Dana Flanders, Holly L. Howe, and John L. Wilhelm. 1996. "Heat-Related Deaths during the July 1995 Heat Wave in Chicago." *New England Journal of Medicine* 335(2):84–90.

Sherraden, Michael, Nancy Morrow-Howell, James Hinterlong, and Philip Rozario. 2001. "Productive Aging: Theoretical Choices and Directions." Pp. 260–84 in *Productive Aging: Concepts and Challenges*, edited by N. Morrow-Howell, J. Hinterlong, & M. Sherraden. Baltimore: Johns Hopkins University.

Steverink, Nardi and Siegwart Lindenberg. 2006. "Which Social Needs are Important for Subjective Well-Being? What Happens to Them with Aging?" *Psychology and Aging* 21:281–290.

Stine-Morrow, Elizabeth A. L., Jeanine M. Parisi, Daniel G. Morrow, Jennifer Greene, and Denise C. Park. 2007. "An Engagement Model of Cognitive Optimization through Adulthood." *Journals of Gerontology: Social Sciences,* 62B:Special Issue 62–9.

Sugihara, Yoko, Hidehiro Sugisawa, Hiroshi Shibata, and Ken Harada. 2008. "Productive Roles, Gender, and Depressive Symptoms: Evidence from A National Longitudinal Study of Late-Middle-Aged Japanese." *Journal of Gerontology: Psychological Sciences* 63B:P227–34.

Sun, Rongjun and Yuzhi Liu. 2006. "Mortality of the Oldest Old in China: The Role of Social and Solitary Customary Activities." *Journal of Aging and Health* 18:37–55.

Tan, Erwin J., George W. Rebok, Qilu Yu, Constantine E. Frangakis, Michelle C. Carlson, Tao Wang, Michelle Ricks, Elizabeth K. Tanner, Sylvia McGill, and Linda P. Fried. 2009. "The Long-Term Relationship between High-Intensity Volunteering and Physical Activity in Older African American Women." *Journal of Gerontology: Social Sciences* 64B:304–11.

Tang, Fengyan. 2006. "What Resources are Needed for Volunteerism? A Life Course Perspective." *The Journal of Applied Gerontology* 25:375–90.

———. 2009. "Late-Life Volunteering and Trajectories of Physical Health." *Journal of Applied Gerontology* 28:524–533.

Tang, Fengyan, Nancy Morrow-Howell, and Song-Iee Hong. 2008. "Inclusion of Diverse Older Populations in Volunteering: The Importance of Institutional Facilitations." *Nonprofit and Voluntary Sector Quarterly* 38(5): 810–27.

Thoits, Peggy A. and Lynda N. Hewitt. 2001. "Volunteer Work and Well-Being." *Journal of Health and Social Behavior* 42:115–131.

Toppe, Christopher and Joe Galaskiewicz. 2006. *Measuring volunteering: Committee report*. Retrieved January 2, 2010 (http://archive.pointsoflight.org/downloads/pdf/resources/research/CommitteeReport.pdf).

Uslaner, Eric M. 2004. "Trust, Civic Engagement, and the Internet." *Political Communication* 21(2):223–42.

Vail, Julia. 2008, November 12. Virtual Volunteering Widens Opportunity. *Philanthropy Journal*. Retrieved January 2, 2010 (http://www.philanthropyjournal.org/resources/ special-reports/volunteers-boards/virtual-volunteering-widens-opportunity).

Van Willigen, Marieke. 2000. "Differential Benefits of Volunteering Across the Life Course." *Journal of Gerontology: Social Sciences* 55B:S308–S18.

Wheeler, Judith A., Kevin M. Gorey, and Bernard Greenblatt. 1998. "The Beneficial Effects of Volunteering for Older Volunteers and The People They Serve: A Meta-Analysis." *International Journal of Aging and Human Development* 47:69–79.

World Health Organization. 2002. *Active Ageing: A Policy Framework.* Geneva, Switzerland: World Health Organization.

Wilson, John. 2000. "Volunteering." *Annual Review of Sociology* 26:215–40.

Wilson, John and Marc A. Musick. 1999. "Attachment to Volunteering." *Sociological Forum* 14(2):243–272.

Wilson, Robert S., Julie A. Schneider, Steven E. Arnold, Julia L. Bienias, and David A. Bennett. 2007. "Conscientiousness and the Incidence of Alzheimer Disease and Mild Cognitive Impairment." *Archives of General Psychiatry* 64:1204–12.

Windsor, Timothy D., Kaarin J. Anstey, and Bryan Rodgers. 2008. "Volunteering and Psychological Well-Being among Young-Old Adults: How Much Is Too Much?" *The Gerontologist* 48:59–70.

Wymer, Walter W. 1999. "Understanding Volunteer Markets: The Case of Senior Volunteers." *Journal of Nonprofit & Public Sector Marketing* 6(2/3):1–23.

Yoshida, Hitomi, Daryl Gordon, and Nancy Henkin. 2008. *Community Treasures: Recognizing the Contributions of Older Immigrants and Refugees*. Philadelphia, PA: Temple University Center for Intergenerational Learning.

Zedlewski, Sheila R. 2007, December. *Will Retiring Boomers Form A New Army Of Volunteers?* (The Retirement Project, Perspectives on Productive Aging No. 7). Washington, D.C.: The Urban Institute.

Zedlewski, Sheila R. and Barbara A. Butrica. 2007. *Are We Taking Full Advantage of Older Adults' Potential?* Washington, D.C.: The Urban Institute.

Zedlewski, Sheila R. and Simone G. Schaner. 2005, July. *Older Adults' Engagement Should Be Recognized and Encouraged* (The Retirement Project, Perspectives on Productive Aging No. 1). Washington, D.C.: The Urban Institute.

———. 2006, May. *Older Adults Engaged as Volunteers* (The Retirement Project, Perspectives on Productive Aging No. 5). Washington, D.C.: The Urban Institute.

Chapter 22
Business and Aging: The Boomer Effect on Consumers and Marketing

Janice I. Wassel

This chapter provides a foundation for understanding businesses' emerging interest in the older adult as a consumer and introduces the silver market phenomenon dedicated to the mature market. We present the influence of consumerism on the life courses of American baby boomers and the importance of considering the life course perspective for researching older adult consumers. Finally, we discuss the potential of aging boomers and older adults as valuable economic contributors.

With increasing numbers of people currently living to old age and the first American baby boomer reaching the traditional retirement age of 65 years in 2011, we experience the demographic shift from America being a young country to an aging country. The success of longevity over the past 50 years spawns major challenges for vast and complex social and economic systems on which government agencies, businesses, and individuals depend. In response to this demographic shift, attention has been focused on the projected costs and issues of an aging society. Recurrent themes characterize older adults as unproductive members of society and "passive consumers of public services, especially of health care and pensions" (Moody 2008:4).

The meaning of old age rooted in social and cultural contexts of classic age-graded roles diminishes with the democratization of old age and the democratization of good health at older ages (Cutler 2002). In the past, little attention was paid to the nearly 80% of adults aged 65 and over with no limitations in movement, sensory, or cognition as consumers (National Center for Health Statisics 2009). Expectations of societal disengagement after orderly transitions from birth to death through education, family, and work stages to retirement and death no longer holds for engaged older adults and aging boomers (Dannefer and Miklowski 2006). Businesses and nonprofits now find that older adults continue to be productive and active consumers. As the sizeable boomer cohort approaches the traditional retirement age of 65, there is increased interest in the mature market, a traditionally neglected group. The United States lags behind aging countries such as Japan and Germany in valuing the older consumer.

Macrolevel economics of aging focus on retirement saving and policy, labor market behavior, sustainability of Social Security, and relationship of health and economic circumstances. Aging and business concentrate on developing and marketing products for older adults (Coulmas 2008). Aging and business address the opportunity for profits while providing services and products and meeting the needs of the growing population of older adult consumers. Neither is the economics of aging and business mutually exclusive, but complementary in seeking to better the lives of older adults.

J.I. Wassel (✉)
University of North Carolina, Greensboro, NY, USA

R.A. Settersten, Jr. and J.L. Angel (eds.), *Handbook of Sociology of Aging*, Handbooks of Sociology and Social Research, DOI 10.1007/978-1-4419-7374-0_22, © Springer Science+Business Media, LLC 2011

Consumers Through the Life Course Lens

The richness of the life course perspective provides a method to examine dynamic influences on lives at earlier years to explain outcomes at later years. The life course perspective measures continuity and change sequences as individuals move through complex roles and the influence that the sequencing and timing these roles have on later outcomes (Elder 2006). Moreover, life course analysis may be used at a population level to examine the cumulative effect of social and historical change resulting in intracohort and intercohort variation. For example, the life course perspective helps understand the *cumulative advantage/disadvantage* of inequality as a cohort ages in boomer's wealth. Recently, the life course analysis serves as a paradigm for family physicians understanding individual patients' health in context over time (Daaleman and Elder 2007; Dannefer and Miklowski 2006; Hughes and O'Rand 2004).

Moschis (2007) proposes that the life course perspective is an appropriate methodology with which to study consumer behavior. Current consumer research has been dominated by cross-sectional studies observing different age groups and then comparing consumption behaviors for different age groups. Studies have not considered consumption patterns over the individual's life course or how past experiences within the historical or social context may influence consumption later in the life course (p. 295). For instance, while recent studies conclude that stressful family events during adolescence affect the development of materialism and compulsive consumptions, Moschis suggests that conclusions would be richer if the timing and duration of the stressful events were measured. Television's role or peer influence on consumer behavior might be advanced if individual exposure and timing was integrated into the research model. Although marketers use role transitions or events as criteria to target certain consumption behaviors, they do not consider how life events and historical consumption patterns contribute to consumer behavior.

As individuals age and move through complex roles within social and historical time, patterns of consumption and saving emerge. Modigliani and Brumberg (1954) developed the "Life Cycle Theory of Consumption" whereby individuals make rational consumption choices fluctuating by individuals' age. Based on total expected lifetime earnings, individuals at different ages adjust personal spending. Savings behavior, they conclude, is driven by the desire to prepare for retirement. They observed that consumption and income often are unequal throughout the life cycle. Younger individuals' consumption usually exceeds income. By middle age, consumption tends to level or decrease while earnings, on average, increase resulting in higher savings levels. At older ages and in retirement when incomes declined, individuals use savings for consumption. This model of savings and consumption assumes a static and traditional life course from education to work and family to retirement. Yet, the role of consumption and savings throughout the life course may have changed since the 1950s as individuals' life courses have changed.

In societies that value consumption, older adults with resources and with a life history of consumption are increasingly important. Rather than devaluation at retirement as modernization theory suggests, older consumers' status should increase as valuable economic contributors (Blaikle 2006). Older adults' consumption may be constrained by health status and resources, but it may also be biased by the perception of them as poor consumers. The demographics of the baby boom and their role as consumers throughout their lives should influence businesses to see older adults as a viable and profitable market.

The American Baby Boom: A Business Opportunity

> Your future is great in a growing America. Every day 11,000 babies are born in America. This means new businesses, new jobs, new opportunities" read public service posters hanging in the New York Subway in the 1950s (Jones 1980:37).

Boomers have been a business phenomenon since they were babies and will continue through old age. The baby boom that followed World War II was a "boom" for businesses, marketers, and the economy (Lindop and DeCapua 2010). Beginning in 1946 and lasting through 1964, America experienced unparalleled and unanticipated fertility. Historically a young nation, with each baby boomer born America became even younger. The median age of the general population dropped from 30.2 years in 1950 to 26.3 years by 1970 (U.S. Census Bureau 2003). The economic and public spotlight on the thousands of babies increased the status of children. Children became a focal point for families, government policy, and business. Public service announcements posted in the New York subway stations during the 1950s advertised innovative and ever-increasing business possibilities this surge of babies and expanding families brought to the American market (Jones 1980). This flood of babies both shaped and was shaped by the culture and times in which the baby boomers were born. The life course perspective is constructive to understand how boomers influenced businesses since infancy and businesses marketed to boomers as they aged.

The boomers are the first generation of children cultivated as consumers through a new technology, television. As the baby boomers grew into kids, teens, and young adults, they set the fads, spent the money (even when their parents paid), and popularized new products and services (Jones 1980). In the 1960s, youth spending grew the economy, on average $12 billion yearly. Fifty-three percent of boomer teens purchased movie tickets, 43% records, and 20% of high school seniors drove their own cars (Hamilton et al. 2000). Never before in American history had youth influenced the economy as the uninterrupted economic growth of the 1960s surpassed the prosperous 1950s. Yet disparity persisted through this economic growth. Nearly 20% of Americans in 1964 lived in poverty leaving many boomers unable to purchase the products they saw advertised on television (DeNavas-Walt et al. 2006).

Overlooked in the social construction of Baby Boom Generation is their diversity. Whether in news articles or conversation, regularly boomers are lumped into a single entity. Yet, as the last boomers (born in 1964) entered kindergarten, the first boomers (born in 1946) might be fighting in Vietnam. Described both as selfish and negative or idealistic and positive, highly educated, affluent, innovative, and "The Me Generation" (Hughes and O'Rand 2004; Moody 2008), this image clouds disparities in household incomes, home ownership, and net worth of current middle-aged and older boomers. Hughes and O'Rand found increased disparity in affluence and poverty between the leading-edge boomers (those born 1946–1954) and late boomers (born 1955–1964) with late boomers significantly at higher risk of near poverty at middle age.

Late boomers comprise the largest population share of all adults aged 18 and older (26%) and are 73% larger than the leading-edge boomers who are only 15% of the adult population. Identified as Generation Jones by Jonathan Pontell, the "Gen Jonesers" grew up in the shadows of early boomers (Sheridan 2008). Unlike the leading boomers, Gen Jonesers in the 1970s found an economy shaken by the Vietnam War, the OPEC price hikes and an already boomer saturated job market. Television was not just entertainment, but brought Watergate into family living rooms. The 1980s economy made home ownership increasingly difficult for Gen Jonesers (Hamilton et al. 2000).

Experiences of early childhood and adolescence for late boomers differed from the robust 1950s and 1960s economy perhaps contributing to lack of boomer identity by late boomers. A 2008 nationally representative sample of 500 U.S. adults born in 1961 were asked: "Do you consider yourself to be a member of the Baby Boom Generation, Generation X, or a lost generation in-between (usually called Generation Jones)?" Only 22% of respondents considered themselves to be part of the Baby Boom Generation, while 57% identified with Generation Jones (Sheridan 2008). Generalizations of the Baby Boomer Generation obscure possible early life course effects on early and late boomers' intracohort differences whether in purchasing power or differential job market experiences.

Historical Beginnings of the Aging and Business Industry

The earliest business endeavors for older adults were that of long-term care or supportive services. These businesses and industries specifically targeting the older adult market have been named "Silver Industries." Silver Industries refers to "businesses that create, produce, market and sell goods and services" to older adults (Cutler 2004–05:6).

As early as the 1600s, England established regulations and laws governing the institutionalization of elderly who were sick and poor in the first long-term care industry (Dennis 2004–05). Following the English model for institutional care of the elderly into the New World, the American Colonies established laws requiring families, including grandchildren, to financially support their sick and poor elderly family members (Farkas 2001). Rapid industrialization splintered many of the established norms of family care for elders, resulting in increasing pressure for long-term care institutions. A supportive service sector for the elderly developed. By the 1950s, the influx of Social Security to older adults' income increased the demand for long-term care. Private sector investors, encouraged by attractive government loans, began building the long-term care housing industry (Dennis 2004–05). In general, the interest in aging and business did not advance into other business sectors and remained focused on frail elderly long-term care.

Beginning with Social Security in 1935 and later the 1965 Older Americans Act creation of the Administration on Aging (AoA), the aging network became the "major vehicle for the organization and delivery of social and nutrition services" to older adults, aged 60 and older (Administration on Aging 2009). Providing the majority of supportive services and nutrition programs to those aged 60 and older, AoA monopolized the mature market. Identified in 1979 as the aging enterprise, Estes envisioned the public aging network of hospitals, care facilities, and services coordinated by area agencies on aging working in partnerships with private businesses for the good of older adults (Moody 2004–05). However, progress in this direction has moved at glacial-like speed obstructed by bureaucratic structure, programs that treated all older people exactly the same and the socially constructed negative stereotypes of older adults.

As organizations evolve into mammoth bureaucracies, multiple layers and appendages are constructed. Bureaucracies develop rigid policies and rules and assign responsibilities to specific departments or programs (specialization of tasks) for efficiency. Fragmented and not rewarded for efficiencies, large organizations have difficulty coordinating and cooperating within, let alone with external competitors. Self-preservation is a strong incentive for monopolizing the market (Thompson 1967). AoA and its aging network retain many bureaucratic characteristics. This does not discredit the significant advancements by the AoA and the aging network for older adults; it does, however, help explain the monopolization of the older adult market.

A commonly held perception is that the profit essential to the private sector outweighs businesses' concerns for the older adults; therefore, older adults need to be protected from private sector businesses. Stereotyping the private sector as predatory businesses exploiting the elderly solidified the aging networks' power and control over the older adult market as the aging network strategically worked to monopolize the products and services provided to older adults (Estes 1969). With growing numbers of current and future older adults, many of whom are wealthier and healthier than previous cohorts, the older population is outpacing public providers' budgets. The projections of having more consumers than public resources as boomers age are forcing a reexamination of resources allocated based on age regardless of personal circumstances. And, it should be a call for revisiting of the public–private partnership proposed by Estes in 1969 but considering boomers' consumer preferences while recognizing their intracohort differences.

Ageism may account for lack of products and advertising to old adult consumers until recently. Invisible to most markets, health aide marketing played on stereotypes of frail and dependent elderly. Hendricks (2005) noted ageism is "woven into the wool of our social fabric." Early constructs of negative stereotyping of aging and seniors in sitcoms, dramas, and commercials of the 1950s targeted younger audiences and used older adults as comic relief by displaying elders as physically or mentally incompetent (Montepare and Zebrowitz 2002). Later, the 1960s' public service announcements of compassionate ageism depicted the elderly as frail, poor, senile, and "deserving" aired on television to earn public support for older adult programs. By the late 1970s, Americans "had learned the catechism of compassionate ageism well" (Binstock 2005:73). In the 1980s, a new ageism emerged, that of older Americans burdening society, self-serving, and depleting the future of America. In response, Longino (2005) presented a more positive view of aging noting private businesses development of products and services encouraging health and autonomy for older adults, together with their higher education and income levels, could over time decrease costs associated with aging.

The New Mature Market: Age of Aquarius

As leading-edge boomers' 60th birthdays grew near, the media renewed its interest in them. Stories about the new "well-elderly" and the wealth and preferences of boomers, together with the massive numbers of boomers soon eligible for early retirement, increased awareness of aging. Interest in the older consumer and mature market grew. While not all boomers are wealthy, the potential customer market of approximately 76 million wallets is a gigantic and attractive economic opportunity for businesses.

A contributing factor to increased interest in the boomers' aging, according to Wolfe (2004–05:91), is their occupying the psychological center of gravity (PCG). The PCG principle is that those "who are within 5 years of the adult median age have a disproportionate influence on society's *zeitgeist,* the spirit of the times." In 2005, when the adult median age was 45, the PCG bracket ranged from ages 40–50, or approximately 40 million boomers. With boomers both comprising the PCG and with their numeric size, Wolfe observed that it should be no surprise that the boomers are "the most influential source of cultural and consumer behaviour trends" (p. 16). An interesting example of the boomers' PCG influence is found by reviewing trade marketing books focusing on boomers. Using the keywords "marketing" and "boomers," we find 43 trade marketing books published between January 2000 and December 2009 available at the online bookstore Amazon.com. Of those, 31 (70%) were published after January 2005.

Another factor in the growing interest of business opportunities for older adults may be the aging of the workforce and the parallel aging of the private business owners and entrepreneurs. Suddenly, the old middle-aged business boomer rejects the ageist stereotype of older adults. As aging consumers themselves, boomers and older adults envision business potential in products and services that are not patronizing or offensive to the older consumer. The data provides support for this assumption. A 2009 study from the Kauffman Foundation found those aged 55–64 to have the highest entrepreneurial activity since 1996 forward. The Kauffman Foundation projects an "entrepreneurial boom" in part as the result of the boomers' aging (Stangler 2009). Later-life entrepreneurs are more apt to have the means or borrowing power for business start-up, a lifetime of work experiences on which to draw, and the desire for social good (Rogoff 2009). Zhang (2008) concludes that the knowledge economy together with an aging workforce offers more opportunities for older adults to be entrepreneurs finding that being aged 62 or older increases the likelihood of entrepreneurship by about 11%.

The convergence of demographics, boomers' controlling PCG, interest in private–public partnerships, increased resources of older adults, and the aging of the workforce and entrepreneurs elevate the older consumer market into the business spotlight. When businesses first envisioned opportunities in the boomers, they were babies; now the opportunities lay in silver industries and silver markets.

Growth of Silver Industries

Silver industries create, produce, market, and sell goods and services to older adults. These include products adapted from existing products and made attractive and useful for older adults and innovative new products or businesses targeting older consumers (Cutler 2004–05). Yet, the most "successful and innovative silver products are those that attract across the life course: ageless or age-neutral with value to older adults and attractive to younger customers. Silver industries are those developing and marketing transgenerational or universal solutions that can be used regardless of age, yet connect and integrate multiple generations" (Kohlbacher 2008:18). Industries targeting silver consumers have been an economic boom for Japan, a superaging society where one out of four Japanese will be aged 60 or older by 2015 (Vogt 2008).

The United States lags behind in innovation, design, and product development for the older consumer market when compared with other superaging countries. Leading innovations and businesses for older Japanese adults, Japan's business community and government both embrace and promote "silver" for economic development (Yashiro 2008; Vogt 2008). The use of "silver" to modify industries and markets targeting older adults evolved from a simple act during 1973, Respect for the Elderly Day, when the Japanese National Railway introduced "silver seats" for the silver-haired seniors and the disabled. Since then silver has become a "common epithet for all sorts of institutions, activities and products designed for and catering to senior citizens" (Coulmas 2008, p. vi). In Japan, it is common for products to include silver when naming products and services, such as the "silver pass" for transit.

Japan's jump on the silver market may in part be contributed to governmental old-age policies encouraging product development for the care and well-being of Japanese elders. As a superaging society faces extreme shrinking of both population and the labor force, new forms of caregiving and products are needed to care for the growing numbers of elders (Yashiro 2008). In addition, with average Japanese senior households controlling nearly half of the entire national savings, lack of spending by elders became an economic concern (Usui 2008). Producing attractive products for older Japanese increases consumption returning capital to the economy which would otherwise be tied up in senior's savings.

Examples of the emerging silver industries span low to high technology. Travel, a traditional mature market product of low technology, reaches new silver markets when agents specialize in "grandtravel packages" for intergenerational travel experiences. Opportunities may be a low-cost investment such as grocery stores that widen aisles, lower shelves, and increase lighting to attract the older customer. Gerontechology combines the use of technology with an understanding of gerontology to promote autonomy and lifestyle preferences into old age (International Society for Gerontechnology 2009). The U.S. Jitterbug and the Japanese Raku-Raku cell phones designed for those with cognitive impairment are examples of midtechnology. Motion sensors in continuing care retirement communities, such as Samsung's Noble County, support autonomy by providing unobtrusive fall monitoring involving all technology levels. Humanoid robots, such as the Japanese Paro, a fuzzy seal-like robot that mimics facial expressions found to increase social interaction for those with dementia, are excellent examples of high technology silver industries (Paro Robots Website, 2009). A further example is mobile robotic guides activated through voice commands being tested in Japanese nursing homes to assist residents' movements (Montemerlo et al. 2002). One of the most complex robotic devices is the supportive HAL Bionic Suit developed at Tsukuba University, designed to assist in movement of limbs and carrying of heavy objects (Kohlbackher 2008). Massachusetts Institute of Technology's AgeLab is studying multiple systems at all levels of technology from health to housing to transportation bringing technological advancements for old adults to businesses. New technologies of all complexity that support autonomy, provide physical care, encourage leisure activities, and combat isolation are a burgeoning area of research and of interest to venture capitalists.

Silver industry opportunities are limitless as long as there is the demand by older adults for products that promote independence, healthy lifestyles, sustained mobility, and financial independence and are transgenerational. Japan's advanced development of silver industries was propelled by the demographic consequences of the Japanese superaging society, existing technology expertise, and their long tradition of honoring elders. While the United States is an aging society invested in technology development, there is no strong tradition of honoring elders in a society that continues to revere youth. Perhaps aging boomers themselves, who refuse to age, will add to the lagging technological development in silver industry production in the United States.

The Future of the Mature Market: Aging is Good Business

The silver market is complex and reaches beyond segmentation by age, product, or service category. The drawback to innovative strategies for the silver market has been a static perspective and undervaluing of older adult consumers. In reality, silver consumers comprise the most diverse market of all age groups. Identified as those aged 50 and older, the silver consumer market includes people from age 50 to centenarians. Silver consumers have very different needs, desires, and lifestyles determined in part by their varied life courses. Some are wealthy, others middle class or poor. Some very fit, others unhealthy or frail. While many remain in the workforce, others have retired (Kohlbackher 2008). The limited success of some products designed and marketed to older adult might be attributed to the lack of understanding of the diversity of the mature market and the life course influences on consumption.

Furlong (2007) proposes that the power to reach the older adult consumer lies in understanding the life stages and events boomers and older adults experience as they age. Once considered an uneventful stage of life, middle-age and older adults "move through more transitions in their 50s and 60s than any other phase of life" (Furlong 2007:11). Each event can generate a business opportunity by marketing to "life-transitions." Examples include family transitions (becoming empty nesters, paying for kids' college, parental deaths, and birth of grandchildren); health transitions (onset of heart disease, health clubs, sexual enhancement drugs, as well as traditional medical services); housing transitions (relocations, downsizing, building, refitting for ageless living); work transitions (increased attention to financial planning, retirement, civic engagement); and daily living transitions (increased leisure, volunteerism, personal activities).

American society has progressed through decades of demographic change, yet a myopic youth-centered marketing focus nurtured by an earlier social construct that "old people" are poor, frail, demented, sexless, or pathetic remains. A rich market of middle-aged and older consumers cultivated since babies await discovery. The concept that aging is good business is novel and perhaps even offensive to some. Over the past decade, astute businesses and marketers have begun, albeit slowly, to appreciate the profitability of the middle-aged and older adult consumer market. There are potential social and economic consequences for ignoring business opportunities that come with an aging society.

The future of good business practices in providing products and services for an aging society lies in understanding the diversity of the boomers and how boomers differ from their parents and within themselves. Boomers were cultivated as consumers throughout their lives. With their PCG influence, there is every reason to expect change for older consumers. Recognizing the influence of the life course on boomers' consumption preferences and needs has positive economic potential for the U.S. economy and the status of older adults. Ruffenach of the *Wall Street Journal* wonders why gerontologists aren't knocking at the doors of companies and businesses to help them understand the needs of the older consumer (2004–05). After decades of productive work, savings, and investment in America, boomers may remain the productive economic driver as consumers at older ages. Recently, the struggling economy has affected Americans of all ages. In 2001, those aged 50 and older held 69% of America's total net worth, an increase of 13% over 20 years for those ages (Peters and Barletta 2005). The 2008 collapse of the housing bubble and the continuous volatile stock market since have implications for baby boomers' wealth and their housing equity (Baker and Rosnick 2009). Notwithstanding, unless the U.S. economy experiences even greater downturns, those currently aged 50 and older will continue to control the majority of wealth and purchasing power as they age.

In a society like the United States that values consumption, older adults with savings and resources together with a life history of consumption will be increasingly important. In this emerging area of aging and business, understanding the distinct boomers' preferences has great potential for producing new and innovative products and linking generations. It is an area calling for research to understand how past life trajectories and current life stages will impact boomer consumer behavior at older ages, the status of older adults, and the economy. Rather than viewing aging as a taboo subject, opportunities arise from recognizing that the aging consumer can make aging "hip." Every day, nearly 11,000 boomers are reaching retirement age. This means new businesses, new jobs, and new opportunities. Those New York subway public service posters from the 1950s can be *recycled* to read: "Your future is limitless in an aging America."

References

Administration on Aging. 2009. *A Profile of Older Americans: 2009*. Retrieved February 23, 2010 (http://www.aoa.gov/AoARoot/Aging_Statistics/Profile/2009/docs/2009profile_508.pdf).

———. 2010. *Older Americans Act and Aging Network*. Retrieved January 30, 2010 (http://www.aoa.gov/AoARoot/AoA_Programs/OAA/Introduction.aspx).

Baker, Dean and David Rosnick. 2009. *The Wealth of the Baby Boom Cohorts after the Collapse of the Housing Bubble*. CEPR Reports and Issue Briefs 2009–07, Center for Economic and Policy Research Retrieved May 5, 2009 (http://www.cepr.net/documents/publications/baby-boomer-wealth-2009-02.pdf).

Binstock, Robert. 2005. "Old-Age Policies, Politics, and Ageism." *Generations* 29(3):73–8.

Blaikle, Andrew. 2006. "Visions of Later Life: Golden Cohort to Generation Z." Pp. 12–21 in *Aging in the New Millennium*, edited by C. Phillipson. Thousand Oaks, CA: Sage.

Coulmas, Florian. 2008. "Looking at the Bright Side of Things." Pp. v–vi in *The Silver Market Phenomenon: Business Opportunities in an Era of Demographic Change*, edited by F. Kohlbacher and C. Herstatt. Berlin: Springer.

Cutler, Neal. 2002. *Advising Mature Clients: The New Science of Wealth Span Planning*. New York, NY: John Wiley.

———. 2004–05. "Introduction." *Generations* 28(4):–7.

Daaleman, Timothy and Glen Elder, Jr. 2007. "Family Medicine and the Life Course Paradigm." *Journal of the American Board of Family Medicine* 20(1):85–91.

Dannefer, Dale and Casey Michalowski. 2006. "Three Developments in the Life Course." Pp. 30–9 in *Aging in the New Millennium*, edited by C. Phillipson. Thousand Oaks, CA: Sage.

Dennis, Helen. 2004–05. "Evolution on the Link between Business and Aging." *Generations* 28(4):8–14.

DeNavas-Walt, Carmen, Bernadette Proctor, and Cheryl Lee. 2006. *Income, Poverty, and Health Insurance Coverage in the United States: 2005*. Current Population Reports, P60–231. Washington D.C.: U.S. Government Printing Office.

Elder, Glen 2006. "The Life Course and Human Development." Pp. 665–714 in *Handbook of Child Psychology*, Volume 1*: Theoretical Models of Human Development*, edited by W. Damon and R. Lerner. Hoboken, NJ: John Wiley.

Estes, Carroll. 1969. *The Aging Enterprise: A Critical Examination of Social Policies and Services for the Aged*. San Francisco, CA: Jossey-Bass, Inc.

Farkas, Janice. 2001. "Grandparents." Pp. 320–24 in *Boyhood in America*, Vol. 1, edited by P. F. Clement and J. S. Reinier. Santa Barbara, CA: ABC-CLIO.

Furlong, Mary. 2007. *Turning Silver into Gold: How to Profit in the New Boomer Marketplace*. Saddle River, NJ: FT Press.

Hamilton, Neal, Jon Brunelle, Beth Scully, and Rebecca Sherman. 2000. *Atlas of the Baby Boom Generation*. New York, NY: Macmillan Library Reference.

Hendricks, Jon. 2005. “Ageism: Looking Across the Margin in the New Millennium.” *Generations* 29(3):4–7.

Hughes, Mary and Angela O’Rand. 2004. *The Lives and Times of the Baby Boomers*. New York, NY: Russell Sage Foundation and Washington DC: Population Reference Bureau.

International Society for Gerontechnology. 2009. Retrieved November 4, 2009 (http://www.gerontechnology.info/Journal/).

Jones, Landon. 1980. *Great Expectations: America and the Baby Boom Generation*. New York: Coward, McCann & Geoghegan.

Kohlbackher, Florian. 2008. *Business Opportunities in an Era of Demographic Change*. TUJ ICJS Business Economics Seminar. Retrieved October 7, 2009 (http://www.tuj.ac.jp/icas/pdf/icjs_080924.pdf).

Lindop, Edmund and Sarah DeCapua. 2010. *America in the 1950s*. Minneapolis, MN: Lerner.

Longino, Jr., Charles. 2005. “The Future of Ageism: Baby Boomers at the Doorstep.” *Generations* 29(3):79–83.

Modigliani, Franco and Richard Brumberg. 1954. “Utility Analysis and the Consumption Function.” Pp. 338–438 in *Post-Keynesian Economics*, edited by K. Kurihara. New Brunswick, N.J.: Rutgers University Press.

Montepare, Joann and Leslie Zebrowitz. 2002. “A Social-Developmental View of Ageism.” Pp: 77–125 in *Ageism: Stereotyping and Prejudice Against Older Persons*, edited by T. Nelson. Cambridge, MA: MIT Press.

Montemerlo, Michael, Joelle Pineau, Nicholas Roy, Sebastian Thrun, and Vandana Varma. 2002. “Experiences with a Mobile Robotic Guide for the Elderly.” Pp. 587–92 in *Proceedings of the International Conference on Artificial Intelligence* (AAAI 2002). Edmonton, AB. Retrieved October 4, 2009 (http://mapleleaf.csail.mit.edu/~nickroy/papers/AAAI02.pdf).

Moschis, George. 2000. “Consumer Behavior in Later Life: Multidisciplinary Approaches and Methodological Issues.” Research in Consumer Behavior. 9:103–128.

———. 2007. “Life Course Perspectives on Consumer Behavior.” *Journal of the Academy of Marketing Science*. 35(3):295–307.

Moody, Harry R. 2004–05. “Silver Industries and the New Aging Enterprise.” *Generations* 28(4):75–9.

———. 2008. *The New Aging Enterprise*. Washington DC: AARP Academic Affairs.

National Center for Health Statisics. 2009. *Health, United States, 2009: With Special Feature on Medical Technology*. Washington D.C.: Government Printing Office. Retrived April 1, 2010 (http://www.cdc.gov/nchs/data/hus/hus09.pdf).

Paro Robots Website. 2009. *Paro Therapeutic Robot*. Retrieved October 4, 2009 (www.parorobots.com).

Peters, Tom and Martha Barletta. 2005. *Trends: Recognize Analyze Capitalize*. New York: Dorling Kindersley.

Rogoff, Edward. 2009. “Entrepreneurship Strategies in Later Life.” Pp. 33–44 in *Boomer Bust?: Economic and Political Issues of the Graying Society*, edited by R. Hudson. Westport, CT: Praeger.

Ruffenach, Glenn. 2004–05. “Why Gerontologists Should Join Forces with the Business Community,” *Generations* 28(4):57–63.

Sheridan, Edward. 2008. “New Poll Shows Those Born in ‘61 – Like Obama – Don’t Think of Themselves as Baby Boomers.” New York, NY: ThirdAge, Inc. Retrieved May 3, 2010 (http://www.thirdage.com/travel-pastimes/birthday-gift-to-obama-your-generational-identity-revealed).

Stangler, Dane. 2009. The Coming Entrepreneurship Boom. Ewing Marion Kauffman Foundation. Retrieved January 15, 2010 (http://www.kauffman.org/uploadedFiles/the-coming-entrepreneurial-boom.pdf).

Thompson, James. 1967. *Organizations in Action: Social Science Bases of Administration Theory*. New York: McGraw-Hill.

U.S. Census Bureau. 2003. *Statistical Abstract of the United States: 2003* (No. HS-3). *Population by Age: 1900 to 2002*. Retrieved December 28, 2009 (http://www.census.gov/statab/hist/HS-03.pdf).

Usui, Chikako. 2008. “Japan’s Demographic changes: Social Implications, and Business Opportunities.” Pp. 78–84 in *The Silver Market Phenomenon: Business Opportunities in an Era of Demographic Change*, edited by F. Kohlbacher and C. Herstatt. Berlin: Springer.

Vogt, Gabriele. 2008. “Talking Politics: Demographic Variables and Policy Measures in Japan.” Pp. 17–29 in *The Silver Market Phenomenon: Business Opportunities in an Era of Demographic Change*, edited by F. Kohlbacher and C. Herstatt. Berlin: Springer

Wolfe, David. 2004–05. “The Changing Psychology of the Older Consumer: The Myth of Aging Boomers’ Difference from Their Parents.” *Generations* 28(4):15–19.

Yashiro, Naohir. 2008. “The Silver Markets in Japan through Regulatory Reform.” Pp. 31–40 in *The Silver Market Phenomenon: Business Opportunities in an Era of Demographic Change*, edited by F. Kohlbacher and C. Herstatt. Berlin: Springer.

Zhang, Ting. 2008. *Elderly Entrepreneurship in an Aging US Economy: It’s Never Too Late*. NJ: World Scientific.

Chapter 23
Consumption and Aging

Chris Gilleard and Paul Higgs

Over the past 30 years, consumption has become a major aspect of the new dynamics of later life. Where the social reproduction of old age was once viewed as an issue of providing basic subsistence for those at risk of absolute poverty, today the pressing issue has become one of maintaining lifestyles into later life. As a 2009 report from the United Nations points out: "*older persons constitute an increasing consumer group with specific needs and significant aggregate purchasing power*" (UNECE 2009:1). This transformation in the significance of consumption for the social reproduction of later life in the most affluent global economies is the focus of this chapter.

Before addressing this transformation, however, it is necessary to outline first the principal ways in which the term "consumption" is used within sociology and within the study of economic processes, and secondly to locate current interest in consumption in later life within the broader development of a sociology of consumption. Four terms are commonly used to encompass the principal phenomena of consumption. *Consumer behavior* focuses on how and why people consume, focusing upon individual acts of consumption. *Consumption expenditure* addresses the question of what and how much households and householders spend their money on, focusing upon aggregate or collective consumption. Terms such as *consumer society* and *consumer culture* are concerned with the social and cultural dimensions of consumption and consumer practices. These latter terms can be interpreted as constituting "an institutional field, i.e. a set of interconnected economic and cultural institutions centered on the production of commodities for individual demand" (Zukin and Maguire 2004:175). Collectively these terms demonstrate the potential inter-disciplinary framework surrounding consumption; from the economic decisions of households to the social psychology of consumers. Consumer activity occupies an increasing amount of people's time and resources and the structures organizing consumption exercise a correspondingly wider influence over social life, turning individuals into "citizen consumers" throughout the life course (Cohen 2003).

However, despite Adam Smith's admonition that "Consumption is the sole end and purpose of all production; and the interest of the producer ought to be attended to only so far as it may be necessary for promoting that of the consumer" (Smith 1936:625[1776]), it has taken some two centuries for consumption to feature significantly in sociology. The classical sociologists writing in the midst of the industrialization of the nineteenth and early twentieth century Europe – most notably Karl Marx, Emile Durkheim, and Max Weber – centered their concerns upon the work process and occupational structures. The lives of most adults were seen to be organized not around the consumption of the goods and services needed to maintain their own lives but around their role in their production. Although there were other voices drawing attention to the social significance of consumption practices, such as Georg Simmel's work on the importance of fashion and Thorsten

C. Gilleard (✉)
Division of Research Strategy, University College London Medical School, London, UK
e-mail: rmhacgi@ucl.ac.uk

R.A. Settersten, Jr. and J.L. Angel (eds.), *Handbook of Sociology of Aging*, Handbooks of Sociology and Social Research, DOI 10.1007/978-1-4419-7374-0_23, © Springer Science+Business Media, LLC 2011

Veblen's ideas of conspicuous consumption (Veblen 1912/1959; Simmel 1904), it was not until the 1960s that consumption began to emerge as an important theme in the social sciences in both the United States and Europe (Miller et al. 1998).

Part of the explanation for this shift in sociological interest toward consumption was because such practices expanded in an era of generalized mass production and relative freedom from scarcity (Bell 1973; Ransome 2005). After the First World War, the time spent working or engaging in household maintenance fell while the resources for "surplus" consumption increased (Gershuny 2000:63). Sociologists began to recognize this shift in observing the "roaring" 1920s and the rise of the "flapper" generation. Writing in the *American Journal of Sociology* in 1922, Lloyd suggested that America was experiencing a new "third age of leisure," where "the liberation of men [sic] from some condition of subjection and distribution of the leisure attending liberation to all in some portion and to as many as possible in large portion" (Lloyd 1922:167). But it was not until the second half of the twentieth century that *mass consumption* properly arrived with the spread of commodities and commodification into most spheres of social activity. The sheer scale with which this happened led to both a quantitative and qualitative shift in the nature of industrialized societies. As choice and consumption began to play an increased role in people's lives, consumption was transformed into "consumerism" (Campbell 2004).

The development and evolution of consumer society became a topic of sociological interest in the 1960s (Riesman 1989 [1961]; Bell 1973). Central to this transformation was the growth in leisure (more "free" time to spend), aggregate consumption expenditure (purchasing a wider range of goods and services), and the complexity of consumption (moving from simple, relatively unmediated acts of consumption to acts "which are deemed by the actor to provide satisfaction of something more than basic needs") (Ransome 2005:67). This fostered not just a widening choice in the type of goods and services that could be consumed but also a broader interpretive framework governing consumer behavior. Marketing played a particularly important part in helping shape that framework (Brown 1995). In order to sell a wider range of goods and services to more people, during the 1950s and 1960s market researchers developed the concept of market segmentation (Cohen 2003:295). This idea – that products should cater to and be marketed for particular segments of the population, defined by "age, affluence, and lifestyle" – led to a growing interactive relationship between business studies, marketing, and sociology.

Since the 1960s, there has been a growing sociological interest in the study of consumption, consumerism, and consumer "lifestyles," often providing an explicit contrast to the earlier dominance of educational and occupational stratification in understanding and analyzing society. This "cultural turn" linked together consumption, lifestyle, and leisure and provided some tentative bridges between business and the social sciences (cf. Lazer 1963). It also saw the identification of consumption and consumer society as ways of explaining the social whole (Baudrillard 1998 [1970]; Miller 1998), and of understanding the practices of social distinction through culture (Warde and Bennet 2008), food (Warde 1996), and fashion (Lipovetsky 1994).

Consumption and Later Life: Some Early Starts

Consumption and the study of later life began to be linked quite early in the post-Second World War development of American sociology. The increase in the numbers of people retiring led some sociologists to suggest that a new "leisure class" was appearing in American society (Michelon 1954). Michelon's study of retired trailer park residents was prescient if somewhat optimistic for the time. It potentially resonates more strongly now as the retired population has continued to increase, while poverty in later life has diminished (Iceland 2003). As a result, perhaps, this term is now employed more critically and "retired Americans are being quietly reframed as the [new] leisure class living off the younger generation" (*The Boston Globe*, June 26, 2005, E. Goodman).

These early attempts to frame later life as a field of study not just for health and disability but for leisure and lifestyle were handicapped by the evident poverty among the retired population and the consequent weakness of older people's consumer power. As one researcher noted, throughout the 1950s and 1960s "*the aged segment of the market*" continued to allocate "*their limited resources [to] the point of view of survival (food, housing and medical care)*" (Goldstein 1968:67). At this time, the vast majority of men retiring did so out of necessity rather than from a desire to or expectation of living a life of leisure (Costa 1998:133). The dominant assumption was that older people were uninvolved in the post war consumer boom: "teenagers" or "youth" were seen as the leading age group developing consumer-oriented lifestyles and subcultures that were designed in part in order that they would be distinct from the older generation (Hebdige 1979).

The tendency, at least until recently, of most American sociologists generally to underplay the topic of consumption and more particularly the participation of older people in "complex consumption" has meant that both empirical and theoretical studies of consumption, leisure, and lifestyle in later life have been limited. This is so, despite the early, prescient interest shown by researchers such as Goldstein, Havighurst, and Michelon. That consumption may come to play an active part in later life was however anticipated, when Goldstein wrote "in the future an increasing number of older people will probably receive larger incomes through a widening participation in pension plans, social security and other benefits [and] as a result they may be able to remain important consumers for longer... and at a level which is above a mere subsistence standard" (Goldstein 1968:69). How well those predictions have been realized provides an important backdrop for this chapter. We aim first to provide an overview of consumer expenditure in later life, drawing upon current and historical household expenditure survey data to show how older households' "complex" consumption has increased relative to other age groups and earlier cohorts. Such empirical data provide the necessary context for understanding how the shift from consumption to consumerism has affected older households particularly over the last 3 decades. We then consider some of the evidence that older people are indeed significant actors, participating as knowing subjects within contemporary consumer culture. This evidence, we argue, indicates that the majority of older people are behaving more like the "leisure class" that was advocated nearly three quarters of a century ago, when the Townsend movement for pension reform articulated the slogan "Youth for work – age for leisure" (in 1934 Graebner 1980:194). We consider the limits to the full realization of that sentiment, both now and in the future before concluding the chapter with some thoughts on future directions for this field and its significance in understanding the new dynamics of aging of the twenty first century.

Household Consumption Expenditure in Later Life: National Survey Data

Most nation states in the prosperous parts of the world conduct regular surveys of household expenditure as part of their processes for determining rates of inflation. These surveys provide useful sources of information on the relative expenditure of later life households on a wide variety of goods and service as well as serving as indicators of well-being and quality of life (UNECE 2009:2; Hurd and Rohwedder 2005). In addition to the use of such surveys to calculate overall consumer expenditure, the nature of household expenditure can also be used to provide a broad distinction between expenditure on basic household welfare directed at supplying necessities such as clothing, food, housing, and fuel/energy and expenditure directed toward nonessential goods and services such as leisure/recreation, personal care, luxury household goods, travel, and holidays – the former representing "simple consumption," the latter "complex consumption" (Ransome 2005:79). This distinction is important both theoretically charting the expansion of "complex" consumption or consumerism and empirically demonstrating differential and often contrasting trends in consumption.

What follows are a series of case examples illustrating contemporary patterns and recent trends in later life household consumption in the United States and in Europe. Although there now exists a growing body of national statistics on household expenditure, we have limited our focus to France and Germany because these European economies yield comparable data over a similar time frame to that of the United States (Casey and Yamada 2002). Our first example is drawn from French household expenditure for 2006. The various categories of household expenditure are summed into indicators of "simple" vs. "complex" consumption, broken down by age group. As Table 23.1 shows, necessary household welfare expenditure (simple consumption) now constitutes less than half the expenditure of all but the very oldest French households (those aged 75 years and over). Historical analyses reported by the French National Institute of Statistics and Economic Studies (INSEE) also indicate that, between 1985 and 2005, the gap between retired householders' expenditure and that of the average French household has narrowed while the consumption expenditure of those in their thirties, forties, and fifties has moved closer to the average (INSEE 2008 Figure VII:1:77).

Data from the United States Consumer Expenditure Survey over a similar time period (1985–2008) demonstrate similar trends (see Fig. 23.1). The proportion of household expenditure devoted to simple consumption (spending on housing, fuel, energy, food, and clothing) by people aged 65–74 years has remained relatively constant (compared to the average U.S. household) while the proportion spent on complex consumption (entertainment, leisure, personal care products, and services) has risen from two thirds to nearly 90% of the national average. In Britain, a similar trend has been observed with spending on leisure goods and services rising by 47 and 75% among pensioner households compared with only 10 and 58% for nonpensioner households, between 1995 and 2007 (Leicester et al. 2009:Table 3.7a, 27).

Incorporated into most surveys of consumption expenditure are reports on household ownership of key consumer durable goods. These illustrate the relative rates of "penetration" of household (domestic) technology into later life households. Such goods also serve as useful indicators of engagement with leisure-oriented consumer culture, since they provide opportunities or serve as resources for leisure activities by members of the household. Data from the 2008 German Household Expenditure Survey (Table 23.2) illustrate rates of domestic technology ownership by age group.

Over three quarters of all German households currently own a bicycle, car, and cell-phone, over two thirds own a CD player, and between one half to one third own a satellite/cable TV. This is the case for all age groups. The majority of households (except the oldest) also own personal computers (PC) while between one third to a quarter of all households (except the youngest) own some type of sports equipment. The household ownership of key items of such "second generation" domestic technology among older age groups is of relatively recent origin. The ownership of microwave

Table 23.1 Household expenditure patterns by age of household respondent: France, 2006

Category	25–34 years (%)	35–44 (%)	45–54 (%)	55–64 (%)	65–74 (%)	75+ (%)
Food	13	16	17	18	21	22
Housing, fuel, utilities	19	15	14	14	17	21
Clothing	9	9	8	7	6	4
Health	3	3	3	4	5	6
Simple consumption	44	43	42	43	49	53
Transport	16	17	17	17	13	8
Eating out	6	6	5	4	3	3
Leisure and culture (excl. holiday travel)	7	8	8	8	8	7
Holiday travel	1	1	1	2	2	1
Hairdressing/personal care	1	2	2	2	2	2
Other expenditures	25	24	24	25	25	26
Complex consumption	56	57	58	57	51	47

Source: Dépense annuelle moyenne par ménage selon l'âge de la personne de référence. L'INSEE 2008

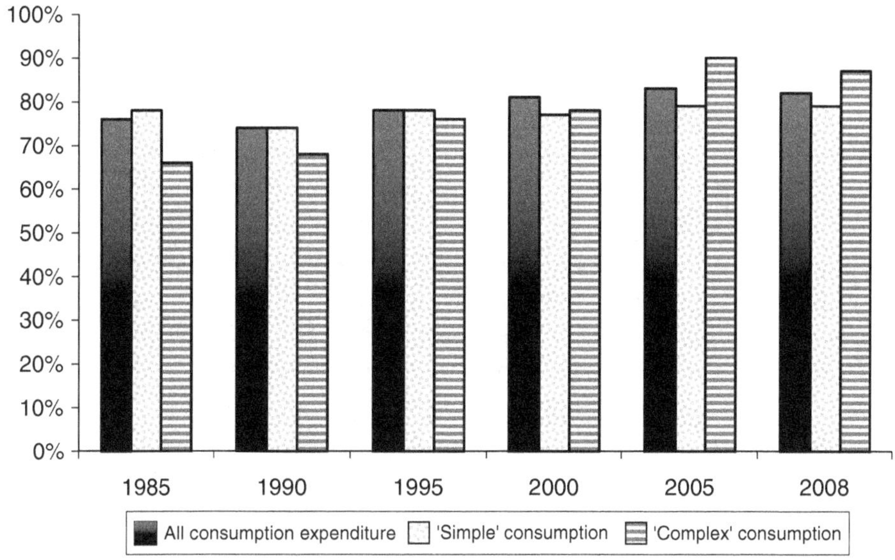

Fig. 23.1 Simple, complex, and overall consumption in U.S. households headed by people aged 65–74 years as a proportion of the national average: 1985–2008. Source: U.S. Consumer Expenditure Surveys, 1985–2008

Table 23.2 Ownership of key "domestic technology" goods in German households according to age of the household reference person (2008)

Consumer durables	25–35 years (%)	35–45 years (%)	45–55 years (%)	55–65 years (%)	65–70 years (%)	70–80 years (%)
Automobile	78	84	82	79	76	72
Bicycle	83	89	86	79	76	67
Home trainer	21	28	31	31	31	28
DVD player	87	86	79	67	52	39
CD player	87	91	87	79	73	62
Satellite TV	35	46	42	40	37	35
PC	91	92	87	76	61	44
Mobile phone	97	95	92	86	78	70
Camera	46	62	64	67	70	69
Digital camera	75	74	67	56	43	30

Source: Statistisches Bundesamt, Statistisches Jahrbuch 2009, Table 22.1.2 (p. 550)

ovens, dishwashers, VCRs, and PCs by U.S. households headed by people aged 65–74 years of age has shown a steady increase in penetration over the last 2 decades from the point when they were absent from or owned by only a small minority of such households to the point where the majority of households now own these goods (Fig. 23.2).

Data on household expenditure show clearly how, over the last quarter century, older age households have become much more engaged with "complex" consumption. Although overall household expenditure tends to decline after retirement, the extent and nature of this decline is no longer as steep nor as excluding as it was once. As Hurd and Rohwedder have observed, most "households seem to be pleasantly surprised by their level of resources… [and] the adequacy of [their] retirement income" (Hurd and Rohwedder 2005:17) with many retired households consuming significantly more than they had anticipated (Ameriks et al. 2002:28). Arguably, the conditions for consumerism in later life are now being met.

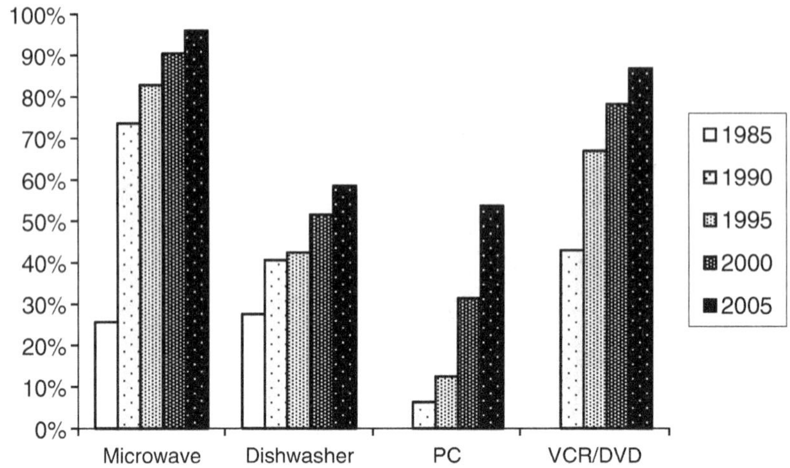

Fig. 23.2 Trends in the ownership of selected items of domestic technology by U.S. households for respondents aged 65–74 years (1985–2005). Source: Division of Consumer Expenditure Surveys, Office of Prices and Living Conditions, U.S. Department of Labor, 1985–2005

Nevertheless, there is still a long way to go and many householders of all ages remain at least partially excluded from consumer citizenship. Consumer expenditure among the oldest old – people in their eighties and nineties – is still dominated by expenditure on basic household welfare. Health status, gender, and ethnicity influence household expenditure, while period and cohort effects also play an important part in accounting for later life consumer inequalitis (Keister 2000; Rentz, Reynolds and Stout 1983). Effects arising from the lifelong inequalities associated with ethnicity may however be less open to change than those associated with either education or gender, even if the overall level and form of consumption inequalities within the older population continue to diminish (Newman 2003). We shall return to this issue of expenditure inequalities later in the chapter.

Market Segmentation and the Dissolution of Old Age

Of course, inequalities in consumption are not the same as differences and distinctions in consumer behavior. Since the universalization of retirement and the expansion of compulsory education, older people (those over age 60) have gained more "free" time with which to engage in consumption than all other age groups. But until relatively recently, members of this age group have not enjoyed an equivalent increase in levels of disposable income to realize the opportunity to do so. Throughout the 1950s and 1960s, older people were significantly over-represented among the poor (Townsend 1979; Pampel 1998:5). By the 1980s, this had changed and in North America, Australasia, and Western Europe, the income of the retired population was rising faster than that of the working age population (Gilleard and Higgs 2005) while the numbers in poverty had dropped below that of the under 65s (Hourriez et al. 2001; Pampel 1998). In France, for example, the purchasing power of retired people increased more than sixfold between 1945 and 2000; that of workers only fourfold (Juvin 2005:43). By the end of the twentieth century, as we have tried to show in the previous section, the conditions for a mass "leisure class" in later life were beginning to materialize.

Despite the growing significance of older households as sites of "complex consumption" the sociological analysis of older people as consumers remains extremely limited (Higgs et al. 2009). One area where research has been more actively undertaken has been in business and marketing studies.

During the late 1970s, market researchers in the United States began to take an active interest in later life and retirement, both as sources of market expansion and segmentation (Burnett 1989; Lazer 1986). This field has continued to grow significantly, and age and generation have become important market factors (Bone 1991; Moschis 1996; Stroud 2005). Within this area, a number of themes have emerged that are of particular interest for the sociology of later life. These include studies emphasizing the salience of "generation" or "cohort" influences as potential sources of market differentiation (Moschis 2009), the heterogeneity of consumption-oriented lifestyles in later life (Stroud 2005:124 ff), the contextual nature of market segmentation within discrete areas of consumption (Moschis and Unal 2008; Shoemaker 2000), and the role of later life status change in stimulating consumption practices (Mathur et al. 2008). Though socioeconomic status continues to have an important role in influencing the consumer power of older people, this traditional structuring of the population is less evident in market research studies than that of generation, life stage, and lifestyle.

Such studies demonstrate the heterogeneity in consumption and consumerism within an older population, the limited significance of chronological age *per se* on consumption and consumer orientation, the relative absence of obvious "conservatism" in consumer choice, and the signal failure of age and agedness to sell products (Gunter 1998; Moschis 2009; Stroud 2005). Ruiz et al. (2004), for example, asked shoppers in a Canadian shopping mall about the various activities that they engaged in at the mall during their visit. From the responses they constructed four "clusters" of shopper – "recreational," "full experience," "browsers," and "mission" shoppers. People aged 65 and over were evenly spread across all categories, although people aged over 65 were more often identified as "recreational shoppers" than other age groups. Similar conclusions have been reached in relation to travel and tourism consumer profiles within the older population (Horneman et al. 2002).

The existence of distinct cohort or generational differences within the consumption patterns and consumerist orientations of the older population has been propounded by several advocates of market segmentation (Cravit 2008; Smith and Clurman 1997). Much of this "generational" focus has been upon the so-called "boomer" generation who as a cohort, some argue, "have played a central role in influencing… corporations and organizations to buy into the market model as well as instilling… a consumer mentality" (Bibby 2006:60). Evidence of such distinct cohort effects on consumption is limited but not completely lacking. Warde (1996), for example, in an analysis of food consumption from the British Family Expenditure Survey, showed how "a particular age cohort, [those] born in the 1940s" brought about an "abrupt transformation in… food culture" (op cit., p. 170). Gilleard et al. (2005), have demonstrated significant "birth cohort" effects on the ownership of key consumer goods in late life British households, a finding that has been replicated in a larger national sample (Higgs et al. 2009). These and similar studies provide some albeit limited empirical support for generational marketing strategies and generational consumer "habitus." Arguably, generational marketing is itself fostering a new generational consciousness within the older population (Gilleard and Higgs 2005).

This segmentation of later life into distinct and distinguishable consumer groups can be linked conceptually to the cultural sociology of Pierre Bourdieu. Bourdieu was interested in how taste formed a new site for class distinction. In doing so he utilized the idea of a cultural or symbolic capital that was aligned with, yet distinct from economic capital (Bourdieu 1987). Drawing upon extensive analysis of preferred food, music, reading, and entertainment, Bourdieu demonstrated how disparate features of consumer choice could be subsumed within an implicit cultural structure that organized preference along a status hierarchy that mapped onto the class divisions of French society. Few attempts have been made to apply Bourdieu's methodology to studies of age or cohort-based distinctions in people's consumer choices. Bourdieu's emphasis upon social class rather than generation or birth cohort as a source of consumer-based distinctions has found echoes in Formosa's recent study of social class in later life in Malta (Formosa 2009). He found that while the processes of class have become more individualized, definite patterns of distinction based upon consumption practices still were evident between working and middle class older Maltese people, most notably in relation to home-based consumption.

Lifestyle and Identity Among Senior Citizen Consumers

While generational marketing strategies have focused upon the significance of cohort and life stage for consumer behavior and consumer choice in later life, other studies have examined consumption as a significant element in maintaining identity, lifestyle, and social differentiation in later life. As pointed out above this theme can be traced back to the work of Veblen and Simmel and their focus upon "conspicuous consumption" and "status differentiation" (Veblen 1912; Simmel 1904), but it was Baudrillard who first made popular the argument that modern consumer society and the complex consumption it supports flourishes through a system of consumption-mediated "symbolic" exchange. He argued that the objects of consumption are embedded within a network of signs that are created, manipulated, or reframed through the media, through advertisements and marketing. For Baudrillard, consumption is both a process of classification and social differentiation and a process of signification and communication – a system of symbolic exchange (Baudrillard 1998:60). As McCracken (1987) notes this, "cultural turn" recognizes that complex, nonutility-focused consumption is more than just a means of indicating status as Simmel and Veblen implied. The goods and services individuals purchase signify more than just an individual's wealth. They respond as much, if not more, to what an individual desires as to what an individual simply needs. They convey a message and a meaning that is shaped by the market, the media, and the social and cultural contexts within which individuals spend their time and money. Traditionally, older people have been considered to be preoccupied with simple, necessary consumption (Goldstein 1968). The goods and services they purchased were seen to privilege their use value, their capacity to satisfy older people's "basic" needs – for food, fuel, shelter, and clothing – not to reflect or reinforce their desires. In the 1950s, the British Ministry of Labour and National Service specifically excluded the consumption expenditure of pensioner households in calculating the cost of living because of their then very limited finances (Ministry of Labour and National Service 1957).

The subsequent shift toward older people's engagement with complex consumption has facilitated what McCracken (1990) described as a "North American" style of cultural categorization that is more indeterminate, elective, and dynamic than the earlier status-based consumption patterns. This socio-economic and cultural transformation of later life is only now being studied. A small but growing number of studies are beginning to illustrate how consumption impacts upon the social processes of being or becoming "old." These studies indicate how complex consumption in later life is creating new sources of differentiation that fragment what was previously considered a homogenous "subculture" of aging (Rose 1965). Increasing numbers of older people can and do engage in consumer lifestyles that ignore, avoid, or deny the identification of self with a social and cultural identity conferred by a chronologically defined agedness. The consumption and identity work that was located in the teenage and early adult years during the "cultural revolution" of the 1960s has since spread upward and outward across the lifecycle. Propelling this lifestyle agenda is a plethora of lifestyle manuals directing the "mature" consumer toward a variety of seemingly "ageless" lifestyles (Gilleard 1996).

Empirical research on consumption, identity, and lifestyle in later life is beginning to show how older people – singly, collectively, or as couples – actively use the purchase of leisure goods and services to frame, change, or expand their identities. Most of this research focuses upon complex consumption – and in particular the consumption of leisure goods and services such as travel, tourism, and entertainment. Therkelsen and Gram (2008), for example, have examined how mature couples use holidays "as a vehicle of identity construction" for themselves as a couple as well as individuals in their own right. Schau et al. (2009) describe a process of "identity renaissance" in retirement in their multidisciplinary, multi-method qualitative study of people aged 61–83 years old. From interviews, questionnaires, and focus groups, they conclude that "[c]onsumer identity renaissance… takes place in a cultural context that favors individuality and where identity play is permissible, where retirement and leisure consumption ideologies are in flux and where there is a multitude of available and attractively priced options" (Schau et al. 2009:272). Studies of attitudes toward and experiences of travel and tourism indicate the particular role that these experiences can play in conferring "identity" and "meaning" (Nimrod 2008; Russell 2008; Sedgeley et al. 2006) and in the process diluting or dissolving the significance of chronological age. These and similar studies indicate how "older consumers [are] becoming increasingly ageless in their consumption behavior" (Szmigin and Carrigan 2000:505).

Other, less direct approaches to exploring the identity and lifestyle influences derived from consumption have used the images of later life present in advertisements and the media. In a study of newspaper articles about retirement and retired people, Katz and his colleague analyzed 138 published personal profiles and biographical narratives (Katz and Laliberte-Rudman 2005) from which they identified two types of lifestyle. The first was an historical identity they called "retired worker" based upon past working roles, while the other was what they describe as that of the "opportunity-seeking consumer." This latter "type" is "a person who chooses specific lifestyle options in the consumer market that promise to transform retirement into an opportunity and a 'second chance' for self-development and social success" (Katz and Laliberte-Rudman 2005:149). Other studies have examined the role of advertisements in shaping the cultural identities of older people as potential consumers of a variety of goods and services. Williams et al. (2007) showed how a marketing strategy for margarine using images of older people evolved over a 7-year period. They found a shift from an initial emphasis upon the agedness/longevity of the characters toward a later emphasis upon the exercise of choice and engaging in activities not usually associated with old people. This shift in focus, the authors suggest, "may apply particularly to the baby boomer generation" (op cit.:17).

Older people's "complex consumption" is not confined to purchasing leisure goods and services. A broad array of financial services have become available such as reverse mortgages, long term care insurance, investment portfolios, and so forth – many of which address later life income security (Brennan and Ritch 2010). Ekerdt and Clark (2001) have explored the role of retirement financial planning advertisements in shaping the meaning(s) of retirement. They found that most of the advertisements conveyed an image of retirement in positive terms, as an emancipatory life stage of active leisure, achieved by self-reliant, financially sophisticated men. Exactly how, if at all, such advertisements shape expectations and interpretations of later life is of course difficult to gauge, but the companies that commission them are counting on the fact that they can and they do.

Older People as Marginalized Consumers

Arguably most advertising and related market research are geared toward selling goods and services to people with the money to buy them. It is a field which tends to emphasize the better off among the older population (Roberts and Zhou 1997). Research in marketing, leisure, and tourism studies and related social research has followed such trends and studies of consumption and consumer behavior among "poor old people" are consequently uncommon. This lack of research is partly the result of, and partly a contributor to the more general assumption that older people are still largely a residual group. For example John Vincent writes:

> Older people's access to consumption opportunities is limited, as is that of disabled people, by the physical structure of retailing. Out of town supermarkets effectively discriminate against noncar users. Not only are older people less likely to drive or to have a car; but the alternative outlets, the corner shops and city center stores are being driven out by competition. There are other retail outlets in addition to shops. Mail order and telephone shopping, for example, can facilitate consumption choices by older consumers. However, these remote systems lose much of the valued social contact which local shopping can facilitate, and older people tend to have less access to the latest communications technologies (Vincent 1999:151).

Such assumptions are made with little reference to empirical observation or interrogation of older people themselves. Such research that has been carried out suggests that these assumptions of marginality significantly underplay the significance of shopping and leisure consumption as sources of positive well-being among older pensioners (Szmigin and Carrigan 2001). In one study conducted in two London, England shopping centers particular focus was given to those groups of elderly persons "stranded in council estates" and displaying signs of "disability, poverty, and loneliness" (Miller et al. 1998:79). The authors observed that although assumptions "that the elderly [sic] are largely a nostalgic and sentimental community whose ties lie firmly attached to the traditions of their own childhood and that they would therefore ideally prefer a sepia tinted image of 'olde worlde' shopping" (Miller et al. 1998:83) were supported in the focus group discussions (in a context that drew attention to the members' collective agedness and limited means) when the individuals from these focus groups were accompanied on their shopping trips – where the focus was on the activity of shopping – the evidence suggested that shopping was in fact associated with a "sense of enjoyment" and that these particular elderly people "showed a marked preference for large supermarkets because of their desire for bright modern looks and access to the large gamut of new produce" (op. cit. 1998:84). Similar assumptions can be discerned in studies of casino gambling, where the perceived "vulnerability" of older gamblers is belied by empirical research. As one study observed "there was no evidence to support the idea that casino gambling activities threatened these older adults in any way" (Hope and Havir 2002:195). For most people, it seemed that they were operating within clear financial limits – and in the process, having fun.

Noteworthy too are the results of research conducted in non-Western "middle income" countries. Although limited in number, such studies suggest that active consumerist attitudes and habits exist among significant numbers of older people. In a survey of older consumers in Malaysia, Ong and Phillips observed that "contrary to the common image of older persons as 'invisible consumers' … older consumers in Malaysia are in many ways 'canny consumers' who know what they want and how to get it" (Ong and Phillips 2007:113). Related research conducted in China (a country with the largest number of mature consumers in the world) also indicates that older consumers share many similarities in consumer choice and participation with the younger generation, with a general preference for "foreign" rather than "domestic" products and similar levels of overall consumption (Yang et al. 2005).

As noted, social science researchers have tended to cling to the assumption that modern developments in leisure, shopping, and the "new" consumerism are "dangerous" or "bad" for more peripheral and marginal groups, risking excluding them still further (Miller et al. 1998:88). Such evidence as has been gathered suggests that this is not the case. The preference of many, if not most older people, is generally to enjoy the opportunities that shopping, leisure, and lifestyle consumption offer. These findings signal an important point for the sociology of consumption – namely the role that consumption plays not only in reproducing and refashioning identities and lifestyles but also in offering opportunities for sharing, for socializing, and for having fun. And in the process, the oppressiveness of chronological age-based identities may be lightened.

Difference, Diversity, and Inequality Among Older Consumers

While we have been at pains to demonstrate that later life has become a site of complex consumption and that older age groups appear to be as active consumers as much as any other age group, it would be a mistake to ignore the differences between groups of older consumers. These differences are reflected both in market segmentation and diversity of lifestyles as well as in socio-economic inequalities. As with any age group, differences in late life income and resources create different constraints on people's capacity to consume and exercise active consumer choice (Peterson 2007). Many of these differences do not emerge specifically in later life. Some reflect continuing inequalities associated with education, ethnicity, and gender that persist over the life course while others reflect discontinuities arising from the contingencies of life in late modernity.

While there exists a considerable body of literature addressing income inequalities in later life (Price and Ginn 2006; Scharf 2009), much less has been written about inequalities in consumption. Since marriage conceals much of the gender imbalance in lifetime earnings and pension entitlements, household expenditure data tell us little about how gender affects men and women's individual choices. Changing rates of divorce, remarriage, separation, and widowhood across cohorts and over time affect the composition of later life households, making it likely that the influence of gender on consumption will vary as a complex function of age, period, and cohort. Given current differences in longevity, the longer women live, the more likely they are to live alone and households headed by single older people spend consistently less than married households. Ethnicity compounds the effects of gender. Older black and Hispanic women are twice as likely to be poor as older black and Hispanic men and older white women and *5 times* as likely to be poor as older white men (Street and Wilmoth 2001:131–2).

Income poverty and consumption poverty are not of course identical, and income poverty generally exceeds consumption poverty for all age groups. Among the over 65-year-old U.S. population, just over 16% are income poor but only around half that number are consumption poor. Some 6% are both income and consumption poor. Within this most marginalized group, according to Fisher and his colleagues, a disproportionate number are "women, widowed, high school drop outs, Blacks, and renters" (Fisher et al. 2009:765). Although detailed research into the sources and consequences of such consumption differentials in later life is lacking, the limited spending power of such groups seems likely to lead significant numbers of old, widowed, and black women into the enforced status of "flawed consumers" (Bauman 1998:38). However, even in circumstances of poverty, later life identity construction and social relatedness through consumption remain possibilities, as the British research study already mentioned has shown.

New Directions for the Study of Aging and Consumption

One of our intentions in writing this chapter has been to revisit Goldstein's prediction, made over 40 years ago, that in the future older people would play a more significant role as "important consumers." Our conclusion is that older people have indeed become more active participants in consumption and consumerism, moving in the words of one recent book "from passive to active consumers" (Jones et al. 2008). What are the consequences of this shift for the sociology of later life? What directions does it suggest for future research and crucially what remains unexplored and unrecognized?

The theme of consumption (and leisure) has been present since the origins of social gerontology, but it has always operated in the background. Like many related areas, it has been obscured by the more dominant focus on health, poverty, and social care. The relative lack of attention to consumption is equally evident in the sociology of old age, where its absence also reflects a broader neglect of the topic within sociology as a whole (Zukin and Maguire 2004). That it is now receiving attention is not only a reflection of the growing importance of later life within consumer society but also a consequence of changes in intellectual orientation within sociology itself. This is not just a consequence of the demographic weight of the number of retirees in the population, nor is it connected solely to the rising standards of living of retirees – though both are important. The significance of consumption in later life is also the story of consumption as a key cultural field within society, creating new sources of distinction as well as new forms of collective identity (Hechter 2004). It is here that aging constitutes an important aspect of the development of consumption. Much has been written about the impact of the aging of the baby boom cohorts in the more prosperous nations in the world as they move from middle age to retirement. What is often overlooked is the role played out by the distinct "generational habitus" (Gilleard and Higgs 2005) exhibited by these new cohorts of retirees in the development of consumption and consumerism.

Many of the important tropes of modern consumer lifestyles have their origins in the "generational schism" of the late 1960s with its focus on identity, self-expression, and social differentiation. What has sustained and developed this new "generational habitus" are the new and enlarged markets for fashion, leisure, and domestic technologies and the declining costs of leisure that have contributed to the shift from consumption to mass consumerism. Any development in the field of consumption and later life must take into consideration the historical emergence of this new consumerist habitus within earlier youth-oriented cohorts and their various subcultures. If people in later life continue to participate in and extend the scope of consumerism within retirement, it is important that research integrates the emerging "cultures of aging" into the core assumptions of both social gerontology and the sociology of aging.

As we have noted several times, there is still a dearth of sociologically informed quantitative and qualitative work on almost all aspects of consumption and consumerism in later life. What work has been done often throws up unexpected results – particularly when older people are not treated as if they are outside the practices shared by the rest of the population, or when consumerism in later life is not treated as confined to the practices of a small and privileged elite within the retired population. More research into the consumer practices and preferences of older people is needed – not in order to promote strategies for further market segmentation but to challenge assumptions that consumption in later life remains a marginal or risky activity that older people cannot afford to engage in.

A focus on the cultures of consumption in later life can also develop a unique perspective in relation to the contradictory representations of agedness in what is, in many parts of the world, a very age sensitive and in some ways "ageist" society. The fact that aging consumer societies remain strongly orientated toward the idea of youth rather than old age has meant that part of the engagement of older people with consumer culture is through the denial, refusal, rejection, or reconstruction of "old age" as a dominant structural identity. As noted earlier, market research has made clear that agedness does not attract sales. Resistance to old age is more common, and consumer products that offer such age-proofing sell. The target for such "antiaging" products is typically the middle-aged consumer, but sales of antiaging products and procedures are rising among both younger and older age groups. Studies of "age-conscious consumption" suggest that later life consumer choice can sometimes focus upon maintaining a "nonaged" identity (Hurd Clarke et al. 2009). Such explorations of consumer choice indicate both the emancipatory potential of consumption alongside the perceived imperative to choose not to be old; or perhaps more accurately to choose to be old in a new and improved manner. Breaking the taboo that such studies of "age conscious" consumption are trivial and inconsequential provides considerable scope to explore the identity work of older consumers of clothing, fashion, and self-care products in the presentation of self as aging in the right way, taking account of the potential significance of health status and gender.

An area for future research that also deserves more exploration is the role of consumption in shaping and sustaining social relationships in later life. Household consumption, for example, is not principally individualized consumption. Mature couples' consumption preferences serve many functions, which may reflect and reinforce their sense of identity and solidarity or which may create and widen schisms within the partnership. Buying and exchanging presents across the generations may provide a less evident means for sustaining the engagement of older people in cultural and technological developments that are "taken up" first by their children or grandchildren, as for example, when adult children buy their parent a mobile phone. Tours, travel, and visits to the shops may offer opportunities for older people to socialize with friends and share and develop their recreational pleasures. The state provision of consumption-oriented benefits to older people – such as free or subsidized transport – may extend such opportunities across a wider segment of the older population, as may "senior" discounts on eating out, personal care, and shopping. Consumption may also involve buying memories and memorabilia that link past and present – maintaining or constructing continuities with one's earlier adult life. Digitalized media now make it possible to "rejuvenate" old records and old movies, for example.

In short, it is important to recognize the significance of consumption in understanding the new dynamics of later life, the changes in the organization of the life course and the significance of age as an organizing principle for determining status and position within society. As complex consumption overshadows basic household expenditure and leisure time exceeds the time required for self and household maintenance and outside employment, older people's participation in consumption and consumer society is growing in size and significance. Earlier socialization into the consumerist habitus of mass consumer society is evident for each successive birth cohort since the 1930s while the mass media provide increasing opportunities for people of all ages to be exposed to developments in lifestyle culture and technology. Such a combination of cohort and period effects suggest that the old age of first modernity will be progressively undermined by a more contingent, conflictual, and effortful later life. Studies of later life consumption are important in charting this process of change.

References

Ameriks, John, Andrew Caplin, and John Leahy. 2002. *Retirement Consumption: Insights from a Survey. NBER Working Paper No. 8735.* Cambridge MA: National Bureau of Economics.

Baudrillard, Jean. 1998. *The Consumer Society.* London: Sage Publications.

Bauman, Zigmund. 1998. *Work, Consumerism, and the New Poor.* Buckingham: Open University Press.

Bell, Daniel. 1973. *The Coming of Post-Industrial Society.* New York: Basic Books.

Bibby, Reginald W. 2006. *The Boomer Factor.* Toronto: Bastion Books.

Bone, Peter F. 1991. "Identifying Mature Segments." *Journal of Consumer Marketing* 8:19–32.

Bourdieu, Pierre 1987. *Distinction: A Social Critique of Judgment and Taste.* Cambridge MA: Harvard University Press.

Brennan, Carol and Elaine Ritch. 2010. "Capturing the Voice of Older Consumers in Relation to Financial Products and Services." *International Journal of Consumer Studies* 34:212–8.

Brown, Stephen. 1995. *Postmodern Marketing.* London: Routledge.

Burnett, J.J. 1989. "Retirement vs. the efficacy of Retirement as a Segmentation Variable." *Journal of the Academy of Marketing Sciences* 17:333–43.

Campbell, Colin. 2004. "I Shop Therefore I Know That I Am: The Metaphysical Basis of Modern Consumerism." Pp. 27–34 in *Elusive Consumption,* edited by K. Ekstrom and H. Brembeck. London: Berg.

Casey, Bernard and Yamada, Atsuhiro. 2002. Getting Older, Getting Poorer? A Study of the Earnings, Pensions, Assets and Living Arrangements of Older People in Nine Countries. *OECD Labour Market and Social Policy Occasional Papers*, no. 60. Paris: OECD.

Cohen, Lizabeth. 2003. *A Consumers' Republic: The Politics of Mass Consumption in Post-War America.* New York: Knopf.

Costa, Dora L. 1998. *The Evolution of Retirement: An American Economic History, 1880–1990.* Chicago: University of Chicago Press.

Cravit, David. 2008. *The New Old: How the Boomers are Changing Everything – Again.* Toronto: ECW Press.

Ekerdt, David J. and Evelyn Clark. 2001. "Selling Retirement in Financial Planning Advertisements." *Journal of Aging Studies* 15:55–68.

Fisher, Jonathan D., David S. Johnson, Joseph T. Marchand, Timothy M. Smeeding, and Barbara B. Torrey. 2009. "Identifying the Poorest Older Americans." *Journals of Gerontology, Social Sciences* 64B:S758–66.

Formosa, Marvin. 2009. *Class Dynamics in Later Life: Older Persons, Class Identity, and Class Action in Malta.* Berlin: Lit Verlag.

Gershuny, Jonathan. 2000. *Changing Times: Work and Leisure in Postindustrial Society.* Oxford: Oxford University Press.

Gilleard, Chris. 1996. "Consumption and Identity in Later Life: Toward a Cultural Gerontology." *Ageing & Society* 16:489–98.

Gilleard, Chris and Paul Higgs. 2005. *Contexts of Ageing: Class, Cohort, and Community.* Cambridge: Polity Press.

Gilleard, Chris, Paul Higgs, Martin Hyde, Richard Wiggins, and David Blane. 2005. "Class, Cohort and Consumption: The UK Experience of the Third Age." *Journals of Gerontology, Social Sciences* 60B:S305–10.

Goldstein, Sydney. 1968. "The Aged Segment of the Market, 1950 and 1960." *Journal of Marketing* 32:62–8.

Goodman, Ellen. 2005. "The New 'Leisure Class'." *The Boston Globe*, June 26, 2005.

Graebner, William. 1980. *A History of Retirement: The Meaning and Function of An American Institution, 1885–1978.* New Haven CT: Yale University Press.

Gunter, Barrie (1998) *Understanding the Older Consumer.* London: Routledge.

Hechter, Michael. 2004. "From Class to Culture." *American Journal of Sociology* 110:400–45.

Hebdige, Dick. 1979. *Subculture: The Meaning of Style*. London: Methuen.

Higgs, Paul, Martin Hyde, Chris J. Gilleard, Christina. R Victor, Richard D. Wiggins, and Ian R. Jones. 2009. "From Passive to Active Consumers? Later Life Consumption in The UK from 1968–2005." *The Sociological Review* 57:102–24.

Horneman, Louise, R. W. Carter, Sherry Wei, and Hein Ruys. 2002. "Profiling the Senior Traveller: An Australian Perspective." *Journal of Travel Research* 41:23–37.

Hope, Janet and Linda Havir. 2002. "You Bet They're Having Fun: Older Americans and Casino Gambling." *Journal of Aging Studies* 16:177–97.

Hourriez, Jean-Michel, Legendre, Nadine, and Le Verre, Robert. 2001. *Plus de Ménages Pauvres Parmi les Salariés, Moins Chez les Retraités*. INSEE Première, no. 761. Paris: Institut National de la Statistique et des Études Économiques.

Hurd Clarke, Laura, Meredith Griffin, and Catherine Maliha. 2009. "Bat Wings, Bunions, and Turkey Wattles: Body Transgressions and Older Women's Strategic Clothing Choices." *Ageing & Society* 29:709–26.

Hurd, Michael and Rohwedder Susann. 2005. The Retirement-Consumption Puzzle: Anticipated and Actual Declines in Spending at Retirement. *RAND Labor and Population working papers*, WR-242.

Iceland, John. 2003. *Poverty in America: A Handbook*. Los Angeles: University of California Press.

INSEE 2008. *Cinquante Ans de Consommation en France*. Paris: Institut National de la Statistique et des Études Économiques.

Jones, Ian R., Victor Hyde, Christina Martin, Dick Wiggins, Christopher Gilleard, and P. Higgs. 2008. *Ageing in a Consumer Society: From Passive to Active Consumption in Britain*, Bristol: Policy Press.

Juvin, Hervé. 2005. *L'Avènement du Corps*. Paris: Editions Gallimard.

Katz, Steven and Debbie Laliberte-Rudman. 2005 "Exemplars of Retirement: Identity and Agency between Lifestyle and Social Movement." Pp.140–59 in *Cultural Aging: Life Course Lifestyle and Senior Worlds,* edited by S. Katz. Toronto: Broadview Press.

Keister, Lisa A. 2000. *Wealth in America: Trends in Wealth Inequality*. Cambridge: Cambridge University Press.

Lazer, W. 1963. "Life Style Concepts and Marketing". Pp. 140–151 in *Toward Scientific Marketing*, edited by S. Greyser. Chicago: American Marketing Association.

Lazer, W 1986. "Dimensions of the Mature Market." *Journal of Consumer Marketing* 3:23–34.

Leicester, Andrew, Cormac O'Dea, and Zoe Oldfield. 2009. *The Expenditure Experience of Older Households*. London: Institute for Fiscal Studies.

Lipovetsky, Giles. 1994. *The Empire of Fashion: Dressing Modern Democracy*, trans. Catherine Porter, Princeton, NJ: Princeton University Press.

Lloyd, Alfred H. 1922. "Ages of Leisure." *American Journal of Sociology* 28:160–78.

Mathur, Anil, George Moschis, and Euehan Lee. 2008. "A Longitudinal Study of the Effects of Life Status Changes on Changes in Consumer Preferences." *Journal of the Academy of Marketing Science* 36:234–46.

McCracken, Grant. 1987. "Culture and Consumption among the Elderly: Three Research Objectives in an Emerging Field." *Ageing and Society* 7:203–24.

———. 1990. *Culture & Consumption.* Bloomington: Indiana University Press.

Michelon, L.C. 1954. "The New Leisure Class." *American Journal of Sociology* 59:371–8.

Miller, Daniel. 1998. *A Theory of Shopping.* Cambridge: Polity Press.

Miller, Daniel, Peter Jackson, Nigel Thrift, Beverley Holbrook, and Michael Rowlands. 1998. *Shopping, Place and Identity.* London: Routledge.

Ministry of Labour and National Service. 1957. *Report of an Enquiry into Household Expenditure, 1953–54.* London: HMSO.

Moschis, George. 1996. *Gerontographics*. Newport, CT: Quorum.

———. 2009. "Generational Marketing." Pp. 149–70 in *Consumption & Generational Change: The Rise of Consumer Lifestyles,* edited by I.R. Jones, P. Higgs, and D. J. Ekerdt. New Brunswick, NJ: Transaction Publishers.

Moschis, George and Belgin Unal. 2008. "Travel and Leisure Services Preferences and Patronage Motives of Older Consumers." *Journal of Travel & Tourism Marketing* 24:259–69.

Newman, Katherine S. 2003. *A Different Shade of Gray: Midlife and Beyond in the Inner City*. New York: The New Press.

Nimrod, Gulit. 2008. "Retirement and Tourism: Themes in Retirees' Narratives." *Annals of Tourism Research* 35:859–78.

Ong, Fon Sim and David R. Phillips. 2007 "Older Consumers in Malaysia." *International Journal of Ageing and Later Life* 2:85–117.

Pampel, Fred C. 1998. *Aging, Social Inequality, And Public Policy.* Thousand Oaks CA: Pine Forge Press.

Peterson, M. 2007. "Effects of Income, Assets and Age on the Vacationing Behavior of U.S. Consumers." *Journal of Vacation Marketing* 13:29–43.
Price, Debora and Ginn, Jay. 2006. The Future of Inequalities in Retirement Income". Pp. 76–84 in The Futures of Old Age. Edited by J. Vincent, C. Phillipson and M. Downs. London: Sage Publications.
Ransome, Paul. 2005. *Work, Consumption, and Culture: Affluence and Social Change in the Twenty-First Century.* London: Sage.
Reisman, David. 1989 (1961) *The Lonely Crowd. A Study of the Changing American Character.* New Haven CT: Yale University Press.
Rentz, Joseph O, Reynolds, Fred D, and Stout, Roy G. 1983. Analyzing Changing Consumption Patterns with Cohort Analysis. *Journal of Marketing Research* 20:12–20.
Roberts, Scott and Nan Zhou. 1997. "The Fifty and Older Characters in the Advertisements of Modern Maturity: Growing Older, Getting Better?" *Journal of Applied Gerontology* 16:208–20.
Rose, Arnold M. 1965. "The Subculture of the Aging: A Framework for Research in Social Gerontology." Pp. 3–18 in *Older People and Their Social World*, edited by A.M. Rose and W. A. Peterson. Philadelphia, Pa., F.A. Davis Company.
Ruiz, Jean-Paul, Jean-Charles Chebat, and Pierre Hansen. 2004. "Another Trip to the Mall: A Segmentation Study of Customers Based on Their Activities." *Journal of Retailing and Consumer Services* 11:333–50.
Russell, Dale W. 2008. "Nostalgic Tourism." *Journal of Travel and Tourism Marketing* 25:103–16.
Scharf, Thomas. 2009. "Too Tight To Mention: Unequal Income in Older Age." Pp. 25–52 in *Unequal Ageing: The Untold Story of Exclusion in Old Age,* edited by P. Cann and M. Dean. Bristol: Policy Press.
Schau, Hope Jensen, Mary Gilly, and Mary Wolfinbarger. 2009. "Consumer Identity Renaissance: The Resurgence of Identity-Inspired Consumption in Retirement." *Journal of Consumer Research* 36:255–76.
Sedgeley, Diana, Annette Pritchard, and N. Morgan. 2006. "Understanding Older Women's Leisure: The Value of Biographical Methods." *Tourism* 54:43–52.
Shoemaker, Stowe 2000. "Segmenting the Mature Market – 10 Years Later." *Journal of Travel Research* 39:11–26.
Simmel, Georg. 1904. "Fashion." *International Quarterly* 10:130–55.
Smith, Adam. 1936 [1788]. *An Inquiry into the Nature and Causes of the Wealth of Nations.* New York: Modern Library.
Smith, J.W. and A. Clurman. 1997. *Rocking the Ages: The Yankelovich Report On Generational Marketing.* New York: HarperCollins.
Statistisches Bundesamt. 2009. *Statistisches Jahrbuch 2009.* Wiesbaden, Germany: Statistisches Bundesamt.
Street, Debra and Janet Wilmoth. 2001. "Social Insecurity? Women and Pensions in the U.S." Pp.120–141 in *Women, Work, and Pensions: International issues and prospects,* edited by J. Ginn, D. Street, and S. Arber. Buckingham: Open University Press.
Stroud, Dick. 2005. *The 50-Plus Market.* London. Kogan Page.
Szmigin, Isabelle and Marilyn Carrigan. 2000. "The Older Consumer as Innovator: Does Cognitive Age Hold the Key?" *Journal of Marketing Management* 16:505–27.
———. 2001. "Learning to Love the Older Consumer." *Journal of Consumer Behaviour* 1:22–34.
Therkelsen, Annette and Gram, Malene. 2008. "The Meaning of Holiday Consumption: Construction of Self Among Mature Couples." *Journal of Consumer Culture* 8:269–92.
Townsend, Peter. 1979. *Poverty in the United Kingdom.* Harmondsworth: Penguin Books.
United Nations Economic Commission for Europe. 2009. *Older Persons as Consumers* UNECE Policy Brief on Ageing No 3: Geneva, UNECE.
Veblen, Thorsten. 1959[1912]. *The Theory of the Leisure Class.* New York: Viking.
Vincent, John. 1999. "Consumers, Identity and Old Age." *Education and Ageing* 14:141–52.
Warde, Alan. 1996. *Consumption, Food and Taste.* London: Sage.
Warde, Alan and Tony Bennett. 2008. "Culture in Common: The Cultural Consumption of the UK Managerial Elite." *Sociological Review* 56:240–56.
Williams, Angie, Virpi Ylanne, and Paul Wadleigh. 2007. "Selling the 'Elixir of Life': Images of the Elderly in an *Olivio* Advertising Campaign." *Journal of Aging Studies* 21:1–21.
Yang, Zhilin, Nan Zhou, and Jie Chen. 2005. "Brand Choice of Older Chinese Consumers." *Journal of International Consumer Marketing* 17:65–81.
Zukin, Sharon and Jennnifer S. Maguire. 2004. "Consumers and Consumption." *Annual Review of Sociology* 30:173–97.

Part VII
Social Vulnerabilities and Aging

Chapter 24
Planning for Old Age

Debra Street and Sarah Desai

For many individuals, planning for old age is somewhat analogous to generals' propensity to fight the last war. What is to be planned for, and when, depends on how typical experiences of old age are perceived, whether an individual expects to fit within the typical mold, and which goals for later life seem to have been attainable in the recent past. The looming "silver tsunami" of Baby Boomer retirees, who transitioned to adulthood during expansive educational opportunities, maturing welfare states and booming economies, reflects a challenge for aging societies now confronting very different contemporary socioeconomic conditions. From the mid to late twentieth century (the Baby Boomers' temporal frame of reference for what to expect in old age) most individuals' eventual retirement in developed countries was taken for granted and pension systems appeared relatively stable. However, both prospects – universal retirement and secure pensions – are far less certain for cohorts contemplating old age at the outset of the twenty-first century.

The choices and risks associated with later life planning reflect opportunities and behaviors constrained by particular national social arrangements, employment markets, and welfare states. In fact, Ulrich Beck (1992, 1999) characterizes modern nation states as *risk societies* that reflect cultural standards for tolerable and intolerable risk at different points in the life course, including the risks associated with old age. Just as individuals adapt personal goals and plans in response to changing conditions, welfare states enact and adjust social policies in nation-specific ways that are consistent with meeting needs of aging populations, maintaining cultural standards of acceptable risk and responsibility (Beck 1999), and reflecting the ideological currents of the time (see Hacker 2006; Béland and Gran 2008). The interplay between structure and agency and cohort-specific experiences of aging through time make consideration of planning for later life a moving target. Broad social and cultural change, the political context of modern welfare states, evolving family structures and health institutions, and transformed labor and investment markets shape the opportunity structures and constraints that individuals planning for later life must consider in their assessments of how best to plan for their futures. The influences of social networks within which individuals are embedded and make decisions add another layer of complexity to the mix.

The idea of planning for a good life in old age is universally appealing but conceptually imprecise. Certainly, no one sets out to accomplish bad outcomes in old age, although many people experience them. What a good old age might entail encompasses a range of diverse outcomes across many life domains. Further, a "good enough" old age – a period at the end of the lifecourse savored

D. Street (✉)
Department of Sociology, State University of New York, Buffalo, NY, USA
e-mail: dastreet@buffalo.edu

R.A. Settersten, Jr. and J.L. Angel (eds.), *Handbook of Sociology of Aging*, Handbooks of Sociology and Social Research, DOI 10.1007/978-1-4419-7374-0_24, © Springer Science+Business Media, LLC 2011

as enjoyable and satisfying – could vary so widely across individuals that there is no widespread agreement among sociologists about what or how to study planning, nor among individuals about what exactly is to be planned for, beyond doing one's best to secure adequate income and avoid frailty and ill health. One candidate outcome associated with planning for a good old age is the psychosocial notion of expressing generativity in midlife that reflects individuals' sense of responsibility toward other generations, lives invested with meaning, and optimism about the future of humanity, leading to ego identity vs. despair in old age (Erickson 1959; McAdams and St. Aubin 1998). The actual expression of life-enhancing midlife generativity may (or may not) be tightly linked to identifiable planning processes for old age.

For the purposes of this chapter, we characterize planning for old age as the range of activities individuals deliberately pursue with a goal of creating desired outcomes in later life. Such planning could involve a range of strategies, from picking a career to becoming parents, from buying a vacation home for future use to investing in the stock market, from purchasing long-term care insurance to exercising regularly, from changing careers to pursuing a hobby, and from having cosmetic surgery to simply making new friends. We highlight a selection of sociological literature that bears on issues associated with individual planning for old age, particularly the well-developed body of research on retirement planning, both at the level of national pension systems and for individuals. However, sociological research that focuses explicitly on planning for old age more generally (and the activities and processes associated with it) is relatively sparse. This is not surprising, given an array of factors that make theorizing and empirical research on planning for old age particularly challenging. Among the challenges are the following:

- Hyper focus on economic aspects of financial planning for retirement, to the exclusion of research on other important aspects of planning for later life.
- Evolving meanings of retirement and old age in youth-dominated cultures.
- The challenge of linking contemporaneous individual preferences and behaviors to outcomes that will not be experienced for decades, if ever.
- The sheer scope of activities that could be planned in ways that impact later life experiences (and unplanned counterpart activities that also influence how individuals experience their lives at advanced ages).
- The pace of social and policy change associated with aging societies.

Social scientific literature that bears on planning underscores how a variety of activities associated with daily life in younger adulthood are major contributors to outcomes in old age, regardless of whether individuals consciously or explicitly include later life goal-oriented planning in their decision-making calculus. For example, while research on goal-setting by young adults indicates that setting goals in the first place is a predictor of the speed and efficiency of achieving positive outcomes in early adulthood, research in this vein typically treats planning and associated outcomes as processes of transitions to early and middle adulthood, rather than an explicit effort to plan for old age (Clausen 1991; Reynolds et al. 2007). While early life decisions set the path toward a range of possible outcomes in later life, members of different cohorts experience similar transitions under different conditions that structure how and what they can reasonably plan (Shanahan et al. 1997). Membership in a cohort that enters their employment years during periods of high labor demand, like current Baby Boomers, represents luckier timing than for cohorts who make such transitions when unemployment is high, as do young adults in 2010. Planning can do only so much to compensate for such socio-historical differences that shape cohort possibilities and constraints. We touch on several areas beyond retirement income where planning for later life is important and discuss potential areas for future sociological research that incorporate shifting meanings and expectations associated with old age, early twenty-first century risks associated with aging, and identifying pathways for individuals in different social positions toward good lives in old age.

Planning

Planning for old age falls within the scope of processes that cognitive psychologist Kathleen Galotti (2002) characterizes as making "decisions that matter." The complexities associated with real-life planning are monumental. Along with selecting a goal (among competing ones) having various levels of "goodness," gathering (often complex) relevant information, structuring decisions to account for the different criteria and options that could maximize the likelihood of goal achievement, making a decision about what set or sets of options to pursue, then finally evaluating progress and strategies associated with goal attainment that might signal need for readjustment (Galotti 2002). Planning and pursuing goals under the complexities of real world experiences is fraught with multiple influences, signaling the need to study planning in multiple dimensions and across levels of analysis. Despite the complexities, individuals often do exert planful competence (Clausen 1991; Shanahan et al. 2003), including deliberate actions to address predictable risks of later life, such as low income, social isolation, and declining health.

While psychologists and economists focus predominantly on individuals, more sociological approaches emphasize that planning does not occur in a social vacuum, but rather is structured by the particular arrangements that shape expectations and experiences of old age within specific societies. There are both individuals who do not plan much and desired outcomes that are unplanned, no matter what level of analysis sociologists consider. Welfare states, other social institutions, and organizations are implicated in the ways individuals try to optimize future outcomes. The institution of retirement, the dominant marker for old age in developed countries for much of the twentieth century, requires a predictable income stream to support it. Pension systems in modern welfare states are an essential structural piece of the old age planning puzzle; consequently pension issues dominate the research literature. This underscores two issues that link sociological research and planning for old age. First, the age of eligibility for public pensions is the ubiquitous social science proxy used to distinguish old age. Second, social gerontologists regard income *adequacy* as an essential component of managing in later life, and it is a centerpiece of their research and theorizing about planning for it. Despite a relative dearth of explicit analytic attention to planning for later life outcomes more generally, a rich body of sociological literature documents later life outcomes for different groups associated with the mixed economies of retirement income systems. In fact, research is often restricted to analysis of the retirement decision or saving for retirement as proxies for whether any planning has occurred at all.

Planning for Retirement and Later Life Income Security

Pension systems are important structuring institutions, and the expansion of public pension systems throughout the mid-twentieth century steadily improved the stability and value of retirement income, leading to normative expectations of a tripartite lifecourse of education, work, and retirement (Kohli 1986). Retirement systems, for the first time, gave ordinary workers the right to "stop work before wearing out" in what John Myles (1989) called *old age welfare states* (Myles 1989). By 1980, however, most developed countries had outrun the postwar expansion that marked the "golden age" of welfare states (Esping-Andersen 1999) and the normative (for men) tripartite mid-twentieth century life course became blurred by the end (Mutchler et al. 1997). Women's employment exploded, welfare states retrenched, transitions between work and retirement became less crisp, and employers scaled back or changed the rules of the game for occupational pensions (Guillemard and Rein 1993; Shuey and O'Rand 2004). Disrupted domestic labor markets, modifications in employee benefits, transformations in welfare states (Quadagno, Chap. 20), and other social institutions have undermined confidence in the predictability of prospects for retirement as an institution and for conditions of later life more generally.

Sociologists in the 1980s and early 1990s researched how occupation and pension benefits influenced labor exit, noting that by the mid-1990s the decades' long trend toward early retirement for men reversed. Early retirement research had focused mainly on men's experiences, but women's unprecedented surge into paid work soon attracted analytic attention. Women's shorter average work histories, differential entitlement to pension benefits, and the early influences of the transformation from defined benefit to defined contribution pensions inspired researchers to investigate how retirement-related processes were different for women (see, e.g., O'Rand 1996; Ginn et al. 2001). Growing diversity in aging cohorts and variations in later life experiences for different race and ethnic groups also began to attract research attention in the 1980s and 1990s.

Income security in old age varies internationally, depending on the size and structure of public pension systems and routine access to private retirement income (see, e.g., Ginn et al. 2001; Gran 2008). In the United States, later life income security typically requires access to private pensions, individual savings and wealth to top up income received from Social Security. Women, members of minority demographic groups, unmarried people, and individuals with lifetime low wage employment are less economically secure (Angel et al. 2007; Harrington Mayer and Herd 2007; Shuey and O'Rand 2004; Street and Wilmoth 2001). Over the past 30 years, employers have replaced defined benefit pensions and their guaranteed retirement income with defined contribution savings plans that shift savings and investment risk from large institutions onto the shoulders of individuals (see Schulz and Borowski 2006; O'Rand 2003; Shuey and O'Rand 2004) at a time when individuals appear to be ill-equipped to handle the risk of uncertain investments in volatile markets.

Baby Boomers entered a workforce that seemed to hold the promise for secure retirement at ever younger ages, so early sociological research on aging often focused on predictors of "early" and "on time" retirement. By the early 1990s, research had evolved to account for the ways that discernible shifts in particular institutional realms – such as women's labor market participation, changing family forms, and the transformation of occupational retirement systems – influenced current and future retirees. For example, researchers documented patterns of the pace and timing of dual worker couple retirement (Henretta et al. 1993; Elder and Pavalko 2003; Moen 2003; Behringer et al. 2005). However, the pace and timing of retirement transitions in previous decades seems an unlikely roadmap for increasingly variable and dynamic transitions now and in the future. In the aftermath of the financial market collapse, more men than women have lost jobs and the relatively uninterrupted trend toward increased equity in pension savings in private financial markets ended in fall 2008. Older workers are remaining employed, postponing anticipated retirements to stabilize their pension prospects. Such secular circumstances, and others, work their way into younger adults' expectations and planning for old age. A need to systematically study and document how planning might contribute to a good old age in life domains that extends beyond knowledge about income security becomes even more important when economic uncertainty abounds.

Why Is Planning so Difficult to Study?

With the exception of research on retirement planning, planning for old age writ large is difficult to observe contemporaneously (see Pearlin 2010:211) and usually studied post hoc under the broader umbrella of sociological research on outcomes in old age. There are myriad reasons. Modern societies are dominated by youth culture, fear of aging, age denial (e.g., Blaikie 1999; Calasanti and Slevin 2001), differences in individual planning horizons and propensities, the rapid pace of social change that alters expectations, and structured inequalities that influence the capacity to choose goals and plan in the first place. For some people, whether because the struggle to keep their heads above water in current circumstances precludes planning, or because some individuals are not very planful in the first place, old age is simply another experience that unfolds in its own way—"old age is just something that happened to me."

Sociologist David Ekerdt observes potential for an opposite trend, what he calls "foreshortening of the lifecourse." A steady drumbeat of media messages reminds adults that they must "do" something now to make retirement happen later. He argues that retirement has "colonized adulthood" (Ekerdt 2004:5) by penetrating the consciousness of even very young adults with the need to plan. This foreshortened lifecourse could lead to foresight in planning for later life, or the specter of planning for the end over an entire adult lifecourse may provoke anxiety and create perverse outcomes rather than instill peace of mind (Ekerdt 2004).

Planning for a Different Time

What old age and retirement might mean in individual futures has always been and remains something of a moving target (see, e.g., Achenbaum 1978; Cole 1992; Ekerdt 2010; Hardy 2002; Phillipson 1998; Quadagno 1988; Weiss 2005), with the position of future goalposts moving faster than individuals can readily incorporate into planning. Change in nearly all institutional realms is so brisk – from education, to families and work (see Harrington-Meyer, Chap. 17), to social policies, to the state of the economy – that perceiving the structural landscape for old age a decade in the future, never mind several, makes planning a daunting task. Traditionally, old age in the United States was conveniently marked by the 65th birthday – the age when individuals could retire with full Social Security benefits. Even that is changing. The age for entitlement to full Social Security retirement is increasing gradually, to 67 for individuals born in 1960 or later (Social Security Administration [SSA] 2010). Sixty-seven-year-old Americans, not 65-year-olds, will soon be the administratively (and social scientifically) "old," and the United States is not alone in recently increasing national retirement ages. The phrase "sixty is the new forty" has invaded popular culture and media reports on aging, spawning websites with the slogan emblazoned on gifts, greeting cards, and t-shirts. Yet in the current recession, older workers who have lost their jobs are the age group most likely to be long-term unemployed (BLS 2010), take jobs at substantially lower pay, and perceive age discrimination in their efforts to be reemployed (NYT 2009). Despite popular support for the program, confidence in the future of Social Security benefits has dwindled since the 1980s (Quadagno 1996) and many younger Americans doubt whether Social Security benefits will be available at all when they hope to retire (Gallup 2010). Logically, this *may* make individual planning more salient than ever for younger workers (even if they are wrong in their policy prognostications), but it does not necessarily mean that most younger workers *can* or *will* behave planfully.

For researchers who tracked savings for prospective retirees (often taken as the most tangible evidence of planning for old age), the ramifications across life domains of the loss in value when financial markets plunged in 2007 and 2008 are still unfolding, although there has been a chilling effect on plans to retire (EBRI 2010). Adding fuel to the retirement savings loss fire, the sub-prime mortgage collapse hit individuals who had planned to cash out soaring home equity for income in later years. From a sociological perspective, it is not just that individuals have less, on average, to fund their retirements. These shocks manifest differently for individuals in different structural locations and undoubtedly change planning orientations. Some individuals were scarcely affected; others experienced an unexpected and profound change in life course trajectories that required adjustment across institutional realms of family, work, and retirement.

Individuals who planned carefully by saving and investing for old age and retirement over the past few decades could have wildly different outcomes for similar efforts. An individual who retired in early 2007 could have converted assets into retirement annuities, sold a home and downsized at the peak of the housing market, creating a comfortable income stream for old age. The downturn of housing and equities markets less than a year later meant that precisely the same kind of planning portfolio and behavior could yield very different outcomes, for no other reason than because retirement conversions were a few months apart. By late 2008, growing numbers of prudent planners realized that the collapse of the value of retirement savings or housing assets – the largest single asset for most American families (U.S. Census Bureau 2001) – jeopardized retirement plans.

Only 16% of workers who responded to the *2010 Retirement Confidence Survey* (Employee Benefits Research Institute [EBRI] 2010) were very confident that they would have enough money to retire comfortably. On average, the value of retirement investments for *Retirement Confidence Survey* (hereafter *Confidence Survey*) respondents declined across the board from 2007 to 2010. The percentage who said they expected to work past 65 increased from 11% in 1991, to 19% in 2000, to 33% by 2010 (EBRI 2010:28). Workers who said they were saving at all for retirement declined from 65% in 2009 to 60% in 2010 and half of workers reported they were "not too" or "not at all" confident they could accumulate enough savings for retirement (EBRI 2010). Twenty-seven percent of working respondents had savings and investments (excluding value of primary residence and defined benefit pension) of less than $1,000, up from 20% in 2009. In 2002, half of respondents to the *Confidence Survey* reported having investments and savings of $25,000 or less, which declined to 48% by 2007. By 2010, however, nearly two thirds of respondents reported their savings and investments were worth less than $25,000 and nearly 30% of workers said they would delay retirement, mainly because they could not afford to (EBRI 2010). The link between retirement and old age experiences, the emerging contours of future public pensions and private retirement systems, an uncertain employment market during the preretirement years, combined with the volatility and individual risk of retirement investments, make understanding the right steps for planning for later life difficult to discern.

What to Plan for Old Age?

Uncertainties aside, some processes that lead to desired or negative outcomes in old age are neither inevitable nor immutable, making better theoretical and empirical understanding of the opportunities and constraints different groups encounter in planning for old age a fertile area for sociological research. The emphasis and importance placed on planning in different sociological frameworks varies across theoretical and research approaches and is often implicit in the models rather than explicitly analyzed. Taken for granted in most sociological work on later life outcomes are differing propensities to plan or save, based on assumptions of individual behavior borrowed from economics. Economists generally take an individualist stance and explain planning for old age using one of two broad theoretical approaches. One emphasis is the life-cycle saving hypothesis that posits that individuals borrow, save, and dissave to smooth consumption over the life course (Ando and Modigliani 1963). The other emphasis is the theory of planned behavior, which assumes that norms, attitudes, and perceived control will increase the likelihood of planning and that successful planning experiences contribute to future planning behavior (Ajzen 1991).

Sociologists have not yet fully articulated planning for old age into sociological theories or empirical work (see Dannefer and Uhlenberg 1999; Marshall 2000; Hitlin and Elder 2007a, b). Rather, sociology of aging and the life course has tended to analyze a range of outcomes associated with aging rather than focusing on particular choices individuals make in the moment and how processes unfold as they prepare for their futures. One influential research track in that vein explores successful (Rowe and Kahn 1997 1998) and/or productive aging (Bass et al. 1993; Bass and Caro 2001), deploying a rich array of quantitative datasets (cross-sectional and longitudinal), indicators, and statistical analyses to demonstrate the associations between individual characteristics and positive late life outcomes. The influential Rowe and Kahn (1997:439) definition of *successful aging* includes "avoidance of disease and disability, maintenance of high physical and cognitive function, and sustained engagement in social and productive activities." Researchers in the *productive aging* tradition have elaborated the benefits of continued engagement in productive activities (including caregiving, volunteering, civic engagement, and employment) for improved physical and cognitive functioning and greater life satisfaction (e.g., Burr et al. 2002; Ekerdt 2010; Morrow-Howell et al. 2001; Morrow-Howell et al. 2003; O'Reilly and Caro 1994). Although criticized because the term *successful* overly pathologizes routine aspects of aging (Kane 2005) and *productive* aligns with economic activity (Herzog et al. 1989), the *successful/productive aging* research stream provides empirical evidence that social activities enhance later life.

The theoretical importance of planfulness in these frameworks is apparent, even if not elaborated: optimal goals can be "attained through individual choice and effort," emphasizing the importance of planning for later life (Rowe and Kahn 1998:37). Successful and/or productive outcomes associated with aging are highly correlated with SES indicators (making retirement planning that contributes to higher SES important), but outcomes are not entirely dependent on them. *Successful/productive* aging frameworks are concerned with later life outcomes that result from individual choices (similar to economic theories), tend to be relatively decontextualized, and focus on outcomes emphasized within neo-liberal ideological frameworks (Estes and Mahakian 2001; Katz 2000; Taylor and Bengtson 2001). Compared to its *successful aging* cousin, *productive* aging is more attuned to opportunities and constraints – more sociological – that influence individual capacities to succeed or produce (Taylor and Bengtson 2001), but processes of planning for old age are central analytic concerns for neither theoretical approach. Researchers highlight positive outcomes for individuals who age optimally within the conceptualizations of each framework, but how or when it is important to plan or the ways social location and life course together influence capacities to plan are generally not incorporated into the models.

McLaughlin and her colleagues (2010), analyzing data from four waves of the Health and Retirement survey, determined that the constraints of successful aging models are so stringent that only 11.9% aged successfully and that rates of successful aging were declining over time. Their findings point to a fundamental flaw in paradigm that otherwise has attracted widespread attention and yielded influential empirical insights. An overwhelming proportion of older Americans fail to age successfully, at least when stringent operational definitions of variables are used (McLaughlin et al. 2010). Whether it is through bad planning, lifestyle choices, cohort-specific life course constraints, or bad luck that individuals fail to hit successful aging markers, it would be implausible to imagine that merely 12% of later lives are "good" ones. Factors such as birth cohort, age, gender, race, ethnicity, and life experiences influence beliefs about what constitutes success in terms of aging (Holstein and Minkler 2003; Jackson et al. 1993; Phelan and Larson 2002). This highlights an obvious criterion for meaningful sociological conceptualization of good lives in old age and how individuals could plan for those: definitions must connect in some tangible way with lived experiences of individuals in different cohorts and what people do and value in their own lives.

From a more experiential, critical, and social relational stance, sociologists skeptical of the productive/successful aging paradigms envision a good life in old age more pragmatically, as individuals making the most out of what is available in the later years, given the contextual circumstances of aging they experience (see, e.g., Holstein and Minkler 2003; Kane 2005; Phillipson 1998; Street et al. 2007). This includes individuals adapting to social losses alongside physical and cognitive declines often associated with aging. While the capacity to age successfully or productively is not available for every individual, accomplishing a good life in old age – getting the most out of life as it comes – is more widely obtainable, particularly if social institutions like workplaces and welfare states also make planful adjustments to help individuals meet their needs. Research on planning that promotes skills in habituation (Kastenbaum 1980) to identify and enact processes that foster adaptation, tend social relationships, and create and/or preserve an array of evolving capacities that enhance well-being and happiness in later years, (despite the age-related onset of limitations most individuals will almost inevitably experience) has received relatively less attention (but see Holstein and Minkler 2003).

Challenges in Planning for Old Age

Women and men enhance or undermine opportunities for good old age through an almost endless range of activities, from accomplishing training or education that leads to lucrative employment, whether they avoid smoking, embrace healthy diets and exercise, or decide to form families or life partnerships, and to be childfree or raise a family. Mundane activities bear inexorably on the outcomes in later life, and while the quotidian choices of individual behavior represent agency, they are not planning for old age (see Hitlin and Glenn 2007a; Pearlin 2010). What distinguishes planning for old age in particular from either planning in general (which has a future goal orientation, but not necessarily one associated with old age) or the minutiae of everyday choice is the goal of the activity, its intended consequences.

Conceptually, planning for old age captures a series of tensions: between human agency and the constraints social structures impose on individual lives; between plans and actions associated with short-term goals and contemporaneous choices vs. strategies enacted for the longer range; between current and delayed consumption; and, between the relative shares of responsibility for the state, employers, and individuals in planning and providing for old age. Culturally and ideologically, planning for old age in modern societies involves deliberate action to accomplish something many younger individuals fear – growing old – and that many individuals want to deny – being old. Understanding what characteristics, expectations, risk perceptions, and opportunities inspire some individuals to plan for old age, and what exactly they plan for, presents a series of puzzles for sociologists of aging and the life course. As mentioned earlier, sociological research on these planning puzzles has usually been linked to outcomes associated with either retirement (income or transition planning) or health promotion for later life.

Over the past 3 decades, a substantial body of social science research has focused on how working-aged individuals in developed countries like the United States plan to avoid or minimize poverty and poor health, two of the major risks associated with old age. The dominant stream of research associated with planning for old age has treated one outcome – having sufficient financial resources in retirement, whether from pensions, individual savings, or wealth – as the implicit proxy for successful planning, if not successful aging. Consequently, analysis has often focused quite narrowly on economic issues at the expense of noneconomic ones, with well-developed bodies of literature in behavioral economics, cognitive psychology, and sociology centered on individual determinants of investment and retirement income planning. Sociologists have generally been more attentive than researchers in other disciplines to the ways that social inequalities are structured and expressed in the sources and amounts of later life incomes for different demographic groups, but economic outcomes are still the most usual proxy for successful planning. A second stream of sociological research has explored more directly the processes of retirement planning and decision-making for individuals and within families under changing conditions of work and retirement. How individual workers decide, and how spouses and partners together negotiate work/life transitions are also important topics of sociological research (Henretta 1992; Moen 2003; Szinovacz and DeViney 2000). A third research trajectory associated broadly with individual planning for old age (although more usually couched in terms of health behaviors rather than "planning") explores the role health behaviors and healthy lifestyles (another set of deliberate activities) play in later life healthfulness (see Ferraro, Chap. 29; Haber 2003; but also see Holstein 1998; Minkler 1999).

Even so, planning *qua* planning for old age, beyond issues associated explicitly with work-to-retirement transitions (Kim and Moen 2001) and retirement income planning, is seldom the focus of sociological research on preparations individuals make for their futures (although see Denton et al. 2004 and Moen et al. 2005 for exceptions). Planning that pays off for health and income security provides a material foundation that permits individuals to savor later life across multiple dimensions. Because highly valued but less tangible goals for old age – like happiness, life satisfaction, and well-being – are subjective, the role planning may play in their pursuit is intrinsically even more difficult to theorize, operationalize, and observe than the planful components of economic security or health behavior. This, no doubt, is why several sociologists observe that agency is more often asserted or assumed than analyzed empirically in life course research (Hitlin and Elder 2007a; Marshall 2000; Pearlin 2010; but see Denton et al. 2004; Hitlin and Elder 2007b; McMullin and Marshall 1999). Although planning and agency are not precisely analogous conceptually or in ways that could be operationalized for research, life course agency (Hitlin and Elder 2007a) operates, in part, through planning that individuals can accomplish within the opportunity structures they encounter.

Cultural roadmaps provide age-related guideposts for sequencing adults' education and employment transitions (Settersten and Hagestad 1996), but simultaneously invoke several planning black boxes of their own. Age norms evolve in ways that are neither entirely predictable nor synchronized across important life domains. Further there is really no age normative guidepost for when in the adult life course deliberate planning for later life should begin, although recently the dominant message is sooner rather than later (Ekerdt 2004). Cohort membership and social location also mediate individuals' perceptions of risk (Shuey and O'Rand 2004) and influences planning for an increasingly diverse aging population. The onset and types of later life planning must also differ across and within groups, with individuals of different demographic characteristics incorporating different types of strategies, goals and priorities (Angel and Angel 2006; Angel et al. 2007; Calasanti and Slevin 2001; Ginn et al. 2001; Holstein 1998), and time horizons (Ekerdt and Hackney 2002; Denton et al. 2004; Hitlin and Elder 2007a; Moen 2003). For some individuals, planning seldom happens at all (Denton et al. 2004). Although the ideal age for planfulness for later life lacks a normative age marker – except, perhaps, in the aspirations of financial planners – there is an age-based pattern bolstered by both logic and evidence that, as adults get older, the salience and propensity to prepare for later life increases.

Structured Inequalities and the Life Course Lens

As life course theorists have emphasized, later lives are the sum of prior statuses, earlier choices and past experiences (Elder 1994, 1998; Elder and Johnson 2003), whether enacted as planful for old age, or not. A tenet of life course research is the variation in individual capacities (and by extension, later life chances for a good life) structured by earlier opportunities and constraints, including choices individuals make (see, e.g., Lutfey and Mortimer 2006; Roberts and Bengtson 1999; Settersten 2003). Life chances are determined by family statuses and origins; gender, race, and ethnicity; the communities where people grow up and grow old; by educational and employment experiences expressed over adult lifetimes; and the socioeconomic and historical circumstances of the time. A strength of sociological approaches for studying old age is the capacity to take into account the way opportunities for planning and choice for individuals are socially structured by cohort and social location. One gap, however, is that the actual processes of planning are rarely modeled in research, but rather are implicit in the models and inferred from outcomes (Marshall 2000; Dannefer and Uhlenberg 1999; Hitlin and Elder 2007a). This is likely because planfulness overlaps with the "problem" of agency in life course research, and is most often treated as the residual that is not explained within analysis of normative patterns (Marshall 2000).

In general, sociologists using the life course framework have indirectly taken some aspects of planning for later life into account through outcomes research. The largest body of literature using a life course framework has focused on *cumulative advantage and disadvantage* (hereafter CAD) as a mechanism by which structural opportunities and risks bestowed by initial social location magnify advantage and disadvantage, translating into later life inequalities (Dannefer 1987, 2003; DiPrete and Eirich 2006; O'Rand 1996, 2003; Ferraro 2007). Advantaged structural locations – high levels of education, employment in benefit-rich employment sectors, high income jobs – provide some individuals with better arrays of choices and capacities to exercise life course agency and plan. Typically, CAD analyses demonstrate categorical differences in SES, gender, race/ethnicity, and other social markers that reflect structured (dis)advantages in early life and manifest in tandem across multiple life domains. Most evidence in the CAD framework demonstrates that individuals with greater income, wealth, and education face circumstances offering more choices and fare significantly better in old age than their less educated, less wealthy, lower income counterparts (see, e.g., O'Rand 1996; Willson et al. 2007; Ekerdt 2010; Burge and Street 2010) . CAD approaches can highlight intergroup differences within cohorts (Dannefer 2003), but the varying capacities for agency that influence intragroup variation are rarely explicitly analyzed (Marshall 2000; Hitlin and Elder 2007a).

Ferraro and his colleagues (2009) have recently explicated *cumulative inequality* theory (hereafter CI) that extends and refines attention to the accumulation of risk and opportunity across multiple life course domains. CI emphasizes that later life variation may arise from disadvantage (higher levels of exposure to risk) or advantage (more exposure to opportunity) in one life domain, but that neither advantage nor disadvantage in one domain is inevitably associated with advantage or disadvantage in another. While research in the CAD tradition focuses on the structured bifurcation and accumulation of advantage and disadvantage over the life course, processual permutations associated with the accumulation process are seldom specified (DiPrete and Eirich 2006; but see Willson et al. 2007). By contrast, the CI approach gives more analytic weight to *potentially* asynchronous (dis)advantage and to choice/human agency across life domains. Using CI, Ferraro and his colleagues argue that life course trajectories are shaped by "the accumulation of risk, available resources, and human agency" (Ferraro et al. 2009:423).

Theoretically, within the CI framework individuals may be structurally disadvantaged in earlier life or one particular life domain, but can potentially compensate by exercising agency – planning and implementing plans – in another. An individual may experience low SES over the life course, yet simultaneously compensate in another life domain. A low income elder may experience substantial well-being and happiness in non-SES domains through planning and behaviors that capitalize on advantages in their psychosocial domain – perhaps by nurturing family relationships or developing a large network of friends for social support. Outcomes across every life domain need not vary in the same direction (Ferraro et al. 2009; Ferraro and Kelley-Moore 2003). Although CI shares with CAD an analytic capacity to examine the opportunity structures for later life income security or healthfulness, CI redirects attention to inequalities experienced differently across life domains. Large pensions or great wealth do not necessarily guarantee good health or well-being. Consequently, CI's attention to distinctive accumulations of advantage and disadvantage represents a recent theoretical innovation in sociology of aging and the life course that may offer more analytic attention to human agency and, by extension, planning processes. This offers a theoretical and analytic approach that departs from dominantly deterministic or economic views of later life outcomes.

Middle-range perspectives like exchange theory (Dowd 1975) provide viewpoints that received more attention in earlier decades of research on the sociology of aging, but which could potentially be retooled to offer additional analytic traction and alternative lenses into planning processes. For example, rational choice or exchange perspectives emphasize actions that weigh the costs and benefits of interpersonal relationships, where issues of time, money, and concern come into play (Silverstein 2006). Exchange theory assumes a certain calculus of reciprocity, including interpersonal investments early in life with the expectation of returns in old age, which may be construed as "planning." For example, having a large family or sending children to expensive private schools could be interpreted as a fallback plan to protect oneself from social isolation or financial fragility in old age. For the most part, however, current behavior is not treated as planning for the future. Rather these "investments" represent a collection of subtle and time-specific cultural expectations about how others one provides for – emotionally or materially – will interact with someone in the future. While it does not make sense to interpret attending every one of a child's soccer games as "planning" in the same way as contributing to a 401k, from an exchange perspective attention to children is a type of investment. These investments, like financial ones, may not have predictable returns due to the pace of social change. Whether a set of current expectations about old age, likely to be affected by social change, can be considered as deliberate planning is an open question, but also a potential starting point using exchange theory concepts and models for research on planning for later life.

Planning for older ages represents an evocative example of gaps and frictions between individuals' life courses and the practices and processes influenced by a range of social institutions. *Structural lag* – the gaps between individuals' contemporary experiences and the inertia of institutions and organizational forms designed to meet the needs and normative expectations of earlier times (Riley and Riley 1994) – makes planning difficult. To optimize life chances and plan effectively, individuals need to enrich their contemporaneous lives across a range of domains and over time. Predictability and stability make planning easier, because it is clearer what needs to be done, and when. Especially during periods of rapid socioeconomic change, for example, the current recession in the United States and abroad, gaps between what individuals need and wherewithal to accomplish those goals become more apparent. Economic turmoil has highlighted the extent of risk and fragility of planning for later life. Recent changes in welfare states suggest that the risks individuals must plan for will increase rather than decline (Esping-Andersen 1999; Phillipson 1998; Estes and Mahakian 2001). Current institutional forms, socioeconomic opportunities, and the expectations and needs of individuals are desynchronized within the structures of the social programs, institutions, and organizations that do not yet take new preferences and risks into account. Yet the ingredients for a good old age – having adequate resources to meet needs (and some wants), relatively good physical and mental health, the capacity to engage in meaningful activities and to enjoy social relationships, enough help when it is needed, a modicum of happiness – are timeless. An ongoing challenge for sociologists is to enrich systematic understanding about how the complexities of twenty-first century institutions endow life courses with capacities for planning for old age, and how those capacities are deployed across different groups.

Where Next?

The large body of sociological research that emphasized retirement and income planning to the exclusion of other planning considerations is understandable for several reasons. The sheer size of national expenditures on state pensions throughout the developed world and the thirst of investment markets for the vast pools of capital that pension and retirement savings provide represent one impetus to pension planning research. Although not without its data challenges, the relative ease and ubiquity of obtaining quantitative indicators have also made issues surrounding pensions and retirement income among the easiest topics associated with planful aging to analyze. Patterns in pension planning and retirement saving can be assessed readily using standard metrics of program participation, scope of coverage, monetized portfolio or benefit value, all data routinely collected in several high-quality, nationally representative datasets. Exclusively economic models of later life planning employ well-developed theoretical frameworks and reflect the economic colonization of so many aspects of human experience. Another reason for this focus arises from the "usual science" pursuits of sociologists, given the attention that SES inequalities receive in life course research. For example, Veenhoven (2006) argues that the discipline is so problem-focused that sociologists seldom pay attention to the distribution of important life experiences, like happiness, in their research repertoires. Yet happiness, and figuring out ways to enhance it, might be the quintessential planning goal for old age.

Studying how retirement became institutionalized and widespread in western democracies spawned a wealth of theorizing on the welfare state, but often failed to connect the macro processes that institutionalized retirement to the micro level experiences of individual women and men. Nonetheless, macro and comparative research at the end of the "golden age of the welfare state" provided a lens through which sociologists could explore how societies shaped the experiences of women and men in time and place. Less available were studies that explored how timing, linked lives, human agency, and lifelong development, the other tenets of life course research, unfolded for individuals planning under the umbrella of the broader political economy.

Researchers in the CAD tradition and life course research pioneered studies that exploited the rich data in several longitudinal data sets to understand how advantage and disadvantage structured planning potential for individuals in different groups. Recent developments of CI theory (Ferraro et al. 2009) and the reflexive planning model (Denton et al. 2004) offer rich analytic platforms for sociologists to more fully incorporate the role human agency plays in planning processes within the structural contexts of different life domains. The challenge for CI, as for CAD research in particular and life course research more generally, will be to settle on conceptual and operational definitions for planning and exploiting longitudinal data sets that can live up to the potential of the theoretical framework. The challenge for the reflexive planning framework will be to extend its analytic reach to larger, more representative populations.

What is a good old age and how can individuals plan to attain them in the future? These are at once philosophical questions, a series of processes and accomplishments, and a set of puzzles that sociologists have yet to adequately define, operationalize, and study. There is no firm agreement on what "good" old age is, what exactly successful aging looks like, or how planning is implicated in the outcomes. This has, so far, left little choice but to study what is most readily understood and measured – often income and occasionally health status, because those are the variables for which we have the best indicators over the longest periods of time – as the proxies for the planning outcomes.

While some workers appear to be very planful in terms of preparation for old age (through retirement savings at least), others have few tools with which to plan. Ekerdt and Hackney (2002) found that, even among workers who had pensions, many did not know when they would become eligible for benefits or the value of their pension income on retirement. These earlier findings are reinforced by current data from the latest *Retirement Confidence Survey* (EBRI 2010). Its executive summary offered bullet points that compared 2010 respondents to respondents in earlier years of the survey:

- Preparations [for retirement] are still eroding (fewer saving for retirement)
- More people have no savings at all (27% have $1,000 or less)
- Clueless about savings goals (most do not know how much they need to retire)
- Expecting to work longer (EBRI 2010)

The *Retirement Confidence Survey* is one barometer of how Americans perceive prospects for retirement, and the main conclusions of the report are that planning for retirement is more disrupted than it has been in decades. Similar effects associated with economic insecurity may also be manifested in other life domains. Alternatively, planning in other life domains may inoculate planners against some consequences of economic disadvantage or uncertainty by magnifying advantage in other facets of life. This suggests fertile ground for sociological research on planning for later life more broadly, with potential for both empirical advances and theoretical ones.

Old age is the historically contingent culmination of individual lives in particular times, viewed from the perspective of planning through the lens of expectations formed in the present for an unknown and (to a large extent) unknowable future. From this vantage point, individual decisions early in life (such as whether to form a family, pursue higher education, or settle for a second-best job) or cohort-related circumstances within strained political economies (e.g., entering the job market or losing employment during a deep recession, rehousing "boomerang" kids, experiencing failure of a retirement investment) represent critical turning points in earlier adulthood that create life course discontinuities and differentiate subpopulations whose transformative early experiences vary. Research has started to document ways time is expressed differently in partnered women's and men's life experiences (Moen 2003), but little yet about how such temporal experiences systematically influence the onset of planning for old age, whether setbacks at critical junctures in the life-course inspire or frustrate planning, nor how much planning involves a series of individual choices or the depth and influence of consultations with others (nor who, besides spouses and partners, those other social relational influences are).

Do young families consider their own old age when they contemplate having children, and if so how and which families? How do parents decide to weigh the trade-offs between investing in the education of their children and cultivating their own retirement prospects? Do individuals with family histories of disease or disability (chronic or otherwise) plan for old age differently from others whose contextual backdrops differ? Such variations in family circumstances and individual experiences all have plausible influences on the capacity to plan, underscoring the potential for turning points associated with planning for old age in one domain of life to be quite separate from, and perhaps to magnify or mitigate risks in another. However, research that could test processual turning points or adjustments in life trajectories associated with deliberate planning for old age remains a task for the future.

One challenge for future sociological research on planning for later life involves reclaiming the analytic import of the structural realities within which planning occurs, often attributed to error terms in economics, but from a sociological perspective the social status and relationships that give meaning to later life. Surely, economic security and good health are important outcomes, attributable both to good planning and good luck over the life course and in old age. Being able to distinguish which process is in play, what matters most (luck or planning, structure or agency) and when, are important potential contributions of future sociological research. The systematic and predictable inequalities associated with income insecurity and poor health in later life are well documented by sociologists. But systematic insights into other less tangible or easily measured components of daily life that may also arise from planning for old age are not.

What exactly needs to be planned for old age? The possibilities are vast and evolving quickly in the twenty-first century. Planning happens more easily when the future seems predictable, yet unpredictability seems to be the watchword for the foreseeable future. For retirement to persist, an income stream for material support is necessary, so saving and investing for that possibility remains an important planning priority. However, instability in financial markets and the potential for policy changes in public retirement systems could have different planning outcomes for individuals in different cohorts or structural positions. Some may read the uncertain fiscal tea leaves as an imperative to save as much and as early as possible. Others may interpret the possible benefits of planning in a very different way, presuming that change is so rapid and outcomes so uncertain that too much planning makes little sense at all. Others may want to plan for future income security but lack the resources necessary to do it effectively because of competing current demands or inadequate income; some may have plenty of resources but lack a sense of urgency or desire to plan, preferring current over delayed consumption. It may be that financial planning seems fraught with peril, but other life domains offer more leverage for planning that could yield predictable outcomes. Setting and acting on goals that optimize social relationships and tweak conditions in nonfinancial realms of life may be the arena for planning more activities for future old age. The laundry list of planning orientations, other life domains and possible later life outcomes, and how individuals might plan (or not) within them is long. A few broad areas for additional research on the way planning may influence later life outcomes include the following:

- How to look and feel in old age: whether an individual attempts to roll back the appearance of aging by such activities as coloring hair or having cosmetic dentistry or surgery; whether to embrace the potential outcome of generativity and achieve ego identity or to pursue activities that stave off a sense of old age and mortality; preserving health and mobility to the extent possible.
- Where to live in old age: planning to move closer to adult children who have moved away, to downsize into a smaller house in the same community, to be a snowbird or a sunbird, to move into an age-segregated retirement community, to age in place.
- When old age is experienced: continuing as many roles from younger adulthood for as long as possible, avoiding activities associated with aging and old age; resisting stereotypes associated with aging.
- Who matters in later life: maintaining social relationships with friends and family, preserving contacts with the broader community, making new friends in new settings, anticipating care giving and receiving relationships and planning for those.

These and many more outcomes may be positively affected by planning, or negatively impacted by its absence. However, knowing when and for whom planning matters most requires data that go far beyond indicators of income and health status. Understanding the role planning plays in these later life processes requires research that considers how inequalities in several facets of life may interact to mitigate some inequalities and magnify others.

One recent example of how casting a broader planning net could be studied is an empirical project by Margaret Denton and her colleagues (2004) who developed a *reflexive planning* framework to study planning for old age in a small convenience sample of midlife and older Canadians. They note that as uncertainty increases around the ability of the welfare state, financial markets, and families to meet individual needs during later life, people reflect on their life situation and make choices accordingly. While four-fifths of individuals in the sample engaged in reflexive planning, a small group lived day-by-day, with little thought about the future. *Planners* in the study had a positive outlook and a strong sense of agency, in contrast to individuals who coped *day-by-day* and had a fatalistic perspective and weak sense of agency. Those who avoided planning were not simply disorganized or spontaneous. Most of the nonplanners were divorced or separated women with low incomes and few resources available for planning. Social structural factors interfered with the luxury of planning for later life (Denton et al. 2004). This research underscores the connections between lived experiences and structural constraints, how adaptation to current circumstances shapes planning and the interpretation of contemporary experiences, and could be extended to many realms of later life experiences and different cohorts of individuals.

Survey research can identify outcomes that may be associated with planning, but is so far poorly suited to understand processes of goal-setting, decision-making, and planning for old age across an adult lifetime. This argues for several different approaches to pry open some of the black boxes associated with planning for later life across multiple life domains. Exploratory studies that encompass the life experiences of diverse groups would be an important first step, including multimethod approaches used to explore the complexities of planning in real world circumstances (see, e.g., Moen 2003 and Denton et al. 2004). Quantitative components of such research with validated metrics for assessing planning activities could foster comparison across group experiences of planning if appropriate measures beyond income and health status can be devised. Qualitative research of processes associated with planning is also essential to build theory and empirical confidence that the range of activities associated with planning – or not – for old age are systematically identified and investigated. Qualitative studies of planning for different demographic groups, on varied life trajectories and at different turning points and transitions across the adult life course, could offer a grounded approach for theory-building, a foundation for broader study, and better insights into planning processes beyond the current focus on health and retirement. Such research would offer insights into processes associated with planning for older ages across successive cohorts, diverse groups, and multiple life domains, to build confidence that larger, more generic national studies have merit and utility for the discipline.

At the beginning of the twentieth century, contemplating routine survival into old age accompanied by a period of leisure at the end of life was beyond the imaginations of most individuals. Economic progress, social and cultural changes, increased life expectancies, and a set of structuring institutions moved potential old age beyond the realm of imagination to realization. But the current pace of social change and the restructuring of social institutions requires reimagination of old age for this century. There is substantial potential for sociological research to explore more carefully and creatively how planning might influence later life outcomes beyond the important considerations of income and health. Processes and outcomes surrounding those two material foundations of good old age have been carefully researched by sociologists over the past several decades, although economic turmoil and health care reform will keep social science research on planfulness for health and income security important areas of study. Understanding how planning directs and redirects life course trajectories that influence social support and friendship, outlooks on life, sense of fulfillment and happiness – the very things each of us would plan for old age if we could – are understudied areas that provide fertile ground for future sociological research.

Acknowledgment We appreciate helpful comments from Jacqui Angel, Stephanie Burge, Jeralynn Cossman, and Rick Settersten.

References

Achenbaum, W. Andrew. (1978). *Old Age in the New Land: The American Experience since 1790*. Baltimore, MD: The Johns Hopkins University Press.

Ajzen, Icek. 1991. "The Theory of Planned Behavior." *Organizational Behavior and Human Decision Processes* 50(2):179–211.

Ando, Albert and Franco Modigliani. 1963. "The Life Cycle Hypothesis of Saving: Aggregate Implications and Tests." *American Economic Review* 53:55–84.

Angel, Jacqueline L., Maren A. Jiménez, and Ronald J. Angel. 2007. "The Economic Consequences of Widowhood for Older Minority Women." *The Gerontologist* 47(2):224–34.

Angel, Ronald J. and Jacqueline L. Angel. 2006. "Diversity and Aging in the United States." Pp. 94–110 in *Handbook of Aging and the Social Sciences*, 6th edition, edited by R. H. Binstock and L. K. George. San Diego, CA: Academic Press.

Bass, Scott and Francis G. Caro. 2001. "Productive Aging: A Conceptual Framework." Pp. 37–79 in *Productive Aging: Concepts and Challenges*, edited by N. Morrow-Howell, J. Hinterlong, and M. Wayne Sherraden. Baltimore, MD: The Johns Hopkins University Press.

Bass, Scott A., Francis G. Caro, and Yun-Ping Chen, editors. 1993. *Achieving a Productive Aging Society*. Westport, CT: Auburn House.

Beck, Ulrich. 1992. *Risk Society: Towards a New Modernity*. London, UK: Sage.

———. 1999. *World Risk Society*. Cambridge, UK: Polity Press.

Behringer, Autumn, Carolyn C. Perrucci, and Richard Hogan. 2005. "Dual-Earner Couples' Expectations for Joint Retirement: A Study of Typical and Atypical Congruent and Non-Congruent Couples." Advances in Gender Research 9:111–132.

Béland, Daniel and Brian Gran, eds. 2008. *Public and Private Social Policy: Health and Pension Policies in a New Era*. Basingstoke UK: Palgrave Macmillan.

Blaikie, Andrew. 1999. *Ageing and Popular Culture*. Cambridge, UK: Cambridge University Press.

Bureau of Labor Statistics [BLS]. 2010. *Issues in Labor Statistics Summary 10–4*, March 2010. Retrieved March 30, 2010 (http://www.bls.gov/opub/ils/pdf/opbils81.pdf).

Burr, Jeffrey A., Francis G. Caro, and Jennifer Moorhead. 2002. "Productive Aging and Civic Participation." *Journal of Aging Studies* 16(1):87–105.

Burge, Stephanie Woodham, and Debra Street. 2010. "Advantage and Choice: Social Relationships and Staff Assistance in Assisted Living." *Journal of Gerontology Series B: Social Sciences* 65B: 358–69.

Calasanti, Toni and Katherine Slevin. 2001. *Gender, Social Inequalities, and Aging*. Walnut Creek, CA: AltaMira Press.

Clausen, John S. 1991. "Adolescent Competence and the Shaping of the Life Course." *American Journal of Sociology* 96:805–42.

Cole, Thomas. R. 1992. *A Journey of Life: A Cultural History of Aging in America*. New York: Cambridge University Press.

Dannefer, Dale. 1987. "Aging as Intracohort Differentiation: Accentuation, the Matthew Effect, and the Life Course." *Sociological Forum* 2:211–37.

———. 2003. "Cumulative Advantage/Disadvantage and the Life Course: Cross-Fertilizing Age and Social Science Theory." *Journal of Gerontology* B 58:S327–S337.

Dannefer, Dale and Peter Uhlenberg. 1999. "Paths of the Life Course: A Typology." Pp. 306–326 in *Handbook of Theories of Aging*, edited by V. L. Bengtson and K. W. Schaie. New York: Springer.

Denton, Margaret A., Candace L. Kemp, Susan French, Amiram Gafni, Anju Joshi, Carolyn J. Rosenthal, and Sharon Davies. 2004. "Reflexive Planning for Later Life." *Canadian Journal on Aging Supplement/La Revue canadienne du vieillissement* Supplément: S71–S82.

DiPrete, Thomas A., and Gregory M. Eirich. 2006. "Cumulative Advantage as a Mechanism for Inequality: A Review of Theoretical and Empirical Developments." *Annual Review of Sociology* 32:271–97.

Dowd, James J. 1975. "Aging as Exchange: A Preface to Theory." *Journal of Gerontology* 30(5): 584–94.

Ekerdt, David J. 2004. "Born to Retire: The Foreshortened Life Course." *The Gerontologist* 44(1):3–9.

———. 2010. "Frontiers of Research on Work and Retirement." *The Journals of Gerontology* 65B(1):69–80.

Ekerdt, David J. and Jennifer Kay Hackney. 2002. "Workers' Ignorance of Retirement Benefits." *The Gerontologist* 42(4):543–51.

Elder, Glen H., Jr. 1994. "Time, Human Agency, and Social Change: Perspectives on the Life Course." *Social Psychology Quarterly* 57:4–15.

———. 1998. *Children of the Great Depression: Social Change in Life Experience*, 25th Anniversary Edition. Boulder, CO: Westview Press.

Elder, Glen H., Jr., and Monica Kirkpatrick Johnson. 2003. "The Life Course and Aging: Challenges, Lessons, and New Directions." Pp. 49–84 in *Invitation to the Life Course: Toward New Understandings of Later Life,* edited by R. A. Settersten, Jr. Amityville, N.Y.: Baywood.

Elder, Glen H., Jr. and Pavalko, Eliza K. 2003. "Work Careers in Men's Later Years: Transitions, Trajectories, and Historical Change." *Journal of Gerontology* 48(4):S180–91.

Employee Benefit Research Institute [EBRI]. *The 2010 Retirement Confidence Survey*, Issue Brief No. 340. Retrieved March 25, 2010 (http://www.ebri.org/publications/ib/index.cfm?fa=ibDisp&content_id=4488).

Erikson, Erik H. 1959. *Identity and the Life Cycle*. New York: International Universities Press.

Esping-Andersen, Gøsta. 1999. *Social Foundations of Postindustrial Economies*. Oxford, UK: Oxford University Press.

Estes, Carroll L. and Jane L Mahakian. 2001. "The Political Economy of Productive Aging," Pp. 197–213 in *Productive Aging: Concepts and Challenges*, edited by N. Morrow-Howell, J. Hinterlong, and M.-W. Sherraden. Baltimore, MD: The Johns Hopkins University Press.

Ferraro, Kenneth F. 2007. "The Gerontological Imagination." Pp. 325–42 in *Gerontology: Perspectives and Issues*, edited by J. M. Wilmoth and K. F. Ferraro. New York: Springer.

Ferraro, Kenneth F. and Jessica A. Kelley-Moore. 2003. "Cumulative Disadvantage and Health: Long-term Consequences of Obesity?" *American Sociological Review* 68:707–29.

Ferraro, Kenneth F., Tetyana P. Shippee, and Markus H. Schafer. 2009. "Cumulative Inequality Theory for Research on Aging and the Life Course." Pp. 413–434 in *Handbook of Theories of Aging*, 2nd ed., edited by V. L. Bengtson, D. Gans, N. M. Putney, and M. Silverstein. New York: Springer.

Galotti, Kathleen M. 2002. *Making Decisions that Matter: How People Face Important Life Choices*, Mahwah, NJ: Lawrence Erlbaum.

Gallup. 2010. *Six in 10 Workers Hold No Hope of Receiving Social Security*, July 20, retrieved February 5, 2011, http://www.gallup.com/poll/141449/six-workers-hold-no-hope-receiving-social-security.aspx.

Ginn, Jay, Debra Street, and Sara Arber, eds. 2001. *Women, Work and Pensions: International Issues and Prospects*. Buckingham, UK: Open University Press.

Gran, Brian. 2008. "Shifts in Public-Private provision of Retirement Income: A Four-Country Comparison." *Research on Aging* 30(5):507–42.

Guillemard, Anne-Marie and Martin Rein. 1993. "Comparative Patterns of Retirement: Recent Trends in Developed Societies." *Annual Review of Sociology* 19:469–503.

Haber, David. 2003. *Health Promotion and Aging: Practical Applications for Health Professionals*, 3rd edition. New York: Springer Publishing Company.

Hacker, Jacob S. 2006. *The Great Risk Shift: The Assault on American Jobs, Families, Health Care, and Retirement and How You Can Fight Back.* New York: Oxford University Press.

Hardy, Melissa A. 2002. "The Transformation of Retirement in Twentieth Century America: From Discontent to Satisfaction." *Generations* 25:9–16.

Harrington-Mayer, Madonna, and Pamela Herd. 2007. *Market Friendly or Family Friendly? The State and Gender Inequality in Old Age.* New York: Russell Sage Foundation.

Henretta, John C. 1992. "Uniformity and Diversity: Life Course Institutionalization and Late Life Work Exit." *The Sociological Quarterly* 33:2:265–79.

Henretta, John C. Angela M. O'Rand, and C. Chan. 1993. "Joint Role Investments and Synchronization of Retirement: A Sequential Approach to Couples' Retirement Timing." *Social Forces* 981–1000.

Herzog, A. Regula, Robert L. Kahn, James N. Morgan, James S. Jackson, and Toni Antonucci. 1989. "Age Differences in Productive Activities." *Journal of Gerontology* 44(4): S129–38.

Hitlin, Steven and Glen H. Elder, Jr. 2007a. "Time, Self, and the Curiously Abstract Concept of Agency." *Sociological Theory* 25(2):170–91.

———. 2007b. "Agency: An Empirical Model of an Abstract Concept." Pp. 33–67 in *Constructing Adulthood: Agency and Subjectivity in Adolescence and Adulthood. Advances in Life Course Research, Vol. 11,* edited by R. Macmillan. New York: Elsevier.

Holstein, Martha. 1998. "Women and Productive Aging: Troubling Implications." Pp. 359–73 in *Critical Gerontology: Perspectives on Political and Moral Economy*, edited by M. Minkler and C. L. Estes. Amityville, NY: Baywood.

Holstein, Martha B. and Meredith Minkler. 2003. "Self, Society and the 'New Gerontology'." *The Gerontologist* 43(6):787–96.

Jackson, James S., Toni C. Antonucci, and Rose C. Gibson. 1993. "Cultural and Ethnic Context of Aging Productively over the Life Course: An Economic Network Framework." Pp. 249–68 in *Achieving a Productive Aging Society*, edited by S. A. Bass, F. G. Caro, and Y.-P. Chen. Westport CT: Auburn House.

Kane, Rosalie. 2005. "Connecting the Dots: Public Health, Health Care, Health Policy, and Successful Aging." *The Gerontologist* 45(2):274–9.

Kastenbaum, Robert J. 1980. "Habituation as a Model of Human Aging." *International Journal of Aging and Human Development* 12(3):159–70.

Katz, Stephen. 2000. "Busy Bodies: Activity, Aging and the Management of Everyday Life." *Journal of Aging Studies* 14(2):135–52.

Kim, Jungmeen E. and Phyllis Moen. 2001. "Moving into Retirement: Preparation and Transitions in Late Midlife." Pp. 487–527 in *Handbook of Midlife Development*, edited by M. E. Bachmann. Hoboken, NJ: John Wiley & Sons Inc.

Kohli, Martin. 1986. "Social Organization and the Subjective Construction of the Life Course." Pp. 271–92 in *Human Development and the Life Course: Multidisciplinary Perspectives*, edited by A. B. Sørensen, F. E. Weinert, and L. R. Sherrod. Hillsdale, NJ: Lawrence Erlbaum.

Lutfey, Karen and Jeylan T. Mortimer. 2006. "Development and Socialization throughout the Adult Life Course." Pp. 183–202 in *Handbook of Social Psychology*, edited by J. Delamater. New York: Springer.

Marshall, Victor W. 2000. "Agency, Structure, and the Life Course in the Era of Reflexive Modernization." Presented at the annual meeting of the American Sociological Association. *The Life Course in the 21st Century*. Washington, DC. Retrieved February 15, 2010 (http://www.aging.unc.edu/infocenter/resources/2000/marshallv.pdf).

McAdam, Dan P. and Ed de St. Aubin. 1998. *Generativity and Adult Development: How and Why We Care for the Next Generation.* Washington, D.C. APA Press.

McLaughlin, Sara J., Cathleen M. Connell, Steven G. Heeringa, Lydia W. Li, and J. Scott Roberts. 2010. "Successful Aging in the United States: Prevalence Estimates from a National Sample of Older Adults." *Journal of Gerontology: Social Sciences* 65B(2):216–26.

McMullin, Julie A. and V. W. Marshall. 1999. "Structure and Agency in the Retirement Process: A Case Study of Montreal Garment Workers." Pp. 305–38 in *Self and Society in Aging Processes*, edited by C. D. Ryff and V. W. Marshall. New York: Springer.

Minkler, Meredith. 1999. "Personal Responsibility for Health? A Review of the Arguments and the Evidence at Century's End." *Health Education and Behavior* 26:121–40.

Moen, Phyllis. 2003. *It's About Time: Couples and Careers*. Ithaca, NY: Cornell University Press.

Moen, Phyllis, Stephen Sweet, and Raymond Swisher. 2005. "Embedded Career Clocks: The Case of Retirement Planning." The Structure of Life Course Research: Standardized? Individualized? Differentiated? *Advances in Life Course Research* 9:237–65.

Morrow-Howell, Nancy, Jim Hinterlong, Philip Rozario, and Fengyan Tang. 2003. "Effects of Volunteering on the Well-being of Older Adults." *Journal of Gerontology: Social Sciences* 58B:S137–45.

Morrow-Howell, Nancy, James Hinterlong, and Michael Wayne Sherraden. 2001. *Productive Aging: Concepts and Challenges*. Baltimore, MD: The John Hopkins University Press.

Mutchler, Jan E., Jeffrey A. Burr, Amy Pienta, and Michael P. Massagli. 1997. "Pathways to Labor Force Exit: Work Transitions and Work Instability." *Journal of Gerontology: Social Sciences* 52B(1):S4–S12.

Myles, John. 1989. *Old Age in the Welfare State: The Political Economy of Public Pensions*. Lawrence KS: University Press of Kansas.

New York Times [NYT]. 2009. "Longer Unemployment for Those 45 and Older." April 30.

O'Rand, Angela M. 1996. "The Precious and the Precocious: Understanding Cumulative Disadvantage and Cumulative Advantage over the Life Course." *The Gerontologist* 36:230–38.

———. 2003. "Cumulative Advantage Theory in Life Course Research." Pp. 14–30 in *Annual Review of Gerontology and Geriatrics: Focus on Economic Outcomes in Later Life*, edited by S. Crystal and D. F. Shea. New York: Springer.

O'Reilly, Patrick, and Francis G. Caro. 1994. "Productive Aging: An Overview of the Literature." *Journal of Aging and Social Policy* 6(3):39–71.

Pearlin, Leonard I. 2010. "The Life Course and the Stress Process: Some Conceptual Comparisons." *Journal of Gerontology: Social Sciences*, 65B(2):207–15.

Phelan, Elizabeth A., and Eric B. Larson. 2002. "'Successful Aging' – Where Next?" *Journal of the American Geriatrics Society* 50:1306–82.

Phillipson, Chris. 1998. *Reconstructing Old Age: New Agendas in Social Theory and Practice*. London, UK: Sage.

Quadagno, Jill. 1988. *The Transformation of Old-Age Security*. Chicago, IL: University of Chicago Press.

———. 1996. "Social Security and the Myth of the Entitlement 'Crisis'." *The Gerontologist* 36(3):391–9.

Reynolds, John, Stephanie Burge, Emily Boyd, Cheryl Robbins, and Brandy Harris. 2007. "Mastery and the Fulfillment of Occupational Expectations by Midlife." *Social Psychology Quarterly* 70:366–83.

Riley, Matilda White and John W. Riley Jr. 1994. "Age Integration and the Lives of Older People." *The Gerontologist* 34:110–15.

Roberts, Robert E. L. and Vern L. Bengtson. 1999. "The Social Psychology of Values: Effects of Individual Development, Social Change, and Family Transmission over the Life Span." Pp. 453–82 in *The Self and Society in Aging Processes*, edited by C. D. Ryff and V. W. Marshall. New York: Springer.

Rowe, John W. and Robert L. Kahn. 1997. "Successful Aging." *The Gerontologist* 37:433–40.

———. 1998. *Successful Aging*. New York: Random House.

Schulz, James H. and Allan Borowski. 2006. "Economic Security in Retirement: Reshaping the Public-Private Pension Mix." Pp. 360–79 in *Handbook of Aging and the Social Sciences*, 6th edition, edited by R. H. Binstock and L. K. George. San Diego, CA: Academic Press.

Settersten, Richard A. Jr. 2003. "Age Structuring and the Rhythm of the Life Course." Pp. 81–102 in *Handbook of the Life Course*, edited by J. T. Mortimer and M. J. Shanahan. New York: Kluwer.

Settersten, Richard A. Jr. and Gunhild O. Hagestad. 1996. "What's the Latest? Cultural Age Deadlines for Educational and Work Transitions." *The Gerontologist* 36(5):602–13.

Shanahan, Michael J., Glen H. Elder, Jr. and Richard A. Miech. 1997. "History and Agency in Men's Lives: Pathways to Achievement in Cohort Perspective." *Sociology of Education* 70:54–67.

Shanahan, Michael J., Scott M. Hofer, and Richard A. Miech. 2003. "Planful Competence, the Life Course, and Aging: Retrospect and Prospect." Pp. 189–211 in *Personal Control in Social and Life Course Contexts*, edited by S. H. Zarit, L. I. Pearlin, and K. W. Schaie. New York: Springer.

Shuey, Kim M. and Angela M. O'Rand. 2004. "New Risks for Workers: Pensions, Labor Markets, and Gender." *Annual Review of Sociology* 30:453–77.

Silverstein, Merrill. 2006. "Intergenerational Family Transfers in Social Context." Pp. 164–80 in *Handbook of Aging and the Social Sciences*. 6th ed., edited by R. H. Binstock and L. K. George. San Diego, CA: Academic Press.

Social Security Administration [SSA]. 2010. *Retirement Benefits by Year of Birth*. Retrieved January 9, 2010 (http://www.ssa.gov/retire2/agereduction.htm).

Street, Debra and Janet Wilmoth. 2001. "Social Insecurity: Women and Pensions in the US." Pp. 120–41 in *Women, Work and Pensions: International Issues and Prospects*, edited by J. Ginn, D. Street, and S. Arber. Buckingham, UK: Open University Press.

Street, Debra, Stephanie Woodham Burge, Jill Quadagno and Anne Barrett. 2007. "The Salience of Social Relationships for Resident Wellbeing in Assisted Living." *Journal of Gerontology*, Series B: Social Sciences 62:S129–34.

Szinovacz, Maximiliane E. and Stanley DeViney. 2000. "Marital Characteristics and Retirement Decisions." *Research on Aging* 22(5):470–98.

Taylor, Brent A. and Vern L. Bengtson. 2001. "Sociological Perspectives on Productive Aging." Pp. 120–44 in *Productive Aging: Concepts and Challenges*, edited by N. Morrow-Howell, J. Hinterlong, and M. Wayne Sherraden, Baltimore, MD: The Johns Hopkins University Press.

U.S. Census Bureau. 2001. *Did You Know? Homes Account for 44 Percent of All Wealth: Findings from the SIPP, Current Population Reports*. Series P70, No. 75. Household Economic Studies. Washington, DC: U.S. Bureau of the Census.

Veenhoven, Ruut. 2006. "Unseen Happiness: Why Sociologists Fail to Acknowledge Findings on this Matter." Presentation at the 16th World Congress of Sociology Working Group 06 Social Indicators, Durban, South Africa, 23–29 July 2006. Retrieved January 20, 2010 (http://www.soc.duke.edu/resources/sinet/papers06/Veenhoven.pdf).

Weiss, Robert S. 2005. *The Experience of Retirement*. Ithaca, NY: Cornell University Press.

Willson, Andrea E., Kim M. Shuey, and Glen H. Elder, Jr. 2007. "Cumulative Advantage Processes as Mechanisms of Inequality in Life Course Health." *American Journal of Sociology* 112(6):1886–1924.

Chapter 25
Responses of the Long-Term Care System to Recent Natural Disasters

Sarah B. Laditka, James N. Laditka, and Dena Shenk

The number of people affected by disasters has grown steadily since the mid-1970s. The number of deaths due to natural disasters world-wide between 1970 and 2006 is estimated at nearly 2.8 million (EM DAT 2009). Recent natural disasters include the tsunami in South Asia, and major earthquakes in Haiti, Japan, Pakistan, China, and Turkey. In the United States, four powerful hurricanes struck Florida in 2004, and Hurricanes Katrina and Rita caused major destruction in the Gulf Coast states in 2005. These events highlight serious and fatal health consequences following major disasters. In the United States, more than 45 million people live in the coastal region of the Atlantic and Gulf Coast, which are exposed to considerable risk of hurricanes (U.S. Census Bureau 1997); the risk of hurricanes has increased by nearly 40% in the past 15 years, due in large part to rising sea temperatures (Pew Center on Global Climate Change 2009). More than 12% of the population of U.S. coastal states will be aged 65 and above by 2025. Large heavily populated areas of the United States are also at risk of major earthquakes, including much of the West Coast, the Central Mississippi River Valley, and the coastal Southeast (U.S. Geological Survey 2009). The growing numbers of natural disasters, as well as the threat of man-made emergencies, underscore the importance of improved planning at the individual and community levels to address the disaster preparedness and recovery needs of older people and their families.

In the United States, the population aged 65 and above will increase to 82 million by 2050, from 35 million in 2000. About four million older people in the United States live in nursing homes or residential care facilities (Association for Homes and Services for the Aged 2009; Centers for Disease Control 2008). The vast majority of older people live in the community. Many of the negative consequences of natural disasters for older people are due to exacerbations of chronic diseases, which disproportionately affect older people (Fernandez et al. 2002; Miller and Arquilla 2008). Over 70% of people aged 80 and older have a disability; over 40% need help to perform basic activities of daily living (U.S. Census Bureau 2002). Effects of Hurricane Katrina underscored the need to improve preparedness for older people, both those living in the community and those in long-term care residential facilities. Hurricane Katrina devastated many communities in the Gulf Coast in August of 2005. Over 1,300 people died, the majority being older people (Simerman et al. 2005). The response to a disaster by emergency management organizations, public health agencies, and medical providers depends in large part on advance planning. This planning must consider the needs of special populations, such as frail older people and disabled individuals (Johnson et al. 2006; Saliba et al. 2004; U.S. Office of Inspector General 2006). There is considerable consensus among public health experts and researchers that medical systems and emergency management in the U.S. and in many other countries are not prepared for public health disasters (Johnson et al. 2006; Miller and

S.B. Laditka (✉)
Public Health Sciences, University of North Carolina, Charlotte, NC, USA
e-mail: sladitka@uncc.edu

R.A. Settersten, Jr. and J.L. Angel (eds.), *Handbook of Sociology of Aging*, Handbooks of Sociology and Social Research, DOI 10.1007/978-1-4419-7374-0_25, © Springer Science+Business Media, LLC 2011

Arquilla 2008). Their lack of preparation is particularly acute regarding needs of older populations (Fernandez et al. 2002; Johnson et al. 2006; Miller and Arquilla 2008; Saliba et al. 2004; U.S. Office of Inspector General 2006).

In this chapter, we first describe preparedness challenges and opportunities affecting nursing homes. Next, we consider the preparedness needs of organizations that provide health and other supportive services to older people living in the community. Our focus in the first two sections is on long-term care organizations in the United States. Next, we focus on preparedness responses and needs of older individuals. In that section, we broaden our lens to include experiences and studies conducted in a number of counties following recent disasters, including Turkey, China, and Japan. We consider responses and perspectives from a life course perspective. Some researchers view disasters as a combination of populations and potentially destructive natural or man-made events, where the combination involves a socially produced condition of vulnerability (Hoffman and Oliver-Smith 1999). Such conditions arise, for example, when large cities grow in areas that are subject to intense hurricanes or massive earthquakes. A disaster creates an immediate need to respond to new and unique conditions. Individuals' responses to such rapid change are influenced by their conceptual frameworks, which are a function of their life experience (Hoffman and Oliver-Smith 1999). Thus, disasters may be symptomatic of strategies societies use to adapt in the context of their social, economic, and built environments (Hoffman and Oliver-Smith 1999).

We turn next to implications for research and practice. To help guide future research, we present an integrated model of long-term care preparedness, The Socio-Ecological Model of Individual and System Preparedness in Long-Term Care. This model applies elements of the socio-ecological model of health to disaster preparedness, including individuals, families, neighborhoods and communities, and political perspectives. Next, we consider how the life course perspective applies to preparedness for older people. Then, implications for practice are considered, including needs for communication and training.

Definition of Disasters and Emergencies

Most studies examining preparedness use the terms "emergency" and "disaster" interchangeably, to refer to life-threatening unexpected events (e.g., Laditka et al. 2008a). These events can encompass a relatively localized "emergency," such as a tornado, or a more massive "disaster," such as Hurricane Katrina. We recognize that the government, communities, social service, and health care organizations play a role in preparing older people for local events affecting a small number of individuals, such as a house fire. However, in this chapter, we use the terms emergency and disaster to refer to widespread events that disrupt community infrastructure over a broad geographic area. This definition of an emergency or disaster is consistent with well-recognized published guidelines (Task Force of Quality Control of Disaster Management 2002). The terms "emergency" and "disaster" relate to the sociology of aging in that these are categories of events for which individuals and society can help older people and the organizations that serve them to prepare, where that preparation includes advanced planning for major unexpected widespread disruptive events and their aftermath.

Disaster Preparedness and the Special Needs of Older Populations

The special needs of older populations in disaster preparedness need to be considered by practitioners, policymakers, and planners, to enhance geriatrics disaster training and education. The Bioterrorism and Emergency Preparedness in Aging Committee, formed in 2002, addressed these areas by developing and disseminating education and training materials to improve preparedness for older people residing in the community and in residential facilities (Johnson et al. 2006). The Bioterrorism and Emergency Preparedness in Aging Committee consisted of members of the National Association of Geriatric Education Centers. Six of the 46 Geriatric Education Centers in the U.S. took the lead in this initiative. The Bioterrorism and Emergency Preparedness in Aging framework conceptualizes the special needs of older populations as stemming from: (1) compromised immune systems, which make older people more susceptible to stress induced by extremes of temperature or biological agents; (2) failure of major body systems such as renal failure, and comorbidities such as diabetes, which put older individuals at higher risk of adverse outcomes following a disaster; (3) declines in functional and cognitive status, limitations in mobility, and loss of vision and hearing, which result in special communication needs and additional preparation time required by older populations; and (4) other factors related to lower socioeconomic status, racial and ethnic disparities, social isolation, disability, and ageism (Johnson et al. 2006).

Preparedness in Nursing Homes

More than 1.5 million older Americans reside in nursing homes (Centers for Disease Control 2008). Despite a declining prevalence of disability, their number will grow due to population aging (Laditka 1998). In this section, we extend the discussion introduced by Laditka et al. (2008b, 2009). Nursing homes are vulnerable to many wide-spread disasters, including hurricanes, earthquakes, floods, outbreaks of infectious disease, terrorism, prolonged loss of power, as well as other challenges involving ice, snow, fire, or wind. Disasters raise particularly difficult challenges for preparedness, because these events substantially disrupt infrastructures of emergency response and regional economies, often affecting the availability of electrical power, communications, gasoline and other fuels, transportation, food, medicines and other supplies, and staff (Laditka et al. 2008b, 2009). Relatively little research examined disaster preparedness in nursing homes before Hurricane Katrina.

The four hurricanes that struck Florida in a 44-day period in 2004 – Charley, Frances, Ivan, and Jeanne – prompted a recently published study of 291 nursing homes. Researchers found that major concerns were transportation and long lasting power outages (Hyer et al. 2009). There was evidence that transportation agreements were not upheld; generators were not adequate to support air conditioning and laundry services. A number of studies were conducted in the aftermath of Hurricane Katrina. One identified perceptions about disaster preparedness among nursing home administrators in South Carolina in the period immediately before and after Hurricane Katrina (Laditka et al. 2007a). Although South Carolina was not directly affected by Katrina, nursing home administrators in the state followed the events surrounding Katrina closely because South Carolina is at risk for severe Hurricane damage. Administrators expressed concerns about their ability to care for evacuees from other nursing homes in the event of a disaster, and about their lack of appropriate transportation (Laditka et al. 2007a, b). In a response to a brief survey of administrators immediately following Hurricane Katrina, the majority of respondents said that they were rethinking their preparedness plans for transportation, supplies, staffing, and communication (Laditka et al. 2007a). In a study of nursing homes primarily in Mississippi that sheltered nursing home residents evacuated from areas with the most damage from Hurricane Katrina, nursing home administrators also raised concerns about communication, transportation, supplies, and staffing (Laditka et al. 2008a). Laditka and colleagues (2008b) found reports of long-term mental health needs among both residents and staff following Katrina. Administrators said that nursing homes were not included in community preparedness planning (Laditka et al. 2008b).

Administrators of nursing homes in New Orleans who evacuated or sheltered affected individuals following Hurricanes Katrina and Rita said major challenges were lack of appropriate transportation, staffing shortages, and a perception of abandonment by state and federal response agencies (Dosa et al. 2007). A study of hospitals in New Orleans serving older patients in skilled nursing facilities and hospice units found that loss of power, shortages of staff and supplies, and extreme heat were major challenges following Katrina (Gray and Hebert 2007).

Another study following Hurricane Katrina focused on responses of 38 staff in four nursing homes in Mississippi that sheltered frail evacuees (Laditka et al. 2009). Staff emphasized the need to provide emotional support to evacuees as well as physical care. Many staff said caring for evacuees was difficult because they were anxious and in poor physical condition after the evacuation. A major challenge in caring for evacuees was communicating with evacuees' families: landline and cell phones did not operate for weeks following the hurricane; further, families were spread by the evacuation (Laditka et al. 2009). Staff also stressed challenges associated with preventing dehydration, lack of food and personal hygiene supplies, and staff exhaustion. At the same time, many described caring for evacuees as "a blessing," saying the experience helped them to bond with residents, evacuees, and other staff. Staff emphasized the importance of teamwork, community help, and the need to have a well-organized disaster plan, extra supplies, and dependable staff (Laditka et al. 2009).

A recent study examined emergency evacuation plans for about 2,100 nursing homes in the U.S. (Castle 2008). Although most facilities were relatively well prepared to shelter in place, most did not include plans to evacuate. Summit meetings of key stakeholders, including leaders in long-term care, transportation, emergency management, federal and state agencies, and nursing homes, were convened in 2007 to discuss challenges and opportunities in disaster preparedness for nursing homes (Hyer et al. 2006). Participants recommended enhancing transportation and communication resources, improving coordination between nursing homes and local emergency preparedness systems, refining disaster preparedness guidelines, and conducting emergency drills (Hyer et al. 2006).

Taken as a whole, research and recent meetings of key stakeholders identified a number of common problems faced by nursing homes following disasters: loss of power; lack of sufficient or appropriate transportation for evacuation; wide-spread disruption of communication systems, with breakdowns in landline and cell phone service; lack of food, water, gasoline, medications, and other medical and general supplies; and lack of adequate staff (Dosa et al. 2007; Hyer et al. 2006; Laditka et al. 2007a, 2008b, 2009; Saliba et al. 2004). Findings suggest that nursing homes receive substantially less support than hospitals from local, state, and federal response agencies, and that this shortcoming affects them before, during, and following disasters (Brown et al. 2007; Dosa et al. 2007; Hyer et al. 2006; Laditka et al. 2008b). For example, state and federal laws do not require that nursing homes receive priority for power restoration following a disaster (Brown et al. 2007).

Summary of Disaster Preparedness Lessons Learned from Nursing Home Studies

There were a number of important "lessons learned" from nursing home-related research conducted following Hurricane Katrina. Laditka et al. (2008b) adapted six of the training and practice domains described by The Bioterrorism and Emergency Preparedness in Aging Committee (Johnson et al. 2006) to describe care, practice, and policy implications for nursing homes and other residential facilities serving older people. The six domains described by Laditka et al. (2008b) are: (1) maintain core functions, i.e., the ability to maintain normal day-to-day operations following a disaster, with a focus on ensuring that there are sufficient supplies for residents and families of staff and their pets; (2) respond to needs of a diverse group of stakeholders, e.g., provide culturally sensitive care to residents; (3) apply geriatric-specific protocols in patient and resident care, including triage and medication management; (4) address mental health needs of residents and staff, recognizing that older people and staff are vulnerable to depression after a disaster and may have long term mental health needs; (5) ensure transportation is available if evacuation is needed, and that transportation services meet the special needs of older people (e.g., heating, cooling, accommodate wheelchairs); and (6) ensure communication systems are in place with backup systems for landline and cell phones and the internet. Laditka et al. (2008b) also introduced two new preparedness domains for long term care populations: (1) ensure that needs of nursing homes are addressed in community preparedness planning, e.g., lifelines such as power are restored promptly; and (2) recognize nursing homes provide valuable health care resources to the community, including oxygen, trained health care personnel, and supplies. The preparedness concepts in the domains described by Laditka et al. (2008b) can be extended to or adapted for many types of residential communities for older people, including independent living communities, assisted living communities, and continuing care retirement communities. These preparedness domains can also be applied to health and social services agencies providing care to older people living in the community, which are described in the section that follows.

Preparedness Among Agencies Providing In-Home Care to Older People in the Community

In this section, we turn to preparedness among agencies providing care to older people in their homes, drawing on the discussion introduced by Laditka et al. (2008a). More than eight million vulnerable older Americans receive long term care services in their homes (Hughes and Renehan 2005). Support for home care will grow, as a continuing response to the Supreme Court's 1999 Olmstead Decision (United States Supreme Court 1999). Most recipients of home care are over age 75 and live alone. These older adults are at high risk of rapid physical decline, mental disorientation, emotional trauma, and death; few would be able to care for themselves during and after a disaster. Thus, agencies providing care in the home serve as a critical component in the long-term care continuum.

Home health agencies and personal care agencies provide a broad array of medical and support services for older people in their homes. Home health agencies provide more highly skilled nursing care, whereas personal care agencies provide basic assistance with bathing, dressing, meal preparation, and similar activities (Hughes and Renehan 2005). The federal Centers for Medicare and Medicaid Services requires home health agencies to have disaster plans. No specific rules govern the types or regularity of training or the content of those plans. Most states require home health agencies to be certified to obtain Medicare reimbursement (Hughes and Renehan 2005). Improving federal and state regulations could substantially enhance preparedness among home health agencies. Home health agencies can also seek accreditation from the Joint Commission or the Community Health Accreditation Program. Since 2006, the Joint Commission standards require home health agencies to conduct one disaster drill per year (Joint Commission 2007).

No federal regulations govern disaster preparedness for personal care agencies. Preparedness among these agencies varies substantially by state. We provide one case example using South Carolina. Most personal care agencies in South Carolina have a contractual agreement to provide services to clients enrolled in the Community Long-term Care Program. This is a Medicaid home- and community-based waiver program serving approximately 12,500 clients who qualify for both nursing home placement and Medicaid (Pande et al. 2007). Case managers are employed by the Community Long-term Care Program. Case managers are required to ensure that clients of personal care agencies have disaster plans (South Carolina Department of Environmental Control, SC DHEC 2005). Case managers develop an "emergency preparedness checklist" and an "emergency telephone list" with clients and/or designated caregivers (SC DHEC 2005). Clients' needs are assessed when they are enrolled in the Program, and updated every 30 days (SC DHEC 2005). Nurse consultants evaluate clients and identify those who should have priority status during disasters. With clients' permission, names are shared with emergency agencies.

Research examining disaster preparedness in agencies providing services to older and/or disabled clients in their homes is sparse. Until recently, almost all studies have been limited to narrative reports about the impact of a disaster on clients of home care services (Riddix and Dellar 2001) or ways home health care nurses can help clients to prepare (Sienkiewicz et al. 2007). There are few empirical studies in this area. Using qualitative methods, researchers evaluated how five home health agencies in Orleans Parish, Louisiana, responded to Hurricane Katrina (Kirkpatrick and Bryan 2007). All of the agencies had preparedness plans. However, the breakdown of communications was widespread due to the loss of landline and cell phone operability. Results showed a lack of coordination among government and home health agencies (Kirkpatrick and Bryan 2007). The researchers recommended additional drills to train agency staff, better ways to identify clients who are reluctant to evacuate, enhanced communications and transportation, and early evacuation.

Another recent study examined preparedness among health care and personal care agencies in South Carolina (Laditka et al. 2008a). Telephone interviews were conducted with administrators of 16 agencies providing in-home personal care to 2,147 clients, and five agencies providing in-home health care to 2,180 clients. For agencies in both categories, findings suggest a lack of preparedness: in identifying clients at high risk and assisting them in planning, providing written materials and/or recommendations for clients, training staff, and coordinating disaster planning and response across agencies (Laditka et al. 2008a). Although home health agencies were better prepared than personal care agencies, a number of home health administrators commented that they were unsure how well their plans would work. Most administrators said that better coordination and/or more preparedness training is needed. The findings support incorporating disaster planning in certification requirements for home health agencies, and developing additional educational materials for administrators and staff and for clients (Laditka et al. 2008a).

Older Adults' Experiences and Responses to Disasters

In this section, we focus on responses to disasters and needs of older people. Responses are considered using a life course perspective, building on the recent review by Shenk et al. (2009). We include studies conducted in the United States and in other countries.

Coping Approaches

From a life course perspective, memories and previous experiences help shape how older people experience and cope with disasters (Krause 1987). Disasters often evoke one or more coping strategies. Two broad coping styles used by older adults have been identified: emotion-focused coping and problem-focused coping.

In emotion-focused coping, older people often turn to activities or emotional states that keep them from directly confronting stressful events (Yeung and Fung 2007). In this strategy emotional support, humor, and disengagement are often used by older people to mediate responses to stressful events. Some responses may be positive. For example, older adults may help others affected by the disaster. This response can distract older individuals from their own problems, and provide a greater sense of emotional control (Heller et al. 2005). Emotion-focused coping can also lead to negative responses. For example, one study suggested that talking about an earthquake caused older people to relive the disaster experience and resulted in increased stress (Alea et al. 2004).

Problem-focused coping involves efforts to reduce and alleviate stressful situations (Yeung and Fung 2007). An example of problem-focused coping is improved preparation for another disaster among people who have experienced a previous disaster (Ecevit and Kasapoglu 2002). Findings in this area are mixed. A lifetime of experiences with life changes – including but not limited to disaster-related changes – may enable older adults to better cope with change (Heller et al. 2005; McMillen et al. 1997; Norris and Murell 1988). Some research suggests that experiencing a disaster prompts older people to recognize the need to prepare (Heller et al. 2005). In one actual experience, however, older people who experienced an earthquake and recognized a need to prepare for a future earthquake did relatively little to prepare (Heller et al., 2005). A recent study examined preparedness among 547 older people in Florida who responded to a survey while visiting an ambulatory Veterans Administration clinic (Cherniack et al. 2008). Nearly 84% reported that they had lived in a hurricane prone area of Florida for more than 10 years; about 31% reported experiencing 5–10 hurricanes, nearly 38% reported experiencing more than 10 hurricanes. Yet, most respondents did not understand common warning terminologies, e.g., "hurricane watch" or "hurricane warning," or what steps to take if they heard these warnings and only 56% had an evacuation plan. About 30% said they had an electric generator, but of these, only 46% said they knew how to operate it (Cherniack et al. 2008). These findings suggest that preparation for hurricanes was lacking even among an older population with substantial disaster experience.

Level of Coping Self Efficacy

According to Bandura (1997), coping self-efficacy is defined as self-reported ability to cope with the effects of a disaster or trauma. Individuals can often recognize their relationship to the environment, and thus anticipate future adaptive responses. Through this process, older people consider possibilities for their future, and attempt to control their future life course. This approach can help define the older person's feelings, thoughts, motivations, and behaviors. Levels of coping self-efficacy are related to the individual's personality and life experience. Those with high coping self-efficacy have better abilities and are more likely to cope effectively with stresses during and after a disaster (Bandura 1997).

Psychological effects: Negative psychological effects of disasters can include anger, fear and shock (Yeung and Fung 2007), sadness (Alea et al. 2004), and emotional numbness and social isolation (Ecevit and Kasapoglu 2002). Depression following disasters varies depending on individual characteristics. For example, among older people who survived the Loma Prieta earthquake in San Francisco, greater exposure to the damage and destructive effects of the earthquake increased depression overall, and exacerbated depressive symptoms of older people with depression (Nolen-Hoeksema and Morrow 1991). Although often negative, psychological effects after a disaster can be positive. Positive effects include personal growth and reassessment of life priorities (McMillen et al. 1997). Surviving such an experience sometimes causes survivors to articulate community needs, and can help to inform improvements in infrastructure, services, and policy (McMillen et al. 1997).

Social support: During and after disasters, social support is provided through formal groups and structures and informal systems of family, friends, and neighbors. Organizations and the community provide resources and services. Support is also needed to address emotional needs. Social workers often fulfill essential roles following disasters, thanks to their skills in providing information and services to older individuals in the community (Javadian 2007) as well as in long-term care facilities

(Laditka et al. 2008b, 2009). In many instances, older people have been living in the disaster area for a longer time (Goto et al. 2006), and have established better social networks than younger residents, factors that can help older people to cope following disasters (Kato et al. 1996). As suggested in our review of recent research on nursing homes and organizations providing services to frail older people in the community, establishing strong community linkages and networks well in advance of a disaster is an important characteristic of organizations that respond more successfully to disasters (Laditka et al. 2008b, 2009). An analogous benefit may accrue to older individuals who have strong social networks.

Life experiences and responses to disasters: As a whole, evidence suggests that older people have greater coping abilities than younger adults. Some research attributes stronger coping abilities to more life experiences (Melick 1985), and experiences with coping with stressful events (Heller et al. 2005; Thompson et al. 1993). A lifetime of experiences with many life changes may help some older people cope with disasters reasonably well (Norris and Murrell 1988; Thompson et al. 1993). Many older adults have experienced a previous disaster (Kato et al. 1996), and may therefore be more confident in their ability to manage a new one (Phifer and Norris 1989). There is evidence that older persons who experienced a disaster previously may experience less stress after another disaster (Goto et al. 2006; Melick 1985; Phifer and Norris 1989). There may be analogous beneficial effects of having experienced a lifetime of lesser emergencies, challenges, and traumatic events, as they can help to prepare the individual for new challenges even if they might be unexpected. Studies of older persons' emotional health have identified individual characteristics that promote successful adaptation, variously described as resilience, mastery, self-efficacy, or emotional vitality (Hendrie et al. 2006). Analyzing data from the Berlin Aging Study, Baltes and Baltes (1990) proposed that accumulated life experiences for some older persons result in "wisdom," the ability to exercise good judgment about important but uncertain matters, which enables successful adaptation. It seems likely that those older persons who have developed this ability to make good judgments in the face of uncertainty would be better adapted to deal with disaster experiences successfully.

Implications for Research and Practice

The Socio-Ecological Model of Individual and System Preparedness in Long Term Care

Drawing on previous research at both the organizational and individual levels, we offer an initial theory of preparedness for older people in long term care. A useful starting point for conceptualizing a framework of preparedness for long term care is the Vulnerability Perspective (Blaikie et al. 1994). Thomas and Soliman (2002) adapted the Vulnerability Perspective to examine factors that contributed to death among older people in heat waves the United States during a 20-year period. As adapted by Thomas and Soliman (2002), the Vulnerability Model suggests that the progression of vulnerability is influenced by "root causes" such as limited resources, "dynamic pressures" such as lack of local institutions, and "unsafe conditions" such as lack of preparedness to cope with disasters and hazards (Thomas and Soliman 2002, p. 27). The Vulnerability Perspective offers a framework for how various external factors, such as social, economic, and political institutions increase or decrease vulnerability to disasters among certain groups of people. The model suggests that personal characteristics such as age, gender, and level of disability interact with the external environment and organizational systems to affect the ability of individuals and organizations serving older people to plan for disasters, and to recover from them. According to this model, risk arises as a combination of vulnerability and hazard; that is, the interaction of these factors increases individuals' susceptibility to risk.

Building on the Vulnerability Perspective, we suggest a framework for conceptualizing preparedness for older people, The Socio-Ecological Model of Individual and System Preparedness in Long Term Care. The model extends the Vulnerability Perspective framework by incorporating domains of preparedness identified in previous studies of disaster preparedness among organizations providing health, medical, residential, and other supportive services to older people (Cherniack et al. 2008; Dosa et al. 2007; Hyer et al. 2006, 2009; Johnson et al. 2006; Kirkpatrick and Bryan 2007; Laditka et al. 2007b, 2008b, 2009; Saliba et al. 2004), and older adults' experiences and responses to disasters (Shenk et al. 2009). Figure 25.1 shows a simplified version of the model. Using this model, the older person's risk can be viewed as a function of vulnerability plus individual characteristics, and organizational and system level preparedness factors. As shown in Fig. 25.1, risk for older people is at the center of the model. Risk is affected by three sets of factors, represented by three ovals, extending out from the center in the following order: (1) individual characteristics, such as health status, mobility, education, location of residence (e.g., rural or urban, coastal or noncoastal, earthquake prone or nonearthquake-prone), disability status, income, access to transportation and information/resources, ability to communicate, social support (e.g., presence of a spouse and/or adult children), and life course experiences; (2) organizational characteristics, including staff number and type, type and number of older people served by the organization, location factors analogous to individual location characteristics, training, communication resources, and transportation resources; and (3) system factors, including health care resources, emergency systems resources, technology resources, transportation network, and communications network. Thus, the model incorporates domains previously identified as important in long-term care preparedness studies at the individual, organizational, and system levels. These factors interact to affect risk among older people from natural or man-made disasters.

Many older people live alone, some with limited resources and few social contacts. Many will not be able to engage in active information gathering during a disaster (Slaughter et al. 2005). An integrated theory of disaster preparedness in long term care will anticipate this need for information, strengthen multilevel communication efforts, and provide specific recommendations for the preparedness needs of older adults at the individual, organizational, and system levels, integrating an understanding of the life course perspective.

Although the Vulnerability Perspective framework and the Socio-Ecological model provide useful conceptual frameworks for disaster preparedness, these models have limitations. They are not causal models. Further, they do not allow researchers to identify certain factors and characteristics that play more (or less) prominent roles in preparedness for older people. It would be useful for future research to identify linkages among sets of factors and causal pathways and to determine factors and characteristics that play a key role in preparedness for older individuals. These steps will help to develop an integrated theory of preparedness in long-term care. We suggest that an integrated theory should present the current state of knowledge regarding such preparedness, depict areas in which revisions of current knowledge, practice, or policy may be desirable, inform future research into emergency preparedness among organizations serving older populations, and lay out directions for future research in the field.

Implications for Life Course Research

The strength of the life course perspective lies in its ability to integrate the context of individuals' lives, including the current phase of life with its previous experiences and consequences, the time and events both before and after the point being studied, and the ways in which the individual's life

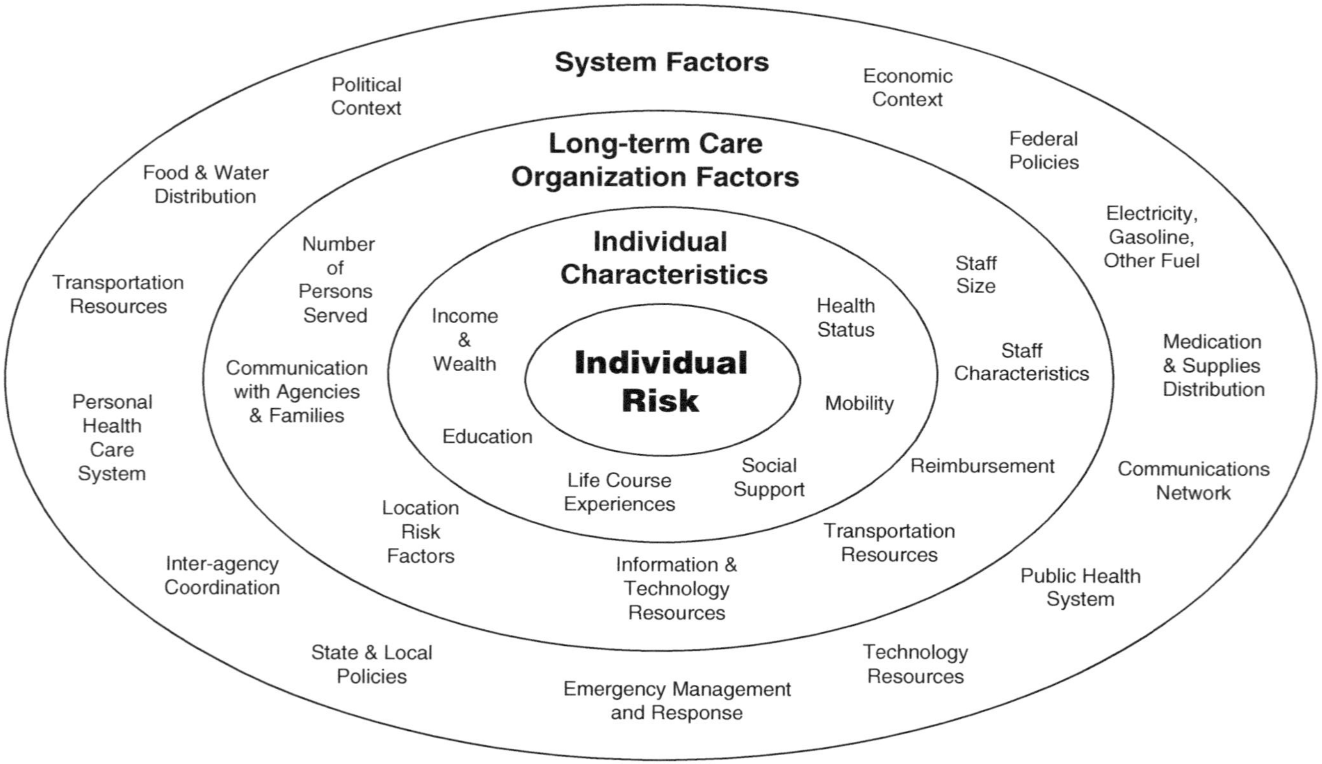

Fig. 25.1 The socio-ecological model of individual and system preparedness in long-term care

is intertwined with others (Shenk et al. 2009). The past helps to shape the individual's orientation to the present and the future so that life before the experience of the disaster and life after the disaster exist in the context of the individual's personal experiences and environment. Thus, there is interplay between biography and history, or self and society. The biography of the older adult interacts with the historical encounter of the disaster experience. It is from this interaction that the life course perspective extends itself to conceptualization and understanding of individual action.

Central to the concept of disaster studies are the obvious effects a disaster brings, including vulnerability, loss of lives, and destruction or loss of material possessions. Following a disaster, everyone is likely to experience losses and be affected. However, more severe and longer lasting adverse effects occur among those who belong to the most vulnerable groups. This includes those with lower socioeconomic status, those most affected by health disparities associated with race or ethnicity, individuals with limited social support, and those with limited ability to access resources to rebuild and recover. Among older adults, their frailty and vulnerability may often be compounded by misconceptions about their abilities and needs. Older people can be further victimized by oversimplified generalizations about their abilities to respond in the event of a disaster.

The experiences of disasters by older adults influence the rest of their lives and the consequences are likely to bring behavioral changes. Their life orientations and intrinsic features can also set them apart from other disaster victims. As noted earlier, previous experiences with a disaster can improve confidence in their ability to manage the situation; longer length of residence in the disaster area may have led to development of stronger social networks (e.g., Kato et al. 1996). The number of years lived with the range of experiences they are likely to have encountered can make older adults emotionally stronger and more mature in handling the stresses of a disaster encounter (e.g., Heller et al. 2005). The emotional competency that derives from these intrinsic features is often affected by the physiological changes that aging often brings. Though older adults are often viewed as vulnerable and frail, these characteristics are countered by the heightened emotional stability they possess (e.g., Hendrie et al. 2006). Given these skills, older people can often serve as sources of information for disaster preparation and recovery.

The best way to mitigate the unfavorable outcomes of a disaster experience is to be prepared. Although researchers have identified many negative effects of disasters, the interpretation of the encounter varies across individuals. Capitalizing on strengths of older adults as expressed in the richness of their lifetime of experiences can make a significant contribution to disaster research. Disaster studies can help to inform life course research. Disaster research provides a vivid example of how the life course perspective can help us to understand the complexities of interactions of history and memory and of individuals' experiences within a historical context. The challenge is to balance strengths and vulnerabilities of older adults as survivors of disasters who can aid in preparing for future disasters.

Disaster and Emergency Communication Needs

Preparedness communication is a key factor in disaster preparedness (Kreps et al. 2005). Unfortunately, preparedness information may not reach older adults or their families, due to limited access to information, confusing communication of messages that are not well designed for their intended audience, and poor understanding of these messages by older adults (Cherniack et al. 2008). At all ages, people increasingly are turning to the internet for information about health risk and disease. Facing a pending disaster or emergency, many older people rely on the internet to provide accurate, current, and easily understood information. A recent study evaluated the readability and suitability (e.g., content, layout, cultural appropriateness) of 50 top websites with information about disasters and preparedness aimed at the general public in the U.S. (Friedman et al. 2008).

The researchers found that most websites required a high reading level, nearly grade 11, and were rated difficult or very difficult. Most were rated as either below average or adequate in suitability. These results suggest that there is a need for web-based preparedness resources that can be more easily understood.

Public and governmental health organizations, and organizations providing long term health care and support services, can help prevent excess morbidity and mortality among older people by communicating about the need for preparedness and its desired characteristics, and about disasters that are predictable in the short term. Community, public health, and organizational responsibilities may include helping older people to prepare for unexpected events, helping them to shelter-in-place or assisting with evacuation and/or transportation to shelters if necessary, and helping them to recover from disasters (Johnson et al. 2006). In all of these areas, clear, targeted messages will result in better attention to the information, and a greater probability of understanding the need to take action when required (Kreuter et al. 1999).

Preparedness Training Needs Among Students of Health Administration and Health Professions

There is an urgent need to enhance disaster preparedness training for students of public health, gerontology, health administration, nursing, social work, and other health-related fields. One recent study used a train-the-trainer process and experiential learning to teach doctoral students to serve as preparedness educators for students of public health and health administration (Laditka et al. 2007a). After reviewing preparedness instructional materials, students worked in small groups of 3 or 4 students to develop teaching materials for master's level students. The results of eight paired Likert scale questions on pre and postassignment surveys showed significant and meaningful improvement in students' perceptions. The results showed markedly improved knowledge about disaster preparedness for vulnerable groups, and in the students' confidence in their ability to develop and teach disaster preparedness (both $p<0.05$) (Laditka et al. 2007a). This type of experiential approach can be adapted by faculty to incorporate a "stand alone" assignment in a course for students, including students of public health, health administration, gerontology, social work, and nursing. It can also be adapted to train leaders in the private and public sectors, as well as health care professionals, to train health care and public health workers and volunteers to plan for disasters and respond to them.

Additional Areas for Future Research in Long-Term Care Preparedness

A growing number of studies have examined preparedness in nursing homes, in response to the aftermath of Hurricane Katrina. In the past 20 years there has been a huge growth in other residential options for older people in the United States, including assisted living facilities, continuing care retirement communities, and retirement communities. To our knowledge, no research has examined preparedness plans in these long-term care organizations and communities in which most residents are older. Thus, little is known about procedures and processes to help these older people during and after disasters. The eight training and practice domains described by Laditka et al. (2008b) could be adapted for research in these residential facilities and communities. In another area, research has not typically distinguished between reactions to natural disasters and reactions to man-made disasters. Reactions to the latter, particularly terrorist events, may differ qualitatively from reactions to natural disasters. They may also result in different life course effects, including, for example, enhanced feelings of patriotism or vulnerability, either of which could influence behaviors throughout the ensuing life course.

Research that explicitly examines cohort effects in preparedness would also be useful. Results of a recent study using nationally representative data for Vietnam era veterans ($n=7,914$) from the 2001 National Survey of Veterans illustrates the usefulness of considering cohort effects in disaster preparedness among older populations. Brooks et al. (2008) compared the treatment of posttraumatic stress disorder and that for other mental health conditions of veterans of the Vietnam War era who served in Vietnam and those who served elsewhere, stratifying the analysis by age (<60, ≥60). Most mental health measures were significantly poorer among veterans who served in Vietnam than those who served elsewhere; veterans who served in Vietnam were also significantly more likely to receive treatment for posttraumatic stress disorder (Brooks et al. 2008). A cohort effect was observed: the negative effect of service in the war zone on mental health measures was substantially worse among those who experienced the war at younger ages; they were dramatically more likely to be treated for posttraumatic stress disorder than were those who experienced the war at older ages (Brooks et al. 2008). These findings highlight the complex interplay of cohort, age, and shared life-altering experiences. There may be similar differential disaster-related effects.

In addition, research has examined the role of race and ethnicity in vulnerability to disasters in the U.S. (for a review of earlier studies, see Fothergill et al. 1999). Most of these studies have been conducted after Hurricane Katrina (e.g., Chen et al. 2007; Elliott and Pais 2006; Lee et al. 2009; White et al. 2007). Taken as a whole, there is substantial evidence that African Americans are disproportionately adversely affected after a disaster, compared with non-Hispanic whites (Chen et al. 2007; Elliott and Pais 2006; Lee et al. 2009; Mills et al. 2007; White et al. 2007). Further, there is evidence that African American women may have the greatest vulnerability for long-term mental health problems after a disaster (Chen et al. 2007; Mills et al. 2007). The experiences of African American women following Hurricane Katrina are the focus of a recently conducted study (Laditka et al. 2010).

Conclusions

In this chapter, we described challenges and opportunities of disaster preparedness in organizations providing long term care for older people. A life course approach was used to consider older adults' experiences and responses to disasters. We also proposed a framework for long-term care preparedness. We encourage researchers, policy makers, practitioners, and administrators to continue to improve knowledge of how to help older people prepare for disasters, and to recover from them. Older adults bring a lifetime of experience to disasters. Their special needs and coping styles need to be addressed in disaster planning and recovery. It is important to recognize the vulnerabilities and special challenges that older adults experience following a disaster. At the same time, the history and memories of older people can prepare them to draw valuable lessons from past disasters and challenging experiences, and often also to cope with present disasters. Public health and community professionals who specialize in preparedness, together with administrators of organizations that provide health and social services to older people, would benefit from incorporating the valuable perspectives of older individuals in disaster planning.

Acknowledgements We are grateful to Daniela Friedman, PhD, and Janice Probst, PhD, for valuable comments about this research. We thank Karel Kalaw for her excellent research assistance.

References

Alea, Nicole, Susan Bluck, and Angeleria B. Semegon. 2004. "Young and Older Adults' Expression of Emotional Experience: Do Autobiographical Narratives Tell a Different Story?" *Journal of Adult Development* 11(4):235–50.

Association for Homes and Services for the Aged. *Aging Services: The Facts. 2009*. Retrieved October 11, 2009 (http://www.aahsa.org/article.aspx?id=74).

Baltes Paul B. and Margaret M. Baltes. 1990. *Successful Aging: Perspectives from the Behavioral Sciences*. Cambridge, England: Cambridge University Press.

Bandura, Albert. 1997. *Self-Efficacy: The Exercise of Control*. New York: W.H. Freeman.

Blaikie, Piers, Terry Cannon, Ian Davis, and Ben Wisner. 1994. *At Risk: Natural Hazards, People's Vulnerability & Disasters*. New York: Routledge.

Brooks, Matthew S., Sarah B. Laditka, and James N. Laditka. 2008. "Long Term Effects of Military Service on Mental Health among Veterans of the Vietnam Era." *Military Medicine,* 173(6):570–5.

Brown, Lisa S., Kathryn Hyer, and LuMarie Polivka-West. 2007. "A Comparative Study of Laws, Rules, Codes and Other Influences on Nursing Homes' Disaster Preparedness in the Gulf Coast states." *Behavioral Science and the Law* 25(5):655–75.

Castle, Nicholas G. 2008. "Nursing Home Evacuation Plans." *American Journal of Public Health* 98:1235–40.

Centers for Disease Control. *National Nursing Home Survey.* 2008.

Chen, Angela Chia-Chen, Verna M. Keith, Chris Airriess, Wei Li, and Karen J. Leong. 2007. "Economic Vulnerability, Discrimination, and Hurricane Katrina: Health among Black Katrina Survivors in Eastern New Orleans." *Journal of the American Psychiatric Nurses Association* 13(5):257–66.

Cherniack, E. Paul, Lauran Sandals, Linda Brooks, and Michael J. Mintzer. 2008. "Trial of a Survey Instrument to Establish the Hurricane Preparedness of and Medical Impact on a Vulnerable, Older Population." *Prehospital and Disaster Medicine* 23(3):242–9.

Dosa, David M., Nancy Grossman, Terrie Wetle, and Vincent Mor. 2007. "To Evacuate or Not To Evacuate: Lessons Learned from Louisiana Nursing Home Administrators Following Hurricanes Katrina and Rita." *Journal of the American Medical Directors Association* 8(3):142–9.

Ecevit, Megnet and Aytul Kasapoglu. 2002. "Demographic and Psychosocial Features and Their Effects on the Survivors of the 1999 Earthquake in Turkey." *Social Behavior and Personality* 30:195–202.

Elliott, James R. and Jeremy Pais. 2006. "Race, Class, and Hurricane Katrina: Social Differences in Human Responses to Disaster." *Social Science Research* 35:295–321.

EM-DAT: The OFDA/CRED International Emergency Disasters Data Base. Université Catholique de Louvain, Brussels, Belgium. Retrieved on September 15, 2009 (http://www.em-dat.net/).

Fernandez, Lauren S., Deana Byard, Chien-Chih Lin, Samuel Benson, and Joseph A. Barbera. 2002. "Frail Elderly as Disaster Victims: Emergency Management Strategies." *Prehospital Disaster Med*icine 17:67–74.

Friedman, Daniela B., Manju Tanwar, and Jane V.E. Richter. 2008. "Evaluation of Online Disaster and Emergency Preparedness Resources." *Prehospital and Disaster Medicine* 23(5):438–46.

Fothergill, Alice, Enrique G.M. Maestas, and JoAnne DeRouen Darlington. 1999. "Race, Ethnicity, and Disasters in the United States: A Review of the Literature." *Disasters* 23(2):156–73.

Goto, Toyomi, John P. Wilson, Boaz Kahana, and Steve Slane. 2006. "The Miyake Island Volcano Disaster in Japan: Loss, Uncertainty, and Relocation as Predictors of PTSD and Depression." *Journal of Applied Social Psychology* 36:2001–26.

Gray, Bradford H. and Kathy Hebert. 2007. "Hospitals in Hurricane Katrina: Challenges Facing Custodial Institutions in a Disaster." *Journal of Health Care for the Poor and Underserved* 18(2):283–98.

Heller, Kenneth, Douglas B. Alexander, Margaret Gatz, Bob G. Knight, and Tara Rose. 2005. "Social and Personal Factors as Predictors of Earthquake Preparation: The Role of Support Provision, Network Discussion, Negative Affect, Age, and Education." *Journal of Applied Social Psychology* 35:399–422.

Hendrie, Hugh, C., Marilyn S. Albert, Meryl A. Butters, Sujuan Gao, David S. Knopman, Lenore J. Launer, Kristine Yaffe, Bruce N. Cuthbert, Emmeline Edwards, and Molley V. Wagster. 2006. "The NIH Cognitive & Emotional Health Project: Report of the Critical Evaluation Study Committee." *Alzheimer's & Dementia, 2*(1):12–32.

Hoffman, Susanna M. and Anthony Oliver-Smith. 1999. "Anthropology and the Angry Earth: An Overview." Pp. 1–16 in *The Angry Earth*, edited by Anthony Oliver-Smith and Susanna M. Hoffman. New York: Routledge.

Hughes, Susan L. and Megan Renehan. 2005. "Home Health." Pp. 87–111 in *The Continuum of Long-Term Care*, edited by Connie J. Evashwick. Clifton Park, NY: Delmar Learning.

Hyer, Kathryn, Lisa S. Brown, Janelle J. Christensen, and Kalie S. Thomas. 2009. "Weathering the Storm: Challenges to Nurses Providing Care to Nursing Home Residents during Hurricanes." *Applied Nursing Research* 4:e9-14.

Hyer, Kathryn, Lisa S. Brown, Amy Berman, and LuMarie Polivka-West. 2006. "Establishing and Refining Hurricane Response Systems for Long-Term Care Facilities." *Health Affairs* 5:w407–11.

Javadian, Reza. 2007. "Social Work Response to Earthquake Disasters: A Social Work Intervention in Bam, Iran." *International Social Work* 50:334–46.

Johnson, Arleen, Robert E. Roush, Jr., Judith L. Howe, Margaret Sanders, Melen R. McBride, Andrea Sherman, Barbara Palmisano, Nina Turmosa, Elyse A. Perweiler, and Joan Weiss. 2006. "Bioterrorism and Emergency Preparedness in Aging (BTEPA): HRSA-Funded GEC Collaboration for Curricula and Training." *Gerontology and Geriatric Education* 26:63–86.

Joint Commission. 2006–2007 Standards for Home Health, Personal Care, Support Services, and Hospice. Available at: http://www.jcrinc.com/3235/. Accessed July 23, 2007.

Kato, H., N. Asukai, Y. Miyake, K. Minakawa, and A. Nishiyama. (1996). "Post-Traumatic Symptoms among Younger and Elderly Evacuees in the Early Stages Following the 1995 Hanshin-Awaji Earthquake in Japan." *Acta Psychiatry Scandinavia* 93:477–81.

Kirkpatrick, Dahlia D. and Marguerite Bryan. 2007. "Hurricane Emergency Planning by Home Health Providers Serving the Poor." *Journal of Health Care for the Poor and Underserved* 18:299–314

Krause, Neal. 1987. "Exploring the Impact of a Natural Disaster on the Health and Psychological Well-Being of Older Adults." *Journal of Human Stress* 13:61–9.

Kreps, Gary L., Kenneth Alibek, Linda Neuhauser, Katherine E. Rowan, and Lisa Sparks. 2005. "Emergency/Risk Communication to Promote Public Health and Respond to Biological Threats." Pp. 349–62 in *Global Public Health Communication: Challenges, Perspectives, and Strategies,* edited by Muhiuddin Haider. Sudbury MA: Jones and Bartlett Publishers.

Kreuter, Matthew W., Victor J. Strecher, and Bernard Glassman. 1999. "One Size Does Not Fit All: The Case for Tailoring Print Materials." *Annals of Behavioral Medicine* 21:276–83.

Laditka, Sarah B. 1998. "Modeling Lifetime Nursing Home Use under Assumptions of Better Health." *Journal of Gerontology: Social Sciences* 53B(4):S177–87.

Laditka, Sarah B., James N. Laditka, Carol B. Cornman, Courtney B. Davis, and Maggi J. Chandlee. 2008a. "Disaster Preparedness for Vulnerable Persons Receiving In-Home, Long-term Care in South Carolina." *Prehospital and Disaster Medicine* 23:133–42.

Laditka, Sarah B., James N. Laditka, Carol B. Cornman, Courtney B. Davis, and Jane V.E. Richter. 2009. "Resilience and Challenges among Staff of Gulf Coast Nursing Homes Sheltering Frail Evacuees Following Hurricane Katrina, 2005: Implications for Planning and Training." *Prehospital and Disaster Medicine* 24(1): 54–62.

Laditka, Sarah B., Louise M. Murray, and James N. Laditka (2010). "In the Eye of the Storm: Vulnerability and Resilience among African American Women in the Wake of Hurricane Katrina." *Health Care for Women International* 31(11):1013–1027.

Laditka, Sarah B., James N. Laditka, Margaret M. Houck, Carol B. Cornman, and Courtney B. 2007a. "Ethical Issues in Training for Disaster Preparedness to Serve Frail Older People: An Innovative Approach to Graduate Health Management Education." Pp. 49–68 in *Business Ethics in Focus*, edited by Laura A. Parrish. New York: Nova Science Publishers.

Laditka, Sarah B., James N. Laditka, Sudha Xirasagar, Carol B. Cornman, Courtney B. Davis, and Jane V.E. Richter. 2007b. "Protecting Nursing Home Residents during Disasters: An Exploratory Study from South Carolina." *Prehospital and Disaster Medicine* 22(1):42–8.

———. 2008b. "Providing Shelter to Nursing Home Evacuees in Disasters: Lessons from Hurricane Katrina." *American Journal of Public Health* 98:1288–93.

Lee, Eun-Kyoung Othelia, Ce Shen, and Thanh V. Tran. 2009. "Coping with Hurricane Katrina: Psychological Distress and Resilience among African American Evacuees." *Journal of Black Psychology* 35(1):5–23.

McMillen, J. Curtis, Elizabeth M. Smith, and Rachel H. Fisher. 1997. "Perceived Benefit and Mental Health after Three Types of Disaster." *Journal of Consulting and Clinical Psychology* 65:733–9.

Melick, Mary Evans. 1985. "The Effect of Disaster on the Health and Well-Being of Older Women." *International Journal of Aging and Human Development* 21:27–38.

Miller, Andrew C. and Bonnie Arquilla. 2008. "Chronic Diseases and Natural Hazards: Impact of Disasters on Diabetic, Renal, and Cardiac Patients." *Prehospital and Disaster Medicine* 23(2):185–94.

Mills, Mary Alice, Donald Edmonson, and Crystal L. Park. 2007. Trauma and Stress Response among Hurricane Katrina Evacuees. *American Journal of Public Health* 97:S116–23.

Nolen-Hoeksema, Susan and Jannay Morrow. 1991. "A Prospective Study of Depression and Posttraumatic Stress Symptoms after a Natural Disaster: The 1989 Loma Prieta Earthquake." *Journal of Personality and Social Psychology* 6:115–21.

Norris, Fran H. and Stanley A. Murrell. 1988. "Prior Experience as a Moderator of Disaster Impact on Anxiety Symptoms in Older Adults." *American Journal of Community Psychology* 16:665–83.

Pande, Archana, Sarah B. Laditka, James N. Laditka, and Dorothy R. Davis. 2007. "Aging in Place? Evidence that a State Medicaid Waiver Program Helps Frail Older Persons. *Home Health Care Services Quarterly* 26:39–60.

Pew Center on Global Climate Change. *Hurricanes and Global Warming FAQs*. Retrieved October 1, 2009 (http://www.pewclimate.org/hurricanes.cfm#freq).

Phifer, James F. and Fran H. Norris. 1989. "Psychological Symptoms in Older Adults Following Natural Disaster: Nature, Timing, Duration, and Course." *Journal of Gerontology, Social Sciences* 44:S207–17.

Riddix Louis and Uno Dellar. 2001. "The Ice Storm in Eastern Canada 1998 KAMDO-Report No. 74." *Prehospital and Disaster Medicine* 16:50–2.

Saliba, Debra, Joan Buchanan, and Raynard S. Kington. 2004. "Function and Response of Nursing Facilities during Community Disaster." *American Journal of Public Health* 94:1436–41.

Shenk, Dena, Blanca Ramos, Karel J. Kalaw, and Ismail Tufan. 2009. "History, Memory, and Disasters among Older Adults: A Life Course Perspective." *Traumatology* 15:20–30.

Sienkiewicz, Josephine, Ginny Wilkinson, and Betsy Cubbage. 2007. "A Patient Classification System for Emergency Events in Home Care." *Home Health Care Nurse* 25:378–85.

Simerman, J., D. Ott, and T. Mellnik. 2005. "Early Data Challenge Assumptions about Katrina Victims." Austin American Statesman, 30 December 2005.

Slaughter, Laura, Alla Keselman, Andre Kushniruk, and Vimla L. Patel. 2005. "A Framework for Capturing the Interactions between Laypersons' Understanding of Disease, Information Gathering Behaviors, and Actions Taken during an Epidemic." *Journal of Biomedical Infomatics* 38:298–313.

South Carolina Department of Health and Environmental Control. Division of Community Long Term Care. *Elderly/Disabled Policy and Procedure Manual*, 2005.

Task Force of Quality Control of Disaster Management. 2002. "Health and Disaster Management Guidelines for Evaluation and Research in the Utstein Style." Prehospital and Disaster Medicine 17(S3):144–67.

Thomas, Norma D. and Hussein Soliman. 2002. "Preventable Tragedies: Heat Disaster and the Elderly." *Journal of Gerontological Social Work* 38:53–66.

Thompson, Martie P., Fran H. Norris, and Betty Hanacek. 1993. "Age Differences in the Psychological Consequences of Hurricane Hugo." *Psychology and Aging* 8:606–16.

US Census Bureau: *Current population reports: Population projections: States 1895–2025*. 1997. Retrieved October 1, 2008 (http://www.census.gov/prod/2/pop/p25/p25-1131.pdf).

———. 2002. *Americans with Disabilities: 2002. Table1: Prevalence of Disability by Age, Sex, Race, and Hispanic Origin*.

U.S. Geological Survey. *National Seismic Hazard Mapping Project*. Retrieved September 30, 2009 (http://earthquake.usgs.gov/hazards).

U.S. Office of Inspector General. *Nursing Home Emergency Preparedness and Response during Recent Hurricanes*. Department of Health and Human Services, OEI-06000020. Retrieved August 18, 2006 (http://oig.hhs.gov).

United States Supreme Court. *Olmstead v. L.C.* 1999.

White, Ismail, Tasha S. Philpot, Kristin Wylie, and Ernest McGowen. 2007. "Feeling the Pain of My People: Hurricane Katrina, Racial Inequality, and the Psyche of Black America." *Journal of Black Studies* 37(4):523–38.

Yeung, Danni Yuen-Lan and Helene H. Fung. 2007. "Age Differences in Coping and Emotional Response toward SARS: A Longitudinal Study of Hong Kong Chinese." *Aging and Mental Health* 11:579–87.

Chapter 26
Elder Mistreatment

Sonia Salari

Quality of life has emerged as a strong theme in the study of sociology of aging and life course. Victims of elder abuse and neglect experience low quality of life, along with a greater risk of morbidity and mortality (CDC 2009; Dong et al. 2009; Lachs et al. 1998). Evidence suggests elder mistreatment has existed historically in various forms (Pillemer and Wolf 1986), and current estimates suggest between one and two million older Americans suffer physically, psychologically, and/or financially at the hands of a trusted caregiver, spouse, or other family member.

At the societal level there has been increasing awareness and response to elder abuse and neglect over the past three decades. In the 1970s, public attention focused mostly on child abuse and neglect. The 1974 Child Abuse Prevention and Treatment Act (CAPTA) included an official definition of mistreatment and a mandatory reporting requirement for suspected child abuse. During this time, the first federal domestic violence shelter for victims of intimate partner violence (IPV) was established in California. The decade was also known for the beginning of the deinstitutionalization movement, where disabled persons were integrated more heavily into community life. Institutional settings were widely considered abusive. In 1975, Adult Protective Services (APS) legislation was funded under the Title XX Social Security Act, with an eye toward mandatory reporting standards. APS units were created in states to investigate reports of maltreatment of vulnerable adults (Fulmer et al. 2002). Later that decade, Claude Pepper chaired the widely publicized House Select Committee on Aging hearings and coined the term "elder abuse" (Bonnie and Wallace 2003).

The 1980s are commonly cited as the decade of greatest public attention to the problem and prevention of elder abuse and neglect. Initially, funding was sparse and research seemed stunted by methodological difficulties. Initially, policymakers and APS workers saw no harm in simply adapting child abuse patterns and solutions to the plight of the abused older person. Pillemer (2005) has critiqued early decades, when elder abuse "knowledge" was not empirically tested and tended to involve repeated claims describing elder abuse victims as highly dependent on overburdened caregivers. Eventually, the repeated claims came to be accepted as truth, with the perspective influencing policy programming. Improvement began in the late 1980s when the Amendments to the Older American's Act, adopted in 1987, defined elder abuse categories (physical, sexual, emotional/psychological, neglect, self-neglect, abandonment, and financial exploitation) (Teaster 2004). The first large-scale incidence study was published by Pillemer and Finkelhor (1988) using surveys from the Boston Metropolitan Area. Early investigators devoted enormous efforts toward establishing the abuse prevalence, and bringing the dialog into the sights of national funding agencies (Wolf and Pillemer 1989). By the late 1980s, the late Rosalie Wolf introduced the *Journal of Elder Abuse and Neglect*, which is still the premier journal for the study of mistreatment in later life.

S. Salari (✉)
Department of Family and Consumer Studies, University of Utah, Salt Lake City, UT, USA
e-mail: sonia.salari@fcs.utah.edu

R.A. Settersten, Jr. and J.L. Angel (eds.), *Handbook of Sociology of Aging*, Handbooks of Sociology and Social Research, DOI 10.1007/978-1-4419-7374-0_26,

Over the past three decades, experts have come to understand the challenges to studying later life mistreatment, and they acknowledge research, policy, and practice in the field lags decades behind other areas of family violence, such as child abuse and IPV (Dong et al. 2009; McNamee and Murphy 2006). By 2003, the National Research Council reported no major foundation had identified elder mistreatment as a priority and federal funding for this problem had been modest at best (Bonnie and Wallace 2003). Since that time, small-scale federal efforts have funded incidence studies and there has been some progress developing accepted definitions.

Recent trends in elder mistreatment knowledge have focused on determining prevalence from a variety of sources which include bureaucratic reports from organizations such as APS. APS substantiated reports of elder abuse in 50 states show most cases took place in domestic settings (89.3%) and victims were typically female (65.7%), Caucasian (77.1%), and disproportionately over the age of 80 (42.8%). Self-neglect was the most frequently reported form of mistreatment (Fulmer et al. 2005; Teaster et al. 2006) and has been shown to be a significant predictor of mortality (Dong et al. 2009). Limitations of bureaucratic measurement include the need for consistent definitions across states and the reliance on reports to authorities, leaving much of the incidence unmeasured. APS units have defined *vulnerable adults* as those who are mistreated or in danger of mistreatment, without the ability to protect themselves due to advanced age or functional disability. Victim age categories vary across states, so groups 18–59 with disabilities, those 60 (55 or 65) and over have all been used – making cross-state comparisons challenging (Teaster 2004).

Social science research has attempted to determine prevalence through primary health care interviews, telephone surveys (Bonomi et al. 2007; Laumann et al. 2008), caregiver self-reports, and content analysis of television and print news stories (Malphurs and Cohen 2002; Salari 2007). Limitations include the biases of media and self-reports related to victimization or perpetration, and variations based on operational definitions. The National Research Council developed a definition of elder mistreatment as "intentional actions that cause harm or create serious risk of harm … to a vulnerable elder by a caregiver or other person who stands in a trust relationship to the elder or failure by a caregiver to satisfy elder's basic needs to protect … from harm" (Bonnie and Wallace 2003:1). These guidelines provide a helpful framework but tend to leave many facets of elder mistreatment unmeasured.

The social construction of elder abuse and mistreatment has expanded in recent decades with a widespread recognition of new categories such as self-abuse and neglect. In addition, material abuse has ballooned in recent years, as evidenced by the reports listed in the National Council of Elder Abuse (NCEA) daily news feed listserv. Evidence suggests greater rates of financial exploitation are related to new vulnerabilities, such as identity theft and investment scams. The recent economic downturn has worsened the rate of material abuse suffered in later life (Metlife 2009). Attention to nonfamily perpetrators now includes strangers who target community dwelling elderly persons as well as negative treatment one might receive in institutional settings. Recently, the "victim's rights" movement has spilled over to survivors of elder abuse, recognizing the need for age-appropriate support services and shelters.

The social construction of elder mistreatment is related to cohort aging and period effects. The current generation of older persons has been socialized to keep hardships to themselves and family life private. Many of the violent or traumatizing incidents have remained hidden, with a high rate of under reporting. Elderly victims of detected mistreatment have a small group of devoted advocates such as researchers, human service personnel, medical staff, and those in law enforcement. Survivors of abuse are often left in precarious positions, as the available support services may not be age appropriate or responsive to the needs of an older person. As the Baby Boom cohort ages, significant increases are expected in the number of vulnerable elders, which is likely to be dealt with more publically. A stronger advocacy lobby is expected to develop to effectively mobilize forces to prevent and respond to violence.

Researchers now understand that child abuse models of dependency are not appropriate to explain elder victimization. However, in order to raise awareness and address the service and policy needs of this population, the scholarship will need to move beyond the focus on individual and family dynamics alone, and recognize the role of the larger community and sociological forces.

Ecological Iceberg of Elder Mistreatment

My experience in the field includes domestic violence victim advocate, university professor (taught courses in aging, public policy, and family violence), and researcher in the area of family violence (Salari 2007, 2009; Salari and Baldwin 2002) and elder mistreatment (Salari 2002, 2006). Based on these roles, I envision an ecological framework which examines the micro (individual), meso (family), exo (community), and macro (societal) level conditions influencing the social construction of elder mistreatment. The first two levels have received a great deal of attention in the literature, namely the individual and family culture perspectives. The vulnerabilities of victims (health, cognitive functioning, isolation, and gender) and the psychological characteristics of perpetrators are considered the *individual* level. *Family* dynamics include maladaptive dependency, violence socialization, unbalanced reciprocity, and poor coping skills. The two macro-level influences have been neglected. Fewer perspectives have focused on *community norms* about the nature of public and private relationships or the effects of the *economic and social structure* as they relate to elder maltreatment. All four ecological levels can be recognized and combined within an "iceberg" framework, where the unmeasured incidence is acknowledged, realizing the true size and shape of the phenomenon (under the surface) are difficult to envision. The social construction of mistreatment is likely to change with cohort aging over time (Fig. 26.1).

Why is there so much unmeasured elder mistreatment? Elderly persons have greater potential isolation than other age groups because they may be homebound and not be expected to appear for

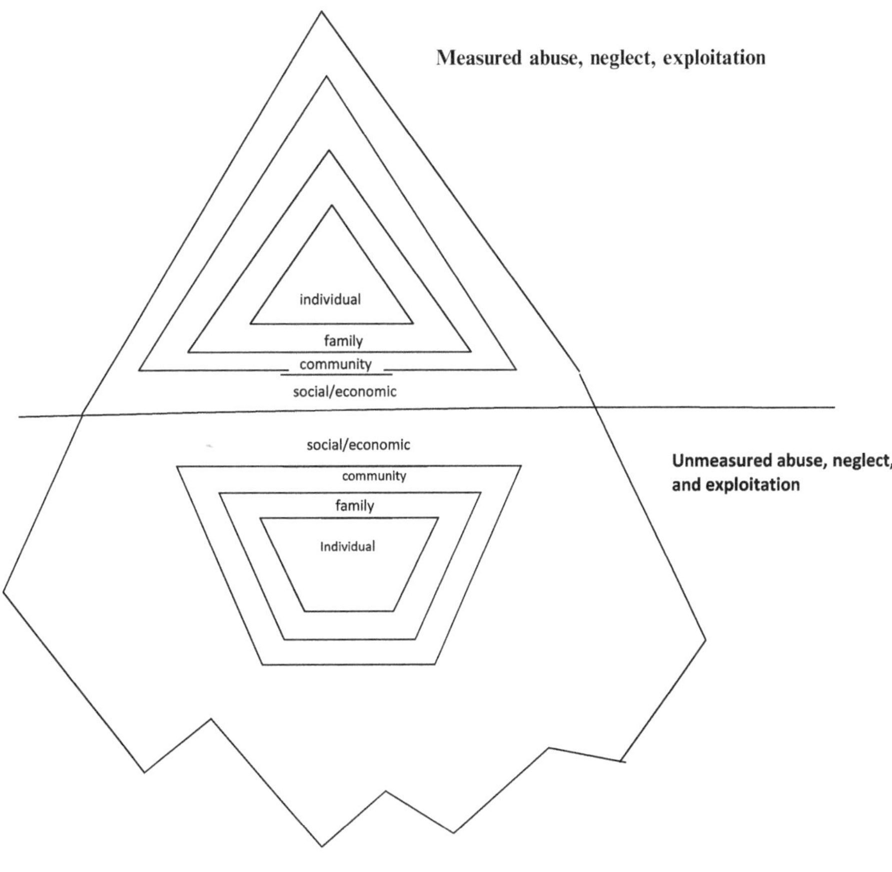

Fig. 26.1 Ecological iceberg of elder mistreatment

work, school, or other regular social obligations. Abused elders with physical evidence of injury are less visible to authorities and service personnel. Detection of abuse is further hampered by factors such as disability, cognitive limitation, depression, technological challenges, and perceived stigma. These characteristics make elderly individuals more vulnerable in families with abusive perpetrators. Relatives who abuse may be overburdened with caregiver responsibilities, or may be manipulative and dependent on the victim and exploit the situation. Abusive spousal behaviors are another source of potential threat to well-being. Intimate partner perpetrators range from psychologically aggressive and isolating to homicidal, and may be motivated by the perceived need for power and control.

Community culture and public vs. private family norms exist to either encourage intervention or turn a blind eye to violence. Denial is a powerful force in individualistic societies where family interactions are considered private business. Under these conditions, the mechanisms of power and control used by members are known only internally and not open to public intervention. In the case of elder mistreatment, the power and control mechanisms may take the form of physical abuse, neglect, psychological maltreatment, and/or financial exploitation. Norms of individualism increase potential for self-abuse and neglect to occur in the home – without a chance of community intervention. Social structural and economic conditions influence the well-being of communities, families, and individuals. Larger influences such as gender oppression, ageism, racism, public policy, and access to resources cannot be ignored with regard to the social construction and experience of mistreatment in later life.

Theoretical and Empirical Perspectives of Elder Mistreatment

Elder mistreatment is not caused by one easily identified element or characteristic, so theoretical perspectives guide our understanding of the potential conditions and outcomes, without definitive proof to support one viewpoint over the other. It is likely that the perspectives described here only begin to explain the true causes and relationships associated with known patterns of abuse and neglect, but much remains unmeasured. The *social-ecological model* supports the notion of individual, interpersonal, and societal factors linking elder mistreatment to "broader social issues," such as the economy and aging services (Dahlberg and Krug 2002; Wolf 2000). Rosalie Wolf (2000) expressed enthusiasm for this framework in her later essays. This section will loosely follow this ecological model to help organize the theoretical literature in the field and identify areas in need of further research.

Individual factors include historical, health, and biological indicators that potentially put a person at greater risk for involvement in violence. An example might include psychological issues related to trauma, mental health, or substance abuse problems. Other issues may include vulnerabilities related to disability, cognitive function, age, race, and gender. For victims, repeated exposure to physical and psychological trauma has been linked to responses, such as learned helplessness, where the sense of empowerment has been lost and individuals must re-learn independent thinking patterns to leave an abusive situation (Walker 2009). Personality types and the need for power and control are individual traits related to perpetrator motives. The intimate partner relationship that previously had little or no aggression, but developed "late onset of abuse" could signify changes related to functional disability or cognitive disease state in the perpetrating individual. Alzheimer's disease has been known to decrease normal inhibitions and increase aggressiveness (Steinmetz 2005).

Family, spousal, and caregiver relationships involve a variety of dynamics and patterns related to quality of life outcomes. Steinmetz (2005) points out that most forms of abuse are defined by the relationship between the abused and the abuser (such as IPV or child abuse). However, elder abuse has only age as a defining characteristic and the relationship can vary.

Social Learning Theory posits those with negative experiences in childhood imitate the behavior they witnessed and continue the cycle (Wolf 2000). This "intergenerational transmission" of abuse is learned within the family unit so that neglecting, angry, or violent reactions are based on previous behaviors or witnessing. Children who grow up under these circumstances might be expected to abuse their children or even aged parents. With few other skills for coping with frustration, family members are sensitized to violence which is condoned in the family culture. Support for this model was found when severely neglected elders treated in emergency departments were more likely to tolerate the lack of care they received at home if they were abused or neglected as children (Fulmer et al. 2005). Wolf and Pillemer (1989) dispute the validity of social learning theory, because elder abuse perpetrators are not necessarily from abusive homes themselves. In addition, this perspective discounts the fact that humans have "free agency" to make their own choices about their behavioral conduct in the family setting.

The overburdened caregiver perspective attributes greater risk of elder abuse and neglect to the overwhelming stress felt by adult children or other relatives who care for a dependent elderly person (Steinmetz 2005). Exchange theory suggests abuse potential increases with a perceived imbalance of reciprocity. Relationships are considered ideal when the benefits outweigh the costs. A dependent older person may balance the relationship by giving financial gifts or emotional support (Wallace 2005: Wolf 2000). Laws of distributive justice suggest that later-life disabilities could tip the scale – causing the older person to lose status or experience abuse, neglect, or exploitation in the family (Pillemer and Wolf 1986).

Another at-risk relationship involves abusers who are dependent on the older victim, such as an adult child or grandchild who is substance abusing, mentally ill, or manipulative. In a twist on the exchange theory perspective, Pillemer, Wolf, and others using control groups, found abused elderly were not more dependent than their nonabused counterparts (Pillemer 1986; Pillemer and Finkelhor 1989). In addition, abusers were actually more likely to be dependent on the elderly person and the aggression may represent attempts by someone with few resources to gain power (Pillemer and Wolf 1986), and/or abuse may be a symptom of addiction or mental illness (Pillemer 2005; Wolf and Pillemer 1989; Wolf 2000).

Spousal or IPV can be exhibited as either a continuation of previous abuse or a pattern with late onset. The CDC (2006) defines IPV as occurring between two people in a close relationship, which is perpetrated by a current or former spouse or dating partner. Regardless of age, spousal or partner violence has been described as the most common form of family violence (Gelles and Strauss 1988), and mid- and later-life abuse may closely resemble violence in younger age categories. Physical size and strength differences, as well as access to resources, influence the IPV violence outcomes between men and women. Women are more likely seriously injured or killed in an IPV event and recent research has focused specifically on the plight of older women. Consequences of IPV can include health problems, depression, trauma, suicidal ideation, disability, and death.

A recent telephone interview survey of 370 women aged 65 and over found 26.5% reported a lifetime prevalence of IPV. Physical or sexual abuse (or both) was reported by 18.4% of the sample and 21.9% reported psychological abuse. Narrowing the timeframe of recollection, 3.5% of women reported IPV in the past 5 years and 2.2% in the past year. Of those with IPV experience, there exists an overlap of exposure, with most women reporting more than one type (Bonomi et al. 2007). IPV is not limited to female victimization in later life, and at least one major survey found wives more often perpetrated (36%) compared to husbands (22%) in partner abuse (Pillemer and Finkelhor 1988).

Not all IPV is equally harmful or potentially deadly. Johnson's research over the past 15 years has helped make distinctions recognizing the most harmful categories (i.e., homicidal). These distinctions serve as a guide to researchers, service professionals, law enforcement, and policymakers who can respond accordingly with appropriate use of public resources. Johnson (2008) describes *intimate terrorism* as an intense pattern of "coercive control" with a campaign that isolates and endangers the victim. Intimate terrorists are most likely to be homicidal, compared to other types of perpetrators. In my content analysis study of mid- and later-life intimate partner homicide suicide (IPHS) events ($n=403$), intimate terrorists could be identified among some perpetrators in the fatal events. For example, a perpetrator (65), upset about a break up, came to his former girlfriend's (54) workplace and chased her, fatally shooting her in the back and face. He then killed himself nearby. Before the incident, the victim had been threatened and stalked. Her daughter said "He was obsessed with her. He would not leave her alone" (Salari 2007).

In contrast, *situational couple violence* is mutual conflict that escalates into only minor aggression, without either party exhibiting coercive control or risk of homicide. *Violent resistance* is usually perpetrated in self-defense by victims, as they contemplate leaving the abusive relationship. This form of "fighting back" should be recognized as part of the leaving process – with implications for law enforcement. Distinctions are important, so that danger can be assessed, and resources can be allocated wisely to punish the most dangerous perpetrators.

The duration of IPV is relevant to later-life abuse. Late onset violence in those with a long history together might suggest a change in the organic brain functions of the perpetrator or may reflect an aggressive response to caregiver burden. In contrast, long-term IPV known as "spouse abuse grown old," describes partners together over time with continuous levels of abuse in the relationship. The dynamic tends to be cyclical with an increase in frequency of "tension," "crisis," and "calm." The calm phase eventually disappears, leaving the repeated high stress tension and the acute battering stages (Walker 2009). The CDC (2006) warns that long-term IPV can be particularly harmful to victims both physically and psychologically. Repeated trauma over time may lead to *Post-Traumatic Stress Disorder* (PTSD), which is associated with elevated anxiety, fear, depression, and difficulty coping. It is not uncommon for victims to become passive with an inability to see a viable way out. This explains the common pattern where a victim exits several times before they go permanently (Walker 2009).

Social networks are key to the abuse survivor. Unfortunately, the family and friends who once supported the victim tend to drop out as they become frustrated with the inability to make a permanent break from the abuser. Victim advocates are beginning to recognize that the abuse victim usually understands best how to manage terminating the relationship. Danger of serious injury or death escalates when victims are attempting to leave or have left an intimate terrorist. Perpetrators manufacture isolation over time, by interfering with family relationships and cutting victims off from potential social supports. In my IPHS research, neighbors sometimes commented that they did not know the deceased couple, even after living next door for 25 or 30 years. One neighbor believed the husband (85) would not allow his wife (81) to socialize. She commented, "I tried to talk to her; she would smile and then turn her face. She didn't dare" (Salari 2007:448).

Extended kin networks from the family of origin may serve to prevent violence or serve as a buffer between a perpetrator and a victim. Perpetrators who perceive they are being monitored by the victim's family members may resist using violent control tactics. A colleague and I (Salari and Baldwin 2002) examined over 4,000 dyads from the National Survey of Families and Households and found that injurious aggression was more prevalent in those couples who had little contact with the woman's family of origin. Those who are motivated to perpetrate family violence will often wage a campaign to isolate their victims from family support networks so that violence or financial exploitation can go undetected. Another point to consider, we must acknowledge, not all extended family networks are protective. Those with a culture of violence and oppression against women may condone abusive behavior and encourage dominant displays of power and control, even within kinship units.

Elder mistreatment perspectives that consider community level influences are sparse. Communities are involved in the experience of violence, through the degree to which they see family violence as a public vs. private matter. Do the members of the community feel comfortable intervening in domestic disputes or providing mediation? Or, do they turn a blind eye to the perpetration?

Some Native American tribes perceive elder abuse as a community problem with the need for community solutions (Maxwell and Maxwell 1992). In more individualistic societies, having laws against violence is not enough if there is no enforcement and perpetrators have no costs associated with harmful behavior (Dahlburg and Krug 2002). Levinson's (1989) study of 90 nonwestern societies found family violence was not universal and the structure of the society served to promote or prevent abuse. Societies with low divorce rates and greater gender equality had low levels of family violence. In addition, family aggression was minimal in societies with norms of public intervention, where extended family and neighbor networks felt it was their *obligation* to mediate in a dispute. Rules against abuse may exist in society, but lack of enforcement can cause serious harm. Perpetrators who abuse and are not punished may scoff at laws when they are only symbolic. Fewer negative repercussions make vulnerable family members "safe" targets for a family system or individual who is abusive. To apply this philosophy to elder abuse, the problem becomes critical as you add vulnerabilities of older victims with complications of aging and disease, cognitive and physical limitations, generational differences in norms of reporting, and ageism. The solution would involve public recognition of family vulnerabilities that encourage mediation, shelter, and enforcement of effective punishments for abuse perpetrators. In addition, community solutions must balance the needs of choice and autonomy with safety and security. Adult abuse victims have the right to self-determination and the right to be free from mistreatment.

In addition, to violent family relationships, there can also be elder mistreatment that occurs in institutional settings. Goffman's work found total institutions have goals which strip residents of their self-identity and require adaptation strategies for psychological survival (see Salari 2006). Institutional abuse at the hands of licensed professionals in health care facilities and nursing homes is now labeled as "malpractice" (Fulmer, Paveza, and Guadagno 2002). All forms of abuse and neglect can also take place in an institutional setting, and facility workers have been known to perpetrate serious crimes against residents. Prevention efforts have included the new requirements of criminal background checks on facility employees, which is only effective if potential employees have previously been *convicted* of abuse crimes. The vast majority of sexual predators are not convicted – so background checks may only serve a false sense of security.

The majority of the theoretical perspectives to date focus on individual (micro) and family level (meso) dynamics to help us understand what makes some vulnerable to abuse and neglect, while others are at lesser risk. Community norms and larger social and economic influences have been largely left out of the explanations. What happens to victims and perpetrators when law enforcement is ineffective, discriminatory, or lacking altogether? Is elder mistreatment worse in periods of economic downturn? How do policies play a role in prevention and treatment of abuse and neglect? Once victims are removed from dangerous situations in families, are they in a better place? Do communities have appropriate services and shelters for an older person? How does the social construction of mistreatment change with new cohorts entering later life? The future of aging and life course scholarship in this area needs to include greater attention to investigations of the larger sociological influences on quality of life in old age.

Types of Elder Mistreatment

Types of elder abuse have been identified to include physical, sexual, emotional/psychological, and financial exploitation. Neglect can sometimes include abandonment and can be perpetrated against oneself. Suicide is the most severe form of self-abuse, and is common among elderly white men. *Physical abuse* refers to the infliction of pain or injury, in the form of bruises, burns, fractures, lacerations, abrasions, hair loss, and injury to sexual organs (Wallace 2005), Result may include health problems, psychological harm, disability, or death. Some definitions of physical abuse include the use of chemical (drug-induced sedation) and physical restraints. *Sexual abuse* includes nonconsensual sexual contact. Persons who are physically or cognitively incapacitated and/or unconscious are unable to consent and are at increased vulnerability to sexual attack. The lack of ability to communicate may exacerbate the trauma of the attack and hinder adequate reporting. Traditional rape kit testing has often been omitted for elderly victims of such attacks (Burgess and Hanrahan 2006).

Emotional/psychological abuse often accompanies physical and other types of abuse. It is commonly described as infliction of mental anguish. Weiss (2003) introduced a more intense interpretation of "psychological abuse" to describe something different from a bad relationship, a fight, a hot temper, or poor communication skills. Psychological abuse is defined as a "campaign" to harm another. Perpetrators have "bad empathy," where they zero in on their victim's vulnerabilities and use weaknesses against them to cause harm. Normal relationships attempt to shelter loved ones from harm and criticism. An example is infantilization, the psychological mistreatment aimed at older persons that attacks their independence and adult status by subjecting them to treatment normally reserved for children (Salari 2006).

Material abuse/financial exploitation. The greatest wealth in American society is typically accumulated by mid- to later-life, which increases vulnerability to financial exploitation. Modern crime perpetration includes identity theft and breaches to financial privacy. Older generations were socialized during a time when cultural norms prescribed free exchange of Social Security numbers and financial information, so strangers were often entrusted with vital account information. A recent survey of APS reports found financial exploitation (unauthorized use of resources) to be the third most common category of investigated and substantiated reports (Teaster et al. 2006). Recently, the 2006 amendments to the Older American's Act officially defined exploitation as fraudulent, illegal, unauthorized, or improper act of a caregiver or fiduciary that uses the resources of an older individual for monetary or personal benefit, profit or gain, etc., depriving the older person of rightful access to or use of benefits, resources, belongings, or assets (OAA 2006). There is an estimated $2.6 billion annual loss from financial elder abuse (Metlife Study 2009), and it seems to be increasing over time. People who are financially hurt by the current economic downturn are experiencing a desperate state – increasing perpetration and/or targets of exploitation. Those with considerable wealth (e.g., Brooke Astor) are also vulnerable to abusive family or nonfamily perpetrators (Ramnarace 2009), who may feel entitled to take inheritance early.

Neglect can be active (an action) or passive (a failure to act) on the part of a caretaker. It includes failure to care (feed, clothe, bathe, diet, supervise), misuse of medications (including overmedication to sedate), and a failure to seek medical care when needed (Wallace 2005). Pressure sores are considered a sign of neglect and are usually preventable with regular movement. According to a national survey of APS, caregiver neglect is the second-most substantiated case (21%) (Teaster et al. 2006). A recent study of adults recruited through the use of four urban emergency departments found older adults diagnosed as neglected are sicker, have fewer financial resources, and less support in the home (Fulmer et al. 2005).

Abandonment is defined as a caregiver's leaving a dependent elderly person alone and no longer providing care for him or her (CDC 2009). Prevalence of this form of mistreatment, sometimes called "granny dumping" is actually not very high among reported APS cases, but may receive disproportionate publicity. Newsweek (1991) stated 38% of hospitals sampled had reported at least one case of elder abandonment, and hospital staff sometimes empathized with overburdened caregivers.

Self-abuse and neglect. Self-neglect is actually the most often substantiated allegation of maltreatment according to APS reports (Teaster et al. 2006). However, not all states include it as an official form of elder mistreatment. Older American's Act amendments of 2006 define "self-neglect" as an adult's inability, due to physical or mental impairment or diminished capacity, to perform essential self-care tasks such as obtaining food, clothing, shelter, medical care, goods and services to maintain health, mental health or safety, and manage one's own financial affairs (OAA 2006). Hoarders are sometimes included in the self-neglect category, as they have a compulsion to collect animals and/or save items to the point of excess – often creating dangers to health and fire safety. Reported elder self-neglect and abuse are significantly associated with an increased risk of mortality (Dong et al. 2009).

Self-abuse can include intentionally using too much medication, or perhaps even suicide – which obviously causes harm to older victims. It is important to realize that suicidal self-abusers can also be harmful to other people. Murder suicides involve both fatal domestic violence and suicide, and Malphurs and Cohen (2002) argue these events are disproportionately perpetrated in later life and increasing over time. Sometimes the primary motivation of the perpetrator is homicide and in other instances it is suicide, where the suicidal person makes a decision to commit homicide first. IPHS is among the most tragic forms of family violence, and in later life there is more often primarily suicidal motive, when compared to younger age categories (Salari 2007, 2009).

Research Methodology and Challenges to Estimates of Prevalence

Research methodology in the area of interpersonal violence has been challenging because detection in caregiver and family relationships is elusive and known cases may just represent the "tip of the iceberg" (Branch 2008). Underreporting is a persistent limitation plaguing studies of elder mistreatment. Rarely do perpetrators admit on a survey that they abuse vulnerable people. Current generations of elderly people are likely to keep victimization to themselves for a variety of reasons (CDC 2009). In addition, victims may feel protective of adult child or grandchild perpetrators and remain guarded about disclosing bad experiences or enduring stigma of "airing dirty laundry." For those living with abuse, humiliation, or fear of retaliation may factor into their lack of disclosure. Older victims remain hidden as they are less likely to seek shelter, make fewer calls to police, and rarely bring legal actions against perpetrators, compared to younger counterparts. Elderly individuals may also find it difficult to report or disclose abuse due to fear of potential institutionalization.

Elderly persons who are "shut in" may be more severely isolated and hidden from professionals in the medical and social services field who could detect and prevent their abuse and neglect. In addition, mental and physical disabilities may intensify the vulnerability in later life. It is possible for older persons to be mistreated without cognitive awareness of the situation. On the other hand, Steinmetz (2005) describes another problem where some elderly persons perceive that they are abused, but instead are actually suffering from paranoia. Cognitive limitations often create difficulty securing informed consent, thus making a high-risk population a challenge to study. Some researchers find it troubling that much of the work in this area depends on reports from health and human service professionals who work with vulnerable populations, rather than victims and families themselves (Pillemer 2005). These ethical issues have implications for Institutional Review Boards as well as potential funding sources (Bonnie and Wallace 2003; McNamee and Murphy 2006).

One of the most persistent difficulties surrounding elder mistreatment research has to do with the definition of mistreatment, including who can be a perpetrator (family, institution, strangers, self). What is the age range for victims (disabled adults 18–64, elders 55 and over, 60+, 65)? What type of abuse or neglect is counted in the prevalence report (physical, psychological, sexual, neglect, abandonment, exploitation)? Prevalence studies have varied widely in these definitions – making a definitive estimate nearly impossible. Even from a policy and practice perspective, there are state-by-state differences in definitions and responses to elder abuse (Teaster 2004). State units have taken a lead in detection, response, and enforcement, but confusion exists when comparing patterns in reporting and outcome (Bonnie and Wallace 2003). Federal attention has mandated reporting and provided some consistent definitions, but not everyone is satisfied with the standards developed. In addition, APS reports represent only the most severe forms of mistreatment reported to authorities, and actual cases are undercounted (Steinmetz 2005).

Prevalence estimates vary based on the types of mistreatment included and the method of measurement. Wolf's (2000) review article estimated between 4 and 6% of elders experienced domestic elder abuse, neglect, and financial exploitation. The National Research Council estimated those 65 and over who had suffered mistreatment by a caregiver or close contact between 0.5 and 10% (Bonnie and Wallace 2003). The most common ranges are measured between 4 and 10% (Costa 1993).

Pillemer and Finkelhor's (1988) widely cited stratified probability sample of community elders aged 65 and older in the Boston Metropolitan area ($n = 2{,}000$) suggested an overall prevalence of 3.2%, but the study was limited to physical (conflict tactics scale), psychological (insulted, sworn at, or threatened), and neglect (deprivation of assistance) types but did not include others such as sexual abuse, financial exploitation, self-abuse, and neglect. Adding those, the estimate is expected to be higher.

A recent random digit dialing nationally representative phone survey of 5,777 older adults found the following past year prevalence of mistreatment: physical 1.6%, emotional 4.6%, sexual 0.6%, potential neglect 5.1%, and current financial exploitation by family 5.2%. Lifetime financial exploitation by a nonfamily member was estimated at 6.5%. Proxy reports were not helpful to identify mistreatment (except for financial exploitation), and they call for alternative methods to capture the abuse rates of those with severe cognitive and physical disability (Acierno et al. 2009).

In another recent study, the National Social Life, Health, and Aging Project (NSHAP) interviewed 3,005 community-dwelling, cognitively aware individuals aged 57–85 in 2004. Respondents were asked directly if there was anyone who insults them or puts them down (verbal), anyone who takes their money or belongings without their permission or prevents getting when they ask (financial), or anyone who hits, kicks, slaps, or throws things at them (physical mistreatment). Findings suggest that verbal mistreatment was the most common form reported (9%). The verbal abuser was most often someone other than an immediate family member (56%), a spouse or intimate partner (26%), or last, an adult child (15%). For financial mistreatment, the highest reported category of offender was someone other than a family member (56%), followed by an adult child, and – rarely – a romantic partner or spouse. Physical mistreatment was the least-frequently reported type (1%). Of those who reported physical abuse, nonrelatives were most often perpetrators (55.6%), followed by an adult child (24.8%), and lastly, an intimate partner or spouse (19.6%).

Interestingly, reports of financial mistreatment were higher among African-Americans compared to their white counterparts, but Latinos reported less financial and verbal mistreatment than others. Running counter to the "vulnerability" hypothesis that suggests more frailty would lead to greater victimization, these authors found respondents in their late 50s and 60s reported higher rates of elder mistreatment than did older persons. Results must be taken in context, as older persons over the age of 85, those in nursing homes, and the cognitively impaired were not included in this study (Laumann et al. 2008).

Outcomes of elder maltreatment can range from psychological distress (PTSD) to loss of life. The CDC points out that physical injury from abuse can lead to premature death or make existing health problems worse (CDC 2009). Several researchers have pointed to the mortality risk of elder mistreatment. Lachs et al. (1998) reported results of a 9-year observational cohort study of 2,812 community dwelling older persons published in the *Journal of the American Medical Association*. They found risk of death among reported and corroborated elder abuse (odds ratio 3.1) and self-neglect (odds ratio 1.7) led to shorter survival after adjusting for demographics, disease, social networks, cognitive status, and depressive disorders. Dong and colleagues 2009 found self-neglect had an alarming mortality risk within the first year after the report. Elder abuse reports were associated with a 40% increased risk of overall mortality – including cardiovascular mortality.

Medical examiners rarely consider elder mistreatment as a cause of death because it is difficult to assess. There is a call for greater training of forensic specialists in the area of geriatric medicine (Dyer et al. 2008) to recognize abuse and neglect – as it differs from accidents, medication side effects, disease, or results of aging. Elder abuse forensics is estimated to be about where child abuse detection was three decades ago and IPV was 10–15 years ago (Dyer et al. 2008). In one extreme instance, dozens of victims in several institutions were killed by a nurse who injected them with lethal doses of medication, but none of these cases were recognized or reported by medical examiners. Attitudes about nursing home deaths among a focus group of forensic specialists indicated disturbing ageism, as nursing home residents were considered a "waste of their time ... because most ... would die eventually anyway" (McNamee and Murphy 2006).

The most severe form of family violence involves multiple fatalities at once – the commonly known murder-suicide or IPHS. Patterns of IPHS in later life show a primary motivation toward suicide for the older man, who makes a decision to simultaneously kill his wife or partner (who is usually unaware of her fate). The consensual suicide pact is very rare (1% in middle aged couples and 4% of elderly). The CDC definition of "mercy killing" is also typically not the motive, as consent on the part of the terminal or hopeless patient is required (see Salari 2007). As a researcher, I have found that some of the mystery of studying family violence is resolved in these cases. Incidents are countable; we know that fatal domestic violence and self-abuse have happened on the scene. There is no potential harm to human subjects – and it does not depend on the ability of the victim to answer questions with credible information. There are variations that allow the researcher to "make distinctions" (See Johnson 2008) between perpetrator/victim relationships. So, for example, some IPHS perpetrators seem focused on homicidal acts, with intimate terrorist qualities and others who are primarily suicidal, with a decision made to kill their partner or ex-partner before they suicide. These distinctions provide us with a chance for more targeted prevention efforts (Salari 2007). From a societal perspective, IPHS incidents seem to take on different patterns across cultures. For example, those countries with restrictive gun control laws have fewer IPHS incidents. In the events, other methods are used, such as carbon monoxide poisoning – but the incident is more likely a suicide pact, rather than one where a partner is caught off guard with a fatal gunshot.

Modern Developments in Research and Policy

What is the prototype for elder mistreatment and why does it matter for policy? Wolf (2000) explained that the typical image of a functionally impaired older person with a well-intentioned, but overburdened, caregiver (similar to the child abuse model) influenced legislation in more than three quarters of the states. Reduction of caregiver burden and stress was a major focus of many of the interventions, and some have argued, older persons were often infantilized, including potential civil rights infringements (Pillemer 2005). Once empirical studies included comparison groups, Wolf and others became convinced the dependency child-abuse model was not appropriate and needed to be modified for the specific experiences of older persons. The "family stress theory" (Wallace 2005) or "situational model" were not supported by the evidence, and elder abuse seemed more similar to domestic violence models than to child abuse. Dependency had a different relationship than expected – with dependent, mentally ill, and/or substance-addicted family abusing older persons (Pillemer 2005; Wolf 2000).

The 1990s were significant because large datasets were used to quell the earlier emphasis on "sandwich generation" stress. Logan and Spitze's (1996) well-known work determined it was rare for the middle generation to be in the position of caring for both elders and dependent children at the same time. Until age 75, older persons give significantly more downward support than they receive and then it equalizes. Modern researchers tend to acknowledge that dependency and stress may be a strong contributing factor – but not the only family elder-abuse dynamic. Steinmetz identified the *perception* of the caregiver's burden was more important than actual level of dependency. Gaining or maintaining control was a primary motivation for this type of abuse perpetrator (Steinmetz 2005).

During the 1990s, technology changed the landscape of safeguarding personal information and property. Widespread use of the internet increased identity theft and financial scams on vulnerable persons who became targets of sophisticated financial schemes by both kin and nonkin perpetrators.

Legislatively, the Americans with Disabilities Act helped bring freedoms and open public places to those with assistive devices. In 1992, Congress passed and funded the Older American's Act Title VII Chap. 3 for prevention of abuse, neglect, and exploitation. Vulnerable Elder Rights Protection included provisions for state legal assistance and Long-Term Care Ombudsmen programs. This legislation was aimed at curbing institutional mistreatment of elderly and disabled persons with assistance and advocacy.

With the murder of Nicole Brown Simpson, public attention was focused on IPV. IPV research made theoretical progress, as Johnson and colleagues identified distinctions among types of victims and perpetrators – with the recognition that not all were the same or deserved the same response (see Johnson 2008). The 1994 Violence against Women Act focused on the needs of violence survivors and represented a step toward public recognition that social structure, funding for shelters, and training of law enforcement and prosecutors play an important role in violence prevention and treatment. Initially, the new legislation focused mostly on the perspectives of younger victims.

Since 2000, family violence, including elder mistreatment, has been viewed as a public health and human rights issue – evident in work from the World Health Organization (WHO), Centers for Disease Control and Prevention (CDC), and *Journal of the American Medical Association* (special issue 5 August 2009). Progress has been made regarding a widely cited definition of elder mistreatment, as well as more inclusive estimates from APS. Recommendations of the National Research Council urged scholars to focus on more studies, and included methodological and logistical suggestions needed to create a solid research base addressing elder mistreatment (Bonnie and Wallace 2003).

The 2000 and 2005 reauthorizations of VAWA included attention to the plight of the older and disabled woman. There was a greater recognition of the special needs of these violence survivors, who might require an alternative physical environment and greater sensitivity training among police personnel and courts. As Baby Boomers age, domestic violence shelters with provisions for an aging population will be required. It is unlikely that an older victim would choose to escape abuse if she would be placed in a dormitory setting with a number of young children, staircases, and other barriers to good quality of life. The VAWA reauthorization act of 2000 provided $5 million per year to STOP grants and training programs encouraging law enforcement, prosecutors, and court officers to recognize, address, investigate, and prosecute instances of elder abuse, neglect, exploitation, and violence against individuals with disabilities, including domestic violence and sexual assault against older or disabled individuals. By 2005 reauthorization, there was an expansion to $10 million per year FY 2007–FY 2011 for domestic-violence programs that deal with domestic and dating violence, sexual assault, and stalking against victims age 50 or older (National Coalition against Domestic Violence 2006).

During the decade since 2000, there has also been a large emphasis on "Victim's Rights," which is reflected in the VAWA reauthorizations. There is new concern over how older and disabled survivors are treated during the process of the report, sheltering, protecting, and prosecuting. What happens to victim's rights in the process of criminally punishing perpetrators? Unfortunately, much of the problem of family violence policy is that it focuses on sheltering survivors, but not getting to the root of the problem … stopping the perpetrators. Abusers are often smart, "slippery," and manipulative, so professionals often advise against couple's counseling, where the therapist may be drawn into the dynamic. Individual counseling for each party is recommended instead (Weiss 2003). As a Victim's Advocate, I often saw a victim escape, and another woman would take her place as the focus of a specific abuser. Perpetrator programs were developed for abusers to talk about their perspectives, learn anger management skills, and attempt to reform. In small doses, this can be effective if they take responsibility for their actions and commit to change. However, once the judicial system began assigning perpetrators to these programs in mass numbers they often did not accept responsibility and the programs were overwhelmed and ineffective (U.S. Department of Justice 2003).

The majority of states do not maintain an "abuse registry" to track alleged elder abuse perpetrators (Teaster et al. 2006). Savvy perpetrators are free to target new victims and move around from state to state to stay ahead of law enforcement. VAWA enacted laws for computerized systems for "full faith and credit" that assure protective orders are enforced across state lines. It seems that similar programs could be applied to combat perpetration of elder mistreatment across the nation.

The 2006 amendments to the Older American's Act Title VII include services and allotments for detection, assessment, intervention, investigation, and response to elder abuse. Responses include public education/outreach about financial exploitation, elder abuse shelters and safe havens, multidisciplinary elder-justice activities, services to underserved populations, and development of accountability measures. Several definitions were provided, including "elder justice" which includes efforts to prevent, detect, treat, intervene in, and respond to elder abuse, neglect, and exploitation, and to protect older individuals with diminished capacity while maximizing their autonomy and self-determination. An older individual has recognized rights, including right to be free from abuse, neglect, and exploitation (OAA 2006).

APS has undergone a transformation over time, to more adequately respect the right of older persons to shape their own lives. Those who are being abused or neglected by others may refuse services and intervention. There is a greater concern that disrupting the family system, even if it is abusive, may cause more harm to the older victim than good. If intervention is undertaken, it is often considered the state's responsibility to improve the situation for the elderly victim and many service providers agree that it may be difficult to accomplish with limited resources. Currently, social service agencies that focus on protection work to investigate the abuse and most often provide referrals to helpful resources in the community. Refusal of services to remedy hoarding and medical neglect creates controversy. Of reported APS cases of abuse, 11% of vulnerable adults refused services (Teaster 2004).

APS is currently limited in its ability to function effectively, due to the economic downturn and a lack of adequate funds to keep up with the demands created by mandatory reporting laws (Fulmer et al. 2002). Certain types of elder mistreatment have been exacerbated by the recent widespread economic instability. Older persons were already the age group most at risk of material abuse, but the crisis conditions seem to have increased their vulnerability. In order to prevent widespread losses, the banking industry has taken an interest in monitoring the spending and withdrawal patterns of older account holders to identify and combat financial abuse – include the freezing of assets when exploitation is suspected. Law scholars have pointed to abuse potentials in guardian and conservator arrangements with the acknowledgment that some vulnerable adults are unable to make decisions in their own best interest.

Elder abuse is defined, counted, and treated differently depending on the specific laws of each state. According to a comparison done between 1986 and 2000, reports to APS agencies of elder/adult abuse increased from 117,000 to 472,813 (Teaster 2004). By 2003, 50 states (plus Guam and Washington, DC) reported a 19.7% increase with a total of 565,747 reports. Despite these substantial increases in reports, states have faced budget shortfalls in the difficult economic conditions, which have often led to program reductions (Teaster et al. 2006). This trend is dangerous, considering that many APS agencies are performing on a bare bones budget and must refer clients out to community services.

Discussion and Future Agenda

Family violence has been recognized as a major cause of injury and a serious threat to public health (CDC 2009; WHO 2010). The consequences of violence and neglect for the elderly are serious because of the greater potential for damage to skin, organs, and bones, as well as the longer recovery time, compared to younger victims (WHO 2010). Elder mistreatment is severely under-reported, and attempts at a more accurate count are thwarted by methodological issues related to measurement, disclosure, cognitive ability, consent and proxy status.

Elder mistreatment literature has developed over the past 30 years, but many of the early review articles focused on the methodological dilemmas, variation in definitions, and the quest for an accurate count, with little attention to theory construction. Some aspects of the phenomenon that are known and a great deal that is still unclear. The ecological iceberg conceptualization recognizes that there are micro, meso, exo, and macro systems that can be measured and that an entire mass that exists below the surface that eludes detection. It is unclear what size or shape represents the lower level of the iceberg, but multiple levels of influence exist in those layers. The theoretical perspectives to date tend to dwell on the micro-level role of individuals and family dynamics, with less attention focused on the larger community, the economy, and the social structure.

A great deal of mystery surrounds the incidence and nature of elder abuse and neglect at these macro levels. Community and cultural beliefs about the public vs. private nature of intimate partnerships and family relationships play a role in the social construction of what is perceived as abuse and what is disclosed to outsiders. What is the role of community intervention? What happens when there is no access to shelters? If laws exist to prohibit mistreatment, what happens with a lack of enforcement? How do vulnerable elderly persons fare with societal oppression against women or minorities, or in a poor economy? When public services for victims are cut back, what is the consequence to public health and abuse prevention?

Social construction of elder mistreatment varies across conditions and cultures. These issues are not only relevant to the United States but have vast global significance. Older people in the developing world have great vulnerability, especially when they are impoverished. According to the WHO (2010) the risk of property seizure and abandonment of older widowed women is greater in countries with severe gender oppression. Other conditions of hardship may exist where healthcare, food, and medicine are rationed to the old. For example, in the recent 2010 earthquake in Haiti, nursing home residents were originally passed over with food and water supplies. The title of the Associated Press article said it best "Abandoned, 85 elderly Haitians await death: nursing home residents, barely a mile from the center of the aid effort, left waiting." Several of the elderly residents of this nursing home died of dehydration and lack of food, and those who survived described being visibly neglected and robbed of their money and possessions as they lay injured in the street (De Montesquiou 2010). Economic and social conditions play a role in whether mistreatment is detected, perpetrators punished, and victims sheltered. The macrosystem is the stage where social problems are identified, policy is made (or neglected), and the rules are then enforced – either effectively or ineffectively.

In the United States, substantiated reports to APS seem to give us a good idea of the nature of abuse, but there is an enormous rate of under-reporting. It is important to recall that the APS reports represent only the most severe cases, and in reality there is more of a continuum. Elder abuse cuts across SES levels, but access to resources might help an individual escape harm without the use of public services. Shelters house those with no other options for escape, and the residents do not represent a cross-section of the abused population. Few older victims use shelter services, possibly because of cohort differences in help-seeking behaviors. Recent policy changes suggest more attention is needed to make services appropriate for the needs of elderly clientele.

Family violence at any age is a human rights issue and it significantly reduces the quality of life. Whatever the rate, it is clear that we need an understanding of why older persons are targeted and how family members and other abusers isolate victims. Advocacy movements and public awareness campaigns are important factors for change. Several recent studies have helped to ease paralysis in the field, and provide a greater understanding of prevalence and undercount, so that a new focus can be placed on federal funding, theoretical advancement, and policy recommendations. Future cohorts including the Baby Boom Generation will have a differing social construction of elder mistreatment and the societal responses necessary for prevention. Based on previous activism exhibited by this cohort, one might expect greater attention to transparency and victim advocacy.

References

Acierno, Ronald, Melba Hernandez-Tejada, Wendy Muzzy, and Kenneth Steve. 2009. *The National Elder Mistreatment Study.* Washington, D.C.: National Institute of Justice.

Bonnie, Richard J. and Robert B. Wallace, eds. 2003. *Elder Mistreatment: Abuse, Neglect, and Exploitation in an Aging America.* Washington D.C.: National Academics Press.

Bonomi, Amy E., Melissa L. Anderson, Robert J. Reid, David Carrell, Paul A. Fishman, Frederick P. Rivara, and Robert S. Thompson. 2007. Intimate Partner Violence in Older Women. *The Gerontologist* 47(1):34–41.

Burgess, Ann and Nancy Hanrahan. 2006. *Identifying Forensic Markers in Elder Sexual Abuse.* National Institute of Justice, Washington, D.C.

Branch, Laurence. 2008. *Florida Elder Abuse Study.* Washington, D.C.: National Institute of Justice.

Centers for Disease Control and Prevention. 2006. *Understanding Intimate Partner Violence. Factsheet 2006.* Retrieved (http://www.cdc.gov/violenceprevention/pdf/IPV_factsheet-a.pdf).

———. 2009. *Understanding Elder Maltreatment Factsheet 2009.* Retrieved (www.cdc.gov/violenceprevention/pdf).

Costa, A. J. 1993. Elder Abuse. *Primary Care* 20(2):375–89.

Dahlberg Linda L. and Etienne G. Krug. 2002. Violence – a Global Public Health Problem. Pp. 1–56 in *World Report on Violence and Health*, edited by Etienne Krug, Linda L. Dahlberg, James A. Mercy, Anthony B. Zwi, and Rafael Lozano. Geneva, Switzerland: World Health Organization.

De Montesquiou, Alfred. 2010. Abandoned, 85 Elderly Haitians Await Death. Associated Press January 17, 2010.

Dong, XinQi., Melissa Simon, Carlos Mendes de Leon, Terry Fulmer, Todd Beck, Liesi Hebert, Carmel Dyer, Gregory Paveza and Dennis Evans. 2009. Elder Self-Neglect and Abuse and Mortality Risk in a Community Dwelling Population. *Journal of American Medical Association* 302(5):517–26.

Dyer, Carmel B., Luis Sanchez, Lucia Kim, Jason Burnett, Stacy Mitchell, Barbara Reilley, Sabrina Pickens, and Michelle Mehta. 2008. *Factors that Impact the Determination by Medical Examiners of Elder Mistreatment as a Cause of Death in Older People.* Washington, D.C.: National Institute of Justice.

Fulmer, Terry, Gregory Paveza, and Lisa Guadagno. 2002. Elder Abuse and Neglect: Policy Issues for Two Very Different Problems. *Public Policy and Aging Report* 12(2):15–18.

Fulmer, Terry, Gregory Paveza, Carla VandeWeerd, Sharon Fairchild, Lisa Guadagno, Marguarette Bolton-Blatt, and Robert Norman. 2005. Dyadic Vulnerability and Risk Profiling for Elder Neglect. *The Gerontologist* 45(4):525–34.

Gelles, R. J. & Straus, M. A. (1988). Intimate Violence. New York: Simon and Schuster.

Johnson, Michael P. 2008. A Typology of Domestic Violence: Intimate Terrorism, Violent Resistance, Situational Couple Violence, and Mutual Violent Control. Lebanon, NH: Northeastern University Press.

Lachs, Mark S., Christianna S. Williams, Shelley O'Brien, Karl A. Pillemer, and Mary E. Charlson. 1998. The Mortality of Elder Mistreatment. *Journal of the American Medical Association* 280(5):428–32.

Laumann, Edward O., Sara A. Leitsch, and Linda J. Waite. 2008. Elder Mistreatment in the United States: Prevalence Estimates From a Nationally Representative Study. *Journal of Gerontology: Social Sciences* 63B(4):S248–54.

Levinson, David. 1989. *Family Violence in Cross Cultural Perspective.* Newbury Park CA. Sage.

Logan, John and Glenna Spitze. 1996. *Family Ties: Enduring Relations between Parents and Their Adult Children.* Philadelphia: Temple University Press.

McNamee, Catherine C., and Mary B. Murphy. 2006. *Elder Abuse in the United States* (Journal No. 255). Washington, D.C.: National Institute of Justice.

Malphurs, Julie E. and Donna Cohen. 2002. A Newspaper Surveillance Study of Homicide-Suicide in the United States. *American Journal of Forensic Medicine and Pathology* 23(2):142–48.

Maxwell Elanor K. and Robert J. Maxwell. 1992. Insults to the Body Civil: Mistreatment of Elderly in Two Plains Indian Tribes. *Journal of Cross-Cultural Gerontology* 7:3–22.

Metlife Mature Market Institute Study. 2009. *Broken Trust: Elders, Family, and Finances. March, 2009*. Retrieved October 27, 2009 (http://www.maturemarketinstitute.com).
National Coalition against Domestic Violence. 2006. *Comparison of VAWA 1994, VAWA 2000, and VAWA 2005 Reauthorization Bill*. Washington, D.C.
Newsweek. 1991. A Dumping Ground for Granny: Weary Families Drop Her in the Emergency Room. December 23, 1991.
Older American's Act Amendments. 2006. Administration on Aging, Department of Health and Human Services. Retrieved (http://www.aoa.gov/aoaroot/aoa_programs/OAA/oaa_change.aspx).
Pillemer, Karl. 1986. Risk Factors in Elder Abuse: Results from a Case-Control Study. Pp. 236–263 in *Elder Abuse: Conflict in the Family*, edited by K. Pillemer and R. S. Wolf. Dover, MA: Auburn House.
———. 2005. Elder Abuse is Caused by the Deviance and Dependence of Abusive Caregivers. Pp. 207–20 in *Current Controversies on Family Violence,* 2nd ed., edited by D. Loseke, R. Donileen, R. J. Gelles, and M. M. Cavanaugh. Newbury Park, CA: Sage.
Pillemer, Karl and David Finkelhor. 1988. The Prevalence of Elder Abuse: A Random Sample Survey. *The Gerontologist* 28:51–7.
———. 1989. Causes of Elder Abuse: Caregiver Stress vs. Problem Relatives. *American Journal of Orthopsychiatry* 59:179–87.
Pillemer, Karl and Rosalie S. Wolf. 1986. *Elder Abuse: Conflict in the Family*. Dover, MA: Auburn House.
Ramnarace, Cynthia. 2009. How Can We Protect Against Elder Abuse During a Down Economy? *AARP Bulletin,* July 17.
Salari, Sonia. 2002. Intergenerational Partnerships in Aging Services: Importance of Age-Appropriate Environments and Behaviors. *The Gerontologist* 42(3):321–33.
———. 2006. Infantilization as Elder Maltreatment: Evidence from 5 Adult Day Centers. *Journal of Elder Abuse and Neglect* 17(4):53–91.
———. 2007. Patterns of Intimate Partner Homicide Suicide in Later Life: Strategies for Prevention. *Clinical Interventions in Aging*, volume 3. Memphis, TN: Dove Press.
———. 2009. Neighborhood and Community Trauma: Responses to Mid and Later Life Intimate Partner Homicide Suicide. Paper Presentation Gerontological Society of America, Atlanta GA, Nov 21.
Salari, Sonia and Bret Baldwin. 2002. Verbal, Physical, and Injurious Aggression among Intimate Couples over Time. *Journal of Family Issues* 23(4):523–50.
Steinmetz, Suzanne. 2005. Elder Abuse is Caused by the Perception of Stress Associated with Providing Care. Pp. 191–205 in *Current Controversies on Family Violence*, 2nd ed., edited by D. R. Loseke, R. J. Gelles, and M. M. Cavanaugh. Newbury Park, CA: Sage.
Teaster, Pamela. 2004. *A Response to the Abuse of Vulnerable Adults: The 2000 Survey of State Adult Protective Services*. Report to the National Center on Elder Abuse, Administration on Aging, Washington, D.C.
Teaster, Pamela B., Tyler D. Dugar, Joanne M. Otto, Marta S. Mendiondo, Erin L. Abner, and Kara A. Cecil. 2006. *The 2004 Survey of State Adult Protective Services: Abuse of Adults 60 Years of Age and Older*. Report to the National Center on Elder Abuse, Administration on Aging, Washington, D.C.
U.S. Department of Justice. 2003. *Do Batterer Intervention Programs Work? Two Studies*. Washington, D.C.: U.S. Department of Justice.
Walker, Lenore E.A. 2009. *The Battered Woman Syndrome*, 3rd ed. New York: Springer.
Wallace, Harvey. 2005. *Elder Abuse. Family Violence: Legal, Medical and Social Perspectives*, 3rd ed. Boston, MA: Allyn and Bacon.
Weiss, Elaine. 2003. *Family and Friends' Guide to Domestic Violence: How to Listen, Talk, and Take Action When Someone You Care About is Being Abused*. Volcano, CA: Volcano Press.
Wolf, Rosalie. 2000. The Nature and Scope of Elder Abuse: Changes in Perspective and Response over the Past 25 Years. *Generations* 24(2):6–12.
Wolf, Rosalie S. and Karl A. Pillemer. 1989. *Helping Elderly Victims: The Reality of Elder Abuse*. New York: Columbia University Press.
World Health Organization. 2010. *Abuse of the Elderly. Chapter 5. World Health Report on Violence and Health*. Retrieved on February 17, 2010 (http://www.who.int/violence_injury_prevention/violence/global_campaign/en/chap5.pdf).

Chapter 27
Crime, the Law, and Aging

Duane Matcha

While much has been made for older American's decreasing rates of poverty, improved health status, continued connection to the worlds of work and community service as well as continued strong family relations, their relationship to the criminal justice system as victims and perpetrators of crime remains less well known. Although research explores facets of this relationship, such efforts generally draw little attention because of older American's limited involvement as either perpetrators or victims of crime.

What we know is that the relationship between older Americans and the criminal justice system is multifaceted. For example, while much has been made of the difference between the reality and perception of criminal victimization of older Americans. Similarly, older Americans are victims of various forms of abuse at the hands of family members or care providers that are occurring within a variety of environmental locations ranging from independent living arrangements (Fisher and Regan 2006; Lachs et al. 2006) to institutional settings such as nursing homes, assisted living facilities (Wood and Stephens 2003), and prisons (Stojkovic 2007). Conversely, older Americans are also perpetrators of crime. Although the United States in not experiencing a "geriatric crime wave," older Americans do commit crimes and are being incarcerated. According to Feldmeyer and Steffensmeier, arrest rates of older Americans between 1980 and 2004 increased in the categories of simple assault, drug law violations, and OBT (Other But Traffic) (Feldmeyer and Steffensmeier 2007). In other categories, arrest rates have remained stable or declined during that period of time. Nevertheless, state and federal prisons continue to experience an increasing percentage of older inmates. The increasing number of elderly inmates and their attendant consequences such as higher costs of incarceration and extensive health problems raises serious questions regarding the future of the prison system.

What are the implications of crime by and against older Americans? It is within this context that issues such as age-related treatment within the criminal justice system as well as long-term cost of incarceration exist. If older Americans are identified as a "special category" of criminal offenders, what are the consequences? Should age-segregated prisons be constructed to ensure the safety of older-aged prisoners? Additionally, the number of older women in prison raises issues of specific care as well as their marginalized position within the larger prison population (Krabill and Aday 2005).

Victimization of older Americans elicits similar policy questions. Here, questions of the social and physical environment within which the elderly lives and functions become central; creating a safer living environment, monitoring impediments to mobility, and improving access to services necessary for independent living. More specifically, does knowing the risk factors for victimization among older Americans provide sufficient reason for the establishment of age-specific services to

D. Matcha (✉)
Professor of Sociology, Siena College, Loudonville, NY
e-mail: matcha@siena.edu

R.A. Settersten, Jr. and J.L. Angel (eds.), *Handbook of Sociology of Aging*, Handbooks of Sociology and Social Research, DOI 10.1007/978-1-4419-7374-0_27, © Springer Science+Business Media, LLC 2011

limit their impact? What, if any, criminal justice policies are essential if older Americans are more likely to die or experience medical problems as a result of victimization?

Such policy questions underscore the impact of age as a social category in the criminal justice system. Although the sociology of aging has been interested in crime relative to the aging population for decades, more recent efforts have attempted to diversify the focus. While interest was initially on the victimization of older Americans and their fear of crime, recent research has broadened its efforts to disaggregate older Americans and explores the impact of race, sex, and social class on victimization as well as reasons for perpetrating crime. More recent research within the sociology of aging also addresses multilayered issues such as the social, medical, ethical, and economic consequences of incarceration of older Americans. These efforts are forcing a deeper thinking of the implications of the aging process and its relation to criminal activity.

While the role of older Americans in the criminal justice system is understood as being victim or perpetrator, their role within the broader American legal system is less clear. This lack of certainty is grounded in the ongoing debate over what obligations society defines as necessary toward older Americans. Given that demographic projections indicate continued expansion of the elderly population throughout the first half of the twenty-first century, the ramifications of legal uncertainty regarding what rights older Americans should have is problematic.

This chapter begins with an historical overview of aging and the law, then transitions to coverage of aging and crime. The application of an historical context throughout provides the means by which changes and challenges can best be understood. In the second section, elderly victimization focuses on the ongoing debate over perception vs. reality of crime against the elderly as well as the extent and consequences of elderly abuse. In the third section, the role of the elderly as the perpetrator of crime and its consequences are addressed. Finally, the last section examines how crime and the law have framed the lives of older Americans to the present and prognostications for the future.

Aging and the Law

The law represents the foundation upon which social contracts are created and enforced. As Cohen states: "The law affecting the elderly is a reflection of how society regards older people, the relationships that exist between and among generations, the views people have about growing old, and the passage of time in another age" (Cohen 1990:184). Questions concerning the role of older Americans relative to the legal system offer insight into the extent that chronological age impacts the legal process. As Eglit notes, "age bias is obviously present in the interactions that, in the aggregate, make up the American legal system" (Eglit 2004:5). The concern of the legal system is the balance between creating public policy that endows a specific population with necessary services and public policy based on equality across the population.

Some 3 decades ago, Cain (1976) addressed the relationship between age and the law. Beginning with definitions of age (chronological, functional, and social) as well as the emergence of aging within the field of law, he focused much attention on the post-World War II period in which "laws have contributed to the creation of a distinctive status for the elderly" (Cain 1976:356–7). Although the focus at the time was that of mandatory retirement, he argued that the emergence of laws that provide the legal basis for programs designed for the elderly had inadvertently reinforced their dependence on those who interpret and administer the programs. As a result, he suggested that "a major challenge facing gerontologists is to move beyond advocacy for particular measures to a perspective which calls for consistency and coherence of status, for programs which can be comprehended and managed more fully by the elderly themselves" (Cain 1976:366).

Similarly, Kapp examined the relationship between aging and the law by assessing the type of law and its relevance to an aging population. The importance of this approach is found in his exploration of the influence of law on the elderly, particularly in terms of government intervention. The fundamental question addressed by Kapp is whether the elderly enjoy equality or special treatment as a result of the law. In terms of equality, the Americans with Disabilities Act and the Age Discrimination in Employment Act are examples in which older Americans are treated the same as other age groups. On the other hand, laws based on special treatment of the elderly include Social Security, Medicare, and government housing for the elderly (Kapp 1996, 2006).

While the debate continues over whether the elderly receive equality or special treatment under the law, it is important to understand the changes that have occurred in the law over the past 30 years. In the mid-1970s, equality under the law did not exist for older Americans. Mandatory retirement rules forbade older Americans from remaining on the job, regardless of their qualifications as chronological age was the primary characteristic defining when a person could or could not work. In the first decade of the twenty-first century, mandatory retirement rules are virtually nonexistent although age discrimination remains a constant struggle for some mature workers. Similarly, equality under the law is assumed under the Americans with Disability Act. However, if application of the law within the criminal justice system is generally viewed as an accommodation of older, rather than younger, criminals, its application becomes viewed as special treatment of the elderly.

As a result of the need for accommodation of older criminals and victims within the criminal justice system, the legal debate is primarily over special treatment under the law on the one hand, and age bias on the other. In terms of special treatment, legal questions regarding the right of the elderly to receive a disproportionate share of government funding and continued protection under the law in entitlement programs such as Medicare and Social Security have become of primary importance. These programs return us to the heart of the argument over social contracts that involve intergenerational transfers of funds from younger to older generations. The long-term issue is whether American society is willing to increase the transfer of funds from younger generations to older as well as the youngest generation that is truly important. Although the elderly population will increase as a percentage of the total population, it is also true that the youngest generation will decrease as a percentage of the total population. In 2000, the total dependency ratio of 51 was comprised of a child ratio of 33 and an elderly ratio of 19. By 2050, the total dependency ratio is expected to reach 61 and be comprised of a child ratio of 28 and an elderly ratio of 33 (United Nations 2004). Furthermore, the argument that programs such as Medicare advances the health of the elderly at the expense of younger generations is evidence that the problem lies not with special treatment of the elderly, but rather with social policy unwilling to provide commensurate health coverage to others generations of Americans.

Relative to crime and the law, the creation of a separate justice system for older Americans does not garner overwhelming support from the public (Preston 2003). However, Rothman and Dunlop suggest that as legal issues of a growing older-aged population increase, "it is important to understand how courts, as an institution, judges, and court administrators perceive this development and what policies or practices they should be developing to address it" (Rothman and Dunlop 2006:32). Addressing the role of the court, Manning, Carroll, and Carp report that "older judges may show greater affinity to plaintiffs in age bias claims" (Manning et al. 2004:16).

Equal treatment of older Americans under the law is also problematic. As Preston and others report, efforts to require prison compliance with the ADA have met with some resistance. Although prison inmates regardless of age are covered under ADA regulations if they are disabled, older prison inmates may face additional hardship (Preston 2003). According to Morton, some prisons require inmates to pay for medical care visits. While the reason for inmates to participate in the cost of medical care may be understood as a method to control utilization rates, the result may have a deleterious impact on older inmates. The problems of copayment programs for older inmates include difficulty paying for services that they are more likely to be in need of. If they are unable to pay for medical services, they are less likely to seek treatment, thus exacerbating the medical condition (Morton 2001).

The question of special vs. equal treatment of older Americans under the law remains central to the discussion of aging, crime, and the law. While special treatment is generally decried as favoring older citizens over the needs of other age groups, equal treatment, as presented by Morton (2001), can result in bias against the elderly. Thus, questions of the impact of equal treatment are of similar importance as questions of special treatment as they relate to older Americans.

Victimization of the Elderly

This section examines the perceptions and realities of older Americans as a victim of criminal activity. The question of perceived victimization is followed by the changing nature of crime against older-aged Americans. Finally, an historical analysis of elder abuse offers evidence of continued physical, social, and financial trauma experienced by older Americans at the hands of friends, family, and strangers.

Perceptions and Realities of Victimization

Thirty years ago, older Americans identified fear of crime as their most important problem (Harris 1976). Research throughout the 1980s and 1990s in the United States has consistently reported that fear of crime among older Americans is not as pervasive as predicted earlier (Yin 1982; LaGrange and Ferraro 1987; Ferraro and LaGrange 1992; Ferraro 1995). Hale in an international review of the fear of crime offers the following assessment that presents a more complex portrait: "Fear of crime is influenced by a general sense of vulnerability, by signs of physical and social decay in the neighborhood and by lack of community cohesion or inter-group conflict" (Hale 1996:131).

In part, the problem between reality and perception of crime is the result of research efforts that are not always measuring what they purport to measure (Ferraro and LaGrange 1987). As Ferraro points out, concepts such as *fear* and *perceived risk* must be understood as identifying fundamentally different experiences (Ferraro 1995). As a result, Ferraro argues that "perceived risk is correlated with both official crime risk and fear; actors are more afraid when they sense a greater likelihood of potential criminal risk" (Ferraro 1995:120). Furthermore, as Yin points out, fear of crime is influenced by ones age, sex, and perceived health status; in other words, the most vulnerable (Yin 1985). The elderly are likely to know the perpetrator of the crime, more likely to defend themselves, and to report a crime. Furthermore, while we now know that initial research findings regarding elderly concerns of crime were inaccurate, questions remain. For example, Ferraro and LaGrange question why the public and mass media continue to believe that the elderly are fearful of crime when in fact they generally are not (Ferraro and LaGrange 1992). It is this effort to engage in a deeper understanding of the relationship between aging and criminal victimization that will move gerontologists in new research directions.

Although research has consistently demonstrated that older Americans are much less likely to be victimized, the perception continues to frame the relationship between aging Americans and crime. The reality of victimization by age is shown in Table 27.1. Statistics from the U.S. Department of Justice demonstrate that older Americans experienced 9.1 violent crimes per 1,000 persons in their age group in 1973. By comparison, younger Americans experienced violent crime rates that were 9 times greater. The differential in violent crime rates between older and younger Americans has continued throughout the remainder of the twentieth century and into the first decade of the twenty-first century. For example, in 1983, the violent crime rate against older Americans decreased to 5.9 per 1,000 persons aged 65 and above while rates among younger age Americans vacillated. By 1993,

Table 27.1 Violent victimization rates by age, 1973–2005 (violent crimes included are homicide, rape, robbery, and both simple and aggravated assault)

Violent crime rate per 1,000 persons in age group

	Age of Victim						
Year	12–15	16–19	20–24	25–34	35–49	50–64	65+
1973	81.8	81.7	87.6	52.4	38.8	17.2	9.1
1974	77.5	90.6	83.5	58.6	37.5	15.5	9.5
1975	80.3	85.7	80.9	59.5	36.9	17.8	8.3
1976	76.4	88.8	79.7	61.5	35.9	16.1	8.1
1977	83.0	90.2	86.2	63.5	35.8	16.8	8.0
1978	83.7	91.7	91.1	60.5	35.8	15.0	8.4
1979	78.5	93.4	98.4	66.3	38.2	13.6	6.2
1980	72.5	91.3	94.1	60.0	37.4	15.6	7.2
1981	86.0	90.7	93.7	65.8	41.6	17.3	8.3
1982	75.6	94.4	93.8	69.6	38.6	13.8	6.1
1983	75.4	86.3	82.0	62.2	36.5	11.9	5.9
1984	78.2	90.0	87.5	56.6	37.9	13.2	5.2
1985	79.6	89.4	82.0	56.5	35.6	13.0	4.8
1986	77.1	80.8	80.1	52.0	36.0	10.8	4.8
1987	87.2	92.4	85.5	51.9	34.7	11.4	5.2
1988	83.7	95.9	80.2	53.2	39.1	13.4	4.4
1989	92.5	98.2	78.8	52.8	37.3	10.5	4.2
1990	101.1	99.1	86.1	55.2	34.4	9.9	3.7
1991	94.5	122.6	103.6	54.3	37.2	12.5	4.0
1992	111.0	103.7	95.2	56.8	38.1	13.2	5.2
1993	115.5	114.2	91.6	56.9	42.5	15.2	5.9
1994	118.6	123.9	100.4	59.1	41.3	17.6	4.6
1995	113.1	106.6	85.8	58.5	35.7	12.9	6.4
1996	95.0	102.8	74.5	51.2	32.9	15.7	4.9
1997	87.9	96.3	68.0	47.0	32.3	14.6	4.4
1998	82.5	91.3	67.5	41.6	29.9	15.4	2.8
1999	74.4	77.5	68.7	36.4	25.3	14.4	3.8
2000	60.1	64.4	49.5	34.9	21.9	13.7	3.7
2001	55.1	55.9	44.9	29.4	23.0	9.5	3.2
2002	44.4	58.3	47.6	26.4	18.2	10.7	3.4
2003	51.6	53.1	43.5	26.5	18.6	10.3	2.0
2004	49.7	46.0	43.2	23.8	18.0	11.0	2.1
2005	44.0	44.3	47.1	23.7	17.6	11.4	2.4

Note: because of changes made to the victimization survey, data prior to 1992 are adjusted to make them comparable to data collected under the redesigned methodology. Estimates for 1993 and beyond are based on collection year while earlier estimates are based on data year. Because of changes in the methods used, these data differ from earlier versions. For additional information about the methods used, see *Criminal Victimization 2005*

Original source: rape, robbery, and assault data are from the National Crime Victimization Survey (NCVS). The homicide data are collected by the FBI's Uniform Crime Reports (CUR) (Supplementary Homicide Reports) from reports from law enforcement agencies

Source: U.S. Department of Justice 2009. Office of Justice Programs. Bureau of Justice Statistics

the violent crime rate against older Americans remained steady at 5.9 per 1,000 persons. However, violent crime rates among younger Americans increased significantly. Among 12–15-year-olds, there were 115.5 violent crimes per 1,000 persons in that age category. Similarly, teenagers aged 16–19 experienced 114.2 violent crimes per 1,000 persons in that age category. Ten years earlier, the respective rates were 75.4 and 86.3. By 2006, violent victimization rates declined for all age categories. Among older Americans, the rate of violent victimization declined to 2.4 per 1,000 persons in that age category. Similarly, the rates among younger citizens declined to 44.0 for those aged 12–15 and 44.3 for teenagers aged 16–19 (Bureau of Justice Statistics 2009).

While the elderly are less likely to be a victim of crime than other age groups, there are particular crimes such as burglary and being the target of con artists that older Americans are more likely to experience. Older Americans, particularly men, are also more likely to be robbed in private locations whereas assaults against older women are most likely to occur in a private location. In addition, the use of a gun was more likely to be used against older men during a robbery and against women during an assault (Bachman et al. 1998). LaGrange and Ferraro (1987) and others (for example, Yin 1980; Hirschel and Rubin 1982; Chu and Kraus 2004) address a range of consequences that crime can have on elderly victims. Consequences include increased medical care, greater likelihood of death, and limited mobility because of injury, longer recuperation periods, as well as increased social isolation (Peguero and Lauck 2008). It is this vulnerability that makes crime against the elderly much more problematic. Because most elderly are on fixed incomes, are more likely to experience health problems, and are limited in their choice for residential location, the fear of losing money and property that potentially cannot be regained through time and work makes the loss much more difficult for older victims.

Victims of Elder Abuse

A particularly devastating area of elderly victimization is that of elder abuse. In this case, the perpetrator of the offense is generally the spouse although adult children and other family members also rank as likely perpetrators. Although the United States was in the forefront of the developed world in identifying elder abuse, the extent of the problem remains relatively unknown. Data from the National Center on Elder Abuse (1997a) indicate a 150% increase in the number of reported cases between 1986 and 1996 and that the number of reported cases is significantly underreported.

According to the National Center on Elder Abuse (1997b), the primary sources of reporting elder abuse were health care and service providers followed by family. Further complicating the reporting process is the variation in state level legislation regarding the reporting and investigation of suspected elder abuse (Jogerst et al. 2003). Brinig et al. (2004) also found that state laws regarding elder abuse are not uniform in their content or application by social service agencies. These discrepancies facilitate the continued growth of elder abuse.

According to Carp, there were 206,000 substantiated cases of elder abuse in 1991 (Carp 2000). However, if the belief that there are 14 cases of elder abuse that exist for every one that is reported, then there were over 2.8 million cases of elder abuse in 1991. By 1999, there were 242,430 investigations of domestic elder abuse (Brinig et al. 2004). Wolf estimated that some 4.5 million "persons ages 60 and over living in domestic settings were abused, neglected, or exploited" (Wolf 2000:8). A nationally representative study of elder mistreatment reported that "1 in 10 NSHAP respondents answered positively to questions on verbal mistreatment, and 3.5% said that they had suffered financial mistreatment" (Laumann et al. 2008:S252).

According to the National Center on Elder Abuse (1999), neglect is the primary cause of mistreatment of the elderly (55.0%), followed by physical abuse (14.6%), financial/material exploitation (12.3%), emotional (7.7%), all other types (6.1%), unknown (4.0%), and sexual (0.3%). Although availability of data on elder abuse in institutional settings is limited, Wolf cites Pillemer and Moore (1990) who reported over one third of nursing home personnel in one state witnessed an incident of physical abuse and over 80% witnessed a case of psychological abuse in the previous year (Wolf 2000).

The consequences of elder abuse on its victims are many but are not clearly understood. According to Wolf, efforts to sort through the impact of aging, abuse and other factors are difficult (Wolf 2000). Research findings do suggest that abuse has a deleterious effect on the physical well-being of the victim resulting in increased mortality rates. Additional research by Fisher and Regan found that elderly victims who experienced psychological/emotional abuse "reported significantly more health conditions, on average, than did women who had not been abused in this manner" (Fisher and Regan 2006:207).

The recent past offers lessons for the future of elder abuse and its victims. As the demographic shift continues throughout the first 3 decades of the twenty-first century, elder abuse and its consequences will remain an underlying crime against older Americans. Efforts to standardize data collection across states as well as funding for research into causes and cures should remain a priority in the twenty-first century (Jogerst et al. 2003).

The Elderly as Perpetrators of Crime

Historically, we know that the elderly are the least likely to commit a crime. However, they are an increasing percentage of those housed in state and federal prisons in the United States. An explanation of these seemingly inconsistent trends is presented below.

Examining changing crime rates among the elderly between 1980 and 2004, Feldmeyer and Steffensmeier found that overall crime committed by the elderly has decreased. This decrease has been fueled by significantly lower arrest rates for alcohol-related crimes. Increased rates of arrests were noted for simple assaults, OBT (Offenses But Traffic), and drug law violations (Feldmeyer and Steffensmeier 2007). In terms of arrest rates for index crimes, larceny-theft continues to be the primary category among older Americans.

Addressing historical patterns of elder criminality, Flynn identified those categories with the most frequent arrest rates. Thus, in 1965, drunkenness, disorderly conduct, and other but traffic represented the three primary arrest categories. In 1984, the primary arrest categories were driving while intoxicated, drunkenness, and larceny-theft. By 1995, gambling, sex offenses, and driving while intoxicated represented the three primary reasons for arrests of elderly Americans (Flynn 2000).

In 1978, the arrest rate of males aged 65 and above was 1.1 per 1,000. Arrest rates among other age groups ranged from 3.5 for those 55–64 years of age to 32.3 for those 18–24 years old (U.S. Bureau of the Census 1978). Recent research indicates that the elderly represented 0.7% of the total number of arrests for 2008 (Federal Bureau of Investigation 2008). More specifically, Table 27.2 identifies arrest rates of older Americans by type of offense.

The data in Table 27.2 illustrate the changes in arrest rates among older Americans. Rates of violent and property crimes decrease significantly in almost all categories. Overall, violent crimes committed by older Americans declined by 20% between 1993 and 2001 and property crimes by 61%. Among violent crimes, older Americans were less likely to be arrested for murder, forcible rape, robbery, and aggravated assault. Among property crimes, there was less change over time, but continued downward shift in arrest rates in almost all categories. The only exceptions were embezzlement which remained static at 0.2 between 1993 and 2001 and drug abuse violations in which the arrest rate increased 34%. Nevertheless, older Americans remained the least likely to be arrested for violent or property crimes.

Reasons for the low arrest rates among the elderly are varied. According to Flynn, one reason is an "aging-out effect" in which biological, individual, and social factors frame the explanation (Flynn 2000). For example, as a person ages, his or her physical and mental conditions may make it more difficult to engage in those activities associated with criminal intent. Furthermore, the "aging-out effect" argues that as a person ages, his or her level of responsibility in adulthood increases, thereby

Table 27.2 Age-specific arrest rates, age 65 and above, United States

	1993	1994	1995	1996	1997	1998	1999	2000	2001
Violent crime	15.5	15.1	16.1	14.7	14.4	14.1	13.3	13.4	12.3
Murder	0.7	0.5	0.6	0.6	0.5	0.5	0.5	0.4	0.4
Forcible rape	1.0	1.0	0.9	0.7	0.7	0.8	0.7	0.7	0.6
Robbery	0.6	0.6	0.6	0.4	0.4	0.4	0.6	0.4	0.4
Aggravated assault	13.1	13.1	14.0	12.9	12.8	12.4	11.5	11.8	10.9
Property crime	51.7	46.0	40.7	36.9	32.9	28.9	27.6	23.9	20.0
Burglary	1.8	1.6	1.7	1.6	1.4	1.5	1.4	1.6	1.1
Larceny-theft	49.1	43.5	38.4	34.6	31.0	26.9	25.6	21.7	18.4
Motor vehicle theft	0.8	0.8	0.6	0.7	0.5	0.5	0.6	0.5	0.4
Arson	0.4	0.3	0.4	0.3	0.3	0.3	0.2	0.2	0.2
Forgery and counterfeiting	1.0	1.1	1.0	1.0	0.9	1.1	0.9	1.0	0.8
Fraud	7.2	7.9	8.2	8.5	8.5	7.8	8.2	8.0	6.7
Embezzlement	0.2	0.3	0.2	0.2	0.3	0.1	0.2	0.2	0.2
Stolen property	1.1	1.2	1.0	1.1	1.1	0.9	0.9	1.0	0.7
Weapons violations	5.8	4.9	4.8	4.5	4.6	3.8	3.6	3.1	3.0
Sex offenses	7.2	7.2	6.5	6.2	6.6	6.2	6.2	5.9	5.1
Gambling	2.7	2.3	2.0	2.2	2.1	2.0	2.1	2.0	1.7
Drug abuse violations	6.4	7.0	7.7	7.9	8.1	8.7	10.0	11.5	8.6

Age-specific arrest rates and race-specific arrest rates for selected offenses, 1993–2001. Uniform crime reports, November 2003. Federal Bureau of Investigation

reducing interest in criminal activity. While this argument provides reasons why the elderly as a group are less likely to commit a crime, it does not address why some elderly engage in criminal activity. Additional explanations for the low rates of criminal behavior among older Americans include a life course perspective in which Sampson and Laub argue that diminished engagement in criminal activity is tied to social connections in adulthood (Sampson and Laub 1990).

Incarceration

While arrest rates of older Americans have been stable or declining, the elderly are increasing in number and as a percentage of the overall prison population. Here, it is important to note that research on prison inmates utilizes a different designation of old. Most of the research on those incarcerated uses age 50 or 55 and above as elderly. The reason for a younger elderly age is that the health status of a 50-year-old prison inmate approximates that of a nonincarcerated elderly person some 10–15 years older (Reimer 2008).

In 1985, there were less than 10,000 prison inmates aged 55 and above in state prisons. By 2010, that number is expected to surpass 60,000. The growth in the number of federal prisoners aged 55 and above has also been increasing but at a much lower rate (Yates and Gillespie 2000). Overall, the elderly represent the fastest growing age group in American prisons (Rothman et al. 2000). This increase is the result of a number of factors including growth in long-term incarceration (5–10 years or more) and older Americans being sentenced for longer periods compared with other age categories (Aday 2003). In addition, prison recidivists, chronic offenders, and problems in retirement may lead to increased chances for arrest and incarceration (Reimer 2008). Among elderly prison inmates, the majority were convicted of nonviolent offenses, particularly those currently in federal prisons (National Center on Institutions and Alternatives 1997). Although in the minority, the number of elderly prison inmates under sentence of death continues to increase. In 1996, 1.8% of the prisoners under sentence of death were 60 years and older. By 2007, the elderly represented 5.8% of prisoners under sentence of death (Pastore and Maguire N.d.). Demographically, elderly prison inmates are predominantly non-Hispanic whites (Dunlop et al. 2000) and as with the general prison population, are overwhelmingly male with generally low levels of education and job skills (Aday 2003). Table 27.3 illustrates the current breakdown of state and federal prisoners by age and sex. While the data demonstrate that relative to other age groups, older Americans are significantly less likely to be incarcerated in state or federal facilities, the number also reflects the level of difference between older male and female incarceration rates. Furthermore, the role of race and/or ethnicity reflects incarceration rates. Blacks, non-Hispanic men, and Hispanic women report the highest rates while white males and females the lowest. Among incarcerated elderly, however, is the increased potential

Table 27.3 Estimated number and rate (per 100,000 U.S. resident population in each group) of sentenced prisoners under jurisdiction of State and Federal correctional authorities

By sex, race, Hispanic origin, and age group, United States, 2007

	Male				Female			
Age group	Total	White, non-Hispanic	Black, non-Hispanic	Hispanic	Total[a]	White, non-Hispanic	Black non-Hispanic	Hispanic
Number, total	1,427,300	471,400	556,900	301,200	105,500	50,500	29,300	17,600
18–19 years	23,700	6,400	10,200	5,000	1,000	400	300	200
20–24 years	207,900	58,800	84,200	49,600	11,600	5,400	3,000	2,300
25–29 years	246,000	65,300	102,100	61,500	16,000	7,300	4,400	3,200
30–34 years	237,200	69,800	96,200	55,300	18,400	8,900	5,000	3,200
35–39 years	225,000	74,200	89,200	46,600	20,900	9,900	6,000	3,300
40–44 years	201,700	74,800	76,700	36,100	16,100	8,700	5,200	2,700
45–49 years	135,200	52,400	50,500	22,800	10,700	5,200	3,100	1,500
50–54 years	75,000	31,000	26,600	12,500	5,000	2,400	1,400	700
55–59 years	38,600	18,600	11,700	6,200	2,100	1,200	500	300
60–65 years	18,900	10,500	4,600	2,900	900	600	200	100
65 years and older	15,500	9,000	3,600	2,200	600	400	100	100
Rate, total[b]	955	481	3,138	1,259	69	50	150	79
18–19 years	539	238	1,561	656	24	16	47	28
2–24 years	1,915	887	5,580	2,507	114	86	203	134
25–29 years	2,256	1,025	7,256	2,624	154	117	303	172
30–34 years	2,385	1,214	8,166	2,500	191	157	386	175
35–39 years	2,113	1,124	7,215	2,344	199	151	434	191
40–44 years	1,859	1,044	6,106	2,111	166	121	364	176
45–49 years	1,196	658	4,013	1,619	93	65	214	113
50–54 years	719	404	2,422	1,164	46	31	108	66
55–59 years	432	274	1,337	787	22	17	47	36
60–64 years	266	188	771	526	12	10	26	16
65 years and older	95	68	294	200	3	2	5	7

Note: these data are compiled by a yearend census of prisoners in State and Federal correctional institutions. Data collection is sponsored by the U.S. Department of Justice, Bureau of Justice Statistics. Sentenced prisoners are defined as those serving sentences of more than 1 year under the jurisdiction of State and Federal correctional authorities. For information on methodology and definitions of terms, see appendix 15

[a]Includes American Indians, Alaska Natives, Asians, Native Hawaiians, other Pacific Islanders, and persons identifying with two or more races

[b]Based on estimates of the U.S. population on Jan. 1, 2008

Source: U.S. Department of Justice, Bureau of Justice Statistics, *Prisoners in 2007*. Bulletin NCJ 224280 (Washington, D.C.: U.S. Department of Justice, December 2008), p. 19, Appendix Tables 8 and 9

Source: Pastore and Maguire 2010

for death. According to Pastore and Maguire (N.d.), 168 elderly prisoners died in local jails in 2006 compared to 115 in 2000.

Once incarcerated, those over the age of 60 represent a greater expense than younger prisoners. One study estimates that prisoners over the age of 60 cost the state of California 2–3 times more than prisoners under the age of 60 (GW News Center 2003). Nationally, cost differences are estimated at $69,000 for inmates aged 50 and above and $22,000 for all other inmates (McMahon 2003). Reasons for the increased cost include the emergence of chronic health conditions that may have precipitated the commission of a crime or for long-term prison inmates, the development of poor health outcomes as a result of imprisonment. According to Kerbs, the need for alternative housing arrangements for elderly prisoners is a necessity in order to lower the overall cost of incarceration (Kerbs 2000). On the other hand, Kerbs and Jolley point out that as age increases, the rate of recidivism decreases (Kerbs and Jolley 2009).

Because of the increasing numbers of elderly prison inmates, some states are constructing prisons specifically for the elderly. The question of age segregation vs. age integration in the prison system is open for debate. Arguments in favor of age segregation include concern for frail elderly prisoners in a population of younger, stronger prisoners. Here, the issue is one of intimidation and control by younger-aged prisoners. In addition, age-segregated prisons could exclude older prisoners from vocational and other training programs. Conversely, proponents of age integration in prisons argue that older inmates can function as a role model for younger prisoners. Perhaps, most important is that older prison inmates are not a homogeneous group with similar needs, thus supporting the belief that elderly inmates represent a range of abilities and needs within the prison system (Yates and Gillespie 2000).

On a more controversial note is the effort to apply the Americans with Disabilities Act to the prison system. Although some prisons adhere to the requirements of the law, others claim that the ADA does not apply to the prison system. As Preston points out, "the disabled (including the elderly) face numerous challenges, in the prison system especially, that may defeat some of the manifest purposes of our justice system" (Preston 2003:324).

The past 30 years provides a rich history within which to assess the level of change in the relationship between aging, the law, and crime. The remaining sections address the broader implications associated with this relationship and offer possible future directions.

Intellectual Frontiers

The issue of crime and the law relative to older Americans requires forward thinking. Over the past 30 years, researchers have attempted to clarify the relationship between age and fear of crime and the social and environmental context in which it occurs. The future must move beyond crime alone and explore more deeply the relevance of other characteristics relative to this relationship. The fact that such a belief persists after 30 years indicates the need for new intellectual frontiers. Here, the emergence of the baby boom generation and their increasing predilection for urban living may offer new hope for rethinking the intersection of crime and aging. In addition, the work of Ferraro and LaGrange provides a beginning point for future gerontological examination of the role of the media relative to aging and the fear of crime. To what extent do the media reinforce negative stereotypes of older Americans as being frail and incapable of providing for their physical well-being, particularly within the urban settings? In addition, the role of new media and the construction of social networking within defined urban environments could facilitate the emergence of interaction and support systems among the elderly and nonprofit organizations. Such support systems could provide assistance to the elderly concerned with issues of crime and safety.

The intellectual frontier of exploding incarceration costs of elderly criminals lies again in technology. With low recidivism rates, it would be more appropriate to provide elderly criminals with less expensive home environment settings rather than institutional arrangements. For those elderly inmates with medical problems, monitoring systems would allow medical personnel the ability to treat them in a home environment much as other elderly are encouraged to age in place with the assistance of routine medical and social visits.

More broad is the issue of age discrimination. Here, the need to rethink age-based laws and its consequences will offer fresh insight into questions of retirement and redefined roles within the world of work. The legal task is to restructure the social contract between society and the elderly. This effort should be based on the best interests of the elderly, not on expediency or economic limitations. The outcomes need to be intra- as well as inter-generational in nature.

Implications for Social Policy

Although older Americans continue to represent the age group least likely to engage in crime or to be a victim of a criminal act, their involvement in the criminal justice system is central to the restructuring of social policy. Sociologically, the role of the elderly within the criminal justice system creates a dynamic interplay between three interrelated foci: special vs. equal treatment under the law, age-based sentencing guidelines, and incarceration.

The following statement illustrates the inadequacies of current social policy toward the elderly, crime, and the law:

> (T)here has been little analysis of whether age-based distinctions intended to aid or assist older people have the unintended effect of further stigmatizing them. Laws that increase penalties for crimes committed against older people, for example, may reflect society's collective judgment that elders should be afforded a positive status and be protected by stiffer penalties for crimes against them. On the other hand, these laws may perpetuate the inaccurate perception that older people, by and large, are weak, helpless, and vulnerable to victimization. ... Furthermore, to single out older people for differential or specialized treatment based simply on generalized or stereotypic ideas about their characteristics creates a risk of paternalism (Dunlop et al. 2000:353–4).

An example of a social policy directed toward reducing the cost of incarceration of older inmates is called the Project for Older Prisoners or POPS. Effectively, this program is designed to allow older prisoners to receive an early release from prison if they have "served the average sentence for their offense and are unlikely to repeat their crimes" (Yates and Gillespie 2000:172). Beyond release, the program offers job and housing assistance as well as support obtaining Social Security if available. The program is voluntary with support from law schools whose students screen possible candidates.

Rikard and Rosenberg offer a number of examples of how a gerontological perspective can assist in rethinking criminal justice policies. For example, rather than basing programs or resources on chronological age, institutions could utilize functional age as a determinant of a person's ability to perform a task. They also suggest explaining differences between age, period, and cohort effects as a tool for understanding what institutional changes are necessary and why. Utilizing indices such as Activities of Daily Living (ADL) and Instrumental Activities of Daily Living (IADL), institutions could better assess the needs of older inmates. Finally, they suggest application of the Person-Environment Model for the purpose of "teaching new behaviors or problem-solving skills by modifying the residential prison environment to reduce its press on aging inmates" (Rikard and Rosenberg 2007:158).

As evidenced, the sociology of aging must explore current social policy to better meet the needs of older Americans when they come into contact with the criminal justice system. The discipline needs to engage in a multilayered analysis that goes beyond how older people should be treated when victimized to how and when to incarcerate, for how long, and guidelines outlining what constitutes special treatment under the law.

Barriers

The barriers to an improved sense of security among the elderly remain. Here, the need to create environments within which older Americans feel secure remains an unfunded barrier. Other barriers include an increasingly negative attitude toward an older age population. With diminishing economic opportunities for younger age citizens, intergenerational divides are likely to generate increased competition for limited public resources. Furthermore, as entitlement programs such as Medicare and Social Security continue to experience funding deficits, questions of their long-term viability will remain an issue. Among the younger age population, the viability of such programs may raise questions of fairness of cost and special treatment under the law. Among older Americans, these programs provide a much needed economic and medical support.

Conclusion

The past 30 years have witnessed significant changes within the elderly population. The demographic shift that continues to result in an increase in the number of elderly Americans, their relationship with the criminal and legal systems continues to evolve. We know that over the past 30 years, the elderly remain the least victimized and the least likely to commit crime. At the same time, their fear of victimization remains an issue. Although research has addressed this inconsistency and locates explanations for the fear in research methodologies, elderly concerns persist. We also know that the percentage of elderly prison inmates continues to increase because of the number of inmates serving longer prison sentences. The aging of prison inmates is beginning to strain state and federal budgets and creating the need for creative housing of elderly inmates.

The relationship between the elderly and the law is more difficult. Here, the evolving social contract has created social policies that vary between treating the elderly as equal with other age groups and as a special age category. As the elderly population increases as a percentage of the population, concerns have arisen over the potential economic consequences of special treatment. These concerns are framed as an intergenerational transfer that provides a disproportionate share of the resources to the elderly.

The future of aging relative to crime and the law is filled with considerable uncertainty and opportunity. The task will be to frame the relationship so as to ensure that the needs and well-being of aging Americans are adequately addressed within a social environment that does not limit the significance of aging.

References

Aday, Ronald H. 2003. *Aging Prisoners: Crisis in American Corrections.* Westport CT: Praeger.

Bachman, Ronet, Heather Dillaway, and Mark S. Lachs. 1998. "Violence against the Elderly: A Comparative Analysis of Robbery and Assault across Age and Gender Groups. *Research on Aging* 20(2):183–98.

Brinig, Margaret F., Gerald Jogerst, Jeanette Daly, Gretchen Schmuch, and Jeffrey Dawson 2004. "The Public Choice of Elder Abuse Law." *The Journal of Legal Studies* 33(2):517–49.

Bureau of Justice Statistics. 2009. *Violent Victimization Rates by Age, 1973–2005.* Key Facts at a Glance. Washington, D.C.

Cain, Leonard D. 1976. "Aging and the Law." Pp. 342–368 in *Handbook of Aging and the Social Sciences*, edited R. H. Binstock and E. Shanas. New York: Van Nostrand Reinhold Company.

Carp, Francis Merchant. 2000. *Elder Abuse in the Family: An Interdisciplinary Model for Research.* New York: Springer.

Chu, Lawrence D. and Jess F. Kraus. 2004. "Predicting Fatal Assault among the Elderly Using the National Incident-Based Reporting System Crime Data." *Homicide Studies* 8(2):71–95.

Cohen, Elias S. 1990. Realism, Law and Aging. *Law, Medicine, & Health Care* 18(3):183–92.

Department of Justice. 2009. *Key Facts at a Glance. Violent Victimization Rates by Age, 1973–2005*. Retrieved (http://www.oip.usdoj.gov/bjs/glance/tables/vagetab.htm). Accessed 9/27/2009.

Dunlop, Burton D., Max B. Rothman, and Pamela Entzel. 2000. "Epilogue: Policy Implications for the 21st Century." Pp. 331–358 in *Elders, Crime, and the Criminal Justice System: Myths, Perceptions, and Reality in the 21st Century*, edited by. M. B. Rothman, B. D. Dunlop, and P. Entzel. New York: Springer.

Eglit, Howard C. 2004. *Elders on Trial: Age and Ageism in the American Legal System*. Gainesville FL: The University Press of Florida.

Federal Bureau of Investigation. 2008. *Arrests. Table*. Retrieved October 22, 2001 (38.http://www.fbi.gov/urc/cius2008/data/table_38.html. Accessed 10/22/09).

Feldmeyer, Ben and Darrell Steffensmeier. 2007. "Elder Crime: Patterns and Trends, 1980–2004." *Research on Aging* 29(4):297–322.

Ferraro, Kenneth F. 1995. *Fear of Crime: Interpreting Victimization Risk*. Albany NY: State University of New York Press.

Ferraro, Kenneth F. and Randy LaGrange. 1987. "The Measurement of the Fear of Crime." *Sociological Inquiry* 57(1):70–101.

———. 1992. "Are Older People Most Afraid of Crime? Reconsidering Age Differences in Fear of Victimization." *Journal of Gerontology* 47(5):S233–44.

Fisher, Bonnie S. and Saundra L. Regan. 2006. "The Extent and Frequency of Abuse in the Lives of Older Women and Their Relationship with Health Outcomes." *The Gerontologist* 46(2):200–9.

Flynn, Edith Elisabeth. 2000. "Elders as Perpetrators." Pp. 43–83 in *Elders, Crime, and the Criminal Justice System: Myths, Perceptions, and Reality in the 21st Century*, edited by Max B. Rothman, B. D. Dunlop, and P. Entzel. New York: Springer.

Jogerst, Gerald J., Jeanette M. Daly, Margaret F. Brinig, Jeffery D. Dawson, Gretchen A. Schmuch, and Jerry G. Ingram. 2003. "Domestic Elder Abuse and the Law". *American Journal of Public Health* 93(12):2131–6.

GW News Center. 2003. (February 24, 2003). GW Professor Jonathan Turley to Testify on California's Prison Overcrowding and Budgetary Crisis before the State Senate February 25. Retrieved (http://www.gwu.edu). Cited p. 93 in Vierck, Elizabeth and Kris Hodges. 2005. Aging: Lifestyles, Work, and Money. Westport, CT: Greenwood Press.

Hale, C. 1996. "Fear of Crime: A Review of the Literature." *International Review of Victimology* 4(2):79–150.

Harris, Louis, 1976. *The Myth and Reality of Aging in America*. Washington, D.C.: National Council on the Aging.

Hirschel, J. David and Karen B. Rubin. 1982. "Special Problems faced by the Elderly Victims of Crime." *Journal of Sociology and Social Welfare* 9(2):357–74.

Kapp, Marshall B. 1996. "Aging and the Law." Pp. 467–79 in *Handbook of Aging and the Social Sciences*, 4th ed., edited by R. H. Binstock and L. K. George. San Diego, CA: Academic Press.

———. 2006. "Aging and the Law." Pp. 419–35 in *Handbook of Aging and the Social Sciences*, 6th ed., edited by R. H. Binstock and L. K. George. San Diego, CA: Academic Press.

Kerbs, John J. 2000. "Arguments and Strategies for the Selective Decarceration of Older Prisoners." Pp. 229–52 in *Elders, Crime, and the Criminal Justice System: Myths, Perceptions, and Reality in the 21st Century*, edited by M. B. Rothman, B. D. Dunlop, and P. Entzel. New York: Springer.

Kerbs, John J. and Jennifer M. Jolley. 2009. "A Commentary on Age Segregation for Older Prisoners: Philosophical and Pragmatic Considerations for Correctional Systems." *Criminal Justice Review* 34(1):119–39.

Krabill, Jennifer J. and Ronald H. Aday. 2005. "Exploring the Social World of Aging Female Prisoners." *Women & Criminal Justice* 17(1):27–53.

Lachs, Mark, Ronet Bachman, Christianna S. Williams, Alice Kossack, Carolyn Bove, and John O'Leary. 2006. "Violent Crime Victimization Increases the Risk of Nursing Home Placement in Older Adults." *The Gerontologist* 46(5):583–89.

LaGrange, Randy L. and Kenneth F. Ferraro. 1987. "The Elderly's Fear of Crime: A Critical Examination of the Research." *Research on Aging* 9(3):373–91.

Laumann, Edward O., Sara A. Leitsch, and Linda J. Waite. 2008. "Elder Mistreatment in the United States: Prevalence Estimates from a Nationally Representative Study." *The Journals of Gerontology*, 63B(4):S248–54.

Manning, Kenneth L., Bruce A. Carroll, and Robert A. Carp. 2004. "Does Age Matter? Judicial Decision Making in Age Discrimination Cases." *Social Science Quarterly* 85(1):1–18.

McMahon. P. 2003. "Aging Inmates Present Prison Crisis". *USA Today*, p. A03 (August 11). Cited p. 204 in Reimer, Glenda. 2008. "The Graying of the U.S. Prison Population." *Journal of Correctional Health Care* 14(3): 202–8.

Morton, Joann Brown. 2001. "Implications for Corrections of an Aging Prison Population." *Corrections Management Quarterly* 5(1):78–88.

National Center on Elder Abuse. 1997a. *Trends in Elder Abuse in Domestic Settings.* Elder Abuse Information Series No. 2. Washington, D.C.

———. 1997b. *Reporting of Elder Abuse in Domestic Settings.* Elder Abuse Information Series No. 3. Washington, D.C.

———. 1999. *Types of Elder Abuse in Domestic Settings.* Elder Abuse Information Series No. 1. Washington, D.C.

National Center on Institutions and Alternatives. 1997. *Elderly Study* Retrieved (http://NCIANET.org). Cited p. 93 in Vierck, Elizabeth and Kris Hodges. 2005. *Aging: Lifestyles, Work, and Money.* Westport, CT: Greenwood Press.

Pastore, Ann L. and Kathleen Maguire, eds. Sourcebook of Criminal Justice Statistics, (Table 6.33) [online]. Retrieved January 15, 2010 (http://www.albany.edu/sourcebook/pdf/t6332007.pdf).

Peguero, Anthony A. and Amanda M. Lauck. 2008. "Older Adults and Their Vulnerabilities to the Exposure of Violence." *Sociology Compass* 2(1):62–73.

Pillemer, Karl and David W. Moore. 1990." Highlights from a Study of Abuse of Patients in Nursing Homes." *Journal of Elder Abuse & Neglect* 2(1/2):5–30.

Preston, Pamela. 2003. "Should there be Separate Justice Systems for Special Needs Populations? Results from the Penn State Public Opinion Poll." *Criminal Justice Policy Review* 14(3):322–38.

Reimer, Glenda. 2008. "The Graying of the U.S. Prison Population." *Journal of Correctional Health Care* 14(3):202–8.

Rikard, R. V. and Ed Rosenberg. 2007. "Aging Inmates: A Convergence of Trends in the American Criminal Justice System." *Journal of Correctional Health Care* 13(3):150–62.

Rothman, Max B., Burton D. Dunlop, and Pamela Entzel. 2000. *Elders, Crime, and the Criminal Justice System: Myths, Perceptions, and Reality in the 21st Century.* New York: Springer.

Rothman, Max B. and Burton D. Dunlop. 2006. "Elders and the Courts: Judicial Policy for an Aging America." *Journal of Aging & Social Policy* 18(2):31–46.

Sampson, Robert J. and John H. Laub. 1990. "Crime and Deviance over the Life Course: The Salience of Adult Social Bonds." *American Sociological Review* 55(5):609–27.

Stojkovic, Stan. 2007. "Elderly Prisoners: A Growing and Forgotten Group within Correctional Systems Vulnerable to Elder Abuse." *Journal of Elder Abuse & Neglect* 19(3/4):97–117.

United Nations. 2004. *World Population Prospects: The 2004 Revision Highlights.* Population Division of the Department of Economic and Social Affairs of the United Nations Secretariat: New York.

U.S. Bureau of the Census. 1978. *Statistical Abstract of the United States: 1978.* 99th ed. Washington, D.C., 1978.

Wolf, Rosalie. 2000. "The Nature and Scope of Elder Abuse. *Generations* 24(2):6–12.

Wood, Stacey and Mary Stephens. 2003. "Vulnerability to Elder Abuse and Neglect in Assisted Living Facilities." *The Gerontologist* 43(5):753–7.

Yates, Jeff and William Gillespie. 2000. "The Elderly and Prison Policy." *Journal of Aging and Social Policy* 11(2/3):167–75.

Yin, Peter. 1980. "Fear of Crime among the Elderly: Some Issues and Suggestions." *Social Problems* 27(4):492–504.

———. 1982. "Fear of Crime as a Problem for the Elderly." *Social Problems* 30(2):240–5.

———. 1985. *Victimization and the Aged.* Springfield IL: Charles C. Thomas.

Chapter 28
Aging Veterans: Needs and Provisions

Janet M. Wilmoth and Andrew S. London

Current knowledge about aging is primarily based on cohorts that were born during the early part of the twentieth century. A substantial percentage of men in these cohorts, and subsequent cohorts who are currently middle aged, served in the military during war, peace, Cold War, or some combination thereof. Consequently, veterans are a sizeable demographic group in the United States. In 2000, over 26 million Americans were veterans, representing approximately 12.7% of those aged 18 years or older (U.S. Census Bureau 2003). Military service is particularly prevalent among older cohorts who served in World War II (WWII) and the Korean War (Hogan 1981); almost 9.2 million men age 65 years and older were veterans in 2000, which represents 64% of men in this age group (Interagency Forum on Aging-Related Statistics 2008). In addition, participation in the military has increased substantially among women; in 2000, nearly 1.6 million American women were veterans (U.S. Census Bureau 2003).

Military service and war are persistent features of Americans' lives. Because of the high prevalence of military service among men who were young adults during World War I (WWI), World War II (WWII), and the Korean War and the necessary concentration on these cohorts in earlier, large-scale data collection efforts that focus on aging Americans, Settersten and Patterson (2007:5) have argued that "...wartime experiences may be important but largely invisible factors underneath contemporary knowledge about aging." We contend that the role of military service in shaping age-related outcomes has more generally been under-acknowledged in gerontological research on men, as well as women who sometimes served in the military service roles that were made available to them historically, but more often lived their lives linked to men who served in the military. At the same time, the seminal work of Glen H. Elder, Jr. and his colleagues on the lives of the children of the Great Depression, who primarily served during WWII, has elucidated the impact of military service on men's lives, while also providing strong theoretical foundations for life-course scholars to use in studies of the effects of military service on the lives of men, women, and children.

In this chapter, we discuss the extant research that has used life-course perspectives to examine military service, paying particular attention to the theoretical contributions of this work to broader life-course studies. We argue that this area of research is unique in its careful articulation of within- and between-cohort variation, and then use that approach to discuss the characteristics of current and future aging veterans. We also consider the needs of and provisions to aging veterans by focusing on current patterns of benefit and service use, and linking this discussion to how needs for benefits and services may change as younger veterans with different historical experiences and demographic characteristics age. We are careful to distinguish between the needs and provisions

J.M. Wilmoth (✉)
Department of Sociology, Gerontology Center and Center for Policy Research,
Syracuse University, Syracuse, NY, USA
e-mail: jwilmoth@maxwell.sys.edu

R.A. Settersten, Jr. and J.L. Angel (eds.), *Handbook of Sociology of Aging*, Handbooks of Sociology and Social Research, DOI 10.1007/978-1-4419-7374-0_28,

that are rooted in general aging processes and those that are unique to the experiences of particular veteran cohorts. The chapter concludes with a discussion of the types of theoretical and empirical work that are necessary to understand the changing role of military service in the life course.

Life-Course Studies on Military Service

Over the past 30 years, a growing body of research has drawn on life-course perspectives to investigate the influence of military service on various life-course trajectories and outcomes. This research has demonstrated that the U.S. military is a critical social institution that can (re)shape educational, occupational, income, marital/family, health, and other life-course trajectories (London and Wilmoth 2006; MacLean and Elder 2007; Mettler 2005; Modell and Haggerty 1991; Settersten 2006). Life-course researchers who have addressed questions related to military service recognize that the military has the potential to transform the lives of those who serve in it for better or worse, but also that the transformative potential of the military varies across individual characteristics, the timing of military service in the life course, service experiences, and historical periods (Angrist 1990; Angrist and Krueger 1994; Gimbel and Booth 1994, 1996; Teachman and Call 1996; Teachman 2004, 2005; Teachman and Tedrow 2004). These insights serve as the foundation for much theorizing about how military service (re)shapes life course outcomes, as well as empirical life-course studies that have examined the role of military service in men's and, to a lesser extent, women's lives. This body of exemplary life-course research aims, in the words of Modell and Haggerty (1991) "to connect the micro- and macro-levels of analysis, thus connecting the soldier's story to that of his *[or her]* changing society" (p. 205).

Studying the role of military service in the life course provides researchers with an amazingly rich opportunity to investigate each of the five major principles of the life-course paradigm: human agency; location in time and place; timing; linked lives; and lifelong development (Elder and Johnson 2002; Giele and Elder 1998). Glen H. Elder, Jr. is responsible for the most-influential, early theorizing about the role of military service in the life course (Elder 1986, 1987) and, with a range of colleagues, for demonstrating the importance of military service for life-course studies (Clipp and Elder 1996; Dechter and Elder 2004; Elder and Bailey 1988; Elder and Clipp 1988, 1989; Elder et al. 1991; 1994, 1997; Pavalko and Elder 1990). This seminal body of research articulates each of the five principles of the life-course paradigm and, by focusing on the timing of military service in the life course, motivates and interrogates two corollary hypotheses that have wide-reaching implications for life-course scholarship: the military as turning point and the life-course disruption hypotheses.

The *military as turning point hypothesis* focuses on young age at entry into the military because it maximizes the chances for redirection of the life course and minimizes disruption to established life-course trajectories. Elder (1987) argues that early entry into the military represents a social and psychological moratorium, which both delays the transition to adulthood and allows for the maximal utilization of service benefits. Early entry may reflect selection of the most disadvantaged, who see military service as a route out of difficult life circumstances. While this would suggest that early entrants would have worse life-course trajectories and outcomes, it is theorized that they are precisely the persons who may benefit most from the range of benefits and services available to veterans. The *life-course disruption hypothesis* is a corollary hypothesis, which posits that relatively late entry into the military has the potential to disrupt established marital, parenting, and occupational trajectories, which may have consequences for the subsequent patterning of the life course and later-life outcomes. Later entrants often come from more advantaged backgrounds and may have already completed their educations. Because of the timing of military service in their lives, later entrants have less opportunity to take advantage of educational benefits for veterans. Consequently, the gains that accrue to more disadvantaged, earlier entrants through access to G.I. Bill educational benefits may not materialize to the same degree or with the same effects in their lives. The psychological effects of military service, especially, but not exclusively, among veterans who have experienced combat, may intersect with disrupted occupational roles in ways that increase strains within families and influence the risk of marriage and family disruption.

Both of these hypotheses emphasize the potential of military service to produce discontinuity in the life course and remind life-course scholars of how participating in social institutions during young adulthood is influenced by and influences the process of cumulative inequality. In that regard, Elder's work laid a theoretical foundation for understanding the role of other institutions, including education, criminal justice, marriage, and family, in shaping life-course outcomes. His work has also spurred other scholars to elaborate the role of military service in the lives of cohorts who served during different historical time periods. This work is concerned with specifying between-cohort variation in employment, earnings, marital, and health outcomes and emphasizing within-cohort variation in experience based on gender and race/ethnicity, and to a lesser extent, sexual orientation.

Characteristics of Aging Veterans

Before turning to a discussion of the needs of aging veterans and the provisions available to them, we now provide a portrait of the current population of older veterans and how this population is projected to change between now and 2030. We believe that it is essential to ground a discussion of needs and provisions within a life-course perspective that takes historical context and individual characteristics into account.

2010

As shown in Fig. 28.1, in 2010, 4% of 25–44 year olds, 11% of 45–64 year olds, and 23% of adults aged 65 and older are veterans. The total number of veterans by age group and period of service is given in Table 28.1, which indicates that the current number of veterans increases from approximately 4.1 million for the 25–45 age group to over 9 million for both the 45–64 year old and 65 year or older age groups. The youngest group served primarily during the ongoing Gulf War: with the largest among them serving prior to 9/30/2001 (48%). The middle-aged group served primarily during the Vietnam Era (53%) or between the Vietnam Era and the Gulf War period (32%).

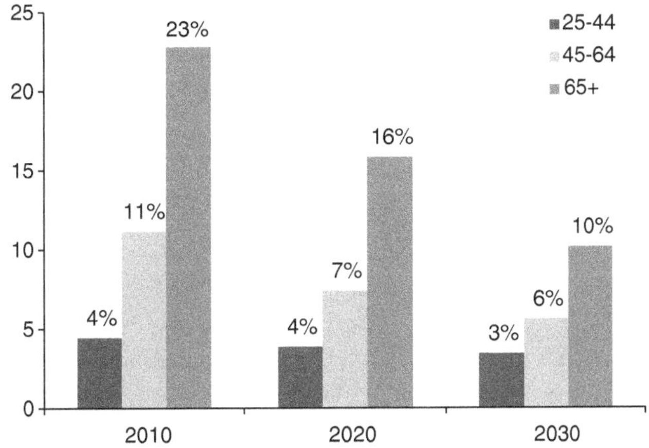

Fig. 28.1 Projected percent of veterans in the total U.S. population by age, 2010, 2020, 2030 (Authors' calculations based on VetPop2007 data (U.S. Department of Veteran Affairs 2008))

Table 28.1 Projected number and percent of veterans by age and period of service

	25–44 Years		45–64 Years		65+ Years	
	Number	%	Number	%	Number	%
2010						
World War II (WWII)	0		0		1,981,216	21.61
Pre-Korean War	0		0		115,440	1.26
Korean War	0		0		2,228,333	24.31
Between Korean War and Vietnam Era	0		17,593	0.19	2,247,980	24.52
Vietnam Era	0		4,756,230	52.59	2,515,536	27.44
Between Vietnam Era and Gulf War	455,173	11.00	2,889,455	31.95	66,683	0.73
Gulf War (Pre-9/30/2001)	1,990,312	48.09	962,681	10.64	9,884	0.11
Gulf War (Post-9/30/2001)	1,692,761	40.90	417,827	4.62	1,209	0.01
Post-Gulf War	761	0.02	0	0.00	0	0.00
Total	4,139,006	100	9,043,786	100	9,166,281	100
2020						
World War II (WWII)	0		0		269,721	3.11
Pre-Korean War	0		0		33,259	0.38
Korean War	0		0		804,947	9.28
Between Korean War and Vietnam Era	0		0		1,311,460	15.12
Vietnam Era	0		173,818	2.79	5,531,833	63.79
Between Vietnam Era and Gulf War	0		2,699,361	43.27	507,485	5.85
Gulf War (Pre-9/30/2001)	409,960	11.86	2,284,136	36.62	165,895	1.91
Gulf War (Post-9/30/2001)	2,080,399	60.18	1,079,714	17.31	47,441	0.55
Post-Gulf War	966,617	27.96	1,051	0	4	0.00
Total	3,456,976	100	6,238,080	100	8,672,045	100
2030						
World War II (WWII)	0		0		7,636	0.10
Pre-Korean War	0		0		3,087	0.04
Korean War	0		0		95,708	1.31
Between Korean War and Vietnam Era	0		0		388,026	5.31
Vietnam Era	0		0		3,341,176	45.70
Between Vietnam Era and Gulf War	0		417,839	8.86	2,313,522	31.64
Gulf War (Pre-9/30/2001)	0	0.00	1,865,246	39.53	756,667	10.35
Gulf War (Post-9/30/2001)	627,962	19.32	2,368,375	50.19	405,007	5.54
Post-Gulf War	2,622,466	80.68	67,065	1.42	169	0.00
Total	3,250,428	100	4,718,526	100	7,310,999	100

Source: Authors' calculations based on VetPop2007 data (U.S. Department of Veteran Affairs 2008). Percentages may not add to exactly 100% due to rounding error.

Notes: The VetPop2007 data draws information from the Census Bureau, the Defense Manpower Data Center, and the Department of Defense's Office of Actuary GORGO data.

Veterans are assigned to the period of war in which they first served. For example, a veteran who served in World War II (WWII) and the Korean War is assigned to World War II. (WWII) Similarly, a veteran from the Pre-9/30/2001 stage of the Gulf War may or may not have continued on to serve in the Post-9/30/2001 stage of the Gulf War.

The dates associated with the periods of service are as follows:

World War II (WWII)– September 1941 to July 1947;

Pre-Korean War – August 1947 to May 1950;

Korean War – June 1950 to January 1955;

Between Korean War and the Vietnam Era – February 1955 to July 1964;

Vietnam Era – August 1964 to April 1975;

Between Vietnam Era and Gulf War – May 1975 to July 1990;

Gulf War – August 1990 through a date to be prescribed by Presidential proclamation or law

Post-Gulf War – Hypothetical period of service after the Gulf War has ended, based on assumptions regarding continued stable enlistment rates.

The oldest group contains a mix of veterans who served in various mid-twentieth century wars prior to the establishment of the All-Volunteer Force (AVF) at the end of conscription in 1973. Among this group, the oldest-old, whose members are in their 80s, served in WWII (22%); the next oldest subgroup, whose member are at least 68 years old, served during Korea (24%) or between Korea and Vietnam (25%); and the youngest subgroup, whose member are mostly in their mid- to late-60s, served in Vietnam (28%).

Table 28.2, which presents the age, sex, and race/ethnicity distribution of veterans by time period, shows that the overwhelming majority of these older veterans are white men (86%). This reflects the gender and racial/ethnic composition of veterans who served during the pre-AVF era (see Lutz 2008 for a detailed review of historical changes in the racial/ethnic characteristics of those serving in the U.S. military). Despite their underrepresentation in terms of percentages, a substantial number of racial and ethnic minorities served in the military during the first half of the twentieth century. For example, it is estimated that over one million African Americans and one-half million Hispanics served in WWII (Lutz 2008; Allsup 1982). Research suggests that serving in the military during this time period profoundly affected the identities and lives of these veterans, and that their

Table 28.2 Projected number and percent of veterans by age, sex, and race/ethnicity

	25–44 Years		45–64 Years		65+ Years	
	Number	%	Number	%	Number	%
2010						
White Men	2,423,350	58.55	6,294,529	69.60	7,759,261	84.65
Black Men	508,852	12.29	1,106,493	12.23	574,052	6.26
Hispanic Men	349,703	8.45	508,853	5.63	314,039	3.43
Other Race Men	180,169	4.35	322,454	3.57	217,301	2.37
White Women	403,010	9.74	546,435	6.04	253,881	2.77
Black Women	161,999	3.91	171,601	1.90	24,816	0.27
Hispanic Women	69,677	1.68	51,281	0.57	11,839	0.13
Other Race Women	42,248	1.02	42,141	0.47	11,092	0.12
Total	4,139,006	100	9,043,786	100	9,166,281	100
2020						
White Men	2,002,274	57.92	3,786,323	60.70	6,865,405	79.17
Black Men	392,946	11.37	871,628	13.97	744,754	8.59
Hispanic Men	306,976	8.88	434,089	6.96	377,277	4.35
Other Race Men	160,652	4.65	246,575	3.95	259,140	2.99
White Women	353,103	10.21	565,166	9.06	316,765	3.65
Black Women	134,996	3.91	215,868	3.46	63,996	0.74
Hispanic Women	67,111	1.94	68,799	1.10	23,512	0.27
Other Race Women	38,918	1.13	49,632	0.80	21,195	0.24
Total	3,456,976	100	6,238,080	100	8,672,045	100
2030						
White Men	1,918,465	59.02	2,758,092	58.45	5,101,586	69.78
Black Men	364,037	11.20	583,849	12.37	851,643	11.65
Hispanic Men	273,266	8.41	386,591	8.19	394,535	5.40
Other Race Men	152,443	4.69	203,520	4.31	251,547	3.44
White Women	327,307	10.07	469,556	9.95	482,735	6.60
Black Women	119,693	3.68	189,939	4.03	147,817	2.02
Hispanic Women	59,425	1.83	78,434	1.66	44,579	0.61
Other Race Women	35,792	1.10	48,545	1.03	36,557	0.50
Total	3,250,428	100	4,718,526	100	7,310,999	100

Source: Authors' calculations based on VetPop2007 data (U.S. Department of Veteran Affairs 2008). Percentages may not add to exactly 100% due to rounding error.

honorable service helped pave the way for desegregation within the military and throughout the broader society (MacGregor 1981; Moskos and Butler 1996; Rivas-Rodriguez 2005).

Among the more than 70% of veterans aged 65 years and older who served prior to the current AVF era, the oldest served during WWII and are members of the birth cohorts that experienced higher rates of economic deprivation as children of the Great Depression, but went on to benefit from the post-WWII economic boom. The younger veterans in this group, who served between WWII and the Korean War or during the Korean War, were part of a relatively smaller set of birth cohorts that experienced WWII as children and came of age during the height of the Cold War. Military service provided men from these pre-Vietnam Era cohorts access to on-the-job training and higher education through generous G.I. Bill benefits. This had positive impacts on these veterans' subsequent life-course outcomes, including employment, earnings, occupational status, and marital stability, particularly among those from disadvantaged backgrounds, those who were officers, those who did not experience combat, and those who were white men (Bound and Turner 2002; Dechter and Elder 2004; Mettler 2005; Turner and Bound 2003). However, there is evidence that older pre-AVF veterans have higher mortality risk than nonveterans (London and Wilmoth 2006), which may be due in part to the pro-tobacco military policy (i.e., free and reduced price distribution of cigarettes) during WWII (Bedard and Deschênes 2006) and exposure to atomic radiation associated with the detonation of nuclear testing devices and the occupation of Hiroshima and Nagasaki, Japan (Bice-Stephans and Wynona 2000).

It is important to note that the G.I. Bill was not available to those who served from 1955 to 1965. Research indicates that during this time period men who were drafted were less likely to attend college than nonveterans and those who were not drafted, and that military service redirected academically ambitious men away from college attendance (MacLean 2005). This suggests that veterans who served during the period between the Korean War and the Vietnam Era might not have reaped the same benefits of service as those who served during WWII and the Korean War. In addition, the benefits of service were not as likely to be experienced by men who entered military service late. A series of studies indicate that late entry reduced the economic and job benefits associated with military service, increased the risk of life-course disruption, and resulted in poorer physical health trajectories over the life course (Elder et al. 1994, 1997; MacLean and Elder 2007).

2020

As shown in Table 28.1, it is projected that by 2020 there will be approximately 3.5 million veterans in the 25–45 year old age group, 6.2 million in the 45–64 year old age group, and 8.7 million in the 65 year and older age group. The youngest group of veterans will increasingly include individuals who served during the post-9/30/2001 period of the Gulf War (60%), which encompasses the Global War on Terror, including Operation Enduring Freedom (OEF) and Operation Iraqi Freedom (OIF). Vietnam Era veterans will begin to move out of the middle age group and be replaced by the cohorts who served during the initial, peaceful years of the AVF era (43%) and the first 10 years of the Gulf War (37%). The oldest age group will overwhelmingly contain veterans from the Vietnam Era (64%), although the oldest and frailest among this group will be the nearly 1.1 million surviving veterans of WWII (3%) and the Korean War (9%). Although it is a small group, some existing research documents the specific needs of and provisions for current centenarian veterans (Selim et al. 2005). Table 28.2 demonstrates that the shift to veterans from the Vietnam Era is accompanied by increasing racial and ethnic diversity among veterans ages 65 years and older, as well as a notable increase in the proportion of older veterans who are women, from 3% in 2010 to 5% in 2020. By 2020, white men will represent 79% of all older veterans. The percentage of this older age group comprised of black men, Hispanic men, and men of other races/ethnicities will increase to almost 8, 4, and 3%, respectively.

The Vietnam Era veterans will have several unique characteristics compared to the cohorts that preceded them. These veterans are part of the large Baby Boom cohort that experienced increased competition for jobs during early adulthood, the economic downturn of the early 1970s, dropping rates of marriage and fertility, and rising divorce rates. In addition, there is evidence that the Vietnam War was more disruptive to the lives of veterans than previous wars (Frey-Wouters and Laufer 1986; Kulka et al. 1990) and that the economic outcomes of Vietnam veterans are lower than veteran cohorts who served earlier in the twentieth century (Angrist 1990; Cohen et al. 1986; Teachman and Call 1996). Due to advances in medical treatment, more wounded and injured soldiers survived, creating unique long-term health care needs due to war-related disability. Vietnam veterans also faced increased cancer risk due to exposure to Agent Orange (Pavuk et al. 2005). In addition, the nature of combat engagement, drug use during service, the relatively quick return to civilian life after assignments in combat zones, and antiwar sentiment in the United States contributed to the emergence of distinct mental health and substance abuse issues for Vietnam veterans.

2030

The projected profile of veterans in 2030 looks quite different than the one that exists today. As shown in Fig. 28.1, a smaller percentage of the total population will be a veteran in 2030: approximately 3, 6, and 10% of the 25–44 year old, 45–64 year old, and 65 year and older age groups, respectively. However, Table 28.1 indicates that the number of veterans will continue to be high: 3.2 million 25–44 year olds, 4.7 million 45–64 year olds, and 7.3 million 65 years and older. The majority of the youngest 25–44 year olds will be veterans the Post-Gulf War group (81%), which is a hypothetical cohort based on assumptions by the Department of Veterans Affairs regarding the end of the Gulf War and stable future enlistment rates. Nearly all of the middle-aged group will be veterans from the Gulf War: 40% will be from the Pre-9/30/2001 period and 50% will be from the Post-9/30/2001 period. The oldest veterans will represent three sets of veterans with distinct service histories. The oldest and largest group will be the surviving veterans of the Vietnam War (47%). There will also be a large number of veterans who served during the early years of the AVF Era that occurred after the end of the Vietnam Era (32%). The leading edge of the Pre-9/30/2001 Gulf War veterans will just be reaching retirement ages by 2030 (10%). The gender and racial/ethnic diversity of older veterans will become increasingly salient as the AVF-era veterans enter later life (see Table 28.2). White men will comprise 70% of veterans and the percentage of black and Hispanic men will increase to approximately 12 and 5%, respectively. The percentage of older veterans who are white and black women will increase to 6 and 2%, respectively.

In 2030, the characteristics of older Vietnam Era veterans are likely to dominate concerns about aging veterans. Not only will these veterans be among the oldest and frailest, but they are likely to have more needs relative to the slightly younger veterans who served during the early AVF era. These AVF-era veterans include the end of the Baby Boom and the beginning of the Baby Bust cohorts who came of age between the Vietnam Era and the start of the Gulf War. Young adults coming of aging during this period of relative peace experienced economic recession in the early 1980s, the rise of service sector employment, delayed ages of first marriage and childbearing, and continued high divorce rates. The experience of serving in the military also changed; as fewer young adults opted to enlist, military service became a less normative part of early adulthood, and more of those who enlisted chose military careers. Fewer veterans from this era took advantage of G.I. Bill benefits, in part because veterans became required to voluntarily contribute some of their own earnings in order to access the military's portion of the benefit. But, there is evidence that military service is associated with higher earnings for AVF-era veterans than nonveterans, particularly for those from disadvantaged backgrounds (Teachman and Tedrow 2007; Angrist 1998; MacLean and Elder 2007). One recent study suggests that the downsizing of the military in the early 1990s had little effect on men's employment, but was associated with substantial increases in college attendance, especially among black men (Kleykamp 2009). In addition, the emerging literature on military service, race, and marriage (Lundquist 2004, 2006; Lundquist and Smith 2005; Teachman 2007; Usdansky et al. 2009), which posits a range of arguments to support the hypothesis that the American military became a relatively "pro-family," "pro-marriage" institution partly to recruit and retain personnel, suggests AVF-era veterans might not experience as much disruption in marital and family relationships as pre-AVF cohorts and that military service might be associated with increased marriage rates particularly among blacks.

Needs of and Provisions for Aging Veterans

The needs of aging veterans are shaped by their earlier premilitary, military, and postmilitary experiences, as well as the provisions they are able to make for themselves privately and obtain from family members, universal and means-tested social welfare programs, and benefits available through the Veterans Administration/Department of Veterans Affairs (VA). Veterans who experience significant life course disruption may find themselves aging without adequate economic and social resources to sustain themselves in the community. Those veterans who experience substantial physical and mental health problems as a result of combat experiences, being a prisoner of war, a sexual assault, or a service-related accident may need special health care services, as well as economic supports as they age.

A considerable literature documents the co-occurring struggles many veterans face with respect to mental health and substance abuse problems (Gatnache et al. 2000; Kang and Hyams 2005; Schinka et al. 1998; Sher 2009; Tessler et al. 2005), employment problems (Cohany 1990; Rosenheck and Mares 2007), incarceration, and homelessness (Kasprow et al. 2000; McGuire 2007; O'Connell et al. 2008; Tessler et al. 2003), as well as programs that aim to address these problems among veterans. As Vietnam Era veterans, whose military service is thought to have been particularly disruptive to their lives and who may have relatively high rates of midlife mental health, substance abuse, employment, and housing problems, become a larger proportion of the older veteran population in the next two decades, these earlier life-course experiences will have to be taken into account. For example, the lifetime prevalence of posttraumatic stress disorder (PTSD) (an anxiety disorder that was formally defined in 1980 in the American Psychological Association's third edition of the *Diagnostic and Statistical Manual of Mental Disorders*) is 30% among Vietnam veterans compared to 8% in the general U.S. population (Kulka et al. 1990; Kessler et al. 1995). In addition, among Vietnam Era veterans, those who served in Vietnam have poorer mental and self-rated health, higher cancer risk, and more treatment for specific health conditions than those who served elsewhere (Brooks et al. 2008a, b). This suggests that geriatric health care providers will need to become increasingly responsive to mental health conditions like PTSD and the unique health needs of veterans who served in different geographic venues as the veteran composition of the older adult population changes. Veterans whose transitions into and out of the military were less disruptive may have been able to capitalize on a range of educational and training benefits that led them to have better socioeconomic and more stable family outcomes. Such stability and status attainment may enhance their ability to care for themselves and obtain informal care from family members as they age.

The increasing gender and racial/ethnic diversity among older veterans may pose unique and complex challenges in terms of service provision. For example, there is some evidence that women and racial/ethnic minorities from disadvantaged socioeconomic backgrounds are more likely to join the military. Given this, some of that early life disadvantage might be carrying through the life course and increasing the later-life needs of these veterans. However, among men, there is evidence that military service actually offsets early-life socioeconomic disadvantage, at least in terms of health outcomes (Parker, Wilmoth, and London 2009). Therefore, among older adults from disadvantaged backgrounds, those who are veterans may be doing better in later life than those who are not veterans. Still, female and racial/ethnic minority veterans face unique challenges as they age.

There is evidence that female veterans have comparable or worse health than male veterans, and more physical and mental health problems than female nonveterans (Frayne et al. 2006; Skinner et al. 1999; Skinner and Furey 1998). In addition, women who have served in the military often have experiences that are related to negative health outcomes, such as sexual assault during military service and weak social ties after military service (Frayne et al. 2006; Suris and Lind 2008; Turner et al. 2004). These negative outcomes appear to be more prevalent among younger female veterans who served during the Vietnam and AVF Eras (Cotton et al. 2000; Wolfe et al. 2000). However, female veterans are less likely than male veterans to have a service-connected disability, which may place them at a disadvantage in terms of accessing VA health care services because veterans with conditions related to or aggravated by military service receive priority for enrollment into the VA health care system. This potential disadvantage is offset to some extent by the expansion of VA services that aim to meet the specific needs of women veterans. Despite the existence of such services, older female veterans' knowledge about VA benefits and services is low, particularly their knowledge of VA aging-related services, such as long-term care (Silverstein and Moorhead 2001). This suggests that more needs to be done to inform older female veterans of the VA services available to them and to educate care providers in the community about these options for older female veterans. There is evidence that female veterans are more likely to use VA outpatient services than male veterans, particularly among those with medical and mental health conditions, which has implications for the necessity of providing integrated delivery of those VA services (Frayne et al. 2007). Integrated VA service delivery is also likely to be increasingly important as relatively advantaged older cohorts of female veterans are replaced by more disadvantaged younger cohorts of female veterans, who are more likely to be racial/ethnic minorities, have lower levels of education, and earn less income (Wolfe et al. 2000).

Thus, when considering the needs of and provisions for aging veterans, it is important to keep in mind that veterans and those to whom their lives are linked are a heterogeneous group. The composition of the veteran population is partly shaped by individual decisions about serving in the military, historical context, and military policies, including the size of, and restrictions on serving in, particular branches of the military. These restrictions on serving are particularly consequential for women, who have been barred from certain types of occupational specialties (Manning 2005; Segal and Segal 2004). These factors, in turn, influence combat exposure and service-related experiences among veterans with different sociodemographic characteristics, which shape the needs of and provisions to particular groups of veterans. Those needs for and provision of benefits to aging veterans are also shaped by the historical contexts in which veterans lived their lives outside of the military and the provisions made available to veterans from the time they re-enter civilian life through their later years. Such benefit and service provision will undoubtedly need to change as the size and composition of the older veteran population changes. Coordinating care for older veterans residing in rural areas will continue to be a specific challenge (Fortney et al. 2005; Ritchie et al. 2002), as will attending to the specific needs of persons with particular health conditions (Hwang et al. 2004; Trudel et al. 2007; U.S. Department of Veterans Affairs 2009c), women and racial/ethnic minority veterans (Frayne et al. 2006, 2007; Silverstein and Moorhead 2001; Skinner et al. 1999; Washington 2004), and LGBT-identified veterans. Coordinating care for veterans' family members will also pose significant challenges.

Provisions from the Department of Veterans Affairs most directly aim to address service-related needs. These provisions try to mitigate some of the disruption that military service can cause in certain circumstances, such as during war-time mobilization, compensate and care for persons who have been harmed in the course of their service, as well as their dependents, and generally reward those who have taken risks and made personal sacrifices of various kinds in service to their country. The tradition of providing benefits to individuals who have served in the armed forces has roots in the founding of the United States (U.S. Department of Veteran Affairs 2009a). However, the social contract shaping benefits provision did not take its current form until after the establishment of the Veterans Administration in 1930 (which became known as the Department of Veterans Affairs when the agency was elevated to Cabinet-level status in 1989). Over time, provisions for veterans have expanded, but the basic types of provisions have remained the same since WWII.

VA benefits and services address the needs of aging veterans from two approaches. Some are designed to meet the general aging-related needs of veterans that are similar to the needs of any subgroup of the population that is aging. These benefits work in tandem with social insurance programs, such as Social Security and Medicare, which are available to all qualifying members of the U.S. population, and social assistance programs, such as Supplemental Security Income and Medicaid, which are used by a subset of the U.S. population with demonstrated need. Other aspects of these VA benefits and services are designed to accommodate the unique needs of specific subgroups of aging veterans, such as veterans with service-connected PTSD and disabilities, veterans from specific wars, and other veterans with unique service-related experiences. Currently, there are nine main categories of VA benefits and services: health care; service-connected disability compensation; pensions; education and training; home loan guaranty; life insurance; burial and memorial benefits; transition assistance, including vocational rehabilitation and employment; and dependent and survivors benefits (U.S. Department of Veterans Affairs 2009b). To qualify for VA benefits, the service member must have been other than dishonorably discharged from full-time active duty service, and in some cases must have served during wartime. Members of Reserve and National Guard qualify for these benefits under certain conditions. Special provisions are made for other historically relevant groups including WWII Merchant Marine Seamen, WWI and WWII Allied Veterans, and WWII Filipino Veterans, and 33 specific civilian groups who provided military-related services during WWI and WWII, such as Women Air Force Service Pilots (WASPs), and U.S civilians of the American Field Service, who served overseas under U.S. armies and U.S. Army groups in WWII (U.S. Department of Veterans Affairs 2009c).

Veterans with service-connected disabilities are given priority in access to benefits and premiums in resource allocations. For example, veterans with service-connected disabilities are given the highest priority for enrollment in the VA health care system (U.S. Department of Veterans Affairs 2009c). The VA's integrated health care system is the nation's largest with more than 1,400 sites of care, including hospitals, community clinics, community living centers, domiciliaries, readjustment counseling centers, and other types of facilities. Veterans are enrolled in priority groups that the VA uses to balance available resources with demand for enrollment; changes in available resources can lead to reductions in the number of priority groups that can be enrolled, with those in higher-rated priority groups retaining access. The top three of the eight priority groups that are currently defined are comprised of veterans with combat-related disabilities: Group 1 includes "veterans with service-connected disabilities rated 50 percent or more and/or veterans determined by VA to be unemployable due to service-connected conditions"; Group 2 includes "veterans with service-connected disabilities rated 30 or 40 percent"; and Group 3 includes "veterans with service-connected disabilities rated 10 and 20 percent, veterans who are former Prisoners of War (POW) or were awarded a Purple Heart medal, veterans awarded special eligibility for disabilities incurred in treatment or participation in a VA Vocational Rehabilitation program, and veterans whose discharge was for a disability incurred or aggravated in the line of duty" (U.S. Department of Veterans Affairs 2009c:2). Additionally, the compensation given to veterans with service-connected disabilities vary from $123/month in 2009 for veterans with a 10% VA disability rating to $376, $770, and $2,673 for veterans with 30, 50, and 100% VA disability ratings, respectively (U.S Department of Veterans

Affairs 2009c). Those with a VA disability rating of 30% or more are eligible for additional allowances for dependents, including spouses, minor children, children between the ages of 18 and 23 years who are attending school, children who are permanently incapable of self-support due to a disability arising before the age of 18 years, and dependent parents. Accounting for both the service member's disability and the access to additional benefits for dependents is important for life-course studies that focus on those whose lives are linked to veterans.

In addition to veterans with service-connected disabilities, other subpopulations of veterans are highlighted in policies governing benefits and services. For example, there are special programs that target homeless veterans, and those at risk of homelessness, including veterans who are re-entering public life after a spell of incarceration (U.S. Department of Veterans Affairs 2009c). Women veterans are eligible for the same VA benefits as men; however, comprehensive health services to address women's specific health care needs are available, including the management of acute and chronic illnesses, preventive care, mental health care, contraceptive services, Pap smears and mammography, gynecological and maternity care, and infertility evaluation at VA health care facilities; referral to community-based providers; and special initiatives. Certain groups of veterans participate in health registries in order to obtain free medical examinations and diagnostic tests. These include the Gulf War Registry, the Depleted Uranium Registries for veterans who served, respectively, in the Gulf War and OIF and those who served elsewhere, including Bosnia and Afghanistan; the Agent Orange Registry for veterans possibly exposed to dioxin or other toxic substances in herbicides used during the Vietnam War, while serving in Korea in 1968 or 1969, or as a result of testing, transporting, or spraying herbicides for military purposes; and the Ionizing Radiation Registry for veterans possibly exposed to atomic radiation across a broad range of historical circumstances (U.S. Department of Veterans Affairs 2009c).

Given that veterans have access to public social insurance and assistance programs, and that some VA programs are only available to veterans who meet certain criteria, not all veterans use VA benefits and services. Of the approximately 23 million currently living veterans, 36% (8,493,700) received at least one VA benefit or service in fiscal year 2008 (U.S. Department of Veteran Affairs 2009d). Among those who received a benefit or service, 68% (5.76 million) received only one, while 32% (2.74 million) used more than one. More veterans used VA health care than any other benefit or service: 61% of veterans who used benefits or services used VA health care; 33% (2.78 million) used VA health care only; and 28% (2.36 million) used VA health care and at least one other benefit or service. Forty percent of veterans receiving VA disability compensation did not use VA health care. Patterns of nonhealth single benefit or service use in declining order of prevalence are loan guaranty (11%), compensation (10%), insurance (9%), burial services (2%), education (2%), pension (1%), and vocational rehabilitation (0%). Patterns of multiple benefit use in declining order of prevalence are health and compensation (46%); health, compensation, and insurance (8%); health and pension (8%); health and insurance (6%); health, compensation, and loan guaranty (5%); compensation and insurance (5%); and compensation and loan guaranty (5%). Seventeen percent use some other combination of VA benefits and services.

The pattern of using VA benefit and services varies by gender and age. Overall, 87% of those who received a benefit or service were male and 81% were 45 years old or older, which indicates that the majority of benefits and services are currently used by older male veterans. Of the veterans who received at least one benefit or service, approximately 37% (3.14 million) were between the ages of 45 and 54 years and 44% (3.75 million) were aged 65 years or older. There is an interesting difference in the age patterns of single versus multiple benefit and service use, which suggests that benefit and service use changes dynamically over the life course. Focusing first on those who only used one benefit or service, 34% (1.93 million) were between the ages of 45 and 64 years, while 48% (2.75 million) were 65 years old or older. Among those using more than one service or benefit, 44% (1.21 million) were in the 45–64 year age category, while only 36% (1.00 million) were 65 years or older.

Although it is only part of the story that is relevant to tell in relation to the needs and provisions for aging veterans, careful consideration of the VA benefits and services available to and used by veterans in the past and present is essential. The snapshot of VA benefits and service use by veterans in one recent year that is provided above suggests considerable use by a sizeable portion of the veteran population, and that benefit and service use is concentrated among older veterans. As the population of aging veterans changes over the next two decades, adjustments to benefits and services will need to be made. Such adjustments should take into account VA benefit and service use histories of aging veterans across the life course, as well as the barriers to benefit and service use that older veterans with particular characteristics experience, such as veterans living in rural areas, homebound veterans, the oldest-old veterans, veterans with histories of homelessness, and veterans with psychiatric and/or substance abuse disorders. Efforts to determine unmet needs for benefits and services and underserved populations are also warranted. For example, even though most of the approximately 12,600 lesbian, gay, and bisexual service members dismissed from the military during the 15 years that Don't Ask, Don't Tell has been in effect were discharged honorably and maintain eligibility for VA benefits, it is unclear to what extent they know they are eligible for VA benefits and services, feel comfortable using them, and do use them. Even if they use benefits and services, they may withhold specific kinds of information about behaviors and social relationships from service providers and have unmet needs because of their sexual orientation. To address these issues and concerns, new data collection efforts will be necessary that allow us to more fully understand how institutionalized military practices, historical context, and human agency jointly affect needs for and use of benefits and services among groups of veterans with different social locations and experiences of cumulative inequality.

Future Directions for Research on Military Service and the Life Course

As we look toward the future of research in this area, we have to acknowledge the importance of the past in shaping what is known and what needs to be known. By necessity, the study of military service and the life course is grounded in historical periods of war and peace. Given this, the research in this area focuses on a classic concern of life-course scholarship: how lives unfold for individuals who experience certain historical times from the vantage point of particular social locations. The next generation of studies on military service and the life course must pay renewed attention to understanding the nexus of historical circumstance and individual biography. Of particular concern is how the structure of institutionalized practices related to military personnel and veterans influences individual choices and chances in ways that shape cumulative inequality. As noted by Ferraro et al. (2009:423), "life course trajectories are shaped by the accumulation of risk, available resources, and human agency." From this perspective, the individual decision to join the military is a critical turning point in early adulthood that creates life-course discontinuity by differentiating a subpopulation who experiences a powerful, potentially transformative, and resourceful social institution from a subpopulation who does not accrue a military service history or the possibility of enduring connections to the military through the use of the benefits and services offered by the VA to veterans and their dependents. This turning point has the potential to mitigate or accelerate unfavorable trajectories (and conversely accelerate or undermine favorable trajectories), depending on the timing, intensity, and duration of the resources provided to military personnel while they are serving and to veterans after being discharged from service. Thus, social structure–in this case represented by the policies and programs of the Department of Defense and Department of Veterans Affairs–acts in tandem with human agency–in this case embodied by individual decisions related to military service–to shape life-course trajectories.

Research on military service and the life course must continue to articulate the interplay between social structure and human agency in the lives of military personnel and veterans, as well as those whose lives are linked to them. Pressing questions include: how historical conditions and personal circumstances influence individual decisions regarding the timing and duration of military service; how military policies shape the timing and sequence of other events during the demographically dense period of early adulthood; and how veterans mobilize the resources provided through VA programs and services over the life course. As the demographic profile of veterans becomes more diverse, it becomes increasingly important to recognize the ways in which intersectionality influences experiences during and after military service. It is also crucial that we develop a better understanding of the lives of those who are linked to veterans, including spouses/partners, children, and parents. All of this work must articulate clearly its contributions to understanding variation between cohorts whose members served during different historical time periods and/or variation within cohorts whose members served in a given period of war or peace.

The primary challenge to research on military service and the life course lies not in developing its theoretical underpinnings, but in collecting appropriate data. There are ample surveys of military personnel and veterans who use VA health care facilities. While these surveys are rich in the information they contain about military service experience, their exclusion of nonveterans precludes making comparisons on the basis of veteran status. Such comparisons, with controls for the selectivity of military service, are essential for understanding how the lives of those who participate in the military are different from the lives of those who do not serve in the military. There are also a number of under-researched nationally representative datasets that contain measures of veteran status. These datasets allow for direct comparisons between veterans and nonveterans, but, often, there are an insufficient number of female and racial/ethnic minority veterans to examine issues related to intersectionality. In addition, these nationally representative surveys typically have very limited information about military service experiences or the timing of military service in relation to other life-course trajectories, such as educational attainment or family formation. Often, all that is known is whether the respondent is a veteran and, if so, the time period of his or her service. Information regarding preservice circumstances, duration of service, branch of service, rank, exposure to combat, and service-related disability is infrequently gathered. The most promising avenues for advancing the scholarship in this area involve merging military and VA benefit usage records with prospective, longitudinal studies of samples that include both veterans and nonveterans.

Data collection efforts will also need to focus on gathering detailed information on veterans' use of VA and non-VA benefits and services across the life course in order to make it possible to determine how such resources shape the life course. It is clear from available data that patterns of VA benefit and service use are complex, vary with age and other characteristics, and are established shortly after deployment ends. But extant research tells us little about changes in use or cumulative use over the life course and how that varies between cohorts with different histories of service or among individuals within the same cohort who entered the military from and exited it to different social locations. This is an important gap in the extant literature and will affect what we know moving forward in the absence of efforts to collect better data. For example, descriptive data suggests that, compared to previous veteran cohorts, recent veterans may be using more benefits and services in early- and mid-adulthood. Specifically, a higher percentage of OEF/OIF veterans are using benefits and services relative to all veterans, and a higher percentage is using multiple benefits and services (U.S. Department of Veterans Affairs 2009d). This is noteworthy given the younger ages of OEF/OIF veterans relative to all veterans. This might reflect needs that are related to postservice adjustment. But, it could also be indicative of a higher pattern of service use that might persist as this cohort ages. These patterns of benefit and service use also reflect early- to mid-life course access to resources that have the potential to affect later life-course trajectories and outcomes as these veterans age. Given the relatively large percentage of OEF/OIF veterans who are women, analysis of how patterns of benefit and service use vary by sex, and sex differences in the effects of benefit and service use as veterans age, will be important topics for future data collection, research, and policy.

References

Allsup, Carl. 1982. *The American G.I. Forum: Origins and Evolution*. Austin, TX: University of Texas Press.

Angrist, Joshua D. 1990. "Lifetime Earnings and the Vietnam Era Draft Lottery: Evidence from Social Security Administrative Records." *American Economic Review* 80:313–36.

Angrist Joshua D., and Alan B. Krueger. 1994. "Why Do World War II (WWII) Veterans Earn More than Nonveterans?" *Journal of Labor Economics* 12:74–97.

Angrist, Joshua D. 1998. "Estimating the labor market impact of voluntary military service using social security data on military applicants." *Econometrica* 66(2):249–288.

Bedard, Kelly and Oliver Deschênes. 2006. "The Long-Term Impact of Military Service on Health: Evidence from World War II (WWII) and Korean War Veterans." *American Economic Review* 96:176–94.

Bice-Stephans, W. M. and M. Wynona. 2000. "Radiation Injuries from Military and Accidental Explosions: A Brief Historical Review." *Military Medicine* 165(4):275–7.

Bound, John and Sarah Turner. 2002. "Going to War and Going to College: Did World War II (WWII) and the G.I. Bill Increase Educational Attainment for Returning Veterans?" *Journal of Labor Economics* 20:784–815.

Brooks, Matthew S., Sarah B. Laditka, and James N. Laditka. 2008a. "Long-Term Effects of Military Service on Mental Health Among Veterans of the Vietnam War." *Military Medicine* 173:570–5.

———. 2008b. "Evidence of Greater Health Care Needs among Veterans of the Vietnam War." *Military Medicine* 173:715–20.

Clipp, Elizabeth Colerick and Glen H. Elder, Jr. 1996. "The Aging Veteran of World War II (WWII): Psychiatric and Life Course Insights." Pp. 19–51 in *Aging and Post-Traumatic Stress Disorder*, edited by P. E. Ruskin and J. A. Talbott. Washington, DC: American Psychiatric Press.

Cohany, Sharon R. 1990. "Employment and Unemployment among Vietnam-Era Veterans." *Monthly Labor Review* 113(4):22–9.

Cohen, Jere, David R. Segal, and Lloyd V. Temme. 1986. "Military Service was an Educational Disadvantage to Vietnam-Era Personnel." *Sociology and Social Research* 70 (April):206–8.

Cotton, Shelia, Katherine Skinner, and Lisa Sullivan. 2000. "Social Support among Women Veterans." *Journal of Women and Aging* 12:39–62.

Dechter Aimée R., and Glen H. Elder Jr. 2004. "World War II (WWII) Mobilization in Men's Work Lives: Continuity or Disruption for the Middle Class? *American Journal of Sociology* 110(3):761–93.

Elder, Glen H., Jr. 1986. "Military Times and Turning Points in Men's Lives." *Developmental Psychology* 22(2):233–45.

———. 1987. "War Mobilization and the Life Course: A Cohort of World War II (WWII) Veterans." *Sociological Forum* 2(3):449–72.

Elder, Glen H., Jr., and Susan L. Bailey. 1988. "The Timing of Military Service in Men's Lives. Pp. 157–74 in *Social Stress and Family Development*, edited by J. Aldous and D. M. Klein. New York: Guilford.

Elder, Glen H. Jr. and Elizabeth C. Clipp. 1988. "Wartime Losses and Social Bonding: Influences Across 40 Years in Men's Lives." *Psychiatry* 51:177–98.

———. 1989. "Combat Experience and Emotional Health: Impairment and Resilience in Later Life." *Journal of Personality* 57(2):311–41.

Elder, Glen H., Jr., Cynthia Gimbel, and Rachel Ivie. 1991. "Turning Points in Life: The Case of Military Service and War." *Military Psychology* 3(4):215–31.

Elder, Glen H., Jr., Michael J. Shanahan, and Elizabeth C. Clipp. 1994. "When War Comes To Men's Lives: Life-Course Patterns in Family, Work, and Health." *Psychology and Aging* 9(1):5–16.

———. 1997. "Linking Combat and Physical Health: The Legacy of World War II (WWII) in Men's Lives." *American Journal of Psychiatry* 154:330–6.

Elder, Glen H., Jr., and Monica Kirkpatrick Johnson. 2002. "The Life Course and Aging: Challenges, Lessons, and New Directions." Pp. 49–81 in *Invitation to the Life course: Toward New Understandings of Later Life*, edited by R. A. Settersten, Jr. Amityville, NY: Baywood.

Ferraro, Kenneth F., Tetyana Pylypiv Shippee, and Markus H. Schafer. 2009. "Cumulative Inequality Theory for Research on Aging and the Life Course." Pp. 413–343 in *Handbook of Theories of* Aging, 2nd ed., edited by V. L. Bengtson, D. Gans, N. M. Putney, and M. Silverstein. New York: Springer.

Fortney, John C., Matthew L. Maciejewski, James J. Warren, and James F. Burgess, Jr. 2005. "Does Improving Geographic Access to VA Primary Care Services Impact Patients' Patterns of Utilization and Costs?" *Inquiry* 42:29–42.

Frayne, Susan M., V.A. Parker, C.L. Christiansen, S. Loveland, M.R. Seaver, L.E. Kazis, and K.M. Skinner. 2006. "Health Status among 28,000 Women Veterans. The VA Women's Health Program Evaluation Project." *Journal of General Internal Medicine* 21(3):S40–6.

Frayne, Susan M., Wei Yi, Elizabeth Yano, Lakshmi Ananth, Samina Iqbal, Ann Thraikill, and Ciaran Phibb. 2007. "Gender and Use of Care: Planning for Tomorrow's Veterans Health Administration." *Journal of Women's Health* 16(8):1188–99.

Frey-Wouters, Ellen and Robert S. Laufer. 1986. *Legacy of a War: The American Soldier in Vietnam*. Armonk, NY: M.E. Sharpe, Inc.

Gatnache, Gail, Robert A. Rosenheck, and Richard Tessler. 2000. "Factors Predicting Choice of Provider among Homeless Veterans with Mental Illness." *Psychiatric Services* 51(8):1024–8.

Giele, Janet Z. and Glen H. Elder, Jr., eds. 1998. *Methods of Life Course Research: Qualitative and Quantitative Approaches*. Thousand Oaks, CA: Sage.

Gimbel, Cynthia, and Alan Booth. 1994. "Why Does Military Combat Experience Adversely Affect Marital Relations?" *Journal of Marriage and the Family* 56:691–703.

———. 1996. "Who Fought in Vietnam?" *Social Forces* 74:1137–57.

Hogan, Dennis P. 1981. *Transitions and Social Change: The Early Lives of American Men*. New York: Academic Press.

Hwang, Shirley, Victor T. Change, Janet Cogswell, Yvette Alejandro, Pamela Osenesko, Erma Morales, Shanti Crinivas, and Basil Kasimis. 2004. "Study of Unmet Needs in Symptomatic Veterans with Advanced Cancer: Incidence, Independent Predictors, and Unmet Needs Outcome Model." *Journal of Pain and Symptom Management* 28(3):421–32.

Interagency Forum on Aging-Related Statistics. 2008. *Older Americans 2008: Key Indicators of Well-being, Population Indicator 5*. Retrieved (http://www.agingstats.gov/agingstatsdotnet/Main_Site/Data/2008_Documents/Population.aspx; last accessed on October 27, 2009).

Kang, Han K. and Kenneth C. Hyams. 2005. "Mental Health Care Needs among Recent War Veterans." *The New England Journal of Medicine* 352(13):1289.

Kasprow, Wesley J., Robert A. Rosenheck, and Linda Frisnan. 2000. "Referral and Housing Processes in a Long-Term Supported Housing Program for Homeless Veterans." *Psychiatric Services* 51(8):1017–23

Kleykamp, Meredith. 2009. "Where did the Soldiers Go? The Effects of Military Downsizing on College Enrollment and Employment." *Social Science Research* 39(3):477–90.

Kessler, Ronald C., Amanda Sonnega, Evelyn Bromet, Michael Hughes, and Christopher Nelson. 1995. "Posttraumatic Stress Disorder in the National Comorbidity Survey." *Archives of General Psychiatry* 52(12):1048–60.

Kulka, Richard A., William E. Schlenger, John A. Fairbank, Richard L. Hough, B. Kathleen Jordan, Charles R. Marmar, David S. Weiss, and David A. Grady. 1990. *Trauma and the Vietnam War Generation*. New York: Brunner/Mazel.

London Andrew S., and Janet M. Wilmoth. 2006. "Military Service and (Dis)Continuity in the Life Course: Evidence on Disadvantage and Mortality from the HRS and AHEAD." *Research on Aging* 28(1):135–59.

Lundquist, Jennifer Hickes. 2004. "When Race Makes No Difference: Marriage and the Military." *Social Forces* 83(2):731–57.

———. 2006. "The Black-White Gap in Marital Dissolution among Young Adults: What Can a Counterfactual Scenario Tell Us?" *Social Problems* 53(3):421–41.

Lundquist, Jennifer Hickes, and Herbert L. Smith 2005. "Family Formation among Women in the U.S. Military: Evidence from the NLSY." *Journal of Marriage and Family* 67(1):1–13.

Lutz, Amy C. 2008. "Who Joins the Military? A Look at Race, Class, and Immigrant Status. *Journal of Political and Military Sociology* 36:167–88.

MacGregor, Morris, J. Jr. 1981. *Integration of the Armed Forces: 1940–1965*. Washington, DC: Center of Military History, United States Army.

MacLean Alair, and Glen H. Elder Jr 2007. "Military Service in the Life Course." *Annual Review of Sociology* 33:175–96.

Manning, Lori. 2005. *Women in the Military: Where They Stand*, 5th ed. Washington, DC: Women's Research and Education Institute.

McGuire, James. 2007. "Closing a Front Door to Homelessness among Veterans." *Journal of Primary Prevention* 28:389–400.

Mettler, Suzanne. 2005. Soldiers to Citizens: *The G.I. Bill and the Making of the Greatest Generation*. New York: Oxford University Press.

MacLean, Alair. 2005. "Lessons from the Cold War: Military service and college education." *Sociology of Education* 78(3):250–266.

Modell, John and Timothy Haggerty. 1991. "The Social Impact of War." *Annual Review of Sociology* 17:205–24.

Moskos, Charles C. and John Sibley Butler. 1996. *All That We Can Be: Black Leadership and Racial Integration the Army Way*. New York: Basic Books.

O'Connell, Maria, Wesley Kasprow, and Robert A. Rosenheck. 2008. "Rates and Risk Factors for Homelessness after Successful Housing in a Sample of Formerly Homeless Veterans." *Psychiatric Services* 59(3):268–75.

Parker, Wendy M., Janet M. Wilmoth, and Andrew S. London. 2009. "Does Military Service Offset the Effect of Childhood SES Disadvantage on Men's Later Life Physical Functioning?" Presented at the Gerontological Association of America, Annual Meeting, Atlanta, GA.

Pavalko, Eliza K. and Glen H. Elder, Jr. 1990. "World War II (WWII) and Divorce: A Life Course Perspective." *American Journal of Sociology* 95(5):1213–34.

Pavuk, M, J.E. Michalek, A. Schecter, N.S. Ketchum, F.Z. Akhtar, and K.A. Fox. 2006. "Did TCDD exposure or service in southeast Asia increase the risk of cancer in air force Vietnam veterans who did not spray agent orange?" Journal of Occupational and Environmental Medicine 27(4):335–343.

Rivas-Rodriguez, Maggie, ed. 2005. Mexican Americans and World War II (WWII). Austin: University of Texas Press.

Ritchie, Christine, Darryl Weiland, Chris Tully, Joseph Rose, Richard Sims, and Eric Bodner. 2002. "Coordination and Advocacy for Rural Elders (CARE)" A Model for Rural Case Management with Veterans." *The Gerontologist* 42(3):299–305.

Rosenheck, Robert A. and Alvin S. Mares. 2007. "Implementation of Supported Employment for Homeless Veterans with Psychiatric or Addition Disordered: Two-Year Outcomes." *Psychiatric Services* 58(3):325–33.

Schinka, John A. Elie Francis, Patrick Hughes, Leif LaLone, and Christopher Flynn. 1998. "Comparative Outcomes and Costs of Impatient Care and Supportive Housing for Substance-Dependent Veterans." *Psychiatric Services* 49(7):946–50.

Silverstein, Nina M. and Jennifer L. Moorhead. 2001. "Responding to Social Service and Health Care Needs of Aging Women Veterans." *Journal of Women and Aging* 13(2):39–56.

Segal, David R. and Mady W. Segal. 2004. "America's Military Population." *Population Bulletin* 59(4).

Selim, Alfredo J., Graeme Fincke, Dan R. Berlowitz, Donald R. Miller, Shirley X. Qian, Austin Lee, Zhongxiao Cong, William Rogers, Bernardo J. Selim, Xinhua S. Ren, Avron Spiro III, and Lewis E. Kazis 2005. "Comprehensive Health Status Assessment of Centenarians: Results from the 1999 Large Health Survey of Veteran Enrollees." *Journal of Gerontology: Medical Sciences* 60A(4):515–19.

Settersten, Richard A., Jr. 2006. "When Nations Call: How Wartime Military Service Matters for the Life Course and Aging." *Research on Aging* 28:12–36.

Settersten, Richard A., Jr., and Robin S. Patterson. 2007. "Military Service, the Life Course, and Aging: An Introduction." *Research on Aging* 28:5–11.

Sher, Leo 2009. "A Model of Suicidal Behavior in War Veterans with Posttraumatic Mood Disorder." *Medical Hypotheses* 73:215–19.

Skinner, Katherine and J. Furey. 1998. "The Focus on Women Veterans Who Use Veterans Administration Health Care: The Veterans Administration Women's Health Project." *Military Medicine* 163 (11) (November): 761–766.

Skinner, Katherine, Lisa M. Sullivan, Rara J. Tripp, Nancy, R. Kressin, Donald R. Miller, Lewis Kasis, and Virginia Casey. 1999. "Comparing the Health Status of Male and Female Veterans Who Use VA Health Care: Results of from the VA Women's Health Project." *Women and Health* 29(4):17–33.

Suris, Alina and Lisa Lind. 2008. "Military sexual trauma - A review of prevalence and associated health consequences in veterans." Trauma, Violence, and Abuse 9(4):250–2699.

Teachman, Jay D. 2004. "Military Service During the Vietnam Era: Were There Consequences for Subsequent Civilian Earnings?" *Social Forces* 83:709–30.

———. 2005. "Military Service in the Vietnam Era and Educational Attainment." *Sociology of Education* 78:50–68.

———. 2007. "Race, Military Service, and Marital Timing: Evidence from the NLSY-79." Demography 44(2):389–404.

Teachman, Jay D., and Vaughn R. A. Call. 1996. "The Effect of Military Service on Educational, Occupational, and Income Attainment." *Social Science Research* 25:1–31.

Teachman, Jay D., and Lucky M. Tedrow. 2004. "Wage, Earnings, and Occupational Status: Did World War II (WWII) Veterans Receive a Premium?" *Social Science Research* 33:581–605.

Teachman, Jay D., and Lucky Tedrow. 2007. "Joining up: Did military service in the early all volunteer era affect subsequent civilian income?" *Social Science Research* 36(4):1447–1474.

Tessler, Richard, Robert Rosenheck, and Gail Gamache. 2003. "Homeless Veterans of the Mil-Volunteer Force: A Social Selection Perspective." *Armed Forces and Society* 29(4):509–24.

Tessler, Richard, Robert Rosenheck, and Gail Gamache. 2005. "Declining Access to Alcohol and Drug Abuse Services among Veterans in the General Population." *Military Medicine* 170(3):234–8.

Trudel, Tina M., F. Don Nidiffer, and Jeffrey T. Barth. 2007. "Community-Integrated Brain Injury Rehabilitation: Treatment Models and Challenges for Civilian, Military, and Veteran Populations." *Journal of Rehabilitation Research and Development* 44(4):1007–16.

Turner, Sarah, and John Bound. 2003. "Closing the Gap or Widening the Divide? The Effects of the G.I. Bill and World War II (WWII) on the Educational Outcomes of Black Americans." *Journal of Economic History* 63:145–77.

Turner, Carole, Susan Frayne, & Editors. 2004. "Veterans Health Initiative: Military Sexual Trauma." TRACE Code: 03.VHI.SH&T.P.A. Independent Study Course: Released January 2004. Washington, DC: Department of Veterans Affairs. http://www.publichealth.va.gov/docs/vhi/military_sexual_trauma.pdf

United States Census Bureau 2003. *Veterans: 2000*. Census 2000 Brief. Retrieved October 27, 2009 (http://www.census.gov/prod/2003pubs/c2kbr-22.pdf).

United States Department of Veterans Affairs. 2008. *Veteran Population Model, VetPop2007*. Retrieved October 29, 2009 (http://www1.va.gov/vetdata/page.cfm?pg=15).

———. 2009a. *History of the Department of Veterans Affairs*. Retrieved October 25, 2009 (http://www1.va.gov/opa/feature/history/index.asp).

———. 2009b. *General Factsheet*. Retrieved October 25, 2009 (http://www.vba.va.gov/VBA/benefits/factsheets/general/21-00-1.pdf).

———. 2009c. *Federal Benefits for Veterans, Dependents, and Survivors*. 2009 Edition. Retrieved October 25, 2009 (http://www1.va.gov/opa/vadocs/current_benefits.asp).

———. 2009d. *Analysis of Unique Veterans Utilization of VA Benefits & Services*. National Center for Veterans Analysis and Statistics (008A3), Office of Policy and Planning. Retrieved October 21, 2005 (http://www1.va.gov/vetdata/docs/uniqueveteransMay.pdf).

Usdansky, Margaret L., Andrew S. London, and Janet M. Wilmoth. 2009. "Race, Veteran Status, and Marriage among Fragile Families." *Journal of Marriage and Family* 71:768–86.

Washington, Donna L. 2004. "Challenges to Studying and Delivering Care to Special Populations: The Example of Female Veterans." *Journal of Rehabilitation Research and Development* 41(2):7–9.

Wolfe, Jessica, Amy Stern, Jennifer Daley, Alan Zaslavsky, Shelly Forsorn Roper, and Kimberly Wilson. 2000. "Changing Demographic Characteristics of Female Veterans: Evidence from a National Sample." *Military Medicine* 165:773–80.

Part VIII
Public Health and Aging

Chapter 29
Health and Aging: Early Origins, Persistent Inequalities?

Kenneth F. Ferraro

A sociological consideration of health and aging can be viewed primarily from one of two lenses. One focuses on the health of older people, emphasizing the centrality of health to the daily experiences and longevity of older people. For the most part, these studies examine older people only. Given that "older people" (65+) typically refers to a very wide cohort, the enormous health variability within this population spanning three or more decades merits systematic consideration.

A second lens focuses on the aging process and draws heavily from the life course perspective in order to understand how health variability plays out over time and culminates in the older adult population. It is also concerned with health variability among older adults, but emphasizes the life course processes leading to such variability. There is a clear need for both approaches; each is a valued intellectual endeavor with profound policy implications. Moreover, there is a need for integrative analyses that identify points of convergence and divergence.

Although both perspectives are needed and will continue to flourish, one of the great changes during the past three decades is a greater appreciation within sociology for the second approach emphasizing the aging process. This perspective is not new to the field of gerontology, but sociologists have played a critical role in promoting its utility. This has been seen most clearly in sociological theories of age, aging, and the life course (Elder 1974, 1998; Riley 1987; Settersten 2003). At the same time, the proliferation of longitudinal data, coupled with methodological innovations to analyze such data, has enabled a new and exciting genre of research on health and aging.

In this chapter, I recognize and draw from both lenses but emphasize the second – studying health processes over the life course. My aim is to highlight the scientific utility of this perspective to better understand health inequality in later life while reviewing some of the major findings during recent decades. Ultimately, I contend that greater appreciation for life course processes of health will aid our understanding of the health of older people, thereby enabling the design of more effective interventions.

Life Course Approach to Health and Aging

Sociologists, demographers, and psychologists have drawn important conceptual distinctions between individual, cohort, and population aging in order to better understand inequalities in physical and mental health (Alwin et al. 2006; Sliwinski and Mogle 2008). At the same time, a subfield of epidemiology known as "life course epidemiology" has emerged that shares many conceptual

K.F. Ferraro (✉)
Department of Sociology and Center on Aging and the Life Course, Purdue University, West Lafayette, IN, USA
e-mail: ferraro@purdue.edu

R.A. Settersten, Jr. and J.L. Angel (eds.), *Handbook of Sociology of Aging*, Handbooks of Sociology and Social Research, DOI 10.1007/978-1-4419-7374-0_29, © Springer Science+Business Media, LLC 2011

frameworks and methods of analysis with the life course sociological approach to health and aging (Kuh and Ben-Shlomo 1997; Lynch and Smith 2005; Wadsworth 1997).

In epidemiology's conceptualization of the interrelationships between host, agent, and environment, there is a premium placed on identifying exposure to potential risks. What is distinctive about life course epidemiology, however, is studying how these risks accumulate over time. When does exposure to a risk first occur? For how long is the host exposed to the risk? Once exposed to a risk, what mechanisms lead to compromised health? These and related questions have led to a pursuit of important periods of risk exposure that can be especially consequential to health. Although most sociologists are skeptical of complex stage theories of human development, they nevertheless recognize the importance of selected periods of the life course for shaping health outcomes (Ferraro and Shippee 2009). As Berkman (2009:30) described in a recent essay, "life course issues have recently come to permeate thinking about a broad number of exposures in public health. It is now commonplace to think of critical or sensitive periods in exposure to risk as well as to understand dynamics related to cumulative exposure."

If aging is a lifelong process from birth to death, then birth may play a pivotal role in life chances. This is seen in sociological analyses of the long-term effects of low birth weight on health and well-being, and this literature carries a heavy emphasis on socioeconomic differentials in birth weights and well-being (e.g., Conley and Bennett 2000). Stratification has a clear impact on the likelihood of a low birth weight, which, in turn, may shape socioeconomic status (SES) and life chances.

Birth weight is an outcome closely linked to maternal health, and British epidemiologists have pioneered a research program on the fetal origins of adult disease. Fetal origins of *childhood* disease is not a surprising link, but Barker's (1997, 2001) research has sought to identify how and why fetal development is related to *adult* health. In this sense, gestation may be a critical period for understanding health in later life. Such a notion was largely foreign to social gerontology three decades ago, but it is now considered quite plausible. Barker's findings that low birth weight is associated with elevated risk of obesity, diabetes mellitus, hypertension, and heart disease during adulthood have sparked other studies about whether diseases are programmed in utero. An alternative explanation to fetal programming is that the accelerated weight gain after a low birth weight is the mechanism for heightened risk of obesity and cardiovascular diseases (Singhal and Lucas 2004). In this sense, it may be the metabolic discontinuity (or whiplash) that links low birth weight babies with adult diseases. Whatever the case, diseases in adulthood and later life may have their origins in the womb. This realization is still permeating the gerontological community, but life course sociologists have widely acknowledged the importance of this critical period in shaping health outcomes.

Childhood is widely noted in psychology as a critical period for personality development, but its significance for physical health is equally important. During the past 30 years, we have witnessed a proliferation of studies on the long-term consequences of childhood conditions or experiences, including: adversity (e.g., Shaw and Krause 2002), health (e.g., Blackwell et al. 2001), or "disadvantage" (i.e., typically measured as parental social class, Smith et al. 2009; Willson et al. 2007; Yi et al. 2007). Many of these studies show that childhood conditions and experiences can have long-term negative consequences on health in later life, and this line of research has come from sociology (Blackwell et al. 2001; Preston et al. 1998), epidemiology (Singh-Manoux and Marmot 2005; Surtees and Wainwright 2007), economics (Case et al. 2002), and medicine (Dube et al. 2001; Felitti 2002).

At the same time, investigations of the link between childhood conditions and adult health also show the power of compensatory or neutralizing forces to reduce or eliminate the negative effects. Psychological, economic, and social resources in adulthood may reduce the noxious effects of early adversity (Irving and Ferraro 2006; Smith et al. 2009), and this line of research is very important for the development of effective interventions (Schafer et al. 2009).

Research on the link between health during childhood and adulthood is especially intriguing because of the mechanisms involved. Many studies reveal that poor health during childhood is associated with poor health during adulthood (Case et al. 2002; Haas 2008), but Preston et al. (1998) identify an alternative outcome that drives different etiologic processes: acquired immunity. For instance, age differences in the recent prevalence of H1N1 infection reveal the highest rates among persons 5–24 years of age. Indeed, the Centers for Disease Control and Prevention (2009) suggest that "older people may have pre-existing immunity to the novel H1N1 virus." What this suggests is that the age differences in H1N1 are actually cohort differences due to historical differences in prior exposure. Aging does not reduce risk for H1N1; rather, life experiences of the earliest cohorts resulted in a protective effect via acquired immunity.

Another mechanism linking health in childhood with later life may be illustrated with a viral disease. Although a vaccine for chicken pox was approved for use in the United States in 1995, millions of adults contracted chicken pox during childhood. The uncomfortable sores associated with chicken pox typically disappeared within a week, but the virus that caused them – herpes zoster – remains in the person's body. Shingles in adulthood and later life is a rash that results from reactivation of the herpes zoster virus, and several studies reveal that stress can cause the reactivation of herpes zoster (Schmader et al. 1998). The stress process is usually identified as leading to the development of chronic disease, but it may also compromise immune responses, leading to heightened susceptibility to or reactivation of a virus. Thus, using a life course lens for studying health in later life may enable investigators to identify early and meaningful antecedents of both chronic and infectious diseases in later life.

Health Dynamics in Social Context: Selected Findings and Research Frontiers

Research on health and aging has proliferated over the past 30 years, and several findings have been transformative on the field. One of the major streams of research during the past three decades was sparked by Fries's (1980) discussion of the compression of morbidity. Fries envisioned a delay in the onset of chronic disease, enabling many people to live a larger proportion of their lives free from disease and disability. Sociological study of the compression of morbidity has given fairly limited attention to disease onset, but warmly embraced efforts to examine whether disability has been compressed. This emphasis is not surprising given that sociologists have long emphasized the functional consequences of disease in terms of social interaction and organization.

In the mid-1990s, two studies of disability were especially influential in shaping the field and discussions of the compression of morbidity. Verbrugge and Jette (1994) outlined the disablement process, by which disease led to functional disability. For older people, the expectation was that the disablement process had a strong gravitational pull – once begun, there seems little hope to escape from its forces, only ways to minimize or stall the decline in functional ability. The authors identified factors that might "speed or slow disablement," with little attention given to reversing the process (Verbrugge and Jette 1994:1). About this time, however, research by Manton et al. (1993:S194) reported that disability actually *declined* among "chronically disabled community-dwelling and institutionalized elderly populations." This sparked a number of studies by sociologists, demographers, and epidemiologists investigating whether this finding could be replicated. The scholarly paradigm at the time accepted slowing down the disablement process, but the notion that we might observe a decline in disability sparked scores of studies to determine whether this actually occurred (Crimmins et al. 2009; Freedman et al. 2004).

Three conclusions emerge from this literature. First, there is considerable evidence that disability among older adults, measured via activities of daily living (ADL), actually declined beginning in the mid-1990s (Freedman et al. 2004; Manton 2008). Second, the meaning of this finding is important: ADL disability taps whether or not a person is functionally capable of independent living.

This measurement of disability is focused on more advanced forms of disability, and some surveys actually limit their samples to persons who were screened to have some disability (e.g., components of the National Long Term Care Survey). In such surveys, the decline in disability, although important, needs to be interpreted as a decline among persons who previously reported some ADL disability. In short, the decline is a real but proverbial tip of the iceberg in the study of disability among adults (Verbrugge 1986). Third, in addition to the observed decline in ADL disability, alternative measures of health and functioning reveal notable improvements; and these improvements are due in part to cohort differences in disease prevalence and physical functioning (Manton et al. 2008; Martin et al. 2007, 2009).

Research on the compression of morbidity spawned a number of studies that elucidate health dynamics in later life. Many of these recent studies highlight the fact that health changes are often episodic and nonlinear. The bulk of our analytic methods, however, are premised on linear change. Disability, in particular, does not change in ways that are easily predicted with linear models. Rather, disability, and health status more generally, may increase, decrease, or plateau – and these transitions are frequently episodic. Cross-sectional point estimates of health status are helpful in some respects, but cutting-edge research on health has moved to dynamic models, especially growth curve models to capture oscillations in health over time.

Although there is excitement that we have recently witnessed notable health improvements for current cohorts of older adults, this satisfaction is tempered by a concern that the rising prevalence of obesity may undo these precious public health advances. Many scholars contend that the growing rate of obesity may well halt any further compression of morbidity (Manton 2008). The rising prevalence of obesity, and severe obesity, will likely yield an epidemic of diabetes mellitus (Ogden et al. 2006). Given that obesity is a potent risk factor for disability (Ferraro and Kelley-Moore 2003), an increase in the prevalence of obesity may compromise further improvements in population health.

We have also witnessed in the past 30 years a growing interest in ecological determinants of health. Although contextual analysis has been used for decades, both theories and methods for studying "clustered observations" have advanced considerably in recent years. Most theories in sociology have some component that addresses multiple levels of analysis such as macro-, meso-, and microstructures and processes. Thus, studying ecological antecedents of health dynamics in later life can be beneficial for both theoretical development and empirical generalization.

Sociologists are well aware of the power of social context in shaping health across the life course, but the past three decades have accentuated this tenet in new ways. Research has identified important influences on physical and mental health due to social capital (Snelgrove et al. 2009), neighborhood characteristics (Lee and Ferraro 2007), and even architectural features of one's residence that facilitate social interaction (Brown et al. 2009). These are especially important findings when considering the health of older people: Limited mobility means that the importance of local social ties is magnified. When social capital is low, older adults may be particularly vulnerable as illustrated by Klinenberg's (2002) finding that mortality was highest for older adults facing a Chicago heat wave. Also, older adults are less likely than younger adults to relocate to independent residences. As such, the irony is that their residential *stability* is often associated with neighborhood *change*. Thus, whether the person moves or the community changes, a frontier for this genre of research is to examine how both social context and social change influence health dynamics (Wheaton and Clarke 2003).

Health Inequality

One of the major conclusions from the study of health dynamics is that there is substantial variability in the pace of change experienced by persons. Some older people are experiencing health declines, while others are experiencing improvements or seasons of stability. As such, sociologists systematically study *health inequality* by identifying the processes leading to advantaged or disadvantaged health status over the life course. What factors mitigate the anticipated health declines associated with aging? What social forces benefit or harm people's life chances? And are the effects of these forces more or less important with aging?

Among sociologists, the influence of SES has been singularly important for identifying health inequality. The influence of SES on health status is pervasive and substantial (Link and Phelan 1995). For instance, there is evidence that the compression of morbidity discussed earlier may be limited to persons of middle- or upper-class standing; it does not extend to lower-status persons (House et al. 2005). Coupled with the study of the early antecedents of health inequality, others have uncovered that there are long-term effects of SES, especially education, on health in later life (Elo 2009; Haas 2008; Ross and Wu 1996).

For decades, the focus of studies on the SES/health relationship was on the distinct disadvantage due to poverty, but there is more to the SES/health relationship than the profound disadvantage experienced by persons in poverty. Rather, what Marmot (2003) has referred to as the "social gradient" in health means that there are noticeable differences in health by SES among persons at each level of social class (Adler et al. 2008). Stated differently, if one were to exclude persons in poverty from the analysis, there is still a striking advantage for higher SES persons.

Beyond the influence of SES on an individual's health, there is also the question of the contribution of parental SES and intergenerational mobility on health. Studies of low birth weights revealed that parental SES is consequential (Conley and Bennett 2000), but, additionally, recent research reveals that intergenerational mobility is consequential to women's mortality (Tiikkaja et al. 2009). In many of these studies, the question is often asked which element of SES is most consequential. Although measurement of each element remains fairly crude, most scholars agree that education is generally the most important predictor of health and mortality. Interestingly, when Herd et al. (2007) differentiated between the onset and progression of disease and functional limitations, they found that education was better than income for predicting onset, but that the opposite occurred for predicting the progression of both outcomes.

Stratification involves overlapping systems of differentiation, and both race and ethnicity are closely related to social class in many modern societies. During the past three decades, numerous studies have documented the antecedents and extent of the health gap between racial and ethnic groups in America. Given that many Asian-American groups have better health and live longer than the White population, the focus of the research during the past three decades has been on Black and Hispanic Americans.

Acknowledging that the racial gap exists at birth (e.g., prevalence of low birth weight), scholars have sought to identify whether the Black/White disparity continues across adulthood and into later life. Studies of older adults only tend to show that the Black/White differences observed are relatively stable during later life – persistent inequality (Kelley-Moore and Ferraro 2004). And a few studies of older adults show that the racial gap decreases or even manifests a crossover after age 75 or 85 on selected outcomes (Johnson 2000). By contrast, more cohort-inclusive studies (i.e., adults of all ages) generally reveal a widening of the gap in health between White and Black Americans (Ferraro and Farmer 1996; Haas and Rohlfson 2010), accompanied by higher mortality for Black adults (Geronimus et al. 2006; Warner and Hayward 2006). Thus, findings on the reduction in the racial disparity in advanced ages must be couched in the context of lifelong disadvantage for the majority of the population. Given the higher mortality risk for Black Americans, the big-picture view is a major racial gap in health from birth into old age, resulting in an exceptionally selective group of Black adults at ages 75 and higher. Indeed, selective survival may give the appearance of decreasing inequality (Ferraro and Shippee 2009).

A major idea that has permeated much of the research on Hispanic Americans is the "Hispanic paradox" (Markides and Coreil 1986). It is deemed paradoxical because mortality risk is *lower* for Hispanic Americans than for many White and African-Americans of comparable socioeconomic standing. Explanations for this paradox have centered on social and cultural practices that have salutary effects on health and the relationship between migration and health. It is possible that the healthy migrant tendency and return migration (i.e., "salmon bias") are two selection processes that may give the appearance of a paradox.

Others have questioned the concept of an epidemiologic paradox on both theoretical and empirical grounds. According to Palloni and Morenoff (2001), documenting a paradox is contingent on three items: a health outcome, a target population, and a contrast population. Thus, rather than an omnibus Hispanic paradox, there could be dozens of paradoxes – or none – and the authors claim that the empirical evidence to support a paradox is quite limited. If a Hispanic paradox exists, it is probably a fairly recent phenomenon and refers to Mexican Americans only, not all Hispanic Americans (Palloni and Arias 2004).

Regardless of the ethnic group considered, research on health and aging is evolving beyond simple contrasts between social categories to studying the underlying processes associated with health inequality (Whitfield and Morgan 2008). In reviewing the literature, I identify four major streams of research that are advancing the cutting edge of minority health research across the life course. First, research on the relationship between discrimination and health is a major innovation for the field. Perceived discrimination negatively affects the physical and mental health of African, Hispanic, and Asian Americans, but according to Williams et al. (2003:200), there is a need to study how "the association between discrimination and health unfolds over the life course" (see also Noh et al. 2007).

Second, for most ethnic groups, information on immigration and nativity is critical for understanding health disparities (Angel et al. 2001). For groups that have recently immigrated, it is essential to measure duration of residence and generational status because residential history and acculturation may shape health inequality.

Third, there is an emergent body of research on the effects of racial/ethnic segregation on health, but the findings are inconsistent. Although most studies of segregation and health reveal a detrimental effect (e.g., Acevedo-Garcia 2001; Jackson et al. 2000), some studies of Hispanic communities reveal a beneficial effect on selected outcomes – what some refer to as the *barrio* benefit (Eschbach et al. 2004; Lee and Ferraro 2007).

Fourth, more studies are integrating information on medical care use over the life course in studies of health. This type of research is clarifying the processes leading to and consequences of primary care (e.g., Decker et al. 2009) and hospitalization (e.g., Ferraro and Shippee 2008).

Health inequality is substantial in many modern societies. Although we can appreciate the progress made to understand the extent and sources of SES and ethnic health inequality, we need to turn to investigating the exposures and available resources over the life course in order to break the chains of risk that generate the inequalities.

Cumulative Inequality in Health

The inequalities discussed in this chapter have stimulated research on how to capture the many insults experienced by persons over the life course. In this regard, theories of cumulative disadvantage (Dannefer 2003; O'Rand 1996) have garnered considerable interest for understanding these life course processes. More recently, cumulative inequality theory has been advanced as a middle-range theory for the study of health and aging (Ferraro and Shippee 2009), integrating elements from cumulative disadvantage, life course (Elder 1998), and stress process theories (Pearlin 1989). Pertinent to this chapter, two elements of the cumulative inequality theory are especially relevant: (a) life course trajectories are shaped by the accumulation of risk, available resources, and human agency and (b) trajectories are affected by the onset, duration, and magnitude of exposures (Ferraro et al. 2009).

As has become clear in this chapter, the literature on health and aging is moving beyond point estimates of exposures to risks and resources to measuring the onset, duration, and magnitude of such exposures. Many scholars use the term "accumulation," but there is precious little specificity in the use of the term. In its simplest form, cumulative exposure often refers to a sum of exposures. Scholars routinely add together events or experiences, and this is a very meaningful first step. Nevertheless, there are alternative ways of conceptualizing and measuring accumulation.

To illustrate different forms of accumulation, consider previous research on the link between childhood adversity and health. If the aim is to study the health consequences of multiple adversities, there are many ways to specify the multiple adversities in analytic models. Figure 29.1 presents four heuristic models for the accumulation of such adversity.

First, some scholars examine the unique effects of given adversities on adult health, addressing questions such as whether material adversity (family poverty) is more or less consequential than insults related to family organization (Kuh et al. 2002). I refer to this as the *unique adversities* approach, depicted in Fig. 29.1 as Model 1. The letters A through K represent separate adverse events or experiences such as parent death or family poverty.

Other investigators conceptualize the effects of childhood adversity as a dose–response function, what I refer to as *additive adversity* (Model 2). They often do this by adding adversities into an overall measure of cumulative adversity, arguing that there is a need to study joint or *cumulative* effects of the adversity (Felitti 2002; Turner and Lloyd 1995). One might also test polynomial forms of this variable to capture nonlinearities in the relationship between childhood adversity and adult health.

Third, still other scholars conceive of clusters of adversity, whereby some adversities tend to co-occur with others (O'Rand & Hamil-Luker 2005). *Adversity clustering* is depicted in Model 3.

Finally, it is possible that some early adversities may be endogenous to other adversities, even during childhood. Such a conception of *pathway adversity* specifies that selected adversities heighten the risk of other adversities or experiences (Model 4 in Fig. 29.1). For an example of Model 4, early adversity may increase the risk of smoking initiation in adolescence (Anda et al. 1999), which may lead to other adversities.

The four models outlined in Fig. 29.1 are a sample of the ways in which previous investigators have used extant measures of adversity or disadvantaged status. One can readily specify additional models such as those that combine one or more elements of these four.

In thinking of how to better study accumulation processes, there are also some phenomena that have received little or no attention in prior studies of health. I contend that we need better measures of some of our familiar concepts as well as new measures to effectively analyze cumulative inequality in health. At a minimum, it may be useful to consider two dimensions of accumulation that merit greater attention: (1) temporal and (2) perceptual.

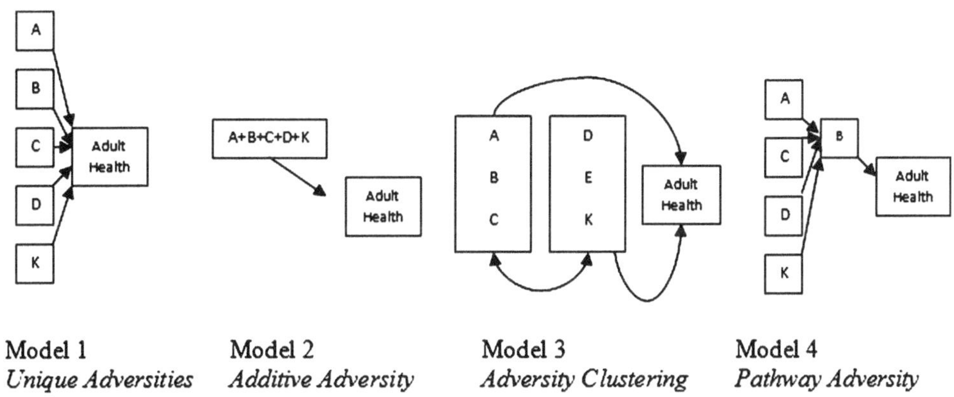

Fig. 29.1 Heuristic models for the accumulation of adversity

First, in studies of SES and health, we all too often resort to a single measure such as education at a given time. In studies of the link between religion and health, the predominant approach is measuring a religious phenomenon at a single point in time. In order to advance the science of the social antecedents of health, we will need more than a single slice in time to adequately capture exposure to risks or resources (Berkman 2009). Ideally, we want to know something about the onset of the exposure and its duration as well as interruptions. Did the accumulation begin during a critical period in the life course? The timing of accumulation matters, a fact easily illustrated with investment practices (i.e., contrast the consequences of investing the same amount of money but on two different time schedules). We are only beginning to see studies capture life course information on exposures and how they affect health outcomes (Schafer and Ferraro 2007; Willson et al. 2007), but they are needed to advance our understanding of health and aging.

Second, a good portion of the literature on cumulative inequality gives little attention to how the actor perceives his or her trajectory in a given life domain. Although the terms *adversity* and *disadvantage* are often used as synonyms, I draw a distinction between the two. Adversity refers to specific events that are perceived to be unfortunate or undesired, but disadvantage relates to a condition – a structural position – in some type of social hierarchy.

Cumulative advantage/disadvantage theory emphasizes the structural positions, but cumulative inequality theory emphasizes the dialectic between the structural positions and the actor's perception of experiences associated with these positions (Schafer et al. 2009). Not all positions, such as limited education, are regarded as a form of adversity. An actor may favorably judge what many consider to be an unfavorable position or vice versa.

Beyond a single judgment of a position, event, or experience, there also is the actor's evaluation of the accumulation of such experiences. Actors seek to make sense of their lived experiences. Thus, scientists may have a better understanding of why certain exposures do or do not have an anticipated effect by asking respondents about their experiences. As an exemplar for this type of research, Surtees and Wainwright (2007) asked respondents to assess the impact of various experiences on their lives. Life course epidemiology would benefit from tapping the actor's perceptions and definitions of life situations rather than presume what is in the actor's worldview. More generally, one could say that there is a subjective life course that needs to be considered in the study of aging and health.

In conclusion, the systematic study of accumulation is an important frontier in research on health and aging. If we can identify the accumulation processes related to health, we are better positioned to prevent disease, disability, and premature death. Moreover, as sociologists increasingly incorporate biomeasures into their analyses, earlier detection of health inequality should be beneficial to health for persons of all ages.

References

Acevedo-Garcia, Dolores. 2001. "Zip Code-Level Risk Factors for Tuberculosis: Neighborhood Environment and Residential Segregation in New Jersey, 1985–1992." *American Journal of Public Health* 91:734–41.

Adler, Nancy, Archana Singh-Manoux, Joseph Schwartz, Judith Stewart, Karen Matthews, and Michael G. Marmot. 2008. "Social Status and Health: A Comparison of British Civil Servants in Whitehall-II with European- and African-Americans in CARDIA." *Social Science and Medicine* 66:1034–45.

Alwin, Duane F., Scott M. Hofer, and Ryan J. McCammon. 2006. "Modeling the Effects of Time: Integrating Demographic and Developmental Perspectives." Pp. 20–38 in *Handbook of Aging and the Social Sciences,* Sixth Edition, edited by R. H. Binstock and L. K. George. Burlington, MA: Academic Press.

Anda, Robert F., Janet B. Croft, Vincent J. Felitti, Dale Nordenberg, Wayne H. Giles, David F. Williamson, and Gary A. Giovino. 1999. "Adverse Childhood Experiences and Smoking during Adolescence and Adulthood." *Journal of the American Medical Association 282*:1652–8.

Angel, Jacqueline L., Cynthia J. Buckley, and Arthur Sakamoto. 2001. "Duration or Disadvantage? Exploring Nativity, Ethnicity, and Health in Midlife." *Journal of Gerontology: Social Sciences* 56:S275–84.

Barker, David J. 1997. "Maternal Nutrition, Fetal Nutrition, and Disease in Later Life." *Nutrition* 13:807–13.

Barker, David J. P., ed. 2001. *Fetal Origins of Cardiovascular and Lung Disease*. New York: Marcel Dekker.

Berkman, Lisa F. 2009. "Social Epidemiology: Social Determinants of Health in the United States: Are We Losing Ground?" *Annual Review of Public Health* 30:27–41.

Blackwell, Debra L., Mark D. Hayward, and Eileen M. Crimmins. 2001. "Does Childhood Health Affect Chronic Morbidity in Later Life?" *Social Science and Medicine* 52:1269–84.

Brown, Scott C., Craig A. Mason, Joanna L. Lombard, Frank Martinez, Elizabeth Plater-Zyberk, Arnold R. Spokane, Frederick L. Newman, Hilda Pantin, and José Szapocznik. 2009. "The Relationship of Built Environment to Perceived Social Support and Psychological Distress in Hispanic Elders: The Role of 'Eyes on the Street'." *Journal of Gerontology: Social Sciences* 64B:234–46.

Case, Anne, Darren Lubotsky, and Christina Paxson. 2002. "Economic Status and Health in Childhood: The Origins of the Gradient." *American Economic Review* 92:1308–34.

Centers for Disease Control and Prevention. 2009. "Novel H1N1 Flu: 2009 H1N1 Early Outbreak and Disease Characteristics." Retrieved October 24, 2009 (http://www.cdc.gov/H1N1FLU/surveillanceqa.htm).

Conley, Dalton and Neil G. Bennett. 2000. "Is Biology Destiny? Birth Weight and Life Chances." *American Sociological Review* 65:458–67.

Crimmins, Eileen M., Mark D. Hayward, Aaron Hagedorn, Yasuhiko Saito, and Nicolas Brouhard. 2009. "Change in Disability-Free Life Expectancy for Americans 70-Years-Old and Older." *Demography* 46:627–46.

Dannefer, Dale. 2003. "Cumulative Advantage/Disadvantage and the Life Course: Cross-Fertilizing Age and Social Science Theory." *Journal of Gerontology: Social Sciences* 58B:S327–37.

Decker, Sandra L., Susan M. Schappert, and Jane E. Sisk. 2009. "Use of Medical Care for Chronic Conditions." *Health Affairs* 28:26–35.

Dube, Shanta R., Robert F. Anda, Vincent J. Felitti, Daniel P. Chapman, David F. Williamson, and Wayne H. Giles. 2001. "Childhood Abuse, Household Dysfunction, and the Risk of Attempted Suicide throughout the Life Span: Findings from the Adverse Childhood Experiences Study." *Journal of the American Medical Association* 286:3089–96.

Elder, G. H., Jr. 1974. *Children of the Great Depression: Social Change in Life Experience*. Chicago: University of Chicago Press.

Elder, Glen H., Jr. 1998. "The Life Course as Developmental Theory." *Child Development* 69:1–12.

Elo, Irma T. 2009. "Social Class Differentials in Health and Mortality: Patterns and Explanations in Comparative Perspective." *Annual Review of Sociology* 35: 553–72.

Eschbach, Karl, Glenn V. Ostir, Kushang V. Patel, Kyriakos S. Markides, and James S. Goodwin. 2004. "Neighborhood Context and Mortality among Older Mexican Americans: Is There a Barrio Advantage?" *American Journal of Public Health* 94:1807–12.

Felitti, Vincent J. 2002. "The Relationship between Adverse Childhood Experiences and Adult Health: Turning Gold into Lead." *The Permanente Journal* 6:44–7.

Ferraro, Kenneth F. and Melissa M. Farmer. 1996. "Double Jeopardy to Health Hypothesis for African Americans: Analysis and Critique." *Journal of Health and Social Behavior* 37:27–43.

Ferraro, Kenneth F. and Jessica A. Kelley-Moore. 2003. "Cumulative Disadvantage and Health: Long-Term Consequences of Obesity?" *American Sociological Review* 68:707–29.

Ferraro, Kenneth F. and Tetyana Pylypiv Shippee. 2008. "Black and White Chains of Risk for Hospitalization Over 20 Years." *Journal of Health and Social Behavior* 49:193–207.

———. 2009. "Aging and Cumulative Inequality: How Does Inequality Get Under the Skin?" *The Gerontologist* 49:333–43.

Ferraro, Kenneth F., Tetyana Pylypiv Shippee, and Markus H. Schafer. 2009. "Cumulative Inequality Theory for Research on Aging and the Life Course." Pp. 413–33 in *Handbook of Theories of Aging*, 2nd. Ed., edited by V. L. Bengtson, d. Gans, N. M. Putney, and M. Silverstein. New York: Springer Publishing.

Freedman, Vicki A., Eileen Crimmins, Robert F. Schoeni, Brenda C. Spillman, Hakan Aykan, Ellen Kramarow, Kenneth Land, James Lubitz, Kenneth Manton, Linda G. Martin, Diane Shinberg, and Timothy Waidmann. 2004. "Resolving Inconsistencies in Trends in Old-Age Disability: Report from a Technical Working Group." *Demography* 41:417–41.

Fries, James F. 1980. Aging, Natural Death, and the Compression of Morbidity. *New England Journal of Medicine* 303:130–5.

Geronimus, Arline T., Cynthia G. Colen, Tara Shochet, Lori Barer Ingber, and Sherman A. James. 2006. "Urban-Rural Differences in Excess Mortality among High-Poverty Populations: Evidence from the Harlem Household Survey and the Pitt County, North Carolina Study of African American Health." *Journal of Health Care for the Poor and Underserved* 17:532–58.

Haas, Steven. 2008. "Trajectories of Functional Health: The 'Long Arm' of Childhood Health and Socioeconomic Factors." *Social Science and Medicine* 66:849–61.

Haas, Steve and Leah Rohlfsen. 2010. "Life Course Determinants of Racial and Ethnic Disparities in Functional Health Trajectories." *Social Science and Medicine* 70:240–50.

Herd, Pamela, Brian Goesling, and James S. House. 2007. "Socioeconomic Position and Health: The Differential Effects of Education versus Income on the Onset versus Progression of Health Problems." *Journal of Health and Social Behavior* 48:223–38.

House, James S., Paula M. Lantz, and Pamela Herd. 2005. "Continuity and Change in the Social Stratification of Aging and Health Over the Life Course: Evidence From a Nationally Representative Longitudinal Study From 1986 to 2001/2002 (Americans' Changing Lives Study)." *Journal of Gerontology: Social Sciences* 60:S15–26.

Irving, Shalon M. and Kenneth F. Ferraro. 2006. "Reports of Abusive Experiences during Childhood and Adult Health Ratings: Personal Control as a Pathway?" *Journal of Aging and Health* 18:458–85.

Jackson, Sharon, Roger Anderson, Norman Johnson, and Paul D. Sorlie. 2000. "The Relation of Residential Segregation to All-Cause Mortality: A Study in Black and White." *American Journal of Public Health* 90:615–7.

Johnson, Nan E. 2000. "The Racial Crossover in Comorbidity, Disability, and Mortality." Demography 37:267–83.

Kelley-Moore, Jessica A., and Kenneth F. Ferraro. 2004. "The Black/White Disability Gap: Persistent Inequality in Later Life?" *Journal of Gerontology: Social Sciences* 59B:S34–43.

Klinenberg, Eric. 2002. *Heat Wave: A Social Autopsy of Disaster in Chicago.* Chicago: University of Chicago Press.

Kuh, Diana and Yoav Ben-Shlomo. 1997. "Introduction: A Life Course Approach to the Aetiology of Adult Chronic Disease." Pp. 3–14 in *A Life Course Approach to Chronic Disease Epidemiology,* edited by D. Kuh and Y. Ben-Shlomo. New York: Oxford University Press.

Kuh, Diana, Rebecca Hardy, Claudia Langenberg, Marcus Richards, and Michael E. J. Wadsworth. 2002. "Mortality in Adults Aged 25–64 Related to Socioeconomic Conditions in Childhood and Adulthood: Post War Cohort Study." *British Medical Journal* 325:1076–80.

Lee, Min-Ah and Kenneth F. Ferraro. 2007. "Neighborhood Residential Segregation and Physical Health among Hispanic Americans: Good, Bad, or Benign?" *Journal of Health and Social Behavior* 48:131–48.

Link, Bruce G. and Jo Phelan. 1995. "Social Conditions as Fundamental Causes of Disease." *Journal of Health and Social Behavior* 35:80–94.

Lynch, John and George Davey Smith. 2005. "A Life Course Approach to Chronic Disease Epidemiology." *Annual Review of Public Health* 26:1–35.

Manton, Kenneth G. 2008. "Recent Declines in Chronic Disability in the Elderly U.S. Population: Risk Factors and Future Dynamics." *Annual Review of Public Health* 29:91–113.

Manton, Kenneth G., Larry S. Corder, and Eric Stallard. 1993. "Estimates of Change in Chronic Disability and Institutional Incidence and Prevalence Rates in the U.S. Elderly Population from 1982, 1984, and 1989 National Long Term Care Survey." *Journal of Gerontology: Social Sciences* 48:S153–66.

Manton, Kenneth G., XiLiang Gu, and Gene R. Lowrimore. 2008. "Cohort Changes in Active Life Expectancy in the U.S. Elderly Population: Experience from the 1982–2004 National Long-Term Care Survey." *Journal of Gerontology: Social Sciences* 63:S269–81.

Markides, Kyriakos and J. Coreil. 1986. "The Health of Hispanics in the Southwestern United States: An Epidemiological Paradox." *Public Health Reports* 101:253–65.

Marmot, Michael G. 2003. "Understanding Social Inequalities in Health." *Perspectives in Biology and Medicine* 46:S9–23.

Martin, Linda, Vicki Freedman, Robert Schoeni, and Patricia Andreski. 2009. "Health and Functioning Among Baby Boomers Approaching 60." *Journal of Gerontology: Social Sciences* 64B:369–77

Martin, Linda G., Robert F. Schoeni, Vicki A. Freedman, and Patricia Andreski. 2007. "Feeling Better? Trends in General Health Status." *Journal of Gerontology: Social Sciences* 62:S11–21.

Noh, Samuel, Violet Kaspar, and K. A. S. Wickrama. 2007. "Overt and Subtle Racial Discrimination and Mental Health: Preliminary Findings for Korean Immigrants." *American Journal of Public Health* 97:1269–74.

Ogden, Cynthia L., Margaret D. Carroll, Lester R. Curtin, Margaret A. McDowell, Carolyn J. Tabak, and Katherine M. Flegal. 2006. "Prevalence of Overweight and Obesity in the United States, 1999–2004." *Journal of the American Medical Association* 295:1549–55.

O'Rand, Angela M. 1996. "The Precious and the Precocious: Understanding Cumulative Disadvantage and Cumulative Advantage over the Life Course." *The Gerontologist* 36:230–8.

O'Rand, Angela M. and Jennifer Hamil-Luker. 2005. "Processes of Cumulative Adversity: Childhood Disadvantage and Increased Risk of Heart Attack across the Life Course." *Journal of Gerontology: Social Sciences* 60B:117–24.

Palloni, Alberto and Elizabeth Arias. 2004. "Paradox Lost: Explaining Hispanic Adult Mortality." *Demography* 41:385–415.

Palloni, Alberto and Jeffrey D. Morenoff. 2001. "Interpreting the Paradoxical in the Hispanic Paradox: Demographic and Epidemiologic Approaches." *Annals of the New York Academy of Sciences* 954:140–74.

Pearlin, Leonard I. 1989. "The Sociological Study of Stress." *Journal of Health and Social Behavior* 30:241–56.

Preston, Samuel H., Mark E. Hill, and Greg L. Drevenstedt. 1998. "Childhood Conditions that Predict Survival to Advanced Ages among African-Americans." *Social Science and Medicine* 47:1231–46.

Riley, Matilda W. 1987. "On the Significance of Age in Sociology." *American Sociological Review* 52:1–14.

Ross, Catherine E., and Chia-Ling Wu. 1996. "Education, Age, and the Cumulative Advantage in Health." *Journal of Health and Social Behavior* 37:104–20.

Schafer, Markus H. and Kenneth F. Ferraro. 2007. "Obesity and Hospitalization over the Adult Life Course: Does Duration of Exposure Increase Use?" *Journal of Health and Social Behavior* 48:434–49.

Schafer, Markus H., Tetyana Pylypiv Shippee, and Kenneth F. Ferraro. 2009. "When Does Disadvantage Not Accumulate? Towards a Sociological Understanding of Resilience." *Swiss Journal of Sociology*: 35:231–51.

Schmader, Kenneth E., Linda K. George, Bruce H. Burchett, John D. Hamilton, and Carl F. Pieper. 1998. "Race and Stress in the Incidence of Herpes Zoster in Older Adults." *Journal of the American Geriatrics Society* 46:973–7.

Settersten, Richard A., Jr., ed. 2003. *Invitation to the Life Course: Toward New Understandings of Later Life*. Amityville, NY: Baywood Publishing.

Shaw, Benjamin A. and Neal Krause. 2002. "Exposure to Physical Violence during Childhood, Aging, and Health." *Journal of Aging and Health* 14:467–94.

Singh-Manoux, Archana and Michael Marmot. 2005. "Role of Socialization in Explaining Social Inequalities in Health." *Social Science and Medicine* 60:2129–33.

Singhal, Atul and Alan Lucas. 2004. "Early Origins of Cardiovascular Disease: Is There a Unifying Hypothesis?" *The Lancet* 363:1642–5.

Sliwinski, Martin and Jacqueline Mogle. 2008. "Time-Based and Process-Based Approaches to Analysis of Longitudinal Data." Pp. 477–91 in *Handbook of Cognitive Aging: Interdisciplinary Perspectives,* edited by S. M. Hofer and D. F. Alwin. Thousand Oaks, CA: Sage Publications.

Smith, Ken R., Geraldine P. Mineau, Gilda Garibotti, and Richard Kerber. 2009. "Effects of Childhood and Middle-Adulthood Family Conditions on Later-Life Mortality: Evidence from the Utah Population Database, 1850–2002." *Social Science and Medicine* 68:1649–58.

Snelgrove, John W., Hynek Pikhart, and Mai Stafford. 2009. "A Multilevel Analysis of Social Capital and Self-Rated Health: Evidence from the British Household Panel Survey." *Social Science and Medicine* 68:1993–2001.

Surtees, Paul G. and Nicholas W. J. Wainwright. 2007. "The Shackles of Misfortune: Social Adversity Assessment and Representation in a Chronic Disease Epidemiological Setting." *Social Science and Medicine* 64:95–111.

Tiikkaja, Sanna, Örjan Hemström, and Denny Vågerö. 2009. "Intergenerational Class Mobility and Cardiovascular Mortality among Swedish Women: A Population-Based Register Study." *Social Science and Medicine* 68:733–39.

Turner, R. Jay and Donald A. Lloyd. 1995. "Lifetime Traumas and Mental Health: The Significance of Cumulative Adversity." *Journal of Health and Social Behavior* 36:360–76.

Verbrugge, Lois M. 1986. "From Sneezes to Adieux: Stages of Health for American Men and Women." *Social Science and Medicine* 22:1195–1212.

Verbrugge, Lois M. and Alan M. Jette. 1994. "The Disablement Process." *Social Science and Medicine* 38:1–14.

Wadsworth, M. E. J. 1997. "Health Inequalities in the Life Course Perspective." *Social Science and Medicine* 44:859–69.

Warner, David F. and Mark D. Hayward. 2006. "Early-Life Origins of the Race Gap in Men's Mortality." Journal of Health and Social Behavior 47:209–26.

Wheaton, Blair and Philippa Clarke. 2003. "Space Meets Time: Integrating Temporal and Contextual Influences on Mental Health in Early Adulthood." *American Sociological Review* 68:680–706.

Whitfield, Keith and Adrienne Aiken Morgan. 2008. "Minority Populations and Cognitive Aging." Pp. 384–97 in *Handbook of Cognitive Aging: Interdisciplinary Perspectives,* edited by S. M. Hofer and D. F. Alwin. Thousand Oaks, CA: Sage Publications.

Williams, David R., Harold W. Neighbors, and James S. Jackson. 2003. "Racial/Ethnic Discrimination and Health: Findings from Community Studies." *American Journal of Public Health* 93:200–8.

Willson, Andrea E., Kim M. Shuey, and Glen H. Elder, Jr. 2007. "Cumulative Advantage Processes as Mechanisms of Inequality in Life Course Health." *American Journal of Sociology* 112:1886–1924.

Yi, Zeng, Danan Gu, and Kenneth C. Land. 2007. "The Condition of Childhood Socioeconomic Conditions with Healthy Longevity at the Oldest-Old Ages in China." *Demography* 44:497–518.

Chapter 30
Mental Health and Aging: A Life-Course Perspective

David Mechanic and Donna D. McAlpine

Much written about aging focuses on the problems and challenges faced by elderly people. However, aging is a lifelong process in which physical, psychological, and social capacities and resources are acquired and modified. Understanding psychological function and mental health in later life requires a life-course perspective that takes account of biological predispositions, socioeconomic status (SES) and environmental exposures, changing social conditions, and interpersonal networks (George 1999).

Such a life-course perspective highlights that one's mental well-being at any particular part of life is shaped by earlier experiences, the historical period in which one lives, and the social context. One of the most remarkable changes in how we have come to think of aging and health over the past 30 years has been the distinction between aging as a biosocial process and disease occurrence. While it is obvious that ill health accumulates over the life course, and that the prevalence of chronic disease and serious comorbidities are found more commonly with advancing age, the link of later stages of the life course to disease and loss of function is hardly inevitable and many people live their lives free of serious morbidity well into old age. The relative role of predisposition and experience in establishing these patterns remains uncertain although we increasingly recognize that biological predisposition is itself shaped in interaction with the environment from fetal development and early life.

Research Perspective

In the field of mental health, much of the research on the determinants of health has been aimed at trying to identify the life experiences that account for later well-being. By mid-life, there is already extraordinary heterogeneity of capacity and function among persons in the same age cohort reflecting the cumulative influence of life experiences as they interact with biological endowments and as they come to vary among persons of different socioeconomic circumstances, race, ethnic and religious background, and the influence of the particular historical and contextual experiences of their age cohort. Even such factors as changing rates of marriage and fertility and the changing size of particular age cohorts set the stage for the types of opportunities and constraints people experience as they age. Thus, any serious examination of the sociology of aging must take account of the heterogeneity of experience and outcomes, a type of understanding very much enhanced by a life-course perspective. The appeal of life-course theory is tempered by the difficulty of fully applying

D. Mechanic (✉)
Institute for Health, Health Care Policy
and Aging Research, Rutgers University, New Brunswick, NJ, USA
e-mail: mechanic@rci.rutgers.edu

R.A. Settersten, Jr. and J.L. Angel (eds.), *Handbook of Sociology of Aging*, Handbooks of Sociology and Social Research, DOI 10.1007/978-1-4419-7374-0_30, © Springer Science+Business Media, LLC 2011

it in empirical research. As Crosnoe and Elder (2002:309) have noted, "Studying the aging process as multifaceted and embedded in the life course is a demanding enterprise that requires long-term longitudinal data on multiple contexts. Except for some pioneering longitudinal studies, few data sets can meet such requirements."

Carrying out single studies of the life course is almost impossible although there have been some studies that cover much of the life span of selected cohorts with repeated measures over many years. While we have benefitted greatly from such studies, they are limited by the selectivity of the populations and the sample attrition that occurs over the years. Also, with new knowledge and theory, often the measures that we become interested in over time are quite different from those that were initially seen as most important, and the older measures themselves often are deficient from the perspective of current measurement standards. Thus, life-course studies, in reality, represent more a perspective and way of looking at human development than a particular type of study methodology. Using a life-course perspective makes possible synthetic approaches that combine data from varying ages, cohorts, and historical periods that illuminate our understanding of the variety of human experience.

Serious consideration of aging in a social context must substantially deal with issues of values and preferences and the cultural systems within which people function. Examples include attitudes toward household composition, work and retirement, personal independence, obligations to give and receive support and assistance among kin, and many related issues. These cultural attitudes are affected, as well, by such broader social patterns as geographic mobility within families, the role of gender in family caregiving, the increased employment of women, and changing educational preparation and economic capacities. For example, as persons enter older ages, becoming frailer and losing spouses and friends, they often have more need for assistance and social support than is available to them. Such persons might reasonably benefit from home sharing arrangements with relatives or other elderly persons in comparable situations. Yet increased economic independence allows many elders to retain single-person households which they prefer to protect their sense of independence or to avoid placing burdens on family members or friends. Thus, acting out one's values and preferences, and the opportunity to do so, may not provide the optimal social context for dealing with increased frailty.

One of the most challenging problems confronting life-course research is untangling the effects of age, cohort, and period when studying various health outcomes. Age refers to the period of time lived while cohort refers to the unique experiences of persons who share the same age at a particular point of time. Period refers to the historical period in which the person lives. Persons who make up a defined cohort, such as those born in a particular year, are exposed to the same historical events (period effects), but other age cohorts are, as well, and the influence of an environmental event will impact cohorts of various ages differently. Thus, period effects are aggregated among diverse cohorts that experience the event. Since the concepts of age, cohort, and period are interrelated, it is virtually impossible to assess the effects of each separately and independent of the others. This presents significant challenges to interpreting the prevalence of various mental disorders or the relative influence of risk factors over the life course.

Age and Risk of Mental Disorder

Definitions of mental disorder in the United States are based on the clinical criteria in the Diagnostic and Statistical Manual of Mental Disorders (DSM) published by the American Psychiatric Association (current version DSM-IV-TR) (American Psychiatric Association 2000). The DSM categorizes disorders on a number of different axes, with Axis I describing common clinical disorders such as depression and anxiety, and Axis II describing personality disorders and mental retardation (although the next version of the DSM may drop this distinction). The DSM is symptom based. For example, in DSM-IV, the diagnosis of major depression is based on patients having five symptoms that are present during the same 2-week period and reflect a change in prior function. At least, one of the symptoms must be either depressed mood or loss of interest or pleasure, but the remaining four may be any of seven others listed. Thus, the symptoms are largely treated as interchangeable although they vary from such symptoms as insomnia nearly every day, weight gain or loss not connected to a diet, fatigue or loss of energy every day as compared to a suicide attempt or a specific plan for committing suicide or psychomotor agitation or retardation nearly every day observable to others.

Early Onset Affecting the Life Course: Childhood and Adolescence

The DSM includes a specific category of disorders that typically have first onset in childhood and adolescence such as pervasive developmental disorders, which includes autism. The clinical definition of autism requires, for example, that onset is before the age of three. However, for the most part, the DSM is relatively indifferent to age. That is, the DSM is based on the assumption that the presentation of symptoms is similar across the life course, and that the essential features of the disorders are the same regardless of whether they are experienced in early or later life. This assumption has been largely untested in research and belies evidence that the presentation of symptoms of many disorders does vary over the life course and may be quite different for older compared to younger adults (Jeste et al. 2005).

The best estimates of the prevalence of various disorders across the life course come from community studies. Characteristics of treated samples may be systematically different in terms of demographics, health status, and help-seeking behavior than the majority of persons with a mental health problem who are not treated. While there are a number of nationally representative community mental health studies of the adult population in the United States, there are no comparable studies of childhood or adolescence.

Estimates of the prevalence of disorder in childhood and adolescence vary widely across samples and the measures employed. Autism is a good example of the difficulty in drawing any firm conclusions about the prevalence of mental illness and points to the importance of understanding the social construction of disorder. The Centers for Disease Control and Prevention (CDC) recently estimated that about one in 110 children has an autism spectrum disorder (ASD) (CDC 2009). These and similar high estimates of the prevalence of autism have supported the popular perceptions that rates have increased, and have generated considerable research and speculation in the popular media about the probable causes; the most controversial and highly unlikely causes involve the link with childhood vaccines. Despite the attention that the "autism epidemic" has received, there continues to be serious debate about whether the rates of autism have really increased. Indeed, there is some evidence that findings in support of increasing prevalence result from expanding the scope of behaviors that constitute autism, increased awareness of the disorder, or changes in diagnostic practices (Nassar et al. 2009; Rutter 2009). There is considerable evidence, for example, that cases previously diagnosed as mental retardation are now commonly diagnosed as ASD. There have been even more intense debates over estimates of the prevalence of attention deficit/hyperactivity disorder (Mayes et al. 2008).

Despite serious questions about the validity of DSM criteria for many mental disorders, a number of small community studies of children and adolescents have used these criteria to estimate the prevalence of disorder in children as young as preschool age. Estimates of disorder vary widely, but anxiety and conduct disorders appear to be the most common disorders among children age five through 17 years of age (Costello et al. 2005), with overall about 1 in 8 experiencing serious emotional disturbance. Egger and Angold (2006) argue that rate of serious emotional disorder among preschoolers (12.1%) is quite similar to that observed for adolescents.

Later Onsets

Risk for common disorders, such as depression and anxiety, and severe mental illnesses, such as schizophrenia, increase from childhood to adolescence into young adulthood, but patterns of risk over the course of adulthood are not well understood. Probably, the clearest evidence for an effect of aging on risk of mental disorders concerns cognitive impairment and other dementias, including Alzheimer's disease, where aging is associated with greater risk. The most recent national data on the prevalence of dementia in the United States suggest that dementia among persons over 70 is about 14%, while the prevalence of Alzheimer's disease is about 9.7% (Plassman et al. 2007). The rates increase with age; for example, almost 37% of the population over the age of 90 years meet the criteria for dementia (Plassman et al. 2007). Cognitive impairment without dementia is more prevalent, affecting about 22% of persons over the age of 70 and 39% of those over the age of 90 (Plassman et al. 2008). It is widely accepted that the aging process contributes to the increased prevalence of cognitive impairment and dementias.

The effect of aging on risk for other types of mental disorders is less clear. The most recent community data in the United States come from the National Comorbidity Survey Replication (NCS-R) and indicate that the 12-month prevalence of common mental disorders is about 32% for the total adult population, but only 16% for persons age 60 years and older (National Comorbidity Study Replication 2007).

The most commonly studied mental disorder in later life is depression because it is prevalent, disabling, often co-occurs with other important chronic diseases of the elderly, and because of the persistent belief among clinicians that depression is especially under-recognized and under-treated in elderly populations. Geriatricians and primary care physicians see many very elderly patients with complex comorbidities and it is hardly surprising that they see much depression as well. Community studies, however, do not bear out the perceptions of clinicians. Estimates from NCS-R, for example, suggest that 1-month, 12-month, and lifetime prevalence measured cross-sectional are lower among older persons compared to younger persons (Kessler et al. 2010).

Schizophrenia is also commonly studied; although the prevalence is much lower than more common disorders such as depression, the associated disability is substantial and commonly chronic. There are few national community studies that examine aging and schizophrenia, and there is much legitimate skepticism of estimates of schizophrenia using interview schedules in community studies. The Epidemiological Catchment Area (ECA) study estimated the overall prevalence rates of schizophrenia for 1 year and lifetime for the population as 1 and 1.5%. Comparable estimates for the population 65 and older were 0.2 and 0.3% (Keith et al. 1991).

Conceptual Issues in Interpreting Aging and Risk

The fact that lifetime prevalence of many common mental disorders appears to be lower for older adults compared to younger adults suggests that there may be important cohort differences. Some studies suggest that recent cohorts of younger persons have increased risk for disorder (Kessler et al. 2005). If correct, we would expect that in future decades the prevalence of mental disorder among older persons will be higher than it is today.

Conceptual and methodological issues make it difficult to draw firm conclusions about the association between aging and risk of disorder or to separate out age from cohort effects. First, it is not clear that the nosological categories and diagnostic criteria used in population surveys apply equally well to persons at older ages and particularly to the old-old. In addition, the low prevalence rates of mental illness at older ages may be a result of problems in memory and other distortions. Using ECA data, investigators examined the test–retest reliability of retrospective symptom reports of depression. They found that approximately 40% of those who reported a lifetime history of depressed mood in 1981 did not report such a history when re-interviewed in 1994 (Thompson et al. 2004). There are various interpretations of such low concordance over time including the fact that respondents who are not depressed at time of interview are less likely to recall a previous episode. In any case, there is limited value in lifetime estimates using retrospective reports from persons in varying age groups.

There are a variety of other interpretations of the lower prevalence rates in the elderly. Persons with serious mental illness have significantly reduced longevity than others in the population and selective mortality obviously affects morbidity observed in the later years. Also, studies that focus on the mental health of the elderly tend to be small and nonrepresentative or have limited and/or inadequate measures. Further, population studies of mental illness typically exclude persons in residential facilities such as nursing homes who are disproportionately old, poor, and disabled. Such persons often have secondary diagnoses of depression although the validity of these diagnoses is not rigorously assessed (Mechanic and McAlpine 2000).

Many studies have focused on the prevalence of psychological distress or depressive symptoms in contrast to clinical depression. Mirowsky and Ross (1999) combined the data from three national surveys and one state survey in Illinois that asked respondents how many days per week they were depressed. The measure used represents general well-being or distress rather than clinical disorder. These data cover some 18,450 respondents and while each survey has somewhat different results they show a comparable pattern. They find an approximation of a U-shaped curve with a relatively high average of depressed days among the age group 18–22 that falls to its lowest levels in the middle years, around ages 40–60. At the end of the middle years, reports of depressed days begin rising and continues with increasing age reaching its highest levels in the age group 88–99.

The matter of the true association between aging and psychological distress or well-being is by no means settled. In an analysis of data from over 70 countries and multiple years which allows for looking at cohort and aging effects, Blanchflower and Oswald (2008) demonstrate an opposite U-shape pattern to that observed by Mirowsky and Ross. They report that happiness, life satisfaction, and psychological well-being are at minimum in middle age; they also find some evidence of a cohort difference in the U.S. data, with later cohorts reporting less happiness than earlier cohorts.

Risk of mental disorder across the life course is associated with social stratification – most importantly gender and social class. Women have a lower lifetime risk of impulse control and substance-related disorders and a higher risk for anxiety and depression (Kessler et al. 2005). It is unclear whether gender differences in the prevalence of specific types of disorders reflect differences in genetic vulnerability, exposure to risk factors, the availability of protective resources, or represent functionally equivalent ways of expressing psychological distress.

SES – commonly measured as education, occupational attainment, and income – is negatively associated with many measures of health, including psychological health, although the direction of the associations varies by type of disorder in the case of major mental illnesses (Muntaner et al. 2004). Low SES appears to have a causal role in depression. In contrast, schizophrenia results in downward social mobility or failure to keep up with peers (Dohrenwend et al. 1992). Understanding the casual mechanisms that link social position to health requires examining the timing of first onset and subsequent changes in SES position.

A Life-Course Perspective

Epidemiological research on the prevalence of various mental disorders for persons of different ages, and from different social statuses, is intriguing and important to our understanding of well-being over the life course. While a considerable amount of research has been devoted to trying to explain the observed patterns, much of this work is cross-sectional and does not really take an explicit life-course approach. However, the principles of life-course theory offer a useful organizing framework for synthesizing this research. The core principles of life-course theory include the importance of (1) developmental transitions, (2) one's location in time and place, (3) human agency played out in the context of social structure, and (4) the concept of linked lives or the interdependence of individuals enmeshed in social networks (Elder 2000). These principles help us interpret observed epidemiological patterns of prevalence and risk.

Developmental Perspective

An essential feature of life-course theory is its emphasis on the timing of events or transitions in life for understanding later trajectories or outcomes. Simply put, "the developmental impact of a succession of life transitions or events is contingent on when they occur in a person's life" (Elder 1998:3). There are critical periods for certain developmental and social maturation phases but firm evidence is stronger in the biological than social areas. For some neurobiological functions there seem to be narrow early stimulative periods required for the system to function, such as the visual cortex (Cynader and Frost 1999). Similar arguments have been made for language acquisition and facility, attachment, and other psychosocial development processes but it is more contested as to whether critical periods exist in the same sense. Nevertheless, there is much evidence that early stimulation and experience help shape subsequent social and biodevelopmental outcomes. Increasingly, we learn that genes themselves are activated or modified by early environmental exposures.

Research indicates that the conditions of prenatal and early life, the nutritional history of the mother, and early growth and development are associated with chronic disease and mortality in middle and later life (Mechanic 2007). Although this work has focused for the most part on cardiovascular disease and diabetes, studies of early nutritional deprivation in famines have also found links with neurodevelopment and more specifically with schizophrenia (St. Clair et al. 2005; Susser et al. 1998). These pathways to causation are very complex and associated with SES. Low SES parents are more likely to have children of low birth weight and who have nutritional deficiencies, and these children have lower levels of educational and occupational attainment (Barker et al. 2001). There is also growing evidence that these influences are intergenerational in that the health status of grandmothers influences the subsequent health status of their children and grandchildren (Mechanic 2007). These outcomes are shaped by both family SES and nutritional deficiencies and their interactions. But they are then further moderated by the levels of education attained and associated factors. Thus, the story, like much else, involves complex and continuing interactions between biology and environment. Poverty itself is not a static state. Families and children move in and out of poverty over time as personal circumstance, economic conditions, and social policies change. Children exposed to more persistent lives of poverty do less well on cognitive and health indicators.

Timing is critical in many developmental and social processes. Three issues of timing are particularly important for understanding mental health across the life course: the timing of first onset of disorder, the timing of events that increase risk of disorder, and the timing of development of personal resources that affect adjustment to the disorder.

Timing of first onset of disorder: Most psychiatric disorders begin relatively early in the lifespan, typically around late adolescence and early adulthood but may not be recognized and treated without long delays. Even after initial recognition of symptoms, delays of 10 years or more are common before people receive any formal assistance for their mental health problems (Wang et al. 2005). However, many early symptoms of disorder are self-limited and resolve themselves without interventions. Much depends on the nature of the problem, the persistence of aggravating circumstances, and the evolving life circumstances of the individual including neglect and abuse, poverty and deprivation, disruption of stable living conditions and social networks, and the like. Biological propensities interact with environmental exposures in ways that make individuals more or less susceptible to later risk of disorder or exacerbation of symptoms. Some disorders are substantially shaped by early biological incapacities that then carry through the entire lifespan such as mental retardation or autism. Others such as schizophrenia usually first become evident in late adolescence and early adulthood with the course of the disorder depending on intrinsic biological factors and environmental exposures, still poorly understood. Serious conduct disorder among children which disproportionately occurs in deprived SES contexts, while socially shaped to a considerable degree, tends to be associated with a negative trajectory involving cumulative failures and troubles that persist over the life course (Robins 1966, 1983).

Evidence of the importance of timing of first onset of mental illness on later outcomes comes from community studies that ask respondents to recall their first experience of serious symptoms of disorders such as depression. This research suggests that early onset of some disorders, particularly externalizing behaviors, is associated with many later life problems (Breslau et al. 2008).

Stronger evidence comes from cohort studies that follow persons over considerable lengths of time. There have been many studies of the outcomes of conduct disorder in children and adolescents beginning with the classic studies of Robins (1966). A more recent report based on the British 1946 birth cohort (children in the United Kingdom born during 1 week in March) and followed for 40 years is indicative of findings more generally (Colman et al. 2009). In this study, 3,652 children were assessed by their teachers for symptoms of externalizing disorders (disobedience, truancy, lying, etc.) at ages 13 and 15. Those characterized as having severe externalizing problems at baseline were significantly more likely to be male, to come from lower social class families, to have poorer cognitive abilities at age 8, and more depression and anxiety at ages 13 and 15. They were likely to become parents before age 20, had one or more divorces, were unhappy with their family lives, reported problems in relationship to others, lacked educational qualifications, had low status manual jobs, had financial difficulties, and high life adversity (Colman et al. 2009). They had more difficult lives as measured by outcomes at ages 36, 43, and 53. Those with severe externalizing problems as adolescents were also more likely to report symptoms of depression/anxiety or a history of nervous problems 40 years later.

Using the same survey, Colman et al. (2007) have also examined outcomes for children who showed evidence of internalizing disorders (fearful, timid, anxious, sad, etc.) at age 13 or 15. They found that while a single episode of problems (at age 13 or 15) in adolescence was not associated with poor mental health outcomes into adulthood and middle age, persistent problems during adolescence predicted more negative outcomes. However, other studies have found that internalizing symptoms early in life are much less damaging in adulthood than externalizing problems (Robins 1983).

We know too little about outcomes for persons with mental illness as they enter older age, although some studies of persons with severe mental illness such as schizophrenia paint a much more optimistic picture than one might suspect given the disability associated with disorders such as schizophrenia and clinical impressions. There are a number of long-term longitudinal studies of patients with schizophrenia that found that in later life the highly disturbing positive symptoms of schizophrenia (e.g., delusions and hallucinations) abate and that these patients are able to adapt to community life but at relatively modest levels of function and quality of life (Mechanic 2008:30–31). Cohen et al. (2009), for example, measured a wide variety of outcomes from recovery through community integration for persons over the age of 55 with schizophrenia. They report that about 17% of persons with schizophrenia had recovered and almost 50% had achieved symptom remission after the age of 55. However, outcomes related to community integration and successful aging (defined as absence of disease and disability, high cognitive and physical functioning, and engagement with family and friends and feeling useful to others) were less positive. Overall this literature contradicts the impression of schizophrenia following an inevitable downward course.

Timing of events and risk of mental disorder: A developmental perspective also calls attention to the importance of the timing of events in the life course that may increase risk of mental disorder. There is a substantial body of research that shows that stressful life events and chronic strains are associated with psychological health during all stages of life (Pearlin et al. 2005), with much of the research focusing on depression or psychological distress. Moreover, exposure to social stressors is not random; for example, women and persons from lower SES statuses are more likely to experience stressful events compared to men and individuals who are more advantaged (Turner et al. 1995). However, few researchers have taken a truly life-course perspective and tried to understand how the timing of events during the life cycle shapes later life transitions and trajectories and ultimately mental health outcomes (Pearlin et al. 2005).

The most intensive and sustained research program on the role of stress has been the work of George Brown, Tirril Harris, and their co-workers in the United Kingdom and collaborators in several other countries (Brown and Harris 1978, 1989; Harris et al. 1987). These investigators, in contrast to many epidemiological survey investigators, use a very intensive interview instrument to obtain rich data. Reported symptoms, life events, difficulties, and other measures are then derived using independent ratings that may be different from the subjective reports provided by the respondent. The purpose of these independent measures is to assess meaning of experiences without the biases characteristic of respondents' ratings of their experience. The model Brown and Harris present conceptualizes stressful life events as provoking agents, and one of their early findings that highlight the importance of life-course work was that early loss of a mother was associated with depression in adulthood. They found that such loss is associated with lack of adequate care, which contributed to a trajectory of premarital pregnancy, less effective coping, and often early and unsuitable marriages (Bifulco et al. 1987; Harris et al. 1987). Although these patterns occur more commonly in the lower classes, the investigators distinguish among the circumstances of lower class families that more or less contribute to depression. The larger theoretical frame characteristic of life-course analysis assumes that life experience at any point derives in important ways from earlier influences and choices that either open or constrain subsequent opportunities. Choices made about schooling, jobs, marriage, childbearing, divorce, retirement and much more, and their timing, establish the conditions for subsequent transitions (Brown 1986).

In addition to loss of a parent, other childhood adversities and traumas have also been shown to be associated with mental health much later in life. Horwitz et al. (2001) compare mental health outcomes in adulthood among persons who had been documented to have suffered sexual abuse, physical abuse, or neglect before the age of 11 with a matched sample who did not experience such abuse. Men and women who had been abused were more likely than the comparison group to have dysthymia and antisocial personality disorder 20 years later and women were also more likely to have alcohol use problems.

The effects of early life adversities extend into later life, but the pathways from childhood abuse to disorder much later in life are not clear. Horwitz et al. (2001) find that individuals who were abused or neglected as children also report more lifetime stressful events such as job loss, financial problems, and marital troubles and that accounting for these events reduces the association between childhood abuse and lifetime prevalence of mental disorder to nonsignificance. Similarly, Brown et al. (2008) argue that maltreatment in childhood is linked to depression in adulthood when it is associated with the greater exposure and vulnerability to severe life events more proximate to onset of depression.

Life-course theory also suggests that the timing of specific events is important for later development. For example, studies of men who joined the military service during World War II in their thirties had more negative outcomes into old age than men who joined the services earlier. Men who joined the service early in their life trajectories suffered less life disruption and related costs (MacLean and Elder 2007).

Timing of development of personal resources: Resources developed earlier in life also affect individuals' adjustment to patienthood and the subsequent course of disorder. When disruptions in educational acquisition, occupational advancement, family formation, childbearing, and many other important functions occur, they significantly shape subsequent outcomes and their impact. The occurrence of a serious mental illness well prior to full educational attainment vastly limits the life-course post illness as compared with its occurring after significant educational and occupational achievement. Although gaining employment is one of the largest problems faced by persons who have had serious mental illnesses, having higher educational attainment is one of the best predictors of workforce participation (Mechanic et al. 2002). It should be noted, however, that factors such as the manifestations and severity of the disorder may affect both educational attainment and subsequent occupational success and this is difficult to account for adequately in studies.

There are many contingencies that moderate outcomes. Whether a person with mental impairments can find work depends substantially on the labor market where toleration of impairment depends on the alternative workers available. Similarly, even late entrants who were generally disadvantaged by service in the armed services who had the prerequisite skills and who served as officers had opportunities for leadership and further skill acquisition that served them well when they returned to civilian life (Dechter and Elder 2004; MacLean and Elder 2007). Also, of course, the provisions that society makes for returning veterans such as the GI Bill significantly shapes future opportunities and trajectories.

The Importance of Historical Times

Life-course theory draws our attention to the fact that how we live out our lives is substantially affected by the time and place. For example, the life experience of persons currently over the age of 65 is quite different from their counterparts a decade or two ago. Demographic trends in marriage, education, mobility, fertility, labor markets, and the like shape individuals' experiences, life chances, and opportunities. Cultural changes in norms and attitudes, for example those related to the role of women or the meaning of race and ethnicity, affect the choices we can make about essential features of daily life. Other changes in health and disability policies and health technology mean that each cohort will confront different understandings of what constitutes illness and face different consequences of illness. It is not surprising, therefore, that the experience of mental health or illness may be quite different for older persons now than it was in the past, and we can expect it to be even more different for future cohorts of older persons.

One of the most dramatic changes observed in the past several decades in mental health policy was the rapid deinstitutionalization of the mentally ill whereby the number of persons in long-term public mental hospitals dropped from over 500,000 in the mid-1950s to less than 60,000 today (Grob 2008; Mechanic 2008). The promise of community care was never realized. As Grob (2008:98) explains, "The ideology of community mental health and the facile assumption that residence in the community would promote adjustment and integration was illusory and did not take into account the extent of social isolation, exposure to victimization, inducement to substance abuse, homelessness, and criminalization of persons with mental disorders." The failure of community care means that many persons with severe mental illnesses who once would have resided in mental institutions may now go through extended periods of time without treatment, have repeated admissions in short-stay hospitals and periods of homelessness, victimization, arrest, imprisonment, and social isolation – experiences that will shape the trajectories of their later lives. Others have benefitted from changes by avoiding the dysfunctions and disabilities associated with long-term institutional care (Mechanic 2008).

Changes in long-term care policies also shape the chances that individuals will stay in the community as they age or whether they live in nursing homes and other institutional settings. In the United States, most public care comes through the Medicaid program for persons who meet eligibility criteria based on disability and inadequate resources or who spend down their resources to attain eligibility. Medicaid has thus been the de facto American long-term care public system. There has been great reluctance on the part of the public sector to take on responsibilities commonly associated with informal care but rather to function more as backup when the informal sector can cope no longer. Although nursing homes remain the largest component of Medicaid long-term care, many states have made efforts to provide community-based long-term care services to persons who could meet the threshold for nursing home admission. In recent years, we have seen more assisted care communities, home care, and foster care arrangements for persons who previously had no alternative to nursing home admission.

Despite increases in the proportion of persons over the age of 65 and in life expectancy, the demand for nursing homes has declined (Manton 2003). The reduction in demand for residential care may be partially accounted for by the fact that successive cohorts attaining particular ages report less disability than previous cohorts at comparable ages. One view is that improved economic and social circumstances over the life course have allowed the delay of disability closer to the end of life (Manton 2003). More microanalysis suggests, however, that most of the gains in disability status have been in the realm of instrumental activities such as using the telephone, paying bills, shopping, and food preparation and less in activities of daily life such as bathing and physical mobility (Freedman et al. 2002). The ability to engage in instrumental activities of daily life has been much facilitated by new technologies and new products that make life easier for persons with increasing fragility. These include communication media such as the internet, technologies with voice commands, instruments offering enhanced visibility, delivery services for prepared meals, home banking, and much more. Mobility has also been enhanced by changing attitudes and laws and environmental modifications that make public places and public transportation more available for persons with impairments. Changes over time in the degree of disability associated with various conditions mean that each historical period brings differences in the chances of people being able to maintain their independence and continue to live in the community as they age.

Historical period also substantially shapes the chances that persons with mental illness in the community will receive care and the type of care received. Over the past few decades, the proportion of persons with a mental illness who receive some treatment has increased (Wang et al. 2005). But there have continued to be substantial barriers to care for older persons, including financial disincentives to seeking care. Medicare, the dominant insurance plan for persons over the age of 65 in the United States, has not covered psychiatric care to the same extent it covered general medical care. However, with the passage of The Medicare Improvements for Patients and Providers Act of 2008, over time coverage for outpatient care for psychiatric illness will achieve parity with care for physical problems. Thus, future cohorts of older persons with mental illness may face fewer barriers than the current cohort.

At the same time, historical period shapes the types of diagnoses received. For example, there has been a rapid increase in diagnosis of bipolar disorder for young children, with little evidence of an actual increase in prevalence and continued debate about whether bipolar disorder exists in children (Duffy 2007). Similarly, disorders once thought to be primary problems of childhood have expanded into adulthood as shown by the increasing numbers of adults receiving diagnoses of ADHD (Conrad and Potter 2000). Fundamental conceptualizations of mental health and illness are the product of time and place. With each revision of the DSM, the number of diagnoses has grown from 106 in DSM-I to 297 in DSM-IV (Mayes and Horwitz 2005).

In its efforts to gain greater precision and reliability, DSM moved to a diagnostic system substantially based on symptom checklists in which clinical diagnoses were determined when patients met specified symptomatic criteria. Using the case of depression, Horwitz and Wakefield (2007) note that throughout centuries depressive symptoms have typically been evaluated in relation to patients' social contexts and life events. But the DSM, in seeking greater reliability, has neglected social context, resulting in defining many patients as having major depression when they are responding in normally expected ways to noxious life events. While the DSM makes exception for bereavement, Horwitz and Wakefield (2007) argue that other life circumstances are comparable and that distress reactions to such circumstances are "normal sadness." By "normal" they mean that persons' affective systems respond as programmed by evolution. They accept that such patients could benefit from help but argue that they are not clinically disordered. They estimate that about a quarter of such diagnoses represent over extensions of a true concept of major depressive disorder (Wakefield et al. 2007).

These debates have important relevance for depression in the elderly because such distress is often manifested in ways different than among younger persons and is commonly linked to physical comorbidities, greater use of medications, loss of physical function, loss of loved ones, and erosion of social networks. Moreover, depression in the elderly, more than among the young, is expressed in physical as compared with psychological complaints and the elderly have greater reluctance to seek and accept specialty mental health services. Definition of what is or is not "disorder" according to the DSM shape what patient and providers attend to, what is or is not covered by health plans, and access to disability programs.

Human Agency and Social Structure

Awful things happen to people not of their own doing; not infrequently they result from social and historical circumstances that expose persons to significant trauma. These include early loss of parents, physical abuse and victimization, rape and other sexual abuse, extended combat and other extreme exposures, life-threatening disease and handicap, famine and hunger, and extreme economic adversity. We know that such traumas increase risk of mental health problems even years after the occurrence of the trauma. But negative consequences are not inevitable. Depending on what people bring to these situations, some may gain personal strength and new capacities in coping with these adversities when they can be overcome.

People in difficult situations not infrequently have "second chances" – opportunities to reverse a negative trajectory. Second chances may occur in a variety of ways, such as developing an intimate relationship with a highly supportive other, having opportunities for educational advancement and for developing new skills, getting a new job that offers satisfactions and reinforcing experiences, or becoming part of a supportive friendship group. Studies of the military, for example, found that service in World War II was often a "second chance" for young men who had disadvantaged backgrounds or had problems with the law (MacLean and Elder 2007). Many social intervention programs have as their goal developing "second chances" for individuals who have had a difficult life trajectory, who lack confidence and self-esteem, and who aspire to a stronger sense of mastery over their life situations.

Individuals may overcome even the most severe events that happened in early life. Studies of survivors of the Holocaust, for example, show that while they continued to have elevated symptoms of psychological distress and posttraumatic distress decades later compared to their counterparts, the differences between groups were modest, and most appeared to show favorable adjustments as measured by marriage, children, and strong families in later life (Kahana et al. 2005).

Studies of Holocaust survivors, like all long-term studies of the effects of traumatic events, may exaggerate the resiliency of the people who experienced the events because the sickest or least resilient are likely to die earlier, leaving the most healthy to survive to old age. Despite this methodological problem, studies of long-term outcomes of traumatic events do show the remarkable resiliency of some people, even after experiencing extraordinary trauma. The challenge is to identify the characteristics of individuals and environments that lead to such resiliency.

Self-efficacy, or "people's assessments of their effectiveness, competence and causal agency" (Gecas 1989:292), is important to health across the life course. In the context of examining the effects of stressful life events on risk of mental disorder, a sense of control over one's life (mastery) may help individuals be active agents in confronting stressful life events to both cope with them and deliberately seek out solutions (Thoits 2006). Some studies suggest that a sense of mastery declines with age (Schieman and Turner 1998) and the experience of negative life events in late life appears to undermine levels of mastery (Pearlin et al. 2007). Moreover, past experiences also influence feelings of control during old age. Pearlin et al. (2007) hypothesize that exposure to stressors over the life course and status (both achieved such as education and ascribed such as gender and race) influences "life course mastery," or the degree that individuals feel that they had control over their life circumstances. Early hardships influence individuals' capacities to be active agents in shaping their own lives as they age.

Sociological work has long recognized that social agency is constrained by social structure, with most attention being devoted to the role of social class or its typical proxy socioeconomic status. While social class is not causative of disorder in any direct sense, it is commonly associated with many factors more proximate to the occurrence of disorder. Social class, or its SES proxies, is associated with almost every indicator of health including infant mortality, adult mortality, chronic disease prevalence, impairment and disability, and psychological distress. Families at the lowest rungs of the SES distribution, as compared with those higher in the SES hierarchy, are more likely to experience economic hardships, deprivations and household disruptions, intrafamily violence and abuse, child neglect, and disruptions in the normal course of development of their children. Children are more likely to do poorly in school, have difficulties with the law, engage in delinquency and substance abuse, leave school early, develop poorly considered and exploitative relationships, and bear children early. Children faced with these adversities have diminished life chances relative to peers from higher SES circumstances and are more likely to follow a life trajectory associated with many adversities. Of course, none of this is inevitable, and many low SES families provide their children the nurturance, encouragement, and coping skills that result in satisfying and productive lives and often high achievement. Moreover, the challenges of overcoming adversities associated with economic and social deprivations allow some to develop perspectives, motivations, and coping capacities that make for impressive success in later life.

Studies of the effect of SES on health over the life course have examined the impact of cumulative advantage or disadvantage. Early SES disadvantage in one's family of origin presages later disadvantages such as early school leaving and poor health which then are the precursors of later disadvantages. Miech and Shanahan (2000) argue that the negative relationship between lesser education and depressive symptoms is stronger in older age than before because older persons with less education have more physical health problems which contribute to depression. As people age and physical health problems become more common, the association between education and depression also strengthens, suggesting a cumulative disadvantage for those with less education over the life course.

Linked Lives

At the simplest level, the concept of "linked lives" in life-course theory recognizes that individuals are embedded in social relationships and social networks. The nature of these relationships affects many aspects of life, and not surprisingly health and well-being. Much of the research has focused on the social support function of our social ties and examined the positive effects of both emotional and instrumental aid. Feeling loved and valued and having people to count on for assistance may directly affect health, and also indirectly promote well-being by buffering the effects of negative life events (Cohen 2004).

Others have examined characteristics of social networks beyond the support they provide. Characteristics of networks can be defined by their size, density, proximity, and closeness (Berkman et al. 2000). In addition to be sources of support, social networks may affect health behaviors, coping, self-efficacy, and physiological functioning, all of which promote health. Social networks can also have negative effects through their norms and influences, the demands and obligations they promote, and the noxious behaviors they may encourage. The culture of one's networks is all important in understanding their effects.

The concept of linked lives highlights that what happens in the lives of others in an individual's social network matters for health and well-being. For example, among older persons, having a spouse hospitalized increases risk of mortality in the surviving spouse (Christakis and Allison 2006). Moreover, risk of mortality is highest for persons whose spouse was hospitalized for a psychiatric problem or dementia. Much of the work on the interdependence of social ties and health has focused on care giving. The chances of having to assume such responsibilities are greater among women than among men, and increases risk of depressive symptoms, especially for women, those caring for a spouse, and those caring for persons with dementia (Pinquart and Sörensen 2003).

Recent studies of social contagion offer further intriguing insights into the importance of social networks for health. Christakis and Fowler (2009) have shown how health behaviors and subjective aspects of well-being, such as happiness, also spread within social networks. For example, if you have a happy friend who lives nearby, it increases the likelihood that you will be happy. Similarly, they find that loneliness (conceptualized as perceived social isolation) is contagious. The mechanisms through which subjective states spread through social networks are not clear, nor is it clear whether network effects may differ with stage of the life course. However, together this work points to the importance of linked lives for understanding health outcomes.

While early accounts of aging portrayed it as a process of increasing isolation – we now know that this holds for only a minority of persons (Cornwell et al. 2008). In a large community sample of persons 57–85 years of age in the United States, Cornwell et al. (2008) find that while the size of social networks and the diversity of ties tend to decrease with age, the frequency of connections with social ties, religious involvement, and volunteer activity increase with age. The only measure of social integration that appears weaker with age is closeness to networks. The life-course transitions associated with aging, such as death of a spouse or retirement, may challenge social integration, but Cornwell et al. suggest that older persons may adjust to these transitions by becoming more involved in community activities.

The experience of mental illness over the life course is played on in the context of family. Familial expressed emotion, or overinvolvement and criticism, is linked to increased risk of relapse for people with depression and schizophrenia (Butzlaff and Hooley 1998). The relationships within families are reciprocal and one member's health can influence the health of others. For example, among older persons, having a spouse who is depressed increases one's chances of becoming depressed (Tower and Kasl 1996). The interdependence of family members also influences caregiving roles. The responsibility for caring for a family member with a severe mental illness brings different burdens and benefits depending on the stage of the life course the illness started, and the age of the care giver (Cook et al. 1997).

Conclusions

One of the challenges of life-course research is to capture life holistically but in a way that takes advantage of rigorous social science methodology. While life narratives offer the richest context for understanding how an individual life unfolds, researchers must seek generalizations across lives. For the study of mental illness, much of the research has focused on correlates of risk. As we noted, however, most mental illnesses occur first relatively early in life. Thus, it is difficult to untangle the precursors of first onset of disorder from the correlates of re-occurrence in research lacking data from over the life course.

Insel (2009:130) sees cardiovascular epidemiology as a useful model for understanding future approaches to mental health intervention. For heart disease, like mental illness, some precursors are already evident in the womb and in early life, and other risk factors over the life span are readily identified, but cardiovascular disease itself may not become evident until late middle and early old age. While cardiovascular understanding is much advanced beyond psychiatry, some believe that it will soon be possible to reshape the course of conditions such as schizophrenia and related psychoses (McGorry et al. 2008). One approach of increasing interest is to identify prodromal attributes that predict subsequent development of the condition (Cannon et al. 2008). For example, unusual thought content, suspicion/paranoia, perceptual anomalies, grandiosity, and disorganized communication may not reach a diagnostic threshold but may be indicative of later disorder. Success in defining appropriate measures would allow early intervention but there are also obvious dangers of false positives in prediction and the possible harm of intervention and labeling.

Very little in life is so constrained as to make different life trajectories impossible. Certainly, there are congenital and genetic factors at their extreme that close out many options, but even here, under extremely favorable conditions, better outcomes than anyone can imagine occur. Preventing the negative sequelae that often follow early disadvantage and trauma and result in mental health problems is challenging but possible. At the individual level people draw on their own strengths and capacities and make choices that may turn a negative trajectory around and prevent mental health problems. They also can draw on social networks to find aid and support. But social policy also plays a large role in enhancing mental health throughout the life course and into old age. The role of policy in a wide variety of arenas from education to housing and disability to long-term care shapes the opportunities and constraints people face as they age.

References

American Psychiatric Association. 2000. "Diagnostic and Statistical Manual of Mental Disorders, 4th ed. Text Revision (DSM-IV-TR)." Washington, D.C.: American Psychiatric Association.

Barker, David J.P., Tom Forsén, Antti Uutela, Clive Osmond, and Johan G. Eriksson. 2001 "Size at Birth and Resilience to Effects of Poor Living Conditions in Adult Life: Longitudinal Study." *British Medical Journal* 323(7324):1273–6.

Berkman, Lisa F., Thomas Glass, Ian Brissette, and Teresa E. Seeman. 2000. "From Social Integration to Health: Durkheim in the New Millennium." *Social Science and Medicine* 51:843–57.

Bifulco, Antonia T., George W. Brown, and Tirril O. Harris. 1987. "Childhood Loss of Parent, Lack of Adequate Parental Care and Adult Depression: A Replication." *Journal of Affective Disorders* 12:115–28.

Blanchflower, David G. and Andrew J. Oswald. 2008. "Is Well-Being U-Shaped Over the Life Cycle?" *Social Science and Medicine* 66:1733–1749.

Breslau, Joshua, Michael Lane, Nancy Sampson, and Ronald C. Kessler. 2008. "Mental Disorders and Subsequent Educational Attainment in a US National Sample." *Journal of Psychiatric Research* 42:708–16.

Brown, George. 1986. "Mental Illness." pp. 175–203 in *Applications of Social Science to Clinical Medicine and Health Policy*, edited by L.H. Aiken and D. Mechanic. New Brunswick, NJ: Rutgers University Press.

Brown, George W. and Tirril O. Harris, eds. 1978. *Social Origins of Depression: A Study of Psychiatric Disorder in Women*. London: Tavistock.

Brown, George W. and Tirril O. Harris. 1989. *Life Events and Illness*. New York: The Guilford Press.

Brown, George W., Tom K.J. Craig and Tirril O. Harris. 2008. "Parental Maltreatment and Proximal Risk Factors Using the Childhood Experience of Care & Abuse (CECA) Instrument: A Life-Course Study of Adult Chronic Depression." *Journal of Affective Disorders* 110:222–33.

Butzlaff, Ronald L. and Jill M. Hooley. 1998. "Expressed Emotion and Psychiatric Relapse: A Meta-Analysis." *Archives of General Psychiatry* 55:547–52.

Cannon, Tyrone D., Kristin Cadenhead, Barbara Cornblatt, Scott W. Woods, Jean Addington, Elaine Walker, Larry J. Seidman, Diana Perkins, Ming Tsuang, Thomas McGlashan, and Robert Heinssen. 2008. "Prediction of Psychosis in Youth at High Clinical Risk: A Multisite Longitudinal Study in North America." *Archives of General Psychiatry* 65:28–37.

Centers for Disease Control and Prevention. 2009. *Prevalence of Autism Spectrum Disorders-Autism and Developmental Disabilities Monitoring Network, United States, 2006. Surveillance Summaries. Morbidity and Mortality Weekly Report, December 18, 2009*, 58:(SS10):1–20. Retrieved February 3, 2011. (http://www.cdc.gov/mmwr/preview/mmwrhtml/ss5810a1.htm).

Christakis, Nicholas A. and James H. Fowler. 2009. *Connected: The Surprising Power of Our Social Networks and How They Shape Our Lives*. New York: Little, Brown and Company.

Christakis, Nicholas A. and Paul D. Allison. 2006. "Mortality after the Hospitalization of a Spouse." *The New England Journal of Medicine* 354:719–30.

Cohen, Sheldon. 2004. "Social Relationships and Health." *American Psychologist* 59:676–684.

Cohen, Carl I., Richa Pathak, Paul M. Ramirez, and Ipsit Vahia. 2009. "Outcome Among Community Dwelling Older Adults with Schizophrenia: Results Using Five Conceptual Models." *Community Mental Health Journal* 45:151–6.

Colman, Ian, Joseph Murray, Rosemary A. Abbott, Barbara Maughan, Diana Kuh, Tim J. Croudace, and Peter B. Jones. 2009. "Outcomes of Conduct Problems in Adolescence: 40 Year Follow-Up of National Cohort." *British Medical Journal* 338:a2981.

Colman, Ian, Michael E.J. Wadsworth, Tim J. Croudace, and Peter B. Jones. 2007. "Forty-Year Psychiatric Outcomes Following Assessment for Internalizing Disorder in Adolescence." *American Journal of Psychiatry* 164:126–33.

Conrad, Peter and Deborah Potter. 2000. "From Hyperactive Children to ADHD Adults: Observations on the Expansion of Medical Categories." *Social Problems* 47:559–82.

Cook, Judith A., Bertram J. Cohler, Susan A. Pickett, and Jeff A. Beeler. 1997. "Life-Course and Severe Mental Illness: Implications for Caregiving within the Family of Later Life." *Family Relations* 46:427–36.

Cornwell, Benjamin, Edward O. Laumann, and L. Philip Schumm. 2008. "The Social Connectedness of Older Adults: A National Profile." *American Sociological Review* 73:185–203.

Costello, E. Jane, Helen Egger, and Adrian Angold. 2005. "10-Year Research Update Review: The Epidemiology of Child and Adolescent Psychiatric Disorders: I. Methods and Public Health Burden." *Journal of the American Academy of Child and Adolescent Psychiatry* 44:972–86.

Crosnoe, Robert and Glen H. Elder, Jr. 2002. "Successful Adaptation in the Later Years: A Life Course Approach to Aging." *Social Psychology Quarterly* 65:309–28.

Cynader, Max S. and Barrie J. Frost. 1999. "Mechanisms of Brain Development: Neuronal Sculpting by the Physical and Social Environment." pp. 153–184 in *Developmental Health and the Wealth of Nations: Social, Biological, and Educational Dynamics*, edited by D.P. Keating and C. Hertzman. New York: Guilford Press.

Dechter, Aimee R. and Glen H. Elder, Jr. 2004. "World War II Mobilization in Men's Work Lives: Continuity or Disruption for the Middle Class." *American Journal of Sociology* 110:761–793.

Dohrenwend, Bruce P., Itzhak Levav, Patrick E. Shrout, Sharon Schwartz, Guedalia Naveh, Bruce G. Link, Andrew E. Skodol, and Ann Stueve. 1992. "Socioeconomic Status and Psychiatric Disorders: The Causation-Selection Issue." *Science* 255:946–52.

Duffy, Anne. 2007. "Does Bipolar Disorder Exist in Children? A Selected Review." *The Canadian Journal of Psychiatry* 52:409–17.

Egger, Helen Link and Adrian Angold. 2006. "Common Emotional and Behavioral Disorders in Preschool Children: Presentation, Nosology, and Epidemiology," *Journal of Child Psychology and Psychiatry* 47:313–37.

Elder, Glen H., Jr. 1998. "The Life Course as Developmental Theory." *Child Development* 69:1–12.

———. 2000. "The Life Course." pp. 1614–1622 in *Encyclopedia of Sociology*, 2nd ed., Vol. 3, edited by E.F. Borgatta and R.J.V. Montgomery. New York: Macmillan Reference USA.

Freedman, Vicki A., Linda G. Martin and Robert F. Schoeni. 2002. "Recent Trends in Disability and Functioning Among Older Adults in the United States: A Systematic Review." *Journal of the American Medical Association* 288:3137–46.

Gecas, Viktor. 1989. "The Social Psychology of Self-Efficacy." *Annual Review of Sociology* 15:291–316.

George, Linda K. 1999. "Life-Course Perspectives on Mental Health." pp. 565–83 in *Handbook of the Sociology of Mental Health*, edited by C.S. Aneshensel and J.C. Phelan. New York: Kluwer Academic/Plenum.

Grob, Gerald N. 2008. "Mental Health Policy in the Liberal State: The Example of the United States." *International Journal of Law and Psychiatry* 31:89–100.

Harris, Tirril O., George W. Brown, and Antonia Bifulco. 1987. "Loss of Parent in Childhood and Adult Psychiatric Disorder: The role of Social Class Position and Premarital Pregnancy." *Psychological Medicine* 17:163–83.

Horwitz, Allan V. and Jerome C. Wakefield. 2007. *The Loss of Sadness: How Psychiatry Transformed Normal Sorrow into Depressive Disorder*. New York: Oxford University Press.

Horwitz, Allan V., Cathy S. Widom, Julie McLaughlin, and Helene R. White. 2001. "The Impact of Childhood Abuse and Neglect on Adult Mental Health: A Prospective Study." *Journal of Health and Social Behavior* 42:184–201.

Insel, Thomas R. 2009. "Translating Scientific Opportunity into Public Health Impact: A Strategic Plan for Research on Mental Illness." *Archives of General Psychiatry* 66:128–33.

Jeste, Dilip V., Dan G. Blazer, and Michael First. 2005. "Aging-Related Diagnostic Variations: Need for Diagnostic Criteria Appropriate for Elderly Psychiatric Patients." *Biological Psychiatry* 58:265–71.

Kahana, Boaz, Zev Harel, and Eva Kahana. 2005. *Holocaust Survivors and Immigrants: Late Life Adaptations.* New York: Springer.

Keith, Samuel J., Darrel A. Regier, and Donald S. Rae. 1991. "Schizophrenic Disorders." pp. 33–52 in *Psychiatric Disorders in America: The Epidemiologic Catchment Area Study*, edited by L.N. Robins and D.A. Regier. New York: The Free Press

Kessler, Ronald C., Howard Birnbaum, Evelyn Bromet, Irving Hwang, Nancy Sampson, and Vicki Shahly. 2010. "Age Differences in Major Depression: Results from the National Comorbidity Survey Replication (NCS-R)." *Psychological Medicine* 40:225–37.

Kessler, Ronald C., Patricia Berglund, Olga Demler, Robert Jin, Kathleen R. Merikangas, and Ellen E. Walters. 2005. "Lifetime Prevalence and Age-of-Onset Distributions of DSM-IV Disorders in the National Comorbidity Survey Replication." *Archives of General Psychiatry* 62:593–602.

Mayes, Rick, Catherine Bagwell, and Jennifer Erkulwater. 2008. "ADHD and the Rise in Stimulant Use among Children." *Harvard Review of Psychiatry* 16:151–66.

Mayes, Rick and Allan V. Horwitz. 2005. "DSM-III and the Revolution in the Classification of Mental Illness." *Journal of the History of Behavioral Sciences* 41:249–67.

MacLean, Alair and Glen H. Elder, Jr. 2007. "Military Service in the Life Course." *Annual Review of Sociology* 33:175–96.

Manton, Kenneth G. 2003. "Forecasting the Nursing Home Population." *Medical Care* 41:21–4.

McGorry, Patrick D., Alison R. Yung, Andreas Bechdolf, and Paul Amminger. 2008. "Back to the Future: Predicting and Reshaping the Course of Psychotic Disorder." *Archives of General Psychiatry* 65:25–7.

Mechanic, David. 2007. "Population Health: Challenges for Science and Society." *Milbank Quarterly* 85:533–59.

———. 2008. *Mental Health and Social Policy: Beyond Managed Care*, 5th ed. Boston: Pearson.

Mechanic, David, Scott Bilder, and Donna D. McAlpine. 2002. "Employing Persons with Serious Mental Illness." *Health Affairs* 21:242–53.

Mechanic, David and Donna D. McAlpine. 2000. "Use of Nursing Homes in the Care of Persons with Severe Mental Illness: 1985 to 1995." *Psychiatric Services* 51:354–8.

Miech, Richard Allen and Michael J. Shanahan. 2000. "Socioeconomic Status and Depression over the Life Course." *Journal of Health and Social Behavior* 41:162–76.

Mirowsky, John and Catherine E. Ross. 1999. "Well-Being across the Life Course." pp. 328–47 in *A Handbook for the Study of Mental Health*, edited by A.V. Horwitz and T.L. Scheid. Cambridge: Cambridge University Press.

Muntaner, Carles, William W. Eaton, Richard Miech, and Patricia O'Campo. 2004. "Socioeconomic Position and Major Mental Disorders." *Epidemiologic Reviews* 26:53–62.

Nassar, Natasha, Glenys Dixon, Jenny Bourke, Carol Bower, Emma Glasson, Nick de Klerk, and Helen Leonard. 2009. "Autism Spectrum Disorders in Young Children: Effect of Changes in Diagnostic Practices." *International Journal of Epidemiology* 38:1245–54.

National Comorbidity Study-Replication. 2007. "Table 2. 12-Month Prevalence of DSM-IV/WMH CIDI disorders by Sex and Cohort." Retrieved February 2, 2011 (http://www.hcp.med.harvard.edu/ncs/ftpdir/table_ncsr_12monthprevgenderxage.pdf).

Pearlin, Leonard I., Kim B. Nguyen, Scott Schieman, and Melissa A. Milkie. 2007. "The Life-Course Origins of Mastery among Older People." *Journal of Health and Social Behavior* 48:164–79.

Pearlin, Leonard I., Scott Schieman, Elena M. Fazio, and Stephen C. Meersman. 2005. "Stress, Health, and the Life Course: Some Conceptual Perspectives." *Journal of Health and Social Behavior* 46:205–19.

Pinquart, Martin and Silvia Sörensen. 2003. "Differences between Caregivers and Noncaregivers in Psychological Health and Physical Health: A Meta-Analysis." *Psychology and Aging* 18:250–67.

Plassman, Brenda L., Kenneth M. Langa, Gwenith G. Fisher, Steven G. Herringa, David R. Weir, Mary Beth Ofstedal, James R. Burke, Michael D. Hurd, Guy G. Potter, Willard L. Rodgers, David C. Steffens, Robert J. Willis, and Robert B. Wallace. 2007. "Prevalence of Dementia in the United States: The Aging, Demographics, and Memory Study." *Neuroepidemiology* 29:125–32.

Plassman, Brenda L., Kenneth M. Langa, Gwenith G. Fisher, Steven G. Herringa, David R. Weir, Mary Beth Ofstedal, James R. Burke, Michael D. Hurd, Guy G. Potter, Willard L. Rodgers, David C. Steffens, John J. McArdle, Robert J. Willis, and Robert B. Wallace. 2008. "Prevalence of Cognitive Impairment without Dementia in the United States." *Annals of Internal Medicine* 148:427–34.

Robins, Lee N. 1966. *Deviant Children Grown Up: A Sociological and Psychiatric Study of Sociopathic Personality.* Baltimore: Williams & Wilkins.

Robins, Lee N. 1983. "Continuities and Discontinuities in the Psychiatric Disorders of Children." pp. 195–219 in *Handbook of Health, Health Care, and the Health Professions*, edited by D. Mechanic. New York: Free Press.

Rutter, Michael. 2009. "Commentary: Fact and Artefact in the Secular Increase in the Rate of Autism." *International Journal of Epidemiology* 38:1238–9.

Schieman, Scott and Heather Turner. 1998. "Age, Disability, and the Sense of Mastery." *Journal of Health and Social Behavior* 39:169–86.

St. Clair, David, Mingqing Xu, Peng Wang, Yaqin Yu, Yourong Fang, Feng Zhang, Xiaoying Zheng, Niufan Gu, Guoyin Feng, Pak Sham, and Lin He. 2005. "Rates of Adult Schizophrenia Following Prenatal Exposure to the Chinese Famine of 1959–61." *Journal of the American Medical Association* 294:557–62.

Susser Ezra, Hans W. Hoek, and Alan W. Brown. 1998. "Neurodevelopmental Disorders After Prenatal Famine: The Story of the Dutch Famine Study." *American Journal of Epidemiology* 147:213–16.

Thoits, Peggy A. 2006. "Personal Agency in the Stress Process." *Journal of Health and Social Behavior* 47:309–23.

Thompson, Richard, Hillary R. Bogner, James C. Coyne, Joseph J. Gallo, and William W. Eaton. 2004. "Personal Characteristics Associated with Consistency of Recall of Depressed or Anhedonic Mood in the 13-Year Follow-up of the Baltimore Epidemiologic Catchment Area Survey." *Acta Psychiatrica Scandinavica* 109:345–54.

Tower, Roni Beth and Stanislav V. Kasl. 1996. "Depressive Symptoms across Older Spouses: Longitudinal Influences." *Psychology and Aging* 11:683–97.

Turner, R. Jay, Blair Wheaton, and Donald A. Lloyd. 1995. "The Epidemiology of Social Stress." *American Sociological Review* 60:104–25.

Wakefield, Jerome C., Mark F. Schmitz, Michael B. First, and Allan V. Horwitz. 2007. "Extending the Bereavement Exclusion for Major Depression to Other Losses: Evidence from the National Comorbidity Survey." *Archives of General Psychiatry* 64:433–40.

Wang, Philip S., Patricia Berglund, Mark Olfson, Harold A. Pincus, Kenneth B. Wells, and Ronald C. Kessler. 2005. "Failure and Delay in Initial Treatment Contact after First Onset of Mental Disorders in the National Comorbidity Survey Replication." *Archives of General Psychiatry* 62:603–13.

Chapter 31
Aging with HIV/AIDS

Allen J. LeBlanc

When the first small clusters of previously healthy, young gay men began to dramatically and mysteriously fall ill in the nation's urban centers toward the end of the 1970s, few might have guessed that the United States was facing an emerging epidemic that would shake our confidence in modern medicine and revive " … fears at least as old as the medieval plagues" (Institute of Medicine 1988:1; Shilts 1987). Tragically, over the past three plus decades such an epidemic has unfolded not only in this country, but worldwide. Indeed, human immunodeficiency virus (HIV) – the virus believed to cause acquired immunodeficiency syndrome (AIDS) – has become a global pandemic, one comprised of thousands of lesser epidemics with unique epidemiologic shapes and social contexts across the world (Mann et al. 1993).

Individuals can live – knowingly or unknowingly – with HIV for long periods of time without manifesting symptoms of AIDS. In 2007, there were approximately 33 million people living with HIV worldwide. While the global *percentage* of individuals with HIV has stabilized since 2000, the overall *number* of people who are seropositive has increased due to new infections and the beneficial impact of more widely available treatments that reduce morbidity and extend life. Sixty-seven percent of the world's HIV-positive population is concentrated in Sub-Saharan Africa, where heterosexual intercourse is the epidemic's driving force, women and children – especially girls – are more heavily affected. In virtually all other regions of the world, HIV disproportionately affects men who have sex with men, injection drug users (IDUs), and sex workers. Areas of specific concern due to increasing rates of HIV infection include Indonesia, the Russian Federation, and some high-income countries. There were an estimated two million people living with HIV in North America, Western and Central Europe – as a region – in 2007. Approximately 1.2 million of those individuals were living in the United States (UNAIDS 2008).

Gross aggregate statistics such as these readily demonstrate the magnitude and reach of HIV in the world today. They also clearly portray the reality that, more and more, people across the globe and from all walks of life are living – and in greater numbers than ever before, growing old – with HIV. Older HIV populations may be emerging more slowly in developing nations where treatments are less available (UNAIDS 2008), and relevant research lags behind. For example, data on the prevalence and incidence of HIV among older persons in developing countries are presently unavailable (Schmid et al. 2009). Therefore, existing data largely concern the experiences of people in developed areas of the world. Moreover, available reports generally rely on recent data, which require continuous monitoring to identify emerging and long-term trends.

A.J. LeBlanc (✉)
Department of Sociology and the Health Equity Institute,
San Francisco State University, San Francisco, CA, USA
e-mail: aleblanc@sfsu.edu

R.A. Settersten, Jr. and J.L. Angel (eds.), *Handbook of Sociology of Aging*, Handbooks of Sociology and Social Research, DOI 10.1007/978-1-4419-7374-0_31, © Springer Science+Business Media, LLC 2011

U.S. Population Estimates: Older People with HIV

The Centers for Disease Control (CDC) has produced statistics on the numbers of Americans aged 50 and over living with HIV/AIDS since 1981. In that time, the prevalence of cases in that age group – a symbolic indicator of "aging" in the HIV context – has increased steadily, in part due to the greater overall prevalence of HIV, which is largely accounted for by the fact that infected individuals are living longer. In short, better treatment and prevention of opportunistic infections associated with HIV – in particular the introduction of highly active antiretroviral therapy (HAART) in 1996 – are extending life, and improving its quality. In addition, the incidence of HIV in this age group has also continued to rise (Simone and Appelbaum 2008).

People aged 50 and older accounted for 15% of newly diagnosed HIV/AIDS cases and 25% of all cases in the United States in 2005, reflecting a rise from 17% in 2001 (Simone and Appelbaum 2008). In addition, persons aged 40–49 comprised the largest proportion of newly diagnosed HIV/AIDS cases among Americans in 2007, illustrating that a significant portion of the HIV-positive population will shortly age into the "older" category (Centers for Disease Control and Prevention [CDC] 2009). According to data compiled in high-prevalence cities, for example New York City, it is predicted that in the next decade the majority of people with HIV in that city will be over age 50 (Karpiak et al. 2006). Moreover, it may be that current statistics underestimate the true prevalence of HIV among older adults as many physicians do not perceive them to be at risk. Consequently, they are less likely to be tested and misdiagnosis is common (Casau 2005). Indeed, the CDC recommends testing people aged 65 and older for HIV only if they have known risk factors (Lee 2009).

Describing the Older HIV-Positive Population

Until recently, efforts to describe the older HIV-positive population in the United States and elsewhere have typically been based on samples recruited through clinics, hospitals, or other HIV-related service providers. These studies offer important descriptive data on local populations; and they also help to identify specific sub-population vulnerabilities and weaknesses in the local service systems. Studies conducted in the UK and Italy are recent examples of such work (Elford et al. 2008; Orchi et al. 2008; Sherr et al. 2009). Ongoing research based in New York City is another (Karpiak et al. 2006).

Nationally representative data for the United States are available. The HIV Cost and Services Utilization Study (HCSUS) – fielded in 1996 – has produced profiles of middle-aged and older (i.e., age 50 and older) Americans with HIV (Crystal et al. 2003; Joyce et al. 2005). According to HCSUS data, this population is perhaps best understood by comparing groups based on mode of HIV transmission: men who have sex with men, a group mainly composed of "gay men"; IDUs; and "other," including individuals whose mode of transmission was heterosexual contact, blood transfusion, and cases in which the source of exposure was unknown. It was estimated that these three groups account for cases of HIV among older persons as follows: gay men (40%); IDUs (23%); and "other" (37%) (Crystal et al. 2003).

With HCSUS data, it has been estimated that: The vast majority of older gay men with HIV are white (92%). Sizeable proportions of this group have a college education (49%), worked in the paid labor force full- or part-time (38%), earned more than $25,000 per year (48%), and had some form of health insurance (92%). In comparison, older IDUs with HIV – 17% of whom are women – are relatively more disadvantaged and vulnerable. Only small proportions have a college education (7%), are employed (11%), or earn more than $25,000 per year (10%). About one-fifth lack health insurance coverage of any kind. Over half (56%) of older IDUs with HIV are African-American. Finally, the "other" category, which contains a greater proportion of women (30%), is also predominantly African-American (64%) and of exceptionally low socioeconomic status (e.g., 14.2% with a college education, 22.3% employed, 16.7% earning more than $25,000 annually, and 17.6% uninsured) (Crystal et al. 2003).

In sum, available data illustrate that the older HIV-positive population in the United States is sociodemographically diverse, which suggests a wide range of experiences and needs. Among older adults with HIV, gay men (a predominantly white group) have greater socioeconomic status – via their higher levels of education, employment, income, and access to insurance coverage – than persons who were infected through IDU, heterosexual sex, or other means (groups that include women and greater proportions of racial/ethnic minority persons).

As a whole, the older HIV-positive population is socially and economically disadvantaged in relation to the general population aged 50–61. For example, in analyses comparing HCSUS data with Health and Retirement Survey data – a longitudinal study representative of U.S. adults born between 1931 and 1941 – it was found that regardless of gender or race/ethnicity, older adults with HIV are much less likely to be employed, have employment-based health insurance, and have far less income and assets than their non-HIV-infected counterparts. Indeed, there are significant numbers of HIV-positive individuals over the age of 50 who are incarcerated or homeless, or moving between those states (Joyce et al. 2005). Even among older HIV-positive white men who are more likely to have a college degree than older men in the general population, total household income and assets are smaller in comparison (Kushel et al. 2005).

Although these studies are relatively recent, it is important to note that the HCSUS data were collected in 1996, which was a breakthrough year in terms of HIV treatments with the introduction of HAART. Consequently, it is essential that researchers continue ongoing examinations of the changing sociodemographic profiles of HIV-infected populations in a range of geographic locales – focusing on populations ranging from local to national. In addition, such work must begin in developing countries.

HIV-Related Risk Behavior Among Older People

A growing body of research suggests that older adults do not engage in sexual and drug-related behaviors that carry risk for HIV at the same rate as younger populations. However, general population studies demonstrate that among adults aged 50 and over who do have known risk factors, only small percentages adopt safer sexual practices or undergo HIV testing (Mack and Ory 2003; Stall and Catania 1994).

Research carried out with targeted samples is beginning to illustrate unique HIV-related risks for subpopulations of midlife and older persons. For instance, data collected from mid-life and older, urban men who have sex with men shows that they engage in risky sexual and drug-related behaviors through the sixth decade of life, at high levels (Dolcini et al. 2003; Stall and Purcell 2000; Koblin et al. 2006). Research on risky substance use among gay men includes some focus on the use of sildenafil, which is intended to treat erectile dysfunction typically associated with aging (Cooperman et al. 2007). In addition, studies of mid-life and older women suggests they face risk for HIV exposure by virtue of their relationship dynamics with male partners, sexual behaviors, and substance use, although these women are particularly unlikely to perceive of themselves as at-risk (Corneille et al. 2008; Neundorfer et al. 2005; Sormanti and Shibusawa 2007). Finally, available data regarding older IDUs suggest they are significantly less likely to share needles and are less active sexually than younger IDUs (Kwiatkowski and Booth 2003). These kinds of age-based differences in HIV-related risk warrant further investigation to articulate their social and cultural determinants (cf. Schensul et al. 2003), and future research must address additional populations.

Clinical Studies

Although the focus on this chapter is on the social and psychological aspects of aging with HIV, it is important to briefly recognize the implications of aging seropositive populations for clinical research and practice. At this point in time, clinical researchers are only just beginning to identify and understand age-based differences in the manifestations of HIV disease and responses to treatment. There has historically been a disproportionate focus on younger generations when it comes to HIV because they have been at greatest risk of infection, and this focus to some extent remains. For example, older patients (and those with comorbidities) are routinely excluded from clinical trials evaluating the efficacy of emerging HIV medications, which in itself is an important reminder of how little is known about the clinical issues faced by people aging with HIV (Gebo 2009; Martin et al. 2008).

In the United States and other developed nations, there is a small, but growing clinical research literature on the prevention, diagnosis, and management of HIV/AIDS among older adults (Martin et al. 2008; Simone and Appelbaum 2008; Stoff et al. 2004). It seems there are a number of interrelated challenges regarding HIV and aging that the clinical scientific community must address. For instance, researchers are attempting to more fully characterize the ways in which the natural history and symptom manifestations of HIV infection may differ between younger and older adults. Available data suggest that, relative to their younger counterparts, older adults living with HIV tend to be diagnosed later. In addition, they may experience a more severe HIV disease course, and scientists are striving to identify the underlying mechanisms that might explain such differences. Moreover, there is also considerable focus on potential interactions among concomitant and overlapping conditions in the older population, including HIV disease, comorbid conditions (HIV-related and non-HIV-related), normal aging, age-related medical challenges, and age-related diseases (e.g., Alzheimer's disease, osteoporosis, adult-onset diabetes, cancer, and hypertension) (Stoff et al. 2004). There is also now a growing literature on the relationship between HIV, reproductive aging, and health among mid-life and older women (Conde et al. 2009).

In short, the overall profile of HIV-associated health conditions and comorbidities is changing as the HIV-positive population ages. Consequently, the fundamental clinical and clinical research challenges will continue to revolve around developing better strategies for the long-term treatment and management of HIV as a chronic condition – often a comorbid one – that can be effectively controlled in most patients as they age into later life. Data from the CDC (2007) have been used to begin the estimation of survival after a diagnosis of AIDS. For example, over 80% of individuals receiving an AIDS diagnosis in 2002 have survived more than 36 months postdiagnosis. Survival appears to decrease as age at diagnosis increases, and is lowest among IDUs, a population with relatively few socioeconomic resources (Crystal et al. 2003).

Living, and Now Aging, with HIV

Receiving an HIV diagnosis – and/or an AIDS diagnosis – is a traumatic life event that alters one's life course. For many, in particular those who were diagnosed in the first waves of the pandemic, there was little time for reflection about the future, only the anticipation of rapidly approaching illness and death. Indeed, historically only limited aging with HIV has been possible. However, while the struggles of infected individuals worldwide continue – especially those without access to quality care and emerging treatments – the future outlook has changed dramatically for the better. Consequently, it is useful to examine key studies that attempted to understand the HIV experience through individual narratives. It is also helpful to distinguish between those conducted before and after the introduction of effective treatments.

Before Effective Treatments: Uncertainty and Adaptive Strategies

A powerful and recurring theme in research on the experience of life with HIV is that of uncertainty, in particular uncertainty about the future. Uncertainty is a central concern for all seriously ill persons, and it has been – and remains – a critical and multifaceted challenge for people living with HIV.

In one study conducted prior to the introduction of HAART, Weitz (1989) found that gay and bisexual men with AIDS struggled to cope with overwhelming uncertainty in their lives, which began with uncertainties about whether they were at risk of contracting HIV, and subsequently about emerging symptoms that may or may not be signs of AIDS. These uncertainties, sometimes coupled with a heightened sense of guilt about the behaviors that led to exposure, were found to drain their emotional and physical resources months and even years before they became ill. Once they had received a confirmed diagnosis of AIDS, many were then compelled to find answers to their questions about how they became infected – which were sometimes unanswerable – and faced further uncertainty regarding whether they would be able to function in the short term, whether their illness would take their lives, and whether they would be allowed to live – and if death was unavoidable to die – with dignity.

This study also illustrated some of the divergent strategies these men employed to manage uncertainty and assert control over their lives, including thoughts and actions reflecting both avoidance (e.g., avoiding knowledge about their serostatus) and vigilance (e.g., seeking knowledge and developing a realistic picture of what they could expect in the future) (Weitz 1989). For those infected early in the history of HIV, like Weitz's sample, learning to manage uncertainty and live with AIDS was all too often a tragically short process overwhelmed by the onset of serious illness and pending death.

In another pre-HAART study – but in recognition of the fact that the trajectory of life with HIV was steadily being prolonged by the first breakthroughs in antiretroviral treatment – Siegel and Krauss (1991) described a number of adapt strategies adopted by gay men to help them adapt to the challenges of living with HIV. Their sample was comprised of men who were HIV-positive, but did not have an AIDS diagnosis. This work identified three broad adaptive challenges, which cut across a number of adaptive tasks: (1) Dealing with the possibility of a curtailed life span, which included feeling a sense of urgency to attain life goals and deciding to what extent to invest in the future; (2) Dealing with reactions to a stigmatizing illness, which included making decisions about disclosing their illness to others and feelings of shame and contamination; and (3) Developing strategies for maintaining physical and emotional health, which included decisions about how to maintain a sense of control over – and an appropriate level of vigilance about – their health and treatment.

Like people facing other serious and chronic illnesses, some of these men worked to achieve purpose and control by reviewing and constructing their personal biographies. In other words, they engaged in biographical work, with many attempting to create a script for a more complete life. Career-related concerns became central to the script because issues relating to career success are key identity issues for people in their 20s, 30s, and 40s – like the men in their sample. Such an emphasis on personal accomplishment offered some not only a source of financial support and meaning, but also an opportunity to assert a normalized or relatively "disease-free" area of their existence (Siegel and Krauss 1991).

In addition, many of the men in Siegel and Krauss's sample described their attempts to avoid the pain of stigmatization, with some opting to primarily affiliate with people who were also seropositive, and not – or only selectively – disclosing their HIV status to others. Finally, the attentiveness – in many cases vigilance – that some exhibited in effort to maintain their emotional and physical health reflected efforts to distance themselves from further impact of HIV, and in cases where it was possible, to reinforce their experience of their selves as "not sick."

In sum, the adaptive strategies of these men reflected their underlying efforts to put their lives back together and resist being defined – by themselves or others – entirely through the lens of their HIV status (i.e., as a master status). Nonetheless, such strategies can become tenuous as physical symptoms emerge. For example, definitions of portions of one's life as "disease-free" or the self as "at risk" (i.e., "not sick") may yield in the face of contradictory sensory evidence. Indeed, Siegel and Krauss (1991) found that the occurrence of symptoms marked a significant turning point for these men in terms of how they viewed themselves in relation to their disease. For those men who did experience significant symptoms, feelings of shame and contamination were common, potentially signaling their inability to insulate themselves from AIDS, the disease, or its social and psychological impacts.

After Effective Treatments: HIV as a Chronic Condition

Much has changed since the earliest and darkest days of the pandemic, especially in terms of advances in treatment, which have provided reason for great optimism and hope in the battle against HIV. With the introduction of HAART, many long-term AIDS patients realized dramatic improvements – sometimes called Lazarus Phenomena (Thompson 2003) – and many more recently infected individuals have been able to strengthen their bodies' capacities to keep opportunistic infections at bay indefinitely. People living with HIV today can enjoy virtually normal lives and pursue personal and work-related goals with a sense that "the future" – in many respects a full and bright one – awaits them. In short, in the past two decades HIV has been transformed from a "death sentence" to a chronic condition not unlike others – manageable with proper care and adherence to treatment.

However, current treatments can have side effects that affect both health and appearance, and studies of seropositive gay men have examined the effects of body image on a range of mental health outcomes – and on adherence with treatment regimens (cf. Sharma et al. 2007). Moreover, treatments do not take away the reality that HIV-positive individuals remain infectious, and they must therefore continue to negotiate their romantic desires and intimate relationships with care. For instance, they are faced with a range of difficult decisions concerning if and when to disclose HIV status, as well as the safety of various sexual practices. In particular, there is increasing focus on the role of serostatus – and the practice of serosorting – in sexual decision making, dating, and relationship formation (Frost et al. 2008; Xia et al. 2006). Moreover, if a woman with HIV is pregnant or if HIV-affected individuals or couples wish to have children, numerous additional issues are raised in that regard as well (Schönnesson 2002).

In addition, as emerging treatment breakthroughs have been heralded in the media, met with excitement, relief, and hope; they simultaneously place an expectation on patients that they should be able to rid themselves of earlier HIV-related traumas and readily adapt to living with what is now *merely* "a chronic illness" (Catalan et al. 2000). In reality, however, making such a transition can be very difficult, and moving through it successfully requires an exceptional capacity to tolerate ambiguity and manage new uncertainties. Moreover, for long-term patients it requires a reconstruction of the future, perhaps especially in relation to issues of work/disability and intimate relationships (Thompson 2003).

From a qualitative study with HIV-positive men and women, Merriam et al. (2001) described this reconstruction process using a developmental perspective of the life course and different dimensions of time. By interviewing seropositive individuals before and after the development of HAART – in late 1995 and early 1998 – they documented a relationship between advances in HIV treatment – which happen in "historical time" – and the efforts of people living with HIV to recapture their ability to see their lives in "social time." This return to social time, which is based in the developmental patterns and ordering of events typically associated with adult life, made possible a view of a future that did not exist just a few years before. This reorientation appears to carry some ambivalence about aging. Consider, for example, the words of one 29-year-old woman – who had lived for about 8 years before beginning effective treatment:

> You know, I don't know what it is like to be twenty-something and not acutely aware of my own mortality. [She goes on to say that she used to think of 30 as] kind of the edge of the world. Now, I'm approaching it and I'm finding out that the world isn't flat; that's an amazing, wonderful thing, but it's scary too. It's about redefining who I want to be and everything (Merriam et al. 2001:179).

In short, evolving treatments are offering people living with HIV the opportunity to envision a future that accompanies a full life. For those who have already experienced considerable illness, reconstructing their futures presents them with new uncertainties, and the need to manage them. Current cohorts of people aging with HIV are historically unique in that they will be the first to grow old. In some senses, they are living their lives "on the boundaries," for example: the boundaries between having a terminal and chronic disease, between illness and health, between despair and hope, and ultimately, between death and the existential concerns about their futures (Schönnesson 2002; Schönnesson and Ross 1999). Much can be learned from their examples.

Studies that have attempted to look "inside" life with HIV as it continually changes in relation to scientific discovery and clinical progress are useful for situating the experiences of aging with HIV within a broad social and psychological context that incorporates personal biography, challenging experiences (both eventful and chronic), and efforts to manage them. Undoubtedly, more work of this kind is needed. Existing studies have disproportionately focused on the experiences of gay or bisexual men in developed countries. Therefore, it is important that future research examines the experiences of diverse samples of men and women – in different social settings – as they grow old with HIV. In the context of aging, such studies must investigate how individuals think about "time" and navigate the events and transitions of later life (e.g., retirement, grandparenthood) that have been previously unforeseen by and for people with HIV. Lessons from life course scholarship will be useful in this regard (Settersten 1999).

Additional Lenses and Frames of Reference

Our evolving understanding of living and aging with HIV is not solely driven by medical advances that improve disease management. HIV and AIDS have been and remain, in significant degree, conditions whose meanings are socially constructed. For instance, as a mysterious new threat, it was initially termed Gay Related Immune Deficiency (GRID) in the United States (Shilts 1987), although worldwide HIV has always disproportionately affected heterosexuals. Institutional response to emerging HIV-related issues – and seropositive individuals who are public about their status – in the social worlds of religion, politics, law, education, entertainment, and sports will continue to influence public discourse surrounding, and personal experience of, HIV.

In turn, social and cultural events and shifts will continue to contribute in important ways to ongoing reconstructions of HIV as a chronic condition, one that individuals can manage into later life. Thus, future social scientific research must strive to better understand how these forces – in addition to future scientific breakthroughs concerning HIV exposure, infection, and disease progression – are shaping life with HIV. Lessons learned from this work have the potential to inform theories about successful aging, as well as policy and practice regarding the needs of aging populations.

Stigma and Discrimination: "Not Like a Lot of Other Diseases"

While HIV has clearly become part of the social fabric of modern life, people living today with HIV experience many of the same negative social reactions that they did early on in the epidemic. Research consistently documents alarming levels of stigma and discrimination, for example, in studies concerning attitudes and beliefs about HIV and AIDS, HIV-positive individuals, and those thought to be at risk for HIV (e.g., Herek 1999; Herek et al. 2002).

To illustrate, although some overt expressions of HIV-related stigma have declined with time, nationally representative data suggest that inaccurate beliefs about risks from routine social interactions have increased (LeBlanc 1993; Herek et al. 2002), as has the belief that individuals with HIV deserve their illness. For instance, data collected in 1999 suggest that one in five Americans are "afraid" of individuals with HIV, and large proportions feel there is some risk of infection associated with being coughed or sneezed on by a person who is seropositive (50%) or simply using a public toilet (41%). In addition, one in four people in the general population endorses the sentiment that people infected with HIV through sex or drug use "have gotten what they deserve" (Herek et al. 2002). Finally, research has shown that negative attitudes toward people with HIV – and the groups most heavily affected by HIV – are associated with support for name-based HIV surveillance policies (Herek et al. 2003). Such findings underscore the reality that serostatus often plays a critical role in the formation and maintenance of social relationships of all kinds.

HIV-related stigma and discrimination also remain evident in high rates of violence and victimization associated with HIV-status, particularly for women (Zierler et al. 2000). Indeed, violations of human rights related to HIV are well documented, as are growing efforts to combat them (Gostin 1990; UNAIDS 2005), and HIV-based discrimination is observable in all areas of social life, including employment and health care settings (Schuster et al. 2005; Studdert 2002). Institutionalized policies may be especially slow to change. For example, it was only in October 2009 that plans were announced to lift the formal U.S. ban on HIV-positive immigrants and visitors, a policy established in 1987 (Preston 2009). Moreover, in many states, people with HIV may be subjected to criminal penalties for sexual activity; even when sexual activity is consensual, involves little or no risk of transmission, there is no intent to transmit, and the activity does not result in transmission (Lamda Legal 2007; Lange 2003; Patterson and London 2002).

The quote below, from a 39-year-old gay man who participated in a study published in 1991, illustrates the powerful social stigma that surrounds HIV as he experienced it:

> I do feel stigmatized. I feel as though somehow I have this dirty disease that everyone has learned to hate … I'm a carrier. At times I feel as though I bear all of the guilt and the responsibility for the negative comments we hear all the time. It's all on me, it's all in me. It's coursing through my veins. Sometimes I feel dirty because of it … And it is not in the same category as cancer. It is not in the same category as any so-called clean disease. The fact that I might be dying of HIV infection is not in the same category as anyone who innocently has a heart attack or any other long term chronic [disease like] diabetes (Siegel and Krauss 1991:24).

Many would argue that this quote applies as much today as it did nearly two decades ago.

Moreover, with the aging of the HIV population, there is increasing attention on the "double jeopardy" created by felt and enacted stigma and discrimination related to both HIV and aging. In a qualitative study of men and women over 50 and living with HIV or AIDS, Emlet (2006) identified stigma-related themes that reflected the experiences of HIV-associated stigma and ageism, as they occur independently and as they are interrelated. He has also developed a multi-dimensional HIV-stigma scale that has been examined in research with older adults (Emlet 2007).

For these older persons, HIV stigma included experiences related to: (1) fear of contagion; (2) homophobia; (3) violations of confidentiality; and (4) protective silence. Such themes are reminiscent of descriptions of stigma from earlier work based on younger samples. A substantial number of Emlet's respondents described others' irrational fears of contracting HIV from them, and several reported feeling stigmatized by the persistent homophobia that disproportionately links HIV with homosexuality. Violations of confidentiality, which is legally protected in medical settings and should never be compromised, were also present. One study participant described a particularly painful experience:

> So when I got there [to the medical clinic] they hand me my chart and I'd go see the blood pressure nurse, then I'd go see the nurse that draws blood, then I'd go see the doctor. I'd walk around with this big chart with

this big sticker on it – HIV Positive. Everyplace I went I was carrying this. I felt, why don't they just tattoo my forehead (Emlet 2006:787).

Thus, it is not surprising that people with HIV of all ages practice "protective silence" – or non-disclosure of their serostatus to others – as a way to manage the fear of anticipated stigma. One woman who lived with her elderly parents chose not to tell them of her condition. One man expressed his fear that his church group of friends would cut both him and his wife off if they learned of his diagnosis. Several described their fears about telling potential sexual partners, with one stating: It is a very difficult disease to have when you still want to have a partner-companion. "It is a very difficult disease to have, not like a lot of other diseases" (Emlet 2006:787).

Experiences at the ageism–HIV stigma intersection are especially illuminating, and included the following themes: (1) rejection; (2) stereotyping; and (3) feeling "separate or alone." Rejection by others was a unifying experience, reported by the majority of study participants as experienced from multiple sources, such as service providers, friends, family, church members, and potential sexual partners. Rejection was not always solely based on HIV status, but often included an element of rejection based on age. This appears to be a typical experience of older, gay men living with HIV – particularly those who are not healthy – in part because of the gay community's heavy emphasis on youth, vitality, and masculinity (cf. Kimmel and Mahalik 2005; Scrimshaw and Siegel 2003). One seropositive gay man stated that ageism "... is a far mightier sword than HIV" (Emlet 2006:785), in keeping with what Green (2008) describes as a "sexual status order" – based in part on attractiveness – that privileges younger and healthier gay men. Such disparities create significant stressors for gay men, particularly if they are aging with HIV. Similar research on women and straight men aging with HIV is much needed.

Stereotypical attitudes about aging and sexuality, or assumptions about how one becomes infected, were also challenging for individuals aging with HIV. For instance, respondents described circumstances where it was clear that society expects older people to be more knowledgeable about the disease simply by virtue of their age. One woman expressed her belief that: "[I]n some cases ... older people are held to a different standard ... [because] ... well, for crying out loud, [they] should know better" (Emlet 2006:786). One man shared: "There's a lot of people that go 'you're awfully old to have this disease'" (Emlet 2006:786). Another man, who is heterosexual, described how women around his age whom he would like to date assume that he must be gay or bisexual because of his HIV status (Emlet 2006).

Finally, older persons with HIV described an internalized stigma that Emlet termed "separate or alone," a fitting label for a variety of experiences that keep them segregated from the mainstream of social life. One single man described his feelings of extreme isolation: "I don't have a partner, you know. I could drop dead in my house and no one might know about it for three weeks" (Emlet 2006:786). Another shared that he felt "radioactive" (Emlet 2006:786), a powerful word that simultaneously portrayed his loneliness and his sense that he is perceived to be a threat to others.

In another study, Siegel et al. (1998) highlighted perceived disadvantages of age in the context of HIV, including: being more socially isolated; receiving less sympathy and being judged more harshly; and feeling that medical providers set higher goals when treating younger patients. Such findings highlight important avenues for additional, and more nuanced, study regarding social relationships for older people with HIV.

In sum, more than three decades into the pandemic, pervasive stigma and discrimination surrounding HIV remains, and we are beginning to understand the unique experiences of seropositive individuals who are now entering later life. Additional research based on individual narratives, of diverse samples across a range of social contexts, will build on this small but important literature.

In addition, future studies can further illuminate HIV-related stigma and discrimination as sources of social stress that can directly and indirectly affect the well-being of people aging with HIV. Existing theoretical frameworks found in the minority stress (Meyer 1995, 2003) and stress process (Pearlin 1999; Pearlin et al. 1997) literatures could provide useful models for such work. Moreover, as important lessons are learned about HIV-related stigma and discrimination, stress, aging, and well-being; applied researchers can work toward the identification of more effective means of addressing stigma and discrimination (cf. Chesney and Smith 1999).

Finally, studies of HIV stigma and discrimination have typically focused on the perceptions and experiences of individuals, either as perpetrators or victims, reflecting the use of individualistic models in approaching this topic. However, social scientists should also attempt to examine HIV-related stigma and discrimination as social processes that are linked to the reproduction of inequality and exclusion – rooted in gender, race/ethnicity, and class as they interact with age. Such a focus may be particularly useful for work focused on the resistance of stigmatized individuals and community mobilization for creating rights-based approaches to transforming social climates in which stigmatization and discrimination are routinely tolerated (Parker and Aggleton 2003).

Social Networks and Social Support

As detailed above, nationally representative data from the United States portray significant social and economic vulnerabilities of the older HIV-positive population. Older individuals living with HIV are more likely than their younger counterparts to live alone, and they are also more likely to report conflict with family, friends, and others in their social networks. In addition, among older people with HIV, racial/ethnic minority persons and older individuals who were exposed to HIV through IDU experience relatively greater economic disadvantage, and consequently are more likely to lack access to supportive services (Crystal et al. 2003; Joyce et al. 2005). It has been suggested, for example, that IDUs are especially likely to exhaust their social resources as drug use lengthens and friends and family become disaffected with their problems (Levy 1998).

Smaller scale studies of the social networks and social support of older HIV-infected populations have also emphasized their high risk of social isolation, due to the lack of traditional supports and the effects of HIV-related stigma and ageism (Emlet 2006; Shippy and Karpiak 2005a). In a study of older people with HIV in New York City, it was estimated that only a third have a partner and nearly three-fourths (71%) live alone (Shippy and Karpiak 2005a).

Moreover, research has suggested that sources of social support vary by sexual orientation, with sexual minority persons reporting greater support from friends, which often comprise their "families of choice" (Weston 1991). Certainly, the large AIDS caregiving literature demonstrates the historically unprecedented rallying of support among persons unrelated by ties of blood or marriage (Turner et al. 1994). Moreover, resilient HIV-positive networks of friendship and support in some instances take the place of support lost to stigma and rejection. For some older people living with HIV, however, friendship networks have been diminished due to AIDS-related deaths and HIV-related morbidity (Scrimshaw and Siegel 2003; Shippy and Karpiak 2005a; Poindexter and Shippy 2008). Increasing bereavement is of course a natural part of the aging process, but unlike their contemporaries, long-term survivors of HIV have more often experienced losses of peers – often to HIV – who were only just entering young adulthood or middle age (cf. Martin and Dean 1993).

Further research into the ways in which older people with HIV seek and receive assistance – or not – from within their varying familial and social networks is needed. In particular, it is important to examine how social and cultural factors may affect support seeking, or lack thereof. For instance, Heckman et al. (2000) focused on differences in sources of social support between older African-American and white men, and found that African-American men received relatively more support from family members, and were simultaneously less likely to disclose their HIV status to close friends. It is also important to better understand the value of various types of social support – for instance, emotional versus instrumental – to people aging with HIV, particularly with changes in their health over time. For example, existing research finds that those experiencing worse health were more likely to report having adequate emotional and instrumental support within their informal support networks, which conversely suggests that those in relatively good health are more apt to perceive inadequate support (Shippy and Karpiak 2005b).

In addition, in keeping with the longstanding literature demonstrating the health benefits of social integration and support (Thoits 1995; Turner and Turner 1999), more work is needed to understand the unique roles various types of social support – for instance, emotional and instrumental support from different sources – play as determinants of both physical and mental health among individuals aging with HIV. Indeed some research of this type has been conducted with HIV-positive populations, and a recent study of seropositive men suggests that the moderating effect of support on health may be especially important to older men (Chesney et al. 2003).

Lastly, this emerging literature suggests high levels of unmet need for informal social support among HIV-positive elders (Shippy and Karpiak 2005b), and we can therefore expect to witness expanding needs for ongoing assistance from the formal social and health care systems in the future as this population continues to grow. Unfortunately, we can also anticipate a high probability of witnessing more cases where individuals fall through the cracks of both the formal systems, and informal networks, of support, and are left to fend for themselves in later life.

HIV, Aging, and Well-Being

Clinical studies tend to neglect the aspects of quality of life (QOL) and health – well-being for short – as they are subjectively experienced. QOL measures typically encompass several dimensions, including patient-reported: physical function and symptoms, performance of social roles, emotional status, cognitive functioning, and feelings about his or her health. The first HIV-related QOL studies coincided with the advent of the first promising antiretroviral therapies in the late 1980s, and since that time QOL data have complemented findings from treatment trials, sometimes in support – and sometimes at odds with – clinical findings (Wu 2000).

QOL has been examined among people with HIV more frequently over time. Generic QOL measures have been adapted to address aspects of living with HIV that can differ from living with other chronic conditions, and new HIV-specific QOL measures have been developed as well (Avis and Smith 1998). The more prominent generic QOL instruments have been implemented with HIV populations to facilitate comparisons in QOL for people living with HIV and QOL for individuals living with other conditions, including breast cancer, clinical depression, and diabetes (Avis and Smith 1998; Hays et al. 2000). There have also been studies examining, within HIV populations, the influence of HIV-specific symptomatology on self-perceived QOL (Lorenz et al. 2001); in recognition of the fact that disease trajectories are diverse, and fluctuate over time.

Although researchers have also begun to examine how self-rated QOL varies by age, studies of aging and QOL in HIV populations are quite rare. Working with data collected between 1988 and 1990, Piette et al. (1995) concluded that older persons with HIV reported worse QOL on multiple dimensions than their younger counterparts. Studies of mental health outcomes among persons aged 50 and older with HIV illustrate the serious emotional difficulties faced by this population. One study estimated that one-fourth experiences "moderate" to "severe" depression, which was related to greater HIV-related stress, less support from friends, and reduced access to services due to HIV-related stigma (Heckman et al. 2002). Another found that just over one-fourth of a sample of middle-aged and older persons with HIV had considered taking their own life in the preceding week (Kalichman et al. 2000).

Findings based on self-report measures of general health and emotional well-being from the HCSUS study suggest that older gay men may enjoy some health-related advantages over their younger counterparts (cf. Bybee et al. 2009), although older persons infected through IDU do not demonstrate such advantages, and the data regarding older people infected in other ways are mixed (Crystal et al. 2003). These better health perceptions reported by older gay men defy simple interpretation. For instance, it may be that they are, at least in part, merely comparing themselves with an older reference group with more health problems, or that with age comes some psychosocial advantages (e.g., wisdom, patience, and a relatively diminished sense of feeling "cheated" by their condition) (Siegel et al. 1998). Alternatively, this difference might also reflect the more consistent adherence to treatment regimens that has been observed in older HIV populations (Casau 2005; Crystal et al. 2003; Sherr et al. 2009).

In sum, additional studies of HIV, aging, and well-being – as assessed along a wide range of dimensions – are greatly needed. Studies that apply a life course framework (Settersten 1999) may be especially useful. For example, as the current research literature expands, it is important for investigators to take into account patient age at time of exposure (e.g., at birth through later life), in order to gain critical insights regarding the full diversity of HIV biographies and trajectories (cf. Avis and Smith 1998). Moreover, the available studies, however limited, serve as a reminder that the emerging data on disease manifestations and clinical markers – which garner disproportionate scientific and media attention – must be viewed alongside more subjective indicators of wellness, which stand to teach us important lessons regarding the subjective experiences of HIV as people age. Such work should focus not only on hardships associated with HIV and aging, but also positive changes that have liberating or transforming effects.

Finally, emerging research on mental health among older people living with HIV may offer especially useful insights for building stronger theoretical frameworks for studying HIV, aging and well-being. Stress process models, which consider how stress begets stress, appear especially relevant (Pearlin 1999) (Pearlin 1999). For instance, attention paid to the subjective experiences of social stress by individuals with HIV, as well as their efforts to access psychosocial resources (e.g., a sense of mastery, social support) to cope with challenges, highlights key psychological and social processes influencing well-being (cf. Goodkin et al. 2003). One developing body of research is focusing on the role of spirituality and religiosity in the lives of people aging with HIV (Vance et al. 2007; Vance 2006). The finding that suicide ideation among midlife and older persons with HIV is associated with greater levels of emotional distress, poor health-related QOL, and the use of escape and avoidance coping strategies (as opposed to positive-reappraisal coping) is one example of research reflecting the relevance stress process frameworks (Kalichman et al. 2000).

HIV-Affected Older Populations

The study of HIV and aging also inherently concerns the lives of older persons who are affected by, but not infected with, HIV. Certainly the longstanding AIDS caregiving literature has drawn attention to the experiences of aging care providers (cf. Wight et al. 1998; Wight 2000). For example, Wight et al. (2008) highlighted how uncertainty about the future is filled with shared experiences, in this case between people with HIV and their caregivers. Indeed, lives affected by HIV – like all others – are "linked lives" in multiple and complex ways, and should be examined as such by focusing on relationship dynamics for people aging with HIV. Moreover, a focus on HIV-affected – in addition to HIV-infected – older persons is perhaps especially relevant and useful in developing nations where at the current historical moment rates of HIV-related morbidity and mortality are relatively high (Williams, Knodel and Lam 2010; Knodel et al. 2010).

It is also essential that researchers simultaneously consider the extent to which older people with HIV are themselves sources of social support for one another (LeBlanc and Wight 2000), and for others in need of support. Indeed, many are primary caregivers to parents, spouses/partners, and children, as well as for friends (cf. Wight 2000), and as they continue realizing improvements in their well-being, the obligations of such roles and obligations will only expand.

Discussion

More than three decades into the global HIV pandemic – and more than two decades after the social scientific community began writing about aging in the context of HIV (Riley et al. 1989) – it is clear that a diverse and growing HIV-positive population is aging into later life. Social science has a significant part to play in the production of knowledge that will facilitate better understandings of their experiences, as well as inform the design of better interventions and policy to support their unique needs.

Current research has accomplished a good deal toward developing sociodemographic profiles of aging seropositive populations. However, this work – like virtually all research on HIV and aging – pertains almost exclusively to populations in developed nations. It is clear that the significant diversity reflected in these initial studies will become even more apparent as similar data are collected across more locales and updated over time. Indeed, much more of this basic descriptive study is required – in developed and developing countries and local settings – because it provides the foundation for both basic research that attempts to examine HIV and aging in greater depth and applied research that strives to address related social problems.

Past studies have produced useful qualitative understandings of what it is like for people to live and age with HIV, in terms of their existential concerns as well as in regard to their social interactions. Because AIDS has taken the lives of so many young adults, existing research has privileged the impact of "disease" over "age" in thinking about the experiences of people aging with HIV. However, with the diverse and rapidly expanding population of people growing old with HIV, it is time for more thoughtful examinations of the unique effects of HIV-status and aging – and their combined effects – on well-being. It is also apparent that this work must be conducted across a wide range of settings with people from all backgrounds. To date, research on the day-to-day experiences of women and racial/ethnic minority persons with HIV – of all ages – is especially lacking (Brown and Sankar 1998; Zablotsky 1998). Life course perspectives – drawn from multiple disciplines – may particularly facilitate this work by situating it within the broader social and cultural contexts of time and place (Settersten 1999).

Many of the relevant studies to date have been necessarily concerned with how the introduction of effective treatment has dramatically altered the experience of living and aging with HIV, at least in populations with access to care. In short, new treatments have made aging a possibility for many. However, it is time for the field to move beyond period effects associated with scientific advances in the war against HIV, paying attention also to the social and cultural changes that also determine the contextual landscape and personal experience of HIV. Moreover, life course researchers must also focus on the effects of cohort. For instance, as the first generation to do so, older people with HIV are now entering later life without role models. Their younger counterparts will benefit from their examples, but they will also navigate a larger proportion of the life course with HIV, managing more of life's important events and transitions along the way. For instance, in the existing literature, there appear to be significant cohort differences in the well-being of gay men with HIV, with the older generation reporting some health-related advantages over their younger counterparts (Crystal et al. 2003); despite facing what the clinical literature is also beginning to suggest, a more severe disease course. Deeper qualitative understandings of such differences stand to teach us important lessons about successful aging with chronic conditions.

There are a range of social and psychological factors that must be considered in relation to this growing understanding of HIV as a chronic illness that can be experienced over the course of a long life. Issues relating to the still pervasive HIV-related stigma and discrimination are of paramount concern. Existing studies poignantly illustrate their demeaning, demoralizing, and isolating powers. Research also suggests that HIV-related stigma overlaps in some respects with ageism, which points to the potential use of further conceptual research regarding the intersection of HIV-related stigma and other forms of marginalization and oppression. Moreover, the recognition that stressful experiences and conditions emanate from HIV- and age-related stigma and discrimination draws our attention to current models of minority stress (Meyer 1995, 2003) and stress process (Pearlin 1999), both of which provide useful approaches for understanding the relationship between social stress and health. Indeed, new frameworks that integrate the two may be particularly promising in the study of HIV and aging.

The emerging literature on the present – and likely growing – unmet need for social support among people aging with HIV also addresses critical issues of interest. Available studies portray differences in the ways in which older people with HIV define their familial and social networks, as well as in whether and how they seek and receive social support. There is also diversity in the networks and supports available to them. As a whole, this population experiences relative socioeconomic disadvantage, with many enduring severe hardship. Available studies document the unique vulnerabilities of sexual minority and IDU support networks as well. Future study must continue to explore for these kinds of differences in order to facilitate a more refined understanding of the potential role of social support in the lives – and well-being – of the aging HIV-positive population.

The recognition of the unmet needs for social support among individuals aging with HIV – viewed against the backdrop of persistent stigma and discrimination – offers a useful vantage point for considering future applied research that informs policy. Certainly, the aging HIV-positive population will continue to expand, and we are only beginning to understand age-based differences in the experience of HIV, not only in terms of symptomatology but also with regard to the social interactions involved in seeking informal and formal supports, as well as in accessing medical care. Thus, it is time for more in-depth research on the support and care needs of diverse, aging seropositive populations. As more about such differences becomes known, appropriate social and medical services can be developed and made available (cf. Emlet et al. 2009; Poindexter and Shippy 2008). This work will also inform the creation of better policy for people entering late life with HIV, as they face personal, social, and economic challenges – similar and unique from those facing the general aging population – associated with medical care, disability, retirement, independent living, and long-term care. In the United States, for example, we might anticipate interesting policy research questions concerning the ways in which steadily growing numbers of HIV-positive individuals aging into Medicare eligibility could create age-based disparities in access to treatment.

Ultimately, success in the battle against HIV is understood by the well-being – along a range of social and health-related dimensions – of people who live with the virus. Amidst so much change in the evolution of our understandings of HIV as a medical condition, research on the subjective well-being of infected individuals has lagged behind. Studies of well-being among older HIV populations are exceptionally rare. As discussed above, research to date points to the need to locate future research within the theoretical frames of the life course and stress research in order to better explain social determinants of well-being among individuals aging with HIV (Pearlin et al. 2005). We stand to learn a great deal from people who are now aging in later life with HIV. They are the first cohort in the history of the pandemic to grow old.

References

Avis, Nancy E. and Kevin W. Smith. 1998. "Quality of Life in Older Adults With HIV Disease." *Research on Aging* 20:822–45.

Brown, Diane R. and Andrea Sankar. 1998. "HIV/AIDS and Aging Minority Populations." *Research on Aging* 20:865–84.

Bybee, Jane A., Eric L. Sullivan, Erich Zielonka, and Elizabeth Moes. 2009. "Are Gay Men in Worse Mental Health than Heterosexual Men? The Role of Age, Shame and Guilt, and Coming Out." *Journal of Adult Development* 16:144–54.

Casau, Nathalie C. 2005. "Perspectives on HIV Infection and Aging: Emerging Research on the Horizon." *Clinical Infectious Diseases* 41:855–63.

Catalan, J., J. Meadows, and A. Douzenis. 2000. "The Changing Pattern of Mental Health Problems in HIV Infection: The View from London, UK." *AIDS Care* 12:333–41.

Centers for Disease Control and Prevention. 2007. *HIV/AIDS Surveillance Report, Vol. 19*. Atlanta, GA: U.S. Department of Health and Human Services, Centers for Disease Control and Prevention. Retrieved (http://www.cdc.gov/hiv/topics/surveillance/resources/reports).

Centers for Disease Control and Prevention. 2009. *HIV/AIDS in the United States*. CDC HIV/AIDS Facts, Retrieved August 2009 (http://www.cdc.gov/hiv/resources/factsheets/PDF/us.pdf).

Chesney, Margaret A., Donald B. Chambers, Jonelle M. Taylor, and Lisa M. Johnson. 2003. "Social Support, Distress, and Well-Being in Older Men Living With HIV Infection." *JAIDS Journal of Acquired Immune Deficiency Syndromes* 33:S185–93.

Chesney, Margaret A. and Ashley W. Smith. 1999. "Critical Delays in HIV Testing and Care: The Potential Role of Stigma." *American Behavioral Scientist* 42:1162–1174.

Conde, Marques Délio, Emanuela T. Silva, Waldemar N. Amaral, Marta F. Finotti, Rui G. Ferreira, Lúcia Costa-Paiva, and Aarão M. Pinto-Neto. 2009. "HIV, Reproductive Aging, and Health Implications in Women: A Literature Review." *Menopause: The Journal of the North American Menopause Society* 16:199–213.

Cooperman, Nina A., Julia H. Arnsten, and Robert S. Klein. 2007. "Current Sexual Activity and Risky Sexual Behavior in Older Men with or at Risk for HIV Infection." *AIDS Education and Prevention* 19:321–33.

Corneille, Maya A., Linda E. Zyzniewski, and Faye Z. Belgrave. 2008. "Age and HIV Risk and Protective Behaviors among African American Women." *Journal of Black Psychology* 34:217–33.

Crystal, Stephen, Ayse Akincigil, Usha Sambamoorthi, Neil Wenger, John A. Fleishman, David S. Zingmond, Ron D. Hays, Samuel A. Bozzette, and Martin F. Shapiro. 2003. "The Diverse Older HIV-Positive Population: A National Profile of Economic Circumstances, Social Support, and Quality of Life." *JAIDS Journal of Acquired Immune Deficiency Syndromes* 33:S76–83.

Dolcini, M. Margaret, Joseph A. Catania, Ronald D. Stall, and Lance Pollack. 2003. "The HIV Epidemic among Older Men Who Have Sex With Men." *JAIDS Journal of Acquired Immune Deficiency Syndromes* 33:S115–S21.

Elford, J., F. Ibrahim, C. Bukutu, and J. Anderson. 2008. "Over Fifty and Living with HIV in London." *Sexually Transmitted Infections* 84:468–72.

Emlet, Charles A. 2006. "You're Awfully Old to Have This Disease: Experiences of Stigma and Ageism in Adults 50 Years and Older Living with HIV/AIDS." *The Gerontologist* 46:781–90.

———. 2007. "Experiences of Stigma in Older Adults Living with HIV/AIDS: A Mixed-Methods Analysis." *AIDS Patient Care and STDs* 21:740–752.

Emlet, Charles, A., Amanda Gerkin, and Nancy Orel. 2009. "The Graying of HIV/AIDS: Preparedness and Needs of the Aging Network in a Changing Epidemic." *Journal of Gerontological Social Work* 52:803–14.

Frost, David M., Michael J. Stirratt, and Suzanne C. Ouellette. 2008. "Understanding Why Gay Men Seek HIV-Seroconcordant Partners: Intimacy and Risk Reduction Motivations." *Culture, Health & Sexuality* 10:513–27.

Gebo, Kelly A. 2009. "Editorial: HIV Infection in Older People." *British Medical Journal* 338:b1460.

Gooding, Karl, Timothy Heckman, Karolynn Siegel, Nathan Linsk, Imad Khamis, Diana Lee, Robert Lecusay, Cynthia C. Poindexter, Sally J. Mason, Paola Suarez, and Carl Eisdorfer. 2003. "'Putting a Face' on HIV Infection/AIDS in Older Adults: A Psychosocial Context." *Journal of Acquired Immune Deficiency Syndromes* 33:S171–84.

Gostin, Lawrence O. 1990. "The AIDS Litigation Project: A National Review of Court and Human Rights Commission Decisions, Part II: Discrimination." *Journal of the American Medical Association* 263:2086–93.

Green, Adam Isaiah. 2008. "Health and Sexual Status in an Urban Gay Enclave: An Application of the Stress Process Model." *Journal of Health and Social Behavior* 49:436–51.

Hays, Ron D., William E. Cunningham, Cathy D. Sherbourne, Ira B. Wilson, Albert W. Wu, Paul D. Cleary, Daniel F. McCaffrey, John A. Fleishman, Stephen Crystal, Rebecca Collins, Ferd Eggan, Martin F. Shapiro, and Samuel A. Bozzette. 2000. "Health-Related Quality of Life in Patients with Human Immunodeficiency Virus Infection in the United States: Results from the HIV Cost and Services Utilization Study." *The American Journal of Medicine* 108:714–22.

Heckman, Timothy G., B. D. Heckman, Arlene Kochman, Kathleen J. Sikkema, J. Suhr, and Karl Goodkin. 2002. "Psychological Symptoms among Persons 50 Years of Age and Older Living with HIV Disease." *Aging & Mental Health* 6:121–8.

Heckman, Timothy G., Arlene Kochman, Kathleen J. Sikkema, Seth C. Kalichman, James Masten, and Karl Goodkin. 2000. "Late Middle-Aged and Older Men Living with HIV/AIDS: Race Differences in Coping, Social Support, and Psychological Distress." *Journal of the National Medical Association* 92:436–44.

Herek, Gregory M. 1999. "AIDS and Stigma." *American Behavioral Scientist* 42:1106–16.

Herek, Gregory M., John P. Capitanio, and Keith F. Widaman. 2002. "HIV-related Stigma and Knowledge in the U.S.: Prevalence and Trends." *American Journal of Public Health* 92:371–7.

———. 2003. "Stigma, Social Risk, and Health Policy: Public Attitudes toward HIV Surveillance Policies and the Social Construction of Illness." *Health Psychology* 22:533–40.

Institute of Medicine. 1988. *Confronting AIDS: Update 1988.* Washington, DC: National Academy Press.

Joyce, Geoffrey F., Dana P. Goldman, Arleen A. Leibowitz, Abby Alpert, and Yuhua Bao. 2005. "A Socioeconomic Profile of Older Adults with HIV." *Journal of Health Care for the Poor and Underserved* 16:19–28.

Kalichman, Seth C., Timothy Heckman, Arlene Kochman, Kathleen Sikkema, and Jo Bergholte. 2000. "Depression and Thoughts of Suicide among Middle-Aged and Older Persons Living With HIV-AIDS." *Psychiatric Services* 51:903–7.

Karpiak, Stephen E., R. Andrew Shippy, and Marjorie Cantor. 2006. *Research on Older Adults with HIV.* AIDS Community Research Initiative of America, New York.

Kimmel, Sara B. and James R. Mahalik. 2005. "Body Image Concerns of Gay Men: The Roles of Minority Stress and Conformity to Masculine Norms." *Journal of Consulting and Clinical Psychology* 73:1185–90.

Knodel, John, Nathalie Williams, Sovan Kiry Kim, Sina Puch, and Chanpen Saengtienchai. 2010. "Community Reaction to Older Age Parental AIDS Caregivers and Their Families: Evidence from Cambodia." *Research on Aging* 31:122–151.

Koblin, Beryl A., Marla J. Husnik, Grant Colfax, Yijian Huang, Maria Madison, Kenneth Mayer, Patrick J. Barresi, Thomas J. Coates, Margaret A. Chesney, and Susan Buchbinder. 2006. "Risk Factors for HIV Infection among Men Who Have Sex with Men." *AIDS* 20:731–9.

Kushel, Margot B., Judith A. Hahn, Jennifer L. Evans, David R. Bangsberg, and Andrew R. Moss. 2005. "Revolving Doors: Imprisonment Among the Homeless and Marginally Housed Population." *American Journal of Public Health* 95:1747–52.

Kwiatkowski, Carol F. and Robert E. Booth. 2003. "HIV Risk Behavior among Older American Drug Users." *Journal of Acquired Immune Deficiency Syndromes* 33(supp. 2):S131–7.

Lange, Tamara. 2003. "HIV and Civil Rights: A Report from the Frontlines of the HIV/AIDS Epidemic." ACLU AIDS Project, New York. Retrieved (http://www.aclu.org/pdfs/hivaids/hivcivilrights.pdf).

LeBlanc, Allen J. 1993. "Examining HIV-Related Knowledge among Adults in the U.S." *Journal of Health and Social Behavior* 34:23–36.

LeBlanc, Allen J. and Richard G. Wight. 2000. "Reciprocity and Depression in AIDS Caregiving." *Sociological Perspectives* 43:631–49.

Lamda Legal. 2007. *The State of HIV Stigma and Discrimination in 2007: An Evidence Based Report.*

Lee, Marlene. 2009. HIV/AIDS and Older Adults in the United States. *Today's Research on Aging, Issue 18.* Washington, DC: Population Reference Bureau. Retrieved (http://www.prb.org/pdf09/TodaysResearchAging18.pdf).

Levy, Judith A. 1998. "AIDS and Injecting Drug Use in Later Life." *Research on Aging* 20:776–97.

Lorenz, Karl A., Martin F. Shapiro, Steven M. Asch, Samuel A. Bozzette, and Ron D. Hays. 2001. "Associations of Symptoms and Health-Related Quality of Life: Findings from a National Study of Persons with HIV Infection." *Annals of Internal Medicine* 134:854–60.

Mack, Karin A. and Marcia G. Ory. 2003. "AIDS and Older Americans at the End of the Twentieth Century." *Journal of Acquired Immune Deficiency Syndromes* 33:S68–75.

Mann, Jonathan, Daniel J.M. Tarantola, and Thomas W. Netter. 1992. *AIDS in the World.* Cambridge, MA: Harvard University Press.

Martin, Clifford P., Mindy J. Fain, and Stephen A. Klotz. 2008. "The Older HIV-Positive Adult: A Critical Review of the Medical Literature." *The American Journal of Medicine* 121:1032–7.

Martin, John L. and Laura Dean. 1993. "Effects of AIDS-related Bereavement and HIV-related Illness on Psychological Distress among Gay Men: A 7-Year Longitudinal Study, 1985–1991." *Journal of Consulting and Clinical Psychology* 61:94–103.

Merriam, Sharan B., Bradley C. Courtenay, and Patricia M. Reeves. 2001. "Time and Its Relationship to Development in the Life Course: Some Reflections from a Study of HIV-Positive Adults." *Journal of Adult Development* 8:173–182.

Meyer, Ilan H. 1995. "Minority Stress and Mental Health in Gay Men." *Journal of Health and Social Behavior* 36:38–56.

———. 2003. "Prejudice, Social Stress, and Mental Health in Lesbian, Gay, and Bisexual Populations: Conceptual Issues and Research Evidence." *Psychological Bulletin* 129:674–97.

Neundorfer, Marcia M., Phyllis Braudy Harris, Paula J. Britton, and Delores A. Lynch. 2005. "HIV-Risk Factors for Midlife and Older Women." *The Gerontologist* 45:617–25.

Orchi, N., R. Balzano, P. Scognamiglio, A. Navarra, G. De Carli, P. Elia, S. Grisetti, A. Sampaolesi, M. Giuliani, A. De Filippis, V. Puro, G. Ippolito, E. Girardi, and on behalf of the SENDIH group. 2008. "Ageing with HIV: Newly Diagnosed Older Adults in Italy." *AIDS Care* 20:419–25.

Parker, Richard and Peter Aggleton. 2003. "HIV and AIDS-related Stigma and Discrimination: A Conceptual Framework and Implications for Action." *Social Science & Medicine* 57:13–24.

Patterson, D. and L. London. 2002. "International Law, Human Rights, and HIV/AIDS." *Bulletin of the World Health Organization* 80:964–9.

Pearlin, Leonard I. 1999. "The Stress Process Revisited: Reflections on Concepts and Their Interrelationships." pp. 395–415 in *Handbook on the Sociology of Mental Health*, edited by Carol S. Aneshensel and Jo Phelan, New York: Plenum.

Pearlin, Leonard I., Carol S. Aneshensel, and Allen J. LeBlanc. 1997. "The Forms and Mechanisms of Stress Proliferation: The Case of AIDS Caregivers." *Journal of Health and Social Behavior* 38:223–36.

Pearlin, Leonard I., Scott Schieman, Elena M. Fazio, and Stephen C. Meersman. 2005. "Stress, Health, and the Life Course: Some Conceptual Perspectives." *Journal of Health and Social Behavior* 46:205–19.

Piette, John, Tom J. Wachtel, Vincent Mor, and Kenneth Mayer. 1995. "The Impact of Age on the Quality of Life in Persons with HIV Infection." *Journal of Aging and Health* 7:163–78.

Poindexter, Cynthia and R. Andrew Shippy. 2008. "Networks of Older New Yorkers with HIV: Fragility, Resilience, and Transformation." *AIDS Patient Care and STDs* 22:723–33.

Preston, Julia. 2009. "Obama Lifts a Ban on Entry Into U.S. by H.I.V.-Positive People." *The New York Times*, October 30, 2009, p. A9. Retrieved (http://www.nytimes.com/2009/10/31/us/politics/31travel.html).

Riley, Matilda White, Marcia G. Ory, and Diane Zablotsky. 1989. *AIDS in an Aging Society: What We Need to Know*. New York: Springer.

Schensul, Jean J., Judith A. Levy, and William B. Disch. 2003. "Individual, Contextual, and Social Network Factors Affecting Exposure to HIV/AIDS Risk among Older Residents Living in Low-Income Senior Housing Complexes." *JAIDS Journal of Acquired Immune Deficiency Syndromes* 33:S138–52.

Schmid, George P., Brian G. Williams, Jesus Maria Garcia-Calleja, Chris Miller, Emily Segar, Monica Southworth, David Tonyan, Jocelyn Wacloff, and James Scott. 2009. "The unexplored story of HIV and ageing." *Bulletin of the World Health Organization* 87:162.

Schönnesson, Lena Nilsson. 2002. "Psychological and existential issues and quality of life in people living with HIV infection." *AIDS Care* 14:399–404.

Schönnesson, Lena Nilsson, and Michael W. Ross. 1999. *Coping with HIV Infection: Psychological and Existential Responses in Gay Men*. New York: Kluwer Academic/Plenum.

Schrimshaw, Eric W. and Karolynn Siegel. 2003. "Perceived Barriers to Social Support from Family and Friends among Older Adults with HIV/AIDS." *Journal of Health Psychology* 8:738–52.

Schuster, Mark A., Rebecca Collins, William E. Cunningham, Sally C. Morton, Sally Zierler, Myra Wong, Wenli Tu, and David E. Kanouse. 2005. "Perceived Discrimination in Clinical Care in a Nationally Representative Sample of HIV-Infected Adults Receiving Health Care." *Journal of General Internal Medicine* 20:807–13.

Settersten, Richard A., Jr. 1999. *Lives in Time and Place: The Problems and Promises of Developmental Science*. Amityville, NY: Baywood.

Sharma, A., A. A. Howard, R. S. Klein, E. E. Schoenbaum, D. Buono, and M. P. Webber. 2007. "Body Image in Older Men with or at-Risk for HIV Infection." *AIDS Care* 19:235–41.

Sherr, L., R. Harding, F. Lampe, M. Johnson, J. Anderson, S. Zetler, M. Fisher, G. Arthur, S. Norwood, H. Leake-Date, and S. Edwards. 2009. "Clinical and Behavioural Aspects of Aging with HIV Infection." *Psychology, Health and Medicine* 14:273–279.

Shilts, Randy 1987. *And The Band Played On: Politics, People and the AIDS Epidemic*. New York: St. Martin's Press.

Shippy, R. Andrew and Stephen E. Karpiak. 2005a. "The Aging HIV/AIDS Population: Fragile Social Networks." *Aging & Mental Health* 9:246–54.

———. 2005b. "Perceptions of Support Among Older Adults With HIV." *Research on Aging* 27:290–306.

Siegel, Karolynn and Beatrice J. Krauss. 1991. "Living with HIV Infection: Adaptive Tasks of Seropositive Gay Men." *Journal of Health and Social Behavior* 32:17–32.

Siegel, Karolynn, Victoria Raveis, and Daniel Karus. 1998. "Perceived Advantages and Disadvantages of Age among Older HIV-Infected Adults." *Research on Aging* 20:686–711.

Simone, Mark J. and Jonathan Appelbaum. 2008. "HIV in Older Adults." *Geriatrics* 63:6–12.

Sormanti, Mary and Tazuko Shibusawa. 2007. "Predictors of Condom Use and HIV Testing Among Midlife and Older Women Seeking Medical Services." *Journal of Aging and Health* 19:705–19.

Stall, Ron and Joe Catania. 1994. "AIDS Risk Behaviors among Late Middle-Aged and Elderly Americans: The National AIDS Behavioral Surveys." *Archives of Internal Medicine* 154:57–63.

Stall, Ron and David W. Purcell. 2000. "Intertwining Epidemics: A Review of Research on Substance Use Among Men Who Have Sex with Men and Its Connection to the AIDS Epidemic." *AIDS and Behavior* 4:181–92.

Studdert, David M. 2002. "Charges of Human Immunodeficiency Virus Discrimination in the Workplace: The Americans with Disabilities Act in Action." *American Journal of Epidemiology* 156:219–29.

Stoff, David M., Jag H. Khalsa, Andrew Monjan, and Peter Portegies. 2004. "Introduction: HIV/AIDS and Aging." *AIDS* 18:S1–2.

Thoits, Peggy. 1995. "Stress, Coping, and Social Support Processes: Where are We? What Next?" *Journal of Health and Social Behavior* Extra Issue: 53–79.

Thompson, B. 2003. "Lazarus Phenomenon: An Exploratory Study of Gay Men Living with HIV." *Social Work in Health Care* 37:87–114.

Turner, Heather A., Joseph A. Catania, and John Gagnon. 1994. "The Prevalence of Informal Caregiving to Persons with AIDS in the United States: Caregiver Characteristics and Their Implications." *Social Science and Medicine* 38:1543–52.

Turner, R. Jay and J. Blake Turner. 1999. "Social Integration and Support." in *Handbook of the Sociology of Mental Health*, edited by C. S. Aneshensel and J. C. Phelan. New York: Kluwer Academic/Plenum.

UNAIDS. 2005. HIV-Related Stigma, Discrimination, and Human Rights Violations: Case Studies of Successful Programmes. *UNAIDS Best Practice Collection*. Retrieved (http://data.unaids.org/publications/irc-pub06/JC999-HumRightsViol_en.pdf).

———. 2008. Report on the Global AIDS Epidemic 2008: Executive Summary. Joint United Nations Program on HIV/AIDS (UNAIDS), Geneva, Switzerland. Retrieved (http://www.unaids.org/en/KnowledgeCentre/HIVData/GlobalReport/2008/2008_Global_report.asp).

Vance, David E. 2006. "Spirituality of Living and Aging with HIV: A Pilot Study." *Journal of Religion, Spirituality & Aging* 19:57–74.

Vance, David E., Thomas C. Struzick, and Toya V. Russell. 2007. "Spiritual and Religious Implications of Aging with HIV: A Conceptual and Methodological Review." *Journal of Religion, Spirituality & Aging* 19:21–42.

Weitz, Rose. 1989. "Uncertainty and the Lives of Persons with AIDS." *Journal of Health and Social Behavior* 30:270–81.

Weston, Kath. 1991. *Families We Choose: Lesbians, Gays, Kinship*. New York: Columbia University Press.

Wight, Richard G. 2000. "Precursive Depression among HIV Infected AIDS Caregivers over Time." *Social Science and Medicine* 51:759–70.

Wight, Richard G., Carol S. Aneshensel, Allen J. LeBlanc, and Kristin P. Beals. 2008. "Sharing an Uncertain Future: Improved Survival and Stress Proliferation among Persons Living with HIV and Their Caregivers." pp. 369–97 in Stress Processes Across the Life Course, edited by Heather A. Turner and Scott Schieman, San Diego, CA: Elsevier, JAI Press.

Wight, Richard G., Allen J. LeBlanc, and Carol S. Aneshensel. 1998. "AIDS Caregiving and Health among Midlife and Older Women." *Health Psychology* 17:130–7.

Williams, Nathalie, John Knodel, and David Lam. 2010. "HIV/AIDS and Older Persons: Shifting the Focus From the Infected to the Affected." *Research on Aging* 32:3–18.

Wu, Albert W. 2000. "Quality of Life Assessment Comes of Age in the Era of Highly Active Antiretroviral Therapy." *AIDS* 14:1449–51.

Xia, Qiang, Fred Molitor, Dennis H. Osmond, Maya Tholandi, Lance M. Pollack, Juan D. Ruiz, and Joseph A. Catania. 2006. "Knowledge of Sexual Partner's HIV Serostatus and Serosorting Practices in a California Population-Based Sample of Men Who Have Sex with Men." *AIDS* 20:2081–89.

Zablotsky, Diane L. 1998. "Overlooked, Ignored, and Forgotten: Older Women at Risk for HIV Infection and AIDS." *Research on Aging* 20:760–75.

Zierler, Sally, William E. Cunningham, Ron Andersen, Martin F. Shapiro, Sam A. Bozzette, Terry Nakazono, Sally Morton, Stephen Crystal, Michael Stein, Barbara Turner, and Patti St. Clair. 2000. "Violence Victimization After HIV Infection in a US Probability Sample of Adult Patients in Primary Care." *American Journal of Public Health* 90:208–15.

Chapter 32
Obesity: A Sociological Examination

Christine L. Himes and Valerie Episcopo

Over the past several decades, the prevalence of obesity has increased across all ages and socio-economic groups in the United States. Obesity has been described as one of the most significant public health issues of the new century (National Institutes of Health 2004). The associated health problems have raised concerns about the costs of lost productivity, increased medical care needs, and the possible effects on longevity (Cutler et al. 2003; Finkelstein et al. 2005; Olshansky et al. 2005). Because of these concerns, researchers are interested in understanding the factors behind the increase and the potential for changing the trend. In this chapter, we will examine the patterns, explanations, and consequences of the increased prevalence of obesity for the older population. We will end with a discussion of the social implications of increasing obesity and social policy options for addressing the issue.

Patterns

Obesity

At the most basic level, obesity is the result of an energy imbalance, individuals taking in more calories than they expend. The equation for energy balance has not changed over time, so the increased prevalence of obesity in the population must be the result of changes in caloric intake, energy expenditure, or both. These changes do not have to be large; researchers have concluded that the observed increase in mean BMI in the United States could be accounted for by an excess of 100–150 calories per day (Hill et al. 2003).

There is evidence that energy intake has gone up over the last 30 years, more calories are consumed per capita. Working at the macrolevel, economists comparing national measures of food consumption concluded that the trend in per capita daily total energy supply is responsible for the trend for increased obese and overweight individuals (Silventoinen et al. 2004). The Centers for Disease Control and Prevention (CDC) agree with this analysis (CDC 2004a). They found that between 1971 and 2000 the mean energy intake in the United States increased from 2,450 to 2,618 kcal for men and from 1,542 to 1,877 kcal for women. The CDC goes on to show that fat and protein intake decreased in the same time period while carbohydrate consumption increased. The CDC suggested that this increase in caloric intake was due to foods consumed away from the home:

C.L. Himes (✉)
School of Behavioral and Social Science, St. Edwards University, Austin, TX, USA
e-mail: clhimes@syr.edu

R.A. Settersten, Jr. and J.L. Angel (eds.), *Handbook of Sociology of Aging*, Handbooks of Sociology and Social Research, DOI 10.1007/978-1-4419-7374-0_32, © Springer Science+Business Media, LLC 2011

salty snacks, soft drinks, pizza, and increased portion sizes. As we will discuss later, the types of food consumed, irrespective of caloric count, may also contribute to obesity.

Food consumption has increased, but what has happened to the other side of the energy balance equation, energy use? The measurement of energy expenditure starts with the basil metabolic rate (BMR). Since the BMR must be measured under strict conditions (before a person gets out of bed and after a 12 h break from all physical activity, consuming any food or beverage, or inhaling any nicotine), it is rarely used (Mahan and Escott-Stump 2000). Instead, an alternative measure, resting energy expenditure (REE), is more common. REE can be calculated by direct calorimetry, in which heat output is measured by, for instance, submerging a person in water and measuring the change in the temperature of the water. Alternatively, indirect calorimetry measures oxygen consumption. For indirect calorimetry, a person is asked to breath into a tube that measures the levels of oxygen and carbohydrates in the air before the person inhales them and after he or she exhales, several times. REE accounts for 65–70% of daily energy expenditures and is thought to have remained relatively constant over time (Mifflin et al. 1990).

Another factor of energy expenditure that has not changed in the last 30 years is the thermic effect of food (TEF), the energy used to digest food. TEF varies by the types of food that an individual consumes and usually is approximately 10% of total energy expenditure (TEE) (Mifflin et al. 1990). The shift in energy consumption noted by the CDC, above, might suggest that an increase in carbohydrate intake with a decrease in fat and protein intake would result in lower TEF, since sugars dissolve in water. It is however important to remember that fruits and vegetables are forms of high caloric foods that often require more energy to digest.

The final piece of the energy expenditure equation is the energy used in activity. Physical activity expenditures account for the remaining 20–30% of daily energy expenditure (Mifflin et al. 1990). This is the element of the equation which has very likely changed during the last 30 years. Activity factors were developed to estimate the energy used in physical activity. A sedentary person is assigned an activity factor between 1.0 and 1.4. Someone who participates in low levels of activity has an activity factor between 1.4 and 1.6. An active person burns 1.6–1.9 times his or her resting metabolic rate, and a very active person can spend up to two and a half times his or her REE. Current guidelines call for a minimum of 150 min of moderate-intensity physical activity a week, like brisk walking, water aerobics, slow bicycling, playing doubles tennis, ballroom dancing, and general gardening (USDHHS 2008). Approximately, 25% of adults receive no leisure time physical activity, and this percentage has been decreasing approximately 1% each year (CDC 2004b). Nearly one-half of those over the age of 65 receive no leisure time physical activity, and 80% participate in no vigorous leisure time activity (NCHS 2009).

The relative contribution of the increased intake of calories and decreased physical activity to the increased prevalence of obesity is unclear. The relationship between energy intake and expenditure is complex with many compensatory loops. Recent research has shown, for instance, that weight loss rarely occurs due to exercise alone (Church et al. 2009). Bleich et al. (2008) conclude that the primary cause of the observed weight gain in the population is overconsumption. Some researchers argue that food intake is more important for changing weight while activity is more important for maintaining weight (USDHHS 2008). Before exploring some of the reasons that both energy intake and expenditure may have changed over the past 30 years, we will look at the overall patterns of obesity in the population.

Measurement

Ideally, the percentage of body fat is determined at the individual level by directly measuring lean and fat body masses. Methods such as dual energy X-ray absorptiometry (DEXA scans) allow the precise measurement of lean soft tissue and total body fat. This, and similar methods, are very accurate; however, they require specialized equipment and trained administrators. Bioelectrical impedance is another alternative for measuring body composition and is based on the differing electrical impedance of water and fat tissues in the body. It is not considered a gold standard method for measuring body composition, but is widely available and relatively simple to use.

Most population surveys rely on the body mass index (BMI) for measuring obesity. Calculation of BMI requires only current height and weight. In most surveys this information is self-reported, but some national health surveys, most notably the National Health and Nutrition Examination Survey (NHANES), clinically measure both height and weight. Individuals with a BMI of 30.0 or higher are considered obese, and those between 25.0 and 29.9 are considered overweight. Within the category of obesity further distinctions are made, having those with a BMI between 30 and 35 are considered Class 1 obese, those with BMIs between 35 and 40 are considered Class 2 obese, and those with BMI of 40.0 or higher are considered Class 3, or morbidly, obese. These levels were selected to parallel the types of health problems associated with different levels of BMI (NIH 1998).

BMI as a measure of adiposity suffers from several flaws. It is imperfectly correlated to body fat, since individuals with muscular bodies can have identical weights to those with more fatty tissue at the same height (Prentice and Jebb 2001). This problem becomes increasingly important with age, as older adults tend to have less fat than younger individuals of the same weight (Gallagher et al. 1996). The distribution of body fat changes with age and BMI becomes increasingly inaccurate as a measure of adiposity (Kuk et al. 2009). At very old ages, underweight and weight loss, rather than obesity, are significant health problems and are often associated with increased frailty (Wallace et al. 1995). The self-reported nature of height and weight is also problematic. Individuals tend to underreport weight and overreport height (Burkhauser and Cawley 2008).

Some sociologists and policy analysts argue that the over reliance on BMI as a definition of obesity reflects current social trends and attitudes, rather than true health problems, and is associated with a moral, rather than medical, stance (Kwan 2009; Oliver 2005). As noted above, BMI is only a proxy measure for body composition and the definition of obesity; a BMI of 30.0 or higher is an arbitrary point selected to reflect the level at which health problems are believed to begin (NIH 1998). Using BMI as the sole measure of body composition will undoubtedly lead to misclassification of some individuals, particularly those with large muscle mass or the elderly (Prentice and Jebb 2001).

Alternatives to BMI include waist circumference (WC), waist to hip ratio (WHR), waist to stature ratio (WSR), and skin fold thickness. In an analysis of the relationship of BMI, WC, and WSR to body fat, Flegal et al. (2009) found that those three indicators were more closely related to each other than to body fat. For any individual, these measures may be inaccurate indicators of body fat, but they correspond fairly well to categories of body fat (such as overweight and obese). There is evidence that abdominal fat is more closely related to cardiovascular disease than BMI. Measurement incorporating WC may identify persons who are at increased risk for obesity-related cardiometabolic disease, above and beyond the measurement of BMI (Klein et al. 2007).

Trends

The proportion of the American population that is overweight and obese has risen dramatically for all ages although there is evidence that this trend may be slowing. Between 1960 and 2004, the proportion of overweight adult men rose from 50 to 71%, while the proportion of women who were overweight rose from 40 to over 62%. The percentage of men who were obese rose from 10 to 31%, and the proportion of women who were obese rose from 15 to about 33% in the same 44-year period (Ogden et al. 2006). However, between 2004 and 2006 there was no significant change in obesity prevalence for either men or women in the United States (Ogden et al. 2007).

The increase of obesity hides an important trend in the overall distribution of BMI. While the overall distribution of body sizes shifted to the right, the shift was greatest among those in the upper percentiles of BMI (Ogden et al. 2007). This shift indicates that while the entire population is heavier, those who are the heaviest have become much heavier over time. Analysis of data from the Behavioral Risk Factor Surveillance System (BRFSS) shows that between 2000 and 2005 the prevalence of a BMI over 40 increased twice as fast as the prevalence of a BMI over 30 (Sturm 2007). The rapid increase of those in the morbidly obese category, for whom the health consequences are most severe, contradicts the past understanding that individuals in this group suffered from a rare pathological condition.

Obesity rates vary across racial, ethnic, and socioeconomic groups. Although the prevalence of obesity is greatest among those at the lowest levels of educational attainment, the prevalence of obesity has increased for all educational groups (Himes and Reynolds 2005; Truong and Sturm 2005). A similar pattern is found by relative income: Those in the highest income group have lower prevalence of obesity, but the increase – particularly in the last 20 years – is similar to the increase in prevalence seen in the lowest income group (Truong and Sturm 2005). Among men, there are no significant racial or ethnic differences in obesity prevalence. This is not the case for women. Non-Hispanic black women and Mexican American women are significantly more likely to be obese compared to non-Hispanic white women (Ogden et al. 2006).

Obesity and the Life Course

Understanding the changing prevalence of obesity across age and time provides insights into the possible explanations for the increase and its effects. The transition from normal weight to overweight to obese tends to occur gradually over the life course, sometimes starting in childhood, but often beginning in middle age. Because obesity is difficult to reverse, understanding the timing of the transition and the lifetime consequences of obesity is vital. A life course perspective underscores particular concern over the increasing rate of obesity among children and adolescents (U.S. Surgeon General 2001). The health problems associated with obesity affect children, as well, and are difficult to reverse. For example, there is a strong relationship between childhood obesity and the development of diabetes in adulthood (Bloomgarden 2004). There is evidence that individuals who are obese as children are more likely to be obese as adults; about one third of obese preschoolers and half of obese school age children are obese as adults (Serdula et al. 1993). Ferraro et al. (2003) find that overweight children are at significantly greater risk of becoming severely obese (BMI≥35) compared to normal weight children. In some cases, obesity in childhood is related to parental obesity. The familial transmission of obesity is poorly understood and could be the result of genetics, learned behaviors of eating and exercising, or other socio-economic factors.

The standards for measuring obesity are slightly different for children and are age and sex dependent (Barlow 2007). About 10% of children aged 2–5 are at or above the 95th percentile BMI for age. This nearly doubles to 19.6% of children age 6–11. This represents an almost tripling in prevalence rates since 1976. Among adolescents age 12–19, the proportion considered obese is 18.1% (Ogden et al. 2010). Below age 6, girls are slightly more likely to be obese, but from age 6 onward, boys have higher levels of obesity, measured by BMI. The racial and ethnic patterns of children mimic those of adults; obesity is more common among non-Hispanic blacks and Mexican Americans, compared to non-Hispanic whites, with differences significantly larger for girls (Ogden et al. 2008).

Weight gain is steady through early adulthood (Baum and Ruhm 2009). After adjusting for the secular trend in obesity, average BMI rose from 24.3 to 27.3 between the ages of 18 and 40 in the National Longitudinal Survey of Youth (NLSY) cohort. Similarly, the prevalence of obesity generally increases with age through adulthood (Cook and Daponte 2008). People between the ages of

18 and 32 have the lowest average BMI (Cook and Daponte 2008) and BMI increases steadily until about age 75, when there is a small drop (Flegal et al. 1998). In cross-sectional studies, peak values of BMI are observed in the age range 50–59 in both men and women, with gradual declines in BMI after age 60 (Flegal et al. 1998; Hedley et al. 2004; Ogden et al. 2006).

However, premature mortality of the obese may influence these cross-sectional relationships. In a 10-year follow-up study, individuals under age 55 exhibited a greater tendency to gain weight, with the magnitude of increase decreasing with age (Williamson 1993). Rates of overweight and obesity in longitudinal studies generally increase with age until age 75, when there is a small drop (Ferraro et al. 2003; Flegal et al. 1998; Must and Strauss 1999). Adults who are obese are more likely to remain obese until death than to leave the classification of obese (Ferraro et al. 2003). The loss of weight in late adulthood is often considered a sign of underlying chronic health problems or increased frailty (Kuk et al. 2009). This loss of weight, however, may mask an increase in adiposity since the amount of fat an individual has for a given weight may change as lean muscle mass is lost (Carmelli et al. 1991).

The observed increase in obesity over time appears to have affected all ages and cohorts. Three studies (Cook and Daponte 2008; Reither et al. 2009; Reynolds and Himes 2007) use age-period-cohort analyses to examine the reasons for the observed increase. Age-period-cohort analyses are useful for determining if an observed trend in the population is the result of a particular cohort's experience, of changes in one age group, or if the observed trend is seen across ages and cohorts at a particular time. The overwhelming conclusion is that the observed increase in obesity prevalence occurred at all ages and for all cohorts, and that period effects are principally responsible for the increased obesity prevalence. Cook and Daponte (2008) note that the increases in obesity prevalence are fastest at youngest adult ages; indicating that more adolescents are reaching adulthood already obese. Within the middle and late adulthood years, however, the rate of increase in obesity has been similar. Older adults were not immune to the obesity "epidemic."

Explanations

A wide variety of explanations have been put forth to explain the increased prevalence of obesity in the population. Some focus on individual psychosocial factors, such as self-control. Other explanations look at genetic influences. Still another set of explanations focuses on cultural, environmental, and economic factors.

Psychosocial Factors

Obesity is often considered a measure of personal failing, an indication of an inability to exercise self-control. This explanation for obesity puts the responsibility on the individual. Research suggests that it is easier for individuals to gain weight than to lose it (Berthoud and Morrison 2008). This would imply that the control of appetite and eating is not symmetrical (Blundell 2002). Although eating is controlled according to biological need, eating is also subject to environmental constraints. Since eating can be consciously controlled, as in hunger strikes or dieting, the assumption is that limiting food intake can be controlled in order to control weight or limit weight gain. Individuals are bombarded with seductive messages designed to increase the consumption of less healthy, high calorie foods. The expectation that individuals can counteract these messages may be unrealistic.

Another psychosocial explanation advanced to explain individual obesity is stress. Stress is thought to influence eating behavior. Individuals experiencing stress may over- or under-eat

(Oliver and Wardle 1999; Popper et al. 1989). Although the evidence is mixed, acute stressors tend to result in a lower food intake while those experiencing chronic stress show a preference for foods high in sugar and fat and a greater food intake (Torres and Nowson 2007). These effects are thought to operate because stress alters the control of cortisol and insulin leading to a disregulation of appetite and fat distribution (Adam and Eppel 2007). The effects can be long term, as at least one study shows a relationship between stress in childhood (measured by exposure to family violence) and obesity in adulthood (Greenfield and Marks 2009).

Genetic Factors

Genetic influences cannot explain the increased obesity of the past 3 decades. However, genetic influences set the stage for the current rapid increase in body weight. Evolutionary influences created a human body with a powerful defense against under-nutrition, the capacity to store excess energy as fat. However, this advantage in times of scarcity is not well adapted to an environment in which food is plentiful. The theory of the "thrifty gene" emerged in the 1960s advanced by James Neel to explain the prevalence of Type II diabetes and obesity in the population (Neel 1962). In its most general form, the theory posits that genotypes enabling the deposition of fat during times of abundance would be an evolutionary advantage. Over time, however, this genotype advantage would be lost as food sources became more abundant. As a result, individuals are predisposed to deposit fat, regardless of the need to store such reserves for times of famine. This hypothesis, although later rejected by Neel himself (Neel 1989), gained popularity as an explanation for high rates of obesity among particular groups exposed to Western diets.

The thrifty gene hypothesis cannot explain all of the increase in obesity prevalence or the variation within population groups. It is more likely that a wide range of genetic influences affect weight gain. Research has shown that there are individual differences, based on biology, in the predisposition to gain weight (Bouchard 2007; Farooqi and O'Rahilly 2007). Genes have been identified associated with poor regulation of satiety and appetite, diminished ability to use dietary fats as fuel, and easily stimulated ability to store body fat (de Krom et al. 2009). The value of pursuing a biological – or genetic predispositional – approach to obesity lies in its ability to contribute to pharmaceutical development and behavioral or dietary strategies for weight loss (Farooqi and O'Rahilly 2007). The interaction of genetic predisposition with shifting environments is likely to be the explanation of increased obesity.

Cultural Factors

Genetic explanations, while useful for understanding the long-term trends in body size, and for understanding the biological framework upon which the current obesity levels can be viewed, cannot explain the very recent changes observed. Similarly, individual level psychosocial factors cannot explain the long-term trend. For this, we must look more closely at social and cultural factors, which have changed in the past 40–50 years and may have contributed to the increased body size seen in most developed nations.

One cultural change often associated with the rise of obesity is the increased participation of women in the labor force. The argument is that as more women have begun working full time, families have tended to eat out more and to prepare more convenience foods at home – foods that, in general, have higher fat contents than foods prepared from fresh ingredients. Anderson et al. (2003) advanced this argument as a partial explanation for the rising obesity among children. Chou et al. (2004)

show that as the value of women's time has increased with labor force participation, the amount of time spent in the home is reduced, increasing the reliance on convenience foods. However, in their analysis Cutler et al. (2003) conclude that there is no relationship between female labor force participation and obesity trends.

Another area of cultural change has been the decline in smoking. In general, smoking is associated with lower body weight. This effect might be particularly strong in determining the initiation of smoking behavior. Young women who are overweight or trying to lose weight are more likely to begin smoking (Cawley et al. 2004). Chou et al. (2004) argue that as the cost of smoking has increased, smoking rates have declined, and there has been a concomitant increase in weight. In their research, Cawley et al. (2004) found that this price effect was strong for young men, but not for young women.

The use of leisure time has changed as well. Over the last few decades more recreational time, especially among children, is spent on television watching, computer use, and video games. Based on data from the National Time Use Survey, Americans spent about 3 h a day watching television, compared to less than 1 h/day in leisure time physical activities, like sports or exercising (American Time Use Survey 2008). Although national estimates of the time spent on the computer are not available, in the latest data available, 2005, over 60% of households had access to the Internet at home (U.S. Bureau of the Census 2007).

Data from the National Health Interview Survey indicate that 32% of American adults engage in some regular leisure time physical activity, defined as light-moderate activity at least 5 times a week for 30 min or vigorous activity at least 3 times a week for 20 min (NCHS 2009). This prevalence declines steadily with age, to about 26% of Americans age 65–74 and 18% of Americans age 75 and older. At all ages, the rates are higher for men than for women. There has been little change in the population level of physical activity since 1997, with the annual age adjusted percentages ranging from 29.6 to 32.8 with little consistent trend evident.

Environmental Factors

A variety of environmental factors may also play a role in the rise of obesity prevalence. Environmental changes have affected both diet and levels of physical activity. One dietary change suspected of contributing to weight gain is the increased use of high fructose corn syrup (HFCS). An extensive review of epidemiologic and clinical evidence finds that this hypothesis can be neither supported nor refuted (Forshee et al. 2007). HFCS is used more extensively in the United States than in other countries where sucrose continues to be the primary caloric sweetener. The link between HFCS and weight gain is indirect. Some research has shown that HFCS is sweeter than sucrose and leads to greater consumption of calories. There is also evidence that increased levels of HFCS have adverse metabolic consequences due to changes in the fructose:glucose (F:G) ratio. HFCS may decrease production of the satiety hormone leptin and increase levels of ghrelin, associated with appetite stimulation. In addition, the use of HFCS may be related to an increased consumption of liquid calories, primarily in the form of carbonated soda drinks. Liquid calories tend to be less satiating than calories obtained through solid food consumption.

Environmental changes may have contributed to a decrease in physical activity, and, consequently to an increase in weight. One such factor is the fear of crime. In areas where crime rates are high, children and adults may not feel safe walking in their neighborhoods or using playgrounds. This effect might be particularly strong for girls (Gomez et al. 2004). In addition, in many urban areas, sidewalks and public parks have not been maintained, making them both unsafe and unattractive for use. In other areas, sidewalks may not exist at all. Schools devote less time to physical activity, either through free play time, like recess or structured physical education courses (Story et al. 2006).

Another environmental change is the changing nature of work and work-related tasks. As labor has become less physically demanding, the amount of energy expended in paid employment has declined (Philipson and Posner 2003). However, evidence from time diary studies indicates that the majority of the declines in energy expenditure occurred between 1965 and 1975 (Cutler et al. 2003). In addition, the shift to cars rather than walking to work may have contributed to a decline in physical activity. However, both of these changes occurred before the obesity increase, making them less likely as explanations.

Economic Factors

One economic explanation for the obesity increase focuses on the relative prices of food items. Lakdawalla and Philipson (2009) argue that technological advances in agriculture have caused food prices to fall. This decline in real prices of groceries caused a surge in caloric intake. Cawley (1999) finds that BMI is negatively related to prices. In addition, if the prices of calorie-dense foods (e.g., foods sold by fast food restaurants) fall faster than those for less calorie-dense foods (e.g., vegetables), consumers will tend to shift consumption to the cheaper alternatives (Finkelstein et al. 2005).

It is unlikely that any one explanation can account for the recent dramatic increases in BMI. The evolutionary predisposition toward storing energy in the form of fat and increasing appetite in times of plenty has created a situation in which imbalances in caloric intake and energy expenditure will lead to excessive fat. In addition, cultural and environmental changes have made such an imbalance more likely to occur. The increased availability and reliance on energy dense foods at the same time as physical activity declined have made the energy balance more precarious. Given the increased levels of overweight and obesity, attention has focused on the potential negative consequences of this trend.

Consequences and Harm

Physical Health

It is well established that overweight and obesity are significant risk factors for developing several chronic health conditions in later life, including diabetes, high blood pressure, high cholesterol, coronary heart disease, arthritis, and certain types of cancer (Mokdad et al. 2003; Paul and Townsend 1995; Wolf and Colditz 1998; Villareal et al. 2005). These effects are found across the life course, beginning in childhood and persisting into later life (Ferraro et al. 2003; Gregg et al. 2005; Koplan et al. 2005; Whitmer et al. 2005). Most generally, obesity has been shown to be related to an overall decline in health-related quality of life (Ford et al. 2001). Most of the increased risk for chronic diseases is a direct physical result of overweight and obesity, but several authors suggest that the stigma of being overweight is so pronounced in the medical field that many avoid going to the doctor, complicating the diagnosis of medical conditions and their care (Puhl and Brownell 2001, 2003; Schwartz et al. 2003).

The diseases most closely associated with obesity are Type II diabetes and coronary heart disease. Since the likelihood of developing diabetes increases significantly as body fatness increases, many have attributed the increased prevalence of Type 2 diabetes to the increased obesity of the world's population. According to the *World Health Report 2002*, approximately 58% of diabetes globally is attributable to excess weight (WHO 2002), while in the United States the prevalence of diabetes rose 132% between 1980 and 2006 (CDC 2008). Diabetes in midlife is associated with a variety of complications later in life. Those with diabetes are more likely to develop heart problems or suffer from strokes (see http://diabetes.niddk.nih.gov/dm/pubs/stroke/index.htm).

The effects of obesity on mortality are less definitive. Although researchers agree that there is some effect, the overall magnitude and age gradient is less clear (Flegal et al. 2005; Fontaine et al. 2003; Olshansky et al. 2005). The risk is highest for those who have been overweight for longer periods of time and decreases if one does not become overweight or obese until after age 50 (Flegal et al. 2005). Adults under the age of 50 show the clearest association between obesity and increased mortality (Stevens et al. 1998; Thorpe and Ferraro 2004). In longitudinal analyses, obesity in middle adulthood (ages 30–49) has been shown to be associated with an approximately 6 year lesser life expectancy when compared to normal weight individuals (Peeters et al. 2003). At very old ages, higher BMIs may be associated with lower mortality risks and the BMI associated with the lowest mortality appears to increase compared to younger age groups (Heiat et al. 2001). The lack of a clear relationship between obesity and mortality at older ages has been labeled by some as the "obesity paradox" (Beddhu 2004; Uretsky et al. 2007).

The reasons for the observation that obese subjects tend to fare better than, or at least as well as, their normal weight peers may be related to an unobserved underlying factor, nutritional reserves, or medical therapy (Osher and Stern 2009). Variations in weight, including weight loss, gain, and weight cycling, among the elderly may signal underlying health problems associated with higher mortality (Arnold et al. 2010). The effect of obesity on mortality may be changing over time as well. The introduction of better drugs for treating high cholesterol and hypertension appears to be reducing the disease risks of the obese, at least with respect to cardiovascular disease (Gregg et al. 2005). Although opinions differ, the long-term impact of these interventions on mortality rates and life expectancy may counteract the increased risks associated with obesity, at least at BMI levels below 35.0 (Olshansky et al. 2005; Preston 2005; Reuser et al. 2008).

Perhaps because of the association between obesity and disease, obese individuals tend to have an increased prevalence of functional limitations (Ferraro et al. 2002; Himes 2000; Jenkins 2004). Ferraro and Booth (1999) summarize three pathways through which obesity may impact physical functioning. First, excess weight adds stress to the skeleton and weight-bearing joints, increasing the likelihood of arthritis and joint problems. In addition, excessive weight contributes to insulin resistance, which may damage connective tissues. Finally, they argue, weight may lead to a pattern of blood lipids that leads to atherogenesis and decreased functioning.

Obesity may also limit physical activity, depriving individuals of the benefits of exercise resulting in decreased cardiovascular fitness and muscle strength. This would be particularly relevant for the development of mobility limitations. Obese individuals have more trouble walking, climbing stairs, and getting in and out of bed compared to those who are not obese (Himes 2000; LaCroix et al. 1993; Launer et al. 1994). Jenkins (2004) examines individuals 70 and older over a 3-year time period and finds that obese are more likely to suffer the onset of functional limitations (loss of lower body mobility, strength, or ADL limitations) than those who are not obese. Ferraro et al. (2002) find similar longitudinal results across a wider age range, and also note that obesity is associated with a more rapid increase in disability over time. This limitation in mobility may be particularly problematic for older individuals who may need assistance to carry out daily tasks.

Numerous cross-sectional studies have examined the relationship between body weight and a variety of measures of functioning and disability. In general, greater levels of impairment are found at the extremes of the body size distribution (Galanos et al. 1994; Okoro et al. 2004).

Cross-sectional studies also have shown that excess weight has a negative effect on lower body functioning when using both direct measures of functioning (Apovian et al. 2002) and self-reports of limitations (Himes 2000). Data from the NHANES III survey show that for women aged 70 and older there is a strong relationship between obesity and mobility-related functional limitations. Obese women were two times as likely to report a functional limitation as normal weight women.

This relationship is weaker for men, but obese men were still 1.5 times as likely to report a limitation as their normal weight counterparts (Davison et al. 2002).

There is a documented negative effect of both obesity and overweight on ADL and IADL functioning. Sturm et al. (2004) find that for older women the probability of ADL limitations doubles for those with moderate obesity and quadruples for those with severe obesity. Using data from the Medicare Current Beneficiary Survey, Lakdawalla et al. (2005) estimate that obese 70-year-olds can expect to spend 40% more time in disability than their normal weight counterparts. In addition, the obese spend more time in the most disabled groups (at least three ADL limitations or institutionalized) than those of normal weight. At the same time, they find few differences in life expectancy or disability free life expectancy when comparing the overweight and normal weight groups.

In terms of cognitive functioning, BMI may operate differently at different stages of the life course. Higher BMI at midlife is associated with higher risk of dementia at older ages (Whitmer et al. 2005). One reason for this relationship may be the increased prevalence of diabetes among the obese in midlife. Diabetes at midlife has been linked to dementia more than 3 decades later (Beeri et al. 2004). At the same time, independent of disease status, there is a link between weight loss and Alzheimer's disease in later life (Atti et al. 2008; Buchman et al. 2005). Unexplained weight loss among the elderly may be an indicator of incipient cognitive decline in late life (Atti et al. 2008).

Combining the effects of mortality and morbidity, researchers have compared estimates of active life expectancy between obese and nonobese adults (Lakdawalla et al. 2005; Peeters et al. 2004; Reynolds et al. 2005). All find that because of the increased risks of disability and mortality, obese populations have significantly lower life expectancy free of disability than the normal weight population. The explanation for this difference may be due to a similar number of years of disability lived between normal weight and obese populations, but shorter life expectancies (Peeters et al. 2004) or due to similar overall life expectancy, but differences in the number of years spent disabled (Lakdawalla et al. 2005; Reynolds et al. 2005).

Mental Health

Obesity is considered one of the strongest stigmas in American society. Obesity is still viewed by many as a personal failing, a sign of character weakness. Common stereotypes of the overweight include laziness, self-indulgence, impulsivity, and incompetence (Paul and Townsend 1995; Roehling 1999; Rothblum 1992). People hold those who are overweight responsible for their own condition and often think that if they simply had more willpower, they would reduce their food intake and, thus, lose weight (Puhl and Brownell 2003; Rothblum 1992). These stereotypes persist despite evidence that overweight people generally do not have higher caloric intakes than those of average weight (Rothblum 1992). Even overweight and obese people themselves have been found to subscribe to these stereotypical anti-fat attitudes, suggesting that they have internalized the social stigma of being overweight (Wang et al. 2004). The stigma of obesity is so strong, that one survey found that 24% of women and 17% of men would choose to live three or more years less if they could be their desired weight (Puhl and Brownell 2003). Being overweight at older ages appears to have somewhat fewer social consequences than does at younger ages. There is significantly more pressure on younger adults to be thin and fit. Moreover, older people are significantly less likely to be in either the marriage or job markets, thus they are less likely to be subjected to sanctions on the basis of body size. Thus being overweight may not be as stigmatizing for the elderly as for those at younger ages.

The relationship between weight and psychiatric disorders is complex. There is little evidence that obesity is associated with anxiety disorders (Bjerkeset et al. 2007; Williams et al. 2009). However, obesity may be a clinical condition that predisposes an individual to depressive disorders

(Williams et al. 2009). Weight stigmatization may be one factor increasing the vulnerability to depression among obese individuals. Recent analyses of a community-based sample in Norway found higher body mass to be associated with increased risk of depression for both men and women (Bjerkeset et al. 2008). Studies of obese patients find a relationship between childhood teasing based on appearances and adult depression (Jackson et al. 2000). This supports the finding that it is not obese itself which contributes to depression, but the interpersonal mistreatment due to weight (Carr et al. 2007).

Individual-Level Economic Consequences

As a result of both the physical consequences and stigmatization, obesity has been shown to be related to generally negative economic outcomes at the individual level. For example, in the United States white females have both lower rates of obesity and higher average income than African-American or Hispanic women. Chang and Lauderdale (2005) find strong negative correlations between income and BMI for white and African-American women, with a weaker correlation for Mexican-American women. Among men, they find a weak negative correlation between income and BMI for white men, but for Mexican-American and African-American men they find a positive correlation between income and BMI. Hedley and Ogden (2006) find that within each race-gender group, the prevalence of obesity rose across all income categories in recent decades, and that the differences in the prevalence of obesity by income (within race-gender group) have decreased over time.

The lower earnings of obese individuals have been attributed to several factors. Obese adolescents, especially women, are less likely to graduate from high school and less likely to attend college than their average weight peers (Crosnoe 2007). This effect is stronger if obesity is normative in the school (Crosnoe and Muller 2004). Obese employees also suffer from discriminatory practices (Puhl and Brownell 2001). In addition, obese workers are more likely to have moderate to high levels of absenteeism from work than their nonobese counterparts (Wolf and Colditz 1998). In addition, Cawley (2004) finds that heavier women are significantly more likely than women at the recommended weight to report employment disability. The effects of lower earnings may persist into later life. Many retirement benefits, including Social Security, are determined at least in part by lifetime earnings. Lower earnings can translate into lower incomes and increased economic insecurity in later life. In fact, older women who are obese have lower net worth than their normal weight peers (Fonda et al. 2004). This relationship appears not to hold true for men, however, indicating a difference in the effects of body size on earnings and asset accumulation.

Societal Economic Impact

The impacts of obesity on individual health, well-being, and economic situation are not the only considerations. The high prevalence of obesity in the adult population also has broader economic implications. Given the public financing of many health care services, especially Medicaid and Medicare, the medical costs of obesity are important to consider. Estimates are that from 5 to 7% of all annual health care expenditures are attributable to obesity (Finkelstein et al. 2005). As noted above, the effects of obesity are much stronger for morbidity than mortality, particularly at old ages. The use of simulation models of the health care costs associated with obesity. Lakdawalla et al. (2005) estimate that from age 70 onward Medicare spends 35% more on an obese person than one of normal weight. Underweight individuals also have high Medicare costs relative to normal weight, but these costs are incurred for fewer years. In similar analyses, Daviglus et al. (2004) have shown that obesity is linked to higher lifetime health care costs.

Health care costs are not the only economic considerations. Most studies show that overweight, and particularly obese, individuals miss more days of work. Obese employees are more likely to have moderate to high levels of absenteeism from work than their nonobese counterparts. Wolf and Colditz (1998) estimate that in 1988 alone, over 52 million work days were lost to obesity, while in 1994 another 58 million work days were lost, at annual price tags in excess of $5 billion. Burton et al. (1998, 1999) found that as BMI increases, so do the number of sick days and absenteeism from work. Tucker and Freidman (1998) studied over 10,000 employees and found that obese workers were 2 times as likely to have high-level absenteeism, defined as seven or more absences in the past 6 months, and 1.5 times as likely to have moderate absenteeism, defined as 3–6 absences in the past 6 months, than lean employees. The higher rates of absenteeism and missed work days contribute to lower overall productivity and reduced public revenue.

The individual and societal costs of obesity are numerous. Individuals who are obese have poorer physical health, lower educational attainment, and lower earnings. As a country, an increasing proportion of health care dollars is spent treating and managing the effects of obesity. Despite these costs, the proportion of the population considered obese has increased. If efforts to date have done little to stop or reverse the trend, what new approaches to the problem can be considered?

New Frontiers

The majority of American adults are now overweight and over a third are obese. Within specific subpopulations, the prevalence is even higher. Current approaches to curbing obesity focus on individual change. However, as the preceding discussion makes clear, the obesity problem has sources outside of the control of individuals. Even individual behaviors like diet and exercise patterns are embedded in social relationships and culture. The recent emphasis on social network analysis to examine the spread of obesity through social relations highlights the complex social nature of the factors leading to weight gain (Christakis and Fowler 2007). To expect that individuals can, on a widespread scale, overcome those cultural strictures is unrealistic. The focus on individual self-control for weight control cannot be effective. Like other public health problems throughout history, obesity needs to be addressed on a societal level. The complex causes and consequences of obesity point to a multidisciplinary and multilevel research approach. If obesity is considered as a social problem, with multiple causes outside of the control of individuals, new considerations can be given to a wide range of approaches (Nestle and Jacobson 2000).

Although obesity is linked to a variety of health problems, the advantages of weight loss at older ages are unclear. Among those with diabetes, a modest weight loss of 5–10 lb has been shown to significantly improve control of glucose levels and a reduction in cardio-vascular risk factors (Espelande et al. 2007). Weight loss is recommended for obese patients suffering from high blood pressure, heart disease, and arthritis (Osher and Stern 2009). The value of weight loss among the elderly population without chronic conditions, however, is unclear. The BMI associated with lower mortality is higher for those over age 65 compared to those under 65 (Janssen and Mark 2007). In fact, weight loss is associated with excess mortality among normal, overweight, and mildly obese among those aged 50–70 (Myrskala and Chang 2009). This suggests that the benefits of lower BMI may not hold true late in life. Still, there is considerable evidence that entering later life obese increases the chances that an individual will have chronic health problems and an increased chance of experiencing limitations in activity (Osher and Stern 2009). Focusing on addressing obesity at younger ages is likely to result in a healthier older population (Ferraro et al. 2003; Olshansky et al. 2005).

An individual focus often leads to medical solutions. For instance, the rates of bariatric surgery have increased dramatically in recent years (Santry et al. 2005). Such surgery can be effective for some individuals; however, long-term results are difficult to maintain (Shah et al. 2006). Other medical interventions are based on pharmacological treatments options. There is great interest in studying the hormonal and neural mechanisms involved in the control of appetite in hopes of identifying methods of controlling eating. But current drug treatments for obesity are limited and the associated weight loss is small, usually not more than 10% (Berthoud and Morrison 2008).

What types of societal interventions are possible? Fiscal and regulatory policies can be implemented at the federal, state, or local level. Two widely discussed options are tax policies that would influence food prices and food labeling policies in restaurants that would provide consumers with nutritional information. Other, broader, options focus on making changes in the built environment which would encourage physical activity and influence children's behavior through school-based programs.

Over the past several years, there have been many attempts to regulate food prices through differential tax policies. These efforts are modeled after the established "sin taxes" in place of tobacco and alcohol products. Proponents argue that taxes on certain foods would limit their consumption and result in desirable health effects. The item most often targeted is sugared beverages. As noted above in the discussion of HFCS, but even more generally of all sugared beverages, their consumption is linked to increased body weight, poor nutrition, and increased risk for diabetes (Nielsen and Popkin 2004). Limiting their consumption, through increased prices, could be one factor in obesity prevention. One study has suggested that for every 10% increase in price of sugared beverages, consumption decreases by 7.8% (Brownell and Frieden 2009). Objections to this type of tax focus on the unfairness of targeting one type of food; that such a limited approach would not solve the obesity problem. However, if obesity is seen as a multileveled problem, requiring multiple approaches, tax policy could be one part of such an approach.

Food labeling policies may influence food choices through better information on the nutritional value of specific foods. One study of the effects including calorie information on menus showed that consumers altered their behavior only when additional information on the daily recommended allowance of calories was included (Roberto et al. 2010). New York City recently implemented a menu labeling policy requiring fast food restaurants to post calorie labels on their menu boards and sit down restaurants to include calorie information on the printed menu. Analysis of the implementation of this policy found limited effects at the population level on the total number of calories purchased (Elbel et al. 2009). One conclusion is that labeling must be accompanied by widespread educational campaigns to be successful in altering behavior (Nestle 2002). As with tax policies, one single intervention is unlikely to have a large impact on behavior, and a multifaceted approach has greater chances of success.

Environmental changes that would increase physical activity are another approach to creating more balance in the energy equation. Federal funding has focused on the development of interstate highways and the facilitation of motorized transportation. Prompted by both health and environmental concerns, emphasis is shifting to the construction of pedestrian and bicycle pathways as alternatives (Handy et al. 2009). This focus on "active travel" can influence behavior. An overview of the evidence demonstrates that changes in the built environment can increase physical activity levels (Transportation Research Board and Institute of Medicine 2005). One study of residents in neighborhoods with high walkability (well-connected streets, sidewalks, and nonresidential destinations) got 40–50 min more exercise per week than their counterparts in low-walkability areas of the same cities (Sallis et al. 2009).

The best programs may involve multiple levels of involvement. An example of a successful partnership between school and community is the Shape Up Somerville program in Somerville, Massachusetts (Economos et al. 2007). Started in 2002 the program involved children attending grades one through three in public elementary schools in three participating communities. The intervention involved increased physical activity and increased availability of healthful foods. Most notably, the efforts extended beyond the school to include changes in home and community environments.

The success of the program, measured by lower BMI increases in children at high risk for obesity, demonstrates the potential value of a multi-pronged approach.

All of these options – tax policy, food labeling, changes in the built environment, and education – can address one piece of the energy imbalance. Just as the increased prevalence of obesity has no one identifiable cause, it has no one identifiable solution. Efforts to limit the increasing body size of the population cannot focus solely on individual and medical intervention. Bariatric surgery can be beneficial for individuals facing severe health consequences, but reliance on surgical and pharmaceutical solutions ignores the societal structure in which individuals will continue to live. With obesity, social scientists have the opportunity to apply their understanding of the social world to address a pressing social problem.

References

Adam, Tanja C. and Elissa S. Eppel. 2007. "Stress, Eating and the Reward System." *Physiology & Behavior* 91:449–58.

American Time Use Survey. 2008. "Time Spent in Detailed Primary Activities, and Percent of the Civilian Population Engaging in Each Detailed Activity Category, Averages per Day by Sex." Retrieved (http://www.bls.gov/tus/tables/a1_2008.pdf). Accessed on 2/7/2011.

Anderson, Patricia M., Kristin F. Butcher, and Philip B. Levine. 2003. "Maternal Employment and Overweight Children." *Journal of Health Economics* 22:477–504.

Apovian, Caroline M., Carolin M. Frey, G. Craig Wood, Joanne Z. Rogers, Christopher D. Still, and Gordon L. Jensen. 2002. "Body Mass Index and Physical Function in Older Women." *Obesity Research* 10:740–7.

Arnold, Alice M., Anne B. Newman, Mary Cushman, Jingzhong Ding, and Stephen Kritchevsky. 2010. "Body Weight Dynamics and Their Association with Physical Function and Mortality in Older Adults: The Cardiovascular Heart Study." *Journal of Gerontology: Biological Sciences and Medical Sciences* 65:63–70.

Atti, Anna Rita, Katie Palmer, Stefano Volpato, Bengdt Winblad, Diana DeRonchi, and Laura Fratiglioni. 2008. "Late-Life Body Mass Index and Dementia Incidence: Nine-Year Follow-Up Data from the Kungsholmen Project." *Journal of the American Geriatrics Society* 56:111–6.

Barlow, Sarah E. 2007. "Expert Committee Recommendations Regarding the Prevention, Assessment, and Treatment of Child and Adolescent Overweight and Obesity: Summary Report." *Pediatrics* 120:S164–92.

Baum, Charles L. and Christopher J. Ruhm. 2009. "Age, Socioeconomic Status, and Obesity Growth." *Journal of Health Economics* 28:635–48.

Beddhu, Srinivisan. 2004. "The Body Mass Index Paradox and an Obesity, Inflammation, and Artherosclerosis Syndrome in Chronic Kidney Disease." *Seminars in Dialysis* 17:229–32.

Beeri, Michal Schnaider, Uri Goldbourt, Jeremy M. Silverman, Shlomo Noy, James Schmeidler, Ramit Ravona-Springer, A. Sverdlick, and Michael Davidson. 2004. "Diabetes Mellitus in Midlife and the Risk of Dementia Three Decades Later." *Neurology* 63:1902–7.

Berthoud, Hans Rudolf, and Christopher Morrison. 2008. "The Brain, Appetite, and Obesity." *Annual Review of Psychology* 59:55–92.

Bjerkeset, Ottar, Pal Romunstad, Jonathan Evans, and David Gunnell. 2007. "Association of Adult Body Mass Index and Height with Anxiety, Depression, and Suicide in the General Population." *American Journal of Epidemiology* 167:193–202.

Bleich, Sara N., David Cutler, Christopher Murray, and Alyce Adams. 2008. "Why Is the Developed World Obese?" *Annual Review of Public Health* 29:273–95.

Bloomgarden, Zachary T. 2004. "Type 2 Diabetes in the Young." *Diabetes Care* 27:998–1010.

Blundell, John E. 2002. "A Psychobiological System Approach to Appetite and Weight Control." Pp. 43–49 in *Eating Disorders and Obesity*, edited by C.G. Fairburn and K.D. Brownell. New York: Guilford Press.

Bouchard, Claude. 2007. "The Biological Predisposition to Obesity: Beyond the Thrifty Gene Hypothesis." *International Journal of Obesity* 31:1337–9.

Brownell, Kelly D. and Thomas R. Frieden. 2009. "Ounces of Prevention: The Public Policy Case for Taxes on Sugared Beverages." *New England Journal of Medicine* 360:1805–8.

Buchman, Aron S., Robert S. Wilson, Julia L. Bienias, Raj C. Shah, Denis A. Evans, and David A. Bennett. 2005. "Change in Body Mass Index and Risk of Incident Alzheimer Disease." *Neurology* 65:892–7.

Burkhauser, Richard V. and John Cawley. 2008. "Beyond BMI: The Value of More Accurate Measures of Fatness and Obesity in Social Science Research." *Journal of Health Economics* 27:519–29.

Burton, Wayne N., Chen Yu Chen, Alyssa B. Schultz, Dee W. Edington. 1998. "The Economic Costs Associated with Body Mass Index in a Workplace." *Journal of Occupational and Environmental Medicine* 40:786–92.

Burton, Wayne N., D.J. Conti, Chen Yu Chen, Alyssa B. Schultz, Dee W. Edington. 1999. "The Role of Health Risk Factors and Disease on Worker Productivity." *Journal of Occupational and Environmental Medicine*. 41:863–77.

Carmelli, Dorit, M.R. McElroy, R. H. Rosenman. 1991. "Longitudinal Changes in Fat Distributions in the Western Collaborative Group Study: A 23 Year Follow-Up." *International Journal of Obesity* 15:67–74.

Carr, Deborah, Michael A. Friedman, and Karen Jaffe. 2007. "Understanding the Relationship Between Obesity and Positive and Negative Affect: The Role of Psychosocial Mechanisms." *Body Image* 4:165–177.

Cawley, John. 1999. *Rational Addiction, the Consumption of Calories, and Body Weight*. Ph.D. dissertation. University of Chicago, Chicago, IL.

———. 2004. "The Impact of Obesity on Wages." *Journal of Human Resources* 39:451–474.

Cawley, John, Sara Markowitz, and John Tauras. 2004. "Lighting Up and Slimming Down: the Effects of Body Weight and Cigarette Prices on Adolescent Smoking Initiation." *Journal of Health Economics* 23:293–311.

Centers for Disease Control and Prevention. 2004a. "Trends in Intake of Energy and Macronutrients: United States, 1971–2000." *Morbidity and Mortality Weekly Report* 53:80–82.

———. 2004b. "Prevalence of No Leisure-Time Physical Activity: 35 States and the District of Columbia, 1988–2002." *Morbidity and Mortality Weekly Report* 53:82–86.

———. 2008. "Crude and Age-Adjusted Percentage of Civilian, Noninstitutionalized Population with Diagnosed Diabetes, United States, 1980–2006." http://www.cdc.gov/diabetes/statistics/prev/national/figage.htm

Chang, Virginia W. and Diane S. Lauderdale. 2005. "Income Disparities in Body Mass Index and Obesity in the United States, 1971–2002." *Archives of Internal Medicine*, 165:2122–8.

Chou, Shin Yi, Murray Grossman, Henry Saffer. 2004. "An Economic Analysis of Adult Obesity: Results from the Behavioral Risk Factor Surveillance System." *Journal of Health Economics* 23:565–87.

Christakis, Nicholas A. and James H. Fowler. 2007. "The Spread of Obesity in a Large Social Network over 32 Years." *New England Journal of Medicine* 357:370–9.

Church, Timothy S., Corby K. Martin, Angela M. Thompson, Conrad P. Earnest, Catherine R. Mikus, Steven N. Blair. 2009. "Changes in Weight, Waist Circumference, and Compensatory Responses with Different Doses of Exercise among Sedentary, Overweight Postmenopausal Women." *PLoS One* 4:e4515.

Cook, Andrew and Beth Daponte. 2008. "A Demographic Analysis of the Rise in the Prevalence of the US Population Overweight and/or Obese." *Population Research and Policy Review* 27:403–26.

Crosnoe, Robert. 2007. "Gender, Obesity, and Education." *Sociology of Education* 80:241–60.

Crosnoe, Robert and Chandra Muller. 2004. "Body Mass Index, Academic Achievement, and School Context: Examining the Educational Experiences of Adolescents at Risk of Obesity." *Journal of Health and Social Behavior* 45:393–407.

Cutler, David M., Edward L. Glaeser, and Jesse M. Shapiro. 2003. "Why Have Americans Become More Obese?" *Journal of Economic Perspectives* 17:93–118.

Daviglus, Martha L., Kiang Liu, Lijing L. Yan, Amber Pirzada, Larry Manheim, Willard Manning, Daniel B. Garside, Renwai Wang, Alan R. Dyer, Philip Greenland, and Jeremiah Stamler. 2004. "Relation of Body Mass Index in Young Adulthood and Middle Age to Medicare Expenditures in Older Age." *Journal of the American Medical Association* 292:2743–49.

Davison, Kirsten K., Earl S. Ford, Mary E. Cogswell, and William H. Dietz. 2002. "Percentage of Body Fat and Body Mass Index Are Associated with Mobility Limitations in People Aged 70 and Older from NHANES III." *Journal of the American Geriatric Society* 50:1802–9.

de Krom, Mariken, Florianne Bauer, David Collier, R.A.H. Adan, Susanne E. laFleur. 2009. "Genetic Variation and Effects of Human Eating Behavior." *Annual Review of Nutrition* 29:283304.

Economos, Christina D., Raymond R. Hyatt, Jeanne P. Goldberg, Aviva Must, Elena N. Naumova, Jessica J. Collins, and Miriam E. Nelson. 2007. "A Community Intervention Reduces BMI z-score in Children: Shape Up Somerville First Year Results." *Obesity* 15:1325–36.

Elbel, Brian, Rogan Kersh, Victoria L. Brescoll, and L. Beth Dixon. 2009. "Calorie Labeling and Food Choices: A First Look at the Effects on Low-Income People in New York City." *Health Affairs* 28:w1110–21.

Espeland, M., et al. 2007. "Reduction in Weight and Cardiovascular Disease Risk Factors in Individuals with Type 2 Diabetes–One-year Results of the Look AHEAD trial. *Diabetes Care* 30:1374–1383.

Farooqi, I. Sadaf and Stephen O'Rahilly. 2007. "Genetic factors in Obesity." *Obesity Reviews* 8:37–40.

Ferraro, Kenneth F., Roland J. Thorpe Jr., and Jody A. Wilkinson. 2003. "The Life Course of Obesity: Does Childhood Overweight Matter?" *Journal of Gerontology: Social Sciences* 58b:S110–9.

Ferraro, Kenneth F., Ya-ping Su, Randy J. Greteback, D. Randy Black, and Stephen F. Badylak. 2002. "Body Mass Index and Disability in Adulthood: A 20-Year Panel Study." *American Journal of Public Health* 92:834–40.

Ferraro, Kenneth F. and Tara L. Booth. 1999. "Age, Body Mass Index, and Functional Illness." *Journal of Gerontology: Social Sciences* 54B:S339–48.

Finkelstein, Eric A., Christopher Ruhm, and Katherine Kosa. 2005. "Economic Causes and Consequences of Obesity." *Annual Review of Public Health* 26:239–57.

Flegal, Katherine M., Barry I. Graubard, David F. Williamson, and Mitchell H. Gail. 2005. "Excess Deaths Associated with Underweight, Overweight, and Obesity." *Journal of the American Medical Association* 293:1861–7.

Flegal, Katherine M., Margaret D. Carroll, Robert J. Kuczmarski, and Clifford L. Johnson. 1998. "Overweight and Obesity in the United States: Prevalence and Trends, 1960–1994." *International Journal of Obesity and Related Metabolic Disorders* 22:39–47.

Flegal, Katherine M., John A. Shepherd, Anne C. Looker, Barry I. Graubard, Lori G. Borrud, Cynthia L. Ogden, Tamara B. Harris, James E. Everhart, and Nathaniel Schenker. 2009. "Comparisons of Percentage Body Fat, Body Mass Index, Waist Circumference, and Waist-Stature Ratio in Adults." *American Journal of Clinical Nutrition* 89:500–8.

Fonda, Stephanie J., Nancy H. Fultz, Kristi Rahrig Jenkins, Laura M. Wheeler, and Linda A. Wray. 2004. "Relationship of Body Mass and Net Worth for Retirement-Aged Men and Women." *Research on Aging* 26:153–76.

Fontaine, Kevin R., David T. Redden, Chenxi Wang, Andrew O. Westfall, and David B. Allison. 2003. "Years of Life Lost to Obesity." *JAMA* 289:187–193.

Ford, Earl S., David G. Moriarty, Matthew M. Zack, Ali H. Mokdad, and Daniel P. Chapman. 2001. "Self-Reported Body Mass Index and Health Related Quality of Life: Findings from the Behavioral Risk Factor Surveillance System." *Obesity Research* 9:21–31.

Forshee, Richard A., Maureen L. Storey, David B. Allison, Walter H. Glinsman, Gayle L. Hein, David R. Lineback, Sanford A. Miller, Theresa A. Nicklas, Gary A. Weaver, and John S. White. 2007. "A Critical Examination of the Evidence Relating High Fructose Corn Syrup and Weight Gain." *Critical Reviews in Food Science and Nutrition* 47:561–82.

Galanos, Anthony N., Carl F. Pieper, Joan C. Cornoni-Huntley, Connie W. Bales, and Gerda G. Fillenbaum. 1994. "Nutrition and Function: Is There a Relationship between Body Mass Index and the Functional Capabilities of Community-Dwelling Elderly?" *Journal of the American Geriatrics Society* 42:368–73.

Gallagher, Dympna, Marjolein Visser, Dennis Sepulveda, Richard N. Pierson, Tamara Harris, and Steven B. Heymsfield. 1996. "How Useful is BMI for Comparison of Body Fatness across Age, Sex and Ethnic Groups?" *American Journal of Epidemiology* 143:228–39.

Gomez, Jorge E., Beth Ann Johnson, Martha Selva, and James F. Sallis. 2004. "Violent Crime and Outdoor Physical Activity among Inner-City Youth." *Preventive Medicine* 39:876–81.

Greenfield, Emily A. and Nadine F. Marks. 2009. "Violence from Parents in Childhood and Obesity in Adulthood: Using Food in Response to Stress as a Mediator of Risk." *Social Science and Medicine* 68:791–8.

Gregg, Edward W., Yiling J. Cheng, Betsy L. Cadwell, Giuseppina Imperatore, Desmond E. Williams, Katherine M. Flegal, K.M. Venkat Narayan, and David F. Williamson. 2005. "Secular Trends in Cardiovascular Disease Risk Factors According to Body Mass Index in U.S. Adults." *JAMA* 293:1868–74.

Handy, Susan L., Barbara McCann, Linda Bailey, Michelle Ernst, Lanier McRee, Emily Meharg, Reid Ewing, and Kate Wright. 2009. *The Regional Response to Federal Funding for Bicycle and Pedestrian Projects*. Institute of Transportation Studies, University of California, Davis, Research Report UCD-ITS-RR-09-15

Hedley, Allison A., and Cynthia L. Ogden. 2006. "More for the Money? Differences in the Prevalence of Adult Obesity in the U.S. by Income Level – 1999–2002." Paper presented at the 2006 Population Association of America Conference. Los Angeles, California.

Hedley, Allison A., Cynthia L. Ogden, Clifford L. Johnson, Margaret D. Carroll, Lester R. Curtin, and Katherine M. Flegal. 2004. "Prevalence of Overweight and Obesity among US Children, Adolescents, and Adults, 1999–2002." *JAMA* 291:2847–50.

Heiat, Asefeh, Viola Vaccarino, and Harlan M. Krumholz. 2001. "An Evidence Based Assessment of Federal Guidelines for Overweight and Obesity as They Apply to Elderly Persons." *Archives of Internal Medicine* 161:1194–203.

Hill, James O., Holly R. Wyatt, George W. Reed, and John C. Peters. 2003. "Obesity and the Environment: Where Do We Go from Here?" *Science* 299:853–5.

Himes, Christine L. 2000. "Obesity, Disease, and Functional Limitation in Later Life." *Demography* 37:73–82.

Himes, Christine L. and Sandra L. Reynolds. 2005. "The Changing Relationship between Obesity and Educational Status." *Gender Issues* 22:45–57.

Jackson, Tamara D., Carlos M. Grilo, and Robin M. Masheb. 2000. "Teasing History, Onset of Obesity, Current Eating Disorder Psychopathology, Body Dissatisfaction, and Psychological Functioning in Binge Eating Disorder." *Obesity Research* 8:451–8.

Janssen, Ian and Amy E. Mark. 2007. "Elevated Body Mass Index and Mortality Risk in the Elderly." *Obesity Reviews* 8:41–59.

Jenkins, Kristi R. 2004. "Obesity's Effects on the Onset of Functional Impairment among Older Adults." *The Gerontologist* 44:206–16.

Klein, Samuel, David B. Allison, Steven B. Heymsfield, David E. Kelley, Rudolph L. Leibel, Cathy Nonas, and Richard Kahn. 2007. "Waist Circumference and Cardiometabolic Risk: A Consensus Statement from Shaping America's Health: Association for Weight Management and Obesity Prevention; NAASO, The Obesity Society; the American Society for Nutrition; and the American Diabetes Association." *Obesity* 15:1061–7.

Koplan, J.P., Catharyn T. Liverman, and Vivica L. Kraak, eds. 2005. *Preventing Childhood Obesity: Health in the Balance*. Washington DC: National Academy Press.

Kuk, Jennifer L., Travis J. Saunders, Lance E. Davidson, and Robert Ross. 2009. "Age-Related Changes in Total and Regional Fat Distribution." *Aging Research Reviews* 8:339–48.

Kwan, Samantha. 2009. "Framing the Fat Body: Contested Meanings between Government, Activists, and Industry." *Sociological Inquiry* 79:25–50.

LaCroix, Andrea Z., Jack M. Guralnik, Lisa F. Berkman, Robert B. Wallace, and Suzanne Satterfield. 1993. "Maintaining Mobility in Late Life: Smoking, Alcohol Consumption, Physical Activity, and Body Mass Index." *American Journal of Epidemiology,* 137:858–69.

Lakdawalla, Darius N., Dana P. Goldman, and Baoping Shang. 2005. "The Health and Cost Consequences of Obesity among the Future Elderly." *Health Affairs,* 24:W5R30–41.

Lakdawalla, Darius N. and Tomas J. Philipson. 2009. "The growth of obesity and technological change." *Economics and Human Biology* 7:283–93.

Launer, Lenore J., Tamara Harris, Catherine Rumpel, and Jennifer Madans. 1994. "Body Mass Index, Weight Change, and Risk of Mobility Disability in Middle-Aged and Older Women." *Journal of the American Medical Association* 271:1093–8.

Mahan, L. Kathleen and Sylvia Escott-Stump. 2000. *Krause's Food, Nutrition, and Diet Therapy*. Philadelphia PA, W. B. Saunders.

Mifflin, Mark D., Sachiko T. St Jeor, Lisa A. Hill, Barbara J. Scott, Sandra A. Daugherty, and Young O. Koh. 1990. "A New Predictive Equation for Resting Energy Expenditure in Healthy Individuals." *American Journal of Clinical Nutrition* 51:241.

Mokdad, Ali H., Earl S. Ford, Barbara A. Bowman, William H, Dietz, Frank Vinicor, Virginia S. Bales, and James S. Marks. 2003. "Prevalence of Obesity, Diabetes, and Obesity-Related Health Risk Factors, 2001." *JAMA* 289:76–9.

Must, Aviva and Richard S. Strauss. 1999. "Risks and Consequences of Childhood and Adolescent Obesity." *International Journal of Obesity* 23:s2–11.

Myrskala, Mikko and Virginia W. Chang. 2009. "Weight Change, Initial BMI, and Mortality among Middle- and Older-Aged Adults." *Epidemiology* 20:840–8.

National Center for Health Statistics. 2009. Early Release of Selected Estimates Based on Data From the 2008 National Health Interview Survey. http://www.cdc.gov/nchs/data/nhis/earlyrelease/200906_07.pdf

National Institutes of Health. 1998. *Clinical Guidelines on the Identification, Evaluation, and Treatment of Overweight and Obesity in Adults*. Bethesda, MD: National Institutes of Health.

———. 2004. *Strategic Plan for NIH Obesity Research*. National Institutes of Health: U.S. Department of Health and Human Services.

Neel, James V. 1962. "Diabetes Mellitus: A 'Thrifty' Genotype Rendered Determined by Progress?" *American Journal of Human Genetics* 14:353–63.

———. 1989. "The Study of Natural Selection in Primitive and Civilized Human Populations-Update" *Human Biology* 61:811–23.

Nestle, Marion. 2002. *Food Politics: How the Food Industry Influences Nutrition and Health*. Berkeley and Los Angeles: University of California Press.

Nestle, Marion and M.F. Jacobsen. 2000. "Halting the Obesity Epidemic: A Public Health Policy Approach." *Public Health Reports* 115:12–24.

Nielsen, Samara Joy and Barry J. Popkin. 2004. "Changes in Beverage Intake between 1977 and 2001." *American Journal of Preventive Medicine* 27:205–210.

Ogden, Cynthia, Margaret D. Carroll, and Katherine M. Flegal. 2008. "High Body Mass Index for Age among US Children and Adolescents, 2003–2006." *JAMA* 299:2401–5.

Ogden, Cynthia L., Margaret D. Carroll, Lester R. Curtin, Margaret A. McDowell, Carolyn J. Tabak, and Katherine M. Flegal. 2006. "Prevalence of Overweight and Obesity in the United States, 1999–2004." *JAMA* 295:1549–55.

Ogden, Cynthia L., Margaret D. Carroll, Margaret A. McDowell, and Katherine M. Flegal. 2007. "Obesity among Adults in the United States – No Change since 2003–2004." *NCHS Data Brief, No. 1*. Hyattsville, MD: National Center for Health Statistics.

Ogden, Cynthia L., Margaret D. Carroll, Lester R. Curtin, Molly M. Lamb, Katherine M. Flegal. 2010. "Prevalence of High Body Mass Index in US Children and Adolescents, 2007–2008." *JAMA,* 303:242–9.

Okoro, Catherine A., Jennifer M. Hootman, Tara W. Strine, Lina S. Balhuz, and Ali H. Mokdad. 2004. "Disability, Arthritis, and Body Weight among Adults 45 years and Older." *Obesity Research* 12:854–61.

Oliver, J. Eric. 2005. *Fat Politics: The Real Story behind America's Obesity Epidemic*. New York: Oxford University Press.

Oliver, Georgina and Jane Wardle. 1999. "Perceived Effects of Stress on Food Choice." *Physiology & Behavior* 66:511–5.

Olshansky, S. Jay, Douglas J. Passaro, Ronald C. Hershow, Jennifer Layden, Bruce A. Carnes, Jacob Brody, Leonard Hayflick, Robert N. Butler, David B. Allison, and David S. Ludwig. 2005. "A Potential Decline in Life Expectancy in the United States in the 21st Century." *New England Journal of Medicine* 352:1138–45.

Osher, Etty and Naftali Stern. 2009. "Obesity in Elderly Subjects: In Sheep's Clothing Perhaps, But Still a Wolf!" *Diabetes Care* 32:S398–402.

Paul, R.J. and J.B. Townsend. 1995. "Shape Up or Ship Out? Employment Discrimination against the Overweight." *Employee Responsibilities and Rights Journal* 8:133–45.

Peeters, Anna, Jan J. Barendregt, Frans Willekens, Johan P. Mackenback, Abdullah Al Mamun, and Luc Bonneux. 2003. "Obesity in Adulthood and Its Consequences for Life Expectancy: A Life-table Analysis." *Annals of Internal Medicine* 138:24–32.

Peeters, Anna, Luc Bonneux, Wilma J. Nusselder, Chris De Laet, and Jan J. Barendregt. 2004. "Adult Obesity and the Burden of Disability throughout Life." *Obesity Research* 12:1145–51.

Philipson, Tomas J. and Richard A. Posner. 2003. "The Long-Run Growth in Obesity as A Function Of Technological Change." *Perspectives in Biology and Medicine* 46:S87–107.

Popper, R., G. Smits, Herbert L. Meiselman, Edward Hirsch. 1989. "Eating in Combat: A Survey of U.S. Marines." *Military Medicine* 154:619–23.

Prentice, Andrew M. and Susan A. Jebb. 2001. "Beyond Body Mass Index." *Obesity Reviews* 2:141–7.

Preston, Samuel H. 2005. "Deadweight? The Influence of Obesity on Longevity." *New England Journal of Medicine* 352:1135–7.

Puhl, Rebecca M. and Kelly D. Brownell. 2001. "Bias, Discrimination, and Obesity." *Obesity Research* 9:788–805.

———. Brownell. 2003. "Psychosocial Origins of Obesity Stigma: Toward Changing a Powerful and Pervasive Bias." *Obesity Reviews* 4:213–27.

Reither, Eric N., Robert M. Hauser, and Yang Yang. 2009. "Do Birth Cohorts Matter? Age-Period-Cohort Analyses of the Obesity Epidemic in the United States." *Social Science and Medicine* 69:1439–48.

Reuser, Mieke, Luc Bonneux, and Frans Willekens. 2008. "The Burden of Mortality of Obesity at Middle and Old Age is Small: A Life Table Analysis of the US Health and Retirement Survey." *European Journal of Epidemiology* 23:601–7.

Reynolds, Sandra L. and Christine L. Himes. 2007. "Cohort Differences in Adult Obesity in the U.S.:1982–1996." *Journal of Aging and Health* 19:831–50.

Reynolds, Sandra L., Y. Saito, and Eileen M. Crimmins. 2005. "The Impact of Obesity on Active Life Expectancy in Older American Men and Women." *The Gerontologist* 45:439–44.

Roberto, Christina, Peter D. Larsen, Henry Agnew, Jenny Baik, and Kelly D. Brownell. 2010. "Evaluating the Effects of Menu Labeling on Food Choices and Intake." *American Journal of Public Health* 100:312–8.

Roehling, Mark V. 1999. "Weight-based Discrimination in Employment: Psychological and Legal Aspects." *Personnel Psychology* 52:969–1016.

Rothblum, Esther D. 1992. "The Stigma of Women's Weight: Social and Economic Realities." *Feminism and Psychology* 2:61–73.

Sallis, James F., Brian E. Saelens, Lawrence D. Frank, Terry L. Conway, Donald J. Slymen, Kelli L. Cain, James E. Chapman, and Jacqueline Kerr. 2009. "Neighborhood Built Environment and Income: Examining Multiple Health Outcomes." *Social Science and Medicine* 68:1285–93.

Santry, Heena P., Daniel L. Gillen, and Diane S. Lauderdale. 2005. "Trends in Bariatric Surgical Procedures." *JAMA* 294:1901–17.

Schwartz, Marlene B., Heather O'Neal Chambliss, Kelly D. Brownell, Steven N. Blair, and Charles Billington. 2003. "Weight Bias among Health Professionals Specializing in Obesity." *Obesity Research* 11:1033–9.

Serdula Mary K., D. Ivery, Ralph J. Coates, David S. Freedman, David F. Williamson, and Tim Byers. 1993. "Do Obese Children Become Obese Adults? A Review of the Literature." *Preventive Medicine* 22:167–77.

Shah, Meena, Vinaya Simha, and Abhimanyu Garg. 2006. "Review: Long-term Impact of Bariatric Surgery on Body Weight, Comorbidities, and Nutritional Status." *Journal of Clinical Endocrinology and Metabolism* 91:4223–31.

Silventoinen, Karri, Susana Sans, Hanna Tolonen, D. Monterde, Kari A. Kuulasmaa, Hugo E.C. Kesteloot, and J.O. Tuomilehto. 2004. "Trends in Obesity and Energy Supply in the WHO MONICA Project." *International Journal of Obesity* 28:710–8.

Stevens, June, Jianwn Cai, Elsie R. Pamuk, David F. Williamson, Michael J. Thun, and Joy L. Wood. 1998. "The Effect of Age on the Association between Body-mass Index and Mortality." *New England Journal of Medicine* 338:1–7.

Story, Mary, Karen M. Kaphingst, and Simone French. 2006. "The Role of Schools in Obesity Prevention." *The Future of Children* 16:109–42.

Sturm. Roland. 2007. "Increases in Morbid Obesity in the USA: 2000–2005." *Public Health* 121:492–6.

Sturm, Roland, Jeanne S. Ringel, Tatiana Anderyeva. 2004. "Increasing Obesity Rates and Disability Trends." *Health Affairs* 23:199–205.

Thorpe, Roland and Kenneth Ferraro. 2004. "Aging, Obesity, and Mortality: Misplaced Concern about Obese Older People?" *Research on Aging* 26:108–29.

Torres, Susan J., and Caryl A. Nowson. 2007. "Relationship between Stress, Eating Behavior, and Obesity." *Nutrition* 23:887–94.

Transportation Research Board and institute of Medicine. 2005. *Does the Built Environment Influence Physical Activity? Examining the Evidence.* Special Report 282. Washington, D.C.: National Academies Press.

Truong, Khoa Dang, and Roland Sturm. 2005. "Weight Gain Trends across Socioeconomic Groups in the United States." *American Journal of Public Health* 95:1602–16.

Tucker, Larry A., and Glenn M. Friedman. 1998. "Obesity and Absenteeism: An Epidemiologic Study of 10,825 Employed Adults." *American Journal of Health Promotion* 12:202–7.

Uretsky, Seth, Franz H. Messerli, Sripal Bangalore, Annette Champion, Rhonda M. Cooper-Dehoff, Qian Zhou, and Carl J. Pepine. 2007. "Obesity Paradox in Patients with Hypertension and Coronary Artery Disease." *American Journal of Medicine* 120:863–870.

U.S. Bureau of the Census. 2007. Computer and Internet Use in the United States: October 2007. http://www.census.gov/population/www/socdemo/computer/2007.html

U.S. Department of Health and Human Services. 2008. *Physical Activity Guidelines Advisory Committee Report.* Washington, DC:U.S. Government Printing Office.

U.S. Surgeon General. *Overweight and Obesity: Health Consequen*ces. Retrieved ([http://www.surgeongeneral.gov/topics/obesity/calltoaction/fact_consequences.htm]. Rockville: MD. 2001).

Villareal, Dennis T., Caroline M. Apovian, Robert F. Kushner, and Samuel Klein. 2005. "Obesity in Older Adults: Technical Review and Position Statement of the American Society for Nutrition and NAASO, The Obesity Society." *American Journal of Clinical Nutrition* 82:923–34.

Wallace, Jeffrey I., R.S. Schwartz, A.Z. LaCroix, Richard F. Uhlmann, and Robert A. Pearlman. 1995. "Involuntary Weight Loss in Older Outpatients: Incidence and Clinical Significance." *Journal of the American Geriatrics Society* 43:329–37.

Wang, Shirley S., Kelly D. Brownell, and Thomas A. Wadden. 2004. "The Influence of the Stigma of Obesity on Overweight Individuals." *International Journal of Obesity* 28:1333–1337.

Whitmer, Rachel A., Erica P. Gunderson, Elizabeth Barrett-Connor, Charles P. Quesenberry, Jr., and Kristine Yaffe. 2005. "Obesity in Middle Age and Future Risk of Dementia: A 27 Year Longitudinal Population Based Study." British Medical Journal, 330:1360.

Williams, Lana J., Julie A. Pasco, Margaret J. Henry, Felice N. Jacka, Seetal Dodd, Geoffrey C. Nicholson, Mark A. Kotowicz, and Michael Berk. 2009. "Lifetime Psychiatric Disorders and Body Composition: A Population Based Study." *Journal of Affective Disorders* 118:173–9.

Williamson, David F. 1993. "Descriptive Epidemiology of Body Weight and Weight Change in U.S. Adults." *Annals of Internal Medicine* 119:646–9.

Wolf, Anne M., and Graham A. Colditz. 1998. "Current Estimates of the Economic Cost of Obesity in the United States." *Obesity Research* 6:97–106.

World Health Organization. 2002. *World Health Report: 2002.* Geneva, Switzerland: World Health Organization.

Chapter 33
Religious Involvement, Health Status, and Mortality Risk

Terrence D. Hill, Amy M. Burdette, and Ellen L. Idler

Religious involvement – indicated by observable feelings, beliefs, activities, and experiences in relation to spiritual, divine, or supernatural entities – is a prevalent and powerful force in the lives of older adults. Despite evidence of secularization and the declining significance of religious institutions (Chaves 1994), elderly Americans continue to exhibit high rates of religious involvement. According to national estimates from the 2008 General Social Survey, a large percentage of adults aged 65 and older affiliate with religious groups (93%), characterize their affiliations as strong (56%), attend religious services weekly or more (42%), pray at least once per day (68%), and believe that the Bible is the actual word of God and is to be taken literally (37%). These figures are remarkable in their own right. They also inspire countless questions concerning the consequences of religious involvement in late life.

This chapter provides an overview and critical examination of published research on the health consequences of religious involvement in the elderly population. After describing general religious variations in health and mortality risk, we present major theoretical and empirical explanations for these patterns. We conclude with a detailed discussion of the limitations of previous work and note viable avenues for future research. In the sections that follow, we emphasize seminal research articles, recent studies, and notable reviews of empirical evidence. Although we intend for this research collection to be representative of the field, we do not consider it to be exhaustive.

Health and Mortality Patterns

In this section, we briefly summarize known religious variations in health and mortality risk. We first highlight patterns across several indicators of mental and physical health. We then consider the evidence from studies of all-cause and cause-specific mortality.

Mental Health

Numerous studies show that religious involvement is associated with favorable mental health outcomes in the elderly population. This general pattern is remarkably consistent across indicators of mental health, including anger (Carr 2003), depression (Braam et al. 2004; Idler 1987; Idler and

T.D. Hill (✉)
Department of Sociology, Florida State University, Tallahassee, FL, USA
e-mail: thill@fsu.edu

R.A. Settersten, Jr. and J.L. Angel (eds.), *Handbook of Sociology of Aging*, Handbooks of Sociology and Social Research, DOI 10.1007/978-1-4419-7374-0_33, © Springer Science+Business Media, LLC 2011

Kasl 1997a; Idler et al. 2009; Jang et al. 2005; Law and Sbarra 2009; Norton et al. 2008; Strawbridge et al. 1998), anxiety (Cicirelli 2002; Krause 2005), nonspecific psychological well-being (Fry 2001; Idler and Kasl 1997a), life satisfaction (Levin et al. 1996; Krause 2003a, 2005), and cognitive functioning (Hill et al. 2006; Reyes-Ortiz et al. 2008; Van Ness and Kasl 2003). Although the majority of studies are cross-sectional, longitudinal evidence also suggests that religious involvement is associated with favorable mental health trajectories in late life (Braam et al. 2004; Hill et al. 2006; Law and Sbarra 2009; Norton et al. 2008; Reyes-Ortiz et al. 2008; Van Ness and Kasl 2003).

Physical Health

Surprisingly few studies have considered the association between religious involvement and physical health in the elderly population. Nevertheless, studies indicate that religious involvement is associated with favorable physical health outcomes. This pattern is consistent across a limited range of physical health indicators, including self-rated health (Idler et al. 2009; Krause 1998, 2006a; Musick 1996), functional status (Benjamins 2004; Idler 1987; Idler and Kasl 1997b; Park et al. 2008), and stroke (Wolinsky et al. 2009). Once again, available longitudinal evidence demonstrates that religious involvement is associated with favorable physical health trajectories in late life (Benjamins 2004; Idler and Kasl 1997b; Krause 1998; Park et al. 2008; Wolinsky et al. 2009).

Mortality Risk

If religious involvement favors mental and physical health in the elderly population, evidence of a survival advantage should come as no surprise. Indeed, several longitudinal studies indicate that religious involvement (typically indicated by attendance at religious services) is associated with lower risk of all-cause mortality (Dupre et al. 2006; Ellison et al. 2000; Gillum et al. 2008; Helm et al. 2000; Hill et al. 2005; Hummer et al. 1999; Koenig et al. 1999; Krause 2006b; la Cour et al. 2006; Lutgendorf et al. 2004; Musick et al. 2004; Oman and Reed 1998; Oman et al. 2002; Strawbridge et al. 1997). There is also some evidence to suggest that religious involvement is associated with lower risk of mortality from specific causes, including circulatory diseases, respiratory diseases, and other causes (Hummer et al. 1999; Oman et al. 2002).

Major Theoretical and Empirical Explanations

In this section, we consider major theoretical and empirical explanations for why religious involvement might favor health and longevity. Several significant articles and books have addressed this issue (e.g., George et al. 2002; Idler 2004; Koenig et al. 2001; Seeman et al. 2003). Drawing on this body of work and relevant empirical evidence, we discuss several potential social, psychological, behavioral, and biological mechanisms.

Social Resources

Social isolation and its health consequences are major concerns in the elderly population (Gray 2009; Kobayashi et al. 2009). As adults transition to old age, they retire from socially engaging work roles, their children grow up and move away, and health conditions increasingly limit social activities, create significant caregiving responsibilities, and enhance the probability of spousal loss.

Religious involvement may help to counteract the risk of social isolation and promote health and longevity by enhancing a range of social resources.

Studies show that religious involvement (especially religious attendance) is associated with larger and more diverse social networks, more contact with network members, more stable marriages and extensive family ties, more types of social support received and perceived (both instrumental and emotional), and greater civic participation, including memberships to religious and secular organizations and groups (Ellison and George 1994; Gray 2009; Idler and Kasl 1997a; Krause 2008; McIntosh et al. 2002; Strawbridge et al. 1997). Recent work by Kobayashi et al. (2009) demonstrates that regular religious attendance (at least monthly) reduces the odds of social isolation among older adults by 128%, even with adjustments for health status, chronic conditions, and activity limitations.

Public involvement in religious communities clearly influences the size and nature of social networks in late life. Strong religious beliefs concerning the sanctification of the family could encourage older adults to maintain strong family ties, although some research suggests that religious beliefs can also contribute to conflict between older parents and adult children (Clarke et al. 1999). Divine relations may also play a significant role in promoting health and well-being when older adults believe that they are constantly supported and loved by a supernatural entity (Harvey and Silverman 2007; Van Ness et al. 2008).

Several studies have formally tested whether social resources help to explain why religious involvement might favor health and longevity in old age; however, many of these tests are difficult to interpret because multiple mediators are entered simultaneously. Research suggests that leisure social activities and, to a lesser extent, number of close kin help to explain favorable disability trajectories (Idler and Kasl 1997b). Other studies show no support for social interaction and interaction satisfaction as links to self-rated health (Musick 1996) or marital status, number of children, and financial support as mechanisms to functional health (Benjamins 2004). With respect to mortality outcomes, some research confirms the mediating influences of marital status, social connections, social activity, and the receipt and provision of social support (Ellison et al. 2000; la Cour et al. 2006; Strawbridge et al. 1997), while others show no evidence of mediation for number of confidants, frequency of social contact, and perceived social support (Hill et al. 2005; Koenig et al. 1999; Musick et al. 2004).

Psychological Resources

Old age is also characterized by the loss of important psychological resources, including self-esteem, the sense of control, optimism, and meaning and purpose (Koenig et al. 1988). In addition to deficiencies in health, valued social roles, and sources of love and support, many elderly persons are especially vulnerable to financial hardship and numerous stressors in the environment (e.g., poor neighborhood conditions). As older adults struggle with multiple adverse conditions, they are forced to question themselves in very fundamental ways: Who controls my life? Am I valuable? Is my life meaningful? Am I loved or even worthy of love? Am I safe? Religious involvement may favor health and longevity by helping to answer some of these questions by reinforcing important psychological resources.

Studies show that religious involvement is positively associated with key psychological resources, including self-esteem (Krause 1992, 2003a, 2005), self-control (McCullough and Willoughby 2009), personal control or mastery (Dillon and Wink 2007; Jang et al. 2003; Schieman et al. 2005), optimism (Ai et al. 2002; Krause 2003a, 2005), and meaning and purpose (Ardelt and Koenig 2006; Krause 2003a). Religious involvement may enhance psychological resources and positive coping by promoting social connections, social activities, and specific religious beliefs.

Religious attendance provides people with the opportunity to interact with those who hold similar values and beliefs, and these interactions can be important for self-esteem because they reinforce positive role identities and role expectations. Ellison (1993) notes that active religious participants are often valued for skills and abilities that are uniquely connected with church-related activities (e.g., singing in choir, participation in religious discussion groups, and praying for others), respected for service to others in the community (e.g., volunteering and specific leadership roles), and admired for personal spiritual qualities (e.g., wisdom and morality). Positive self-perceptions gained through religious involvement may be especially valuable for the elderly who lack important socially valued roles.

Religious involvement is characterized by social control and self-regulation. Within the context of religious communities, there are social (and perceived divine) sanctions associated with conformity to and deviance from established religious standards (e.g., behavioral and ritual standards and expectations). Religious involvement contributes to self-control by building generic self-regulatory strength over the life course (McCullough and Willoughby 2009). Because religion is, in many respects, a routine practice of constraint and restraint, older religious adults are more likely to believe that they can control their emotions and behavior. A strong sense of divine control may also help to promote a sense of personal control or mastery over various aspects of life when older adults trust that anything is possible through faith and a strong partnership with God (Harvey and Silverman 2007; Schieman et al. 2005).

Religious involvement may contribute to hope and optimism by fostering positive self-conceptions and control beliefs. When older adults are faced with adversity, confidence derived from self-esteem and a sense of control can be instrumental in solving problems. It can be useful to think of life as unfolding according to some divine plan. Beliefs such as these promote a sense of meaning and purpose, which helps to buffer appraisals of difficult life conditions (Harvey and Silverman 2007; Pargament 1997).

Although psychological resources are theoretically viable explanations for why religious involvement might favor health and longevity in old age, very few studies have formally examined this class of mechanisms. Nevertheless, research suggests that self-esteem (Krause 1992) and a sense of meaning and purpose (Ardelt and Koenig 2006; Krause 2003a) help to explain the association between religious involvement and mental health.

Healthy Behaviors

When older adults experience the loss of health and valuable social and psychological resources, they often struggle to maintain certain healthy behaviors. Healthy behaviors like exercise require a certain degree of health. Proper adherence to complex medical regimens also calls for a strong sense control. Religious involvement could contribute to health and longevity in old age by promoting healthy behaviors.

Studies show that religious involvement is associated with a wide range of healthy behaviors in the elderly population, including lower levels of smoking (Idler and Kasl 1997a; Klemmack et al. 2007; Koenig et al. 1998a) and drinking (Idler and Kasl 1997a; Krause 2003b), higher levels of exercise (Idler and Kasl 1997a; Oman and Reed 1998; Strawbridge et al. 1997), greater use of preventive health care services (Benjamins 2005, Benjamins and Brown 2004), and more rigid adherence to medication regimens (Koenig et al. 1998b). There is even some evidence to suggest that older religious adults exhibit a stronger preference for and willingness to undergo life-sustaining treatments (Cohen-Mansfield et al. 1992; Van Ness et al. 2008). Body mass is one possible exception to the healthy lifestyle profile of religious adults. Studies clearly demonstrate that older religious adults tend to weigh more, not less, than their less religious counterparts (Idler and Kasl 1997a; Oman and Reed 1998; Strawbridge et al. 1997). However, research also suggests that religious adults are less likely to be underweight (Musick et al. 2004), which is especially relevant in old age.

There are several compelling explanations for why religious involvement is associated with so many healthy behaviors in old age. First and foremost, religious involvement exposes individuals to moral directives that are supported by the authority of longstanding religious traditions and sacred texts. With prolonged exposure (through religious involvement), individuals may internalize religious messages that discourage specific health-relevant behaviors (e.g., biblical proscriptions against intoxication). Specific religious proscriptions may help to explain why religious individuals might avoid particular behaviors (e.g., heavy drinking), but they cannot account for the effects of religious involvement on health-relevant behaviors that are unspecified in religious scripture (e.g., smoking and use of preventive health care services).

Interestingly, many religious groups adhere to general religious principles that sanctify the body and promote the instrumental importance of physical health as a means to greater spiritual commitment and involvement (e.g., 1 Corinthians 3:16–17; 1 Corinthians 6:19–20). Mahoney et al. (2005) refer to sanctification as a process through which objects are infused with divine or spiritual significance. 1 Corinthians (6:19–20) provides an especially direct example of the sanctification of the body: "…your body is the temple of the Holy Spirit who is in you…therefore glorify God in your body, and in your spirit, which are God's." Many religious groups use this passage to promote the body as a sacred object and to discourage a wide range of health-relevant behaviors, including, for example, alcohol consumption, tobacco smoking, illicit drug use, risky sexual behaviors, and even body piercing and tattooing.

Religious involvement could also contribute to healthy behaviors by encouraging deference to authority and conformity to rules and laws (Welch et al. 2006). Numerous biblical passages counsel adherents to submit to various "authorities" and "ordinances" (e.g., Hebrews 13:17; Peter 2:13–14; Romans 13:1–7). For instance, Romans (13:1–2) advises: "Let every soul be subject to the governing authorities. For there is no authority except from God, and the authorities that exist are appointed by God. Therefore whoever resists the authority resists the ordinance of God, and those who resist will bring judgment on themselves." Welch et al. (2006) explain that religious involvement may favor conformity through fear of divine retribution, internalized moral codes, guilt avoidance, and the social context of similarly obedient peer networks. If religious individuals are more sensitive to authority, which is often sanctified, they may be more likely to trust physicians, adhere to recommended medical regimens, and avoid risky health-related activities. Indeed, many religious older adults view physicians and institutions of medicine as instruments through which God heals (King et al. 2005).

Healthy behaviors are the most intuitive explanations for the effects of religious involvement on health and longevity, and several studies have formally examined these links. With respect to disability status, studies support the mediating influences of smoking (Benjamins 2004) and exercise (Benjamins 2004; Idler and Kasl 1997b); however, Idler and Kasl (1997b) also show no evidence of mediation for alcohol consumption, smoking, and body mass. Musick (1996) tested smoking and alcohol consumption as links to self-rated health, but found no support for these mechanisms. Some mortality studies indicate that smoking (Dupre et al. 2006; Ellison et al. 2000; Hummer et al. 1999; Strawbridge et al. 1997), body mass – especially underweight (Dupre et al. 2006; Hummer et al. 1999; Musick et al. 2004; Strawbridge et al. 1997), exercise (Musick et al. 2004; Strawbridge et al. 1997), and alcohol consumption (Strawbridge et al. 1997) are important explanatory factors, while others show little to no mediating influence for smoking (Helm et al. 2000; Koenig et al. 1999; la Cour et al. 2006), body mass (Helm et al. 2000; Koenig et al. 1999; la Cour et al. 2006), and alcohol consumption (Dupre et al. 2006; Ellison et al. 2000; Hill et al. 2005; Hummer et al. 1999).

Biological Markers

Biological markers or biomarkers are objective indicators (derived from independent assessments like blood and saliva, not self-reports) of physiological functioning (e.g., cardiovascular and immune functioning) that are known to predict health and mortality risk. Like most health outcomes, biomarkers are not randomly distributed in society. They are shaped by repeated and patterned social, psychological, and behavioral processes. Does religious involvement favor healthier biomarker profiles in the elderly population? If so, do biomarkers help to explain any health and mortality advantages associated with religious commitment?

To this point, very few studies have considered biomarkers as outcomes of religious involvement. Nevertheless, research shows that various indicators of religious involvement are associated with biomarkers across sympathetic nervous, hypothalamic–pituitary–adrenal, cardiovascular, immune, and metabolic systems (Seeman et al. 2003; Seybold 2007). When we limit the evidence to older adults, we find that religious involvement is associated with lower levels of blood pressure (Koenig et al. 1998b; Krause et al. 2002), c-reactive protein (Gillum et al. 2008; King et al. 2001, 2002), interleukin-6 (Koenig et al. 1997; Lutgendorf et al. 2004), white blood cells (King et al. 2001), and cortisol (Ironson et al. 2002).

How might religious involvement get "under the skin" to contribute to favorable biomarker profiles? Religious involvement (e.g., religious meaning systems) may help to buffer appraisals of stressful life conditions and, by extension, their physiological consequences (Seybold 2007). Religious involvement might also support healthy biological functioning indirectly by promoting important social, psychological, and behavioral resources. For example, instrumental support, the sense of control, and moderate drinking practices could help older adults to avoid stressful life conditions (events and appraisals) and chronic activation of the physiological stress response (i.e., allostatic load). In the event of stressful life conditions (and the activation of sympathetic systems), religious beliefs and practices, supportive relationships, strong self-concepts, and healthy lifestyles may also favor healthy coping strategies (and efficient activation of parasympathetic systems and various growth responses).

Given the limited amount of research in this area, it should come as no surprise that very few studies have formally tested whether biomarkers help to mediate or explain the effects of religious involvement on health and longevity. Lutgendorf et al. (2004) demonstrate that the inverse association between religious attendance and all-cause mortality risk in older adults is fully mediated by lower levels of interleukin-6, a biomarker implicated in the development of heart disease, cancer, osteoporosis, frailty, and functional limitations. Although Gillum et al. (2008) report a similar pattern for c-reactive protein, these results are unclear because several potential mediators were entered simultaneously.

Research Limitations and Future Directions

The research reviewed in this chapter is characterized by several limitations. In this section, we review these shortcomings and highlight important directions for future research and policy initiatives.

Measurement Issues

Religious involvement is a multidimensional phenomenon (Idler et al. 2003; Krause 1993); however, most studies employ only one or two single-item indicators of religious involvement (typically religious attendance or some measure of religious salience). Single items prevent a comprehensive understanding of religious involvement as a multidimensional phenomenon and are less reliable than would be ideal. Several multi-item indices have been developed to assess multiple dimensions of religious involvement (e.g., Frey et al. 2005; Idler et al. 2003; Krause 1993). To this point, however, limited and discipline-specific applications of these instruments undermine the potential for comparisons across studies (and disciplines).

Health Outcomes

Although previous studies have considered a wide range of health and mortality outcomes, this body of literature is heavily focused on mental health. Because most studies of mental health emphasize depression, it is important for future research to examine outcomes like anger and anxiety in the elderly population. Studies of physical health and mortality are usually limited to self-rated health and all-cause mortality, respectively. Additional research is needed to establish the effects of religious involvement on specific physical health conditions (e.g., heart disease, diabetes, cancer, and sensory impairments), biomarker profiles, and cause-specific mortality outcomes.

Indirect Effects

Previous research is also limited by theoretical models that overemphasize the main or direct effects of religious involvement on health and longevity. Although studies often speculate as to why religious involvement might favor health and longevity, empirical support for these explanations is sorely lacking. We have noted limited empirical support for several classes of mediators. It is important for future research to confirm these patterns and to examine understudied mechanisms for mental health (e.g., social resources, lifestyle factors, and physical health), physical health (e.g., psychological resources, biomarker profiles, and mental health), and mortality risk (e.g., psychological resources, mental health, and physical health). While it is important to establish individual mechanisms, it is time to focus more on developing and testing elaborate theoretical models with multiple mediators and complex causal chains.

Healthy Lifestyles

The idea that religious involvement might contribute to health and longevity by promoting healthy behaviors is perhaps the most intuitive and most widely accepted explanation in the religion-health literature (and the popular mind). Religious involvement is clearly associated with a wide range of healthy practices, and these healthy practices are strong predictors of health and longevity. This process should work theoretically, but, as discussed above, empirical support is limited and mixed. Given that religious involvement is associated with so many healthy behaviors, it makes sense to think less in terms of individual health behaviors and more in terms of healthy lifestyles (Hill et al. 2007). Are older religious adults healthier simply because they avoid smoking and heavy drinking or because they are also more likely to exercise, use preventive health care services, and follow medication regimens simultaneously? Adjusting for an index of healthy behaviors would directly test the mediating influence of the clustering of healthy behaviors within individuals and groups. This re-conceptualization could be the key to finding consistent empirical support for this traditional explanation.

Subgroup Variations

It is also often unclear whether the effect of religious involvement varies according to theoretically relevant subgroups. In other words, under which social, psychological, and physiological conditions is religious involvement more or less protective? Some studies show that religious involvement is more beneficial to the mental health of older blacks (Krause 2003a, 2005) and older women (Idler 1987; Jang et al. 2005). Research also suggests that the positive association between religious involvement and self-rated health is more pronounced among individuals with functional limitations (Idler et al. 2009; Musick 1996). There is even some indication that the association between religious involvement and longevity is stronger for older women (Koenig et al. 1999; la Cour et al. 2006; Strawbridge et al. 1997) and weaker among elderly individuals in the oldest ages (Dupre et al. 2006; Ellison et al. 2000; Musick et al. 2004), with functional limitations (Helm et al. 2000), and low levels of social support (Oman and Reed 1998). Additional research is needed to confirm these patterns and to consider new and understudied subgroup variations. We also need better theoretical explanations for these variations *a priori*.

Stress Moderation

While numerous studies emphasize how religious involvement might influence health and mortality risk directly or, to a lesser extent, indirectly, less attention has been devoted to religious involvement as a moderator or buffer of stressful life conditions. Studies show that religious involvement is protective against the mental health consequences of spousal loss (Carr 2003; Coleman et al. 2007; Fry 2001), financial hardship (Strawbridge et al. 1998), neighborhood disorder (Strawbridge et al. 1998), hospitalization (Koenig et al. 2004), disability (Braam et al. 2004; Strawbridge et al. 1998), other physical health conditions (Braam et al. 2004; Strawbridge et al. 1998; Wink et al. 2005), and terminal illness (Ardelt and Koenig 2006). Research also indicates that religious involvement is protective against the physical health consequences of poor neighborhood conditions (Krause 1998, 2006a). Our review of the religion-mortality literature revealed only one test of stress-moderation. In a national study of older adults, Krause (2006b) reports that the positive association between financial hardship and mortality risk is attenuated by the provision and receipt of church-based social support. More work is needed to establish these patterns and to comprehensively examine the extent to which religious involvement protects older adults against the health consequences of stressful life conditions.

The "Dark-Side" of Religious Involvement

Although the evidence reviewed thus far clearly demonstrates a salubrious role of religious involvement in the elderly population, this research does not exclude the possibility that religious involvement might also undercut health and longevity. Studies show that religious involvement can undermine psychological well-being when older adults have religious doubts (Krause et al. 1999), feel abandoned by God (Allen et al. 2008), exhibit high levels of extrinsic religiosity (Ardelt and Koenig 2006), and combine strong religious beliefs with low religious attendance (Dillon and Wink 2007). Strawbridge et al. (1998) report that religious involvement can exacerbate the mental health consequences of intimate partner violence, marital problems, caregiving, and problematic parent–child relations. Research also indicates that religion can contribute to poor physical health outcomes when older adults suppress religious doubts (Krause and Ellison 2009). One study of elderly hospital patients demonstrates that various indicators of religious struggle – wondering whether God has abandoned you, questioning God's love, and attributing poor health conditions to the devil – can actually elevate the risk of mortality (Pargament et al. 2001). Although studies confirm that there is a distinct "dark side" of religious involvement, explanations for these processes are not as thoroughly developed as explanations for the healthful consequences of religious involvement. More research is needed to explore and explain these processes.

Alternative Explanations

There are three main alternative explanations for why religious involvement might favor health and longevity: health selection, personality selection, and social desirability. Although longitudinal evidence is growing, most mental and physical health studies are limited to cross-sectional data. Because cross-sectional studies are unable to establish the causal order of any observed associations, it is often unclear why religious people appear healthier. We assume that religious involvement predicts health status, but health status might also predict religious involvement. For example, studies show that physical health problems, including broken hips, disability, cancer, and stroke, can undermine or limit public religious activities in old age (Benjamins et al. 2003; Kelley-Moore and Ferraro 2001). In the absence of longitudinal designs and adequate controls for baseline health status, certain indicators of religious involvement (especially indicators of public religious activities) "select" healthier older adults into religious activities. This pattern can be seen in various mortality studies when associations with religious attendance are noticeably attenuated or even eliminated with comprehensive adjustments for baseline physical health and functioning (e.g., Ellison et al. 2000; Hill et al. 2005; Hummer et al. 1999; Musick et al. 2004).

Research clearly demonstrates that a significant portion of the association between religion and health is produced by health selection processes in old age. Researchers also argue that individuals with certain personality traits are selected into religious involvement. Some even view religious involvement as a strong indicator of personality. Personalities are patterned ways of thinking, feeling, and behaving. Studies show that religious involvement is reliably associated with several personality characteristics, including lower levels of psychoticism (risk-taking and lack of responsibility) and higher levels of agreeableness (friendly and helpful to others), conscientiousness (dependability and self-discipline), and cooperativeness (Koenig forthcoming). Because studies of religion and health rarely (if ever) adjust for personality, there is little to no evidence for personality selection. Psychoticism could reasonably select individuals into risky lifestyle patterns and, as a consequence, out of religious institutions. Agreeableness could even draw individuals into religious communities through various social aspects of public religious activities. Although there appears to be considerable continuity between religious involvement (rituals and expectations) and conscientiousness (especially the component of self-discipline), it is important to acknowledge that religious involvement can also foster self-regulatory strength over the life course. If any of these processes are at work, personality selection could account for at least some of the effect of religious involvement through social and behavioral mechanisms.

Finally, there is the potential for bias due to social desirability. The idea is that some older religious adults might falsely respond to health-related questions (e.g., risky health behaviors and depressive symptoms) in order to protect their religious beliefs, values, and identities. Since studies generally fail to control for social desirability, reports may, to some extent, exaggerate associations between religious involvement and self-reported health outcomes. To this point, however, there is little to no empirical support for social desirability as an alternative explanation for religious influence.

Policy Implications

This chapter clearly demonstrates that religious involvement favors health and longevity in the elderly population, but what are the policy implications of this research? Neither scholars nor practitioners can ethically or practically prescribe religious practice, much less belief. Are there other possibilities? Physicians routinely take note of employment status and marital status because these social characteristics provide health-relevant information. If religious involvement is associated with health outcomes and health-relevant decisions and practices, it may be important for physicians and other health professionals to take such socio-cultural information into account. As discussed above, religious involvement may influence how well older adults follow medical regimens. Religious beliefs and practices may also disagree with specific treatments. For example, religious beliefs surrounding substance use and mental illness could affect the use of medications like pain relievers and antidepressants, respectively. Religious norms concerning dietary restrictions and fasting may also play various roles. When older adults experience health problems and marital loss, social isolation becomes a major cause for concern. Under these conditions, older religious adults may become disconnected from their religious communities. Programs or interventions designed to help the elderly to maintain their religious involvement (e.g., ride programs) may uniquely contribute to the health and well-being of older religious adults.

Given that a substantial portion of older adults are involved in religious institutions, the congregation is an ideal setting for addressing health maintenance and disease prevention, as well as mobilizing support for those elderly with health problems in need of care (Koenig and Brooks 2002). Parish nursing is one example of programs linking religious institutions with health care systems. A parish nurse is a registered nurse who is a member of the congregation and often provides medical services on a voluntary basis. A parish nurse usually does not provide much of the hands-on care needed by sick members of the congregation, but instead acts as a health educator (teaching classes on diet and exercise, medication usage, and disease management), medical interpreter, and trainer of volunteers that may help with these activities (Koenig and Brooks 2002). Similarly, the Faith-Based and Community Initiative launched by President George W. Bush developed a number of programs linking religion with health care. Faith-based programs provide a range of services aimed at the elderly, ranging from disease prevention and treatment to outreach programs for those with dementia. Such programs may be particularly relevant for delivering services too hard to reach elderly populations (e.g., racial/ethnic minorities and rural residents) (Brooks and Koenig 2002).

Conclusion

Our overview and critical examination of published research suggests that older religious adults are happier and healthier and tend to live longer than their less religious counterparts. These general patterns are remarkably consistent across studies of various health and mortality outcomes. Even more impressive is the consistency of findings across disciplines, including sociology, religious studies, gerontology, psychology, public health, medicine, and geriatrics. Having said this, additional research is needed to establish: (1) valid and reliable measures of religious involvement, (2) understudied health outcomes, especially in the areas of physical health and cause-specific mortality, (3) social, psychological, behavioral (lifestyle), and biological explanations, (4) subgroup variations, (5) and patterns of life stress moderation. It is also important for future studies to: (6) thoroughly explore the "dark side" of religion, (7) formally test alternative explanations like health and personality selection and social desirability, and (8) consider the policy ramifications of the religion–health connection. Research along these lines will no doubt contribute to a better and more practical understanding of how and why religious involvement might favor health and longevity in the elderly population.

References

Ai, Amy, Christopher Peterson, Steven Bolling, and Harold Koenig. 2002. "Private Prayer and Optimism in Middle-Aged and Older Patients awaiting Cardiac Surgery." *The Gerontologist* 42:70–81.

Allen, Rebecca, Laura Phillips, Lucinda Roff, Ronald Cavanaugh, and Laura Day. 2008. "Religiousness/Spirituality and Mental Health among Older Male Inmates." *The Gerontologist* 48:692–97.

Ardelt, Monika and Cynthia Koenig. 2006. "The Role of Religion for Hospice Patients and Relatively Healthy Older Adults." *Research on Aging* 28:184–215.

Benjamins, Maureen. 2004. "Religion and Functional Health among the Elderly: Is there a Relationship and is it Constant?" *Journal of Aging and Health* 16:355–74.

———. 2005. "Social Determinants of Preventive Service Utilization: How Religion Influences the use of Cholesterol Screenings in Older Adults." *Research on Aging* 27:475–97.

Benjamins, Maureen, Marc Musick, Deborah Gold, and Linda George. 2003. "Age-Related Declines in Activity Level: The Relationship between Chronic Illness and Religious Activities." *Journal of Gerontology: Social Sciences* 58B:S377–85.

Benjamins, Maureen and Carolyn Brown. 2004. "Religion and Preventative Health Care Utilization among the Elderly." *Social Science & Medicine* 58:109–18.

Braam, Arjan, Erick Hein, Dorly Deeg, Jos Twisk, Aartjan Beekman, and Willem Tilburg. 2004. "Religious Involvement and 6-year Course of Depressive Symptoms in Older Dutch Citizens." *Journal of Aging and Health* 16:467–89.

Brooks, Robert and Harold Koenig. 2002. "Having Faith in an Aging Health System: Policy Perspectives." *Public Policy and Aging Report* 12:23–6.

Carr, Deborah. 2003. "A 'Good Death' for Whom? Quality of Spouse's Death and Psychological Distress among Older Widowed Persons." *Journal of Health and Social Behavior* 44:215–32.

Chaves, Mark. 1994. "Secularization as Declining Religious Authority." *Social Forces* 72:749–74.

Cicirelli, Victor. 2002. "Fear of Death in Older Adults: Predictions from Terror Management Theory." *Journal of Gerontology: Psychological Sciences* 57B:P358–66.

Clarke, Edward, Mar Preston, Jo Raskin, and Vern Bengtson. 1999. "Types of Conflicts and Tensions between Older Parents and Adult Children." *The Gerontologist* 39:261–70.

Cohen-Mansfield, Jiska, Janet Droge, and Nathan Billig. 1992. "Factors Influencing Hospital Patients' Preferences in Utilization of Life-Sustaining Treatments." *The Gerontologist* 32:89–95.

Coleman, Peter, Fionnuala McKiernan, Marie Mills, and Peter Speck. 2007. "In Sure and Uncertain Faith: Belief and Coping with Loss of Spouse in Later Life." *Ageing and Society* 27:869–90.

Dillon, Michele and Paul Wink. 2007. *In the Course of a Lifetime: Tracing Religious Belief, Practice, and Change.* Berkley, CA: University of California Press.

Dupre, Matthew, Alexis Franzese, and Emilio Parrado. 2006. "Religious Attendance and Mortality: Implications for the Black-White Mortality Crossover." *Demography* 43:141–64.

Ellison, Christopher. 1993. Religious Involvement and Self-perception among Black Americans. *Social Forces* 71:1027–55.

Ellison, Christopher and Linda George. 1994. "Religious Involvement, Social Ties, and Social Support in a Southeastern Community." *Journal for the Scientific Study of Religion* 33:46–61.

Ellison, Christopher, Robert Hummer, Shannon Cormier, and Richard Rogers. 2000. "Religious Involvement and Mortality Risk among African American Adults." *Research on Aging* 22:630–67.

Frey, Bruce, Timothy Daaleman, and Vicki Peyton. 2005. "Measuring a Dimension of Spirituality for Health Research: Validity of the Spirituality Index of Well-being." *Research on Aging* 27:556–77.

Fry, P.S. 2001. "The Unique Contribution of Key Existential Factors to the Prediction of Psychological Well-Being of Older Adults Following Spousal Loss." *The Gerontologist* 41:69–81.

George, Linda, Christopher Ellison, and David Larson. 2002. "Explaining the Relationships between Religion and Health." *Psychological Inquiry* 13:190–200.

Gillum, R., Dana King, Thomas Obisesan, and Harold Koenig. 2008. "Frequency of Attendance at Religious Services and Mortality in a U.S. National Cohort." *Annals of Epidemiology* 18: 124–9.

Gray, Anne. 2009. "The Social Capital of Older People." *Ageing & Society* 29:5–31.

Harvey, Idethia and Myrna Silverman. 2007. "The Role of Spirituality in the Self-Management of Chronic Illness among Older African and Whites." *Journal of Cross-Cultural Gerontology* 22:205–20.

Helm, Hughes, Judith Hays, Elizabeth Flint, Harold Koenig, and Dan Blazer. 2000. "Does Private Religious Activity Prolong Survival? A Six-year Follow-up Study of 3,851 Older Adults." *Journal of Gerontology: Medical Sciences:* 55A:M400–05.

Hill, Terrence, Jacqueline Angel, Christopher Ellison, and Ronald Angel. 2005. "Religious Attendance and Mortality: An 8-year Follow-Up of Older Mexican Americans." *Journal of Gerontology: Social Sciences* 60B:S102–9.

Hill, Terrence, Amy Burdette, Jacqueline Angel, and Ronald Angel. 2006. "Religious Attendance and Cognitive Functioning among Older Mexican Americans." *Journal of Gerontology: Psychological Sciences* 61B:P3–9.

Hill, Terrence, Christopher Ellison, Amy Burdette, and Marc Musick. 2007. "Religious Involvement and Healthy Lifestyles: Evidence from the Survey of Texas Adults." *Annals of Behavioral Medicine* 34:217–22.

Hummer, Robert, Richard Rogers, Charles Nam, and Christopher Ellison. 1999. "Religious Involvement and U.S. Adult Mortality." *Demography* 36:273–85.

Idler, Ellen. 1987. "Religious Involvement and the Health of the Elderly: Some Hypotheses and an Initial Test." *Social Forces* 66:226–38.

———. 2004. "Religious Observance and Health: Theory and Research." Pp. 20–43 in *Religious Influences on Health and Well-Being in the Elderly*, edited by K. Schaie, N. Krause, and A. Booth. New York: Springer.

Idler, Ellen and Stanislav Kasl. 1997a. "Religion among Disabled and Nondisabled Persons I: Cross-Sectional Patterns in Health Practices, Social Activities, and Well-Being." *Journal of Gerontology: Social Sciences:* 52B:S294–S305.

———. 1997b. "Religion among Disabled and Nondisabled Persons II: Attendance at Religious Services as Predictor of the Course of Disability." *Journal of Gerontology: Social Sciences:* 52B:S306–16.

Idler, Ellen, Marc Musick, Christopher Ellison, Linda George, Neal Krause, Marcia Ory, Kenneth Pargament, Lynda Powell, Lynn Underwood, and David Williams. 2003. "Measuring Multiple Dimensions of Religion and Spirituality for Health Research: Conceptual Background and Findings from the 1998 General Social Survey." *Research on Aging* 58:64–74.

Idler, Ellen, Julie McLaughlin, and Stanislav Kasl. 2009. "Religion and the Quality of Life in the Last Year of Life." *Journal of Gerontology: Social Sciences* 64B:S528–37.

Ironson, Gail, George Solomon, Elizabeth Balbin, Conall O'Cleirigh, Annie George, Mahendra Kumar, David Larson, and Teresa Woods. 2002. "The Ironson-Woods Spirituality/ Religiousness Index is associated with Long Survival, Health Behaviors, Less Distress, and Low Cortisol in People with HIV/AIDS." *Annals of Behavioral Medicine* 24:34–48.

Jang, Yuri, Amy Borenstein-Graves, William Haley, Brent Small, and James Mortimer. 2003. "Determinants of a Sense of Mastery in African American and White Older Adults." *Journal of Gerontology: Social Sciences* 58B:S221–4.

Jang, Yuri, Amy Borenstein, David Chiriboga, and James Mortimer. 2005. "Depressive Symptoms among African American and White Older Adults." *Journal of Gerontology: Psychological Sciences* 60B:P313–19.

Kelley-Moore, Jessica and Kenneth Ferraro. 2001. "Functional Limitations and Religious Service Attendance in Later Life: Barrier and/or Benefit Mechanism." *Journal of Gerontology: Social Sciences* 56B:S365–73.

King, Dana, Arch Mainous, Terrence Steyer, and William Pearson. 2001. "The Relationship between Attendance at Religious Services and Cardiovascular Inflammatory Markers." *International Journal of Psychiatry in Medicine* 31:415–25.

King, Dana, Arch Mainous, and William Pearson. 2002. "C-reactive Protein, Diabetes, and Attendance at Religious Services." *Diabetes Care* 25:1172–6.

King, Sharon, Elisabeth Burgess, Makungu Akinyela, Margaret Counts-Spriggs, and Nesonya Parker. 2005. "'Your Body is God's Temple': The Spiritualization of Health Beliefs in Multigenerational African American Families." *Research on Aging* 27:420–46.

Klemmack, David, Lucinda Roff, Michael Parker, Harold Koenig, Patricia Sawyer, and Richard Allman. 2007. "A Cluster Analysis Typology of Religiousness/Spirituality among Older Adults." *Research on Aging* 29:163–83.

Kobayashi, Karen, Denise Cloutier-Fisher, and Marilyn Roth. 2009. "Making Meaningful Connections: A Profile of Social Isolation and Health among Older Adults in Small Town and Small City, British Columbia." *Journal of Aging and Health* 21:374–97.

Koenig, Harold, Mona Smiley, and Jo Ann Gonzales. 1988. *Religion, Health, and Aging: A Review and Theoretical Integration*. Westport, CT: Greenwood.

Koenig, Harold, Harvey Cohen, Linda George, Judith Hays, David Larson, and Dan Blazer. 1997. "Attendance at Religious Services, Interleukin-6, and other Biological Parameters of Immune Function in Older Adults." *International Journal of Psychiatry in Medicine* 27:233–250.

Koenig, Harold, Linda George, Harvey Cohen, Judith Hays, David Larson, and Dan Blazer. 1998a. "The Relationship between Religious Activities and Cigarette Smoking in Older Adults." *Journal of Gerontology: Medical Sciences*: 53A:M426–34.

Koenig, Harold, Linda George, Judith Hays, David Larson, Harvey Cohen, and Dan Blazer. 1998b. "The Relationship between Religious Activities and Blood Pressure in Older Adults." *International Journal of Psychiatry in Medicine* 28:189–213.

Koenig, Harold, Judith Hays, David Larson, Linda George, Harvey Cohen, Michael McCullough, Keith Meador, and Dan Blazer. 1999. "Does Religious Attendance Prolong Survival? A Six-year Follow-Up Study of 3,968 Older Adults." *Journal of Gerontology: Medical Sciences:* 54A:M370–6.

Koenig, Harold, Michael McCullough, and David Larson. 2001. *Handbook of Religion and Health.* New York: Oxford University Press.

Koenig, Harold and Robert Brooks. 2002. "Religion, Health and Aging: Implications for Practice and Public Policy." *Public Policy and Aging Report* 12:13–19.

Koenig, Harold, Linda George, and Patricia Titus. 2004. "Religion, Spirituality, and Health in Medically Ill Hospitalized Older Patients." *Journal of the American Geriatrics Society* 52: 554–62.

Koenig, Harold. Forthcoming. "Personality and Personality Disorder." *Handbook of Religion and Health,* 2nd Ed. New York: Oxford University Press.

Krause, Neal. 1992. "Stress, Religiosity, and Psychological Well-Being among Older Blacks." *Journal of Aging and Health* 4:412–39.

———. 1993. "Measuring Religiosity in Later Life." *Research on Aging* 15:170–97.

———. 1998. "Neighborhood Deterioration, Religious Coping, and Changes in Health during Late Life." *The Gerontologist* 38:653–64.

———. 2003. "Religious Meaning and Subjective Well-Being in Late Life." *Journal of Gerontology: Social Sciences:* 58B:S160–70.

———. 2003b. "Race, Religion, and Abstinence from Alcohol in Late Life." *Journal of Aging and Health* 15:508–33.

———. 2005. "God-Mediated Control and Psychological Well-Being in Late Life." *Research on Aging* 27:136–64.

———. 2006a. "Gratitude toward God, Stress, and Health in Late Life." *Research on Aging* 28:163–83.

———. 2006b. "Church-Based Social Support and Mortality." *Journal of Gerontology: Social Sciences* 61B:S140–6.

———. 2008. *Aging in the Church: How Social Relationships Affect Health.* West Conshohocken, PA: Foundation Press.

Krause, Neal, Berit Ingersoll-Dayton, Christopher Ellison, and Keith Wulff. 1999. "Aging, Religious Doubt, and Psychological Well-Being." *The Gerontologist* 39:525–33.

Krause, Neal, Jersey Liang, Benjamin Shaw, Hidehiro Sugisawa, Hye-Kyung Kim, and Yoko Sugihara. 2002. "Religion, Death of a Loved One, and Hypertension among Older Adults in Japan." *Journal of Gerontology: Social Sciences* 57B:S96–S107.

Krause, Neal and Christopher Ellison. 2009. "The Doubting Process: A Longitudinal Study of the Precipitants and Consequences of Religious Doubt in Older Adults." *Journal for the Scientific Study of Religion* 48:292–312.

la Cour, Peter, Kirsten Avlund, and Kirsten Schultz-Larsen. 2006. "Religion and Survival in a Secular Region. A Twenty Year Follow-Up of 734 Danish Adults Born in 1914." *Social Science & Medicine* 62:157–64.

Law, Rita and David Sbarra. 2009. "The Effects of Church Attendance and Marital Status on the Longitudinal Trajectories of Depressed Mood among Older Adults." *Journal of Aging and Health* 21:803–23.

Levin, Jeffrey, Kyriakos Markides, and Laura Ray. 1996. "Religious Attendance and Psychological Well-Being in Mexican Americans: A Panel Analysis of Three Generations Data." *The Gerontologist* 36:454–63.

Lutgendorf, Susan, Daniel Russell, Philip Ullrich, Tamara Harris, and Robert Wallace. 2004. "Religious Participation, Interleukin-6, and Mortality in Older Adults." *Health Psychology* 23:465–75.

Mahoney, Annette, Robert Carels, Kenneth Pargament, Amy Wachholtz, Laura Leeper, Mary Kaplar, and Robin Frutchey. 2005. "The Sanctification of the Body and Behavioral Health Patterns of College Students." *The International Journal of the Psychology of Religion* 15:221–8.

McIntosh, William, Dianne Sykes, and Karen Kubena. 2002. "Religion and Community among the Elderly: The Relationship between Religious and Secular Characteristics of their Social Networks." *Review of Religious Research* 44:109–25.

McCullough, Michael and Brian Willoughby. 2009. "Religion, Self-Regulation, and Self-Control: Associations, Explanations, and Implications." *Psychological Bulletin* 135:69–93.

Musick, Marc. 1996. "Religion and Subjective Health among Black and White Elders." *Journal of Health and Social Behavior* 37:221–37.

Musick, Marc, James House, and David Williams. 2004. "Attendance at Religious Services and Mortality in a National Sample." *Journal of Health and Social Behavior* 45:198–213.

Norton, Maria, Archana Singh, Ingmar Skoog, Christopher Corcoran, JoAnn Tschanz, Peter Zandi, John Breitner, Kathleen Welsh-Bohmer, and David Steffens. 2008. "Church Attendance and New Episodes of Major Depression in a Community Study of Older Adults: The Cache County Study." *Journal of Gerontology: Psychological Sciences* 63B:P129–37.

Oman, Douglas and Dwayne Reed. 1998. "Religion and Mortality among the Community-Dwelling Elderly." *American Journal of Public Health* 88:1469–75.

Oman, Douglas, John Kurata, William Strawbridge, and Richard Cohen. 2002. "Religious Attendance and Cause of Death Over 31 Years." *International Journal of Psychiatry in Medicine* 32:69–89.

Park, Nan Sook, David Klemmack, Lucinda Roff, Michael Parker, Harold Koenig, Patricia Sawyer, and Richard Allman. 2008. "Religiousness and Longitudinal Trajectories in Elders' Functional Status." *Research on Aging* 30:279–98.

Pargament, Kenneth. 1997. *The Psychology of Religion and Coping.* New York: Guilford.

Pargament, Kenneth, Harold Koenig, Nalini Tarakeshwar, and June Hahn. 2001. "Religious Struggle as a Predictor of Mortality among Medically Ill Elderly Patients: A 2-year Longitudinal Study." *Archives of Internal Medicine* 161:1881–5.

Reyes-Ortiz, Carlos, Ivonne Berges, Mukaila Raji, Harold Koenig, Yong-Fang Kuo, and Kyriakos Markides. 2008. "Church Attendance Mediates the Association between Depressive Symptoms and Cognitive Functioning among Older Mexican Americans." *Journal of Gerontology: Medical Sciences* 63A:480–6.

Schieman, Scott, Tetyana Pudrovska, and Melissa Milkie. 2005. "The Sense of Divine Control and the Self-Concept." *Research on Aging* 27:165–96.

Seeman, Teresa, Linda Dubin, and Melvin Seeman. 2003. "Religiosity/Spirituality and Health: A Critical Review of the Evidence for Biological Pathways." *American Psychologist* 58:53–63.

Seybold, Kevin. 2007. "Physiological Mechanisms Involved in Religiosity/Spirituality and Health" *Journal of Behavioral Medicine* 30:303–9.

Strawbridge, William, Richard Cohen, Sarah Shema, and George Kaplan. 1997. "Frequent Attendance at Religious Services and Mortality over 28 Years." *American Journal of Public Health* 87:957–61.

Strawbridge, William, Sarah Shema, Richard Cohen, Robert Roberts, and George Kaplan. 1998. "Religiosity Buffers Effects of Some Stressors on Depression but Exacerbates Others." *Journal of Gerontology: Social Sciences* 53B:S118–26.

Van Ness, Peter and Stanislav Kasl. 2003. "Religion and Cognitive Dysfunction in an Elderly Cohort." *Journal of Gerontology: Social Sciences:* 58B:S21–9.

Van Ness, Peter, Virginia Towle, John O'Leary, and Terri Fried. 2008. "Religion, Risk, and Medical Decision Making at the end of Life." *Journal of Aging and Health* 20:545–59.

Welch, Michael, Charles Tittle, and Harold Grasmick. 2006. "Christian Religiosity, Self-Control and Social Conformity." *Social Forces* 84:1605–23.

Wink, Paul, Michele Dillon, and Britta Larsen. 2005. "Religion as a Moderator of the Depression-Health Connection: Findings from a Longitudinal Study." *Research on Aging* 27:197–220.

Wolinsky, Frederic, Suzanne Bentler, Elizabeth Cook, Elizabeth Chrischilles, Li Liu, Kara Bright, John Geweke, Maksym Obrizan, Claire Pavlik, Robert Ohsfeldt, Michael Jones, Robert Wallace, and Gary Rosenthal. 2009. "A 12-year Prospective Study of Stroke Risk in Older Medicare Beneficiaries." *BMC Geriatrics* 9:1–11.

Part IX
Care Arrangements and Aging

Chapter 34
Civil Society and Eldercare in Posttraditional Society

Ronald J. Angel

My maternal grandmother died when I was very young so I never knew her well. What I remember most clearly about her, though, is that whenever I saw her she was living in one of my aunts' homes. For a while she lived with us. My grandfather built the house in which my grandmother gave birth to thirteen children, many of whom died in infancy, and in which she raised the survivors once she became a young widow. Without a husband she raised her children by herself with whatever income she could earn, primarily as a janitor at the local school. When the children had grown and left home and when grandmother became too frail to take care of herself, my parents, my sister and I moved into the old house with her. Four of my aunts lived nearby. Even though they grew up in poverty my mother and three of her sisters managed to attend university and become teachers. They were ahead of their time in terms of modern gender roles since they worked full-time while raising children and taking care of their own homes, but they were traditional in that they combined efforts to care for their mother, even as they continued to work and raise their own families. After several years of living with one or another of her daughters, my grandmother died in her own bed at the age of 87 in the house in which she had raised her children.

Unlike my grandmother, when my mother died at the age 96 it was in a long-term care facility far from the city in which she was born. Since her only daughter had died at 50, she had only one surviving son. Although she was far more financially secure than her own mother, she had no large group of family caregivers who could combine eldercare with work and other family responsibilities. Formal long-term care was the only realistic option. In only one generation, then, my family had changed dramatically in terms of its eldercare capacity. But my family's story is not unique. Changes in family size, the migration of children away from their place of birth, longer lives, and much more have changed the caregiving capacities of families generally. Today, with clear exceptions, the family is less able to provide all of the care that frail elderly parents need. Other alternatives must be explored. In this chapter I investigate the possibilities for civil society to provide at least some of the day-to-day care that the family can no longer provide. By civil society I refer to the wide range of nongovernmental organizations (NGOs), from faith-based organizations and community groups to formal charitable organizations like Meals on Wheels. In the modern world such organizations are becoming an increasingly important part of the social world and in the future they may assume a greater role in eldercare.

For individuals born at the end of the nineteenth or early in the twentieth centuries the family was the major source of care and support when they became old and frail. With the passage of the Social Security Act in 1936 and the introduction of Medicare in 1965 the State has assumed a large part of the financial and long-term care responsibility for elderly citizens. Income supports and

R.J. Angel (✉)
Department of Sociology, University of Texas at Austin, Austin, TX, USA
e-mail: rangel@austin.utexas.edu

R.A. Settersten, Jr. and J.L. Angel (eds.), *Handbook of Sociology of Aging*, Handbooks of Sociology and Social Research, DOI 10.1007/978-1-4419-7374-0_34, © Springer Science+Business Media, LLC 2011

medical care consume a large fraction of the budgets of most developed nations. Yet the State faces limits in what it can provide in terms of instrumental care. Many of the needs of frail elders are routine and personal and require close and frequent contact. This sort of care is most easily and efficiently provided by family members. The state might pay for such services, but there are practical limits as to what the State can afford. In this new context organized voluntary action represents at least a partial solution. The challenge for public policy is to determine how best to combine the efforts of nongovernmental actors and organizations with those of formal governmental programs.

The New Demographic and Social Reality

Today longer life spans, migration, lower fertility, increased female labor force participation and more have contributed to a rapidly changing long-term care reality for the elderly (Haber 2006; Quadagno 1982; Thane 2005). During the nineteenth century fertility declined significantly in the United States, but families remained large by today's standards even at the turn of the twentieth century when the average female had nearly four children (Coale and Zelnik 1963; Himes 1989; Tolnay et al. 1982). Family members for the most part stayed close to home and older parents could call upon at least some of their children who lived nearby to provide basic care when they became frail. Nobody, or almost nobody, entered a long-term care facility other than perhaps a poorhouse or an insane asylum (Grob 1983; Wagner 2005). That remains true today in most parts of the world, although the situation is changing even in countries like India where the number of long-term care facilities is increasing (Liebig 2003).

The reality of the situation, then, is that given the demographic and social changes that have occurred over the last two generations traditional family caregivers find it increasingly difficult to combine eldercare with the demands of work, child rearing, and other duties. Today in Western Europe fertility rates have reached historic lows (Sobotka 2006). In Italy, Spain, and other European countries where almost everyone married and had large families just a generation ago, fertility rates among the native populations are below replacement. The fertility rate in Italy and Spain is barely 1.3, far below the 2.1 that is necessary to assure a stable population (Central Intelligence Agency 2009). In the absence of immigration these countries would have shrinking populations. In the United States the total fertility rate has remained higher, at about 2.0 since the 1980s. To a large degree this reflects the higher fertility of Hispanics (Hamilton et al. 2009). These differential fertility rates suggest significant ethnic and racial group differences in family eldercare capacity in years to come (Angel and Angel 1997).

Growing older populations and lower fertility even affect the developing world. South American populations are aging rapidly and will continue to age well into the twenty-first century (Chamie 2005; Palloni et al. 2002). This growth in older populations in both developed and developing countries will in all likelihood be accompanied by a continuing fiscal austerity that will strain the ability of governments to provide for all of their citizens' needs (Pierson 2001). This new demographic reality of aging populations and lower fertility, in combination with other social transformations such as increasing divorce rates and the decline of the traditional family requires novel approaches to the care of older adults. For most of human history tradition has largely determined who would care for the frail elderly, but tradition no longer provides firm answers and we are in a new world that will require new institutional arrangements including a greater role for civil society and NGOs. Understanding how nonfamily and nongovernmental sources of elder support might be employed in eldercare represents a core intellectual and practical question for future research and policy design.

Posttraditional Society and Old Age Support Systems

We might usefully ask what the term "tradition" refers to and how has it changed with reference to the care of the elderly. Tradition lies at the core of human social interaction and defines our major social institutions like marriage and the family. Tradition, often codified in law, provides the rules that determine inheritance and dictate the nature and extent of responsibility among generations. Even today tradition dictates that family members assume primary responsibility for the care of aging parents (Bullock et al. 2003; Chatters et al. 1986; Dilworth-Anderson et al. 1999; Jolicoeur and Madden 2002; Stone 2006). In the absence of informal family care greater public expenditures for formal care would be necessary (Bolin et al. 2007; Hayward et al. 2004; Himes and Reidy 2000). Yet when basic structures change radically traditions cannot remain unchanged. Given limited time and energy budgets, there may be an inevitable tradeoff for potential caregivers between caring for older family members and other commitments, including employment. Some studies find that employed women provide less informal care than unemployed women (Fast et al. 1999; Kotsadam 2009). Caregiving responsibilities can interfere with the ability of caregivers to look for work (Bullock et al. 2003).

Tradition, by its very nature resists change since it is undone by change. But change occurs regardless of tradition's dictates. Today in the developed world we are living in what Anthony Giddens has described as "posttraditional" society (Giddens 1991, 1994). Giddens and others use this term to emphasize the fact that traditional norms and practices have weakened and that individuals today cannot, or have the option of choosing not to, look to tradition for the basic instructions of life. Traditions clearly still exist and remain powerful. Because of the traditional social value placed in the institution or marriage, as well as other practical advantages, same-sex couples wish to partake of the tradition. The fact that marriage can be construed as a union between same-sex individuals underscores the reality that the rules governing the social institution of marriage are in the process of change. Today individuals choose and reinterpret traditions to reflect changing circumstances and newer social realities. Daughters and daughters-in-law are not constrained by strong community expectations to stay at home to care for their aging parents or in-laws.

Traditions and norms do not exist independent of history, and in the domain of eldercare history has moved on. Structurally, though, there remain a limited number of possibilities in terms of the long-term care of the elderly. Financial, instrumental, and emotional support can be provided by some combination of the family, the State, and civil society. By civil society I refer to the entire range of nongovernmental groups and organizations that provide support and services to older individuals. These include faith-based organizations and congregations, NGOs like On Lok in San Francisco, HelpAge India, and many others, some of which I mention below. Unfortunately, there is little systematic research on the role of NGOs and other civil society organizations in the support of the elderly. In what follows, then, I offer a theoretical framework for understanding the potential role of these organizations in eldercare and review the relatively sparse empirical data available. I conclude by proposing a research agenda focused on the potential role of such organizations in the mix of care providers. The core question relates to the extent to which civil society organizations, including NGOs, can complement the State in providing support without further eroding the role of the family or contributing to the State's neglect of its basic responsibility for guaranteeing the basic social rights of vulnerable elderly citizens.

Addressing this question requires an examination of the role of civil society organizations in terms of the mix of elderly support services. Despite the changes the family has undergone family members continue to provide much of the care older parents need. One might justifiably conclude that the family is the most desirable source of care given the lifelong emotional bonds that exist between parents and children. As desirable as family care might be, though, in posttraditional society children, and especially those with higher levels of education, move away from their parents for career reasons and cannot assume the daily responsibility for dealing with their parent's immediate needs (Rogerson et al. 1993). Adult children can, of course, provide financial support, and if they or their parents can afford the cost they may chose to move their aging parents to assisted living arrangements nearer by (Silverstein 1995; Silverstein and Angelelli 1998). For adult children with demanding careers having an older parent move in to their homes, especially once the parent suffers serious functional and cognitive decline, may not be realistic.

In addition to the family, or perhaps in place of the family, today the State assumes a major role in the financial care of the elderly. Public pensions, health care, and other supports for the elderly largely define the welfare state. In the United States Social Security, Medicare, and Medicaid have basically socialized the financial responsibility for the elderly. What remains, then, is to examine the actual and potential roles of the nongovernmental and voluntary sector in eldercare.

The New Focus on Nongovernmental Solutions to Social Problems

In order to begin to develop a better sense of the role of NGOs and similar organizations in eldercare it is necessary to briefly review the growing importance of NGOs in the modern world. Although NGOs have been involved in various social issues since the nineteenth century and even earlier, the period since the Second World War has seen a rapid growth in their number. Today NGOs are involved in health, education, human rights, women's issues, the environment, sustainable development, and much more (Boli and Thomas 1997, 1999; Carroll 1992; Hudock 1999; Keck and Sikkink 1998; Robinson and Riddell 1995; Salamon 2003; Salamon and Anheir 1996). They have assumed an important and growing role as advisors to the United Nations (Global Policy Forum 2010). Many explanations for this growth have been offered and a large literature focused on NGOs as expressions of new social movements has appeared (Albrow et al. 2008; Boli and Thomas 1999; Keck and Sikkink 1998; Pichardo 1997). Some observers view the rapid growth of NGOs in the post-WWII era as a reflection of the spread of a global culture that is supranational in character and largely Western in values (Boli and Thomas 1997). Others see it as reflecting the global demand by poor countries for greater global social justice and a new collective consciousness (Bendaña 2006; Grzybowski 2000). Whatever the causes of the increase in the number and role of NGOs they are more salient actors internationally and locally and their presence calls for a more complex approach to understanding how society operates at all levels (Fisher 1997; Ruggie 1998).

The domains in which NGOs and other civil society organizations operate differ greatly along many dimensions. NGOs that deal with environmental issues and those dealing with human rights, women's rights, and the rights of native peoples differ in mission and structure from those that provide medical, educational, and emergency services to specific groups. Understanding the potential utility of such organizations requires not only a focus on their structure and organization, but also an understanding of the needs of the populations they serve. In terms of addressing the needs of the elderly the relevant questions relate to how well such organizations function as advocates and service providers.

NGO activities can be characterized in terms of two broad categories, advocacy and service delivery (Pereira and Angel 2009). Advocacy refers to activities that are aimed at improving the situation of a particular group though legislation, public awareness campaigns, and other activities that draw attention to the structural and system factors that produce and perpetuate disadvantage. Organizations like Amnesty International that focus on human rights or organizations like Greenpeace that oppose the exploitation of animal species engage primarily in advocacy. Other organizations like Habitat for Humanity, Doctors without Borders, Oxfam, and the Red Cross focus more on service delivery and the response to crises. Many, if not most, NGOs engage in both activities to one degree or another, providing services while they engage in advocacy and political action aimed at changing laws and practices to encourage democracy, safeguard civil rights, or develop local capacities to deal with longer-term needs.

One of the reasons for the renewed interest in NGOs as service providers is that they often have certain advantages in dealing with the more routine and manageable needs of specific populations (Pereira and Angel 2009; Pereira et al. 2007). For example, although complex and expensive high-tech medicine can only be paid for or provided by the State, routine and relatively inexpensive services, such as basic primary care, assistance with activities of daily living, and companionship are often more effectively and economically provided by NGOs and other local groups. Ideally in the case of eldercare such groups can complement the State in supporting older individuals in the community and enhancing the quality of their lives. Because of their more detailed knowledge of the legal, transportation, nutritional, and other needs of the elderly they could potentially act as both service providers and advocates.

Relatively little research has been carried out focused on the role of civil society organizations in eldercare. A large body of research documents the importance of social networks and social support in later life and offers indirect evidence that nonfamily support could be important in assuring the well-being of isolated older persons (Krause 2006; Moren-Cross and Lin 2006). It is clear that social engagement and human contact are vital to the well-being of the elderly and the potential of nonfamily and nongovernmental sources to provide such contact may be high. Studies of volunteerism, including a recent study in Mexico, reveal that in situations of high need many individuals provide significant amounts of support and assistance to needy neighbors and others either through formal groups, such as churches or voluntary organizations, or informally as personal assistance (Burcher 2008).

Service to the Elderly

The organizations and affiliative groups that make up civil society not only help individuals in need, but they also provide their members opportunities to engage in pro-social activities ultimately strengthening communities (Etzioni 1993, 1995). Churches and religious groups have historically cared for the poor and infirm and as part of that mission they provide much needed companionship and care to the elderly (Idler 2006). Faith-based international NGOs such as *CARITAS, Catholic Charities, and Lutheran Social Services* provide assistance to the elderly as part of their general missions. The Red Cross and many of the other major international and local relief agencies identify the elderly as a vulnerable population in need of special attention in their relief and support missions. A quick perusal of NGO directories on the Web yields hundreds of such organizations in every country with some focus on the elderly as a vulnerable population.

In addition to NGOs that offer assistance to elderly persons with special needs or in times of crisis, many other international, national, and local organizations of differing sizes and reach provide care to the elderly. Again their number is too large for even a partial enumeration but some examples are informative. What follows is clearly not a systematic sample of NGOs that deal with the elderly. Rather the cases I present represent very well-known organizations or those that illustrate the point that eldercare and issues related to the elderly are drawing greater attention from civil society organizations. For the elderly, of course, companionship and visitation are important needs that volunteers can readily provide. Companionship and the prevention of isolation are major objectives of these organizations.

The *Meals on Wheels Association of America* (http://www.mowaa.org) is the oldest and best known nongovernmental nutrition program for the elderly in the country. Founded during World War II the organization has grown in size and mission and today is dedicated to ending hunger among the elderly. In addition to providing nutritious meals, volunteers provide clearly important human contact to older persons. The less well-known *Little Brothers – Friends of the Elderly (LBFE)* (http://www.littlebrothers.org), an international network of nonprofit, volunteer-based organization with branches in the U.S. provides companionship to elderly people to reduce isolation and loneliness. The organization is a member of a larger international organization, the Fédération Internationale des petits frères des Pauvres (International Federation of Little Brothers of the Poor, http://www.petitsfreres.org). Another international organization is the *Fédération Internationale des Associations de Personnes Agées* (International Federation of Associations of Older Persons, FIAPA: http://www.fiapa.org) headquartered in Paris. The organization's mission is also to prevent isolation and improve the quality of life (QOL) for older individuals.

Many other examples of specific NGO activities in various countries are easily found on the internet. Three examples illustrate the potential of NGOs in eldercare in less developed countries. India, like most of the rest of the world, is facing a serious problem related to the care of an older population. Even as developing nations with high fertility rates remain comparatively young, their older populations are growing in absolute and relative size. In India NGOs are important advocates for and service providers to the elderly (Sawhney 2003). *Dignity Foundation* (http://www.dignityfoundation.com), a member of the American Association of Retired Persons (AARP) Global Network, provides housing, companionship, recreation, and other services to elderly individuals in several Indian cities. *HelpAge India* (http://www.helpageindia.org) has a similar service mission. This NGO provides financial, medical, and emotional support to poor elderly Indians. The organization has introduced new programs and is extending its services to previously underserved areas. One example highlighted on the organization's website is a Mobile Medicare Unit (MMU) program that provides basic health care and is introducing new initiatives such as disability aids, shelter assistance, yoga, specialized home visits, and psychological therapy.

The cases of Dignity Foundation and HelpAge India are examples of eldercare NGOs moving into resource starved areas in which formal supports are rare. Another example in a nation that is far more developed relates to the eldercare functions of *Hogar de Cristo* (Christ's Home: http://www.hogardecristo.cl) in Chile (Pereira et al. 2007). Hogar de Cristo is a Catholic organization in a highly Catholic country, a fact that no doubt has contributed to its success. Begun in 1944 by a Catholic priest named Alberto Hurtado, from its founding the organization has focused on the needs of poor Chileans. Given the specific vulnerabilities of the elderly, especially the seriously curtailed social services that were part of the neoliberal reforms introduced by the Pinochet dictatorship, Hogar's mission has expanded to provide the full range of services to poor elderly individuals. These services include day care, nutritional programs, and even housing. In the absence of an adequate old-age welfare state Hogar de Cristo fills a void that is created by limited government commitments or capacities to address serious social problems.

In the United States a similar comprehensive care approach named *On Lok*, a Cantonese term which means "peaceful, happy abode," was begun in the early 1970s in San Francisco to provide services to frail Asian elderly individuals in certain Bay area communities in order to allow them to remain in their own homes (http://www.onlok.org) (Bodenheimer 1999). This success of the program led to its formal adoption by Congress as a model for the *PACE* program (Program of All-inclusive Care for the Elderly: http://www.cms.hhs.gov/pace), which provides comprehensive services paid for primarily by Medicare and Medicaid to high-need frail elderly individuals (Gross et al. 2004). The On Lok experience serves as an example of how private nongovernmental initiatives can serve as laboratories in which best practices related to the care of older persons can be tried and eventually inform State initiatives. Currently 70 PACE programs employ interdisciplinary teams of care providers who develop care plans for each individual and monitor their progress with the objective of allowing them to enjoy the highest possible QOL. In addition to primary care the programs offer specialist care, home health aides, transportation, recreation and companionship (Gross et al. 2004).

Hundreds of examples similar to those I have presented can be found in all nations of the world and it would be impossible to summarize the activities of even a few. As the PACE example shows, there is often a blurring of the definition of nongovernmental. PACE programs rely heavily on Medicaid and Medicare for financing their operations. The category of nongovernmental, therefore, includes many different degrees of government/civil society cooperation. As I mentioned before, in addition to organizations focused specifically on the needs of the elderly, many other nongovernmental and faith-based groups include assistance to the elderly as part of their missions. Service delivery, though, is not the only mission of NGOs. Adequate pensions and other legal guarantees require changing laws and that objective requires different approaches and organization than short-term crisis interventions.

Advocacy for the Elderly

As important as services are for vulnerable elderly individuals especially in developing nations, basic assistance does not change the fundamental vulnerabilities that undermine the welfare of older persons. While basic assistance with food, medical care, and housing might alleviate some of the most immediate problems that older individuals face they are no substitutes for more comprehensive and continuous social security programs (HelpAge International 2009; Willimore 2006). In addition to service delivery, then, a major role of NGOs is advocacy for the elderly. In the United States the *AARP* is undoubtedly the most well-known and effective advocate for its membership (Binstock 2004). Other advocacy organizations include the *National Committee to Preserve Social Security and Medicare* (NCPSSM: http://www.ncpssm.org/), the *Alliance for Retired Americans* (ARA: http://www.retiredamericans.org), and the *National Hispanic Council on Aging* (NHCOA: http://www.nhcoa.org/).

According to its Web page NHCOA's mission includes advocacy, the support of research, the funding of community-based projects, as well as the creation of support networks, capacity-building in Hispanic communities and the support and strengthening of Hispanic community-based organizations. The organization's core objective is to "empower Hispanic community organizations and agencies, as well as Hispanic older adults and their families."

The organization offers educational programs focused on the major health risks to Hispanics, like diabetes and it has developed an e-course on cultural competence that educates health care professionals concerning the culture of their patients (http://edu.nhcoa.org).

Smaller local organizations, such as Family Eldercare of Austin, Texas (http://www.familyeldercare.org), the author's home town, provide important legal services and perform what are basically case-management services in coordinating a wide range of services that the organization's poor and largely minority clientele needs in terms of housing, instrumental support, and more. The organization participates in a summer fan drive which collects fans and money to purchase them to provide vital cooling to older people who live in a part of the country in which the heat of summer can be life threatening. These activities are replicated in various forms by any number of NGOs all over the country.

Potential Downsides to Nongovernmental Approaches

It is quite possible, of course, to romanticize civil society and nongovernmental approaches to complex social problems or to overestimate the ability of voluntary collective action to empower the powerless. Local approaches to the care and support of vulnerable individuals are appealing because they appear to be more personal and on more human scale than services provided by large impersonal bureaucracies. The new Neoliberalism that has informed public policy in recent decades reflects a deep distrust of big government and the welfare state, a large part of which consists of old-age supports (Binstock 2004). Although support for public pensions remains high in most countries, fiscal constraints and the recent economic crisis have furthered efforts to control the growth of such public programs (Chap. 20). Thatcherism in Great Britain, the Regan revolution in the United States, and new Third Way philosophies in other developed nations reflect a growing anti-welfare state rhetoric and the adoption of policies intended to cap or even reduce social expenditures and the paternalistic role of the state (Andrews 1999; Faux 1999). Public antiwelfare state sentiment and neoliberal social and economic policies have serious implications for the material and social welfare of the elderly, and especially those with limited resources including the elderly, women, African-Americans, and Hispanics (Angel and Angel 1997, 2009).

There can be little doubt that the elderly benefit from local programs that provide companionship and assistance with activities of daily living. The On Lok model and the PACE program clearly demonstrate the benefits of intensive involvement in providing frail older individuals with limited resources the full range of services and supports they need in order to remain in the community. Yet there are serious dangers in placing too much faith in the ability of local communities and organizations to provide all of the supports that older citizens need. Social Security and Medicare are fundamental and indispensable support programs without which millions of older individuals would fall into poverty and be denied the medical care they need. As we attempt to mobilize family and local resources for the care of the elderly we confront the very real danger that the supposed superiority of nongovernmental approaches will be used to justify excessive and harmful cuts in Social Security, Medicare, and Medicaid. Among the big losers would be single and poor older African-American and Hispanic females and other vulnerable individuals. The life-long labor force disadvantages faced by minority Americans leaves them particularly dependent on publicly funded programs in old age. If an unrealistic belief in the support capacities of local organizations and community groups were to encourage reductions in funding for formal services the results could be devastating.

As appealing and noncontroversial as civil society approaches to dealing with human needs might seem at first glance, then (after all who can argue with generosity and civic engagement), there are serious criticisms of NGOs and the nongovernmental approach generally that potentially apply to civil society solutions to dealing with the elderly. In the area of development and assistance to poor countries NGOs have been criticized for often being too apolitical and unwilling or unable to deal with basic social structural factors that account for the powerlessness and exclusion of particular groups. They are seen as part of the neoliberal agenda of privatizing what are in fact public responsibilities (Clarke 1998; Sangeeta 2004; Srinivas 2009). Development NGOs are criticized for having an excessively narrow focus on specific projects and fostering dependency rather than furthering sustainable development. They are criticized for being too accountable to their funders and less concerned with the perspectives of those they supposedly intend to help.

NGOs and the civil society focus more generally are also criticized for their often narrow and excessive focus on subgroups defined in terms of gender, race, ethnicity, disability status, specific illnesses, and more. An excessively narrow focus on particular subgroups or problem areas undermines efforts to address collective political issues. Supporters of the international labor movement, for example, continue to believe that the major locus of disadvantage and struggle in the world that in principle unifies workers in developed and developing nations is that between labor and capital. From this perspective an excessive focus on gender, race, or other narrow interests undermines the possibility for the sort of collective political action necessary to better the lot of all citizens (Roman 2004). Whatever the merits of the various criticisms of a civil society approach in general it seems clear that the core danger to the interests of the elderly lies in the potential justification of the devolution of federal responsibilities for income support and health care or drastic reductions in funding for such programs on the basis of the supposed superiority of local nongovernmental approaches.

A Research Agenda Focused on Civil Society and Eldercare

NGOs have become major international players in developing and implementing environmental policy, furthering women's rights, redefining development policy, and much more. They are major players at local levels in dealing with problems of poverty, homelessness, teenage pregnancy among others. In this chapter I have posed the question as to whether civil society organizations, including NGOs might meaningfully supplement the family and the State in addressing the needs of growing older populations. It is clear that solutions to the major social problems that developed and developing societies face are extremely complex and that civil society is at best a complement to the State, which remains as important as ever in guaranteeing basic political and social rights to all citizens. Indeed it is probably not possible for NGOs to operate at all in the absence of an effective and strong State. Unfortunately, there is little published empirical research on the potential effectiveness of civil society activity in providing services to or advocating for elderly groups. This lack of research is accompanied by a lack of theory to assist in understanding the complex interaction of various layers of government, private enterprise, the family, and the nongovernmental sector in addressing the needs of aging populations.

At the most general, level the most theoretically and practically important questions relate to the actual possibilities for civil society action in eldercare and advocacy and the forms that they might take. Specifically it is important to know how the role of civil society organizations in eldercare service provision and advocacy is constrained by state policy and funding sources at multiple levels in different political, economic, and cultural contexts. Addressing these questions requires a comparative approach based on national, state, provincial, municipal, and other differences of significance in policy design and implementation. Such analyses, for example, could examine differences in elder support policies in countries like Argentina, Chile, and Mexico which differ in terms of degree of national federalization, public policies, political structures, and more. Such research could provide useful information on how levels of development interact with state policies and civil society activity to influence policy to affect the comparative welfare of the elderly. It would also answer questions as to the range of possibilities for dealing with generational equity in a time of universal fiscal austerity.

Other theoretically important questions relate to the potential of eldercare and elder advocacy to take on characteristics of a social movement that may even be transnational in character. Other questions entail the complex and difficult assessment of the potential negative aspects of civil society involvement. As I mentioned earlier, to the extent that the devolution of services and supports results in State abandonment of its basic responsibilities and shifting of those responsibilities to levels of government and the nongovernmental sectors with inadequate capacities the results may well be negative. These overarching theoretical questions call for several dissertations and serious focused theorizing and empirical research.

At a more practical level the questions of interest relate to a better understanding of how formal State supports for the elderly can be coordinated with civil society efforts to more efficiently and economically address the needs of the elderly and insure their optimal QOL. The objective of a research agenda with such a focus would be to identify the best organizational forms and best administrative practices for insuring older individuals' personal autonomy, dignity, social participation, and life satisfaction. Since the potential utility and success of civil society approaches depends on the nature of the social problems they address an assessment of the role of NGO's in elder support must begin with a detailed understanding of the clientele and its needs, including its diversity in terms of race, ethnicity, social class, gender, religious orientation, immigration status and more. Since NGOs and other civil society organizations, to the extent they are more than just casual associations, have some sort of organizational form it would be useful to ask what organizational forms are associated with the greatest success. Of course, defining success represents an important part of the research agenda in and of itself.

Let me suggest a limited set of general questions that might form part of a larger research agenda focused on NGOs as organizations and their roles as advocates for and service providers to the elderly. There are, of course, many more questions and I invite the reader to criticize these and propose others. The list implies a qualitative approach, which is necessary to understand specific organizations and their cultures and internal workings. It would be useful of course to do larger-scale comparative and quantitative analyses to compare different organizations. The questions I propose include the following:

- How significant is NGO and other civil society activity in advocacy or service provision to the elderly? How many organizations are there with significant elder agenda and how rapidly has this movement grown?
- To what extent does the entry of new organizations into the field or the redirection of the missions of older organizations toward elder advocacy and service represent a new social movement? What are the demographic, social, and political forces that might propel such a movement and how might these contribute to the definition of the objectives of specific governmental and nongovernmental actors, including those in the for-profit sector?
- How is the mission of the movement as a whole and the missions of individual organizations involved in eldercare or advocacy determined and by whom? Are the clients of these organizations passive recipients of services or are they involved in framing the mission statement, monitoring progress, and evaluating effectiveness?
- How is effectiveness defined and measured, and by whom? Are serious QOL criteria applied in the evaluation of effectiveness?
- How are individual organization's missions "framed" for presentation to larger audiences, including funders? Does the framing of issues related to eldercare and support truly reflect the needs of this population or subpopulations?
- How do the political, legal, and regulatory environments in which various civil society organizations operate affect their operation and effectiveness? This question is particularly relevant in comparative research.
- How are the target populations of specific organizations defined and identified? What sort of outreach efforts, if any, do various organizations employ? Are issues of access and payment addressed adequately?
- What are the major funding sources for different organizations? To what extent do specific organizations rely on grants, public fundraising drives, or client fees? How do different sources of funding affect accountability and the targeting of services? What is their mix of service provision and advocacy for the elderly in specific organizations? Is the advocacy role of specific organizations or the movement as a whole effective or is it constrained or co-opted by political and financial factors?
- To whom do specific organizational leaders feel the organization is accountable? How are funders' and clients' differing conceptions of what the clients need reconciled? Does the need to be accountable to funders distort the intended mission of these organizations?
- How do organizations' basic structures, in terms of information flows, gender, racial, and ethnic composition of staff and clientele, decision making, the role of board members, etc. influence the ability of organizations to identify and respond to the needs of their clienteles?
- How professionalized is the organization and the movement as a whole? What is the mix of paid staff to volunteers and how does this mix relate to the operation of the organization? Do volunteers have meaningful input into the definition of the mission and the design of service delivery?
- Does the organization communicate and coordinate its efforts with other organizations engaged in similar activities?

There are of course many additional questions that we might ask concerning all aspects of the eldercare movement and the social, political, and economic forces that determine its organizational expression.

Addressing these questions would provide useful theoretical and practical knowledge concerning the potential role of NGOs and other civil society answers in dealing with aging populations in the U.S. and elsewhere.

Conclusion

For the majority of human history change has occurred relatively slowly and one generation's life course and the traditions it followed were similar to those of generations that had come before and those that came after. That world of stability and tradition has passed in a relatively short time in historical terms. Baby boomers lives have been governed by a very different set of traditions than those that constrained their grandparent's lives. Their children's lives will no doubt be more different in ways that we cannot yet predict. It is in this context of demographic and social change that old age must be reconceptualized and new social arrangements suited to the new posttraditional order developed. Most likely that menu of options will contain many variants of traditional institutional arrangements and some groups, especially those with high fertility or those that reject what they see as the moral decay of the modern world, may continue in ways that look very similar to the past.

There is no doubt, though, that the dramatic decline in fertility and the decline of the traditional family though late marriage and divorce mean that families will simply not have the personnel to take care of aging parents, especially as lifespans increase. Some people in second and third marriages may find themselves having to deal with several aging parents at the same time. As baby boomers age they have already found out from their parent's experience that early old age can be a very independent time of life if one is in good health and has sufficient income and wealth. Older boomers, though, are already finding out that when physical and cognitive capacities decline the care burden for older parents can become overwhelming. Some arrangement other than caring for those parents oneself often becomes necessary. Assisted living in one form or another is a possibility if a family has sufficient resources. With the high cost of assisted living that option eludes many families. For those families Medicaid will become even more important than it is today and the questions addressed in this chapter become especially salient: How can civil society organizations complement Medicaid-financed community or institutional care to optimize the QOL for frail older individuals with limited resources while not breaking the bank? There are probably several answers to the basic question, but the only way to begin to address it is to develop a more sophisticated and informed understanding of the role of civil society organizations in the mix of help providers and advocates for the elderly.

The question is particularly salient in a period that is likely to be characterized by protracted fiscal austerity, limitations on State budgets in all developing and developed nations, and the continued inroads of neoliberal policy aimed specifically at social security programs. This new research agenda must be informed by a sophisticated understanding of the fact that the elderly are not a homogeneous population either in terms of social characteristics, needs, or political profiles. The elderly differ in terms of such politically and economically relevant dimensions as gender, race, ethnicity, immigration status, rural/urban residence, and more. Even as we advocate for the rights of the elderly as a group, more targeted civil society approaches allow us to deal with the great heterogeneity that characterizes aging in the world today.

References

Albrow, Martin, Helmut Anheier, Marlies Glasius, Monroe Price, and Mary Kaldor, eds. 2008. *Global Civil Society 2007/8*. Thousand Oaks, CA: Sage.

Andrews, Geoff. 1999. "New Left and New Labour." *Soundings* 13:14–24.

Angel, Ronald J. and Jacqueline L. Angel. 2009. *Hispanic Families at Risk: The New Economy, Work, and the Welfare State*. New York: Springer.
———. 1997. *Who Will Care for Us? Aging and Long-Term Care in Multicultural America*. New York: New York University Press.
Bendaña, Alejandro. 2006. "NGOs and Social Movements: A North/South Divide?" edited by United Nations Research Institute for Social Development. New York: United Nations.
Binstock, Robert H. 2004. "Advocacy in an Era of Neoconservatism: Responses of National Aging Organizations." *Generations* 28:49–54.
Bodenheimer, Thomas. 1999. "Long-Term Care for Frail Elderly People: The On Lok Model." *New England Journal of Medicine* 341:1324–8.
Boli, John G. and George M. Thomas. 1997. "World Culture in the World Polity: A Century of International Non-Governmental Organization." *American Sociological Review* 62:171–90.
———. 1999. *Constructing World Culture: International Nongovernmental Organizations Since 1875*. Stanford, CA: Stanford University Press.
Bolin, K., B. Lindgren, and Peter Lundborg. 2007. "Informal and Formal Care Among Single-Living Elderly in Europe." *Health Economics* 17(3):393–409.
Bullock, Karen, Sybil L. Crawford, and Sharon L. Tennstedt. 2003. "Employment and Caregiving: Exploration of African American Caregivers." *Social Work* 48:150–62.
Burcher, Jacqueline, ed. 2008. *México solidario: Participación cudadana y voluntaiado*. Mexico, DF: Limusa.
Carroll, Thomas. 1992. *Intermediary NGOs: The Supporting Link in Grassroots Development*. West Hartford, CN: Kumarian Press.
Central Intelligence Agency. 2009. "Spain." In *The World Factbook* (online). Washington, DC: Central Intelligence Agency. Available at: https://www.cia.gov/library/publications/the-world-factbook/geos/sp.html.
Chamie, J. 2005. *World Population Ageing: 1950–2050*. New York: Population Division, Department of Economic and Social Affairs, United Nations Secretariat.
Chatters, Linda M., Robert Joseph Taylor, and James S. Jackson. 1986. "Aged Blacks' Choices for an Informal Helper Network." *Journal of Gerontology* 41:94–100.
Clarke, Gerard. 1998. "Non-Governmental Organizations (NGOs) and Politics in the Developing World." *Political Studies* 46:36.
Coale, Ansley Johnson and Melvin Zelnik. 1963. *New Estimates of Fertility and Population in the United States*. Princeton, NJ: Princeton University Press.
Dilworth-Anderson, Peggye, Sharon Wallace Williams, and Theresa Cooper. 1999. "Family Caregiving to Elderly African Americans: Caregiver Types and Structures." *Journal of Gerontology* 54B:S237–41.
Etzioni, Amitai. 1993. *The Spirit of Community: Rights, Responsibilities, and the Communitarian Agenda*. New York: Crown Publishers.
———. 1995. *New Communitarian Thinking: Persons, Virtues, Institutions, and Communities*. Charlottesville: University Press of Virginia.
Fast, Janet E., Deanna L. Williamson, and Norah C. Keating. 1999. "The Hidden Costs of Informal Care." *Journal of Family and Economic Issues* 20:301–26.
Faux, Jeff. 1999. "Lost on the Third Way." *Dissent* 46:67–76.
Fisher, William F. 1997. "Doing Good? The Politics and Antipolitics of NGO Practices." *Annual Review of Anthropology* 26:439–64.
Giddens, Anthony. 1991. *Modernity & Self-Identity: Self and Society in the Late Modern Age*. Stanford, CA: Stanford University Press.
———. 1994. *Beyond Left and Right: The Future of Radical Politics*. Oxford, UK: Polity Press.
Global Policy Forum. 2010. "Paper on NGO Participation at the United Nations." New York: United Nations.
Grob, Gerald N. 1983. *Mental Illness and American Society*. Princeton, NJ: Princeton University Press.
Gross, Diane L., Helena Temkin-Greener, Stephen Kunitz, and Dana B. Mukamel. 2004. "The Growing Pains of Integrated Health Care for the Elderly: Lessons from the Expansion of PACE." *The Milbank Quarterly* 82:257–82.
Grzybowski, Candido. 2000. "We NGOs: A Controversial Way of Being and Acting." *Development in Practice* 10:436–444.
Haber, Carole. 2006. "Old Age Through the Lens of Family History." Pp. 59–75 in *Handbook of Aging and the Social Sciences*, edited by R. H. Binstock and L. K. George. New York: Academic.
Hamilton, Brady E., Joyce A. Martin, and Stephanie J. Ventura. 2009. "Births: Preliminary Data for 2007." Washington, DC: National Center for Health Statistics.
Hayward, Lynda, Sharon Davies, Roberta Robb, Margaret Denton, and Greg Auton. 2004. "Publicly Funded and Family-Friend Care in the Case of Long-Term Illness: The Role of the Spouse." *Canadian Journal on Aging Supplement* 23 Supplement:S39–S48.
HelpAge International. 2009. "Working for life: making decent work and pensions a reality for older people." London, UK: HelpAge International.

Himes, Christine L. and Erin B. Reidy. 2000. "The Role of Friends in Caregiving." *Research on Aging* 22:315–36.
Himes, Michael R. 1989. "American Fertility in Transition: New Estimates of Birth Rates in the United States, 1900–1910." *Demography* 26:137-48.
Hudock, Ann C. 1999. *NGOs and Civil Society: Democracy by Proxy?* Cambridge, England: Polity Press.
Idler, Ellen. 2006. "Religion and Aging." Pp. 277–300 in *Handbook of Aging and the Social Sciences*, edited by R. H. Binstock and L. K. George. New York: Academic.
Jolicoeur, Pamela M. and Teresa Madden. 2002. "The Good Daughters: Acculturation and Caregiving among Mexican-American Women." *Journal of Aging Studies* 16:107–20.
Keck, Margaret and Kathryn Sikkink. 1998. *Activists Beyond Borders: Advocacy Networks in International Politics*. Ithaca, NY: Cornell University Press.
Kotsadam, Andreas. 2009. "Effects of Informal Eldercare on Female Labor Supply in Different European Welfare States." Göteborg, Sweden: Institutionen för Nationalekonomi med Statistik, Handelshögskolan vid Göteborgs Universitet.
Krause, Neal. 2006. "Social Relationships in Later Life." Pp. 181–200 in *Handbook of Aging and the Social Sciences*, edited by R. H. Binstock and L. K. George. New York: Academic.
Liebig, Phoebe S. 2003. "Old-Age Homes and Services: Old and New Approaches to Aged Care." Pp. 159–78 in *An Aging India: Perspectives, Prospects, and Policies*, edited by P. S. Liebig and S. I. Rajan. New York: The Hayworth.
Moren-Cross, Jennifer L., and Nan Lin. 2006. "Social Networks and Health." Pp. 111–26 in *Handbook of Aging and the Social Sciences*, edited by R. H. Binstock and L. K. George. New York: Academic.
Palloni, Alberto, Guido Pinto-Aguirre, and Martha Pelaez. 2002. "Demographic and Health Conditions of Ageing in Latin America and the Caribbean." *International Journal of Epidemiology* 31:762–71.
Pereira, Javier and Ronald Angel. 2009. "From Adversary to Ally: The Evolution of Non-Governmental Organizations in the Context of Health Reform in Santiago and Montevideo." Pp. 97-114 in *Social Inequality and Public Health*, edited by S. Babones. Bristol, UK: Polity Press.
Pereira, Javier, Ronald J. Angel, and Jacqueline L. Angel. 2007. "A Case Study of the Elder Care Functions of a Chilean Non-Governmental Organization." *Social Science and Medicine* 64:2096–2106.
Pichardo, Nelson A. 1997. "New Social Movements: A Critical Review." *Annual Review of Sociology* 23:411–30.
Pierson, Paul. 2001. *The New Politics of Welfare*. New York: Oxford University Press.
Quadagno, Jill. 1982. *Aging in Early Industrial England*. New York: Academic.
Robinson, M. and R. Riddell. 1995. *Non-Governmental Organizations and Poverty Alleviation*. New York: Oxford University Press.
Rogerson, Peter A., Richard H. Weng, and Ge Lin. 1993. "The Spatial Separation of Parents and Their Adult Children." *Annals of the Association of American Geographers* 83:656–71.
Roman, Joseph. 2004. "The Trade Union Solution or the NGO problem? The Global Fight for Labour Rights." *Development in Practice* 14:100–09.
Ruggie, John Gerard. 1998. "What Makes the World Hang Together? Neo-Utilitarianism and the Social Constructivist Challenge." *International Organization* 52:855–85.
Salamon, Lester M. 2003. *The Resilient Sector: The State of Nonprofit America*. Washington, DC: Brookings Institution Press.
Salamon, Lester M., and Helmut K. Anheir. 1996. *The Emerging Sector: An Overview*. Manchester, UK: Manachester University Press.
Sangeeta, Kamat. 2004. "The Privatization of Public Interest: Theorizing NGO Discourse in a Neoliberal Era." *Review of International Political Economy* 11:155–76.
Sawhney, Maneeta. 2003. "The Role of Non-Governmental Organizations for the Welfare of the Elderly: The Case of HelpAge India." Pp. 179-92 in *An Aging India: Perspectives, Prospects, and Policies*, edited by P. S. Liebig and S. I. Rajan. Binghamton, NY: The Hayworth.
Silverstein, Merril. 1995. "Stability and Change in Temporal Distance between the Elderly and Their Children." *Demography* 32:29–45.
Silverstein, Merril and Joseph J. Angelelli. 1998. "Older Parents' Expectations of Moving Closer to Their Children." *Journal of Gerontology 53*B:SI53–63.
Sobotka, Tomáš. 2006. "Is Lowest-Low Fertility in Europe Explained by the Postponement of Childbearing?" *Population and Development Review* 30:195–220.
Srinivas, Nidhi. 2009. "Against NGOs? A Critical Perspective on Nongovernmental Action." *Nonprofit and Voluntary Sector Quarterly* 38:614–26.
Stone, Robyn. 2006. "Emerging Issues in Long-Term Care." Pp. 397–418 in *Handbook of Aging and the Social Sciences*, edited by R. H Binstock and L. K. George. New York: Academic.
Thane, Pat, ed. 2005. *A History of Old Age*. Los Angeles: The J. Paul Getty Museum.
Tolnay, Stewert E., Stephen N. Graham, and Avery M. Guest. 1982. "Own-Child Estimates of U.S. White Fertility, 1886–99." *Historical Methods* 15:127–38.
Wagner, David. 2005. *The Poorhouse: America's Forgotten Institution*. Lanham, MA: Roman & Littlefield.
Willimore, Larry. 2006. "Universal Age Pensions in Developing Countries: The Example of Mauritius." *International Social Security Review* 59:67–89.

Chapter 35
Population Aging, Health Systems, and Equity: Shared Challenges for the United States and Canada

Amélie Quesnel-Vallée, Jean-Simon Farrah, and Tania Jenkins

In the wake of growing concern with social determinants of health, health systems and their potential contribution to health disparities have often been overlooked. Yet, as the WHO Commission on Social Determinants of Health (CDSH 2008) report compellingly reminds us, health systems' organization likely play a role in the development and the persistence of social inequalities in health. These effects are exacerbated among populations with frequent need for care: children, individuals with chronic diseases, and the elderly, and where service provision and funding are fragmented in a mix of public and private sources.

In this chapter we show how population aging creates common challenges for the U.S., an entrepreneurial health system, and Canada, a national health system. We will make this argument on the basis of cross-national comparison of three categories of health services that are most prominently used, or have seen their growth most affected by aging populations: prescription drugs, long-term care (LTC), and end-of-life (EOL) care.

As the United States considers the implementation of a plan for mandated health insurance coverage, and while Canada ponders carving out a greater responsibility of private health insurance in its national health insurance system, it may not be surprising to observe convergence between these two countries in coming years. Yet, the convergence in policies that affect most the aging population has deeper roots, which we examine with a particular focus on the considerations for equity that are raised by this comparison. More specifically, we will show that the organization of these services poses equity challenges for both countries in terms of geographical disparities and even within jurisdictions, in terms of socioeconomic inequalities in access and outcomes.

In the next section we will present a brief comparison of the aging of the Canadian and U.S. populations, followed by an overview of these health systems. We will then look in turn at the situation concerning prescription drugs, LTC, and EOL care, contrasting for each the Canadian and U.S. specificities and drawing out common challenges. Finally, we conclude with broader concerns about the impact of these trends on socioeconomic equity in access to care.

Population Aging and Intergenerational Equity

In a comparison of population aging in Canada and the U.S., LaPierre and Hughes (2009) show that those populations are currently quite similar in terms of their age distribution. However, due to lower fertility rates in the past couple of decades (and the associated decline in the proportion under 15),

A. Quesnel-Vallée (✉)
Department of Sociology, Epidemiology, Biostatistics and Occupational Health, McGill University, Montreal, Canada
e-mail: amelie.quesnelvallee@mcgill.ca

R.A. Settersten, Jr. and J.L. Angel (eds.), *Handbook of Sociology of Aging*, Handbooks of Sociology and Social Research, DOI 10.1007/978-1-4419-7374-0_35,

projections suggest that Canada could experience a slightly steeper rate of increase in population aging that would result in a greater proportion of 65+ individuals in this country starting in the 2030s (23.4% in Canada vs. 19.7% in the U.S.). Interestingly though, these trends do not seem to differentially affect the proportion of the oldest-old, those aged 85 years and over, which remains fairly similar, and increases at the same pace in both countries well into the 2050s (LaPierre and Hughes 2009).

The growth in this group of people, which is projected to reach about 5% of the total population in both Canada and the United States by 2050 (in contrast with less than 2% today), is a source of concern for policymakers across the developed world (Lafortune et al. 2007). Indeed, given that 20% of Americans aged 85 years and over report some limitation to their activities of daily life (ADL) – as opposed to only 3% of those aged 65–74 the ensuing multiplication of needs may require profound changes in the way services are provided to these populations (Lafortune et al. 2007).

As we will see in the next section both Canadian and U.S. systems share the historical heritage of Liberal Welfare States whereby the elderly (workers) were designated as a group particularly worthy of social protection. But this perspective emerged in a very different demographic context where the working-age population far outnumbered the elderly, thus allowing for the sustainability of a social insurance system. Similarly, as medical interventions were fairly limited, so were the associated costs (to the system, not to individuals, for whom they were always prohibitively high). While population aging per se is not the main driver of health care costs, the fact that most of an individual's lifetime health care costs are incurred in the last 6–12 months of life entails that the bulk of health care spending goes to elderly individuals. Any consideration of health systems, the elderly and equity must therefore first address the issue of intergenerational equity.

With the aging of their population and parallel (though not unrelated) changes in medical technology and patient demands, concerns are periodically raised regarding the financial sustainability of both Canadian and U.S. health systems, and particularly, about the risks for intergenerational equity and solidarity (Thurow 1996). However, while these "calls to arms" have not ceased, their dire predictions have not materialized either. Foner (2000) argues that this lack of age-based schism stems first from two essential features of social life: (1) The most basic social structure that individuals belong to, namely families, are age-heterogeneous, which tends to promote intergenerational solidarity; (2) Aging is an inevitable process, and thus programs directed at the elderly may still receive support from the working-age population that stands to eventually benefit from them. Moreover, public subsidies to the elderly can benefit younger family members through direct transfers, and indirectly, by removing the obligation to provide financial support to elderly relatives. Finally, as it is typically the family who cares for elderly relatives, this ensures an intimate, emotional awareness of aging-related issues and the availability (or need) of public support to surmount those challenges.

Therefore, Foner (2000) argues, the concerns with "age wars" are overstated, and serve to mask the more important social rift of social class, where public programs are supported by the working and middle class, but increasingly less so by more fortunate classes. While we do not adopt a class perspective in this paper, we follow in Foner's (2000) perspective to argue that the major challenge that both Canadian and U.S. health systems tackle with regards to eldercare does not so much have to do with public support sustainability, but rather with eroding social equity in access to services.

In the next section we replace health systems and notions of the "deservingness" of the elderly in the broader context of the development of the Welfare State (WS). This will provide the context for our contrast of the Canadian and U.S. health systems organization, and the more in-depth examination of the three cases of prescription drugs, LTC, and EOL care.

Welfare States, Health Systems, and the Elderly

Health systems are perhaps one of the most emblematic symbols of WS provision in the twentieth century. They are indeed commanding massive amounts of human and financial resources, and often represent the first and most obvious point of contact of the individual with the social system.

Following Briggs's (2000) seminal historical account, we can trace this preeminent position back to the very origins of the WS. Indeed, Briggs argues that WS are organizations whose primary function is the deliberate use of organized power (through politics and administration) in an effort to modify the play of market forces. This has historically been done through a guarantee of minimum income; by limiting insecurity in the face of both unexpected and unavoidable risks (such as sickness, unemployment, and old age); and by making the *best standards* (and not merely a minimum standard) of the above available to all citizens (within a certain agreed upon range of options, of course). Organized modern health systems – largely publicly funded and under some form of governmental regulation if not provision – obviously play a large role in fulfilling the second objective of WS.

Thus, the development of modern health systems closely mirrors that of the WS in Western societies, starting with the establishment in 1883 of the first government-mandated social health insurance program by Otto Von Bismarck, first Chancellor of the German Empire. In this landmark piece of legislation that was subsequently widely copied and adopted around the world, Bismarck enacted a law requiring employer (one-third) and employee (two-thirds) contributions to a sickness fund (World Health Organization 2000). The moral underpinnings of this legislation was that workers, or perhaps more cynically, their work capacity, was most deserving of social insurance. Thus, while there is strong evidence that this policy was passed at least in part to appease social tensions resulting from the woes of industrialization (and to stave off the growth of socialist ideology), this was also explicitly done in the name of insuring national productivity (Briggs 2000).

Old age insurance was soon to follow in the German Empire (in 1889), but as the age limit of 70 years substantially exceeded the average life expectancy of 41 years old at the time (Gapminder Foundation 2010; Social Security Online 2010), we can safely assume that this program was obviously not designed to pay pension annuities to a large segment of the population, or for lengthy periods of time into retirement. Nevertheless, this legislation stemmed from a slightly different moral imperative than the first, where the elderly and disabled (as a result of work) were explicitly designated as deserving of social protection from the state, ostensibly by virtue of their past contribution to the country's productivity.

While the first part of the twentieth century is marked by the extension of similar programs to larger segments of the population within countries and to more countries around the world, the second part of this past century sees the emergence of new principles of social protection. Often linked to the seminal Beveridge (1942) report, reforms that followed broadened the focus of state protection from the worker to all citizens, and "from cradle to grave."

The Canadian health system is generally thought to reflect a Beveridgean approach to social protection following the orienting principle of equity, while the U.S. health system is generally more reflective of a Bismarckian perspective upholding the values of free choice and individual responsibility (Roemer 1991). However, because of historical contingencies in the development of these systems (Maioni 1998), as we will see later, this overall assessment is not entirely reflective of eldercare in those two countries. Indeed, in Canada, some of the services most crucially needed by the elderly were not explicitly listed under the Canada Health Act of 1984, and have thus remained unevenly covered by a mix of public and poorly regulated private sources. In contrast, in the U.S., the elderly are eligible for universal social security coverage under Medicare, though not without out-of-pocket expenses, also with a substantial contribution from private sources.

Thus, while at *prima facie* one might expect more substantial differences with regards to these countries' health systems' capacity for handling these changes equitably, we will show below that the challenges they face are in fact very similar, though they emanate from radically different orienting principles. In the following section we provide a brief overview of these systems' organization, with a focus on a comparative perspective and on those services that are most relevant to an aging population.

Canadian and U.S. Health Systems in Comparative Perspective

The Canadian Health System(s)

Canada's "health system" in fact consists of multiple provincial and territorial single-payer national health systems (Marchildon 2005). Indeed, the Canadian constitution dictates that health matters are primarily under the jurisdiction of the provinces and thus hampers the federal government's capacity to intervene in the delivery of health care services.

There are, however, some commonalities from coast to coast. In the 1950s, in an effort to standardize plans across the country, the federal government pledged to subsidize half of each province's total costs in establishing universal health care systems around a consistent set of guidelines (Maioni 1997). As this initiative still left some substantial variation across the country, more recently, the Canada Health Act was passed in 1984 to consolidate (reduced) federal grants to provincial health care systems on the condition that the provinces and territories comply with five basic principles: universality, public administration, comprehensiveness, portability, and accessibility (Canadian Healthcare Association 2009). However, these conditions are only imposed on "medically necessary hospital, physician and surgical-dental services," thus leaving a substantial range of health services unregulated at the federal level. This has resulted in, at best, provincial variations in the populations covered and the extent to which this coverage is free, and, at worse, substantial proportions of the Canadian population being uninsured across all provinces.

In particular, the Canadian system developed during a period where population aging was not yet a preoccupation, and this is reflected in the publicly insured basket of services, which is inadequate to address the needs of this population. Moreover, Canada also faces a common challenge of national health systems, which is a lack of responsiveness to patients' needs. Indeed, a centralized organization and a single-payer funding scheme are typically ill-suited to fulfill individualized health service needs, which are often the hallmark of elderly care. For instance, prescription drugs were left out of the Canada Health Act in 1984, because it was deemed these costs would quickly rise out of control due to the substantial demand and high potential for individual access to these services (Boychuk and Banting 2008).

As we will see later, in some instances, as for prescription medication, a supplementary private health insurance market has developed to patch these holes in public coverage. Yet in other cases, the insurance market is still only at the incipient stages of development, with very little public regulation of the offer. This situation leaves these areas open to private speculation and very muchvulnerable to the development of socioeconomic inequalities, which is the case for LTC and EOL services.

The U.S. Health System

As a classic example of an entrepreneurial health system (Roemer 1991), the U.S. health care system is diametrically opposed to Canada's in many ways, as it does not mandate health insurance coverage to its general population, let alone fund care through a single-payer system. However, these differences are much less significant when it comes to the elderly population, which is covered through Medicare (Maioni 1997; Wiener and Tilly 2002). This program, which covers most expenses for seniors' acute needs, is primarily financed through social security funds collected at the federal level, and to which every American worker contributes. Also, Medicaid, a state–federal public program, targets the needs of the poor and the severely disabled according to a minimum standard set by the federal government. Beyond this minimum level, states are entitled to extend coverage as they see fit. Therefore, the federal government fulfills the crucial role of maintaining the public sector so as to protect the neediest Americans.

The postwar public health care system was designed to respond to the needs of seniors by endowing the federal government with the legal responsibility of regulating this sector. However, as more Americans are getting older and living longer and healthier lives, seniors' needs are also changing. Thus, the prevailing approach of acute care adopted thus far by Medicare and Medicaid may no longer constitute an appropriate response. Yet, Washington is mainly preoccupied with funding issues as the central administration is slowly realizing that Medicare's financing is going to be difficult to sustain in the short term. Financial constraints, then, are one of the major roadblocks in adapting the public system to the needs of an older population, along with organizational constraints stemming from a fragmented system. At the state level, pressure is mounting to provide more assistance through Medicaid and to develop initiatives within their jurisdictions that will allow older Americans to remain at home later in life.

Having very summarily introduced the particular challenges faced by these systems as a result of their idiosyncratic organization, we will now turn to the similar challenges they face with regards to three categories of services: prescription drugs, LTC, and EOL care.

Prescription Drug Coverage for Seniors in Canada and the United States

Prescription drug coverage for seniors is an issue of growing importance for both Canadian and American policymakers, as evidenced by dramatic reforms in this area in recent years. Indeed, drug expenditures have risen dramatically in the past decades across the world. The United States in particular constituted over 50.2% of the global market sales of drugs in 2004, whereas Canada only accounted for 2.9% (Gray 2005). This translates into per capita expenditures of $752 (US PPP) in the United States compared to $559 (US PPP) in Canada (OECD 2004). Expenses on drugs for seniors have grown exponentially in the United States, from $10 billion in 1987 to over $52 billion in 2004 (Hartman et al. 2008) – an increase which has been attributed in the literature to changes in that country's demographic characteristics (i.e., the aging population) as well as to technology (Shah et al. 2006). In contrast, Morgan et al. (2004) have found that between 1991 and 2001, prescription drug expenditures for the Canadian elderly have increased from $149 to $320 million – but that less than half this variation can be explained by increases in the number of seniors or the rate at which they use these drugs. Instead, studies have found that increases in expenditures are best explained by changes in the types of therapies employed (Morgan 2006; Morgan et al. 2004).

What follows is an overview of prescription drug insurance coverage in both countries, as well as a discussion on rising drug expenditures and some of the techniques employed to curb those costs. We will then examine some of the repercussions associated with rendering these plans cost-effective, namely the issue of cost-related noncompliance (CRNC) amongst seniors, and finally conclude with some recommendations for both countries.

Prescription Drug Coverage for Seniors in Canada

Interestingly, despite universal access to health care in Canada and the Canada Health Act's guarantee to cover medically necessary services, public coverage for prescription drug medication varies wildly from province to province and by individual characteristics (Anis et al. 2001; Demers et al. 2008; Grootendorst 2002). For example, Demers et al. (2008) found that the same elderly patient would pay anywhere between 0 and 100% of the costs of their medications, depending on the province in which they reside. This means that an elderly man, earning an average income with an annual prescription burden of $1,283, would pay anywhere between $60 for those prescriptions in New Brunswick to $1,332 in Manitoba, when the cost of yearly premiums is factored in (Demers et al. 2008).

Not all Canadian seniors are eligible for public coverage, either. Seniors in Saskatchewan, Manitoba, and Newfoundland are only covered if they have low income, whereas Ontario, Quebec, and Nova Scotia offer reimbursements proportional to costs and income level (Demers et al. 2008). Quebec is the only province in the country to have what is known as mandated health insurance for prescription drugs (Pomey et al. 2007). As such, it requires all residents to be covered by either a private plan (which is usually obtained through an employer) or the Régie de l'assurance maladie du Québec (RAMQ) public plan. Low-income seniors who receive the Canadian government's Guaranteed Income Supplement (along with children and individuals receiving social welfare) are not required to pay a premium to this public plan (Régie de l'assurance maladie du Québec 2009).

In sum, prescription drug insurance coverage in Canada is characterized by a fragmented system both across the various provinces and within them, with coverage and costs varying substantially according to individual income and supplementary private health insurance coverage.

Prescription Drug Coverage for Seniors in the United States

Prior to 2006, American seniors had to obtain their prescription medication by either purchasing private health insurance (independently or through a retirement benefit plan), paying out-of-pocket, or receiving some form of supplemental Medicare coverage (Frencher and Glied 2006). However, as a result of ever-growing expenditures on drugs, uneven coverage across social groups, impressive lobbying efforts on behalf of various public actors, and the population's general discontent with the status quo, the Bush administration introduced the Medicare Prescription Drug, Improvement, and Modernization Act in December of 2003 (Pomey et al. 2007). This resulted in the development of Medicare Prescription Drug Coverage (also known as Medicare Part D) in 2006, the country's first drug plan universally available to all seniors, regardless of income or health status (Centers for Medicare and Medicaid Services and U.S. Department of Health and Human Services 2009). Over 25 million Medicare beneficiaries now enjoy Medicare drug coverage (Hoadley et al. 2008).

There is still substantial variation in plan offerings, however, such that individuals can select among a total of 1,824 stand-alone plans in all 50 states which can vary in coverage and cost (Hoadley et al. 2008). That said, the federal government sets a minimum standard level for all plans, such that the balance of coverage and beneficiary contributions is somewhat standard across individuals (Centers for Medicare and Medicaid Services and U.S. Department of Health and Human Services 2009). This means that virtually all plans (with some exceptions for low-income individuals) include user contributions in the form of monthly premiums, yearly deductibles, copayments or coinsurance, and a coverage gap.

In addition, many Medicare plans in the United States include a so-called "doughnut hole" or $3,454 coverage gap, which requires beneficiaries to pay 100% of their drug costs once they've already spent $2,700 in total drug expenditures for the year (Kaiser Family Foundation 2009). After $6,154 dollars in total annual prescription costs, seniors then can benefit from what is known as catastrophic coverage, whereby they only contribute a 5% co-payment and Medicare and their insurance plan cover the other 95% of the cost. The coverage gap does not apply to everyone; however, eligible low-income individuals are entitled to coverage regardless of year-to-date costs, and some plans even offer limited coverage during the gap, but these often charge higher premiums to enrollees (Centers for Medicare and Medicaid Services and U.S. Department of Health and Human Services 2009). In recognition of the financial and health risks that this coverage gap entailed, President Obama's Affordable Care Act proposed to fill this "donut hole" (U.S. Department of Health and Human Services 2011). However, as these initiatives were only implemented starting in 2010 and are set to progressively address this gap in coverage until 2020, it is still too early to evaluate the impact of this policy.

Thus, Medicare drug coverage requires some out-of-pocket expenditures for most people (Frencher et al. 2006). Some individuals can help offset the costs of these expenses through supplementary private cover. For others, publicly funded "extra help" is now available for eligible low-income seniors to offset some of the costs which fall under the beneficiary's responsibility. Seniors with incomes below $21,855 and resources less than $25,010 in 2009 are entitled to assistance with paying their drug plan's monthly premium, yearly deductible, coinsurance, or copayments, and are not subject to coverage gaps. Yet, many individuals who cannot afford private insurance but are not needy enough to be eligible for "extra help" are at risk of incurring substantial out-of-pocket costs related to their prescription drug needs, despite this universal drug coverage (Hoadley et al. 2008).

Implications of Out-of-Pocket Expenses: Cost-Related Drug Noncompliance

In sum, governments in both countries expect seniors to pay out-of-pocket expenses for prescription drugs. One important consequence of requiring seniors to contribute to their drug insurance plans is CRNC. Despite almost universal drug coverage in the United States, older patients with chronic diseases are still susceptible to cost-related nonadherence to prescription medication, even when enrolled in a prescription insurance plan (Briesacher et al. 2007; Maio et al. 2005). Low-income, higher out-of-pocket costs, lower drug benefits, and poorer health are all associated with higher risks of CRNC (Kennedy et al. 2004; Poisal and Murray 2001; Rector and Venus 2004). As a result, Medicare's gap in coverage may have a particularly alarming effect on those individuals who suffer from chronic illnesses but do not have a low enough income to qualify for assistance, as they may be left with little incentive to take newly prescribed medications which would push them over the edge of the "doughnut hole" in terms of year-to-date expenditures (Gellad et al. 2006). Gellad et al. (2006) also found that the near-poor who do not quite qualify for subsidies but do not have private drug coverage tend to pay more for their prescriptions than wealthier seniors because of the gaps in coverage. And since individuals who do not comply with their medication are more likely to experience worse health and visit the emergency department more frequently, this perversely generates higher medical costs in general (Heisler et al. 2004; Olshaker et al. 1999).

Americans of all ages are consistently less likely to comply with their drug regimens because of cost than Canadians (Kennedy and Morgan 2006, 2009). Thanks to the recent changes in Medicare part D; however, seniors in both countries are now less likely than nonseniors to experience CRNC (9.2 vs. 25.8% in the United States, $p<0.001$ and 4.6 vs. 8.7% in Canada, $p<0.001$) (Kennedy and Morgan 2009).

Long-Term Care

LTC refers to a broad range of services encompassing skilled nursing care, assisted-living facilities, home care, hospice care, respite care, adult day care, and different in-home living arrangements (Sultz and Young 2009). These services provide a long-term response to chronic increases in disability brought about by functional limitations inherent in the process of individual aging (Feder 2000).

Even in the absence of comprehensive publicly funded LTC services, this sector has experienced constant growth since the Second World War (Grabowski 2008). However, while the formal provision of these services was once concentrated in institutions, demand appears to be shifting towards assistance aimed at remaining in the community as long as possible: from 1994 to 2005, the 85+ age group showed a decrease of 9% in the report of at least one ADL limitation for those living in institutions, while an increase of 3% took place among those living in the community (Lafortune et al. 2007). This suggests that there is an increasing proportion of those aged 85 and over who are in need of LTC services and now remain in the community.

This trend is probably not unrelated to the fact that governments across the board in developed countries have been, since the 1970s, championing a view of seniors as being more independent, more involved in their community and living as long as possible outside of institutional settings (World Health Organization 2002). However, this demand and the funding for it have so far been borne mostly by families and informal caregivers. Therefore, as we will see below, disparities in access to these services will be increasingly highlighted as the oldest cohorts grow in number and governments have to face issues of social justice in the provision of LTC services (World Health Organization 2002).

Long-Term Care in Canada

In Canada (as in the United States), local authorities are recognized as having a better perception of the LTC services and funding required in their area (Liebig 1993). Moreover, Canadian public LTC services are available to *complement* families' role in caregiving (Chappell et al. 2004). Requests for assistance thus usually stem from the family unit – a process that is often far from straightforward – and local health authorities deal with each case depending on available resources. Consequently, there is a great variation in access to publicly funded LTC services between and within provinces and territories in Canada.

Thus, the Federal government's influence on LTC remains minimal and is currently limited (at best) to promoting inter-provincial dialog and horizontal emulation of best LTC practices in the federation. As such, the potential for innovation at the sub-national level is greater than in the United States because provinces have complete freedom in the allocation of resources to LTC services. However, a standardization of practices may be harder to achieve across all provinces since the central government – assuming it had sufficient unifying strength *and* the will to act on this issue – currently lacks formal means through which it could exert influence on LTC-related policy.

Long-Term Care in the United States

In the United States, Medicare does not cover LTC services, with the exception of home care needed in the weeks following an episode of acute care (rehabilitative care). Thus, Medicare contributed only 19% to total spending for LTC for the elderly in 2004 (Komisar and Thompson 2007). In contrast, Medicaid, a federal–state shared program, does target the LTC needs of the poorest elderly Americans, accounting for 49% of national LTC expenditures in 2004 and is, therefore, one of the main players in the field (Komisar and Thompson 2007). States are responsible for implementing (and are free to expand) Medicaid services as they wish, but remain constrained by federal guidelines in doing so.

Typically, this program has tended towards the institutionalization of patients in Medicaid-approved facilities for LTC services. In fact, most states still spend 70–80% of their LTC funds in nursing facilities (Polivka and Zayac 2008). This fact led many scholars to criticize the "institutional bias" of Medicaid programs on the basis that it does not respond to growing demands from seniors to receive community and home care, or simply put, to facilitate aging in the community.

Thus, in contrast to Canada, the U.S. federal government has high stakes in the development of LTC services, since a large share of the formal LTC system operated in the country is funded, regulated, and assessed at the state and federal governmental levels through Medicare and Medicaid. Still, because of its financially shared nature, Medicaid has led to important geographical variations in coverage; states that are more conservative and less open to public coverage of LTC will simply meet the minimum requirements of the program, whereas those that are more progressive (and potentially those that contain a larger proportion of voting seniors) may extend the coverage to citizens who are further away from the poverty-line threshold, for example (Feder 2000).

Challenges to Equity in Long-Term Care in North America

The current state of the delivery of LTC services in both Canada and the United States tends to reflect the contemporary conception of individual aging in the private market; indeed, this is where elderly people, as consumers, can spend considerable amounts of money to receive the necessary support which enables them to remain active. The shift from an institutionalization-centered medical model of LTC to a patient-directed model has yet to fully penetrate the public system in North America, which constitutes one of the greatest challenges to policy, and particularly to Medicaid, in coming decades (Robinson and Reinhard 2009).

In fact, to this day, most long-term caregiving is still undertaken by informal providers and paid for out-of-pocket, particularly with regards to care in the community. With the implementation of a senior-oriented WS in the United States and a universal public health care system in Canada, this informal care network came to be complemented by formal caregiving options, either through scarce and means-tested publicly subsidized services and/or a plethora of private providers (Wiener and Tilly 2002). Drawbacks to such a loose health sector include lack of information-sharing on available public services, an unstable workforce, little follow-up on elderly people's health and social issues, and greater inequities regarding access to these services (Rantz et al. 2000).

This generates unmet needs which, at the individual level, accelerate the deterioration of health and promote institutionalization which could have been otherwise prevented with efficient and integrated LTC coordination (Keefe et al. 2007). In the United States, a Commonwealth Fund study reported that 58% of the elderly respondents of a 1999 cross-state survey declared having unmet needs (Komisar et al. 2005). In Canada, a Statistics Canada study drawing on the 2003 wave of the Canadian Community Health Survey showed that 42% of seniors requiring help with moving about in their home did not receive any help and that even among those who received a mix of formal and informal home care, 19% of seniors expressed unmet home care needs (Carriere 2006). This sheds light on the fact that even limited and fragmented public provision of LTC services may not be sufficient to ameliorate or even maintain patients' health status.

Many scholars and practitioners recognized the need to change this and designed experimental integrated care systems for the elderly as a way of influencing policy-making (Hollander et al. 2007). So far, efforts to generalize these local experiments to a larger policy scope have largely failed. Another avenue for reform may come from LTC insurance, which exists in both countries, but whose growth has been relatively limited. At the American federal level, the Health, Education, Labor and Pensions Committee (HELPC) has been active in testing ideas underlying a hypothetical national LTC insurance scheme. Known as the CLASS Act sponsored by the late Senator Edward Kennedy (former HELPC chairman), this federal policy would automatically enroll workers in the insurance scheme, with the option of opting out (Gleckman 2009). If enacted as part of the Obama administration's health care reform bill, this program could potentially provide all Americans with one alternative to Medicare and Medicaid in accessing LTC services (O'Malley Watts 2009). Currently, only a handful of insurers in Canada offer these services, but these offerings may grow in the future, as they are being heavily marketed to both employers and young senior citizens by the industry.

Finally, there is substantial evidence from both countries showing that elderly people and their families – especially those from lower socioeconomic classes without an informal caregiving network – do not possess sufficient resources or control over their LTC needs (Feder 2000; Messinger-Rapport 2009). Indeed, while a vast array of private care options exist, ranging from home care services to high-end retirement communities or elder care centers (Grabowski 2008), many elderly persons lack the means to consume the variety of LTC products offered by the market and yet fall below the stringent requirements to access publicly covered, means-tested LTC services. It is difficult to envision how such strongly ingrained views that see LTC as an entitlement reserved to the poor in the United States, and as a family responsibility in Canada (Chappell et al. 2004), could be circumvented to ensure care for those lower middle-class elderly people who, for various reasons (geographical mobility or smaller families, for example), lack informal care support. Yet, along with the issues of integrating LTC for individual patients, and providing alternatives to institutionalization, this is without a doubt a challenge that both countries will be facing in the years to come.

End-of-Life Care in Canada and the United States

As seniors continue to live ever longer, there is a growing need for what has come to be known as "end-of-life" (EOL) care, or "the services that may be needed by dying persons and their family or friends in the last year of life, as well as bereavement services following death" (Wilson et al. 2008:323). It seems that there have been changes in the locus of EOL care in recent decades in both Canada and the United States, away from hospitals and towards the home or LTC facilities. Let us take a closer look at the broader trends in EOL care in Canada and the United States as well as the trends regarding the expenditures associated with this type of care. We will then examine how socioeconomic status (SES) and place of death are closely related, in the context of two healthcare systems where health care proffered outside of an institution is scarcely covered by health insurance.

End-of-Life Care Trends in Canada

Across Canada, hospital deaths have declined from 77.7% in 1994 to 60.6% in 2004 (Wilson et al. 2009). In a separate study of Canadians who died during 2001–2002, it was found that 49.2% of those sampled died in hospitals, 30.5% in LTC facilities, 9.6% died at home while receiving home care, and 10.7% died at home without any home care (Motiwala et al. 2006). Place of death is known to be determined by various factors. Certain morbidities have been found to be the strongest predictors of location of death, such that cancer for example, increased chances of home death, while dementia made death in a LTC facility more likely, and major acute conditions increased the probability of death in a hospital (Motiwala et al. 2006). Regarding home care more specifically, other researchers have found that older individuals, women, people residing in urban regions, and individuals of higher SES are all more likely to receive care at home during their last year of life (Brackley and Penning 2009). Similarly, Wilson et al. (2008) have noted that access to hospice and palliative care programs is unevenly distributed across Canada because these services tend not to be covered by provincial health plans, and as such, not all Canadians have equal opportunity when it comes to choosing their place of death.

End-of-Life Care Trends in the United States

In trends that parallel those in Canada, Flory et al. (2004) found that from 1980 to 1998, the percentage of Americans dying in hospitals decreased from 54 to 41%, whereas home deaths increased from 17 to 22%. Similarly, nursing home deaths increased from 16 to 22% from 1990 to1998. Parallel to this, Dy et al. (2007) showed that the proportion of Medicare recipients hospitalized during their last year of life remained stable from 1989 to 1999, but that the use of institutionalization in LTC facilities increased. They also found that changes in the place of death had occurred during this period, towards increases in LTC facilities and decreases in the proportion of those dying in the hospital. Fried et al. (1999) have also noted this diversification in places of death in their study on dying at home. They found that 49% of people in their sample died in hospital, 21% at home, and 27% in a LTC facility, numbers that essentially mirror the Canadian ones.

Expenditures on End-of-Life Care

In sum, it seems that both countries have experienced a general trend towards the deinstitutionalization of death in recent decades – a policy which seems to have been adopted almost internationally (Merlis 2000). This shift could reflect a more humane approach to dying, as well as respond to patient preferences; Buntin and Huskamp (2002) indeed found that if the care were available or accessible, many more patients would choose to die at home than elsewhere. However, a closer look at expenditures in EOL care suggest that this qualitative improvement in services places a higher cost burden on older decedents and their families, and is not distributed evenly in the population.

Indeed, despite the aging of the population, governmental expenditures on EOL care have remained stable over time in both countries (Buntin and Huskamp 2002; Canadian Institute for Health Information 2007; Emanuel 1996; Hogan et al. 2001). In the United States, approximately 25% of all Medicare spending occurs during the last year of life (Buntin and Huskamp 2002; Hogan et al. 2001), whereas in both countries, between 10 and 12% of all health care expenditures are dedicated to EOL care (Buntin and Huskamp 2002; Canadian Institute for Health Information 2007; Emanuel 1996).

Expenditures on EOL care in both countries generally decrease as age of death increases. In the United States, for all sexes, races, levels of co-morbidity, type of care, and cause and site of death, Medicare expenditures decreased consistently with age at death (Bird et al. 2002; Levinsky et al. 2001; Lubitz and Riley 1993). In explaining this pattern, Levinsky et al. (2001) estimated that declining aggressiveness of care in the United States accounted for approximately 80% of the decrease in medical expenditures with age. These trends were found in Canada as well (Demers 1998).

However, in all these cases, the focus is on measurable *government* spending during an individual's last year of life. Thus, some authors have argued that these figures exclude many services which are not covered through government health plans such as home or hospice care, and may thus be underestimated (Buntin and Huskamp 2002; Scitovsky 1994). Indeed, Hoover et al. (2002) confirmed that EOL expenditures covered by Medicare were lower for those dying at older ages, but also that nonMedicare-covered expenditures were in fact higher for those dying at older ages than at younger ages. Similarly, Scitovsky's review of the literature (1994) finds that although the aggressiveness of care decreases with age of death, these savings are often offset by rising LTC costs. So it appears that while governments spend less on the oldest decedents, individuals and their families are bearing a relatively greater burden than for younger decedents. As we have shown above that one of the greatest (shared) challenges for both countries will be the significant increase in the oldest-old (85+), these patterns beg the question of who will bear the additional burden of these older decedents.

Inequalities in Place of Death

Moreover, health coverage during the last year of life affects patients' "trajectories of death" (Buntin and Huskamp 2002), and nowhere is this relationship clearer than when examining the link between SES and place of death. As Gruneir et al. (2007) found in the United States, ethnic minority status, lower levels of educational attainment, and less access to resources and social support were associated with a higher probability of hospital death, despite patient preference to die at home. In Canada, higher SES was also found to increase chances of dying at home (Motiwala et al. 2006). More specifically, it has been noted that the expenses incurred by Canadian families at the end of life are quite substantial, and that as a result, a two-tiered system of EOL care has developed, such that only those with the most resources can afford to die at home (Chochinov and Janson 1998). These social inequalities in death may in fact be only the tip of the iceberg of an emerging trend: in the United States in 1980, there were no racial differences between the proportion of whites and blacks dying in hospital. In 1998, however, fewer whites died as inpatients in hospital than did African Americans (Flory et al. 2004).

Several studies have specifically examined the relationship between government policies and social inequalities in EOL care. For example, high government investment in LTC, measured by regional density of nursing home beds and state Medicaid payment rate, was associated with higher probability of nursing home death in the United States (Gruneir et al. 2007). In another American study, Muramatsu et al. (2008) found that state spending on home and community-based services indirectly increased the chances of dying at home by reducing the risk of nursing home relocation near death. In Canada, however, this relationship is not quite as clear. Despite specific policy reforms designed to make home and community care more accessible in British Columbia in the 1990s, the results of one study suggest that equity in access to this type of care did not improve during that period, nor did home care service utilization increase (Brackley and Penning 2009). So despite deinstitutionalization efforts on behalf of the government, it may be because community care requires out-of-pocket resources that fewer people are opting for home deaths than expected.

Common Challenges for Equity: Different Solutions for the U.S. and Canada?

There is often an expectation that the Canadian system is inherently more equitable than the American one (World Health Organization 2000). Yet, in our examination of three types of services that are particularly salient to aging populations, namely prescription drugs, LTC, and EOL care, we have shown that, due to various historical contingencies in the development of the Canadian and U.S. health systems, population aging poses comparable challenges to equity on both sides of the border. Below we summarize the main points of common challenges shared by Canada and the United States.

Prescription Drugs Coverage

In an interesting departure from most cross-national trends between these two countries, prescription drug insurance coverage for the elderly in the United States is in fact more homogenous than in Canada, thanks to the recently implemented Medicare Part D. Indeed, as this was never a category of services covered under Canada's national insurance plan (deliberately so; see Boychuk and Banting 2008), coverage evolved into a fragmented patchwork system of public and private coverage with strong inter-provincial variations in both levels and mechanisms of coverage. As a result, levels of out-of-pocket contributions vary widely in Canada, depending on a number of factors including an individuals' income level. Of course, variations in coverage are endemic in the United States as well, with more than 1,824 stand-alone plans offered (Hoadley et al. 2008). This fragmentation may at least partly explain why American seniors still experience far greater CRNC than Canadians (25.8% in the United States vs. 4.6% in Canada), though these data were gathered only a year after the implementation of Medicare Part D, and thus may transient health inequalities between the two nations that will diminish over time (Kennedy and Morgan 2009). Another contributor probably lies in the so-called "doughnut-hole" that leaves some seniors vulnerable to all but the most catastrophic medication expenses. One unintended consequence of these out-of-pocket payments in both countries could be additional costs to the health care system, since medication noncompliance is known to lead to worse health and increased emergency department visits.

Long-Term Care

Two main parallel trends are unavoidable for students of health care systems in our graying countries: a fairly high demand in LTC services on the one hand, and an increase in service fragmentation within and between states and provinces – as opposed to an integrated system carefully monitored by governments – on the other. Most explanations of these two opposite trends tend to point to the fact that the increasingly rapid aging of the population tests the institutional limits established at the inception of both Canadian and American Medicare programs. Therefore, LTC largely falls outside the purview of these programs.

Perhaps more than any other type of health service, the delivery of LTC (and particularly home care) must contend with highly individual issues of human dignity and privacy, as well as the social consequences of disability for caregivers and the broader community. Moreover, because of the multifactorial nature of disability, these services are difficult to standardize according to a unique clinical protocol, and are most efficacious when the response varies according to individual needs. This complexity in service delivery means that a social consensus has yet to be reached on achieving equitable access to and better quality of LTC, as this entails both integrating diverse resources (a number of them falling outside of the purview of the medical system) to provide a continuum of care to elderly patients and enhancing support for informal caregivers (Merlis 2000).

Some degree of fragmentation may be unavoidable, considering the needs unique to each patient's idiosyncratic experience of aging; however, experiments in both countries *facilitating* a continuum of care across diverse LTC service providers have yielded interesting results, though they are still confined to local settings, such as the SIPA and PRISMA programs (Béland et al. 2006; Hébert et al. 2003). A large part of the current demand for LTC has so far been absorbed by informal caregivers, and funded through out-of-pocket payments. However, the stakes are likely to grow in the face of baby boomers' increasing demand on these services, longer life expectancy, and smaller families (with the commensurate decreases in informal caregivers this may entail). Thus, the calls for reform are likely to grow as disparities among elderly individuals with disabilities will become an apparent consequence of the steady rise in unmet LTC needs in this population.

End-of-Life Care

Our review indicated that not only are trends in EOL changing in both countries, but so are out-of-pocket expenditures for services not covered by public health insurance. In both countries, these expenses tend to be offset by family members, which means that EOL care in general and the location of death in particular are now becoming socially stratified. This should serve as a reminder that social inequalities in health exist not only in life, but death as well. In light of these findings, it may be useful to draw from Wilson et al.'s (2008) recommendations for best practices in EOL care. In their exhaustive review of the data and literature, these authors identified four essential components of an ideal EOL model of care which include: universality, care coordination, assured access to a broad range of basic and advanced EOL services, and EOL care provision regardless of care setting. By being mindful of the potential inequalities that can be caused by governmental policies such as deinstitutionalization, we can grow towards greater equity in public health – even in death.

Avenues for Policy Reform

The examples above illustrate that, while stemming from very different institutional sources, we observe similar challenges across Canada and the United States with regards to equity in prescription drug coverage, LTC, and EOL care.

In Canada these challenges are born out of a national health insurance system that was developed around the insurance and provision of physician and hospital services. Services falling out of this strict biomedical purview have been left to be funded and provided by the private market for the most part, with the concomitant fragmentation and inequities in access that ensued. This, despite being a universal health system as far as hospitalization and physician services are concerned, Canada leaves domains that are not covered by these categories essentially unregulated, and open to private markets. This tends to disproportionately affect the elderly, as many of the services most salient for this population are either highly individualized – and thus poorly addressed by central organizations – or fall in a gray zone that bridges the social and medical policy realms.

In the United States a social consensus around the importance of providing care for the elderly was reached, in part thanks to the notion that this population "deserves" these services for having paid into the system for most of their working lives. This has led to the coverage of the elderly population through a public system that is funded through social security, and is therefore congruent in its design with the moral underpinning that justified it. Because they share many of the characteristics of a single-payer model, Medicare and Medicaid must also address the issues of responsiveness that were highlighted above for Canada. Of course, because of the overall orientation of the U.S. health system, the private market is also by design never very far, and low levels of public regulation contribute to inequity among the elderly population, as with LTC and EOL. Yet, where the private market is publicly regulated, as with prescription drug coverage, the United States has the potential to reach a higher uniformity of coverage than in Canada.

The challenges of a centralized organization of services for LTC and EOL we outlined earlier, along with the sometimes difficult to reconcile social and medical goals of these services may complicate the extension of these services under a strictly publicly funded model. In other words, the private offering of these services is likely to persist, and perhaps for good reasons. However, governments can and must intervene to regulate these markets to ensure equitable access and fairness of financial contribution, two of the primary goals of high-performing health systems (World Health Organization 2000). To be clear, we are not arguing here necessarily for increased public funding (although that would certainly help) but rather for greater public regulation of the existing private markets.

Thus, both countries could benefit from a greater emphasis towards a publicly regulated, mandated insurance market for the services we studied here. The mandate for individual insurance (i.e., that all elderly must obtain insurance of some form) is an important condition here, as it ensures adequate pooling of (financial) risk between those who need the least and those who need the most care. Furthermore, a necessary condition to attaining the goals stated above would be to ensure greater public regulation of private contracts. Notably, as health conditions can deteriorate rapidly from one year to the next among the elderly, this would entail public protection from frequent renewals of contracts (and the afferent hike in rates as actuarial risk increases), or even lifetime cover as is currently being considered in Australia (Palangkaraya and Yong 2007). Finally, public insurance options that offer more flexibility than the existing public services, and that would be made available on a means-tested or even universal (for the whole population of the elderly) basis constitute a pillar of a responsive policy aimed at equitable access.

In terms of the likelihood of these reforms, Canada and the United States both suffer from inequitable regional variations and benefit from the fragmentation of power between the federal and sub-national units. Indeed, while this fragmentation can bring challenges in reaching a uniform national policy, it also allows for innovation and later diffusion (as was the case in Canada with national insurance, for instance). Comparatively, as Medicare Part D shows, the United States may have more structures already in place to enact a sweeping reform, such as the progressive "closing of the doughnut hole" that was indeed proposed in the Affordable Care Act passed in March 2010 (U.S. Department of Health and Human Services 2011). However, as the epic challenges waged against this bill since it was passed tellingly remind us the U.S. political system in and of itself can also constitute a significant roadblock that can trump the implementation of evidence-based policies.

Thus, beyond organizational capacity and political context, perhaps the most important criteria for reform in these countries would be a social consensus around the need for more equitable access and fairness of financing of those services for the elderly. At a cultural level, stronger public regulation of private contracts (for the elderly) should be fairly viable in both countries. A mandate for insurance may be slightly more contentious in the United States than in Canada, but given the existence of Medicare (in itself a mandated program), it may be more palatable among the elderly than the working-age population. Finally, while a public option has met strong resistance from the working-age population in the U.S., again, the existence of Medicare in the United States and of a national health system in Canada may provide the necessary cultural and organizational foundations for this in those countries. The case of Quebec, a province which requires mandatory health insurance coverage for prescription drugs and consequently has the lowest CRNC rate in both countries, shows to what extent mandatory insurance can help mitigate social inequalities (Kennedy and Morgan 2009).

In terms of the challenges specific to each country, Canada probably needs to supplement its general population health and equity perspective with a stronger focus on certain underserved subpopulations, including the elderly (Frohlich and Potvin 2008). Indeed, provisions are already made in a number of provinces to protect this population from out-of-pocket expenses for prescription drugs for instance, but the basket of services offered by the public medical system still overwhelmingly consists of curative medicine, which is not optimal for a population dealing primarily with chronic diseases, and managing disability. In turn, the United States has covered, and keeps covering, the elderly on the premise that they deserve this support (while working-age individuals do not deserve such support, as per the principle of individual responsibility), but there, the population needs to realize that health must be also nurtured across the life course, and not just starting at 65.

What Next for Sociological Research?

One of the main challenges to studying the impact of policy is lack of variation in the population's exposure to national legislation. Yet, by virtue of the fragmentation of our health systems, natural policy experiments are happening everywhere in Canada and the United States. The shared challenges we have uncovered in this chapter mean that we can learn from one another, and that cross-nationally comparative research would prove to be a particularly advantageous strategy here.

Moreover, as sociologists, we are uniquely positioned to make a significant contribution in this field. Indeed, the focus in evaluating the impact of those policy interventions (to the extent that there is any evaluation at all) is often towards health outcomes, which obscures the existence, persistence, or even development of social inequalities. Yet, as Foner (2000) reminds us, sociologists are uniquely poised (and tooled) to uncover those processes.

Thus, through these very different institutional pathways, both of these countries find themselves with very similar needs for adaptation posed by this common demographic process of population aging. The ball is now in our camp to evaluate ongoing interventions and their impact on equitable access and fairness of financing among the elderly so that evidence-based policies that seek those goals may eventually prevail.

References

Alfons Palangkaraya & Jongsay Yong, 2007. "How effective is "lifetime health cover" in raising private health insurance coverage in Australia? An assessment using regression discontinuity," *Applied Economics, Taylor and Francis Journals*, vol. 39(11), pages 1361-1374.

Anis, Aslam H., Daphne Guh, and Xiao-Hua Wang. 2001. "A Dog's Breakfast: Prescription Drug Coverage Varies Widely Across Canada." *Medical Care* 39:315–26.

Béland, Francois, Howard Bergman, Paule Lebel, Luc Dallaire, John Fletcher, André Pierre Contandriopoulos, and Pierre Tousignant. 2006. "Integrated Services for Frail Elders (SIPA): A Trial of a Model for Canada." *Canadian Journal on Aging-Revue Canadienne Du Vieillissement* 25:5–42.

Beveridge, William. 1942. "Social Insurance and Allied Services." HMSO, Cmd. 6404, London.

Bird, Chloe E., Lisa R. Shugarman, and Joanne Lynn. 2002. "Age and Gender Differences in Health Care Utilization and Spending for Medicare Beneficiaries in Their Last Years of Life." *Journal of Palliative Medicine* 5:705–12.

Boychuk, Gerard W. and Keith G. Banting. 2008. "The Canadian Paradox: The Public-Private Divide in Health Insurance and Pensions." Pp. 140–93 in *Public and Private Social Policy: Health and Pension Policies in a New Era*, edited by D. Béland and B. Gran. New York: Palgrave Macmillan.

Brackley, Moyra E. and Margaret J. Penning. 2009. "Home-Care Utilization Within the Year of Death: Trends, Predictors and Changes in Access Equity During a Period of health policy reform in British Columbia, Canada." *Health & Social Care in the Community* 17:283–94.

Briesacher, Becky A., Jerry H. Gurwitz, and Steven B. Soumerai. 2007. "Patients At-Risk for Cost-Related Medication Nonadherence: A Review of the Literature." *Journal of General Internal Medicine* 22:864–71.

Briggs, Asa. 2000. "The Welfare State in Historical Perspective." Pp. 16–29 in *The Welfare State Reader*, edited by C. Pierson and F. G. Castles. Cambridge: Polity.

Buntin, Melinda Beeuwkes and Haiden Huskamp. 2002. "What Is Known About the Economics of End-of-Life Care for Medicare Beneficiaries?" *Gerontologist* 42:40–8.

Canadian Healthcare Association. 2009. "Home Care in Canada: From the Margins to the Mainstream" *Policy Brief.*, Retrieved 30 October, 2009 (http://www.cha.ca/index.php?option=com_content&view=article&id=185&Itemid=101).

Canadian Institute for Health Information. 2007. *Health Care Use at the End of Life in Western Canada*. Ottawa: CIHI.

Carriere, Gisele. 2006. *"Seniors" Use of Home Care,"* Health Reports 17:4, pp. 43–47 [Statistics Canada Cat. 82-003].

CDSH. 2008. *Closing the Gap in a Generation: Health Equity Through Action on the Social Determinants of Health. Final Report of the Commission on Social Determinants of Health*. World Health Organization, Geneva.

Centers for Medicare and Medicaid Services and U.S. Department of Health and Human Services. 2009. *Medicare & You* 2009. Washington D.C.: Department of Health and Human Services.

Chappell, Neena L., Betty H. Dlitt, Marcus J. Hollander, Jo Ann Miller, and Carol McWilliam. 2004. "Comparative Costs of Home Care and Residential Care." *Gerontologist* 44:389–400.

Chochinov, Harvey M. and Krist Janson. 1998. "Dying to Pay: The Cost of End-of-Life Care." *Journal of Palliative Care* 14:5–15.

Demers, Marie 1998. "Age Differences in the Rates and Costs of Medical Procedures and Hospitalization During the Last Year of Life." *Canadian Journal on Aging-Revue Canadienne Du Vieillissement* 17:186–96.

Demers, Virginie, Magda Melo, Cynthia Jackevicius, Jafna Cox, Dimitri Kalavrouziotis, Stéphane Rinfret, Karen H. Humphries, Helen Johansen, Jack V. Tu, and Louise Pilote. 2008. "Comparison of Provincial Prescription Drug Plans and the Impact on Patients' Annual Drug Expenditures." *Canadian Medical Association Journal* 178:405–9.

Dy, Sydney M., Jennifer L. Wolff, and Kevin D. Frick. 2007. "Patient Characteristics and End-of-Life Health Care Utilization Among Medicare Beneficiaries in 1989 and 1999." *Medical Care* 45:926–30.

Emanuel, Ezekiel J. 1996. "Cost Savings at the End of Life: What Do the Data Show?" *Journal of the American Medical Association* 275:1907–14.

Feder, Judith. 2000. "Long-Term Care in the United States: An Overview." *Health Affairs* 19:40–56.

Flory, James, Yinong Young-Xu, Ipek Gurol, Norman Levinsky, Arlene Ash, and Ezekiel Emanuel. 2004. "Place of Death: U.S. Trends Since 1980." *Health Affairs* 23:194–200.

Foner, Anne. 2000. "Age Integration or Age Conflict as Society Ages?" *Gerontologist* 40:272–76.

Frencher, Stanley K. and Sherry Glied. 2006. "The Medicare Prescription Drug Improvement and Modernization Act: Prescription Drugs and Academic Medicine." *Academic Medicine* 81:812–6.

Fried, Terri R., Daphna M. Pollack, Margaret A. Drickamer, and Mary E. Tinetti. 1999. "Who Dies at Home? Determinants of Site of Death for Community-Based Long-Term Care Patients." *Journal of the American Geriatrics Society* 47:25–9.

Frohlich, Katherine L. and Louise Potvin. 2008. "The Inequality Paradox: The Population Approach and Vulnerable Populations." *American Journal of Public Health* 98:216–21.

Gapminder Foundation. 2010, "Life Expectancy at Birth", Retrieved February 10, 2010 (www.gapminder.org).

Gellad, Walid F., Haiden A. Huskamp, Kathryn A. Phillips, and Jennifer S. Haas. 2006. "How the New Medicare Drug Benefit Could Affect Vulnerable Populations." *Health Affairs* 25:248–55.

Gleckman, Howard. 2009, "Will Long-Term Care Be Included In Health Reform?" *Kaiser Health News*. Retrieved July 29, 2010 (http://www.kaiserhealthnews.org/Columns/2009/September/091409Gleckman.aspx).

Grabowski, David C. 2008. "The Market for Long-Term Care Services." *Inquiry* 45:58–74.

Gray, Jean 2005. "Experience with Providing Drugs for Seniors in Canada." *American Journal of Geriatric Cardiology* 14:313–6.

Grootendorst, Paul. 2002. "Beneficiary Cost Sharing Under Canadian Provincial Prescription Drug Benefit Programs: History and Assessment." *Canadian Journal of Clinical Pharmacology* 9:79–99.

Gruneir, Andrea, Vincent Mor, Sherry Weitzen, Rachael Truchil, Joan Teno, and Jason Roy. 2007. "Where People Die: A Multilevel Approach to Understanding Influences on Site of Death in America." *Medical Care Research and Review* 64:351–78.

Hartman, Micah, Aaron Catlin, Davis Lassman, Jonathan Cylus, and Stephen Heffler. 2008. "U.S. Health Spending By Age, Selected Years Through 2004." *Health Affairs* 27:w1–12.

Hébert, Réjean, Pierre J. Durand, Nicole Dubuc, André Tourigny, and PRISMA Group. 2003. "PRISMA: A New Model of Integrated Service Delivery for the Frail Older People in Canada." *International Journal of Integrated Care*, 3: 1–8.

Heisler, Michele, Kenneth M. Langa, Elizabeth L. Eby, A. Mark Fendrick, Mohammed U. Kabeto, and John D. Piette. 2004. "The Health Effects of Restricting Prescription Medication Use Because of Cost." *Medical Care* 42:626–34.

Hoadley, Jack, Elizabeth Hargrave, Juliette Cubanski, and Tricia Neuman. 2008. "Medicare Prescription Drug Plans in 2008 and Key Changes Since 2006: Summary of Findings." *Kaiser Family Foundation*. Retrieved on October 12, 2009 (www.kff.org/medicare/7762.cfm).

Hogan, Christopher, June Lunney, Jon Gabel, and Joanne Lynn. 2001. "Medicare Beneficiaries' Costs of Care in the Last Year of Life." *Health Affairs* 20:188–95.

Hollander, Marcus J., Neena L. Chappell, Michael J. Prince, and Evelyn Shapiro. 2007. "Providing Care and Support for an Aging Population: Briefing Notes on Key Policy Issues." *Healthcare Quarterly* 10:34–45.

Hoover, Donald R., Stephen Crystal, Rizie Kumar, Usha Sambamoorthi, and Joel C. Cantor. 2002. "Medical Expenditures during the Last Year of Life: Findings from the 1992–1996 Medicare Current Beneficiary Survey." *Health Services Research* 37:1625–42.

Kaiser Family Foundation. 2009. "The Medicare Prescription Drug Benefit – An Updated March 2009 Fact Sheet.", Retrieved September 3, 2009 (http://www.kff.org/medicare/7044-02.cfm).

Keefe, Janice, Jacques Légaré, and Yves Carrière. 2007. "Developing New Strategies to Support Future Caregivers of Older Canadians with Disabilities: Projections of Need and Their Policy Implications." *Canadian Public Policy* 33:65–80.

Kennedy, Jae, Joseph Coyne, and David Sclar. 2004. "Drug Affordability and Prescription Noncompliance in the United States: 1997–2002." *Clinical Therapeutics* 26:607–14.

Kennedy, Jae and Steve Morgan. 2006. "A Cross-National Study of Prescription Nonadherence due to Cost: Data from the Joint Canada-United States Survey of Health." *Clinical Therapeutics* 28:1217–24.

———. 2009. "Cost-Related Prescription Nonadherence in the United States and Canada: A System-Level Comparison Using the 2007 International Health Policy Survey in Seven Countries." *Clinical Therapeutics* 31:213–9.

Komisar, Harriet L. and Lee Shirley Thompson. 2007. "National Spending for Long-Term-Care." Georgetown University Long-Term Care Financing Project, Washington D.C. Retrieved July 29, 2010 (http://ltc.georgetown.edu).

Komisar, Harriet L., Judith Feder, and Judith D. Kasper. 2005. "Unmet Long-Term Care Needs: An Analysis of Medicare-Medicaid Dual Eligibles." *Inquiry* 42:171–82.

Lafortune, Gaetan, Gaelle Balestat, and Disability Study Expert Group Members. 2007. "Trends in Severe Disability Among Elderly People: Assessing the Evidence in 12 OECD Countries and the Future Implications." in *OECD Health Working papers*: OECD. Retrieved October 14, 2009 (http://www.oecd.org/dataoecd/13/8/38343783.pdf).

LaPierre, Tracey A. and Mary Elizabeth Hughes. 2009. "Population Aging in Canada and the United States." Pp. 191–230 in *International Handbook of Population Aging*, edited by P. Uhlenberg. New York: Springer.

Levinsky, Norman G., Wei Yu, Arlene Ash, Mark Moskowitz, Gail Gazelle, Olga Saynina, and Ezekiel J. Emanuel. 2001. "Influence of Age on Medicare Expenditures and Medical Care in the Last Year of Life." *Journal of the American Medical Association* 286:1349–55.

Liebig, Phoebe. 1993. "Federalism and Suitable Housing for the Frail Elderly: A Comparison of Policies in Canada and the United States." *Housing Policy Debate* 4:199–237.

Lubitz, James D. and Gerald F. Riley. 1993. "Trends in Medicare Payments in the Last Year of Life." *New England Journal of Medicine* 328:1092–6.

Maio, Vittorio, Laura Pizzi, Adam R. Roumm, Janice Clarke, Neil I. Goldfarb, David B. Nash, and David Chess. 2005. "Pharmacy Utilization and the Medicare Modernization Act." *Millbank Quarterly* 83:101–30.

Maioni, Antonia. 1997. "Parting at the Crossroads: The Development of Health Insurance in Canada and the United States, 1940–1965." *Comparative Politics* 29:411–31.

———. 1998. *Parting at the Crossroads: The Emergence of Health Insurance in the United States and Canada*. Princeton, NJ: Princeton University Press.

Marchildon, Gregory P. 2005. *Health Systems in Transition: Canada*. WHO Regional Office for Europe on Behalf of the European Observatory on Health Systems and Policies, Copenhagen.

Merlis, Mark. 2000. "Caring for the Frail Elderly: An International Review." *Health Affairs* 19:141–9.

Messinger-Rapport, Barbara. 2009. "Disparities in Long-Term Healthcare." *The Nursing clinics of North America* 44:179–85.

Morgan, Steven G. 2006. "Prescription Drug Expenditures and Population Demographics." *Health Services Research* 41:411–28.

Morgan, Steve G., Jonathan D. Agnew, and Morris L. Barer. 2004. "Seniors' Prescription Drug Cost Inflation and Cost Containment: Evidence from British Columbia." *Health Policy* 68:299–307.

Motiwala, Sanober S., Ruth Croxford, Denise N. Guerriere, and Peter C. Coyte. 2006. "Predictors of Place of Death for Seniors in Ontario: A Population-Based Cohort Analysis." *Canadian Journal on Aging-Revue Canadienne Du Vieillissement* 25:363–71.

Muramatsu, Naoko, Ruby L. Hoyem, Hongjun J. Yin, and Richard T. Campbell. 2008. "Place of Death Among Older Americans: Does State Spending on Home- and Community-Based Services Promote Home Death?" *Medical Care* 46:829–38.

O'Malley Watts, Molly. 2009. "The Community Living Assistance Services and Supports (CLASS) Act." in *Focus on Health Reform*: Kaiser Family Foundation. Retrieved October, 19, 2009 (http://www.kff.org/healthreform/upload/7996.pdf).

OECD. 2004, "Drug Expenditure per Capita, Public and Private expenditure, OECD Countries, 2004." Retrieved October 19, 2009 (http://www.oecd.org/dataoecd/5/24/36984917.pdf).

Olshaker, Jonathan S., Robert A. Barish, Jerome F. X. Naradzay, David A. Jerrard, Ellen Safir, and Larry Campbell. 1999. "Prescription Noncompliance: Contribution to Emergency Department Visits and Cost." *Journal of Emergency Medicine* 17:909–12.

Poisal, John A. and Lauren Murray. 2001. "Growing Differences Between Medicare Beneficiaries With and Without Drug Coverage." *Health Affairs* 20:74–85.

Polivka, Larry and Helen Zayac. 2008. "The Aging Network and Managed Long-Term Care." *Gerontologist* 48:564–72.

Pomey, Marie Pascale, Pierre Gerlier Forest, Howard A. Palley, and Elisabeth Martin. 2007. "Public/Private Partnerships for Prescription Drug Coverage: Policy Formulation and Outcomes in Quebec's Universal Drug Insurance Program, With Comparisons to the Medicare Prescription Drug Program in the United States." *Milbank Quarterly* 85:469–98.

Rantz, Marilyn J., Karen D. Marek, and Mary Zwygart-Stauffacher. 2000. "The Future of Long-Term Care for the Chronically Ill." *Nursing Administration Quarterly* 25:51–8.

Rector, Thomas S. and Patricia J. Venus. 2004. "Do Drug Benefits Help Medicare Beneficiaries Afford Prescribed Drugs?" *Health Affairs* 23:213–22.

Régie de l'assurance maladie du Québec. 2009. "Prescription Drug Insurance," Retrieved 1 October 5, 2009 (http://www.ramq.gouv.qc.ca/en/citoyens/assurancemedicaments/regimepublic/lescouts_laprime.shtml).

Robinson, Karen M. and Susan C. Reinhard. 2009. "Looking Ahead in Long-Term Care: The Next 50 Years." *The Nursing Clinics of North America* 44:253–62.

Roemer, Milton Irwin. 1991. *National Health Systems of the World,* Vol. 1. Oxford: Oxford University Press.

Scitovsky, Anne A. 1994. "The High Cost of Dying Revisited." *Milbank Quarterly* 72:561–91.

Shah, N., J. Mullahy, M. A. Smith, and D. J. Vanness. 2006. "Determinants of Growth in Prescription Drug Expenditures." *Paper presented at the annual meeting of the Economics of Population Health: Inaugural Conference of the American Society of Health Economists, TBA, Madison, WI, USA*. Retrieved May 25, 2009 (http://www.allacademic.com/meta/p93362_index.html).

Social Security Online. 2010, "Age 65 Retirement." Retrieved February 10, 2010 (http://www.ssa.gov/history/age65.html).

Sultz, Harry A. and Kristina M. Young. 2009. "Long-Term Care." Pp. 279–315 in *Health Care USA: Understanding its Organization and Delivery,* Sudbury, Mass.: Jones and Bartlett Publishers.

Thurow, Lester C. "The Birth of a Revolutionary Class." *New York Times Magazine*, May 19, 1996, pp. 46–7.

U.S. Department of Health and Human Services. 2011. "Understanding the Affordable Care Act: About the Law." Retrieved: February 8, 2011 (http://www.healthcare.gov/law/about/index.html).

Wiener, Joshua W. and Jane Tilly. 2002. "Population Ageing in the United States of America: Implications for Public Programmes." *International Journal of Epidemiology* 31:776–81.

Wilson, Donna M., Stephen Birch, Sam Sheps, Roger Thomas, Christopher Justice, and Rod MacLeod. 2008. "Researching a Best-Practice End-of-Life Care Model for Canada." *Canadian Journal on Aging-Revue Canadienne Du Vieillissement* 27:319–30.

Wilson, Donna M., Corrine D. Truman, Roger Thomas, Robin Fainsinger, Kathy Kovacs-Burns, Katherine Froggatt, and Christopher Justice. 2009. "The Rapidly Changing Location of Death in Canada, 1994–2004." *Social Science & Medicine* 68:1752–8.

World Health Organization. 2000. *The World Health Report 2000 – Health Systems: Improving Performance*. Geneva: World Health Organization.

———. 2002. *Ethical Choices in Long-Term Care: What Does Justice Require?* Geneva: World Health Organization.

Chapter 36
Long-Term Care: Tradition and Innovation

Eva Kahana, Loren Lovegreen, and Boaz Kahana

The goal of this chapter is to explore both conceptual and empirical advances in understanding the nature, determinants, social processes, and impact of long-term care (LTC) provided to elders in the United States. We call attention to relevant conceptual and research literature that can help readers connect current ideas and approaches to prior seminal research and theoretical formulations.

It is well established that most older persons prefer to age in place (Wagnild 2001) and enter institutional facilities only when other options are unavailable (Kane 2001). LTC institutions serve an important function by offering medical and residential services to older adults who suffer from chronic illnesses, accompanied by physical, sensory, or cognitive impairments (Stone 2006). Such care becomes necessary when informal assistance by family (or friends and neighbors) is not available or is insufficient to meet the care needs of an older adult (Charles and Sevak 2005). The concept of LTC designates services for meeting the ongoing needs of older adults for assistance with activities of daily living. Such services are not limited to residential care but can also include home care or day care. However, the focus of this chapter is on institutional LTC which is the predominant approach in meeting the needs of frail elders in the US (Kane 2009).

With the rapid aging of the population and increases in the ranks of the very old, demand for LTC is accelerating (Johnson et al. 2007) and the delivery of such care poses a major challenge to social policy and practice in the United States and around the world (Kane 2009; Settersten 2007). Key unresolved policy issues in this field include the balance between institutional and noninstitutional alternatives, coordination of state with federal financing of care, coordination between acute care and LTC, and providing access without excessive burdens to consumers (Feder et al. 2000). In addition, a major challenge relates to finding mechanisms for ensuring provision of high-quality care. Such care allows for maintenance of good quality-of-life among residents, even in the face of their physical and/or cognitive limitations (Kane 2003). We focus on social processes that shape the daily lives of elderly persons who are care recipients along the spectrum of LTC settings.

Our discussion calls attention to important early work that can serve as scaffolding for future research and practice to insure that foundational sociological approaches are not obscured or forgotten. We consider perspectives of diverse stakeholders in LTC, ranging from families of elders to direct service providers, professional staff, and policy makers. Even as we attempt to take a broad view of LTC, our central interest remains with the older care-receiver whose lived experience is most fundamentally shaped by LTC (Kahana et al. 2009). Taking a sociological approach, we highlight the complexities, the human dimensions, and the countervailing values involved in caring for frail older adults.

E. Kahana (✉)
Department of Sociology, Case Western Reserve University, Cleveland, OH, USA
e-mail: eva.kahana@case.edu

R.A. Settersten, Jr. and J.L. Angel (eds.), *Handbook of Sociology of Aging*, Handbooks of Sociology and Social Research, DOI 10.1007/978-1-4419-7374-0_36,

One useful approach for organizing our understanding of LTC for the aged has been provided by Stone (2006). She refers to the triple knot of LTC – financing, type of setting, and work force issues – as the defining features of care delivery. A second useful organizing framework has been offered in the work of Kane (2001) who considers important linkages between quality of LTC and residents' quality-of-life. She relates structural and organizational features of LTC to indicators of multidimensional quality-of-life articulated in the gerontological literature (Abeles et al. 1994). Although we recognize the value of structural approaches that contribute to quality of care, we direct our major attention to the social processes that link quality of care to residents' quality-of-life that are rooted in prior sociological theorizing (Goffman 1961; Sommer and Osmond 1961) but have been less central in recent treatises on LTC.

Important emerging themes in the literature on LTC reflect efforts at reform, directed at enhancing both the quality of care and quality of resident life. The most prominent among the recently implemented approaches is the culture change movement (for a review and critique see Rahman and Schnelle 2008). There is also a growing literature focused on more homelike alternatives in LTC such as assisted living facilities (ALFs) (Kane and Cutler 2009; Zimmerman et al. 2001) and continuing care retirement communities (CCRCs) (Sherwood 1997; Shippee 2009).

The theme of our discussion is *tradition and innovation*, as we explore the impact of LTC on care-receivers' quality-of-life. The structural model guiding this discussion is depicted in Fig. 36.1. In our model, we relate personal characteristics of the resident (component A) to the type of LTC settings in which the older adult resides (component B). We also briefly consider the environmental context of LTC settings (component C) that shape the quality of care delivered (component D). Based on our interest in social processes that define residents' lived experience, we next focus on linkages between quality of care (component D) and resident quality-of-life (component E). We argue that the type of LTC setting influences the quality of care provided, indirectly via the

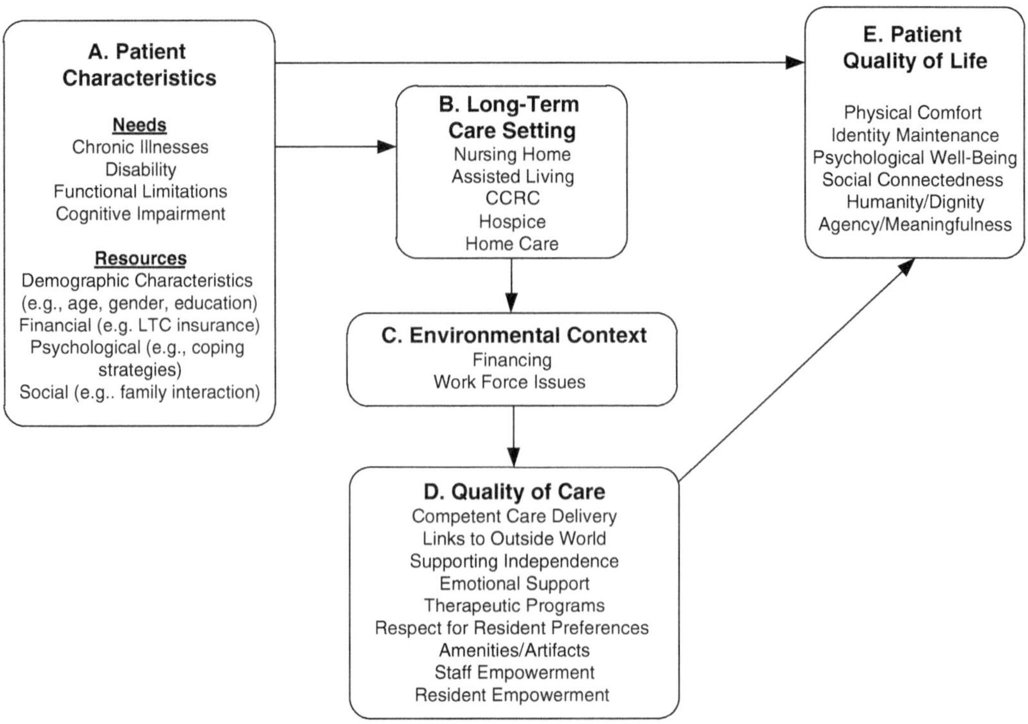

Fig. 36.1 Structural model of long-term care impact

environmental and social milieu (e.g., financing, workforce issues). Our structural model also recognizes that beyond residents' quality of care, their quality-of-life is also affected by their own needs and resources. Thus, component A has a direct impact on component E. These characteristics cut across diverse theoretical models and may receive greater or lesser emphasis in alternative conceptualizations. Our model depicts a range of LTC settings including nursing homes, along with newer LTC options of ALF, CCRCs, and hospice programs. Notably, the majority of work focusing on LTC from the perspective of sociology relates to nursing home care (Goffman 1961; Gubrium 1975; Kahana 1973).

In the discussion of our structural model (Fig. 36.1) we offer a brief description of alternative LTC settings and the key environmental characteristics of financing of care and workforce issues. We then turn to a review of social processes that link quality of care with quality-of-life in LTC facilities. The latter discussion is organized utilizing a concept map presented in Table 36.1. This map specifies theoretical orientations used in the study of social processes in LTC primarily in sociology, but also includes relevant models from other disciplines. We also identify programs, policies, and practices associated with each model that proponents of conceptual models or others have developed.

Table 36.1 Concept map: processes of quality of care and quality-of-life

Theoretical/conceptual model	Quality of care: environmental and social processes	Quality-of-life: impact on resident's physical and psychosocial well-being and identity	Program, policy, and practice implications
1. Medical Model			
The Sick Role (Parsons 1951, 1975)	Efficient physical care delivery Ensure patient safety	"Patient" identity Conformity	Patient assessment Patient monitoring Reporting of adverse events
2. Institutional Analysis Models			
Total Institution (Goffman 1961; Sommer and Osmond 1961; Kleemeier 1963)	Congregate Segregate Institutional control	Mortification of self Stripping De-individuation	Homelike environment Age integration Humanizing care
Dependency Inducing Environments (Langer 1975; Baltes et al. 1980)	Reinforcement of dependency Staff control of environment	Learned helplessness Loss of control	Culture change movement Environmental flexibility Support personhood Aid self-determination Enhance patient control
3. Ecological Models			
Stress Model (Schooler 1982)	Threats and hazards	Diminished health and well-being	Reduce stressors Improve coping skills Increase social support
Environment Press/Adaptation (Lawton 1982)	Adaptive capacities of resident taxed by environment	Environmental docility	Therapeutic programs to enhance competence Environmental design
Congruence Model of Person Environment Fit (P–E Fit) (Kahana 1982)	Lack of P–E Fit	Low morale Residential dissatisfaction	Increase P–E Fit
4. Ethnographic Model			
Social Construction of Everyday Life (Clark and Anderson 1967; Henry 1963; Gubrium 1975)	Indignities of daily life Assaults on the self Front stage and back stage of care	Suffering Negative transformations of identity Broken social connections	Life review to restore selfhood Community building in LTC Personal testimony to effect legislation

(continued)

Table 36.1 (continued)

Theoretical/conceptual model	Quality of care: environmental and social processes	Quality-of-life: impact on resident's physical and psychosocial well-being and identity	Program, policy, and practice implications
5. Political Economy Model			
Structures of Domination (Estes 1979)	Commodification of care Exploitation of staff Patient abuse/neglect	Physical harm Alienation	Staff empowerment Collective bargaining Regulation
6. Advocacy Models			
Consumerism/Self-advocacy (Haug and Lavin 1983)	Demand patient rights Self-care	Responsive care getting	Skill-building for self-advocacy
Empowerment/Family Advocacy (Cox and Parsons 1994)	Family involvement Accountability in care	Self-esteem Self-efficacy	Family/staff collaboration Family/staff confrontation
Activism/Community Advocacy (Kane 2009)	Community presence	Self-actualization	Action group alliances Stakeholders demand legislative changes

Long-Term Care Settings

Our discussion covers key LTC settings noted in our structural model. However, additional LTC options include adult foster care (Kane et al. 1991), small board and care homes (Carder et al. 2008), and home-sharing arrangements (Pritchard 1983). Furthermore, naturally occurring retirement communities (NORCs) have also received attention as growing options in LTC (Bookman 2008; Hunt and Gunter-Hunt 1985). Even though different types of LTC settings are often treated as distinct entities, there is growing concern about the overlap between traditional nursing homes and ALF settings, with the latter assuming more institutional properties (Kane and Cutler 2009). Moreover, LTC settings are increasingly diversified in the types of care provided. For example, within each type of setting there may be special units, such as dementia care, rehabilitation, or hospice that constitute distinct entities.

Nursing Homes

Nursing homes have served as the primary source of institutional care for older adults since the 1930s (Mollica 2001). They provide a broad spectrum of services for people who often have multiple disabilities and who suffer from physical, sensory, and cognitive impairments. These residents also need extensive assistance with activities of daily living (e.g., bathing, eating, and dressing) (Jones et al. 2009). Typically those with severe dementia are now cared for in special units that provide environmental support and therapeutic programs (Zeisel et al. 2003). Nursing homes also provide short-term rehabilitation for discharged hospital patients (Kane 2005).

In 2004, there were 16,100 nursing homes in the United States, with a majority operating as a for-profit organization (Jones et al. 2009). Rates of institutionalization for the very old are high, with 17% for those aged 75 years and older residing in nursing homes in 2005 (Jones et al. 2009). However, the overall proportion of elderly persons (over age 65) in nursing homes has declined in recent years, from 7.5% in 1982 to 4% in 2005 (Manton et al. 2006). This may reflect decreasing disability rates among older adults and a wider range of LTC options. Based on 2008 U.S. Census data, 68% of nursing home residents are Caucasian women and the median age is 83.2 years (Jones et al. 2009). The under-representation of minority elders in diverse LTC facilities reflects structural constraints in access to care (Quadagno 1999).

Assisted Living Facilities

Since the 1990s assisted living has become an increasingly popular alternative in LTC in the United States for older adults with financial means (Zimmerman et al. 2001). At present there are approximately 39,500 ALFs in the U.S (AAHSA 2009). The rise of ALFs is largely a response to the desire among older adults to avoid nursing home placement and reflects the medical model of care (Kane and Cutler 2009). ALFs, at least in principle, support an "aging in place" philosophy. Resident autonomy and privacy within a homelike setting are two central goals of ALFs (Hawes et al. 1999). Housing and nonmedically oriented services are provided to mildly or moderately frail and/or cognitively impaired older adults in ALFs. Older adults typically expect to continue living in these facilities for an extended period of time, even as their needs for care may change (Golant and Hyde 2008). However, restrictive policies at some ALFs require relocation by frailer residents to nursing home settings.

Continuing Care Retirement Community

CCRCs offer a broad range of services within a single setting. Most residents enter into independent living while they are still in good health. They live in private apartments and are provided with meals and recreational services. They are provided with more extensive services sponsored by the same facility when greater service needs arise (Sherwood 1997). Residents with some ADL and IADL needs typically first transfer to ALFs and later, those with more extensive needs, enter the nursing home within the same comprehensive facility (Shippee 2009). Many well-functioning older adults are attracted by the independent living options of CCRCs, while planning ahead for increasingly complex services without burdening family members as they become frailer (Lovegreen, 2010). As of 2009, approximately 1,850 CCRCs were operating in the United States with most (82%) operating as not-for-profit organizations (AAHSA 2009).

Hospice

Hospice care is aimed at meeting the needs of patients (including the elderly) close to the end of life (Saunders 1978). Developed originally in England, the hospice philosophy embraces patient-centered, compassionate care. Incorporating principles of palliative care, the tenets of hospice care include comfort, dignity, meaning, and connectedness for the dying person and their family. Pain management, emotional and spiritual support, family education, counseling, advocacy and bereavement support, are all central goals of hospice care (Saunders 1978). Hospice was originally focused on terminally ill cancer patients; though its clientele has been expanded over time and now includes patients with varying diagnoses including Alzheimer's disease and Parkinson's (Han et al. 2006). However, hospice services are still underutilized, particularly in LTC institutional settings (Miller et al. 2002).

Environmental Context: Financing of LTC

The LTC settings reviewed reflect growing options for older adults requiring assistance with activities of daily living. However, access to such options is often based on financial resources (Stone 2006). States also play a major role in access to LTC through a variety of control mechanisms.

These include regulation of the number of nursing home beds allowed at a given time and reimbursement rates paid to nursing homes for Medicare-eligible residents. States also exert control over home health services (e.g., setting certification criteria for reimbursement), which influences the need for nursing home alternatives. However, state and federal coordination of LTC financing has been limited. As in 2004, Medicaid served as the major source (64%) of LTC financing for nursing home residents (Mollica and Johnson-Lamarche 2005). Many of these individuals initially financed their own care, and Medicaid assumed responsibility only after personal funds were "spent down." Medicaid finances care from the outset for those who completely lack resources. It is important to recognize that the homelike options of CCRCs and ALFs are not readily available to older adults of limited financial means. Those elders with the greatest need and most limited resources have narrower choices and rely predominately on nursing home care (Stone 2006).

Environmental Context: Work Force Issues

The quality-of-life of residents in LTC facilities is largely shaped by the quality of care that they receive from staff working in such settings (Stone 2006). The adequacy of staff has been recognized to play a key role in insuring high quality of LTC. The Institute of Medicine (IOM) report (2001) on improvements in quality of LTC calls for improved compensation and training of staff in LTC facilities. The report emphasizes staff supervision and training, working conditions, and good management practices as important influences on quality of care to residents.

Registered nurses serve primarily in supervisory roles in LTC, whereas paraprofessionals provide hands on care. Many of the challenges for providing good quality of care in LTC facilities have been associated with problems facing direct care workers (DCWs) (Ejaz et al. 2008). In the sociological literature, low job satisfaction and high turnover of DCWs have been associated with the feminization of poverty (Armstrong et al. 2008). Direct care is provided primarily by underpaid, poorly educated women (Harris-Kojetin et al. 2004). Personal stressors experienced by workers, along with job-related stressors and lack of social support, are among the primary factors accounting for low job satisfaction in this group (Ejaz et al. 2008).

Nevertheless, there have been some noteworthy advances in workforce development within LTC, attributable to the increased professionalization of nursing. Nurses have assumed key leadership roles in developing research and training programs relevant to LTC. Indeed, much of the research on specific interventions to enhance caring and to reduce patient suffering has appeared in the nursing literature (Watson 2005). In terms of management of patients in LTC, nursing professionals have been charged with individualized care plans that address both physical and social needs. Care plans are designed to maximize patient functioning through therapeutic programs (Kane 2005).

Processes Linking Quality of Care and Quality-of-Life

In Table 36.1 we provide a concept map that serves as a guide to understanding alternative theoretical formulations about social processes linking quality of care and resident quality-of-life in LTC. It is useful to recognize that this work spans disciplinary boundaries. While our major interest is in sociological orientations, we recognize that theoretical models used in the field of LTC often draw on traditions of other disciplines, including medicine, psychology, anthropology, and nursing. Different orientations are illustrated by the divergent terms used for labeling care-receivers. They are alternatively referred to as the patient (medical model), client (social work/psychology), resident (social models), or consumer (business and advocacy models). The assisted living movement assigns the label of resident as a defining aspect of their form of care. In our recent work we identify the older care recipient as a "care-getter" emphasizing the role of agency in marshaling support in late life (Kahana et al. 2009).

Our discussion of environmental and social processes that influence quality of care in LTC and the resulting impact of residents' quality-of-life follow our concept map. In Table 36.1 we present six broad theoretical models which have shaped our understanding of social processes in delivery of LTC in the United States. The six models include: (1) Medical Model, (2) Institutional Analysis Model, (3) Ecological Model, (4) Ethnographic Tradition (5) Political Economy Model, and (6) Advocacy Model. These are broad models reflecting distinctions in theory, methods, and/or disciplinary orientation. In some cases, related conceptualizations are noted within the broader model. We also briefly note implications of each model for practice, policy, program development, or improvement of care. We recognize that there is some overlap among models, and that some program initiatives such as culture change may address quality of care issues identified in several models. The references in our concept map identify each tradition through some of the leading proponents or foundational scholars who are associated with conceptualizing the model. Influential theorists, whose work paved the way for different models, outside of aging, medical sociology, or LTC are noted in the text describing each model. We also reference contributions to research and program initiatives in the text rather than the concept map.

The Medical Model

The medical model is the most prevalent paradigm in the delivery of LTC. The nursing home industry has modeled care delivery to patients on the acute care medical model. Indeed, many nursing home residents are transferred to LTC after hospitalization (Kane 2005). Although the medical model of nursing home care mimics many aspects of hospital care, it is noteworthy that physicians are largely absent in LTC facilities (Katz and Karuza 2005). Nurses who are linked to the very identity of the nursing home are also in short supply and are largely responsible for administrative and supervisory functions. Nursing assistants represent the key direct caretakers in LTC facilities (Ejaz et al. 2008).

Parsons' formulations regarding the sick role represent a sociological foundation for the medical model (Parsons 1951). These propositions about the sick individual's obligation to assume a "patient" identity and to seek competent professional care also have implications for responsibilities of care settings to offer good care. The medical model in general, and Parsons' sick role model in particular (1975), have been primarily associated with acute illness and situations where a cure is possible. This model has been critiqued as having only partial applicability to chronic illness and to LTC (Haug and Lavin 1983). In considering assumptions about older patients in the medical model, it is also possible to draw linkages to disengagement theory (Cumming and Henry 1961). The expected withdrawal of elderly persons from society proposed by this theory is consistent with limited interest in patients' social needs by caretakers espousing the medical model.

Sociological analyses linking quality of care and quality-of-life in LTC seldom explicitly discuss the medical model. Yet, this model is implicitly criticized for neglecting the social needs of elderly patients (Cockerham and Scambler 2010). Based on the medical model, the goals of high-quality care in nursing homes include efficient physical care of patients and avoidance of accidents and of iatrogenic illnesses. To guarantee patients' safety, regulations must be carefully followed by staff and compliance by patients is also required. Any efforts to attend to the psychosocial needs of residents represent superficial enhancements to the basic bed and body work characterizing the culture of patient care (Henderson 1995). The emphasis of the medical model on safety often deters staff from respecting resident's desires or preferences and offers few opportunities for self-directed activities (Baltes et al. 1980).

An elaborate system of regulation has evolved to ensure uniform delivery of care and to guard against medical mistakes. This regulatory system is based on mandatory assessments of residents' physical condition and functional abilities, such as the minimum data set (MDS) (Hawes et al. 1999). The resultant "report cards" note each facility's adherence to expected standards. This information may be utilized by prospective patients and their families in choosing a nursing home. By 2005, in order to overcome adverse medical events in nursing homes, interventions were developed to reduce infections, pressure sores, and incontinence (Mody et al. 2005). However, the lack of attention to social needs of patients and to the maintenance of personhood that inheres in the medical model limits the impact of programs to improve the quality of patient life (Kahana et al. 1999).

Institutional Analysis Model

Total Institution

Sociological understandings of LTC have a strong theoretical anchor in the incisive analyses of Erving Goffman (1961) and his exposition of the total institution. Goffman's original work on asylums outlined a framework, based on mental hospitals, that has provided a prototype for characterizing residential care in nursing homes. Asylums were described by Goffman as highly restrictive and even as punitive social milieus. The impact of totalistic environments on residents is described by Goffman as including stripping, mortification of self, and de-individuation. The negative impact of institutionalization in such facilities was further detailed by Sommer and Osmond (1961) who noted processes of de-individuation, disculturation, psychological damage, estrangement, isolation, and stimulus deprivation.

Such total institutions inflicted assaults on the self, based on congregate, segregate, and institutionally controlled features of the environment (Kleemeier 1961). Nursing homes, where physically frail and cognitively impaired elders are cared for today, possess many of the environmental features of the total institution. Furthermore, nursing home residents typically exhibit even greater vulnerability than the mentally ill, as many suffer from both physical frailty and dementia (Kane 2001).

Concerns about social processes in nursing home care were explored by early gerontological researchers (Townsend 1962; Bennett 1963; Kleemeier 1963). Research by Coe (1965) found that commercial facilities exhibited more totalistic features than nonprofit ones. Furthermore, higher institutional totality was related to withdrawal and depersonalization by residents. Gerontologists called for humane treatment of older people in institutions through reducing totalistic features that create barriers between residents and the outside world (Kahana 1973; Vladek 1980).

Dependency Inducing Environments

During the 1970s, social psychologists became intrigued by the mechanisms contributing to individuals' sense of control and dependency (Langer 1975).This orientation was consistent with views of Seligman (1975), based on studies of younger persons, suggesting that environments can create learned helplessness. The mechanisms by which care processes foster dependency among aged residents of LTC were identified by Baltes et al. (1980). They found that independent behavior of residents was ignored by nursing home staff, while dependent behavior was supported (Baltes et al. 1980). Successful interventions were designed by Langer and Rodin (1976) to enhance personal control of residents through provision of meaningful choices and responsibility.

A desire to restore control and self-determination (Ryan and Deci 2000) to nursing home residents became the conceptual foundations of the culture change movement (Thomas 1999). We will return to a more detailed discussion of this important movement that has become a foundation for nursing home reform. Conceptualizations of control and helplessness in LTC have also been linked to broader ecological issues of person–environment transactions (Kahana et al. 1989). These issues are discussed next.

Ecological Model

There is a rich theoretical tradition applicable to the study of LTC environments that relates to person and environment transactions in the context of aging. Gerontologists interested in the physical and social environments have developed creative frameworks, guided by the intellectual leadership of Powell Lawton. These diverse ideas were first published in *Aging and the Environment, Theoretical Approaches* (Lawton et al. 1982). This book presents three conceptual frameworks that are of particular interest for understanding social processes in LTC. These include: Kermit Schooler's stress theoretical perspective, Powell Lawton's environmental press and adaptation model, and Eva Kahana's congruence model of person–environment interaction. We will briefly review each of these models as they relate to quality of care and its impact on quality-of-life in LTC settings.

Stress Model

Schooler's (1982) stress theoretical model provides a framework for understanding threats and hazards posed by factors such as forced relocation, lack of privacy, and crowding associated with life in LTC facilities. The stress model, earlier articulated by Lazarus (1966) has been further developed in the field of sociology by Pearlin (1989). It associates adverse social and environmental characteristics with subsequent loss of morale and other poor quality-of-life outcomes. The model also recognizes the potential benefits of effective coping strategies and of social supports in ameliorating adverse effects of stress on quality-of-life. Stress models have guided both subsequent research and interventions in LTC settings. Instrumental coping strategies were found to be useful for promoting well-being of institutionalized aged (Kahana et al. 1987). The effectiveness of social supports offered by family involvement has also been noted (Lieberman and Tobin 1983). However, LTC institutions have not always welcomed family participation (Kahana 2007). Important work in evaluating residential facilities using stress-based theoretical frameworks was conducted by Moos and Lemke (1996).

Environmental Press/Adaptation Model

Lawton's (1982) model is concerned with environmental demands that are placed on the older individual in relation to personal competence to handle those demands. When environmental demands exceed personal capacities, adverse consequences to quality-of-life occur. Lawton's concept of environmental docility proposes that the frailest elderly residents are at greatest risk for adverse quality-of-life outcomes in institutions. Lawton also recognized the dual roles of the individual as both a responder and an initiator in interacting with the environment (Lawton and Nahemow 1973).

Lawton's work has led to improvements in architectural design of LTC settings. These designs help overcome environmental docility of frail and cognitively impaired older adults by focusing on negotiability, personalization, orientation, and safety (Calkins 2003). Lawton's recognition of the need for enhancing personal competence is consistent with therapeutic programs that aim to enhance cognitive functioning (Lichtenberg 1990).

Congruence Model of Person–Environment Interaction

Kahana's (1982) congruence model of person environment interaction has its roots in Murray's (1938) "need-press" formulation regarding human behavior. The model recognizes individual differences in both needs and preferences of older adults and the consequent importance of matching LTC environments to the unique needs of individual residents. It is hypothesized that in institutional settings, congruence is particularly important along segregate, congregate, and institutional control dimensions. These dimensions are based on Goffman's (1961) and Kleemeier's (1961) concepts of institutional totality. Additional dimensions include stimulation and tolerance of affective expression. Poor quality of care, according to the congruence model, would be demonstrated in LTC institutions that disregard resident preferences. Lack of person environment fit is expected to result in diminished resident morale. Research by Kahana et al. (1980) confirmed the contribution of selected aspects of congruence to morale among residents in homes for the aged. Recent approaches of the culture change movement that tailor schedules and programming to individual preferences of residents are consistent with the congruence model (Thomas 1999).

All three of the ecological models reviewed share in common an interest in both physical and social environmental features and in quantitative orientations to assessing resident psychological well-being. Each of these models has been influential in offering conceptual scaffolding for later research and intervention.

Ethnographic Model

Ethnographic models may be classified as methodological rather than conceptual orientations. Yet, the focus of ethnographers on micro-processes involved in the social construction of everyday life offers unique perspectives and raises conceptual issues that cannot be readily subsumed by other models. It is also notable that books representing ethnographic studies of nursing homes are among the most widely disseminated works on LTC by sociologists. This tradition originates in anthropological research focusing on the lived experience of older persons (Clark and Anderson 1967). The ethnographic tradition is based on participant observations and on detailed unstructured interviews with informants that allow for understanding quality of care and quality-of-life as defined by the lived experience of key participants: residents and staff. The complex realities of daily life appreciated from the perspectives of patients, and nursing assistants (as well as visiting family members and volunteers), allow the ethnographer to describe the inter-subjective reality that is referred to as culture (Henderson and Vesperi 1995). By offering multiple perspectives, ethnographers gain a more textured understanding that allows for an appreciation of both patients' and staff members' predicaments.

A pioneering, ethnographic study of nursing home life is offered in Jules Henry's classic (1963) book, *Culture against Man*. Henry eloquently describes the valiant efforts of nursing home residents to defend and sustain their core identity in the face of indignities suffered. This theme of the nursing home residents' struggle against assaults on their sense of self is consistent with the symbolic interactionist framework in sociology (Mead 1934). It allows for use of the interpretive tradition that considers nursing home residents as social actors who live in social worlds that are characterized by different "front stage" and "back stage" interaction patterns (Goffman 1961).

The perspectives of ethnographers reflect a diversity of connections to the nursing homes they describe and study. Savishinsky (1991) initiated his study as a volunteer helping with a pet program that was aimed to humanize nursing home care. Diamond (1992) made his incisive observations about disempowered staff while working as a nursing assistant in a LTC facility. This tradition is also exemplified in *Limbo*, the autobiography of Carolbeth Laird (1979), an anthropologist, who writes about her first-hand experiences in living in a nursing home. Similarly Kahana (2007) articulated perspectives of a daughter spending time daily with her mother who was living in a nursing home.

One of the most important works of modern nursing home ethnography is Gubrium's (1975), *Living and Dying in Murray Manor*. This book reflects orientations of the new ethnography that considers residents' narratives not only as the reflections of the social realities of nursing home life, but views residents as active participants who construct the realities of their worlds (Gubrium and Holstein 1997). This orientation foreshadows later developments in participatory action research with nursing home residents (Dannefer et al. 2008). Such research views narratives as a valuable tool for bringing about policy changes that may improve quality of care and quality-of-life in nursing homes and other health care settings.

Political Economy: Structures of Domination

LTC has been viewed as a confluence of failures; on the part of the older person (e.g., being frail rather than vibrant), family members (e.g., placing their loved one in a nursing home rather than caring for them at home), and on the part of LTC staff (e.g., working for low wages) (Armstrong et al. 2008). Sociologists have attributed these failures to larger forces operating in society (i.e., historical, economic, political, and social). Using this lens, political economists have sought to understand and identify sources of oppression, conflict, and power imbalance in LTC settings (Estes 1979). Work by Close and Estes (1994), Diamond (1992), and Harrington (1991) exemplify research on LTC in a political economy tradition and call our attention to how structures of domination, based on class, gender, ethnicity, and race permeate the delivery and receipt of LTC. These influences are generally discussed by proponents of the political economy model, in terms of the commodification of care and workforce issues.

The concept of the commodification of care is rooted in Marxist theory (Marx 1904) and relates to historical shifts from caring at home for the old to placement in an institution. As a result, the nature of caring changed, as it was no longer a part of daily life where the ebb and flow of living and dying was connected to home and hearth. This shift had a negative impact on both the delivery and receipt of care in LTC facilities. In care work, social and affective aspects (e.g., nurturing, compassion, camaraderie) are discouraged and replaced by the dispensing of medical work units that can be timed, measured, and observed (Estes 1979). Staff may become helpless cogs in a hierarchical and inflexible health care machine. Diamond uses the example of staff having to give cold showers to residents, when hot water is unavailable, during their specified bath time (Diamond 1992).

With the commodification of care, corporate LTC institutions strive for efficiency and cost-containment strategies, at the expense of staff, and ultimately of residents (Close and Estes 1994). The political economy model has informed studies linking staff exploitation in LTC facilities to abuse and neglect of patients (Pillemer 1991) and to residents' poor quality-of-life (IOM 2003). The effectiveness of regulatory mechanism for improving quality of care has not been clearly demonstrated (IOM 2003). There has been particular concern about the success of LTC in meeting needs and preferences of increasingly diverse populations being served (Quadagno 1999). Proponents of the political economy orientation generally serve as social critics; however, specific suggestions for change that could be implemented in the current political framework of care delivery are not generally offered. Staff empowerment may be the most concrete direction for social action within this framework. This may be accomplished through labor organizing and collective bargaining approaches (Dodson and Zincavage 2007). The improvements for working conditions for staff, in turn, would be expected to translate into better quality-of-life for residents.

Advocacy Model

There has been growing interest in considering the role of consumers and their advocates in influencing quality of care and quality-of-life in LTC facilities. This orientation has intellectual roots in sociological work on consumerism in medicine, articulated in the work of Haug and Lavin (1983). It also shares common ground with the broader consumer movement (Nader 1976), leading to changes in social policy that resulted in the Patient Self Determination Act (PSA 1990). Another relevant social movement, laying the groundwork for patient advocacy is presented by disability activism (DeJong 1984), that has culminated in the Americans with Disabilities Act (ADA 1990). A related paradigm, focused on professional advocacy, is posed by the ideology of empowerment originally articulated in the social work literature (Abramson 1985). Finally, families also serve as natural advocates of the institutionalized aged (Richard 1986). We briefly review implications of these orientations to advocacy that reflect human agency within institutional structure for linking quality of care and quality-of-life in LTC facilities.

The health care consumerism paradigm (Haug and Lavin 1983) calls for increased power, wrested by patients from representatives of the health care system. Based on an approach of "doubt and caution rather than faith and trust" patients are urged to resist coercive components of care delivery through lack of compliance with medical regimens and alternative focus on self-care (Haug and Lavin 1983:10). Independence and self-confidence, resulting from consumerism, may benefit both the health status and self-concept of the patient. It is important to note, however, that many elderly persons receiving LTC may lack the personal and social resources needed for engaging in self-care or in active consumerism.

Family may serve as advocates for LTC residents by challenging authority when they observe poor quality of care by lodging complaints (Kahana 2007). When family advocates find staff to be unreceptive, they may enlist formal advocates (e.g., ombudsmen) or they may file lawsuits to improve quality-of-life of a loved one. However, advocacy may be limited by fear of retaliation against the family member living in a LTC facility (Nelson 2000). It is useful to note that patient and family advocacy need not be adversarial and can also succeed through building alliances with staff.

Community presence can also serve useful advocacy functions in LTC (Wiener 2003). The nursing home reform act of OBRA (1987) established the right of residents and family members to organize resident councils in LTC. Consumer advocacy groups have been successful in generating state-level policy changes to benefit LTC residents, but often lack the organizational infrastructure to maintain long-term operations (Phillips et al. 2008). The nursing home ombudsmen program created by the Older Americans Act represents the most successful public initiative to improve the quality of care and quality-of-life in nursing homes (Kahana 1994).

In addition to the role of advocacy in improving quality of care in existing LTC facilities, more dramatic paradigm changes through advocacy have been recently proposed by gerontologists (Haber 2009). These bold ideas recognize the potential of new cohorts of older adults to advocate for social policies that could minimize placement in LTC facilities following the models of disability activism. Elders could thus purchase alternative community-based and self-directed housing services (Kane 2009). Some critiques of "agency" based approaches attribute self-serving motives to proponents of advocacy (King and Calasanti 2009). Nevertheless, social movements that wrest power from existing hierarchies of LTC are among the major practical solutions toward the improvement of quality-of-care for frail elders. The Pioneer Network represents a promising social movement working toward a humane, meaningful, and life-affirming orientation to LTC (Pioneer Network 2008).

Synergy Among Models and Directions for Change: From Tradition to Innovation

In the final section of the paper we challenge ourselves to consider implications of the alternative models we presented for understanding likely future directions linking quality of care and quality-of-life in nursing homes. In so doing, we consider divergence as well as common ground across sociological and other disciplinary orientations that could lead to improvement in quality of LTC. We also explore linkages between tradition and innovation.

A perusal of the six models of care that we presented reveals some noteworthy similarities as well as differences. Three models (the medical model, the ecological model, and the ethnographic model) reflect primarily micro level phenomena. Focus is on individual actors and interactions between individual patients and staff. The institutional analysis model is a mezzo level model reflecting Goffman's interests in the reality of the total institution. The advocacy model incorporates both micro and macro components. On the level of individual consumerism and family empowerment it is a micro model. However, its focus on community activism calls for legislative and society-level change. Finally, the political economy model is a macro level model that represents a critique of societal forces beyond the nursing home.

The diverse models presented also reflect different disciplinary orientations. They underscore that it is useful to have multiple disciplines engaged with issues linking quality of care and quality-of-life with the goal of improving LTC. This synergy is well illustrated as we consider the institutional analysis model. We consider Goffman, a sociologist, to be the founder of this approach, but acknowledge that his research was based on anthropological traditions of ethnographic analysis. Goffman's (1961) work was further elaborated and brought into mainstream gerontology by Kleemeier, a psychologist specializing in aging (1963). Institutional control components of the total institution became influential concepts in the field of social psychology in the 1970s. New work on personal and environmental control was published in prestigious social psychology journals by Langer (1975). Concepts of learned helplessness (Seligman 1975) and self-determination theory (Ryan and Deci 2000) bolstered initiatives based on these concepts to improve quality-of-life of nursing home residents. Some of these initiatives were conducted within the framework of an experimental paradigm by Langer and Rodin (1976). The understanding of underlying social processes was further advanced by the work of German psychologist Margaret Baltes and colleagues. (1980). They identified behavioral patterns by staff that reinforced resident dependency in nursing homes settings and called for both environmental and behavioral interventions to break maladaptive patterns.

Subsequently, in the 1980s, there were fledgling social movements developing to address, in a more holistic way, needs for fundamental change in the culture of nursing home care. Work of the "nursing home pioneers" a group of utopian reformers, converged with the vision of geriatrician and family physician William Thomas (1999), who designed the "Eden Alternative" for nursing care reform and culture change. There is a clear connection between the holistic approaches of culture change and Goffman's brilliant critique of organizations that stripped away the very humanity of their inmates. The attention by these reformers to counteract indignities suffered by patients in every mundane aspect of their daily lives echoes observations of ethnographers about the importance of everyday interactions that define the social world of the individual who lives in an institution (Gubrium 1975).

The new movement of culture change thus becomes part of continuing efforts by gerontologists to offer policies and practices that reduce totalistic features of institutions. The nursing home reform movement also follows notable early work conducted by Coons and Reichel (1988) to create homelike residences for elderly patients in mental hospitals. Ecological principles have also been advocated and implemented in the work of proponents of the Eden Alternative (Bergman-Evans 2004).

The environmental side of the culture change movement is an important area for future innovation. Innovative building design may be allied with changes in federally mandated life safety code regulations. It is recognized that striving for safety must be counterbalanced with consideration of resident preferences for homelike and independence supporting environments (Bowman 2008). Consequently, there are continuing efforts to create "home" in the nursing home. Thus, we see promising government and industry alliances in promoting culture change. However, grass-roots efforts by recipients of care and their families are also needed for enhancing options in LTC. One promising direction can be seen in the growing NORC movement. NORCs are rental or condominium housing sites that were not specifically planned and designed for older adults, but over time, as residents age in place acquire features of retirement communities (Hunt and Gunter-Hunt 1985:4). Services to support care needs that typically arise in such settings tend to be resident-generated. Advocates of this approach state that participation in the development of services provides an opportunity for older residents to build social capital, allowing elders to become empowered "community citizens" rather than passive and dependent "clients" (Bookman 2008). As we consider other avenues for grass-roots initiatives by older adults, the world of technology also offers creative alternatives to traditional LTC options. For example, frail elders are increasingly open to installing intelligent home monitoring systems or "smart home" technology in their homes as a means maintaining independent living arrangements (Mahmood et al. 2008).

Other proponents of nursing home reform, based on culture change, call for realigning structures of nursing homes so as to improve the life experience of both staff and residents (Fagan et al. 1997). These staff empowerment initiatives that are aimed at removing hierarchical features of staff organization are consistent with aspects of the political economy model. The reform movement that created the Pioneer Network in Long Term Care was noted as an example of community action in our discussion of the Advocacy Model.

The culture change movement calls for offering nursing home environments that reflect life-supporting habitats, incorporating animals and plants into a more livable environment (Thomas 1999). This innovative approach goes beyond orientations of the ecological model and is uniquely attributable to the culture change movement. While this focus was not listed in our concept map, we acknowledge its importance through inclusion in our structural model where we have added "artifacts" and "amenities" as key components of quality of care. Culture change also offers greater resident participation and control (Langer 1975). Furthermore, by involving family members and other community representatives, they reduce the isolation of residents from the outside world (Goffman 1961). Approaches to culture change in nursing homes reflect a humanistic orientation and aim to maintain the personhood and identity of frail residents (Kahana 1973). Their holistic efforts of redesign hold many promising insights. The evaluation of the success of such efforts, however, is limited, and is largely based on qualitative and ethnographic studies conducted in not-for-profit facilities (Rahman and Schnelle 2008).

As the culture change movement has gained momentum in the United States, the federal government is now facilitating the implementation of culture change (CMS 2008). This is part of broader initiatives exemplified in efforts to restructure Medicare in order to improve chronic care (Wolff and Boult 2005). It is noteworthy that sociologists are playing an important role in conducting research based on the culture change movement. They have enriched the movement by introducing the concept of participatory action research, which enhances resident control and self-determination (Dannefer et al. 2008).

Consideration of totalistic features of LTC had long evoked indignation and calls for reform by gerontologists. These efforts primarily aim to reduce institutional control features. Congregate and segregate features of institutional care are more entrenched and are more difficult to change. Consequently, these features are likely to continue to characterize institutional life even when good quality of care is provided. It is for this reason that beyond efforts to counteract institutional totality through culture change, activists call for social movements that can lead to deinstitutionalization of LTC residents (Kane 2009).

Promising initiatives are also focused on changing the managerial culture of nursing homes and point to the value of labor–management partnerships in facilitating person-centered care (Leutz et al. 2009). Our focus on the culture change movement has served to illustrate the synergies that often go unrecognized, between tradition and innovation. Those working at present, within the boundaries of any given conceptual orientation, may be too close to their own work; and may overlook linkages to the past and alternative orientations that reflect competing ideologies.

It is useful to consider recent developments in program innovation as an alternative pathway for analyzing the intellectual origins of creative new approaches. Thus, recent approaches to translational research advocating the rapid adoption of innovative programs deserve attention (Bradley et al. 2004). Indeed, federal agencies have shown a growing interest in supporting translational initiatives to improve quality in LTC. Examples of such initiatives include the Wellspring model for improving the quality of clinical care in nursing homes (Kehoe and Heesch 2003). This is accomplished by multidisciplinary care resource teams that include nursing assistants who are empowered to develop and carry out interventions (Bradley et al. 2004). Other successful innovations include communication interventions that involve cooperation between nursing home staff and family members, yielding positive outcomes for both family and staff (Pillemer et al. 2003).

There are many useful approaches that we could not include in our discussion, as we focused primarily on sociological traditions. Readers could create concept maps that unpack contributions in the field of organizational behavior and innovation as they relate to LTC (Anderson et al. 2003). Similarly, there are many promising developments in evidence-based interventions that can also lead to enhancing the lives of LTC residents (Cummings and Kropf 2009). Promising insights could also be gained by focusing on diversity as a lens for understanding staff and resident experiences of life in LTC (Ball et al. 2009). There is a growing literature on palliative and end-of-life care, anchored largely in the nursing literature that offers lessons about beneficence and caring (Watson 2005).

Scholarly work that promotes the understanding of social factors that can humanize LTC has exciting traditions in sociology. Yet, there is much room for innovation in this field that can make life better for each of us who will eventually cross the threshold of a LTC facility.

References

Abeles, Ronald P., Helen C. Gift, and Marcia G. Ory. 1994. *Aging and Quality of Life*. New York: Springer.

Abramson, Marcia. 1985. "The Autonomy-Paternalism Dilemma in Social Work Practice." *Social Casework* 66:387–93.

Americans with Disabilities Act (ADA). 1990. "ADA Compliance Guide." *ADA Compliance Bulletin* 1(5):1–8.

Anderson, Ruth A., L. Michele Issel, and Reuben R. McDaniel. 2003. "Nursing Homes as Complex Adaptive Systems: Relationship between Management Practice and Resident Outcomes." *Nursing Research* 52:12–21.

American Association of Homes and Services for the Aging (AAHSA). 2009. "Aging Services: The Facts." Retrieved November 27, 2009 (http://www.aahsa.org).

Armstrong, Hugh, Pat Armstrong, and Krista Scott-Dixon. 2008. *Critical to Care: Women's Ancillary Work in Health Care*. Toronto: University of Toronto Press.

Ball, Mary M., Molly M. Perkins, Carole Hollingsworth, Frank J. Whittington, and Sharon V. King. 2009. "Pathways to Assisted Living: The Influence of Race and Class." *Journal of Applied Gerontology* 28(1):81–108.

Baltes, Margaret M., Robert L. Burgess, and Robert B. Stewart. 1980. "Independence and Dependence in Self-Care Behaviors in Nursing Home Residents: An Operant-Observational Study." *International Journal of Behavioral Development* 3:489–500.

Bennett, Ruth. 1963. "The Meaning of Institutional Life." *The Gerontologist* 3:117–25.

Bergman-Evans, Brenda. 2004. "Beyond the Basic: Effects of the Eden Alternative Model on Quality of Life Issues." *Journal of Gerontological Nursing* 30(6):27–34.

Bookman, Ann. 2008. "Innovative Models of Aging in Place: Transforming our Communities for an Aging Population." *Community, Work, and Family* 11:419–38.

Bowman, Carmen S. 2008. "The Environmental Side of the Culture Change Movement: Identifying Barriers and Potential Solutions to Furthering Innovation in Nursing Homes." Pre-symposium background paper presented at Creating Home in the Nursing Home: A National Symposium on Culture Change and the Environment Requirements, April 3, Washington, DC.

Bradley Elizabeth H., Tashonna R. Webster, Dorothy Baker, Mark Schlesinger, Sharon K. Inouye, Michael C. Barth, Kate L. Lapane, Debra Lipson, Robyn Stone, and Mary Jane Koren. 2004. "Translating Research into Practice: Speeding the Adoption of Innovative Health Care Programs." *Issue Brief: Commonwealth Fund* 724:1–12.

Calkins, Margaret P. 2003. "Powell Lawton's Contributions to Long-Term Care Settings." *Journal of Housing for the Elderly* 17(1–2):67–84.

Carder, Paula C., Leslie A. Morgan, and J. Kevin Eckert. 2008. Pp. 143–68 in *Assisted Living Residences: Visions for the Future,* edited by S. M. Golant and J. Hyde. Baltimore: Johns Hopkins University Press.

Centers for Medicare and Medicaid Services (CMS). 2008. "Action Plan for Further Improvement of Nursing Home Quality." Retrieved December 1, 2009 (http://www.cms.hhs.gov/CertificationandComplianc/Downloads/2008NHActionPlan.pdf).

Charles, Kerwin Kofi and Purvi Sevak. 2005. "Can Family Caregiving Substitute for Nursing Home Care?" *Journal of Health Economics* 24:1174–90.

Clark, Margaret and Barbara Gallatin Anderson. 1967. *Culture and Aging: An Anthropological Study of Older Americans*. Springfield, IL: Charles C Thomas.

Close, Liz and Carroll L. Estes. 1994. "A Political Economy Perspective on Frontline Workers in Long-Term Care." *Generations* 18(23):23–7.

Cockerham, William C. and Graham Scambler. 2010. "Medical Sociology and Sociological Theory." Pp. 3–26 in *The New Blackwell Companion to Medical Sociology,* edited by W. C. Cockerham. Malden, MA: Blackwell.

Coe, Rodney M. 1965. "Self-Conceptions and Institutionalization." Pp. 224–44 in *Older People and Their Social World,* edited by A. M. Rose and W. A. Peterson. Philadelphia: FA Davis.

Coons, Dorothy H. and William Reichel. 1988. "Improving the Quality of Life in Nursing Homes." *American Family Physician* 37:241–8.

Cox, Enid Opal and Ruth J. Parsons. 1994. *Empowerment Oriented Social Work Practice with the Elderly.* Belmont, CA: Brooks Cole.

Cumming, Elaine and William Earl Henry. 1961. *Growing Old: The Process of Disengagement.* New York: Basic Books.

Cummings, Sheryl M. and Nancy P. Kropf. 2009. Handbook of Psychosocial Interventions with Older Adults: Evidence-Based Approaches. New York: Hawthorne.

Dannefer, Dale, Paul Stein, Rebecca Siders, and Robin Shura Patterson. 2008. "Is That All There Is? The Concept of Care and the Dialectic of Critique." *Journal of Aging Studies* 22(2):101–8.

DeJong, Gerben. 1984. "Independent Living: From Social Movement to Analytic Paradigm." Pp. 39–64 in *The Psychological and Social Impact of Physical Disability,* edited by R. P. Marinelli and A. E. Dell Orto. New York: Springer.

Diamond, Timothy. 1992. *Making Gray Gold: Narratives of Nursing Home Care*. Chicago: University of Chicago.

Dodson, Lisa and Rebekah M. Zincavage. 2007. "'It's Like Family': Caring Labor, Exploitation, and Race in Nursing Homes." *Gender & Society* 21:905–28.

Ejaz, Farida K., Linda S. Noelker, Heather L. Menne, and Joshua G. Bagaka. 2008. "The Impact of Stress and Support on Direct Care Workers' Job Satisfaction." *The Gerontologist* 48:60–70.

Estes, Carroll L. 1979. *The Aging Enterprise.* San Francisco: Jossey-Bass.

Fagan, Rose Marie, Carter Catlett Williams, and Sarah Greene Burger. 1997. *Meeting of Pioneers in Nursing Home Culture Change.* Rochester, NY: Lifespan of Greater Rochester.

Feder, Judith, Harriet L. Komisar, and Marlene Niefeld. 2000. "Long-Term Care in the United States: An Overview." *Health Affairs* 19(3):40–57.

Goffman, Erving. 1961. *Asylums: Essays on the Social Situation of Mental Patients and other Inmates.* New York: Doubleday.

Golant, Stephen M., and Joan Hyde, eds. 2008. *The Assisted Living Residence: A Vision for the Future*. Baltimore: The Johns Hopkins University Press.

Gubrium, Jaber F. and James A. Holstein. 1997. *The New Language of Qualitative Methods*. New York: Oxford University Press.

Gubrium, Jaber F. 1975. *Living and Dying at Murray Manor*. New York: St. Martin's Press.

Haber, David. 2009. "Gerontology: Adding a New Empowerment Paradigm." *Journal of Applied Gerontology* 28:283–97.

Han, Beth, Robin E. Remsburg, William J. McAuley, Timothy J. Keay, and Shirley S. Travis. 2006. "National Trends in Adult Hospice Use: 1991–1992 to 1999–2000." *Health Affairs* 25(3):792–9.

Harrington, Charlene. 1991. "The Nursing Home Industry: A Structural Analysis." Pp. 153–64 in *Critical Perspectives on Aging: The Political and Moral Economy of Growing Old*, edited by M. Minkler and C. L. Estes. Amityville, NY: Baywood.

Harris-Kojetin, Lauren, Debra Lipson, Jean Fielding, Kristen Kiefer, and Robyn I. Stone. 2004. *Recent Findings on Frontline Long-term Care Workers: A Research Synthesis 1999–2003*. Washington, DC: Office of the Assistant Secretary for Policy and Evaluation, DHHS.

Haug, Marie R. and Bebe Lavin. 1983. *Consumerism in Medicine: Challenging Physician Authority*. Beverly Hills: Sage.

Hawes, Catherine, Miriam Rose, and Charles D. Phillips. 1999. *A National Study of Assisted Living for the Frail Elderly: Results of a National Survey of Facilities*. Beachwood, OH: Meyers Research Institute.

Henderson, J. Neil. 1995. "The Culture of Care in a Nursing Home: Effects of a Medicalized Model of Long Term Care." Pp. 37–54 in *The Culture of Long Term Care: Nursing Home Ethnography*, edited by J. N. Henderson and M. D. Vesperi. London: Bergin & Garvey.

Henderson, J. Neil, and Maria D. Vesperi, eds. 1995. *The Culture of Long Term Care: Nursing Home Ethnography*. London: Bergin & Garvey.

Henry, Jules. 1963. *Culture against Man*. New York: Random House.

Hunt, Michael E. and Gail Gunter-Hunt. 1985. "Naturally Occurring Retirement Communities." *Journal of Housing for the Elderly* 3(3/4):3–21.

Institute of Medicine (IOM). 2001. *Improving the Quality of Long-Term Care*. Washington, DC: National Academies Press.

———. 2003. *Unequal Treatment: Confronting Racial and Ethnic Disparities in Health Care*. Washington, DC: National Academies Press.

Johnson, Richard W., Desmond Toohey, and Joshua M. Wiener. 2007. *Meeting the Long-Term Care Needs of the Baby Boomers: How Changing Families Will Affect Paid Helpers and Institutions*. Discussion Paper 07-04. Washington, DC: The Urban Institute.

Jones, Adrienne L., Lisa L. Dwyer, Anita R. Bercovitz, and Genevieve W. Strahan. (2009). "The National Nursing Home Survey: 2004 Overview." *National Center for Health Statistics. Vital Health Statistics* 13:167.

Kahana, Eva. 1973. "The Humane Treatment of Old People in Institutions." *The Gerontologist* 13(3):282–9.

———. 1982. "A Congruence Model of Person-Environment Interaction." Pp. 97–121 in *Aging and the Environment: Theoretical Approaches,* edited by M. P. Lawton, P. G. Windley, and T. O. Byerts. New York: Springer.

———. 2007. "Dreams of Hot Toast and Smiling Nurses: Toward a Model of Patient-Responsive Care in Nursing Homes." Pp. 109–33 in *Research in the Sociology of Healthcare,* edited by J. J. Kronenfeld. Bingley: Emerald Group Inc.

Kahana, Eva, Boaz Kahana, May Wykle, and Diana Kulle. 2009. "Marshalling Social Support: Active Roles in Care-Getting for Cancer Patients throughout the Life Course." *Journal of Family Social Work* 12:168–93.

Kahana, Eva, Boaz Kahana, and Heidi T. Chirayath. 1999. "Innovations in Institutional Care from a Patient Centered Perspective." Pp. 249–75 in *Innovations in Practice and Service Delivery across the Lifespan,* edited by D. E. Biegel and A. Blum. Oxford: Oxford University Press.

Kahana, Eva, Jersey Liang, and Barbara J. Felton. 1980. "Alternative Models of Person-Environment Fit: Prediction of Morale in Three Homes for the Aged." *Journal of Gerontology* 35:584–95.

Kahana, Eva, Boaz Kahana, and Kathy Riley. 1989. "Person-Environment Transactions Relevant to Control and Helplessness in Institutional Settings." Pp. 121–53 in *Psychological Perspectives of Helplessness and Control in the Elderly,* edited by P. S. Fry. Amsterdam: North Holland.

Kahana, Eva, Boaz Kahana, and Rosalie Young. 1987. "Strategies of Coping and Post-Institutional Outcomes." *Research on Aging* 9:182–99.

Kahana, Jeffrey S. 1994. "Reevaluating the Nursing Home Ombudsman's Role with a View toward Expanding the Concept of Dispute Resolution." *Journal of Dispute Resolution* 2:217–33.

Kane, Robert L. 2005. "Changing the Face of Long Term Care." *Journal of Aging and Social Policy* 17(4):1–18.

———. 2009. "Empowerment of Seniors and Paradigm Change: A Commentary on Haber's Empowerment Paradigm." *Journal of Applied Gerontology* 28:298–303.

Kane, Rosalie A. 2001. "Long-term Care and a Good Quality of Life: Bringing them Closer Together." *The Gerontologist* 41:293–304.

———. 2003. "Definition, Measurement, and Correlates of Quality of Life in Nursing Homes: Toward a Reasonable Practice, Research, and Policy Agenda." *The Gerontologist* 43(Special Issue II):28–36.

Kane, Rosalie A. and Lois J. Cutler. 2009. "Promoting Homelike Characteristics and Eliminating Institutional Characteristics in Community-Based Residential Care Settings: Insights from an 8-State Study." *Seniors Housing & Care Journal* 17(1):15–38.

Kane, Rosalie A, Robert L. Kane, Laurel Hixon Illston, John A. Nyman, and Michael D. Finch. 1991. "Adult Foster Care for the Elderly in Oregon: A Mainstream Alternative to Nursing Homes?" *American Journal of Public Health* 81:1113–20.

Katz, Paul R. and Jurgis Karuza. 2005. "Physician Practice in the Nursing Home: Missing in Action or Misunderstood." *Journal of the American Geriatric Society* 53:1826–8.

Kehoe, Mary Ann and Betsy Van Heesch. 2003. "Culture Change in Long Term Care: The Wellspring Model." Pp. 159–222 in *Culture Change in Long Term Care,* edited by A. S. Weiner and J. L. Ronch. New York: The Haworth Social Work Practice Press.

King, Neal and Toni Calasanti. 2009. "Aging Agents: Social Gerontologists' Imputations to Old People." *International Journal of Sociology and Social Policy* 29(1/2):38–48.

Kleemeier, Robert Watson. 1961. *Aging and Leisure: A Research Perspective into the Meaningful Use of Time.* New York: Oxford University Press.

Kleemeier, Robert Watson. 1963. "Attitudes toward Special Settings for the Aged." Pp. 101–21 in *Processes of Aging,* by Richard Hays Williams. New York: Atherton Press.

Laird, Carolbeth. 1979. *Limbo: A Memoir about Life in a Nursing Home by a Survivor.* Novato, CA: Chandler & Sharp.

Langer, Ellen J. 1975. "The Illusion of Control." *The Journal of Personality and Social Psychology* 32:311–28.

Langer, Ellen J. and Judith Rodin. 1976. "The Effects of Choice and Enhanced Responsibility: A Field Experiment." *Journal of Personality and Social Psychology* 34:191–8.

Lawton, Mortimer Powell and Lucille Nahemow. 1973. "Ecology and the Aging Process." Pp. 619–674 in *The Psychology of Adult Development and Aging,* edited by M. P. Lawton. Washington, DC: American Psychological Association.

Lawton, Mortimer Powell. 1982. "Competence, Environmental Press, and the Adaptation of Older People." Pp. 5–16 in *Aging and the Environment: Theoretical Approaches,* edited by M. P. Lawton, P. G. Windley, and T. O. Byerts. New York: Springer.

Lawton, Mortimer Powell, Paul G. Windley, and Thomas O. Byerts. 1982. *Aging and the Environment: Theoretical Approaches.* New York: Springer.

Lazarus, Richard S. 1966. *Psychological Stress and the Coping Process.* New York: McGraw Hill.

Leutz, Walter, Christine E. Bishop, and Lisa Dodson. 2009. "Role for a Labor-Management Partnership in Nursing Home Person Centered Care." *The Gerontologist* 53:340–51.

Lichtenberg, Peter A. 1990. "Reducing Excess Disabilities in Geropsychiatric In-Patients: A Focus on Behavioral Problems." Pp. 65–77 in *Mental Health in the Nursing Home,* edited by T. L. Brink. New York: Hawthorne.

Lieberman, Mortin A. and Sheldon S. Tobin. 1983. *The Experience of Old Age: Stress, Coping, and Survival.* New York: Basic Books.

Lovegreen, Loren. 2010. "Coping with Health Stressors and Normative Losses Associated with Aging Proactively: Residential Relocation in Later Life." *Open Longevity Journal Special Issue: Enhancing Disability-Free Life Expectancy through Proactive Lifestyles.* 10–10\9.

Mahmood, Atiya, Toshiko Yamamoto, Megan Lee, and Carmen Steggell. 2008. "Perceptions and Use of Gerotechnology: Implications for Aging in Place." *Journal of Housing for the Elderly* 22(1&2):104–26.

Manton, Kenneth G., XiLiang Gu, and Vicki L. Lamb. 2006. "Change in Chronic Disability from 1982 to 2004/2005 as Measured by Long-Term Changes in Function and Health in the U.S. Elderly Population." *Proceedings of the National Academy of Science* 103:18374–9.

Marx, Karl. 1904. *A Contribution to the Critique of Political Economy.* Chicago: Charles H Kerr Company.

Mead, George Herbert. 1934. *The Mind, Self, and Society: From the Standpoint of a Social Behavioralist.* Chicago: University of Chicago Press.

Miller, Galen W, J. Richard Williams, David J. English, and Jonathan Keyserling. 2002. *Delivering Quality Care and Cost-Effectiveness at the End-of-Life: Building on the Twenty-Year Success of the Medicare Hospice Benefit.* Retrieved February 25, 2010 (http://www.nhpco.org/public/articles/).

Mody, Lona, Kenneth M. Langa, Sanjay Saint, and Suzanne F. Bradley. 2005. "Preventing Infections in Nursing Homes: A Survey of Infection Control Practices in Southeast Michigan. *American Journal of Infection Control* 33:489–92.

Mollica, Robert. 2001. "The Evolution of Assisted Living: A View from the States." *Caring* 20:24–6.

Mollica, Robert and Heather Johnson-Lamarche. 2005. *State Residential Care and Assisted Living Policy: 2004*. Portland, ME: National Academy for State Health Policy.

Moos, Rudolph H. and Sonne Lemke. 1996. *Evaluating Residential Facilities: The Multiphasic Environmental Assessment Procedure*. Thousand Oaks, CA: Sage.

Murray, Henry Alexander. 1938. *Explorations in Personality*. New York: Oxford University Press.

Nader, Ralph. 1976. "Consumerism and Legal Services: The Merging of Movements." *Law & Society Review* 11:247–56.

Nelson, H. Wayne. 2000. "Injustice and Conflict in Nursing Homes: Toward Advocacy and Exchange." *Journal of Aging Studies* 14(1):39–62.

Omnibus Budget Reconciliation Act (OBRA). 1987. *OBRA '87 Summary*. Retrieved December 16, 2009 (http://www.allhealth.org/briefingmaterials/OBRA87Summary-984.pdf).

Parsons, Talcott. 1951. *The Social System*. Glencoe, IL: Free Press.

———. 1975. "The Sick Role and the Role of the Physician Reconsidered." *Milbank Memorial Fund Quarterly/Health and Society* 53(3):257–78.

Patient Self Determination Act (PSA). 1990. *United States Code* 42:1395cc, 1396a.

Pearlin, Leonard I. 1989. "The Sociological Study of Stress." *Journal of Health and Social Behavior* 30:241–56

Phillips, Charles D., Anne-Marie Kimbell, Catherine Hawes, Janet Wells, Jean Badalamenti, and Mary Jane Koren. 2008. "It's a Family Affair: Consumer Advocacy for Nursing-Home Residents in the United States." *Ageing and Society* 28(1):67–84.

Pillemer, Karl and Ronet Bachmen-Prehn. 1991. "Helping and Hurting: Predictors of Maltreatment of Patients in Nursing Homes." *Research on Aging* 13(1):74–95.

Pillemer, Karl, J., Jill Suitor, Charles R. Henderson, Rhoda Meador, Leslie Schultz, Julie Robison, and Carol Hegeman. 2003. "A Cooperative Communication Intervention for Nursing Home Staff and Family Members of Residents." *The Gerontologist* 43(Special Issue II):96–106.

Pioneer Network. 2008. "The Pioneer Network: About Us." Retrieved December 16, 2009 (http://www.pioneernetwork.net/AboutUs/).

Pritchard, David C. 1983. "The Art of Matchmaking: A Case Study in Shared Housing." *The Gerontologist* 23(2):174–9.

Quadagno, Jill. 1999. "Review of the Book Shifting the Color Line: Race and the American Welfare State." *Journal of Sociology & Social Welfare* 26:184–5.

Rahman, Anna N. and John F. Schnelle. 2008. "The Nursing Home Culture-Change Movement: Recent Past, Present, and Future Directions for Research." *The Gerontologist* 48:142–8.

Richard, Michel P. 1986. "Goffman Revisited: Relatives vs. Administrators in Nursing Homes." *Qualitative Sociology* 9:321–38.

Ryan, Richard M. and Edward L. Deci. 2000. "Self-Determination Theory and the Facilitation of Intrinsic Motivation, Social Development, and Well-Being." *American Psychologist* 55:68–78.

Saunders, Cicely. 1978. "Hospice Care." *American Journal of Medicine* 65:726–8.

Savishinsky, Joel S. 1991. *The Ends of Time: Life and Work in a Nursing Home*. New York: Bergin and Garvey.

Schooler, Kermit K. 1982. "Response of the Elderly to the Environment: A Stress Theoretical Perspective." Pp. 80–96 in *Aging and the Environment: Theoretical Approaches*, edited by M. P. Lawton, P. G. Windley, and T. O. Byerts. New York: Springer.

Seligman, Martin E.P. 1975. *Helplessness: On Depression, Development, and Death*. New York: WH Freeman.

Settersten, Richard A. Jr. 2007. "Ten Reasons why Shake-Ups in the Life Course Should Change Approaches to Old-Age Policies." *Aging and Public Policy Report* 17(3):21–7.

Sherwood, Sylvia. 1997. *Continuing Care Retirement Communities*. Baltimore, MD: Johns Hopkins University Press.

Shippee, Tetyana Pylypiv. 2009. "But I Am Not Moving: Residents' Perspectives on Transitions within a Continuing Care Retirement Community." *The Gerontologist* 49(3):418–27.

Sommer, Robert and Humphry Osmond. 1961. "Symptoms of Institutional Care." *Social Problems* 8:254–63.

Stone, Robyn I. 2006. "Emerging Issues in Long-Term Care." Pp. 397–418 in *Handbook of Aging and Social Sciences*, R. H. Binstock and L. K. George. Amsterdam: Elsevier Press.

Thomas, William H. 1999. Eden Alternative Handbook: The Art of Building Human Habitats. Sherburne, NY: Summer Hill Co.

Townsend, Peter. 1962. Last Refuge: A Survey of Residential Institutions and Homes for the Aged in England and Wales. London: Routledge.

Vladek, Bruce C. 1980. *Unloving Care: The Nursing Home Tragedy*. New York: Basic Books.

Wagnild, Gail. 2001. "Growing Old at Home." Pp. 71–84 in *Housing Choices and Well-Being of Older Adults: Proper Fit*, edited by L. A. Pastalan and B. Schwarz. New York: Hawthorn.

Watson, Jean. 2005. "Caring Science: Belonging Before Being as Ethical Cosmology." *Nursing Science Quarterly* 18(4):4–5.

Wiener, Joshua M. 2003. "An Assessment of Strategies for Improving Quality of Care in Nursing Homes." *The Gerontologist* 43(2):19–27.

Wolff, Jennifer L. and Chad Boult. 2005. "Moving Beyond Round Pegs and Square Holes: Restructuring Medicare to Improve Chronic Care." *Annals of Internal Medicine 143*:439–45

Zimmerman, Sheryl, Phillip D. Sloane, and J. Kevin Eckert, Editors. 2001. *Assisted Living: Needs, Practices, and Policies in Residential Care for the Elderly.* Baltimore, MD: Johns Hopkins University Press.

Zeisel, John, Nina M. Silverstein, Joan Hyde, Sue Levkoff, M. Powell Lawton, and William Holmes. 2003. "Environmental Correlates to Behavioral Health Outcomes in Alzheimer's Special Care Units." *The Gerontologist* 43:697–711.

Chapter 37
Caregiving and the Life Course: Connecting the Personal and the Public

Eliza K. Pavalko

At some point in our adult lives, most of us will face challenges of caregiving. For many, the issue will come up several times, either when we need to be cared for by others or when a parent, spouse, sibling or adult child becomes ill or disabled and needs care. National studies estimate that currently 44 million, or 21% of the adult population, representing 21% of U.S. households, are providing unpaid care to an ill or disabled adult (AARP 2005). The need for informal, unpaid care work has grown in recent decades and is expected to continue to grow with an aging population. For example, between 1987 and 1998 the population 65 and older increased from approximately 28 to 34 million. Corresponding with this increase, the number of individuals providing care rose from 7.8 to 22% of adults (Wagner 1997). The need for care has been tempered by declining rates of disability, but there is some evidence that disability rates may be on the upswing because of increasing rates of obesity (Lakdawalla et al. 2004). Informal (and unpaid) care is also a critical component of the U.S. health care system, particularly long-term care. Among those needing long-term care, 40% rely solely on informal, unpaid care and 25% use a combination of informal and formal care (Uhlenberg and Cheuk 2008).

Caregiving is an ideal illustration of C. Wright Mills's (1959) distinction between personal troubles and public issues, and the importance of using the sociological imagination to connect the two. Caregiving is first and foremost a personal issue that challenges individuals and families as they balance providing care with other work and family roles, and much of the research on caregiving has necessarily focused on providing information that both documents and informs these personal challenges. Understanding who provides care, the effects of care work on the health and well-being of caregivers, and developing strategies for reducing these stresses are critical for understanding care work. Bringing a sociological imagination to caregiving issues, however, seeks to understand and locate the personal struggles and rewards of caregiving within the broader context of the organization and financing of health care, the organization of work, and the role of gender, race and class in framing both public issues and personal lives. Linking the personal and public, biography and history sheds light on why concerns about providing care are becoming more acute. Individual illness and disability is not a new phenomenon, nor is reliance on family members to provide that care. It is the larger institutional and historical context, including demographic changes in family size, social changes in patterns of labor force participation, and the organization of work and health care systems that increasingly make caregiving a public issue as well as personal trouble.

Social scientists have been at the forefront of efforts to document, understand, and inform caregiving issues. In this chapter, I draw on several decades of social science research on care work to examine the intersection of caregiving as a personal and public issue, and to identify areas where

E.K. Pavalko (✉)
Department of Sociology, Indiana University, Bloomington, IN, USA
e-mail: epavalko@indiana.edu

R.A. Settersten, Jr. and J.L. Angel (eds.), *Handbook of Sociology of Aging*, Handbooks of Sociology and Social Research, DOI 10.1007/978-1-4419-7374-0_37,

more research is needed. I use the terms "caregiving" and "care work" (Harrington Meyer 2000) interchangeably to convey the complex nature of informal care work as potentially rewarding and something people want to provide, but also work, albeit unpaid. A major strength of caregiving research is that it has been informed by interdisciplinary and cross-disciplinary research, and that caregiving concerns span the entire life course, from concerns about caring for young children who may or may not have severe health problems to caring for ill or disabled parents. This breadth provides not only an opportunity to identify the unique challenges and opportunities of different types of care work, but also the potential to inform broader questions about family and gender relationships across the life course.

I organize research on care work into three broad areas. *First*, much research on caregiving documents the extent of the problem, including identifying how many people are providing informal care, rates and changes in disability, and the extent to which informal care is being used by persons who are ill or disabled (AARP 2005; Manton and Gu 2001). I begin by examining in more detail why care work is a growing public issue in the United States, tracing the roots of the problem not only to the changing demographics of an aging society, but also to changes in family structure, patterns of paid work, structural lag in the organization of work, and changes in the health care system. I argue that the intersection of these demographic and social trends not only heighten the personal challenges of care work but also create a public issue – a "caregiving squeeze" – where we have more persons who need care but fewer with time to provide that care.

A *second* somewhat smaller body of research examines individual and family decisions about who provides care. Theories of gender, family dynamics and rational choice inform and are informed by the unique complexities of caregiving. I thus review some of the main findings and theoretical perspectives relevant to questions about who provides care.

Third, the largest body of research on caregiving focuses on the short- and long-term consequences of care work for the caregiver. Much of our understanding of the personal consequences of caregiving has been shaped by stress theory, but role theory and political-economy and gender perspectives have also been important for our understanding of the consequences of caring. In the remaining sections, I review each of these three domains of research on caregiving in more detail and discuss what different theoretical perspectives add to our understanding of who provides care and the short- and long-term consequences of doing so. I conclude the chapters by proposing several directions for future research on caregiving that connect the public and personal issues of care work.

The Caregiving Squeeze: Intersecting Demographic and Social Trends

Changing demographics, rates of disability, patterns of work, and health care all influence the number of people needing care and the number of persons providing care. A recent study by the National Alliance for Caregiving (NAC) and the American Association of Retired Persons (AARP) estimates that 21% of the U.S. population, or 44.4 million people, provide unpaid help to an adult on at least one activity of daily living (help getting in or out of bed or a chair, getting dressed, getting to and from the toilet, bathing, eating, etc.) or one instrumental activity of daily living (help with finances, grocery shopping, medications, housework, etc.) (AARP 2005). This study estimates that 61% of caregivers are women; the average caregiver age is 46, and 37% report a child under 18 living in the household. Most caregivers are employed, with 48% employed full time and 11% employed part-time. More than half (52%) of the caregivers surveyed reported spending 9 or more hours per week providing care (AARP 2005).

The NAC survey provides one of the best recent profiles of caregivers in the United States, but we know less about trends in caregiving and informal care over time. Recent trends are important, however, as they are our best indicator of future directions of caregiving. The National Long Term Care Surveys document trends in disability among Medicare beneficiaries 65 and older and provide information on types of care used by the disabled elderly and how patterns of care have changed since the mid-1980s (Spillman and Black 2005; Spillman and Pezzin 2000). One consistent finding is that the disabled elderly rely heavily on informal family care. Through the 1980s and 1990s, more than 90% of long-term care users relied on at least some informal care, either solely or in combination with paid formal care. However, from 1984 to 1999 the percent of disabled elderly relying on any informal care dropped from 95% in 1984 to 91.5% in 1999. At the same time, and particularly from 1994 to 1999, the use of formal care also declined, resulting in an increase in the percent of care recipients relying entirely on informal care. By 1999, 66% of care recipients relied entirely on informal care (Spillman and Black 2005).

Trends in care receipt among the disabled elderly and in care work are complex because they are influenced by demographic and social trends that shape both the need for informal care and the supply of family members who can provide that care. Despite this complexity, one thing is clear. Even with some encouraging trends suggesting a decline in disability, as we look forward to future levels of demand for and supply of care, the intersection of an aging population, changes in family size, our health care system and patterns of work suggest we will be increasingly facing a "caregiving squeeze" – an increasing need for informal care at the same time that there is a declining supply of persons who can provide that care. In order to understand that squeeze, and ultimately to minimize its impact, we first review changes influencing the need for care and then those most likely to affect the supply of available caregivers.

Changes in the Need for Care

A major, and often cited, factor fueling a growing need for care is the aging of the population, driven largely by the aging of the baby boom cohort. In 1995, 33.5 million people, or 12.8% of the U.S. population was 65 and older. By 2050, the Census Bureau estimates that 79 million Americans, or one-fifth of the population, will be 65 or older (Day 1996). The largest increases in the 65 and older group will begin in 2010 and continue to 2030. Expected growth in the population 85 and older is especially dramatic both because of the aging of the population and increases in life expectancy. In 1995, 3.6 million Americans were 85 or older with an expected increase to 18.2 million by 2050 (Day 1996). The growth in the oldest old suggests an increase in need for care since it is this segment of the population most in need of acute and long-term care for illness and disability. Among those aged 65–74 only about 7% receive care for an ADL or IADL disability, compared to 63% of those 90 or older who need care (Uhlenberg and Cheuk 2008).

While trends in the aging of the population are clear, the extent to which population aging will translate into increased need for care also depends on trends in disability among the elderly. Recent trends in disability have suggested some decline in rates of disability among the elderly (Freedman et al. 2004). In general, in the mid to late 1990s, the percent of persons 70 and older who had difficulty with daily activities such as bathing, eating, or walking across a room declined 1–2.5% per year (Freedman et al. 2004). Any decline in disability rates among the elderly will offset some of the demand for care. There are a number of reasons to be optimistic that reductions in disability and need for care will continue, particularly as levels of education rise among more recent cohorts of elderly and availability and use of assistive technology expands (Freedman et al. 2004; Uhlenberg and Cheuk 2008). However, other trends suggest less optimism in disability projections because of increases in rates of obesity (Reynolds et al. 2005). Lakdawalla et al. (2004) report, for example, that during the 1980s and 1990s when disability was declining among the elderly, it was increasing for adults under age 60 and these are also the cohorts that have experienced large increases in obesity. These findings suggest that we cannot assume that recent declines in disability among elderly Americans will continue for future cohorts.

In addition to population aging and disability, the need for care is also shaped by the structure and financing of health care. Changes in the availability and cost of formal care alter reliance on informal care. Length of hospital stays have declined dramatically, from an average of 7.5 days in 1980 to 4.7 days in 2006 (National Center for Health Statistics 2009). Declines in hospital and other institutional care increase reliance on informal caregivers for skilled nursing tasks, such as changing IVs and catheters (Glazer 1990). Likewise, increases in the number of persons who are uninsured or underinsured, restrictions on covered services, and increases in out of pocket expenses are all likely to increase the reliance on unpaid and informal care. For example, Spillman and Black (2005) find a significant decline in use of formal care and an increase in use of informal care from 1994 to 1999, a change they attribute in part to the changes in Medicare home health payments after the Balanced Budget Act of 1997. As health care reform is implemented, increased access to care may reduce some pressure on need for informal care. However, the countervailing pressure to reduce costs of health care are likely to offset these gains as we increasingly look to informal care to provide care for those who are ill or disabled.

Changes in the Supply of Informal Care

The caregiving squeeze is not only created by trends in the need for care, but also changes that reduce the available supply of people to provide that care. The decline in family size is a major factor influencing the supply of informal caregivers. Compared to earlier cohorts, the baby boom cohort significantly reduced their average number of children. Unlike their parents' generation which had a fertility rate of over 3.0, baby boom cohorts had a total fertility rate of 2.0 (National Center for Health Statistics 2009). As noted by Uhlenberg and Cheuk (2008), "when mothers of the baby boom are in greatest need of caregiving (i.e., when they are over age 80 around 2020), more than a fourth of them will have at least four children. However, 20 years later when baby boomers are over 80, only 10% of them will have four or more children" (p. 27). Smaller families not only reduce the number of available children to provide informal care to elderly adults, but also reduce the number of siblings and other extended kin who might provide care. Uhlenberg and Cheuk argue, however, that the decline in family size, while important, should not be overstated. Much of the decline in family size is due to a drop in large families and the bigger risk to informal care is among older adults who are childless. While there projected an increase in women 80 and older who are childless between 2020 and 2040, the proportion of women 80 and older who are childless is expected to drop between 2000 and 2020. By 2040, the proportions of childless elderly women is thus expected to be similar to levels seen in 2000 (Uhlenberg and Cheuk 2008).

A much more significant challenge to the available supply in informal caregivers is the well documented change in women's labor force participation. Between 1980 and 2000, the proportion of women aged 45–54 in the labor force increased from 60 to 77, and it is expected to remain at roughly that level through 2016 (U.S. Census Bureau 2008). As of 2007, 55% of married couples in the United States were a part of dual-earner households. Increases in dual-earner households shrink the reserve of nonemployed family members who can step in to provide care and increase the pool of men and women who face caregiving decisions when they are employed. The National Alliance/AARP 2004 survey of U.S. caregivers estimates that 59% of caregivers are employed, up from 52% in a similar study conducted in 1997 (National Alliance for Caregiving and American Association of Retired Persons (NAC/AARP) (1997)). At least among earlier cohorts, when employed women were faced with an ill or disabled family member who needed care, they were not less likely to than the nonemployed to provide that care, but doing so did put them at greater risk of reducing or stopping work (Pavalko and Henderson 2006; Pavalko and Artis 1997). Although we know far less about men's employment and caregiving, a study of care work among married men in Massachusetts found that neither their own employment, nor that of their wives, affected the amount or kind of help they provided to parents, in-laws, adult children or friends (Gerstel and Gallagher 2001).

Confounding the increase in employment among potential caregivers is the limited availability of workplace or government policies that support workers needing flexibility to balance care needs with employment. We will return to this issue below. It remains unknown whether we will see similar choices among more recent cohorts of women and men as gender attitudes and patterns labor force participation change. However, the more consistent labor force attachment among women with young children in recent years suggests that more recent cohorts may be less likely to take on intensive caregiving that would require them to reduce or stop work.

In sum, we have a range of demographic and social trends that suggest we will see an increasing need for informal care in the coming decades at the same time we have a decreasing supply of available caregivers. This caregiving squeeze has implications for the quality and type of care received by those who are ill or disabled. Those who have fewer sources of informal care are more likely to be cared for in institutions and to rely on paid care (Uhlenberg and Cheuk 2008). At the same time, as potential care providers, family members are more likely to face difficult choices about who will provide care and the challenges of balancing employment with care work. The caregiving squeeze is also problematic for employers who will have larger numbers of employees caring for an ill or disabled family member.

Who Cares?

Demographic and social changes in social roles, disability and health care resources suggest a pending caregiving squeeze, but how that squeeze plays out will depend on how individuals, families and communities negotiate who will provide that care. As noted earlier, the majority of long-term care is informal care provided in the community, and the bulk of that care is provided by adult children and spouses (Johnson 2008; Spillman and Pezzin 2000). We also know that in the short-run, employed women are not less likely to provide care (Pavalko and Artis 1997), although if they do provide care they are more likely to stop employment (Pavalko and Henderson 2006). However, to date we know little about whether broader cultural and economic pressures have altered the ability of more recent cohorts of men and women to provide care.

Research investigating who provides care most commonly views the provision of care as a choice made by individuals and families. In line with this assumption, a vast body of research seeks to better understand how men and women make these difficult choices about care. The relationship between the care recipient and care provider is a key factor in these decisions, and for both men and women there appears to be a care hierarchy (Gerstel and Gallagher 2001). Spouses are generally most likely to care for one another, and they tend to continue providing care until they can no longer do so because of their own health problems (Johnson 2008; Stoller and Miklowski 2008). Care for children is also high in men and women's care hierarchy, followed by care for extended kin, and finally care for friends is generally last (Gerstel and Gallagher 2001).

Beyond one's relationship to the care recipient, a *rational choice perspective* theorizes that when a family member needs care, decisions about who will provide that care are made rationally by families as various family members weigh the emotional and financial costs and incentives for providing care (Becker 1991). Economically, families are theorized to weigh the opportunity costs of each potential caregiver and the likely wages they would lose if not caregiving, along with other factors such as geographical proximity to the care recipient. Empirical findings suggest that wage opportunity considerations are important in how families choose to provide long-term care. For example, in analyses of data from the 2002 wave of the Health and Retirement Study, Johnson (2008) finds that "frail older adults received less unpaid help from their children and more unpaid help from paid sources, both at home and in nursing homes, when all of their offspring could earn relatively high wages in the labor market than when some had relatively poor labor market prospects." Economic considerations may be part of the decision process in the type of support children provide to parents. For example, in one study, children earning high wages were more likely to give parents money but spend less time providing care work (Couch et al. 1999). However, there are alternative noneconomic influences, such as cultural differences across social class that could also account for many of these patterns (Johnson 2008). Finally, we know little about how (or even whether) the larger structural context, such as workplace or community structures that are more supportive of work–family balance, influence care decisions. To the extent that care decisions are based even partially on economic opportunity costs, increases in women's labor force participation and the increasing movement of women into higher paying professional and managerial positions will have widespread implications for those who provide care.

While economic opportunity costs may be part of the decision process, it is clear that a number of other influences also play a role in care decisions. *Gender theories* point to the complex cultural and emotional pressures and rewards for women to provide care, and there is little question that gender plays a central role in the amount and type of care provided. Although the proportion of male caregivers has increased in recent years, women continue to provide the majority of unpaid care. While opportunity costs, and particularly women's lower pay and labor force participation, account for some of the gender disparities in care, it does not account for the consistent finding that gender differences in care remain even as women's labor force participation and occupational status rival that of men. For example, in a large, multigenerational study, Silverstein et al. (2008) find that, when comparing the amount of care provided for mothers within families, daughters continue to provide more care than sons. Notably, these gender differences remain after controlling for each sibling's income, and indeed income itself is not a significant predictor of care in these intra-family comparisons.

An underlying assumption of many choice-based models of caregiving, including those weighing opportunity costs, is that potential caregivers substitute for one another in a zero-sum fashion. Thus, as one family member increases care, others will reduce their share. In this substitution model, as opportunities of women's paid work increase their opportunity costs, men would be expected to increase their involvement in care work. Both in the case of care for children and care for ill or disabled family members, this does not always appear to be the case (Gerstel and Gallagher 2001). An alternative is that, especially within families, men's and women's care work is complimentary, and that women pull men into caregiving rather than being replaced by them (Gerstel and Gallagher 2001). In a study of married couples, Gerstel and Gallagher (2001) find that the strongest factor increasing men's help to in-laws and adult children was a wife who also provided care. In contrast, wives' contributions to family income and overlap in work schedules made little difference in how much care men provided. However, the same study did find a substitution effect between men and their sisters. Men with more sisters spend less time providing parental care, although brothers had no effect. These findings suggest that while wives pull men into parental care, sisters replace care that otherwise would be provided by men.

Network theories also enlighten the complexity of care decisions by examining how the allocation of care within families operates as a series of interrelated decisions made by a network of potential care providers. Care networks may expand and contract over time and different members of the care network may provide different types of support. While network approaches to care work are in their infancy, attention to caregiving networks and efforts to unravel their changes can provide unique insights into why caregiving networks change and how those changes unfold. Early research on care networks confirms that changes in these networks over a 2-year period were more common than not.

Interestingly, change in networks is more common if that network includes male caregivers, particularly if that network included both male and female caregivers (Davey and Szinovacz 2008). Attention to network changes also provides potential insights into broader questions about how gender, job opportunity and other factors influence which network members provide care and the factors prompting changes in the caregiving network.

Understanding who provides care and why they do so, or why they do not, is essential for projecting how changes in work and family life may impact the availability of caregivers for future generations. At the same time the complexity of care decisions offers ample opportunities for research on caregiving to inform fundamental questions about family dynamics, gender, and economic choice.

Consequences of Caring

The consequences of caregiving have been widely studied and consistently documented. There is little doubt that caregiving affects the health and well-being of the caregiver. Studies from a wide range of samples and a numerous levels and types of care have found that caregivers have higher rates of depression and psychological distress than noncaregivers (Schulz et al. 1990, 1995). However, there are also notable benefits to caring, such as opportunities for personal growth and enhanced relationships with others (Abel 1986; Chesley and Moen 2006; Tarlow et al. 2004). The effect of caregiving on physical health is more varied, but studies have found that caregivers have more disease symptoms, physical limitations, chronic conditions, or poorer immune function (Kiecolt-Glaser et al. 1991; Pavalko and Woodbury 2000; Seltzer et al. 2009). There is growing evidence that, at least for women, there are significant short- and long-term economic consequences of caring (Pavalko and Henderson 2006; Wakabayashi and Donato 2006). A number of theoretical perspectives are useful for understanding why caregiving has such broad implications for the caregiver, and the mechanisms through which caregiving influences unfold. While these theories are not mutually exclusive, each does identify and highlight different dimensions of the caregiving experience.

Stress theory has served as the major orientating framework for understanding the consequences of caring for health (Aneshensel et al. 1995; Pearlin 1989). Stressors, such as caring for an ill or disabled family member, are problematic or difficult circumstances that have the potential to affect one's health and well-being. The stress process model theorizes that these effects are not automatic or experienced in the same way for all people, in large part because these effects are moderated by social, personal and material resources, such as economic resources that may allow one to purchase more or better health care services, educational resources that provide more knowledge about disease processes or their management, and social-psychological resources that provide more avenues for social support (Pearlin 1989). Attention to stress mediators is particularly important because it locates the stress process within a larger social-structural context and helps identify interventions that might reduce the impact of stressors (Aneshensel 2009). The stress process model also distinguishes between primary and secondary stressors associated with caregiving (Aneshensel et al. 1995). Primary stressors are those associated with objective conditions surrounding caregiving, such as managing the care recipient's behavioral problems or helping with physical or instrumental tasks. Secondary stressors stem from problems in other roles, such as work and family, produced because of the temporal, physical and emotional demands of caring. Thus, caregivers may experience stress proliferation as primary and secondary stressors accumulate over time (Pearlin et al. 1990). Moderators may influence the extent to which both primary and secondary stressors translate into adverse health outcomes. Research over the last several decades has documented the overall impact of caregiving stressors and a wide range of mediators in caregiving stress processes.

One direction in more recent research has been to identify how caregiving stressors "get under the skin" by influencing various influence on various biomarkers such as cortisol and telomere shortening (Epel et al. 2004; Kiecolt-Glaser et al. 1991; Seltzer et al. 2009). For example, Seltzer et al. (2009) found that parents of children with disabilities had elevated levels of stress and different diurnal rhythms of cortisol expression than a matched comparison groups, particularly on days when they were spending more time with their children. Alternatively, researchers have also begun to identify distinct stressors associated with having a parent who is ill or disabled as well as those stemming from the stressors of providing care (Amirkhanyan and Wolf 2006).

The importance of stress theory for our understanding of caregiving experiences has largely emphasized the negative consequences of caring, but a growing body of research has expanded the traditional stress-coping model to assess the positive effects of care work. While the specific challenges and benefits of caring vary by the type of illness or disability and the relationship between the caregiver and care recipient, a wide range of caregivers report benefits of caring. For example, research based on a national study of Alzheimer's caregivers found that they frequently reported that caring for someone with dementia made them feel needed and good about themselves, and that caring increased their ability to appreciate life more and strengthened their relationships with others (Tarlow et al. 2004). Likewise, a study of mothers caring for children with disabilities found that most mothers perceived benefits such as a greater awareness of what is important in life and a greater sense of personal strength because of their parenting experiences (Green 2007). There is also growing evidence of gender differences in benefits of caring. Several studies suggest that men experience more positive benefits of caring, such as increased sense of personal growth, than do women (Chesley and Moen 2006; Marks 1998).

Role theory has also been used to understand, and been informed by research on caregiving. While stress theory views roles and role conflicts as an important part of the stress process, it largely assumes that multiple roles contribute to the stress process. In contrast, role theory examines the ways in which the multiplicity of social roles may conflict or enhance one another. Early research on multiple roles questioned whether role combinations such as wife, parent, and worker were harmful or enhanced health (Thoits 1987), while more recent work documents how the strains and benefits of role combinations are contextualized by the nature of those roles and the larger structural conditions in which they are imbedded (Moen et al. 1995; Waldron et al. 1998).

One important line of research on caregiving that has been informed by role theory has been attention to the "sandwiched generation." The sandwiched generation refers to the challenges faced by full-time workers who have the responsibility of caring for both elderly parents and minor children (Ingersoll-Dayton et al. 2001; Spillman and Pezzin 2000). While there is some debate over how many persons are faced with simultaneously caring for young children and aging parents and whether this has increased over time, because care work is most likely to fall during midlife, there is little question that caregivers are often squeezed by conflicting roles from both work and family.

Role theory has also been valuable for framing our understanding of how, or whether, role conflicts, particularly between caregiving and employment, affect health. Most caregivers are employed and employment does not appear to reduce the likelihood that women will take on caregiving tasks (Pavalko and Artis 1997). Despite many concerns about role conflicts, a considerable body of research indicates that multiple roles combined with caregiving enhance rather than hurt health (Hong and Seltzer 1995; Pavalko and Woodbury 2000). Employment appears to be a particularly protective role for caregivers, with most studies finding that employed caregivers fare better than nonemployed caregivers. However, it remains unclear the extent to which this is a protective effect of employment and how much it reflects the selection of those with the greatest caregiving demands out of the labor force. We know that employed women who take on caregiving tasks are at greater risk of leaving the labor force (Pavalko and Henderson 2006; Pavalko and Artis 1997) and that women who remain employed while caregiving differ in several ways from those who stop work while caring (Pavalko and Woodbury 2000).

Understanding the role conflicts and enhancements of caregiving with other roles highlights an important lesson for role theory. Combining employment and caregiving is neither uniformly positive nor uniformly negative, in part because people are not just passive recipients of roles. Individuals are "role makers" who actively manage and negotiate roles to maximize their benefits and minimize costs (Stryker and Statham 1985; Thoits 2006; Thoits and Hewitt 2001). Caregiving differs from many other roles because it is, on average, shorter duration than most work and family roles, but decisions whether or when to start or stop caregiving are often beyond an individual's immediate control. Caregiving thus offers a unique lens for understanding the health implications of adding or losing a major role and for observing the agency individuals bring in taking on or shedding roles (Pavalko and Woodbury 2000).

Role making and stress proliferation have contributed to another theoretical development important for understanding care work, the *caregiving career* (Aneshensel et al. 1995; Pavalko and Woodbury 2000). Attention to the caregiving career emphasizes the dynamic nature of caregiving experiences and incorporates insights into how people actively manage their roles, even when faced with role demands over which they have little control, such as the health of a family member. Caregiving is dynamic in many respects. On average, spells of caregiving last 2 years or less (Moen et al. 1994) and the experiences of caregiving are likely to vary as caregivers adjust and adapt to the role, as the care needs of the care recipient change, and as caregivers take on and shed other roles. While a handful of studies have mapped health changes across the caregiving career, some find that initial periods of caregiving stress plateau or even drop as caregiving continues (Whitlatch et al. 1997) while others find a steadier pattern of stress proliferation (Pearlin et al. 1997). Pavalko and Woodbury (2000) found that psychological distress increased as women moved into caregiving and also for those who had been caregiving and continued to do so. Caregiving had less of an influence on changes in physical health, but women who did experience an increase in physical limitations when caregiving were more likely to stop providing care. Caregiver health across the caregiving career also differed depending on whether or not they were also employed. Employment appears to offer a greater buffer against health decline in the early stages of the caregiving career, but there was also evidence that combining caregiving and employment was unhealthy for some women, prompting them to stop employment.

Stopping or reducing hours of employment is a strategy for managing the demands of caregiving employed by significant numbers of women. Employed women who take on caregiving are more likely to stop work (Pavalko and Henderson 2006; Pavalko and Artis 1997). While this may help manage role demands in short run, it produces another notable consequence of caring – short- and long-term economic risk. Stopping employment to provide care reduces immediate income and access to employer benefits such as health insurance, but it also risks the long-term economic security of the caregiver. For example, spells out of employment affect primary insurance amounts, which serve as the basis for determining Social Security benefits in later life (Kingson and O'Grady-LeShane 1993). It is thus not surprising that a recent study finds that caregivers who spent 20 hours per week assisting parents with personal care were subsequently 25% more likely to live in poverty, 27% more likely to receive public assistance, and 46% more likely to receive Medicare (Wakabayashi and Donato 2006).

A *political economy perspective* calls attention to the economic consequences of caring but also notes that neither the health nor the economic consequences of caring occur in a vacuum. A growing body of research has begun to identify workplace policies that are effective in helping employees remain employed when they take on caregiving tasks. Three policies – family leave, sick and vacation days and flexible hours – show the most promise for improving work–family balance, labor market outcomes and work commitment (Chesley and Moen 2006; Kossek 2005; Pavalko and Henderson 2006). In one recent study, women working at jobs with flexible hours, unpaid family leave and paid sick and vacation days were more likely to remain employed and maintain their prior level of work hours even if they were not currently caregiving (Pavalko and Henderson 2006). Women who had begun caregiving in the previous 2 years and who had access to unpaid family leave were far more likely to remain employed than caregivers without family leave. A study of dual-earner couples reports that women caregivers with flexible work arrangements report better emotional health (Chesley and Moen 2006).

Theoretically, these studies of workplace policies suggest that how individuals manage roles and the consequences of doing so are contextualized by structural supports and constraints that make it more or less difficult to combine roles. More immediate policy contexts, such as whether a workplace is "family friendly" and has policies available for workers such as flexible hours and paid leave is also contextualized by broader policy and cultural contexts. These broader contexts, such as whether a country locates responsibility for work–family policies in the workplace, as is the case in the United States, or with the state, as in many Western European countries, most likely has implications for how people combine work and family roles and the consequences of doing so for health and economic well-being. Cultural contexts which define appropriate strategies balancing work and family roles are also likely to be important for shaping many of the stressors that arise from caregiving. The political economy perspective can thus be valuable for contextualizing the individual decisions and consequences about care work within the broader political and economic context.

New Directions for Understanding Care Work

Questions about family dynamics, gender inequities, and the division of labor, as well as those about roles and stress processes, continue to be informed by the unique context of caregiving. While the study of caregiving has much to offer these theories, there remains much that we do not understand about care work. At its core, care work is a personal issue faced by individuals and families with an ill or disabled family member but it is also a public issue. Individuals and families have always been faced with illness and disability and the personal challenges for families to provide that care are certainly not unique to the twenty-first century. What is unique is the particular intersection of demographic trends, institutional structures and social trends which make caregiving a public issue of concern to health care providers, employers and communities as well as individuals and families facing an immediate need for care. The public issue – a growing "caregiving squeeze," – means that increased needs for care occur at a time when we have fewer people with time to provide that care. While research on caregiving has done much to inform the personal experiences of care and to identify the public issues of care work, our understanding of the *intersection* of personal and public is more limited. Comparative research, whether the comparisons are across personal contexts such as racial/ethnic groups, temporal, institutional, and even country-level contexts, can provide unique insights into the public dimensions of care. However, each of these comparisons offers different opportunities to improve our understanding of how care needs, decisions and consequences vary across contexts. Below I highlight several contextual comparisons that may be particularly fruitful in broadening our understanding of the personal and institutional contexts of care work.

One important, but understudied, context of care work is race/ethnicity. Much of what we know about caregiving has been based on the experiences of racial/ethnic majority families in the United States, but there are a number of reasons to expect that the need for care, the decisions about care providers, and the consequences of care work varies across racial/ethnic groups. Variations in life expectancy, morbidity and disability across racial/ethnic groups are well documented. African-Americans generally have higher rates of mortality and morbidity than whites, but U.S.-born Hispanics and Asian-Americans are generally healthier and live longer than U.S.-born whites (Singh and Siahpush 2002). Immigrants to the United States, including foreign-born blacks, Hispanics, and Asian/Pacific Islanders, however, have lower mortality and morbidity rates than U.S.-born whites (Singh and Siahpush 2002). While greater longevity is clearly a positive goal, it does increase the probability that especially women will develop osteoporosis, which places them at a greater risk of disability in later life (Angel and Angel 2006). Greater longevity may thus increase the need for some types of care, which are further complicated by racial/ethnic differences in access to health care and health insurance. Health care reform in the United States will likely reduce some of the more immediate disparities in access to health care, but it will take far longer to reduce disparities that may have accumulated from differential access to care across the life course. Current health care reforms that increase access to insurance and health also exclude immigrants from much of the coverage. All of these changes suggest a changing landscape of racial/ethnic disparities in the need for care. Family decisions and resources for providing care also vary across racial/ethnic groups, as U.S.-born and immigrant racial/ethnic groups differ in family size, cultural norms about family support and gender roles. Comparative research across racial/ethnic groups will thus be increasingly important for understanding the caregiving squeeze and the varying cultural contexts of care decisions within families.

We also know surprisingly little about how historical contexts shape both decisions and consequences of care work. Given rapid changes in women's employment, the role of opportunity costs and employment in care decisions among contemporary cohorts of men and women may differ from that found among cohorts born even a decade or two before. Likewise, the gendered nature of care decisions is likely to differ as attitudes about gender, family and work change and as more women are engaged in full-time employment. However, we know relatively little about how decisions about care by men and women vary across birth cohorts or the life course (but see Moen et al. 1994; Robison et al. 1995). It is plausible that more recent cohorts of men and women may make different decisions, or face different consequences, of care work than prior cohorts and these differences could have consequences for the extent of thc caregiving squeeze. However, to date we know relatively little about cohort and historical shifts in care work.

A third important point of comparison is the institutional context in which care work takes place. These types of comparisons are particularly valuable for assessing influences of policies and programs on care decisions and consequences. Local communities are one source of institutional context, but few communities, at least in the United States, provide comprehensive resources for long-term care. Workplaces provide another institutional context, allowing us to assess whether company policies affect caregiving decisions and consequences for employed caregivers. For example, do employed caregivers with access to "family friendly" policies such as paid leave, sick and vacation time, or flexible hours have less psychological distress or are they more likely to remain in the labor force than employed caregivers who do not have access to these policies? Recent evidence suggests that they may (Chesley and Moen 2006; Pavalko and Henderson 2006). An innovative assessment of the effects of flexible hours on employees uses an experimental design to assess the effects of a new policy implemented by a large employer that gives workers control over when, where, and how they work (i.e., flexibility and control). Preliminary results suggest that after adopting this policy, workers report fewer turnover intentions, more organizational commitment, and greater job satisfaction than those in the control group (Moen and Kelly 2007). While these results suggest the power of workplace policies for reducing the stresses of care work, more work is needed to identify which policies have the greatest impact and whether the effects vary across different types of workplaces or occupations. These comparisons will be particularly valuable for understanding whether specific policies or institutional contexts influence the personal experience and effects of care.

Finally, cross-national comparisons provide a more macro lens for assessing the institutional contexts surrounding personal and public dimensions of care. There is little question that the institutional and policy context of caregiving in the United States differs from that found in other western industrialized countries. Although policy-makers often invoke language of family values, policies that make it possible for workers to balance work and provide care for family members are far more scattered and modest in the United States than other industrialized countries (Gornick and Meyers 2003).

While the United States only recently mandated that workers receive 12 weeks *unpaid* family leave with the FMLA act passed in 1993, most industrialized countries provide at least 2–3 months of *paid* leave after childbirth (Ruhm 2005). In contrast, most of the protections for family care available to workers in the United States are provided by individual employers, thus making those protections far more variable and highly dependent on specific employment contexts. Cultural norms about family care and gender roles also vary across countries. Cross-national comparisons of caregiving decisions, experiences, and short- and long-term effects on income and employment could provide a much broader scope for understanding how the personal experience of caregiving is, and is not, shaped by cultural and institutional contexts in which the care takes place.

At some point in our lives, most of us will face personal challenges of care work. Many of us will face it multiple times, both as someone in need of care and as potential providers of care for an ill or disabled family member. Sociologists have an important role to play in preparing individuals and communities for the emerging challenges associated with care and assessing ways that care work can most effectively be balanced with other work and family roles. Bringing a sociological imagination to these personal challenges will be essential for understanding how and why contemporary challenges of care work may differ from those faced by previous and future generations. Attention to both the personal and the public dimensions of care work is also essential for imagining, and thus creating, the contexts where care work is more likely to enhance rather than compete with other work and family roles.

References

AARP. 2005. "Caregiving in the United States." AARP and National Alliance for Caregiving. Washington, D.C.

Abel, Emily. 1986. "Adult Daughters and Care for the Elderly." *Feminist Studies* 12:479–98.

Amirkhanyan, Anna A. and Douglas A. Wolf. 2006. "Parent Care and the Stress Process." *Journal of Gerontology: Social Sciences* 61B:S248–55.

Aneshensel, Carol S. 2009. "Toward Explaining Mental Health Disparities." *Journal of Health and Social Behavior* 50:377–94.

Aneshensel, Carol S., Leonard I. Pearlin, Joseph T. Mullan, Steven H. Zarit, and Carol J. Whitlatch. 1995. *Profiles in Caregiving: The Unexpected Career*. San Diego, CA: Academic.

Angel, Ronald J. and Jacqueline L. Angel. 2006. "Diversity and Aging in the United States." Pp. 94–110 in *Handbook of Aging and the Social Sciences*, edited by R. H. Binstock and L. K. George. New York: Academic.

Becker, Gary S. 1991. *A Treatise on the Family*. Cambridge, MA: Harvard University Press.

Chesley, Noelle and Phyllis Moen. 2006. "When Workers Care: Dual-Earner Couples' Caregiving Strategies, Benefit Use, and Psychological Well-Being." *American Behavioral Scientist* 49:1248–69.

Couch, K.A., M.C. Daly, and Douglas A. Wolf. 1999. "Time? Money? Both? The Allocation of Resources to Older Parents." *Demography* 36:219–32.

Davey, Adam and Maximiliane E. Szinovacz. 2008. "Division of Care Among Adult Children." Pp. 133–59 in *Caregiving Contexts: Cultural, Familial, and Societal Implications*, edited by M. E. Szinovacz and A. Davey. New York: Springer.

Day, Jennifer Cheeseman. 1996. *Population Projections of the United States, by Age, Sex, Race, and Hispanic Origin: 1993–2050*. Current Population Reports, Series P25-1104. U.S. Census Bureau, Washington, D.C.

Epel, Elissa S, Elizabeth H. Blackburn, Jue Lin, Firaus S. Dhabhar, and Richard M. Cawthon. 2004. "Accelerated Telomere Shortening in Response to Life Stress." *Proceedings of the National Academy of Sciences* 101:17312–15.

Freedman, Vicki A., Eileen Crimmins, Robert Schoeni, Brenda Spillman, Hakan Aykan, Ellen Kramarow, Kenneth C. Land, James Lubitz, Kenneth G. Manton, Linda G. Martin, Diane Shinberg, and Timothy Waidmann. 2004. "Resolving Inconsistencies in Trends in Old-Age Disability: Report from a Technical Working Group." *Demography* 41:417–414.

Gerstel, Naomi and Sally K. Gallagher. 2001. "Men's Caregiving: Gender and the Contingent Character of Care." *Gender & Society* 15:197–217.

Glazer, Nona Y. 1990. "The Home as a Workshop: Women as Amateur Nurses and Medical Care Providers." *Gender & Society* 4:479–99.

Gornick, Janet C. and Marcia K. Meyers. 2003. *Families that Work: Policies for Reconciling Parenthood and Employment*. New York: Russell Sage Foundation.

Green, Sara Eleanor. 2007. "We're Tired, Not Sad: Benefits and Burdens of Mothering a Child with a Disability." *Social Science and Medicine* 64:150–63.

Harrington Meyer, Madonna. 2000. *Care Work: Gender, Class, and the Welfare States*. New York: Routledge.

Hong, Jinkuk and Marsha Mailick Seltzer. 1995. "The Psychological Consequences of Multiple Roles: The Nonnormative Case." *Journal of Health and Social Behavior* 36:386–98.

Ingersoll-Dayton, B, Margaret B Neal, and Leslie B. Hammer. 2001. "Aging Parents Helping Adult Children: The Experience of the Sandwiched Generation." *Family Relations* 50(3):262–71.

Johnson, Richard W. 2008. "Choosing Between Paid Elder Care and Unpaid Help From Adult Children: The Role of Relative Prices in the Care Decision." Pp. 35–70 in *Caregiving Contexts: Cultural, Familial, and Societal Implications*, edited by M. E. Szinovacz and A. Davey. New York: Springer.

Kiecolt-Glaser, J. K., J. R. Dura, C. E. Speicher, O. J. Trask, and R. Glaser. 1991. "Spousal Caregivers of Dementia Victims: Longitudinal Changes in Immunity and Health." *Psychosomatic Medicine* 53:345–62.

Kingson, Eric R. and Regina O'Grady-LeShane. 1993. "The Effects of Caregiving on Women's Social Security Benefits." *The Gerontologist* 33:230–9.

Kossek, Ellen Ernst. 2005. "Workplace Policies and Practices to Support Work and Families." Pp. 97–115 in *Work, Family, Health and Well-Being*, edited by S. Bianchi, L. Casper, and R. King. Mahwah, New Jersey: Lawrence Erlbaum.

Lakdawalla, D., J. Bhattacharya, and D. P. Goldman. 2004. "Are the Young Becoming More Disabled? Rates of Disability Appear to be on the Rise Among People Ages Eighteen to Fifty-Nine, Fueled by a Growing Obesity Epidemic." *Health Affairs* 23:168–76.

Manton, Kenneth G. and XiLiang Gu. 2001. "Changes in the Prevalence of Chronic Disability in the United States Black and Non-Black Population Above Age 65 from 1982 to 1999." *Proceedings of the National Academy of Sciences* 98:6354–9.

Marks, Nadine F. 1998. "Does it Hurt to Care? Caregiving, Work-Family Conflict, and Midlife Well-Being." *Journal of Marriage and the Family* 60:951–66.

Mills, C. W. 1959. *The Sociological Imagination*. New York: Oxford University Press.

Moen, Phyllis and Erin L. Kelly. 2007. "Flexible Work and Well-Being Study, Final Report." University of Minnesota, Minneapolis, MN.

Moen, Phyllis, Julie Robinson, and Donna Dempster-McCain. 1995. "Caregiving and Women's Well-Being: A Life Course Approach." *Journal of Health and Social Behavior* 36:259–73.

Moen, Phyllis, Julie Robison, and Vivian Fields. 1994. "Women's Work and Caregiving Roles: A Life Course Approach." *Journal of Gerontology: Social Sciences* 49:S176–86.

National Alliance for Caregiving and American Association of Retired Persons (NAC/AARP). 1997. "Family Caregiving in the U.S.: Findings from a National Survey." National Alliance for Caregiving and American Association of Retired Persons (NAC/AARP), Bethesda, MD.

National Center for Health Statistics. 2009. *Health, United States, 2008 With Special Feature on the Health of Young Adults*. National Center for Health Statistics, Hyattsville, MD.

Pavalko, Eliza K. and Kathryn A. Henderson. 2006. "Combining Care Work and Paid Work: Do Workplace Policies Make a Difference?" *Research on Aging* 28:359–74.

Pavalko, Eliza K. and Julie E. Artis. 1997. "Women's Caregiving and Paid Work: Causal Relationships in Late Midlife." *Journal of Gerontology: Psychology and Social Sciences* 52:S170–9.

Pavalko, Eliza K. and Shari Woodbury. 2000. "Social Roles as Process: Caregiving Careers and Women's Health." *Journal of Health and Social Behavior* 41:91–105.

Pearlin, Leonard I. 1989. "The Sociological Study of Stress." *Journal of Health and Social Behavior* 30:241–56.

Pearlin, Leonard I., Carol S. Aneshensel, and Allen J. LeBlanc. 1997. "The Forms and Mechanisms of Stress Proliferation: The Case of AIDS Caregivers." *Journal of Health and Social Behavior* 38(3):223–36.

Pearlin, Leonard I., Joseph T. Mullan, Shirley J. Semple, and Jeremy Skaff. 1990. "Caregiving and the Stress Process: An Overview of Concepts and Their Measures." *The Gerontologist* 30:583–91.

Reynolds, S. L., Y. Saito, and Eileen M. Crimmins. 2005. "The Impact of Obesity on Active Life Expectancy in Older American Men and Women." *The Gerontologist* 45:438–44.

Robison, J., Phyllis Moen, and D. Dempster-McCain. 1995. "Women's Caregiving: Changing Profiles and Pathways." *Journal of Gerontology: Social Sciences* 50B:S362–73.

Ruhm, Christopher J. 2005. "How Well Do Government and Employer Policies Support Working Parents?" Pp. 313–325 in *Work, Family, Health and Well-Being*, edited by S. M. Bianchi, L. M. Casper, and R. B. King. Mahwah, NJ: Lawrence Erlbaum.

Schulz, R, A. T. O'Brien, J. Bookwala, and K. Fleissner. 1995. "Psychiatric and physical Morbidity Effects of Dementia Caregiving: Prevalence, Correlates, and Causes." *The Gerontologist* 35:771–91.

Schulz, Richard, Paul Visintainer, and Gail M. Williamson. 1990. "Psychiatric and Physical Morbidity Effects of Caregiving." *Journal of Gerontology: Psychological Sciences* 45:P181–91.

Seltzer, Marsha M., David M. Almeida, Jan S. Greenberg, Jyoti Savla, Robert S. Stawski, Jinkuk Hong, and Julie L. Taylor. 2009. "Psychosocial and Biological Markers of Daily Lives of Midlife Parents of Children with Disabilities." *Journal of Health and Social Behavior* 50:1–15.

Silverstein, Merril, Stephen J. Conroy, and Daphna Gans. 2008. "Commitment to Caring: Filial Responsibility and the Allocation of Support by Adult Children to Older Mothers." Pp. 71–91 in *Caregiving Contexts: Cultural, Familial, and Societal Implications*, edited by M. E. Szinovacz and A. Davey. New York: Springer.

Singh, Gopal K. and Mohammad Siahpush. 2002. "Ethnic-Immigrant Differentials in Health Behaviors, Morbidity, and Cause-Specific Mortality in the United States: An Analysis of Two National Data Bases." *Human Biology* 74:83–109.

Spillman, Brenda C and Kirsten J. Black. 2005. "Staying the Course: Trends in Family Caregiving." AARP Public Policy Institute, Washington, D.C.

Spillman, Brenda C and Liliana E. Pezzin. 2000. "Potential and Active Family Caregivers: Changing Networks and the 'Sandwich Generation'" *The Milbank Quarterly* 78:347–74.

Stoller, Eleanor Palo and Casey Schroeder Miklowski. 2008. "Spouses Caring for Spouses: Untangling the Influences of Relationship and Gender." Pp. 115–31 in *Caregiving Contexts: Cultural, Familial, and Societal Implications*, edited by M. E. Szinovacz and A. Davey. New York: Springer.

Stryker, Sheldon and Anne Statham. 1985. "Symbolic Interaction and Role Theory." Pp. 311–78 in *The Handbook of Social Psychology*, edited by G. Lindzey and E. Aronson. New York: Random House.

Tarlow, Barbara J., Stephen R. Wisniewski, Steven H. Belle, Mark Rubert, Marcia G. Ory, and Dolores Gallagher-Thompson. 2004. "Positive Aspects of Caregiving." *Research on Aging* 26:429–53.

Thoits, Peggy. 2006. "Personal Agency in the Stress Process." *Journal of Health and Social Behavior* 47:309–23.

Thoits, Peggy A. 1987. "Negotiating Roles." Pp. 11–22 in *Spouse, Parent, Worker: On Gender and Multiple Roles*, edited by F. J. Crosby. New Haven, CT: Yale University Press.

Thoits, Peggy A. and Lyndi N. Hewitt. 2001. "Volunteer Work and Well-Being." *Journal of Health and Social Behavior* 42:115–31.

U.S. Census Bureau. 2008. *Statistical Abstract of the United States: 2009*. Washington.

Uhlenberg, Peter and Michelle Cheuk. 2008. "Demographic Change and the Future of Informal Caregiving." Pp. 9–34 in *Caregiving Contexts: Cultural, Familial, and Societal Implications*, edited by M. E. Szinovacz and A. Davey. New York: Springer.

Wagner, Donna L. 1997. "Comparative Analysis of Caregiver Data for Caregivers to the Elderly 1987 and 1997." Bethesda, MD: National Alliance for Caregiving.

Wakabayashi, Chizuko and Katherine M. Donato. 2006. "Does Caregiving Increase Poverty Among Women in Later Life? Evidence from the Health and Retirement Survey." *Journal of Health and Social Behavior* 47:258–74.

Waldron, Ingrid, C.C. Weiss, and Mary Elizabeth Hughes. 1998. "Interacting Effects of Multiple Roles on Women's Health." *Journal of Health and Social Behavior* 39:216–36.

Whitlatch, Carol J., Lynn F. Feinberg, and Douglas S. Sebesta. 1997. "Depression and Health in Family Caregivers." *Journal of Aging and Health* 9:222–43.

Part X
Sociological Lives: Personal Reflections on the Sociology of Aging

Chapter 38
Gerontology with a "J": Personal Reflections on Theory-Building in the Sociology of Aging

Vern L. Bengtson

Geh-Ron-Tology or Jeh-Ron-Tology?

My introduction to the field of gerontology was not auspicious. I was sitting in the office of Bernice Neugarten at the University of Chicago, scared silly. I was a senior in college applying to graduate schools and Dr. Neugarten (I still can't bring myself to call her "Bernice") was interrogating me. Why should I, a philosophy major from a very small and undistinguished college, feel myself qualified to join the Human Development and Aging program at the great University of Chicago? "Well," I began, "I think I would be interested in *Geh*-ren-tology, pronouncing the 'G' as in 'guest.'" "*NO!*," she shouted. "It is pronounced *Jeh*-ren-tology!" This was the first time I had heard anyone *say* the word. The study of aging was very new in 1963.

Successful Aging: Activity or Disengagement or Both?

Theories of aging have a very short history compared to theories in other areas of science. I arrived at Chicago after the start of the first big theoretical controversy in what later came to be called the sociology of aging. In the previous decade, Robert Havighurst and his students had collected survey data that linked "successful aging," indicated by high life satisfaction scores, to the maintenance of high levels of activity and social engagement (Havighurst and Albrecht 1953). This came to be known as the "activity theory" of aging. It became immediately popular with gerontological practitioners working with the aged, because it justified what they had been doing to keep elderly patients active.

Then to Chicago came two young and ambitious researchers, Elaine Cumming and Bill Henry, who proposed exactly the opposite in their "disengagement theory of aging" (Cumming and Henry 1961). This theory boldly proposed that withdrawal was the normal pattern of aging, and that both the individual and the society benefited when aging persons disengaged from roles and activities in which they had previously been engaged. Such disengagement is universal as well as functional: satisfying for the aging individual, because it enhances life satisfaction by freeing him/her from increasingly unrealistic demands of mid-life and useful for the social system, because it helps the group maintain equilibrium while preparing younger members to fit into the retiree's position. However, disengagement theory very quickly became unpopular (Achenbaum and Bengtson 1994). The advocates of "activity theory" felt disengagement theory was overgeneralized and deterministic

V.L. Bengtson (✉)
School of Gerontology, University of Southern California, Los Angeles, CA, USA
e-mail: Bengtson@usc.edu

R.A. Settersten, Jr. and J.L. Angel (eds.), *Handbook of Sociology of Aging*, Handbooks of Sociology and Social Research, DOI 10.1007/978-1-4419-7374-0_38, © Springer Science+Business Media, LLC 2011

(Rose 1964). Furthermore gerontological practitioners, therapists, and nursing home administrators were alarmed because it undercut activity-based therapies. And then there was the sense that advocating disengagement from productive activities was, in a vague sort of way, un-American.

Into this theoretical fray I sailed with my dissertation research, an attempt to test (and, I thought, to refute) the postulate of "universality" in the theory of disengagement. The data, from middle-aged and retired steel workers and teachers in several nations, did not provide any support for the majority of disengagement theory's postulates. Nor did the data support activity theory. In fact, high life satisfaction varied wildly with social engagement across the two occupational groups and the eight national settings involved in the study, in no consistent "pattern of aging" (Bengtson 1969). While the retired Americans showed a high correlation between activity and life satisfaction in ways supporting activity theory, the Dutch and the Italians showed low activity and high satisfaction – supporting disengagement theory. Later, in my first research project following graduate school I developed a more systematic test of activity theory (Lemon et al. 1972). The data provided little support for the axiom that high activity produces greater well-being in aging. There was just too much variation between individuals, income groups, or men and women to support such a sweeping, one-size-fits-all theory.

I learned some important lessons about theory from my first attempts to grapple with explanations of aging, lessons I have tried to pass on to my students. The first was that a good theory must account for variability as well as for modal patterns. In aging there are not only individual differences but social and structural differences that must be taken into account by the theory. Second, a theory is in danger if it attempts to explain too much, if it over-generalizes. A one-size-fits-all theory of aging will not last very long. A third lesson (I thought) was that a theory not supported by data will wither and die. A theory must be testable and when data show the theory to be disconfirmed we must modify the theory or develop a new one.

By these three criteria, the activity theory of aging should now be encountered only in the footnotes of gerontology textbooks. But still, to this date, activity theory lives on, 60 years later, as the implicit theoretical base for a recent best-seller, *Successful Aging* (Rowe and Kahn 1998), touted as the prescription for a good old age. The activity-is-good formula is presented (sadly) without any indication that the authors were aware of the earlier incarnation of activity theory, or that they were conscious of the individual variations and social structural constraints that critics noted must be taken into account, or that they were aware of the *lack of empirical evidence to support activity theory* (Schmeeckle and Bengtson 1999). I had thought that such one-size-fits-all, grand theories of "successful aging" that ignore diversity, personality preferences, or structural inequalities had been put to rest a half-century ago. It seems to me that theoretical indifference has been one of the greatest limitations to the development of cumulative knowledge about aging.

The Generation Gap: Family Solidarity and the Generational Stake Theory

My first job after Chicago was with James Birren at his new Institute on Aging at the University of Southern California. I had also been recruited by John Clausen (with Glen Elder) at the University of California Berkeley, but USC got me by offering a magnificent $500 per year more than Berkeley. Those were frugal years. They were also the years of protest, with the Civil Rights Movement and the Student Rights Movement and the Anti-Vietnam War Movement and the Counter-Culture Movement each gaining more publicity month by month. It appeared that the "generation gap" was wide and deeply rooted in radically different political and economic values between youth and their elders. I saw it differently. In a review of the literature on generational differences I argued that generational contrasts were more apparent than real (Bengtson 1970), and that there were far greater differences *within* generations than *between* them. This was brought home to me 40 years later during the heated 2010 elections with the "Tea Party" protesters, a political movement led by mostly 60ish activists. It struck me that these people are of the same "generation" that produced the Anti-War and Students Rights protests of the 1960s and 1970s. They are the *other* members of this generation, the group in their age cohort that *did not* protest back then.

I began a study to test the extent of generational differences with a study of 2,000 grandparents, parents, and young adults that would examine values, attitudes, and relationships across generational lines. I had not the foggiest notion that this study would go on for another 35 years as the Longitudinal Study of Generations (Glass et al. 1986; Bengtson et al. 2002) and that in 2005 the Wave 8 data would be collected under Merril Silverstein (Silverstein and Bengtson 1997), the new principal investigator.

In our first look through the data we did note distinct generational differences (Bengtson 1975). These were not, however, in political or social or moral perceptions or even (much to our surprise) in attitudes toward sex. They were rather in *perceptions of intergenerational relationships*, what we called "family solidarity" (Bengtson et al. 1976; Bengtson and Schrader 1982). We found differences in the perception of family solidarity that showed a consistent *generational bias* – parents rated their relationship as more warm and cohesive, for example, than the children rated it, and parents also reported more frequent rates of interaction than did their offspring. This was true whether the children were 18 or 48 and whether the parents were 48 or 88. What would explain this?

I developed the "generational stake" theory with Joe Kuypers, my buddy from graduate school (he got the Berkeley job I turned down for $500), to account for the consistent bias that continually colored the two generations' perceptions of their common relationship. We argued (Bengtson and Kuypers 1971) that each generation had an investment, in terms of emotion, time, and money (depending on the life stage) in the other generation; these investments were inevitably unequal. The parents' stake or investment in their adolescent and young adult children was huge, involving enormous commitments of time, money, love, and support in them. This investment prompted many parents to justify it by perceiving a warm and cohesive response from their child, a high degree of value similarity and frequent interaction – in other words, high solidarity. The child's lesser investment in the relationship, and higher investment in individuation, caused him/her to minimize the cohesion, similarity, and contact. Their differing developmental stakes caused a systematically different perception of their common relationship. It was an explanation that many family theorists found useful in explaining family dynamics. But it was also a theory that questioned the validity of survey data that was based on the report – potentially biased by the generational stake – of one family member alone; thus it was not nearly as popular with researchers using large-scale survey data. To adequately compensate for the generational stake bias would require collecting data from at least one other family member, an expense most surveys are not willing to undertake.

The generational stake theory was an attempt to explain differences in the perception of one age group by another in the context of the family. What about the larger society? What explains differences in the perception of age groups, and what are the consequences as an individual moves into advanced age?

Linking Macro and Micro: The Social Breakdown and Reconstruction Theories of Aging

The years from 1965 to 1980 were an amazingly innovative period in the sociology of aging. Social interactionism inspired some innovative perspectives on aging, such as Rose's (1962) subculture of aging theory, which explained (and justified?) the initial development of Leisure World retirement communities across America. Modernization theory was proposed by Cowgill and Holmes (1972) to explain the devalued status of the elderly, also called "ageism." Riley et al. (1971) presented a theory of age stratification that linked societal change and individual change over time to processes of cohort flow and individual aging. A political economy approach that linked social structures to the experiences of the aged was outlined by Estes (1979).

Kuypers and I, with the audacity of youth, wanted to try to connect these perspectives together in a theory that linked macro- and micro-social levels. We came up with the Social Competence vs. Breakdown theory (Kuypers and Bengtson 1972), which started with symbolic interactionist concepts (what today might be called the interpretive perspective) and ended with something vaguely resembling the political economy of aging. We argued that an older individual's sense of self, his/her ability to mediate between self and society, and the society's willingness to acknowledge his/her competence are related to the kind of social labeling and valuing s/he experiences in aging.

The Social Breakdown model of aging begins (Stage 1) with the vulnerability in health or economic status that characterizes many older persons – for example, an older woman who is a new widow, experiencing failing health, and who has lost her job to a younger worker. While dealing with these personal crises at the micro-social level, she is also feeling the victimizations of ageism, a cultural or macro-social condition reflecting devaluation of elderly persons in an America that values youth over age and equates an individual's worth with his or her productivity or wealth. Then, for this woman, a vicious spiral downward is set in motion (Stage 2) with additional micro-social crises having to do with role losses (involuntary retirement, loss of social contacts through illness, loss of intimacy through widowhood), leading her to feel confused about what she should do, where to turn for support, and how to behave in responding to these losses in her life. Unless she finds new age-appropriate reference groups, she may lack social feedback concerning who she is, what roles and behaviors she can usefully perform, and what value she is to her social world. Resulting from this feedback vacuum is (Stage 3) dependence on external sources of self-labeling and identification, many of which are stereotypical images of the elderly as useless and obsolete, a characterization common in a society which places so much importance on productivity. As one woman put it, "I feel old, sick, dumb, and ugly." The individual who accepts such negative labeling is then inducted into the negative, dependent situation of the final stage of life (Stage 4), learning to act like old people are supposed to act. This reduces her social and psychological competence even more, with a corresponding atrophy of skills involving social and cognitive coping (Stage 5). This, in turn, makes her even more vulnerable to even greater debilitation, thus setting the stage for still another round of the vicious spiral of the Social Breakdown Syndrome.

How might the cycle of Social Breakdown be broken? The efforts of practitioners and policy-makers to meet the needs of an aging population represent attempts to make beneficial inputs to this end. We proposed a Social Reconstruction theory with inputs at each stage. To ameliorate problems in each stage efforts could be made to (1) liberate the individual from an age-inappropriate view of status, the functional ethic; (2) improving social services to elderly persons or lessening the debilitating environmental conditions faced by elderly persons. We also applied this model to family problems involving the elderly (Bengtson and Kuypers 1985) with what I think was a practical list of interventions for therapists and practitioners working with families and elders: (1) clarify the nature of the crisis event; (2) suggest roles of short-term involvement for each potential family caregiver; (3) discuss moralisms, unrealistic expectations, and guilt; (4) program some quick short-term successes; and (5) follow-up and develop external supports.

This attempt at linking micro- and macro-theory was very well received by gerontological social workers but virtually ignored by sociologists in the field of aging. Why? I learned some valuable lessons from this failure. The theory, while innovative and quite unique in the field of aging, tried to account for too much. It was bewilderingly eclectic, borrowing ideas from symbolic interactionism and phenomenological psychology and Marx and political economy theory. It attempted to both explain and prescribe, linking theory with its practical application (which, according to social work textbooks, it succeeded in doing). And the lesson that unfortunately impressed me the most was the one least relevant to substantive theory: We did not publish it in the right place. We should have published in a sociology journal or the *Journal of Gerontology* instead of in human development journals and as chapters in books. As I have emphasized to my students since then, too many sociologists will not take seriously anything that is not published in a sociological journal.

Race, Ethnicity, and Differences (or Similarities) in Aging

As an assistant professor I suddenly became, by default, principal investigator of a huge National Science Foundation study of black, Mexican-American, and Anglo Los Angeles middle-aged and elderly people. This project took several unexpected turns. One, happily, was an Oscar in 1977 in the Best Picture Documentary category for *Number Our Days*. This was based on ethnographic research by Barbara Meyerhoff (1978), a member of our NSF research team, of elderly Jews in Venice, California. These older individuals had created their own culture of aging far from memories of their Holocaust losses and far from their upwardly mobile children.

Also unexpected was the suspension of the project by NSF because of charges of racism. A group of minority community activists protested that there were no minorities represented on the research staff and there were no provisions for benefits to come back to the communities that furnished the data – the USC research project was a classic case of white elites exploiting minority groups. As a result the project was re-organized, the original principal investigator resigned (replaced by a woefully inexperienced new PI, me) and recruitment began for minority staff and graduate students. A Community Research Planning Committee (note that it was not an *advisory* committee) was established. Their mandate was to vote on actions at each step of the research process, from designing the survey instrument to submitting papers for journal publication. After months of heated meetings and mutual mistrust, the minority groups and the academic researchers finally got to work designing the community survey.

The initial protests and conflicts led eventually to a third unexpected outcome of the project. We *did* end up working together. Moreover, the research that resulted turned out to be *better* because of it. The Community Planning Committee found areas in the survey we were proposing that were poorly framed or not covered at all. My favorite example is the family relationship section and the questions we had drafted about children and grandchildren, brothers and sisters. A member of the Black Caucus asked, "But what about children you raise that aren't your children?" Reluctantly, and complaining inwardly about the cost of each additional question in a 1,200-respondent survey, we added some questions about "children you have raised that are not your own." When the survey results came in we found that *over 50%* of the black respondents reported they had raised children who were not biologically theirs but who they considered family (Burton and Bengtson 1982). Had we omitted this question we would not have learned about support networks outside the children usually asked about in surveys, nor about the investments, emotional and financial, that middle-aged and elderly minority respondents were making to children not their own.

This project was the most memorable experience of my research career, and I learned more from it than from any other. This awkward experiment in community-academic collaboration proved useful for other researchers (Bengtson et al. 1977) and today community involvement in social research is taken for granted. I should add a comment about the personal support I received from the Community Research Planning group. These were difficult times for me personally because my wife had been diagnosed with leukemia and my girls were 3 and 6 years old at the time. Several members of the black and Mexican American caucuses came to her funeral a year later. I will never forget what one member, Carmella Banks, said to me: "Vern, you and your daughters will *always* be members of *our* community."

A fourth set of unexpected consequences involved the research results of the project and their implications for theory. At that time the dominant orientation was what came to be known as the multiple jeopardy theory, in which the disadvantages of, say, an elderly black woman was explained by her cumulative minority status of being black, old, poor, and a woman. I wondered about this, thinking that by contrast there may be resources available to this woman that may buffer her disadvantages – particularly family support, in contrast to white elderly's lack of family support. I wanted, with the audacity of youth, to challenge the multiple jeopardy theory. So we began analyzing the data expecting to find marked differences between the black, Mexican-American, and Anglo individuals reflecting social aspects of aging. For example, we "knew" that minority groups placed higher emphasis on familistic values and that whites' higher mobility led their elders to feel lonely and often abandoned by their families. But the data did *not* support the hypotheses of either racial–ethnic advantage *or* disadvantage in family support for the aged. We did find that Mexican Americans had higher rates of contact with and affection toward family members than blacks and Anglos, but the differences were less than expected. We found that Anglos *expected* more assistance from their children than blacks or Mexican Americans, but there were insignificant differences in the *actual* provision or receipt of care. The most important finding was this: there were more differences *within* the three groups – black, Mexican American, and Anglo – in family relations and other aspects of aging than there were differences *between* the groups (Bengtson et al. 1990). We found little evidence in the data of this cross-ethnic comparative study to support the multiple jeopardy theory of aging; but we saw little evidence of data to disconfirm it either.

The research on aging in minority communities caused me to rethink several sociological generalizations I had accepted about social inequality. Could it be that we sociologists tended to maximize *between*-group differences – at the level of race, gender, or wealth – while minimizing, or perhaps even missing, differences *within* these social categories? Then there was another thing I became aware of, something even more relevant today than it was then. In contemporary sociology we are preoccupied with social inequality, with power, conflict, empowerment, and hegemony the social forces to be examined or critiqued. But there are other social forces as well: cohesion, cooperation, supportiveness, solidarity. My hope is that in the next decade we will get back to studying what binds people together as well as what drives them apart, and that sociologists will look at aging individuals for the resilience their lives represent as much as the social inequalities that make them vulnerable.

The Future of Theory in the Sociology of Aging

Since I can claim almost 50 years of experience doing research in the sociology of aging, I think I am allowed to offer some observations about the future of our field. First, in the sociology of aging today we appear to be drifting away from science and careful observation as practiced in related fields such as the psychology and biology of aging and toward the interpretive methods of history and the study of literature. Some among us have felt that this is a welcome trend and have vigorously advocated that we move away from the reductionism of positivism and move toward the freedom and questioning of critical analysis. But such polarization is unfortunate, I think; we need both. For one thing, I think we can profit greatly by following the model of cumulative knowledge building through observation, classification, prediction, and intervention or control, a model that has characterized science and has led to so many developments for the betterment of humankind's health and well-being. At the same time, however, I believe that we who do research in aging have responsibility to work for social justice, to advocate for dependent elders, and above all to add knowledge that can be useful for interventions. To accomplish this I believe that the process of developing theory – *explanations* – is the best way to develop more effective *interventions* to improve the lives of older people, interventions such as more effective health practices and more equitable social policies regarding aging.

Second, we need to develop more cross-disciplinary dialogs. Successful collaboration across fields to forge multidisciplinary models of aging and intervention is all too rare for those of us doing research in the sociology of aging. In part that may be because of an increasingly insular or antiscientific culture in sociology at large, which our colleagues in medicine, psychology, biology, and economics simply cannot understand. We need to recognize that aging is a multifaceted dynamic and requires multiple research perspectives to understand it. We need to believe in ourselves and that we have something useful to offer our research colleagues in medicine, biology, and psychology.

My third concern is that in the sociology of aging today we seem to be forgetting our history and our past attempts, whether successful and unsuccessful, to explain and to understand. Some of these old theories have been useful, others not; but each one of the attempts to build explanations, particularly those that have failed to be supported by subsequent data, can teach us a good deal about age and society or social processes of aging. Knowing their history can protect us against making their mistakes.

Finally, I believe we have an intellectually exciting future before us because of the creative cohorts of younger researchers that I see entering the sociology of aging. This is evident from the papers they present at the Gerontological Society of America and the American Sociological Association meetings. We have a responsibility to help them as they develop their professional competence and humanitarian concern. We need to encourage them to propel their research toward theory, beyond mere description and classification. We need to remind them that such explanations can be used to change things for the better, to produce more effective social policy about aging and more effective programs and health interventions for older people with problems. We need to remind them that they will do a better job of theorizing about age and aging than those of us who were in the first generation of the sociology of aging were able to do. This is because they will be able to learn from our mistakes and our successes. And because they will already know how to say "Gerontology" with a "J."

Acknowledgments I am grateful to Norella Putney, Victor Marshall, and Hannah Bengtson for their comments on earlier drafts of this chapter.

References

Achenbaum, Andrew and Vern L. Bengtson. 1994. "Re-Engaging the Disengagement Theory of Aging: On the History and Assessment of Theory Development in Gerontology." *The Gerontologist* 34:756–63.

Bengtson, Vern L. 1969. "Cultural and Occupational Differences in Level of Present Role Activity in Retirement." Pp. 25–53 in *Adjustment to Retirement: A Cross-National Study*, edited by R. J. Havighurst, E. P. M. Joep, B. Neugarten, and H. Thomae. Assen, Netherlands: Van Gorkum.

———. 1970. "The 'Generation Gap': A Review and Typology of Social-Psychological Perspectives." *Youth and Society* 2(2):7–32.

———. 1975. "Generation and Family Effects in Value Socialization." *American Sociological* Review 40:358–71.

Bengtson, Vern L., Timothy J. Biblarz, and Robert E. E. Roberts. 2002. *How Families Still Matter: A Longitudinal Study of Youth in Two Generations.* New York: Cambridge University Press.

Bengtson, Vern. L., Eugene Grigsby, Elaine M. Corry, and Manuel Hruby. 1977. "Relating Academic Research to Community Concerns: A Case Study in Collaborative Effort." *Journal of Social Issues* 33(4):75–92.

Bengtson, Vern. L. and Joseph A. Kuypers. 1971. "Generational Difference and the 'Developmental Stake.'" *Aging and Human Development* 2(1):249–60.

———. 1985. "The Family Support Cycle: Psychosocial Issues in the Aging Family." Pp. 61–77 in *Life Span and Change in a Gerontological Perspective*, edited by J.M.A. Munnichs, P. Mussen, and E. Olbriich. New York: Academic Press.

Bengtson, Vern L., Edward B. Olander, and Anees A. Haddad. 1976. "The 'Generation Gap' and Aging Family Members: Toward a Conceptual Model." Pp. 237–63 in *Time, Roles and Self in Old Age*, edited by J. E. Gubrium. New York: Human Sciences Press.

Bengtson, Vern L., Carolyn J. Rosenthal, and Linda M. Burton. 1990. "Families and Aging: Diversity and Heterogeneity." Pp. 263–87 in *Handbook of Aging and the Social Sciences (3rd edition)*, edited by R. Binstock and L. George. New York: Academic Press.

Bengtson, Vern L. and Sandra S. Schrader. 1982. "Parent-Child Relations." Pp. 115–85 in *Handbook of Research Instruments in Social Gerontology, Vol. 2*, edited by D. Mangen and W. Peterson. Minneapolis: University of Minnesota Press.

Burton, Linda M. and Vern L. Bengtson. 1982. "Research in Elderly Minority Communities: Problems and Potentials." Pp. 215–22 in *Minority Aging*, edited by R. C. Manuel. Westport, CN: Greenwood Press.

Cowgill, Donald A. and Lowell H. Holmes, 1972. *Aging and Modernization.* New York: Appleton-Century Crofts.

Cumming, Elaine M. and William T. Henry. 1961. *Growing Old: The Process of Disengagement.* New York: Basic Books.

Estes, Carol. 1979. *The Aging Enterprise.* San Francisco: Jossey-Bass.

Glass, Jennifer, Vern L. Bengtson, and Charlotte C. Dunham. 1986. "Attitude Similarity in Three-Generation Families: Socialization, Status Inheritance, or Reciprocal Influence." *American Sociological Review* 51:685–98.

Havighurst, Robert J. and Ruth Albrecht. 1953. *Older People.* New York: Longmans.

Kuypers, Joseph A. and Vern L. Bengtson. 1973. "Social Breakdown and Competence: A Model of Normal Aging." *Human Development* 16:181–201.

Lemon, Bruce L., Vern L. Bengtson, and James A. Peterson. 1972. "An Exploration of Activity Theory of Aging: Activity Types and Life Satisfaction among In-Movers to a Retirement Community." *Journal of Gerontology* 27:511–23.

Meyerhof, Barbara. 1978. *Number our Days.* New York: Simon and Schuster.

Riley, Matilda W., Marilyn Johnson, and Anne Foner. 1971. *Aging and Society, Vol. 3, Sociology of Age Stratification.* New York: Russell Sage Foundation.

Rose, Arnold M. 1962. "Subculture of the Aged: A Topic for Sociological Research." *The Gerontologist* 2:123–7.

———. 1964. "A Current Theoretical Issue in Social Gerontology." *The Gerontologist* 4:238–45.

Rowe, John W. and Robert L. Kahn. 1998. *Successful Aging.* New York: Pantheon.

Schmeeckle, Maria and Vern L. Bengtson. 1999. "Successful Aging. Conclusions from a Longitudinal Study: Cross National Perspectives. Review of J. W. Rowe and R. L. Kahn, *Successful Aging*." *Contemporary Gerontology* 5(3):87–90.

Silverstein, Merril and Vern L. Bengtson. 1997. "Intergenerational Solidarity and the Structure of Adult Child-Parent Relationships in American Families." *American Journal of Sociology* 103:429–60

Chapter 39 The Sociology of Aging and the Life Course Comes of Age

Stephen J. Cutler

As I thought about this chapter and the occasion it presented to reflect on my years as a sociologist and social gerontologist, two trends significant to my career repeatedly surfaced. The first is the growing legitimization of aging as a topic of investigation, and the second is the development of a research infrastructure supporting the study of aging. I make no claim that these two represent *the* most significant developments. I single them out only because of their personal importance and because of my belief that they represent critical indicators of how the field of aging has evolved over the past 4 decades. In what follows, I will place these trends in an autobiographical context.

My Michigan Years: The Study of Aging in the Mid to Late 1960s

I have often been asked how I became interested in aging and what led to a career-long interest in gerontology. (This question was especially prevalent when I was a younger sociologist; it has tapered off somewhat in my older years as the apparent connection between my personal biography and the field of study has become closer and as the field has gained legitimacy.) I expect many people thought my response would have something to do with warm, loving grandparents. I did have warm, loving grandparents, but the answer to the question is quite different (and quite possibly a disappointment to many people).

My interest in the sociology of aging dates precisely to a particular lecture in my first semester as a graduate student at the University of Michigan in the fall of 1964. I was enrolled in a course on population taught by the eminent demographer Ronald Freedman. At one point in the course, he was discussing the end of the Baby Boom – after all, it was 1964, the date generally acknowledged as the end of the baby boom era in the United States – and he was pointing out the demographic implications. He noted that the declining birth rate associated with the end of the Baby Boom would contribute to population aging. Fewer births meant a shrinking base of the population pyramid and a greater proportion of people at the older ages. Gains in life expectancy would also play a role in population aging, as would the eventual flow of Baby Boom cohorts into the older age strata, but declining fertility would be an early and critical component of the aging of our population. However, it was his next comment that caught my attention. Sociologists, he said, knew very little about aging. But between the lines he also seemed to be suggesting that sociologists cared little about aging. (In retrospect, I believe he was right on both scores.) And that was my "light bulb" moment! Something in his spoken and unspoken words clicked, and a huge intellectual gap and challenge presented itself.

S.J. Cutler (✉)
Department of Sociology, University of Vermont, Burlington, VT, USA
e-mail: Stephen.Cutler@uvm.edu

R.A. Settersten, Jr. and J.L. Angel (eds.), *Handbook of Sociology of Aging*, Handbooks of Sociology and Social Research, DOI 10.1007/978-1-4419-7374-0_39, © Springer Science+Business Media, LLC 2011

From that lecture given by Professor Freedman on an autumn afternoon 45 years ago, the seed that grew into my career-long interest in aging was planted.

The University of Michigan in the mid- to late-1960s, however, was far from the hotbed of gerontological activity that it is today. Any graduate student in sociology who wanted to do work on aging – although I am pretty sure I was the only one – had to carve relevant topics out of courses that had little if anything to do with aging. My term paper in a course on deviance, for instance, was on the topic "The Organizational Structure of the Mental Hospital and Geriatric Withdrawal." As best I could determine, and despite the presence of Wilma Donahue on campus, only one course on aging was offered in the entire university and that was in the School of Social Work. Taught by Professor Sheldon Siegel, the course brought together an eclectic group of graduate students from biology, psychology, social work, sociology, and other disciplines. There, among a group of students with a passionate but somewhat offbeat interest in aging, I got my first taste of the excitement of the multidisciplinary nature of gerontology.

As my time at Michigan progressed, I was faced with the task of developing a dissertation topic. My interest in aging had continued, and I designed what I thought would be a fascinating and significant research project, one I believed to be in the best tradition of Michigan sociology at the time (a la Amos Hawley, Otis Dudley Duncan, and others). What I proposed to study was how community functional specialization affected the prestige of older persons. Building on Burgess's notion of the "roleless" role of old people and notions about the declining status of older people with industrialization and modernization, my project viewed the status and prestige of elders not as givens but as variables, and it sought to examine how status and prestige varied as a function of the nature of community specialization. I wrote up the proposal, made an appointment with our department chair Albert J. Reiss to discuss it, and then witnessed first-hand the accuracy of Professor Freedman's assessment of the state of sociological interest in topics having to do with aging. After the requisite chit-chat, Professor Reiss said he had read the proposal, thought it was all well and good, but told me that what I really needed to do was something much more in the sociological mainstream. Apparently, aging was not. The handwriting was unmistakably on the wall. It was clear that my gerontological bent was judged to be somewhat misguided, and it was equally evident that I was not about to receive much, if any, support from the University of Michigan's Department of Sociology to carry out dissertation research on a topic having to do with aging.

Having had the error of my ways pointed out to me, I put my gerontological interests on the back burner, availed myself of some excellent data from the Detroit Area Study, and for my dissertation did a secondary analysis of the implications of voluntary association memberships for mass society. Yet, sometimes obstacles do pay off. Anyone who is at all familiar with my work will see there the beginning of two major threads to my research: first, a career-long interest in correlates and consequences of voluntary association memberships and, second, a heavy, though not exclusive, reliance on secondary analyses of survey data.

My Oberlin Years: 1969–1984

My first faculty appointment was as an Assistant Professor of Sociology-Anthropology at Oberlin College. I had made a conscious decision in my last year of graduate school that I wanted to begin my professional career at an undergraduate liberal arts college. This was a reflection of the nurturance and support I had received as an undergraduate at Dartmouth College in the early 1960s, especially from Professor Robert Sokol. The following is an excerpt from a letter I wrote in October, 1996 on the occasion of retirement festivities in Professor Sokol's honor:

> Long before the term was fashionable, you were a mentor in the very best sense of the word. You took me under your intellectual wing and gave me opportunities that few other undergraduates had. It sounds like ancient history now, but those many hours I spent running IBM cards through the counter-sorter using data sets you

> provided and calculating percentages and chi-squares on rotary calculators gave me a firm and enduring foundation for understanding the logic of sociological research. You found opportunities for me to do some interviewing and—I hope I'm not betraying anything here—you even let me try my hand at constructing multiple choice questions for an exam. As I became more and more interested in the possibility of graduate work in sociology, you gently encouraged me to test my interest by taking advantage of summer opportunities in related fields. And when I did decide that I wanted to go on to graduate school, you were there to help me think through the graduate programs that were best suited for me. I could go on at length, giving many more examples of your generosity, but the important point is how you consistently went out of your way to nurture my intellectual and professional development. In doing so, you served as a model for me and as an influence on my career in yet another way. My decision to go to Oberlin, where I taught for 15 years after getting my PhD, was a direct result of the undergraduate experience I had at Dartmouth and of which you were such a large part. I wanted to be in an environment that encouraged and valued the same type of student-faculty relationship, and I wanted to be able to work closely with undergraduates as you did. Being at Oberlin gave me the chance to work with exceptional undergraduate students, an opportunity in a real sense to carry on your legacy.

Several events during my early years at Oberlin brought my gerontological interests back to the forefront. First, in what Harvey Stearns at the University of Akron once referred to as the "Dracula complex" – or the incessant search for new blood! – the then Gerontological Society sponsored an Institute for Research and Teaching in Gerontology in August, 1970. Imagine having an opportunity to spend a week studying and talking gerontology with the likes of George Maddox, Bernice Neugarten, Ethel Shanas, Jack Botwinick, Carl Eisdorfer, and Juanita Kreps, among many others. If anything could reignite the smoldering intellectual embers, that week did.

Second, senior colleagues at Oberlin, especially George Simpson and Milton Yinger, were supportive – or, at least tolerant – of my interests in aging, even going so far as to permit me to offer a seminar on the sociology of aging every 2 or 3 years.

Third, a neighbor who was on the local Oberlin Health Commission and aware of my interests in aging informed me that the Commission was interested in doing a study of the needs and status of older persons in the community. They were looking for someone to conduct the survey and, as it turned out, I was looking for an opportunity for my research methods students to get some first-hand exposure to the survey research process. That fortunate combination of circumstances resulted in the 1970 Oberlin Survey of the Aged, a longitudinal follow-up supported by an NIMH grant 2½ years later, and several of my early presentations and articles on prestige loss, transportation, and voluntary association memberships.

Of course, time and resource constraints made it very difficult to mount major survey research projects from a setting such as Oberlin. *N*s of 170 and 106 did not go far when running first- and second-order tabular analyses (state-of-the-art at the time), and there were valid questions about the sample representativeness and generalizability of the Oberlin data. Nevertheless, over the many years I taught research methods at Oberlin, it was my belief that students profited from "doing" research. Social science data analysis packages were in their infancy, and runs still were submitted via IBM cards at input–output windows. But eager and energetic students seemed perfectly willing to try their hand at data analysis, so I soon began requiring my methods students to do an empirical research paper for their final project. (Still later I began to give my introductory sociology students the option of doing an empirical paper – somewhat streamlined, but still a chance for them to get their hands dirty.) It was because of this bent that I greeted so enthusiastically the arrival in 1972 of James Davis's brainchild, the National Opinion Research Center's (NORC) General Social Surveys (GSS). For the bargain-basement price of $100, students could access large-scale, nationally representative surveys of the adult population of the United States on topics of central sociological significance. In a review symposium on the NORC GSSs in 1978, I wrote:

> Most teachers have experienced the great pleasure of writing a letter of recommendation for that one student who comes along every five or ten years: head and shoulders above others, unsurpassed performance, and immense potential. So it is with the NORC General Social Surveys....Having used them for a variety of instructional purposes since they first became available in 1972, I have come to consider them as the single most important data resource for the teaching or undergraduate sociology (Cutler 1978:541).

But their value extended well beyond their instructional uses. For minimal cost, they gave those of us without access to on-campus survey research centers the possibility of working with an exceptionally rich source of current, nationally representative data. Moreover, an additional feature proved to be of inestimable value. From the outset, many of the questions included in the GSSs were replications of items asked in earlier surveys. Coupled then with the growing availability and accessibility of data archives (e.g., the Roper Center, NORC, the Inter-University Consortium for Political and Social Research [ICPSR]), the GSSs lent themselves to various sorts of diachronic analyses. For example, along with Robert Kaufman (an exceptional undergraduate student and now Chair of the Department of Sociology at Temple University), we were able to pair measures of tolerance of ideological nonconformity in the 1972 GSS with the very same measures in Samuel Stouffer's 1954 surveys of attitudes about civil liberties. The availability of those two sets of items asked over an 18-year period allowed us to conduct a cross-sequential analysis of cohort changes in attitudes about civil liberties (Cutler and Kaufman 1975), the first in a series of several studies showing both the occurrence of attitude change among older cohorts as well as shifts over time that often were in a liberal direction. Both findings countered prevalent stereotypes of aging as being a time of growing attitudinal rigidity or a time of adopting increasingly conservative social and political attitudes. Other collaborative studies with outstanding undergraduates took advantage of this replicated item design feature (e.g., Cutler et al. 1980, in which we examined cohort changes in attitudes about the legalization of abortion using NORC data from 1965 and six GSSs between 1972 and 1977). Still other projects began to take advantage of the cumulative aspect of the GSSs themselves. Because of the initial annual and subsequent biannual administration of the surveys, the GSSs alone increasingly provided the basis for trend analyses. The most recent example of this is a comprehensive examination of change in sociopolitical attitudes using 33 items asked in anywhere from 14 to 21 of the surveys conducted between 1972 and 2004 (Danigelis et al. 2007).

Although Oberlin College emphasized undergraduate instruction, it recognized that an active program of scholarly engagement would enhance the instructional experience. Not only did the college have a generous sabbatical program but it also had a limited number of "research status awards" which, when combined with a one semester sabbatical, gave the recipient a full year off at full pay. I was fortunate enough to receive one of the research status awards and spent the 1975–1976 academic year at the Center for the Study of Aging and Human Development at Duke University. There I had the opportunity to spend a year in the company of George Maddox, George Myers, Erdman Palmore, Gerda Fillenbaum, Linda George, Dick Campbell, and others, to avail myself of the rich set of postdoctoral and other program activities of the Center, and to have access to data from the Duke Longitudinal Studies to use in conjunction with data from the Oberlin surveys (e.g., Cutler 1977). This extraordinary intellectual and networking opportunity more than compensated for the gerontological isolation I experienced at Oberlin.

Other developments during this period speak to the growing legitimacy of aging as a field of inquiry and have proven to have had a lasting and significant impact. In 1974, the National Institute on Aging was established as part of the National Institutes of Health. Under the visionary leadership of the founding director, Dr. Robert Butler, NIA from the outset proved to be receptive to and encouraging of sociological research on aging. In fact, the first and long-time director of the Behavioral and Social Research Program was Matilda White Riley, herself a sociologist and a vocal and effective proponent of the value of social and behavioral research.

And within sociology, of course, the Section on Aging was established in 1979. The culmination of efforts spearheaded by Harold Orbach and others gave an ASA organizational imprimatur to the growing interest among sociologists in the wide array of fascinating theoretical, methodological, and substantive issues surrounding late life and the life course.

The late 1970s and early 1980s also saw efforts to facilitate access to the aging research infrastructure. A noteworthy example was the establishment of the National Archive of Computerized Data on Aging (NACDA) as a unit of the University of Michigan's ICPSR. With funding from NIA,

NACDA has become the principal repository of easily accessible, thoroughly documented, and analytically ready aging-related data sets. Recognizing the value of ultimately making such data available to the wider research community, beginning in 2003 NIH has required investigators to detail plans for archiving and distributing data gathered with NIH funding.

My Vermont Years: 1984–2009

My move from Oberlin to the University of Vermont (UVM) in 1984 reflects still another indicator of the growing visibility and legitimacy of aging as a significant focus of social and behavioral education and research. Beginning in the 1970s with Title IV funding from the Administration on Aging (AoA), colleges and universities started developing centers, institutes, and programs on aging, as well as courses, minors, and certificate programs; master's degrees and even PhD programs in gerontology would eventually appear. UVM had an active but informal interest group (the Multidisciplinary Committee on Aging) and a smattering of courses on aging around the university, but no officially organized programs. In an effort to catalyze and coordinate aging-related activities on the UVM campus, an endowed chair was created, the Bishop Robert F. Joyce Distinguished University Professorship of Gerontology. According to Bell (1986), this was 1 of 29 endowed positions in gerontology and geriatrics in the mid-1980s, 80% of which had been created since 1980. In an effort to align my academic and curricular responsibilities with my major scholarly focus (aging), I applied for the position and became the first incumbent of the Joyce Professorship, holding the position until my retirement in 2009. Curricular and programmatic activities were proposed, approved, and initiated, and UVM joined the large and growing list of institutions of higher education that recognized the significance of aging as an area of study.

The move to Vermont not only provided me with a group of local colleagues who shared my interest in aging from a variety of disciplinary perspectives but it also provided me with the encouragement, support, and indeed the expectation to immerse myself in the burgeoning professional activities in aging and gerontology. If my entry into aging and gerontology in 1964 was on the "ground floor," the Vermont years beginning 2 decades later allowed me to be part of the explosive growth and maturation of the field from a variety of professional vantage points: as Chair of the Behavioral and Social Sciences Section of GSA (1987), as a member and chair of NIH's Human Development and Aging Study Section (1988–1992), as Editor of the *Journal of Gerontology: Social Sciences* (1989–1993), as Chair of ASA's Section on the Sociology of Aging (1994), and as President of GSA (1998).

A Last Look Back

If I think back to that gerontological awakening some 45 years ago in Professor Freedman's population course and reflect on the development of the field and on my involvement in it, two unmistakable achievements stand out. First, the sociological study of aging has come a long distance since Professor Reiss suggested that my dissertation topic needed to be more mainstream. The sociological study of aging and the life course has, indeed, become mainstream. Witness the occasion for this set of essays: the 30th anniversary of the establishment of the Section on Aging (and the Life Course) and the regular appearance of articles on aging and life course topics in the major sociological journals. Second, a research infrastructure that both supports and reflects the growth of the field of aging is now well established. Witness the public investment of financial resources in large-scale, multi-wave data collection efforts (e.g., Michigan's Health and Retirement Survey, Chicago's

National Social Life, Aging and Health Project, Duke's National Long Term Care Survey, etc.), and the theoretical and methodological advances that guide us in formulating the questions, interpreting the findings, and analyzing the data gathered by such studies. From these and the many other perspectives explored in this volume, the sociology of aging and the life course has come of age.

References

Bell, William G. 1986 "Endowed Positions in Gerontology and Geriatrics in Higher Education: Results from a Preliminary Survey." *Educational Gerontology* 12:507–18.

Cutler, Stephen J. 1978 "Instructional Uses of the General Social Surveys." *Contemporary Sociology* 7:541–5.

———. 1977 "Aging and Voluntary Association Participation." *Journal of Gerontology* 32:470–9.

Cutler, Stephen J. and Robert L. Kaufman. 1975. "Cohort Changes in Political Attitudes: Tolerance of Ideological Nonconformity." *Public Opinion Quarterly* 34:69–81.

Cutler, Stephen J., Sally Ann Lentz, Michael J. Muha, and Robert N. Riter. 1980 "Aging and Conservatism: Cohort Changes in Attitudes about Legalized Abortions." *Journal of Gerontology* 35:115–23.

Danigelis, Nicholas L., Melissa Hardy, and Stephen J. Cutler. 2007 "Population Aging, Intracohort Aging, and Sociopolitical Attitudes." *American Sociological Review* 72:812–30.

Chapter 40
Long Time Coming, Not Here Yet: The Possibilities of the Social in Age and Life Course Studies

Dale Dannefer

The advance of scientific knowledge and understanding proceeds through a continuous dialectic of multiple intellectual tensions – between theory and research, and sometimes practice, and between multiple and often competing conceptual and methodological paradigms within and across disciplines. At its best, scientific work entails an ongoing discourse of discovery and interpretation guided by logic and evidence, and yet it is also a discourse that is shaped unavoidably by the biographical experience and social location of individual researchers. "Making science" is a human process, located in a specific sociohistorical space, within the broader everyday life processes through which society is continuously reconstituted. Thus, it is epistemically incumbent on the researcher to consider carefully the impact of her own life history and social location upon her work. The editors of this volume are to be especially thanked for inviting essays that encourage such reflexive work – combining a view of the field of aging and life course studies with autobiographical particularity, which is what I shall attempt in this essay.

Although I had the great benefit during the course of my graduate training of working and studying with Matilda White Riley (including taking her very first *Sociology of Age* course at Rutgers in the mid-1970s), I had no particular interest in the study of age while in graduate school. My doctoral research was informed by interests in the sociology of knowledge, culture, meaning, and identity in late modernity. My first real substantive connection with the study of age and the life course came as a result of empirical studies of those issues via intensive qualitative interviews with targeted groups of respondents as they revealed that meaning-making is a lifelong process that can only be apprehended as part of the actor's broader biographical experience or life narrative.

With that realization, I returned to the sociology of age and discovered the life course, as well as the fields of lifespan psychology and adult development, as I searched for frameworks and concepts that could help organize and offer insights for dealing with the rich and complex qualitative interview data I was amassing. These literatures – serving up terms and topics such as *life history, seasons, cohorts, stages and transitions, fallacies, age structures and allocation,* and *change vs. constancy* – represented several distinct paradigms and offered a provocative range of multiple and competing approaches to understanding human development and aging over the life course, often leavened by lively debate and a sense of discovery.

Nevertheless, what I found most striking in this array of approaches to aging and development over the life course was not their *differences* but two underlying *similarities* that characterized most of them, and that many had in common: (1) a predilection to envision the basic contours of aging primarily as an organismically grounded individual process and a corresponding lack of recognition of the magnitude of the importance of social forces to account for age-related patterns and outcomes

D. Dannefer (✉)
Selah Chamberlain Professor of Sociology Chair, Department of Sociology,
Case Western Reserve University, 10900 Euclid Avenue, Cleveland, OH 44106, USA
e-mail: dale.dannefer@cwru.edu

R.A. Settersten, Jr. and J.L. Angel (eds.), *Handbook of Sociology of Aging*, Handbooks of Sociology and Social Research, DOI 10.1007/978-1-4419-7374-0_40, © Springer Science+Business Media, LLC 2011

and (2) a lack of attention to the role of institutionalized power relations in society in shaping both the life course and its study, inviting an implicit assumption of a generally benign relationship between social structures and dynamics and individual needs and aging.

Thus, rather than providing the framework I was seeking for my own research, what I found in this set of literatures was a cluster of theoretical assumptions and substantive assertions that themselves needed to be scrutinized and critically analyzed. As a consequence, I began to focus my scholarly work on the implications of social theory and sociological knowledge for aging and the life course (e.g., Dannefer 1984, 1987), and I have yet to finish the empirical studies that sent me to these literatures in the first place.

I begin with this account not for the sake of reviewing the past but because the field of aging and life course studies continues to be constrained by these same problems. The tendency toward life course reductionism, which leaves the explanatory potential of social forces unrecognized in studies of age, has consistently been – and remains – an enduring tension in the field of aging and life course studies. Moreover, even when the reach of social influence is more fully recognized (as in constructivist work), the systemic nature of social structures and processes in shaping of the life course is generally neglected. Taken together, these two problems represent what I have called a *functional-developmental symbiosis* that sustains, albeit often in disguised or subdued form, a "normal aging" view of the individual life course. The general problem is, of course, most readily visible in work that explicitly presents aging as a highly general or universal process (e.g., Gutmann 1987; Levinson 1994). However, on careful inspection it becomes clear that even many approaches that attempt to emphasize the role of social context continue to rely on a view of age does not venture far from a "normal aging" view of the individual, relying implicitly if not explicitly on an individualized and organismic view of change across the life course.

As I have suggested elsewhere (Dannefer 2011), the overall problem can be clarified by contrasting two fundamentally different intellectual postures – or *heuristic positions* – that provide alternative paradigmatic orientations to the subject matter and that are the basis for reacting to and interpreting findings. I call these the heuristic of *containment* and the heuristic of *openness*. Both of these heuristic positions recognize the challenges posed to the normal aging paradigm by cohort analysis, constructivist approaches, and related breakthroughs of the late twentieth century that demonstrated diversity in patterns of aging and the relevance of context in understanding such diversity. Where they differ is in their response to these developments. The *heuristic of containment* describes a defensive move on behalf of the "normal aging" paradigm – an intellectual predisposition to recognize and give credit to, but then promptly limit, the encroachments of external forces on the normal aging paradigm. The *heuristic of openness*, by contrast, describes a logic that imposes no preconceived limits on external influences and embraces the attendant complexities and challenges for measurement, analysis, and theorizing. Thus, openness implies a view that countenances the unrealized and even unrecognized possibilities for the future of the life course and of the social institutions that organize it. However, it also means that the task of the science of aging is much more daunting. If the tasks of science are to understand, bring order to, and make predictions about empirical phenomena, the heuristic of openness acknowledges that the phenomenon age and the life course are less predictable than the dominant heuristic of containment assumes them to be. And when a phenomenon is as rich, multivariate, and contingent as is human aging, it will not advance understanding to force it to conform to a paradigmatic framework selected *a priori* that, no matter how appealing, does not fit the nature of the phenomena under investigation. As Herbert Blumer (1969) frequently emphasized, the first task of science is to be true to its subject matter.

In this brief essay, I can only discuss a couple of examples of this persistent tension in aging and life course studies. I will describe one of its significant early manifestations in social science research, (cohort analysis), followed by a discussion of a contemporary instance (nursing home research and reform), which is one of several domains in which an opportunity currently exists to move beyond the heuristic of containment. This discussion will also require consideration of the special place of the life course as a social institution both in the past, and in contemplating future possibilities of this paradigmatic tension.

Cohort analysis may seem an unlikely target for this complaint. As is well known, cohort analysis was central to the successful challenge posed in the late 1960s and 1970s to the widespread assumption that cross-sectional age patterns could be equated with actual life course trajectories of aging. It demonstrated that aging is contingent on historical circumstance, and that how individuals mature and age will depend on the kind of world they experience. It posed a major challenge to a universalistic view of aging and was rightly seen as a threat by those who argued for a view of age as a transcultural, transhistorical process (Gutmann 1987; see also Dannefer and Perlmutter 1990).

By posing this challenge, cohort analysis opened a "Pandora's box" of logical possibilities, inviting consideration of other forms of external and social influence upon aging beyond those indexed by social change. However, as I have noted elsewhere, the implications of this opening were hardly embraced. Instead, a primary mode of response to cohort analysis encouraged a tendency to equate and conflate the scope of social influence with social change, assuming that social forces are mainly visible or of interest during periods of rapid change (Hagestad and Dannefer 2001). The paradigmatic form of cohort analysis probably remains the intercohort comparison of mean trajectories, so that each cohort itself is implicitly treated as having its own normally distributed pattern of aging. This is more than curious since it takes only a moment's thought to realize that if social influence creates change in patterns of aging, it may also be at least partly responsible for stability in patterns of aging and for the variation in age-related characteristics among the members of a cohort, as well as between cohorts.

That is why in my work I began to focus on *intracohort variability* and on evidence that variability and inequality tend to increase with age, which seemed to me to demonstrate clearly the operation of life course processes of social stratification and exclusion, and cumulative dis/advantage (CDA). Nevertheless, reviewers of my first submitted paper on this topic (Dannefer 1987) criticized it for lack of attention to individual-level processes, suggesting that the psychosocial accentuation of personality or other individual characteristics could be the primary factor accounting for divergence with age (see also Clausen 1993:521). Despite such efforts at *containment* of the complications social forces, I regard the CDA perspective as one domain of scholarship in which the expansion of the sociological imagination is quite manifest (e.g., Crystal and Shea 2002; Dannefer 2003; Ferraro and Shippee 2009).

In contrast to the dominant "containment" approach, the *heuristic of openness* recognizes that the discovery of cohort differences represents only the beginning of the problem of specifying external effects on individual aging, not the sum total of it. Cohort differences thus appear as only one among many potential kinds of evidence showing that individuals "age differently" (Maddox 1987), due in substantial part to the nature of their daily life experience, and life experience is inevitably socially organized and regulated.

The same tension applies in the treatment of numerous other topics in social gerontology and related literatures (e.g., lifespan theory, activity theory, family solidarity theory, cumulative dis/advantage, gene-environment interaction), as I have discussed elsewhere. In the space remaining, I focus on a significant contemporary site of the tension between these two heuristics that has real-life relevance – the topic of long-term care research, reform, and culture change. To approach this problem it is important first, however, to introduce explicitly the second major analytical fulcrum of life course studies, which is a focus on the life course as a social institution.

A concept of the age and the life course as elements of social structure is clearly present in the initial formulations of Cain (1964) and Riley and associates (1972), but it was most systematically elaborated in the work of Martin Kohli (1986) and others in the 1980s, which revealed the significance of the life course as a social institution. The crucial importance of this perspective is that it makes explicit the power of age, not as an inherent feature of individual organisms but as having its own independent reality as a constructed, yet potent, feature of social order. Age is thus a property of social structure. Increasingly, age is relied upon in late modern societies as a central basis of social organization and reality construction.

Given the recency of this historical development (e.g., Chudacoff 1989; Kett 1977), its taken-for-grantedness is remarkable. It reflects a *naturalization* of age, as there is a widespread popular, professional, and academic assumption that the "three boxes of life" (school, work, and retirement) and the observed empirical trajectories of development and aging represent the "normal" human life course to which the age-graded institutions of school, work, and retirement serve as *accommodations* (Dannefer 2008), meeting the age-graded needs of the normal life course. The acceptance of these institutional forms as a response to human needs and of positive benefit to human interests reflects the heuristic of containment. So unrealistic and destructive is the logic of containment when applied to institutions (and not just individuals) that it prompted Matilda Riley (among whose best friends were counted the leading functionalists Parsons and Merton) to declare the three boxes of life a reflection of "society's failure to provide meaningful roles" in the subtitle of her book *Age and Structural Lag* (Riley et al. 1994).

Institutions of long-term care that are associated with advancing age – assisted living and especially nursing homes – can be thought of as a "fourth box." This fourth box is officially defined as a further accommodation to the normal life course to some imagined final life stage of social and existential, as well as physical withdrawal and decrepitude, a legitimate and professionally and scientifically sanctioned form of the systematic social exclusion of elders. The resilience of this widely accepted exclusionary practice is a reflection of the depth and resilience of ageism in late modernity. It is noteworthy that it ordman nursing home life appears as a shock to frontline caregivers in the United States, who are themselves from traditional societies (e.g., West Africa or the Caribbean), in whose eyes the circumstances faced by elders appear cruel and barbaric abandonment. Their culturally grounded perspectives on aging involve hope and expectation of more for elders, in terms of quality of life and social engagement, than does the ageist culture of late modernity.

Why, then, have the devastating and well-founded critiques of the prototypic institutional forms of long-term care – as warehouses that require the "elderly" to adjust to the logic of a total institution (Boyd 1994; Vladeck 1980), as regimes that impose "helplessness, boredom, and loneliness" (Thomas 1996) and "depersonalization" and "infantilization" (Kayser-Jones 1990) upon their residents – come mainly from practitioners and professionals working in the field and not from social and behavioral scientists?

Over the past 15 years, a national movement to promote a fundamental *culture change* of rehumanization of long-term and other elder care (Fagan 2003) and to resist ageism more generally (Barkan 1995; Thomas 2004) has grown rapidly, with goals squarely in line with the intellectual concerns of established developmental theory (e.g., Deci and Ryan 1985) and progressive social gerontology. It is a movement that sees its mission precisely as combatting an especially insidious manifestation of what Matilda Riley called structural lag.

Long-term care institutions that have sought to apply the vision of this movement to change their structures and practices to be more responsive to elder residents have had some, if limited, success (Dannefer and Stein 2000; Kane et al. 2007; Weiner and Ronch 2003). In virtually every case, the critical analysis of both the contradictions of institutional practices and their adverse effects on human development and human potential came from the field and from critical practitioners while the published discourse of social and behavioral scientists on nursing homes largely contented itself to consider topics such as incontinence, caregiver burden, decontextualized recreational "therapies," regulatory change, and a host of other topics that largely left the medical-model structure of the traditional nursing home unquestioned and therefore legitimated.

Why has the cutting edge of analysis of institutional structures and practices – and hence the call for reform – come, with few exceptions, not from social gerontology but from imaginative frontline leaders who have derived their own critiques of ageist assumptions about the diminishing lives of elders from the manifest human contradictions and destructiveness with which the empirical reality of everyday life confronted them? These are legitimate questions for social and behavioral scientists to ponder. Such unquestioned acceptance of the status quo is, of course, a reflection of the power of the heuristic of containment.

In the opening paragraph of this essay, I emphasized the importance of scientific reflexivity. When considering the current movement of long-term care reform, one obvious, overarching question is, "*Where were we?*" Where were social and behavioral scientists in analyzing the damage done by institutions and considering the possibilities for change, instead of largely accepting their basic form? I suggest that, overall, the problem can be diagnosed as a lack of development of the heuristic of openness and a reliance on the logic of containment.

In the reflexive moment, these considerations raise yet other questions. Culture change and other reform initiatives have highlighted and sought to change some adverse aspects of the treatment of elders, but they represent only a few halting steps in making institutions more responsive and in nurturing late-life development. What arenas of needed change are not yet visible? What additional frontiers remain to be explored that hold possibilities for upgrading expectations and opportunities for improvements in engagement, quality of life, and further personal development for frail and impaired elders (see Baker 2007; Siders et al. 2006)?

This brief essay cannot, of course, even enumerate the array of other questions about the tensions of long-term care reform that confront us as scholars – questions of balancing stability and change and viable aspects of existing structures with possibly risky reforms; questions of co-optation of rhetoric and trappings of reform to make unwarranted claims of transformational success; and of sustaining celebrated and publicized change initiatives after the researchers and media crews have left.

To identify and pursue the promising questions, it is clear that a rigorous and imaginative advance in considering the maximal possibilities for elders is needed. As reformer Barry Barkan (1995) puts it, "If someone has only 3% of their functioning left, we try to make it 5%." In my view, the potential contributions of the social and behavioral sciences to understanding the realities and untapped possibilities of aging will be meaningfully advanced to the extent that we adopt a similar posture that reflects an underlying heuristic of openness.

Acknowledgments I would like to thank Elaine Dannefer, Jessica Kelley-Moore, and Setari Parsa for comments on an earlier version of this paper, and Setari Parsa for research assistance.

References

Baker, Beth. 2007. *Old Age for a New Age: The Promise of Transformative Nursing Homes.* Nashville: Vanderbilt University Press.

Barkan, Barry. 1995. "The Regenerative Community: The Live Oak Living Center and the Quest for Autonomy, Self-Esteem, and Connection in Elder Care." In *Enhancing Autonomy in Long-Term Care,* edited by L. G. Semradek and E.M. Tornquist. New York: Springer Publishing.

Blumer, Herbert. 1969. *Symbolic Interactionism: Perspective and Method.* Englewood Cliffs: Prentice-Hall.

Boyd, Charlene. 1994. "Residents First." *Health Progress* (September).

Cain, Leonard D. 1964. "Life Course and Social Structure." Pp. 272–309 in *Handbook of Modern Sociology,* edited by R.E.L. Faris. Chicago: Rand McNally.

Chudacoff, Howard P. 1989. *How Old Are You? Age Consciousness in American Culture.* Princeton, NJ: Princeton University Press.

Clausen, John A. 1993. *American Lives: Looking Back at the Children of the Great Depression.* New York: Free Press.

Crystal, Stephen and Dennis Shea. 2002. *Annual Review of Gerontology and Geriatrics.* Vol. 22, *Economic Outcomes in Later Life.* New York: Springer Publishing.

Dannefer, Dale. 1984. "Adult Development and Social Theory: A Paradigmatic Reappraisal." *American Sociological Review* 49:100–16.

———. 1987. "Aging as Intracohort Differentiation: Accentuation, the Matthew Effect, and the Life Course." *Sociological Forum* 2:211–36.

———. 2003. "Cumulative Advantage/Disadvantage and the Life Course: Cross-Fertilizing Age and Social Science Theory." *Journal of Gerontology: Social Sciences* 58:S327–37.

———. 2008. "The waters we swim: Everyday social processes, macrostructural realities and human aging." Pp. 3-22 in *Social structures and aging individuals: Continuing challenges*, edited by K. W. Schaie and R. P. Abeles. New York: Springer.
———. 2011. "Aging, the Life Course, and the Sociological Imagination: Prospects for Theory." Pp. 1-26 in *The Handbook of Aging and the Social Sciences,* 7th ed., edited by R. Binstock and L.K. George. New York: Academic Press.
Dannefer, Dale and Marion Perlmutter. 1990. "Development as a Multidimensional Process: Individual and Social Constituents." *Human Development* 33:108-37.
Dannefer, Dale and Paul Stein. 2000. *From the Top to the Bottom, from the Bottom to the Top: Systemically Changing the Culture of Nursing Homes*. Final Report. New York: van Ameringen Foundation.
Deci, Edward and Richard Ryan. 1985. *Intrinsic Motivation and Self-Determination in Human Behavior.* New York: Plenum.
Fagan, Rose Marie. 2003. "The Pioneer Network: Changing the Culture of Aging in America." Pp. 125–39 in *Culture Change in Long-Term Care*, edited by A. S. Weiner and J. L. Ronch. New York: Routledge.
Ferraro, Kenneth F. and Tetyana P Shippee. 2009. "Aging and Cumulative Inequality: How Does Inequality Get Under the Skin?" *The Gerontologist* 49(3):333-43.
Gutmann, David. 1987. *Reclaimed Powers: Toward a New Psychology of Men and Women in Later Life*. New York: Basic Books.
Hagestad, Gunhild and Dale Dannefer. 2001. "Concepts and Theories of Aging." In *Handbook of Aging and the Social Sciences,* 5th ed., edited by R. Binstock and L. George. New York: Academic.
Kane, Rosalie A., Terry Y. Lum, Lois J. Cutler, et al. 2007. "Resident Outcomes in Small-House Nursing Homes: A Longitudinal Evaluation of the Initial Green House Program." *Journal of the American Geriatrics Society* 55:(June)832–39.
Kayser-Jones, Jeanie. 1990. *Old, Alone and Neglected: Care of the Aged in Scotland and the United States.* Berkeley: University of California Press.
Kett, Joseph F. 1977. *Rites of Passage: Adolescence in America, 1790 to the Present*. New York: Basic Books, Inc.
Kohli, Martin. 1986. "Social Organization and Subjective Construction of the Life Course." In *Human Development and the Life Course: Multidisciplinary Perspectives*, edited by A.B. Sorensen, F.E. Weinert and L.R. Sherrod. Hillsdale, NJ: Erlbaum.
Levinson, Daniel J. 1994. *The Seasons of a Woman's Life*. New York: Random House.
Maddox, George. 1987. "Aging Differently." *The Gerontologist 27*:557–64.
Riley, Matilda W., Marilyn E. Johnson, and Anne Foner. 1972. *Aging and Society: Vol. 3. A Sociology of Age Stratification*. New York: Russell Sage Foundation.
Riley, Matilda W., Robert L Kahn, and Anne Foner. 1994. *Age and Structural Lag: Society's Failure to Provide Meaningful Opportunities in Work, Family, and Leisure*. New York: Wiley.
Siders, Rebecca, Robin S. Patterson, and Dale Dannefer. 2006. "Learning from Those Who Know: Action Research as a Mode of Culture Change in LTC Settings and as an Educational Tool for LTC Professionals, Academics, and Student Researchers." Presented at the annual meeting of the Gerontological Society of America, Dallas, TX (Nov).
Thomas, William. 1996. *Life Worth Living: How Someone You Love Can Still Enjoy Life in a Nursing Home*. Acton, MA: Van der Wyk & Burnham.
———. 2004. *What Are Old People For? How Elders Will Save the World.* New York: Van der Wyke & Burnham.
Weiner, Audrey S. and Judah L. Ronch, eds. 2003. *Culture Change in Long-Term Care*. New York: Routledge.
Vladeck, Bruce. 1980. *Unloving Care: The Nursing Home Tragedy*. New York: Basic Books.

Chapter 41
Looking Back: My Half Century as a Sociologist of Aging and Society

Anne Foner

The 1960s do not seem to have been a propitious time to embark on a major project exploring social aspects of aging in the United States. In those days, when the baby boom cohorts were about to come of age, public attention was on young people rejecting traditional ways and in apparent revolt against those over 30 – not on aging, which was generally thought of primarily as growing old. Yet it was at that time that Matilda White Riley received a major grant from the Ford Foundation and the Russell Sage Foundation to pull together, evaluate, and analyze social science knowledge about the middle and later years. It was then, in the initial stages of the project, that I was invited to join in what turned out to be a lifelong career studying age and aging and a decades long association, collaboration, and friendship with Matilda Riley, one of the leading scholars in the field.

The sociology of age and aging as a field of study was relatively undeveloped in the 1960s, although there were important foundations. At the theoretical level, as early as 1928, Karl Mannheim ([1928] 1952) had written about the problem of generations; in the 1940s Ralph Linton (1942), Talcott Parsons (1942), and Pitirim Sorokin ([1947] 1969) had analyzed age as an element in the social structure, and in the 1960s Leonard Cain (1964) focused on the life course and social structure, and Norman Ryder (1965) called attention to successive cohorts and social change. But there was no broad conceptual scheme tying these important strands together. On a substantive level there was a substantial body of interdisciplinary research and a treasure trove of data, much of it collected by government bodies such as the US Bureau of the Census. But often these were focused on particular research objectives. At the institutional level, the field was not a specialty in many sociology departments. And the Section on Aging and the Life Course (earlier, Section on the Sociology of Aging) had yet to be established in the American Sociological Association.

In setting about tackling some of these gaps and to carry out the mission of codifying and analyzing existing knowledge about age and aging as social phenomena, Matilda Riley recruited a group of scholars, many of whose main interest had been in other fields. Matilda Riley herself had made her mark as a methodologist and had done major substantive research on social communication and adolescent–parent relationships. At the time, I was relatively new to sociology and my interests were in theory, class, and social conflict. Whatever our particular previous interests, the small group of us who did the initial spade work for what ended up as the three-volume *Aging and Society* came to the project with fresh eyes, not tied to any specific approach to understanding this key topic. On a personal level, I was returning to complete graduate study and looking ahead to entering the workforce after a hiatus devoted to family obligations – part of an early wave of women "returnees" in their 30s and 40s. Thus, in our own lives, as we broke with previous traditions about the appropriate roles for women, we exemplified an important principle: that life course patterns are subject to change.

A. Foner (✉)
Department of Sociology, Rutgers University, New Brunswick, NJ, USA
e-mail: FONERAM@aol.com

R.A. Settersten, Jr. and J.L. Angel (eds.), *Handbook of Sociology of Aging*, Handbooks of Sociology and Social Research, DOI 10.1007/978-1-4419-7374-0_41, © Springer Science+Business Media, LLC 2011

The first volume of *Aging and Society: An Inventory of Research Findings* (Riley and Foner 1968) involved more than collecting and organizing the existing literature on aging. It also entailed evaluating the research. In the process, the volume set forth a preliminary conceptual scheme for understanding age as an element in social structure and social change as both a process that proceeds from birth to death and a process that itself changes as each new cohort proceeds through a unique segment of history.

This scheme was further elaborated in *Aging and Society vol. 3, A Sociology of Age Stratification* (Riley et al. 1972), whose subtitle captured key aspects of the approach. Indeed, the term "age stratification" became synonymous with the conceptual scheme being proposed. Although heretofore the focus of much scholarly work in the field had been on the individual, the age stratification perspective also called attention to the age structure, which consists of society-wide age strata – for example, the youth, middle-aged, and older strata – existing apart from the particular individual incumbents, with rules and roles governing opportunities, attitudes, and behavior of strata "members." (Later the term "age stratification" was dropped and the approach relabeled the "Aging and Society Paradigm" in order to capture the broad range of issues not clearly suggested by the earlier title.)

Given my interest in social structure and social stratification, my own early efforts at developing the broad age stratification conceptual scheme were directed at further exploring age as a basis of social stratification. How good an analogy to social stratification in general and class structure in particular was the age stratification approach? Was there a theoretical payoff to pursuing this idea?

First, there was the issue of social inequality, a fundamental feature of social stratification. I noted that age inequalities are pervasive and they are built into the social structure. Access to social roles that are unequally rewarded with wealth, power, and prestige is governed by age criteria that persist beyond the incumbency of given individuals – in short, constituting a system of structured social inequality based on age.

Of course, age-based structures of inequality differ from class systems and other forms of social inequality in important ways, including the dynamisms of individual aging over the life course and the succession of cohorts through different periods of history. I sought to pursue the implications of both the similarities with and the differences between class- and age-based systems of inequality by addressing questions fundamental to analysis of class but often neglected in the study of age: How do advantaged or disadvantaged positions influence individual behavior and attitudes? To what extent does common social location lead to within-group solidarity and awareness? Or to conflict or interdependence with other strata?

In looking into these issues I noted that there were some parallels with class patterns and some intriguing differences. Much like class patterns in modern American society, studies conducted in the 1960s and 1970s showed that there were, for example, age differences in political and social attitudes, life style, organizational attachments, and subjective reactions to their economic status, family situation, and life in general. But as compared to class patterns, fewer of these age differences could be clearly attributed to social location in the age-based stratification system. For example, despite many deprivations, older people were more likely than those younger than they to report satisfaction with their jobs, their marriages, and life in general. To understand such apparently anomalous reactions, various hypotheses pointed to the role of dynamic age-related processes. The aging process itself, as people experience and deal with successive changes over their lifetimes, can lead older people to accommodate to new changes they undergo. Moreover, the cohorts under study may not have had high expectations for their lives in the later years because these expectations were influenced by the social environment in which they spent their formative years, an environment with lower life expectancy, poorer health, and greater economic risks for the older population than existed in the 1960s and 1970s.

Given the undoubted influence of such dynamic processes on the behavior and attitudes of older people in this period, I suggested that future cohorts of older people might not react in similar ways.

It is an open question whether subsequent cohorts of older people socialized in their youth to have relatively high expectations for their future lives will be likely to accept setbacks in old age with the equanimity that appeared to characterize the elderly born in the early twentieth century – surely a question for scholars to revisit in future times of economic troubles.

Going beyond age patterns in individual reactions to their age status a fundamental question for sociologists is the impact of social location on group life. Age as a basis of within group cohesion and intergroup relations was an issue that was particularly highlighted in the United States of the 1960s and 1970s by the collective actions of the post-World War II baby-boom cohorts. Taking off from the specific age-based struggles in those years I sought to look at the general issue of age conflicts and age solidarity (e.g., Foner 1974).

Several factors, I argued, encouraged a sense of common fate among age peers. At any one period they are likely to have many similar roles and share the satisfactions and pains of current life tasks. As they grow up and grow older they face the problems of making major role transitions at the same stages of their lives. Because they were born at the same time, they move together through and are influenced by the events of a particular slice of history. In addition, especially in highly complex and developed societies like the United States, various structural factors reinforce a sense of *we-ness*: age grading in schools, mass media programming organized to appeal to different age groups, a host of age-graded voluntary associations, age-segregated residences for many young people in their college years, and increasing numbers of retirement communities for the old.

Whether or not people of similar age do develop a degree of cohesiveness and sense of common interests has implications for relations with other strata. With the student rebellion in the 1960s and 1970s receiving much public attention at the time I was beginning to write about these issues, I wanted to explore conditions that might lead to societal-wide age conflicts.

As I noted in several essays (e.g., Foner 1974), various factors appear to trigger cleavages among the age strata. Power, status, and material inequalities endemic within age systems can motivate the disadvantaged to oppose existing arrangements. In addition, members of successive cohorts whose collective histories differ widely are likely to have divergent values and world views, life styles, and cultural tastes. When they feel they are without the institutionalized power to impose their views, the powerless – often the young – can feel impelled to challenge other age groups in order to assert their "rights."

In fact, tensions between the young, the middle-aged, and the old *are* fairly common, but these tend to play out within particular institutional spheres such as the family and the work place, and the issues at stake differ from institution to institution. Societal-wide conflicts such as the youth rebellion of the 1970s have been less frequent; when they have occurred, they have typically been initiated by the young, over "ideal" issues such as war and peace, justice for all, and social morals. I argued that such ideal issues are less susceptible to conflict-reducing mechanisms than material issues dividing age strata. In the first place, economic issues often activate class interests that create divisions within age strata and supersede age-based interests. In addition, material issues are more subject to compromise than so many struggles over ideal "all or nothing" issues. Futhermore because of the inevitability of aging, young people's discontent with their relatively poor economic status can be allayed as they look forward to future gains.

As I write in the first decade of the twenty-first century, these general principles may soon be tested. The United States is in the midst of a deep economic recession, and class divisions appear to have come to the fore as public policies are debated about how to deal with the economic crisis. At the same time, however, there is talk of coming age-based struggles among the young, middle-aged, and old over shoring up or diluting public entitlement programs such as Social Security for those 65 and over. Such predictions are not new; they were made in the 1980s, another period when the solvency of old age programs in the United States was of public concern. At that time, age-based clashes did not erupt; indeed, there were coalitions of organizations with constituents crossing age lines to save, not weaken, *Social Security* as a social program. As for the coming years, whether or not age conflicts over age-based entitlement programs do break out across the society, the underlying issues – understanding the nature of group life and intergroup relationships among age strata – will continue to be of fundamental importance to sociologists.

My discussion of age stratum cohesiveness or interstrata conflict suggests that the age stratification analogy has had relevance for understanding key aspects of age structures. Nevertheless, exploring the applicability of the stratification paradigm points to other important themes, not clearly derived from the stratification analogy, that await in-depth analyses. In my own work I considered some of these other themes such as life course transitions, age-related aspects of the family and political life, and changing age structures.

Life course transitions, stages when individuals give up familiar roles and take on new ones, punctuate the aging process in all societies. According to long-held views, such transitions have been seen as particularly likely to be troubled in modern societies such as the United States, in contrast to smoother transitions in less complex and less developed societies. In several articles on life course transitions (e.g., Foner and Kertzer 1978), David Kertzer (an anthropologist) and I showed that these ideas did not hold up in a particular case, a comparison of life course transitions in the United States with those in 21 African age-set societies, a group of preliterate, small-scale societies where age is a major organizing principle. Despite the vastly different societal contexts, in both settings, transitions were associated with tensions for the individual. At the same time, in both contexts there were mechanisms – although particular to each environment – that served to ease some of these difficulties. We also found that as both individuals and societies adapted to transitional problems and to social and environmental transformations, the transition processes themselves were subject to change.

Turning from these general conclusions to a specific and major life course transition – retirement – Karen Schwab and I examined the retirement process in the contemporary United States, dealing with issues on the individual level as well as societal-wide problems associated with retirement (Foner and Schwab 1981). Our analysis was consistent with the general principles noted above: that the retirement process involves an interplay between the individual and the society and that this interplay is likely to put a new face on the process and provide new challenges in the future.

The theme of social change also runs through much of my other work. For example, my articles dealing with intergenerational relations in the family noted the continuing vitality of intergenerational connections. They also discussed the ways these relationships have changed historically, as when independent living among older people became the norm. In writing about age-related aspects of political life, I took note of the influence of succeeding cohorts on political events. As each new cohort experiences and deals with a unique economic, social, and political environment, its members often react with a new "take" on political affairs and help to influence new political alignments in the society.

More generally, I suggested that age structures are subject to change both from forces outside the age systems – such as the creation of a national old age pension program that helped provide the means for older people to establish new patterns of residential independence – and forces intrinsic to the age system – such as new norms reflecting values and world views introduced by succeeding cohorts. In addition, I proposed that there was frequently a tension between the structure of the age system and the continual flow of new cohorts into that system. For if the existing norms governing an age system are not appropriate for oncoming cohorts of different size or composition, there is often a push for change in these norms.

In short, while we may not be able to predict the exact nature of change in age systems, transformations are inevitable. This leads to a few final words about future work in the field. For students of age and aging as social phenomena there will be opportunities to reexamine "old" issues under changed circumstances as well as discover new developments for a research agenda. While our studies must deal with the particulars at hand, our full understanding of age and aging will be enhanced as scholars try to extract and articulate the general principles underlying these analyses. And in so doing they will be contributing also to sociological theory in general.

References

Cain, Leonard D. Jr. 1964. "Life Course and Social Structure." Pp. 272–309 in *Handbook of Modern Sociology*, edited by R.L. Faris. Chicago: Rand McNally.

Foner, Anne. 1974. "Age Stratification and Age Conflict in Political Life." *American Sociological Review* 39:187–96.

Foner, Anne and David Kertzer. 1978. "Transitions over the Life Course: Lessons from Age-Set Societies." *American Journal of Sociology* 83:1081–1104.

Foner, Anne and Karen Schwab. 1981. *Aging and Retirement*. Belmont, CA: Wadsworth.

Linton, Ralph. 1942. "Age and Sex Categories," *American Sociological Review* 7:589–603.

Mannheim, Karl. [1928] 1952. "The Problem of Generations." Pp. 276–322 in *Essays in the Sociology of Knowledge*, edited and translated by P. Kecskemeti. London: Routledge and Kegan Paul.

Parsons, Talcott. 1942. "Age and Sex in the Social Structure of the United States." *American Sociological Review* 7:604–16.

Ryder, Norman B. 1965. "The Cohort as a Concept in the Study of Social Change." *American Sociological Review* 30:843–86.

Riley, Matilda White and Anne Foner. 1968. *Aging and Society vol. 1 An Inventory of Research Findings*. New York: Russell Sage.

Riley, Matilda White, Marilyn Johnson, and Anne Foner. 1972. *Aging and Society vol. 3 A Sociology of Age Stratification*. New York: Russell Sage.

Sorokin, Pitirim. [1947] 1969. *Society, Culture, and Personality*. New York: Cooper Square.

Chapter 42
As Time Goes By: Gerontological and Life Course Musings

Linda K. George

When I was a senior in high school, everyone in my class was asked to submit his or her aspiration, to be printed under our pictures in the year book. While my classmates were submitting such aspirations as "work for a tire company," "be a beautician," and "drive a truck," I submitted the ambition "to be happy." Mrs. Ludwig, the English teacher supervising the year book, said that my aspiration was "unsuitable" and told me to come up with an acceptable alternative. But I was not to be dissuaded and this can be verified by looking at the 1965 Highland High School year book where, under my smiling picture, are the words "to be happy." I had no inkling then that I would spend about equal portions of my life course being happy and studying happiness.

During both undergraduate and graduate school, one topic of lively intellectual debate revolved around questions of whether older adults are typically satisfied with their lives and, if so, what life circumstances underlie happiness in later life. Disengagement theory (Cumming and Henry 1961) was well-established and posited that the key to a comfortable old age was to give up the roles that had structured life earlier in adulthood and to live in a smaller, less demanding world. By the time that I was in graduate school, disengagement theory was under attack on multiple fronts. George Maddox, who was one of my primary graduate school mentors, used longitudinal data from the Duke University Normal Aging Study to demonstrate that the strongest predictors of life satisfaction in later life were such high levels of social activities as spending informal time with others and participating in organizations (Maddox and Douglas 1974). Research by my undergraduate mentor, Robert Atchley, generally supported the conclusion that high levels of social engagement were the foundations of life satisfaction in later life. Atchley added the proviso, however, that the activities that promoted well-being in later life were those that reinforced pre-existing identities and emotional commitments – a perspective that he labeled continuity theory (Atchley 1989). By the mid-1970s, ground-breaking empirical data based on a representative sample of American adults documented that not only were the vast majority of older adults satisfied with their lives, they were more satisfied with their lives than young and middle-aged adults (Campbell et al. 1976).

These debates had repercussions for the field of aging far beyond specific questions about the extent to which later life is satisfying versus dissatisfying or happy versus unhappy. Before these debates, a "crisis orientation" characterized the field – a sense that older adults were a semi-invisible minority group who were systematically, if perhaps inadvertently, barred from the rewards due to them in an affluent society. And on purely objective grounds, a case could be made for that perspective. During the 1960s and first half of the 1970s, older adults were more likely to live in poverty, lack access to medical care, have disabling illnesses, and report being socially isolated than young and middle-aged adults. Research on life satisfaction among older adults did not challenge these

L.K. George (✉)
Department of Sociology, Center for the Study of Aging, Duke University, Durham, NC, USA
e-mail: lkg@geri.duke.edu

R.A. Settersten, Jr. and J.L. Angel (eds.), *Handbook of Sociology of Aging*, Handbooks of Sociology and Social Research, DOI 10.1007/978-1-4419-7374-0_42, © Springer Science+Business Media, LLC 2011

objective facts, but it did challenge – indeed, debunk – the notion that these deprivations severely compromised perceptions of life quality.

The objective life conditions of old age began to improve substantially by the end of the 1970s. Legislation was passed that guaranteed cost-of-living adjustments to Social Security benefits. The first cohorts in which large proportions of members had private pensions – and most were defined-benefit pensions – transitioned into later life. As a result of these changes, within a decade the elderly moved from being the most likely age group to live in poverty to being the least likely to do so. Though initiated in the 1960s, it was not until a decade later that Medicare's contributions to promoting the health and well-being of older citizens became visible. Improved medical technology also disproportionately benefited older adults. It was in the mid to late 1970s that CT scans first allowed physicians to distinguish between strokes caused by blockages and those caused by bleeding in the brain – a critical distinction because one required thinning the blood and the other required clotting it. At this time, a variety of interventions, including bypass surgery, stents, and pacemakers also became routine weapons against cardiovascular disease, the largest cause of death for older adults. In addition to improvements in income security and medical care, new cohorts entered late life better off in education, health, wealth, and social resources than their predecessors.

As a result of all these changes, the crisis orientation that had been the foundation of aging theory and research in the social and behavioral sciences became difficult to justify. Indeed, by the 1980s, public discourse included references to older adults as "greedy geezers," and inaccurate but nonetheless fervent predictions of large-scale intergenerational battles in which economically strapped young and middle-aged adults revolted against older adults who spent their time in luxurious leisure were rampant. Although the "greedy geezer" image was obviously inaccurate, it was clear that traditional stereotypes of older adults as poor, needy, and disabled were too. But I believed then – and still do – that the aging field lost much of its vigor, integration, and sense of purpose once it was clear that most older adults are not "underdogs."

Beginning in the late 1970s, subjective well-being (SWB) became the "umbrella term" used to denote positive orientations toward life (i.e., life satisfaction, happiness, and morale). A prominent theme that characterized research on SWB from the late 1970s through the end of the century involved questions about the extent to which SWB rests on objective life conditions. As would be expected, strong and significant relationships between objective life conditions – including education, income, being married, friendships, and health – and SWB were observed. The surprising finding, however, was that objective life conditions explained less than half the variation in SWB. That is, mismatches between objective indicators of life quality and subjective assessments of life quality were exceedingly common. Many individuals who were relatively deprived of the objectively good things in life reported high levels of SWB and, conversely, many people who seemed on objective grounds to be advantaged reported low SWB. My contribution to this body of research was to demonstrate that the relationships between objective life conditions and SWB were smaller for older adults than they were for young and middle-aged adults (George et al. 1985). In short, older adults were less dependent on health, wealth, and social relationships for happiness and life satisfaction than were their younger counterparts.

A second, and related, theme of research on SWB at this time was the search for explanations that could account for the frequent mismatches between individuals' levels of objective resources and their SWB. The primary theory that emerged to account for these mismatches was *aspiration theory*. The basic tenet of aspiration theory is that people report that their lives are satisfying and of high quality when the discrepancy between their aspirations and their achievements is small. Conversely, when the discrepancy between individuals' aspirations and achievements is large, they view their lives as less rewarding and satisfying. Thus, we would expect financially advantaged, healthy, socially connected persons to be unhappy or dissatisfied when they aspire to even higher levels of the "good things" in life. And relatively disadvantaged persons should report high levels of SWB if their aspirations are met.

Aspiration theory received strong support in multiple studies. I was able to show that older adults (1) had, on average, smaller discrepancies between their aspirations and achievements than young and middle-aged adults, (2) had generally lower aspirations than young and middle-aged adults, (3) but were not advantaged, relative to their younger counterparts, on objective resources. Indeed, older adults reported the lowest levels of education and, especially, health. With regard to marital status (i.e., being married) and income, older adults scored approximately equal to young adults and significantly lower than middle-aged adults (George 1992).

Although it received less attention, *equity theory* also appears to partially explain mismatches between objective life conditions and SWB, especially for older adults. In brief, equity theory posits that individuals will be satisfied with the rewards that they receive if they perceive that the distribution of rewards is fair. I was able to show that the perceived fairness by which economic rewards (income and wealth) are distributed was a significant predictor of SWB for older adults but was non-significant for young and middle-aged adults (George 1992).

Social comparisons also help explain the high levels of SWB among older adults, many of whom are not socially or economically advantaged. The basic tenet of *social comparison theory* is that we evaluate our own qualities and status by comparing ourselves to relevant others. In general, *upward comparisons*, in which we compare ourselves to those who are more advantaged than we are, generate relatively low levels of satisfaction. In contrast, *downward comparisons*, in which we compare ourselves to those who are less advantaged, promote a sense of satisfaction. A number of researchers reported that most older adults use downward comparisons, thus promoting SWB.

I interpreted the support of these theories as demonstrating the power of social–psychological processes to produce either the sense that life is satisfying or feelings of dissatisfaction – and these processes are at least as strong as objective life conditions. These findings also suggest that the capacity of older adults to establish aspirations that are closely linked to their achievements, to perceive the distribution of societal resources as fair, and to compare themselves to those who are less advantaged are tremendously adaptive, permitting them to sustain a sense of SWB despite loss of resources and age-related limits on their ability to increase their achievements.

The publication of *Children of the Great Depression* by Glen Elder in 1974 introduced the life course perspective and opened a new era of theory and research on aging. As described there, as well as in subsequent writings by Elder and others, the life course perspective rests on four primary propositions. First, and probably most important, the life course perspective rests on the assumption that lives unfold over time – that to understand the present, one must understand the past. Consequently, a requisite for life course research is longitudinal data covering long periods of time. Second, the life course perspective focuses on the intersection of history and personal biography – that is, the ways that historical conditions shape life course trajectories, providing unique opportunities for and constraints upon human actors. Third, the life course perspective pays attention to human agency – to the consequences of individual decisions. Social determinism certainly is not ignored by life course researchers, but it is balanced by a focus on human agency. Fourth, and finally, the life course perspective focuses on linked lives – on ways an individual's life course is intertwined with those of others.

I believe that the life course perspective energized aging research in a way that we had not seen since the demise of the crisis orientation. It provided a paradigm that is broadly applicable, regardless of the specific outcomes of interest. It is as applicable, for example, to health outcomes as to economic outcomes. The life course perspective also focuses on heterogeneity and recognizes the myriad of pathways that can lead to the same (or different) outcomes. Failure to acknowledge and investigate heterogeneity was perhaps the fatal flaw of the crisis orientation. By promoting images of older adults as poor, sick, and socially isolated, the crisis orientation failed to recognize the heterogeneity of late life – heterogeneity that recent theories, such as cumulative advantage theory, and research results demonstrate to be greater in late life than at earlier ages. When, as a result of social change, the crisis orientation described fewer and fewer older adults, it was certain to be reputed.

There are still, of course, older adults who are poor, sick, and/or socially isolated. The life course perspective, however, highlights the fact that this is just one of the many destinations observed in later life.

Empirically applying the life course perspective is methodologically demanding. Longitudinal data are required that span large proportions of the life course. Although the number and quality of data sources suitable to life course analysis have grown exponentially in the last decade or so, a number of topics have not been examined from a life course perspective because of lack of sufficient data. SWB is one such topic. A few studies have predicted SWB in late life using information collected at earlier points in time. None of these, however, have addressed fundamental issues about SWB across the life course, including the dynamics of SWB over time. I hypothesize, for example, that one common trajectory of SWB – probably the most common one – is increasing SWB as individuals move from young adulthood to middle-age to later life. Cross-sectional studies suggest this kind of aging effect, but they inevitably confound age changes with cohort differences. I also hypothesize that aspirations in multiple life domains – health, income, and amount of social contact – decline from middle-aged to late life. Again, this hypothesis is supported by cross-sectional research findings, but it is unclear whether this pattern is observed over time as individuals age.

Beyond the methodological issues involved in distinguishing between age changes and cohort differences, there are logical reasons for suspecting that cohort differences account for at least part of the age differences observed in cross-sectional data. Because succeeding cohorts have consistently entered old age (as it is conventionally defined) in better health, and with more education, wealth, and social resources than their predecessors, it is possible that these "resource-rich" cohorts will be less willing – or have less need – to lower their aspirations. Rather than not declining at all, however, it is likely that these cohorts will reduce their aspirations at later ages than earlier cohorts.

It is also possible that the Baby Boomers will behave differently from earlier cohorts. Although research evidence is scant and inconsistent, public discourse and the popular media claim that Baby Boomers will be more active and have higher expectations for their later years than earlier cohorts. If this proves to be true, it is possible that we also will observe a decline in SWB among Baby Boomers as they reach their late 70s and beyond, as age-related losses inevitably occur.

I believe that life course research requires multiple observations of the *same* individuals over long periods of time – in other words, true panel data. Although we lack suitable panel data for life course studies of SWB, some authors have used repeated cross-sectional data to approximate the "natural histories" of cohorts. Such analyses are very useful for understanding the "average" behavior of a cohort. They cannot, however, reveal the heterogeneous ways that lives unfold over time. Nonetheless a recent study of SWB in the US adult population, based on repeated cross sections over more than 30 years, provides a rich description of the average effects of age and cohort on SWB.

Yang (2008) used data from the General Social Survey to perform an age-period-cohort (APC) analysis of happiness among representative samples of US adults from 1972 to 2004. Yang also used a new methodology for APC analysis. Using this analytic technique, age effects are estimated with cohort effects statistically controlled and vice-versa. Yang observed age effects over the 32 years of data that mirrored those reported in cross-sectional studies: increasing happiness over the life course. There were significant cohort effects as well, with the Baby Boomers reporting lower levels of happiness at all ages than the cohorts that precede and follow them.

SWB also is now accepted by social indicators researchers as a valid indicator of the extent to which societies meet the needs of their members (Veenhoven 2009). Researchers in that field recognize that there are subjective as well as objective bases for individuals' reports of SWB and that these reports reflect comparisons to both absolute and relative standards of what constitutes the "good life." Although I continue to await the long-term panel study that will permit examination of the multiple trajectories of SWB over the life course, there is ample evidence that older adults in the USA are highly satisfied with the quality of their lives.

And what about my youthful ambition to be happy? Well, overall I am. I've had some pretty tough obstacles to get past, but so do we all. I also seem to be taking the path that research suggests. As I approach the conventional definition of late life, I am more satisfied than I can remember being at any other time of my life. With regard to aspirations, I'm not sure whether they have declined or simply become largely irrelevant. I just don't spend any time these days comparing my professional accomplishments or personal life to other people's or to some abstract notion of what constitutes "making it." I still have goals, of course, but they tend to be short-term goals about things such as finishing a paper, shampooing my carpet, or weeding my garden. I'm almost ready to believe that these really are "golden years."

References

Atchley, Robert C. 1989. "A Continuity Theory of Normal Aging." *The Gerontologist* 29:183–90.

Campbell, Angus, Philip Converse, and Willard Rodgers. 1976. *The Quality of American Life.* New York: Russell Sage Foundation.

Cumming, Elaine and William Henry. 1961. *Disengagement: The Process of Growing Old.* New York: Basic Books.

Elder, Glen H., Jr. 1974. *Children of the Great Depression.* Chicago: University of Chicago Press.

George, Linda K., Morris A. Okun, and Richard Landerman. 1985. "Age as a Moderator of the Determinants of Life Satisfaction." *Research on Aging* 7:209–33.

George, Linda K. 1992. "Economic Status and Subjective Well-Being." Pp. 233–64 in *Age, Money, and Life Satisfaction*, edited by N. E. Cutler, D. W. Gregg, and M. P. Lawton. New York: Springer.

Maddox, George L. and Elizabeth B. Douglass. 1974. "Aging and Individual Differences: A Longitudinal Analysis of Social, Psychological, and Physiological Indicators." *Journal of Gerontology* 29:555–63.

Veenhoven, Ruut. 2009. "Well-Being in Nations and Well-Being of Nations: Is There a Conflict between Individual and Society?" *Social Indicators Research* 91:5–21.

Yang, Yang. 2008. "Social Inequalities in Happiness in the United States, 1972 to 2004: An Age-Period-Cohort Analysis." *American Sociological Review* 73:204–26.

Chapter 43
Studying Age Across Borders

Gunhild O. Hagestad

Life Travels

When I was approached about writing this piece, there was talk of a "senior in the field." My first reaction was, "A senior? *Me*?" There are two bases for my puzzlement. The first is an experience many of us share: problems with accepting and internalizing such labels as "old, elderly, senior." A few years ago, I was in a long security line at an airport. People were tense, worried about missing their flights. A security officer was trying to speed up things and loudly announced, "The shoes coming through on the belt belong to the elderly woman." I looked around, but saw no older woman! I would like to believe that today I would respond differently. But reality is that it takes a while to live with such labels. We are not ready to become *geronts*, presumably the objects of gerontology. I return to that topic below.

A quite different side to my reaction was that I do not see myself as a senior in the field. I do not feel like a giant on whose shoulders others can stand, to paraphrase Robert Merton a bit. One key reason for this feeling is that I have been fortunate enough to know true giants, trailblazers in the study of age and aging. Some of them are no longer among us, but their work is very much alive. The list shows how fortunate I have been: Paul Baltes, Leonard Cain, Henning Friis, Bill Henry, Peter Laslett, Powell Lawton, Ursula Lehr, George Maddox, Bernice Neugarten, Michel Philibert, Matilda Riley, Leopold Rosenmayr, Irv Rosow, Warner Schaie, Ethel Shanas, Gordon Streib, and Hans Thomae. As the reader will notice, the list includes primarily psychologists, some sociologists, a philosopher (Philibert) and a historian/journalist (Laslett).

Among the people I mentioned, I worked most closely with Bernice Neugarten. She hired me in 1972 when she was head of the University of Chicago's Committee on Human Development. After studying sociology at the University of Oslo and the University of Minnesota, this was my first regular academic job.

As a student in Norway, I was interested in social change, but I also discovered George Herbert Mead! In the Twin Cities, I had several professors with a strong "symbolic interaction" perspective, among them Arnold Rose. As part of my doctoral work at the University of Minnesota, I did a minor in philosophy. Courses on pragmatism helped further my understanding of Mead and his circle. Thus, I got extensive discussion of *meaning*, its social production and reproduction. Since my minor was in the philosophy of science, I also became acquainted with two powerful scholars, May Brodbeck and Herbert Feigl. Both had spent a good deal of time pondering the issue of *levels* and units of analysis in science.

G.O. Hagestad (✉)
NOVA/Norwegian Social Research, Oslo, Norway
e-mail: gohagestad@gmail.com

R.A. Settersten, Jr. and J.L. Angel (eds.), *Handbook of Sociology of Aging*, Handbooks of Sociology and Social Research, DOI 10.1007/978-1-4419-7374-0_43, © Springer Science+Business Media, LLC 2011

My graduate courses did not teach me much about age and aging. After arriving at the University of Chicago, I had a pretty steep learning curve and was very pleased when a chapter on Age and the Life Course, coauthored with Bernice, appeared in the 1976 edition of *Handbook of Aging and the Social Sciences.* There was a lot to cover because so much had happened from the mid-1960s to the mid-1970s. A major, comparative study of aging in three industrial societies appeared in 1968. The international group of authors, led by Ethel Shanas, stated quite clearly that social *integration versus segregation* was the key issue facing the sociology of aging. Between 1968 and 1972, Matilda Riley and her colleagues published landmark volumes on *Aging and Society,* summarizing existing knowledge and presenting a model of age stratification. Age was here linked to social differentiation and the division of labor. The 1970s also marked the start of Glen Elder's groundbreaking work on the life course. By 1980, it was clearly time for ASA to have a section on aging. But it took until the late 1990s to add the life course to the name!

One scholar who, in my view, has not been given enough attention within sociological discussions of age is Leonard Cain. His 1964 chapter on life course and social structure in the *Handbook of Modern Sociology* was a pioneering overview of issues which have continued to challenge students of life course and age stratification. Already in 1959 he wrote a state-of-the-art article on the sociology of aging, covering work from several countries. Here, he made the somewhat prophetic statement that "ameliorative gerontology" might overpower theoretical sociology!

What was Bernice's disciplinary anchor? Throughout her professional life, she belonged to APA as well as ASA. Her research topics spanned from personality to age norms to policies for aging societies. For her doctoral work at the University of Chicago, her adviser was Lloyd Warner, famous for his work on inequality, but also a student of age classes and age sets. My impression was, however, that although she had worked closely with a scholar focused on social systems, Bernice anchored much of her thinking in individuals and their personal experiences. Towards the end of her career, she would often comment on presentations by asking, "Where is *the person* here?" There were times when I wanted to pose a corresponding question: "Where is *the society* here?"

In the villa that houses Human Development, there were many intense scholarly discussions. Nearly always, someone would suggest that we try to move up a notch in the level of abstraction: "What is the bigger question here?" A thoughtful, widely traveled colleague, anthropologist Robert Le Vine, always placed the meanings of age in a broad comparative context. As Ralph Linton did in the 1940s, Bob also reminded us that men and women have different life scripts. He also stressed the degree of individualism/collectivism in different societies. So did another anthropologist who came to visit from the University of Illinois: David Plath. He used Japan as his main basis of comparison and in the process gave us new insights on lives in our own society. His beautifully written book *Long Engagements* deserves to be a classic. Another person who came to visit, offering important *diachronic* comparisons, was historian Tamara Hareven, who later spent considerable time in Japan. She highlighted the need to consider time dimensions on three levels: individual, family, and history. Glen Elder spent time in Human Development when he was trying to finish "the book," juggling lots of information about individual development, family, and historical context. *Children of the Great Depression* appeared in 1974. To me, the greatest contribution of that classic volume will always be its powerful illustrations of how levels matter and must be taken into account. Within the context of a national economic crisis, the financial stability or instability of *families* made the difference in determining future lives of children and youth. The lesson is that the meso-level of family functions as a "lens" between history and individual lives. Consequently, cohort contrasts entail intracohort differentiation. In some cases, the lens intensifies effects of historical circumstances; in other cases, it deflates and minimizes historical impact.

Ever since I first heard Glen talk about his book, I have wondered about *personal meaning* and historical change. If people are "carriers" of marked cohort effects which show up when we follow them across time, are they themselves aware of it? Is it part of their life story? I kept going back to Karl Mannheim's essay on generations, in which he argued that true generational units have *generational awareness* – a sense of their "location" that they have in common. But then there was a conversation, many years ago, with life course researcher Karl Ulrich Mayer, who has provided striking examples of how history left footprints in the lives of German cohorts. He also has authored articles on how states structure life course patterns, showing policy shifts change lives. When I raised the awareness issue with him, Uli responded, "Gunhild, that is for psychologists to ponder; *we* are sociologists!"

Differences between psychological and sociological eyes; micro- versus macro views also marked ongoing discussions in my next place of work, the Department of Individual and Family Studies at Pennsylvania State University. I joined the interdisciplinary faculty in the late 1970s. Again, I was privileged to be around scholars who helped define the field of aging research: Paul Baltes, John Nesselroade and K. Warner Schaie, a significant triumvirate in laying conceptual and methodological foundations for the study of adulthood from a life span perspective. Paul was head of the department. He later became the director of the Max Planck Institute for Human Development in Berlin. We shared the experience of being "bi-continental," which has both positive and negative aspects. While belonging in two places represents a strength, it also raises issues of marginality. I believe we both experienced being seen as quite European when in the United States and quite American when in Europe. For me personally, it has always been difficult to understand and accept that many of my American colleagues do not seem to be familiar with relevant and important European research, even though it is readily available in English.

As is well-known, Warner Schaie was among the first to seriously address the age-period-cohort conundrum in the study of development. It might be the lenses of a sociologist again, but I detected a shift in his thinking over the years. As a psychologist, he initially seemed to think of cohort effects as bothersome "noise" messing up our understanding of "pure development," especially in the cognitive domain. Over the years, he admitted that the noise might actually be the most interesting thing to focus on!

After nearly a decade at Penn state, I moved back to the Chicago area, taking a position in a program founded by Bernice: Human Development and Social Policy at Northwestern University. At that time, I became truly bicontinental, dividing my year between Northwestern and the Institute for Social Research in Oslo. For a sociologist, a focus on connections between structural conditions and lives offered exciting possibilities, but most of my colleagues were psychologists. My main conversation partner was Fay Cook, who was deeply engaged in issues related to aging, posing research questions about political debates and policy issues. She later served as president of the Gerontological Society of America.

My life was profoundly changed by a cancer diagnosis in 1993. The illness put me totally "out of commission" for a year and a half, an experience I have described as *falling out of time*. Because long-term effects of treatment and a systemic infection greatly reduced my work capacity for nearly a decade, I decided to take early retirement in 2004. Since then, new medical intervention increased my energy and optimism, allowing me to go back to research and writing, as well as some teaching. I have collaborated with colleagues in the United States and a number of other countries. There have also been some intense, interesting meetings organized by the United Nations. The biggest marker for my "second chance" was an invitation to deliver a keynote address for the UN International Year of Older Persons. I was asked to reflect on the theme of the year: *Towards a Society for All Ages*. Work on the speech rekindled my interest in issues of age segregation. I was reminded of how I read the 1968 book by Shanas and her coauthors with excitement and optimism. Thinking about it 30 years later made me discouraged, but collaboration with Peter Uhlenberg has helped me move further in my thinking about age segregation and integration in some papers we have written together. Very few sociologists have actually addressed these issues. There was a major work by Irv Rosow, published in 1967, which did not get the attention it deserved. Was he too gerontological for the sociologists and too sociological for the gerontologists? Why do we have so few contemporary scholars who use aging as a domain for raising core sociological questions? Has the study of aging helped make sociology wiser? Or is "ameliorative gerontology" still a risk, as Cain saw it 50 years ago? I must admit that when I read GSA journals, there is very little that tickles my sociological imagination!

In a speech delivered in early 1994, Bernice said, "I am greatly tempted to predict that the field of gerontology is going to disappear over the next couple of decades." Why did she make this prediction? And why does it seem that it was false? She argued that we have come to realize that aging is a lifelong process and lives are indivisible. Furthermore, as she put it, "need without age must override age without need" in policies and programs. Quite different age groups, she argued, have similar needs. Recently, this latter argument has been voiced by several of us, for instance in pointing to communalities between our youngest and our oldest.

A realistic view of current political debates and social planning could easily make us pessimistic. Across countries, there seems to be a hegemonic paradigm for discussing old age, one emphasizing dependency, needs, and burden. Recently, I had a strong impression of ruling paradigms when I participated in a working conference funded by the EU. Although my subgroup had been asked to discuss old people as care providers, it was nearly impossible to get people in the group to think of examples of old people providing care, to consider care other than physical care, or to admit that care provision is not always a source of stress. I came away with a strong discomfort about how research position, clout, and funding may require buying into – and thereby reproducing – a construction of aging that we know to be fundamentally flawed. I keep coming back to French sociologist Pierre Bordieu's reminder: We must never forget that we are engaged in a battle over common sense!

Now I am back to the point where I started, the *geronts*. As I get older, I am increasingly struck by how much "othering" is going on, even in scholarly discussions. It is both comical and sad to see a group of gerontologists, all well over 60, *talk about them* – the old. We talk as if we concur with Mary Pipher, who calls old age "a different country," a place where *we* do not reside. Would we make more progress if we said *we* and *us?*

I have tried to briefly sketch how thinking about age has been a part of my academic life across phases of my adulthood, starting as a mother in her 30s, through my middle years, to my current life as an older person and a grandmother of two rapidly developing grandchildren (who really want me to finish this piece so that I can join them on a long hike and even some bouldering!). Over those years, I have had extremely rich opportunities for learning. What is my hope for the future? That the study of age and aging can cross borders – across national research communities, across levels of human contexts, and across life phases.

For members of my own generation of sociologists, I wish a rich dialogue about our own aging, our own understanding of networks across age boundaries, in different cultures and in contrasting policy contexts.

Chapter 44
Living the Gendered Life Course in Time and Space

Phyllis Moen

I have always been interested in the life course, with age performing as a social marker related to the time and timing of events. My father's tales about the Great Depression were stories about the way historical and biographical events shape life chances and life quality depending on when they occur. (In his mid-20s, he had worked for years to save money to go to college and lost it all. But he then found a sense of purpose as a soldier in the Second World War.) My own life too has been a study of transitions and timing in context. I was widowed at age 32 and recoiled in horror when 70- or 80-something women would come up and hug me and say they were widows, too. I knew our age difference made me not "like them" but couldn't really formulate that notion until as a graduate student I read work by Bernice Neugarten. I never met Bernice, but her insights into time and timing norms became a key theme in my own work.

Going back to school as a young widow in the 1970s underscored this lesson. I was only 32, but life experience (two daughters and a husband who had passed away from cancer) made me feel out of sync with fellow graduate students even though the age difference was not as large as I felt it to be. Graduate school in the 1970s also brought the historical period, and especially the Women's Movement, into sharp relief. I was refused a credit card – and asked why. Specifically, I asked was it because I was a single parent? A woman? A widow? A graduate student? They gave me the credit card.

I learned about the life course as a subject of study when graduate students in the Family Development Center at the University of Minnesota read *Children of the Great Depression*. I was so impressed by the book that I lobbied to have Glen Elder invited to come give a talk. I picked him up at the airport but didn't recognize him as the person I came to meet. I told him that he was much younger than I had expected. This was in 1977 or 1978. He asked why. Was it because of his name, "Elder"? I said no, it was because he seemed so young to have written such an important book.

I drove Glen in my pickup to Reuben Hill's house for a party and got lost. This turned out to be great because I used the time to talk to and learn from him. I think he must have thought I was a bit of a crazy driver in my (late) husband's pickup and with no knowledge of Minneapolis streets. I was fortunate enough to share time with Glen at Cornell University in 1979. Even though he was my colleague then, I still sat in on his graduate classes, still learned much from him, and have continued to learn from his work.

Another Cornell colleague, Robin M. Williams, Jr. and my "first" graduate student, Donna Dempster-McClain, led me to the sociology of aging, but quite unwittingly. Donna was taking a class from Robin, where he mentioned a never-studied survey he had undertaken many years back, in 1956, on women's roles in Elmira, New York (less than an hour from Ithaca, where Cornell is

P. Moen (✉)
Department of Sociology, University of Minnesota,
South Minneapolis, MN, USA
e-mail: phylmoen@umn.edu

R.A. Settersten, Jr. and J.L. Angel (eds.), *Handbook of Sociology of Aging*, Handbooks of Sociology and Social Research, DOI 10.1007/978-1-4419-7374-0_44, © Springer Science+Business Media, LLC 2011

located). Donna was intent on recapturing and doing something with Robin's data, as a now historical investigation of women's lives before the Women's Movement. Somehow, we got it into our heads that we would find these women and reinterview them 30 years later.

At first we couldn't get funding because people thought we would never find these women. Then our team (Donna, I, Dikkie Schoggen, and Melanie Miller) set out to prove them wrong. Melanie Miller was a young graduate student, and looked like someone's granddaughter or daughter as she knocked on doors in Elmira, asking about women who had lived there 30 years earlier. Neighbors helped. The town's historical records helped. Trips to view gravestones in the cemetery helped. (Remember there was no internet then, this was all done laboriously.) But we *found* them – all but two.

We wrote and received funding for a proposal to reinterview these women who were young mothers in the middle of the 1950s and their (now adult) daughters. Reading and working with the 1956 data had kept us thinking about them as they were in their twenties and thirties. When we went to interview the mothers, we found, to our amazement, that they were old – 30 years older than when interviewed the first time. And that is how I got interested in aging. But I always thought of them (and other older people I have studied) as young adults who have grown older.

In fact, I wasn't interested in old people. I was interested in how these Elmira women's lives played out over a 30-year period. And I continue in thinking of older people as people who have grown older, not as a separate category of "elders," the "aged," or the "young-old" or the "old old." This put me at odds with what was much of gerontology in the 1980s and 1990s. But it contributed to my interest in the sociology of age and the life course. And I am happy to report that gerontology has changed to become more "life coursey" over the years.

On the other side of the spectrum were my colleagues in human development (I had a joint appointment in human development and in sociology at Cornell), who saw "age" as going from zero to six, or at most to eighteen. I had a hard time getting students interested in what I was increasingly interested in – adults at all ages and stages. It helped that I studied women, since most women were mothers and, hence, were "mothering" some child at some point in their life course. I was ecstatic when Karl Pillemer and Elaine Wethington came to Cornell. We started the Cornell Gerontology Institute as a Roybal Center, and studied the second half of the life course in earnest.

Matilda White Riley was another pivotal force in my education in aging and the life course. Her ideas about structural lag permeated all of my work. But it was also her personality that touched my life. Watching her as the head of the ASA section on Aging and the Life Course, as a senior scientist at the NIH, was a wonderful chance to see age integration as she studied and lived it. My own enthusiasm for the research process – and belief in the importance of this topic – was reinforced by Matilda. The significance of what we do was also a lesson learned from Linda George over the years. And Martin Kohli's inspired work on the institutionalization of the lock-step life course into a tripartite of first education, then a lifetime of continuous, full-time paid work, then punctuated by continuous leisure in retirement, became key to my own thinking, since it so obviously excluded women's diverse and very contingent experiences.

Urie Bronfenbrenner always said that "we are the people in our lives." I couldn't agree more. Urie was another moving force in shaping my intellectual thinking, as he began to incorporate time and the life course into his ecology of human development approach, and I, influenced by him, attended ever more to the multilayered ecological contexts of lives.

I have been fortunate that my own biographical and intellectual journey has occurred in tandem with the development of the aging and life course field. Glen Elder has been a prime actor in moving the life course to center stage in the study of lives, highlighting historical events as potential turning points and introducing such important concepts as "linked lives" and "life course dynamics" in the form of interlocking transitions and trajectories. Bernice Neugarten played a key role in specifying the normative temporal contexts of behavior and decision-making, with her focus on being "on" or "off" time in one's actions. (I, for example, got married at age 18, which was "on" time in the South in the early 1960s, but appallingly early from the vantage point of contemporary norms.) And Matilda Riley along with Martin Kohli supplied the policy contexts of the life course, even as Linda George contributed important insights in role theory and subjective meaning throughout the life course, and Urie Bronfenbrenner emphasized the developing individual embedded in multiple ecologies.

Where will the field move in the future? This is an exciting time to study aging and the life course, given the striking demographic changes, medical and technological advances, labor market, and economic dislocations converging to shape lives in new and unexpected ways. Suddenly age and the life course are high on both research and policy agendas, as the boomers move into their 50s and 60s and nations recognize the mismatch (structural lag) between existing policies and practices and emergent realities of an aging workforce, an aging society, a growing retired force – all overlaid by disparities along the fault lines of gender, race and ethnicity, occupation, and education.

Married women's movement into the workforce was the "story" of the last half of the twentieth century, bringing into sharp relief the need to study women's as well as men's life course paths. The "story" of the first half of twenty-first century will be the aging of the population and what it means for the care of infirm family members and assistance to the next generation, the (very probably multiple) retirement passages of men, women, and couples and the restructuring of retirement as we know it, and the health and well-being of adults in the second half of the life course. This is further problematizing the taken-for-granted lock-step (what I call the career and retirement mystiques), institutionalized based on the experiences of middle-class or unionized blue-collar male breadwinners with full-time homemaking wives in the middle of the last century and perpetuating a gendered life course.

The challenge for scholars is nothing less than to capture contemporary patterns of lives as they play out in distinctive ways over time and in multilayered contexts. We don't know, for example, whether the existing evidence on retirement (based on prior cohorts) applies to those now confronting this status passage and doing so on a moving platform of risk and change. As I write this, I am spending a month as a visiting professor at the Social Science Research Institute (WZB) in Berlin, which points to the importance of comparative cross-cultural analysis. By documenting a range of patterned experiences and expectations and their impacts, scholars can illuminate the complexity and the varieties of the contemporary life course, including the importance of cultural and regulatory contexts.

Moving to the University of Minnesota in 2003 after many years at Cornell opened up an opportunity to study (with Erin Kelly) actual changes taking place in the way work is accomplished in organizations, underscoring the importance of the normative clockworks of work (and retirement) as taken-for-granted, even invisible, forms of structural lag. Studying the links between policies and practices (and changes in them) and their impacts in shaping life courses is important future work. Another promising avenue of work is investigating time allocations by men and women at different points in their life course. By pursuing these and other fruitful research directions, scholars can also assist policymakers in the fashioning of structures conducive to multiple alternative pathways to gender and other forms of equality, as well as to optimal health and life quality at all ages and life stages.

Part XI
The Future of the Sociology of Aging

Chapter 45
Sociology of Aging in the Decade Ahead

Jacqueline L. Angel and Richard A. Settersten, Jr.

As the chapters of the *Handbook of Sociology of Aging* have shown, many remarkable and ground-breaking strides have been made in the field over the past 30 years. The histories and current state of scholarship in each of the topics represented in the handbook provide a solid foundation on which to build future research agendas. In this final chapter, we identify some of the most provocative or pressing topics that might be nurtured in the decade ahead and how sociology can play a role in understanding them.

We have chosen some because they have great social import and others because they show great promise for advancing theories, methods, and data. We begin with issues that seem relatively universal in their potential or their challenge, widely shared by people and governments around the world. These include the need for physical care and housing, issues related to financial support and inheritance, and the new reality of diverse family forms and networks. As the chapter progresses, we touch on other important issues that apply to more specific groups of individuals, families, or cohorts. Of course, the topics we cover throughout the chapter represent only a limited number of many possibilities and, for each, our comments are necessarily brief. It is our aim to simply turn attention to them. In addition, the significance of some of these topics may not now be immediately apparent but seem certain to emerge in the future.

Who Will Care For Older People and How

The physical environment and conditions of life affect the pace at which individuals and populations age. These are, of course, profoundly affected by social factors (e.g., social class, behaviors, and networks) and require social institutions to address them. But in developing and highly developed societies alike, it is within the context of families that the care of needy family members is negotiated, and it is here that commitments to providing care, especially at home, will be sorely tested in the coming years.

The challenges of family life and care in the later years raises difficult but important questions about the expectations and responsibilities of siblings, children, grandchildren, and other relationships in a "post-traditional" society. The concept of a post-traditional world refers to a new social order, one in which rapid social change and modernization have left people with more choices and

J.L. Angel (✉)
Department of Sociology and LBJ School of Public Affairs,
University of Texas at Austin, Austin, TX, USA
e-mail: angel@austin.utexas.edu

R.A. Settersten, Jr. and J.L. Angel (eds.), *Handbook of Sociology of Aging*, Handbooks of Sociology and Social Research, DOI 10.1007/978-1-4419-7374-0_45, © Springer Science+Business Media, LLC 2011

fewer constraints on their lives – but also with more risks, as the choices they make are perceived to be of their own making and there are fewer safety nets to protect them. Unlike traditional societies, in which values, practices, and social institutions more heavily structure opportunities and decisions, post-traditional societies lack well-established social norms and rules for expectations and obligations and, consequently, place greater emphasis on individual action and choices.

The freedoms of a post-traditional society have in part created the extraordinary breadth in family forms that exists today. But we know little about what the strengths and vulnerabilities of these family forms means for the aging of their members, or what normative expectations and obligations their members have for giving and receiving care. Traditional definitions of family, especially the two-parent nuclear family, have become outdated in the face of a wide variety of family or family-like forms and relationships. This variety is driven by many factors, including multiple divorces and remarriages across generations, multi-partner fertility and nonmarital fertility, nontraditional partnerships, the co-survival of multiple generations over many decades, and the presence of four and even five generations at any single moment in family life. The growing physical distance from immediate family members also creates role ambiguities about filial obligations and consequently strains support systems. Together, these and other changes might lead us to envision societies that are relatively kinless, for lack of a better phrase, characterized by fragmented or superficial family ties and limited or uncertain support to family members in the face of poor health, economic distress, and other needs. Of course, we might equally envision that the kin and kin-like relationships that are actively chosen and voluntarily nurtured within these very same conditions might be stronger and more meaningful than those in more traditional societies. Either way, it is unclear how these new or complex configurations of relationships are supposed to function or feel, and who is responsible to whom and for what.

Such role ambiguity, for example, may carry significant consequences for the quality of relationships as family members grow old. Who should provide care, or is expected to provide care, when family relationships are tenuous or fragile – and even when they are not? Curiously, we lack rigorous multi-method studies about how adult children and aging parents make decisions regarding health care, living arrangements, and financial management. Many factors clearly come into play in these decisions, including the number and gender of siblings, their history of relationships with one another and with their parents, work-related issues, other family demands, and so on. Systems theories are but one example of the sort of heuristic device that can help us understand decision-making processes in families.

For these reasons, it will become necessary to determine how informal mechanisms of support might be augmented (and in an extreme scenario, replaced) by formal sources designed to ease the potential dependency burden of care and to improve the quality of life for everyone involved. As of yet, however, the cost-effectiveness or sustainability of such programs, and their ability to keep infirm older persons out of nursing homes, has not been demonstrated. Rather than serving as a substitute for nursing home care, home health care evidently taps a large reservoir of otherwise unmet need in the community. Under current arrangements, many infirm older individuals who might seriously benefit from assistance in the community simply do without it.

As nursing home costs continue to rise, home-based programs seemingly hold even greater potential to support individuals with intermediate levels of need in the United States. Research should more carefully document this need, especially in determining the extent to which individual services reduce the risk of institutionalization. This will also permit better targeting of services to the very frail and vulnerable, as impaired older clients at risk of dependency show greatest potential for long-term care cost reductions. Research should also project the potential relative costs and benefits – social, economic, and human, and to families and society – of caring for older parents and populations with severe dementia and other serious chronic illnesses.

Providing care to an increasingly diverse older population presents an entirely different set of challenges. What does it now mean to grow old in the face of the dramatically shifting racial and ethnic composition of a society? In the United States, for example, these population changes will have implications of great consequence for the nation as ethnic minorities, particularly Hispanics, and their children assume a greater role in the formal and informal labor market. Members of these groups are also among the most vulnerable in our population with respect to health outcomes and access to health care and insurance. Much remains to be learned about their health care needs as they age, the specific social contexts in which those needs are not met, and the potential role of families, the government, elder care organizations, and businesses in meeting them.

More broadly, it is important to design and evaluate new institutional arrangements and services that better take into account the needs of older people – a priority that has only been heightened by dramatic changes in the health care industry in the past 20 years. Consider, for example, the factors that influence when an older person and family caregiver come into contact with the acute and chronic health care system. These have far reaching implications for the outcomes of the encounter as well as for how satisfied patients and caregivers are with treatments and services. It is also vital to ascertain the role of the profession of medicine in treating – and creating – the health care problems of older people in different settings. What are the pathways of elder health care service use in ambulatory settings? How does acute care articulate with chronic caregiving systems? The more we know about how health care organizations affect the health and illness experiences of older adults, the better positioned we will be to improve the geriatric health workforce of tomorrow. More attempts, however, must be made to understand help-seeking processes for both formal and informal types of support.

Role Inversion and Parenting

Another interesting question at the intersection of sociology of aging and family life pertains to a sort of role inversion: children who, because of their special needs, are unable to live independent lives, and whose parents will not have the experience of fully launching them. This is a growing social challenge in part because medical care and technological advances have improved the infant and child survival rates of those who are medically fragile (e.g., infants who are extremely premature, very low birth weight, or drug-addicted) or developmentally disabled. The growing incidence and prevalence of attention, hyperactivity, and autism spectrum disorders among children has similarly created new challenges for parents. What is the experience of aging like for parents as they are responsible for children with intensive special needs for extended periods, and even their lifetimes? How do these provisions affect the social, psychological, and financial well-being of parents as they age? What is the fate of that child's aging?

These "role inversion" challenges do not only apply to a small slice of the population. Because the transition to adulthood is a more prolonged period today, parents across all social classes are providing much more support to their adult children, and for longer periods of time, than ever before. As a result, the "empty nest" today is one that comes later and remains open or cluttered as children make their way into adulthood and as the achievement of adulthood has become more elusive. This may bring serious "ripple effects" in the lives of aging parents, affecting their financial, emotional, and other resources at midlife and beyond, even prompting the need to revisit plans for retirement. Of course, new generations of parents produce new kinds of children, and there is evidence that parents and their young adult children are now closer and more connected today, even if they experience bumps as children forge new pathways into adulthood. One begins to wonder whether the greater investments that parents are now making in children from infancy well into their 30s will come with greater support when parents are old. Will these children feel stronger obligations to reciprocate support to aged parents? Will aged parents have stronger expectations that it be provided?

Social Sources and Consequences of Intergenerational Inequalities

The old-age dependency ratio, typically defined as the number of working-age people aged 15–64 relative to the population 65 and over, is beginning to present major challenges for governments in guaranteeing both the financial and health security of older people in the developed and developing world. In the United States, for example, there is a vastly unequal distribution of retirement income in old age among the baby boomers, yet few investigators have examined options for improving overall income adequacy in later life for vulnerable groups. What will reductions in, or the fading away of, pensions mean for people in the middle of their careers? How will working families with low incomes and few or no benefits afford to grow old? It will be crucial, therefore, to understand how old age policies and programs, especially social security, will impact intergenerational equity in the distribution of benefits – across generations in the family and across cohorts in society – as well as the financial well-being of future generations of workers and retirees. Similar challenges exist for many European nations and Japan.

Serious and controversial questions about intergenerational inequality demand closer scrutiny and the contributions of sociologists. The 1990s brought much attention to this topic, but gerontologists then largely dismissed charges that growing attention to the needs of the old would necessarily drive inequalities or fuel resentments from other age groups. But there is reason to believe that these heated debates will again emerge in future social and political conversation. For instance, to what extent will the future generation of workers receive a fair rate of return on their investment based on their lifetime contributions to the social security system? Cohorts of men who are now old were able to count on continuous work histories in well-paid and well-protected jobs, and cohorts of women were able to count on long-lived marriages – both of which are assumed in determining access to social security in the United States. How might more fragmented and unprotected work and marriage experiences today create serious risks in old age for future cohorts?

Sociologists of aging are well positioned to inform policymaking in an aging society, particularly in determining how inequalities in income and wealth are connected to age and cohort, gender, and race and ethnicity and in anticipating the consequences of alternative policy options on various groups. Consider, for example, the raising of the age-of-eligibility for retirement pensions and social security. What are the changing age norms for retirement in the United States and elsewhere? What are the various possibilities for strengthening economies to support aging societies? Who will be protected, weak, or fall through the cracks? How do retirement policies perpetuate privilege or further disadvantage women and members of other minority groups whose work lives cannot or will not match policy criteria? What conditions are necessary for generating broad public support for social security reform?

Thus far, we have focused on intergenerational issues across cohorts and on variability within cohorts. But we must also consider how social security retirement policies and programs affect the likelihood of intergenerational financial transfers within *families*. One of the most fascinating issues related to retirement security is the extent to which patterns of *inter vivos* ("between the living") gifts and bequests from older parents to adult children are connected to aging policies and programs. For example, is it the provision of social security and Medicare to old parents that makes it possible for parents to transfer wealth upon their death to their children? How is inheritance affected when the provisions of the government are made more generous or frugal? One window into these questions is cross-national: How do social security policies and programs affect intergenerational financial transfers in different national contexts? Most countries face the same problem of caring for a large retirement-age population, yet little is known about why welfare systems are different or how they result in meaningful differences in the well being of older people or the nation. What factors produce or explain the similarities and differences among countries in the developed or developing worlds?

Beyond *inter vivos* transfers and bequests, there is the grave problem of how individuals and employers will fund retirement plans, including defined contribution plans like 401(k)s. This is a crisis for municipal and state governments, let alone the federal government, which has made retirement fund commitments that far exceed what they could have ever realistically funded, especially to labor unions. This retirement tsunami raises crucial questions about who will pay for the oncoming wave of boomer retirement, especially early retirees from cohorts who have long life expectancies. What sort of structures should be in place for the old and the young, and for the poor?

This begs the question of what social consequences might result from these over-commitments when resources are limited. The aging of the baby boomer cohorts will increase competition for funding among the various recipient groups, and carry the potential for inter-cohort conflict, especially in times of significant economic retrenchment. Because of limited federal and state funding, the needs of impoverished women and children compete with those of the disabled and the elderly: Although the majority of enrollees in Medicaid are children, the majority of funding is spent on care for the disabled and the elderly. Almost half of long-term care is paid for by Medicaid. Medicaid long-term care for chronically ill elderly persons consumes about 17% of the average state budget, with spending estimated at $1.6 trillion dollars over the next two decades. The need for long-term care will only increase because longer lives mean that a large fraction of these cohorts will spend-down to Medicaid eligibility.

The new health care reform bill has clearly changed the health care financing landscape, but how it will directly affect Americans, and older Americans in particular, is as yet unknown. While it represents the only realistic means of covering any substantial fraction of the uninsured, there are numerous uncertainties about the specifics of the new program for younger and older Americans alike. It is unclear how health care coverage exchanges will work, and there is also the real possibility of repeal. As national health care expenditures grow toward 20% of GDP, cost controls are also inevitable. Medicare enrollment is anticipated to be the major force that drives public health care spending in the future fueled largely by a new prescription drug benefit and the rapid growth of the baby boomer population reaching retirement age. If Medicaid continues to pay less than other forms of insurance, providers may not participate unless they are forced. Still, Medicaid expenditures will increase as well.

Sources of Social Support

The stresses and strains of living with and caring for chronically ill people are well documented in the literature – whether the chronically physically ill, the severely and persistently mentally ill, or the infirmed. Arguably, the health consequences of extended life expectancy for older men and especially women will create difficult challenges for families and other societal institutions. The role of social support in preserving the physical and mental health of individuals who are old or very old, then, is clearly worth our attention.

Despite growing literature that documents the significance of social support for the overall well-being of the aged, there is relatively little good comparative data on how social support differentially enables some groups of people, including men and women, to better cope with the declines in resources (time, people, money, and power) that come with aging. The nature of relationships between parents and children has clearly changed as the result of shrinking family size, geographic mobility, and technological change – and yet we have so much to learn about how exactly those support systems have changed, and whether or how they have the capacity to provide emotional and instrumental support.

Studies of older minorities and immigrants suggest that there are reasons to suspect that cultural groups differ in the availability of social support, whether emotionally, financially, or otherwise.

How do culturally influenced differences in levels and types of social support affect the health of older minority group members? It is especially important to understand Latino cultures, which are large and rapidly growing segments of the aging population in the United States. Research indicates that Hispanics generally, and Mexican Americans particularly, benefit from aspects of their culture which are shown to be socially protective. This cultural protection operates primarily through strong family traditions and support systems.

Yet disparities in the health outcomes of different racial and ethnic groups also mean that aging itself is likely to be quite different for individuals and family members in specific groups. For example, despite a favorable mortality regime, the socioeconomic disadvantages among aging Mexican Americans and late-life immigrants point to a potentially high dependency burden placed on their families – a burden about which we know very little. For older African-Americans, multiple health vulnerabilities undermine attempts to age independently, and consequently may increase reliance on adult children and relatives who are ill equipped to provide adequate instrumental support because of other work and family obligations.

Emotionally supportive intimate relationships allow individuals to develop a strong self-identity that gives life meaning and purpose. Yet we have much to learn about how social environments influence the life satisfaction and morale of older people. The excessive focus on the individual factors that create vulnerability – vs. those in the social systems outside of individuals – has prompted research emphases that pay too much attention to depression per se rather than the social conditions that arise from or exacerbate it (e.g., role strain) or those that inhibit or diminish it (e.g., supportive relationships or services). Developing new research approaches to understand the protective role of social supports in late life is vitally important because social isolation, loneliness, and depression have been shown to increase the risk of death and physical or other mental illnesses. Other stressors, such as economic losses associated with the current "Great Recession," for instance, may also create new risks for older people who would otherwise not be at risk, or elevate levels of risk or distress for those who already are. A socially supportive network is crucial in helping older individuals cope with health challenges or economic losses.

The mental health benefits of social support suggest that it is not the number of people in one's social network that matters most, with more being better, but rather a step function in which the presence of at least one close confidant provides near total protection. While age-graded life events like widowhood or the death of an adult child punctuate the need for social support, they often come at a time when the social networks themselves are unraveling, often through the deaths of longstanding friends and peers. Social support is crucial in these situations and other harmful life events by preventing a severe stressful response and increasing the capacity to continue living.

Besides the clear practical benefits of social integration, the availability of relatives and friends with whom one interacts has a direct effect on mental health and morale as individuals grow older. Loneliness is a major health risk for the old and the socially isolated, and in particular for elders who lack English language proficiency. Studies frequently show that integration into a cohesive social network appears to protect individuals from many kinds of negative health outcomes. These findings, if replicated on ethnically-diverse populations, will bring us closer to demonstrating just how significant intimate relationships are in preserving healthful aging – or, likewise, to show just how significant their *absence* is in jeopardizing it.

There is also great need to develop a qualitatively-driven research agenda on how social environments support or fracture the potential to improve or protect health and quality of life for people and whole communities. Specifically, qualitative studies are needed to address the health consequences of the nature, degree, and quality of social support available to individuals across the multiple settings they inhabit. Detailed and highly textured analyses are also needed on the health consequences of emotional and social support provided at home and work. These settings, along with other nontraditional and emergent settings (e.g., internet cafes, social media sites, naturally occurring retirement communities, petting zoos), will provide fresh and valuable insights into contemporary contexts.

Social Dimensions and Determinants of Successful Aging

The field of life course studies has at its core two propositions for which there is an inherent tension: one emphasizing that the life course is the product of social forces (broadly construed as "social structure"), and the other emphasizing individual capacities and effort (broadly construed as "human agency"). A vital feature of an emphasis on aging is that people actively and regularly evaluate their behavior and make decisions about how to act in new situations depending on their past experiences. It also assumes that the choices people make – whether regarding education, occupation, marriage, and fertility or other major experiences – bring serious and enduring consequences for well-being in later life. Years of socialization and accumulated advantages and disadvantages equip individuals with a repertoire of social skills, liabilities, and resources that affect how they age and meet the realities of later life.

This theoretical point about the sociology of autonomy and aging underscores the complexity of the concept of successful aging. What does it mean to age well (as opposed to pathologically)? There has in the last two decades been an explosion of interest in "successful aging," which appears routinely in our lexicon. But what is this term or social label ultimately meant to capture? The literature suggests that it implies a certain amount of money, personal prestige and respect, and power that assures one can remain as independent and autonomous as possible. The social–psychological aspects of successful aging tend to overemphasize life satisfaction and happiness. But research must go beyond prescriptive definitions of successful aging toward an understanding of what the concept means for assuring the greatest personal autonomy and adequate security in old age. At the same time, the need for autonomy should not negate the need for social support – both are important parts of the human experience. The voices of sociologists are essential for broadening the lens of successful aging so that it is not as exclusively focused individuals and more often makes visible how particular aspects of particular social environments nurture or inhibit successful aging.

Changing Work and Leisure Roles

It is also important for sociologists to address how work-related experiences, roles, and expectations for them have changed in recent decades. New models of work and retirement have emerged as the manufacturing sector has given way to the emergence of the knowledge and service sector, as "lifetime" models of work have eroded, as the contract between employers and employees has weakened, and as employment has become more discontinuous. These changes have brought both new possibilities and new risks for aging.

At the broadest level, the meanings of "work" and "retirement" are being called into question and recreated in our postindustrial world. People with flexible work and ample resources are able to shift careers in midlife or reduce or step away from work in order to manage family obligations. Many retirees are seeking life-altering experiences and opportunities to reclaim old interests, beginning second careers, or volunteering. But what do "work" and "retirement" mean for the less fortunate among the old? And just as important, what is the fate of future cohorts, whose work lives will be more fragmented than those who are now moving into and through old age? What will work and retirement mean to them in their later years, how will they experience work and retirement, and how will our society in turn be affected?

We also know little about what work and retirement will mean to women, minorities, and immigrants in their later years, and how secure their experiences will be. For example, because of traditional gender roles and the gendered division of labor among couples, especially in pre-baby boom cohorts, women never or rarely worked, or did so only part time. The greater labor force commitment of women in the latter half of the twentieth century, coupled with the high degree of marital instability, however, mean that women are increasingly responsible for their own retirement security and many are at risk of economic insecurity in old age. Are women's work trajectories likely to get longer and stronger in the years ahead – or will they continue to be more discontinuous then men's, regardless of the gains they have made, and at risk, especially in the face of shorter marriages?

Furthermore, there is broad public concern about whether retirement as we now know it will be possible or sustainable for future cohorts, whether because of their own diminishing resources, freedom from the three-box lockstep life, or the diminishing resources of the state and old age policies, especially as (or after) the boomers retire. The growing diversity of successive cohorts also means that disadvantages in the work opportunities for women, racial and ethnic minorities, and other groups will also diminish their retirement security and material and social well-being in old age.

Coping with Change in an External World

Late-life displacement is an emerging topic that encompasses aging at the nexus of social change and disorganization in the physical environment. For example, global climate change is an environmental threat and it, like hurricanes, earthquakes, and other natural disasters, poses a challenge for individuals, families, and the state in caring for older people. As keen observers of the social world, sociologists are especially equipped to shed light on how aging experiences are affected by a constantly and often rapidly changing social world.

To this end, we must reveal how particular cohorts, given their unique histories and position in life, might be differentially able to adjust to social change. In the event of a crisis, how might, say, being in a cohort that grew up in the Great Depression or wartime leave people more or less resilient in the face of hardships in late life? How might being old – that is, having lived a long life – leave people in stronger or weaker positions in weathering social change? Scant information is available on the resilience of special populations to late-life displacement due to political upheaval, economic depression, natural disaster, and the like, and how age interacts with other social statuses or roles to place people at greater or lesser risk (e.g., those who are old *and* poor, or old *and* a member of a minority group). Or is it "something else" that is instead the major factor?

We also have much to learn about how social institutions can be designed so that, when unexpected changes occur, older people are best prepared to handle those changes. Research on victims of natural disasters, for example, highlights the significance of research along these lines. There is a need for local studies that provide empirical verification of how nursing homes plan to carry out complex recovery efforts for elders in the wake of natural disasters. Researchers and government officials alike will find research like this useful as they recognize the daunting challenges linked to the mitigation of future hazards and unspeakable catastrophes.

Besides environmental disasters, late-life displacement includes disorganization brought about by economic decline, including the "Great Recession" that began in 2008. Economic downturns of this type can dramatically alter the realities of old age for many people and their families as they struggle with unemployment, face home foreclosures, and come to grips with lost savings and pensions, which shatter plans for retirement. The deficits worldwide are staggering.

As a result, governments in the developed world are wrestling with the plight of the unemployed, and especially the long-term jobless. In the United States, for example, political fears concerning the nation's financial health and growing budget deficit have dimmed hopes for extending unemployment insurance benefits. Will the productivity of the future labor force depend on continued and increased federal support to compensate for the new economic realities? Might employment programs to employ the unemployable, like those that existed during the Great Depression, be reconsidered today, and how might they capitalize on the skills and experiences of older workers? Finally, how will the current economic crisis affect the magnitude of public transfers, such as social security, or private transfers within families, for future generations?

In addressing issues like these, sociologists can begin to uncover the lifelong effects of economic decline and growth on an aging population and society and provide information to help determine how available resources can be allocated in a socially productive and responsible way.

Legacies of Immigration

Immigration and the income and health care needs of the older population will be among the most pressing and interrelated issues the United States will face in the near future. Historically, wars, natural disasters, political conflicts, and other major societal-level events have always displaced large numbers of people. But the need to respond to the needs of large numbers of refugees presents governments and civil society organizations with new and significant challenges. Of course, so much of the current political dialog is not about old immigrants, but about working-age adults, and often young parents and their children. There are, however, sizable numbers of immigrants now in their later years – and the experiences of being an old immigrant are surely different from being a young one, and conditioned by the age of the person at the point of immigration. To immigrate as an old person is different from being an old immigrant who arrived decades ago. The surges in the young immigrant population will also have effects on our society decades from now, as they grow older and eventually reach late life.

Other questions about immigration and aging also need to be addressed. For example, aging in another country a brings several health and social welfare challenges and opportunities issues for migrants in specific U.S. cities and those who return to specific foreign cities. This research could provide valuable information into how the migration process interacts with experiences in the labor market to affect general health and well-being. What impact does large-scale migration have on national and state-level health policies and health delivery systems on access to care in the United States and elsewhere? How is rapid population aging and migration in the United States and other developed nations affecting the structure of communities and even entire regions? In countries and states of high immigration and emigration, families, communities, schools, workplaces, and governments are being transformed. Transnational families are an important part of that process.

Late-Life Loss

While the experience of loss in later life would seem normative (e.g., widowhood for women), many aspects of loss are poorly understood. Aging into the eighth, ninth, and tenth decades of life brings with it the inevitable losses of spouses and partners, members of the extended family, and friends and neighbors. Although some of these losses may be normal and expected, and therefore may be prepared for, they are clearly not easy to live through. These aging-related losses often push the limits of human coping, adaptation, and resilience, and the normative physical and cognitive declines that come with aging may make that process more difficult.

Other losses in roles and social relationships are unexpected and even the result of an extended life span. For example, given the new certainties of a long life and the predictability of death in old age rather than earlier in life, it is natural for parents to expect to out-survive their children. But as both parents and children can now jointly survive into old age, the experience of the death of a child may become more likely. The process by which older parents cope with the loss of a child who is middle-aged or even older is not one that is acknowledged by physicians, policy makers, or family members. Understandably, stories of the loss of a child, regardless of how old that child is, portray family tragedy beyond comprehension and words. So, too, are painful stories of being the sole survivor among siblings or longstanding friendship groups – what Bernice Neugarten once called the "costs of survivorship" in advanced old age. Yet research on the complexity and intensity of the emotions that accompany this kind of grieving in late life is missing in the literature. Sociologists have important work to do in conducting deeply textured research on how people make sense of and deal with the impact of the loss in life's final decades.

Moral and Ethical Dilemmas of the End of Life

We have much to learn about the moral dimensions of end-of-life care – for example, who should help and be helped in the end of life, how they should be helped, and who should pay. There is a need for studies to assess barriers to health care service use, especially those that stem directly from policies. An aging society simultaneously brings an increased need for hospital and home health care services, but also to find innovative ways to contain the growth in expenditures for the care of older people and maximize the use of less expensive community supports. These innovations must take into account a range of cultural differences, social trends, and moral issues, as well as changing economic realities.

In addition, research needs to examine the ethical considerations of competing costs for families and societies. When should one cease heroic life supports, and who makes that decision? Is it based on cost or on quality of life? The practical issues of curbing rising health care costs raises a serious moral dilemma related to the rationing of expensive life-saving interventions. How much should be spent on older people, given that public funding is a finite pie and given the sheer size of the baby boomer cohort? Who will win and lose in the game of support? What are the tradeoffs of supporting the old versus the young as the ratio of old-to-young increases? The complex family structures of baby boomers also seem likely to heighten questions of who is responsible for whose care, but few answers exist about the expectations and obligations attached to that responsibility. The ethical and moral dimensions of caregiving have been in large part glossed over, leaving a window of new opportunities wide open for fresh ideas to address the support systems of an aging society for generations to come.

These are just a few of the important topics we hope will be nurtured in the future. While they represent what seem like formidable challenges, to meet in the future, these and other advances will also be aided by some basic principles for guiding the generation and use of theories, questions, data, and methods. We now turn to these parting thoughts.

Basic Principles for Moving Forward

First, there is a natural synergy between aging and the life course, a point that was reinforced when the American Sociological Association's Section on Aging renamed itself in 1997 to the Section on Aging and the Life Course, as discussed in Chapter 1. Greater attention to the life course has yielded many new insights. Yet as attention to the life course has grown exponentially – and will continue to do so in the decade ahead – it is our hope that this will not compromise the scope and clarity of the sociology of *aging*. There are questions about aging that do not entail the life course, and many more questions about the life course that do not entail aging. Both are naturally treated in our work, but how we put the two together has tremendous implications for future scholarship. The life course perspective has grabbed hold of us, offered valuable insights, and transformed scholarship in our field. But it is important for researchers to consider what is gained and what is lost or put at risk in our understanding of the sociology of aging if too great an emphasis is placed on the life course.

In pursuing an understanding of "aging and the life course," does our field become too big and too broad to manage – a field of every possible age, and every possible transition, in every possible domain of life? As the life course perspective gains prominence in other subfields of sociology, these fields, too, begin to overlap with our own – indeed, the "life course" tag can be found with great regularity in the sociology of family, education, work, health, and criminology. What is it, then, that leaves our own subject matter distinct? Sociologists of aging must continue to wrestle with these difficult and critical questions in the years to come.

Second, and related, it is also important not to lose sight of the sociology of *age* (rather than aging), which was a central point of inquiry in the early years of our field. This scholarship, for example, examined how social roles and activities are allocated based on age, how age underlies the organization of social institutions, how age structures legal rights and responsibilities or is used to determine eligibility in social policies, how age is used to determine expectations of the self and others, or how it enters into social interactions. The salience of age in our scholarly lenses has diminished over time. This has occurred in tandem with what seems to be a growing *denial* of age – that age is something that can be defied or transcended – and an accompanying emphasis on *successful* aging among gerontologists and in our society. It is good that gerontology has, in the last few decades, seriously challenged the belief that old age is a dark period of irrecoverable physical and cognitive declines. Yet with the pervasive sense of optimism in aging research (and its emphases on successful, productive, and optimal development and on positive processes and outcomes), we must ask whether we do ourselves and old people a great disservice in the process. These emphases threaten to obscure from our scientific lenses the real underbellies of aging and old age that must be acknowledged if they are to be dealt with effectively. Here, we are thinking of even normative changes and vulnerabilities in physical health and cognition, and of the compression of illness and disease during the final years. When we deny the realities of aging and emphasize the differences among old people to the exclusion of the things they share, we jeopardize the political activities and policy agendas that serve elders' interests and needs.

Third, this demands that we renew attention to the things that people in a given life period have in common, alongside our investments in understanding difference. Sociological research on aging routinely references the high degree of variability among old people, consciousness about which has also been promoted by popular theories of cumulative advantage and disadvantage. But the field is in need of comprehensive empirical treatments of variability and examinations of its *social* sources and its *social* consequences. And because we now assume that variability is a key hallmark of old age, only rarely do we consider the things that old people have in common. This trend, coupled with more attention to the whole life course, make it increasingly important *and* difficult to clarify how old age is distinct from periods before it. When we lose sight of commonness, we lose sight of the things that make our subject matter distinct. To what extent are the challenges of old age simply those of earlier periods that are prolonged or revisited? To what extent does old age pose unique developmental challenges and opportunities? What are the markers that define entry into old age and movement through the "young-old," "old-old," and "oldest-old" periods so commonly cited in research on aging?

Fourth, it is important to reclaim attention to the macro phenomena that preoccupied sociological attention to age and aging in the early years. We are beginning to live in the top-heavy aging population that prompted so much reflection and concern, and drew attention to our field a few decades ago. It is important that scholars do not emphasize the individualization of aging so much that we lose sight of social structure – one of the very things that defines us as sociologists. We have an obligation as sociologists to keep social forces and factors front and center in our inquiry. It is especially important to keep in focus the implications of our aging society – and the demographic parameters that produce it (mortality, morbidity, and fertility) – *for* our society, families, and individuals.

The need to keep social factors in focus is also heightened by reductionist tendencies in science and the obsession with genomes and genetics. As we travel further out into social spaces and attempt to take them into account, our empirical work becomes more difficult. Yet the tendency in individual-based disciplines, such as psychology or biology, to dismiss external forces as being too unwieldy to measure or as already represented in lower-order measures, means that we must, as sociologists, make them visible, which will also require advances in our measures and methods. This may be why biomarkers seem so attractive today, especially in the areas of abilities and health, for the further we drill down into biology and put our money on those factors, the more quickly we simplify the complexity posed by the social world. There is comfort lurking there, it seems, if only the right markers can be connected to the right outcomes. But these connections are rarely clear, and our theories do not get down that far or that specific to make meaningful connections. Worse still, in emphasizing what lies within individuals over what lies outside of them, we laud people for positive outcomes with little regard to the ways in which social environments bring them about, and we blame individuals for problems or failures that have little to do with their own actions and more to do with how their social worlds constrain their possibilities or squash them out. It is our responsibility as sociologists to show the way. New research focusing on the interplay of genetics, the social environment and aging creates a new opportunity to learn about the social implications of genetic research.

Fifth, it is important that our work become more anticipatory. Virtually everything that is known about aging is bound to cohorts born in the first few decades of the twentieth century. We do not know how much of our current knowledge base about aging will apply to future cohorts whose characteristics and experiences have been very different. Later cohorts have also grown up and older with different resources, expectations, and needs, and they have been subject to different constellations of government programs and policies, with different types and levels of support. We do not need to wait to see how they are different. These cohorts are the future of aging, which can be understood by getting more intimately acquainted with them and by *making history visible and tracing its legacy in their lives.* Understanding the middle aged is particularly important, for they are next in the queue and include boomers; understanding young people matters too, for their transition into adult life looks dramatically different from what we have seen before.

Sixth, we must remember that there are people behind the numbers, and that we must do more to bring the person back into research. As sociologists, we have an obligation to understand whole people, and to understand them in relation to the multiple social contexts in which they live. We also hope that in the face of increasingly specialized and fine-grained empirical work, we put more value to the task of integration, synthesis, reflection, and theoretical development. Both of these points speak to the need to develop more holistic views of people and knowledge, and to develop a science of greater meaning, for the people and things we study are rich and complex.

Finally, if we are to do right by our subject matter, we must do right by ourselves – to become more self-reflective and self-critical, to think more seriously about why we are drawn to our topics, to unearth the values that lie beneath our own work and the commitments of our profession, and to become more conscious of the things we are invested in promoting, accepting, or denying. One thing will not change: Most people who create expectations, conduct research, make policies, engage in practice related to old people and old age *are not themselves old*. We are outsiders to the very people and phenomena we hope to understand. Our values and assumptions affect what we do (or do not do) with and for old people. This predicament creates challenges for what we know, how we know it, and what we do with it. We hope that in the future we will have a stronger science of action, one that will permit us to more seriously wrestle with the moral or political imperatives we have to advocate for or intervene on behalf of the populations we study, and to feel compelled toward action and the improvement of the greater social good. In the *sociology* of aging, we must grapple with social problems, solutions, and interventions, for it is here that some of our most important contributions are to be found and realized.

Index

R.A. Settersten, Jr. and J.L. Angel (eds.), *Handbook of Sociology of Aging*, Handbooks of Sociology and Social Research, DOI 10.1007/978-1-4419-7374-0, © Springer Science+Business Media, LLC 2011

D

M

CPSIA information can be obtained
at www.ICGtesting.com
Printed in the USA
LVHW06s1407210818
587646LV00012B/383/P

9 781461 440956

ELSEVIER'S DICTIONARY OF GEOGRAPHY

ELSEVIER'S DICTIONARY OF GEOGRAPHY

in

English, Russian, French, Spanish and German

compiled by

V.M. KOTLYAKOV and A.I. KOMAROVA
Moscow, Russia

ELSEVIER

Amsterdam – Boston – Heidelberg – London – New York – Oxford
Paris – San Diego – San Francisco – Singapore – Sydney – Tokyo

Elsevier
Radarweg 29, PO Box 211, 1000 AE Amsterdam, The Netherlands
The Boulevard, Langford Lane, Kidlington, Oxford OX5 1GB, UK

First edition 2007

Library of Congress Cataloging-in-Publication Data
A catalog record for this book is available from the Library of Congress

British Library Cataloguing in Publication Data
A catalogue record for this book is available from the British Library

ISBN-13: 978-0-444-51042-6
ISBN-10: 0-444-51042-7

Computer layout of the Dictionary is prepared by V. Kotlyakov and M. Kunakhovich

For information on all Elsevier publications
visit our website at books.elsevier.com

Transferred to Digital Print 2008
Printed and bound by CPI Antony Rowe, Eastbourne

CONTENTS

ABOUT AUTHORS

Vladimir Kotlyakov is a famous scientist, specialist in geography and glaciology. He has graduated from Lomonosov Moscow State University and now works as a Director of the Institute of Geography, Russian Academy of Sciences; he is also the Honorary President of the Russian Geographical Socety.

Vladimir Kotlyakov worked in the Northern Island of Novaya Zemlya, spent more than a year in Antarctica, worked on glaciers in Pamirs, Caucasus and Tien Shan and participated in many other field works. He is the author of more than 20 books in different fields of geography, his main contribution is the World Atlas of Snow and Ice Resources published in 1997. Collecting material for this Atlas Vladimir Kotlyakov visited many countries and all continents. He is a member of the Academy Europaea, French and Georgian Academies of Sciences; he is also a foreign member of the American, Mexican, Italian, Georgian and Estonian Geographical Societies.

Vladimir Kotlyakov is the author of Elsevier which published in 1990 his Dictionary of Glaciology.

Anna Komarova is Doctor of Philology, brilliant teacher of foreign languages, specialist in linguistics, terminology and lexicography. She graduated from Lomonosov Moscow State University in geography, then turned to linguistics and now works as the Professor of the Faculty of Foreign Languages at Moscow State University, Head of the Department of Foreign Languages for Geographers.

Anna Komarova teaches foreign languages (English, Spanish and Portugese) and linguistics at Lomonosov Moscow State University. She is the author of a number of textbooks on English and Spanish (among them "English for Geographers" published in 2005) as well as scientific articles and books on linguistics and philology. She is a member of International Geographical Union and National Association of Applied Linguistics taking an active part in the conferences and congresses worldwide. She is professionally interested in the issues of Geographical Education – the role of foreign languages for the education of Russian Geographers and specialists in Tourism, and in the problems of functional stylistics, lexicology, lexicography and terminography.

TERMINOLOGY OF GEOGRAPHY

Geography as a multidisciplinary science

Geography is a system of highly developed sciences about the environment. Geographical science embracing the study of the Earth's physical phenomena, people and their economic activities has always been in need of an extensive terminology. Geographical terms are related to the terms of natural sciences (physics, chemistry, biology, geology, etc.) and humanities (history, economics, sociology, etc.) since geography is based on these fundamental subjects.

Like mathematics, history or philosophy, geography is one of the oldest sciences in the world. It originated in ancient times and over many centuries the main purpose of geography was to discover and describe "new" lands and oceans. The tendency to record individual phenomena led to the formation of chorology and regional geography. At the same time scientists strove to find and explain the differences and the similarities between the geographical phenomena and started to organize and classify geographical facts creating general or systematic geography.

At the later stages of the evolution of geography such sciences as geology, biology, etc. separated from it and became independent. Gradually the sphere of scientific interests of geography had been narrowing down, in the middle ages it became dominated by the clerical ideology which resulted in the statical and schematic character of geography of those times. The Great Discoveries have expanded the horizons of scientific thought and confirmed the idea of the integrity of the world. In the 18th and early 19th centuries a considerable progress in the understanding of the Earth's physical phenomena was achieved.

In the second half of the 19th century the ideas of geographical determinism became very popular. The concept of geographical determinism stated that the physical factors of the environment dominated and even determined the pattern of human life and human behaviour, that the development of certain peoples and countries was largely conditioned by the environmental factors. However in the 20th century because of the increase of human impact on the environment these ideas lost their popularity and today remain only in the concept of environmentalism. On the turn of the 19th and 20th centuries there appeared a concept of geographical possibilism, which stated the existence of a passive uniform environment and various forms of the interaction between humans and nature.

In the course of the world development the most general features of the geographical thought were determined by the greatest achievements of natural sciences and the invention of new technologies, cyclic evolution of human society, political events of global scale, the logic of the development of science and its branches, the change of paradigms in the course of scientific revolutions. The character of the development of geographical science was determined mainly by the traditions of national scientific schools like French school of human geography with its steady social trend; German school having the traditions of a deep theoretical analysis, regional planning and geopolitics;

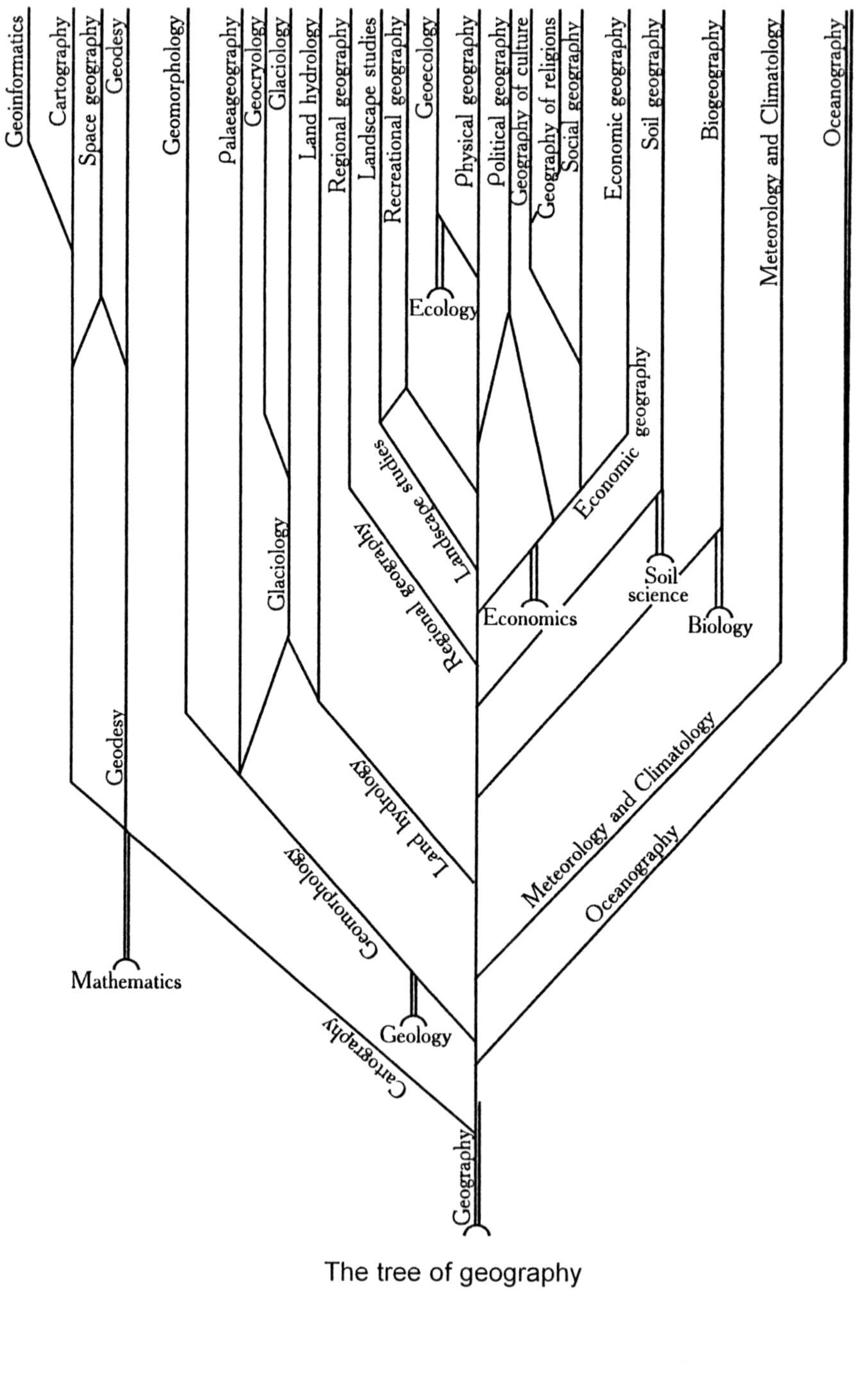

The tree of geography

Anglo-American and Swedish schools of theoretical geography which widely used the quantitative methods of research.

The whole complex of geographical disciplines closely related to a certain space of land or water, have been divided into two main parts: general geography and regional geography. These both parts of geography reveal the integrity of geography, since they deal with the relations and interaction of a number of physical and social components of the environment. Studying social and physical spheres, geography does not oppose them, but rather take them in combination and interrelation. Its two main branches – physical geography and human geography – coexist enjoying equal rights, interlace and interact within the limits of one integrated science – Geography.

All the processes occurring in the environment have common energy basis which is the light and thermal energy of the Sun, the internal heat of the Earth, gravitational and tectonic forces. The components of the environment and their spatial combinations determine the global zonal and regional regularities, but each territory is unique and requires individual investigation.

Geography includes a number of disciplines and subdivisions which appeared along with the development of the science. In spite of being very different geographical disciplines have some common tools of investigation which is maps, comparative method of exploration, remote sensing, geoinformation systems.

In the study of the Earth's physical environment geography has very close connections with physics, chemistry, biology, geology, in the study of human geography it is related with regional economics, demography, sociology. Geography has actively used the methods of other fundamental sciences, such as historical, physical, mathematical, geochemical and other methods. However, it has its own subject and methods and consequently its own integrated system of scientific terms well-distinguished among the terminologies of other fundamental and applied sciences.

Today along with the progress of scientific knowledge and techniques the word-stock of geography grows exponentially, terminology of different subfields constantly develop and many new scientific terms appear, various kinds of complex databases are being created in many areas, as well as highly sophisticated geoinformational systems. They need a very well organized scientific terminology both in national languages and at the international level.

Geographical and linguistic principles of compiling the Dictionary

Today very well developed terminologies of all the specialist fields of geography and related subjects exist in the main world languages. However, they are not always well-correlated. Nowadays geographical terminology requires unification and international correlation more than ever before. Hence the idea of compiling a multilingual polydisciplinary dictionary.

The Dictionary consists of the basic table of terms arranged according to the order of the English alphabet with each term numbered. Each entry consists of the term in English and its equivalents in Russian, French, German, and Spanish. Short definitions of terms are given in English and in Russian. The terms are supplied with the necessary grammar labels, such as gender of nouns, plural number, etc.

The Dictionary combines two functions: that of a defining dictionary and that of a bilingual dictionary. These two functions are basically contradictory because usually the defining dictionary is aimed at giving one meaning of the word which is the main and essential one, while the bilingual dictionary tries to give different equivalents of a given word in the other language in order to supply the user with maximum possible translations, differing in the shades of meanings, thus giving him the possibility to choose the appropriate word. But in our Dictionary we intentionally decided to combine the two functions – defining and multilingual, because a short definition of the term and equivalents in other languages help to achieve our main aim which consists in showing the basic geographical terminology and harmonizing it in several languages. Having this into consideration we deliberately mixed two types of dictionaries in one.

The dictionary presents geography as an integrated science covering a wide spectrum of geographical disciplines and subdisciplines: general geography; physical geography and landscape study; palaeogeography; geomorphology; soil geography; biogeography; meteorology and climatology; hydrology; glaciology; geocryology; oceanography; geodesy, cartography and geoinformatics; geoecology; human geography and also terms of relatively new branches, such as political geography; geography of tourism and leisure, resource management; medical geography, etc. In the Russian scientific tradition these disciplines have always been regarded as indispensable elements which form part of geography, rather than related subjects.

Each discipline is represented in the Dictionary by a limited but sufficient number of essential terms, describing the basic notions of the subject. We selected and included into the Dictionary the terms which are commonly used in geography and are generally recognized. For the actual choice of terms for definition we have been guided by some strict lexicographic principles confirmed by the scientific experience of the authors of the Dictionary and the help of experienced colleagues who are experts in separate fields of geography.

Terms are words and word-combinations used to express concepts and categories of a given subject, since concepts and categories form a system of knowledge of the subject – the terms also form a system, called terminological system or terminosystem. Scientific terms are considered to be linguistic units of a specific type – rational and easy to be classified. However, terminology changes in time, has different sources of origin, various ways of borrowing which lead to some "linguistic imperfection" of the terms. Because basically terms are the units of natural human language, often they behave as ordinary words, i.e. develop synonyms, homonyms, different meanings, etc. These features are very typical for the words of general language and provide the richness and diversity of any natural language. But from the point of view of rational systems, as it should be terminology, those features are imperfections and shortcomings, which spoil terminology and cause troubles in using it in scientific discourse and makes it difficult to register terms in dictionaries.

The additional difficulty consists in the fact, that different national languages divide the reality in a different way, so, the equivalents in different languages are far from being in one-to-one correspondence. But we tried to harmonize them as much as we could. While analyzing and comparing terminologies in English, Russian, French, German and Spanish we noticed that those branches which emerged relatively recently when computers and information networks started to be widely used (such as geoinformatics, remote sensing, environmental management) have much better coordinated terminologies, than the terminologies of traditional branches of geography, such as geomorphology or cartography which developed in less international contacts or even in isolation.

In our work of compiling the Dictionary we followed a certain sequence of stages elaborated and accepted in modern lexicography. At the first stage the type of the future dictionary and its main characteristics (lexicographic parameters) have been determined: the purpose and functions of the dictionary, the scope of the lexical units devoted to description and the way they would be described, the approximate volume and the principles of selecting the words for the dictionary, the information that would be given in the entries and the structure of the entry as well as the whole structure of the dictionary. At the second stage the list of head-words have been made: terms (words and word-combinations with their preliminary definitions) have been selected to be included in the dictionary. Initially, the lists of head-words have been compiled separately for each subdiscipline and submitted to reviewing. At the third stage the dictionary entries have been formed: the terms have been supplied with short definitions in Russian and in English, the equivalents in French, German and Spanish have been found and necessary grammatical notes have been supplied. Separate thematic lists have been united in the basic table of the terms arranged according to the order of the English alphabet, with each term numbered. At the forth stage the final reviewing and editing have been realized, the alphabetical indexes for Russian, French, German, Spanish terms have been made, the cross-checking have been done to make sure that every technical term used in the definitions had been itself defined in the dictionary. Then the camera-ready copy of the dictionary has been prepared.

In order to select the essential vocabulary for the Dictionary first we made the so-called thematic or conceptual lists of head-words, showing clearly and logically the structure of each subdiscipline which form part of Geography. The "conceptual word list" indicated how the terms are grouped into "areas of meaning" or "topic areas". Within each thematic group of terms there have been established the main categories and oppositions. Thus the category of *environmental pollution* is represented by two opposed elements *natural pollution* and *antropogenic pollution* distinguished on the basis of the main factors causing pollution either a physical factor or human activity. Another category is related to the type of a substance polluting the environment: *chemical pollution, noise pollution, thermal pollution, visual pollution.* Setting up similar classification and oppositions was also helpful for other subdisciplines. Thus for example: *biotic factors – abiotic factors, freezing front – thawing front* (*in frozen ground*), *renewable resources – non-renewable resources;* types of permafrost occurrence: *continuos permafrost – discontinuous permafrost – sporadic permafrost,* different kinds of protected areas: *reserve, biosphere reserve, naatural monument, sanctuary, national park, natural-historic park, marine national park.* By applying this method it was possible to establish semantic links between the terms inside each thematic group without missing important elements, while the terms denoting too small elements of geographical objects or too detailed concepts have been excluded.

Another big problem has been solved by making conceptual lists of head-words - they helped to exclude the terms which do not belong to geography, such as general scientific words, interdisciplinary words, terms belonging to other scientific disciplines. Thus there is no place in conceptual lists for such words as *refraction,* for example, since it refers to physical process and belongs to absolutely different system of oppositions of light effects *refraction – reflection – diffraction – interference,* while the term *wave refraction* (*The process by which water wave crests change direction as they approach a shoreline, owing to the shallowing of water*) belongs to oceanography and finds its place in the system of geographical terms together with the other terms describing coastal processes.

This example shows that along with occupying its place in the conceptual list it is very important for a term to pass the so-called "definability test" , i.e. its definition should refer the term to geography. There is such a physical term *diffusion* (*The process of slow mixing of liquids, gases, or solids which intermingle as the result of molecular interpenetration*). It must not be included into our dictionary, because it belongs to the system of concepts of physics. But at the same time there exist the term *diffusion of innovations* (*The widespread dispersal, the dissemination of cultural, social, economic and other phenomena, innovation, idea, and information from the sources where they initially concentrated*) which has geographical meaning and can be included into our Dictionary.

In order to draw the demarcation line between the terms of geography and those of other sciences we had to treat them not as separate words, but as lexical representatives of underlying concepts.

Similarly the conceptual word-lists made it possible to separate geographical terms from "words of general language". For example, *depression* is a term having two meanings relevant for geography: *1*) *Any hollow or relatively sunken ares, especially one enclosed by higher area; 2*) *Region of the atmosphere where the atmospheric pressure is lower than that of its surroundings, i.e. low pressure system* (*cyclone*). We did not include in our dictionary the nominative meaning of the word *depression* which is used in the general language (*A feeling of being extremely unhappy, a great sadness and disappointment*) but included the economic meaning *depression* (*economic recession*): *A slow-down or sudden decline of economic activity accompanied by reduced production that leads to economic stagnation.*

The word *accident* (*A sudden breakdown of a technological process or other economic activity resulting in material losses and often environmental damage*) is a word of general language while the term *accident emission* (*A release of pollutants into the environment as a result of an accident or breakdown in a technological process*) is a geographical term pertaining to geoecology and forming part of the whole system.

The word *absolute* has no specific geographical meaning, being the word of the general language but such terms as *absolute humidity* is definitely the meteorological term being in opposition to *relative humidity* and deserves to be included in the Dictionary. The word *monitoring* is widely used in many spheres of modern life having the meaning (*regular checking or examining of the progress of something*) is not included in the Dictionary, while *environmental monitoring* is a full-fledged compound geographical term consisting of two words.

Besides, it is very important to distinguish between a compound term and free word-combination with a terminological word. Let's take examples to clarify the point. The terminological adjective *autochtonous* is applied to *material and processes which originated in its present position, and is not the product of transportation being brought in by outside influences* (*allochtonous*). It is obvious that *autochtonous* and *allochtonous* are both geographical terms in its own right, while the units like *autochtonous substance, autochtonous sediments, autochtonous rock, autochtonous process, etc.* and respectively, *allochtonous substance, allochtonous rock, allochtonous sediments, allochtonous process* must be regarded as free word-combinations denoting in scientific discourse a multitude of specific objects and processes of *autochtonous* and *allochtonous* character. Such lexical units are not included in the Dictionary because their meaning is formed by a simple sum of the meanings of its constituent words.

On the contrary, *autochtonous river* (*A river the basin of which is located completely in one and the same geographical region and the stream-flow regime reveals the local nature peculiarities*) must be included into the dictionary as a compound terms proper, characterized by the globality of nomination since the whole multiword unit is not only the sum of the meanings of separate words that form part of it but it has its individual global meaning.

On the other hand, lexical units are so intricate, that sometimes being absolutely the same in form they may be different in content. Thus, such units as *geographical education, biological education, mathematical education* are free word-combinations meaning *"professional training (instruction) acquired by a learner in this or that science or technology"*. They cannot be included in the terminological geographical dictionary as headwords because they are specifications of the general word *education*. But in connection to the unit *ecological education* we cannot say that it is only professional training or instruction in the field of ecology. The lexical unit *ecological education* must be regarded as primary term because the individual words forming part of it are closely tied together having global idiomatic meaning (*Public education and cultivating in people the respect and care of nature and delicate use and protection of the environment*).

The classic examples of such terms having global meaning independent of the meanings of the constituent parts are *diamond dust* (*The precipitation of fine columnar ice crystals or their fragments under the weather conditions of cloudless sky or high clouds*) and *water pocket* (*A film of water enveloping air bubbles in temperate glaciers*). Both word-combinations are geographical terms. They have nothing to do with a *diamond* or *pocket* and their globality of nomination is obvious. However, a lexicographer encounters in its work with many less obvious cases and has to make decision for each controversial unit whether to include it into the dictionary or not.

On the basis of conceptual word-lists we tried to produce classificatory definitions in order to show the hierarchy and factual relations between the notions. Each term included into the dictionary is supplied with a concise scientific definition. We did not try to write encyclopedic definitions which usually contain comprehensive information and give explanations in sufficient detail. Our definitions give a short information of the most characteristic features of a given object or phenomenon which would help a user of the dictionary to understand the essence of the notion and if necessary to find a more exact equivalent term in his own language (especially when his mother tongue is not English). Each definition contains a succinct description of the object or phenomenon and in some cases explains the factors which cause it. The definitions are based on systemic relationships of scientific concepts which in its turn reflect in underlying reality. They are written in a clear, simple, easy-to-understand style. We have avoided to explain terms by using other terms or highly technical words but where their use is unavoidable we have ensured that they are clearly defined at their own entry in the Dictionary.

It is very important that the definitions are formulated from the general geographic point of view, rather than from the viewpoint of a particular subdiscipline. That is why, for example, the definitions of biogeographical terms are not overloaded with biological details as it should be for the biogeographical dictionary or the terms of geomorphology are deprived of geological details, they are all written in a general geographic "foreshortening".

Those classificatory definitions helped to solve the problem of synonyms. When definitions did not allow to semantically separating the terms they were regarded as synonyms. In general language synonyms are different words which mean one and the same thing, i.e. they are close in their main semantic content but differ in the shades of meaning or stylistic connotations. In terminology people are interested in the equivalence and substitutability, in how far the definition will be applicable not only to term *A*, but also to its synonyms. The words which coincide completely in their meanings are called «doublets» (triplets, etc.). The difficulty of the task here is to decide whether it is enough grounds to regard two given word as individual terms or it is better to find the common core of meaning and regard them as synonyms. For example: *pollution* = *contamination*; *maximum permissible emission* = *threshold emission*; *environmental impact* = *environmental load* = *environmental pressure* = *anthropogenic impact.*

Although the semantic structure of the different terms may be different, and they may be different words, but if their definitions are similar or close, i.e. they denote the same "object" they are regarded terminological synonyms and are described in one entry. We regard *atmospheric pollution* and *air pollution* as synonyms, although there could be found a slight difference between them. Then we understood that the semantic difference between them is very slight and it would be only right not to present them separately in their individual entries.

The problem of synonyms in terminology is complicated by the fact that in real scientific texts people use not only strict terms but very often substitute the terms by the so-called "free descriptive equivalents of terms" which are speech variants of the term such as for example in Spanish it is used the word *ambiente* instead of the term *medio ambi-*

ente (environment); in English the multitude of words and phrases is used to denote *maximum permissible environmental impact – critical environmental load, threshold impact.*

Of course if there is such a concept in geoecology as "human influence on the environment" which can be somehow measured and is being investigated by geographers (then there must be an appropriate terminological appellation) a strict term. In scientific texts we come across many descriptive fragments of speech of different length to denote one and the same thing. It is the task of the compilers of the dictionary to reduce them to one or two terms which are generally accepted and have an optimal linguistic form (compact, consisting of one or two words) to be included into the dictionary. In the example given above those basic and optimal terms are *environmental impact.*

Here it is very helpful to draw parallels between language in general and terminology. No natural language can be uniform since lexical units have territorial and social varieties. Nevertheless it is always possible to ignore some unimportant details in order to concentrate on the core features. This is the objective basis of the existence of any language although each language exists in different styles and dialects. And this is especially true of terminology.

Fourteen branches of geography in the Dictionary

According to the history of development and structure of the geographical science and the peculiarities of the terminological systems of its branches we have distinguished 14 geographical disciplines: General Geography (324 terms), Physical Geography (167), Palaeogeography (146), Geomorphology (1039), Soil Geography (349), Biogeography (545), Meteorology and Climatology (702), Hydroloogy (674), Glaciology (688), Geocryology (133), Oceanology (428), Geodesy, Cartography and Geoinformatics (374), Geoecology (343), Human geography (959). The total number of terms described in the Dictionary is 6871. To this number should be added about 2400 main synonyms for each language which form a part of the basic table. Thus, the number of terms defined in the Dictionary for each language is 9300, the total number of terms in five languages being about 46000.

General Geography includes terms denoting the basic concepts of geographical science as it is understood in its classical sense from the ancient times. In its core being the concepts of the Earth globe and its relations with the celestial sphere, cardinal points and the sphericity of the Earth, the rotation of the Earth around the axis and the movement of the planet Earth on its orbit, the related change of day and night, the characteristics of gravity and magnetism. The terms denoting the main methods of geographical investigations, the main geographical trends such as geographical determinism, geographical possibilism, catastrophism, etc. have been included. The concepts about the environment, geographical boundaries, spatial interactions have also been explained.

Physical Geography includes the terms related to natural complexes, natural zones, vertical zonality. Special attention is given to the types of landscapes, their structure and functioning.

In a relatively small section called **Palaeogeography** such terms as ice age, glacial and interglacial epochs, stadial and interstadial, methods of the investigation of the history of the environment, previous geological periods, natural events in the Pleistocene and the stages of the development of humans have been defined. The central place here occupy the notions related to the change of ice ages which have a detailed terminology which differs in many languages. In order to correlate them we made a table showing the events in the Pleistocene represented by the terms used in Western Europe, Central Russia and North America (see Table 2 on pages 350-351).

The biggest thematic group of terms in the Dictionary is **Geomorphology**. It includes terms denoting different kinds of rocks and sediments, basic types of tectonic and volcanic processes, earthquakes, folding processes and elements of rock bedding. Also rock formation processes, types of relief, mountainous and plain landforms, such as river valleys, eolic, coastal and karst forms and processes have been described. The multiplicity of geomorphological terms can be explained by the fact that the landforms are the outlines of the Earth's surface familiar to people since long ago and are very well seen. At the same time many words denoting too small details of the landforms and their geological features have been excluded.

A very important although not very numerous is the group of **Soil Geography** terms. The terms here are related to soil fertility and agriculture. The terms of different properties of soils, soil formation processes, the structure of soil profile and the role of soils in agriculture have been included. The main characteristic feature of soil geography is the words denoting various genetic and natural types of soils. Out of a large number of these terms we have selected for the Dictionary only those types and classes of soils which are dominant in this or that geographical zone or typical for the zonal landscapes.

A very specific group is the terms of **Biogeography** which contains the names of flora and fauna typical of this or that geographical conditions. In the short definitions of these terms the natural conditions in which a living thing usually dwells have been explained. There are also terms denoting various types of ecosystems, biogeographical processes, productivity and succession of ecosystems. Terrestrial and aquatic types of flora and fauna and types of vegetation, especially the types of forest and grassland have been defined.

Meteorology and Climatology are very ancient geographical disciplines having a very well developed and numerous terminology involving the elements of the atmosphere, characteristics of temperature and humidity, insolation, input and output of solar energy, physical state and processes in the atmosphere, its circulation, cyclones, anticyclones, clouds, precipitations, and meteorological phenomena, winds. Many terms designate types of climate, characteristic features of climates, its forecast and modelling. Here some terms of agrometeorology have also been included.

The thematic group **Hydrology** contains an extensive terminology because water serves as a basis of life on the Earth and the main agent of weathering and erosion being the main factor of the shaping of land surface. Over more than 300 years of the development of hydrology as a geographical discipline a huge group of terms denoting various types of rivers, lakes, swamps and other bodies of water have appeared. The names of the types of rivers are related to different conditions of climate and relief in which they run. Similarly, the types of lakes are conditioned by the physical, especially climatic conditions of the environment where the lakes originated and exist. The terms of the characteristics of water and its properties, water processes, hydrological regime of the territory, drainage, runoff, water streams, loads and sediments, morphology of water objects, river and drainage system; types, structure and stratification of lakes, channel processes, ground waters as well as the terms on hydrochemistry, hydrobiology, hydroenergetics, and the use of water in agriculture have been defined.

An extremely numerous group of terms is **Glaciology.** It includes terms denoting many natural processes, related to snow and ice objects such as snow cover, avalanches, marine ice, mountain glaciers, ice sheets. There are terms of the elements of geomorphology of ice and snow cover – glacial landforms and elements of ice morphology; of climatology – characteristics of climatic conditions in the polar regions; of meteorology – atmospheric ice; of hydrology – snowmelt, snow and glacial runoff, river and lake ice cover; of oceanography – ice regime of seas and marine ice; of palaeogeography – glaciation of the past epochs, their spreading and degradation. The terms of such natural dynamic phenomena as snow drifting, mudflows, aufeises, as well as glacial sediments, deposits and seasonal behaviour of glaciers have also been defined.

There are two reasons to explain why the group of terms of **Geocryology** is not very numerous. The main subject of study of geocryology (permafrost) is not very widely spread on the Earth's surface, it is found in a relatively limited area. Besides, the scientific investigation of permafrost began quite recently mainly in Russia and Canada. The terms of different phenomena in permafrost, ground ice, permanently frozen grounds, permafrost landforms (frost mounds, polygonal tundra) have been explained. Some of the terms of geocryology, such as pingo, taryn, baidzherakh, etc. derive from the of local languages of peoples inhabiting the regions of permafrost.

A specific place is occupied in the geographical terminosystem by **Oceanography,** because oceanography has diverged from the classical geography but still remaining the member of the family of geographical disciplines. It includes the terms of ocean dynamics, sea geomorphology, physical processes in sea water, waves, currents, tides, coastal processes, marine sediments, etc.

Another thematic group unites the terms of **Geodesy, Cartography and Geoinformatics**. It includes the terms of different types of maps, cartographic projections, properties and elements of geographical maps, as well as methods of making digital maps and geoinformation systems.

A very important role in the terminological system of modern geography plays **Geoecology.** The inclusion of the group of geoecological terms in the Dictionary has been accompanied with a number of difficulties. It was necessary to draw the line between the numerous terms of ecology proper which function today practically in all the branches of natural and social sciences. The criterion was the relation of the terms of geoecology directly with the geosphere and its possible degradation as a result of human impact. This group of terms includes words related to the state of the environment, conservation of nature, resources use, environmental management, as well as term of medical geography, recreational geography, tourism, pollution of the environment and waste disposal, the changes of the environment caused by human impact on nature, and the legislation of human-environment interaction.

A very huge thematic group of terms is **Human Geography.** The boundaries of this geographical subject with economics, sociology, demography, geopolitics, and engineering are not sharp. Human geography operate some of the terms of these sciences but many of the terms form part of geography proper, e.g. humanity, civilization, state, colony, metropolis, as well as the terms of political geography, geography of population, energetics, resources, geography of world economy, industry, agriculture, communication, transport and trade.

One of the characteristic features of the geographical terminology is the presence of local terms, which exist practically in all the geographical disciplines. Many of these local terms are now very widespread. Especially numerous they are in geomorphology (local landforms, such as bolson, hammada, sebkha, etc.), meteorology (names of local winds, such as chinook, mistral, sirocco, etc.), hydrology (creek, loch, pororoca, etc.), physical geography (cerrado, dasht, shott, etc.). In order to include these terms in the final word list we had to analyze their usage and the extent to which they are spread. As a result of this study only about 200 terms of this kind have been selected (mainly those which are widely used out of the locality where they originated) to be included in the main word-list of the Dictionary.

We are fully aware of the fact that geography, in addition to its own diversity, is dependent on the flow of concepts and techniques from other fundamental disciplines such as physics, chemistry, biology, mathematics as well as more specialized areas as medicine, engineering, informatics. That is why a lot of terms of those sciences are often used in geographical texts. But it would be wrong and impossible to include all of them into our dictionary, however useful they may be for geographical discourse and however frequently they are used in geographical texts, because it would destroy the unity of the subject and overload the dictionary.

Conclusion

In compiling the Dictionary we proceed from the assumption that the terms only embody the classification of geographical facts, their hierarchy and oppositions, thus forming a close-knit system called terminosystem. Some authors of the dictionaries think that the task of a terminological dictionary is descriptive, not prescriptive: it records words and expressions that are in current use and explains the meanings attached to them, but in no sense it is intended to be a textbook in its own right. Our position is different. By including or excluding this or that term we show our understanding of the subject, we express our position with regard to the scope of Geography. In this sense we do not regard ourselves as simply recorders, who express no opinions. We definitely express our opinion by selecting for the Dictionary the concepts, and therefore the terms, we believe, pertaining to the sphere of modern geography.

This terminological dictionary has been compiled by making preliminary conceptual lists of headwords which made it possible: 1) to single out the central, core vocabulary which corresponds to central concepts by selecting for the Dictionary the terms belonging only to a certain "scale" or hirachical level by excluding very detailed concepts or too small elements of geographical objects or phenomena; 2) to give to the terms scientifically objective classificatory definitions which establish the relation between generic terms and specific terms; 3) to solve the problem of synonyms and decide whether a term has the same meaning and can be qualified as a synonym and should be included in the dictionary or one term is preferable to the other and should be excluded from the word-list; 4) to exclude terms which do not belong to geography, because there is no place for this or that term in the conceptual system; 5) to separate terms from "words of general language" since it is impossible to find a place for the general word in the terminological system and it has no any specifically geographical meaning.

We designed our terminological dictionary not as a mere collection of terms but as a coherent body of geographical thought, though presented in the form of isolated entries – terms. At the same time we understand that it is not possible to eliminate the subjective element from dictionaries and no grouping of words can be ideal or the only possible. We agree with the opinion that lexicography is about "exercise of judgment". In order the Dictionary be what we designed it – to be compact, with wide coverage of subjects, integrated, and coordinating terms in 5 languages – we had to make hard but well-founded choices based on the strict lexicographic principles and our knowledge of geography.

The Dictionary is intended for a wide variety of users: professional geographers, under-graduate and post-graduate students and teachers of colleges and universities, nature-lovers and travelers. The Dictionary aims to be useful to those who are interested in environmental problems or deal with the world resources and their use. It should serve the readers to see geography as an integrated academic discipline with a wide scope of its modern interests and directions. The coordination between the geographical terminologies of geography in five languages is the call of the day caused by the ever growing contacts between geographers of different countries, their joint research and international projects. And by our Dictionary we try to contribute to the improvement of international understanding and cooperation in the sphere of geographical science and related disciplines.

We have been working on the Dictionary for more than five years since the year 2000. We also know that terminology is always one jump ahead of lexicography. And since always there is room for improvement, constructive criticism and suggestions will be of great value. We'll be very grateful if users send their comments, corrections, and amendment to the authors by e-mail:

direct@igras.geonet.ru

komarova@ffl.msu.ru

USED SOURCES

In English, French, Spanish and German

A Glossary of Geographical Terms / Ed. Stamp L.D. New York, John Wiley and sons, 1962, 539 p.

Baions Liris, 2000, 190 p.

Clark A.N. The Penguin Dictionary of Geography. Third edition. Penguin Books, 2003. 468 p.

Clark A.N. Longman Dictionary of Geography: human and physical Longman, 1985, 724 p.

Das Gesicht der Erde. Leipzig, 1956, 627 S.

Diccionario de Geografia. Madrid, ANAYA, 1986, 480 p.

Diccionario de Geografia. Bajo la direccion de Pierre George. Madrid, Akal, 1991, 622 p.

Diccionario Terminologico de Biologia. La Habana, Editorial Cientifico-Tecnica, 1982, 262 p.

DIERCKE-Wörterbuch Allgemeine Geographie. München, 2001, 1037 S.

George P., Verger F. Dictionnaire de la geographie. 7e edition. Paris, Presses universitaires de France, 2000, 501 p.

Gregory D., Smith D.M. The Dictionary of Human Geography. 2nd edition. Oxford, Dlackwell Reference, 1986, 576 p.

Gudelis V. Juros krantotyros terminu (odinas. Vilnius, «Academia», 1993, 408 p.

Huschke R.E. Glossary of Meteorology. American Meteorological Society, 1980, 638 p.

International meteorological vocabulary. Second edition. WMO, No. 182. Secretariat of the World Meteorological Organization, Geneva, Switzerland, 1992, 784 p.

Kotlyakov V.M., Smolyarova N.A. Elsevier's Dictionary of Glaciology. Elsevier, Amsterdam–Oxford–New York–Tokyo, 1990, 336 p.

Lozet L., Mathieu C. Dictionnaire de science du sol. Technique et Documentation–Lavoisier, 1990, 398 p.

Mathew S. A Dictionary of Geography. 2nd edition. Oxford, New York, Oxford University Press, 1997, 460 p.

Multi-language glossary of permafrost and related ground-ice terms (in 12 languages). Compiled and edited by R.O. van Everdingen. Calgary, 1998.

Multilingual dictionary of technical terms in cartography. Wiesbadem/Germany, Franz Steiner Verlag GMBH, 1973, 573 p.

Permafrost terminology. Biuletyn Peryglacjalny, ' 32. Wroc(aw, Warszawa, Krakow, 1992, 176 p.

Schreiner J. Praxis-Wörterbuch Umwelt, Naturschutz und Landnutzungen. Wissenschaftliche Verlagsgesellschaft mbH, Stuttgart, 2004, 556 S.

Small J., Witherick M. A Modern Dictionary of Geography. London, Edward Arnold, 1986, 233 p.

The Encyclopedic Dictionary of Physical Geography / Ed. Goudie A. Blackwell Reference, 1985, 528 p.

Whittow J. The Penguin Dictionary of Physical Geography. 2nd edition. Penguin Books, 1999, 590 p.

In Russian

Алаев Э.Б. Социально-экономическая география. Понятийно-терминологический словарь. М., 1983, 350 с.

Англо-русский гидрологический словарь. М., «Советская энциклопедия», 1966, 299 с.

Англо-русский океанографический словарь. М., Госуд. изд-во технико-теоретич. литературы, 1957, 292 с.

Быков Б.А. Геоботанический словарь, 2-е изд. Алма-Ата, «Наука»,1973, 214 с.

Быков Б.А. Экологический словарь. Алма-Ата, «Наука», 1988, 245 с.

Викторов Д.П. Краткий словарь ботанических терминов. М.–Л., «Наука», 1964, 177 с.

Географический энциклопедический словарь. Понятия и термины. М., «Советская Энциклопедия», 1988, 432 с

Геоинформатика. Толковый словарь основных терминов / Под ред. А.М. Берлянта и А.В. Кошкарева. М., 1999, 203 с.

Гляциологический словарь / Под ред. В.М. Котлякова. Л., Гидрометеоиздат, 1984, 528 с.

Гребенщиков О.С. Геоботанический словарь. М., «Наука», 1963, 227 с.

Дедю И.И. Экологический энциклопедический словарь. Кишинев, 1989, 406 с.

Зайцева Н.А., Раввина К.Б. Англо-русский метеорологический словарь. Санкт-Петербург, Гидрометеоиздат, 2003, 515 с.

Зенин А.А., Белоусова Н.В. Гидрохимический словарь. Л., Гидрометеоиздат, 1988, 239 с.

Клубов С.В., Прозоровский Л.Л. Геоэкология: Русско-английский понятийно-терминологический словарь. М., 1994, 104 с.

Конищев В.Н., Розенбаум Г.Э., Романовский Н.Н. Русско-английский словарь по геокриологии и смежным наукам. М., Изд-во МГУ, 1999, 102 с.

Красильников П.В. Почвенная номенклатура и корреляция. Петрозаводск, 1999, 435 с.

Краткий топографо-геодезический словарь / Под ред. Б.С. Кузьмина. М., «Недра», 1979, 312 с.

Лобжанидзе А.А., Манусов В.М. Эколого-экономический словарь-справочник. М., 1990, 252 с.

Морская геоморфология. Терминологический справочник. Береговая зона: процессы, понятия, определения / Под ред. В.П. Зенковича и Б.А. Панова. М., «Мысль», 1980, 280 с.

Немецко-русский океанографический словарь. М., Госуд. изд-во технико-теоретич. литературы, 1957, 240 с.

Океанографическая энциклопедия. Пер. с англ. Л., Гидрометеоиздат, 1974, 631 с.

Охрана ландшафтов. Толковый словарь, М., «Прогресс», 1982;272 с.

Пармузин Ю.П., Карпов Г.В. Словарь по физической географии. М., «Просвещение», 1994, 367 с.

Перелет Р. Экономика и окружающая среда. Англо-русский словарь-справочник. Гарвардский ин-т международного развития, 1996, 120 с.

Прох Л.З. Словарь ветров. Л., Гидрометеоиздат, 1983, 312 с.

Словарь общегеографических терминов / Под ред. Л.Д. Стампа. Пер. с англ., т. 1-2, М., 1975-1976.

Соловьев А.И., Карпов Г.В. Словарь-справочник по физической географии. М., 1983, 224 с.

Тимашев И.Е. Геоэкологический русско-английский словарь-справочник. М., Издат. Дом «Муравей–Гайд», 1999, 168 с.

Тимофеев Д.А. Терминология аридного и эолового рельефообразования. М., «Наука», 1980, 164 с.

Тимофеев Д.А. Терминология поверхностей выравнивания. М., «Наука», 1974; 87 с.

Тимофеев Д.А. Терминология денудации и склонов. М., «Наука», 1978; 242 с.

Тимофеев Д.А. Терминология флювиальной геоморфологии. М., «Наука», 1981;268 с.

Тимофеев Д.А., Втюрина Е.А. Терминология перигляциальной геоморфологии. М., «Наука», 1983; 233 с.

Тимофеев Д.А., Дублянский В.Н., Кикнадзе Т.З. Терминология карста. М., «Наука», 1991, 260 с.

Тимофеев Д.А. Маккавеев А.Н. Терминология гляциальной геоморфологии. М., «Наука», 1986, 256 с.

Тимофеев Д.А., Уфимцев Г.Ф., Онухов Ф.С. Терминология общей геоморфологии. М., «Наука», 1977; 200 с.

Толковый словарь по почвоведению. М., «Наука», 1975, 286 с.

Уфимцев Г.Ф., Онухов Ф.С., Тимофеев Д.А. Терминология структурной геоморфологии и неотектоники. М., «Наука», 1979, 253 с.

Французско-русский геологический словарь. М., Изд-во физ.-мат. литературы, 1958, 406 с.

Хромов С.П., Мамонтова Л.И. Метеорологический словарь, 3-е изд. Л., Гидрометеоиздат, 1974; 568 с.

Чеботарев А.И. Гидрологический словарь, 3-е изд. Л., Гидрометеоиздат, 1978, 308 с.

Четырехъязычный энциклопедический словарь терминов по физической географии / Под ред. И.С. Щукина. М., «Советская энциклопедия», 1980, 703 с.

Экологический энциклопедический словарь. М., Издательский дом «Ноосфера», 1999, 930 с.

The following sources were used to prepare illustrations for the Dictionary:

materials of V.M. Kotlyakov – Figures 12, 28, 85, 127, 132, 157, 158, 160, 163, 172, 198, 214, 218, 297, 298, 330, 347;
Clark, 2003 – Figures 33, 35, 46, 54, 83, 95, 109, 111, 113, 124, 210, 224, 225, 228, 243, 251, 259, 314, 326;
Das Gesicht der Erde, 1956 – Figures 13, 60;
Diercke, 2001 – Figures 7, 22, 34, 47, 48, 56, 68, 87, 88, 98, 116, 121, 122, 174-177, 182, 189, 197, 202, 212, 213, 235, 262, 267, 274, 276, 291, 292, 308, 313, 327, 328, 338, 340, 341, 368;
Mathew, 1997 – Figures 38, 40, 72, 94, 215, 270, 317, 359;
Whittow, 1999 – Figures 3, 8, 41, 50, 53, 58, 65, 91, 104, 129, 133, 136, 140, 142, 143, 150, 164, 167, 168, 170, 171, 173, 178, 186-188, 192, 201, 203, 207, 208, 211, 226, 229, 232, 239, 242, 250, 253, 261, 280, 282, 290, 303, 311, 329, 333, 335-337, 342;
Большая Советская Энциклопедия, 3-е издание – Figures 30, 49, 61, 75, 79, 86, 102, 205, 206, 233, 256, 263;
Географический энциклопедический словарь, 1988 – Figures 1, 11, 14, 17, 23, 29, 52, 69, 78, 114, 117, 119, 149, 155, 193, 209, 279, 294;
Дедю, 1989 – Figure 321.
Краткий топографо-геодезический словарь, 1979 – Figures 67, 284;
Пармузин, Карпов, 1994 – Figures 15, 141, 169, 306, 331;
Прох, 1983 – Figures 44, 118, 216, 307, 361;
Соловьев, Карпов, 1983 – Figures 63, 126, 153, 207, 236, 238, 273, 277, 287, 289, 348, 351, 353, 355, 364;
Хромов, Мамонтова, 1974 – Figure 144;
Экологический энциклопедический словарь, 1999 – Figures 42, 55, 110, 154, 278.

Photographies published are from archives:

V.M. Kotlyakov – Figures 2, 12, 20, 24, 26, 32, 39, 62, 72, 73, 94, 97, 100, 106, 109, 125, 130, 145, 146, 148, 156, 161, 162, 176, 183, 184, 186, 194, 200, 204, 222, 230, 231, 237, 240, 245, 249, 264, 272, 280, 206, 288, 295, 315, 322, 323, 325, 339, 357, 365, 367;
S.A. Bulanov – Figures 31, 45, 74, 76, 77, 82, 96, 98, 103, 105, 119, 138, 147, 179, 180, 195, 196, 234, 241, 252, 254, 284, 304, 309, 310, 316, 319, 320, 343, 344, 349, 360, 362;
L.V. Desinov – Figures 6, 37, 43, 80, 81, 101, 107, 107, 123, 151, 159, 166, 191, 220, 221, 223, 227, 271, 281, 283, 293, 298, 299, 324, 331, 356;
space images from the Russian segment of the International Orbital Station – Figures 4, 19, 23, 59, 64, 66, 78, 99, 128, 185, 217, 219, 260, 266, 312, 346, 350, 352, 254.

Acknowledgements

The vocabulary for the Dictionary, compiled by authors, have been reviewed by some specialists. We thank very much all colleagues who checked the vocabularies for different disciplines and gave their remarks and additions: Kirill Dyakonov (general and physical geography), Andrei Velichko (palaeogeography), Sergei Bulanov, Bruno Messerli (geomorphology), Sergei Goryachkin (soil geography), Arkady Tishkov (biogeography), Vera Moiseeva (soil geography and biogeography), Andrei Shmakin (climatology), Nikolai Koronkevich (hydrology), Andrei Glazovskiy (glaciology), Nina Zaitseva (meteorology, climatology, glaciology and oceanography), Roger Barry, Vyacheslav Konishev (geocryology), Yuri Dolotov, Viktor Neiman (oceanography), Alexander Berlyandt, Nikolai Komedchikov, Alexander Khropov (cartography), Nikita Glazovskiy, Kim Losev (geoecology), Yuly Lipets, Vladimir Streletskiy, Alexander Gorkin, Vladimir Kolosov (human geography).

HOW TO USE THE DICTIONARY

1. The English terms in the basic table of the Dictionary appear in alphabetic order; in the compound terms (consisting of two or more words), the order of words is not inverted, but direct e.g. "ocean deep", not "deep, ocean".

2. The noun headwords are given in the singular, only some terms according to tradition are given in the plural.

3. Each term has its number; synonyms are referred to the appropriate main term.

4. The entries contain definitions in which some words are printed in italics. This points the reader to cross-references.

5. The reference to the illustration has the symbol: → Fig. with the indication of the page, if an illustration placed not on the same page.

6. When meanings are very different, they are shown as separate senses with numbers 1, 2, 3; when meanings are closely related, their definitions are separated by a semi-colon.

7. The terms have grammar labels when it is adjective (*adj*), adverb (*adv*), gender of nouns (*m*, *f*, *n*) and plural number of nouns (*mpl*, *npl*, *fpl*).

8. In the Indexes in Russian, French, German, Spanish each term (including synonyms) has the indication to the number of the appropriate headword in the basic table.

9. Some headwords in the basic table and in the indices have small numbers at the top end or at the bottom end; the former indicates that these are homonyms – different words although spelt the same, the latter indicates the sense number when the headword has more than one meaning defined as separate senses in the same entry.

Basic Table

A

1 **aa-lava**
A rough scoriaceous lava flow.
r аа-лава
Застывшая базальтовая лава с шероховатой поверхностью.
f aa-lave *f*
s aa-lava *f*
g Aa-Lava *f*

2 **aberration of light**
Apparent displacement of a celestial body due to the effect of annual motion of the Earth round the Sun (annual aberration) and its daily rotation round its axis (daily aberration) upon the results of measurements.
r аберрация света
Искажение видимого положения небесного светила из-за того, что на результаты измерений оказывает влияние годичное движение Земли вокруг Солнца (годичная аберрация) и суточное вращение ее вокруг своей оси (суточная аберрация).
f aberration *f* de la lumière
s aberración *f* de luz
g Lichtaberration *f*

3 **abiotic environment; non-living environment**
A complex of non-living conditions and elements of the environment including the atmosphere, hydrosphere and lithosphere.
r абиотическая среда
Совокупность неживых условий обитания организмов и элементов окружающей среды, включая атмосферу, гидросферу и литосферу.
f environnement *m* abiotique
s medio *m* ambiente abiótico
g abiotische Umwelt *f*; unbelebte Umwelt *f*

4 **ablation**
The wastage or removal of snow and ice by such processes as sublimation, evaporation, melting including the losses by wind action and breaking-off of icebergs.
r абляция
Уменьшение массы снега и льда за счет таяния, испарения, сдувания снега ветром, обвалов льда и откалывания айсбергов.
f ablation *f*; ablation *f* glaciare
s ablación *f*; ablación *f* glaciar
g Ablation *f*; Abtrag *m*

5 **ablation area; ablation zone**
The portion of a glacier below the equilibrium line where the ablation exceeds accumulation (the opposite of accumulation area).
r область абляции ледника
Нижняя часть ледника, где в течение балансового года масса ледника уменьшается вследствие таяния и испарения снега и льда.
f zone *f* d'ablation
s zona *f* de ablación
g Ablationsgebiet *n*; Zehrgebiet *n*

6 **ablation microrelief**
Topography of a glacier surface formed by uneven melting of ice.
r абляционный микрорельеф
Рельеф поверхности ледника, формирующийся в результате неравномерного таяния льда.
f microrelief *m* d'ablation
s microrelieve *m* de ablación
g Ablationsmikrorelief *n*

7 **ablation moraine**
A type of *deposited moraine* resulted from melting-down of surface and englacial moraines of glaciers.
r абляционная морена
Отложенная морена, формирующаяся в результате вытаивания поверхностной и внутренней морены ледников.
f moraine *f* d'ablation
s morrena *f* de ablación
g Ablationsmoräne *f*

8 **ablation period**
The period of a balance year when ablation of the glacier exceeds its accumulation.
r период абляции
Часть балансового года, в течение которой расход вещества с ледника превышает его приход.
f période *f* d'ablation
s período *m* de ablación
g Ablationsperiode *f*

* **ablation zone → 5**

* **aboriginal → 4045**

* **aborigine → 4045**

9 **abrasion**
The process of wearing down or wearing away of rocks by the hard particles carred by rivers, glaciers, wind and waves.
r абразия
Разрушение твердых горных пород под воздействием скальных частиц, переносимых текущей водой, ледниками, ветром или волнением на море (озере).
f abrasion *f*
s abrasión *f*
g Abrasion *f*; Abrieb *m*; Abschliff *m*

10 **abrasion arch**
A narrow rocky projection on the sea coast in which waves have cut an opening.
r абразионная арка
Узкий скалистый выступ берега, в котором прибоем выбито сквозное отверстие.

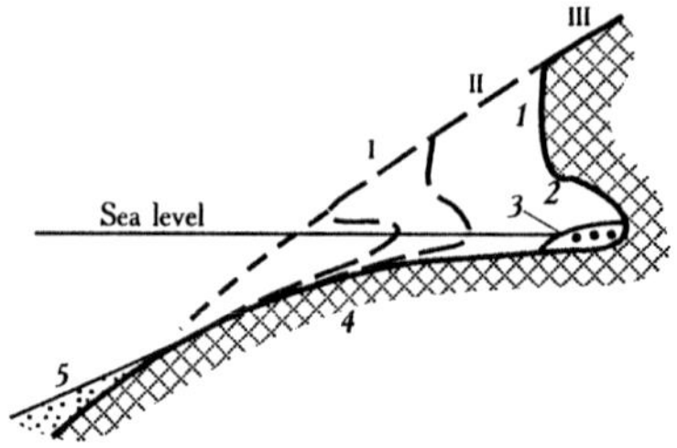

Fig. 1. **Abrasion shore**: I, II, III – stages of the shore retreat, 1 – cliff, 2 – notch, 3 – beach, 4 – bench, 5 – abrasion platform

f arc *m* d'abrasion
s arco *m* de abrasión
g Brandungsgewölbe *n*

11 abrasion platform; strandflat; wave-cut platform; shore platform
A gently sloping marine or lake platform cut by waves at a shoreline.
r абразионная платформа; подводная терраса; стрэндфлет
Полого наклоненная, прибрежная часть поверхности дна озера, моря или океана, выработанная абразией.
f plate-forme *f* d'abrasion; strandflat *m*; terrasse *f* sous-marine
s plataforma *f* de abrasión; terraza *f* submarina; strandflat *m*; plataforma *f* litoral
g Brandungsplatte *f*; submarine Terrasse *f*; Strandplatte *f*

12 abrasion shore
A high steep sea or lake shore cut by breaking waves (swash). → Fig. 1
r абразионный берег
Высокий крутой берег моря или озера, разрушаемый действием прибоя. → Рис. 1
f littoral *m* d'abrasion
s litoral *m* de abrasión; costa *f* de abrasión
g Abrasionsküste *f*; Steilküste *f*

13 abrasion terrace; water-cut bench; marine-built terrace; water-built terrace
A marine erosion plane formed at the base of a sea cliff, sloping down towards the sea, in some cases merging into a submarine terrace.
r абразионная терраса; волноприбойная терраса
Ровная площадка у основания морского клифа, наклоненная в сторону моря, в некоторых случаях незаметно переходящая в подводную террасу.
f terrasse *f* d'abrasion; gradin *m* d'abrasion
s terraza *f* de abrasión; grada *f* de abrasión
g Wellenstufe *f*; Schorre *f*

* **abrazem → 4985**

14 absolute age; radiometric age
The age of rocks expressed in units of astronomical time (years, thousands and millions of years) identified on the basis of radioactive decay of some chemical elements.
r абсолютный возраст; радиометрический возраст
Возраст горных пород в единицах астрономического времени (годы, тысячи и миллионы лет), устанавливаемый на основе радиоактивного распада некоторых химических элементов.
f âge *m* absolu; âge *m* radiométrique
s edad *f* absoluta; edad *f* radiométrica
g absolutes Alter *n*; radiometrisches Alter *n*

* **absolute elevation → 227**

15 absolute extremes
The maximum or minimum value of some meteorological or hydrological element observed in a long-term data series.
r абсолютные экстремумы
Наибольшее или наименьшее из наблюдавшихся в течение длительного периода времени значение какого-либо метеорологического или гидрологического элемента.
f extrêmes *mpl* absolus
s extremos *mpl* absolutos
g absolute Extremwerte *mpl*

16 absolute geochronology; radiometric dating
The dating of absolute age of rocks by using the method of radioactive decay of some chemical elements going on with a constant rate.
r абсолютная геохронология; радиохронология; изотопное датирование
Установление абсолютного возраста горных пород с использованием закономерностей радиоактивного распада некоторых химических элементов, идущего с постоянной скоростью
f géochronologie *f* absolue; datation *f* au isotope; radiochronologie *f*
s geocronología *f* absoluta; datación *f* por radio-isótopos; radiocronología *f*
g absolute Geochronologie *f*; Isotopendatierung *f*; Radiochronologie *f*

17 absolute humidity; vapour concentration
Mass of water contained in a unit volume of moist air.
r абсолютная влажность; концентрация водяного пара
Масса воды в единице объема влажного воздуха.
f humidité *f* absolue
s humedad *f* absoluta
g absolute Luftfeuchtigkeit *f*; absolute Feuchtigkeit *f*

18 absolute maximum

The greatest value of some meteorological or hydrological element changing in time at a certain point or area.

r абсолютный максимум
Наибольшее значение какого-либо изменяющегося во времени метеорологического или гидрологического элемента в данном пункте или области.

f maximum *m* absolu
s máximo *m* absoluto
g absolutes Maximum *n*

19 absolute minimum
The minimal value of some meteorological or hydrological element changing in time at a certain point or area.

r абсолютный минимум
Наименьшее значение какого-либо изменяющегося во времени метеорологического или гидрологического элемента в данном пункте или области.

f minimum *m* absolu
s mínimo *m* absoluto
g absolutes Minimum *n*

20 absolute monarchy
A monarchy headed by a supreme ruler having unlimited legislative and executive power.

r абсолютная монархия; абсолютизм
Монархия, в которой глава государства обладает неограниченной законодательной и исполнительной властью.

f monarchie *f* absolue
s monarquía *f* absoluta
g absolute Monarchie *f*

21 absolute temperature scale; Kelvin scale
Thermodynamic scale of temperature where 0°K=–273.16°C (1K=1°C).

r абсолютная температурная шкала; температурная шкала Кельвина; шкала Келвина
Термодинамическая шкала температуры (TK), в которой 0°K=–273.16°C (1K=1°C).

f échelle *f* de température absolue; échelle *f* de température Kelvin
s escala *f* de temperatura absoluta; escala *f* de temperaturas Kelvin
g absolute Temperaturskala *f*; Kelvin-Temperaturskala *f*

22 absolute zero
The point (0°K) at which thermal molecular movement ceases (the Kelvin temperature scale is based on absolute zero which corresponds to –273.16±0.01°C).

r абсолютный нуль
Предельно низкая температура, при которой прекращается тепловое движение молекул, в шкале Кельвина абсолютный нуль (0°K) соответствует –273,16±0,01°C.

f zéro *m* absolu
s cero *m* absoluto
g absoluter Nullpunkt *m*

23 absorbed solar radiation
A part of total solar radiation retained by the Earth's surface or the atmosphere.

r поглощенная солнечная радиация
Часть суммарной солнечной радиации, поглощенная земной поверхностью или атмосферой.

f rayonnement *m* solaire absorbé
s radiación *f* solar absorbida
g absorbierte Sonnenstrahlung *f*

24 absorbing capacity of soil
Capacity of soil for taking up dissolved substances which are in contact with its solid components.

r поглотительная способность почв
Свойство почв задерживать растворенные вещества, соприкасающиеся с ее твердой частью.

f capacité *f* d'absorption du sol
s capacidad *f* de absorción del suelo
g Absorptionskapazität *f* des Bodens

* **absorbing complex → 25**

25 absorbing soil complex; absorbing complex
The whole complex of components capable of keeping (absorbing) some elements which play a leading role in the exchange processes in soil, e.g. clayed minerals and humus.

r поглощающий комплекс почвы
Совокупность компонентов, способных удерживать в почве (поглощать) некоторые элементы, играющие ведущую роль в процессах обмена, к ним относятся прежде всего глинистые минералы и гумус.

f complexe *m* absorbant du sol
s complejo *m* absorbente del suelo
g Absorptionskomplex *n* des Bodens

26 absorption
The process by which the atmosphere or Earthrs surface retain radiant energy (heat- and light-waves) instead of reflecting it, or the soil takes in and holds the dissolved substances.

r абсорбция
Физический процесс поглощения радиационной энергии атмосферой или земной поверхностью, либо процесс задерживания растворенных веществ, как это происходит в почве.

f absorption *f*
s absorción *f*
g Absorption *f*

27 absorption coefficient in the atmosphere
A measure of the amount of normally incident radiant energy absorbed by a unit distance or by a unit mass of the atmosphere.

r коэффициент поглощения в атмосфере

Количество поглощенной энергии нормально падающего потока на единицу расстояния или на единицу массы атмосферы.
f coefficient *m* d'absorption
s coeficiente *m* de absorción
g Absorptionskoeffizient *m*

* **abyss → 4209**

28 **abyssal** ***adj***
Applied to the deepest regions (and associated phenomena) of the ocean or the ocean floor in general, as well as to the deepest regions of a lake.
r абиссальный
Относящийся к самой глубокой области (и соответствующим явлениям) океана или в целом океанского ложа, а также к глубинам озера.
f abyssal *adj*
s abisal *adj*; abismal *adj*
g abyssal *adj*

* **abyssal basin → 4209**

29 **abyssal cone**
A cone-like deposit of terrigenous material accumulated at a considerable depth in the ocean near the mouths of many large rivers.
r абиссальный конус выноса
Конусовидные отложения терригенных осадков вблизи устьев большинства крупных рек, располагающиеся на значительных океанических глубинах.
f cône *m* de déjection abyssale
s cono *m* de evacuación abisal
g Abyssalschüttkegel *m*; Abyssalschüttfächer *m*

30 **abyssal deposits; abyssal sediments**
Modern and ancient marine sediments (mainly carbonate and siliceous oozes, red clay), accumulated at the depths of 3,000 m and below.
r абиссальные отложения; глубоководные отложения
Современные и древние морские и океанические отложения (в основном карбонатные и кремнистые илы, красная глина) на глубинах 3000 м и более.
f dépôts *mpl* abyssaux
s depósitos *mpl* abisales
g Tiefseeablagerungen *fpl*

31 **abyssal fauna**
Marine animals living in complete darkness in soft substrate of deep ocean and having uniformly greyish or black colouring.
r абиссальная фауна
Глубоководные животные, обитающие в темноте, в спокойной среде, на мягком грунте, имеющие одинаковую окраску, чаще всего сероватую или черную.
f faune *f* abyssale
s fauna *f* abisal
g Tiefseefauna *f*

32 **abyssal plain**
A rather flat deep-sea plain, a very large, relatively level area of the deep ocean floor or marginal sea.
r абиссальная равнина
Ровная глубоководная равнина, типичная для котловин ложа океана и впадин краевых морей.
f plaine *f* abyssale
s llanura *f* abisal
g Abyssalebene *f*

* **abyssal sediments → 30**

33 **abyssal zone**
The bottom zone of the ocean from 3,700 to 6,000 m characterized by a slow water motion, constant low temperatures (below 0°C), absence of the sunlight, and specific fauna.
r абиссальная зона; абиссаль; глубоководная зона
Зона глубин ложа океана от 3700 до 6000 м, отличающаяся слабой подвижностью воды, постоянными температурами ниже 0°C, отсутствием солнечного света и специфическим животным миром.
f zone *f* abyssale; étage *m* abyssal
s zona *f* abisal; piso *m* abisal; abismo *m*
g Abyssal *n*; Abyssus *m*; Abyssalzone *f*

34 **accelerated erosion**
Intensive destruction of soil and underlying rock by surface waters mainly caused by human activity.
r ускоренная эрозия
Интенсивное разрушение почвы и подстилающих горных пород поверхностными водами, вызванное главным образом нерациональным ведением хозяйства.
f érosion *f* accélérée
s erosión *f* acelerada
g beschleunigte Erosion *f*

35 **accessibility**
The ease to get to a place normally measured in the terms of the distance travelled, the time taken, the cost of travel and the location of one place in relation to transportation routes.
r доступность
Удобство достижения какого-либо пункта из данной точки, оцениваемое обычно необходимым для этого временем, а также положением пункта по отношению к транспортным коммуникациям.
f accessibilité *f*
s accesibilidad *f*
g Erreichbarkeit *f*; Zugänglichkeit *f*

36 **accident emission**
A release of pollutants into the environment

as a result of an accident or breakdown in a technological process.
r аварийный выброс
Поступление загрязняющих веществ в окружающую среду в результате аварии или нарушения технологического процесса.
f émission *f* accidentée
s expulsión *f* accidental; emisión *f* accidental
g Notausstoss *m*

37 accidental avalanche; accidental triggering of avalanche
The release of avalanche caused by unintended human intervention, e.g. by a skier.
r непреднамеренный спуск лавины
Спуск лавины, инициированный непреднамеренными действиями человека, например, лыжником.
f avalanche *f* accidentelle
s avalancha *f* accidental
g Unfalllawine *f*

* **accidental triggering of avalanche → 37**

38 acclimatization
The action of becoming used to a different sort of environment, usually to a change in climate.
r акклиматизация
Приспособление популяции (в том числе человека) к новым условиям среды, в которые она попала в результате переселения.
f acclimatation *f*
s aclimatación *f*; aclimatamiento *m*
g Akklimatisation *f*; Anpassung *f*; Eingewöhnung *f*

* **accordant summits → 1186**

* **accordant valley → 1216**

39 accretion
1. The growth of cloud elements or precipitation particles due to collision and freezing together of ice particles and supercooled liquid droplets. 2. The accumulation of mineral material in a paricular locality or environment, e.g. lacustrine muds in a lake basin.
r аккреция; нарастание
1. Рост элементов облаков или осадков вследствие столкновения и смерзания ледяных частиц с переохлажденными каплями. 2. Отложение минеральных частиц на небольшом участке местности, например, прибрежного ила в озере.
f accrétion *f*
s acreción *f*
g Akkretion *f*

40 accretion coast; accumulation coast
Bank of a water body or watercourse growing due to the accumulation of loads brought by waves, flow or surf.
r аккумулятивный берег
Берег водоема или водотока, нарастающий за счет наносов, приносимых волнами, течением или прибоем.
f côte *f* d'accumulation
s costa *f* de accumulación
g Akkumulationsküste *f*

41 acculturation
A complete or partial adoption by a nation of a new culture.
r аккультурация
Восприятие одним народом полностью или частично культуры другого народа.
f acculturation *f*
s aculturación *f*
g Akkulturation *f*

42 accumulation
1. Deposition of loose mineral or organic material on the Earth's surface. 2. The process of snow and ice nourishment including precipitation, rime ice, hoar-frost, avalanche material and drifting snow.
r аккумуляция
1. Накопление рыхлого минерального материала или органических остатков на поверхности Земли. 2. Накопление на леднике твердых атмосферных осадков в результате выпадения снега, метелевого переноса, а также схода снежных и ледяных лавин.
f accumulation *f*
s acumulación *f*
g Akkumulation *f*

43 accumulation area of glacier; accumulation zone; surplus area
The portion of a glacier above the equilibrium line where the accumulation exceeds ablation (the opposite of ablation area).
r область аккумуляции ледника
Верхняя часть ледника, где в течение балансового года преобладает аккумуляция.
f zone *f* d'accumulation de glacier
s zona *f* de acumulación de glaciar
g Akkumulationsgebiet *n* des Gletschers; Nährgebiet *n*

44 accumulation area ratio
The ratio of accumulation area to the total area of a glacier.
r доля области питания
Отношение площади области аккумуляции ко всей площади ледника.
f pourcentage *m* de zone d'accumulation
s porcentaje *m* de zona de acumulación
g Prozentverhältnis *n* des Akkumulationsgebiets

* **accumulation coast → 40**

45 accumulation period of glacier
The period of a balance year when accumulation of the glacier exceeds its ablation.
r период аккумуляции ледника

Часть балансового года, в течение которой приход вещества на ледник превышает его расход.
f période *f* d'accumulation de glacier
s período *m* de acumulación de glaciar
g Akkumulationsperiode *f* des Gletschers

* **accumulation zone → 43**

46 **accumulative relief**
Landforms created by the processes of the deposition of material.
r аккумулятивный рельеф
Рельеф, созданный процессами аккумуляции.
f relief *m* d'accumulation
s relieve *m* de acumulación
g Akkumulationsrelief *n*

47 **acid lava**
Lava rich in silica having a high melting-point (850°), that is why flowing slowly in a stiff or viscous stream only short distances from the volcanic vent.
r кислая лава
Лава, богатая кремнеземом и имеющая высокую температуру плавления (около 850°), что ведет к ее быстрому затвердеванию недалеко от очага извержения.
f lave *f* acide
s lava *f* ácida
g saure Lava *f*

* **acid precipitation → 48**

48 **acid rain; acid precipitation**
The deposition of atmospheric precipitations containing sulphates or other acidic compounds released by industrial plants and transported long distances by atmospheric processes.
r кислотный дождь; кислотные осадки
Отложения кислотных веществ с осадками (дождем, снегом) в результате переноса на дальние расстояния загрязнений, выброшенных в атмосферу предприятиями металлургической и химической промышленности.
f pluie *f* acide
s lluvia *f* ácida
g Säureregen *m*

49 **acid rock**
Igneous rock containing more than 10% of free quarz and more than 66% of silica.
r кислая порода
Изверженная порода, содержащая более 10% чистого кварца и более 66% кремния.
f roche *f* acide
s roca *f* ácida
g saures Gestein *n*

50 **acid soil**
A low-fertile soil with a pH less than 7, possible due to the parent material or to leaching of the soluble basis (especially sodium and calcium).
r кислая почва
Малоплодородная почва, имеющая pH меньше 7, формирующаяся на кислой материнской породе или характеризующаяся промывным режимом, приводящим к выщелачиванию.
f sol *m* acide
s suelo *m* ácido
g saurer Boden *m*

51 **acid waters**
Natural waters with a pH of less than 0.5 which results in increase of their destructive action on rocks.
r кислые воды
Природные воды с водородным показателем (pH) менее 0,5, что ведет к увеличению их растворяющего воздействия на горные породы.
f eau *f* acide
s agua *f* ácida
g Sauerwasser *n*

52 **acidophile**
Organisms that torelate and flourish on acid soils, e.g. some plants of raised bogs.
r ацидофил
Организмы, способные жить и развиваться в условиях значительной кислотности среды, например, некоторые растения верховых болот.
f acidiphile *m*; acidophile *m*
s acidófilo *m*
g Azidophil *m*

53 **acrisol**
A soil having a horizon of low-active clay accumulation and characterized by low saturation with bases which develops in tropical and subtropical regions where intensive weathering on ancient rocks is typical.
r акрисоль
Почва, имеющая горизонт аккумуляции малоактивной глины и низкую насыщенность основаниями, формируется в тропических и субтропических районах с интенсивным выветриванием на древних породах.
f acrisol *m*
s acrisol *m*
g Acrisol *m*

54 **actinometry**
A branch of *meteorology* studying solar, terrestrial and atmospheric radiation whithin the atmosphere and the techniques of their measurement.
r актинометрия
Раздел *метеорологии*, изучающий солнечное, земное и атмосферное излучение в условиях атмосферы и методы его измерений.
f actinométrie *f*

s actinometría *f*
g Aktinometrie *f*; Strahlenmessung *f*

55 activated sludge
A substance formed due to biological treatment of sewage and containing active microorganisms, which oxydize organic matter intensively and speed up the purification process.
r активный ил
Вещество, образующееся при биологической очистке сточных вод, содержащее большое количество микроорганизмов, которые ускоряют процесс очистки.
f boue *f* activée
s limo *m* activo
g aktivierter Schlamm *m*; belebter Schlamm *m*

56 active *adj*
Effective or energetic, as opposed to *passive*, quiet, dormant or dead.
r активный
Относящийся к эффективным и быстро протекающим природным процессам и явлениям, в противоположность *пассивным* и мертвым.
f actif *adj*
s activo *adj*
g aktiv *adj*

57 active cirque
Cirque or corrie situated higher than the present-day snow line, filled with firn and continuing its morphological development.
r деятельный цирк
Цирк или кар, расположенный выше современной снеговой линии, заполненный фирном и продолжающий морфологическое развитие.
f cirque *m* fonctionnel; cirque *m* vif
s circo *m* activo
g aktiver Zirkus *m*

58 active drainage area
Part of a catchment area the water flowing from which determines the formation of water discharge at a certain time.
r действующая площадь водосбора
Часть площади водосбора, с которой стекает вода, формирующая расход в данный момент времени.
f surface du bassin versant actif
s cuenca *f* vertiente activa
g Abflusswirksame Fläche eines Einzugsgebietes

59 active glacier
A glacier with a high rate of accumulation and ablation, it flows very rapidly and transports big amount of ice and debris. → Fig. 2
r активный ледник
Ледник с высокой скоростью аккумуляции и абляции, который быстро движется и переносит много обломочного материала. → Рис. 2
f glacier *m* actif

Fig. 2. **Active** (right) and **passive** (left) **glaciers** at the Pamirs

s glaciar *m* activo
g aktiver Gletscher *m*

60 active layer
The top layer of ground the temperature of which fluctuates during the day and season.
r деятельный слой
Верхний слой горных пород, в пределах которого происходят суточные и сезонные колебания температуры.
f couche *f* active
s capa *f* activa
g aktive Schicht *f*; Auftauboden *m*

61 active layer of bog
A layer of active water exchange in a swamp, which is transitional from peat deposits to the living vegetative cover.
r деятельный горизонт болота
Слой активного водообмена в болоте, переходный от торфяной залежи к поверхности живого растительного покрова.
f nappe *f* active de tourbière
s capa *f* activa de turbera
g aktive Schicht *f* eines Sumpfes

62 active layer of ocean
The uppermost sea water layer which directly interacts with the atmosphere.
r деятельный слой моря
Поверхностный слой морской воды, находящийся в непосредственном взаимодействии с атмосферой.
f couche *f* active de l'océan
s capa *f* activa de océano
g aktive Schicht *f* des Ozeans

63 active layer of permafrost; seasonally thawed layer
The top layer of permafrost where it is subject to annual thawing and freezing.
r деятельный слой многолетнемерзлых пород; сезонноталый слой
Верхний слой многолетнемерзлых пород, где они подвергаются сезонному промерзанию и протаиванию.
f couche *f* active de pergélisol
s capa *f* activa de permafrost

g aktive Schicht *f* des Permafrosts; saisonale Auftauschicht *f*

64 active population; working population
People with independent sources of subsistence (wages, scholarships, etc.).
r экономически активное население; самодеятельное население
Совокупность лиц, имеющих самостоятельный источник средств существования: заработную плату, стипендии и пр.
f population *f* active
s población *f* activa
g erwerbstätige Bevölkerung *f*

65 active porosity
Total volume of pores and other cavities along which ground waters can easily move in rocks.
r активная пористость
Совокупность пор и других пустот, по которым грунтовые воды могут свободно перемещаться в горных породах.
f porosité *f* active
s porosidad *f* activa
g aktive Porosität *f*

66 active temperature
The temperature of air higher than the biological minimum, determined for a given phase of the development of a cultural plant or for the whole period of vegetation.
r активная температура
Температура воздуха выше биологического минимума, установленного для данной фазы развития сельскохозяйственной культуры или для всего периода вегетации.
f température *f* active
s temperatura *f* activa
g aktive Temperatur *f*

67 active volcano
A volcano characterised by constant or recurrent eruptions at present or in the historical past or by solfatares.
r действующий вулкан
Вулкан, которому свойственны постоянные или периодические извержения в настоящее время или в историческом прошлом, либо характерны сольфатары.
f volcan *m* en activité
s volcán *m* activo
g tätiger Vulkan *m*

68 activity index; energy of glacierization
The degree of the activity of glaciers expressed in vertical gradient of glacier mass balance that is the sum of gradients of the annual increment and loss of ice.
r энергия оледенения
Мера активности ледников, которая выражается вертикальным градиентом их баланса массы, представляющим собой сумму градиентов годового прироста и убыли льда.
f coefficient *m* d'activité
s índice *m* de actividad
g Vergletscherungsenergie *f*

69 actualism
A comparative-historical method in geology and geography according to which the interpretations of the Earth's history can be based on the present-day evidence of natural processes.
r актуализм
Сравнительно-исторический метод в геологии и географии, согласно которому, изучая современные природные процессы, можно судить об аналогичных процессах в прошлом.
f actualisme *m*
s actualismo *m*
g Aktualismus *m*

* **adamantine braize → 1585**

* **adaptability → 70**

70 adaptation; adaptability
The ability of living organisms and their populations to change in morphology and function to fit new environmental conditions and withstand the impact of biotic and abiotic factors.
r адаптивность; адаптация
Морфологическая и функциональная приспособляемость организмов и их популяций к местным условиям существования и воздействию биотических и абиотических факторов.
f adaptation *f*
s adaptación *f*
g Adaptativität *f*; Anpassungsfähigkeit *f*

71 adfreeze; adfreezing
The hardening and fusion of ground due to the freezing of present water; the process by which two objects are bonded together by ice formed between them.
r смерзание
Цементирование грунта благодаря переходу находящейся в нем влаги в лед; соединение двух предметов с появлением льда между ними.
f géli-adhérence *f*
s congelamiento *m* adherente
g Anfrieren *n*

* **adfreezing → 71**

72 adhesion
A process in which dry sand blown on to a damp surface is held by the surface tension of water rising between the grains by capillarity.
r слипание
Процесс, при котором сухой песок, переносимый над влажной поверхностью, задерживается силами поверхностного натяжения капиллярной воды, находящейся между зернами.

f adhésion *f*
s adhesión *f*
g Adhäsion *f*

73 adhesion ripple
A small (30-40 cm long) irregilar sand ridge with the windward side steeper than the lee side and transverse to wind direction which forms when dry sand is moved across a smooth moist surface.
r смоченная рябь
Небольшие (длиной 30-40 см) песчаные гряды с более крутой наветренной стороной, формирующиеся, когда сухой песок перемещается в направлении господствующего ветра по гладкой слегка смоченной поверхности.
f rides *fpl* de adhésion
s arrugas *fpl* de adhesión
g Adhäsionsrippeln *n*

74 adiabatic
Referring to the changes of temperature and volume which occur in a mass of air when it is compressed, heated or expanded, cooled without the aid of any external sources of heating or cooling.
r адиабатический
Относящийся к изменениям температуры и объема, происходящим в воздушной массе, когда она сжимается (прогревается) или расширяется (охлаждается) без участия каких-либо внешних источников тепла или холода.
f adiabatique *adj*
s adiabático *adj*
g adiabatisch *adj*

75 adiabatic compression
The compression of air without any external exchange of heat.
r адиабатическое сжатие
Уменьшение объема воздуха без теплообмена с внешней средой.
f compression *f* adiabatique
s compresión *f* adiabática
g adiabatische Kompression *f*

76 adiabatic cooling
The cooling of air occurring when the air volume is allowed to expand without any external exhange of heat (it takes place in nature in upward convective air flows).
r адиабатическое охлаждение
Охлаждение воздуха при расширении его объема без притока тепла извне (наблюдается в природе в восходящих конвективных потоках воздух).
f refroidissement *m* adiabatique
s enfriamiento *m* adiabático
g adiabatische Abkühlung *f*

77 adiabatic equivalent temperature
Temperature that an air parcel would have after dry adiabatic expansion until being saturated, followed by pseudo-adiabatic expansion until all the moisture is precipitated, followed by dry adiabatic compression to the initial pressure.
r адиабатическая эквивалентная температура
Температура, которую принял бы объем воздуха после сухоадиабатического расширения до состояния насыщения, а затем псевдоадиабатического расширения – до тех пор пока вся влага не выпадет из него в виде осадков с последующим сухоадиабатическим сжатием до исходного давления.
f température *f* adiabatique équivalente
s temperatura *f* adiabática equivalente
g adiabatische äquivalente Temperatur *f*

78 adiabatic expansion
The expansion of air without any external exchange of heat.
r адиабатическое расширение
Увеличение объема воздуха без теплообмена с внешней средой.
f détente *f* adiabatique
s ampliación *f* adiabática
g adiabatische Ausdehnung *f*

*** adiabatic gradient → 79**

79 adiabatic lapse rate; environmental lapse rate; adiabatic gradient
The rate at which the temperature of an air mass, ascending or descending, changes in response to the expansion or compression of the air mass.
r адиабатический градиент температуры
Скорость изменения температуры в массе воздуха при ее адиабатическом перемещении по вертикали как реакция на расширение или сжатие этой воздушной массы.
f gradient *m* adiabatique
s gradiente *m* adiabático
g adiabatischer Temperaturgradient *m*

80 adiabatic process
The change of temperature and volume of air without any external exchange of heat, the opposite to *diabatic process*.
r адиабатический процесс
Изменение термодинамического состояния воздушной массы, протекающее без обмена теплом с окружающей ее средой, противоположен *диабатическому процессу*.
f processus *m* adiabatique
s proceso *m* adiabático
g adiabatischer Prozess *m*

*** adiabatic trail → 1191**

81 adiabatic warming
The warming of air occurring when the air is compessed without any outer heat excange (it takes place in nature under the air subcidence in anticyclones).

r адиабатическое нагревание
Нагревание воздуха при сжатии его объема без притока тепла извне (наблюдается при оседании воздуха в антициклоне).
f réchauffement *m* adiabatique
s calentamiento *m* adiabático
g adiabatische Erwärmung *f*

82 adit; gallery
A nearly horizontal passage or working dug into a hill side from the surface for the purpose of extracting minerals.
r штольня
Горизонтальная или наклонная горная выработка с выходом на земную поверхность.
f galèrie *f* de mine
s galería *f* minera; socavón *m*
g Stollen *m*

83 administrative centre
A city, town or settlement with the functions of government and control over a certain territory.
r административный центр
Город или поселок, обладающий функциями административного управления на данной территории.
f centre *m* administratif
s centro *m* administrativo
g Verwaltungszentrum *n*

*** administrative frontier → 4035**

84 administrative methods of management
The management of national economy and social life through reglamentation by directives, orders, rules and norms.
r административные методы управления
Руководство народным хозяйством и жизнью населения посредством властных распоряжений: приказов, предписаний, норм, обязательных для выполнения.
f méthodes *fpl* administratives de géstion
s métodos *mpl* administrativos de gestión
g administrative Methoden *fpl* der Leitung

85 administrative region; district$_2$
The smallest territorial unit in the administrative division of certain countries.
r административный район
Низшая территориальная ячейка в административном делении некоторых стран.
f région *f* administrative; district$_2$ *m*
s región *f* administrativa; distrito$_2$ *m*
g Verwaltungsbezirk *m*; Verwaltungskreis *m*

86 administrative-territorial division
A hierarchical arrangement of territorial organization under which a state is divided into several regions, provinces, and districts in accordance to which the bodies of state power and administration function.
r административно-территориальное деление
Иерархическая система территориальной организации государства, предусматривающая разделение территории на ряд областей, провинций и районов, на основе которой функционируют органы государственной власти и управления.
f division *m* administrative
s división *f* administrativa
g administrative Gliederung *f*; territoriale Gliederung *f*

87 adret
The sunny, warm slope of a hill or valleyside, as opposed to the *ubac* or shady side.
r солнечный склон
Освещенный солнцем, теплый склон холма или речной долины, в противоположность *теневому склону.*
f adret *m*; soulane *f*; solane *f*
s solana *f*
g Sonnenseite *f*

*** advance of glacier → 2505**

88 advanced dune
A small sand dune formed round an obstacle and lying in front, i.e. to windward, of a largar dune which is separated by an eddying wind.
r передовая дюна
Небольшая песчаная дюна, формирующаяся на наветренной стороне крупной дюны, где порывистый ветер затухает.
f avant-dune *f*
s ante-duna *f*
g Vordüne *f*

*** advanced technology → 2767**

89 advection
The horizontal movement of masses of air and water, in contrast to vertical movement (*convection*).
r адвекция
Перенос водной или воздушной масс в горизонтальном направлении, в отличие от переноса в вертикальной плоскости (*конвекции*).
f advection *f*
s advección *f*
g Advektion *f*

90 advection fog; sea fog
Fog formed whem a warm, moisture-laden air stream moves horizontally over a cooler surface, so that its lower layers are chilled below dewpoint.
r адвективный туман; морской туман
Туман, возникающий вследствие натекания влажной воздушной массы на более холодную подстилающую поверхность земли или воды, так что нижние слои натекающего воздуха охлаждаются ниже точки росы.
f brouillard *m* d'advection; brouillard *m* marin

s niebla *f* por advección; niebla *f* marina
g Advektionsnebel *m*; Meeresnebel *m*; Seerauch *m*

91 **advection frost**
Frost caused by the advection (penetration) of cold air.
r адвективный заморозок
Заморозок, обусловленный адвекцией (натеканием) холодного воздуха.
f gelée *f* d'advection
s helada *f* por advección
g Advektivfrost *m*

92 **advection inversion**
The *inversion of temperature* connected with advection of warm air upon colder underlying surface.
r адвективная инверсия
Инверсия температуры, связанная с адвекцией (переносом) теплого воздуха на более холодную подстилающую поверхность.
f inversion *f* d'advection
s inversión *f* advectiva
g Advektionsinversion *f*

93 **advective thunderstorm**
A thunderstorm caused by unstable stratification of the atmosphere occurred due to advection.
r адвективная гроза
Гроза, вызванная неустойчивой стратификацией атмосферы, возникшей вследствие адвекции.
f orage *m* d'advection
s tormenta *f* por advección
g Advektionsgewitter *n*; advektives Gewitter *n*

94 **adventitious organism**
A plant or animals deliberately or accidentally brought by man from other climatic zones to a new environment.
r адвентивный организм
Растение или животное, преднамеренно или случайно завезенное человеком из других климатических зон в новую для них область и условия среды.
f organisme *m* adventice
s organismo *m* adventicio; organismo *m* adventizo
g adventiver Organismus *m*

* **adventive cone → 4417**

* **adventive crater → 4418**

* **adventure tourism → 2081**

95 **adyrs**
A hilly piedmont fringing the mountain ranges in Central Asia.
r адыры
Холмисто-увалистые предгорья, окаймляющие хребты в Центральной Азии.
f adyrs *mpl*
s adyres *mpl*
g Adyren *mpl*

* **aeolian abrasion → 1470**

96 **aeolian *adj*; eolian *adj***
Related to the wind action.
r эоловый
Обусловленный деятельностью ветра.
f éolien *adj*
s eólico *adj*
g äolisch *adj*; Wind-

* **Aeon → 1947**

97 **aerated ice**
Ice artificially saturated with air bubbles.
r аэрированный лед
Лед, искусственно насыщенный пузырьками воздуха.
f glace *f* bulleuse
s hielo *m* con burbujas
g belüftetes Eis *n*; mit Luft angereichertes Eis

* **aerated zone → 6858**

98 **aeration**
Any process by which a substance becomes impregnanted with air.
r аэрация
Процесс, при котором вещество насыщается воздухом.
f aération *f*; aérage *m*
s aeración *f*
g Belüftung *f*; Durchlüftung *f*

99 **aeration of sewage; sewage aeration**
The aturation of sewage with air resulting in a quick oxidation of organic matter.
r аэрация сточных вод
Насыщение сточных вод воздухом, вызывающее быстрое окисление содержащихся в них органических веществ.
f aérage *m* des eaux d'égout
s aeración *f* de aguas residuales
g Abwässerbelüftung *f*

100 **aeration of water; water aeration**
The enrichment of water with oxygen in order to improve its quality.
r аэрация воды
Обогащение воды кислородом с целью улучшения ее качества.
f aération *f* de l'eau
s aeración *f* del agua
g Wasserbelüftung *f*

101 **aerial photograph**
Any photograph taken from the air and used for the map-making. → Fig. 3, p. 14
r аэрофотоснимок; аэроснимок
Фотография местности, сделанная с воз-

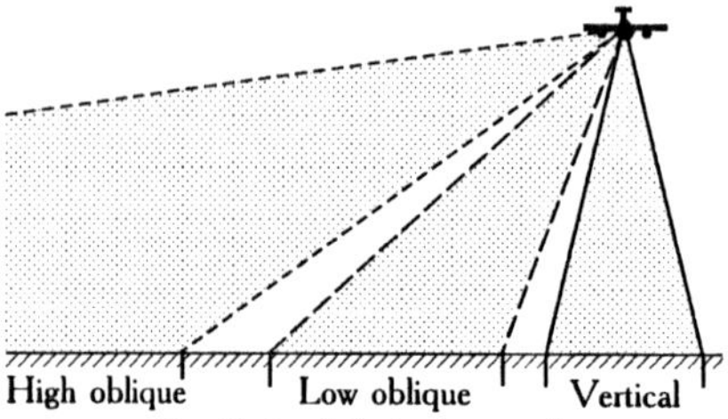

Fig. 3. **Aerial photograph**

духа и используемая при составлении карт. → Рис. 3
f photo *f* aérienne; cliché *m* aérien
s fotografía *f* aérea; clisé *m* aéreo; aerofoto *f*
g Luftbild *n*; Luftaufnahme *f*

102 aerial photography
The obtaining of photographs of the surface of the Earth taken from aircrafts or flying vehicles in different spectrum zones with the help of photocamera systems.
r аэрофотосъемка
Фотографирование местности с летательных аппаратов в разных зонах спектра с помощью фотокамер.
f photographie *f* aérienne
s fotografía *f* aérea
g Luftbildaufnahme *f*; Luftvermessung *f*

103 aerobic *adj*
Applied to living organisms that require free oxygen to live, and to processes that occur only in the presence of oxygen.
r аэробный
Относящийся к организмам, способным существовать только при наличии свободного кислорода, а также к процессам, происходящим в присутствии кислорода.
f aérobie *adj*
s aeróbico *adj*; aerobio *adj*
g aerob *adj*

104 aerobic process
Microbiological, biochemical and chemical transformations which occur under the presence of free oxygen in contrast to *anaerobic process*.
r аэробный процесс
Совокупность микробиологических, биохимических и химических трансформаций, протекающих, в отличие от *анаэробного процесса*, в присутствии свободного кислорода.
f processus *m* aérobie
s proceso *m* aeróbico
g aerober Prozeß *m*

105 aeroclimatology
Climatology of the free atmosphere, i.e. the study of climatic conditions in the troposphere and stratosphere up to 20-25 km.
r аэроклиматология
Климатология свободной атмосферы, т.е. изучение климатических условий в тропосфере и стратосфере до высоты 20-25 км.
f aéroclimatologie *f*
s aeroclimatilogía *f*
g Aeroklimatologie *f*

106 aerology
The study of the free atmosphere and methods of upper-air observations.
r аэрология
Изучение процессов, протекающих в свободной атмосфере.
f aérologie *f*
s aerología *f*
g Aerologie *f*; Höhenwetterkunde *f*

107 aeronomy
The study of the physics and chemistry of the high atmosphere (mesosphere and ionosphere).
r аэрономия
Изучение физических и химических свойств верхней атмосферы (мезосферы и ионосферы).
f aéronomie *f*
s aeronomía *f*
g Aeronomie *f*

108 aerophototopography
Making of topographic maps by materials of aerial photograpy.
r аэрофототопография
Создание топографических карт по материалам аэрофотосъемки.
f aérophototopographie *f*
s fototopografía *f* aérea
g Aerophototopographie *f*; Luftbildvermessung *f*

109 aerophyte
A plant inhabiting mainly the trunks, branches, and leaves of trees and getting nutrients from air, e.g. some lichens.
r аэрофит
Растение, обитающее в основном на стволах, ветвях, листьях деревьев и получающее все основные питательные вещества только из воздуха, например, некоторые лишайники.
f aérophyte *m*
s aerofita *m*
g Aerophyt *m*

110 aeroplankton
Minute organisms (spores, bacteria, etc.) floating freely in the atmosphere.
r аэропланктон
Мельчайшие организмы (споры, бактерии и т.п.), свободно парящие в атмосфере.
f aéroplancton *m*; plancton *m* aérien
s aeroplanctón *m*
g Luftplankton *n*; Aeroplankton *n*

111 aerosol pollutant; tropospheric aerosol

Suspended minute particles (solid or liquid) of dust, sea salt and, in urban environments, carbon, lead and aluminium compound produced by man's combustion of fuels, appearing in the troposphere as mist, fog, etc.

r тропосферный аэрозоль
Мельчайшие рассеянные (твердые и жидкие) частицы пыли, морских солей, а на урбанизированных территориях также частицы углерода, свинца и алюминия, возникающие от сгорания углеводородного топлива, присутствующие в тропосфере в виде тумана, дымки и пр.

f aérosol *m* troposphérique
s aerosol *m* troposférico
g tropospherischer Aerosol *m*

112 aerospace research
Related to investigations and map-making of the Earth's surface with the use of flying vehicles.

r аэрокосмические исследования
Исследования и картографирование земной поверхности с помощью летательных аппаратов.

f recherche *f* aérospatiale; recherche *f* aérocosmique
s investigación *f* aeroespacial; estudio *m* aerocósmico
g aerokosmische Erforschungen *fpl*

113 aesthetic geography
Geography of beauty considering the impact of natural landscapes upon the feelings of people.

r эстетическая география
География прекрасного, рассматривающая природные ландшафты как объекты природы, воздействующие на чувства людей.

f géographie *f* esthétique
s geografía *f* estética
g ästhetische Geographie *f*

114 aestivation
The state of inactivity adopted by desert animals during the hottest and driest seasons in order to survive.

r летняя спячка
Отсутствие активности у некоторых пустынных животных в самые жаркие и сухие периоды, позволяющее им выжить в эти неблагоприятные сезоны.

f estivation *f*
s estivación *f*
g Sommerruhe *f*

* **affluent → 6347**

115 afforestation
The establishment of forest by the planting of trees on land where they did not grow formerly.

r облесение
Разведение леса путем посадки деревьев в местах, где раньше они не произрастали.

f afforestation *f*
s repoblación *f* forestal
g Bewaldung *f*; Aufforstung *f*

* **afoliage → 333**

116 afterglow; evening glow
A faint radiance occasionally visible in the western sky just after sunset, caused by the scattering effect of atmospheric dust-particles on sunlight.

r вечерняя заря
Слабое свечение в западной части неба сразу после захода Солнца, вызываемое эффектом рассеяния солнечных лучей атмосферными пылевыми частицами.

f lueurs *fpl* du soleil couchant
s crepúsculo *m* vespertino; luz *f* del sol poniente
g Abendrot *n*

117 aftermath
The grass stand recovered after beeing mowed or grazed when the damaged and new sprouts start to grow.

r отава
Травостой, восстановленный в результате отрастания поврежденных побегов и роста новых (из спящих почек) после скашивания или стравливания.

f regain *m*
s retallo *m*
g Grummet *n*; Öhmd *n*

118 aftershock
Underground shocks following an earthquake and connected with it genetically.

r афтершок
Подземные толчки, следующие за землетрясением и генетически с ним связанные.

f ébranlement *m* consécutif
s sacudida *f* consecutiva; temblor *m* consecutivo
g Nachbeben *n*

119 age of landscape
The period of time which has passed from the beginning of formation or change of the main features of the modern landscape caused by climate change, geological processes, human activity, etc.

r возраст ландшафта
Время, прошедшее с момента формирования (перестройки) основных черт современной структуры конкретного ландшафта, обусловленной изменением климата, геологическими процессами, деятельностью человека и пр.

f âge *m* de paysage
s edad *f* de paisaje
g Alter *n* der Landschaft

120 age of relief
Time period from the begging of formation of the main elements of the contemporary relief till the present time.

r возраст рельефа
Промежуток от начала формирования основных элементов современного рельефа до настоящего времени.
f âge *m* du relief
s edad *f* de relieve
g Alter *n* des Reliefs

* **agent → 2087**

* **ageostrophic wind component → 121**

121 **ageostrophic wind; ageostrophic wind component**
Vector difference between the actual (observed) wind and *geostrophic wind.*
r агеострофический ветер; агеострофическая составляющая ветра
Векторная разность между фактическим (наблюденным) ветром и *геострофическим ветром.*
f vent *m* agéostrophique; composante *m* agéostrophique du vent
s viento *m* ageostrófico; componente *m* ageostrófico del viento
g ageostrophischer Wind *m*; ageostrophische Windkomponente *f*

* **age-sex pyramid → 4700**

122 **agglomerate**
Loose deposits of angular large fragments of sedimantary or volcanic genesis.
r агломерат
Рыхлые отложения неокатанного крупнообломочного материала осадочного или вулканического происхождения.
f agglomérat *m*
s aglomerado *m*
g Agglomerat *n*

123 **agglomeration economies; localization economies**
For an industry, the benefits of locating in a densely peopled and highly industrialized situation where a large market is concentrated in a relatively small area and transport costs are therefore low.
r агломерационный эффект; агломерационная экономия
Снижение себестоимости продукции отдельных предприятий, работающих рядом друг с другом или имеющих общий рынок на компактно расположенной территории, что приводит к снижению транспортных и иных издержек.
f effet *m* d'agglomération
s efecto *m* de aglomeración
g Agglomerationseffekt *m*

* **agglomeration of enterprises → 6111**

124 **aggradation**
A process in which a land surface is build up by deposition and accumulation of detritus, rock waste, sand, alluvium, etc.
r аградация
Процесс формирования земной поверхности путем отложения и переотложения осадочных пород, что в береговой полосе приводит, в частности, к возникновению намывных берегов, островов и пр.
f aggradation *f*
s agradación *f*
g Anschwemmung *f*; Anlandung *f*

125 **aggradation terrace; alluvial terrace**
A terrace composed of alluvium, marine or lake sediments.
r аккумулятивная терраса; аллювиальная терраса
Терраса, сложенная аллювием, морскими или озерными отложениями.
f terrasse *f* alluviale
s terraza *f* alluvial
g Aufschüttungsterrasse *f*; Alluvialterrasse *f*

126 **aggradational soils**
Soils forming under the intensive accumulation of mineral soil-forming materials, and occurring mainly at the foot slopes, on bottoms of ravines and gullies.
r намытые почвы
Почвы, формирующиеся в условиях активного накопления минерального почвообразующего материала, встречающиеся чаще всего у подножья склонов, на днищах балок и оврагов.
f sols *mpl* d'aggradation
s suelos *mpl* de agradación
g Anschwemmungsböden *mpl*

127 **aggraded plain; plain of aggradation**
A plain formed as a result of the accumulation of loose deposits over a lengthy period of time.
r аккумулятивная равнина
Равнина, образовавшаяся в результате длительного накопления толщ рыхлых отложений.
f plaine *f* d'accumulation
s llanura *f* de acumulación
g Akkumulationsebene *f*

* **aggregate map → 1170**

* **aggregate of soil → 5587**

128 **aggregation**
In biogeography a localized group composed of many individuals of a population.
r агрегация
В биогеографии локализованная группа особей одной популяции.
f agrégation *f*
s agregación *f*
g Aggregation *f*

129 **aggressive water**

Water which is able to destroy different material (concrete, metal, etc.) by the chemical action of substances.

r агрессивная вода
Вода, способная, благодаря растворенным в ней веществам, разрушать путем химического воздействия различные материалы: бетон, металл и пр.

f eau *f* aggressive
s agua *f* agresiva
g schade Wasser *n*; agressives Wasser *n*

130 agrarian *adj*; agricultural *adj*
Related to land, especially to its cultivation, management, and landed property.

r аграрный; сельскохозяйственный
Относящийся к сельскохозяйственному производству и собственности на землю.

f agraire *adj*; agricole *adj*
s agrario *adj*; agrícola *adj*
g agrarisch *adj*; agrarwirtschaftlich *adj*

131 agrarian crisis
An economic crisis of overproduction in agriculture that shows itself in the growth of stocks of agricultural produce, fall in prices, decrease of land under crops and decrease of production.

r аграрный кризис
Экономический кризис перепроизводства в сельском хозяйстве, проявляющийся в росте нереализуемых запасов сельскохозяйственных товаров, падении цен, сокращении посевных площадей и объемов производства.

f crise *f* agraire
s crisis *m* agrario
g Agrarkrise *f*

132 agrarian reform; land reform
State measures aimed at changing and improving rural life and agricultural efficiency.

r аграрная реформа; земельная реформа
Государственные мероприятия по преобразованию системы землевладения и землепользования.

f réforme *f* agraire
s reforma *f* agraria
g Agrarreform *f*; Bodenreform *f*

133 agribusiness complex
A large-scale farming business run along the lines of a conventional economy, often involving the growing, processing, packaginig and sale of farm products.

r продовольственный комплекс
Совокупность отраслей (предприятий), производящих продукты питания из сельскохозяйственного сырья.

f complexe *m* agro-alimentaire
s complejo *m* agro-alimentario
g Lebensmittelkomplex *m*

*** agricultural *adj* → 130**

*** agricultural area → 136**

134 agricultural climatology; agroclimatology
The study of those aspects of climate which are relevant to agricultural problems.

r сельскохозяйственная климатология; агроклиматология
Раздел *климатологии*, изучающий воздействие климата на сельское хозяйство.

f agroclimatologie *f*
s agroclimatología *f*
g Agrarklimatologie *f*

135 agricultural effluent; farm runoff
Sewage waters coming from agricultural, mainly from cattle-breeding farms.

r сельскохозяйственные стоки
Сточные воды сельского хозяйства, преимущественно животноводческих комплексов.

f effluents *mpl* agricoles; déchets *mpl* agricoles
s aguas *fpl* residuales agrícolas
g landwirtschaftliche Abwässer *npl*

*** agricultural geography → 2428**

136 agricultural land; farmland; agricultural area
An area of land used for farming, including arable land, land under tree crops (orchards, etc.), fallow, hay fields, pastures.

r сельскохозяйственные угодья; культивируемые земли
Участки земли, используемые в сельскохозяйственном производстве: пахотные земли, многолетние насаждения (сады и пр.), залежи, сенокосы и пастбища.

f terre *f* agricole; surface *f* agricole utile
s tierra *f* agrícola; superficie *f* agrícola útil
g landwirtschaftliche Fläche *f*

137 agricultural landscape; agrolandscape
A man-made landscape in which natural vegetation is largely replaced by agrocenoses such as fields, pastures, grasslands, perennial crops.

r агроландшафт
Антропогенный ландшафт, естественная растительность которого на подавляющей части территории заменена агроценозами - полями, пастбищами, сенокосами, многолетними посадками культур.

f paysage *m* agraire; paysage *m* rural
s paisaje *m* agrário; paisaje *m* rural
g Agrarlandschaft *f*; Ackerlandschaft *f*

138 agricultural meteorology; agrometeorology
Branch of *meteorology* studying atmosphere conditions important for agriculture as well as interaction between hydrometeorological conditions and agriculture.

r сельскохозяйственная метеорология; агрометеорология

Раздел *метеорологии*, изучающий атмосферные условия, имеющие значение для сельского хозяйства, а также взаимодействие между гидрометеорологическими факторами и сельским хозяйством.
f agrométéorologie *f*
s agrometeorología *f*
g Agrarmeteorologie *f*

139 agricultural plants
Varieties of plants developed through selection and grown for foodstuffs, industrial raw materials and fodder.
r сельскохозяйственные культуры
Выведенные человеком виды и сорта растений, возделываемые для получения продуктов питания, производства промышленного сырья и корма для сельскохозяйственных животных.
f cultures *fpl* agricoles
s cultivos *mpl* agrícolas
g landwirtschaftliche Kulturen *fpl*

140 agricultural region
An economic region in which agricultural production is the main human activity.
r сельскохозяйственный район
Экономический район с сельскохозяйственным производством в качестве важнейшей отрасли производственной специализации.
f région *f* agricole
s región *f* agrícola
g Agrargebiet *n*; landwirtschaftliche Region *f*

141 agricultural revolution
Periods of deep changes in the world agriculture: the replacement of fallow with forage crops at the end of the XVIII century, mechanization of agriculture and growing high-yield grain crops in the middle of the XX century, biotechnology at the end of the XX century.
r сельскохозяйственная революция
Этапы резких изменений в мировом сельском хозяйстве, связанных с происшедшей в конце XVIII в. заменой залежей возделыванием кормовых культур, в середине XX в. механизацией сельского хозяйства и внедрением высокоурожайных зерновых культур, в конце XX в. применением новых биотехнологий.
f révolution *f* agricole
s revolución *f* agrícola
g landwirtschaftliche Revolution *f*

* **agricultural science → 151**

142 agriculture; husbandry
Cultivation of land, including horticulture, fruit growing, crop and seed growing, dairy farming and livestock breeding in order to get foodstuff and commodities for industry.
r сельское хозяйство
Возделывание сельскохозяйственных культур и разведение сельскохозяйственных животных для получения продовольствия и сырья для промышленности.
f agriculture *f*
s agricultura *f*
g Landwirtschaft *f*

143 agrocenology
The science which studies agrocoenoses and their features conditionned by plant species under cultivation.
r агроценология
Наука, изучающая агроценозы и их особенности, определяемые возделываемой культурой.
f agrocénologie *f*
s agrocenología *f*
g Agrozönologie *f*

144 agrocenosis; agroecosystem; agrophytocenosis
A biotic community created and regularly maintained by people in order to get the produce of one or several species of agricultural plants or animals.
r агроценоз; агросистема; агробиоценоз
Биотическое сообщество, созданное человеком и регулярно им поддерживаемое для получения продукции одного или нескольких избранных видов (сортов, пород) растений или животных.
f agrocénose *f*; agrosystème *m*
s agrocenosis *f*; agrosistema *m*
g Agrozönose *f*; Agrarbiozönose *f*; Agrosystem *n*

145 agroclimatic index
An index relating some particular agricultural activity with one or more elements of the local climate, e.g. with temperature, humidity, etc.
r агроклиматический индекс
Индекс, выражающий связь какого-либо вида сельскохозяйственной деятельности с одной или несколькими климатическими характеристиками, например, с показателями температуры, влажности и пр.
f indice *m* agroclimatique
s índice *m* agroclimático
g agroklimatischer Index *m*

146 agroclimatic resources
Resources of natural heat- and water supply of agriculture at a certain region.
r агроклиматические ресурсы
Ресурсы естественной тепло- и влагообеспеченности земледелия в данном регионе.
f ressources *fpl* agroclimatiques
s recursos *mpl* agroclimáticos
g agroklimatische Ressourcen *fpl*

147 agroclimatic zone
A climatic zone based on the effect of climate on agricultural activities.
r агроклиматическая зона

Климатическая зона, выделяемая по характеру влияния климатических условий на земледелие.
f zone *f* agroclimatique
s zona *f* agroclimática
g agroklimatische Zone *f*

* **agroclimatology → 134**

* **agroecosystem → 144**

148 **agroforestal** ***adj***
Related to the use of land for a combination of agriculture and forestry.
r агролесной
Относящийся к комбинированному использованию земли и для сельского и для лесного хозяйства.
f agroforestière *adj*
s agroforestal *adj*
g agroforstwirtschaftlich *adj*

149 **agro-forestry**
A land-use system in which trees are deliberately left (e.g. as shade), planted or encouraged (e.g. for fuel and food) on land where animals are grazed or crops grown.
r агролесное хозяйство
Система землепользования, в которой на землях, где пасутся животные или выращиваются сельскохозяйственные культуры, также сохраняются деревья (в качестве тени, а также для топлива или еды).
f agroforesterie *f*
s agrosilvicultura *f*
g Agroforstwirtschaft *f*

150 **agro-industrial complex**
A group of sectors (enterprises) involved directly or indirectly in the production of food or other products out of agricultural raw materials.
r аграрно-промышленный комплекс; агропромышленный комплекс; АПК
Совокупность отраслей (предприятий), деятельность которых прямо или косвенно направлена на производство продуктов питания или другой продукции, вырабатываемой из сельскохозяйственного сырья.
f complexe *m* agro-industriel
s complejo *m* agro-industrial
g agro-industrialer Komplex *m*

* **agrolandscape → 137**

* **agrometeorology → 138**

151 **agronomy; agricultural science**
The theory and practice of crop production and the scientific management of soils.
r агрономия
Теория и практика сельскохозяйственного растениеводства и научные основы использования почв.
f agronomie *f*; agrologie *f*
s agronomía *f*; agrología *f*
g Agronomie *f*; Ackerbau *m*

152 **agrophyte**
Alien (introduced) plant species, cultivated in agrocenoses.
r агрофит
Чужеродные (интродуцированные) виды растений, культивируемые в агроценозах.
f agrophyte *m*
s agrofita *m*
g Agrophyt *m*

* **agrophytocenosis → 144**

153 **agrozem**
A soil the profile of which consists of agriculturally transformed horizon over 25 cm thick which sharply changes into mineral soil layer or into soil-forming rock.
r агрозем
Почва, профиль которой состоит из агрогенно преобразованного горизонта мощностью более 25 см, резко сменяющегося минеральным почвенным горизонтом или почвообразующей породой.
f agrosol *m*; sol *m* agricole
s agrosuelo *m*; suelo *m* agrário
g Agrarboden *m*

154 **aheliophyte**
Plants which do not need sunlight, e.g. fungi.
r агелиофит
Растения, не нуждающиеся в солнечном освещении, например, грибы.
f ahéliophyte *m*
s aheliófita *m*
g Aheliophyt *m*

* **A-horizon → 1875**

155 **air**
Natural mixture of gases which composes the Earth's atmosphere.
r воздух
Естественная смесь газов, образующая земную атмосферу.
f air *m*
s aire *m*
g Luft *f*

156 **air corridor**
A strip of territory belonging to one country used for international flights to reach other part of the world.
r воздушный коридор
Полоса суверенной территории страны, используемая для международных полетов в другие страны.
f corridor *m* d'air
s corredor *m* aéreo
g Luftkorridor *m*

157 **air diffusion**
The random mixing of air bodies brought about by a slow process of mixing (termed

molecular diffusion) or by turbulence (eddy diffusion).

r диффузия воздуха
Перемешивание воздуха, вызываемое молекулярными процессами (молекулярная диффузия) или турбулентными движениями (вихревая диффузия).

f diffusion *f* de l'air
s difusión *f* de aire
g Luftdiffusion *f*

* **air flow → 164**

* **air frost → 2327**

158 air hole
An abrupt change of the air density causing a sudden drop down of an aircraft.
r воздушная яма
Резкое изменение плотности воздуха, вызывающее неожиданный бросок самолета в полете.
f fosse *f* aérienne
s foso *m* aéreo
g Luftloch *n*

* **air humidity → 2837**

159 air mass
An almost homogeneous mass of air of great lateral extent (generally hundreds or thousand of kilometres) bounded by atmospheric fronts.
r воздушная масса
Относительно однородная масса воздуха, распространенная на сотни и тысячи километров, ограниченная атмосферными фронтами.
f masse *f* d'air
s masa *f* del aire
g Luftmasse *f*

160 air navigation chart
A geographical map intended for flights of aircrafts.
r аэронавигационная карта
Специальная географическая карта, предназначенная для выполнения полетов самолетов.
f carte *f* aéronautique
s carta *f* aeronáutica
g Luftfahrtkarte *f*

* **air pollution → 437**

* **air pressure → 438**

161 air quality
The conformity of air to health of people and technological requirements determined by a number of indices.
r качество воздуха
Степень соответствия воздуха потребностям людей и технологическим требованиям, определяемая по ряду показателей.
f qualité *f* d'air
s calidad *f* de aire
g Luftqualität *f*

162 air regime of soil
Content and composition of soil air and their time variation.
r воздушный режим почвы
Содержание и состав почвенного воздуха и его изменение во времени.
f régime *m* aérien de sol
s régimen *m* aéreo del suelo
g Bodenlufthaushalt *m*

163 air space
Part of the atmosphere lyingover the land and water territory of a state and being under its jurisdiction.
r воздушное пространство
Часть атмосферы над сухопутной и водной территорией государства, находящаяся под юрисдикцией данной страны.
f espace *m* aérien
s espacio *m* aéreo
g Luftraum *m*

164 air stream; air flow
A system of winds which has a distinctive source or place of origin but which is not necessarily homogeneous.
r воздушное течение
Система ветров, не всегда однородная, но имеющая ясный источник или место формирования.
f courant *m* aérien
s corriente *f* de aire
g Luftstrom *m*

165 air transport
A type of transport moving passengers and goods by air.
r воздушный транспорт
Вид транспорта, занятый перевозкой пассажиров и грузов воздушным путем.
f transport *m* par avion;
transport *m* aérien
s transporte *m* aéreo
g Luftverkehr *m*

166 aircraft industry; aviation industry
A branch of machine-building industry which produces airplanes and helicopters.
r авиационная промышленность
Отрасль машиностроения, производящая самолеты и вертолеты.
f industrie *f* aéronautique
s industria *f* aeronáutica;
industria *f* de aviones
g Flugzeugindustrie *f*

167 airglow; nightglow
A weak and variable radiation of light from the upper atmosphere producing the faint light of the night sky, even when moonless, it is a result of the radiation emitted by the ionized products of ultraviolet waves during the day.

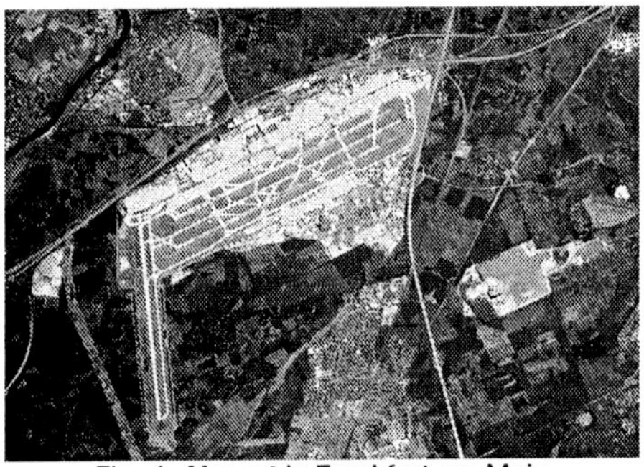

Fig. 4. **Airport** in Frankfurt-on-Main

r ночной отблеск
Слабый поток света из верхней атмосферы, приводящий к слабому свечению ночного неба, даже в отсутствии Луны, вызывается свечением ионизированных продуктов ультрафиолетового излучения, полученного днем.
f luminescence *f* du ciel
s luminiscencia *f* de cielo; luminosidad *f* del cielo
g Himmelsleuchten *n*

168 air-mass clouds
Clouds which occur within a given unstable air mass and not caused by the passage of a front.
r внутримассовые облака
Облака, возникающие в неустойчивой воздушной массе и не связанные с прохождением фронта.
f nuages *mpl* de masse d'air
s nubes *fpl* de masa de aire
g Luftmassenwolken *fpl*

169 air-mass thunderstorm
Any non-frontal thunderstorm caused by great instability of the air mass.
r внутримассовая гроза
Гроза, не связанная с фронтальной деятельностью, а вызванная сильной неустойчивостью состояния атмосферы.
f orage *m* de masse d'air
s tormenta *f* de masa de aire
g Innermassen-Gewitter *n*

170 airport
A transportation complex fully-equipped to be used as a base for the transport of passengers, cargoes and post by air. → Fig. 4
r аэропорт
Транспортный комплекс, обеспечивающий регулярные перевозки пассажиров, грузов и почты средствами авиации. → Рис. 4
f aéroport *m*
s aeropuerto *m*
g Flughafen *m*

171 ajacent sea; marginal sea; semi-closed sea
A part of an ocean adjoining continent and partially isolated from an ocean by peninsulas, islands or bottom sills, e.g. Sea of Japan.
r окраинное море; полузамкнутое море
Часть океана, примыкающая к материку и частично обособленная от открытого океана полуостровами, островами или донными порогами, например, Японское море.
f mer *f* bordiére; mer *f* demi-fermée
s mar *m* costero; mar *m* semicerrado
g Randmeer *n*; halbgeschlossenes Meer *n*; Nebenmeer *n*

172 Akchaghylian basin
A sea basin which in the late Pliocene occupied the area of the contemporary Caspian Sea exceeding it size.
r Акчагыльский бассейн
Морской бассейн, находившийся в позднем плиоцене на месте современного Каспийского моря и значительно превосходивший его по площади.
f bassin *m* akchaghylien
s cuenca *f* de Akchagil
g Aktschaghyl-Becken *n*

173 aksu
A mountain river in Eurasia carrying transparent water or glacier milk.
r аксу
Горная река в Евразии, несущая прозрачную воду или ледниковое молоко.
f aksu *m*
s aksu *m*
g Aksu *n*

174 alas
A large flat-bottomed depression found in the permafrost zone and formed by the melting of ground ice and subsequent land subsidence.
r алас
Обширная плоскодонная котловина в области многолетнемерзлых пород, возникшая в результате протаивания подземного льда и просадки дневной поверхности.
f alass *m*
s alass *m*
g Alas *m*

175 alatau
A high mountain range in Siberia or Central Asia with well defined vertical zonality and landscape characterised by the combination of snow patches, glaciers, rockfields, forests and meadows.
r алатау
Высокие хребты в горах Сибири и Средней Азии с хорошо выраженной высотной поясностью и сочетанием в ландшафтах снежников, ледников, каменных россыпей и участков, покрытых лесами и лугами.
f alatau *m*
s alatau *m*
g Alatau *n*

176 albedo
The ratio of the solar radiation falling on the surface of land to the amount reflected from it, expressed as a percentage or decimal (the average albedo of the Earth is equal to 33% or 0.33). → Fig. 5
r альбедо
Отношение солнечной радиации, отраженной от поверхности земли, к интенсивности радиации, падающей на нее, выражается в процентах или десятичных долях (среднее альбедо Земли равно 33%, или 0,33). → Рис. 5
f albédo *m*
s albedo *m*
g Albedo *f*

177 albeluvisol
An acid soil that has a bleached horizon deposited on a horizon of clay accumulation and penetrating into it in the form of tongues, it is widespread in north-eastern Europe, north-western Asia and south-western Canada.
r альбелювисоль
Кислая почва, имеющая отбеленный горизонт, залегающий на горизонте аккумуляции глины и проникающий в него в виде языков, альбелювисоли широко распространены на север-востоке Европы, северо-западе Азии и юго-западе Канады.
f albeluvisol *m*
s albeluvisol *m*
g Albeluvisol *m*

178 aleurite
Loose fine sedimentary rock consisting mainly of mineral grains from 0.01 to 0.1 mm in size.
r алеврит
Рыхлая мелкообломочная осадочная порода, состоящая преимущественно из минеральных зерен размером от 0,01 до 0,1 мм.
f aleurite *f*
s aleurita *f*
g Schluff *m*; Aleurit *m*

179 aleurolite
Coherent sedimentary rock composed of more than 50% of aleurite particles.
r алевролит
Сцементированная осадочная порода, сложенная более чем наполовину алевритовыми частицами.
f aleurolithe *f*
s aleurolita *f*
g Aleurolith *m*; chluffstein *m*

180 alfisol
A saturated or poorly saturated soil characterized by well-developed illuvial-clay horizon and moistening during the vegetation period, it occurs on all continents and is intensively used in agriculture.
r альфисоль
Насыщенная или слабонасыщенная почва с иллювиально-глинистым горизонтом, увлажненная во время вегетационного периода, распространена на всех континентах и интенсивно используется в сельском хозяйстве.
f alfisol *m*
s alfisol *m*; alfisuelo *m*
g Alfisol *m*

181 algae
Simple autotrophic (photosynthetic) plants containing chlorophyll and living mainly in water.
r водоросли
Низшие автотрофные растения, содержащие хлорофилл и живущие преимущественно в воде.
f algues *fpl*
s algas *fpl*
g Algen *fpl*

* **algae blooming → 6642**

182 algal mat
Accumulation of algae in the upper beach part of surf zone or in near shore shelf depressions.
r водорослевый мат
Скопление водорослей на пляже в верхней части зоны прибоя или на прибрежных участках дна в его депрессиях.
f paillet *m* d'algues
s pallete *m* de algas
g Algenteppich *m*

183 algotrophism
The nourishment of living organisms by means of algae which participate in the symbiosis (e.g., hydra through the association of chlorella, or fungus inside lichen).
r альготрофизм
Питание растений и животных при по-

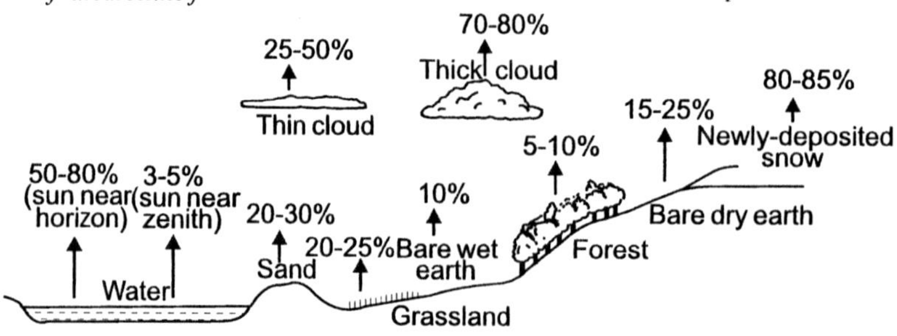

Fig. 5. **Albedo** of different kinds of the Earth's surface

средстве водорослей – участников симбиоза, например, гидры посредством хлореллы, или гриба внутри лишайника.

f algotrophisme *m*
s algotrofismo *m*
g Algotrophismus *m*

* **alien → 2054**

* **alimentation → 184**

184 **alimentation of glacier; glacial alimentation; alimentation**
The combined processes which serve to increase the mass of a glacier or snowfield.
r питание ледника
Поступление на ледник новых масс снега и льда.
f alimentation *f* de glacier
s alimentación *f* de glaciar
g Akkumulation *f* auf dem Gletscher; Gletscherspeisung *f*

185 **alimentation of water bodies; feeding of water bodies**
The inflow of water into a water body from different sources such as rain, snow, glaciers, ground water or a combination of these sources.
r питание водных объектов
Поступление воды в водный объект от источника питания, может быть дождевым, снеговым, ледниковым, подземным, смешанным.
f alimentation *f* des objets hydrologiques
s alimentación *f* hidrológica
g Speisung *f* der Gewässer

186 **alisol**
A soil of tropical and subtropical areas that has a horizon of accumulation of active clay and absorbing complex saturated with aluminium.
r алисоль
Почва тропических и субтропических областей, имеющая горизонт аккумуляции активной глины и поглощающий комплекс, насыщенный алюминием.
f alisol *m*
s alisol *m*; suelo *m* alítico
g Alisol *m*

* **alkali flat → 5287**

187 **alkali flat**
A dried-out floor of a lake where the intense evaporation has left behind a level plain of alkaline sediments, e.g. in the Jordan valley of Israel.
r содовая поверхность
Днище высохшего озера, покрытого содовой коркой, образованной интенсивным испарением в пустынных условиях, как, например, в долине р. Иордан в Израиле.
f superficie *f* alkaline
s superficie *f* alkalina
g Langenoberfläche *f*

188 **alkali soil**
A soil having a high degree of alkalinity (pH>8.5) and/or having a high exchangeable sodium content.
r солонцовая почва
Почва с высокой степенью щелочности (pH>8,5) и с большой изменчивостью содержания натрия.
f sol *m* alkali
s suelo *m* alkali
g alkalischer Boden *m*

189 **alkaline soil**
A soil of arid areas of the temperate belt that contains exchange sodium which gives it specific properties such as clay accumulation, columnar-prismoidal structure, etc.
r солонцеватая почва
Почва засушливых областей умеренного пояса, содержащая обменный натрий, который придает ей специфические свойства: накопление глины, столбчато-призматическую структуру и пр., имеет щелочную реакцию, как правило, pH>7,0.
f sol *m* alcalin
s suelo *m* alcalino
g alkaliner Boden *m*

190 **alkalinity of soil**
The increased content of alkali salts in soil, mainly sodium or magnesium carbonates which favours the development of sodium carbonate salinization which is toxic for many plants.
r щелочность почвы
Повышенное содержание в почве щелочных солей, главным образом карбонатов натрия или магния, что способствует развитию содового засоления и токсично для большинства растений.
f alcalinité *f* du sol
s alcalinidad *f* del suelo
g Bodenalkalinität *f*

191 **alkalinization of soil**
The increase of exchange sodium content in soil by way of colloidal fixation.
r ощелачивание почвы
Процесс увеличения содержания в почве обменного натрия путем его фиксации коллоидами.
f alcalinisation *f* du sol
s alcalinización *f* del suelo
g Bodenalkalisierung *f*

192 **allelopathy**
The release into the environment by organisms of a chemical substance that acts as the growth inhibitor to other organisms, it may even lead to the transformation of a biomass into mortmass.
r аллелопатия
Влияние организмов одной популяции на организмы другой с выделением в окру-

жающую среду продуктов жизнедеятельности и превращением биомассы в мортмассу.
f allélopathie *f*
s alelopatía *f*
g Allelopathie *f*

193 Alleroed
The phase of the late glacial warming of climate (at the end of Valdai Ice Age) dividing the phases of cooling in the middle and late Dryas, approximately from 11400 to 10800 years ago.
r аллерёд
Фаза позднеледникового потепления климата (в конце валдайской ледниковой эпохи), разделяющая фазы похолодания среднего и позднего дриаса, приблизительно от 11400 до 10800 лет назад.
f Alleröd *m*
s Alleröd *m*
g Allerödzeit *f*; Alleröd *m*

194 allitic soil
A deeply weathered soil which lost a considerable amount of silica and is enriched with aluminium.
r аллитная почва
Глубоко выветрелая почва, утратившая значительное количество кремнезема и обогащенная алюминием.
f sol *m* allitique
s suelo *m* alítico
g allitischer Boden *m*

195 allitization
Type of chemical weathering of rocks that is characteristic of humid tropics and subtropics and that results in the formation of thick waste mantles and allitic soils.
r аллитизация
Тип химического выветривания горных пород, свойственный влажным тропикам и субтропикам, в результате которого формируются мощные коры выветривания и аллитные почвы.
f allitisation *f*
s alitización *f*
g Allitisation *f*

* **allocation → 1330**

196 allochore
A plant spreading its fruits, spores, seeds by the action of some external factors such as wind (anemochores), water (hydrochores), animals (zoochores), etc.
r аллохор
Растение, распространяющее плоды, споры, семена с помощью какого-либо внешнего фактора: ветра (анемохоры), воды (гидрохоры), животных (зоохоры) и пр.
f plante *f* allochore
s planta *f* alocora
g Allochor *m*; Fremdverbreiter *m*

197 allochthonous *adj*
Applied to material and features which did not originate in its present position (e.g. till formed from sediments transported by glacier) but introduced by some processes from outside, as opposed to autochthonous.
r аллохтонный
Относящийся к материалу (веществу), принесенному в данный природный объект извне (например, валунная глина, сформировавшаяся из материала, перенесенного ледником), в отличие от автохтонного, образовавшегося здесь же.
f allochtone *adj*
s alóctono *adj*
g allochthon *adj*

198 allochthonous species
A plant or animal species originated not in the area it inhabits, e.g. opossum of North America which came there from South America, as distinct from the *autochtonous species.*
r аллохтонный вид
Виды растений и животных, возникшие в процессе эволюции не в той местности, где они обитают в настоящее время, например, опоссум Северной Америки, расселившийся сюда из Южной Америки, в противоположность *автохтонным видам.*
f espèce *f* allochtone
s especie *f* alóctona
g allochthone Art *f*

199 allochtonous river
Part of a river the water regime of which is formed in the conditions different from those through which it flows.
r транзитная река
Участок реки, водный режим которой сформирован в условиях, отличных от района, по которому она протекает.
f fleuve *m* allogène
s río *m* alóctono
g allochthoner Fluß *m*; Fremdlingsfluß *m*

200 alluvial *adj*
Related to sediments deposited by running water of streams.
r аллювиальный
Относящийся к отложениям, образованным водными потоками.
f alluvial *adj*
s aluvial *adj*
g alluvial *adj*; Schwemm-

201 alluvial cone; debris fan
An accumulation of loose debris material deposited by a permanent or temporary stream at the lower end of a gulch or valley where a sharp decrease of kinetic energy of the stream occurs.
r конус выноса; аллювиальный конус
Скопление рыхлого обломочного материала, отложенного постоянным или временным водотоком у нижнего конца ов-

рага или долины, где происходит резкое уменьшение живой силы потока.
f cône *m* alluvial;
cône *m* de déjection torrentiel
s cono *m* aluvial;
cono *m* de deyección torrencial
g Schwemmfächer *m*; Schwemmkegel *m*;
Schuttfächer *m*

* **alluvial deposits → 207**

202 alluvial fan
A fan-shaped mass of a coarse alluvium (sand and gravel), deposited by a stream where it emerges from a narrow valley at a mountain front and flows on to a plain or into a wide trunk valley.
r пролювиальный конус выноса; веер
Конусообразная масса грубого материала (песка и гравия), отложенная потоком на выходе его из узкой горной долины на равнину или в широкую главную долину.
f éventail *m* alluvial
s abanico *m* aluvial
g Schwemmfächer *m*

203 alluvial fan deposit
Terrestrial accumulations of loads by temporary streams in the mouths of their valleys.
r пролювий; пролювиальные отложения
Скопления наносов временных водотоков в устьевых частях долин, в местах слияния образующих пролювиально-аллювиальные шлейфы.
f alluvion *f* torrentielle
s aluvión *m* torrentoso
g Proluvium *n*; proiluviale Ablagerungen *fpl*

204 alluvial plain; washplain
A plain formed as a result of the accumulative action of a large river.
r аллювиальная равнина;
намывная равнина
Равнина, образовавшаяся в результате аккумулятивной деятельности крупной реки.
f plaine *f* alluviale
s llanura *f* alluvial
g Alluvialebene *f*; Schwemm(land)ebene *f*

205 alluvial soil
A fertile soil usually forming in river flood-plains and deltas under the conditions of periodical flooding and accumulation of alluvium.
r аллювиальная почва; пойменная почва
Плодородная почва, формирующаяся в речных поймах и дельтах в условиях периодического затопления и аккумуляции наносов.
f sol *m* alluvial
s suelo *m* aluvial
g Alluvialboden *m*; Schwemmlandboden *m*;
Aueboden *m*

* **alluvial terrace → 125**

206 alluvial water
Water contained in alluvial deposits of the contemporary and ancient river valleys.
r аллювиальная вода
Вода, залегающая в аллювиальных отложениях современных и древних речных долин.
f eau *f* alluviale
s agua *f* aluvial
g Alluvialwasser *n*

* **alluvian coast → 6730**

* **alluvian island → 6729**

207 alluvium; fluvial deposits; alluvial deposits; fluvial sediments
Unconsolidated, loose material (not only silt but also the gravel and sand) brought down and deposited by running water in the river beds, flood-plains and terraces of river valleys.
r аллювий; речные наносы; речные отложения; аллювиальные отложения
Отложения русловых водных потоков, слагающие поймы и террасы речных долин.
f alluvion *f*; sédiments *mpl* fluviaux;
dépots *mpl* fluviaux
s aluvión *m*; depósitos *m* aluviales;
sedimentos *mpl* fluviales
g Alluvium *n*; Flußsedimente *mpl*;
alluviale Ablagerungen *fpl*;
Flußablagerungen *fpl*

208 alluvium sheet
A thin cover of loose sediments deposited on watersheds and slopes.
r плащ наносов[1]
Тонкий покров рыхлых отложений на междуречьях и склонах.
f nappe *f* alluviale; manteau *m* de débris
s capa *f* aluvial; manto *m* de derrubios
g Schwemmschicht *f*

209 alp
The high-altitude bench or shoulder of a mountain slope lying usually above the tree line and covered with snow in winter, which provides valuable summer pastures for grazing animals.
r альпийское пастбище
Сравнительно пологий склон, лежащий выше границы леса и зимой покрытый снегом, представляющий плодородное летнее пастбище для травоядных животных.
f alpage *m*; alpe *m*
s alpe *m*
g Alm *f*; Bergweide *f*

210 alpine *adj*
Related to any high mountain terraines and more specifically to mountain plants, climate, glaciers, soils, etc.

r альпийский
Относящийся к природным особенностям высокогорья, высокогорной растительности, климата, ледников, почв и т.п.
f alpin *adj*; alpestre *adj*
s alpino *adj*; alpestre *adj*
g Alpen-; alpin *adj*

211 alpine belt
An altitudinal belt in mountains of humid climate in temperate and subtropical latitudes found between subalpine and niveo-glacial belts, it is characterized by cold high mountain climate, abundant snowfalls and high solar radiation.
r альпийский пояс
Высотный пояс в хорошо увлажняемых горах умеренных и субтропических широт, расположенный между субальпийским и нивально-гляциальным поясами, отличается холодным высокогорным климатом, обильными снегопадами и значительным приходом солнечной радиации.
f étage *m* alpin
s piso *m* alpino
g alpine Stufe *f*

212 alpine desert
A type of cold desert forming on high plateaus with cold dry climate (typical of Eastern Pamir, Tibet, some parts of the Andes).
r высокогорная пустыня
Разновидность холодной пустыни, формирующаяся на высоких нагорьях с холодным сухим климатом, характерна для Восточного Памира, Тибета, некоторых частей Анд.
f désert *m* d'altitude
s desierto *m* de altitud
g Hochgebirgswüste *f*

*** alpine folding → 216**

213 alpine glacier
A simple valley glacier descending off a mountain glacial cirque along just one valley.
r альпийский ледник
Простой долинный ледник, спускающийся за пределы горного цирка по одной долине.
f glacier *m* alpin
s glaciar *m* alpino
g Hochgebirgsgletscher *m*;
Gletscher *m* von alpinem Typ

214 alpine glow
A series of phenomena observed over mountains around sunrise and sunset when the sun is 2° above the horizon and snow-covered peaks assume yellow-pink-purple tints.
r альпийская заря; пурпурный свет
Природное явление, наблюдаемое в горах при восходе и заходе Солнца, когда оно находится на 2° над горизонтом и окрестные снежные пики приобретают желто-розово-пурпурный оттенок.

Fig. 6. **Alpine meadow**

f lueurs *fpl* crépusculaires
s resplandor *m* alpino;
luzes *fpl* crepusculares
g Alpenwiderschein *m*; Alpenglühen *n*

215 alpine meadow
Short-grass meadows which occur in the mountains above the subalpine belt. → Fig. 6
r альпийский луг
Низкотравные луга, распространенные в горах выше субальпийского пояса. → Рис. 6
f prairie *f* alpine
s prado *m* alpino; pradera *f* alpina
g Alpenwiese *f*; Matte *f*

216 alpine orogeny; alpine folding
A mountain-building movement which culminated in mid-Tertiary, having created the Alpine ranges which stretch from the Pyrenees, Atlas and Alps in Western Europe and Northern Africa through the Caucasus to the Hymalayas of Asia.
r альпийская складчатость
Совокупность процессов складчатости и горообразования, происходивших в кайнозойскую эру и приведших к формированию альпийских хребтов, протягивающихся от Пиренеев, гор Атласа и Альп к Кавказу и Гималаям.
f plissement *m* alpin
s plegamiento *m* alpino
g alpine Faltung *f*; Alpenfaltung *f*

217 alpine relief
Deeply dissected relief with steep rocky slopes, sharp crests and tops, and with the traces of the latest glaciation.
r альпийский рельеф
Резко расчлененный рельеф с крутыми скалистыми склонами, острыми гребнями и вершинами, со следами недавнего оледенения.
f relief *m* alpin
s relieve *m* alpino
g alpines Relief *n*

218 alpine tundra
A vegetation belt in the mountains above the treee-line characterized by lichens, mosses, cold-resistant grasses and shrubs.

r горная тундра
Пояс растительности в горах, располагающийся выше границы леса и характеризующийся господством лишайников, мхов и некоторых холодостойких трав и кустарников.
f toundra *f* de montagnes; toundra *f* alpine
s tundra *f* de montañas; tundra *f* alpina
g Gebirgstundra *f*

219 **alpine vegetation**
The vegetation of the alpine belt represented by short grass "carpet" meadows and dwarf shrub brushes.
r альпийская растительность
Растительность альпийского пояса гор, представленная низкотравными «ковровыми» лугами и зарослями кустарничков.
f végétation *f* alpine
s vegetación *f* alpina
g alpine Vegetation *f*

* **alpinism → 3994**

220 **alternative energy; green energy**
Electricity produced from renewable sources of energy, such as wind power, tidal power, etc.
r альтернативная энергия
Электрическая и иные виды энергии, получаемые от использования нетрадиционных источников, таких как энергия ветра, приливов и т.п.
f énergie *f* alternative
s energía *f* alternativa
g alternative Energie *f*; grüne Energie *f*

221 **alternative sources of energy; renewable energy sources**
Generic term for such non-conventional renewable sources of energy as wind, tides, geothermal heat, etc. in contrast to conventional ones such as coal, oil, gas, etc.
r альтернативные источники энергии; возобновляемые источники энергии
Собирательное понятие для нетрадиционных возобновляемых источников энергии (ветра, приливов, геотермического тепла) в отличие от традиционных (нефти, газа, угля и пр.).
f ressources *fpl* énergétiques alternatives; sources *fpl* énergetiques alternatives; energie *f* renouvelable
s recursos *mpl* energéticos alternativos; energía *f* renovable; fuentes *mpl* energéticos alternativos
g alternative Energieresourcen *fpl*; erneuerbare Energie *f*

222 **alternative technology**
The technology concerned with the technical processes that make use of renewable rather than non-renewable natural resources and which are therefore regarded as ecologically sustainable.
r альтернативная технология
Технология, основанная на производственных процессах, использующих в основном возобновляемые ресурсы и поэтому рассматриваемая как экологически устойчивая.
f technologie *f* alternative
s technología *f* alternativa
g alternative Technologie *f*

223 **altimetric frequency curve**
A curve constructed to show generalized altitudes in an area that is made by dividing a map into small squares and taking the highest or lowest points in each.
r частотная кривая рельефа
Кривая, показывающая генерализованные высоты в данном районе, построенная по карте, разбитой на небольшие квадраты, в которых вычисляются наивысшие или наинизшие высоты.
f courbe *f* de fréquence altimétrique
s curva *f* de frecuencia altimétrica
g Häufigkeitskurve *f* des Reliefs

224 **altiplanation**
The flattening of relief in high mountainous and subpolar areas which takes place under the impact of solifluction, frost weathering, nivation, and other cryogenic processes.
r альтипланация
Выравнивание рельефа в высокогорных и субполярных областях, происходящее под действием солифлюкции, морозного выветривания, нивации и прочих криогенных процессов.
f altiplanation *f*
s altíplanación *f*
g Altiplanation *f*

225 **altiplanation terrace; solifluction terrace; goletz terrace**
A terrace formed on a hill slope by a combination of freeze-thaw processes and solifluction under periglacial conditions.
r нагорная терраса; солифлюкционная терраса; гольцовая терраса
Терраса, формирующаяся на склоне под действием процессов замерзания-таяния и солифлюкции в перигляциальных условиях.
f terrasse *f* d'altiplanation; terrasse *f* de goletz; terrasse *f* de solifluxion
s terraza *f* de altiplanicie; terraza *f* de solifluxión; terraza *f* de goletz
g Gebirgsterrasse *f*; Goletzterrasse *f*

226 **altiplano**
An elevated plateau in the Andes.
r альтиплано
Высокогорное плато, нагорная равнина в Андах.
f altiplano *m*
s altiplano *m*
g Altiplano *m*

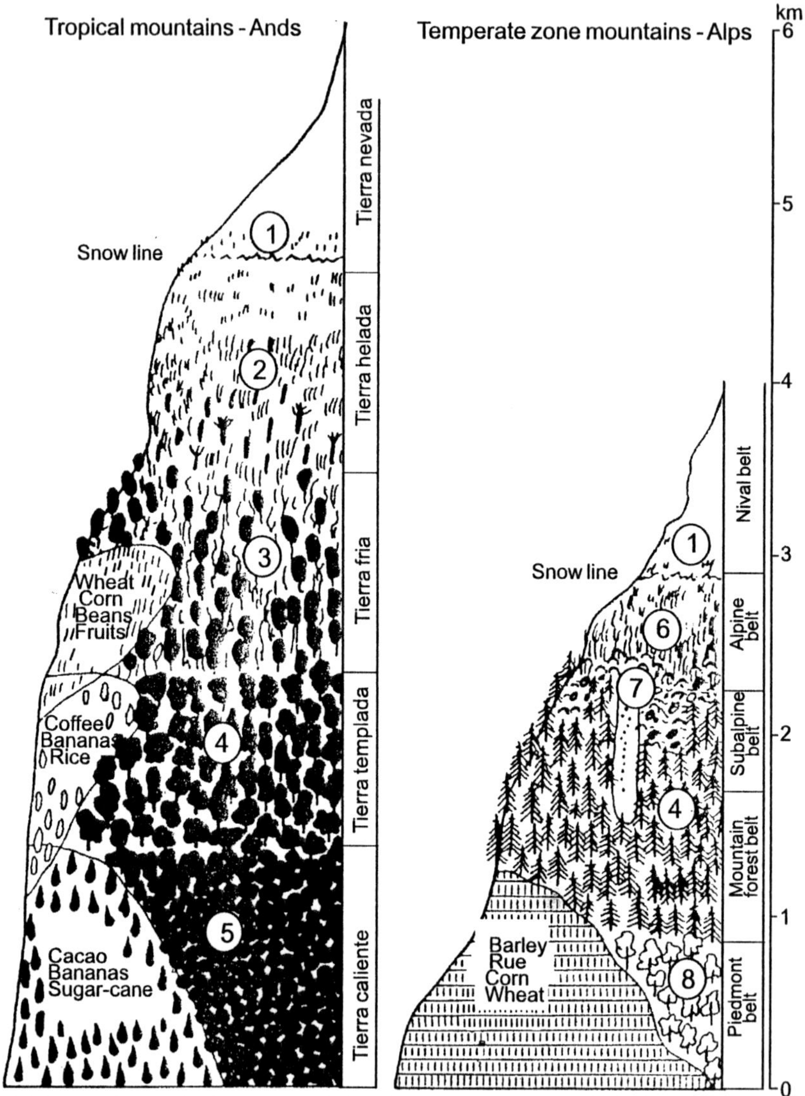

Fig. 7. **Altitudinal belts.** 1– pioneer vegetation, 2 – paramos, 3 – cloud forests, 4 – mountain forest, 5 – rain forest, 6 – carpe-like communities, 7 – shrub, 8 – broad-leaved forest

227 altitude; absolute elevation
The height of a given point above mean sea level, measured vertically
r абсолютная высота; альтитуда
Расстояние по вертикали от среднего уровня поверхности океана до данной точки земной поверхности.
f altitude *f* absolue
s altitud *f* absoluta
g absolute Höhe *f*; Höhenlage *f*

228 altitudinal belt; vertical zone; altitudinal zone
Landscape areas in the mountains with more or less homogeneous natural conditions which lie, sometimes interrupted, one above another. → Fig. 7
r высотный пояс
Сравнительно однородные по природным условиям, иногда прерывистые полосы ландшафтов в горах, расположенные одна над другой. → Рис. 7
f étage *m* altitudinal
s piso *m* altitudinal
g Höhenstufe *f*; Höhengürtel *m*

229 **altitudinal vegetation belt**
Strips of vegetation changing successively with height in the mountains varying in the slopes.
r высотный пояс растительности
Последовательно сменяющиеся с высотой типы растительности, образующие сплошные или прерывистые полосы, не одинаковые на склонах разной экспозиции.
f étage *m* de végétation
s piso *m* de vegetación
g Vegetationshöhenstufe *f*

230 **altitudinal zonation; vertical zonation**
A successive change of landscapes in the mountains connected with the change of climatic conditions (the decrease of heat, change of the amount of precipitation and air humidity) with height above sea level.
r высотная поясность; вертикальная поясность
Последовательная смена ландшафтов в горах, связанная с изменением климата (уменьшением тепла, изменением количества осадков и влажности воздуха) в зависимости от высоты над уровнем моря. .
f zonalité *f* altitudinale; zonalité *f* verticale; étagement *m* des paysages
s zonalidad *f* altitudinal; escalonamiento *m*; zonación *f* vertical
g Höhenstufung *f*; Höhengliederung *f*

* **altitudinal zone → 228**

231 **altocumulus**
A type of cloud (symbol: Ac) of middle altitudes (2000-7000 m), white and grey in colour, occurring in a patch, sheet or r layer, generally casting shade to the surface (typical medium-level clouds).
r высококучевые облака
Тип облаков (символ: Ac) на высотах 2000-7000 м белого или серого цвета, появляющихся в виде пелены или облачного слоя, обычно дающих тени на земной поверхности (типичные облака среднего яруса).
f altocumulus
s altocumulus
g Altocumulus

232 **altostratus**
A type of cloud (symbol: As) of middle altitudes (2000-7000 m), greyish and bluish in colour, occurring in sheets or layers of fibrous or homogeneous structure (it is associated with rainy weather).
r высокослоистые облака
Тип облаков (символ: As) на высотах 2000-7000 м сероватого или синеватого цвета, появляющихся в виде пелены или слоя волокнистой или однородной структуры (эти облака сопровождаются дождливой погодой).
f altostratus
s altostratus
g Altostratus

233 **alvar**
A landscape in Scandinavia characterized by a bare or treeless limestone terrain with a thin layer of soil.
r альвар
Ландшафт в Скандинавии, характеризующийся голой или покрытой тонким слоем почвы безлесной поверхностью, сложенной известняками.
f alvar *m*
s alvar *m*
g Alvar *m*

234 **ambes *pl***
Flat tops of Ephiopian Highlands composed of horizontally bedded rock strata dissected by deep valleys.
r амбы
Плоские вершины Эфиопского нагорья, сложенные горизонтально залегающими пластами горных пород и разделенные глубокими долинами.
f ambes *fpl*
s ambes *fpl*
g Amben *fpl*

235 **amenity**
An object of the environment which is perceived as pleasant and attractive.
r привлекательный объект; аттрактивный объект
Объект окружающей среды, приятный для восприятия людей.
f aménité *m*
s atracción *f* turística
g Anmut *f*

236 **amictic lake**
A lake permanently covered with ice, in which no seasonal changes of vertical circulation of water occur.
r амиктическое озеро
Озеро, постоянно покрытое льдом, в водной толще которого отсутствуют сезонные изменения вертикальной циркуляции.
f lac *m* amictique
s lago *m* amíctico
g amiktischer See *m*

237 **amorphous ice**
Ice having no definite crystalline structure (does not occur in nature).
r аморфный лед
Лед, не имеющий кристаллического строения (в природных условиях не встречается).
f glace *f* amorphe
s hielo *m* amorfo
g amorphes Eis *n*

238 **amphiatlantic distribution**
A kind of the distribution of sea organisms in the northern part of the Atlantic Ocean when the species inhabiting the littoral zone of

Europe and North America are not found in the open ocean, e.g. cod.

r амфиатлантическое распространение
Разорванное распространение морских организмов в северной части Атлантического океана, когда виды, встречающиеся у берегов Европы и Северной Америки, отсутствуют в открытой части океана, например, треска.

f distribution *f* amphiatlantique
s dispersión *f* anfiatlántica
g amphiatlantische Verbreitung *f*

239 amphibians
The vertebrates adapted both to terrestrial and aquatic conditions.

r земноводные; амфибии
Животные, приспособленные к жизни как в воде, так и на суше.

f amphibies *fpl*
s anfibios *mpl*
g Amphibien *fpl*; Lurche *mpl*

240 amphiboreal distribution
A kind of the distribution of sea organisms when one and the same species lives both in theb temperate and cold belts of the Atlantic and Pacific Oceans, but is not found in the seas of the north of Asia and North America, e.g. dolphins and seals.

r амфибореальное распространение
Разорванное распространение морских организмов, при котором один и тот же вид обитает в умеренном и на окраине арктических поясов Атлантического и Тихого океанов, но отсутствует в морях, омывающих Северную Азию и север Северной Америки, например, дельфины и обыкновенный тюлень.

f distribution *f* amphiboréale
s dispersión *f* anfiboreal
g amphiboreale Verbreitung *f*

241 amphidromic point
A point on certain hydrographic charts from which cotidal lines radiate and no tidal fluctuations of sea level are observed. → Fig. 8

r амфидромическая точка
Точка, из которой расходятся котидальные линии (линии равных приливов) и где приливные колебания уровня моря отсутствуют. → Рис. 8

f point *m* amphidromique
s punto *m* anfidrómico
g amphidromischer Punkt *m*; Knotenpunkt *m*

242 amphidromic system
A tidal system in which the high water rotates round a central point (the nodal or amphidromic point), the rotation being caused by the Coriolis force of the Earth's rotation. → Fig. 8

r амфидромическая система
Система приливо-отливных течений, в ко-

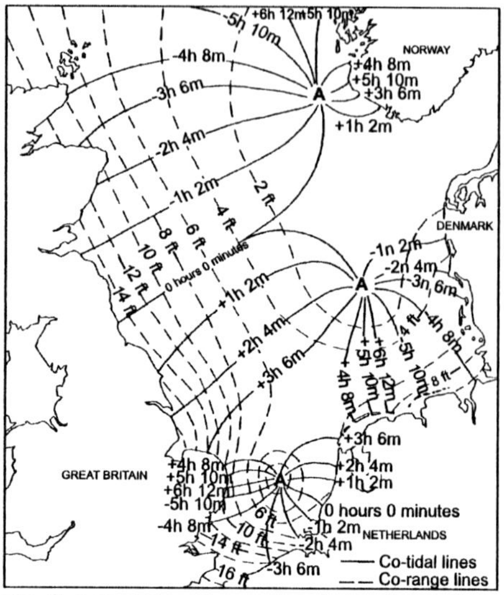

Fig. 8. **Amphidromic system** in the North Sea. A – amphidromic point

торой высокая вода вращается вокруг центральной (амфидромической) точки, что вызвано действием силы Кориолиса. → Рис. 8

f système *m* amphidromique
s sistema *m* anfidrómico
g amphidromisches System *n*

243 amphipacific distribution
A kind of the distribution of sea organisms in the northern part of the Pacific Ocean, when one and the same species inhabits the littoral zone of both Asia and North America, but is not found in the open ocean.

r амфипацифическое распространение
Разорванное распространеие морских организмов в северной половине Тихого океана, встречающихся в водах у побережья Азии и Северной Америки, но отсутствующих в открытой части океана.

f distribution *f* amphipacifique
s dispersión *f* anfipacífica
g amphipazifische Verbreitung *f*

244 amphiphyte
A water plant rooted into the ground of a water body with only the leaves and flowers located above the water surface.

r амфифит; земноводное растение
Водное растение, прикрепляющееся к грунту водоема, листья и цветы которого в основном расположены над водой.

f amphiphyte *m*
s anfifita *m*
g Amphiphyt *m*

245 amphitheatre
A semicircular depression of glacial, landslide or volcanic genesis which resembles an ancient Greek theatre.

r амфитеатр
Полукруглое замыкание склонов ледникового, оползневого или вулканического происхождения, по форме напоминающее древний греческий театр.
f amphithéâtre *m*
s anfiteatro *m*
g Amphitheater *n*

246 anabatic front
Front in which the warm air ascends along the frontal surface.
r фронт восходящего скольжения; анафронт
Атмосферный фронт с восходящим скольжением теплого воздуха над фронтальной поверхностью.
f front *m* anabatique
s frente *m* anabático
g Anafront *f*

247 anabatic wind
An upslope wind formed due to convection when air on hill-sides is heated by insolation more rapidly and to a greater extent than air lying above the valley floor. → Fig. 178, p. 395
r восходящий ветер
Ветер, дующий вверх по склону, вызванный ковекцией, когда воздух на склоне холма нагревается благодаря инсоляции быстрее и в большей степени, чем воздух над дном долины. → Рис. 178, с. 395
f vent *m* anabatique
s viento *m* anabático
g Hangaufwind *m*; anabatischer Wind *m*

248 anabiosis
Temporary state of a plant or animal when the metabolism slows down considerably, and all the visible manifestations of life are absent.
r анабиоз
Временное состояние растительного или животного организма, при котором значительно снижается обмен веществ и отсутствуют все видимые проявления жизни.
f anabiose *f*
s anabiosis *f*
g Anabiose *f*

* **anabranch → 779**

249 anabranching
The condition when a river is divided into two or more channels by islands which have a width greater than three times the main channel width.
r разветвление реки
Разделение русла реки на два и более крупных потоков с островами между ними, ширина которых по крайней мере втрое превышает ширину вновь образованных рукавов.
f formation *f* des bras; diffluence *f*
s formación *f* de brazos; difluencia *f*
g Verzweigung *f* des Flusses; Flußverzweigung *f*

250 anaerobic *adj*
Applied to organisms that do not need free oxygen for existence, and receive the energy to live by decomposing organic and inorganic substances, as well as to processes that occur without free oxygen.
r анаэробный
Относящийся к организмам, способным существовать при отсутствие свободного кислорода, за счет энергии, получаемой путем расщепления органических и неорганических веществ, а также к процессам, протекающим без кислорода.
f anaérobie *adj*
s anaerobio *adj*; anaeróbico *adj*
g anaerob *adj*

251 anaerobic process
Microbiological, biochemical and chemical transformations going on under deficit or absence of oxygen, they are typical of closed water bodies, bogs, waterlogged grounds and soils (favours peat formation).
r анаэробный процесс
Совокупность микробиологических, биохимических и химических трансформаций, протекающих при недостатке или отсутствии кислорода, широко распространен в замкнутых водоемах, болотах, переувлажненных грунтах и почвах, способствует образованию торфа.
f processus *m* anaérobie
s proceso *m* anaeróbico
g anaerober Prozeß *m*

* **anaglyph → 252**

252 anaglyphical map; anaglyph
A map printed in two mutually complementary colors (green and red) with parallax shift so as two images form a stereo pair, which permits to get a three-dimensional image.
r анаглифическая карта
Карта, отпечатанная двумя взаимно дополняющими цветами (зеленым и красным) с параллактическим смещением, так что оба изображения образуют стереопару, дающую объемное изображение.
f anaglyphe *m*
s mapa *m* anaglífico
g Anaglyphenbild *n*; Anaglyphe *f*

253 anaglyphy
A method of obtaining a three-dimensional image of topography by viewing two adjoining aerial photographs that have been printed in two complementary colours, e.g. red and green.
r анаглифический метод
Метод получения стереоскопической модели рельефа путем просматривания двух смежных аэрофотоснимков, напечатанных двумя дополнительными цветами, например, зеленым и красным.
f méthode *f* des anaglyphes

s método *m* anáglifo
g Anaglyphenverfahren *n*

254 anahalinity
The increase of the sea water salinity from the bottom to the surface.
r анахалинность
Возрастание солености морской воды от дна к поверхности.
f anahalinité *f*
s anahalinidad *f*
g Anahalinität *f*

255 analogue landscapes
Geographical landscapes in different locations but with similar features, e.g. delta landscapes or desert landscapes of different zones.
r ландшафты-аналоги
Географические ландшафты, разобщенные территориально, но сходные по основным признакам, например, дельтовые или пустынные ландшафты различных зон.
f paysages-analogies *mpl*
s paisajes *mpl* análogos
g Landschaften-Analogen *npl*

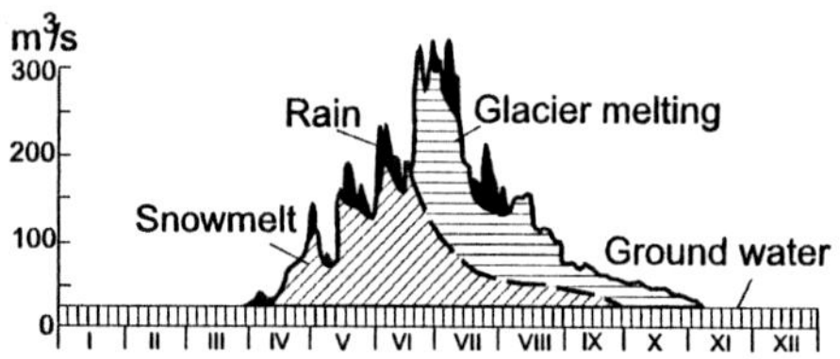

Fig. 9. The **analysis of a hydrograph**

256 analysis of hydrograph
Graphical singling out of water volumes on a hydrograph formed by different sources of alimentation. → Fig. 9
r расчленение гидрографа
Графическое выделение на гидрографе объемов воды, сформированных различными источниками питания. → Рис. 9
f analyse *f* de courbe hydrométrique
s análisis *m* de curva hidrométrica
g Gliederung *f* der Abflußkurve

257 analythical map
A geographical map showing the characteristics of a phenomenon (e.g. air temperature) which are not well generalised, or only separate features of an object (e.g. slope aspect).
r аналитическая карта
Географическая карта, показывающая малообобщенные показатели какого-либо явления (например, температуру воздуха) или только отдельные стороны объекта (например, экспозицию склонов).
f carte *f* analytique
s mapa *m* analítico
g analytische Karte *f*

* **anamorphoid → 258**

258 anamorphotic map; anamorphoid
Topological cartographic image in which the composing parts of a territory are proportional to the mapped index, e.g. population.
r анаморфированная карта; картоид
Топологическое картографическое изображение, в котором составляющие части территории пропорциональны картографируемому показателю, например численности населения.
f carte *f* anamorphotique
s mapa *m* anamorfótico; anamorphosis *m*
g Kartenanamorphote *f*; verzerrte Karte *f*

* **anastomosing channel → 776**

* **anastomosis → 777**

259 anchor ice; bottom ice
Ice which forms on the bed of a body of water and attached to the bottom of a stream, lake or shallow sea.
r донный лед; якорный лед
Лед, образующийся на дне рек, озер, мелких морей и скрепленный с грунтом.
f glace *f* de fond
s hielo *m* de fondo
g Grundeis *n*; Bodeneis *n*

260 ancient glaciation
Glaciations of the past geological epoques.
r древнее оледенение
Оледенения прошлых геологических эпох.
f glaciation *f* ancienne
s glaciación *f* antigua
g alte Vereisung *f*; ehemalige Vereisung *f*

261 ancient shoreline
A shoreline of a water body found above or below the present zone of breaking of waves because it formed due to the vertical movement of land or change of sea level.
r древняя береговая линия
Береговая линия водоема, находящаяся выше современной полосы прибоя или ниже нее (под водой) вследствие поднятия или опускания суши, подъема и опускания уровня водоема.
f ligne *f* littorale ancienne;
ligne *f* de rivage ancien
s línea *f* de costa antigua
g alte Küstenlinie *f*; alte Strandlinie *f*

262 ancient valley
An old valley which existed in the area before the present valley.
r прадолина
Древняя долина, предшествовавшая ныне существующей в данном месте.
f vallée *f* ancienne
s valle *m* antiguo
g Urtal *n*

263 Ancylus lake
A closed fresh water basin which existed in

boreal period (about 7.5-6 thousand years ago) in the area of the contemporary Baltic Sea.

r Анциловое озеро
Замкнутый пресноводный бассейн, существовавший в бореальное время (около 7,5-6 тыс. лет назад) на месте современного Балтийского моря.

f lac *m* à Ancylus

s lago *m* de Ancylus

g Ancylus-See *m*; Ancylussee *m*

264 andosol; volcanic soil
A soil with dark surface layers rich in organic substances that develops on volcanic ash in the regions of volcanic activity, especially along the coast of the Pacific Ocean.

r вулканическая почва; андосоль
Молодая пеплово-вулканическая почва, встречается в районах повышенной вулканической активности, особенно по побережью Тихого океана.

f sol *m* volcanique

s suelo *m* volcánico; suelo *m* ándico; andosuelo *m*

g vulkanischer Aschenboden *m*; Andosol *m*

265 anemochore
Plants whose fruits, seeds and spores are disseminated by wind.

r анемохор
Растения, у которых плоды, семена и споры распространяются ветром.

f anémochore *m*

s anemócoro *m*

g Anemochor *m*; Windwanderer *m*

* **anémophile *m* → 266**

266 anemophyte; anémophile *m*
Plants which are pollinated by wind.

r анемофит
Растения, обладающие способностью к опылению путем переноса пыльцы ветром.

f anémophyte *m*

s anemófita *m*; anemófilo *m*

g Anemophyt *m*

267 angle of repose; angle of rest
The maximum slope gradient at which a mass of unconsolidated material will remain stable.

r угол покоя
Максимальный уклон, при котором массы рыхлого материала остаются неподвижными на склоне.

f angle *m* de repos

s ángulo *m* de reposo

g Winkel *m* der Ruhelage

* **angle of rest → 267**

* **angular alteration → 268**

268 angular distortion; angular alteration
The inequality of angles on a geographical map and corresponding angles on the surface of the Earth's ellipsoid, measured by the difference between them. → Fig. 10

r искажение углов
Неравенство углов на географической карте и соответствующих им углов на поверхности земного эллипсоида, измеряемое разницей между ними. → Рис. 10

f altération *f* angulaire

s deformación *f* angular

g Winkelverzerrung *f*

269 angular uncomformity
The difference in the inclination of two succesive strata which indicates that in the past the sedimentation occurred in different condition.

r угловое несогласие
Прилегание двух последовательных толщ осадочных пород под углом друг к другу, что свидетельствует об изменении в прошлом условий их напластования.

f discordance *f* angulaire

s discordancia *f* angular

g Winkeldiskordanz *f*

* **animal husbandry → 3537**

* **animal sanctuary → 5193**

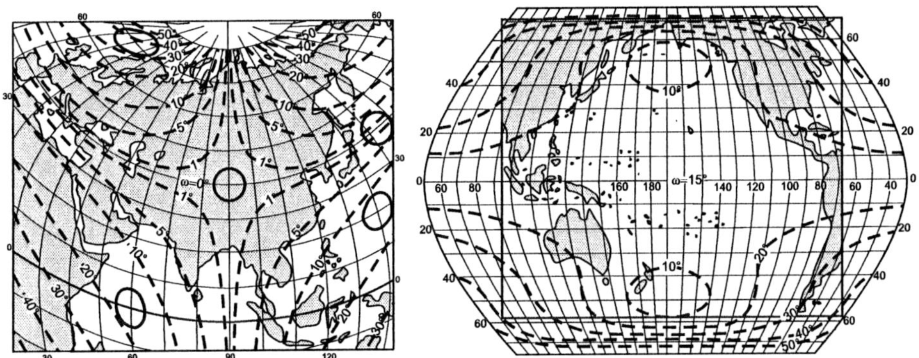

Fig. 10. **Angular distortion** (isolines) at the pseudoconical equivalent (left) and pseudocylindrical (right) cartographic projections

270 **animal trail**
A relatively permanent path made by animals, especially in dense thickets and through deep snow in winter.
r звериная тропа
Относительно постоянный путь передвижения животных, проторенный одним или несколькими видами, особенно в густых зарослях, а зимой по глубокому рыхлому снегу.
f sentier *m* des bêtes
s senda *f* de fieras
g Tierpfad *m*

271 **animated map; animation**
Screen program-controlled map, whichhas the effect of motion and transfer of the whole image or separate objects over the screen during demonstration.
r анимационная карта; динамическая карта
Экранная программно-управляемая карта, воспроизводящая при демонстрации эффект движения, перемещения всего изображения или отдельных объектов по экрану.
f carte *f* d'animation
s mapa *m* de animación
g Animationskarte *f*

* **animation → 271**

272 **annexation**
Taking possession or incorporation of a country or other territory within the domain of a state or keeping it within the limits of another state by force.
r аннексия
Насильственное присоединение (захват) всей или части территории другого государства или народа, а также насильственное удержание народа в границах другого государства.
f annexion *f*
s anexión *f*
g Annexion *f*; Annektierung *f*

273 **annual *adj***
Occurring annually or once a year.
r годичный; однолетний; годовой
Случающийся ежегодно или раз в год.
f annuel *adj*
s anual *adj*
g jährlich *adj*; annuell *adj*; einjährig *adj*

274 **annual course**
Change of a meteorological element during a year.
r годовой ход
Изменение метеорологического элемента в течение года.
f évolution *f* annuelle
s curso *m* anual
g Jahresgang *m*

275 **annual growth rings; annual ring; tree ring**
The annual increase of wood added during a single growth period, seen as concentric rings in a cross-section of the stem or trunk of woody plants.
r годичные кольца древесины
Ежегодно формирующиеся кольца на древесном стволе, легко видимые на спиле.
f couche *f* annuelle
s anillo *m* de crecimiento del árbol
g Jahrring *m*

276 **annual layer**
A layer of ice in a glacier or sediments on a lake bottom deposited during one year.
r годовой слой
Горизонт в ледниковой или озерной толще, отложенный за год.
f couche *f* annuelle
s lámina *f* annual
g Jahresschicht *f*

277 **annual plant**
A plant whose life cycle (germination, flowering, fruiting) takes place within a period of one year or only a growing season.
r однолетнее растение
Растение, развитие которого от прорастания до созревания семян проходит в течение одного года, а иногда лишь одного вегетационного периода.
f plante *f* annuelle
s planta *f* anual
g Einjahrespflanze *f*

278 **annual range**
A measure of the difference between the maximum and minimum average monthly values of a meteorological element, recorded whithin a period of one year.
r годовая амплитуда
Разность наибольшего и наименьшего средних месячных значений какого-либо изменяющегося в течение года метеорологического элемента.
f amplitude *f* annuelle
s amplitud *f* annual
g Jahresamplitude *f*

* **annual ring → 275**

279 **annual runoff**
The total water volume flowing over a year from a drainage area usually referred to the outlet of a drainage basin.
r годовой сток
Общий объем воды, стекающей за год, обычно отнесенный к замыкающему створу водосборного бассейна.
f écoulement *m* annuel
s escorrentía *f* annual
g jährlicher Abfluß *m*; Jahreabfluß *m*

280 **annular drainage pattern**
A drainage pattern in which the consequent streams radiate, as a spokes of a wheel, and

the subsequent streams eroding their valleys in the weaker strata following an underground circular structure (e.g. dome-like igneous intrusion, batholith) make a series of concentric circles. → Fig. 95, p. 202

r кольцевая речная сеть
Речная сеть, обычно развивающаяся на склонах батолита, когда субсеквентные реки следуют в плане дугообразно вплоть до впадения в консеквентную реку. → Рис. 95, с. 202

f réseau *m* hydrographique annulaire
s red *f* hidrográfica anular
g ringförmiges Flußnetz *n*

281 anomalous watershed
A watershed in a mountain region which does not run along the crest of the highest range of a mountain chain.

r аномальный водораздел
Водораздел в горном районе, протягивающийся не по гребню высочайшего хребта в горной цепи.

f ligne *f* anomale de partage des eaux
s línea *f* anómala de división de aguas
g anomale Wasserscheide *f*

282 anoxic zone
Anaerobic zone of a water body where the concentration of organic matter is high and the velocity of oxygen consumption exceeds that of its income.

r аноксичная зона
Анаэробная зона водоема, где высока концентрация органических веществ, а скорость потребления кислорода превышает скорость его поступления.

f zone *f* dépourvu d'oxygéne
s zona *f* desprovista de oxígeno
g anoxische Zone *f*

283 antarctic ***adj***
Related to the polar region of the Southern Hemisphere.

r антарктический
Относящийся к полярному региону Южного полушария Земли.

f antarctique *adj*
s antártico *adj*
g antarktisch *adj*

*** antarctic air → 284**

284 antarctic air mass; antarctic air
Cold air masses formed over the Antarctic continent and Southern Ocean, having lower levels which are extremely cold

r антарктическая воздушная масса; антарктический воздух
Исключительно холодная воздушная масса, особенно в нижних уровнях, формирующаяся над Антарктическим континентом и Южным океаном.

f masse *f* d'air antarctique; air *m* antarctique
s masa *f* de aire antártica; aire *m* antártico
g antarktische Luftmasse *f*; antarktische Luft *f*; Antarktikluft *f*

285 Antarctic circle
The polar circle (latitude 66°33′ S) in the Southern Hemisphere.

r Южный полярный круг
Полярный круг, расположенный в Южном полушарии, проходит по 66°33′ ю.ш.

f cercle *m* antarctique
s círculo *m* antártico
g südlicher Polarkreis *m*; antarktischer Kreis *m*; Südpolarkreis *m*

286 Antarctic Circumpolar Current; West Wind Drift
The ocean current with the largest volume transport, and the swiftest velocity of water which flows from west to east through all oceans around the Antarctic continent.

r южное океаническое кольцо; течение западных ветров
Течение в Мировом океане, опоясывающее Южное полушарие между Антарктидой и южными оконечностями трех других материков, где результирующий перенос вод направлен на восток.

f courant *m* circumpolaire antarctique; dérive *f* due aux vents d'ouest
s corriente *f* circumpolar antártica
g antarktische Zirkumpolarströmung *f*; Westwinddrift *f*

287 antarctic convergence
The well-defined narrow zone in the South Ocean at which denser and colder antarctic waters sink under subantarctic waters almost without mixing with them.

r антарктическая конвергенция
Четко выраженная узкая зона в Южном океане, на которой более плотные и холодные воды антарктического происхождения погружаются, почти не смешиваясь, под субантарктические воды.

f convergence *f* antarctique
s convergencia *f* antártica
g antarktische Konvergenz *f*

288 antarctic desert zone
A natural zone which comprises Antarctica and the neighbouring islands, and is characterized by severe Antarctic climate and poor polar vegetation.

r зона антарктических пустынь
Природная зона, включающая Антарктиду и близлежащие острова, имеет суровый антарктический климат и скудную полярную растительность.

f zone *f* du désert antarctique
s zona *f* de desierto antártico
g Zone *f* der antarktischen Wüsten

289 antarctic front
The front of great extent, occurring in high southern latitudes, which separates relatively

cold Antarctic air from relatively warmer polar air, and on which atmospheric waves form.
r антарктический фронт
Протяженный фронт в высоких южных широтах, отделяющий холодный антарктический воздух от более теплой полярной воздушной массы, на этом фронте нередко возникают атмосферные волны.
f front *m* antarctique
s frente *m* antártico
g antarktische Front *f*

290 **antarctic oasis**
A patch of land free of ice in the coastal zone of Antarctica.
r антарктический оазис
Участок прибрежной зоны Антарктиды, свободный от ледникового покрова.
f oasis *m* antarctique
s oasis *m* antártico
g antarktische Oase *f*

291 **antarctic oscillation**
Variation of the atmospheric pressure between the middle and the high latitudes of the Southern Hemisphere, the greatest in October-December (symbol: AAO).
r антарктическое колебание
Колебания разности атмосферного давления между средними и высокими широтами Южного полушария, особенно значительные в октябре–декабре.
f oscillation *f* antarctique
s oscilación *f* antártica
g antarktische Oszillation *f*

292 **antarctic zone**
The most southerly geographical region of the Earth covering Antarctica, the neighbouring islands and adjoining sea which is characterized by severe climate with frosts even in summer.
r антарктический пояс
Самый южный географический пояс Земли, включающий Антарктиду, близлежащие острова и прилегающие акватории, характеризующийся суровым климатом с морозами даже в летние месяцы.
f zone *f* antarctique
s zona *f* antártica
g antarktische Zone *f*

293 **antecedent river**
A river flowing along a deep narrow valley across a mountain chain.
r антецедентная река
Река, текущая в глубокой узкой долине, пересекающей поперек горную цепь.
f fleuve *m* antécédent
s río *m* antecedente
g antezedenter Fluß *m*

294 **antecedent valley**
A deep narrow river valley that cuts across the uplifting mountains and uplands. → Fig. 11

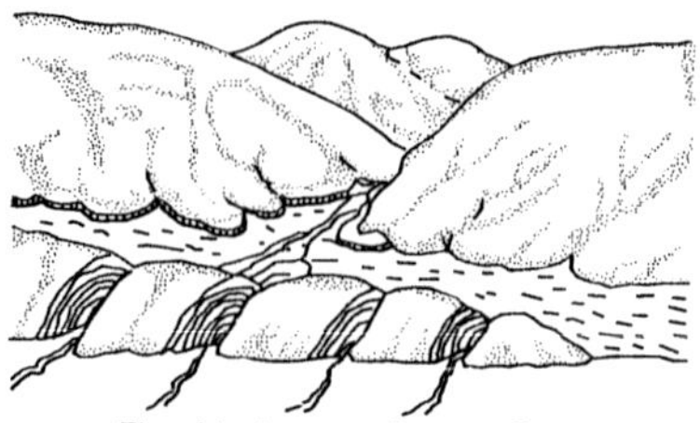

Fig. 11. **Antecedent valleys**

r антецедентная долина
Глубокая и узкая сквозная долина реки, прорезающая поднимающиеся горы и возвышенности вкрест их простирания. → Рис. 11
f vallée *f* antécédente
s valle *m* antecedente
g antezedentes Tal *n*

295 **anteclise**
A gentle uplift of the sedimentary cover of a platform, several hundreds kilometres across.
r антеклиза
Пологое поднятие осадочного чехла платформы поперечником в несколько сотен километров.
f antéclise *m*
s anteclisa *f*
g Anteklise *f*

296 **anthropocentrism**
An antiecological doctrine that opposes man against nature and consideres human beings to be the most significant entities of the universe.
r антропоцентризм; антропизм
Учение, противопоставляющее человека природе и считающее его высшей целью созидания, отличается антиэкологической направленностью.
f anthropocentrisme *m*
s antropocentrismo *m*
g Anthropozentrismus *m*

297 **anthropochore**
A plant the seeds and spores of which are spread due to human activity (weeds and ruderal plants).
r антропохор
Растение, распространяющееся при непроизвольном участии человека, преимущественно сорные, мусорные и тому подобные растения.
f anthropochore *m*
s antropócoro *m*
g Anthropochor *m*

* **Anthropogene → 4853**

298 **anthropogenic *adj*; man-made *adj*; human-induced *adj***
Applied to man-made objects, phenomena or processes or resulting directly or indirectly from human activity.

r антропогенный
Относящийся к объектам, явлениям или процессам, вызванным прямым или косвенным влиянием человека или созданным им.
f anthropique *adj*; anthropogène *adj*
s antropógeno *adj*; antrópico *adj*
g anthropogen *adj*

* **anthropogenic contamination → 302**

299 **anthropogenic desert**
A barren land formed as a result of direct or indirect human impact on an area which has totally degraded as habitat.
r антропогенная пустыня
Пустынный тип ландшафта, возникший вследствие прямого или косвенного воздействия человека на какой-либо территории, которая практически полностью деградировала как среда жизни.
f désert *m* anthropique
s desierto *m* antropógeno
g antropogene Wüste *f*

300 **anthropogenic environment; technosphere**
The part of the physical environment created or modified by human action.
r антропогенная среда; техносфера
Окружающая среда, отдельные элементы которой сильно изменены хозяйственной деятельностью людей.
f environnement *m* anthropogène
s medio *m* ambiente antropógeno
g menschliche Umwelt *f*; antropogene Umwelt *f*

* **anthropogenic impact → 1932**

301 **anthropogenic landscape; man-made landscape**
A natural landscape changed by human economic activity.
r антропогенный ландшафт
Природный ландшафт, измененный в процессе хозяйственной деятельности.
f paysage *m* anthropique
s paisaje *m* antropógeno
g anthropogene Landschaft *f*

302 **anthropogenic pollution; man-made pollution; anthropogenic contamination**
Pollution of the environment caused by human activity (as opposed to natural pollution).
r антропогенное загрязнение
Загрязнение окружающей среды, вызванное деятельностью людей (в отличие от природного загрязнения).
f pollution *f* anthropique; contamination *f* anthropique
s polución *f* antropógena; contaminación *f* antropógena
g anthropogene Verschmutzung *f*; anthropogene Verunreinigung *f*

303 **anthropogenic relief**
The relief forms changed or created as a result of human activity.
r антропогенный рельеф; техногенный рельеф
Совокупность форм рельефа, измененных или созданных в результате хозяйственной деятельности человека.
f relief *m* anthropogénétique
s relieve *m* antropógeno
g anthropogenes Relief *n*

304 **anthropogenic soils; man-made soils**
Soils greatly transformed by human economic activity both improved for cultivation or reclamed soils and spoiled soils as a result of change of soil-forming factors (e.g. salinization).
r антропогенные почвы
Почвы, сильно преобразованные в результате деятельности человека, улучшенные (окультуренные почвы) или испорченные (засоленные и пр.) в результате хозяйствования.
f sols *mpl* anthropogènes
s suelos *mpl* antropogenos
g anthropogene Böden *mpl*

305 **anthropogenic succession**
The sequential change of vegetation caused by the economic activity of people.
r антропогенная сукцессия
Смена растительности, возникающая в результате хозяйственной деятельности человека.
f succession *f* anthropique
s sucesión *f* antropogénica
g anthropogene Sukzession *f*

306 **anthropogenic vegetation**
The vegetation planted or transformed by people.
r антропогенная растительность
Растительность, выращеннная или измененная человеком.
f végétation *f* anthropique
s vegetación *f* antropogena
g anthropogene Vegetation *f*

307 **anthropogeography; geography of human being**
The study of environmental conditions directly influencing human life and activities.
r география человека; антропогеография
Изучение особенностей и объектов окружающей среды, непосредственно воздействующих на человека и его деятельность.
f anthropogéographie *f*; géographie *f* de l'homme
s antropogeografía *f*; geografía *f* de hombre
g Anthropogeographie *f*; Geographie *f* des Menschen

308 **anthropology**
The study of human beings (physical characteristics, social relations and culture) and the

origin and distribution of human races in relation to the environment.
r антропология
Изучение происхождения и эволюции человека и становления человеческих рас, в частности, в связи с окружающей средой.
f anthropologie *f*
s antropología *f*
g Anthropologie *f*

309 anthropophyte
Plants growing permanently in man-made habitats or introduced by man into the local flora (cultural and ruderal plants).
r антропофит
Растение, постоянно растущее на созданных человеком местообитаниях или вошедшее в местную флору благодаря человеку.
f anthropophyte *m*
s antropofita *m*
g Anthropophyt *m*

310 anti-avalanche structures
Various structures designed for protection against avalanches. → Fig. 12
r противолавинные сооружения
Разнообразные сооружения, предназначенные для защиты от лавин. → Рис. 12
f ouvrages *mpl* pare-avalanche
s barrera *f* antiavalancha
g Lawinenverbauung *f*; Lawinenschutzwerk *n*

311 anti-avalanche wedge
A wedge-like structure, making an avalanche to flow it over, slowing the avalanche down and reducing its destructive force.
r лавинорез
Клиновидная инженерная конструкция, вынуждающая снежную лавину обтекать сооружение и, таким образом, снижать ее скорость и разрушительную силу.
f étrave *f* pare-avalanche
s cuña *f* contra avalanchas; quebranta *f* de avalanchas
g Lawinenbrecher *m*

* **anticlinal fold → 314**

312 anticlinal ridge
A mountain ridge that coincides with the anticlinal structure of rock strata.
r антиклинальный хребет
Горный хребет, совпадающий с антиклинальной структурой залегания горных пород.
f cordillère *f* anticlinale
s cordillera *f* anticlinal
g antiklinaler Gebirgszug *m*; antiklinaler Rücken *m*

313 anticlinal valley
A river valley which runs along the axis of an anticlinal fold.
r антиклинальная долина

Fig. 12. **Anti-avalanche structures** in Swiss Alps (above). 1, 2 – avalanche dam, 3 – avalanche gallery, 4 – snow-remaining structures

Речная долина, проходящая вдоль свода антиклинали.
f vallée *f* anticlinale
s valle *m* anticlinal
g Antiklinaltal *n*

314 anticline; anticlinal fold
A fold in geologic strata that is convex upward like an arch, formed by the compression of rocks in the Earth's crust. → Fig. 13
r антиклиналь
Складка слоев горных пород, обращенная выпуклостью вверх, с более древними отложениями в центральной части. → Рис. 13
f pli *m* anticlinal
s pliegue *m* anticlinal
g Antiklinale *f*; Sattel *m*

315 anticlinorium
A great arch of strata, stretching for dozens and hundreds kilometres, composed of numerous subordinate folds.
r антиклинорий
Сложный комплекс складок земной коры, протягивающийся на десятки и сотни километров.
f anticlinorium *m*

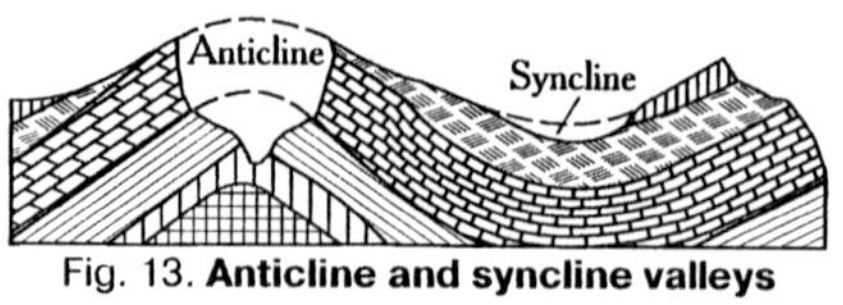

Fig. 13. **Anticline and syncline valleys**

s anticlinorio *m*
g Antiklinorium *n*

316 anticyclogenesis
The process of origin and development of anticyclonic activity.
r антициклогенез
Процесс образования или усиления интенсивности антициклонической деятельности в атмосфере.
f anticyclogénèse *f*
s anticiclogénesis *f*
g Antizyklogenese *f*

317 anticyclolysis
The process of weakening, or ceasing, of anticyclonic activity.
r антициклолиз
Процесс ослабления или окончательного прекращения антициклональной деятельности.
f anticyclolyse *f*
s anticiclólisis *f*
g Antizyklolyse *f*

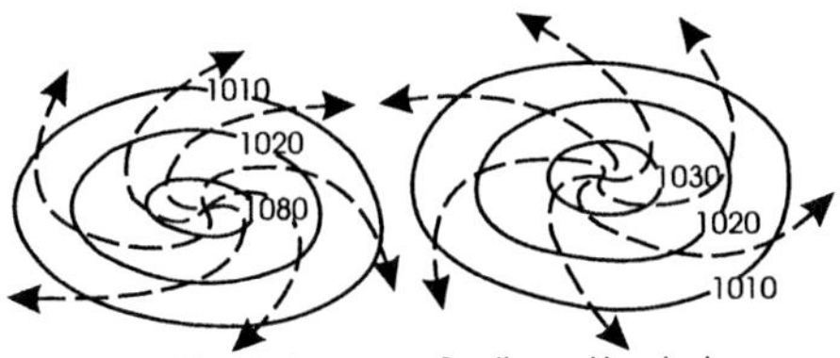

Fig. 14. **Anticyclone**

318 anticyclone; high
A system in the atmosphere where the pressure is high relative to the surrounding air at the same level and air moves around the center (counter clockwise in the Southern Hemisphere, and clockwise in the Northern Hemisphere). → Fig. 14
r антициклон
Область повышенного атмоферного давления в тропосфере с максимальным давлением в центре и барическими градиентами, направленными от центра к периферии (против часовой стрелки в Южном полушарии и по часовой в Северном). → Рис. 14
f anticyclone *m*
s anticiclón *m*; alta presión *f*; alta *f*
g Antizyklone *f*; Hoch *n*

319 anticyclonic circular current
A circular current in seas and oceans directed clockwise in the Northern Hemisphere, and counter-clockwise in the Southern Hemisphere.
r антициклонический круговорот
Круговое течение воды в морях и океанах, направленное в Северном полушарии по часовой стрелке, а в Южном – против часовой стрелки.
f courant *m* de circulation anticyclonique
s corriente *f* de circulación anticiclónica
g antizyklonale zirkulare Strömung *f*

320 anticyclonic eddy; anticyclonic vortex
A whirl in the ocean with vertical or inclined axis having the clockwise rotation in the Northern Hemisphere, and the counter-clockwise rotation in the Southern Hemisphere.
r антициклонический вихрь
Вихрь в океане с вертикальной или наклонной осью, вращающийся в Северном полушарии по часовой стрелке, а в Южном - против часовой стрелки.
f tourbillon *m* anticyclonique
s torbellino *m* anticiclónico
g antizyklonaler Wirbel *m*

* **anticyclonic vortex → 320**

321 antidune
A ripple on the bed of a stream or river made up of bed load which migrates against the direction of the flow (upstream).
r антидюна; донная гряда
Гряда донных наносов, возникающая в потоках и перемещающаяся вверх по течению.
f antidune *f*
s antiduna *f*
g Antidüne *f*

322 antiglacialism
A scientific view of the Quaternary geology and paleogeography denying the existence of ancient ice sheets in the temperate and subarctic zones.
r антигляциализм
Научная доктрина в четвертичной геологии и палеогеографии, отрицающая существование древних покровных оледенений на равнинах умеренного и субарктического поясов.
f théorie *f* antiglaciaire
s antiglacialismo *m*; teoría *f* antiglaciar
g Antiglazialismus *m*

323 antiglobalism
A public movement protesting against the globalization of the larger part of the world.
r антиглобализм
Общественное движение, протестующее против происходящей на большей части мира глобализации.
f antiglobalism *m*
s antiglobalismo *m*
g Antiglobalismus *m*

324 antipodes
Places on the Earth's surface diametrically opposite to each other, in Great Britain this term is applied to Australia and New Zealand which are situated on the opposite side of the globe with regard to British Islands.
r антиподы

Места на земной поверхности, диаметрально противоположные друг другу, в Великобритании этим термином называют Австралию и Новую Зеландию, лежащие на противоположной стороне земного шара по отношению к Британским островам.
f antipodes *mpl*
s antípodas *mpl*
g Antipoden *npl*

325 anti-trade; anti-trade wind
Air current with a westerly component which blows in subtropical regions of both hemispheres above the trade winds.
r антипассат
Воздушный поток с преимущественно западной составляющей, который иногда возникает над пассатами в субтропических широтах обоих полушарий.
f contre-alizé *m*
s contralisio *m*
g Antipassat *m*

* **anti-trade wind → 325**

326 antitriptic current
The settled staight-line water movement in the ocean formed when the pressure gradient of water counterbalances by frictional force and the horizontal components of Coriolis force absent.
r антитриптическое течение
Установившееся прямолинейное движение воды в океане, когда градиент давления воды уравновешивается силой трения, а горизонтальные составляющие силы Кориолиса отсутствуют.
f courant *m* antitriptique
s corriente *f* antitriptica
g antotriptische Strömung *f*

327 anti-uafeis constructions
Engineering structures intended to protect against harmful effects of aufeis.
r противоналедные устройства
Инженерные сооружения, предотвращающие вредное влияние наледей.
f ouvrages *mpl* de défense contre les givrages
s defensas *fpl* contra icing
g Aufeisschutzwerke *npl*

328 apartheid; racial segregation
The separation of racial groups inside a country accompanied by the discrimination of any of the ethnic groups.
r апартеид; расовая сегрегация
Принудительно поддерживаемое разделение расовых групп населения страны, сопровождаемое дискриминацией той или иной национальной группы.
f apartheid *m*; ségrégation *f* raciale
s apartheid *m*; segregación *f* racial
g Apartheid *f*; Rassensegregation *f*

329 apchipelago
A group of islands located close to each other and having the same origin, e.g. archipelagos of continental (Canadian Arctic Archipelago), coral (Marshall Islands) and volcanic (Hawaiian Islands) origin.
r архипелаг
Группа островов, лежащих на небольших расстояниях друг от друга и рассматриваемых как одно целое, различают архипелаги материкового (Канадский Арктический архипелаг), кораллового (Маршалловы острова) и вулканического (Гавайские острова) происхождения.
f archipel *m*
s archipiélago *m*
g Archipel *m*

330 aphelion
The furthest point of elliptical orbit of a planet, comet or some other body of the solar system which is the most remote from the Sun.
r афелий
Наиболее удаленная от Солнца точка эллиптической орбиты планеты, кометы или какого-либо другого тела Солнечной системы.
f aphélie *f*
s afelio *m*
g Aphel *n*; Sonnenferne *f*

331 aphotic zone; disphotic zone
A deep-water area of marine ecosystems below 200 m depth where the amount of light is not enough for photosynthesis.
r афотическая зона
Область океанических глубин ниже 200 м, освещенность которых недостаточна для фотосинтеза.
f couche *f* aphotique; étage *m* aphytal; étage *m* disphotique
s capa *f* afótica; piso *m* afital; piso *m* afótico
g aphotische Zone *f*

332 aphyllous plant; leafless plant
A plant adapted to dry climate with no leaves which reduces the loss of moisture while photosynthesis goes on in green stems, e.g. cactus.
r безлистное растение
Растение сухого климата, лишенное листьев, что ограничивает потерю влаги из-за уменьшения испаряющей поверхности, а фотосинтез происходит в зеленых стеблях, как, например, у кактуса.
f plante *f* aphylle
s planta *f* áfila
g blattlose Pflanze *f*; Nacktpflanze *f*

333 aphylly; afoliage
A complete reduction of leaves in plants due to extremely dry conditions and high temperatures of the environment.
r афиллия
Отсутствие листьев у растений, обуслов-

ленное высокой степенью сухости и высокими температурами окружающей среды.
f aphyllie *f*
s afilía *f*
g Aphyllie *f*

* **apiculture → 636**

* **apogean tide → 4079**

334 apogee
The point furthest from the Earth in the orbit of a planet (especially of the Moon) or a satellite.
r апогей
Наиболее удаленная от Земли точка орбиты Луны или искусственного спутника Земли.
f apogée *m*
s apogeo *m*
g Erdferne *f*; Apogäum *n*

335 appalachian relief
A type of *inverted relief* characteristic of relatively low, old mountains with complex fold structure.
r аппалачский рельеф
Тип *инверсионного рельефа*, свойственный невысоким древним горам, обладающим сложной складчатой структурой.
f relief *m* appalachien
s relieve *m* apalachense
g Appalachen-Relief *n*

* **apparent time → 5651**

336 apparent dip
The dip as seen by the observer and not necessarily the angle of maximum slope.
r видимое падение
Падение пласта, как оно видится наблюдателю и не обязательно совпадающее с максимальным углом склона.
f pendage *m* apparent
s buzamiento *m* aparente
g scheinbares Einfallen *n*

* **apparent horizon → 5466**

337 appearance of ice
The beginning of water freezing in the upper layer of water bodies.
r появление льда
Начало процесса ледообразования на водных объектах.
f apparition *f* de glace
s aparición *f* de hielo
g Eisentstehen *n*

338 applied geography
The application of geographical knowledge and techniques to the solution of economic and social problems on a local to a world scale, in the sphere of regional planning, land use, land-reclamation, etc.
r прикладная география
Географические знания, исследования и методы, направленные на решение конкретных экономических и социальных проблем в области регионального планирования, землепользования, мелиорации земель и т.п.
f géographie *f* appliquée
s geografía *f* aplicada
g angewandte Geographie *f*

339 applied geomorphology; engineering geomorphology
The application of geomorphological methods of survey and analysis to the solution of such problems as flood control, stabilization of slopes, land reclamation etc.
r прикладная геоморфология; инженерная геоморфология
Применение геоморфологических методов исследования для решения проблем использования окружающей среды в хозяйственных целях, например, регулирование паводков, стабилизация склонов, мелиорация земель и т.п.
f géomorphologie *f* appliquée
s geomorfología *f* aplicada
g angewandte Geomorphologie *f*

340 apposed glacier
A glacier formed when two glaciers coalesce.
r слившийся ледник
Ледник, формирующийся при слиянии двух отдельных ледников.
f glacier *f* de jonction
s glaciar *m* de conexión; glaciar *m* de junción
g zusammengießender Gletscher *m*

* **appraisal → 3373**

341 appropriate technology
The technology that is suited to the local environment, usually involving skills or materials that are easily available locally, and uses the resources with the maximum care and efficiency.
r приемлемая технология; ресурсосберегающая технология
Технология, при которой наиболее бережно и эффективно используются природные и общественные ресурсы, включая землю, рабочую силу, а также местные легко доступные материалы.
f technologie *f* d'épargne
s tecnología *f* de ahorro; tecnología *f* propia
g annehmbare Technologie *f*

342 appropriative economy
The way of life of primitive people, dominant in the Palaeolithic, when food was obtained in nature, mainly by hunting and gathering.
r присваивающий способ хозяйства
Господствовавший в палеолите способ существования первобытных людей, когда продукты питания добывались в природе,

прежде всего с помощью охоты и собирания плодов диких растений.

f économie *f* d'appropriation
s economía *f* de apropiación
g Aneignungswirtschaft *f*

343 **apron; bahada**
A belt of adjacent alluvial fans deposited by temporary or permanent streams at the mountain foot which form a gently sloping plain.
r пролювиально-аллювиальный шлейф
Полоса слившихся конусов выноса временных и постоянных водотоков, образующая у подошвы гор наклонную равнину.
f glacis *m* d'accumulation alluviale
s glasis *m* de acumulación aluvial
g Schwemmlandstreifen *m*

344 **Apsheronian basin**
A salty-water basin which existed at the end of the late Pliocene in the area of the contemporary Caspian Sea.
r Апшеронский бассейн
Солоноватоводный бассейн, существовавший в конце позднего плиоцена на месте современного Каспийского моря.
f bassin *m* apchéronien
s cuenca *f* de Apcheron
g Apscheron-Becken *n*

345 **aquaculture**
Growing water plants, breeding and rearing fish, shellfish and other animals in aquatic environments under human control for commercial and aestetic purposes.
r аквакультура; марикультура
Разведение и выращивание в контролируемых человеком условиях водных растений и животных, представляющих экономическую или эстетическую ценность.
f aquaculture *f*; aquiculture *f*
s aquacultura *f*; acuicultura *f*
g Aquakultur *f*; Teichwirtschaft *f*

* **aquafer → 352**

346 **aquanaut**
Underwater explorer specially trained and capable of staying in depths for a long time in special deep-diving submersibles.
r акванавт
Исследователь морских глубин, получивший специальную подготовку и способный длительное время находиться под водой в специальных глубоководных аппаратах.
f aquanaute *m*
s acuanauta *m*
g Aquanaut *m*

347 **aquatic *adj***
Living or growing in the water conditions, related to the water objects.
r водный
Живущий или растущий в водных условиях, относящийся к водным объектам.
f aquatique *adj*
s acuático *adj*
g Wasser-

* **aquatic animals → 348**

* **aquatic ecosystem → 350**

348 **aquatic fauna; aquatic animals**
Species of animals permanently living in water.
r водная фауна; водные животные
Животные, постоянно обитающие в воде.
f faune *f* aquatique;
animals *mpl* aquatiques
s fauna *f* acuática; animales *mpl* acuáticos
g Wasserfauna *f*; Wassertiere *npl*

349 **aquatic flora; aquatic plants**
Species of plants living in water.
r водная флора; водные растения
Растения, населяющие водоемы.
f flore *f* aquatique; plantes *fpl* aquatique
s flora *f* acuática; plantas *fpl* acuáticas
g Wasserflora *f*; Wasserpflanzen *fpl*

350 **aquatic landscape; aquatic ecosystem**
Geographical landscape and ecosystem, in the structure and functioning of which water ecosystems prevail.
r аквальный ландшафт; водная экосистема
Географический ландшафт и экосистема, структура и функционирование которых определяются водой, служащей средой обитания организмов.
f paysage *m* aquatique;
écosystème *m* aquatique
s paisaje *m* acuático; ecosistema *m* acuático
g aquatische Landschaft *f*; aquatisches Ökosystem *n*; Wasserökosystem *n*

* **aquatic plants → 349**

351 **aqueduct**
An artificial channel built to carry water from one place to another, especially such a channel supported on a series and spanning a valley or river.
r акведук
Сооружение в виде моста или эстакады с лотком или трубой, служащее для пропуска воды через овраги, каналы, дороги и пр.
f aqueduc *m*
s acueducto *m*
g Aquädukt *m*; Wasserleitungbrücke *n*

* **aqueous vapour → 6694**

352 **aquifer; aquifer horizon; aquafer; water-bearing layer**
A water-bearing stratum of rock, sufficiently porous to carry the water and permit its use, lying between two impermeable layers or above the underlying imperable layer.

r водоносный горизонт; водоносный пласт
Толща горной породы, насыщенная водой, залегающая между двумя водоупорными пластами или лежащая на водоупорном пласте.
f nappe *f* aquifère
s capa *f* acuífera; capa *f* freática; aguífero *m*
g Grundwasserschicht *m*; Grundwasserträger *m*; Grundwasserleiter *m*; wasserführende Schicht *f*

* **aquifer horizon → 352**

* **arable farming → 1343**

353 arable horizon of soil
The surface layer of soil rich in humus and other organic substances and periodically turned over in ploughing, it is characterized by a loose, cloddy structure and a high water holding capacity.
r пахотный слой
Поверхностный слой почвы, богатый гумусом и иной органикой, периодически обрабатываемый сельскохозяйственными орудиями, отличается рыхлой, комковатой структурой и значительной водоудерживающей способностью.
f couche *f* arable du sol; couche *f* labourée du sol
s capa *f* labrada de suelo
g Ackerkrume *f*

354 arable land; cropland; tillage$_2$; ploughland
Land used for growing crops.
r пахотная земля; пашня
Земля, систематически обрабатываемая для получения урожая сельскохозяйственных культур.
f champ *m* labouré; terre *f* travailée; terre *f* labourée; terrain *m* cultivé
s campo *m* labrado; tierras *fpl* de cultivo; labrado *m*; campo *m* de cultivo; tierras *fpl* arables
g Ackerland *n*; Acker *m*

355 arboretum
A place set aside for the cultivation, display or study of trees, including rare species.
r дендрарий
Участок территории, где в открытом грунте выращиваются древесные и кустарниковые растения, размещаемые по систематическим признакам и используемые для научных исследований и демонстрации публике.
f arboretum *m*
s arboretum *m*
g Arboretum *n*; Baumgarten *n*

* **arboriculture → 4911**

356 arch^1
A curved opening formed in a rock mass by different processes, e.g. by the collapse of the roof in a limestone cavern, by marine erosion of projecting rocks (abrasional arch), or by weathering of a relatively soft layer in a rock mass.
r арка
Закругленное отверстие в скалах, возникшее под влиянием разных процессов, например, при обрушении крыши полости в известняках, при морской абразии в выступающих береговых скалах (абразионная арка) или при выветривании относительно рыхлого слоя в горных породах.
f arc *m*
s arco *m*
g Bogen *m*

357 arch^2
A dome-like uplift of a platform basement of about a hundred kilometres across.
r свод
Куполовидное поднятие фундамента платформы с поперечником в несколько десятков или сотню километров.
f voûte *f*
s bóveda *f*; domo *m*
g Beule *f*

358 archaean
The lower of the two largest parts of the Precambrian with the duration up to two billion years and the upper boundary of about 2.5 billion years ago. → Table 1, p. 284
r архей
Нижнее из двух крупнейших подразделений докембрия продолжительностью до 2 млрд лет и верхним рубежом около 2,5 млрд лет назад. → Табл. 1, с. 284
f achéen *m*
s arcaico *m*
g Archäikum *m*

359 archean foldings
The earliest (over 2.6 billion years ago) epochs of folding in the Earth's history characterised by the deep metamorphism and granitisation of rocks, typical of Baltic and Canadian shields.
r архейские складчатости
Наиболее ранние (более 2,6 млрд лет назад) эпохи складчатости в истории Земли, отличающиеся глубоким метаморфизмом и гранитизацией горных пород, характерны для Балтийского и Канадского щитов.
f plissements *mpl* d'Archéen
s plegamientos *mpl* arcáicos
g archäische Faltungen *fpl*

360 archeological culture
Archaeological monuments belonging to the same period of time and located in the same region.
r археологическая культура
Общность археологических памятников,

относящихся к одному времени и сосредоточенных на одной территории.
f culture *f* archéologique
s cultura *f* arqueológica
g archäologische Kultur *f*

361 archeological dating
The age of archaeological finds and deposits containing them, determined on the basis of the concept of the chronologial order of succession of archaeological cultures.
r археологическое датирование
Возраст археологических находок и вмещающих их отложений, получаемый на основе концепции последовательной смены во времени археологических культур.
f âge *m* archéologique
s edad *f* arqueológica
g archeologisches Alter *n*

362 archeological site
The place of permanent or temporal settlement of primitive man.
r стоянка
Место поселения (кратковременного или долговременного) первобытного человека.
f camp *m*
s estación^2 *f*
g Wohnstätte *f*

363 archeophyte
Weeds brought to cultivated lands in prehistoric time.
r археофит
Сорные растения, занесенные на культурные земли еще в доисторическое время.
f archéophyte *m*
s arqueofita *m*
g Archäophyt *m*

364 arctic *adj*
Related to the polar region of the Northern Hemisphere.
r арктический
Относящийся к полярному региону Северного полушария Земли.
f arctique *adj*
s ártico *adj*
g arktisch *adj*

* **arctic air → 365**

365 arctic air mass; arctic air
An exceedingly cold air mass originating over the Arctic Ocean.
r арктическая воздушная масса; арктический воздух
Исключительно холодная воздушная масса, зарождающаяся над Северным Ледовитым океаном.
f masse *f* d'air arctique; air *m* arctique
s masa *f* de aire ártica; aire *m* ártico
g arktische Luftmasse *f*; Arktikluft *f*

366 Arctic circle
The polar circle (latitude 66°33′ N) in the Northern Hemisphere.
r Северный полярный круг
Полярный круг, расположенный в Северном полушарии, проходит по 66°33′ ю.ш.
f cercle *m* arctique; cercle *m* polaire arctique
s círculo *m* ártico; círculo *m* polar ártico
g nördlicher Polarkreis *m*; Nordpolarkreis *m*; arktischer Kreis *m*

367 arctic desert zone
A natural region of the Arctic zone, it comprises islands and the northern margins of continents, and is characterized by numerous glaciers, permafrost and poor grass vegetation.
r зона арктических пустынь
Природная зона арктического пояса, включает острова и северные окраины материков, для которых характерны многочисленные ледники, многолетнемерзлые породы и скудная травянистая растительность.
f zone *f* du désert arctique
s zona *f* de desierto ártico
g Zone *f* der arktischen Wüsten

368 arctic front
The front of great extent occurring in high northern latitudes, which separates relatively cold Arctic air from relatively warm polar air, and on which atmospheric waves form.
r арктический фронт
Протяженный фронт в высоких северных широтах, отделяющий холодный арктический воздух от более теплой полярной воздушной массы, на этом фронте нередко возникают атмосферные волны.
f front *m* arctique
s frente *m* ártico
g Arktikfront *f*

369 arctic oscillation
Variation of the atmospheric pressure between middle and high latitudes of the Northern Hemisphere (symbol: AO).
r арктическое колебание
Колебания разности атмосферного давления между средними и высокими широтами Северного полушария.
f oscillation *f* arctique
s oscilación *f* ártica
g arktische Oszillation *f*

* **arctic smoke → 2337**

370 arctic soils
Soils of high latitudes the profile of which is constituted by thin biogenic horizon and mineral mass underlaid at the depth of about 0.5 m by permafrost.
r арктические почвы
Почвы высоких широт, состоящие из маломощного органогенного горизонта и минеральной массы, подстилаемой на

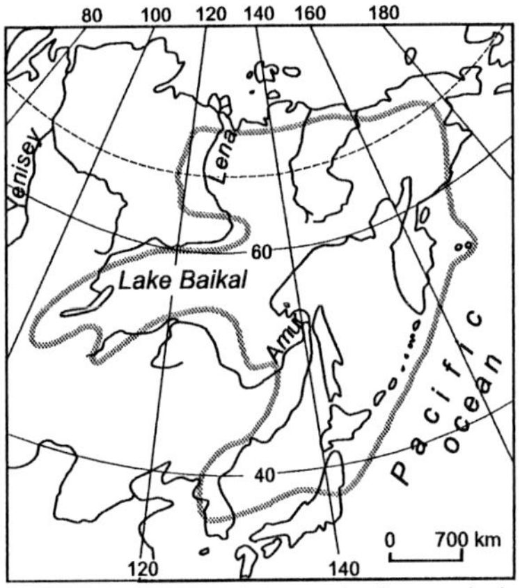

Fig. 15. The **area** of the cedar krummholz

глубине около 0,5 м многолетнемерзлыми породами.

f sols *mpl* arctiques
s suelos *mpl* árticos
g arktische Böden *mpl*

371 arctic tundra
The most northerly subzone of tundra in which frost mounds and polygons are typical, and vegetation occupies up to 60% of the surface.
r арктическая тундра
Самая северная подзона тундр, для которой характерны бугры пучения и полигональные образования, растительностью занято до 60% поверхности.
f toundra *f* arctique
s tundra *f* ártica
g arktische Tundra *f*

372 arctic zone
The most northerly geographical region of the Earth which includes the greater part of the Arctic (its boundary is drawn along the isotherm of +5°C of the warmest month).
r арктический пояс
Самый северный географический пояс Земли, включающий большую часть Арктики (его границу проводят по изотерме +5°C самого теплого месяца).
f zone *f* arctique
s zona *f* ártica
g arktische Zone *f*

373 arcuate delta
A delta in which the outermost margin exhibits an arc-life form, convex towards the sea, e.g. Nile delta.
r дугообразная дельта
Дельта с внешним краем в форме арки, выпуклой в сторону моря, например, дельта Нила.
f delta *m* arqué; delta *m* arrondi
s delta *m* arqueado; delta *m* redondeado
g bogenförmiges Delta *n*; Bogendelta *n*

374 area; geographic range; range$_2$
A space of existence of a phenomenon, e.g. an ecosystem or a region which a species of living thing linhabits. → Fig. 15
r ареал
Площадь распространения какого-либо явления, например, экосистемы, или область обитания какого-либо вида растений или животных. → Рис. 15
f aire *f*
s área *f*
g Areal *n*; Ausbreitung *f*

* **area deformation → 375**

375 area distortion; area deformation
Differences of scale of areas in different places of a geographical map, measured by the ratio of the local scale of areas to the base scale. → Fig. 16
r искажение площади
Различия масштаба площадей в разных местах географической карты, измеряется отношением частного масштаба площадей к главному масштабу. → Рис. 16
f altération *f* des surfaces
s deformación *f* de superficies
g Flächenverzerrung *f*

* **area of internal drainage → 1064**

* **area studies → 4980**

* **areal differentiation → 993**

* **areic** ***adj*** **→ 378**

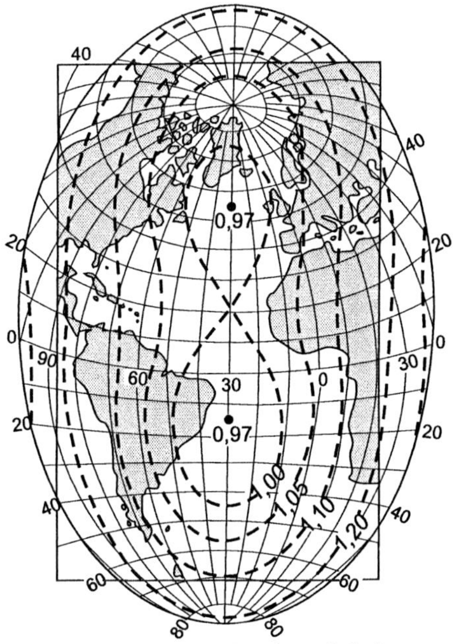

Fig. 16. **Area distortion** shown in isolines

376 arenosol
A poorly developed sandy non-alluvial soil in which there are no pronounced diagnostic horizons.
r ареносоль
Слаборазвитая песчаная неаллювиальная почва, в которой нет выраженных диагностических горизонтов.
f arénosol *m*
s arenosol *m*
g Arenosol *m*

377 areography
The study of territorial distribution of areas of animals and plants.
r ареография
Изучение пространственного распространения ареалов животных и растений.
f airegraphie *f*
s areografía *f*
g Areographie *f*

378 aretic *adj*; areic *adj*; arheic *adj*
Applied to desert regions where the rainfall is so slight that it sinks into the ground, or evaporates, so that there are no flowing streams.
r нестекающий
Применяется к пустынным территориям, где редкая дождевая влага просачивается в грунт или испаряется и не образует постоянных водотоков.
f aréique *adj*
s arréico *adj*
g areisch *adj*

* **aretic depression → 1065**

* **aretic drainage → 1902**

379 argillaceous *adj*; clayey *adj*; argillic *adj*
Containing clay, or clay-like in composition and texture, applied especially to a sedimentary rock in which clay minerals predominate.
r глинистый
Содержащий глину или глиновидные частицы в составе и текстуре осадочных пород, где подобный состав преобладает.
f argileuse *adj*
s arcilloso *adj*
g Lehm-; lehmhaltig *adj*; lehmig *adj*

* **argillic *adj* → 379**

380 argillic desert
A desert formed on clay deposits of the arid zone where solonchaks with halophytes and takyrs with almost no vegetation develop.
r глинистая пустыня
Пустыня, развитая на глинистых отложениях аридной зоны, на ее поверхности формируются солончаки с галофитами и такыры, почти лишенные растительности.
f désert *m* argileux
s desierto *m* arcilloso
g Tonwüste *f*

* **argillic horizon → 3029**

381 argillic horizon; clayey subsoil
A soil horizon (commonly the B horizon) in which clay minerals have accumulated by illuviation.
r глинистый горизонт
Почвенный горизонт (обычно горизонт B), в котором аккумулируются глинистые частицы, поступающие сюда в результате вмывания.
f horizon *m* argileux
s horizonte *m* arcilloso
g Tonhorizont *m*

382 argillite
A solid dense clay-like rock, often finely laminated, which does not get sodden in water.
r аргиллит
Твердая плотная глинистая горная порода, не размокающая в воде, часто со слабо выраженной сланцеватостью.
f argilite *f*
s argilita *f*
g Argillit *m*

* **arheic *adj* → 378**

* **arid *adj* → 1727**

383 arid climate
A climate having insufficient precipitation to support vegetation where evaporation of moisture exceeds considerably its income with precipitation.
r аридный климат
Климат пустынь и полупустынь, когда величина испаряемости сильно превышает годовую сумму выпадающих осадков.
f climat *m* aride
s clima *m* árido
g Tockenklima *n*; arides Klima *n*

384 arid desert; hot desert
A desert in the zone of trade winds of subtropical or tropical belts in both hemispheres, characterized by insufficient irregular precipitation.
r аридная пустыня
Пустныня в зоне пассатных ветров субтропического или тропического поясов обоих полушарий, для которой характерно ничтожное количество атмосферных осадков, выпадающих крайне нерегулярно.
f désert *m* aride; désert *m* tropical et subtropial; désert *m* vrai
s desierto *m* árido; desierto *m* tropical y subtropial; desierto *m* cálido
g aride Wüste *f*

385 arid landforms
The type of relief characteristic of deserts, semideserts and dry steppes formed by the action of wind, weathering, and temporary watercourses.

r аридный рельеф
Тип рельефа, свойственный пустыням, полупустыням и сухим степям, формирующийся под влиянием эоловой деятельности, выветривания, эрозии временных водотоков.
f relief *m* des zones arides
s relieve *m* de zonas áridas
g arides Relief *n*

386 arid soils; aridisol; aridosol
Soils developing in arid climate of deserts, semi-deserts, dry steppes and desertified savannas, with or without argillic horizon, in which an accumulation of soluble salts or carbonates is common.
r аридные почвы; аридисоль
Почвы, возникающие в условиях засушливого климата пустынь, полупустынь, сухих степей и опустыненных саванн, отличающиеся сильно минерализованным профилем, где нередко присутствует глинистый горизонт, обычны растворимые соли и карбонаты.
f sols *mpl* arides; aridisol *m*
s suelos *mpl* áridos; aridisuelos *mpl*; aridisol *m*
g trockene Böden *mpl*; aride Böden *mpl*; Aridisol *m*

387 arid zone
A geographical zone with arid climate characterized by deserts and semi-deserts.
r аридная зона
Географическая зона с аридным климатом, в которой господствуют пустыни и полупустыни.
f zone *f* aride
s zona *f* árida
g Trocksngebiet *n*

* **aridisol → 386**

388 aridity
Very dry, deficient of moisture, desert natural conditions.
r аридность
Сухие, засушливые, пустынные природные условия.
f aridité *f*
s aridez *f*
g Aridität *f*; Trockenheit *f*; Wasserarmut *f*

* **aridity factor → 389**

389 aridity index; index of aridity; aridity factor
A measure expressing the degree of dryness of a climate as a combination of various climatic factors.
r индекс аридности; индекс сухости
Параметр, выражающий степень засушливости климата как функцию различных климатических факторов.
f indice *m* d'aridité
s índice *m* de aridez

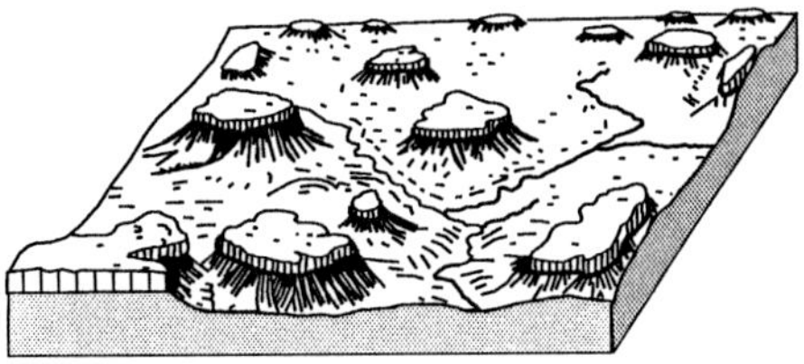

Fig. 17. **Armoured landforms**

g Ariditätsindex *m*; Trockenheitsindex *m*

* **aridosol → 386**

* **arm of a river → 779**

* **armament industry → 3858**

390 armoured karst
A type of karst developing under the layer of solid rocks: basalt, sandstone, etc.
r бронированный карст
Тип карста, развивающийся под покровом твердых пород: базальтов, песчаников и пр.
f karst *m* cuirassé; karst *m* barré
s karst *m* acorazado; karst *m* barré
g gepanzerter Karst *m*

391 armoured landforms
A part of a tableland, a plateau or a slope with the surface composed of solid rock beds overlying less-resistant rocks and protecting them from destruction. → Fig. 17
r бронированный рельеф
Участки плоскогорий, плато и склонов, сложенные с поверхности пластами твердых пород, предохраняющими нижележащие менее стойкие породы от разрушения. → Рис. 17
f relief *m* cuirasse
s relieve *m* acorazado; relieve *m* encostrado
g gepanzertes Relief *n*

392 artefact; artifact
A simple object made by human workmanship, especially one related to prehistoric time, e.g. a stone tool.
r артефакт
Любой предмет, изготовленный первобытным человеком, отличающийся от природного объекта.
f artefact *m*
s artefacto *m*
g Artefakt *n*

* **artery → 3644**

393 artesian basin
A depression in the Earthr's crust in which water-bearing strata lie between impermeable strata, the whole being folded to form the synclinal basin. → Fig. 18, p. 48
r артезианский бассейн
Бассейн напорных подземных вод, приуроченный к тектонической впадине, выполненной преимущественно осадочными слоистыми породами. → Рис. 18, с. 48

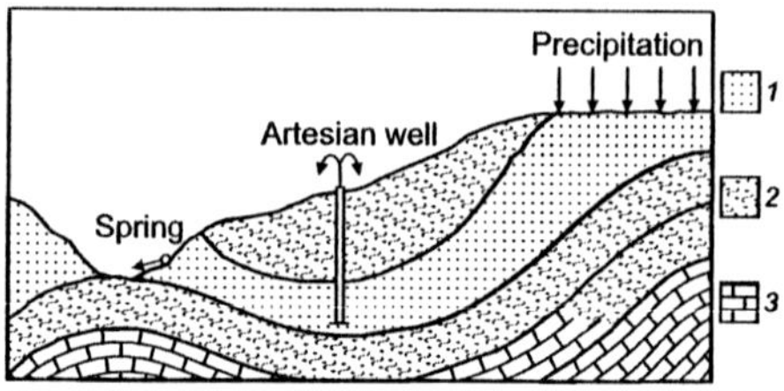

Fig. 18. **Artesian basin**: 1– sands (aquifer), 2 – permeable rock, 3 – limestone

f bassin *m* artésien
s cuenca *f* artesiana
g artesisches Becken *n*

394 artesian water
Ground waters lying between impermeable strata and having its own pressure hence when open they outflow on the surface.
r артезианские воды
Подземные воды, залегающие между водоупорными пластами и имеющие напор, вследствие чего, будучи вскрытыми, они изливаются на поверхность.
f eau *f* artésienne; nappe *f* artésienne
s agua *f* artesiana; capa *f* artesiana
g artesisches Wasser *n*

395 artesian well
A perpendicular boring sunk through the upper impermeable layer in an artesian basin to reach the aquifer and get artesian water which rises to the surface under a head pressure. → Fig. 18
r артезианский колодец; фонтанирующий колодец
Вертикальная скважина, пробуренная через верхние водонепроницаемые слои в артезианском бассейне и вошедшая в водоносный пласт, из которого напорная артезианская вода поднимается к поверхности под давлением. → Рис. 18
f puits *m* artésien
s pozo *m* artesiano
g artesischer Brunnen *m*

*** artifact → 392**

396 artificial avalanche
The release of avalanche initiated by means of artificial modification of the snow cover.
r искусственный спуск лавины
Лавина, инициированная целенаправленным активным воздействием.
f avalanche *f* artificielle
s avalancha *f* artificial; desencadeno *m* de alud
g Lawinenauslösung *f*

397 artificial satellite
A space vehicle, housing various scientific instruments, that is placed into orbit around the Earth or other planets in order to collect and transmit data.
r искусственный спутник
Космический аппарат, снабженный различным научным оборудованием, отправленный на орбиту вокруг Земли или других планет для сбора и передачи данных.
f satellite *m* artificiel
s satélite *m* artificial
g künstlicher Satellit *m*

398 artificial snow
A snow produced by a *snow cannon* by spraying frozen water droplets and used mainly to prepare the runs for skiing.
r искусственный снег
Снег, получаемый из *снежной пушки* и используемый в основном для устройства лыжных трасс на спортивных соревнованиях.
f neige *f* artificielle
s nieve *f* artificial
g Kunstschnee *m*

399 aryk
A canal of irrigation system in Central Asia.
r арык
Канал оросительной системы в Средней Азии.
f aryk *m*
s arik *m*
g Aryk *m*

400 ash; volcanic ash
The unconsolidated pyroclastic material consisting of finely comminuted fragments of rock and lava which have been ejected explosively from a volcano and solidified in the air.
r вулканический пепел
Мелкие (до 2 мм) частицы, образующиеся в результате застывания в воздухе мелких капель лавы и дробления твердых частиц.
f cendre *f* volcanique
s ceniza *f* volcánica
g Vulkanasche *f*; vulkanische Asche *f*

401 ash cone
A volcanic cone formed by ash ejected during eruption.
r пепловый конус
Вулканический конус, сложенный рыхлыми пирокластическими продуктами – пеплами.
f cône *m* de cendre
s cono *m* de ceniza
g Aschenkegel *m*

402 aspect$_{\text{I}}$
Seasonal state of vegetation, e.g. spring aspect.
r аспект
Сезонное состояние растительного сообщества, например, весенний аспект.
f aspect$_{\text{I}}$ *m*
s aspecto$_{\text{I}}$ *m*
g Aspekt *m*

* **aspect$_2$ → 2066**

403 **aspen forest**
Deciduous small-leaved forest with the dominance of aspen, often grows as a secondary forest that grows in place of broad-leaved and coniferous forests after tree felling or forest fires.
r осиновый лес; осинник
Летнезеленый мелколиственный лес с господством осины, вырастает на месте широколиственных и хвойных лесов после порубок или пожаров.
f tremblaie *f*; peupleraie *f*
s pobeda *f*
g Espenwald *m*

* **asrect *m* de terrain → 3396**

404 **assimilation**
The process by which various groups (nations or communities) in society merge, lose their distinguishing characteristics, and become more similar and culturally homogeneous.
r ассимиляция
Процесс слияния народов, включая язык, культуру, традиции и приемы хозяйствования.
f assimilation *f*
s asimilación *f*
g Assimilation *f*

405 **assimilation capacity of the environment**
The amount of parmful substance which can be absorbed or removed from a land area or water body without any harmful impact.
r ассимиляционный потенциал окружающей среды
Количество вредного вещества, которое может быть поглощено или устранено на данной территории или акватории без вредного воздействия.
f capacité *f* d'assimilation de l'environnement
s capacidad *f* de asimilación de medio ambiente
g Assimilationskapazität *f* der Umwelt

406 **astatic lake; basinal lake**
A lake which has no permanent outflow from it neither by river nor underground. flow. → Fig. 19

Fig. 19. **Astatic lake** in the Tibet Mountains

r бессточное озеро
Озеро, из которого нет постоянного и речного, и подземного стока. → Рис. 19
f lac *m* sans émissaire
s lago *m* sin emisario
g abflußloser See *m*

407 **asteroid**
One of several thousand small planets moving in a stable orbit around the Sun in the Solar system, occurring mainly in the asteroid belt between Mars and Jupiter.
r астероид
Одна из тысяч малых планет, движущихся по стабильным орбитам вокруг Солнца, в основном в астероидном поясе Солнечной системы между Марсом и Юпитером.
f astéroïde *m*
s asteroide *m*
g Asteroid *m*

408 **asthenosphere**
A layer of low viscosity, the zone of hot rock in a plastic condition in the upper mantle of the Earth.
r астеносфера
Слой пониженной вязкости в верхней мантии Земли.
f asthénosphère *f*
s astenosfera *f*
g Astenosphäre *f*

409 **astrobleme**
A crater-like depression on the Earth's surface (usually an ancient meteorite crater) edged with a ridge.
r астроблема
Замкнутая округлая впадина земной поверхности, окаймленная кольцевым валом, обычно древний метеоритный кратер.
f astroblème *m*
s astroblema *f*
g Astroblem *n*

410 **astronaut**
One who works in space outside the Earth's atmosphere, a space traveller (in Russia it is called cosmonaut).
r космонавт; астронавт
Специалист, работающий на космическом аппарате вне пределов земной атмосферы.
f astronaute *m*
s astronauta *m*
g Astronaut *m*; Raumfahrer *m*

411 **astronomical coordinates**
System of geographical coordinates (points on the Earth's surface) in which longitude and latitude are identified by observing the stars the co-ordinates of which are known.
r астрономические координаты
Система географических координат (точек на земной поверхности), в которых долгота и широта определяются путем наблюдений звезд с известными координатами.

f coordonnées *fpl* astronomiques
s coordenadas *fpl* astronómicas
g astronomische Koordinaten *fpl*

412 **astronomical observation**
The determination of geographical co-ordinates of a ship by observing the stars the coordinates of which are known.
r астрономическая обсервация
Определение географических координат судна по наблюдениям звезд с известными координатами.
f observation *f* astronomique
s observación *f* astronómica
g astronomische Observation *f*

413 **astronomical twilight**
Interval of time in the morning and in the evening, until sunrise and after sunset, when the centre of Sun's disc is between 12 and 18° below the horizon.
r астрономические сумерки
Промежуток времени утром и вечером, перед восходом и после заката Солнца, в течение которого центр солнечного диска находится от 12 до 18° ниже горизонта.
f crépuscule *m* astronomique
s crepúsculo *m* astronómico
g astronomische Dämmerung *f*

* **asylum seeker → 4958**

414 **asymmetrical fold**
Curve of rock layers with walls falling at different angles. → Fig. 20
r асимметричная складка
Изгиб слоев горных пород с крыльями, падающими под разными углами. → Рис. 20
f pli *m* dissymétrique
s pliegue *m* disimétrico; pliegue *m* inclinado
g asymmetrische Falte *f*

415 **asymmetrical valley**
A river valley that has one side steeper and shorter than the other, or one side straight and the other with terraces.
r асимметричная долина

Fig. 20. **Asymmetrical fold**

Речная долина с неодинаковыми склонами: одним крутым и коротким, другим пологим и длинным, либо одним прямым, а другим террасированным.
f vallée *f* dissymétrique
s valle *m* disimétrico
g asymmetrisches Tal *n*

416 **asymmetry of relief**
A valley, a ridge, etc., with the sides of different steepness and length.
r асимметрия рельефа
Сопряженные склоны долин, гряд, хребтов и пр., имеющие разную крутизну и длину.
f asymétrie *f* du relief
s disimetría *f* de relieve
g Asymmetrie *f* des Reliefs; Reliefasymmetrie *f*

417 **Atlantic**
The period of the maxmimum warming in the Holocene which occurred 8-5 thousand years ago.
r атлантическое время; климатический оптимум голоцена
Время наибольшего потепления в голоцене, относящееся к 8-5 тыс. лет назад.
f Atlantique *m*
s Atlántico *m*
g Atlantikum *n*

418 **atlantic lande**
Flat monotonous sandy plains in the south-western France (along the Bay of Biscay) once covered by heath and marshes, but reclaimed now and planted with maritime pine; secondary heaths covered with evergreen dwarf shrubs and grasses in Western Europe.
r ланды
Ландшафт пустошей в приморских областях Западной Европы с влажным климатом; в более узком понимании – полоса дюн и болот вдоль побережья Бискайского залива во Франции.
f lande *f* atlantique
s landa *f* atlántica
g Landes *fpl*

419 **atlas; geographic atlas**
A systematic collection of maps designed in accordance with a single program and kept (bound or loose) in a volume.
r атлас; географический атлас
Систематическое собрание карт, выполненных по общей программе, изданное в виде книги или набора листов.
f atlas *m*; atlas *m* géographique
s atlas *m*; atlas *m* geográfico
g Atlas *m*; Kartenwerk *n*

420 **atlas cartography**
A branch of *cartography* concerned with geographical atlases.
r атласная картография

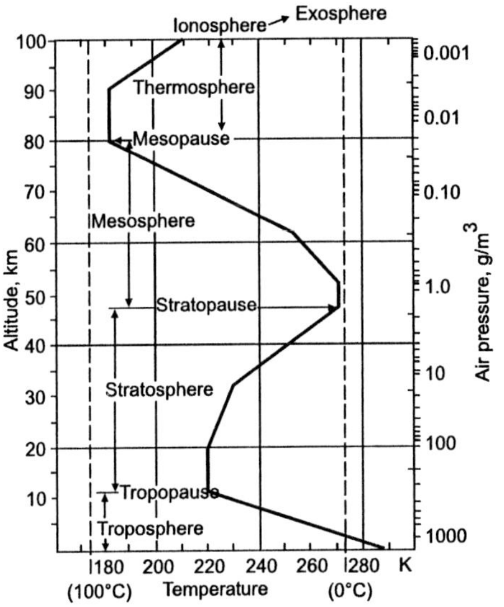

Fig. 21. The structure of the **atmosphere** and vertical distribution of temperature

Раздел *картографии*, предметом которого служат географические атласы.
f cartographie *f* de atlas
s cartografía *f* de atlas
g Atlaskartographie *f*

* **atmogenic ice → 430**

421 atmosphere
A layer of gases (air) and dust surrounding the Earth attracted to it by the Earth's gravitational pull, and subdivided into concentric sublayers. → Fig. 21
r атмосфера
Воздушная оболочка Земли, связанная с ней силой тяжести и принимающая участие в ее суточном и годовом вращении, подразделяется на концентрические слои. → Рис. 21
f atmosphère *f*
s atmósfera *f*
g Atmosphäre *f*; Lufthülle *f*

422 atmospheric absorption
Transformation of radiant energy into thermal, mechanical, electric, etc. energy because it is retained by atmospheric gases and aerosols.
r поглощение радиации в атмосфере
Превращение солнечной энергии в тепловую, механическую, электрическую и иные виды энергии при ее поглощении атмосферными газами и аэрозолями.
f absorption *f* atmosphérique; absorption *f* du rayonnement solaire
s absorción *f* atmosférica; absorción *f* de la radiación solar
g atmosphärische Strahlungsabsorption *f*

423 atmospheric aerosol; atmospheric dust
The smallest solid or liquid particles of natural origin suspended in the upper layers of the atmosphere.
r атмосферный аэрозоль; атмосферная пыль
Мельчайшие твердые или жидкие частицы естественного происхождения, взвешенные в верхних слоях атмосферы.
f aérosol *m* atmosphérique
s aerosol *m* atmosférico
g atmosphärisches Aerosol *n*

424 atmospheric attenuation; attenuation of solar radiation
The loss of intensity of direct solar radiation while passing through the atmosphere due to absorption or scattering by the atmospheric constituents.
r ослабление радиации в атмосфере
Уменьшение интенсивности прямой солнечной радиации при прохождении ее сквозь атмосферу, обусловленное поглощением или рассеянием радиации атмоферными газами и примесями в воздухе.
f atténuation *f* atmosphérique; atténuation *f* du rayonnement solaire
s atenuación *f* atmosférica; atenuación *f* de la radiación solar
g atmosphärische Strahlungsminderung *f*

425 atmospheric cell
A vertical circulation section in the tropopause associated with meridional circulation of the atmosphere.
r атмосферная ячейка
Секция вертикальной циркуляции в тропопаузе, связанная с меридиональной циркуляцией атмосферы.
f cellule *f* atmosphérique
s célula *f* atmosférica
g Luftzelle *f*

426 atmospheric chemistry
Chemical composition, chemical and photochemical processes that occur in the atmosphere, as well as their scientific study.
r химия атмосферы
Химический состав атмосферного воздуха и происходящие в атмосфере химические и фотохимические процессы, а также отрасль метеорологии, их изучающая.
f chimie *f* de l'atmosphère
s química *f* de la atmósfera
g Chemie *f* der Atmosphäre; Atmosphärenchemie *f*

427 atmospheric circulation; circulation of the atmosphere
Planetary system of air flow patterns over the whole globe and its complete statistical description. → Fig. 22, p. 52
r атмосферная циркуляция; циркуляция атмосферы

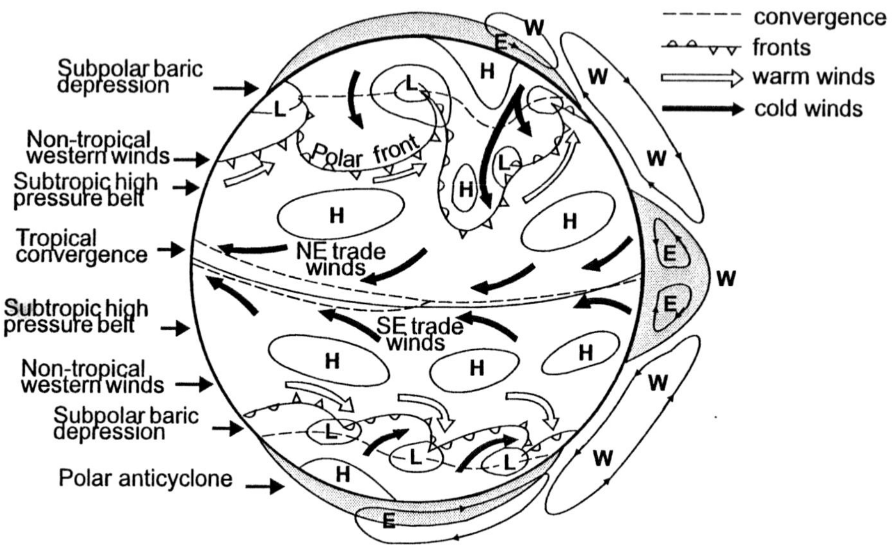

Fig. 22. **Atmospheric circulation**

Планетарная система воздушных течений, охватывающая весь земной шар, и ее полное статистическое описание. → Рис. 22

f circulation *f* atmosphérique; circulation *f* planétaire
s circulación *f* atmosférica; circulación *f* de la atmósfera
g atmosphärische Zirkulation *f*; Zirkulation *f* der Atmosphäre

* **atmospheric dust → 423**

428 atmospheric electricity
Various electrical phenomena which occur naturally in the atmosphere.
r атмосферное электричество
Совокупность различных естественных электрических явлений в атмосфере.
f électricité *f* atmosphérique
s electricidad *f* atmosférica
g Luftelektrizität *f*

429 atmospheric front
Surface of separation of two air masses with different physical properties.
r атмосферный фронт
Поверхность раздела между двумя воздушными массами с разными физическими свойствами.
f front *m* atmosphérique
s frente *m* atmosférico
g Wetterfront *f*

430 atmospheric ice; atmogenic ice
Ice formed in the air or on various surfaces by the sublimation from the air.
r атмосферный лед
Лед, образующийся в воздухе или на различных поверхностях из воздуха.
f glace *f* atmosphérique
s hielo *m* atmosférico
g atmosphärisches Eis *n*

431 atmospheric instability
A physical condition of the atmosphere in which a parcel of air rises and expands because it is warmer and therefore less dense than the air immediately above it.
r неустойчивость атмосферы
Физические условия в атмосфере, вызывающие подъем и расширение некоторого объема воздуха – более теплого, а потому менее плотного, чем расположенный непосредственно выше.
f instabilité *f* de atmosphère
s inestabilidad *f* de atmósfera
g Luftinstabilität *f*

433 atmospheric opacity; atmospheric turbidity
The loss of transparency of the atmosphere due to the scattering and absorption of light by solid and liquid particles suspended in the atmosphere.
r мутность атмосферы
Уменьшение прозрачности воздуха, обусловленное рассеянием и поглощением света взвешенными в атмосфере твердыми или жидкими частицами.
f opacité *f* atmosphérique
s opacidad *f* atmosférica
g atmosphärische Trübung *f*

434 atmospheric optics
Optical properties and optical phenomena in the atmosphere caused by scattering, absorption, refraction, diffraction, and polarization of light in the air, as well as their scientific study.

r атмосферная оптика
Оптические свойства атмосферы и оптические явления в атмосфере, вызываемые рассеянием, поглощением, преломлением, дифракцией и поляризацией света в воздухе, а также отрасль метеорологии, их изучающая.
f optique *f* atmosphérique
s óptica *f* atmosférica
g atmosphärische Optik *f*

435 **atmospheric perturbations**
Any disturbance in a stable zonal flow in the atmosphere.
r возмущение атмосферы
Любое отклонение в устойчивом зональном потоке атмосферного воздуха.
f perturbation *f* atmosphérique
s perturbación *f* atmosférica
g Luftunruhe *f*

436 **atmospheric phenomenon**
A physical phenomenon visually abserved at a meteorological station and around: hydrometeors, thunderstorm, fog, squall, etc.
r атмосферное явление
Физическое явление, визуально наблюдаемое на метеостанции и в ее окрестностях: гидрометеоры, гроза, шквал, туман и пр.
f phénomène *m* atmosphérique
s fenómeno *m* atmosférico
g atmosphärische Erscheinung *f*

* **atmospheric physics → 4555**

437 **atmospheric pollution; air pollution**
Pollution of the atmospheric air with solid and liquid particles as well as harmful gases.
r загрязнение атмосферы; атмосферное загрязнение; загазованность
Загрязнение атмосферного воздуха твердыми и жидкими частицами, а также газами, не входящими в состав воздуха.
f pollution *f* atmosphérique; contamination *f* atmosphérique
s polución *f* atmosférica; contaminación *f* atmosférica
g Luftverschmutzung *f*; Verunreinigung *f* der Atmosphäre

438 **atmospheric pressure; air pressure**
The pessure exerted by the weight of the atmosphere on the Earth's surface.
r атмосферное давление; давление воздуха
Давление, оказываемое весом атмосферы на земную поверхность.
f pression *f* atmosphérique
s presión *f* atmosférica
g Atmosphärendruck *m*; Luftdruck *m*

439 **atmospheric refraction**
The bending of light rays passing through the atmosphere due to differences in the refractive indices of adjacent layers associated with differences in density.
r атмосферная рефракция
Искривление световых лучей в земной атмосфере из-за изменения показателей преломления в слоях атмосферы, имеющих разную плотность.
f réfraction *f* atmosphérique
s refracción *f* atmosférica
g atmosphärische Refraktion *f*

440 **atmospheric stability; stable atmospheric equilibrium**
The state of the atmosphere where the environmental lapse rate of an air mass is less than the dry adiabatic lapse rate and vertical air movement cannot be sustained.
r устойчивое равновесие атмосферы
Состояние атмосферы, когда вертикальный градиент температуры воздуха меньше сухого адиабатического градиента и не происходит вертикального движения воздуха.
f équilibre *m* stable de l'atmosphère
s equilibrio *m* estable de atmósfera
g stabiles Gleichgewicht *n* der Atmosphäre

441 **atmospheric tide; solar atmospheric tide**
Periodical (diurnal, semi-diurnal) oscillations of the atmosphere caused by the gravitational attraction of the Moon and Sun.
r атмосферный прилив; солнечный атмосферный прилив
Периодические (суточные, полусуточные) колебания атмосферы, вызываемые притяжением Луны и Солнца.
f marée *f* atmosphérique; marée *f* atmosphérique solaire
s marea *f* atmosférica; marea *f* atmosférica solar
g atmosphärische Gezeiten *pl*

442 **atmospheric transparency**
The property of the atmosphere which enables a flux of solar radiation to pass through it.
r прозрачность атмосферы
Способность атмосферы пропускать поток солнечной радиации.
f transparence *f* de atmosphère
s transparencia *f* de atmósfera
g Durchsichtigkeit *f* der Atmosphäre

* **atmospheric turbidity → 433**

443 **atmospheric water**
Water in the atmospheric air in the form of vapour or suspended products of condensation such as drops, crystals, etc.
r атмосферная вода
Вода, находящаяся в атмосферном воздухе в виде водяного пара или взвешенных продуктов конденсации: капель, кристаллов.
f eau *f* atmosphérique
s agua *f* atmosférica
g atmosphärisches Wasser *n*

444 atmospheric wave
A wave in the horizontal flow o air in the atmosphere, e.g. Rossby waves.
r атмосферная волна
Волна в системе горизонтального потока воздуха в атмосфере, например, волны Россби.
f onde *f* atmosphérique
s onda *f* atmosférica
g Luftwelle *f*; atmosphärische Welle *f*

445 atmospheric window; window
Those parts of the long-wave, infrared radiation spectrum, with wavelengths of 8.5-14 microns, in which radiation is not absorbed by water vapour and CO_2, and thus when there is little cloud cover, terrestrial radiation of this wavelength is lost into space, allowing a cooling of the Earth's surface through these windows.
r атмосферное окно
Полоса спектра длинноволновой инфракрасной радиации с длиной волны 8,5-14 микрон, в которой радиация не абсорбируется водяными парами и углекислым газом, и поэтому при незначительной облачности земное излучение в этом «окне» уходит в мировое пространство, тем самым охлаждая Землю.
f fenêtre *f* atmosphérique
s ventana *f* atmosférica
g atmosphärisches Fenster *n*; Atmosphärenfenster *n*

446 atoll
A ring-shaped island reef that encircles a central lagoon. → Fig. 23
r атолл
Коралловое островное сооружение в форме сплошного или разорванного кольца, окружающего мелководную лагуну. → Рис. 23
f atoll *m*; récif *m* annulair
s atolón *m*; arrecife *m* anular
g Riffinsel *f*; Atoll *n*

447 atrio
A circular depression separating the central, younger cone of a volcano from the surrounding crest of the old crater (*somma*).
r атрио
Кольцевая впадина, отделяющая центральный, более молодой конус вулкана от окружающего гребня старого кратера (*соммы*).
f atrio *m*
s atrio *m*
g Atrio *n*

448 attached dune
A sand dune formed around an obstacle lying in the path of blown sand in arid lands, the obstacle forming the nucleus that «fixes» the dune.
r прилегающая дюна
Песчаная дюна, формирующаяся вокруг

Fig. 23. **Atoll** in Polynesia (above); the formation of an atoll

препятствия на пути перевеваемого песка в аридных районах, причем препятствие служит ядром, фиксирующим дюну.
f dune *f* attachèe; dune *f* d'obstacle
s duna *f* adyacente; duna *f* de obstáculo
g Nachbardüne *f*

449 attached island
An island jointed to a continent or a larger island by loose aggradation in the form of *tombolo*.
r причлененный остров
Остров, причлененный к берегу материка или более крупного острова полосой рыхлых морских наносов в виде *томболо* (перемы).
f île *f* attachée
s isla *f* adyacente
g angeschlossene Insel *f*

450 attached water; constitution water; fixed water
Motionless underground waters kept in rock by the cohesion with ground particles.
r связанная вода
Неподвижные подземные воды, удерживаемые в горной породе за счет ее сцепления с частицами грунта.
f eau *f* liée; eau *f* fixe
s agua *f* de constitución
g gebundenes Wasser *n*; Haftwasser *n*

* **attenuation of solar radiation → 424**

451 attribute
In geoinformatics, qualitative or quantitative feature characterizing a spatial object (but not connected with indication of place) and associated with its unique number.
r атрибут
В геоинформатике непозиционное свойство пространственного объекта (в отличие от его координат), его качественная или количественная характеристика.

f attribut *m*
s atributo *m*
g Attribut *n*; Merkmal *n*

452 attrition
The mechanism by which the particle size of any material is reduced by friction during transport – by ocean waves, wind or running water.
r измельчение
Уменьшение частиц любого материала вследствие трения при их переносе, возникающего от воздействия морских волн, ветра или текущей воды.
f attrition *f*
s desgaste *m*
g Mahlung *f*

453 aufeis; naled
A surface accumulation of ice on the ground when groundwater freezes seasonally where it repeatedly emerges from springs and when rivers flood surrounding countryside and engineering structures during the period of freezing. → Fig. 24
r наледь
Масса льда на поверхности земли, ледяного покрова (на реке) или на поверхности инженерных сооружений, образующаяся при замерзании периодически изливающихся природных или техногенных вод. → Рис. 24
f glaçage *m*; naledj *m*
s naledj *m*
g Aufeis *n*

454 aufeis blister
A convex part of an aufeis mass formed by the frost heaving.
r наледный бугор пучения
Выпуклая часть наледного массива, образовавшаяся в результате пучения льда.
f dôme *m* de glaçage
s domo *m* de naledj
g Aufeisblase *f*

455 aufeis flood-plain
Part of a river valley with a running water stream covered with the body of aufeis.
r наледная пойма
Часть речной долины со сквозным водным потоком, перекрываемая толщей наледного льда.
f plaine *f* inondable de glaçage
s llanura *f* de inundación de naledj
g Hochflutbett *n* mit Aufeis; Eismulde *f*

Fig. 24. **Aufeis** at a tributary of Indigirka river

456 aufeis glade
Part of a gentle slope or a valley bottom kept clear of trees and shrubs by the annual occurrence of aufeis.
r наледная поляна
Безлесная часть пологого склона или дна долины, в пределах которых ежегодно формируется наледь.
f zone *f* de glaçage
s clara *f* de naledj
g Aufeis-Delle *f*

457 aufeis resources
Total resources of ice in aufeises formed during a certain time interval over some territory.
r наледные ресурсы
Суммарные запасы наледного льда, образующегося за конкретный период времени на какой-либо территории.
f ressources *fpl* en glaçage
s recursos *mpl* de naledj
g Aufeisquellen *fpl*

458 aureole
A poorly developed corona, characterised by a bluish-white disk seen immediately around the luminary reddish-brown outer edge.
r ореол; венец
Плохо развитая корона вокруг Солнца или Луны в виде голубоватого диска, красновато-коричневая на внешнем крае.
f auréole *f*
s aureola *f*
g Aureole *f*; Korona *f*

459 aurora; aurora australis; aurora borealis
The luminous effect in the high atmosphere, electro-magnetic origin, seen near the horizon in the night sky in high latitudes as streamers, veils, sheets and rays of white, red green colour (the terms aurora borealis or Northern Lights areis used in the Northern Hemisphere and aurora australis in the Southern Hemisphere).
r полярное сияние; северное сияние
Свечение газов в верхних слоях атмосферы, вызываемое взаимодействием атомов и молекул с заряженными космическими частицами, видимое в ночном небе высоких широт в виде занавесей, лент и отдельных лучей белого, красного или зеленого цвета.
f aurore *f*; aurore *f* polaire
s aurora *f*; aurora *f* polar; aurora *f* borealis
g Polarlicht *n*; Nordlicht *n*; Aurora *f* Borealis

* **aurora australis → 459**

* **aurora borealis → 459**

460 **autarchy; autarky**
National economic isolation of a country achieved creating a self-sufficient independent economy.
r автаркия
Хозяйственное обособление страны, создание самообеспеченной, замкнутой экономики.
f autarcie *f*
s autarcía *f*; autarquía *f*
g Autarkie *f*

* **autarky → 460**

461 **autochore**
A plant that spreads its fruits, spores or seeds without the action of external factors, such as wind.
r автохор
Растение, распространяющееся без содействия внешних факторов, например, ветра.
f plante *f* autochore
s planta *f* autocora
g Autochor *m*; Selbstverbreiter *m*

462 **autochthonous *adj***
Applied to material and features which originated in its present position (e.g. the plant material in a deposit, such as peat, which actually grew where it is found), rather than being brought in from outside, as opposed to *allochthonous*.
r автохтонный
Относящийся к материалу (веществу), образующемуся на месте своего первоначального произрастания или залегания (например, торф, сформировавшийся из растительной массы, выросшей в данном месте), в отличие от *аллохтонного*, принесенного извне.
f autochtone *adj*
s autóctono *adj*
g autochthon *adj*

463 **autochthonous species; native species; indigenous species**
A species of living things originating in a particular area (not introduced) and inhabiting this area for a long time (e.g. duck-billed platypus and eucalyptus in Australia), as distinct from the *allochthonous species*.
r автохтонный вид; аборигенный вид
Вид живых организмов, возникший или с древних времен обитающий на данной территории (например, утконос и эвкалипт в Австралии), в отличие от *аллохтонных видов*.
f espèce *f* autochtone; espèce *f* aborigène
s especie *f* autóctona; especie *f* aborigena; especie *f* nativa
g autochthone Art *f*; einheimische Art *f*

464 **autocracy; dictatorship**
A form of government in which undivided and uncontrolled absolute power is concentrated in one dictorial person or a small clique, e.g. despotic states of Ancient Orient, absolute monarchies, etc.
r автократия; диктатура
Форма государственного правления с неограниченным бесконтрольным полновластием одного лица (деспотии Древнего Востока, абсолютные монархии и др.).
f autocratie *f*
s autocracia *f*; dictadura *f*
g Autokratie *f*; Diktatur *f*

465 **autoecology**
The study of the interactions of individual organisms, and populations of species with the environment.
r аутоэкология
Изучение взаимоотношения отдельных особей, популяций видов с окружающей средой, в отличие от изучения сообществ.
f autoécologie *f*
s autoecología *f*
g Autökologie *f*

* **autogeneous *adj* → 466**

466 **autogenic *adj*; autogeneous *adj***
Produced by an organism itself, not by an external influence.
r автогенетический
Сформированный самим организмом, без внешнего влияния.
f autogénétique *adj*
s autógeno *adj*
g autogenetisch *adj*

467 **automated cartography**
The process of drawing maps with the aid of such computer-driven devices as plotters.
r автоматизированная картография
Процесс создания карт с помощью компьютерных средств, например, плоттера.
f cartographie *f* automatique
s cartografía *f* automática
g automatisierte Kartographie *f*

468 **automated mapping system**
Software that includes technical, program, information, and linguistic means providing input of data into computer, processing and storage of this information, compilation and updating of maps.
r автоматическая картографическая система
Компьютерная программа составления карт, включающая технические, программные, информационные и лингвистические средства, обеспечивающие ввод в ЭВМ данных, обработку и хранение этой информации, составление и обновление карт.
f système *m* de cartographie automatique

s sistema *m* de cartografía automática
g EDV-gestütztes kartographisches System *n*; automatisiertes Kartierungssystem *n*

469 automobile industry; motor-car industry
A branch of machine-building industry which produces lorries, cars, and buses.
r автомобильная промышленность
Отрасль машиностроения, производящая грузовые, легковые автомобили и автобусы.
f industrie *f* automobile
s industria *f* automovilística; industria *f* de automóvil
g Fahrzeugbau *m*; Straßenfahrzeugbau *m*; Automobilindustrie *f*

470 autonomous *adj*
Usually applied to a state which is not a sovereign state or absolutely independent, but one which is self-governing in home affairs while being under the control of a larger (and sovereign) state.
r автономный
Обычно применяется к государству или части другого государства, которые имеют определенную долю самостоятельности, но находятся под контролем основного государства или государства-суверена.
f autonome *adj*
s autónomo *adj*
g autonome *adj*

471 autonomous landscape; eluvial landscape
A geographical landscape of flat interfluves which do not influence the biological cycle of matter.
r автономный ландшафт; элювиальный ландшафт
Географический ландшафт, приуроченный к плоским водоразделам, не оказывающим заметного влияния на биологический круговорот веществ.
f paysage *m* autonome; paysage *m* de régime éluvial
s paisaje *m* autónomo
g autonome Landschaft *f*; eluviale Landschaft *f*

472 autonomy
A form of government or a state which is not sovereign or absolutely independent, but is self-governing in home affairs while being under the control of a larger state.
r автономия
Форма правления или часть государства, обладающая некоторыми особенностями национального состава и быта, закрепленными в общегосударственной конституции.
f unité *f* autonome; région *f* autonome; autonomie *f*
s unidad *f* autónoma; región *f* autónoma; autonomía *f*
g Autonomie *f*; Selbstverwaltung *f*

473 auto-oscillations; self-induced oscillations
Continuous oscillations in a natural object, the period and amplitude of which are determined by properties of the object itself.
r автоколебания
Незатухающие колебания в природном объекте, например в леднике, период и амплитуда которых определяются свойствами самой объекта.
f autofluctuations *fpl*
s autofluctuaciónes *fpl*
g Autooszillatione *npl*

* **autotroph → 4776**

474 autotrophic soil formation
The processes of soil formation going on mainly on flat interfluve areas under the atmospheric moistening, without any ground water input.
r автоморфное почвообразование
Процессы почвообразования, протекающие преимущественно в плакорных условиях при поступлении влаги только с атмосферными осадками, без участия грунтовых вод.
f formation *f* du sol autotrophe
s formación *f* del suelo autotrofa; génesis *f* del suelo autotrofa
g autotrophe Bodenbildung *f*

475 autumn; fall[2]
The season of the year between summer and winter, the period astronomically defined from the autumnal equinox to the winter solstice (from September 23 to December 21-22 in the Northern Hemisphere, from March 20-21 to June 21-22 in the Southern Hemisphere).
r осень
Время года, переходный сезон между летом и зимой, в астрономическом понимании – промежуток времени от момента осеннего равноденствия до зимнего солнцестояния (с 23 сентября до 21-22 декабря в Северном полушарии, с 20-21 марта до 21-22 июня в Южном).
f automne *m*
s otoño *m*
g Herbst *m*

476 autumnal equinox
The moment when the Sun crosses the celestial equator during its apparent annual movement from the north to south on September 23, which is the beginning of the astronomic autumn in the Northern Hemisphere. → Fig. 25, p. 58
r осеннее равноденствие
Момент времени, когда Солнце пересекает небесный экватор при своем видимом движении из Северного полушария в Южное 23 сентября, что служит началом

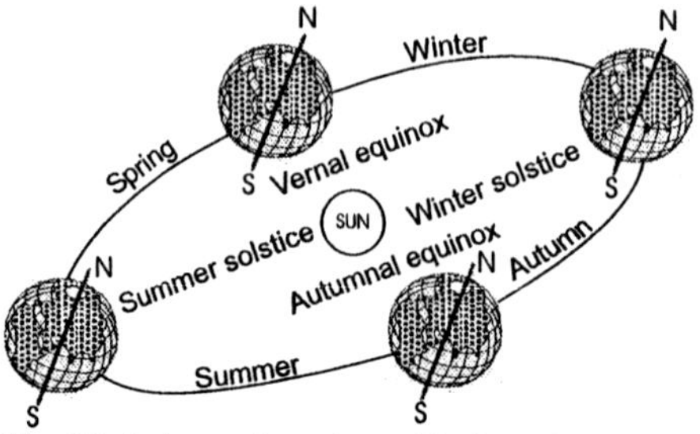

Fig. 25. **Autumnal equinox**. Moving of the Earth on its orbit and changes of seasons

астрономической осени в Северном полушарии. → Рис. 25

f équinoxe *m* d'automne
s equinoccio *m* de otoño
g Herbstäquinoktrum *m*; Herbsttagundnachtgleiche *f*

477 auxiliary industry
A manufacturing industry or infrastructure that provide the main specialized industries of a given region with semi-finished products and services.
r вспомогательная отрасль хозяйства
Производство или инфраструктура, обеспечивающие полуфабрикатами, услугами, фондами основные отрасли специализации данного района.
f industrie *f* auxiliaire
s industria *f* auxiliar
g Hilfzweig *m*; Zulieferindustrie *f*

478 avalanche; snow avalanche
A rapid gravitational movement of snow and ice down steep slopes. Fig. 26
r лавина; снежная лавина
Масса снега, движущаяся с большой скоростью вниз по горному склону. Рис. 26
f avalanche *f*; avalanche *f* de neige
s avalancha *f*; alud *m*; avalancha *f* de nieve
g Schneelawine *f*; Lawine *f*

* **avalanche barrier → 483**

479 avalanche basin; avalanche catchment
A part of a mountain side and valley bottom where a snow avalanche arises, moves and stops.
r лавиносбор; лавинный бассейн
Участок горного склона и дна долины, на котором образуется, движется и останавливается снежная лавина.
f bassin *m* d'avalanche
s cuenca *f* de avalancha
g Lawineneinzugsgebiet *n*

480 avalanche body
A snow mass that forms an avalanche.
r тело лавины
Снежная масса, приобретающая определенную форму в процессе движения лавины.
f corps *m* d'avalanche; culot *m* d'avalanche
s cuerpo *m* de avalancha; culote *m* de avalancha
g Lawinenkörper *m*

* **avalanche catchment → 479**

481 avalanche channel
A trough on a slope, along which a channelled avalanche moves down.
r лавинный лоток; лавинный лог
Желоб на склоне, по которому сходит лотковая лавина.
f couloir *m* d'avalanche
s pasillo *m* de avalancha
g Lawinengleitbahn *f*; Lawinengang *m*

482 avalanche cone; avalanche deposit; avalanche debris
A mass of snow deposited by an avalanche at its runout zone and the debris material remained after the avalanche melts out. → Fig. 28, p. 60
r конус выноса лавины; лавинный мусор; лавинный конус; лавинные отложения
Отложения снега, вынесенного лавиной, а также обломочного материала, сохраняющегося после стаивания лавинного снега. → Рис. 28, с. 60
f cône *m* d'avalanche; dépôts *mpl* d'avalanche; éboulis *m* d'avalanche
s cono *m* de avalancha; depósitos *mpl* de avalancha
g Lawinenkegel *m*; Ablagerungszone *f* einer Lawine

483 avalanche dam; avalanche barrier; avalanche wall
A stationary construction used to stop a falling down avalanche and protect the adjoining areas.
r противолавинная дамба; противолавинная стенка
Стационарная конструкция, используемая для остановки лавины и защищающая прилегающую к ней территорию.
f digue *m* d'arrêt de l'avalanche
s dique *m* de frenado de avalancha

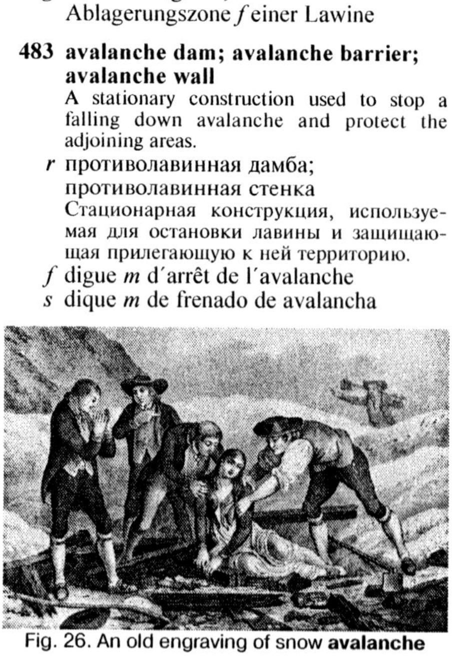

Fig. 26. An old engraving of snow **avalanche**

Fig. 27. **Avalanche dog**

g Lawinendamm *m*

* **avalanche danger → 487**

* **avalanche debris → 482**

* **avalanche deposit → 482**

* **avalanche deposition zone → 491**

484 avalanche dog
A dog specially trained for rescuing people burried by a snow avalanche. → Fig. 27
r лавинная собака
Специально обученная собака для поиска людей, засыпанных снежной лавиной. → Рис. 27
f chien *m* d'avalanche
s perro *m* de avalancha
g Lawinenhund *m*

485 avalanche frequency
Average number of avalanches annually released upon a specific path in a given year.
r частота схода лавин
Среднее количество снежных лавин, ежегодно сходящих в конкретном лавиносборе.
f fréquence *f* des avalanches
s frecuencia *f* de avalanchas
g Lawinenhäufigkeit *f*

486 avalanche gallery
A reinforced concrete construction in the form of a gallery protecting avalanche-hazardous sections of mountain roads and railways against avalanches.
r противолавинная галерея
Железобетонная конструкция (типа галереи), защищающая лавиноопасный участок горной дороги от воздействия лавин.
f galérie *f* pare-avalanche
s galería *f* contra avalanchas
g Lawinengalerie *f*

487 avalanche hazard; avalanche danger
The possibility of avalanching in a region characterized by typical features and conditions of snow accumulation favourable for the avalanching.
r лавинная опасность
Возможность схода лавин на местности, отличающейся характерными признаками лавинной деятельности и условиями снегонакопления, благоприяными для образования лавин.
f risque *m* d'avalanche
s riesgo *m* de avalancha; riesgo *m* de aludes
g Lawinengefährdung *f*

488 avalanche landscape
An area with typical features of the avalanche activity, manifested in the landforms, vegetation, soils, etc.
r лавинный ландшафт
Участок территории с типичными признаками лавинной деятельности, выраженными в рельефе, растительности, почвах и пр.
f terrain *m* avec des traces d'avalanche
s paisaje *m* de avalanchas
g Lawinenlandschaft *f*

* **avalanche path → 496**

489 avalanche research station
Research station in mountains intended for regular observations of a snow cover and the development of avalanches.
r снеголавинная станция
Исследовательская станция в горах, предназначенная для наблюдений за снежным покровом и условиями схода снежных лавин.
f station *f* de surveillance des avalanches
s estación *f* de observación de avalanchas
g Lawinenforschungsstation *f*

490 avalanche runout distance
The distance between the place where an avalanche starts to slide and the place where stop its frontal part.
r дальность выброса лавины
Расстояние от зоны отрыва лавины до места остановки ее фронта.
f distance *f* parcourue par avalanche
s distancia *f* de desencadenamiento de avalancha
g Reichweite *f* einer Lawine

491 avalanche runout zone; avalanche deposition zone
The part of an avalanche path where the avalanche begins to break, stops and snow is deposited.
r зона выброса лавины
Участок лавиносбора, на котором происходит торможение лавины и отлагается лавинный снег.
f zone *f* d'arrêt d'avalanche
s zona *f* de llegada de avalancha

g Auslaufstrecke *f*

492 avalanche site; starting zone of avalanche
The upper part of an avalanche basin where avalanches start to slide. → Fig. 28
r лавинный очаг; зона зарождения лавины
Верхняя часть лавиносбора, где происходит зарождение снежной лавины. → Рис. 28
f zone *f* de départ d'avalanche
s zona *f* de partida de avalancha
g Lawinenanrisszone *f*

493 avalanche snow-patch
A snow-patch which is a remain of an avalanche.
r лавинный снежник
Снежник, представляющий собой остаток лавины.
f névé *m* d'avalanche
s nevero *m* de avalancha
g Lawinenschneefeld *n*

494 avalanche studies
The study of structure, origin and dynamics of snow avalanches, including the ways to prevent their release.
r лавиноведение
Изучение строения, механизма формирования и схода, а также способов предупреждения снежных лавин и борьбы с ними.
f étude *f* des avalanches
s estudio *m* de avalanchas
g Lawinenkunde *f*

495 avalanche swath
A vertical treeless path on a forest-covered mountain slope, because the trees have been destroyed by an avalanche.
r лавинный прочес
Вертикальная полоса на залесенном горном склоне, лишенная древесной растительности, выбитой лавинами.
f laie *f* d'avalanche
s entresaca *f* de avalancha
g Lawinenwaldsteig *m*

496 avalanche track; avalanche path
The middle part of an avalanche below the avalanche site and above the runout zone. → Fig. 28
r зона транзита лавины
Средняя часть пути движения снежной лавины ниже зоны зарождения и выше ее конуса выноса. → Рис. 28
f zone *f* d'écoulement d'avalanche
s zona *f* de trayecto de avalancha
g Lawinensturzbahn *f*

497 avalanche trail
The rear part of an avalanche body where the velocity of a snow mass movement is smaller and its surface is gradually flattened.
r шлейф лавины

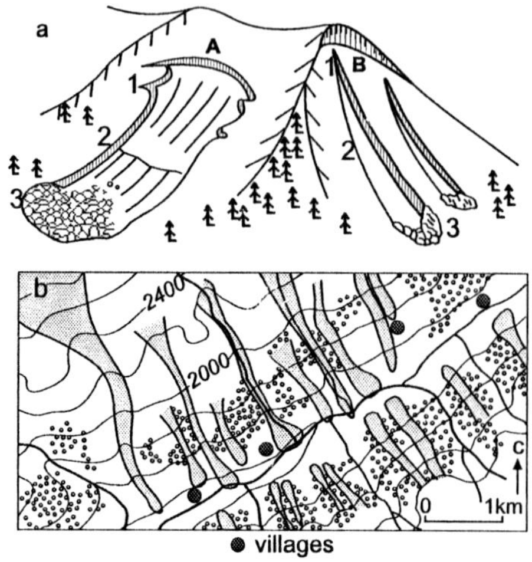

Fig. 28. **Avalanche tracks**: a – avalanche starting from a line (A) or a point (B); 1 – avalanche site, 2 – avalanche track, 3 – avalanche cone. b – avalanche tracks on the geographical map

Хвостовая часть тела лавины, где скорость движения снежных масс меньше и поверхность лавины постепенно выполаживается.
f traîne *f* d'avalanche
s cola *f* de avalancha
g Lawinenschleppe *f*

* **avalanche wall → 483**

498 avalanche wind
The blast of air which preceedes an avalanche and runs beyond the visible extent of the powder avalanche and often causes great damage.
r воздушная волна лавины; лавинный ветер
Ударная волна, возникающая при падении пылевой лавины и приводящая к разрушениям вне зоны отложения основной массы лавинного снега.
f souffle *m* d'avalanche
s viento *m* de avalancha
g Lawinenedruckwelle *f*; Lawinenwind *m*

499 avalanche-hazardous period
Time interval during which avalanches can occur in a given region.
r лавиноопасный период
Интервал времени, в течение которого условия снегонакопления и характер механической устойчивости снега на склонах могут привести к сходу лавин.
f période *f* d'avalanches
s período *m* de riesgo de avalanchas
g lawinengefährdete Zeit *f*

500 avalanche-hazardous slope
A mountain slope where avalanches can potencially occur.

r лавиноопасный склон
Горный склон, с которого возможен сход лавин.
f pente *f* d'avalanches
s vertiente *m* peligroso de avalanchas
g lawinengefährdeter Hang *m*

501 avalancher
A compressed gas gun which launches projectiles onto potential by avalanche-hazardous slopes.
r аваланчер
Небольшая пушка, действующая на сжатом газе, предназначенная для обстрела потенциально лавиноопасных склонов.
f avalancheur *m*
s cañón *m* de desencadenamiento de avalanchas
g Avalancher *m*

* **aven → 4726**

* **aviation industry → 166**

* **axial rift → 5083**

* **axial trough → 5083**

502 axis of a fold
The central line of a fold, from the crest of which strata dip downwards and away in an aticline, or the central line of the foot of the trough from which strata rise in opposite directions in a syncline.
r ось складки
Центральная линия складки, от гребня которой пласты падают вниз и во вне в антиклинали, или центральная линия подошвы впадины, от которой пласты поднимаются в противоположных направлениях в синклинали.
f axe *m* de pli
s eje *m* de pliegue
g Achse *f* der Falte

503 axis of the celestial sphere
An imaginary straight line parallel to the axis of the Earth rotation around which the visible diurnal rotation of the celestial sphere takes place (the axis of the celestial sphere crosses the celestial sphere in celestial poles).
r ось мира
Воображаемая прямая, параллельная оси вращения Земли, вокруг которой происходит видимое суточное вращение небесной сферы (ось мира пересекает небесную сферу в полюсах мира).
f axe *m* du monde
s eje *m* del mundo
g Weltachse *f*; Himmelsachse *f*

504 axis of the Earth
The diameter between the North pole and South pole tilted at an angle of about 60°30′ to the plane of the Earth's orbit, around which the Earth rotates (anti-clockwise) once in every 24 hours.
r земная ось; ось вращения Земли
Диаметр, соединяющий Северный и Южный полюса Земли, наклоненный под углом 60°30′ к плоскости земной орбиты, вокруг которого происходит суточное вращение Земли против часовой стрелки.
f axe *m* de la Terre
s eje *m* de la Tierra
g Erdachse *f*

505 azimuth; bearing$_2$
An angle in the horizontal plane read clockwise from the northern direction up to the direction to a certain object. → Fig. 29
r азимут
Угол в горизонтальной плоскости, отсчитываемый по часовой стрелке от северного направления до направления на заданный объект. → Рис. 29
f azimut *m*
s acimut *m*; acimut *m* verdadero; azimut *m*
g Azimut *m*

506 azimuth of dip
The angle between the projection of line of a dip (of a stratum, fault plane, etc.) on a horizontal plane and the local meridian. → Fig. 90, p. 192
r азимут падения
Угол между проекцией линии падения (пласта, поверхности сброса и пр.) на горизонтальную плоскость и меридианом местности. → Рис. 90, с. 192
f azimut *m* du pendage
s acimut *m* de pendiente estratigráfica; acimut *m* de pendiente estructural; acimut *m* de buzamiento
g Einfallsrichtung *f*

507 azimuth of strike; trend azimuth
The angle between a strike line (of a stratum, fault plane, etc.) and the local meridian.
r азимут простирания
Угол между линией простирания (пласта, поверхности сброса и пр.) и меридианом местности.
f azimut *m* de la direction
s acimut *m* de la línea de pendiente
g Streichwert *m*; Streichwinkel *m*

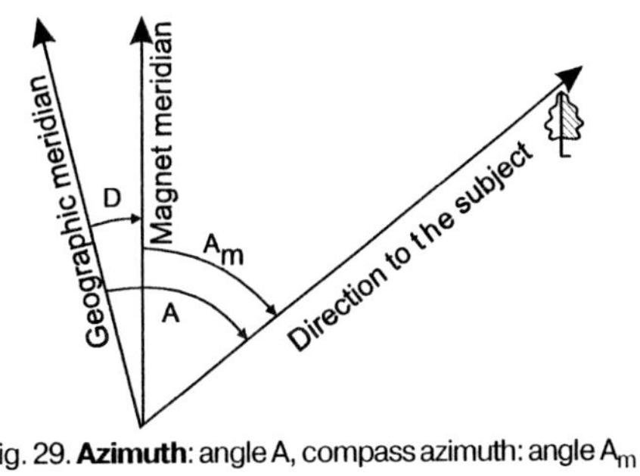

Fig. 29. **Azimuth**: angle A, compass azimuth: angle A_m

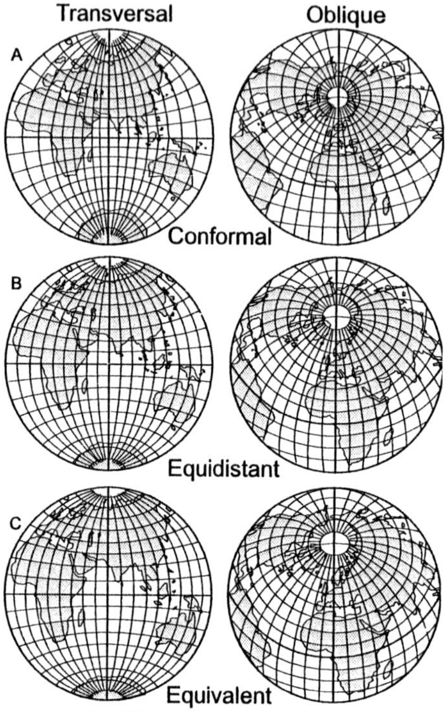

Fig. 30. **Azimuthal projections**

508 azimuthal projection; zenithal projection

A map projection in which parallels are indicated by concentric circles, and meridians – by their radiuses the angles between which are equal to the corresponding differences of longitudes. → Fig. 30

r азимутальная проекция

Картографическая проекция, в которой параллели изображаются концентрическими окружностями, а меридианы – их радиусами, углы между которыми равны соответствующим разностям долгот. → Рис. 30

f projection *f* azimutale
s proyección *f* acimutal; proyección *f* central
g azimutaler Kartennetzentwurf *m*

509 azoic *adj*

Without life, hence applied to the Precambrian rocks reflected the Earth's history before organic life, as well as those parts of the oceans where organisms cannot exist.

r азойский

Не содержащий органических остатков, поэтому термин применяется к докембрийским отложениям, в которых не встречаются следы органики, а также к глубоководной части океана, где организмы не могут существовать.

f azoïque *adj*
s azoico *adj*
g azoisch *adj*

510 azonal *adj*

Related to natural phenomena which is not determined by the zonal features of a given area.

r азональный; экстразональный

Относящийся к природным явлениям, не имеющим связи с зональными особенностями данной территории.

f azonal *adj*
s azonal *adj*
g azonal *adj*

* **azonal landscape → 512**

511 azonal soils

Soils that have slightly pronounced or not pronounced features of zonal soil formation and retain their properties in different zones.

r азональные почвы

Почвы со слабовыраженными или невыраженными чертами зонального почвообразования, сохраняющие свои свойства в разных природных зонах.

f sols *mpl* azonaux
s suelos *mpl* azonales
g azonale Böden *mpl*

512 azonal vegetation; azonal landscape

The plant communities and accompanying features of landscape, which do not belong solely to a certain geographical zone but occur in many zones, e.g. bogs, rock vegetation, etc.

r азональная растительность; азональный ландшафт; интразональная растительность

Растительные сообщества и сопутствующие черты ландшафта, не образующие самостоятельной зоны, но встречающиеся во многих зонах, например, болота, растительность скал и пр.

f végétation *f* azonale; paysage *m* azonal
s vegetación *f* azonal; paisaje *m* azonal
g azonale Vegetation *f*; azonale Landschaft *f*

B

513 back radiation; downward terrestrial radiation; counterradiation
The terrestrial infrared (long-wave) radiation absorbed and re-emitted by the atmosphere that is the principal factor in the greenhouse effect.
r встречное излучение; противоизлучение атмосферы
Длинноволновая (тепловая) радиация Земли, поглощенная атмосферой и переизлученная ею назад к поверхности Земли, основной механизм парникового эффекта.
f rayonnement *m* de retour; contre-rayonnement *m*
s radiación *f* de retorno; contrarradiación *f*
g Gegenstrahlung *f*; rückwärtige Strahlung *f*

514 back swamp; valley bog
Lowland eutrophic swamp located in a river valley on a floodplane and covered with the vegetation of reed and sedge.
r пойменное болото
Низинное болото, лежащее в речной долине, заросшее тростником и осокой.
f marais *m* de plaine d'inondation
s turbera *f* de lecho mayor; turbera *f* lateral
g Sumpf *n* in der Flußniederung

515 background; background concentration; background level
General level of noise, pollution, concentration of a substance, etc. in an area which is determined by global or macroregional natural processes.
r природный фон; фоновая концентрация; фоновый уровень
Показатели содержания веществ в воздухе или воде, отвечающие средним условиям, характерным для данной территории или акватории, которые определяются глобальными или макрорегиональными природными процессами.
f fond *m*; concentration *f* de fond; niveau *m* de fond naturel
s fondo *m*; concentración *f* de fondo; nivel *m* de fondo natural
g Hintergrund *m*; natürliche Grundbelastung *f*; Hintergrundkonzentration *f*

* **background concentration → 515**

* **background level → 515**

516 background radiation
General level of radiation in the area which comes from natural sources such as rocks, earth or atmosphere, and not from a man-made source.
r природный радиационный фон
Естественный уровень содержания радиоактивных веществ в окружающей среде.
f niveau *m* de fond radioactif
s nivel *m* de fondo radiactivo
g Hintergrundstrahlung *f*; natürliche Erdstrahlung *f*

517 backing of wind
The change of the wind in a cyclonic direction, i.e. anti-clockwise from the north through the west to the south in the Northern Hemisphere, in contrast to veering wind (in the Southern Hemisphere backing wind is equivalent to the veering wind in the Northern Hemisphere).
r левое вращение ветра
Изменение ветра в циклоническом (против часовой стрелки) направлении, т.е. с севера через запад на юг (в противоположность вращению ветра по часовой стрелке) в Северном полушарии, а в Южном полушарии такой поворот ветра эквивалентен вращению ветра в Северном полушарии.
f vent *m* lévogyre
s viento *m* con giro antihorario
g Windinversion *f*

518 backshore
The part of a beach that is above the level of normal high tides and flooded during severe storms.
r тыловая часть пляжа
Часть пляжа, лежащая выше береговой линии в период прилива и заливаемая водой при сильных штормах.
f arrière-plage *f*; plage *f* soulevée
s playa *f* trasera; playa *f* elevada
g Hinterstrand *m*

519 backslope
The gentler slope of a cuesta.
r тыловой склон
Более пологий склон куэсты.
f versant *m* cataclinal; reverse *m*
s vertiente *f* cataclinal; reverso *m*; vertiente *f* reverso
g Hinterhang *m*

520 backwash; backwash flow
The seaward flow of water moving down the slope after a wave has broken on the beach, in contrast to swash.
r откат волны; обратный поток заплеска
Поток морской воды, движущийся под действием силы тяжести по направлению наибольшего уклона пляжа, после того как приходящая волна разбивается о пляж.
f retrait *m* d'eau; jet *m* du retrait
s retirada *f* de agua; flujo *m* de retirada
g Rückwasserstrom *m*; Wellenrücklauf *m*; rückläufige Welle *f*

* **backwash flow → 520**

* **backwasting → 5054**

521 **backwater**
The rise of water level in a stream channel accompanied by the decrease of flow velocity in a certain sector of the channel.
r подпор
Повышение уровня воды в речном русле, сопровождаемое уменьшением скорости течения на определенном участке реки.
f remous *m*
s remanso *m*
g Altwasser *n*; Staugewässer *n*; Rückstau *m*

522 **back-water**
A small part of a river with a slow flow, which gets backward in flood, and has no flow in low-water.
r заводь
Небольшой речной залив с медленным течением, которое в половодье становится обратным, а в межень совсем отсутствует.
f anse *f*
s ancón *m*; ensenada *f*
g kleine Bucht *f*

523 **backwater area of groundwater**
The area where the level of groundwaters while the reservoir is created.
r зона подпора вод; зона подтопления
Область повышения уровня грунтовых вод в результате подпора речных вод при создании водохранилища.
f aréa *f* de remous; aréa *f* d'engorgement
s área *f* de remanso;
área *f* de encharcamiento
g Rückstauzone *f*

524 **backwoods**
Inward, wild part of a forest far from human settlement.
r лесная глушь; чащоба
Девственный нерасчищенный лес вдалеке от населенных мест.
f épaisseur *f*; fourré *m*
s espesura *f*; bosque *m* frondoso
g Walddickicht *n*

525 **badland**
Deeply and intricately dissected areas in arid or semiarid regions where erosion resulted in a land of narrow ravines, sharp crests and pinnacles, devoid or almost devoid of vegetation and unsuitable for agriculture. → Fig. 31
r бедленд; дурные земли
Резко и сложно расчлененный преимущественно низкогорный рельеф в областях с семиаридным и аридным климатом, лишенный растительности и непригодный для земледелия. → Рис. 31
f mauvaise terre *f*; badland *m*
s tierra *f* mala; malpaís *m*; bad-land *m*
g Badland *n*

526 **Baer's mounds**
Parallel sublatitudinal ridges and oval hills, 10-25 m high, 200-300 m wide and up to 25 km long, on Caspian Lowland.
r бэровские бугры
Параллельные субширотные гряды и холмы овальной формы высотой 10-25 м, длиной до 25 км и шириной 200-300 м, сложенные главным образом песками и плотными глинами, на Прикаспийской низменности.
f buttes *f* de Baer
s bultos *m* de Baer
g Baerische Hügel *m*

527 **baguio**
Tropical cyclones observed on Philippine Islands from July to November.
r багио
Тропические циклоны, наблюдающиеся на Филиппинских островах с июля по ноябрь.
f baguio *m*
s baguio *m*
g Baguio *m*

* **bahada → 343**

528 **baidzherakh**
A mound several metres high found in the permafrost zone formed as the residual melted-out core through the melting of an ice wedge in polygonal ground.
r байджерах
Холм высотой несколько метров в области многолетнемерзлых пород, представляющий собой остаточное грунтовое ядро после вытаивания повторно-жильного льда криогенного полигона.
f butte *f* de thermokarst

Fig. 31. **Bad land** in the Kopetdag Mountains

s montículo *m* de termocarst
g Thermokarst-Hügel *m*

529 Baikalien folding
The era of mountain formation at the end of the Precambrian and the early Cambrian, as a result of which the mountains of Eastern Sayan, Baikal region, Arabian Peninsular, etc. appeared.
r Байкальская складчатость
Эра горообразования в конце докембрия и начале кембрийского периода, в результате которой возникли горы Восточного Саяна, Прибайкалья, Аравийского полуострова и др.
f plissement *m* baikalien
s plegamiento *m* baikaliano
g baikalische Faltung *f*

530 bairak
A dry gully or ravine covered with broad-leaved trees in the zone of steppe and forest-steppe of the European Russia.
r байрак
Сухой овраг или балка в лесостепных и степных районах Европейской России, зарастающие широколиственным лесом.
f bairak *m*
s bairak *m*
g Bairak *m*

531 bajir
A sand ridge, ouval, and inter-ridge depression in deserts and semi-deserts of Central Asia.
r баир
Песчаная гряда, увал и межгрядовое понижение в пустынных и полупустынных районах Центральной Азии.
f bair *m*
s bair *m*
g Bajir *m*

532 Baku basin
A low-salty basin which in the early Pleistocene occupied the area of the contemporary Caspian Sea.
r Бакинский бассейн
Слабосоленый водоем, находившийся в раннем плейстоцене на месте современного Каспийского моря.
f bassin *m* de Bakou
s cuenca *f* de Baku
g Baku-Becken *n*

533 balance curves
The graphs showing the relation of specific mass balance of a glacier and its components (accumulation and ablation) to the altitude.
r балансовые кривые
Графики зависимости от высоты удельного баланса массы ледника и его составляющих – аккумуляции и абляции.
f courbes *fpl* des bilans glaciers
s curvas *fpl* de balance glaciar
g Kurven *fpl* der Massenbilanz

534 balance methods
Methods of investigation of natural dynamic phenomena by calculation of substance and energy fluxes.
r балансовые методы
Методы исследования динамичных природных явлений путем расчета потоков вещества и энергии.
f méthodes *fpl* de bilan
s métodos *mpl* de balance
g Bilanzmethoden *fpl*

* **balance of nature → 1921**

535 balance of payments
The difference between the amount of money going out for the goods and services a country imports, and the amount of money coming into the country from the goods and services it exports during a certain period (usually a year).
r платежный баланс
Баланс, отражающий соотношение всех платежей, произведенных страной за границу, со всеми поступлениями, полученными из-за границы за определенный (обычно годовой) период.
f balance *f* des paiements
s balance *m* de pagos
g Zahlungsbilanz *f*

536 balance of trade
The relationship between a nationrs total visible imports and exports, i.e. goods.
r торговый баланс
Соотношение между стоимостью экспорта и импорта страны за определенный период, как правило, за год.
f balance *f* de commerce
s balance *m* de comercio
g Handelsbilanz *f*

* **balance year → 812**

537 balanced rock
A big block of rock which rests on the ground with its small part.
r качающийся камень
Каменная глыба значительных размеров, опирающаяся на поверхность земли очень небольшой своей частью.
f bloc *m* branlant
s bloque *m* balanceador
g Wackelstein *m*

538 bald mountain
An isolated hill or mountains with rounded top in Kazakhstan, Transbaikalia, in the Russian Far East.
r сопка
Изолированный холм или гора с округлой вершиной в Казахстане, Забайкалье, на российском Дальнем Востоке.

f mont *m* chauve
s monte *m* pelad
g K

539 balka
At erosion valley dry or with a temporary watercourse, found in steppes and forest-steppes of Russia and Ukraine.
r балка; суходол; лощина; лог
Эрозионная долина, сухая или с временным водотоком, в степных и лесостепных районах России и Украины.
f balka *f*; ravin *m*
s barranco *m*
g Balka *f*; Gerinne *n*; Trockental *n*

540 balkanization
The division of an area into small states, sometimes implying mutual hostility among such units, or between those units and surrounding states.
r балканизация
Разделение территории на сравнительно маленькие государственные образования, нередко вызывающие обоюдную враждебность и конфликтные ситуации как между этими территориями, так и с окружающими регионами.
f balkanisation *f*
s balcanización *f*
g Balkanisierung *f*

541 ball lightning
A rare form of lightning having a shape of a reddish, luminous ball which can move rapidly along solid objects or remain floating in the air.
r шаровая молния
Редкий вид молнии, имеющий форму красноватого светящегося шара, который может быстро передвигаться или зависать в воздухе.
f éclair *m* en boule
s relámpago *m* esférico
g Kugelblitz *m*

542 balme
Overhanging rocky slope created by differential weathering of rock layers different in density.
r бальма
Нависающий скалистый склон, формирующийся при неравномерном выветривании разных по плотности слоев горных пород.
f abri *m* sous roche; encorbellement *m* d'une corniche rocheuse
s abrigo *m* rocoso; saledizo *m* de cornisa rocosa
g Überhang *f*

543 balneology
The study of the origin, physical and chemical properties of natural mineral waters and their use for medical purposes.
r бальнеология
Изучение происхождения и физико-химических свойств природных минеральных вод, методов их использования в лечебных целях.
f balnéologie *f*
s balneología *f*
g Balneologie *f*; Bäderkunde *f*

544 Baltic Ice Lake
A fresh water basin which existed by the edge of the retreating ice sheet in the final phase of the last glaciation (11-10 thousand years ago) in the area of the contemporary Baltic Sea.
r Балтийское ледниковое озеро
Пресноводный бассейн, существовавший у края отступавшего ледникового покрова в заключительной фазе последнего оледенения (11-10 тыс. лет назад) на месте современного Балтийского моря.
f lac *m* glaciaire Baltique
s lago *m* glacial Báltico
g Baltischer Eisstausee *m*

545 bank; river bank
The rising land bordering a river, stream or lake.
r берег
Приподнятая земная поверхность, ограничивающая по бокам реку, ручей или озеро.
f rive *f*; rive *f* de rivière
s orilla *f*; orilla *f* del río
g Ufer *n*; Flußufer *n*

546 bank; swell$_2$
An elevated area of considerable areal extent.
r вал
Относительно узкая длинная и невысокая форма рельефа.
f rempart *m*; bombement *m*
s caballón *m*
g Wall *m*; Schwelle *f*

547 bank fixation
The protection of seashores and river banks from the destruction of waves by reinforcing and making them more resistant.
r крепление берега
Комплекс инженерных мер по защите от разрушения берегов морей, рек и водоемов вследствие размыва потоком или волнения.
f fixation *f* de rivage
s encofrado *m* de orilla
g Ufersicherung *f*

* **bank reef → 5922**

548 bank scour
A sudden destruction of river banks undercut by the stream due to their sliding caused by sharp fluctuations of river water level.
r дейгиш
Внезапное разрушение подмываемых потоком берегов русел в результате их спол-

зания, вызванного резкими колебаниями уровня воды в реке.
f affouillement *m*
s derrubio *m* de ribera
g Ufereinsturz *m*

549 bank stream flow control
Infiltration of river waters into banks at the initial stage of high water and their return back to the river with flood recession.
r береговое регулирование стока
Естественная фильтрация речных вод в берега в начальной фазе половодья и возврат их в реку при спаде половодья.
f régulation *f* des débits par rives
s regulación *f* de caudal por las orillas; regularización *f* de caudal por las orillas
g Abflußpedelung *f* durch Ufers; Abflußregulierung *f* durch Ufer

550 bankfull stream
A stage when the river channel is completely filled with water, from bank to bank, as it happens right before the overflow.
r переполненная река
Состояние реки, когда ее русло целиком, т.е. от берега до берега, заполнено водой, и вот-вот начнется наводнение.
f stade *m* de surcharge; débit *m* à plein bord
s estadio *m* de repleción; caudal *m* a bordo pleno
g überfüllter Fluß *m*

551 banking; endigage *m*; diking
The building of walls of earth in order to protect a land from flooding.
r обвалование
Ограждение местности земляными дамбами от затопления.
f terre-plein *m* terrassement; endiguement *m*
s terraplenamiento *m*; abancalado *m*
g Eindeichung *f*

552 banner cloud
Stationary orographic cloud which forms next to a mountain crest or peak and takes the shape of a banner streaming upwind from the mountain peak. → Fig. 32

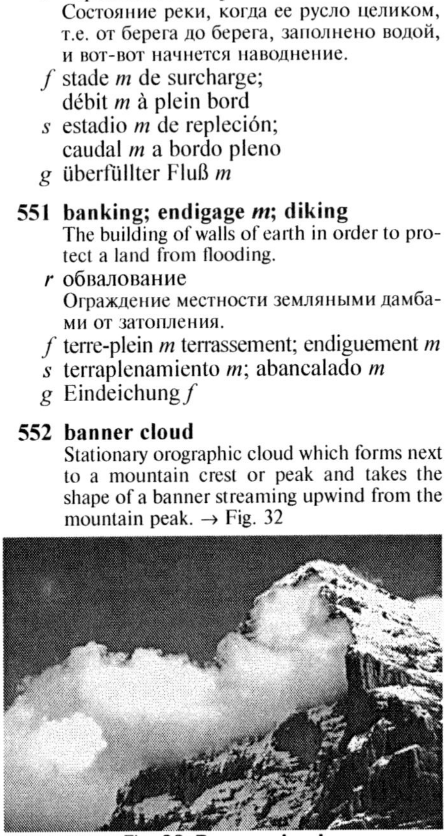

Fig. 32. **Banner cloud**

r облачный флаг; облачный вымпел
Стационарное орографическое облако, образующееся на горном хребте или вершине натекающим потоком воздуха и имеющее форму вымпела. → Рис. 32
f nuage *m* en banderole
s nube *f* en banderola
g Wolkenfahne *f*

553 bar; longshore bar$_2$
A narrow belt of land laid down by marine sediments which runs parallel to the coast (*barrier-beach*) or a sand bank across an estuary (*mouth bar*).
r бар; отмель
Узкая, вытянутая вдоль берега наносная полоса суши, отделяющая от моря лагуну (*береговой бар*); песчаный подводный вал в прибрежной полосе морского дна перед устьем реки (*приустьевой бар*).
f barre *f*
s barra *f*
g Barre *f*; Bank *f*

* **bar scale (*amer*) → 3511**

554 barbed drainage pattern
A pattern of drainage caused by the capture of a river, when tributaries meet the main stream at such angles that their flow appears to be directed to the source of the main stream. → Fig. 95, p. 202
r остроугольная речная сеть
Речная сеть, сформированная в результате перехвата реки, когда притоки впадают в главную реку под таким углом, что кажется, будто их течение направлено к истоку основной реки. → Рис. 95, с. 202
f réseau *m* hydrographique acutangle
s red *f* hidrográfica acutangular
g spitzwinkliges Flußnetz *n*

555 barbotage; sparging
The saturation of water with oxygen by way of passing through it a strong airflow, this method is used for eutrophication control of water bodies, preservation of the patches of open water in ice cover, etc.
r барботаж
Насыщение воды кислородом путем пропускания через нее сильной струи воздуха, применяемое для борьбы с евтрофированием водоемов, сохранения полыней и пр.
f barbotage *m*
s barbotaje *m*
g Durchblasen *n*

556 barchan
A crescent-shaped mobile dune created by wind accumulation in a sand desert or semi-desert, with the «horns» pointing downwind. → Fig. 33, p. 68
r бархан
Форма рельефа песчаных пустынь и полупустынь, образованная ветровой аккуму-

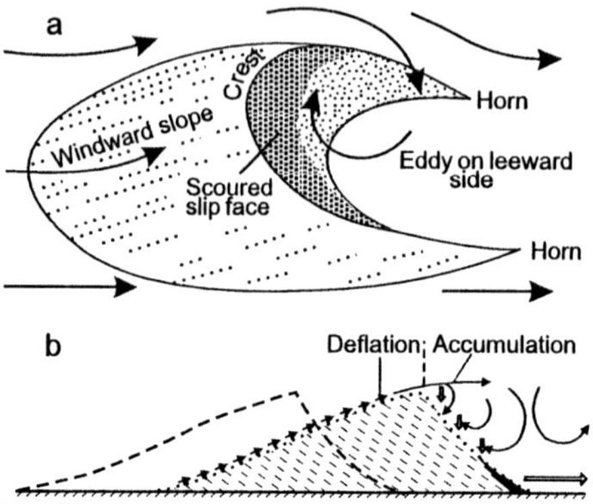

Fig. 33. **Barchan**: a – elements of a barchan, b – formation and moving of the barchan

ляцией: скопления песка серповидной формы, обычно движущиеся, с рогами по направлению ветра. → Рис. 33
f barkhane *m*
s barkhana *f*; barjana *f*
g Barchan *m*

557 barchan chain
Accumulation of sand not fixed by vegetation consisting of line of joint barchans, stretched transverse to prevailing winds and slowly moving in the direction of winds.
r барханная цепь
Скопление незакрепленного растительностью песка, состоящее из слившихся барханов, вытянутое поперек господствующих ветров и медленно продвигающееся по ветру.
f chaîne *f* de barkhanes
s cadena *f* de barkhanas
g Barchankette *f*

558 barchan relief
A typical relief of sand deserts in which the moving barchans and dunes occupy the better part of the area.
r барханный рельеф
Типичный рельеф песчаных пустынь, когда большую часть территории занимают движущиеся барханы и дюны.
f relief *m* des barkhanes
s relieve *m* de barkhanas
g Relief *m* der Barchane

559 bare fallow
A fallow land left without a crop for a vhole season so that the nutrients can build up again in the soil and water can accumulate.
r чистый пар; черный пар
Поле севооборота, которое с весны до осени остается свободным от растительности и подвергается ряду профилактических работ с целью повысить плодородие земли и аккумулировать в ней влагу.
f terre *f* en jachère
s barbecho *m* negro; barbechera *f*
g Schwarzbrache *f*

560 bare karst
Karstic surfaces with no soil and vegetation cover typical for the Mediterranean region.
r голый карст
Карстующиеся поверхности, лишенные почвенно-растительного покрова, характерные для Средиземноморья.
f karst *m* nu
s karst *m* desnudo
g nackter Karst *m*

561 bare minimum flow
The amount of water which should be left in a water body used for economic purposes to secure its natural regime.
r минимально допустимый сток
Количество воды, которое необходимо оставить в водотоке для обеспечения природного режима водного объекта в условиях водопользования.
f décharge *f* minimum admissible
s desagüe *f* mínima admisible
g Mindestabfluß *m*

562 baric col
A region of almost uniform pressure which appears between two depressions and two anticyclones arranged alternately in a cross.
r барическая седловина
Область слабоградиентного поля давления, образующаяся между двумя циклонами и двумя антициклонами, расположенными в шахматном порядке.
f col *m* barométrique
s ensillada *f* barométrica
g barometrischer Sattel *m*

563 baric field
A spatial distribution of the atmospheric pressure at sea level.
r барическое поле
Пространственное распределение атмосферного давления на уровне моря.
f champ *m* barométrique
s campo *m* barométrico
g Druckfeld *n*

564 baric topography
General geometrical configuration of the distribution of geopotential height on an isobaric surface.
r барическая топография
Общее геометрическое распределение геопотенциальных высот на изобарической поверхности.
f topographie *f* barométrique
s topografía *f* bárica
g barometrische Topographie *f*

565 baric trough
A part of the atmosphere in which the pressure is low regarding to the surrounding regions at the same level, represented on a

synoptic chart by a system of nearly parallel isobars or contours, of approximately V-shape which are concave towards the depression.

r барическая ложбина
Область атмосферы, давление в которой ниже, чем в окружающих областях на том же высотном уровне, на синоптической карте представлена системой почти параллельных изобар приблизительно V-образной формы, выпуклой по отношению к циклону.

f creux *m* barométrique
s vaguada *f* barométrica
g Tiefdruckrinne *f*

566 baric wave
A free long wave appearing in the ocean and atmosphere under the impact of sharp changes of horizontal gradients of the atmospheric pressure.

r барическая волна
Свободная длинная волна, возникающая в океане и атмосфере под влиянием резких изменений горизонтальных градиентов атмосферного давления.

f vague *f* barométrique
s ola *f* bárica
g barische Welle *f*

567 baric wind law; Buys-Ballot's law
A law describing the relatioship of the directed wind in the atmosphere to the distribution of pressure, according to which the wind deviates to the right from the baric gradient in the Northern Hemisphere and to the left in the Southern Hemisphere.

r барический закон ветра; закон Бейс-Балло
Закон, описывающий связь направленного горизонтального ветра и распределения давления в атмосфере, согласно которому отклонение ветра от барического градиента в Северном полушарии происходит вправо, а в Южном - влево.

f régle *f* de Buys Ballot
s ley *f* de Buys Ballot
g barisches Windgesetz *n*

568 barochore
A plant in which the ripe heavy seeds or fruits fall under the force of gravity without any action of other factors.

r барохор
Растение, у которого зрелые тяжелые семена или плоды опадают под влиянием силы тяжести, без участия других факторов.

f barochore *m*
s barócoro *m*
g Barochor *m*

569 barocline
Sea water layer in which both isobaric and isosteric surfaces are not parallel.

r бароклин
Слой морской воды, в котором изобарические и изостерические поверхности не параллельны друг к другу.

f couche *f* barocline
s baroclino *m*
g barokline Zone *f*

570 baroclinic instability
Atmospheric instability arising from the horizontal temperature gradient and thus strong thermal wind.

r бароклинная неустойчивость
Неустойчивость атмосферы, возникающая в результате горизонтального градиента температуры и, соответственно, термического ветра.

f instabilité *f* barocline
s inestabilidad *f* baroclínica
g barokline Instabilität *f*

* **baroclinicity → 571**

571 baroclinity; barocliny; baroclinicity
The state of the atmosphere whereby surfaces of constatnt pressure intersect surfaces of constant density. → Fig. 34

r бароклинность
Состояние атмосферы, при котором поверхности постоянного давления пересекаются с поверхностями постоянной плотности. → Рис. 34

f baroclinie *f*; baroclinicité *f*
s baroclinicidad *f*; baroclinia *f*
g Baroklinität *f*

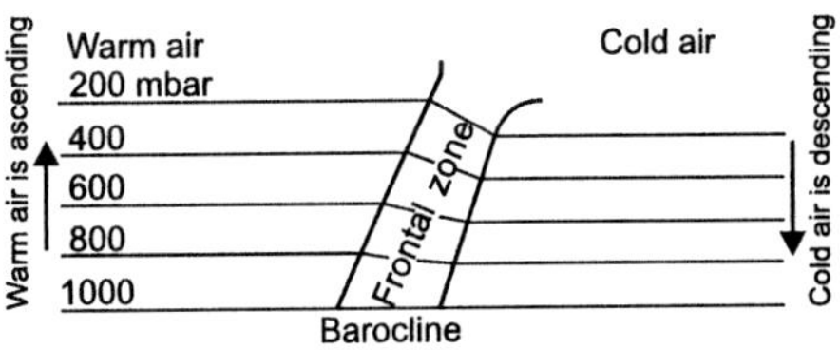

Fig. 34. **Baroclinity**

* **barocliny → 571**

572 barogradient current
Water motion caused by sea surface slope occuring due to the difference of the atmospheric pressure over different parts of the ocean.

r барградиентное течение; бароклинное течение
Движение воды, вызываемое наклоном уровенной поверхности за счет разности атмосферного давления над различными областями океана.

f courant *m* de barogradient
s corriente *f* de barogradiente
g Barogradientströmung *f*; barokline Strömung *f*

573 barometric depression; depression$_2$
A low pressure system, a region of the atmosphere where the atmosrheric pressure is lower

than that of its surroundings, it is represented on a synoptic chart by a system of isobars which enclose relatively low values of pressure. → Fig. 35

r барическая депрессия
Область атмосферы, давление в которой ниже, чем в окружающих областях на том же уровне, на синоптической карте представлена системой изобар, которые очерчивают область низкого давления. → Рис. 35

f dépression *f* barométrique; dépression$_2$ *f*

s depresión *f* barométrica; depresión$_2$ *f*

g barometrische Depression *f*; Tiefdruckgebiet *n*

574 barometric gradient; pressure gradient
The change in atmospheric pressure between two points, perpendicular to an isobaric line or surface and directed toward low pressure.

r барометрический градиент; барический градиент
Изменение атмосферного давления между двумя точками перпендикулярно изобарической линии или поверхности и направленное в сторону убывания давления.

f gradient *m* barométrique; gradient *m* de pression

s gradiente *m* barométrico; gradiente *m* de presión

g barometrischer Gradient *m*; Druckgradient *m*

575 barometric levelling
The levelling based on the dependence between the absolute altitude and atmospheric pressure.

r барометрическое нивелирование
Нивелирование, основанное на зависимости между абсолютной высотой и атмосферным давлением.

f nivellement *m* barométrique

s nivelación *f* barométrica

g Barometerhöhenmessung *f*; barometrische Höhenbestimmung *f*

576 barometric step; pressure step
A vertical distance in meters corresponding to the change of the atmospheric pressure by 1 hPa.

r барическая ступень
Расстояние по вертикали в метрах, соответствующее изменению атмосферного давления в 1 гПа.

f gradin *m* barométrique d'altitude

s escalón *m* barométrico

g barometrische Höhenstufe *f*

577 barotropy
The state of the atmosphere whereby surfaces of constant pressure are parallel to the surfaces of constant density.

r баротропность; баротропия
Состояние атмосферы, при котором поверхности постоянного давления параллельны поверхностям постоянной плотности.

f barotropie *f*

s barotropía *f*

g Barotropie *f*

578 barrage; dyke
An artificial construction or natural obstacle in a river channel and valley damming its stream, like a wall, a ridge, an embankment, a ditch.

r запруда; барраж
Искусственное сооружение или естественное препятствие в русле и долине реки, подпруживающее ее течение.

f barrage *m*

s barrera *f*; represa *f*; presa *f*

g Deich *m*; Wehr *n*

579 barranco
A deep radially developing erosion furrows on the slopes of a volcano.

r барранкос
Глубокие радиально расходящиеся эрозионные борозды на склонах вулканов.

f barranco *m*

s barranco *m*

g Barranco *m*

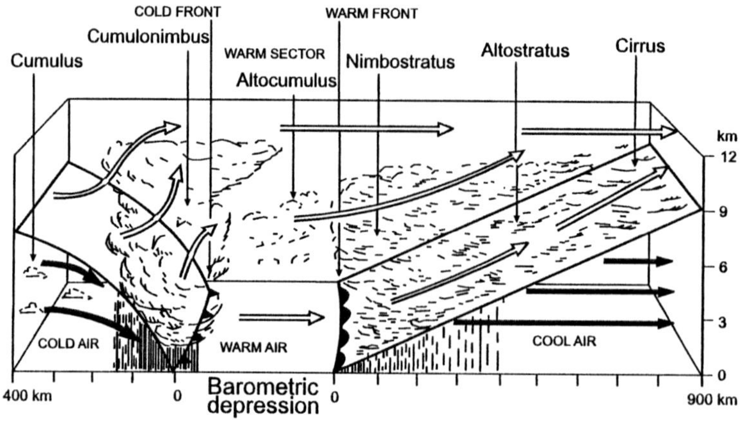

Fig. 35. Scheme of a **barometric depression**

*** barren lands → 6628**

*** barrier bar → 580**

580 barrier beach; barrier bar
A long, narrow, sandy ridge lying above high tide level parallel to the coast and separating a lagoon.
r береговой бар
Длинная и узкая, песчаная вдольбереговая гряда, протягивающаяся параллельно берегу и обычно отделяющаяся от него лагуной.
f barre *f* de côte
s barra *f* costera
g Küstenbarre *f*

*** barrier iceberg → 6038**

581 barrier island
An elongated sand or shingle bank which lies parallel to the coastline and not submerged by the tide, covered with vegetation and having swampy areas occurring on the lagoon side.
r барьерный остров
Вытянутая вдоль берега песчаная или галечная гряда, не заливаемая приливными водами, с растительностью и болотистыми участками на стороне, обращенной к лагуне.
f île *f* de barrière
s isla *f* de barrera
g Barriereinsel *f*

582 barrier reef
A ridge of coral constructions stretching parallel to the shore at some distance from it and separating a shallow lagoon from the sea.
r барьерный риф; окаймляющий риф
Гряда коралловых построек, протягивающаяся параллельно берегу на некотором рассстоянии от него и отгораживающая от моря мелководную лагуну.
f récif *m* à barrière; récif-barrière *f*
s arrecife *m* barrera; arrecife *m* lineal
g Barriereriff *n*; Wallriff *n*; Saumriff *n*

*** barrow → 6403**

583 barter
A form of trade where goods or services are echanged for a certain amount of other goods or services, i.e. with no money involved in the transaction.
r бартер
Обмен определенного количества товара на эквивалентное по стоимости количество других товаров, т.е. без использования денежных средств.
f troc *m*; échange *m*
s trueque *m*; bárter *m*; comercio *m* trueque
g Bartergeschäft *n*; Tauschgeschäft *n*

584 basal *adj*; basic *adj*
Pertaining to, situated at, forming the basement of relief or rock deposits.
r базальный
Относящийся к основанию рельефа или к свите отложений.
f basal *adj*
s basal *adj*
g basal *adj*; grundlegend *adj*; Grund-

*** basal complex → 597**

585 basal horizon
A layer of the stony channel deposits which cannot be worn out by the present flow of running water.
r базальный слой
Слой наиболее крупных русловых отложений, которые не поддаются размыву современным потоком.
f horizon *m* basal
s horizonte *m* basal
g Basalschicht *f*

586 basal moraine
Material carried in the lower (basal) layers of the glacier debris-containing ice, a part of the *transported moraine*.
r придонная морена
Переносимый ледником грубообломочный материал в слоях придонного мореносодержащего льда, часть *влекомой морены*.
f moraine *f* basale
s morrena *f* basal
g intraglaziale Grundmoräne *f*

*** basal sliding → 587**

587 basal slip; basal sliding
The sliding movement of a glacier over its rock-floor when the whole mass of glacier or its portions move relative to its bed, walls or other parts of the glacier causing no deformation inside the ice blocks.
r глыбовое движение льда; блоковое движение льда
Движение всей массы ледника или отдельных его частей относительно ложа, бортов и других частей ледника без деформаций внутри глыб льда.
f glissement *m* en bloc de glace
s deslizamiento *m* en bloque de hielo
g Eisblockschollenbewegung *f*

588 basalt
A fine-grained and dark-coloured effusive rock which sometimes forms hexagonal closely adjoined columns.
r базальт
Эффузивная горная порода черного цвета, иногда образует шестигранные столбы, тесно примыкающие друг к другу.
f basalte *m*
s basalto *m*
g Basalt *m*

*** basaltic lava → 598**

589 **basaltic layer**
The lower layer of the Earth's crust lying between the granite layer and the upper Earthrs mantle.
r базальтовый слой
Нижний слой земной коры, расположенный между гранитным слоем и верхней мантией Земли.
f couche *f* basaltique
s capa *f* basáltica
g Basaltschicht *f*

590 **base alimentation; base flow**
Inflow of ground waters to the surface watercourses and water bodies.
r подземное питание
Приток подземных вод в поверхностные водотоки и водоемы.
f alimentation *f* par les eaux souterraines
s alimentación *f* por las aguas subterráneas
g Speisung *f* durch unterirdische Gewässer

* **base flow → 590**

591 **base level of accumulation**
The level above which accumulation can not occur.
r базис аккумуляции
Уровень, выше которого аккумуляция происходить не может.
f niveau *m* de base d'accumulation
s nivel *m* de base de accumulación
g Akkumulationsbasis *f*

592 **base level of denudation**
The level below which the products of degradation of rocks cannot be transported.
r базис денудации
Уровень, ниже которого продукты разрушения горных пород переместиться не могут.
f niveau *m* de base de dénudation
s nivel *m* de base de denudación
g Denudationsbasis *f*

* **base level of erosion → 595**

593 **base level of glaciation**
The lowest limit of possible advance of the glacier tongues corresponding to the altitude of the valley bottom in a given region.
r базис оледенения
Нижняя граница возможного сползания языков ледников, соответствующая высоте днищ долин в данном районе.
f niveau *m* inférieur de glaciation
s nivel *m* inferior de glaciación
g untere Vergletscherungsgrenze *f*

594 **base map; outline map**
Type of base maps containing elements of geographical basis and coordinate grid and intended for teaching.
r контурная карта
Вид бланковых карт, содержащих элементы географической основы и координатную сетку и предназначенных для обучения.
f carte *f* muette; carte *f* de base
s mapa *m* mudo; mapa *m* de base
g Umrißkarte *f*

* **base of bed → 3602**

595 **base-level; base level of erosion**
A level down to which but not below to which a running stream can erode its bed under stable conditions of the Earth's crust and below which a land surface cannot be reduced by running water.
r базис эрозии
Уровень поверхности, на котором водный поток теряет свою силу и не может далее углубить свое ложе.
f niveau *m* de base; base *f* de l'érosion
s nivel *m* de base; base *f* de erosión
g Erosionsbasis *f*

596 **baseline**
An accurately measured line on the Earth's surface which is used as a base in triangula rsurvey, and in topographic map construction.
r базисная линия
Линия на земной поверхности, аккуратно измеренная и используемая как базис для тригонометрических наблюдений и, следовательно, для картографирования суши.
f ligne *f* de base
s línea *f* de base
g Basislinie *f*

597 **basement; basal complex**
Old folded and metamorphic rocks that make up the lower part of a platform.
r фундамент
Комплекс древних, обычно интенсивно складчатых и метаморфических горных пород, слагающих цоколь платформы.
f soubassement *m*
s basamento *m*; basamiento *m*
g Grundgebirge *n*; Fundament *n*

* **basic *adj* → 584**

598 **basic lava; basaltic lava**
The most common type of lava, molten igneous rock, high in metallic element content, low in silica, and with a low melting point, which pours easily and quietly from a volcanic vent, spreading widely before hardening to form a shield volcano or broad plateau.
r основная лава; базальтовая лава
Наиболее частый тип лавы, расплавленная вулканическая порода с высоким содержанием металлических компонентов и малым содержанием кремнезема, легко плавится и потому распространяется далеко от очага извержения, формирует щи-

товой вулкан или обширное вулканическое плато.
f lave *f* basaltique; lave *f* basique
s lava *f* basáltica; lava *f* básica
g Basaltlava *f*

599 basin
Any natural or artificial depression, varying in extent, in the Earth's crust.
r бассейн
Естественная или искусственная депрессия в земной коре.
f bassin *m*
s cuenca *f*
g Becken *n*

* **basin cultivation → 600**

600 basin irrigation; basin cultivation
A type of irrigation in which floodwater from a river that annually and for a short time overflows on to its floodplain is led off into prepared basins of different size, and separated from one on other by bankingю
r бассейновое орошение; бассейновая культивация
Тип орошения, при котором воды ежегодного половодья, на короткое время заливающего пойму, собираются в заранее приготовленные бассейны, имеющие разные размеры и отделенные друг от друга земляными валами.
f irrigation *f* par bassin
s irrigación *f* por embalse
g Beckenbewässerung *f*; Bassinbewässerung *f*

601 basin landscape
A landscape typical of karst and arid areas characterized by depressions which are not or poorly linked together by the network of valleys.
r ванновый ландшафт
Ландшафт, в рельефе которого господствуют впадины-ванны, не связанные или слабо связанные долинной сетью, особенно типичен для карстовых и сухих областей.
f paysage *m* des bassins fermés
s paisaje *m* de cubetas; paisaje *m* de cuencas cerradas
g Wannenlandschaft *f*

602 basin of ground waters
A land form or underground structure which can accumulate and discharge water due to its shape and properties of composing rocks, and serve as a source of base alimentation of a certain river.
r подземный водосбор; бассейн подземных вод
Форма рельефа или подземная структура, способные накапливать и разгружать воду благодаря своей форме и свойствам слагающих пород и служащие источником подземного питания данной реки.
f bassin *m* souterrainé
s cuenca *f* subterranea
g Grundwassereinzugsgebiet *n*

* **basinal lake → 406**

603 basin-and-range terrain
A landscape characterized by a series of asymmetrical ridges separated by basins, the ridge consisting of titled fault blocks. → Fig. 36
r вереница бассейнов
Ландшафт, характеризующийся серией асимметричных увалов, разделенных речными бассейнами, причем увалы представляют собой опрокинутые сбросовые блоки. → Рис. 36
f terrain *m* de rang des bassins
s terreno *m* de fila de cuencas
g Landschaft *f* der Beckenreihe

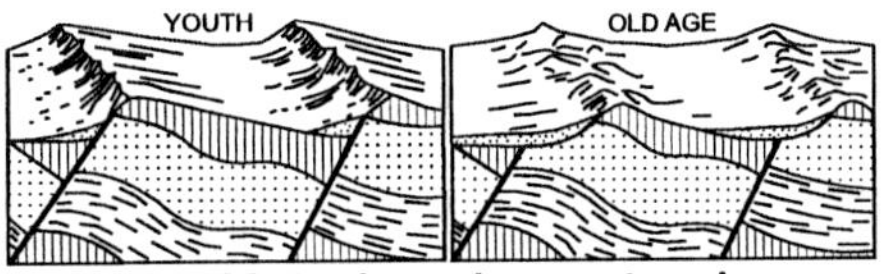

Fig. 36. **Basin-and-range terrain**

* **basin-cooler → 1286**

604 basiphyte
Plants growing on alkali soils, among these are steppe, desert and some mountain plants (plants which inhabit carbonate soils are called calciphytes, and plants which grow on alkaline soils are halophytes).
r базифит
Растения, обитающие на щелочных почвах (карбонатных – кальцефиты, солонцеватых – галофиты), это преимущественно степные, пустынные и некоторые горные растения.
f basiphyte *m*; basophile *m*; basiphile *m*
s basifita *m*; basófilo *m*; basifilo *m*
g Basiphyt *m*

605 batholith; bathylith
A very large dome-shaped mass of igneous rock (granite) formed by a large-scale, deep-seated intrusion of magma into sedimentary rocks.
r батолит
Интрузивное тело площадью более 200 км^2 неправильных очертаний, залегающее среди осадочных толщ складчатых областей.
f batholite *m*
s batolito *m*
g Batholith *m*

606 bathyal *adj*
Applied or pertaining to the deeper parts of the ocean, i.e. continental slope between the continental shelf and abyssal zone.
r батиальный
Относящийся или принадлежащий к глубокой части океана, т.е. лежащий на кон-

тинентальном склоне между материковым шельфом и абиссальной зоной.
f bathyal *adj*
s batial *adj*
g bathyal *adj*

607 bathyal deposits; bathyal sediments
Deep-water deposits accumulated at a continental slope at the depths from 200 to 3,000 m and consisting mainly of terrigenous material and silts.
r батиальные отложения
Умеренно глубоководные отложения, образовавшиеся на материковом склоне на глубинах от 200 до 3000 м и представленные главным образом терригенными осадками и илами.
f dépôts *mpl* bathyaux
s depósitos *mpl* batiales
g bathyale Ablagerungen *fpl*

* **bathyal sediments → 607**

608 bathyal zone
The zone in the ocean between sublittoral and abyssal zones approximately corresponding to a continental slope (at the depths from 200 down to 3,000 m).
r батиальная зона; батиаль
Зона в океане между сублиторалью и абиссалью, сооответствующая материковому склону (на глубинах от 200 до 3000 м).
f zone *f* bathyale
s zona *f* batial
g Bathyalzone *f*; Bathyal *n*

* **bathylith → 605**

* **bathymetric contour → 3223**

609 bathymetric curve
A plot showing the percentage of the ocean depth levels.
r батиграфическая кривая
Часть или разновидность гипсографической кривой, показывающая процентное соотношение разных ступеней глубин дна океана или морей.
f courbe *f* bathymétrique
s curva *f* batimétrica
g bathymetrische Kurve *f*

610 bathymetric map
A geographical map showing the underwater relief with isobaths and depth marks.
r батиметрическая карта
Географическая карта, отображающая подводный рельеф при помощи изобат, дополняемых отметками глубин.
f carte *f* bathymétrique
s mapa *m* batimétrico
g bathymetrische Karte *f*

611 bathymetry
The measurement of the depth of the ocean from the water surface.
r батиметрия
Измерение глубины морского дна с водной поверхности.
f bathymétrie *f*
s batimetría *f*
g Tiefenmessung *f*; Bathymetrie *f*

612 bathyorographical map
A map showing both the relief of the land and the depth of the ocean.
r батиорографическая карта
Карта, показывающая и рельеф земной поверхности, и глубины в океане.
f carte *f* bathyorographique
s mapa *m* batiorográfico
g bathyorographische Karte *f*

* **bathyscap → 613**

613 bathyscaph; bathyscap; bathyscaphe
A deep-diving autonomous manned vehicle for oceanographic investigations.
r батискаф
Глубоководный обитаемый автономный аппарат для океанологических исследований.
f bathyscaphe *m*
s batiscafo *m*
g Bathyskaph *m*

* **bathyscaphe → 613**

614 bathysphere
A solid ball-shaped camera equipped for underwater observations, and connected with a ship by a cable.
r батисфера
Прочная камера в форме шара с аппаратурой для подводных наблюдений, связанная кабель-тросом с надводным судном.
f bathysphère *m*
s batisfera *f*
g Bathysphäre *f*

615 bay; embayment
An open, wide curved indentation of the sea or of a lake into the land.
r бухта; губа
Небольшая часть моря или озера, вдающаяся в сушу, иногда ограниченная от открытых вод мысами, островами или косами.
f baie *f*
s bahía *f*
g Bucht *f*

616 bay bar; bay barrier
An elongated bank of sand, shingle or mud which links two headlands across the mouth of a bay.
r банка в мористой части залива
Вытянутый риф из песка, гальки или ила, соединяющий два выступа суши на выходе из залива и прикрывающий залив с моря.
f barre *f* de baie
s barra *f* de bahía
g Nehrung *f*

Fig. 37. **Bazaar** in Turkmenistan

* **bay barrier → 616**

617 bazaar; bazaar economy
Trade in foodstuffs and fast moving consumer goods brought by their producers to cities. → Fig. 37
r базар
Торговля продуктами питания и ходовыми товарами, привозимыми в город, как правило, самими производителями, восточный тип рынка. → Рис. 37
f bazar *m*; marché *m* public
s bazar *m*; mercado *m* público
g Basar *m*

* **bazaar economy → 617**

618 beach
A part of the sea coast above the present sea-level which forms in the zone of upswash.
r пляж
Надводная часть современной береговой зоны, образующаяся в зоне прибойного потока.
f plage *f*; estran *m*
s playa *f*; estrán *m*
g Strand *m*; Küste *f*

619 beach cusps
A crescent-shaped accumulation of sand or shingle surrounding a semicircular depression on a beach alternating with bay-like depressions, resulting from the swash and backwash of waves breaking at right angles to the coast. → Fig. 38
r пляжевые фестоны
Чередование на пляже миниатюрных треугольных мысов и разделяющих их вогнутостей берегового контура, формирующихся под действием наката и отката волн, разбивающихся под определенным углом к берегу. → Рис. 38

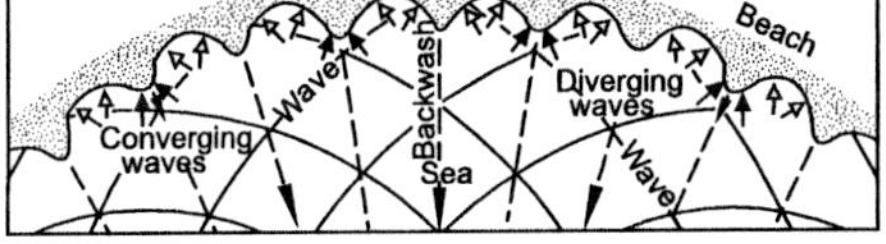

Fig. 38. **Beach cusps**

f festons *mpl* de plage
s festónes *mpl* de playa; playa *f* en arco
g Strandfestonen *npl*

* **beach ridge → 4237**

620 beach scarp; wave-cut platform
Abrasion terrace of small height (a few meters or even less a meter) in loose rocks, eroded by waves in the coastal zone.
r уступ размыва
Абразионный уступ малой высоты (первые метры или доли метра) в рыхлых породах, эродированный волнами в береговой зоне.
f berge *f*
s talud *m* de costa; entalladura *f*; estrán *m*
g Strandkante *f*

* **beach tourism → 5277**

621 beach-rock
A littoral deposit of consolidated sand which can develop in the intertidal zone of certain beaches, it is extremely resistant to erosion and its reef-like form acts as a barrier to movement of sediment across the beach.
r бич-рок
Сцементированные известью прибрежно-морские отложения, формирующиеся в приливной зоне некоторых пляжей, они особо устойчивы к эрозии и их рифоподобная форма служит барьером на пути движения отложений вдоль пляжа.
f grés *m* de plage
s arenisca *f* de playa
g Beachrock *m*

* **beacon → 3491**

622 beaded drainage pattern
A pattern of drainage where small pools joined by streams have been caused by the melting of the ground surface in permafrost regions.
r нанизанная речная сеть
Речная сеть, в которой имеются небольшие водоемы на пути речного потока, образованные в результате протаивания грунта в районах многолетней мерзлоты.
f réseau *m* hydrographique d'enfilage
s red *f* hidrográfica de ensarte
g aufgereichtes Flußnetz *n*

623 beaded valley
A transverse valley in which wider parts alternate with narrower linking sections, when the valley crosses the rocks of different resistance.
r чётковидная долина
Поперечная долина, состоящая из чередующихся широких и узких участков, возникающих при пересечении долиной горных пород разной устойчивости к размыву.
f vallée *f* en chapelet
s valle *m* en forma de rosario
g perlschnutförmiges Tal *n*

624 **bearing$_1$; orientation; taking a bearing**
The determination of an exact position with regard to cardinal points by using a compass, map or aerial image, as well as finding out an approximate location by local landmarks, position of the Sun, the Moon or stars, or with the help of radio-, light and sound signals, and GPS.
r ориентирование на местности
Определение своего местоположения относительно сторон горизонта с помощью компаса, карты, аэроснимка или приближенное ориентирование по местным ориентирам: положению Солнца, Луны или звезд, с помощью радио-, световых и звуковых сигналов, ГПС и др.
f orientation *f*; positionnement *m*
s posicionamiento *m*; orientación *f* sobre el terreno; orientación *f*
g Orientierung *f*

* **bearing$_2$ → 505**

* **beast of burden → 4365**

625 **Beaufort scale**
Wind force scale, based on the state of the sea, expressed in numbers from 0 to 12.
r шкала Бофорта
Шкала для определения силы ветра, основанная на визуальной оценке состояния моря, выражается в баллах от 0 до 12.
f échelle *f* Beaufort
s escala *f* de Beaufort
g Beaufortskala *f*; Beaufort-Skala *f*

626 **bed^1**
The floor of a water body, e.g. ocean, lake or river.
r ложе
Дно водоема или водотока (океана, озера, реки).
f lit *m*
s lecho *m*
g Bett *n*

* **bed^2→ 5851**

627 **bed load; load of a stream; traction load**
Solid material, e.g.sand, graval and sometimes large boulders, transported by flow in the bottom layer by way of dragging or rolling over the bottom, as well as *saltation*.
r влекомые наносы
Обломочный материал, переносимый потоком в придонном слое путем волочения или перекатывания по дну, а также *сальтацией*.
f charge *f* de lit; matériaux *mpl* de fond
s carga *f* de fondo; corrimiento *m* de fondo
g Geschiebe *n*

* **bed roughness → 5139**

628 **bedding$_1$**
The disposition of a rock stratum, horizontal or tilted, e.g. the relationship of a bedding plane to the horizontal plane in terms of its dip and strike.
r залегание
Пространственное положение горных пород в земной коре, определяемое падением и простиранием.
f disposition *f* des couches
s disposición *f* de capas
g Lagerung *f*

* **bedding$_2$ → 5838**

629 **bedding cave**
A wide, low horizontal cave formed in a crack in the bedding or in a slate interbed in the rock.
r пластовая пещера
Широкая и низкая горизонтальная пещера, выработанная в трещине напластования или в сланцевом прослое среди скальных пород.
f caverne *f* de lit
s caverna *f* de capas
g Schichthöhle *f*

630 **bedding plane; stratification plane**
The surface separating the successive distinctive layers of sedimentary rock.
r стратиграфическая граница; плоскость напластования; поверхность напластования
Линия или поверхность, разделяющая два смежных согласно залегающих в разрезе слоя осадочных пород.
f plan *m* de stratification; joint *m* de stratification
s plano *m* de estratificación
g stratigraphische Grenze *f*; Schichtfläche *f*

631 **bedforms**
Features developed by fluid flow over bottom alluvial sediments.
r донный рельеф
Формы поверхности на дне водоемов и водотоков, создаваемые деятельностью движущейся воды.
f formes *fpl* de relief de lit
s relieve *m* de lecho; formas *fpl* de relieve del fondo
g Grundrelief *n*; Bodenrelief *n*

632 **bedouin**
A nomadic Arab herdsman (mainly herding camels) in the Middle East and Northern Africa.
r бедуин
Кочевой араб-скотовод (преимущественно верблюдовод) в Передней Азии и Северной Африке.
f bédouin *m*
s beduino *m*
g Beduine *n*

633 **bedrock coast**
A river bank composed of older solid rocks compared with the alluvial deposits in the valley.
r коренной берег
Берег реки, сложенный коренными породами, более древними по отношению к речным отложениям в долине.
f côte *f* de roche de fond
s costa *f* de roca de fondo
g Felsküste *f*

634 **bedrock[1]**
The rocks more ancient with regard to the relief under consideration.
r коренная порода
Породы, более древние по отношению к рассматриваемому рельефу.
f roche *f* de fond
s roca *f* de fondo; lecho *m* de roca; roca *f* originaria
g Grundgestein *n*

635 **bedrock[2]**
Upper bedding plane of bedrock stratum, often dislocated, lying at the bottom of a river valley or lake and covered with younger river or lake deposits.
r коренное ложе
Кровля горных пород, часто дислоцированных, выстилающих долину либо озерную ванну и покрытых современными осадочными отложениями реки или озера.
f lit *m* primaire
s lecho *m* primario
g primäres Bett *n*

636 **beekeeping; apiculture**
Breeding bees for honey, wax, apitoxin, etc.
r пчеловодство
Отрасль животноводства, занимающаяся разведением пчел для получения меда, воска, пчелиного яда и пр.
f apiculture *f*
s apicultura *f*; cría *f* de abejas
g Bienenzucht *f*; Imkerei *f*

* **behavioural approach → 637**

637 **behaviourism; behavioural approach**
The study of peoplers perception of their natural and social environment within certain regions and areas, their understanding of the optimal conditions of living and preferences when selecting places for habitation.
r бихевиоризм; поведенческий подход
Изучение особенностей восприятия людьми окружающей природной и социальной среды в пределах определенных районов и местностей, их представлений об оптимальных условиях обитания и предпочтений в выборе мест проживания.
f béhaviorisme *m*
s behaviourismo *m*; conductismo *m*
g Behaviorismus *m*

* **beheading of river → 871**

638 **belligerent landscape**
A geographical landscape formed owing to military activities (ancient fortress ramparts, watch barrows, trenches, shell-holes, caponiers, firing grounds, etc.).
r беллигеративный ландшафт
Географический ландшафт, обязанный своим возникновением военной деятельности (старинные крепостные валы, сторожевые курганы, траншеи, нынешние воронки взрыва, капониры, полигоны для стрельбы и пр.).
f paysage *m* belligérant
s paisaje *m* beligerante
g Kriegslandschaft *f*

639 **bench**
A sea or lake terrace cut by ocean waves in the bedrock at the base of a cliff.
r бенч
Морская или озерная терраса, выработанная абразией в коренных породах у основания клифа.
f plate-forme *f* d'abrasion marine
s terraza *f* de abrasión marina
g Bank *f*

640 **bench mark; survey mark**
A sign fixed in the walls of long-term constructions or in the ground, the altitude of which is determined by high-precision levelling.
r репер
Закрепляемый в стенах долговременных сооружений или в грунте знак с известной абсолютной высотой, которая определена высокоточным нивелированием.
f repère *m*
s punto *m* de referencia de nivel; marca *f* de referencia
g Höhenmarke *f*; Pegelfestpunkt *m*

641 **benefication *f*; concentration of minerals; dressing of minerals**
The process of removal of a commercially valuable mineral from the rock surrounding it after extraction from the ground (by means of crushing, magnetic separation, flotation) carried out at or near the site of the mine or other working.
r обогащение руды
Процесс выделения полезного минерала из горной породы, окружающей его после выемки из грунта (путем размельчения, магнитной сепарации и флотации), производимый на месте или недалеко от места добычи.
f enrichissement *m*
s enriquecimiento *m* de minerales
g Erzaufbereitung *f*

642 **benthal *adj***
Related to the bottom of bodies of water: lakes, oceans, seas, and to organisms dwelling in the and near the bottom.

r придонный
Относящийся к донным частям озер, морей и океанов и к организмам, обитающим на грунте или в придонной толще.
f bentique *adj*
s bental *m*; béntico *adj*
g benthal *adj*

643 benthic division; benthic zone
One of the two chief divisions of the aquatic environment based on depth of water (the other being the *pelagic* division), consisting of all the floor of the ocean or lake, where benthos live.
r бентальная зона; бенталь
Одно из двух главных подразделений водной окружающей среды (другое подразделение носит название *пелагиаль*), представляющее собой придонную часть морей и озер, где обитает бентос.
f étage *m* bentique
s piso *m* béntico
g Benthalzone *f*; Benthal *n*

* **benthic zone → 643**

644 benthos
All the plant and animal organisms that live on or near the bottom of seas and lakes irrespective to its depth, they are subdivided into phyto- and zoobenthos.
r бентос
Совокупность растительных и животных организмов, обитающих на дне и в грунте морских и материковых водоемов, подразделяется на фито- и зообентос.
f benthos *m*
s bentos *m*
g Benthos *n*

645 bergschrund
A rupture of a firn mass at the foot of a slope which separates the moving ice of a glacier from a relatively immobile ice (ice apron) on the slope. → Fig. 39
r бергшрунд; подгорная трещина
Разрыв толщи фирна у подошвы склона, отделяющего неподвижную толщу на склоне от движущегося льда в области питания горного ледника. → Рис. 39
f rimaye *f* du névé
s rimaya *f* de nevero
g Bergschrund *m*

Fig. 39. **Bergschrund** at a glacier in the Central Pamirs

* **bergy bits → 2667**

646 bergy waters
Ice-free navigable waters where icebergs are likely to occur.
r айсберговые воды
Судоходное водное пространство, где нет морского льда, но встречаются айсберги.
f eaux *fpl* avec icebergs
s aguas *fpl* con icebergs
g Gewässer *n* mit Eisbergen

647 Beringia
Region between north-eastern coastal areas of Asia and north-western coasts of North America which in past glacial epochs became land due to regression of the ocean and served as a bridge for flora, fauna and people to be exchanged between two continents.
r Берингия
Регион между северо-восточными прибрежными участками Азии и северо-западным побережьем Северной Америки, который в прошлые ледниковые эпохи в результате морской регрессии становился сушей и служил мостом в расселении флоры, фауны и человека между двумя континентами.
f Beringie *f*
s Beringia *f*
g Beringia *f*

648 berm
1. A narrow ledge, shelf or terrace formed by material thrown up on the beach by storm waves to make a horizontal shelf above the foreshore. 2. A narrow bank of a river or canal situated opposite a tow-path.
r берма
1. Узкая насыпь или терраса, сформированные материалом, выброшенным штормовыми волнами на фронтальную зону пляжа или отлогого песчаного берега. 2. Узкий берег (ступень) реки или канала, расположенный напротив бечевника.
f gradins *mpl* de plage
s gradas *fpl* de playa
g Stufe *f*

* **B-horizon → 3029**

649 bid price curve; bid rent curve
A curve relating price (rent) to distance, showing the price a landuser would be willing to pay for a given area of land at various distances from a given point, especially from a city centre. → Fig. 40, p. 79
r кривая «цена-расстояние»
Кривая зависимости цены (земельной ренты) от расстояния, показывающая,

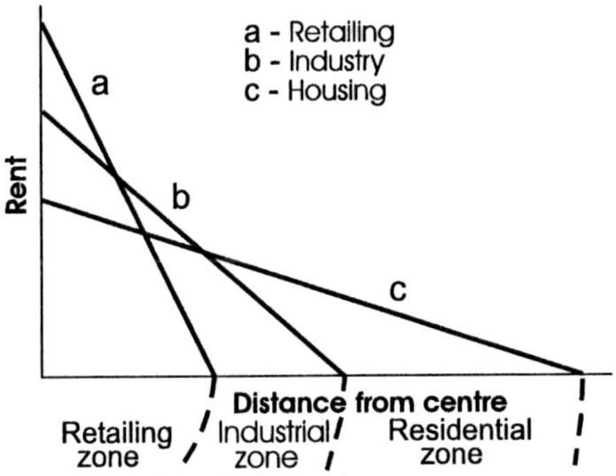

Fig. 40. **Bid price curve**

сколько пользователь готов заплатить за данный участок земли в зависимости от его положения по отношению, например, к центру города. → Рис. 40

f coubre *f* offre-prix
s curva *f* oferta-precio
g Preisangebotkurve *f*

* **bid rent curve → 649**

650 biennial plant
A plant which completes its life cycle (germination, flowering, fruiting) over a period of two years.
r двухлетнее растение
Растение с жизненным циклом, укладывающимся в два года: на первом году у него развивается розетка листьев, а на втором оно цветет, плодоносит, после чего погибает.
f plante *f* bisannuelle
s planta *f* bienal
g zweijährige Pflanze *f*; bienne Pflanze *f*

651 biennial wind oscillation; quasi-biennial oscilation
Alternation of easterly and westerly wind regimes in the stratosphere (23-30 km) within about 12° of the equator, with a period varying from about 24 to 30 months.
r двухгодичное колебание ветра; квазидвухлетняя осцилляция
Смена преобладающих восточных и западных ветров в стратосфере (23-30 км) в зоне примерно 12° от экватора с периодичностью от 24 до 30 месяцев.
f oscillation *f* biennale du vent; oscillation *f* quasi biennale
s variación *f* bienal del viento
g quasizweijährige Oszillation *f*

652 bifurcation of stream; stream bifurcation
Division of a valley and river channel into two branches, which further do not converge and form independent mouths.
r бифуркация реки
Разделение долины и русла реки на две ветви, которые в дальнейшем не сливаются и образуют самостоятельные устья.
f bifurcation *f* de rivière
s bifurcación *f* del río
g Flußgabelung *f*; Stromgabelung *f*

653 bifurcation ratio
The quantitative relationshiop between the different orders of magnitute of streams within a drainage basin, expressed as the number of streams of one order divided by the number of streams of the next highest order, which they create through their confluence.
r коэффициент разветвления рек
Численное отношение количества речных потоков разного порядка в пределах одного бассейна, выражаемое как число рек одного порядка, деленное на число рек следующего по величине порядка.
f coefficient *m* de bifurcation
s coeficiente *m* de bifurcación
g Verzweigungskoeffizient *m*

654 bight
A large-scale indentation of a coastline, generally of continental propotion, normally much wider than a bay, e.g. Great Australian Bight, Bight of Benin in Africa.
r открытый залив
Очень крупная, вплоть до континентальных пропорций, часть океана, вдающаяся в сушу, гораздо более широкая, чем обычный залив, например, Большой Австралийский залив.
f baie *f*
s caleta *f*; ensenada *f*; cala *f*
g Bucht *f*

655 billow cloud
A series of parallel regular cloud rolls separated one from another by clear sky bands, generally of the same thickness to the cloud bands, they occur usually at 6-8 km above the surface due to the increase of the speed of the wind with height.
r волнистые облака
Серия параллельных и одинаковых облачных валов, разделенных такими же по размеру полосами чистого неба, возникающая на высоте 6-8 км в результате резкого возрастания скорости ветра с высотой.
f nuages *mpl* ondulés
s nubes *fpl* onduladas
g Wellenwolken *fpl*

656 binodal seiche
Standing wave when the water level rises in the middle of a water body (usually lake) and sinks near its shores.
r двухузловая сейша
Стоячая волна, при которой уровень воды в середине водоема (обычно озера) поднимается, а у краев опускается.
f seiche *f* binodale
s seiche *f* binodal
g binodale Schwingung *f*

657 **biochemical oxidation**
Process of bacterial decomposition and oxidation of organic substances in water.
r биохимическое окисление
Процесс бактериального разложения и окисления органических веществ в водной толще.
f oxydation *f* biochimique
s oxidación *f* bioquímica
g biochemische Oxydation *f*

* **biochemical oxygen demand → 683**

658 **biochore**
A group of similar habitats situated in similar climatic conditions and having approximately one and the same composition of organisms.
r биохора
Группа сходных местообитаний, расположенных в однотипных климатических условиях и имеющих близкий состав организмов.
f biochore *m*
s biocora *f*
g Biochor *f*; Biochore *f*

659 **bioclymatology**
The study of climate as it affects the life and health of animals, plants and people.
r биоклиматология
Изучение воздействий климата на жизнь и здоровье растений, животных и людей.
f bioclimatologie *f*
s bioclimatología *f*
g Bioklimatologie *f*

660 **biocoenology**
The study of biotic communities – biocoenoses, their systematization, origin, structure, functioning, development, and spatial distribution.
r биоценология; биогеоценология
Наука, изучающая биоценозы, их систематику, происхождение, структуру, функционирование, развитие и распределение в пространстве.
f biocénologie *f*
s biocenótica *f*; biocenología *f*
g Biozönologie *f*

661 **biocoenosis; community$_2$**
The community of plants or animals living in the same relatively small area of land or water, plant communities (*phytocoenosis*) and animal communities (*zoocoenosis*) are distinguished. → Fig. 321, c. 718
r биоценоз; сообщество
Взаимосвязанная совокупность всех живых организмов, населящих более или менее однородный участок суши или водоема, выделяют сообщества растений (*фитоценоз*) и животных (*зооценоз*). → Рис. 321, с. 718
f biocénose *f*; communauté$_2$ *f*
s biocenosis *f*; comunidad$_2$ *f*
g Biozönose *f*; Lebensgemeinschaft *f*

662 **biocycle**
The largest ecological division of the biosphere, such as land, sea, estuaries and inland water bodies (each biocycle is divided into *biochores*, consisting of a considerable number of *biotopes*).
r биоцикл
Высшая единица экологического подразделения биосферы: суша, море, эстуарии и внутренние водоемы (каждый биоцикл подразделяется на *биохоры*, включающие значительное число *биотопов*).
f biocycle *m*
s biociclo *m*
g Biozyklus *m*; Lebenskreis *m*

663 **biodegradable *adj***
Applied to a substance or material that can be decomposed by living organisms.
r биоразложимый
Относящийся к веществам или материалам, которые могут разлагаться под воздействием живых организмов.
f biodégradable *adj*
s biodegradable *adj*
g biodegradiert *adj*

664 **biodegradation; biodeterioration**
The breaking down of a substance or material by the action of living organisms, mainly by aerobic bacteria.
r биодеградация
Разложение веществ или материалов под воздействием живых организмов, главным образом аэробных бактерий.
f biodégradation *f*
s biodegradación *f*
g Bioabbau *m*; biologischer Abbau *m*

* **biodeterioration → 664**

665 **biodiversity**
The variety of different species of plants and animals inhabiting a certain area: alpha-diversity is the variety of species, beta-diversity is the variety in composition of plant or animal communities, gamma-diversity is the variety both of species and communities.
r биоразнообразие; биологическое разнообразие
Разнообразие живых организмов, обитающих на определенной площади: альфа-разнообразие - это видовое разнообразие, бета-разнообразие - это разнообразие состава растительных или животных сообществ, гамма-разнообразие - это разнообразие и видов и сообществ.
f biodiversité *f*
s biodiversidad *f*
g Biodiversität *f*; biologische Vielfalt *f*

666 **biofiltration**
Separation of suspended particles from water and their concentration in the form of large lumps (*pellets*).

r биофильтрация
Отделение взвешенных частиц от воды и их концентрация в виде крупных комков (*пеллет*).
f biofiltration *f*
s biofiltración *f*
g Biofiltration *f*; Tropfkörperbehandlung *f*

667 biofuel; biomass fuel
Fuel from organic domestic waste and other sources which does not use up fossil fuels, organic matters used to produce energy.
r биотопливо
Горючий материал, получаемый из бытовых отходов, сельскохозяйственных культур и т.п.
f biocombustible *m*
s biocombustible *m*
g Biobrennstoff *m*; Biokraftstoff *m*

*** biogenic deposit → 671**

668 biogenic landforms
The landforms created by the activity of animals and plants, such as coral reefs, termite hills, molehills, etc.
r органогенный рельеф
Формы рельефа, возникшие в результате деятельности животных и растений – коралловые сооружения, термитники и др.
f relief *m* biogénique; formes *fpl* de relief biogéniques
s relieve *m* biógeno; formas *fpl* de relieve biogenas
g biogenes Relief *n*

669 biogenic migration
The movement of chemical elements in nature as a result of life activity of plants, animals, microorganisms.
r биогенная миграция
Перемещение химических элементов в процессе жизнедеятельности растений, животных, микроорганизмов.
f migration *f* biogéne
s migración *f* biogena
g biogene Migration *f*

*** biogenic reef → 2314**

670 biogenic rocks
Sedimentary rocks consisting of products of life activity of animals and plants or their not decayed remains.
r органогенная горная порода; биогенная горная порода
Осадочные породы, состоящие из продуктов жизнедеятельности животных и растений или их неразложившихся остатков.
f roches *f* biogénique
s roca *f* biogena; roca *f* de origen orgánico
g biogenes Gestein *n*

671 biogenic sediment; organic deposit; biogenic deposit
A sediment formed as a result of life activity of organisms or consisting of remains of dead organisms.
r биогенные отложения; органогенные отложения
Отложения, образующиеся в результате жизнедеятельности организмов, либо отмершие остатки самих организмов.
f sédiment *m* biogénique
s sedimento *m* biógeno; sedimento *m* orgánico
g biogene Sedimente *npl*; biogene Ablagerungen *fpl*

672 biogenic sedimentation
Deposition of organic substances in water.
r биогенная седиментация
Осаждение органических веществ в водной среде.
f sedimentation *f* biogénique
s sedimentación *f* biógena
g biogene Sedimentation *f*

673 biogeochemical cycle
Mechanisms by which chemicals such as carbon, oxygen, phosphorus, nitrogen and water are moved through the ecosphere to be renewed over and over again (the major cycles are gaseous, sedimentary and hydrologic). → Fig. 41, p. 82
r биогеохимический цикл
Трансформация соединений химических элементов в земной коре, обусловленная климатическими, гидрологическими и биологическими процессами, нередко приводящая химические соединения к исходному состоянию. → Рис. 41, с. 82
f cycle *m* biogéochimique
s ciclo *m* biogeoquímico
g biogeochemischer Zyklus *m*

*** biogeocoenosis → 1838**

674 biogeography
The study of the distribution of plants, animals and their communities over the Earth's surface.
r биогеография
Изучение распространения на Земле растений, животных и образуемых ими сообществ.
f biogéographie *f*
s biogeografía *f*
g Biogeographie *f*

675 biogeosphere
The outer part of the lithosphere down which there is no organic life.
r биогеосфера
Внешняя часть литосферы, ниже которой органическая жизнь отсутствует.
f biogéosphère *f*
s biogeosfera *f*

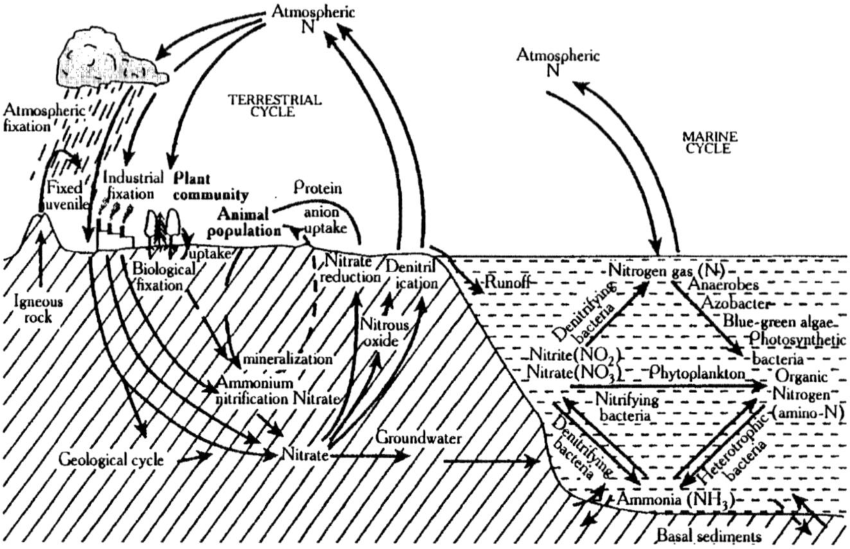

Fig. 41. **Biogeochemical cycles**

g Biogeosphäre *f*

676 bioherm; reef-knoll
A calcareous outgrowth on a basin bottom formed by fastened organisms depositing lime and remaining in this place after death.
r биогерм; старый коралловый риф
Известковый нарост на дне водоема, образованный прикрепленными организмами, отлагающими известь и сохраняющими после своей смерти прижизненное положение.
f bioherme *m*
s bioherm *m*
g Kuppe *f* des Meeresbodens; Bioherm *n*

677 bioindication
Estimation of water quality by change of the biocenosis in the conditions of pollutants impact.
r биоиндикация
Оценка качества воды по изменению характера биоценоза в условиях воздействия загрязняющих веществ.
f bioindication *f*
s bioindicación *f*
g Bioindikation *f*

678 biological activity of soil
All biological processes occurring in soil such as "breathing of soil" (oxygen consumption and carbon dioxide release), the release of thermal energy by soil organisms, etc.
r биологическая активность почвы
Совокупность биологических процессов в почве, таких как «дыхание почвы» (потребление кислорода и выделение углекислоты), выделение тепловой энергии организмами почвы и др.
f activité *f* biologique du sol
s actividad *f* biológica del suelo
g biologische Bodenaktivität *f*

679 biological amplification; biological magnification
The increase of the concentration of toxic substances (pesticides, herbicides, etc.) that are often non-biodegradable in an ecosystem; the accumulate of the toxins at succesive tropic levels of the food chain.
r биотическое усиление
Увеличение токсичных веществ (пестицидов, гербицидов и т.п.), которые нередко не разлагаются в экосистемах; концентрация организмом и как следствие увеличение количества устойчивого химического вещества на каждом трофическом уровне пищевой цепи.
f amplification *f* biologique
s amplificación *f* biológica
g biologische Amplifikation *f*; biologische Verstärkung *f*

680 biological desert
An area where there is no life because of heavy pollution, for example in the bottom of a contaminated lake.
r биологическая пустыня
Территория, на которой отсутствует жизнь из-за сильного загрязнения окружающей среды, например, на дне очень загрязненного озера.
f désert *m* biologique
s desierto *m* biológico
g biologische Wüste *f*

* **biological factor → 700**

681 biological indicator; indicator species
Organisms or species whose presence in the area shows the state of natural processes and, which being very sensitive to environmental conditions, can indicate the changes of nature.
r индикаторный организм; биоиндикатор
Организмы или сообщества организмов, присутствие и деятельность которых служат показателями естественных условий или антропогенных изменений природной среды.
f organisme *m* indicateur; indicateur *m* biologique
s indicador *m* biológico; organismo *m* indicador
g Zeigerorganismus *m*; Bioindikator *m*

* **biological magnification → 679**

682 biological minimum of temperature
The low temperature limit for the active vegetation of agricultural plants for each phase of their growth.
r биологический минимум температуры
Нижний предел температуры для активной вегетации сельскохозяйственных растений в той или иной фазе их развития.
f minimum *m* biologique de température
s mínimo *m* biológico de temperatura
g biologisches Temperaturminimum *n*

* **biological oceanography → 3687**

683 biological oxygen demand; biochemical oxygen demand
The amount of dissolved oxygen which aquatic organisms need for the decomposition of organic matter in polluted water (symbol: BOD).
r биологическая потребность в кислороде
Количество растворенного кислорода, необходимое водным организмам для разложения органического вещества в загрязненной воде.
f demande *f* biologique d'oxygène
s demanda *f* biológica de oxígeno
g biologischer Sauerstoffbedarf *m*

684 biological pollution
The introduction into the ecosystem and reproduction of alien living organisms.
r биологическое загрязнение
Привнесение в экосистему и размножение здесь чуждых ей видов живых организмов.
f pollution *f* biologique; contamination *f* biologique
s polución *f* biológica; contaminación *f* biológica
g biologische Verschmutzung *f*; biologische Verunreinigung *f*

685 biological production; biological productivity
The total amount of organic matter produced by organisms of a particular community, calculated per unit area or unit volume over a certain period of time (usually over a year). → Fig. 42
r биологическая продукция; биологическая продуктивность
Общее количество органического вещества, произведенного организмами, входящими в состав того или иного сообщества, на единицу площади или объема за единицу времени (обычно за год). → Рис. 42

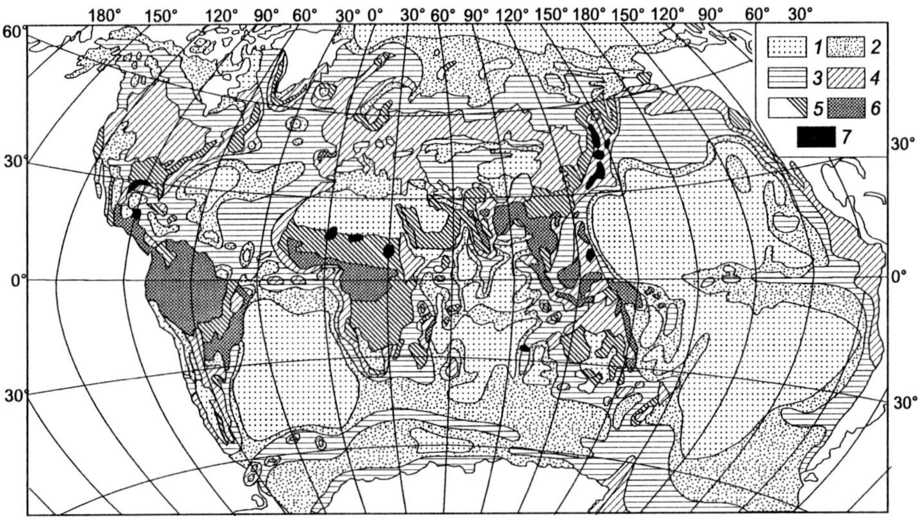

Fig. 42. **Biological production** of the Earth's biota in a live weight, kg/m^2 per year: 1 – 0.25; 2 – 0.25-1.0; 3 – 1.0-2.0; 4 – 2.0-3.75; 5 – 3.75-7.5; 6 – 7.5-12.5; 7 – more than 12.5

f production *f* biologique; productivité *f* biologique
s productividad *f* biológica
g biologische Produktion *f*; biologische Produktivität *f*

* **biological productivity → 685**

686 **biological purification; biopurification**
The removal of impurities from the environment and its components by means of living things.
r биологическая очистка; биоочистка
Очистка от посторонних примесей окружающей среды и ее компонентов с помощью живых организмов.
f purification *f* biologique; épuration *f* biologique
s purificación *f* biológica; depuración *f* biológica
g biologische Reinigung *f*

687 **biological resources**
The sum total of living organisms which have real or potential value for humans.
r биологические ресурсы
Совокупность живых организмов, имеющих реальную или потенциальную ценность для людей.
f ressource *f* biologique
s recursos *mpl* biológicos
g biologische Resourcen *fpl*

688 **bioluminescence**
The luminescence of the upper sea water layer caused by chemical reactions running in the bodies of some sea organisms under a mechanical impact upon the water.
r свечение моря; биолюминесценция
Свечение верхнего слоя морской воды «холодным» светом, вызываемое химическими реакциями, происходящими в теле некоторых морских организмов при механическом воздействии на воду.
f bioluminescence *f*
s bioluminicencia *f*
g Biolumineszenz *f*

689 **biomass**
The total weight of organisms living in a unit area or volume of the habitat.
r биомасса
Суммарный вес живых организмов, приходящийся на единицу площади или объема местообитания.
f biomasse *f*
s biomasa *f*
g Biomasse *f*

* **biomass fuel → 667**

690 **biome**
The largest community of living organisms in their interaction with the environment corresponding to major regions, e.g. natural zone: desert biome, tundra biome, etc.
r биом
Совокупность живых организмов, обитающих на какой-либо крупной территории, например, на территории природной зоны: биом пустыни, биом тундры и т.п.
f biome *m*
s bioma *m*
g Biom *n*

691 **biometry**
Measurements of any elements of the vegetation cover, such as the age of trees in the forest, the canopy density, the turf cover, etc.
r биометрия
Измерения элементов растительного покрова, таких как возраст деревьев в лесу, сомкнутость полога, задернованность и пр.
f biométrie *f*
s biometría *f*
g Biometrie *f*

* **biopurification → 686**

692 **bioremediation**
The use of microorganisms and plants to detect, degrade or remove environmental pollutants from soil, water or air.
r биокорректировка
Использование микроорганизмов и растений для обнаружения, разложения и удаления загрязнителей из почвы, воды и воздуха.
f bioremède *m*
s biomediación *f*
g Bioremediation *f*; mikrobielle Sanierung *f*

693 **biosphere**
The zone at the interface of the Earth's crust, ocean, and atmosphere where life is found.
r биосфера
Оболочка земного шара на контакте литосферы, приземного слоя атмосферы и гидросферы, в которой сконцентрировано живое вещество планеты.
f biosphère *f*
s biosfera *f*
g Biosphäre *f*

694 **biosphere reserve**
A nature reserve which is included into the international system of protected natural areas created for ecological monitoring.
r биосферный заповедник
Природный заповедник, который входит в международную систему биосферных резерватов, выполняющих глобальный экологический мониторинг.
f réserve *f* biospherique
s reserva *f* biosférica
g Biosphärenreservat *n*

695 **biostasis; biostasy**
A state of biological equilibrium between soil, vegetation and climate.

r биостазия
Состояние биологического равновесия между почвой, растительностью и климатом.
f biostasie *f*
s biostasía *f*
g Biostasie *f*

* **biostasy → 695**

696 biota
A community of flora, fauna and microorganisms inhabiting some territory.
r биота
Исторически сложившаяся совокупность флоры, фауны и микроорганизмов, населяющих какую-либо территорию.
f biote *m*
s biota *f*
g Biota *f*

697 biotic barrier
Natural conditions hindering the spreading of organisms to the neighbouring territory.
r биогеографический барьер; биотический барьер
Природные условия, препятствующие распространению организмов на соседнюю территорию.
f barrière *f* biotique
s barrera *f* biótica
g biotische Schranke *f*

698 biotic continuum
A number of gradually changing habitats and related communities which reveal the continuity of life on the planet.
r биотический континуум
Ряд постепенно изменяющихся местообитаний и связанных с ними сообществ, что служит проявлением непрерывности жизни на планете.
f continuum *m* biotique
s continuo *m* biótico
g biotisches Kontinuum *n*

699 biotic environment
A complex of living organisms effecting other organisms and non-living environment.
r биотическая среда
Совокупность живых организмов, оказывающих своей жизнедеятельностью влияние на окружающую среду.
f environnement *m* biotique
s medio *m* ambiente biótico
g biotische Umwelt *f*

700 biotic factor; biological factor
The impact upon the environment of activity of living organisms, including people, as distinct from such factors as climatic and geological.
r биотический фактор
Воздействие на окружающую среду деятельности живых организмов, включая человека, в противоположность влиянию таких факторов, как климатические или геологические.
f facteur *m* biotique
s factor *m* biológico
g biotischer Faktor *m*

701 biotic index
Scale for showing the quality of an environment (e.g. how clean a river is) by indicating the types of organisms present in it.
r биотический индекс
Показатель природных условий данной территории, определяемый через состав и состояние обитающих здесь организмов.
f index *m* biotique
s índice *m* biótico
g biotischer Index *m*

702 biotic potential
Maximum rate at which a population can increase under ideal conditions.
r биотический потенциал
Способность популяции к размножению, выживанию и развитию при оптимальных условиях окружающей среды.
f potentiel *m* biotique
s potencial *m* biótico
g biotisches Potential *n*

703 biotic pyramid
A graph, shaped like a pyramid, showing the number of individuals at each trophic level in a stable food chain, from the primary producers, to the primary, secoindary and tertiary consumers.
r биотическая пирамида
График, имеющий форму пирамиды, показывающий количество особей на каждом трофическом уровне устойчивой пищевой цепи, от первичных продуцентов к первичным, вторичным и третичным консументам.
f piramide *f* biotique
s pirámide *f* biótica
g Nahrungspyramide *f*

704 biotic regulation
The concept which states that biota has a mechanism providing the stability of climate and environment as a whole.
r биотическая регуляция
Концепция, согласно которой биота обладает механизмом, обеспечивающим устойчивость климата и окружающей среды в целом.
f régulation *f* biotique
s regulación *f* biótica
g biologische Regulation *f*

705 biotic weathering
Mechanical and chemical destruction of rocks as a result of life activity of plants, animals and microorganisms.
r биологическое выветривание
Механическое и химическое разрушение

горных пород в результате жизнедеятельности растений, животных и микроорганизмов.
f altération *f* biologique
s alteración *f* biológica
g biologische Verwitterung *f*

706 biotope
A piece of land or water with characteristic uniformity of environmental conditions that is occupied by one biocenosis.
r биотоп
Однородный в экологическом отношении участок земной поверхности (территории или акватории), занятый одним биоценозом.
f biotope *f*
s biotopo *m*
g Biotop *n*; Lebensraum *m*

707 bioturbation
The mixing of the surface layers of soil or of marine and lacustrine sediments by the action of organisms or microorganisms, occurring naturally (e.g. by earthworms) or by human activity (e.g. ploughing).
r биотурбация
Перемешивание поверхностных слоев почвы или морских и озерных отложений организмами и микроорганизмами, происходящее естественным путем (например, с помощью земляных червей) или благодаря действиям людей (например, вспашке земли).
f bioturbation *f*
s bioturbación *f*
g Bioturbation *f*

708 bipolar *adj*
Pertaining to, occurring at, associated with, two poles or with the two polar regions.
r биполярный; двухполюсный
Относящийся к двум географическим полюсам или двум полярным регионам.
f bipolaire *adj*
s bipolar *adj*
g bipolar *adj*

709 bipolar distribution
A kind of the distribution of organisms when one and the same species inhabits middle latitudes of the Northern and Southern hemispheres, but is not found in the tropical belt.
r биполярное распространение; биполярность
Разорванное распространение организмов, при котором один и тот же вид обитает в умеренных широтах Северного и Южного полушарий, но отсутствует в тропическом поясе.
f distribution *f* bipolaire
s distribución *f* bipolar
g bipolare Verbreitung *f*

710 bipolar world
Geopolitical arrangement of the world with two main centres of economic and political influence (the US and the USSR in the past).
r двухполярный мир
Геополитическое устройство мира, в котором существуют два основных центра экономического и политического влияния (в недалеком прошлом США и СССР).
f monde *m* bipolaire
s mundo *m* bipolar
g bipolare Welt *f*; zweipolige Welt *f*

711 birch forest; birch wood
Deciduous small-leaved forest of secondary or primary origin in which different species of birch dominate.
r березовый лес; березняк
Летнезеленый мелколиственный лес вторичного или реже первичного происхождения с господством разных видов березы.
f boulaie *m*; bétulaie *m*
s bosque *m* de abedul; abedular *m*
g Birkenwald *m*

* **birch wood → 711**

* **birdfoot delta → 3544**

* **birds of passage → 3853**

712 birth control
The policy of limiting family size in poor and densely populated countries by the use of contraception and sterilization.
r контроль рождаемости
Политика ограничения численности населения экономическими средствами, а также путем контрацепции и стерилизации.
f contrôle *m* de natalité;
contrôle *m* des naissances
s control *m* de natalidad
g Geburtenzahlkontrolle *f*

713 birth rate; fertility
The number of births in a year per one thousand of the population.
r рождаемость; коэффициент рождаемости
Число рожденных за определенный период (обычно за год) в расчете на 1000 человек, проживающих на данной территории.
f natalité *f*; taux *m* de natalité
s natalidad *f*; tasa *f* de natalidad
g Geburtenziffer *f*; Geburtenzahlrate *f*

714 bituminous coal
A carbonaceous fossil fuel, a brownish-black or black combustible mineral substance found in beds or seams in sedimentary rocks and derived from vegetable material growing in the Carboniferous period.
r каменный уголь
Карбонатное ископаемое топливо, горючее минеральное вещество коричнево-черного или черного цвета, залегающее в слоях осадочных пород и состоящее из

растительных остатков, произраставших на Земле в каменноугольный период.

f houille *f*; charbon *m*

s carbón *m* de piedra; carbón *m* mineral; hulla *f*

g Steinkohle *f*

* **black cotton soil → 4991**

715 black earth

A general term covering chernozem and the dark plastic clays of tropical regions.

r черная земля

Общий термин, включающий чернозем и темные пластичные глины тропических регионов.

f terre *f* noire

s tierra *f* negra

g schwarzes Erdboden *m*

716 black economy; informal economy; shadow economy

The part of the job market which is not reported to the tax authorities.

r теневая экономика

Экономическая деятельность, оплачиваемая наличными или товарами, не декларируемыми налоговой службе.

f économie *f* informelle

s economía *f* informal

g unformale Wirtschaft *f*

717 black ice

A layer of ice formed on surface of roads which is transparent and practically invisible hence extremely dangerous to traffic.

r черный лед

Слой гололеда на дорогах, состоящий из очень прозрачного, а потому невидимого льда, создающий крайнюю опасность при движении на дороге.

f glace *f* noire

s hielo *m* negro

g schwarzes Eis *n*

718 black smoker

A plume of hydrothermal fluid containing finely dispersed particles of black sulphides issuing from a submarine vent on a mid-oceanic ridge.

r черный курильщик

Источник высокотемпературных гидротермальных растворов, извергающихся в виде черного дыма в рифтовых зонах океанского дна.

f fumeur *m* noir

s fumador *m* negro

g schwarzer Raucher *m*

719 blackening of glacier

The artificial spreading of dust on the surface of a glacier in order to decrease the surface albedo and stimulate the melting of ice and increasing melt water runoff.

r зачернение ледника

Искусственнное добавление пыли на ледники с целью уменьшения альбедо их поверхности, усиления таяния льда и увеличения стока.

f nourrisson *m* de glacier

s ennegrecimiento *m* de glaciar

g Gletscherbeschwarzung *f*

720 blanket bog

A bog occurring on a relatively horizontal land surface in the regions of high rainfall and low evaporation, and covering the countryside like a blanket, except on steep slopes and rock outcrops.

r покровное болото

Болото, формирующееся на относительно горизонтальной поверхности в районах с большим количеством влаги и сплошь покрывающее поверхность, за исключением крутых склонов и выходов скал.

f marais *m* de couverture

s ciénaga *f* de manto

g Deckenmoor *n*

721 blasting; sandblasting

The impact of the fine material carried by wind or water on a bare exposured surface.

r истирание

Разрушение предметов под воздействием мельчайших частиц, переносимых ветром или водой.

f broyage *m*

s cherro *m* de arena

g Abreibung *f*; Abnützung *f*

* **blighted area → 1532**

* **blind drainage basin → 1065**

722 blind mouth

A section of a river channel where a stream loses all water due to strong evaporation or infiltration into soil before entering another water body or river; a river which has no mouth because the stream disappears underground into karst hollows.

r слепой конец

Место, где поток из-за сильного испарения или просачивания теряет всю воду, не доходя до водоема или другой реки; река, не имеющая устья из-за поглощения водотока карстовыми пустотами.

f embouchure *f* aveugle

s desembocadura *f* ciega

g blinde Mündung *f*; Talschluß *m*

723 blind valley

A type of valley, the head of which begins with the amphitheatre of steep slopes, i.e. wall that served as a limit for the periglacial regressive erosion.

r слепая долина

Долина, начинающаяся в верховьях амфитеатром крутых склонов, т.е. уступом, до которого распространялась доледникоавая регрессивная эрозия.

f vallée *f* aveugle
g Blindtal *n*
s valle *m* ciego

724 **blistering of ice**
The ratio of the volume of air bubbles enclosed in ice to its total volume expressed as a percentage.
r пузырчатость льда
Объем замкнутых воздушных пузырьков во льду в процентах от его общего объема.
f teneur *f* en bulles de la glace
s cantidad *f* de burbujas en hielo
g Luftblasengehalt *m* im Eis

725 **blizzard**
A violent blowing snow accompanied by intense cold.
r близзард
Интенсивная низовая метель при сильном морозе.
f blizzard *m*
s blizzard *m*
g Blizzard *m*; Schneegestöber *n*

726 **block diagram**
A three-dimensional sketch joining a perspective image of the object with its longitudinal and transverse vertical profiles showing the structure of the layers.
r блок-диаграмма
Трехмерный рисунок, совмещающий перспективное изображение объекта с продольными и поперечными вертикальными разрезами, показывающими строение толщи.
f bloc-diagramme *m*
s bloque-diagrama *m*
g Blockbild *n*

727 **block mountains**
The mountains formed by block-faulting, i.e. by vertical movement of separate blocks along the faults.
r глыбовые горы; сбросовые горы
Горы, сформированные разнонаправленными движениями блоков земной коры по разломам.
f montagnes-blocs *fpl*
s montañas *fpl* de bloques
g Schollengebirge *n*

728 **block of the Earth's crust**
A part of the Earth's crust, considerable smaller than a tectonic plate, limited by fault lines which may be stable or move with all its mass.
r блок земной коры
Участок земной коры, значительно меньший, чем тектоническая плита, стабильный или движущийся всей массой и ограниченный разрывами.
f bloc *m* de l'écorce terrestre
s bloque *m* de la corteza terrestre
g Scholle *f* der Erdkruste; Erdkrustenblock *m*

729 **block stream**
A linear accumulation of stone blocks and rubble that slowly moves in a narrow gully on a mountain slope due to gravity and solifluction.
r каменная река
Линейно вытянутое скопление глыб и щебня, медленно движущееся в неглубокой ложбине на горном склоне под действием силы тяжести и процессов солифлюкции.
f rivière *f* des pierres
s río *m* de piedra
g Blockstrom *m*

730 **block structure**
The pattern of the Earth's crust divided by a series of fractures into a number of blocks.
r блоковая структура
Структура участка земной коры, рассеченного системой разрывов на ряд блоков.
f structure *f* de blocs
s estructura *f* de bloques
g Schollenstruktur *f*

* **blockage → 4199**

731 **blocked delta**
A river delta, closed from the sea by the bar which is growing due to the waves of one direction.
r блокированная дельта
Речная дельта, отгороженная от моря приустьевой косой, растущей в результате преобладания волнений одного направления.
f delta *m* bloqué
s delta *m* bloqueado
g gesperrtes Delta *n*

732 **block-faulting**
Vertical movements on a section of the Earth's crust in which parallel faults produce a series of individual aligned blocks.
r блоковые движения
Вертикальные движения участка земной коры, в котором параллельные разломы создают отдельные блоки.
f bloc-faille *f*
s bloque-falla *f*
g Blockbewegung *f*

733 **blockfield**
An area covered with stone blocks chaotically spread on mountain slopes and flat tops formed due to frost weathering, solifluction and other slope processes.
r каменное море; каменная россыпь
Хаотическое нагромождение каменных глыб на горных склонах и плоских вершинах, образовавшееся в результате морозного выветривания, солифлюкции и других склоновых процессов.
f mer *f* de rocher; champ *m* de pierres
s mar *f* de piedra; campo *m* de piedras
g Felsenmeer *n*; Steinfeld *n*; Blockhalde *f*

734 blocking action
A process in the atmosphere which makes the air to move in the meridional direction as a result of which the normal zonal current in temperate latitudes becomes interrupted.
r блокирование
Процесс в атмосфере, вызывающий движение воздуха в меридиональном направлении, в результате чего нормальный зональный перенос средних широт надолго прерывается.
f action *f* de blockage
s acción *f* de bloqueo
g Blockierungeffekt *m*

735 blocking anticyclone
A slow-moving anticyclone remaining in the middle latitudes for a long time which blocks the normal westerlies.
r блокирующий антициклон
Малоподвижный антициклон, долго находящийся в средних широтах и блокирующий западный перенос в тропосфере.
f anticyclone *m* de blockage
s anticiclón *m* de bloqueo
g blockierendes Hoch *n*

* **blockstripes → 5826**

736 blow-hole
An inclined chamber in an abrasion sea coast which appears when the waves break down.
r канал продувания
Наклонный канал в породах, слагающих морские абразионные берега, выработанный мощным воздушным потоком при ударах волн.
f trou *m* souffleur
s agujero *m* de soplado
g Blasloch *n*

737 blowing snow
The transport and removal by the wind of a previously deposited snow.
r низовая метель; дефляционная метель
Ветровой перенос отложенного ранее снега.
f chasse-neige *f* basse; neige *m* soufflée
s ventisca *f* de nieve baja; arrastre *m* eólico bajo nieve
g Schneetreiben *n*; Schneefegen *n*

738 blow-out
Streched along wind a trough-shaped depression on a snow or sand surface formed by the wind whirls near a single obstacle.
r желоб выдувания; борозда выдувания; выдуй
Вытянутое по ветру желобообразное понижение снежной или песчаной поверхности около одиночного препятствия, возникающее в результате завихрений ветра.
f sillon *m* de déflation
s surco *m* de deflación
g Windfurche *f*

739 blue ice
A pure transparent ice containing no air bubbles and acquiring a bluish tint of colour.
r голубой лед
Прозрачный лед, лишенный пузырьков воздуха, в массиве приобретающий голубоватый оттенок.
f glace *f* bleue
s hielo *m* azul
g blaues Eis *n*

740 blue ice bands
Bands and lenses of blue ice, almost totally deprived of any air inclusions, typical for near-bottom layers of glaciers.
r голубые ленты льда
Слои и линзы голубого льда, почти полностью лишенного включений, характерные для придонных частей ледников.
f bandes *fpl* bleues de glace
s bandas *fpl* azules de hielo
g blaue Eisbänder *npl*; blaue Eisstreifen *mpl*

741 blue sky
The apparent colour of the cloudless sky in daylight, due to the scattering of sunlight by obstructing molecules in the air.
r голубое небо
Видимый цвет безоблачного неба в дневное время, вызванный рассеянием солнечных лучей молекулами воздуха.
f ciel *m* bleu
s cielo *m* azul
g blauer Himmel *m*

742 blue-green algae
One-celled, multicellular, and colonial algae found mainly in fresh water, sometimes in seas, in soils, on tree bark, etc., which often cause algae bloom of water.
r синезеленые водоросли
Одноклеточные, многоклеточные и колониальные водоросли, распространенные преимущественно в пресных водах, но иногда в морях, на почве, коре деревьев и т.п., и нередко вызывающие цветение воды.
f algues *fpl* bleu-vertes; Chlorophycées *fpl*
s algas *fpl* azul-verdes
g blaugrüne Algen *fpl*

743 bluff; high bank
A high wall at the edge of a former flood plane, in particular a river-cut cliff or steep slope on the outside of a meander. → Fig. 201, p. 446
r нагорный берег; яр; обрыв
Высокий склон на краю поймы, в частности, обрыв крутого склона на внешней стороне речной излучины, подмываемой рекой. → Рис. 201, с. 446
f escarpement *m*
s escarpe *m*; ribazo *m*
g Steilufer *n*

744 boat-yard
A long stagnant river inlet formed of an

Fig. 43. **Boat-yard** near Moscow in the winter

oxbow, arm of the river or separated from the main channel by a spit and often used for keeping and repairing vessels. → Fig. 43

r затон
Длинный непроточный речной залив, образованный из старицы, протоки или отделенный от основного русла косой, нередко используемый для отстоя и ремонта судов. → Рис. 43

f anse *f*
s ensenada *f*
g Einbuchtung *f*

745 **bocca**
A hole in a crater bottom or on a slope of a volcano from where eruption products are emitted.

r бокка
Отверстие на дне кратера или на внешнем склоне вулкана, откуда выходят продукты извержения.

f bocca *f*
s boca *f* volcánica
g Bocca *f*

746 **bodden coast**
Widely branching inlets in the coast of the Baltic and Northern seas deeply cut inland and separated from the sea by islands and peninsulas. → Fig. 60, p. 130

r боддены
Широко ветвящиеся заливы на юге Балтийского и Северного морей, глубоко вдающиеся в сушу и отчлененные от моря островами и полуостровами. → Рис. 60, с. 130

f côte *f* à bodden; bodden *m*
s costa *f* de bodden; bodden *m*
g Boddenküste *f*

* **body of water → 6643**

747 **body of water; water body**
An accumulation of water on the earth's surface or underground having a certain hydrological regime.

r водный объект
Скопление природных вод на земной поверхности или в верхних слоях земной коры, обладающих определенным гидрологическим режимом, например, река, озеро, водохранилище.

f objet *m* hydrologique
s objeto *m* hidrológico; objeto *m* acuático
g Wasserköpfer *m*

748 **Boelling**
The interstadial characterised by the warming of climate which divided the cold phases of the low and middle Dryas at the end of the Late Pleistocene of Europe (approximately 12.8-12.3 thousand years ago).

r бёллинг
Межстадиал, характеризующийся потеплением климата и разделяющий холодные фазы позднего и среднего дриаса в конце позднеплейстоценовой ледниковой эпохи Европы (приблизительно 12,8-12,3 тыс. лет назад).

f Bölling *m*
s Bölling *m*
g Bölling *m*

* **bog → 6009**

749 **bog landscape; fenland**
Landscapes having excessive moisture which include bogs and adjoining waterlogged areas and found mainly in temperate and equatorial belts.

r болотный ландшафт
Избыточно увлажненный ландшафт, включающий собственно болота и прилегающие заболоченные территории, распространен главным робразом в умеренном и экваториальном поясах.

f paysage *m* de marécage
s paisaje *m* palustre; pantanal *m*
g Moorlandschaft *f*

750 **bog soils; pen soils; peat soils**
Soils of bogs or greatly swamped forests that form under stagnant moistening by atmospheric, surface or ground waters, the upper part of peat deposits of bogs.

r болотные почвы; торфяные почвы
Почвы болот или сильно заболоченных лесов, формирующиеся в условиях застойного увлажнения атмосферными, поверхностными или грунтовыми водами, верхняя часть торфяных залежей болот.

f sols *mpl* de marais tourbeux
s suelos *mpl* de turbera
g Moorböden *mpl*; Torfböden *mpl*; Sumpfböden *mpl*

751 **bolson**
An enclosed flat-bottomed depression in the mountainous arid regions in the south-west of North America.

r больсон
Межгорная плоскодонная, обычно бессточная впадина в засушливых районах на юго-западе Северной Америки.

f bolson *m*
s bolsón *m*
g Bolson *m*

752 **Bond cycles**

Late Pleistocene climate oscillations that average 10-15 thousand years in duration, they correspond to glacial stages and interstades, and their terminations in North Atlantic were marked by formation of specific layers of glaciomarine deposits known as *Heinrich events*.

r циклы Бонда
Осцилляции климата позднего плейстоцена периодичностью в 10-15 тыс. лет, соответствовавшие ледниковым стадиям и межстадиалам и в Северной Атлантике завершавшиеся формированием особых прослоек ледниково-морских отложений, соответствующих *событиям Хайнриха*.

f cycles *mpl* de Bond
s ciclos *mpl* de Bond
g Bond-Zyklen *mpl*

753 bonitet
An index used to evaluate the quality of arable lands, soils, forests with the aim of their rational use.

r бонитет
Условный показатель, применяемый для оценки качества земельных угодий, почв, лесов с целью их рационального использования.

f classe *f* de fertilité
s clase *f* de fertilidad
g Bonität *f*

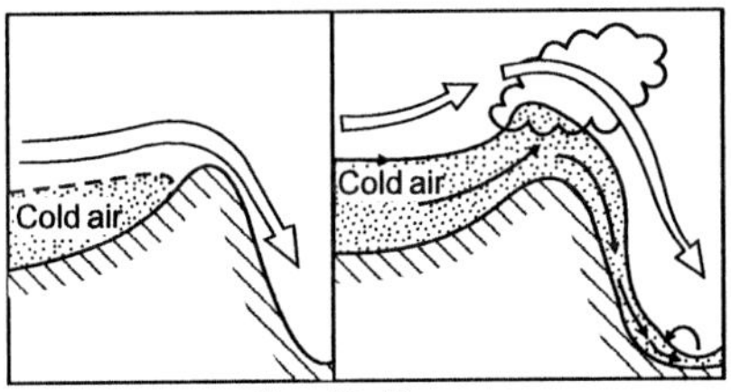

Fig. 44. **Bora**

754 bora
A strong very dry north-easterly wind which blows in cold gusts from the Balkan mountains down to the eastern shore of the Adriatic and in the Black sea, most commonly in winter. → Fig. 44

r бора
Сильный и порывистый, большей частью холодный ветер, направленный вниз по склону и дующий с невысоких гор, характерен для восточного побережья Адриатического и Черного морей. → Рис. 44

f bora *f*
s bora *f*
g Bora *f*

755 border; boundary; frontier
The limit of any territorial unit be it political, economic, social or physical.

r граница
Территориальный рубеж, разделяющий местности с разными природными, экономическими, социальными или политическими условиями.

f limite *f*; frontière *f*
s límite *m*; frontera *f*
g Grenze *f*

756 border mountains
The mountains edging highlands and separating them from lower-lying and plain territories.

r окраинные горы
Горы, окаймляющие нагорья и отделяющие их от более низких и равнинных территорий.

f montagnes *fpl* bordiéres
s montañas *fpl* marginales
g Randgebirge *n*

*** bore → 6206**

757 Boreal
The initial stage of the Holocene (9.5-7.5 thousand years ago) in the Northern Europe characterised by temperate and dry climate.

r бореальное время
Начальная стадия голоцена (9,5-7,5 тыс. лет назад), характеризовавшаяся в Северной Европе умеренным и сухим климатом.

f Boréal *m*
s Boreal *m*
g Boreal *n*

758 boreal *adj*
Applied to natural conditions of the Northern Hemisphere between 40 and 60° N where short hot summer and long cold winter are typical.

r бореальный
Относящийся к природным условиям Северного полушария между 40 и 60° с.ш. с коротким жарким летом и длительной холодной зимой.

f boréal *adj*
s boreal *adj*
g boreal *adj*; nördlich *adj*

759 boreal climate
A relatively cold climate of middle latitudes with clearly manifested seasons.

r бореальный климат; субполярный климат
Умеренно холодный климат средних широт с ясно выраженными временами года.

f climat *m* boréal
s clima *m* boreal
g boreales Klima *n*

*** boreal forest → 6041**

760 boreal transgression
The transgression of the Arctic ocean towards the north of Eurasia during the Miculino interglacial 125-130 thousand years ago.

r бореальная трансгрессия
Трансгрессия Северного Ледовитого океана на север Евразии во время микулинского (эемского) межледниковья 125-130 тыс. лет назад.

f transgression *f* boréale
s transgresión *f* boreal
g Borealtransgression *f*

* **borehole → 1713**

* **boring → 1712**

* **bornhardt → 3110**

* **botanic garden → 761**

761 botanical garden; botanic garden
A place where plants are grown as a collection in order to show them to the public and for scientific study.
r ботанический сад
Место, где выращиваются коллекции растений для демонстрации публике и научных исследований.
f jardin *m* botanique
s jardín *m* botánico
g botanischer Garten *m*

762 bottom; floor
The bed of a natural depression or body of water such as the bottom of a stream, lake, valley, etc.
r дно; днище
Нижняя поверхность природной депрессии или водоема, например, дно реки, озера, долины и пр.
f fond *m*
s fondo *m*
g Grund *m*

763 bottom countercurrent; bottom counterflow
The movement of the lower water layers compensating the surface wind-driven current.
r донное противотечение
Течение в нижних слоях воды, компенсирующее поверхностное ветровое течение.
f contre-courant *m* de fond
s contra-corriente *f* de fondo
g Grundgegenströmung *f*

* **bottom counterflow → 763**

764 bottom current; bottom flow
The motion of water along the bottom of the water body.
r придонное течение
Движение вод в придонном пограничном слое водоема, где существенную роль играет трение о поверхность дна.
f courant *m* de fond
s corriente *f* del fondo
g Grundströmung *f*; Bodenstrom *m*

* **bottom flow → 764**

* **bottom ice → 259**

765 bottom sediments
Mineral and organic material sank to the bottom of a water body and being consolidated.
r донные отложения; донные осадки
Осевшие из воды на дно водоема и уплотнившиеся с течением времени минеральные и органические вещества.
f sédiments *mpl* de fond
s sedimentos *mpl* del fondo
g Bodensedimente *npl*; Bodenablagerungen *fpl*

* **bottom soil → 5937**

766 bottom water; deep watermass
The lowest water layer in an ocean just above the sea floor, especially in the abyss, which is relatively cold and dense with temperature only a little above freezing point.
r придонная вода
Слой воды в океанах и морях, расположенный непосредственно над поверхностью их дна и отличающийся относительно низкой температурой и высокой плотностью.
f eau *f* profonde
s agua *f* profunda
g Tiefseewasser *n*

767 boudinage
The breaking down of a layer or dyke into lenses connected with pinches or even separated from each other.
r будинаж
Тектоническое разлинзование, т.е. распадение пласта или дайки на линзы, соединенные пережимами или вовсе отделившиеся друг от друга.
f boudinage *m*
s boudinage *m*
g Boudinage *n*

768 boulder; cobble
A large rounded fragment of rock of different origin exceeding 10 cm in diameter.
r валун
Окатанный обломок горной породы разного происхождения поперечником более 10 см.
f caillou *m* roulé
s canto *m* rodado
g Rollstein *m*; Findling *m*

769 boulder balk
The boundary of a plot of land in the form of a stone wall made of glacial boulders.
r валунная межа
Граница земельного участка, обнесенного каменной стенкой из ледниковых валунов.
f enclos *m* de pierre
s pared *f* de piedra; cerca *f* de piedra
g Geschieberain *m*

770 boulevard
Originally a broad tree-lined street but now also applied to various wide new roads, e.g. a ring road inside a city, or a scenic highway.
r бульвар
Первоначально широкая улица, засаженная деревьями, но ныне также применя-

ется к разным новым широким дорогам, таким как кольцевая дорога внутри города, или живописное шоссе.
f boulevard *m*
s bulevar *m*
g Boulevard *m*

* **boundary → 755**

771 boundary layer; surface layer of air; ground layer of air
The lower part of the boundary layer of the atmosphere adjacent to the Earth's surface, about 10 to100 m thick.
r приземный слой воздуха
Нижняя часть пограничного слоя атмосферы, непосредственно примыкающая к земной поверхности до высоты 10-100 м.
f couche *f* limite de surface; couche *f* superficielle
s capa *f* de aire límite superficial; capa *f* de aire superficial; capa *f* de aire cerca del suelo
g Bodenschicht *f*

772 boundary layer of the atmosphere; friction layer of the atmosphere; planetary boundary layer
The lowest layer of the troposphere, from 300-400 up to 1500-2000 m having the properties which are significantly effected by the Earth's surface.
r пограничный слой атмосферы
Нижний слой тропосферы толщиной от 300-400 до 1500-2000 м, свойства которого определяются в основном взаимодействием с земной поверхностью.
f couche *f* limite planetaire de atmosphère; couche *f* de frottement; couche *f* limite atmosphérique
s capa *f* limite planetaria de l'atmósfera; capa *f* límite atmosférica; capa *f* de fricción
g atmosphärische Grenzschicht *f*

773 boundary strip
The boundary of a plot of land in the form of a narrow strip of uncultivated land.
r межа
Граница земельных владений в виде узкой полосы необрабатываемой земли.
f dérayure *f*; lisière *f*
s linde *m*; surco *m*; lindero *m*; lomo *m* divisorio
g Rain *m*

* **boundary waves → 3158**

774 box canyon
A valley with a wide flat-floor and steep slopes.
r ящикообразная долина
Долина с широким, плоским днищем и крутыми склонами.
f vallée *f* à fond plat
s valle *m* de fondo plano
g Kastental *n*

775 brackish *adj*
Applied to any water which exhibits slight salinity intermediate between sea water and fresh water.
r солоноватый
Применяется к любой воде небольшой солености, промежуточной между пресной и морской водой.
f peu salé *adj*
s poco salado *adj*
g salzig *adj*

* **brackish groundwater → 5939**

* **brackish lagoon → 3495**

* **brackish lake → 5167**

* **brackish water → 5170**

* **braided channel → 776**

776 braided river; braided channel; anastomosing channel
A river having wide shallow channel within which it forms a network of interlaced branches separated from each other by bars and islands.
r блуждающая река; разветвленное русло
Река, имеющая мелкое русло, в пределах которого формируется сеть меняющих свое положение рукавов, отделенных друг от друга отмелями и островами.
f rivière *f* divagante; chenal *m* divagate; lit *m* d'un cours d'eau anastomosé
s río *m* divagante; curso *m* anastomosado; lecho *m* divagante
g verwildeter Fluß *m*; Umlagerungsstrecke *f*; wandernder Fluß *m*

777 braiding; anastomosis
The process of splitting of a stream with a wide shallow channel into many small shallow interlaced channels separated by bars of alluvial material. → Fig. 45
r блуждание реки
Процесс разветвления реки с широким и

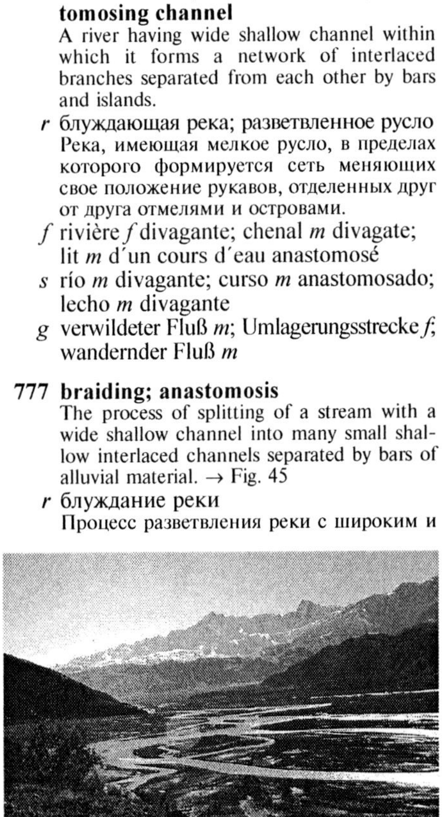

Fig. 45. **Braiding** of the Surkhob river in the Pamirs Mountains

мелким руслом на несколько рукавов, отделяющихся друг от друга отмелями, заполненными аллювиальным материалом. → Рис. 45, с. 93
f divagation *f* de rivière
s divagación *f* de río
g Verwilderung *f* des Flusses; Flußverwilderung *f*

778 brain drain
The movement of the most capable, highly-skilled, technical and professional people from the country where they have been trained and gained their first work experience to another country which offer better carrier opportunities and/or higher rewards.
r утечка мозгов
Выезд наиболее способных, технически и профессионально подготовленных людей из страны, где они получили образование, в поисках возможностей лучшей карьеры и более адекватной оплаты своего труда.
f fuite *f* des cerveaux
s fuga *f* de cerebros
g Brine-Drain *n*

779 branch of a river; anabranch; distributary$_2$; arm of a river
A stream which flows away from the main river and may not return to it, as in a delta.
r рукав
Боковое русло реки, отделяющееся на некоторых участках от главного потока, в дельте нередко самостоятельно впадает в водоем.
f bras *m* d'un fleuve; défluent *m*
s brazo *m* del río; defluente *m*
g Flußarm *m*; Nebenarm *m*

780 branch of industry; industrial branch
A structural part of an industry that includes all the enterprises producing similar products or rendering similar services.
r отрасль промышленности
Структурное подразделение промышленности, охватывающее предприятия, которые производят однотипную продукцию или оказывают однотипные услуги.
f branche *f* industrielle
s rama *f* industrial
g Zweig *m*

781 brash ice; ice gruel
A loose accumulation of the fragments of sea ice or fresh-water ice (each piece is not more than 2 m across).
r ледяная каша; тертый лед
Скопление обломков плавучего льда до 2 м в поперечнике.
f brash *m*; masse *f* de glace flottante
s masa *f* de hielo flotante
g Trümmereis *n*; Matscheis *n*

* **Brave West Winds → 6751**

782 break of slope
A sudden change of the steepness of a slope.
r перегиб склона
Резкое изменение крутизны склона.
f rupture *f* de pente
s ruptura *f* de pendiente
g Hangknick *m*

783 break off of slope
A slow gravity process of the detachment and downslope movement of large blocks of rock which leads to the formation of cracks up to 20 m deep and 100 m wide, that extend for several hundred metres parallel to the slope.
r отседание склона
Медленный гравитационный процесс отрыва и соскальзывания крупных блоков прочных горных пород, приводящий к появлению глубоких (до 20 м) и широких (до 100 м) трещин (рвов отседания), которые простираются параллельно склону на несколько сотен метров.
f détachement *m* de versant
s desprendimiento *m* de vertiente
g Abtrennung *f* eines Hangs

784 breaker
A mass of turbulent water and foam, breaking violently against a rocky shore or passing over a reef or shallows, formed when a heavy ocean wave rushes from deep to shallow water, so that its crest steepens, rolls over, and breaks.
r бурун
Пенистая масса воды, образующаяся на гребне волны при его разрушении над подводными валами, рифами, отмелями.
f ressac *m*
s resaca *f*
g Brecher *m*; Sturzwelle *f*

785 breaker zone; wave splash; breaking zone
A coast belt being under the impact of wave-swash flow.
r прибойная зона; зона заплеска волн
Полоса побережья, в пределах которой действует волноприбойный поток.
f zone *f* du déferlement
s zona *f* de rompiente
g Brandungszone *f*

* **breaking up of ice → 787**

* **breaking zone → 785**

786 break-of-bulk point
The point in a transport system where cargo is transferred from one mode of transport to another, e.g. at a railway station, port or airport.
r перевалочный пункт
Пункт в транспортной системе, где груз перегружается с одного вида транспорта на другой, например, на железнодорожной станции, в порту или аэропорту.

f point *m* de transbordement
s punto *m* de transbordo
g Umschlagplatz *m*

787 breakup of ice; breaking up of ice
The destruction of the ice cover on rivers during the spring thaw when the solid sheet of ice on rivers breaks into ice-floes moving downstream.
r вскрытие ледяного покрова
Интенсивное разрушение ледяного покрова на реках вследствие уменьшения его толщины и прочности от таяния и воздействия механических сил.
f débâcle *m*
s rotura *f* del hielo
g Eisaufbruch *m*; Aufbrechen *n* der Eisdecke

788 breakwater; groyne
An artificial construction built out into the sea from a shoreline to protect coastal areas, harbour waters, entrances of canals and locks from waves by damping or reflecting them.
r волнолом; волнорез
Гидротехническое сооружение для защиты от волнения (гашения или отражения волн) береговых участков моря, акватории порта, подходов к каналам и шлюзам.
f brise-lames *m*
s rompeolas *m*; escollera *f*
g Wellenbrecher *m*

789 breastwall reservoir
A reservoir created by means of a breastwall which holds mass of water behind it.
r подпорное водохранилище
Водохранилище, созданное с помощью подпорной плотины, удерживающей находящуюся за ней массу воды.
f réservoir *m* de remous
s embalse *m* de remanso
g Staudammsee *m*

790 breccia
A rock consisting of angular fragments of other rocks generally more than 10 mm in size cemented together by some finer material.
r брекчия
Крупнообломочная горная порода, состоящая из сцементированных угловатых обломков разных пород размером более 10 мм.
f brèche *f*
s brecha *f*
g Brekzie *f*; Trümmergestein *m*; Breccie *f*

791 breckland
A tract of land covered with vegetation, which is cleared for cultivation from time to time, but allowed to revert.
r временно культивируемая земля
Участок земли, покрытый растительностью, которая время от времени очищается для культивации, а затем возвращается в прежнее состояние.

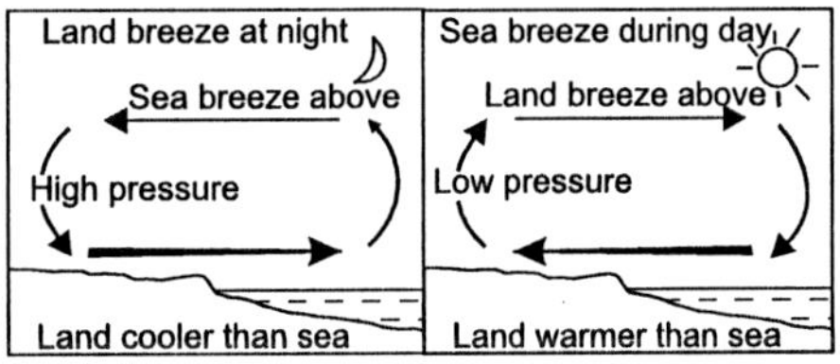

Fig. 46. Land and sea **breezes**

f terres *fpl* de culture temporaire
s tierras *fpl* de cultivo temporal
g Brachland *n*

792 breeding bottom; spawning ground
Area of water where fish come each year to produce their eggs.
r нерестилище
Место икрометания (нереста) рыб.
f frayére *f*
s desovadero *m*
g Laichplatz *m*

793 breeze
A light or moderate local wind occurring on shores of seas and lakes which blows during the day time from water surface to the land, and in the opposite direction at night. → Fig. 46
r бриз
Слабый или умеренный местный ветер на берегах морей и больших озер, который днем дует в сторону суши, а ночью наоборот. → Рис. 46
f brise *f*
s brisa *f*
g Brise *f*

794 bridlepath; bridleway
A path fit for the passage of a horse or a pedestrian, but not for a vehicle.
r вьючная тропа; верховая тропа
Тропа, по которой возможно передвижение на лошадях или пешком, но невозможен проезд автомобилей.
f sentier *m* de bât; draille *f*
s senda *f*; cañada *f*
g Saumpfad *m*

* **bridleway → 794**

795 brine
Natural water almost saturated with salt with the concentration of salts over 50‰; found in water bodies, underground cavities and pores in the bottom sediments of salt lakes.
r рассол; рапа; тузлук
Природные воды с минерализацией свыше 50‰; насыщенный соляной раствор в водоемах, подземных пустотах и порах донных отложений солеродных (самосадочных) озер носит название рапа, или тузлук.
f saumure *f*
s salmuera *f*
g Sole *f*; Salzlösung *f*

796 **brine ice**
Ice with mineralization over 50 ‰.
r рассоловодный лед
Лед с минерализацией более 50 ‰.
f glace *f* saumâtre
s hielo *m* asalmuerado
g Salzeis *n*

797 **brine pocket**
A small hole in a growing sea ice filled with a residual liquid which is salter than the sea water around.
r солевой карман
Небольшая выемка в растущем морском льду, содержащая жидкость, более соленую по сравнению с исходной морской водой.
f poche *f* de saumure
s bolsillo *m* de salmera
g Salztasche *f*

798 **broad-leaved forest**
The forest composed of trees having broad leaves (oak, beech, maple, linden, hornbeam, etc.), occurs mainly in the temperate zone of the Northern Hemisphere.
r широколиственный лес
Лес, состоящий из мезофильных древесных пород с широкими листовыми пластинками (дуб, бук, клен, липа, граб и др.), распространен преимущественно в умеренном климатическом поясе Северного полушария.
f forêt *f* latifoliée
s bosque *m* foliáceo
g breitblätteriger Wald *m*

799 **broad-leaved forest zone**
A natural zone of temperate forest of the Northern and Southern hemispheres found towards the equator from the mixed forest zone, they are mainly developed in the west of Eurasia and to the south-east of Great Lakes in North America, in the Southern Hemisphere it includes hemihylaea.
r зона широколиственных лесов
Природные зоны умеренных поясов Северного и Южного полушарий, расположенные ближе к экватору от зон смешанных лесов, в основном развиты на западе Евразии и к юго-востоку от Великих озер в Северной Америке, в Южном полушарии включает гемигилею.
f zone *f* des forêts latifoliées
s zona *f* de bosques foliáceos
g Zone *f* breitblättriger Laubwälder

800 **Brocken spectrum**
An optical phenomenon occurring in the mountains: when an observerrs shadow is seen on a water cloud or fog, sometimes surrounded by coloured rings (glories) around the shadow of the head.
r брокенский призрак
Оптическое явление в горах: тень наблюдателя на жидко-капельном облаке или тумане, иногда с цветными кольцами (глориями) вокруг тени головы.
f spectre *m* du Bpocken
s espectro *m* de Brocken
g Brockengespenst *n*

* **broken ice → 5493**

801 **Bronze Age**
An era in human development extending from 4000 to 2000 B.C. (Subboreal) when bronze was used to make tools and weapons, writing and arithmetic developed, animals for riding and pulling came to be used, towns were formed.
r бронзовый век
Археологическая эпоха в развитии человечества, для которой характерно изготовление бронзовых изделий наряду с каменными и керамическими, относится к концу 4-го – 2-му тысячелетию до н.э. (суббореальное время).
f âge *m* de bronze
s edad *f* de bronce
g Bronzezeit *f*

802 **brook; runnel; draw (*amer*); rivulet; rill**
Small permanent or intermittent watercourse.
r ручей
Небольшой постоянный или временный водоток.
f ruisseau *m*
s arroyo *m*
g Bach *m*; Gerinne *n*

803 **brow**
The point where a steep slope changes into a gentle slope.
r бровка
Резкий перелом склона, отделяющий вышележащую пологую его часть от расположенного ниже более крутого участка.
f bord *m*
s borde *m*
g Kante *f*

* **brown coal → 3494**

* **brown earth → 804**

804 **brown forest soil; brown earth**
A grey-brown soil developing under broad-leaved and mixed forests of moderately-warm humid climate which is common Mountains, Appalachians, China, in the south of Chile and New Zealand.
r бурозем; бурая лесная почва
Почва серо-коричневого цвета, образующаяся под широколиственными и смешанными лесами в условиях умеренно-теплого влажного климата, распространена на Кавказе, в Крыму, Карпатах, Аппалачах, на юге Чили и Новой Зеландии.
f sol *m* brun forestier
s suelo *m* pardo forestal

g Waldbraunerde *f*; Burosem *m*; Braunerde *f*; brauner Waldboden *m*

805 **brown semidesert soil**
A soil that develops in the semidesert zone under a thinned vegetation of moderately hot climate which is common in Central Asia and the Caspian Region.
r бурая полупустынная почва
Почва, развивающаяся в полупустынной зоне под разреженной растительностью в условиях умеренно-жаркого климата, широко распространена в Центральной Азии и Прикаспии.
f sol *m* brun subaride
s suelo *m* pardo de semidesierto
g brauner Halbwüstenboden *m*

806 **brown soils**
Soils that form in cold, temperate and tropical zones under where no leaching at all or a weak leaching of clays and iron occur, and either the whole profile of the soil or its upper horizon does not contain carbonates.
r бурые почвы
Почвы, формирующиеся в холодных, умеренных и тропических зонах, с отсутствием или слабым выщелачиванием глин и железа, не содержащие карбонатов по всему профилю или по крайней мере в поверхностном горизонте.
f sols *mpl* bruns
s suelos *mpl* pardos; tierras *fpl* pardas
g Braunböden *mpl*

807 **browse**
Tender shoots, twigs, and leaves of trees and shrubs that provide food for animals.
r потрава
Поедание животными мягких почек, веток и листьев деревьев и кустарников.
f dégât *m*
s holladura *f*
g Feldfrevel *m*; Abweiden *n*

* **Brückner cycle → 808**

* **Brückner period → 808**

808 **Brueckner period; Brückner cycle; Brückner period**
A cycle of climatic change with irregular alternations of cold, wet periods with warm, dry periods, which occur with a periodicity of 35 years, have been identified by German scientist E. Brueckner.
r брикнеров цикл
Установленное немецким ученым Э. Брюкнером чередование холодных влажных периодов с теплыми сухими, наступающее в среднем через 35 лет.
f période *f* de Brückner
s período *m* de Brückner
g Brücknerische Periode *f*

809 **brunification**
Soil processes such as soil weathering and transformation of minerals of iron oxides resulting in the appearance of brown colour in soil horizon.
r брюнификация
Почвенные процессы, приводящие к появлению бурой окраски в почвенном горизонте в результате внутрипочвенного выветривания и трансформации минералов – оксидов железа.
f brunification *f*
s brunificación *f*
g Verbrauung *f*

810 **brunizem; prairie soil**
A greatly humuficated chernozem-like soil of North American prairies that occupies a considerable part of "the grain belt" in the USA and Canada.
r брюнизем; почва прерий
Сильно гумусированная черноземовидная почва североамериканских прерий, занимающая значительную часть «зернового пояса» в США и Канаде.
f brunizem *m*; sol *m* brun de prairie
s suelo *m* pardo de pradera
g Brunisem *m*; brauner Prärieboden *m*

* **brushfires → 6770**

811 **brushwood; thicket; dense bushes**
Dense growth of shrubs and small trees, e.g. brushwood of bamboo, blackberry, fern, etc., that may hinder the development of other species of plants.
r заросли
Густые кустарниковые, древесные или смешанные сообщества, входящие в состав подлеска и обычно препятствующие проникновению растений других видов, например, заросли бамбука, ежевики, папоротника и др.
f broussailles *fpl*; hallier *m*
s maleza *f*; breña *f*
g Gestrüpp *n*

812 **budget year; balance year**
The one-year period with the start of the accumulation season and extending through the following summerr's ablation season.
r балансовый год
Интервал времени, в течение которого последовательное преобладание процессов аккумуляции и абляции образует замкнутый цикл массообмена ледника.
f année *f* budgétaire
s año *m* de balance
g Massenhaushaltsjahr *n*; Bilanzjahr *n*

813 **buffer zone of a reserve**
A territory adjacent to a reserve where any activity disturbing nature, e.g. cutting of trees, is prohibited.
r буферная зона заповедника; охранная зона заповедника

Территория, прилегающая к заповеднику, в пределах которой запрещены любые действия, нарушающие характерный для нее природный режим, например, вырубка деревьев.
f zone *f* de protection de réserve naturelle
s zona *f* de protección de reserva natural
g Pufferzone *f* des Naturschutzgebiets

* **built-up area → 814**

814 built-up estate; built-up area
The part of a town where the land is covered with buildings and roads very densely, so that there is space for further development only if existing structures are demolished.
r застроенная территория
Часть города, тесно застроенная зданиями и дорогами, так что новое строительство возможно лишь при сносе прежних сооружений.
f terrain *m* à bâtir; espace *m* à bâtir
s terreno *m* de edificación
g Bebauungsgelände *n*; bebautes Gelände *n*

* **bulgunyakh → 4571**

815 bulk cargo
Materials such as coal, grain, construction materials, etc. transported without packaging.
r насыпные грузы; массовые грузы
Грузы, состоящие из однородной сыпучей массы, перевозимые без упаковки: уголь, зерно, строительные материалы и пр.
f marchandises *fpl* pondérales
s cargamento *m*; mercancías *fpl* masivas
g Massengut *n*; Schüttgut *n*

816 buoy; float
A floating device used as a navigational aid to mark a channel or hazard as well as to study surface and deep-sea currents and other oceanographic parameters.
r буй
Плавучий маяк (поплавок) для обозначения фарватера или местонахождения какого-либо предмета, используется также для измерения морских течений и других океанографических параметров.
f bouée *f*
s boya *f*
g Schwimmer *m*; Boje *f*

817 buoyancy; floatage
The ability of an object where specific weight is smaller than that of water, to stay on the water surface.
r плавучесть; сила плавучести
Способность предмета, удельный вес которого меньше, чем у воды, держаться на водной поверхности.
f flottabilité *f*
s flotabilidad *f*
g Schwimmfähigkeit *f*; Schwimmkraft *f*

818 buran; poorga; purga
A strong northeast wind blowing in Central Asia at all seasons but most frequently and fiercely in winter when it lifts and carries the snow, and ice particles (in the tundra, especially in Siberia, it is termed purga).
r буран; пурга
Сильный северо-восточный ветер, дующий в Азии во все сезоны года, но особенно жестокий в зимнее время, когда он переносит большое количество снега (в северной России, главным образом в тундре, носит название пурга).
f bourane *m*; pourga *f*; poudrerie *f*
s bouran *m*; borrasca *f* de nieve; pourga *f*
g Schneesturm *m*

* **burial mound → 6403**

* **burial site → 5019**

819 buried glacier
A glacier covered with a thick moraine mantle.
r погребенный ледник
Ледник, забронированный мощным моренным чехлом.
f glacier *m* fossile
s glaciar *m* fósil
g bedeckter Gletscher *m*

820 buried ice
An ice mass formed or deposited on the Earth's surface and later covered with mineral sediments.
r погребенный лед
Лед, сформировавшийся на поверхности земли и впоследствии погребенный под толщами минеральных отложений.
f glace *f* enterrée; glace *f* enfouie
s hielo *m* enterrado
g bedecktes Eis *n*; Steineis *n*; subterranes Eis *n*

821 buried ice vein
An ice vein which ceased to grow and lying inside the permafrost layer below its boundary with the seasonally thawed ground.
r погребенная ледяная жила
Жила льда, прекратившая рост и залегающая в толще многолетнемерзлых пород ниже их контакта с сезонноталыми породами.
f glace *f* enfouie
s vena *f* de hielo enterrada
g subterranes Eiskeil *m*

* **buried karst → 4384**

822 buried landforms
An old land surface covered by sedimentary or volcanogenic deposits.
r погребенный рельеф
Древний рельеф, перекрытый толщей осадочных или вулканогенных отложений.
f relief *m* fossile; relief *m* enfoui
s relieve *m* fósil; relieve *m* fosilizado

g begrabenes Relief *n*

* **buried soil → 4393**

823 buried terrace
A terrace covered with a layer of sedimentary or volcanic rocks and not visible in the open relief.
r погребенная терраса
Терраса, погребенная под толщей осадочных или вулканогенных пород и не видимая в открытом рельефе.
f terrasse *f* fossile
s terraza *f* fósil
g begrabene Terrasse *f*

824 burn
A site where the vegetation is destroyed by fire and secondary pyrogenic communities (e.g. birch and aspen) start to form.
r гарь
Участок растительности, погибшей в результате пожара, впоследствии зарастающий вторичными пирогенными сообществами.
f brûlé *m*; brûlis *m*
s chamicera *f*
g Brandstelle *f*

825 burning
A controlled burning of trees or grasses to improve herbage or to clear land for agriculture, it is one of the methods of slash and burn cultivation.
r пал
Выжигание древесной или степной растительности для улучшения травостоев и расширения пахотных земель, один из методов подсечно-огневого земледелия.
f brûlage *m*; écobuage *m*
s quemada *f*; quema *f*; roza *f*
g Abbrennen *n*; Ausbrennen *n*

* **bursting of a lake → 2493**

* **bush → 5422**

* **bushfires → 6770**

* **bushwood-road → 3571**

* **butte → 2102**

826 butte; knob
A small rounded hill of various origin with relatively steep slopes.
r бугор
Округлое возвышение самого разного происхождения со сравнительно крутыми склонами небольшой высоты.
f butte *f*
s cerro *m*; cerillo *m*
g Hügel *m*; Anhöhe *f*

827 buttress
A rugged, projecting part of a hill or a mountain composed of undistorted rocks.
r контрфорс
Выступающая вперед часть холма или горы, сложенная ненарушенными породами.
f contrefort *m*
s contrafuerte *m*
g Strebepfeiler *m*

* **Buys-Ballot's law → 567**

828 bypass
A road which skirts a place, especially a road designed to divert through-traffic from roads in a congested area, e.g. a town centre.
r объездная дорога; объезд
Дорога, проложенная таким образом, чтобы облегчить проезд через перегруженные и перенаселенные районы, в особенности через центр большого города.
f détour *m*
s camino *m* de rodeo
g Umgehungsstraße *f*

829 bypassing of sediments
The transport of coastal sediments from the windward side of a bay to its leeward side and vice versa.
r байпассинг наносов
Перемещение береговых наносов от наветренной стороны залива к его подветренной стороне и наоборот.
f bypassing *m* de dépôts
s bypassing *m* de depósitos
g Bypassing *n* der Sedimenten

830 by-product
A secondary product obtained during a specific production process, of greater or less value than the product which is the primary objective of the operation.
r побочный продукт
Вторичный продукт, получаемый в результате какого-либо производственного процесса, обладающий определенной ценностью наряду с основной продукцией данного процесса.
f sous-produit *m*
s producto *m* colateral; subproducto *m*
g Nebenprodukt *n*

C

831 caatinga
A tropical deciduous scrubly forests composed of stunted thorny xerothytic trees and shrubs occurring in arid regions of South America.
r каатинга
Характерное для аридных районов Южной Америки тропическое листопадное редколесье из низкорослых деревьев и кустарников, обильны суккуленты и колючие растения.
f caatinga *f*
s caatinga *f*
g Caatinga *f*

832 cable car; cableway
A kind of transportation route in mountains consisting of a cable fixed on high poles, cabins (aerial gondola) or seats (chair lift) hanging on the cable.
r канатная дорога
Способ доставки пассажиров и грузов в горах с помощью кабеля на высоких мачтах, к которому крепятся кабины (маятниковая и гондольная дороги) или кресла (кресельная дорога).
f téléphérique *m*
s funicular *m* aéreo; telecabina *f*
g Seilbahn *m*; Luftseilbahn *m*

*** cableway → 832**

833 cabotage; coastwise navigation
Navigation between ports of one country, large cabotage is coastal shipping between ports of several seas and small cabotage is navigation between ports of the same sea.
r каботаж
Судоходство между портами одной страны (различают большой каботаж, т.е. перевозки между портами разных морей, и малый каботаж - перевозки между портами одного и того же моря).
f cabotage *m*
s cabotaje *m*
g Küstenschiffahrt *f*

834 cadastral map
A large scale map showing the ownership of real estate of land.
r кадастровая карта
Крупномасштабная карта, показывающая земельную собственность на данной территории.
f carte *f* cadastrale
s mapa *m* catastral
g Katasterkarte *f*

835 cadastre
A record of the area, boundaries, location, value, and ownership of land, e.g. land cadastre, forest cadastre.
r кадастр
Количественная и качественная опись земель, включающая их положение, границы, качество, права собственности и т.п., например, земельный кадастр, лесной кадастр, кадастр рекреационных ресурсов.
f cadastre *m*
s catastro *m*
g Kataster *m*

836 Cainozoic; Cenozoic
The youngest geological era in the Earth's history, it began 67 million years ago and continues up to now. → Table 1, p. 284
r кайнозой
Самая верхняя (молодая) геологическая эра в истории Земли, началась 67 млн лет назад и продолжается поныне. → Табл. 1, с. 284
f Cainozoïque *m*; Cénozoïque *m*
s era *f* cenozoica; Cenozóica *f*
g Känozoikum *m*

837 cake ice; ice-cake
A fragment of sea ice smaller than 20 m across.
r мелкобитый лед
Отдельные льдины с относительно ровной поверхностью, менее 20 м в поперечнике.
f petits glaçons *mpl*
s bloques *mpl* pequeños de hielo
g Eisschollenbruchstücke *npl*; Feinscholleneis *m*

*** calcareous tuff → 6333**

*** calcicole → 838**

*** calcifuge → 839**

838 calciphile; calcicole
Plant species inhabiting alkaline soils rich in lime and carbonates, and chalk outcroppings, e.g. grape, sesame.
r кальцефил
Виды растений, обитающих на солонцеватых, богатых известью и карбонатами почвах и меловых обнажениях, например, виноград, кунжут.
f calciphile *m*; calcicole *m*
s calcífilo *m*; calcícolo *m*
g Kalkpflanze *f*

839 calciphobe; calcifuge
Plant species thriving in relatively acid environment and avoiding carbonate soils and waters, e.g. plants of sphagnum bogs.
r кальцефоб
Растения, избегающие почву и воды, насыщенные карбонатами, например, растения сфагновых болот.
f calcifuge *m*
s calcífugo *m*
g kalkfliehende Pflanze *f*; kalkmeidende Pflanze *f*

840 calcisol
A soil in the profile of which there is a horizon of secondary calcite accumulation which is characteristic of the Mediterranean region and semiarid subtropical climate zone.
r кальцисоль
Почва с горизонтом накопления вторичного кальцита в профиле, характерна для областей со средиземноморским и семиаридным субтропическим климатом.
f calcisol *m*
s calcisol *m*
g Ñalcisol *m*

841 calcite
Calcium carbonate, the main rock-forming mineral of limestones, marbles, carbonates.
r кальцит
Карбонат кальция, главный породообразующий минерал известняков, мраморов, карбонатов.
f calcite *f*
s calcita *f*
g Kalzit *m*; Kalkspat *m*; Calcit *m*

*** calculating chart → 4147**

842 caldera
A broad, roughly circular volcanic crater formed by the blowing off of the top of the crater by explosive eruption or by subsidence. → Fig. 47
r кальдера
Широкая округлая впадина на вершине вулкана с крутыми склонами, возникшая в результате обрушения его верхней части во время извержения. → Рис. 47
f caldeira *f*
s caldera *f*
g Caldera *f*

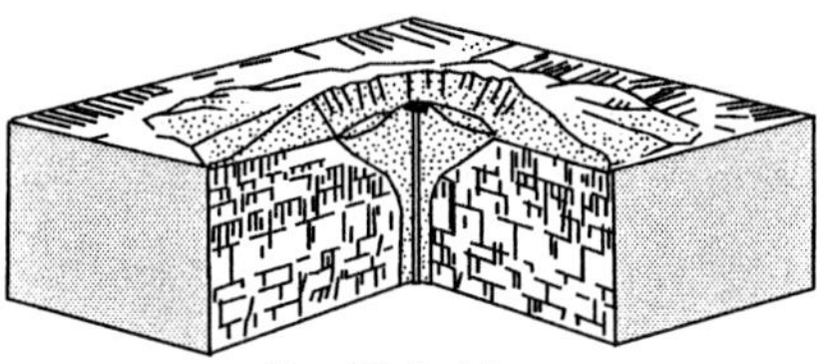

Fig. 47. **Caldera**

843 caldera glacier
A glacier lying inside a volcanic crater or caldera, sometimes with one or more short outlet tongues.
r кальдерный ледник
Ледник, располагающийся внутри кратера или кальдеры вулкана, иногда с одним или несколькими короткими выводными языками.
f glacier *m* de caldeira
s glaciar *m* de caldera
g Kaldergletscher *m*

844 Caledonian folding
The era of mountain formation in the Ordovician and Silurian periods, as a result of which the mountains of British Islands, north-western part of Scandinavia, western part of Central Kazakhstan, etc. appeared.
r каледонская складчатость
Эра горообразования в ордовикском и силурийском периодах, когда возникли горы Британских островов, северо-западной части Скандинавии, западной части Центрального Казахстана и др.
f plissement *m* calédonien
s plegamiento *m* caledoniano
g kaledonische Faltung *f*

845 calendar
A system of dividing time into fixed periods, initially a system of counting long time intervals based on periodical of natural phenomena connected with the movement of celestial bodies.
r календарь
Система счета длительных промежутков времени, основанная на периодичности явлений природы, связанных с движением небесных светил.
f calendrier *m*
s calendario *m*
g Kalender *m*

846 calendar day; meansolar day
The interval equal to the period of the Earth's rotation around its axis, i.e. 24 hours.
r сутки
Промежуток времени, равный периоду вращения Земли.
f jour *m* et nuit *f*
s día *m*; día *m* civil; día *m* solar medio
g Tag *m* und Nacht *f*

847 calendar year
The time interval that is equal to 365 days for usual years and 366 days for leap-years.
r календарный год
Промежуток времени, составляющий 365 суток для простых годов и 366 суток для високосных.
f année *f* civile
s año *m* civil; año *m* gregoriano; año *m* calendar
g Kalenderjahr *n*

848 caliche
More or less cemented calcareous, sometimes dolomitic crust which accumulates in soils in arid and semiarid conditions as a result of the evaporation of soil water, it is common in the Atacama desert of Peru and Chile.
r каличе
Поверхностный, более или менее сцементированный известковый, иногда доломитовый слой, формирующийся над почвой в аридных и семиаридных условиях в результате испарения почвенной влаги, широко распространен в Перу и Чили.

f caliche *m*; croûte *f* calcaire
s caliche *m*; costra *f* calcárea
g Caliche *m*

849 calm
A windless weather; the state of a sea when no wind waves occur on the surface.
r штиль; безветрие
Безветренная погода; состояние моря, когда на его поверхности отсутствуют ветровые волны.
f calme *m*
s calma *f*
g Windstille *f*; Kalme *f*

850 calving
Breaking-off of icebergs from ice shelves and from glaciers terminating in the sea or a lake.
r облом айсберга; откалывание айсберга; отёл ледника
Откалывание айсбергов от шельфовых ледников или от наземных ледников, оканчивающихся в море или озере.
f vêlage *m*
s desprendimiento *m* de icebergs; velage *m*
g Eisbergkalben *n*; Eisbergabstoßen *n*

851 calving bay
A deep-water bay where icebergs are frequently formed be breaking off from a seaward edge of the land ice which moves to the sea.
r бухта отела
Глубоководный залив, благоприятный для откалывания айсбергов от ледника, сползающего к морю.
f baie *f* de vêlage
s bahía *f* de desprendimiento de icebergs; bahía *f* de velage
g Kalbungsbucht *f*

852 cambisol
An immature soil in which the soil formation reveals mainly in the change of colour and structure of horizons, it is common on young surfaces of temperate boreal areas, mountainous and arid regions.
r камбисоль
Неполноразвитая почва, в которой почвообразование проявляется преимущественно в изменении окраски и структуры горизонтов, распространена на молодых поверхностях умеренных бореальных областей и горных районов, а также в аридных областях.
f cambisol *m*
s cambisol *m*
g Ñambisol *m*

853 Cambrian
The first geological period of the Palaeozoic era, which began 570 million years ago and lasted for about 70 million years. → Table 1, p. 284
r кембрий
Первый период палеозойскй эры геологической истории Земли, начался 570 млн лет назад и продолжался около 70 млн лет. → Табл. 1, с. 284
f Cambrien *m*
s Cámbrico *m*
g Kambrium *n*

854 campo
A savanna vegetation in central Brazil where the ratio of trees to grass allows a division into: «campo cerrado» which has scattered trees, and «campo limpo» which has tall grass and virtually no trees.
r кампос
Ландшафты высокотравных или мелкотравных (в зависимости от обилия осадков) степей и саванн, распространенных в Бразилии, различаются кампосы без деревьев (campo limpo) и с участками деревьев и кустарников (campo cerrado).
f campo *m*
s campo *m*
g Campo *mpl*

855 canal
An artificial watercourse made by people for navigation, drainage or irrigation, in which water flows without pressure.
r канал
Искусственное русло с безнапорным течением воды, предназначенное для навигации, дренажа или ирригации.
f canal *m*
s canal *m*
g Kanal *m*

*** canal river → 960**

856 canopy; crown cover
The high, leafy continuous layer in a forest community that is formed by the crowns of the trees of approximately the same hight.
r полог леса; лесная крона
Высокий лиственный слой в лесном сообществе из крон деревьев близкой высоты.
f couvert *m* des couronnes; couvert *m* forestier
s fronda *f* del bosque; cubierta *f* forestal
g Kronendach *n*; Bläteerdach *n*

857 canopy density
The condition in which the crowns of individual trees overlap to form a virtually continuous layer.
r сомкнутость полога
Смыкание крон отдельных деревьев в единый непрерывный слой.
f densité *f* des couronnes; fermeture *f* des couronnes
s densidad *f* de fronda del bosque
g Bestandesschluß *m*

*** canyon → 2584**

858 capacity of a stream; carrying capacity of a stream

The maximum load of stones, pebbles, sand, etc. a stream can carry, as measured in grams per second.

r предельная нагрузка потока
Максимальное количество твердых частиц, которое способен переносить поток (воды или ветра) за данный отрезок времени.
f capacité *f* de torrent
s capacidad maxima *f* de torrente; carga *f* límite
g Geschiebetransportkapazität *f*; Stromkapazität *f*

859 capacity of snow drifting
The maximum amount of snow that can be transported by a snow-storm.
r транспортирующая способность метели
Предельная масса снега, которая может переноситься метелью.
f capacité *f* de transport du chasse-neige
s capacidad *f* de transporte de ventisca
g Frachtvermögen *n* der Schneegestöber

*** capacity of territory → 4725**

860 cape; headland; promontory
A land area projecting into the sea, lake or river.
r мыс; нос
Участок суши, вдающийся в море, озеро или реку.
f cap *m*; promontoire *m*
s cabo *m*; promontorio *m*
g Kap *n*; Nase *f*

861 capillarity
The ability of the ground and soil to keep and pass water via the hairlike diameter spaces of the capillary tubes.
r капиллярность
Способность грунта и почвы содержать и пропускать воду через тончайшие волосяные пространства.
f capillarité *f* du sol
s capilaridad *f* del suelo
g Bodenkapillarität *f*

862 capillary fringe
A layer of soil lying immediately over the water table in which water drawn up from the ground water level is held in capillary spaces and available to plant roots.
r капиллярная кайма; капиллярная зона
Зона капиллярного поднятия воды в пористом грунте или почве, разделяющая зоны аэрации и насыщения.
f frange *f* capillaire
s franja *f* capilar
g Kapillarzone *f*; Kapillarsaum *m*; Saugraum *m*

*** capillary moisture → 863**

863 capillary water; capillary moisture
Underground water kept in pores and capillary cavities of rocks by surface tension force.
r капиллярная вода
Подземная вода, удерживаемая в порах и капиллярных пустотах горных пород силами поверхностного натяжения.
f eau *f* capillaire
s agua *f* capilar
g Kapillarwasser *n*; Porensaugwasser *n*

864 capital
The main city, political and administrative centre of a state or province.
r столица
Главный город, политико-административный центр государства.
f capitale *f*
s capital *f*
g Hauptstadt *f*; Kapitale *f*

865 capital goods
The machinery and equipment, and the primary and partly processed raw materials used in the manufacture of other goods, contrasting with consumer goods.
r основные фонды
Основное оборудование и производственные помещения, а также сырье и материалы, используемые в производстве товаров.
f biens *mpl* équipées
s bienes *mpl* de equipo
g Hauptfonds *mpl*

866 capital investment
Input of labour, material and financial resources into the reproduction of *capital goods.*
r капитальные вложения
Затраты трудовых, материальных ресурсов и денежных средств на вопроизводство *основных фондов.*
f capital *m* de investissement
s inversiones *fpl* básicas; inversión *f* de capital
g Anlagekosten *pl*

867 capital-intensive *adj*
Needing a large investment of capital for higher earnings or increased productivity, as opposed to *labour-intensive.*
r капиталоемкий
Нуждающийся в больших инвестициях капитала для получения большей прибыли или увеличения выпуска продукции, в противоположность *трудоемкому* производству.
f capital-intensif *adj*
s capital-intensivo *adj*
g investionsintensiv *adj*; anlageintensiv *adj*

868 capitalism
An economic system characterized by private ownership of the production of goods, private enterprise, competition, profit-making and a market economy.
r капитализм
Экономическая система, основанная на частной собственности на средства производства, частных предприятиях, конку-

ренции, извлечении прибыли и рыночной экономике.
f capitalisme *m*
s capitalismo *m*
g Kapitalismus *m*

869 cap-rock
A layer of hard, resistant rock which forms the flat summit of a hill and protects the underlying, less-resistant strata; an impermeable stratum overlying an aquifer, or salt dome.
r кепрок
Слой твердой горной породы, формирующий плоскую вершину холма и предохраняющий от разрушения более рыхлые нижележащие слои; водонепроницаемый слой, покрывающий водоносный горизонт или соляной купол.
f chapeau *m* de roche
s chapa *f* de roca
g Gesteinshut *m*

870 captation
Devices making it possible to collect and remove ground waters to the earth surface.
r каптаж
Устройства, позволяющие собирать и выводить подземные воды на земную поверхность.
f captage *m*
s instalación *f* para captar aguas subterráneas; captaje *m*
g Fassen *n*; Wasserfassung *f*

871 capture of river; river piracy; river capture; beheading of river
The action by which a river, by a rapid headward erosion, captures and diverts into its own channel the headwaters of another stream. → Fig. 48
r перехват реки
Захват одной рекой в результате пятящейся эрозии стока соседней реки. → Рис. 48
f capture *f* de fleuve
s captura *f* fluvial
g Flußanzapflung *f*

872 carbon capacity
The amount of carbon taken from the environment to produce an item of any product.
r карбоноемкость
Количество углерода, забираемого из окружающей среды на производство единицы той или иной продукции.
f capacité *f* de carbone

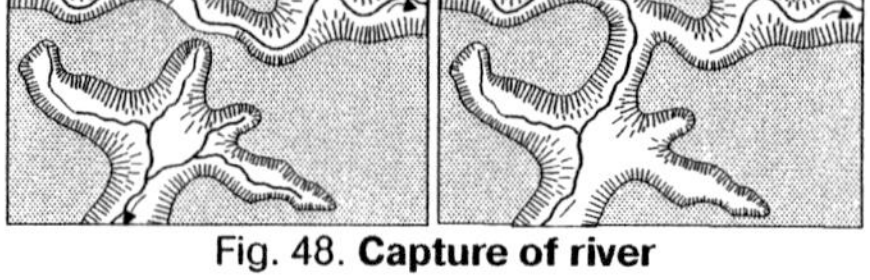

Fig. 48. **Capture of river**

s capacidad *f* de carbono
g Kohlenstoffkapazität *f*

873 carbon credit
A credit provided by international organizations or a state for measures aimed at the reduction of carbon dioxide emissions.
r углеродный кредит
Кредит, выдаваемый международными организациями или государством на мероприятия по снижению выбросов углекислого газа.
f crédit *m* de carbone
s crédito *m* por la emisión de CO_2; crédito *m* de carbono
g Kredit *m* für CO_2-Emissionsreduzierung

874 carbon cycle
Succesive transformation of carbon in the biosphere from its form of carbon dioxide into biomass and its return to a gaseous form by respiration and decay processes.
r углеродный цикл
Последовательное превращение углерода в биосфере от двуокиси углерода в биомассу и до его возвращения в газообразную форму в результате процессов дыхания и разложения.
f cycle *m* du carbone
s ciclo *m* del carbono
g Kohlenstoffzyklus *m*

* **carbon dioxide → 879**

875 carbon sink
The consept that large areas of forest, by photosynthesis, may significantly contribute to the reduction of carbon dioxide in the atmosphere.
r углеродный сток
Концепция, согласно которой большие территории, покрытые лесами, благодаря фотосинтезу, вносят существенный вклад в уменьшение содержания углекислого газа в атмосфере.
f écoulement *m* de carbone
s vertimiento *m* de carbono
g Kohlenstoffsenke *f*

876 carbon tax
The amount of money added by a government to the price of fuel in an attempt to reduce greenhouse gas emissions.
r углеродный налог
Государственный налог на нефтяное топливо с целью сократить чрезмерные выбросы углекислого газа в атмосферу.
f impôt *m* sur le CO_2; taxe *f* sur le rejet de CO_2
s impuesto *m* sobre la emisión de CO_2
g Kohlenstoffsteuer *n*

877 carbonaceous *adj*
Containing carbon, applied to rocks (e.g. coal, shale) or other sedimentary material

(e.g. peat) consisting largely of carbon which has organic origin.

r углеродистый
Содержащий углерод, относящийся к породам (например, уголь или сланец) или осадочному материалу (например, торф), содержащим значительную массу углерода, имеющего органическое происхождение.

f carbonè *adj*
s carbonado *adj*
g kohlenstoffhaltig *adj*; Kohlenstoff-

878 carbonate horizon; lime horizon
A soil layer developing under insufficient moistening and containing carbonates, often in the form of impregnation, veins, and concretions.

r карбонатный горизонт
Почвенный горизонт, развивающийся в условиях недостаточного увлажнения и содержащий карбонаты, часто в виде пропитки, прожилок, конкреций.

f horizon *m* carbonaté
s horizonte *m* de carbonatación; horizonte *m* carbonatado
g Karbonathorizont *m*; Kalkhorizont *m*

879 carbonic acid; carbon dioxide
A colourless, heavy gas (CO_2), present in the atmosphere (about 0.03%) and in solution in the hydrosphere, it gets into the atmosphere as a result of the burning of organic matter, evapotranspiration of plants and breathing of animals, it is effective greenhouse gas.

r углекислый газ; двуокись углерода
Бесцветный газ, CO_2, составляющий около 0,03% состава земной атмосферы, поступает в воздух при сгорании органических веществ и при дыхании животных и растений, обладает парниковым эффектом.

f gaz *m* carbonique; acide *m* carbonique
s gas *m* carbónico; ácido *m* carbónico
g Kohlensäuregas *n*

880 Carboniferous
The fifth geological period of the Palaeozoic era which began 345 million years ago and lasted for about 65 million years. → Table. 1, p. 284

r карбон; каменноугольный период
Пятый период палеозойской эры геологической истории Земли, начался 345 млн лет назад и продолжался около 65 млн лет. → Табл. 1, с. 284

f Carbonifère *m*
s Carbonífero *m*
g Karbon *n*

881 cardinal direction
The angle counted off from northern or southern direction of magnetic or true meridian with the indication of a corresponding quarter (NE, SE, SW, NW) before the degree value.

r румб
Угол, отсчитываемый от северного или южного направления магнитного или истинного меридиана вправо или влево до данного направления с указанием перед градусной величиной соответствующей четверти: СВ, ЮВ, ЮЗ, СЗ.

f rhumb *m*
s rumbo *m*; cuarta *f*; cuarta *f* de rosa náutica
g Kompaßstrich *m*

882 cardinal points
The four main points of the compass: North, South, East and West and corresponding quarters of the horizon (northern, southern, eastern and western).

r страны света; стороны горизонта; главные румбы
Четыре главные точки горизонта: север, юг, восток и запад (иногда так же называют соответствующие четверти горизонта: северную, южную, восточную и западную).

f points *mpl* cardinaux
s puntos *mpl* cardinales
g Himmelsrichtungen *fpl*

883 cargo turnover; fright turnover; turnover of goods
A characteristic of the transportation of goods calculated as the freight volume multiplied by the distance to which it is moved.

r грузооборот
Характеристика грузовых перевозок, выражаемая произведением количества перевезенного груза и расстояния, на которое он перемещен.

f trafic *m* des marchandises; trafic *m* marchand; circulation *f* des marchandises
s circulación *f* de mercancías; movimiento *m* de mercancías
g Güterumschlag *m*

884 carnivore; predator
An animal feeding on other animals.

r хищное животное; плотоядное животное; хищник
Животное, питающееся другими животными.

f carnassier *m*; rapace *m*
s carnívoro *m*; carnicero *m*
g Raubtier *n*; Raubvogel *m*

885 carpet-like communities; mat-vegetation
Short grass lands that occur in the alpine belt of mountains, in the regions with abundant snowfall, especially in the mountains of central and southern Europe.

r матты
Низкотравные луга в альпийском поясе гор, в районах с обильным выпадением снега, особенно характерные для гор Центральной и Южной Европы.

f pelouse *f* alpaine «matte»
s pradera *f* alpina «matte»
g Matte *npl*

* **carrying capacity of a stream → 858**

886 **carrying capacity of area; profitable capacity**
The population size, both the number of species or the number of people, which the resources of a particular environment can maintain without a tendency to decrease or increase.
r емкость территории
Количество людей или животных особей на данной территории, наиболее оптимальное при существующих здесь природных ресурсах и не испытывающее тенденций к росту или сокращению естественным путем.
f capacité *f* de terrain
s capacidad *f* del terreno
g Umschlagkapazität *f*

887 **carrying capacity of ice**
The ability of a fast ice or an ice field to carry loads on their surface.
r несущая способность ледяного покрова
Способность припая и ледяных полей выдерживать нагрузку на их поверхность.
f trafficabilité *m* de la banquise; capacité *m* de charge de la banquise
s capacidad *f* de carga de banquisa
g Tragfähigkeit *f* der Eisdecke

888 **carrying capacity of transport**
The maximal weight of goods a means of a transport can carry at one and the same time.
r грузоподъемность
Максимальная масса груза, который транспортное средство способно в один раз поднять и перевезти.
f capacité *f* de charge
s capacidad *f* de carga
g Ladekapazität *f*; Tragfähigkeit *f*

889 **carrying intensity**
The index of intensity of a transport network expressed as the ratio between cargo turnover of the transport network and its length.
r грузонапряженность
Показатель интенсивности работы транспортной системы, выражаемый отношением грузооборота транспортной сети к ее длине.
f charge *f* de transport
s cargamento *m* de transporte
g Gütertransportintensität *f*

* **cartage → 891**

890 **cartel**
An association of producers and suppliers that controls prices, output, selling arrangements, and recruitment.
r картель
Форма экономического объединения, участники которого, сохраняя коммерческую и производственную самостоятельность, заключают соглашение о регулировании объемов производства, условий сбыта продукции, найма рабочей силы.
f cartel *m*
s cartel *m*
g Kartell *n*

891 **carting; cartage**
A type of land transport that uses animals (horses, oxen, camels, reindeer, dogs, etc) as the moving force.
r гужевой транспорт
Вид наземного транспорта, в котором в качестве тяги используется сила животных: лошадей, волов, верблюдов, оленей, собак и др.
f transport *m* hippomobile; charroi *m*
s transporte *m* hiromóvil
g Wagentransport *m*; Fuhrwerktransport *m*

892 **cartogram; chorogram**
A way of cartographic representation of some statistical data (index, information) of a phenomenon by hatching or colouring the area of the territorial unit on a map. → Fig. 49
r картограмма
Способ изображения штриховкой или закраской какого-либо относительного статистического показателя в пределах нанесенных на карту территориальных единиц. → Рис. 49
f cartogramme *m*
s cartograma *m*
g Kartogramm *n*

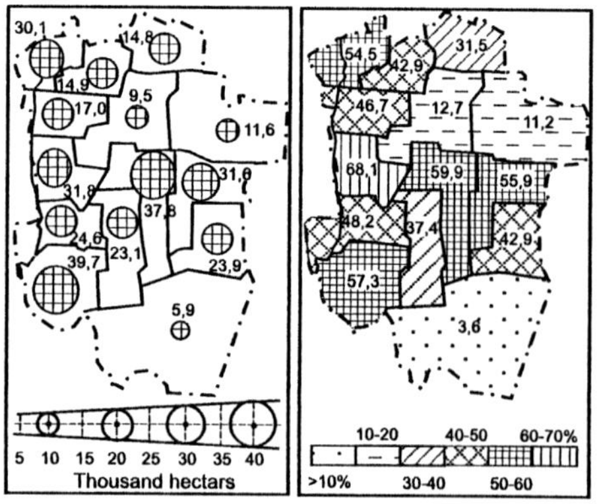

Fig. 49. **Cartogram** (right) showing the share of the arable land, and **diagram map** (left) showing the area of the arable land

* **cartographic alteration → 1646**

* **cartographic data base → 893**

893 **cartographic databank; cartographic data base**
A set of cartographic data arranged by certain rules establishing the general principles of their description, storage and use.

r картографический банк данных; картографическая база данных
Совокупность картографиченских данных, организованных по определенным правилам, устанавливающим общие принципы их описания, хранения и использования.
f base *f* des données cartographiques
s base *f* de datos cartográficos
g kartographische Datenbank *f*

* **cartographic distortion → 1646**

894 **cartographic format**
The size of a geographical map sheet, and for a relief map its maximum height by vertical as well.
r формат карты
Размер листа географической карты, а для рельефной карты – также наибольшая ее высота по вертикали.
f format *m* de la carte
s formato *m* de un mapa
g Kartenformat *n*

895 **cartographic generalization**
Selection and simplified representation of geographical objects in accordance with the purpose, scale, and theme of the map, and the peculiarities of the mapped area.
r картографическая генерализация
Отбор и обобщение изображаемых на карте объектов в соответствии с назначением, масштабом, тематикой карты и особенностями отображаемой территории.
f généralisation *f* cartographique
s generalización *f* cartográfica
g Generalisierung *f*

896 **cartographic pattern**
Spatial combination of cartographic symbols apprehended by the reader of a geographical map or by a recognizing device.
r картографический образ
Пространственная комбинация картографических знаков, воспринимаемая читателем географической карты или распознающим устройством.
f forme *f* cartographique
s forma *f* cartográfica
g kartigraphisches Gefüge *n*

* **cartographic projection → 3670**

897 **cartographical modelling**
Construction and analysis of mathematical models by the data read from maps; compilation of new maps on the basis of mathematical models.
r картографическое моделирование
Построение и анализ математических моделей по данным, снятым с карт; создание новых карт на основе математических моделей.
f modelage *m* cartographique
s modelación *f* cartográfica
g kartographische Modellierung *f*

898 **cartographical animation**
A dynamic succession of electronic maps which being demonstrated on the screen of a computer produce the effect of motion.
r картографическая анимация
Динамическая последовательность электронных карт-кадров, создающих при их демонстрации на экране компьютера эффект движения.
f animation *f* cartographique
s animación *f* cartográfica
g karographische Animation *f*

899 **cartographical browser**
A program which enables to see a cartographic image on a computer screen.
r картографический браузер
Программа просмотра картографического изображения на экране компьютера.
f browser *m* cartographique
s browser *m* cartográfico
g kartographischer Browser *m*

900 **cartography; mapping**
The art, science and technology of making maps and atlases, together with their study as scientific documents.
r картография
Область науки, техники и производства, охватывающая изучение, создание и использование географических карт и атласов.
f cartographie *f*
s cartografía *f*
g Kartographie *f*; Kartenkunde *f*

901 **cartology**
A branch of *cartography* studying maps, their elements, properties, types, history of development, ways of using.
r картоведение
Раздел *картографии*, изучающий карты, их элементы, свойства, виды, историю развития, способы использования.
f cartologie *f*
s cartología *f*
g Kartenkunde *f*

902 **cartometry**
Measuring and calculation of numerical values of objects from maps, together with their graphic presentation.
r картометрия
Измерение по картам размеров объектов и их иных количественных характеристик.
f cartométrie *f*
s cartometría *f*
g Kartometrie *f*

903 **cartouche**
An embellishment of a map, often in the form of scroll, which encloses a title, legend or name.
r картуш
Заголовок и другие данные, помещенные на карте в рамке с орнаментом.

f cartouche *m*
s orla *f*
g Kartusche *f*

904 cascade
Natural or artificial waterfall in which the water tumbles down a series of steps (ledges).
r каскад
Естественный или искусственный водопад, в котором вода скатывается вниз по серии уступов.
f cascade *m*; caída *f*
s cascada *f*
g Kaskade *f*; Fallstufe *f*

905 cascade of reservoirs
A series of reservoirs created in one river.
r каскад водохранилищ
Комплекс водохранилищ, построенных на одной и той же реке.
f cascade *f* des réservoirs
s cascada *f* de embalses
g Stauseekaskade *f*

906 cash crop
A crop grown primarily for sale as contrasted with a *subsistence crop* grown for the use of the grower.
r коммерческий урожай
Урожай, выращиваемый в основном для продажи, в отличие от *потребительского урожая*, потребляемого самим производителем.
f récolte *f* d'argent
s cosecha *f* de mercado
g Währungsernte *f*

907 caste
A closed hierarchically arranged group of people separated from others due to its specific social functions, inherited occupations and trades often associated with ethnic or religious affiliation and connected with certain territories.
r каста
Замкнутая группа людей, обособившаяся вследствие выполнения специфических социальных функций, наследственных занятий и профессий, что часто связано с принадлежностью к определенной этнической, а иногда и религиозной общности, и имеющая отношение к определенной территории.
f caste *f*
s casta *f*
g Kaste *f*

908 cataclasis
Deformation of rocks, including ice, accompanied by the fracturing and rotation of mineral grains or their aggregates due to tectonic processes.
r катаклаз
Деформация горных пород, включая лед, сопровождающаяся дроблением и вращением минеральных зерен или их агрегатов под влиянием тектонических процессов.
f cataclase *f*
s cataclasa *f*; cataclastía *f*
g Kataklase *f*

909 catacombs
Underground cavities often in the form of galleries and labirinths of natural and man-made origin, e.g. the Roma Catacombs.
r катакомбы
Подземные полости, нередко образующие лабиринты, искусственного (каменоломни) или естественного происхождения, например, Римские или Одесские катакомбы.
f catacombes *fpl*
s catacombas *fpl*
g Katakomben *fpl*

910 catalogue of glaciers
Making inventories of glaciers, containing systematic data on a location, dimension, type and regime of glaciers grouped according to river basins.
r каталогизация ледников
Составление каталогов ледников, где приводятся систематические сведения о положении, размерах, типах и некоторых чертах режима ледников, сгруппированных по речным бассейнам.
f catalogage *m* des glaciers; inventaire *m* glaciaire
s catalogación *f* de glaciares
g Gletscherinventarisation *f*; Gletscherkatalogisierung *f*

911 cataract
A large waterfall in which vertically falling water forms one perpendicular flow above a precipice.
r катаракт
Большой водопад, в котором вертикально падающий поток образует над пропастью одну отвесную струю.
f cataracte *f*
s catarata *f*
g Katarakt *m*

912 catastrophic avalanche; climax avalanche
A rarely occuring large avalanche traveling far beyond the limits of the detrital cone previously formed by regular avalanches, or a snow avalanche that has tremendous destructive force.
r катастрофическая лавина
Лавина редкой повторяемости, распространяющаяся далеко за пределы минерального конуса выноса, сформированного обычными лавинами, либо большая лавина, обладающая значительной разрушительной силой.
f avalanche *f* catastrophique
s avalancha *f* catastrófica
g Schadenlawine *f*

913 catastrophism
A largerly recognized XIX century belief that geological deformations of the Earth's crust, as well as changes of composition of organic world happen by sudden, isolated, and forceful events (such as Noahrs flood), divided by much more longer periods of rest, rather than by slow process.
r катастрофизм
Концепция XIX века, согласно которой преобразования земной коры (ее деформации, а также изменения состава органического мира) происходят катастрофически быстро, в течение коротких периодов геологического времени, разделенных гораздо более длительными периодами покоя.
f catastrophisme *m*
s catastrofismo *m*
g Katastrophismus *m*; Kataklysmentheorie *f*

914 catch crop
A fast-maturing crop grown between two main crops in a rotation, or between the rows of a main crop.
r дополнительный урожай
Быстро созревший урожай, выращенный в промежутке между двумя главными урожаями, в том числе при ротации земель.
f récolte *f* de prise
s cosecha *f* de premio
g Preisernte *f*

* **catchment → 1690**

* **catchment area → 1690**

915 catena
A connected series of soil varieties along the horizontal profile of a slope, from well-drained relatively flat upper surfaces down along a steeper part to low-lying flat surface at the foot of the slope. → Fig. 50
r катена
Взаимосвязанная почвенная серия на склоне от его верхней относительно плоской и хорошо дренируемой части вниз по более крутому откосу к выполаживающемуся подножью. → Рис. 50
f catena *f*; chaîne *f* de sols
s catena *f*; cadena *f* de suelos
g Katena *f*

* **cattle driving → 6302**

916 cattle-breeding; stock-raising
A branch of agriculture engaged in raising horned cattle to produce milk, meat and hides.

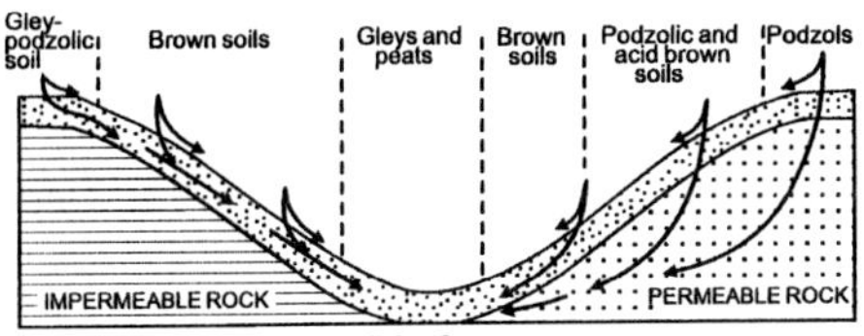

Fig. 50. **Catena**

r скотоводство
Отрасль животноводства, занимающаяся разведением крупного рогатого скота для получения молока, мяса и кожевенного сырья.
f élevage *m* bovin
s ganadería *f* bovina
g Viehzucht *f*

917 cave
A hollow in the upper part of the Earth's crust with one or several openings to the surface.
r пещера; полость
Полость в верхней толще земной коры, открывающаяся на земную поверхность одним или несколькими входными отверстиями.
f caverne *f*
s caverna *f*; cueva *f*
g Höhle *f*

918 cave glacier
A big accumulation of cave ice of different origin which slowly moves in a definite direction.
r пещерный ледник
Наиболее крупное смешанное по генезису скопление пещерного льда, обладающее самостоятельным движением.
f glacier *m* de caverne
s glaciar *m* de caverna
g Höhlengletscher *m*

919 cave hoarfrost
Large ice crystals formed by sublimation of moisture in still air inside caves.
r пещерная изморозь
Крупные кристаллы льда, формирующиеся в спокойном воздухе ледяных пещер вследствие сублимации парообразной влаги.
f givre *m* de caverne
s escarcha *f* de caverna
g Höhlenreif *m*

920 cave ice
Ice formations of various shapes and structure, occuring in underground cavities and hollows in permafrost or in temporary freezing rocks.
r пещерный лед
Разные по форме и строению ледяные образования, возникающие в подземных полостях и пустотах мерзлых и промерзающих горных пород.
f glace *f* de caverne
s hielo *m* de caverna
g Höhleneis *n*

921 cave lake; underground lake; karst lake
An individual water body in a cave or a pool in an underground karst river.
r пещерное озеро; подземное озеро
Самостоятельный водоем в пещере или местный разлив подземной карстовой реки.

f lac *m* de caverne; lac *m* karstique
s lago *m* de caverna; lago *m* kárstico
g Höhlensee *m*; Karstsee *m*; unterirdischer See *m*

922 cavern
A hollow space in rock several millimetres to several centimetres in diameter.
r каверна
Пустота, полое пространство в горной породе поперечником от нескольких миллиметров до нескольких сантиметров.
f petite caverne *f*
s caverna *f* pequeña
g Höhlung *f*; Kaverne *f*

923 cavitation
A process in which air bubbles in the water are formed due to the sudden increase of velocity in a fast-running stream that results in a more intensive erosion of rocks which is characteristic of a waterfall, rapids or a cataract.
r кавитация
Формирование пузырьков в воде, происходящее в случаях резкого увеличения скорости текущего потока на водопадах, перекатах и пр., что вызывает повышенную эрозию горных пород.
f cavitation *f*
s cavitación *f*
g Kavitation *f*

*** CBD → 1013**

924 Celcius scale
A temperature scale (t°C) proposed by the Swedish astronomer A. Celsius, which divides the interval between the freezing-point and the boiling point of water into 100 parts, so the freezing point of water at standard atmospheric pressure is 0°C and the corresponding boiling point is 100°C.
r температурная шкала Цельсия
Температурная шкала (t°C), предложенная шведским астрономом А. Цельсиусом, которая делит интервал между точкой замерзания и точкой кипения воды на 100 частей, так что точка замерзания воды при стандартном атмосферном давлении равна 0°C, а соответствующая точка кипения равна 100°C.
f échelle *f* de température Celsius
s escala *f* de temperaturas Celsius
g Celsius-Temperaturskala *f*

925 celestial *adj*
Related to the sky, heaven, or space.
r небесный
Относящийся к небесной сфере или космическому пространству.
f céleste *adj*
s celeste *adj*
g Himmels-

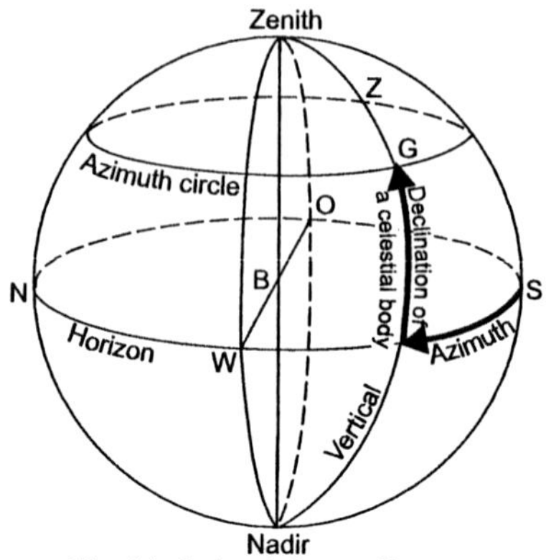

Fig. 51. **Celestial coordinates**

926 celestial coordinates
Angles which serve to identify the locations of heavenly bodies on the celestial sphere, they are measured by arcs of large circles or angles near centre of the celestial sphere, expressed in degrees or hour measure at 1 hour = 15°. → Fig. 51
r небесные координаты
Углы, с помощью которых определяют положение небесных светил на небесной сфере, измеряются дугами больших кругов или углами при центре небесной сферы, выражаемыми в градусах или в часовой мере из расчета 1 час = 15°. → Рис. 51
f coordonnées *fpl* célestes
s coordenadas *fpl* celestes
g Himmelskoordinaten *fpl*

927 celestial equator; equinoctial
An imaginary line in the sky over the Earth's equator formed where the celestial sphere is crossed by a plane passing through its centre perpendicular to the axis of the celestial sphere, it divides the celestial sphere into Northern and Southern hemispheres.
r небесный экватор
Воображаемая линия в небе над земным экватором, представляющая собой линию пересечения небесной сферы плоскостью, проходящей через ее центр перпендикулярно оси мира, делит небесную сферу на Северное и Южное полушария.
f équateur *m* céleste
s ecuador *m* celeste
g Himmelsäquator *m*

928 celestial meridian
A large circle of the celestial sphere passing through celestial poles, zenith and nadir.
r небесный меридиан
Большой круг небесной сферы, проходящий через полюсы мира, зенит и надир.
f méridien *m* céleste
s meridiano *m* celeste
g Himmelsmeridian *m*

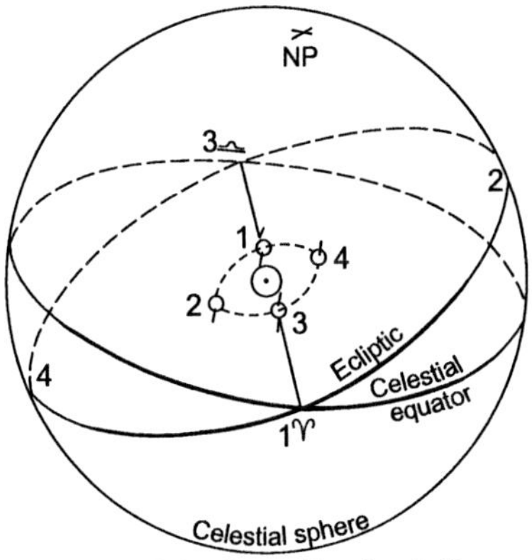

Fig. 52. **Celestial sphere**. The Sun in the centre, NP - North Pole. Four positions of the Earth on its orbit: 1, 3 - equinox, 2, 4 - solstices

929 celestial pole
A point of crossing of the celestial sphere by its axis (near the North celestial pole the Polaris is situated).
r полюс мира
Точка пересечения небесной сферы осью мира (около Северного полюса мира расположена Полярная звезда).
f pôle *m* céleste
s polo *m* celeste
g Himmelspol *m*

930 celestial sphere
An imaginary sphere of infinite radius with the Earth at its centre against which celestial bodies appear to be projected. → Fig. 52
r небесная сфера
Воображаемая сфера произвольного радиуса, на которую проецируют небесные тела. → Рис. 52
f sphère *f* céleste
s esfera *f* celeste
g Himmelskugel *f*; Himmelsphäre *f*

931 cellular sands
Accumulation of sand, usually stabilized, shaped as rounded honeycombs up to 15 m deep in sand deserts.
r ячеистые пески
Обычно неподвижные скопления песков в песчаных пустынях в виде ячеек закругленной формы глубиной до 15 м.
f sables *m* cellulaires
s arenas *f* celulares
g Zellsände *mpl*

932 cellulate sand relief
A variety of barchan relief of deserts when a hole forms in front of the leeward side of a barchan, and rake-like ridges that form, can reach several tens of metres in height.
r лунковый рельеф
Разновидность барханного рельефа пустынь, когда перед откосом осыпания у бархана образуется выемка (лунка), а формирующиеся граблевидные гряды достигают высоты в несколько десятков метров.
f relief *m* alvéolaire; topographie *f* alvéolaire
s relieve *m* alveolar; topografía *f* alveolar
g Alveolenrelief *n*

933 cementation
The process of consolidation of loose weathered rocks which leads to the formation of a solid mass, e.g.sands are converted into sandstone, rubble into breccia, and pebbles into conglomerate.
r цементация
Скрепление в прочную массу рыхлых продуктов выветривания горных пород: песков – в песчаник, щебня – в брекчию, галечника – в конгломерат.
f cémentation *f*
s cementación *f*
g Zementation *f*

*** Cenozoic → 836**

934 census
A periodic counting of the population and gathering of demographic, economic and social information carried out by government that describes people living in one country or on any other territory for a particular point in time.
r перепись населения
Сбор демографических, экономических и социальных данных, характеризующих жителей страны или иной территории на единый момент времени.
f recensement *m* de population
s censo *m*
g Volkszählung *f*

935 central business district; downtown (*amer*)
The main buisness district of a town or city (symbol: CBD).
r даунтаун
Главный деловой район города.
f downtown *m*; quartier *m* d'affaires
s downtown *m*; barrio *m* de negocios; centro *m* cívico
g stadteinwärts adw; Stadtmitte *f*

936 central eruption
A volcanic eruption from a single vent or from a tight group of vents, producing a cone, in contrast to a *fissure eruption.*
r центральное извержение
Вулканическое извержение из единственного жерла вулкана или из компактной группы нескольких выходов лавы, формирующих вулканический конус, в противоположность *трещинному извержению.*
f éruption *f* centrale
s erupción *f* central
g Zentraleruption *f*

937 **central meridian**
Meridian which is the axis of symmetry of the grid on a certain part of latitudinal arch.
r осевой меридиан
Меридиан, являющийся осью симметрии градусной сети на данном участке широтной дуги.
f méridien *m* central d'une projection
s meridiano *m* principal de proyección
g Hauptmeridian *m*

938 **central place**
A settlement or nodal point which provides goods, services, administrative functions for the consuming population of the surrounding area.
r центральное место; опорный центр расселения
Населенный пункт или узловая точка, выполняющие административные функции, а также снабжающие товарами и обслуживающие население окружающей территории.
f lieu *m* central
s lugar *m* central; asentamiento *m* central
g Zentralort *m*

939 **central place system**
The spatial distribution of a number of central places of different size and spacing which satisfy the daily, weekly, monthly and yearly needs of the general consuming population.
r система центральных мест
Пространственное распределение серии центральных мест разного размера и разного положения, которое удовлетворяет ежедневные, еженедельные, ежемесячные и круглогодчиные потребности населения, проживающего на данной территории.
f système *m* de lieu central
s sistema *m* de asentamiento central
g System *n* der Zentralstellen

940 **central place theory**
The theory which states that the numbers, sizes, and patterns of spatial distribution of central places can be explaned by the operation of the forces of supply and demand.
r теория центральных мест
Теория, утверждающая, что количество, размеры и основные черты пространственного распределения центральных мест могут быть объяснены действием соотношения спроса и предложения.
f théorie *f* de lieux centraux
s teoría *f* del lugar central
g Zentralorttheorie *f*

941 **centrality**
In central place theory the relative importance of a place with regard to surrounding area, or the degree to which a centre serves its surrounding area.
r центральное положение
В теории центральных мест важность данного места по отношению к окружающей территории или то положение, которое оно занимает в обслуживании окружающей его территории.
f centralité *f*
s centralidad *f*
g Zetralität *f*

942 **centralization**
The tendency towards localization of economic activity in and around a relatively small number of urban centres.
r централизация
Тенденция сосредоточения экономической активности вокруг относительно небольшого количества городских центров.
f centralisation *f*; centralité *f*
s centralización *f*; centralidad *f*
g Zentralisierung *f*; Zentralisation *f*

* **centrally planned economy → 1143**

943 **centre of atmospheric action**
An extensive and almost stationary cyclone or anticyclone which controls the movement of atmospheric disturbances over a large region.
r центр действия атмосферы
Обширный и почти стационарный циклон или антициклон, оказывающий воздействие на циркуляцию атмосферы на значительной территории.
f centre *m* d'action atmosphérique
s centro *m* de acción atmosférico
g Aktionszentrum *n* der Atmosphäre

944 **centre of origin**
An area of genetic diversity of cultivated plants found mainly in the mountainous regions of tropics, subtropics and in temperate belt (Russian scientist N.I. Vavilov distinguished eight centres of origin).
r центр происхождения
Географические центры генетического разнообразия культурных растений, расположенные преимущественно в горных районах тропиков, субтропиков и отчасти умеренного пояса (русский ученый Н.И. Вавилов открыл восемь таких центров).
f centre *m* d'origine
s centro *m* de origen
g Entwicklungszentrum *m*

945 **centre of population density**
An index describing population distribution across a country or region, calculated in the way the centre of gravity in mechanics is calculated.
r центр населенности
Показатель, характеризующий размещение населения страны или района, вычисляется аналогично центру тяжести в механике.
f barycentre *m* caché
s centro *m* de gravedad de población
g Besiedlungsgewichtszentrum *n*

946 **centripetal drainage pattern**
A drainage pattern in which the streams drain radially inwards, either towards a single trunk river which drains the basin, or to a lake which may or not have an outlet.
r центростремительная речная сеть
Речная сеть, когда реки со многих направлений стремятся к главной реке, протекающей на данной территории, либо текут в озеро, которое может иметь или не имеет стока.
f réseau *m* hydrographique centripète
s red *f* hidrográfica centrípeta
g zentripetales Flußnetz *n*

947 **cereals**
Cultivated grasses which produce grains used for human or animal food and to produce baking flour or fodder.
r зерновые культуры
Злаки, возделываемые для получения зерна, идущего в основном на выпечку хлеба и корм скоту.
f céréales *fpl*
s cereales *mpl*
g Getreidekultur *f*

948 **cerrado**
A type of savanna found in Brazil in which trees 9-15 m in height are intermixed with tall grassland.
r серрадо
Тип саванны, распространный в Бразилии, где деревья высотой 9-15 м перемежаются с высоким травостоем.
f cerrado *m*
s cerrado *m*
g Cerrado *m*

949 **cession**
The transition of all sovereign rights over a certain territory of one state to another by an agreement, e.g. the selling of Alaska to the United States of America in 1867 by Russia or the selling of Western Louisiana to the U.S. by Napoleon in 1803.
r цессия
Передача всех суверенных прав над определенной территорией одним государством другому по соглашению между ними (например, продажа Россией в 1867 г. Соединенным Штатам Америки Аляски, или продажа Наполеоном в 1803 г. тем же США Западной Луизианы).
f cession *f*
s cesión *f*
g Zession *f*

950 **chaco**
A landscape of a vast almost level alluvial plain with poor soil supporting grassy savanna or deciduous scrub woodland north of the pampas of Argentina between Andes and the river Paraguai, where heavy rainfall is usual, and the chaco is marshy nearly everywhere.
r чако
Ландшафт широкой и почти плоской аллювиальной равнины, покрытой травянистой саванной или кустарником, к северу от аргентинской пампы между Андами и рекой Парагвай, где обычны сильные дожди, и поэтому чако почти всюду заболочена.
f chaco *m*
s chaco *m*
g Ñhaco *m*

951 **chain migration**
Migration in which the individual worker moves and his/her dependants follow once any initial loan given for travel has been repaid and the worker has found a secure job, and than other members of the family follow, using as a base the home of the first migrant.
r цепная миграция
Миграция, при которой отдельный рабочий находит работу в другом месте, и как только им/ею полученная на переезд ссуда выплачена, иждивенцы и другие члены семьи также переезжают на новое место.
f migration *f* à chaîne
s migración *f* en cadena
g Kettenmigration *f*; Kettenwanderung *f*

952 **chain of hills**
A linear stretch of uplands with rounded tops and a well-defined foot line.
r кряж
Линейно вытянутая возвышенность с мягкими округлыми очертаниями вершин и отчетливой линией подножья.
f chaîne *f* de basses montagnes; chaîne *f* des collines
s ramal *m* corto; cadena *f* de colinas; cadena *f* de montañas bajas
g Gebirgskette *f*

* **channel → 5836**

953 **channel deformation**
Changes of dimensions and form of channels under the action of running water, erosion, transport and accumulation of material.
r русловые деформации
Изменение размеров и формы русел под действием водного потока, эрозии, транспорта и аккумуляции наносов.
f deformation *f* de lit
s deformación *f* del lecho
g Verformung *f* des Flußbettes

954 **channel flow; riverbed flow**
Surface runoff occurring along the channels of rivers and temporary watercourses.
r русловой сток
Поверхностный сток, происходящий по руслам рек и временных водотоков.
f écoulement *m* de lit
s escorrentía *f* de lecho
g Abflus *m* im Flussbett

955 channel processes
The interactions of a stream and its bed resulting in constant changes in the morphological structure of the channel.
r русловые процессы
Взаимодействия потока и его ложа, вызывающие постоянные изменения морфологии русла.
f processus *mpl* de lit
s procesos *mpl* de lecho
g Prozesse *mpl* im Flußbett

956 channel reach; cross section of a stream
A transverse crossing of a stream by a plane perpendicular to the mean direction of the flow, it is limited by the channel profile and the water surface from above.
r живое сечение реки; поперечное сечение реки; сечение русла; водное сечение
Сечение реки плоскостью, перпендикулярной среднему направлению течения, оно ограничено профилем русла, а сверху уровнем воды.
f section *f* mouillée
s sección *f* mojada
g Flußquerschnitt *n*

* **channel resistance → 5139**

957 channel storage
The change of river water level at the outlet section due to the sharp deepening of the channel and its further discharge resulting in the water level lowering.
r русловое регулирование
Изменение уровня воды в замыкающем створе реки вследствие резкого углубления русла, что приводит к спаду уровня.
f régulation *f* du lit
s regulación *f* de lecho
g Flußbettregulierung *f*

958 channelled avalanche
An avalanche that moves along a narrow depression on the slope, such as a ravine, gully, etc.
r лотковая лавина
Лавина, движущаяся по узкой депрессии склона – ложбине, логу и пр.
f avalanche *f* en colée
s avalancha *f* en colada
g Runsenlawine *f*; Furchlawine *f*

959 channelled glacier; wall-sided glacier
A long narrow glacier occurring in a gully on a steep slope or in a narrow depression on a gentle slope.
r русловой ледник
Узкий длинный ледник, залегающий в эрозионной ложбине на крутом склоне или в узком углублении на пологом склоне.
f glacier *m* canalisé
s glaciar *m* canalizado; glaciar *m* de curso
g Rinnengletscher *m*

960 channelled river; canal river
A river or its part which is shaped into a canal (bounded on the bottom and sides or streightened), or a canal arranged in a river channel.
r канализованная река
Река или ее участок, руслу которой придан вид канала, либо канал, устроенный в русле реки.
f rivière *f* canalisée; fleuve *m* canalisé
s río *m* canalizado
g kanalisierter Fluß *m*

961 chapada
A flat-topped areas composed of sandstone steeply descending to the crystal basement, common on Plateau of Brazil.
r шапада
Плато на Бразильском плоскогорье, сложенные песчаником, круто обрывающиеся к кристаллическому основанию.
f chapada *f*
s chapada *f*
g Chapada *f*

962 chaparral
Natural subtropical vegetation composed of broad-leaved evergreen shrubs and small trees, which form dense forests, common in California and Mexico, similar to maquis in the Mediterranean area.
r чапарраль
Субтропические вечнозеленые кустарниковые сообщества, характерные для Калифорнии и Мексики, аналог маквиса в Средиземноморье.
f chaparral *m*
s chaparral *m*
g Chaparral *m*

* **characteristic sheet → 1279**

* **characteristic species → 1851**

* **charriage → 5383**

963 chart datum; datum plane; reference plane
The level or plane to which all depths on a nautical chart are related (in the seas with tides it is the lowest sea level).
r нуль глубин
Условная поверхность, от которой отсчитываются отметки глубин на морских навигационных картах (в морях с приливами это наинизший возможный уровень моря).
f zéro *m* hydrographique; zéro *m* des profondeurs
s cero *m* de profundidades
g Normalnull *f*; Kartennull *f*

964 charter
The process of hiring a boat, plane or bus, especially for use by a group of people and to movement of loads and people by air or water by a rented (completely or partially) boat, plane or bus.

r чартер
Наем судна, самолета или автобуса для специальной перевозки людей или грузов с предоставлением всего транспортного средства или его части.
f charter *m*
s charter *m*
g Charter *m*

* **chatter mark → 2496**

965 **«cheeks»**
A Russian term for steep rocky river banks in narowed parts of the river valley.
r щёки
Широко распространенное в России название скалистых обрывистых берегов реки в местах сужения ее долины.
f «joues» *fpl*
s «mejillas» *fpl*
g «Backen» *fpl*

966 **chemical abrasion**
The destruction of shores and a bottom of a water course or body of water as a result of dissolution and washing-out of rocks.
r химическая абразия
Разрушение берегов и дна водоемов и водотоков в результате растворения и вымывания горных пород.
f abrasion *f* chimique
s abrasión *f* química
g chemische Abrasion *f*

967 **chemical deposit**
Sediments formed by way of deposition of material on the bottom of water bodies due to chemical and biochemical reactions or water evaporation.
r хемогенные отложения
Отложения на дне водоемов, возникающие путем осаждения материала из растворов в результате химических и биохимических реакций или испарения воды.
f dépôt *m* chimique
s depósito *m* químico
g chemische Ablagerungen *fpl*

968 **chemical industry**
A group of industries including mining of chemical minerals, production of mineral fertilizers, polymers, household chemicals, etc.
r химическая промышленность
Комплекс отраслей, включающих горнохимическую промышленность, производство минеральных удобрений, полимерных материалов, бытовой химии и пр.
f industrie *f* chimique
s industria *f* química
g chemische Industrie *f*

* **chemical oceanography → 971**

969 **chemical pollution**
Pollution of the environment with chemical substances not present there before, or changing their natural concentration that exceeds the usual level.
r химическое загрязнение
Загрязнение окружающей среды химическими веществами, ранее здесь отсутствовавшими, либо изменяющими свою естественную концентрацию до уровня, превышающего обычную норму.
f contamination *f* chimique
s contaminación *f* química
g chemische Verschmutzung *f*

* **chemical runoff → 5185**

970 **chemical weathering; corrosion**
The wearing away of solid rocks by means of chemical reactions and the removal of the material in aqueous solution.
r химическое выветривание; коррозия; выщелачивание
Разрушение горных пород вследствие их растворения химическим воздействием воды и выноса веществ в водном растворе.
f météorisation *f* chimique; corrosion *f*; décomposition$_2$ *f*; altération *f*; érosion *f* chimique
s meteorización *f* química; decomposición$_2$ *f*; alteración *f*; corrosión *f*; erosión *f* química
g chemische Verwitterung *f*; Korrosion *f*; Auswaschung *f*

971 **chemistry of ocean; chemical oceanography**
Chemical properties of sea water and processes determining its chemical composition, and their study.
r химия моря
Химические свойства морской воды и процессы, определяющие ее химический состав, а также их изучение.
f chimie *f* de mer
s química *f* de mar
g Meereschemie *f*

972 **chemocline**
The layer in a lake having the maximum vertical gradient of salinity: the surface water layer being fresh and deep water layer being denser and with higher salinity.
r хемоклин
Горизонт с максимальным значением вертикального градиента солености в озере от пресного поверхностного слоя воды к более плотным глубинным слоям с повышенной соленостью.
f chimiocline *f*; couche *f* de discontinuité chimique
s quimoclina *f*
g Chemokline *f*

973 **chemosphere**
A part of the high atmosphere in which molecular dissociation and recombination take place under the influence of ultra-violet radiation.

r хемосфера
Область верхней атмосферы, в которой происходят молекулярное разложение и рекомбинация под влиянием ультрафиолетовой радиации.
f chimosphère *f*; chimiosphère *f*
s quimosfera *f*
g Chemosphäre *f*

974 **chemosynthesis**
The process by which bacteria produce organic compounds from inorganic substances using the energy of chemical reactions.
r хемосинтез
Процесс синтеза органических веществ из неорганических путем химических реакций, производимых некоторыми бактериями в процессе их жизнедеятельности.
f chimiosynthèse *f*
s quimosíntesis *f*
g Chemosynthese *f*

975 **chemotrophic *adj***
Applied to an organism that obtains energy from a source other than light, i.e. by chemical reactions, in contrast to phototrophic.
r хемотрофический
Относящийся к организмам, получающим энергию не от света, а от химических реакций, в отличии от фототрофических.
f chimotrophique *adj*; chimiotrophique *adj*
s quimotrófico *adj*
g chemotroph *adj*

976 **chern (Abies sibirica dominated taiga)**
Dense coniferous taiga consisting of fir and spruce typical of Siberia.
r чернь; черневая тайга
Густые хвойные пихтово-еловые леса с примесью осины и березы в Сибири.
f chern *m*
s chern *m*
g Tschern *f*

977 **chernozem**
A soil with the thick, dark, grainy humus and illuvial carbonate horizons which forms under perennial herbaceous vegetation of steppes and forest-steppes and has non-leaching or periodically leaching water regime.
r чернозем
Почва, формирующаяся под многолетней травянистой растительностью степей и лесостепей в условиях непромывного или периодически промывного режима, имеющая мощный темный зернистый гумусовый и иллювиальный карбонатный горизонты.
f chernozem *m*
s chernozem *m*; chernoziom *m*
g Tschernosem *m*; Schwarzerde *f*

978 **chernozem-like soil**
A soil similar to chernozem in that it has the horizon of dark humus, but it devoid of some other typical horizons such as illuvial or carbonate horizon.
r черноземовидная почва
Почва, сходная с черноземом по наличию темного гумусового горизонта, но отличающаяся, например, отсутствием иллювиального карбонатного горизонта.
f sol *m* de type du chernozem
s suelo *m* de tipo de chernozem
g Tschernosemartiger Boden *m*

979 **chestnut soil**
A soil of dry steppes of temperate arid belt that has thick, dark-brown horizon, rich in organic matter, under which there are horizons rich in calcium carbonate or gypsum.
r каштановая почва; каштанозем
Почва сухих степей умеренно засушливого пояса, имеющая мощный, богатый органическим веществом темно-коричневый горизонт, под которым залегают горизонты, богатые карбонатом кальция или гипсом.
f sol *m* châtain
s suelo *m* castaño
g kastanienfarbiger Boden *m*; Kastanosem *m*

* **chill factor → 6773**

980 **chimney**
A narrow vertical fissure or cleft in a rock-wall which is used by mountaineers for climbing to almost inaccessible rock-walls.
r камин
Узкая вертикальная трещина-расселина на горном склоне, нередко используемая альпинистами для подъема по труднодоступным скальным склонам.
f chemineé *f*; couloir *m* étroit
s chimenea *f*; pasillo *m* estrecho
g Kamin *m*

981 **chimney rock; rock pillar**
A tall, figure rock column composed of resistant rock.
r каменный истукан; каменный столб
Изолированный фигурный столб, сложенный устойчивыми к разрушению горными породами.
f roche *m* façonnée; aiguille *f* rocheuse
s aguja *f* rocosa
g Felsfigur *f*; Steinsäule *f*

982 **chink**
A steep escarpment of a plateau or table upland in Central Asia.
r чинк
Обрывистый уступ, ограничивающий плато или столовую возвышенность в Казахстане и Средней Азии.
f chink *m*
s chink *m*
g Tschink *m*

983 **chinook**

A warm dry, south-west wind similar to foehn which blows down on the eastern side of the Rocky Mountains in North America.

r чинук

Теплый и сухой юго-западный ветер, пдобный фёну, дующий на восточных склонах Скалистых гор в Северной Америке.

f chinook *m*
s chinook *m*
g Chinook *m*

984 chionophile

Organisms adapted to a high snow cover in winter.

r хионофил

Организмы, способные существовать в условиях мощного снежного покрова зимой.

f chionophile *m*
s kionófilo *m*
g Chionophil *m*

985 chionophite

Plants that live on the surface of snow or ice cover, e.g. some algae.

r хионофит

Растения, развивающиеся на поверхности снега или льда, например, некоторые водоросли.

f chionophyte *m*
s nievefita *m*; kionófito *m*
g Chionophyt *n*

986 chionophob

Animals which avoide areas with deep snow cover, such as some ungulates, and plants that are not snow-resistant, such as citric trees, cotton, rice, etc.

r хионофоб

Животные, избегающие местностей с глубоким снежным покровом (некоторые копытные), и растения, не обладающие снеговыносливостью (цитрусовые, хлопок, рис и пр.).

f chionophobe *m*
s kionófobo *m*
g Chionophob *m*

987 chionosphere

The regions of the Earth's land surface where permanent snow and glaciers can originate and exist in the appropriate land forms.

r хионосфера

Область поверхности суши, где в благоприятных формах рельефа возможно зарождение и существование многолетних снежников и ледников.

f chionosphère *f*
s kionósfera *f*
g Chionosphäre *f*

988 chop; choppi sea

A chaotic waving on the water surface of the sea where the waves of different directions meet.

r толчея

Беспорядочное волнение на водной поверхности моря, образуемое волнами разных направлений.

f ressac *m*
s resaca *f*
g Kabbelsee *f*

* **choppi sea → 988**

989 choppi wind

Wind with strong fluctuation of velocity and direction induced by the turbulence of air flows.

r порывистый ветер

Ветер с сильной пульсацией по скорости и направлению, связанной с турбулентностью воздушных течений.

f vent *m* à rafales
s viento *m* huracanado
g böiger Wind *m*

990 C-horizon

The lowest layer of the soil, underlying the B-horizon (illuvial horizon), the little-altered horizon of the parent material from which the true soil is derived.

r горизонт C

Самый нижний слой почвы, подстилающий горизонт B (иллювиальный горизонт), содержит лишь слегка измененный материал материнской породы.

f horizon C *m*
s horizonte C *m*
g Horizont C *m*

991 chorochromatic map

A type of thematic map in which qualitative spatial distributions are depicted by colour-tinting.

r хорохроматическая карта

Тип тематической карты, когда пространственное распределение количественных характеристик изображается путем послойного раскрашивания.

f carte *f* chorochromatique
s mapa *m* corocromático
g chorochromatische Karte *f*

* **chorogram → 892**

992 chorographic map; general map

Any comprehensive map representing large territories at a small scale.

r обзорная карта

Общегеографическая карта какой-либо территории в мелком масштабе.

f carte *f* chorographique
s mapa *m* corográfico
g Übersichtskarte *f*

* **chorography → 993**

993 chorology; areal differentiation; chorography

The establishment and description of geographical regions.

r хорология; пространственная дифференциация
Изучение областей распространения (ареалов) отдельных видов, семейств и других систематических групп растений и животных; установление и описание географических районов.
f chorologie *f*
s corología *f*
g Chorologie *f*

994 choropleth map
A type of thematic map in which qualitative spatial distributions are depicted from data that have been calculated from mean values, e.g. density of population shown within local, regional or national administrative areas.
r хороплетная карта
Тип тематической карты, когда пространственное распределение количественных характеристик вычисляется на основе средних величин, отнесенных к какой-либо площади, например, плотность населения в пределах местных, региональных или национальных административных границ.
f carte *f* des choroplèthes
s mapa *m* de coropletas
g Choroplethenkarte *f*

995 chronostratigraphy
A branch of *stratigraphy* studying the age of layers and their correlation in time.
r хроностратиграфия
Раздел *стратиграфии*, исследующий возраст слоев и их соотношения во времени.
f chronostratigraphie *f*
s cronostratigrafía *f*
g Chronostratigraphie *f*

* **chute → 4906**

996 Cimmerian folding
The era of mountain formation at the end of the Triassic and Jurassic periods within geosynclinal areas which developed in the late Palaeozoic and early Mesozoic, which led to the formation of the mountains of Verkhoyansk-Chukcha area, some parts of Cordilleras of North America, etc.
r киммерийская складчатость
Эра горообразования в конце триасового и в юрском периоде в пределах геосинклинальных областей, которые развивались в позднем палеозое и раннем мезозое, в результате возникли горы Верхояно-Чукотской области, части Кордильер Северной Америки и др.
f plissement *m* cimmérien
s plegamiento *m* cimmeriano
g kimmerische Faltung *f*

997 cinder cone
A cone formed round the vent of a volcano by debris, usually of volcanic origin, cast up during eruption.
r шлаковый конус
Конус, формируемый вокруг жерла вулкана обломками, обычно вулканического происхождения, сохраняющийся лишь до следующего сильного извержения.
f cône *m* de scories
s cono *m* de escorias
g Schlackenkegel *m*

998 cinnamonic soil
A soil coloured brown which gets lighter downwards, that forms in semiarid subtropical Mediterranean and monsoon climate where xerophytic trees and shrubs are typical.
r коричневая почва
Почва, имеющая коричневую окраску, светлеющую книзу, образующаяся в условиях семиаридного субтропического средиземноморского и муссонного климата, где развита ксерофитная древесно-кустарниковая растительность.
f sol *m* marron
s suelo *m* marrón
g zimtfarbiger Boden *m*

999 circulation
The movements of air masses, or ocean currents.
r циркуляция
Движение воздушных масс и морские течения.
f circulation *f*
s circulación *f*
g Zirkulation *f*

1000 circulation epoch
A period with duration from 10 years and over which is characterised by exceeding the norm by zonal or meridional types of the atmosphere circulation (by Russian scientists B. Dzerdzeevsky and Ya. Vangengeim).
r циркуляционная эпоха
Период длительностью от 10 лет и более, характеризующийся превышением над нормой зональных или меридиональных типов циркуляции атмосферы (по Б. Дзердзеевскому и Я. Вангенгейму).
f époque *f* de circulation
s época *f* de circulación
g Zirkulationepoche *f*

* **circulation of the atmosphere → 427**

1001 circumpolar diseases
The diseases spread in the areas of cold climate and caused mainly by cooling of human organism, e.g. non-specific pneumonias.
r циркумполярные болезни
Болезни, распространенные в странах с холодным климатом, вызываемые в основном переохлаждением организма человека, например, неспецифические пневмонии.
f maladies *fpl* circumpolaires
s enfermedades *fpl* circumpolares
g zirkumpolar Kkrankheiten *fpl*

1002 circumpolar distribution
The distribution of species of living things inhabiting the area in the polar regions either around the North Pole or around the South Pole, e.g. Arctic fox and reindeer dwell in the Arctic, the Weddell seal and the Adelie penguin in Antarctica.
r циркумполярное распространение
Ареалы видов животных и растений, охватывающие кольцом одну из полярных зон земного шара, например, в Арктике – ареалы песца, северного оленя, в Антарктике – тюленя Уэдделла, пингвина Адели.
f distribution *f* circumpolaire
s dispersión *f* circumpolar
g zirkumpolare Verbreitung *f*

1003 circumpolar vortex; polar vortex
The large-scale cyclonic circulation in the middle and upper troposphere in the polar regions.
r циркумполярный вихрь; полярный вихрь
Крупномасштабная циклоническая циркуляция в средней и верхней тропосфере полярных районов.
f tourbillon *m* circumpolaire
s vórtice *m* circumpolar
g circumpolarer Jetstream *m*

1004 circumtropical distribution
The distribution of species of living things inhabiting the tropical belt of the globe, e.g. monkeys, crocodiles, etc.
r циркумтропическое распространение
Ареалы видов растений и животных, опоясывающие тропики земного шара, например, ареалы обезьян, крокодилов.
f distribution *f* circumtropicale
s dispersión *f* circumtropical
g circumtropische Verbreitung *f*

1005 cirque; corrie; kar
A half-closed bowl-like deep depression worked out by a glacier at the head of a mountain valley on a mountain slope, in some cases contains a small lake. → Fig. 53

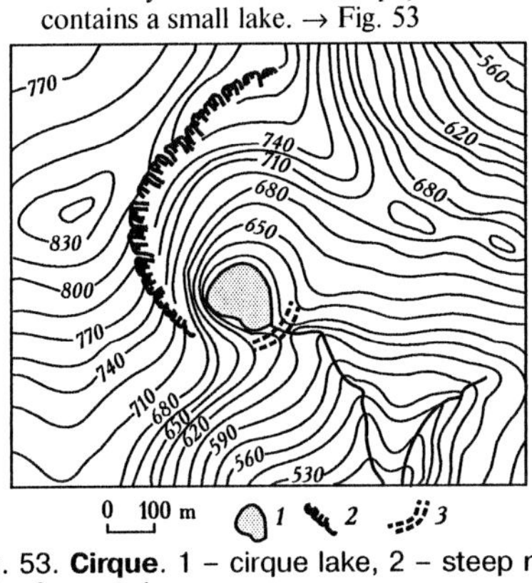

Fig. 53. **Cirque**. 1 – cirque lake, 2 – steep rock face, 3 – moraine

r кар; ледниковый цирк; цирк
Глубокая полузамкнутая чашеобразная впадина, выработанная ледником в привершинной части склона ледниковых высокогорий, на ее дне нередко находится озеро. → Рис. 53
f cirque *m*; cirque *m* glaciare
s circo *m*; circo *m* glaciar
g Kar *n*; Zirkus *m*

* **cirque glacier → 1308**

1006 cirque lake
A small lake occupying the bottom of a steep-walled depression formed by the glacier. → Fig. 53
r каровое озеро
Небольшое ледниковое озеро, занимающее дно кара. → Рис. 53
f lac *m* de cirque
s lago *m* de circo
g Karsee *m*

1007 cirque platform
A terrace on a slope of a ridge which formed when the bottoms of adjacent cirques joined together as they enlarged.
r каровая терраса
Террасовидная площадка на склоне хребта, возникшая от соединения днищ смежных ледниковых каров при их расширении в прошлом.
f plate-forme *f* de cirque
s plataforma *f* de circo
g Karterrasse *f*

1008 cirque steps
A series of glacier cirques situated one above the other one like giant steps.
r каровая лестница
Ряд ледниковых цирков, расположенных в несколько ярусов один над другим.
f cirque *m* en escalier
s circo *m* en escalera
g Kartreppe *f*; Stufenkar *n*

1009 cirrocumulus
A type of high clouds occurring above 6,000 m (symbol: Cc), white in colour, formed in thin sheets or layers without shading but in more-or-less regular ripples and grains.
r перисто-кучевые облака
Слои или гряды тонких белых облаков верхней тропосферы (выше 6000 м) без теней, состоящие из мелких элементов, имеющих вид хлопьев или ряби (символ: Cc).
f cirrocumulus
s cirrocumulus
g Cirrocumuluswolken *fpl*; Cirrocumulus; Zirrokumuluswolken *fpl*; Zirrokumulus

1010 cirrostratus
A type of high clouds occurring above 6,000 m (symbol: Cs), whitish in colour, formed in transparent sheets of smooth appearance and

producting solar or lunar halo phenomena, heralding the approach of a warm front.

r перисто-слоистые облака
Один из основных типов облаков верхнего яруса (выше 6000 м), имеющий вид белесоватой полупрозрачной пелены, обычно волокнистой, может покрывать все небо и часто вызывает явление гало, предшествуют прохождению теплого фронта (символ: Cs).

f cirrostratus
s cirrostratus
g Cirrostratuswolken *fpl*; Cirrostratus; Zirrostratus; Zirrostratus-Wolken *fpl*

1011 cirrus
A type of high clouds occurring above 6,000 m (symbol: Ci), whitish in colour, formed as silky, fibrous bands or wisp, detached from each other, usually associated with fair weather.

r перистые облака
Основной тип облаков верхнего яруса (выше 6000 м), они состоят из отдельных перистообразных элементов в виде тонких белых нитей или клочьев и вытянутых узких гряд, обычно ассоциируются с плохой погодой (символ: Ci).

f cirrus
s cirrus
g Cirrus-Wolken *fpl*; Cirrus; Zirrus-Wolken *fpl*; Cirruswolken *fpl*; Zirrus

1012 city
A large urban place with the population engaged mainly in non-agricultural occupations.

r город
Значительный населенный пункт с населением, занятым преимущественно в несельскохозяйственных сферах деятельности.

f ville *f*
s ciudad *f*
g Stadt *f*

* **city business district → 1013**

1013 city centre; city business district; CBD
The central part of a city with wide avenues and squares where many buisness, culture buildings and athority offices are located.

r центр города
Центральная часть города с широкими проспектами и площадями, в которой расположены многие деловые, культурные и властные учреждения.

f centre *m* de la ville; centre *m* des affaires; centre *m* des affaires; quartier *m* des affaires
s centro *m* cívico-comercial; distrito *m* comercial
g Stadtmitte *f*; Stadtzentrum *m*

* **city expansion → 6506**

1014 city region
A city or large town and the surrounding area which are being so functionally linked that the two are interdependent and function together as one unit.

r городской район
Город и окружающая его территория, функционально тесно связанные и поэтому представляющие по существу одну территориальную единицу.

f région *m* de ville
s región *m* de la ciudad
g Stadtgebiet *n*

1015 city-state
A city which has sufficient power to constitute an independent, sovereign state, usually exerting its authority over the surrounding region, e.g. Athens, being one of the city states of clssical Greece, or modern Singepore.

r государство-город
Город, имеющий достаточную силу и авторитет для сохранения своей государственной независимости, обычно распространяющий свой суверенитет на окружающий регион, например, Афины в эллинской Греции, или современный Сингапур.

f état-ville *f*
s estado-ciudad *f*
g Stadtstaat *m*

1016 civil twilight
The lightest part of evening and morning twilight during which the centre of the Sun's disc is below or above the horizon not more than by 6°.

r гражданские сумерки
Наиболее светлая часть вечерних и утренних сумерек, в течение которой центр диска Солнца погружен под горизонт не более чем на 6°.

f crépuscule *m* civil
s crepúsculo *m* civil
g bürgerliche Dämmerung *f*

1017 civilization
A human society which has its own highly developed social organization, culture, and way of life which makes it distinct from other societies, the largest civilizations being Western, Slavic-Christian Orthodox, Islamic, Confucian, and Latin American, etc.

r цивилизация
Человеческое общество, обладающее собственной высокоразвитой социальной организацией, культурой и образом жизни, отличающими его от других подобных обществ (крупнейшие цивилизации – западная, славяно-православная, исламская, конфуцианская, латиноамериканская и др.).

f civilisation *f*
s civilización *f*
g Zivilisation *f*

1018 class of landscapes
The highest taxonomic unit in the classification of landscapes which unites groups of

landscapes (class of plain landscapes and mountain landscapes) on the basis of megarelief.

r класс ландшафтов
Высшая таксономическая единица в систематике ландшафтов, объединяющая группы ландшафтов по мегарельефу (класс равнинных и горных ландшафтов).

f classe *f* des paysages
s clase *f* de paisajes
g Landschaftsklasse *f*

* **clastic rock → 1561**

1019 clathrate
Natural formations found in the deep layers of glaciers in which the molecules of gas enter into the crystal structure of water molecules.

r клатрат
Образования, обнаруживаемые в глубоких слоях ледников, в которых молекулы газа внедряются в пустоты кристаллических структур, составленных из молекул воды.

f clathrate *m*
s clatrate *m*
g Klathrate *n*

1020 clay
A fine-textured sedimentary rock composed of particles of aluminium silicates and other minerals less than 0.01 mm in size, which becomes plastic and impermeable when wet.

r глина
Пластичная во влажном состоянии осадочная горная порода, состоящая из частиц глинистых минералов размером менее 0,01 мм.

f argile *f*
s arcilla *f*
g Ton *m*; Lehm *m*

1021 clay loam
A loam which contains between 27 and 40% clay and between 20 and 45% sand, this mixture produces a relatively permeable and friable soil with particles of different sizes.

r жирный суглинок
Суглинок, содержащий 27-40% глинистых и 20-45% песчаных частиц, что формирует относительно рыхлую и проницаемую почву с частицами разного размера.

f limon *m* argileux
s limo *m* arcilloso
g lettiger Lehmboden *m*

* **clayey *adj* → 379**

1022 clayey karst
Pseudokarst formations created by weathering and suffosion which resemble true karst, they form in clays due to the underground washout along the cracks.

r глинистый карст
Псевдокарстовые образования, внешне сходные с карстовыми, формирующиеся в глинистых породах за счет подземного размыва по трещинам, возникшим в результате выветривания и суффозии.

f karst *m* argileux
s karst *m* arcilloso
g Lehmkarst *m*

* **clayey subsoil → 381**

1023 claypan
A hardened soil horizon with a considerable accumulation of clays carried from the upper soil horizons in the process of leaching or formed in situ during pedogenesis which hampers the air and water circulation and growth of plant roots.

r клейпен
Уплотненный почвенный горизонт со значительной аккумуляцией глин, вынесенных в процессе выщелачивания из поверхностных горизонтов или образовавшихся in situ в процессе почвообразования и затрудняющих циркуляцию воздуха, воды и рост корней растений.

f claypan *m*
s horizonte *m* arcilloso
g Lehmpfanne *f*; verdichtete Tonschicht *f*

* **clean production → 1816**

1024 clean water
Water free of pollution.

r чистая вода
Вода, не содержащая загрязнений.

f eau *f* pure
s agua *f* pura
g Reinwasser *n*

1025 clear water
1. A water surface free of underwater and above-water navigational obstacles. 2. A large water area on the verge of ice cover which is free of ice.

r свободная вода$_1$
1. Свободная от подводных и надводных рифов и отмелей акватория моря, безопасная для судоходства. 2. Большое пространство свободной ото льда воды, расположенное на границе с ледяным покровом.

f eau *f* libre
s agua *f* limpia; agua *f* libre
g freies Wasser *n*

* **cleared path → 5074**

1026 clearing; opening
A tract of forest where the trees were completely cut down or vigorously thinned.

r вырубка
Сильно прореженный или полностью вырубленный участок леса.

f taillis *m*
s lugar *m* talado; lugar *m* de tala
g Aushau *m*; Holzschlag *m*

1027 cleavage; schistosity
The ability of rocks to split into thin plates and prisms in the direction quite different from that of the original bedding.

r кливаж
Способность горных пород раскалываться на тонкие пластинки и призмы по системе параллельных поверхностей, секущих слоистость или согласных с ней.
f clivage *m*; schistosité *f*; feuilletage *f*
s laminamiento *m*; esquistosidad *f*; foliamiento *m*
g Schieferung *f*

*** cleft → 4267**

1028 clevage of ice; ice foliation
The secondary texture of ice characterised by a system of thin closed parallel fissures and other planes of weakening which do not coincide with those of the initial texture.
r кливаж льда; сланцеватость льда
Вторичная текстура льда, для которой характерны системы тонких закрытых параллельных трещин и других поверхностей ослабления, не совпадающие с первичной текстурой.
f clivage *m* de la glace
s clivaje *m* de hielo
g Schieferung *f* des Eises; Spaltbarkeit *f* des Eises

1029 cliff
A vertical or stepped escarpment formed as a result of the degradation of a high bedrock coast by breaking waves, which sometimes may sharply fall down into the sea for some tens meters. → Fig. 54
r клиф; береговой уступ
Отвесный или ступенчатый обрыв, образовавшийся в результате разрушения высокого коренного берега действием прибоя, нередко продолжающийся на десятки метров под водой. → Рис. 54
f falaise *f*
s acantilado *m*
g Kliff *m*; Uferabhang *m*; Felswand *f*

1030 climate
The average weather conditions in a given area, considered over many years (usually 30-35 years) on the basis of long-term statistics of the meteorological elements in that area.
r климат
Обобщенная характеристика условий погоды данного региона, описываемая многолетними статистическими значениями метеорологических элементов.
f climat *m*
s clima *m*
g Klima *n*

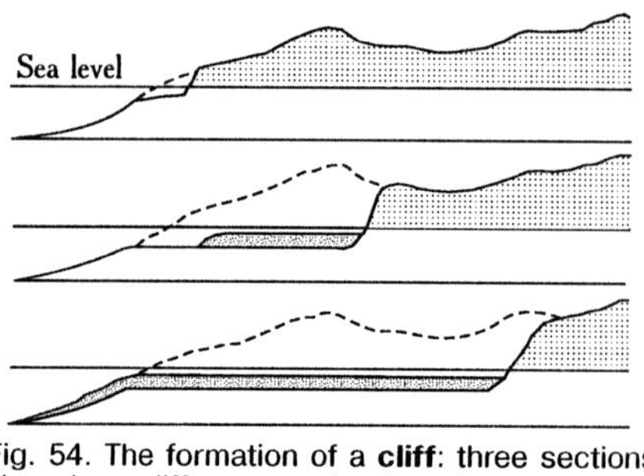

Fig. 54. The formation of a **cliff**: three sections show how cliffs are cut back and an abrfasion platform formed

1031 climate change; climatic change
The slow variations on climate characteristics considered in different time scales from decades to millions of years.
r изменение климата
Изменения климата Земли в разных временных масштабах от десятилетий до миллионов лет.
f changement *m* climatique
s cambio *m* climático
g Klimaveränderung *f*

1032 climate feedback
The interaction between the processes within a climate system when one of the processes causes changes in the other one and in its turn effects the first process.
r климатические обратные связи
Взаимодействие между процессами в климатической системе, когда один из них вызывает изменения во втором, в свою очередь влияющие на первый.
f feed-back *m* climatique
s feed-back *m* climático
g Klimarückkopplung *f*

1033 climate model
The representation of the climate system by the mathematical equations which describe physical processes in the atmosphere and interactions between the meteorological elements.
r модель климата
Описание климатической системы математическими уравнениями, характеризующими физические процессы в атмосфере и взаимодействия метеорологических элементов.
f modèle *m* de climat
s modelo *m* climático
g Klimamodell *n*

1034 climate scenario; climatic forecast
A model of the future climate based on the probable external, including anthropogenic, impact upon it.
r климатический сценарий; климатический прогноз; сценарий изменения климата
Представления о будущем климате, основанные на предполагаемых внешних, прежде всего антропогенных, воздействиях на него.
f scénario *m* climatique; prévision *f* de climat
s escenário *m* climático; predicción *f* de clima
g Klimaszenario *n*; Klimaprognose *f*

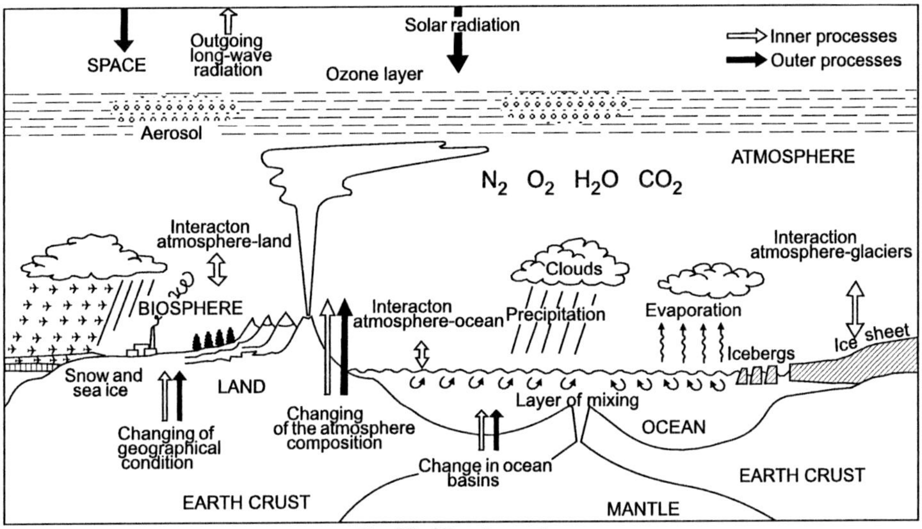

Fig. 55. **Climate system**

1035 climate sensitivity
A possible equilibrium change of the average *global temperature* near the Earth's surface under a unit change of the solar radiation.
r чувствительность климата
Равновесное изменение средней *глобальной температуры* у поверхности Земли при единичном изменении радиационного воздействия.
f sensibilité *f* de climat
s sensibilidad *f* de clima
g Klimaempfindlichkeit *f*

1036 climate system
A system consisting of the atmosphere, hydrosphere, cryosphere, surface lithosphere, and biosphere which under the impact of solar radiation, received by the Earth, forms the Earth's climate. → Fig. 55
r климатическая система
Система, состоящая из атмосферы, гидросферы, криосферы, поверхностной литосферы и биосферы, которая под воздействием солнечного излучения, падающего на Землю, определяет климат Земли. → Рис. 55
f système *m* climatique
s sistema *m* climático
g Klimasystem *n*

1037 climate system stability
The retention of climate character during a long period determined by the climate feedback.
r устойчивость климатической системы
Сохранение характера климата в течение достаточно длительного времени, обусловленное системой обратных связей между протекающими в ней процессами.
f stabilité *f* de système climatique
s estabilidad *f* de sistema climático
g Stabilität *f* des Klimasystems

1038 climatic ***adj***
Applied to natural processes and phenomena related to climate, e.g. climatic anomaly, climatic zone, etc.
r климатический
Относящийся к природным процессам и явлениям, связанным с климатом, например, климатическая аномалия, климатическая зона и др.
f climatique *adj*
s climático *adj*
g klimatisch *adj*

1039 climatic anomaly
The deviation the value of a climatic element from its normal value.
r климатическая аномалия
Отклонение значения климатического элемента от его нормального значения.
f anomalie *f* climatique
s anomalía *f* climática
g Klimaanomalie *f*

* **climatic change → 1031**

1040 climatic characteristics; climatic element
The numerical values of meteorological elements calculated on the basis of observations made over many years.
r климатические характеристики; климатический элемент
Числовые характеристики метеорологических элементов, вычисленные на основе многолетних рядов наблюдений.
f indices *mpl* climatiques; élément *m* climatique
s índices *mpl* climáticos; elemento *m* climático
g Klimakennziffern *fpl*; Klimaelement *n*

1041 climatic cycle
The changes of climatic characterisctics (temperature and humidity) which occur at relatively regular intervals.
r климатический цикл
Достаточно регулярные колебания климата (температуры или увлажнения) близкие по амплитуде и продолжительности.
f cycle *m* climatique
s ciclo *mpl* climático
g klimatischer Zyklus *m*; Klimazyklus *m*

* **climatic element → 1040**

1042 climatic factors
Combination of geographical conditions (latitude, elevation, topography, distribution of land and sea, ocean currents, etc.) which control the climate.
r климатические факторы
Комплекс географических условий (широта, высота, рельеф, распределение суши и моря, морские течения и пр.), определяющих характер климата.
f facteurs *mpl* climatiques
s factores *mpl* climáticos
g klimatische Faktoren *mpl*

* **climatic fluctuations → 2212**

* **climatic forecast → 1034**

1043 climatic health resort
A type of resort with climatic conditions favourable for health and having medicinal properties.
r климатический курорт
Курорт, основным достоинством которого служат благоприятные для лечения климатические условия.
f station *f* de cure climatique; station *f* climatique
s balneario *m* climático
g Luftkurort *m*

1044 climatic macrocycle
A large-scale (about 100 thousand years) global climatic change during which it passes through glacial and interglacial stages.
r климатический макроцикл
Долговременное (около 100 тыс. лет) глобальное колебание климата, в течение которого происходит смена ледниковых и межледниковых эпох.
f macrocycle *m* climatique
s ciclo *m* macroclimático
g klimatischer Makrozyklus *m*

1045 climatic minimum
The phase of glacial epoch with the maximum fall of the average global temperature.
r климатический минимум
Фаза ледниковой эпохи с максимальным понижением среднеглобальной температуры.
f minimum *m* climatique
s mínimum *m* climático
g Klimaminimum *n*

1046 climatic optimum
The period within an interglacial epoch characterised by the most favourable combination of temperatures and humidity.
r климатический оптимум
Отрезок времени внутри межледниковой эпохи, характеризующийся наиболее благоприятным сочетанием тепла и влаги.
f optimum *m* climatique
s optimum *m* climático
g Klimaoptimum *n*

1047 climatic region
A part of the climatic zone with a fairly uniform climate, e.g. Amazon climatic region with its hot and humid climate.
r климатическая область
Часть климатической зоны, обладающая относительно однородным климатом, например, Амазония с ее жарким и влажным климатом.
f région *f* climatique
s región *f* climática
g Klimagebiet *n*; Klimaregion *f*

1048 climatic resources
Renewable natural resources that include solar energy, moisture and wind energy.
r климатические ресурсы
Неисчерпаемые природные ресурсы, включающие солнечную энергию, влагу и энергию ветра.
f ressources *fpl* climatiques
s recursos *mpl* climáticos
g klimatische Ressourcen *fpl*; Klimaressourcen *fpl*

1049 climatic snow line

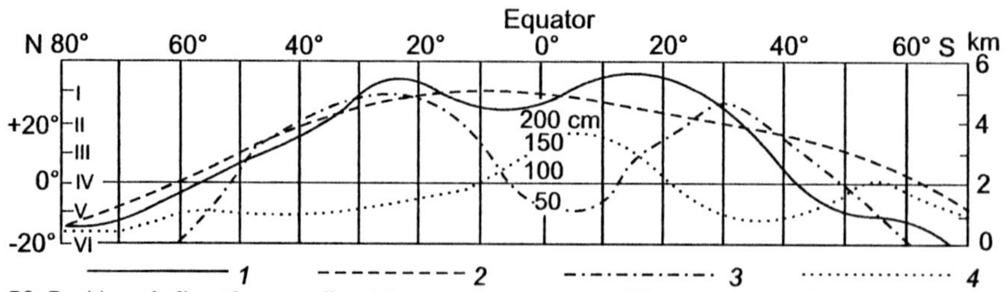

Fig. 56. Position of **climatic snow line** (1), annual temperature (2), annual sum of precipitation (3) and index aridity (4) versus geographical latitude

The altitude above which a flat surface (fully exposed to sun, wind, and precipitation) would experience a net accumulation of snow over an extended period of time. → Fig. 56, c. 124

r климатическая снеговая линия
Условный уровень в горах, выше которого часть твердых атмосферных осадков, выпавших на горизонтальную незатененную поверхность, не успевает стаять и испариться за теплый период года. → Рис. 56, с. 124

f limite *f* climatique des neiges permanentes
s límite *m* climático de nieves perpetuas
g klimatische Schneegrenze *f*

1050 climatic standard normal
Average values of climatological data calculated for the succesive 30-year periods: 1 January 1901 - 31 December 1930, 1 January 1931 - 31 December 1960, etc.

r климатическая стандартная норма
Средние климатологические данные, рассчитанные для последовательных 30-летних периодов: с 1 января 1901 г. по 31 декабря 1930 г., с 1 января 1931 г. по 31 декабря 1960 г. и т.д.

f normale *f* climatologique standard
s normal *f* climatológica estandart
g klimatische Standardnorm *f*

1051 climatic trend
A climatic change characterised by a smooth, monotonous increase or decrease of the average values of meteorological elements in the period of records.

r климатический тренд
Изменение климата, характеризуемое плавным монотонным увеличением или уменьшением средних значений метеорологических элементов за период наблюдений.

f tendance *f* climatique; tendance *f* du climat
s tendencia *f* climática
g Klimatrend *m*

1052 climatic type
A regional climate characterised by a certain series of means and annual variations of basic climatological characteristics in any classification of climate, e.g. the Mediterranean climate.

r тип климата
Региональный климат, характеризуемый определенным сочетанием средних и годовых колебаний основных климатических характеристик в той или иной классификации климатов, например, средиземноморский климат.

f type *m* de climat; type *m* climatique
s tipo *m* de clima
g Klimatyp *m*

1053 climatic zonation
The division of the Earth's surface into belts, zones, areas and smaller regions by climatic parameters.

r климатическое районирование
Разделение местности на пояса, зоны, области и более мелкие регионы по климатическим признакам.

f régionalisation *f* climatique
s división *f* en regiones climáticas
g klimatische Zonierung *f*

1054 climatic zone
A zone distinguished by the latitudinal distribution of the climatic elements, these are polar, temperate, subtropical, tropical and equatorial climatic zones. → Fig. 57

r климатическая зона
Зона, характеризуемая распределением климатических элементов вдоль широты

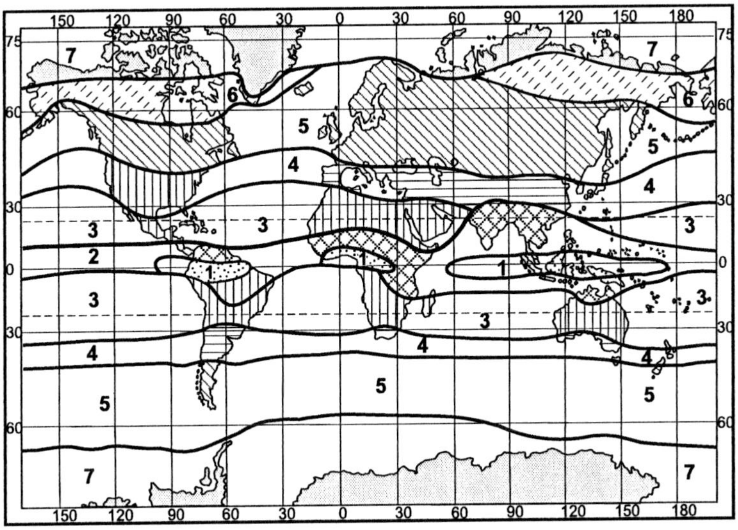

Fig. 57. **Climatic zones** (after B. Alisov): 1 – equatorial, 2 – equqtorial monsoon, 3 – tropical, 4 – subtropical, 5 – temperate, 6 – subarctic, 7 – Arctic (Antarctic)

(выделяют полярную, умеренную, субтропическую, тропическую и экваториальную климатические зоны). → Рис. 57, с. 126
f zone *f* climatique
s zona *f* climática
g Klimazone *f*

1055 **climatogram**
A diagram showing the combinations of the main climatic characteristic typical of a certain natural phenomenon, e.g. the distribution of a certain species or genus of organisms within a given area, or the degree of comfort of geographical conditions for people, etc.
r климатограмма
Диаграмма, демонстрирующая сочетания основных климатических параметров, характерных для данного природного явления, например, распространения определенного вида или рода организмов в пределах их ареала, либо комфортности географических условий для проживания людей и пр.
f climatogramme *m*
s climograma *m*
g Klimagramm *n*

1056 **climatography**
The quantative discription of climate by charts, tables, diagrams, texts.
r климатография
Совокупность сведений о климатах земного шара в форме карт, таблиц, графиков, текстов.
f climatographie *f*
s climatografía *f*
g Klimatographie *f*

1057 **climatological *adj***
Applied to observations, records and analysis of climatic characteristics.
r климатологический
Относящийся к наблюдениям и записям климатических характеристик и их анализу.
f climatologique *adj*
s climatológico *adj*
g klimatologisch *adj*

1058 **climatological front**
Mean or characteristic geographical position of fronts in a given region of the Earth, closely related to the mean position of the centres of atmospheric action.
r климатологический фронт
Среднее или характерное географическое положение фронтов в конкретном регионе, тесно связанное со средним положением центров действия атмосферы.
f front *m* climatologique
s frente *m* climatológico
g klimatologische Front *f*

1059 **climatological times**
Fixed hours of meteorological observations (in average local time), used to calculate the climatic characteristics.
r климатологические сроки
Фиксированные сроки метеорологических наблюдений (по среднему местному времени), используемые для вычисления климатических характеристик.
f heures *fpl* climatologiques
s horas *fpl* climatológicas
g klimatologische Zeiten *fpl*

1060 **climatology**
A science studying climates of the Earth: their types, factors of formation, geographical distribution and changes in time.
r климатология
Наука о климатах земного шара: их типах, факторах формирования, географическом распространении и изменениях во времени.
f climatologie *f*
s climatología *f*
g Klimatologie *f*; Klimakunde *f*

1061 **climax; climax community**
The final stage in the possible development of the natural vegetation of a locality or region, when the composition of the plant community is relatively stable and in equilibrium with the existing environmental conditions.
r климакс; климаксная растительность
Конечная стадия возможного развития естественной растительности в данном районе, когда состав растительного сообщества относительно устойчив и находится в равновесии с условиями окружающей среды.
f climax *m*; climax *m* végétal
s clímax *m*; clímax *m* vegetal; comunidad *f* vegetal climax
g Klimaxvegetation *f*; Klimaxgesellschaft *f*

* **climax avalanche → 912**

* **climax community → 1061**

1062 **cloak of sediments**
A bottom cluster of mobile alluvium of irregular shape usually near a dissected shore.
r покров наносов; плащ наносов$_2$
Донное накопление подвижных наносов неправильных очертаний, обычно у расчлененного берега.
f manteau *m* de sédiments
s capa *f* de sedimentos
g Sedimentdecke *f*

1063 **closed canopy; continuous canopy**
A continuous leafy layer in a forest which does not give for sunny light to reach the earth surface.
r сомкнутый полог леса; сомкнутая крона
Сплошной лиственный слой в лесу, почти не пропускающий солнечные лучи к земной поверхности.
f couronne *f* de bois dense; couvert *m* des couronnes fermées

s frondas *fpl* unidas; cubierto *m* de arboles cerrado; copas *fpl* cerradas
g geschlossenes Kronendach *n*

1064 closed drainage; area of internal drainage
The area of inland runoff which is not connected with the ocean through river systems.
r бессточная область; область внутреннего стока
Область внутриматерикового стока, лишенная связи через речные системы с Мировым океаном.
f région *f* aréique; drainage *m* endoréique
s drenaje *m* endorreico
g abflussloses Gebiet *n*; endorheisches Gebiet *n*

1065 closed drainage basin; blind drainage basin; aretic depression
A river or lake basin located within the limits of an internal drainage in the areas of arid climate.
r бессточный бассейн; бессточная впадина; замкнутый бассейн
Бассейн реки или озера, расположенный в пределах бессточной области, лишенный связи с океаном через речные системы, особенно характерен для областей аридного климата.
f bassin *m* sans émissaire superficiel; bassin *m* aréique; depression *f* de décantation
s cuenca *f* cerrada; depresión *f* de decantación; depresión *f* arrética
g abflußloses Becken *n*

1066 closed economy
The economy of a country that relies solely on its own resources and restricts foreign trade and relations.
r закрытая экономика
Экономика страны, ориентированная исключительно на собственные силы и ограничивающая внешние связи и элементы внешней торговли.
f économie *f* fermée; autarcie *f* economique
s economía *f* cerrada; autarquía *f* económica
g autochtone Wirtschaft *f*

* **closed forest → 1511**

1067 closed plant community
A plant community in which no other plant species can be admitted and constant species composition is retaining.
r закрытое растительное сообщество; замкнутый фитоценоз
Растительное сообщество, недоступное для внедрения в него иных растений и тем самым сохраняющее постоянный видовой состав.

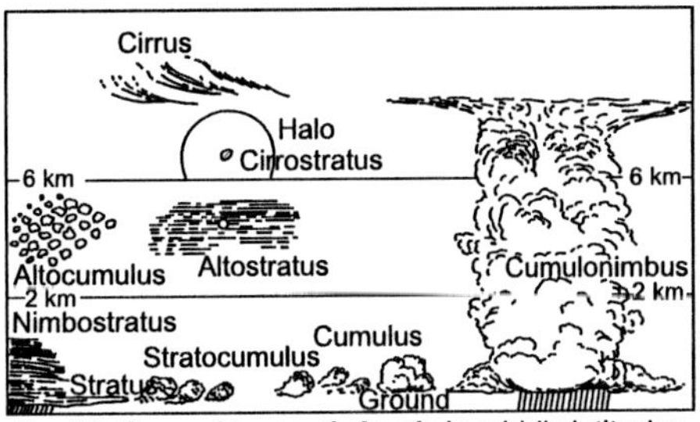

Fig. 58. General types of **clouds** in middle latitudes

f communauté *f* végétale fermé
s comunidad *f* vegetal cerrada
g geschlossene Pflanzengesellschaft *f*

1068 closeness of ice
The ratio of the area covered by drifting ice to the total area of a water body.
r сплоченность льда
Отношение площади, занятой дрейфующим льдом, к общей площади данного участка водного объекта.
f consolidation *f* de glace
s consolidación *f* de hielo
g Eisgeschlossenheit *f*

* **closing error → 3879**

1069 cloud
A visible mass of minute particles of liquid water or ice, or both, suspended in the free atmosphere and usually not touching the Earth's surface. → Fig. 58
r облако
Видимая совокупность мельчайших частиц воды и/или льда, возникающих в результате конденсации и сублимации водяного пара, находится во взвешенном состоянии в свободной атмосфере и обычно не опускается к Земле. → Рис. 58
f nuage *m*
s nube *f*
g Wolke *f*

* **cloud amount → 1079**

1070 cloud base
The lowest level of a cloud or cloud layer.
r основание облака; нижняя граница облака
Нижняя кромка облака или облачного слоя.
f base *f* d'un nuage
s base *f* de nube
g Wolkenuntergrenze *f*

* **cloud cover → 1079**

1071 cloud étage; cloud level
A tropospheric layer with clouds of certain genera.
r облачный ярус
Слой тропосферы с облаками определенных форм.

f étage *m* des nuages
s piso *m* de las nubes
g Wolkenhöhenlage *f*; Wolkenetage *f*

1072 cloud forest; mist forest
The usually broad-leaved evergreen forest which grows in the mountains between 1000 and 2500 m where clouds form (e.g. in the Andes) and give sufficient moisture to support a substantial vegetation formed by short and crooked trees with abundant lianas, mosses and well developed grasses.
r туманный лес
Тропический вечнозеленый лес, расположенный на склонах гор в полосе конденсации туманов (например в Андах), образован невысокими деревьями с обильными лианами, мхами и хорошо развитым травостоем.
f forêt *f* des nuages
s bosque *m* nublado
g Nebelwald *m*; Wolkenwald *m*

* **cloud form → 1077**

1073 cloud formation
The process by which various types of clouds are formed which involves adiabatic cooling of ascending humid air resulting in the water vapour condensation.
r облакообразование
Процесс образования облаков разных типов, как правило, в результате адиабатического охлаждения поднимающегося влажного воздуха, что приводит к конденсации водяного пара.
f formation *f* des nuages
s formación *f* de nubes
g Wolkenbildung *f*; Wolkenformation *f*

* **cloud level → 1071**

1074 cloud seeding; seeding of clouds
A weather modification technique in which an active substance or agent are introduced into clouds in order to change the phase state of clouds and stimulate precipitation.
r засеивание облаков; засев облаков
Метод активных воздействий на погоду, при котором в облака вносятся активные вещества или реагенты с целью изменения фазового состояния облаков и вызывания осадков.
f ensemencement *m* des nuages
s siembra *f* de las nubes
g Wolkenimpfung *f*

1075 cloud system
A well distinct and lasting mass of clouds united by the same process of cloud formation.
r облачная система
Легко различимая в течение длительного времени совокупность облаков, объединенных общим процессом облакообразования.
f système *m* nuageux
s sistema *m* nuboso
g Wolkensystem *n*

1076 cloud system of a cyclone; cyclonic cloud system
Any cloud system which is typical for a depression and presented by different combinations of the frontal clouds
r облачная система циклона; циклоническая облачная система
Облачная система, типичная для области низкого давления и представленная различными сочетаниями фронтальных облаков.
f système *m* nuageux dépressionnaire
s sistema *m* nuboso de una depresión
g zyklonisches Wolkensystem *n*

1077 cloud type; cloud form
Three main forms of clouds (cirrus, cumulus and stratus) generated by the vertical motion of air associated with convection and frontal processes.
r тип облаков
Три основные группы облаков (перистые, кучевые и слоистые), вызванных вертикальными движениями воздуха, обусловленными конвекцией и фронтальными процессами.
f type *m* des nuages
s tipo *m* de nubes
g Wolkentyp *m*

1078 cloudburst
A violent torrential shower of rain or hail caused by violent upward and downward air streams, generally short-lived but capable of cosiderable damage because of the large volume of water precipitated.
r шквал осадков
Сильнейший проливной дождь или град, вызванный резким подъемом или опусканием воздушных потоков, обычно скоротечный, но способный вызвать существенные разрушения из-за большого количества выпавшей влаги.
f rafale *f* nuageuse
s ráfaga *f* de precipitación
g Wolkenbruch *m*

1079 cloudiness; cloud amount; cloud cover
The coverage of the sky covered by clouds of a certain genesis, type, cloud йtage, or of a certain combination of clouds which is measured in tenths according to the international conventions or in eights (oktas) in the UK. → Fig. 59, p. 129
r облачность; количество облаков
Степень покрытия небосвода облаками определенного происхождения, яруса, формы или сочетания разных облаков, согласно международным конвенциям, выражается в десятых или (в Великобритании) в восьмых долях (октах). → Рис. 59, с. 129

Fig. 59. **Cloudiness** above the globe. View from the orbital station

f nébulosité *f*; couverture *f* nuageuse
s nubosidad *f*
g Bewölkung *f*; Himmelsbedeckung *f*; Wolkendecke *f*

1080 cloudy ice
An ice body containing numerous air inclusions, and having a milk-white colour.
r мутный лед
Лед, богатый воздушнымии включениями, в массиве приобретающий молочнобелый оттенок.
f glace *f* sale
s hielo *m* turbio
g weisses Eis *n*; trübes Eis *n*

1081 cluster
A group of natural similar things growing or gathered together (e.g. trees) which can be shown on a map by means of symbols or dots.
r кластер
Группа однородных природных объектов, растущих или собранных вместе (например, группа деревьев), которая может быть показана на карте с помощью символов или точек.
f claster *m*
s claster *m*
g Haufen *m*

* **coadaptation → 1103**

1082 coagulation
1. A process of clumping together of tiny particles dissolved or suspended in water by the force of molecular attraction. 2. A growth of cloud droplets due to their collision and coalescence.
r коагуляция
1. Процесс взаимного слипания мельчайших растворенных частиц в воде под действием сил молекулярного притяжения. 2. Укрупнение облачных капель вследствие их столкновения и слияния.
f coagulation *f*
s coagulación *f*
g Koagulation *f*; Koagulieren *n*; Flockung *f*

1083 coal
A carbonaceous fossil fuel, a brownish-black or black combustible mineral substance found in sedimentary rocks and derived from vegetable material growing in the Carboniferous period.
r уголь
Горючая ископаемая порода бурого или черного цвета, залегающая в осадочных отложениях карбона, представляющая собой остатки растительности, произраставшей в каменноугольный период.
f charbon *m*
s carbón *m*
g Kohle *f*

1084 coalfield
A tract of land underlain by coal, in many cases the workable coal is covered by younger deposits (concealed coalfield).
r угольное месторождение
Участок земли с выходами угольных пластов, во многих случаях прикрытых более молодыми отложениями.
f champ *m* de charbon
s campo *m* de carbón
g Kohlefeld *n*

1085 coalition
A political union, joint activity of two or more states directed against any country or group of countries.
r коалиция
Политический союз, совместная деятельность двух или нескольких государств, направленная против какого-либо государства или группы государств.
f coalition *f*
s coalición *f*
g Koalition *f*

1086 coarse-grained
Consisting of granules and particles of some millimetres in size.
r крупнозернистый
Состоящий из гранул и зерен размером в несколько миллиметров.
f à gros grain
s de grano grueso
g grobkörnig *adj*

1087 coast; shore
A zone of interaction between land and a water body or watercourse. → Fig. 60, p. 130.
r берег; побережье
Полоса взаимодействия между сушей, с одной стороны, водоемом и водотоком, с другой. → Рис. 60, с. 130.
f côte *f*; rivage *m*; littoral *m*
s costa *f*; litoral *m*; ribera *f*
g Küste *f*; Ufer *n*

1088 coast of Calatype
The indented abrasion coast with numerous small bays separated by pointed headlands.
r берег балеарского типа
Изрезанный абразионный берег с много-

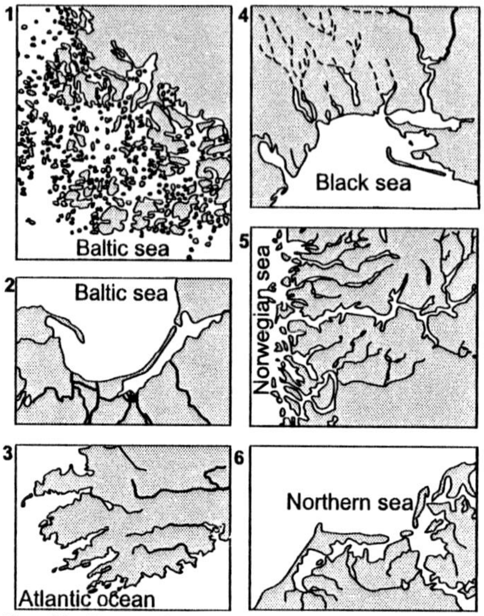

Fig. 60. Types of **coasts**: 1 – skerries, 2 – bays with spits, 3 – rias, 4 – brackish lagoon, 5 – fjords, 6 – bodden coasts

численными небольшими полукруглыми бухтами, разделенными острыми мысами.
f côte *f* du type Cala
s costa *f* de tipo Baleares
g balearische Küste *f*

* **coast of emergence → 1880**

1089 coast of thermal abrasion
A coast created in permafrost by thawing due to the thermal activity of water on ice.
r термоабразионный берег
Берег, образованный многолетнемерзлыми горными породами, протаявшими в результате термического воздействия воды на лед.
f côte *f* d'abrasion thermique
s costa *f* de abrasión térmica
g Thermoabrasionsküste *f*

1090 coast protection structures
Engineering constructions (walls, dams, etc.) used to protect a shore against the action of sea waves. → Fig. 61
r берегоукрепительные сооружения; береговая защита
Инженерные сооружения (стены, дамбы и пр.), используемые для защиты берега от воздействия волнения. → Рис. 61
f constructions *fpl* de consolidation de côte
s obras *fpl* para consolidación de costa
g Uferschutzeinrichtungen *fpl*; Küstenschutz *m*

1091 coastal *adj*
Applied to the zone of indeterminate width where land and sea or other extensive tract of water meet.
r береговой; прибрежный
Относящийся к области неопределенной ширины, где соприкасаются суша и значительные участки воды (море, озера, река).
f littoral *adj*; côtier *adj*
s litoral *adj*; costero *adj*
g Küsten-

* **coastal area → 1097**

1092 coastal desert; cold-water desert
A desert along the western coast of a continent in the zone of trade-winds where the climate is influenced by cold sea currents flowing towards the equator, common of Africa and South America, e.g. Namib and Atacama deserts.
r береговая пустыня
Пустыня у западного берега материка в зоне действия пассатов, где на климат также воздействует холодное морское течение, текущее в сторону экватора, наиболее распространены в Африке и Южной Америке, например, пустыни Намиб и Атакама.

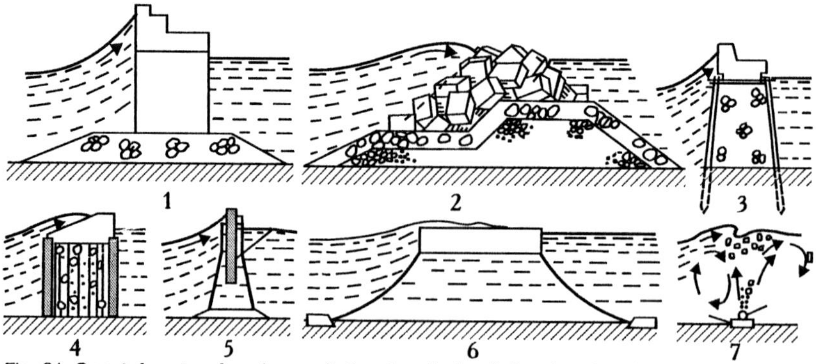

Fig. 61. **Coastal protecrion**: 1 – gravitational vertical wall, 2 – slope break water, 3 – vertical piles, 4 – vertical wall with cylindrical jacket, 5 – through breakwater, 6 – floating breakwater, 7 – pneumatic breakwater

f désert *m* littoral; désert *m* côtier
s desierto *m* litoral; desierto *m* costero
g Küstenwüste *f*

* **coastal dune → 2256**

1093 coastal effect
The change of a wind velocity when it blows across a shoreline due to a different character of underlying surface.
r береговой эффект
Изменение скорости ветра при переходе воздушного потока через береговую зону вследствие изменения характера подстилающей поверхности.
f effet *m* côtier
s efecto *m* costero
g Küsteneffekt *m*

1094 coastal lake
A shallow area of sea water near the coast completely or partly separated from the open sea by accumulations of sediments.
r прибрежное озеро
Краевые части морей, отчлененные наносными образованиями, бывают двух типов: лагуны и лиманы.
f lac *m* littoral
s lago *m* litoral
g Strandsee *m*

* **coastal lowland → 1095**

1095 coastal plain; coastal lowland
A plain which borders the landward side of a coastal zone.
r береговая равнина
Равнина, окаймляющая со стороны суши береговую зону.
f plaine *f* littorale; plaine *f* de niveau de base
s llanura *f* litoral; llanura *f* de nivel de base
g Küstenebene *f*

1096 coastal processes
Geomorphological processes occurring in coastal zone which result from the combined action of wind and currents, include abrasion, transport, attrition, sorting and accumulation of material.
r береговые процессы
Процессы, происходящие в береговой зоне под действием волн и течений и включающие абразию, перемещение, истирание, сортировку и аккумуляцию наносов.
f processus *mpl* littoraux
s procesos *mpl* litorales
g Küstenvorgänge *mpl*

1097 coastal zone; coastal area
A near shore part of the shelf experiencing the impact of sea waves.
r береговая зона; прибрежная полоса
Прибрежная часть шельфа, постоянно испытывающая воздействие морского волнения.
f zone *f* littorale
s zona *f* litoral
g Strandzone *f*; Küstenstreifen *m*; Küstenvorfeld *n*

* **coastline → 5411**

* **coastwise navigation → 833**

1098 coating; revestment
Deposits of mineral matter on the surface of soil structural jointings.
r кутана
Отложение минерального вещества на поверхности структурных отдельностей почвы.
f revêtement *m*
s revestido *m*
g Verblendung *f*

* **cobble → 768**

1099 cockpit
A deeply incised hollow in tropical karst region, which have been developed from alluvial-floored dolines due to the fluctuating water-table in the limestone.
r кокпит
Глубоко врезанная крутосклонная впадина в районах тропического карста, возникающая на аллювиальном дне дóлины вследствие колебаний водного зеркала в известняках.
f cockpit *m*
s cockpit *m*
g Kockpit *m*

1100 cockpit karst
A type of tropical karst characterized by numerous deep (up to 100-150 m), steep-sided hollows which develop in cracked limestones where water table is lying quite low.
r котловинный карст
Разновидность тропического карста, отличающаяся обилием близко расположенных друг к другу глубоких (до 100-150 м) крутосклонных котловин, развивающихся в сильно трещиноватых известняках при глубоком стоянии подземных вод.
f karst *m* à dolines
s karst *m* de dolinas
g Kesselkarst *m*

1101 codominant species
Two or more plant and animal species simultaneously dominating in a particular community.
r кодоминантный вид; кодоминант
Два и более видов растений и животных, одновременно доминирующих в данном сообществе.
f espèce *f* codominante
s especie *f* codominante
g kodominante Art *f*

1102 coefficient of slope stability
The capacity of a slope to resist displacement (the ratio of the shear strength of the rock to the applied shear stresses along the most dangerous landslide surface).

r коэффициент устойчивости склона
Способность склона противостоять сдвигу (отношение сил сопротивления сдвигу к сумме сдвигающих сил по наиболее опасной поверхности оползневого скольжения).
f coefficient *m* de stabilité de versant
s coeficiente *m* de estabilidad de vertiente
g Koeffizient *m* der Hangstabilität

1103 coevolution; coadaptation
Mutual adaptation of species to each other in the course of evolution, for example insects to pollination of plants and plants to pollination by insects.
r коэволюция; коадаптация
Взаимное приспособление видов друг к другу в ходе эволюции, например, насекомых к опылению растений и растений к опылению насекомыми.
f coévolution *f*; coadaptation *f*
s coevolución *f*; coadaptación *f*
g Koevolution *f*; Koadaptation *f*

1104 coevolution of society and environment
Mutual and interconnected evolution of human society and nature when the impact of people upon nature is deliberately limited.
r коэволюция общества и природы
Совместная и взаимосвязанная эволюция общества и природы, когда воздействие человека на природу сознательно ограничивается.
f coévolution *f* de société et environnement
s coevolución *f* de sociedad y medio ambiente
g Koevolution *f* der Gesellschaft und Umgebung

1105 cogwheel railway; mountain railway
A narrow-gauge railway having the additional rail and toothwheel which enable the train to run up rather deep slopes, (up to 25°) it is very widely developed in Switzerland and some other mountainous countries. → Fig. 62
r зубчатая железная дорога
Узкоколейная железная дорога с третьим зубчатым рельсом, позволяющим поезду двигаться на достаточно крутых склонах (до 25°), широко распространены в Швейцарии и некоторых других горных странах. → Рис. 62
f train *m* à crémaillère
s tren *m* de montaña
g Zahnradbahn *m*; Bergbahn *m*

Fig. 62. **Cogwheel railway** in Switzerland

1106 coherent *adj*
Related to the rock or ground which gets consolidated in the process of transformation from loose sediments.
r сцементированный
Относящийся к горной породе или почве, уплотнившейся в процессе преобразования первоначально рыхлых осадков.
f cohérent *adj*
s coherente *adj*
g kohärent *adj*

*** coherent rock → 1221**

1107 cohesion of snow
Sticking of snow as a result of the appearance of crystalline bond caused by the molecular attraction between the snowflake surface and thin film of melt water.
r сцепление снега
Слипание снега в результате возникновения межкристаллических связей, обусловленных молекулярными взаимодействиями поверхности снежинок с тонкими пленками талой воды.
f cohésion *f* de la neige
s cohesión *f* de nieve
g Schneekohäsion *f*; Schneehaftung *f*

1108 colatitude
The complement of the latitude, i.e. the difference between 90° and the latitude.
r дополнение широты
Дополнение географической широты, т.е. разница между 90° и широтой данной точки.
f colatitude *f*
s colatitud *f*
g Kobreite *f*

1109 cold air mass
An air mass which is colder than the surrounding air masses, or an air mass advancing on a warmer underlying surface, as opposed to warm air mass.
r холодная воздушная масса
Масса воздуха с более низкой температурой, чем соседние воздушные массы, или масса воздуха, надвигающаяся на более теплую подстилающую поверхность, в противоположность теплой воздушной массе.
f masse *m* d'air froid
s masa *f* del aire frio
g kalte Luftmasse *f*; Kaltluftmasse *f*

1110 cold current; cool current
A sea or ocean stream carring water which is colder than the surrounding water masses.

r холодное морское течение
Течение, в котором средняя температура воды ниже средней температуры окружающих вод.
f courant *m* froid
s corriente *f* fría
g kalte Meeresströmung *f*

1111 cold desert
An arctic or high mountain region where poor vegetation cover is caused by low temperatures and physiolodical drought rather than the lack of moisture, as opposed to *arid* (hot) *desert.*
r холодная пустыня
Арктические или высокогорные области, в которых скудость растительного покрова определяется в первую очередь низкими температурами, а не отсутствием влаги, противопоставляется *аридной пустыне.*
f désert *m* froid
s desierto *m* frío; desierto *m* helado; desierto *m* polar y subpolar

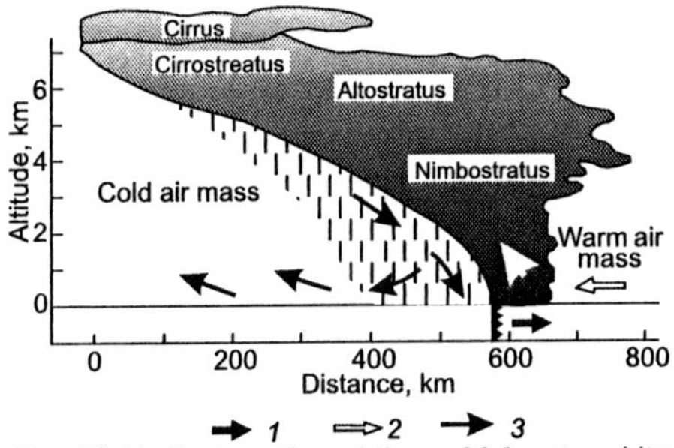

Fig. 63. Vertical section of the **cold front** and its cloud system. 1 – direction of the front movement, 2, 3 – circulation of warm and cold air

g Kältewüste *f*

1112 cold front
The front boundary of a mass of advancing cold air where it undercuts and replaces the warmer air. → Fig. 63
r холодный фронт
Фронтальная граница холодной воздушной массы, вытесняющей теплый воздух. → Рис. 63
f front *m* froid
s frente *m* frío
g Kaltfront *f*

1113 cold glacier; polar glacier
A glacier in which cold temperatures dominate below the active layer and no melting occurs.
r холодный ледник; полярный ледник
Ледник, во всей толще которого ниже деятельного слоя господствует отрицательная температура и таяния не бывает.
f glacier *m* froid; glacier *m* polaire
s glaciar *m* frío; glaciar *m* polar
g kalter Gletscher *m*; Polargletscher *m*

* **cold lake → 4649**

* **cold low → 4650**

1114 cold occlusion
An occlusion in which the advancing mass of cold air has lower temperatures than the air mass in front.
r холодная окклюзия
Окклюзия циклона, в которой натекающая масса холодного воздуха имеет более низкую температуру, чем воздушная масса на атмосферном фронте.
f occlusion *f* froide
s oclusión *f* fría
g kalte Okklusion *f*

1115 cold pole
The point on the Earth's surface with the lowest mean annual temperature (it is located in Antarctica, near Russian station Vostok, -89.2°C).
r полюс холода
Область в данном полушарии или на всем земном шаре, где наблюдаются самые низкие среднегодовые значения температуры воздуха у земной поверхности (полюс холода Земли находится в Антарктиде в районе российской станции Восток, -89,2°C).
f pôle *m* de froid
s polo *m* de frío
g Kältepol *m*

1116 cold wave
The fall of temperature caused by the sudden burst of cold air which flows in.
r волна холода
Резкое понижение температуры воздуха, или вторжение (адвекция) холодной воздушной массы на обширную территорию.
f vague *f* de froid
s ola *f* de frío
g Kältewelle *f*

* **cold-blooded animal → 4634**

* **cold-water desert → 1092**

1117 coli-index
A degree of the bacterial pollution of water expressed as the number of coliform bacilli in 1 dm^3 of water.
r коли-индекс
Показатель бактериального загрязнения воды, соответствующий числу кишечных палочек в 1 дм^3 воды.
f coli-indice *m*
s coli-índice *m*
g Coli-Index *m*

1118 coli-titer
A degree of bacterial pollution of water expressed as the volume of water per one coliform bacillus.

r коли-титр
Показатель бактериального загрязнения воды, соответствующий объему исследуемой воды, приходящемуся на одну кишечную палочку.
f coli-titre *m*
s coli-título *m*
g Colititer *m*

1119 collapse earthquake
An earthquake caused by the roof collapse of a subterranean cave in soluble rocks.
r обвальное землетрясение
Землетрясение, вызванное провалом свода полости в растворимых горных породах.
f séisme *m* d'éboulement
s seísmo *m* de derrumbe
g Einsturzbeben *n*

1120 collapse polje
A small hollow (polje) which is formed by the collapse of the roof of an underground river tunnel.
r провальное полье
Полье небольших размеров, образовавшееся путем провала свода туннеля подземной реки.
f poljé *m* d'éffondrement
s polje *m* de desmoronamiento
g Einsturzpolje *f*

1121 collapse structures
Geological or geomorphological structures formed by gravity collapse, such as the break-up of the uppermost bed of sedimentary rocks, the roof collapse of a subterranean cave (polje), etc.
r обвальные структуры
Геологические или геоморфологические структуры, образованные посредством гравитационных обвалов: обрушения верхнего слоя осадочных пород, провала крыши карстовой пещеры (полье) и пр.
f structures *fpl* d'éboulement
s estructuras *fpl* de derrumbe
g Einsturzstruktur *f*

1122 collective consumption
Consumption of the services which can be consumed only collectively and are thus provided by the state, e.g. public transport, defence services.
r коллективное потребление
Потребление услуг, которые предоставляются государством для всего общества, например, общественный транспорт, органы защиты населения.
f consommation *f* collective
s consumo *m* colectivo
g Kollektivverbrauch *m*

1123 collective farm; kolkhoz
A type of socialist farm organization originating in the USSR when peasants work collectively on state-owned land.
r колхоз
Объединение крестьян для совместного труда в сельскохозяйственном производстве, одна из форм организации труда в СССР.
f kolkhoze *m*
s koljós *m*; koljoz *m*
g Kolchos *m*

1124 collective farming
Agricultural organization in which the farms are subjected to collectivization, they are acquired and amalgamated by a village (as in the kibbutz of Israel) or by the state (as in the kolkhoz of the former USSR).
r коллективное хозяйство
Сельскохозяйственная организация, при которой отдельные хозяйства коллективизированы, т.е. объединены деревней (киббуцы в Израэле) или государством (колхозы в бывшем СССР).
f exploitation *f* collective; économie *f* collective
s hacienda *f* colectiva; economía *f* colectiva
g Kollektivwirtschaft *f*

1125 collectivization
The transformation of private peasant households into large collective farms through cooperation.
r коллективизация
Преобразование частных крестьянских хозяйств в крупные общественные хозяйства путем кооперирования.
f collectivisation *f*
s colectivización *f*
g Kollektivierung *f*

1126 collision zone
In plate tectonics, a zone where converging plates carrying continental crust, meet with the result that the edge of one plate dives under the other but the rocks pile up, crushed, buckled and mixed with material swept up from the oceanic floor.
r зона столкновения
В тектонике плит зона, где встречаются сходящиеся плиты континентальной земной коры, в результате чего край одной плиты погружается под другую, а горные породы нагромождаются, дробятся, сгибаются и смешиваются с материалом, поднимающимся с океанического дна.
f zone *f* de collision
s zona *f* de colisión
g Kollisionszone *f*

1127 colluvium; slope-wash
The accumulation of rock debris of varied origin at the foot of a slope as a result of the slope-wash and mass movement processes.
r коллювий
Скопления всех смещаемых вниз по склону продуктов разрушения горных пород в шлейфах, прислоненных к нижним частям склонов.

f colluvion *m*; dépôt *m* colluvial
s coluvión *m*; accumulación *f* coluvial
g Kolluvium *m*; kolluviale Ablagerungen *fpl*

1128 colmatage
Sedimentation of clay or silt particles on the surface and in the pores of permeable rock resulting in the decrease of its water conveying capacity; artificial filling of pores with clay particles in order to decrease the filtration of water.
r кольматаж; кольматация
Отложение глинистых или илистых частиц на поверхности и в порах водопроницаемой породы, в результате чего снижается ее водопропускная способность; искусственное заполнение пор грунта глинистыми частицами для снижения фильтрации воды.
f colmatage *m*
s colmación *f*
g Kolmation *f*

1129 colonial animal
An animal which is a member of an association of incompletely separated individuals, e.g. coral.
r колониальное животное
Животное, являющееся членом ассоциации связанных между собой индивидуумов, например, коралл.
f animal *m* colonial
s animal *m* colonial
g Kolonietier *n*

1130 colonial empire
A group of countries subordinated economically and politically to the parent state and usually less developed than it in social and economic respect.
r колониальная империя
Группа стран, как правило, менее развитых в социально-экономическом отношении, политически и экономически подчиненных метрополии.
f empire *m* colonial
s imperio *m* colonial
g Kolonialoreich *n*

* **colonial system → 1131**

1131 colonialism; colonial system
A system of interstate relations and a corresponding international division of labor under which certain countries are ruled by other countries and are deprived of political and economic independence.
r колониализм; колониальная система
Система межгосударственных отношений и соответствующего международного разделения труда, при которой отдельные страны находятся под властью другого государства (метрополии) и лишены политической и экономической самостоятельности.
f colonialisme *m*; système *m* colonial
s colonialismo *m*; sistema *m* colonial
g Kolonialismus *m*; Kolonialsystem *n*

1132 colonization
1. The establishing of settlements and economic development of marginal areas of one's own country (internal colonization) and organization of settlements in another territory (external colonization). 2. The occupation by an organism of a new habitat, thereby extending its ecologicalor geographical range.
r колонизация
1. Процесс заселения и хозяйственного освоения пустующих окраинных земель своей страны (внутренняя колонизация) и основание поселений за пределами своей страны (внешняя колонизация). 2. Освоение организмом нового местообитания, т.е. расширение своего ареала.
f colonisation *f*
s colonisación *f*
g Kolonisation *f*; Kolonisierung *f*

1133 colony
1. A country or a territory ruled by the parent state and deprived of political and economic independence. 2. A human settlement formed in a territory by people from other territory, usually from another country, e.g. German settlements in Russia. 3. A group of closely associated organisms, such as corall, social insects.
r колония
1. Страна или территория, находящаяся под властью иностранного государства (метрополии) и лишенная политической и экономической самостоятельности. 2. Поселок, состоящий из выходцов из другой страны, например, немецкие колонии в России. 3. Группа животных, растений или микроорганизмов, живущих вместе на определенной территории (колонии кораллов, птичьи базары и пр.).
f colonie *f*
s colonia *f*
g Kolonie *f*

1134 colour model
A proof defining the tints to be reproduced in printing and delimiting in the map the areas they are to cover.
r красочный макет
Пробный оттиск, показывающий цвета, площади и границы фоновых элементов карты, которые должны печататься разными красками.
f modèle *m* des teintes
s modelo *m* de colores; modelo *m* de tintas
g Farb(platten)vorlage *f*

1135 colour of water
A visible colour of the upper water layer in a water body, an important index of water quality.
r цвет воды
Видимая окраска верхнего слоя воды в водоеме, важный показатель качества воды.

f color *m* de l'eau
s color *m* de agua
g Wasserfarbe *f*

1136 colour proof
A proof combining all constituent colours, which is made for comparison with the original to check colour rendering.
r красочная проба
Пробный совмещенный оттиск со всех печатных форм, дающий полное представление об окончательном виде карты и служащий эталоном при печати тиража.
f épreuve *f* de couleur
s prueba *f* en color
g Farbenandruck *m*

1137 coloured snow
A snow which is coloured by some mineral or organic admixtures.
r цветной снег
Снег, окрашенный какими-либо минеральными или органическими примесями.
f neige *f* colorée
s nieve *f* coloreada
g gefärbter Schnee *m*

1138 columnar jointing
Parallel prismatic columns in basaltic lavas and other effusive and intrusive rocks.
r столбчатая отдельность
Параллельные призматические столбы в базальтовых лавах и в иных эффузивных и интрузивных породах.
f prisme *m* disjoncté; colonne *f*
s juntura *f* columnar; columna *f* disyunctada
g Säulenabsonderung *f*; säulige Ablagerung *f*

1139 columnar structure
A type of jointing of rocks comprising hexagonal columns formed at the cooling of igneous rocks, especially of basalt; a similar structure of smaller scale is formed in snow cover under the process of recrystallization.
r столбчатая текстура
Геологическая структура, представляющая собой гексагональные призмы, образующиеся при остывании изверженных пород, особенно базальта; подобная же структура небольших размеров в снежном покрове, формирующаяся в процессе его перекристаллизации.
f structure *f* de colonne; disjonction *f* en prismes
s estructura *f* columnar; disyunción *f* columnar
g Pfostenstruktur *f*

1140 comet
A celestial body consisting of a gaseous cloud enveloping a bright nucleus, moving around the Sun in a parabolic or elliptical orbit so that some comets are nearing the Earth and some escape from the solar system.
r комета
Небесное тело, состоящее из газового облака вокруг яркого ядра, движущееся вокруг Солнца по параболической или эллиптической орбите так, что некоторые кометы иногда приближаются к Земле.
f comète *m*
s cometa *f*
g Komet *m*

1141 comfort of environment
Optimal conditions of the environment which secure good human health and capacity for work.
r комфортность среды
Оптимальное для человека состояние окружающей среды, обеспечивающее здоровье и работоспособность.
f confort *m* de l'environnement
s conforte *m* de medio ambiente
g Umweltkomfort *m*

1142 comfort zone
A range of temperature, humidity and wind under which the majority of people feel the most comfortable.
r зона комфорта
Интервал комбинаций температуры, влажности и ветра, при которых большинство людей чувствуют себя наиболее комфортно.
f zone *f* de confort
s zona *f* de conforte; intervalo *m* de conforte
g Komfortzone *f*

1143 command economy; centrally planned economy
An economic system in which all the buisness activity and industry are ruled by the government.
r плановая экономика
Социально-экономическая система, в которой государство планирует всю деловую активность.
f économie *f* planifiée
s economía *f* planificada; economía *f* dirigida
g staatliche Planwirtschaft *f*

1144 commensalism; mutualism
A relationship between two types of organism living in the same environment without harm to either species and in which one or both may obtain sustenance, protection or other benefits.
r комменсализм
Связь организмов двух видов, живущих в аналогичных условиях окружающей среды без ущерба для обоих видов, когда оба они получают пищу, защиту или другую выгоду от совместного обитания.
f commensalisme *m*
s comensalismo *m*
g Kommensalismus *m*

1145 commensurability of symbols

A conventional correspondence between the size of cartographic symbols and numerical parameters of represented features.

r соизмеримость знаков
Условное соответствие размеров условных знаков количественным показателям отображаемых объектов.

f échelle *f* des symboles
s escala *f* cuantitativa de signos; conmensurabilidad *f* de signos
g Kartenzeichenmaßstab *m*

* **commerce → 6280**

1146 **commercial economy; commodity economy; goods economy**
A type of economy when the production is oriented on the market excange while the relations between producers and consumers are realized through trade.

r товарное хозяйство
Тип хозяйства, в котором производство ориентируется на рынок, а связь производителей и потребителей продуктов осуществляется через куплю-продажу товаров.

f économie *f* marchande
s economía *f* mercantil
g Warenwirtschaft *f*

1147 **commercial game**
The wild animals which are allowed to be hunted or which are bred for commercial purposes.

r промысловая фауна
Дикие животные, на которых разрешено охотиться или которых разводят в коммерческих целях.

f gibier *m* commercial
s volatería *f* comercial
g Handelswild *n*

1148 **commodity; goods and services**
A manufactured product or a material that is bought and sold.

r товар
Продукт труда, произведенный для обмена или продажи, которые определяются общественным и географическим разделением труда.

f marchandise *f*; commodité *f*
s mercancía *f*
g Ware *f*

* **commodity economy → 1146**

1149 **commodity flow**
The volume of goods transported along the lines of transport network and between cooperating enterprises and regions.

r грузопоток
Объем перевозок грузов по участкам и направлениям транспортной сети, а также между взаимодействующими предприятими и районами.

f flux *m* des marchandises
s flujo *m* de mercancías
g Güterfluß *m*

1150 **common rights; rights of common**
The right of several people to use common land.

r совместное право
Общее право нескольких лиц на совместное использование общинных земель.

f droit *m* en commun
s derecho *m* común
g gemeinsames Recht *n*

1151 **commonwealth**
A loose association of autonomous states under a common allegiance, e.g. the association consisting of Britain and the states that belonged formerly to the British Empire.

r содружество наций
Объединение стран и территорий, входивших ранее в Британскую империю.

f communauté *f*
s comunidad *f*
g Gemeinschaft *f* der Nationen

1152 **communal field**
A form of land ownership when the land is owned by a community.

r общинные земли
Форма земельной собственности, когда определенный земельный участок принадлежит крестьянской общине.

f terres *fpl* communales; ejido *m*
s tierras *fpl* comunales; tierras *fpl* colectivas; ejido *m*
g Gemeindeländereien *fpl*

1153 **commune**
In some countries the smallest administrative-territorial unit consisting of several countryside settlements.

r коммуна
В ряде стран низшая административно-территориальная единица, состоящая из нескольких сельских поселений.

f commune *f*
s comuna *f*; comunal *m*
g Kommune *f*

* **communication network → 6318**

1154 **communism**
A theory advocating elimination of private property.

r коммунизм
Теория, в основе которой лежит отрицание права частной собственносити.

f communisme *m*
s comunismo *m*
g Kommunismus *m*

1155 **community$_1$**
A primary form of social organization, an integrating group of people living in a partic-

ular area often sharing a common faith, culture, professional or life-style.

r община

Первичная форма социальной организации людей, нередко связанная с конкретной территорией, например, городские коммуны или сельские землячества.

f communauté$_1$ *f*

s comunidad$_1$ *f*

g Gemeinde *f*

* **community$_2$ → 661**

* **commuter village → 1678**

1156 **commuting; labour linkages**

Daily movements of people to the place of work from suburban or rural locations and back.

r маятниковая миграция; трудовые связи; челночная миграция

Ежедневные перемещения людей на работу из пригорода в город с возвращением к месту проживания.

f migration *f* quotidienne; navettes *fpl*; migration *f* pendulaire; migration *f* alternante

s migración *f* cotidiana; migración *f* pendular; migración *f* alternante; ligazón *f* de trabajo

g Pendelwanderung *f*; Arbeitsverbindungen *fpl*; Pendelverkehr *m*

1157 **compact soils; vertisols**

Dark-coloured clay soils that develop on clay rocks in equatorial, tropical and temperate belts under periodical excessive moistening and strong drying up (the high content of swelling clays in these soils leads to the formation of deep cracks during the dry season).

r слитые почвы; вертисоли

Темноокрашенные глинистые почвы, формирующиеся в экваториальном, тропическом и умеренном поясах на глинистых породах в условиях периодического переувлажнения и сильного иссушения, набухают при увлажнении и растрескиваются при высыхании.

f sols *mpl* compacts; vertisols *mpl*

s suelos *mpl* compactos; vertisols *mpl*; vertisuelos *mpl*

g kompakte Böden *mpl*; Vertisole *mpl*

1158 **compaction of rock**

A gradual increase of density of rocks, including snow and ice.

r уплотнение горной породы

Постепенное увеличение плотности горной породы, включая снег и лед.

f compactage *m* de roche

s compactación *f* de roca; consolidación *f* de roca

g Gesteinsverdichtung *f*

1159 **compaction of soil; consolidation of soil**

The pressing together of soil particles and a gradual reduction in volume of a soil mass due to the increased compression or drying up.

r уплотнение почвы; усадка почвы

Уменьшение порового пространства в почве, вызванное потерей влаги при ее высушивании или под действием внешнего давления.

f compactage *m* du sol

s compactación *f* del suelo; consolidación *f* del suelo; asiento *m* del suelo

g Bodenverdichtung *f*

1160 **comparative method of exploration**

An identification of similarity and difference of processes, properties and states of some geographical objects – both similar and situated close to each other at the same time, and different, but comparable by structure and functions, remote in space and time.

r сравнительный метод исследования

Выявление сходства и различия процессов, свойств и состояний нескольких географических объектов – как сходных и рядом расположенных, существующих в одно и то же время, так и различных, но сопоставимых по структуре и функциям, отдаленных в пространстве и времени.

f méthode *f* comparative de l'exploration

s método *m* de exploración comparativo

g vergleichende Forschungsmethode *f*

1161 **compass**

An instrument used to find direction to the Earth's magnetic poles and serving to take bearing regarding the cardinal points.

r компас; буссоль

Прибор, указывающий направление на магнитные полюса Земли и служащий для ориентирования относительно сторон горизонта.

f boussole *f*; compas *m*

s brújula *f*

g Kompaß *m*

1162 **compass azimuth; magnetic azimuth**

The horizontal angle measured clockwise from the north indicated by the magnetic needle of a compass to the given direction.

r магнитный азимут

Угол в горизонтальной плоскости, отсчитываемый от северного направления магнитной стрелки компаса по часовой стрелке до заданного направления.

f azimut *m* magnétique

s acimut *m* magnético; azimut *m* magnético

g magnetischer Azimut *m*

* **compass declination → 3633**

1163 **compensating current**

A horizontal water current compensating the outflow of water from some part of a basin, e.g. equatorial counter currents in oceans.

r компенсационное течение

Горизонтальный перенос вод, возмещающий отток из какой-либо части водоема, например, межпассатные противотечения в океанах.
f courant *m* de compensation
s corriente *f* de compensación
g Kompensationsstrom *m*

1164 compensation flow
A legally established minimum rate of river discharge below a water intake in order to maintain the water resource of a given region or catchment.
r компенсационный сток
Предельная величина расхода воды ниже водозабора с целью сохранения водных ресурсов в данном районе или на данном водосборе.
f écoulement *m* compensateur
s desagüe *m* compensatório; desagüe *m* de compensación
g Kompensationabfluss *m*

1165 competence of a stream
The ability of a stream to move particles of different size: sand, gravel, pebbles or boulders.
r транспортирующая способность речного потока
Способность текучей воды или реки переносить твердый материал: песок, гравий, а иногда и крупные валуны.
f capacité *f* de charge de flux
s capacidad *f* de carga del flujo
g Geschiebetransport *m*

1166 competition
1. The interaction between individuals of the same species (intraspecific competition) or between different species (interspecific competition) when they strive for the same resources of the environment, such as food, habitat, etc. 2. The fight of the producers for the markets which leads to the rise of the quality of a product and lowering of its price, it is the basis of market economy.
r конкуренция
1. Взаимоотношения между организмами одного или того же вида (внутривидовая конкуренция) или между разными видами организмов (межвидовая конкуренция), когда они являются антагонистами в борьбе за пищу, места обитания и другие необходимые для жизни условия. 2. Соревнование, а иногда борьба производителей за рынки сбыта, сопровождающаяся повышением качества товаров и снижением их себестоимости, основа рыночной экономики.
f compétition *f*
s competencia *f*
g Wettbewerb *n*; Konkurenz *f*

1167 compilation plot; original plot
An exact original of a geographical map prepared by the author.
r авторский оригинал; составительский оригинал
Точный оригинал географической карты, изготовленный автором как образец ее содержания и оформления.
f minute *f* d'auteur
s minuta *f* de autor
g Autorenoriginal *n*; Minute *f*

1168 complex atlas
Geographical atlas containing maps which represent different aspects of a given area.
r комплексный атлас
Географический атлас, содержащий карты, характеризующие изображаемую территорию или акваторию с разных сторон.
f atlas *m* complexe; atlas *m* d'inventaire
s atlas *m* complejo; atlas *m* de inventario
g komplexer Atlas *m*

1169 complex climatology
A method of climatological analysis of meteorological observations when separate meteorological elements are combined into complexes called weather types.
r комплексная климатология
Метод климатологического анализа, при котором отдельные метеорологические элементы объединяются в комплексы, называемые типами погоды.
f climatologie *f* intégrée
s climatología *f* integrada
g komplexe Klimatologie *f*

*** complex cross-section → 1171**

1170 complex map; aggregate map
A geographical map showing together several different interconnected phenomena (or several properties of one phenomenon) but each in its system of indices.
r комплексная карта
Географическая карта, показывающая совместно несколько разных взаимосвязанных явлений (или несколько свойств одного явления), но каждое в своей системе показателей.
f carte *f* d'inventaire
s mapa *m* de inventario
g komplexe Karte *f*

1171 complex profile; complex cross-section
A profile showing several parameters, e.g. hypsometric profile together with geological section or with the graph of hydroclimatic parameters, etc.
r комплексный профиль
Профиль, совмещающий разные показатели, например, гипсометрический профиль вместе с геологическим разрезом, графиком гидроклиматических показателей и т.п.
f profil *m* conjugué
s perfil *m* complejo
g Komplexprofil *n*

1172 component of nature; landscape component
The main structural parts of landscape which are fragments of the Earth's spheres: lithosphere, hydrosphere, atmosphere, biosphere.
r природный компонент; компонент ландшафта
Основные структурные части ландшафта, представленные фрагментами отдельных земных сфер: литосферы, гидросферы, атмосферы, биосферы.
f composante *m* du paysage
s componente *m* de paysaje; componente *m* de naturaleza
g Landschaftsbestandteil *m*

1173 composite coast
A coast that has resulted from the alternating spatial subsidence and rising of coastal tracts along the faults that run transverse to the coastline itself.
r смешанный берег
Побережье, сформировавшееся в результате чередования опускания и подъема прибрежных участков вдоль линеаментов (нередко разломов и сбросов), протягивающихся поперек береговой линии.
f côte *f* composé
s costa *f* compuesta
g Mischstrand *m*

1174 composite map
A type of map, in which information from several other maps is brought together and displayed to facilitate comparison between different sets of data.
r композитная карта
Тип карты, при составлении которой объединяется информация, взятая с других карт, в целях сравнения разных наборов данных.
f carte *f* composite
s mapa *m* de composición
g Bestandkarte *f*

* **composite volcano → 5850**

1175 composition of the population
Distribution of population across the country or any territory according to the national (ethnic) affiliation, sex, age , etc.
r состав населения
Распределение населения страны или какой-либо территории по ряду признаков: национальности, полу, возрасту и др.
f composition *f* de population
s composición *f* de la población
g Zusammensetzung *f* der Bevölkerung

1176 compost
A soil conditioner and fertilizer produced by the planned decomposition of organic material, such as vegetable remains.
r компост

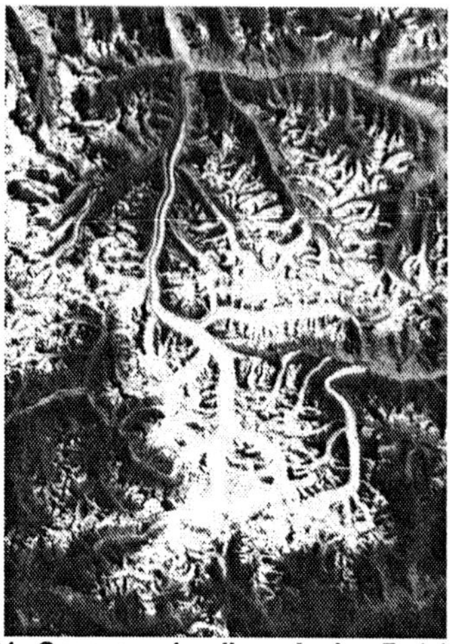

Fig. 64. **Compound valley glacier** Fedchenko, 77 km long, in the Central Pamirs

Удобрение, получаемое в результате разложения разных органических веществ под влиянием деятельности микроорганизмов.
f compost *m*; fumure *m* organique naturelle
s compost *m*; estiércol *m*; fertilizante *m* orgánico compuesto
g Kompost *m*; Mischdünger *m*

1177 compound valley glacier; dendritic glacier
A valley glacier consisting of several tributary glaciers with independent accumulation areas flowing down into the glacier trunk. → Fig. 64
r сложный долинный ледник; дендритовый ледник
Долинный ледник, состоящий из ряда притоков разного порядка с самостоятельными областями питания, вливающихся в основной ствол ледника. → Рис. 64
f glacier *m* de vallée composé; glacier *m* dendritique; glacier himalayen
s glaciar *m* de valle compuesto; glaciar *m* dendriforme; glaciar *m* himaláyico
g zusammengezetzter Talgletscher *m*; baumartiger Gletscher *m*

1178 comprehensive atlas
Geographical atlas consisting mainly of comprehensive maps.
r общегеографический атлас
Географический атлас, состоящий преимущественно из общегеографических карт.
f atlas *m* géographique général
s atlas *m* geográfico general
g allgemein-geographischer Atlas *m*

1179 comprehensive map
A map presenting the main features of the area: relief, drainage network, forests, settlements, roads, etc. in approximately equal detail.
r общегеографическая карта

Карта, отображающая примерно с одинаковой подробностью основные элементы местности: рельеф, реки, леса, населенные пункты, дороги и пр.
f carte *f* géographique générale
s mapa *m* geográfico general
g allgemein-geographische Karte *f*

* **compression fussure → 1265**

1180 concave slope
A slope the steepness of which declines downslope.
r вогнутый склон
Склон с крутой верхней частью, постепенно выполаживающейся книзу.
f pente *f* concave
s pendiente *f* cóncava
g konkaver Hang *m*

* **concentration of minerals → 641**

1181 concentration of production
The centralization of production and labour force in large enterprises in order to increase the scale and efficiency of production of similar or technologically related units.
r концентрация производства; укрупнение производства
Сосредоточение производства и рабочей силы на крупных предприятиях в целях увеличения эффективности выпуска однородной или технологически связанной продукции.
f concentration *f* de production
s concentración *f* de producción
g Produktionskonzentration *f*

1182 concern; corporation
An alliance of industrial, transport, trade, banking enterprises, research and educational institutions on the basis of combination and diversification, while the finances, investment and research get centralized.
r концерн; корпорация
Объединение предприятий промышленности, транспорта, торговли, банков, научных и учебных центров на базе комбинирования и диверсификации производства при централизованных финансах, капиталовложений и научно-исследовательских разработок.
f conzern *m*; corporation *f*
s corporación *f*; cuerpo *m*
g Konzern *m*; Korporation *f*; Körperschaft *f*

1183 concession
A right given to a state or an individual, allowing them to use or operate an enterprise, land, mineral deposit or other natural resources, during a certain period or certain terms.
r концессия
Передача в эксплуатацию на определенный срок и определенных условиях государству или частному лицу предприятий, земельных угодий, месторождений полезных ископаемых и других природных богатств.
f concession *f*
s concesión *f*
g Konzession *f*

* **concordance → 1203**

1184 concordant *adj*; conformable *adj*
Pertaining to the harmonious correspondence between a morphological feature and its underlying geological structure.
r согласный
Расположенный или протекающий параллельно структурным линиям рельефа, согласующийся с общим направлением напластования.
f concordant *adj*
s concordante *adj*
g konkordant *adj*

* **concordant bedding → 1203**

1185 concordant coast; longitudinal coast
A coast which lies parallel to the geological ctructure of the coastal region.
r согласный берег
Берег, общее направление которого совпадает с простиранием геологических структур прибрежной суши.
f côte *f* concordante
s costa *f* concordante
g konsequente Küste *f*; Längsküste *f*

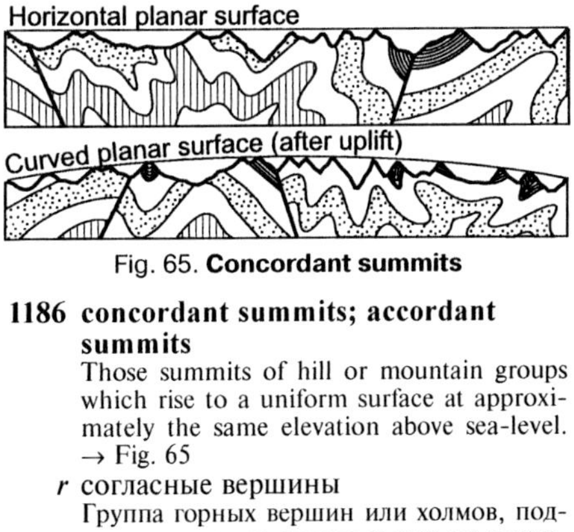

Fig. 65. **Concordant summits**

1186 concordant summits; accordant summits
Those summits of hill or mountain groups which rise to a uniform surface at approximately the same elevation above sea-level. → Fig. 65
r согласные вершины
Группа горных вершин или холмов, поднимающихся до одного и того же высотного уровня. → Рис. 65
f sommets *fpl* concordant
s cumbres *fpl* concordantes
g konkordante Gipfel *fpl*

1187 concretion; nodule
A solid mineral formation of rounded or irregular shape appearing mainly in sedimentary rocks and soils under the diagenesis, weathering and other processes.
r конкреция

Минеральное образование округлой или неправильной формы, возникающее в осадочных породах и почвах при диагенезе, выветривании и других процессах.
f concrétion *f*
s concreción *f*
g Konkretion *f*

1188 **condensation level**
The level at which an ascending air reaches the state of saturation due to expansion and adiabatic cooling.
r уровень конденсации
Высота, до которой должен подняться воздух, чтобы содержащийся в нем водяной пар достиг состояния насыщения вследствие адиабатического охлаждения.
f niveau *m* de condensation
s nivel *m* de condensación
g Kondensationshöhe *f*

1189 **condensation nucleus**
A liquid or solid particle suspended in the air upon which condensation of water vapour begins in the atmosphere.
r ядро конденсации
Жидкая или твердая частица, взвешенная в воздухе, на которой начинается конденсация водяного пара в атмосфере.
f noyau *m* de condensation
s núcleo *m* de condensación
g Kondensationskern *m*

1190 **condensation of water vapour**
The physical process by which water vapour is transformed into dew, fog or cloud droplets.
r конденсация водяного пара
Физический процесс, при котором происходит превращение водяного пара в капельки росы, тумана или облака.
f condensation *f* de vapeur d'eau
s condensación *f* de vapor de agua
g Wasserdampfkondensation *f*

1191 **condensation trail; contrail; adiabatic trail**
A cloud of artificial origin, resembling cirrus, which forms behind the aircraft when the air at the flight level is sufficiently cold and moist. → Fig. 66
r конденсационный след; адиабатический след
Искусственное облако, напоминающее перистое, возникающее за самолетом, если атмосфера на уровне полета достаточно холодная и влажная. → Рис. 66
f traînée *f* de condensation; traînée *f* adiabatique
s estela *f* de condensación; estela *f* adiabática
g Kondensationsspur *f*

1192 **conditional instability**
The state of an air mass with an environmental lapse rate greater than the saturated adiabatic lapse rate but less than the dry adiabatic lapse rate, the instability depending on the amount of water vapour held in the air mass.
r условная неустойчивость
Состояние воздушной массы, когда адиабатический градиент температуры больше, чем влажноадиабатический градиент, но меньше сухоадиабатического градиента, неустойчивость зависит от количества водяных паров, содержащихся в воздушной массе.
f instabilité *f* conventionelle
s inestabilidad *f* convencional
g konventionelle Instabilität *f*

1193 **conditionally clean water**
Sewage waters the discharge of which into a water object after purification or without it does not break the norms of water quality.
r условно чистая вода
Сточные воды, спуск которых в водный объект после очистки или без очистки не нарушает нормы качества воды.
f eau *f* pure conventionnelle
s agua *f* condicionalmente pura
g Reinabwasser *n*

1194 **condominium**
A territory under joint possession and administration of two or more states, e.g. Anglo-Egyptian Sudan in 1899-1955.
r кондоминиум
Совместное владение и управление одной и той же территорией двумя (или более) государствами, например, Англо-Египетский Судан в 1899-1955 гг.
f condominium *m*
s condominio *m*
g Kondominium *n*

* **conducting channel → 2215**

* **conduit → 4576**

1195 **cone of depression; cone of exhaustion; cone of water table depression**
The funnel-shaped depression in the water table surface around a well caused by the extraction of water, by pumping or artesian

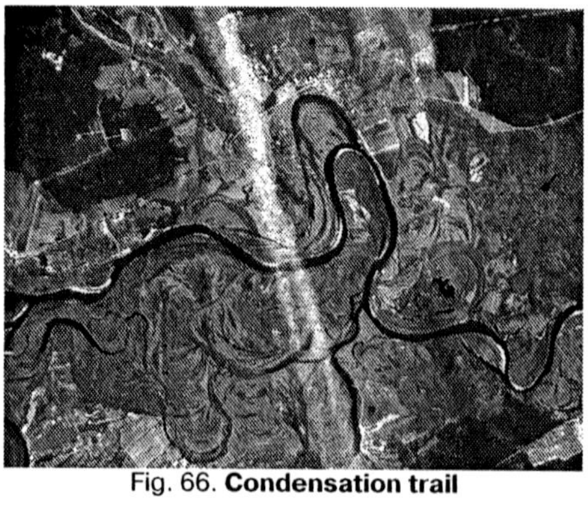

Fig. 66. **Condensation trail**

flow, at a rate grater than that at which the aquifer is being recharged.

r депрессионная воронка
Пьезометрическая поверхность напорных вод или свободная поверхность безнапорных вод, приобретающие форму воронки в месте откачки воды (скважины, шахты, колодца).

f entonnoir *m* de dépression de la nappe

s embudo *m* de depresión

g Depressionstrichter *m*

* **cone of exhaustion → 1195**

* **cone of water table depression → 1195**

1196 **confederation**
A form of state organization in which the states that form confederation maintain their independence and set up common organs of power and administration for military and foreign policy purposes.

r конфедерация
Форма государственного устройства, при которой государства, образующие конфедерацию, полностью сохраняют свою независимость, но создают объединенные органы власти и управления в военных и внешнеполитических целях.

f confédération *f*

s confederación *f*

g Konföderation *f*; Staatenbund *m*

1197 **configuration map**
A simplified map on which major crests, ridges and spurs are represented by a skeleton of single lines.

r орографическая схема
Схематическая карта с изображением рельефа скелетными линиями.

f croquis *m* de l'ossature du relief; carte *f* orographique

s croquis *m* configurativo; mapa *m* orográfico

g Geripplinienskizze *f*

1198 **confined aquifer; confined water**
Ground waters lying between the impermeable layers and being under hydrostatic pressure.

r напорные воды; межпластовые воды
Подземные воды, заключенные между водоупорными слоями и находящиеся под гидростатическим давлением.

f nappe *f* captive; nappe *f* aquifère captive

s capa *f* acuífera cautiva

g gespanntes Grundwasser *n*

* **confined meander → 3052**

* **confined water → 1198**

1199 **confluence; junction$_2$**
The point at which two streams meet including the inflowing of tributaries into the main river; a joint movement of two or several glaciers or a convergence of glacier tongues having separate accumulation areas.

r слияние; конфлюэнция
Соединение речных потоков в один, в том числе слияние притоков с главной рекой; совместное движение двух или нескольких ледников, слияние ледниковых языков, имеющих обособленные области питания.

f confluence *f*

s confluencia *f*

g Konfluenz *f*; Zusammenfluß *m*

1200 **confluence step**
A transversal scarp in the longitudinal profile of a glacial valley formed as a result of its overdeepening down the place of confluence of ancient glaciers.

r устьевая ступень; конфлюэнтная ступень
Поперечный уступ в продольном профиле ледниковой долины, возникший в результате ее переуглубления ниже места слияния древних ледников.

f gradin *m* de confluence

s grada *f* de confluencia

g Konfluenzstufe *f*; Mündungsstufe *f*

1201 **confluent glacier**
A glacier formed by the convergence of several glaciers' tongues.

r конфлюэнтный ледник
Ледник, образованный слиянием окончаний нескольких ледников.

f glacier *m* de confluence

s glaciar *m* de confluencia

g Konfluenzgletscher *m*

* **conformable *adj* → 1184**

1202 **conformal projection; equal-angle projection**
A map projection in which there is no angular distortions.

r равноугольная проекция
Картографическая проекция, в которой нет искажений углов.

f projection *f* conforme

s proyección *f* conforme

g winkeltreuer Kartennetzentwurf *m*

1203 **conformity; concordance; concordant bedding**
The disposition of rock strata formed by the layers lying parallel to each other and having no gaps in sedimentation.

r согласное залегание
Залегание горных пород, отличающееся параллельностью слоев и непрерывностью разреза.

f concordance *f*

s concordancia *f*

g konkordante Lagerung *f*

* **congelation → 2299**

1204 **congelation ice; hydrogenic ice**
Ice formed by freezing of liquid water.

r конжеляционный лед; гидрогенный лед
Лед, формирующийся при замерзании жидкой воды.
f glace *f* de congélation
s hielo *m* de cogelación
g Kongelationseis *n*

1205 congelation ice formation
The process of ice formation from water mass.
r конжеляционное льдообразование
Образование ледяной породы из водной массы.
f formation *f* de la glace par congélation
s formación *f* de hielo por congelación
g Eisbildung *f* durch Gefrieren

1206 congelifluction; congelifluxion
A flow of earth (solifluction) in conditions of a permanently frozen ground.
r конжелифлюкция
Течение грунта (солифлюкция) в условиях многолетней мерзлоты.
f congelifluxion *f*
s congelifluxión *f*
g Kongelifluxion *f*

* **congelifluxion → 1206**

* **congelifraction → 2339**

* **congeliturbation → 1357**

1207 conglomerate; pudding rock
A detrital rock consisting of coherent shingle beds with the admixture of thinner material.
r конгломерат
Обломочная горная порода, представляющая собой сцементированный галечник с примесью более тонкого материала.
f conglomérat *m*
s conglomerado *m*
g Konglomerat *n*

1208 congress tourism
A type of tourism that includes a participation in the scientific or buisness congresses, symposia, etc.
r конгрессный туризм
Вид туризма, предусматривающий участие в научных или деловых конгрессах, симпозиумах и пр.
f tourisme *m* de congrès
s turismo *m* de congreso
g Kongreßtourismus *n*

1209 conical projection
A map projection in which parallels are represented by arches of concentric circles, and meridians are their radiuses, the angles between which are proportional to corresponding differences of longitudes. → Fig.67
r коническая проекция
Картографическая проекция, в которой параллели изображаются дугами концентрических окружностей, а меридианы – прямыми, расходящимися из общего центра параллелей, углы между которыми равны разностям их долгот. → Рис. 67
f projection *f* conique
s proyección *f* cónica
g Kegelentwurf *m*

1210 conical-summit glacier
A glacier covering all slopes of a single mountain peak.
r ледник конической вершины
Ледник, покрывающий со всех сторон склоны отдельной вершины.
f glacier *m* de sommet conique
s glaciar *m* de cumbre cónico
g Kegelberggletscher *m*

1211 coniferous forest; needle forest
The forest composed of evergreen trees with needle-shaped leaves that occurs naturally under many conditions of soil, climate and elevation (the largest continuous areas of coniferous forest is the boreal forest also called taiga, it occurs in the subtropical and tropical belts).
r хвойный лес
Лес, образуемый хвойными породами, характерен для северной части умеренных

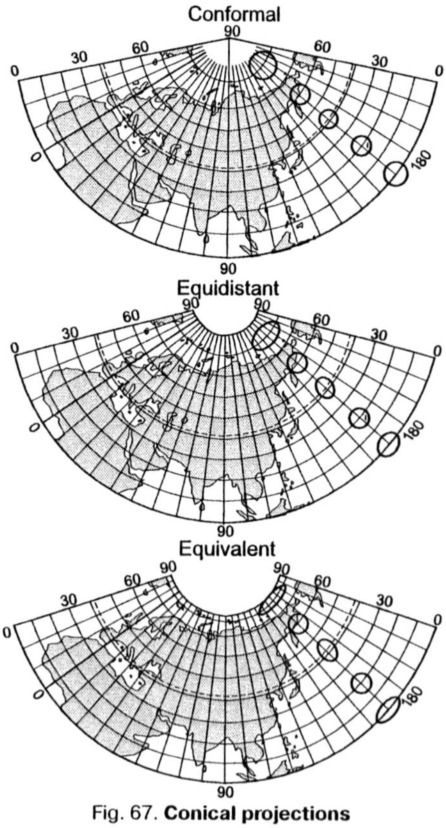

Fig. 67. **Conical projections**

широт Северного полушария, встречается также в субтропическом и тропическом поясах.
f forêt *f* des conifères; forêt *f* résineuse
s bosque *m* conífero; bosque *m* de coníferos
g Nadelwald *m*

1212 **coniferous plant**
Mostly evergreen trees and shrubs with needles unlike the leaf-bearing plants.
r хвойное растение
Преимущественно вечнозеленые деревья и кустарники с хвоей, в отличие от лиственных растений.
f plante *f* conifère
s planta *f* conífera
g Nadelpflanze *f*

1213 **conjunction**
The position of two planets or other celestial bodies which, when viewed from the Earth, lie in one line in the same direction (when the Sun and the Moon are in the conjunction, it induces strong tidal forces).
r сближение двух небесных тел
Положение двух планет или иных небесных тел на одной линии и в одном направлении при взгляде с Земли, что вызывает действие более сильных приливных сил.
f conjunction *f*
s conjunción *f*
g Konjunktion *f*

1214 **connate water; fossil water**
Ocean water or fresh water which has become trapped in the interstices of a sedimentary rock during the time when the material was being deposited.
r реликтовая вода
Океанская или пресная вода, которая была вкраплена в узкие расщелины осадочной породы во время ее формирования.
f eau *f* innée
s agua *f* innata
g Reliktwasser *n*

1215 **connecting bar; tombolo**
A spit resulting from longshore drift which joints an offshore island to the mainland.
r томболо; намывная коса; перейма

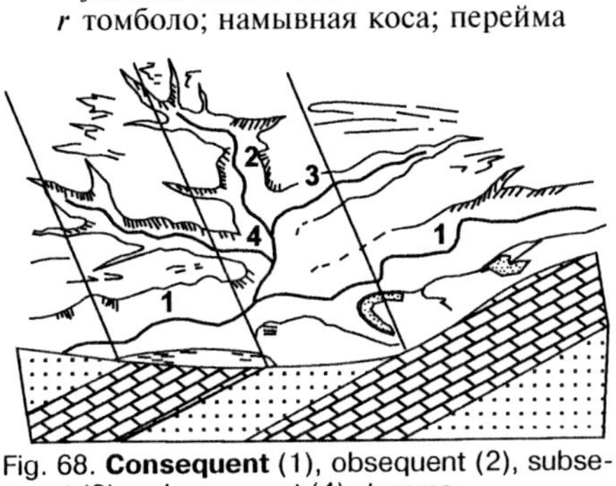

Fig. 68. **Consequent** (1), obsequent (2), subsequent (3) and resequent (4) streams

Коса из намытого морем рыхлого материала, причленяющая бывший остров к берегу ближайшей суши.
f tombolo *m*; flèche-isthme *m*
s tómbolo *m*; flecha-istmo *m*
g Inselnehrung *f*; Tombolo *n*; Hals *m*

1216 **consequent stream; accordant valley**
The main river running in a valley whose direction conforms to the slope of the land surface which often coincides with the dip of the rock strata. → Fig. 68
r консеквентная река; консеквентная долина
Главная река, текущая в долине, направление которой соответствует уклону поверхности, часто совпадающему с падением пластов горных пород. → Рис. 68
f fleuve *m* conséquent; vallée *f* conséquente; vallée *f* cataclinale
s río *m* consecuente; valle *m* consecuente; valle *m* cataclinal
g consequenter Fluß *m*; konsequentes Tal *n*

* **conservation → 4797**

1217 **conservation map**
A map which shows the spatial distribution of some measures aimed at the protection, rational use and restoration of natural resources and the environment.
r карта охраны природы
Географическая карта, показывающая систему мероприятий, направленных на сохранение, рациональное использование и возобновление природных ресурсов и окружающей среды.
f carte *f* de protection de l'environnement
s mapa *m* de protección de medio ambiente; mapa *f* de conservación de naturaleza
g Naturschutzkarte *f*

* **conservation movement → 1802**

* **conservationism → 1802**

1218 **conservationist; environmentalist**
A participant of a public movement or a political party who works to protect the environment from damage or destruction.
r защитник окружающей среды; зеленый
Участник общественного движения или партии, выступающей за сохранение окружающей природной среды.
f conservationiste *m*
s ecologista *m*
g Naturschützer *m*; Umweltschützer *m*

1219 **consociation**
Any unit of vegetation which is dominated by a single species.
r консоциация
Любая растительная ассоциация, в которой доминирует единственнный вид.
f consociation *f*

s consociación *f*
g Konsoziation *f*

* **consolidated drifting ice → 1220**

1220 **consolidated pack ice; consolidated drifting ice**
An ice field formed by freezing together of ice floes of different shapes and age.
r сморозь
Ледяное поле, образовавшееся от смерзания льдин разного возраста и формы.
f pack *m* consolide; couverture *f* complète
s banco *m* de hielo consolidado
g konsolidierendes Packeis *n*

1221 **consolidated rock; coherent rock**
Dense rock the particles or fragments of which are tightly connected.
r сцементированная горная порода
Плотная горная порода, частицы или обломки которой прочно связаны между собой.
f roche *f* cohérente
s roca *f* coherente
g Festgestein *n*

1222 **consolidation of ice**
1. The increase of cohesion between the constituent parts of ice or snow cover in the course of its metamorphism, can be induced artificially in order to improve the mechanical properties of ice or snow by the introduction of admixtures or by additional packing. 2. The increase of the closeness of sea ice under the effect of winds and currents.
r упрочение льда$_1$; сплочение льда$_2$
1. Увеличение сил сцепления между составными частями ледяного или снежного покрова в процессе его метаморфизма, может вызываться искусственно для улучшения механических свойств льда или снега посредством введения добавок или уплотнения. 2. Увеличение сплоченности морского льда под влиянием ветров и течений.
f consolidation *f* de glace$_1$; jointage *m* de la glace$_2$
s consolidación *f* de hielo$_1$; apretamiento *m* de hielo$_2$
g Eisfestigung f_1; Eisschollenzusammenschluß m_2

* **consolidation of soil → 1159**

1223 **consortium**
A temporary association of several banks or companies for a common purpose, such as to underwrite an issue or carry out a capital-intensive industrial project.
r консорциум
Временное соглашение между несколькими банками или промышленными предприятиями для совместного размещения займа или выполнения единого капиталоемкого промышленного проекта.
f consortium *m*
s consorcio *m*
g Konsortium *n*

* **constancy → 1226**

1224 **constant pressure surface; isobaric surface**
Surface in the atmosphere with the same value of atmospheric pressure at every point.
r изобарическая поверхность; поверхность постоянного давления
Поверхность в атмосфере, на которой атмосферное давление во всех точках одинаково.
f surface *f* à pression costante; surface *f* de isobare
s superficie *f* de presión constante; superficie *f* isobárica
g isobare Fläche *f*

1225 **constant slope**
The part of a slope profile which stands above the waning slope and below free face and waxing slope, its gradient is determined by the angle of repose of the fallen debris.
r стабильный склон
Часть профиля склона, сохраняющегося выше отступающей части склона и ниже наступающей, его уклон определяется углом покоя залегающих здесь рыхлых обломочных отложений.
f versant *m* constant
s vertiente *f* constante
g konstanter Abhang *m*

1226 **constant species; constancy**
A species of wide ecological amplitude present in all, or almost all, biotops occupied by a particular community.
r константный вид; константность
Вид с широкой экологической амплитудой, присутствующий во всех или почти всех биотопах данного растительного сообщества.
f espèce *f* constante; constance *f*
s especie *f* constante; constancia *f*
g Stetigkeit *f*; konstante Art *f*

1227 **constellation**
A group of stars in the sky named after the pattern they form, which may be used to take bearing.
r созвездие
Группы звезд в небе (всего их 88), выделенные для удобства ориентировки на небесной сфере и иногда используемые для ориентирования по странам света.
f constellation *f*
s constelación *f*
g Sternbild *n*

* **constitution water → 450**

1228 **constitutional ice**
The ice formed when wet soils and rocks freeze.

r конституционный лед
Лед, образующийся при замерзании увлажненных почв и горных пород.
f glace *f* constitutionnelle
s hielo *m* constitucional
g Konstitutionseis *n*

1229 **constitutional monarchy**
A monarchy in which the power of the supreme ruler is limited by constitution, i.e. the legislative power belongs to the parliament and executive power belongs to the government.
r конституционная монархия
Монархия, где власть монарха ограничена конституцией, т.е. законодательные функции переданы парламенту, а исполнительные – правительству.
f monarchie *f* constitutionnelle
s monarquía *f* constitucional
g konstitutionelle Monarchie *f*

1230 **construction inustry**
A branch of industry engaged in the realisation of all kinds of construction.
r строительная промышленность
Отрасль промышленности, занимающаяся всеми видами строительства.
f industrie *f* du bâtiment; industrie *f* de construction
s industria *f* de construcción
g Bauindustrie *f*

* **construction materials industry → 3078**

1231 **constructive geography**
The study of the problems of nature transformation for the purpose of the complex use of natural resources (in the middle of the XX century it meant transition from descriptive geography to experimental geography on the basis of mathematical, geophysical, geochemical, aerospace and other methods of investigation).
r конструктивная география
Разработка проблем преобразования природы в целях рационального и комплексного использования природных ресурсов (в середине XX в. означала переход от описательной географии к экспериментальной на базе внедрения математических, геофизических, геохимических, аэрокосмических и других методов исследования).
f géographie *f* constructive
s geografía *f* constructiva
g konstruktive Geographie *f*

1232 **constructive wave**
A water wave breaking on a beach, with a powerful swash and a weaker backwash, so that it moves material up the slope of the beach.
r волна намыва
Волна, разбивающаяся о берег с мощным накатом и более слабым откатом, что приводит к перемещению материала вверх по склону пляжа.
f vague *f* constructive
s ola *f* constructiva
g konstruktive Welle *f*

1233 **consumer; heterotroph; heterotrophic organism**
Organisms that feed on living or dead organic material produced by other organisms.
r консумент; гетеротроф
Организмы, использующие в качестве источника питания органические вещества, произведенные другими организмами.
f consommateur *m*; hétérotrophe *m*
s consumidor *m*; heterotrofo *m*
g Konsument *m*; Heterotroph *m*

1234 **consumer goods**
The products and services intended for personal consumption, such as foods, clothing, housing, etc., which directly satisfy the needs and desires of the individual person.
r предметы потребления
Продукты труда и услуги, предназначенные прежде всего для личного потребления (продукты питания, одежда, жилища и т.п.).
f biens *mpl* de consommation
s bienes *mpl* de consumo
g Gebrauchsartikel *mpl*; Konsumartikel *mpl*

1235 **consumption**
The use of the product, worked out at the final stage of the production process.
r потребление
Использование продукта, произведенного на заключительной стадии производственного процесса.
f consommation *f*
s consumo *m*
g Konsum *m*; Verbrauch *m*

1236 **container; container transport**
A metal packing case of standard size used for the transport of cargo by road, rail, or water and suitable for repeated use and for mechanical handling.
r контейнерный транспорт; контейнер
Металлический ящик стандартного размера, используемый для грузовых перевозок по автомобильным, железным дорогам и водным путям, приспособленный для механизированных перегрузок.
f transport *m* par conteneurs; container *m*
s transporte *m* en containeres; contenedor *m*
g Containertransport *m*; Container *m*

* **container transport → 1236**

1237 **containerization**
The transfer to a more economical system of trancporting goods by using containers that make it possible to move loads easily and quickly from one mode of transport to another.
r контейнеризация
Переход на экономичную контейнерную систему перевозки грузов, позволяющую

использовать разные виды транспорта для бестарной транспортировки грузов.
f conteneurisation *f*
s containerisación *f*
g Einführung *f* des Containerverkehrs

* **contamination → 4669**

1238 **contamination coefficient of the environment**
The amount of pollutants formed when one unit of final goods produced with the use of existing technology.
r коэффициент загрязнения окружающей среды
Количество загрязнителей на единицу конечной продукции, образующихся при использовании существующей технологии.
f coefficient *m* de contamination de l'environnement
s coeficiente *m* de contaminación de medio ambiente
g Verschmutzungsfaktor *m* der Umwelt

1239 **continent; landmass; mainland**
One of the large continuous masses of land on the Earth's surface, the major part of which is lying above the sea level, and the periphery extends below its level (seven continents are usually distinguished: North and South America, Europe, Asia, Africa, Australia and Antarctica, while some authors insist that Eurasia is a single continent).
r континент; материк
Крупный массив земной коры, большая часть которого выступает над уровнем Мирового океана, а периферия находится ниже его уровня (обычно различают семь континентов: Северная и Южная Америка, Европа, Азия, Африка, Австралия и Антарктида, некоторые авторы считают Евразию единым континентом).
f continent *m*
s continente *m*; masa *f* continental
g Kontinent *m*; Erdteil *m*

1240 **continental** ***adj***
Related to a very large land masses, continents.
r континентальный; материковый
Относящийся к значительным пространствам суши, континентам.
f continental *adj*
s continental *adj*
g kontinental *adj*; Festlands-

* **continental air → 1241**

1241 **continental air mass; continental air**
An air mass, usually of low humidity, formed over a continental interior where high atmospheric pressure prevails.
r континентальная воздушная масса; континентальный воздух
Воздушная масса, обычно с низкой влажностью, формирующаяся над внутренними частями континента, где преобладает высокое атмосферное давление.
f masse *f* d'air continentale; air *m* continental
s masa *f* de aire continental; aire *m* continental
g kontinentale Luftmasse *f*; kontinentale Luft *f*

1242 **continental anticyclone**
An anticyclone travelling over a continent in the cold season.
r континентальный антициклон
Зимняя область высокого атмосферного давления над континентом.
f anticyclone *m* continental
s anticiclón *m* continental
g kontinentales Hoch *n*; kontinentale Antizyklone *f*; Kontinentalhoch *n*

* **continental apron → 1249**

1243 **continental climate**
The climate associated with the interior of a continent and characterised by great extreems of temperature between day and night and between summer and winter, low relative humidity and moderate or low and irregular precipitation.
r континентальный климат
Климат, характерный для внутренних частей континентов, отличающийся большими годовыми и суточными колебаниями температуры, низкой относительной влажностью и умеренными или редкими осадками.
f climat *m* continental
s clima *m* continental
g Landklima *n*

1244 **continental crust**
The Earth's crust 35-45 km thick (maximum up to 75 km) underlying continents and consisting of sedimentary, granitic and basaltic layers.
r континентальная земная кора
Земная кора мощностью 35-45 км (максимально до 75 км), распространенная под континентами и состоящая из осадочного, гранитного и базальтового слоев.
f écorce *f* terrestre continentale
s corteza *f* terrestre continental
g kontinentale Erdkruste *f*

1245 **continental deposits; continental sediments**
All the deposits formed on land including those laid down on the bottom of inland water bodies.
r континентальные отложения
Отложения, образовавшиеся на суше, в том числе на дне внутриматериковых водоемов.
f dépôts *mpl* continentaux; sediments *mpl* continentaux

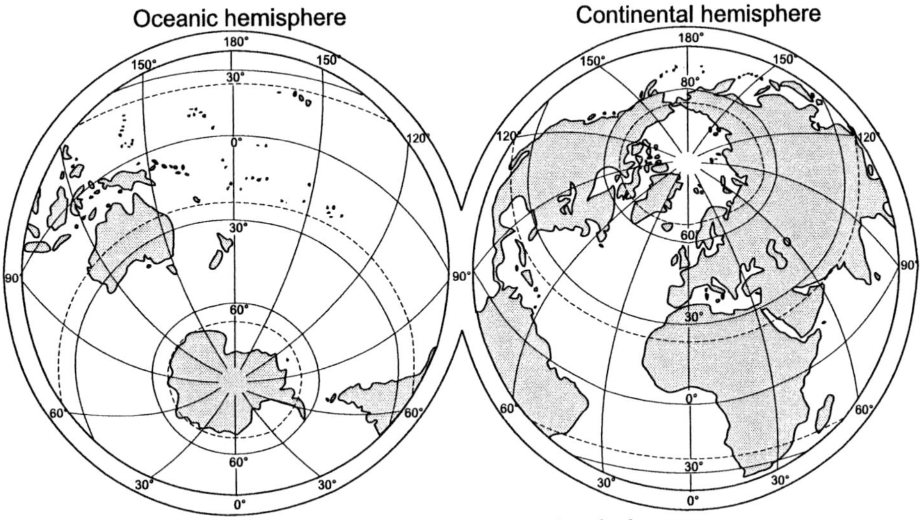

Fig. 69. **Continental** and **oceanic hemispheres**

s depósitos *mpl* continentales; sedimentos *mpl* continentales
g kontinentale Ablagerungen *fpl*; kontinentale Sedimente *npl*

1246 continental divide; continental watershed
A line or watershed area on a continent dividing the basins of the rivers flowing to different oceans.
r континентальный водораздел; материковый водораздел
Линия или водораздельное пространство на континенте, разделяющие бассейны рек, текущих в разные океаны.
f partage *m* continental des eaux
s divisoria *f* continental de aguas
g kontinentale Wasserscheide *f*; Kontinentalwasserscheide *f*

1247 continental drift
The movement of continents relative to each other across the Earth's surface and with respect to the poles which led to the present distribution of the continental masses.
r дрейф материков
Большие, до нескольких тысяч километров, горизонтальные перемещения материков относительно друг друга и по отношению к полюсам в течение геологического времени.
f dérive *f* des continents
s deriva *f* de los continentes
g Kontinentalverschiebung *f*; Kontinentaldrift *f*

* **continental drift theory → 4607**

1248 continental dune
An accumulation of sand found not in a desert but another sand space away from the coast in a sandur or a river valley.
r материковая дюна
Дюны, приуроченные к внепустынным песчаным пространствам, удаленным от побережий: дюны зандров, речных долин.
f dune *f* continentale
s duna *f* continental
g Kontinentaldüne *f*; kontinentale Düne *f*

1249 continental foot; continental apron; continental rise
The external part of an underwater margin of a continent which adjoins the base of a continental slope forming a deep depression at the junction of continental and oceanic crust filled with a thick layer of loose deposits.
r материковое подножие
Внешняя часть подводной окраины материка, примыкающая к основанию материкового склона и представляющая собой глубокий прогиб на стыке континентальной и океанической коры, заполненный мощной толщей рыхлых осадков.
f glacis *m* précontinental; marge *f* continentale; pied *m* du talus continental
s glacis *m* precontinental; pie *m* de talud continental; margen *f* continental
g Kontinentalfuß *m*

* **continental glacier → 3100**

1250 continental hemisphere
The hemisphere of the globe where the continents are concentrated occupying 47% of its area, the centre of it being in the south-west of France. → Fig. 69
r материковое полушарие
Полушарие Земли, в пределах которого суша развита наиболее широко, занимает

47% площади, центр такого полушария расположен на юго-западе Франции. → Рис. 69, с. 149
f hémisphère *m* continental
s hemisferio *m* continental
g kontinentale Halbkugel *f*

* **continental ice → 3374**

1251 **continental island**
A large island that separated from the continent as a result of abrasion or sinking of a part of land below the sea level.
r материковый остров
Крупный остров, отделившийся от материка вследствие абразии или опускания части суши ниже уровня моря.
f île *f* continentale
s isla *f* continental
g kontinentale Insel *f*

1252 **continental margin**
A rim part of the oceanic bottom with continental structure of the crust, it is subdivided into shelf, continental slope and continental foot.
r подводная окраина материка
Периферическая часть дна Мирового океана с материковым типом строения земной коры, по особенностям рельефа подразделяется на шельф, материковый склон и материковое подножье.
f marge *f* continentale
s margen *m* continental
g submariner Kontinentalrand *m*

* **continental rise → 1249**

* **continental sea → 1893**

* **continental sediments → 1245**

1253 **continental shelf; shelf**
A gently sloping submarine plain forming a border to nearly every continent, stretchng from the coast to the continental slope.
r шельф; материковая отмель; континентальный шельф
Мелководная часть подводной окраины материков и островов, имеющая относительно ровную поверхность и незначительные уклоны дна.
f plate-forme *f* continentale
s plataforma *f* continental
g Kontinentalschelf *m*

1254 **continental slope**
The marked slope lying between the edge of the continental shelf and the continental foot.
r материковый склон
Часть подводной окраины материка между шельфом и материковым подножьем, отличающаяся большими уклонами поверхности и резкой расчлененностью рельефа.
f talus *m* continental; pente *f* continentale
s talud *m* continental; pendiente *m* continental
g Kontinentalabfall *m*; kontinentaler Abfall *m*

1255 **continental socle**
A part of a continent situated lower than the mean depth of a shelf edge.
r материковый цоколь
Часть материка, лежащая ниже средней глубины бровки шельфа.
f cocle *m* continental
s zócalo *m* continental
g Kontinentalsockel *m*

1256 **continental waters**
Water located on land and contained in rivers, lakes, reservoirs, bogs, glaciers as well as ground waters.
r воды суши
Воды рек, озер, водохранилищ, болот, ледников, а также подземные воды.
f eaux *fpl* des continents; eaux *fpl* continentale
s aguas *fpl* continentales
g Binnengewässer *pl*

* **continental watershed → 1246**

1257 **continentality**
The degree to which a climate is affected by continental influence, in contrast to *oceanity*.
r континентальность климата
Степень, до которой климат подвержен влиянию континента, противоположно *океаничности*.
f continentalité *f*
s continentalidad *f*; continentalización *f*
g Klimakontinentalität *f*

1258 **continentality index**
Index which characterizes the degree of continentality of a place, generally based on the annual range of temperature.
r индекс континентальности
Параметр, характеризующий степень континентальности климата местности, обычно основанный на годовой амплитуде температуры.
f indice *m* de continentalité
s índice *m* de continentalidad
g Kontinentalitätsindex *m*

1259 **continuous area**
An uninterrupted extension of the area occupied by a plant or animal species.
r сплошной ареал
Целостная площадь распространения вида растений или животных, на которой он встречается во всех типичных для него местообитаниях.
f aire *f* continue
s área *f* continua
g kontinierliches Areal *n*

* **continuous canopy → 1063**

1260 **continuous permafrost**
A zone in which permafrost occurs almost everywhere beneath the land surface. → Fig. 247, p. 539
r сплошная вечная мерзлота
Область почти сплошного, без перерывов, распространения многолетнемерзлых пород. → Рис. 247, с. 539
f pergélisol *m* continu
s permafrost *m* continuo
g kontinuierlicher Permafrost *m*

1261 **continuous precipitation; continuous rain; steady precipitation**
Long-lasting precipitation in the form of rain or snow of almost uniform intensity falling simultaneously over a vast area.
r обложные осадки; обложной дождь
Длительные осадки достаточно равномерной интенсивности в виде дождя или снега, одновременно выпадающие на значительной площади.
f précipitations *fpl* continue; pluie *f* continue
s precipitaciones *fpl* continuos; lluvia *f* continua
g Dauerniederschläge *fpl*; Landregen *m*

* **continuous rain → 1261**

* **contorted till → 1632**

1262 **contour; isohypse; form-line**
An imaginary line on a map joining all the points on the ground that are at the same height above sea-level or, for submarine contours, below sea-level.
r горизонталь; изогипса
Линия на карте, соединяющая все точки поверхности с одной и той же абсолютной высотой.
f courbe *f* de niveau; isohypse *f*; courbe *f* hypsométrique
s curva *f* de nivel; curva *f* hipsométrica; isohipsa *f*
g Höhenlinie *f*; Isohypse *f*

1263 **contour interval; index contours**
The difference in height represented by adjacent contours shown on a topographic or geographic map.
r высота сечения рельефа; сечение рельефа
Разность высот между смежными горизонталями на топографической или географической карте.
f équidistance *f* des courbes de niveau
s equidistancia *f* de curvas de nivel
g Höhenlinienabstand *m*; Schichthöhe *f*

1264 **contour ploughing**
The farming practice of cutting furrows across a hillslope, following the contours, to reduce soil erosion caused by the run-off of rainwater.
r распашка вдоль склона
Практика пахоты на склонах холмов приблизительно вдоль горизонталей с тем, чтобы уменьшить эрозию, вызываемую стоком дождевых вод по распаханным бороздам.
f labourage *m* en landes de niveau; labourage *m* en niveau de contour
s labranza *f* a lo largo de curvas de nivel
g horizontales Aufpflügen *n*; Konturpflügen *n*

1265 **contraction fissure; compression fussure**
A fissure appearing in soil or rocks as a result of their contraction under the changes of temperature and water content, e.g. frost fissure, desiccation crack, etc.
r контракционная трещина; усадочная трещина
Трещина, возникающая в грунтах или горных породах в результате их сжатия при изменениях температуры и влажности, например, морозобойные трещины, трещины усыхания и др.
f fissure *f* de retrait
s fisura *f* de contracción
g Kontraktionsspalte *f*

1266 **contraction hypothesis**
Suggestion that as the Earth's interior cooled, so it would contract resulting the compression of the Earth's crust and creation of fold mountains.
r контракционная гипотеза
Концепция, согласно которой складчатость слоев горных пород и горообразование происходят в результате охлаждения Земли и сокращения ее объема, радиуса и площади земной поверхности.
f hypothèse *f* de contraction; théorie *f* de la contraction
s hipótesis *f* de contracción; teoría *f* de la contracción
g Kontraktionstheorie *f*

* **contrail → 1191**

1267 **contrastive relief**
The alteration of positive and negative landforms with a big difference of heights on a particular area of the Earth's surface.
r контрастность рельефа
Частая смена положительных и отрицательных форм рельефа с большими амплитудами высот на смежных участках земной поверхности.
f relief *m* vigoureux; relief *m* différencié
s relieve *m* contrastivo; relieve *m* diferencial
g Reliefgegensätze *mpl*

1268 **conurbation; metropolitan area**
A group of cities forming a continuous built-up area as a result of the expansion and consequent coalescense of previously separate

urban areas, none of which dominates the others (e.g. Ruhr basin in Germany).
r конурбация
Группа близко расположенных и связанных друг с другом городов, ни один из которых не доминирует над другими, например, скопление городов в Рурском бассейне Германии.
f conurbation *f*
s conurbación *f*
g Konurbation *f*

1269 convection
The process of the vertical heat transfer from place to place caused by the differences in temperature and hence in density of the fluid itself and the pull of gravity.
r конвекция
Процесс вертикального переноса тепла от места к месту, вызываемый различиями в температуре и плотности воды или воздуха.
f convection *f*
s convección *f*
g Konvektion *f*

1270 convection cloud; convective cloud
A cumuliform cloud formed by atmospheric convection as a result of the adiabatic cooling of ascendng air and condensation of water vapour in it.
r конвективное облако; облако конвекции
Кучевообразное облако, образующееся при атмосферной конвекции в результате адиабатического охлаждения поднимающегося воздуха и конденсации в нем водяного пара.
f nuage *m* de convection; nuage *m* convectif
s nube *f* de convección; nube *f* convectiva
g Konvektionswolke *f*

1271 convection in the atmosphere
Vertical air motions caused by irregular distribution of temperature of underlying surface. → Fig. 70
r атмосферная конвекция
Вертикальные движения воздуха, вызываемые неравномерным распределением температуры подстилающей поверхности. → Рис. 70

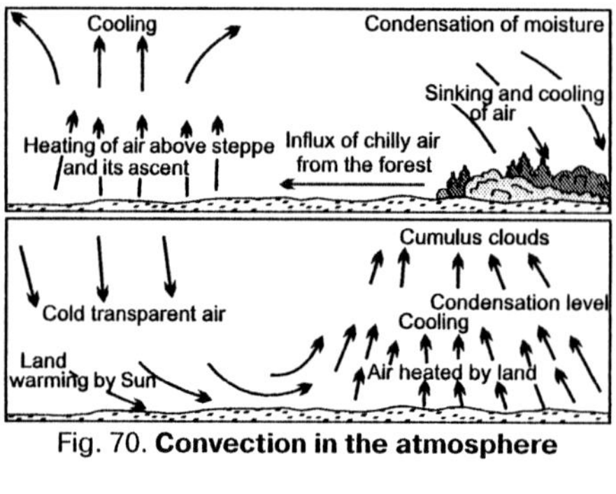

Fig. 70. **Convection in the atmosphere**

f convection *f* atmosphérique
s convección *f* atmosférica
g atmosphärische Konvektion *f*

1272 convection level
A height at which the temperature of ascending air becomes equal to that of the surrounding air, and the convection attenuates.
r уровень конвекции
Уровень, на котором температура восходящего воздуха выравнивается с температурой окружающей воздушной среды и конвекция затухает.
f niveau *m* de convection
s nivel *m* de convección
g Konvektionshöhe *f*

1273 convection rain; convectional rain
Liquid precipitation caused by the warming of air by thermal conduction from the heated land surface (moist with water is taken up from the ground and its vegetation).
r конвективный дождь
Жидкие осадки, вызванные прогреванием воздуха путем теплопроводности от нагретой земной поверхности, при этом влага берется из грунта и покрывающей его растителььности.
f pluie *f* de convection
s lluvia *f* de convección; lluvia *f* convectiva
g Konvektionsregen *m*

* **convectional rain → 1273**

* **convective cloud → 1270**

1274 convective condensation level
The altitude in the atmosphere with corresponding air pressure at which condensation and cumuliform cloud formation occurs.
r уровень конвективной конденсации
Высота над земной поверхностью с соответствующим атмосферным давлением, на которой происходит конденсация влаги и формируются облака кучевых форм.
f niveau *m* de condensation par convection
s nivel *m* de condensación convectiva
g konvektive Kondensationsniveau *n*

1275 convenience good
A good which needs to be bought frequently but for which people are not willing to travel a great distance.
r товар первой необходимости
Товар повседневного спроса, который люди стараются покупать недалеко от места своего проживания или работы.
f article *m* de première nécessité
s artículo *m* de primera necesidad
g Massenbedarfsartikel *pl*

1276 conventional fuel
A unit used in technical and economic calculations to compare heat value of various types of organic fuels.

r условное топливо
Принятая при технико-экономических расчетах единица, служащая для сопоставления тепловой ценности разных видов органического топлива.
f combustible *m* convenu
s combustible *m* convencional; combustible *m* condicional
g konventioneller Brennstoff *m*

1277 conventional name; exonym
The name of towns, localities, physical features, given by the people of one linguistic group or nation to the place-names etc. of another.
r экзоним
Название городов, местностей и других природных особенностей в местах проживания одного народа, данное представителями иной лингвистической группы, например, город Любек на севере Германии.
f exonyme *m*
s exónimo *m*
g Exonym *n*

1278 conventional projection
A map projection the calculation of which is determined by some conditions, such as type of a cartographic grid, form of parallels and meridians, the maximum values of distortions of angles and areas, etc.
r условная проекция
Картографические проекции, при расчете которых исходят из каких-либо условий: вида картографической сетки, формы параллелей и меридианов, предельных значений искажения углов и площадей и т.п.
f projection *f* aphylactique
s proyección *f* convencional
g vermittelnde Kartennetzentwurf *m*

1279 conventional sign; characteristic sheet
A standard symbol, used on a map and explaned in the legend, to show a location, shape, qualitative and quantitative parameters etc.
r условный знак; условное обозначение
Графические обозначения, при помощи которых на картах показывают местоположение, очертания, качественные и количественные характеристики объектов.
f signe *m* conventionnel; symbole *m* d'une carte
s signo *m* convencional; símbolo *m* de un mapa
g Kartenzeichen *n*

1280 convergence
The movement of sea waters of different salinity and temperature to meet at a very defined line and sink of surface water to the deep layers; a type of horizontal air-movement in which inflow exceeds outflow, thereby creating a tendency for air to accumulate at a particular place in the atmosphere.
r конвергенция
Схождение поверхностных течений и опускание вод на глубину; горизонтальное движение воздуха, при котором его приток превышает отток, в результате чего воздух скапливается в определенном месте атмосферы.
f convergence *f*
s convergencia *f*
g Konvergenz *f*

1281 convergence of landforms
The external similarity, often temporary,of landforms of different genesis, e.g. volcanic and meteorite craters.
r конвергенция форм рельефа
Внешнее, часто временное, сходство форм рельефа разного происхождения, например, вулканических и метеоритных кратеров.
f convergence *f* des formes du relief
s convergencia *f* de formas de relieve
g Konvergenz *f* der Reliefformen

1282 converging terraces
River terraces with the elevation above each other decreasing downstream.
r сходящиеся террасы
Речные террасы с превышениями друг над другом, уменьшающимися вниз по долине.
f terrasses *fpl* convergentes
s terrazas *fpl* convergentes
g konvergierende Terrassen *fpl*

1283 convex slope
A slope that gets steeper downhill and has a well-defined foot.
r выпуклый склон
Склон с пологой верхней и крутой нижней частями и с резко выраженной подошвой.
f pente *f* convexe
s pendiente *f* convexa
g konvexer Hang *m*

1284 conveyance distance
The distance between the place of loading and unloading of cargoes or from boarding and unloading of passengers.
r дальность перевозки
Расстояние от места погрузки до места выгрузки или перевалки груза, или посадки и высадки пассажиров.
f distance *f* de transport
s distancia *f* de transporte
g Güterbeförderungszone *f*

1285 convolute bedding
The appearance of a crumbled or highly disturbed layer in a well-bedded sedimentary succession.
r изогнутое залегание
Появление раздробленного или сильно нарушенного слоя в правильной последовательности осадочных отложений.
f disposition *f* convolutée des couches; stratification *f* convolutée

s disposición *f* de capas convoluta; estratificación *f* convoluta
g gebogenes Lagerung *n*

* **cool current → 1110**

1286 cooling water pool; basin-cooler; water reservoir-cooler
Natural or artificial water bodies used to cool the warm water discharged from cooling systems.
r бассейн-охладитель; водоем-охладитель
Естественный или искусственный водоем для охлаждения сбрасываемых в них теплых вод из систем охлаждения.
f bassin *m* de refroidissement
s embalse *m* enfrigerante; estanque *m* de refriamiento
g Kühlbecken *n*

* **co-operation → 1287**

1287 cooperation; co-operative; cooperative; co-operation
A combination of individual farmers in a group for mutual benefit in collective tilliage and agricultural production or combination of small private enterprises in order to share by the members the costs and profits of produce and marketing.
r кооперация
Объединение мелких сельских производителей для совместной обработки земли и производства сельскохозяйственной продукции, либо мелких частных предприятий для улучшения снабжения и сбыта своих товаров.
f coopération *f*; coopérative *f*
s cooperativa *f*; cooperación *f*
g Kooperation *f*; Zusammenwirken *n*

1288 co-operation of production; production co-operation
The process of organized interaction between specialized enterprises which provides the basis of production specialization.
r кооперирование производства
Организованная взаимосвязь специализированных предприятий, служащая основой специализации производства.
f coopération *f* de production
s cooperación *f* de producción
g Kooperation *f* der Produktion

* **cooperative → 1287**

* **co-operative → 1287**

* **coordinate grid → 2635**

* **coppice → 5496**

* **coprolite → 1776**

1289 coprophagous animal
Animals feeding on excrements.
r копрофаг
Животные, питающиеся экскрементами.
f animal *m* coprophage
s animal *m* coprófago
g Koprophag *m*

* **copse → 5496**

1290 coral
Sedentary marine animals that live in colonies and have calcareous skeletons of different shapes, which fused together form the basis of coral reefs.
r коралл
Морское животное, образующее известковый скелет разнообразной формы, служащий основой коралловых рифов.
f corail *m*
s coral *m*
g Koralle *f*

1291 coral bleaching
The loss of coralrs colourful symbiotic algae, usually leading to death of the coral reef.
r обесцвечивание кораллов
Потеря кораллами водорослей, живущих с ними в симбиозе, что приводит к потере цвета и омертвению кораллового рифа.
f blanchiment *m* des coraux; décolorisation *f* des coraux
s blanqueo *m* de corales
g Korallenbleichen *n*

1292 coral coast
A coast formed by coral constructions and found in tropical seas.
r коралловый берег
Берега в тропических морях с коралловыми постройками.
f côte *f* corallienne
s costa *f* coralina
g Riffküste *f*; Korallenküste *f*

1293 coral island
An island formed from the products of destruction of corals or composed of coral limetones.
r коралловый остров
Остров, образованный из продуктов разрушения коралловых построек, реже сложенный коралловыми известняками.
f île *f* corallienne
s isla *f* coralina
g Koralleninsel *f*

1294 coral limestone
Hard porous lime rock composing coral reefs.
r коралловый известняк
Твердая пористая известковая порода, слагающая коралловые рифы.
f calcaire *m* corallien
s caliza *f* coralina
g Korallenkalk *m*

1295 coral mud

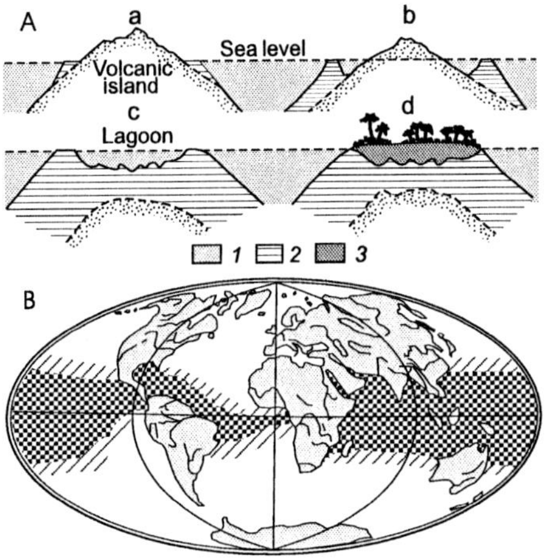

Fig. 71. **Coral reef**. A. Successive formation of reef: a – border reef, b – barrier reef, c – atoll, d – reef island; 1 – sea water, 2 – reef rock, 3 – reef sediment. B. Distribution of coral reefs over the world

An accumulation of very fine coral fragments which lie near to coral reef.

r коралловый ил
Современный осадок, состоящий в основном из обломков кораллов.

f vase *f* corallienne
s limo *m* coralino
g Korallenschlick *m*

1296 coral reef
A submarine lime construction formed along the shallow (from 1 to 50 m) shores of some tropical areas as a result of life activity of colony corals, calcareous algae, pearlweeds and other reef-forming organisms. Fig. 71

r коралловый риф
Подводное известняковое сооружение, формирующиеся в тропических морях на небольшой глубине (от 1 до 50 м) колониальными кораллами, известковыми водорослями, мшанками и другими рифообразующими организмами. → Рис. 71

f récif *m* corallien; corallien *m*; récif *m*; récif *m* de corail
s arrecife *m* coralino; arrecife *m*
g Korallenriff *n*

1297 coral sand
An accumulation of fine white sand composed of comminuted coral fragments bordering a coral reef.

r коралловый песок
Отложения белого мелкозернистого песка, состоящего из измельченных фрагментов кораллов и ограничивающего коралловый риф.

f sable *m* corallien
s arena *f* coralina
g Korallensand *m*

* **corded lava → 5132**

1298 core$_1$
1. The central part of planet Earth with the radius approximately 3500 km, probable consisting of a dense nickel-iron alloy with temperature estimated at about 2,700°C. 2. The centre of the core-periphery model.

r земное ядро
Центральная часть планеты Земля радиусом приблизительно 3500 км, вероятно, состоящая из плотного железо-никелевого сплава с температурой около 2700°С.

f noyau *m*
s núcleo *m*
g Kern *m*

1299 core$_2$; drill core; core sampling
A cylindrical rock column got as a result of drilling which serves as a material for various laboratory analyses; wood sample 5-7 mm in diameter taken perpendicularly to the centre of the tree trunk, used for the purpose of dendroindication of various natural events.

r керн
Цилиндрическая колонка горной породы, получаемая в результате бурения и служащая материалом для разнообразных лабораторных анализов; образец древесины, диаметром 5-7 мм, взятый перпендикулярно к центру ствола дерева и используемый в целях дендроиндикации различных природных событий.

f carotte *f*; carottage *f*
s testigo *m* de perforación; testigo *m* cilíndrico de sondeo
g Bohrkern *m*

1300 core of fold
The internal part of a fold in anticlines composed of more ancient deposits than with the external part of the fold, and in synclines it is composed of younger deposits.

r ядро складки
Внутренняя часть складки, сложенная в антиклиналях более древними отложениями по сравнению с внешней частью складки, а в синклиналях – более молодыми.

f noyau *m* de pli; coeur *m* de pli
s núcleo *m* de pliegue; corazón *m* de pliegue
g Faltenkern *m*

* **core sampling → 1299**

1301 core-periphery model
A model of spatial organization of human activity in which centres or industrially developed cores and resource-frontier regions or peripheral zones are in constant functional cooperation.

r модель «центр-периферия»
Модель пространственной организации хозяйственной деятельности, в которой промышленно развитые центры, с одной стороны, и окраинные ресурсо-производящие территории, с другой, находятся в постоянном функциональном взаимодействии.

f modèle *m* «centre-périphérie»
s modelo *m* «centro-periferia»
g Zentrum-Peripherie-Modell *n*

1302 Coriolis force; deviating force
A force produced by the Earth's rotation which effects every particle moving relative the Earth, deviating it to the right in the Northern Hemisphere, and to the left in the Southern Hemisphere.
r сила Кориолиса
Отклоняющая сила вращения Земли, воздействующая на любую частицу, движущуюся относительно Земли, и отклоняющая ее движение вправо в Северном полушарии и влево в Южном.
f force *f* de Coriolis; force *f* deviante
s fuerza *m* de Coriolis; fuerza *m* desviadora
g Corioliskraft *f*

1303 corona
The coloured rings, typically from blue inside to red outside, that surround the Sun or Moon; the fringe of radiant light seen around the dark rim of the Moon during a total lunar eclipse, or around the dark rim of the Sun during a total solar eclipse.
r корона
Цветные кольца, обычно от голубого внутри до красного снаружи, иногда появляющиеся вокруг Солнца или Луны; бахрома радиальных световых лучей, видимая вокруг темного обода Луны во время полного лунного затмения или вокруг темного обода Солнца во время полного солнечного затмения.
f couronne *f*
s corona *f*
g Korona *f*

* **corona discharge → 5749**

* **corporation → 1182**

1304 corrasion
The process of mechanical impact of rock debris transported by running water, by glaciers, by wind, or by gravity upon bedrock.
r корразия
Механическое воздействие на коренные горные породы твердого обломочного материала, влекомого водой, льдом, ветром или движущегося под действием силы тяжести.
f corrasion *f*
s corrasión *f*
g Korrasion *f*; Abschleifung *f*

1305 correction sheet
A geographical map used to record the changes happened in the area, this information should be taken into account while updating the existing and compiling the new maps.
r дежурная карта
Географическая карта, на которой ведется регистрация изменений местности, подлежащих учету при исправлении существующих и составлении новых карт.
f modéle *m* de tenue à jour
s hoja *f* de correcciones
g Korrekturblatt *n*

1306 correlate deposits
Geological sediments belonging to one and the same chronostratigraphic subdivision.
r коррелятные отложения
Геологические отложения, принадлежащие к одному и тому же хроностратиграфическому подразделению.
f dépôts *mpl* corrélatives
s depósitos *mpl* correlativos
g korrelative Ablagerungen *fpl*

1307 corridor
A strip of territory belonging to one country but running through the territory of another country in order to give the country owning the strip access to an international waterway, the sea, etc.
r территориальный коридор
Сравнительно узкая полоса территории, принадлежащая данной стране, но проходящая в окружении территории другого государства, предоставленная в качестве выхода страны к международным водам, морю и т.п.
f corridor *m* territorial
s corredor *m* territorial
g Territorialkorridor *m*

* **corrie → 1005**

1308 corrie glacier; cirque glacier
A short-tongued small glacier which fills a

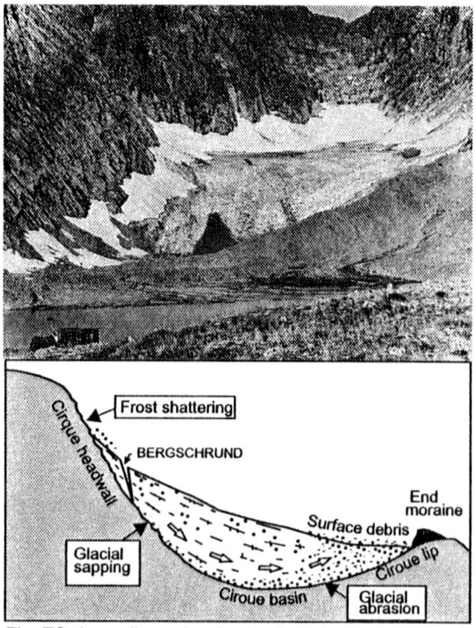

Fig. 72. A small **corrie glacier** in Polar Urals (above); scheme of the corrie glacier (down)

separate, rounded basin glacially eroded on a mountain slope. → Fig. 72, p. 156
r каровый ледник
Небольшой ледник с коротким языком, лежащий в чашеобразном углублении горного склона. → Рис. 72, с. 156
f glacier *m* de cirque
s glaciar *m* de circo
g Kargletscher *m*

1309 corrie shoulder
Escarp of a corrie wall.
r плечо кара
Уступ, образуемый стенкой кара.
f balcon *m* de cirque
s balcón *m* de circo
g Karwandstufe *f*

1310 corrie-valley glacier
A corrie glacier moving down to the valley below for a distance not longer than one or two thirds of the glacier total length.
r карово-долинный ледник
Каровый ледник, спускающийся в нижележащую долину на расстояние, не превышающее одну-две трети общей длины ледника.
f glacier *m* de cirque avec une langue
s glaciar *m* de circo y valle
g Kartalgletscher *m*

* **corrosion → 970**

1311 corvée
Work which had to be done by peasants without payment under the feudal system.
r барщина
Форма земельной ренты, даровой принудительный труд зависимого крестьянина, работающего собственным инвентарем в хозяйстве феодала.
f corvée *f*
s angaria *f*
g Fron *f*; Frondienst *m*

1312 cosmic *adj*; extraterrestrial *adj*; spatial$_2$ *adj*
Applied to phenomena of space outside of the Earth's atmosphere.
r космический
Относящийся к явлениям, происходящим в мировом пространстве, вне атмосферы Земли.
f cosmique *adj*
s cósmico *adj*; extraterrestre *adj*
g kosmisch *adj*; Raum-; Weltraum-

1313 cosmic dust; meteoric dust
Very small particles of solid matter present in the atmosphere derived from extraterrestrial sources and falling on to the Earth's surface.
r космическая пыль
Очень маленькие частицы твердого вещества, присутствующие в мировом пространстве и выпадающие на Землю.
f poussière *f* cosmique
s polvo *m* cósmico
g kosmischer Staub *m*

* **cosmic radiation → 1314**

1314 cosmic rays; cosmic radiation
A flow of energetic particles from outside the solar system and the products of their collision with atmospheric gases.
r космические лучи; космическое излучение
Поток частиц высоких энергий, приходящих на Землю из мирового пространства, и продуктов их взаимодействия с атмосферными газами.
f rayons *mpl* cosmiques
s rayos *mpl* cósmicos
g kosmische Strahlung *f*

1315 cosmography; cosmology
The science concerned with the structure of the Earth and the universe, including astronomy, geography and geology (historically, a synonym for physical geography).
r космография
Наука о структуре Земли и Вселенной, включает астрономию, географию и геологию, в прошлом синоним физической географии.
f cosmographie *f*
s cosmografía *f*
g Kosmographie *f*; Weltbeschreibung *f*

* **cosmology → 1315**

* **cosmopolite → 6435**

* **cosmos → 6466**

1316 costs; expenditures
The total amount of materialized and live labour, including losses caused by imperfect organization of labour, production process, transportation of products, etc. (on the level of an enterprise the term means production costs).
r затраты; издержки
Все объективно необходимые вложения (расходы) овеществленного и живого труда, вместе с потерями, возникающими в результате несовершенной организации труда, производства, транспортировки продукции и пр. (на уровне предприятия это себестоимость продукции).
f expenditures *fpl*; frais *mpl* généraux; coût *m* de production
s gastos *mpl* generales; coste *m* de producción
g Gesamtkosten *pl*

1317 costs of transport; transport costs
All the expenses involved in the moving of goods and people from one place to another, including inventory, paperwork, handling, packaging, insurance, freight rates, temporary warehousing en route, etc.

r транспортные затраты
Затраты, связанные с перемещением грузов, людей из одного пункта в другой с использованием транспортных средств, включая оформление и переработку грузов, их упаковку, стахование, стоимость фрахта, временную перегрузку в пути и пр.
f frais *mpl* de transport; coût *m* de transport
s gastos *mpl* de transporte; coste *m* de transporte
g Thansportkosten *pl*

1318 co-tidal line
A line drawn on a tidal chart joining points at which high water (or low water) occurs at the same time, co-tidal lines radiate from an amphidromic point.
r котидальная линия; линия равных приливов
Линия на карте приливов, соединяющая точки одновременного наступления полной (или малой) воды, эти линии отходят радиально от амфидромической точки.
f ligne *f* cotidale
s línea *f* cotidal
g Flut(stunden)linie *f*

1319 cottage industry; domestic industry
Manufacturing wholly or partly carried on in the home of the worker.
r кустарная промышленность; надомное производство
Изготовление товаров, полностью или частично выполняемое рабочим в своем доме.
f entreprise *f* à domicile
s industria *f* casera; industria *f* domiciliaria
g hausgewerbliche Industrie *f*; Hausindustrie *f*; Heimindustrie *f*

1320 couloir
A deep narrow gully on a steep mountain slope devoid of vegetation.
r кулуар
Глубокая ложбина на крутых, лишенных растительности склонах гор.
f couloir *m*
s pasillo *m*
g Couloir *m*

1321 countercurrent
Sea current directed opposite to winds prevailing in a given region or to other permanent surface currents.
r противотечение; вторичное течение
Морское течение, направленное против преобладающих в данном районе ветров или против движения других устойчивых поверхностных течений.
f contre-courant *m*
s contra-corriente *f*
g Gegenströmung *f*

* **counterradiation → 513**

* **counter-urbanisation → 1564**

1322 counter-urbanization
The movement of people and industry away from major towns and sities.
r контрурбанизация
Перемещение людей и промышленности из главных городов на периферию.
f contre-urbanisation *f*
s contraurbanización *f*
g Kontraurbanisierung *f*

* **country → 5778**

* **country road → 1769**

* **country study → 4976**

1323 countryside; rural area
Populated territory outside urban settlements with natural and cultural landscapes, population and rural settlements.
r сельская местность
Обитаемая территория, находящаяся вне городских поселений, с ее естественными и антропогенными ландшафтами, населением и сельскими населенными пунктами.
f campagne *f*; zone *f* rurale; espace *m* rural
s campo *m*; zona *f* rural
g Landesgebiet *n*

1324 cover loam
A thin layer of loam without boulders covering different relief elements in the area of the Pleistocene glaciation.
r покровный суглинок
Маломощный слой безвалунного суглинка, покрывающий различные элементы рельефа в области плейстоценового оледенения.
f limon *m* de couverture
s limo *m* de manta
g Decklehm *m*

1325 covered karst
A type od karst in which karstic rocks are buried beneath a cover of loose insoluble material of various genesis.
r покрытый карст
Тип карста, в котором карстующиеся породы прикрыты рыхлыми нерастворимыми отложениями разного генезиса.
f karst *m* couvert
s carso *m* cubierto
g bedeckter Karst *m*

1326 crack in ice
A rupture in drifting ice not spreading outwards.
r трещина во льду
Разрывы плавучего льда, которые не разошлись в стороны.
f fissure *f* dans la glace
s grieta *f* en hielo
g Treibeisspalte *mpl*

1327 crag
An isolated, high rugged mass of resistant rock.

r скала; утес
Изолированная часть горного отрога или одиноко стоящая каменная глыба, сложенная устойчивыми к выветриванию породами.
f rocher *m*
s peña *f*; peñasco *m*; roca *f*
g Fels *m*

1328 **crater**
A bowl-shaped depression at the top or on the flanks of a volcanic cone from dozens meters to several kilometres in diameter formed as a result of explosive eruption.
r кратер
Округлое углубление на вершине или склоне вулканического конуса диаметром от десятков метров до нескольких километров, возникшее в результате эксплозивного (взрывного) извержения.
f cratère *m*
s cráter *m*
g Krater *m*

* **crater lake → 6594**

* **craton → 4610**

1329 **creationism**
The concept of species invariability considering the diversity of organic world as a result of its creation by God.
r креационизм
Концепция постоянства видов, рассматривающая многообразие органического мира как результат его сотворения Богом.
f creationisme *m*
s creacionismo *m*
g Kreationismus *m*

1330 **credit; allocation**
Lending money and goods with interest, there are bank, consumer, international and other kinds of credit.
r кредит
Предоставление денег или товаров в долг и, как правило, с уплатой процентов, бывает банковский (денежные ссуды банков) и коммерческий (отсрочка платежа продавцом товара) кредиты, а также международный (предоставляемый государствами и международными банками).
f crédit *m*
s crédito *m*
g Kredit *m*

1331 **creek**
An intermittent watercourse in Australia; an estuary of a small river or rill inflowing into the sea on a low coast of the USA.
r крик
Периодически пересыхающая река или водоток в Австралии; эстуарий небольшой речки или ручья, впадающий в море на низменном побережье США.
f crique *f*
s caleta *f*
g Creek *n*; Priel *m*

1332 **creep**
The slow downslope displacement of loose rocks resulting from periodic changes of their volume due to different causes, including the alternate freezing and thawing.
r крип; оползание грунта; десерпция
Медленное сползание рыхлых пород по склону вследствие периодического изменения их объема из-за разных причин, в числе которых промерзание и оттаивание.
f reptation *f*
s reptación *f*
g Gekriech *n*; Abgleiten *n*

1333 **crepuscular rays**
The rays of sunlight which diverge upwards immediately following the setting of the Sun, they represent true sunlight from below the horizon and are caused by interruptions of the sun light by hills and by cloud masses.
r сумеречные лучи
Лучи солнечного света, которые появляются из-под горизонта сразу после захода Солнца, они вызываются препятствиями на пути света в виде неровностей рельефа и облачных масс.
f rayons *mpl* crépusculaires
s rayos *mpl* crepusculares
g Dämmerungsstrahlen *mpl*

1334 **crescent moon**
The figure of the Moon at any stage between new Moon and first quarter or between last quarter and the succeeding new Moon.
r лунный серп; месяц^2
Фигура Луны в любой момент между молодой Луной и первой четвертью или между последней четвертью и следующей молодой Луной.
f croissant *m*
s luna *f* menguante
g Mondsichel *f*

1335 **crest**
The highest part of the summit ridge of a mountain; the highest point of an anticline; the highest point of an ocean wave; the maximun stage of a flood as it moves down river; the summit ridge of a sand dune; the highest point of a ripple.
r гребень
Высочайшая часть горного хребта; самая высокая точка антиклинали, океанской волны, песчаной дюны, ветровой ряби; максимальная стадия паводка, движущегося по реке.
f crête *f*
s cresta *f*
g Kamm *m*

1336 **crest of wave**
The highest point of a wave.

Fig. 73. Glacier **crevasse** in the Pamir mountains

r гребень волны
Высшая точка волны.
f crête *f* de vague
s cresta *f* de ola
g Wellenkamm *m*; Wellenkopf *m*

1337 Cretaceous
The third (the latest) geological period of the Mesozoic era which began 135-137 million years ago and lasted for about 70 million years. → Table 1, p. 284
r мел
Верхний (последний) период мезозойской эры геологической истории Земли, начался 135-137 млн лет назад и продолжался около 70 млн лет. → Табл. 1, с. 284
f Crétacé *m*
s Cretácico *m*
g Kreide *f*

1338 crevasse; glacier crevasse; glacier fissure
A vertical or steeply inclined crack in the glacier body resulted from the breaking-up of ice in the course of glacier flow over the irregular earth surface. → Fig. 73
r ледниковая трещина
Вертикальный или круто наклонный разлом в леднике, вызванный разрывом льда в процессе его движения. → Рис. 73
f crevasse *f* glaciaire; fissure *f* glaciaire
s fisura *f* glaciar; crevasse *f* glaciar
g Gletscherspalt *m*; Gletscherriss *m*

* **critical environmental load → 3733**

1339 critical flow velocity
Mean flow velocity under which laminar flow changes to the turbulent one and vise versa.
r критическая скорость потока
Средняя скорость потока, при которой происходит смена ламинарного течения на турбулентное и наоборот.
f vélocité *f* de courant critique
s velocidad *f* de corriente crítica
g kritische Strömungsgeschwindigkeit *f*

1340 critical level of groundwater
The depth of the unconfined groundwater bedding above which the capillary salt-bearing solutions reach the soil surface causing the salinisation of soil.
r критический уровень грунтовых вод
Глубина залегания поверхности грунтовых вод, выше которой капиллярные соленосные растворы достигают поверхности почв, вызывая их засоление.
f niveau *m* critique d'eau souterraine
s nivel *m* crítico de agua subterránea
g höchstzulässiger Grundwasserstand *m*

1341 critical temperature
1. A temperature of vital importance to plants, i.e. the temperature below which growth cannot take place (in midlatitudes 6°C for most food crops). 2. The highest temperature at which a liquid and its vapour can co-exist.
r критическая температура
1. Важнейший температурный рубеж для растений, ниже которого рост не происходит (в средних широтах для большинства сельскохозяйственных растений он равен 6°С). 2. Наивысшая температура, при которой данное вещество может находиться одновременно и в жидком и в газообразном состояниях.
f température *f* critique
s temperatura *f* crítica
g kritische Temperatur *f*

1342 crop; yield; harvest
The annually or seasonally harvested produce resulting from the cultivation of grain, grass, fruit, etc; cultivated produce while growing.
r урожай
Ежегодное или сезонное производство сельскохозяйственных культур: зерновых, фруктов, овощей и пр.
f récolte *f*
s cosecha *f*; cultivo *m*
g Ernte *f*; Ernteertrag *m*

1343 crop farming; arable farming; crop-raising
A system of methods of cultivating the land for growing crops such as cereals, vegetables etc. on arable land.
r земледелие
Система приемов воздействия на почву для выращивания сельскохозяйственных культур и получения высоких, устойчивых урожаев.
f agriculture *f* arable
s agricultura *f* de cultivos
g Ackerbau *m*; Feldbau *m*

* **crop rotation → 3379**

* **cropland → 354**

* **crop-raising → 1343**

1344 **cross profile of a valley**
A transverse profile of a valley drawn approximately at right angles to the stream flowing through it.
r поперечный профиль долины
Поперечное сечение долины примерно под прямым углом к протекающему здесь потоку.
f profil *m* transversal de vallée
s perfil *m* transversal de valle
g Talquerschnitt *m*

1345 **cross section; transverse section; cross-profile**
A profile of a natural object along the line perpendicular to its longitudinal axis.
r поперечный профиль
Профиль природного объекта по линии, перпендикулярной его продольному направлению.
f profil *m* à traversal; section *f* transversale
s perfil *m* transversal; sección *f* transversal
g Querprofil *n*

* **cross section of a stream → 956**

* **cross-bedding → 2100**

1346 **cross-circulation in rivers**
Flows in a plane of cross section of a stream having closed character.
r поперечная циркуляция в реках
Течения в плоскости поперечного сечения речного потока, имеющие замкнутый характер.
f circulation *f* transversale
s circulación *f* transversal
g Querzirkulation *f*

* **cross-over → 5080**

* **cross-profile → 1345**

* **crown cover → 856**

* **crude oil → 4524**

* **cruise → 1347**

1347 **cruise tourism; cruise**
A sea or river trip along a closed or nearly closed circuit for tourist purposes normally with short excursions from ports of call further into the countries visited.
r круизный туризм; круиз
Морское или речное туристское путешествие по замкнутому или почти замкнутому маршруту обычно с короткими радиальными поездками из портов во внутренние районы посещаемых стран.
f croisiére *f*
s crucero *m*
g Kreuzfahrt *f*

1348 **crumb**
A naturally formed rounded, porous aggregate of soil particles, up to 10 mm in maximum dimension.
r крупица
Округлый, пористый агрегат почвенных частиц размером до 10 мм, формирующийся в процессе почвообразования.
f grain *m* du sol
s grano *m* del suelo
g Krümel *m*

1349 **crumb structure**
A soil in which the constituent particles are aggregated into crumbs, up to 10 mm in size, which permit the percolation of air and water between them.
r крупитчатая структура
Строение почвы, когда отдельные частицы соединяются в крупинки размером до 10 мм, дающие возможность циркуляции воздуха и воды между ними.
f structure *f* grenue
s estructura *f* granoulosa
g Krümelstruktur *f*

* **crust of the Earth → 1773**

1350 **crustification**
Ther accumulation of concentric or parallel mineral layers on the walls of hollows and open fissures, and on the surface of rock fragments.
r крустификация
Нарастание концентрических или параллельных минеральных слоев на стенках пустот и открытых трещин, а также на поверхности отдельных обломков.
f crustification *f*
s crustificación *f*
g Krustenbildung *f*

* **cryergic *adj* → 1356**

1351 **cryobiont**
An organism living on ice or in snow, e.g. snow fleas or Chlamidomonas nivalis.
r криобионт
Организмы, обитающие на льду или в снегу, например, глетчерные блохи или снежная хламидомонада.
f cryobionte *m*
s criobionte *m*
g Kryobiont *m*

1352 **cryoconite**
A minute dust material formed by the weathering of moraine deposits, and accumulated on surfaces of glaciers and snow patches.
r криоконит
Мельчайший пылевой материал, образующийся в результате выветривания моренных отложений и скапливающийся на поверхности ледников и снежников.
f cryoconite *m*
s crioconita *m*
g Kryokonit *n*

* **cryoconite hole → 2519**

1353 **cryodiagenesis**
The transformation of rock including the formation of ice and mineral composition change due to its freezing.
r криодиагенез
Преобразование горной породы под влиянием ее промерзания, включая образование льда и изменение минерального состава.
f cryodiagenèse *f*
s criodiagénesis *f*
g Kryodiagenese *f*

1354 **cryofront**
The boundary between the cryotic and noncryotic ground as indicated by the position of the 0°C isotherm in the ground.
r фронт охлаждения
Граница между мерзлым и немерзлым грунтами, демонстрируемая положением нулевой изотермы в грунте.
f cryofront *m*
s criofrente *f*
g Kryofront *f*

1355 **cryogenesis**
A series of natural processes occurring at the surface of the Earth in the cryosphere and leading to ice formation.
r криогенез
Совокупность природных процессов, происходящих в криосфере и сопровождающихся льдообразованием.
f cryogenèse *f*
s criogénesis *f*
g Kryogenese *f*

1356 **cryogenic *adj*; cryergic *adj***
Caused by the processes related to low temperatures, to freeze-thaw processes, i.e. frost action.
r криогенный
Обусловленный процессами, связанными с низкими температурами и фазовыми переходами воды.
f cryogénique *adj*
s criogénico *adj*
g kryogen *adj*

1357 **cryogenic dislocation; congeliturbation; cryoturbation**
Dislocations of rocks caused by their freezing and pressure due to ice formation in ground waters.
r криогенные дислокации; криотурбации
Нарушения горных пород в результате их промерзания и давления, вызванного образованием льда в грунтовых водах.
f géliturbation *f*; cryoturbation *f*
s crioturbación *f*; turbación *f* periglaciar
g Kryoturbation *f*

1358 **cryogenic head; freezing pressure**
The pressure occurring as a result of alternate freezing and thawing of water in the ground.
r криогенный напор; криогенное давление
Давление, возникающее в результате замерзания и оттаивания воды в грунтах.
f pression *f* de gel
s presión *f* de congelamiento
g Gefrierdruck *m*

1359 **cryogenic heaving; frost heaving**
The upward movement of separate portions of ground under the effects of internal stresses related to the expansion of water in lenses and layers upon freezing.
r криогенное пучение
Выдавливание отдельных частей грунта под влиянием внутренних напряжений, связанных с промерзанием линз и пластов воды.
f gonflement *m*; soulèvement *m* gélival
s levantamiento *m* por congelamiento
g Eisauftreiben *n*; Frosthebung *f*

1360 **cryogenic polygon; ice polygon; frost polygon; stone polygon; polygon**
Closed, multi-sided, roughly equidimensional macro- and meso-scale features bounded by more or less straight sides and resulting from frost fracturing of fine-grained ground.
r криогенный полигон
Округлые или многоугольные формы микро- и мезорельефа, возникающие при морозном растрескивании тонкодисперсных грунтов.
f polygone *m* cryogénique
s polígono *m* criogénico; polígono *m* de tierra; polígono *m* de tundra
g Strukturbödenpolygon *n*

1361 **cryogenic relief**
Land forms created or changed by the processes of ground freezing and thawing.
r криогенный рельеф; мерзлотный рельеф
Рельеф, созданный или преобразованный под воздействием процессов промерзания и оттаивания грунтов.
f relief *m* cryogénique
s relieve *m* criogénico
g kryogenetisches Gelände *n*

1362 **cryogenic soil; gelisol**
A mineral soil that forms in cryogenic areas and has permafrost within the upper 1-2 m of its profile and also indications of cryoturbations.
r криосоль; криозем; гелисоль
Минеральная почва, формирующаяся в мерзлотных областях, имеет в профиле многолетнюю мерзлоту в пределах верхних 1-2 м, а также признаки криотурбаций.
f cryosol *m*; gélisol *m*
s criosol *m*; gelisol *m*
g Ñryosol *m*; Frostboden *m*

1363 **cryogenic stage of Pleistocene**
The coldest period over the last 245 million years embracing 25-10 thousand years ago when due to vast glaciation the sea level low-

ered by 130 m, and permafrost area was much larger than the contemporary one.

r криогенный этап плейстоцена
Самый холодный период за последние 245 млн лет, охватывавший 25-10 тыс. лет назад, когда из-за обширного оледенения уровень моря понижался на 130 м, а область многолетней мерзлоты была намного больше современной.
f étape *f* cryogénique de Pléistocène
s etapa *f* criogénica de Pleistocene
g kryogene Periode *f* des Pleistozäns

1364 cryogenic structure
The character of the fine-dispersed and biogenic permanently frozen ground formed by the segregated ice, pore ice and other types of constitutional ice.
r криогенная текстура
Текстура тонкодисперсных и органогенных многолетнемерзлых пород, образованная сегрегационным льдом, льдом-цементом и другими видами конституционного льда.
f texture *f* cryogénique
s textura *f* criogénica
g kryogene Struktur *f*

1365 cryohydrate
A fine dispersed mixture of minute crystals of ice and some salt, which melts producing a solution of the same composition.
r криогидрат
Тонкодисперсная смесь мельчайших кристаллов льда и какой-либо соли, которая плавится с образованием раствора того же состава.
f cryohydrate *m*
s criohidrato *m*
g Kryohydrat *n*

1366 cryolithogenesis
A complex of the processes of ice-bearing ground formation.
r криолитогенез
Совокупность процессов образования горных пород, содержащих в своем составе лед.
f cryolithogenèse *f*
s criolitogénesis *f*
g Kryolithogenese *f*

1367 cryolithology
A branch of *geocryology* that deals with lithological and geomorphological manifestations resulting from the freezing and thawing of the ground.
r криолитология
Раздел *геокриологии*, изучающий литологические и геоморфологические последствия промерзания и протаивания грунтов.
f cryolithologie *f*
s criolitología *f*
g Kryolithologie *f*

1368 cryology
The study of natural phenomena and processes occurring in the cryosphere.
r криология
Наука о природных объектах и процессах, происходящих в криосфере.
f cryologie *f*
s criología *f*
g Kryologie *f*; Frostkunde *f*

* **cryopedology → 4509**

1369 cryopeg
A body of liquid saline water or brine below 0°C located inside permafrost but not containing ice.
r криопег
Природные солоноватые, соленые и рассольные воды в многолетнемерзлых породах, имеющие отрицательную температуру, но не содержащие льда.
f cryopeg *m*
s criopeg *m*
g Kryopeg *n*

1370 cryophile; thermophobe
Organisms adapted to permanently low temperatures.
r криофил; термофоб
Организмы, обитающие в условиях постоянных низких температур.
f cryophile *m*; thermophobe *m*
s criófilo *m*; termofobo *m*
g Kryophil *m*; Thermophob *m*

* **cryophyte → 1375**

1371 cryoplanation
The flattering and lowering of a terrain by the processes related to the action of frost.
r криопланация
Выравнивание рельефа под влиянием криогенных процессов, в частности, вследствие термоденудации.
f cryoplanation *f*; planation *f* thermique
s crioplanación *f*
g Kryoplanation *f*

1372 cryoplankton
Microorganisms, mainly algae that inhabit thawed puddles on the surface of snow and ice in polar areas and high mountains.
r криопланктон
Микроорганизмы, преимущественно водоросли, населяющие талые лужи на поверхности снега и льда в полярных областях и высокогорьях.
f cryoplancton *m*
s crioplancton *m*
g Kryoplankton *m*

1373 cryosphere
The interface of the atmosphere, hydrosphere and lithosphere, in which negative or zero temperatures prevail and water is present in solid or supercooled state.

r криосфера
Область взаимодействия атмосферы, гидросферы и литосферы, где господствуют отрицательные или нулевая температуры и вода в твердой фазе или переохлажденном состоянии.
f cryosphère *f*
s criosfera *f*
g Kryosphäre *f*

1374 **cryotic ground**
Soil or rock at temperatures of 0°C or lower (this term refers solely to the temperature of the rock or soil material independent of its water or ice content.
r отрицательно-температурный грунт
Почва или грунт при температуре 0°C или ниже, не обязательно содержащий воду, т.е. не мерзлый.
f sol *m* cryotique
s suelo *m* criótico
g kryotischer Boden *m*

* **cryoturbation → 1357**

1375 **cryovegetation; cryophyte**
Plants and plant communities, especially algae, mosses and lichens, which have adapted to live in environments of permanent snow and ice.
r холодолюбивая растительность
Растения и растительные сообщества, особенно водоросли, мхи и лишайники, которые прекрасно себя чувствуют в условиях постоянного снега и льда.
f cryovégétation *f*
s criovegetación *f*; criofitas *fpl*
g Kryovegetation *f*

1376 **cryoxerophyte**
The plants of dry and cold habitats, such as steppes and deserts of the mountains in Northern, Eastern and Central Asia.
r криоксерофит
Растения сухих и холодных местообитаний, например, степей и пустынь гор Северной, Восточной и Центральной Азии.
f cryoxérophyte *m*
s crioxerofita *m*
g Kryoxerophyt *m*

1377 **cryptopodzolic soil**
An acid podzolic soil forming on sand deposits, it has an acid reaction and no morphologically pronounced bleached horizon (the podzolization of this soil may be identified only analytically).
r скрытоподзолистая почва
Подзолистая почва на песчаных отложениях, обладающая кислой реакцией, в ней отсутствует морфологически выраженный осветленный горизонт, а оподзоливание обнаруживается лишь аналитически.
f sol *m* cryptopodzolique

Fig. 74. **Cuesta**

s suelo *m* criptopodzólico
g Kryptopodsolboden *m*

1378 **crystallization of water**
The process by which ice crystals are formed in water due to the replacement of the irregular arrangement of molecules by the regular arrangement one.
r кристаллизация воды
Образование в воде кристаллов льда в результате смены неупорядоченного расположения молекул на упорядоченное.
f cristallisation *f* de l'eau
s cristalización *f* de agua
g Wasserkristallisation *f*

1379 **cuesta**
An assymetrical monoclinal ridge in which the long, gentle slope coincides with the dip of the resistant strata. → Fig. 74
r куэста
Моноклинальная гряда или хребет с асимметричным поперечным профилем, где пологий склон соответствует полого падающим пластам прочных горных пород. → Рис. 74
f cuesta *f*
s cuesta *f*
g Cuesta *f*; Schichtkamm *m*

1380 **cultivation of land**
Preparing and using land for growing crops.
r культивация земель
Возделывание сельскохозяйственных земель для получения урожая.
f culture *f* des terres
s cultivo *f* de tierras; labrado *m* de suelo
g Bodenkultivierung *f*

* **cultural area → 1385**

1381 **cultural geography; geography of culture**
The study of the spatial distridution of the elements of the material and spiritual culture of the peoples: traditions, ways of life, beliefs, arts, ways of interaction with nature, etc.
r география культуры
Изучение пространственного разнообразия элементов материальной и духовной культуры народов: обычаев, верований, искусств, взаимоотношений с окружающей средой и пр.

f géographie *f* culturelle
s geografía *f* cultural
g Kulturgeographie *f*

1382 cultural hearth
The centre of a culture or of a cultural group from which, in favourable conditions, the successful culture spread, effecting the changes in the natural landscape.
r очаг культуры
Центр, из которого в благоприятных условиях распространяется соответствующая культура, что, в частности, приводит к ее воздействию на естественные ландшафты.
f foyer *m* de la culture
s foco *m* de la cultura
g Kulturstätte *f*

1383 cultural landscape
A geographical landscape changed by economic activity of people for some specific purpose, e.g. an oasis on irrigated lands in a desert.
r культурный ландшафт
Географический ландшафт, целесообразно измененный хозяйственной деятельностью, например, оазис на орошаемых землях в пустыне.
f paysage *m* cultural
s paisaje *m* cultural
g Kulturlandschaft *f*

1384 cultural layer; cultural level
A fossil layer found in the places of ancient people settlements which contains the remains that serve as the evidencies of human activity.
r культурный слой
Ископаемый слой земли на местах поселений древних людей, содержащий следы их деятельности.
f couche *f* archéologique
s capa *f* arqueológica
g Kulturschicht *f*

*** cultural level → 1384**

1385 cultural region; cultural area
A region identified by the existence within it of a single, distinctive culture or of cultures similar to one other.
r область единой культуры
Район, отличающийся распространением характерных черт одной культуры или нескольких близких друг другу культур.
f région *f* de culture
s región *f* de cultura
g Kulturgebiet *n*

1386 cultural tourism
The touristic visits of historic and cultural monuments, sites of natural and cultural heritage.
r культурно-познавательный туризм
Туристские посещения памятников истории и культуры, достопримечательностей и объектов культурного наследия.
f tourisme *m* culturel
s turismo *m* cultural
g kulturorientirter Tourismus *n*; Kulturturismus *n*; Bildungstourismus *n*

1387 cultural vegetation
The plants grown by man.
r культурная расттительность
Растения, разводимые человеком.
f vegetation *f* cultivée; vegetation *f* sative
s vegetación *f* de cultivo
g Kulturpvegetation *f*

1388 culture
The collective mental and spiritual manifestations (customs, beliefs, ideas, morals, values, aesthetic perception, the forms of behaviour and social structures) together with material and artistic manifestations (tools, buildings, works of art, etc.) formulated and created by people according to the conditions of their lives.
r культура
Совокупность жизненных форм и идей, включающих обычаи, верования и обряды, знания и умения, язык и искусство, строительство и технологии, сформулированные и созданные людьми в соответствии с их историческим опытом и условиями жизни.
f culture *f*
s cultura *f*
g Kultur *f*

1389 culture contact
The meeting or mingling of groups with differing cultural traditions, brought about by migration, displacement, or enlargement of territory, etc. that may result in the emergence of a new society with its own particular cultural traditions.
r смешение культур
Встреча или смешение групп людей с разными культурными традициями, происходящее вследствие миграции, переселения или расширения территории, что в конечном счете нередко приводит к формированию общества со своими особенностями культурных традиций.
f contact *m* de cultures
s contacto *m* de culturas
g Kulturbegegnung *f*

1390 culvert siphon
A pressure conduit built under the channel of a river, canal, etc. in order to let the crossing watercourse pass. → Fig. 75
r дюкер
Напорный водовод, прокладываемый под руслом реки, канала и пр. для пропуска пересекающего их водотока. → Рис. 75
f aqueduc-siphon *m*

Fig. 75. **Culvert siphon** under a channel

Fig. 76. **Cumulus**

s sifón *m* invertido
g Düker *m*

* **cumulation of snow → 5504**

1391 cumuliform cloud
A cloud having a bulging appearance of a cumulus.
r кучевообразное облако
Облако, имеющее характерный «клубящийся» вид кучевого.
f nuage *m* cumuliforme
s nube *f* cumuliforme
g cumuliförmige Wolke *f*

1392 cumulonimbus
A heavy and dense cloud (symbol: Cb), with a considerable vertical extent up to 14 km, in the shape of a mountain or huge towers, usually associated with heavy rain, hail and thunder-storm.
r кучево-дождевые облака
Плотные и мощные облака с сильным развитием по вертикали до высоты 14 км в форме гор или огромных башен, дающие обильные ливневые осадки с грозовыми явлениями (символ: Cb).
f cumulonimbus
s cumulonimbus; cumulonimbo *m*
g Cumulonimbuswolken *fpl*; Kumulonimbus; Kumulonimbuswolken *fpl*; Cumulonimbus

1393 cumulus
A detached cloud (symbol: Cu), generally dense and with sharp outlines, developing vertically in the form of growing mounds, domes or towers. → Fig. 76
r кучевое облако
Отдельное плотное ярко-белое облако с резкими очертаниями, развивающееся вверх в форме растущих холмов, куполов или башен (символ: Cu). → Рис. 76
f cumulus
s cumulus; cúmulo *m*
g Cumuluswolken *fpl*; Quellwolken *fpl*; Cumulus; Kumulus; Kumuluswolken *fpl*

1394 current; flow$_2$
A body of water or air moving vertically or horizontally in a definite direction, including convection current, movement of water in certain channel of a river, permanent or semi-permanent movement of water in the ocean.
r течение
Массы воды или воздуха, перемещающиеся горизонтально или вертикально в определенном направлении, включая движение в результате конвекции, перемещение воды в реках, постоянное или квазипостоянное движение воды в океане и т.п.
f courant *m*
s corriente *f*
g Strömung *f*

* **current ripple → 5094**

1395 cushion plant
A plant life form found in tundra and arid mountain deserts where plants with short dense sprouts form tight hummocks having its own microclimate. → Fig. 77
r подушковидное растение
Густые скопления растений, образующих подушку, внутри которой сохраняется собственный микроклимат, распространены главным образом в тундрах и горных пустынях с аридным климатом. → Рис. 77
f coussinet *m*
s almohadilla *f*
g Polsterpflanze *f*

* **cuspate delta → 6342**

1396 custom duty
The money which is collected for nthe goods brought into a country.
r таможенная пошлина
Денежный сбор, взыскиваемый государством с товаров, провозимых через его границу.
f droit *m* de douane
s derecho *m* arancelario; derecho *m* de aduana
g Zollgebühr *f*; Zollabgabe *f*

Fig. 77. **Cushion plant**

* **customs barrier → 6058**

1397 **customs union**
An agreement between two or more states on eliminating tariff barriers between a territory with common customs tariffs.
r таможенный союз
Соглашение двух или нескольких государств об упразднении таможенных границ между ними и образовании территории с единым таможенным тарифом.
f alliance *f* douanière; union *f* douanière
s alianza *f* aduanera; unión *f* aduanera
g Zollunion *f*

1398 **cut-off**
A new channel formed in a river meander neck; a newly formed branch in a river delta; a place where a spit is attached to the see shore is broken by water. → Fig. 201, p. 446
r прорва; проран
Новое русло, образовавшееся в шейке речной излучины; вновь образовавшийся рукав в дельте реки; место размыва песчаной косы. → Рис. 201, с. 446
f coupure *f* de méandre; percement *m*
s corte *m* de meandro; corte *m*
g Durchbruch *m*

1399 **cut-off meander**
A meander broken by a river along a shorter way which becomes the main river channel.
r прорванная излучина
Излучина, прорванная рекой по более короткому пути, который становится основным речным руслом.
f méandre *m* recoupé
s meandro *m* recordato
g durchschnittener Hals *m* eines Mäanders

* **cutting → 5074**

* **cutting area → 2124**

1400 **cycle**
A succession of stages which occur repeatedly in some order of a sequence of events which return to its starting point, e.g. hydrological cycle.
r цикл
Последовательность стадий какого-либо природного процесса или события, повторяющихся в определенном порядке и в конечном счете возвращающихся к стартовой точке, например, гидрологический цикл.
f cycle *m*
s ciclo *m*
g Zyklus *m*

* **cycle of energy → 1906**

1401 **cycle of erosion**
The successive stages of the development of fluvial landforms. There are three types of the cycle of erosion: normal, arid and glacial.
r эрозионный цикл; цикл денудации
Совокупность последовательных стадий развития водно-эрозионных и водно-аккумулятивных форм рельефа, различают три типа эрозионного цикла: нормальный, аридный и гляциальный.
f cycle *m* d'érosion
s ciclo *m* de erosión
g Erosionszyklus *m*

1402 **cycle of matter**
A series of transformations and transport of substances in nature repeating many times, in which the process is completed and the substance is returned to its starting point.
r круговорот веществ
Многократно повторяющиеся процессы превращения и перемещения веществ в природе, имеющие разный масштаб и циклический характер.
f cycle *m* de la matière
s ciclo *m* de materia
g Kreislauf *m* der Stoffe; Stoffkreislauf *m*

* **cycle of nutrients → 4190**

* **cyclic migration → 4499**

1403 **cyclicity**
Presence or manifestation of cycles of different duration in natural processes.
r цикличность
Наличие или проявление циклов разной продолжительности в природных процессах.
f caractère *m* cyclique
s caracter *m* cíclico
g Zyklizität *f*

1404 **cyclogenesis**
The processes by which any cyclonic circulation is initiated and develops.
r циклогенез; циклонообразование
Процесс возникновения или развития циклонической циркуляции.
f cyclogénèse *f*
s ciclogénesis *f*
g Zyklogenese *f*

1405 **cyclolysis**
The weakening or decay of a cyclonic circulation.
r затухание циклона; циклолиз
Процесс ослабления или прекращения (затухания) циклонической джеятельности.
f cyclolyse *f*
s ciclolisis *f*
g Zyklolyse *f*

1406 **cyclone; depression$_2$**
A system in the atmosphere where the pressure is low relative to the surrounding air at the same level and air moves around the center (counter clockwise in the Northern Hemisphere, and clockwise in the Southern Hemisphere). → Fig. 78, p. 168
r циклон

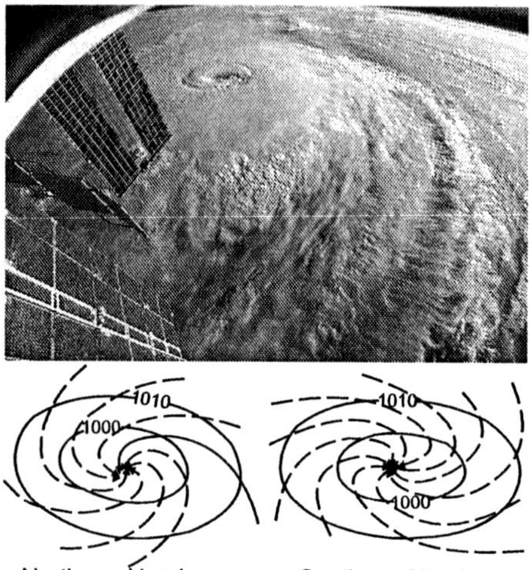

Fig. 78. **Cyclone**: space image made at the orbital station (above); scheme of a cyclone (down)

Область пониженного атмоферного давления в тропосфере с минимальным давлением в центре и барическими градиентами, направленными от периферии к центру (против часовой стрелки в Северном полушарии и по часовой в Южном). → Рис. 78

f cyclone *m*
s ciclón *m*
g Zyklone *f*; Tief *n*; Tiefdruckgebiet *n*

1407 cyclonic circulation

Atmospheric circulation associated with the moving of cyclones.

r циклоническая циркуляция
Циркуляция атмосферы в условиях преобладания областей пониженного давления.

f circulation *f* cyclonique
s circulación *f* ciclónica
g zyklonale Zirkulation *f*

*** cyclonic cloud system → 1076**

1408 cyclonic eddy; cyclonic vortex

A whirl in the ocean with vertical or inclined axis having counter-rotation in the Northern Hemisphere and clockwise in the Southern Hemisphere, in the centre of which the upwelling of cold waters rich in biogenic substances takes place.

r циклонический вихрь
Океанический вихрь с вертикальной или наклонной осью, вращающийся против часовой стрелки в Северном полушарии и по часовой стрелке в Южном, в центре которого происходит подъем с глубин холодных вод, богатых биогенными веществами.

f tourbillon *m* cyclonique
s torbellino *m* ciclónico
g zyklonaler Wirbel *m*

*** cyclonic precipitation → 2322**

*** cyclonic rain → 2322**

*** cyclonic vortex → 1408**

*** cyclonic wave → 2324**

1409 cylindrical projection

A map projection constructed as if the globe is inserted within a cylinder and meridians and parallels are drawn as straght lines intersecting each other at right angles. → Fig. 79

r цилиндрическая проекция
Картографическая проекция, построенная как будто земной шар помещен внутри цилиндра так, что параллели и меридианы представляют собой прямые линии, пересекающие друг друга под прямыми углами. → Рис. 79

f projection *f* cylindrique
s proyección *f* cilíndrica
g Zylinderentwurf *m*

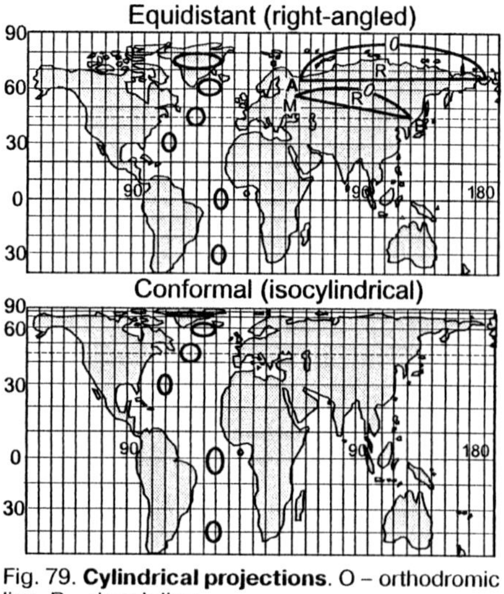

Fig. 79. **Cylindrical projections**. O – orthodromic line, R – rhumb line

D

1410 daban
A mountain pass, or saddle, in a mountain range in Siberia and Central Asia.
r дабан
Горный перевал, седловина в хребте Сибири и Центральной Азии.
f daban *m*
s daban *m*
g Daban *m*

*** dairy farm → 1411**

1411 dairy farming; dairy farm
Farming devoted primarily to the keeping of cows for their yield of milk and milky products such as butter, cheese and others.
r молочное хозяйство
Отрасль животноводства, специализирующаяся на разведении коров для производства молока и молочных продуктов: масла, сыра и пр.
f laiterie *f*; élevage *m* laitier
s hacienda *f* lechera; ganadería *f* lechera
g Molkerei *f*; Milchwirtschaft *f*

1412 dale
A linear flat-bottomed depression with a rudimentary channel, or a broad open valley characteristic of northern England and the regions of permafrost and is formed due to solifluction (the term is commonly used in place names).
r делль
Линейно вытянутая плоскодонная ложбина с зачаточным руслом, чаще безрусловая, характерна для районов распространения многолетнемерзлых пород, формируется в результате солифлюкции.
f vallon *m* de solifluxion
s vallejo *m* de solifluxión
g Delle *f*

1413 Dalmatian coast
A partially drowned coastal area resulting in a coastline of narrow peninsulas, lengthy gulfs and channels and a number of linear islands running parallel with the lineament of the topography and underlying geological structures.
r далматинский берег
Подтопленное морем побережье, состоящее из узких полуостровов, длинных узких заливов и вытянутых островов, протягивающихся параллельно наземным формам рельефа и нижележащим геологическим структурам.
f côte *f* de type dalmate
s costa *f* de tipo dálmata
g Dalmatische Küste *f*

Fig. 80. **Dam** of Nurek hydroelectric power station in the Pamirs Mountains

1414 dam
A construction in the form of a barrier made of aggregated ground (earth dam), stone, concrete (concrete dam) across the course of a river to block it or regulate the water supply, preserving river banks and sea shores from wash-out and submergence. → Fig. 80
r плотина; дамба
Гидротехническое сооружение из намывного грунта (земляная плотина), камня, бетона (бетонная плотина), предохраняющее берега рек и морей от размыва и затопления, а также создающее подпор на водохранилищах. → Рис. 80
f barrage *m*; digue *f*
s presa *f*; dique *m*
g Damm *m*; Staudamm *m*

1415 damage by grazing cattle; overbrowsing
A pasture on which the grass cover is completely destroyed as a result of overgrazing.
r сбой; потрава
Пастбище, на котором в результате неумеренного выпаса травяной покров полностью уничтожен.
f dégât *m*
s holladura *f*
g zerstampftes Feld *n*

1416 dambo
The floodplain of a river in central Africa, swampy in the wet season, but dry for most of the year, supporting long grass; a shallow depression, lacking distinct drainage channels, at the head of a drainage system.
r дамбо
Речная пойма в Центральной Африке, болотистая в сезон дождей, но сухая большую часть года, зарастающая пышной травой; узкая котловина с еле заметными руслами в верховье речной системы.
f dambo *m*
s dambo *m*
g Dambo *m*

1417 dam-break; dam-failure

Fig. 81. Sarez **dammed lake** in the Central Pamirs

A hole, which appeared in the body of a dam under its breaking by a water stream.

r проран
Отверстие, образовавшееся в теле земляной плотины при прорыве ее водным потоком.

f rapture *f* du barrage
s brecha *f* del dique; rotura *f* del dique
g Dammbruch *m*

* **dam-failure → 1417**

1418 dammed lake
A lake formed when the river valley is blocked by mountain landslides, lava streams, etc. → Fig. 81

r завальное озеро; плотинное озеро
Озеро, возникшее в результате перекрытия речной долины горными обвалами, оползнями, потоками лавы. → Рис. 81

f lac *m* barragé
s lago *m* de dique
g Bergsturzsee *m*

1419 damming of glacier
The process of breaking of a glacier flow due to an obstacle resulting in the elevation of the glacier surface.

r подпруживание ледника
Препятствие движению льда, приводящее к поднятию ледниковой поверхности.

f barrage *m* de glacier
s contención *f* de glaciar
g Gletscheraufstau *m*

1420 damp snow
A snow containing an imperceptible amount of water and sticking together when pressed.

r влажный снег
Снег, содержащий незаметное для глаза количество воды, слипается при сжатии.

f neige *f* humide
s nieve *f* húmeda
g feuchter Schnee *m*; Pappschnee *m*

1421 Daniglacial
An initial phase of the retreat of the last glaciation in Western Europe when the edge of the ice sheet covered the surface of contemporary Danmark.

r даниглациал
Начальная фаза отступания последнего оледенения Западной Европы, когда край ледникового покрова находился в пределах современной Дании.

f stade *m* daniglaciaire
s estadio *m* daniglaciar
g Daniglazial *n*

1422 Dansgaard-Oeschger oscillations
Periodic changes of the Late Pleistocene climate averaging 1000-3000 years in duration as revealed from isotopic profiles of Greenland ice sheet and deep-sea sediments of the North Atlantic.

r осцилляции Дансгора-Эшгера
Периодические изменения климата позднего плейстоцена продолжительностью в 1000-3000 лет, выявляемые по изотопным профилям гренландского льда и глубоководных осадков Северной Атлантики.

f oscillations *fpl* de Dansgaard-Oeschger
s oscilaciónes *fpl* de Dansgaard-Oeschger
g Dansgaard-Oeschger-Oszillations *fpl*

* **daoism → 6055**

1423 darja
Term which is a part of compound names of a number of Central Asia rivers indicating their abundance-of-water character.

r дарья
Термин, входящий в состав названий ряда рек Центральной Азии, указывающий на их полноводность.

f darja *f*
s darya *f*
g Darja *f*

1424 dark coniferous forest
Taiga dominated by spruce, larch and Siberian pine, found in the middle latitudes of the Northern Hemisphere.

r темнохвойный лес
Тайга с господством ели, пихты, сосны сибирской, развита в умеренных широтах Северного полушария.

f forêt *f* sombre résineuse
s taiga *f* oscura
g dunkler Nadelwald *m*

1425 dark soils
Soils with a dark humus horizon that occur as isolated tracts among podzolic soils in the temperate belt, they are fertile and ploughed since early times.

r темноцветные почвы
Почвы с темным гумусовым горизонтом, распространенные в умеренном поясе отдельными массивами среди подзолистых почв, плодородны и издавна распахиваются.

f sols *mpl* foncés
s suelos *mpl* oscuros
g dunkel Böden *mpl*

1426 Darwinism
The theory of evolution by natural selection as stated by Charles Darwin.
r дарвинизм
Теория эволюции органического мира путем естественного отбора, разработанная Чарльзом Дарвином.
f darwinisme *m*
s darwinismo *m*
g Darwinismus *m*

1427 dasht
Desert plains on Plateau of Iran and in Central Asia with arid rubble and pebble surfaces, solonchaks.
r дашт
Пустынные равнины на Иранском нагорье и в Центральной Азии с безводными щебнистыми и галечными поверхностями и солончаками.
f dasht *m*
s dasht *m*
g Dasht *m*

1428 dating
The determining of dates for structures, events, artifacts using dendrochronology, lichenometry, radiocarbon, varve, palaeomagnetism investigation, etc.
r датирование
Определение возраста геологических отложений, структур, событий, артефактов и пр., с помощью дендрохронологии, лихенометрии, исследований радиоуглерода, ленточных глин, палеомагнетизма и пр.
f datation *f*
s datación *f*
g Datierung *f*

1429 datum level; datum line; datum plane
The fixed starting point for the measurements of heights.
r исходный уровень; условный уровень
Исходная нулевая точка, используемая для измерения высот.
f niveau *m* de référence
s nivel de referencia
g Höhenniveau *n*; Höhenbezugsniveau *n*

* **datum line → 1429**

* **datum plane → 963**

1430 dawn
The first appearance of light in the eastern sky before sunrise.
r утренняя заря
Первое появление света в восточной части неба перед восходом Солнца.
f aube *f*
s alba *f*
g Morgenrot *n*

1431 day
The time between sunrise and sunset during which the Sun is above the horizon, the opposite to night.
r день
Промежуток времени между восходом Солнца и его заходом, в течение которого Солнце бывает над горизонтом.
f jour *m*
s día *m*
g Tag *m*

1432 daya
Flat depressions with no outflow in Algerian Sahara in which water accumulates from time to time.
r дайя
Плоская бессточная впадина в Алжирской Сахаре, в которой временами может накапливаться вода.
f daya *f*
s daya *f*
g Daya *f*

1433 daylight-saving time
The time in an area advanced, usually by one hour, in relation to standard time of that area in order to extend the period of daylight at the end of a normal working day (symbol: DST).
r летнее время
Вводимое в ряде стран летом поясное время, переведенное на один час вперед с целью более рационального использования светлой части суток.
f heure *f* d'été
s hora *f* estival
g Sommerzeit *f*

1434 dead ***adj***
Having no life; inactive, in contrast to *active*.
r мертвый
Лишенный жизни; относящийся к неактивным природным процессам и явлениям, в противоположность *активным*.
f morte *adj*
s muerto *adj*
g tot *adj*

* **dead fall → 1439**

1435 dead ice; stagnant ice
The residual parts of a glacier that has stopped its motion.
r мертвый лед; застойный лед
Остатки ледника, прекратившего свое движение.
f glace *f* morte
s hielo *m* muerto
g Toteis *n*

* **dead ice hollow → 3313**

* **dead lake → 3956**

* **dead neap → 4079**

1436 dead storage o

The mass of water accumulated in a reservoir and can not be removed under normal conditions.

r мертвый объем водохранилища
Объем воды, накопленный в водохранилище, который не может быть спущен при обычных условиях.

f reserve *f* inutilisable de réservoir
s capacidad *f* inutilizada de embalse
g unbenutzbares Volumen *n* eines Stausees

1437 dead swell
A very long and flat swell waves running over a large distance in the open ocean from the area of its initiation under the conditions of calm.

r мертвая зыбь
Длинная зыбь исключительно правильной формы и периодичности, распространяющаяся на большие расстояния в условиях полного безветрия.

f mer *f* calme
s mar *m* de calma; marea *f* muerta
g Dünung *f*

1438 dead water
Part of water river section where water flowing is not observed.

r мертвое пространство
Часть водного сечения реки, где течения воды не наблюдается.

f eau *f* morte
s agua *f* muerta
g Totwasser *n*

1439 dead wood; dead fall; wood debris
Big parts of trees (dry branches, trunks, etc.) both fallen on soil surface and standing dead, which form a part of mortmass. → Fig. 82

r сухостой; отпад
Мертвые, остающиеся на корню деревья и мертвые части растений: стволы, сухие вершины и ветви, образуют часть мортмассы. → Рис. 82

f broussailles *fpl*; bois *m* mort
s árboles *mpl* secos; bosque *m* seco
g Trockenholz *n*; Trockenstand *m*; Dürrholz *n*

Fig. 82. **Dead wood**

1440 dealkalization of soil
The change of the composition of absorbed cations and the chbange of the properties of alkali soils that occurs naturally or may be caused by the improvement of soils.

r рассолонцевание
Процесс изменения состава поглощенных катионов и свойств солонцовых почв, происходящий естественным путем или вызванный мелиоративными мероприятими.

f déalcalinisation *f* du sol
s dealcalinización *f* del suelo
g Dealkalisierung *f*

* **death-rate → 3955**

1441 debacle
The displacement of broken sea ice one relative to another resulting in the formation of hummocks and clearings; the breakup of river ice before the beginning of ice drift.

r подвижка льда
Смещение льдин дрейфующего морского льда относительно друг друга, приводящее к формированию полос тертого льда и разводий; вскрытие речного льда перед началом ледохода.

f avance *f* de la glace
s desplazamiento *m* del hielo
g Eisaufbruch *m*; Eisschub *m*

* **debris → 1563**

1442 debris avalanche; debris flow
A rapid flowage of debris due to saturation when the mixture becomes extremely fluid and flows downslope considerable distances, sometimes with catastrophic results.

r поток рыхлых отложений
Быстрое движение обломков вниз по склону, нередко вместе с жидкой поверхностной водой, в результате чего эта смесь становится исключительно текучей и стекает на значительное расстояние, иногда с катастрофическими последствиями.

f torrent *m* de débris; flux *m* de débris
s torrente *m* de detrito; flujo *m* de escombros
g Trümmerstrom *m*; Schuttfließen *n*

1443 debris cone; sand cone; dirt cone
A conical mound, several meters high, on a glacier tongue covered with a detrital or earthy material.

r грязевой конус; муравьиная куча
Куполовидное или конусовидное возвышение на поверхности ледникового языка высотой в несколько метров, покрытое каменистым или землистым чехлом.

f cône *m* de boue
s cono *m* de barro
g Sandkegel *m*

1444 debris cover of a glacier

A degree or extent of the total coverage of the ablation area of a glacier by all types of surface moraine material.

r замореннность ледника
Степень суммарного покрытия области абляции ледника всеми видами поверхностной морены.

f degré *m* de couverture morainique
s grado *m* de recumbrimiento morrénico
g Schuttbedeckung *f* auf einem Gletscher

1445 debris fall
The fall of rock masses of variable size from a steep slope under the force of gravity.

r камнепад
Падение обломков горных пород и крупных каменных масс с крутых склонов под действием силы тяжести.

f chute *f* de pierres
s pedrero *m*
g Steinschlag *m*; Bergsturz *m*

* **debris fan → 201**

* **debris flow → 1442**

1446 decalcification
The removal of calcium carbonate from rocks, soil, etc. by the proces of leaching by ground waters.

r декальцификация
Удаление (выщелачивание) подземными водами карбоната кальция из горных пород, почвы и пр.

f décalcification *f*
s decalcificación *f*
g Entkalkung *f*

1447 decarbonation
The dissolving of carbonates of the surface horizons of soil and of carbonate parent rock by rainwater and their transport outside the soil horizon or the whole soil profile.

r декарбонатизация
Растворение карбонатов поверхностных почвенных горизонтов и карбонатной материнской породы дождевой водой и вынос их за пределы почвенного горизонта или всего почвенного профиля.

f décarbonisation *f*
s decarbonatación *f*
g Dekarbonatisierung *f*

* **decay ooze → 5211**

1448 decentralization; deconcentration
The distribution of economic or political factors evenly over the country as opposed to their concentration and centralization.

r децентрализация
Рассредоточение экономических или политических центров влияния по территории страны, в противоположность их концентрации при централизации.

f décentralisation *f*
s decentralización *f*
g Dezentralisierung *f*

1449 deciduous ***adj***
Applied to trees and bushes that ahed their leaves at the end of the growing season, as opposed to evergreen.

r листопадный
Относящийся к деревьям и кустарникам, сбрасывающим листья на неблагоприятный для их развития период года.

f caducifolié *adj*
s caducifolio *adj*
g laubabwerfend *adj*

1450 deciduous forest; leaf-bearing forest
The forest formed by decidious trees, which shed their leaves in the end of the growing season (there are temperate deciduous forests in middle-latitude regions and dry deciduous forest in tropical areas).

r лиственный лес; листопадный лес
Лес, образованный деревьями, сбрасывающими листья в неблагоприятный для вегетации период: холодный во внетропических или сухой в тропических областях.

f forêt *f* décidue; forêt *f* caducifoliée; estisilve *f*
s bosque *m* caducifolio; bosque *m* de hoja caduca; estisilva *f*
g Laubwald *m*; laubabwerfender Wald *m*

1451 declination of a celestial body; elevation of a celestial body
The angle between the direction to a celestial body and the plane of horizon, one of the coordinates in horizontal system of celestial coordinates: read off from the equator northwards (positive declination) or southwards (negative declination).

r склонение небесного светила; высота небесного светила
Угол между направлением на светило и плоскостью истинного горизонта (одна из координат в горизонтальной системе небесных координат, измеряется в градусах от плоскости наблюдателя: к северу – положительное склонение, к югу – отрицательное склонение).

f déclinaison *f* d´un astre; hauteur *f* d´un astre
s declinación *f* de cuerpo celeste; elevación *f* de cuerpo celeste
g Deklination *f* eines Gestirns; Höhe *f* eines Gestirns

1452 décollement
The deformation of superficial rocks by folding and faulting without the rocks at depth being affected.

r разрывное отслаивание
Деформации поверхностных слоев горных пород при образовании складок и сбросов, не затрагивающие нижележащих горных пород.

f décollement *m*
s degollamiento *m*
g Abplatzen *n*

1453 **decolonization**
Transformation of the low-developed marginal areas of a country or its former colonies into fast-developing areas or free countries.
r деколонизация
Превращение отсталых в прошлом окраин страны или ее бывших колоний в полноправные быстро развивающиеся регионы или страны, теряющие свой периферийный и колониальный облик.
f décolonisation *f*
s decolonisación *f*
g Dekolonisation *f*

1454 **decomposer**
Organisms, mainly bacteria, that in an ecosystem break down dead organic matter into mineral substances.
r редуцент
Организмы, в основном бактерии, занятые в экосистеме разложением мертвого органического вещества до минерального.
f décomposeur *m*
s reductor *m*; desintegrador *m*; descomponedor *m*
g Reduzent *m*

1455 **decomposition**
1. The breaking-down of organic matter into simpler compounds. 2. The weakening and breaking-down of rocks by chemical weathering.
r распад; разложение
1. Разложение органического вещества на его более простые составляющие. 2. Ослабление и разрушение горных пород вследствие химического выветривания.
f décomposition$_1$ *f*
s descomposición$_1$ *f*
g Dekomposition *f*

* **deconcentration → 1448**

1456 **decree time; legal time**
A system in which time in an area is advanced by one hour, in relation to the standard time of that area, in order to extend the period of daylight and use it more rationally.
r декретное время
Система, в которой время в данном часовом поясе переведено на 1 час вперед с целью более рационального использования светлой части суток.
f heure *f* décrétée
s hora *f* de decreto
g gesetzliche Zeit *f*

1457 **deep**
A hollow formed in the bed of a stream channel and constantly filled with water.
r бочаг
Углубление дна в русле реки или ручья, постоянно заполненное водой.
f fossé *f*
s poza *f*
g Kolk *m*

1458 **deep; ocean trench**
Deep (5-11 km) narrow depression in the ocean floor with steep slopes stretched out for several thousand kilometres.
r глубоководный желоб; абиссальный желоб; океанический желоб
Глубокие (5-11 км) понижения океанического дна, вытянутые на несколько тысяч километров с крутыми склонами и узким днищем.
f fosse *f* océanique; fosse *f* abyssale; ravin *m* abyssal
s fosa *f* oceánica; fosa *f* abisal; barranco *m* abisal
g Tiefseerinne *f*; Tiefseegraben *m*

1459 **deep drilling**
The drilling of boreholes over 6000 m deep for the purpose of studying the structure of the Earth's crust and the upper mantle of the Earth.
r сверхглубокое бурение
Бурение скважин глубиной более 6000 м в целях изучения строения земной коры и верхней мантии Земли.
f forage *m* profond
s perforación *f* superprofunda
g Tiefbohrung *f*

1460 **deep fault**
A narrow linear zone of ruptures that goes through the Earth's crust and penetrates into the upper mantle of the Earth.
r глубинный разлом
Узкая, линейно вытянутая зона нарушения сплошности геологических пластов, пронизывающая земную кору и проникающая в верхнюю мантию Земли.
f fracture *f* profonde
s fractura *f* profunda
g Tiefenbruch *m*

1461 **deep karst**
Karst cavities penetrating deep under the surface, sometimes below the water-table, due to the tectonic sinking of the region.
r глубокий карст
Карстовые полости, проникающие глубоко под землю, иногда ниже уровня грунтовых вод вследствие тектонического прогибания территории.
f karst *m* profond
s karst *m* profundo; karst *m* interno
g Lehmkarst *m*

1462 **deep ocean**
Ocean or sea in which surface the waves do not effect the formation of sea-floor forms.
r глубокий океан

Океан или море, в котором поверхностные волны не оказывают влияния на формирование рельефа дна.
f océan *m* profond
s océano *m* profundo
g Tiefsee *f*

* **deep scattering → 5674**

* **deep watermass → 766**

1463 **deep weathering**
A type of weathering of rocks that extends to great depth (30 cm) below surface in the humid tropics, where the process reaches its maximum efficiency.
r глубокое выветривание
Тип выветривания, захватывающий значительный слой горной породы (до 30 см от поверхности), характерный для влажных тропиков, где он достигает максимальной эффективности.
f météorisation *f* profonde
s meteorización *f* profunda
g Tiefenverwitterung *f*

* **deepening → 1464**

1464 **deepening of cyclone; deepening**
A decrease in atmospheric pressure at the centre of a cyclone which generally leads to the worsening of weather, as opposed to *filling of cyclone.*
r углубление циклона
Уменьшение атмосферного давления в центре циклонической системы, что ведет к ухудшению погоды, противоположно *заполнению циклона.*
f creusement *m* de dépression
s crecimiento *m* de depresión
g Zyklonvertiefen *n*

1465 **deep-freezing; flash-freezing**
A longterm storage at temperatures below freezing point; many crops, such as peas and beans (are grown for the freezer market).
r глубокое замораживание
Долговременное удержание температуры ниже нуля, используемое для сохранения части урожая с последующей реализацией его на рынке в замороженном состоянии.
f congélation *f* profonde
s congelación *f* profunda
g Tieffrosten *n*

1466 **deep-sea benthos**
Plant and animal organisms living at the bottom of deep seas and oceans.
r глубоководный бентос
Совокупность донных организмов, обитающих в глубоких морях и океанах.
f benthos *m* abyssal
s bentos *m* de aguas profundas
g Tiefseebenthos *n*

1467 **deep-water animals**
The animals living in lakes, seas and oceans at depths of over 500 m, they are subdivided into bathyal, abyssal, and ultraabyssal fauna.
r глубоководные животные
Животные, обитающие в океанах, морях и озерах на глубинах более 500 м, подразделяются на батиальную, абиссальную и ультраабиссальную фауну.
f animals *mpl* d'abyssal
s animales *mpl* de aguas profundas
g Tiefseetiere *npl*

* **deepwater current → 5940**

* **deep-water wave → 5414**

1468 **deer-husbandry**
Breeding deer for foodstuffs (reindeer husbandry in Russia, Finland, Sweden, Norway, Canada, Alaska) and raw materials for pharmacentical industry (deer are bred for antlers in China, Korea, Mongolia, and Kazakhstan).
r оленеводство
Отрасль животноводства, занимающаяся разведением оленей с целью получения продуктов питания (северное оленеводство: Россия, Финляндия, Швеция, Норвегия, Канада, Аляска) и сырья для лекарств (пантовое оленеводство: Китай, Корея, Монголия, Казахстан).
f élevage *m* des rennes
s cría *f* de ciervos; cría *f* de renos
g Rentierzucht *f*

1469 **deficiency of rainfall; deficient humidity; lack of moisture; insufficient moistering**
Natural conditions when the value of potential evaporation exceeds the amount of precipitation on the average over a year.
r недостаточное увлажнение
Природные условия, при которых величина испаряемости в среднем за год превышает количество выпадающих осадков.
f humidité *f* déficiente
s humedad *f* deficiente
g ungenügende Feuchtigkeit *f*

* **deficient humidity → 1469**

1470 **deflation; aeolian abrasion**
The destruction of rocks, soils, deposited snow by the action of wind and the removal of solid particles by air currents sometimes over considerable distances.
r дефляция; ветровая абразия; развевание
Разрушение горных пород, почв, снежного покрова под действием ветра и вынос твердых частиц воздушным потоком, иногда на значительные расстояния.
f déflation *f*
s deflación *f*
g Deflation *f*; Abwendung *f*

1471 deflation hollow
A depression formed as a result of the removal of unconsolidated material by the wind, especially when the protective vegetation cover has been removed or destroyed.
r дефляционная впадина
Котловина на земной поверхности, возникшая в результате местного выноса ветром рыхлого материала, чему способствует удаление растительности, прежде защищавшей рыхлые отложения от развеивания.
f cuvette *f* de déflation; cuvette *f* hydro-éolienne
s depresión *f* de deflación; depresión *f* hidro-eólica
g Deflationsbecken *n*

1472 deflation lake
A lake in a desert the basin of which was formed by the action of wind (deflation) occurring in the depressions between barchans and dunes.
r дефляционное озеро
Озеро в пустыне в понижениях между барханами и дюнами, коловина которого создана дефляцией.
f lac *m* de déflation
s lago *m* de deflación
g Deflationssee *m*

1473 defluxion
Slow pressing-out of slightly humidified grounds under soil and vegetation cover.
r дефлюкция
Медленное выдавливание слабо увлажненных грунтов под почвенно-растительным покровом.
f defluction *f*
s deflucción *f*
g Defluktion *f*

1474 defoliant
Chemical used for defoliation of plants, i.e. applied to plants to cause the leaves to drop off prematurely.
r дефолиант
Химическое вещество, используемое для удаления листьев живых растений.
f défoliant *m*
s defoliante *m*
g Entlaubangsmittel *n*

* **defoliation → 3455**

1475 deforestation
The disappearance of forests due to natural causes or as a result of economic activity of people.
r обезлесение
Исчезновение леса по естественным причинам или в результате хозяйственной деятельности человека.
f déboisement *m*; déforestation *f*
s deforestación *f*; desmonte *m*
g Entwaldung *f*

1476 deformation
Any change in the shape, volume or structure of rock masses produced by tectonic forces.
r деформация
Любое изменение формы, объема или структуры горных пород тектонического происхождения.
f déformation *f*
s deformación *f*
g Deformation *f*

1477 deformational eustatism
A global change of ocean level due to a variation in the capacity of ocean beds, caused by filling in by sedimentation or movements of the ocean floors.
r деформационная эвстазия
Планетарные изменения уровня моря вследствие колебаний вместимости океанского ложа, вызванных донной седиментацией или движениями океанического дна.
f eustatisme *m* de déformation
s eustatismo *m* de deformación
g Deformationseustatismus *m*

* **defreezing of ground → 6138**

1478 deglaciation; shrinking of glaciers; degradation of glaciers; deglacierization; glacial recession; recession of glaciers
The general decrease of glaciation caused by the long-term climate warming.
r дегляциация; деградация ледников; деградация оледенения
Общее убывание и отмирание оледенения при длительном потеплении климата.
f déglaciation *f*; récession *f* des glaciers; dégradation *f* des glaciers
s deglaciación *f*; degradación *f* de glaciares; recesión *f* de glaciares
g Entgletscherung *f*; Degradation *f* der Vereisung

* **deglacierization → 1478**

1479 deglomeration
The setting-up of productive enterprises at a distance from the centre of an established agglomeration, occurring when the advantages or economies of the agglomeration begin to diminish with its increasing size and land values, etc.
r рассредоточение агломерации
Перемещение продуктивных предприятий на некоторое расстояние от центра основной агломерации в тех случаях, когда уменьшается эффективность предприятия из-за чрезмерного расширения территории, подорожания земли и пр.
f déglomération *f*
s deglomeración *f*
g Deglomeration *f*

* **degradated soil → 1521**

1480 **degradation**
Gradual destruction of and lowering of a land surface by endogenetic and exogenetic forces, as opposed to aggradation.
r деструкция
Совокупность процессов разрушения земной поверхности экзогенными и эндогенными силами.
f destruction *f*
s destrucción *f*
g Destruktion *f*; Abbau *m*

* **degradation of glaciers → 1478**

1481 **degradation of soil; soil structural decline; soil degradation; soil degradation**
Decrease of soil fertility caused by deterioration of its properties (destruction of structure, eluviation of nutrients, etc.) as a result of natural change of conditions responsible for soil formation or of human economic activity.
r деградация почвы
Снижение плодородия почвы, вызванное ухудшением ее свойств (разрушением структуры, вымыванием питательных веществ и пр.) в результате естественного изменения условий почвообразования или хозяйственной деятельности человека.
f dégradation *f* du sol
s degradación *f* del suelo
g Bodendegradierung *f*

* **degradation sheet → 1519**

1482 **degree**
1. The measure of temperature of any thermometric scale. 2. The unit of angular measurement of latitude and longitude (symbol: °).
r градус
1. Единица измерения температуры в любой термометрической шкале. 2. Угловая мера широты и долготы (символ: °).
f degré *m*
s grado *m*
g Grad *m*

1483 **degree of adaptability**
The extent to which an organism can change to fit new conditions of the environment.
r степень адаптивности; экологическая амплитуда
Степень приспособления организмов к новым условиям или меняющимся факторам окружающей среды.
f degré *m* d'adaptation
s grado *m* de adaptación
g Anpassungsgrad *m*; ökologische Amplitude *f*

1484 **degree of glaciation**
The ratio of the area of glaciers to the total area of the glacierized catchment or the region under consideration.
r степень оледенения
Соотношение площади ледников и общей площади ледникового бассейна или рассматриваемого района.
f degré *m* de glaciation
s grado *m* de glaciación
g Vergletscherungsanteil *m*

* **degree square → 4844**

1485 **dehumification**
The loss of soil humus as a result of the change of natural processes of soil formation, in particular when reclaiming virgin lands and pastures are used for farming without taking necessary measures to restore soil humus.
r дегумификация
Потеря почвой гумуса при изменении естественного хода почвообразования, в частности, при освоении целинных земель и пастбищ под земледелие без необходимых мероприятий по поддержанию гумусового состояния почвы.
f dehumification *f*
s dehumificación *f*
g Dehumifikation *f*

1486 **dehydration**
The process of removal of water from minerals, rocks and soils.
r дегидратация; обезвоживание
Процесс обезвоживания, т.е. выделения воды из минералов, горных пород и почв.
f dehydratation *f*
s dehidratación *f*
g Dehydratation *f*; Entwässerubg *f*

1487 **deindustrialization**
A decline of large-scale machine building in some countries during certain periods of their history (destructive wars, the cultural revolution in China, etc.); the process in which there is marked movement of employment away from the production of goods usually to the provision of services.
r дезиндустриализация; деиндустриализация
Процесс сокращения крупного машинного производства в некоторых странах, происходящий на отдельных этапах их истории (разрушительные войны, культурная революция в Китае и другие события); происходящий в современную эпоху заметный переход от производства товаров к предоставлению разнообразных услуг.
f désindustrialisation *f*
s desindustrialización *f*
g Deindustrialisierung *f*

1488 **delimitation**
The determination of the position of a frontier on a geographical map, which is a component of a corresponding agreement between contiguous states.
r делимитация границы

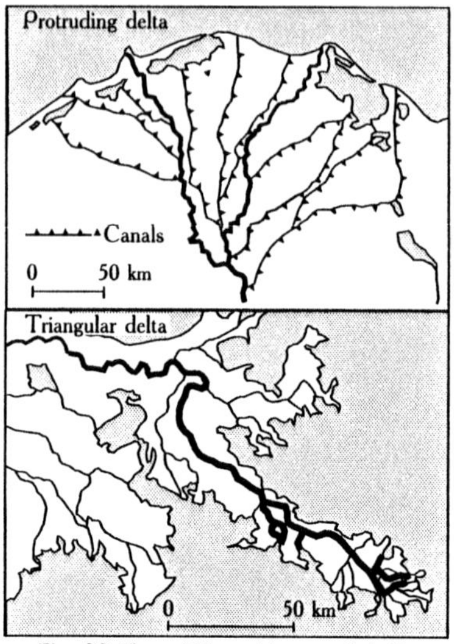

Fig. 83. Two different types of **delta**

Проведение государственной границы на географической карте, которая является составной частью соответствующего соглашения между сопредельными государствами.

f délimitation *f*
s delimitación *f*
g Abgrenzung *f*

1489 delta
An alluvial plain formed when a river deposits solid material at its mouth at a rate faster than that by which it can be moved by tidal and other currents, as deposits grow the river splits into channels and distributaries. → Fig. 83
r дельта
Равнина в низовьях реки, сложенная речными наносами и расчлененная разветвленной сетью рукавов и проток. → Рис. 83
f delta *m*
s delta *m*
g Delta *n*; Deltamündung *f*

1490 delta arm
A branch of a river or distributary in delta the of large rivers inflowing into the Black Sea and the Sea of Azov.
r гирло
Рукав или протока в дельтах крупных рек, впадающих в Черное и Азовское моря.
f grau *m*
s garganta *f*
g Durchfluß *m*

1491 delta coast
A low-lying, often fan-like, flat coast formed of loose layers of sediment carried by a river.
r дельтовый берег
Низкий, равнинный, часто лопастной берег, сложенный рыхлыми слоистыми наносами, принесенными рекой.
f rivage *m* deltaïque
s costa *f* deltáica
g Deltaküste *f*

1492 delta head
A place up to which the seawater intrudes into a river in the conditions of tides and surges at low-water river flow.
r вершина дельты
Предел максимального распространения в реку воды при приливах и нагонах в условиях меженного речного стока.
f faîte *m* de delta
s cima *f* de delta
g Deltaspitze *f*

1493 delta lake; deltaic lake
Lakes with slowed-down water exchange, formed among unevenly growing delta lobates (ilmens on the Volga, brackish lagoons on the Kuban, laydas on the Lena and the Yana).
r дельтовое озеро
Водоемы с замедленным водообменом, формирующиеся среди неравномерно нарастающих лопастей дельты (ильмени на Волге, лиманы на Кубани, лайды на Лене и Яне).
f lac *m* de delta
s lago *m* de delta
g Deltasee *m*

1494 delta-front
The underwater sloping continuation of a delta.
r авандельта
Пологий подводный склон, примыкающий к дельте и продолжающий ее надводную часть.
f avant-delta *m*
s antedelta *m*
g Vordelta *n*

*** deltaic lake → 1493**

1495 deluvial train
A gently sloping surface at the foot of a slope which is formed by the deposition of the material washed off the slope (deluvium).
r делювиальный шлейф; гласис
Пологонаклонная поверхность, формирующаяся у подножья склона в результате накопления смываемого со склона материала – делювия.
f traîne *f* deluviale; glacis *m*
s cola *f* deluvial; glacis *m*
g Glacis *m*

1496 deluvium
The deposits of temporary streams appearing on slopes as a result of the accumulation of loose weathered material, washed off by rain and snowmelt waters.
r делювий; делювиальные отложения
Отложения временных водотоков, возникающие на склонах в результате накопления рыхлых продуктов выветривания, смытых дождевыми и талыми снеговыми водами.

f deluvium *m*; dépôt *m* deluvial
s deluvium *m*; depósito *m* deluvial
g Deluvium *n*; deluviale Ablagerungen *fpl*

* **delve → 5084**

1497 **demand**
One of two (supply and demand) market forces that govern prices, operating freely in a market economy, apparent through the price mechanism as responces in quantities that cosumers are prepared to buy when the market price changes. → Fig. 317, c. 713
r спрос
Одна из двух (предложение и спрос) категорий товарного производства, соотношение которых – главный ценообразующий фактор, спрос представляет собой общественную потребность, выраженную в денежной форме и обеспеченную платежными средствами. → Рис. 317, c. 713
f demande *f*
s demanda *f*
g Nachfrage *f*

1498 **demarcation**
Establishing of the frontier line in the terrain with the use of posts, signs, etc.
r демаркация границ
Установление линии государственной границы на местности с использованием столбов, указателей и пр.
f démarcation *f*
s demarcación *f*
g Demarkation *f*; Grenzmarkierung *f*; Demarkierung *f*

1499 **democracy**
A form of government in which political power lies in the hands of the people through elected representatives; a state so governed.
r демократия
Форма государственного устройства общества, в котором политическая власть находится в руках народа, избирающего своих представителей в органы власти различных уровней.
f démocratie *f*
s democracia *f*
g Demokratie *f*

1500 **demographic crisis; demographic decline**
A dramatic decline in population reproduction manifested, in particular, through the depopulation.
r демографический кризис
Резкое ухудшение воспроизводства населения и его условий, проявляющееся, в частности, в депопуляции.
f crise *f* démographique; déclin *m* démographique
s crisis *f* demográfica
g demographische Krise *f*

* **demographic decline → 1500**

1501 **demographic explosion**
A sudden increase in the number of people in a country induced by the improvement of economic, political and environmental conditions.
r демографический взрыв
Резкое увеличение народонаселения в данной стране, связанное с улучшением экономических, политических и экологических условий.
f explosion *f* démographique
s explosión *f* demográfica
g demographishe Explosion *f*

1502 **demographic indicators**
Quantitative indices describing the processes of population reproduction: birth rate, mortality, population growth, etc.
r демографические показатели
Количественные показатели, характеризующие процессы воспроизводства населения: коэффициенты рождаемости, смертности, прироста населения и мн. др.
f indices *mpl* démographiques
s índices *mpl* demográficos
g demographische Kennziffern *fpl*

1503 **demographic potential**
The number of population of a country, region, settlement etc, seen from the point of view of population reproduction and the potentially possible number of people employed in the national economy.
r демографический потенциал
Численность населения страны, района, населенного пункта и пр., рассматриваемая с точки зрения воспроизводства населения и потенциально возможного числа занятых в отраслях народного хозяйства и др.
f potentiel *m* démographique
s potencial *m* demográfico
g demographisches Potenzial *m*

1504 **demographic transition**
The change of population growth from the high birth and death rates to low birth and death rates.
r демографический переход
Изменение естественного движения населения от высоких показателей рождаемости и смертности к более низким их значениям.
f transition *f* démographique
s transición *f* demográfica
g demographischer Übergang *m*

1505 **demography**
The study of the dynamics of reproduction of human population as a whole and its particular groups by analyzing population growth, migration patterns and changes in age-sex, marriage and family patterns.
r демография

Изучение динамики воспроизводства населения и отдельных его групп путем анализа естественного движения населения и изменения его возрастно-половой, брачной и семейной структур.
f démographie *f*
s demografía *f*
g Demographie *f*; Bevölkerungslehre *f*

* **demoiselle → 1768**

1506 **denationalization; privatization**
The removal of state control or ownership of an activity (including land) in order to allow it to pass into private hands, the reverse of nationalization.
r денационализация; приватизация
Переход средств производства (включая землю), банков и отраслей экономики из собственности государства в частную собственность.
f dénationalisation *f*; privatisation *f*
s denacionalización *f*
g Entnationalisierung *f*; Privatisierung *f*

1507 **dendritic drainage pattern**
A drainage pattern resembling the branchibg of a tree, developed usually on a gentle, nearby uniform plain surface. → Fig. 95, p. 202
r дендритовая речная сеть
Речная сеть, напоминающая в плане ветвящееся дерево, обычно развита на пологой, почти однородной равнинной поверхности. → Рис. 95, с. 202
f réseau *m* hydrographique dendritique
s red *f* hidrográfica dendrítica
g dendritisches Flußnetz *n*

* **dendritic glacier → 1177,** Fig. 84

1508 **dendrochronology**
The dating and reconstruction of natural events by counting and studying the annual growth rings of trees.

Fig. 84. **Dendritic glacier** in Alaska

r дендрохронология
Датирование и реконструкция природных явлений на основе анализа годичных колец древесины.
f dendrochronologie *f*
s dendrocronología *f*
g Dendrochronologie *f*

1509 **dendroclimatology**
The study of the variations of local climates in the past by comparing the annual growth rings of perennial trees.
r дендроклиматология
Изучение изменений местных климатов в исторический период времени преимущественно по толщине годичных колец у многолетних древесных растений.
f dendroclimatologie *f*
s dendroclimatología *f*
g Dendroklimatologie *f*

1510 **denitrification**
Reduction of nitrates through the agency of anaerobic microorganisms in soils of humid and cold climate that results in the escape of gaseous nitrogen into the air and impoverishment of soil with nitrogen compounds.
r денитрификация
Сокращение содержания нитратов в почвах влажного и холодного климата под действием анаэробных микроорганизмов, ведущий к потере газообразного азота и обеднению почвы соединениями азота.
f dénitrification *f*
s denitrificación *f*
g Denitrifikation *f*

* **dense bushes → 811**

1511 **dense forest; closed forest**
A forest with the dense undergrowth.
r густой лес; чаща
Заросли леса с плотным подлеском.
f forêt *f* dense; fourré *m*
s bosque *m* cerrado; bosque *m* denso
g dichter Wald *m*; Dickicht *n*

1512 **dense phytocoenosis**
The condition in which the crowns or roots of individual plants in a community overlap to form a virtually continuous layers.
r сомкнутый фитоценоз
Растительное сообщество, в котором кроны или корневые системы растений сомкнуты в одном или двух ярусах.
f phytocénose *f* dense
s fitocenosis *f* densa
g zugedeckte Phytozönose *f*

1513 **density**
The concentration of matter or mass at a specified temperature and pressure, expressed in terms of mass per unit volume, it can refer to the density of air and other gases, to all types of solid materials and to water and other fluids.

r плотность
Концентрация вещества или массы при определенных температуре и давлении, выраженная как отношение массы к занимаемому ею объему (относится к плотности воздуха и других газов, ко всем типам твердых материалов, к воде и другим жидкостям).
f densité *f*
s densidad *f*
g Dichte *f*

1514 density current
An ocean current resulting from the differences in water density caused by variations of salinity and temperature.
r плотностное течение; плотностной поток
Течения в морях и океанах, возбуждаемые горизонтальными градиентами плотности воды, вызванными различиями температуры и солености.
f courant *m* de densité
s corriente *f* de densidad
g Dichteströmung *m*

1515 density of rock
The mass of rock per unit volume.
r плотность породы
Масса горной породы в единице объема.
f densité *f* de roche
s densidad *f* de roca
g Gesteinsdichte *f*

1516 denudation
The incovering of underlying rocks or strata by the removal of overlying material by any agent of erosion, weathering and mass movement (by water, wind, ice, direct impact of gravity).
r денудация
Снос и перенос (водой, ветром, льдом, непосредственным действием силы тяжести) продуктов выветривания горных пород.
f dénudation *f*
s denudación *f*
g Abtragung *f*; Denudation *f*

1517 denudation landforms
A type of relief formed by denudation processes, i.e. removal of overlying material by any agent of erosion, weathering and mass movement.
r денудационный рельеф
Рельеф, образовавшийся в результате процессов денудации, т.е. сноса продуктов выветривания под действием экзогенных процессов.
f relief *m* de dénudation; formes *fpl* de denudation
s relieve *m* de denudación; formas *fpl* de denudación
g Abtragungsrelief *n*

1518 denudation plain; destructional plain
A planed terrain formed as a result of gradual lowering of high landforms by denudation.
r денудационная равнина; остаточная равнина
Выровненная поверхность, образовавшаяся в результате разрушения возвышенного рельефа длительно протекающими процессами денудации.
f plaine *f* d'érosion; plaine *f* de dénudation
s llanura *f* de denudación; llanura *f* de erosión
g Fastebene *f*; Rumpffläche *f*

1519 denudation sheet; degradation sheet
The thickness of the rock layer removed by denudation processes.
r денудационный срез
Толщина слоя горных пород, удаленных процессами денудации.
f pellicule *f* détritique
s película *f* detrítica
g abgetragene Schichtsdicke *f*

1520 denudation slope
A type of slope from which material is being lost by erosion and weathering, leading to a net loss of ground.
r денудационный склон
Склон, образованный в результате процессов денудации, т.е. сноса материала путем выветривания и эрозии.
f versant *m* de dénudation; glacis *m* de dénudation
s vertiente *f* de denudación; glacis *m* de denudación
g Abtragungshang *m*

1521 depleted soil; degradated soil; poor soil; impoverished soil
A soil that has lost most of its available nutrients, either by the bad management, such as over-cropping, or by intense leaching.
r истощенная почва; бедная почва
Почва, потерявшая значительную часть плодородия из-за ее неправильного использования или интенсивного выщелачивания.
f sol *m* dégradé; sol *m* appauvri
s suelo *m* degradado; suelo *m* empobrecido
g erschöpfter Boden *m*

1522 depletion of natural resources; natural resources depletion
The intensive use of natural resources which exceeds the level providing the sustainability of the environment.
r истощение природных ресурсов
Изъятие природного ресурса, превышающее нормы, обеспечивающие устойчивость окружающей среды.
f épuisement *m* de ressource naturelle
s agotamiento *m* de recursos naturales
g Erschöpfung *f* der Naturschätze

1523 depletion of natural waters

The decrease of the minimum permissible surface waters runoff, water body volume or ground water reserves under which the typical regime of the water body becomes disturbed.

r истощение природных вод
Уменьшение минимально допустимого стока поверхностных вод, объема водоема или запаса подземных вод, при котором нарушается обычный режим водного объекта.

f épuisement *m* des eaux naturelles
s agotamiento *m* de las aguas naturales
g Erschöpfung *f* der Gewässer

1524 depletion of soil; impoverishment of a soil; soil depletion; overcropping
The overplanting and taking of too many harvests from the soil without restoring its fertility, thus leading to the creation of poor soils which lack sufficient plant nutrients.

r истощение почвы
Обеднение почвы питательными веществами в результате неправильной агротехники, длительного выращивания сельскохозяйственных культур без внесения удобрений или при недостаточном их внесении.

f épuisement *m* du sol; appauvrissement *m* du sol
s agotamiento *m* del suelo; esterilización *f*; empobrecimiento *m* del suelo
g Bodenerschöpfung *f*; Bodenmüdigkeit *f*

1525 depletion of the ozone layer
The destruction of ozone at the heights of 10-50 km under the impact of industrial release into the atmosphere of freons and other halogen-containing hydrocarbons.

r истощение озонового слоя
Разрушение озона на высотах 10-50 км под воздействием производственных выбросов в атмосферу фреонов и других галогеносодержащих углеводородов.

f épuisement *m* de couche d'ozone
s agotamiento *m* de capa de ozono
g Ozon(schicht)abbau *m*

1526 depopulation
A decline of the total population of an area brought about by a fall in fertility, exessive mortality or out-migration.

r депопуляция
Уменьшение численности населения вследствие суженного воспроизводства, когда последующие поколения численно меньше предыдущих, а также из-за значительной миграции из данной страны или региона.

f dépeuplement *m*; depopulation *f*
s despoblamiento *m*; despoblación *f*
g Depopulation *f*

1527 deposit; sediment
Unconsolidated particles or grains of rocks laid down by a river, ocean, ice, wind, a natural accumulation of material.

r отложения; осадки$_2$
Несцементированные частицы или зерна горных пород, отлагающиеся в водной среде (в реках, озерах и морях), а также на суше в результате действия ледниковых и эоловых процессов.

f dépôt *m*; sédiment *m*
s depósito *m*; sedimento *m*
g Ablagerung *f*; Sedimente *npl*

1528 deposited and metamorphosed ice
Ice formed on the Earth's surface from the accumulation of solid precipitation which has undergone recrystallisation and metamorphism (glaciers consist mainly of this type of ice).

r осадочно-метаморфические льды
Льды, образовавшиеся на земной поверхности из скопления твердых атмосферных осадков, подвергшихся перекристаллизации и метаморфизму (ледники состоят главным образом из этого типа льдов).

f glaces *fpl* déposées et métamorphosées
s hielos *mpl* sedimentarios y metamórficos
g abgelagertes und metamorphisiertes Eis *n*

1529 deposited moraine
A complex of ridges and mounds related to the irregular accumulation of unsorted, unstratified coarse detritus of glacial origin. → Fig. 85

r отложенная морена
Комплекс холмистых и грядовых форм рельефа, связанных с неравномерной аккумуляцией ледникового материала, представляющих собой несортированные, неслоистые грубообломочные отложения ледникового происхождения. → Рис. 85

f moraine *f* déposée
s morrena *f* depositada
g abgelagerte Moräne *f*

1530 deposition; sedimentation
The act and process of deposition of material from mobile, suspended or dissolved state in water or air and its transformation into a sediment.

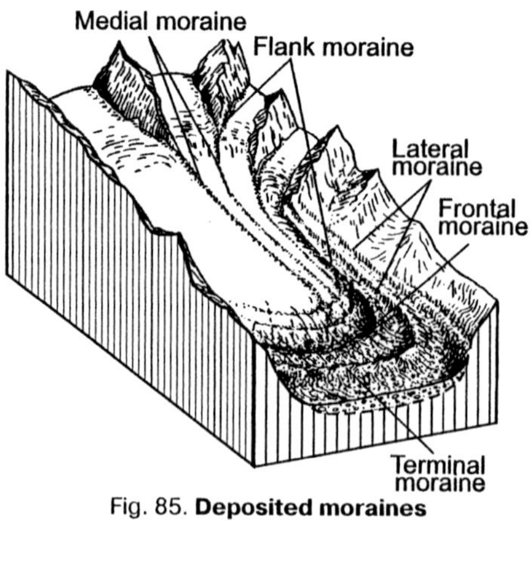

Fig. 85. **Deposited moraines**

r осадконакопление; седиментация; отложение
Процесс образования отложений путем перехода осаждаемого материала из подвижного, взвешенного или растворенного состояния в водной или воздушной среде в неподвижное – наносы.
f sédimentation *f*
s sedimentación *f*
g Absetzen *n*; Ablagerung *f*; Sedimentation *f*

* **depressed area → 1532**

1531 depression
1. Any hollow or relatively sunken ares, especially one enclosed by higher area. 2. Region of the atmosphere where the atmospheric pressure is lower than that of its surroundings, i.e. low pressure system (*cyclone*).
r депрессия
1. Любое понижение или низко расположенная территория, окруженная более высокими пространствами. 2. Область атмосферы с пониженным давлением воздуха по сравнению с соседней воздушной массой, т.е. система низкого давления (*циклон*).
f dépression *f*
s depresión *f*
g Depression *f*

1532 depressive region; depressed area; blighted area
An economic region that in the past demonstrated relatively high economic development rates while now is gradually declining.
r депрессивный район
Район, в недавнем прошлом демонстрировавший относительно высокие темпы экономического развития, но затем пришедший в упадок.
f région *f* dépressive; région *f* déprimée
s región *f* depresiva
g depressive Region *f*

* **depth contour → 3223**

* **depth current → 5940**

* **depth dump → 5388**

1533 depth hoar; sugar hoar
Large skeletal (goblet-shaped) ice crystals formed inside the snow mass in the layers of sharp temperature changes.
r глубинная изморозь
Крупные скелетные (бокалообразные) ледяные кристаллы, образующиеся в толще снега в слоях температурных скачков.
f givre *m* de profondeur
s escarcha *f* de profundidad
g Schwimmschnee *m*; Tiefenreif *m*

1534 depth of friction
A depth in water at which the direction of wind-driven current changes by 180° owing to the Coriolis force.
r глубина трения
Глубина в водной толще, на которой направление скорости ветрового течения под влиянием силы Кориолиса изменяется на 180°.
f profondeur *f* de frottement
s profundidad *f* de frotamiento
g Reibungstiefe *f*

* **depth of frictional influence → 1864**

1535 depth of measurement
A depth at which oceanographic characteristics are measured or water samples are taken.
r горизонт наблюдений
Глубина, на которой проводятся измерения океанографических характеристик или отбор проб.
f profondeur *f* d'observation
s profundidad *f* de observación
g Beobachtungstiefe *f*

1536 depth of runoff; water equivalent
Amount of water flowing from a catchment over a time interval, equal to the thickness of the layer evenly distributed over this catchment area.
r слой стока
Количество воды, стекающее с водосбора за какой-либо интервал времени, равное толщине слоя, равномерно распределенного по площади этого водосбора.
f indice *m* d'écoulement
s capa *f* de escorrentía
g Abflußhöhe *f*

1537 depth of snow; snow depth; snow thickness
Vertical distance between the top surface of a snow layer and the underlying surface of ground or ice.
r толщина снежного покрова; высота снежного покрова; глубина снежного покрова
Вертикальное расстояние между внешней поверхностью слоя снега и подстилающей поверхностью (почвой или льдом).
f hauteur *f* de la couche de neige
s altura *f* de la capa de nieve
g Schneehöhe *f*; Mächtigkeit *f* der Schneedecke

1538 deranged drainage
A confused drainage system which produces a mosaic of small lakes, streams, marshes, small islands (as in Finland) caused by the haphazard distribution of glacial drift.
r беспорядочная речная сеть
Речная сеть, представляющая собой мозаику из небольших озер, болот, рек, островов (как в Финляндии), сформированную случайным распределением ледниковых потоков.
f réseau *m* hydrographique sans suite
s red *f* hidrográfica desordenada
g ungeordnetes Flußnetz *n*

1539 **derelict land**
Land damaged by some process (e.g. by extractive or other industry) or neglect, abandoned and left to fall to ruin, incapable of being used in its present condition.
r бесхозная земля
Земля, испорченная каким-либо хозяйственным процессом (например, добычей полезных ископаемых или каким-либо индустриальным использованием) и брошенная из-за этого, т.е. оставленная вне хозяйственного использования в существующих условиях.
f terre *f* abandonnée
s tierra *f* abandonada
g herrenloses Land *n*

1540 **derivation channel**
A canal aimed at the removal of water from the river channel with the purpose of carrying out hydrotechnical activities or because of the threat of flooding.
r водоотводный канал; отводной канал
Канал для временного освобождения от воды русла реки с целью проведения в нем гидротехнических работ или в связи с угрозой затопления.
f canal *m* de dérivation; canal *m* de fuite
s canal *m* de derivación
g Ableitungskanal *m*

1541 **derivation system**
A system of constructions aimed at taking water from a water object and conducting it to a pumping plant or hydroelectric plant. → Fig. 86
r деривационная система
Система гидротехнических сооружений для отвода воды из водного объекта и ее транспортировки к насосной станции или ГЭС. → Рис. 86
f système *m* de dérivation
s sistema *m* de derivación
g Wasserableitungssystem *n*

1542 **derived map**
A geographical map compiled, usually at a reduced scale, from existing maps.
r производная карта
Географическая карта, обычно более мелкого масштаба, составленная по имеющимся картам.
f carte *m* dérivée

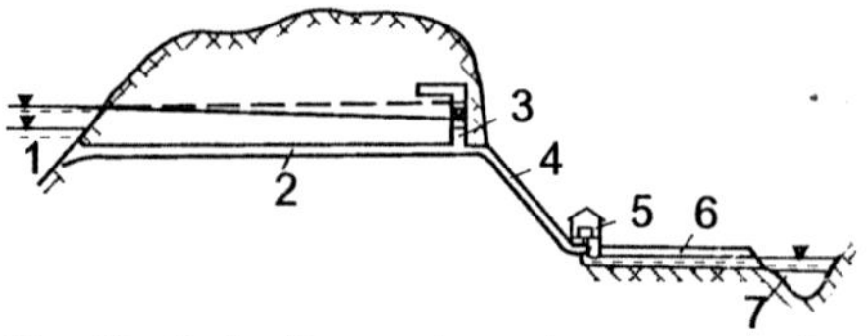

Fig. 86. **Derivation system**: 1 – reservoir, 2 – pressure tubing, 3 – equalizing reservoir, 4 – turbine tubing, 5 – hydropower station, 6 – derivation kanal, 7 – river

s mapa *m* derivado
g abgeleitete Karte *f*

1543 **desalinization of ice**
The extraction of salt from sea ice caused by the downward movement of the brine cells due to the temperature gradient and gravity.
r рассоление льда; опреснение льда
Выпадение соли из морского льда вследствие движения вниз солевых ячеек, обусловленного температурным градиентом и силой тяжести.
f dessalement *m* de la glace
s desalación *f* de hielo
g Entsalzung *f* von Meereis; Eisaussussung *f*

1544 **desalinization of soil**
Natural or man-induced loss of salts from soils by percolating waters under the increased precipitation or irrigation rate, as well as under the improvement of drainage.
r рассоление почвы
Естественный или искусственный процесс вымывания солей из почвы при увеличении количества осадков и оросительной нормы, а также при увеличении дренированности.
f désalinisation *f* du sol
s desalinización *f* del suelo
g Absalzung *f* des Bodens; Entsalzung *f* des Bodens; Bodenentsalzung *f*

1545 **desalinization of water**
The process of removal of ions, salts and other minerals from sea water to produce fresh water, or the dilution of strongly mineralised water with fresh water.
r опреснение воды
Процесс удаления из воды ионов, солей и других минеральных веществ, или разбавление сильно минерализованных вод водами меньшей минерализации,
f dessalage *m* d'eau; desaumurage *m* d'eau
s desalinización *f* de agua; desalación *f* de agua
g Wasserentsalzung *f*; Verdünnung *f* des Salzwassers

1546 **descending spring**
A spring of free ground waters moving from the nourishment area of the aquifer downward to the place in which the spring water emerges.
r нисходящий источник
Источник безнапорных подземных вод, движущихся сверху вниз от области питания водоносного слоя к месту выхода вод.
f source *f* de débordement
s manantial *m* de desbordamiento
g absteigende Quelle *f*

1547 **desert**
A type of landscape that develops under the lack of moisture (arid deserts) or lack of heat (cold deserts), and vegetations is scanty or absent, unless specially adapted.

r пустыня
Тип ландшафта, развивающийся в условиях недостатка влаги (аридные пустыни) или недостатка тепла (холодные пустыни).
f désert *m*
s desierto *m*
g Wüste *f*

1548 **desert crust**
A hard cemented layer on the rock or soil surface forming in a semi-arid desert in which the annual distribution of precipitation is uneven.
r пустынная корка
Плотная корка на поверхности горных пород или почв, образующаяся преимущественно в семиаридных пустынях с неравномерным распределением в году атмосферных осадков.
f croûte *f* désertique
s costra *f* desértica
g Wüstenkruste *f*

1549 **desert fauna**
Animals adapted to live in deserts by coping with the insufficient moisture, e.g. rodents, ungulates, lizards, scorpions, and etc.
r пустынная фауна
Совокупность видов животных, приспособившихся к обитанию в пустынях, характерны грызуны и копытные, вараны, фаланги, скорпионы и мн. др.
f faune *f* désertique
s fauna *f* desértica
g Wüstenfauna *f*

* **desert mosaic → 1550**

1550 **desert pavement; desert mosaic**
An extensive area in a desert covered with large rock fragments and bare pebbles created by weathering and wind removal of the finer dust and sand material.
r пустынная мостовая
Скопление на поверхности пустыни крупных обломков горных пород, образовавшееся вследствие выветривания и лишенное благодаря дефляции мелкозема.
f pavage *m* désertique
s pavimento *m* desértico
g Steinsohle *f*

1551 **desert soils**
Soils of arid regions where there is a net deficiency of rainfall, hence a lack of vegetation and a thin or discontinuous organic layer, and the leached layer being only about 15 cm thick, underclaimed by a carbonate horizon.
r пустынные почвы
Почвы аридных регионов с недостатком влаги, имеющие слабый растительный покров и тонкий прерывающийся органический горизонт, а также выщелоченный слой толщиной до 15 см, подстилаемый карбонатным горизонтом.
f sols *mpl* désertiques
s suelos *mpl* desérticos
g Wüstenböden *mpl*

1552 **desert varnish**
A thin shiny, almost black coating that commonly occurs on exposed rock surfaces in the desert.
r пустынный загар
Тонкая (до 5 мм) блестящая почти черная корка на поверхности хорошо освещаемых солнцем горных пород в пустынях.
f vernis *m* désertique
s tostado *m* desértico
g Wüstenlack *m*

1553 **desert zone**
Natural regions with desert landscapes which is found in the temperate zone of the Northern Hemisphere, and subtropical and tropical zones of the Northern and Southern hemispheres.
r пустынная зона
Природная зона, в естественных ландшафтах которой преобладают пустыни, распространены в умеренном поясе Северного полушария, субтропическом и тропическом поясах Северного и Южного полушарий.
f zone *f* des déserts
s zona *f* de desiertos
g Wüstenzone *f*

1554 **desertification; desiccation**
The extension of typical desert landscapes and landforms to areas where they did not occur in the recent past, the increase the processes by which soil degrades becoming a desert (looses fertility, productivity and biodiversity) due to climate change or due to human activity, such as intensive farming, overgrazing, deforestation, etc.
r опустынивание; аридизация
Наступание пустыни (уменьшение биомассы, продуктивности, видового разнообразия) в результате направленных климатических изменений или интенсивной хозяйственной деятельности.
f désertification *f*
s desertificación *f*; expansión *f* de desierto
g Verwüstung *f*

* **desiccation → 1554**

1555 **desiccation crack; shrinkage crack**
A fissure appearing in the fine-grained deposits under their drying and decrease of the volume.
r трещина усыхания
Трещина, возникающая в тонкозернистых осадках при их высыхании и уменьшении объема.
f fissure *f* de dessiccation
s fisura *f* de desecación
g Trockenriß *m*

1556 **desilication**
The solution and removal of silica from rocks generally by chemical weathering.
r десиликация
Растворение и вынос кремнезема в основном при химическом выветривании горных пород.
f désilification *f*
s desilicificación *f*
g Desilifizierung *f*

1557 **desintegration**
A gradual weakening of ties in the economic complex of the formerly connected national economies of two or more states leading to isolation of their economies from the world economic community.
r дезинтеграция
Постепенное ослабление связей в экономическом комплексе ранее связанных национальных хозяйств двух и более государств, ведущее к изоляции их экономики от мирового экономического сообщества.
f désintégration *f*
s desintegración *f*
g Desintegration *f*

* **desquamation → 2047**

1558 **destratification**
A deliberate destruction of the discontinuity layer in a lake by way of mechanic mixing of water or its aeration.
r дестратификация
Искусственное разрушение слоя скачка в водоеме путем механического перемешивания воды или аэрации.
f destratification *f*
s destratificación *f*
g Destratifikation *f*

1559 **destruction of ice**
Gravitational crushing and rounding of the ice fragments during ice falls and avalanches from termini of hanging glaciers.
r деструкция льда
Гравитационное дробление и окатывание обломков льда при ледяных обвалах и лавинах с концов висячих ледников.
f désagrégation *f* de la glace
s destrucción *f* de hielo
g Eisdestruktion *f*

* **destructional plain → 1518**

* **destructive margin → 5891**

* **destructive plate margin → 5891**

1560 **destructive wave**
One of a rapid succession of strong storm waves which drop almost vertically on to the beach, the backwash of the wave being so much more powerful than the swash that the beach material is dragged seaward.
r волна размыва
Одна из серии быстро движущихся мощных штормовых волн, падающая почти вертикально на берег, вследствие чего ее откат гораздо сильнее наката, так что материал пляжа стаскивается в море.
f vague *f* destructive
s ola *f* destructiva
g destruktive Welle *f*

* **detached mass → 1996**

1561 **detrital rock; clastic rock**
A rock composed of broken fragments of preexisting rocks converted into a consolidated mass.
r обломочная горная порода; кластическая горная порода
Осадочная порода, состоящая из обломков различных минералов и пород, превратившихся в консолидированную массу.
f roche *f* détritique; roche *f* clastique
s roca *f* detrítica; roca *f* clástica
g Trümmergestein *n*; klastisches Gestein *n*

* **detritivore → 1562**

1562 **detritophage; detritivore; detritus feeder**
Organisms feeding on detritus which is dead or partially decomposed organic matter.
r детритофаг
Организмы, питающиеся детритом – мертвым или частично разложившимся органическим веществом.
f détritophage *m*
s detritofago *m*
g Detritophag *m*

1563 **detritus; debris**
1. An accumulation of rock waste consisting of disintegrated rocks, sand, clay, moved from their place of origin and redeposited. 2. Organic and mineral paricles formed as the product of desintegration of biotic substances.
r детрит
1. Обломки горной породы, перемещенные от места своего происхождения и переотложенные. 2. Остатки органических и минеральных частиц биогенного происхождения.
f détritus *m*; debris *m*
s detrito *m*; derrubio *m*; escombros *mpl*; detritus *m*
g Detritus *m*; Trümmer *pl*

* **detritus feeder → 1562**

1564 **deurbanisation; disurbanisation; counter-urbanisation**
The movement of population and economic activity away from urban areas.
r дезурбанизация; деурбанизация
Сокращение населения в крупных городах.

f désurbanisation *f*; contre-urbanisation *f*
s desurbanización *f*; contraurbanización *f*
g Desurbanisierung *f*; Desurbanisation *f*

1565 **developed country**
A country with a high level of economic development, culturally and socially advanced and which plays an important role in the world politics.
r развитая страна
Страна с высоким уровнем экономического, культурного и социального развития, играющая важную роль в мировой политике.
f pays *m* développé
s país *m* desarrollado
g hochentwickeltes Land *n*

1566 **developing country**
A country with a relatively low level of economic development and multi-structural economy mainly providing raw-material to the developed countries.
r развивающаяся страна
Страна со сравнительно низким уровнем развития и многоукладной экономикой, служащая в основном поставщиком сырьевых ресурсов для развитых стран.
f pays *m* en voie de développement; pays *m* en développement
s país *m* en vías de desarrollo; país *m* subdesarrollado
g Entwicklungsland *n*

1567 **development of territory**
The inclusion of a territory having unchanged natural landscapes into economic activities by carrying out the building, engineering, mining, etc.
r освоение территории
Включение в хозяйственную деятельность новой территории, до этого имевшей неизмененный природный ландшафт.
f mise *f* en valeur du territoire; développement *m* du territoire
s desarrollo *m* del territorio
g Erschließung *f* des Gebiets

* **deviating force → 1302**

1568 **deviation**
1. In statistical calculations, the difference between an oserved value and a calculated value. 2. In geodesy, the angle between the geoid and spheroid of the Earth.
r отклонение; девиация
1. В статистических вычислениях, разница между наблюденной и вычисленной величиной. 2. В геодезии, угол между геоидом и сфероидом Земли.
f déviation *f*
s desviación *f*
g Deviation *f*

1569 **deviation of plumb line**
The angle between a vertical line coinciding with the direction of gravity at a certain point and the perpendicular to the surface of the Earth ellipsoid at the same point.
r уклонение отвесной линии
Угол между отвесной линией, совпадающей с направлением силы тяжести в данной точке, и перпендикуляром к поверхности земного эллипсоида в той же точке.
f déviation *f* de la verticale
s desviación *f* de la vertical
g Lotabweichung *f*

1570 **devolution**
In regional politics, the process of handling down power from a cenral authority to local administrative centres.
r передача функций
В региональной политике процесс передачи полномочий от центральной власти в местные административные центры.
f dévolution *f*
s devolución *f*
g Devolution *f*

1571 **Devonian**
The fourth geological period of the Palaeozoic era which began 395 million years ago and lasted for about 50 million years. Table 1, p. 284
r девон
Четвертый период палеозойской эры геологической истории Земли, начался 395 млн лет назад и продолжался около 50 млн лет. Табл. 1, с. 284
f Dévonien *m*
s Devónico *m*
g Devon *n*

1572 **dew**
Droplets of water deposited on the surface of vegetation and other cool objects due to direct condensation of water from the atmospheric air, especially at night after a hot day.
r роса
Отложение водяных капель на поверхности растений и других холодных предметов вследствие конденсации водяного пара из воздуха в результате радиационного выхолаживания, особенно ночью после жаркого дня.
f rosée *f*
s rocío *m*
g Tau *m*

1573 **dew point**
The temperature at which air on cooling becomes saturated with water vapour and below which condensation begins and dew forms. → Fig. 87, p. 188
r точка росы
Температура, при которой охлаждающийся воздух насыщается водяным паром и начинается его конденсация с образованием росы. → Рис.87, с. 188

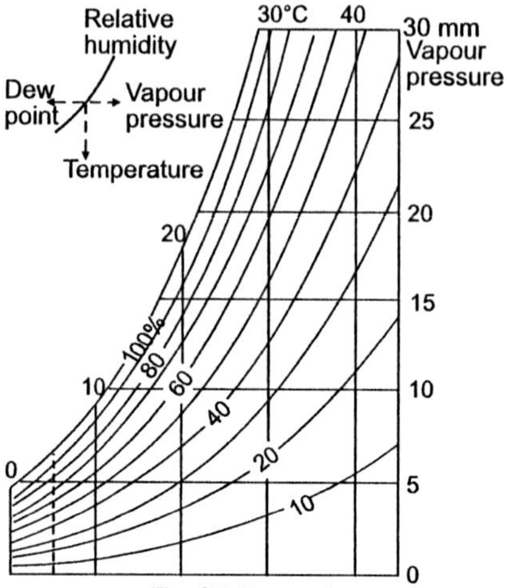

Fig. 87. **Dew point**

f point *m* de rosée
s punto *m* de rocío
g Taupunkt *m*

1574 dew point depression; dew-point deficit
The difference between the air temperature and the dew point.
r дефицит точки росы
Разность между температурой воздуха и точкой росы.
f dépression *f* du point de rosée
s depresión *f* del punto de rocío
g Taupunktdifferenz *f*

* **dew-point deficit → 1574**

1575 diabatic *adj*; non-adiabatic *adj*
Related to the thermodynamic process in which a mass of air gains or loses heat from/to external sources, the opposite of *adiabatic*.
r неадиабатический
Относящийся к термодинамическому процессу, в котором происходит обмен теплом между данной массой воздуха и внешними источниками тепла, противоположен термину *адиабатический*.
f diabatique *adj*
s diabático *adj*
g diabatisch *adj*

1576 diabatic process; non-adiabatic process
The change of temperature and volume of air with the external exchange of heat, the opposite to *adiabatic process*.
r диабатический процесс
Изменение термодинамического состояния воздушной массы, сопровождающееся обменом теплом с окружающей ее средой, противоположен *адиабатическому процессу*.
f processus *m* diabatique
s proceso *m* diabático
g diabatischer Prozess *m*

1577 diagenesis
Processes of physical and chemical transformation of rocks in the upper parts of the Earth's crust and on its surface under the impact of the atmosphere, hydrosphere and living organisms.
r диагенез; гипергенез
Совокупность физико-химических процессов преобразования рыхлых осадков на дне водоемов и на суше в осадочные породы в условиях верхней зоны земной коры.
f diagenèse *f*
s diagénesis *f*
g Diagenese *f*

1578 diagenesis of snow cover
The transformation of snow cover without the influence of high temperature and pressure that does not result in the conversion of snow into firn.
r диагенез снежного покрова
Преобразование снежного покрова без участия высокой температуры и давления, не ведущее к превращению снега в фирн.
f diagenèse *f* de manteau nival
s diagénesis *f* de manto nivoso
g Diagenese *f* der Schneedecke

1579 diagnostic horizon
A strictly defined soil horizon which clearly shows the soil-forming processes at work in an area, and is used as a basis for classifying soils in a soil classification.
r диагностический горизонт
Строго определенный почвенный горизонт, ясно показывающий почвообразовательные процессы на исследуемой территории и служащий основой для классификации почв.
f horizon *m* diagnostique
s horizonte *m* diagnóstico
g diagnostischer Horizont *m*

1580 diagonal lamination
The initial inclined bedding of interlayers and lenses inside sedimentary rocks.
r диагональная слоистость; косая слоистость
Первичное наклонное залегание прослоев и линз внутри осадочных горных пород.
f litage *m* diagonal
s litaje *m* diagonal
g Diagonalschichtung *f*; Schrägschichtung *f*; Kreuzschichtung *f*

1581 diagonal scale
A linear scale in form of a graph for precise measurement and setting of distances on a geographical map.
r поперечный масштаб

Линейный масштаб в виде графика, предназначенный для точных измерений и откладывания расстояний на географических картах.
f régle *f* á vernier
s regla *f* con nonio
g Transversalmaßstab *m*

1582 diagonal valley
A river valley which cuts at some angle across the direction of folds.
r диагональная долина
Долина реки, заложенная под некоторым углом к линии простирания складок горных пород.
f vallée *f* diagonale
s valle *m* diagonal
g Diagonaltal *n*

1583 diagram
A graphic representation of numerical data (a drawing or design) that shows the arrangement and relations, and commonly used on maps.
r диаграмма
Графическое изображение ряда сопоставляемых величин, используемое, в частности, при картографировании.
f diagramme *m*
s diagrama *m*
g Schaubild *n*; Diagramm *n*

1584 diagram map
A representation on a geographical map of the phenomena which have a continuous or stripe arrangement, by graphs and diagrams placed at the points of the measurement of these phenomena. → Fig. 49, p. 106
r способ локализованных диаграмм; картодиаграмма
Изображение на географической карте явлений, имеющих сплошное или полосное простирание, с помощью графиков и диаграмм, помещенных в пунктах измерения этих явлений. → Рис. 49, с. 106
f carte-diagramme *f*
s mapa-diagrama *m*
g Kartodiagramm *n*; Kartodiagrammethode *f*

1585 diamond dust; powedery snow; adamantine braize
The precipitation of fine columnar ice crystals or their fragments under the weather conditions of cloudless sky or high clouds.
r алмазная пыль; снежная пыль
Выпадение мелких столбчатых кристаллов льда или их обломков при безоблачном небе или высокой облачности.
f poudre *f* de diamand
s polvo *m* diamantino; nieve *f* pulverulenta; polvo *m* de diamante
g Diamantstaub *m*

1586 diaphtoresis of snow cover

Fig. 88. **Diapir**

The loosening of a snow cover as a result of metamorphism.
r диафторез снежного покрова
Разрыхление снега в результате метаморфизма.
f diaphtorése *f* de couverture *f* nivale
s diaftoresis *f* de manto nival
g Diaphthorese *f* der Schneedecke

1587 diapir
A dome-shaped anticlinal fold which has resulted from the upward movement of mobile rocks such as halite, lying beneath the more competent strata. → Fig. 88
r диапировая складка; диапир
Куполообразная антиклинальная складка со смятым ядром, возникающим вследствие выдавливания снизу высокопластичных пород (соль, глины). → Рис. 88
f pli *m* diapirique; diapir *m*
s pliegue *m* diapír; diapiro *m*
g Diapirfalte *f*

1588 diaspora
An ethnic group of people living outside its country of origin.
r диаспора
Этническая группа, живущая вне страны своего происхождения.
f diaspora *f*
s diáspora *f*
g Diaspora *f*

1589 diastrophism
Tectonic processes which produce dramatic changes in the shape of the Earth's surface, such as orogenies, faulting and folding.
r диастрофизм
Интенсивное проявление тектонических движений и деформаций земной коры.
f diastrophisme *m*
s diastrofismo *m*
g Diastrophismus *m*

1590 diatom
A type of unicellular microscopic algae having cell walls composed of silica and containing brown pigment (diatomin), they occur in salt and fresh water, in soils, on rocks, tree bark, and on sea ice.
r диатомовые водоросли
Одноклеточные водоросли, содержащие

твердый скелет и бурый пигмент диатомин, распространены в соленых и пресных водах, в почве, на скалах, коре деревьев, на морских льдах.

f Diatomées *fpl*
s diatomeas *fpl*
g Kieselalgen *fpl*; Diatomeen *fpl*

1591 dichothermy
The distribution of temperature in a water body when the temperature minimum is located at a certain depth.

r дихотермия
Распределение температуры воды в водоеме, при котором ее минимум находится на некоторой глубине от поверхности.

f dichothermie *f*
s dicotermia *f*
g Dichothermie *f*

* **dictatorship → 464**

1592 difference of glacierization
The difference between the altitudes of the highest point of the glacier basin and the valley bottom where the glacier moves down.

r разность оледенения
Разница высотных отметок наивысшей точки в бассейне ледника и днищ долин, в которые спускаются языки ледников.

f dénivellation *f* d'un glacier
s diferencia *f* de glaciación
g Vergletscherungsdifferenz *f*

1593 differential ablation
The melting of a glacier surface at different rates to produce variable landforms on it.

r дифференциальная абляция
Неодинаковая скорость таяния на разных участках ледниковой поверхности, что приводит к формированию на ней неровностей.

f ablation *f* différentielle
s ablación *f* diferencial
g Differentialablation *f*

1594 differential erosion
The selective desintegration and removal of rocks by the action of running water in the places which are the least resistant.

r избирательная эрозия
Неравномерный размыв горных пород в местах наименьшего сопротивления под действием текучей воды.

f érosion *f* différentielle
s erosión *f* selectiva; erosión *f* diferencial
g selektive Erosion *f*

1595 differential rent
A form of ground rent that brings additional profit due to the difference in labour productivity on average and better lands, or due to additional capital investment.

r дифференциальная рента
Форма земельной ренты, представляющая добавочную прибыль, возникающую как разница в производительности труда при равновеликих затратах на средних и лучших землях или при добавочных вложениях капитала на одном и том же земельном участке.

f rente *f* différentielle
s renta *f* diferencial
g Differentialrente *f*

1596 differential weathering
The relatively greater rate of rock desintegration in certain parts of a rock mass than in others which may be due to difference in porosity, grain size or mineral composition.

r избирательное выветривание
Относительно большая скорость разрушения горных пород в определенных местах, возникающая вследствие неодинаковых пористости, размеров зерен или минерального состава пород.

f météorisation *f* différentielle
s meteorización *f* selectiva
g selektive Verwitterung *f*

1597 diffluence of glacier; glacial diffluence
The formation of a lateral branch of a glacier or separation of a glacier when the ice flows into a nedighbouring valley over a watershed.

r диффлюэнция ледника
Боковое ответвление или разделение ледникового потока со стеканием льда через водораздел в соседний бассейн.

f diffluence *f* de glacier
s difluencia *f* de glaciar
g Diffluenz *f* eines Gletschers

1598 diffraction of a wave; wave diffraction
Bending of the wave front due to its collision against an above-water obstacle or when it is running through a narrow passage.

r дифракция волн
Изгиб фронта волны вследствие столкновения с надводным препятствием или при проходе ее через узкое пространство.

f diffraction *f* des vagues
s difracción *f* de olas
g Wellendifraktion *f*

1599 diffuse front
The front which is difficult to determine from study of the meteorologocal elements since the contrast between the elements near the front is small and the frontal area is wide.

r размытый фронт
Трудно различимый атмосферный фронт, поскольку контрасты метеоэлементов вблизи фронта малы, а ширина фронта достаточно велика.

f front *m* diffus
s frente *m* difuso
g unscharfe Front *f*

1600 diffuse solar radiation; sky radiation
Indirect scattered and reflected solar radiation, coming from the sky with the exeption.

r рассеянная солнечная радиация
Нисходящая рассеянная и отраженная солнечная радиация, поступающая на земную поверхность от небесного свода.
f rayonnement *m* solaire diffus; rayonnement *m* du ciel
s radiación *f* solar difusa
g diffuse Sonnenstrahlung *f*; Himmelsstrahlung *f*

* **diffusion → 1601**

1601 **diffusion of innovations; diffusion**
The widespread dispersal, the dissemination of cultural, social, economic and other phenomena, innovation, idea, information from the sources where they initially concentrated.
r диффузия нововведений
Распространение культурных, социальных, экономических и других явлений, новшеств, идей, информации, первоначально концентрирующихся в одном или нескольких очагах-источниках.
f diffusion *f* des innovations; diffusion *f*
s difusión *f* de innovaciones; difusión *f*
g Innovationsdiffusion *f*; Ausbreitung *f* der Neuerungen

1602 **digital cartography; digital mapping**
A branch of *cartography* dealing with the theory and practice of creation and use of digital maps and models of the Earth's surface.
r цифровая картография; цифровое картографирование
Раздел *картографии*, охватывающий теорию и практику создания и использования цифровых карт и моделей земной поверхности.
f cartographie *f* digitale
s cartografía *f* digatal
g Digitalkartographie *f*; Digitalkartierung *f*

1603 **digital elevation model**
A digital representation of three-dimensional spatial objects as a set of elevations or depths, normally at joints of the regular grid.
r цифровая модель рельефа
Объемное представление трехмерных пространственных объектов как совокупности отметок высот или глубин, обычно в узлах регулярной сети.
f modèle *m* de relief digital
s modelo *m* de relieve digital
g digitales Höhenmodell *n*

1604 **digital map; electronic map**
Cartographic product in electronic form presenting digital data together with software for their visualisation.
r цифровая карта; электронная карта
Географическая карта или набор слоев ГИС, визуализированные или подготовленные к визуализации на видеоэкране компьютера или другого устройства.
f carte *f* digitale; carte *f* électronique

Fig. 89. **Dike**

s mapa *m* digital; mapa *f* electrónico
g digitale Karte *f*; Elektronenkarte *f*

* **digital mapping → 1602**

1605 **dike; dyke**
A sheet-like body of rock which cuts through the host rock and often rises above the surface like a wall. → Fig. 89
r дайка
Плитообразное геологическое тело, секущее геологические породы и нередко возвышающееся над поверхностью в виде стены. → Рис. 89
f dyke *m*; filon-couche *m*; filon *m* éruptif; sill *m*
s dique *f*; filón-capa *f*; filón *m*; sill *m*
g Eruptivgang *m*

* **diking → 551**

1606 **dilution of natural waters**
The mixing of polluted natural water with natural non-polluted water which improves the state of the body of water.
r разбавление природных вод
Смешение загрязненных природных вод с природными незагрязненными водами, что улучшает состояние водного объекта.
f dilution *f* d'eau
s delución *f* de aguas
g Verwässerung *f*

1607 **diluvium**
An obsolete term used widely by XIX century, originally given to material supposedly deposited by the Noachian Flood described in the Bible, then for glacial drift and finally to describe any flood deposit.
r дилювий
Устаревший термин, широко использовавшийся в XIX в. для описания предполагаемых отложений Библейского потопа, позже его применяли для описания отложений ледниковых, а затем и любых других паводков.
f diluvium *m*
s diluvium *m*
g Diluvium *m*

1608 **dimictic lake**
A lake having two periods of the total vertical water circulation a year, e.g. in temperate cli-

mate the convective circulation occurs in spring and autumn.

r димиктическое озеро
Озеро с двумя ежегодными периодами полной вертикальной циркуляции воды, например, в умеренном климате конвективная циркуляция происходит весной и осенью.

f lac *m* dimictique
s lago *m* dimíctico
g dimiktischer See *m*

1609 dingja
A flat volcano with gentle slopes in Island which forms after repeated lava-flows.

r дингья
Пологоплоский вулкан в Исландии, образующийся при повторных излияниях жидкой лавы.

f dingja *f*
s dingja *f*
g Dingja *f*

1610 dip; true dip
The maximum inclination of a stratum of rocks determined relative to a horizontal plane (angle of dip) and relative to the meridian of the place (direction of dip). → Fig. 90

r падение; линия падения
Наибольший наклон пласта, определяемый относительно горизонтальной плоскости (угол падения) и меридиана местности (азимут падения). → Рис. 90

f pendage *m*
s buzamiento *m*
g Einfallen *n*; Schichteinfallen *n*

1611 direct solar radiation
The solar radiation coming from the Sun's disk without being absorbed.

r прямая солнечная радиация
Солнечная радиация, поступающая непосредственно от солнечного диска.

f rayonnement *m* solaire direct
s radiación *f* solar directa
g direkte Sonnenstrahlung *f*

Fig. 90. **Dip**

1612 direct-current water supply
The water supply from a water body to a consumer and its discharge after the use back into the water body with or without the preliminary purification.

r прямоточное водоснабжение
Подача воды из водного объекта потребителю и сброс ее после использования в водный объект с очисткой или без очистки.

f alimentation *f* en eau direct
s abastecimiento *m* de agua directo
g direkte Wasserversorgung *f*

*** dirt cone → 1443**

1613 dirt ice bands
Bands or lenses of ice mixed with mud and debris material (fragments of rocks).

r грязные ленты льда
Линзы льда, засоренного дисперсной породой, а иногда и обломками горных пород.

f bandes *fpl* de boue en glace
s bandas *fpl* de barro en hielo
g Eisstreifen *mpl* mit Moräneneinschlüsse

*** dirt road → 1769**

1614 disappearing stream; influent stream
A river in a karst area which suddenly falls down into the karst cavity and continues its flow in the underground conditions. → Fig. 91, p. 193

r исчезающая река
Река в карстовой области, внезапно проваливающаяся в карстовой воронке или полости и продолжающая свое течение под землей. → Рис. 91, с. 193

f rivière *f* disparaissante
s río *m* desapareciente
g verschwindender Fluß *m*

1615 disastrous flood
An especially high flood of rare frequency (the discharge openings of hydraulic constructions are rated at this amount of flood water).

r катастрофический паводок
Выдающийся по величине паводок редкой повторяемости, на пропуск которого рассчитываются водосбросные отверстия гидротехнических сооружений.

f crue *f* catastrophique
s crecida *f* inundante; crecida *f* catastrófica
g Flutkatastrophe *f*; Katastrophenhochwasser *n*

1616 discharge basin
A cirque-shaped depression on a slope of an ice cap or ice sheet serving as the head of an outlet glacier.

r бассейн истечения
Циркообразная впадина на склоне ледникового щита или купола, служащая верховьем выводного ледника.

f bassin *m* d'écoulement
s cuenca *f* de desagüe
g Ausfliessbecken *n*

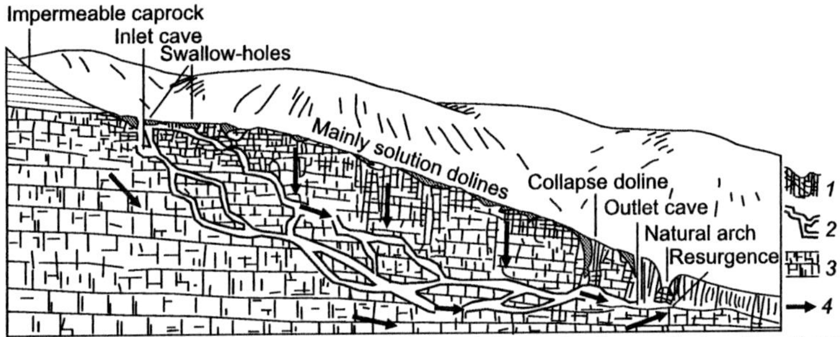

Fig. 91. **Disappearing stream**: 1 – zone of percolation, solution and collapse, 2 – location of stream-water corrosion and corrasion, 3 – zone of phreatic water solution, 4 – dominant direction of water movement

1617 discharge current; hydraulic current
Horizontal movement of waters caused by the sea surface slope occurring due to its local rise or drop.
r стоковое течение
Горизонтальное движение вод, вызванное наклоном уровня моря под влиянием его местного повышения или понижения.
f courant *m* de décharge
s corriente *f* de descarga
g Abflußströmung *f*

1618 discharge fluctuations
The change of the discharge of stream due to alternation of abundance of water and little water periods.
r колебания стока; колебания расхода воды
Изменения водности рек, проявляющиеся в последовательной смене многоводных и маловодных циклов.
f variation *f* des débits
s variación *f* de escorrentias; variación *f* de caudales
g Schwankungen *fpl* der Wassermenge

1619 discharge of spring
The volume of water incoming from a spring in a given unit of time.
r дебит источника
Количество воды, поступающей из источника в единицу времени.
f débit *m* de source
s caudal *m* de manantial
g Quellergiebigkeit *f*

1620 discharge rating curve
Graph of relationship between the height of water level and its discharge for a given section of a watercourse over a certain period of time.
r кривая расхода воды
График связи между высотой уровня воды и ее расходом для данного сечения водотока за определенный период времени.
f courbe *f* des débits
s curva *f* de caudal
g Abflußkurve *f*

* **discharge site → 2892**

1621 dischargeof a stream
The volume of water flowing through a cross section of a stream in a given unit of time.
r расход воды
Объем воды, протекающий через живое сечение потока в единицу времени.
f débit *m*
s caudal *m*; aforo *m*
g Abflußmenge *f*

1622 disclimax; plagioclimax
A stable state of the environment which has been caused by human intervention, such as a desert caused by deforestation (much of the worldrs vegetation today can be regarded as plagioclimax).
r дисклимакс
Стабильное состояние окружающей среды, вызванное воздействием человека, например, возникновение пустыни на месте некогда сведенного леса.
f disclimax *m*
s disclímax *m*
g Disklimax *n*

* **disconformity → 1625**

1623 discontinuous area; disjunctive area
The area occupied by a species consisting of several separate territories located far from each other so that the excange of seeds and spores between plants or migration of animals becomes impossible (for example, area of arctic-alpine plants, broad-leaved forests in Northern Eurasia, etc.).
r разорванный ареал; дизъюнктивный ареал; прерывистый ареал
Пространство обитания, распадающееся на несколько разобщенных территорий, удаленных друг от друга до такой степени, что обмен семенами/спорами между растениями или миграции животных невозможны (например, ареал аркто-альпийских растений, широколиственных лесов в Северной Евразии и пр.).
f aire *f* disjointe
s área *f* disjunta
g disjunktes Areal *n*

1624 discontinuous permafrost
A peripheral zone of permanently frozen ground interrupted by non-frozen ground. → Fig. 247, p. 539
r прерывистая вечная мерзлота
Краевая часть области многолетнемерзлых пород, местами прерывающихся немерзлыми отложениями. → Рис. 247, с. 539
f pergélisol *m* discontinu
s permafrost *m* discontinuo
g diskontinuierlicher Permafrost *m*

*** discontinuous stream → 3168**

1625 discordance; discordant bedding; unconformity; disconformity
A lack of parallelism between contiguous rock strata; the disposition of younger deposits on the surface of more ancient rocks with the interval in accumulated sediments. → Fig. 92
r несогласное залегание
Залегание более молодых отложений на размытой поверхности более древних пород, свидетельствующее о перерыве в накоплении осадков. → Рис. 92
f discordance *f*
s discordancia *f*
g diskordante Lagerung *f*

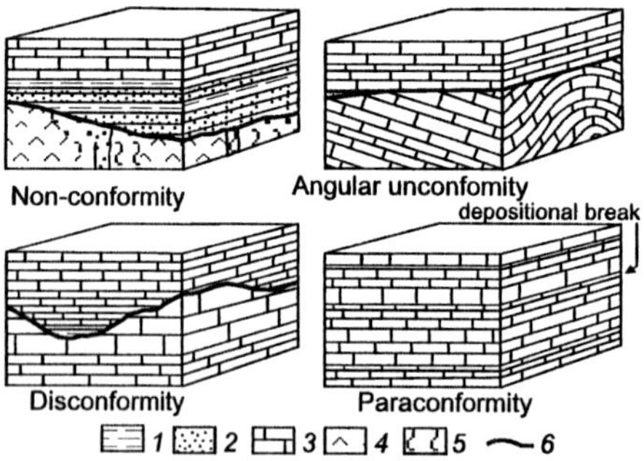

Fig. 92. Four types of **discordance**. 1 – shale, 2 – sandstone, 3 – carbonate, 4 – igneous rock, 5 – metamorphic rock, 6 – erosion surface

1626 discordance surface
The surface of contact of two rock strata of different age, in particular along a tectonic fracture.
r поверхность несогласия
Поверхность соприкосновения двух разновозрастных толщ горных пород, в частности, по тектоническому разрыву.
f surface *f* de discordance
s superficie *f* de discordancia
g Diskordanzfläche *f*

1627 discordant *adj*
Any landform that bears little or no relationship to the geological structure of the «core» of the region.
r несогласный
Любая черта рельефа, которая не согласуется с основной геологической структурой данного региона.
f discordant *adj*
s discordante *adj*
g diskordant *adj*

*** discordant bedding → 1625**

1628 discordant coast; transverse coast
A coastline which cuts across the geological structure of the coastal region.
r поперечный берег
Берег, общее направление которого сечет под прямым или близким к нему углом простирание геологических структур прибрежной суши.
f côte *f* transversale
s costa *f* transversal
g Querküste *f*

*** diseases of civilization → 1798**

1629 disgarmonic folding
A group of folds formed simultaneously along the whole thickness of layers but having different forms at different levels of stratigraphic section.
r дисгармоничная складчатость
Группа складок, образующихся одновременно по всей толще слоев, но имеющих неодинаковые формы на разных уровнях стратиграфического разреза.
f plissement *m* disgarmonique
s plegamiento *m* disarmonioso
g disgarmonische Faltung *f*

1630 disjunction of areas
Separation of plants or animal areas caused by the drift of continents, lowering of land areas, plants migration, change of climatic conditions, extinction of some species, etc.
r дизъюнкция ареалов
Разъединение ареалов растений или животных, обусловленное дрейфом континентов, опусканием участков суши, миграцией растений, изменением климатических условий и вымиранием отдельных видов и пр.
f disjonction *f* des aires
s disjunción *f* de áreas
g Arealdisjunktion *f*

*** disjunctive area → 1623**

1631 dislocated bed stratum
A rock layer, contorted, curved, ruptured or shifted as a result of geological processes.
r дислоцированный слой
Слой горных пород, смятый, изогнутый, разорванный или смещенный в результате геологических процессов.
f couche *f* disloquée
s capa *f* dislocada; estrato *m* dislocado
g gestörte Schicht *f*

1632 **dislocated moraine; contorted till; disturbation till**
A moraine crumpled by a moving glacier into well pronounced folds.
r дислоцированная морена
Морена, смятая движущимся ледником в заметно выраженные складки.
f moraine *f* disloquée; moraine *f* feuilletée
s morrena *f* dislocada; morrena *f* moldurada
g Strauchmoräne *f*; Staumoräne *f*

1633 **dislocation**
Disturbance of forms of bedding of a geological body caused by tectonic motions or processes inside rocks, compressive dislocation (bending of layers and folds) and disjunctive dislocation (fracture with shift) are distinguished.
r дислокация
Нарушение формы залегания геологического тела, вызываемое тектоническими движениями или процессами внутри горных пород, различают пликативные (смятие пластов и складок) и дизъюнктивные (разрыв со смещением) дислокации.
f dislocation *f*
s dislocación *f*
g Störung *f*; Dislokation *f*

1634 **dispatched production**
A statistical index representing the money which enterprises receive or will receive for the goods produced or services rendered (when products have been shipped on conditions of subsequent cover).
r отгруженная продукция
Статистический показатель, представляющий собой полученные и подлежащие получению (при отгрузке на условиях последующей оплаты) денежные поступления предприятий от продажи продукции или оказания услуг.
f production *f* expédiée
s producción *f* expedida
g Verladungsproduktion *f*

1635 **dispersed ground**
Ground consisting of small homogeneous particles, often wet.
r дисперсный грунт
Грунт, состоящий из мелких однородных частиц, нередко смоченных водой.
f sol *m* despersu
s suelo *m* disperso
g disperser Boden *m*

1636 **dispersed settlement**
A settlement pattern in which farmhouses and rural dwellings are scattered instead of being grouped together.
r рассредоточенное поселение
Особенности сельского поселения, когда отдельные крестьянские фермы и жилые дома расположены на значительном расстоянии друг от друга, не образуя компактных населенных пунктов.
f peuplement *m* dispersé
s poblamiento *m* disperso
g verstreute Siedlung *f*

* **disphotic zone → 331**

1637 **dissected plain**
A plain cut by a network of valleys, gullies and gulches.
r расчлененная равнина
Равнина, изрезанная долинно-балочной и овражной сетью.
f plaine *f* ravinée
s llanura *f* barrancal
g gegliederte Ebene *f*

1638 **dissected volcano**
A volcanic cone with the slopes deeply dissected by radial erosion furrows.
r эродированный вулкан
Вулканический конус, склоны которого глубоко изрезаны радиальными эрозионными бороздами – барранкосами.
f volcan *m* écossais
s volcán *m* escocés
g zerschnittener Vulkan *m*

1639 **dissection; dissection of relief**
The process by which a land surface is cut up by eroding streams into numerous valleys.
r расчленение рельефа
Процесс разъединения единой до этого формы рельефа главным образом путем эрозии.
f dissection *f*; dissection *f* de relief
s disección *f*; disección *f* de relieve
g Reliefgliederung *f*

* **dissection of relief → 1639**

1640 **dissipation of wave energy**
The loss and dispersion of the wave energy mainly due to the interaction of waves with sea bottom as they break.
r диссипация энергии волн
Рассеивание волновой энергии, главным образом при взаимодействии волн с морским дном, когда волны разбиваются.
f dissipation *f* de l'énergie de vagues
s disipación *f* de energía de ondas
g Dissipation *f* der Wellenenergie

1641 **dissolved load**
The organic and inorganic material in solution carried by a stream, as distinct from the bed load and the suspended load.
r растворенные наносы
Растворенный в воде органический и неорганический материал, переносимый рекой, в отличие от влекомых и взвешенных наносов.
f charge *f* en dissolution
s carga *f* de disolución
g lösliche Anschwemmung *f*

1642 distal *adj*
External or remote with regard to the place of origin.
r дистальный
Внешний или дальний по отношению к месту происхождения.
f distal *adj*
s distante *adj*
g distal *adj*

1643 distance distortion
The differences of scale of lengths in different places of a geographical map and in different directions in a given point of the map.
r искажение длины
Различия масштаба длин в разных местах географической карты и по разным направлениям в данной ее точке.
f altération *f* linéaire
s distorsión *f* lineal
g Längenverzerrung *f*

1644 distance of snow transport
The distance over which snow particles can be transported by a snow drifting before they settle on snow cover, get destroyed or evaporate.
r дальность переноса снега
Расстояние, на которое могут быть перенесены отдельные метелевые частицы, прежде чем они закрепятся в снежном покрове, разрушатся или испарятся.
f distance *f* de transport de la neige
s distancia *f* de transporte de nieve
g Reichweite *f* der Schneeverfrachtung

1645 distortion isogram
A line along which the values of distortion of a cartographic projection are equal. → Fig. 93
r изокола
Линия равных искажений в картографической проекции. → Рис. 93
f ligne *f* d'égale altération
s línea *fpl* de igual distorsión
g Äquideformate *n*

1646 distortion of map projection; cartographic alteration; cartographic distortion
Distortions of distances, areas, angles, and shapes when the surface of the ellipsoid or globe is represented on a map.
r искажение в проекциях
Искажение длин, площадей, углов и форм при изображении поверхности эллипсоида или шара на карте.
f altération *f* du projection
s deformación *f* cartográfica; distorsión *f* en proyecciónes cartográficas
g Netzverzerrung *f*

1647 distributary$_1$
A branch canal distributing water from the main canal in an irrigation system.
r распределительный канал
Ответвляющийся канал, служащий для распределения воды из главного канала в оросительную систему.
f distributeur *m*
s distribuidor *m*
g Verteilungskanal *m*

* **distributary^2 → 779**

1648 district$_1$; urban district
A clearly identifiable section of a city, having a distinct image and geographical extent.
r район города
Ясно различимая часть города, имеющая характерный облик и занимающая определенную территорию.
f district$_1$ *m*; quartier *m*
s distrito$_1$ *m*
g Bezirk *m*

* **district^2 → 85**

1649 distrophic lake
A lake having a distorted cycle of nutrients in the water ecosystem.
r дистрофное озеро
Озеро с нарушенным круговоротом питательных веществ в водной экосистеме.
f lac *m* distrophe
s lago *m* distrófico
g distropher See *m*

* **disturbation till → 1632**

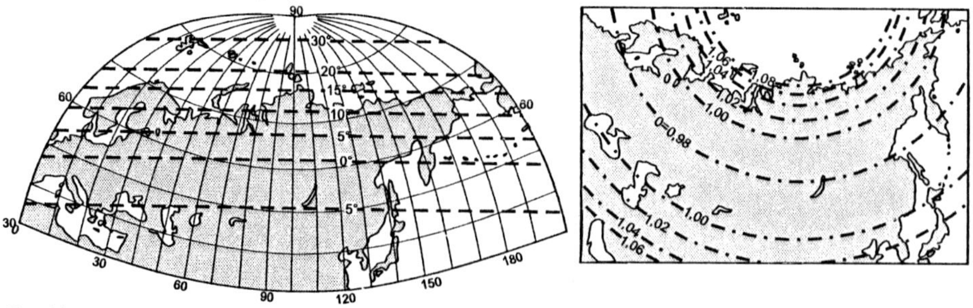

Fig. 93. **Distortion isograms** of an area scale in an equidistant conical cartographic projection (right) and of angles in an oblique perspective cylindrical projection (left)

* **disurbanisation → 1564**

* **ditch → 5084**

1650 **diurnal** ***adj***
Related to each day, completed once a day.
r дневной
Относящийся к конкретному дню, выполненный за день, каждодневный.
f diurne *adj*
s diurno *adj*
g Tages-; täglich *adj*

1651 **diurnal range; diurnal variation**
A measure of the difference between the minimum and maximum values of a meteorological element, recorded whithin a 24-hours period, e.g. as applied to air temperature.
r дневной ход; суточный ход
Различия за 24 часа между минимальным и максимальным значениями какой-либо величины, например, температуры.
f variation *f* diurne; amplitude *f* diurne
s variación *f* diurna; amplitud *f* diurna
g Tagesbereich *m*; Tagesgang *m*

1652 **diurnal tide**
A type of tidal pattern found in a few areas of the world where, because of the configuration of the coastline (e.g. the Philippine Islands) only one high tide and one low tide occurs every 24 hours (each lunar day).
r суточный прилив
Тип приливов, происходящих в некоторых районах земного шара, где благодаря конфигурации береговой линии (например, на Филиппинских островах) высокая и низкая вода отмечаются с циклом, близким к лунным суткам - примерно 24 час. 49 мин.
f marée *f* diurne
s marea *f* diurna
g eintägige Gezeiten *fpl*

* **diurnal variation → 1651**

1653 **divagating lake**
A lake changing its location from year to year.
r блуждающее озеро
Озеро, меняющее свое положение от года к году.
f lac *m* divagante
s lago *m* divagante
g wandernder See *m*

1654 **divagating meander**
A meander of a river which is liable to variation from time to time because the surface on which it occurs approaches the condition of a peneplain.
r блуждающая излучина
Речная излучина, время от времени подверженная перемещениям, так как поверхность, по котороой течет река, близка к условиям пенеплена.
f méandre *m* divagant; méandre *m* libre
s meandro *m* divagante; meandro *m* libre
g wandernder Mäander *m*

1655 **divergence**
The movement of surface masses of sea water move away as a result of wind-drift, leading to upwelling of deep water; discrepancy of air flows in the direction of its general motion.
r дивергенция
Расхождение поверхностных морских течений или линий тока воздуха в направлении общего переноса.
f divergence *f*
s divergencia *f*
g Divergenz *f*

1656 **diverging terraces**
River terraces with the elevation above each other and above the river level gradually increasing downstream.
r расходящиеся террасы
Речные террасы с превышениями друг над другом и над уровнем реки, постепенно увеличивающимися вниз по течению.
f terrasses *fpl* divergentes
s terrazas *fpl* divergentes
g divergierendeTerrassen *fpl*

1657 **diversification**
Widening of industrial branches, the range of products or servises of a given area.
r диверсификация
Увеличение разнообразия отраслей хозяйства, номенклатуры их продукции, расширение видов предоставляемых услуг и др. на данной территории.
f diversification *f*
s diversificación *f*
g Diversifizierung *f*; Diversifikation *f*

1658 **diversification index**
A measure of economic variety of a given area.
r индекс диверсификации
Мера разнообразия отраслей хозяйства на данной территории.
f indice *m* de diversification
s índice *m* de diversificación.
g Diversifikationsindex *m*

1659 **diversified economy**
The economy of a state or region which has a wide range of different sectors of industry, agriculture and services.
r многоотраслевая экономика
Экономика страны или региона, в которой представлен широкий спектр отраслей промышленности, сельского хозяйства и сферы услуг.
f économie *f* diverse
s economía *f* diversa
g diverse Ökonomik *f*

* **divide → 6705**

1660 divided circle diagram; pie graph; wheelgraph; pie diagram
A diagram in which a circle, representing the total of the values, is divided into sectors, each sector being proportional to the value it represents.
r круговая диаграмма
Диаграмма, в которой круг, представляющий всю величину, разделен на сектора, каждый из которых пропорционален изображаемой здесь величине.
f diagramme *m* circulaire divisible
s diagrama *m* circular divisorio
g Runddiagramm *n*

1661 division of labour
The process of development of productive forces when different types of labour become isolated while productive units and regions become highly specialized and exchange products among themselves.
r общественное разделение труда
Процесс развития производительных сил, заключающийся в обособлении разных видов трудовой деятельности, специализации отдельных производящих единиц и регионов, обмене между ними продуктами труда.
f division *f* du travail
s división *f* de trabajo
g Arbeitsteilung *f*

1662 division of water; water parting
The redistribution of waters between neighbouring drainage basins usually occurring in headwaters of plain rivers without the formation of branches.
r деление вод
Перераспределение вод между соседними водосборами, происходящее обычно в верховьях равнинных рек и не сопровождающееся разветвлением реки.
f division *f* des eaux; partage *m* des eaux
s división *f* de aguas
g Wasserteilung *f*

1663 dock
An enclosure or artificial basin, fitted with floodgates, in a harbor or river, in which vessels are loaded, unloaded, refitted, repaired.
r док
Замкнутый или специально построенный бассейн в гавани или на реке, где происходит погрузка, разгрузка, ремонт и переоборудование кораблей.
f dock *m*
s dock *m*; dársena *f*
g Dock *n*

1664 dockland
The district around the docks in a large port together with dock basins, adjoining wharfs, warenhouses, workshops and yards often regenerated in modern towns to become zones of entertainment and tourism, as in San Francisko.
r район доков
Территория вокруг доков в большом порте, включающая сами доки с бассейнами, верфи, мастерские и склады, в наше время нередко перестраиваемая в зоны отдыха и туризма, как, например, в Сан-Франциско.
f dockland *m*
s dockland *m*
g Dockland *n*

1665 doldrums; equatorial calms
The equatorial belt of calms or light variable winds lying between the two trade wind belts.
r зона затишья; экваториальная зона затишья
Зона слабых переменных ветров и штилей вблизи экватора, лежащая между поясами пассатов обоих полушарий.
f calmes *mpl* équatoriaux; pot-au-noir *m*
s calmas *fpl* ecuatoriales
g Kalmengürtel *m*; äquatoriale Kalmen *fpl*

1666 doline
A relatively small enclosed hollow of karst origin, often with a *ponor* in the floor. → Fig. 94, p. 199
r до́лина; карстовая воронка
Замкнутая впадина сравнительно небольшого размера карстового происхождения, часто с *понором* на дне. → Рис. 94, с. 199
f doline *f*
s dolina *f*
g Doline *f*; Karsttrichter *m*

1667 dolomite
A sedimentary limestone rock composed mainly of the mineral of calcium magnesium carbonate.
r доломит
Осадочная горная порода, состоящая в основном из одноименного минерала – двойного карбоната кальция и магния.
f dolomite *f*; dolomie *f*
s dolomita *f*; dolomía *f*
g Dolomit *m*

1668 dome
1. A dome-like landform or rock massif. 2. A zone in the body of water where the thermocline rises towards the surface.
r купол
1. Любая куполообразная форма рельефа или массива горных пород. 2. Район подъема термоклина в толще воды в океане к поверхности: в Северном полушарии в районе антициклонических вихрей, в Южном – циклонических.
f coupole *f*; dôme *m*
s cúpula *f*; bóveda *f*; domo *m*
g Dom *m*; Kuppe *f*

1669 dome-shaped iceberg; rounded iceberg
An iceberg with a flat rounded top, often highly crevassed, broken off from an outlet glacier.

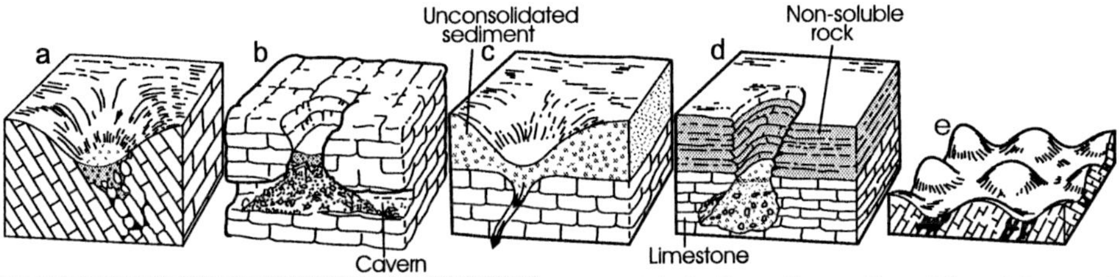

Fig. 94. **Doline**. Above: Five different types of limestone solution hollow: a – solution doline produced by widening of joints; b – collapse doline produced by a cave roof falling in; c – subsidence doline produced by removal of unconsolidated sediment down a fissure of joint; d – subjacent karst collapse doline; e – cockpits (intersecting star-shaped dolines), typical of tropical areas. Left: Doline at the top part of the Yaila Mountains in Crimea

r куполообразный айсберг
Айсберг с вершиной в форме пологого купола, как правило, со множеством трещин, отколовшийся от выводного ледника.
f iceberg *m* en coupole
s iceberg *m* de cúpula
g Kuppeleisberg *m*

1670 domestic *adj*
Belonging or related to native land or nation, situated within a country, as opposed to international.
r внутренний; отечественный
Принадлежащий или относящийся к дому, родной стране или нации, расположенный внутри страны.
f domestique *adj*
s doméstico *adj*
g innerlich *adj*; heimlich *adj*; Innen-

1671 domestic animal
A tame animal living under and dependent on human care.
r домашнее животное
Прирученное животное, живущее рядом с человеком и используемое им в разных целях.
f animal *m* domestique
s animal *m* doméstico
g Haustier *n*

* **domestic industry → 1319**

1672 domestic sewage
Sewage from peoplers homes.
r бытовые стоки
Сточные воды, образующиеся в результате повседневной деятельности людей и на месте их проживания.
f effluents *mpl* domestiques
s aguas *fpl* residuales domésticas
g Haushaltsabwässer *npl*

1673 domestic tourism
Travelling within a country of residents of this country.
r внутренний туризм
Путешествие в пределах страны лиц, постоянно проживающих в данной стране.
f tourisme *m* intérieur; tourisme *m* domestique; tourisme *m* national
s turismo *m* interior; turismo *m* nacional; turismo *m* doméstico
g innerer Tourismus *m*; Innlandstourismus *m*

* **domestic trade → 3156**

1674 domestic waste; household waste
Waste material from peoplers homes.
r бытовые отходы
Отходы от бытовой деятельности людей.
f déchets *mpl* domestiques
s desechos *mpl* domésticos; residuos *mpl* domésticos
g Haushaltsabfälle *mpl*

* **dominant → 1675**

1675 dominant species; dominant
Plant or animal species which is the most abundant in different layes of a community.
r доминантный вид; доминант
Преобладающие в разных ярусах сообщества виды растений и животных.
f espèce *f* dominante
s especie *f* dominante
g dominante Art *f*; herrschende Art *f*

1676 dominion
A self-governing nation of the Commonwealth other than the United Kingdom (existed until 1947) headed by the British monarch represented by governor-generals.
r доминион
Самоуправляемые государства британско-

го Содружества вне Великобритании (существовавшие до 1947 г.), главой которых был английский король, представляемый генерал-губернатором.
f dominion *m*
s dominio *m*
g Dominion *n*

* **dormancy → 2760**

1677 dormant volcano
A volcano having a long period on inactivity, which has not erupted in historic time, but is not regarded as extinct.
r уснувший вулкан
Вулканический конус, в течение длительного времени не проявляющий активности.
f volcan *m* en repos; volcan *m* assoupi; volcan *m* endormi
s volcán *m* latente; volcán *m* dormido
g ruhender Vulkan *m*

1678 dormitory village; residential district; residential area; commuter village
Area of land with buildings specially designed and constructed for residential use; a region from which residents travel daily to work in an accessible nearby large town or conurbation.
r селитебная зона; селитьба; спальный район
Территория, предназначенная для постройки жилых зданий или занятая жилыми зданиями; жилой район, жители которого ежедневно ездят на работу в ближайший большой город или конурбацию.
f ville *m* residentielle; zone *f* résidentielle; quartiers *mpl* résidentiels
s ciudad *f* residential; zona *f* residencial; barrio *m* residencial; ciudad *f* dormitorio
g Schlafstadt *f*; Siedlungsraum *m*; Siedlungsgebiet *n*

1679 dot method
Representation on a geographical map of the spatial distribution of a phenomenon by the use of dots, usually of uniform size, each representing a specific number of the objects concerned. → Fig. 206, p. 456
r точечный способ
Изображение на географической карте массового рассеянного явления с помощью точек, каждая из которых обозначает некоторое количество показываемых объектов. → Рис. 206, с. 456
f méthode *f* des points
s método *m* de puntos
g Punktmethode *f*

1680 double coast line
A lagoon shoreline consisting of the external line formed by shoal heads and spits, and the internal coastline limiting the bedrock coast.
r двойная береговая линия
Береговая линия лагун, состоящая из внешней линии, образованной косами и пересыпями, и внутренней, ограничивающей коренной берег.
f ligne *f* de rivage lagunaire
s línea *f* de costa albufera
g Doppelküste *f*

1681 double tide
A tidal regime in which there is a double high tide (the first falling a little before rising again to a second maximum) or a double low tide, both due to the effects of the shape of the coast or of shallow water.
r двойной прилив; четвертьсуточный прилив
Режим прилива, при котором образуется двойная высокая (первая слегка спадает и поднимается вновь как второй максимум) или двойная низкая вода, в обоих случаях этому способствует конфигурация берега или малая глубина.
f marée *f* double
s marea *f* doble
g Doppelflut *f*

1682 downcutting
The action of a stream as it lowers its bed by fluvial erosion, generally as a result of increased energy due to tectonic uplift.
r глубинная эрозия; врезание; линейная эрозия
Углубление водным потоком своего русла, главным образом в результате увеличения энергии воды при тектоническом воздымании территории.
f érosion *f* linéaire; encaissement *m*; érosion *f* verticale
s erosión *f* lineal; erosión *f* vertical; encajonamiento *m*
g Tiefenerosion *f*; Einschneidung *f*; lineare Erosion *f*

1683 downdraught
A powerfull downward movement of air within cumulonimbus cloud, during a thunderstorm, caused by the evaporation cooling of the air beneath the cloud due to falling rain.
r нисходящий поток воздуха
Сильное движение вниз воздуха внутри кучево-дождевого облака во время грозы, вызванное охлаждением воздуха в результате испарения в нижней части облака благодаря выпадающему дождю.
f abaissement *m*; courant *m* dessendant
s corriente *f* descendente
g Luftsenken *n*

1684 downland
An open expanse of gently undulatuing, elevated land usually of chalk and supporting pasture.
r даунленд
Слегка всхолмленное открытое пространство возвышенностей, сложенных известняками, используемое как пастбище.
f downland *m*

s downland *m*
g Downland *m*

1685 **downscaling**
The way of getting meteorological parameters on a detailed grid on the basis of their values shown on a gross grid by using certain rules and taking into account the specific features of the region.
r даунскейлинг
Процедура получения метеорологических параметров на мелкой сетке по данным на грубой сетке в соответствии с определенными правилами и учетом природных условий данного региона.
f downscaling *m*
s downscaling *m*
g Downscaling *n*

1686 **downstream *adv*, *adj***
In the direction of the flow of a stream and towards the riverr's mouth from the point further up the river.
r вниз по течению
Указывает направление от истока к устью реки или вдоль течения реки.
f aval *adv*
s aguasabajo *adv*; a favor de la corriente
g stromab *adv*; stromabwärts *adv*

* **downtown (*amer*) → 935**

* **downward terrestrial radiation → 513**

* **downwearing → 5479**

1687 **downwelling; sinking**
The process of slow moving down of a big water mass from the surface of ocean as a result of the increase of its density during mixing which occurs in convergence and frontal zones or due to the cooling of sea water and consequent increase of salinity.
r даунвеллинг; синкинг
Процесс медленного погружения большой массы воды с поверхности океана в результате увеличения ее плотности при перемешивании в зонах конвергенции и во фронтальных зонах или при понижении температуры и повышении солености.
f abaissement *m* des eau
s descendimiento *m* de agua
g Downwelling *n*; Sinking *n*

1688 **drain$_1$**
A pipe or cavity placed in the ground in order to collect the excess of surface and soil water.
r дрена
Труба или полость, закладываемая в грунте с целью сбора почвенно-грунтовых вод.
f drain *m*
s dren *m*
g Drän *m*

* **drain$_2$ → 3460**

1689 **drainage**
1. The natural (surface and underground) runoff of water from an area by streams, rivers, etc., or by sheet flow. 2. The removal of the excess surface or ground waters from the land by artificial channels or by pumping-out.
r дренирование; дренаж
1. Естественный (поверхностный и подземный) сток воды с какой-либо территории (водосборного или речного бассейна). 2. Удаление поверхностных или грунтовых вод с данной территории самотеком или откачкой.
f drainage *m*
s drenaje *m*
g Dränierung *f*; Dränung *f*

1690 **drainage area; catchment; drainage basin; watershed (*amer*); catchment area**
That part of the land surface, including water-bearing rocks from where waters flow down to a river, lake or sea.
r водосборный бассейн; водосбор
Часть земной поверхности, включающая толщу водоносных пород, откуда воды стекают в реку, озеро или море.
f bassin *m* versant; bassin *m* hydrographique; bassin *m* de réception
s cuenca *f* vertiente; cuenca *f* hidrográfica
g Einzugsgebiet *n*; Abflußgebiet *n*

* **drainage basin → 1690**

* **drainage density → 1693**

1691 **drainage lake**
A lake having intensive water exchange when one of the inflowing rivers brings the amount of water equal to the outflowing water discharge.
r проточное озеро
Озеро с интенсивным водообменом, когда один из его притоков по водоносности сравним с рекой, вытекающей из озера.
f lac *m* avec émissaire
s lago *m* con emisario
g durchflossener See *m*

1692 **drainage network**
A system of all rivers on a certain area.
r речная сеть
Совокупность всех рек, находящихся в пределах какой-либо территории.
f réseau *m* fluvial
s red *f* fluvial
g Flußnetz *n*

1693 **drainage network density; drainage density**
The ratio of the total length of watercourses in a river basin to its area.
r густота речной сети
Отношение суммарной длины водотоков в речном бассейне к его площади.

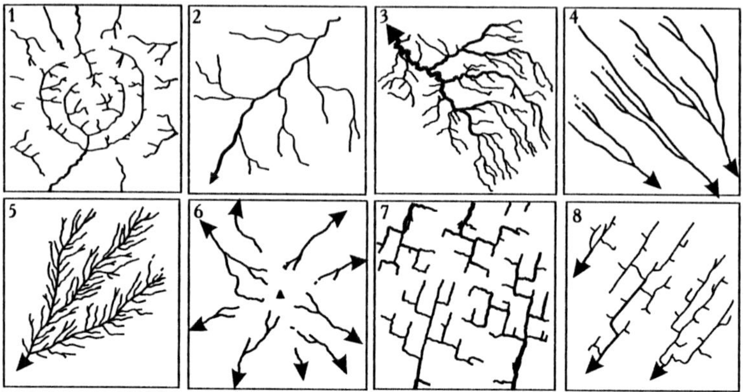

Fig. 95. **Drainage pattern**: 1 – annular, 2 – barbed, 3 – dendritic, 4 – parallel, 5 – pinnate, 6 – radial, 7 – ectangular, 8 – trellis

f densité *f* de réseau hydrographique; densité *f* de réseau fluvial
s densidad *f* de red hidrográfica; densidad *f* de red fluvial
g Gewässernetzdichte *f*; Wasserlaufdichte *f*

1694 drainage of confined waters
A natural outflow of water from the confined aquifers by river vallies and other land surface depressions.
r разгрузка напорных вод
Естественный дренаж напорных водоносных горизонтов долинами рек и другими понижениями рельефа.
f drainage *m* d'eau souterraine
s drenaje *f* de agua subterránea
g Dränung *f* des gespannten Wasser

1695 drainage of glacier
Water discharge from a glacier through a system of channels, cracks, etc., on the glacier surface, inside and beneath its body.
r дренаж ледника
Спуск жидкой воды из ледника по системе каналов, трещин и полостей на поверхности ледника, внутри него и под ним.
f drainage *m* de glacier
s drenaje *m* de glaciar
g Entwasserung *f* der Gletscher

1696 drainage pattern
The arrangement of the main river and its tributaries. → Fig. 95
r рисунок речной сети
Форма и расположение в плане речных русел. → Рис. 95
f dessin *m* de réseau hydrographique; dessin *m* de réseau fluvial
s diseño *m* de la red hidrográfica; diseño *m* de la red fluvial
g Zeichnung *f* eines Flußnetzes

1697 drainage rate
Calculated level of unconfined groundwater lowering on the artificially drained territory.
r норма осушения
Расчетная величина понижения уровня грунтовых вод на осушаемой территории.
f norme *f* de drainage
s norma *f* de drenaje
g Draingerate *f*; Entwässerungrate *f*

1698 drainage runoff
Runoff collected and removed by collectors or drainage systems from irrigated and drained lands.
r дренажный сток
Сток, собираемый и отводимый коллекторами и дренажными системами с орошаемых и осушаемых земель.
f écoulement *m* de drenage
s escorrentía *f* de drenaje
g Dränageabfluß *m*; Drainageabfluß *m*

1699 drainage system
A complex of channels cut in naturally wet lands constructed to remove excess water.
r осушительная система; дренажная сеть
Совокупность гидротехнических и агромелиоративных средств, направленных на устранение естественного или вторичного переувлажнения почв.
f système *m* de drainage; réseau *m* de drainage
s sistema *m* de drenaje; red *f* de drenaje
g Dränsystem *n*; Entwässerungssystem *n*

1700 drainage water
Surface or ground waters collected and removed from a certain area and discharged into water bodies.
r дренажные воды
Поверхностные и подземные воды, собираемые и отводимые с данной территории и сбрасываемые в водные объекты.
f eau *f* de drenage
s aguas *pl* de drenaje
g Dränwasser *n*; Drainagewasser *n*

1701 **draining spot; drying area**
1. A part of coastal zone undergone to alternate flooding and drying during tides and under wind effected phenomena. 2. A belt of a reservoir bed drying up when some water is consumed and the level of water in the reservoir lowers down.
r зона осушки
1. Часть береговой зоны, подвергающаяся попеременному затоплению и осушению во время приливов и при сгонно-нагонных явлениях. 2. Полоса ложа водохранилища, обсыхающая при понижении уровня воды, вызванном расходованием ее запасов.
f zone *f* d'assèchement
s zona *f* de secamiento
g Trockungszone *f*

* **draw (*amer*) → 802**

1702 **drawdown; drawdown of a water table**
The act of lowering a water table by artificial means, isually by pumping or by artesian flow.
r снижение зеркала вод
Искусственное понижение уровня грунтовых вод, обычно путем откачивания воды или артезианским истечением.
f baisse *f* de plan d'eau
s bajada *f* de espejo del agua
g Absenkung *f* des Wasserspiegels; Wasserspiegelsenkung *f*

* **drawdown of a water table → 1702**

1703 **dredger**
A ship equipped with a scoring device to take ground samples out of a water reservoir floor or to deepen a river channel.
r драга; землечерпательный снаряд
Судно, оснащенное черпаковым устройством для извлечения грунта со дна водоема и используемое для сбора донных проб или для углубления дна.
f drague *f*
s draga *f*
g Bagger *m*; Nassbagger *m*; Schwimmbager *m*

1704 **dredging**
The deepening and broadening of a river or water body by removing ground with the help of mechanical means; sampling bottom sediments and rocks and producing mineral resources by means of dredge or from board a ship.
r дноуглубление; землечерпание; драгирование
Увеличение глубины и ширины водного пути на водоемах и водотоках путем удаления грунта механическими средствами; процесс взятия проб донных отложений водоема, либо извлечения полезных ископаемых из подводных месторождений с борта судна или с помощью плавучей *драги*.
f dragage *m*
s dragado *m*
g Ausbaggern *n*; Nassbraggerung *f*; Baggearbeit *f*

* **dressing of minerals → 641**

1705 **dried ice**
A bare floating ice from the surface of which standing water and slush have disappeared due to melt water discharge through cracks and leads.
r обсохший лед
Плавучий лед, с поверхности которого исчезли снежницы в результате стекания талой воды через трещины и прогалины.
f glace *f* asséchée
s hielo *m* resecado
g getrocknetes Meereis *n*

1706 **drift**
1. A passive movement of an object (floe, ship) in water under the effect of wind or current. 2. The movement and accumulation of loose material such as snow (snow drift) or sand (sand drift) caused by wind.
r дрейф
1. Пассивное перемещение по воде какого-либо предмета (льдины, судна) под влиянием ветра или течения. 2. Перенос и аккумуляция рыхлого материала – снега или песка, вызванные ветром.
f dérive *f*
s deriva *f*
g Drift *f*

* **drift$_3$ → 1707**

1707 **drift current; wind drift; wind-driven current; drift$_3$**
The motion of ocean water generally at a low velocity under the influence of prevailing winds (e.g. North Atlantic Drift and Gulf Stream).
r дрейфовое течение
Медленное движение морской воды, возникающее в результате поверхностного трения при постоянно дующих ветрах (например, Северный Атлантический Дрейф и Гольфстрим).
f courant *m* de dérive; courant *m* dus aux vents; courant *m* d'impulsion
s corriente *f* de deriva; corriente *f* de impulsión; corriente *f* debido a los vientos
g erzwungene Strömung *f*; Driftströmung *f*

1708 **drift theory**
A theory suggested by Ch. Lyell according to which erratic boulders found in the Quaternary deposits had been transported by icebergs floating over a hypothetical sea.
r дрифтовая теория
Выдвинутая Ч. Лайелем теория, согласно которой эрратические валуны, встречающиеся в четвертичных отложениях, раз-

носились айсбергами, плававшими по гипотетическому морю.

f theorie *f* du drift
s teoría *f* de deriva
*g*締Drifttheorie *f*

1709 drifted snow
Dense snow redeposited by strong wind.
r метелевый снег
Плотный снег, состоящий из мелких обломков кристаллов, переотложенных сильной низовой метелью.
f neige *f* soufflée
s nieve *f* arrancada
g Schneedriftablagerungen *fpl*

* **drift-ice → 2184**

* **drifting ice → 2184**

* **drifting ice station → 1710**

1710 drifting station; drifting ice station
A research station established on a floe of the Arctic Ocean.
r дрейфующая станция
Научно-исследовательские станции, создаваемые на плавучих льдах Арктического бассейна.
f station *f* dérivante
s estación *f* derivante
g Driftstation *f*

1711 driftwood
Wooden remains brought to the shore by sea or to a river bank. → Fig. 96
r плавник
Стволы деревьев и обломки деревянных построек, выброшенные прибоем на берег моря или течением на берег реки. → Рис. 96
f nageoire *f* flottante
s madera *f* flotante
g Schwimmflosse *f*

* **drill core → 1299**

* **drill hole → 1713**

1712 drilling; boring
The process of making boreholes into rocks and ice in order to study the internal structure and composition of rocks and take samples for the analysis.
r бурение
Проходка скважин в горных породах и во льду, как правило, с отбором керна, с целью изучения внутреннего строения и состава исследуемой толщи и анализа условий ее формирования.
f forage *m*; perforation *f*
s perforación *f*
g Bohrung *f*; Durchbohrung *f*

Fig. 96. **Driftwood**

1713 drilling hole; borehole; drill hole; well$_2$
An underground working of a cylindrical form made in rocks by drilling.
r буровая скважина
Горная выработка цилиндрической формы, пройденная в горных породах с помощью бурения.
f trou *m* de forage; puits *m* de forage
s pozo *m* de perforación
g Bohrloch *n*; Sonde *f*

1714 drinking water
Epidemically safe water, harmless by chemical composition and having favourable organoleptic properties.
r питьевая вода
Вода, безопасная в эпидемическом отношении, безвредная по химическому составу и имеющая благоприятные органолептические свойства.
f eau *f* potable
s agua *f* potable
g Trinkwasser *n*

1715 drip
Falling of water in drops from the roof and walls of a subterranean cavity.
r капёж; dripping water
Падение капель воды в подземных полостях с кровли и со стенок.
f suintement *m* d'eau
s goteo *m*
g Tröpfeln *m*

* **driplet cone → 5707**

1716 drizzle
A very fine rainfall the droplets of which are less than 0.5 mm in diameter, falling from a cloud like a fine spray.
r морось
Равномерные осадки в виде очень мелких капель воды диаметром менее 0,5 мм, выпадающие из облака.
f bruine *f*
s llovizna *f*
g Sprühregen *m*; Nieselregen *m*

1717 dropstone
Erratic fragment of boulder- or pebble-size released from floating ice and dropped into thin-laminated soft sediments on the ocean, sea or lake bottom.

r дропстон
Эрратический обломок валунной или галечной размерности, выпавший из плавучего льда в тонкослоистые осадки дна океана, моря или озера.
f dropstone *m*
s dropston *m*
g Dropstone *m*

1718 **drought**
A prolonged continuous period during which no significant precipitation is recorded.
r засуха
Период длительного отсутствия или значительного недостатка атмосферных осадков.
f sécheresse *f*
s sequía *f*
g Dürre *f*; Trockenheit *f*

1719 **drought index**
Measure of potential drought intensity expressed by different combinations of climatic and soil characteristics.
r индекс засушливости
Мера потенциальной интенсивности засух, выражаемая разными комбинациями климатических и почвенных показателей.
f indice *m* de sécheresse
s índice *m* de sequía
g Dürreindex *m*

1720 **drought resistance**
The ability of plants to withstand the lack of moisture (dry air or dry soil) and to recover quickly after wilting.
r засухоустойчивость
Способность растений переносить недостаток влаги (атмосферную и почвенную засуху), легко оправляться после завядания.
f résistance *f* au sécheresse
s resistencia *f* a sequía
g Dürreresistenz *f*; Trockenresistenz *f*

1721 **drowned valley**
A valley which was excavated in a land surface but owing to a change in sea level has been partly or wholly drowned by the sea.
r затопленная долина
Долина, выработанная в рельефе, но затем вследствие подъема уровня моря оказавшаяся частично или полностью затопленной морской водой.
f vallée *f* inondable; vallée *f* d'inondation
s valle *m* de inundación; valle *m* inundable
g überschwemmtes Tal *n*

1722 **drumlin**
A smooth low hill resulted from the ancient glacial accumulation composed mainly of the moraine material and extended in the direction of the movement of the ice that deposited it. → Fig. 97
r друмлин
Холм, возникший в результате древнеледниковой аккумуляции и вытянутый в направлениии движения ледника, сложен преимущественно моренным материалом. → Рис. 97
f drumlin *m*
s drumlin *m*
g Drumlin *m*; Schillberg *m*

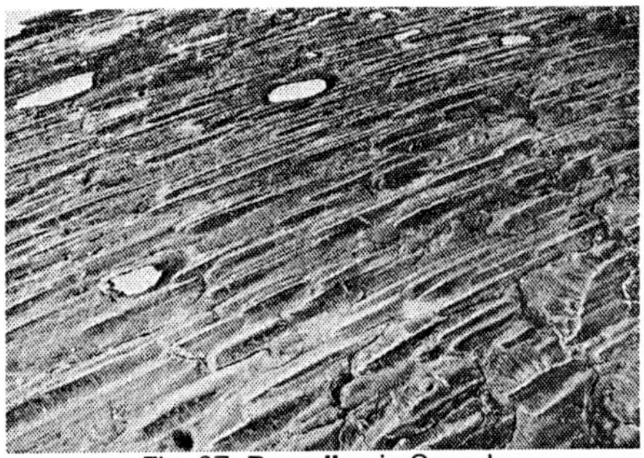
Fig. 97. **Drumlins** in Canada

1723 **drunken forest**
A forest dominated by the trees with bent trunks inclined in different directions mainly due to uneven cryogenic heaving and subsidence of ground through thawing.
r пьяный лес
Лес с изогнутыми и наклоненными в разные стороны стволами деревьев, возникающий в результате неравномерного криогенного пучения и оседания пород при протаивании.
f forêt *f* ivre
s bosque *m* ebrio
g Törkelwald *m*; betrunkener Wald *m*

1724 **druse**
A cavity in a rock, into which crystals grow.
r друза
Полость в коренных породах, где вырастают кристаллы.
f druze *f*
s drusa *f*
g Druse *f*

1725 **dry adiabatic process**
An adiabatic process in a system of dry air.
r сухоадиабатический процесс
Адиабатический процесс в объеме сухого воздуха.
f processus *m* adiabatique sec
s proceso *m* adiabático seco
g trockenadiabatischer Prozess *m*

1726 **dry adiabatic lapse rate**
The rate of loss of temperature with increasing height occurring in an unsaturated body of air as it ascends adiabatically, about 1°C in 100 m of ascent.
r сухоадиабатический градиент
Вертикальный адиабатический градиент температуры сухого или влажного ненасыщенного воздуха, равен примерно 1°C на 100 м высоты.

f gradient *m* adiabatique sec
s gradiente *m* adiabático seco
g trockenadiabatischer Gradient *m*

1727 dry *adj*; arid *adj*
Referring to a season or climate when evaporation exceeds precipitation.
r сухой; аридный
Относящийся к сезону или климату, когда испарение превышает атмосферные осадки.
f sec, sèche *adj*; aride *adj*
s seco *adj*; árido *adj*
g trocken *adj*; arid *adj*

1728 dry channel
Hollow along which a stream runs seldom and not for long.
r сухое русло
Ложбина, по которой вода реки течет редко и непродолжительное время.
f chenal *m* sec
s lecho *m* seco
g trockenes Strombett *n*; trockenes Flußbett *n*

1729 dry deposition
Precipitation of considerable dust masses from air in the absence of condensation and liquid precipitation, mainly as a result of gravity.
r выпадение пыли
Выпадение значительных масс пыли в отсутствии конденсации и осадков, главным образом под действием силы тяжести.
f chute *f* de poussière
s caída *f* de polvo
g Staubniederschlag *m*

1730 dry farming
Growing crops without irrigation in the zone of irrigated farming (in foothills and on the fringes of oases), that involves special treatment of the land to overcome the shortage of water is to sow the land only one of two years and then leave fallow.
r богарное земледелие; богара; аридное земледелие
Земли в зоне орошаемого земледелия, как правило, в предгорьях и на окраинах оазисов, где сельскохозяйственные культуры возделывают без полива, для чего практикуется технология, когда поле поочередно один год распахивается и засевается, а на другой год остается под залежь.
f culture *f* sèche
s cultivo *m* de secano
g Aridlandwirtschaft *f*

1731 dry flat-floor valley
A gently inclined linear depression with a wide bottom, gentle slopes and usually without a permanent stream.
r дол
Длинная ложбина с небольшим уклоном, широким днищем, пологими склонами, обычно без водотока.
f vallée *f* sèche à fond plat
s valle *m* seco de fondo plano; vallejo *m* en cuna; vallejo *m* de fondo plano
g Mulde *f*

1732 dry ice
The solid form of carbon dioxide when used in the process of the cloud seeding in order to produce rainfall.
r сухой лед
Сухая форма углекислоты, используемая для засева облаков с целью вызывания дождя.
f glace *f* sèche
s hielo *m* seco; hielo *m* carbónico
g Trockeneis *n*

1733 dry meadow
A meadow growing on interfluves in the forest zone of the temperate belt in deforested areas.
r суходольный луг
Луг, встречающийся на междуречьях в лесной зоне умеренного пояса на месте сведенных лесов.
f pré *m* sec; pelouse *f*
s prado *m* seco
g Trockenwiese *f*

1734 dry season
A period of the year with small amount or almost complete absence of rainfall which is common in the low latitude regions.
r сухой сезон
Ежегодно повторяющийся период с малым количеством осадков или без них, характерный для низких широт.
f saison *m* sèche
s estación *f* seca
g Trockenzeit *f*; Trockenperiode *f*

1735 dry snow
A type of powdery snow containing no wet or liquid particles, in which the very cold crystals do not bond together and have charp contours, it sparkles in the sunlight and crunches under pressure.
r сухой снег
Снег, который не имеет жидкой воды, его кристаллы сохраняют резкие очертания, он искрится на солнце и скрипит при надавливании на него.
f neige *f* sèche
s nieve *f* seca
g trockener Schnee *m*

1736 dry snow zone; recrystallization zone
The zone on a glacier where no melting occurs and the ice formation brought about completely by settlement and recrystallization.
r зона сухого снега; снежная зона
Зона на леднике, где таяние отсутствует и льдообразование происходит целиком путем оседания и перекристаллизации.

f zone *f* de recristallisation
s zona *f* de recristalización
g Schneezone *f*

1737 dry valley
A valley originally carved by water, especially in chalk and limestone, which no longer has a running stream for the better part of the year.
r сухая долина
Долина, первоначально выработанная водой в меловых и известняковых породах, но теперь бульшую часть года лишенная водного потока.
f vallée *f* sèche
s valle *m* seco
g Trockental *n*

1738 dry wind; sukhovey
A hot and dry wind causing an abrupt decrease of the air humidity in the area where it comes to (it has different names depending on the location).
r суховей; сухой ветер
Сухой жаркий ветер, вызывающий резкое уменьшение влажности воздуха в местности, куда он приходит (имеет разные названия в зависимости от места возникновения).
f vent *m* sec
s viento *m* seco
g Trockenwind *m*; Suchowej *n*

1739 Dryas
Three phases of cooling at the end of the last glaciation in the Late Pleistocene in Europe 13-10 thousand years ago alternated with the periods of warming – the Bølling and Alleroed.
r дриас
Три фазы похолодания в конце последнего позднеплейстоценового оледенения в Европе 13-10 тыс. лет назад, разделенные периодами потепления – бёллинг и аллерёд.
f Dryas *m*
s Dryas *m*
g Dryas *f*

*** drying area → 1701**

1740 drying up of rivers
The cease of runoff in a river channel in the long periods with no rains and the depletion of ground water reserves.
r пересыхание рек
Прекращение стока в русле реки в периоды длительного отсутствия дождей и истощения запасов подземных вод.
f tarissement *m*
s desecación *f*; agotamiento *m* del río
g Austrocknung *f* der Flüsse

1741 dual economy; economic dualism
1. The economy which combines commodity production with subsistence agriculture, typical for developing countries. 2. The condition in the economy of a country where a relatively small group of socially and economically advanced elite live in the central city or in the larger towns, where the economy is dynamic and growing, while the majority of the population (who are much poorer) live in the countryside where the economy is static.
r экономический дуализм; двойственная экономика
1. Характерное для экономики развивающихся стран сочетание товарного сектора с мелкими натуральными хозяйствами. 2. Особенности экономики, когда относительно небольшая группа социально и экономически продвинутой элиты живет в центральном городе или больших городах, где экономика динамична и неуклонно растет, в то время как большинство населения (которое гораздо беднее) живет вне этих городов, где экономика развивается медленнее.
f dualisme *m* économique; economie *f* duale
s dualismo *m* económico; economía *f* dual
g wirtschaftlicher Dualismus *m*; Dualwirtschaft *f*

*** dump → 3875**

1742 dump; waste site; dumping ground
A place where solid waste is thrown away.
r свалка
Специально отведенная территория для складирования твердых бытовых и промышленных отходов.
f voirie *f*; décharge *f*
s basurero *m*
g Abladestelle *f*; Mülldeponie *f*

1743 dumped moraine; moraine line
A relatively narrow, long and rather low range in a mountain valley composed of the moraine material.
r моренный вал
Относительно узкая, длинная и невысокая гряда, протягивающаяся вдоль горной долины, сложенная моренными отложениями.
f rempart *m* de moraine; rognon *m*; cordon *m* morainique
s cordón *m* morrénico; riñon *m*
g Wallmoräne *f*

1744 dumping
Selling goods in a foreign market at prices lower than the domestic ones which is a way to compete for foreign markets.
r демпинг
Продажа товаров на внешних рынках по ценам более низким, чем на внутреннем рынке, одно из средств конкурентной борьбы за внешние рынки.
f dumping *m*
s dumping *m*
g Dumping *n*; Schleuderkonkurrenz *f*

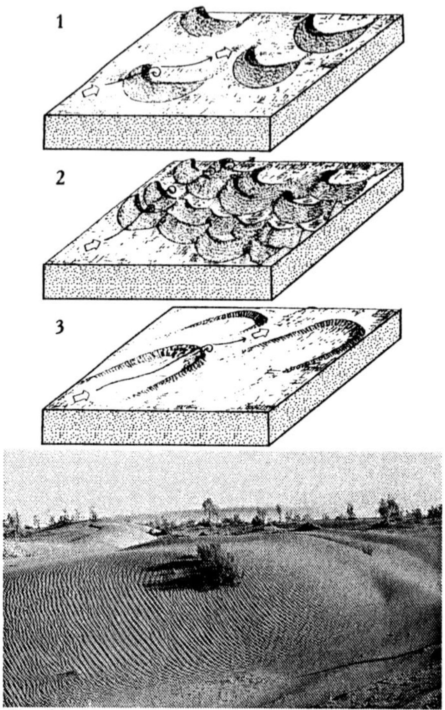

Fig. 98. Above: Different forms of sand **dunes**. 1 – barchans, 2 – dune ridges, 3 – longitudinal dunes (above). Down: An overgrown dune

* **dumping ground → 1742**

1745 dune; sand dune
A ridge of sand formed by wind and formed in deserts or on the low-lying coasts under different climates. → Fig.98
r дюна
Песчаная форма рельефа, созданная деятельностью ветра в пустынях, на низких морских берегах в разных климатических условиях → Рис. 98
f dune *f*
s duna *f*
g Düne *f*

1746 dune lake
A lake formed in a depression between the coastal sand dunes and the near-by hills.
r дюнное озеро
Озеро во впадине между береговыми дюнными грядами и окружающими холмами.
f lac *m* des dunes
s lago *m* de dunas; estanque *m* costero
g Dünensee *m*

1747 dune ridges
A system of coastal dunes stretched perpendicular to seasonal winds, and at the initial stages of formation stretchng parallel to the coast line.
r дюнные цепи
Система прибрежных дюн, вытянутых перпендикулярно сезонным ветрам, а на начальных стадиях – параллельно берегу.
f chaînes *fpl* de dunes; alignements *mpl* dunaires; dunes *fpl* en vagues; alignements *mpl* des dunes
s cadenas *fpl* de dunas; dunas *fpl* en olas; alineamientos *mpl* de dunas
g Dünenzüge *mpl*

1748 dune slack
A deflation depression formed between longitudinal sand ridges and having dimentions up to several kilometres wide and dozens kilometres long.
r межгрядовое понижение
Дефляционное понижение между продольными песчаными грядами шириной до нескольких километров и длиной в десятки километров.
f couloir *m* interdunaire
s pasillo *m* entre dunas
g Dünental *n*; Dünensenke *f*

1749 duopoly
The exclusive control of the supply of a product or service in a particular market by two suppliers, who thus dominate the market and control prices supplied goods.
r двухполия
Исключительный контроль поставок продуктов на конкретный рынок и обслуживания двумя поставщиками, которые доминируют на данном рынке и тем самым определяют цены и масштаб поставок.
f duopole *m*
s duopolio *m*
g Duopol *n*

1750 durable goods; durables
Consumer goods (e.g. vehicles and household appliances) expected to have a long lifetime.
r товар длительного пользования
Потребительские товары (например, автомобили или предметы домашнего хозяйства), которые служат продолжительное время.
f marchandise *f* d'une longue durée
s mercancía *f* de larga vida; mercancía *f* de larga duración
g langlebice Ware *f*

* **durables → 1750**

1751 duration of flood tide
Time interval between culminations of sea level at a given place at low water (*ebb tide*) and subsequent high water (*flood tide*).
r время прилива; длительность прилива
Период времени между кульминациями уровня воды в данном месте при отливе и последующем приливе.
f durée *f* de la montée
s duración *f* de marea creciente
g Flutdauer *f*

1752 duricrust
A hard crust covering a relatively soft soil surface in semi-arid, flat areas with a short rainy season and a long, hot dry season, consisting of minerals drawn to the surface by capillary during the dry season.
r тяжелая корка
Твердый слой, покрывающий относительно мягкую поверхность почвы в плоских семиаридных районах с коротким сезоном дождей и длинным сухим сезоном, состоит из минералов, поднятых к поверхности по капиллярам в сухой сезон.
f croûte *f* dure; dalle *m*
s costra *f* densa; lamina *f*
g schwere Kruste *f*

1753 durisol
A soil which has a coherent horizon of accumulation of secondary silica, common in the regions of arid and semiarid climate, less typical for the regions of Mediterranean climate.
r дурисоль
Почва с цементированным горизонтом накопления вторичного кремнезема, распространена в районах с аридным и семиаридным, реже средиземноморским климатом.
f durisol *m*
s durisol *m*
g Durisol *m*

1754 dust
Tiny particles of solid matter (less than 0.6 mm in diameter) occurring anywhere in the atmosphere and light enough to be carried in suspension, by the wind, sometimes for vast distances around the globe before falling to earth.
r пыль
Мельчайшие твердые частицы (диаметром менее 0,6 мм), плавающие в атмосфере и достаточно легкие, чтобы переноситься ветром в суспензии, иногда на огромные расстояния, прежде чем они упадут на землю.
f poussière *f*
s polvo *m*
g Staub *m*

1755 dust avalanche
The sliding of a mass consisting of loose volcanic ash, which has not yet become coherent into dense tuff.
r сухая лавина
Стекание по склону масс рыхлого вулканического пепла, не успевшего еще сцементироваться в плотный вулканический туф.
f avalanche *f* sèche
s avalancha *f* seca
g Staublawine *f*

* **dust devil → 5202**

1756 dust storm; sand storm
A storm in arid or semi-arid regions which transports great amounts of sand and dust picked up by strong dry wind from the soil surface having no vegetation. → Fig. 99
r пыльная буря; черная буря; песчаная буря
Перенос сильным сухим ветром больших количеств песка и пыли, поднятых с поверхности почвы, лишенной растительности. → Рис. 99
f tempête *f* de poussière; tempête *f* de sable
s tempestad *f* de polvo; tempestad *f* de arena
g Staubsturm *m*; Sandsturm *m*

Fig. 99. **Dust storm** above the drying Aral sea

* **dust whirlwind → 5202**

1757 duty
The money collected under tariffs imposed on the importation, exportation, or consumption of goods and services.
r пошлина
Денежный сбор, взимаемый государственными органами с граждан, учреждений и организаций за определенные виды услуг.
f droits *mpl*
s derechos *mpl*
g Zoll *m*

1758 dwarf plant; dwarfish formation
Short plants growing in unfavourable conditions of the environment.
r карликовое растение
Низкорослое растение, развивающееся при неблагоприятных условиях окружающей среды.
f formes *fpl* naines; plante *f* naine
s planta *f* enana; formación *f* enana
g Zwergpflanze *f*

* **dwarfish formation → 1758**

* **dwarfshrub → 5930**

* **dyke → 578**

1759 dynamic axis of stream
A line connecting points inside the water flow with the highest velocities.
r динамическая ось потока
Линия, соединяющая точки в речном потоке с самыми высоким скоростями течения.
f axe *m* dynamique de fleuve
s eje *m* dinámico del río

Fig. 100. Sequence of the **dynamomethamorphism**: tectonic structures of ice

g dynamische Achse *f* des Flußes

1760 dynamic climatology
The study of the dynamic processes of the general and local circulation of the atmosphere.
r динамическая климатология
Отрасль климатологии, изучающая особенности климата, обусловленные процессами общей и региональной циркуляции атмосферы.
f climatologie *f* dynamique
s climatología *f* dinámica
g dynamische Klimatologie *f*

1761 dynamic equilibrium
A state in which balance is maintained despite continual change, e.g. on a slope where the rate of weathering of the rock is balanced by the rate of removal of the weathered rock material.
r динамическое равновесие
Состояние, при котором сохраняется баланс, несмотря на непрерывно идущие изменения, например, на склоне, где скорость выветривания горных пород сбалансирована скоростью выноса материала выветривания.
f équilibre *f* dynamique
s equilibrio *m* dinámico
g dynamisches Gleichgewicht *n*

1762 dynamic geomorphology
A branch of *geomorphology* studying the processes and mechanisms of the formation of landforms and their dynamics.
r динамическая геоморфология; геодинамика
Раздел *геоморфологии*, изучающий динамику рельефа и рельефообразующие процессы.
f géomorphologie *f* dynamique
s geomorfología *f* dinámica
g dynamische Geomorphologie *f*

* **dynamic head → 6543**

1763 dynamic meteorology
A branch of *meteorology* studying the dynamics (mechanics, thermodynamics) of the atmosphere.
r динамическая метеорология
Раздел *метеорологии*, изучающий динамику (механику, термодинамику) атмосферы.
f météorologie *f* dynamique
s meteorología *f* dinámica
g dynamische Meteorologie *f*

1764 dynamical topography of sea surface
A cartographic representation by a system of different horizontals of sea surface which constantly changes under the influence of external and internal factors.
r динамическая топография поверхности моря
Картографическое изображение с помощью системы разных горизонталей морской поверхности, постоянно изменяющейся под влиянием внешних и внутренних факторов.
f topographie *f* dynamique de mer
s topografía *f* dinámica de mar
g dynamische Meerestopographie *f*

1765 dynamometamorphism
The alteration of rocks, including natural ice, by intense pressure associated with earth movement or caused by the motion of ice. → Fig. 100
r динамометаморфизм
Деформации горных пород, включая природный лед, в результате интенсивного давления и движений земной коры, а также перемещения льда. → Рис. 100
f dynamométamorphisme *m*
s dinamometamorfismo *m*
g dynamische Metamorphose *f*

E

* **eagre → 6206**

1766 **Earth**
The planet of the solar system, the third in order from the Sun, it is studied by many sciences: geography, geology, geophysics, etc.
r Земля
Планета Солнечной системы, третья по порядку от Солнца, изучается географией, геологией, геофизикой и другими науками.
f Terre *f*
s Tierra *f*
g Erde *f*

1767 **Earth ellipsoid; ellipsoid of reference**
The shape of planet Earth expressed as a ratio of ellipticity (Earth ellipsoid has semimajor axis = 6378160 m, semiminor axis = 6356775 *m*, and flattening = 0,003353).
r земной эллипсоид
Форма планеты Земля, выраженная через ее эллиптичность, т.е. сжатие Земли (принятый стандарт земного эллипсоида: R=6378160 м, r=6356775 м, сплюснутость = 0,003353).
f ellipsoïde *m* terrestre; ellipsoïde *m* référence
s elipsoide *m* terrestre; elipsoide *m* de referencia
g Erdellipsoid *n*

1768 **earth pillar; demoiselle; pedestral rock; pedestral**
A narrow cone, column or pyramid up to15 m high capped by a boulder formed as a result of the washout of loose unsorted material by rainwater. → Fig. 101

Fig. 101. **Earth pillar** in the West Pamirs

r земляная пирамида
Узкий конус, столб или пирамида высотой до 15 м с крупным валуном на вершине, образующийся в результате размыва дождевыми водами отложений рыхлого несортированного материала. → Рис. 101
f cheminée *f* de fée; demoiselle *f* coiffée
s chimenea *f* de hada; demoiselle *f* coifée
g Erdpyramide *f*; Erdpfeiler *m*

1769 **earth road; earthen highway; unpaved road; country road; dirt road**
A local road, usually not paved, connecting small settlements and farms.
r грунтовая дорога; проселок; проселочная дорога
Дорога, пригодная для движения колесного транспорта, но не имеющая твердого дорожного покрытия, а потому временами непроезжая.
f chemin *m* vicinal
s camino *m* natural
g Feldweg *m*; unbefestigte Landstraße *f*

1770 **Earth sciences; geosciences**
A collective term for all studies concerned with the physical characteristics of planet Earth, which includes such disciplines as geology, geophysics, geochemistry, physical geography, geodesy, etc.
r науки о Земле
Науки, изучающие физические аспекты земного шара, к ним относятся геология, геофизика, геохимия, физическая география, геодезия и др.
f géosciences *fpl*; sciences *fpl* de la Terra
s geociencias *fpl*; ciencias *fpl* de la Tierra
g Erdwissenschaften *fpl*; Geowissenschaften *fpl*

1771 **earth tremor; jolt**
A small, low intensity earthquake.
r сотрясение земли
Землетрясение слабой интенсивности.
f ébranlement *m* de terre
s conmoción *f* de tierra
g Erderschütterung *f*

1772 **Earth´s corona**
The external layer of the Earth′s atmosphere lying above the exosphere and extending from the height of 1-2 up to 20 thousand kilometres.
r земная корона
Внешняя область земной атмосферы, расположенная над экзосферой и простирающаяся от 1-2 до 20 тыс. км.
f couronne *f* de la terre
s corona *f* de la tierra
g Geokorona *f*

1773 **Earth´s crust; crust of the Earth**
The upper layer of the rigid Earth which has a thickness from 5-10 km under oceans up to 35-70 km under continents, separated from

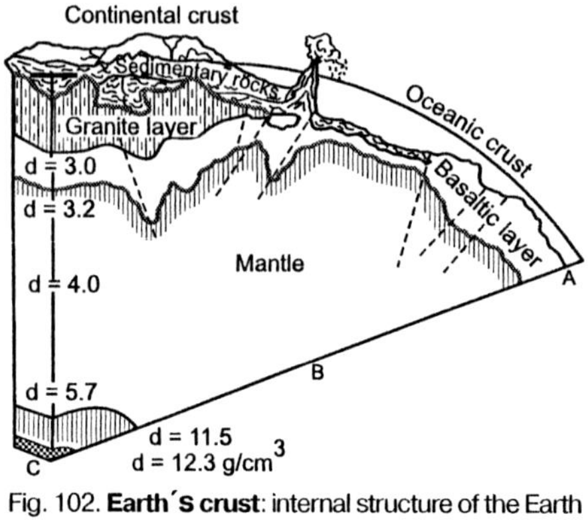

Fig. 102. **Earth´s crust**: internal structure of the Earth

the underlying mantle by the Mohorovičič discontinuity. → Fig. 102

r земная кора
Верхняя оболочка «твердой Земли» толщиной от 5-10 км под океанами до 35-70 км под материками, отделенная от нижележащей мантии поверхностью Мохоровичича. → Рис. 102

f écorce *f* terrestre; croûte *f* terrestre
s corteza *f* terrestre
g Erdkruste *f*; Erdrinde *f*

* **Earth´s magnetic field → 6104**

* **Earth´s mantle → 3652**

* **Earth´s physical environment → 3398**

* **earthen highway → 1769**

1774 earthflow; earthslide
A gravitational movement of soil and loose surface material downslope when saturated by water.

r оплывина
Смещение вниз по склону под действием силы тяжести маломощного слоя почвы или грунта при насыщении их водой.

f coulée *f* boueuse
s colada *f* de fango
g Herabfließen *n*; Fließerde *f*

1775 earthquake; temblor (*amer*)
Underground shocks and shakings of the Earthrs surface caused by deep-seated tectonic disturbances.

r землетрясение
Подземные удары и колебания поверхности Земли, вызванные главным образом тектоническими причинами.

f secousse *f* sismique; tremblement *m* de terre; séisme *m*
s terremoto *m*; seísmo *m*; sismo *m*; temblor *m* de tierra; sacudida *f* sísmica
g Erdbeben *n*

* **earthquake waves → 5327**

* **earthslide → 1774**

1776 earthworm cast; coprolite
Petrified excrements of soil worms and other soil invertebrates that contain a great amount of mineral particles and form structural aggregates with high stability to the water impact.

r копролит
Окаменевшие экскременты почвенных червей и других почвенных беспозвоночных, содержащие большое количество минеральных частиц и образующие в почве структурные агрегаты, обладающие большой устойчивостью к водному воздействию.

f coprolite *m*
s coprolito *m*
g Koprolith *m*

1777 easement
Land on both sides of railways and roads belonging to transport organizations and not used for other economic purposes.

r полоса отчуждения
Территория по обе стороны железной или автомобильной дороги, принадлежащая транспортными организациями и не используемая для другой хозяйственной деятельности.

f bande *f* d´aliénation
s zona *f* de alienación
g Schutzstreifen *m*

1778 east
One of the four cardinal points, the direction at which the sun rises at the equinox, lying on the right of a person facing due north.

r восток
Одна из четырех главных сторон горизонта, расположенная справа от наблюдателя, стоящего лицом к северу, точка восхода Солнца в период равноденствия.

f est *m*; orient *m*
s este *m*; oriente *m*
g Ost(en) *m*

1779 east *adj*; oriental *adj*; eastern *adj*
Relating to, belonging to, coming from or situated towards the east, e.g. winds blowing from that direction.

r восточный
Относящийся, расположенный или приходящий с востока, например, ветер, дующий с этого направления.

f d´est; oriental *adj*
s del este; oriental *adj*
g Ost-; östlich *adj*; orientalisch *adj*

1780 east wind transport; easterlies
A sufficiently stable movement of air masses from east to west occurring in tropics (trade winds) or in polar regions along the peripheries of stable anticyclones.

r восточный перенос

Достаточно устойчивый перенос воздушных масс с востока на запад в тропиках (пассаты) или в полярных широтах вдоль периферий устойчивых антициклонов.
f vents *m* d'est constants
s vientos *m* del este constantes
g Ostwinddrift *f*

* **easterlies → 1780**

* **eastern *adj* → 1779**

1781 **e a s t i n g**
In cartography, a linear distance eastwards from the north-south grid line on a map which passes through the true origin of a grid.
r ордината
В картографии координата, отсчитываемая по направлению, перпендикулярному осевому меридиану.
f ordonnée *f*
s ordenada *f*
g Ordinate *f*

1782 **ebb; ebb tide; low tide**
The retreating tide or the outgoing tidal stream, immediately following the period of high tide, the opposite to the high tide (flood tide). → Fig. 103
r отлив
Состояние моря, противоположное кульминации прилива, спад уровня моря, следующий непосредственно за приливом. → Рис. 103
f reflux *m*; jusant *m*; marée *f* descedante
s marea *f* menguante; reflujo *m*
g Ebbe *f*; Niedergang *m*

Fig. 103. **Ebb**

1783 **ebb and flow**
The backward and forward movement of tidal water in relation to the shore.
r приливо-отливные течения
Попеременный накат и откат приливной воды по отношению к берегу.
f flot *m* et jusant *m*; marée *f* haute et basse
s flujo *m* y reflujo *m*
g Gezeitenströme *mpl*

* **ebb current → 1784**

* **ebb tide → 1782**

1784 **ebbing current; ebb current**
Horizontal movement of water into the sea occurring in the phase of the ebb tide (but not always coinciding with it), it moves from the shore or in the direction opposite to the tidal wave propagation.
r отливное течение
Горизонтальное движение воды в море, возникающее в фазе отлива (однако не всегда с ним совпадающее), направлено от берега или в направлении, противоположном распространению приливной волны.
f courant *m* de jusant
s corriente *f* de reflujo
g Ebbstrom *m*

1785 **eccentricity of orbit**
An element of the elliptical Earth's orbit, at present equal to 0.0167.
r эксцентриситет орбиты
Элемент эллиптической орбиты Земли, в современную эпоху равный 0,0167.
f excentricité *f* de orbite
s excentricidad *f* de órbita
g Bahnexzentrizität *f*

1786 **ecesis**
The establishment of a plant or animal in the new environment.
r эцезис
Процесс завоевания растениями или животными свободного пространства.
f écèse *m*
s ecesis *f*
g Einbürgerung *f*

1787 **echelon section**
The parallel stretching of mountain ranges and spurs or at an acute angle.
r кулисы
Простирание горных хребтов и их отрогов в виде параллельных отрезков, либо образующих в плане острый угол.
f dissection *f* en échelon
s disección *f* sectoral
g kulissenartige Gliederung *f*

1788 **echo-sounder; fathometer; sonic depth finger; sonar**
A hydroacoustic instrument used for measuring the depth of water in a sea or lake by sound vibrations generated at the surface to return after being reflected from the sea floor.
r эхолот
Гидроакустический прибор для измерения глубин в море или озере путем возбуждения на поверхности вибраций, которые достигают дна, отражаются от него и фиксируются прибором.
f sonde *f* acoustique
s eco-sonda *f*; sonda *f* acústica
g Echolot *n*

1789 **ecistics**

A theory concerned with the process of formation and evolution of human settlements, it was elaborated in the 1950-60s to deal with the problems of settlement patterns in a comprehensive and global way.

r экистика
Теория формирования и эволюции человеческих поселений, разработанная в 1950-60-х годах, стремящаяся охватить проблемы расселения целостно и глобально.

f équistique *f*
s equística *f*
g Ökistik *f*

1790 eclipse
The total or partial cutting off of light received by one celestial body from another by the interception of a third celestial body in passing between the other two, e.g. solar or lunar eclipse.

r затмение
Полное или частичное отсечение света, получаемого одним небесным телом от другого при прохождении третьего небесного тела между ними, например, солнечное или лунное затмение.

f éclipse *f*
s eclipse *m*
g Finsternis *f*

1791 ecliptic
The large circle of the celestial sphere along which the apparent annual movement of the Sun relative to the fixed stars takes place, the plane of ecliptic is inclined to the plane of celestial equator at an angle of 23°27′ crossing it in two points which determines the change of seasons.

r эклиптика
Большой круг небесной сферы, по которому происходит видимое годичное движение центра Солнца среди звезд, плоскость эклиптики наклонена к плоскости небесного экватора под углом 23°27′, пересекая его в двух точках, что определяет смену времен года.

f écliptique *f*
s eclíptica *f*
g Ekliptik *f*; Sonnenbahn *f*

1792 ecocide
Delibarate destruction of the environment by human activity.

r экоцид
Осознанное разрушение окружающей среды деятельностью человека.

f écocide *m*
s ecocidio *m*
g Ökozid *n*

1793 ecogenesis
The development of relations between organisms and the environment in the course of evolution.

r экогенез
Развитие отношений между организмами и окружающей средой в процессе эволюции.

f écogenèse *f*
s ecogénesis *f*
g Ökogenesis *f*

* **ecogeography → 2408**

* **ecological *adj* → 1920**

1794 ecological alarmism
A public movement putting the emphasis on the catastophic impact of people on the environment.

r экологический алармизм
Общественное движение, акцентирующее внимание на катастрофичности антропогенного воздействия на природу.

f alarmisme *m*
s alarmismo *m*
g Alarmismus *m*

* **ecological balance → 1921**

* **ecological catastrophe → 1797**

1795 ecological certificate
A document containing data on the use of natural resources by an enterprise and its impact upon the environment.

r экологический паспорт
Нормативно-технический документ, где представлены данные о предприятии, отражающие уровень использования им природных ресурсов и степень воздействия на окружающую среду.

f certificat *m* environnemental; certificat *m* écologique
s certificado *m* medioambiental; certificado *m* ecológico
g Umweltzertifikat *n*

* **ecological charges → 1933**

1796 ecological crisis
A tense state of the relationships between society and nature when the development of society does not correspond to ecological potential of the biosphere.

r экологический кризис
Напряженное состояние взаимоотношений общества и природы, когда развитие человеческого общества не соответствует экологическим возможностям биосферы.

f crise *f* écologique; crise *f* environnementale
s crisis *f* ecológica; crisis *f* medioambiental
g Umweltkrise *f*

1797 ecological disaster; ecological catastrophe
The irreversible environmental changes occured on a certain territory as a result of natural disasters or human impact.

r экологическое бедствие; экологическая катастрофа

Необратимые изменения окружающей среды на определенной территории в результате опасных природных явлений или антропогенного воздействия.
f catastrophe *f* écologique; catastrophe *f* environnementale
s catástrofe *f* ecológica; catástrofe *f* medioambiental
g Umweltkatastrophe *f*

1798 **ecological diseases; diseases of civilization**
Diseases associated with the development of industry and technology and consequent pollution of the environment, e.g. cancer caused by radioactive contamination of the area.
r экологические болезни; болезни цивилизации
Болезни человека, возникшие в результате издержек научно-технического прогресса и вызванных им деформаций окружающей среды, например, раковые заболевания в результате радиационного заражения местности.
f maladies *fpl* écologiques; maladies *fpl* de civilisation
s enfermedades *fpl* de civilización
g ökologische Krankheiten *fpl*; Zivilisationskrankheiten *fpl*

1799 **ecological economics; environmental economics**
The aim to conserve, maintain, use and reuse natural resources so that the quality of life is retained without excessive waste.
r экологическая экономика
Рациональное использование природных ресурсов в хозяйственных целях с целью минимизации отходов, выбрасываемых в окружающую среду.
f économie *f* écologique
s economía *f* medioambiental
g ökologische Ökonomie *f*; Umweltökonomie *f*

1800 **ecological education**
Public education and cultivating in people the respect and care of nature and delicate use and protection of the environment.
r экологическое образование
Просвещение и воспитание населения в духе бережного отношения к природе.
f éducation *f* écologique
s educación *f* medioambiental; educación *f* ecológica
g Umweltbildung *f*

* **ecological emergency → 1925**

1801 **ecological foundation**
An organization intended for channelling purpose-oriented money to nature protection and conservation of the environment.
r экологический фонд
Организация, предназначенная для направления целевых денежных поступлений на охрану и улучшение окружающей среды.
f fonds *m* écologique
s fondo *m* medioambiental; fondo *m* ecológico
g Umweltschiftung *f*; Umweltfonds *m*

* **ecological imbalance → 6486**

* **ecological law → 1934**

1802 **ecological movement; conservation movement; green movement; conservationism**
A public movement or political party which supports the protection of the environment.
r движение в защиту окружающей среды; движение зеленых
Общественное движение или политическая партия, выступающие за сохранение природной среды.
f mouvement *m* écologiste; mouvement *m* vert
s movimiento *m* verde
g Grünebewegung *f*

1803 **ecological niche; niche$_2$**
The optimum habitat in an environment for a specific species.
r экологическая ниша
Совокупность условий окружающей среды, в которых данный вид чувствует себя лучше всего.
f niche *f* écologique; niche$_2$ *f*
s nicho *m* ecológico; nicho$_2$ *m*; compartimiento *m* ecológico
g ökologische Nische *f*; Nische$_2$ *f*

1804 **ecological optimism**
The conviction that it is possible to prevent the ecological catastrophe by way of co-evolution of human society and nature.
r экологический оптимизм
Убеждение в возможности предотвращения экологической катастрофы, обеспечения коэволюции общества и природы.
f optimisme *m* écologique
s optimismo *m* ecológico
g ökologischer Optimismus *m*

1805 **ecological optimum**
The environmental conditions which are the most favourable for the existence of a certain species, community or ecosystem.
r экологический оптимум
Условия окружающей среды, наиболее благоприятные для существования данного вида, сообщества или экосистемы в целом.
f optimum *m* écologique
s óptimo *m* ecológico
g ökologisches Optimum *n*

1806 **ecological pessimism**
The conviction that the ecological catastro-

phe and its tragic consequences for humanity are inevitable.

r экологический пессимизм
Убеждение в неизбежности экологической катастрофы и ее трагических последствий для человечества.

f pessimisme *m* écologique

s pesimismo *m* ecológico

g ökologischer Pessimismus *m*

1807 ecological realism
The use of current problems and former positive experience to regulate the human impact on the environment and the consumption of natural resources.

r экологический реализм
Использование текущих проблем и позитивного опыта прошлого для регулирования воздействия человека на окружающую среду и потребления природных ресурсов.

f réalisme *m* écologique

s realismo *m* ecológico;
realismo *m* medioambiental

g ökologischer Realismus *m*;
Umweltrealismus *m*

1808 ecological refugees; eco-refugees
People who have left their homeland to escape drought, desertification, soil erosion and other irreversible environmental changes.

r экологические беженцы
Беженцы с территорий, подвергшихся необратимым экологическим изменениям: засухам, опустыниванию, почвенной эрозии и пр.

f réfugiés *mpl* écologiques

s refugiados *mpl* ecológicos

g ökologische Flüchtlinge *mpl*

1809 ecological regulation
A system of measures, laws and rules aimed at the protection of the environment along with its intensive economic use.

r экологическое регулирование
Система мер и законов, направленная на сохранение окружающей среды при активном ее использовании.

f réglage *m* écologique

s regulación *f* medioambiental

g Umweltregulierung *f*

1810 ecological risk; environmental risk
A potential harmful impact upon the environment of the economic projects which use natural resources and conditions of the territory.

r экологический риск
Вероятность неблагоприятных последствий для окружающей среды от принятых решений при использовании природных условий и ресурсов территории.

f risque *m* écologique

s riesgo *m* ecológico;
riesgo *m* medioamdiental

g Umweltrisiko *n*

1811 ecological risk assessment
The evaluation of the increasing danger of pollution and other impact upon the environment.

r оценка экологического риска
Определение растущей опасности загрязнения и иных воздействий на окружающую среду.

f évaluation *f* de risque écologique

s valoración *f* de riesgo ecológico

g Naturrisikoeinschätzung *f*

1812 ecological safety
The conditions of being protected (of any person, society, state) from the excessive and undesirable changes of the environment.

r экологическая безопасность
Состояние защищенности человека, общества, государства от чрезмерных и нежелательных изменений окружающей среды.

f sécurité *f* écologique

s seguridad *f* medioambiental

g Umweltsicherheit *f*;
ökologische Sicherheit *f*

1813 ecological series
The ranking of ecosystems in accordance with the gradient of environmental conditions (humidity, mechanical composition and fertility of soils, ground salinity, etc), usually along a selected profile (e.g. along a mountain slope).

r экологический ряд
Распределение организмов и сообществ в соответствии с градиентом элементов окружающей среды: влажности, механического состава и богатства почв, солености грунтов и т.п.

f série *f* écologique

s serie *f* ecológica

g ökologische Reihe *f*

* **ecological system → 1838**

1814 ecological tension
The degradation of the environmental components which threatens the normal functioning of other natural systems and processes.

r экологическая напряженность
Деградация элементов природной среды, создающая угрозу нормальному функционированию других природных систем и процессов.

f tension *f* écologique

s tensión *f* ecológica;
tensión *f* medioambiental

g ökologische Gespannheit *f*

* **ecological tourism → 1841**

1815 ecology
The study of interrelationships between organisms and their environment, in the broader sense – the study of the interactions between society and nature.

r экология
Изучение взаимоотношений организмов и окружающей среды, в широком смысле – общества и природы.
f écologie *f*
s ecología *f*
g Ökologie *f*

1816 ecolologically friendly technology; wholesome production; clean production
Production of commodities and services accompanied with a small impact upon the environment and its minimum pollution.
r экологически чистое производство
Производство товаров и услуг, сопровождаемое малым воздействием на окружающую среду и минимальным ее загрязнением.
f technologie *f* écologiquement pure; production *f* innofensive
s technología *f* ecológicamente pura
g umweltfreundliche Technologie *f*

* **economic activity → 2827**

1817 economic *adj*
Related to the economy of a given country or region.
r экономический
Относящийся к экономике данной страны или региона и к хозяйственным взаимоотношениям.
f économique *adj*
s económico *adj*
g ökonomisch *adj*; Wirtschafts-; wirtschaftlich *adj*

1818 economic aid
Material resources (money, food supplies, etc.) or other help that the government of one country, a group of countries or international organization give to another country which needs assistance because of a war, economic crisis or natural disaster.
r экономическая помощь
Передача стране, нуждающейся в помощи вследствие войны, экономического кризиса или стихийного бедствия, материальных средств и ресурсов от другой страны, группы стран или международных организаций.
f aide *f* au développement
s ayuda *f* económica
g Wirtschaftshilfe *f*

1819 economic crisis
One of the stages of the market economy development formerly characterized by the decrease of production and today having the form of a financial crisis.
r экономический кризис
Одна из фаз развития экономики в рыночных условиях, выражавшаяся в прошлом в падении производства, а ныне – в форме валютно-финансового кризиса.
f crise *f* économique
s crisis *f* económica
g ökonomische Krise *f*; Wirtschaftskrise *f*

* **economic depression → 4916**

1820 economic development
Improvements in economy accompanied by increased labour productivity, production efficiency, etc., which characterize a group of countries, country or region that reflects their economic development and hierarchical (ranking) position.
r экономическое развитие
Качественные изменения хозяйства, сопровождающиеся ростом производительности труда, повышением эффективности производства и пр., характеризует группу стран, страну или район и позволяет определить достигнутую ступень экономического развития и их место в иерархическом ряду сравниваемых территорий.
f développement *m* économique
s desarrollo *m* económico
g wirtschaftliche Entwicklung *f*

1821 economic distance
The distance a commodity may travel before its value is exceeded by the transport costs.
r экономичное расстояние
Расстояние, на котором при перевозке товара его ценность не будет перекрыта транспортными расходами.
f distance *f* économe
s distancia *f* económica
g wirtschaftliche Entfernung *f*

* **economic dualism → 1741**

1822 economic efficiency
Making the best use of limited natural resources.
r эффективность производства; экономическая эффективность
Наиболее рациональное использование в производстве ограниченных природных ресурсов.
f efficacité *f* économique
s eficacia *f* económica
g ökonomische Effizienz *f*; effiziente Nutzung *f*

1823 economic geography
The study of the spatial distribution and territorial organization of economy which includes the use of resources, industrial and agricultural production, consumption of goods and services, transport and other types of infrastructure.
r экономическая география
Изучение территориальной организации хозяйства, включая использование ресурсов, производство, потребление и сопутствующую инфраструктуру.
f géographie *f* économique

s geografía *f* económica
g Wirtschaftsgeographie *f*

1824 **economic growth**
The increase in real national income expressed as the growth of production of goods and services over a long period of time shown as a percentage per capita.
r экономический рост
Увеличение национального дохода, выраженное в росте производства товаров и услуг за длительный период времени в расчете на душу населения.
f croissance *f* économique
s crecimiento *m* económico
g Wirtschaftswachstum *n*

1825 **economic integration**
Gradual unification of economic complexes of two or more countries, in particular, through removing national economic barriers and creating uniform conditions for international exchange within this group of countries.
r экономическая интеграция
Постепенное объединение в экономическом комплексе национальных хозяйств двух и более государств, в частности, путем устранения национальных экономических барьеров и создания однородных условий международного обмена в данной группе стран.
f intégration *f* économique
s integración *f* económica
g internationale Wirtschaftsintegration *f*

1826 **economic mechanism**
The organization and control of a territorial economic system by administrative, public and economic methods.
r хозяйственный механизм
Управление территориально-экономической системой путем административных, экономических и общественных методов.
f mécanisme *f* économique
s mecanismo *m* económico
g Volkswirtschaftsmechanismus *m*

1827 **economic methods of management; mechanism of economic management**
The management of national economy and life of people by creating the conditions and norms profitable for both producers and customers.
r экономические методы управления
Руководство народным хозяйством и жизнью населения путем создания таких условий и нормативов, выполнение которых приносит выгоду исполнителям и потребителям.
f méthodes *fpl* économiques de géstion
s métodos *m* económicos de gestión
g Wirtschaftsmethoden *fpl* der Leitung

1828 **economic potential**
The capacity of the economies and industries of a group of countries, a country or region to provide industrial and agricultural production, capital construction, transportation of cargoes and render various services.
r экономический потенциал
Совокупная способность отраслей хозяйства группы стран, страны или района производить промышленную и сельскохозяйственную продукцию, вести капитальное строительство, выполнять перевозки грузов, оказывать разнообразные услуги.
f potentiel *m* économique
s potencial *m* económico
g wirtschaftliches Potential *n*

* **economic recession → 4916**

1829 **economic region**
An area with economic features which give it a measure of unity and make it differ from the areas surrounding it, i.e. territorially and economically integral part of a country or a large region with particular natural and economic conditions, historically formed economic specialization based on the territorial division of labour.
r экономический район
Территориально и экономически целостная часть страны или крупного региона, характеризующаяся своеобразием природных и экономических условий, исторически сложившейся специализацией хозяйства на основе территориального разделения труда.
f région *f* économique
s región *f* económica
g Wirtschaftsregion *f*; Wirtschaftsgebiet *n*

1830 **economic regionalization**
The identification of a system of related economic regions of a country that objectively reflects the existing territorial division of labour.
r экономическое районирование
Выделение системы соподчиненных экономических районов страны или крупного региона, объективно отражающее сложившееся территориальное разделение труда.
f régionalisation *f* économique
s regionalización *f* económica; división *f* en regiones económicas
g ökonomische Regionalisation *f*; wirtschaftsräumliche Gliederung *f*

1831 **economic sector; sector of economy**
A part of economic system that includes groups of closely related branches: agriculture, industry, transport, and the services, as well as different sectors - primary, secondary and others.
r сектор экономики; отрасль хозяйства
Часть экономической системы, включающая группы близких по специализации отраслей, различают сельское хозяйство, промышленность, транспорт, сферу услуг, а также первичный, вторичный и иные секторы экономики.

f secteur *m* économique; secteur *m* d'activité
s sector *m* económico
g Ökonomiksektor *m*; Wirtschaftssektor *m*

1832 economic stagnation
The state of an economic object or region in which no production growth or development has been registered for a long time.
r экономическая стагнация; стагнация; застой
Состояние экономического объекта или района, при котором в течение длительного времени отсутствует рост производства и тем более его развитие.
f stagnation *f* économique
s estancamiento *m* económico
g ökonomische Stagnation *f*; Stillstand *m*

1833 economics
The study of the production, distribution, exchange, consumption of goods and services and the way money is managed.
r экономика$_2$
Наука, изучающая производство, распределение, продажу, использование и потребление товаров и услуг.
f économie *f*
s economía *f*
g Ökonomik *f*; Wirtschaftswissenschaft *f*; Wirtschaftslehre *f*

1834 economy
1. The system of production, exchange, distribution and consumption of goods and services and management of money. 2. Efficient and sparing use of material and non-material resources which leads to saving money.
r экономика$_1$; экономия$_2$
1. Совокупность общественных отношений в сфере производства, обмена и распределения продукции, а также народное хозяйство данной страны или его часть, включающая определенные отрасли и виды производства. 2. Эффективное и бережливое использование материальных и иных ресурсов, что ведет к их сохранению.
f économie *f*
s economía *f*
g Wirtschaft *f*; Einsparung *f*

1835 economy of scale
The lowering of unit costs achieved by increasing the scale of production.
r экономия от укрупнения производства
Снижение себестоимости продукции за счет сосредоточения производства на предприятиях, выпускающих крупные партии однородных и технологически связанных товаров.
f economies *fpl* d'échelle
s economías *fpl* de escala
g Ausmaßökonomie *f*

1836 ecopolis
An urban settlement planned with due account for man's ecological requirements that combines adequate architectural forms, green zones and other features of the territory.
r экополис
Городское поселение, спланированное с учетом комплекса экологических потребностей человека, сочетающее соразмерность архитектурных форм, озеленения и других особенностей территории.
f écopolis *f*
s ecopoli *f*
g Ökopole *f*

* **eco-refugees → 1808**

1837 ecosphere
The biosphere and all the ecological factors which affect organisms.
r экосфера
Взаимосвязанная система биосферы и всех экологических факторов, воздействующих на организмы.
f écosphère *f*
s ecosfera *f*
g Ökosphäre *f*

1838 ecosystem; biogeocoenosis; ecological system
A system which includes all the organisms of an area and the environment in which they live i.e. it consists of living and nonliving parts interacting to form a system.
r экосистема; биогеоценоз
Совокупность живых организмов и окружающей их среды во взаимодействии.
f écosystème *m*; biogéocénose *f*
s ecosistema *m*; biogeocenosis *f*
g Ökosystem *n*; Biogeozönose *f*

1839 ecotone
A transitional zone between two bordering ecosystems where they blend together characterized by increased density and diversity of organisms, e.g. forest edge.
r экотон
Переходная зона между экосистемами, где происходит их взаимопроникновение, отличается ростом разнообразия и плотности организмов на границе двух соседствующих биогеоценозов, например, опушка леса.
f écotone *m*
s ecotono *m*
g Ökoton *n*

* **ecotope → 2688**

1840 ecotopology
The study of interrelations between plant communities and their habitats in different parts of landscape.
r экотопология
Учение о взаимосвязанности растительных сообществ с местообитаниями в разных частях ландшафта.

f ecotopologie *f*
s ecotopología *f*
g Ökotopologie *f*

1841 ecotourism; ecological tourism
Environmentally responsible travel to natural areas in order to enjoy and appreciate nature that promotes conservation and sustains the well being of local people.
r экотуризм; экологический туризм
Туризм, направленный на лучшее понимание и бережное отношение людей к природе и побуждающий местное население сохранять природные ресурсы.
f écotourisme *m*; tourisme *m* vert
s ecoturismo *m*
g Ökotourismus *m*

* **écoulement → 5122**

* **ecumene → 4235**

1842 edaphic; pedogenic *adj*
Related to the soil or having characteristics due to the nature of the soil.
r эдафический; почвенный
Относящийся к почве и протекающим в ней процессам.
f édaphique *adj*; pédologique *adj*
s edáfico *adj*; edafológico *adj*
g edaphisch *adj*; Boden-

1843 edaphic climax
A relatively stable state of soil which attained its equilibrium with the vegetation growing there.
r почвенный климакс
Относительно устойчивое состояние почвы, достигшее своего равновесия с произрастающей здесь растительностью.
f climax *m* édaphique
s clímax *m* edáfico; pedoclímax *m*
g Bodenklimax *f*; edaphische Klimax *f*

1844 edaphology
The study of the relationships between the soil and soil organisms, including the use made of land by humans.
r эдафология
Изучение связей между почвой как природным телом и обитающими в ней организмами, включая хозяйственное вмешательство людей.
f édaphologie *f*; microbiologie *f* du sol
s microbiología *f* del suelo
g Edaphologie *f*

1845 edaphon
A community of living organisms, mainly microflora, micro- and mesofauna, inhabiting soil.
r эдафон
Совокупность живых организмов, главным образом микрофлоры, микро- и мезофауны, населяющих почву.
f édaphon *m*
s edáfono *m*
g Edaphon *n*

* **eddy → 6606**

1846 eddy coefficients; exchange coefficients
Coefficients of eddy flux (e.g., of momentum, heat, water vapour, etc.) in turbulent flow, expressing the degree of turbulence development in the atmospheric boundary layer.
r коэффициенты турбулентного обмена
Коэффициенты интенсивности вихревого потока (т.е. количества движения, тепла, водяного пара и т.п.) в турбулентном течении, отражающие степень развития турбулентности в пограничном слое атмосферы.
f coefficients *mpl* de tourbillon de turbulence
s coeficientes *mpl* de torbellino turbulento
g Koeffizienten *mpl* der Wirbelaustausch

1847 eddy formation
The occurrence of whirls of different origin in the water or air.
r вихреобразование
Формирование вихрей разного происхождения в водной или воздушной среде.
f formation *f* de tourbillon
s formación *f* de torbellino
g Wirbelbildung *f*

1848 eddy structure
A system of sea water circulation consisting of a number of cyclonic and anticyclonic whirls contacting one another.
r вихревая структура
Система циркуляции морской воды, состоящая из нескольких циклонических и антициклонических вихрей, граничащих между собой.
f structure *f* de tourbillon
s estructura *f* de torbellino
g Wirbelstruktur *f*

1849 edge city
A business and trade centre located in the suburbs of a large city.
r периферийный центр
Деловой и торговый центр на окраине крупного города.
f edge cité *f*
s centro *m* periférico
g Peripheriezentrum *n*

1850 edge of water; water edge
Line of intersection of free water surface with land surface.
r урез воды
Линия пересечения свободной поверхности воды с поверхностью суши.
f bord *m* d'eau
s borde *m* de agua
g Uferlinie *f*

1851 **edificator; characteristic species**
A species that occurs solely or mainly in a certain plant community and determines its status, e.g. trees in forest biocenoses, grasses in savanna and steppes.
r эдификатор; характерный вид
Вид, встречающийся исключительно или преимущественно в определенном растительном сообществе и определяющий ее статус, например, деревья в лесных биоценозах, злаки в саванне и степях.
f plante *f* édificatrice
s planta *f* edificadora
g charakteristische Art *f*; umweltbildende Pflanze *f*

1852 **edoma**
Flat hills or low ridges covered with sparse low-growing wood and found among the alases on tundra plains actively influenced by thermokarst.
r едома
Плосковерхие холмы или невысокие гряды с мелким лесом среди аласов на равнинах тундр, подверженные активному воздействию термокарста.
f edoma *f*
s edoma *f*
g Edoma *f*

1853 **effective capacity of water reservoir**
The mass of water accumulated in a reservoir and intended for the purpose of its economic use and river runoff regulation.
r полезный объем водохранилища
Объем воды, накопленный в водохранилище и предназначенный для ее хозяйственного использования и регулирования стока реки.
f capacité *f* effective de réservoir
s capacidad *f* efectiva de embalse
g effektive Kapazität *f* des Staubeckens

1854 **effective fertility**
The fertility of soil resulting from the land cultivation, improvement, application of fertilizers, etc.
r эффективное плодородие
Плодородие почвы, обусловленное влиянием обработки, мелиорации, удобрений и пр.
f fertilité *f* efficace du sol
s fertilidad *f* efectiva del suelo
g effektive Fruchtbarkeit *f*

1855 **effective precipitation**
That part of the precipitation which reaches stream channels as runoff; in irrigation, that part of the total precipitation which remains in the soil and is available for vegetation.
r эффективные осадки
Часть атмосферных осадков, образующая сток; в ирригации, часть осадков, которая остается в почве и доступна растениям.
f précipitation *f* efficace
s precipitación *f* efectiva
g effektive Niederschläge *mpl*

1856 **effective radiation**
The difference between the terrestrial radiation and back radiation absorbed by the Earth's surface.
r эффективное излучение
Разность собственного излучения земной поверхности и поглощенного ею встречного излучения атмосферы.
f rayonnement *m* effectif
s radiación *f* efectiva
g effektive Ausstrahlung *f*

1857 **effective temperature**
The active temperature minus biological minimum, i.e. the low limit of the temperature for the active vegetation of a given cultural plant at this or that phase of development.
r эффективная температура
Активная температура, уменьшенная на величину биологического минимума температуры, т.е. нижнего предела температуры для активной вегетации сельскохозяйственных растений в той или иной фазе развития.
f température *f* effective
s temperatura *f* efectiva
g effektive Temperatur *f*

1858 **efflorescence**
The beginning of the period of flowering in plants.
r зацветание
Распускание цветов на растениях в период вегетации.
f épanouissement *m*; effloraison *f*
s florecimiento *m*; floración *f*; florescencia *f*
g Aufblühen *n*

1859 **effluent$_1$**
A stream flowing out of a lake or out of a reservoir.
r вытекающая река
Речной поток, вытекающий из озера или водохранилища.
f émissaire *m*
s emisario *m*
g Ausfluß *m*

* **effluent$_2$ → 5369**

* **effluent river** → 3359

1860 **effluent standard; emission standard**
The amount of sewage or pollutant which is allowed to be released into the environment.
r разрешаемый сброс; норма выброса
Количество отходов, жидких или газообразных, которые разрешено сбрасывать в окружающую среду.
f émission *f* standardisée; norme *f* de rejet; norme *f* d'émission
s norma *f* de emisión; emisión *f* admisible; efluente *m* admisible

g Einleitungsstandard *m*

* **effluent stream → 3359**

1861 efforation
The dissolution and washout of limestones by water running under pressure.
r эффорация
Растворение и размыв известняков водой, текущей под давлением.
f efforation *f*
s eforación *f*
g Efforation *f*

1862 effusion
The process of quiet, nonexplosive outpouring of lava to the Earth's surface.
r эффузия
Процесс спокойного, без сильных взрывов, излияния жидкой лавы на поверхность Земли.
f effusion *f*
s efusión *f*
g Effusion *f*

1863 effusive rock
An igneous rock formed as a result of a nonexplosive outpouring of molten rock to the Earth's surface which quickly solidified in the form of lava flows or lava sheets.
r изверженная горная порода; эффузивная горная порода
Магматическая горная порода, образовавшаяся в результате излияния магмы на земную поверхность и быстро отвердевшая в виде лавовых потоков или покровов.
f roche *f* effusive
s roca *f* efusiva
g Effusivgestein *n*; Erstarrungsgestein *n*; Ergussgestein *n*

1864 Ekman spiral; depth of frictional influence
A mathematical model of the wind distribution with height and a theoretical model which explains the currents that would result from a steady wind blowing over an ocean of unlimited depth and extent. → Fig. 104
r спираль Экмана
Математическая модель распределения ветра с высотой и теоретическая модель, которая объясняет морские течения, возникающие в результате устойчивых ветров над океаном неограниченной глубины и протяжения. → Рис. 104
f spirale *f* d'Ekman
s espiral *f* de Ekman
g Ekman-Spirale *f*

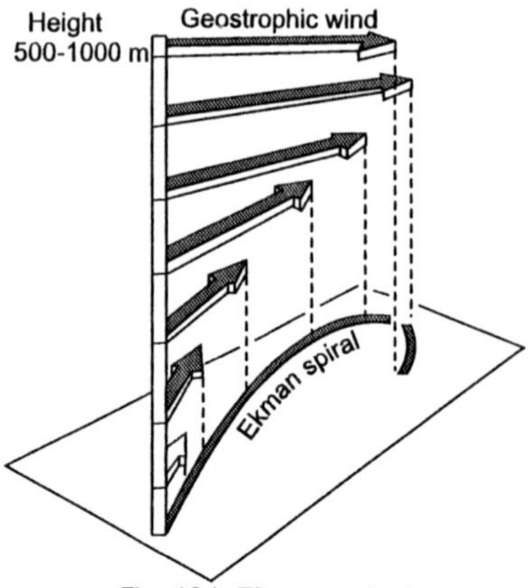

Fig. 104. **Ekman spiral**

1865 El Niño; southern oscillation
The abnormal warming of ocean water off the west coast of South America replacing the cold Humbolf current which brings heavy rainfall in the coastal region of Peru and Chile, which occurs irregularly and results from the weakening of the southeast trade winds in the Pacific.
r эль-ниньо; южное колебание
Аномальное потепление воды в океане у западного побережья Южной Америки, замещающее холодное течение Гумбольдта, что приносит сильные ливни в прибрежных районах Перу и Чили и случается время от времени в результате воздействия юго-восточных пассатов в Тихом океане.
f El Niño *m*; oscillation *f* australe
s El Niño *m*; oscilación *f* del sur
g El Niño *n*; Südoszillation *f*; El Ninjo *n*

1866 elan'
Forest meadows or glades with dense grasses common in middle Russia and in the Altai region.
r елань
Лесные луга или поляны с густым травостоем, распространенные в Средней России и на Алтае.
f élan' *f*
s elan' *f*
g Jelan *f*

1867 electoral geography
The study of political preferences of population expressed through elections at different regional levels.
r электоральная география
Изучение территориальных особенностей политических предпочтений населения, выражаемых во время выборов разных уровней.
f géographie *f* électorale
s geografía *f* electoral
g Geographie *f* der Wahlen; Elektoralgeographie *f*

1868 electrical industry
A branch of machine-building industry that produces electrical machinery for the power generation, supply, and transformation into other types of energy.

r электротехническая промышленность
Отрасль машиностроения, выпускающая продукцию для производства электроэнергии, передачи ее потребителям и преобразования в другие виды энергии.
f industrie *f* électrotechnique
s industria *f* electrotécnica; industria *f* electromecánica
g elektrotechnische Industrie *f*

1869 electronic atlas
A set of electronic maps.
r электронный атлас
Набор электронных карт.
f atlas *m* électronique
s atlas *m* electrónico
g Elektronenatlas *m*

* **electronic map → 1604**

1870 electronics industry
A branch of machine-building industry engaged in the design and production of electronic equipment, materials and apparatus.
r радиоэлектронная промышленность; micro-electronics
Отрасль машиностроения, занимающаяся разработкой и производством электронного оборудования, материалов и аппаратуры.
f industrie *f* électronique
s industria *f* electrónica
g elektronische Industrie *f*

1871 elevated reef
A dead coral reef raised above the sea level by tectonic movements of the Earth's crust.
r поднятый коралловый риф
Отмерший коралловый риф, поднятый тектоническими движениями земной коры выше уровня моря.
f récif *m* soulevé
s arrecife *m* levantado
g gehobenes Riff *n*

* **elevation → 4996**

* **elevation of a celestial body → 1451**

1872 elevation
1. The vertical distance above a specific level, e.g. above sea-level, or between a hill base and a hill top 2. The vertical angle between the horizontal and a high point, e.g. between the horizon and a star.
r высота$_1$; склонение$_2$
1. Расстояние по вертикали над заданным уровнем, например, над уровнем моря, или между подножьем и вершиной холма 2. Вертикальный угол между горизонтальной плоскостью и лежащей выше нее точкой, например, между горизонтом и небесным светилом.
f hauteur *f*
s elevación *f*
g Höhe *f*

1873 elfin woodland; elfinwood; kampzone; gnarled forest
Low trees (up to 10 m high) with gnarled trunks and branches curved in the lower part, characteristic of forest-tundra of the sea coast, and mountains at the upper tree-line.
r криволесье
Заросли низкорослых (до 10 м) деревьев с искривленными внизу стволами и ветвями, характерны для лесотундры близ морских побережий и в горах у верхней границы леса.
f forêt *f* rabougrie; forêt *f* basse
s bosque *m* enano; bosque *m* raquítico
g Kampzone *f*

* **elfinwood → 1873**

1874 ellipse of distorsion; Tissot's indicatrix
In map projections, an infinitesimal ellipse at any point of a map which represents an infinitesimal circle on the Earth's ellipsoid or sphere.
r эллипс искажений; индикатриса Тиссо
В картографических проекциях бесконечно малый эллипс в любой точке карты, отображающий бесконечно малую окружность на поверхности земного эллипсоида или шара.
f indicatrice *f* de Tissot
s elipse *f* indicatriz; indicatriz *f* de Tissot
g Verzerrungsellipse *f*; Tissot'sche Indikatrix *f*

* **ellipsoid of reference → 1767**

* **eluvial deposits → 1876**

1875 eluvial horizon; A-horizon
The top horizon in a soil profile from which leaching removes soluble salts and colloids but where an admixture of mineral and organic matter is retained.
r элювиальный горизонт; горизонт А; горизонт вымывания
Горизонт в верхней части почвенного профиля, из которого вынесены в более глубокие горизонты или за пределы профиля растворимые вещества и илистые частицы.
f horizon *m* éluvial; horizon *m* lessivé
s horizonte *m* eluvial; horizonte *m* lavado; horizonte *m* A
g Eluvialhorizont *m*; Auslaugungshorizont *m*

* **eluvial landscape → 471**

1876 eluvium; eluvial deposits
Weathered material formed by the break-up of solid rock, components of which have remained at or near to their point of formation.
r элювий; элювиальные отложения
Отложения, состоящие из продуктов выветривания горных пород, остающихся на месте своего образования.

f éluvium *m*
s eluvión *m*
g Eluvium *n*; eiluviale Ablagerungen *fpl*

1877 **embankment**
A raised structure to hold back water or to carry a roadway or railway.
r насыпь
Линейное сооружение из горной породы, предназначенное для строительства дороги, защиты от наводнений или иных инженерных целей.
f remblai *m*
s terraplén *m*
g Damm *m*

1878 **embargo**
A government order that forbids the buying or selling of particular goods.
r эмбарго
Наложение государством запрета на ввоз другими странами или вывоз золота или иностранной валюты, либо отдельных видов товаров.
f embargo *m*
s embargo *m*
g Embargo *n*; Handelssperre *f*

1879 **embayed coast**
An indented coast with alternating bays and headlands.
r бухтовый берег
Расчлененный, изрезанный берег с чередованием бухт и мысов.
f rivage *m* dentelé
s litoral *m* dentado
g gebuchtete Küste *f*; Buchtenküste *f*

* **embayment → 615**

1880 **emergent shoreline; coast of emergence**
Any type of former coastline that once marked the juxtaposition of land and ocean but which is now either raised above present high-water mark or located at some distance inland from the present coastline.
r проявляющаяся береговая линия
Любой тип береговой линии, которая когда-то обозначала соседство суши и моря, но теперь или поднялась над отметкой нынешнего уровня высокой воды, или удалена в глубь суши на некоторое расстояние от современной береговой линии.
f ñôte *f* d'emersion
s costa *f* de emersión
g Auftauchküste *f*

1881 **emigrant**
A person who migrates (moves) voluntary from the home country to another country to take up permanent residence.
r эмигрант
Человек, выехавший из своей страны в другую страну на постоянное жительство или на более или менее длительный срок.
f émigrant *m*; émigré *m*
s emigrante *m*
g Emigrant *m*; Auswanderer *m*

1882 **emigration**
The process of leaving one country to go to another country for the purpose of permanent residence.
r эмиграция
Выезд граждан из своей страны в другую страну на постоянное жительство или на более или менее длительный срок.
f émigration *f*
s emigración *f*
g Emigration *f*; Auswanderung *f*

1883 **emission**
The release of any pollutant into the environment.
r выброс; эмиссия
Поступление в окружающую среду любых загрязнителей.
f émission *f*
s expulsión *f* contaminante; emisión *f* contaminante
g Emission *f*; Ausstoss *m*

1884 **emission quotas**
Fixed by law or an international agreement the maximum amount of a substance which is allowed to be emitted into the environment.
r квотирование выбросов
Установленный законодательно или в результате международного соглашения лимит выбросов определенного вещества в окружающую среду.
f quotas *mpl* d'émission
s cuotas *fpl* de emisión
g Emissionsquoten *npl*

1885 **emission rate**
The amount of a pollutant released into the atmosphere per time unit.
r мощность выброса; интенсивность выброса
Масса загрязняющего вещества, выделяемая в атмосферу в единицу времени.
f intensité *f* de l'émission
s intensidad *f* de emisión
g Emissionsrate *f*; Emissionswerf *m*

1886 **emission source**
An object which produces pollutants and releases them into the environment.
r источник выброса; источник загрязнения
Объект, распространяющий вещества, загрязняющие окружающую среду.
f source *f* de contamination; source *f* de pollution
s fuente *f* de contaminación; fuente *f* de emisión contaminante
g Emissionsquelle *f*; Emittent *m*

1887 **emission standard**
The maximum amount of a pollutant that is permitted to be discharged from a single source of pollution.
r допустимый выброс
Максимальное количество загрязняющего вещества, которое разрешено к выбросу из одного источника загрязнения.
f émission *f* standard
s emisión *f* standard
g zulässig Emission *f*

1888 **empire**
The territory of a number of countries or peoples ruled by one authority, such as the British Empire, the Russian Empire, etc.
r империя
Территория, включающая несколько стран или народов, управляемая из одного центра, например, Британская империя, Российская империя и др.
f empire *m*
s imperio *m*
g Keiserreich *n*; Reich *n*

* **employment level → 1889**

1889 **employment rate; employment level**
The proportion of the total population of working age who are engaged in regular paid work in the national economy.
r занятость населения
Доля населения трудоспособного возраста, занятая в народном хозяйстве.
f emploi *m* de population
s empleo *m* de población
g Beschäftigung *f*

1890 **empoldering**
The reclamation of land from the sea, lake, river delta by the creation of polders.
r польдеризация
Мелиорация земель у моря, озера и в речных дельтах путем устройства польдеров.
f poldérisation *f*
s polderización *f*
g Polderisation *f*

1891 **enclave**
A territory or part of the territory of one state lying within the territory of another state, e.g. San Marino.
r анклав
Территория или часть территории одного государства, окруженная со всех сторон территорией другого государства, например, Сан-Марино.
f enclave *f*
s enclave *m*
g Enklave *f*

1892 **enclosed landscape**
A cultural landscape, in which fields and meadows are separated by hedges, i.e. stripes of forests and shrubs.
r сельскохозяйственный ландшафт; бокаж
Культурный ландшафт, в котором поля и луга разделены живыми изгородями, т.е. полосами лесов и кустарников.
f bocage *m*
s campos *mpl* cercados; boñage *m*; paisaje *m* de bocage
g Bocage *n*

1893 **enclosed sea; continental sea**
A sea deeply cut inland and connected with an ocean or adjoining sea through channels or straits, e.g. Black Sea.
r внутреннее море
Море, глубоко вдающееся в сушу и сообщающееся с океаном или с прилегающим морем через проливы, например, Черное и Мраморное моря.
f mer *f* intérieure
s mar *m* interior
g Mittelmeer *n*

1894 **enclosure**
Open land or land formerly worked in common cut up into individual fields surrounded by fences, walls, hedges, etc., forming a typical enclosed landscape.
r огораживание
Разделение открытых в прошлом полей или общинных земель на отдельные участки, огороженные стенками или живыми изгородями, что создает типичный сельскохозяйственный ландшафт.
f enclosures *fpl*
s enclosures *fpl*
g Einhegen *n*

* **end moraine → 6091**

1895 **endangered species; threatement species; species at risk of extinction**
An animal or plant species with population numbers so reduced (often by human activities) that their future existence seems uncertain.
r исчезающий вид
Вид животных или растений, находящийся под угрозой полного вымирания, когда сохранившаяся численность особей вида недостаточна для самоподдержания популяций в естественных условиях размножения.
f espèce *f* ménacée; espèce *f* en voie de disparition
s especie *f* en peligro de extinción; especie *f* en vías de extinción; especie *f* amenazada
g verschwindende Art *f*; gefährdete Art *f*

* **endemic → 1897**

1896 **endemic** ***adj***
Applied to the species of living organisms, the distribution of which is confined to a limited area within one region and which do not occur in other places and also to a disease that is normally confined and always likely to

occur in certain areas, sometimes reaching epidemic propotions.
r эндемичный
Относящийся к видам животных и растений, имеющим ограниченный ареал в пределах одной какой-либо области и в других местах не встречающимся, а также к болезням, распространенным в конкретных районах и нередко приобретающим характер эпидемий.
f endémique *adj*
s endémico *adj*
g endemisch *adj*

1897 **endemic disease; endemic**
A disease registered in a restricted area and caused by the local environmental and social conditions.
r эндемическое заболевание; эндемия
Заболевание, регистрируемое на ограниченной территории и обусловленное местными природными и социальными условиями.
f affection *f* endémique; endémie *f*
s enfermedad *f* endémica; endemia *f*
g endemische Krankheit *f*; Endemie *f*

1898 **endemism**
The property of some species to have a restricted geographic distribution, i.e. to be confined to a certain area and never occur in other places with similar conditions.
r эндемизм
Характерная для некоторых видов способность распространяться только на определенной территории или акватории (например, на острове, озере и пр.) при отсутствии в других аналогичных регионах.
f endémisme *m*
s endemismo *m*
g Endemismus *m*

* **endigage *m* → 551**

1899 **endodynamic succession**
The sequential change of vegetation that results from the change of environmental conditions caused directly by life activity of the plants.
r автогенная сукцессия; эндогенетическая сукцессия
Смена растительности, обусловленная влиянием самого растительного сообщества в результате изменения им среды обитания.
f succession *f* endodynamique
s sucesión *f* endodinámica
g autogene Sukzession *f*

1900 **endogenic *adj***
Related to the tectonic processes originating from within the Earth by its internal energy, gravity and forces caused by the rotation of the planet.
r эндогенный
Относящийся к процессам, происходящим в недрах Земли и обусловленным ее внутренней энергией, силой тяжести и силами, возникающими при вращении планеты.
f endogène *adj*
s endógeno *adj*
g endogene *adj*

* **endoreic *adj* → 1901**

1901 **endorheic *adj*; endoreic *adj***
Applied to desert and semi-desert regions where the ephemeral or seasonal rivers flow within the limits of closed drainage basin.
r частично стекающий
Применяется к пустынным территориям, где дождевая влага после редких дождей не только просачивается в грунт, но и дает временный сток в пределах бессточного бассейна.
f endoreique *adj*
s endorreico *adj*
g endoreisch *adj*

1902 **endorreic drainage; aretic drainage**
Land water runoff into continental basins.
r внутренний сток
Сток вод суши во внутренние замкнутые бассейны.
f endoréisme *m*
s endorreísmo *m*
g Endorheismus *m*

* **end-product → 2153**

1903 **energy balance; fuel-energy balance**
The balance of production, conversion and consumption of energy.
r топливно-энергетический баланс
Баланс получения, преобразования и потребления всех видов энергии.
f bilan *m* énergétique
s balance *m* energético
g Energiebilanz *f*

1904 **energy consumption per unit; power consumption per unit**
The amount of energy needed to produce a unit of production or service (one of the major indexes of resource intensity).
r энергоемкость
Количество энергии, необходимое для получения единицы производимого продукта или услуги, один из основных показателей ресурсоемкости.
f consommation *f* d'énergie
s consumo *m* de energía
g Energiekapazität *f*

1905 **energy crisis**
A structural economic crisis caused by the growing shortage of fuel and energy resources, mainly oil (the major element of the raw-materials crisis which took place in the 1970s).
r энергетический кризис

Структурный экономический кризис, вызванный возросшим дефицитом топливно-энергетических ресурсов, главным образом нефти (важнейшая часть сырьевого кризиса, имевшего место в 1970-х годах).
f crise *f* énergétique
s crisis *f* energética
g Energirkrise *f*

1906 energy cycle; cycle of energy; energy flow
A cycle consisting of the absorption of solar energy by plants and its conversion into carbonhydrate compounds, which are a food source for animals.
r круговорот энергии
Круговорот, состоящий в поглощении солнечной энергии растениями и преобразовании ее в углеродные соединения, служащие пищей для животных.
f cycle *m* de énergie
s ciclo *m* de energía
g Energiezyklus *m*

*** energy flow → 1906**

*** energy of glacierization → 68**

1907 energy of relief
The degree of the relief dissection showing the potential intensity of relief formation processes.
r энергия рельефа
Размах рельефа, т.е. степень его расчлененности, показывающая возможность и силу проявления рельефообразующих процессов.
f énergie *f* de relief
s energía *f* de relieve
g Reliefenergie *f*

1908 energy resources; fuel-and-energy resources
Non-renewable mineral resources, renewable organic resources, and some natural processes (the energy of flowing water, wind, tides, etc.) used to produce energy.
r энергетические ресурсы; топливно-энергетические ресурсы
Невозобновляемые минеральные вещества, возобновляемые органические ресурсы и ряд природных процессов (энергия текущей воды, ветра, приливов и пр.), используемые для получения энергии.
f ressources *fpl* énergétiques
s recursos *mpl* energéticos; reservas *fpl* energéticas
g Energieressourcen *fpl*

1909 energy savings
A system of measures resulting the maximum diminution of the energy consumption in the production and conditions of life.
r энергосбережение
Система мер, призванных максимально сократить потребление энергии на производстве и в быту.
f économies *fpl* d'énergie
s economía *f* de energía
g Energieerhaltung *f*

*** energy system → 4736**

*** engineering geomorphology → 339**

1910 englacial ablation
The melting of ice inside the body of a glacier caused by the friction of separate ice layers, and also by water and air circulation in the cavities and fractures of the glacier.
r внутриледниковая абляция
Таяние льда внутри ледника в результате трения отдельных слоев льда, циркуляции воды и воздуха в полостях и трещинах ледника.
f ablation *f* interne
s ablación *f* interior; ablación *f* interna
g innere Ablation *f*; interglaziale Ablation *f*

1911 englacial *adj*
Related to the processes occurring inside the body of a glacier.
r внутриледниковый
Относящийся к природным явлениям внутри ледника.
f intraglaciaire *adj*
s intraglaciar *adj*
g intraglazial *adj*

*** englacial debris → 1914**

1912 englacial lake; glacier lake
A temporary lake lying within or upon a glacier.
r наледниковое озеро
Временное озеро на поверхности ледника.
f lac *m* surglaciaire
s lago *m* sobreglaciar
g Schmelzwassersee *m* an der Gletscheroberfläche

1913 englacial mass-and-energy exchange
Mass and energy transfer inside the body of a glacier.
r внутренний массоэнергообмен ледника
Перенос массы и энергии внутри ледника.
f échange *m* interne de masse d'énergie
s cambio *m* interno de energía y masa de un glaciar
g interglazialer Massen- und Energieaustausch *m*

1914 englacial moraine; englacial debris
Fragments of rock enclosed in the mass of a glacier and carried inside its body.
r внутренняя морена
Обломки горных пород, переносимые внутри ледника.

f moraine *f* interne; moraine *f* intraglaciaire
s morrena *f* interna; morrena *f* intraglaciar
g Innenmoräne *f*

1915 englacial water
Liquid water contained inside the body of a glacier.
r внутриледниковые воды
Жидкая вода внутри ледника.
f eau *f* intraglaciaire
s agua *f* intraglaciar
g intraglaziales Wasser *n*

1916 enterprise; firm
An industrial or commercial organization which is a primary territorial unit of territorial division of labour.
r предприятие; фирма
Самостоятельный хозяйственный субъект, первичное территориальное звено территориального разделения труда.
f entreprise *f*; firme *f*
s empresa *f*; compañía *f*; firma *f*
g Betrieb *m*; Firma *f*; Unternehmen *n*; Geschäftshaus *n*

1917 entisol
A poorly developed mineral soil having no developed horizons.
r энтисоль
Слаборазвитая минеральная почва, не имеющая сформированных горизонтов.
f entisol *m*
s entisol *m*; entisuelo *m*
g Entisol *m*

1918 entomochorous plant
Plants whose fruits, seeds and spores are disseminated by insects.
r энтомохор
Растения, у которых плоды, семена и споры распространяются насекомыми.
f plante *f* entomochore
s planta *f* entomócora
g Entomochor *m*

* **entrepot → 5829**

1919 environment; natural environment
The complete range of external conditions (physical and biological) with which people interact in their life and economic activities.
r окружающая среда; природная среда
Совокупность природных (биотических и абиотических) объектов, явлений и процессов, внешних по отношению к человеку, с которыми он взаимодействует в процессе своей деятельности.
f environnement *m*; milieu *m* naturel
s medio *m* ambiente; entorno *m* natural
g Umgebung *f*; Umwelt *f*

1920 environmental *adj*; ecological *adj*
Referring to the environment and nature conservation.
r экологический
Относящийся к окружающей среде, как правило, с оттенком сохранения ее природных качеств и роли в жизни людей.
f écologique *adj*; écologiste *adj*; environnemental *adj*
s medioambiental *adj*; ecológico *adj*; ecologista *adj*
g Umwelt-; ökologisch *adj*

1921 environmental balance; ecological balance; balance of nature
The ability of the environment to keep its qualities by the gradual readjustments in the components of nature in response to different influences which provides long-term stability of the biosphere.
r природное равновесие; экологическое равновесие
Способность окружающей среды сохранять свои природные качества и возможность воспроизведения своих компонентов, обеспечивая длительное устойчивое состояние биосферы.
f équilibre *m* écologique
s equilibrio *m* ecológico; equilibrio *m* de la naturaleza
g Umweltbilanz *f*

* **environmental conservation → 4797**

* **environmental contamination → 4669**

1922 environmental crime
A socially dangerous act infringing the environment and stipulated for by criminal legislation or international agreements.
r экологическое преступление
Предусмотренное уголовным законодательством или международными соглашениями общественно опасное деяние, посягающее на окружающую среду.
f crime *m* écologique
s crimen *m* medioambiental
g Umweltkriminalität *f*

1923 environmental damage
Economic and social losses due to the upset of the environmental balance as a result of economic activity.
r экологический ущерб
Экономические и социальные потери общества из-за нарушения устойчивости окружающей среды в результате хозяйственной деятельности.
f dommage *m* environnemental
s daño *m* al medio ambiente
g Umweltschaden *m*

1924 environmental degradation; environmental deterioration
Deterioration of the environment as a result of natural phenomena or economic activity of people including its pollution and loss of aesthetic properties.

r деградация природной среды
Ухудшение природной среды в результате естественных явлений или хозяйственной деятельности людей, включая ее загрязнение и потерю эстетического вида.
f dégradation *f* d'environnement
s degradación *f* de medio ambiente
g Umweltdegradation *f*

* **environmental deterioration → 1924**

* **environmental economics → 1799**

1925 **environmental emergency; ecological emergency**
A dangerous deviation from the normal state of the environment caused by a natural disaster or economic activity of people.
r чрезвычайная экологическая ситуация
Резкое и опасное отклонение от обычного состояния окружающей среды, вызванное стихийным бедствием или хозяйственной деятельностью человека.
f émergence *f* environnementale; situation *f* d'émergence environnementale
s emergencia *f* ecológica; emergencia *f* medioambiental
g Umweltunfall *m*

1926 **environmental engineering**
Applied discipline elaborating the system of measures for preservation of the environment under economic activities.
r экологическая инженерия
Прикладная дисциплина, разрабатывающая систему мероприятий для обеспечения сохранности окружающей среды в условиях хозяйственной деятельности.
f génie *m* écologique
s ingeniería *f* ecológica; ingeniería *f* medioambiental
g Umwelttechnik *f*

1927 **environmental ethics**
Care of nature which is acquired, in particular via ecological education.
r экологическая культура
Культура бережного отношения к природе, приобретаемая, в частности, с помощью экологического образования.
f éthique *f* écologique
s ética *f* ecologista; ética *f* medioambiental
g Umweltethik *f*

1928 **environmental examination; environmental impact assessment**
Evaluation of the effect upon the environment of a development project: construction programme, drainage system, etc.
r экологическая экспертиза; географическая экспертиза
Всесторонняя оценка вероятных воздействий предлагаемого проекта на окружающую среду.
f contrôle *m* de l'environnement; expertise *f* écologique
s peritación *f* ecológica; peritaje *m* medioambiental
g Umweltbewertung *f*

1929 **environmental forecast**
Forecasting the development of the environment in a given area under theb changing natural or man-made conditions.
r экологический прогноз
Предсказание будущего развития окружающей среды на данной территории в изменяющихся природных и техногенных условиях.
f prévision *f* environnementale; prévision *f* écologique
s previsión *f* ecológica; previsión *f* medioambiental
g Umweltprognose *f*

1930 **environmental geomorphology**
The study of the influence of landforms and geomorphological processes over living organisms including humans.
r экологическая геоморфология
Изучение влияния рельефа и рельефообразующих процессов на живые организмы, включая человека.
f géomorphologie *f* écologique
s geomorfología *f* medioambiental
g Umweltgeomorphologie *f*

1931 **environmental hazard; environmental risk**
A risk (usually to human being) associated with the natural hazards, such as flash-flood, landslide, volcanic eruption, etc. or produced by environmental impact.
r риск природных катастроф
Риск для людей, связанный со стихийными бедствиями, такими как наводнение, оползни, вулканические извержения и т.п., а также с природными бедствиями, вызванными техногенной нагрузкой на окружающую среду.
f risque *m* environnemental
s riesgo *m* medioambiental
g Umweltrisiko *n*

1932 **environmental impact; environmental load; environmental pressure; anthropogenic impact**
The degree of direct or indirect influence of human economic activity upon nature and its componens, it is subdivided into permissible and ecologically dangerous impact.
r воздействие на окружающую среду; техногенная нагрузка на окружающую среду
Прямое и косвенное влияние хозяйственной деятельности человека на природу и ее отдельные компоненты, подразделяется на допустимое и экологически опасное.

f impact *m* anthropique
s impacto *m* ecológico; impacto *m* medioambiental; carga *f* medioambiental; presión *f* medioambiental; carga *f* ecológica
g anthropogene Belastung *f*; Umweltbelastung *f*

* **environmental impact assessment → 1928**

1933 **environmental investments; ecological charges**
The cost of measures providing environmental protection.
r природоохранные затраты; экологические платежи
Стоимость мероприятий, обеспечивающих охрану окружающей среды и здоровье населения.
f dépenses *fpl* de l'environnement; paiements *mpl* écologiques
s gastos *mpl* medioambientales; gastos *mpl* para la conservación; pagos *mpl* ecológicos
g Umweltkosten *pl*; Umweltzahlungen *fpl*

* **environmental lapse rate → 79**

1934 **environmental legislation; ecological law**
Established legal norms and rules in the sphere of nature conservation, as well as the responsibility for their violation.
r природоохранное законодательство; экологическое право
Установленные юридические нормы и правила, а также ответственность за их нарушение в области охраны природы.
f législation *f* de l'environnement; droit *m* écologique; législation *f* environnementale
s legislación *f* de medio ambiente; derecho *m* medioambiental; legislación *f* sobre conservación; legislación *f* medioambiental
g Umweltrecht *n*

* **environmental load → 1932**

1935 **environmental management; rational environmental management**
Great variety of methods and tgechniques of the rational use of natural resources and measures aimed at environmental protection.
r природопользование; рациональное природопользование; адаптивное управление
Совокупность способов использования природных ресурсов и мер по их сохранению.
f management *m* de l'environnement; gestion *f* de l'environnement
s gestión *f* de medio ambiente; gestión *f* medioambiental
g Umweltgestaltung *f*

1936 **environmental monitoring**
A system of regular and long-term observations over the environment used to assess its present state and forecast its development in the future.
r мониторинг окружающей среды
Система регулярных и длительных наблюдений за состоянием окружающей среды с целью оценки настоящего и прогноза возможных изменений в будущем.
f monitoring *m* d'environnement; surveillance *f* de l'environnement
s control *m* medioambiental; observación *f* de medio ambiente
g Umweltmonitoring *n*; Umweltbeobachtung *f*

1937 **environmental policy; green politics**
A set of plans and actions of a society aimed at environmental protection.
r экологическая политика
Целенаправленное воздействие общества на природу с целью предупреждения, минимизации или ликвидации отрицательных для человека и природы последствий такого воздействия.
f politique *f* environnementale
s política *f* medioambiental
g Umweltpolitik *f*

* **environmental pollution → 4669**

* **environmental pressure → 1932**

* **environmental quality → 4848**

* **environmental quality standards → 5775**

* **environmental risk → 1810**

1938 **environmental services**
A system of measures aimed at the maintenance of life quality of people in the areas of impact upon the environment.
r экологические услуги
Система мероприятий, направленных на поддержание качества жизни в условиях активного воздействия на окружающую среду, например, очистка городов, разбивка парков и т.п.
f services *mpl* environnementaux
s servicios *mpl* medioambientales
g Umweltdienstleistungen *fpl*

1939 **environmental studies**
The study of the state of the environment, of the environmental management, monitoring and protection.
r изучение окружающей среды
Исследования состояния окружающей среды, возможностей рационального природопользования, средств мониторинга и охраны природы.
f études *fpl* environnementales
s estudios *mpl* medioambientales
g Umweltkunde *f*

1940 **environmental tax**

A payment for pollution or any other impact upon the environment included into taxes and used for nature conservation.

r экологический налог
Включение в состав налогов платы за загрязнение и любое иное воздействие на окружающую среду с последующим использованием этих налогов на природоохранные цели.

f impôt *m* écologique; taxe *f* écologique
s impuesto *m* medioambiental
g Umweltsteuer *f*

* **environmentalist → 1218**

* **environmentally sensitive area → 6614**

1941 environmetalism
A theory that views natural environment as the important factor in human development, life and activity of people.

r энвайронментализм
Философская концепция, подчеркивающая значение влияния окружающей природной среды на жизнь и деятельность человека.

f environnementalisme *m*
s ecologismo *m*; medioambientalismo *m*
g Ökologismus *m*; Environmentalismus *m*

1942 environs
The area adjoining a geographical object, or surrounding the place.

r окрестность
Местность, прилегающая к какому-либо географическому объекту, окружающее его пространство.

f alentours *mpl*; environs *mpl*
s alrededores *m pl*
g Umgebung *f*

1943 Eocene
The middle epoch of the Palaeogene period which began 55 million years ago and lasted about 17 million years. → Table. 1, p. 284

r эоцен
Средняя эпоха палеогенового периода геологической истории Земли, началась 55 млн лет назад и продолжалась около 17 млн лет. → Табл. 1, с. 284

f Éocène *m*
s Eoceno *m*
g Eozän *n*

* **eolian *adj* → 96**

1944 eolian deposits
Accumulations of fine loose material (sands, loess) brought by wind.

r эоловые отложения
Накопления тонкого рыхлого материала (песков, лёсса), принесенного ветром.

f sédiments *mpl* éoliens; dépôt *m* éolien
s sedimentos *mpl* eólicos; depósito *m* eólico
g äolische Ablagerungen *fpl*; äolische Sedimenten *npl*

1945 eolian erosion
The destruction and formation of landforms due to the action of wind.

r золовая эрозия; ветровая эрозия
Разрушение и формирование рельефа в результате деятельности ветра.

f érosion *f* éolienne
s erosión *f* eólica
g äolische Erosion *f*; Winderosion *f*

1946 eolian landforms
The landforms created by the action of wind, most common in arid climate.

r эоловый рельеф
Формы рельефа, возникающие под действием ветра, наиболее характерны для аридного климата.

f relief *m* éolien; modelé *m* éolien
s relieve *m* eólico; modelado *m* eólico
g äolisches Relief *n*

* **eolic park → 6778**

1947 Eon; Aeon
Geochronological subdivision uniting several geological eras, e.g. the Phanerozoic Eon. → Table 1, p. 284

r эон
Геохронологическое подразделение, объединяющее несколько геологических эр, например фанерозойский эон. → Табл. 1, с. 284

f Eon *m*
s Eon *m*
g Eon *m*

* **epeiric sea → 1954**

* **epeirogenesis → 1948**

1948 epeirogenic movements; epeirogenesis
The type of mass earth movement which results in changes of level over large areas of the Earth's crust, e.g. continental uplift and depression, without intense folding, in contrast to orogenic movement.

r эпейрогенические движения; эпейрогенез
Медленные и очень продолжительные движения земной коры, как восходящие, так и нисходящие.

f mouvement *m* épeirogénique; épirogénique *f*; épirogenése *f*
s movimiento *m* epirogénico; epirogénesis *f*
g epirogenetische Bewegungen *fpl*; Epirogenese *f*

1949 ephemer; ephemeroid
A short-lived annual (ephemer) or perennial (ephemeroid) plant, the life cycle of which from germination to production of seeds takes several weeks when conditions are favourable.

r эфемер; эфемероид
Однолетнее (эфемер) или многолетнее (эфемероид) растение, цикл развития которого

Fig. 105. **Ephemeral lake**

от прорастания семян до плодоношения происходит в течение нескольких недель.

f éphémère *m*; éphémérophyte *m*; éphéméroïde *m*
s efímero *m*; efimerófita *m*; efimeróido *m*
g Ephemere *n*; Ephemeroide *n*

1950 ephemeral ***adj***
1. Applied to a plant with a short life cycle. 2. Applied to a water body or a stream which flows at the surface only periodically.
r эфемерный
1. Относящийся к растениям с коротким жизненным циклом. 2. Относящийся к водоемам или речным потокам, текущим по поверхности лишь периодически.
f éphémère *adj*
s efímero *adj*
g ephemerisch *adj*

1951 ephemeral lake
A water body in arid areas appearing in certain seasons after heavy rains. → Fig. 105
r пересыхающее озеро; эфемерное озеро
Временный водоем в засушливых областях, возникающий в определенные сезоны после больших дождей. → Рис. 105
f lac *m* éphémère
s lago *m* efímero
g periodischer See *m*

* **ephemeral stream → 3149**

1952 ephemeris
Tables showing the position of celestial bodies at different times during the year.
r эфемериды
Таблицы, содержащие заранее вычисленные значения положения небесных светил на выбранные даты и моменты суток.
f éphémérides *fpl*
s efemérides *fpl* astronómicas
g Ephemeriden *fpl*

* **ephemeroid → 1949**

1953 epicentre
The point on the Earth's surface immediately above the centre of origin (seismic focus) of the earthquake.
r эпицентр
Точка на земной поверхности непосредственно над очагом землетрясения.
f épicentre *m*
s epicentro *m*
g Epizentrum *n*

1954 epicontinental sea; epeiric sea
A shallow sea resulting from the transgression of the World Ocean caused by the lowering of the marginal part of the continent or due to the raise of the ocean level e.g., Sea of Okhotsk or Baltic Sea.
r эпиконтинентальное море; мелководное внутреннее море
Море, образующееся при трансгрессии Мирового океана в результате опускания окраинной части материка или повышения уровня океана, например, Охотское и Балтийское моря.
f mer *f* épicontinentale
s mar *m* epicontinental
g epikontinentales Meer *n*; Schelfmeer *n*

1955 epidemic
A disease that becomes widespread in a particular place at a particular time.
r эпидемия
Болезнь, широко распространившаяся в определенное время в конкретной местности.
f épidémie *f*
s epidemia *f*
g Epidemie *f*; Seuche *f*

1956 epigenesis; epigenetic process
Any geological or rock-forming process which takes place upon the Earth's surface and leads to different changes of minerals and rocks.
r эпигенез; эпигенетический процесс
Вторичные процессы, ведущие к любым изменениям минералов и горных пород.
f épigénie *f*
s epigenia *f*
g Epigenese *f*

1957 epigenetic ice
A type of ice lens formed under conditions of permafrost in the subsoil.
r эпигенетический лед
Небольшие линзы льда, формирующиеся в подпочве при близко расположенных многолетнемерзлых породах.
f glace *f* épigénétique
s hielo *m* epigenético
g epigenetisches Eis *n*

1958 epigenetic ice wedge
A thin ice wedge formed after the deposition of the surrounding material.
r эпигенетическая ледяная жила
Ледяная жила небольшой вертикальной мощности, образовавшаяся после накопления вмещающих осадков.
f coin *m* de glace épigénétique; fente *f* de glace épigénétique

s cuña *f* de hielo epigenética
g epigenetischer Eiskeil *m*

* **epigenetic process → 1956**

1959 epigenetic valley
A river valley which cut newer cover rocks and overlaid older buried relief.
r эпигенетическая долина
Сквозная долина реки, прорезавшая поверхностные пласты и наложенная на древний, погребенный рельеф.
f vallée *f* épigénique
s valle *m* epigénico
g epigenetisches Tal *n*

* **epigleyic podzol → 2562**

1960 epilimnion
The topmost water layer in lakes, reservoirs and ocean when they have direct thermal stratification.
r эпилимнион
Верхний слой воды в озерах и водохранилищах при наличии прямой температурной стратификации.
f épilimnion *m*
s epilimnión *m*
g Epilimnion *n*; Deckschicht *f*

* **epiphyllous plant → 1961**

1961 epiphyte; epiphyllous plant
Plants that grow on other plants but use them only for support and not for food, e.g. algae, mosses, lichens.
r эпифит; эпифилл
Растительные организмы, обитающие на других растениях и получающие питательные вещества из окружающей среды, т.е. не являющиеся паразитами, к ним относятся, например, водоросли, мхи, лишайники.
f épiphyte *m*; plante *f* épiphylle
s epifita *m*; planta *f* epifila
g Epiphyt *m*; Epiphyll *m*

1962 epiplatformian mountains
The mountains formed by the latest tectonic movements on the place of old mountain regions (peneplains).
r возрожденные горы
Горы, возникшие в результате новейших тектонических движений на месте древних, пенепленизированных горных областей.
f montagnes *fpl* rajeunies de plate-forme; massifs *mpl* anciens rajeunis
s montañas *fpl* de rejuvenecimiento; macizos *mpl* antiguos rejuvencidos
g Bruchgebirge *n*

* **equal-angle projection → 1202**

* **equal-area projection → 1986**

1963 equator
An imaginary great circle round the Earth in a plane perpendicular to the axis of the Earth's rotation, lies midway between the two poles, thus dividing the Earth into the Northern and Southern hemispheres, its latitude is 0°.
r экватор
Линия сечения поверхности Земли плоскостью, проходящей через ее центр перпендикулярно оси вращения, имеет широту 0° и делит земной шар на Северное и Южное полушария.
f équateur *m*
s ecuador *m*
g Äquator *n*

1964 equatorial *adj*
Related to an area of the latitudinal zone between approximately 10° N and 10° S and to the natural phenomena (climate, forest, etc.) characteristic for these latitudes.
r экваториальный
Относящийся к региону, расположенному примерно между 10° с.ш. и 10° ю.ш. и к природным явлениям (климат, лес и т.п.), распространенным в этих широтах.
f équatorial *adj*
s ecuatorial *adj*
g äquatorial *adj*

1965 equatorial air
Air masses with high moisture content and high temperature formed over the regions lying close to the equator.
r экваториальный воздух
Воздушные массы с высокими влагосодержанием и температурой, сформированные над регионами вблизи экватора.
f air *m* équatorial
s aire *m* ecuatorial
g Äquatorialluft *f*

* **equatorial belt → 1973**

* **equatorial calms → 1665**

1966 equatorial climate
The type of hot and humid climate which occurs in low latitudes in the zone of equatorial depression (5-10° northwards and southwards from the equator), characterized by constant high temperature of about 27°C and high humidity with little range throughout day and night and the year.
r экваториальный климат
Тип жаркого и влажного климата, охватывающего полосу экваториальной депрессии (5-10° к северу и югу от экватора), характеризующегося постоянными высокими температурами около 27°C и высокой влажностью с малыми различиями днем и ночью и в течение года.
f climat *m* équatorial
s clima *m* ecuatorial
g Äquatorialklima *n*

* **equatorial countercurrent → 1971**

1967 **equatorial current; trade-wind current**
The surface movement of ocean water currents in tropical and equatorial latitudes caused by trade winds.
r пассатные течения; экваториальная циркуляция
Поверхностные течения Мирового океана в тропических и экваториальных широтах, вызываемые господствующими здесь пассатами.
f courants *mpl* équatoriaux; circulation *f* équatoriale
s corrientes *fpl* ecuatoriales; circulación *f* ecuatorial
g Passatströme *fpl*; äquatoriale Zirkulation *f*

1968 **equatorial forest; equatorial rain-forest; selva**
The specific forest mostly of evergreen hydrophillous species, growing at lower latitudes in a belt from approximately 7° N to 7° S in area of equatorial climate, particularly in the Amazon and Congo (Zaire) basins.
r экваториальный лес; сельва; экваториальный дождевой лес
Богатый лес, состоящий в основном из вечнозеленых гидрофитов, произрастающий в низких широтах в поясе между 7° с.ш. и 7° ю.ш., в области экваториального климата, преимущественно в бассейнах Амазонки и Конго (Заир).
f forêt *f* équatoriale; selva *f*
s bosque *m* ecuatorial; selva *f*
g äquatorialer Wald *m*; Selva *f*

1969 **equatorial forest zone**
Natural zones of the equatorial belt on both sides of the equator in South America, Africa, South-Eastern Asia, on the islands of Oceania, it has constant high temperatures (26-28°C) and abundant precipitation, covered with equatorial forests.
r зона экваториальных лесов
Природные зоны экваториального пояса по обе стороны от экватора в Южной Америке, Африке, Юго-Восточной Азии, на островах Океании, имеют постоянные высокие температуры (26-28°C) и обильное увлажнение с преобладающей растительностью экваториальных лесов.
f zone *f* des forêts équatoriales
s zona *f* de bosques ecuatoriales
g Zone *f* der äquatorialen Wälder

* **equatorial front → 1970**

* **equatorial rain-forest → 1968**

1970 **equatorial trough; equatorial front**
The belt of relatively low atmospheric pressure which lies at the equator between the subtropical high pressure belts of the two hemispheres.
r экваториальная депрессия
Полоса пониженного давления, охватывающая земной шар вблизи экватора между субтропическими зонами высокого давления обоих полушарий.
f dépression *f* equatoriale
s vaguada *f* ecuatorial
g äquatoriale Tiefdruckrinne *f*

1971 **equatorial undercurrent; equatorial countercurrent**
Along-equator water flows moving to the east in undersurface layers and compensating wind-induced surge near the western shores of oceans by the trade-wind currents (e.g. Lomonosov Stream in the Atlantic Ocean, Tareev Stream in the Indian Ocean, and Kromwell Stream in the Pacific Ocean).
r экваториальное противотечение
Вдольэкваториальные потоки воды, направленные на восток в подповерхностных слоях и компенсирующие нагон воды у западных берегов океанов пассатными течениями (к ним относятся течение Ломоносова в Атлантике, течение Тареева в Индийском океане, течение Кромвеля в Тихом океане).
f contre-courant *m* équatorial
s contra-corriente *f* ecuatorial
g äquatorialer Gegenstrom *m*

1972 **equatorial westerlies zone; equatorial west-wind zone**
A latitudinal belt of westerly winds at the equator between the zone of trade winds or between the belt of trade winds and pre-monsoon variable winds especially over the eastern part of the Indian Ocean and over western part of the Pacific.
r экваториальная зона западных ветров
Широтная полоса западных ветров у экватора между пассатами или между пассатом и предмуссонными переменными ветрами, особенно над восточной частью Индийского океана и над западной частью Тихого океана.
f zone *f* équatoriale des vents d'ouest
s zona *f* ecuatorial de vientos del oeste
g äquatoriale Westwindzone *f*

* **equatorial west-wind zone → 1972**

1973 **equatorial zone; equatorial belt**
The latitudinal zone between 5-8° N and 4-14° S, the area with equatorial climate with constant high temperatures and abundant precipitation evenly distributed through the year.
r экваториальный пояс
Географический пояс, расположенный между 5-8° с.ш. и 4-14° ю.ш., с постоянно высокими температурами и обильными осадками, равномерно распределенными в течение всего года.
f zone *f* équatoriale
s zona *f* ecuatorial

Fig. 106. **Equilibrium line** in the upper part of a small glacier at the Polar Urals

g äquatoriale Zone *f*; Äquatorialzone *f*

* **equiareal projection → 1986**

1974 **equidistant projection**
A map projection that maintains its scale in one of the main directions by spacing the meridians and parallels equidistantly outwards from the centre of the map.
r равнопромежуточная проекция
Картографическая проекция, сохраняющая свой масштаб длин по одному из главных направлений путем размещения параллелей и меридианов на равном расстоянии от центра карты.
f projection *f* équidistante
s proyección *f* equidistante
g abstandstreuer Kartennetzentwurf *m*

1975 **equilibrium line**
The boundary between the accumulation area and the ablation area of a glacier. → Fig. 106
r граница питания ледника
Линия, разделяющая область аккумуляции и область абляции ледника. → Рис. 106
f ligne *f* d'équilibre
s línea *f* de equilibrio
g Gleichgewichtslinie *f*

1976 **equilibrium models**
The reconstruction of any natural phenomenon which by its parameters completely corresponds to the conditions pre-determining its state.
r равновесные модели
Воспроизведение тех или иных природных явлений, которые по своим параметрам пришли в полное соответствие с условиями, предопределяющими их состояние.
f modèles *mpl* équilibrés
s modelos *mpl* equiponderantes
g Gleichgewichtsmodelle *npl*

1977 **equilibrium of landscape; landscape stability**
The ability of a landscape to maintain its structure and features over a long period by external (including anthropogenic) factors, or as a result of self-regulation processes.
r равновесие ландшафта; устойчивость ландшафта
Способность ландшафта длительно сохранять свою структуру и функциональные особенности под влиянием внешних (включая антропогенные) факторов или в результате процессов саморегулирования.
f équilibre *m* de paysage; stabilité *f* de paysage
s equilibrio *m* de paisaje; estabilidad *f* de paisaje
g Gleichgewicht *n* der Landschaft; Landschaftsstabilität *f*

1978 **equilibrium prices**
The prices that establish a balance between supply and demand on the market (they are lower when the market operates in the conditions of free competition and are higher if the market is monopolized).
r цены равновесия
Цены, при которых объем спроса на рынке равен объему предложения (устанавливаются в ходе действия рыночного механизма на низком уровне при свободной конкуренции и на высоком – при монополизации рынка).
f prix *mpl* d'équilibre
s precios *mpl* de equilibrio
g Eqiillibriumpreise *mpl*

1979 **equilibrium profile of river**
A longitudinal profile of a channel formed by a river under stable erosion basis.
r профиль равновесия реки
Продольный профиль русла, выработанный рекой при стабильном базисе эрозии.
f profil *m* d'équilibre de fleuve
s perfil *m* de equilibrio del río
g Gleichgewichtsprofil *n* des Flußes

* **equinoctial → 927**

1980 **equinoctial rains**
Rainy periods usually associated with the equinoxes in many equatorial regions.
r равноденственные дожди
Дождливые периоды, обычно выпадающие в дни равноденствий во многих экваториальных районах.
f pluies *fpl* équinoxiales; pluies *fpl* d'équinoxe
s lluvias *fpl* equinocciales
g Äquinoktialregen *mpl*

1981 **equinox**
The moment of time when the Sun's centre crosses the celestial equator during its apparent annual movement at the point of the spring (vernal) equinox (March 21-22) or the autumnal equinox (September 22-23) in the Northem Hemisphere, when the day and night are of equal lengh.
r равноденствие
Момент времени, когда центр Солнца при его годичном движении по небесной

сфере пересекает небесный экватор в точке весеннего (21-22 марта) или осеннего (22-23 сентября) равноденствия в Северном полушарии, в это время день и ночь имеют одинаковую продолжительность.
f équinoxe *m*
s equinoccio *m*
g Äquinoktium *m*; Tagundnachtgleiche *f*

1982 equiplain
A denudation surface formed as a result of the development and subsequent merging of cirques formed on the opposite slopes of mountains.
r эквиплен
Денудационная поверхность, образующаяся в результате развития и последующего слияния каров противоположных склонов возвышенности.
f équiplaine *f*
s equiplano *m*
g Equiplain *f*

1983 equiplanation
The processes of glacial denudation above the snow line leading to the planation of the relief irrespectively of the base level.
r эквипланация
Процессы ледниковой денудации выше снеговой линии, ведущие к выравниванию рельефа вне зависимости от базиса эрозии.
f équiplanation *f*
s equiplanación *f*
g Equiplanation *f*

1984 equipment of map; map elements
Graphical elements and explanations placed on a geographical map for its easier use: coordinate grids, map's legend, scale of horizontal equivalent, numerical and linear scales of a map, mathematical base of the map and reliability diagram.
r оснащение карты; элементы карты
Графические элементы и пояснения, помещаемые на географической карте для облегчения ее использования: координатные сетки, легенда карты, шкала заложений, численный и линейный масштабы карты, элементы математической основы и зарамочные сведения.
f équipement *m* de carte; éléments *mpl* de carte
s equipamiento *m* de mapa; elementos *mpl* de mapa
g Kartenausstattung *f*; Kartenelemente *n*; Kartenbestandteile *mpl*

1985 equiscalar surfaces
The surfaces of equal values of some characteristics showing their spatial distribution: isothermal surface for temperature, isopycnic surface for water density, isobaric surface for pressure, isosteric surface for volume.
r изоповерхности в океане
Поверхности равных величин каких-либо характеристик, дающие представление об их пространственном распределении: изотермическая поверхность – для температуры, изопикническая – для плотности воды, изобарическая – для давления, изостерическая – для объема.
f surfaces *fpl* équivalentes
s superficies *fpl* equivalentes
g Isoflächen *fpl*

1986 equivalent projection; equal-area projection; equiareal projection
A map projection in which there is no distortion of areas.
r равновеликая проекция
Картографическая проекция, в которой нет искажения площадей.
f projection *f* équivalente
s proyección *f* equivalente
g flächentreuer Kartennetzentwurf *m*

1987 era
Subdivision of geochronological scale corresponding to large stage of geological history and development of life on the Earth; Phanerozoic Palaeozoic, Mesozoic and Cainozoic eras are singled out. → Table 1, p. 284
r эра
Подразделение геохронологической шкалы, отвечающее крупному этапу геологической истории и развития жизни на Земле; в фанерозое выделяются палеозойская, мезозойская и кайнозойская эры. → Табл. 1, с. 284
f ère *f*
s era *f*
g Zeitalter *n*

1988 erg
Sand accumulations in wide basins in Northern Africa.
r эрг
Песчаные массивы Северной Африки, обычно расположенные в понижениях рельефа.
f erg *m*
s erg *m*
g Erg *m*

1989 erik
A linear hollow of intermittent watercourses formed on flood-plains while the river is flooding, common in the south of European Russia and the Ukraine.
r ерик
Ложбины временных водотоков, образующиеся на пойме при разливах реки на юге Европейской России и Украины.
f erik *m*
s erik *m*
g Erik *n*

1990 erosion
The wearing away of soil and rocks by natural agents such as running water, glaciers, wave action, wind, and the transport of the destroyed material. → Fig. 107, p. 237

Fig. 107. **Erosion** in the Kafirnigan river valley in Tadjikistan

r эрозия
Разрушение почв и горных пород текучими водами, ледниками, действием волн, ветром, а также вынос продуктов разрушения. → Рис. 107
f érosion *f*
s erosión *f*
g Erosion *f*

1991 **erosion landforms**
An irregular degradational surface formed by the eroding action of permanent or temporarystreams and other agents of erosion, in contrast to a level surface formed by deposition.
r эрозионный рельеф
Совокупность форм рельефа, созданных в основном эродирующей деятельностью постоянных и временных водотоков и других агентов эрозии, в противоположность формам рельефа, сформированным при отложении материала.
f relief *m* d´érosion; modelé *m* d´érosion; formes *fpl* d´érosion
s relieve *m* de erosión; modelado *m* de erosión; formas *fpl* de erosión
g Erosionsrelief *n*

1992 **erosion mountains**
The mountains formed as a result of the deep erosion of elevated plate-like areas of the Earth's surface with horizontal geological ctructure, e.g. the Colorado valley in USA.
r эрозионные горы
Горы, возникшие в результате глубокого эрозионного расчленения высоко поднятых платообразных областей земной поверхности с горизонтальной геологической структурой, например, бассейн реки Колорадо в Северной Америке.
f montagnes *fpl* d´érosion
s montañas *fpl* de erosión
g Erosionsgebirge *n*

1993 **erosion of sea floor**
Linear washout of sea floor produced by bottom currents resulting in the formation of different underwater relief.
r эрозия морского дна; субаквальная эрозия
Линейный размыв поверхности морского дна, обусловленный придонными течениями, в результате чего образуются различные формы подводного рельефа.
f érosion *f* du fond marine
s erosión *f* de fondo marino
g Meeresbodenerosion *f*

* **erosion remnant → 3912**

* **erosion surface → 4586**

1994 **erratic** ***adj***
Deposited in an area remote from its place of origin.
r эрратический
Отложенный в районе, далеком от места его образования.
f erratique *adj*
s errático *adj*
g erratisch *adj*

1995 **erratic boulder**
Boulders of hard rock, their petrography is different from the underlying substratum, as they were transported by a glacier or floating ice over long distances from the bedrock. → Fig. 108
r эрратический валун; руководящий валун
Валуны и глыбы прочных пород, петрографически отличных от подстилающего субстрата, перенесенные ледником или плавучим льдом на значительные расстояния от выходов коренных пород. → Рис. 108
f cailloux *m* erratique
s canto *m* rodado errático
g erratischer Block *m*; Erratiken *n*

1996 **erratic mass; detached mass; outlier**
A block of sedimentary rock that had been transported by a glacier for some distance and then became a component of glacial deposits but having retained a character of the primary bedding.
r отторженец
Блок осадочных пород, который перемещен ледником и стал составной частью

Fig. 108. Giant **erratic boulder** in the Tien Shan Mountains

ледниковых отложений, сохранив характер первичных напластований.
f bloc *m* erratique
s bloque *m* errático
g erratischer Block *m*; Auslieger *m*

* **eruption → 6591**

1997 eruption centre
The mouth part of a volcanic vent through which lava and gases come out from the magma chamber.
r эруптивный центр
Устьевая часть жерла вулкана, по которому лава и газы поступают из магматического очага.
f centre *m* éruptif
s centro *m* eruptivo
g Eruptionszentrum *n*

* **escape → 5722**

* **escarpment → 5240**

1998 esker
A narrow, sinuous ridge composed of stratified false-bedded sand, gravel and pebble of glaciofluvial origin. → Fig. 109
r оз; эскер
Линейно вытянутые гряды водно-ледникового происхождения, сложенные косослоистыми песками, гравием и галькой. → Рис. 109
f ôsar *m*; esker *m*
s esker *m*
g Esker *m*; Os *m*; Ås *m*

1999 espinal
Communities of tropical and subtropical evergreen thorny shrubs and low trees in the Chilean Andes and on the Gran Chaco plain in Argentina, similar to *maquis* and *chaparral*.
r эспиналь
Сообщества тропических и субтропических вечнозеленых колючих кустарников в

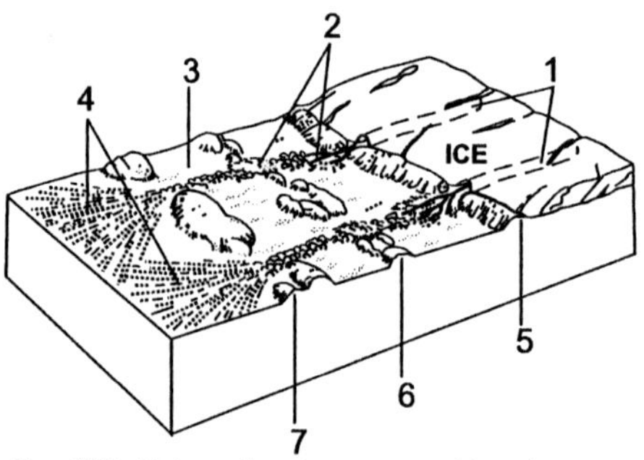

Fig. 109. **Esker**: 1 – ice caves with sub-glacial streams, 2 – esker «beaded» at still stand posiitons, 3 – esker marking course of former sub-glacial stream, 4 – fans of outwash sand, 5 – recessional moraine marking still stand in ice retreat, 6 – minor recessional moraine, 7 – terminal push-moraine

Чилийских Андах и на равнине Гран-Чако в Аргентине, аналог *макписа* и *чапараля*.
f espinal *m*; épinaie *f* sempervirente
s espinal *m*
g Espinal *m*

* **estancia → 2000**

2000 estate; hacienda; estancia; manor
An area of land forming a landed property that is passed by inheritance.
r поместье; имение; усадьба
Земельное имение, собственность на землю, переходящая по наследству.
f hacienda *f*; estancia *f*; fazenda *f*
s hacienda *f*; fazenda *f*; estancia *f*
g Landgut *n*; Gutshof *m*

2001 estavelle
A funnel-like hole on the floor of a polje in a karst area which at one moment absorbs water as ponor, at another ejects it as spring.
r эставелла
Воронкообразное отверстие на дне полья в карстовых областях, периодически действующее то как поглотитель воды (понор), то как источник, выбрасывающий воду под гидростатическим напором.
f estavelle *f*
s estavela *f*
g Estavelle *f*

2002 estuarian ***adj***
Related to an estuary and to the environmental conditions of an estuary.
r эстуарный
Относящийся к эстуарию или к природным условиям, характерным для эстуария.
f estuarien *adj*
s de estuario; estuarino *adj*
g ästuarisch *adj*

2003 estuarian circulation
Water circulation in estuaries caused by the interaction of river flow, ocean and tidal currents as well as by local features of water stratification.
r эстуарная циркуляция
Циркуляция вод в эстуариях, обусловленная взаимодействием речного стока, морских (в том числе приливных) течений и особенностями стратификации вод.
f circulation *f* estuarienne; courant *m* estuarien
s circulación *f* de estuario; corriente *f* de estuario
g Zirkulation *f* in Ästuaren

2004 estuary
A widened tidal mouth of a river where the saltwater of the tide meets the freshwater of the river current.
r эстуарий
Расширенное приморское устье реки, в пределах которого речная вода смешивается с морской.

f estuaire *m*
s estuario *m*
g Ästuar *n*

* **eternal snow → 4513**

2005 etesian wind
A strong north to northwesterly wind blowing in summer in the eastern Mediterranean Sea and the Aegean Sea.
r этезии
Ветры преимущественно северо-западных и северных румбов, дующие летом в восточной части Средиземного моря и в Эгейском море.
f vent *m* étésien
s viento *m* etesiano
g Etesien *f*

2006 ethnic *adj*
Relating to people unified by common geographic origin, cultural, religious or linguistic ties.
r этнический
Относящийся к группе людей, объединенных общим географическим происхождением, культурными, религиозными и языковыми связями.
f ethnique *adj*
s étnico *adj*
g ethnisch *adj*; Volks-

* **ethnic cleansing → 2384**

2007 ethnic geography
The study of the peoples and nationalities of the world, countries and regions and their spatial distribution over time.
r этническая география; этногеография
Изучение особенностей расселения (в прошлом и настоящем) народов мира, отдельных стран и регионов.
f géographie *f* éthnique
s geografía *f* étnica
g Ethnogeographie *f*; ethnische Geographie *f*

2008 ethnic group
A historically formed group of people with common self-identity and self-name, common origin, culture and often language, at early stages the group has been connected with a certain territory.
r этнос; этническая общность
Исторически сложившаяся группа людей, обладающих общим самосознанием и самоназванием, общностью происхождения и культуры, чаще всего языка, при своем возникновении связан с определенной территорией.
f ethnie *f*; groupe *f* ethnique
s grupo *m* étnico; etnia *f*
g ethnische Gruppe *f*; Ethnie *f*

2009 ethnic minority; minority ethnic group
A small ethnic group with the same culture and traditions commonly living in a place where most people have a different culture and traditions, the minority group is often economically and politically oppressed by the dominant nation.
r национальное меньшинство
Национальная группа населения, находящаяся на данной территории в численном меньшинстве и нередко экономически и политически притесняемая основной нацией.
f minorité *f* ethnique; minorité *f* nationale
s minoría *f* nacional; minoría *f* étnica
g etnische Minderheit *f*; Minorität *f*

2010 ethnicity
The fact that someone belongs to a particular ethnic group.
r этническая принадлежность
Тот факт, что данный человек относится к определенной этнической группе.
f ethnicité *f*
s etnicidad *f*
g Ethniezität *f*

2011 ethnocentric *adj*
Regarding oners own ethnic group as superior and showing a failure to recognize that other peoplers cultures are also important and valuable.
r этноцентрический
Признающий свою собственную этническую группу выдающейся и не способный понять ценность и важность других культур.
f ethnocentrique *adj*
s etnocéntrico *adj*
g ethnozentrisch *adj*

2012 ethnocentrism
The state or quality of being ethnocentric, i.e. regarding oners own ethnic group as being of superior, or supreme, importance, while the human attributes of other groups are oppressed (an extreme form of ethnocentrism is exhibited in *racism*).
r этноцентризм
Состояние общества, когда абсолютно превалирует одна этническая группа, в то время как жизненные атрибуты всех других групп всячески подавляются (крайняя форма этноцентризма выражается в *расизме*).
f ethnocentrisme *m*
s etnocentrismo *m*
g Ethnozentrismus *m*

2013 ethnocultural diversity
The variety of ethnic groups and their cultures.
r этнокультурное разнообразие
Существующее разнообразие этносов и их культур.
f diversité *f* éthnique et culturelle
s diversidad *f* étnica y cultural
g ethnische und kulturelle Diversität *f*

2014 **ethnogenesis**
The process of formation of an ethnic group on the basis of varied natural, historical and cultural components.
r этногенез
Процесс становления этнической общности людей на основе различных компонентов природы, истории и культуры.
f ethnogénèse *f*
s etnogénesis *f*
g Ethnogenese *f*

2015 **ethnography; ethnology**
The study of different ethnic groups and the ways in which different societies and cultures have developed.
r этнография; этнология
Изучение бытовых и культурных особенностей этносов и народов, их происхождения, расселения и культурно-исторических взаимоотношений.
f ethnographie *f*; ethnologie *f*
s etnografía *f*; etnología *f*
g Ethnographie *f*; Völkerkunde *f*

* **ethnology → 2015**

* **euphotic zone → 4534**

2016 **eurybathic animal**
Aquatic animals with a wide range of vertical distribution.
r эврибатное животное
Водные животные с широким диапазоном вертикального распространения.
f animal *m* eurybathique
s animal *m* euribático
g eurybathisches Tier *n*

2017 **euryhaline animal**
Aquatic and soil animals capable of living in a broad range of salinity conditions.
r эвригалинное животное
Водные или почвенные животные, способные существовать при больших перепадах солености окружающей среды.
f animal *m* euryhalin
s animal *m* eurihalino
g euryhalines Tier *n*

2018 **euryhygrobiont**
Organisms capable of living under sharp fluctuations of air humidity in contrast to *stenohygrobionts*.
r эвригигробионт
Организмы, способные существоаать при резких колебаниях влажности воздуха, в противоположность *стеногигробионтам*.
f euryhygrobionte *m*
s eurihigrobionte *m*
g Euryhygrobiont *m*

2019 **euryoxybiont**
Aquatic organisms capable of surviving under some fluctuations of oxygen content of water.
r эвриоксибионт
Водные организмы, способные переносить значительные колебания содержания растворенного в воде кислорода.
f euryoxybionte *m*
s eurioxibionte *m*
g Euryoxybiont *m*

2020 **euryphagous animal**
Animals that feed on a wide variety of food.
r эврифаг; полифаг
Всеядные животные, поедающие разнообразные виды корма растительного и животного происхождения.
f animal *m* euryphage; polyphage *m*
s animal *m* eurifago; polifago *m*
g Euryphag *m*; Polyphag *m*

2021 **eurythermobiont**
Organisms capable of living under great temperature fluctuations, e.g. animals and plants of the temperate zone.
r эвритермобионт
Организмы, способные переносить большие колебания температуры, например, животные и растения умеренного пояса.
f eurythermobionte *m*
s euritermobionte *m*
g Eurythermobiont *m*

* **eurytopic organism → 6435**

2022 **eustasy; eustatism; eustatic oscillations**
General slow changes of the World Ocean level caused by tectonic processes on the ocean bed and resulting in transgressions and regressions on its coasts.
r эвстатические колебания
Повсеместные медленные изменения уровня Мирового океана, вызываемые тектоническими процессами на дне океана и ведущие к трансгрессиям и регрессиям на его берегах.
f mouvements *m* eustatiques; eustatisme *m*
s movimiento *m* eustático;
eustasia *f*; eustatismo *m*
g eustatische Schwankungen *pl*

* **eustatic oscillations → 2022**

* **eustatism → 2022**

2023 **eutrofication**
Excessive growth of plants and microorganisms in water promoted by the accumulation of organic and mineral nutrients which leads to the lack of oxygen to support animal life, which becomes depleted and destroyed. → Fig. 110, p. 241
r евтрофикация; евтрофирование
Увеличение концентрации в водоеме питательных веществ, вследствие чего усиливается продукция биомассы и одновременно ухудшаются условия для жизни фауны, испытывающей недостаток кислорода. → Рис. 110, р. 241

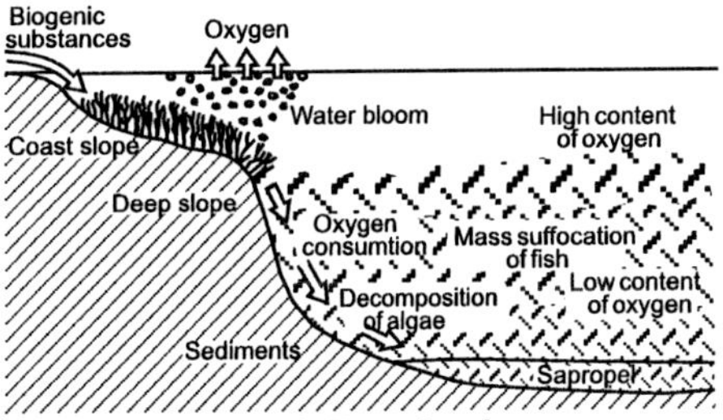

Fig. 110. **Eutrofication**

f eutrophisation *f*
s eutrofización *f*
g Eutrophierung *f*

2024 eutrophic ***adj***
Applied to a water body when it has an excess of organic and mineral nutrients usually derived from agricultural fertillizers, from farming actvities and from other human activities.
r евтрофный
Относящийся к водоему, содержащему излишек органических и минеральных питательных веществ, что обычно вызывается поступлением удобрений с полей, вообще сельскохозяйственной и иной деятельностью людей на окружающей территории.
f eutrophe *adj*
s eutrofo *adj*; eutrófico *adj*
g eutrophe *adj*

* **eutrophic fen → 2125**

2025 eutrophic lake
A shallow (10-15 m deep) lake well warmed in summer and having a very high content of nutrients in water.
r евтрофное озеро
Неглубокое (10-15 м) и хорошо прогреваемое летом озеро с большим содержанием питательных веществ.
f lac *m* eutrophe
s lago *m* eutrofo
g eutropher See *m*

2026 eutrophic plant; megatrophic plant
Plants growing on rich soils, especially in chernozem steppes, broad-leaved forests, flood-plain meadows.
r евтрофное растение
Растения, обитающие на богатых питательными веществами почвах, особенно в черноземных степях, широколиственных лесах, пойменных лугах.
f plante *f* eutrophique;
plante *fpl* mégatrophique
s planta *f* éutrofa; planta *f* eutrófica
g eutrophe Pflanze *f*

2027 evaporation; evaporation of water
The emission of water vapour into atmosphere by the free surface of liquid water.
r испарение воды
Поступление водяного пара в атмосферу со свободной поверхности жидкой воды.
f évaporation *f*; évaporation *f* de l'eau
s evaporización *f*; evaporización *f* del agua
g Wasserverdunstung *f*;
Wasserverdampfung *f*

* **evaporation fog → 5790**

2028 evaporation of ice
The conversion of ice directly to the gaseous state (water vapour) with no liquid state.
r возгонка льда; испарение льда
Переход льда в газообразное состояние (пар), минуя стадию таяния.
f évaporation *f* de la glace
s evaporación *f* de hielo
g Eisverdunstung *f*; Verdunstung *f* von Eis

* **evaporation of water → 2027**

2029 evaporative capacity; potential evaporation
The highest amount of water that can be evaporated from a given area under the existing climatic conditions and the unlimited water supply.
r испаряемость
Потенциально возможное испарение с единицы территории в данной местности при существующих климатических условиях и допущении неограниченного запаса влаги.
f évaporation *f* potentielle
s evaporación *f* potencial
g Verdunstungsfähigkeit *f*;
Verdunstungsvermögen *n*

* **evaporativity → 4721**

2030 evaporite
A water-soluble sedimentary rock composed of minerals that have been deposited by precipitation from saline water, as a result of evaporation, especially in the bottom of lagoons or in salt lakes.
r эвапорит
Осадочная порода, формирующаяся на дне лагун и соленых озер в результате выпаривания воды и ее перенасыщения солями.
f évaporite *m*
s evaporita *f*
g Evaporit *m*

2031 evapotranspiration
The loss of moisture from the Earth's surface by means of direct evaporation allied with transpiration from vegetation.
r эвапотранспирация; суммарное испарение; валовое испарение
Полное количество воды, возвратившейся с земной поверхности в атмосферу, включая испарение с почвы и транспирацию с растительного покрова.
f évapotranspiration *f*

s evapotranspiración *f*
g Evapotranspiration *f*; Gesamtverdunstung *f*

2033 evasion of gases
The emission of gases from the surface layer of water into the atmosphere.
r эвазия
Выделение газов из поверхностного слоя воды в атмосферу.
f évasion *f*
s evasión *f*
g Evasion *f*

* **evening glow → 116**

2034 everglade
A low marshy region, usually under water, with hummocks and small islands, and overgrown with tall grass, cans and trees, occupies large areas in Florida where National Park of the Everglades is located.
r эверглейд
Пространства маршей, на большей части залитые водой, с небольшими холмами и островами, покрытыми высокой травой, тростником и деревьями, занимают обширные площади во Флориде, где находится национальный парк Эверглейд.
f everglade *f*
s everglade *f*
g Everglade *f*

* **evergreen → 2035**

2035 evergreen plant; evergreen
A plant that keeps its leaves through the year instead of losing them seasoally, common mainly in the equatorial, tropical and subtropical regins with humid climate.
r вечнозеленое растение
Растения, сохраняющие листву круглый год, распространены главным образом в экваториальных, тропических и субтропических областях с влажным климатом.
f plante *f* sempervirente
s planta *f* de verdor perenne; planta *f* perennifolia
g immergrüne Pflanze *f*

2036 evolution
The gradual process by which living organisms change from more primitive forms and slowly evolve into species different from a common ansector; in the general meaning it is the gradual change and development of nature and its components.
r эволюция
Постепенный процесс изменений живых организмов от примитивных форм к более совершенным, сопровождающийся формированием биологических видов, отличных от некогда общего предка; в общем смысле это постепенное развитие и изменения природы и ее составляющих.
f évolution *f*
s evolución *f*
g Evolution *f*

2037 evolutionary geography
The study of the contemporary state of the environment on the basis of the knowledge of its evolution and by taking into account the impact upon the environment of human activity.
r эволюционная география
Изучение современного состояния окружающей среды на основе познания ее эволюции, с учетом влияния деятельности человека.
f géographie *f* évolutionniste
s geografía *f* evolutiva
g Evolutionsgeographie *f*

2038 evorsion
Local erosion which scours out cavities in the bed of a stream, on a seashore, and on the bed of a glacier caused by the rotational action of vertically falling water.
r эворзия
Локальная эрозия, при которой в результате придонного вращения вертикально падающей воды вырабатываются полости в русле быстро текущей реки, на морских берегах и у днищ ледников.
f évorsion *f*
s evorsión *f*
g Evorsion *f*

2039 exaration
The action by which glacier ice, unladen with debris, carries out erosion of the bedrock by quarring or joint-block removal.
r экзарация; ледниковое выпахивание
Разрушение ледникового ложа непосредственно движущимся льдом, в котором отсутствуют каменные обломки, которые могли бы производить абразивное действие.
f exaration *f*
s exaración *f*
g Exaration *f*

2040 excavated waste rock; Haldenbaden *m*; mining waste
Rocks excavated from the lithosphere in mining operations.
r отвальный грунт
Пустые породы, изъятые из литосферы при горных выработках.
f déblai *m*
s escombrera *f*; vaciadero *m*
g Halde *f*

2041 excess ice
The volume of ice in the ground which exceeds the total pore volume that the ground would have under the natural unfrozen conditions.
r избыточное льдосодержание
Количество подземного льда, по объему превышающее пористость грунта до промерзания.

f glace *f* en excès; glace *f* excessive
s hielo *m* en exceso; hielo *m* excesivo
g Überschußeis *n*

2042 excessive humidity
Natural conditions when the amount of precipitation over a year exceeds the average evaporation and infiltration of water into soil.
r избыточное увлажнение
Природные условия, когда количество выпавших за год атмосферных осадков в среднем превышает испарение и просачивание воды в почву.
f humidité *f* excessive
s humedad *f* excesiva
g überschissige Feuchtigkeit *f*

2043 exchange capacity of soil
Soil capability of absorbing different cations and anions from soil solution, emitting the equivalent amount of other cations and anions into the solution.
r обменная способность почвы
Свойство почвы поглощать из почвенного раствора различные катионы и анионы, выделяя при этом в раствор эквивалентное количество других катионов и анионов.
f capacité *f* d'échange du sol
s capacidad *f* de intercambio del suelo
g Austauschncapazität *f* des Bodens

* **exchange coefficients → 1846**

* **exclusionary enclave → 2368**

2044 exclusive economic zone
Marine regions outside territorial waters and adjacent islands.
r экономическая зона
Морские районы, находящиеся за пределами территориальных вод государства, и прилегающие к ним острова.
f zone *f* économique exclusive
s zona *f* económica exclusiva
g exklusive ökonomische Zone *f*

2045 excursion
A short journey in order to visit places of interest.
r экскурсия
Кратковременное путешествие с посещением достопримечательных объектов.
f excursion *f*
s excursión *f*
g Exkursion *f*; Ausflug *m*

2046 exfoliation; onion-skin weathering
A hydration process of breaking, splitting and peeling-off of outer rock layers caused by the expansion of salt crystals within the interstices of the rock surface.
r отслаивание
Процесс дробления, расщепления и отслаивания внешних слоев горных пород, вызванный проникновением кристаллов соли в расщелины скальной поверхности.
f exfoliation *f*
s exfoliación *f*
g Abblättern *n*

2047 exfoliation; onion-weathering; desquamation
A weathering process of peeling off of rounded scales from exposed surfaces of rocks due to sharp, mainly daily temperature fluctuations.
r десквамация
Отслаивание тонких округлых чешуй с обнаженных скалистых поверхностей горных пород под влиянием резких, главным образом суточных, колебаний температуры.
f exfoliation *f*
s exfoliación *f*
g Abschuppung *f*; Desquamation *f*

2048 exhumation of relief
The exposing by denudation of a land surface or landform that had previously been buried by deposition.
r эксгумация рельефа
Процесс обнажения ранее погребенного рельефа в результате денудации земной поверхности.
f exhumation *f* de relief
s exhumación *f* de relieve
g Reliefexhumierung *f*

2049 exhumed relief
The land surface features exposed by denudation that had previously been buried by deposition.
r откопанный рельеф
Формы погребенного рельефа, вскрытые процессами денудации.
f relief *m* exhumé;
formes *fpl* de relief exhumées
s relieve *m* exhumado;
formas *fpl* de relieve exhumadas
g ausgegrabenes Relief *n*

2050 exodynamic succession
The sequential change of vegetation caused by the impact of external factors.
r аллогенная сукцессия;
экзогенетическая сукцессия
Смена растительности, вызванная воздействием внешних факторов.
f succession *f* exodynamique
s sucesión *f* exodinámica
g allogene Sukzession *f*

2051 exogenetic *adj*; exogenic *adj*
Related to processes occurring at or near the Earth's surface created by the solar energy, gravity and life activity of organisms, and to the landforms produced by such processes.
r экзогенный
Относящийся к процессам, происходящим на поверхности Земли или на небольшой глубине в земной коре, обуслов-

ленным энергией солнечного излучения, силой тяжести и жизнедеятельностью организмов.
f exogène *adj*
s exógeno *adj*
g exogene *adj*

* **exogenic** ***adj*** **→ 2051**

* **exonym → 1277**

2052 **exoreic drainage; external drainage**
The area from which rivers flow outside carrying water to the seas connected with the ocean.
r область внешнего стока
Часть суши, с которой реки несут воду в моря, соединенные с Мировым океаном.
f drainage *m* exoréique
s drenaje *m* exorreico
g Außenabflußgebiet *n*; exorheisches Gebiet *n*

2053 **exosphere**
A layer of the atmosphere above 600-800 km from which the lighter gases escape to the space.
r экзосфера
Внешняя часть атмосферы (выше 600-800 км), из которой частицы легких газов рассеиваются в мировое пространство.
f exosphère *f*
s exosfera *f*
g Exosphäre *f*

* **exotherm → 4634**

2054 **exotic species; introduced species; alien; invasive species**
An organism which is not indigenous and is subject of the deliberate or accidental introduction by humans into an area outside the range of their natural distribution.
r чужеродный вид; экзотический вид; чужестранец
Организм, намеренно или случайно интродуцированный людьми в регионы, выходящие за пределы их естественного распространения.
f espèce *f* exotique; étranger *m*
s especie *f* exótica; especie *f* ajena; especie *f* introducida; exótico *m*
g exotische Art *f*; Ausländer *m*; gebietsfremde Art *f*

2055 **expanded foot glacier; wall-sided glacier**
A glacier spreading out from the mouth of a valley to form a broad tongue of ice on the plain, a small piedmont glacier.
r ширококонечный ледник
Ледник с широким языком при выходе из горной долины на равнину, небольшой предгорный ледник.
f glacier *f* dilatante
s glaciar *m* dilatado
g Gebirgsfußgletscher *m*; Lobusgletscher *m*

2056 **expatriate**
One who voluntary chooses to live and in some cases work away from his/her native country, but not necessarily permanently.
r экспатриант
Тот, кто добровольно предпочитает жить, а нередко и работать, вне своей страны, но не обязательно постоянно.
f expatrié *m*
s expatriado *m*; expatriante *m*
g Expatriierte *m*; Auswanderer *m*

2057 **expatriation**
Voluntary or compulsory expulsion from native land.
r экспатриация
Добровольное или принудительное выселение за пределы родины.
f expatriation *f*
s expatriación *f*
g Ausbürgerung *f*; Expatriation *f*

* **expenditures → 1316**

2058 **explerent**
Plants with low competitiveness but rapid reproduction and growth which allow them to invade free spaces in a short time, but not for long.
r эксплерент
Растения, которые в борьбе за существование мало конкурентоспособны, но, благодаря быстрому размножению и росту, способны захватывать освободившуюся территорию, хотя и на непродолжительный срок.
f explerent *m*
s explerente *m*
g Explerent *m*

2059 **exploitation**
Withdrawal of a part of biomass (hunting, fishing, felling, etc.) for economic purposes, sometimes with its restoration (fish farming, forest-planting).
r промысел
Изъятие в хозяйственных целях (охота, лов, рубка и пр.) части биомассы в виде полезной продукции, иногда с искусственным восстановлением (рыборазводные станции, лесонасаждения), различают охотничий, рыбный, в прошлом китобойный промыслы.
f exploitation *f*
s explotación *f*
g Gewerbe *n*; Erwerbszweig *m*

2060 **explosive eruption; volcanic explosion**
Eruption of a volcano as a result of explosion of magma gases accompanied by outburst of great masses of loose products from a volcanic vent funnel.

r взрывное извержение; вулканический взрыв; эксплозивное извержение
Извержение вулкана в результате взрыва магматических газов, сопровождаемое выбросом из жерла вулкана больших масс рыхлых продуктов.
f éruption *f* explosive; explosion *f* volcanique
s erupción *f* explosiva; explosión *f* volcánica
g explosiver Ausbruch *m*; vulkanische Explosion *f*

2061 **export**
The action of selling and transporting goods or capital to a foregn country.
r экспорт
Вывоз товаров или капиталов за границу.
f exportation *f*
s exportación *f*
g Export *m*; Ausfuhr *f*

2062 **export duty**
The money which is collected for the goods exported from a country.
r экспортная пошлина
Денежный сбор, взыскиваемый государством с товаров, вывозимых за границу.
f droits *mpl* d'exportation
s derechos *mpl* de exportación
g Ausfuhrzoll *f*

2063 **export economy**
An economy which is mainly oriented to foreign market rather than domestic market.
r экспортная экономика
Экономика, в которой основная часть производства ориентирована на зарубежный, а не национальный рынок.
f économie *f* d'exportation
s economía *f* de exportación
g exportorientierte Wirtschaft *f*

2064 **export of capital**
Taking capital out of the country by private firms and by the state in the form of foreign investments and loans.
r вывоз капитала; экспорт капитала
Вывоз капитала частными фирмами и государством, происходящий в двух формах: предпринимательского капитала (вложения в разные отрасли хозяйства) и ссудного капитала (в виде займов).
f sortie *f* de capital
s salida *f* de capital
g Kapitalausfuhr *f*

2065 **export quota**
A limit on the amount of a particular commodity to be exported from one country to another.
r экспортная квота
Разрешение на экспорт определенного объема товара.
f contingent *m* d'exportation; quota *f* d'exportation
s contingente *m* de exportación; cupo *m* de exportación; cuota *f* de exportación
g Exportsquote *f*

2066 **exposure; slope exposure; aspect$_2$**
The position of mountain slopes and any other landform with regard to cardinal points and to climatic influences such as prevailing winds, etc.
r экспозиция; экспозиция склона
Ориентировка склонов гор и любых форм рельефа по отношению к странам света и к преобладающим воздушным течениям.
f exposition *f*; aspect$_2$ *m*; exposition *f* de versant
s exposición *f*; exposición *f* de pendiente; aspecto$_2$ *m*
g Exposition *f*; Abhangexposition *f*

2067 **extension of drainage network**
The total length of all rivers within a basin or some area.
r протяженность речной сети
Сумма длин всех рек в пределах бассейна или какой-либо территории.
f extension *f* de réseau hydrographique
s extención *f* de la red hidrográfica
g Gesamtlänge *f* eines Flußnetzes

2068 **extensive agriculture; extensive farming**
A farming system characterized by large areas of cultivated land, low inputs of labour and minimal production costs.
r экстенсивное сельское хозяйство
Система хозяйствования, при которой обрабатываются большие площади земли при минимальных затратах труда и производственных издержках.
f agriculture *f* extensive
s agricultura *f* extensiva
g extensive Landwirtschaft *f*

2069 **extensive economy**
An economic system based on the expansion of production that makes little use of innovations, opposite to ***intensive economy***.
r экстенсивное хозяйство
Система хозяйствования, основанная на расширении производства, мало связанном с введением инноваций (противоположно ***интенсивному хозяйству***).
f économie *f* extensive
s economía *f* extensiva
g extensive Wirtschaft *f*

* **extensive farming → 2068**

2070 **external costs; externalities**
The effects of economic activity such as environmental pollution that create costs not added to production cost of goods or services, these costs are borne by those affected by the results of the economic activity (consumers, population, and other countries in case of transborder pollution).

r внешние издержки
Издержки хозяйственнной деятельности, такие как загрязнение окружающей среды, которые не учитываются в себестоимости продукции или услуг и которые оплачивают те, кто оказался вовлеченным в последствия данной хозяйственной деятельности (потребители, население, другие страны в случае трансграничного переноса).
f coûts *mpl* externes
s gastos *mpl* externos
g Außenkosten *pl*

* **external drainage → 2052**

2071 external perspective projection
A map projection, a kind of perspective azymuthal projection, in which the perspective centre lies outside the surface of the general globe, it is used for processing space images.
r внешняя проекция
Картографическая проекция, вид перспективной азимутальной проекции, получаемой проектированием из точки, расположенной вне сферы на конечном расстоянии, применяется для обработки космических снимков.
f projection *f* perspective externe
s proyección *f* perspectiva exterior
g äußerer perspektiver Kartennetzentwurf *m*

* **external trade → 3163**

2072 external water circulation
A hydrological cycle where the water vapour evaporated from a sea surface condenses in the form of precipitation which falls on the continent.
r внешний влагооборот
Такой влагооборот, когда водяной пар, испаряющийся с поверхности моря, конденсируется в форме осадков, которые затем выпадают на континент.
f cycle *m* hydrologique externe
s ciclo *m* hidrológico externo
g äußere Wasserzirkulation *f*

* **externalities → 2070**

2073 extinct volcano
A volcano which functioned in the distant geological past and has no longer any volcanic activity.
r потухший вулкан
Вулкан, не проявляющий активности на протяжении исторического времени.
f volcan *m* éteint
s volcán *m* apagado; volcán *m* extinguido
g erloschener Vulkan *m*

2074 extinction of species
The disappearance of a species in an ecosystem due to the influence of natural conditions or under the impact, direct or indirect, of human activity.
r вымирание вида
Исчезновеие вида в процессе развития живой природы в результате эволюции или воздействия человека, прямого или опосредованного.
f extinction *f* des espéces
s extinción *f* de especies
g Aussterben *n* der Arten

2075 extractive agriculture
An agricultural practice which is more concerned with immediate short-term benefits than with conserving resources for future agricultural use, opposed to the *sustainable agriculture.*
r расточительное сельское хозяйство
Сельскохозяйственная практика, ориентированнная скорее на быстрое извлечение выгоды, чем на сохранение ресурсов для будущего сельскохозяйственного использования, в противоположность *устойчивому сельскому хозяйству.*
f agriculture *f* extractive
s agricultura *f* extractiva
g extraktive Landwirtschaft *f*

* **extractive industry → 3876**

* **extraglacial deposits → 4494**

2076 extralayer plants
The plants which do not form their own layers, but inhabit the layers of trees, shrubs and undershrubs, they include epiphytes and lianas.
r внеярусные растения
Растения, не образующие самостоятельных ярусов, а населяющие ярусы высших растений лесной, кустарниковой и полукустарниковой растительности, преимущественно эпифиты и лианы.
f plantes *fpl* hors-étagés
s plantas *fpl* extrapiso
g Pflanzen *fpl* ausserhalb der Schichten

* **extraterrestrial *adj* → 1312**

2077 extratropical cyclone; extratropical depression
Cyclone which originates and develops in the westerlies zone of middle and polar latitudes.
r внетропический циклон
Циклон, возникший и развивающийся в зоне западного переноса воздушных масс в умеренных и полярных широтах.
f cyclone *m* extratropical
s ciclón *m* extratropical
g aussertropisches Tief *n*

* **extratropical depression → 2077**

2078 extratropical monsoon
Monsoon of middle and high latitudes resulted from a stable winter anticyclone and summer cyclone over Asia.
r внетропический муссон
Муссон умеренных и высоких широт, обусловленный устойчивым зимним антициклоном и летним циклоном над Азией.

f mousson *m* extratropicale
s monzón *m* extratropical
g aussertropischer Monsun *m*

* **extrazonal landscape → 2079**

2079 extrazonal vegetation; extrazonal landscape
Plant communities and accompanying features of landscape of a certain geographical zone found within other zones, e.g. patches of forest islands in steppe.
r экстразональная растительность; экстразональный ландшафт
Растительные сообщества и сопутствующие черты ландшафта, характерные для определенной географической зоны, располагающиеся в пределах других зон, например, островки леса в степи.
f végétation *f* extrazonale; paysage *m* extrazonal
s vegetación *f* extrazonal; paisaje *m* extrazonal
g extrazonale Vegetation *f*; extrazonale Landschaft *f*

2080 extreme continental climate
A climate with the greatest range of average daily and monthly temperatures, small precipitation and cloudiness.
r резко континентальный климат
Климат с большими колебаниями средних суточных и месячных температур, малым количеством атмоферных осадков и малой облачностью.
f climat *m* hyper-continental
s clima *m* hypercontinental
g extremes Landlklima *n*

2081 extreme tourism; adventure tourism
Tourism organised at high altitudes, in the open ocean, in uninhabited lands, etc. having the prime motivation of physical exercise and challenging situations in natural environments.
r экстремальный туризм; приключенческий туризм
Туризм, организуемый в экстремальных природных условиях: на больших абсолютных высотах, в открытом океане, на необитаемых землях и пр.
f tourisme *m* extrême; tourisme *m* d'aventure
s turismo *m* extremo; turismo *m* de aventura
g Extremtourismus *m*

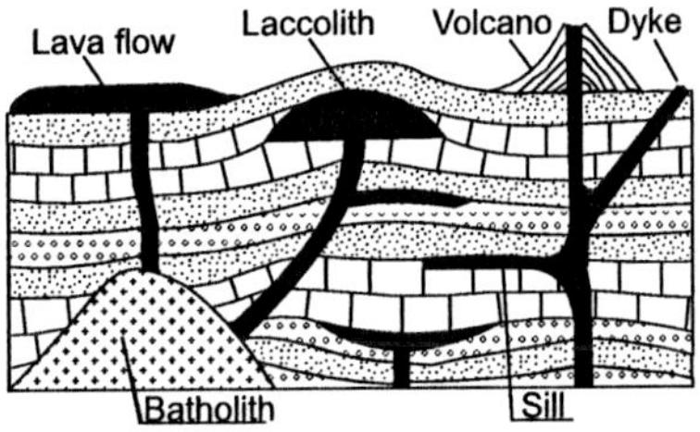

Fig. 111. **Extrusive** and **intrusive rocks**

2082 extrusion
The discharge of solid, liquid or gaseous material to the Earth's surface during a volcanic eruption as well as a geological body produced by rocks which have been deposited at the Earth's surface after eruption.
r экструзия
Выжимание вязкой лавы из жерла вулкана и образование экструзивного купола над ним, а также геологическое тело, формирующееся в результате этих процессов.
f extrusion *f*
s extrusión *f*
g Extrusion *f*

2083 extrusive rock
An ingeous rock formed as a result of queezing up of viscous lava from a volcanic vent at the Earth's surface. → Fig. 111
r экструзивная горная порода
Магматическая горная порода, образовавшаяся в результате выжимания вязкой лавы из жерла вулкана и образования над ним экструзивного купола. → Рис. 111
f roche *f* extrusive
s roca *f* extrusiva
g Extrusivgestein *n*

2084 eye of a tropical cyclone; storm eye
The still, calm area of the central part of a tropical cyclone, generally between 16 and 24 km in diameter, characterized by clear or almost clear sky, light winds and the lowest atmospheric ptressure. → Fig. 340, p. 759
r глаз бури
Область в центре тропического циклона с диаметром от 16 до 24 км и с ясным или почти ясным небом, слабыми ветрами и самым низким атмосферным давлением. → Рис. 340, с. 759
f oeil *m* du cyclone
s ojo *m* del ciclón
g Auge *n* eines Wirbelsturms

F

2085 facet of fall
A horizontal or inclined part of a plane surface limited by the margins of other crossed surfaces.
r фасет
Горизонтальный или наклонный участок поверхности, ограниченный гранями других пересекающихся поверхностей.
f facette *f* de faille
s faceta *f* de falla
g Facette *f*

2086 facies
The total character of any part of a geological formation, shown by the fossils it contains, the composition, colour, texture, form of stratification and nature of the constituent rocks, or by other special features.
r фация
Общий характер любой части геологической формации, отличающейся содержащимися в ней окаменелостями, составом, цветом, текстурой, формой пластов и природой вмещающих пород и пр.
f faciès *m*
s facies *m*
g Fazies *f*

2087 factor; agent
An agent or influence contributing to a result, effect or natural condition.
r фактор
Природный процесс или явление, создающие определенный эффект или воздействующие на окружающую среду.
f facteur *m*; agent *m*
s factor *m*; agente *m*
g Faktor *m*

2088 factory; plant
A place where industrial or manufacturing processes are carried out and power or manufactured goods are produced.
r завод; фабрика
Промышленное предприятие с механизированными процессами производства.
f usine *f*; établissement *m* industriel
s fábrica *f*; unidad *f* de producción; planta *f*; usina *f*
g Werk *n*; Fabrik *f*

2089 Fahrenheit temperature scale
A temperature scale in which the freezing point of water is 32°F and the boiling point of water is 212°F, so 1o Celsius equals 1.8° Farenheit [the Celcius temperature is calculated by the formula:C=(F-32)5/9].
r температурная шкала Фаренгейта
Температурная шкала с точкой замерзания воды -32°F и точкой кипения -212°F [перевод в температурную шкалу Цельсия (C) делается по формуле: C=(F-32)5/9].
f échelle *f* de température Fahrenheit
s escala *f* de temperaturas Fahrenheit
g Fahrenheit-Temperaturskala *f*

2090 fairway; channel
A navigable channel in a river, bay or harbour.
r фарватер
Безопасный в навигационном отношении проход по водному пространству.
f chenal *m* passe
s canal *m* de paso
g Fahrrinne *f*; Fahrwasser *n*

2091 fall[1]
Downhill rolling and sliding of rock fragments of different size on a denued slope.
r осыпание
Скольжение и скатывание вниз по оголенному склону обломков разного размера.
f éboulement *m* continu
s derrumbamiento *m* continuo
g Abbröckeln *n*

* **fall[2] → 475**

2092 fall velocity; sinking velocity
The velocity of uniform falling of solid particles in still water or air.
r гидравлическая крупность
Скорость равномерного падения твердых частиц в неподвижной воде или воздухе.
f vitesse *f* de la cute des sédiments dans l´eau
s velocidad *f* de asentamiento
g Sinkgeschwindigkeit *f*

2093 fall wind
A strong, cold, downslope wind, which differs from a foehn in that the air is initially cold enough so that it remains relatively cold despite adiabatic warming upon descent, *bora* is an example of a fall wind.
r падающий ветер; «сваливающийся» ветер; нисходящий ветер
Сильный холодный ветер, «падающий» вниз по горному склону, отличается от фёна тем, что первоначально воздух, перетекающий через вершину или хребет, достаточно холодный и не успевает нагреться за счет адиабатического нагревания при опуснании, *бора* – пример падающего ветра.
f vent *m* de gravité
s viento *m* de gravitación
g Fallwind *m*

* **fall zone → 2094**

2094 fall-line; fall zone
A line or narrow zone where a number of nearly parallel rivers plunge over the edge of a plateau to the lowland below, forming a series of waterfalls on the deep slope of a plateau.

r зона водопадов
Линия или узкая зона, где ряд почти параллельных рек стремительно падает с края плато на нижележащую равнину, образуя серию водопадов на крутом склоне плато.
f zone *f* des chutes
s zona *f* de cascadas
g Zone *f* der Wasserfälle

2095 fallout; radioactive fallout
Radioactive material which falls from the atmosphere with rain, snow or as dust, especially harmful after a nuclear explosion in the atmosphere.
r радиоактивные осадки
Радиоактивные вещества, выпадающие из атмосферы на поверхность Земли вместе с дождем или снегом, либо в сухом виде, особенно опасные после ядерных взрывов в атмосфере.
f retombées *fpl* radioactives
s lluvia *f* radioactiva
g radioaktive Niederschläge *mpl*

2096 fallow; ley; bare fallow; lea
A ploughed or cultivated land which is allowed to rest, uncropped or partially cropped during the vegetation period in order to restore its fertility and moisture.
r пар
Поле севооборота, не занимаемое посевами в течение всего вегетационного периода или части его и содержащееся в рыхлом и чистом от сорняков состоянии с целью повышения плодородия почвы и накопления в ней влаги.
f jachère *f*
s barbecho *m*
g Brache *f*

2097 fallow land
Lands left untilled for a long time (15-20 years) so that their vegetation cover hardly differs from that of virgin lands.
r залежные земли; залежь
Участок давно (15-20 лет) не паханной земли с растительным покровом, мало отличающийся от целины.
f jachère *f* forestière
s barbecho *m* forestal
g Brachland *n*

* **false color image → 2098**

2098 false colour composite; false color image
Photograph made by way of combining images got in several wavelands of multiband photography.
r спектрозональный снимок; ложноцветной снимок
Фотоснимок, созданный путем комбинирования изображений, полученых в нескольких каналах многоспектральной съемки.
f image *m* en couleur fausse
s imagen *m* espectrozonal; imagen *m* de satélite en color falso
g Falschfarbenkomposition *f*; Falschfarb-Luftbild *n*

2099 false terrace
A planed, horizontal or gently inclined surface on a slope the formation of which was different from the formation of river terraces.
r псевдотерраса
Выровненная, горизонтальная или слабонаклонная поверхность на склоне, не обязанная причинам, вызывающим формирование речных террас.
f pseudo-terrasse *f*; replat *m*
s pseudoterraza *f*; rellano *m*
g Pseudoterrasse *f*

2100 false-bedding; cross-bedding
The lamination, especially in sandstones, deposited under the influence of chainging currents in shallow water bodies parallel to each other for short distances, but inclined at varying oblique angles to the main stratificatioon of the bedding planes.
r ложное залегание
Слоистость, особенно в песчаниках, сформированная под влиянием разных потоков в мелководных бассейнах, параллельная на коротких расстояниях, но лежащая под разными углами к основному напластованию.
f stratification *f* entrecroisée; litage *m* faux
s estratificación *f* cruzada; sedimentación *f* falsa
g falsche Lagerung *f*

2101 fan of wandering
A system of alternating bow-shaped floodplain ridges and depressions which formed due to the shift of a river bed and the growth of meanders.
r веер блужданий
Система дугообразно изогнутых пойменных гряд и разделяющих их понижений, возникших в результате смещения русла и роста меандр.
f éventail *m* de divagation; éventail *m* divagant
s abanico *m* de divagación; abanico *m* divagante
g Fächer *m* der Wanderungen

2102 farewell rock; zeugen; butte; mushroom rock
A small isolated, flat-topped hill, the remnant

Fig. 112. **Farewell rock**

of a higher surface lowered by denudation. → Fig. 112, p. 249

r останец; гора-свидетель
Изолированная возвышенность, остаток некогда более высокой поверхности, сохранившийся после денудации. → Рис. 112, с. 249

f butte-témoin *f*

s cerro-testigo *m*

g Restberg *m*; Auslieger *m*; Zeugenberg *m*

2103 farm; homestead
An individual agricultural enterprise on privately owned or leased land, used for growing crops or raising animals, it includes a living house and ajacent buildings.

r ферма; усадьба
Индивидуальное сельскохозяйственое предприятие, ведущееся на собственой или арендованной земле, включает жилой дом и сопутствующие строения.

f ferme *f*

s granja *f*; finca *f*

g Farm *f*; Bauernhof *m*

* **farm runoff → 135**

* **farming system → 6033**

* **farmland → 136**

2104 farmstead
The land and buildings of a small farm.

r хутор
Земля и постройки небольшой фермы, в России обособленная крестьянская усадьба на собственном земельном участке.

f ferme *f* isolée; hameau *m*

s granja *f* aislada; finca *f* aislada

g Vorwerk *n*; Einzelhof *m*; Gehöft *n*; Einödhof *m*

2105 fast ice; ice foot; landfast ice
Any type of sea, river, or lake ice attached to the shore.

r припай
Лед водоемов и водотоков, примороженный к берегу.

f banquise *f*; glace *f* fixe; pied *m* de glace

s banquisa *f*; pie *m* de hielo

g Standeis *n*; Festeis *n*; Eisfuß *m*

2106 fata morgana
An impressive multiple mirage when distant objects are repeatedly observed with various distortions.

r фата-моргана
Яркий множественный мираж, когда отдаленные предметы видны многократно и с разнообразными искажениями.

f fata Morgana *f*; fée *f* Morgane

s fata Morgana *f*; hada Morgana *f*

g Fata Morgana *f*

2107 fathogram

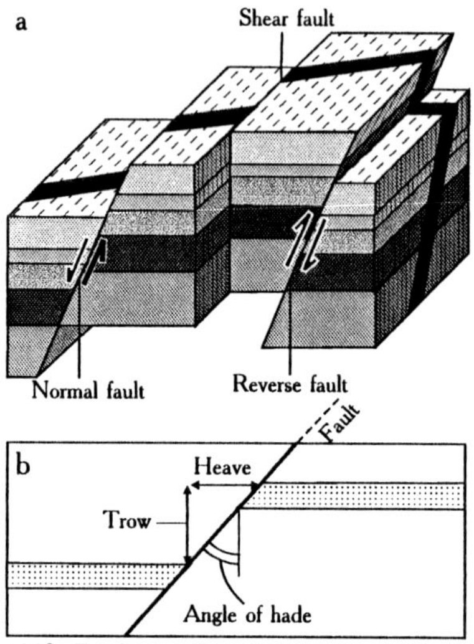

Fig. 113. **Fault**. a – Diagram showing the direction of movement at some types of fault in the rock of the Earth's crust; b – Elements of a fault

The profile of the sea-floor recorded on a chart by an echo sounder in order to ascertain the depth of sea water.

r батиграмма; эхограмма
Нанесенный на карту профиль морского дна, полученный с помощью эхолота.

f fondogramme *m*

s fondograma *f*; curva *f* de profundidades

g Tiefenkurve *f*

* **fathometer → 1788**

2108 fault
A fracture or break in a series of rocks along which there has been vertical or lateral movement or both as a result of exessive strain. → Fig. 113

r разлом
Крупное разрывное нарушение большой протяженности и ширины, прослеживающееся на значительную глубину. → Рис. 113

f faille *f*

s falla *f*

g tectonischer Bruch *m*

2109 fault block
A block of a terrain, often of considerable size, bounded by faults, in many cases resulting from vertical movements in the Earth's crust.

r сбросовый массив
Крупный блок горных пород, ограниченный сбросами, возникший, как правило, в результате вертикальных движений земной коры.

f bloc *m* de faille; bloc *m* faillé

s bloque *m* de falla; bloque *m* fallado

g Bruchblock *m*

2110 fault breccia
A rock composed of shattered angular fragments resulting from movement along a fault.
r сдвиговая брекчия
Горная порода, состоящая из раздробленных угловатых обломков, сформированная в результате движений вдоль разлома.
f brèche *f* de faille
s brecha *f* de falla
g Bruchbrekzie *f*

2111 fault plane
The surface of movement of a fault where material may be shattered and broken off to form fault breccia.
r плоскость сброса
Поверхность движения в разломе, где происходит раздробление материала и образуется сдвиговая брекчия.
f plan *m* de faille
s plano *m* de falla
g Verwerfungsfläche *f*

2112 fault scarp; scarp
A steep face of rock, up to 10 m in height, resulting from the uplift along one side of recently active fault.
r сбросовый уступ
Крутой склон, обычно до 10 м высотой, возникший в результате поднятия вдоль одной из сторон современного разлома.
f escarpement *m* de faille
s escarpe *m* de falla
g Abstufung *f*; Absturz *m*; Bruchstufe *f*

2113 fault zones
Bow-shaped zones in mid-ocean ridges dividing them into series of large blocks displaced relative to one another by many kilometres.
r зоны разломов
Дугообразно изогнутые зоны в срединно-океанических хребтах, разбивающие их на серии крупных блоков, смещенных друг относительно друга на многие километры.
f zones *fpl* des failles
s zonas *fpl* de fallas
g Bruchzonen *fpl*

2114 fauna
The animal life considered collectively which is typical for an area or a geological period.
r фауна; животный мир
Животные, свойственные какой-либо местности или геологической эпохе.
f faune *f*
s fauna *f*
g Fauna *f*; Tierwelt *f*

2115 faunal realm
The largest biogeographical subdivision that contains a fauna of similar origin, age of characteristic animal groups and peculiarities of their distribution, four faunal realms are recognized: Notogaea, Neogaea, Palaegaea and Arctogaea. → Fig. 114
r фаунистическое царство
Наиболее крупное биогеографическое подразделение суши, основанное на сходстве происхождения, древности характерных групп животных и особенностях их распространения, различают четыре фаунистических царства: Нотогея, Неогея, Палеогея и Арктогея. → Рис. 114
f règne *m* faunistique

Fig. 114
Faunal realms

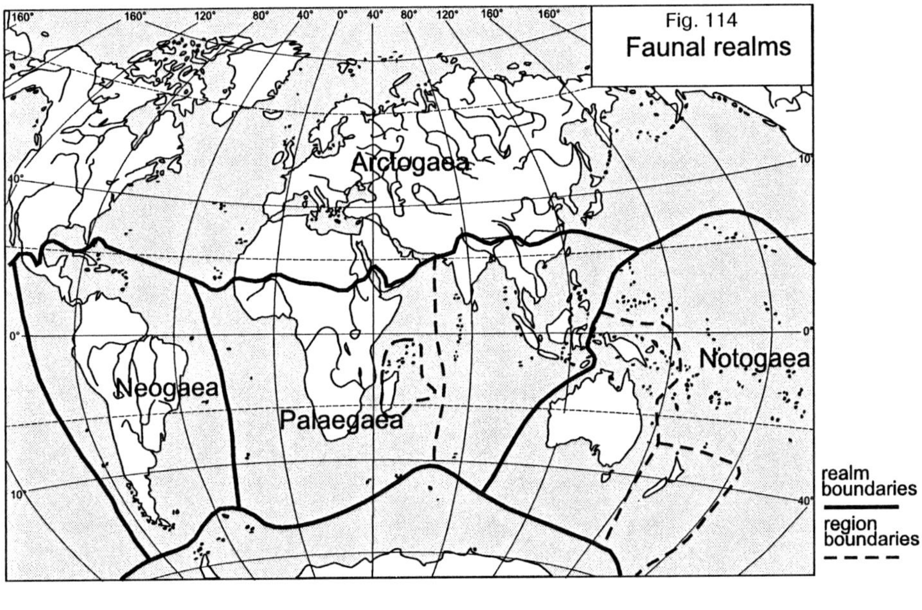

s reino *m* fáunico
g Tierreich *n*

2116 faunal region
A large biogeographical area (land or water) that contains a fauna more or less peculiar to it in terms of its composition and origin, a part of the faunal realm.
r фаунистическая область
Крупный регион земной поверхности (территории или акватории), выделяемый по сходству состава и генезиса фауны, часть фаунистического царства.
f région *f* faunistique
s región *f* fáunica
g Faunengebiet *n*

2117 faunistic complex
A group of animal species that originated in the same region and have similar habitats.
r фаунистический комплекс
Группа видов животных, возникших в одном географическом регионе и имеющих сходные местообитания.
f complexe *m* faunistique
s complejo *m* fáunico
g Faunenkomplex *m*

2118 federal *adj*
Related to the central government of a federation as distinguished from those of constituent units.
r федеральный; федеративный
Относящийся к федеративному государству в целом.
f fédéral *adj*
s federal *adj*; federativo *adj*
g Bundes-; Föderal-

2119 federal district
A district set apart as the seat of a federal government, e.g. District of Columbia (for Washington) in USA.
r федеральный округ
Особая территория, предназначенная для размещения правительства страны, как, например, округ Колумбия близ Вашингтона в США.
f district *m* fédéral
s distrito *m* federal
g Bundesbezirk *m*

* **federal state → 2122**

2120 federal tax
A tax collected by the central authorities for the state budget to cover federal expenses.
r федеральный налог
Налог, взимаемый центральными органами власти в государственный бюджет для покрытия федеральных расходов.
f taxe *f* fédérale
s impuesto *m* federal
g Bundessteuer *f*

2121 federalism
A system of government, a political system with a clear and constitutionally agreed division of powers between the different authorities, each of which is sovereign in its own particular sphere of responsibility.
r федерализм
Форма государственного устройства, предполагающая значительные права у отдельных территориальных образований, существующих в пределах данной страны.
f fédéralisme *m*
s federalismo *m*
g Föderalismus *m*

2122 federation; federal state
A country made up of individual states with the power to make their own decisions, in which the main authoritative control is usually divided between a central or national government and a regional government of federal units (republics, states, cantons, lands) which have a certain degree of independence while are kept together by common federal institutions, citizenship, etc. (examples are Russia and Germany).
r федеративное государство; федерация
Государство, в которое входят федеративные единицы (республики, штаты, кантоны, земли), сохраняющие некоторую долю самостоятельности, но существуют единые федеральные органы, единое гражданство и пр., например, современная Россия или Германия.
f fédération *f*; etat *m* fédéral
s federación *f*; estado *m* federal
g Föderation *f*; Bund *m*

* **feeding by snowmelt → 5506**

2123 feeding ground; sea pastureland
A sea area abundant with food used by fish and other sea animals.
r зона нагула; морское пастбище
Участок моря, богатый пищей, используемой рыбами и другими морскими животными.
f pâturage *m* de mer
s pasto *m* marino
g Seegraswiese *f*

* **feeding of water bodies → 185**

* **felling → 3570**

2124 felling area; cutting area
An area of mature forest intended for felling or already felled (the basic concept of forestry).
r лесосека
Отведенный для вырубки или уже вырубленный участок спелого леса, основное понятие лесного хозяйства.
f coupe *f*
s tallar *m*; monte tallar *m*
g Holzschlag *m*; Kahlschlag *m*

* **fen → 6009**

2125 fen; eutrophic fen
A lowland covered wholly or partly with shallow water with waterloving vegetation which decays to form peat, normally it is formed along the river banks, springs and ponds filled with vegetation.
r низинное болото; евтрофное болото
Болото с богатым минеральным питанием, формирующееся при зарастании водоемов, по берегам рек и в местах выхода ключей.
f marais *m* eutrophe; tourbière *f* basse
s turbera *f* eutrofa; turbera *f* baja
g eutrophes Moor *n*; Niedermoor *n*

* **fenland → 749**

2126 feral *adj*
Applied to a once domesticated animal or to a once cultivated plant which has now established itself in the wild.
r одичавший
Применяется к ранее одомашненным животным или ранее культивированным растениям, которые ныне вернулись к самостоятельной жизни в природе.
f ensauvagè *adj*
s cimarrón *adj*
g verwildert *adj*

2127 ferralitization
The weathering of parent rock in hot and humid climate that results in almost complete decomposition of silicates, release of iron and aluminum oxides and accumulation of kaolinite.
r ферралитизация
Выветривание материнской породы в условиях жаркого и влажного климата, приводящее почти к полному разложению силикатов, высвобождению оксидов железа и алюминия и накоплению каолинита.
f ferralitisation *f*
s ferralitización *f*
g Ferralitisierung *f*

2128 ferrallitic soil; latosol
A red or yellow-red soil with a high content of iron and aluminium in fine silty fraction which forms in equatorial, tropical and subtropical belts.
r ферралитная почва; латосоль
Почва с высоким содержанием железа и алюминия в тонкой илистой фракции, как правило, красного или желто-красного цвета, встречается в экваториальном, тропическом и субтропическом поясах.
f sol *m* ferrallitique
s suelo *m* ferralítico
g ferrallitischer Boden *m*

2129 ferralsol
A thick, greatly weathered soil of tropical and humid climate with heavy texture and low absorption capacity of clays due to the presence of kaolinite, common mainly in Africa and Brazil.
r ферральсоль
Мощная, сильно выветрелая почва тропического и влажного климата с тяжелым механическим составом и низкой поглотительной способностью глин из-за присутствия каолинита, распространена в основном в Африке и Бразилии.
f ferralsol *m*
s ferralsol *m*
g Ferralsol *m*

2130 ferrolysis
The process of releasing, migration and decomposition of iron compounds and clayed minerals in acid medium under periodical water logging of soils.
r ферролиз
Процесс высвобождения, перемещения и разрушения соединений железа и глинистых минералов в условиях кислой среды и периодического переувлажнения почв.
f ferrolyse *f*
s ferrolisis *f*
g Ferrolysis *f*

* **ferrous metallurgy → 5791**

2131 ferruginous crust; laterite
A continuous horizon consisting of residual products of rock decay which are strongly compacted and cemented with residual or input sesquioxides of iron and/or aluminium (seldom of magnesium).
r железистый панцирь; латеритный панцирь
Сплошной горизонт, состоящий из продуктов выветривания горных пород или отложений, сильно уплотненных и сцементированных остаточными или привнесенными окислами железа и/или алюминия (реже магния).
f cuirasse *m* ferrugineuse; cuirasse *m* latéritique
s coraza *f* ferruginosa; costra *f* ferralítica; coraza *f* laterítica; costra *f* ferruginosa
g Lateritpanzer *m*

2132 ferruginous soil
A zonal soil developed in warm temperate climates without a dry season or in the tropical savanna and bushlands.
r железистая почва
Зональная почва, развивающаяся в теплом умеренном климате, где отсутствует сухой сезон, либо в саванне или кустарниковом сообществе.
f sol *m* ferrugineux
s suelo *m* ferruginoso
g eisenhaltiger Boden *m*

2133 ferry
A floating construction or a vessel used to carry people or goods across a body of water (e.g. a river).
r паром

Плавучее сооружение или судно для регулярной переправы сухопутных транспортных средств, людей и грузов через водную преграду.
f bac *m*
s transbordador *m*
g Fähre *f*

2134 fersiallitic soil
A soil coloured red, devoid of carbonates, enriched with bases and formed as a result of rubefaction, it is characteristic of Mediterranean climate.
r ферсиаллитная почва
Почва, характерная для средиземноморского климата и формирующаяся в результате рубефикации, имеет красный цвет, лишена карбонатов и обогащена основаниями.
f sol *m* fersiallitique
s suelo *m* fersialítico
g fersiallitischer Boden *m*

* **fertility → 713**

* **fertility of soil → 5608**

2135 fertilization
The process of application of special natural or synthetic substances to soil to conserve or to increase its fertility.
r удобрение почвы
Мероприятия по сохранению или увеличению продуктивности почвы путем внесения специальных природных или синтетических веществ.
f fertilisation *f*; apport *m* des engrais
s abonado *m*; aporte *m* de abonos
g Düngung *f*; Düngen *n*

2136 fertilizer
Chemical or natural substance which is added to the soil to enrich it and make it more fertile.
r удобрение
Химическое или природное вещество, которое вносят в почву для увеличения ее плодородия.
f engrais *m*; fertilisant *m*
s abono *m*; fertilizante *m*
g Dünger *m*

2137 fertilizing soil with clay; marling
A method of improvement of water-retaining properties of sandy soils by way of applying clay into them, it is usually used in irrigated oases to decrease the irrigation water discharge.
r глинование почвы
Метод улучшения водно-физических свойств песчаных почв путем внесения в них глины, обычно применяется в орошаемых оазисах и приводит к снижению расходов оросительных вод.
f argilisation *f* du sol
s arcillado *m* del suelo
g Lehmdüngung *f*

2138 fetch; length of wave run-out
The distance that a sea wave has travelled from its initiation to the coast where it breaks.
r область разгона волн; разгон волны; длина разгона волн
Расстояние между местами зарождения ветровой волны и ее полного развития при заданной скорости ветра.
f aire *f* de génération des vagues
s área *f* de generación de ondas
g Wellenanlauflänge *f*

* **fiard → 2169**

2139 fibric layer
A layer of fibrous soil material in which the weakly decomposed organic material can be easily recognized and the separate species identified.
r волокнистый слой
Слой волокнистого почвенного материала, в котором легко различимы разлагающиеся органические остатки и отдельные экземпляры растений.
f couche *f* fibreuse
s capa *f* fibrosa
g Faserschicht *f*

2140 field inspection of aerial photograph
Comparison of aerial photographs with the real natural landscape directly in the field.
r полевое дешифрирование
Сравнение аэрофотоснимков с натурой непосредственно в полевых условиях.
f interprétation *f* de photo aérienne sur la terrain
s interpretación *f* de foto aérea sobre el terreno; restitutión *f* de fotos aéreas sobre el terreno
g Luftbildauswertung *f* im Feld

2141 field sketching
A simplified topographic survey made by simplest instruments for getting sketch plan of an area.
r глазомерная съемка
Упрощенная топографическая съемка с помощью простейших приборов для получения приближенного плана местности.
f levé *m* visuel; levé *m* à vue
s levantamiento *m* visual
g Krokieren *n*

2142 fill
Accumulation in a channel of loose material brought by the river due to the decreased load transportation capacity at a certain part of the stream.
r намыв
Аккумуляция наносов в русле из-за пониженной транспортирующей способности реки на данном участке русла.
f alluvion *f*
s aluvión *f*

g Ablagerung *f*

* **filling → 2144**

2143 filling in of a lake; plantgrowing of waterbody; vegetation infill of a lake
Gradual filling of the volume or only the surface of a lake or reservoir with macroscopic water plants (macrophytes).
r зарастание водоема
Постепенное заполнение объема или только поверхности водоема макроскопическими водными растениями (макрофитами).
f couverture *f* de plantes
s cubrimiento *m* de plantas
g Verlandung *f* eines Gewässers

2144 filling of cyclone; filling; filling of depression
The increasing atmospheric pressure near the centre of a depression, which ultimately leads to its disappearance (opposite to *deepening*).
r заполнение циклона
Увеличение атмосферного давления в центре циклона, что в конце концов приводит к его исчезновению (противоположный процесс носит название *углубление циклона*).
f remplissage *m* de dépression
s llenado *m* de depresión
g Auffüllung *f*

* **filling of depression → 2144**

2145 filling up of a water body
Gradual filling of a water body or its surface with vegetation, it is accompanied by the formation of a floating mat, accumulation of sapropel and extension of the coastal belt of macrophytes.
r зарастание водоема
Постепенное заполнение всего объема или поверхности водоема растительностью, происходит с образованием сплавины, накоплением сапропеля и расширением прибрежной полосы макрофитов.
f encombrement *m* de lac par végétaux aquatiques
s cubrimiento *m* de agua por vegetales acuáticos
g Zuwachsen *n* eines Gewässer

2146 film water; pellicular water
A thin layer of water surrounding soil particles above hygroscopic water and not available for a plant growth.
r пленочная вода
Тонкий слой воды, обволакивающий частицы почвы или грунта поверх гигроскопической воды и недоступный для использования этой воды растениями.
f eau *f* pelliculaire
s agua *f* pelicular
g Haftwasser *n*

* **filtration beds → 2147**

2147 filtration fields; filtration beds
Territory intended for natural biological treatment of sewage water through the natural purification of organic substances.
r поля фильтрации
Территория, предназначенная для биологической очистки сточных вод путем естественного обезвреживания органических соединений.
f champs *mpl* de filtration; lits *mpl* filtrants
s campos *mpl* de estercolamiento; campos *mpl* de filtraje
g Rieselfelder *npl*

2148 financial resources
Money stock which could be invested in the development of production or used for other economic and social purposes.
r финансовые ресурсы
Фонды денежных средств, которые могут вложены в развитие производства или для достижения экономической или социальной целей.
f ressources *fpl* financières
s recursos *mpl* financieros
g Finanzressource *f*

2149 fine material; fine-grained sediment
Soil and rock particles less than 1 mm in size.
r мелкозем
Частицы почв и горных пород размером менее 1 мм.
f sable *m* fin; roche *f* à grain fin
s arena *f* fina; roca *f* de grano fina
g feiner Grund *m*; feine Erde *f*

2150 fine-grained *adj*; small-grained *adj*
Consisting of granules and particles less than 1 mm in size.
r мелкозернистый
Состоящий из гранул и зерен размером менее 1 мм.
f à grain fin
s de grano fino; microgranulado *adj*
g feinkörnig *adj*

* **fine-grained sediment → 2149**

2151 finger lake; trough lake
A narrow, linear body of water occupying a glacially overdeepened valley and sometimes impounded by a morainic dam.
r пальцеобразное озеро; троговое озеро
Узкий, линейно вытянутый водоем, занимающий переуглубленную гляциальную долину и иногда подпруженный моренным валом.
f lac *m* digitiforme
s lago *m* digitiforme de artesa glaciar
g fingerförmiger See *m*

2152 Finiglacial

The final phase of the retreat of the last glaciation in Europe during which the glaciation did not come beyond the limits of Finland.

r финигляциал
Заключительная фаза сокращения последнего оледенения Европы, в течение которой ледник не выходил за пределы Финляндии.

f stade *m* finiglaciaire
s estadio *m* finiglaciar
g Finiglazial *n*

2153 finished product; end-product
The final article produced by the manufacturing process for industrial or personal consumption.

r готовая продукция
Продукция, выпущенная предприятием или отраслью материального производства, предназначенная для производственного потребления, накопления и личного потребления.

f production *f* finie
s producción *f* final; producto *m* final; bienes *mpl* de consumo finales
g fertige Produktion *f*

* **firm → 1916**

2154 firn; nevé
A porous grainular ice mass, transitional phase between snow and glacier ice.

r фирн
Пористая зернистая ледяная порода, переходная форма между снегом и ледниковым льдом.

f névé *m*
s neviza *f*
g Firn *m*

* **firn basin → 2155**

2155 firn field; firn basin
A widened part of a mountain glacier lying above the firn line.

r фирновое поле; фирновый бассейн
Расширенная часть горного ледника, лежащая выше фирновой линии.

f champ *m* de névé
s campo *m* de neviza
g Firnbecken *n*

2156 firn line
The lower boundary of the firn field on a glacier.

r фирновая линия
Нижняя граница области фирнового питания на леднике.

f ligne *f* de névé
s línea *f* de neviza
g Firngrenze *f*

2157 firnification
The transformation of snow into firn.

r фирнизация
Превращение снега в фирн.

f névéfication *f*
s nevización *f*
g Verfirnung *f*

2158 first-year ice; white sea ice; one-year ice
Floating ice of not longer than one winter age.

r однолетний лед; белый лед
Плавучий лед, просуществовавший не более одной зимы.

f glace *f* de l'année; glace *f* de l'hiver
s hielo *m* de un año; hielo *m* de invierno
g einjähriges Meereis *n*; Wintereis *n*

* **fish breeding → 4578**

2159 fish farm
A farm where fish are bred under carefully controlled conditions in tanks or in protected parts of coastal waters, lakes, rivers and estuaries.

r рыбная ферма
Ферма, где в тщательно контролируемых условиях на продажу разводят рыбу, используя специальные емкости, а также защищенные части прибрежных вод, озера, реки и эстуарии.

f ferme *f* de poisson
s granja *f* de pesca; granja *f* pesquera
g Fischfarm *f*; Fischzuchtbetrieb *m*

* **fish farming → 4578**

* **fishery → 2160**

2160 fishing; fishery
Catching fishs, marine mammals, molluscs and marine plants for commercial purposes (especially well-developed in Japan, China, Russia, the U.S., Chile, and Norway).

r рыболовство
Отрасль хозяйства по добыче рыбы, морских млекопитающих, моллюсков, водной растительности, особенно развито в Японии, Китае, России, США, Чили, Норвегии.

f pêche *f*; pêcherie *f*
s industria *f* pesquera; pesquería *f*
g Fischerei *f*; Fischfang *m*

2161 fissure; joint; fracture
A crack in a rock following a dominant direction along the line of weakness, but without dislocation.

r трещина
Разрыв сплошных горных пород без смещения разобщенных частей относительно друг друга.

f fissure *f*; fracture *f*
s fisura *f*; fractura *f*
g Spalt *m*; Kluft *f*; Spalte *f*; Riss *m*; Schrund *m*

2162 fissure eruption; linear eruption
A steady, non-explosive outpouraing of lava (mainly basaltic lava) along linear fissures of the Earth's crust or through a large crack rather than through a pipe or vent.

r линейное извержение; трещинное извержение
Излияния жидкой, главным образом, базальтовой лавы по линейным трещинам земной коры.
f éruption *f* linéaire; éruption *f* fissurale
s erupción *f* lineal; erupción *f* fisural
g Spalteneruption *f*

2163 fissure volcano
A volcano sitting on a fracture in the Earth's crust along which the eruptions occur.
r трещинный вулкан
Вулкан, находящийся на трещине в земной коре, вдоль которой нередко и происходят извержения.
f volcan *m* fissural
s volcán *m* fisural
g Spaltenvulkan *m*

2164 fissuring; fracturing
Totality of fissures dissecting rocks.
r трещиноватость
Совокупность трещин, рассекающих горные породы.
f fissurité *f*
s fisuración *f*; fracturidad *f*
g Klüftung *f*

2165 fiumare
Small rivers in the Mediterranean which dry up in summer and have rough floods in the winter rainy season.
r фиумары
Небольшие реки в Средиземноморье, пересыхающие летом и имеющие бурные паводки в период дождей зимнего полугодия.
f fiumare *m*
s fiumara *f*; rambla *f*
g Fiumara *f*

2166 fixation
The process occurring in a soil whereby certain chemical elements essential for plant growth are converted from an exchangeable form to less soluble and non-exchangeable forms.
r фиксация
Почвенный процесс, при котором некоторые химические вещества, существенные для роста растений, становятся мало растворимыми, не способными к изменениям.
f fixation *f*
s fijación *f*
g Fixierung *f*

2167 fixation of sands
Preventing the mobility of shifting sand by planting vegetation or by way of mechanical means in extra-arid conditions.
r закрепление песков
Борьба с подвижностью песков с помощью посадок растительности, а в экстрааридных условиях – механическими способами.
f fixation *f* de sable
s fijación *f* de arenas
g Verfestigung *f* von Sand; Sandbefestigung *f*

* **fixed point → 2406**

* **fixed water → 450**

2168 fixism
A group of tectonic hypotheses based on the ideas of stability of continents and the importance of the vertical tectonic movements.
r фиксизм
Группа тектонических гипотез, основанных на представлениях о незыблемости континентов на поверхности Земли и о решающей роли вертикальных тектонических движений.
f fixisme *m*
s fijismo *m*
g Fixismus *m*

2169 fjard; fiard
A relatively narrow inlet of the sea in low rocky coasts being a part of half-drowned glaciated lowland but without the deep glacial troughs of a fjord.
r фиард
Сравнительно узкий залив низменного скалистого побережья, представляющий собой часть полузатопленной ледниковой долины.
f fjärd *m*
s fjard *m*
g Fjärd *m*

2170 fjel; fill
A plateau-like top surface of a mountain massif in Scandinavia covered by glaciers or tundra.
r фьельд
Платообразная вершинная поверхность горных массивов в Скандинавии, покрытые ледниками или тундрой.
f fjell *m*; fjeld *m*
s fjell *m*; fjeld *m*
g Fjeld *m*

2171 fjord
A long narrow and deep sea inlet, sometimes branching, with high rocky parallel sides formed by the submergence of glacially overdeepened valleys due to a rising sea-level after the melting of the Pleistocene ice sheets. → Fig. 115

Fig. 115. **Fjord**

r фьорд
Длинные, узкие и глубокие, иногда ветвящиеся заливы и проливы, расчленяющие гористые побережья областей покровного оледенения, воникшие в результате поднятия уровня моря после стаивания плейстоценовых ледниковых покровов. → Рис. 115, с. 257
f fjord *m*
s fiordo *m*; fjordo *m*; fjord *m*
g Fjord *m*

2172 flank moraine; side moraine
Rock material transported by a glacier at a lateral part of its tongue. → Fig. 85, p. 182
r боковая морена
Морена, переносимая ледником в боковой части языка. → Рис. 85, с. 182
f moraine *f* de flanc
s morrena *f* de flanco
g Randmoräne *f*; Ufermoräne *f*

2173 flash-flood
A sudden violent flood caused by exceptionally heavy rain in a normally dry valley in a semi-arid area and other factors, the torrential stream sometimes being laden with debris.
r внезапный паводок
Неожиданный резкий подъем воды на реке, вызванный исключительно сильным дождем в обычно сухой семиаридной местности и иными причинами, нередко переносящий большое количество обломочного материала.
f crue *f* d'averse
s crecida *f* de chubasco
g plötzliches Hochwasser *n*; plötzliche Überschwemmung *f*

* **flash-freezing → 1465**

* **flashy flow → 6257**

* **flat → 4582**

2174 flat bank
A coast with a very gentle wide underwater slope.
r отмелый берег
Берег, имеющий очень пологий и широкий подводный склон.
f rivage *m* à pente douce
s ribera *f* de banco de arena
g flache Küste *f*

2175 flat interfluve
A flat or slightly inclined watershed area which separates two drainage areas.
r плакор
Плоское или слабонаклонное приводораздельное пространство, разделяющее два отдельных речных бассейна.
f interfluve *m* plat
s interfluvio *m* plano
g flaches Zwischenstramland *n*

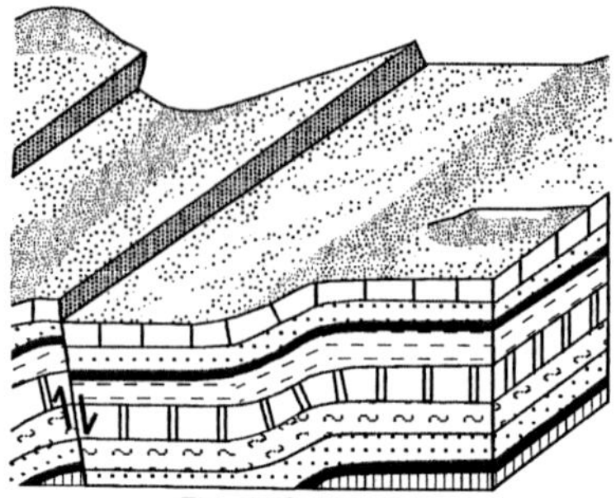
Fig. 116. **Flexure**

* **flat-bottom steppe depressions → 4371**

2176 flattering of the flood wave
The decrease of height and increase of length of a flood wave as well as the decrease of the maximum water discharge downstream during the period of flood.
r распластывание паводка; трансформация паводка
Уменьшение высоты и увеличение длины волны паводка, а также уменьшение максимального расхода воды вниз по реке за время прохождения паводка.
f aplatissement *m* de crue
s aplanamiento *m* de crecida
g Verflachung *f* der Hochwasserwelle

2177 flat-top glacier
A slightly convex glacier covering the gently sloping individual tops and ridges.
r ледник плоской вершины
Ледник в форме слегка выпуклого купола, покрывающий отдельные слабонаклонные вершины и гребни.
f glacier *m* de sommet plat
s glaciar *m* de cumbre plano
g Plattberggletscher *m*

* **flaw lead → 5408**

2178 flexure
A step-like curve of horizontally bedded deposits, i.e. a type of fold where bedded rocks have slipped over each other. → Fig. 116
r флексура
Ступенеобразный изгиб горизонтально залегающих отложений, т.е. тип складки, когда пласты пород сползли один над другим. → Рис. 116
f flexure *f*
s flexura *f*; flexión *f*
g Flexur *f*

2179 flint
A very hard, grey or grey-black quarz material (a variety of silica) found especially in chalk or limestone, which has a widespread use by prehistoric man of the Stone Age who manufactured many tools and weapons, when struck with steel it emits a spark, hence its use in a flintlock gun.

r кремень
Очень твердый, серый или темносерый кварцевый материал (разновидность кремнезема), особенно часто встречающийся в меловых породах и известняке, широко использовавшийся доисторическими людьми в каменном веке, которые изготавливали из него орудия труда и защиты.
f silex *m*
s sílex *m*; pedernal *m*; sílice *m*
g Kiesel *m*; Flint *m*; Kieselstein *m*

* **float → 816**

2180 floatable river; floatable way
River used for logging timber by scattering without making rafts.
r сплавная река; сплавной путь
Река, используемая для сплава леса россыпью, без сплачивания в плоты.
f rivière *f* flottable; voie *m* de flottage
s río *m* flotable; vía *f* flotable; vía *f* de flotamiento
g flößbarer Fluß *m*; Flößbarstraße *f*

* **floatable way → 2180**

* **floatage → 817**

2181 floating; rafting$_3$
Transportation of timber along rivers by using its floating capacity.
r лесосплав
Транпортировка лесоматериалов по воде с использованием плавучести древесины.
f flottage *m*
s conducción *f* de maderadas; flotamiento *m*
g Flößerei *f*; Holzflößen *n*

2182 floating bog; floating mat
A layer of water and swamp vegetation growing from the shore to the center of a water body as well as a layer of slightly decomposed peat emerged to the surface.
r сплавина; зыбун
Слой водной и болотной растительности, нарастающий со стороны берега к центру водоема; всплывший на поверхность слой слаборазложившегося торфа, толщиной до 2 м.
f marais *m* flottant
s ciénaga *f* flotante
g schwimmende Scholle *f*

* **floating glacier → 2183**

2183 floating glacier tongue; floating glacier
The frontal part of an outlet glacier advancing into the sea and being afloat.
r плавающий ледниковый язык
Выдвигающаяся в море и находящаяся на плаву часть выводного ледника.
f langue *f* de glacier flottant
s lengua *f* de glaciar flotante
g schwimmende Gletscherzunge *f*

2184 floating ice; drifting ice; drift-ice
Ice blocks or ice fields floating in river or sea water by wind or current.
r плавучий лед; дрейфующий лед
Всякий лед, плавающий на воде и находящийся под воздействием ветров и течений.
f glace *f* flottante; glace *f* de dérive
s hielo *m* flotante; hielo *m* a la deriva; deriva *f*
g schwimmendes Eis *n*; Drifteis *n*; Treibeis *n*

2185 floating ice studies
The study of origin, growth and melting of sea ice, lacustrine and river ice, its physical and mechanical properties and possibilities of artificial destruction.
r ледоведение водоемов и водотоков
Изучение возникновения, развития и исчезновения морских, речных и озерных льдов, их физико-механических свойств и возможностей искусственного разрушения.
f étude *f* des glaces flottantes
s estudio *m* de hielos flotantes
g Eiskunde *f* von Gewässen

* **floating layer → 4856**

* **floating mat → 2182**

2186 floating snow
Loose, coarse-grained snow with vertical pillar texture including numerous crystals of depth hoar.
r снег-плывун
Рыхлый крупнозернистый снег с вертикально-столбчатой текстурой, включающий множество кристаллов глубинной изморози.
f neige *f* coulante; neige *f* flottante
s nieve *f* movediza; nieve *f* flotante
g Schwimmschnee *m*; Tiefenreif *m*

2187 flocculation
The process in soils by which fine-grained particles come into contact and gather together to form crumbs, thereby improving the soil texture.
r аггрегирование
Процесс в почвах, при котором мелкозернистые частицы соединяются и формируют крупитчатую текстуру, в результате чего заметно улучшается строение почвы.
f floculation *f*
s floculación *f*
g Flockung *f*

2188 flock
A group of birds or insects feeding and travelling together.
r стая
Группа птиц или насекомых, обитающих на какой-либо территории и совершающих совместные перелеты.
f volée *f*
s banda *f*

g Schwarm *m*; Vogelschwarm *m*

* **flock of sheep → 6762**

2189 floe; ice-cake; ice-pan; ice-floe
A fragment of floating ice of tens to hundreds meters across, sometimes it is smaller than 1 m in diameter.
r льдина
Обломок плавучего льда размером от десятков или сотен метров до долей метра в поперечнике.
f glaçon *m* flottant; floe *m*; banc *m* de glace
s banco *m* de hielo; floe *m*
g Eisscholle *f*

2190 floeberg
A mass of hummocked ice formed by the piling up of many ice floes by lateral pressure.
r несяк
Большой торос или группа смерзшихся торосов, образующих отдельную льдину, находящуюся на плаву.
f floeberg *m*
s floeberg *m*
g Scholleneisberg *m*; Packeisberg *m*

* **flood → 3182**

2191 flood; high water; spate
The rise of water level in a river as a result of quick snowmelt while thaw, heavy rains, release of reservoirs, etc.
r паводок; половодье
Поднятие уровня воды в реке в результате быстрого таяния снега при оттепели, обильных дождей, сработки водохранилища.
f crue *f*
s crecida *f*
g Hochwasser *n*

2192 flood control
The system of measures to prevent inundations of the land by the building of river barrages, the constructing of sea walls to prevent inundations by exceptionally high tides, etc.
r борьба с паводками
Система мер по предупреждению возможных наводнений путем создания временных водохранилищ, построения отбойных стенок для предохранения от высоких приливов и т.п.
f contrôle *m* des crues
s control *m* de crecidas
g Flutkontrolle *f*

2193 flood flow
Water volume flowing through a cross section of a stream during the flood period of high water.
r паводочный сток; максимальный сток
Объем воды, протекающей через живое сечение реки за время паводка.
f écoulement *m* de crue; débit *m* maximum
s escorrentía *f* de crecida
g Hochwasserabfluß *m*; Höchstabfluß *m*

* **flood tide → 2768**

2194 flood wave
Wave appearing by a sharp rise of water level while passing flood and high water.
r паводочная волна; волна паводка
Волна, возникающая при резком подъеме уровня воды во время прохождения половодья и паводков.
f flot *m* de crue
s oleada *f* de crecida
g Hochwasserwelle *f*

2195 flood zone; inundation zone
A land temporarily covered with water when water discharge exceeds a channel capacity and the river overflows, or when a dam is built and reservoir is created.
r зона затопления
Территория, временно покрываемая водой, когда расход воды превышает пропускную способность русла и вода выходит из берегов, или оказавшаяся под водой в результате строительства плотины и образования водохранилища.
f zone *f* d'inondation
s zona *f* de inundación; zona *f* anegada
g Hochwassergebiet *n*; Überflutungsgebiet *n*

2196 flood-control storage capacity
A part of a reservoir capacity intended to be filled with water at maximum inflow which happens seldom.
r резервный объем водохранилища
Часть емкости водохранилища, предназначенная для заполнения водой при максимальном притоке, случающемся очень редко.
f volume *m* en réserve de réservoir
s volumen *m* de reserva de embalse
g eiserner Bestand *m* des Stausees

2197 floodplain
The relatively level part of a river valley adjacent to the channel over which a river flows in times of floods or high water and which is covered with contemporary alluvium.
r пойма
Относительно ровная часть дна речной долины, выстланная современным аллювием, затопляемая в половодье или во время паводков.
f plaine *f* inondable; lit *m* majeur; plaine *f* d'inondation
s llanura *f* de inundación; lecho *m* mayor
g Hochwasserbett *n*; Aue *f*; Flußniederung *f*

2198 floodplain deposits
Alluvial sedimentary material deposited on a flood-plain in times of a highwater.
r пойменные отложения
Аллювиальные отложения поймы, отлагающиеся здесь рекой во время половодья.

f dépôts *mpl* alluviaux du lit majeur
s depósitos *mpl* aluviales de lecho mayor
g Überschwemmungsablagerungen *fpl*; Auenablagerungen *fpl*

2199 floodplain meadow; water meadow
A zone of grassland along the floodplain of a river where the growth of herbage is stimulated by perodic flooding, by either natural and artificial means.
r заливной луг; займище; пойменный луг
Луговая полоса в пойме реки, ежегодно затопляемая весенними полыми водами, способствующими, наряду с другими природными и антропогенными воздействиями, обильному росту трав.
f prairie *f* inondable
s prado *m* inundable; anegadizo *m*; pradera *f* inundable
g Auenwiese *f*; Aue *f*; Überschwemmungswiese *f*; Flussau *f*

2200 floodplane alluvium; overbank deposits
Alluvial deposits of a river flood-plain forming during high water.
r пойменные отложения
Аллювиальные отложения речной поймы, формирующиеся здесь во время половодья, годовой слой таких отложений носит название наилок.
f dépôts *mpl* alluviaux du lit majeur
s depósitos *mpl* aluviales de lecho mayor
g Überschwemmungsablageungen *fpl*; Auenablagerungen *fpl*

2201 floodplane island
An island created in the floodplain of a meandering river when a narrow meander neck gets broken by waters of the stream.
r пойменный остров
Остров в пойме меандрирующей реки, возникший при прорыве узкой шейки излучины.
f île *f* en plaine inondable
s isla *f* en llanura inundable
g Aueninsel *f*

* **floor → 762**

* **floor of a valley → 6521**

2202 flora
The plant life considered collectively which is typical for an area or geological epoch.
r флора
Совокупность видов растений, свойственных какой-либо местности или геологической эпохе.
f flore *f*
s flora *f*
g Flora *f*

2203 floriculture
Production of ornamental plants, flowers and pot plants.
r цветоводство
Выращивание декоративных растений для парков, цветов на срезку, растений в горшках для украшения помещений.
f floriculture *f*; culture *f* des fleurs
s floricultura *f*
g Blumenzucht *f*

2204 floristic element
A group of plant species having similar origin, natural habitat and the time when they started to inhabit a particular territory.
r флористический элемент
Группа видов растений, объединяемых по сходству их ареалов, времени появления на данной территории и по происхождению.
f élément *m* floristique; cortège *m* floristique
s elemento *m* florístico; cortejo *m* florístico
g Florenelement *n*

2205 floristic kingdom; floristic realm
The largest biogeographical subdivision identified on the basis of the peculiarities of its floristic composition and distribution of plant especies, Holarctic (Holarctica), Palaetropical, Neotropical, Cape, Australian, Holantarctic floral kingdoms are recognized. → Fig. 117, p. 262
r флористическое царство
Наиболее крупное биогеографическое подразделение, выделяемое по особенностям флористического состава и видов растений, различают Голарктическое (Голарктика), Палеотропическое, Неотропическое, Капское, Австралийское, Голантарктическое царства. → Рис. 117, с. 262
f règne *m* floristique
s reino *m* florístico
g Florenreich *n*

2206 floristic province
Biogeographical subdivision identified on the basis of the peculiarities of floristic composition, particularly with regard to the presence of endemic species, a part of the floral kingdom, there are about 150 floristic provinces on the Earth.
r флористическая провинция
Биогеографическое подразделение, основанное на особенностях флористического состава, в первую очередь на присутствии эндемичных видов, часть флористического царства, на Земле известно около 150 флористических провинций.
f province *f* floristique; domaine *m* floristique
s provincia *f* florística
g Florenprovinz *f*

* **floristic realm → 2205**

2207 flotation

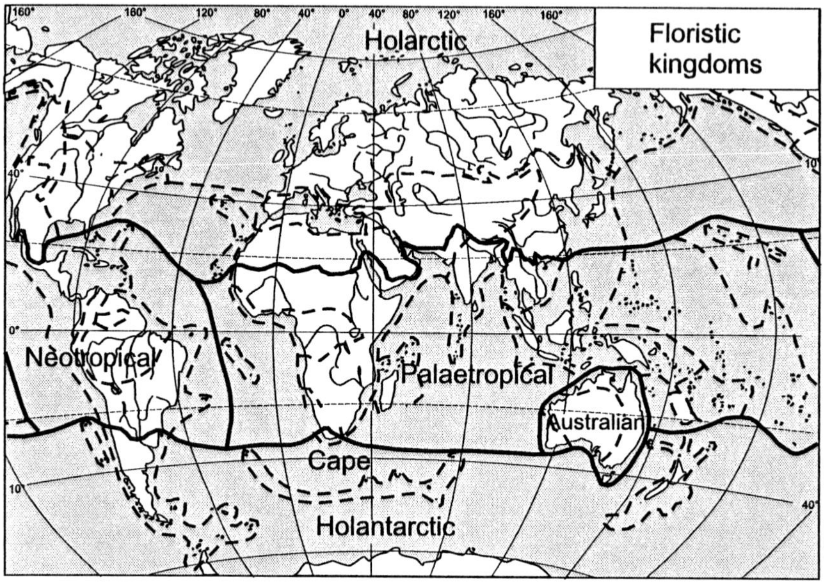

Fig. 117

kingdom boundaries

region boundaries

A method of extraction of insoluble impurities, molecules and ions from aqueous solutions by rising to the surface air bubbles adsorbed on their surface.

r флотация
Извлечение из воды нерастворимых примесей всплывающими пузырьками воздуха, адсорбирующимися на их поверхности.

f flottation *f*
s flotación *f*
g Flotation *f*

2208 flow$_1$
The smooth movement of a fluid or plastic mass of rocks with a continuous change in shape, e.g. flow of lava, or of a stream confined between banks, or of solid rocks under stress, or of ice in a glacier.

r поток
Медленное движение жидкости или пластичных горных пород, сопровождающееся непрерывным изменением формы движущейся массы, например, течение магмы, водного потока, ограниченного берегами, горных пород под большим давлением, льда в ледниках.

f flux *m*
s flujo *m*
g Strom *m*

* **flow$_2$ → 1394, 5148**

* **flow line → 5859**

2209 flow loss
Difference between the amount of precipitation in a certain territory and the amount of water run off from this area through rivers.

r потери стока
Разность между количеством осадков, выпаших на данной территории, и количеством воды, стекшей с этой площади через реки.

f déficit *m* d'écoulement;
déficit *m* de escorrentía
s déficit *m* de desagüe
g Abflußverlust *m*

2210 flowing avalanche
An avalanche flowing over the ground or another snow layer and having dense core at its base.

r текучая лавина
Снежная лавина, движущаяся по грунту или снежному слою с плотным ядром в основании.

f avalanche *f* fluide
s avalancha *f* de flujo
g Fliesslawine *f*

* **flowstone → 5451**

2211 fluctuation of glacier;
glacier fluctuations
The changes in dimensions and shapes of glaciers usually associated with glacier advance or glacier retreat.

r колебания ледника
Изменения размеров и формы ледников, обычно связанные с их наступанием или отступанием.

f fluctuation *f* glaciaire;
oscillation *f* glaciaire
s fluctuación *f* glaciar; oscilación *f* glaciar
g Gletscherschwankungen *fpl*

2212 fluctuations of climate; climatic fluctuations
Quasi-periodic or cyclic variations of climate.
r колебания климата
Квазипериодические или циклические изменения климата.
f fluctuations *fpl* de climat
s fluctuaciones *fpl* de clima
g Klimaschwankungen *fpl*

2213 fluidity of ice
A property of ice to undergo plastic deformations under permanent stress.
r текучесть льда
Способность льда пластически деформироваться при неизменяющейся нагрузке.
f fluidité *f* de la glace
s fluidez *f* de hielo
g Fliessfähigkeit *f* des Eises

2214 fluidization
A process in volcanic activity by which hot gases, moving through fine-grained material, cause it to flow, thereby cracking and enlarging existing cracks in the rock by physical and chenical action.
r флюидизация
Процесс, сопровождающий вулканические извержения, когда горячие газы, двигаясь через мелкозернистый материал, заставляют его течь, разбивают на трещины или увеличивают уже существующие трещины в горной породе путем физического и химического воздействия.
f fluidification *f*
s fluidificación *f*
g Fluidisation *f*

2215 flume; conducting channel; conduit
A construction for passing water to the place of its consumption.
r водовод
Сооружение для пропуска воды к месту ее потребления.
f canal *m* d'amenée
s canal *m* de conducción
g Wasserleitung *f*

2216 flush
A sudden rise of water in a stream, creating a rush of water that does not quite overflow the banks.
r приток воды
Внезапное увеличение объема потока, вызывающее подъем уровня воды в реке, которая, однако, не выходит из берегов.
f afflux *m*
s aflujo *m*
g Zufluß *m*

2217 flushed soil
A soil enriched by transported materials, either dissolved mineral salts or rock particles.
r обогащенная почва
Почва с большим содержанием принесенного материала в виде растворенных минеральных солей или отдельных минеральных частиц.
f sol *m* enrichi
s suelo *m* enriquecido
g bereicherter Boden *m*

* **fluted moraine → 2218**

2218 fluting-moraine; fluted moraine
Parallel morainic ridges, several meters high and up to 100 m wide, linearly extended in the direction of a glacier movement.
r флютинг-морена
Линейно вытянутые в направлении движения ледника параллельные моренные гряды высотой в несколько метров и шириной до 100 м.
f fluting-moraine *f*
s fluting-morrena *f*
g Fluting-Moräne *f*

* **fluvial abrasion → 6636**

2219 fluvial *adj*; riverine *adj*; fluviatile *adj*
Applied to rivers, river action, to organisms which are found in rivers and to anything produced by river action.
r речной; флювиальный
Относящийся к реке или речным процессам, к организмам живущим в реках, а также ко всему, что служит результатом воздействия рек.
f fluvial *adj*
s fluvial *adj*
g Fluß-; fluvial *adj*; Strom-

* **fluvial deposits → 207**

2220 fluvial hydraulics
Water motion in river flows, transport of sediments and the processes of channel formation.
r речная гидравлика
Движение воды в речных потоках, перемещение ими наносов и процессы формирования русла.
f hydraulique *f* fluviale
s hidráulica *f* fluvial
g Flußhydraulik *f*

2221 fluvial landform
A landform created by the action of temporary or permanent streams.
r флювиальный рельеф
Формы рельефа, создаваемые деятельностью постоянных и временных водных потоков.
f relief *m* fluvial; modelé *m* fluviatile
s relieve *m* fluvial; modelado *m* fluvial
g fluviales Relief *n*

* **fluvial sediments → 207**

2222 fluvial terrace above floodplain

The terrace above a flood-plain separated by a scarp.

r надпойменная терраса
Терраса, возвышающаяся над поймой и отделенная от нее уступом.

f terrasse *f* fluviatile
s terraza *f* por encima de llanura de inundación
g Auenterrasse *f*

* **fluviatile** ***adj*** **→ 2219**

2223 fluvioglacial ***adj*****; glaciofluvial** ***adj***
Related to the action of meltwater streams flowing from a glacier and the landforms created by the glacial meltwater.

r флювиогляциальный
Относящийся к процессам и формам рельефа, своим происхождением обязанным воздействию талых ледниковых вод.

f fluvio-glaciaire *adj*
s fluvio-glaciar *adj*
g fluvioglazial *adj*

* **fluvio-glacial deposits → 2547**

2224 fluvisol
Young soils on recent fluvial deposits (alluvial soils, soils of marshes and mangroves) which have a characteristic lamination and uneven decrease of content of organic carbon along the profile.

r флювисоль
Молодые почвы на современных флювиальных отложениях: пойменные почвы, почвы маршей и мангров, отличающиеся характерной слоистостью и неравномерным убыванием содержания органического углерода по профилю.

f fluvisol *m*
s fluvisol *m*
g Fluvisol *m*

2225 flysch
A sedimentary rock consisting of thick succession of alternations of the sediments ranging from marine sandstones, shales, marls to clays deposited in a deep water environment.

r флиш
Осадочная порода, состоящая из ритмического чередования отложений морских песков, глинистых сланцев, мергелей и глин, образовавшихся в водной среде.

f flysch *m*
s flysch *m*
g Flysch *m*

2226 foam ice
An artificially frozen mass of air bubbles and water.

r пенолед
Искусственно замороженная водно-воздушная пена.

f mousse *f* de glace
s espuma *f* de hielo
g Gischteis *n*

2227 fodder; forage
Any food for cattle, horses, sheeps or other domestic animals, including annual or perennial plants grown to be used for livestock in fresh, dried or fermented (silage) form.

r фураж; кормовые культуры; фуражные культуры
Корм для крупного рогатого скота, лошадей, овец и прочих домашних животных, включая однолетние и многолетние растения, выращиваемые для скармливания сельскохозяйственным животным в свежем, сухом или консервированном (как силос) виде.

f fourrage *m*; culture *f* fourragère
s forraje *m*; cultivo *m* forrajero
g Futter *n*; Futterkultur *f*

2228 fodder field
A plot of land used for grazing and for growing food for cattle (fodder).

r кормовое угодье
Земельные участки, используемые для выпаса скота и производства кормов.

f terres *fpl* fourragères; fourragère *f*
s tierras *fpl* forrajeras
g Futtergelände *n*

2229 foehn; föhn
A warm and dry strong descending wind usually blowing on the leeward side of a mountain ridge as a result of adiabatic air heating during its descent down a high mountain slope. → Fig. 118

r фён
Теплый и сухой сильный нисходящий ветер, обычно дующий на подветренной стороне горы или горного хребта в результате адиабатического нагревания воздуха во время его опускания по склону. → Рис. 118

f foehn *m*
s foehn *m*
g Föhn *m*

2230 fog
The obscurity exhibited by the surface layers of the atmosphere owing to the presence of suspendede very small, usually microscopic water droplets (or ice crystals) reducing the horizontal visibility at the earth`s surface to less than 1 km.

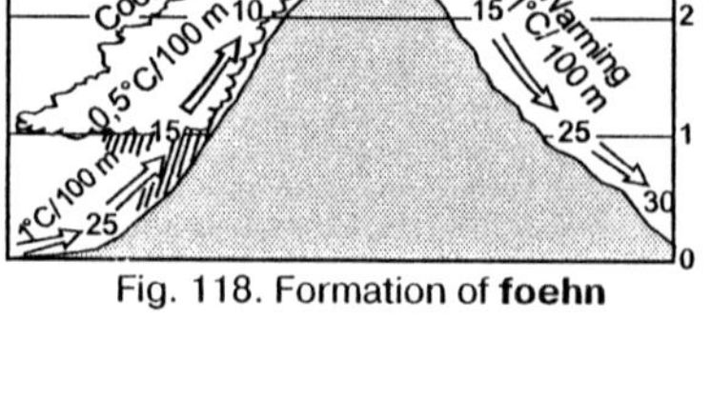

Fig. 118. Formation of **foehn**

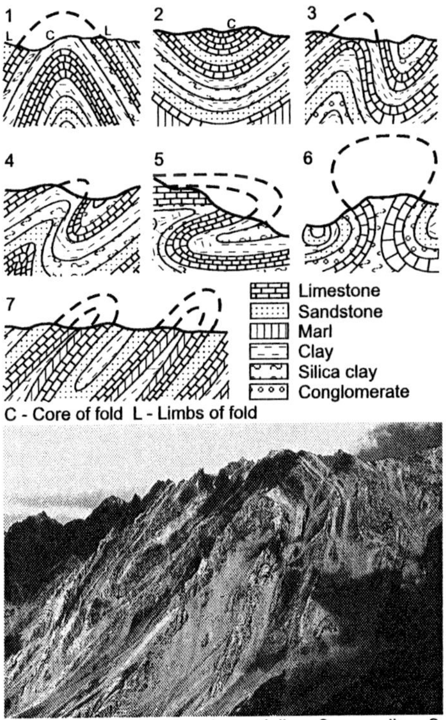

Fig. 119. **Folds**. Above: 1 – anticline, 2 – syncline, 3 – asymmetric, 4 – inclined, 5 – lying, 6 – fan-shaped, 7 – isoclinal folds. Down: Natural fold in Swiss Alps

r туман
Скопление мелких водяных капель или ледяных кристаллов в приземном слое атмосферы, снижающее горизонтальную видимость у поверхности земли до 1 км.
f brouillard *m*
s niebla *f*
g Nebel *m*

2231 fogbow; white rainbow
A virtually colourless arc (with a radius of about 40°) of brighter light, similar in form to a rainbow, seen in a fog when the Sun is behind the observer.
r туманная радуга; белая радуга
Практически бесцветная дуга (с радиусом около 40°) яркого света, похожая по форме на радугу, видимая в тумане, когда Солнце находится позади наблюдателя.
f arc *m* dans le brouillard; arc-en-ciel *m* blanc
s arco *m* iris blanco de la niebla; arco *m* iris blanco
g Nebelbogen *m*

* **föhn → 2229**

2232 fold
A bend of a rock strata or planes of bedding not disturbing the continuity of rocks, and resulting from the Earth's crust compression (there are many types of folds, depending on the intensity of compression and on the character of the rock). → Fig. 119
r складка
Изгиб или искривление слоя горных пород, плоскостей напластования (вплоть до обратного падения), не нарушающие сплошности составляющих их пород, возникают под действием напряжений в земной коре и в зависимости от силы напряжений и характера горных пород могут приобретать разные формы. → Рис. 119
f pli *m*
s pliegue *m*
g Falte *f*

2233 fold mountains
An upland area formed by the folding of the Earth's crust.
r складчатые горы
Горы, возникающие в результате смятия толщ осадочных горных пород в складки.
f montagnes *fpl* plissée
s montañas *fpl* de plegamiento; relieve *m* de plegamiento
g Faltengebirge *n*

2234 fold system
A large area of the Earth's crust folds of clear strike formed on the place of a geosynclinal system or at the margin of a platform.
r складчатая система
Крупная система складок земной коры выдержанного простирания, образующаяся на месте геосинклинальной системы или на краю платформ.
f système *m* de plis
s sistema *f* de pliegues
g Faltungssystem *n*

2235 fold-block mountains
The mountains formed by both folding and block-faulting.
r складчато-глыбовые горы
Горы, формирующиеся при совместном действии складчатых и глыбовых тектонических процессов.
f montagnes *fpl* plissées faillées; montagnes *fpl* des pis et blocs
s montañas *fpl* de pliegues y fallas; montañas *fpl* de pliegues y bloques
g Falten- und Schollengebirge *n*

2236 folding
The process of flexure or bendings of the rocks and formation of fold areas and fold systems, as well as the tectonic phase characteristic by the formation of complexes of folds in the Earth's crust.
r складчатость; складкообразование
Процесс образования складчатых облас-

тей и складчатых систем, а также формирования складок в горных породах и совокупность складок того или иного участка земной коры.
f plissement *m*
s plegamiento *m*
g Faltung *f*

2237 folding zone
Planetary system of fold mountains formed in the place of a geosynclinal belt between two platforms or a continental platform and the ocean.
r складчатый пояс
Планетарная система складчатых горных сооружений, возникшая на месте геосинклинального пояса между двумя платформами или континентальной платформой и океаном.
f zone *f* de plissement
s zona *f* de plegamiento
g Faltungszone *f*

2238 foliage
The leaves of a plant or tree.
r листва
Листья кустарника или дерева.
f feuillage *m*
s follaje *m*
g Krone *f*; Blätterwerk *n*

2239 foliation
A laminated or banded structure within a metamorphic rock, caused by the segregation of different minerals into parallel layers following the schistosity of the rock; the stratiform structure of ice in the lower layers of a glacier, propably the result of shearing of one part over another.
r сланцеватость
Полосчатая структура метаморфической горной породы, вызванная расслоением различных минералов на параллельные слои; полосчатое строение ледниковой толщи в нижней части ледника, возникающее, вероятно, в результате сдвига одной части льда по другой.
f foliation *f*
s foliación *f*
g Foliation *f*; Schieferung *f*

2240 folk ecology; popular ecology
The skills and habits acquired by people in the process of the centuries-long experience to cope with wild nature and use it for meeting the versatile requirements of humans.
r народная экология
Навыки, сложившиеся в процессе многовекового опыта обживания дикой природы, ее использования для удовлетворения многосторонних потребностей населения.
f écologie *f* populaire
s ecología *f* popular
g Volksökologie *f*

* **food and biverage industry → 2242**

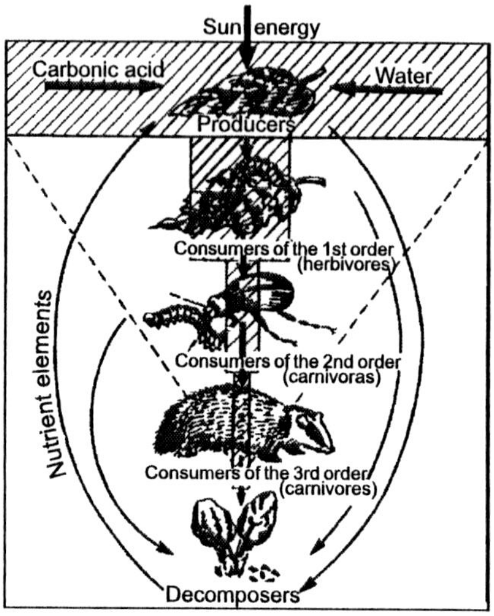

Fig. 120. **Food chain**

2241 food chain; trophic chain; food web
A series of organisms (producers, consumers and decomposers) existing in any natural community, along which energy passes from one to another as each provides food for next. → Fig. 120
r трофическая цепь; цепь питания
Система организмов (консументов, продуцентов и редуцентов), последовательно взаимодействующих в процессе питания, проявляется в питании особей одного вида за счет живых особей другого вида, продуктов их жизнедеятельности или их мертвых остатков. → Рис. 120
f chaîne *f* trophique; chaîne *f* alimentaire
s cadena *f* trófica; cadena *f* de alimentación; cadena *f* alimentaria
g Ernährungskette *f*

2242 food industry; food and biverage industry; food processing industry
A branch of industry producing foodstuffs, beverages, tobacco, perfumes and cosmetics.
r пищевкусовая промышленность
Отрасль промышленности, производящая пищевые продукты, напитки, табачные изделия, парфюмерию и косметику.
f industrie *f* alimentaire
s industria *f* alimenticia; industria *f* de alimentos y bevidas; industria *f* de alimentación
g Nahrungs- und Genußmittelindustrie *f*; Lebensmittelindustrie *f*

* **food processing industry → 2242**

2243 food security
The provision of a country with the main types of foodstuffs, taking into account the national interests and social requirements.

r продовольственная безопасность
Уровень обеспеченности страны важнейшими видами продовольствия с позиции национальных интересов и общественных потребностей.
f sécurité *f* alimentaire
s seguridad *f* alimentaria
g Lebensmittelsicherheit *f*

* **food web → 2241**

2244 **foot of slope**
A narrow transitional strip of land which separates a mountain slope from the adjacent plain.
r подошва склона; подножие склона
Узкая переходная полоса, ограничивающая склон и отделяющая его от прилегающей равнины.
f pied *f* de versant
s pie *f* de vertiente
g Abhangfuß *m*

2245 **foothills**
The lower marginal part of mountains with hilly or low-mountain relief.
r предгорья
Пониженные окраинные части горных стран, отличающиеся холмистым или низкогорным рельефом, прилавки на Северном Тянь-Шане.
f pays *m* situé au pied d´un massif
s zona *f* premontañosa;
llanura *f* al pie de una montaña
g Vorgebirge *n*; Vorland *n*

* **footloose activity → 2246**

2246 **footloose industry; footloose activity**
A mobile enterprise, usually light industry, which, by the nature of the raw materials used and its labour, market and transport needs, is not tied to a particular location.
r местная промышленность
Предприятия, обычно относящиеся к легкой промышленности, которые практически могут быть размещены в любой местности, что связано с доступностью сырья, рабочей силы, транспорта и рынка.
f industrie *f* nomade; industrie *f* «footloose»
s industria *f* «peregrina»; industria *f* «transeunte»; industria *f* de tipo footloose
g unbegrenzte Industrie *f*

* **forage → 2227**

2247 **forb steppe**
A plant community dominated by steppe herbs with a closed grass cover, it is characteristic of Eurasian steppe.
r разнотравная степь
Растительное сообщество с господством степного разнотравья с сомкнутым травяным покровом, характерна для степей Евразии.
f steppe *f* à herbes diverses
s estepa *f* herbosa; estepa *f* de hierbas diferentes
g Krautsteppe *f*; Grassteppe *f*

2248 **forb-meadow steppe**
A plant community with considerable domination of steppe herbs, it is characteristic of meadow and true steppe.
r разнотравно-луговая степь
Растительное сообщество со значительным участием в растительном покрове степного разнотравья, характерна для луговых и настоящих степей.
f steppe-prairie *f* à herbes diverses
s estepa *f* de hierbas y gramíneas
g Gras- und Wiesesteppe *f*

2249 **forced convection**
The convection of air caused by mechanical forces, when the moving air comes into contact with a rough earthrs surface of a slope.
r вынужденная конвекция
Конвекция, вызванная механическими причинами, например, при движении воздуха над шероховатой поверхностью или вдоль склона.
f convection *f* forcée
s convección *f* forzada
g erzwungene Konvektion *f*

2250 **forced migrants**
Population of other regions made to migrate to a given territory for economic or ecological reasons.
r вынужденные переселенцы
Жители других районов, вынужденные мигрировать на данную территорию в связи с экономическими или экологическими обстоятельствами.
f migrants *mpl* forcés
s migrantes *mpl* forzados
g Zwangsübersiedler *mpl*

2251 **ford**
A part of a river channel shallow enough to be crossed on foot (hence common in riverside place-names, e.g. Oxford).
r брод
Участок речного русла, достаточно мелкий для перехода через реку в этом месте.
f gué *m*
s vado *m*
g Furt *f*

2252 **Fordism**
A system of industrial organization of line production and social relations that emerged in the U.S. in the first quarter of the XX century at Henri Ford's plants.
r фордизм
Система организации поточного производства и социальных отношений, возникшая в США в первой четверти XX в. на автомобильных заводах Г. Форда.
f Fordisme *m*

s fordismo *m*
g Fordismus *m*

2253 forecast
A prediction based on scientific observation, on experience, or by an estimate of probability, as in a weather forecast.
r прогноз
Предсказание, основанное на научных наблюдениях, на жизненном опыте или на оценке выроятности событий, как в случае прогноза погоды.
f prévision *f*
s previsión *f*
g Prognose *f*

2254 forecast map
Geographical map representing scientifically grounded development and distribution of objects and phenomena in a certain time interval in the future.
r прогнозная карта
Географическая карта, отображающая научно обоснованное развитие и распространение объектов и явлений через определенный промежуток времени в будущем.
f carte *f* prévisionnel
s mapa *m* previsivo; mapa *m* pronóstico
g Vorhersagekarte *f*

* **foredeep → 3679**

2256 foredune; coastal dune
A dune or line of dunes nearest to the sea in a system of dunes on a sandy shore, comparable with an advanced dune in a hot desert.
r авандюна; береговая дюна
Дюна, расположенная недалеко от моря в системе дюн на песчаном побережье, аналогичная передовой дюне в жаркой пустыне.
f dune *f* bordière; avant-dune *f*
s duna *f* de bordo; duna *f* lateral; ante-duna *f*
g Küstendüne *f*

2257 foreign economic connections; foreign economic relations
Economic cooperation of states based on the international division of labour, it involves the movement of goods and services, capital and work force, technology, and international tourism.
r внешнеэкономические связи
Хозяйственное взаимодействие государств, в основе которого лежит международное разделение труда, включающее движение товаров и услуг, финансовых и трудовых ресурсов, технологий, международный туризм.
f relations *fpl* économiques exterieures
s relaciones *fpl* económicas exteriores
g Außenwirtschaftsbeziehungen *fpl*

* **foreign economic relations → 2257**

* **foreign exchange crisis → 3915**

* **foreign trade → 3163**

2258 foreign trade balance; foreign trade turnover
The difference in value between the exports and imports of a country during a certain period, usually a year.
r внешнеторговый баланс; внешнеторговый оборот
Баланс, отражающий соотношение стоимости экспорта и импорта страны за определенный период (обычно за год).
f balance *f* commerciale
s balance *m* comercial
g Außenhandelbilanz *f*; Außenhandelumsatz *m*

* **foreign trade turnover → 2258**

2259 foreland
1. Land or territory lying in front of something, in particular, a promontery of land jutting out from a coastline or a stable continental mass on the margin of a geosyncline. 2. The seaward trading area connected with the port through maritime organization. → Fig. 121
r форланд
1. Любое пространство, лежащее впереди чего-либо, в частности, устойчивая масса континентальной суши на краю геосинклинали. 2. Морская акватория, регулярно используемая для получения и отправки торговых грузов и включающая водное пространство, которое связывает основной порт и его аванпорт. → Рис. 121
f avant-pays *m*; foreland *m*
s antepaís *m*; foreland *m*
g Vorland *m*

* **foreland basin → 3679**

2260 forerange
A mountain range on the periphery of the mountains lying along the adjoining plain.
r передовой хребет
Горный хребет, протягивающийся на периферии горной страны вдоль прилегающей равнины.

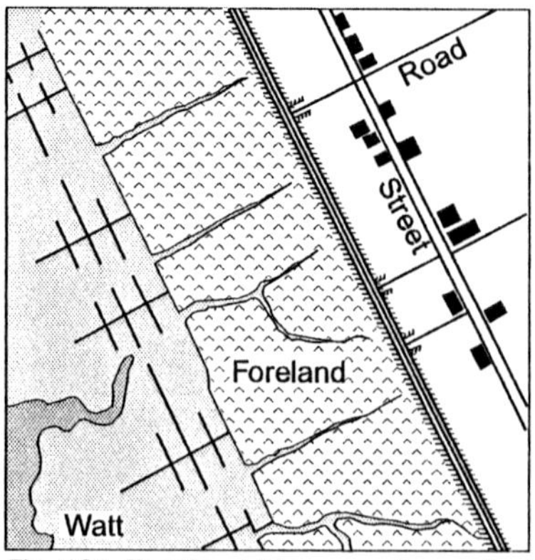

Fig. 121. **Foreland** (the second meaning)

f chaîne *f* d'avant-pays; cordillère *f* d'avant-pays
s cadena *f* de antepaís; cordillera *f* de antepaís
g Vorrücken *m*

2261 foreshock
A weak underground shock sometimes preceding the main earthquake by short time interval.
r форшок
Слабые подземные толчки, иногда предшествующие сильному землетрясению.
f frémissement *m* préliminaire; tremblement *m* de terre initial; ébranlement *m* initial
s sacudida *f* inicial; temblor *m* preliminar
g Vorbeben *n*

2262 foreshore; offshore[2]
The lower zone of a beach, extending between low-water spring-tide level and high-water spring-tide level, i.e. the part of the beach covered by spring tide.
r фронтальная часть пляжа
Нижняя часть пляжа, протягивающаяся между нижним и верхним уровнями самого высокого прилива, т.е. полоса пляжа, затапливаемого сизигийным приливом.
f avant-plage *f*; avant-côte *f*
s anteplaya *f*; antecosta *f*
g Vorstrand *m*; Küstenvorland *n*

2263 forest; wood
A type of vegetation, the dominating layer of which is formed by trees of one or several species, the crowns of which touch, so forming a continuous canopy, grasses, shrubs, lianas, subshrubs, mosses, lichens are also typical. → Fig. 122
r лес
Тип растительности, где господствуют деревья одного или нескольких видов с сомкнутыми кронами, характерны также травы, кустарники, лианы, кустарнички, мхи, лишайники. → Рис. 122
f forêt *f*; bois *f*
s bosque *m*
g Wald *m*; Forst *m*

2264 forest assessment; forest inventory; forest valuation
Estimation of potential productivity of forests: growth and timber stock, number of trees, etc.

Fig. 122. Some types of **forest**

r таксация лесов
Определение потенциальной продуктивности леса: прироста и запаса древесины, объема деревьев и пр.
f évaluation *f* de forêt
s valoración *f* de bosque
g Waldabschätzung *f*

2265 forest cadastre
A part of the land cadastre that contains information on forest resources.
r лесной кадастр
Составная часть земельного кадастра, содержащая сведения о лесных ресурсах.
f cadastre *m* forestier
s catastro *m* forestal
g Forstkataster *m*

2266 forest cover
The persentage of the surface of a region which is covered by forest.
r лесистость
Отношение площади, покрытой лесом, к общей площади данной территории, выражается в процентах.
f pourcentage *m* de terre boisée
s porcentaje *m* de tierra boscosa; boscosidad *f*
g Waldanteil *m*; Waldbedeckung *f*; Walddecke *f*

2267 forest damaged by snow
Destruction and breaking of trees and their branches caused by heavy snow load, resulting from snowfalls and hoarfrost deposit.
r снеголом; снеговал
Повреждение и поломка деревьев в результате большой снеговой нагрузки, возникающей после обильных снегопадов и отложений изморози.
f forêt *f* endommagée par la neige
s bosque *m* destruido por la nieve
g Baumbeschädigung *f* infolge Schneelast

2268 forest edge
A narrow transitional strip on the border of the forest and adjoining communities which is covered with bushes, lianas, undergrowth of the trees dominating in the forest.
r опушка леса
Край неровной границы леса, узкая переходная полоса к соседнему сообществу, здесь обычно растут кустарники, лианы, подрост доминирующих в лесу деревьев.
f lisière *f*; orée *f*
s lindero *m*
g Waldrand *m*; Waldsaum *m*

2269 forest fauna
The animals adapted to the forest conditions.
r лесная фауна
Совокупность видов животных, приспособившихся к обитанию в лесах.
f faune *f* de forêt
s fauna *f* de bosque; fauna *f* forestal

g Waldfauna *f*

* **forest fire → 6770**

* **forest gallery → 2355**

2270 **forest improvement; forest reclamation**
A controlled transformation of environment by way of reafforestation, change of species and age composition of forests, improvement of proportion between forest and woodless areas, etc.
r лесомелиорация
Направленное изменение окружающей среды через улучшение лесных насаждений путем лесовозобновления, изменения породного и возрастного состава лесов, улучшения соотношения лесных и безлесных площадей и т.п.
f amélioration *f* de la forêt
s bonificación *f* de bosque; mejoramiento *m* de bosque
g Waldmelioration *f*; Forstmelioration *f*

* **forest inventory → 2264**

2271 **forest management; forest use**
Forms and methods of forest utilization and protection regulated legally and economically.
r лесопользование
Формы и методы использования и охраны лесов, регламентируемые юридически и экономически.
f usage *m* de forêt
s uso *m* de bosque
g Forstnutzung *f*; Waldnutzung *f*

* **forest reclamation → 2270**

2272 **forest resources**
Timber and non-timber materials, hunting, food, and medicinal resources of forests and their aesthetic and recreational value.
r лесные ресурсы
Древесные запасы леса, недревесные технические, кормовые, охотничье-промысловые, пищевые, лекарственные ресурсы леса, а также его оздоровительное и эстетическое значение.
f ressources *fpl* forestières
s recursos *mpl* forestales
g Forestressourcen *fpl*; Forestvorräte *mpl*

* **forest stand → 5768**

* **forest strip → 5391**

* **forest use → 2271**

* **forest valuation → 2264**

2273 **forestal *adj***
Related or inhabited in a forest.
r лесной
Относящийся к лесу, обитающий в лесу.
f forestier *adj*; de bois; sylvestre *adj*; de forêt
s forestal *adj*; de bosque; silvestre *adj*
g Wald-; Forst-

2274 **forested lowland**
A wooded sandy lowland found in the marginal zone of the Pleistocene glaciation, often swampy or waterlogged.
r полесье
Залесенная песчаная низина в окраинной полосе плейстоценового материкового оледенения, часто заболоченная или переувлажненная.
f plaine *f* basse boisée
s llanura *f* baja boscosa
g Polessje *n*; Waldland *n*

2275 **forest-industrial complex; timber-industry complex**
A group of timber enterprises on one site with an adjoining timber-felling site.
r лесопромышленный комплекс
Совокупность лесоперабатывающих предприятий, расположенных на одной промышленной площади с тяготеющим к ней лесным массивом, в котором ведутся лесозаготовительные работы.
f industrie *f* forestière
s complejo *m* forestal-industrial
g forstwirtschaftlicher Komplex *m*; Forstwirtschaft *f*

2276 **forestry**
Planting, cultivating and managing forests.
r лесное хозяйство; лесоводство
Отрасль хозяйства, занимающаяся учетом, воспроизводством, охраной и защитой лесов, а также регулированием их использования.
f foresterie *f*; sylviculture *f*
s cultivo *m* de bosques; silvicultura *f*
g Forstwirtschaft *f*

2277 **forest-steppe; forest-steppe zone**
A zonal type of landscapes characterized by alternating forest and steppe areas on watersheds, common in temperate and subtropical belts.
r лесостепь; лесостепная зона
Зональный тип ландшафта, характеризующийся чередованием на водоразделах лесных и степных участков, распространен в умеренном и субтропическом поясах.
f steppe *f* boisée; forêt-steppe *f*; zone *f* des forêt-steppes
s bosque-estepa *f*; zona *f* de bosque-estepa
g Waldsteppe *f*; Waldsteppenzone *f*

* **forest-steppe zone → 2277**

2278 **forest-tundra; forest-tundra zone**
A zonal type of landscape characterized by alternating of tundra and forest areas on watersheds, common southwards of the tundra zone in the north of Eurasia and North America. → Fig. 123, c. 271

Fig. 123. **Forest-tundra** in the Kola Peninsula

r лесотундра; лесотундровая зона
Зональный тип ландшафта, характеризующийся чередованием на водоразделах участков тундры и леса, распространен южнее тундровой зоны на севере Евразии и Северной Америки. → Рис. 123
f toundra *f* arborée; zone *f* de toundra arborée
s tundra *f* arbolada; bosque-tundra *f*; zona *f* de tundra arbolada; zona *f* de bosque-tundra
g Waldtundra *f*; Waldtundra-Zone *f*

* **forest-tundra zone → 2278**

* **formal symmetrical park → 4988**

* **form-line → 1262**

2279 fossil
The remains or traces of plants or animals, preserved in the Earth's crust by natural processes, usually in sedimentary rocks (sometimes in metamorphic rocks if the metamorphism has not been too violent).
r ископаемые остатки
Остатки или следы растений и животных, сохранившиеся в земной коре от прошлых эпох, обычно в осадочных породах (иногда и в метаморфических, если метаморфизм не был слишком глубоким).
f fossile *m*
s fósil *m*
g fossile Versteinerung *f*; fossile Überreste *mpl*

2280 fossil fuel
Organic combustible materials found underground (coal, oil, natural gas) which were formed in the past geological epochs from the remains of animals or plants and which are used to produce energy at thermic power plants and in car engines.
r ископаемое топливо; энергоносители
Ископаемое органическое топливо (уголь, нефть, природный газ), образовавшееся из животных и растительных остатков прошлых геологических эпох, ныне используюшееся для получения энергии на тепловых электростанциях и в двигателях машин.
f combustible *m* fossil
s combustible *m* fósil
g fossiler Brennstoff *m*

* **fossil ice → 5007**

* **fossil magnetism → 4388**

* **fossil soil → 4393**

* **fossil water → 1214**

2281 foul air
Air which has been circulated in a building or mine without being changed.
r грязный воздух; спертый воздух
Загрязненный несменяемый воздух, циркулирующий внутри зданий и шахт.
f air *m* sale; air *m* vicié
s aire *m* sucio
g mattige Luft *f*

2282 foundering
The formation of fractures and sinking of earlier formed tectonic structures along them.
r обрушение
Процесс возникновения разломов и опускания по ним частей ранее сформированных тектонических структур.
f effondrement *m*; éboulement *m* catastrophique
s derrumbamiento *m* catastrófico
g Abbruch *m*; Einsturz *m*

* **fracture → 2161**

2283 fracture; rupture
Any disturbance of the rock integrity: shear-fault, gapping fault, fault, reversed fault, overthrust, etc.
r разрыв
Всякое нарушение сплошности горных пород: сдвиг, раздвиг, сброс, взброс, надвиг и пр.
f fracture *f*; cassure *f*; rupture *f*
s fractura *f*; ruptura *f*; diaclasa *f*
g Bruch *m*

2284 fracture line of an avalanche
The breakaway wall at the top of the wind slab where snow avalanching starts.
r линия отрыва лавины
Сводовая стенка, вдоль которой происходит обрушение снежной доски и начинается сход лавины.
f ligne *f* de rupture d'une avalanche
s línea *f* de ruptura de avalancha
g Bruchlinie *f* einer Lawine

* **fracturing → 2164**

2285 fractus

Clouds (symbol: fra) in the form of irregular shreds which have a clearly ragged appearance, indicating strong winds and stormy conditions in the upper atmosphere (the prefix fracto- is used to indicate a similar ragged condition in other cloud forms, e.g. fractocumulus, fractonimbus, fractostratus, etc).
r разорванные облака
Облака (символ: fra) в форме беспорядочных клочьев резко разорванного вида, свидетельствующие о сильных ветрах и штормовых условиях в верхних слоях атмосферы.
f fractus
s fractus
g Fractus

2286 fragipan
A deep soil horizon that is characterized by medium granulometric composition, increased volume weight as compared to adjoining horizons, and polygonal pattern of fissures, its origin is timed to the Wьrm Ice Age.
r фраджипен
Глубокий почвенный горизонт со средним гранулометрическим составом и повышенной относительно окружающих горизонтов плотностью, включает трещины, образующие полигональный рисунок, его происхождение связано с вюрмским оледенением.
f fragipan *m*
s fragipan *m*
g Fragipan *n*

2287 fragmentation of a farm
The division of the land of a farm into separate, isolated parts according to the law of equal inheritance or caused by special circumstances, e.g. the construction of a motorway cuts through the farm.
r раздел фермы
Разделение земли, принадлежащей одной ферме, на изолированные участки в соответствии с законом о равном наследовании земли, или вызванный особыми обстоятельствами, например, проведением автострады, разрезающей поля фермы.
f fragmentation *f* de ferme
s fragmentación *f* de granja
g Einteilung *f* einer Farm

2288 framework of settlement pattern
The network of the largest settlements on a definite territory and transportation routes that connect them.
r опорный каркас расселения
Сеть наиболее значительных поселений на определенной территории и соединяющих их транспортных коммуникаций.
f armature *f* urbaine
s armazón *m* urbano
g Grundgerüst *n* der Ansiedelung

*** frazil → 5425**

2289 frazil ice; frazil
Primary ice crystals in the form of minute needles which develop into spongy accumulations in the body of water below freezing-point.
r внутриводный лед
Скопления первичных ледяных кристаллов в форме мельчайших игл, формирующихся в толще воды и на дне.
f glace *f* frasil
s hielo *m* intraquático; hielo *m* frazil
g freischwebende Eiskristalle *mpl*

*** freak wave → 6086**

2290 free atmosphere
That part of the atmosphere situated above the friction layer in which the air motion is only slightly affected by the surface friction.
r свободная атмосфера
Часть атмосферы выше слоя трения, в которой движение воздуха мало подвержено воздействию трения о поверхность Земли.
f atmosphère *f* libre
s atmósfera *f* libre
g freie Atmosphäre *f*

*** free city → 2292**

2291 free economic zone; free zone
A region in a country where enterprises are exempt from taxes paid in the rest of the territory.
r свободная экономическая зона
Область внутри государства, где предприятия не облагаются теми налогами, которые действуют на остальной территории.
f zone *f* franche
s zona *f* de franquicia
g freie Wirtschaftszone *f*

*** free economy → 3709**

*** free groundwaters → 6442**

*** free market economy → 3709**

2292 free port; free city
A port or an area of a port in which imported goods can be held or processed free of customs duties before reexport.
r порто-франко; вольный город; вольная гавань
Охраняемый порт или часть порта, где можно получать и отправлять товары без таможенной пошлины и без соблюдения большинства таможенных правил.
f port *m* franc; ville *f* franche; franches *fpl*
s puerto *m* franco; ciudades *f* francas; francas *fpl*
g Freihafen *m*

2293 free solifluction
The solifluction occurring on bare slopes deprived of vegetation or the solifluction that destroys the plant cover.
r свободная солифлюкция

Солифлюкция на лишенных растительности склонах, либо разрушающая растительный покров.
f solifluxion *f* libre
s solifluxión *f* libre
g freie Solifluktion *f*

2294 free trade
The commercial interchange between countries in which there are no tariff barriers or other protective measures such as quota restrictions.
r свободная торговля
Обмен товарами между государствами или экономической группой и другими партнерами в условиях отмены таможенных барьеров и свободы импорта и экспорта между ними.
f libre-échange *m*
s cambio *m* libre
g Freihandel *m*

2295 free trade area; free trade zone
An area where there are no trade restrictions between member countries (i.e. where tariff barriers and quotas do not apply), e.g. European Economic Community where the states agreed to operate according to the free trade rules while each member maintains its separate trade politics with the states outside the area.
r зона свободной торговли
Территория, где действует соглашение между государствами о политике свободной торговли между ними (например, внутри Европейского союза), в то время как каждая страна сохраняет свою политику в области торговли со странами за пределами этой территории.
f zone *f* de libre échange
s zona *f* de libre cambio
g Freihandelzone *f*

* **free trade zone → 2295**

* **free water → 2616**

* **free zone → 2291**

2296 freeze-free period
Period of the year free of subfreezing temperatures at ground level or in the air, as specified.
r безморозный период
Период года, свободный от понижения температуры ниже 0°C на уровне почвы или в воздухе.
f période *f* sans gel
s período *m* sin heladas
g frostfreie Zeit *f*; frostfreie Periode *f*

* **freeze-thaw action → 2339**

2297 freeze-thaw cycle
Annual, seasonal or daily cycles of freezing and thawing of the ground resulting in frost weathering and other cryogenic processes.
r цикл промерзания-протаивания
Годовой, сезонный или суточный циклы промерзания и протаивания пород, ведущие к морозному выветриванию и другим криогенным процессам.
f cycle *m* gel-dégel
s ciclo *m* de congelamiento y descongelamiento
g Gefrier- und Auftauzyklus *m*

2298 freeze-up
The process of formation or existence of a solid stable ice cover in the surface of a body of water.
r ледостав
Процесс образования, а впоследствии и наличие на поверхности реки или водоема неподвижного ледяного покрова.
f gel *m* des rivières et lacs; prise *f* en glace
s formación *f* de la capa de hielo en el río or lago
g Eisstand *m*; Gefrieren *n*; Eisbildung *f*

* **freezing → 3489**

2299 freezing; congelation
The natural process when water changes or becoming frozen from liquid to solid state as the temperature falls below the freezing point.
r замерзание
Естественный процесс, когда охлаждение жидких веществ ниже их точки замерзания приводит к образованию льда.
f congélation *f*; gel *m*
s congelación *f*; helamiento *m*; hielo *m*
g Gefrieren *n*; Einfrieren *n*

2300 freezing crust
An ice layer, up to 15 cm thick, covering the floor, walls or ceiling of an ice cave or an old working, which is formed by the freezing of water seeping through fractures.
r кора обледенения
Слой натечного льда толщиной до 15 см, образовавшийся на дне, стенах или потолке в ледяной пещере или горной выработке при замерзании воды, просачивающейся сверху по трещинам.
f croûte *f* gelée
s corteza *f* helada
g Eisrinde *f*; Eiskruste *f*

2301 freezing drizzle; freezing rain
Supercooled droplets, falling from a relatively warm air above a temperature inversion, which freeze when theytouch a cold surface below 0°C to form the glaze.
r ледяной дождь; переохлажденный дождь
Переохлажденные капли воды, падающие из относительно теплого воздуха выше температурной инверсии и замерзающие при соприкосновении с холодной, ниже 0°C, поверхностью, что формирует гололед.
f bruine *f* glacée

s llovizna *f* de hielo
g gefrorener Staubregen *m*

2302 **freezing front**
The boundary between frozen ground and the non-frozen ground which moves downward and upward through the earth.
r фронт промерзания
Граница между мерзлым и немерзлым грунтом, продвигающаяся при промерзании грунта.
f front *m* de gel
s frente *m* de congelación; frente *m* de hielo
g Gefrier-Front *f*

2303 **freezing nucleus**
Any particle on which a water droplet freezes to form an ice crystal.
r ядро замерзания
Частица в переохлажденной воде или переохлажденном облаке, вызывающая замерзание воды.
f noyau *m* de congélation
s núcleo *m* de congelación
g Gefrierkern *m*

2304 **freezing point**
The temperature at which a liquid substance changes to the solid state, e.g. at which fresh water changes to ice, 0°C (32°F).
r точка замерзания
Температура, при которой вещество, находящееся в жидком состоянии, становится твердым, например, пресная чистая вода превращается в лед при 0°C (32°F).
f point *m* de congélation
s punto *m* de congelación
g Gefrierpunkt *m*

* **freezing pressure → 1358**

* **freezing rain → 2301**

2305 **freezing through of river**
A cessation of a river flowing at its shallow sections due to a very thick ice or freezing of water up to the bottom.
r промерзание реки
Прекращение руслового стока на мелких участках из-за образования мощного ледяного покрова или промерзания до дна.
f congélation *f* de rivière
s congelación *f* de río
g Zugefrieren *n* von Flußen

2306 **freezing-out; upfreezing**
A method of production of fresh water from a salt water by its freezing.
r вымораживание
Метод получения пресной воды из соленой путем ее замораживания.
f dessalement *m* par congélation
s desalación *f* por congelación
g Wasserentsalzung *f* durch Ausfrieren

2307 **freight**
The service of carrying goods, originally by ship but now extended to land vehicles and aircraft.
r фрахт
Услуги по перевозке товаров и пассажиров разными видами транспорта, в прошлом морским и речным, а ныне также наземным и воздушным.
f fret *m*
s flete *m*
g Fracht *f*

2308 **freight rates**
The money paid to a carrier for the loading, transporting and unloading of goods.
r цена фрахта
Деньги, уплачиваемые перевозчику за погрузку, перевозку и разгрузку товара.
f prix *m* de transport
s pago *m* el flete
g Frachtgebühr *f*

2309 **fresh ice; fresh-water ice**
Natural ice produced by freezing of fresh water, with the content of salts less than 1 mg/l.
r пресный лед; пресноводный лед
Природный лед, образующийся из пресной воды, с минерализацией до 1 мг/л.
f glace *f* douce
s hielo *m* dulce
g Süßeis *n*

2310 **fresh water**
Natural water with the concentration of salts less than 1 mg/l.
r пресная вода
Природная вода с минерализаций до 1 мг/л.
f eau *f* douce
s agua *f* dulce
g Süßwasser *n*

2311 **freshwater *adj*; limnic *adj*; limnetic *adj***
Pertaining to organisms which live in fresh water lakes and ponds.
r пресноводный
Относящийся к организмам, живущим в пресной воде.
f dulçaquicole *adj*; limnetique *adj*; dulcicole *adj*
s dulceacuícola *adj*; de agua dulce; limnético *adj*
g Süßwasser-; limnisch *adj*

2312 **freshwater environment**
A water mass of rivers and freshwater lakes inhabitated by the specific flora and fauna.
r пресноводная среда
Водная масса рек и пресных озер со специфической флорой и фауной.
f milieu *m* dulçaquatique
s medio *m* dulceacuícolo
g Süßwasserumwelt *f*

* **fresh-water ice → 2309**

2313 **freshwater lake**
A lake containing fresh water.
r пресное озеро
Озеро с пресной водой.
f lac *m* à l'eau douce; lac *m* doux
s lago *m* de agua dulce; lago *m* dulce
g Süßwassersee *m*

* **friction coefficient of river bed → 5138**

* **friction layer of the atmosphere → 772**

* **fright turnover → 883**

* **fringing forest → 2355**

2314 **fringing reef; biogenic reef**
Coastal shoal with coral constructions directly adjoining a continental or insular land.
r береговой риф; биогенный риф
Прибрежная отмель с коралловыми сооружениями, непосредственно примыкающая к материковой или островной суше.
f récif *m* frangeant; récif *m* biogénétique
s arrecife *m* frangeante; arrecife *m* biogénico
g Küstenriff *n*

2315 **front; frontal zone; frontal surface**
A surface of separation between two large air or water masses that exhibit markedly different physical parameters such as temperature, humidity, salinity, etc.
r фронт; фронтальная зона
Поверхность раздела двух больших воздушных или океанических масс с резко различающимися физическими параметрами, такими как температура, влажность, соленость и др.
f front *m*; zone *f* frontale
s frente *m*; zona *f* frontal
g Front *m*; Frontalzone *f*

2316 **front of growing activity**
The lowest limit of the active ice in a glacier tongue moving forward as a prominent swelling while the glacier is surging.
r фронт активизации
Нижняя граница активного льда, продвигающегося в виде выпуклого вала вниз по языку ледника во время его пульсации.
f front *m* d'activation
s frente *m* de activación
g Aktivierungsfront *f*

2317 **front of snow melting**
The boundary that separates the area of snow melting from that where no melting has started.
r фронт снеготаяния
Граница, отделяющая площадь, охваченную снеготаянием, от территории, где таяние еще не начиналось.
f front *m* de fusion de la neige
s frente *m* de fusión de la nieve
g Schneeschmelzfront *f*

2318 **frontal**
1. Related to an atmospheric front. 2. Related to the forward part of any object.
r фронтальный
1. Относящийся к атмосферному фронту. 2. Относящийся к передней части какого-либо объекта.
f frontal *adj*
s frontal *adj*
g Frontal-; frontal *adj*

2319 **frontal cloud system**
A combination of clouds of certain type and sequence formed on an atmospheric front.
r облачная система фронта
Совокупность облаков определенного вида и последовательности, сформировавшаяся на атмосферном фронте.
f système *m* nuageux de front
s sistema *m* nuboso de frente
g frontale Wolkensystem *n*

2320 **frontal fog**
Fog formed when very fine warm drizzle resulting from the passing of the warm front of a depression encounters cold air near the ground, which becomes saturated.
r фронтальный туман
Туман, формирующийся при прохождении теплого фронта циклона, сталкивающегося с холодным воздухом вблизи поверхности, который быстро насыщается влагой.
f brouillard *m* frontal
s niebla *f* frontal
g Frontalnebel *m*

2321 **frontal moraine**
Transversal bow-shaped bands of moraine deposits on the surface of the forward part of a glacier. → Fig. 85, p. 182
r фронтальная морена
Поперечные дуги моренных отложений на поверхности фронтальной части ледника. → Рис. 85, с. 182
f moraine *f* frontale
s morrena *f* frontal
g Frontalmoräne *f*; Stirnmoräne *f*

* **frontal occlusion → 4202**

2322 **frontal precipitation; cyclonic precipitation; frontal rainfall; cyclonic rain**
Precipitation associated with the passage of an atmospheric depression in middle and high latitudes, as the warm moist air mass of the depression meets and overrides colder, heavier air.
r фронтальные осадки
Осадки, связанные с прохождением циклона в средних и высоких широтах, когда циклоническая теплая и влажная воздушная масса встречается и натекает на более холодный и тяжелый воздух.
f précipitation *f* frontale

s precipitación *f* frontal
g Frontalniederschläge *mpl*

* **frontal rainfall → 2322**

* **frontal surface → 2315**

2323 **frontal thunderstorm**
A thunderstorm resulting from the passing of the atmospheric front.
r фронтальная гроза
Гроза, связанная с прохождением атмосферного фронта.
f orage *m* frontal
s tormenta *f* frontal
g Wirbelgewitter *n*

2324 **frontal wave; cyclonic wave**
A wave disturbance on an atmospheric front giving origin to the first stage of development of an extratropical depression.
r фронтальная волна;
циклоническая волна
Волновое возмущение на атмосферном фронте, дающее начало первой стадии образования и развития внетропического циклона.
f onde *f* frontale; onde *f* cyclonique
s onda *f* frontal; ola *f* frontal; onda *f* ciclónica
g Frontalwelle *f*

* **frontal zone → 2315**

* **frontier → 755**

2325 **frontogenesis**
The process of formation or intensification of an atmospheric front due to physical (e.g. radiation) or kinematic influences.
r фронтогенез
Процесс образования или обострения атмосферного фронта вследствие физического (радиационного) или кинематического воздействия.
f frontogénèse *f*
s frontogénesis *f*
g Frontogenese *f*

2326 **frontolysis**
The process of weakening or disappearance of an atmospheric front due to physical (e.g. radiation) or kinematic influences.
r размывание фронта; фронтолиз
Процесс размывания или окончательного прекращения фронтальной деятельности вследствие физического (радиационного) или кинематического воздействия.
f frontolyse *f*
s frontolisis *f*
g Frontolyse *f*

2327 **frost; air frost**
The state of freezing or becoming frozen, when the air temperature falls to or below the freezing point.
r мороз
Состояние замерзания, когда температура на уровне около 1 м падает до 0°C и ниже.
f gelée *f*; froid *m*; gel *m*
s helada *f*; fuerte helada *f*; frío *m*
g Frost *m*

* **frost blister → 2878**

* **frost boundary → 2333**

* **frost climate → 4518**

* **frost crack → 2329**

2328 **frost danger of the ground**
The degree of impact on engineering installations of the changes in the state of the ground caused by its freezing and thawing.
r морозоопасность грунтов
Степень влияния изменчивости грунтов на инженерные сооружения при их промерзании и протаивании.
f danger *m* gélival
s peligro *m* de congelación;
peligro *m* de helada
g Frostgefahr *f*

2329 **frost fissure; frost crack**
A fissure in soil or rocks formed by their contraction brought about by frost action.
r морозобойная трещина
Трещина в почве или горной породе, образующаяся в результате их сжатия при сильных морозах.
f fissure *f* de gel;
fissure *f* de refroidissement
s fisura *f* de hielo; fisura *f* de enfriamiento
g Frostspalte *f*

2330 **frost flowers; window frost$_2$**
1. Agglomerations of freezing sea salts which are formed on the snow accumulated above sea ice surface near polynyas. 2. A thin deposit of hoarfrost often found on interior surfaces of windows in winter which get strikingly beautiful design (frost flowers form by a direct sublimation on the glass, while window ice forms from freezing of liquid water that had condensed on the glass).
r морозные узоры
1. Разросшиеся «кустики» замерзших морских солей на поверхности снега, лежащего на морском льду вблизи полыней. 2. Тонкие отложения инея на внутренней стороне окон зимой, нередко приобретающие красивые очертания, образующиеся путем непосредственной сублимации на стекле, в то время как оконный лед формируется при замерзании жидкой воды, которая конденсируется на стекле.
f fleurs *f* de givre
s flores *pl* frios
g Eisblumen *f*

2331 **frost fracturing**

The formation of cracks, fractures and fissures in soil and rocks due to their contraction by the action of hard frosts.

r морозное трещинообразование
Возникновение морозобойных трещин в почвах и горных породах в результате их сжатия при сильных морозах.

f fragmentation *f* par le gel
s fragmentación *f* por enfriamiento; gelivación *f*
g Spaltenfrost *m*

* **frost heaving → 1359**

* **frost hillock → 2973**

* **frost hollow → 2334**

2332 **frost jacking; upfreezing**
A gradual upward displacement of blocks of rock or stones embedded in a seasonally frozen ground caused by frost-heaving.

r выморaживание грунта
Постепенное смещение к дневной поверхности находящихся в толще сезонно-мерзлых грунтов валунов и щебня, вызванное криогенным пучением.

f éjection *f* gélivale
s extrusión *f* criogénica; extrusión *f* gelival
g Auffrieren *n*

2333 **frost line; frost boundary**
A boundary delimiting areas with cold climate characterized by perpetual frost (i.e. no month with a mean temperature above freezing point).

r граница мороза
Граница территорий с холодным климатом, характеризующимся постоянным морозом, когда средняя температура любого месяца не поднимается выше 0°C.

f limite *f* des gelées
s límite *m* de heladas
g Frostgrenze *f*

* **frost mound → 2973**

2334 **frost pocket; frost hollow**
A small, low-lying area into which cold air rolls down by gravity from a hillslope and may stay at freezing point while the surrounding slopes remain frost-free.

r линза холодного воздуха
Небольшая котловина, куда под действием силы тяжести скатывается холодный воздух с окружающих склонов.

f poche *f* de gelée
s bolsa *f* de heladas
g Kaltluftsenke *f*

* **frost polygon → 1360**

2335 **frost resistance; frost stability**
The power of plants, animals and grounds to remain unharmed or unaffected by hard frosts.

r морозостойкость
Способность грунтов, растений и животных противостоять сильным морозам.

f résistance *f* à gelée; stabilité *f* à froid
s resistencia *f* a heladas; resistencia *f* al frío
g Frostfestigkeit *f*

2336 **frost resistant ground**
The ground with the little ability to heave during freezing.

r морозоустойчивый грунт
Грунт, незначительно вспучивающийся в условиях низких температур.

f sol *m* non-gélif
s suelo *m* estable a congelación
g froststabiler Boden *m*

* **frost shattering → 2339**

2337 **frost smoke; arctic smoke**
Fog occurring in high latitudes when an ice-cold air from land passes over a warmer air lying above the water surface and causes the water in the air to condence.

r морозный туман; арктическая дымка; морозная дымка
Туман, случающийся в высоких широтах, когда очень холодный воздух с суши оказывается над более теплым воздухом, лежащим над водным пространством, что вызывает конденсацию влаги в холодном воздухе.

f fumée *f* de mer arctique; voile *m* froid
s humo *m* de mar *m* ártico; humo *m* frío
g Polarnebel *m*; Frostnebel *m*; Eisnebel *m*; Frostrauch *m*

2338 **frost sorting**
The differential movement of stones on the surface in periodically freezing and thawing ground, which causes the formation of stone borders along cryogenic polygons.

r морозная сортировка; мерзлотная сортировка
Перемещение камней на поверхности периодически замерзающего и оттаивающего грунта в соответствии с их размерами, что приводит к образованию каменных бордюров у криогенных полигонов.

f triage *m* gélival
s selección *f* por congelación
g frostbedingte Sortierung *f*

* **frost stability → 2335**

2339 **frost weathering; frost shattering; gelivation; congelifraction; freeze-thaw action**
A form of weathering in periglacial areas where the temperature hovers around freezing point, below which frost breaks up the rock and above which the ice melts, so that water flows and carries away the rock fragments.

r морозное выветривание; криогенное выветривание

Процесс разрушения горных пород в условиях частого замерзания и оттаивания при колебаниях температуры около 0°C.
f cryoclastie *f*; météorisation *f* gélivale; gélifraction *f*
s crioclastia *f*; criometeorización *f*; gelifracción *f*
g Frostspendung *f*; Frostverwitterung *f*

2340 **frost wedging**
A kind of frost weathering in which the fractured rocks are broken up and split by the pressure of water frozen in cracks and fissures acting like a wedge.
r криогенное расклинивание
Тип морозного выветривания, когда растрескивающаяся порода взламывается и смещается льдом, действующим как клин.
f délogement *m* gélival
s crioclastismo *m*
g Frostspaltung *f*

* **frosts → 3489**

2341 **frozen ground**
Ground or rock having temperatures at or below 0°C and containing water in the solid state.
r мерзлый грунт; мерзлая горная порода
Грунт с отрицательной температурой и влагой в замерзшем состоянии.
f gélisol *m*
s suelo *m* congelado
g Frostboden *m*

2342 **frozen snow crust; wind slab snow**
A hard firn layer formed on the surface of snow when a steadily blowing moist wind makes the snow to freeze.
r наст; ветровая корка
Уплотненная фирновая корка, образующаяся путем смерзания снега на поверхности под действием устойчивого влажного ветра.
f croûte *f* de neige
s costra *f* de nieve
g Schneekruste *f*; Windkruste *f* im Schnee

2343 **frozen waterfall**
Ice columns on a steep mountain slope formed by the freezing of a falling water flow.
r замерзший водопад
Массив льда на крутых и обрывистых склонах гор, образующийся в результате замерзания падающих водяных струй.
f chute *f* d'eau congelée
s salto *m* de agua congelada
g gefrorener Wasserfall *m*

* **fuel-and-energy resources → 1908**

* **fuel-energy balance → 1903**

* **full depth avalanche → 2646**

2344 **full Moon**
A phase of the Moon when the Earth lies between the Moon and the Sun, so that viewed from Earth the whole face of the Moon is illuminated by light from the Sun (in the sky the full Moon is situated near the point opposite to the Sun).
r полнолуние
Фаза Луны, когда она находится близ точки неба, противоположной Солнцу, обращена к Земле всем своим освещенным полушарием и имеет вид полного диска.
f pleine lune *f*
s luna *f* llena; plenilunio *m*
g Vollmond *m*

2345 **full-time employment**
Working population engaged in his regular paid work whole working week.
r полная занятость
Трудоспособное население, имеющее работу полную рабочую неделю.
f plein emploi *m*
s plena ocupación *f*
g Vollbeschäftigung *f*

2346 **fully mature shoreline**
A levelled shoreline the bays of which are filled with sea and river silt or separated from the open sea by spits, while the capes are cut by surf impact.
r зрелая береговая линия
Выровненная береговая линия, бухты которой заполнены морскими и речными наносами или отделены от открытого моря пересыпями, а мысы срезаны деятельностью прибоя.
f ligne *f* de rivage mûr
s línea *f* de costa madura
g reife Küstenlinie *f*; reife Strandlinie *f*

2347 **fumarole**
Small fissures and canals in the volcanic crater, from where the effusion of hot volcanic gases and water vapours are ejected, often with considerable force in the form of jets.
r фумарола
Небольшие отверстия и трещины, по которым в виде струй поднимаются горячие вулканические газы и пары воды, выделяющиеся из магмы, неостывших лавовых потоков и пирокластических отложений.
f fumerolle *f*
s fumarola *f*
g Fumarole *f*

2348 **funicular**
A cablerailway in which two small cars or carriages move up and down on the same cable counterbalancing one another.
r фуникулёр
Кабельно-рельсовая дорога на склоне, по которой два небольших вагона двигаются вверх и вниз на едином кабеле, уравновешивая друг друга.
f funiculaire *m*

s funicular *m*; ferrocarril *m* funicular
g Drahtseilbahn *f*

2349 funnel cloud; tornado cloud
A typical funnel- or trunk-shaped cloud in the lower part of a cumulonimbus cloud, which may develop into a spout or tornado.
r воронкообразное облако; воронка смерча; облачная воронка
Свисающая в нижней части кучево-дождевого облака «воронка» или «хобот», возникающие в начальной фазе развития торнадо или смерча.
f nuage *m* en entonnoir
s nube *f* de embudo
g Trichterwolke *f*

2350 fur farm
A farm concerned with the rearing of certain fur-bearing animals under controlled conditions with the objective of producing fur pelts for sale.
r пушная ферма
Ферма, на которой в определенных условиях содержатся животные, которые используются для продажи меха.
f ferme *f* de fourrures
s granja *f* de pieles;
granja *f* de animales de piel fina
g Pelzfarm *f*

2351 fur-farming
Rearing of fur-bearing animals for fur pelts (especially developed in the U.S., Canada, Russia, Northern and Central Europe).
r пушное звероводство; звероводство
Отрасль животноводства, занимающаяся разведением в неволе ценных пушных зверей для получения шкурок (пушнины), особенно развито в США, Канаде, России, странах Северной и Средней Европы.
f élevage *m* des animaux fourrures
s cría *f* de animales de piel fina
g Pelztierzucht *f*

2352 furrow
A narow trench made in the earth by a plough, into which seed is sown; a relatively small linear depression made on the ground by running water.
r борозда
Мелкий ров, сделанный плугом при распашке земли, куда помещают зерна при засеве; небольшая линейная рытвина, размытая текущей водой.
f sillon *m*
s surco *m*; ranura *f*
g Furche *f*; Ackerfurche *f*

G

2353 gafe
A special form of a gulf in river mouths of the Southern Hemisphere resembling a freshwater lagoon separated from the sea by islands or sandy bars.
r гафы
Особая форма залива в устьях рек Южного полушария в виде пресноводного лимана, отделенного от моря островами или песчаными грядами.
f gafe *m*
s gafe *m*
g Haffe *pl*; Nebenarm *m*

2354 galaxy
One of the great number of systems in the universe consisting of stars, planets and other celestial bodies (the Earth lies in a spiral galaxy called The Milky Way, or The Galaxy).
r галактика
Одна из огромного числа звездных систем во Вселенной, состоящих из звезд, планет и других небесных тел (Земля принадлежит к Галактике, называемой также Млечный путь).
f galaxie *f*
s galaxia *f*
g Galaxis *f*; Galaxie *f*

* **galeria forest → 2355**

* **gallery → 82**

2355 gallery forest; fringing forest; galeria forest; forest gallery
The flood plain forest fringing both river banks in otherwise open areas in savannas of Africa and America, the foliage of trees may meet in an arch over narrow streams forming a tunnel (in Central Asia, they are called *tugai*).
r галерейный лес
Пойменный лес по берегам рек, текущих в безлесных областях, в саваннах Африки и Америки (в Средней Азии их называют *тугаями*).
f forêt-galérie *f*; galérie *f* forestière
s bosque-galería *f*; galería *f* forestal
g Galeriewald *m*

2356 game reserve
A protected area set aside for the acclimatization and intensive reproduction of some game animal species where strictly regulated hunting is allowed.
r заповедно-охотничье хозяйство
Охраняемая территория, на которой акклиматизируются и воспроизводятся охотничьи виды животных и разрешена строго регулируемая охота.
f réserve *f* de chasse
s reserva *f* de caza
g Jagdschutzgebiet *n*; Jagdreservat *n*

2357 gap
A deep, flat through depression, a break in a belt of hills or in a mountain ridge of tectonic genesis about ten or more kilometres wide (if the break is just an indent and no stream is present, it is termed a dry gap or *wind-gap*, if the cutting is deep and a stream flows through it, it is termed a *water-gap*).
r горный проход
Сквозное глубокое и плоскодонное понижение в горном хребте тектонического происхождения шириной до десяти и более километров (если в этом проходе нет реки, он называется *сухой проход*, а если течет река – это *сквозная долина*).
f défilé *m*
s desfiladero *m*; paso *m*
g Gebirgspaß *m*; Engpaß *m*

2358 gara
An isolated rock in a desert shaped as a mushroom by the wind action.
r гара
Грибовидная скала в пустынях, сформированная ветровой абразией.
f gara *f*; butte *m* témoin rocheuse
s gara *f*; cerro *m* testigo rocoso
g Gara *f*

* **garbage → 5143**

2359 garden city
A planned industrial-residential settlement, rural in character, with open spaces and trees, catering for inhabitatnts of various grades of society.
r город-сад
Хорошо спланированное промышленное и жилое поселение, имеющее сельский облик, со значительными открытыми пространствами и участками леса, предназначенное для представителей самых разных слоев общества.
f cité-jardin *m*
s ciudad jardín *m*
g Gartenstadt *f*

* **gardening → 2824**

2360 garigue
An open scrubland of low evergreen xerophytic shrubs, mainly dwarf oak (Quercus coccifera) or dwarf palm (Chamaeros humilis) covering stony and rocky slopes of the Mediterranean region (it is anthropogenic origin resulting from the effect of human settlement, fire, browsing and grazing).
r гаррига
Сообщество низкорослых разреженных вечнозеленых ксерофитных кустарников, главным образом кустарникового дуба (Quercus coccifera) или карликовой пальмы

(Chamaeros humilis) на каменистых и скалистых склонах Средиземноморья.

f garrigue *f*
s garriga *f*
g Garrigue *f*

2361 **garúa**
A dense fog and/or drizzle falling from low stratus on the western coast of South America (Peru, Ecuador and Chile).
r гаруа
Плотный моросящий туман, оседающий зимой и весной на пустынных побережьях Эквадора, Перу и Чили.
f garua *f*
s garúa *f*
g Garua *f*

2362 **gas; natural gas**
A substance or mixture of substances in a gaseous state trapped in pore spaces in rocks with or without liquid petroleum used as a fuel, for provision of light and heat, and as raw material for the chemical industry for making plastics, detergents, fertilizers, etc.
r газ; природный газ
Вещество или смесь веществ в газообразном состоянии, вкрапленные в порах горных пород вместе с жидкой нефтью или без нее, используемое в качестве топлива, а также как сырье для производства пластических масс, моющих средств, удобрений и т.п.
f gaz *m* naturel
s gas *m* natural
g Erdgas *n*; Naturgas *n*

2363 **gas hydrate**
A special form of chemical compounds formed at temperatures below 0°C or slightly above 0°C and high pressure which mainly consist of solid water molecules and guest molecules are those of varios gases.
r газогидрат
Особая форма химических соединений (образующихся при отрицательных и низких положительных температурах и повышенном давлении), в которых основную массу составляют молекулы твердой фазы воды, а инородные молекулы представлены в основном газами.
f hydrate *m* gazeux
s hidrato *m* de gas
g Gashydrat *n*

2364 **gas-pipeline**
A pipeline used to transport inflammable gases from the places of extraction to areas of use.
r газопровод
Трубопровод для транспортировки горючих газов от мест их добычи или производства к районам потребления.
f gazoduc *m*
s gasoducto *m*; conducción *f* de gas
g Gasleitung *f*

2365 **gästarbeiter; guest worker**
A foreign migrant worker usually engaged in manual labour.
r гастарбайтер
Иностранный мигрант, как правило, занятый неквалифицированным трудом.
f hôte-ouvrier *m*
s huésped-obrero *m*
g Gästarbeiter *m*

2366 **gat; ship canal**
An inshore channel or strait dividing offshore islands from the mainland, as well as an artificially deepened passage in a sea bay connecting port constructions with the open sea.
r морской канал
Пролив между песчаными отмелями, а также искусственно углубленный проход в морском заливе, соединяющий портовые сооружения с открытым морем.
f canal *m* navigable
s canal *m* navegable
g Meereskanal *m*

2367 **gate**
A narrowed part of a river valley with high steep slopes.
r ворота
Суженная часть речной долины, окаймленная по бокам высокими крутыми склонами.
f porte *f*
s puerta *f*
g Pforte *f*

2368 **gated community; exclusionary enclave**
A group of houses or flats in an area surrounded by a fence or wall that can be entered only by people who live their their guests.
r замкнутое сообщество
Территория богатой застройки, изолированная от окружающего пространства с помощью забора и контролируемых входов, доступная лишь для узкого круга проживающих здесь людей и их гостей.
f communauté *f* cloitrée
s comunidad *f* cerrada; comunidad-fortaleza *f*
g geschlossene Gesellschaft *f*

2369 **gateway city**
A settlement, usually occurring a favourable, commanding site, which acts as a link between two areas and often becomes the main city in the region.
r пограничный город
Поселение, обычно расположенное в удобном месте и служащее связующим звеном между двумя районами, часто становящееся главным городом на данной территории.
f cité *f* de porte
s ciudad *f* de puerta
g Grenzstadt *f*

2370 **gathering**

An economy activity of low level where people collected elible plants and materials.

r собирательство

Примитивная экономическая деятельность людей, когда они собирают легко доступные растения и материалы.

f cueillette *f*

s recolección *f*

g Aufsammeln *n*

* **gauge line → 2892**

* **gauging rod → 5275**

* **gauging station → 2885**

2371 **Gauss-Krüger coordinates**

A system of flat rectangular geodetic co-ordinates where the central meridian of a given territory is considered as X-axis, and the equator as Y-axis.

r координаты Гаусса-Крюгера

Система плоских прямоугольных геодезических координат, где за ось X принимается осевой меридиан рассматриваемой территории, а за ось Y – экватор.

f coordonnées *fpl* de Gauss-Krüger

s coordenadas *fpl* de Gauss-Krüger

g Gauss-Krüger Koordinaten *fpl*

2372 **gazetteer**

An alphabetic list of place-names and geographical features, usually with a reference to their location by coordinates and sometimes with descriptive information.

r указатель географических названий; газетир

Алфавитный перечень географических названий с указанием местоположения объектов, к которым эти названия относятся и нередко с дополнительной информацией о них.

f index *m* des noms géographiques

s índice *m* de nombres geográficos

g Namenverzeichnis *n*

* **GDP → 2639**

2373 **geest**

A landscape of flat sandy lowlands (sandur) in Germany and the Netherlands near the coast of the North Sea.

r геесты

Ландшафт плоских песчаных низменностей (зандровых равнин) в Германии и Нидерландах близ побережья Северного моря.

f geest *m*

s geest *m*

g Geest *m*

2374 **gelifluction**

The slow downslope flow of unfrozen earth materials on a frozen substrate, a type of solifluction implying the presence of either seasonal frost or permafrost.

r желифлюкция

Медленное движение вниз по склону рыхлого грунта на мерзлом субстрате, вид солифлюкции на сезонном или постоянно мерзлом основании.

f gélifluxion *f*

s gelifluxión *f*

g Gelifluktion *f*

* **gelisol → 1362**

* **gelivation → 2339**

2375 **gendarme**

A sharp-pointed rock on a narrow mountain ridge, usually close to the highest top of a ridge.

r жандарм

Остроугольный выступ скального гребня, обычно по соседству с высшей точкой хребта.

f gendarme *m*

s gendarme *m*

g Bergzinne *f*

2376 **gender**

The different parts played by males and females in the social and economic life of a society, often determined by custom and tradition.

r пол

Разная роль, которую играют мужчины и женщины в социальной и экономической жизни общества, нередко определяемая обычаями и традициями.

f genre *m*

s genero *m*

g Gens *f*

2377 **gender geography**

A branch of human geography that concentrates on the gender differences in society, particularly access to employment, social services etc. and the extent to which women are disadvantaged.

r гендерная география

Отрасль социальной географии, изучающая роль женщины в общественной жизни, особенно их занятость в трудовом процессе, социальных службах и пр., а также причины, вызывающие неравновное положение женщины в разных странах.

f geographie *f* du genre

s geografía *f* genérica; geografía *f* de genero

g feministische Geographie *f*

2378 **gene pool**

The total amount of genes or the amount of genetic information possessed by all the species of living things inhabiting the Earth.

r генетический фонд; генетические ресурсы; генофонд

Совокупность генов или генетической информации, которой обладают все биологические виды, населяющие Землю.

f banque *f* génétique; fonds *m* génique

s banco *m* genético

g Genbestand *m*; Genbank *f*; Genpool *m*

2379 **general cargo**
Packaged and individual cargoes carried in batches as distinct from *bulk cargoes.*
r генеральные грузы
Тарные (в упаковке) и штучные грузы, перевозимые партиями, в отличие от *насыпных грузов.*
f marchandises *fpl* générales
s cargamento *m* general; mercancías *fpl* generales
g Generalgut *n*

2380 **general geography**
The study of the Earth's physical environment in the most general features of its composition, structure, development, dynamics, and functioning.
r землеведение
Изучение географической оболочки Земли и наиболее общих закономерностей ее состава, структуры, развития, динамики и функционирования.
f géographie *f* générale
s geografía *f* general
g Erdkunde *f*

2381 **general geomorphology**
The study concerned with landforms, especially morphological features, genesis, evolution and processes involved in the formation of the surface forms of the Earth both on land and in oceans.
r общая геоморфология
Наука, изучающая рельеф земной поверхности на суше, дне океанов и морей – его внешние признаки, происхождение и законы развития.
f géomorphologie *f* générale
s geomorfología *f* general
g allgemeine Geomorphologie *f*

2382 **general hydrology**
The science dealing with the Earth's water (the origin, distribution and circulation) as well as phenomena and processes in the hydrosphere.
r гидрология
Наука, изучающая природные воды (их происхождение, распространение, циркуляцию), а также явления и процессы, протекающие в гидросфере.
f hydrologie *f*
s hidrología *f*
g Hydrologie *f*; Gewässerkunde *f*

* **general map → 992**

2383 **generation of waves**
Any disturbance caused by the action of wind, pressure gradient, etc. in the water or air medium in the formation of a wave motion.
r генерирование волн
Возмущение водной или воздушной массы, вызванное внешними воздействиями: колебаниями атмосферного давления, ветрами и пр.
f génération *f* des vagues
s generación *f* de olas
g Wellenanregung *f*

2384 **genocide; ethnic cleansing**
The deliberate murder or persecution of people for their belonging to a definite group or for their ethnic origin, the unexpressed agreement to hold guilty people on the strength of their belonging to a definite social, racial, cultural or biological group.
r геноцид; этническая чистка
Уничтожение или преследование людей по признаку определенной общности или происхождения, негласное признание виновности людей в принадлежности к той или иной социальной, культурной или биологической группе.
f génocide *m*
s genocidio *m*
g Völkermord *m*; Genozid *n*

2385 **gentle slope**
A slope smoothed down and turned to the slightly inclined plane.
r пологий склон
Мягкий, выровненный склон, постепенно переходящий в слабо наклонную равнину.
f pente *f* douce
s pendiente *f* suave
g sanfer Hang *m*

2386 **gently rolling plain; rolling plain**
A plain not deeply dissected by erosion, with a gently rising and falling landscape.
r полого-волнистая равнина; волнистая равнина
Равнинные пространства, подвергшиеся неглубокому эрозионному расчленению.
f plaine *f* ondulée
s llanura *f* ondulada
g wellige Ebene *f*

2387 **gentrification**
The rebuilting, renewing, and rehabilitation of depressed areas of the inner city as more affluent families seek to live near the city centre, trading off space and quiet for access to the goods and services of the city centre.
r джентрификация
Возрождение центрального района города, улучшение в нем качества окружающей среды, сопровождающееся подъемом здесь деловой активности, приходит на смену субурбанизации и дезурбанизации.
f gentrification *f*
s gentrificación *f*
g Gentrifikation *f*

2388 **geoarcheology**
A conjugated analysis of cultural, economic, geological and paleogeographical data for certain archaeological sites in order to find out the relations between the existed human society and the environment.

Table 1. **Geochronology** of the Earth

Time	Eon 1)	Era	Period		Epoch
0	PRESENT TIME				
0.1	PHANEROZOIC	CAINOZOIC	QUARTERNATY		HOLOCENE
1.8					PLEISTOCENE
7			TERTIARY	NEOGENE	PLIOCENE
26					MIOCENE
38				PALAEOGENE	OLIGOCENE
55					EOCENE
65					PALAEOCENE
136		MESOZOIC	CRETACEOUS		3)
195			JURASSIC		
225			TRIAS		
280		PALAEOZOIC	PERMIAN		
345			CARBONIFEROUS		
395			DEVONIAN		
440			SILURIAN		
500			ORDOVICIAN		
570			CAMBRIAN		
2500	PRECAMBRIAN (CRYPTOZOIC)	PROTEROZOIC			
4600		ARCHAEAN			

1)In the column time of beginning of a period or epoch is given (in mln years ago).
2)History of the Earth is divided into two eons – Cryptozoic (eon of latent life) and Phanerozoic (eon of real life).
3)Epochs of these periods have no special names, and are indicated as early, middle and late epochs of a corresponding period (e.g. early Triassic).

r геоархеология
Сопряженный анализ культурно-хозяйственных, геологических и палеогеографических данных для конкретных археологических местонахождений с целью установления причинных связей между существовавшим человеческим обществом и окружающей средой.
f géoarchéologie *f*
s geoarqueología *f*
g Geoarchäologie *f*

2389 geocentric *adj*
Relating to the Earth as a centre, e.g. referring to the motion of the Moon or an artificial satellite that has the Earth as centre of mass.
r геоцентрический
Относящийся к Земле как центру пространства, например, к движению Луны или искусственного спутника Земли, которые вращаются вокруг Земли как центра массы.
f géocentrique *adj*
s geocéntrico *adj*
g geozentrisch *adj*

2390 geochemical barrier
A part of the environment where the migration of chemical elements and compounds slows down and their accumulation takes place.
r геохимический барьер
Участок природной среды, на котором резко снижается интенсивность миграции химических элементов и соединений и происходит их концентрация.
f barrière *f* géochimique
s barrera *f* geoquímica
g geochemische Sperre *f*

2391 geochemical endemia
Diseases of living organisms, including human being, caused by the geochemical properties of the area.

r геохимическая эндемия
Заболевание живых организмов, включая человека, вызванное геохимическими особенностями территории.
f endémie *f* géochimique
s endemia *f* geoquímica
g geochemische Endemie *f*

2392 **geochemical landscape**
An area which joins several structural landscape elements, and where the migration of chemical elements of the atmosphere, hydrosphere, lithosphere and living matter takes place.
r геохимический ландшафт
Участок территории, объединяющий несколько элементов ландшафта, где происходит миграция химических элементов атмосферы, гидросферы, литосферы и живого вещества.
f paysage *m* géochimique
s paisaje *m* geoquímico
g geochemische Landschaft *f*

2393 **geochemistry**
The study of the chemical composition of the Earth, the distribution of the Earth's chemical elements and the way they react in natural processes.
r геохимия
Изучение химического состава Земли, закономерностей распределения химических элементов и их изотопов, особенностей их проявления в природных процессах.
f géochimie *f*
s geoquímica *f*
g Geochemie *f*

2394 **geochron**
A time interval equivalent to a stratigraphic unit in rocks which may vary between different locations, depending on the age of the rock unit.
r геохрон
Интервал времени, эквивалентный стратиграфической единице в горных породах, разнящийся в разных местах в зависимости от возраста пород.
f géochron *m*
s geocron *m*
g Geochron *m*

2395 **geochronology; geological timescale**
1. The order in which the rocks making up the Earthrs crust formed. 2. The determination of the age of rocks by using geological data. → Table 1, p. 284
r геохронология
1. Хронологическая последовательность формирования горных пород, слагающих земную кору. 2. Определение возраста горных пород геологическими методами. → Табл. 1, с. 284
f géochronologie *f*
s geocronología *f*
g Geochronologie *f*

2396 **geocoding**
The method and process of positioning of objects with regard to a system of coordinates.
r геокодирование
Метод и процесс позиционирования пространственных объектов относительно некоторой системы координат.
f géocodage *m*
s geocodificación *f*
g Geokodierung *f*

2397 **geocratic period**
A period in the geological history of the Earth when there was a considerable extension of land surface, as opposed to a *thalassocratic period*.
r геократический период
Период в геологической истории Земли, когда происходило значительное расширение площади суши, в противоположность *талассократическому периоду*.
f période *f* géocratique
s período *m* geocrático
g geokratische Phase *f*

* **geocryology → 4509**

* **geodesic control point → 2406**

* **geodesic line → 2402**

2398 **geodesy**
The science of measurements of the size and shape of the Earth, together with the determination of the exact position of particular points on its surface.
r геодезия
Совокупность средств и методов измерений и вычислений пространственного положения объектов земной поверхности, размера и формы Земли и изменений этих параметров во времени.
f géodésie *f*
s geodesia *f*
g Geodäsie *f*; Erdmessung *f*

2399 **geodetic base line**
The distance between two fixed points measured with high accuracy for the determination of the lengths of lines of the geodetic net.
r геодезический базис
Расстояние между двумя закрепленными на местности точками, измеренное с высокой точностью и используемое для вычисления длин сторон геодезической сети.
f base *f* géodésique
s base *f* geodésica
g Basisgrundlinie *f*

2400 **geodetic control**
A set of geodetic data necessary for making maps and determining the location of objects on a map by latitude, longitude and altitude.
r геодезическая основа карты

Совокупность геодезических данных, необходимых для создания карты и определяющих положение объектов на карте по широте, долготе и абсолютной высоте.
f canevas *m* géodésique
s retícula *f* geodésica
g geodätische Grundlage *f*

2401 geodetic coordinates
The latitude and longitude of a point of the Earthrs surface determined by way of geodetic measurements of distances and directions from a point with known geographic coordinates, as well as the elevation of the point with regard to the surface of the Earth ellipsoid.
r геодезические координаты
Широта и долгота точки земной поверхности, определенные путем геодезических измерений расстояния и направления от точки с известными географическими координатами, а также высота точки относительно поверхности земного эллипсоида.
f coordonnées *fpl* géodésiques
s coordenadas *fpl* geodésicas
g geodätische Koordinaten *fpl*

2402 geodetic distance; geodetic line; geodesic line
The shortest distance between any two points on the Earth's surface (it has an arc shape of a large circle on the sphere, and a straight line on a plane surface).
r геодезическая линия
Линия кратчайшего расстояния между двумя пунктами на поверхности, в том числе на эллипсоиде (на сфере – это дуга большого круга, на плоскости – прямая).
f ligne *f* géodésique
s línea *f* geodésica
g geodätische Linie *f*

* **geodetic framework → 2405**

2403 geodetic framework of astronomical points
System of interconnected surface geodetic points situated at a distance of about 70-100 km from each other, the location of which is determined by astronomic methods.
r астрономо-геодезическая сеть
Система связанных между собой наземных геодезических пунктов, положение которых определено астрономическими способами, расположенных друг от друга на расстоянии порядка 70-100 км.
f réseau *m* astrogéodésique
s red *f* astronómico-geodésica
g geodätisches Netz *n* von astronomischen Punkten

2404 geodetic intersection
The way of determination of coordinates of a point by measuring the of parameters at it or at the initial points with known coordinates.
r геодезическая засечка; засечка
Способ определения координат точки измерением параметров на ней или на исходных пунктах с известными координатами.
f intersection *f* géodésique; intersection *f*
s intersección *f* geodésica; intersección *f*
g geodätisches Einschneiden *n*

* **geodetic line → 2402**

* **geodetic net → 2405**

2405 geodetic network; geodetic net; geodetic framework
Net of interconnected points fixed on the Earth's surface, the position of which is determined in the common system of co-ordinates.
r геодезическая сеть
Сеть взаимосвязанных пунктов, закрепленных на земной поверхности, положение которых определено в общей для них системе координат.
f réseau *m* géodésique
s red *f* geodésica
g geodätisches Netz *n*

2406 geodetic point; geodesic control point; fixed point
A point on the Earth's surface for which its position in general system of coordinates is determined and which serves as an element of geodetic network.
r геодезический пункт; опорная точка
Точка на земной поверхности, для которой определено ее положение в общей системе координат и которая служит составляющей геодезической сети.
f point *m* géodésique; point *m* de repère
s punto *m* geodésico; punto *m* de referencia
g geodätischer Punkt *m*; Festpunkt *m*

2407 geodetic satellite
An artificial satellite intended to determine the position of particular points on the Earth's surface as well as to measure the form and dimensions of celestial bodies.
r геодезический спутник
Искусственный спутник Земли, предназначенный для определения положения географических объектов на земной поверхности, а также для определения формы и размеров космических тел.
f satellite *m* géodésique
s satélite *m* geodésico
g geodätischer Satellit *m*

2408 geoecology; ecogeography
A branch of geography studying the interactions between humans and landscapes of different hierarchical levels including the biosphere.
r геоэкология
Раздел географии, исследующий взаимодействия человека и ландшафтов на разных иерархических уровнях, вплоть до биосферы.

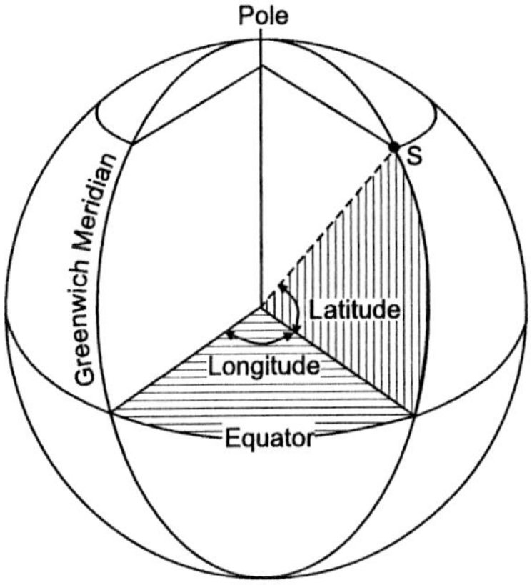

Fig. 124. **Geographic coordinates** of point S

f geoécologie *f*; écogéographie *f*
s geoecología *f*; geografía *f* medioambiental
g Geoökologie *f*

2409 geoeconomics
The study developing in the interface of economics, economic geography and geopolitics dealing with the interactions of large geographical regions, states, transnational corporations, etc. on a global scale.
r геоэкономика
Научная дисциплина на стыке экономической науки, экономической географии и геополитики, изучающая действия субъектов экономического развития (крупные географические районы, государства, транснациональные корпорации и пр.) на глобальном уровне.
f géoéconomie *f*
s geoeconomía *f*
g Geoökonomie *f*

* **geographic *adj* → 2413**

* **geographic atlas → 419**

2410 geographic coordinates
Linear and angular quantities determining the location of a point on the Earth's surface (latitude and longitude). → Fig. 124
r географические координаты
Величины, определяющие положение точки на земной поверхности (географическая широта и географическая долгота). → Рис. 124
f coordonnées *fpl* géographiques
s coordenadas *fpl* geográficas
g geographische Koordinaten *fpl*

* **geographic information system → 2436**

* **geographic map → 3653**

* **geographic object → 2418**

2411 geographic position; geographical location; geographical situation; geographic situation
Characteristics of the situation of an object on the Earth's surface in relation to other geographical objects and cardinal points.
r географическое положение
Характеристика расположения объекта на земной поверхности относительно других географических объектов и стран света.
f situation *f* géographique
s ubicación *f* geográfica; posición *f* geográfica; situación *f* geográfica
g geographische Lage *f*

* **geographic range → 374**

* **geographic situation → 2411**

2412 geographic space
A territory with interconnected geographic objects situated on it and developing in time.
r географическое пространство; геопространство
Территория с расположенными на ней взаимосвязанными географическими объектами, развивающимися во времени.
f espace *m* géographique
s espacio *m* geográfico
g geographischer Raum *m*

* **geographical name → 6252**

2413 geographical *adj*; geographic *adj*
Relating to objects and phenomena in the environment, including formed by human activity, which have the spatial distribution-and which can be represented on a map.
r географический
Относящийся к объектам и явлениям окружающей среды, в том числе вызванным деятельностью людей, имеющим пространственное распространение и отраженным на географической карте.
f géographique *adj*
s geográfico *adj*
g geographisch *adj*

2414 geographical belt
The largest latitudinal subdivision of the landscape geosphere which is characterized by the common values of incoming solar radiation, circulation of air masses, rhythm of biochemical processes, etc., it is subdivided into zones and latitudinal subzones.
r географический пояс
Наиболее крупное широтное подразделение географической оболочки, характеризующееся общими особенностями поступления тепла, циркуляции воздушных масс, определенной ритмикой биогеохимических процессов и пр., подразделяется на географические зоны и подзоны.
f ceinture *f* géographique
s cinturión *m* geográfico

g geographischer Gürtel *m*

2415 geographical boundary
Not a very sharp line or transitional belt marking the limits of a unit of land, a geographical region, etc., e.g. fault line, shoreline, boundaries between geographical zones.
r географическая граница
Переходная полоса (как правило, не резкая), разделяющая смежные пространственные (географические) образования, существенно различающиеся между собой, например, линия сброса, береговая линия, зональные границы.
f limite *f* géographique
s límite *m* geográfico
g geographische Grenze *f*

2416 geographical description
A traditional method of the geographical investigation and presenting of its results consisting in the generalization of materials collected during the field investigation of a territory.
r географическое описание
Традиционный метод географического исследования и изложения его результатов, заключается в обобщении материалов, собранных во время полевого обследования территории.
f description *f* géographique
s descripción *f* geográfica
g geographische Beschreibung *f*

2417 geographical determinism
The theory that states that the environment (particularly its physical factors) dominates, even determines, the pattern of human life and human behaviour and that people are largely conditioned by environmental factors.
r географический детерминизм; вульгарный географизм
Концепция, признающая преобладание или даже доминирование природных факторов в жизни и поведении людей, зависимость человеческого общества от окружающей среды.
f déterminisme *m* géographique
s determinismo *m* geográfico
g geographischer Determinismus *m*

* **geographical distribution → 3562**

* **geographical division of labour → 6115**

2418 geographical feature; geographical object; geographic object
A natural or man-made integral geographical formation with a certain stable position on the Earth's surface.
r географический объект
Естественный или созданный человеком целостный и относительно стабильный объект с определенным постоянным положением на поверхности Земли.
f objet *m* géographique
s objeto *m* geográfico
g geographisches Objekt *n*; Geoobjekt *n*

* **geographical horizon → 4054**

2419 geographical inertia
The tendency of a place with established installations and services to maintain its size and its importance as a focus of activity after the conditions originally influencing its development have appreciably altered, ceased to be relevant, or disappeared.
r географическая инерция
Тенденция определенного места со сложившимся размещением и обслуживанием сохранять некоторое время свои размеры и значение как фокус активности после того, как условия, влияющие на его развитие, заметно изменились или исчезли.
f inertie *f* géographique
s inercia *f* geográfica
g geographische Trägheit *f*

2420 geographical landscape; landscape
A territorial unit with a common origin and history of development, geological basement, relief and climate, a uniform combination of soils, biocoenoses, and structure which serves as the main unit of physical-geographical regionalization.
r географический ландшафт
Конкретная территориальная единица, однородная по происхождению и истории развития, обладающая единым геологическим фундаментом, однотипными рельефом, климатом, единообразным сочетанием почв, биоценозов и определенной присущей ей структурой (основная единица физико-географического районирования).
f paysage *m* géographique; paysage *m*
s paisaje *m* geográfico; paisaje *m*
g geographische Landschaft *f*

* **geographical location → 2411**

* **geographical market → 3712**

2421 geographical modelling
The description of natural processes and phenomena with a set of mathematical equations with corresponding boundary conditions and algorithm for their numerical solution.
r географическое моделирование
Отображение природных процессов и явлений с помощью системы математических уравнений с соответствующими краевыми условиям и алгоритмом их численного решения.
f moulage *m* géographique
s modelación *f* geográfica
g geographische Modellierung *f*

* **geographical object → 2418**

2422 geographical pathology
A branch of medical geography studying the

deviation from normal health in humans, plants and animals caused by the environment.

r географическая патология
Отрасль медицинской географии, изучающая отклонения от нормы состояния здоровья человека, животных и растений в связи с особенностями окружающей среды.

f pathologie *f* géographique
s patología *f* geográfica
g geographische Pathologie *f*

* **geographical pole → 4660**

2423 geographical possibilism; possibilism
A concept which considers geographical environment as a factor limiting and changing the human activity, at the same time accepts in contrast to *determinism* that within those limits there is a set of opportunities which offer humans freedom of choice of action.

r географический поссибилизм; поссибилизм
Концепция, рассматривающая географическую среду как фактор, ограничивающий и изменяющий деятельность людей, но признающая (в отличие от *географического детерминизма*) определенную роль исторических и иных условий.

f possibilisme *m* géographique
s posibilismo *m* geográfico
g geographischer Possibilismus *m*; Possibilismus *m*

2424 geographical prognosis; geographical prognostication
A prediction based on scientific observation, experience, and an estimate of probability, of how natural and socioeconomic systems will develop in future on global, regional and local levels.

r географический прогноз
Научно обоснованное суждение о состоянии и тенденциях развития природных и природно-хозяйственных систем в будущем (прогнозы могут быть отраслевыми и комплексными, а по территориальному охвату – глобальными, региональными и локальными).

f prévision *f* géographique
s previsión *f* geográfica
g geographische Prognose *f*

* **geographical prognostication → 2424**

* **geographical situation → 2411**

2425 geographical subzone
A part of a geographical zone with distinctive local features found in a zonal type of landscapes, e.g. in the steppe zone of Eurasia – the subzone of northern steppes on chernozems and the subzone of southern steppes on dark-chestnut soils.

r географическая подзона
Часть географической зоны с достаточно выраженными местными особенностями в зональном типе ландшафтов, например, в зоне степей Евразии – подзона северных степей на черноземах и подзона южных степей на темно-каштановых почвах.

f subzone *f* latitudinale
s subzona *f* latitudinal
g geographische Subzone *f*

2426 geographical zone
A latitudinal strip of land or the World Ocean which has homogeneous thermal characteristics and atmospheric moisture, and relatively similar landscape elements (it forms part of a geographical belt).

r географическая зона; природная зона; ландшафтная зона
Широтная природная полоса суши или Мирового океана, имеющая однородные термические условия и атмосферное увлажнение, и соответственно относительное однородные элементы ландшафтов, является составной частью географического пояса Земли.

f zone *f* latitudinale
s zona *f* latitudinal
g geographische Zone *f*

2427 geography
The science studying the phenomena which make up the Earth's physical environment together with people and human economic activities, their spatial distribution, relationships and changes over time.

r география
Комплекс наук, изучающих природные и антропогенные процессы и явления на поверхности Земли и в окружающих ее сферах, их пространственное распределение и изменения во времени.

f géographie *f*
s geografía *f*
g Geographie *f*

2428 geography of agriculture; agricultural geography
The study of spatial distribution of agricultural activity and its links with other branches of economy.

r география сельского хозяйства
Изучение территориальной дифференциации сельскохозяйственного производства и его взаимосвязей с другими отраслями экономики.

f géographie *f* de l'agriculture; géographie *f* agricole
s geografía *f* de agricultura; geografía *f* agrícola
g Landwirtschaftsgeographie *f*; Agrargeographie *f*

* **geography of culture → 1381**

* **geography of human being → 307**

* **geography of industry → 3067**

* **geography of population → 4697**

2429 geography of religion
The study of the spatial distribution of various confessions and beliefs, their local peculiarities and the relations with the social structure and traditional ways of life.
r география религий
Изучение распространения различных конфессий и верований, их местных особенностей, связей с социальной структурой и традиционным укладом жизни населения.
f géographie *f* de religion; géographie *f* religieuse
s geografía *f* de religión; geografía *f* religiosa
g Religionsgeographie *f*

2430 geography of tourism and leisure; recreation geography
A branch of geography studying regional features of people's activity connected with recreation, tourism, in particular the intensity of *tourist flows*.
r рекреационная география; география отдыха и туризма
Отрасль географии, изучающая территориальные закономерности деятельности людей, связанные с отдыхом, туризмом, в частности, направления и интенсивность *туристских потоков*.
f géographie *f* de la récréation; géographie *f* de loisir et tourisme
s geografía *f* de recreación; geografía *f* de ocio y turismo
g Rekreationsgeographie *f*; Geographie *f* der Freizeit und des Tourismus

2431 geography of trade
The study of commodities, their ways of distribution and consumption.
r география торговли
Изучение путей распределения и потребления товаров.
f géographie *f* de commerce
s geografía *f* de comercio
g Handelsgeographie *f*

* **geography of transport → 6316**

2432 geography of world economy
The study of interconnected national economies and their development in the course of international division of labour.
r география мирового хозяйства
Изучение взаимосвязанных национальных хозяйств стран мира и их развития в ходе международного разделения труда.
f géographie *f* de l'économie mondiale
s geografía *f* de economía mundial
g Weltwirtschaftsgeographie *f*

2433 geoiconics
Elaboration of general theory of georepresentations, methods of their analysis, transformation and use.
r геоиконика
Разработка общей теории геоизображений, методов их анализа, преобразования и использования.
f géoiconique *f*
s geoicónica *f*
g Geoikonik *f*

2434 geoid
The Earth's figure formed by level surface coinciding with the world ocean surface (in the state of rest and equilibrium) and continued under continents.
r геоид
Фигура Земли, образованная уровенной поверхностью, совпадающей с поверхностью Мирового океана (в состоянии покоя и равновесия) и продолженной под материками.
f géoïde *m*
s geoide *m*
g Geoid *n*

2435 geoinformatics
Science and technology of projecting, creation and use of geoinformation systems.
r геоинформатика
Наука и технология проектирования, создания и эксплуатации геоинформационных систем.
f géo-informatique *f*
s geoinformática *f*
g Geoinformatik *f*

2436 geoinformation system; geographic information system; GIS
An information system providing the collection, storage, processing, access, representation and dissemination of data on geographical objects and processes related to a certain system of coordinates.
r геоинформационная система; ГИС
Информационная система, обеспечивающая сбор, хранение, обработку, доступ, отображение и распространение пространственно-координированных данных о земных объектах и процессах.
f système *m* d'information géographique
s sistema *m* de información geográfica
g geographisches Informationsystem *n*

2437 geoinformational *adj*
Related to the information on terrestrial objects and processes presented in various forms: descriptive, measuring, numerical.
r геоинформационный
Относящийся к информации о земных объектах и процессах, представленной в самой разнообразной форме: описательной, измерительной, цифровой.
f géo-informatique *adj*
s geoinformático *adj*
g geoinformatik *adj*

2438 geoinformational mapping

Automated compilation and use of maps on the basis of geoinformation technologies and bases of geographical knowledge.

r геоинформационное картографирование
Автоматизированное составление и использование карт на основе геоинформационных технологий и баз географических знаний.

f cartographie *f* de géoinformation
s cartografía *f* de geoinformación
g Geoinformationskartierung *f*

2439 **geological age**
The time which has passed since the formation of rocks or lahdforms (absolute age and relative age are distinguished).

r геологический возраст
Время, прошедшее с момента образования горных пород или форм рельефа (различают абсолютный возраст и относительный возраст).

f âge *m* géologique
s edad *f* geológica
g geologisches Alter *n*

2440 **geological inversion**
Strata which have the normal sequence of layers but in reverse order, i.e. upside down, as in the lower limb of an *overfold.*

r геологическая инверсия
Пласт горной породы, который имеет обычную последовательность слоев, но они лежат в обратном порядке, как это бывает в нижнем крыле *опрокинутой складки.*

f inversion *f* géologique
s inversión *f* geológica
g geologische Inversion *f*

2441 **geological record**
The history of the Earth's development based on the conclusions derived from the study of stratigraphic succession and arrangement of rock strata.

r геологическая летопись
История развития Земли, основанная на выводах, связанных с изучением стратиграфической последовательности напластования горных пород.

f notation *f* géologique
s notas *fpl* geológicas
g geologische Aufzeichnung *f*

2442 **geological section**
A graphic representation of geological structure of the Earth's crust revealed by a cut made vertically through rocks.

r геологический разрез
Графическое изображение геологического строения земной коры в вертикальном сечении.

f section *f* géologique
s sección *f* geológica
g geologisches Schnitt *m*

2443 **geological time**
The chronology of the history of the Earth revealed by its rocks.

r геологическое время
Хронология истории Земли, обнаруживаемая в горных породах.

f temps *m* géologique
s tiempo *m* geológico
g geologische Zeit *f*

* **geological timescale → 2395**

2444 **geology**
The scientific study of the composition, structure and history of the development of the Earth's crust and other parts of the Earth's interior.

r геология
Комплекс наук о составе, строении и истории развития земной коры и более глубоких сфер Земли.

f géologie *f*
s geológía *f*
g Geologie *f*

* **geomagnetism → 6104**

* **geomorphic *adj* → 2446**

2445 **geomorphic levels**
The levels of existing or existed earlier (at present dissected) surfaces formed in the conditions of long tectonic stabilization or slow upward movements.

r геоморфологические уровни
Уровни существующих или существовавших ранее (ныне расчлененных) выровненных поверхностей, которые формировались в условиях длительной тектонической стабилизации или слабых восходящих движений.

f niveaux *mpl* géomorphologiques
s niveles *mpl* geomorfológicos
g geomorphologische Ebenen *fpl*

2446 **geomorphological *adj*; geomorphic *adj***
Related to the landforms or solid surface features of the Earth, or concerned with the geomorphology.

r геоморфологический
Относящийся к формам рельефа или общим чертам земной поверхности.

f géomorphologique *adj*
s geomorfológico *adj*
g geomorphologisch *adj*

2447 **geomorphological profile**
The shape of the relief shown in outline where the plane of a section cuts vertically the surface of the ground.

r геоморфологический профиль
Графическое изображение сечения рельефа какого-либо участка земной поверхности вертикальной плоскостью.

f profil *f* géomorphologique

s perfil *m* geomorfológico
g geomorphologisches Profil *n*

2448 **geomorphological survey**
A direct field observation and examination of landforms carried out to gather data in order to make geomorphological maps.
r геоморфологическая съемка
Непосредственные наблюдения в поле за формами рельефа, необходимые для составления геоморфологических карт.
f levé *m* géomorphologique
s levantamiento *m* geomorfológico
g geomorphologische Aufnahme *f*

2449 **geomorphology**
A branch of *general geomorphology* studying the shape, form, nature of continental landforms.
r геоморфология суши
Раздел *геоморфологии*, изучающий рельеф суши.
f géomorphologie *f*
s geomorfología *f*
g Geomorphologie *f*

2450 **geophil**
Organisms inhabiting soil, silt and denser substrates, e.g. earthworms, some molluscs, etc.
r геофил
Организмы, обитающие в почве, илах и более плотных породах, например, дождевые черви, некоторые моллюски и др.
f géophile *m*
s geofilo *m*
g Geophil *m*

2451 **geophysical satellite**
An artificial satellite aimed at the investigation of geophysical parameters of the Earth: density of the atmosphere, geomagnetic, radiation fields, aurora polaris, etc.
r геофизический спутник
Искусственный спутник Земли, предназначенный для исследования геофизических параметров Земли: плотности атмосферы, геомагнитного, радиационного полей, полярных сияний и др.
f satellite *m* géophysique
s satélite *m* geofísico
g geophysischer Satellit *m*

2452 **geophysics**
The study of the physical phenomena relating to the Earth's structure, physical forces and processes in the lithosphere, hydrosphere, atmosphere and in the Earth's interior.
r геофизика
Изучение строения и физических свойств Земли в целом, а также физических процессов, происходящих в ее твердых сферах, в гидросфере и атмосфере.
f géophysique *f*
s geofísica *f*
g Geophysik *f*

2453 **geophyte**
A perennial plant whose regeneration buds are hidden at some depth in soil.
r геофит
Многолетнее растение, у которого почки возобновления скрыты на некоторой глубине в почве.
f géophyte *m*
s geofita *m*
g Geophyt *m*

2454 **geopolitics**
The study of the influence of the environment, economics and demography on politics of states.
r геополитика
Изучение влияния окружающей среды, экономики и демографии на политику государств.
f géopolitique *f*
s geopolítica *f*
g Geopolitik *f*

2455 **geopotential**
That potential with which the Earth's gravitational field is associated, it is equivalent to potential energy of unit mass relative to mean sea level and is equal to work which would be done against gravity in raising the unit mass from mean sea-level up to the level under cosideration.
r геопотенциал
Потенциал поля земной силы тяжести, равный потенциальной энергии массы относительно уровня моря, или работе, которую надо совершить для поднятия единицы массы от среднего уровня моря до рассматриваемого уровня.
f géopotentiel *m*
s geopotencial *m*
g Geopotential *n*

2456 **geopotential altitude; geopotential height**
The height of a given point or a surface in the atmosphere in relation to the potential energy of unit mass (*geopotential*) at this height, it is calculated with consideration for a change of the gravity acceleration depending on a latitude and the dynamic height, this term is used in aerological reports.
r геопотенциальная высота
Высота точки или поверхности в атмосфере по отношению к потенциальной энергии единицы массы (*геопотенциал*) на этой высоте, вычисляется с учетом изменения ускорения свободного падения в зависимости от широты и высоты над уровнем моря, используется в аэрологии.
f altitude *f* géopotentielle
s altitud *f* geopotencial
g geopotentielle Höhe *f*

* **geopotential height → 2456**

2457 georepresentation; graphic image
Any generalised model of terrestrial objects or processes presented in iconic figurative form (can be flat, volumetric, dynamic).
r геоизображение; графический образ
Любая генерализованная модель земных объектов или процессов, представленная в иконической образной форме (может быть плоской, объемной, динамической).
f géoreprésentation *f*; image *f* graphique
s georepresentación *f*; imagen *f* gráfica
g Georepräsentation *f*; graphische Gestalt *f*

* **geosciences → 1770**

2458 geospheres
Continuous or intermittent layers surrounding the Earth, differing from each other by chemical composition, physical state and properties (the magnetosphere, atmosphere, lithosphere, hydrosphere and the landscape geosphere as the interface of some spheres of the Earth have been distinguished).
r геосферы
Концентрические, сплошные или прерывистые оболочки, образованные веществом Земли, отличающиеся друг от друга по химическому составу, агрегатному состоянию и физическим свойствам (выделяют магнитосферу, атмосферу, гидросферу и литосферу, а также географическую оболочку как область взаимодействия многих геосфер).
f géosphères *fpl*
s geosferas *fpl*
g Geosphären *fpl*

2459 geostationary satellite
An artificial satellite the angular velocity of which is equal to the velocity of the Earth's rotation due to which it moves synchronically with a subsatellite point and as if «hanging» over it.
r геостационарный спутник
Искусственный спутник Земли, угловая скорость движения которого равна скорости вращения Земли, благодаря чему он движется синхронно с подспутниковой точкой и как бы «висит» над ней.
f satellite *m* géostationnaire
s satélite *m* geoestacionario
g geostationärer Satellit *m*

2460 geostrategy
A branch of geopolitics that deals with the role of a state's territory in the formation of its military power, also used practically in the location of transnational corporations.
r геостратегия
Отрасль геополитики, изучающая роль территории в формировании военной мощи государства, используемая также при размещении транснациональных корпораций.
f géostratégie *f*
s geostrategia *f*
g Geostrategie *f*

2461 geostrophic current
An ocean current caused by the pressure gradient driving force which is balanced by the Coriolis force.
r геострофическое течение
Движение вод, вызываемое градиентом давления, уравновешиваемым силой Кориолиса.
f courant *m* géostrophique
s corriente *f* geostrófica
g geostrophische Strömung *f*

* **geostrophic flow → 2462**

2462 geostrophic wind; geostrophic flow
Theoretical wind blowing parallel to the isobars, the wind is a resultant of the opposing forces: the pressure gradient or pressure force in one direction and the deflecting force (Coriolis force) in the other.
r геострофический ветер
Теоретический ветер, являющийся результатом равновесия силы горизонтального градиента давления и отклоняющей силы вращения Земли в предположении, что на воздух действуют только эти две силы, направлен параллельно прямолинейным изобарам или изогипсам.
f vent *m* géostrophique
s viento *m* geostrófico
g geostrophischer Wind *m*

2463 geosyncline
A very large depression, perhaps several hundred kilometres across and up to ten kilometres deep, filled with a deep layer of sediments which develops along a contonental margin gradually turning into a folded mountain construction.
r геосинклиналь
Длинный, протягивающийся на многие десятки и сотни километров, относительно узкий и глубокий прогиб земной коры, возникающий на дне морского бассейна и постепенно превращающийся в складчатое горное сооружение.
f géosynclinal *m*
s geosinclinal *m*
g Geosynklinale *f*

2464 geosystem
Geographical formation of different scale (from an individual landscape to the landscape geosphere) which consists of a number of interrelated and interacting components of the environment and includes human component.
r геосистема
Географическое образование разного масштаба (от географического ландшафта до географической оболочки), состоящее из взаимосвязанных и взаимодействующих компонентов окружающей среды, включая социально-экономическую составляющую.
f géosystème *m*

s geosistema *m*
g Geosystem *n*

2465 geotechnical system
A complex of natural objects and engineering structures which are in close interaction, e.g. a river dam, water reservoir and the adjacent area.
r геотехническая система
Совокупность природных объектов и технических сооружений, находящихся в тесном взаимодействии, например, плотина на реке, водохранилище и окружающая территория.
f système *m* géotechnique
s sistema *m* geotécnico
g geotechnisches System *n*

2466 geothermal energy; geothermal heat
The energy, in the form of heat derived from the hot rocks in the Earth's crust, some of which emitted as hot water and steam to the Earth's surface by thermal springs and geysers.
r геотермальная энергия
Энергия тепла Земли в местах повышенного геотермического градиента, вызывающая появление горячей воды и пара из недр Земли, т.е. термальных источников или гейзеров.
f énergie *f* géothermique
s energía *f* geotérmica
g geothermische Energie *f*; Geothermalenergie *f*

2467 geothermal gradient
The degree of rise in temperature in the rocks of the Earth's crust with increasing depth.
r геотермический градиент
Величина, на которую повышается температура горных пород с увеличением глубины.
f gradient *m* géothermique
s gradiente *m* geotérmico
g geotermischer Gradient *m*

*** geothermal heat → 2466**

2468 geothermal power station
A power station that converts the heat of the Earth's interior (the energy of hot springs) into electrical energy.
r геотермальная электростанция
Электростанция, преобразующая внутреннее тепло Земли (энергию горячих пароводяных источников) в электрическую энергию.
f centrale *f* électrique géothermale
s central *f* geotérmica
g geothermales Kraftwerk *n*

2469 geothermal resources
The heat from the Earth's interior which can be used for heating or to generate electricity.
r геотермальные ресурсы
Запасы внутреннего тепла Земли, которые могут быть использованы для получения тепла или выработки электроэнергии.

Fig. 125. **Geyser** in Iceland which gave a name to these phenomena

f ressource *f* géothermique
s recursos *mpl* geotérmicos
g geotermische Resourcen *fpl*

*** geothermal spring → 6163**

2470 geothermic step
An interval of the Earth's crust depth at which the temperature rises by 1°C (it ranges depending on the depth and location from 5 to 150 m).
r геотермическая ступень
Интервал глубины земной коры, на котором температура повышается на 1°C, колеблется в зависимости от глубины и местоположения от 5 до 150 м.
f gradin *m* géothemique
s grado *m* geotérmico
g geothermische Tiefenstufe *f*

2471 geyser
A hot spring that intermittently, sometimes at regular intervals, throws up a jet of hot water, steam, etc. in areas that are or were volcanic. → Fig. 125
r гейзер
Источник, в основном в областях современного вулканизма, периодически выбрасывающий фонтаны горячей воды и пара. → Рис. 125
f geyser *m*
s géiser *m*; geyser *m*
g Geiser *m*; Springquelle *f*

2472 ghetto
A part of a city occupied by an ethnic minority group; historically the enforced concentration of Jews into specific residential areas in European cities from the Middle Ages.
r гетто
Район города, в котором селятся национальные меньшинства; исторически территория, где в европейских городах еще со средних веков селилось еврейское население.
f ghetto *m*
s ghetto *m*
g Getto *m*

2473 ghost town
A town, usually once a mining town, now wholly abandoned, or inhabited by a few people if the town is now a tourist attraction.
r город-призрак
Город или поселок, обычно горнодобывающий прииск, ныне полностью оставленный людьми, либо служащий теперь туристским центром.
f ville *f* fantôme
s ciudad *f* fantasma
g Geisterstadt *f*

2474 giant's pavement
A natural surface of bare lava plateaus broken by vertical fissures into basaltic prisms up to 1.5 m in diameter, looking like giant cobbles.
r мостовая гигантов
Обнаженная поверхность некоторых лавовых плато, разбитая вертикальными трещинами на базальтовые призмы с вершинами поперечником до 1,5 м, имеющая вид гигантской булыжной мостовой.
f chaussée *f* des géants
s pavimento *m* de gigantes
g Fussweg *m* der Riesen

2475 gilgaï; gilgai soil; melonhole
An undulating microrelief of clay soils characterized by the alternating microridges and depressions (1 m deep) which becomes saturated and expands with rainfall (it is typical of areas of seasonal rainfall and *vertisols*).
r гильгаи
Типичный микрорельеф глинистых почв, представляющий собой чередование повышений и понижений с разницей высот до 1 м и обладающий повышенной способностью к набуханию (характерен для климатических условий с контрастным режимом осадков, типичен для *вертисолей*).
f mottureau *m*
s mottureau *m*
g Gilgai *n*

* **gilgai soil → 2475**

* **GIS → 2436**

2476 GIS technologies
Technology of making of geoinformation systems, which provide collection, storage, processing, transformation and dissemination of spatial information.
r геоинформационные технологии
Технологии создания геоинформационных систем, обеспечивающие сбор, хранение, обработку, преобразование и распространение информации о географических объектах.
f technologies *fpl* de géoinformation
s tecnologías *fpl* de geoinformación; tecnologías *fpl* geoinformaticas
g GIS-Technologien *fpl*

2477 GIS-based analysis
The analysis of distribution, structure of interrelation and dynamics of objects and phenomena using methods of spatial analysis and spatial modelling.
r геоинформационный анализ
Анализ размещения, структуры взаимосвязей и динамики объектов и явлений с использованием методов пространственного анализа и геомоделирования.
f analyse *f* de géoinformation
s análisis *m* geoinformático
g Geoinformationsanalyse *f*

2478 GIS-data
In Geographical Information Systems, data in the form of points, lines, polygons, or a mixtre of these.
r ГИС-данные
Данные в географических информационных системах, представленные в виде точек, линий, полигонов или комбинации этих данных.
f données *fpl* des images
s datos *mpl* de imagen
g Gestaltungsdaten *pl*

2479 glacial abrasion
The mechanical destruction, polishing and grinding of rock surfaces by the friction of hard debris transported by glaciers.
r ледниковая абразия
Механическое разрушение, шлифование, соскабливание или истирание поверхности горных пород при трении и столкновении с твердыми обломками, переносимыми ледником.
f abrasion *f* glaciaire
s abrasión *f* glaciar
g Gletscherabrasion *f*

2480 glacial accumulation
The process of deposition of rock debris carried by glaciers on subglacial and periglacial ground, i.e. transformation of transported moraine into a deposited moraine.
r ледниковая аккумуляция
Отложение обломков горных пород, переносимых ледниками, на подледниковый и приледниковый грунт, т.е. преобразование влекомой морены в отложенную морену.
f accumulation *f* glaciaire
s accumulación *f* glaciar
g Gletschersedimentation *f*

2481 glacial *adj*
Belonging to a glacier or directly associated with it (glacial cycle, glacial mudflow).
r ледниковый; гляциальный
Принадлежащий леднику (ледниковый лед) или непосредственно связанный с ним (ледниковый цикл, гляциальный сель).
f glaciaire *adj*; glaciel *adj*
s glaciar *adj*; glacial *adj*
g Gletscher-; glazial *adj*

*** glacial age → 2924**

*** glacial alimentation → 184**

2482 glacial anticyclone
Area or system of high atmospheric pressure resulting from the cooling effect of large ice sheets.
r ледниковый антициклон
Область повышенного атмосферного давления, возникающая в результате охлаждающего действия крупных ледниковых покровов.
f anticyclone *m* glaciaire
s anticiclón *m* glaciar
g glaziale Antizyklone *f*

2483 glacial cycle
The time interval between two successive peaks of the glacial advancing (glacial maximum) as well as sequential changes of relief under the prevailing effect of ice and snow.
r ледниковый цикл; цикл оледенения
Интервал времени между двумя последовательными пиками разрастания оледенения, а также совокупность последовательных изменений рельефа в процессе его формирования под преобладающим воздействием льда и снега.
f cycle *m* glaciaire
s ciclo *m* glaciar
g glazialer Kreislauf *m*; Vereisungszyklus *m*

2484 glacial denudation
The removal and transportation by a glacier of the products of rock destruction.
r ледниковая денудация
Снос и перенос ледником продуктов разрушения горных пород.
f dénudation *f* glaciaire
s denudación *f* glaciar
g glaziale Abtragung *f*; glaziale Denudation *f*

2485 glacial deposit
Any type of sediment formed as a result of the activity of a glacier and melt water streams on land, as well as the sediments brought by icebergs and deposited in the sea.
r ледниковые отложения
Отложения, образующиеся в результате деятельности ледника и водных потоков при его таянии на суше, а также осадки, перенесенные айсбергами и отложенные в море.
f dépôts *mpl* glaciaires;
alluvion *m* glaciaire
s depositos *mpl* glaciares;
aluviones *mpl* glaciares
g Glazialablagerungen *fpl*;
glaziale Sedimente *npl*

*** glacial diffluence → 1597**

2486 glacial divergence; glacial diversion
The blocking of a river drainage system by an advancing glacier or ice sheet, leading to the change of course of the established drainage pattern.
r гляциальное отклонение
Возникновение препятствия в существующей дренажной системе, вызванного наступающим ледником или ледниковым покровом, что ведет к изменению рисунка речной сети.
f divergence *f* glaciaire
s divergencia *f* glaciar
g Glazialdivergenz *f*

*** glacial diversion → 2486**

2487 glacial dynamics
The study of the structure of contemporary and ancient glaciers, differentiation of their margins, rates of stadial advance and retreat, direction of movement, etc.
r гляциодинамика
Изучение структуры современных и древних ледников, дифференциации их краев, скоростей стадиального наступания и отступания, направлений движения и т.п.
f dynamique *f* glaciaire
s dinámica *f* glaciar
g Gletscherdynamik *f*

*** glacial epoch → 2924**

2488 glacial erosion; plucking
The action of breakdown and removal of the bedrock by a glacier or an ice sheet.
r ледниковая эрозия
Разрушение и снос ледником или ледниковым покровом горных пород ложа.
f érosion *f* glaciaire
s erosión *f* glaciar
g Gletschererosion *f*; glaziale Abtragung *f*

2489 glacial flood; glacial outburst
A flood occurring in a river which is caused by the intensive melting of a glacier or a sudden release of melt water from a glacier dammed lake, often resulting in disastrous inundations.
r ледниковый паводок; ледниковое наводнение
Паводок в реке за счет воды, поступающей с ледников, в частности, внезапный, часто ежегодный прорыв талой воды из запруженного ледником озера, нередко приводящий к катастрофическому затоплению.
f crue *f* glaciaire
s crecida *f* glaciar
g glaziales Hochwasser *n*;
Gletscherhochwasser *n*

2490 glacial lake
A small lake between the margin of a glacier and its valley wall or associated moraine; a lake the origin of which is connected with the past activity of a glacier (morainic lake and corrie lake).
r ледниковое озеро

Озеро между краем ледника и бортом ледниковой долины или моренным валом; озеро, возникновение которого связано с прошлой деятельностью ледника (моренное озеро, каровое озеро).
f lac *m* glaciaire
s lago *m* glaciar
g Glazialsee *m*; Gletschersee *m*

* **glacial landforms → 2495**

2491 glacial maximum
The largest extent achieved by ice-sheets during the Pleistocene ice age, in the narrow sence the position achieved by a valley glacier at its greatest extent.
r ледниковый максимум
Наибольшее распространение ледниковых покровов во время плейстоценового оледенения, в более узком смысле положение долинного ледника при его максимальном продвижении вниз по долине.
f maximum *m* glaciaire
s máximo *m* glaciar
g glaziales Maximum *n*

* **glacial monadnock → 3912**

2492 glacial mudflow
A mudflow of high destructive force caused by some glacier-related event, such as active melting of snow and ice, outburst of temporary glacier-dammed lakes or by the instability of morainic deposits.
r гляциальный сель
Селевой поток большой разрушительной силы на горных реках, сформированный в результате интенсивного таяния снега и льда, прорыва вод временных подпруженных ледником озер и нарушения устойчивости моренных отложений.
f coulée *f* de boue glaciaire
s colada *f* fangosa glaciar
g glazigener Schlammstrom *m*

* **glacial oscillation → 4317**

* **glacial outburst → 2489**

2493 glacial outburst; outburst of glacial lake; bursting of a lake
A flood caused by the sudden release of an ice-dammed or englacial lake, often with disastrous consequences.
r прорыв ледникового озера
Паводок, вызванный быстрым опорожнением озера, подпруженного ледником в боковой долине, нередко с разрушительными последствиями.
f crue *f* de lac glaciaire;
rupture *f* de barrage de glace
s ruptura *f* de lago glaciar;
ruptura *f* de barrera de hielo
g Gletscherseedurchbruch *m*

* **glacial period → 2924**

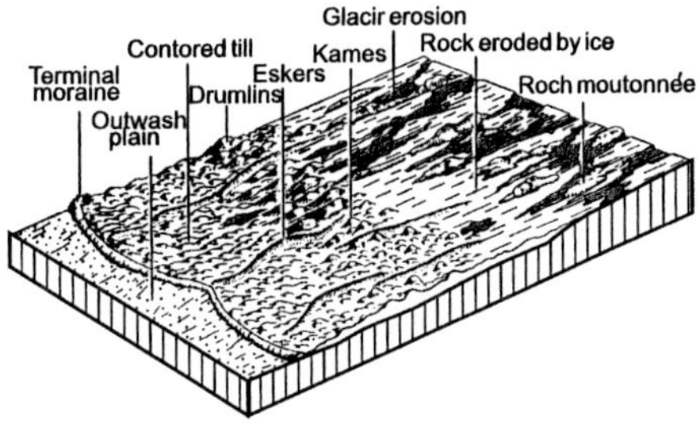

Fig. 126. **Glacial relief**

2494 glacial polish
Wearing away and grinding of rock surfaces by a moving glacier with morainic material transported by it.
r ледниковая шлифовка; ледниковая полировка
Обтачивание поверхности горных пород движущимся ледником с помощью переносимого им моренного материала.
f polissage *m* glaciaire
s pulimento *m* glaciar; pulido *m* glaciar
g Gletscherabschliff *m*; Detersion *f*

* **glacial recession → 1478**

2495 glacial relief; glacial landforms
The relief formed by the action of ice sheets and glaciers, and meltwater running from them. → Fig. 126
r ледниковый рельеф
Рельеф, создаваемый деятельностью покровных и горных ледников и стекающими с них талыми водами. → Рис. 126
f relief *m* glaciaire
s relieve *m* glaciar
g glaziales Relief *n*

2496 glacial scar; glacial stria; chatter mark; glacial scratch
A tiny groove or furrow on a rock surface made by a rock fragment transported in the base of a glacier.
r ледниковый шрам; ледниковая борозда
Выбоина или борозда, оставленная на поверхности горной породы обломком, переносимым в придонном слое ледника.
f cannelure *f* glaciaire; coup *m* de gouge
s trozado *m* glaciar; acanaladura *f* glaciar
g Gletscherstrich *m*; Gletscherfurche *f*

* **glacial scoring → 5863**

* **glacial scratch → 2496**

2497 glacial spillway; overflow channel
Discharge channel of a glacier-dammed lake.
r спиллвей
Канал сброса ледниково-подпрудного озера.
f chenal *m* d'écoulement;
chenal *m* proglaciaire
s canal *m* de desagüe; canal *m* proglaciar
g Entwässerungskanal *m*

2498 **glacial stage**
An interval in a glacier development accompanied with its stagnation both in the phase of advance and in the phase of degradation when stadial terminal moraines form (glacial stages are interrupted by the epochs of warming, i.e. *interstadials*). → Table 2, p. 350-351.
r ледниковая стадия
Интервал в развитии ледника, сопровождающийся его стагнацией как в фазы наступания, так и деградации, когда формируются стадиальные конечные морены (ледниковые стадии разделяются эпохами потеплений – *межстадиалами*). → Табл. 2, с. 350-351.
f stade *m* glaciaire
s estadio *m* glaciar
g Glazialstadium *n*

2499 **glacial stairway**
A glaciated valley the floor of which rises upstream in a series of rock steps and alternating, glacially excavated rock basins, often infilled with lakes.
r ледниковая лестница
Ледниковая долина, дно которой представляет собой серию ступеней – отдельных выработанных льдом локальных бассейнов, нередко занятых озерами.
f escalier *m* glaciaire
s escalera *f* glaciar
g Gletschertreppe *f*

2500 **glacial stream**
A flow of melt water on the surface of a glacier or a river nourished with melt waters of glaciers.
r ледниковая река
Поток талых ледниковых вод на поверхности ледника, либо река, питающаяся талыми водами ледников.
f fleuve *m* glaciaire
s río *m* glaciar
g Gletscherfluß *m*

* **glacial stria → 2496**

* **glacial striation → 5863**

* **glacial transfluence → 6297**

* **glacial valley → 6382**

* **glacialism → 2537**

* **glaciar *m* de piamonte → 4563**

* **glaciated valley → 6382**

2501 **glaciation**
1. The occupation of an area by glaciers or by an ice sheet ice both at present and in the past, e.g. the Pleistocene glaciation. 2. The transformation of droplets of water to ice crystals at the top of a developing cumulus cloud.
r оледенение
1. Сильное разрастание ледников и других многолетних льдов, отражающее общеклиматические изменения. 2. Замерзание облачных капель.
f glaciation *f*
s glaciación *f*
g Vereisung *f*; Vergletscherung *f*

2502 **glaciation boundary**
A line showing the limit of the maximum spreading of ancient mountain glaciers or ice sheets.
r граница оледенения
Граница максимального распространения древних горных ледников или ледниковых покровов.
f limite *f* de glaciation
s límite *m* de glaciación
g Glazialgrenze *f*; Eisrandlage *f*

* **glaciation level → 2503**

2503 **glaciation limit; glaciation level**
A statistical calculation referring to the altitude above which mountain glaciers should occur; the level separating mountain peaks and slopes covered with snow and ice from subjacent Alpine meadows or forested slopes.
r предел оледенения
Статистически рассчитанный уровень, выше которого должны существовать горные ледники; уровень, отделяющий покрытые снегом и льдом горные вершины от нижележащих склонов, покрытых лесом или альпийскими лугами.
f niveau *m* de glaciation
s nivel *m* de glaciación
g Vergletscherungsgrenze *f*

2504 **glacier**
An extensive body of land ice which forms by the recrystallization of firn and moves slowly downslope under the influence of gravity from the accumulation area to the ablation area. → Fig. 127, p. 298
r ледник
Масса наземного льда, образованная последующей перекристаллизацией фирна, медленно текущая (в настоящее время или в прошлом) из области питания (накопления льда) в область абляции. → Рис. 127, с. 298
f glacier *m*
s glaciar *m*
g Gletscher *m*

2505 **glacier advance; advance of glacier**
The increase of a glacier mass leading to a thickening of the ice and an advance of its front when accumulation exceeds ablation.
r наступание ледника
Увеличение массы ледника, вызывающее продвижение его фронта.
f avancée *f* du glacier; progression *f* de glacier
s avance *m* de glaciar; progresión *f* de glaciar

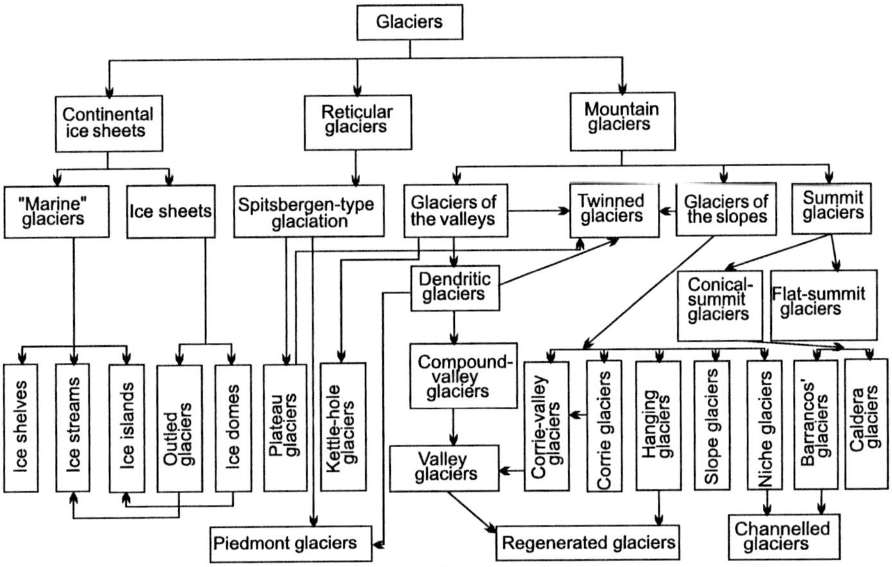

Fig. 127. Morphological types of **glaciers**

g Gletschervorstoß *m*

2506 glacier alimentation; glacier nourishment
Water incoming to rivers due to melting of glaciers.
r ледниковое питание
Питание рек за счет воды, поступающей с ледников.
f alimentation *f* glaciaire
s alimentación *f* glaciar
g Speisung *f* durch Gletscherschmelzwasser

2507 glacier basin
The whole catchment area embracing a glacier and its tributaries together with surrounding slopes from which snow comes to the glacier. → Fig. 128
r бассейн ледника
Территория, охватывающая ледник с подчиненными ему притоками, а также окружающие склоны, с которых снег поступает на ледник. → Рис. 128
f bassin *m* du glacier
s cuenca *f* de glaciar
g Einzugsbecken *n* eines Gletschers

2508 glacier bed
The floor, the land at the bottom of a glacier.
r ложе ледника
Поверхность горных пород, на которых лежит ледник.
f lit *m* glaciaire
s lecho *m* glaciar
g Gletscherbett *n*

2509 glacier breeze; glacier wind
A cold wind blowing, in day time and at night, down a valley of a glacier as a result of cooling of the air in contact with the ice.
r ледниковый ветер
Ветер в приземном слое, дующий вниз по склону ледника днем и ночью.
f brise *f* de glacier
s brisa *f* de glaciar
g Gletscherwind *m*

2510 glacier catchment
A part of the catchment area with glaciers where a considerable part of the moisture cycle runs through a solid phase.
r ледниковый бассейн
Часть водосборного бассейна, в пределах которого имеются ледники и значительная доля влагооброта проходит через твердую фазу.
f bassin *m* glaciaire
s cuenca *f* glaciar

Fig. 128. **Glacier basin** in the Nanga Parbat region, Hymalaya

g Gletscherbassin *n*

2511 glacier coefficient
The ratio of the accumulation area to the ablation area of a glacier.
r ледниковый коэффициент
Отношение площадей области аккумуляции и области абляции ледника.
f coefficient *m* glaciaire
s coeficiente *m* glaciar
g Gletscherkoeffizient *m*

2512 glacier complex
1. A system of glaciers including different morphological types connected by continuous processes running within them. 2. Combination of regularly located glacial landforms and glacial deposits connected with the tongue of a valley glacier or with individual lobe of the ice sheet edge.
r ледниковый комплекс
1. Единое ледниковое тело, состоящее из ледников разных морфологических типов, связанных непрерывностью происходящих в них гляциологических процессов. 2. Сочетание закономерно расположенных гляциальных форм рельефа и ледниковых отложений, связанных с концевой частью горно-долинного ледника или с отдельной лопастью края ледникового покрова.
f complexe *m* glaciaire$_1$; assemblage *m* topographique glaciaire$_2$
s complejo *m* glaciar
g Gletscherkomplex m_1; glazialer Formenschatz m_2

* **glacier crevasse → 1338**

2513 glacier desert
The surface of a glacier with no vegetation due to permanent low air temperatures.
r ледниковая пустыня
Пустынный ландшафт поверхности ледников, характеризующийся постоянными низкими температурами воздуха и отсутствием растительности.
f désert *m* de glace; désert *m* glaciaire
s desierto *m* de hielo; desierto *m* glaciar
g Eiswüste *f*

* **glacier fissure → 1338**

2514 glacier flow; movement of glacier
The complex mechanisms of downslope movement of an ice body including basal sliding, gliding along internal slip-planes, plastic deformations, etc. → Fig. 129
r движение ледника; течение ледника
Сложный механизм перемещения ледникового тела вниз по склону, включает базальное скольжение, смещение льда вдоль плоскостей скольжения, пластические деформации и мн. др. → Рис. 129
f écoulement *m* du glacier; cours *m* de glacier

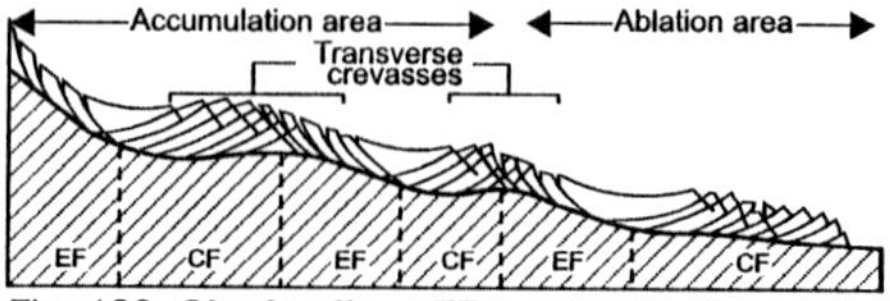

Fig. 129. **Glacier flow**: EF – extending flow, CF – compressing flow

s arroyada *f* de glaciar; corriente *f* de glaciar; movilidad *f* de glaciar; fluencia *f* de glaciar
g Gletscherfliessen *n*; Gletscherströmen *n*

* **glacier fluctuations → 2211**

2515 glacier fountain
A jet of water ejected from a glacier crevasse under lateral thrust.
r ледниковый фонтан
Струя воды, выбрасываемая под напором из ледниковой трещины.
f fontaine *f* glaciaire
s fuente *m* glaciar
g Gletscherfontäne *f*

2516 glacier front; glacier snout; glacier terminus
The lower boundary of a glacier, the end of a glacier tongue.
r фронт ледника; конец ледника
Нижняя граница, низшая точка ледника.
f front *m* de glacier; front *m* glaciaire
s frente *m* glaciar
g Gletscherfront *f*; Gletschertor *n*; Gletscherende *n*

2517 glacier funnel
A small rounded depression with gentle slopes, having the diameter up to 200 m and the depth up to 20 m, occupying a part of a glacier tongue devoid of crevasses.
r ледниковая воронка
Округлая впадина с пологими склонами диаметром до 200 м и глубиной до 20 м на участке ледникового языка, лишенном трещин.
f entonnoir *m* glaciaire
s embudo *m* glaciar
g Eistrichter *m*; Eiswanne *f*

* **glacier gate → 2518**

2518 glacier grotto; glacier gate
A cavity at the front of a glacier from which melt waters flow out in the form of a stream.
r ледниковый грот; ледниковые ворота
Полость у нижнего конца ледника, из которой вытекают в виде небольшой реки талые подледниковые воды.
f grotte *f* de glacier
s gruta *f* glaciar
g Gletscherhöhle *f*; Gletschergrotte *f*

2519 glacier hollow; cryoconite hole

Vertical cylindrical hollow in ice, 3-5 cm wide and 10-40 cm deep, arising from the ice fusion caused by the dark inclusions heated by the sun.

r ледниковый стакан
Вертикальные цилиндрические углубления шириной 3-5 см и глубиной 10-40 см, проплавляемые во льду нагретыми солнцем темными включениями.

f trou *m* de cryoconite
s vaso *m* de crioconita
g Kryokonitloch *n*

* **glacier hydrology → 2549**

2520 glacier ice
A huge monolithic ice mass of a glacier.

r ледниковый лед
Монолитная ледяная порода, слагающая ледник.

f glace *f* de glacier
s hielo *m* de glaciar
g Gletschereis *n*

2521 glacier karst
The irregular relief and honeycombed character of a stagnant ice mass in which features analogous to those of true limestone karst or thermokarst are exibited.

r ледниковый карст
Неровный рельеф и сотовый характер поверхности мертвого льда, в котором встречаются формы, аналогичные обычному карсту или термокарсту в подземных льдах.

f karst *m* glaciaire
s karst *m* glaciar
g Gletscherkarst *m*

2522 glacier knot
Concentration of numerous glaciers at the crossing of two or several mountain ridges.

r ледниковый узел
Сосредоточение множества ледников в местах пересечения двух или нескольких горных хребтов.

f noeud *m* des glaciers
s nudo *m* de glaciares
g Gletschersendverwachsung *f*

* **glacier lake → 1912**

2523 glacier lobe
A periphery convex part of an ancient or contemporary glacier formed in a favourably oriented land depression.

r лопасть ледника
Периферическая выпуклая часть древнего или современного ледника, образующаяся в благоприятно ориентированной депрессии рельефа.

f lobe *m* de glacier; lobe *m* glaciaire
s lóbulo *m* glaciar
g Gletscherlappen *m*

2524 glacier margin; ice edge$_2$; ice margin
A side boundary of a glacier or ce sheet terminating on land.

r край ледника
Боковая граница ледника либо ледникового покрова, оканчивающаяся на суше.

f bout *m* de glacier
s canto *m* de glaciar
g Gletscherrand *m*

2525 glacier mass balance; glacier mass budget
The difference between the increase and decrease in mass of a glacier over a certain time interval.

r баланс массы ледника
Соотношение прихода и расхода массы снега и льда на леднике за определенное время.

f bilan *m* de masse du glacier; bilan *m* glaciaire
s balance *m* de masa de glaciar; balance *m* glaciar
g Massenbilanz *f* des Gletschers

* **glacier mass budget → 2525**

2526 glacier milk
A milk-white turbid meltwater issuing from a glacier which is characterized by a high percentage of suspended fine alluvial material.

r ледниковое молоко; ледниковая муть
Мутно-белая (от взвешенных наносов) вода приледниковых потоков, вытекающих из ледниковых гротов.

f lait *m* de glacier
s leche *f* de glaciar
g Gletschermilch *n*

2527 glacier mill; glacier well; moulin
A deep vertical well in a glacier tongue, several meters in diameter carved by swirling melt water as it pours down from the surface.

r ледниковая мельница; ледниковый колодец
Глубокий спиральный колодец на языке ледника диаметром несколько метров, пропиленный в трещине при стекании в нее воды из наледникового ручья.

f moulin *m* glaciaire; puits *m* glaciaire; trou *m* glaciaire
s molino *m* glaciar; pozo *m* glaciar
g Gletschermühle *f*; Gletscherloch *n*

* **glacier nourishment → 2506**

* **glacier pothole → 3312**

2528 glacier regime
The overall state of a glacier in relation to its mass balance, i.e. a set of processes by which the glacier is nourished, expanded and ultimately reduced by melting.

r режим ледника
Состояние ледника, определяемое балансом его массы, т.е. совокупность процес-

сов, происходящих на поверхности и в толще ледника.
f régime *m* glaciaire
s régimen *m* glaciar
g Gletscherregime *n*

2529 glacier retreat; glacial recession; retreat of a glacier
The reduction in both the area and the thickness of a glacier or ice sheet when its front is receding although the body of the ice itself is still moving forward (it happens when the rate of ablation exceeds that of accumulation).
r отступание ледника
Уменьшение массы ледника, приводящее к сокращению его длины.
f recul *m* du glacier
s retroceso *m* de glaciar
g Gletscherrückzug *m*

2530 glacier runoff
All types of melt water runoff from a glacier.
r ледниковый сток
Сток талых вод сезонного снега, фирна и льда, а также жидких осадков, поступающих в речную сеть с поверхности ледника.
f écoulement *m* glaciaire
s escorrentía *f* glaciar
g Gletscherabfluß *m*

* **glacier snout → 2516**

2531 glacier stability
The ability of a glacier to keep its size for a long time, even under changing conditions of accumulation and ablation of the glacier.
r устойчивость ледника
Способность ледника к длительному сохранению размеров, даже в изменившихся условиях аккумуляции и абляции.
f équilibre *m* glaciaire
s equilibrio *m* glaciar
g Gletscherstabilität *f*

2532 glacier studies
The branch of *glaciology* studying glaciers and ice sheets.
r ледниковедение
Отрасль *гляциологии*, изучающая ледники и ледниковые покровы.
f étude *f* des glaciers
s ciencia *f* de glaciares
g Gletscherkunde *f*

2533 glacier surge; surge
A sudden acceleration of a glacier flow, followed by a rapid dramatic advancing movement of the glacier terminus caused by unstable dynamic constraints in its thickness but with no change in the total mass of the glacier ice.
r подвижка ледника; сёрдж; пульсация ледника
Резкое ускоренное движение ледника, сопровождающееся быстрым продвижением его конца, возникающее из-за нестационарности динамических связей в теле ледника без изменения общей массы льда.
f crue *f* glaciaire; chute *f* du glacier
s crecida *f* glaciar
g Gletschervorstoß *m*

Fig. 130. **Glacier table**

2534 glacier swamp
A gentle depression on a glacier surface filled with the mixture of snow and water.
r ледниковое болото
Плоский участок на леднике со снежным покровом, пропитанным водой.
f marécage *m* glaciaire
s ciénaga *f* glaciar
g Gletschersumpf *m*

2535 glacier system
A complex of glaciers found in a certain region and having common interactions with the environment.
r ледниковая система
Совокупность ледников, объединенных территорией и общими взаимосвязями с окружающей средой.
f système *m* glaciaire
s sistema *m* glaciar
g Gletschersystem *n*

2536 glacier table
A large stone block supported by a relatively thin ice column or pedestal, 1-1.5 m high, on the surface of a glacier tongue. → Fig. 130
r ледниковый стол; ледниковый гриб
Крупная каменная глыба на языке ледника, покоящаяся на сравнительно тонкой ледяной колонне высотой 1-1,5 м. → Рис. 130
f table *f* glaciaire
s mesa *f* glaciar
g Gletschertisch *m*

* **glacier terminus → 2516**

2537 glacier theory; glacialism
A system of ideas about the ancient glaciations of the Earth, according to which the cold periods of different duration repeated regularly in the history of the Earth leading to the development of large ice sheets.
r ледниковая теория; гляциализм
Система научных представлений о древних оледенениях Земли, согласно которой

в истории Земли неоднократно повторялись холодные периоды разной длительности, способствовавшие развитию больших ледниковых покровов.
f théorie *f* glaciaire
s teoría *f* glaciar
g Gletschertheorie *f*

2538 glacier tongue
A long narrow part of a glacier extending below the equilibrium line, sometimes extending to the sea and afloat. → Fig. 131
r язык ледника
Узкая часть ледника, расположенная ниже границы питания, иногда в плавучем состоянии. → Рис. 131
f langue *f* glaciaire
s lengua *f* glaciar
g Gletscherzunge *f*

2539 glacier trunk
The central part of a compound valley (dendritic) glacier which is usually the longest and the thickest part of the glacier.
r ствол ледника
Главная, наиболее длинная и мощная составляющая сложного долинного (дендритового) ледника.
f tronc *m* du glacier
s tronco *m* de glaciar
g Hauptaufstrom *m* eines Gletschers

*** glacier well → 2527**

*** glacier wind → 2509**

2540 glacierization
The occupation of an area by a glacier or ice sheet at the present time.
r современное оледенение
Распространение ледников и ледниковых покровов в настоящее время.
f glaciation *f* actuelle
s glaciación *f* moderna
g heutige Vergletscherung *f*

2541 glaciers of barrancos
Long and narrow glaciers moving in the deep eroded gullies which radially diverge downslope on a volcanic cone.
r ледники барранкосов
Узкие длинные ледники в радиально расходящихся эрозионных бороздах на склонах вулканического конуса.
f glaciers *mpl* de barrancos
s glaciares *mpl* barrancosos
g Barrancosgletscher *mpl*

*** glacifluvial relief → 2546**

2542 glacioclimatology; glaciometeorology
The study of interaction of glacio-nival systems with the atmosphere.
r гляциоклиматология; гляциометеорология
Изучение взаимодействия нивально-гляциальных систем с атмосферой.
f climatologie *f* glaciaire; météorologie *f* glaciaire
s glacioclimatología *f*; glaciometeorología *f*
g Glazialklimatologie *f*; Glazialmeteorologie *f*

2543 glaciodislocation
The disturbance of the bedding of rocks caused by the action of glaciers.
r гляциодислокация
Нарушения в залегании горных пород, вызванные воздействием ледников.
f glaciodislocation *f*; distorsion *f* glaciotectonique
s glaciodislocación *f*

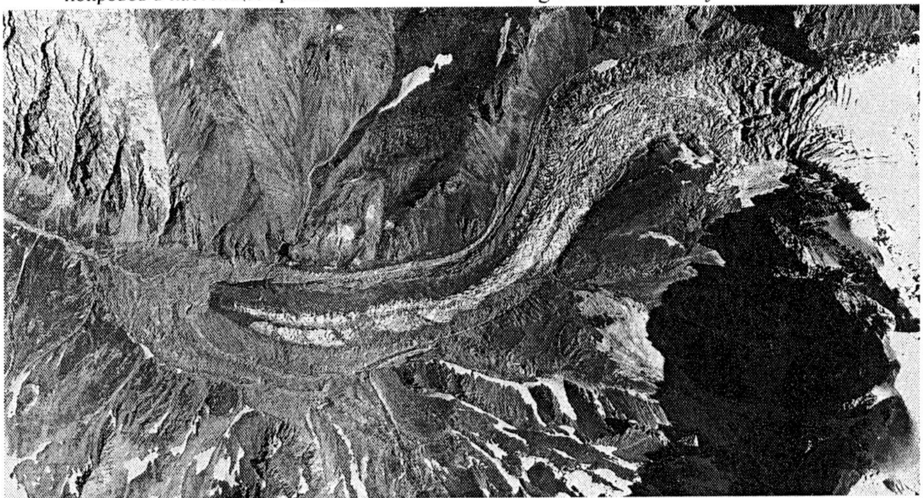

Fig. 131. Aerial photography of the **glacier tongue** which flows from the Elbrus Mountain

g glazigene Verschiebung *f*; Glazialstörung *f*; Eisstauchung *f*

2544 glacio-eustasy; glacio-eustatism
A global change of ocean level owing to the growth and destruction of ice sheets during ice ages.
r гляциоэвстазия
Глобальные изменения уровня Мирового океана, вызванные ростом и разрушением ледниковых покровов во время ледниковых эпох.
f glacio-eustatisme *m*
s glacio-eustatismo *m*
g Glazioeustasie *f*; Glazialeustasie *f*

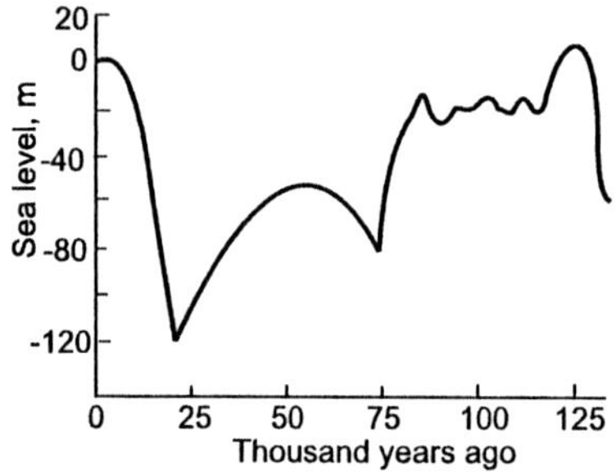

Fig. 132. **Glacio-eustatic curve**

2545 glacio-eustatic curve
A line showing the temporal sea-level variations caused by the changes of ice sheets on the continents. → Fig. 132
r гляциоэвстатическая кривая
Линия, показывающая изменения во времени уровня Мирового океана из-за колебания массы льда на суше. → Рис. 132
f courbe *f* glacio-eustatique
s curva *f* glacio-eustática
g glazialeustatische Kurve *f*

* **glacio-eustatism → 2544**

* **glaciofluvial** *adj* **→ 2223**

2546 glaciofluvial landforms; glacifluvial relief
Landforms related to the action of meltwater streams running from a glacier on land beyond the edge of the glacier or ice sheet. → Fig. 133
r флювиогляциальный рельеф
Формы рельефа, (гряды и ложбины) созданные под действием талых ледниковых вод на участках суши, непосредственно примыкающих к краю ледников. → Рис.133
f modelé *m* fluvio-glaciaire
s modelado *m* fluvio-glaciar
g fluvioglaziales Relief *n*

2547 glaciofluvial sediments; outwash sands and gravels; fluvio-glacial deposits
Stratified, mainly coarse deposits accumulated by streams of water derived from the melting of the ice of a glacier or the margin of an ice sheet.
r флювиогляциальные отложения; ледниково-речные отложения
Слоистые, преимущественно грубые по составу отложения, аккумулированные потоками талых ледниковых вод перед краем ледников.
f sédiments *mpl* fluvio-glaciaires
s depositos *mpl* fluvio-glaciares; acumulación *f* fluvio-glaciar; aluviones *mpl* fluvio-glaciares
g fluvioglaziale Ablagerungen *fpl*

2548 glaciofluvial terrace
A terrace on a slope of a river valley composed of the products of the washout and redeposition of glacial deposits.
r флювиогляциальная терраса
Терраса на склоне речной долины, сложенная продуктами размыва и переотложения ледниковых наносов.
f terrasse *f* fluvio-glaciaire
s terraza *f* fluvio-glaciar
g fluvioglaziale Terrasse *f*

2549 glaciohydrology; glacier hydrology
The study of the formation and regime of liquid water in glaciers and glacio-nival systems.
r гляциогидрология; гидрология ледников
Изучение формирования и режима жидкой воды в ледниках и нивально-гляциальных системах.
f hydrologie *f* glaciaire
s hidrología *f* glaciar; glaciohidrología *f*
g Gletscherhydrologie *f*; Glazialhydrologie *f*

2550 glacio-isostasy; glacioisostatic movement
Vertical movements of the Earth's crust in areas of the present-day and past glaciation caused by the excessive weight of an ice sheet and its decrease as the ice melts.
r гляциоизостазия; гляциоизостатические движения
Вертикальные движения земной коры в областях современного и плейстоценового оледенения, вызываемые нагрузкой ледниковых покровов и ее исчезновением при их стаивании.
f glacio-isostasie *f*; glacio-isostatisme *m*
s glacio-isostasía *f*; glacio-isostatismo *m*

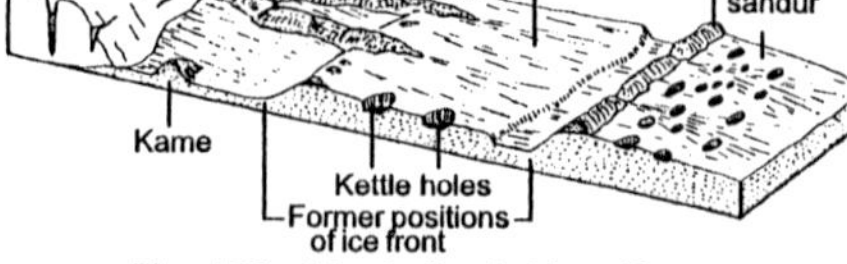

Fig. 133. **Glaciofluvial landforms**

g Glazioisostasie *f*; glazialisostatische Bewegungen *f*

* **glacioisostatic movement → 2550**

2551 **glaciokarst**
Landforms of karstic character formed by a combination of glacial processes acting upon an area of massive limestone, together with the features produced by glacial erosion and deposition.
r гляциокарст
Формы рельефа карстового характера, образованные комбинацией гляциальных процессов, действующих в области распространения массивных известняков, вместе с ледниковой эрозией и гляциальными наносами.
f glaciokarst *m*
s glaciokarst *m*
g Glaziokarst *m*

2552 **glaciolacustrine sediments**
The deposits of periglacial lakes, formed by the deposition of suspended material, carried by streams of glacial melt water, they have a clear seasonal and annual lamination.
r ледниково-озерные отложения
Отложения приледниковых озер, сформированные осаждением взвешенного материала, выносимого потоками талых ледниковых вод, имеющие четкую сезонную и годовую слоистость.
f sédiments *mpl* glaciolacustres
s depósitos *mpl* glaciolacustres
g glazilimnische Ablagerungen *fpl*

2553 **glaciology**
The study of snow and ice on the Earth's surface, in the atmosphere, hydrosphere and lithosphere, their regime and dynamics, interaction with the environment, the role of ice in the evolution of the Earth.
r гляциология
Наука, изучающая природные льды на поверхности земли, в атмосфере, гидросфере и литосфере – режим и динамику их развития, взаимодействие с окружающей средой, роль льда в эволюции Земли.
f glaciologie *f*
s glaciología *f*
g Glaziologie *f*

2554 **glacio-marine deposits; iceberg deposits**
Marine deposits, containing mainly terrigenous fragmental material brought by icebergs.
r ледниково-морские отложения; айсберговые отложения
Морские отложения, в составе которых преобладает терригенный обломочный материал, связанный с перемещением айсбергов.
f sédiments *mpl* glacio-marins
s depósitos *mpl* glaciomarinos
g glazimarine Ablagerungen *fpl*; Eisbergablagerungen *fpl*

* **glaciometeorology → 2542**

2555 **glacio-nival system**
A natural system with the dominating role of snow and ice in its composition and processes running within it.
r нивально-гляциальная система
Природная система с ведущей ролью снежного покрова и льда в ее вещественном составе и процессах.
f système *m* nivo-glaciaire
s sistema *m* nivoglacial
g nival-glaziales System *n*

2556 **glaciosphere**
Snow and ice formations on the Earth's surface.
r гляциосфера
Совокупность снежно-ледовых образований на поверхности Земли.
f sphère *f* glaciaire; glaciosphère *f*
s esfera *f* glacial
g Glaziosphäre *f*

2557 **glaciotectonics**
The study of dislocation of rocks caused by the action of glaciers.
r гляциотектоника
Изучение нарушений в залегании горных пород, вызванных действием ледников.
f tectonique *f* glaciaire
s glaciotectónica *f*
g Glazialtektonik *f*

2558 **glade**
An open space in a wood or forest covered with grass and surrounded by trees.
r поляна
Заросший травой участок, расположенный среди леса.
f clairière *f*
s calvero *m*; calva *f*; claro *m*
g Waldlichtung *f*; Waldwiese *f*

* **glaze → 6545**

* **glazed frost → 6545**

* **glei soils → 2561**

* **gleization → 2559**

* **glen → 5837**

2559 **gley formation; gleization; gleying; gleyification**
The development of gley soil horizon (coloured blue-grey, pale blue or blue) as a result of reducing reactions going on under waterlogging conditions and insufficient for oxygen.
r глееобразование; оглеение
Процесс образование глеевого горизонта

почвы (окрашенного в сизый, голубой или синий цвет) в результате восстановительных реакций в условиях переувлажнения и затрудненного доступа кислорода.

f formation *f* de gley
s formación *f* de gley
g Gleybildung *f*

2560 gley horizon
A soil layer coloured blue-grey, blue or dark-blue, formed as a result of the reduction processes occurring under the lack of oxygen in periodically or permanently waterlogged soils.

r глеевый горизонт
Слой почвенного профиля сизого, голубого или синего цвета, образующийся в результате восстановительных процессов при затрудненном доступе кислорода в условиях избыточного увлажнения.

f horizon *m* à gley
s horizonte *m* de gley
g Gleyhorizont *m*

2561 gley soils; gleysol; glei soils
Soils forming under a constant excessive moistening, i.e. affected by periodic or permanent saturation by water in the absence of effective drainage, they have a gley horizon coloured bluish-grey, or are gleyed through the whole profile.

r глеевые почвы; глейсоли; глееземы
Почвы, формирующиеся в условиях постоянного избыточного увлажнения, содержат глеевый горизонт, окрашенный в холодных сизых тонах, или оглеены во всем профиле.

f sols *mpl* à gley
s suelos *mpl* de gley
g Gleyböden *mpl*

* **gleyification → 2559**

* **gleying → 2559**

2562 gley-podzolic soil; epigleyic podzol
Podzolic soils with the gleyed upper mineral bleached horizon laying under the forest litter involved in peat formation, they are widespread in the northern taiga sub-zone.

r глее-подзолистая почва
Подзолистые почвы с проявлением оглеения в верхнем минеральном осветленном горизонте, лежащем под оторфованной лесной подстилкой, распространены в подзоне северной тайги.

f sol *m* podzolique à gley
s suelo *m* podzólico de gley
g gleyiger Podsolboden *m*

* **gleysol → 2561**

2563 glimmer ice
A thin ice accumulated on the surface of water above the older floating ice at its melting.

r наслуд
Тонкий лед на воде, скопившейся на более старом плавучем льду, или образовавшийся в разводьях, полыньях, снежницах.

f glace *f* éclatante
s hielo *m* resplandeciente
g abgelagertes Dünneis *n*

2564 glint
An extended steep slope, steep edge of a structural plateau formed by denudation and abrasion.

r глинт
Протяженный крутой обрыв структурного плато, образованный денудацией или абразией.

f glint *m*
s glint *m*
g Klint *m*

2565 global change
Considerable changes of the environment taking place over the whole globe at present time which are induced mainly by the changes of climate and human impact on the environment.

r глобальные изменения
Заметные изменения окружающей среды на всем земном шаре, происходящие в настоящее время и обусловленные главным образом изменениями климата и антропогенным воздействием.

f changement *m* global
s cambio *m* global
g Globalveränderung *f*

* **global city → 6815**

* **global contamination → 2567**

2566 global ocean conveyor belt
A system of large-scale oceanic fluxes providing the global water mass and heat exchange between the oceans. → Fig. 134

r глобальный океанский конвейер
Система взаимосвязанных крупномасштабных потоков в Мировом океане, обеспечивающая глобальный тепло- и водообмен между океанами. → Рис. 134

f bande *f* convoyeuse globale

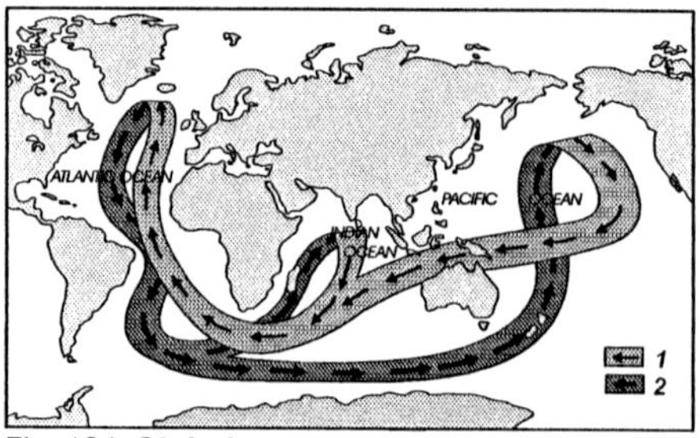

Fig. 134. **Global ocean conveyor belt**: 1 – warm shallow current, 2 – cold and salty deep current

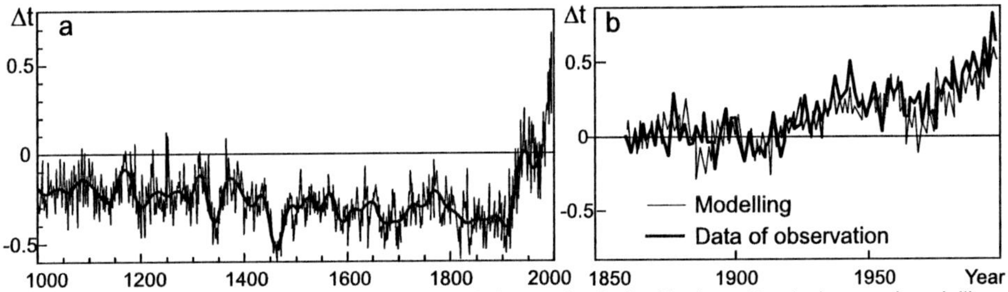

Fig. 135. **Global warming** as shown by the data of observation in the Northern Hemisphere and modelling: a – the last thousand years, b – the last 150 years

s circulación *f* global de cinta transportadora
g Globalfließband *n*

2567 global pollution; global contamination
Pollution found in different places of the Earth far from the sources of pollution.
r глобальное загрязнение
Загрязнение, обнаруживаемое в разных точках Земли далеко от источников возникновения.
f pollution *f* globale; contamination *f* globale
s polución *f* global; contaminación *f* global
g globale Verschmutzung *f*; globale Verunreinigung *f*

2568 global positioning system; GPS
Technological complexes intended for the determination of three-dimensional co-ordinates of the Earth's objects by using simple earthbased instruments that receive signals from two or three geostationary satellites.
r глобальные системы позиционирования
Технологические комплексы, предназначенные для определения трехмерных координат земных объектов, использующие наземные инструменты, которые получают сигналы с двух или трех геостационарных спутников.
f système *m* de positionnement globale
s sistema *f* global de posicionamiento
g globales Positionierungssystem *n*

2569 global solar radiation
A sum of the direct solar radiation and reflected solar radiation.
r суммарная солнечная радиация
Сумма прямой и рассеянной радиации.
f rayonnement *m* global
s radiación *f* global
g Globalstrahlung *f*

2570 global temperature
The air temperature averaged for one (more often for the Northern Hemisphere) or two hemispheres taking into account the areas of geographical zones, it is a measure of the thermal state of the global climatic system.
r глобальная температура
Температура воздуха, осредненная по одному (чаще Северному) или двум полушариям с учетом площадей широтных зон, мера термического состояния планетарной климатической системы.
f temperature *f* globale
s temperatura *f* global
g Globaltemperatur *f*

2571 global warming
The rise of mean global temperature on the Earth observed in the XX century and at the beginning of the XXI century. → Fig. 135
r глобальное потепление
Повышение средней глобальной температуры на Земле, наблюдающееся в XX и начале XXI вв. → Рис. 135
f réchauffement *m* global; réchauffement *m* climatique
s calentamiento *m* global
g Erderwärmung *f*

2572 globalization
Economic and social processes leading to the development of a single global economy and culture as a result of improved technology and communications and the influence of very large multinational companies.
r глобализация
Экономические и социальные процессы, ведущие к сближению разных культур и цивилизаций, вызываемые развитием новых технологий и коммуникаций, а также интернационализацией деятельности экономических корпораций и фирм.
f mondialisation *f*
s globalisación *f*; mundialisación *f*
g Globalisierung *f*; Globalisation *f*

2573 globe
A reduced model of a celestial body (usually the Earth) or of the celestial sphere made with the use of cartographic signs, i.e. a map depicted on a sphere.
r глобус
Уменьшенная шаровидная модель Земли, небесного тела или небесной сферы с нанесенным картографическим изображением.
f globe *m* terrestre; globe *m*
s globo *m* terráqueo
g Erdball *m*; Globus *m*

2574 globigerine silt
A light sea sediment found at depths down to 5000 m that consists of the foraminifera plankton lime shells .
r глобигериновый ил
Светлый морской осадок, встречающийся на глубинах до 5000 м и состоящий из известковых раковинок планктонных фораминифер.
f boue *f* à Globigérines
s fango *m* globigerino
g Globigerinenschlamm *m*

2575 globular projection
A cartographic projection which is commonly used in paires to illustrate each hemisphere, where hemispheric graticules are enclosed by a circle, thereby resembling an azimuthal projection, but which do not represent the transverse aspects of the azimuthal projection.
r сферическая проекция
Картографическая проекция, используемая в паре для изображения каждого полушария, градусная сеть на ней ограничивается кругом и поэтому она имеет сходство с азимутальной проекцией, но не отражает ее поперечных свойств.
f projection *f* sphérique
s proyección *f* esférica
g Kugelprojektion *f*

2576 glory
An optical phenomenon in the atmosphere seen as a coloured ring around the shadow of an observer's head.
r глория
Оптическое явление в атмосфере, представляющее собой цветной круг вокруг тени головы наблюдателя.
f gloire *f*
s gloria *m*; circulo *m* de Ulloa; corona *f* de Ulloa
g Glorie *f*

2577 glowing avalanche
A volcanic eruption similar to a nuée ardente but which is propelled more by gravity than by the force of the explosion.
r раскаленная лавина
Вулканическое извержение типа раскаленной тучи, при котором вулканический материал движется в большей мере под действием гравитационных сил, чем под действием сил взрыва.
f avalanche *f* brûlant
s avalancha *f* candente
g glühende Lawine *f*

*** GMT → 2628**

*** gnarled forest → 1873**

2578 gnomonic projection
A map projection of perspective azymuthal projection in which the type produced by projecting from the centre of the generating globe. → Fig. 136

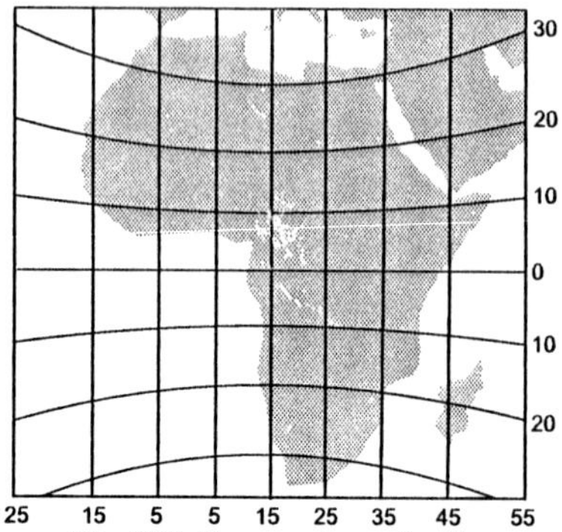

Fig. 136. **Gnomonic projection**

r гномоническая проекция
Картографическая проекция, вид перспективной азимутальной проекции, получаемой при расположении точки зрения в центре шара. → Рис. 136
f projection *f* gnomonique
s proyección *f* gnomónica
g gnomonische Projektion *f*

*** GNP → 2641**

2579 goat-husbandry
Breeding goats for foodstuffs and raw materials for light and food industries, most developed in Russia and Central Asia.
r козоводство
Отрасль животноводства, занимающаяся разведением коз с целью получения продуктов питания и сырья для легкой и пищевкусовой промышленности, наиболее развито в России и Средней Азии.
f élevage *m* des chèvres; élevage *m* cappin
s cría *f* de cabras
g Ziegenzucht *f*

2580 gold-field
A place normally in a river valley where gold is mined manually or with the help of small-scale mechanization.
r прииск
Место, как правило, в долине реки, где добывают золото ручным способом или с помощью малой механизации.
f mine *f* d'or
s mina *f* de oro
g Fundgrube *f*; Goldgrube *m*

2581 goletz
A flattered or rounded mountain top in Siberia which rises above the tree-line and is almost devoid of vegetation.
r голец
Горная вершина уплощенной или округлой формы в Сибири, поднимающаяся выше границы леса и почти лишенная растительности.

f goletz *m*
s goletz *m*
g Golez *m*

2582 goletz relief
The relief of levelled interfluves and slopes above the tree-line created by cryogenic and nival processes and extremes of continental climate in middle latitudes.
r гольцовый рельеф
Рельеф сглаженных междуречий и склонов выше границы леса в горах, формируется криогенными и нивальными процессами в условиях резко континентального климата умеренных широт.
f relief *m* des goletz
s relieve *m* de goletz
g Golezrelief *n*

* **goletz terrace → 225**

* **Gondwana → 2583**

2583 Gondwanaland; Gondwana
Hypothetical continent which existed during the greater part of the Paleozoic and at the beginning of the Mesozoic in the Southern Hemisphere and included some parts of the contemporary continents of South America, Africa, Asia, Australia and, probably, Antarctica.
r Гондвана
Гипотетический материк, существовавший в течение большей части палеозоя и в начале мезозоя в Южном полушарии и включавший части современных материков Южной Америки, Африки, Азии, Австралии и, возможно, Антарктиды.
f continent *m* de Gondwana; Gondwana *f*
s continente *m* de Gondwana
g Gondwanaland *n*

* **goods and services → 1148**

* **goods economy → 1146**

2584 gorge; canyon
A deep valley with steep or sheer walls and a narrow bottom, usually completely occupied by a river channel. → Fig. 138
r ущелье; каньон
Глубокая долина с крутыми или отвесными склонами и узким дном, как правило, полностью занятым руслом реки. → Рис. 138
f gorge *f*; canyon *m*
s garganta *f*; cañón *m*
g Schlucht *f*; Klamm *f*

* **Götiglacial → 2585**

2585 Götiglacial stage; Götiglacial
The phase of the retreat of the last glaciation in Northern Europe when the edge of continental ice sheet was situated within the limits of southern Sweden.
r готигляциал
Фаза сокращения ледников последней ледниковой эпохи Северной Европы, когда край материкового ледникового покрова находился в пределах Южной Швеции.
f stade *m* gotiglaciaire
s estadio *m* gotiglaciar
g Götiglazial *n*

* **GPS → 2568**

* **GPS measurement → 4711**

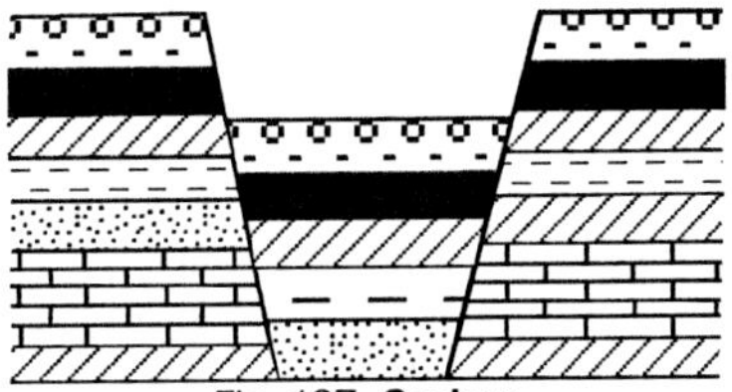

Fig. 137. **Graben**

2586 graben
A stretched block of land downthrown between parallel faults. → Fig. 137
r грабен
Вытянутый участок земной коры, опущенный относительно окружающих участков по крутым разрывам – сбросам. → Рис. 137
f graben *m*; fossé *f* tectonique
s graben *m*; fosa *f* tectónica
g Graben *m*

2587 graded bedding; graded sediments
Loose or cemented sediments with particles sorted by natural means, according to size.
r сортированное напластование
Рыхлые или сцементированные отложения, состоящие из частиц, отсортированных естественным путем в соответствии с их размером.

Fig. 138. **Gorge**

f sédimentation *f* gradée
s sedimentación *f* ordenada; estratificación *f* granulométrica
g sortierte Ablagerung *f*

2588 graded river; graded stream
A river channel which has the equilibrium between the erosion and deposition and the rate of its flow is exactly sufficient for the carrying of the material supplied by the drainage basin.
r сбалансированная река
Река, в русле которой сложилось нулевое соотношение эрозии и отложения наносов и скорость течения соответствует той, что необходима для переноса всего материала, попадающего в речной бассейн.
f rivière *f* graduée
s río *m* gradual
g Ausgleichsfluß *m*

* **graded sediments → 2587**

2589 graded shore
The type of a shore formed due to the straightening of a submerged (ingression) irregular coast by abrasion and aggradation, when the headlands are cut by marine erosion while the bays are enclosed by the building of bars across the entrance.
r выровненный берег
Тип берега, образовавшегося в результате выравнивания ингрессионного берега под воздействием абразионно-аккумулятивной деятельности моря, когда выступы берега срезаются морской эрозией, а у входа в бухты вырастают отмели.
f côte *f* régularisée
s costa *f* regularizada
g Ausgleichsküste *f*

* **graded stream → 2588**

2590 gradient
1. A change or variation in the value of a variable, especially that relating to climate, e.g. temperature gradient, barometric gradient. 2. The steepness of a slope expressed in degrees, percentage, or as a ratio.
r градиент
1. Изменение значения любой переменной величины, чаще всего используется в метеорологии, например, градиент температуры, барометрический градиент. 2. Крутизна склона, выраженная в градусах, процентах или как отношение.
f gradient *m*
s gradiente *m*
g Gradient *m*

* **gradient² → 3054**

2591 gradient observations
Simultaneous measurements of the main meteorological elements at several levels within near-surface or boundary layer.
r градиентные наблюдения
Одновременные измерения основных метеорологических элементов на нескольких уровнях в приземном или пограничном слое воздуха.
f observations *fpl* de gradient
s observacións *fpl* de gradiente
g Gradientbeobachtungen *fpl*

2592 gradient wind
The wind which results from a balance between the horizontal pressure force, the Coriolis force, and the centrifugal force due to air motion on a curved path, it blows parallel to the curved isobars.
r градиентный ветер
Теоретический ветер, возникающий в результате равновесия между силой горизонтального барического градиента, горизонтальной составляющей силы Кориолиса и центробежной силой, обусловленной криволинейным движением воздуха, движется вдоль криволинейных изобар или изогипс.
f vent *m* du gradient; vent *m* gradiental
s viento *m* del gradiente
g Gradientenwind *m*

2593 gradualism
A scientific concept which sees the evolution of species taking place gradually.
r градуализм
Научная концепция, доказывающая постепенную эволюцию отдельных видов.
f gradualisme *m*
s gradualismo *m*
g Gradualismus *m*

2594 graduation
A graphic representation of the succession of colours, intensity, quantitative characteristics of conventional signs.
r цветовая шкала на карте
Графическое изображение последовательности изменения цвета, насыщенности, количественных характеристик условных обозначений.
f échelle *f* des couleurs
s escala *f* graduada; escala *f* de colores
g Teilung *f*

2595 grain
1. A very small hard particle, especially of a mineral or soil. 2. The seed of a sereal, or harvested cereal crops in general.
r зерно
1. Небольшая твердая частица, компонент минералов или почвы. 2. Семя любой зерновой культуры, в широком смысле – урожай зерновых культур.
f grain *m*
s grano *m*
g Korn *n*

2596 grain farm; grain growing
Growing cereal and leguminous crops, producing and selling grain.

r зерновое хозяйство
Производство и сбыт зерна, выращенного при возделывании зерновых и бобовых культур.
f production *f* des céréales; économie *f* céréalière
s producción *f* de cereales; economía *f* cerealista
g Getreideerzuugung *f*; Getreideanbau *m*

* **grain growing → 2596**

2597 **grained snowflake**
A snowflake covered by minute ice crystals or frozen droplets.
r обзерненная снежинка; снежные зерна; круповидная снежинка
Снежинка диаметром менее 1 мм, беспорядочно обросшая мелкими кристаллами инея или замерзшими капельками.
f flocon *m* givré
s copo *m* de nieve granulado; nieve *f* granular
g verkörnte Schneeflocke *f*; Griesel *pl*

2598 **gramineous grassland; gramineous steppe**
The drought-resistant steppe with gramineous plants, widespread in Eurasia (*steppe*), North America (*prairies*) and South America (*pampa*).
r злаковая степь
Засухоустойчивая степь с господством в растительном покрове злаков и ковылей, широко распространена в Евразии (*степи*), Северной Америке (*прерии*) и Южной Америке (*пампа*).
f steppe *f* à graminées
s estepa *f* de gramíneas
g Getreidesteppe *f*

* **gramineous steppe → 2598**

2599 **granite**
A coarsely-crystalline light-coloured acid igneous rock the most common and widespread in the crust of continents formed by the slow cooling of a large intrusion of magma, its hardness makes it an excellent building stone.
r гранит
Полнокристаллическая интрузивная кислая светлоокрашенная горная порода – самая распространенная в земной коре континентов, широко применяется в строительстве благодаря высокой твердости.
f granite *m*
s granito *m*
g Granit *m*

2600 **granitic layer**
The layer between sedimentary rocks and basaltic layer typical of the Earth's crust of continental type.
r гранитный слой
Промежуточный слой между осадочным (сверху) и базальтовым (снизу), типичен для земной коры континентального типа.
f couche *f* granitique
s capa *f* granítica
g Granitschicht *f*

2601 **granular disintegration**
One of the main types of mechanical weathering of rocks, caused either by the freezing of pore water and the dislodgement of particles, or by differential expansion and contraction of rock due to insolation, leading to the destruction of the grains.
r зернистое выветривание
Один из основных типов механического выветривания горных пород, возникающего из-за замерзания воды в порах и смещения частиц, либо в результате дифференциального расширения и сужения частиц при проникновении инсоляции, вызывающего разрушение зерен.
f désagrégation *f* granuleuse
s disgregación *f* granulosa
g körnige Desintegration *f*

2602 **granular ice**
An ice formed from the frozen water drops dispersed in the cold air.
r гранулированный лед
Лед, образованный из замерзших капель воды, диспергированной в холодный воздух.
f glace *f* granulée; glace *f* granulaire
s hielo *m* granulado
g Eisgranulat *n*; Eiskörner *npl*

2603 **granulometric composition**
The content of grains of different size in a rock or soil expressed as a percentage of the mass or the number of grains in a sample under investigation.
r гранулометрический состав
Содержание в горной породе или почве зерен разного размера, выраженное в процентах от массы или количества зерен исследованного образца.
f composition *f* granulométrique
s composición *f* granulométrica
g Korngrössenzusammensetzung *f*

* **graphic image → 2457**

* **graphic industry → 4772**

2604 **graphic variables**
Graphic means used for some cartographic symbols, symbol systems, graphic images.
r графические переменные; графические знаки
Графические средства, используемые для построения отдельных картографических знаков, знаковых систем, графических образов.
f variables *fpl* graphiques; représentation *f* graphique
s variables *fpl* gráficos; representación *f* grafica; lenguaje *m* gráfico
g graphische Variablen *fpl*

* **graphical chart → 4147**

2605 **graphite**
A soft, grey, laminar form of carbon, the most widespread and stable in the Earth's crust, because of its very high melting-point, excellent mechanical strenhgt and good thermal and electrical conductivity it has many industrial uses.
r графит
Мягкий, серый минерал, наиболее распространенная и устойчивая в земной коре разновидность углерода, благодаря большой механической прочности, хорошей теплопроводности и электропроводности находит широкое применение в хозяйстве.
f graphite *m*
s grafito *m*
g Graphit *m*

* **grass → 2748**

2606 **grass savanna; savanna grassland**
A type of African savanna with tall water-resistant gramineous plants (up to 3,5 m).
r злаковая саванна
Тип африканской саванны с высоким (до 3,5 м) и густым влаголюбивым злаковым травостоем.
f savane *f* des graminées
s sabana *f* de gramíneas
g Grassavanne *f*

* **grass stand → 2751**

2607 **grass swamp**
A type of a lowland fen with the predominance of herbaceous vegetation.
r травяное болото
Тип низинного болота с преобладанием луговой растительности.
f marais *m* tourbeux;
tourbière *f* basse herbacée
s turbera *f* baja herbácea
g Grasmoor *n*

2608 **grassland; steppe zone; steppe**
A zonal type of landscape with xerophylous grass vegetation on chernozem and chestnut soils, it is common in the temperate and subtropical zones of the Northern and Southern hemispheres.
r степь; степная зона
Зональный тип ландшафта, представленный травянистой ксерофитной растительностью с сообществами преимущественно дерновинных злаков и разнотравья на черноземных и каштановых почвах, образует природные зоны умеренных и субтропических поясов Северного и Южного полушарий.
f steppe *f*; zone *f* des steppes
s estepa *f*; zona *f* de estepas
g Steppe *f*; Steppenzone *f*

2609 **graticule**
A network of meridians and parallels on geographical maps and globes serving to make an exact reference to any point of the Earth's surface (longitude and latitude) or mapping the objects by their coordinates.
r градусная сеть; географическая сетка; картографическая сетка
Система меридианов и параллелей на географических картах и глобусах, служащая для отсчета географических координат точек земной поверхности (долгот и широт) или нанесения на карту объектов по их координатам.
f réseau *m* géographique; réseau *m* des coordonnées géographiques
s red *f* geográfica
g Gradnetz *n*; Kartennetz *n*

2610 **gravel**
Loose sediments composed of rounded stones 1-10 mm in size.
r гравий
Рыхлая осадочная порода, сложенная окатанными обломками горных пород размером 1-10 мм.
f gravier *m*; gravillon *m*
s grava *f*; gravilla *f*
g Kies *m*; Geröll *n*; Schotter *m*

* **graveyard → 5019**

2611 **gravimetry**
The measurements of the Earth's gravitation, and the use of the results to determine the Earth's form and to study its internal structure.
r гравиметрия
Измерение величин, характеризующих гравитационное поле Земли, и использование их для определения фигуры Земли и изучения ее внутреннего строения.
f gravimétrie *f*
s gravimetría *f*
g Schwerkraftmessung *f*; Gravimetrie *f*; Schweremessung *f*

2612 **gravitation; gravity; gravitational force; gravitational pull**
The attraction between bodies by a force proportional to their masses, this force determines the spherical form of the Earth, many features of relief of the Earth's surface, rivers flow, glaciers movement, etc.
r гравитация; тяготение; сила тяжести
Свойство тел притягивать друг друга силой, зависящей от их масс, действиями этой силы обусловлена сферическая форма Земли, многие черты рельефа земной поверхности, течение рек, движение ледников и мн. др.
f gravitation *f*; gravité *f*
s gravitación *f*; gravedad *f*
g Gravitation *f*; Schwerkraft *f*

* **gravitational force → 2612**

2613 **gravitational model; gravity model**
Used in human geography mathematical

model of spatial interconnection of settlements depending on their sizes and distances between them that creates people, commodity and information flows.

r гравитационная модель
Применяемая в социально-экономической географии математическая модель пространственного взаимодействия между населенными пунктами в зависимости от их размеров и расстояния между ними, что приводит к образованию потоков людей, грузов, информации.

f modèle *m* de gravitation; modèle *m* de gravité

s modelo *m* de gravitación; modelo *m* de gravedad

g Gravitationsmodell *n*

* **gravitational pull → 2612**

* **gravitational water → 2616**

* **gravity → 2612**

2614 gravity anomaly
The difference between an observed gravity value of rocks and its normal (theoretical) value due to the existence of rocks of unexpectedly different densities or magnetic characters in the Earthrs crust.

r гравитационная аномалия; аномалия силы тяжести
Разность между наблюдаемой силой тяжести и ее нормальным (теоретическим) значением в той же точке, вызванная резкими различиями плотности горных пород или особенностями магнетизма в земной коре.

f anomalie *f* gravimétrique

s anomalía *f* gravitacional; anomalía *f* de gravedad

g Gravitationsanomalie *f*; Schwerkraftsanomalie *f*

* **gravity model → 2613**

2615 gravity processes
The slope processes that act under gravitational force.

r гравитационные процессы
Процессы на склонах, идущие под действием силы тяжести.

f processus *mpl* de gravitation

s procesos *mpl* de gravitación

g Gravitationsprozesse *mpl*

2616 gravity water; free water; gravitational water
The water which is filling pores, cavities and fissures in soil and rocks and moving by the force of gravity towards the water table.

r гравитационная вода; свободная вода$_2$
Вода, заполняющая поры, пустоты и трещины в почвах и горных породах, перемещающаяся под действием силы тяжести к водному зеркалу.

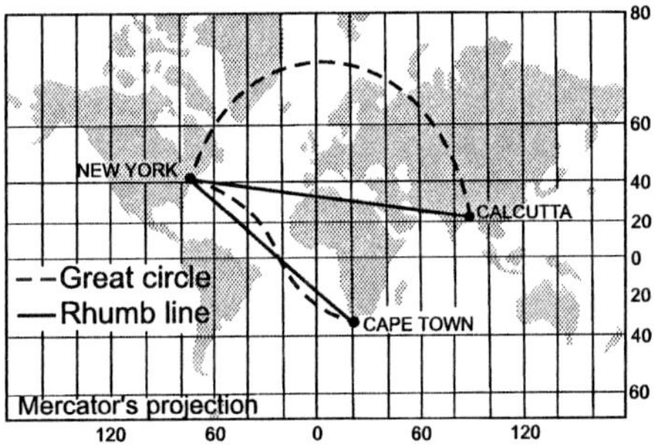

Fig. 139. **Great circle**

f eau *f* de gravité

s agua *f* gravitacional

g Gravitationswasser *n*; Sickerwasser *n*

* **gravity wind → 3306**

* **grazing → 4441**

2617 grease ice; ice fat; lard ice
A thick layer of ice crystals in the sea that gives to the water surface a greasy appearance.

r ледяное сало
Густой слой ледяных кристаллов на поверхности воды, придающий ей маслянистый вид.

f glace *f* pelliculaire

s hielo *m* pelicular

g Eisschlamm *m*

2618 great circle
Any hypothetical circle on the Earth's surface the plane of which passes through the Earth's centre, and thus divides the Earth into two hemispheres (the equator is a great circle, and each meridian is half of a great circle). → Fig. 139

r большой круг
Любой гипотетический круг на земной поверхности, плоскость которого проходит через центр Земли и, таким образом, он делит Землю на два полушария (экватор представляет собой большой круг, а каждый меридиан – половину большого круга). → Рис. 139

f cercle *m* grand

s círculo *m* grande

g großer Kreis *m*; Großkreis *m*

2619 Great Discoveries
The epoch since the middle of the XV century to the middle of the XVII century when the European travellers found new lands not known for them before.

r великие географические открытия
Эпоха крупнейших открытий новых земель, сделанных европейскими путешественниками с середины XV до середины XVII вв.

f grandes découvertes *fpl* géographiques

s grandes descubrimientos *mpl* geográficos

g die großen geographischen Entdeckungen *fpl*

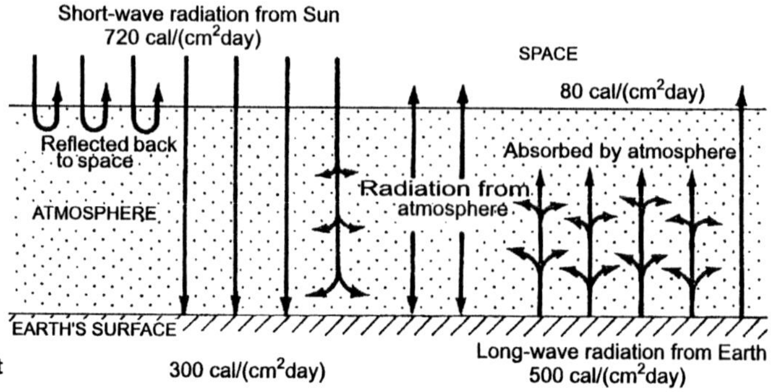

Fig. 140. **Greenhouse effect**

2620 **green belt**
A tract of open recreational areas, woodland or parks lying around cities and inside urban areas protected from further building in order to prevent urban srrawl and aimed at performing ecological functions.
r зеленая зона; зеленый пояс
Территория вокруг и внутри городской застройки, занятая лесами, лесопарками и другими озелененными участками, выполняющая экологические функции и используемая для отдыха населения.
f ceinture *f* verte
s cinturón *m* verde; cinturón *m* forestal
g Grünanlagenzone *f*; Grüngürtel *m*; Grünzug *m*

* **green energy → 220**

2621 **green flash; green ray**
A brilliant green colouration of the upper limb of the Sun occasionally observed just as the Sunrs apparent disk is about to sink below a distant clear horizon.
r зеленый луч
Последний луч заходящего Солнца под горизонтом, обычно морским, или первый луч восходящего Солнца, окрашенный в изумрудно-зеленый цвет.
f éclat *m* vert; rayon *m* vert
s destello *m* verde; rayo *m* verde
g grüner Strahl *m*; grüner Blitz *m*

2622 **green manure**
A green crop, especially one rich in nitrogen, e.g. clover, which is, while still green, ploughed directly into the soil to increase soil fertility and improve structure.
r зеленое удобрение
Зеленые растения, особенно богатые питательными веществами, например, клевер, которые, будучи еще зелеными, перепахиваются прямо в почву с тем, чтобы увеличить ее плодородие и улучшить структуру.
f engrais *m* vert
s estiércol *m* verde
g Gründünger *m*

* **green movement → 1802**

* **green politics → 1937**

* **green ray → 2621**

2623 **Green Revolution**
A considerable increase of food production, mainly in developing countries since the mid 1960s due to the development of new forms of cereal plants and the use of more powerful fertilizers which give much higher yields.
r зеленая революция
Значительный рост с середины 1960-х годов во многих, в основном в развивающихся странах, производства продовольственных, главным образом зерновых культур.
f révolution *f* verte
s revolución *f* verde
g grüne Revolution *f*

2624 **«green» gross domestic product**
Gross national product including environmental investments in nature protection.
r «зеленый» валовой внутренний продукт
Валовой внутренний продукт с учетом всех затрат на природоохранные мероприятия.
f produit intérieur brut *m* vert
s producto interior bruto *m* verde
g grünes Bruttoinlandprodukt *n*

2625 **greenhouse effect**
The warming of the lower layers of the atmosphere due to its different absorption properties for long- and short-wave radiation. → Fig. 140
r парниковый эффект
Нагревание нижних слоев атмосферы в результате разных поглощающих свойств атмосферного воздуха для длинно- и коротковолновой (солнечной) радиации. → Рис. 140
f effet *m* de serre
s efecto *m* de invernadero
g Treibhauswirkung *f*; Treibhauseffect *m*

2626 **greenhouse gases**
Atmospheric gases, including water vapour and carbon dioxide, the molecules of which absorb terrestrial radiation and form the greenhouse effect.
r парниковые газы
Атмосферные газы, включая водяной пар, молекулы которых поглощают тепловое из-

лучение Земли и вызывают явление парникового эффекта.
f gazes *mpl* de serre
s gases *mpl* de invernadero
g Treibhausgase *npl*

2627 Greenpeace
An international pressure group, which takes physical action to prevent the pollution of the environment.
r Гринпис
Общественное движение, оказывающее открытое давление на власти для предотвращения загрязнения окружающей среды.
f Greenpeace *m*
s Greenpeace *m*
g Greenpeace *n*

2628 Greenwich mean time; GMT; Greenwich universal time
The local time at the meridian of Greenwich, England (0° longitude), the time from which most of the countries calculate their standard time (Symbol: GMT).
r среднее гринвичское время
Местное время на Гринвичском меридиане в Англии (0° долготы), время, от которого отсчитывается поясное время во многих странах.
f temps *m* moyen de Greenwich; temps *m* universel
s hora *f* universal; hora *f* del meridiano de Greenwich
g mittlere Greenwich-Zeit *f*

* **Greenwich universal time → 2628**

2629 gregarious animal
Animals which have the tendency to live in groups or communities or to group for a period of time.
r стадное животное
Объединение животных одного вида на определенный период в стада или группы, характерно для рыб, ракообразных, китообразных, копытных.
f animal *m* grégaire
s animal *m* gregario
g Herdentier *m*

2630 Gregorian calendar
A system of chronology in which years divisible by 100, but not divisible by 400 (e.g. 1700, 1800, 1900) are not leap years (in contrast with less precise *Julian calendar*), it was introduced by Pope Gregory XIII in 1582.
r Григорианский календарь; новый стиль
Система летоисчисления, в которой (в отличие от менее точного *Юлианского календаря*) не считаются високосными годы с номерами, делящимися на 100, но не делящимися на 400 (например, 1700, 1800, 1900), введен папой Григорием XIII в 1582 г.
f calendrier *m* grégorien
s calendario *m* gregoriano
g Gregorianischer Kalender *m*

2631 grey brown soil
A soil that forms in the regions of dry subtropical climate under xerophytic grasses and shrubs, it has light-grey humus horizon, clayed horizon and illuvial-carbonate horizon.
r серо-коричневая почва
Почва, формирующаяся в районах с сухим субтропическим климатом под ксерофитной травянистой и кустарниковой растительностью, имеет светло-серый гумусовый, оглиненный и иллювиально-карбонатный горизонты.
f sol *m* gris-marron
s suelo *m* gris-marrón
g graubrauner Boden *m*

* **grey desert soil → 5434**

* **grey earth → 5434**

2632 grey-brown desert soil
A soil that forms in deserts of the temperate belt under a thinned halophyte vegetation, it has porous crust, pale-yellow and grey horizon as well as packed brown horizon with new formations of carbonate and gypsum.
r серо-бурая пустынная почва
Почва, формирующаяся в пустынях умеренного пояса под разреженной галофитной растительностью, имеет ноздреватую корку, палево-серый и уплотненный бурый горизонты с карбонатными и гипсовыми новообразованиями.
f sol *m* brun-gris désertique
s suelo *m* pardo-gris de desierto
g graubrauner Wüstenboden *m*

2633 grey-brown podzol
An acidic soil, transitional between a podzol and brown forest soil, less leached and with a greater organic content than a podzol, it occurs mainly in western Europe and northeast USA.
r серо-коричневый подзол
Кислая почва, переходная между подзолом и коричневой лесной почвой, менее выщелоченная и с большим содержанием органики, чем подзол, встречается преимущественно в Западной Европе и на северо-востоке США.
f podzol *m* gris-brun
s podzol *m* gris-marrón
g graubrauner Podzol *m*

2634 grey-forest soil
A soil forming under broad-leaved and mixed forests of continental climate under periodical leaching regime, it usually forms on carbonate loams and has dark humus horizon bleached in the lower part.
r серая лесная почва
Почва, формирующаяся под широколиственными и смешанными лесами конти-

нентального климата в условиях периодического промывного режима, обычно образуется на карбонатных суглинках и имеет темный гумусовый горизонт, осветленный в нижней части.
f sol *m* gris forestier
s suelo *m* gris forestal
g grauer Waldboden *m*

2635 grid; coordinate grid; reference grid
The system of lines on a geographical map serving to determination of co-ordinates of objects, their plotting and search by co-ordinates, orientation, routing (geographical, rectangular, kilometre grids are distinguished).
r координатная сетка
Система линий на географической карте, служащая для определения координат объектов, ориентирования, прокладки маршрутов (различают географическую, прямоугольную, километровую сетки).
f réseau *m* cartographique; canevas *fpl*
s retícula *m* de un mapa; canevás *mpl*
g Koordinatennetz *n*

* **grid azimuth → 2636**

* **grid bearing → 2636**

2636 grid bearing; grid bearing; grid azimuth
Angle on a plane counted at a given point off from a line parallel to the northern direction of central meridian clockwise till the given direction.
r дирекционный угол
Угол между северным направлением вертикальной линии координатной сетки и направлением на предмет, отсчитываемый по часовой стрелке.
f angle *m* de plane
s angulo *m* del plano; azimut *m* decluadriculado
g Gitterazimut *m*

2637 grid sell
The element of the division of a map (image) by the lines of a regular or irregular grid.
r ячейка
Элемент разбиения поля карты (снимка) линиями регулярной или нерегулярной сети.
f maille *f*
s malla *f*
g Masche *f*

2638 groin; groyne
A coastal protecting structure built across the shoreline aimed at the accumulation of beach material and protection of the coast from washing it out.
r буна
Поперечное берегозащитное сооружение, предназначенное для накопления пляжа, предохраняющего берег от размыва.
f traverse *f*
s espigón *m*
g Wuhre *f*

2639 gross domestic product; GDP
The annual value of goods produced and services provided by a country (whithout an income from abroad).
r валовой внутренний продукт; ВВП
Совокупная стоимость конечных товаров и услуг, произведенных на территории данной страны, в рыночных ценах (без доходов из-за границы).
f produit *m* intérieur brut; PIB
s producto *m* interior bruto; PIB
g Bruttoinlandprodukt *n*

2640 gross income
Money earned by industrial and agricultural enterprises for their products and services minus their material expenditures.
r валовой доход
Денежная выручка предприятий промышленности и сельского хозяйства, полученная от реализации продукции и услуг, за вычетом материальных затрат.
f revenu *m* brut
s renta *f* bruta; ingreso *m* bruto
g Bruttoeinkommen *n*

2641 gross national product; GNP
The annual total value of goods produced and services provided by a country plus the total income from abroad.
r валовой национальный продукт; совокупный общественный продукт; валовой общественный продукт; ВНП
Совокупная стоимость конечных товаров и услуг, произведенных в данной стране, в рыночных ценах, в отличие от валового внутреннего продукта, включает и сумму чистых доходов из-за границы.
f produit *m* national brut; PNB
s producto *m* nacional bruto; PNB
g Bruttosozialprodukt *n*

2642 gross product
A part of the gross national product produced by an enterprise or a branch of economy in a given time period.
r валовая продукция
Часть валового национального продукта, произведенного на предприятии или в отрасли за определенный период, выраженная в оптовых ценах.
f produit *m* brut
s producto *m* bruto
g Bruttowertschöpfung *f* der Produktion

2643 gross regional product
The annual total value of goods produced and services provided by a region, it is a part of the *gross national product* that shows the economic development level of the region.

r валовой региональный продукт
Суммарное годовое количество товаров и услуг, произведенных в данном регионе, это часть *валового национального продукта*, характеризующая уровень экономического развития отдельных районов и центров.
f produit *m* régional brut
s producto *m* regional bruto
g Bruttoregionalprodukt *n*

2644 grotto
A cave with a wide opening and a low arched roof as well as a widening of a cave after a narrow passage.
r грот
Широко открытая пещера с невысоким сводчатым потолком, а также расширение пещеры после узкого прохода.
f grotte *f*
s gruta *f*
g Grotte *f*

2645 ground
Rocks occurring in the zone of weathering of the Earth's crust and which are being used in engineering activity and building or which serves as a substratum.
r грунт
Горные породы, залегающие в зоне выветривания земной коры и служащие объектом инженерно-строительной деятельности человека.
f terrain *m*
s terreno *m*; suelo *m*
g Grund *m*; Erde *f*

2646 ground avalanche; full depth avalanche
A spring wet snow avalanche which incorporates the entire snowpack depth and moves over ground or soil surface.
r грунтовая лавина; лавина полной глубины
Влажная весенняя лавина, сдирающая весь снежный покров и движущаяся по грунту или почве.
f avalanche *f* de fond
s avalancha *f* de fondo
g Grundlawine *f*

* **ground fog → 4868**

2647 ground freezing
The change of water into ice in soil, ground or rock.
r промерзание грунта
Переход влаги, содержащейся в грунте, из жидкого состояния в твердое.
f congélation *f* de sol
s congelamiento *m* del suelo
g Gefrieren *n*

2648 ground frost
A frost occurring on the surface of the ground or in the upper soil layer when the ground surface temperature drops to -1°C, although the air temperature may remain above freezing-point.

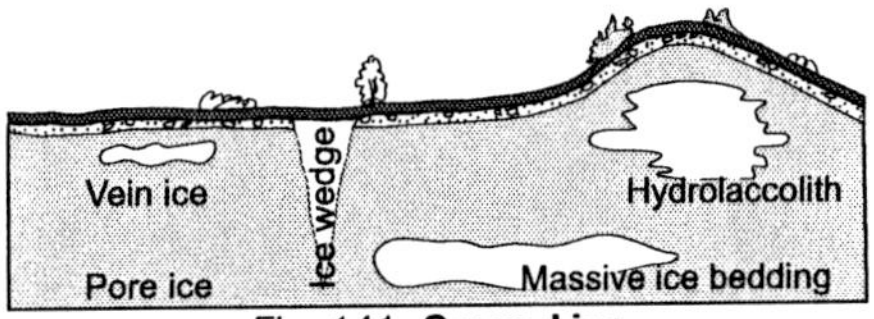

Fig. 141. **Ground ice**

r заморозок на почве
Заморозок на поверхности грунта или в верхнем слое почвы, когда температура поверхности грунта падает до -1°C, хотя температура воздуха может оставаться выше точки замерзания.
f gelée *f* du sol
s helada *f* sobre suelo
g Bodenfrost *m*

2649 ground ice
Ice of any origin occurring under the surface of the ground. → Fig. 141
r подземный лед; грунтовый лед
Лед любого генезиса, находящийся под поверхностью земли. → Рис. 141
f glace *f* souterraine; glace *f* de sol
s hielo *m* de suelo
g Bodeneis *n*; Grundeis *n*

2650 ground information
In remote sensing, information concerning the state of the physical environment derived from aerial photographs and measurements of soil moisture, temperature, biomass, etc. (ideally this information is recorded at the time when a remote sensor is passing over the area concerned).
r наземная информация
В дистанционном зондировании, информация о физическом состоянии окружающей среды, получаемая из аэрофотоснимков и наземных измерений влажности почвы, температуры, биомассы и пр. (эту информацию получают, когда дистанционный прибор находится над изучаемой территорией).
f information *f* terrestre
s información *f* terrestre
g Grundinformation *f*

2651 ground inversion; surface inversion
The inversion of the temperature in the air layer close to the earthrs surface, often caused by nocturnal net radiation.
r приземная инверсия
Инверсия температуры непосредственно над земной поверхностью, часто вследствие охлаждения воздуха путем эффективного излучения.
f inversion *f* au sol; inversion *f* en surface
s inversión *f* a partir del suelo; inversión *f* en superficie
g Bodeninversion *f*

* **ground layer of air → 771**

* **ground mist → 4868**

2652 **ground moraine; subglacial moraine**
Debris carried in the base of a glacier or ice sheet and deposited during the ice melting.
r донная морена; основная морена
Отложение под ледниками грубообломочного материала, переносимого в составе *придонной морены.*
f moraine *f* de fond
s morrena *f* de fondo; morrena *f* inferior
g Grundmoräne *f*; Untermoräne *f*

2653 **ground rent**
A regular income of a landowner received from a leaseholder who is engaged in agriculture or mining.
r земельная рента
Регулярный доход собственника земли, получаемый им с непосредственного производителя в сельском хозяйстве и отраслях добывающей промышленности, использующих эту землю.
f rente *f* foncière
s renta *f* del suelo
g Bodenrente *f*; Grundrente *f*

* **ground resolution → 5706**

2654 **ground subsidence**
A downward sinking of a land surface due to the thawing of frozen ground.
r просадка грунта
Оседание поверхности в результате протаивания мерзлых пород.
f abaissement *m*
s hundimiento *m* del suelo
g Absenkung *f*

2655 **ground swell**
The deep, slow rolling of the sea (sometimes caused by a distant storm or a seismic disturbance) which in passing into shallow water raises the height of the waves.
r придонное волнение
Придонное, медленное вращение морской воды (иногда вызванное отдаленным штормом или сейсмическими смещениями), которое при переходе на мелководье увеличивает высоту волн.
f vagues *fpl* de fond
s olas *fpl* del fondo
g Grundwelle *f*

2656 **ground vein**
A wedge-shaped geological body formed as a result of filling up of surface fissures with a fine material.
r грунтовая жила
Клиновидное в поперечном разрезе геологическое тело, образующееся в результате заполнения поверхностных трещин мелкоземом.
f coin *m* de sol; filon *m* de sol
s filón *m* de suelo; cuña *f* de suelo
g Grundader *f*

* **ground water → 2661**

2657 **ground water balance**
The relation of the amount of water got into a water-bearing layer of rock (aquifer) and discharged from it over a certain period of time.
r баланс подземных вод
Соотношение количества воды, поступившей в водоносную толщу и израсходованной из нее за определенный период времени.
f bilan *m* des eaux souterraines
s balance *m* de aguas subterráneas
g Groundwasserbilanz *f*

2658 **ground water lens**
A limited volume of ground saturated with free (gravity) water, located between two bedding plains.
r скопление подземных вод; линза подземных вод
Небольшой по площади объем горных пород, насыщенных свободной (гравитационной) водой, находящийся между двумя поверхностями напластования.
f nappe *f* captive fossile
s capa *f* cautiva fósil; agua *f* fósil
g Grundwasserlinse *f*

2659 **grounded hummock; stranded hummock**
Isolated ice blocks being aground.
r стамуха
Отдельные глыбы льда высотой до 10 м и более, стоящие на мели.
f hummock *m* échoué
s banco *m* de hielo encallado
g verankerter Presseishügel *m*; Eishügel *n*

2660 **grounding line**
Boundary on a glacier bed separating the grounded portions of glaciers from those afloat. → Fig. 163, p. 359
r линия налегания
Граница на ледниковом ложе, разделяющая сидящие на грунте и плавающие части ледников. → Рис. 163, с. 359
f ligne *f* de mise á flot; ligne *f* d'échouage
s línea *f* de encalladura
g Gründungslinie *f*

2661 **groundwater; ground water**
Water lying under the surface of the ground in the rocks and in soil in its liquid, solid or gaseous state.
r подземные воды; грунтовые воды
Вода, находящаяся ниже земной поверхности в толще горных пород и в почве в любых физических состояниях.
f eau *f* souterraine
s agua *f* subterránea
g Grundwasser *n*; unterirdisches Wasser *n*

2662 **groundwater flow; subsurface flow**

The movement of ground water due to hydraulic pressure head or gravity.
r подземный сток
Перемещение подземных вод под действием гидравлического напора или силы тяжести.
f écoulement *m* souterrain
s escorrentía *f* subterránea
g unterirdischer Abfluß *m*

2663 groundwater stock
The total volume of water in a groundwater deposit.
r запасы подземных вод
Общий объем воды в месторождении подземных вод.
f capacité *f* souterraine; réserve *f* d`eau souterraine
s reservas *fpl* de agua subterránea
g Grundwasservorräte *mpl*

2664 grove; coppice
A small, usually secondary, leaf-bearing wood (e.g. oak, olive, palm groves occur), or an area that is composed entirely of timber trees without underwood.
r роща
Небольшой массив леса, обычно вторичного лиственного (например, дубовая, оливковая, пальмовая рощи), иногда лишенного подлеска.
f bosquet *m*
s arboleda *f*; bosquete *m*
g Hain *m*; Gehölz *n*; Wäldchen *n*

2665 growing delta
A river delta increasing its area due to the accumulation of sediments by the river.
r растущая дельта
Речная дельта, увеличивающая свою площадь в результате аккумулирующей деятельности реки.
f delta *m* en formation
s delta *m* en formación; delta *m* creciente
g wachsendes Delta *n*

2666 growing season; vegetation season
The period of the year when meteorological conditions are favourable to the growth of plants.
r вегетационный период
Период года, на протяжении которого метеорологические условия благоприятны для роста и развития растений.
f saison *f* de la végétation; saison *f* de la croissance
s período *m* de vegetación; período *m* de crecimiento
g Vegetationszeit *f*; Vegetationsperiode *f*

2667 growler; bergy bits; iceblock
A piece of glacier ice detached directly from a glacier or produced by the breakdown of an iceberg.
r айсблок
Отколовшийся от ледника или айсберга плавучий кусок льда высотой над водой в несколько метров.
f bloc *m* de glace
s bloque *m* de hielo
g Eishügel *m*

2668 growth area; growth pole; growth centre
An industrial centre or a district capable of accelerating the development of the surrounding area, spreading prosperity from the core to the periphery.
r центр роста; полюс роста
Промышленный центр или целый район, призванный дать импульс территориальному развитию в окружающем регионе, распространяя свое влияние от центра к периферии.
f centre *m* de croissance; pôle *m* de croissance; pôle *m* de développement
s centro *m* de crecimiento; polo *m* de desarrollo; polo *m* de crecimiento
g Entwicklungzentrum *m*; Wachstumspol *m*

* **growth centre → 2668**

* **growth pole → 2668**

* **groyne → 2638**

* **gruss → 5137**

2669 guano
The excreta of sea birds decomposed under the arid climate found in the coastal zone and the islands of Chile, Peru, South Africa, and the Caribbean Sea, where there is little or no rain to wash it away, gathered and used as fertilizer.
r гуано
Разложившийся в условиях сухого климата помет морских птиц, используемый в качестве ценного удобрения, распространен на островах близ побережья Чили, Перу, Южной Африки, в Карибском море.
f guano *m*
s guano *m*
g Guano *m*

* **guest worker → 2365**

* **gulch → 2671**

2670 gulf
A large indentation of the sea cutting into the land more deeply than a bay, being more enclosed by the coast and may itself contain one or more bays.
r залив
Значительная часть океана, моря или озера, довольно глубоко вдающаяся в сушу, может включать в свою акваторию одну или несколько бухт.
f golfe *m*
s golfo *m*
g Busen *m*; Golf *m*

2671 gulley; gulch; ravine; gully
A small narrow channel with steep sides formed by the action of water, especially resulting from a heavy downpour of rain but not permanently occupied by stream, it is not used for cultivation.
r овраг; эрозионная ложбина
Крутосклонная сильно разветвленная ложбина, образованная деятельностью временных водотоков, которую невозможно использовать при возделывании земли.
f ravin *m*; vallon *m*
s barranco *m*; vallejo *m*
g Wasserriß *m*; Erosionschlucht *f*

* **gully → 2671**

2672 gully erosion; gullying
The linear washout of slopes and valley floors leading to the formation of gullies that dissect the surface.
r овражная эрозия
Линейный размыв временными водотоками поверхности склонов и днищ долин, приводящий к развитию оврагов и расчленению ими террритории.
f érosion *f* ravinée; ravinement *m*
s erosión *f* barrancal; abarrancamiento *m*
g Schluchterosion *f*; Rinnenerosion *f*

* **gully terrain → 2673**

* **gullying → 2672**

2673 gullying; gully terrain; ridge-and-ravine terrain
A type of erosion relief with gulches and small flat-bottom valleys that dissect flat or convex interfluves.
r овражно-балочный рельеф
Тип эрозионного рельефа с преобладанием оврагов и балок, рачленяющих плоские или выпуклые междуречья.
f relief *m* raviné
s relieve *m* barrancal
g Schluchtrelief *n*

2674 gust
A sudden, strong, brief increase of the wind speed over its mean value.
r порыв ветра
Внезапное, кратковременное возрастание скорости ветра по сравнению с ее средней величиной.
f rafale *f*
s ráfaga *f*; racha *f* de viento
g Bö *f*; Windstoß *m*

2675 gust-and-glaze loading
The loading on wires and structures produced by the joint action of glaze and dynamic pressure of wind.
r гололедно-ветровая нагрузка
Нагрузка на провода и сооружения от совместного воздействия гололеда и скоростного напора ветра.
f charge *f* de givre et de vent
s carga *f* de helada y viento
g Eis- und Windbelastung *f*

2676 gustiness component
The ratio of wind velocity in a gust over 10-minute interval to the average wind velocity taken for the same time interval.
r коэффициент порывистости ветра
Отношение скорости ветра в порыве за 10-минутный интервал к средней скорости ветра, взятой за тот же интервал времени.
f composante *f* de la turbulence du vent
s coeficiente *m* de rafagosidad del viento
g Windturbulenzkomponente *f*; Böigkeitkomponente *f*

2677 Gutenberg Discontinuity
The discontinuity occurring between the lower surface area of the mantle and the core of the Earth, where the material of the Earth's interior is in transition between the solid and plastic state, and where the speed of the P-waves of earthquakes slackens and S-waves are absent.
r поверхность Гутенберга
Граница между областью нижней поверхности мантии и ядром Земли, где внутренний материал находится в переходной стадии между твердым и пластичным состоянием и скорость P-волн землетрясений становится меньше, а S-волны отсутствуют.
f discontinuité *f* de Gutenberg
s discontinuidad *f* de Gutenberg
g Gutenberg-Diskontinuität *f*

2678 guttation
The process by which the water transpired from vegetation is unable to evaporate into the air because it is already saturated with water vapour.
r гуттация
Процесс, при котором капли воды, образовавшиеся из растения, не могут испариться в окружающий воздух, потому что он уже насыщен водяным паром.
f guttation *f*
s gutación *f*
g Tropfenbildung *f*

2679 guyot
An isolated volcanic underwater flat-topped mountain. → Fig. 142, p. 321
r гайот
Плосковершинная вулканическая подводная гора. → Рис. 142, с. 321
f guyot *m*
s guyot *m*
g Guyot *m*

2680 gypsic horizon
A soil layer accumulating gypsum found in the middle or lower part of soil profile, it contains gypsum in the form of crystals of different size.

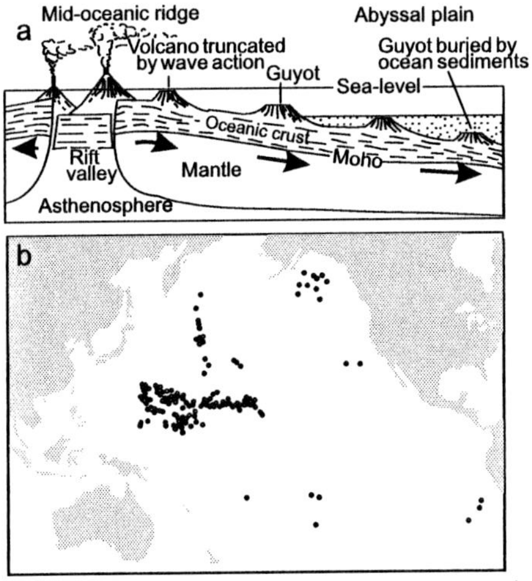

Fig. 142. Formation of a **guyot** (a) and the distribution of guyots in the Pacific (b)

r гипсовый горизонт;
гипсиковый горизонт
Слой аккумуляции гипса в почве, находящийся обычно в средней или нижней части почвенного профиля, содержит гипс в виде отдельных кристаллов разной величины.
f horizon *m* gypsique
s horizonte *m* yesoso
g Gipshorizont *m*

2681 gypsisol
Soils of the most arid areas with accumulation of gypsum, they are widespread mainly in deserts of Africa and Australia.
r гипсисоль
Почвы наиболее аридных областей с аккумуляцией в них гипса, распространены в основном в пустынях Африки и Австралии.
f gypsisol *m*
s gipsisol *m*; gipsosuelo *m*
g Gypsisol *m*

2682 gypsophyte
Plants that grow on soils rich in gypsum.
r гипсофит
Растения, обитающие на почвах, богатых гипсом.
f gypsophite *m*
s gipsofita *m*; gipsófilo *m*
g Gypsophyt *m*

2683 gypsum
Sedimentary rock and mineral hydrated calcium sulphate in which often karst phenomena develop.
r гипс
Осадочная порода и минерал – водный сульфат кальция, нередко область развития карста.
f gypse *m*
s yeso *m*
g Gips *m*

2684 gypsum crust
A dense coherent cover, hard when dry, cemented by gypsum, sometimes mixed with carbonates, it is characteristic of the extremely arid desert regions.
r гипсовая корка
Плотный или с кавернами, твердый в сухом состоянии покров, сцементированный гипсом, иногда в смеси с карбонатами, характерный для самых безводных районов пустынь.
f croûte *f* de gypse
s costra *f* de yeso
g Gipskruste *f*

2685 gypsuming of soil
The application of gypsum to the upper horizons of solonetz and solonetz-like soils in order to increase their fertility.
r гипсование почвы
Повышение плодородия солонцов и солонцеватых почв путем внесения гипса в их верхние горизонты.
f plâtrage *m* du sol;
introduction *f* de gypse à sol
s abonado *m* del suelo con yeso
g Gipsdüngung *f*

2686 gyre
The circulation of surface water in closed circulatory systems around the subtropical high-pressure systems at latitudes 20°-30° N and S.
r круговое течение
Циркуляция поверхностных вод в замкнутых циркуляционных системах вокруг субтропических областей высокого давления на широтах 20°-30° в обоих полушариях.
f giration *f*
s giración *f*
g Kreislauf *m*

2687 gyttja
A grey-brown to blackish soil, rich of organic matter, consisting mainly of plant and animal residues precipitated from stagnant water.
r гиттия
Почва, от серо-коричневого до темносерого цвета, богатая органическим веществом, состоящим в основном из остатков растений и животных, осаждающихся в застойной воде.
f gyttja *f*
s gyttja *f*
g Gyttja *f*

H

2688 **habitat; ecotope**
A piece of land or water body where an organism or population of a species lives, and which provides all biotic and abiotic environmental conditions for its life.
r местообитание; экотоп; среда обитания
Участок суши или водоема, занятый организмом или его популяцией и обладающий необходимыми для их существования условиями обитания, включая живую и неживую природу.
f habitat *m*; environnement *m* vital; écotope *m*; milieu *m* vital
s hábitat *m*; ecotopo *m*; ambiente *m* habitado
g Lebensort *m*; Lebensumwelt *f*; Ökotop; Lebensraum *m*; Habitat *n*; Standort *m*

2689 **habitat loss; habitat reduction**
A permanent disappearance of, or decrease in, the amount of suitable environment available to an organism.
r сокращение местообитания
Временное исчезновение или ухудшение условий окружающей среды, способствовавших ранее существованию в данном месте определенных организмов.
f réduction *f* de l'habitat
s reducción *f* de hábitat
g Habitatverlust *m*

* **habitat reduction → 2689**

2690 **hachures**
Short lines drawn on a map to run in the direction of maximum slope, which indicate the relief by their thickness and spacing. → Fig. 143
r гашюры; штрихи крутизны
Графическое средство для изображения рельефа на картах с помощью штрихов, передающих крутизну склонов. → Рис. 143
f hachures *fpl*
s hachuras *fpl*
g Schraffen *fpl*; Böschungsschraffen *pl*

* **hacienda → 2000**

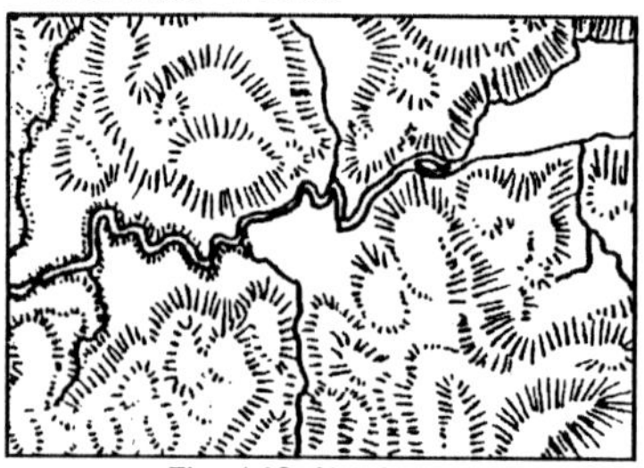

Fig. 143. **Hachures**

* **hadal zone → 6438**

2691 **hade**
The angle made by a mineral vein or fault plane with the vertical, i.e. the slope of the fault as contrasted with the dip of the beds.
r отклонение от вертикали
Угол склонения минеральной жилы или плоскости сдвига по отношению к вертикали, т.е. склон сброса, отличающийся от падения пласта.
f écart *m* de faille
s desviación *f* vertical
g Lotabweichung *f*

2692 **Hadley cell**
A thermally driven vertical circulation cell forming the part of the general circulation of the atmosphere lying between the equator and approximately 30° N and 30° S.
r ячейка Хедли
Термически обусловленная ячейка вертикальной циркуляции воздуха как часть общей циркуляции атмосферы между экватором и приблизительно 30° северной и южной широты.
f cellule *f* de Hadley
s célula *f* de Hadley
g Hadley-Zelle *f*

2693 **haff**
A coastal lagoon of fresh and brackish water, fed by a stream which is blocked by a spit (nehrung), through which it is linked to the sea by a channel, typical of the Baltic coast of Germany, Poland and Russia.
r гафф
Прибрежная лагуна с пресной и солоноватой водой, образованная ручьем или небольшой рекой, отгороженная косой, но имеющая связь с морем через естественный канал, такие лагуны типичны для балтийского побережья в Германии, Польше и России.
f haff *m*
s haff *m*
g Haff *n*

2694 **hail**
A type of precipitation of compact irregular pieces of ice (hailstones) which usually fall from cumulonimbus clouds in the warm season at the passing of a cold front.
r град
Частицы плотного льда неправильной формы, выпадающие в теплое время года из мощных кучево-дождевых облаков при прохождении холодного фронта.
f grêle *f*
s granizo *m*; pedrisco *m*
g Hagel *m*

2695 **hail damage**
The damage to crops, vegetation, buildings, vehicles, etc, which is a direct result of hail.

r градобитие
Ущерб, наносимый посевам, плантациям, зданиям и т.п., являющийся прямым следствием выпадения града.
f dommage *m* occasionné par la grêle
s destrucción *f* causada por el granizo
g Hagelschlag *m*; Hagelschaden *m*

2696 halmyrolysis
Chemical and mineralogical transformation of primary sediments on the ocean floor under the effect of dissolution, oxidation, etc.
r гальмиролиз; подводное выветривание
Химико-минералогическое преобразование отложений на дне моря под влиянием процессов растворения, окисления и пр.
f halmyrolyse *f*
s halmirólisis *f*
g Halmyrolyse *f*; submarine Verwitterung *f*

2697 halo
A ring of light arcs, pillars or bright spots around the Sun or Moon, produced by the refraction or reflection of light by ice crystals and water drops within the cloud. → Fig. 144
r гало
Световое кольцо вокруг Солнца или Луны, возникающее в результате преломления и отражения света в ледяных кристаллах, к гало относятся окрашенные круги, дуг, столбы, пятна и т.п. → Рис. 144
f halo *m*
s halo *m*
g Lichthof *m*; Halo *m*

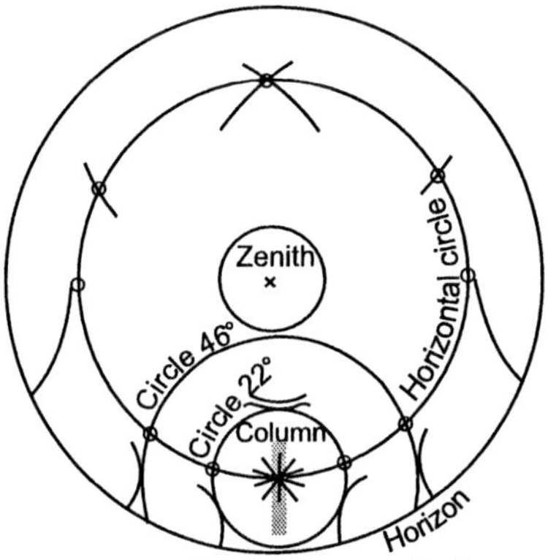

Fig. 144. Principal forms of **halo**

* **haloclasty → 5190**

2698 halocline
Sea water layer with pronounced increase of salinity.
r галоклин
Слой воды в океане (море) с резким ростом солености.
f halocline *f*
s haloclina *f*
g Salzgehaltssprungschicht *f*; Halokline *f*

2699 halogenic soil
A soil formed mainly as a result of migration of easily soluble salts (*solonchaks*).
r галогенная почва; галоморфная почва
Почвы, в формировании которых существенную роль играют процессы миграции легкорастворимых солей (к ним относятся в первую очередь *солончаки*).
f sol *m* halogène
s suelo *m* halógeno; suelo *m* halomorfo
g halogener Boden *m*; halomorpher Boden *m*

2700 halophile
Any organism which can tolerate saline conditions, e.g. sea shore, salt deserts, salt marsh (plants that prefer saline soils are called halophytes).
r галофил
Организмы, предпочитающие соленую среду обитания, к ним относятся радиолярии, кораллы, большинство иглокожих, а также многие ракообразные, насекомые и др. (растения, приспособленные к обитанию на засоленных почвах, называют галофитами).
f halophile *m*
s halófilo *m*
g Halophil *m*

* **hamada → 2701**

* **hamlet → 6563**

2701 hammada; hamada
A stony desert in the Middle East and Northern Africa.
r гамада
Каменистая пустыня в странах Ближнего Востока и Северной Африки.
f hammada *f*
s hammada *f*
g Hammada *f*

2702 handicrafts
Items made by hand or with simple equipment.
r ремесленные товары
Изделия, выполненные мастером вручную или на личном, весьма простом оборудовании.
f artisanat *m*
s artesania *f*; artesanado *m*
g Handwerksarbeit *f*

2703 hanging glacier
A small glacier on such a precipitous slope that its snout is constantly breaking away as ice avalanches. → Fig. 145, p. 324
r висячий ледник
Небольшой ледник, лежащий на обрывистом склоне так, что лед на его конце регулярно обламывается и падает вниз в виде ледяной лавины. → Рис. 145, с. 324
f glacier *m* suspendu
s glaciar *m* suspendido
g Gehängegletscher *m*

Fig. 145. **Hanging glaciers**

2704 hanging valley
A tributary valley which descends abruptly towards the floor of a larger valley and the lower end of which is well above the bed of the main valley from which a waterfall may descent.
r висячая долина
Боковая долина, лежащая значительно выше главной долины и обрывающаяся уступом к ее днищу, с которого нередко вода спускается в виде водопада.
f vallée *f* suspendue
s valle *m* colgado
g Hängetal *n*

2705 hanging-up water
Water that is durably kept by soil almost without movement and devoid of hydraulic connection with water-bearing horizons.
r подвешенная влага
Влага, длительно удерживаемая в почве практически в неподвижном состоянии и не имеющая гидравлической связи с водоносным горизонтом.
f eau *f* suspendue
s agua *f* suspensa; aguas *fpl* de retención capilar; humedad *f* suspensa
g Hangwasser *n*

2706 haploidization
The process of soil formation that does not lead to the differentiation of the initial homogenous soil material into distinct soil horizons, it is characteristic of vertisols.
r гаплоидизация
Почвообразовательный процесс, не приводящий к дифференциации исходного гомогенного почвенного материала на четко выраженные горизонты, характерен для вертисолей.
f haploïdisation *f*
s haploidización *f*
g Haploidisierung *f*

2707 harbour
A stretch of water, close to and sheltered by the shore and protected from the sea and swell by natural or artificial walls for the anchorage of vessels which enter through a narrow entrance (the harbour mouth).
r гавань
Прибрежная часть акватории, естественно или искусственно защищенная от волн, ветров и течений для стоянки судов, которые попадают в гавань через узкий проход.
f port *m*
s puerto *m*
g Hafen *m*

* **harbour bar → 3998**

2708 harbour oscillation; range action
The resonance waving of water in ports, bays and harbours in the form of horizontal and vertical water movements.
r тягун
Резонансные волновые колебания воды в портах, бухтах и гаванях в виде горизонтальных и вертикальных движений вод.
f agitation *f* intérieure
s agitación *f* interior
g Brandungsschlag *n*

2709 hard water
Water having a high concentration (more than 60 mg/l) of dissolved calcium, magnesium and iron compounds derived from the rocks over or through which the water has passed.
r жесткая вода
Вода, содержащая повышенное (более 60 мг/л) количество солей щелочных металлов, преимущественно кальция, магния и железа, извлеченных ею из горных пород, сквозь которые она протекает.
f eau *f* dure
s agua *f* dura
g hartes Wasser *n*

2710 hard-leaved forest; sclerophyllous forest
A subtropical forest consisting of evergreen species with hard thick, leathery leaves that prevent moisture loss during summer heat, it is widespread in the Mediterranean region, America (the USA, Chile), Australia.
r жестколистный лес
Субтропический лес, состоящий из пород с жесткими листьями и прутьевидными стеблями, обладающих способностью сохранять влагу во время летней жары, распространен в Средиземноморье, Америке (США, Чили), Австралии.
f forêt *f* sclérophylle; durisilve *f*
s bosque *m* esclerófilo; durisilva *f*
g Hartlaubwald *m*

2711 hardness of rock
The resistance of rock to scratching, cutting, cracking, i.e. to the penetration of another body.
r твердость породы
Способность горной породы оказывать сопротивление проникновению другого тела.
f dureté *f* de roche
s dureza *f* de roca
g Gesteinshärte *f*

2712 hardpan
A hard (indurated or cemented) layer of soil of varying character found at different distances below, or sometimes at the upper surface of the soil.
r твердая корка
Затвердевший или сцементированный почвенный почвы разного характера и происхождения на разной глубине, а иногда и на поверхности почвы.
f encroûtement *m*; croûte *f*
s encostramiento *m*; costra *f*
g Verhärtungsschicht *f*

2713 hardsetting
The compaction and destruction of the initial soil structure that results in the formation of lumpy horizons which are characterized by the plasticity when wet, and high hardness and fissuring when dry.
r слитизация
Уплотнение и разрушение исходной структуры почвы, приводящие к образованию глыбистых горизонтов, отличающихся пластичностью во влажном состоянии и высокой твердостью и трещиноватостью в сухом.
f fusion *f*
s fusion *f*
g Slitisation *f*

2714 hardwood
The timber of all those broad-leaved trees which comprise the deciduous woodlands, which is relatively hard and durable, in contrast to that of the coniferous softwoods.
r твердая древесина
Древесина всех широколиственных деревьев, составляющих листопадные леса, которая относительно тверже и долговечнее по сравнению с мягкой древесиной хвойных деревьев.
f bois *m* dur
s madera *f* brava; madera *f* dura
g Hartholz *n*; Laubholz *n*

2715 harmattan
A very dry, dust-laden wind which blows from the Sahara to the north-west African coast over west Africa.
r харматан
Зарождающийся в Сахаре сухой пыльный восточный или северо-восточный ветер над Западной Африкой.
f harmattan *m*
s harmattan *m*
g Harmattan *m*

* **harmful algae blossoming → 4942**

* **harvest → 1342**

* **Hawaiian eruption → 3441**

2716 hay
The stems and leaves of grasses, cut and dried, and used as fodder.

Fig. 146. **Hazard of aufeis**: Destruction of a bridge by the river aufeis at a tributary of Amur river

r сено
Стебли и листья травы, скошенные и высушенные, используемые как корм для скота.
f foin *m*
s heno *m*
g Heu *n*

2717 hayfield
Agricultural land used to grow grass for hay.
r сенокос
Сельскохозяйственное угодье, используемое для выращивания трав на сено.
f fenaison *f*
s segazón *f*
g Heuschlag *m*

2718 hazard of aufeis
A hazard to people and structures occurring at pouring-out and successive freezing of water. → Fig. 146
r наледная опасность
Угроза людям или инженерным сооружениям, возникающая при излиянии и последующем замерзании наледеобразующих вод. → Рис. 146
f risque *m* de glaçage
s riesgo *m* de icing
g Aufeisgefahr *f*

2719 haze
The obscurity in the lower layers of the atmosphere resulting from the suspension in the air of extremely small, dry particles of dust or smoke, in particular caused by the smoke of forest or peat fires.
r мгла; сухая дымка
Помутнение воздуха в атмосфере, вызванное большим содержанием в нем взвешенных мелких сухих частиц, в частности, дымом от лесных или торфяных пожаров.
f brume *f* séche
s calina *f*; bruma *f* seca
g trockener Dust *m*

* **head → 4753**

2720 head percolation
The downward seeping of water into soil and rock under the pressure exceeding the atmospheric pressure.

r напорная фильтрация
Фильтрация в подземных горизонтах при давлении, превышающем атмосферное.
f percolation *f* de pression
s percolación *f* aprisionada
g Perkolation *f* unter Druck

* **headland → 860**

2721 headward erosion
The action of a river in cutting back upstream.
r регрессивная эрозия; пятящаяся эрозия; попятная эрозия
Распространение глубинной эрозии вверх по течению водотока.
f érosion *f* régressive
s erosión *f* regresiva; erosión *f* remontante
g rückschreitende Erosion *f*

* **headwater avalanche → 6754**

2722 headwaters; upper course; source area
The upper parts of a river drainage system.
r верховье реки
Верхние участки речной системы.
f cours *m* supérieur
s curso *m* superior; curso *m* alto
g Oberlauf *m*; Quellgebiet *n*

2723 health resort
A place with natural medicinal resources (climate, mineral water, etc.) and approrpriate facilities.
r лечебный курорт
Местность, обладающая природными лечебными свойствами (климат, минеральные воды и пр.) и необходимыми условиями для их использования.
f ville *f* d´eaux; station *f* de cure
s balneario *m*
g Kurort *m*

2724 health tourism
Tourism aimed at medical treatment.
r лечебно-оздоровительный туризм
Туризм с целью отдыха и лечения.
f tourisme *m* de santé; tourisme *m* médicinal
s turismo *m* medicinal
g Gesundungstourismus *n*; Kurtourismus *n*

2725 heartland
In geopolitics, an area encompassing the rich interior lowlands of Eurasia that covers a considerable part of Northern Eurasia (Russia included) which lies out of reach of the navies of maritime powers.
r хартленд
В геополитике сердцевинный регион земного шара, охватывающий значительную часть Северной Евразии (включая Россию), недоступную судам морских держав.
f heartland *m*
s heartland *m*
g Heartland *m*

2726 heat balance of the atmosphere
Algebraic sum of incoming and outgoing heat flows in the atmosphere.
r тепловой баланс атмосферы
Алгебраическая сумма потоков тепла, поступающих в атмосферу и уходящих из нее как радиационным, так и нерадиационным путем.
f bilan *m* thermique de l´atmosphère
s balance *m* térmico de la atmósfera
g Wärmebilanz *f* der Atmosphäre

2727 heat balance of the Earth surface
Algebraic sum of heat flows coming to the Earthґs surface and radiated by it.
r тепловой баланс земной поверхности
Алгебраическая сумма потоков тепла, приходящих к земной поверхности и излучаемых ею.
f bilan *m* thermique de la surface terrestre
s balance *m* térmico de la superficie terrestre
g Wärmebilanz *f* der Erdoberfläche

* **heat equator → 6155**

2728 heat exchange in soil
Heat transfer from soil surface to deeper layers or inversely.
r теплообмен в почве
Передача тепла от поверхности почвы к более глубоким слоям или в обратном направлении.
f échange *m* thermique dans le sol
s cambio *m* térmico en el suelo
g Wärmeaustausch *m* im Boden

2729 heat flow
Process of heat transfer together with river waters.
r тепловой сток
Процесс переноса тепла вместе с речными водами.
f écoulement *m* thermique
s escorrentía *f* termal
g Wärmeabfluß *m*

2730 heat flux
Quantity of heat transferred through isothermal surface per a time unit.
r поток тепла
Количество теплоты, переданное через изотермическую поверхность (не меняющую при этом температуры) в единицу времени.
f flux *m* de chaleur
s flujo *m* del calor
g Wärmefluß *m*

* **heat haze → 3588**

2731 heat island
A zone of warm and polluted air which covers a large urban area and in which the temperature is higher than in the surroundings due to the cityґs industries, buildings, traffic, etc.

r остров тепла
Купол теплого и загрязненного воздуха над центральной частью большого города, в котором температура воздуха выше, чем вокруг, из-за городской индустрии, застройки, транспорта и т.п.
f îlot *m* de chaleur; îlot *m* thermique
s isla *f* de calor; islote *m* térmico
g Wärmeinsel *f*

2732 heat lightning
A distant lightning flash observed as a short illumination of the sky or a cloud.
r зарница
Отдаленная вспышка молнии, наблюдаемая как короткое свечение неба или облака.
f éclair *m* de chauleur
s relámpago *m* de calor
g Wetterleuchten *n*

2733 heat wave
A drastic warming up of air, or the invasion of very warm air over a large area.
r волна тепла
Резкое повышение температуры воздуха, или вторжение теплой воздушной массы на обширную территорию.
f vague *f* de chaleur
s ola *f* de calor
g Wärmewelle *f*

* **heath → 2734**

2734 heath land; heather moor; heath
A vegetation community and type of landscape formed by dwarf evergreen shrubs of the heath family, grasses and mosses, developing on acidic podzolized soils and typical of cool and humid oceanic climate.
r верещатник
Тип растительности и своебразный ландшафт, представленный вечнозелеными кустарничками из семейства вересковых, травами и мхами, распространенный в условиях прохладного и влажного океанического климата.
f bruyère *f*; brande *f*; touya *f*
s brezal *m*; tuya *f*
g Heide *n*; Heideland *n*

* **heather moor → 2734**

2735 heave of a fault
The forward, lateral displacement of strata on an inclined normal fault, expressed as a horizontal distance between the ends of the surfaces of the displaced strata.
r горизонтальное смещение пород
Направленное вперед смещение пласта по наклонной плоскости сдвига, выраженное в горизонтальном расстоянии между концами плоскости смещенной области.
f déplacement *m* de faille
s desplazamiento *m* de falla
g Verschiebung *f* der Erdschichten

2736 heavy industry
A group of industries producing fuel, ferrous and non-ferrous metals, large machines and mechanisms.
r тяжелая промышленность
Комплекс отраслей, занимающихся производством топлива, черных и цветных металлов, крупных машин и механизмов.
f industrie *f* lourde
s industria *f* pesada; industria *f* básica
g Schwerindustrie *f*

* **heavy land → 2737**

2737 heavy soil; heavy land
A soil having high percentage of fine particles, especially clay, making it difficult to cultivate.
r тяжелая почва
Почва с высокой долей мелких частиц, особенно глинистых, что делает ее трудной для обработки.
f sol *m* massif; sol *m* épais
s suelo *m* masizo; suelo *m* grueso
g schwerer Boden *m*

2738 heavy water
A natural isotope variety of water in which one or both of the hydrogen atoms in the molecule are deuterium.
r тяжелая вода
Встречающаяся в природе изотопная разновидность воды, в состав которой вместо обычного водорода входит его тяжелый изотоп дейтерий.
f eau *f* lourde
s agua *f* pesada
g schweres Wasser *n*; Schwerwasser *n*

2739 hedge
A row of dense trees and bushes planted along roads and water bodies, and along the boundaries of estates, to provide a barrier.
r живая изгородь; живая защита
Густая полоса деревьев или кустарников, посаженных вдоль дорог, на границе земельных владений, вдоль водоемов.
f haie *f*
s seto *m*
g Hecke *f*

2740 Heinrich events
Cataclysmic discharges (outbursts) of icebergs, resulting from ice-sheet collapses and surges, recorded by the layers of ice-rafted detritus in the bottom sediments of the North Atlantic (they appear coincident with the ends of the climatic *Bond cycles*).
r события Хайнриха
Выбросы в океан гигантских масс айсбергов, связанные с коллапсами ледниковых покровов или их частей, установленные по слоям детрита в донных осадках Северной Атлантики (эти события случаются на конечных стадиях *циклов Бонда*).

f événements *mpl* de Heinrich
s eventos *mpl* de Heinrich
g Heinrich-Ereignisse *npl*

* **hekistothermal plant → 4820**

2741 heliogeophysics
The impact of the Sun's activity upon the physical processes of the Earth, mainly upon its magnetosphere and atmosphere and the study of these phenomena.
r гелиогеофизика
Влияние солнечной активности на физические процессы Земли, главным образом на ее магнитосферу и атмосферу и исследование этих явлений.
f héliogéophysique *f*
s heliogeofísica *f*
g Heliogeophysik *f*

2742 heliophobe
An organism that theives in a well-shaded environment, e.g. the inhabitants of caves, deep seas and lakes.
r гелиофоб
Организмы, избегающие света, например, обитатели пещер, глубоководных зон морей и озер.
f héliophobe *m*
s heliófobo *m*
g Heliophob *m*

* **heliophobic plant → 5247**

2743 heliophyte
A plant adapted to life in full sunlight in sunlit habitats, e.g. pine, larch, etc.
r гелиофит; светолюбивое растение
Растения, предпочитающие ярко освещенные солнцем местообитания, например, сосна и лиственница.
f héliophyte *m*; héliophile *m*
s heliófita *m*; heliófilo *m*
g Heliophyt *m*; Sonnenpflanze *f*

2744 heliothermal water body
A lake or reservoir having two water layers of different salinity and where the lower salter and denser layer is heated by the solar radiation to higher temperatures.
r гелиотермический водоем
Озеро или искусственный водоем с двумя слоями воды разной минерализации, нагреваемый солнечной радиацией сильнее в нижнем более соленом и плотном слое.
f étendue *f* d'eau héliothermique
s extención *f* de agua heliotérmica
g heliothermisches Gewässer *n*

2745 helophyte
A plant typical of salt water habitats (marshy or lake-edge environments) in which shoot buds lie in soil or mud below the water level.
r гелофит
Болотные растения, почки возобновления которых находятся в почве под водой, характерны для маршей или озерно-прибрежной среды.
f hélophyte *m*; hélobio *m*
s helófito *m*; helobio *m*
g Helophyt *m*; Sumpfpflanze *f*

2746 hemihylaea
An evergreen forest that grows in the moist subtropical climate, in the south-west of the USA, south of Plateau of Brazil, south-east of Africa, and in New Zealand.
r гемигилея
Вечнозеленый влажный лес, формирующийся в условиях субтропического климата, распространен на юго-западе США, юге Бразильского плоскогорья, юго-востоке Африки, в Новой Зеландии.
f hémihyléa *f*
s hemihylea *f*
g Hemihyläa *f*

2747 hemisphere
A half of the spherical surface of the Earth distinguished by some feature, e.g. the Northern and Southern hemispheres divided by the equator, or the Eastern and Western hemispheres divided by a meridian, continental and oceanic hemispheres.
r полушарие
Половина сферической поверхности Земли, разделенной по какому-либо признаку, например, Северное и Южное полушария, разделяемые экватором, Западное и Восточное полушария, разделяемые одним из меридианов, материковое и океаническое полушария.
f hémisphère *m*
s hemisferio *m*
g Erdhalbkugel *f*; Erdhemisphäre *f*

2748 herb; grass; herbaceous plant
A small, non-woody plant in which all the aerial parts die back at the end of each growing season, herbaceous plants have worldwide distribution and provide valuable food for people and their livestock.
r травянистое растение; трава
Однолетнее или многолетнее растение, надземные побеги которого не одресневают и отмирают в неблагоприятный для развития сезон, широко распространены на земном шаре и служат прекрасной пищей для людей и животных.
f herbe *f*
s hierba *f*
g Gras *n*; Kraut *n*

2749 herb layer
In the stratified structure of vegetation the lower layer composed of grasses in forest and shrub communities and the upper layer in herbaceous communities.
r травяной ярус

Слагающийся из трав нижний ярус в лесных и кустарниковых растительных сообществах и верхний – в травянистых сообществах.
f étage *m* herbacé
s piso *m* herbáceo
g Krautschicht *f*

2750 **herbaceous *adj***
Pertaining to a herb, a plant that does not develop permanent woody tissul and dies at the end of a growing season.
r травянистый; травяной
Относящийся к траве, травянистым растениям.
f herbacé *adj*
s herbáceo *adj*
g Gras-; Kraut-

* **herbaceous plant → 2748**

2751 **herbage; grass stand**
The above-surface part of grasses in a plant community; herbs when they are considered as an agricultural crop or used for grazing.
r травостой
Наземная часть трав в растительном сообществе; это также в значительной степени сельскохозяйственный урожай либо главная составляющая пастбища.
f peuplement *m* herbacé
s herbaje *m*
g Grasbestand *m*

2752 **herbarium**
A collection of dried up and systematically arranged plant specimens as well as an institution where they are kept.
r гербарий
Коллекция специально собранных и засушенных растений, а также учреждение, в котором они хранятся.
f herbier *m*
s herbario *m*
g Herbarium *n*

2753 **herbicide**
A chemical used in agriculture to destroy or inhibit plant growth, especially weeds and other pest plants.
r гербицид
Вещества, используемые в сельском хозяйстве для уничтожения сорняков и другой нежелательной растительности.
f herbicide *m*
s herbicida *m*
g Herbizide *pl*

2754 **herbivore; herbivorous animal; plant-eating animal**
A plant-eating animal, a primary consumer in the food chain.
r травоядное животное
Травоядное животное, первичный потребитель в пищевой цепи.
f herbivore *m*; consommateur *m*
s herbívoro *m*
g Pflanzenresser *m*

* **herbivorous animal → 2754**

2755 **Hercynian folding; Hercynian orogeny; Variscian folding**
The era of mountain formation within geosynclinal areas at the end of the Palaeozoic when the mountains of Western Europe, Urals, Tien Shan, Altai, Kunlun, etc. appeared.
r герцинская складчатость
Эра горообразования в пределах геосинклинальных областей, сформировавшихся в конце палеозоя, в результате возникли горы Западной Европы, Урала, Тянь-Шаня, Алтая, Куньлуня и др.
f plissement *m* hercynien
s plegamiento *m* herciniano
g herzynische Faltung *f*

* **Hercynian orogeny → 2755**

2756 **herd**
A group of wild or domestic animals, either feeding and travelling together or kept together.
r стадо; отара; табун
Группа диких или домашних животных, обитающих на какой-либо территории, например, стадо оленей или яков.
f troupeau *m*
s rebaño *m*
g Herde *f*

2757 **heterochronism**
The development of two similar geological and palaeogeographical formations not simultaneously although the conditions of their formation were the same.
r гетерохронность
Явление, когда два аналогичных геологических и палеогеографических образования развиваются неодновременно, хотя условия их формирования были одни и те же.
f hétérochronisme *m*
s heterocronismo *m*
g Heterochronismus *m*

2758 **heterosphere**
The layers of the atmosphere above about 100-110 km in which the atmospheric air composition varies considerably as a result of photo-dissociation or diffusive separation of some of its gaseous constituents.
r гетеросфера
Слои атмосферы выше 100-110 км, в которых состав воздуха значительно изменяется в результате фотодиссоциации молекул и диффузного разделения некоторых газовых составляющих.
f hétérosphère *f*
s heterosfera *f*
g Heterosphäre *f*

* **heterotroph → 1233**

2759 **heterotrophic** ***adj***
Applied to an organism which needs an organic food supply from the outside in order to produce its own constituents and energy.
r гетеротрофный
Относящийся к организмам, нуждающимся в органической пище для поддержания своей жизни и производства необходимой им энергии.
f hétérotrophe *adj*
s heterotrófico *adj*; heterotrofo *adj*
g heterotroph *adj*

* **heterotrophic organism → 1233**

* **hglacial trough → 6382**

2760 **hibernation; dormancy**
The way to survive unfavourable winter period by animals of temperate and cold belts, when the metabolic rate and activity in the organism reduce to a minimum and the animal uses up the stored body reserves being in deep sleep.
r зимняя спячка
Способ сохранения жизненных функций в неблагоприятный зимний период животными умеренного и холодного поясов, когда активность организма снижается до минимума, и животное использует накопленные в теле резервы, погружаясь в глубокий сон.
f hibernation *f*
s hibernación *f*
g Winterschlaf *m*; Winterruhe *f*

* **high → 318**

2761 **high altitude sickness; mountain sickness**
An unhealthy state arising in humans at high altitudes due to low atmospheric pressure and lack of oxygen.
r горная болезнь
Болезненное состояние, возникающее при подъеме на большие высоты вследствие пониженного атмосферного давления и недостатка кислорода.
f mal *m* des montagnes
s mal *m* de montaña
g Bergkrankheit *f*; Höhenkrankheit *f*

* **high bank → 743**

* **high cloud → 2771**

2762 **high latitudes**
Subpolar areas of the globe limited approximately by 65° north and south.
r высокие широты
Условное название приполярных областей земного шара, ограниченных примерно 65° северной и южной широты.
f hautes latitudes *fpl*
s latitudes *fpl* altas
g hohe Breiten *fpl*

2763 **high mountain climate**
The climate at high levels in mountains which is characterised by a low atmospheric pressure, temperature and air humidity, and high intensity of solar radiation.
r высокогорный климат
Климат на высоких уровнях в горах, отличающийся низкими атмосферным давлением, температурой и влажностью воздуха, высокой интенсивностью солнечной радиации.
f climat *m* des hautes montagnes
s clima *m* alpino; clima *m* de alta montaña
g Höhenklima *n*

* **high mountain relief → 2764**

2764 **high mountains; high mountain relief**
The type of mountain relief with the absolute height of over 3,000 m, depth of dissection over 1,000 m, slopes with the gradient of over 25° and many traces of modern and ancient glaciation: pointed peaks, steep slopes, kars, troughs, snow patches and glaciers.
r высокогорье; высокогорный рельеф
Тип горного рельефа с абсолютными отметками вершин более 3000 м, глубиной расчленения более 1000 м и склонами средней крутизны более 25°, для него типичны следы современного и древнего оледенения: остроконечные вершины, крутые склоны, кары, троги, снежники и ледники.
f haute montagne *f*; relief *m* des hautes montagnes; hautes terres *fpl*
s alta montaña *f*; relieve *m* de alta montaña; tierras *fpl* altas
g Hochgebirge *n*; Hochgebirgsrelief *n*

2765 **high oblique aerial photograph**
Aerial or space image made with a considerable deflection of the axis of photographing from a vertical position. → Fig. 3, p. 14
r перспективный аэрофотоснимок
Аэрофото- или космический снимок, сделанный со значительным отклонением оси фотографирования от отвесного положения. → Рис. 3, с. 14
f photographie *f* aérienne oblique
s fotografía *f* aérea de perspectiva
g Schrägbild *n*

2766 **high plain**
A plain lying at the elevation above some 600 m.
r высокая равнина
Равнина, лежащая на высотах свыше 600 м.
f plaine *f* élevée; haute plaine *f*
s llanura *f* elevada; llanura *f* alta
g Hochebene *f*

* **high sea → 4269**

2767 **high technology; advanced technology**
Technology that involves the most modern or advanced methods, especially electronic equipment and computers.

r высокие технологии
Технологии, имеющие дело с разработкой в основном компьютерного и электронного оборудования.
f technologie *f* de pointe
s tecnología *f* de punta
g hohe Technologie *f*

2768 high tide; rising tide; flood tide
The opposite to ebb tide, the rising of water or inflowing tidal stream immediately following the period of low water.
r состояние прилива
Состояние моря, противоположное кульминации отлива, подъем уровня моря, следующий непосредственно за отливом.
f marée *f* montante
s marea *f* creciente; flujo *m* altamar
g Steigen *n*

* **high water → 2191**

2769 high water; high-water mark$_2$
The highest water level reached during a tide cycle.
r полная вода; кульминация прилива
Максимальная высота подъема уровня моря в приливном цикле.
f étale *f* de haute mer; haute mer *f*; niveau *m* de haute mer; pleine mer *f*
s pleamar *f*; nivel *m* pleamar; estado *m* de pleamar
g Tidehochwasser *n*

2770 high-altitude frontal zone
Zone of increased horizontal temperature and pressure gradients in the middle and upper troposphere with great extension in middle and subtropical latitudes.
r планетарная высотная фронтальная зона
Зона увеличенных горизонтальных градиентов температуры и давления в средней и верхней тропосфере с большим протяжением в умеренных и субтропических широтах.
f zone *f* frontale à haute altitude
s zona *f* frontal de altura
g Frontalzone *f* der Höhen

* **highland → 4608**

2771 high-level cloud; high cloud
Clouds of the high level: Cirrus, Cirrocumulus and Cirrostratus. → Fig. 147
r высокие облака; облака верхнего яруса
Наиболее высокие облака в тропосфере: перистые, перисто-слоистые, перисто-кучевые. → Рис. 147
f nuage *m* de l'étage supérieur; nuage *m* élevé
s nube *f* del piso superior; nube *f* alta
g hohe Wolken *fpl*

Fig. 147. **High-level clouds**

2772 high-level waste
The waste which emits strong radiation and is very difficult to dispose of safely.
r высокоактивные отходы
Токсичные отходы с высокой радиоактивностью, представляющие большие сложности для захоронения.
f déchets *mpl* de haute radioactivité; déchets *mpl* d'hautement radioactif
s desechos *mpl* de alta radioactividad; residuos *mpl* de alta radioactividad
g hochaktive Abfälle *mpl*

* **high-tech industry → 2773**

2773 high-technology industry; high-tech industry
Any industry concerned with advanced technology, such as biotechnology, computers, microprocessors, telecommunications, pharmacenticals, fibre optics, etc.
r высокотехнологичное производство
Производство информационных и компьютерных изделий и приборов, продуктов биоинженерии, телекоммуникаций и т.п.
f industrie *f* de pointe; industrie *f* high-tech
s industria *f* de punta; high-tech industria *f*
g Informations- und Kommunikations-mittelindustrie *f*

2774 high-water mark$_1$
The mark left on land by the high water level.
r метка высоких вод
След, оставляемый на местности высоким уровнем воды.
f marque *f* des hautes eaux
s marca *f* de aguas altas
g Hochwassermarke *f*

* **high-water mark$_2$ → 2769**

* **highway → 3963**

2775 highway; motorway
A wide road built for fast travel between towns and cities.
r шоссе; автомагистраль
Дорога, предназначенная для автомобильного движения.

f autoroute *f*; autostrade *f*
s carretera *f*; autopista *f*
g Autobahn *f*; Shaussee *f*

* **hiking → 6616**

2776 hill
A rounded natural elevation of land up to 200 m high.
r холм
Небольшое (до 200 м относительной высоты) возвышение рельефа с мягкими очертаниями.
f colline *f*
s colina *f*; lomo *m*
g Hügel *m*

2777 hill shading; illuminated relief
A type of shading employed to create a three-dimensional impression of relief on a map applying either vertical or oblique illumination of the ground (the darker the shading the steeper the slope).
r отмывка рельефа
Изображение на карте рельефа местности методом теневой пластики при косом, реже отвесном освещении, что создает иллюзию объемности изображаемого рельефа.
f estompage *m*
s sombreado *m* del relieve
g Reliefschummerung *f*

* **hill-shading → 3824**

* **hillslope → 5478**

* **hilly lands → 3324**

2778 hinge of fold
The line of intersection of the axial plane with the surface of any layer forming a fold.
r шарнир складки
Линия пересечения осевой плоскости с поверхностью любого из слоев, образующих складку.
f charnière *f* de pli
s charnela *f* de pliegue; bisagra *f* de pliegue
g Faltenscharnier *n*

2779 hinterland
1. The area served by a port, in particular, the area where storage facilities and warehouses are situated. 2. The area influenced by a settlement.
r хинтерланд
1. Вся территория, обслуживаемая портом, и, в частности, пространство на суше, где расположены склады и пакгаузы. 2. Территория, на которую распространяется влияние какого-либо поселения.
f hinterland *m*; arrière-pais
s hinterland *m*; interior de país
g Hinterland *m*

2780 histogram
A two-dimentional representation of data showing a frequency distribution by means of retangles (bars).
r гистограмма
Изображение статистических распределений какой-либо величины по количественному признаку в виде столбчатой диаграммы.
f histogramme *m*
s histograma *m*
g Histogramm *n*

2781 historical area; historical region
A region with specific features of historical development, formerly being a state or other territorial unit, e.g. Bretagne or Bourgogne in France.
r историческая область
Регион, имеющий общие черты исторического развития и в прошлом представлявший собой государственное или административно-территориальное образование, например, Бретань или Бургундия во Франции.
f région *f* historique; province *f* historique
s provincia *f* histórica; región *f* histórica
g historisches Gebiet *n*; historische Region *f*

2782 historical geography
A study of the state of the physical environment (over historical period in the past) and its changes due to both natural processes and human economic activity.
r историческая география
Изучение былых состояний географической среды (за исторический период) и их изменений, в том числе антропогенных, с течением времени.
f géographie *f* historique
s geografía *f* histórica
g historische Geographie *f*

2783 historical geology
A branch of *geology* studying the history of the Earth's development since the period of the Earth's crust formation till its contemporary state.
r историческая геология
Раздел *геологии*, изучающий историю развития Земли с момента образования земной коры до современного ее состояния.
f géologie *f* historique
s geología *f* histórica
g historische Geologie *f*

* **historical region → 2781**

2784 history of geography
The development of geographical ideas, concepts and knowledge at different stages of the history of humanity, it includes the history of geographical investigations and discoveries, development of geographical ideas and methods of investigation.
r история географии
Развитие географических представлений и знаний на разных этапах истории человече-

ства, включает историю географических исследований и открытий, развитие географических идей и методов исследований.
f histoire *f* de géographie
s historia *f* de geografía
g Geschichte *f* der Geographie; Geographiegeschichte *f*

2785 **histosol**
A soil with thick biogenic horizon developed mainly by the accumulation of organic matter in a waterlogged area (the typical soils of peat bogs, moors and swamps).
r гистосоль
Почва с мощным органогенным горизонтом, формирующаяся при обилии влаги и богатых остатков растительности, наиболее типична на болотах.
f histosol *m*
s histosol *m*; histosuelo *m*
g Histosol *m*

2786 **hoar; hoarfrost; hoar frost; white frost**
A deposit of interlocking ice crystals formed by a direct sublimation, i.e. directly from the cooling of water vapour on cold surfaces of objects, usually those of small diameter freely exposed to the air, such as tree branches, plant stems, leaf edges, wires, poles, etc.
r изморозь
Рыхлые отложения ледяных кристаллов, покрывающие ветви деревьев, провода, выступы предметов при тумане в морозную погоду, образованные путем прямого осаждения водяного пара на холодные поверхности.
f givre *m*
s escarcha *f*
g Nebelfrost *m*

* **hoar frost → 2786**

* **hoarfrost → 2786**

2787 **hobby farming**
Growing fruit, vegetables and flowers on personal plots of land for pleasure.
r дачное земледелие
Устройство сада, огорода и цветников на приусадебных и дачных участках, как правило, для удовольствия.
f agriculture *f* de loisir; hobby farming *m*
s agricultura *f* de ocio; hobby farming *m*
g Sommerfrischeackerbau *m*

2788 **Holarctic**
A formerly unified circumpolar biogeographic region enbracing the nontropical part (i.e. cold and temperate belts) of the Northern Hemisphere and characterized by the great floral and faunal similarities between North America, Europe and Asia.
r голарктика
Биогеографический регион, включавший в прошлом внетропическую часть (т.е. холодный и умеренный пояса) Северного полушария и характеризовавшийся значительной общностью флоры и фауны Северной Америки, Европы и Азии.
f région *f* holarctique
s region *f* holártica
g Holarktis *f*

2789 **holding company**
A company that owns the majority of the shares in another company.
r холдинг
Компания, владеющая контрольным пакетом акций других компаний.
f holding *m*
s holding *m*
g Holding *n*

2790 **holism**
The view that the environment can be understood only by looking at it as a complex system of interrelated parts and wholes, which are more than simply the sum of their parts.
r холизм
Концепция, утверждающая, что окружающая среда может быть понята только как комплексная система взаимосвязанных частей и целого.
f holisme *m*
s holismo *m*
g Holismus *m*

2791 **hollow**
Any hollow in the ground surface enclosed by higher land or open from one or two opposite sides, a closed depression often filed with water.
r котловина; впадина; ванна
Понижение земной поверхности, замкнутое со всех сторон или открытое с одного или двух противоположных направлений, нередко заполненное водой.
f combe *f*; creux *m*
s hondonada *f*; hueco *m*
g Mulde *n*; Vertiefung *f*; Senke *f*

2792 **Holocene**
The contemporary geological epoch which has been lasting for already 11 thousand years, the youngest part of the Quaternary.
r голоцен
Современная геологическая эпоха, продолжающаяся уже 11 тыс. лет и представляющая последний отрезок четвертичного периода геологической истории Земли.
f Holocéne *m*
s Holoceno *m*
g Holozän *n*

2793 **holokarst**
The total entity of karstic landforms in a region.
r развитый карст
Глубоко зашедшее в данной метности преобразование рельефа карстовыми процессами.
f holokarst *m*
s holokarst *m*
g Holokarst *m*

2794 **holomictic lake**
A lake in which water circulation periodically occurs in the whole water mass from the surface to the bottom.
r голомиктическое озеро
Озеро, в котором циркуляция периодически охватывает всю толщу воды от поверхности до дна.
f lac *m* holomictique
s lago *m* holomíctico
g holomiktischer See *m*

2795 **holorganic horizon**
A soil horizon composed of peat deposits.
r органогенный горизонт
Почвенный горизонт, в основе которого находятся торфяные отложения.
f horizon *m* holorganique
s horizonte *m* holorgánico
g organogener Horizont *m*; holorganischer Horizont *m*

* **home trade → 3156**

2796 **homeostasis; selfregulation**
The tendency of a natural system to resist change and maintain itself in a state of stable equilibrium.
r гомеостаз; саморегулирование окружающей среды; авторегуляция
Свойство природной системы противостоять изменениям внешней среды и сохранять устойчивость внутренних свойств.
f homéostasie *f*; autoréglage *m*
s homeostasis *m*; autorregulación *f*
g Homöostasis *f*; Homöostaisse *f*

* **homestead → 2103**

2797 **hominidae**
A family of primates comprising recent man (Homo sapiens) and his immediate ancestors: Pithecanthropus, Neanderthal man, Australopithecus, etc.
r гоминиды
Собирательное название приматов, включая человека современного типа (Homo sapiens) и ископаемых людей: питекантропов, неандертальцев, австралопитеков и др.
f hominidés *mpl*
s homínidos *mpl*
g Hominiden *pl*

2798 **homogeneous atmosphere**
A theoretical atmosphere, whose depth is about 8 km, in which the air density is assumed to be constant at all levels and the pressure at sea-level is approximately the same as that of the real atmosphere.
r однородная атмосфера
Теоретическая атмосфера высотой приблизительно 8 км, в которой плотность воздуха принимается постоянной на всех уровнях, а давление на уровне моря близко к наблюдаемому в реальной атмосфере.
f atmosphère *f* homogène
s atmósfera *f* homogénea
g homogene Atmosphäre *f*

2799 **homogeneous series**
Data consisting of a sequence of values of a meteorological element which have been observed for a long time at a given point with the same or similar equipment.
r однородный ряд наблюдений
Ряд данных о метеорологическом элементе, полученных одинаковыми методами наблюдений и приборами без переноса мест наблюдений.
f série *f* homogène
s serie *f* homogénea
g homogene Reihe *f*

* **homoiotherm → 2800**

2800 **homoiothermic animal; homoiotherm**
A warm-blooded animal with constant body temperature despite variations in temperature of their environment, among these are birds and mammals.
r гомойотермное животное; теплокровное животное
Животное с постоянной, устойчивой температурой тела, почти не зависящей от температуры окружающей среды, сюда относятся птицы и млекопитающие.
f animal *m* homoiotherme; animal *m* à sang chaud
s animal *m* homeotermo
g homoiothermes Tier *n*

2801 **homopause**
Boundary between the homosphere and heterosphere at the height of about 100 km.
r гомопауза
Граница между гомосферой и гетеросферой на высоте примерно 100 км.
f homopause *f*
s homopausa *f*
g Homopause *f*

2802 **homosalinity**
The homogeneous vertical distribution of sea water salinity.
r гомохалинность
Однородное распределение солености морской воды по глубине.
f homohalinité *f*
s homohalinidad *f*
g Homohalinität *f*

* **homoseism → 3249**

* **homoseismal line → 3249**

2803 **homosphere**
A part of the atmosphere lying between the Earth's surface and the height of 100-110 km, in which the gas composition of the air is constant.

r гомосфера
Часть атмосферы, простирающаяся от поверхности Земли до высоты 100-110 км и характеризующаяся постоянством относительного содержания основных газов (азота, кислорода, аргона).
f homosphère *f*
s homosfera *f*
g Homosphäre *f*

2804 homothermy
The state of a lake when the temperature is the same in the whole water body, it occurs usually in spring (spring homothermy) or in autumn (autumn homothermy) when the water temperature is about 4°C.
r гомотермия
Состояние водоема, при котором температура одинакова во всем его объеме воды, наблюдается, как правило, весной (весенняя гомотермия) или осенью (осенняя гомотермия), когда температура равна около 4°C.
f homéothermie *f*
s homotermia *f*
g Homothermie *f*

2805 honeycomb relief
Barchan sands consisting of deflation hollows surrounded by ramparts.
r ячеистый рельеф
Барханные пески, включающие дефляционные понижения и окружающие их валы насыпания.
f modelé *m* cellulaire
s modelado *m* celular
g Zellenrelief *n*

2806 honeycomb rocks
A rough rock surface which assumes the appearance of honeycomb with a network of rounded hollows created by chemical weathering and deflation of less resistant material.
r сотовые скалы
Шероховатая поверхность скал в пустынях и полупустынях, испещренная округлыми углублениями, вызванными химическим выветриванием и дефляцией.
f rochers *mpl* cellulaires
s rocas *fpl* celulares
g Zellensteine *mpl*

2807 honeycomb snow
A type of snow with small holes on the surface resembling a honeycomb, caused by melting.
r ячеистый снег
Округлые, похожие на соты углубления на поверхности снега, возникающие в результате таяния.
f neige *f* cellulaire
s nieve *f* celular
g Wabenschnee *m*; Zellenschnee *m*

2808 honeycomb weathering

Fig. 148. **Honeycomb weathering** in an Antarctic oasis

A type of weathering which causes a rock surface to resemble that of a honeycomb, the reason is that hard material filling some of the joints in the rock is left standing as softer material is worn down more speedily by wind and water. → Fig. 148
r ячеистое выветривание
Тип выветривания, при котором скальная поверхность становится похожей на пчелиные соты, так как более твердые участки скал разрушаются медленнее, чем более рыхлые, частички которых уносятся ветром и водой. → Рис. 148
f météorisation *f* cellulaire
s meteorización *f* cecular
g Wabenverwitterung *f*

2809 horizon
1.The line limiting a part of the Earth's surface visible by an observer in an open place, i.e. the apparent junction of earth and sky. 2. Genetically related and relatively homogenious layer of soil. 3. The smallest layer (or stratum) of a stratified sedimentary rock which can be divided from its adjacent layers, both above and below, by a bedding plane.
r горизонт
1. Линия, ограничивающая часть земной поверхности, видимой наблюдателем на открытой местности. 2. Относительно однородный по строению и происхождению слой почвы. 3. Элементарный слой стратифицированных осадочных пород, который может быть отделен и сверху и снизу стратиграфической границей.
f horizon *m*
s horizonte *m*
g Horizont *m*

* **horizon² → 5611**

2810 horizontal barometric gradient
The rate of the atmospheric pressure change along the normal to an isobar per distance unit (100 km) or in 1° of meridian.
r горизонтальный барический градиент
Изменение атмосферного давления по нормали к изобаре на единицу расстояния (100 км) или на 1° меридиана.

f gradient *m* barique horizontal
s gradiente *m* horizontal barométrico
g horizontaler barometrischer Gradient *m*; horizontaler Druckgradient *m*

2811 horizontal coordinates
A system of celestial coordinates in which mathematical or true horizon is the main circle and zenith is a pole, in this system the coordinates are azimuth and altitude (or zenith distance).
r горизонтальные координаты
Система небесных координат, в которой основным кругом служит астрономический, или истинный, горизонт, а полюсом – зенит, в этой системе координаты – это азимут и высота (или зенитное расстояние).
f coordonnées *fpl* horizontales
s coordenadas *fpl* horizontales
g Horizontalkoordinaten *fpl*

2812 horizontal dissection of mountains
The plain-view pattern of mountain ridges, river valleys, and other orographic elements of mountains.
r горизонтальное расчленение гор; густота расчленения
Рисунок в плане, образуемый хребтами разного порядка, речными долинами и другими орографическими элементами горных стран.
f dissection *f* horizontale
s disección *f* horizontal
g horizontale Gliederung *f*

2813 horizontal equivalent
The distance on a map between two successive contour lines, it depends on the steepness of the slope and contour interval.
r заложение горизонталей
Расстояние на карте между двумя последовательными горизонталями, определяемое крутизной склона и высотой сечения.
f écarpement *m* des courbes de niveau
s escalón *m* de curvas de nivel; distancia *f* entre horizontales; equivalente *m* horizontal
g Horizontalabstand *m*

2814 horizontal lamination
Stratification of rocks formed by the layers lying parallel to each other.
r горизонтальная слоистость; горизонтальное залегание
Слоистость горных пород, образованная напластованиями, параллельными друг другу.
f litage *m* horizontal
s litaje *m* horizontal
g waagerechte Schichtung *f*

2815 horizontal scale
The ratio of a distance on a geographical map to the corresponding horizontal distance in nature.
r горизонтальный масштаб
Отношение длины линии на географической карте к длине соответствующего расстояния на местности.
f échelle *f* des longueurs
s escala *f* horizontal
g horizontaler Maßstab *m*; Längemaßstab *m*

2816 horizontal temperature gradient
Rate of air temperature change in horizontal direction (along the normal to an isotherm) per distance unit (100 km).
r горизонтальный градиент температуры
Изменение температуры воздуха в горизонтальном направлении (по нормали к изотерме) на единицу расстояния (100 км).
f gradient *m* horizontal de température
s gradiente *m* horizontal de temperatura
g horizontaler Temperaturgradient *m*; horizontales Temperaturgefälle *n*

2818 horn
A steep-sided isolated mountain summit formed when several cirques cut into it, back to back, leaving a high central mass with marked faces and sharp ridges.
r пирамидальный пик
Ярко выраженная вершина в горном хребте, возникающая, когда несколько цирков сближаются своими задними стенками.
f corne *f*
s cuerno *m*
g Horn *n*

2819 horn-peak
A sharp cogged crest or a pyramidal peak located in the ice-covered or glaciated highlands.
r карлинг
Острый зубчатый гребень или пирамидальный пик в ледниковом высокогорье.
f pyramide *f* de recoupement des cirques
s pico *m* piramidal
g Karling *m*; Zwischenkarscheide *f*

2820 horse latitudes
Either two belts of calms, light variable winds and descending air, the subtropical belts of high atmospheric pressure lying north and south of the equator between about 30 and 35°.
r конские широты
Два пояса штилей или слабых переменных ветров и опускающегося воздуха во внутренних частях субтропических океанических антициклонов в Северном и Южном полушариях на широтах между 30 и 35°.
f calmes *mpl* subtropicaux; hautes pressions *fpl* subtropicaux
s latitudes *fpl* de los caballos; calmas *fpl* subtropicales
g Rossbreiten *fpl*

2821 horse-husbandry
The breeding off horses for stock, racing, and, rarely, meat and milk.
r коневодство

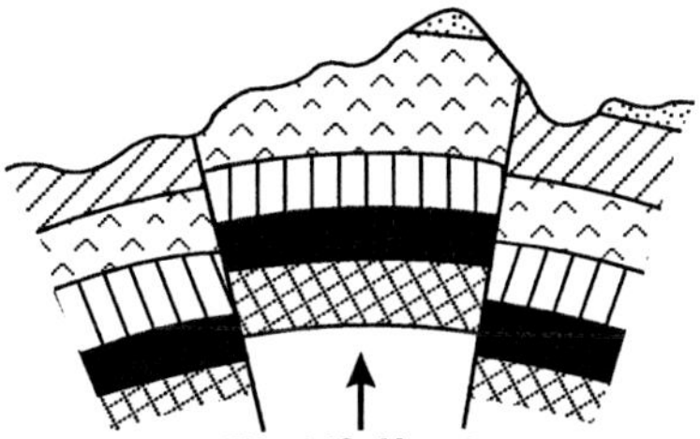

Fig. 149. **Horst**

Отрасль животноводства, направленная на племенное, спортивное и реже мясо-молочное разведение и использование лошадей.
f élevage *m* chevalin
s cría *f* de caballos; hipotecnia *f*; ganadería *f* equina
g Pferdezucht *f*

* **horses → 6762**

2822 horseshoe glacier
A group of small corrie glaciers located around the main cirque in the form resembling a horseshoe.
r подковообразный ледник
Группа небольших каровых ледников, расположенных в виде подковы в карах по периферии главного цирка.
f glacier *m* en fer á cheval
s glaciar *m* de herradura
g Hufeisengletscher *m*

2823 horst
An elevated, usually stretched, block of land lying between two fault blocks. → Fig. 149
r горст
Приподнятый и, как правило, вытянутый участок земной коры, ограниченный круто наклоненными разрывами – сбросами. → Рис. 149
f horst *m*
s horst *m*
g Horst *m*

2824 horticulture; truck farming (*amer*); gardening
The intensive cultivation of fruits and ornamental plants.
r садоводство
Выращивание садовых культур и декоративных растений.
f jardinage *m*
s horticultura *f*; jardinería *f*
g Gartenbau *m*; Gärtnerei *f*

* **hot desert → 384**

2825 hot spring
A continuous flow of hot water with the temperature of 50-90°C from the ground, usually associated with present or former volcanic activity.
r горячий источник
Постоянное излияние горячих вод с температурой 50-90°C из грунта, обычно связанное с современными или недавними вулканическими проявлениями.
f source *f* chaude
s fuente *f* caliente
g heisse Quelle *f*

* **household waste → 1674**

2826 hum
An isolated residual hill on the floor of a polje.
r хум
Останцовое возвышение среди днища полья или по его окраине.
f hum *m*
s hum *m*
g Hum *m*

2827 human activity; economic activity
A various activity of the mankind which is one of the agents of global changes.
r хозяйственная деятельность
Разнообразная деятельность людей, оказывающая влияние на окружающую среду и глобальные изменения природы.
f activité *f* humaine; activité *f* économique
s actividad *f* humana; actividad *f* económica
g Menschentätigkeit *f*; Wirtschaftstätigkeit *f*

2828 human development index
The measure of country-by-country comparison of life quality comprising three aspects: expected life-span, population literacy, and level of income.
r индекс человеческого развития
Сравнительный показатель качества жизни в данной стране, слагающийся из трех оценок: ожидаемой продолжительности жизни, грамотности населения и уровня доходов.
f indice *m* de développement humain
s índice *m* de desarrollo humano
g Indikator *m* der Menschenentwicklung

2829 human ecology
The study of man and communities of people, the place which they occupy in the natural world, and the ways in which they adapt to or change the einvironment.
r экология человека
Изучение взаимодействия людей с природными, социальными и производственными объектами, и их воздействия на состояние и здоровье человека.
f écologie *f* humaine
s ecología *f* humana
g Humanökologie *f*

2830 human geography; socioeconomic geography
One of the two main parts of geography (the other being *physical geography*) studying the spatial and temporal distribution of population, their activities, social organization and culture on a local to global scale.

r общественная география; социально-экономическая география
Отрасль географии, изучающая пространственную организацию общества, включая расселение людей, хозяйство, культуру и политическое устройство, вместе с *физической географией* образует единую систему географических наук.
f géographie *f* humaine; géographie *f* sociale et économique; géographie *f* socio-économique
s geografía *f* social y económica; geografía *f* humana
g Sozialwirtschaftsgeographie *f*

* **human-induced** *adj* → **298**

2831 **human-induced hazard; man-made disaster**
An accident resulting serious or even catastrophic environmental changes on a certain territory, e.g. the destruction of a dam or explosion at a nuclear reactor, etc.
r техногенная катастрофа; антропогенная катастрофа
Техногенная авария, повлекшая за собой серьезные, а иногда катастрофические изменения окружающей среды на определенной территории, например, разрушение или обрушение плотин, взрыв на атомной электростанции и т.п.
f catastrophe *f* technique; décastre *m* artificiel
s catástrofe *f* antropógena
g anthropogene Gefahr *f*

2832 **humic acid**
A complex organic acid found in soils and resulting from the partial decay of organic matter, it is characteristic of chernozems, chestnut and other soils.
r гуминовая кислота
Высокомолекулярная органическая кислота, возникающая при частичном распаде органического вещества почв, характерна для черноземов и каштановых почв.
f acide *m* humique
s ácido *m* húmico
g Huminsäure *f*

2833 **humic horizon; humus horizon**
The most fertile, dark-coloured surface layer of soil that stands out among other horizons of soil profile as containing the greatest amount of organic matter.
r гумусовый горизонт
Наиболее плодородный приповерхностный почвенный слой, имеющий темную окраску и содержащий больше всего органического вещества по сравнению с другими горизонтами почвенного профиля.
f horizon *m* humifère; horizon *m* humique
s horizonte *m* de humus; horizonte *m* húmico
g Humushorizont *m*

* **humic soils** → **2844**

2834 **humid climate**
Climate with abundant moistening of the area when the annual amount of precipitation exceeds the water which evaporates and percolates into the soil over the same period.
r гумидный климат
Климат с избыточным увлажнением, когда годовая сумма осадков превышает количество воды, которое за то же время испаряется и просачивается в грунт.
f climat *m* humide
s clima *m* húmedo
g humides Klima *n*

* **humid tropicality** → **2835**

2835 **humid tropics; humid tropicality**
The climatic condition whereby relative humidity exceeds 65% and mean temperature remains above 20°C, the term is applied to tropical areas in which such conditions prevail for a minimum of nine months per year.
r влажные тропики
Климатические условия, при которых относительная влажность превышает 65%, а средняя температура 20°C, и такие условия преобладают минимум 9 месяцев в году.
f tropiques *mpl* humides
s trópicos *mpl* húmedos
g feuchte Tropen *pl*

2836 **humidification**
An artificial supplying of moisture under conditions of its lack.
r увлажнение
Внесение дополнительной влаги в почву искусственным путем в условиях недостатка влаги.
f humidification *f*
s humidificación *f*
g Befeuchtung *f*

2837 **humidity; air humidity**
Water vapour content of the air.
r влажность воздуха
Содержание водяного пара в воздухе.
f humidité *f*
s humedad *f*
g Luftfeuchtigkeit *f*; Feuchte *f*; Luftfeuchte *f*; Feuchtigkeit *f*

2838 **humification**
The transformation of organic material into humus by slow decomposition and oxidation.
r гумификация
Процесс превращения растительных и животных остатков в гумус под влиянием распада и окисления.
f humification *f*
s humificación *f*
g Humifikation *f*; Humusbilding *f*

2839 **hummock**

1. A low mound of earth, rock or ice. 2. A conglomeration of ice floes piled up by lateral pressure of floes.

r торос; ледяной торос
1. Небольшое возвышение грунтовой поверхности или льда. 2. Нагромождение льдин (до 10-20 м высотой), образовавшееся в результате бокового сжатия ледяных полей.

f hummock *m*; toross *m*; entassement *m* de glace

s hielo *m* de cresta; amontonamiento *m* del hielo; toross *m*

g Packeishügel *m*

* **hummocked ice → 4754**

2840 **hummocked ice row; pressure ridge**
Extended heaps of broken ice blocks and fragments formed by lateral pessing.

r гряда торосов
Вытянутое нагромождение льда, образующееся в результате сжатия и раздробления льда.

f chaîne *f* de torosses

s fila *f* de torosses

g Packeiskette *f*; Eisbankkette *f*

* **hummocky ice → 4754**

2841 **hummocky moraine**
A type of moraine characterized by considerably undulating terrain owing to the presence a large number of kettle holes and crevasses usually formed in dead ice.

r бугристая морена
Моренные отложения, отличающиеся исключительно неровной поверхностью с множеством исполиновых котлов и трещин, они возникают, как правило, на участках мертвого льда.

f moraine *f* mamelonnée; moraine *f* perturbée

s morrena *f* quebrada; morrena *f* perturbada

g hügelige Moräne *f*

2842 **hummocky terrain**
A combination of hills and ridges up to 100 m high scattered in disorder, with flat basins and valleys.

r мелкосопочник
Сочетание беспорядочно разбросанных холмов и гряд высотой до 100 м, чередующихся с плоскими котловинами и долинами.

f relief *m* collinèen; relief *m* des collines

s relieve *m* de colinas

g Kleinhügelland *n*

2843 **humus**
A black or dark-brown soil organic matter formed as a result of decomposition and biochemical transformation of organic residues.

r гумус; перегной
Совокупность органических веществ почвы черного или темнокоричневого цвета, образующихся в результате разложения и биохимического превращения органических остатков.

f humus *m*

s humus *m*

g Humus *m*

* **humus horizon → 2833**

2844 **humus-accumulating soils; humic soils**
A group of soils the humus horizon of which is well pronounced (among these are chernozems, chestnut and dark compact soils as well as their analogues transformed by human activity).

r аккумулятивно-гумусовые почвы
Группа почв, в которых хорошо выражен гумусовый горизонт, объединяет черноземы, каштановые и темные слитые почвы, а также их аналоги, преобразованные человеком.

f sols *mpl* d'accumulation de l'humus

s suelos *mpl* de accumulación de humus

g akkumulative Humusböden *mpl*

2845 **hunting**
An economic activity of low level where people killed wild animals for food.

r охота
Примитивная деятельность людей, когда они убивают диких животных для своего пропитания.

f chasse *f*

s caza *f*

g Jagd *f*

2846 **hurricane**
1. Violent wind blowing with a mean velocity exceeding 30 *m*/s or 120 km per hour. 2. Tropical revolving storm of the Caribbean and Gulf of Mexico. → Fig. 150

r ураган
1. Ветер разрушительной силы, дующий со скоростью более 30 м/с. 2. Тропический циклон, зарождающийся в районе Карибских островов и Мексиканском заливе. → Рис. 150

f ouragan *m*

s huracán *m*

g Orkan *m*; Hurrikan *m*; Wirbelsturm *m*

* **husbandry → 142**

2847 **hydatophyte**
Hydrophyte that is completely or almost complitely submerged.

r гидатофит

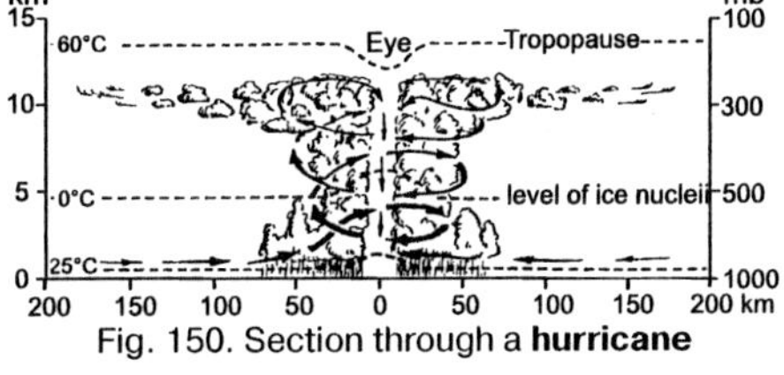

Fig. 150. Section through a **hurricane**

Водное растение, целиком или большей своей частью погруженное в воду.

f hydatophyte *m*
s hidatofita *m*
g Hydatophyt *m*

2848 hydration
In the mechanical weathering of rocks, the process, in which minerals combine with water and expand, thereby exerting pressure within the rock pores.

r гидратация
Вид механического выветривания горных пород, когда при наличии воды внутри минералов она расширяет поры и приводит к разрушению породы.

f hydratation *f*
s hidratación *f*
g Hydratation *f*

2849 hydraulic ***adj***
Related to water in motion or to the pressure exerted by water when carried through pipes, or to mechanical devices operated by moving fluids.

r гидравлический
Относящийся к текущей воде или к давлению, вызываемому водой, когда она проходит по трубам, или к механическим приборам, регулируемым с помощью движущихся жидкостей.

f hydraulique *adj*
s hidráulico *adj*
g hydraulisch *adj*

2850 hydraulic constructions; hydraulic works
Constructions intended for use of water resources or prevention of harmful impact of water upon the environment, e.g. dam, water reservoir, etc.

r гидротехнические сооружения
Сооружения, предназначенные для использования водных ресурсов или предотвращения вредного воздействия воды на окружающую среду, например, плотина, водохранилище.

f installations *fpl* hydrauliques; aménagement *m* hydraulique
s instalaciónes *fpl* hidráulicas
g Wasserbauten *fpl*

* **hydraulic current → 1617**

* **hydraulic engineering → 6656**

2851 hydraulic force
The eroding power of water exerted on rocks, by turbulence, eddying, wave action.

r гидравлическая сила
Эродирующая сила воды, воздействующая на горные породы путем турбулентного, вихревого и волнового движений.

f force *f* hydraulique
s fuerza *f* hidráulica
g hydraulische Kraft *f*

2852 hydraulic gradient
The rate of change in hydraulic head with distance.

r гидравлический уклон
Градиент изменения гидравлического напора с расстоянием.

f gradient *m* hydraulique
s gradiente *m* hidráulico
g hydraulischer Gradient *m*

2853 hydraulic head; head; piezometric head
The pressure exerted by the weight of water above a given point.

r гидравлический напор; пьезометрический напор
Давление, вызываемое весом водяного столба над данной точкой.

f chute *f* piézométrique
s caída *f* piezométrica
g piezometrische Druckhole *f*

2854 hydraulic jump
A sudden change in the discharge along an open river channel, which alters as the flow moves from a state of low velocity in deep water to one of high velocity in shallow water or vice versa.

r гидравлический прыжок
Внезапное изменение расхода воды в открытом речном русле, когда медленное течение в глубоком русле резко сменяется быстрым течением над мелководьем и наоборот.

f ressaut *m* d´exhaussement
s salto *m* hidráulico
g Wechselsprung *m*; hydraulischer Sprung *m*

* **hydraulic mean depth → 2855**

2855 hydraulic radius; hydraulic mean depth
The ratio of the cross-sectional area of flow to the wetted perimeter of a channel, in rivers is practically equal to the mean stream depth.

r гидравлический радиус
Отношение площади поперечного сечения потока к смоченному периметру русла, на реках практически равен средней глубине потока.

f rayon *m* hydraulique
s radio *m* hidráulico
g hydraulischer Radius *m*

2856 hydraulic structures
Different types of constructions such as dams, canals, pipelines, hydroelectric plants, etc., aimed at the use of water resources and fight against the destructive force of water.

r гидротехнические сооружения
Разные типы сооружений (плотины, каналы, трубопроводы, гидроэлектростанции и др.), предназначенные для использования водных ресурсов и борьбы с разрушающим действием водной стихии.

f installations *fpl* hydrauliques
s instalaciónes *fpl* hidráulicas
g Wasserbauwerke *npl*

* **hydraulic works → 2850**

2857 **hydraulics of glaciers**
The dynamics of water and its equilibrium inside glaciers.
r гидравлика ледников
Процессы движения и равновесие воды внутри ледников.
f hydraulique *f* glaciaire
s hidráulica *f* de glaciares
g Gletscherhydraulik *f*

2858 **hydrobiological *adj***
Relating to the water flora and fauna and biological processes in water bodies and water courses.
r гидробиологический
Относящийся к живым организмам, обитающим в воде, и к биологическим процессам, протекающим в водоемах и водотоках.
f hydrobiologique *adj*
s hidrobiológico *adj*
g hydrobiologisch *adj*

2859 **hydrobiology**
The science studying aquatic flora and fauna, their interrelations with the environment, biological productivity of oceans, seas, and inland water bodies.
r гидробиология
Наука, изучающая растительный и животный мир вод, его взаимоотношение с условиями обитания, биологическую продуктивность океанов, морей и внутренних водоемов.
f hydrobiologie *f*
s hidrobiología *f*
g Hydrobiologie *f*

* **hydrobiont → 2860**

2860 **hydrobiose; hydrobiont**
An organism (animal, plant, fungi or microorganism) living in water environment.
r гидробиос; гидробионт
Живой организм (животное, растение, гриб, микрорганизм), обитающий в водной среде.
f hydrobiose *f*; hydrobionte *m*
s hidrobios *f*; hidrobionto *m*
g Hydrobiose *f*; Hydrobiont *m*

2861 **hydrocharite**
An aquatic plant attached to the ground, e.g. reed.
r гидрохарит
Водное растение, прикрепленное к грунту, например, тростник.
f hydrocharite *m*
s hidrocarito *m*
g Hydrocharit *m*

2862 **hydrochemical *adj***
Relating to the chemical composition of natural waters and chemical processes occuring in water.
r гидрохимический
Относящийся к химическому составу воды и протекающим в ней химическим процессам.
f hydrochimique *adj*
s hidroquímico *adj*
g hydrochemisch *adj*

2863 **hydrochore**
A plant whose fruits, seeds or spores are disseminated by water.
r гидрохор
Растение, у которого плоды, семена или споры переносятся водой.
f hydrochore *m*
s hidrocoro *m*
g Hydrochor *m*

2864 **hydrodynamic *adj***
Related to the motion of water, the pressures created by water turbulence and its flow through channels, conduits, pipes and over rivers.
r гидродинамический
Относящийся к движению воды, а также давлению, вызываемому турбулентностью потока и его течению в каналах, трубах и речных руслах.
f hydrodynamique *adj*
s hidrodinámico *adj*
g hydrodynamisch *adj*

2865 **hydroelectric *adj***
Related to the production of electricity by waterpower.
r гидроэнергетический
Относящийся к производству электроэнергии за счет гидроресурсов.
f hydro-électrique *adj*
s hidroeléctrico *adj*
g Wasserkraft-

* **hydroelectric energy → 6655**

2866 **hydroelectric engineering; hydropower engineering**
A branch of power engineering that uses the power of water to produce electricity at hydroelectric power stations.
r гидроэнергетика
Отрасль энергетики, использующая энергию вод для выработки электроэнергии на гидроэлектростанциях.
f hydroélectricité *f*
s energía *f* hidroeléctrica; hidroelectricidad *f*
g Wasserkraftwirtschaft *f*; Hydroenergiewirtschaft *f*

* **hydroelectric plant → 2868**

2867 **hydroelectric potential**

Fig. 151. **Hydroelecytric power ststion** at the Dniepr river

The total volume of hydroelectric resources that can be used at a certain level of economic and technical development of a country.

r гидроэнергетический потенциал; гидроэнергоресурсы; гидроэнергетические ресурсы
Общий объем гидроэнергоресурсов, который может быть освоен на данном уровне технико-экономического развития страны.

f potentiel *m* hydro-électrique; ressources *fpl* hydriques; ressource *f* hydroénergétique

s potencial *m* hidroeléctrico; recursos *mpl* hidroeléctricos; recursos *mpl* hidráulicos

g Wasserkraftspotential *n*; Wasserkraftressourcen *fpl*

* **hydroelectric power → 6655**

2868 hydroelectric power station; hydroelectric plant
A power station where mechanical energy of running water is converted into electrical energy. → Fig. 151

r гидроэлектростанция; ГЭС
Электростанция, преобразующая механическую энергию потока воды в электрическую энергию. → Рис. 151

f centrale *f* hydroélectrique; usine *f* hydroélectrique; centrale *f* hydraulique

s central *f* hidroeléctrica; central *f* hidráulica

g Wasserkraftwerk *n*; Wasserkraftanlage *f*

2869 hydroelectric scheme; hydropower construction
A set of hydraulic constructions united by the arrangement and conditions of their joint work at a water object.

r гидроузел
Комплекс гидротехнических сооружений, предназначенных для выработки электроэнергии и объединенных по расположению и условиям их совместной работы на водном объекте.

f complexe *m* hydroénergétique; centre *m* hydraulique

s complejo *m* hidroeléctrico; centro *m* hidráulico; complejo *m* hidroenergético

g Wasserzentrale *f*

* **hydrogenic ice → 1204**

* **hydrogenic soil → 2896**

2870 hydrogenic succession
The sequential change of vegetation caused by the change of ground waters level or temporal flooding.

r гидрогенная сукцессия
Последовательная смена растительности, вызванная изменением уровня грунтовых вод или временным затоплением.

f succession *f* hydrogène

s sucesión *f* hidrogénica

g hydrogene Sukzession *f*

2871 hydrogeological *adj*
Relating to ground waters, their origin, distribution and relationship with rocks.

r гидрогеологический
Относящийся к подземным водам, их происхождению, распространению и взаимосвязям с горными породами.

f hydrogéologique *adj*

s hydrogeológico *adj*

g hydrogeologisch *adj*

2872 hydrogeology
The study of ground waters, their origin, composition, properties, distribution and motion as well as relationship with rocks.

r гидрогеология
Наука о подземных водах, их происхождении, составе, свойствах, распространении и движении, а также о взаимосвязях с горными породами.

f hydrogéologie *f*

s hidrogeología *f*

g Hydrogeologie *f*

2873 hydrograph
A graph showing the change in time of water discharge in a river or an aquifer over a period of time at a given point. → Fig. 152

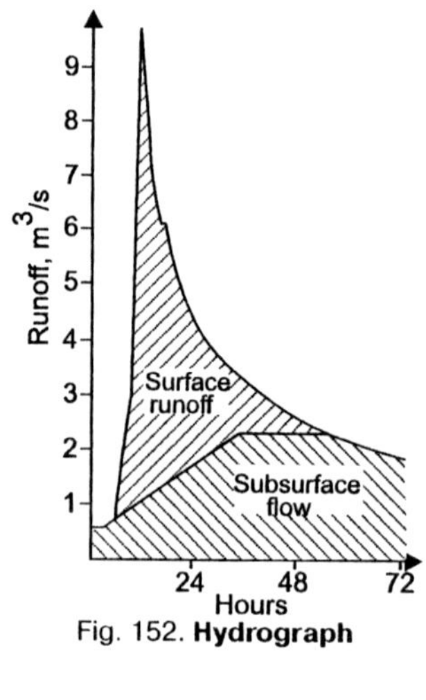

Fig. 152. **Hydrograph**

r гидрограф
График изменения во времени расходов воды в реке за год или за часть года. → Рис. 152, с. 342
f courbe *m* hydrométrique; hydrogramme *m*
s curva *f* hidrométrica; hidrograma *m*
g Abflußganglinie *f*; Abflußkurve *f*

2874 hydrographic *adj*
Relating to the description of bodies of water, including the measurement of their size and characteristics of regime.
r гидрографический
Относящийся к описанию водных объектов, включая количественные характеристики их режима и размеров.
f hydrographique *adj*
s hidrográfico *adj*
g hydrographisch *adj*

2875 hydrographic name
A geographical name of a water object: sea, lake, river, etc.
r гидроним
Географическое название водного объекта: реки, озера, моря и т.п.
f hydronyme *m*
s hidrónimo *m*
g Gewässername *m*

2876 hydrographic network
A system of water bodies and water courses on a certain area.
r гидрографическая сеть
Совокупность водоемов и водотоков на какой-либо территории.
f réseau *m* hydrographique
s red *f* hidrográfica
g Gewässernetz *n*

2877 hydrography
The description and study of any body of water and its parts, including their measurement and characteristics of regime as well as their mapping, especially charting oceans and seas, their beds and coast lines, and the tides, currents, winds with particular regards to safe navigation.
r гидрография
Описание и изучение водных объектов и их отдельных частей, включая измерения размеров и характеристик режима, а также их картографирование, с особым вниманием к океанам и морям (их глубинам, береговой линии, приливам, течениям, ветрам), имея в виду в первую очередь безопасность мореплавания.
f hydrographie *f*
s hidrografía *f*
g Hydrographie *f*

2878 hydrolaccolith; frost blister
A kind of frost mound having a dome shape with the height of 30-40 m and horizontal dimension of 300-400 m consisting of a core of massive ice produced due to freezing of injected water. → Fig. 153
r гидролакколит
Один из видов бугров пучения – куполообразное возвышение высотой 30-40 м и диаметром 300-400 м с ледяным ядром. → Рис. 153
f pingo *m* en system ouvert
s pingo *m* de sistema abierto
g Offner-System-Pingo *m*

*** hydrologic *adj* → 2881**

2879 hydrologic front
An interface zone between two different water masses.
r гидрологический фронт
Граница между соседними водными массами с разными физическими и химическими свойствами.
f front *m* hydrologique
s frente *m* hidrológico
g hydrologische Front *f*

2880 hydrologic series
The data of oceanographic observations at a certain point in ocean or sea by means of bathometers (water samplers).
r гидрологическая серия
Данные многочасовых или многодневных наблюдений на океанографической станции, включающие отбор проб воды с помощью батометров и измерение ее температуры.
f série *f* des mesures limnimétriques
s serie *f* de mensuras hidrológicas
g hydrologische Station *f*

2881 hydrological *adj*; hydrologic *adj*
Relating to natural water objects, their physical and chemical properties, processes and phenomena which are determined by liquid water.
r гидрологический
Относящийся к природным водным объектам, их физическим и химическим ха-

Fig. 153. **Hydrolaccolith**: 1 – strata of active layer, 2 – ice, 3 – fissures with water

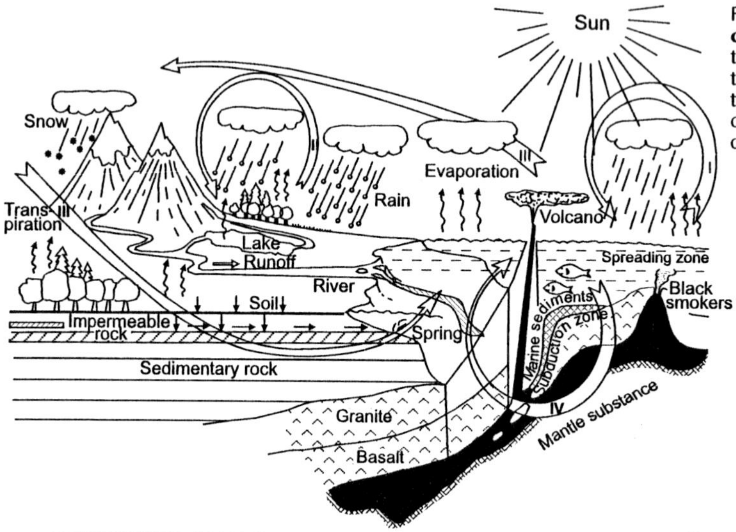

Fig. 154. **Hydrological cycle**: I – above the ocean, II – above the land, III – between the land and ocean, IV – geological part

рактеристикам, процессам и явлениям, режим которых определяется жидкой водой.
f hydrologique *adj*
s hidrológico *adj*
g hydrologisch *adj*

2882 hydrological cycle; water cycle
The continious transformation of water between the gaseous, liquid, and solid forms and its movement or circulation from the Earth's surface to the atmosphere and back to the Earth's surface, brought about by solar radiation and gravity force. → Fig. 154
r круговорот воды; гидрологический цикл; водный цикл; влагооборот
Непрерывная трансформация воды на Земле в ее газообразной, жидкой и твердой формах, движение или циркуляция воды от земной поверхности к атмосфере и обратно к земной поверхности, происходящее под влиянием солнечной радиации и под действием силы тяжести. → Рис. 154
f cycle *m* hydrologique; cycle *m* de l'eau
s ciclo *m* del agua
g Wasserkreislauf *m*

2883 hydrological regime; water regime
Regularly recurring natural changes of state of a water body.
r гидрологический режим; водный режим
Закономерно повторяющиеся природные изменения уровней, объемов и других показателей состояния водных объектов.
f régime *m* hydrologique
s régimen *m* hidrológico; régimen *m* fluvial
g hydrologisches Regime *n*

2884 hydrological season
Part of an annual cycle during which water or ice regime at rivers and lakes is characterized by similar features of its formation and manifestation.
r гидрологический сезон
Часть гидрологического года, в течение которого водный или ледовый режим рек и озер характеризуется общими чертами его формирования и проявления.
f saison *f* hydrologique
s estación *f* hidrológica[1]
g hydrologische Jahreszeit *f*

2885 hydrological station; gauging station
A point on a water body or on bank of a stream where hydrological observations are carried out in order to obtain continuous record of stream-flow and discharge or to measure groundwater level.
r гидрологическая станция; гидрологический пост; водомерный пост
Место на акватории или на берегу водного объекта, где проводятся гидрологические наблюдения с целью получения непрерывных записей стока и расхода воды или измерения уровня грунтовых вод.
f station *f* hydrologique
s estación *f* hidrológica[2]
g hydrologische Station *f*; Pegelstation *f*

2886 hydrological year
Continuous 12-month period selected in such a way that all solid and liquid precipitation and their runoff took place during this period.
r гидрологический год
Непрерывный 12-месячный период, выбранный таким образом,чтобы все твердые и жидкие осадки, выпавшие за этот интервал времени, попали в сток.
f année *f* hydrologique; année *f* pluviométrique
s año *m* hidrológico
g hydrologisches Jahr *n*; Abflußjahr *n*

2887 hydrology

The study of the distribution, movement, properties and use of continental water (before it reaches the ocean or before it evaporates into the air).

r гидрология суши
Наука, изучающая свойства, распределение, движение и использование природных вод на материках (пока они не достигнут океана или испарятся в атмосферу).

f hydrologie *f* continentale; hydrologie *f* des eaux des continents
s hidrología *f* continental; hidrología *f* de las aguas continentales
g Binnengewässerkunde *f*

2888 hydrology of bogs
The study of water regime of bogged up areas covered by natural peat sediments.

r гидрология болот
Водный режим заболоченных территорий, покрытых естественными торфяными отложениями, и их изучение.

f hydrologie *f* des tourbières
s hidrología *f* de turberas
g Moorhydrologie *f*

2889 hydrology of lakes
Water regime and hydrological processes in lakes and their scientific study .

r гидрология озер
Водный режим и гидрологические процессы в озерах и их изучение.

f hydrologie *f* lacustre
s hidrología *f* lacustre
g Hydrologie *f* der Seen

2890 hydrology of reservoirs
Hydrological regime of reservoirs and their scientific study.

r гидрология водохранилищ
Гидрологический режим водохранилищ и их изучение.

f hydrologie *f* des réservoirs
s hidrología *f* de embalse
g Hydrologie *f* der Stauseen

2891 hydrometeor
Any product of condensation or sublimation of atmospheric water vapour, whether formed in the free atmosphere and at the Earth's surface, such as cloud, dew, hail, rain, snow as well as any water particle blown by the wind from the earth's surface.

r гидрометеор
Любой продукт конденсации или сублимации атмосферного водяного пара, образующийся в свободной атмосфере или на поверхности Земли (облако, роса, град, дождь, снег), а также любая водная частица, унесенная ветром с земной поверхности.

f hydrométéore *m*
s hidrometeoro *m*
g Hydrometeore *n*

2892 hydrometric section; measuring section; discharge site; gauge line
Conventional cross section of a watercourse or water body where a series of hydrological obsrvations are carried out.

r створ; гидрометрический створ
Условное поперечное сечение водотока или водоема, где проводится комплекс гидрометрических наблюдений.

f ligne *f* du limmimètre; exutoire *m*
s sección *f* de sondeo; línea *f* de sondeo
g Pegelprofil *n*

2893 hydrometry
A set of methods and techniques of observation and measurement of water flows: depth, pressure, velocity of the running water.

r гидрометрия
Совокупность методов определения характеристик водных объектов: измерение уровней, глубин, напоров, давлений, скоростей водных потоков.

f hydrométrie *f*
s hidrometría *f*
g Hydrometrie *f*

2894 hydromorphic
Related to areas with waterlogged soils characterized by an insufficient or lacking local drainage which leads to the development of anaerobic organisms and retards humification of organic remains.

r гидроморфический
Относящийся к областям с водонасыщенными почвами, характеризующимися недостаточным дренажем, что приводит к развитию анаэробных организмов и замедляет гумификацию органических остатков.

f hydromorphe *adj*
s hidromorfo *adj*
g hydromorphe *adj*

2895 hydromorphic podzolic soil
Soils of taiga zone that form under periodical, but long-term surface or ground excessive moistening which causes podzolization and gleization.

r болотно-подзолистая почва; подзолисто-болотная почва
Почва таежной зоны, формирующаяся в условиях периодического, но длительного поверхностного или грунтового переувлажнения, вызывающего оподзоливание и оглеение.

f sol *m* podzolique hydromorphe
s suelo *m* de podzólico hidromorfo
g hydromorpher Podsolboden *m*

2896 hydromorphic soil; hydrogenic soil
A soil forming under the moistening of the excessive surface and ground and insufficient aeration, with a consequent gleization, deposition of peat, etc.

r гидроморфная почва

Почва, формирующаяся в условиях избыточного поверхностного и грунтового увлажнения и недостаточной аэрации, что приводит к оглеению, образованию торфа и пр.

f sol *m* hydromorphe; sol *m* hydrogénique
s suelo *m* hidromorfo; suelo *m* hidrogénico
g hydromorpher Boden *m*; hydrogener Boden *m*

2897 hydrophobe
An organism which avoids damp habitats or the excess of water, e.g. cacti, steppe, desert or semi-desert grasses, locusts, scorpions, etc.

r гидрофоб
Организмы, избегающие близости или избытка воды, например, кактусы, степные, пустынные и полупустынные злаки, саранча, скорпионы и др.

f hydrophobe *m*
s hidrofobo *m*
g Hydrophob *m*

2898 hydrophyte
An aquatic (water) plant morphologically and/or physiologically adapted to grow in water or very wet environments, having large floating or submerged leaves, reduced root system, etc.

r гидрофит
Водное растение, морфологически и физиологически приспособленное к росту в воде или на очень мокрых местообитаниях, имеет большие плавающие или погруженные в воду листья, редуцированную корневую систему и др.

f hydrophyte *m*
s hidrofita *m*
g Hydrophyt *m*

2899 hydroponics
The growing of plants in nutrient solution with or without such inert medium as gravel, sand, etc. to provide mechanical support.

r гидропоника
Выращивание культурных растений в условиях замены почвы водным раствором с питательными солями – на гравии, песке, в воде или воздухе.

f hydroponie *f*; culture *f* hors-sol; culture *f* hydroponique
s hidropónica *f*; cultivo *m* hidropónico
g Hydroponik *f*; Hydrokultur *f*

* **hydropower construction → 2869**

* **hydropower engineering → 2866**

2900 hydrosphere
The water sphere, all the water (liquid and solid) of the surface of the Earth collectively, including soil and ground water.

r гидросфера
Прерывистая водная оболочка Земли, располагающаяся между атмосферой и литосферой, представляет собой совокупность вод океанов, морей, поверхностных и подземных вод суши.

f hydrosphère *f*
s hidrosfera *f*
g Hydrosphäre *f*; Wasserhülle *f*

2901 hydrostatic pressure
The pressure exerted by water at rest equally at any point within that body of water due to gravity.

r гидростатическое давление; гидростатический напор
Сжимающее напряжение покоящейся воды, обусловленное действием силы тяжести.

f pression *f* hydrostatique
s presión *f* hidrostática
g hydrostatischer Druck *m*

2902 hydrothermal ***adj***
Relating to or caused by the action of hot water, especially on the Earth's crust.

r гидротермический
Относящийся или вызванный воздействием горячей воды, связанной с земными недрами.

f hydrothermique *adj*
s hidrotérmico *adj*
g hydrothermisch *adj*

2903 hydrothermal coefficient
The ratio of the amount of precipitation over the period of active vegetation to the average daily temperatures over the same period.

r гидротермический коэффициент
Отношение суммы осадков за период активной вегетации к сумме среднесуточных температур за то же время.

f coefficient *m* hydrothermique
s coeficiente *m* hidrotérmico
g hydrothermischer Koeffizient *m*

2904**hydrothermal process**
The process of mineralogical and chemical transformation of rocks on the ocean bottom under the impact of thermal water solutions (hydrothermal vent).

r гидротермальный процесс
Процесс минералогического и химического изменения горных пород на дне океана под воздействием термальных водных растворов (гидротерм).

f processus *m* hydrothermique
s proceso *m* hidrotermal
g hydrothermischer Prozeß *m*

2905 hydrothermal vent
Gases and mineralized water which eject from under ocean floor and form hier thermal water solutions.

r гидротерма
Горячие воды и газы, вытекающие из земной коры на некоторых участках морского дна и создающие здесь термальные водные растворы.

f ouverture *f* hydrothermique
s abertura *f* hidrotermal
g hydrothermische Öffnung *f*

2906 hygrophile
An animal adapted to wet habitats, such as wet shores of water bodies, lower layers of rain forests, etc.
r гигрофил
Животное, предпочитающее влажные места обитания (сырые берега водоемов, нижние ярусы влажных лесов и т.п.).
f hygrophile *m*
s higrofilo *m*
g Hygrophil *m*

2907 hygrophobe
A land animal which avoids the conditions of excessive moisture in the habitat.
r гигрофоб
Наземное животное, избегающее избыточного увлажнения в местах своего обитания.
f hygrophobe *m*
s higrófobo *m*
g Hygrophob *m*

2908 hygrophyte
A plant adapted to conditions of the constant excessive moisture or plentiful supply of water.
r гигрофит
Растение, адаптированное к условиям постоянного поступления значительного количества воды.
f hygrophyte *f*
s higrofita *m*
g Hygrophyt *m*

2909 hygroscopic *adj*
Referring to substances which absorb moisture from the atmosphere and retain it.
r гигроскопический
Относящийся к веществам, которые абсорбируют влагу из атмосферы и удерживают ее.
f hygroscopique *adj*
s higroscópico *adj*
g hygroskopisch *adj*

* **hygroscopic moisture → 2911**

2910 hygroscopic nuclei
Condensation nuclei consisting of a hygroscopic substance (sea salt, combustion products), which tend to attract and condense water vapour.
r гигроскопические ядра
Ядра конденсации, состоящие из гигроскопических веществ (морская соль, продукты сгорания), которые притягивают и конденсируют влагу.
f noyau *m* hygroscopique
s núcleo *m* higroscópico
g hygroskopischer Kern *m*

2911 hygroscopic water; hygroscopic moisture
Water in the soil that is adhered to the ground particles and therefore unavailable for plants.
r гигроскопическая вода
Вода, прочно связанная с поверхностью частиц грунта и поэтому недоступная произрастающим здесь растениям.
f eau *f* hygroscopique
s agua *f* higroscópica
g hygroskopisches gebundenes Wasser *n*; hygroskopisches Wasser *n*

* **hylea → 4885**

2912 hypermarket; superstore
A self-service store much larger than supermarket and provides clothing and consumer goods as well as foodstuffs with the total trade area of over 2500 m^2, usually located in the outskirts of a large city where land is cheaper and parking is easy.
r гипермаркет
Торговый комплекс самообслуживания с суммарной торговой площадью более 2500 м^2, реализующий пищевые товары наряду с одеждой и многими другими товарами, необходимыми покупателю, располагающийся, как правило, на окраине крупного города, где земля дешевле и место для стоянки машин найти проще.
f hypermarché *m*
s hypermercado *m*
g Hypermarket *m*

2913 hyperspectral survey
A kind of remote sensing when the images are taken simultaneously in a great number (over 20) of spectral ranges.
r гиперспектральная съемка
Вид дистанционного зондирования, когда фотосъемку ведут одновременно в большом числе (более 20) спектральных диапазонов.
f levé *m* hyperspectral
s levantamiento *m* hiperespectral
g hyperspektrale Aufnahme *f*

2914 hyperthermal water
Ground and spring waters having the temperature of 42-100°C.
r гипертермальные воды
Подземные воды и воды источников с температурой 42-100°C.
f eau *f* hyperthermale
s agua *f* hipertermal
g Hyperthermalwasser *n*

2915 hyperzonation
The system of natural zonality in which two or more contemporary zones form a single landscape zone, uniform in its main properties.
r гиперзональность
Система природной зональности, в которой на месте двух и более современных зон формируется одна ландшафтная зона, единая по своим основным свойствам.

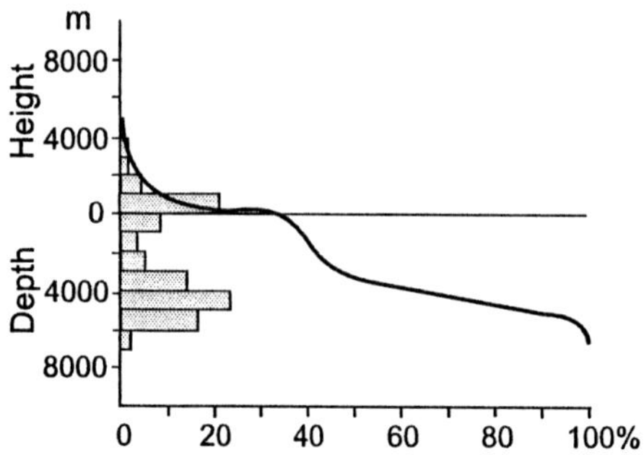

Fig. 155. **Hypsometric curve**

f hyperzonalité *f*
s hiperzonalidad *f*
g Hyperzonalität *f*

* **hypocentre → 5325**

2916 **hypolimnion**
The lower water layer in oceans, lakes and reservoirs stratified in summer which is lying below the layer of temperature jump (metalimnion).
r гиполимнион
Нижний слой воды в стратифицированных летом морях, озерах и водохранилищах, залегающий ниже слоя температурного скачка – металимниона.
f hypolimnion *m*
s hypolimnión *m*
g Hypolimnion *n*; Tiefenschicht *f*

* **hypsographic tinting → 2917**

2917 **hypsography; hypsographic tinting**
The way of mapping the Earth's topography with the help of contour lines connecting the points of equal altitude.
r гипсометрия
Способ изображения рельефа земной поверхности с помощью горизонталей, соединяющих точки с одинаковой абсолютной высотой.
f hypsométrie *f*; hypsographie *f*
s hipsometría *f*; hipsografía *f*
g Hypsometrie *f*; Höhenmessung *f*

2918 **hypsometric curve**
A graph which represents the proportion of the surface of the Earth at various altidudes (on land) or depth (at sea). → Fig. 155
r гипсографическая кривая
Кривая в системе прямоугольных координат, изображающая распространенность на земной поверхности высот (на суше) и глубин (на море). → Рис. 155
f diagramme *m* hypsométrique
s curva *f* hipsométrica
g hypsometrische Kurve *f*

2919 **hypsometric layer step**
The difference in height between two successive grades of hypsometric scale.
r гипсометрическая ступень; высотная ступень
Разность высот двух уровней гипсометрической шкалы.
f intervalle *m* des courbes hypsométrique
s intervalo *m* hipsométrico
g Höhenstufe *f*

2920 **hypsometric map**
A map showing the relief of an area presented by contours with the colouring by altitudinal levels.
r гипсометрическая карта
Карта, основным содержанием которой служит рельеф, изображенный горизонталями с раскраской по высотным ступеням.
f carte *f* hypsométrique
s mapa *m* hipsométrico
g Höhenschichtenkarte *f*; hypsometrische Karte *f*

2921 **hypsometric scale**
A colour wedge in the margin of a map, which indicates absolute heights and depths by graduated shades of colours.
r гипсометрическая шкала
Шкала цветовых тонов и оттенков, применяемых для окраски рельефа на географической карте по ступеням высот.
f échelle *f* hypsométrique
s escala *f* hipsométrica
g Höhenschichtenskala *f*

I

2922 ice; natural ice
The solid form of water formed in nature by the freezing of water as in a river or sea, by the condensation of atmospheric water vapour into ice crystals, by the compaction of snow, etc.
r лед; природный лед
Твердая форма воды, образуемая в природе путем замерзания воды на реках, озерах и морях, конденсации атмосферного водяного пара в ледяные кристаллы, уплотнения снега и т.п.
f glace *f*; glace *f* naturelle
s hielo *m*; hielo *m* natural
g Eis *n*; Natureis *n*

2923 ice adhesion
The sticking of ice to any introduced body due to molecular attraction.
r адгезия льда
Примерзание льда к инородным предметам за счет сил межмолекулярного взаимодействия.
f adhésion *f* de la glace
s adhesión *f* de hielo
g Eisadhäsion *f*

2924 ice age; glacial age; glacial epoch; glacial period
Any period in the Earth's history characterised by the strong cooling of climate and the development of vast continental ice sheets in high and mid latitudes (terminological correlation of ice ages in different territories shown in the Table 2, p. 350-351).
r ледниковая эпоха; ледниковье; ледниковый период; гляциал
Отрезок времени в геологической истории Земли, характеризующийся сильным похолоданием климата и развитием обширных материковых ледниковых покровов не только в полярных, но и в умеренных широтах (терминологическую корреляцию ледниковых эпох на разных территориях см. в табл. 2, с. 350-351).
f époque *f* glaciaire; période *m* glaciaire
s época *f* glaciar; período *m* glaciar
g Eiszeit *f*; Eisperiode *f*; Glazialzeit *f*; Glazialperiode *f*

2925 ice apron
A protective structure on the upstream side of a bridge pier or dam which serves to break up floating ice during the ice drift and to prevent the formation of ice dams in river channels.
r ледорез
Защитное ограждение на опорах мостов, быках и плотинах для защиты их от повреждений льдом во время ледохода и предупреждения образования ледяных заторов.
f éperon *m* d'un pont ou d'un barrage
s espolón *m* rompehielos; guardahielos *m*
g Pfeilerkopf *m*

2926 ice avalanche; ice slide
A rapid gravitational movement of the ice mass down, usually due to the presence of liquid water beneath the ice, it often arises on a hanging glacier or on icefall as a result of the seracs tearing off.
r ледяная лавина; ледяной обвал
Быстрое движение вниз по склону масс льда, обычно связанное с наличием жидкой воды, нередко возникает на висячем леднике и на ледопадах в результате отрыва сераков.
f avalanche *f* de glace
s avalancha *f* de hielo
g Eislawine *f*; Eisabruch *m*

2927 ice barrier; ice front
The seaward, cliff-like edge of a floating glacier or an ice shelf. → Fig. 156
r ледниковый барьер
Обращенный к морю вертикальный обрыв плавучего ледникового языка или шельфового ледника. → Рис. 156
f falaise *f* de glace
s barrera *f* de hielo
g Schelfeisrand *m*; Eisfront *f*

Fig. 156. **Ice barrier** of Shakleton ice shelf in the East Antarctica

2928 ice belt; ice strip
A band of accumulated fragments of sea ice in otherwise open water.
r ледовый пояс; полоса льда
Полоса битого льда на открытой воде.
f ceinture *f* de glace; cordon *m* de glaçons
s cinturón *m* de hielo; franja *f* de hielo
g Eisgürtel *m*; Eisstreifen *m*

2929 ice bight
An indentation in the sea or lake ice where open water comes in.
r залив во льду
Изгиб кромки морского или озерного льда, где чистая вода вклинивается на занятую льдом акваторию.

Table 2. Correlation matrix of glacial events (**ice ages** and **interglacials**) of the Quaternary in the Central Europe, Russia and North America

Общие геохронологические подразделения		Региональные климатохронологические подразделения						Глобальная корреляция	
Между-народные	Россия	Западная Европа: Равнины		Западная Европа: Альпы	Центральная Россия		Северная Америка	Изотопные стадии	Астрономическое время, тыс. л.н.
Holocene	Голоцен	Holocene		Holocene	Голоцен		Holocene	1	
Late Pleistocene	Поздний плейстоцен =(поздний нео-плейстоцен)	Late glacial	Late Weichselian glacial	Main Würm glacial	Поздневалдайское оледенение		Late Wisconsin glacial	2	10,3
		Pleni-glacial	Middle Weichselian Cold stage	Middle Würm Megainterstadial	Средневалдайский мегаинтерстадиал		Middle Wisconsin interstadial	3	25
								4	70-75
		Early Weichselian cold stage		Early Würm	Ранневалдайское оледенение		Early Wisconsin glacial Eowisconsin	5a-d	
		Eemian interglacial		Riss/Würm interglacial	Микулинское межледниковье		Sangamon interglacial	5e	116
Middle Pleistocene	Средний плейстоцен = средний нео-плейстоцен	Saalian glacial	Warthe-stage Schoningen interstadial Drente-stage	Riss glacial	Днепровское оледенение	Московская стадия Костромской интерстадиал (КУРСКАЯ почва) Днепровская стадия	Illinois glacial	6-8	135
			Dömnitz=Wacken interglacial		Каменское межледниковье (Горкинское межледниковье, Чекалинское потепление)				
			Fuhne cold stage		Печорское оледенение (Вологодское оледенение, Калужское похолодание).				

		Holstein interglacial		Mindel/ Riss Interglacial	Лихвинское межледниковье			9?-11?	380
	Ранний плей-стоцен =ранний неоплей-стоцен	Elster glacial 2 Elster glacial 1		Mindel glacial	Окское оледенение		Pre-Illinois glacials: A, B, C		
		Cromer	Cromer IV	Haslach/ Mindel Interglacial	Мучкапское (Рославльское) межледниковье			12-13	
			Cold stage glacial ?	Haslach cold stage	Донское оледенение		Pre-Illinois glacials: D, E, F, G, H, I		
			Cromer III	Günz /Haslach interglacial	Возможные подразделения	Окатовское межледниковье			
			Cold stage, glacial B			Сетуньское оледенение			
			Cromer II			Красиковское межледниковье ?			
			Cold stage, glacial A	Günz glacial		оледенение ?			?
Early Pleistocene	Эо-плейстоцен		Cromer I	Donau interglacial?	Петропавловское (Михайловское, Акуловское) потепление			19-20	? 780
		Bavel warm stage Menap cold stage Waal warm stage Eburon cold stage		Uhlenberg interglacial ?	Ликовское оледенение				
Pliocene	Плиоцен	Tegelen-complex Pretegelen							1800

f baie *f* dans glace
s golfo *m* en hielo
g Eisbucht *f*

2930 ice blink; ice sky
White glare on the underside of low clouds indicating the presence of ice which may be beyond the range of vision.
r ледовое небо; ледовый отблеск
Светлое сияние на нижней поверхности слоистых облаков над ледяным покровом или плавающими льдами в полярных водах.
f réflexion *f* de glace
s resplandor *m* de hielo; reflejo *m* glaciar
g Eisblink *m*

2931 ice bridge
A section of ice cover lifted above the water surface due to the water level drop and resting upon the banks.
r ледяной мост
Участок ледяного покрова, отделившийся от водной поверхности при понижении уровня воды и опирающийся на берега.
f pont *m* de glace
s puente *m* de hielo
g Eisbrücke *f*

2932 ice brittleness
The property of ice to be broken under the load smaller than its elasticity limit.
r хрупкость льда
Способность льда разрушаться при нагрузке, меньшей предела упругости.
f fragilité *f* de la glace
s fragilidad *f* de hielo
g Eissprödigkeit *f*

2933 ice canopy
Pack ice seen from under the water by a submariner.
r ледяной потолок
Дрейфующий лед с точки зрения подводника.
f plafond *m* de glace
s techo *m* de hielo
g Eisdach *n*

2934 ice cap; ice dome
A covering of ice over a tract of land, a dome-shaped glacier that covers a high watershed.
r ледниковый купол; ледниковая шапка
Плоско-выпуклый покровный ледник, занимающий водораздельную часть плосковершинного массива.
f dôme *m* glaciaire
s cúpula *f* glaciar
g Gletscherkappe *f*; Gletscherkuppel *f*

2935 ice cascade
A steplike mass of ice formed on steep slopes of mountains, coastal cliffs, and engineering structures by layered freezing of underground or surface water.
r ледяной каскад
Ступенчатый ледяной массив, формирующийся на крутых склонах и в распадках гор, на береговых обрывах и инженерных сооружениях при послойном замерзании подземных или поверхностных вод.
f cascade *f* de glace
s cascada *f* de hielo
g Eiskaskade *f*; Eistreppe *f*

2936 ice catchment
A part of ice cover bounded by the ice divide, from which the ice moves to a separate outlet glacier or to a certain section of a glacier front.
r ледосборный бассейн
Часть ледникового покрова, оконтуренная ледоразделом, откуда лед стекает в отдельный выводной ледник или на определенный участок фронта ледника.
f bassin *m* versant glaciaire
s cuenca *f* vertiente glaciar
g Eiseinzugsbassin *n*

2937 ice cave
A cave in which the subzero temperatures exist throughout the year making it possible for the ice to form and remain in the cave.
r ледяная пещера; пещера-ледник
Пещера, в которой в течение всего года сохраняется отрицательная температура воздуха и поэтому присутствует лед.
f grotte *f* glaciaire
s gruta *f* glaciar; cueva *f* glaciar; caverna *f* glaciar
g Eishöhle *f*; Frosthöhle *f*

2938 ice chart
A map showing the sea surface on which the information about ice is registered: distribution and closeness of ice, its thickness, etc.
r ледовая карта
Бланковая карта морской акватории с нанесенной на ней ледовой информацией: распространением и сплоченностью льда, его толщиной и пр.
f carte *f* des glaces
s mapa *f* de hielos
g Meereiskarte *f*

2939 ice clearing; lead
A linear opening in sea ice, pack ice, fast ice or in a floe caused by its shifts and deformations.
r разводье
Свободная поверхность воды, возникшая при разломе и разрыве сплоченного льда, припая или ледяного поля в результате подвижек и деформаций.
f clairière *f*
s claro *m* entre los hielos
g Küstenrinne *f*; Tümpel *m*

2940 ice cloud
A cloud composed entirely, or almost entirely, of ice crystals, e.g. Cirrus.

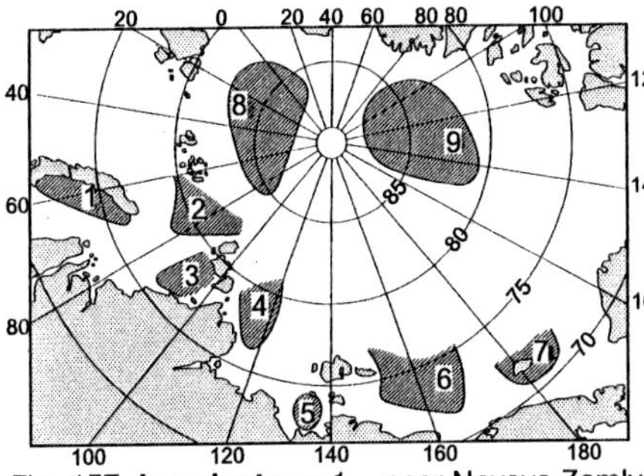

Fig. 157. **Ice clusters**: 1 – near Novaya Zemlya, 2 – north Kara Sea, 3 – near Severnaya Zemlya, 4 – near Taimyr Peninsula, 5 – near Yana mouth, 6 – near Aion island, 7 – near Vrangel island, 8 – in the eastern part of the Arctic Basin, 9 – in the Canadian part of the Arctic Basin

r ледяное облако
Облако, целиком или почти целиком состоящее из кристаллов льда, например, перистое.
f nuage *m* glacé
s nube *f* de hielo
g Eiswolke *f*

2941 ice cluster
A concentration of floating sea ice of very wide extent covering hundreds of square kilometres and remaining in the same region for a long time in summer. → Fig. 157
r ледяной массив; скопление льдов
Скопление морских льдов, занимающее сотни км^2 и сохраняющееся летом длительное время в одном районе. → Рис. 157
f grand champ *m* de glace permanent
s macizo *m* de hielo
g Dauereis *n*; Meerstandeis *n*; Eisanhäufung *f*

2942 ice coast
The coast where the sea or lake washes a glacier.
r ледяной берег
Берег, на котором море или озеро омывают ледник.
f glace *f* littorale
s hielo *m* litoral
g Eisstrand *m*; Strandeis *n*

2943 ice concrete
A frozen mixture of water with stones which has a high mechanical strength.
r льдобетон
Замороженная смесь воды с каменным материалом, обладающая высокой механической прочностью.
f béton *m* de glace
s hormigón *m* de hielo
g Eisbeton *m*

* **ice conditions → 2996**

2944 ice cone
A conical or pyramidal mound of ice on the surface of a glacier tongue which arises due to different rate of ablation of the ice covered by a debris material.
r ледяная пирамида; ледяной конус
Возвышение пирамидальной формы на поверхности ледникового языка, возникающее в результате неравномерного таяния льда, покрытого обломками морены.
f cône *m* couvert de glace
s pirámide *f* de hielo
g Eispyramide *f*

2945 ice content
The amount of ice contained in frozen or partially frozen soil or rock.
r льдистость
Степень насыщения породы льдом вообще или определенным его видом.
f teneur *f* en glace apparante
s contenido *m* de hielo; candidad *f* relativa de hielo
g Eisgehalt *m* der Frostboden; Vereisung *f*

2946 ice core
A cylindrical column of ice obtained by the drilling of glaciers.
r ледяной керн
Цилиндрическая колонка ледяной породы, получаемая в результате бурения ледников.
f carotte *f* de glace
s testigo *m* de hielo
g Eiskern *m*

2947 ice cover
A continuous or broken ice fields on the surface of a sea, lake or river.
r ледяной покров
Сплошные или прерывистые ледяные поля на поверхности моря, реки, озера.
f englacement *m*; manteau *m* de glace; revêtement *m* de glace
s manto *m* de hielo; revestimiento *m* de hielo
g Eisdecke *f*; Standeis *n*

2948 ice coverage
The area covered by the ice of any closeness expressed as a ratio (or percentage) of the ice area to the total area under consideration of water surface.
r ледовитость
Площадь, занятая льдом любой сплоченности, в процентах от общей площади моря или любой другой акватории.
f couverture *f* de glace
s cobertura *f* de hielo
g Eisbedeckung *f*

2949 ice creep
The property of ice to have gradual deformation caused by the internal stresses and accompanied ice by the intergranular motion.
r ползучесть льда

Постепенная деформация льда, вызываемая внутренними напряжениями и сопровождаемая движениями внутри зерен.

f fluage *m* de glace
s fluencia *f* de hielo
g Eiskriechen *n*

2950 ice crust
A thin layer of ice upon the surface of snow formed by the freezing of melt water or rain water.
r ледяная корка
Тонкий слой льда, образующийся при замерзании талой или дождевой воды на поверхности снега.
f croûte *f* de glace
s costra *f* de hielo
g Eiskruste *f*

2951 ice dam$_1$
The accumulation of broken river ice in narrow channels, frequently producing local floods during the spring melting phase and ice drift. → Fig. 158
r затор льда
Нагромождение льдин во время весеннего ледохода в сужениях и излучинах реки, на мелях и в других местах, где проход льдин затруднен, что приводит к резкому подъему уровня воды и вызывает наводнения. → Рис. 158
f embâcle *m* de glace
s dique *m* de hielo
g Eisstau *m*; Eisdamm *m*

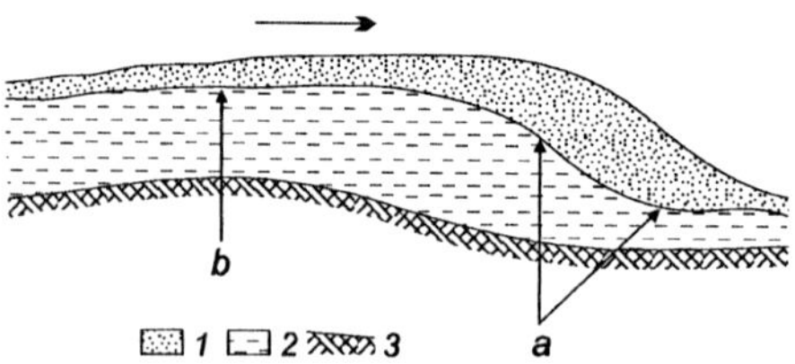

Fig. 158. **Ice dam$_1$**: a – head of ice dam, 2 – tail of ice dam; 1 – river ice, 2 – water, 3 – river bed

2952 ice dam$_2$
1. A glacier blocking a lateral valley which causes the ponding of melt waters to create lake. 2. Congelated blocks of ice in the form of an extended trunkated pyramid used to protect a river bank during spring ice drifts and floods. → Fig. 159
r ледяная дамба; ледяная плотина
1. Ледник, перегораживающий боковую долину, за которым образуется озеро. 2. Намороженный ледяной массив в виде вытянутой усеченной пирамиды, используемый, в частности, в качестве ограждающей конструкции во время весенних ледоходов и паводков. → Рис. 159
f barrage *m* de glace
s espolón *m* de hielo
g Eisbarre *f*; Eisdamm *m*; Eisstauwehr *n*

2953 ice divide
A line that separates two different parts of an inland ice which move in the opposite or different directions.
r ледораздел
Линия, разделяющая две части ледникового покрова, лед которых движется в противоположных или сильно отличающихся направлениях.
f ligne *f* de partage des glaces
s línea *f* divisoria de hielos
g Eisscheide *f*

* **ice dome → 2934**

2954 ice drift
The movement of broken ice-floes in spring and autumn on the rivers and lakes of temperate and high latitudes which have been caried downstream during the seasonal melting phase.
r ледоход
Движение льда весной и осенью на реках и озерах умеренных и высоких широт под действием течения или ветра.
f débâcle *f*
s movimiento *m* de hielo; deshielo *m* de los ríos
g Eisgang *m*; Eistrieb *m*

2955 ice drifting
The movement of ice floes and ice fields due to the action of currents and wind in the seas. → Fig. 160, p. 355
r дрейф льда
Движение льдин и ледяных полей под воздействием течений и ветров в море. → Рис. 160, с. 355
f dérive *f* des banquises
s deriva *f* de hielo
g Eisdrift *f*

2956 ice druse
An aggregation of sublimation ice crystals formed on the surface of some cavity.
r ледяная друза
Совокупность кристаллов сублимационного льда, наросших на поверхность какой-либо полости.
f givre *m*

Fig. 159. **Ice dam$_2$** of Severny Inylchek Glacier in the temporary Merzbacher lake (Tien Shan)

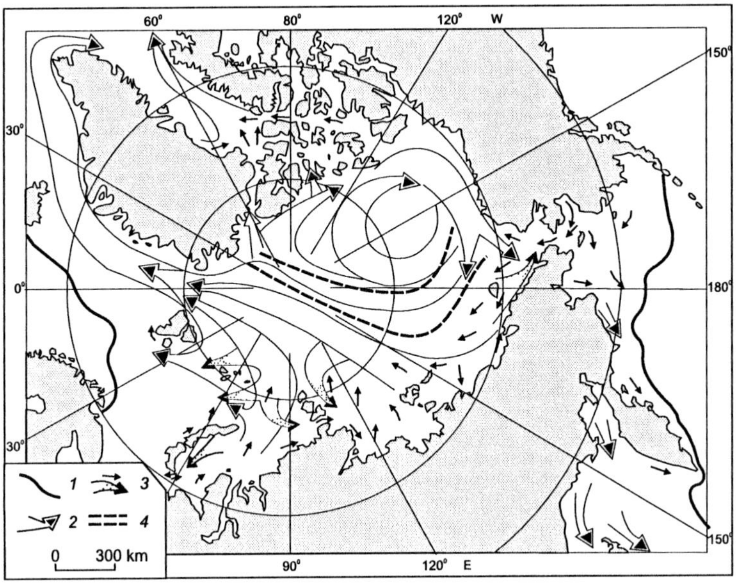

Fig. 160. **Ice drifting**: 1 – margin of the drifting ice maximum, 2 – prevailing direction of the ice drifting, 3 – secondary directions of the ice drifting in spring and autumn, 4 – the boundary between two main systems of the ice drifting in the Arctic Basin

s drusa *f* de hielo
g Eisbarren *m*

2957 ice edge$_1$
The boundary between the open water and a mass of floating sea ice of any kind.
r кромка льда
Линия, разграничивающая чистую воду и морские льды разной сплоченности.
f limite *f* des glaces flottantes
s límite *f* de hielo flotante
g Eisrandlinie *f*

* **ice edge$_2$ → 2524**

* **ice facies → 3017**

* **ice fat → 2617**

2958 ice field
A large area of land ice, especially in Canada, i.e. Columbia Ice Field in Jasper National Park, Alberta.
r ледниковое поле
Обширное пространство наземного льда, в Канаде входит в наименование ледников, например Колумбийское ледниковое поле в провинции Альберта.
f champ *m* glaciaire
s campo *m* glaciar
g Gletscherfeld *n*

2959 ice fog; ice mist
A thin milky veil seen over the land surface and formed by suspended fine particles of ice crystals at very low temperatures.
r ледяной туман; морозный туман; морозная дымка
Помутнение воздуха при сильных морозах, вызываемое наличием в воздухе большого количества мелких ледяных кристаллов.
f brouillard *m* glacé; brume *f* glacée
s velo *m* de hielo; niebla *f* de hielo
g Eisdunst *m*; Eisnebel *m*; Frostbrauch *m*

* **ice foliation → 1028**

2960 ice foot
Sea ice firmly frozen to the shore at high-tide line, and unaffected by tide.
r подошва припая
Лед, крепко смерзшийся с берегом или дном вблизи берега на уровне высокого прилива и поэтому не подвергающийся действию прилива.
f banquette *f* côtière
s banco *m* de hielo costero
g Festeisfuß *m*

2961 ice forecast
The prediction of the ice cover state and duration of ice phenomena in a sea, river or lake.
r ледовый прогноз
Заблаговременное предсказание состояния ледяного покрова или сроков ледовых явлений в море, на реке или озере.
f prévision *f* des glaces flottantes
s previsión *f* de hielos flotantes
g Eisvorhersage *f*

2962 ice formation
The process of formation of ice mass from snow and firn or directly from frozen water.
r льдообразование; ледообразование
Формирование ледяной породы из снежного покрова и фирна или при замерзании воды.
f formation *f* de glace

s formación *f* de hielo
g Eisbildung *f*

* **ice front → 2927**

* **ice ground → 2997**

* **ice gruel → 781**

* **ice heap → 2979**

2963 ice heaving
The pressing-out of individual parts of ice cover caused by the effect of internal stresses due to the freezing-up of ground as well as water lenses or layers.
r пучение льда
Выдавливание отдельных частей ледяного покрова под влиянием внутренних напряжений, связанных с промерзанием грунта, а также линз и пластов воды.
f gonflement *m* de la glace
s levantamiento *m* de hielo
g Eisauftreiben *n*; Eisquellen *n*

2964 ice hillock
A mound of ice up to 6 m high formed around a blowing well or geyser by the layered freezing of the pouring out water.
r ледяной бугор
Ледяной холм высотой до 6 м, сформировавшийся вокруг фонтанирующих скважин при послойном намораживании изливающихся подземных вод.
f butte *f* de glace
s bulto *m* de hielo
g Eiserhebung *f*; Eishügel *m*

2965 ice hummocking
The deformation of ice cover due to lateral stresses leading to the formation of ice hummocks.
r торошение льда
Образование нагромождений и обломков морского льда в процессе его деформации за счет бокового сжатия.
f hummockage *m* de glace; entassement *m* de glace
s amontonamiento *m* del hielo
g Aufpressung *f* des Eises; Eispressung *f*

2966 ice island
A large tabular iceberg floating in the Arctic Ocean.
r ледяной остров
Обширный столовый айсберг, дрейфующий в Северном Ледовитом океане.
f île *f* de glace
s isla *f* de hielo
g Eisinsel *f*

2967 ice jam
A temporary accumulation of small broken ice-floes and frazil (shuga) across a river making an obstruction in autumn which causes the rise of the water level and sometimes flooding of the river banks.
r зажор
Скопление битого льда, вызывающее в период осеннего ледохода стеснение водного сечения реки и связанный с ним подъем уровня воды.
f embâcle *m*
s dique *m* de hielo
g Eisstopfung *f*

2968 ice keel
A wedge-like underwater part of hummocked ice row.
r ледяной киль
Гребень подводной части гряды торосов.
f crête *f* de pression sous-glace
s quilla *f* de hielo submarina
g Eiskiel *m*

2969 ice kernel
A mass of ground ice lying at the foot of a frost mound or buried ice in terminal moraines.
r ледяное ядро
Залежь внутригрунтового льда в основании бугров пучения или погребенного льда в конечных моренах.
f noyau *m* de glace
s núcleo *m* de hielo
g Eiskern *m*

2970 ice lace
Fine horizontal ice plates resting on ice pillars and walls appearing in summer on the surface of a glacier under the action of solar radiation whithin the upper 15-cm layer of ice. → Fig. 161
r ледяные кружева
Тонкие горизонтальные пластинки льда на ледяных столбиках и стенках, образующиеся летом под действием солнечной радиации в верхнем 15-сантиметровом слое льда. → Рис. 161
f dentelle *f* de glace
s encaje *m* de hielo
g Eisspitze *f*

2971 ice lens
An area of ice, often having convex upper and lower surfaces, which exists underground in

Fig. 161. **Ice lace**

periglacial environments or the inclusion of ice in a firn layer resulted from the accumulation and freezing of water above a impremeable layer.

r ледяная линза

Ледяное включение, как правило, с выпуклой верхней и нижней поверхностями, встречающееся в подземном льду в перигляцильной зоне, или включение льда в фирновой толще, образующееся в результате накопления и замерзания воды над водоупорной прослойкой.

f lentille *f* de glace

s lente *f* de hielo

g Eislinse *f*

* **ice margin → 2524**

* **ice mist → 2959**

2972 **ice mooring**

A natural ice wall of a glacier or stranded iceberg, or artificially frozen ice block adapted for the ship mooring.

r ледяной причал

Естественная стенка ледяного обрыва или искусственно намороженный массив льда на берегу, приспособленные для швартовки морских или речных судов.

f quai *m* de glace

s amarradero *m* de hielo; amarradero *m* glaciar

g Eiskai *m*; Eisanlegestelle *f*

2973 **ice mound; frost mound; frost hillock**

Any mound-shaped landform produced by ground freezing combined with the accumulation of ground ice due to the movement of groundwater or the migration of soil moisture. → Fig. 162

r мерзлотный бугор; бугор пучения

Родовой термин для обозначения всех бугристых образований, генетически связанных с мерзлотными процессами (бугры пучения, булгунняхи, пальзы и др.). → Рис. 162

f butte *f* cryogénique

s domo *m* criogénico; domo *m* decongelación

g Frosthügel *m*; Schwellungshügel *m*

Fig. 162. **Ice mound**

2974 **ice mylonite**

A compact ice consisting of fine material that is produced in the zones of differently-directed ice flows whithin glaciers.

r ледяной милонит

Компактный лед, состоящий из тонкоперетертого материала, формирующийся в зонах разнонаправленного движения в ледниках.

f mylonite *f* de glace

s milonita *m* de hielo

g Eismylonit *m*

2975 **ice needles**

Minute slender ice crystals in the near-surface layer of the air under conditions of clear sky or high level cloudness or slender needles and plates of ice suspended in water at the beginning of its freezing.

r ледяные иглы

Очень мелкие удлиненные кристаллы льда, парящие в приземном слое воздуха при безоблачном небе или при высокой облачности, либо тонкие иглы или пластинки льда, взвешенные в воде при начале ее замерзания.

f aiguille *f* de glace

s agujas *fpl* de hielo

g freischwebende Eiskristalle *mpl*; Eisprismen *npl*; Eisnadeln *fpl*

2976 **ice nip**

Pressing of a drifting ship by ice blocks.

r затирание льдами

Сжатие дрейфующего судна льдами.

f prise *f* en glace

s apresamiento *m* en hielo

g Einklemmung *f* im Eis

2977 **ice outburst**

A dramatic growth of ice run-off through outlet glacier into the ocean not compensated by the increase of ice accumulation.

r спуск льда

Резкое, некомпенсированное ростом питания усиление стока льда в океан через выводные ледники.

f évacuation *f* de glace

s evacuación *f* de hielo

g Eisansturz *m*; Eisabfussverstarkung *f*

* **ice pack → 4366**

* **ice pedestal → 5350**

2978 **ice pellets; small hail**

A type of precipitation consisting of transparent particles of ice of 5 mm or less which are spherical or irregular, seldom conical.

r ледяная крупа

Осадки прозрачных частиц льда, имеющих сферическую или неправильную форму, редко коническую, диаметром 5 мм или меньше.

f granules *mpl* de glace

s gránulos *mpl* de hielo
g Eisкügelchen *f*

2979 **ice pile; ice heap**
A mass of rafted ice-floes which pile up along the banks of rivers and shorelines of seasonally frozen lakes and water reservoirs during the ice drift.
r навалы льда
Нагромождения льдин на берегах рек, озер и водохранилищ в период ледохода.
f tas *m* de glace
s montón *m* de hielo
g Eisaufhäufung *f*

2980 **ice plug**
A natural accumulation of ice in a pipe, borehole or tunnel, formed by the freezing of the inflowing surface or underground waters.
r ледяная пробка
Естественное скопление льда в трубе, скважине, туннеле, образующееся путем замерзания затекающих поверхностных или подземных вод.
f bouchon *m* de glace
s tapón *m* de hielo
g Eispropfen *m*

* **ice polygon → 1360**

2981 **ice rafting**
A kind of deformation of floating ice whereby one floe overrides another floe, and both of them remain more or less horizontal.
r наслоение льда
Вид деформации плавучего льда, когда одна льдина надвигается на другую и обе сохраняют более или менее горизонтальное положение.
f chevauchement *m* de glace
s adherencias *fpl* de hielos
g Aufschieben *n* von Eis

2982 **ice ram**
A horizontal underwater protrusion of ice from an iceberg, or an ice barrier.
r ледяной таран
Подводный выступ айсберга или ледникового барьера.
f éperon *m* sous-glace
s esperón *m* de hielo
g Eisspor *m*

2983 **ice rampart; lake rampart**
A small ridge along the shore of a lake formed by the shoreward movement of surface ice during the winter freeze.
r ледяной вал
Небольшой вал вдоль берега озера, образованный при движении озерного льда к берегу во время зимнего ледостава.
f rempart *m* de glace;
empilage *m* de glace
s caballón *f* de hielo
g Eiswall *m*

2984 **ice reconnaissance**
A regular monitoring of the ice conditions carried out from board of a special ship or an aircraft.
r ледовая разведка
Сбор информации о состоянии ледяного покрова на морях, реках и других водных объектах с помощью технических средств.
f reconnaissance *f* des glaces
s reconocimiento *m* de hielos;
prospección *f* glaciar
g Eisdeckeerkundung *f*

2985 **ice regime**
The set of regularly occurring processes of the formation, development and destruction of ice cover on rivers and water bodies.
r ледовый режим
Совокупность закономерно повторяющихся процессов образования, развития и разрушения ледяного покрова водных объектов.
f régime *m* de glace
s régimen *m* de los hielos
g Eisregime *n*

2986 **ice related community**
A community of organisms that inhabits, usually in anabiotic state, the ice cover of water bodies.
r пагон
Сообщество организмов, населяющих поверхность или толщу льда, обычно в состоянии анабиоза.
f pagon *m*
s pagón *m*
g Pagon *n*

2987 **ice remnant**
A residual of aufeis mass up to 2.5 m high and 1-2 m wide covered with a firm mass of soil, plant roots, grass, peat or sand.
r ледяной останец
Остаток наледного льда высотой до 2,5 м и шириной 1-2 м, прикрытый сверху плотной массой почвы, корней растений, травы, торфа или песка.
f témoin *m* de glace
s peñón *m* de hielo
g Aufeisrestberg *m*

2988 **ice reservoir**
A large element of reticular glaciers having massives of ice filling the intermontane basins.
r ледоем
Крупный элемент сетчатого оледенения, представляющий собой массы льда, заполняющие межгорные котловины.
f réservoir *m* de glace
s depósito *m* de hielo
g Eiseinzugsgebiet *n*

2989 **ice rind**
A thin (up to 5 cm) but hard layer of ice in seas, rivers or lakes, formed where fresh water freezes on the surface.

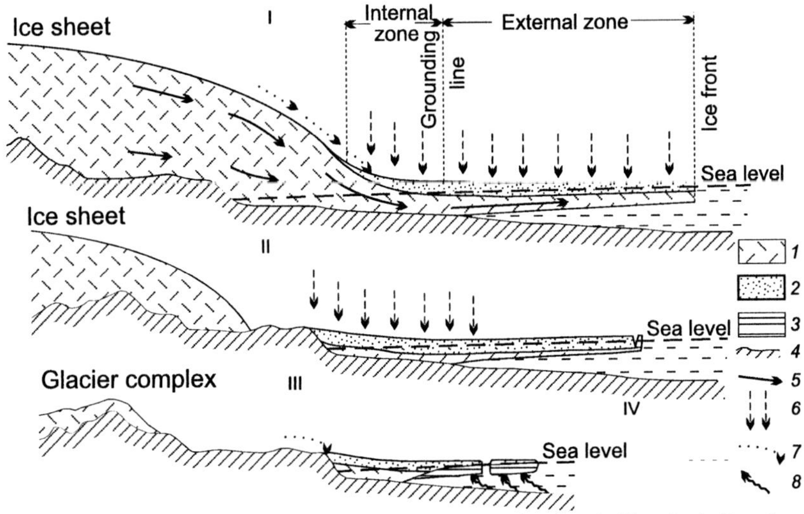

Fig. 163. I, II, III – stages of the reduction of an **ice shelf**; 1 – glacier ice, 2 – shelf ice, 3 – bottom frozen ice, 4 – bedrock, 5 – ice movement, 6 – atmospheric precipitation, 7 – drifting snow, 8 – freezig of ice from sea water

r склянка
Блестящая корка льда толщиной до 5 см, образующаяся на спокойной поверхности пресной или слабо соленой воды.
f glace *f* en feuille
s hielo *m* de hollejo
g Eishaut *f*

2990 ice rise
An ice massif resting on bedrock and surrounded by either an ice shelf or a combination of an open sea and ice-free land.
r ледяное поднятие
Масса льда, лежащая на коренном ложе и окруженная либо целиком шельфовым ледником, либо частично открытым морем и свободной ото льда сушей.
f élévation *f* de glace
s elevación *f* de hielo
g Eiserhöhung *f*

2991 ice rock
A body of ice occurring in the Earth's crust or on its surface.
r ледяная порода
Ледяное тело мономинерального состава, залегающее внутри земной коры или на ее поверхности.
f glacite *m*
s roca *f* de hielo
g Eisgestein *n*; Steineis *m*

2992 ice season
A period during which the ice cover exists on water bodies and watercourses.
r ледовый сезон
Период, в течение которого на акватории водоема или водотока держится ледяной покров.
f saison *f* glaciaire
s estación *f* glaciar
g Eissaison *f*

2993 ice segregation
The formation of discrete bodies of ground ice (lenses or bands of clear ice) in frost-susceptible materials, especially silt, partly by freezing of existing interstitial soil water but largely by freezing of groundwater drawn upwards by capillary action.
r льдовыделение; выделение льда
Формирование линз или полос льда в мелкоземистых отложениях, частично в результате замерзания влаги, находящейся в почве, но в основном за счет замерзания грунтовой воды, поднимающейся вверх по капиллярам.
f ségrégation *f* de glace
s segregación *f* de hielo
g Eissegregation *f*

2994 ice sheet
An area of land ice of considerable thickness spreading over more than 50,000 km^2 forming a continuous ice cover, e.g. currently existing the Antarctic and Greenland ice sheets, or the ice sheets of the Quaternary Age.
r ледниковый щит
Массив наземного или налегающего на континентальный шельф мощного ледяного покрова площадью более 50 тыс. км^2, например, современные Антарктический и Гренландский ледниковые покровы или ледниковые покровы плейстоцена.
f calotte *m* glaciaire
s casquete *m* glaciar
g Eisschild *m*; Inlandeis *n*

2995 ice shelf
A sheet of ice of great thickness and considerable extent with a fairly level surface formed as an extention of land ice sheet over the sea in shallow bays and inlets of polar coasts where it is fastened to the shore and often reaches bottom. → Fig. 163, p. 359
r шельфовый ледник
Мощный плавучий ледник плитообразной формы с довольно ровной поверхностью, образующийся как продолжение материкового ледникового покрова вдоль побережья в полярных районах, в мелких заливах и в шельфовых морях, где он прикреплен к берегу или покоится на мели. → Рис. 163, с. 359
f glacier *m* de plate-form; barrière *f* flottante
s glaciar *m* de plataforma; barrera *f* flotante
g Schelfeis *n*

2996 ice situation; ice conditions
The state of ice cover over a section of sea and at a particular point or on rivers and lakes at a given time.
r ледовая обстановка
Состояние ледяного покрова в море, на каком-либо его участке или в отдельном пункте в конкретный момент времени.
f situation *f* des glaces
s situación *f* de hielos; estado *m* glaciar
g Meereissituation *f*

* **ice sky → 2930**

* **ice slide → 2926**

2997 ice soil; ice ground
A rock, in which the ice content exceeds by volume the total content of all the other components.
r ледогрунт
Горная порода, в которой содержание льда по объему превышает суммарное содержание всех других компонентов.
f sol *m* de glace
s suelo *m* de hielo
g vereister Boden *m*

2998 ice stalactite
An icicle-shaped formation hanging from the ceiling of an ice cave and produced by freezing of water seeping through cracks in the cave arch.
r ледяной сталактит
Натечно-капельное образование в виде сосульки, растущее с потолка ледяной пещеры вследствие замерзания воды, просачивающейся по трещинам в своде пещеры.
f stalactite *f* de glace
s estalactita *f* de hielo
g Eisstalaktit *m*

2999 ice stalagmite
A cone-shaped or pencil-shaped ice formation growing from the floor of an ice cave towards an ice stalactite as the water drops from the ceiling.
r ледяной сталагмит
Натечно-капельное ледяное образование в виде конуса или столба, растущее с пола ледяной пещеры навстречу ледяному сталактиту.
f stalagmite *f* de glace
s estalagmita *f* de hielo
g Eisstalagmit *m*

3000 ice stalagnate
A pillar-shaped formation, produced by the intergrowth of ice stalactites and ice stalagmites.
r ледяной сталагнат
Натечно-капельное образование в виде колонны, возникающее в ледяной пещере при срастании ледяных сталактитов и сталагмитов.
f colonne *f* de glace
s columna *f* de hielo
g Eisstalagnat *m*

3001 ice stream
1. The part of an ice sheet in which there occurs a corridor or stream of faster-moving ice surrounded by large areas of slow-moving or stagnant ice. 2. A band of drifting ice isolated from the main ice cover by wind.
r ледяной поток
1. Коридор в ледниковом щите с быстро двигающимся льдом, окруженный медленно движущимся или неподвижным льдом. 2. Полоса дрейфующего льда, отделившаяся от основной массы, например, под влиянием ветра.
f fleuve *m* de glace; courant *m* de glace
s flujo *m* de hielo
g Eisstrom *m*; Eisflut *f*

* **ice strip → 2928**

3002 ice survey
The measurements of the ice thickness and slush under it and coastal ice on rivers and water reservoirs, as well as the thickness and density of snow on a fast ice.
r ледомерная съемка
Измерение толщины льда и шуги под ним, толщины и плотности снега на морском припае, озерах, реках или водохранилищах.
f levé *f* glaciométrique
s levantamiento *m* glaciométrico
g Eismessungsaufnahme *f*

3003 ice tongue
The projection of an ice edge, up to several kilometers long, into the open water.
r язык льда
Часть кромки льда, выступающая на несколько километров в сторону чистой воды.

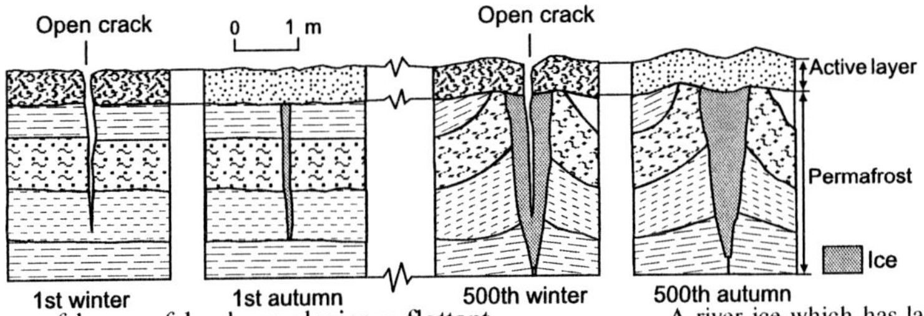

Fig. 164. Formation of **ice veins**

f langue *f* de glace; glacier *m* flottant
s lengua *f* de hielo; glaciar *m* flotante
g Eiszunge *f*

3004 ice trafficability
Limiting thickness of continuous ice cover which can be broken by an icebreaker moving continuously at a rate of one knot.
r ледопроходимость
Предельная толщина сплошного льда, преодолеваемого ледоколом при непрерывном движении со скоростью около одного узла.
f navigabilité *f* de la glace
s navegabilidad *f* de hielo
g Befahrbarkeit *f* des Meereises

3005 ice vein; ice wedge
An inclusion of ice formed as a result of freezing of water in a crack or fissure which cuts through the rocks. → Fig. 164
r ледяная жила; ледяной клин
Клиновидное ледяное тело, образующееся в результате замерзания воды в трещине, которая рассекает лед или другие породы. → Рис. 164
f filon *m* de glace; veine *f* de glace
s vena *f* de hielo; filón *m* de hielo
g Eisader *f*; Eiskeil *m*

3006 ice viscosity
The resistance of ice to the relative motion inside an ice body (deformation).
r вязкость льда
Сопротивление льда относительному движению внутри него (деформации).
f viscosité *f* de glace
s viscosidad *f* de hielo
g Eisviskosität *f*; Eisbindigkeit *f*

3007 ice wall
Ice scarp that is a seaward edge of an ice sheet, ice cap or ice rise.
r ледяная стена
Ледяной обрыв, образующий596 мористый край ледникового покрова, ледникового купола или ледяного поднятия.
f mur *m* de glace
s pared *f* de hielo
g Eiswand *f*

* **ice wedge → 3005**

3008 ice with voids
A river ice which has large air-filled bubbles within or beneath it.
r пустолед
Лед на реках, в толще которого или под ним имеются крупные воздушные полости.
f glace *f* avec cavités
s hielo *m* con cavernas
g Flüßeis *m* mit Lufthöhlen

3009 ice-bearing ground; ice-bounded ground
A frozen dispersed ground that contains many visible inclusions of different types of ice.
r льдистый грунт; льдонасыщенный грунт
Мерзлый дисперсный грунт с включением значительного количества видимых включений льда разного происхождения.
f sol *m* à haute teneur en glace
s suelo *m* rico en hielo
g eisreicher Boden *m*

3010 iceberg
A mass of floating ice that has broken away from a glacier, ice shelf or ice sheet and floats in the sea or a lake. → Fig. 165
r айсберг
Крупная плавучая глыба льда, отколовшаяся от ледника, сползающего в море или озеро. → Рис. 165

Fig. 165. **Iceberg** of middle size with its submarine part which forms more than 80 per cent of the total volume of an iceberg

f iceberg *m*; montagne *f* de glace
s iceberg *m*
g Eisberg *m*

* **iceberg deposits → 2554**

* **iceblock → 2667**

* **ice-bounded ground → 3009**

3011 ice-breaker
A special ship equipped to make and maintain a channel through ice and to pilot other ships.
r ледокол
Специальное судно, предназначенное для проводки других судов через льды, покрывающие поверхность морей, озер и рек.
f brise-glace *f*
s rompehielos *m*; buque *m* rompehielos
g Eisbrecher *m*

* **ice-cake → 4401**

3012 ice-cap island
A polar island composed completely of ice and looking like an ice dome.
r остров – ледяная шапка
Полярный остров, целиком сложенный льдом и представляющий собой ледниковый купол.
f île-calotte *f* glaciaire
s isla-casquete *m* glaciar
g Eiskappe-Insel *f*

* **ice-cement → 4702**

3013 ice-cored moraine
Any morainic mound or ridge marginal to an ice sheet or glacier which encloses buried ice.
r морена с ледяным ядром
Моренный вал у края ледника или ледникового покрова, содержащий внутри мертвый лед.
f moraine *f* à noyau de glace
s morrena *f* con el núcleo de hielo
g Eiskernmoräne *f*

3014 ice-dammed lake
A large water reservoir arising in front of the edge of an ice sheet or in a mountain valley when it is dammed by the valley glacier.
r ледниково-подпрудное озеро
Крупное скопление воды, возникающее перед краем ледникового щита или в горной долине при ее подпруживании долинным ледником.
f lac *m* de barrage glaciaire
s lago *m* de contención de glaciar;
lago *m* de barrera de hielo
g Gletscherstausee *m*; Eisdammsee *m*

3015 icefall
A portion of a glacier where it moves down a steep slope and which is characterized by numerous crevasses and ice pinnacles. → Fig. 166

Fig. 166. **Icefall** from a helicopter window

r ледопад
Участок ледника над *ригелем*, разбитый множеством поперечных трещин на отдельные глыбы. → Рис. 166
f chute *f* de séracs
s salto *m* de hielo
g Eisfall *m*

3016 ice-field
A widespread and continuous area of floating ice, generally greater than 10 km across, defined as large (over 20 km across), medium (15 to 20 km across), or small (10 to 15 km across).
r ледяное поле
Крупное скопление плавучего льда в поперечнике не менее 10 км, различают большое (более 20 км), среднее (15-20 км) и малое (10-15 км) ледяное поле.
f champ *m* de glace; champ *m* de banquise
s campo *m* de hielo; campo *m* de banquisa
g Eisscholle *f*; Treibeisfeld *n*

* **ice-floe → 2189**

3017 ice-formation zone; ice facies
Parts of glaciers where the rates and regime of ice formation are different that results in a non-uniform structure of the body of firn and ice and a different rate of the firn transformation into ice.
r зона льдообразования
Участки на ледниках с различными интенсивностью и ходом льдообразования, что приводит к неодинаковому строению фирново-ледяной толщи и разной скорости превращения фирна в лед.
f zones *fpl* de cryogenèse
s zonas *fpl* de formación de hielo
g Zonen *fpl* der Eisbildung

* **ice-pan → 2189**

3018 iceport

A section of the ice front favourable for the ship mooring.

r шельфовая гавань

Ниша в ледниковом барьере, часто временного характера, где суда могут швартоваться непосредственно у шельфового ледника.

f port *m* de glace

s puerto *m* de hielo

g Eishafen *m*

3019 **ice-quake**

Shaking of ice caused by the processes running in glaciers that is recorded by seismic instruments as a weak earthquake.

r льдотрясение

Сотрясение льда вследствие процессов, происходящих в ледниках, которое фиксируется сейсмическими приборами подобно слабому землетрясению.

f tremblement *m* de glace

s temblor *m* de hielo

g Gletschereisbeben *n*

3020 **ice-survey station**

A site within which visual and instrumental observations of the ice conditions are regularly carried out.

r ледомерный пост

Участок, в пределах которого проводятся визуальные и инструментальные наблюдения за ледовой обстановкой.

f station *f* glaciométrique

s estación *f* glaciométrica

g Eismessstation *f*

3021 **ice-technology**

The development of methods and instruments for the use of ice and protection from its damaging effect.

r ледотехника

Разработка методов и оборудования для использования льда и борьбы с его вредным влиянием.

f cryotechnologie *f*

s criotecnología *f*

g Eistechnik *f*

3022 **icicle**

A pendant of clear ice, generally in the form of a spike, formed by the freezing of water dripping down from the edges of objects, buildings, rocks, etc.

r сосулька

Острый ледяной выступ, образующийся у края нависающих предметов, на склонах и в подземных полостях при послойном намораживании медленно стекающей воды.

f glaçon *m*

s carámbano *m*

g Eiszapfen *m*

3023 **icing**

Any deposit or coating of ice on an exposed object or surface such as leading edges of aircraft wings, when supercooled droplets in a cloud come into contact with the surface.

r обледенение

Любое отложение или окутывание льдом предмета или поверхности в результате выпадения переохлажденных капель.

f givrage *m*

s capa *f* helada

g Vereisung *f*

3024 **icy** ***adj***

Covered with, full of, or consisting of ice.

r ледяной; ледовый

Состоящий изо льда, сложенный льдом (например, ледяной покров), или имеющий отношение ко льду (ледовый режим).

f de glace *adj*

s glaciario *adj*

g Eis- *adj*; eisbedeckt *adj*

3025 **idiomorphism of ice**

The ability of ice to take characteristic crystallographic contours corresponding to the type of its space lattice in the course of its crystallization from water or water vapour as well as in the process of metamorphism.

r идиоморфизм льда

Способность льда принимать при кристаллизации из воды или водяного пара, а также в процессе метаморфизма кристаллографические очертания, соответствующие типу его пространственной решетки.

f idiomorphisme *m* de glace

s idiomorfismo *m* de hielo

g Idiomorphie *f* des Eises

3026 **idle water spill**

The release of water through spillways of hydroelectric power complex under the threat of overfilling of the effective capacity of a reservoir.

r холостой сброс воды

Попуск воды через водосбросы гидроузла при угрозе переполнения полезного объема водохранилища.

f écoulement *m* à vide

s rebosamiento *m* en vacío

g notgedrungenes Ablassen *n*

3027 **igapo**

Landscape of low river flood-plains in Amazon lowland usually inundated for long period of time and covered with *hylea* vegetation.

r игапо

Ландшафт низких длительно затопляемых пойм рек на Амазонской низменности, покрытых *гилеей*.

f igapo *m*

s igapó *m*

g Igapo *m*

3028 **igneous rock**

A rock formed as a result of cooling and solidification of magma coming from the heated lower layers of the Earth's crust.

r магматическая горная порода
Горные породы, образовавшиеся в результате охлаждения и затвердевания магмы (подразделяются на интрузивные и эффузивные).
f roche *f* magmatique; roche *f* ignée
s roca *f* magmática; roca *f* ígnea
g Massengestein *n*; Erstarrungsgestein *n*; Magmagestein *n*

* **illuminated relief → 2777**

3029 **illuvial horizon; B-horizon; argillic horizon**
The layer in a soil profile which lies beneath the eluvial horizon (A-horizon) and may be enriched by the deposition of material washed down from above.
r иллювиальный горизонт; горизонт В; горизонт вмывания
Почвенный горизонт, в котором накапливаются вещества, вынесенные из вышележащих (элювиальных) горизонтов почвы.
f horizon *m* illuvial; horizon *m* d'accumulation
s horizonte *m* iluvial; horizonte *m* B; horizonte *m* de acumulación
g Illuvialhorizont *m*

3030 **illuviation**
The process by which material removed in suspension or solution from the upper part of a soil (the eluvial horizon) is washed down and deposited in the lower layers or illuvial horizons.
r вмывание
Процесс, посредством которого почвенный материал в форме суспензии или в растворе переносится из верхнего (элювиального) горизонта почвы и откладывается в нижних (иллювиальных) горизонтах.
f illuviation *f*
s iluviación *f*
g Illuviation *f*

3031 **ilmen**
A shallow lake (ox-bow lake) in the delta or flood-plain of a river with shores covered with reed and cane.
r ильмень
Мелкое озеро (старица) в дельте или пойме реки с берегами, заросшими тростником и камышом.
f ilmen *m*
s ilmen *m*
g Ilmen *n*

* **image → 5694**

3032 **image enhancement**
The process of improving the visual quality of a remotely sensed image by optical or digital means.
r повышение качества снимка
Процесс улучшения визуального качества дистанционно полученного снимка оптическими или цифровыми средствами.
f amélioration *f* de l'image
s mejoramiento *m* de imagen
g Bildverbesserung *f*

3033 **image interpretation; interpretation of aerial photographs**
The study of territories, water bodies, and the atmosphere by aerial and space images based on the correlation between the properties of identified objects and the character of their representation on images.
r дешифрирование
Изучение по аэрофото- и космическим снимкам территорий, акваторий и атмосферы, основанное на зависимости между свойствами объектов и характером их воспроизведения на снимках.
f photo-interprétation *f*; restitutión *f* photogrammétrique
s interpretación *f* de fotos aéreas; restitución *f* fotogramétrica
g Luftbildinterpretation *f*; Luftbildauswertung *f*

3034 **immature soil**
A soil which exhibits poorly developed horizons or the profile of which has incomplete set of developed genetic horizons characteristic of a given zone.
r неполноразвитая почва
Почва, профиль которой не имеет полного набора развитых генетических горизонтов, характерных для данной зоны.
f sol *m* peu évolué
s suelo *m* poco desarrollado; suelo *m* poco evolucionado
g unreifer Boden *m*

3035 **immigrant**
A person who voluntary comes from the home country to settle in another country for the purpose of permanent or temporal residence.
r иммигрант
Человек, приехавший в чужую страну на постоянное или временное (как правило, длительное) проживание.
f immigrant *m*
s inmigrante *m*; inmigrado *m*
g Immigrant *m*; Einwanderer *m*

3036 **immigration**
1. The movement of a person into a foreign country in order to make up permanent residence. 2. The movement of an organism into a new area.
r иммиграция
1. Въезд в страну на постоянное или временное (как правило, длительное) проживание граждан других стран. 2. Перемещение организма в новый район обитания.
f immigration *f*

s inmigración *f*
g Immigration *f*; Einwanderung *f*

* **impact mitigation → 3882**

* **impact of recreation on the landscape → 4922**

* **impact zone → 3037**

3037 **impacted area; impact zone**
The area under the effect of an industrial plant, building project and other human economic activity.
r импактная зона; зона техногенного воздействия
Территория вокруг промышленного предприятия, строительства или другого объекта хозяйственной деятельности, на которую распространяется его воздействие.
f aire *f* d'impact; zone *f* d'impact
s área *f* de impacto
g Belastungsbereich *m*

3038 **impeded drainage**
The restriction of the downward gravitational movement of water through the soil.
r связанный дренаж почвы
Ограничение направленного вниз гравитационного движения воды в почве.
f drainage *m* embarrassé
s drenaje *m* embarazoso
g unfrei Dränage *f*

3039 **impeded solifluction**
A kind of solifluction occurring on slopes covered with vegetation without disturbing it.
r связанная солифлюкция
Солифлюкция на склонах с растительным покровом и не нарушающая его.
f solifluxion *f* entravée
s solifluxión *f* coherente
g zusammenhängende Solifluktion *f*

3040 **imperialism**
The control of one or a number of countries by a dominant nation, i.e. the creation of the empire through the extension by one sovereign state of its authority over other territories, by military, political and economic means, as well as the policy or doctrine of such an extension of authority.
r империализм
Создание империи путем расширения суверенитета одной страны на другие территории с использованием военной силы, политических и экономических средств, а также политика или доктрина подобного захвата территории.
f impérialisme *m*
s imperialismo *m*
g Imperialismus *m*

* **impermeability factor → 5150**

3041 **impermeable *adj***
Not permitting the passage of fluids, especially of water.
r непроницаемый
Препятствующий проникновению жидкости и, в частности, воды.
f imperméable *adj*
s impermeable *adj*
g impermeabel *adj*

3042 **impermeable rock; impervious rock**
A rock which does not allow water to soak into it and through which oil or gas cannot pass freely, e.g. clay, unfissured granite.
r водонепроницаемая порода; водоупор
Плотная горная порода, не пропускающая подземные воды, газ или нефть, обычно служит водоупором, подстилающим водоносный горизонт.
f roche *f* imperméable; couche *m* imperméable
s roca *f* impermeable; capa *f* impermeable
g wasserundurchlässiges Gestein *n*; Grundwassersohle *f*

* **impervious rock → 3042**

3043 **import**
Goods which are brought into a country by trade from a foreign country and the action of bringing goods into oners own country.
r импорт
Товары, ввозимые в данную страну в результате торговли с другой страной, и сам процесс ввоза товаров в страну из-за границы.
f importation *f*
s importación *f*
g Import *m*; Einfuhr *f*

3044 **import duty**
The money which is collected for the goods brought into a country.
r импортная пошлина
Денежный сбор, взыскиваемый государством с товаров, ввозимых из-за границы.
f droits *mpl* d'entrée; droits *mpl* d'importation
s derechos *mpl* de entrada; derechos *mpl* de importación
g Einfuhrzoll *f*

3045 **import quota**
A limit on the amount of a particular commodity to be imported from one country to another.
r импортная квота
Фиксированная норма товара, разрешенная для ввоза в страну.
f contingent *m* d'importation; quota *m* d'importation
s contingente *m* de importación; cuota *f* de importación
g Einfuhrquote *f*; Importquote *f*

3046 import substitution
The strategy for economic development which encourages the replacement of previously imported goods by home produced goods in order to improve the balance of payments.
r замещение импорта
Стратегия экономического развития, предполагающая замену ранее импортируемых товаров производимыми в этой стране с целью улучшения платежного баланса.
f substitution *f* d'importation
s sustitución *f* de importación
g Importsubstitution *f*

* **impoverished soil → 1521**

* **impoverishment of a soil → 1524**

3047 improved soil
A soil the properties of which are deliberately changed (neutral values of pH, high content of humus) for growing of crops.
r окультуренная почва
Почва, свойства которой направленно изменены для выращивания более высокого урожая, например, получены нейтральные значения pH, высокое содержание гумуса.
f sol *m* cultivé
s suelo *m* cultivado; suelo *m* labrado
g kultivierter Boden *m*

* **improvement of land → 3375**

3048 improvement of soil; soil improvement
The action of making a soil more productive for plant growth by drainage, irrigation, addition of fertilizers, washing irrigation of salty soils, lime pretreatment, adding of gypsum and clay, mud grouting and a number of other methods.
r мелиорация почв
Улучшение свойств почвы с целью повышения ее плодородия, включает осушение, орошение, промывку засоленных почв, внесение удобрений, известкование, гипсование, глинование, кольматаж и ряд других методов.
f amendement *m* du sol; aménagement *m* agricole; améliorations *fpl* foncières
s mejoramiento *m* del suelo; acondicionamiento *m* del suelo
g Bodenmelioration *f*; Bodenverbesserung *f*

3049 improvement of soil by snow
The action of making a soil more productive by the improvement of hydro-thermal regime of soil by the regulation of snow accumulation, duration of snow cover, snow melting and surface snow melt runoff.
r снежная мелиорация
Улучшение водно-теплового режима почв путем регулирования снегонакопления, продолжительности залегания снежного покрова, снеготаяния и поверхностного стока талых вод.
f amélioration *f* par la neige
s bonificación *f* del suelo con nieve; mejora *f* del suelo con nieve
g Schneemelioration *f*

3050 inbound tourism
Tourism involving nonresidents of a country visiting this country for touristic activities.
r въездной туризм
Туризм, относящийся к людям, приезжающим на данную территорию с туристическими целями.
f tourisme *m* réceptif
s turismo *m* receptivo
g Einfahrtstourismus *m*; Einreiseverkehr *m*

3051 inceptisol
In US classification an order of young soils with weakly developed horizons, occurring on newly exposed surfaces recently vacated by an ice or snow cover or on recently accumulated floodplain sediments.
r инсептисоль
В классификации, принятой в США, порядок молодых почв со слабо развитыми горизонтами, появляющихся на недавно обнажившихся поверхностях, которые в настоящее время не покрываются снежным покровом и льдом, либо представляют собой пойменные отложения.
f inceptisol *m*
s inceptisol *m*; inceptisuelo *m*
g Inceptisol *m*

3052 incised meander; confined meander; ingrown meander; infrenched meander
A river meander deeply sunk into the general level of flood-plain and cut into the bedrock shores of the valley.
r врезанная излучина
Речная излучина, углубленная в первичную поверхность поймы и местами врезанная в коренные берега долины.
f méandre *m* encaissé
s meandro *m* encajado
g eingesenkter Mäander *m*; eingetiefter Mäander *m*

3053 incised valley
A river valley whose floor lies much deeper than the original surface on which the river formed.
r врезанная долина
Долина с дном, лежащим значительно ниже первичной поверхности, на которой заложилась река.
f vallée *f* encaissée
s valle *m* encajonado
g eingeschnittenes Tal *n*

3054 inclination; gradient$_2$; slope angle
Angle formed by the direction of a slope at a certain point from a horizontal plane.
r крутизна склона; угол наклона; уклон; крутизна ската

Угол, образуемый направлением склона с горизонтальной плоскостью в данной точке.
f inclinaison *f* de pente; déclivité *f* du versant
s pendiente *m*; angulo *m* de inclinación; inclinación *f*
g Fallwinkel *m*; Abschüssigkeit *f*; Neigung *f*; Hanggefälle *n*

3055 inclined plain
A plain gently inclined towards one side.
r наклонная равнина
Равнина, полого наклоненная в одну сторону.
f plaine *f* inclinée
s llanura *f* inclinada
g schiefe Ebene *f*; schräge Ebene *f*; geneigte Ebene *f*

3056 inconsequent drainage
A drainage pattern not conditioned by the present structure, being either antecedent or superimposed.
r несогласованная речная сеть
Рисунок речной сети, не согласующийся с современной структурой рельефа, так как он связан с предшествующим этапом его развития или является наложенным.
f réseau *m* hydrographique inconséquent
s red *f* hidrográfica discordante; red *f* hidrográfica inconsequente
g inkonsequentes Flußnetz *n*

* **index contours → 1263**

* **index map → 3316**

* **index of aridity → 389**

* **index of sheets → 3316**

3057 Indian summer; St. Martin summer; Old Wive's summer
A period of calm, dry and abnormally warm weather occurring regularly in autumn (fall).
r бабье лето
Сравнительно длительный период сухой, солнечной и необычно теплой погоды осенью.
f été *m* indien; été *m* de Saint Martin
s verano *m* indio; veranillo *m* de San Martin; veranillo *m* de San Miguel
g Indianersommer *m*; Altweibersommer *m*

3058 indicator plant; indicator species; plant indicator
A plant species which shows by its presence in a locality the existence of a particular environmental factor or certain environmental conditions.
r растение-индикатор; индикаторный вид
Виды растений, прочно связанные с определенными экологическими условиями, так что эти условия могут быть распознаны по присутствию данных растений.
f plante *f* indicatrice; espèce *f* indiquée
s planta *f* indicadora; especie *f* indicadora
g Zeigerpflanze *f*; Indikatorart *f*; Indikatorpflanze *f*; Zeigerart *f*

* **indicator species → 681**

3059 indigenous *adj*
Originating in a particular area, not introduced, applied to rocks, plants, animals, human population, not carried in from somewhere else.
r исконный
Происходящий из определенного региона, не интродуцированный в эти места (применяется к горным породам, растениям, животным, местному населению).
f indigène *adj*
s indígeno *adj*
g eingeboren *adj*; einheimisch *adj*

* **indigenous species → 463**

3060 induration
The hardening of a rock by heat, pressure or cementation, or of soil horizons by chemical action.
r отвердение
Затвердевание горной породы под воздействием тепла, давления или цементации, а также почвенных горизонтов под воздействием химических реакций.
f induration *f*
s induración *f*
g Induration *f*; Erhärtung *f*

* **industrial agglomeration → 6119**

* **industrial branch → 780**

3061 industrial centre
A settlement in which industry is the major specialization and the basis for urban development.
r промышленный центр; промышленный узел
Населенный пункт, в котором промышленность служит главной градообразующей базой и основной отраслью специализации.
f centre *m* industriel
s centro *m* industrial
g Industriezentrum *n*

3062 industrial city
A city with a central business district and large industrial periphery, typical of the highly developed contries of today.
r индустриальный город
Город с центральным деловым районом и обширной промышленной периферией, типичный для современных высокоразвитых стран.
f ville *f* industrielle
s ciudad *f* industrial
g Industriestadt *f*

3063 **industrial clouds; plumes**
Clouds of different nature formed above industrial enterprises and regions due to industrial activity.
r индустриальные облака
Облака разной природы, возникающие над промышленными предприятиями и районами в результате хозяйственной деятельности.
f nuages *mpl* industriels
s nubes *fpl* industriales
g Industrewolken *fpl*

* **industrial complex → 6119**

3064 **industrial crop**
A crop grown as a raw material for manufacturing industry, e.g. cotton.
r технические культуры
Растения, возделываемые с целью получения сырья для различных отраслей промышленности, например, хлопок.
f culture *f* technique; culture *f* industrielle
s cultivo *m* técnico; cultivo *m* industrial
g technische Nutzpflanze *f*; Industriepflanze *f*

3065 **industrial district; industrial estate**
A territory with buildings specially designed and constructed for industries with supporting services, often located in suburbs or at the edge of a town or city.
r промышленная зона
Территория внутри или на окраине крупного города со значительной концентрацией промышленных предприятий, общей инфраструктурой и другими внутренними связями.
f zone *f* industrielle; espace *m* industriel
s zona *f* industrial; espacio *m* industrial; polígono *m* industrial
g industrielle Zone *f*; Industriegelände *n*

3066 **industrial effluent; industrial sewage**
Waste waters from industrial processes released into the environment.
r промышленные стоки; промышленный слив
Отработанные производственные воды, спускаемые в естественные или искусственные объекты.
f effluents *mpl* industriels; déchets *mpl* industriels
s aguas *fpl* residuales industriales; effluente *m* industrial
g Industrieabwässer *npl*

* **industrial estate → 3065**

3067 **industrial geography; geography of industry**
The study of the spatial arrangement of manufacturing industry as a whole and as individual manufacturing activities of territorial production complexes at different levels.
r география промышленности
Изучение территориальной структуры промышленного производства в целом и по группам отраслей в составе территориально-хозяйственных комплексов разного уровня.
f géographie *f* industrielle
s geografía *f* industrial
g Industriegeographie *f*

* **industrial landscape → 6063**

3068 **industrial map**
Geographic map showing the location of industrial enterprises of individual branches of industry and their economic ties with allied economic branches.
r отраслевая карта
Географическая карта, отражающая размещение промышленных предприятий конкретной отрасли и экономические связи со смежными отраслями народного хозяйства.
f carte *f* industrielle
s mapa *m* de industria; mapa *m* industrial
g Karte *f* der Brauchen; Karte *f* der Industriezweige

3069 **industrial park**
A tract of land, in some cases park-like, planned and officially designated for the accomodation of clean, relatively small industrial enterprises, which do not pollute the environment.
r индустриальный парк
Участок земли, нередко имеющий облик парка, спланированный и построенный для размещения относительно небольших промышленных предприятий, не загрязняющих окружающую среду.
f parc *m* industriel
s parque *m* industrial
g Industriegelände *n*

3070 **industrial pollution**
Pollution of the environment caused by the activity of industrial enterprises.
r промышленное загрязнение
Загрязнение окружающей среды, вызванное деятельностью промышленных предприятий.
f pollution *f* indusrtielle
s polución *f* industrial
g Industrieverschmutzung *f*

3071 **industrial port complex**
A group of industrial enterprises involved directly or indirectly in the port and adjacent industrial enterprises.
r портово-промышленный комплекс
Совокупность предприятий, деятельность которых прямо или косвенно связана с портом и прилегающими промышленными предприятиями.
f complexe *m* industriel portuaire

s complejo *m* industrial portuario
g port-industrialer Komplex *m*

3072 **industrial region**
An economic region with the domination of industry in its economy.
r промышленный район
Экономический район с преобладанием в производственной специализации промышленных отраслей хозяйства.
f région *f* industrielle
s región *f* industrial
g industrielle Region *f*; Industriegebiet *n*

3073 **industrial revolution**
The change from the cottage industry to the factory system produced by the mechanization which led to the mass production of goods with the consequent great changes in social, economic and technical structures, it began in Britain in the XVIII and early XIX centuries.
r промышленная революция
Исторический период перехода от мануфактуры к машинному производству, начавшийся в Великобритании во второй половине XVIII в. и завершившийся в развитых странах в XIX в.
f révolution *f* industrielle
s revolución *f* industrial
g industrielle Revolution *f*

* **industrial sewage → 3066**

3074 **industrial society**
The stage of the development of a society and social relationships following the industrial revolution when along with raw materials production, the manufacturing industry (secondary industry) formed the basis of economics.
r индустриальное общество
Этап развития общества и общественных отношений, сложившийся после промышленной революции, когда наряду с сырьевыми отраслями основой экономики стали развиваться обрабатывающие отрасли (вторичный сектор экономики).
f société *f* industrielle
s sociedad *f* industrial
g industrielle Gesellschaft *f*

3075 **industrialization**
The process of development of large-scale machine production in countries where people were engaged mainly in agricultural activities, which is accompanied by social changes of the transition from pre-industrial to industrial society.
r индустриализация
Процесс создания крупного машинного производства в странах, где население было занято преимущественно сельскохозяйственной деятельностью, что сопровождалось социальными изменениями в процессе перехода от доиндустриального к индустриальному обществу.
f industrialisation *f*
s industrialización *f*
g Industrialisierung *f*

3076 **industrialized country**
A country in which the contribution of manufacturing industry (secondary industry) and the service industry (tertiary industry) to the economy is greater than that of primary econome, e.g. of agriculture.
r индустриальная страна; промышленно развитая страна
Страна, в которой вклад машинного производства и сферы услуг (вторичный и третичный сектора экономики) в экономику гораздо больше, чем первичный сектор экономики, например, сельское хозяйство.
f pays *m* industriel
s país *m* industrial
g industrielles Land *n*

3077 **industry**
Manufacturing activity as a whole, the largest part of material production divided into the manufacture of means of production and production of consumer goods.
r промышленность; индустрия
Крупнейшая часть материального производства, подразделяющаяся на производство средств производства и производство предметов потребления.
f industrie *f*
s industria *f*
g Industrie *f*

3078 **industry of construction materials; construction materials industry**
A processing industry that produces materials and products for all kinds of construction.
r промышленность строительных материалов
Обрабатывающая промышленность, производящая материалы и изделия для всех видов строительства.
f industrie *f* des matériaux de construction
s industria *f* de materiales de construcción
g Baumaterialindustrie *f*

3079 **inertial current**
Residual water movement in oceans going on after all factors which produce the motion, stop their action.
r инерционное течение
Остаточное движение вод, наблюдающееся после прекращения действия всех возбуждающих движение сил.
f courant *m* de inertia
s corriente *f* de inercia
g Trägheitsströmung *f*

3080 **infanticide**
The act of killing a baby which was practiced in order to eliminate deformed human beings or to regulate the sex proportion of

population, usually it has been faminine infanticide.

r инфантицид; детоубийство
Акт убийства новорожденного, встречающийся в некоторых странах, с целью регулирования соотношения полов в обществе или сохранения единственного наследника.

f infanticide *m*
s infanticidio *m*
g Kindesmord *m*

3081 **infauna**
Aquatic animals living deep in bottom sediment of water bodies, forming part of benthos.
r инфауна
Водные животные, обитающие глубоко в донных отложениях водоемов, составная часть бентоса.
f infaune *f*
s infauna *f*
g Infauna *f*

3082 **infield-outfield farming**
A system of farming which confines intensive cultivation, manuring etc. to enclosed fields near the farm (infield), the outer parts of the farm (outfield) being used for grazing or periodically cropped on a shifting cultivation basis.
r близлежащее земледелие
Система хозяйствования, при которой в основном интенсивно обрабатываются земли, находящиеся недалеко от фермы, в то время как более далекие участки используются под пастбища или для нерегулярного получения урожая.
f proche agriculture *f*
s labranza *f* próxima
g naheliegender Feldbau *m*

3083 **infiltration**
The absorption of atmospheric and surface water by soils and rocks and its downward seeping through capillaries, pores and interstices towards ground waters.
r инфильтрация
Просачивание атмосферной и поверхностной воды в почву, горные породы и движение ее по капиллярам, порам и пустотам к уровню грунтовых вод.
f infiltration *f*; suintement *m*
s infiltración *f*
g Infiltration *f*

3084 **infiltration and recrystallization ice**
Ice formed in accumulation areas of glaciers as a result of infiltrative compaction and recrystallization.
r инфильтрационно-рекристаллизационный лед
Лед, формирующийся в областях питания ледников в результате инфильтрационного уплотнения и перекристаллизации.
f glace *f* d'infiltration et recrystallisation
s hielo *m* de infiltración y recristalización
g Infiltration-Rekristallisation Eis *n*

* **infiltration capacity → 3086**

3085 **infiltration ice**
Ice formed as a result of melt water or rain water entering into a snow or firn cover with its further freezing.
r инфильтрационный лед
Лед, формирующийся в результате просачивания в снег и фирн талой или дождевой воды с последующим ее замерзанием.
f glace *f* d'infiltration
s hielo *m* de infiltración
g Infiltrationseis *n*

3086 **infiltration rate; infiltration capacity**
The maximum rate at which water can enter the porous substance, such as soil or snow, under specified conditions.
r скорость инфильтрации
Максимальная скорость проникновения воды в пористое вещество (почву, снег) при определенных условиях.
f vitesse *f* d'infiltration
s velocidad *f* de infiltración
g Infiltrationgeschwindigkeit *f*

* **infiltration zone → 5490**

3087 **inflow of ice**
Increase of the glacier mass in the ablation area owing to the ice income from the accumulation area.
r привнос льда
Повышение поверхности ледника в области абляции за счет притока льда из области аккумуляции.
f apport *m* de glace
s ingreso *m* de hielo
g Eiszufluß *m*

* **influent → 6347**

* **influent stream → 1614**

3088 **influx**
Inflow of surface waters through fissures, canals and funnels into a cave of karsting rocks.
r инфлюация; исчезновение реки
Втекание поверхностных вод через трещины, каналы и воронки в карстующиеся породы.
f transsudation *f*; rivière *f* perte
s río *m* perdido
g Flußschwinden *n*; Flußversickerung *f*

* **informal economy → 716**

3089 **information flow**
The amount of information sent along a line of communication network during a certain period of time.
r информационный поток
Объем информации, передаваемой по

коммуникационной сети в течение определенного времени.
f flux *m* d'information
s flujo *m* de información *f*
g Informationfluß *m*

* **infralittoral zone → 4100**

3090 infrared photography
A technique used in the remote sensing to record reflected visible and infrared radiation with wavelengths of up to 0.9 micron.
r инфракрасное фотографирование
Метод дистанционного зондирования, при котором съемкой воспроизводится видимое и инфракрасное излучение с длиной волны до 0,9 микрон.
f photographie *f* infra-rouge
s fotografía *f* infraroja
g infrarote Photographie *f*

3091 infrared radiation
Electromagnetic waves in the wavelength range of 0.7 to about 200 microns, invisible, perceived as heat.
r инфракрасная радиация
Невидимое электромагнитное излучение в интервале длин волн приблизительно от 0,7 до 200 микон, ощущаемое как тепло.
f rayonnement *m* infrarouge
s radiación *f* infrarroja
g Infrarotstrahlung *f*; infrarote Strahlung *f*

* **infrared sensing → 3092**

3092 infrared thermography; infrared sensing; thermal sensing
A technique of remote sensing which is concerned with the measurement of the Earth's surface radiant temperature, recorded from the aircraft or a space satellite using the thermal infrared band with wavelengths of 3-20 microns.
r инфракрасная термография; термозондирование
Метод дистанционного зондирования земной поверхности, при котором измеряется ее радиационная температура с борта самолета или искусственного спутника Земли, при этом используется электромагнитное излучение с длиной волны 3-20 микрон.
f thermographie *f* infra-rouge
s termografía *f* infraroja
g infrarote Thermographie *f*

3093 infrastructure
The system of constructions and services which make possible economic development and everyday life of people, e.g. the power and water supplies, the transport and communication facilities, the sewage system, etc.
r инфраструктура
Совокупность сооружений, систем и служб, необходимых для материального производства и обеспечения жизни населения, например, электро- и водоснабжение, транспортная и коммуникационная системы и пр.
f infrastructure *f*
s infraestructura *f*
g Infrastruktur *f*

* **infrenched meander → 3052**

3094 ingression
A gradual intrusion of sea or lake water into coastal depressions as a result of sea level rise or the shore line subsidence.
r ингрессия
Постепенное проникновение морских или озерных вод в понижения рельефа прибрежной суши при повышении уровня водоема или опускании береговой линии.
f ingression *f*
s ingresión *f*
g Ingression *f*

3095 ingression coast
A sea or lake shore with an irregular shoreline which formes as a result of ingression.
r ингрессионный берег
Морской или озерный берег с изрезанной береговой линией, образовавшийся в результате ингрессии.
f côte *f* d'ingression
s costa *f* de ingresión
g Ingressionsküste *f*

3096 ingression sea
A sea resulted from flooding of depressions of a continental plain.
r ингрессионное море
Море, образующееся при затоплении опустившихся частей материковой равнины.
f mer *f* d'ingression
s mar *m* de ingresión
g Ingressionsmeer *n*

3097 ingression strait
A strait formed as a result of the sea ingression into coastal depressions after the sinking of the coast.
r ингрессионный пролив
Пролив, образовавшийся в результате ингрессии моря в понижения рельефа при погружении прибрежной суши.
f détroit *m* d'ingression
s estrecho *m* de ingresión
g Ingressionsstraße *f*

* **ingrown meander → 3052**

* **inherited features → 3098**

3098 initial landform; inherited features
The present relief which retains the major features of the past landforms.
r унаследованный рельеф
Современный рельеф, сохраняющий в целом основные особенности предшествующего рельефа.

f relief *m* hérité; héritage *f* géomorphologique; paléomorphologie *f*; formes *fpl* héritées
s relieve *m* heredado; paleomorfología *f*; formas *fpl* de modelado heredadas
g vererbtes Relief *n*

* **initial meridian → 4767**

3099 initial plain
A low plain formed after the regression of the epicontinental sea and composed of young unmodified marine sediments.
r первичная равнина
Невысокая равнинная суша, образовавшаяся вследствие регрессии эпиконтинентального моря и сложенная молодыми недислоцированными морскими осадками.
f plaine *f* initiale
s llanura *f* inicial
g ursprüngliche Ebene *f*

3100 inland ice; continental glacier; land ice-sheet
A large, continuous layer of land ice of considerable thickness, a continuous system of ice domes, ice streams, outlet glaciers and ice shelves covering extensive areas of land and continental shelves.
r покровное оледенение; континентальный ледник; материковый ледник
Обширный, непрерывный покров наземного льда значительной толщины, система ледниковых куполов, ледяных потоков, выводных ледников, шельфовых ледников, погребающих сушу и континентальные шельфы.
f inlandsis *m*; glacier *m* continental
s inlandsis *m*; glaciar *m* continental
g Inlandeis *n*; Kontinentaleisdecke *f*

3101 inland sea
A large expanse of saline water disposed inside the continent without a link to the open sea, e.g. Kaspian Sea or Aral Sea.
r закрытое море
Крупный водоем с соленой водой (фактически озеро), лежащий внутри материка и не связанный с Мировым океаном, например, Каспийское или Аральское моря.
f mer *f* fermée
s mar *m* cerrado
g Binnenmeer *n*

3102 inlet
A short narrow opening or waterway which connects smaller body of water with a large body.
r узкий залив
Короткое и сравнительно мелкое водное пространство, соединяющее небольшую часть водоема с его основной частью.
f anse *f*
s cala *f*; ensenada *f*
g Bai *f*

3103 inner city
An area at or near the city centre, an important zone characterized by aged, run-down housing occupied by people with low incomes.
r внутренние кварталы
Территория вблизи городского центра или в самом центре города, характеризующаяся старыми обветшавшими зданиями, где живут люди с небольшими доходами.
f aire *f* intra-urbaine
s ciudad *f* interior
g innere Stadtviertel *f*

3104 inner core
The probably solid central part of the Earth's core, below a depth of 5,000 km.
r внутреннее ядро
По всей вероятности твердая часть земного ядра, расположенная ниже 5000 км.
f noyau *m* intérieur
s núcleo *m* interno
g innerer Kern *m*

3105 Inner Crescent
In the geopolitical theory of heartland an inner belt including Western Europe, Turkey, India and China that protects the *heartland* from naval expansion.
r внутренний полумесяц
В геополитической теории хартленда внутренний пояс, включающий Западную Европу, Турцию, Индию и Китай, который служит защитой собственно *хартленда* от военно-морской экспансии.
f croissant *m* intérieur
s creciente *m* interior
g Innere Halbmond *m*; Inner Crescent *n*

3106 inner margin
The inner margin of a terrace surface adjoining the rock slope above, or the scarp of the terrace above.
r тыловой шов
Край поверхности террасы, граничащий с вышележащим коренным склоном или уступом более высокой террасы.
f bout *m* de terrasse
s canto *m* de terraza
g Terrassenkante *f*

3107 innovation
1. The introduction of practices, processes, features, etc. which are new in a practical context. 2. Technological novelties in industry and agriculture that considerably increase productivity and contribute to scientific and technological progress.
r инновации; нововведение
1. Нечто новое, вновь придуманное, получившее широкое распространение в повседневной жизни. 2. Технологические нововведения в промышленности и сельском хозяйстве, существенно повышающие

производительность труда и способствующие научно-техническому прогрессу.
f innovation *f*
s innovación *f*
g Neurung *f*; Innovationen *fpl*

3108 input
The resources, such as land, labour, or raw materials, needed to get output.
r вложение продукции
Исходные ресурсы, такие как земля, труд, сырье, необходимые для получения продукции.
f input *m*; entrée *f*
s input *m*; entrada *f*
g Input *m*

3109 input-output model balance; interindustry balance
The balance between production and distribution of social product between sectors which is used as the basis for analysing and planning in extended reproduction at sector level.
r межотраслевой баланс
Баланс производства и распределения общественного продукта по отраслям, служащий методом анализа и планирования пропорций при расширенном воспроизводстве в отраслевом разрезе.
f bilan *m* entrée-sortie; modèle *m* d'entrée-sortie
s balance *m* insumo-producto
g Verflechtungsbilanz *f*; input-output Bilanz *f*

3110 inselberg; bornhardt
A steep-sided, isolated residual hill or upland rising from a plain which is, in many cases, monotonously flat.
r островная гора
Изолированные холмы или возвышенности, расположенные поодиночке или группами среди обширной равнины.
f inselberg *m*
s monte-isla *m*; inselberg *m*
g Inselberg *m*

3111 insequent valley
A river valley developed in horizontally bedded rocks with homogeneous lithology and is not influenced by the tectonic structure of the area.
r инсеквентная долина
Речная долина, заложенная в литологически однородных горизонтально залегающих породах, ее направление почти не зависит от тектонической структуры территории.
f vallée *f* inséquente
s valle *m* insecuente
g insequentes Tal *n*

3112 insert
An additional map within the margins of the main map which explains or supplements the latter.
r врезка
Дополнительная карта, помещенная в одной рамке с основной картой, необходимая для пояснения или дополнения содержания карты.
f carton *m*; carte *f* complémentaire
s mapa *m* complementario
g Nebenkarte *f*

3113 inset terrace
An aggradation river terrace with the alluvium layer deposited onto the alluvium of an older terrace.
r вложенная терраса
Аккумулятивная речная терраса, толща наносов которой как бы вложена в аллювий более древних террас.
f terrasse *f* emboîtée
s terraza *f* encajada
g eingelegte Terrasse *f*

3114 in-shore station
A point in the littoral zone of a sea where oceanographic observations are regularly carried out.
r рейдовая станция
Точка в прибрежной зоне моря, в которой регулярно выполняются океанографические работы.
f station *f* maritime
s óstación *f* marítima
g Wattstation *f*

3115 insolation
Solar radiation received on the Earth's surface, i.e. the amount of direct solar radiation reaching a horizontal area at a given level. → Fig. 167
r инсоляция
Прямая солнечная радиация, достигающая поверхности Земли, измеряется количеством энергии, падающей на единицу горизонтальной поверхности на данном высотном уровне. → Рис. 167
f insolation *f*
s insolación *f*
g Insolation *f*; Sonneneinstrahlung *f*

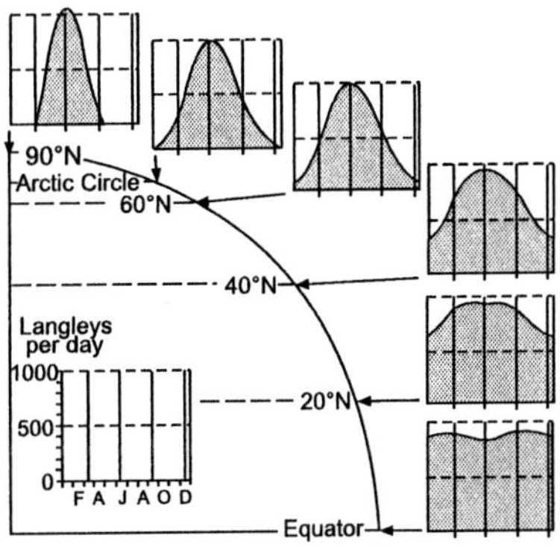

Fig. 167. **Insolation**

3116 insolation avalanche
An avalanche arising due to the snow-melting caused by the solar radiation in spring which is more intense in the slopes of south aspect (in the Northern Hemisphere).
r инсоляционная лавина
Лавина, возникающая в результате таяния снега весной на склонах южной экспозиции (в Северном полушарии).
f avalanche *f* de printemps
s avalancha *f* de primavera
g Sonnenstrahlungslawine *f*

3117 insolation duration; sunshine duration
A time interval during which solar radiation is intensive enough to cast distinct shadows.
r продолжительность солнечного сияния
Время, в течение которого солнечная радиация достаточно интенсивна для создания теней.
f durée *f* d'insolation
s duración *f* de la insolación
g Sonnenscheindauer *f*; Insolationsdauer *f*

3118 insolation weathering; thermal expansion; thermoclasty
The destruction of rocks as a result of their cracking in response to daily temperature changes caused by the differences of solar radiation.
r инсоляционное выветривание
Разрушение горных пород в результате их растрескивания при резких колебаниях температуры, особенно при нагревании солнцем.
f thermoclastisme *m*
s termoclastia *f*
g Insolationsverwitterung *f*

3119 instable humidity
Natural conditions, intermediate between the excessive and deficient humidity, when relative equality of the average annual potential evaporation and precipitation is observed.
r неустойчивое увлажнение
Природные условия, промежуточные между избыточным и недостаточным увлажнением, когда наблюдается относительное равенство средних годовых величин испаряемости и осадков.
f humidité *f* inconstante
s humedad *f* inconstante
g instabile Feuchtigkeit *f*

3120 instantaneous water level
Water level registered simultaneously in a number of points along a river or on the shores of a water body.
r мгновенный уровень воды
Уровень воды, фиксируемый одновременно в ряде пунктов по длине реки или по берегам водоема.
f niveau *m* d'eau instantané
s nivel *m* de agua instantaneo
g momentauer Wasserstand *m*

3121 instructional map
A special geographical map showing a certain generalisation of phenomena presented in accordance with the educational program.
r учебная карта
Специальная географическая карта, содержание которой основано на генерализации изображаемых явлений в соответствии с учебной программой.
f carte *f* didactique
s mapa *m* didáctico; mapa *m* de enseñanza
g didaktische Karte *f*

3122 instrument-making industry
A branch of industry that produces instruments, small mechanisms, clocks and watches.
r приборостроение
Отрасль промышленности, занимающаяся изготовлением приборов, небольших механизмов и часов.
f construction *f* d'appareils de précision
s construcción *f* de aparatos de precisión; industria *f* mecánica de précision
g Gerätebau *m*; Apparátebau *m*

* **insufficient moistering → 1469**

3123 insular *adj*
Relating to an island: inhabiting, situated on, forming, characteristic of.
r островной
Относящийся к острову: расположенный, произрастающий, обитающий на нем и т.п.
f insulaire *adj*
s insular *adj*
g Insel-

3124 insular climate
A climatic regime with little range of temperature between summer and winter, characteristic of many islands and some sheltered coastal areas.
r островной климат
Климатический режим с незначительной разницей температуры зимой и летом, характерный для многих островов и некоторых защищенных прибрежных районов.
f climat *m* insulaire
s clima *m* insular
g Inselklima *n*

3125 integral resources
The sum total of all types of natural resources together with material and labor resources.
r интегральные ресурсы
Совокупность всех частных видов природных ресурсов в сочетании с материальными и трудовыми ресурсами общества.
f ressource *f* intégral
s recursos *mpl* integrales
g Gesamtresourcen *fpl*

3126 integrated plant

A group of industrial enterprises from different sectors of economy where the products of one of them are used as raw materials or semi-finished products by others.

r комбинат
Объединение промышленных предприятий разных отраслей, в котором продукция одного предприятия служит сырьем или полуфабрикатом для другого.

f combinat *m*
s combinado *m*
g Kombinat *n*

3127 integration of landscapes
The process of landscape development when some differences in the structure and functioning of landscapes become less distinct owing to natural or anthropogenic factors.

r интеграция ландшафтов
Процесс развития ландшафтов, при котором под влиянием природных или антропогенных факторов стираются некоторые различия в структуре и характере функционирования ландшафтов.

f intégration *f* de paysages
s integración *f* de paisajes
g Landschaftsintegration *f*

3128 intensification of production
Production growth based on more efficient and rational use of technological, material and labour resources made possible due to scientific progress.

r интенсификация производства
Рост производства, основанный на более полном, эффективном и рациональном использовании технических, материальных и трудовых ресурсов на базе научно-технического прогресса.

f intensification *f* de production
s intensificación *f* de producción
g Intensivierung *f* der Produktion

3129 intensity of earthquake
A strentgh of the manifestation of an earthquake on the Earth's surface expressed by the international scale based on the degree of building destructions, residual ground deformations, etc.

r интенсивность землетрясения
Сила проявления землетрясений на поверхности Земли, измеряемая по 12-балльной международной шкале, основанной на оценке степени повреждения зданий, остаточных деформаций грунта и т.п.

f intensité *f* de séisme
s intensidad *f* de seísmo
g Erdbebenintensität *f*

3130 intensity of rainfall; rainfall intensity
A measure of the rate at which rain or snow falls, expressed in terms of rainfall or snowfall amount related to a specific time period.

r интенсивность осадков;
интенсивность дождя
Мера скорости, с которой выпадает дождь или снег, выраженное как отношение количества выпавших осадков к продолжительности их выпадения.

f intensité *f* de précipitations;
intensité *f* de pluie
s intensidad *f* de precipitaciones;
intensidad *f* de lluvia
g Intensität *f* der Niederschläge;
Regenintensität *f*

3131 intensive agriculture; intensive farming
A farming system characterized by high yields due to better means and methods of production.

r интенсивное сельское хозяйство
Система хозяйствования, при которой получают высокие урожаи благодаря применению более совершенных средств и методов производства.

f agriculture *f* intense
s agricultura *f* intensiva
g intensive Landwirtschaft *f*

3132 intensive economy
An economic system based on strenuous and highly productive labour.

r интенсивное хозяйство
Система хозяйствования, основанная на напряженном, высокопроизводительном труде.

f économie *f* intensive
s economía *f* intensiva
g intensive Wirtschaft *f*

* **intensive farming → 3131**

3133 interactive mapping
Compilation and editing of maps and atlases in the regime of dialogue interaction of a cartographer and computer.

r интерактивное картографирование
Составление и редактирование карт и атласов в режиме диалогового взаимодействия картографа и компьютера.

f cartographie *f* interactive
s cartografía *f* interactiva
g interaktive Kartierung *f*

3134 interbedded ***adj***
Applied to a layer of rock deposited in sequence between two other beds.

r переслоенный
Применяется к слою, залегающему в промежутке между двумя другими слоями.

f interstratifié *adj*
s interstratificado *adj*
g verblätterig *adj*

3135 interbedded layer
A thin rock layer lying between thicker layers of a different composition.

r прослой
Тонкий слой горных пород между более мощными слоями иного состава.

f banc *m* intercalé
s capa *f* intermedia
g Zwischenlage *f*; Ader *f*; Zwischenschicht *f*

3136 interception;
interception of precipitation
Process by which precipitation is caught and retained on vegetation or buildings and subsequently evaporate without reaching the ground.
r перехват осадков
Задержание части выпадающих осадков растительностью в лесу, а также строениями в городе, эти осадки отчасти испаряются, не достигнув почвы.
f interception *f* de précipitation
s intercepción *f* de precipitación
g Auffang *m* von Niederschläge

* **interception of precipitation → 3136**

3137 intercontinental tourism
Travel to other continents for pleasure and interest.
r межконтинентальный туризм
Туристские поездки на другой континент.
f tourisme *m* intercontinental
s turismo *m* intercontinental
g interkontinentaler Tourismus *m*

3138 inter-diurnal variation
Arithmetic difference between the mean diurnal values of meteorological element on two consecutive days.
r междусуточные колебания
Арифметическая разница средних суточных значений метеорологического элемента за двое последовательных суток.
f variation *f* interdiurne
s variación *f* interdiurna
g Tag-zu-Tag-Veränderung *f*

* **inter-flow → 6195**

3139 interfluve
The land between the valleys of adjacent rivers.
r междуречье
Пространство между долинами смежных рек.
f interfluve *m*
s interfluvio *m*
g Zwischenstromland *n*

3140 interglacial
A period of time between two glacials (glacial epochs) of the Quaternary and characterised by the warming of climate up to the contemporary temperatures or even warmer when the ice sheets retreated from mid latitudes (the terminological correlation of interglacials in different territories is shown in the Table 2, p. 350-351).
r межледниковье; интергляциал; межледниковая эпоха
Промежуток времени, разделявший две ледниковые эпохи четвертичного периода, характеризовавшийся потеплением климата, близким к современному или теплее, и освобождением от ледниковых покровов умеренных широт (терминологическую корреляцию межледниковий на разных территориях см. в табл. 2, с. 350-351).
f interglaciaire *m*
s época *f* interglaciar; interglaciar *m*
g Interglazialzeit *f*; Zwischeneiszeit *f*

3141 interglacial ***adj***
Relating to the period dividing two succeeding ice ages in the Pleistocene.
r межледниковый
Относящийя ко времени, разделяющем две последовательных ледниковых эпохи в плейстоцене.
f interglaciaire *adj*
s interglacial *adj*
g interglazial *adj*

* **interindustry balance → 3109**

3142 interior waters
All water objects found within the land area of a given state and within its territorial waters.
r внутренние воды
Все водные объекты, находящиеся в пределах сухопутной территории данного государства и в пределах его территориальных вод.
f eaux *fpl* intérieures
s aguas *fpl* interiores
g Binnengewässer *npl*

3143 inter-island sea
A part of ocean surrounded by more or less close ring of islands, the reefs between which prevent a free water exchange between this sea and the ocean, e.g. seas of Malay Archipelago.
r межостровное море
Часть океана, окруженная более или менее тесным кольцом островов, пороги между которыми препятствуют свободному водообмену этого моря с открытой частью океана, например, моря Малайского архипелага.
f mer *f* interinsulaire
s mar *m* interinsular
g interinsulares Meer *n*

3144 interland sea; mediterranean sea
A sea lying between continents and connected with ocean by one or several straits, e.g. the Mediterranean Sea and the Red Sea.
r межматериковое море;
средиземное море
Море, расположенное между материками и связанное с океаном одним или несколькими проливами, например, Средиземное и Красное моря.
f mer *f* intercontinentale
s mar *m* intercontinental
g Mittelmeer *n*

3145 intermediate contour

A contour line on a map that occurs between index contours.

r полугоризонталь
Горизонталь, проведенная через половину принятой высоты сечения рельефа.
f courbe *f* de niveau intercalaire
s curva *f* de nivel intercalada
g Zwischenhöhenlinie *f*; Halbhöhenlinie *f*

* **intermediate moraine → 3755**

3146 intermediate product
The article produced to be used secondly in the manufacturing process.
r промежуточный продукт
Продукция, предназначенная для вторичного использования в материальном производстве.
f bien *m* intermédiaire
s producto *m* intermedio; productos *mpl* semielaborados
g Zwischenproduktion *f*

3147 intermittent saturation zone
A layer of soil, lying below the surface, which may hold *vadose water* in a period of prolonged rainfall, but which quickly dries out in even a short-lived drought.
r периодически пропитываемая зона
Подповерхностный почвенный слой, наполняемый *вадозными водами* в период продолжительных дождей, но быстро высыхающий даже при короткой засухе.
f zone *f* de saturation intermittente
s zona *f* de saturación intermitente
g periodisch imprägnierte Zone *f*

3148 intermittent spring
A spring which does not flow continuously but dries up from time to time, usually depending on the height of the water table.
r перемежающийся родник
Родник, действующий не в течение всего года, а временами истощающийся.
f source *f* intermittent
s manantial *m* intermitente; fuente *f* intermitente
g intermittierende Quelle *f*

3149 intermittent stream; ephemeral stream
A stream which does not flow continuously but dries up from time to time, it usually flows after a period of heavy rainfall and as a result of snowmelt.
r временный водоток; пересыхающая река
Водоток, в котором движение воды наблюдается время от времени, обычно после сильных дождей и во время снеготаяния.
f cours *m* d'eau temporaire; cours *m* d'eau intermittent
s corriente *f* fluvial temporal; río *m* efímero; corriente *f* fluvial intermitente
g intermittierender Wasserlauf *m*; periodischer Wasserlauf *m*

3150 intermodal transportation
The transportation of goods from the place of shipment (loading) to the place of destination with the use of two or more modes of transport.
r интермодальные перевозки
Перевозки от места отправления (отгрузки) до места назначения, выполняемые двумя или более сопряженными видами транспорта.
f transport *m* intermodal
s transporte *m* intermodal
g intermodale Beförderung *f*

3151 intermontane *adj*
Lying between mountains or mountain ranges, e.g. the high plateaus, lying between the east and west ranges of Andes.
r межгорный
Лежащий между горами или горными хребтами, например, высокие плато, расположенные между западными и восточными цепями Анд.
f intermontagnarde *adj*
s intermontaña *adj*
g zwischengebirgig *adj*

3152 intermontane trough
Tectonic depression between mountain ridges formed as a result of the flexure of the Earth's crust simultaneously with the elevation of the surrounding ridges.
r межгорный прогиб; внутригорная впадина; межгорная впадина
Тектоническая депрессия между горными хребтами, образованная в результате прогибания земной коры одновременно с подъемом окружающих хребтов.
f intra-fosse *f*; dépression *f* intermontagnarde
s intra-fosa *f*; depresión *f* intermontaña
g Zwischensenke *f*

3153 inter-monthly variation
Arithmetic difference between the mean monthly values of meteorological element on two consecutive months.
r междумесячные колебания
Арифметическая разница средних месячных значений метеорологического элемента за два последовательных месяца.
f variation *f* intermensuelle
s variación *f* intermensual
g Monat-zu-Monat-Veränderung *f*

3154 internal accumulation of glacier
A repeated freezing of melt water inside the body of firn and ice.
r внутреннее питание ледника
Повторное замерзание талой воды в толще фирна и льда.
f accumulation *f* interne du glacier
s accumulación *f* interna de glaciar
g innere Akkumulation *f* des Gletschers

3155 internal migration
Movement of population inside the boundaries of a country, e.g. in search of employment.
r внутренняя миграция
Перемещение населения в пределах границ своей страны, например, в поисках работы.
f migration *f* intérieure
s migración *f* interior
g Innermigration *f*

3156 internal trade; domestic trade; home trade
The trade on the market of the given country, as opposed to *international trade.*
r внутренняя торговля
Реализация товарной продукции на внутреннем рынке страны, в противоположность *внешней торговле*
f commerce *m* intérieur
s comercio *m* interior
g Binnenhandel *m*

3157 internal water circulation
Hydrological cycle limited to a definite continental surface: water vapour evaporated from this surface condenses in the form of precipitation within its boundaries.
r внутренний влагооборот
Влагооборот, ограниченный определенным географическим регионом: водяной пар, испаряющийся с поверхности, конденсируется и выпадает в пределах этого региона.
f cycle *m* hydrologique interne
s ciclo *m* hidrológico interno
g innere Wasserzirkulation *f*

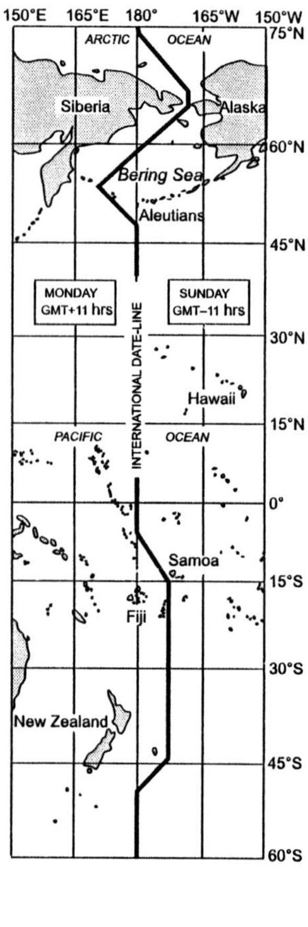

Fig. 168. **International date line**

3158 internal waves; boundary waves
The waves appearing within a water mass at the surface of a divide between water layers of different density while one layer shifts with regard to the other layer.
r внутренние волны
Волны, возникающие внутри водной массы на поверхности раздела слоев воды разной плотности при перемещении одного слоя относительно другого.
f ondes *fpl* intérieures
s olas *fpl* interiores
g innere Wellen *fpl*

3159 international date line
An imaginery line agreed internationally which follows the meridian of 180° and delimits the places having calendar dates differing in one day at the same time. → Fig. 168
r линия перемены даты
Условная линия на земной поверхности для разграничения мест, имеющих в один и тот же момент времени календарные даты, различающиеся на одни сутки. → Рис. 168
f ligne *f* de changement de date
s línea *f* de cambio de fecha
g Datumgrenze *f*

3160 international division of labour
Territorial division of labour between individual countries that exchange products and services.
r международное разделение труда
Территориальное разделение труда между отдельными странами с обменом между ними продукцией и услугами.
f division *f* internationale du travail
s división *f* international del trabajo
g internationale Arbeitsteilung *f*

3161 international monopoly
A large corporation or an alliance of corporations of different countries with assets abroad that dominate in one or several spheres of the world economy.
r международная монополия
Крупнейшая корпорация или союз фирм разной национальной принадлежности с активами за рубежом, устанавливающие господство в одной или нескольких сферах мирового хозяйства.
f monopole *m* international
s monopolio *m* internacional
g internationale Korporation *f*; transnationale Korporation *f*

* **international nautical mile → 4072**

3162 **international tourism**
Travel to any country by residents of a different country with touristic purposes, one of the most dynamic and profitable branches of modern economy.
r международный туризм
Посещение любого государства гражданами других стран с туристскими целями, один из наиболее динамичных и высокодоходных секторов мирового хозяйства.
f tourisme *m* international
s turismo *m* internacional
g internationaler Tourismus *m*

3163 **international trade; overseas trade; external trade; foreign trade**
The excange of products and services of a country with other countries (there are export and import trade).
r внешняя торговля
Торговля одной страны с другими странами, составная часть внешнеэкономических связей (подразделяется на экспорт и импорт).
f commerce *m* extérieur;
commerce *m* international
s comercio *m* exterior
g Außenhandel *m*

3164 **International Trusteeship**
System set up by the UN to rule the former mandate territories of the League of Nations as well as the territories detached from Germany and Italy after World War II.
r международная опека
Созданная ООН система управления бывшими мандатными территориями Лиги Наций и территориями, отторгнутыми от Германии и Италии после второй мировой войны.
f tutelle *f* internationale
s tutoría *f* internacional
g internationales Treuhändersystem *n*

3165 **internationalization of production**
The establishment of closer ties between enterprises of different countries resulting in that the industry of one country becomes a part of the world industrial process.
r интернационализация производства
Установление более тесных связей между предприятиями разных стран, в результате чего производство одной страны становится частью мирового производственного процесса.
f internationalisation *f* de production
s internacionalización *f* de producción
g Internationalisation *f* der Produktion

3166 **internet-GIS**
Geoinformation system using the Internet to transfer data, access to remote databases and program modules with the purpose of carrying out analyses, making decisions and presentation of electronic maps.
r интернет-ГИС
Геоинформационная система, использующая интернет для передачи данных, доступа к удаленным базам данных и программным модулям с целью проведения анализа, принятия решений и презентаций электронных карт.
f internet-SIG *f*
s internet-SIG *f*
g Internet-GIS *n*

* **interpretation of aerial photographs → 3033**

3167 **interregional balance**
The balance between production, consumption and spatial redistribution of social product between economic regions which is used to analyse and plan the extended reproduction at regional level.
r межрайонный баланс
Баланс производства, потребления и территориального перераспределения продукции по экономическим районам для анализа и планирования расширенного воспроизводства по территориям.
f bilan *m* interrégional
s balance *m* interregional
g interregionale Bilanz *f*

3168 **interrupted stream; discontinuous stream**
River with alternating parts of surface flow and underflow.
r прерывистая река
Река, у которой чередуются участки поверхностного и подруслового течения.
f cours *m* d'eau discontinu
s curso *m* de agua interrumpido;
curso *m* de agua discontínuo
g teilweise unterirdischer Fluß *m*

3169 **interstadial**
A period of the slight warming of climate and decrease of glaciers between two stages of their advance during one and the same glacial epoch in the Quaternary.
r межстадиал; интерстадиал
Время слабого потепления климата и сокращения ледников между двумя стадиями их наступания в течение одной и той же ледниковой эпохи в четвертичном периоде.
f interstade *m*
s interestadio *m*
g Interstadial *n*

3170 **interstitial ice**
Ice formed in the cracks of permafrost and frozen ground of any origin.
r трещинный лед
Лед, образовавшийся в любых по генезису трещинах многолетнемерзлых пород и мерзлого грунта
f glace *f* des fissures
s hielo *m* de fisuras

g Spalteneis *n*

3171 intertropical convergence zone
A narrow global belt near the equator where the trade winds of two hemispheres meet (symbol: ITCZ).
r внутритропическая зона конвергенции
Узкая зона вблизи экватора, где встречаются пассаты двух полушарий.
f zone *f* intertropicale de convergence
s zona *f* intertropical de convergencia
g innertropische Konvergenzzone *f*

3172 intrapermafrost water
Ground water occurring or moving within the permafrost or between the layers of permanently frozen rocks.
r межмерзлотные воды; внутримерзлотные воды
Подземные воды, залегающие или перемещающиеся внутри толщи или между слоями многолетнемерзлых пород.
f eau *f* intrapergélisol
s agua *m* de intrapermafrost
g Intrapermafrostwasser *n*

3173 intraregional tourism
Travel to neighbouring countries within one and the same continent for pleasure and interest.
r внутрирегиональный туризм
Туристские поездки в близлежащие страны в пределах одного континента.
f tourisme *m* régional
s turismo *m* regional
g regionaler Tourismus *m*

3174 intrasecular fluctuations of climate
The recurring changes of climate connected with the fluctuations of solar activity, with the rhythm of 11, 35 and 80-90 years.
r внутривековые колебания климата
Периодические изменения климата, связанные скорее всего с колебаниями солнечной активности, с ритмом в 11, 35 и 80-90 лет.
f oscilations *fpl* intraséculaires de climat
s oscilaciónes *fpl* intraseculares de clima
g innersäkulare Klimaschwankungen *fpl*

3175 intrasoil weathering
The destruction, change and formation of minerals in soil which result in the changes of chemical and mineralogical composition of soils as well as crystal-chemical transformations of minerals in the soil cover.
r внутрипочвенное выветривание
Процессы разрушения, изменения и образования минералов, происходящие в почве, включая изменения их химического и минералогического состава, а также кристаллохимические преобразования минералов в почвенном покрове.
f altération *f* du intrasol
s alteración *f* intrasuelo
g interne Bodenverwitterung *f*

3176 intrazonal *adj*
Applied to soils, vegetation and landscapes in the form of inclusions within a geographic zone, e.g. raised bogs in taiga, solonchaks in desert.
r интразональный
Относящийся к участкам почв, растительности и ландшафтов, образующих вкрапления внутри одной зоны, например, верховые болота в тайге, солончаки в пустыне.
f intrazonal *adj*
s intrazonal *adj*
g intrazonal *adj*

3177 intrazonal soils
Soils, the characteristics of which influenced by local factors of parent rocks, relief, salinization of ground waters, etc. rather than climate and vegetation, e.g. solonchaks and solonetz, bog soils, etc.
r интразональные почвы
Почвы, встречающиеся среди зональных почв и отличающиеся от них благодаря действию местных факторов: особых свойств материанской породы, рельефа, засоленности грунтовых вод и пр., к ним относятся солончаки и солонцы, болотные почвы и др.
f sols *mpl* intrazonaux
s suelos *mpl* intrazonales
g intrazonale Böden *mpl*

* **introduced species → 2054**

3178 introduction of species
The process of bringing living organisms accidentally or intentionally by people into a new environment.
r интродукция видов
Переселение человеком непреднамеренно или специально отдельных видов животных и растений за пределы естественнного ареала в места, где они раньше не жили.
f introduction *f* des espèces
s introducción *f* de especies
g Introduktion *f* der Arten

3179 intrusion
The forceful entry of a mass of molten rock (magma) in the pre-existing rock of the Earth's crust as well as a geological body formed after its solidification.
r интрузия
Процесс внедрения магмы в толщу земной коры, а также геологическое тело, образующееся при ее застывании.
f intrusion *f*
s intrusión *f*; intrusiva *f*
g Intrusion *f*

3180 intrusive ice
Ice formed from freezing of water injected into rocks under pressure and spread along the impermeable frozen layers.
r инъекционный лед; интрузивный лед
Лед, возникший в результате замерзания воды, внедрившейся под напором вдоль

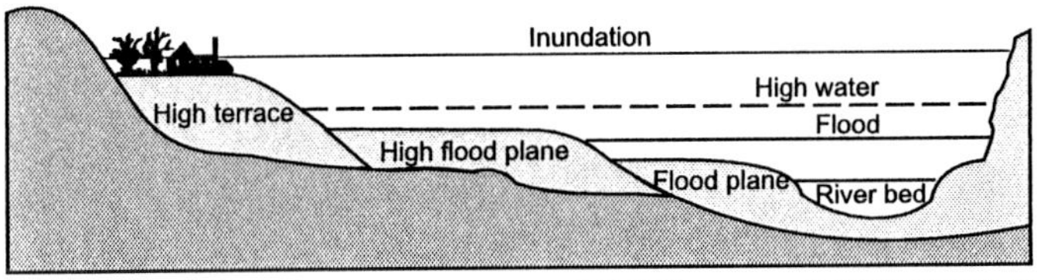

Fig. 169. **Inundation**

водонепроницаемых мерзлых пластов горных пород.
f glace *f* intrusive
s hielo *m* intrusivo
g Intrusionseis *n*

3181 intrusive rock; plutonic rock
An igneous rock formed from the solidation of molten material (magma) which has penetrated into pre-existing solid rocks at depth in the Earth's crust under the conditions of high pressure and slow cooling. → Fig. 111, p. 247
r интрузивная горная порода
Магматическая порода, образовавшаяся при застывании магмы в земной коре в условиях высокого давления и медленного охлаждения. → Рис. 111, с. 247
f roche *f* intrusive
s roca *f* intrusiva
g Intrusivgestein *n*

3182 inundation; spate; flood
The overwhelming of usually dry land by a large amount of water that comes from an overflowing river or lake at high water period, an exceptionally high tide, melting snow, sudden excessive rainfall, or due to construction of dams in a river channel and valley. → Fig. 169
r наводнение; затопление
Заполнение водой территории суши в период половодья или паводков, исключительно высоких приливов, сильных дождей, а также вследствие устройства плотин в русле и долине реки. → Рис. 169
f inondation *f*
s inundación *f*
g Überschwemmung *f*; Flut *f*; Überflutung *f*

* **inundation zone → 2195**

3183 invasion
An accidental getting into a new community of plant or animal species which become very widespread or even domonant.
r инвазия
Естественное (без участия человека) переселение вида растений или животных вне их ареала и включение его в сообщество новых для него видов.
f invasion *f*
s invasión *f*
g Invasion *f*

3184 invasion of air
A rapid and widespread arrival of air masses from the region where they were formed.
r вторжение воздушной массы
Быстрое широкомасштабное вторжение массы воздуха из района его формирования.
f invasion *f* d'air
s invasión *f* de aire
g Einbruch *m* der Luftmasse; Lufteinbruch *m*

3185 invasion of gases
The absorption of gases by the surface layer of water.
r инвазия газов
Поглощение газов поверхностным слоем воды.
f invasion *f* des gazes
s invasión *f* de gases
g Invasion *f* des Gases

* **invasive species → 2054**

3186 inverse moraine cone
A conic or a dome-like hill, up to 10 m high, on the bottom of glacier valley composed of redeposited moraine.
r инверсионный моренный конус
Конический или куполовидный холм высотой до 10 м на дне ледниковой долины, сложенный переотложенной мореной.
f cône *m* morainique d'inversion
s cono *m* morrénico de inversión
g Inversionsmoränenkegel *m*

3187 inversion layer
The atmosphere layer in which the temperature increases with height.
r инверсионный слой
Слой атмосферы, в котором температура растет с высотой.
f couche *f* d'inversion
s capa *f* de inversión
g Inversionsschicht *f*

3188 inversion of precipitation
The phenomenon observed in mountain regions in which the amount of precipitation increases with height up to a certain level and then decreases.
r инверсия осадков
Явление, наблюдаемое в горах, когда количество осадков возрастает с высотой до некоторого уровня, а затем уменьшается.
f inversion *f* de précipitation
s inversión *f* de la precipitación
g Niederschlagsinversion *f*

* **inversion of relief → 3190**

3189 **inversion of temperature; temperature inversion**
A phenomenon in which there is in the air a temperature increase with incfreasing height instead the normal decrease.
r инверсия температуры; температурная инверсия
Повышение температуры с высотой в некотором слое атмосферы вместо обычного ее понижения.
f inversion *f* de température; inversion *f* thermique
s inversión *f* de temperatura
g Temperaturinversion *f*

3190 **inverted relief; inversion of relief**
A relief created by erosion and denudation, in which anticlines form the valleys and the synclines coincide with the mountain ridges. → Fig. 170
r инверсионный рельеф; инверсия рельефа; обращенный рельеф
Эрозионно-денудационный рельеф, формы которого обратны по отношению к геологическим структурам (горные хребты расположены на месте синклиналей, а речные долины – на месте антиклиналей). → Рис. 170
f relief *m* d'inversion; inversion *f* de relief
s relieve *m* de inversión; inversión *f* de relieve
g umgekehrtes Relief *n*; Umkehrung *f* des Reliefs; Inversion *f* des Reliefs

3191 **investment**
Long-term spending money on economy inside and outside the country which can be either financial (in securities) or real (in production).
r инвестиции; инвестирование
Долгосрочные вложения капитала в отрасли экономики внутри страны и за границей, различают финансовые (покупка ценных бумаг) и реальные (вложение капитала в производство) инвестиции.
f investissement *m*
s inversión *f*; inversiones *fpl*
g Investition *f*; Kapitalanlage *f*

3192 **investment climate**
The creation of political, economic and other conditions favourable for the inflow and investment of capital into new branches of economy and modernization of old branches of economy.
r инвестиционный климат

Fig. 170. **Inverted relief**: 1 – limestone, 2 – gritstone, 3 – phyllites, 4 – intrusions

Создание политических, экономических и иных условий для притока капитала и вложения его в новые и модернизацию старых отраслей хозяйства.
f climat *m* d'investissement
s clima *m* de inversiones
g Investitionsklima *n*

3193 **invisible export**
The income-earning items in the international trade of a country, representing not the sale or transfer of goods, but payments made by foreign country to that country, for serving provided, for shipping, air freight, for interest on ivestments, and also expenditure by tourists visiting the country.
r скрытый экспорт
Статьи, приносящие доход в международной торговле страны, представляющие однако не продажу или передачу товаров, но оплату зарубежными партнерами услуг, таких как перевозки по морю, воздуху, проценты от инвестиций, а также расходы туристов, посещающих данную страну.
f exportation *f* invisible
s exportación *f* invisible
g unsichtbarer Export *m*

3194 **invisible import**
The outcome items in the international trade of a country which include payments for serving provided, for shipping, air freight, and also expenditure by tourists outcoming the country, these payments are made by the country to foreign countries.
r скрытый импорт
Статьи расхода в международной торговле страны, включающие оплату зарубежным партнерам услуг, таких как перевозки по морю, воздуху, а также расходы туристов, выезжающих из данной страны.
f importation *f* invisible
s importación *f* invisible
g unsichtbarer Import *m*

3195 **involution**
A contortion of superficial rocky material and soils due to slumping, loading and volume changes in sediments, as well as a deformation of unconsolidated surface materials by periglacial activity.
r смятие пластов
Искривление поверхностных слоев грунта и почв вследствие оползания, нагружения и изменения объема отложений, а также деформации неконсолидированного поверхностного материала из-за перигляциальных процессов.
f involution *f*
s involución *f*
g Involution *f*

* **ion runoff → 5185**

3196 **ionization of atmosphere**

Acquisition of an electronic charge by neutral atmospheric molecules or other suspended particles, mainly as the result of collision with high-energy particles.

r ионизация атмосферы
Образование электрического заряда у нейтральных молекул или взвешенных частиц в атмосфере, главным образом в результате столкновения с частицами, обладающими высокой энергией.

f ionisation *f* atmosphérique
s ionización *f* de atmósfera; ionización *f* atmosférica
g Ionisation *f* der Atmosphäre

3197 ionosphere
A layer of the atmosphere extending from about 70 to 500 km, in which ions and free electrons exist in sufficient quantities to reflect electromagnetic waves.

r ионосфера
Часть атмосферы, простирающаяся приблизительно от 70 до 500 км, в которой плотность ионов и свободных электронов достаточна для отражения электромагнитных волн.

f ionosphère *f*
s ionosfera *f*
g Ionosphäre *f*

*** iridescence → 3199**

3198 iridescent cloud
A cloud with rainbow colouring of edges appearing due to the diffraction of light.

r иризирующее облако
Облако с радужной окраской по краям, обусловленной дифракцией света.

f nuage *m* irisé
s nube *f* iridiscente
g irisierender Wolke *m*

3199 irisation; iridescence
Colour spots appearing on clouds, sometimes mingled, sometimes in the form of bands nearly parallel to the edges of the clouds.

r иризация
Цветные пятна, появляющиеся на облаках, бесформенные или в форме полос, параллельных краю облака.

f irisation *f*
s irisación *f*; iridiscencia *f*
g Irisieren *n*; Schillern *n*

3200 Iron Age
The era in human development succeeding the Bronze Age when iron was smelted and used for tools, utensils and weapons, it began by 1400-1000 B.C.

r железный век
Период в развитии человечества, наступивший с распространением металлургии железа и сменивший бронзовый век, начало датируется 1400-1000 гг. до н.э.

f âge *m* de fer
s edad *f* de hierro
g Eisenzeit *f*

*** iron pan → 4314**

*** iron-and-steel industry → 5791**

3201 irradiance
The amount (W/m^2) of radiant power per unit area that falls upon an object or surface.

r иррадиация
Количество (Вт/м^2) радиационной (электромагнитной) энергии, приходящейся на единицу площади какого-либо объекта или поверхности.

f irradiation *f*
s irradiación *f*
g Irradiation *f*

3202 irretrievable water consumption; irrevocable water consumption
The use of water for economic or domestic purposes when some part of it is not returned to the same or a neighbouring water body.

r безвозвратное водопотребление
Тип водопотребления, при котором часть изъятого на хозяйственные нужды объема воды не возвращается обратно в тот же или соседний водный объект.

f consommation *f* d'eau sans restitution
s consumo *m* de agua sin retorno
g nicht zurückgehender Wasserverbrauch *m*

3203 irreversible change of the environment
Changes of the environment, which can not be indemnified in the course of natural restoration processes including natural succession.

r необратимое изменение окружающей среды
Перемены в окружающей среде, которые не могут быть компенсированы в ходе природных восстановительных процессов, включая естественную сукцессию.

f changement *m* irréversible d'environnement
s cambio *m* irreversible de medio ambiente
g unumkehrbare Umweltveränderung *f*

*** irrevocable water consumption → 3202**

3204 irrigation
The supply of water to the land by means of channels, streams, sprinclers, etc. in order to permit the growth of crops and better yields in arid and semiarid conditions.

r орошение; ирригация; обводнение
Искусственное увлажнение почвы путем подачи воды из водного источника с целью получения большего урожая в аридных и семиаридных условиях.

f irrigation *f*
s irrigación *f*
g Bewässerung *f*; Irrigation *f*

*** irrigation canal → 3205**

3205 **irrigation channel; irrigation canal**
Water conduit intended for water supply from a water source (a river or reservoir) to the irrigation network.
r оросительный канал
Водовод, предназначенный для подачи воды от ее источника до оросительной сети.
f canal *m* d'irrigation
s canal *m* de irrigación; canal *m* de regadío
g Bewässerungskanal *m*

3206 **irrigation farming**
A type of farming when additional water is supplied to crops through a system of man-made canals.
r орошаемое земледелие
Вид земледелия, включающий дополнительную подачу воды с целью обеспечения питательными элементами и влагой сельскохозяйственных культур.
f agriculture *f* irriguée
s agricultura *f* de irrigación; agricultura *f* de regadío
g Bewässerungslandwirtschaft *f*

3207 **irrigation rate**
The amount of water which is necessary to take from the irrigation system to supply a unit of agricultural area to get the projected yield.
r норма орошения; поливная норма
Необходимое количество воды из оросительной системы, используемое на единицу сельскохозяйственной площади, для получения ожидаемого урожая.
f norme *f* d'irrigation
s norma *f* de irrigación
g Bewässerungsrate *f*

* **irrigation scheme → 3208**

3208 **irrigation system; irrigation scheme**
Interconnected constructions supplying water for the irrigation of land.
r оросительная система
Совокупность взаимосвязанных гидротехнических сооружений для орошения земель.
f système *m* d'irrigation
s sistema *m* de irrigación
g Bewässerungssystem *n*

3209 **irrigation water**
Water supplied to the agricultural lands with the purpose of irrigation.
r оросительная вода
Вода, подаваемая на сельскохозяйственные угодья с целью орошения полей.
f eau *f* d'irrigation
s agua *f* de irrigación
g Bewässerungswasser *n*

3210 **irrigation with limans**
The use for irrigation of the snowmelt runoff to create flooded areas (limans).
r лиманное орошение
Использование стока, формирующегося в период снеготаяния, путем создания в удобных местах затапливаемых водой участков (лиманов) для орошения полей.
f irrigation *f* de l'estuaire en voie de comblement
s irrigación *f* con limanes; irrigación *f* de sumersión
g Bewässerung *f* mit Tauwasserbecken

3211 **isallobar**
A line drawn on a map to join places undergoing equal change in atmospheric pressure during a given period, plotted to indicate the development and progress of a pressure system.
r изаллобара
Линия, соединяющая на карте места с одинаковым темпом изменения атмосферного давления за выбранный промежуток времени и показывающая тенденцию развития барической системы.
f isallobare *f*
s isalobara *f*
g Isallobare *f*

3212 **isanomal; isanomalous line**
A line on a map joining points of equal difference from the normal or average of any natural phenomena.
r изаномала
Линия равных значений аномалии природного явления, т.е. отклонений от среднего или нормального их значения.
f isanomale *f*
s isanómala *f*
g Isanomale *f*

* **isanomalous line → 3212**

3213 **island**
An area of land completely surrounded by water.
r остров
Участок суши, окруженный со всех сторон водами океана, моря, озера или реки.
f île *f*
s isla *f*
g Insel *f*

3214 **island arc**
A lengthy chain of oceanic islands formed by a part of a mountain ridge rising above the sea-level which separates a ajacent sea from a deep and forms a curvi linear mountainous archipelago. → Fig. 171, p. 385
r островная дуга
Надводная часть горного хребта, отделяющего котловины окраинных морей от глубоководных желобов, образует линейно вытянутый гористый архипелаг. → Рис. 171, с. 385
f guirlande *f* insulaire; arc *m* insular
s arco *m* insular; guirnalda *f* insular
g Inselbogen *m*; Inselkette *f*

3215 **island fauna**

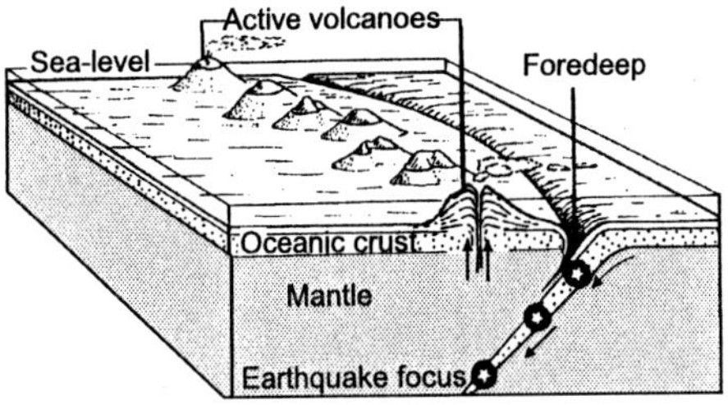

Fig. 171. **Island arc**

All the animals inhabiting islands, characterized by a high proportion of endemics and the absence of some animal species widespread on the nearest continent.

r островная фауна
Совокупность видов животных, обитающих на островах, для которых характерна высокая доля эндемиков, а также отсутствие некоторых видов и родов животных, широко распространенных на ближайшем материке.

f faune *f* insulaire
s fauna *f* insular
g Inselfauna *f*

3216 island flora
The plant life of isolated localities and islands that differs from the continental flora in species composition and origin.

r островная флора
Совокупность видов растений обособленных участков суши и островов, отличающаяся по видовому составу и путям формирования от материковой флоры.

f flore *f* insulaire
s flora *f* insular
g Inselflora *f*

*** island permafrost → 5728**

3217 island state
A state which consists of an island and the ajacent water, e.g. Cyprus, Malta, Tahiti.

r островное государство
Государство, территория которого состоит из острова и прилегающей части моря, например, Кипр, Мальта, Таити.

f état *m* insulaire
s estado *m* insular
g Inselstaat *m*

3218 isoamplitude
A line on a map joining points with equal value of the amplitude of a meteorological element.

r изоамплитуда
Линия на карте, соединяющая точки с одинаковым значением амплитуды метеорологического элемента.

f isoamplitude *f*
s isoamplitud *f*
g Isoamplitude *f*

3219 isoanthe
A line on a map connecting points of the same time of efflorescence of a certain species of plant.

r изоанта
Изолиния, соединяющая на карте пункты с одинаковым сроком зацветания какого-либо вида растений.

f isoanthe *f*
s isoante *f*
g Isoanthe *f*

3220 isoathme
A line on a map joining points with equal value of evaporation.

r изоатма
Линия одинаковой величины испарения.

f isoathme *f*
s isoatma *f*
g Isoathme *f*

3221 isobar
A line on a map joining points of equal barometric pressure on a given surface.

r изобара
Линия, соединяющая точки равного атмосферного давления на данной поверхности.

f isobare *f*
s isobara *f*
g Isobare *f*

*** isobaric surface → 1224**

3222 isobase
A line on geographic map connecting points on the Earth's surface which underwent equal tectonic uplift or subsidence during a specified time period. → Fig. 172, p. 386

r изобаза
Линия на географической карте, соединяющая точки земной поверхности, испытавшие в результате тектонических движений за одинаковый промежуток времени равное поднятие или опускание. → Рис. 172, с. 386

f isobase *f*
s isobase *f*
g Isobase *f*

3223 isobath; bathymetric contour; submarine contour; depth contour
A line joining points on the map of the sea or lake floor situated at an equal vertical distance beneath the surface of water.

r изобата
Линия на карте, соединяющая все точки на дне моря или озера, имеющие одинаковую глубину.

f isobathe *f*
s isobata *f*
g Tiefenlinie *f*; Isobathe *f*

3224 isochion; isochional line
Line on a map joining points of the same altitude of the snow line.

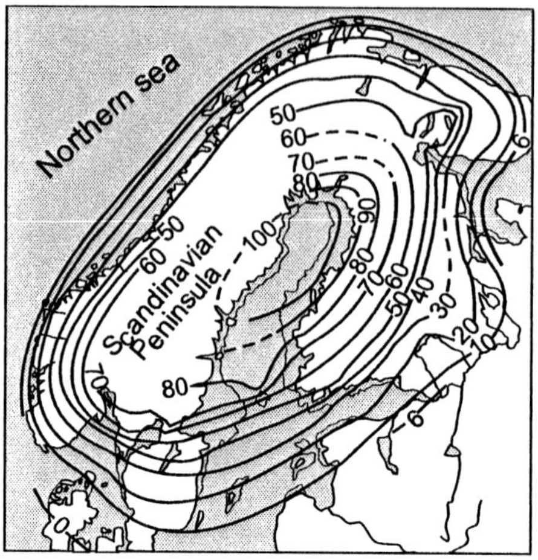

Fig. 172. **Isobases** of glacioisostatic rise (m) in Scandinavia

r изохиона
Изолиния равной высоты снеговой линии.
f isochione *f*
s isokiona *f*
g Isochione *f*

* **isochional line → 3224**

3225 isochron
A line joining points on the Earth's surface which have the same time, including points at which a seismic wave arrives at the same time as well as points where a particular phenomenon or a particular value of a meteorological element occurs simultaneously.
r изохрона
Линия, соединяющая точки на земной поверхности с одинаковым временем, в том числе с одинаковым временем прихода сейсмической волны, а также с одновременным наступлением того или иного метеорологического или фенологического явления.
f isochrone *f*
s isocrona *f*
g Isochrone *f*

3226 isochronous *adj*
Related to a process that is coincident in time with another process.
r изохронный
Относящийся к природному процессу, совпадающему по времени с другим процессом.
f isochrone *adj*
s isócrono *adj*
g isochrone *adj*

3227 isochronous surface
A surface within a body of sediments or within a glacier formed at a common period of time.
r изохронная поверхность
Поверхность внутри осадочных отложений или в леднике, относящаяся к единому времени образования.
f surface *f* de isochrone
s superficie *f* isócrona
g isochrone Oberfläche *f*

3228 isoclinal fold; isocline; isoclinal folding
A fold of any type in which the limbs have parallel dips, i.e. incline in the same direction and approximately the same angle. → Fig. 119, p. 265
r изоклинальная складка
Складка любого типа, крылья которой имеют параллельное падение, т.е. близкий наклон в том же направлении. → Рис. 119, с. 265
f pli *m* isoclinal
s pliegue *m* isoclinal
g isoklinale Falte *f*

* **isoclinal folding → 3228**

* **isoclinal line → 3229**

* **isocline → 3228**

3229 isocline; isoclinal line
A line joining points of equal magnetic inclination, the isoline of zero magnetic inclination being the magnetic equator.
r изоклина
Линия, соединяющая точки с одинаковым магнитным наклонением (изолиния нулевого наклонения – это магнитный экватор).
f isocline *f*
s isoclina *f*
g Isokline *f*

3230 isodapane
A line on a map joining points of equal total transport costs to supply any materials or goods from a certain point.
r изодапана
Линия, соединяющая на карте точки с равной стоимостью перевозки грузов из заданной точки.
f isodapane *f*
s isodapana *f*
g Isodapane *f*

3231 isogonic line
1. A line joining points with the same wind direction. 2. A line drawn on a map joining points of compass declination. → Fig. 173, p. 387
r изогона
1. Линия, соединяющая точки с одинаковым направлением ветра. 2. Линия на карте, соединяющая точки с одинаковым магнитным склонением. → Рис. 173, с. 387
f isogone *f*; ligne *f* isogonique
s isogona *f*; línea *f* isogónica
g Isogone *f*

3232 isogram
A line on a diagram or chart of equal value of a meteorological element.

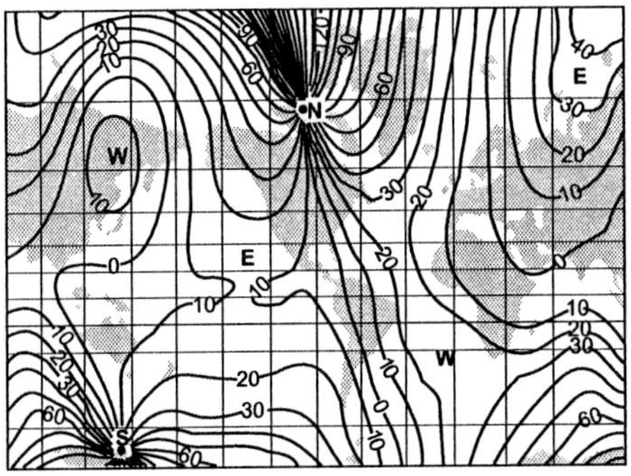

Fig. 173. **Isogonic lines$_2$** on the World map

r изограмма
Линия равных значений метеорологического элемента на диаграмме или карте.
f isogramme *m*
s isograma *f*
g Isogramm *n*

3233 isohaline
A line on a map or water profile connecting points having the same values of salinity.
r изогалина
Изолинии солености вод на карте или профиле акватории.
f isohaline *f*
s isohalina *f*
g Isohaline *f*

3234 isohumic soil
A soil in which organic matter is completely included into mineral base as a result of high biological activity, it is characterised by the gradual decrease of humus content with depth.
r изогумусовая почва
Почва, в которой органическое вещество полностью включено в минеральную основу вследствие высокой биологической активности, поэтому она характеризуется медленным уменьшением содержания гумуса с глубиной.
f sol *m* isohumique
s suelo *m* isohumico
g Isohumusboden *m*

3235 isohyet
A line of equal rainfall amount recorded during a certain time interval.
r изогиета
Линия равных сумм атмосферных осадков за определенный период.
f isohyète *f*
s isoyeta *f*
g Isohyete *f*

* **isohypse → 1262**

3236 isolated population
A population of a species of animals located at a big distance from the major area or when the exchange of individuals between them is very limited.
r изолят
Популяция животных, пространственно изолированная от основного ареала вида или с крайне ограниченным обменом особями между ними.
f isolement *m*; isolation *f*
s población *f* aislada
g Isolat *n*

3237 isolated settlement
A settlement located outside the sphere of influence of a central settlement.
r изолированное поселение
Поселение, лежащее вне сферы влияния центрального поселения.
f peuplement *m* isolé
s asentamiento *m* aislado; población *f* aislada
g Einzelsiedlung *f*

3238 isolationism
A trend in the foreign policy of states oriented mainly on their own economy and on the limitation of international relations.
r изоляционизм
Направление во внешней политике стран, заключающееся в ориентации в основном на собственные силы и ограничении внешних связей.
f isolationnisme *m*
s aislacionismo *m*
g Isolationismus *m*

3239 isoline; isopleth
A line on a map connecting points of equal values, e.g. rainfall, temperature, etc.
r изолиния; изоплета
Линия на карте, проведенная через точки с одинаковым значением той или иной величины.
f isoligne *f*; isoplèthe *f*
s isolínea *f*; isopleta *f*
g Isolinie *f*; Isoplethe *f*

3240 isomorphic *adj*
Having the same or analogous landform, or a similar structure.
r изоморфный
Имеющий ту же или аналогичную форму рельефа, либо структурно близкую.
f isomorphe *adj*
s isomorfo *adj*
g isomorph *adj*

3241 isoneph
A line joining on a chart places of equal cloudiness.
r изонефа
Линия, соединяющая на карте точки с одинаковым количеством облаков.
f isonèphe *f*
s isonefa *f*

g Isonephe *f*

3242 **isooxygene**
A line on a map or chart connecting places of equal concentration of oxygen dissolved in water.
r изооксигена
Линия, соединяющая на карте или схеме водного объекта точки с равной концентрацией растворенного в воде кислорода.
f isooxygéné *f*
s isooxígena *f*
g Isooxygene *f*

3243 **isopach; isopachyte**
A line joining points on the same thickness a rock-stratum.
r изопахита
Изолиния, соединяющая точки на карте с аналогичной мощностью геологического пласта.
f isopaque *f*
s isopaqui *f*
g Isopache *f*

3244 **isopage**
A line on a map showing the same length of time during which the ice cover retains on water bodies.
r изопага
Изолиния на карте, обозначающая продолжительность ледяного покрова на водоемах.
f isopage *f*
s isopaga *f*
g Isopage *f*

3245 **isopectic line**
A line on a map joining points of the same dates of water freezing.
r изопектик
Изолиния на карте, отображающая даты замерзания водных объектов.
f isopectique *f*
s isopéctica *f*
g Isopekte *f*

3246 **isophene**
A line on a chart joining points with the same dates of appearance of a certain stage of plant or animal life.
r изофена
Линия, соединяющая на карте точки одинаковых дат наступления определенных фаз в жизни растений или животных.
f isophène *f*
s isofena *f*
g Isophene *f*

3247 **isophytochrone**
A line on a map connecting points of the same duration of vegetative season.
r изофитохрона
Линия на карте, соединяющая места с равной продолжительностью вегетационного периода.
f isophytochrone *f*
s isofitocrono *f*
g Isophytochrone *f*

* **isopleth → 3239**

* **isopycnic → 3248**

3248 **isopycnic line; isopycnic**
A line joining up points which exhibit the same density of water or air on certain meteorological and oceanographic charts.
r изопикна
Линия, соединяющая на метеорологических и океанографических картах точки с одинаковой плотностью воды или воздуха.
f isopycne *f*
s isopicna *f*
g Isopykne *f*

3249 **isoseist; homoseismal line; homoseism**
A line joining places at which the earthquake-wave arrival time from a particular epicentre is identical.
r изосейста
Линия на карте, соединяющая точки, в которые волна из данного эпицентра землетрясения приходит в одно и то же время.
f isoséiste *f*; ligne *f* isoséiste
s isosista *f*; línea *f* isosista
g Isoseiste *f*

3250 **isostasy of the Earth's crust; isostatic adjustment**
The condition of relative equilibrium of the Earth's crust: if part of the surface is removed by erosion, the continental crust will rise to compensate this erosion while continental crust pushed down by the weight of glacial ice, land masses will rise again as the ice melts (*glacioisostasy*). Fig. → 174
r изостазия
Условие относительного равновесия земной коры, когда при ее разрушении, например, эрозией, континентальная кора поднимается, чтобы компенсировать убыль массы горных пород, и наоборот, при возникновении, например, больших ледниковых покровов континентальная кора опускается (*гляциоизостазия*). → Рис. 174
f isostasie *f*
s isostasia *f*; compensación *f* isostática
g Isostasie *f* der Erdrinde

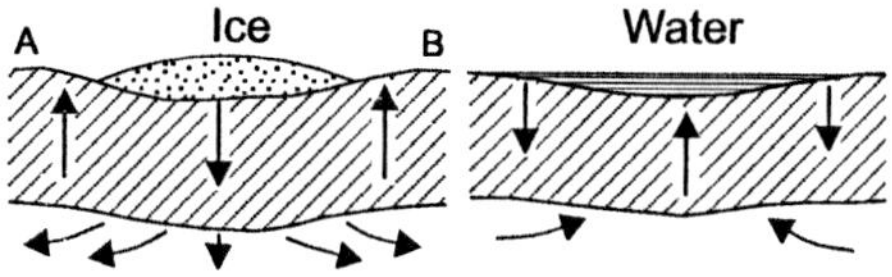

Fig. 174. **Isostasy of the Earth's crust**: A – pressure of an ice sheet, B – rise of the Earth's crust induced by melting of an ice sheet

3251 **isostatic** ***adj***
Relating to, or characterized by isostasy.
r изостатический
Относящийся к природным процессам, связанным с изостазией, или характеризующий равновесие земной коры.
f isostatique *adj*
s isostático *adj*
g isostatisch *adj*

* **isostatic adjustment → 3250**

3252 **isotache**
A line on a chart of equal velocities of any natural or man-made phenomenon.
r изотаха
Линия одинаковых скоростей любого природного или антропогенного явления, изображаемого на карте.
f isotache *f*
s isotaquia *f*; isovela *f*
g Isotache *f*

3253 **isotherm**
A line joining points of equal air or water temperature.
r изотерма
Линия равных значений температуры воздуха или воды.
f isotherme *f*
s isoterma *f*
g Isotherme *f*

3254 **isothermal** ***adj***
Having the same temperature within the whole volume or some layer of water or air, without change of temperature.
r изотермический
Имеющий одинаковую температуру во всей массе или в некотором слое воды или воздуха.
f isothérmique *adj*
s isotérmico *adj*; isotermal *adj*
g isothermisch *adj*

3255 **isothermal atmosphere**
A model atmosphere in which the temperature is assumed to be constant with height.
r изотермическая атмосфера
Модель атмосферы, в которой температура принимается постоянной на всех уровнях.
f atmosphère *f* isotherme
s atmósfera *f* isotérmica
g isothermische Atmosphäre *f*

* **isothermal glacier → 6079**

3256 **isothermal layer; isothermy**
A layer of air in which temperature remain constant and does not change with height.
r изотермический слой; изотермия
Слой воздуха, в котором температура с высотой не меняется.
f couche *f* isotherme; isothérmie *f*
s capa *f* isotérmica; isotermía *f*
g isothermische Schicht *f*; Isothermie *f*

3257 **isothermobath**
A line joining points at a given depth in the sea which exhibit the same temperature.
r изотермобата
Линия, соединяющая точки на заданной глубине в море, имеющие одинаковую температуру.
f isothermobathe *f*
s isotermobata *f*
g Isothermobathe *f*

* **isothermy → 3256**

3258 **isotope**
Different forms of a chemical element having almost the same chemical properties, the same atomic number but different atomic weights, some isotopes are used to date organic material (^{14}C) or determine the paleotemperature in cores of marine sediments and ice sheets ($^{16}O/^{18}O$).
r изотоп
Разные формы химического элемента, имеющие общие химические свойства, тот же атомный номер, но неодинаковый атомный вес, некоторые изотопы используются для датирования органического материала (^{14}C) или для определения палеотемператур в кернах морских осадков и ледниковых щитов ($^{16}O/^{18}O$).
f isotope *m*
s isótopo *m*
g Isotope *pl*

3259 **isotope-oxygen stage**
A period during which the deposits characterised by a certain proportion of $^{16}O/^{18}O$ isotopes have been formed.
r изотопно-кислородная стадия
Временной интервал, в течение которого формируются отложения, характеризующиеся определенным соотношением изотопов $^{16}O/^{18}O$.
f stade *m* isotope-oxygène
s estadio *m* de isótopo-oxígeno
g Isotopen-Sauerstoffstadium *n*

3260 **isthmus; neck**
A narrow strip of land joining two larger areas of land or two continents.
r перешеек
Узкая полоса суши, соединяющая два более значительных массива суши или два материка.
f isthme *m*
s istmo *m*
g Landenge *f*; Isthmus *m*; Hals *m*

J

3261 **jet**
A form of water or air flowing in a surrounding space having the parameters which differ greately from those of the jet.
r струя
Форма течения жидкости или газа, при которой они текут в окружающем пространстве, имеющем отличные от струи параметры.
f jet *m*; filet *m* d'eau
s chorro *m*
g Strahl *m*

3262 **jet stream**
A high aititude, fast-moving air current, a few thousand kilometres in length, a few hundred kilometres wide, a few kilometres in depth, more or less horizontal, occurring in vicinity of the tropopause, usually blowing more strongly in winter than in summer.
r струйное течение
Высотное воздушное течение с большими скоростями, обычно вблизи тропопаузы, длиной в тысячи километров, шириной в сотни километров и глубиной в несколько километров, более или менее горизонтальное, обычно более сильное в зимнее время по сравнению с летним.
f courant-jet *m*
s corriente *f* en chorro
g Strahlstrom *m*

* **jetty → 4566**

3264 **joint**
A crack or fissure in a rock following a dominant direction along a line of weakness, usually transverse to the bedding, and produced by tearing apart under tension or shearing under compression, but whithout dislocation (as in a fault).
r контактная трещина
Трещина в горной породе, протягивающаяся вдоль ослабленного слоя, обычно поперек напластования, вызванная разрывом или сдвигом под напряжением, но без дислокаций, как это случается при образовании складки.
f brisure *f*
s raja *f*; fisura *f* de contacto
g Kluft *f*

3265 **joint venture**
A business activity in which two or more partners work together and jointly invest in a separate business or project.
r совместное предприятие
Форма делового сотрудничества на основе вложения капитала двух или нескольких партнеров для выполнения хозяйственной деятельности.
f entreprise *f* mixte; joint venture *f*; entreprise *f* conjointe
s empresa *f* mixta
g gemeinsamer Betrieb *m*

3266 **jointing**
More or less regular blocks divided by a number of parallel joints formed under the natural splitting of rocks.
r отдельность
Характерная часть горных пород, образующаяся при естественном раскалывании их по определенным поверхностям.
f désagrégation *f*; disjonction *f*
s desagregación *f*; juntura *f*; disyunción *f*
g Absonderung *f*

3267 **joint-stock company**
A form of business organization whose authorized capital is divided into shares.
r акционерное общество
Товарищество, уставной фонд которого разделен на определенное число акций равной стоимости.
f société *f* anonyme; société *f* par actions
s sociedad *f* anónima
g Aktiengesellschaft *f*

3268 **jokullhlaup**
A flood caused by a sudden release of meltwater accumulated as a result of subglacial volcanic activity or a volcanic eruption near a glacier.
r йокульлауп; вулканогенный ледниковый паводок
Мощный паводок и наводнение в результате прорыва талых вод, возникающих при извержении вулкана под ледником или около ледника.
f jökullhlaup *m*
s jökullhlaup *m*
g Jökullhlaup *m*

* **jolt → 1771**

3269 **journey; trip**
A travel from one place to another over a cosiderable distance with different purposes including education, investigation, tourism etc.
r путешествие
Длительное передвижение по какой-либо территории или акватории с общеобразовательными, исследовательскими или туристскими целями.
f voyage *m*
s viaje *m*
g Reise *f*

3270 **Julian calendar**
A system of chronology in which three years out of each four have 365 days and one leap-year has 366 days, the average duration of calendar year is equal to 365.25 average solar

days (it was introduced by Gaius Julius Caesar since January 1, 45 year B.C.).

r Юлианский календарь; старый стиль
Система летоисчисления, в которой из каждых 4 лет три года имеют по 365 суток и один високосный год имеет 366 суток, средняя длина юлианского года равна 365,25 средних солнечных суток (календарь введен Юлием Цезарем с 1 января 45 г. до н.э.).

f calendrier *m* julien
s calendario *m* juliano
g Julianischer Kalender *m*

3271 jumping avalanche
An avalanche that meets a scarp or bench on a slope and thus falls down freely.

r прыгающая лавина
Лавина, которая сваливается с уступа склона и свободно падает.

f avalanche *f* bondissante
s avalancha *f* saltante
g Sprunglawine *f*

3272 junction$_1$
A confluence of routeways where usually motels and serving shops are located.

r дорожная развязка
Соединение двух или нескольких магистральных дорог, где обычно располагаются мотели и центры обслуживания.

f junction *f*
s conexión *f* de carreteras; desvío *m* de carreteras
g Verbindung *f*

* **junction$_2$ → 1199**

3273 jungle
A dense impenetrable tropical forest with tree and shrub vegetation typical of the areas with humid tropical climate, occurs in India, Indochina, on Sunda Isles.

r джунгли
Густые, труднопроходимые древесно-кустарниковые леса в областях с влажнотропическим климатом, распространены в Индии, Индокитае, на Зондских островах.

f jungle *f*
s jungla *f*
g Dschungel *m*

3274 juniper stand
Arid light forest formed mainly by arboraceous forms of juniper.

r арчевник
Ксерофитное редколесье, образованное преимущественно арчой – разными видами древовидного можжевельника.

f genévraie *f*
s enebral *m*
g Wacholderbestand *m*

3275 Jurassic
The middle geological period of the Mesozoic era, it began 190-195 million years ago and lasted for 58-60 million years when sediments of clay were laid down, dinosaurus reached their peak, birds began to appear. → Table 1, p. 284

r юра
Средний период мезозойской эры геологической истории Земли от 190-195 до 58-60 млн лет назад, характеризовался господством динозавров и появлением птиц. → Табл. 1, с. 284

f Jurassique *m*
s Jurásico *m*
g Jura *f*

3276 just-in-time
A system of production whreby the raw materials and assembling parts come directly from the suppliers for the assemblage without any storage.

r точно в срок; точно по графику
Концепция управления, предполагающая поставку ресурса как раз в тот момент, когда его надо использовать; система производства, когда сырье и детали для изготовления продукции поступают прямо в цеха, минуя склады.

f juste-à-temps
s justo a tiempo
g genau zu dem Zeitpunkt

3277 juvenile relief
A relief with steep-sided valleys characteristic of the early stages of cycle of erosion.

r молодой рельеф
Рельеф с крутостенными речными долинами, характерный для начальной стадии эрозионного цикла.

f relief *m* jeune
s relieve *m* jóven
g junges Relief *n*

3278 juvenile valley
A forming river valley with narrow and steep slopes, irregular longitudinal profile, and with no floodplain and terraces.

r молодая долина
Формирующаяся долина реки – с узкими и крутыми склонами, невыровненным продольным профилем, без поймы и террас.

f vallée *f* jeune
s valle *m* jóven
g junges Tal *n*

3279 juvenile water
Ground waters originated from the interior of the Earth reaching the earth's surface as a result of volcanic activity and have not previously participated in the hydrological cycle.

r ювенильные воды
Подземные воды, впервые поступающие в круговорот воды из глубин Земли.

f eau *f* juvénile
s agua *f* juvenil
g juveniles Wasser *n*

K

3280 kalema
A strong surf (from July to September oversplash up to 30 m) near the shores of Gulf of Guinea in West Africa.
r калема
Название сильного прибоя (с июля по сентябрь взброс волны до 30 м) у берегов Гвинейского залива в Западной Африке.
f kalema *f*
s kalema *f*
g Kalema *f*

3281 kaltus
Swamps and swamped glades in the north of the European Russia and in Siberia overgrown by bushes and small forest.
r калтус
Болота и заболоченные поляны на севере Европейской России и в Сибири, зарастающие кустарником и мелким лесом.
f kaltus *m*
s kaltus *m*
g Kaltus *m*

3282 kame
A combination of the steep-sided hills and short ridges usually found in areas of ancient glaciation formed as a result of fluvioglacial and moraine accumulation at places where the dead ice melts. → Fig. 175
r кам
Крутосклонные холмы и короткие гряды в областях древних оледенений, обязанные своим возникновением в основном флювиогляциальной и моренной аккумуляции на участках таяния мертвого льда. → Рис. 175
f kame *m*
s kame *m*
g Kam *m*

3283 kame terrace
A terrace running along the valley side, formed of sand and gravel deposited by a stream of meltwater in the depression between a glacier and the sides of its trough.
r камовая терраса
Терраса, сфомированная песком и гравием, отложенными потоком талых вод в понижении между ледником и склонами его трога.
f terrasse *f* de kame; terrasse *f* externe
s terraza *f* de kame; terraza *f* externa
g Kamenterrasse *f*

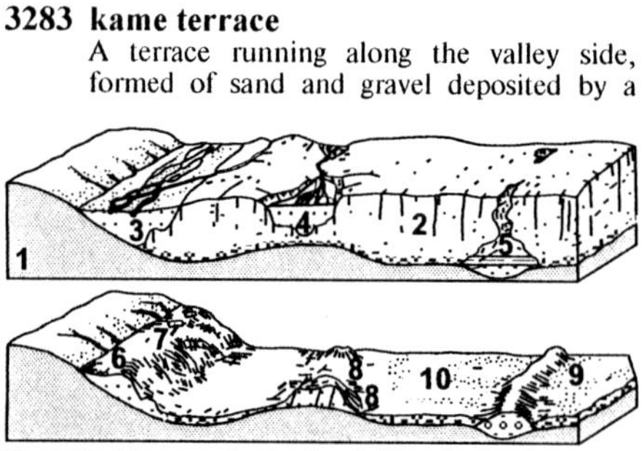

Fig. 175. **Kames**: 1 – bedrock, 2 – glacier, 3 – gravel deposits, 4 – glaciofluvial sediments, 5 – subglacial stream, 6 – kame terrace, 7 – thermokarst funnels, 8 – kames, 9 – os (esker), 10 – moraine landscape

* **kame-and-kettle terrain → 3324**

3284 kamien
A ridge composed of bedrock, or steep rocky river banks in the central and northern parts of the European Russia and in the Urals.
r камень$_2$
Каменистая гряда, сложенная коренными породами, или скалистые обрывы берегов рек в центре и на севере Европейской России и на Урале.
f kamen *m*
s kamen *m*
g Kamen *m*

* **kampzone → 1873**

* **kar → 1005**

3285 karaburan
A strong northeast wind laden with dust and sand blowing in daytime in the arid Tarim basin in Central Asia, darkening the sky, changing river courses by depositing sand, and carrying dust particles at great distances to settle as loess.
r карабуран
Сильный северо-восточный ветер с пылью и песком, который дует в дневное время в бассейне Тарима в Центральной Азии, изменяет течение рек из-за отложений песка и переносит на значительные расстояния твердые частицы, откладывающиеся затем как лёсс.
f karaburan *m*
s karaburan *m*
g Karaburan *m*

3286 karasu
Spring and small permanent stream with ground nourishment and pure transparent water found in Central Asia, in the Altai, the Caucasus and in many countries of Near and Middle East.
r карасу
Родник и небольшая постоянная река грунтового питания в Средней Азии, на Алтае, Кавказе и во многих странах Ближнего и Среднего Востока, несущая чистую, прозрачную воду.
f karasu *m*
s karasu *m*
g Karasu *n*

3287 karez

Horizontal or slightly inclined underground gallery serving for outlet of confined groundwaters to the surface, it is used for irrigation and water supply in many countries Asia and Africa.

r кяриз
Горизонтальная или слабонаклонная подземная галерея, служащая для выхода грунтовых вод на поверхность, которая используется для орошения и водоснабжения во многих странах Азии и Африки.
f karez *m*
s karez *m*
g Karez *m*

3288 karren; lapies
Parallel rows and labyrinths of steep-sloped furrows up to 2 m deep or shallow channels on the surface of bare karst, formed by solution on exposed limestone. → Fig. 176
r карры
Параллельные ряды или лабиринты крутостенных борозд глубиной до 2 м на поверхности голого карста, образованные в результате растворения выходов известняка. → Рис. 176
f lapiés *mpl*; karren *mpl*
s lapiaz *m*; karren *mpl*
g Karren *mpl*

3289 karrenfeld; limestone pavement
An exposed surface of karstifiable rocks dissected by furrows (karrens).
r карровое поле
Обнаженная поверхность карстующихся пород, изборожденная каррами.
f champ *m* de lapiés
s campo *m* de lapiazes
g Karrenfeld *n*

3290 karst; karst relief
The solution of limestone, dolomite, gypsym

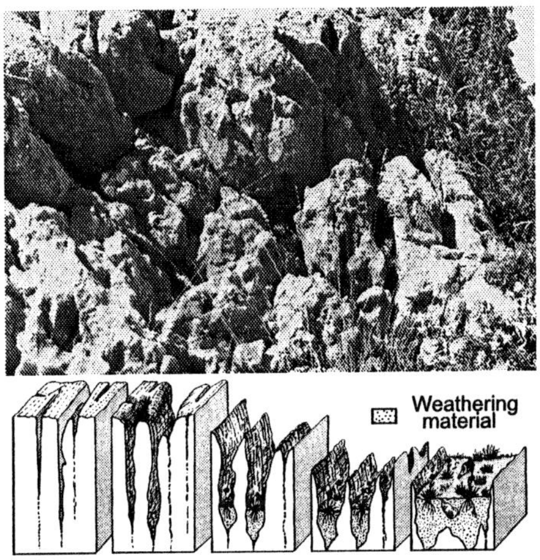

Fig. 176. **Karren**. Down: stages of the karren formation

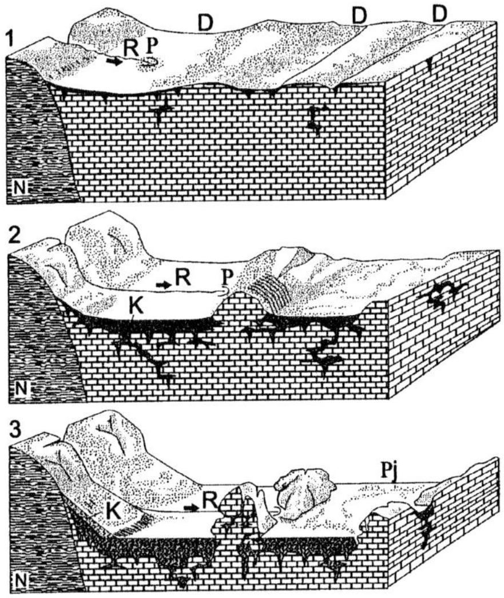

Fig. 177. Successive stages of the **karst** development. R – river, Pj – poliye, P – ponor, K – karstifiable rocks D – dry valleys, N – non karstifiable rocks

and halite by water and the landforms this phenomenon creates, such as bare, limestone ridges, caverns, sinks and underground drainage. → Fig. 177
r карст; карстовый рельеф
Явления, связанные с растворением природными водами горных пород (известняков, доломитов, гипсов, каменной соли), и формы рельефа, образующиеся в областях распространения таких пород: подземные полости, провалы, исчезающие реки и пр. → Рис. 177
f karst *m*; relief *m* karstique
s karst *m*; modelado *m* kárstico; relieve *m* kárstico; relieve *m* cársico; carso *m*
g Karst *m*; Karstrelief *n*

3291 karst base-level
The level down to which water circulation in a karstic terrain is possible, hence the lowest level to which water can erode.
r базис карста
Нижний уровень, до которого возможна циркуляция воды в карстующемся массиве.
f niveau *m* de base karstique
s nivel *m* de base cársico;
nivel *m* de base kárstico
g Karstbasis *f*

* **karst basin → 6517**

3292 karst cavity
Natural shaft, well, cave found in karstifiable rocks.
r карстовая полость
Естественная шахта, колодец, пещера, заложенные в карстующихся породах.
f cavité *f* karstique

s cavidad *f* cársica
g Karstschlotte *f*; Karsthöhle *f*

3293 **karst lake**
A periodic lake formed in a polje or dóline as a result of rising water level.
r карстовое озеро
Временный водоем, возникающий в полье или долине в результате повышения уровня воды.
f lac *m* karstique
s lago *m* cárstico
g Karstsee *m*

3294 **karst landscape; limestone terrain**
The landscape of karst areas: in temperate latitudes it is characterized by rugged terrain with hollows and no surface watercourses, and in subtropical and tropical latitudes – by residual landforms: cones, towers, etc.
r карстовый ландшафт
Ландшафт карстовых областей: в умеренных широтах неровная поверхность с замкнутыми углублениями и отсутствием поверхностных водотоков, в субтропических и тропических широтах останцовые формы рельефа в виде конусов, башен и пр.
f paysage *m* karstique
s paisaje *m* cársico; paisaje *m* kárstico
g Karstlandschaft *f*

3295 **karst pit**
A narrow vertical passage of irregular shape tens of metres deep created by the collapse of the roof of a karst cavity.
r карстовый колодец
Узкий вертикальный канал неправильной формы глубиной в несколько десятков метров, возникающий в результате провала свода карстовой полости.
f puits *m* karstique
s pozo *m* cársico; cenote *m*
g Karstbrunnen *m*

3296 **karst plateau**
A plateau composed of a thick layer of karstifiable rocks with a variety of karstic landforms on the surface of the plateau.
r карстовое плато
Плато, сложенное мощной толщей карстующихся пород, с обилием форм карстового рельефа на поверхности.
f plateau *m* karstique
s meseta *f* cársica; meseta *f* del karst
g Karstplateau *n*

* **karst relief → 3290**

3297 **karst river**
A river originating from a karst spring or running in a karst terrain.
r карстовая река
Река, берущая начало из карстового источника или текущая в карстовом районе.
f rivière *f* karstique
s río *m* kárstico
g Karstfluß *m*

3298 **karst studies**
The study of karst, i.e. processes and landforms developing in the rocks dissolved by natural waters.
r карстоведение
Изучение карста, т.е. процессов и форм рельефа, развивающихся в растворимых природными водами горных породах.
f karstologie *f*
s karstología *f*
g Karstkunde *f*; Karstforschung *f*

3299 **karst valley**
An erosion depression several kilometres long with karst holes and ponors at the bottom.
r карстовая долина
Эрозионное понижение протяженностью несколько километров с карстовыми воронками и понорами на дне.
f vallée *f* karstique
s valle *m* kárstico; valle *m* cársico
g Karsttal *n*

3300 **karst waters**
Ground waters filling cracks and pores, karst cavities and canals in soluble rocks.
r карстовые воды
Подземные воды, заполняющие трещины и поры, карстовые полости и каналы в растворимых горных породах.
f eau *f* karstique
s agua *f* kárstica
g Karstwasser *n*

3301 **karst window**
That part of an underground river in a karst area revealed by the collapse of a section of the roof.
r карстовое окно
Отрезок подземной реки в карстовой области, открывшийся вследствие провала части свода над этой рекой.
f fenêtre *f* karstique
s ventana *f* cársica
g Karstfenster *n*

3302 **karstic** *adj*
Related to the natural phenomena originated by the solution of limestone, dolomite, gypsym and halite by water and the landforms created by this processes.
r карстовый
Относящийся к явлениям, связанным с растворением природными водами известняков, доломитов, гипсов, каменной соли, и формам рельефа, образующимся в результате этих процессов.
f karstique *adj*
s cársico *adj*; kárstico *adj*
g Karst-

3303 **karstic cycle**

The sequential stages of the evolution of karstic landforms.
r карстовый цикл
Совокупность последовательных изменений рельефа, характерных для карстовых областей.
f cycle *m* karstique
s ciclo *m* cársico; ciclo *m* kárstico
g Karstzyklus *m*

3304 **karstifiable rock**
A rock easily dissolved by water which provokes the development of karst.
r карстующаяся порода
Горная порода, легко поддающаяся растворению природными водами, что приводит к развитию в ней карстовых процессов.
f roche *f* karstifiable
s roca *f* carsificada
g verkarstungsfächiges Gesteine *n*

3305 **katabatic front; katafront**
A front in which the warm air descends along the frontal surface.
r фронт нисходящего скольжения; катафронт
Атмосферный фронт с нисходящим движением теплого воздуха вдоль поверхности фронтального раздела.
f front *f* catabatique; catafront *f*
s frente *m* catabático; catafrente *m*
g Katafront *f*

3306 **katabatic wind; gravity wind**
A downslope wind, usually cold, which is caused by ground-surface cooling and the increase of density of the air which then moves downhill by gravity. → Fig. 178
r катабатический ветер; стоковый ветер
Нисходящий ветер, направленный вниз по склону и обусловленный выхолаживанием поверхности склона, благодаря чему плотность воздуха повышается и он начинаете двигаться вниз под влиянием силы тяжести. → Рис. 178
f vent *m* catabatique; vent *m* de gravité
s viento *m* catabático; viento *m* descendente
g katabatischer Wind *m*

* **katafront → 3305**

3307 **katathermy**
Thermal stratification of sea or lake water characterised by the rise of its temperature with depth.

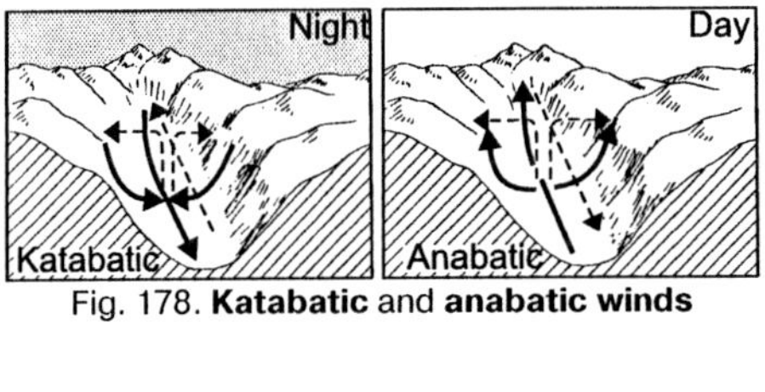

Fig. 178. **Katabatic** and **anabatic winds**

r кататермия
Термическая стратификация морской или озерной воды, характеризующаяся повышением температуры с увеличением глубины.
f catathermie *f*
s catatermía *f*
g Katathermie *f*

3308 **kavir**
A solonchak or salt desert in Iran, the term is often used as part of the geographical names of some depressions or deserts, e.g. Deshte-Kavir.
r кевир
Солончаковые и соляные пустыни Ирана, нередко употребляется как составная часть географических названий отдельных впадин или пустынь, например, Деште-Кевир.
f kévir *m*; kewir *m*; kavir *m*
s kavir *m*; kevir *m*; kewir *m*
g Kavir *m*

3309 **kegelkarst**
A type of tropical karst terrain which is characterized by numerous cone-like limestone hills up to 200 m high, sometimes separated by cockpits, which are enclosed depressions interconnected by sinuous corridors between the cones. → Fig. 335, p. 749
r конический карст
Разновидность тропического карста, представленного массовым развитием известняковых возвышений конической формы высотой до 200 м, нередко разделенных резкими понижениями (*кокпитами*), образующими изогнутые коридоры между конусами. → Рис. 335, с. 749
f karst *m* à pitons; kegelkarst *m*
s karst *m* de pitones; kegelkarst *m*
g Kegelkarst *m*

3310 **kekur**
1. A high rock on the seashore or in the sea near the coast. 2. A ridge or rampart composed of boulders and pebbles which were displaced onto the shore by river ice during a flood.
r кекур
1. Высокая скала-останец на берегу моря или в море близ побережья. 2. Гряда или вал из валунов и галек, вытесненных на берег речными льдами во время половодья.
f kekour *m*
s kekur *m*
g Kekur *m*

* **Kelvin scale → 21**

3311 **kernlose winter**
The singular temperature conditions in winter with a smooth minimum temperature typical of the high Arctic latitudes and long winter night of Central Antarctica.
r безъядерная зима
Зима со сглаженным, слабо выраженным минимумом температуры, характерная

для морского климата в высоких широтах Арктики и для длинной зимней ночи в Центральной Антарктиде.
f hiver *m* non-nucléaire
s invierno *m* sin nucleo
g kernloser Winter *m*

3312 **kettle; glacier pothole; kettle hole**
A bowl-shaped depression in the rocky bed of a glacier or the wall of rocky river bed, usually over 10 m deep and wide, formed at a place of heavy whirlpool or at the foot of a waterfall.
r исполиновый котел; эворзионная ванна
Полусферическое углубление в скальных породах ледникового ложа или на борту долины сброса талых вод глубиной и поперечником более 10 м, образованное на месте сильного водоворота или у основания водопада.
f marmite *f* de géant; cuvette *f* d'evorsion
s marmita *f* de gigante
g Evorsionsbecken *m*; Riesentöpfe *f*; Riesenkessel *m*

* **kettle hole → 3312**

3313 **kettle hole; dead ice hollow**
A circular basin of thermokarst origin some tens of metres in diameter and a few metres deep which is filled with water or peat, formed when a block of dead ice incorporated into sediments finally melts and leaves an enclosed depression in the landscape.
r золль
Округлая впадина-ванна термокарстового происхождения глубиной в несколько метров и диаметром в несколько десятков метров, заполненная водой или торфом, образующаяся на месте растаявшего блока подземного льда.
f sölle *m*
s sölle *f*
g Soll *n*

* **kettle lake → 4727**

3314 **kettle-hole glacier**
A glacier whose accumulation area is situated in a large cirque and its tongue moves down out of the depression for a distance not longer than one to two thirds of its length.
r котловинный ледник
Ледник с областью питания в обширном цирке и языком, спускающимся за пределы котловины не далее одной-двух третей ее длины.
f glacier *m* de dépression
s glaciar *m* de depresión
g Kesselgletscher *m*

3315 **key area; sample plot**
A dictinctive locality, a site selected as a standard for studying a landscape or its components.
r ключевой участок
Характерный участок, выбранный в качестве эталона для изучения ландшафта, либо того или иного его компонента.
f terrain *m* clef
s terreno *m* clave
g Schlüsselbereich *n*

3316 **key map; index of sheets; index map**
A small-scale map or cheme showing the whole territory covered by a multi-sheet geographical map, position, sheet division and numbers of individual sheets as a means of easy reference.
r сборный лист; схема листов карты
Схема, показывающая всю территорию, покрываемую многолистной географической картой, расположение, разграфку и номенклатуру карт отдельных листов.
f schéma *m* d'ensemble de groupe des cartes
s mapa *m* complementario index; mapa-índice *m*; cuadro *m* de alzado
g Blattübersicht *f*

3317 **key stratum**
A layer in rock characterized by some distinctive features and used to compare the contiguous profiles and establish correlations.
r маркирующий горизонт; реперный горизонт
Слой в толще горных пород, выделяющийся по каким-либо признакам и служащий для сопоставления смежных разрезов.
f couche-repère *m*
s capa-referencia *f*
g Leithorizont *m*

3318 **key village**
A minor centre with facilities (e.g. primary school, a village hall) to serve even smaller villages and hamlets, but reliant on major centres for other facilities.
r сельский центр
Небольшое поселение (как правило, с начальной школой и местной администрацией), обслуживающее самым необходимым небольшие окрестные деревни и хутора, но требующее более полного обслуживания (например, больница) в районном центре.
f village *m* de centre
s aldea *f* céntrica
g Grunddorf *n*

3319 **khamsin**
A very hot, dry, and dust-laden wind which blows periodically from the desert in Egypt and over the Red Sea.
r хамсин
Жаркий сухой и пыльный ветер, дующий из пустыни в Египте и над Красным морем.
f khamsin *m*
s khamsin *m*
g Khamsin *m*

3320 Khvalyn transgression
The late Quaternary transgression of the Caspian Sea during which the sea level raised up to the absolute height of 50 m, and it occupied the whole Caspian Lowland.
r Хвалынская трансгрессия
Позднечетвертичная трансгрессия Каспийского моря, во время которой уровень моря поднимался до абсолютной высоты 50 м и оно занимало всю Прикаспийскую низменность.
f transgression *f* khvalynienne
s transgresión *f* de cvalyn
g Chvalyn-Transgression *f*

3321 kibboutz
An agricultural community in the newly developed regions of Israel in which common land is used collectively while houses are privately owned.
r киббуц
Сельская община в районах нового освоения Израиля, совместно эксплуатирующая земли общего пользования при сохранении частного домовладения.
f kibboutz *m*
s kibboutz *m*
g Kibbutz *n*

3322 kinematic wave in a glacier
A disturbance of glacier surface having a form of waves of compression and extension which periodically propagate through the glacier and create alternating mounds and depressions on the glacier surface topography.
r кинематическая волна в леднике
Периодически проходящие по леднику волны сжатия и растяжения, приводящие к чередованию на тех же участках бугров и впадин в рельефе ледниковой поверхности.
f onde *f* cinématique dans les glaciers
s onda *f* cinemática en glaciar
g kinematische Welle *f* in Gletschern

* **klippe → 4326**

3323 knickpoint; rejuvenation head
The break of slope in the long profile of a stream which results from a fall of base-level due to uplift of the land or a fall of sea-level (rejuvenation).
r перегиб русла
Резкая смена длинного профиля речного русла, вызванная падением базиса эрозии из-за воздымания территории или понижения уровня моря.
f point *m* d'entaille
s punto *m* de entalladura
g Kerbepunkt *m*

* **knob → 826**

3324 knob-and-kettle topography; hilly lands; kame-and-kettle terrain
The landscape of hillocks alternating with depressions occupied by lakes or bogs, it is characteristic of periglacial areas.
r холмисто-западинный рельеф; камово-западинный рельеф
Чередование беспорядочно расположенных холмов и впадин, занятых озерами и болотами, характерен для перигляциальных областей.
f morphologic *f* de moraines de fond
s modelado *m* de morrenas de fondo
g kuppige Grundmoränenlandschaft *f*; Kamenkesselgelände *n*

* **knot → 3977**

* **kolkhoz → 1123**

3325 Kondratieff cycles
Long waves of economic development that reflect the cyclical nature of scientific and technological progress (the current, fifth cycle is based on the use of microelectronics, new construction materials, biotechnology, etc.).
r циклы Кондратьева
Длинные волны технико-экономического развития в мировой экономике, отражающие цикличность научно-технического прогресса (текущий, пятый цикл связан с внедрением в производство микроэлектроники, новых конструкционных материалов, биотехнологий и др.).
f cycles *mpl* de Kondratieff
s ciclos *mpl* económicos Kondratieff
g Kondratieff-Zyklen *mpl*

3326 kopje
A small isolated hill in Southern Africa, usually comprising a prominent pile of rocks.
r копье
Небольшой изолированный холм в Южной Африке, состоящий из большой груды горной породы.
f kopje *m*
s kopje *m*
g Kopje *m*

3327 krakatao
A layer of easterly winds over the tropics at the altitude of about 18 to 24 km, more prominent and better defined in the summer hemisphere, it derives its name from the observed behaviour of the volcanic dust carried around the globe after the great eruption of Krakatao in 1883.
r ветры кракатау
Слой восточных ветров в тропиках на высоте от 18 до 24 км, лучше выраженный в летнем полушарии, получил название после известного извержения вулкана Кракатау в 1883 г., когда вулканическая пыль распространялась с этими ветрами вокруг земного шара.
f krakatao *m*
s krakatao *m*
g Krakatau-Winde *mpl*

3328 Kronshtadt sea-gauge
Sea-gauge in the granite abutment of a bridge in Kronshtadt on which the average sea level is marked calculated on the basis of long-term observations (it serves as an initial point of the levelling network of Russia).
r Кронштадтский футшток
Футшток в гранитном устое моста в Кронштадте, на котором отмечен средний уровень моря на базе многолетних наблюдений, служит исходным пунктом нивелирной сети, от которой ведется отсчет высот поверхности в России.
f échelle *f* maritime de Kronchtadt
s mariógrafo *m* de Kronchtadt
g Kronstadt-Pegel *m*

*** krotovina → 3910**

Fig. 179. **Krummholz**

3329 krummholz; prostrate wood; prostrate shrubs
Low trees (birch, spruce, pine, cedar, etc.) with trailing sprouts that grow in the mountains at the tree line and in tundra on plains. → Рис. 179
r стланик
Низкорослые заросли деревьев (кедр, сосна, береза, ель и др.), надземные побеги которых стелются по поверхности почвы, встречаются в горах у верхней границы леса и на равнинах в тундре. → Рис. 179
f forêt *f* rampante; forêt *f* traçante; formation *f* traçante; forêt *f* couchée
s bosque *m* rastrero; formación *f* rastrera
g Knieholz *n*; Krummholz *n*

3330 kum
A sand desert or sand massif in Central Asia.
r кум
Пески и песчаные массивы в Средней Азии.
f koum *m*
s kum *m*
g Kum *m*

3331 kurum
A coarse clastic formation on a slope of 2-3° to 40°, moving downslope mainly due to creep. → Fig. 180

Fig. 180. **Kurum** in Transbaikalye Mountains

r курум
Грубообломочное образование на склоне крутизной от 2 до 40°, медленно движущееся вниз под действием крипа. → Рис. 180
f kurum *m*
s kurum *m*
g Kurum *n*

3332 Kyoto treaty
International treaty signed in 1997, by which goverments agreed to reduce their emissions of greenhouse gases to lower than their 1990 levels by 2010.
r Киотский протокол
Международный договор, подписанный в 1997 г., в котором достигнуто соглашение об уменьшении к 2010 г. выбросов парниковых газов до уровня 1990 г.
f Protocole *m* de Kyoto
s Protocolo *m* de Kyoto
g Kyoto-Protokol *n*

3333 kyr
A flat upland, an area with hummocky terrain, a low ridge, or a plateau with steep slopes rising above the plains in Central Asia, Iran and Turkey.
r кыр
Возвышающаяся над равнинами в Центральной Азии, Иране и Турции плосковерхая возвышенность, участок мелкосопочника, невысокая гряда, а также плато с крутыми обрывистыми склонами.
f kyr *m*
s kyr *m*
g Kyr *m*

L

3334 La Niña
A cooling of ocean water off the west coast of South America, coming after an El Nico.
r ла-нинья
Охлаждение воды в океане у западного побережья Южной Америки, наступающее после периода эль-ниньо.
f La Niña *f*
s La Niña *f*
g La Niña *n*

3335 labour
The manual or intellectual work which is one of the factors of production.
r труд
Физическая или интеллектуальная работа, дающая конкретный результат (продукцию).
f travail *m*
s trabajo *m*; labor *f*
g Arbeit *f*

* **labour force → 3338**

* **labour linkages → 1156**

3336 labour market
The mechanism whereby labour is exchanged for material reward or capital.
r рынок труда
Экономическое пространство, где происходит наем и предложение для выполнения конкретных видов работ, а также определяется их оплата.
f marché *m* de travail
s mercado *m* de trabajo
g Arbeitsmarkt *m*

3337 labour productivity
A measurement of labour efficiency in terms of the input of time to produce a unit of production or in terms of the number of products produced in a time unit.
r производительность труда
Мера эффективности труда, измеряемая количеством времени, затраченным на производство единицы продукции, или количеством продукции, произведенной в единицу времени.
f productivité *f* du travail
s productividad *f* del trabajo
g Arbeitsproduktivität *f*

3338 labour resources; labour force; manpower; workforce; workforce
The total supply of people available for work or service.
r трудовые ресурсы; рабочая сила
Часть населения страны, обладающая необходимым физическим развитием, знаниями и практическим опытом для работы в народном хозяйстве.
f main-d'oeuvre *f* du travail
s mano *m* de obra; fuerza *f* de trabajo
g Arbeitskräfte *fpl*

3339 labour-extensive *adj*
Needing a very small work force to achieve a very high output.
r трудоэкономный
Требующий приложения небольшой рабочей силы, чтобы достичь высокого конечного результата.
f travail-extensif *adj*
s trabajo-extensivo *adj*
g arbeitsextensiv *adj*

3340 labour-intensive *adj*
Needing the efforts of a large work force for increased productivity or higher earnings.
r трудозатратный; трудоемкий
Требующий приложения значительной рабочей силы, чтобы достичь большей продуктивности или более высокой прибыли.
f travail-intensif *adj*
s trabajo-intensivo *adj*
g arbeitsintensiv *adj*

3341 labyrinth
A complicated network of cavities in rocks developed on the same plane or on several levels.
r лабиринт
Сильно разветвленная система полостей в горных породах, развитая в одной плоскости или имеющая этажи.
f labyrinthe *m*
s laberinto *m*
g Höhlenlabyrinth *n*

* **laccolite → 3342**

3342 laccolith; laccolite
A mushroom-like intrusive bodyof igneous rock formed at a small depth while magma is intruded along the bedding planes of sedimentary rocks. → Fig. 181
r лакколит
Грибовидное интрузивное тело, образующееся на небольшой глубине при внедрении магмы в толщу осадочных пород. → Рис. 181
f laccolite *m*
s lacolito *m*
g Lakkolith *m*

* **lack of moisture → 1469**

3343 lacustrine *adj*

Fig. 181. **Laccoliths**

Applied to phenomena and processes related to a lake particularly to deposits laid down in a lake and land forms of its banks.

r озерный
Относящийся к явлениям и процессам, связанным с озером, включая отложения на его дне и рельеф его берегов.

f lacustre *adj*
s lacustre *adj*
g See-; lakustrisch *adj*

3344 lacustrine delta
A delta spreading into a lake, built by a stream flowing into the lake.

r озерная дельта
Дельта, сформированная рекой на берегу озера и постепенно расширяющаяся на его акваторию.

f delta *m* lacustre
s delta *m* lacustre
g Seedelta *n*

3345 lacustrine deposits
Detrital, chemical and biogenic sediments formed on the bottom of lakes.

r озерные отложения
Обломочные, химические и органогенные осадки, образующиеся на дне озер.

f dépôts *mpl* lacustres
s depósitos *mpl* lacustres
g Seeablagerungen *fpl*

* **lacustrine plain → 3358**

* **lacustrine terrace → 3361**

3346 lag; lagged time; time-lag
The interval between an event and the time when its effects are apparent.

r отставание; задержка
Интервал времени между происшедшим природным явлением и некоторым его следствием, например, между возникновением паводка и прохождением его через какой-то створ.

f retard *m*; retardement *m*
s retraso *m*; retardamiento *m*
g Rückstand *m*

* **lagged time → 3346**

3347 lagoon
A shallow water basin separated from the open sea by a bar, spit, or coral reef; the circular body of water partly enclosed within an atoll. → Fig. 71, p. 155

r лагуна
Мелководная часть океана (моря), отделенная от него баром, косой, коралловым рифом и соединяющаяся с ним узким проливом или проливами; округлое пространство воды, частично замкнутое внутри атолла. → Рис. 71, с. 155

f lagune *f*; lagon *m*
s albufera *f*; lagoon *f*; laguna *f* litoral
g Lagune *f*

3348 lagoon coast
The coast where the irregularities of the coastline are cut off from the open sea by sand bars or coral reefs.

r лагунный берег
Тип берега, у которого все неровности береговой линии отгорожены от открытого моря песчаным баром, косой или коралловым рифом и образуют вытянутые вдоль берега цепочки лагун.

f côte *f* lagunaire; côte *f* des lagunes
s costa *f* lagunar; costa *f* de albuferas
g Lagunenküste *f*

3349 lagoon sediment
Sediments deposited in shallow-water closed or semi-closed basins (lagoons).

r лагунные осадки
Отложения мелководных замкнутых или полузамкнутых бассейнов (лагун).

f sédiment *m* de lagune
s sedimento *m* lagunar
g Lagunenablagerungen *fpl*; Lagunensedimente *npl*

3350 lahar; volcanogenic mudflow
A mudflow composed mainly of volcanic ash arising at volcanic eruption as a result of a sudden breakthrough of a crater lake or intensive melting of snow and ice on the slopes.

r лахар; вулканогенный грязевой поток
Селевой поток, возникающий при извержении вулкана вследствие прорыва кратерных озер или интенсивного таяния снега и льда на склонах.

f lahar *m*; lave *f* torrentielle volcanique
s lahar *m*; colada *f* de lava viscosa
g Lahar *m*; vulkanogener Schlammstrom *m*

3351 lake
A natural water body lying in a depression of the Earth's surface having no direct connection with the World Ocean. → Fig. 182

r озеро
Естественный водоем на поверхности суши, не связанный непосредственно с Мировым океаном. → Рис. 182

f lac *m*
s lago *m*
g See *m*

3352 lake breeze

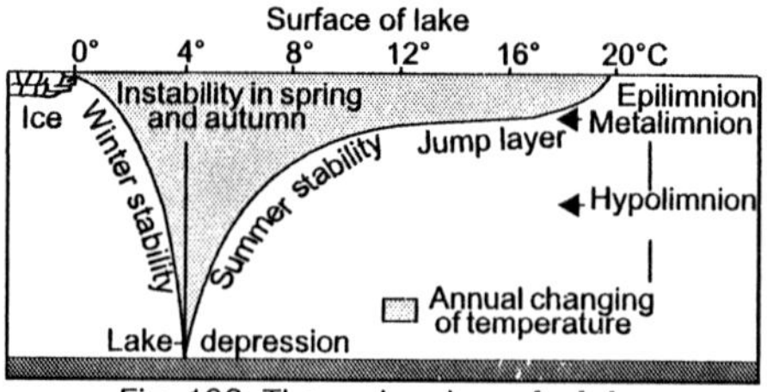

Fig. 182. Thermal regime of a **lake**

Fig. 183. Mountain lake Issyk near Almaty city in Kazakhstan (left) and **lake depression** bared after the violent mudflow happened in the summer of 1963 (right)

A local breeze blowing during the afternoon from a large enclosed body of water during periods of calm weather.

r озерный бриз
Местный бриз, дующий с прилегающего водоема после полудня в период тихой погоды.

f brise *f* de lac
s brisa *f* de lago
g Seebrise *f*

3353 lake depression
A natural hollow on the Earth's surface in which a lake is lying. → Fig. 183

r озерная котловина; озерная ванна; ложе озера
Понижение земной поверхности, служащее вместилищем озера. → Рис. 183

f fossé *f* lacustre
s depresión *f* lacustre; lecho *m* del lago; fosa *f* del lago
g Seewanne *f*

3354 lake ice
The ice formed on lakes and characterised by the vertical structure. → Fig. 184

r озерный лед
Лед, образующийся на озерах и отличающийся вертикальной структурой. → Рис. 184

f glace *f* de lac
s hielo *m* de lago
g Seeeis *n*

3355 lake landscape
A geographical landscape which abounds in lakes, e.g. lakes in the alases of Eastern Siberia, exaration depressions of Finland, etc.

Fig. 184. Structure of **lake ice**

r озерный ландшафт
Географический ландшафт, изобилующий озерами, например, скопления озер в аласах Восточной Сибири, в экзарационных понижениях Финляндии и пр.

f paysage *m* lacustre
s paisaje *m* lacustre
g Seenlandschaft *f*

3356 lake of rift-valley
A lake occupying a depression of rift origin: graben, rift valley, etc. → Fig. 185

r рифтовое озеро
Озеро во впадинах рифтового происхождения: грабенах, рифтовых долинах и др. → Рис. 185

f lac *m* de vallée d'effondrement
s lago *m* de cuenca rift
g Grabenbruchsee *m*

3357 lake percentage
The ratio of the total surface of lakes and reservoirs to the total area under consideration.

r озерность
Отношение суммарной поверхности озер и водохранилищ ко всей площади рассматриваемой территории.

f pourcentage *m* des lacs
s porcentaje *m* de lagos
g Seendichte *f*

3358 lake plain; lacustrine plain
A plain formed on the place of a disappeared lake.

r озерная равнина

Fig. 185. **Lake of rift valley** – Baikal lake

Равнина, образованная на месте исчезнувшего озера.
f plaine *f* d'origine lacustre
s llanura *f* lacustre
g Seenebene *f*

* **lake rampart → 2983**

3359 lake river; effluent stream; effluent river
The river outflowing from a lake or flowing through it, the regime of which is regulated by the lake.
r озерная река
Река, вытекающая из озера или протекающая через него, ее режим регулируется озером.
f rivière *f* du lac
s río *m* del lago
g Seefluß *m*

3360 lake stagnation
The state of a water body when it is stratified into water layers of different density over a long period of time and there is no mixing.
r стагнация озера
Состояние водоема во время длительного расслоения всей его водной толщи на слои воды разной плотности в отсутствии перемешивания.
f stagnation *f* de lac
s estancamiento *m* de lago
g Stagnation *f* des Sees

3361 lake terrace; lacustrine terrace
A terrace which results from the action of waves in a lake.
r озерная терраса
Терраса, обязанная своим происхождением волновой деятельности озера.
f terrasse *f* lacustre
s terraza *f* lacustre
g Seeterrasse *f*

3362 lake-dwelling; pile-dwelling
A dwelling built on piles driven into a marsh or the bed of a shallow lake, common in some tropical areas today, and especially characteristic of certain periods of neolithic times in Central Europe.
r озерное жилище
Жилище, построенное в мелких озерах на сваях, обычное в ряде тропических областей в настоящее время и особенно характерное для некоторых периодов неолита в Центральной Европе.
f habitation *f* lacustre
s vivienda *f* lacustre
g Pfahlbau *m*

3363 lamarckism
Lamarck's theory of evolution stating that environmental changes and the intention of an organism to be perfect cause the evolution of wildlife, and that acquired characteristics are inheritable.
r ламаркизм
Сформулированная Ж.Б. Ламарком теория, согласно которой эволюция живой природы совершается под влиянием окружающей среды и внутреннего стремления всех организмов к усовершенствованию, а приобретенные признаки наследуются.
f lamarcisme *m*
s lamarquismo *m*
g Lamarckismus *m*

3364 lamina
The thinnest layer of stratification in a layered sequence of soils or sedimentary rocks, defined as less than 1 cm in thickness.
r плоскость отслоения
Тончайший слой в почве или осадочных отложениях, обычно толщиной менее 1 см.
f lame *f*
s lamina *f*
g Lamina *f*

3365 laminar flow; parallel flow
A smooth flow of water or air in parallel layers with the same velocity following the direction of the stream with no mixing between adjacent layers, in contrast to *turbulent flow*.
r ламинарное течение
Спокойное, упорядоченное движение воды или воздуха, перемещающихся параллельно направлению течения, в отличие от *турбулентного течения*.
f courant *m* laminaire
s corriente *m* laminar
g Laminarströmung *f*

3366 lamination
In stratigraphy a relatively uniform and frequent alternation of layers differing in some features.
r слоистость; ленточность; полосчатость
Чередование в горных породах сравнительно тонких параллельных слоев, различающихся составом, цветом, структурой, ориентировкой зерен.
f litage *m*
s litaje *m*
g Schichtung *f*; Streifung *f*

3367 land
The area of the Earth's surface lying above sea-level, i.e. continents and islands.
r суша
Часть земной поверхности, не покрытая океанами и морями, т.е. материки и острова.
f terre *f*; terre *f* ferme
s tierra *f*; tierra *f* firme
g Festland *n*

* **land animals → 6103**

3368 land breeze
A wind that blows at night from the land to the sea in coastal regions (or from the land surrounding a large lake to the lake), due to

the differential heating of land and water. → Fig. 46, p. 95

r береговой бриз
Ветер, дующий с суши на море в прибрежных районах (или с суши, окружающей большое озеро, в сторону озера) вследствие разного прогревания земли и воды. → Рис. 46, с. 95

f brise *f* de terre
s brisa *f* de tierra
g Landbrise *f*

3369 land bridge
A hypothetical land connection between continents which served as a migratory route for terrestrial organisms and accounts for the modern distribution of certain flora and fauna.

r материковый мост
Гипотетическая суша, соединявшая континенты и служившая своеобразным мостом для миграции наземных организмов, что привело к современному распределению на Земле флоры и фауны.

f pont *m* de terre
s puente *m* de tierra
g Landbrücke *f*

3370 land cadastre
An official register of the extent and value of real estate: the area, boundaries, location and ownership of land made up for land classification purposes.

r земельный кадастр
Свод систематизированных сведений о природном, хозяйственном и правовом положении земельных угодий какой-либо страны или ее региона.

f cadastre *m* foncier; cadastre *m* des terrains
s catastro *m* parcelário
g Grundbuch *n*; Flurbuch *n*

3371 land capability
The potential of land for agriculture and forestry depending on its physical and environmental qualities, soil type, climate, etc.

r земельный потенциал
Пригодность земли для сельского или лесного хозяйства в зависимости от ее физических и экологических особенностей, характеристик почвы, климата и пр.

f potentiel *m* foncier
s potencial *m* agrario
g Landpotential *n*

* **land classification → 3373**

3372 land disturbance
Reduction in the quality of land caused by economic activity, e.g. changes of soil cover, hydrogeological regime, etc.

r нарушение земель
Качественные изменения земель в результате хозяйственной деятельности, вызывающей нарушение почвенного покрова, гидрогеологического режима и пр.

f dégénération *f* du sol
s degeneración *f* del suelo
g Landbelastung *f*

3373 land evaluation; land classification; appraisal; terrain analysis
Determination of comparative productive ability of soils based upon main natural factors of their fertility (valuation of soils), as well as upon indices of effectiveness of their use (economic estimation of lands).

r оценка земель; бонитировка земель
Установление сравнительной производительной способности земель по основным природным факторам их плодородия (бонитировка почв), а также по показателям эффективности их использования (экономическая оценка земель).

f évaluation *f* des terres; taxation *f* du sol
s valoración *f* de terreno
g Bodenschätzung *f*; Länderbewartung *f*; Bonitierung *f*

3374 land ice; continental ice
Ice formed from solid precipitation and freezing of fresh water lying inland, as opposed to *sea ice*.

r наземное оледенение
Многолетний или сезонный покров снега и льда на поверхности суши, возникающий в результате накопления твердых атмосферных осадков или замерзания пресных вод (в противоположность *морскому льду*, возникающему в основном при замерзании морской воды).

f glace *m* continentale
s hielo *m* continental
g Erdeis *n*

* **land ice-sheet → 3100**

3375 land improvement; improvement of land; land reclamation
A complex of economic and technical measures to make the land useful or more useful for economic (mainly agricultural) purposes, e.g. by drainage of waterlogged land, by irrigation of arid land, by planting vegetation cover in lands subject to erosion, etc.

r мелиорация
Совокупность хозяйственных и технических приемов по улучшению земель, например, с помощью осушения переувлажненных земель, орошения засушливых земель, посадок растительности на землях, подверженных эрозии, и пр.

f amélioration *f*; bonification *f*; amélioration *f* des terres
s mejoramiento *m* del terreno; acondicionamiento *m*; mejoras *f* territoriales
g Bodenverbesserung *f*; Melioration *f*

* **land potential → 4780**

* **land reclamation → 3375**

3376 **land reclamation; reclamation**
Making virgin land, waste sites or land which has already been developed, available for economic, agricultural or social purposes.
r освоение земель
Вовлечение целинных, осушенных и восстановленных земель в хозяйственную деятельность.
f récupération *f* du sol; réhabilitation *f* du sol
s recuperación *f* del suelo; rehabilitación *f* de las tierras
g Erschließung *f* von Land; Urbarmachung *f* von Land

* **land reform → 132**

3377 **land reserve**
The lands of a country or a region classified according to their economic use and legal regime.
r земельный фонд
Совокупность всех земель в пределах страны или какого-либо ее региона, разделяемых по типу хозяйственного использования и правовому режиму.
f réserves *mpl* foncières; fonds *m* foncier
s reservas *fpl* territoriales
g Struktur *f* der Landesfläche

3378 **land resources**
The lands that are used or can be used in different branches of economy including arable and other agricultural lands as well as industrial and housing areas.
r земельные ресурсы
Земли, которые используются или могут быть использованы в разных отраслях народного хозяйства, включая пахотные земли и иные сельскохозяйственные угодья, а также территории под промышленной и жилой застройкой.
f ressources *fpl* foncières
s recursos *mpl* territoriales
g Landressourcen *fpl*

3379 **land rotation; crop rotation; rotation of crops**
A system of land management in which land is cultivated for a few years and then alowed to rest so that soil fertility is not exhausted.
r севооборот; ротация земель
Последовательная смена засеваемых и оставляемых под паром участков земли, определенный порядок смены или последовательности культур на данном участке, позволяющий избежать истощения почвы.
f assolement *m*; rotation *f* culturelle; rotation *f* des cultures
s rotación *f* de cultivos; cultivos *mpl* alternados
g Saatfolge *f*; Anbaufolge *f*; Landesrotation *f*

3380 **land subsidence**
Sinking of the ground due to suffosion, karstic processes, thawing of the ground in a permafrost zone.
r просадка грунта
Оседание грунта, вызываемое суффозией, карстовыми процессами, протаиванием грунта в области многолетнемерзлых пород.
f tassement *m*; soutirage *m*
s hundimiento *m*
g Erdsenkung *f*

3381 **land tenure**
Land ownership on certain legal conditions: possession, use, inheritance, rent.
r землевладение
Обладание землей на определенных правовых основаниях: собственности, пользования, пожизненного наследуемого владения, аренды.
f tenure *f*
s tenencia *f*
g Grundbesitz *m*; Grundeigentum *n*

3382 **land transport**
A type of transport carrying passengers and goods on the land.
r наземный транспорт
Вид транспорта, занятый перевозкой пассажиров и грузов по суше.
f transport *m* terrestre
s transporte *m* terrestre
g Landtransport *m*

3383 **land use; land utilisation**
The use of land resources for economic and other purposes (agricultural, industrial, recreational and residential) based on tradition or regulations.
r землепользование; использование земель
Использование земельных ресурсов в хозяйственных и иных целях на основе обычаев или в установленном законом порядке.
f utilisation *f* du sol
s usos *mpl* del suelo; aprovechamiento *m* del suelo
g Landnutzung *f*

* **land utilisation → 3383**

* **landfast ice → 2105**

3384 **landform**
The shape, form, nature of a specific physical feature of the Earth's surface (e.g. a hill, a plateau) produced by the natural erosive processes or by tectonic processes.
r форма рельефа
Неровность земной поверхности разного размера и положения.
f forme *f* de relief
s forma *f* de relieve
g Reliefform *f*

3385 **landlocked** ***adj***

Applied to an area (particularly a state) which lacks a sea coast and thus does not have direct access to the sea.

r закрытый от моря
Относится к территории (как правило, к государству) без морского побережья и тем самым не имеющей прямого выхода к морю.
f fermé *adj* de land
s cercado *adj* de tierra
g landumschlossen *adj*

* **landmark → 4954**

* **landmass → 1239**

* **landscape → 2420**

3386 landscape aesthetics
Preservation and restoration of a landscape which is based on the view that the beauty of scenery should be preserved to give a person the opportunity to feel the excellence of nature.
r ландшафтная эстетика
Сохранение и восстановление ландшафта как некоего законченного целого, дающего возможность человеку ощутить красоту и совершенство природы.
f esthétique *f* de paysage
s estética *f* de paisaje
g Landschaftsästhetik *f*

3387 landscape aesthetics; landscape esthetics (*amer*)
The scenic beauty of a landscape which affects the aesthetic feelings of people.
r эстетика ландшафта
Зрительная красота ландшафта, оказывающая воздействие на чувственное восприятие людей.
f esthétique *f* de paysage
s estética *f* de paisaje
g Ästhetik *f* der Landschaft

3388 landscape architecture
Combining architectural constructions and landscape elements into a harmonious ensemble, including the design of gardens.
r ландшафтная архитектура
Объединение архитектурных сооружений и элементов ландшафта в единое гармоническое целое.
f architecture *f* de paysage
s arquitectura *f* de paisaje
g Lanaschaftsarchitekteur *f*

3389 landscape capacity; landscape potential; natural potential
The ability of a landscape to provide normal conditions for organisms living in the area or to meet the demands of the economy without any negative effect on nature.
r емкость ландшафта;
потенциал ландшафта;
природный потенциал территории
Способность ландшафта обеспечить нормальную жизнедеятельность организмов на этой территории или удовлетворить разнообразные потребности общества без отрицательных последствий для природы.
f capacité *f* de paysage;
potentiel *m* de paysage
s capacidad *f* de paisaje;
potencial *m* de paisaje
g Landschaftskapazität *f*;
Landschaftspotential *n*

3390 landscape change
Acquiring new or losing some former properties of landscape under anthropogenic the impact or owing to self-development process.
r изменение ландшафта
Приобретение ландшафтом новых или утрата прежних свойств в результате антропогенного воздействия или под влиянием процесса саморазвития.
f changement *m* de paysage
s cambio *m* de paisaje
g Landschaftsveränderung *f*

* **landscape component → 1172**

3391 landscape degradation
Natural or anthropogenic changes of a landscape that cause irreversible deterioration of its structure and functioning.
r деградация ландшафта
Естественные или антропогенные изменения ландшафта, необратимо ухудшающие его структуру и функционирование.
f dégradation *f* de paysage
s degradación *f* de paisaje
g Landschaftdegradation *f*

3392 landscape design
Creating a plan of how to lay out a landscape in order to make it most attractive and suitable for use.
r ландшафтный дизайн
Проектирование фрагментов архитектурно-ландшафтной среды в целях их наилучшего приспособления к потребностям человека.
f dessein *m* de paysage
s desesio *m* de paisaje
g Lanaschaftsdesign *n*

3393 landscape differentiation
The growth of differences in structure and functioning of separate parts of landscape resulting in the isolation of morphological units (facies, stows, etc.) within a homogeneous landscape.
r дифференциация ландшафта
Процесс углубления различий в структуре и функционировании отдельных частей ландшафта, приводящий к обособлению морфологических единиц (фаций, урочищ и пр.) в пределах ранее однородного ландшафта.
f différenciation *f* de paysage

s diferenciación *f* de paisaje
g Landschaftsdifferenzierung *f*; Differentiation *f* der Landschaft

3394 landscape dynamics
Reversible changes of a landscape which are not accompanied by the changes of its structure.
r динамика ландшафта
Обратимые изменения состояния ландшафта, не сопровождающиеся изменениями его структуры.
f dynamique *f* de paysage
s dinámica *f* de paisaje
g Landschaftsdynamik *f*

*** landscape esthetics (*amer*) → 3387**

3395 landscape evaluation
The assessment of landscape as a scenic resource, i.e. from the point of view of its scenic beauty, which is vital for the conservation of high-quality of traditional landscapes, especially in densely populated, industrially developed regions.
r оценка ландшафта
Определение качеств ландшафта как живописного ресурса, т.е. с точки зрения его зрительной красоты, что жизненно важно для сохранения высоких природных качеств ландшафта, особенно в густонаселенных и промышленно развитых регионах.
f évaluation *f* du paysage
s valoración *f* de paisaje
g Landschaftabschätzung *f*

3396 landscape facies; asrect *m* de terrain
An elementary morphological unit of geographical landscape, i.e. an elementary natural complex associated with one element of mesorelief or form of microrelief, with the uniform of moisture and insolation conditions.
r ландшафтная фация
Элементарная морфологическая единица географического ландшафта, т.е. простейший природный комплекс, приуроченный к одному элементу мезорельефа или отдельной форме микрорельефа, обладает однородными условиями увлажнения и инсоляции.
f faciès *m* du paysage
s facies *m* de paisaje
g Landschaftsfazies *f*

3397 landscape geochemistry
Chemical composition of landscape components, the main forms and regularities of the migration of chemical elements in landscapes, and the study of these processes.
r геохимия ландшафта
Химический состав компонентов ландшафта, основные формы и закономерности миграции в ландшафтах химических элементов и изучение этих процессов.
f géochimie *f* de paysage
s geoquímica *f* de paisaje
g Geochemie *f* der Landschaft

3398 landscape geosphere; Earth´s physical environment
A natural complex in which the upper part of the lithosphere, hydrosphere, lower layers of the atmosphere and all living matter on the Earth (biosphere) come into contact and interact, the main subject of physical geography.
r географическая оболочка
Сложный природный комплекс, в котором соприкасаются, взаимно проникают и взаимодействуют верхняя часть литосферы, вся гидросфера, нижние слои атмосферы и все живое вещество на Земле (биосфера), служит основным объектом изучения физической географии.
f sphère *f* géographique
s esfera *f* geográfica
g Landschaftshülle *f* der Erde

3399 landscape inversion
Reverse altitudinal distribution of landscapes not following the usual altitudinal zonaltion, it is caused by the inversion of temperature or different conditions of drainage along a slope.
r инверсия ландшафтов
Обратное распределение ландшафтов по высоте, не соответствующее обычной высотной поясности, вызывается инверсией температуры или неодинаковыми условиями дренажа на разных участках склонов.
f inversion *f* altitudinal des paysages
s inversión *f* altitudinal de paisajes
g Inversion *f* der Landschaft

3400 landscape morphology; landscape structure
The structure of landscape and its sepafrate parts (morphological units) – facies and stows, their interrelationships.
r морфология ландшафта; структура ландшафта
Внутреннее строение ландшафта и отдельных его частей – ландшафтных фаций и урочищ, их связи и взаимоотношения.
f morphologie *f* de paysage; structure *f* de paysage
s morfología *f* de paisaje; estructura *f* de paisaje
g Landschaftsmorphologie *f*; Morphologie *f* der Landschaft; Landschaftsstruktur *f*

3401 landscape park
A park designed to resemble natural land.
r пейзажный парк
Парк со свободной планировкой, как бы воспроизводящей естественный ландшафт.
f parc *m* de paysage
s parque *m* de paisaje
g Landschaftspark *m*

3402 landscape planning
The design of landscape utilization or change of the purposes and methods of its use.

r ландшафтное планирование
Проектирование использования ландшафтов или изменения целей и методов их использования.
f planification *f* de paysage
s planificación *f* de paisaje
g Landschaftsplanung *f*

* **landscape potential → 3389**

3403 landscape recultivation
The restoration of economic, medical, biological and aesthetic values of a disturbed or degraded landscape.
r рекультивация ландшафта
Восстановление хозяйственной, медико-биологической и эстетической ценности нарушенного или деградировавшего ландшафта.
f remise *f* en culture de paysage
s recultivación *f* de paisaje
g Landschaftsrekultivierung *f*

* **landscape science → 3404**

* **landscape stability → 1977**

* **landscape structure → 3400**

3404 landscape study; terrain analysis; landscape science
The study of geographical landscapes, their genesis, evolution, hierarchy, dynamics and functioning, interrelationship and interaction of components, possible economic use.
r ландшафтоведение
Изучение географических ландшафтов, их генезиса, эволюции, иерархической структуры, динамики и функционирования, взаимосвязей отдельных компонентов, возможностей хозяйственного использования.
f étude *f* des paysages; analyse *f* du paysage; science *f* du paysage
s estudio *m* de paisajes
g Landschaftskunde *f*

3405 landscape succession
The sequential changes within a landscape which is taken into consideration during its economic use.
r сукцессия ландшафта
Процесс смены изменяющихся естественных состояний ландшафта, принимаемый во внимание при его хозяйственном использовании.
f succession *f* de paysage
s sucesión *f* de paisaje
g Landschaftssukzession *f*

* **landscape surveying → 3675**

* **landscape unit → 5835**

3406 landslide; slide; slide; landslip
The downslope gravitational movement of a mass of rock or earth where the displaced material retains its form, it may be induced by natural agencies, such as heavy rain, earthquake, or it may be caused by human interference with slope stability. → Fig. 186
r оползень
Движение вниз по склону масс поверхностного материала благодаря его обрушению, иногда в результате появления в почвенном покрове воды и уменьшения трения вдоль плоскости сдвига, нередко вызывается хозяйственной деятельностью, ведущей к потере устойчивости склона. → Рис. 186
f glissement *m* de terrain
s deslizamiento *m* de terreno
g Rutschung *f*; Bergrutsch *m*; Erdrutsch *m*

3407 landslide base; landslide foot
The lower part of a landslide adjoining the surface of a slope.
r подошва оползня
Нижняя часть оползня, прилегающая к поверхности его скольжения.
f sol *m* de glissement
s pie *m* de deslizamiento
g Fuss *m* eines Erdrutsches

3408 landslide body
A sliding rock mass that remains unbroken or collapses into individual blocks.
r тело оползня
Сползшая масса грунта, сохраняющая свою монолитность или распавшаяся на отдельные глыбы.
f masse *f* de glissement
s masa *f* de deslizamiento
g Rutschmasse *f*; Rutschkörper *n*

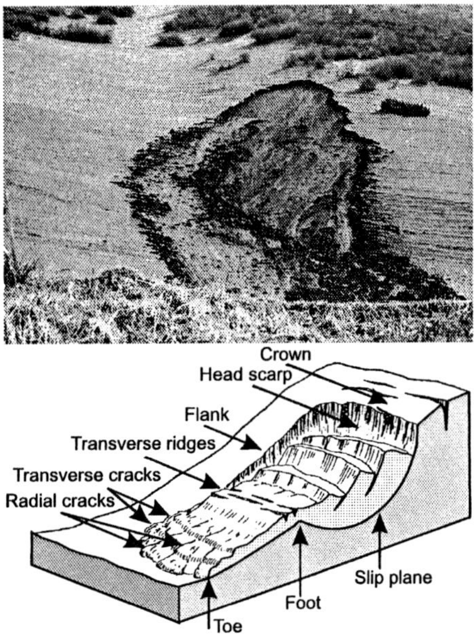

Fig. 186. Natural **landslide** and its elements

* **landslide foot → 3407**

3409 **landslide front**
The front part of a landslide body.
r фронт оползня
Передняя, надвигающаяся часть тела оползня.
f front *m* de glissement
s frente *m* de deslizamiento
g Rutschfront *f*

3410 **landslide lobe**
The lower part of a landslide body.
r язык оползня
Нижняя часть тела оползня.
f loupe *f* de glissement
s lengua *f* de deslizamiento
g Rutschzunge *f*

3411 **landslide slope**
A slope with the morphological traces of a landslide or a series of landslides.
r оползневой склон
Склон с морфологическими следами оползня или серии оползней.
f versant *m* à glissement
s vertiente *f* de deslizamiento
g Rutschhang *m*

3412 **landslide terrace**
A small surface, often inclined towards the slope, which appears after a landslide.
r оползневая терраса
Небольшая площадка, часто с обратным уклоном, возникающая на склоне после схода оползня.
f gradin *m* de glissement de terre; glacis-terrasse *f*
s terraza *f* de deslizamiento; glacis-terraza *f*
g Rutschterrasse *f*

* **landslip → 3406**

3413 **land-surveying; land-use survey; land-use planning**
The demarcation of land for specific uses, based on environmental, social and economis criteria, which takes into account present and possible future needs, and mapping the use of land surface.
r межевание
Подразделение земель для конкретных целей их использования на основе экономических, социальных и экологических критериев, с учетом современного и возможного в будущем использование этих земель, а также картографическое и юридическое оформление этой работы.
f arpentage *m*
s agrimensura *f*; apeo *m*
g Landvermessung *f*; Feldmessung *f*

* **land-use planning → 3413**

* **land-use survey → 3413**

3414 **land-use system**
A system of measures that regulate the use of land as a means of production and its protection.
r землеустройство
Система мероприятий по регулированию земельных отношений и организации охраны и использования земли как средства производства.
f réglementation *f* de utilisation du sol
s reglamentación *f* de usos del suelo; sistema *m* de usos del suelo
g Flurbereinigung *f*

3415 **lane; narrows**
A strait connecting a bay or interior sea with the outer sea in the north of the European Russia.
r горло
Пролив, соединяющий залив или внутреннее море на севере Европейской России с внешним морем.
f goulet *m*
s estrecho *m* garganta
g Gurgel *f*; Enge *f*

* **Langmuir circulation → 3416**

3416 **Langmuir eddy; Langmuir circulation**
A spiral-like whirling of water around a horizontal axis arising in the upper ocean layer due to its interaction with the atmosphere.
r вихрь Лангмюра; циркуляция Лангмюра
Спиралеобразное вихревое движение воды вокруг горизонтальной оси, возникающее в верхнем слое океана вследствие его взаимодействия с атмосферой.
f tourbillon *m* de Langmuir; circulation *f* de Langmuir
s torbellino *m* de Langmuir; circulación *f* de Langmuir
g Langmuir-Wirbel *m*; Langmuirsche Zirkulation *f*

* **lapies → 3288**

3417 **lapilli**
Tiny pieces, between 4 and 32 mm in diameter, of porous lava ejected during the volcanic eruption and solidified in the air.
r лапилли
Кусочки пористой лавы, выбрасываемые при извержении вулкана и затвердевающие в воздухе.
f lapilli *m*
s lapilli *m*
g Lapilli *mpl*

* **lapse rate → 6558**

3418 **lapse-rate process**
The rate of temperature decrease of a parcel of air as it moves upward in the atmosphere, following either the dry adiabatic lapse rate or the saturated adiabatic lapse rate.

r градиентный процесс
Скорость уменьшения температуры в единичном объеме воздуха по мере того как он поднимается в атмосфере, следуя либо сухоадиабатическому, либо влажноадиабатическому градиенту температуры.
f processus *m* de gradient
s proceso *m* de gradiente
g Gradientenprozeß *m*

* **Laramide orogeny → 3419**

3419 **Laramidian folding; Laramide orogeny**
The era of mountain formation at the end of the Cretaceous and beginning of thePalaeogene, fully manifested in the Rocky Mountains of North America.
r Ларамийская складчатость
Эра горообразования в конце мелового — начале палеогенового периодов, наиболее полно проявившаяся в Скалистых горах Северной Америки.
f plissement *m* laramien
s plegamiento *m* laramiano
g laramische Faltung *f*

3420 **larch-tree forest**
A light forest composed of larch trees and having a well developed grass and moss cover, it occurs on permafrost grounds of North-eastern Asia.
r лиственничный лес; лиственничник
Негустой лес с хорошо развитым травянистым и моховым покровом, образованный различными видами лиственницы, широко распространен на многолетнемерзлых грунтах Северо-Восточной Азии.
f forêt *f* de mélèze
s bosque *m* de alerce
g Lärchenwald *m*

* **lard ice → 2617**

3421 **large scale map**
Topographic map on the scale of 1:100,000 and more.
r крупномасштабная карта
Топографическая карта масштаба 1:100 000 и крупнее.
f carte *f* à grande échelle
s mapa *m* a gran escala
g großmaßstäbige Karte *f*

3422 **latent ground ice**
A hidden ground ice which after thawing remains in permafrost in the form of bounded water.
r латентный лед
Скрытый подземный лед, при протаивании многолетнемерзлых пород остающийся в породе в виде связанной воды.
f glace *f* latente
s hielo *m* latente
g Schmelzeis *n*

3423 **latent heat**
The amount of heat absorbed or released when a substance changes its physical state at constant temperature, i.e. without any change of temperature within that body, e.g. from a solid to a liquid at its melting point, or from a liquid to a gas at its boiling point.
r скрытая теплота
Количество тепла, поглощаемое или выделяемое при изменении физического состояния вещества в условиях постоянной температуры, например, при переходе из твердой фазы в жидкую на точке таяния или от жидкой фазы в парообразную на точке кипения.
f chaleur *f* latente
s calor *m* latente
g gebundene Wärme *f*

* **latent overpopulation → 4998**

3424 **lateral accretion**
The built-up of sediments at the slope of a river or during braiding.
r боковой нанос
Отложения на склоне речной долины, возникающие, в частности, в результате блуждания реки.
f accrétion *f* latérale
s acrecimiento *m* lateral
g laterale Akkreszenz *f*

* **lateral crater → 4418**

3425 **lateral erosion**
The action performed by a river when it wears away its banks in a horizontal direction which makes the channel migrate and form meanders.
r боковая эрозия
Усиленный размыв одного из берегов водным потоком, ведущий к смещению русла и образованию меандр.
f érosion *f* latérale
s erosión *f* lateral
g Seitenerosion *f*

3426 **lateral eruption**
The eruption of lava from fissures on the slopes or at the foot of a volcanic cone.
r боковое извержение
Извержение лавы из трещин на склоне или у подножья вулканического конуса.
f éruption *f* latérale
s erupción *f* lateral
g Lateraleruption *f*; Flankenausdruch *m*

3427 **lateral migration**
The gradual shift of a stream channel across a floodplain due to lateral erosion.
r боковая миграция
Постепенный сдвиг речного потока поперек поймы в результате боковой эрозии.
f migration *f* latérale
s migración *f* lateral
g Seitenwanderung *f*

3428 lateral moraine
An elongated range of debris deposited along the mountain glacier sides. → Fig. 85, p. 182
r береговая морена
Моренная гряда, отложенная у бокового края горного ледника. → Рис. 85, с. 182
f moraine *f* latérale
s morrena *f* lateral; morrena *f* ribereña
g Seitenmoräne *f*; Randmoräne *f*

3429 laterite
A red residual deposit created from the weathering of rocks under humid tropical conditions.
r латерит
Красноцветные элювиальные отложения, формирующиеся в результате выветривания горных пород во влажных тропических и субтропических условиях.
f latérite *f*
s laterita *f*
g Laterit *m*

* **laterite soil → 3431**

3430 lateritic *adj*
Related to products (soils, rocks, etc.) of deep and long weathering of alumosilicate rocks in the conditions of humid tropical and subtropical climate.
r латеритный
Относящийся к продуктам (почвам, горным породам и т.п.) глубокого и длительного выветривания алюмосиликатных горных пород в условиях влажного тропического и субтропического климата.
f latéritique *adj*
s laterítico *adj*
g lateritisch *adj*

3431 lateritic soil; laterite soil; latosol
A red and yellow soil of equatorial and humid tropical forests that has laterite horizon and a high content of iron and aluminium, and a low content of silica, therefore it has little agricultural value.
r латеритная почва; латерит
Красные и желтые почвы экваториальных и влажных тропических лесов, имеющие в профиле горизонт латерита и характеризующиеся высоким содержанием железа и алюминия и, наоборот, низким содержанием кремнезема, и поэтому обладающие невысоким плодородием.
f sol *m* latéritique
s suelo *m* laterítico
g Lateritboden *m*

* **lateritisation → 3432**

3432 laterization; lateritisation
The process of deep and long weathering of alumosilicate rocks under humid tropical and subtropical conditions resulting in the formation of laterites or lateritic soils in which little or no humus is found, iron and aluminium compounds accumulate in layers, silica is leached out.
r латеритизация
Процесс глубокого и длительного выветривания алюмосиликатных горных пород в условиях влажного тропического и субтропического климата, приводящий к образованию латеритов или латеритных почв, содержащих малое количество гумуса и, наоборот, значительное количество соединений железа и алюминия.
f lateritisation *f*
s lateritización *f*
g Lateritibildung *f*

3433 latifundium
A large farm or estate mainly engaged in livestock breeding, originally set up as an imperial grant to New World settlers in Latin America.
r латифундия
Крупное земельное владение, на котором преимущественно разводят скот, первоначально они утверждались в качестве грантов для испанских переселенцев в Латинской Америке.
f latifundium *m*; grande propriété *f* foncière
s latifundio *m*; gran propiedad *f* agrária
g Latifundie *f*

3434 latitude
The angular distance of any point on the Earth's surface along a meridian measured north or south from the equator in degrees, minutes and seconds, according to the angle parallels of latitude from 0° to 90°.
r географическая широта; широта местности
Угловое расстояние любой точки на поверхности Земли вдоль меридиана, отсчитываемое к югу и северу от экватора в градусах, минутах и секундах в соответствии с углом данной широтной параллели от 0° до 90°.
f latitude *f*
s latitud *f*
g geographische Breite *f*

* **latitudinal zonation → 6856**

* **latosol → 2128**

3435 Laurasia
The hypothetical continent of the Northern Hemisphere which broke from the supercontinent Pangea by continental drift and existed since the middle of the Palaeozoic being separated by Thetys Ocean from the Southern continent (Gondwanaland).
r Лавразия
Гипотетический материк Северного полушария Земли, который откололся от суперконтинента Пангеи в результате континентального дрейфа и существовал с середины палеозоя, будучи отделенным океаном Тетис от Южного материка – Гондваны.
f Laurasie

s Laurasia; continente *m* de Laurasia
g Laurasia

3436 **Laurentian ice sheet**
Continental ice sheet which existed in the Pleistocene in the north of America having the centre located in the Canadian shield.
r Лаврентийский ледниковый покров
Материковый ледниковый покров, существовавший в плейстоцене на севере Американского континента с центром на Канадском кристаллическом щите.
f bouclier *m* glaciaire laurentienne
s casquete *m* glaciar lauréntico
g laurentischer Eisschild *m*

3437 **lava**
Hot molten rock (magma) that issues during volcanic eruptions to the surface of the ground where it cools and solidifies.
r лава
Огненно-жидкий расплав, изливающийся во время вулканических извержений на земную поверхность.
f lave *f*
s lava *f*
g Lava *f*

3438 **lava blister**
A hollow protuberance on the surface of a lava-flow caused by gas bubbles.
r лавовый пузырь
Пустотелый выброс на поверхности лавововго потока, вызванный пузырями газов.
f ampoule *f* de lave
s ampolla *f* de lava
g Lavablase *f*

3439 **lava cone**
A gentle volcanic cone with low-angled slopes built entirely of lava-flows in which explosive activity was minimal.
r лавовый конус
Пологий вулканический конус, целиком построенный потоками лавы в условиях малой эксплозивной деятельности.
f cône *m* de lave
s cono *m* de lava
g Lavakegel *m*

3440 **lava effusion; volcanic effusion**
The thrusting out of magma from the surface of the Earth during a volcanic eruption.
r излияние лавы
Выход наружу огненно-жидкого расплава во время вулканического извержения.
f épanchement *m* de lave; épanchement *m* volcanique
s derramamiento *m* de lava; derramamiento *m* volcánico
g Effusion *f* der Lava

3441 **lava eruption; Hawaiian eruption**
A non-explosive discharge of volcanic material in the form of effusion of liquid basaltic lava accompanied by only insignificant outburst of loose pyroclastic products. → Fig. 187
r лавовое извержение; гавайский тип извержения
Относительно спокойное извержение вулкана в виде излияния жидкой базальтовой лавы, сопровождающегося незначительным выбросом рыхлых пирокластических продуктов. → Рис. 187
f éruption *f* de lave; éruption *f* lavique; éruption *f* hawaïenne
s erupción *f* de lava; erupción *f* hawaiana; erupción *f* lávica
g Lavaeruption *f*; Lavaausbruch *m*; hawaiiansischer Eruptionstyp *m*

3442 **lava field**
A wide expanse of lava-flows which have coalesced at the foot of one or more volcanic cones to produce a terrain of continuous lava extending over many square kilometres.
r лавовое поле
Широкое распространение лавовых потоков, образующих у подножий одного или нескольких вулканических конусов обширную лавовую поверхность, нередко простирающуюся на много квадратных километров.

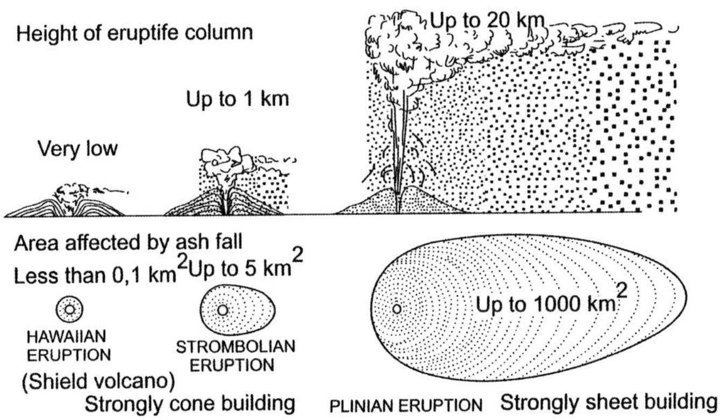

Fig. 187. **Lava eruption**: Hawaiian, Strombolian and Plinian types of eruption

f terrain *m* de lave; champ *m* de lave
s terreno *m* de lava; campo *m* de lava
g Lavafeld *n*

3443 lava lake
A lake of hot liquid lava occupying a volcanic crater or depression in the Earth's surface.
r лавовое озеро
Озеро из огненно-жидкой лавы в вулканическом кратере или углублении земной поверхности.
f lac *m* de lave
s lago *m* de lava
g Lavasee *m*

*** lava plateau → 3444**

3444 lava sheet; lava plateau
A form of bedding of lava effused in great amounts in the conditions of plain relief and often forming a volcanic plateau.
r лавовый покров
Форма залегания лавы, излившейся в больших количествах в условиях равнинного рельефа и нередко формирующей вулканические плато.
f nappe *f* de lave
s capa *f* de lava
g Lavadecke *f*

3445 lava-flow
A narrow and thin stream up to 10 km long of acid lavas and up to 80 km of basaltic lavas effused from a volcano.
r лавовый поток
Узкий и тонкий поток длиной до 10 км у кислых лав и до 80 км у базальтовых, излившийся из вулкана.
f coulée *f* de lave
s colada *f* de lava
g Lavastrom *m*

3446 lavaka
A depression cut into a convex slope of a dome-like upland or into a plateau edge in humid tropics.
r лавака
Впадина на выпуклом склоне куполовидной возвышенности, либо на краю плато во влажных тропиках.
f lavaka *f*
s lavaka *f*
g Lavaka *f*

3447 layda
Coastal swamped meadow on low coasts of European Russia flooded by high tides.
r лайда
Приморский заболоченный луг на низменных побережьях Европейской России, затопляемый высокими приливами.
f lajda *m*
s layda *m*
g Laida *m*

*** layer → 5851**

3448 layer of vegetation
A stratum of a plant community that differs either in height of constituent plants or in the depth of roots growing in soil.
r ярус растительности
Отдельные слои растительного сообщества, выделенные по высоте входящих в них растений или по глубине проникания корней в почву.
f strate *f* de végétation
s estrato *m* de vegetación
g Pflanzenschichten *f*

*** layered freezing of water → 4020**

3449 layered glacier
A glacier formed by the coalescence of two or more glaciers superimposed above each other and having different size, rate of flow, and structure of ice.
r слоистый ледник
Ледник, образованный из слившихся потоков двух или нескольких ледников, наложенных друг на друга и различающихся по величине, скорости движения и структуре льда.
f glacier *m* étagé
s glaciar *m* estratiforme
g geschichteter Gletscher *m*

3450 layering of relief
The successive change of the types of relief in the mountains with their height.
r ярусность рельефа
Последовательная смена типов рельефа в горах по мере возрастания их высоты.
f étagement *m* du relief
s pisos *mpl* de relieve
g Reliefschichtung *f*

*** lea → 2096**

3451 leached soil
A soil in the profile of which there are no easily soluble salts, gypsum and carbonates.
r выщелоченная почва
Почва, в профиле которой нет легкорастворимых солей, гипса и карбонатов.
f sol *m* lessive; sol *m* lévige
s suelo *m* lavado
g ausgelaugter Boden *m*; lessivierter Boden *m*; Fahlerde *f*

3452 leaching
The process by which water moves materials down the soil profile in solution or suspension: humus, soluble bases, clay, etc. in such a way making the upper layer increasingly acid and mineral-deficient while the eluvial horizon develops.
r выщелачивание; иллимеризация почв
Процесс вымывание из почвенного профиля в виде раствора или суспензии частиц гумуса, растворимых оснований, гли-

нистых частиц и т.п., в результате чего в верхних горизонтах возникает дефицит кислотных и минеральных частиц и получает развитие элювиальный горизонт.
f éluviation *f*
s lixiviación *f*; eluviación *f*
g Auslaugen *n*; Lösungsverwitterung *f*

3453 leaching groundwaters
Ground waters formed as a result of an intensive infiltration of atmospheric precipitation and processes of leaching of soils and rocks.
r воды выщелачивания
Подземные воды, образовавшиеся в результате интенсивной инфильтрации атмосферных осадков и процессов выщелачивания ими почв и горных пород.
f eau *f* souterraine de lessivage
s agua *f* subterránea lixiviada
g Auswaschungswasser *n*

*** lead → 2939**

3454 lead; sounding lead
An instrument for measuring water depth from board a ship.
r лот
Прибор для измерения глубины воды с борта судна (различают механические и гидроакустические лоты).
f sonde *f*
s sonda *f*
g Lot *n*

3455 leaf fall; defoliation
Mass shedding of leaves from plants in some seasons such as autumn, dry period of summer, or which may be induced by defoliants
r листопад
Массовая потеря листьев у растений в некоторые сезоны года: осенью, в сухой период лета, может быть вызван искусственно воздействием дефолиантов.
f chute *f* des feuilles; défeuillaison *f*
s deshoje *m*; defoliación *f*; caída *f* de la hoja
g Blattfall *m*; Entlaubung *f*

3456 leaf litter
A layer of dead leaves and broken tree branches covering the soil under the trees in a forest.
r подстилка
Слой опавших листьев и веток, покрывающих почву в лесу.
f litière *f* des feuilles; couverture de feuilles mortes; couverture morte
s lecho *m* de hojas
g Laubstreu *f*

*** leaf-bearing forest → 1450**

3457 leaf-bearing plant
Trees and shrubs composed of broad-leaved, small-leaved or hard-leaved species, found almost in all geographical zones.
r лиственное растение
Деревья и кустарники широколиственных, мелколиственных и жестколистных пород, встречаются почти во всех географических зонах.
f plante *f* feuillue
s planta *f* foliácea
g Blattpflanze *f*

*** leafless plant → 332**

3458 leap-year
A calendar year having 366 days, according to the Gregorian calendar every year divisible by four is a leap year except the years which are divisible by 100, but not divisible by 400.
r високосный год
Календарный год, содержащий 366 дней, по новому стилю високосным является каждый год, кратный четырем, кроме тех годов, числа которых оканчиваются на два нуля, но не делятся на 400.
f année *f* bissextile
s año *m* bisiesto
g Schaltjahr *n*

3459 lease of land
A form of land use in which the owner of land gives it another person (tenant) the wright to use it for a specified period in return for a payment of rent.
r аренда земли
Форма землепользования, когда владелец земельного участка передает его во временное пользование другому лицу (арендатору) за вознаграждение.
f fermage *m*; bail *m*
s arrendamiento *m*; regimen *m* de tenencia
g Pacht *f*

3460 leat; drain$_2$
A small artificial channel, often temporary, constructed to allow the free drainage flow of water from an area.
r канава; лоток
Небольшое искусственное русло, часто временное, предназначенное для отвода по нему вод.
f fossé *m*; rigole *f*
s zanja *f*; foso *m*
g Überleitungskanal *m*

*** lee eddy → 3461**

3461 lee wave; lee eddy
A stationary wave created in a moving air or water stream, which forms on the leeward side of an obstacle such as a hill or mountain range.
r подветренная волна
Стационарная волна в воздушном или водном потоке, образующаяся на подветренной стороне неровности рельефа, например, холма или горного хребта.
f onde *f* sous le vent
s onda *f* a sotavento
g Leewelle *f*

3462 lee-shore
Any shoreline towards which the wind is blowing, thereby creating a danger to shipping.
r наветренный берег
Участок берега, обращенный к господствующим ветрам и тем самым опасный для судоходства.
f côte *f* sou le vent
s sotavento *m*
g Luvküste *f*

3463 leeward *adj*
Turned to the side where wind blows.
r подветренный
Расположенный «по ветру», т.е. обращенный в ту сторону, куда дует ветер.
f sous le vent
s de sotavento *adj*; a sotavento *adj*
g leeseitig *adj*

3464 leeward slope
The slope opposite to the one exposed to the direction from which the prevailing wind blows.
r подветренный склон
Склон, обращенный в сторону, противоположную той, откуда дуют господствующие ветры.
f versant *m* sous-le-vent
s vertiente *f* bajo-el-viento; sotavento *m*
g windabgewandter Hang *m*; Leehang *m*

* **legal time → 1456**

3465 legend; map's legend
A set of conventional signs and explanations of the symbols used on a map.
r легенда карты
Свод условных знаков и пояснений, раскрывающих содержание карты.
f légende *f*
s leyenda *f* de mapa; simbología *f* de mapa
g Zeichenerklärung *f*; Legende *f*

3466 leisure
That time left over after time for work and after obligation.
r досуг; свободное время
Время, остающееся у человека после всякой работы и выполнения обязательств.
f loisir *m*
s ocio *m*
g Mußestunde *f*

3467 length of a river
The distance from the source of a river to the mouth measured by the axis of the channel, and for the major rivers by fairway.
r длина реки
Расстояние от истока до устья реки, измеряемое по геометрической оси русла, а для больших рек по фарватеру.
f longueur *f* de fleuve
s longitud *f* del río
g Lauflänge *f* eines Flußes; Flußlänge *f*

* **length of wave run-out → 2138**

3468 lens
A geological body of lentil-like shape wedging out in all directions.
r линза
Геологическое тело чечевицеобразной формы, выклинивающееся по всем направлениям.
f lentille *f*
s lenteja *f*
g Linse *f*

3469 lenticular cloud
A cloud shaped like a lens, usually produced by an whirlwind in mountains or hills.
r чечевицеобразное облако
Облако, по форме напоминающее линзу, обычно формирующееся под действием воздушного вихря в холмистой или гористой местности.
f nuage *m* lenticulaire
s nube *f* lenticular
g linsenförmige Wolke *f*

3470 leptosol
An extremely thin soil underlain by dense rock or loose stony and crushed rock material formed everywhere on young surfaces, especially in mountains.
r лептосоль
Крайне маломощная почва, подстилаемая плотной породой или рыхлым каменисто-щебнистым материалом, встречается повсюду на молодых поверхностях, особенно в горах.
f leptosol *m*
s leptosol *m*
g Leptosol *m*

3471 lessivage
The removal of clay colloids in a soil with no change in their chemical composition.
r лессиваж
Перемещение глинистых коллоидов в почве, не ведущее к изменению химического состава.
f lessivage *m*
s lavado *m*
g Lessivierung *f*

3472 lettering
All names, terms, explanations, letters and figures appearing on maps.
r надписи на карте
Все названия, термины, пояснения, буквенные и цифровые обозначения, помещаемые на карте.
f écritures *fpl* cartographiques
s escrituras *fpl* cartográficas; escrituras *fpl* en el mapa; menciones *fpl* escritas en mapas
g erläuternde Kartenzeichen *npl*

3473 levada
A type of deciduous forests that occur on river

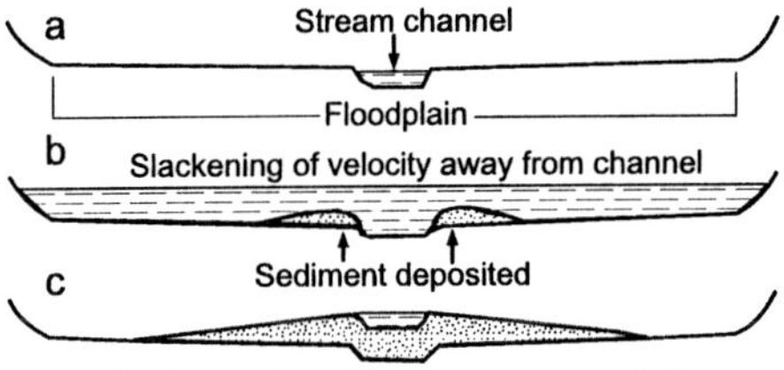

Fig. 188. Formation of **levee**: a – normal character of river valley, b – flood maximum conditions, c – formation of natural levee

flood plains in the south of European Russia, in Romania, Bulgaria, and Greece.

r левада
Название лиственных лесов в поймах рек на юге Европейской России, в Румынии, Болгарии, Греции.

f levada *f*
s levada *f*
g Lewada *f*; Auwald *m*

3474 levante
A humid fresh east or north-east wind blowing over the western Mediterranean area from southern France to the strait of Gibraltar and causing overcast cloudy and rainy weather.

r леванте
Влажный свежий восточный или северо-восточный ветер, дующий над западной областью Средиземноморья от южной Франции до Гиралтара и вызывающий пасмурную и дождливую погоду.

f levante *m*
s levante *m*
g Levante *n*

3475 levee
A natural or artificial enbankment of a river which confines the river within its channel and hinders or prevents flooding. → Fig. 188

r прирусловой вал; намывной вал реки
Естественный (образующийся в результате аккумуляции аллювия) или искусственный береговой вал, который ограничивает течение реки в ее русле и предохраняет пойму от затопления. → Рис. 188

f levée *f*; bourrelet *m* de crue
s levée *f*; caballón *m* de curso; aporte *m*; terraplén *m*
g Deich *m*; Uferwall *m*

3476 level surface
Conventional surface at all point of which gravity has equal value.

r уровенная поверхность
Условная поверхность, во всех точках которой сила тяжести имеет одинаковую величину.

f surface *f* de niveau
s superficie *f* de nivel
g Niveaufläche *f*

3477 levelling
The process to determine height differences between pointsof the Earth's surface in relation to a point of known height or above sea level.

r нивелирование
Определение высот точек земной поверхности относительно некоторой избранной точки или над уровнем моря.

f nivellement *m*
s nivelación *f*; operación *f* de nivel
g Nivellement *n*; Nivellierung *f*; Höhenmessung *f*

3478 levelling network
Network of geodetically connected points on the Earth's surface the absolute height of which is determined.

r нивелирная сеть
Сеть геодезически связанных пунктов, для которых определена абсолютная высота земной поверхности.

f réseau *m* de nivellement
s red *f* de nivelación
g Höhenmessungnetz *n*

* **ley → 2096**

3479 ley-farming; shifting cultivation
An intensive system of farming in which perennial grass or clover are grown on a part of land under crop rotation to restore its fertility.

r травопольная система земледелия
Интенсивная система земледелия, при которой часть площади севооборота занята многолетними бобовыми и злаковыми травами, восстанавливающими плодородие почвы.

f assolement *m* avec praire temporaire; rotation *f* «champ-praire»
s rotación *f* herbácea; rotación *f* «campo-pradera»
g Feldgraswirtschaft *f*

3480 liana; liane; vine
Plant which climbs or winds around other plants for support.

r лиана
Растение с быстро растущими тонкими стеблями, использующее окружающие растения в качестве опоры.

f liane *f*
s liana *f*
g Liane *f*

* **liane → 3480**

3481 lichen
Low plants formed from symbiotic association of algae and fungi, the most diversified in tropics and subtropics, the dominante in tundra and in mountainous regions where few other organisms can survive, (lichens are extremely sensitive to atmospheric pollution and have been used as pollution indicators).

r лишайник
Организмы, образованное симбиозом гриба и водоросли, наиболее разнообразны в тропиках и субтропиках, обильны в тундре и на высокогорьях (многие лишайники особенно чувствительны к атмосферному загрязнению и поэтому используются как индикаторы загрязнения).
f lichen *m*
s líquen *m*
g Flechte *f*

3482 lichen tundra
Dry and stony tundra with surface vegetation cover dominated by lichens.
r лишайниковая тундра
Сухая и каменистая тундра с наземным растительным покровом, в котором преобладают лишайники.
f toundra *f* de lichens; toundra *f* sèche
s tundra *f* liquénica; tundra *f* seca
g Flechtentundra *f*

* **lichenometric dating → 3483**

3483 lichenometry; lichenometric dating
A method of establishing the relative age of deposits based on the rate of growth of lichens.
r лихенометрия
Метод определения относительного возраста отложений на основе определения скорости роста лишайников.
f lichenométrie *f*
s liquenometría *f*
g Flechtemetrie *f*

3484 life expectancy; lifespan
The average number of years which an individual can expect to live in a given society, it is based on statistical probability. → Fig. 189
r продолжительность жизни
Статистический показатель средней длительности жизни населения в данной социальной группе в конкретной стране или регионе. → Рис. 189

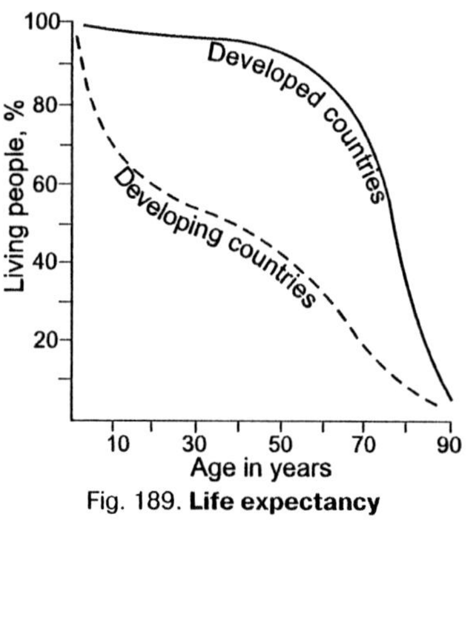

Fig. 189. **Life expectancy**

f espérance *f* de vie
s esperanza *f* de vida; vida *f* mediana o probable
g Lebensdauer *f*; Lebenserwartung *f*

3485 life-cycle
A series of developmental changes such as, birth, growth, maturity, reproduction, gradual decline and death through which an organism goes during its lifetime.
r жизненный цикл
Совокупность фаз развития, пройдя которые организм достигает зрелости, становится способным дать начало потомству, стареет и умирает.
f cycle *m* vital; cycle *m* de vie
s ciclo *m* vital; ciclo *m* de vida
g Lebenszyklus *m*

3486 life-form
A set of features (morphological, behavioural, etc) of living things which result from their adaptations to the environment, e.g. trees, shrubs, ephemers, etc.
r жизненная форма
Совокупность строения, поведения и функционирования животных и растительных организмов (как систематически близких, так и далеких), отражающая их приспособленность к определенным условиям среды, например, деревья, кустарники, эфемеры и пр.
f forme *f* biologique; type *m* biologique
s forma *f* biológica; tipo *m* biológico
g Lebensform *f*

* **lifespan → 3484**

3487 light belts
Latitudinal belts of the Earth differing in ratio of the duration of day and night: equatorial belt, tropical, subtropical, temperate belts, belts of white nights and short winter days, subpolar and polar belts. → Fig. 190, p. 417
r пояса освещенности
Широтные пояса Земли, различаемые по соотношению продолжительности дня и ночи, выделяют экваториальный, тропические, субтропические, умеренные, пояса белых ночей и коротких зимних дней, субполярные и полярные. → Рис. 190, с. 417
f zones *fpl* d'èclairage
s zonas *fpl* de luminosidad
g Gürtel *mpl* der Beleuchtungsstärke

3488 light coniferous forest
Taiga dominated by pine and larch, it grows on poor soils in temperate climate of the Northern Hemisphere.
r светлохвойный лес
Тайга с господством сосны и лиственницы, распространена на бедных почвах в умеренном климате Северного полушария.
f forêt *f* claire résineuse; taïga *f* claire
s taiga *f* clara

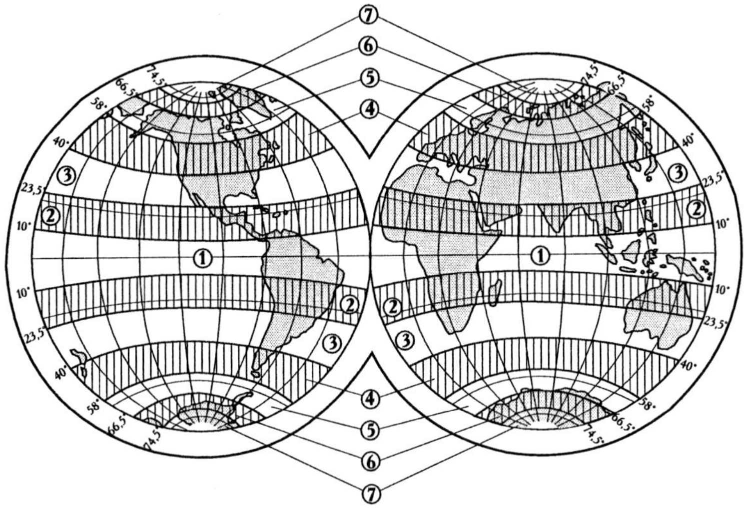

Fig. 190. **Light belts**: 1 – equatorial, 2 – tropical, 3 – subtropical, 4 – temperate, 5 – white summer nights and short winter days, 6 – subpolar, 7 – polar

g lighter Nadelwald *m*

* **light forest → 4271**

3489 **light frosts; frosts; freezing**
The drop of air temperature to 0°C or below, especially the first autumn frosts or night frosts.
r заморозки
Понижение температуры воздуха ниже 0°C, легкий утренний мороз осенью или весной.
f premières gelées *fpl*; gelée *f*; gel *m*
s heladas *fpl*
g Morgenfrost *m*

3490 **light industry**
Branches of industry that produce relatively small consumer goods, e.g. textiles, leather articles, etc., using small amounts of raw materials.
r легкая промышленность
Отрасли промышленности, производящие предметы потребления, прежде всего ткани, кожаное сырье и изделия из них, используя при этом небольшое количество сырья.
f industrie *f* légère
s industria *f* ligera; industria *f* liviana (***Lat. Am.***)
g Leichtindustrie *f*

* **light radiation → 6573**

3491 **lighthouse; beacon**
A towerlike construction equipped with a powerful light or other signalling aids and serving to warn or guide shipping at sea (or for aircrafts in the air).
r маяк
Сооружение башенного типа, оборудованное светооптическими и иными системами и средствами сигнализации, устанавливаемое на мысах, скалах, устьях рек и т.п., служит ориентиром для судов в море или для самолетов при заходе на посадку.
f phare *m*
s faro *m*
g Leuchtturm *m*; Leuchtfeuer *n*

3492 **lightning**
A electrical discharge between clouds or between a cloud and the Earth's surface accompanied by a light flash and thunder.
r молния; грозовой разряд
Внезапная световая вспышка, мгновенный разряд скопившегося атмосферного электричества в воздухе между противоположно заряженными облаками или между облаком и землей.
f éclair *m*
s relámpago *m*; rayo *m*
g Blitz *m*

3493 **lightship**
A moored vessel equipped with a powerful light to warn or guide shipping at sea.
r плавучий маяк
Судно специальной конструкции с маячным оборудованием, устанавливаемое на якорях в отдалении от берега для предупреждения об опасностях на море.
f bateau-phare *m*; bateau-feu *m*
s buque-faro *m*
g Feuerschiff *n*

3494 **lignite; brown coal**
A brown, fibrous deposit containing the plant fragments which have been changed into an amorphous mass and used mainly as a fuel in local thermal power stations.
r бурый уголь
Волокнистые отложения бурого цвета, содержащие остатки растений, превратившиеся в аморфную массу, используется в

основном в качестве топлива на местных теплоэлектростанциях.
f lignite *f*
s lignito *m*; carbón *m* lignitoso
g Braunkohle *f*

3495 liman; brackish lagoon
A wide estuary with low shores nearly closed by coastal bars, typical of the Black Sea.
r лиман
Широкий эстуарий, почти полностью закрытый прибрежными косами, которые превращают его в лагуну, часто встречаются на Черном море.
f liman *m*
s liman *m*
g Liman *m*; Mündungssee *m*

3496 liman coast
A sea coast with no tides which forms by the drowning of river mouths, it is characterised by narrow curvy bays (limans) with steep abrasion shore.
r лиманный берег
Берег бесприливных морей, возникающий при затоплении устьев речных долин и отличающийся извилистыми узкими бухтами (лиманами) с крутыми абразионными берегами.
f côte *f* de marais; côte *f* à liman
s litoral *m* de marisma; costa *f* de liman
g Limanküste *f*

3497 limb of fold
The rock strata on one or the other side of the central line (the axis) of a fold.
r крылья складки
Части складки по обе стороны от ее шарнира.
f flanc *m* de pli
s flanco *m* de pliegue
g Faltenflügel *m*

3498 lime
A caustic, infusible solid consisting essentially of calcium oxide, obtained by heating calcium carbonate, it is used in agriculture and various manufacturing processes.
r известь
Тугоплавкое твердое вещество, получаемое при нагревании карбоната кальция и широко используемое в сельском хозяйстве, строительстве и промышленных процессах.
f chaux *f*
s cal *f*
g Kalk *m*

* **lime horizon → 878**

3499 limes
Formerly the term denoted fortification complexes that separated the Roman Empire from the Barbarian territories; today the term describes boundaries between civilizations as well as frontiers between territories controlled by the state and uncontrolle ones; e.g. boundaries with the Basque country in Spain or the Kurd territories in Turkey.
r лимы
В историческом прошлом комплекс фортификаций на границе, отделяющей владения Римской империи от территории варваров; ныне употребляют для обозначения границ между цивилизациями, а также конфликтных границ между территориями, контролируемыми и не контролируемыми властями, например с областями басков в Испании или курдов в Турции.
f limes *mpl*
s limes *mpl*
g Limes *mpl*

3500 limestone
A sedimentary rock of white or light grey colour, consisting mainly of calcium carbonate or calcite skeletal remains of organisms.
r известняк
Осадочная горная порода белого или светло-серого цвета, состоящая в основном из кальцита или кальцитовых скелетных остатков организмов.
f calcaire *m*
s caliza *f*
g Kalkgestein *m*

* **limestone pavement → 3289**

* **limestone terrain → 3294**

3501 lime-tree forest
Deciduous broad-leaved forest dominated by lime trees.
r липовый лес
Летнезеленый широколиственный лес с господством липы.
f tilaie *f*
s bosque *m* de tilo
g Lindenwald *m*

3502 liming of soil
Applying of lime to acid soils in order to reduce their acidity and thus to increase their fertility.
r известкование почвы
Способ повышения плодородия кислых почв путем внесения в них извести для устранения избыточной почвенной кислотности.
f chaulage *m* du sol
s encalado *m* del suelo
g Kalkdügung *f*

3503 limit of uprush
The limit of a swash run onto a beach.
r граница прибоя
Максимальная дальность наката волны на берег во время прибоя.
f limite *f* de jet de rive
s límite *m* de uprush
g Brandungsgrenze *f*

3504 **limits of growth**
Maximum possible external impact to the biosphere and environment (including population growth, land and natural resources exploitation, etc.) when their structure and functioning systems are destroyed.
r пределы роста
Максимально возможные внешние нагрузки на биосферу и окружающую среду (рост народонаселения, эксплуатация земель и природных ресурсов и т.п.), после чего наступет нарушение их структуры и нормального функционирования.
f limites *fpl* de la croissance
s límites *mpl* de desarrollo
g Wachstumsgrenzen *fpl*

* **limnetic *adj* → 2311**

* **limnetic organism → 3507**

3505 **limnetic zone**
The area of deep water of a fresh lake away from the edge which is well-illuminated zone mainly inhabited by plankton.
r лимническая зона
Толща воды внутреннего пресного водоема до глубины эффективного проникновения света, населенная в основном планктоном.
f zone *f* limnique
s zona *f* límnica
g limnische Zone *f*

* **limnic *adj* → 2311**

* **limnic organism → 3507**

3506 **limnology**
The study of fresh water ponds and lakes, their physical, chemical, hydrological and biological properties.
r озероведение; лимнология
Изучение пресноводных озер и прудов, их физических, химических и биологических свойств.
f limnologie *f*
s limnología *f*
g Limnologie *f*; Seenkunde *f*

3507 **limnophil; limnetic organism; limnic organism**
Organisms inhabiting freshwater lakes and bogs.
r лимнофил
Организмы, обитающие в пресноводных озерах и болотах.
f organisme *m* limnophile; organisme *m* limnétique
s organismo *m* limnofilo; organismo *m* limnético
g Limnophil *m*; limnischer Organismus *m*

* **line of constant bearing → 5066**

3508 **line production**
A series of processes that an item passes through in a factory to convert it into a finished product.
r поточное производство; конвейерное производство
Производственный процесс, разделенный на отдельные, сравнительно короткие операции, выполняемые на специально оборудованных, последовательно расположенных рабочих местах – поточных линиях.
f production *f* sur la chaîne
s producción *f* en cadena
g Fließfertigung *f*; Fließbandproduktion *f*

* **line squall → 5745**

3509 **lineament**
A large-scale linear topographic feature on the land surface, such as trough or ridge, which is structurally controlled.
r линеамент
Линейные или дугообразные структурные элементы рельефа планетарного значения, обычно относящиеся к зонам разломов.
f linéament *m*
s lineamento *m*
g Lineament *n*

* **linear city → 3512**

* **linear eruption → 2162**

3510 **linear feature**
A line on a map that has the length which corresponds to the map scale while its width does not.
r линейный объект
Линия на карте, длина которой соответствует масштабу карты, а ширина не соответствует.
f symbole *m* linéaire
s símbolo *m* lineal
g Linienobjekt *n*

3511 **linear scale; bar scale (*amer*)**
A sub-divided line on a geographical map which shows distances at a given scale.
r линейный масштаб
Графический масштаб на географической карте в виде отрезка прямой, разделенного на равные части с подписанными значениями соответствующих им расстояний на местности.
f échelle *f* graphique linéaire
s escala *f* gráfica lineal
g Maßstabsleiste *f*

* **linear settlement → 3512**

3512 **linear town; linear settlement; linear city**
An elongated urban settlement, developed alongside a major line of transport on account of either the construction of the terrain or the advantages of the ease of transport provided by the routeway.

r линейный город
Растянувшееся городское поселение, сформировавшееся вдоль главного транспортного пути в соответствии с особенностями местности или транспортных путей.
f ville *f* linéaire
s ciudad *f* lineal
g lineare Stadt *f*

3513 liquefaction of soil
The process by which sediments and soil collapse from a sudden loss of cohesion following a loss of shearing resistance.
r разжижение грунта
Процесс, при котором отложения или почва обрушиваются из-за внезапной потери сцепления в результате отсутствия сопротивления сдвигу.
f liquéfaction *f* du sol
s licuación *f* del suelo
g Bodenverdünnung *f*; Bodenverwässerung *f*

3514 liquid deposit
Water deposited from the air on lee sides of cold vertical surfaces under a dull weather or fog.
r жидкий налет
Вода, осаждающаяся из воздуха с наветренной стороны на холодные вертикальные поверхности в пасмурную погоду или при тумане.
f dépôt *m* liquide
s depósito *m* líquido
g flüssiger Niederschlag *m*

3515 liquid precipitation
Liquid water precipitating from clouds, fog or directly from air.
r жидкие осадки
Вода в жидком виде, выпадающая из облаков, тумана или непосредственно выделяющаяся из воздуха.
f précipitations *fpl* atmosphériques liquides
s precipitaciones *fpl* atmosféricas pluviales
g flüssige Niederschläge *pl*

3516 liquid waste
Waste matter, which is liquid, generated in the process of human economic activity.
r жидкие отходы
Всякие жидкие вещества, возникающие в результате хозяйственной деятельности.
f déchets *mpl* liquides
s desechos *mpl* liquidos; residuos *mpl* liquidos
g flüssige Abfälle *mpl*

3517 liquified natural gas
Natural gas that is stored and transported in liquid form.
r сжиженный газ
Природный газ, сохраняемый и транспортируемый в жидкой форме.
f gaz *m* liquéfié
s gas *m* liquidado; gas *m* licuado
g Flüssiggas *n*

* **lithifaction → 3518**

3518 lithification; lithifaction
The processes by which newly deposited sediments are converted into a solid rock by compaction and cementation.
r затвердение
Превращение рыхлых отложений в твердую горную породу путем уплотнения и цементации.
f solidification *f*
s solidificación *f*
g Verhärtung *f*

* **lithogenesis → 5318**

3519 lithogenic landscape
A geographical landscape the essential characteristics of which are connected with the features of the underlying rocks and their effect on relief, water regime and soils, for example limestone-karst landscape, landscape of blown sands.
r литогенный ландшафт
Географический ландшафт, существенные свойства которого связаны с особенностями приповерхностных горных пород и их влиянием на рельеф, водный режим и почвы, например, известняково-карстовый ландшафт, ландшафт развеваемых песков.
f paysage *m* lithogénique
s paisaje *m* litógeno
g lithogene Landschaft *f*

3520 lithology; sedimentology
Science that deals with sedimentary rocks, their composition, structure, genesis, and distribution.
r литология; седиментология
Наука об осадочных горных породах, их составе, строении, происхождении и закономерностях размещения.
f lithologie *f*; sédimentologie *f*
s litología *f*; sedimentología *f*
g Lithologie *f*; Sedimentologie *f*

3521 lithophyte
A plant adapted to grow on rocks and stones, which flourish on a bare rock surface, e.g. lichens.
r литофит
Растение, приспособленное к жизни на скалах и камнях, например, лишайник.
f plante *f* lithophyte; plante *f* saxicole
s planta *f* litofita; planta *f* saxícola
g Lithophyt *m*; Steinpflanze *f*

3522 lithosol
A poorly developed thin stony soil formed on solid rocks mainly in mountains.
r литосоль
Слаборазвитая маломощная сильно каменистая почва, формирующаяся на плотных породах, преимущественно в горах.

f lithosol *m*
s litosol *m*; litosuelo *m*
g Lithosol *m*

3523 lithosphere
The upper layer of the rigid Earth, 50-200 km thick, comprising the Earth's crust and upper part of the upper mantle.
r литосфера
Верхняя оболочка «твердой Земли» мощностью 50-200 км, включающая земную кору и верхнюю часть подстилающей ее верхней мантии Земли.
f lithosphère *f*
s litosfera *f*
g Lithosphäre *f*; Gesteinshülle *f*

* **lithospheric plate → 6071**

3524 lithozem
A soil with organogenic horizon of different nature not more than 30 cm thick, lying directly on dense bedrock.
r литозем
Почва с органогенным горизонтом разной природы мощностью не более 30 см, залегающим непосредственно на плотной коренной породе.
f lithosem *m*
s litozem *m*
g Lithosem *m*

* **litter → 5143**

3525 litter; plant litter; plant residue; leaf litter
An accumulation of dead plant remains on the soil surface or dead leaves lying on the floor of a forest.
r опад; растительные остатки
Скопления различных частей отмерших растений на поверхности почвы или опавших листьев в лесу.
f restes *mpl* végétaux; litière *f*; scorie *f*
s residuos *mpl* de plantas; escoria *f*; pajaza *f*
g Pflanzenreste *mpl*; Streuschicht *f*

3526 little ice age; neoglacial (*amer*)
The epoch of the global climatic cooling on the Earth in the XVII-XIX centuries accompanied with a considerable advance of mountain glaciers.
r малый ледниковый период
Эпоха глобального похолодания на Земле в XVII-XIX вв., сопровождавшаяся значительным разрастанием горных ледников.
f période *m* glaciaire petite
s período *m* glaciar pequeño
g kleine Eiszeit *f*

3527 littoral *adj*
Applied to the shore, whether of sea, lake or river, in marine ecosystems the shore area of intertidal zone.
r литоральный
Относящийся к прибрежной зоне морей, озер и рек, в морских экосистемах – к приливно-отливной зоне.
f littoral *adj*
s litoral *adj*
g litoral *adj*; Ufer-; Küsten-

3528 littoral benthos
Plant and animal organisms living at the bottom of the coastal zone of water bodies.
r прибрежный бентос
Совокупность прибрежных донных организмов, обитающих в морях, озерах, водохранилищах.
f benthos *m* littoral
s bentos *m* litoral; organismos *mpl* litorales
g Litoralbenthos *n*

3529 littoral deposits; littoral sediments
Materials that accumulate between high- and low-water marks, generally comprising sand, shells or shingle.
r литоральные отложения
Отложения на прибрежной полосе приливных морей, т.е. на литорали, содержащие в основном песок, раковины и гальку.
f dépôts *mpl* littoraux
s depósitos *mpl* litorales
g litorale Ablagerungen *fpl*; litorale Sedimenten *npl*

* **littoral drift → 3583**

3530 littoral fauna
All animals inhabiting the coastal area of oceans, seas, and lakes.
r литоральная фауна
Совокупность животных организмов, населяющих береговую зону озер, морей и океанов.
f faune *f* littorale
s fauna *f* litoral
g Litoralfauna *f*

3531 littoral flora
All the plant organisms inhabiting the coastal area of oceans, seas, and lakes.
r литоральная флора
Растительный мир береговой зоны озер, морей и океанов.
f flore *f* littorale
s flora *f* litoral
g Litoralflora *f*

3532 littoral landscape
A geographical landscape which forms in the coastal strip of land.
r литоральный ландшафт
Географический ландшафт, формирующийся в прибрежной полосе суши.
f paysage *m* littoral
s paisaje *m* litoral
g Küstenlandschaft *f*; litorallandschaft *f*

* **littoral sediments → 3529**

3533 **littoral zone**
Area of the sea and shore between the high and low tide; in lakes it is the coastal zone of the lake basin from the edge of waves to the depth of light penetration. → Fig. 244, p. 534
r литораль; береговая отмель; литоральная зона
Зона морского дна, затопляемая во время прилива и осушаемая при отливе; в озерах - прибрежная часть озерной котловины от зоны заплеска волн до глубины проникновения света. → Рис. 244, с. 534
f littoral *m*
s litoral *m*; estrán *m*
g Litoral *n*

3534 **livestock**
Farm animals, such as cattle, sheep, horses, chickens and other animals, raised for milk, wool, meat, eggs, and other animal products.
r скот
Домашние животные, выращиваемые для получения молока, шерсти, мяса, яиц.
f bétail *m*
s ganado *m*
g Vieh *n*

3535 **livestock breeding**
The rearing highly productive animals including the improvement of the existing races and rasing the best strains of cattle.
r племенное животноводство
Выращивание высокопродуктивных животных, включая улучшение существующих и выведение новых пород.
f élevage *m* de race
s ganadería *f* de raza
g Rassenzucht *f*

3536 **livestock capita**
The amount of domestic animals at any farm.
r поголовье скота
Количество домашнего скота в том или ином хозяйстве.
f cheptel *m*
s cabezas *fpl* de ganado; ganado *m* vivo; cabaña *f*
g Viehbestand *m*

3537 **livestock farming; animal husbandry; pastoral farming; pastoralism**
A branch of farming engaged in the rearing of animals for meat, milk, wool, eggs, and other animal products.
r животноводство
Отрасль сельского хозяйства, занимающаяся разведением сельскохозяйственных животных для производства животноводческих продуктов.
f élevage *m*
s ganadería *f*; cría *f* de ganado
g Viehhaltung *f*; Viehzucht *f*

* **living space → 6576**

3538 **living standard; standard of living**
An economic index determined by the level of incomes and consumption of goods and services, normally measured by the per capita income level corrected according to the local currency's purchasing power.
r жизненный уровень
Экономический показатель, определяемый уровнем дохода и размерами потребления товаров и услуг, обычно измеряется уровнем душевого дохода, скорректированного по покупательной способности местной валюты.
f standard *m* de vie
s estandart *m* de vida
g Lebensstandard *m*

3539 **lixisol**
A soil that has a horizon of the accumulation of low-active clay of kaolinite composition which is widespread in the regions of tropical, subtropical and warm temperate climate with marked dry season.
r ликсисоль
Почва, имеющая горизонт накопления малоактивной глины каолинитового состава, распространена в районах тропического, субтропического и теплого умеренного климата с выраженным засушливым сезоном.
f lixisol *m*
s lixisol *m*
g Lixisol *m*

3540 **llanos**
The extensive plains covered with tall grass savannah vegetation with scattered trees and fringing forests along the rivers in the northeast of South America with very wet conditions of soil during the rainy season where swampy grasses and sedges develop.
r льянос
Высокотравные саванны на северо-востоке Южной Америки с густым, главным образом злаковым, покровом и отдельными экземплярами, группами деревьев и галерейными лесами вдоль рек.
f llanos *mpl*
s llanos *mpl*
g Llanos *mpl*

3541 **load**
All the solid matter transported by a river or stream, by being rolled or bounced along its bed, or carried in suspension or in solution.
r нанос
Всякий твердый материал, переносимый речным потоком путем перекатывания или сальтации, либо в растворе или суспензии.
f alluvion *f*
s aluvión *f*

g Anschwemmung *f*

* **load of a stream → 627**

3542 **loam; loamy soil**
Loose continental sediment consisting of particles less than 0.01 mm and detrital material, i.e. clay, silt and sand in a ratio 20:40:40.
r суглинок
Рыхлый континентальный осадок, состоящий из частиц менее 0,01 мм и обломочного материала, например из частиц глины, мелкозема и песка в соотношении 20:40:40.
f limon *m*
s limo *m*
g Lehmboden *m*

3543 **loamy sand**
Loose sediment (ground) consisting of 70-90% of aleurite-sandy material and of 10-30% of clay particles less than 0.01 mm.
r супесь
Рыхлый осадок (грунт), состоящий на 70-90% из алеврито-песчаного материала и на 10-30% из глинистых частиц размером менее 0,01 мм.
f sable *m* argileux; terre *f* sablonneuse
s terreno *m* arenoso; tierra *f* arenosa; greda *f* arenosa
g Lehmsand *m*

* **loamy soil → 3542**

3544 **lobate delta; birdfoot delta**
A delta consisting of branching off narrow borders of sediments projecting seawards in the pattern of a bird's foot and formed by a river carrying a considerable amount of load, e.g. Missisipi delta.
r лопастная дельта
Речная дельта, состоящая из расходящихся узких прирусловых лопастей, формирующихся при значительном потоке речных наносов, например, дельта р. Миссисипи.
f delta *m* digité; delta *m* à patte d'oiseau
s delta *m* digitado; delta *m* de pata de pájaro
g Vogelfußdelta *n*

3545 **lobate frontal moraine**
Ridges of terminal moraine limiting lobes of an ancient glacier.
r лопастная конечная морена
Гряды конечной морены, ограничивающие лопасти древнего ледника.
f moraine *f* frontale lobée
s morrena *f* frontal de lóbulo
g gelappte Frontalmorane *f*; gezackte Frontalmorane *f*

3546 **lobate shore**
A shore of fault origin with projections and bays with angular contours.
r лопастной берег
Берег сбросового происхождения, с выступами и бухтами угловатых очертаний.
f côte *f* des lobes
s costa *f* de lóbulos
g Blattküste *f*

3547 **local** ***adj***
Relating to a particaular place or area, not general widespread.
r местный; локальный
Относящийся к данной конкретной территории.
f local *adj*
s local *adj*
g lokal *adj*; örtlich *adj*

3548 **local anticyclone**
Area of high pressure strongly pronounced in a certain region and visible on mean pressure charts for winter months or for any month of a year [the following local anticyclones are widely known: Asian, Azores, Arctic, Greenland, Antarctic, Hawaiian, North-American (Canadian), South-Atlantic, South-Indian, and South-Pacific].
r местный антициклон
Область высокого давления в конкретном регионе, обнаруживаемая на картах среднего давления за зимние месяцы или за любой месяц года [широко известны азиатский (сибирский), азорский, арктический, гренландский, антарктический, гавайский, североамериканский (канадский), южноатлантический, южноиндийский и южнотихоокеанский антициклоны].
f anticyclone *m* local
s anticiclón *m* local
g lokales Hoch *n*

3549 **local base level**
The base level of erosion of local importance provided at any height.
r местный базис эрозии
Базис эрозии, имеющий местное значение и располагающийся на любой высоте.
f niveau *m* de base local; niveau *m* d'érosion local
s nivel *m* de base local; base *f* de erosión local
g lokale Erosionsbasis *f*

3550 **local circulation**
Circulation of the atmosphere above a relatively small area of the Earth's surface determined by the features of this area.
r местная циркуляция
Циркуляция атмосферы над сравнительно небольшой площадью земной поверхности, обсуловленная особенностями этой территории.
f circulation *f* locale
s circulación *f* local
g Lokalzirkulation *f*

3551 **local climate**
Climate of a relatively small territory, e.g. valley, forest, city etc.
r местный климат
Климат сравнительно небольших территорий, таких как долина, лес, город и т.п.
f climat *m* local
s clima *m* local
g Ortsklima *n*

* **local contamination → 3554**

* **local development → 4973**

3552 **local geographical term**
A folk name of a natural or man-made object having a limited geographical distribution (often local terms form part of place names).
r местный географический термин
Народное название природного и социального объекта, имеющего ограниченное географическое распространение (нередко местные термины входят в состав географических названий данного региона).
f terme *m* géographique local
s termino *m* geográfico local
g lokaler Ortsname *m*

3553 **local geography and ethnography**
The study of nature, population, economy, history and culture of some part of a country, natural region, or human settlement, mainly by local people.
r краеведение
Изучение природы, населения, хозяйства, истории и культуры какой-либо части страны, административного или природного района, населенных пунктов, главным образом силами местного населения.
f ethnologie *f* locale
s etnología *f* local
g Heimatkunde *f*

* **local horizon → 5466**

3554 **local pollution; local contamination**
Pollution of a small area around an industrial plant, settlement, or along a road.
r локальное загрязнение
Загрязнение небольшого района вблизи источника загрязнения: промышленного предприятия, населенного пункта, дороги.
f pollution *f* locale; contamination *f* locale
s polución *f* local; contaminación *f* local
g lokale Verschmutzung *f*;
lokale Verunreinigung *f*

3555 **local self-government**
Independent administration by the population of an administrative region (through elected bodies or directly) within the rights established by the state.
r местное самоуправление
Самостоятельное руководство населением через выборные органы или непосредственно в делах конкретной административного района в пределах прав, установленных государством.
f autogestion *f* locale
s autogestión *f* local
g lokale Selbstverwaltung *f*;
örtliche Selbstverwaltung *f*

3556 **local snow line**
The snow line the altitude of which depends on local factors such as differences in precipitation, aspect, shelter, etc. often changing from place to place over very short distances.
r местная снеговая линия
Снеговая линия, высота которой определяется преимущественно местными факторами, нередко изменяющимися от места к месту на очень коротких расстояниях.
f limite *f* locale des neiges permanentes
s límite *m* local de nieves perpetuas
g örtliche Schneegrenze *f*

3557 **local tax; municipal tax**
A tax collected by local authorities for the local budget.
r местный налог
Налог, взимаемый местными органами власти и поступающий в местный бюджет.
f taxe *f* locale; taxe *f* municipale
s impuesto *m* local; impuesto *m* municipal
g Ortssteuer *f*

3558 **local terrace**
A river terrace formed due to local conditions which affect a relatively small portion of the valley.
r локальная терраса
Тип речной террасы, образование которой обусловлено местными причинами, оказывающими влияние на сравнительно небольшой участок долины.
f terrasse *f* locale
s terraza *f* local
g Lokalterrasse *f*

3559 **local time**
The time determined for a given place on the Earth with reference to the meridian of a given place and it is the same for all the points of that meridian.
r местное время; поясное время
Время, определяемое для данного места на Земле, зависит от географической долготы места и одинаково для всех пунктов, расположенных на одном меридиане.
f heure *f* locale
s hora *f* local
g Ortszeit *f*; lokale Zeit *f*

3560 **local wind**
Wind blowing in a smaller area than planetary winds and caused by local conditions.
r местный ветер
Ветер, создаваемый местными условиями и дующий на ограниченном пространстве.
f vent *m* local

s viento *m* local
g Lokalwind *m*

* **locality → 6101**

3561 localization
The distribution of a geographical object or process in a certain place.
r локализация
Ограничение распространения какого-либо природного явления или процесса пределами определенной территории или акватории.
f localisation *f*
s localización *f*
g Lokalisierung *f*; örtliche Beschränkung *f*

* **localization economies → 123**

* **localization of enterprises → 6111**

3562 location; geographical distribution
The position, arrangement or frequency of occurrence of natural or man-made object or phenomenon, over the Earth's surface.
r размещение; распространение; географическое распределение
Конкретное расположение природных или антропогенных объектов и явлений на поверхности Земли.
f distribution *f*
s distribución *f*
g Standortverteilung *f*

3563 location of industry
The areal distribution of industrial activity.
r размещение промышленности
Пространственное распределение промышленной деятельности.
f localisation *f* de industrie
s instalación *f* de industria
g Standortverteilung *f* der Industrie

3564 loch
A body of water, either an enclosed freshwater lake or long, narrow penetration of seawater extending inland from the ocean, e.g. the lake Lochness in Scotland.
r лох
Водоем, представляющий собой либо пресноводное озеро, либо узкое пространство с проникшей на сушу морской водой, например, озеро Лохнесс в Шотландии.
f loch *m*
s loch *m*
g Loch *m*

3565 lock
An enclosed section of waterway (river, canal) which has gates at each end and in which the water level can be raised or lowered to move boats from one level to another. → Fig. 191
r шлюз
Сооружение для подъема или опускания судов с одного уровня воды (в канале, реке) на другой. → Рис. 191
f écluse *f*
s esclusa *f*
g Schleuse *f*

* **lode → 6540**

3566 loess; löss
Homogeneous, usually non stratified unconsolidated porous sediments composed by silt-sized material, deposited primarily by wind.
r лёсс
Однородные, обычно неслоистые, пористые, слегка сцементированные пылеватые отложения, сформировавшиеся преимущественно с помощью ветра.
f loess *m*
s loess *m*; loss *m*
g Löß *m*

3567 loess loam
A loess-like sediment with great content of clayey particles and the presence of coarse sand or pebbles.
r лёссовидный суглинок
Лёссовидная порода с большим содержанием глинистых частиц и присутствием грубого песчаного и редко галечного материала.
f limon *m* loessique
s limo *m* de loess
g Lößlehm *m*

3568 loess-like deposit
Very fine-grained loams and loamy sands of different genesis resembling loess, but having stratification and other properties which are not typical for loess.
r лёссовидные отложения
Пылеватые суглинки и супеси разного генезиса, внешне напоминающие лёсс, но обладающие слоистостью и другими, не свойственными лёссу особенностями.
f sédiment *m* loessigue
s depósito *mpl* de tipo de loess
g Lössablagerungen *fpl*; Lösssedimente *fpl*

3569 log gorge
The piling appearing on shallow or narrow places of a river channel as a result of accumulation of drifting trees, branches, etc.
r залом

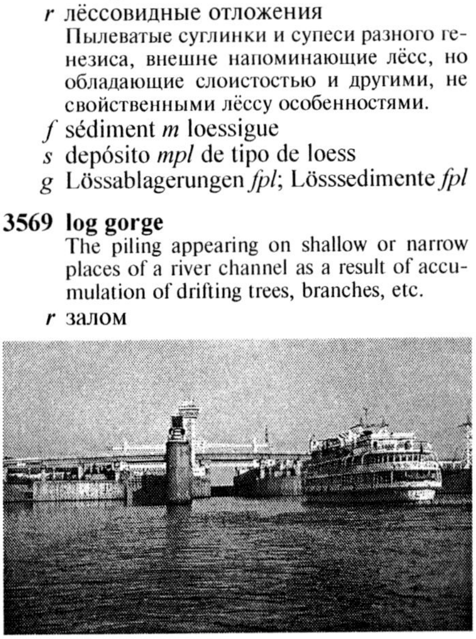

Fig. 191. **Lock** at the kanal Moscow – Volga river

Завал, возникающий на мелководных или узких местах русел рек в результате скопления плывущих по реке деревьев, веток и пр.
f bourrelet *m* détritique
s anillo *m* de troncos; barrera *f* de troncos
g Holzsperre *f*

3570 logging; felling
The work of cutting trees for wood.
r рубка леса
Регулярная промышленная заготовка леса с соблюдением необходимых правил по сохранению и восстановлению лесных массивов.
f abattage *m*
s corta *f*
g Fällen *n*

3571 log-pass; bushwood-road
The fooring made of logs or brushwood for the passage, walking along a bog or swampy place.
r гать
Настил из бревен или хвороста для проезда, прохода через болото или топкое место.
f chemin *m* de fascines
s estriberón *m*
g Faschinenweg *m*; Knüppeldamm *m*

3572 loma
Ephimeral plant community of Pacific coastal deserts in South America, mainly in Peru, where winter mists bring some humidity.
r ломас
Эфемерная растительность формации тихоокеанских береговых пустынь Южной Америки, главным образом в Перу.
f loma *f*
s lomas *f*
g Lomas *fpl*

3573 long fallow
A farm land left unused for 8 to 15 years to restore soil fertility.
r перелог; залежь
Долговременная земельная залежь, когда после снятия нескольких урожаев землю оставляют без обработки на 8-15 лет для восстановления плодородия почвы.
f jachère *f* longue
s barbecho *m* largo;
barbecho *m* de larga duración
g Altbrache *f*

3574 long grass prairie; tall grassland
A kind of prairie, an area of natural vegetation of herbaceous plant communities in the grassland zone in the east of North America.
r высокотравная прерия
Тип прерии, степные травянистые растительные сообщества на востоке Северной Америки.
f prairie *f* à herbes hautes;
prairie *f* altherbeuse
s pradera *f* de hierbas altas
g Hochgrasprärie *f*

* **long profile of a river → 3580**

3575 long wave; shallow-water wave
An ocean wave whose lengh is sufficiently large compared to the water depth (25 or more times the depth).
r длинная волна
Морская волна, длина которой сильно превышает глубину моря, до 25 раз и более.
f vague *f* longue
s ola *f* larga
g langperiodische Welle *f*; Langwelle *f*

3576 longitude
Angular distance between the prime meridian and the meridian of a given point which is measured westwards or eastwards from the prime meridian (from 0° to 180°).
r географическая долгота;
долгота местности
Угловое расстояние от начального меридиана до меридиана в данной точке, отсчитываемое к западу и востоку (от 0° до 180°) от начального меридиана.
f longitude *f*
s longitud *f*
g geographische Länge *f*

3577 longitudinal *adj*
Oriented parallel to the strike of rocks or other elements of the environment, e.g. longitudinal coast. 2. Pertaining to longitude.
r продольный$_1$; долготный$_2$
1. Ориентированный параллельно простиранию горных пород или других элементов окружающей среды, например, продольный берег. 2. Относящийся к географической долготе.
f longitudinal *adj*
s longitudinal *adj*
g längsläufig *adj*; Längs-

* **longitudinal coast → 1185**

3578 longitudinal dune; seif-dune
An accumulation of sand up to hundred metres high, not fixed by vegetation, stretched in the direction of prevailing winds.
r продольная дюна
Скопление незакрепленного растительностью песка высотой до сотни метров, вытянутое вдоль господствующих ветров.
f dune *f* longitudinale
s duna *f* longitudinal
g Strichdüne *f*; Löngsdüne *f*

**3579 longitudinal profile;
longitudinal section**
Profile of a natural object along its strike.
r продольный профиль
Профиль природного объекта вдоль его простирания.
f profil *m* longitudinal

s perfil *m* longitudinal
g Längsprofil *n*

* **longitudinal section → 3579**

3580 longitudinal stream profile; long profile of a river
A section of a river along the river thalweg, showing the slope of the river bed from source to mouth.
r продольный профиль реки
Линия сечения дна реки вертикальной плоскостью, проходящей через тальвег долины.
f profil *m* de fleuve en long; profil *m* de fleuve longitudinal
s perfil *m* del río longitudinal
g Längenprofil *n* eines Flußes

3581 longitudinal valley; trench; strike valley
A river valley with a trend roughly parallel to the strike of the rocks of an area.
r продольная долина
Речная долина, вытянутая в направлении простирания горных пород или основных геологических структур на данной территории.
f vallée *f* longitudinale
s valle *m* longitudinal
g Längstal *n*

* **long-range forecasting → 3586**

* **long-range weather forecast → 3586**

3582 longshore bar$_1$
A bar situated in the intertidal zone and roughly aligned with shoreline.
r береговая отмель
Отмель в приливной зоне, протягивающаяся параллельно береговой лини.
f banc *m* littoral
s banco *m* litoral
g Strandsand bank *f*

* **longshore bar$_2$ → 553**

* **longshore current → 4081**

3583 longshore drift; littoral drift
The movement of material along the beach by the action of waves, which take sediment up and along the coast. → Fig. 192
r поток наносов; вдольбереговой поток наносов
Массовое перемещение обломочного материала прибойным потоком вдоль берега. → Рис. 192

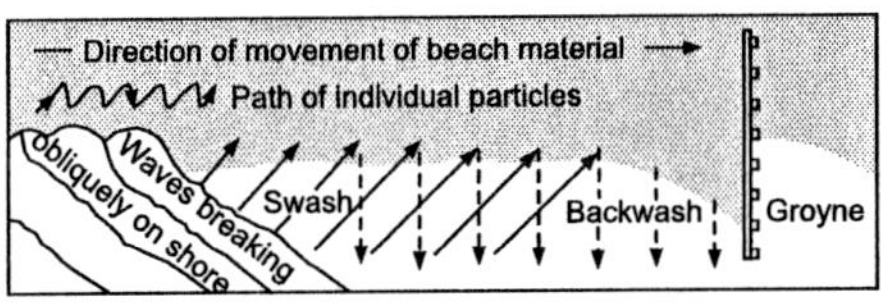

Fig. 192. **Longshore drift**

f dérive *f* littorale; transfert *m* à la côte
s deriva *f* litoral; traslado *m* litoral
g Küstenlängstransport *m*; Uferlängstransport *m*

3584 long-term data set
The results of regular observations over a meteorological element during a long period of time which is used to get climatic characteristics.
r многолетний ряд
Сводка результатов регулярных наблюдений за метеорологическим элементом в течение длительного времени, используемая для получения климатических характеристик.
f serie *f* des plusieurs années
s serie *f* de observaciones de muchos años
g langfristige Reihe *f*

3585 long-term river regulation
A system of measures applied to increase the runoff in years with little water at the expense of the runoff in years with abundant water.
r многолетнее регулирование стока
Система мер, применяемая для повышения стока маловодных лет за счет стока многоводных.
f réglage *m* d'écoulement des plusieurs années
s regulación *f* de escorrentía a largo plazo
g vieljährige Abflußregulierung *f*

3586 long-term weather forecast; long-range forecasting; long-range weather forecast
Weather forecast for a period longer than five days.
r долгосрочнй прогноз погоды
Предсказание погоды на срок более 5 дней.
f prévision *f* météorologique à longue échéance; prévision *f* météorologique à long terme
s predicción *f* meteorológica a largo plazo; pronóstico *m* de tiempo a largo plazo
g Langfristprognose *f*

3587 long-wave radiation; terrestrial radiation
Infrared radiation of the earth surface, the atmosphere and clouds with wavelengths greater than 4 micron.
r длинноволновая радиация
Инфракрасное излучение земной поверхности, атмосферы и облаков в интервале длин волн больше 4 микрон.
f rayonnement *m* de grandes longueurs d'onde
s radiación *f* de onda larga
g langwellige Strahlung *f*; Langwellllenstrahlung *f*

3588 looming; heat haze
State of near-surface layers of air during very

hot weather when hot air streams distort images of distant objects.

r марево

Состояние приземных слоев воздуха в условиях жаркой погоды, когда от нагретой земли поднимаются струи нагретого воздуха, искажающие изображение отдаленных предметов.

f brouillard *m* blafard

s espejismo *m* emergente

g Höhenrauch *m*

* **loose rock → 6444**

3589 loose snow

A snow layer with weak cohesion between the snow particles (occurring on mountain slopes it can cause snow avalanches).

r рыхлый снег; несвязный снег

Снег, лежащий на поверхности и обладающий очень низким сцеплением (на горных склонах служит источником лавин).

f neige *f* sans cohésion; neige *f* incohérente

s nieve *f* sin cohesión; nieve *f* suelta; nieve *f* incoherente

g Lockerschnee *m*

* **loose snow avalanche → 4731**

3590 loosening of snow

A rapid growth of snow crystals within a snow mass as its density decreases or remains the same.

r разрыхление снега

Быстрый рост кристаллов снежной толщи при уменьшении или сохранении ее плотности.

f ameublissement *m* de neige

s mullido *m* de nieve

g Auflockerung *f* des Schnees

3591 loose-snow horizon

A layer of snow in which the process of recrystallization gives rise to the growth of crystals, increase of its porosity and permeability to air.

r горизонт разрыхления снега

Слой снега, в котором в процессе перекристаллизации происходит укрупнение кристаллов, увеличение пористости и воздухопроницаемости.

f couche *f* de neige coulante; couche *f* de neige incoherente

s capa *f* de nieve mullida; horizonte *m* de nieve incoherente

g Wildschneehorizont *m*

3592 lop; overfall

A disturbance and turbulence water, sometimes in a river, more commonly in the sea, occurring when two tidal streams flowing from different directions meet or a strong outflowing surface or near-surface current meets the incoming surf. → Fig. 193

r сулой; смешение волн

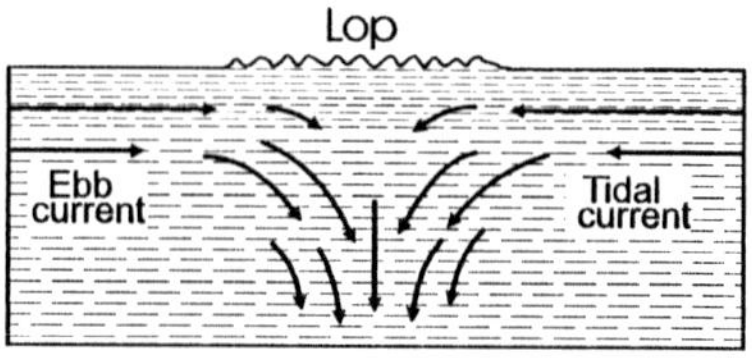

Fig. 193. **Lop**

Турбулентное движение воды, изредка наблюдающееся на реке, но более характерное для моря, возникающее, когда встречаются два приливных потока с разных направлений, или когда сильное приповерхностное течение испытывает действие накатывающегося прибоя. → Рис. 193

f clapotis *m*; clapotis *m* de marée

s chapoteo *m*; chapoteo *m* de marea

g Kabbelung *f*; Wallung *f*

* **löss → 3566**

3593 loss of snow cover

The end of the period of stable snow cover.

r сход снежного покрова

Завершение периода залегания устойчивого снежного покрова.

f fonte *f* du manteau neigeux

s disgregación *f* de manto nivoso

g Schnee(decke)schwinden *n*

3594 low anticyclone

An anticyclone identified only in the baric field of the lower troposphere.

r низкий антициклон

Антициклон, достаточно отчетливо обнаруживаемый лишь в барическом поле нижней тропосферы.

f anticyclone *m* bas

s anticiclón *m* bajo

g niedrieges Hoch *n*

* **low cloud → 3609**

3595 low cyclone

A cyclone identified only in the baric field of the lower troposphere.

r низкий циклон

Циклон, достаточно отчетливо обнаруживаемый лишь в барическом поле нижней тропосферы.

f dépression *f* basse

s depresión *f* baja

g niedrieges Tief *n*

* **low flow → 3601**

3596 low latitudes

The tropical and subtropical areas of the globe lying approximately between the latitudes 40° north and south or between the tropic of Cancer and tropic of Capricorn.

r низкие широты

Условное название тропических и субтро-

пических областей земного шара, расположенных примерно между 40° северной и южной широты или между тропиками Рака и Козерога.
f basses latitudes *fpl*
s baja latitud *f*; latitud *f* baja
g niedrige Breiten *fpl*

3597 low oblique aerial photograph
Aerial photogpaph or space image taken when the optical axis of the camera is inclined (with the deviation more than 3°). → Fig. 3, p. 14
r панорамный аэрофотоснимок
Аэрофото- или космический снимок, сделанный при наклонном (с отклонением более 3°) положении оси фотографирования. → Рис. 3, с. 14
f photographie *f* aérienne panoramique
s fotografía *f* aérea panorámica
g Panorama-Luftaufnahme *f*

3598 low ridge
A low, stretched plain landform of varied size and genesis.
r гряда; грива
Вытянутая, относительно невысокая положительная форма равнинного рельефа разного размера и происхождения.
f bourrelet *m*; grinière *f*
s cadena *f*; anillo *m*
g Landrücken *m*; Hügelkette *f*

*** low tide → 1782**

3599 low tide; low water; lowest ebb
The lowest water level reached during a tide cycle.
r малая вода; низшая точка отлива
Самый низкий уровень моря, достигаемый в течение цикла прилива.
f étale *f* de basse mer; basse mer *f*; niveau *m* de basse mer
s bajamar *f*; nivel *m* bajamar; estado *m* de bajamar
g Niedrigwasser *n*

*** low water → 3599**

3600 low water discharge
The average water discharge in a river in the period of low-water.
r меженный расход
Средний расход воды в реке в период межени.
f débit *m* d'étiage
s caudal *m* de estiaje
g Niedrigwassermenge *f*

3601 low water flow; low flow; minimum flow
The smallest river runoff usually observed in low-water.
r меженный сток; минимальный сток; низкий сток
Наименьший сток рек, обычно наблюдающийся в межень.
f écoulement *m* minimum
s escorrentía *f* mínima
g Mindestabfluß *m*; Niedrigwasserabfluß *m*

3602 lower bedding plane of stratum; base of bed
Stratigraphically the lower surface limiting a stratum as well as a rock directly underlying the stratum.
r подошва пласта
Стратиграфически нижняя поверхность, ограничивающая пласт, а также горная порода, непосредственно его подстилающая.
f plancher *m* de couche; soubassement *m* de couche
s suelo *m* de capa; basamento *m* de capa
g Sohlenfläche *f* einer Schicht

3603 lower course
The lower part of a river with the greatest water-bearing capacity and the smallest inclinations of water surface.
r низовье реки
Нижний участок реки с наибольшей водоносностью и наименьшими уклонами водной поверхности.
f rivière *f* inférieure
s curso *m* inferior
g Unterlauf *m*

3604 lower pool
Part of a river or canal down-stream of a backing construction (dam, lock).
r нижний бьеф
Участок реки или канала ниже подпорного сооружения (плотины, шлюза).
f bief *m* d'aval; bief *m* de fuite
s aguas *fpl* abajo
g Unterhaltung *f*

3605 lower snow drifting
An ensemble of snow particles raised by the wind to small heights above the ground or snow cover with no snowfall. → Fig. 194
r поземок; поземка
Слабая низовая метель, когда снег переносится ветром по поверхности почвы или снежного покрова без снегопада. → Рис. 194

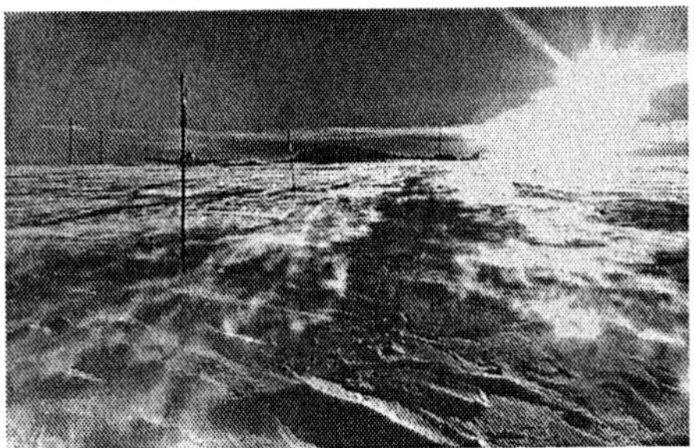

Fig. 194. **Lower snow drifting** near Mirny at the Antarctica

Fig. 195. **Lowland river** Oka in Central Russia

f chasse-neige *f* au sol; chasse-neige *f* basse
s ventisca *f* baja
g niedrige Schneedrift *f*

* **lowest ebb → 3599**

3606 lowland meadow
A meadow that grows in gullies, gulches, and on lowlands in the forest zone.
r низинный луг
Луг, приуроченный к понижениям междуречий, днищам балок и оврагов, незаливаемым низменностям лесной зоны.
f prairie *f* de pays-bas; prairie *f* de bas-fond
s prado *m* de depresión
g Niederungswiese *f*

3607 lowland plain
A vast flat sometimes gently rolling terrain up to 200 m above the sea level.
r низменность
Равнинный, реже холмистый участок суши большой протяженности, расположенный на высоте до 200 м над ур. моря.
f plaine *f* basse
s planicie *f*
g Niederung *f*; Tiefland *n*; Tiefebene *f*

3608 lowland river
A river running in a wide well-developed valley with sluggish flow and small water surface gradient. → Fig. 195
r равнинная река
Река в широкой хорошо разработанной долине с медленным спокойным течением и небольшими уклонами водной поверхности. → Рис. 195
f fleuve *m* de plaine
s río *m* de llanura
g Flachlandfluß *m*

3609 low-level cloud; low cloud
Clouds which are lower than 2 km including Stratus and Startocumulus. → Fig. 196
r нижние облака; облака нижнего яруса
Облака, расположенные ниже 2 км, включают слоистые и слоисто-кучевые облака. → Рис. 196
f nuage *m* de l'étage inférieur; nuage *m* bas
s nube *f* del piso inferior; nube *f* baja

Fig. 196. **Low-level clouds**

g tiefhängende Wolke *f*

3610 low-level waste
The waste which is only slightly radioactive and does not cause problems for disposal.
r низкоактивные отходы
Токсичные отходы с низкой радиоактивностью, не представляющие большой сложности для захоронения.
f déchets *m* de basse radioactivité
s desechos *mpl* de baja radioactividad; residuos *mpl* de baja radioactividad
g niedrigaktive Abfälle *mpl*

3611 low-mountain relief
A type of mountain relief characterised by the absolute heights up to 1,500 m, the depth of dissection up to 500 m, gentle slopes (up to 15°), and poorly expressed altitudinal zonation of landscapes.
r низкогорье; низкогорный рельеф
Тип горного рельефа, характеризующийся абсолютными высотами вершин до 1500 м, глубиной расчленения до 500 м, пологими склонами (в среднем до 15°) с мягкими очертаниями и слабо выраженной высотной поясностью ландшафтов.
f relief *m* des montagnes basses
s relieve *m* de montañas bajas
g Mittelgebirgsrelief *n*

3612 low-shrub tundra
Tundra vegetation with dwarf shrubs, birches and willows with moss and lichen surface cover.
r кустарничковая тундра
Тундровая растительность с преобладанием кустарничков, главным образом карликовых берез и ив с подстилкой из мхов и лишайников.
f toundra *f* à sous-arbrisseaux
s tundra *f* subarbustiva
g Zwergstrauchtundra *f*

* **low-waste technology → 6634**

3613 low-water
The low level of water in a lake or river characterised by small amount of water.
r межень
Низкий уровень воды в реке или озере, сопровождающийся малой водностью.

f étiage *m*
s estiaje *m*
g Niedrigwasserstand *m*; Niedrigwasser *n*

* **loxodrome → 5066**

3614 lull
A momentary decrease of the wind speed below its mean value.
r затишье
Кратковременный спад скорости ветра по сравнению с ее средней величиной.
f accalmie *f*
s calma *f*
g Kalme *f*

* **lumber (*amer*) → 6232**

3615 lunar *adj*
Relating to the Moon or caused by the movement and attraction of the Moon.
r лунный
Относящийся к Луне или связанный с земными процессами, определяемыми движением Луны относительно Земли.
f lunaire *adj*
s lunar *adj*
g Mond(es)-; lunar *adj*

3616 lunar day
The period of time in which the Earth rotates once in relation to the Moon, i.e. interval between two successive crossings by the Moon of the local meridian averaging 24.84 solar hours.
r приливные сутки; лунные сутки
Интервал между двумя последовательными прохождениями Луны через местный меридиан, составляющий в среднем 24,84 солнечных часов.
f jour *m* lunaire
s día *m* lunar
g Mondtag *m*

3617 lunar month; synodic month; lunation
The time interval during which a complete change of Moon phases takes place, i.e. the period from one new Moon to the next, averaging 29.5306 mean solar days – 29 days, 12 hours, 44 min.
r синодический месяц; лунный месяц
Промежуток времени, в течение которого происходит полная смена лунных фаз, его продолжительность непостоянна, в среднем равна 29,5306 средних солнечных суток.
f mois *m* synodique; lunaison *f*
s mes *m* sinódico; mes *m* lunar
g synodischer Monat *m*; Lunation *f*

3618 lunar tide
That portion of a tide which is caused by the tide-producting force of the Moon.
r лунный прилив
Составляющая прилива, вызываемая приливообразующей силой Луны, которая в 2,2 раза больше солнечной составляющей (солнечного прилива).
f marée *f* lunaire
s marea *f* lunar
g Mondtide *f*

* **lunation → 3617**

3619 luvisol
A soil that has a horizon of accumulation of active clay with high capacity of cation exchange and with high saturation with bases, it occurs in the regions of moderately humid, sub-humid, and rarely of subtropical climate.
r лювисоль
Почва, имеющая горизонт накопления активной глины с высокой емкостью катионного обмена и с высокой насыщенностью основаниями, встречается в районах с умеренным гумидным, субгумидным, реже субтропическим климатом.
f luvisol *m*
s luvisol *m*
g Luvisol *m*

3620 lysimeter
A device or (a block of soil, covered with vegetation, placed in a container and replaced in the site from which it came) a piece of equipment which serves to measure the amount of water storage and percolation through soil.
r лизиметр
Прибор или стационарное сооружение (блок почвы, покрытой растительностью, который помещается в контейнер и затем устанавливается на месте взятия этой почвы), используемый для учета и сбора влаги, профильтровавшейся через почву.
f lysimètre *m*
s lisimetro *m*; caja *f* lisimétrica
g Lysimeter *n*

M

3621 maader
An extended part of a wadi occasionally flooded and fit for crops after the flood is over.
r маадер
Расширенная часть русла уэда, заливаемая водой и пригодная для посевов после прохождения паводка.
f maader *m*
s maader *m*
g Maader *n*

3622 maar
An old volcanic craterin form of a funnel-like or cylindrical depression up to 3.5 km in diameter and up to 400 m deep formed at gas explosion usually without lava effusion, often occupied by a lake. → Fig. 197
r маар
Старый вулканический кратер воронкообразной или цилиндрической формы диаметром до 3,5 км и глубиной до 400 м, возникающий при газовом взрыве обычно без излияния лавы, в нем нередко располагается озеро. → Рис. 197
f maar *m*
s maar *m*
g Maar *n*

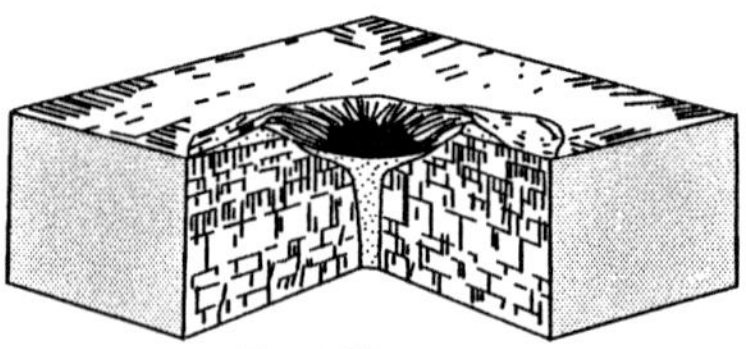

Fig. 197. **Maar**

* **macchia → 3676**

* **machine-building industry → 3753**

* **macquis → 3676**

3623 macroclimate
The climate of large geographical region, or continent.
r макроклимат
Климат обширного географического региона или континента.
f macroclimat *m*
s macroclima *m*
g Makroklima *n*

3624 macroeconomics
Analysis of a nation's economy as a whole, using such aggregate data as price levels, unemployment, inflation, and industrial production.
r макроэкономика
Общие аспекты экономического развития, такие как национальный доход, уровень цен, безработица, инфляция, уровень производства.
f macroéconomie *f*
s macroeconomía *f*
g Makroökonomie *f*

3625 macrophyte
Large water plants with floating leaves, growing in a water body, as opposed to *phytoplankton*.
r макрофит
Растения сравнительно больших размеров с плавающими листьями, произрастающие в водоеме, в отличие от *фитопланктона*.
f macrophyte *m*
s macrofito *m*
g Makrophyt *m*

3626 macrorelief
Large landforms (mountain ridges, plateaus, depressions) created mainly by endogenic processes.
r макрорельеф
Крупные формы рельефа (горные хребты, плато, впадины), созданные главным образом эндогенными процессами.
f macrorelief *m*; relief *m* majeur
s macrorelieve *m*; relieve *m* mayor
g Makrorelief *n*

3627 made land
Any part of the land surface which has been accumulated by the actions of mankind.
r искусственная суша
Любой участок суши, который был искусственно наращен.
f terre *f* artificielle
s tierra *f* artificial
g küstliches Land *n*

3628 magma
Extremely hot viscous, molten rock material lying under great pressure in the Earth's crust and in the upper mantle.
r магма
Горячий силикатный расплав в земной коре и верхней мантии.
f magma *m*
s magma *f*
g Magma *n*

3629 magma chamber; magmatic chamber
Magma reservoir in the Earth's crust and in the upper Earth's mantle from which the nourishment of a volcano occurs.
r вулканический очаг;
магматический очаг
Резервуар магмы в земной коре и верхней мантии Земли, откуда происходит питание вулкана.
f chambre *f* à magma;
réservoir *m* magmatique

s cámara *f* de magma; depósito *m* magmático
g Vulkanherd *m*

* **magmatic chamber → 3629**

3630 **magmatism**
The process of formation, transport and solidification of magma in the Earth's crust and on its surface.
r магматизм
Процесс образования, перемещения и застывания магмы в земной коре и на ее поверхности.
f magmatisme *m*
s magmatismo *m*
g Magmatismus *m*

3631 **magnetic *adj***
Relating to a magnet or magnetism.
r магнитный
Относящийся к магнитному полю или магнетизму.
f magnétique *adj*
s magnético *adj*
g Magnet-; magnetisch *adj*

3632 **magnetic anomaly**
A deviation from the predicted value of the magnetic field of the Earth at a particular point on the ground surface.
r магнитная аномалия
Отклонение магнитного поля Земли в данном месте от его расчетного значения.
f anomalie *f* magnétique
s anomalía *f* magnética
g Magnetanomalie *f*; magnetische Anomalie *f*

* **magnetic azimuth → 1162**

3633 **magnetic declination; compass declination**
The angle between the magnetic and geographical meridians at a given point of the Earth's surface.
r магнитное склонение
Угол между магнитным и географическим меридианами в данной точке земной поверхности.
f déclinaison *f* magnétique
s declinación *f* magnética
g magnetische Deklination *f*; Magnetdeklination *f*

* **magnetic dip → 3635**

3634 **magnetic equator**
A closed line on the Earth's surface where the tension of the magnetic field is parallel to the Earth's surface.
r магнитный экватор
Замкнутая линия на земной поверхности, где напряженность магнитного поля Земли параллельна поверхности.
f équateur *m* magnétique
s ecuador *m* magnético
g magnetischer Äquator *m*

3635 **magnetic inclination; magnetic dip**
The angle between the vector of the magnetic field tension and the plane of horizon.
r магнитное наклонение
Угол между вектором напряженности магнитного поля и плоскостью горизонта.
f inclinaison *f* magnétique
s inclinación *f* magnética
g magnetische Inclination *f*

3636 **magnetic meridian**
A line joining the position of the magnetic poles on the Earth's surface, i.e. the meridian referred to magnetic poles.
r магнитный меридиан
Линия, соединяющая магнитные полюса на земной поверхности, т.е. меридиан по отношению к магнитным полюсам (проекция силовой линии магнитного поля на поверхность Земли).
f méridien *m* magnétique
s meridiano *m* magnético
g magnetischer Meridian *m*

3637 **magnetic needle**
Mobile part of a compass swinging freely in a point, indicating the direction to the northern magnetic pole of the Earth.
r магнитная стрелка; стрелка компаса
Подвижная часть компаса, свободно вращающаяся на острие, указывающая направление на Северный магнитный полюс Земли.
f aiguille *f* aimantée
s aguja *f* imantada; aguja *f* de brújula; aguja *f* magnética
g Magnetnadel *f*

3638 **magnetic pole**
The two specific points on the Earth's magnetic field, where it is directed normally to the Earth's surface.
r магнитный полюс Земли
Точка на земной поверхности, где магнитное поле направлено перпендикулярно к поверхности Земли.
f pôle *m* magnétique
s polo *m* magnético
g Magnetpol *m*

3639 **magnetic reversal**
A 180° swing round of direction of the Earth's magnetic field, causing the magnetic poles to change position, so that the north magnetic pole becomes the south magnetic pole, and vice versa. → Fig. 198, p. 434
r миграция геомагнитных полюсов; инверсия геомагнитного поля
Перемещение геомагнитного полюса Земли из одного полушария в другое, что сопровождается изменением напряженности геомагнитного поля. → Рис. 198, с. 434

f réversion *f* magnétique
s reversión *f* magnética; desplazamiento *m* de los polos magnéticos
g Magnetpolwechsel *m*

3640 magnetic storm
A marked local disturbance of the Earth's magnetic field which lasts from several hours up to several days.
r магнитная буря
Сильное возмущение магнитного поля Земли, продолжающееся от нескольких часов до нескольких суток.
f orage *m* magnétique
s perturbación *f* magnética
g magnetisches Gewitter *n*; magnetischer Sturm *m*

3641 magnetosphere
The region of space surrounding the Earth which is influenced by the Earth's magnetic field.
r маг
Область околоземного пространства, в котором действует магнитное поле Земли.
f magnétosphère *f*
s magnetosfera *f*
g Magnetosphäre *f*

3642 magnitude of earthquake
A measure of the total energy released at a seismic focus of the earthquake and measured on Richter scale.
r магнитуда землетрясения
Условная величина, характеризующая количество энергии, выделившейся в очаге землетрясения, измеряемая по шкале Рихтера.
f magnitude *f* de séisme
s magnitud *f* de seísmo
g Magnitude *f* des Erdbebens; Erdebenmagnitude *f*

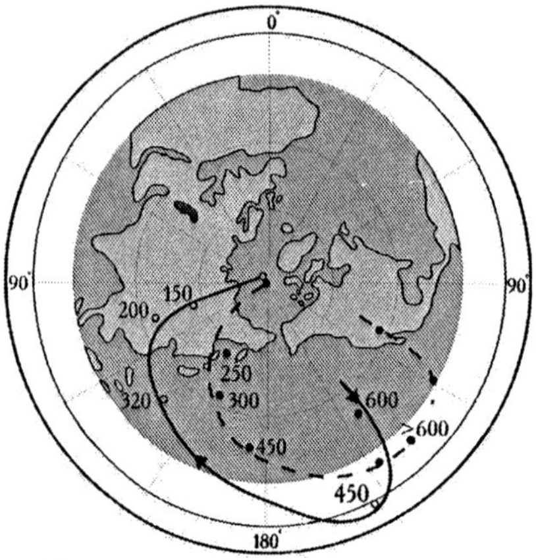

Fig. 198. **Magnetic reversal**: Both lines show possible moving of the North geomagnetic pole in the geological past, figures are past time in millions years before present

3643 main river; trunk stream
The dominant stream of a river system which flows directly into a sea or lake, even if it disappears in sands or swamps.
r главная река
Основной поток речной системы, непосредственно впадающий в море или озеро, а иногда теряющийся в песках и болотах.
f maîtresse; fleuve *m* principal
s río *m* principal
g Hauptfluß *m*; Mutterfluß *m*

3644 main transport line; artery
An especially main channel (e.g. a river or road) of transport or communication that connects two or more districts.
r магистраль
Основная транспортная линия (дорога или река), связывающая два или несколько районов.
f artère *f* de transport; artère *f* principale; voie *f* principale
s vía *f* de transporte principal; arteria *f* de communicación
g Magistrale *f*

* **mainland → 1239**

3645 mainstream
Line of the maximum velocity of the stream surface flow.
r стрежень реки
Линия наибольших поверхностных скоростей течения реки.
f ligne *f* de chenal
s línea *f* de canal; línea *f* de corriente
g Stromstrich *m*

3646 Malthusianism
The system of ideas formulated by T. Malthus according to which the human population, if unchecked by wars, famine, pestilence etc. alwais overgrows the limit of the means of its subsistence such as food and resources.
r мальтузианство
Предложенная Т. Мальтусом система воззрений, согласно которой население Земли, если оно не контролируется войнами, голодом и болезнями, растет быстрее средств существования людей (прежде всего пищи и природных ресурсов).
f malthusianisme *m*
s maltusianismo *m*
g Malthusianismus *m*

3647 mammoth
Existed in the Pleistocene, now extinct animal of elephants family with big spirally curved tusks, dense and long hair, the remnants of mammoths serve as a principal fossil indicator while studying the Quaternary continental deposits of glacial epochs.
r мамонт
Существовавшее в плейстоцене, ныне вы-

мершее животное семейства слонов с большими спирально изогнутыми бивнями, густой и длинной шерстью, остатки мамонтов служат руководящими ископаемыми при изучении четвертичных континентальных отложений, преимущественно ледниковых эпох.
f mammouth *m*
s mamut *m*
g Mammut *n*

3648 mandate; mandate territory
Former German colonies and certain Turkish (Ottoman) territories transferred after the World War I by the Leagues of Nations to the victor-countries for administration.
r мандатная территория
Общее название бывших германских колоний и некоторых османских (турецких) владений, переданных после первой мировой войны по мандату Лиги Наций в управление странам-победительницам.
f territoire *m* sous mandat
s territorio *m* en mandato
g Mandatsgebiet *n*

* **mandate territory → 3648**

* **mangal → 3649**

3649 mangrove; mangal; mangrove forest
A swamp short evergreen forest of saline or brackish water which develops on tropics and subtropics on low, silty sea coasts in tidal zone, on the areas protected from breaking waves (in bays, lagoons, estuaries, etc.), it has a dense tangle of aerating roots projecting above the mud. → Fig. 199
r мангры; мангровые заросли
Тропический тип низкоствольных вечнозеленых лесов и кустарников, развитых на низменных илистых морских побережьях в полосе прилива и отлива, на защищенных от прибоя участках (в заливах, лагунах, эстуариях и т.п.). → Рис. 199
f forêt *f* de mangrove; mangrove *f*
s manglar *m*; mangle *m*; bosque *m* de manglar
g Mangrovenwald *m*; Mangrovendickicht *n*

* **mangrove forest → 3649**

3650 mangrove swamp
Low tropical muddy coast, saturated with sea water, occupied with mangrove vegetation.
r мангровое болото

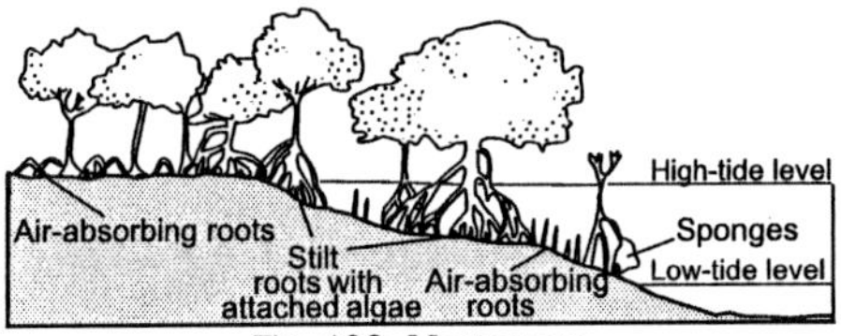

Fig. 199. **Mangrove**

Низменные аккумулятивные тропические побережья с илистым грунтом, пропитанным морской водой, занятые мангровой растительностью.
f marais *m* à palétuviers; marais *m* littoral de mangrove
s marisma *f* manglar
g Mangrovensunpf *m*

* **man-made *adj* → 298**

3651 man-made beach
An artificial band of sand or pebble made to protect the coast against a washout or to widen the beach.
r искусственный пляж
Искусственно намытое или насыпанное образование из песка или гальки для защиты берега от размыва или для расширения пляжа.
f plage *f* artificielle
s playa *f* artificial
g künstlicher Strand *m*

* **man-made disaster → 2831**

* **man-made landscape → 301**

* **man-made pollution → 302**

* **man-made soils → 304**

* **manor → 2000**

* **manpower → 3338**

3652 mantle of the Earth; Earth's mantle
The layer of rigid Earth between the crust and the core, the lower surface lying at 2900 km, the upper surface is the Mohorovičič discontinuity.
r мантия Земли
Оболочка «твердой Земли» между земной корой и ядром Земли, ее верхняя граница проходит по поверхности Мохоровичича, нижняя лежит на глубине около 2900 км.
f manteau *m* de la Terre
s manto *m* de la Tierra
g Erdmantel *m*

* **manufacturing industry → 5295**

3653 map; geographic map
Decreased and generalised flat image of the Earth's surface, some other celestial bodies or heavens constructed in cartographic projection with the use of conventional signs.
r карта; картографическое изображение; географическая карта
Построенное в картографической проекции (на плоскости) с использованием специальных условных знаков, уменьшенное и обобщенное изображение поверхности Земли, других небесных тел или небесной сферы.
f carte *f*; carte *f* géographique
s mapa *m*; mapa *m* geográfico

g geographische Karte *f*

3654 map accuracy
The correspondence to the reality of the location, dimensions, plan outline and altitude of the objects represented on a geographical map.
r точность карты
Соответствие действительности изображенных на географической карте местоположения, размеров, плановых очертаний и высотного положения объектов.
f précision *f* de la carte
s precisión *f* de mapa
g Kartengenauigkeit *f*

3655 map adjustment; map reconciliation
The co-ordination of spatially interconnected and mutually determined elements of the contents in the process of compilation and editing of maps and atlases.
r согласование карт
Увязка пространственно взаимосвязанных и взаимно обусловленных элементов содержания в процессе составления и редактирования карт и атласов.
f règlement *m* des cartes
s coordinación *f* cartográfica
g Kartenabstimmung *f*

3656 map ageing; outdating of a map
The loss of conformity of the whole cartographic representation or some of its elements to the actual reality.
r старение карты
Утрата соответствия с современностью всего картографического изображения или отдельных его элементов.
f vieillissement *m* de la carte
s envejecimiento *m* de mapa
g Kartenalterung *f*

* **map appearance → 3828**

3657 map assemblage; map montage
Determination of the limits of cartographic image, marking map frames and placing legend and additional characteristics with regard to them.
r компоновка карты
Установление границ картографического изображения, нанесение рамок карты и размещение относительно них легенды и дополнительных характеристик.
f assemblage *m* de carte
s ensamblaje *m* de mapa;
montaje *m* de mapa
g Kartenmontage *f*

3658 map clarity
The representation of features on a map in such a way which makes it easy for the users to comprehend the information.
r наглядность карты
Возможность зрительного восприятия пространственных форм, размеров, размещения объектов и другой информации, отображаемой на карте.
f lisibilité *f* de carte
s claridad *f* de mapa
g Anschaulichkeit *f* der Karte

* **map compilation → 3673**

3659 map content
The set of features and information represented on a map.
r содержание карты
Совокупность изображаемых на карте объектов и сообщаемых о них сведений.
f contenu *m* de carte
s contenido *m* de mapa
g Karteninhalt *m*

3660 map coverage
The number or percentage of maps, surveys and cartographic materials, existing for a given area.
r картографическая изученность
Степень покрытия какой-либо территории съемками и картографическими материалами.
f couverture *f* cartographique
s cubrimiento *m* cartográfico
g kartographische Erschließung *f*; Kartenabdeckung *f*

* **map design and production → 3673**

* **map elements → 1984**

3661 map error
The discrepancy between the quantitative or qualitative information shown on a particular map and the actual situation.
r ошибка на карте
Количественное или качественное отклонение картографического изображения от реальной ситуации.
f erreur *f* cartographique
s error *m* del mapa; error *m* cartográfico
g Kartenfehler *m*

3662 map incunabula
Any map issued before the XV century.
r карта-инкунабула
Любая карта, изданная до XV века.
f carte-incunabula *f*; carte *f* ancienne
s mapa-incunabula *f*; mapa *m* antíguo
g Karteninkunabel *f*

3663 map informativity
The amount and value of information which a user can get from a map.
r информативность карты
Степень насыщенности карты содержательным материалом.
f valeur *m* informative de carte
s valor *m* informativo de mapa
g Informationsgehalt *m* der Karte

* **map interpretation → 3671**

3664 **map language; map semiotics**
Conventional signs, ways of cartographic presentation, rules of their construction, use and reading aimed at making and use of geographical maps.
r язык карты;
картографическая семиотика
Условные обозначения, способы картографического изображения, правила их построения, употребления и чтения, применяемые для целей создания и использования географических карт.
f langue *f* de la carte;
sémiologie *f* cartographique
s lenguaje *f* de mapa;
semiología *f* cartográfica
g Kartensprache *f*

3665 **map loading**
The number of conventional symbols and inscriptions on a map expressed in their quantity per unit of the map's area.
r нагрузка карты
Заполнение карты условными знаками и надписями, выражаемая их количеством на единицу площади карты.
f charge *f* de la carte
s cargo *m* del mapa
g Kartenbelastung *f*

3666 **map margin**
The framework, including the neat line, which surrounds a map.
r рамка карты
Линия или совокупность линий, обрамляющих карту.
f filet *m* d'encadrement
s fileta *f* de cuadro; cuadro *m*
g Kartenrahmen *m*

* **map montage → 3657**

* **map of regionalization → 3667**

3667 **map of regions; map of regionalization**
Representation of an area, divided into regions delimited on the basis of one or several features of natural or socioeconomic character.
r карта районирования
Изображение территории, разделенной по одному или по совокупности природных или социально-экономических признаков в соответствии с целевым назначением районирования.
f carte *f* des régions
s mapa *m* de regiones;
mapa *m* de regionalización
g Karte *f* der Regionalisierung;
Karte *f* der Rayonierung

3668 **map of the fields**
A geographical map showing the continuous distribution of natural phenomena or calculated indices presented by isolines.
r карта полей
Географическая карта, на которой посредством изолиний изображаются непрерывные распределения природных явлений или расчетных показателей.
f carte *f* de champs
s mapa *m* de campos
g Feldkarte *f*

3669 **map of the hemispheres**
A cartographic representation of the Earth divided into two hemispheres.
r карта полушарий
Картографическое изображение земного шара, разделенного на два полушария.
f carte *f* des hémisphères
s mapa *f* de hemisferios
g Karte *f* der Erdhalbkugeln

* **map of the world → 4597**

3670 **map projection; projection; cartographic projection**
The method of representing of pert or the whole of the Earth's curved surface on a plane surface of a map, by an orderly system of meridians and parallels.
r картографическая проекция
Математический способ изображения, а также собственно изображение поверхности эллипсоида или шара на плоскости географической карты.
f projection *f* cartographique
s proyección *f* cartográfica
g Kartennetzentwurf *m*

3671 **map reading; map interpretation**
The perception of a geographical map, i.e. the identification of cartographic images, their analysis and understanding of the meanings of map images.
r чтение карты
Восприятие географической карты, связанное с распознаванием картографических образов, истолкованием и пониманием ее содержания.
f interprétation *f* de carte; lecture *f* de carte
s interpretación *f* de mapa; lectura *f* de mapa
g Kartenlesen *n*

* **map reconciliation → 3655**

3672 **map reliability**
Correctness of information presented on a map for a certain date.
r достоверность карты
Правильность сведений, приведенных на карте на определенную дату.
f certitude *f* de la carte
s fiabilidad *f* de mapa;
confiabilidad *f* de mapa
g Kartenzuverlässigkeit *f*

* **map semiotics → 3664**

* **map's legend → 3465**

* **mapping → 900**

3673 **mapping; map design and production; map compilation**
The processes, methods and technologies of making maps, atlases and other cartographic products.
r картографирование; проектирование карты; составление карты
Совокупность методов и технологий создания карт и атласов, а также сам процесс изготовления карты.
f establissement *m* des cartes; compilation *f* cartographique
s construcción *f* de mapas; confección *f* de mapas; producción *f* de mapas; establicimiento *m* de mapas
g Kartierung *f*; Kartieren *n*; Kartenentwerfen *n*

3674 **mapping agency**
A governmental institution responsible for the collection, production, and distribution of maps.
r картографическая служба
Государственное учреждение, задачей которого является накопление, изготовление и распространение карт.
f service *m* cartographique
s servicio *m* cartográfico
g Kartendienst *m*

3675 **mapping of landscape; landscape surveying**
Making a landscape map by means of field survey and by studying key areas.
r ландшафтная съемка
Составление ландшафтной карты с помощью маршрутных исследований и изучения ключевых участков.
f cartographie *f* de paysage
s cartografía *f* de paisajes
g Landschaftsaufnahme *f*

3676 **maquis; macchia; macquis**
Thick evergreen, leathery leaved, sometimes thorny shrubs and small trees in the Mediterranean region, growing in place of cut forests.
r маквис
Сообщество вечнозеленых жестколистных, иногда колючих кустарников и невысоких деревьев в Средиземноморье, возникающее, как правило, на месте сведенных лесов.
f maquis *m*
s maquis *m*; maquia *f*
g Maquis *f*

3677 **margalite soil**
A black tropical soil that develops on carbonate or basalt rocks in regions with distinct dry and wet seasons, when dry it is monolithic, fractured, and when wetted, it is sticky and viscous.
r маргалитовая почва
Черная тропическая почва, формирующаяся на карбонатном или базальтовом основании при чередовании резкого переувлажнения и периодического сильного пересыхания, слитая, трещиноватая, во влажном состоянии клейкая и вязкая.
f sol *m* du margalite
s suelo *m* de margalito
g Margalitboden *m*

* **marginal analysis → 3683**

3678 **marginal channel; periglacial drainage channel**
A valley worked out by the streams of melt water in rocks and loose deposits of the glacier bed and in periglacial areas.
r маргинальный канал; приледниковый дренажный канал
Долина, выработанная в коренных породах и рыхлых отложениях ледникового ложа и приледниковых областей потоками талых ледниковых вод.
f canal *m* marginal
s canal *m* marginal
g Marginalkanal *m*; periglazialer Schmelzwasserkanal *m*

3679 **marginal deep; foreland basin; foredeep**
A relatively narrow, deep, elongated, steep-sided depression in the ocean floor, near or parallel to a mountainous land area, or on the seaward side of an island arc.
r краевой желоб; предгорный желоб
Относительно узкая, глубокая, вытянутая впадина с крутыми стенками на морском дне, расположенная рядом или параллельно гористому берегу, либо с мористой стороны островной дуги.
f fosse *f* marginale
s fosa *f* marginal
g Vorseerinne *f*; marginaler Tiefseegraben *m*

* **marginal fringe → 3682**

3680 **marginal furrow**
A longitudinal depression in a mountain valley occupied by a glacier lying between its bedrock slope and a margin of a bulging surface of the glacier tongue.
r краевая борозда
Продольное понижение в занятой ледником горной долине между ее коренным склоном и краем выпуклой поверхности ледникового языка.
f sillon *m* marginal
s surco *m* marginal
g Randfurche *f*

3681 **marginal glacial landforms**

Landforms formed by the contemporary and ancient glaciers in their periphery zone, which indicate the location of the glacier boundaries and dynamic state of its margin.

r краевые ледниковые образования
Формы рельефа, созданные современными и древними ледниками в их периферической зоне, свидетельствующие о положении ледников и динамическом состоянии их края.

f formes *fpl* de relief glaciaires marginales
s formas *fpl* marginales de relieve glacial
g marginale Gletscherformationen *fpl*

3682 marginal lands; marginal fringe
Land, such as upland, or desert border, which is difficult to calculate and which yields little profit, no more than is sufficient to cover the costs of production, that is why it goes in or out of agricultural use according to fluctuations in economic conditions.

r маргинальные земли
Сельскохозяйственные угодья, средний урожай с которых позволяет покрывать издержки производства, но недостаточен для оплаты ренты, поэтому они вовлекаются в хозяйственный оборот или устраняются из него в зависимости от экономической конъюнктуры.

f terres *fpl* marginales
s tierras *fpl* marginales
g marginale Länder *npl*

* **marginal sea → 171**

3683 marginalism; marginal analysis
Mathematical method in economics and economic geography which widely uses the limit values and differential calculus to analyze economic processes.

r маргинализм
Математический метод в экономике и экономической географии, широко использующий предельные величины и дифференциальное исчисление для анализа экономических процессов.

f marginalisme *m*
s marginalismo *m*
g Marginalismus *m*

3684 mari
A boggy terrain in the permafrost zone of Eastern Siberia and the Russian Far East, thawing only to a depth of 40 cm and covered by sparse larch forest alternating with the patches of treeless bogs and birch communities.

r марь
Заболоченный (протаивающий лишь на глубину до 40 см) редкостойный лиственничный лес в Восточной Сибири и на Дальнем Востоке, перемежающийся с участками безлесных болот и зарослей ерника.

f mari *m*
s mari *m*
g Mar *n*

3685 marigram
A graphic representation of the tidal oscillation at a particular coastal station.

r мариграмма; кривая прилива
График приливных колебаний уровня моря, запись на ленте самописца уровня моря.

f marigramme *m*
s marigrama *m*
g Marigramm *n*

* **marine abrasion → 6636**

3686 marine *adj*; maritime *adj*
Relating to the sea, shipping and navigation.

r морской
Относящийся к морю и судоходству.

f maritime *adj*
s marino *adj*; marítimo *adj*
g maritim *adj*; See-; Meer-

3687 marine biology; biological oceanography
The study of all kinds of marine life found in oceans and seas, their spatial distridution, migration and adaptations to human impact.

r морская биология; биологическая океанология
Изучение распределения и миграции организмов в морской среде, их функционирования, трофической структуры, продуктивности и адаптации к факторам среды и антропогенным воздействиям.

f biologie *f* marine
s biología *f* marina
g marine Biologie *f*; Meeresbiologie *f*

* **marine climate → 3704**

3688 marine cycle of erosion
The evolution of the sea coast and change of coastal landforms due to abrasion and aggradation.

r морской цикл эрозии
Эволюция морского берега и смена комплекса береговых форм рельефа вследствие абразионно-аккумулятивной деятельности моря.

f cycle d'érosion marin
s ciclo *m* de erosión marino
g Meereszyklus *m*

* **marine deposits → 3700**

3689 marine ecology
The study of the relationships between sea organisms and their environment in seas and oceans.

r морская экология
Изучение биоценозов морских организмов и их взаимоотношений с окружающей средой.

f écologie *f* marine
s ecología *f* marina
g marine Ökologie *f*; Meeresökologie *f*

3690 marine ecosystem

A natural complex formed by living organisms and their habitat, the components of which are interconnected by metabolism and energy exchange.

r морская экосистема
Единый природный комплекс, образованный живыми морскими организмами и средой их обитания, компоненты которого связаны между собой обменом вещества и энергии.

f écosystème *m* marin
s ecosistema *m* marino
g marines Ökosystem *n*

3691 marine fauna
All the species of animals inhabiting seas and oceans.

r морская фауна
Совокупность животных, населяющих моря и океаны.

f faune *f* marine
s fauna *f* marina
g Meeresfauna *f*

3692 marine flora
Plants inhabiting seas and oceans, mainly algae, macrophytes and phytoplankton.

r морская флора
Совокупность растений, обитающих в морях и океанах, в основном водорослей, макрофитов и фитопланктона.

f flore *f* marine
s flora *f* marina
g Meeresflora *f*

3693 marine geology
The study of the genesis and geological structure of sea floor.

r морская геология
Изучение геологического строения и динамики дна морей и океанов.

f géologie *f* marine
s geología *f* marina
g marine Geologie *f*; Meeresgeologie *f*

3694 marine geophysics
The study of composition and tectonic structures of sea floor and the Earth's crust below oceans and seas, seismic activity, magnetic and gravitational fields, heat flow and magmatism of the ocean floor.

r морская геофизика
Изучение строения и тектонических структур дна и земной коры под океанами и морями, сейсмичности, магнитного и гравитационного полей, теплового потока и магматизма океанского дна.

f géophysique *f* de mer
s geofisíca *f* de mar
g marine Geophysik *f*

3695 marine hydroacoustics
The study of the processes of propagation, reflection and attenuation of sound waves in sea water.

r морская гидроакустика; акустика океана
Изучение распространения, отражения и затухания звуковых волн в морской среде.

f hydroacoustique *f*
s hidroacústica *f*
g Hydroakustik *f*

* **marine hydrophysics → 4554**

3696 marine ice sheet
Ice sheet overthrusting on the continental shelf and ice shelves dynamically connected with it.

r морской» ледниковый покров
Ледниковый щит, налегающий на континентальный шельф, и динамически связанные с ним шельфовые ледники.

f calotte *m* glaciaire marine
s casquete *m* glaciar marino
g mariner Eisschild *m*

3697 marine meteorology
A branch of *meteorology* studying atmospheric phenomena in the oceanic areas including island and coastal regions.

r морская метеорология
Раздел *метеорологии*, изучающий атмосферные явления над океанами, включая островные и прибрежные территории.

f météorologie *f* marine
s meteorología *f* marina
g marine Meteorologie *f*

3698 marine national park
Protected part of the sea and coast where marine flora and fauna are preserved.

r морской национальный парк
Охраняемая акватория с прилегающей частью побережья, на которой установлен режим охраны морской фауны и флоры.

f parc *m* national marin
s parque *m* nacional marino; reserva *f* nacional marina
g Nationalmeerespark *m*

3699 marine nautical chart
A special geographical map which assists navigation at sea or on other waterways, compiled as a rule in Mercator projection.

r морская навигационная карта
Специальная географическая карта, как правило в проекции Меркатора, предназначенная для прокладки пути корабля.

f carte *f* nautique; carte *f* de navigation
s carta *f* náutica; carta *f* de navegación
g Seekarte *f*

3700 marine sediments; marine deposits
Sedimentary material on the sea floor of oceans not yet transformed into rocks.

r морские отложения; океанические осадки
Осадочные образования на дне современных океанов, еще не превращенные в горные породы.

f sédiments *mpl* marins;
sédiments *mpl* océaniques
s sedimentos *mpl* marinos;
sedimentos *mpl* oceánicos
g Meeresablagerungen *fpl*;
Ozeansedimenten *npl*

3701 marine terrace
A step stretched along a coast either above or below present sea-level which marks the coastline of the past.
r морская терраса
Вытянутая вдоль побережья ступень в рельефе, расположенная выше или ниже современного уровня моря, фиксирующая положение береговой линии в прошлом.
f terrasse *f* marine; terrasse *f* littorale
s terraza *f* marina; terraza *f* litoral
g Meer(es)terrasse *f*

* **marine-built terrace → 13**

3702 marinism
A scientific view assuming that unsorted detrital clays of the northern plains and erratic boulders included in them are the deposits of cold seas and icebergs.
r маринизм
Научное течение, приписывающее несортированные щебнистые глины северных равнин и встречающиеся в них эрратические валуны отложениям холодных морей и айсбергов.
f marinisme *m*
s marinismo *m*
g Marinismus *m*

* **maritime *adj* → 3686**

* **maritime air → 3703**

3703 maritime air mass; maritime air
Air mass which is formed or transformed over the sea, it has a relatively high moisture content, especially in the lower levels.
r морская воздушная масса;
морской воздух
Воздушная масса, сформировавшаяся или трансформировавшаяся над морем и имеющая вследствие этого высокое влагосодержание, особенно в нижних слоях.
f air *m* maritime
s aire *m* marítimo
g Meeresluft *f*

3704 maritime climate; marine climate
The climate of regions adjacent to the sea, and directly influenced by the moderating effect of water mass which causes a small diurnal and annual range of temperature, high relative humidity, much clouds, fairly uniform precipitation.
r морской климат; приморский климат
Климат регионов, близких к морю, отличающийся небольшими суточными и годовыми амплитудами температуры и высокой относительной влажностью, что вызвано смягчающим влиянием водных масс.
f climat *m* maritime; climat *m* marin
s clima *m* marítimo; clima *m* marino
g Seeklima *n*; maritimes Klima *n*

3705 maritime geomorphology;
submarine geomorphology
The study of the structure, origin and development of sea floor topography in seas and oceans.
r морская геоморфология;
геоморфология моря
Изучение строения, происхождения и развития рельефа дна морей и океанов.
f géomorphologie *f* sous-marine
s geomorfología *f* marina
g marine Geomorphologie *f*;
Meeresgeomorphologie *f*

3706 mark; point; number (in a scale)
A conventional unit for quantitative and qualitative estimation of a natural phenomenon (wind force, sea waves, intensity of earthquakes, etc.).
r балл
Условная единица для количественной и качественной оценки природного явления (силы ветра, волнения моря, интенсивности землетрясения и др.).
f point *m*
s grado *m*
g Grad *m*; Punkt *m*; Stufe *f*

3707 marker
A conventional sign in the form of a circle used to indicate human settlements on maps.
r пунсон
Внемасштабный условный знак в виде окружности, используемый для изображения на картах населенных пунктов.
f poinçon *m*
s punto *m*; círculo *m*
g Kreisring *m*; punktförmige Signatur *f*

3708 market
The exchange of goods and services, i.e. demand and supply of commodities, services, financial resources, gold and securities, investments, labour, etc. on the global level (world market), national level (domestic market) and regional level (local market); a public place, a particular site where goods and services are bought and sold.
r рынок
Сфера товарного обмена, т.е. спрос и предложение товаров, услуг, финансовых ресурсов, ценных бумаг и золота, инвестиций, труда и пр. в масштабах мирового хозяйства (мировой, внешний рынок), страны (национальный, внутренний рынок) и ее отдельного региона (местный рынок); специальное место, где продаются и покупаются различные товары и услуги.

f marché *m*
s mercado *m*
g Markt *m*

3709 **market economy; free market economy; free economy**
An economy in which the major parts of production, distribution, and excange are carried out by private individuals or companies rather than by the government, hence there is free enterprise and competition, prices being determined by the free excange of commodities for money (supply and demand).
r рыночная экономика; либеральная экономика; открытая экономика
Социально-экономическая система, развивающаяся на основе частной собственности и товарно-денежных отношений, когда распределение ресурсов, производство, обмен и потребление товаров и услуг определяются главным образом спросом и предложением и не контролируются непосредственно государством.
f économie *f* de marché; économie *f* ouverte
s economía *f* de mercado
g Marktwirtschaft *f*

3710 **market eqiilibrium**
The balance between the prices in supply and demand for all resources, commodities, services and labour.
r рыночное равновесие
Сбалансированность цен предложения и спроса на все ресурсы, товары, услуги и труд.
f équilibre *m* du marché
s equilibrio *m* de mercado
g Marktseqiillibrium *n*

3711 **market of raw materials**
The market that meets the demands of economy in raw materials.
r сырьевой рынок
Рынок, обеспечивающий потребности хозяйства в сырьевых ресурсах.
f marché *m* des matières premières
s mercado *m* de materias primas
g Rohstoffmarkt *m*

3712 **market of sale; geographical market**
A geographical area of any size and rank where particular goods and services are on stable demand.
r рынок сбыта
Географические ареалы любого размера и ранга, в границах которых находят устойчивый спрос данные товары и услуги.
f débouché *m* commercial; débouché *m*
s mercado *m* de venta
g Absatsmarkt *m*

3713 **market zone**
A territory where economic relations are regulated by common laws, taxes, customs and other special rules.
r особая рыночная зона
Территория, где экономические отношения регулируются едиными правовыми, налоговыми, таможенными и другими специальными установлениями.
f zone *f* économique spéciale
s zona *f* económica especial
g Marktzone *f*

3714 **marketing principle**
The sale of production following the main geographical principle, i.e. the supply of goods and services from a central place should be as close as possible to the place supplied.
r сбыт продукции
Реализация готовой продукции в соответствии с основным географическим принципом: предложение товаров и услуг из центрального места дожно быть как можно ближе к этому месту.
f mercatique *f*; commercialisation *f*
s mercado *m* de venta
g Absatz *m*

* **marling → 2137**

3715 **marsh → 3715**
A wet area of mineral soil flooded periodically or at intervals and covered with water-loving vegetation.
r марш
Болотистая равнина с минеральной почвой и влаголюбивой растительностью, периодически или время от времени затапливаемая водой.
f marais *f* littoral; marais *f* maritime
s marisma *f*
g Marsch *f*

* **marsh soil → 3716**

* **marsh-gas → 3821**

3716 **marshy soils; marsh soil**
Soils of different types that form at the margins of bogs or on poorly drained plains which are periodically over-moistened.
r заболоченные почвы
Почвы разных типов, формирующиеся на окраинах болот или на низменных слабо дренированных равнинах, периодически испытывающие переувлажнение.
f sols *mpl* marécageux; sols *mpl* d´engorgement
s suelos *mpl* pantanosos; suelos *mpl* de encharcamiento
g versumpfte Böden *mpl*

3717 **mass movement; mass wasting**
The spontaneous downward movement of surface materials, including solid rock under the influence of gravity, assisted by rainfall or meltwater from snow and ice, classified as: flow, creeping and landslides.
r движение склоновых масс

Стихийное движение вниз по склону горных пород под действием гравитационных сил, а также дождевых или талых вод, различают течение, скольжение и оползни.
f mouvement *m* de masse
s movimiento *m* de masa
g Massenverlagerung *f*

3718 mass production
The continuous production of identical products using a repeated production process and line production.
r массовое производство
Непрерывность изготовления в течение длительного периода одинаковой продукции при строгой повторяемости производственного процесса и поточном производстве.
f production *f* en série
s producción *f* en serie
g Massengüterproduktion *f*

* **mass wasting → 3717**

3719 mass-and-energy exchange
Energy, water, mineral inclusions, and ice exchange between a natural object and its environment.
r массоэнергообмен
Обмен природных объектов теплом, водой, льдом, минеральными включениями с окружающей средой.
f échange *m* d'énergie et de mass
s cambio *m* de energía y masa
g Massen- und Energieaustausch *m*

* **massif → 3980**

3720 massive ice
A large mass of ground ice of any genesis including ice wedges, veins, laminae and blocks deposited in rocks.
r ледяная залежь; залежеобразующий лед
Скопление подземного льда любого генезиса жильной, пластовой и глыбовой формы в горных породах.
f glace *f* massive
s hielo *m* macizo
g Massiveis *n*

3721 massive ice bedding
A massive ice having a quasihorizontal extent according to the stratification of the enclosing rocks.
r пластовая ледяная залежь
Ледяная залежь, залегающая согласно с напластованием вмещающих горных пород.
f glace *f* massive horizontale
s hielo *m* macizo horizontal
g horizontales Massiveis *n*

3722 massive ice vein
A large ice vein formed due to repeated frost fracturing and refilling of fractures with ice. → Fig. 200
r повторная ледяная жила
Ледяная жила, сформировавшаяся в результате многократно повторяющегося морозного растрескивания и заполнения трещин льдом. → Рис. 200
f filon *m* de glace massive
s cuña *f* de hielo secundaria
g sekundärer Eiskeil *m*

* **massive vein ice → 3723**

3723 massive wedge ice; massive vein ice
Ice occurring in massive ice veins and having vertical and subvertical stratification.
r повторножильный лед; полигонально-жильный лед
Лед повторных ледяных жил, отличающийся вертикальной и наклонной слоистостью.
f glace *f* en coin compose; glace *f* filon massive
s hielo *m* de cuña secundaria
g sekundäres Keileis *n*

3724 material production
All types of manufacturing activity that create material wealth and increase its cost in commodity circulation.
r сфера материального производства; материальное производство
Совокупность всех видов производственной деятельности, создающих материальные блага в вещественной форме и увеличивающих их стоимость в сфере обращения.
f production *f* matérielle
s producción *f* material
g materielle Produktion *f*

3725 mathematical base of map
A set of elements determining the accuracy of a geographical map, it includes geodetic basis, scale, and cartographic projection.
r математическая основа карты
Совокупность элементов, определяющих точность географической карты, включает геодезическую основу, масштаб и картографическую проекцию.
f base *m* mathématique de carte
s base *f* matemática de mapa

Fig. 200. **Massive ice veins** in the bank of Indigirka river

g mathematische Grundlage *f* der Karte; mathematische Kartenbasis *f*

3726 mathematical cartography
The working out of mathematical bases of maps and cartographic projections.
r математическая картография
Создание, разработка математической основы карт и картографических проекций.
f cartographie *f* mathématique
s cartografía *f* matemática
g mathematische Karographie *f*

3727 matorral; mattoral
The communities of xerophyllous shrubs that grow in South America in the regions with the Mediterranean climate.
r мато
Заросли кустарников в переменно влажных и засушливых районах Южной Америки.
f matorral *m*
s mattoral *f*
g Mato *m*

* **mattoral → 3727**

3728 mature *adj*
Having reached a stage of full natural development, as applied to landscape, river, shoreline, soil, etc.
r зрелый
Достигший стадии полного естественного развития, применяется к ландшафту, реке, побережью, почве и т.п.
f mûr *adj*
s maduro *adj*
g reif *adj*

3729 mature soil
A soil with a complete set of soil horizons typical for a given geographical zone and substrate; a soil in equilibrium with its environment.
r зрелая почва
Почва с полным набором горизонтов, характерным для данной географической зоны на данном субстрате; почва, находящаяся в равновесии с окружающей средой.
f sol *m* mûr; sol *m* évolué
s suelo *m* maduro; suelo *m* evolucionado
g reifer Boden *m*

* **mat-vegetation → 885**

3730 maximum discharge
The greatest (daily average or instantaneous) water discharge of a flood or high water.
r максимальный расход воды
Наибольший (средний суточный или мгновенный) расход воды половодья или паводка.
f débit *m* maximum; débit *m* de pointe
s caudal *m* máximo; caudal *m* de punta
g Höchstwassermenge *f*

3731 maximum permissible concentration; threshold concentration
The amount of harmful substances which makes the environment unhealthy or dangerous when this concentration is exceeded.
r предельно допустимая концентрация
Количество вредного вещества в окружающей среде, при превышении которого среда или ее элементы непригодны или опасны для использования.
f concentration *f* maximale admissible; seuil *m* de concentration
s concentración *f* umbral; concentración *f* máxima admisible
g höchstzulässige Konzentration *f*

3732 maximum permissible emission; threshold emission
The amount of a pollutant discharged by a source per time unit the exceeding of which causes harmful effect on the environment and human health.
r предельно допустимый выброс
Масса загрязняющего вещества, выбрасываемого отдельным источником за единицу времени, превышение которой ведет к нарушениям окружающей среды и здоровья человека.
f émission *f* limite; seuil *m* d'émission
s emisión *f* umbral; expulsión *f* límite
g höchstzulässige Emission *f*

3733 maximum permissible environmental impact; critical environmental load; threshold impact
The highest level of human effect which does not cause irreversible damage to the environment.
r предельно допустимое воздействие; критическая нагрузка
Максимальная степень воздействия человека на окружающую среду, не приводящая к ее безвозвратному нарушению.
f impact *m* écologique critique; action *f* écologique critique
s impacto *m* ecológico límite; impacto *m* umbral sobre medio ambiente
g höchstzulässige Umweltbelastung *f*

3734 maximum temperature
The highest temperature observed at a certain point or region during a time interval.
r максимальная температура
Наивысшее значение температуры, наблюдавшееся в данном пункте или регионе за конкретный период.
f température *f* maximale
s temperatura *f* máxima
g Maximumtemperatur *f*; maximale Temperatur *f*

3735 maximum water consumption
The total consumption of water taken from a water body including irretrievable water consumption and the water which returns back, i.e. sum of retrievable and irretrievable water consumption.

r полное водопотребление
Общее потребление воды, отобранной из водного объекта, включая безвозвратное водопотребление и ту часть воды, которая возвращается обратно (т.е. сумма возвратного и безвозвратного водопотребления).
f consommation *f* totale d'eau
s consumo *m* total de agua
g totaler Wasserverbrauch *m*; Gesamtwasserverbrauch *m*

3736 **meadow**
A piece pf land permanently covered with dense grassy vegetation that grows on flood plains, watersheds in the forest zone, and in the mountains above timberline (subalpine and alpine meadows).
r луг
Тип сомкнутой травяной растительности, распространенной в поймах рек и на водоразделах в лесной зоне, а также в горах выше границы леса (субальпийские и альпийские луга).
f prairie *f*; pré *m*; pelouse *f*
s prado *m*; pradera *f*
g Wiese *f*

3737 **meadow chernozemic soil**
A soil that forms in the chernozem zone under meadow-steppe vegetation in the conditions of increased moistening, it has a thick humus horizon with a high humus content.
r лугово-черноземная почва
Почва, формирующаяся в черноземной зоне при повышенном увлажнении под лугово-степной растительностью, имеет мощный гумусовый горизонт с высоким содержанием гумуса.
f chernozem *m* de prairie
s suelo *m* chernozémico de pradera
g Wiesen-Tschernosemboden *m*

3738 **meadow soils**
Soils of different natural zones that develop under meadow vegetation in the conditions of increased surface moistening and constant suppl of ground waters.
r луговые почвы
Почвы различных природных зон, развитые под луговой растительностью в условиях повышенного поверхностного увлажнения и постоянной связи с грунтовыми водами.
f sols *mpl* de prairie
s suelos *mpl* de pradera
g Wiesenböden *mpl*

3739 **meadow steppe**
A grass community that is composed mostly of herbs and has solid sodding, it is characteristic of the forest-steppe subzone.
r луговая степь
Травяное сообщество, характерное для подзоны лесостепи, обычно большое участие разнотравья и почти сплошное залужение почвы.
f prairie-steppe *f*
s estepa *f* de pradera
g Wiesensteppe *f*

3740 **meadow-brown semi-desert soil**
A soil that forms among brown semi-desert soils under the additional surface or ground moistening, it is characterized by the increased humus content and gleyzation of lower horizons.
r лугово-бурая полупустынная почва
Почва, формирующаяся среди бурых полупустынных почв при дополнительном поверхностном или грунтовом увлажнении, отличается повышенным содержанием гумуса и оглеением нижних горизонтов.
f sol *m* brun subaride à nappe phréatique
s suelo *m* pardo subárido de pradera
g Wiesen- und brauner Halbwüstenboden *m*

3741 **meadow-chestnut soil**
A soil developing in flat-bottom depressions of dry steppe zone under the additional surface or ground moistening, as compared to chestnut soils, it has greater humus and carbonate content.
r лугово-каштановая почва
Почва, формирующаяся в западинах сухостепной зоны при дополнительном поверхностном или грунтовом увлажнении, по сравнению с каштановыми почвами обладает большим содержанием гумуса и карбонатов.
f sol *m* châtain de prairie
s suelo *m* castaño de pradera
g Wiesen- und kastanienfabriger Boden *m*

3742 **meadow-swamp soil**
A soil that forms in forest-steppe and steppe zones, in the depressions on flat plains of Eurasia and North America and along river terraces under hydrophilous herbaceous vegetation under the conditions of long-term moistening.
r лугово-болотная почва
Почва, формирующаяся в лесостепной и степной зонах под влаголюбивой травянистой растительностью в понижениях на плоских равнинах Евразии и Северной Америки и по террасам рек в условиях длительного увлажнения.
f sol *m* de prairie humide à gley
s suelo *m* de pradera-pantano
g Wiesenmoorboden *m*

* **mean sea-level → 5261**

3743 **mean solar time**
Time that is based on the motion of the mean sun, i.e. a point moving at a certain rate along the celestial equator, making one turn over a year and passing through the point of vernal equinox simultaneously with the real Sun, day being 24 hours.
r среднее солнечное время

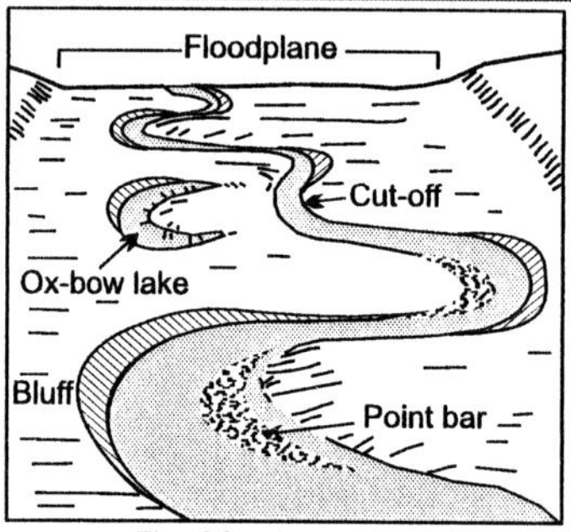

Fig. 201. **Meanders**

Шкала времени, определяемая видимым суточным движением так называемого среднего Солнца, т.е. точки, равномерно движущейся по небесному экватору, делающей один оборот за год и проходящей через точку весеннего равноденствия одновременно с истинным Солнцем.

f temps *m* solaire moyen
s tiempo *m* solar medio; hora *f* media
g mittlere Sonnenzeit *f*

3744 meander
A loop-like bend in the channel of a sluggish river, the radius of which depends on the stream discharge and its flow velocity. → Fig. 201
r излучина; меандра
Плавные изгибы русла равнинной реки, радиус кривизны которых зависит от водоносности реки и скорости ее течения. → Рис. 201
f méandre *m*
s meandro *m*
g Mäander *m*

*** meander bar → 4635**

3745 meander belt
The part of the flat floor of a valley across which a river forms its meanders.
r пояс меандрирования
Участок дна долины, на котором происходит формирование речных излучин.
f lit *m* de méandres
s fondo *m* de meandros
g Mäanderbogen *m*

*** meander core → 3746**

3746 meander lobe; meander core
A large tract of land, often a hill, surrounded on three sides by the river channel bending off the elevation, i.e. a piece of land around which an incised meander swings before breaking through the narrow neck of the meander.
r лука
Полуостров, нередко холм, образуемый большой дугообразной излучиной реки, обтекающей эту возвышенность; изгиб морского берега (морская лука).
f coude *m*; lobe *m* de méandre
s codo *m*; lóbulo *m* de meandro
g Flußkrümmung *f*; Küstenbiegung *f*; Flußbiegung *f*

3747 meander neck
The strip of land located between closely spaced meander loops of a well-developed meander. → Fig. 202
r шейка речной излучины; шейка излучины
Полоса суши, лежащая между обеими сторонами хорошо выраженной излучины. → Рис. 202
f col *m* de méandre
s cuello *m* de meandro
g Mäanderhals *m*

*** meander scroll → 4635**

3748 meander terrace
A terrace formed on one bank of a river as it meanders in its valley and at the same time (due to rejuvenation) erodes the valley floor.
r меандровая терраса
Терраса, формирующаяся на одном берегу реки, по мере того как она меандрирует по долине и в то же время эродирует свое дно.
f terrasse *f* de méandre
s terraza *f* de meandro
g Mäanderterrasse *f*

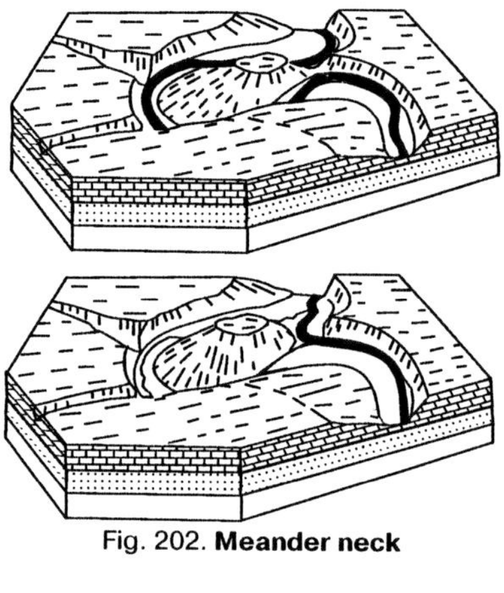

Fig. 202. **Meander neck**

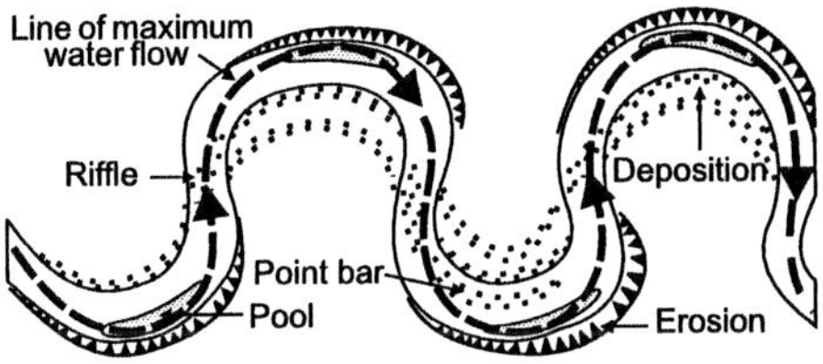

Fig. 203. **Meandering**

3749 **meandering**
The development of pronounced bends or loops in the course of a sluggish river having a wide floodplane. → Fig. 203
r меандрирование
Формирование хорошо выраженных излучин речного русла на широкой пойме. → Рис. 203
f formation *f* des méandres
s ameandramiento *m*; formación *f* de meandros
g Mäanderbildung *f*; Mäanderung *f*

3750 **meandering valley**
A winding valley with large, sweeping curves that cut into the solid rocks.
r меандровая долина
Извилистая речная долина с большими округлыми излучинами, прорезанными в твердых горных породах.
f vallée *f* des méandres; vallée *f* d'auge glaciaire
s valle *m* de meandros; valle *m* meandriforme
g Mäandertal *n*

3751 **means of production; producer goods**
The instruments and objects of labour that are used to manufacture goods and services, i.e. non-human, inputs in production.
r средства производства
Средства и предметы труда, используемые людьми в процессе производства.
f moyens *mpl* de production
s bienes *mpl* de producción
g Produktionsmittel *pl*

* **meansolar day → 846**

3752 **measuring grid**
Grid of parallel or radial lines, squares, hexagons and other geometrical cells plotted on a transparent material and used to take measurements by geographical maps, plans, images.
r палетка
Сетка параллельных или радиальных линий, квадратов, шестиугольников и других геометрических ячеек, нанесенная на прозрачный материал и используемая для измерений по географическим картам, планам, снимкам.
f grille *f* de mesure; réseau *m* pour la mesure
s retícula *f* para medición; rejilla *f* de medida
g Messunggitter *n*

* **measuring section → 2892**

3753 **mechanical engineering; machine-building industry**
A sector of industry that produces motor vehicles (car industry), aircraft (aviation industry), ships (shipbuilding), mashine tools and instruments (machine-tool industry).
r машиностроение
Группа отраслей промышленности, производящая автомобили (автомобильная промышленность), летательные аппараты (авиационная промышленность), речные и морские суда (судостроение), станки и инструменты (станкостроение).
f constructions *fpl* méchaniques
s industria *f* mecánica; construcción *f* de maquinaria
g Maschinenbau *m*

3754 **mechanical weathering; physical weathering**
The disintegration of rocks as a result of their cracking under the expansion of the volume caused by the temperature fluctuations and water freezing in fissures but without any chemical change.
r механическое выветривание; физическое выветривание
Процесс разрушения горных пород в результате их растрескивания при изменении объема, вызванном резкими колебаниями температуры и замерзанием воды в трещинах.
f météorisation *f* mécanique; désagrégation *f*; météorisation *f* physique
s meteorización *f* mecánica; meteorización *f* física; desagregación *f*
g mechanische Verwitterung *f*; physikalische Verwitterung *f*

* **mechanism of economic management → 1827**

3755 **medial moraine; intermediate moraine; median moraine**
A linear accumulation of debris extending along the centre of a glacier tongue from nunataks or the place of coalescence of tributary glaciers. → Fig. 85, p. 182
r срединная морена
Гряды грубообломочного материала, протягивающиеся вдоль ледниковых языков от нунатаков или места слияния главного ледника с ледниками-притоками. → Рис. 85, с. 182
f moraine *f* médiane
s morrena *f* media; morrena *f* mediana; morrena *f* central
g Mittelmoräne *f*

* **median moraine → 3755**

3756 **medical climatology**
The study of the influence of climate on the health and spreading of diseases in human beings.

r медицинская климатология
Отрасль климатологии, изучающая влияние климатических условий на здоровье человека и распространение болезней.
f climatologie *f* médicale
s climatología *f* médica
g medizinische Klimatologie *f*

3757 **medical geographical forecast**
The scientific prediction of the state of health of population in the future depending on the state of the environment.
r медико-географический прогноз
Научно обоснованное суждение о возможном уровне здоровья населения при изменении состояния окружающей среды.
f pronostic *m* géomédical
s pronóstico *m* geomédico
g geomedizinische Prignose *f*

3758 **medical geography; medicinal geography**
The study of the geographical distribution of diseases and the effect of the environment on the health of population.
r медицинская география
Наука, изучающая распространение болезней и закономерности влияния окружающей среды на здоровье населения.
f géographie *f* médicale
s geografía *f* médica
g medizinische Geographie *f*; Geomedizin *f*

* **medicinal geography → 3758**

3759 **medina**
An old Muslim part of a town in the Arabic countries in the north of Africa.
r медина
Старая мусульманская часть города в арабских странах Северной Африки.
f médina *f*
s medina *f*
g Medina *f*

3760 **Mediterranean**
A natural region which comprises the Mediterranean sea with the islands and adjacent territories of Europe, Asia and Africa, evergreen dry forests and bushes on brown forest soils prevail (maquis, garrigue, shiblyak, phrigana).
r Средиземноморье
Природная область, включающая бассейн Средиземного моря с его островами и примыкающими территориями Европы, Азии и Африки вместе с поясом Переднеазиатских нагорий, где преобладают вечнозеленые сухие леса и кустарники на коричневых и бурых лесных почвах (маквис, гарига, шибляк, фригана).
f Méditerranéen *m*
s Mediterráneo *m*
g Mittelmeergebiet *n*; Mittelmeerraum *m*

3761 **Mediterranean climate**
A type of climate characterised by hot, dry summer and warm rainy winter.
r средиземноморский климат
Субтропический климат с жарким сухим летом и прохладной дождливой зимой.
f climat *m* méditerranéen
s clima *m* mediterráneo
g Mediterranklima *n*; Mittelmeerklima *n*

* **mediterranean sea → 3144**

3762 **mediterranean zone**
Natural subzones of tropical zones of the Northern Hemisphere (the Mediterranean basin, California) and the Southern Hemisphere (Middle Chile, South Africa, the south and west of Australia), they are sometimes viewed to as belonging to the forest zones of subtropics.
r средиземноморская зона
Природные подзоны субтропических поясов Северного (бассейн Средиземного моря, Калифорния) и Южного (Среднее Чили, Южная Африка, юг и запад Австралии) полушарий, иногда рассматриваются как отдельные зоны в составе лесных зон субтропиков.
f zone *f* méditerranéenne
s zona *f* mediterránea
g Mittelmeerzone *f*

* **medium cloud → 3764**

3763 **medium scale map**
A geographical map having the scale which ranges from 1:100,000 to 1:1,000,000.
r среднемасштабная карта
Географические карты масштаба от 1:100 000 до 1:1 000 000.
f carte *f* à moyenne échelle
s mapa *m* a mediana escala
g mittelmaßstäbige Karte *f*

3764 **medium-level cloud; medium cloud; middle-level cloud**
Clouds at the elevations from 2 to 4-7 km including Altostratus and Altocumulus.
r облака среднего яруса
Облака на высотах от 2 до 4-7 км, включающие высокослоистые и высококучевые облака.
f nuage *m* de l'étage moyen; nuage *m* moyen
s nube *f* del piso medio; nube *f* media
g mittlelhohe Wolke *f*

* **medium-range weather forecast → 3765**

3765 **medium-term weather forecasting; medium-range weather forecast**
Weather forecast for 3-10 days from the day of issue.
r среднесрочный прогноз погоды
Прогноз погоды на срок от 3 до 10 суток со дня выпуска.
f prévision *f* météorologique à moyenne échéance

s previsión *f* meteorológica a medio plazo; pronóstico *m* a medio plazo; predicción *f* a medio plazo
g Mittelfristprognose *f*

3766 megacity
A very large urban agglomeration with a population exceeding some 25 million.
r мегагород
Крупнейшая городская агломерация с населением более 25 млн человек.
f mégacité *f*
s megaciudad *f*
g Megastadt *f*

3767 megalith
A large stone surved in some cultures as a monument or used in construction.
r мегалит
Крупный камень, использовавшийся в некоторых прошлых культурах в качестве монумента или при строительстве зданий.
f megalithe *m*
s megalito *m*
g Megalith *m*

3768 megalopolis
A very large many-centered, multi-city, dispersed urban area, formed when built-up areas become linked together, as in the seaboard of the USA from Washington to Boston or between Tokyo and Osaka in Japan.
r мегалополис
Срастание большого числа крупных соседствующих городских агломераций в единый урбанизированный ареал, как это имеет место между Вашингтоном и Бостоном в США или от Токио до Осаки в Японии.
f mégalopole *f*; mégalopolis *f*
s megalópolis *f*
g Megalopolis *f*

3769 megarelief
The largest landforms of the Earth: the greatest plains, lowlands, tablelands, mountains.
r мегарельеф
Крупнейшие формы рельефа Земли: величайшие равнины, низменности, плоскогорья, горные страны.
f mégarelief *m*
s megarelieve *m*
g Megarelief *n*

3770 megathermal plant
A plant of tropical rainforest whose annual temperature requirement is that each month should have mean temperature no lower than 18°C.
r мегатермное растение
Растение тропического дождевого леса, существующее в условиях, когда средняя месячная температура в течение всего года не опускается ниже 18°C.
f plante *f* mégatherme
s planta *f* megaterma
g megatherme Pflanze *f*

* **megatrophic plant → 2026**

* **melonhole → 2475**

3771 melt water; meltwater
The water resulting from the melting of snow and ice.
r талая вода
Вода, получившаяся в результате таяния снега и льда.
f eau *f* de fusion; eau *f* de fonte
s agua *f* de fusión
g Schmelzwasser *n*

3772 melting crust
An upper layer of loose ice, 5-10 cm thick, appearing in the ablation area of a glacier due to the effect of a direct solar radiation penetrating into the ice.
r кора таяния
Верхний сильно разрыхленный слой льда толщиной 5-10 см в области абляции ледника, возникающий в результате проникновения в лед прямой солнечной радиации.
f croûte *f* de fusion
s corteza *f* de deshielo; corteza *f* de fusión
g Schmelzharsch *m*; Schmelzkruste *f*

3773 melting of ice
The phase transition of ice from a solid to a liquid state.
r таяние льда
Фазовый переход льда из твердого состояния в жидкое.
f fusion *f* de la glace; dégel *m*
s deshielo *m*; fusión *f* del hielo
g Eisschmelze *f*

3774 melting of snow; snowmelt; snow melting
The process through which the snow particles are converted into water.
r снеготаяние
Процесс превращения снежных кристаллов, из которых состоит снежный покров, в воду.
f fusion *f* de la neige
s fusión *f* de la nieve
g Schneeschmelze *f*

3775 melt-out
The release of objects from ice, snow or frozen ground during their thermal destruction.
r вытаивание
Освобождение предметов из толщи льда, снега или мерзлых пород при их тепловом разрушении.
f décongélation *f*
s decongelación *f*
g Ausschmelzen *n*

* **meltwater → 3771**

3776 melt-water channel
A channel in the solid rock or in drift deposits in areas of former glaciation, unrelated to the

Fig. 204. **Melt-water grooves**

trend of the present drainage pattern, and in some cases cutting across it.

r канал талых вод
Канал в твердой горной породе или в наносах в области прошлого оледенения, не связанный с современной дренажной сетью и в некоторых случаях секущий ее поперек.
f canal *m* de eau de fusion
s canal *m* de agua de fusión
g Schmelzwasserkanal *m*

3777 melt-water grooves
Parallel narrow furrows on a snow surface cut by melt-water streams running downslope. → Fig. 204
r борозды таяния
Вытянутые по склону параллельные бороздки на поверхности снега, образующиеся под воздействием струек талой воды. → Рис. 204
f sillons *mpl* de fusion
s surcos *mpl* de fusión
g Schmelzrinnen *fpl*

3778 Mercalli scale
A scale formerly used in measuring earthquake intensity, based on the observed effects on buildings, etc., ranging from I to XII, and has been later superseded by the Richter scale.
r шкала Меркалли
12-балльная шкала, использовавшаяся в прошлом для измерения интенсивности землетрясений на основе наблюдаемых следов разрушений, впоследствии замененная шкалой Рихтера.
f échelle *f* de Mercalli
s escala *f* de Mercalli
g Mercalli-Skala *f*

3779 mercantilism
A theory popular among European nations during the sixteenth and seventeenth centuries that the economic and political power of a country consists in its acquiring large quantities of gold and silver, to be achieved by restricting imports, developing production for exports, and prohibiting the export of gold and silver.
r меркантилизм
Теория, популярная среди европейских стран в XVI и XVII вв., утверждавшая, что в основе экономического и политического авторитета страны лежит приобретение большого количества золота и серебра, что достигается ограничением импорта, развитием производства продукции на экспорт и запрещением вывоза золота и серебра.
f mercantilisme *m*
s mercantilismo *m*
g Merkantilismus *m*

3780 Mercator map projection
The conformal cylindrical map projection in which all rhumb lines are represented by straight lines. → Fig. 205
r проекция Меркатора
Прямая равноугольная цилиндрическая картографическая проекция, в которой любые локсодромии представлены прямыми линиями. → Рис. 205
f projection *f* de Mercator
s proyección *f* de Mercator
g Mercatorentwurf *m*

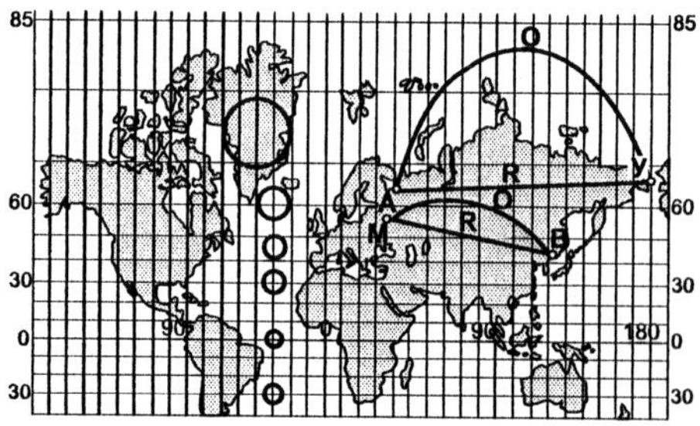

Fig. 205. **Mercator map projection**: O – orthodromic line, R – rhumb line

3781 meridian
An imaginary circle along the Earth's surface from the North pole to the South pole, serving as a line of longitude.
r географический меридиан; меридиан
Условный круг на поверхности земного шара, проведенный через какую-либо точку на земной поверхности от Северного до Южного полюса и являющийся линией географической долготы.
f méridien *m*
s meridiano *m*
g Meridian *m*; Längenkreis *m*

3782 meridian convergence
The angle between true north and grid north on a map.

r сближение меридианов
Угол на карте между меридианом данной точки и линией абсцисс прямоугольной сетки.
f convergence *f* des méridiens
s convergencia *f* de meridianos; angulo *m* de convergencia de meridianos
g Missweisung *f*

3783 **meridian line**
Line along which shadow falls in a true solar noon.
r полуденная линия
Линия, вдоль которой падает тень в истинный солнечный полдень.
f ligne *f* méridienne
s línea *f* de mediodía
g Mittagslinie *f*

* **meridian of Greenwich → 4767**

* **meridian of origin → 4767**

* **meridional circulation → 3784**

3784 **meridional flow; meridional circulation**
Atmospheric circulation in which the dominant airflow follows the meridians, i.e. the air moves from north to south, or from south to north, in contrast to *zonal flow.*
r меридиональная циркуляция; межширотный обмен
Циркуляция атмосферы, при которой потоки воздуха преимущественно следуют вдоль меридианов, т.е. направлены на юг или на север, в противоположность *зональной циркуляции.*
f circulation *f* méridienne
s circulación *f* meridiana
g meridionale Zirkulation *f*

3785 **merokarst**
An imperfect or partially developed karst which occurs where limestones are relatively thin and in areas of impure limestone, often interbedded with marly bands.
r неполноразвитый карст
Частично развитый карст, встречающийся в тех местах, где известняки относительно тонкие или неоднородные, нередко перемежающиеся со слоями мергеля.
f merokarst *m*
s merokarst *m*
g Merokarst *m*

3786 **meromictic lake**
The two-layer lake in which the water of the lower layer is considerably more salty and dense than the water of the upper layer.
r меромиктическое озеро; двухслойное озеро
Двухслойное озеро, в котором вода нижнего слоя существенно более минерализованная и плотная, чем в верхнем слое.
f lac *m* méromictique
s lago *m* meromíctico
g meromiktischer See *m*

3787 **meromixis**
The phenomenon of incomplete, partial water mixing of meromictic lakes developing only in its upper layer (*mixolimnion*).
r меромиксия
Явление неполного, частичного перемешивания вод меромиктических озер, развивающегося только в пределах верхнего слоя (*миксолимниона*).
f méromélange *m*
s meromezcla *f*
g Meromixis *f*

3788 **mesa**
A tableland or an isolated flat-topped hill created by the dissection of vast basaltic plateaus in Spain and Mexico.
r меза
Обособленная столовая возвышенность, результат эрозионного расчленения обширных базальтовых плато в Испании и Мексике.
f mesa *f*
s mesa *f*; meseta *f*
g Mesa *f*; Hochebene *f*

3789 **mesic layer**
A layer of organic material whithin a soil which has reached a stage of decomposition between that of the fibric layer and the humic layer.
r среднеразложившийся слой
Почвенный слой органического материала, достигший средней стадии разложения, промежуточной между волокнистым и гумусовым слоями.
f couche *f* médiale; couche *f* de décomposition médial
s capa *f* medial; capa *f* de descomposición media
g mediale Schicht *f*

3790 **mesoclimate**
The climate of a limited area marked by its deviation from the norms of the macroclimate in which it lies, in area it is larger than a local climate or microclimate.
r мезоклимат
Климат ограниченной территории, демонстрирующий отличия от норм макроклимата, в пределах которого находится, он охватывает бульшую площадь, нежели местный климат.
f mésoclimat *m*
s mesoclima *m*
g Mesoklima *n*

3791 **Mesolithic**
The second period of the Stone Age extending between the Paleolithic to Neolithic (10-6 thousand years ago) when the bow and arrow came into use and small silisic articles

came to be made, its beginning approximately coincides with the *Preboreal.*

r мезолит

Второй период каменного века, переходный от палеолита к неолиту (10-6 тыс. лет назад), характерный мелкими кремневыми изделиями и появлением лука и стрел, его начало примерно совпадает с *пребореальным временем.*

f mésolithique *f*

s mesolítico *m*

g Mesolithikum *m*

3792 mesopause

The upper limit of the mesosphere at about 80-85 km. → Fig. 21, p. 51

r мезопауза

Верхняя граница мезосферы на высоте около 80-85 км. → Рис. 21, с. 51

f mésopause *f*

s mesopausa *f*

g Mesopause *f*

3793 mesophile

Organisms, the optimum development of which occurs under medium conditions: at moderate moisture content and temperatures between 20 and 40°C, they are typical of the temperate zone (terrestrial plants that grow under medium temperatures and moistening are called mesophytes).

r мезофил

Умеренно влаголюбивые организмы, нормально существующие при температурах 20-40°C, преобладают в умеренном поясе (наземные растения, обитающие в условиях устойчивого, но не избыточного увлажнения, называют мезофиты).

f mésophile *m*

s mesófilo *m*

g Mesophil *m*

3794 mesoplankton

Plankton organisms which take the form of plankton for part of their life cycle.

r мезопланктон

Организмы, принимающие форму планктона в части их жизненного цикла.

f mésoplancton *m*

s mesoplancton *m*

g Mesoplankton *n*

3795 mesorelief

Landforms several tens of metres high: moraine hills, guilles, etc.

r мезорельеф

Формы рельефа с амплитудами высот до нескольких десятков метров: моренные холмы, балки.

f mésorelief *m*; relief *m* mineur

s mesorelieve *m*; relieve *m* menor

g Mesorelief *n*

3796 mesoscale eddy; mesoscale vortex

A whirl in the ocean (both cyclonic and anticyclonic), from 10 to 1000 km in diameter, embracing the water layer of several hundreds or thousands meters in depth.

r мезомасштабный вихрь

Вихревое образование в океане (как циклоническое, так и антициклоническое) диаметром от 10 до 1000 км, охватывающее толщу воды в несколько сотен или тысяч метров.

f tourbillon *m* de puissance moyenne

s torbellino *m* de grado medio

g mittelmäßiger Wirbel *m*

*** mesoscale vortex → 3796**

3797 mesosphere

The region of the atmosphere between the stratopause and the mesopause in which the temperature generally decreases with height. → Fig. 21, p. 51

r мезосфера

Область атмосферы между стратопаузой и мезопаузой, в которой температура обычно падает с высотой. → Рис. 21, с. 51

f mésosphère *f*

s mesosfera *f*

g Mesosphäre *f*

*** mesotherm → 3798**

3798 mesothermal plant; mesotherm

Plants found in the areas with the annual average temperature of 15-20°C, such as in the Mediterranean region, e.g. citruses, cypress, etc.

r мезотермное растение

Растения, распространенные в областях со средней годовой температурой 15-20°C, в частности, в Средиземноморье, например, цитрусовые, кипарис.

f plante *f* mésotherme

s planta *f* mesoterma

g mesotherme Pflanze *f*

3799 mesothermy

The distribution of temperature in a water body in such a way that its maximum is located at a certain depth.

r мезотермия

Распределение температуры воды в водоеме, при котором ее максимум находится на некоторой глубине от поверхности.

f mésothermie *f*

s mesotermia *f*

g Mesothermie *f*

3800 mesotrophic *adj*

Applied to a water body or hydrous solution within a soil which is intermediate in character between a euthrophic and oligotrophic states, as far as its nutrients are concerned.

r мезотрофный

Относящийся к водоему или водному раствору внутри почвы, которые в отношении питательных веществ носят промежуточный характер между евтрофным и олиготрофным состояниями.

f mésotrophe *adj*
s mesotrofo *adj*
g mesotrophe *adj*

3801 mesotrophic bog
A bog the character of nourishment and vegetation of which is intermediate between a lowland (eutrophic) and raised (oligotrophic) bog.
r переходное болото; мезотрофное болото
Болото, промежуточное по характеру питания и растительности между низинным (евтрофным) и верховым (олиготрофным) болотами.
f marais *m* mésotrophique
s turbera *f* mesótrofa
g mesotrophes Moor *n*; Übergangsmoor *n*

3802 mesotrophic lake
A lake having a medium content of organic and biogenic mineral substances in water.
r мезотрофное озеро
Озеро с умеренным содержанием в воде органических и биогенных минеральных веществ.
f lac *m* mésotrophe
s lago *m* mesotrofo
g mesotropher See *m*

3803 mesotrophic plant
A plant that grows in soils having the medium content of nutrients, e.g. taiga vegetation.
r мезотрофное растение
Растения, произрастающие на субстратах со средним содержанием питательных веществ, например, растения тайги.
f plante *f* mésotrophique
s planta *f* mesotrofa
g Mesotroph *m*

3804 Mesozoic
The geological era in the Earth's history which began 235 million years ago and lasted for 170 million years. → Table 1, p. 284
r мезозой
Геологическая эра в истории Земли, начавшаяся 235 млн лет назад и продолжавшаяся 170 млн лет. → Табл. 1, с. 284
f Mésozoïque *m*
s era *f* mesozoica; Mesozoico *m*
g Mesozoikum *m*

3805 Mesozoic folding
The epoch of mountain formation of the Mesozoic occurred mainly along the periphery of the Pacific Ocean, the main phases are Cimmerian and Laramian foldings.
r мезозойская складчатость
Эпоха горообразоования, проявившаяся в течение мезозойской эры главным образом по периферии Тихого океана, основные фазы – киммерийская и ларамийская складчатости.
f plissement *m* mésozoïque
s plegamiento *m* mesozoico
g mesozoische Faltung *f*

3806 metalimnion
The intermediate water layer in lakes and reservoirs within which the temperature in summer sharply decreases while water density increases.
r металимнион
Промежуточный слой воды в озерах и водохранилищах, в пределах которого температура летом резко падает, а плотность воды возрастает.
f couche *f* critique
s metalimnión *m*
g Metalimnion *n*; Sprungschicht *f*

* **metallurgic industry → 3807**

3807 metallurgy; metallurgic industry
A branch of heavy industry engaged in producing ferrous and non-ferrous metals and their alloys.
r металлургия
Отрасль тяжелой промышленности, занимающаяся производством черных и цветных металлов и их сплавов.
f métallurgie *f*
s metalurgia *f*; industria *f* metalúrgica
g Metallurgie *f*

3808 metamorphic ice formation
A complex of processes of snow and firn metamorphism that results in the transformation of porous aggregates of ice crystals into the ice rock.
r метаморфическое льдообразование
Комплекс процессов метаморфизма снега и фирна, в результате которого пористые скопления ледяных кристаллов превращаются в ледяную породу.
f formation *f* métamorphique de la glace
s formación *f* metamórfica de hielo
g Eisbildung *f* durch Metamorphose

3809 metamorphic rock
A well-defined new type of rock formed as a result of mineralogical, chemical and structural changes in the pre-existing rock brought about by its contact with a great heated mass of magma or by folding and pressure.
r метаморфическая горная порода
Породы, образовавшиеся в результате минералогических, химических и структурных изменений при соприкосновении с горячими массами магмы или под действием внутреннего давления и сжатия.
f roche *f* métamorphique
s roca *f* metamórfica
g metamorphes Gestein *n*

3810 metamorphism
The change in texture, structure, and mineral composition of rocks under the impact of the high temperature and pressure as a result of which the pre-existing rock is transformed

into a new one, e.g. limestone may be converted to marble, granite to gneiss, etc.

r метаморфизм

Процесс изменения текстуры, структуры и минерального состава горных пород под воздействием высокой температуры и давления, в результате чего существовавшие ранее породы трансформируются в другие, например, известняк превращается в мрамор, гранит – в гнейс и т.д.

f métamorphisme *m*

s metamorfismo *m*

g Metamorphismus *m*

3811 meteor

A solid body moving rapidly in space glowing when it enters the Earth's atmosphere due to the heat generated by friction with the atmosphere as a «shooting star» or «falling star».

r метеор

Твердое тело, быстро двигающееся в мировом пространстве и раскаляющееся при входе в земную атмосферу благодаря выделению тепла от трения о воздух, из-за чего оно носит название «падающей звезды».

f météore *m*

s meteoro *m*

g Meteor *m*

* **meteoric dust → 1313**

3812 meteorite

A solid body (a fragment of a broken asteroid) that manages to reach the Earth's surface despite the frictional effect of the atmosphere which consumes the majority of meteorites.

r метеорит

Твердое тело (фрагмент разрушенного астероида), достигающее земной поверхности несмотря на эффект трения в атмосфере, которое «съедает» большинство падающих на Землю метеоритов.

f météorite *m*

s meteorito *m*

g Meteorit *m*

3813 meteorite crater

A depression of the Earth's surface created by the impact of a meteorite.

r метеоритный кратер

Депрессия на земной поверхности, созданная упавшим на Землю метеоритом.

f cratère *m* de météorite

s cráter *m* de meteorito

g Meteoritenkrater *m*

3814 meteorological conditions

The state of the atmosphere at a concrete place at a given time.

r метеорологические условия

Состояние атмосферы в конкретной местности в заданный момент времени.

f condition *f* météorologique

s condición *f* meteorológica

g Wetterverhältnisse *npl*

3815 meteorological element

Parameters of state of the weather and atmospheric phenomena: temperatura and humidity of the air, speed and direction of wind, precipitation, etc.

r метеорологический элемент

Показатели переменного состояния погоды и атмосферных явлений: температура и влажность воздуха, скорость и направление ветра, осадки и пр.

f élément *m* météorologique

s elemento *m* meteorológico

g meteorologisches Element *n*

* **meteorological forecast → 6740**

3816 meteorological observations

Measurements of meteorological elements and visual qualitative estimation of the atmospheric phenomena.

r метеорологические наблюдения

Измерения метеорологических элементов в сочетании с визуальной качественной оценкой атмосферных явлений.

f observation *f* météorologique

s observación *f* meteorológica

g Wetterbeobachtungen *fpl*

3817 meteorological satellite; weather satellite

An artificial satellite intended for current observation over the distribution of cloud cover and radiant heat of the Earth to get meteorological data for weather forecasting.

r метеорологический спутник

Искусственный спутник Земли, предназначенный для оперативного наблюдения за распределением облачного покрова и теплового излучения Земли с целью получения метеорологических данных для прогноза погоды.

f satellite *m* météorologique

s satélite *m* meteorológico

g Wettersatellit *m*; meteorologischer Satellit *m*

3818 meteorological screen

A white painted wooden box standing on legs, about 1-2 m above the ground, designed to shelter meteorological instruments from strong wind and solar and terrestrial radiation.

r метеорологическая будка

Выкрашенный в белый цвет деревянный ящик, поставленный на подпорки на высоте 1-2 м над поверхностью земли, оборудованный для размещения внутри метеорологических приборов и защиты их от ветра, солнечной и земной радиации.

f cabine *f* météorologique

s cabina *f* meteorológica

g Wetterhäuschen *n*

3819 meteorological station

Place where regular meteorological observations are made.

r метеорологическая станция
Место, где проводятся регулярные метеорологические наблюдения.
f station *f* météorologique
s estación *f* meteorológica
g Wetterwarte *f*; Wetterstation *f*

3820 meteorology
The study of the character of the atmosphere and the processes within it, together with the interaction between the atmosphere and the Earth's surface, in order to use these data for weather forecasting.
r метеорология
Изучение атмосферы Земли и протекающих в ней процессов, а также взаимодействия между атмосферой и земной поверхностью с целью подготовки прогнозов погоды на основе этих данных.
f météorologie *f*
s meteorología *f*
g Wetterkunde *f*; Meteorologie *f*

3821 methane; marsh-gas
A colourless, odourless gas, CH_4, found in swampy, oil-containing sediments and in permafrost, it occurs also in the atmosphere being an effective greenhouse gas.
r метан; болотный газ
Бесцветный газ, CH_4, залегающий в болотных, нефтяных отложениях и в мерзлоте, а также находящийся в атмосфере, где обладает парниковым эффектом.
f méthane *m*
s metano *m*
g Methan *n*; Sumpfgas *n*

3822 method of areas
The representation on a geographical map of an area occupied by an object or phenomenon by way of contouring and colouring or hatching of the area. → Fig. 206, p. 456
r способ ареалов
Изображение на географической карте площади распространения какого-либо объекта или явления путем оконтуривания, раскраски или штриховки этой области. → Рис. 206, с. 456
f méthode *f* des aires
s método *m* de áreas
g Flächenmethode *f*; Arealmethode *f*

3823 method of conventional symbols
The representation of certain objects on a geographical map by the symbols the sizes of which are accepted as constant or vary according to some scale.
r способ внемасштабных знаков; способ значков
Изображение на географической карте отдельных объектов значками, размеры которых принимаются постоянными или меняются по какой-либо шкале.
f méthode *f* des symboles; méthode *f* des signes conventionnels
s método *m* de símbolos; método *m* de signos convencionales
g Kartenzeichenmethode *f*; Signaturenmethode *f*

3824 method of hachures and shading; hill-shading
The representation of topography on a geographical map by way of shading with hachures of the slopes of mountains.
r способ теневой пластики; способ штрихов
Изображение рельефа на географической карте с помощью оттенения штрихами изображенных на карте склонов.
f méthode *f* des hachures
s método *m* de hachuras y sombreado
g Schraffenmethode *f*

3825 method of isolines
The representation of a quantitative characteristic on a geographical map with the help of lines of its equal values. → Fig. 206, p. 456
r способ изолиний
Изображение на географической карте какого-либо количественного показателя с помощью линий его равных значений. → Рис. 206, с. 456
f méthode *f* des isolignes
s método *m* de isolíneas
g Isolinienmethode *f*

3826 method of line symbols
The representation on a geographical map of linear objects, the width of which is not expressed in the map's scale, by the lines of different colour, width, pattern.
r способ линейных знаков
Изображение на географической карте линейных объектов, ширина которых не выражается в масштабе карты, с помощью линий разного цвета, ширины, рисунка.
f méthode *f* des signes linéaires
s método *m* de signos lineales; método *m* de líneas
g Linienkartenzeichenmethode *f*; Linearsignaturenmethode *f*

3827 method of paleoanalogues
The way of estimation of the coming changes in the environment with the help of the inferences of the past conditions which by their initial parameters are close to the expected ones.
r метод палеоаналогов
Способ оценки предстоящих изменений окружающей среды с помощью реконструкций условий прошлого, которые по своим исходным параметрам близки к ожидаемым.
f méthode *f* des paléoanalogies
s método *f* paleoanálogo
g Gleichsetzungsmethode *f*

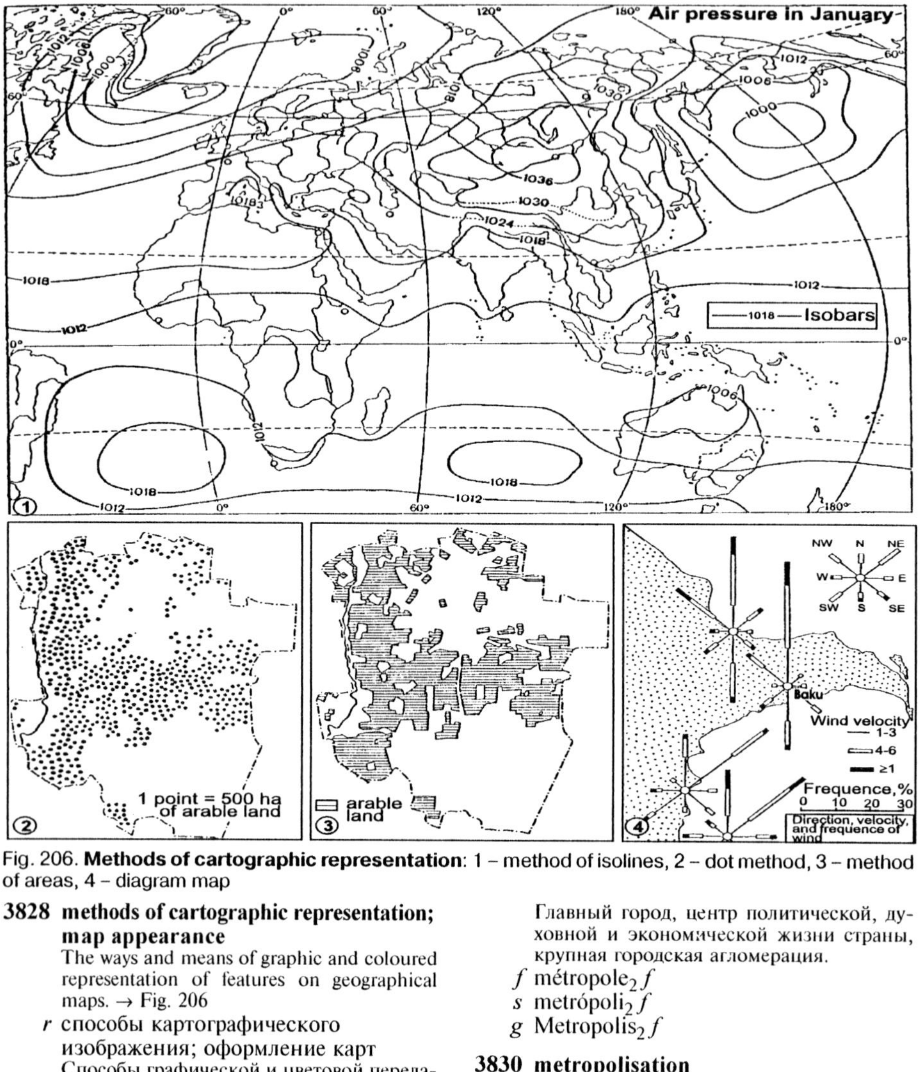

Fig. 206. **Methods of cartographic representation**: 1 – method of isolines, 2 – dot method, 3 – method of areas, 4 – diagram map

3828 methods of cartographic representation; map appearance
The ways and means of graphic and coloured representation of features on geographical maps. → Fig. 206

r способы картографического изображения; оформление карт
Способы графической и цветовой передачи содержания географических карт. → Рис. 206

f méthodes *mpl* de représentation cartographique; présentation *f* des cartes

s métodos *mpl* de representación cartográfica; presentación *f* cartográfica

g kartographische Darstellungsweise *f*; Kartengestaltung *f*

* **metro → 6448**

3829 metropolis
A large urban agglomeration, the main city, being a center of political, spiritual and economic life of a country.

r метрополия$_2$
Главный город, центр политической, духовной и экономической жизни страны, крупная городская агломерация.

f métropole$_2$ *f*

s metrópoli$_2$ *f*

g Metropolis$_2$ *f*

3830 metropolisation
A fast growth of population and concentration of the state's industrial, financial and cultural functions in a single centre, usually in the capital.

r метрополизация
Быстрый рост населения и концентрации промышленных, финансовых и культурных функций государства в единственном центре страны, как правило, в столице.

f métropolisation *f*

s metropolización *f*

g Metropolisation *f*

* **metropolitan area → 1268**

3831 microclimate
The climate conditions within a few metres of the atmosphere in contact with the ground surface and in a relatively small area.
r микроклимат
Климатические условия в пределах слоя в несколько метров на контакте атмосферы и земной поверхности, охватывающие относительно небольшую территорию.
f microclimat *m*
s microclima *m*
g Mikroklima *n*; Geländeklima *n*

3832 microcoenosis
A small community of plants within a biocoenosis, a constituent component of vegetation cover mosaic.
r микроценоз
Небольшое сообщество растений в пределах одного биоценоза, составной элемент мозаичности растительного покрова.
f microcénose *f*
s microcenosis *f*
g Mikrozönose *f*

3833 microconvection
The process of vertical circulation and exchange of air and small masses of water in the surface layer of water bodies.
r микроконвекция
Процесс вертикального перемещения и обмена воздуха, а также небольших объемов воды в поверхностном слое водоемов.
f microconvection *f*
s microconvección *f*
g Mikrokonvektion *f*

3834 microeconomics
The study of the benaviour of particular economic units such as an industry, a firm, a single commodity production as well as a particular market, price, product or a service.
r микроэкономика
Раздел экономики, имеющий дело с отдельными производственными единицами (отраслями, фирмами и пр.), а также с конкретными рынками, ценами, товарами и услугами.
f microéconomie *f*
s microeconomía *f*
g Mikroökonomie *f*

3835 microfauna
All the microscopic animal organisms considered collectively.
r микрофауна
Совокупность микроскопических животных организмов.
f microfaune *f*
s microfauna *f*
g Mikrofauna *f*; Kleinfauna *f*

3836 microflora
Plant microorganisms including microscopic fungi, algae, bacteria considered collectively.
r микрофлора
Совокупность растительных микроорганизмов, в том числе микроскопических грибов, водорослей, бактерий.
f microflore *f*
s microflora *f*
g Mikroflora *f*; Kleinflora *f*

3837 microrelief
Small landforms a few metres in height such as furrows, barchans, zastrugi, etc.
r микрорельеф
Мелкие формы рельефа размером до нескольких метров: борозды, барханы, заструги и пр.
f microrelief *m*
s microrelieve *m*
g Mikrorelief *n*; Kleinrelief *n*

3838 microtripton
Very small inorganic particles, which are in suspended state in lake water.
r микротриптон
Очень мелкие неорганические частицы, находящиеся во взвешенном состоянии в воде озера.
f microtripton *m*
s microtripton *m*
g Mikrotripton *n*

* **mid ocean ridge → 3845**

3839 mid-channel bar
A bar in the middle of a river often dried in low-water and sometimes vegetated.
r осерёдок
Отмель вытянутой формы посредине реки, образованная аллювиальными наносами и иногда закрепленная растительностью.
f barre *f* alluviale
s barra *f* alluvial
g Bank *f* in der Mitte eines Flusses

3840 middle latitudes; midlatitudes
Areas which extend from the Tropics to the Arctic and Antarctic circles.
r средние широты; умеренные широты
Условное название территорий земного шара, расположенных между тропиками Рака и Козерога и Северным и Южным полярными кругами.
f latitudes *fpl* moyennes
s latitudes *fpl* medias; latitudes *fpl* templadas
g mittlere Breiten *fpl*

3841 middle mountains
A type of mountain relief characterised by absolute heights of 1,500-3,000 m, the depth of dissection of 500-1,000 m, average steepness of slopes (15-25°).
r среднегорье; среднегорный рельеф
Тип горного рельефа, характеризующийся абсолютными высотами вершин 1500-

3000 м, глубиной расчленения 500-1000 м, средней крутизной склонов (15-25°).
f montagnes *fpl* moyennes
s montañas *fpl* medianas
g Mittelgebirge *n*

* **middle-level cloud → 3764**

3842 **mid-latitude karst**
A type of karst with isolated depressions found close to each other on the surface, they gradually join together to form larger depressions.
r карст умеренных широт
Тип карста, состоящий из близко расположенных изолированных впадин (воронок, блюдец) на дневной поверхности, постепенно сливающихся и формирующих впадины больших размеров.
f karst *m* des latitudes moyennes
s karst *m* de latitudes templadas; carso *m* de latitudes medias
g Karst *m* der mittleren Breiten

* **mid-latitude zone → 6080**

* **midlatitudes → 3840**

3843 **midnight**
The middle of the night, i.e. the lower culmination of the Sun when in its apparent diurnal movement it crosses the meridian of a given place.
r полночь
Момент нижней кульминации Солнца, когда оно в видимом суточном движении пересекает меридиан места наблюдения.
f minuit *m*
s medianoche *f*
g Mitternacht *f*

3844 **midnight sun**
The sun still to be seen shining above the horizon at midnight in latitudes higher than 63°30′ north or south, in the Northerm Hemisphere between mid-May and the end of July, and in the Southern one between mid-November and the end of January.
r полуночное Солнце
Солнце, находящееся над горизонтом в полночь на широтах более 63°30′, в Северном полушарии с середины мая до конца июля и в Южном с середины ноября до конца января.
f soleil *m* de minuit
s sol *m* de medianoche
g Mitternachtssonne *f*

3845 **mid-oceanic ridge; mid ocean ridge; oceanic ridge**
An underwater mountain ridge occupying the middle-ocean position which is a part of the planetary system of interconnected mountain formations. → Fig. 207
r срединно-океанический хребет
Мощный подводный горный хребет, занимающий срединное положение на дне океанов и прослеживающийся в виде планетарной системы связанных между собою горных сооружений. → Рис. 207
f dorsale *f* médio-océanique
s dorsal *f* oceánica; cadena *f* montañosa submarina
g mittelozeanische Schwelle *f*; Ozeanrücken *m*

3846 **migrant**
A person who migrates (moves crossing borders of territories) in order to change his permanent place of residence forever or for a long period of time.
r мигрант
Человек, совершающий миграцию, т.е. пересекающий границы тех или иных территорий со сменой постоянного места жительства навсегда или на некоторое время.
f migrant *f*
s migrante *m*
g Migrant *n*

3847 **migration**
The act or process of moving of humans, animals, plants, mineral elements etc. from one place to another to settle, extend habitat or accumulate.
r миграция
Процесс перемещения людей, животных, растений, минеральных элементов с одного места на другое для поселения, рас-

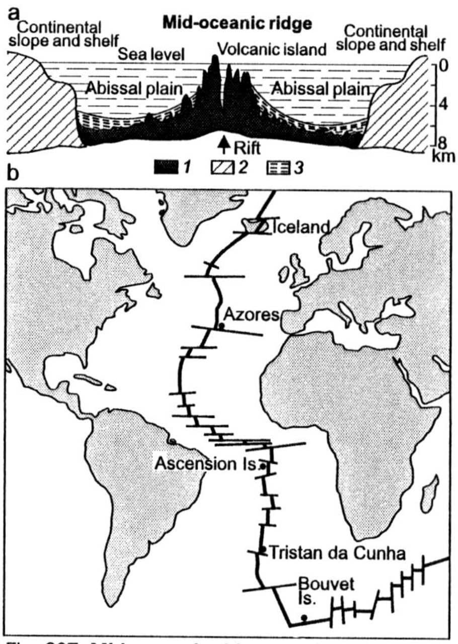

Fig. 207. **Mid-oceanic ridge**. a – Structure of the ridge: 1 – oceanic igneous rock, 2 – bedrock, 3 – sedimentary rock; b – Mid-oceanic ridge in the Atlantic ocean

ширения местообитания или области отложения.
f migration *f*
s migración *f*
g Migration *f*; Wanderung *f*

3848 migration flow
An intensive migration of the people in a certain direction.
r миграционный поток
Интенсивная миграция населения в определенном направлении.
f flux *m* de migration
s flujo *m* migratorio
g Migrationsstrom *m*

3849 migration of animals
Seasonal or spontaneous movements of animals from one region or climate to another which may be caused by changes of life conditions in habitats, overpopulation (lemmings) or associated with their life cycle (ungulates).
r миграция животных
Перемещение животных, вызванное изменениями условий существования в местах обитания, расселением при подъеме численности (лемминги) или связанное с циклом их жизни (копытные).
f migration *f* des animaux; déplacement *m* des animaux
s migración *f* de animales
g Tierwanderung *f*

3850 migration of divide
A change in position on an interfluve owing to the lateral recession of slopes by erosional processes.
r миграция водораздела
Изменение положения водораздела, вызываемое отступанием склонов вследствие эрозионных процессов.
f migration *f* de partage
s migración *f* de línea divisoria de agua
g Wasserscheidenwanderung *f*

3851 migration of elements; migration of substances
The movement and redistribution of chemical elements in the Earth's crust and on the Earth's surface, occurring in liquid (melts, solutions), gaseous and solid states.
r миграция элементов; миграция веществ
Перенос и перераспределение химических элементов в земной коре и на поверхности Земли, происходит в жидкой (расплавах, растворах), газообразной и твердой фазах.
f migration *f* des éléments; migration *f* des substances
s migración *f* de elementos; migración *f* de substancias
g Migration *f* der Elementen; Wanderung *f* der Stoffe

3852 migration of population
Spontaneous or indirectly guided change of residence of people for a relatively long period of time or to settle permanently.
r миграция населения
Стихийная или косвенно управляемая перемена места жительства людей на более или менее длительное время или навсегда.
f migration *f* de population
s migración *f* de población
g Bevölkerungsmigration *f*; Bevölkerungswanderung *f*

* **migration of substances → 3851**

3853 migratory birds; birds of passage
Birds nesting in the countries of temperate or cold belts and regularly migrating in the warmer countries for winter, stopping for a short time before moving on.
r перелетные птицы
Птицы, гнездящиеся в странах умеренного или холодного поясов и совершающие регулярные перелеты на зимовку в более теплые страны.
f oiseaux *mpl* migrateurs; oiseaux *mpl* de passage
s aves *fpl* de paso; aves *fpl* migratórias; aves *fpl* pasajeras
g Zugvögel *mpl*

3854 Milankovitch curve
A graphic representation showing the fluctuations of solar radiation received at certain latitude. → Fig. 208
r кривая Миланковича
Кривая, показывающая колебания прихода солнечной радиации для определенной широты. → Рис. 208
f courbe *f* de Milankowitch
s curva *f* de Milankowitch
g Milankovitch-Kurve *f*

3855 Milankovitch oscillations

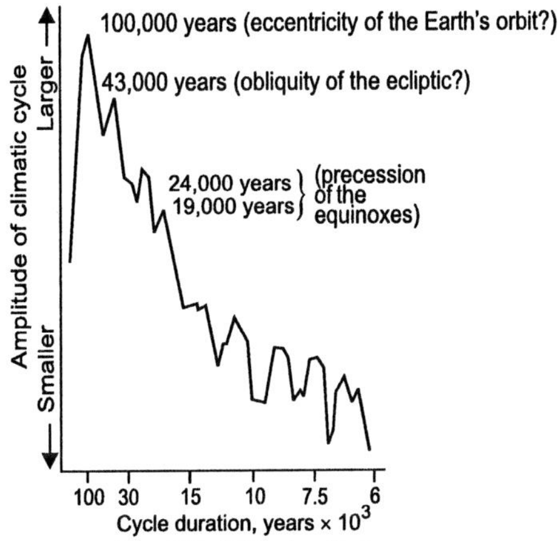

Fig. 208. **Milankovitch curve**

According to calculation of M. Milankovitch, there are three interacting, astronomical cycles in the Earth's orbit around the Sun: in the shape of the elliptical orbit (about 95,000 years); in the axis of rotation (about 42,000 years); and in the time of year when the Earth is clolsest to the Sun (about 21,000 years).

r циклы Миланковича
Согласно вычислениям М. Миланковича, существуют три типа геометрических взаимоотношений Земли и Солнца, носящих циклический характер: 1) эксцентриситет земной орбиты, с периодичностью 95000 лет, 2) отклонение плоскости эклиптики между (40400 лет), 3) предварение равноденствий (21000 лет).

f cycles *mpl* de Milankovitch
s ciclos *mpl* de Milankovitch
g Milankovitch-Zyklen *mpl*

3856 mile; statute mile
A British and USA linear measurement equal to 1609 m.
r миля
Единица длины, принятая в Великобритании и США, равная 1609 м.
f mille *m*
s milla *f*
g Meile *f*

3857 military geography
The study of geographical and socio-economic conditions of possible land and marine theatres of war as well as the efficient supply of military operations with information on the environment.
r военная география
Изучение физико- и экономико-географических условий возможных сухопутных и морских театров военных действий, а также оперативное обслуживание военных операций информацией об окружающей среде.
f géographie *f* militaire
s geografía *f* militar
g Militärgeographie *f*

3858 military industry; armament industry
A wide range of industries that supply the army with weapons and other necessary ammunition and equipment, a part of the military-industrial complex.
r военная промышленность
Широкий спектр отраслей промышленности, снабжающих армию вооружением и иным необходимым снаряжением и оборудованием, часть военно-промышленного комплекса.
f industrie *f* de guerre; industrie *f* militaire; industrie *f* de l'armement
s industria *f* de guerra
g Verteidigungsindustrie *f*

3859 military-economic potential
A part of economic potential of a country or group of countries that can be used to enhance military power or carry out military operations.
r военно-экономический потенциал
Часть экономического потенциала страны или группы стран, которая может быть использована для укрепления оборонной мощи и ведения войны.
f potentiel *m* économique militaire
s potencial *m* militar-económico
g militärwirtschaftlicher Potenzial *n*

3860 military-industrial complex
The combination of military industry, army and the related part of the government and science which usually function independently from local economies.
r военно-промышленный комплекс; ВПК
Совокупность предприятий военной промышленности, армии и связанных с ними части государственного аппарата и науки, как правило, существующих автономно от местного хозяйства.
f complexe *m* militar-industriel
s complejo *m* militar-industrial
g militär-industrieller Komplex *m*

3861 mine
An excavation deep in the ground, made for purpose of extracting minerals, together with its shaft and associated buildings.
r шахта
Глубокая выработка в земной коре, пройденная с целью добычи полезных ископаемых, включая саму шахту и относящийся к ней комплекс зданий и сооружений.
f mine *f*
s mina *f*
g Grube *f*

* **mine digging → 3862**

3862 mine working; mine digging
An excavation of any shape and direction deep in the ground made for the purpose of extraction of minerals.
r горная выработка
Пустота любой формы и направления в земной коре, искусственно созданная при горных работах.
f mine *f*; taille *f*
s mina *f*; minera *f*; excavación *f*
g Schacht *m*; Grube *f*

3863 mineral
A substance usually crystalline with a homogeneous chemical composition and physical properties, which forms part of rocks.
r минерал
Однородное по химическому составу и физическим свойствам обычно твердое тело, служащее составной частью горных пород.

f minéral *m*
s mineral[1] *m*
g Mineral *n*

3864 mineral deposit
The accumulation of a mineral substance naturally occurring on and under the surface of the Earth which is suitable for industrial use due to its size, quality, and conditions of bedding.
r месторождение полезного ископаемого
Скопление минерального вещества на поверхности или в недрах Земли, пригодное по количеству, качеству и условиям залегания для промышленного использования.
f gisement *m*; gîte *m*
s yacimiento *m*
g Lagerstätte *f*

3865 mineral horizon
A soil horizon which consists mainly of inorganic matter and low in humans content.
r минеральный горизонт
Мало плодородный почвенный горизонт, состоящий в основном из неорганического вещества.
f horizon *m* minéral
s horizonte *m* mineral
g Mineralhorizont *m*

3866 mineral reserves; reserves
Part of mineral resources considered to be usable in current conditions of technical skill and economic and social needs, which can be extracted using the prevailing technology.
r запасы полезных ископаемых
Часть минеральных ресурсов, которые могут быть полезны в современных условиях технического развития, экономических и социальных потребностей и могут быть добыты с использованием современных технологий.
f réserves *fpl* des minéraux
s reserva *f* mineral
g Vorräte *mpl* der Bodenschätze

3867 mineral resources; minerals
Naturally occurring mineral substances in the Earth's crust which are used in material production.
r полезные ископаемые; минерально-сырьевые ресурсы; минеральные ресурсы
Природные вещества минерального происхождения, пригодные для получения энергии, сырья и материалов в современных условиях и в перспективе.
f ressources *fpl* minérales
s recursos *mpl* minerales
g Mineralvorräte *mpl*; Mineralien *npl*; Bodenschätze *mpl*

3868 mineral soil
A soil low in humus, consisting mainly of inorganic matter.
r минерализованная почва
Почва, бедная гумусом, состоящая в основном из материала минерального происхождения.
f sol *m* minéral
s suelo *m* mineral
g Mineralboden *m*

3869 mineral spring
A natural spring containing high proportion of mineral salts in water which may be used for medical purposes.
r минеральный источник
Источник воды с повышенным содержанием минеральных солей, которая часто используется в лечебных целях.
f source *f* minérale
s fuente *f* mineral
g Mineralquelle *f*

3870 mineralization of soil
The final stage of biodegradation of humus and vegetal remains by soil microorganisms that results in decomposition of organic compounds into carbon dioxide, water and elementary salts.
r минерализация почвы
Заключительная стадия биодеградации гумуса и растительных остатков под влиянием почвенных микроорганизмов, заключающаяся в распаде органических соединений до углекислоты, воды и простых солей.
f minéralisation *f* du sol
s mineralización *f* del suelo
g Bodenmineralisierung *f*

3871 mineralization of water
The total content of dissolved inorganic substances in water.
r минерализация воды
Суммарное содержание в воде растворенных неорганических веществ.
f minéralisation *f* de l'eau
s mineralización *f* de agua
g Mineralisation *f* des Wassers

3872 mineralogy
The study of minerals, their composition, properties, peculiar features of structure, origin, occurrence and distribution in nature, particularly in soil.
r минералогия
Изучение минералов, их состава, свойств, особенностей строения, условий образования, нахождения и изменений в природе, и, в частности, в почвенном покрове.
f minéralogie *f*
s mineralogía *f*
g Mineralogie *f*; Mineralkunde *f*

* **minerals → 3867**

* **minimum flow → 3601**

3873 **minimum temperature**
The lowest temperature atteined at a certain point or region during a specific time interval.
r минимальная температура
Наименьшее значение температуры, наблюдавшееся в данном пункте или регионе за конкретный период.
f température *f* minimale
s temperatura *f* mínima
g Minimumtemperatur *f*; minimale Temperatur *f*

3874 **mining**
The extraction of non-renewable natural resources, such as coal or metal from under ground.
r добыча полезных ископаемых
Разработка геологических пластов с целью получения невозобновляемых природных ресурсов.
f extraction *f*
s extracción *f*
g Gewinnung *f*; Förderung *f*

3875 **mining dump; slag-heap; spoil heap; dump; spoil bank**
An accumulation of the excavated waste, sub-standard minerals, and other discarded material.
r отвал
Искусственная насыпь из отвальных грунтов, некондиционных полезных ископаемых и других твердых отходов.
f dépôts *mpl* de déblai; remblai *m*
s montón *m*
g Halde *f*

3876 **mining industry; mining; extractive industry**
A group of industries envolved in the extraction and dressing of mineralsg used as sources of energy and raw materials for heavy industry.
r горнодобывающая промышленность; разработка полезных ископаемых; горное дело; добывающая промышленность
Комплекс отраслей, занимающихся добычей и обогащением полезных ископаемых как источников энергии и сырья для тяжелой промышленности.
f industrie *f* minière; industrie *f* extractive
s industria *f* minera; minería *f*; industria *f* extractiva
g Bergbau *m*; extraktive Industrie *f*

* **mining waste → 2040**

* **minority ethnic group → 2009**

3877 **Miocene**
The lower epoch of the Neogene period which began about 26 million years ago and lasted more than 20 million years. → Table 1, p. 284
r миоцен

Fig. 209. **Mirage**

Нижняя эпоха неогенового периода (неогена), началась около 26 млн лет назад и продолжалась более 20 млн лет. → Табл. 1, с. 284
f Miocène *m*
s Mioceno *m*
g Miozän *m*

3878 **mirage**
An optical effect of the atmosphere caused by refraction of light through layers of air, in which the image of an object appears displaced from its true position, it may be seen inverted or suspended in air or give the impression of shimmering pools of water on the ground. → Fig. 209
r мираж
Оптический эффект в атмосфере, вызванный рефракцией, в результате чего изображение предмета смещается относительно его истинного положения и может вызвать впечатление перевернутого или плывущего в воздухе предмета, а также мерцающей водной поверхности. → Рис. 209
f mirage *m*
s espejismo *m*
g Luftspiegelung *f*

* **mire → 4846**

3879 **misclosure; closing error**
Numerical expression of discrepancy between measured and pre-calculated values of distances and angles in the course of a topographic survey.
r невязка
Числовое выражение несоответствия между измеренными и предвычисленными величинами расстояний и углов при топографической съемке.
f écart *m*
s error *m* de cierre; discrepancia *f*
g Widerspruch *m*

3880 **mist**
The suspension in the air of microscopic water droplets or wet hygroscopic particles reducing the visibility at the Earth's surface, it is intermediate between haze and fog.
r влажная дымка; легкий туман
Скопление микроскопических водяных капель или влажных гигроскопических частиц, уменьшающих видимость у поверхности земли, занимает промежуточное положение между сухой дымкой (мглой) и туманом.
f brume *f*
s neblina *f*; bruma *f*; húmeda *f*
g Dunst *m*

* **mist forest → 1072**

3881 **mistral**
A strong cold, dry northerly wind blowing in winter in the south of France from the Alps down the Rhфne valley.
r мистраль
Сильный холодный и сухой северный ветер, дующий на юге Франции со склонов Альп по долине р. Роны.
f mistral *m*
s mistral *m*
g Mistral *m*

3882 **mitigation; impact mitigation**
Making the effects of a natural hazard or the human impact on the envirinment less severe or serious, e.g. flood mitigation, emissions mitigation, wild life mitigation, etc.
r уменьшение воздействия на окружающую среду
Борьба с неблагоприятным воздействием на окружающую среду с целью смягчить отрицательный эффект антропогенного или природного характера, например, борьба с наводнением, обезвреживание выбросов в атмосферу и т.п.
f mitigation *f*; atténuation *f*
s atenuación *f*; mitigación *f*; aminoración *f* del daño erosivo
g Milderung *f*

3883 **mixed alimentation**
The alimentation of water bodies when not a single type of its (snow, rain, underground and glacial) is prevailing.
r смешанное питание
Питание водных объектов, когда ни один из видов питания (снеговое, дождевое, подземное и ледниковое) не является преобладающим.
f alimentation *f* mixte
s alimentación *f* combinada
g kombinierte Speisung *f*

3884 **mixed cloud**
A cloud in which ice particles are intermingled with supercooled droplets of water.
r смешанное облако
Облако, состоящее из смеси снежных кристаллов и переохлажденных капелек воды.
f nuage *m* mixte
s nube *f* mezcla
g Mischwolke *f*

* **mixed economy → 4021**

* **mixed farm → 3885**

3885 **mixed farming; mixed farm**
A type of commercial agriculture concerned with the production of both crops and animals on one farm.
r комплексное хозяйство
Тип рыночного сельскохозяйственного производства, сочетающего и земледелие и животноводство.
f polyculture *f*
s agricultura *f* mixta; policultivo *m*; mixed-farming *m*
g Mischlandwirtschaft *f*

3886 **mixed forest; mixed wood stand**
The forest composed of different tree species, e.g. of coniferous and leaf-bearing trees in the temperate zone and of evergreen and deciduous trees in the tropical zone.
r смешанный лес
Лес с древостоем, образованным деревьями разных пород: в умеренном поясе хвойными и лиственными, в тропическом поясе – вечнозелеными и листопадными.
f forêt *f* mixte; forêt *f* mélangée
s bosque *m* mixto
g Mischwald *m*

3887 **mixed forest zone**
Natural regions of temperate zones of the Northern and Southern hemispheres with the predominance of mixed forests and characterized by cold snowy winter and warm summer.
r зона смешанных лесов
Зоны умеренных поясов в Северном и Южном полушариях с преобладанием ландшафтов смешанных лесов, отличаются холодной снежной зимой и теплым летом.
f zone *f* des forêts mixtes
s zona *f* de bosques mixtos
g Mischwaldzone *f*

3888 **mixed tide**
Irregular, diurnal or semi-diurnal tide, in the daily cycle of which two periods of high water and two periods of low water with different heights of their levels are observed.
r смешанный прилив
Неправильный суточный или полусуточный прилив, в суточном цикле которого отмечаются два периода полной воды и два периода малой воды с разной высотой уровня.
f marée *f* mixte
s marea *f* mixta
g Kombinationstide *f*

* **mixed wood stand → 3886**

3889 **mixing of waters; overturning**
The process of energy and substance exchange between contacting water masses which is determined by vertical and horizontal gradients of the water temperature and density.
r перемешивание вод
Процесс обмена энергией и веществом между соприкасающимися водными массами, обусловленный вертикальными и горизонтальными градиентами температуры и плотности воды.

f mélange *m* des eaux
s mezcla *f* de aguas; agitación *f*
g Vermischung *f* des Wassers; Wasserdurchmischung *f*

3890 mixolimnion
The upper layer of a meromictic (two-layer) lake which is less dense as compared with the lower layer.
r миксолимнион
Верхний слой меромиктического (двухслойного) озера меньшей плотности по сравнению с нижележащим слоем.
f mixolimnion *m*
s mixolimnión *m*
g Mixolimnion *n*

3891 mobile *adj*
Capable of moving or being moved from place to place.
r подвижный; мобильный
Способный двигаться или двигающийся от места к месту.
f mobile *adj*
s móvil *adj*
g beweglich *adj*; mobil *adj*

3892 mobile belt; orogenic belt
A lengthy linear portion of the Earth's crust in whicn intense deformation takes place during an orogeny; mountain construction appeared on the place of a geosyncline at the final stage of its development or on the place of a platform at the study of its activisation. → Fig. 210
r мобильный пояс; подвижный пояс; орогенический пояс

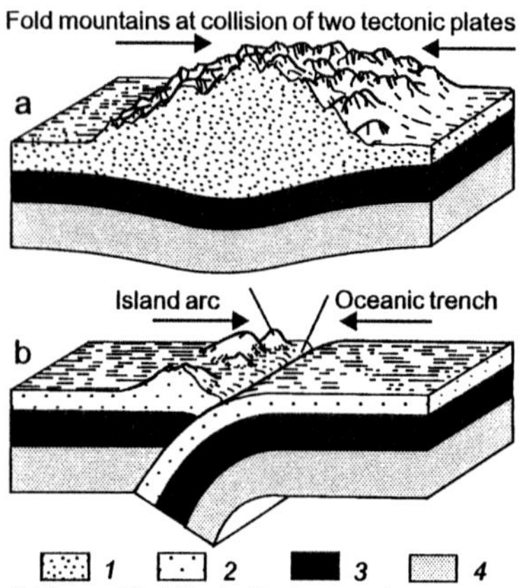

Fig. 210. a – **Mobile belt** at a continent: tectonic plates carrying continental crust, colliding and making fold mountains; b – Mobile belt at an ocean: tectonic plates carrying oceanic crust, converging to form an oceanic trench and a volcanic island arc as one plate dives under the other, the crust being consumed in the mantle; 1 – continental crust, 2 – oceanic crust, 3 – tectonic plate, 4 – mantle

Линейно вытянутая часть земной коры, в которой во время орогенеза происходят интенсивные деформации; горное сооружение, возникшее на месте платформы в стадию ее активизации. → Рис. 210
f ceinture *f* mobile
s cinturón *m* móvil
g Mobilgürtel *m*

3893 mobile dune
An eolian accumulation of sand not fixed by vegetation which moves in the direction of the prevailing wind.
r подвижная дюна
Песчаное эоловое накопление, не закрепленное растительностью и перемещающееся в направлении господствующих в данной местности ветров.
f dune *f* mouvante; dune *f* mobile; dune *f* libre
s duna *f* móvil; duna *f* moviente; duna *f* libre
g Wanderdüne *f*

3894 mobilism
A hypothesis suggested by Alfred Wegener in 1912 that big tectonic plates (up to several thousand km in area) may move slowly around the Earthrs surface.
r мобилизм
Выдвинутая Альфредом Вегенером в 1912 г. геологическая гипотеза, предполагающая большие (до нескольких тысяч километров) горизонтальные перемещения крупных литосферных плит.
f mobilisme *m*
s mobilismo *m*
g Mobilismus *m*

3895 mobility; population mobility
The state or quality of being mobile, e.g. of people moving readily from one place to another and from job to job, in search of employment of higher incomes, or of individuals and households in being able to move between social classes and income groups.
r мобильность; подвижность населения
Состояние или способность быть подвижным, применяется к людям, с готовностью меняющим место жительства или работы в поисках лучшего заработка, либо к конкретному человеку и его домашнему хозяйству при перемене социального и имущественного положения.
f mobilité *f*; mobilité *f* de population
s mobilidad *f*; mobilidad *f* de población
g Beweglichkeit *f*; Bevölkerungsmobilität *f*

* **mock Moon → 4416**

* **mock sun → 4422**

3896 mode of production
The way in which human societies organize their productive activities and reproduce their

social life, for example a small commodity economy in the agrarian sector in the developing countries.

r общественно-экономический уклад; способ производства
Общественная форма производства, образующая систему производственных отношений, например мелкое товарное хозяйство в аграрном секторе ряда развивающихся стран.
f mode *m* de production
s modo *m* de producción
g gesellschaftlich-ökonomische Formation *f*; gesellschaftliche Wirtschaftsweise *f*

3897 modelling of relief
The transformation of relief mainly under the combined impact of exogenetic factors.
r моделировка рельефа
Преобразование рельефа под совокупным воздействием преимущественно экзогенных факторов.
f modelé *m* du relief
s modelado *m* de relieve
g Ausarbeitung *f* des Reliefs

3898 moder
A humus layer intermediate in composition between *mor* and *mull*, occurring where decomposition is greater than in mor, but has not advanced so much as in mull, but where there is some mixing with particles from the underlying mineral soil, due to presence of soil fauna.
r модер
Гумусовый гооризонт, промежуточный между *мором* и *мюллем*, встречающийся там, где разложение больше, чем в горизонте мора, и не так велико, как в горизонте мюлля, но где развито перемешивание с частицами из нижележащего минерального горизонта благодаря присутствию почвенной фауны.
f moder *m*
s moder *m*
g Moder *m*

3899 modernization of economy
Advances in economy based on scientific and technological innovations.
r модернизация экономики
Усовершенствование экономики, введение инноваций на базе научно-технического прогресса.
f modernisation *f* de l'économie
s modernización *f* de economía
g Modernisierung *f* der Wirtschaft

3900 mofette
A jet of carbon dioxide with the admixture of steam and other gases emitting from small canals and fissures on the bottom and slopes of a volcanic crater and on solidified lava flows.
r мофета
Струя углекислого газа с примесью водяного пара и других газов, выделяющаяся из небольших каналов и трещин на дне и склонах кратера вулкана и на застывших лавовых потоках.
f mofette *f*
s mofeta *f*
g Mofette *f*

3901 mogote
A residual mass of karst limestone that rises precipitously from a flat valley floor or plain as an example of *tower karst*.
r могота
Останец карстованного известняка, поднимающийся в виде отвесного холма в днище долины или среди равнины, яркий пример *башенного карста*.
f mogote *m*
s mogote *m*
g Mogote *m*

* **Moho → 3902**

3902 Mohorovičič discontinuity; Moho
The boundary between the Earth's crust and the mantle lying at a depth of some 40 km under the continents and only some 6-10 km under the ocean, it is identified by sharp increase in the rate of travel of seismic waves. → Fig. 211
r поверхность Мохоровичича; граница Мохо
Граница раздела между земной корой и мантией Земли, проходящая на глубине 40 км под континентами и 6-10 км под океанами, устанавливается по резкому увеличению с глубиной скоростей сейсмических волн. → Рис. 211
f discontinuité *f* de Mohorovičič
s discontinuidad *f* deMohorovičič
g Mohorovičič-Diskontinuität *f*

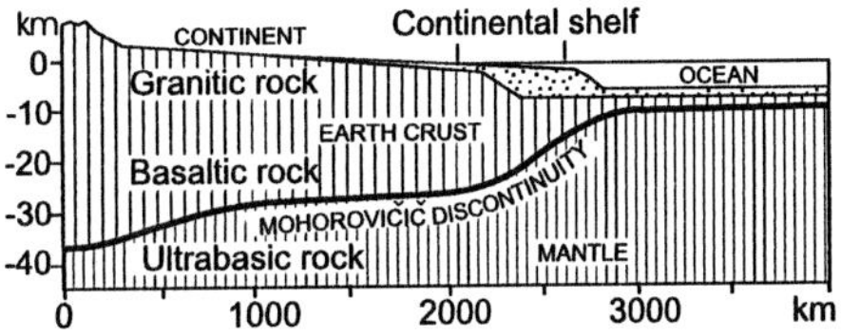

Fig. 211. **Mohorovičič discontinuity**

* **moist adiabatic lapse rate → 5223**

3903 moisture budget
Difference between the amount of precipitation and evaporative capacity over a certain period.
r баланс увлажнения
Разность между количеством осадков и испаряемостью за некоторый период.
f bilan *m* hydrique
s balance *m* de humectación
g Feuchtebilanz *f*

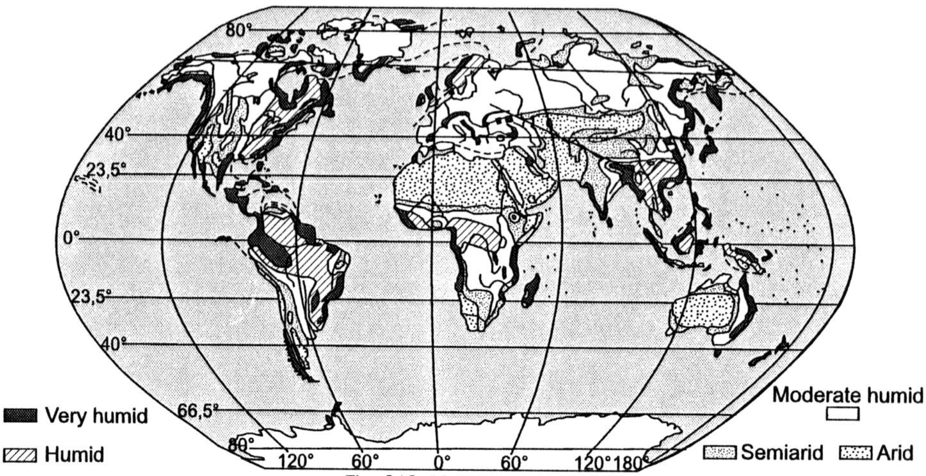

Fig. 212. **Moisture iindex**

3904 moisture capacity; water capacity; specific retention
The ability of rock, including snow, to contain and retain a certain amount of water in pores.
r влагосодержание; влагоемкость
Способность горной породы (включая снег) вмещать и удерживать определенное количество воды в порах.
f capacité *f* de rétention d'eau
s capacidad *f* de retención de agua
g Wasserhaltungsvermögen *n*; Wasserkapazität *f*

3905 moisture capacity of snow
The maximum quantity of water which the snow cover can retain before the gravitational water will flow out.
r водоудерживающая способность снега
Наибольшее количество воды, удерживаемое снежным покровом после смачивания до свободного стекания гравитационной воды.
f teneur *f* en eau liquide de neige
s capacidad *f* de nieve de retener agua
g Wasserhaltekapazität *f* des Schnees

3906 moisture content of soil; water content of soil
The content of liquid, gaseous and solid moisture in soil determined from the mass of dry soil or its volume.
r влажность почвы; влагоемкость почвы
Содержание в почве влаги в твердом, жидком и газообразном состоянии, определяется от массы сухой почвы или от ее объема.
f humidité *f* du sol
s humedad *f* del suelo
g Bodenfeuchtigkeit *f*

3907 moisture deficit
The difference between the total water capacity and the water content in soil.
r дефицит влаги
Разность между полной влагоемкостью и фактической влажностью почвогрунта.
f déficit *m* d'humidité
s déficit *m* de humedad
g Feuchtigkeitsdefizit *n*; Feuchtigkeitsmangel *m*

3908 moisture index
A measure of the water balance of an area in terms of precipitation effectiveness for plant growth. → Fig. 212
r индекс увлажнения
Параметр, показывающий эффективность осадков для роста растений. → Рис. 212
f indice *m* d'humidité
s índice *m* de humedad
g Humiditätsindex *m*; Feuchtigkeitsindex *m*

3909 molasse
Sedimentary rocks accumulated in marginal and intermontagne depressions.
r моласса
Толщи осадочных пород, накапливающихся в краевых и межгорных прогибах.
f molasse *f*
s molasa *f*
g Molasse *f*

3910 molehill; krotovina
An animal burrow that has been filled with a soil material from another horizon and therefore is clearly seen on the soil cross-section.
r кротовина
Небольшая подземная полость, вырытая и переотложенная насекомоядными животными или грызунами, хорошо видимая на почвенном разрезе благодаря заполнению ее почвенной массой иной окраски.
f taupinière *f*
s madriguera *f* de topo
g Maulwurfsgang *m*; Maulwurfshöhle *f*

3911 mollisol
A soil characterized by a very dark humus horizon are formed under grasslands and forests, represent the world's most productive agricultural soils.
r моллисоль
Насыщенные и слабонасыщенные почвы с темным гумусовым горизонтом, широко распространенные в лесах и лугах, весьма продуктивны для земледелия.
f mollisol *m*
s molisol *m*
g Mollisol *m*

3912 monadnock; glacial monadnock; erosion remnant
An isolated hill of small size or a remnant of more resistant rocks left due to denudation, common in the area of former glaciation.
r монаднок
Изолированный останец небольших размеров, сложенный твердыми и более стойкими к выветриванию и денудации горными породами по сравнению с породами, преобладающими на окружающей территории, часто встречается в древнеледниковой области.
f monadnock *m*; relief *m* résiduel
s monadnock *m*; relieve *m* residual
g Monadnock *n*; Härtling *m*

3913 monarchy
A government or state headed by a supreme ruler whose power is inherited.
r монархия
Государство, главой которого является верховный правитель, передающий власть по наследству.
f monarchie *f*
s monarquía *f*
g Monarchie *f*

3914 monetarism
The economic theory which asserts that the level of activity of an economy can be controlled by controlling the money supply.
r монетаризм
Теория, согласно которой уровень экономической активности может контролироваться путем регулирования денежной массы.
f monétarisme *m*
s monetarismo *m*
g Monetarismus *m*

3915 monetary crisis; foreign exchange crisis
The intensification of contradictions in the monetary sphere revealed by the dramatic fluctuations of currency rates, unbalanced payment balances and displacements of loan capital.
r валютный кризис
Обострение противоречий в валютной сфере, проявляющееся в резких колебаниях валютных курсов, несбалансированности платежных балансов, перемещениях ссудного капитала.
f crise *f* monétaire
s crisis *f* monetario
g Währungskrise *f*

3916 monimolimnion
The lower non-circulating layer of a meromictic (two-layer) lake having higher salinity and density, as well as isolated cycle of matter.
r монимолимнион
Нижний неперемешивающийся слой меромиктического (двухслойного) озера, отличающийся повышенными соленостью и плотностью, обособленным круговоротом веществ.
f monimolimnion *m*
s monimolimnión *m*
g Monimolimnion *n*

3917 monk stone
A rock projection or separate stones up to several metres high, formed due to the uneven weathering.
r камни-монахи
Скальные выступы или отдельные камни высотой до нескольких метров, возникающие в результате неравномерного выветривания.
f roche *f* pénitente
s roca *f* penitente
g Büssersteine *m*

* **monocentric world → 3924**

3918 monoclinal bedding
The disposition of rock strata when the layers are inclined into one side approximately at equal angle.
r моноклинальное залегание
Нарушенное залегание горных пород, при котором слои наклонены в одну сторону приблизительно под одинаковым углом.
f gisement *m* monoclinal
s estratificación *f* monoclinal
g monoklinale Lagerung *f*

3919 monoculture
Cultivation in which a single crop is grown successively on the same land without rotating crops, or a single crop grown in a region, e.g. cotton in Central Asia.
r монокультура
Длительное непрерывное возделывание какой-либо одной сельскохозяйственной культуры на одном и том же участке без соблюдения севооборота, либо единственная сельскохозяйственная культура, практикуемая в хозяйстве, как, например, возделывание хлопка в Средней Азии.
f monoculture *f*
s monocultivo *m*
g Monokultur *f*

3920 monoglacialism

A notion widespread at the end of the XIX century of a single glaciation in the Pleistocene with several long and short stages.

r моногляциализм

Представление, господствовавшее в конце XIX в., об однократном оледенении в плейстоцене с несколькими большими и малыми стадиями.

f monoglacialisme *m*

s monoglacialismo *m*

g Monoglazialismus *m*

3921 **monolith**

A large rock, ice or snow sample taken from the natural undisturbed structure to be used for laboratory investigations as well as a vertical section of a soil profile removed from the ground and mounted for display or research purposes.

r монолит

Крупный образец горной породы, почвы, льда или снега, отобранный из естественного залегания с ненарушенным строением и предназначенный для лабораторных исследований.

f monolithe *m*

s monolito *m*

g Monolith *m*

3922 **monomictic lake**

A lake having the total vertical circulation once a year, being warm in winter, and cold in summer.

r мономиктическое озеро

Озеро с периодом полной вертикальной циркуляции один раз в год: теплое зимой и холодное летом.

f lac *m* monomictique

s lago *m* monomíctico

g monomiktischer See *m*

3923 **monomineral rock**

A rock consisting mainly of one mineral with insignificant admixture of other minerals.

r мономинеральная горная порода

Порода, состоящая преимущественно из одного минерала с незначительной примесью других.

f roche *f* monominérale

s roca *f* monomineral

g monomineralisches Gestein *n*

3924 **monopolar world; monocentric world; unipolar world**

Geopolitical arrangement of the world with only one main centre of economic and political influence which is the United States of America.

r однополярный мир

Геополитическое устройство мира, в котором существует один главный центр экономического и политического влияния – Соединенные Штаты Америки.

f monde *m* unipolaire; monde *m* monopolaire

s mundo *m* unipolar; mundo *m* monopolar

g unipolare Welt *f*

3925 **monopoly**

The exclusive ownership or control of something; a situation on the market when a few suppliers can influence on the overall supplies and market prices and thus dominate the market.

r монополия

Исключительное право в какой-либо сфере деятельности государства, организации, фирмы и пр.; ситуация на рынке, характеризующаяся небольшим числом продавцов, каждый из которых способен повлиять на общий объем предложения и на цену товара или услуги.

f monopole *m*

s monopolio *m*

g Monopol *n*

3926 **monsoon**

The winds which blow with considerable regularity at definite seasons of the year due to the seasonal reversal of pressure over land masses and their neighbouring oceans. → Fig. 213

r муссон

Ветры, дующие с определенной регулярностью в разные сезоны года благодаря се-

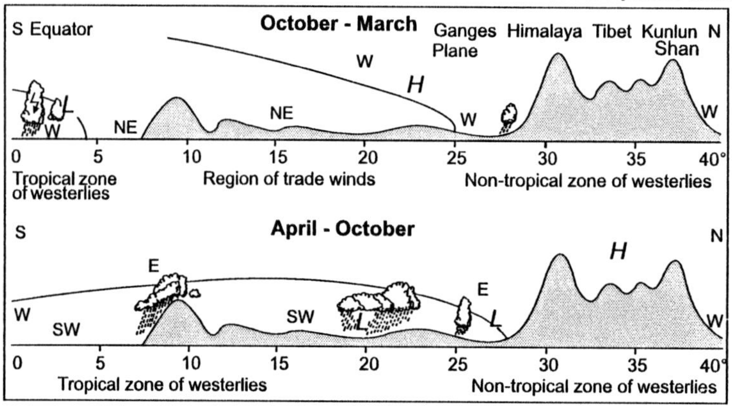

Fig. 213. **Monsoon**. H – high atmospheric pressure, L – low atmospheric pressure

зонному изменению давления над материком и прилегающим океаном. → Рис. 213, с. 468
f mousson *m*
s monzón *m*
g Monsun *m*

3927 monsoon circulation
Part of general atmospheric circulation above a certain area of the globe with regular occurence of monsoons.
r муссонная циркуляция
Часть общей циркуляции атмосферы над определенной областью земного шара с регулярной повторяемостью муссонов.
f circulation *f* de mousson
s circulación *f* de monzón
g Monsunzirkulation *f*

3928 monsoon climate
A type of climate formed in the regions subject to monsoons which is characterised by a dry winter and wet summer.
r муссонный климат
Климат области с муссонной циркуляцией атмосферы, отличающийся сухой зимой и дождливым летом.
f climat *m* des moussons
s clima *m* de monzón
g Monsunklima *n*

3929 monsoon current
Surface (not deeper than 200 m) drift currents in oceans and seas characterized by the seasonal reversal of the direction produced by monsoon winds.
r муссонное течение
Поверхностное (до глубины около 200 м) ветровое течение в океанах и морях с сезонной сменой направления, вызываемое муссонами.
f courant *m* de mousson
s corriente *f* de monzón
g Monsunströmung *m*

3930 monsoon forest
A type of tropical forest with semi-decidious or evergreen trees, found in the regions which are affected by a monsoonal climatic regime showing a marked dry season followed by torrential rain.
r муссонный лес
Леса тропических и субтропических широт, сбрасывающие листву в сухое время года, распространены в основном в юго-восточной Азии, где господствует муссон.
f forêt *f* de mousson
s bosque *m* de monzón
g Monsunwald *m*

3931 monsoon rain
Precipitation, usually heavy, associated with summer monsoon winds.
r муссонный дождь
Осадки, обычно сильные, связанные с летним муссонным ветром.
f pluie *f* de mousson
s lluvia *f* de monzón
g Monsunregen *m*

3932 monsoon season
The period when the summer monsoon blows onto the Asian continent.
r муссонный сезон
Период господства летнего (влажного) муссона в Индии и Южной Азии.
f époque *f* de la mousson
s estación *f* de monzón
g Monsunzeit *f*

3933 monte
A plant formation of xerophite and succulent drought-resistant shrubs that occupies dry high plains in Venezuela and Argentina in which acacia, cacti, etc. are typical.
r монте
Растительная формация ксерофитно-суккулентных кустарников на сухих возвышенных равнинах в Венесуэле и Аргентине, типичны акация, кактусы и др.
f monte *f*
s monte *f*
g Monte *f*

3934 month
The time interval close to the period in which the Moon makes a complete revolution around the Earth, one of 12 unequal parts which a calendar year is divided into.
r месяц[1]
Промежуток времени, близкий к периоду обращения Луны вокруг Земли, одна из 12 неравных частей, на которые разделяется календарный год.
f mois *m*
s mes *m*
g Monat *m*

3935 montmorillonite
A group of clay minerals produced during the alteration of basic rocks under alkaline conditions, they can absorb water and cause soils to swell when wet.
r монтмориллонит
Группа глинистых минералов, формирующихся при метаморфизации коренных пород в щелочных условиях, они обладают способностью абсорбировать влагу и вызывают набухание почвы во влажном состоянии.
f montmorillonite *m*
s montmorillonita *f*
g Montmorillonit *m*

* **moor → 4897**

3936 moorland; moor
A landscape of wet uplands covered with heather and other dwarf shrubs with acid peaty soil, which is not good for pasture, though used as rough grazing, common in Great Britain, northern Germany, etc.

r мур
Ландшафт хорошо увлажняемых, как правило, возвышенных пространств, с кислыми торфянистыми почвами и вересковыми зарослями, нередко используемый как пастбище, широко распространен в Великобритании, северной Германии и пр.
f moorland *m*
s moorland *m*
g Moorlandschaft *f*

3937 mor
A type of humus consisting of died off, partially decomposed organic remains, it is characteristic of soils found under coniferous forests, especially in cold regions and at high altitudes.
r мор
Тип гумуса, состоящего из отмерших, частично разложившихся органических остатков, характерен для почв под хвойными лесами, особенно в холодных районах и на значительных высотах.
f mor *m*; humus *m* brut
s mor *m*; humus *m* bruto
g Rohhumus *m*

* **moraine → 3946**

3938 moraine; till$_2$
An accumulation of unstratified rock debris transported and deposited by a glacier or ice sheet in the form of hills and ridges. → Fig. 214
r морена; тилл$_2$
Скопление обломков горных пород, переносимых ледниками, отложенный ледниками обломочный материал, а также холмы и гряды, образованные в результате аккумуляции ледниковых отложений. → Рис. 214
f moraine *f*
s morrena *f*
g Moräne *f*

3939 moraine amphitheatre
A semicircular rampart of several terminal moraines fringing the basin of an ancient glacier which have moved from the mountains to the adjoining plain.
r моренный амфитеатр
Полукольцевой вал из нескольких гряд конечных морен, обрамляющих бассейны древних горных ледников, выходивших из гор на прилегающую равнину.
f amphithéâtre *m* morainique
s anfiteatro *m* morrénico
g Moränenamphitheater *m*

* **moraine line → 1743**

3940 moraine relief
Rampart-like or hilly ridges of glacially deposited material laid down at the edge of a fluctuating glacier during its stationary phase.
r моренный рельеф
Валообразные или холмистые гряды из морены, отложенной при колебаниях края ледника в период его временного стационарного положения.
f relief *m* morainique
s relieve *m* morrénico
g Moränenrelief *n*

3941 moraine terrace
A lateral moraine of a valley glacier which has been transformed into a terrace on the rock slope after the degradation of the glacier.
r моренная терраса оседания
Боковая морена горно-долинного ледника, после его деградации преобразованная в террасу на коренном склоне долины.
f terrasse *f* de moraine; banquette *f* morainique
s terraza *m* morrénica; banco *m* morrénico
g Moränenterrasse *f*

3942 moraine-containing ice
The ice of basal layers of a glacier with a considerable admixture of rock debris.
r мореносодержащий лед
Лед придонных частей ледников со значительной примесью моренного материала.
f glace *f* chargée de débris

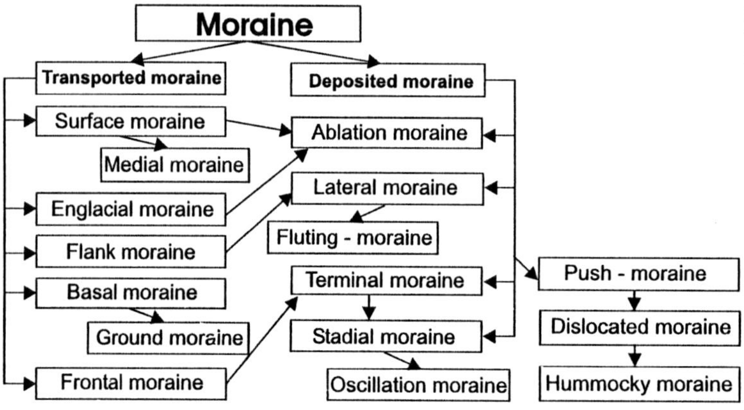

Fig. 214. Types of transported and deposited **moraines**

s hielo *m* cargado de morrena
g moränenhaltges Eis *n*

3943 moraine-covered glacier; turkestan-type glacier
A valley glacier lying on the bottom of narrow and deep valley, and having no firn basin.
r ледник туркестанского типа
Долинный ледник, лежащий на дне глубокой и узкой долины, на большей части засыпанный мореной и лишенный фирнового бассейна.
f glacier *m* recouvert
s glaciar *m* recubierto
g Gletscher *m* vom turkestanischen Typ

3944 morainic lake
A lake occupying a depression between ancient morainic ridges, or formed behind the rampart of a terminal moraine recently left in a valley by a retreating glacier.
r моренное озеро
Озеро, занимающее впадину между древними моренными грядами, либо образующееся за валом конечной морены, недавно оставленной в долине отступающим ледником.
f lac *m* de barrage morainique; lac *m* de moraine
s lago *m* de morrenas
g Moränensee *m*

3945 morainic mantle
A continuous cover of debris on the surface of a glacier tongue or dead ice.
r моренный плащ
Сплошной покров морены на поверхности ледниковых языков или мертвого льда.
f couverture *f* morainique
s capa *m* de morrena; manto *m* morrénico
g Moränenaufschuttung *f*; Deckmoräne *f*

3946 morainic ridge; moraine
A rampart-like accumulation of the rock products of weathering deposited by a glacier.
r моренная гряда
Валообразное скопление продуктов выветривания горных пород, отложенных ледником.
f arc *m* morainique; vallum *m* morainique
s arco *m* morrénico; vallum *m* morrénico
g Moränenwall *m*; Moränenzug *m*

3947 morphogenesis
The creation and evolution of the landforms of the Earth's surface by the endogenetic and exogenetic processes.
r морфогенез; рельефообразование
Возникновение и развитие форм рельефа земной поверхности под влиянием эндогенных и экзогенных процессов.
f morphogenèse *f*; géomorphogenèse *f*
s morfogénesis *f*; geomorfogénesis *f*
g Morphogenese *f*; Reliefbildung *f*

* **morphogenic crisis → 5870**

3948 morphography; orography
A branch of *geomorphology* concerned with the study of land forms of the Earth's surface relief and the systematisation of landforms according to their topographic features.
r орография; морфография
Раздел *геоморфологии*, включающий описание форм рельефа земной поверхности и систематизацию их по внешним признакам, иногда употребляется как синоним понятия рельеф.
f orographie *f*; morphographie *f*
s orografía *f*; morfografía *f*
g Orographie *f*; Morphographie *f*

3949 morphological analysis
The method of reconstruction of the development of the Earth's crust motions on the basis of the study of the external structure of landforms and topographic features resulting from erosion.
r морфологический анализ
Метод восстановления хода и развития движений земной коры на основе изучения внешнего облика земной поверхности и характера экзогенных рельефообразующих процессов.
f analyse *f* morphologique
s análisis *m* morfológico
g morphologische Analyse *f*

3950 morphological map
A map showing the detailed location of the microrelief on a small area by means of standard symbols. → Fig. 215
r морфологическая карта

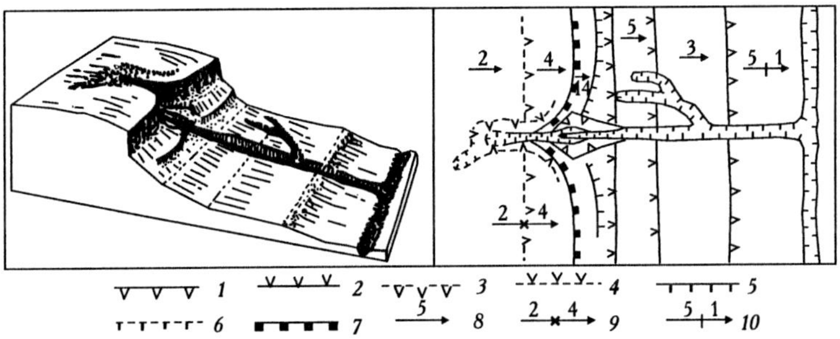

Fig. 215. **Morphological map**: 1, 2 – convex, 3, 4 – concave, 5, 6 – convex and concave too close to permit separate symbols, 7 – cliffs, bedrock, 40° or more, 8 – angle of slope in degrees, 9 – convex slope unit, 10 – concave slope unit

Карта, показывающая детали микрорельефа на небольшой площади посредством стандартизованных символов. → Рис. 215, с. 471

f carte *f* morphologique
s mapa *m* morfológico
g morphologische Karte *f*

3951 morphometric map
A map which represents quantitative characteristics of landforms and types of relief (relative elevations, depth of dissection, etc.).
r морфометрическая карта
Карта, показывающая количественные характеристики форм и типов рельефа: относительные высоты, глубину расчленения и пр.
f carte *f* morphométrique
s mapa *m* morfométrico
g morphometrische Karte *f*

3952 morphometry
The quantitative description of any forms of relief, the measurement of numerical characteristics of landforms: area, length, volume, height, etc.
r морфометрия
Измерение числовых характеристик форм рельефа земной поверхности: площади, длины, объема, высоты, густоты расчленения и пр.
f morphométrie *f*
s morfometría *f*
g Morphometrie *f*

3953 morphosculpture
A relatively small landforms formed mainly by exogenic processes.
r морфоскульптура; скульптурный рельеф
Относительно небольшие формы рельефа, в формировании которых главную роль играют экзогенные процессы.
f morphosculpture *f*
s morfoescultura *f*
g Skulpturformen *fpl*

3954 morphostructure
Predominantly large forms of the Earth's surface created mainly by endogenic processes (their morphology is determined by geological structures exposed by denudation).
r морфоструктура; структурный рельеф
Преимущественно крупные формы земной поверхности, в образовании которых главная роль принадлежит эндогенным процессам (их морфология обусловлена геологическими структурами, отпрепарированными денудацией).
f morphostructure *m*; relief *m* structural
s morfostructura *f*; relieve *m* estructural
g Strukturformen *fpl*; Morphostrukture *f*

3955 mortality; mortality rate; death-rate
The number of deaths in a year per one thousand of the population.
r смертность; коэффициент смертности
Число умерших за определенный период, обычно за год, в расчете на 1000 жителей.
f mortalité *f*; taux *m* de mortalité
s mortalidad *f*; tasa *f* de mortalidad
g Sterblichkeit *f*; Sterblichkeitsrate *f*

* **mortality rate → 3955**

3956 mort-lake; dead lake
A lake where water organisms do not live due to its high mineralisation.
r мертвое озеро
Озеро, в котором вследствие высокой минерализации отсутствуют водные организмы.
f lac *m* mort
s lago *m* muerto
g toter See *m*

3957 mortmass
Dead part of organic matter of biocenosis including dead wood and plant residue, peat, litter and detritus (it is expressed in specific mass).
r мортмасса; мертвый покров
Мертвая часть органического вещества биоценоза, включающая отпад, опад, торф и детрит (подсчитывается в удельных единицах массы).
f masse *f* mort
s masa *f* muerta
g tote Masse *f*

3958 mosaic
In aerial photography, a composite photographic representation of an area obtained by joining together individual photographic prints.
r мозаика
В аэрофотографии составное фотографическое представление территории, полученное путем объединения отдельных фотоотпечатков.
f mosaïque *f*
s mosáico *m*
g Mosaik *n*

3959 mosaic *adj*
Related to a vegetation pattern in which community types are interspersed, it is caused by the differences in the local environmental conditions for plant growth, e.g. hillock bog, grass-forb steppe.
r мозаичный
Относящийся к особенностям растительного покрова, когда разные типы растительности перемежаются друг с другом, что вызывается неодинаковыми условиями существования растительности, например, кочкарное болото, злаково-разнотравная степь.
f mosaïque *adj*
s mosáico *adj*
g mosaikartig *adj*; Mosaik-

3960 moss and lichen layer

The ground layer in the stratified structure of vegetation composed of mosses and lichens that occur in some forest and tundra plant communities.

r мохово-лишайниковый ярус
Напочвенный ярус в некоторых лесных и тундровых растительных сообществах, образованный мхами и лишайниками.

f strate *f* muscinal et des lichens
s estrato *m* muscinal y liquénico
g Moos- und Flechtenschicht *f*

3961 moss and lichen tundra
The tundra, the sod of which consists of green mosses and lichens, and sedge, bluegrass, willow, etc. in the thin upper layer.

r мохово-лишайниковая тундра
Тундра, в растительном покрове которой дернина состоит из зеленых мхов и лишайников, а в разреженном верхнем ярусе растут осока, мятлик, ива и др.

f toundra *f* de mousses et lichens
s tundra *f* de musgos y líquenes; tundra *f* liquénica y muscinal
g Moosflechtentundra *f*

3962 mossy forest
Forest in which the living and dead trees and ground surface support a strong growth of mosses occurring particularly on uplands, in the humid cloudy conditions of the windward side of high oceanic islands in the tropics.

r мшистый лес
Лес, в котором и живые и мертвые деревья, а также поверхность грунта покрыты толстым слоем мхов, часто встречается на плоскогорьях, во влажных облачных условиях на наветренной стороне высоких океанических островов в тропиках.

f forêt *f* moussue
s bosque *m* musgoso
g Mooswald *m*

* **mother country → 4421**

* **mother-of-pearl cloud → 4025**

* **motor-car industry → 469**

* **motorway → 2775**

3963 motorway; highway
A road designed for the movement of heavy volumes of traffic at high speed which has slight gradients and gentle curves.

r автострада
Автомобильная дорога, предназначенная для движения больших потоков транспорта с высокими скоростями, имеющая незначительные уклоны и плавные закругления.

f autoroute *f*
s autopista *f*; autovía *f*
g Autobahn *f*

3964 mottled *adj*

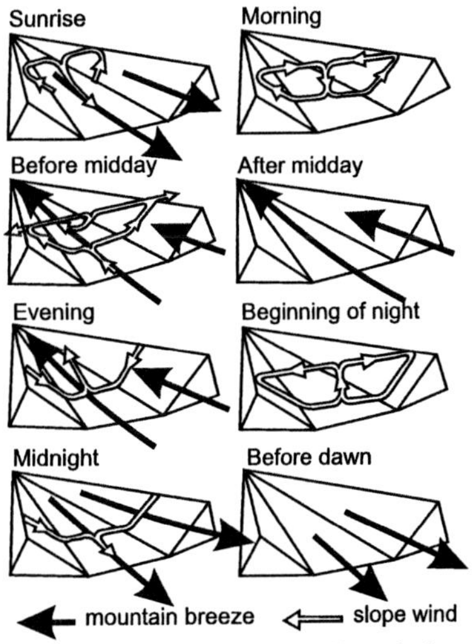

Fig. 216. **Mountain and valley winds**

Marked with blotches, with patches of different colours, e.g. in a gley soil.

r крапчатый
Отличающийся пятнистостью, участками разного цвета, как это имеет место в глеевых почвах.

f tacheté *adj*
s con lunares
g gesprenkelt *adj*

* **moulin → 2527**

3965 mount; mountain
A natural land elevation, higher than a hill rising above its surroundings as a single eminence; a top in the mountains.

r гора
Изолированное крутосклонное поднятие среди относительно ровной местности; вершина в горной стране.

f mont *m*; montagne *f*
s monte *m*; montaña *f*
g Berg *m*

* **mountain → 3965**

3966 mountain and valley winds; valley breeze; valley wind; mountain breeze
Diurnal winds of local circulation in mountain regions associated with diurnal solar heating and nocturnal cooling, which blow along the axis of a valley, uphill or upvalley by day, and downhill or downvalley by night, prevailing mostly in calm, clear weather. → Fig. 216

r горно-долинные ветры; горный ветер; долинный ветер
Ветры местной циркуляции в горных районах, связанные с дневным солнечным нагреванием и ночным охлаждением и поэтому имеющие суточный период: до-

Fig. 217. **Mountain glaciation** near Nanga Parbat Mount in Hymalaya

линный ветер днем дует вверх по склонам гор и дну долины, горный ветер ночью наоборот. → Рис. 216, с. 473

f brise *f* de montagne; brise *f* de vallée
s brisa *f* de montaña; brisa *f* de valle
g Berg- und Talwinde *m*; Bergbrise *f*; Talbrise *f*

* **mountain breeze → 3966**

3967 **mountain chain; mountain range; range₃**
A row or series of ridges and mountain massifs that stretch in one direction, forming a connected system.
r горная цепь
Ряд отдельных хребтов и массивов, тянущихся в одном направлении, иногда соединенных перемычками.
f chaîne *f* de montagnes; ensemble *m* montagneux
s cadena *f* de montaña; conjunto *m* montañoso
g Gebirgskette *f*; Kettengebirge *n*

3968 **mountain crest**
The highest, jagged part of a mountain range or a mountain massif.
r горный гребень; гребень горного хребта
Наиболее высокая острая зубчатая часть горного хребта или массива.
f crête *f* de montagne
s cresta *f* de montaña
g Bergrat *m*; Bergkamm *m*; Gebirgskamm *m*

3969 **mountain desert**
A desert in the medium-height mountains with a poorly developed layer of weathered material.
r горная пустыня
Пустыня в средневысотных горах со слабым развитием коры выветривания.
f désert *m* de montagne
s desierto *m* de montañas
g Gebirgswüste *f*; Bergwüste *f*

3970 **mountain desert-steppe belt**
An altitudinal belt in the mountains of tropical, subtropical and partly temperate belts with desert and steppe vegetation, it occupies the lower parts of mountain slopes with insufficient moisture.
r горный пустынно-степной пояс
Высотный пояс в горах тропического, субтропического и отчасти умеренного поясов с преобладанием пустынной и степной растительности, обычно располагается в нижних частях склонов гор в условиях недостаточного увлажнения.
f étage *m* des déserts-steppes de montagne
s piso *m* de desierto y estepa montañoso; piso *m* pseudo-alpino
g Höhenstufe *f* der Wüsten und Steppen

3971 **mountain fauna**
Animals inhabiting mountain regions.
r горная фауна
Совокупность видов животных, населяющих горные территории.
f faune *f* de montagne
s fauna *f* de montaña
g Gebirgsfauna *f*; Bergfauna *f*

3972 **mountain forest belt**
An altitudinal belt with forest landscapes; it is located above the belt of mountain steppes and forest-steppes or mountain savannas, up to slope it is replaced by subalpine or mountain tundra belts.
r горно-лесной пояс
Высотный пояс с преобладанием лесных ландшафтов; располагается выше пояса горных степей и лесостепей или горных

саванн, выше по склону сменяется субальпийским или горно-тундровым поясами.
f étage *m* montagnard forestier
s piso *m* forestal montañoso; piso *m* de montaña; piso *m* forestal
g Waldstufe *f*

3973 mountain glaciation
A complex of different forms of glaciation developed in the mountain relief. → Fig. 217, p. 474
r горное оледенение
Совокупность разных форм оледенения, приуроченных к горному рельефу. → Рис. 217, с. 474
f glaciation *f* de montagne
s glaciación *f* de montañas
g Gebirgsvergletscherung *f*

3974 mountain glacier
A glacier which moves downslope along a mountain valley. → Fig. 218
r горный ледник
Ледник, залегающий в горах и движущийся вниз по склону по горной долине. → Рис. 218
f glacier *m* de montagne
s glaciar *m* de montañas
g Berggletscher *m*; Hochlandgletscher *m*

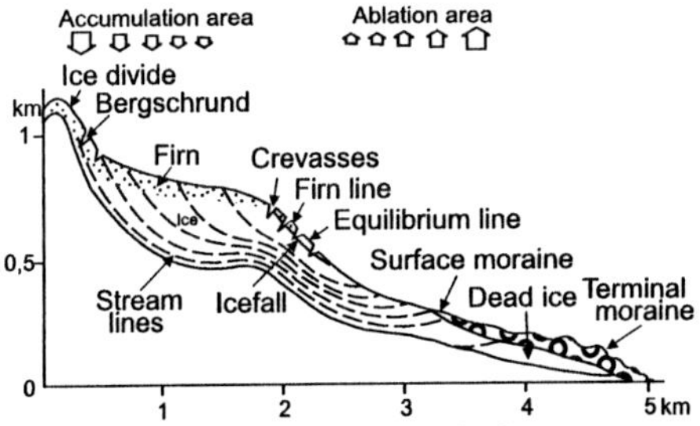

Fig. 218. **Mountain glacier**

3975 mountain group
A part of a mountain system separated by deep valleys, mountain passes or intermountain depressions.
r горная группа
Часть горной системы, обособленная глубоко врезанными долинами, перевалами или внутригорными впадинами.
f groupe *m* des montagnes
s grupo *m* de montañas
g Berggruppe *f*; Gebirgsgruppe *f*

3976 mountain health resort
A resort with favourable mountain climate which has medical effect.
r горный курорт
Лечебный курорт, основным достоинством которого служат благоприятные для лечения горные климатические условия.
f station *f* de cure de montagne
s estación *f* balnearia en montañas
g Gebirgskurort *m*

3977 mountain knot; knot
The crossing-point or meeting place of two or several mountain ranges.
r горный узел
Место пересечения двух или нескольких горных хребтов.
f noeud *m* des montagnes
s nudo *m* montañoso
g Gebirgsknoten *m*

3978 mountain landscape
A landscape which forms in the mountains and has contrastive features caused by the differences of altitude, aspect, geological structure.
r горный ландшафт
Ландшафт, формирующийся в горах и отличающийся контрастностью природных условий, вызванной различиями в высотном положении, экспозиции склонов, геологическом строении.
f paysage *m* montagnard
s paisaje *m* montañoso; paisaje *m* de altitud
g Gebirgslandschaft *f*

3979 mountain livestock farming
Seasonal pasture high in the mountains, usually on the lands belonging to the community.
r высокогорное животноводство
Сезонный выпас скота в высокогорье, как правило, на землях той же горной общины.
f remue *f*
s alpegio *m*; estivaje *f*
g Alpwirtschaft *f*

3980 mountain massif; massif
A more or less isolated section of mountains with approximately equal length and width which is not deeply dissected and has clearly defined boundaries.
r горный массив
Более или менее изолированный участок гор примерно одинаковой длины и ширины, относительно слабо расчлененный и имеющий ясно выраженные границы.
f massif *m* montagneux
s macizo *m* montañoso
g Gebbirgsstock *m*; Bergmassiv *n*; Gebirgsmassiv *n*; Bergstock *m*

3981 mountain meadow belt
Altitudinal belt in the mountains, in the natural landscapes of which meadows prevail, it is located in the alpine and subalpine belts.
r горно-луговой пояс
Высотный пояс в горах, в естественных ландшафтах которого преобладают луга, встречается в пределах альпийского и субальпийского поясов.
f étage *m* des pres de montagne
s piso *m* de prado montañoso
g Wiesenstufen *fpl*

3982 **mountain meadow soil**
Acid soils of subalpine and alpine meadows with thick sod cover and humus horizon gradually turning into rock debris.
r горно-луговая почва
Кислые почвы субальпийских и альпийских лугов с мощной дерниной и гумусовым горизонтом, постепенно переходящим в щебнистую породу.
f sol *m* de près de montagne
s suelo *m* montañoso de prado
g Bergwiesenboden *m*

3983 **mountain meadow steppe soil**
A soil that develops under mountain steppe meadows and has a dark humus horizon grading into the parent rock rich in debris, often carbonate one.
r горная лугово-степная почва
Почва горных остепненных лугов с темным гумусовым горизонтом, переходящим в щебнистую, нередко карбонатную почвообразующую породу.
f sol *m* de près-steppe de montagne
s suelo *m* montañoso de prado y estepa
g Wiesen- und Steppenboden *m* des Gebirges

* **mountain pass → 4432**

3984 **mountain peak**
A pointed prominent top or summit of a mountain.
r пик
Остроконечная горная вершина.
f pic *m*
s pico *m*
g Bergspitze *f*; Pik *m*

* **mountain railway → 1105**

* **mountain range → 3967**

* **mountain ridge → 5075**

3985 **mountain river**
A river with a rapid current usually flowing in a deep poorly developed valley.
r горная река
Река с быстрым течением, протекающая обычно в глубокой слабо разработанной долине.
f fleuve *m* de montagne
s río *m* de montaña
g Gebirgsfluß *m*; Bergstrom *m*

* **mountain sickness → 2761**

3986 **mountain socle**
The folded basement of a mountain system.
r цоколь гор
Складчатое основание горной системы.
f socle *m* de massif montagnex
s zócalo *m* de macizo montañoso
g Gebirgssockel *m*

3987 **mountain soils**
A collective name for the soils found in mountains and having shallow profile (less than 1 m) and high content of rock debris.
r горные почвы
Собирательное название для почв гор, отличающихся небольшой глубиной профиля (менее 1 м) и высокой щебнистостью.
f sols *mpl* de montagne
s suelos *mpl* montañosos
g Bergböden *mpl*

3988 **mountain steppe soil**
Soils that are characteristic of the lower belt of mountains and foothills of steppe and partially semi-desert zones and that are in many respects similar to soils of adjoining plains, they are usually used as pastures.
r горно-степная почва
Почвы, свойственные нижнему поясу гор и предгорьям в степной и отчасти полупустынной зонах, во многом сходные с почвами прилегающих равнин, обычно используются под пастбища.
f sol *m* de steppe de montagne
s suelo *m* de estepa montañoso
g Bergsteppenboden *m*

3989 **mountain system**
A number of mountain ridges that stretch in the same direction and share the common geological structure, genesis and history of development.
r горная система
Совокупность горных хребтов, объединенных общим простиранием, внутренней структурой, генезисом и историей развития.
f système *m* montagneux
s sistema *m* montañoso
g Gebirgssystem *n*

3990 **mountain taiga soil**
A poorly differentiated soil that is found under mountain taiga and has no well-pronounced features of podzolization.
r горно-таежная почва
Слабодифференцированная почва под горной тайгой, не имеющая ярко выраженных признаков подзолистого процесса.
f sol *m* de taïga de montagne
s suelo *m* montañoso de taiga
g Bergtaigaboden *m*

3991 **mountain tundra belt**
An altitudinal belt in the mountains of sub-Arctic and temperate latitudes found above the mountain-forest or alpine belts, it is characterized by severe climate and tundra vegetation with patches of mountain desert.
r горно-тундровый пояс
Высотный пояс в горах Субарктики и умеренных широт, лежащий выше горнолесного или альпийского поясов, обладает суровым климатом и тундровой растительностью с участками горной пустыни.

Fig. 219. **Mountains** of Central Caucasus

f étage *m* de toundra de montagne; toundra *f* alpine
s piso *m* montañoso de tundra; tundra *f* alpina
g Höhenstufe *f* der Tundra

3992 mountain tundra soil
A soil formed under mountain tundra vegetation and characterized by gleization, permafrost-related deformations and often accumulation of matter above permafrost.
r горно-тундровая почва
Почва под растительностью горных тундр, отличающаяся наличием оглеения, мерзлотными деформациями и нередко аккумуляцией вещества над мерзлыми породами.
f sol *m* de toundra de montagne; sol *m* de toundra alpine
s suelo *m* montañoso de tundra; suelo *m* de tundra alpina
g Bergtundraboden *m*

3993 mountain zone
A vast linear region of mountains (it is the largest division in the classification of mountains).
r горный пояс
Обширная линейно вытянутая совокупность последовательно расположенных горных стран, крупнейшая единица районирования горных сооружений.
f zone *f* montagneuse
s zona *f* montañosa
g Gebirgszone *f*; Gebirgsgürtel *m*

* **mountainbuilding → 4301**

3994 mountaineering; alpinism
The technique of climbing mountains and rock faces (may be used in geographical investigations).
r альпинизм
Техника восхождения на труднодоступные горные вершины, может использоваться при географических исследованиях высокогорий.
f alpinisme *m*
s alpinismo *m*; andinismo *m*
g Bergsteigen *n*; Alpinistik *f*

3995 mountainous ***adj***
Related, located, originated in the mountains.
r горный
Относящийся к горам, находящийся в горах, происходящий из гор.
f de montagne
s de montaña; montañoso *adj*
g Gebirgs-; Höhen-; Berg-

3996 mountains
A portion of the land surface rising considerable above the surrounding plains as elevated ranges, chains, massifs, plateaus. → Fig. 219
r горы; горная страна
Часть земной поверхности, существенно поднятая над окружающей равниной и образующая горные хребты, цепи, массивы, плато. → Рис. 219
f montagnes *fpl*
s montañas *fpl*
g Gebirge *n*; Bergland *n*

3997 mountology
The scientific study of physical-geographical characteristics and socioeconomic problems of mountainous regions.
r монтология
Отрасль *географии*, изучающая физико-географические особенности и социально-экономические проблемы горных территорий.
f montelogie *f*
s montelogía *f*
g Bergkunde *f*

* **mouth → 5104**

3998 mouth bar; harbour bar; near-mouth bar
Near shore shallow underwater sand bar in front of a river mouth formed by the deposition of solid river discharge by sea waves.
r приустьевой бар
Песчаный подводный вал, расположенный на небольшой глубине в прибрежной полосе морского дна перед устьем реки, формирующийся в процессе перераспределения морскими волнами твердого речного стока.
f barre *f* d'embouchure
s barra *f* de desembocadura; barra *f* de estuario
g Mündungsbarre *f*

3999 mouth point
A point of intersection of two or more talwegs of river valleys.
r устьевая точка
Место пересечения (схождения) двух или более тальвегов речных долин.
f point *m* d'embouchure; point *m* de confluence
s punto *m* de desembocadura; punto *m* de confluencia
g Mündungspunkt *m*

* **movement of glacier → 2514**

4000 mshari
Raised bogs with sphagnum mosses and peat found in the north and in the centre of the East European Plain.
r мшары
Верховые болота со сфагновыми мхами и торфяниками на севере и в центре Восточно-Европейской равнины.
f mshari *m*
s mshari *m*
g Mschara *n*

4001 mud; ooze; silt
Fine grains of uncosolidated sediments with much water, ranging between clay and sand in size commonly carried and deposited by water in stream channels, lakes and reservoirs.
r ил
Тонкодисперсный водонасыщенный неуплотненный осадок, от глинистой до песчаной размерности, образующийся в руслах рек, на дне озер и водохранилищ.
f boue *f*; limon *m*; vase *f*
s fango *m*; lomo *m*
g Schlamm *m*; Schlick *m*

Fig. 220. **Mud lake**

* **mud flat → 6708**

4002 mud lake
A lake on the bottom of which a thick layer of mud sediments is accumulated. → Fig. 220
r грязевое озеро
Озеро, на дне которого накопился толстый слой иловых отложений. → Рис. 220
f lac *m* à boue
s lago *m* de fango; lago *m* de barro
g Schlammsee *m*

* **mud stream → 4004**

4003 mud volcano
An ejection of hot mud often mixed with water, oil and gas from a volcanic vent to the Earthrs surface building a small ephimeral cone.
r грязевой вулкан
Геологическое образование, постоянно или периодически извергающее на поверхность земли грязевые массы и газы, часто с водой и нефтью.
f volcan *m* de boue; salse *f*
s volcán *m* de barro
g Schlammvulkan *m*

* **mudflat → 6211**

4004 mudflow; mud stream
A moving mass of ground including large blocks of rock made fluid by rain and melting snow or arising suddenly in mountains due to a sudden release of water accumulated above. → Fig. 221
r сель; селевой поток
Обладающий большой разрушительной силой грязевый или грязекаменный бурный поток, внезапно возникающий в го-

Fig. 221. **Mudflow** in the Genaldon river in North Caucasus

рах вследствие прорыва воды, накопленной в результате сильных ливней, интенсивного снеготаяния, прорыва озер или подвижек пульсирующих ледников. → Рис. 221, с. 478
f coulée *f* de boue
s colada *f* fangosa; colada *f* de lodo
g Schlammstrom *m*; Murbruch *m*; Murgang *m*

4005 mudflow basin
A catchment area where mudflows originate and move down the main channel.
r селевой бассейн
Водосборный бассейн, в пределах которого происходит формирование селей и их движение по главному руслу.
f bassin *m* de coulée de boue
s cuenca *f* de colada fangosa
g Schlammstromeinzugsgebiet *n*

* **mudflow centre → 4012**

4006 mudflow channel
A course of periodic mudflows characterized by the alternating sections of scour and sedimentation.
r селевое русло
Русло с признаками периодического схода селей, характеризующееся чередованием участков размыва и отложения рыхлообломочных толщ.
f canal *m* de coulée de boue
s cauce *m* de colada fangosa
g Schlammstrombett *n*

4007 mudflow chute; mudflow gutter
A reinforced concrete trough constructed above a road or channel for the mudflows discharge above the structures to be protected.
r селеспуск
Железобетонный лоток, строящийся над дорожным полотном или над каналом для пропуска селей над защищаемым объектом.
f conduit *m* de coulée de boue
s conducto *m* de colada fangosa
g Schlammstromauslösebahn *f*

Fig. 222. **Mudflow cone** in the Pamirs Mountains

4008 mudflow cone
A detrital fan formed by a mudflow at the end of its channel. → Fig. 222
r конус выноса селя
Форма рельефа, образованная отложениями обломочного материала, вынесенного селевым потоком. → Рис. 222
f cône *m* de dépôt de coulée de boue
s cono *m* de colada fangosa
g Schlammstromablagerungszone *f*; Endschwemmkegel *m*

* **mudflow danger → 4010**

4009 mudflow deposits
The products deposited by mudflows, forming mudflow cones, mudflow ridges and terraces.
r селевые отложения
Продукты выноса селевых потоков, слагающие конусы выноса селя, селевые гряды и террасы.
f dépôts *mpl* de coulée de boue
s depósitos *mpl* de colada fangosa
g Schlammstromablagerungen *fpl*

* **mudflow gutter → 4007**

4010 mudflow hazard; mudflow risk; mudflow danger
Possibility of the occurrance of mudflows in an area characterized by typical features of the mudflow activity, first of all, by the presence of mudflow basins.
r селеопасность
Возможность развития селевых процессов на местности, отличающейся характерными признаками селевой деятельности и прежде всего наличием селевых бассейнов.
f risque *m* de coulée de boue
s riesgo *m* de colada fangosa
g Schlammstromgefahr *f*

4011 mudflow mass
The mixture of detritus, mud, stones, etc. with water, forming the body of a moving mudflow.
r селевая масса
Смесь обломков горных пород, камней, грязи и воды, образующая тело движущегося селевого потока.
f masse *f* de coulée de boue
s masa *m* de colada fangosa
g Schlammstrommasse *f*

* **mudflow risk → 4010**

4012 mudflow site; mudflow centre
The place in the upper parts of a mudflow basin where mudflows originate.
r селевой очаг; зона зарождения селя

Fig. 223. **Mudflow-protective structures**

Участок в верховьях селевого бассейна, где происходит зарождение селевого потока.
f centre *m* actif de coulée de boue
s foco *m* de colada fangosa
g Ursprungszone *f* eines Schlammstroms

4013 mudflow-protective structures; mudflow-retaining constrictions
Protective structures designed to block mudflows. → Fig. 223
r селезадерживающие сооружения; противоселевые сооружения; селенаправляющие сооружения
Инженерные сооружения, построенные с целью задержания селевых потоков. → Рис. 223
f ouvrages *mpl* de défense contre les coulées de boue
s defensas *fpl* contra coladas fangosas
g Schlammstromrückhaltsbauwerke *npl*

*** mudflow-retaining constrictions → 4013**

4014 mulch
A loose layer of organic and inorganic materials (a layer of straw or decaying plant leaves, etc.) that forms naturally or is spread artificially on the ground to protect the soil and the plant roots from the effect of raindrops, evaporation and freezing.
r мульча
Рыхлый слой органических или неорганических материалов, который формируется естественным образом или специально покрывает грунт, чтобы защитить почву и корни растений от прямого воздействия дождевых капель, испарения и промерзания.
f couche *f* protectrice; mulch *m*
s capa *f* protectora; mulch *m*
g Mulch *m*

4015 mulching
A protective covering of soil surface with different materials such as sawdust, compost, etc. in order to reduce evaporation, maintain an even temperature, protect soil structure from destruction, control weeds, etc.
r мульчирование почвы
Покрытие поверхности почвы разными материалами (опилками, компостом и пр.) в целях снижения испарения влаги, регулирования температуры, предохранения почвенной структуры от разрушения, борьбы со всходами сорняков и т.п.
f mulching *m*
s mulching *m*
g Mulchung *f*

4016 mull
A type of humus that rapidly decomposes and becomes well mixed into the mineral soil, so that distinct layers are not apparent, it is characteristic of soils of hardwood forests, deciduous forests, or grasslands of warm, humid climate.
r мюлль
Тип гумуса, для которого характерно полное включение органического вещества в минеральный субстрат, распространен в почвах лиственных лесов и лугов в теплом и влажном климате.
f mull *m*; humus *m* doux
s mull *m*; humus *m* suave
g Müll *m*

*** multinational corporation → 6309**

4017 multinational state
A state which contains some ethnic groups as identified by religion, language or colour, e.g. Spain includes Castillians, Catalans, and Basques, Russia includes dozens different nationalities.
r многонациональное государство
Государство, включающее несколько этнических групп с разной религией, языком или цветом кожи, например, в Испании – кастильцы, каталонцы и баски, в России десятки разных национальностей.
f etat *m* multinational
s estado *m* multinacional
g Nationalitätenstaat *m*

4018 multiple land use
The use of a tract of land for two or more purposes, including recreation (as when an area of forestry).
r многоцелевое использование земли
Использование участка земли для разных целей, в том числе для отдыха.
f utilisation *f* multiple du sol
s uso *m* multiple de terreno
g Mehrzwecklandnutzung *f*

*** multipolar world → 4685**

4019 multispectral *adj*
Related to a survey or sensor that records the images of the Earthrs surface in a number of wavebands.
r многоспектральный
Относящийся к съемке или приборам, делающим снимки земной поверхности одновременно в нескольких диапазонах волн.

f multispectral *adj*
s multispectral *adj*
g Multispectral-

* **multispectral survey → 5716**

4020 **multistratified freezing of water; layered freezing of water**
The formation of ice by the freezing of water slowly which flows down a cold solid surface in the form of drops, streamlets and outflows.
r намораживание; послойное намораживание
Образование льда путем замерзания воды, медленно стекающей по холодному твердому основанию в виде капель, струек и разливов.
f glaçage *m*
s helamiento *m*
g Eiszuwachs *m* durch Gefrieren; Auffrieren *n*

4021 **multi-structural economy; mixed economy**
An economic system which combines private and state-owned productive activities.
r смешанная экономика; многоукладная экономика
Социально-экономическая система, включающая как национализированную промышленность, так и частные предприятия.
f économie *f* mixte
s economía *f* mixta
g Mischwirtschaft *f*

4022 **municipal *adj***
Relating a place that has its own local self-government, especially of a town or city.
r муниципальный
Городской, общественный, относящийся к местному самоуправлению.
f municipal *adj*
s municipal *adj*
g munizipal *adj*; Gemeinde-

* **municipal tax → 3557**

* **mushroom rock → 2102**

4023 **muskeg**
A sphagnum bog in Canada and Alaska, often with tussocks.
r маскег
Сфагновое болото в Канаде или на Аляске, часто с кочковатой поверхностью.
f muskeg *m*
s muskeg *m*
g Muskeg *n*

* **mutualism → 1144**

4024 **mylonitization of ice**
A dislocative metamorphism of ice within zones where movements along internal planes in glaciers have different directions.
r милонитизация льда
Дислокационный метаморфизм льда в зонах разнонаправленного движения по внутренним плоскостям разрывов в ледниках.
f mylonitisation *f* de glace; broyage *m* de glace
s milonitisación *f* de hielo
g Mylonitisation *f* von Eis

N

4025 **nacreous cloud; mother-of-pearl cloud**
A rare, very high (20-30 km) cloud resembling mother-of-pearl because of its iridescence, the most brilliant colours are observed when the Sun is shining from below the horizon.
r перламутровое облако
Очень высокие (20-30 км) и редко наблюдаемые облака, похожие на перистые или высоко-кучевые с очень сильной радужной окраской, напоминающей перламутр, окрас наиболее ярок, когда Солнце находится на несколько градусов ниже горизонта.
f nuage *m* nacré
s nube *m* nacarada
g Perlmutt-Wolke *f*

4026 **nadir**
The point on the celestial sphere directly opposite to the zenith. → Fig. 51, p. 110
r надир
Точка небесной сферы, диаметрально противоположная зениту. → Рис. 51, с. 110
f nadir *m*
s nadir *m*
g Nadir *m*; Fußpunkt *m*

* **naled → 453**

4027 **nalep′**
A cover of wet or frozen snow on trees, pillars, etc.
r налепь
Покров мокрого или замерзшего снега на деревьях, столбах и т.п.
f nalep *m*
s pedadura *f* de nieve
g Nalep *f*

4028 **nanorelief**
Landforms some ten centimetres high, e.g. ripple.
r нанорельеф
Формы рельефа высотой до нескольких десятков сантиметров, например, рябь.
f nanorelief *m*
s nanorelieve *m*
g Nanorelief *n*

4029 **nappe**
A mass of rock that has moved forward by an overthrust for many kilometres from its «roots» covering the formations beneath.
r тектонический покров
Масса горных пород, перемещенная надвигом от места их первичного залегания в горизонтальном направлении на несколько сотен километров.
f nappe *f* de charriage
s manto *m* tectónico; capa *f* de deslizamiento
g tektonische Decke *f*

4030 **narrow gully; runnel**
A general term for shallow (up to several metres) linear depressions with gentle slopes.
r ложбина; лощина
Общее название неглубоких (до нескольких метров) линейно вытянутых понижений рельефа с мягкими пологими склонами.
f vallon *m*
s vallejo *m*; quiebra *f*
g Furche *f*

4031 **narrows**
A constricted passageway in a strait, in part of a river, in a valley, or in a pass.
r теснина
Суженный проход в проливе, речном русле, долине или на перевале.
f col *m*
s garganta *f*
g Engpaß *m*

4032 **nation$_1$**
A historical community of people possessing a specified territory, having common origin, language, traditions, and capable of constituting a nation-state.
r нация
Историческая общность людей, складывающаяся в процессе формирования их территориальных, экономических связей, языка, особенностей культуры и характера.
f nation *f*
s nación$_1$ *f*
g Nation *f*

* **nation^2 → 5778**

* **nation′s economy → 4036**

4033 **national *adj***
Related or characteristic of a nation, or certain country, e.g. national income.
r национальный
Относящийся или свойственный нации, либо конкретной стране, например, национальный доход.
f national *adj*
s nacional *adj*
g national *adj*

4034 **national atlas**
An atlas which contains maps characterising different aspects of nature, population, economics and culture of a certain country.
r национальный атлас
Атлас, содержащий карты с разносторонней характеристикой природы, населения, хозяйства и культуры данной страны.
f atlas *m* national
s atlas *m* nacional
g Nationalatlas *m*

4035 national boundary; administrative frontier; political border
A limit between any states as well as between federation units and other territorial units of a state.
r государственная граница; политико-административная граница
Граница между отдельными странами (государственная граница), а также между субъектами федерации и иными территориальными единицами какого-либо государства (политико-административная граница).
f frontière *f* politique; frontière *f* administrative
s frontera *f* política; límite *m* administrativo
g administrative Grenze *f*; Staatsgrenze *f*

4036 national economy; nation´s economy
The system by which a countryrs trade, industry and money are organized, including the branches of material production and non-material spheres (services).
r народное хозяйство; национальное хозяйство
Совокупность отраслей и сфер экономики страны, взаимосвязанных общественным разделением труда, включает отрасли материального производства и непроизводственной сферы.
f économie *f* nationale
s economía *f* nacional
g Volkswirtschaft *f*

4037 national income
The value of the aggregate national product minus its production costs.
r национальный доход
Стоимость валового национального продукта за вычетом всех материальных затрат на его производство.
f revenu *m* national
s renta *f* nacional
g Volkswirtschaftseinkommen *n*

4038 national income per capita
Mean national income calculated for one citizen of a given country.
r национальный доход на душу населения
Средний национальный доход в расчете на одного жителя страны.
f revenu *m* national per capita
s renta *f* nacional per capita
g Pro-Kopf-Volkswirtschaftseinkommen *n*

4039 national park
A large land or water area containing natural or cultural landscapes of great beauty or scientific interest owned and managed by the government for a strictly regulated recreational and educational use by the public.
r национальный парк
Охраняемая территория и акватория с мало нарушенными ландшафтами и уникальными природными комплексами, предназначенная для строго регулируемого рекреационного использования.
f parc *m* national; parc *m* naturel
s parque *m* nacional; reserva *f* nacional
g Nationalpark *m*; Naturschutzpark *m*

*** national separatism → 5348**

4040 national territory
A part of the Earth´s surface (including inland and territorial waters, mineral resources and the atmosphere) which is under the jurisdiction of any given country.
r государственная территория
Часть земной поверхности, включая внутренние и территориальные воды, недра и атмосферу, находящаяся под юрисдикцией данной страны.
f territoire *m* national
s territorio *m* nacional
g Staatsterritorium *n*

4041 nationalism
Loyalty and devotion to a nation, the wish for political independence of a group of people with the same language, culture, history, etc. within certain geographical territory; the exalting of one nation above all others.
r национализм
Желание группы людей с единым языком, культурой, историей и др. получить политическую независимость в пределах определенной територии; разжигание превосходства одной нации над всеми другими.
f nationalisme *m*
s nacionalismo *m*
g Nationalismus *m*

4042 nationality
A relationship to the particular nation, the nation-state.
r национальность
Принадлежность к определенной нации, национальному государству.
f nationalité *f*
s nacionalidad *f*
g Nationalität *f*; Völkerschaft *f*

4043 nationalization
The act of putting private property (an activity, industry, land) under public property, i.e. under the control of the state, with or without compensation.
r национализация
Переход средств производства (включая землю), банков и отраслей экономики из частной собственности в собственность государства.
f nationalisation *f*
s nacionalización *f*
g Nationalisierung *f*; Verstaatlichung *f*

4044 nation-state
A sovereign state most of the members of which constitute a nation, e.g. Italy.

r национальное государство
Суверенное государство, подавляющее большинство жителей которого составляет единую нацию, например Италия.
f état-nation *f*
s estado *m* national
g Nationalstaat *m*

4045 **native; aborigine; aboriginal**
Original inhabitants of an area occupying it before invasion or colonization.
r абориген; коренной житель
Изначальные жители данной территории, поселившиеся здесь еще до вторжения других людей или ее колонизации.
f aborigène *m*
s aborigen *m*
g Ureinwohner *m*; Urbewohner *m*

* **native species → 463**

4046 **natural** ***adj***
Pertaining to, existing in, or formed by nature and not produced or caused by people.
r природный
Относящийся, существующий или сформированный в природе и не произведенный или вызванный деятельностью людей.
f naturel *adj*
s natural *adj*
g Natur-; natürlich *adj*

4047 **natural and technical system**
A complex of natural and artificial elements making a whole and functioning as a single system, such as a river dam and the reservoir formed by the dam.
r природно-техническая система
Совокупность природных и техногенных элементов, функционирующих как единая система, например, дамба на реке и возникшее в верхнем бьефе водохранилище.
f système *m* naturel et technique
s sistema *m* natural y técnico
g naturtechnisches System *n*

* **natural boundary → 4551**

4048 **natural bridge**
A remnant of a collapsed arch above the channel of an underground river.
r естественный мост
Сохранившаяся часть от обрушившегося свода над каналом подземной реки.
f pont *m* naturel
s puente *m* natural
g natürliche Brücke *f*; Naturbrücke *f*

4049 **natural conditions; physical conditions**
Geographical situation, natural resources and other characteristics of the natural environment of a given place.
r природные условия
Совокупность географического положения, природных ресурсов и других компонентов окружающей среды, характерных для данной территории.
f conditions *fpl* naturelles
s condiciones *fpl* naturales
g Naturbedingungen *fpl*

* **natural contamination → 4058**

* **natural disaster → 4053**

* **natural environment → 1919**

4050 **natural fertility**
The fertility of soil determined by the resources of natural nutrients in soil, natural waters, air and heat regimes.
r потенциальное плодородие
Плодородие почвы, определяющееся природными запасами в почве питательных веществ, естественными водами, воздушным и тепловым режимом.
f fertilité *f* naturel du sol
s fertilidad *f* natural del suelo
g natürliche Fruchtbarkeit *f*

4051 **natural flow**
A river flow having uneven character over a year, i.e. natural discharge fluctuations, heavy floods and low-water periods.
r незарегулированный сток
Естественный сток в реке, характеризующийся неравномерным распределением в течение года: ярко выраженными паводками и меженью.
f débit *m* naturel
s caudal *m* natural
g ungleichmässiger Abfluß *m*; natürlicher Abfluß *m*

4052 **natural focus of disease**
An area with a typical disease transmitted by the wild animals living there.
r природный очаг болезни; природная очаговость
Территория, для которой характерно заболевание, передаваемое обитающими здесь дикими животными.
f foyer *m* naturel de maladie
s foco *m* natural de infección
g natürlicher Krankheitsherd *m*; Naturseuchenherd *m*

* **natural gas → 2362**

4053 **natural hazard; natural disaster**
An unexpected threat to humans and their property caused by an event of natural origin which by its intensity, distribution and duration may affect the environment, people, and economy, e.g. an earthquake, hurricane, tsunami, volcanic eruption, etc.
r стихийное бедствие; природная катастрофа
Природное явление, которое по интенсивности, распространению и продолжительности оказывает разрушительное воз-

действие на окружающую среду, людей и их хозяйство, например, землетрясение, ураган, цунами, извержение вулкана.

f catastrophe *f* naturelle
s catástrofe *f* natural
g Naturkatastrophe *f*

4054 **natural horizon; geographical horizon**
A line at which the earth or sea and sky seem to meet.
r видимый горизонт
Линия, ограничивающая часть земной поверхности, доступной взору наблюдателя.
f horizon *m* naturel
s horizonte *m* natural
g Naturhorizont *m*

* **natural ice → 2922**

* **natural increase of population → 4059**

4055 **natural landscape**
A geographical landscape that has not been effected by the direct human impact.
r природный ландшафт
Географический ландшафт, не испытавший влияния непосредственной человеческой деятельности либо испытавший ее в очень слабой степени.
f paysage *m* naturel
s paisaje *m* natural
g Naturlandschaft *f*

4056 **natural monument**
A unique natural area or separate objects, which is preserved because of its ecological, scientific, cultural and aesthetic value.
r памятник природы
Уникальный, невосполнимый в экологическом, научном, культурном и эстетическом отношениях природный комплекс или отдельный объект.
f monument *m* naturel
s monumento *m* natural; monumento *m* de naturaleza
g Naturdenkmal *n*

* **natural movement of population → 4698**

4057 **natural phenomenon**
A fact or event which occurs in the environment, that can be described and explained scientifically.
r природное явление
Факт или событие, происходящее в природной среде, которое может быть описано и научно объяснено.
f phénomène *m* naturel
s fenómeno *m* de la naturaleza
g Naturerscheinung *f*

4058 **natural pollution; natural contamination**
Pollution of the environment which occurs naturally, and is usually caused by natural disasters (as opposed to man-made pollution).
r природное загрязнение
Загрязнение окружающей среды, вызванное (в отличие от антропогенного загрязнения) естественными, обычно катастрофическими причинами.
f pollution *f* naturelle; contamination *f* naturelle
s polución *f* natural; contaminación *f* natural
g natürliche Verschmutzung *f*; natürliche Verunreinigung *f*

4059 **natural population growth; natural increase of population**
The growth of population in a given region when birth rate exceeds mortality rate.
r естественный прирост населения
Прирост населения в данном регионе за счет превышения рождаемости над смертностью.
f accroissement *m* naturel de population
s crecimiento *m* demográfico natural
g natürlicher Bevölkerungswachstum *m*; natürlicher Bevölkerungszuwachs *m*

* **natural potential → 3389**

* **natural purification of the environment → 5332**

4060 **natural regime**
The fluctuations in natural processes of the environment determined by the geographical conditions of a given area.
r природный режим
Естественный ход процессов в окружающей среде, определяемый существующими географическими условиями данной территории.
f régime *m* natural
s régimen *m* natural
g Naturverhältnisse *pl*

4061 **natural region; physical-geographical region**
A unit of physio-geographical regionalization confining relatively uniform and distinctive natural characteristics (e.g. relief, climate, vegetation) which includes landscapes of similar age and origin although they can be found in different zones and subzones.
r физико-географическая область
Единица физико-географического районирования, обладающая отличительными чертами рельефа, климата, растительности и других компонентов природы и объединяющая ландшафты, близкие по возрасту и происхождению, хотя их территория может находиться в разных зонах и подзонах.
f région *f* naturelle
s región *f* natural
g physiogeographische Region *f*

4062 natural resources
The wealth supplied by nature and available for human use, i.e. the components of the physical environment such as fossil fuels, minerals, forests, water, soil fertility, climate, which can be used commercially to satisfy different needs of people (natural resources may be classified as renewable and non-renewable).
r природные ресурсы
Природные богатства, пригодные для использования человеком, т.е. компоненты окружающей среды, такие как ископаемое топливо, минеральное сырье, леса, воды, плодородие почв, климат, которые могут удовлетворить различные потребности людей (природные ресурсы делятся на возобновляемые и невозобновляемые).
f ressource *f* naturelle
s recursos *mpl* naturales
g Naturressourcen *fpl*; Naturschätze *fpl*

4063 natural resources assessment; natural resources evaluation
Determination of economic, social and ecological value of natural resources by means of economic estimates and sociological surveys.
r оценка природных ресурсов
Определение экономической, социальной и экологической ценности природных ресурсов путем экономических расчетов и социологических опросов.
f évaluation *f* de ressource naturelle
s valoración *f* de recursos naturales
g Resourcenbewertung *f*

* **natural resources depletion → 1522**

* **natural resources evaluation → 4063**

4064 natural selection
The mechanism of evolutionary change suggested by Charles Darwin in his theory of evolution.
r естественный отбор
Механизм эволюционных изменений, предложенный Чарльзом Дарвином в его теории эволюции органического мира.
f sélection *f* naturelle
s selección *f* natural
g natürliche Zuchtwahl *f*; natürliche Selektion *f*; natürliche Auslese *f*

4065 natural shaft
A narrow passage of karst origin which leads from the earth's surface to the depth of 300-400 m.
r естественная шахта; вертикальная пещера
Узкий канал карстового происхождения, начинающийся отверстием на поверхности земли и уходящий на 300-400 м вглубь.
f aven *m*; abîme *m*
s aven *m*; conducto *m* vertical
g Naturschacht *m*; senkrechte Höhle *f*

4066 natural synoptic period
A period of time during which a particular synoptic situation persist over some region.
r естественный синоптический период
Период времени, в течение которого над каким-либо районом сохраняется определенная синоптическая ситуация.
f période *f* sinoptique naturelle
s período *m* sinóptico natural
g natürliche synoptische Periode *f*

4067 natural vegetation
The vegetation which has developed in an area and has not been changed by humans through agriculture, forest clearance, and industry.
r естественная растительность
Растительность, не испытавшая сильного воздействия человека (распашку, вырубку, выпас, сенокос и пр.).
f végétation *f* spontanée; végétation *f* naturelle
s vegetación *f* natural
g natürliche Vegetation *f*

* **natural water → 6635**

4068 natural-historic park
A relatively small protected area with historic and cultural monuments having very close connection with the surrounding natural landscape, e.g. Royal Palaces and Parks of Potsdam (Germany) or historical scenic ensemble of the monastery on the Solovetskiye Islands in the White Sea (Russia).
r природно-исторический парк; музей-заповедник
Сравнительно небольшой участок охраняемой территории, включающий памятники истории и культуры, с которыми неразрывно связаны окружающие их природные ландшафты, например, парк около королевского дворца в Потсдаме (Германия), или монастырь на Соловецких островах (Россия).
f parc *m* naturel-historique
s parque *m* natural-histórico; museo-reserva *f*
g naturhistorischer Park *m*

4069 naturalization
1. The admittance of a foreigner to the citizenship of a country. 2. The introduction of species of an animal or plant to a habitat in which it is not native, but where it can flourish and reproduce.
r натурализация
1. Разрешение иностранцу стать гражданином данной страны. 2. Интродукция видов животного или растения в местность, где они раньше не обитали, но могут нормально существовать и размножаться.
f naturalisation *f*
s naturalización *f*
g Naturalisierung *f*; Einbürgerung *f*

4070 nature; physical environment
The physical conditions of environment including the laws and forces ruling changes made by humans, the subject of natural and socioeconomic studies.
r природа
Совокупность естественных условий существования человеческого общества, а также изменений окружающей среды, вносимых в нее деятельностью людей, объект естествознания и социально-экономических наук.
f nature *f*
s naturaleza *f*
g Natur *f*

* **nature conservation → 4797**

* **nature preserve → 5027**

4071 nautical chart; sea-charts
Map intended for navigation and other activities in ocean representing the sea bottom relief, currents, areas perilous for navigation, etc.
r морская навигационная карта
Карта, предназначенная для мореплавания и других видов деятельности в Мировом океане, показывающая рельеф морского дна, течения, приливы, места навигационной опасности и пр.
f carte *f* marine
s carta *f* marina
g Seekarte *f*

4072 nautical mile; international nautical mile
A unit of distance equal to 1853 m which is average meridian lengh of one minute of latitude (one minute of arc at 48° latitude).
r морская миля; географическая миля
Мера длины, равная 1853 м, что соответствует средней длине одной минуты большого круга Земли (длине одной минуты дуги на широте 48°).
f mille *m* marin
s milla *f* marina
g Seemeile *f*

4073 nautical twilight
A part of evening and morning twilight when the Sun is not more than 12° below the horizon when the light is sufficient to allow vague shapes to be seen (at the lower position of the Sun the navigation without signal lights is impossible).
r навигационные сумерки; морские сумерки
Часть периода вечерних и утренних сумерек, когда Солнце погружено не более чем на 12° под горизонт (при более глубоком погружении Солнца невозможна навигация без сигнальных огней).
f crépuscule *m* nautique
s crepúsculo *m* náutico
g nautische Dämmerung *f*

4074 navigability
The possibility of navigation along narrow waterways (canals, straits, rivers, etc.) or within small closed water bodies (bays, lakes, etc.).
r судоходность
Возможность свободного плавания по узким водным путям (проливам, каналам, рекам и др.) или в небольших замкнутых водоемах (бухтах, озерах и др.).
f navigabilité *f*
s navegabilidad *f*
g Schiffbarkeit *f*

4075 navigable river
River which is used for transport and appropriately equipped.
r судоходная река
Река, используемая для судоходства и соответствующим образом обустроенная.
f fleuve *m* navigable
s río *m* navegable
g schiffbarer Fluß *m*

4076 navigation; shipping
The use of waterways (seas, rivers, lakes, canals, etc.) by vessels; the period of time within which a sea, lake or river can be used for traffic; the science of determining position, course, and distance travelled during a journey and hence advising on the best course to be steered or taken.
r судоходство; навигация
Использование водных путей (морей, рек, озер, каналов и др.) для движения судов; период времени, в течение которого по природным условиям возможно движение судов в данном море, озере или на реке; наука о способах выбора пути и методах вождения судов (морских, воздушных, космических).
f navigation *f*
s navegación *f*
g Schiffahrt *f*; Schiffahrtsperiode *f*; Navigation *f*

4077 navigation observation
The determination of geographical co-ordinates of a point in the sea by observing the objects with the known co-ordinates (satellites, shore stations, signs, etc.).
r навигационная обсервация
Определение географических координат точки на море по наблюдениям объектов с известными координатами (искусственных спутников, береговых станций и знаков и т.п.).
f observation *f* de navigation
s observación *f* de navegación
g Navigationsüberwachung *f*

4078 navolok
Low river bank in Central Russia composed of soft sediments.
r наволок
Низменный берег реки в Центральной России, сложенный рыхлыми отложениями.

f navolok *m*
s navolok *m*
g Navolok *n*

4079 neap tide; apogean tide; dead neap
The tidal condition when the Moon is at its apogee, so that its gravitational pull is reduced, resulting in lower high tides and higher low tides, with a tidal range smaller than is usual.
r квадратурный прилив; апогейный прилив
Прилив во время первой и последней четвертей (квадратур) Луны, когда Луна находится в апогее и ее приливная сила уменьшается, в результате чего высокий прилив понижается, а низкий повышается, и величина прилива становится меньше обычной.
f marée *f* morte
s marea *f* muerta
g Nippflut *f*; Nipptide *f*; taube Flut *f*

4080 near-by thunderstorm
A thunderstorm with lightnings and thunder the interval between which is not longer than 10 seconds.
r близкая гроза
Гроза с молниями и громом, промежуток между которыми не превышает 10 секунд.
f orage *m* proche
s tormenta *f* cercana
g Nahgewitter *n*

* **near-mouth bar → 3998**

4081 nearshore current; longshore current
The movement of water mass along the shore of a sea or lake caused by tides, waves and wind.
r вдольбереговое течение; прибрежное течение
Движение масс воды вдоль берега моря или озера, обусловленное приливом, волнением и ветром.
f courant *m* littoral
s corriente *f* litoral
g Küstenströmung *f*

* **nearshore sediments → 4238**

4082 nebka; nebkha
A small arrow-shaped hummock of sand, less than 1 m high and 1.5 m long, which has collected leeward of a rocky obstacle or clump of vegetation.
r небха
Небольшой стреловидный песчаный бугор высотой менее 1 м и длиной около 1,5 м, образующийся с подветренной стороны скального выступа или изолированного куста растительности.
f nebka *f*; nebkha *f*
s nebka *f*; nebkha *f*
g Nebka *f*; Nebkha *f*

* **nebkha → 4082**

* **neck → 3260**

4083 neck; volcanic neck
An eroded remnant of solidified lava or agglomerate which formerly filled the volcanic vent but which has now been exposed by denudation of the surrounding cone.
r некк
Обнажившееся в результате денудации магматическое тело столбообразной формы, ранее заполнявшее жерло вулкана.
f neck *m*
s neck *m*
g Neck *m*

4084 necrophage; scavenger
Animals feeding on remains of dead animal organisms, e.g. jackals, hyenas.
r некрофаг
Животные, питающиеся остатками умерших животных организмов, например, шакалы, гиены.
f animal *m* nécrophage
s animal *m* necrófago; consumidor *m* terciario
g Nekrophag *m*

4085 necton; nekton
The actively swimming marine organisms inhabiting the open ocean, as distinct from rhe ocean floor.
r нектон
Совокупность активно плавающих организмов, обитающих в толще воды пелагиали, т.е. в среде открытого океана в отличие от океанского ложа.
f necton *m*
s necton *m*
g Nekton *n*

4086 needle
A sharp-pointed steep-sloped mountain peak or the needle-shaped rocks.
r игла
Крутосклонный горный пик или верхняя часть пика.
f aiguille *f* alpine
s aguja *f* alpina
g Nadel *f*

* **needle forest → 1211**

4087 needle ice; pipkrake
Elongated horizontal prismatic ice crystals formed on the calm surface of river water; near-surface formations of *segregation ice* composed of separate or groups of elongated needle-like parallel ice crystals perpendicular to the ground surface lying under a thin layer of turf, soil and stones.
r игольчатый лед
Вытянутые в горизонтальном направлении призматические кристаллы льда, образующиеся в спокойной воде на поверхности реки; приповерхностные образова-

ния *сегрегационного льда* в виде отдельных или групп вертикально удлиненных столбчатых кристаллов под тонким слоем дернины, грунта, камней.
f aiguille *f* de glace; glace *f* aciculaire
s hielo *m* acicular
g Kammeis *n*; Nadeleis *n*

4088 negative difference of glacierization
The difference in the altitude between the snow line and the bottom of the valley where the glacier tongue moved down.
r отрицательная разность оледенения
Разница в отметках снеговой линии и днищ долин, в которые спускаются языки ледников.
f différence *f* de glaciation négative
s diferencia *f* de glaciación negativa
g negative Vergletscherungsdifferenz *f*

4089 negative water setout
The lowering of the water level in a coastal zone, due to the effect of winds blowing off the shore or along it.
r сгон
Понижение уровня воды в береговой зоне, как правило, под действием ветров, дующих с берега или вдоль него.
f mouvement *m* négatif de niveau de mer par le vent
s movimiento *m* negativo del nivel del mar por el viento
g Wasserspiegelsenkung *f*

4090 neighbourhood
A district forming a community within a town or city, where inhabitants recognize each other by sight.
r соседство
Небольшой городской район с неопределенными границами, население которого живет единой социальной жизнью и большинство людей знают друг друга.
f voisinage *m*
s vecindad *f*
g Nachbarschaft *f*

* **nekton → 4085**

4091 neobiont; neophyte
Animal or plant species introduced into a territory relatively recently.
r необионт; неофит
Сравнительно недавно занесенные или интродуцированные на данную территорию виды животных или растений.
f néobionte *m*; néophyte *m*
s neobionte *m*; neofita *m*
g Neobiont *m*; Neophyt *m*

4092 neocolonialism
A system of unequal relations between states under which certain developing countries depend on a more powerful state in the economic, political and military spheres.
r неоколониализм
Система неравноправных отношений между государствами, когда некоторые развивающиеся страны попадают в тесную экономическую, политическую и военную зависимость от иностранного государства.
f néocolonialisme *m*
s neocolonialismo *m*
g Neokolonialismus *m*

4093 Neogene
The second period of the Cainozoic era which began 26 million years ago and lasted for 23.5 million years, it is subdivided into two epochs – Miocene and Pliocene. → Table 1, p. 284
r неоген
Второй период кайнозойской эры геологической истории Земли, начавшийся 26 млн лет назад и продолжавшийся 23,5 млн лет, подразделяется на две эпохи - миоцен и плиоцен. → Табл. 1, с. 284
f Néogène *m*
s Neogeno *m*
g Neogen *n*

* **neoglacial (*amer*) → 3526**

4094 Neolithic
The last period of the *Stone Age* (about 8-3 thousand years ago) when ceramic articles came to be made, grinding and polishing were applied to the making of stone implements, the domestication of animals and the planting of crops began.
r неолит
Завершающий период *каменного века* (около 8-3 тыс. лет назад), с которым связано появление керамических изделий, шлифовки и полировки каменного инвентаря.
f Néolithique *m*
s Neolítico *m*
g Neolithikum *m*

4095 neolithic revolution
The change from the appropriate economy predominant in the Paleolithic to the productive economy that is mainly farming and rearing cattle (about 12-11 thousand years ago).
r неолитическая революция
Переход от присваивающего способа хозяйства, господствовавшего на протяжении всего палеолита, к производящему, прежде всего к началу земледелия и животноводства (около 12-11 тыс. лет назад).
f révolution *f* néolithique
s revolución *f* neolítica
g neolithische Revolution *f*

* **neophyte → 4091**

4096 neotectonics
The study of the structures and history of tectonic development of the Earth's crust since the Oligocene during the Neogene, which relates the crustal movements and their effect on the processes and landforms.

r неотектоника
Раздел тектоники, изучающий структуры и историю тектонического развития земной коры, начиная с олигоцена.
f néotectonique *f*
s neotectónica *f*
g Neotektonik *f*

4097 neptunism
The geological concept of the late XVIII century which considers that all rocks originated from waters of initial ocean, and their formation and transformation occured due to sea sedimentation.
r нептунизм
Геологическая концепция, господствовавшая с конца XVIII в. до 1820-х годов, основанная на представлениях о происхождении всех горных пород из вод первичного океана, о формировании и преобразовании их в результате морского осадконакопления.
f neptunisme *m*
s neptunismo *m*
g Neptunismus *m*

4098 neritic *adj*
Associated with shallow water, especially with shallow coastal water.
r неритический; сублиторальный
Связанный с мелководьем, особенно с прибрежными мелководьями, относящийся к глубинам до 200 м.
f néritique *adj*
s nerítico *adj*
g neritisch *adj*

4099 neritic fauna
Marine animals inhabiting coastal waters, mainly the shelf zone unlike animals living in the open sea.
r неритическая фауна
Морские животные, обитающие в прибрежных водах, главным образом над шельфом, в отличие от животных, обитающих в открытом океане.
f faune *f* néritique
s fauna *f* nerítica
g neritische Fauna *f*

4100 neritic zone; sublittoral zone; infralittoral zone; subtidal zone
A sea-shore zone lying immediately below the littoral (intertidal) zone and extending to a depth of about 200 m or to the edge of the continental shelf. → Fig. 244, p. 534
r неритическая зона; сублитораль; сублиторальная зона
Прибрежная часть океана с глубинами менее 200 м, расположенная между литоралью и краем материкового шельфа. → Рис. 244, с. 534
f étage *m* néritique
s piso *m* nerítico
g neritische Zone *f*

4101 net ablation
The loss of a glacier mass through the evaporation and melting with runoff.
r чистая абляция
Убыль вещества с ледника путем испарения и таяния со стоком.
f ablation *f* nette
s ablación *f* neta
g Nettoablation *f*

* **net radiation → 4866**

4102 network
Any set of interlinking lines (links or routes) that cross or meet one another (at nodes, junctions, terminals) in the manner of those in a net, e.g. a railway network, a grid. → Fig. 224
r сеть
Любой набор линий, будь то проводная связь или дороги, пересекающие или связывающие друг друга (в узлах, дорожных развязках или конечных пунктах) и образующие единую систему, например, железнодорожная сеть. → Рис. 224

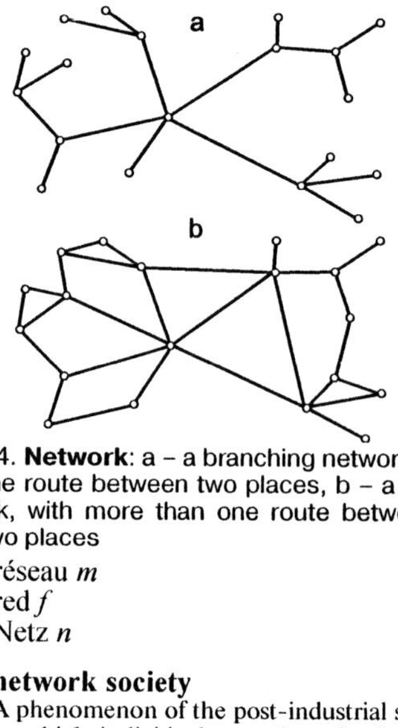

Fig. 224. **Network**: a – a branching network, with only one route between two places, b – a circuit network, with more than one route between at least two places
f réseau *m*
s red *f*
g Netz *n*

4103 network society
A phenomenon of the post-industrial society, in which individuals anywhere in the world who have access to the appropriate technology can communicate with each other, with institutions and with information services.
r коммуникационное общество
Феномен постиндустриального общества, в котором любой человек в мире, имеющий доступ к соответствующим технологиям, может связываться с любым другим индивидуумом, организацией или информационными службами.
f société *f* de communication
s sociedad *f* de comunicación
g Netzgesellschaft *f*

4104 neuston

Organisms inhabiting the surface film of the a water body.

r нейстон

Совокупность организмов, обитающих в поверхностной пленке водной среды водоема.

f neuston *m*

s neuston *m*

g Neuston *n*

4105 neutral shoreline

A coast which is not influenced neither by sinking or rising (typical examples are shorelines of coral reefs, dunes, deltas, marshes, mud-flats, sandur, volcanoes and those produced by faulting).

r нейтральный берег

Берег, основные черты которого не определяются ни его опусканием, ни поднятием (примерами могут служить береговые линии коралловых рифов, дюн, дельт, маршей, зандровых равнин, вулканов и др.).

f ñôte *f* neutre; trait *m* de ñôte neutre

s línea *f* de costa neutral

g neutraler Strand *m*

4106 neutral soil

A soil which neither alkaline nor acid, with a pH value of around 7.0.

r нейтральная почва

Почва, и не щелочная и не кислая, имеющая значение pH около 7,0.

f sol *m* neutre

s suelo *m* neutral

g Neutralboden *m*

4107 neutralism

The type of relationship between species in which they do not affect each other.

r нейтрализм

Тип взаимоотношений между видами, когда они в процессе совместного существования не оказывают влияния друг на друга.

f neutralisme *m*

s neutralismo *m*

g Neutralismus *m*

4108 neutrality

In international law, the status of a state which has a declared policy of noparticipation, i.e. not supporting either of two opposing sides in a war, disagreement, etc.

r нейтралитет

В международном праве, статус страны, декларирующей политику неучастия в военных и иных конфликтах других стран.

f neutralité *f*

s neutralidad *f*

g Neutralität *f*

* **nevé → 2154**

4109 new formation

An accumulation of mineral and organic substances greately different in composition and structure from the surrounding soil or rock.

r новообразование

Скопление минеральных и органических веществ, резко отличающееся по составу и структуре от вмещающей почвы или горной породы.

f néoformation *f*; néogenèse *f*

s neoformación *f*; neogénesis *f*

g Neubildung *f*

4110 new moon

The phase of the Moon when it occurs between the Sun and the Earth so that viewed from the Earth it is not illuminated by the Sun's light, although it may reflect a faint glow of light.

r новолуние

Фаза Луны, когда она располагается между Солнцем и Землей и невидима для земного наблюдателя, так как повернута к нему неосвещенным полушарием.

f nouvelle lune *f*

s luna *f* nueva; novilunio *m*

g Neumond *m*

4111 new settler

A person staying on a new territory for the first few years.

r новосел

Человек в течение первых лет после поселения на новом месте.

f arrivant *m*

s nuevo habitante *m*

g Neusiedler *m*

4112 new snow; newly fallen snow

Loose snow deposited during snowfalls and snow driftings, not accompanied by strong winds.

r свежевыпавший снег

Рыхлый снежный покров, отлагающийся при снегопадах и метелях, не сопровождающихся сильным ветром.

f neige *f* fraîche

s nieve *f* fresca

g gefallener Neuschnee *m*

* **new town → 5221**

4113 New World

General name of a part of the world including North and South America, especially at the time when Europeans first discovered these areas and knew little about them.

r Новый свет

Общее название части света, включающей Северную и Южную Америку.

f Nouveau Monde *m*

s Nuevo Mundo *m*

g Neue Welt *f*

* **newly fallen snow → 4112**

4114 newly industrialized country

A country once designated as less developed, but which has undergone recent rapid and successive industrialization, such as Hong-Kong, Singapore, Malaisia and South Korea, which achieved industrial growth in the 1970s and 1980.

r ныне развивающаяся страна

Еще недавно слаборазвитая страна, которая в настоящее время получила быстрое индустриальное развитие, как это случилось в Гонконге, Сингапуре, Малайзии и Южной Корее в 1970-80-х годах.

f pays *m* nouvellement industrialisé

s país *m* recién industrializado desarrollo

g heutzutage Entwicklungsland *n*

4115 newly-deposited snow

A relatively dense snow cover packed by moderate winds accompanied by snow drifting.

r свежеотложенный снег

Сравнительно плотный снежный покров, отлагающийся при ветрах средней силы, сопровождаемых метелью.

f neige *f* déposée récemment

s nieve *f* recién depositada

g frischabgelagerter Schnee *m*

4116 niche$_1$

A small hollow or shelf on a slope or at its foot.

r ниша

Незначительное углубление на склоне или у его подножья.

f niche$_1$ *f*

s nicho$_1$ *m*

g Nische$_1$ *f*

*** niche$_2$ → 1803**

4117 niche glacier; ural glacier

A small glacier extended on a narrow slope and located under a steep scarp.

r присклоновый ледник

Небольшой ледник, вытянутый в ширину на узкой поверхности склона под крутым уступом.

f glacier *m* de niche

s glaciar *m* de nicho

g Nischengletscher *m*

4118 night

The time of darkness during which the Sun is below the horizon, the opposite of day, the duration of which depends on the geographical latitude and season.

r ночь

Темная часть суток от конца вечерних сумерек до начала утренних, продолжительность ночи зависит от географической широты и времени года.

f nuit *f*

s noche *f*

g Nacht *f*

*** nightglow → 167**

4119 nilas

Thin ice on the surface of a fresh or salt water body which looks dark because of its transparency.

r нилас

Тонкий лед на поверхности пресной или соленой воды в водоеме, имеющий темный вид вследствие его прозрачности.

f nilas *f*

s nilas *f*

g Nilas *f*

4120 nimbostratus

A type of cloud (symbol: Ns), dark grey in colour, thick enough to blot out the Sun and which is associated with more or less continuously falling rain or snow reaching the ground.

r слоисто-дождевые облака

Серый облачный слой (символ: Ns), часто темный, кажущийся размытым из-за более или менее непрерывно выпадающего дождя или снега, которые, как правило, достигают земной поверхности.

f nimbostratus *m*

s nimbostratus *m*

g Nimbuswolken *fpl*; Nimbostratus; Nimbostratus-Wolken *fpl*

4121 nimbus

Any cloud from which rain is falling.

r дождевые облака

Любое облако, из которого идет дождь.

f nimbus *m*

s nimbus *m*

g Nimbus *m*

4122 nitosol

A dark-red or brown coloured soil with poorly pronounced clay transportation, lacking both bleached and dark humus horizons, with well-pronounced nutty structure.

r нитосоль

Почва темно-красного или бурого цвета со слабо выраженными признаками перемещения глины без осветленного и темного гумусового горизонта, с ярко выраженной ореховатой структурой.

f nitosol *m*

s nitosol *m*

g Nitosol *m*

4123 nitrate

A natural salt of nitric acid, well dissoluble in water, and widely spread in soils, which is formed by a biogenic oxydation of the atmospheric nitrogen, used for improving the quality of soil as fertilizer.

r нитрат

Природная соль азотной кислоты, хорошо растворимая в воде и широко представленная в почвах, образуется биогенным путем и в результате окисления азота атмосферы, используется как удобрение.

f nitrate *m*

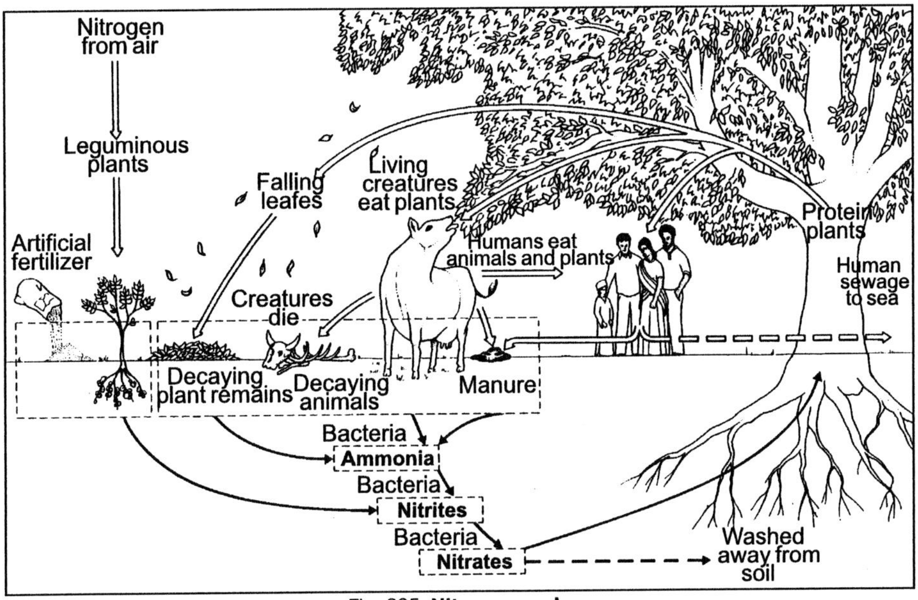

Fig. 225. **Nitrogen cycle**

s nitrato *m*
g Nitrat *n*

4124 nitratophile
Organisms thriving in the environment rich in nitrogen compounds.
r нитрофил
Организмы, предпочитающие среду, богатую соединениями азота.
f nitratophile *m*; nitrophile *m*
s nitrófilo *m*; nitratófilo *m*
g Nitrophil *m*

4125 nitrification
The transformation by aerobic soil bacteria of organic compounds of nitrogen, which cannot be absorbed by plants, into nitrates usable by green plants.
r нитрификация
Процесс превращения почвенными аэробными бактериями азотных органических веществ, которые не могут абсорбироваться растениями, в нитраты, доступные для питания зеленых растений.
f nitrification *f*
s nitrificación *f*
g Nitrifikation *f*

4126 nitrogen
The most abundant gas of the atmosphere and a very important constituent of soil, from which it is utilized by certain bacteria.
r азот
Наиболее распространенный газ в атмосфере и очень важный компонент почвы, который усваивается определенными бактериями.
f azote *m*
s nitrógeno *m*
g Stickstoff *m*

4127 nitrogen cycle
The sequence of biochemical changes by which nitrogen is taken from the atmosphere into the soil by nitrogen-fixing bacteria, making it readily available for assimilation by living organisms which at the end of their lifespan die and are decomposed, thereby returning nitrogen to the atmosphere. → Fig. 225
r азотный цикл
Последовательные биохимические изменения, в процессе которых азот попадает из атмосферы в почву и с помощью азотфиксирующих бактерий становится доступным для его ассимиляции живыми организмами, которые в конце своего жизненного цикла умирают и разлагаются, а азот возвращается в атмосферу. → Рис. 225
f cycle *m* de l'azote
s ciclo *m* de nitrógeno
g Stickstoffzyklus *m*

4128 nitrogen fixation
The process by which atmospheric nitrogen is converted into organic nitrogen compounds by some blue-green algae and by nitrogen-fixing bacteria living in the soil.
r фиксация фзота
Процесс, при котором атмосферный азот превращается в органические азотные соединения с помощью сине-зеленых водорослей и азот-фиксирующих бактерий, живущих в почве.
f fixation *f* de l'azote

s fijación *f* de nitrógeno
g Stickstoffixierung *f*

* **nitrophile** ***m*** **→ 4124**

4129 nival ***adj***
Related to snow cover in origin or natural processes, e.g. nival climate, nival belt.
r нивальный
Связанный со снежным покровом происхождением или природными процессами, например, нивальный климат, нивальный пояс.
f nival *adj*
s nival *adj*
g nival *adj*

4130 nival belt
An altitudinal mountain belt above the climatic snow line, for which snow-patches, glaciers and intensive physical weathering are typical.
r нивальный пояс;
нивально-гляциальный пояс
Высотный пояс гор, расположенный выше климатической снеговой линии, для которого характерны снежники, ледники и интенсивное физическое выветривание.
f étage *m* nival
s piso *m* nival
g nivaler Berggürtel *m*; nivale Stufe *f*

4131 nival region
Area where glaciers, permanent snow, or perennial sea ice occur.
r нивальная область
Область распространения ледников и многолетних снежников, а также многолетних морских льдов.
f zone *f* nivale; chionozone *f*
s zona *f* nival
g nivale Zone *f*

4132 nivation
The effects produced by snow, mostly by permanent snow patches, on the underlying rocks, forming specific landforms and deposits.
r нивация
Воздействие снега и прежде всего снежников на подстилающие горные породы, формирующее специфические формы рельефа и отложений.
f nivation *f*
s nivación *f*
g Nivation *f*; subnivale Denudation *f*

4133 nivation cirque
A large nivation hollow, approaching the scale of a true cirque and possessing its semicircular plan, but lacking its basin- or armchair-shaped form or its rocky «threshold».
r нивальный цирк
Крупная нивальная впадина, достигшая величины настоящего ледникового цирка и полукруглая в плане, но не имеющая кресловидной формы и скального порога.
f cirque *m* de nivation
s circo *m* de nivación
g nivaler Zirkus *m*

4134 noctilucent clouds
Rare clouds, at height between 75 and 90 km, resembling thin Cirrus but usually with a bluish or silverly or sometimes orange to red colour, against the dark night sky, these clouds may be formed of ice crystals or meteoric dust.
r серебристые облака
Облака на высотах 75-90 км, напоминающие перистые, имеющие голубоватую или серебристую окраску, иногда оранжево-красные, они могут формироваться из ледяных кристаллов и метеорной пыли.
f nuages *mpl* nocturnes lumineux
s nubes *fpl* noctilucentes
g nachtleuchtende Wolken *fpl*

4135 nocturnal ***adj***
Relating to night or happening at night (nocturnal animals are active at night rather than during the day).
r ночной
Относящийся к ночному времени (ночные животные активны главным образом ночью).
f nocturne *adj*
s nocturno *adj*
g nächtlich *adj*; Nachts-

4136 nocturnal cooling
The lowering of air temperature caused by the intense effective radiation.
r ночное выхолаживание
Понижение температуры ночью вследствие эффективного излучения.
f refroidissement *m* nocturne
s enfriamiento *m* nocturno
g nächtliche Abkühlung *f*

4137 nocturnal radiation
Long-wave terrestrial radiation at night emitted by the Earth's surface in the absence of solar radiation.
r ночное излучение
Длинноволновое земное излучение в ночное время суток, в отсутствии солнечной радиации.
f radiation *f* nocturne
s radiación *f* nocturna
g nächtliche Ausstrahlung *f*

4138 nodal region
An economic region in which the concentration of economy diminishes from the centre toward periphery.
r узловой район
Экономический район, на территории которого концентрация хозяйства убывает от центра к периферии.
f région *f* nodale

s región *f* nodal
g nodale Region *f*

4139 node
In a network, the point at which routes (edges, links) meet.
r узел
Точка в сети, где встречаются или пересекаются различные связи (коммуникации, дороги и пр.).
f noeud *m*
s nudo *m*
g Knoten *m*

4140 node of seiche
A line of no fluctuations of water level during seiches.
r узел сейши
Линия, вдоль которой отсутствуют колебания уровня воды при сейшах.
f noeud *m* de seiche
s nudo *m* de seiche
g Knoten *m* der Seiche

* **nodule → 1187**

* **noise nuisance → 4141**

4141 noise pollution; noise nuisance
Unpleasant or loud sounds which cause discomfort, annoyance and disturb the hearing.
r шумовое загрязнение
Посторонние звуки в окружающей среде, вызывающие у человека неприятные ощущения или даже разрушения органов слуха.
f pollution *f* bruyante; nuisance *f* sonore; contamination *f* bruyante; pollution *f* sonore
s polución *f* fónica; contaminación *f* fónica; ruido *m*
g Lärmverschmutzung *f*

4142 nomad
Someone who belongs to a group of people moving from place to place, usually seasonally within a well-defined territory in order to find food and water for their grazing animals or themselves.
r кочевник
Представитель народа, передвигающегося в определенные сезоны с места на место в пределах определенной территории в поисках средств к существованию, особенно пастбищ для скота.
f nomade *m*
s nómada *m*; nómade *m*
g Nomade *m*

4143 nomadic *adj*
Related to a type of human life style where cattle-breeders constantly move to change pastures.
r кочевой
Относящийся к образу жизни скотоводов, основанному на миграциях по мере смены пастбищ.
f nomade *adj*
s nómada *adj*; nómade *adj*
g Nomaden-; Wander-; nomadisch *adj*

4144 nomadic birds
The birds migrating from nesting grounds for relatively short distances.
r кочующие птицы
Птицы, откочевывающие от места гнездовья на относительно небольшие расстояния и не совершающие дальних перелетов.
f oiseaux *mpl* nomades; oiseaux *mpl* erratiques
s aves *fpl* nómadas
g Nomadenvögel *mpl*

4145 nomadic camp
A place where nomadic people stay at any given moment of time.
r кочевье
Место пребывания кочевых скотоводов в данный момент времени.
f campement *m* nomade
s campamento *m* de nómadas
g Nomadenlager *n*

4146 nomadism
A type of human life where people and animals move from place to place in search of pasture following seasonal climate fluctuations in the steppes, semi-deserts and deserts, using regular routes.
r кочевничество; номадизм; кочевой образ жизни
Систематическое перемещение населения по мере истощения пастбищ, как правило, вместе со скотом, в степях, полупустынях и пустынях по определенным путям, связанным с сезонными изменениями климата.
f nomadisme *m*; vie *f* nomade
s nomadismo *m*
g Nomadenleben *n*; Nomadismus *m*; Wanderhirtentum *n*

4147 nomogram; graphical chart; calculating chart
Graphical image in the form of straight or curve lines of functional relationship between several variable values, used in the map making.
r номограмма
Графическое изображение в виде прямых или кривых линий функциональных связей между несколькими изменяющимися величинами, применяемое при картографировании.
f nomogramme *m*; abaque *m*
s nomograma *m*; ábaco *m* del nomógrafo
g Nomogramm *n*; nomographische Rechentafel *f*

* **non-adiabatic *adj* → 1575**

* **non-adiabatic process → 1576**

4148 nonaligned *adj*

Applied to a state that does not support the policies of either of two opposed groups of powers, e.g. historically of NATO or of the Warsaw Pact countries.

r неприсоединившийся
Относится к странам, не поддерживающим политику одного из двух противостоящих блоков, например, НАТО и Варшавского договора в прошлом.

f non-aligné *adj*
s no alineado *adj*
g nichtpaktgebunden *adj*

* **non-channeled avalanche → 6443**

4149 non-channeled runoff hollows
Small hollows up to 2 m wide, very closely spaced and well developed in areas with heavy rainfall.

r безрусловые ложбины стока
Плоские, слабовыраженные, линейно вытянутые понижения рельефа без постоянного водотока и следов свежего эрозионного размыва.

f filet *m* de ruissellement
s torrentera *f*
g Abflußvertiefungen *fpl*

4150 non-cryotic ground
Soil or rock at temperatures above 0°C, this term is not synonymous with seasonally thawed layer (which implies an earlier frozen state), nor with unfrozen ground (the temperature of which may be below 0°C).

r положительно-температурный грунт
Почва или грунт при температуре выше 0°C, не является синонимом сезонноталого слоя или не мерзлого грунта, температура которого может быть ниже 0°C.

f sol *m* non cryotique
s terreno *m* no criótico
g akryotischer Boden *m*

4151 non-equivalent exchange
A form of trade when the price of a product differs significantly from its cost and production cost.

r неэквивалентный обмен
Форма товарообмена, при которой цена товара существенно отличается от его стоимости и от цены производства.

f échange *m* inégal
s cambio *m* desigual
g nichtäquivalenter Austausch *m*

4152 non-ferrous metallurgy
A branch of heavy industry engaged in producing non-ferrous metals and their alloys.

r цветная металлургия
Отрасль тяжелой промышленности, занимающаяся производством цветных металлов и их сплавов.

f métallurgie *f* nonferreuse
s metalurgia *f* no ferrosa
g Nicht-Eisen-Metallurgie *f*; NE-Metallurgie *f*; Nichteisenmetallurgie *f*

* **non-frozen ground → 6458**

4153 non-gravity spring
A natural spring having the ascending flow of water under its own pressure.

r восходящий источник
Естественный источник с восходящим движением воды под напором.

f source *f* ascendante
s manantial *m* de emersión
g aufsteigende Quelle *f*

* **non-living environment → 3**

4154 non-paired terrace
A river terrace whose remnants on opposite sides of a valley do not correspond in altitude along a particular stretch of the valley. → Fig. 239, p. 523

r непарная терраса
Речная терраса, остатки которой на противоположных стронах конкретного участка долины не соответствуют друг другу по высоте. → Рис. 239, с. 523

f terrasse *f* formant l'impaire
s terraza *f* impareada
g unpaarige Terrasse *f*

4155 nonperiodic current; nontidal current
A horizontal ocean current produced by different forces, except the tide, i.e. by the wind, differences in water density, and pressure, etc.

r непериодическое морское течение
Горизонтальное движение вод, вызываемое различными силами, кроме приливообразующих, т.е. силой ветра, разностью плотности воды, давления и т.п.

f courant *m* instable; courant *m* non-périodique
s corriente *f* inestable
g instabile Strömung *f*; nichtperiodische Strömung *f*

4156 non-renewable natural resources
Natural resources such as coal, oil or gas which can not be replaced if they are consumed, as opposed to renewable resources.

r невозобновляемые природные ресурсы
Природные ресурсы, не способные к самовосстановлению естественным путем, например, нефть, газ, каменный уголь, в отличие от возобновляемых природных ресурсов.

f ressource *f* naturelle non-renouvelable
s recursos *mpl* naturales no renovables
g nicht-erneuerbare Resourcen *fpl*

* **nontidal current → 4155**

4157 non-washing regime of soil
The moistening which does not exceed the moisture capacity of a soil, as a result of which the drainage of soil water into lower soil horizons is reduced or is completely absent.

r непромывной режим почвы

Слабое поступление воды в почву, не приводящее к превышению ее влагоемкости, вследствие чего сток почвенной воды в нижележащие почвенные горизонты или грунт невелик или полностью отсутствует.
f régime *m* du sol sans lessivage
s régimen *m* del suelo sin lavado
g Bodennichtspülenregime *n*; durchflussloses Regime *n* des Bodens

* **non-waste technology → 6634**

4158 noon
The middle of the day, the moment of the upper culmination of the Sun when it passes the highest point of its diurnal way in the sky with regard to the horizon.
r полдень
Момент верхней кульминации Солнца, когда оно проходит относительно горизонта наивысшую точку своего суточного пути по небу.
f midi *m*
s mediodía *f*
g Mittag *m*

4159 noosphere
A new state of the biosphere when cognitive judicious human activity becomes the main factor of its evolution.
r ноосфера
Новое эволюционное состояние биосферы, при котором разумная деятельность человека становится решающим фактором ее развития.
f noosphère *f*
s noosfera *f*
g Noosphäre *f*

4160 nor
A water body of different origin in Southern Siberia, Middle and Central Asia.
r нур; нор
Водоемы различного происхождения в Южной Сибири, Средней и Центральной Азии.
f nor *m*
s nor *m*
g Nor *n*

4161 Nordic projection
A map projection designed to give optimum representation to Europe and to routes across the Atlantic, Arctic and Indian oceans, it is a modified oblique, equal-area projection with an origin on the Greenwich meridian at 45° N latitude. → Fig. 226
r нордическая проекция
Картографическая проекция, дающая оптимальное представление Европы и основных торговых путей в Атлантическом, Северном Ледовитом и Индийском океанах, это видоизмененная косая, равновеликая проекция, полюс которой расположен на пересечении Гринвичского меридиана и параллели 45° с.ш. → Рис. 226
f projection *f* Nordique
s proyección *f* nórdica
g nordische Projektion *f*

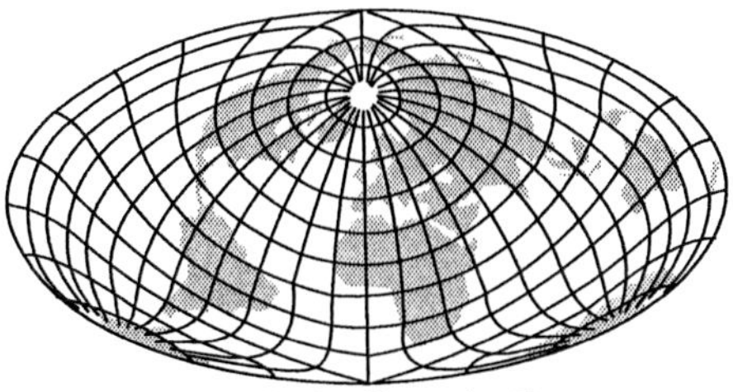

Fig. 226. **Nordic projection**

4162 normal fault; throw
A tectonic displacement where rocks have moved down the slope of the fault. → Fig. 113, p. 250
r сброс
Разрывное тектоническое смещение горных пород по наклонной плоскости, азимут падения которой направлен в сторону лежачего крыла складки. → Рис. 113, с. 250
f faille *f* normale; faille *f* directe
s falla *f* normal; falla *f* directa
g Verwerfung *f*; Abschiebung *f*

4163 normal level; standard level
An average water level in rivers, bays and certain points of sea coast, calculated from a series of continuous records over a long period.
r ординар
Среднее (за много лет) положение уровня воды в реках, заливах и отдельных пунктах морского побережья.
f niveau *m* normal
s nivel *m* de referencia; nivel *m* cero
g Normalwasserstand *m*

4164 normal projection
A map projection where a normal system of polar spherical coordinates is applied, where the projection origin coincides with the geographical pole.
r нормальная проекция
Картографическая проекция, получаемая в случае применения нормальной системы полярных сферических координат, полюс которой совмещен с географическим полюсом.
f projection *f* directe
s proyección *f* directa
g normalachsiger Kartennetzentwurf *m*

4165 normal reservoir level
The highest level at which the dam of a hydroelectric scheme can keep the water level over a very long period.
r нормальный подпорный уровень; проектный уровень водохранилища
Наивысшая отметка, на которой плотина гидроузла может поддерживать уровень воды в течение сколь угодно длительного периода.

f niveau *m* normal de remous
s nivel *m* normal de retención
g Normalstand *m*;
Normalniveau *n* eines Stausees

4166 normal runoff
The mean value of the surface water runoff over a period of many years.
r норма стока
Среднее значение величины поверхностного водного стока за многолетний период.
f écoulement *m* moyen
s escorrentía *f* media
g mittlerer Jahresabfluß *m*; Normalabfluß *m*

4167 normal watershed
A watershed in a montainous region which runs along the crest of the highest range of a mountain chain.
r нормальный водораздел
Водораздел в горном районе, протягивающийся вдоль гребня высочайшего хребта в горной цепи.
f ligne *f* normale de partage des eaux
s línea *f* normal de división de aguas
g normale Wasserscheide *f*

4168 north
One of the four cardinal points, directly opposite the south, lying on the left side of a person facing the east; the northern part, especially of a country.
r север
Одна из четырех главных сторон горизонта, прямо противоположная югу, расположенная слева от наблюдателя, стоящего лицом к востоку.
f nord *m*
s norte *m*
g Norden *m*; Nord *m*

4169 north *adj*; northern *adj*
Relating to, belonging to, situated towards, coming from the north, e.g. of winds blowing from the north.
r северный
Относящийся, расположенный или приходящий с севера, например, ветер, дующий с этого направления.
f du nord; septentrional *adj*; boréal *adj*
s del norte; boreal *adj*; septentrional *adj*; norteño *adj*
g Nord-; nördlich *adj*

4170 North Atlantic oscillation
The dominant pattern of the atmospheric circulation over the Atlantic sector of the Northern Hemisphere (symbol:NAO) expressed by the average pressure difference between the Azores High and Iceland Low.
r северо-атлантическое колебание
Доминирующий механизм атмосферной циркуляции в атлантическом секторе Северного полушария, выраженный стандартизованной разностью давления между азорским максимумом и исландским минимумом.
f oscillation *f* nord-atlantique
s oscilación *f* del Atlántico Norte
g nordatlantische Oszillation *f*

4171 North Pole
The point of crossing of the Earth's rotation axis with the Earth's surface in the Northern Hemisphere.
r Северный полюс
Точка пересечения оси вращения Земли с земной поверхностью в Северном полушарии.
f pôle *m* nord; pôle *m* boreal
s polo *m* norte; polo *m* boreal
g Nordpol *m*

4172 north side
Cold and dry northern or north-eastern wind in mountain regions of France and Switzerland similar to *mistral.*
r бизе; биза
Холодный и сухой северный или северо-восточный ветер в горных районах Франции и Швейцарии, сходный с *мистралем.*
f bise *m*
s bise *m*
g Bise *f*

* **North-East trade wind → 4173**

* **northeast trades → 6281**

4173 north-east trades;
North-East trade wind
A constant north-east wind of the tropics in the Northern Hemisphere.
r северо-восточный пассат
Постоянный северо-восточный ветер тропиков Северного полушария.
f alizé *m* du nord-est
s alisio *m* del nordeste
g Nordostpassat *m*

* **northern *adj* → 4169**

4174 Northern Hemisphere
The half of the Earth north of the equator.
r Северное полушарие
Половина Земли к северу от экватора.
f hémisphère *m* boréal
s hemisferio *m* norte; hemisferio *m* boreal
g nördliche Halbkugel *f*

4175 northing
A linear distance northwards from the east-west grid line on a map which passes through the true origin of a grid.
r абсцисса
В картографии координата, отсчитываемая по направлению осевого меридиана.
f abscisse *f*
s abscisa *f*
g Abszisse *f*

4176 nosogeography
A branch of *medical geography* studying the distribution of human and animal diseases in different geographical zones.
r нозогеография
Раздел *медицинской географии*, изучающий распространение болезней человека и животных в разных географических зонах.
f nosogéographie *f*
s nosogeografía *f*
g Nosogeographie *f*

4177 nosorange
Areas inhabited by humans which are or have recently been active centres of some diseases.
r нозоареал
Совокупность территорий, населяемых человеком, на которых имеются или были в недалеком прошлом активные очаги той или иной болезни.
f nosoaire
s nosoárea *f*
g Nosoareal *n*

4178 notch
A wave-cut hollow at the base of a marine cliff created by upwash. → Fig. 1, p. 4
r волноприбойная ниша
Углубление в основании берегового уступа (клифа), выработанное прибойным потоком. → Рис. 1, с. 4
f encoche *f*
s nicho *m* de oleaje; entalladura *f*
g Brandungskehle *f*; Brandungshöhle *f*

* **nuclear energy → 4179**

4179 nuclear power; nuclear energy
In national economy, a branch of power engineering that uses the energy of nuclear chain reaction as a source of energy; a form of energy which uses nuclear reactions to produce steam to turn generators.
r атомная энергетика; атомная энергия; ядерная энергетика
Отрасль народного хозяйства, использующая энергию цепной ядерной реакции как источник энергии; особая форма энергии, использующая ядерную реакцию для вращения генераторов и получения электроэнергии.
f énergie *f* nucléaire
s energía *f* nuclear
g Kernenergiewirtschaft *f*; Atomenergiewirtschaft *f*

4180 nuclear power station
A power station where atomic (nuclear) energyis converted into electrical energy.
r атомная электростанция; АЭС
Электростанция, на которой атомная (ядерная) энергия преобразуется в электрическую.
f centrale *f* nucléaire
s central *f* nuclear
g Kernkraftwerk *n*

* **nuclear waste → 4873**

4181 nuclear winter
A climatic anomaly with severe wintry conditions and catastrophic effect on ecosystems, which it is thought might occur if sufficient nuclear explosion on Earth created enough dust in the atmosphere to cut off the Sunrs radiation from the Earthrs surface.
r ядерная зима
Климатическая аномалия очень суровых условий, которая, как полагают, может произойти в случае крупного ядерного взрыва на Земле, поднимающего массу пыли в атмосферу, способную затмить Солнце и тем самым резко понизить температуру на Земле.
f hiver *m* nucléaire
s invierno *m* nuclear
g nuklearer Winter *m*

4182 nucleated settlement
A rural settlement comprising a cluster of dwellings, in contrast with a *dispersed settlement.*
r компактное поселение
Сельское поселение с тесным расположением жилых домов, в противоположность *рассредоточенному поселению.*
f peuplement *m* compact
s poblamiento *m* compacto
g dichte Siedlung *f*

4183 nuée ardente
A glowing dense mass of volcanic gases and fragments of solid, partially melted volcanic material forming high hot clouds moving at high speed and overwhelming the surrounding area with devastating effects.
r раскаленная туча
Подвижная взвесь из вулканических газов и обломков твердого, частично расплавленного вулканического материала, образующая высокие горячие тучи, движущиеся с большой скоростью.
f nuée *f* ardente
s nube *f* ardiente
g Glutwolke *f*

4184 nuisance
Natural or man-made phenomenon in the environment which causes discomfort and annoyance.
r раздражитель
Природное или техногенное явление в окружающей среде, вызывающее неудобства или неприятные ощущения у людей.
f nuisance *f*; irritant *m*; incommodité *f*
s molestía *f*; incomodidad *f*
g Belästigung *f*

* **number (in a scale) → 3706**

4185 numerical forecast

Fig. 227. **Nunatak** at a mountain glacier

Forecast of the fields of meteorological elements by numerical solution of the hydrodynamical equations.

r численный прогноз
Прогноз полей метеорологических элементов на основе численного решения гидродинамических уравнений.
f prévision *f* numérique
s predicción *f* numérica
g numerische Wetterprognose *f*

4186 numerical modelling of climate
The description of climatic processes with a set of mathematical equations with corresponding boundary conditions and algorithm for their numerical solution.
r численное моделирование климата
Отображение климатических процессов серией математических уравнений с соответствующими краевыми условиями и с алгоритмом их численного решения.
f moulage *m* numérique de climat
s modelación *f* numérica de clima
g numerische Klimamodellierung *f*

4187 numerical scale
The scale of a geographical map or chart expressed as a fraction or ratio the numerator of which equals one.
r численный масштаб
Масштаб, указанный на географической карте в виде дроби, числитель которой – единица.
f échelle *f* numérique
s escala *f* numérica
g numerischer Maßstab *m*

4188 nunatak
An isolated peak or rock rising above a glacier or ice sheet and may never have been glaciated. → Fig. 227
r нунатак
Изолированный скалистый пик, горный гребень или холм, выступающий над ледниковой поверхностью. → Рис. 227
f nunatak *m*
s nunatak *m*
g Nunatak *m*

4189 nutation
A small (up to 9.2″) oscillation of the Earth's axis with the period of about 18.6 years determined mainly by the non-spherical shape of the Earth and change of mutual situation of planes of the Earth's equator, ecliptic and the Moon's orbit.
r нутация
Небольшие (до 9,2″) колебания оси вращения Земли с периодом около 18,6 года, обусловленные в основном несферичностью Земли и изменением взаимного положения плоскостей земного экватора, эклиптики и лунной орбиты.
f nutation *f*
s nutación *f*
g Nutation *f*

4190 nutrient cycle; cycle of nutrients
The process in which nutriens from living organisms are transferred into the physical environment and back to the organisms through food chains.
r биологический круговорот; биотический круговорот
Природный процесс, при котором питательные вещества от живых организмов переносятся в окружающую среду и возвращаются обратно к организмам через пищевые цепи.
f cycle *m* des nutriments
s ciclo *m* de nutrientes
g biologischer Kreislauf *m*; Nährstoffkreislauf *m*

O

4191 oak forest; oak grove
Deciduous broad-leaved forest consisting of deciduous species of oak widespread in Eurasia and North America.
r дубовый лес; дубрава; дубняк
Летнезеленый широколиственный лес, состоящий из листопадных видов дуба, широко распространен в Евразии и Северной Америке.
f chênaie *f*
s bosque *m* de roble; robledal *m*
g Eichenwald *m*; Eichenhain *m*

* **oak grove → 4191**

4192 oasis
An area in a hot desert where the presence of surface or ground water permits to support plant growth.
r оазис
Место в жаркой пустыне, где благодаря присутствию поверхностых или подземных вод произрастает древесная, кустарниковая и травянистая растительность.
f oasis *f*
s oasis *m*
g Oase *f*

4193 obduction
The process of incorporation of the oceanic crust to the edge of a continental plate.
r обдукция
Надвигание участка океанической коры на край континентальной плиты.
f obduction *f*
s obducción *f*
g Obduktion *f*

4194 oblate *adj*
Relating to a spheroid, flattened at the poles.
r сплюснутый; сжатый
Относящийся к сфероиду, сплюснутый у полюсов.
f aplati *adj*
s aplastado *adj*; achatado *adj*; aplanado *adj*
g abgeplattet *adj*

4195 oblateness of the Earth
The degree of flattening of the Earth expressed as the ratio of the difference between the equatorial and polar radiuses (diameters) of the Earth to the radius (diameter) of the equator.
r сжатие Земли
Отношение разности экваториального и полярного радиусов (диаметров) Земли к радиусу (диаметру) экватора, характеризует отклонение фигуры Земли, принимаемой за эллипсоид вращения, от шарообразной формы.
f aplatissement *m* terrestre
s achatamiento *m* terrestre; aplanamiento *m* terrestre
g Erdabplattung *f*

4196 oblique projection
A map projection where an oblique system of polar spherical coordinates is applied, the origin of which lies between the geographical pole and the equator.
r косая проекция
Картографическая проекция, получаемая в случае, когда применяется косая система полярных сферических координат, полюс которой расположен между географическим полюсом и экватором.
f projection *f* oblique
s proyección *f* oblicua
g schiefachsiger Kartennetzentwurf *m*

4197 obliquity of the ecliptic
The angle between the ecliptic plane and the plane of the celestial equator, equal to 23°77′.
r наклонение эклиптики
Угол между плоскостью эклиптики и плоскостью небесного экватора, равный 23°77′.
f obliquité *f* d'écliptique
s oblicuidad *f* de la eclíptica; inclinación *f* de la eclíptica
g Neigung *f* der Ekliptik

* **obsequent stream → 4198**

4198 obsequent stream; obsequent valley
A small affluent of a subsequent river flowing in the valley running in the direction opposite to the flow of a consequent river or opposite to that of the dip of the rock strata. → Fig. 68, p. 145
r обсеквентная река; обсеквентная долина
Небольшой приток субсеквентной реки, текущий в долине, направленной противоположно течению консеквентной реки, т.е. в обратном направлении по отношению к падению геологических пластов. → Рис. 68, с. 145
f fleuve *m* obséquent; vallée *f* obséquente
s río *m* obsecuente; valle *m* obsecuente
g obsequenter Fluß *m*; obsequentes Tal *n*

* **obsequent valley → 4198**

4199 obstruction; blockage
A heap of unsorted debris appearing as a result of rock fall.
r завал
Груда несортированного обломочного материала, возникающая в результате обвала.
f bourrelet *m* détritique
s obstrucción *f*; escombro *m*
g Verschüttung *f*

4200 occasional rain
Rain which falls occasionally and lasts for some minutes.

r кратковременный дождь
Дождь, выпадающий в течение нескольких минут.
f averse *m*
s chubasco *m*
g kurzdauernder Regen *m*

4201 Occident
Originally Europe as opposed to Asia (the Orient), now extended to the part of the world peopled by Europeans or by those of European descent.
r Запад; западные страны
Первоначально Европа, противопоставляемая Азии, ныне это часть мира, населенная европейцами или выходцами из Европы.
f occident *m*
s occidente *m*
g Okzident *m*

* **occidental *adj* → 6752**

4202 occluded front; frontal occlusion
The front separating two cold air masses which are brought unto contact as a result of the process of occlusion, i.e. when a warm front is lifted and overtaken by a cold front.
r фронт окклюзии
Фронт, разделяющий две холодные воздушные массы, соединенние которых произошло в результате процесса окклюзии, т.е. когда теплый фронт поднимается и натекает на холодный фронт.
f front *m* occlusif; occlusion *f* frontale
s frente *m* ocluido; oclusión *f* frontal
g okkludierte Front *f*

4203 occlusion
The process of progressive decrease of area of warm sector at the Earth's surface, and its ultimate dissapearance, by the junction of the cold air masses which initially precede the warm front and follow the cold front, as well as front. → Fig. 228

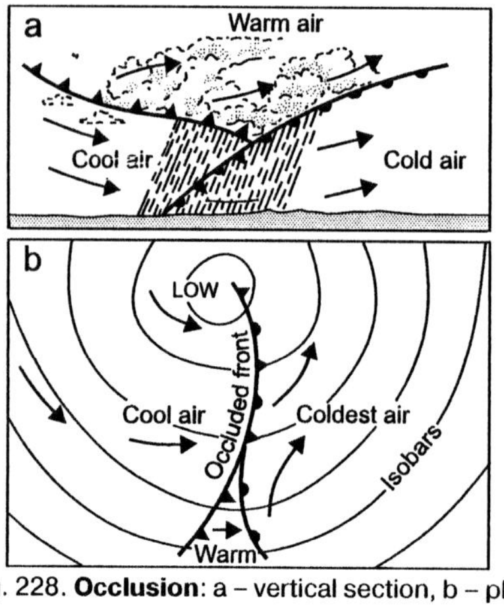

Fig. 228. **Occlusion**: a – vertical section, b – plan

r окклюзия циклона
Процесс последовательного сокращения теплого сектора циклона у земной поверхности и его окончательное вытеснение вверх из-за смыкания холодной воздушной массы, предшествующей теплому фронту, с тыловой, следующей за холодным фронтом воздушной массой. → Рис. 228
f occlusion *f*
s oclusión *f*
g Okklusion *f*

4204 occultation
The impediment for the light to penetrate into water of lakes and seas created by nontransparent hard particles suspended in the water which cause the reflection and absorption of the light rays.
r оккультация
Препятствие проникновению света в воду озер и морей, создаваемое непрозрачными взвешенными в воде твердыми частицами, обуславливающими отражение и поглощение световых лучей.
f occultation *f*
s ocultación *f*
g Okkultation *f*

4205 ocean
The mass of water which fills the ocean basins, i.e. Atlantic, Pacific, Indian, Arctic and Southern oceans.
r океан
Масса воды, заполняющая океанические бассейны, т.е. Атлантический, Тихий, Индийский, Северный Ледовитый и Южный океаны.
f océan *m*
s océano *m*
g Ozean *m*

4206 ocean basin
The deep-water bottom surface mainly located between continental slope and mid-oceanic ridges.
r океаническая котловина
Глубоководная область океанского ложа между материковым подножием и срединно-океаническими хребтами.
f dépression *f* océanique; bassin *m* océanique; bassin *m* maritime; cuvette *f* océanique
s depresión *f* oceánica; cuenca *f* marítima; cuenca *f* oceánica
g ozeanischer Kessel *m*; ozeanisches Becken *n*

4207 ocean circulation
Unified system of main continuous ocean currents determining the general mass and heat transport in the ocean.
r циркуляция вод Мирового океана
Единая система основных устойчивых течений океана, обуславливающая общий перенос и взаимодействие вод.

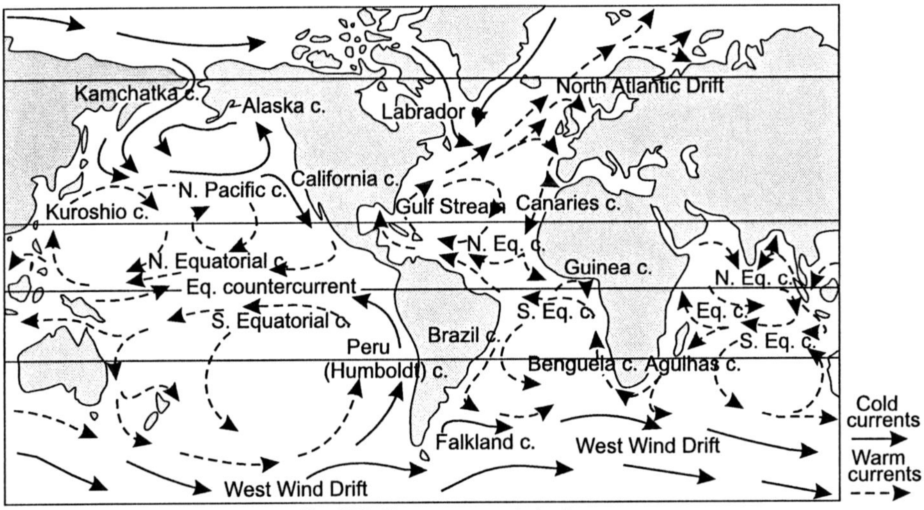

Fig. 229. **Ocean currents** in January

f circulation *f* océanique
s circulación *f* oceánica
g Zirkulation *f* des Weltozeans

4208 ocean current; sea current
A large-scale water movement in the oceans characterized by a regular or definite path, arising from wind stress acting on the surface of the sea, tidal motion and density differencies in sea water caused by the differential heating and cooling, and salinity differencies. → Fig. 229
r морское течение; океанское течение
Регулярное движение воды в океане или море, имеющее циклический или постоянный характер потока, двигающегося вдоль определенной траектории, вызывается воздействием ветра на водную поверхность, приливно-отливными движениями и различиями в плотности воды, связанными с неодинаковым нагреванием и охлаждением воды и разной ее соленостью. → Рис. 229
f courant *m* marin
s corriente *f* marina
g Meeresströmung *f*; ozeanische Strömung *f*

4209 ocean deep; abyss; abyssal basin
A large depression of an ocean floor with depths below 4-6 km, its bottom is flat or slightly hilly.
r глубоководная впадина; глубоководная океаническая котловина
Обширное понижение в пределах ложа океана глубиной 4000-6000 м и более с ровным или холмистым дном.
f dépression *f* abyssale; bassin *m* abyssale
s depresión *f* abismal; cuenca *f* abisal
g Tiefseebecken *n*

4210 ocean dumping
Discharging waste (solid, liquid or radioactive) into the ocean.
r океанский дампинг
Сброс и захоронение отходов в океане.
f dumping *m* océanique; immersion *f* des déchets
s dumping *m* oceánico
g Verklappung *f*

4211 ocean eddy
Complex ring- and ellipse-like horizontal water flow formed as a result of sea current meandering.
r океанский вихрь
Сложное кольце- и эллипсообразное перемещение воды в горизонтальной плоскости, образующееся в результате меандрирования течения.
f tourbillon *m* océanique
s torbellino *m* oceánico
g Ozeanwirbel *m*

4212 ocean floor
The entire sea-bed below low-water mark, i.e. the continental shelf, continental slope and ocean basins, it includes the neritic zone, the bathyal zone and the abyssal zone.
r дно океана; ложе океана
Вся поверхность, на которой лежит толща океанской воды, т.е. континентальный шельф, материковый склон и океанические котловины, дно океана включает неритическую, батиальную и абиссальную зоны.
f fond *m* océanique; plancher *m* des océans; fond *m* sous-marin
s fondo *m* océanico; fondo *m* del océano; suelo *m* de los océanos
g Ozeanboden *m*

* **ocean thermal expansion → 6157**

* **ocean trench → 1458**

4213 **oceanarium**
A large marine aquarium, a water pool with sea water or isolated part of a coastal bay intended for sea animals and to show them to people.
r океанариум
Огромный морской аквариум, наполненный морской водой, или изолированная часть морской бухты, предназначенные для содержания морских животных и показа их публике.
f océanarium *m*
s oceanario *m*
g Meeresaquatorium *n*; Ozeanarium *n*; Meerwasseraquarium *n*

4214 **ocean-atmosphere interaction**
The global processes of energy exchange between the ocean and the atmosphere effecting the Earth's weather and climate.
r взаимодействие океана и атмосферы
Глобальные процессы обмена энергией между океаном и атмосферой, существенным образом влияющие на погоду и климат Земли.
f interaction *f* de l'océan et l'atmosphère
s interacción *f* de océano y atmósfera
g Ozean-Atmosphäre-Wechselwirkung *f*

4215 **oceanic** ***adj***
Pertaining to or occurring in the ocean as well as associated with the deep sea as opposed to neritic.
r океанский; океанический
Относящийся к океану или происходящий в океане, а также связанный с глубинной частью океана (в отличие от неритической), расположенный в окрытом океане (за пределами шельфа) или относящийся к открытому океану.
f océanique *adj*
s oceánico *adj*
g ozeanisch *adj*

4216 **oceanic climate**
The climate characterized by the increased humidity, high cloudiness and a moderate range of temperature in summer and winter owing to the moderating influence of surrounding watres of oceans and big seas.
r океанический климат
Климат, отличающийся повышенной влажностью, большой облачностью, незначительной суточной и годовой амплитудой температуры воздуха, формирующийся под влиянием океанов и больших внутренних морей.
f climat *m* océanique
s clima *m* oceánico
g Meeresklima *n*

4217 **oceanic crust**
The Earth's crust 5-10 km thick underlying deep-water oceanic depressions and mid-oceanic ridges.
r океаническая земная кора
Земная кора мощностью 5-10 км, распространенная под дном глубоководных океанских впадин и в срединно-океанических хребтах.
f écorce *f* terrestre océanique
s corteza *f* terrestre océanica
g ozeanische Erdkruste *f*

* **oceanic farm → 5257**

4218 **oceanic fringing rise**
Elongated arch-like elevatios of the Earth's crust in an ocean fringing deep-sea trenches from the open ocean.
r океанический окраинный вал
Вытянутое сводообразное поднятие земной коры океанического типа, окаймляющее глубоководный желоб со стороны открытого океана.
f bourrelet *m* marginal océanique
s caballón *m* marginal oceánico
g ozeanischer Randrücken *m*

4219 **oceanic front**
An inclined interface or transition zone between two water masses in an ocean within which horizontal gradients of all oceanographic characteristics are maximal.
r океанический фронт
Наклонная плоскость раздела двух различных водных масс в океане, где отмечаются максимальные горизонтальные градиенты всех океанографических характеристик.
f front *m* océanique
s frente *m* oceánico
g ozeanische Front *f*

4220 **oceanic gyration**
A large-scale closed element of the general circulation of the World Ocean waters (more than 1000 km in diameter) spreading over a considerable part of the ocean.
r океанический водоворот
Крупномасштабный замкнутый элемент общей циркуляции вод Мирового океана (более 1000 км в поперечнике), охватывающий значительную часть океана.
f cycle *m* océanique
s ciclo *m* oceánico
g ozeanischer Wirbel *m*; Ozeanwirbel *m*

4221 **oceanic hemisphere**
The hemisphere in which the ocean covers the greatest area while the land occupies only 9% of the area, the centre of it being near New Zealand. → Fig. 69, p. 149
r океаническое полушарие
Полушарие Земли, в пределах которого Мировой океан покрывает наибольшую площадь, а суша занимает лишь 9% пло-

щади, центр такого полушария расположен близ Новой Зеландии. → Рис. 69, с. 149
f hémisphère *m* océanique
s hemisferio *m* oceánico
g ozeanische Erdhalbkugel *f*

4222 oceanic island
An island which rises from the ocean floor or a mid-ocean ridge.
r океанический остров
Остров, расположенный в пределах ложа океана или срединно-океанического хребта.
f île *f* océanique
s isla *f* oceánica; Ozeaninsel *f*
g ozeanische Insel *f*

* **oceanic mud → 4264**

* **oceanic plateau → 5911**

* **oceanic ridge → 3845**

4223 oceanic volcanic ridge
A straight-line chain of underwater mountains connected with tectonic faults and lava flows.
r вулканический океанический хребет
Погруженные в океане вытянутые в одном направлении горы, связанные с тектоническими разломами и излиянием лав.
f crête *f* volcanique sous-marine
s cordillera *f* volcánica submarina; sistema *f* volcánico submarino
g vulkanischer ozeanischer Rücken *m*

* **oceanicity → 4224**

4224 oceanity; oceanicity
Degree in which a climate is affected by maritime influences, opposite to *continentality*.
r океаничность климата
Степень, в которой климат испытывает влияние моря, понятие, противоположное *континентальности*.
f oceanicité *f*; océanité *f*
s oceanicidad *f*; oceanidad *f*
g Maritimität *f*; Ozeanität *f*

4225 oceanographic ***adj***
Pertaining to the investigations in the oceans and seas as well as to the results of those observations.
r океанологический
Относящийся к исследованиям океанов и морей и результатам измерений в океане.
f océanographique *adj*
s oceanográfico *adj*
g ozeanographisch *adj*

4226 oceanographic characteristics
The physical values of the sea water properties, such as temperature, salinity, sea currents, etc.
r океанографические характеристики
Физические значения свойств морской воды, такие как температура, соленость, морские течения и пр.
f charactéristiques *fpl* océanographiques
s características *fpl* oceanográficas
g ozeanographische Charakteristik *f*

4227 oceanographic satellite
Artificial satellite intended for measurements of physical characteristics of ocean surface and surface layers of the ocean as well as identification of the geoid's form.
r океанографический спутник
Искусственный спутник Земли, предназначенный для измерения физических характеристик океанической поверхности и приповерхностных слоев океана, а также определения формы геоида.
f satellite *m* océanographique
s satélite *m* oceanográfico
g ozeanographischer Satellit *m*

4228 oceanographic section
A vertical cross section of ocean (sea) representing the distribution of the sea water characteristics in depth.
r океанографический разрез; океанографический отдел
Вертикальное сечение океана (моря), отражающее распределение каких-либо характеристик морской воды в этой плоскости.
f coupe *f* océanographique
s sección *f* oceanográfica
g ozeanographischer Schnitt *m*

4229 oceanographic station
A point in a sea where oceanographic observations are carried out from the board of a drifting or anchored ship.
r океанографическая станция; морская гидрологическая станция
Точка на поверхности океана, в которой проводятся наблюдения с борта дрейфующего или заякоренного судна.
f station *f* océanographique
s estación *f* oceanográfica
g ozeanographische Station *f*

4230 oceanographic survey
A set of oceanographic measurements carried out by a ship (or a group of ships simultaneously) to obtain data on a spatial distribution of oceanographic characteristics at a certain time.
r океанографическая съемка
Совокупность океанографических измерений, выполняемых судном (или группой судов одновременно) для получения данных о пространственном распределении океанографических характеристик в определенный промежуток времени.
f levé *m* océanographique
s levantamiento *m* oceanográfico
g ozeanographische Aufnahme *f*

4231 oceanographic tables
Tables for calculation of the main characteristics of sea water (mainly salinity and densi-

ty), as well as indices of the main oceanographic processes (currents, waves, tides).

r океанографические таблицы; океанологические таблицы
Таблицы для расчета основных характеристик морской среды (прежде всего солености и плотности), а также показателей главных океанологических процессов (течений, волн, приливов).

f tables *fpl* océanographiques
s tablas *fpl* oceanográficas
g ozeanographische Tabellen *fpl*

4232 oceanography; oceanology
The study of all things relating to the oceans, including the structures of the ocean floor, the character of the sea water, the oceanic flora and fauna, the ocean currents and drifts and the impact of human interference.

r океанология; океанография
Изучение природных явлений, связанных с океаном, включая строение океанского дна, особенности морской воды, океаническую флору и фауну, морские течения, а также антропогенное влияние на моря и океаны.

f océanologie *f*; océanographie *f*
s oceanología *f*; oceanografía *f*
g Meereskunde *f*; Ozeanologie *f*; Ozeanographie *f*

* **oceanology → 4232**

4233 ochreous mud
Terrigenous oxidized brown mud found in seas of the Arctic Ocean.

r коричневый ил
Терригенный окисленный ил коричневого цвета, развитый в морях Северного Ледовитого океана.

f boue *f* brune; boue *f* terrigène
s fango *m* marrón; fango *m* terrígeno
g brauner Schlamm *m*

4234 odour nuisance
Presence in the environment of a smell which is annoying or unpleasant.

r неприятный запах
Раздражающий человека запах в окружающей среде.

f odeur *f* désagréable
s olor *m* desagradable; molestía *f* de olor desagradable
g Geruchbelästigung *f*

4235 oecumene; ecumene
The inhabited part of the Earth including all the areas settled, developed or in some other way involved into social life, for ancient Greeks the ecumene was the inhabited part of the Earth with the centre in Ellas.

r ойкумена; эйкумена
Обитаемая часть суши, включающая все заселенные, освоенные или иным образом вовлеченные в орбиту жизни общества территории, в представлениях древних греков – обитаемая часть Земли с центром в Элладе.

f oecoumène *f*; oekoumène *f*
s ecúmene *m*
g Ökumene *f*

* **offshoot → 5743**

4236 offshore[1]
Narrow shallow belt of seaside from flooded part of a beach towards the edge of the continental shelf.

r предбрежье
Сравнительно плоская мелководная часть шельфа, прилегающая к фронтальной зоне пляжа.

f avant-côte *f*
s antecosta *f*; anteplaya *f*
g Vorküste *f*; Vorstrand *m*

* **offshore[2] → 2262**

* **offshore bar → 4238**

4237 offshore bar; beach ridge
A linear accumulation of shingle at the back of a beach, created by the action of breaking wave parallel with high-water mark, it marks the limit of swash and consists of succesive series of *berms*.

r прибойный вал; береговой вал
Узкая, вытянутая вдоль берега пересыпь, формирующаяся в результате поперечного перемещения морских наносов и их отложения, вал отмечает границу прибоя и состоит из серии последовательных *берм*.

f cordon *m* littoral
s cordón *m* litoral
g Strandwall *m*; Küstendamm *m*; Nehrung *f*

4238 offshore sediments; offshore bar; nearshore sediments
A ridgelike accumulation of sand lying in a semi-submerged position seaward of the breaker zone but not elevated enough to be called barrier-beach.

r прибрежно-морские отложения
Толща наносов в береговой зоне, частично в полосе прибоя, но не выше тыловой части пляжа.

f sédiments *mpl* côtieres
s sedimentos *mpl* costeros
g Küstenablagerungen *fpl*; Küstensedimenten *npl*

4239 offshore zone
A territory that offers privileged conditions (reduced taxes and exemption from foreign exchange regulation, etc.) for financial operations with foreign partners and in foreign currencies.

r офшорная зона
Территория, предоставляющая льготный режим (снижение налогов, освобождение

от валютного контроля и т.п.) для финансово-кредитных операций с иностранными участниками и в иностранной валюте.
f off-shore zone *f*
s off-shore zona *f*
g Off-shore *n*

4240 off-stream storage reservoir
An artificial water body created in a closed drainless basin or low area of a river floodplain the capacity of which is usually increased by banking.
r наливное водохранилище
Искусственный водоем, создаваемый в бессточной котловине или пониженном участке речной поймы, вместимость которого обычно увеличивают обвалованием.
f réservoir *m* remplie
s embalse *m* de rellenamiento; embalse *m* de reserva
g gefülltes Wasserbecken *n*

Fig. 230. **Ogives** at an Caucasian glacier

4241 ogives
Curved bands across the surface of a glacier, a system of wide alternate bands of dark and light ice having a form of arches and stretching right across the glacier tongue. → Fig. 230
r огивы; волновые огивы
Система широких полос льда светлого и темного цвета, чередующихся на поверхности некоторых ледников ниже ледопадов и протягивающихся в виде стрельчатых дуг поперек языков. → Рис. 230
f ogives *fpl*; chevron *m*
s ojivas *fpl*
g Ogiven *fpl*; Sparren *mpl*

*** O-horizons → 4283**

4242 oil dome
An underground, roughly hemispherical structure in gently flexed sedimentary strata containing an accumulation of petroleum.
r нефтесодержащий купол
Подземная, полусферическая структура в мягко изогнутых осадочных отложениях, содержащая залежи нефти.
f dôme *m* pétrolier
s domo *m* petrolero; domo *m* de petróleo
g Erdöldom *m*

4243 oil pipeline
A pipeline for moving crude oil from places of extraction to places of use.
r нефтепровод
Трубопровод для транспортировки сырой нефти от мест ее добычи к районам потребления.
f oléoduc *m*
s oleoducto *m*; conducción *f* de petróleo
g Erdölleitung *f*

4244 oil refining; oil-processing industry; oil-refining industry
A branch of processing industry that produces petrochemicals (petrol, paraffin, diesel oil, lubricants) from crude oil.
r нефтеперерабатывающая промышленность; нефтехимическая промышленность
Отрасль обрабатывающей промышленности, производящая из сырой нефти разнообразные нефтепродукты (бензин, парафин, дизельное топливо, смазочные материалы).
f industrie *f* pétrolière
s industria *f* petrolera; industria *f* petroquímica
g eldölverarbeitende Industrie *f*

4245 oil spill
Escape of crude oil products into the environment due to any accident or disturbance of technological process.
r нефтяной разлив
Попадание в окружающую среду нефти или нефтяных продуктов в результате аварии или нарушения производственного цикла.
f marée *f* noire
s marea *f* negra; vertido *m* de petrólo
g Ölteppich *m*; Ölpest *f*

4246 oil well
A shaft sunk into the ground, in many cases to a great depth, for the purpose of extracting petroleum.
r нефтяная скважина
Скважина в горных породах, иногда пройденная до большой глубины, используемая для добычи нефти.
f puits *m* à pétrole
s pozo *m* petrolífero; pozo *m* de petróleo
g Erdölbohrung *f*

4247 oilfield
An area containing one or more subterranean oil-pools of petroleum associated with one geological phenomenon.
r нефтяное поле
Территория, включающая одно или несколько участков залежи нефти, связанных с одной и той же геологической структурой.
f champ *m* pétrolier
s terreno *m* petrolero

g Erdölfeld *n*

4248 oil-pool
A separate reservoir of petroleum or natural gas occurring, with water, in the pores and fissures of sedimentary rocks.
r нефтяная залежь; залежь нефти
Отдельный резервуар нефти или природного газа, находящихся, вместе с водой, в порах и трещинах осадочных пород.
f gisement *m* pétrolifère
s yacimiento *m* petrolífero; yacimiento *m* de petróleo
g Erdöllagerstätte *f*

* **oil-processing industry → 4244**

* **oil-refining industry → 4244**

* **old fold mountains → 4251**

4249 old ice
Floating sea ice which underwent melting during at least one summer.
r старый лед
Плавучий лед, подвергшийся таянию по крайней мере в течение одного лета.
f vieille glace *f*
s hielo *m* viejo
g Alteis *n*

4250 old irrigated soils
Soils of different types transformed by long-term irrigation, which are found in ancient oases and on the lowlands irrigated from the earliest time.
r культурно-поливные почвы; староорошаемые почвы
Почвы разных типов, измененные многолетним орошением, распространены в древних оазисах и на издавна орошаемых низменностях.
f sols *mpl* vieux d'irrigation
s suelos *mpl* viejos de regardio
g alte Bewässerungsböden *mpl*

4251 old mountains; old fold mountains
The mountains which were created before the Alpine orogeny, their geological structure became more complicated during later orogenic phases.
r древние горы
Горы, геологическая структура которых возникла в одну из доальпийских эпох складчатости и затем неоднократно усложнялась при повторных орогенезах.
f montagnes *fpl* vieilles; massif *m* ancien
s montañas *fpl* viejas; macizo *m* antiguo
g altes Gebirge *n*

4252 old platform
A platform with the fold basement composed of the Precambrian formations and a sedimentary cover of the Phanerozoic deposits.
r древняя платформа
Платформа, складчатый фундамент которой сложен докембрийскими образованиями, а осадочный чехол – фанерозойскими отложениями.
f plate-forme *f* ancienne
s plataforma *f* antigua
g uralte Plattform *f*

4253 old resident
A person living in the same place for a long period of time.
r старожил
Человек, живущий в данном месте уже много лет.
f ancien habitant *m*
s antiguo habitante *m*
g Alteingesessene *m*; Ureinwohner *m*

4254 old snow
Deposited snow in which the original crystalline forms are no longer recognizable
r лежалый снег; старый снег
Перекристаллизованный снег, пролежавший в толще более месяца и потерявший всякое сходство со свежеотложенным снегом.
f vieille neige *f*
s nieve *f* vieja
g Altschnee *m*

* **Old Wive's summer → 3057**

4255 Old World
General name of three parts of the world known in ancient times: Europe, Asia and Africa as opposed to the *New World.*
r Старый свет
Общее название трех известных в античное время частей света – Европы, Азии и Африки, обычно противопоставляется *Новому свету.*
f Ancien Monde *f*
s Viejo Mundo *m*
g Alte Welt *f*

4256 Oligocene
The lower epoch of the Palaeogene period which began 38 million years ago and lasted about 12 million years. → Table. 1, p. 284
r олигоцен
Верхняя эпоха палеогенового периода (палеогена) геологической истории Земли, началась 38 млн лет назад и продолжалась около 12 млн лет. → Табл. 1, с. 284
f Oligocène *m*
s Oligoceno *m*
g Oligozän *m*

4257 oligomictic lake
A lake which is mixed by convection from time to time only in the periods of especially strong cooling of the water surface.
r олигомиктическое озеро
Озеро, редко и нерегулярно перемешиваемое конвекцией до дна лишь в периоды особенно сильного охлаждения поверхно-

сти воды.
f lac *m* oligomictique
s lago *m* oligomíctico
g oligomiktischer See *m*

4258 **oligopoly**
A situation when a few large competing companies dominate on the market.
r олигополия
Ситуация на рынке, когда несколько крупных конкурирующих фирм монополизируют производство и сбыт основной массы продукции в отрасли.
f oligopole *m*
s oligopolio *m*
g Oligopol *n*

4259 **oligotrophic *adj***
Applied to a water body, soil and bog when it has a low nutrient content and is therefore unable to support a large aquatic flora and fauna.
r олиготрофный
Относящийся к водоему, почве или болоту, бедному питательными веществами и поэтому неспособному поддерживать в своей среде богатую речную флору и фауну.
f oligotrophe *adj*
s oligotrofo *adj*
g oligotrophe *adj*

* **oligotrophic bog → 4897**

4260 **oligotrophic lake**
A lake with a low concentration of organic and biogenic mineral substances in water.
r олиготрофное озеро
Озеро с малой концентрацией в воде органических и биогенных минеральных веществ.
f lac *m* oligotrophe
s lago *m* oligotrofo
g oligotropher See *m*

4261 **oligotrophic plant**
Plants growing on poor soils, in particular on sands in pine forests and in raised bogs.
r олиготрофное растение
Растения, обитающие на малоплодородных почвах, в частности, в сосняках на песках и на верховых болотах.
f plante *f* oligotrophique
s planta *f* oligotrofa
g oligotrophe Pflanze *f*

4262 **olive grove**
A place where a lot of olive trees grow.
r оливковая роща
Место, где произрастает множество оливковых деревьев.
f olivière *f*
s olivar *m*
g Olivenhain *m*

4263 **omnivore**
An animal that eats any type of food.
r всеядное животное
Животное, питающееся и травянистой растительностью и другими животными.
f animal *m* omnivore
s animal *m* omnívoro
g Allesfresser *m*

* **one-year ice → 2158**

* **onion-skin weathering → 2046**

* **onion-weathering → 2047**

* **ooze → 4001**

4264 **ooze; oceanic mud**
A fine-textured sediment formed at great depths on the ocean floor far from land, classified as biogenic or non-biogenic according to its origin.
r глубоководный ил; океанический ил
Отложения на океаническом дне вдали от суши, классифицируемые в зависимости от происхождения как биотические и небиотические.
f boue *f* de abyssal; boue *m* océanique
s fango *m* abisal; fango *m* oceánico
g Tiefseeschlamm *m*;
ozeanischer Schlamm *m*

4265 **open burning**
Burning of waste matter in the open air creating pollution with smoke.
r открытое сжигание
Сжигание отходов на открытом воздухе, что вызывает загрязнение атмосферы, распространяющееся вместе с дымом.
f incinération *f* ouverte
s combustión *f* abierta
g offene Verbrennung *f*

* **open channel → 4270**

4266 **open field$_1$**
A distribution of farm land associated with feudalism in Europe, when each manor had two or three large common fields and each farmer was awarded a number of strips within each field.
r открытое поле
Распределение крестьянской земли, связанное еще с феодализмом в Европе, когда каждый помещик имел два или три больших открытых поля, а каждому крестьянину полагалось несколько полос земли внутри такого поля.
f champ *m* ouvert
s campo *m* abierto; openfield *m*
g offenes Feld *n*

* **open field$_2$ → 5867**

* **open forest → 4271**

4267 **open fracture; cleft**

An open fissure in rocks, sometimes expanded by denudation or washed by water.
r открытая трещина; расселина; зияющая трещина
Трещина в скальных породах с раскрытыми стенками, иногда расширенная денудацией или промытая водой.
f fissure *f* ouverte; fente *f*
s fisura *f* abierta; grieta *f*
g offene Spalte *f*; offener Spalt *m*

4268 **open plant community**
A plant community characterized by unstable species composition and inter-influence among the species, it is characteristic of initial stages of succession.
r открытое растительное сообщество; незамкнутый фитоценоз
Растительное сообщество, отличающееся непостоянным видовым составом и взаимовлиянием видов, характерно для начальных стадий сукцессии.
f groupement *m* végétal ouvert
s comunidad *f* vegetal abierta
g offene Pflanzengesellschaft *f*

4269 **open sea; high sea**
Part of the ocean lying outside of territorial waters (or economic zone) of some state and used by all states.
r открытый океан; открытое море
Часть морского пространства за пределами территориальных вод (или экономической зоны) какого-либо государства и используемая всеми государствами.
f mer *f* libre; haute mer *f*
s mar *m* libre; altamar *m*
g offenes Meer *n*; Hochsee *f*

4270 **open surface channel; open channel**
A watercourse the upper surface of which is not isolated from the atmosphere.
r открытое русло
Русловой поток, верхняя поверхность которого не изолирована от атмосферы, например ледяным покровом.
f surface *f* libre de chenal
s superficie *f* abierta del río
g offenes Strombett *n*; offenes Flußbett *n*

4271 **open woodland; light forest; thin forest; open forest**
Low-density spouse growth of trees in the areas of insufficient moisture, e.g. savanna. → Fig. 231
r редколесье; редкостойный лес
Разреженная древесная растительность, развивающаяся в условиях недостаточного увлажнения (например, саванна). → Рис. 231
f forêt *f* claire
s bosque *m* claro
g lichter Wald *m*; offenes Gehölz *n*

* **opencast mining → 4272**

Fig. 231. **Open woodland**

4272 **opencast mining; opencast mining**
The form of excavation used when extensive mineral deposits, e.g. of coal, brown coal or iron ore, lie near the earth's surface and worked mechanically on a large scale.
r открытая добыча
Форма добычи полезных ископаемых, например, каменного угля, бурого угля, железной руды, залегающих вблизи земной поверхности, выполняемая механически и в крупных размерах.
f extraction *f* minière ouverte
s extracción *f* a cielo abierto
g Tagebau *m*

* **opening → 1026**

* **opening in sea-ice → 4683**

4273 **opolje**
An elevated, treeless, well-drained rolling plain in the middle belt of East European Plain.
r ополье
Возвышенная, безлесная, хорошо дренированная слабоволнистая равнина в средней полосе Восточно-Европейской равнины.
f opolje *m*
s opolje *m*
g Opolje *n*

4274 **opposition**
The position of a heavenly body when it is directly opposite another one, e.g. when three celestial bodies share a common line, the two outer bodies are in opposition to each other and to the central body.
r оппозиция небесного тела
Положение небесного тела, когда оно находится непосредственно напротив другого, например, если три небесных тела располагаются на общей линии, говорят, что

два внешних тела находятся в оппозиции друг к другу и к центральному телу.
f opposition *f*
s oposición *f*
g Opposition *f*

4275 optical oceanography
The study of the optical properties of sea water and the spreading in seas the light from different sources.
r оптическая океанография
Изучение оптических свойств морских вод и процессов распространения в море света от разных источников.
f océanographie *f* optique
s oceanografía *f* óptica
g optische Ozeanographie *f*

4276 optical phenomenon in the atmosphere
Visible phenomenon in the atmosphere due to light refraction, reflection, scattering, and diffraction on the atmospheric particles (mirages, halo, dawns, twilight, etc.).
r оптическое явление в атмосфере
Видимое в атмосфере явление, обусловленное преломлением, отражением, рассеянием или дифракцией света на атмосферных частицах (миражи, гало, зори, сумерки и др.).
f phénomène *m* optique en atmosphère
s fenómeno *m* óptico en la atmósfera
g optische Erscheinung *f*

4277 optimal humidity
Natural conditions when the total annual value of precipitation is approximately equal to its possible evaporation.
r оптимальное увлажнение
Природные условия, при которых годовая сумма атмосферных осадков примерно равна величине их возможного испарения.
f humidité *f* optimale
s humedad *f* óptima
g optimale Feuchtigkeit *f*

4278 orbit of the Earth
The counterclockwise closed elliptical path of the Earth around the sun taking 365.26 days to complete.
r орбита Земли
Траектория в форме эллипса, по которой Земля обращается вокруг Солнца по часовой стрелке, совершая полный круг за 365,26 дней.
f orbite *f* terrestre
s órbita *f* de la Tierra
g Erdbahn *f*; Erdorbit *m*

4279 orbital motion of a wave
The circular or near-circular movement of water particles beneath a wave.
r орбитальное движение волны
Круговое или квазикруговое движение водных частиц у подошвы волны.
f mouvement *m* orbital
s movimiento *m* orbital
g Orbitalbewegung *f*

4280 Ordovician
The second geological period of the Palaeozoic era, it began 500 million years ago and lasted for about 60 million years. → Table. 1, p.284
r ордовик
Второй период палеозойской эры геологической истории Земли, начался 500 млн лет назад и продолжался около 60 млн лет. → Табл. 1, с. 284
f Ordovicien *m*
s Ordovícico *m*; Ordovicience
g Ordoviziumn *n*

4281 ore
Any solid naturally-occurring mineral aggregate of economic interest from which one or more valuable constituents (especially metallic minerals) may be obtained by treatment.
r руда
Любой твердый природный минеральный агрегат, из которого путем обогащения могут быть получены один или более полезных компонентов, находящих применение в промышленности.
f minerai *m*
s mineral[2] *m*; mena *f*
g Erz *m*

* **organic deposit → 671**

4282 organic farming
A method of farming which does not involve to use chemical fertilizers or pesticides.
r органическое земледелие
Земледелие, ориентированное на агротехнические методы, не применяющие минеральные удобрения и ядохимикаты.
f agriculture *f* biologique
s agricultura *f* biológica
g organische Landwirtschaft *f*

4283 organic horizons; O-horizons
Certain layers in which humified matter occurs above the mineral soil in a soil profile.
r органические горизонты
Определенные слои почвенного профиля, в которых над минерализованной почвой появляются гумусовые вещества.
f horizons *mpl* organiques
s horizontes *mpl* orgánicos
g organische Horizonte *mpl*

4284 organic soil
A soil consisting mainly of organic material, as distinct from a *mineral soil.*
r органическая почва
Почва, содержащая в основном органический материал, в отличие от *минерализованной почвы.*
f sol *m* organique

s suelo *m* orgánico
g organischer Boden *m*

4285 organic weathering
The disintegration and chemical change of rocks by the biological processes, the organic agents being plant roots, mosses, lichens, the burrowing of animals and products of their decomposition.
r органическое выветривание; биологическое выветривание
Механическое разрушение и химические изменения горных пород под влиянием жизнедеятельности организмов и продуктов их разложения.
f météorisation *f* organique
s meteorización *f* orgánica
g organische Verwitterung *f*

4286 organoleptic properties
Properties of objects which are identified and estimated by using sense organs (e.g. taste, smell, color, etc.).
r органолептические свойства
Свойства объектов, которые выявляются и оцениваются с помощью органов чувств (например, на вкус, запах, цвет и пр.).
f propriétés *fpl* organoleptiques
s propiedades *fpl* organolépticas
g organoleptische Eigenschaften *mpl*

4287 oricanga; solution basin
A small depression, about 1 m in diameter and several centimetres deep, found on level tock surfaces devoid of rock waste.
r орисанга
Замкнутое углубление диаметром около 1 м и глубиной до нескольких десятков сантиметров, встречающееся на пологих, лишенных коры выветривания поверхностях горных пород.
f oriçanga *m*
s orizanga *f*
g Oricanga *f*

4288 Orient
The countries lying to the East of Europe, especially the Far East (the opposite of the *Occident*).
r Восток; восточные страны
Страны, лежащие к востоку от Европы, особенно Дальний Восток, в противоположность *Западу* (западным странам).
f orient *m*
s oriente *m*
g Orient *m*

* **oriental *adj* → 1779**

4289 orientalism
The study of peoples and cultures of the Orient.
r ориенталистика
Изучение народов и культуры Востока.
f orientalisme *m*

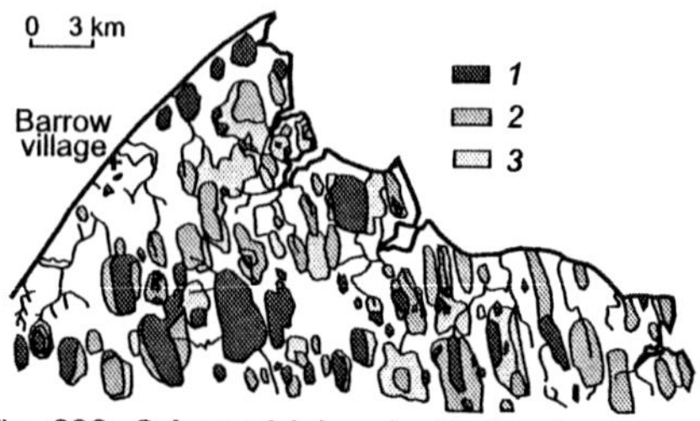

Fig. 232. **Oriented lakes** in Alaska: 1 – present lakes, 2 – recently drained lakes, 3 – ancient drained lakes

s orientalismo *m*
g Orientalismus *m*

* **orientation → 624**

4290 oriented lakes
A group of enclosed water bodies occupying structural hollows picked out by selective glacial erosion or deflation by wind, and exhibiting a parallel alignment. → Fig. 232
r ориентированные озера
Группа водоемов параллельного залегания, занимающих стуктурные впадины рельефа, возникшие благодаря ледниковой экзарации или эоловым процессам. → Рис. 232
f lacs *mpl* orientés
s lagos *mpl* orientados
g orientierte Seen *mpl*

4291 origin
The point defined by the intersection of two axes in a system of coordinates used to work out a map projection.
r полюс проекции
Точка, определяемая пересечением двух осей в системе координат, используемая для построения картографической проекции.
f origine *f*
s origen *m*
g Projektionpol *m*

4292 original bedding
The disposition of rock strata as it was at the time of the formation of sedimentary rock.
r первичное залегание
Залегание горных пород, которое они приобретают в процессе своего формирования.
f stratification *f* initiale
s estratificación *f* primaria
g ursprüngliche Lagerung *f*

4293 original forest; virgin forest; primary forest
Forest which originally covered a region before changes in the environment brought about by people, or the forest that is not affected by direct or indirect economic activity.
r девственный лес; первичный лес
Лес, не тронутый прямой или косвенной хозяйственной деятельностью человека.

f forêt *f* vierge; forêt *f* primaire
s bosque *m* virgen; selva *f* primaria
g Urwald *m*; jungfräulicher Wald *m*; urschprüngliche Wald *m*

4294 original map
Initial copy of a map including all its contents and compiled in established cartographic symbols with prescribed generalisation and accuracy.
r оригинал карты
Первичный экземпляр карты, включающий все ее содержание и составленный в установленных картографических знаках с заданной генерализацией и точностью.
f dessin *m* cartographique
s dibujo *m* cartográfico
g Kartenzeichnung *f*

* **original plot → 1167**

4295 original vegetation; primary vegetation
The vegetation of flat watershed areas existing over a long period of time and correlating with the zonal climatic conditions.
r первичная растительность; коренная растительность
Основная, длительно существующая растительность водораздельных (плакорных) пространств, соответствующая зональным условиям климата.
f végétation *f* originaire
s vegetación *f* originaria; vegetación *f* primaria
g urschprügliche Vegetation *f*

4296 ornithofauna
All bird species which inhabit certain area or lived in some period of geologic history.
r орнитофауна; авифауна
Совокупность видов птиц (главным образом гнездящихся и оседлых), населяющих определенную территорию или встречавшихся в какой-либо отрезок времени в истории Земли.
f ornithofaune *f*
s ornitofauna *f*
g Avifauna *f*; Vogelfauna *f*

4297 ornithogenic soil
A soil that forms in areas of the accumulation of seafowl excrements (guano) with high content of nitrogen and phosphorus.
r орнитогенная почва
Почва, формирующаяся в условиях накопления гуано птиц с высоким содержанием азота и фосфора.
f sol *m* ornithogénique
s suelo *m* ornitogénico
g ornithogener Boden *m*

4298 oroclimatic *adj*
Depending on climate and relief.
r ороклиматический
Находящийся в зависимости от климата и рельефа.
f oroclimatique *adj*
s oroclimático *adj*
g oroklimatisch *adj*

4299 orocline
Any structural mountain chain or orogenic belt, the line of which has been sharply displaced by subsequent horizontal movement.
r ороклин
Горная цепь или орогенический пояс, направление которых резко смещено в результате горизонтальных движений земной коры.
f orocline *f*
s oroclina *f*
g Orokline *f*

4300 orogen
The total mass of rock deformed during an orogeny (mountain-building episode).
r ороген
Горное сооружение, возникшее на месте геосинклинали в завершающую стадию ее развития.
f orogène *m*
s orógeno *m*
g Orogen *n*

4301 orogenesis; orogeny; mountainbuilding; orogenic movements
The event, or mechanism of the construction and uplift of mountain chains, by movements of the Earth's crust.
r орогенез; горообразование; орогенические движения
Совокупность восходящих тектонических движений и денудационных процессов, приводящих к образованию гор.
f orogenése *f*; mouvement *m* orogénique
s orogénesis *f*; movimiento *m* orogénico
g Orogenese *f*; Gebirgsbildung *f*; orogenetische Bewegungen *fpl*

4302 orogenic *adj*
Formed by mountain-building processes, conneced with orogenesis, e.g. orogenic movements.
r орогенический; орогенный
Вызванный процессами горообразования, связанный с формированием гор, например, орогенические движения.
f orogénique *adj*
s orogénico *adj*
g orogenetisch *adj*

* **orogenic belt → 3892**

* **orogenic movements → 4301**

4303 orogenic phase
The geologically short epoch of the acceleration of tectonic movements causing the formation of folds.
r фаза складчатости
Геологически кратковременная эпоха ус-

корения тектонических движений, вызывающих складкообразование.
f phase *f* orogénique
s fase *f* de plegamiento
g Faltungsphase *f*

* **orogeny → 4301**

4304 orographic *adj*
Related to the general features of landforms, or caused by the Earth's relief, e.g. orographic snow line or orographic precipitation.
r орографический
Относящийся к общим чертам земной поверхности или возникающий под влиянием неровностей рельефа, например, орографическая снеговая линия или орографические осадки.
f orographique *adj*
s orográfico *adj*
g orographisch *adj*

4305 orographic cloud
A cloud that results from the uplift of an air stream by a topographic barrier, e.g. banner cloud, lee wave, lenticular cloud.
r орографическое облако
Облако, возникновение и форма которого определяются подъемом воздуха из-за топографического препятствия, например, облачный флаг, подветренная волна, чечевицеобразное облако.
f nuage *m* orographique
s nube *f* orográfica
g orographische Wolke *f*

4306 orographic occlusion
Occlusion forced by the presence of an topographic barriere which breaks the warm frontrs progress, thus allowing the cold front to overtake the warm front.
r орографическая окклюзия
Окклюзия, вызванная наличием орографического препятствия, которое нарушает развитие теплого фронта, из-за чего холодный фронт обгоняет его.
f occlusion *f* orographique
s oclusión *f* orográfica
g orographische Okklusion *f*

4307 orographic precipitation; orographic rain; relief rainfall
Precipitation caused by the ascent of air over which forms when moisture-laden air masses are forced to rise over topographic barriers when the air is cooled, the water vapour codenses and precipitation occurs.
r орографические осадки
Осадки, выпадающие при подъеме воздуха, когда насыщенные влагой воздушные массы вынуждены подниматься над орографическим препятствием и, охлаждаясь с высотой, конденсировать влагу и давать осадки.
f précipitation *f* orographique
s precipitación *f* orográfica
g orographische Niederschläge *mpl*

* **orographic rain → 4307**

4308 orographic snow line
The lower limit at which permanent snow patches form.
r орографическая снеговая линия
Нижняя граница снежников-перелетков.
f limite *f* orographique de la neige
s límite *m* orográfica de nieves perpetuas
g orographische Schneegrenze *f*

4309 orographical name
A proper name of any landform or element of the relief.
r ороним
Собственное имя, название любого элемента земной поверхности.
f oronyme *m*
s orónimo *m*
g orographischer Name *m*

* **orography → 3948**

4310 orthodromic line
Line of the shortest distance between two points of the Earth's surface, the smallest section of an arch of a large circle passing through these points. → Fig. 205, p. 450
r ортодромия
Линия кратчайшего расстояния между двумя точками на поверхности земного шара, наименьший из отрезков дуги большого круга, проходящий через эти точки. → Рис. 205, с. 450
f orthodromie *f*
s ortodrómica *f*
g Orthodromie *f*

* **orthogonal projection → 4311**

4311 orthographic projection; orthogonal projection
A perspective map projection in which the perspective centre is considered to be infinitely far from the centre of the generating globe. → Fig. 233, p. 515
r ортографическая проекция; ортогональная проекция
Перспективная азимутальная картографическая проекция, получаемая при расположении точки зрения на бесконечно большом расстоянии от центра шара. → Рис. 233, с. 515
f projection *f* orthographique; projection *f* orthogonale
s proyección *f* ortográfica
g orthographische Projektion *f*

4312 orthophotomap
A photo map composed of vertical aerial photographs transforned to the orthographic projection.

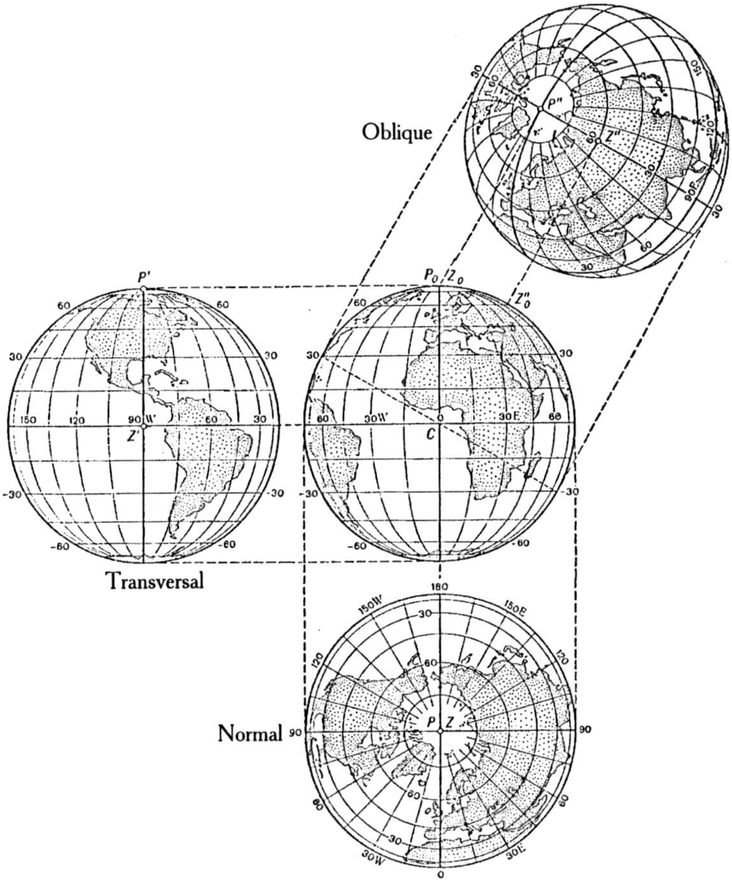

Fig. 233. **Ortographic projection**

r ортофотокарта
Карта, смонтированная из плановых аэрофотоснимков, трансформированных в ортогональную проекцию.
f orthophotocarte *f*
s ortofotomapa *m*
g Orthophotokarte *f*

4313 ortsand
Compacted and cemented band coloured rusty, red-brown or dark brown that occurs in sandy soils.
r ортзанд
Уплотненная сцементированная прослойка в песчаных почвах, ржавого, красно-бурого или кофейного цвета.
f ortsand *m*
s ortsand *m*
g Ortsand *m*

4314 ortstein; iron pan
Rounded ferromanganese soil concretions 1-10 mm in diameter, scattered or forming a dense horizon which develop under varying water-air or oxidation-reduction regimes.
r ортштейн
Округлые железо-марганцевые почвенные конкреции диаметром 1-10 мм, отдельные или составляющие плотный горизонт, образуются в условиях переменного водно-воздушного или окислительно-восстановительного режимов.
f ortstein *m*; alios *m*
s ortstein *m*; alios *m*
g Ortstein *m*

4315 oscillation
A variation above and below a mean value.
r осцилляция
Колебание той или иной величины выше или ниже ее среднего значения.
f oscillation *f*
s oscilación *f*
g Oszilation *f*

4316 oscillation moraine

Fig. 234. **Outcrop** on the Volga river

Ridges of terminal moraines formed during insignificant advances of a glacier which have a local nature.

r осцилляционная морена
Гряды конечных морен, сформированные при незначительных наступаниях ледника, не носящих повсеместного характера.
f moraine *f* d'oscillation
s morrena *f* de oscilación
g Oszillationsmoräne *f*

4317 oscillation of glacier; glacial oscillation
The shifts of the frontal part of a galcier.
r осцилляция ледника
Перемещение конца ледника.
f oscillation *f* glaciaire
s oscilación *f* glaciar
g Gletscheroszillation *f*

* **oued → 6615**

4318 outbound tourism
Tourism involving residents of a country visiting other countries for touristic actvities.
r выездной туризм
Туризм, относящийся к людям, покидающим данную территорию с туристическими целями.
f tourisme *m* émetteur
s turismo *m* externo; turismo *m* emisivo
g Ausfahrtstourismus *m*; Ausreiseverkehr *m*

* **outburst of glacial lake → 2493**

4319 outcrop; rock exposure
A location where bare rock can be seen at the surface, either naturally or by artificial excavation, whether or not it is covered by an overlying superficial cover of drift or detritus. → Fig. 234
r обнажение; выход на поверхность
Участок земной поверхности, на котором естественным образом или при зачистке появляются коренные горные породы, иногда прикрытые тонким чехлом рыхлых поверхностных отложений. → Рис. 234
f affleurement *m*
s afloramiento *m*
g Aufschluss *m*

* **outdating of a map → 3656**

4320 Outer Crescent
In the geopolitical theory of heartland an outer belt that includes Britain, South Africa, both Americas and Japan that encircles the *heartland* and runs a risk of naval expansion.
r внешний полумесяц
В геополитической теории хартленда внешний пояс, включающий Британию, Южную Африку, обе Америки и Японию, который окружает извне собственно *хартленд* и может быть объектом военно-морской экспансии.
f croissant *m* extérieur
s creciente *m* exterior
g Aussere Halbmond *m*; Outer Crescent *n*

4321 outfall offshore
A river delta-front together with the adjacent part of coastal waters.
r устьевое предбрежье; устьевое взморье
Авандельта реки и прилегающая часть прибрежной морской акватории.
f avant-plage *f* d'estuaire
s anteplaya *f* de estuario
g Mündungsvorküste *f*

4322 outflow cave; outlet cave
An effluent cave from out of which an underground karst watercourse flows, or formerly flowed.
r выходная пещера
Карстовая полость, из которой вытекает или вытекал прежде подземный водоток.
f caverne *f* de sortie
s caverna *f* de salida; cueva *f* de salida
g Ausgangshöhle *f*

* **outgoing long-wave radiation → 6105**

4323 outlet
A small course of river connecting two water bodies.
r протока
Небольшой водоток, соединяющий два водоема.
f grau *m*; passe *f*
s canal *m* de enlace; paso *m*
g Verbindungskanal *m*; Durchfluß *m*

* **outlet cave → 4322**

4324 outlet glacier; ice stream
An ice flow rapidly moving down from an ice sheet and carrying the main mass of ice from the ice-catchment basin.
r выводной ледник
Быстро движущийся ледяной поток, переносящий основное количество льда с ледосборного бассейна на ледниковом покрове.
f glacier *m* émissaire
s glaciar *m* de descarga
g Abflussgletscher *m*

4325 **outlet section**
The lowest hydrographical post at the main river of the basin under consideration.
r замыкающий створ
Самый нижний пост на главной реке рассматриваемого бассейна.
f section *f* de jaugeage;
ligne *f* de la section de jaugeage
s sección *f* de salida;
estación *f* de sondeo en el río
g abschliessende Abflußmessstelle *f*

* **outlier → 1996**

4326 **outlier; klippe**
A rock outcrop, a remnant of an ancient nappe.
r клипп
Скала или утес, сохранившиеся от древнего тектонического покрова.
f klippe *f*
s klippe *f*
g Klippe *f*

4327 **outline$_1$ → 4327**
A line, or lines, bounding the outer limits of a discrete area on a geographical map.
r контур
Очертание объекта на географической карте.
f contour *m*
s esquema *m* topográfico
g Umrißlinie *f*; Kontur *m*

* **outline$_2$ → 5462**

* **outline map → 594**

4328 **outport**
A port situated nearer the sea and consequently more accessible to vessels (especially those of greater size and draught) than the main port to which it is subordinate, e.g. Bremerhaven in Northern Germany which is an outport of Bremen or Ouistrïham is the outport for Caen, Normandy.
r аванпорт; внешний порт
Порт, расположенный ближе к морю и тем самым более доступный для судов (особенно имеющих значительный размер и большую осадку), чем главный порт, с которым он тесно связан, например, Бремерхафен в Северной Германии – аванпорт Бремена.
f avant-port *m*
s antepuerto *m*
g Außenhafen *m*; Vorhafen *m*

4329 **output; produce; production**
Mineral, agricultural, or industrial product, a salable or marketable commodity.
r продукция; выход продукции
Сырьевые, сельскохозяйственные или промышленные продукты производства.
f output *m*; produit *m*; production *f*; sortie *f*
s producción *f*; salida *f*; producto *m*
g Output *m*; Produktion *f*; Erzeugung *f*

4330 **outwash plain; sandr *m*; sandur**
Spacious accumulative plains composed of pebbles, gravel and stratified sands washed out from glaciers or ice caps by glacial melt waters.
r зандровая равнина; флювиогляциальная равнина; зандры
Обширные равнинные аккумулятивные поверхности, состоящие из гальки, гравия и косослоистых песков, формирующиеся потоками талых вод перед фронтом ледниковых покровов.
f plaine *f* de lavage proglaciaire; sandur *m*
s llano *m* de lavado postglaciar; sandr *m*
g Schwemmebene *f*; Sander *m*;
Sanderfläche *f*

* **outwash sands and gravels → 2547**

4331 **ouval**
A stretched upland up to 200 m high with a flat or slightly convex top and gentle slopes without a clearly marked foot.
r увал
Вытянутая возвышенность относительной высотой до 200 м с плоской или слегка выпуклой вершиной и пологими склонами без ясно выраженного подножия.
f colline *f* alongée
s colina *f* alargada
g Uval *m*

4332 **overall soil moistening**
The amount of water saturated into soil over a definite period of time, in water balance calculations it is determined as the difference between the amount of precipitation and water loss through surface runoff.
r валовое увлажнение
Количество воды, впитавшейся в почву, за определенный отрезок времени, в воднобалансовых расчетах определяется по разности осадков и поверхностного стока.
f humidification *f* total du sol
s humidificación *f* total del suelo
g Gesamtbodenbefeuchtung *f*

* **overbank deposits → 2200**

4333 **overbank derosit**
Any fluvial material deposited beyond the confines of the channel during the overbank flow of a river.
r внерусловые отложения
Любой материал, который отлагается рекой вне границ русла при выходе реки из берегов.
f dépôt *m* hors de lit de rivière
s depósito *m* fuera de lecho del río
g außerhalb des Strombett Ablagerungen *fpl*

4334 **overbank stage of a stream**
The stage when a river overflows the banks of its normal channel and spreads in flood on to the floodplain.

r выход реки из берегов
Состояние реки, когда она выходит из своего русла и разливается по пойме.
f débordement *m* de rivière
s desbordamiento *m* del río
g Uferübertritt *m*

* **overbrowsing → 1415**

4335 **overburden; overburden rock**
Soil and rock lying on top of geological beds rich in mineral and has to be removed in opencast mining (in the process of mineral extraction it is dug away and turned into waste rock polluting the environment).
r вскрышные породы
Кровля геологических пластов, богатых полезными ископаемыми, которая в процессе их открытой добычи переходит в отвальный грунт, загрязняющий окружающую среду.
f décapage *m*; terres *fpl* de couverture
s montera *f*; terrenos *mpl* de recubrimiento
g Abraum *m*

* **overburden rock → 4335**

* **overcast → 4336**

4336 **overcast sky; overcast**
The sky completely covered with clouds.
r сплошная облачность
Состояние неба, когда оно полностью покрыто облаками.
f nébulosité *f* absolue
s cielo *m* cubierto; nubosidad *f* absoluta
g dichte Wolkendecke *f*; dichte Bewölkung *f*

4337 **overconsolidation of soil**
The change of soil structure caused by the use of heavy machines, overgrazing, and etc. that results in a denser packing of soil particles and decrease of porous space.
r переуплотнение почвы
Изменение сложения почвы под воздействием тяжелой техники, перевыпаса скота и т.п., приводящее к более плотной упаковке почвенных частиц и уменьшению порового пространства.
f sous-consolidation *f* du sol; induration *f* du sol; sous-compactage *f* du sol
s sobrecompactación *f* del suelo; sobreconsolidación *f* del suelo; induración *f* del suelo
g Überverdichtung *f* des Bodens

* **overcropping → 1524**

* **overcrowding → 4349**

4338 **overdeepened valley**
1. A river valley with alluvium deposits thicker than usual, and with the rock floor which lies much lower than the river bottom. 2. The valley of the main trough cut by a glacier deeper than the valleys of the tributary troughs which had smaller glaciers.
r переуглубленная долина
1. Речная долина, у которой мощность аллювия превышает обычную и коренное ложе находится значительно ниже дна реки. 2. Долина главного трога, выпаханная ледником на бóльшую глубину, чем боковые троги, служившие вместилищем ледников меньшей мощности.
f vallée *f* surcreusée
s valle *m* sobreexcavado
g übertieftes Tal *n*

4339 **overdeepening**
The process by which valley floors are excavated by glaciers, sometimes below base-level, so that post-glacial rivers will be forced to infill the rock basins with sediment to restore the valleyrs long profile.
r переуглубление
Процесс, при котором днища долин выпахиваются ледниками, иногда ниже базиса эрозии, так что пост-ледниковые реки заполняют такие долины своими отложениями, что приводит к выравниванию продольного профиля реки.
f superreusement *m*
s sobreexcavación *f*
g Übervertiefung *f*

4340 **overdeepening of channel**
The increase of depth of a stream channel resulting in a serious change of the character of water flow.
r переуглубление русла
Увеличение глубины речного русла, ведущее к заметному изменению характера течения воды в русловом потоке.
f aprofondissement *m* démesuré du lit
s ahondamiento *m* demasiado del lecho; sobreahondamiento *m* del lecho
g Neuanlegung *f* des Flußbettes

* **overfall → 3592**

* **overfault → 6196**

4341 **overflow channel**
A channel, often streamless, cut in solid rocks, having been carved out by the overflow of an ice-dammed lake.
r канал слива
Канал, нередко безводный, выработанный в твердых породах мощным водным потоком при прорыве ледниково-подпрудного озера.
f canal *m* de transvasement
s canal *m* de desbordamiento; canal *m* evacuador
g Abflußkanal *m*

4342 **overflow ice**
A type of congelation ice formed by freezing of drops, water streamlets and unconfined

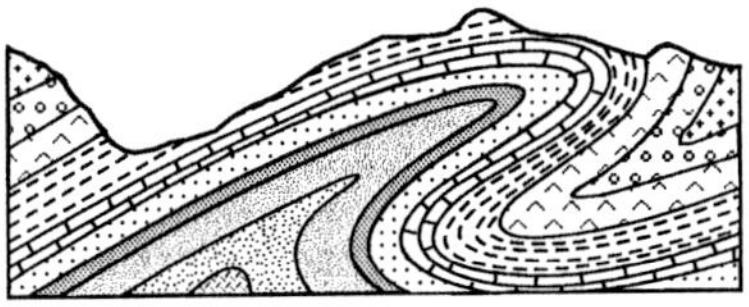

Fig. 235. **Overfold**

streams of water flowing down a cold solid surface.

r натечный лед

Разновидность конжеляционных льдов, формирующаяся при намораживании растекающихся по твердому основанию капель, струек воды или безрусловых потоков.

f concrétion *f* de glace

s concreción *f* de hielo

g Sintereis *n*

4343 overflow lake

A lake having permanent outflow during the whole year.

r сточное озеро

Озеро, имеющее постоянный сток в течение всего года.

f lac *m* à ezutoire

s lago *m* de derramo

g See *m* mit Abfluß

4344 overfold; overturned fold

An asymmetrical anticline in which one limb is inverted and lies beneath the other. → Fig. 235

r опрокинутая складка; перевернутая складка

Асимметричная антиклинальная складка, одно крыло которой перевернуто и лежит ниже другого крыла складки. → Рис. 235

f pli *m* couché; pli *m* retourné

s pliegue *m* tunlado; pliegue *m* acostado

g umkippend Falte *f*

4345 overgrazing

The excessive use of pastures that depletes them by the putting of so many animals on land that the vegetation cover is damaged, sometimes beyond recovery.

r перевыпас

Чрезмерная нагрузка на пастбище, сокращающая его способность производить корма для животных.

f surpâturage *m*

s sobrepastero *m*; pasto *m* excesivo

g Überweiden *n*

4346 overland flow; sheetflow; surface runoff; overland runoff

The unchannelled downslope movement of water produced by sudden, heavy precipitation, often leading to soil erosion when rainfall exceeds infiltration.

r склоновый сток; внерусловой сток; поверхностный сток

Безрусловой сток, формирующийся в пределах склона, вызываемый, как правило, внезапным сильным дождем и нередко приводящий к эрозии почв, если интенсивность дождя превышает фильтрацию воды в почву.

f écoulement *m* en nappe; ruissellement *m*

s escorrentía *f* de vertiente; arroyada *f* superficial

g Oberflächenabfluß *m*; Überlandabfluß *m*

* **overland runoff → 4346**

4347 overlap

The part of an image area lapped over by an adjacent image.

r перекрытие

Часть площади снимка, перекрываемая смежным снимком.

f recoupement *m*

s superposición *f*; zona *f* de superposición

g Überdeckung *f*; Decke *f*

4348 overloaded river

River with a great amount of bedload and suspended load.

r перегруженная река

Река с большим количеством влекомых и взвешенных наносов.

f rivière *f* surchargée

s río *m* sobrecargado

g überlasteter Fluß *m*

4349 overpopulation; overcrowding

The excess of people in an area where the available resources are inadequate for the support of the great number of people, which results in the decline of the standard of living and decrease of per capita income.

r перенаселенность

Избыток людей в районах, испытывающих недостаток рабочих мест, что приводит к снижению уровня жизни и уменьшает душевой доход.

f surpeuplement *m*

s superpoblación *f*

g Überbevölkerung *f*

* **overseas trade → 3163**

4350 overthrust

A type of reverse fault characterised by a low angle of inclination and active movement of the upthrow block. → Fig. 236

r складка-взброс

Смещение горных пород с пологим накло-

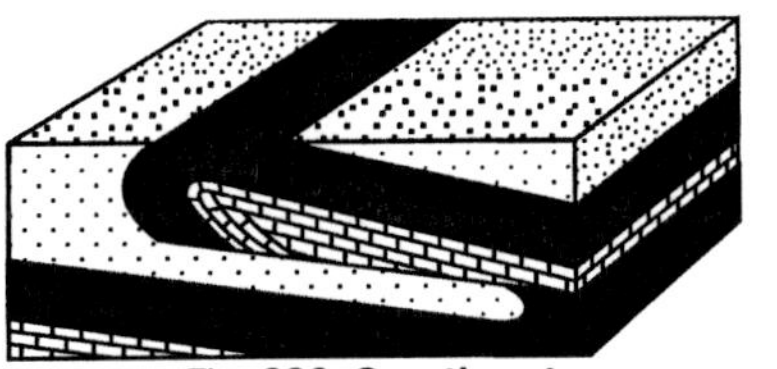

Fig. 236. **Overthrust**

Fig. 237. **Ovring** on the steep slope of the Pyanj river in the Pamirs

ном плоскости и активным движением вышележащего крыла складки. → Рис. 236, с. 519

f faille *f* inverse conforme

s falla *f* inversa conforme

g Aufschiebung *f*

* **overturned fold → 4344**

* **overturning → 3889**

4351 overwetting

Complete saturation of soil with water that results in disturbance of its structure and reduction of aeration, which both cause deterioration of conditions for plant root breathing and often to death of plants.

r переувлажнение

Полное насыщение почвы водой, вызывающее нарушение ее структуры и снижение аэрации, что ведет к ухудшению условий дыхания корней растений и часто к гибели растений.

f sous-humectation *f*; engorgement *m*

s sobrehumedecimiento *m*; encharcamiento *m*; humectación *f* excesiva

g Überfeuchtigkeit *f*

4352 overwintering

Spending the winter and surviving in severe cold conditions.

r зимовка

Жизнь людей, животных или растений в неблагоприятных для них зимних морозных условиях.

f hivernage *m*

s invernada *f*; invernación *f*

g Überwinterung *f*; Überwintern *n*

4353 ovring

Artificially strengthened part of a path made on a steep slope over mountain rivers in Central Asia. → Fig. 237

r овринг

Участок тропы, поддерживаемый специально укрепленным настилом на отвесном склоне над бурной горной рекой в Центральной Азии. → Рис. 237

f ovring *m*

s ovring *m*

g Ovring *m*

4354 ox-bow; oxbow lake

A crescent-shaped lake in a river floodplain formed when the river breaks across the meander neck of a well-developed meander. → Fig. 238

r старица; пойменное озеро

Водоем в пойме реки, возникающий в результате отчленения участка речного русла при спрямлении излучины. → Рис. 238

f bras *m* mort; lit *m* délaisé; mort-bras *m* délaiseé

s bayou *m*; brazo *m* muerto; manga *f*

g Altwasser *n*

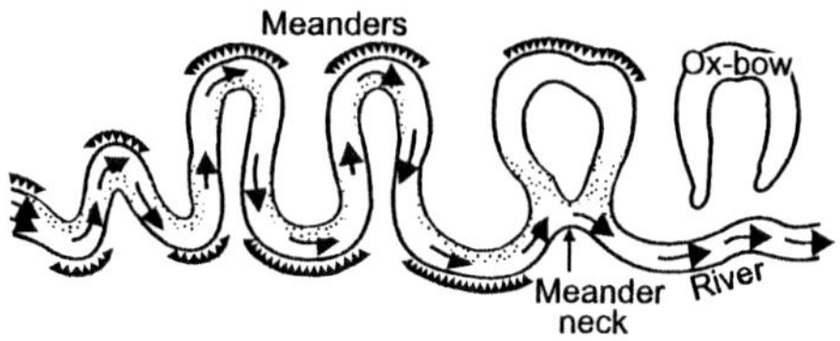

Fig. 238. Formation of an **ox-bow**

* **oxbow lake → 4354**

4355 oxidation

One of the varieties of chemical weathering, particularly in rocks containing iron which leads to a brownish or yellowish staining of the rock surface, which ultimately disintegrates.

r окисление

Одна из разновидностей химического выветривания, особенно в горных породах, содержащих железо, в результате которого скальная поверхность приобретает коричневатый или желтоватый цвет и соответственно разрушается.

f oxydation *f*

s oxidación *f*

g Oxydation *f*; Oxydierung *f*

4356 oxisol

A soil of tropical regions that is characterized by low absorbtion capacity of clays, saturated with water in some periods of most years, the world's two largest areas of this soil type are the Amazon Basin of South America and the Congo Basin of Africa.

r оксисоль

Почва тропических областей с низкой поглотительной способностью глин, насыщенная водой в отдельные периоды большинства лет, широко распространена в бассейнах рек Амазонка и Конго.

f oxysol *m*

s oxisol *m*; oxisuelo *m*

g Oxisol *m*

4357 oxygen

A gaseous element, one of the main constituents of the atmosphere (20.95%), second only to nitrogen in amount.

r кислород
Один из основных составляющих газового состава атмосферы (20,95%), по количеству занимает второе место после азота.

f oxygène *m*
s oxígeno *m*
g Sauerstoff *m*

4358 oxylophile
Organisms preferring rather high concentration of free oxygen in the environment or acid soils (plants preferring acid soil of coniferoys forests or peat bogs, e.g. bilberry, sorrel, are called oxylophytes).

r оксилофил; оксифил
Организмы, предпочитающие достаточно большую концентрацию свободного кислорода в среде или окисленную среду (растения, предпочитающие окисленную среду или торфяные болота, например, черника или щавель, называют оксифитами).

f oxylophile *m*; oxybiont *m*; oxyphile *m*
s oxifilo *m*; oxibionte *m*
g Oxyphil *m*

4359 ozone
A form of oxygen, the proportion of which in the atmosphere is very small, but it is of vital importance in absorbing solar ultraviolet radiation.

r озон
Форма кислорода, доля которой в атмосфере очень мала, но озон имеет жизненное значение для биосферы, так как абсорбирует ультрафиолетовую составляющую солнечной радиации.

f ozone *m*
s ozono *m*
g Ozon *n*

* **ozone depletion → 4360**

4360 ozone hole; ozone depletion
Significant decrease of the atmospheric ozone in polar regions occurring mainly in late winter and early spring.

r озоновая дыра
Наблюдаемое в полярных широтах уменьшение содержания озона в стратосфере в конце зимы или начале весны.

f trou *m* d'ozone
s agujero *m* de ozono; adelgazamiento *m* de la capa de ozono
g Ozonloch *n*

4361 ozone layer; ozonosphere
An ozone-rich band of the atmocphere, at about 10 and 50 km above the Earth, when the ozone layer is depleted, the increased solar ultraviolet radiation reaches the surface of the Earth with consequent damage to human health.

r озоносфера
Слой атмосферы на высотах 10-50 км, в котором сосредоточена основная масса озона атмосферы, при уменьшении озонового слоя увеличенное количество ультрафиолетовой солнечной радиации достигает земной поверхности и оказывает вредное влияние на здоровье людей.

f ozonosphère *f*; couche *m* d'ozone
s ozonosfera *f*; capa *f* de ozono
g Ozonosphäre *f*; Ozoneschicht *f*

* **ozonosphere → 4361**

P

4362 **Pacific folding**
The era of mountain formation in the Mesozoic, occurred in the regions adjoining the Pacific Ocean (Eastern Asia, Cordilleras, Andes).
r тихоокеанская складчатость
Горообразование в мезозойской эре, проявилось в районах, примыкающих к Тихому океану (Восточная Азия, Кордильеры, Анды).
f plissement *m* pacifique
s plegamiento *m* pacífico
g pazifische Faltung *f*

4363 **Pacific Rim**
A geographical region loosely held together by trade and financial links, includes the coastal cities of eastern Asia, Japan, western North America and Australia, together with some Pacific islands.
r Тихоокеанское кольцо
Географический район, объединенный торговыми и финансовыми связями, включает прибрежные города Восточной Азии, Японии, запада Северной Америки и Австралии, а также некоторые острова Тихого океана.
f Rim côte *f* du Pacifique
s anillo *m* de Pacífico; Rim *m* del Pacífico
g Pazifikring *m*

4364 **Pacific-North American oscillation**
Change of difference between the atmosphere pressure above the northern part of the Pacific Ocean and North America (symbol PNA) connected with the southern fluctuation – El-Nico.
r тихоокеанско-североамериканское колебание
Изменение разности между атмосферным давлением над северной частью Тихого океана и Северной Америкой, связанной с южным колебанием – Эль-Ниньо.
f oscillation *f* pacifique-amérique du nord
s oscilación *f* pacífica-norteamérica
g Pazifik-Nordamerika-Oszillation *f*

* **pack → 4366**

4365 **pack animal; beast of burden**
An animal that carried goods, fodder etc., usually loaded on a pack saddle.
r вьючное животное
Животное, используемое для перевозки грузов, которые обычно упаковывают во вьюки.
f animal *m* de bât
s animal *m* de carga
g Packtier *n*; Saumtier *n*

4366 **pack ice; pack; ice pack; packed ice**
Any kind of drifting floating sea ice, in contrast to fast ice.
r пак; паковый лед
Любой дрейфующий лед; многолетний тяжелый морской лед в высоких широтах Арктики, просуществовавший более двух годовых циклов нарастания и таяния.
f pack *m*
s campo *m* de hielo a la deriva; pack-ice *m*; pack *m*
g Packeis *n*

4367 **pack transport**
A type of land transport that moves the loads in packs attached to the backs of such animals as camels, mules, horses, asses, etc.
r вьючный транспорт
Вид наземного транспорта для перевозки грузов во вьюках, которые крепятся на спинах вьючных животных: вюрблюдов, мулов, лошадей, ишаков и др.
f transport *m* de bât
s transporte *m* de carga
g Packtransport *m*

* **packed ice → 4366**

4368 **package tour**
A trip organized by a travel agent, for which one pays a fixed price that includes the cost of transport, accommodation, meals, excursions, etc.
r комбинированный тур; пакетный тур
Туристическая поездка по путевке, обеспечивающей полную организацию тура: размещение в гостинице, транспорт, питание, экскурсии и пр.
f voyage *m* organisé; voyage *m* à forfait; voyage *m* tout compris
s viaje *m* organizado; paquete *m* turístico; viaje *m* programado
g Pauschalreise *f*

4369 **pad'**
1. A valley without a permanent watercourse in the mountains of Siberia and Far East. 2. A swampy depression in a forest on the plains of the Western Siberia.
r падь
1. Долина, не имеющая постоянного водотока, в горах юга Сибири и Дальнего Востока. 2. Заболоченное понижение в лесу на равнинах Западной Сибири.
f pad *m*
s pad *m*
g Pad *f*

4370 **paddock**
An enclosed part of a pasture serving to regulate its use and prevent overgrazing.
r загон
Огороженный участок пастбища, служащий для регулирования его использования и предотвращения перевыпаса.

f parcelle *f* de pâturage
s parcela *f* de pasto; aprisco *m*
g Koppel *f*

4371 padings; flat-bottom steppe depressions
Almost flat, more often rounded depressions occurring individually or in groups on the plain surfaces of interfluves and on the upper river terraces in semi-desert and forest-steppe zones.
r степные блюдца
Почти плоские, чаще округлые, понижения, встречающиеся одиночно или группами на равнинных поверхностях междуречий и на верхних террасах рек в полупустынной, степной и лесостепной зонах.
f soucoupes *fpl*
s platillos *mpl* de estepa
g flache Senke *f* in der Steppe; Steppenpfühle *mpl*

4372 padun
A fall or rapid on a mountainous part of a river in Siberia and in the north of European Russia.
r падун
Водопад или порог на горном участке реки в Сибири и на севере Европейской России.
f padun *m*
s padun *m*
g Padun *n*

* **pahoehoe → 5132**

4373 paired terrace
A river terrace whose remnants on opposite sides of a valley correspond in altitude along a particular stretch of the valley. → Fig. 239
r парная терраса
Речная терраса, остатки которой на противоположных стронах конкретного участка долины соответствуют друг другу по высоте. → Рис. 239
f terrasse *f* formant la paire
s terraza *f* pareada
g paarige Terrasse *f*

4374 Palaeocene
The lower epoch of the Palaeogene period which began 65 million years ago and lasted about 10 million years. → Table 1, p. 284
r палеоцен
Нижняя эпоха палеогенового периода (палеогена) геологической истории Земли, началась 65 млн лет назад и продолжалась около 10 млн лет. → Табл. 1, с. 284
f Paléocène *m*
s Paleoceno *m*
g Paläozän *n*

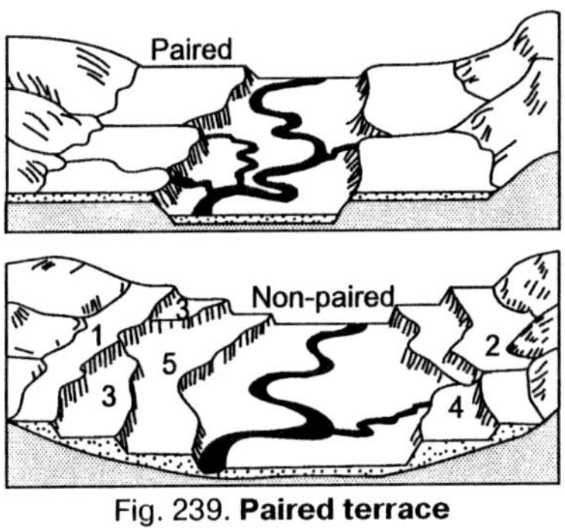

Fig. 239. **Paired terrace**

4375 palaeoclimate
A climate of a prehistoric period the main characteristics of which may be reconstructed from geological and paleobiological evidence: isotope-oxygen records, spores and pollen analysis, etc.
r палеоклимат
Климат доисторического периода, основные характеристики которого определяются по геологическим и палеобиологическим данным: изотопно-кислородным записям, спорово-пыльцевому анализу и т.п.
f paléoclimat *m*
s paleoclima *m*
g Paläoklima *m*

4376 palaeoclimatology
The study of former climates in the geological past based on the investigations of sedimentary rocks, fossils, deep-sea cores, etc.
r палеоклиматология
Изучение климатов геологического прошлого, основанное на исследованиях осадочных пород, ископаемых организмов, глубоководных кернов и пр.
f paléoclimatologie *f*
s paleoclimatología *f*
g Paläoklimatologie *f*

4377 palaeoecology
The study of the relationship between fossil organisms and their palaeoenvironments, achieved by an examination of the geological record using such techniques as pollen analysis and radocarbon dating.
r палеоэкология
Изучение связи между ископаемыми организмами и условиями их обитания, достигнутое на основе исследований геологических записей с использованием методов спорово-пыльцевого анализа и радиоуглеродного датирования.
f paléoécologie *f*
s paleoecología *f*
g Paläoökologie *f*

4378 Palaeogene
The earliest period of the the Cainozoic era which began 65 million years ago and lasted for 40 million years, it is subdivided into three epochs: Palaeocene, Eocene, and Oligocene. → Table 1, p. 284
r палеоген

Самый ранний период кайнозойской эры геологической истории Земли, начался 65 млн лет назад и продолжался около 40 млн лет, подразделяется на три эпохи: палеоцен, эоцен и олигоцен. → Табл. 1, с. 284
f Paléogène *m*
s Paleogeno *m*
g Paläogen *n*

4379 palaeogeographic map
A map representing physical-geographical conditions of the past: the distribution of the land and sea, topography of the continents, landscape zones, etc.
r палеогеографическая карта
Карта, отображающая физико-географические условия прошлого, в частности распределение суши и моря, характер рельефа материков, ландшафтные зоны и пр.
f carte *f* paleogéographique
s mapa *m* paleogeográfico
g paläogeographische Karte *f*

4380 palaeogeography
Branch of geography studying geographical conditions (former land masses, oceans, etc.) which existed on the Earth during the Quaternary including nature conditions of formation and development of humanity.
r палеогеография
Отрасль географии, изучающая географические условия, существовавшие на Земле на протяжении четвертичного периода, включая природные условия становления и развития человечества.
f paléogéographie *f*
s paleogeografía *f*
g Paläogeographie *f*

4381 palaeogeomorphology
A branch of *geomorphology* studying the landforms of past geological periods including buried landforms.
r палеогеоморфология
Раздел *геоморфологии*, рассматривающий рельеф прошлых геологических эпох, нередко погребенный.
f paléogéomorphologie *f*
s paleogeomorfología *f*
g Paläogeomorphologie *f*

4382 palaeoglaciology
The study of natural ice of the past.
r палеогляциология
Изучение природных льдов прошлого.
f paléoglaciologie *f*
s paleoglaciología *f*
g Paläoglaziologie *f*

4383 palaeohydrology
The branch of *hydrology* dealing with rivers and lakes of the past geological epochs.
r палеогидрология
Раздел *гидрологии суши*, изучающий реки и озера в прошлые геологические эпохи.
f paléohydrologie *f*
s paleohidrología *f*
g Paläohydrologie *f*

4384 palaeokarst; buried karst
Ancient karst landforms buried under more recent sediments.
r палеокарст; погребенный карс
Древние карстовые формы рельефа, впоследствии погребенные более молодыми осадками.
f paléo-karst *m*; karst *m* fossile; karst *m* couvert
s paleokarst *m*; karst *m* fósil; karst *m* cubierto
g Paläokarst *m*; begrabener Karst *m*

4385 Palaeolithic
The first period of the Stone Age when humans began to make simple stone tools particularly of flint, began to construct dwellings and use fire, it is subdivided into early (up to 35 thousand years ago) and late (up to 10 thousand years ago) subperiods.
r палеолит
Первый период каменного века, когда происходила эволюция каменных изделий первобытного человека, стали сооружаться жилища и использоваться огонь, подразделяется на ранний (до 35 тыс. лет назад) и поздний (до 10 тыс. лет назад).
f Paléolithique *m*
s Paleolítico *m*
g Paläolithikum *n*

4386 palaeomagnetic chron
Geological interval of time characterised by the prevailing (direct or reverse) direction of magnetic field.
r палеомагнитный хрон
Геологический интервал времени, характеризующийся преимущественным (прямым или обратным) направлением магнитного поля, общеприняты хроны Брюнес, Матуяма, Гаусс и пр.
f chron *m* paléomagnétique
s crono *m* paleomagnético
g paläomagnetische Chron *n*

4387 palaeomagnetic scale
A scale reflecting change of epochs of direct and reverse polarity of the Earthrs magnetic field and the shifting position of the magnetic poles during such epochs.
r палеомагнитная шкала
Шкала, отражающая смену эпох прямой и обратной полярности магнитного поля Земли и его изменений, смещений магнитных полюсов внутри этих эпох.
f échelle *f* paléomagnétique
s escala *f* paleomagnética
g Paläomagnetskala *f*

4388 palaeomagnetism; fossil magnetism

The property of rocks to retain the magnetism present in them at the time of their cooling due to the presence of magnetic minerals, this indicates the direction of the Earth's magnetic field at that time and provides an ancient record of the shifting position of the magnetic poles.

r палеомагнетизм

Свойство горных пород намагничиваться в период своего формирования и сохранять приобретенную намагниченность в последующие геологические эпохи.

f paléomagnétisme *m*

s paleomagnetismo *m*; magnetismo *m* fósil

g Paläomagnetismus *m*

4389 **palaeooceanology**

The study of seas and oceans of the past geological epochs.

r палеоокеанология

Раздел *океанологии*, изучающий моря и океаны в прошлые геологические эпохи.

f paléoocéanographie *f*

s paleooceanografía *f*

g Paläoozeanographie *f*

4390 **palaeopedology**

The study of fossil soils.

r палеопочвоведение

Изучение ископаемых почв.

f paléopédologie *f*

s paleoedafología *f*

g Paläobodenkunde *f*

4391 **palaeovolcanology**

A branch of *volcanology* studying the volcanic activity of the past geologic epochs.

r палеовулканология

Раздел *вулканологии*, изучающий вулканическую деятельность прошлых геологических эпох.

f paléovolcanologie *f*

s paleovulcanología *f*

g Paläovulkanologie *f*

4392 **Palaeozoic**

Geological era in the history of the Earth which began about 570 million years ago and lasted for 340-350 million years. → Table 1, p. 284

r палеозой

Ранняя эра в геологической истории Земли, началась около 570 млн лет назад и продолжалась 340-350 млн лет. → Табл. 1, с. 284

f Paléozoïque *m*

s era *f* paleozoica; Paleozoico *m*

g Paläozoikum *n*

4393 **paleosol; fossil soil; buried soil**

An ancient soil formed under the climatic conditions that differ from the contemporary ones, it may outcrop or be buried by recent deposits.

r палеопочва; погребенная почва; реликтовая почва

Древняя почва, сформированная в условиях климата, отличного от современного, может находиться на поверхности или быть погребенной новейшими отложениями.

f sol *m* fossile; paléosol *m*

s suelo *m* fósil; suelo *m* viejo; paleosuelo *m*

g Paläoboden *m*; Reliktboden *m*

4394 **palsa**

A peat mound, 1-7 m high and 10-50 m across, occurring in an area of permafrost, containing ice lenses composed mainly of mineral material formed as a result of ice segregation, cryogenic heaving and thermokarst.

r пальза; торфяной бугор

Торфяной бугор высотой 1-7 м и диаметром 10-50 м, преимущественно вмещающий минеральный грунт, сформированный в результате льдовыделения, криогенного пучения и термокарста.

f palse *f*

s palsa *f*

g Palsa *f*

4395 **paludal**

Referring to a marshy or swampy environment and to the sediments deposited therein.

r болотный; болотистый

Относящийся к переувлажненному ландшафту и к отложениям, накапливающимся в нем.

f paludéen *adj*; palustre *adj*

s palúdico *adj*; palustre *adj*

g sumpfig *adj*; Sumpf-

4396 **paludification; water-logging**

The gradual lateral expansion of a bog or marsh as peat growth inhibits water drainage, leading to a rise in the water-table.

r заболачивание

Постепенное расширение площади болота по мере роста торфяника, препятствующего стоку воды, что приводит к подъему водного зеркала.

f paludification *f*

s paludificación *f*

g Versumpfung *f*; Vermoorung *f*

4397 **palynology**

The study and analysis of plantsr spores and pollen grains in soil or peat including those the fossil of which helps to reconstruct past climatic conditions preserved and aids to stratigraphic correlation.

r палинология

Изучение спор и пыльцы растений, в том числе ископаемых, что помогает восстанавливать климатические особенности прошлых эпох.

f palynologie *f*

s palinología *f*

g Palynologie *f*

4398 **pampa**

A plain subtropical steppe of South America covered with grassy vegetation (natural grassland of southern Brazil, Argentina and Uruguay).

r пампа; пампасы

Равнинные субтропические степи в Южной Америке, занятые многолетними злаками и разнотравьем (природные степи в южной Бразилии, Аргентине и Уругвае).

f pampa *f*

s pampa *f*

g Pampa *f*

4399 pan

1. Any shallow, generally rounded hollow found in arid and semi-arid regions which may hold water only in the rainy season or may be flooded permanently. 2. A strongly compacted indurated soil horizon or one that is very high in clay content.

r пан

1. Мелкое, как правило, округлое в плане понижение в аридных и семиаридных районах, заполненное водой в период дождей, но иногда и в течение всего года. 2. Сильно затвердевший почвенный горизонт, содержащий большое количество глины.

f pan *m*

s pan *m*

g Pan *m*

4400 Panarctic ice sheet

The hypothetic ice sheet which in the postglacial period occupied the whole Arctic basin and northern parts of Eurasia and North America.

r Панарктический ледниковый покров

Гипотетический ледниковый покров, занимавший в позднеледниковое время весь Арктический бассейн и северные части Евразии и Северной Америки.

f bouclier *m* glaciaire panarctique

s casquete *m* glaciar panártico

g panarktischer Eisschild *m*

4401 pancake ice; ice-cake

Small thin discs of ice, usually less than some meters across, formed during the initial stages of freezing of sears surface. → Fig. 240

r блинчатый лед

Fig. 240. **Pancake ice**

Небольшие округлые пластины молодого плавучего льда, диаметром до нескольких метров. → Рис. 240

f crêpes *fpl* de mer; glace *f* en crêpe

s crepes *fpl* de mar; hielo *m* en crêpe

g Pfannkucheneis *n*; Tellereis *n*

4402 pandemic

A disease so widespread that it affects the whole world or a whole continent.

r пандемия

Болезнь, широко распространившаяся по всему миру или целому континенту.

f pandémie *f*

s pandemia *f*

g Pandemie *f*; Volksseuche *f*

4403 pandemic *adj*

Applied to a disease so widespread that it affects the whole world or a whole continent or a whole country.

r пандемический

Относящийся к болезни, распространившейся по всему миру, по крупной стране или целому континенту.

f pandemique *adj*

s pandémico *adj*

g pandemisch *adj*

4404 Pangaea

The name given by A. Wegener to an enormous land mass, the supercontinent of Pre-Cambrian age, probably split in two parts, Gondwanaland in the south being separated by a vast ocean, Tethis, from Laurasia in the North.

r Пангея

Название, данное А. Вегенером суперконтиненту докембрийского времени, который, возможно, состоял из двух частей - Гондваны на юге и Лавразии на севере, разделенных огромным океаном Тетис.

f Pangaea *f*

s Pangaea *f*

g Pangäa *n*

4405 panoramic photograph

A compound photograph made by phototheodolite or conventional camera from the Earth's surface and covering the whole circle of horizon or its considerable sector.

r панорамный снимок

Составной наземный снимок, сделанный с помощью фототеодолита или обычного фотоаппрата и охватывающий весь круг горизонта или значительный его сектор.

f photographie *f* panoramique

s fotografía *f* panorámica

g Panorama-Aufnahme *f*

4406 pantanal

The savanna-type vegetation occurring along some Brazilian rivers where lengthy dry periods are interrupted by occasional floods.

r пантанал

Растительность саванного типа, распространенная вдоль некоторых бразильских рек, где продолжительные сухие периоды изредка прерываются сильными паводками.
f pantanal *m*
s pantanal *m*
g Pantanal *n*

4407 pantograph
An instrument used to enlarge or reduce the scale of geographical maps.
r пантограф
Прибор, предназначенный для увеличения или уменьшения масштаба географических карт.
f pantographe *m*
s pantógrafo *m*
g Pantograph *m*; Storchschnabel *m*

4408 paper industry
A branch of industry producing paper for different purposes.
r бумажная промышленность
Отрасль промышленности, занимающаяся изготовлением бумаги разного назначения.
f industrie *f* du papier
s industria *f* de papel
g Papierindustrie *f*

4409 parabolic dune
A type of curved sand dune, similar in shape to a barchan but with the horns pointing upwind instead of downwind.
r параболическая дюна
Тип изогнутой песчаной дюны, по форме напоминающей бархан, но с рогами, направленными в сторону господствующего ветра, а не наоборот.
f dune *f* parabolique
s duna *f* parabólica
g Parabeldüne *f*

4410 parallax
The apparent change in position of an object in relation to another when the viewpoint is changed.
r параллакс
Видимое изменение в положении объекта по отношению к другому объекту, когда изменяется точка наблюдения.
f parallaxe *f*
s paralaje *f*; paralaxi *f*
g Parallaxe *f*

4411 parallel
A line of latitude running around the Earth parallel to the equator, on which all points have the same latitude.
r географическая параллель; параллель
Линия сечения поверхности земного шара плоскостью, параллельной экватору, на которой все точки имеют одинаковую географическую широту.
f parallèle *m*
s paralelo *m*
g Breitenkreis *m*

4412 parallel drainage pattern
A drainage pattern in which the channels of streams and their tributaries are almost parallel to one other, due to the underlying structure or a uniform slope gradient. → Fig. 95, p. 202
r параллельная речная сеть
Речная сеть, в которой русла рек и их притоков почти параллельны друг другу в соответствии со структурой коренных пород или однородным падением горных пластов. → Рис. 95, с. 202
f réseau *m* hydrographique parallèle
s red *f* hidrográfica paralela; drenaje *m* paralelo
g paralleles Flußnetz *n*

*** parallel flow → 3365**

*** parallel retreat of slope → 5054**

4413 parametrization
Approximate representation of subgrid-scale process in numerical modelling.
r параметризация
Приближенное представление процессов подсеточного масштаба в численном моделировании.
f paramétrisation *f*
s parámetrización *f*
g Parameterisation *f*

*** paramo → 4414**

4414 paramos; paramo
The bleak and barren plateux of the high Andes, between the snow line and the puna, covered with the high mountain vegetation that occurs at the altitude of 3,200-4,500 m in the Andes and in the mountains of eastern Africa and composed of short grasses and scattered trees 2-5 m high.
r парамос; парамо
Суровые и пустынные плато в Высоких Андах, между снеговой линией и пуной, покрытые высокогорной растительностью на высотах 3200-4500 м над ур. моря в Андах и горах Восточной Африки, представленной низкотравьем с редкими деревьями высотой 2-5 м.
f paramo *m*
s páramo *m*
g Paramo *m*

4415 pararendzina; regosolic forest soil
A soil with dark humus horizon that develops on loose carbonate rocks of humid areas.
r парарендзина
Почва гумидных областей на рыхлых карбонатных породах с темным гумусовым горизонтом.
f pararendzine *m*

s pararendzina *f*
g Pararendzina *f*

4416 **paraselene; mock Moon**
An image of the Moon which is formed in the same way as the *parhelion* but because of its weaker light is less bright than the parhelion.
r ложная Луна
Видимость Луны, которая формируется таким же образом, что и *ложное Солнце*, но из-за более слабого света она видна значительно слабее, чем ложное Солнце.
f fausse lune *f*
s falsa luna *f*; paraselena *f*
g falscher Mond *m*

4417 **parasitic cone; subsidiary volcano; adventive cone; subordinate volcano**
A cone situated at the foot or on a slope of the principal volcano.
r паразитический конус
Конус, расположенный у подножья или на склоне главного вулкана.
f cône *m* adventif
s cono *m* parasitario; cono *m* adventicio; cono *m* secudario
g Parasitärkegel *m*

4418 **parasitic crater; lateral crater; adventive crater**
A small volcanic cone or crater which appears on the flanks of a major volcano.
r паразитический кратер
Побочный кратер, образовавшийся на склоне главного вулканического конуса.
f cratère *m* adventif
s cráter *m* parasitário; cráter *m* lateral; cráter *m* adventivo
g Parasitärkrater *m*; Adventivkrater *m*

4419 **parcel; parcel household**
1. A small plot of land on which a peasant is mainly engaged in subsistence farming. 2. In biogeography, a structural horizontal subdivision of an ecosystem, e.g. separate groups of spruces, pines, aspens in forest communities. → Fig. 321, c. 718
r парцелла; парцеллярное хозяйство
1. Мелкий земельный участок, на котором крестьянин ведет свое (преимущественно натуральное) хозяйство. 2. В биогеографии, структурная часть горизонтального расчленения биогеоценоза, например, обособленные группы елей, сосен, осин в лесных сообществах. → Рис. 321, с. 718
f parcelle *f*; agriculture *f* parcellaire
s parcela *f*; agricultura *f* parcelaria
g Parzelle *f*; Paezellenwirtschaft *f*

* **parcel household → 4419**

* **parent material → 4420**

4420 **parent rock; parent material**
The upper layer of rocks on which the formation of soil takes place, the characteristics of which (e.g. mineral composition, permeability, texture) have an important effect on the soil-forming processses.
r материнская порода
Верхний слой горных пород, на котором происходит образование почвы, его характеристики (например, минеральный состав, воздухопроницаемость, текстура) играют важную роль в почвообразовательном процессе.
f roche mére *f*
s roca *f* madre; roca *f* originaria
g Muttergestein *n*

4421 **parent state; mother country**
The state with territorial possessions: colonies, protectorates, etc.
r метрополия$_1$
Государство, имеющее территориальные владения: колонии, протектораты и пр.
f métropole$_1$ *f*
s metrópoli$_1$ *f*
g Metropole$_1$ *f*

4422 **parhelion; mock sun; sun dog**
An image of the Sun which appears as an intensely bright spot on or near to the solar halo or a photometeor of the halo family (it can be in a form of circle or two luminous spots on both sides of the Sun situated at the same angular height as the Sun), it is caused by the refraction of sunlight by a particular type of ice crystals, especially in Cirrus.
r ложное Солнце
Яркое пятно из семейства оптических явлений гало, появляющееся в форме круга или двух пятен по обе стороны от Солнца на той же угловой высоте, что и Солнце, оно вызывается рефракцией солнечного света в ледяных кристаллах определенного типа, характерных для перистых облаков.
f parhélion *m*; faux soleil *m*
s parhelio *m*; falso sol *m*
g Parhelion *n*; Sonnenhund *m*; Nebensonne *f*

4423 **park; park vegetation**
Originally an enclosed area used for hunting, today an open land covered with trees, shrubs and herbs both planted and natural, that grow on a limited area and are used for recreation in a town or city.
r парк; парковые насаждения
Большой сад или насаженная роща с аллеями, цветниками, т.е. совокупность древесной, кустарниковой и травяной растительности, как саженой, так и естественной, культивируемой на ограниченной площади и предназначенной для отдыха, игр, прогулок.
f parc *m*; plantations *fpl* de parc
s parque *m*; plantaciones *fpl* de parque
g Park *m*; Parkanlagen *fpl*

Fig. 241. **Parkland**

* **park landscape → 4426**

4424 **park savanna**
The type of savanna in which trees are scattered relatively thinly among the grassland, in the manner of a European park.
r парковая саванна
Тип саванны, в которой относительно редкие деревья растут среди травы, напоминая европейские парки.
f savane-parc *m*
s sabana *f* parque
g Parksavanne *f*

* **park vegetation → 4423**

4425 **park-forest**
A woodland in a settlement outskirts that is used for recreation.
r лесопарк
Лесной массив в населенном месте или городе, используемый в рекреационных целях.
f pré-bois *m*; forêt-parc antropique
s prado-bosque *m*;
bosque-parque *m* antrópico
g Waldpark *m*

4426 **parkland; park landscape**
Thin forests of natural origin with scattered trees in different natural zones, e.g. savannah of tropics or forest-tundra. → Fig. 241
r парковый ландшафт
Разреженные леса естественного происхождения в различных природных зонах, например, саванны тропиков или лесотундровое редколесье. → Рис. 241
f forêt-parc *f*
s bosque-parque *f*
g Parklandschaft *f*

4427 **parliamentarianism**
A form of government in which the functions of the legislative and executive powers are strictly divided with the parliament being in a privileged position.
r парламентаризм
Система политической организации государства, при которой четко разграничены функции законодательной и исполнительной власти при привилегированном положении парламента.
f parlementarisme *m*
s parlamentarismo *m*
g Parlamentarismus *m*

4428 **paroxysmal volcanic eruption; Plinian eruption**
A sudden violent eruption of a volcano, usually of one that has been dormant for a long period, e.g. the eruption of Mt St Helens, USA, in 1980.
r пароксизмальное извержение; плинианский тип извержения
Внезапное сильное извержение вулкана, до этого долгое время пребывавшего в спокойном состоянии, например, извержение вулкана Сент-Хеленс в США в 1980 г.
f éruption *f* volcanique paroxysmale
s erupción *f* volcánica paroxismal
g paroxysmale Explosion *f*

4429 **parts of the world»**
A conventional division of large land territories together with the islands lying near them - Asia, America, Antarctica, Africa, Europe, Oceania.
r части света
Условное подразделение крупных территорий суши с расположенными вблизи них островами, к ним относятся Азия, Америка, Антарктида, Африка, Европа, Океания.
f parties *fpl* du monde
s partes *fpl* del mundo
g Weltteile *mpl*

4430 **part-time employment**
Working population engaged in his regular paid work not whole working week.
r частичная занятость
Трудоспособное население, имеющее работу неполную рабочую неделю.
f partiel emploi *m*
s subempleo *m*
g Kurzwochebeschäftigung *f*

4431 **part-time farmer**
A farmer who has a regular occupation, other than that of farming, to provide an income.
r частично занятый фермер
Фермер, который на регулярной основе работает в другом месте, помимо фермы, и доход его складывается из обоих этих занятий.
f fermier *m* occupé partielement
s granjero *m* ocupado parcialmente

g kurzbeschäftigter Farmer *m*

* **parvifoliate forest → 5497**

4432 **pass; mountain pass**
A narrow passage over low ground in a mountain range providing a passageway through the barrier.
r перевал
Глубокая седловина в гребне горного хребта или массива, доступная для его пересечения.
f passe *f*; col *m* de montagne
s paso *m*; puerto *m* de montaña
g Gebirgspass *m*

* **passability → 4737**

4433 **passenger flow**
The number of passengers moving along a transportation line.
r пассажиропоток
Количество пассажиров, перевозимых по транспортной линии.
f flux *m* des passagers
s flujo *m* de pasajeros
g Fahrgaststrom *m*

4434 **passenger intensity**
An index showing the intensity with which a transport system is used, expressed as the ratio between passenger turnover and the length of the way over which the turnover is realized.
r пассажиронапряженность
Показатель интенсивности использования транспортной системы, который выражается отношением пассажирооборота к длине участка дороги, на которой этот пассажирооборот выполнен.
f intensité *f* de trafic des passagers
s intensidad *f* de tráfico de pasajeros
g Passagierverkehrsintensitt *f*

* **passenger traffic → 4436**

4435 **passenger transportation**
Any type of transport that moves passengers.
r пассажирский транспорт
Разные виды транспорта, занятые перевозкой пассажиров.
f transport *m* des voyageurs
s transporte *m* de pasajeros
g Personentransport *m*

4436 **passenger turnover; passenger traffic**
A characteristic of passenger traffic calculated as the product of the number of passengers and the length to which they are transported.
r пассажирооборот
Характеристика пассажирских перевозок, выражаемая произведением количества перевезенных пассажиров и расстояния перевозки.
f trafic *m* des passagers
s circulación *f* de pasajeros; tráfico *m* de pasajeros
g Passagierumschlag *m*

4437 **passive glacier**
A glacier with a low rate of accumulation and ablation, which flows very slowly and transports little debris. → Fig. 2, p. 9
r пассивный ледник
Ледник с невысокой скоростью аккумуляции и абляции, который движется очень медленно и переносит мало обломочного материала. → Рис. 2, с. 9
f glacier *m* passif
s glaciar *m* pasivo
g Passivgletscher *m*

4438 **pastoral *adj***
Pertaining to the care of grazing animals.
r пастбищный; пасторальный
Относящийся к содержанию скота на пастбище.
f de pâturage; de berger
s pastoral *adj*; de pasto
g Weide-; pastoral *adj*

* **pastoral farming → 3537**

4439 **pastoral stock-breeding; grazing**
Raising farm animals (sheep, horses, camels, and yaks) in deserts, semi-deserts and some steppe and mountainous areas on distant pastures.
r отгонное животноводство; пастбищное животноводство
Практикуемое в пустынях, полупустынях и некоторых степных и горных областях содержание животных (овцы, лошади, верблюды и яки) на пастбищах, отдаленных от центрального хозяйства.
f élevage *m* de pâturage; élevage *m* pastoral
s cría *f* de ganado en pastos; pastoreo *m*; ganadería *f* pastoral
g Freilandhaltung *f*

* **pastoralism → 3537**

4440 **pasturage**
Keeping the domestic animals at the land with pasture grasses.
r пастьба
Содержание домашних животных на земельных угодьях с травянистой растительностью (пастбищах).
f pâture *f*
s pastoreo *m*
g Weiden *n*; Weidegang *m*

4441 **pasture; grazing**
Land covered with herbs where livestock can feed on growing grass (there can be natural pastures or semi-natural pastures).
r пастбище; выпас; выгон
Земельные угодья с травянистой растительностью, используемые для пастьбы животных, бывают естественные и сеяные.
f pâturage *m*

s pastos *m*; rebaño *m*; pastadero *m*
g Weide *f*; Viehweide *f*

4442 pasture capacity
A number of domestic animals, which can be fed on a certain pasture during certain time.
r емкость пастбища
Число домашних животных, которых можно прокормить на данном пастбище в течение определенного отрезка времени.
f capacité *f* de pâturage
s capacidad *f* de pastadero
g Weidekapazität *f*

4443 pasture digression
Serious deterioration of a pasture due to overgrazing.
r пастбищная дигрессия; пасторальная дигрессия
Резкое ухудшение состояния пастбища в результате чрезмерного выпаса скота.
f digression *f* de pâturage
s digresión *f* de pastos
g Weidedigression *f*

4444 pasture rotation
Grazing on a part of a pasture and then allowing it «to rest» (normally it is a 3 to 5-year cycle).
r пастбищеоборот
Обычная 3-5-летняя циклическая ротация участков пастбища, заключающаяся в чередовании их использования для пастьбы и «отдыха» земли.
f rotation *f* des pâturages
s rotación *f* de pastos
g Weideumschlag *m*

4445 patch cutting
The forestry practice of felling trees in a defined area within a forest so that the tree cover may regenerate naturally, towards the centre of the cleared area from the surrounding forest.
r пятнистая вырубка
Вырубка деревьев на небольших участках внутри леса таким образом, чтобы древесный покров мог возобновляться естественным путем, распространяясь от опушки леса в сторону вырубки.
f abattage *f* fait avec des cliffons
s tala *f* del bosque parcial
g fleckiger Holzschlag *m*

4446 paternoster lakes
A chain of lakes in a glaciated valley caused by the damming action of morainic ridges or rock bars. → Fig. 242
r чёткообразные озера
Цепь озер в ледниковой долине, образованных вследствие подпруживания моренными грядами или выходами скал. → Рис. 242
f lacs *mpl* à chapelet
s lagos *mpl* de rosario
g rosenkranzförmige Seen *mpl*

4447 path; trail
A track made by the frequent passage of people or animals through a wilderness.
r тропа
Пешеходный или конный путь через местность, не имеющую дорог.
f sentier *m*
s senda *f*
g Fußweg *m*

4448 patterned ground
The forms of the surface created by the sorting of the ground mass subject to the freeze-thaw action.
r структурный грунт
Вид грунта, образующегося в результате сортировки грунтовой массы при ее замерзании и оттаивании.
f sol *m* structuré
s suelo *m* estructurado
g Strukturgrund *m*

4449 paved road
A road covere with stones or asphalt.
r мощёная дорога; дорога с покрытием
Дорога, покрытая асфальтом или камнем.
f route *f* pavée
s camino *m* pavimentado
g gepflasterte Straße *f*

4450 pavement
A smoothed-out surface which most often forms along a river bank on a tow-path, where fine material is carried away by flash-floods, and pebbles, boulders are rounded during ice drift.
r мостовая
Выровненная поверхность, чаще всего возникающая по берегам рек на бечевнике, где мелкозем выносится паводковыми водами, а галька и валуны отшлифовываются льдом во время ледохода.
f pavé *m*
s pavimento *m*
g Pflaster *n*

4451 peasant
A farmer whose activities are dominated by the family group where the family provides all the labour and the produce is for the family as a whole.
r крестьянин
Фермер, в хозяйстве которого преобладает семейный труд, а полученная продукция в основном используется для содержания всей семьи.
f paysan *m*
s campesino *m*

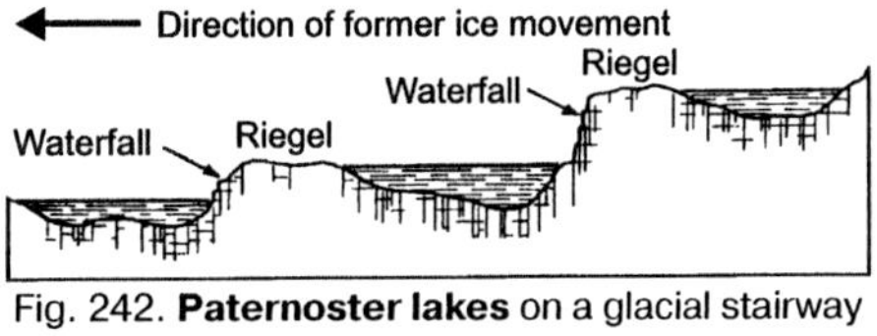

Fig. 242. **Paternoster lakes** on a glacial stairway

g Bauer *m*

4452 peasant farming
A farming activity carried out by a family at or near subsistence level on a small area of land, as distinct from that on a large farm with employed labour, organized primarily for commerce.
r крестьянское хозяйство
Семейное хозяйство на сравнительно небольшом участке земли, служащее основным средством существования, в отличие от крупного рыночного сельского хозяйства, использующего труд наемных рабочих.
f agriculture *f* de paysanne
s agricultura *f* campesina
g Bauernwirtschaft *f*

* **peat → 6417**

* **peat soils → 750**

4453 peatland study
The study of bogs including geological, hydrological, geobotanical and soil aspects.
r болотоведение
Учение о болотах, включающее геологические, гидрологические, геоботанические и почвенные аспекты.
f étude *f* des marais; étude *f* des tourbières
s estudio *m* de ciénaga;
estudio *m* de turberas
g Moorkunde *f*

* **pebble → 5398**

4454 ped
An aggregate of soil particles formed by natural processes.
r пед
Агрегат почвенных частиц, сформированный природными процессами.
f agrégat *m* du sol
s agregado *m* del suelo
g Ped *m*

4455 pedalfer
A soil from which the base compounds such as calcium carbonate and magnesium carbonate have been leached, and which contains the accumulations of aluminium, and iron compounds, such soils occur especially in humid climate.
r педалфер
Почва, из которой были выщелочены карбонаты кальция и магния и, наоборот, она богата накоплениями алюминия и железа, такие почвы особенно часто встречаются во влажном климате.
f pédalfer *m*
s pedalfer *m*
g Pedalfer *m*

* **pedestral → 1768**

* **pedestral rock → 1768**

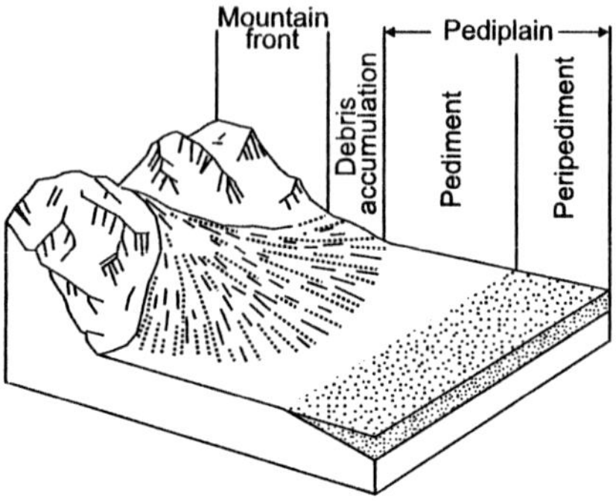

Fig. 243. **Pediment**

4456 pediment
A slightly inclined rocky platform at the foot of abrupt mountain slopes or uplands sloping gently away and gradually becoming featureless. → Fig. 243
r педимент
Слабонаклонная скалистая равнина у подножий крутых склонов гор и возвышенностей, постепенно понижающаяся и выполаживающаяся по мере удаления от них. → Рис. 243
f pédiment *m*
s pedimento *m*
g Pediment *n*

4457 pedimentation
The process of formation of pediments due to the retreat of steep slopes and the formation at their foot of gently inclined denudation plains.
r педиментация
Процесс формирования педиментов за счет отступания крутых склонов и формирования у их подножия пологих наклонных денудационных равнин.
f pédimentation *f*
s pedimentación *f*
g Pedimentierung *f*

4458 pediplain; pediplane
A level, slightly inclined denudation surface formed by merging of several pediments in the course of erosion. → Fig. 243
r педиплен
Выровненная слабонаклонная денудационная поверхность, образующаяся путем слияния педиментов на месте существовавшего ранее возвышенного рельефа. → Рис. 243
f pédiplaine *f*
s pedillanura *f*
g Pediplain *f*

4459 pediplanation
The action of processes which taken together result in the formation of a pediplain.
r педипланация
Группа геоморфологических процессов, приводящих к формированию педиплена.

f pédiplanation *f*
s pediplanación *f*
g Pediplanierung *f*

* **pediplane → 4458**

4460 **pedocal**
A soil, only slightly leached, containing an accumulation of calcium carbonate and magnesium carbonate, and occurring mainly in climates with an annual precipitation below 600 mm, insufficient to remove the soluble constituents.
r педокал
Почва, лишь слегка выщелоченная, содержащая накопления карбонатов кальция и магния, распространена в климатических условиях с годовыми осадками ниже 600 мм, что недостаточно для удаления из почвы растворимых компонентов.
f pédocal *m*
s pedocal *m*
g Pedocal *m*

* **pedogenesis → 5609**

* **pedogenic *adj* → 1842**

* **pedogenic processes → 5609**

4461 **pedolite**
A mixture of soil components and rocks that is formed under an intensive income of mineral materials (as volcanic ash, alluvium, etc.).
r педолит
Смесь компонентов почв и пород, образуемая в результате почвообразования при активном привносе минерального материала (вулканического пепла, аллювия и пр.).
f pédolite *m*
s pedolito *m*
g Pedolith *m*

* **pedology → 5626**

4462 **pedon; soil-individual**
A small ample of soil, an elementary soil column 1-10 m^2 in size that is enough to identify all the soil horizons and characteristics.
r педон; почвенный индивидуум
Элементарный объем почвы площадью 1-10 $м^2$, достаточный для выявления всех почвенных горизонтов и их соотношения.
f pedon *m*; sol *m* individu
s pedón *m*; suelo *m* individuo
g Pedon *n*; Bodenindividuum *n*

4463 **pelagic *adj***
Applied to the environment of the open oceanor a lake as distinct from the *ocean floor.*
r пелагический
Относящийся к окружающей среде открытого океана, в отличие от океанского ложа.
f pélagique *adj*
s pelágico *adj*
g pelagisch *adj*

4464 **pelagic deposit**
Sediments accumulated on a continental slope and ocean floor consisting of the remains of plankton organisms, finest mineral particles brought from land, space dust as well as the products of chemical processes in the sea.
r пелагические отложения
Осадки материкового склона и ложа океанов, состоящие из остатков планктонных организмов, тончайших минеральных частиц, приносимых с суши, космической пыли, а также продуктов химических процессов, протекающих в море.
f dépôts *mpl* pélagiques
s depósitos *mpl* pelágicos
g pelagische Sedimenten *npl*; pelagische Ablagerungen *fpl*

* **pelagic division → 4466**

4465 **pelagic organisms**
The animal and plant organisms occurring in the mass of open water of the oceans, seas, and lakes, i.e. unlike benthos are not connected with the bottom or littoral of a water body.
r пелагические организмы
Животные и растительные организмы, обитающие в толще водной массы океанов, морей и озер, в отличие от бентоса не связанные с дном водного бассейна или его побережьями.
f organismes *mpl* pélagiques
s organismos *mpl* pelágicos
g pelagische Organismen *mpl*

4466 **pelagic zone; pelagic division**
One of the two chief divisions of the aquatic environment based on depth of water consisting of the whole mass of open water (the other being the *benthic division*). → Fig. 244, p. 534
r пелагиаль; пелагическая зона
Одно из двух главных подразделений водной окружающей среды, представляющее собой всю водную массу, за исключением придонной части (в отличие от *бентали*). → Рис. 244, с. 534
f domaine *m* pélagique
s dominio *m* pelágico
g Pelagial *n*; Tiefsee *f*

4467 **Pelean eruption**
A type of volcanic eruption characterized by the extrusion of a very viscous lava which tends to consolidate as a solid plug or a volcanic spine, formed as a result of pressing it out from a vent.
r пелейский тип извержения
Извержение очень вязкой лавы, застывающей до выхода из жерла и образующей пробку, которая возвышается над вулканическим конусом в виде монолитного обелиска.
f éruption *f* péléenne
s erupción *f* peleana

g peleanischer Eruptionstyp *m*

4468 pellets
Large lumps of biogenic particles sedimenting in water.
r пеллеты
Крупные комки биогенных частиц, осаждающиеся в водной среде.
f pellets *mpl*
s pelletas *fpl*
g Pellets *mpl*

*** pellicular water → 2146**

4469 pelosol
A heavy clay soil without distinct genetic horizons that crack deeply in dry seasons in some periods free water circulates on its surface and in fissures.
r пелосоль
Тяжелая глинистая почва без выраженных генетических горизонтов, в некоторые периоды на ее поверхности и в трещинах циркулирует свободная вода.
f pélosol *m*
s pelosol *m*
g Pelosol *m*

*** pen soils → 750**

4470 peneplain; peneplane
An almost plain, featureless, gently undulating land surface reducing dislocated rocks of different genesis to base level.
r пенеплен
Слабо волнистая денудационная равнина, срезающая дислоцированные породы разного генезиса под один общий уровень.
f pénéplaine *f*
s penillanura *f*
g Peneplain *f*; Schnittfläche *f*; Festebene *f*

4471 peneplanation
The process of forming peneplains due to denudation and weathering resulting in the formation of level surfaces under the conditions of long stable position of the base level of erosion.
r пенепленизация
Выравнивание рельефа в результате процессов денудации и выветривания, приводящее в условиях длительного стабильного положения базиса эрозии к образованию пенепленов.
f pénéplanation *f*
s peneplanación *f*
g Bildung *f* einer Festebene; Bildung *f* einer Peneplain

*** peneplane → 4470**

4472 peninsula
An area of land almost surrounded by water or projecting far into the sea or lake from the mainland or an island.
r полуостров
Участок суши, резко и далеко выступающий от края материка или острова в море или озеро.
f presqu'île *f*; péninsule *f*
s península *f*
g Halbinsel *f*

4473 peninsular *adj*
Pertaining or belonging to a peninsula.
r полуостровной
Относящийся или принадлежащий к полуострову.
f péninsulaire *adj*
s peninsular *adj*
g Halbinsel-

4474 penitents; snow and ice penitent
Spikes of ice formed by melting on the surface of firn and ice which are inclined in the direction to the midday sun. → Fig. 245, p. 535
r кающиеся; снега кающихся
Остроконечные образования на поверхности фирна и льда, наклоненные в направлении на полуденное положение солнца. → Рис. 245, с. 535
f pénitents *mpl* de neige
s penitentes *mpl* de nieve
g Büßerschnee *m*; Zackenfirn *m*

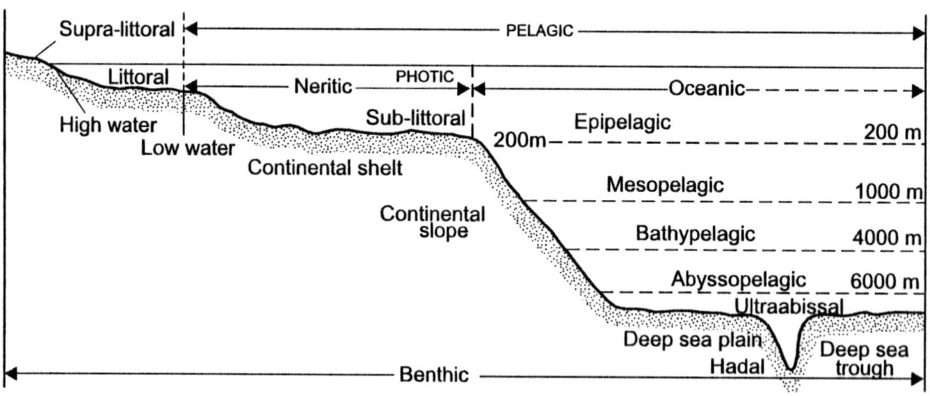

Fig. 244. Different zones of an ocean, including divisions of the **pelagic zone**

Fig. 245. **Penitents** in the high altitude of the Pamirs

4475 **peonage**
Debt dependence of a peasant (peon) on the landowner under which he has to repay his debt by working on the owner's farm.
r батрачество
Долговая зависимость крестьянина от помещика, обязывающая должника отработать свой долг в хозяйстве помещика-латифундиста.
f péonage *m*
s peonaje *m*
g Peonage *f*

4476 **people**
A body of persons that are united by a common culture; the citizens of a state.
r народ
Общность людей, объединенных общей культурой и историей; граждане определенной страны.
f peuple *m*
s pueblo[1] *m*
g Volk *n*

4477 **perched groundwater; perched water-table**
A temporary accumulation of ground waters in the zone of aeration formed due to the percolation of snow melt and rain waters, some height above the normal water-table, from which it is separated by an impermeable rock.
r верховодка
Временное скопление подземных вод в зоне аэрации, образовавшихся от просачивания талых снеговых или дождевых вод, обычно выше нормального водного зеркала, от которого отделяется водоупорным пластом.
f nappe *f* suspendue temporaire
s capa *f* temporal suspendida
g gespanntes Grundwasser *n*

* **perched water-table → 4477**

4478 **percolating waters**
Ground waters formed as a result of infiltration of rain, melt and river waters from the Earth's surface through pores of rocks.
r инфильтрационные воды
Подземные воды, образующиеся в результате просачивания через поры горных пород с земной поверхности дождевых, талых и речных вод.
f eaux *fpl* d'percolation; eaux *fpl* d'infiltration
s aguas *fpl* de percolación; aguas *fpl* de infiltración
g Perkolationswasser *n*; Infiltrationswasser *n*

4479 **percolation**
Tha downward movement of surface water through cracks, joints and pores in soil or rock to deeper layers where ground water forms, sometimes accompanied by the extraction of soluble substances.
r просачивание; перколяция; фильтрация
Движение воды через почву или горную породу (нередко сопровождаемое извлечением из них растворимых компонентов) в более глубокие слои, где формируются грунтовые воды.
f percolation *f*
s percolación *f*
g Versickerung *f*; Durchsickerung *f*; Perkolation *f*

4480 **percolation zone**
The zone on a glacier where melting constitutes a part of snow deposited over the year, and the ice formation proceeds mainly by the snow settlement and recrystallization.
r зона просачивания; фирновая зона
Зона на леднике, где таяние охватывает лишь часть отложенного за год снега и льдообразование идет в основном путем оседания и перекристаллизации.
f zone *f* de percolation
s zona *f* de percolación
g Perkolationzone *f*

4481 **percolative regime of soil**
Regular or periodical excessive moistening of a soil exceeding its moisture capacity, as a result of which some part of the soil water drains into lower soil horizons or to the ground.
r промывной режим почвы
Регулярное или периодическое поступление воды в почву, превышающее ее влагоемкость, что обеспечивет сток части почвенной воды в нижележащие почвенные горизонты или грунт.
f régime *m* de lessivage du sol
s régimen *m* lavado del suelo
g Bodenspülenregime *n*; Durchflussregime *n* des Bodens

4482 **percoline**
A soil horizon along which water seeps laterally.
r перколин
Почвенный горизонт, в который вода просачивается со стороны.
f percoline *m*

s percoline *m*
g Perkolin *m*

4483 pereletok
A frozen layer at the base of the active layer in the soil which remains unthawed for one or two summers.
r перелеток
Мерзлый слой в основании деятельного слоя многолетней мерзлоты, который не тает в течение одного или двух летних сезонов.
f péréletok *m*
s pereletok *m*
g Pereletok *m*

4484 perennial *adj*
Applied particularly to plants that continue in growth from year to year.
r многолетний
Относящийся чаще всего к растениям, продолжающим свой рост из года в год.
f pérennant *adj*; vivace *adj*
s perennal *adj*; perenne *adj*; vivaz *adj*
g perennierend *adj*

4485 perennial aufeis
An aufeis that does not melt for the summer period.
r многолетняя наледь
Наледь, не успевающая стаять за летний период.
f glaçage *m* permanent
s naledj *m* permanente
g vieljähriges Aufeis *n*; Daueraufeis *n*

4486 perennial ice
Sea ice, which existed longer than two annual cycles of growth and melting, being often up to 3 m thick.
r многолетний лед
Морской лед, просуществовавший более двух годовых циклов нарастания и таяния, нередко достигающий толщины 3 м.
f glace *f* de plusieurs années; glace *f* polaire
s hielo *m* de muchos años; hielo *m* polar
g mehrjähriges Alteis *n*

4487 perennial irrigation
A system of irrigation in which water is artificially made available to plants, not only seasonally, but throughout the year.
r круглогодичное орошение
Система орошения, при котором вода искусственно поступает к растениям не сезонно, а круглый год.
f irrigation *f* pérenne
s irrigación *f* perenne
g ganzjährige Bewässerung *f*

4488 perennial plant
A plant which lives for a long time, and its life cycle takes more than two years.
r многолетнее растение
Растение, жизненный цикл которого длится более двух лет.
f plante *f* vivace; plante *f* pérenne
s planta *f* vivaz; planta *f* perenne; planta *f* perennal
g perennierende Pflanze *f*; Dauerpflanze *f*; ausdauernde Pflanze *f*

4489 perennial snow patch; permanent snow stretch
An accumulation of snow and ice which retains unthawed for the whole warm season.
r снежник-перелеток
Снежник, сохраняющийся в течение всего теплого времени года.
f plaque *f* de neige pérenne
s lugar *m* de nieve perenne; parcela *f* de nieve perenne
g perennierendes Firnfeld *n*; Dauerfirnfeld *n*

4490 perennial stream; permanent stream
A watercourse which flows throughout the year, in contrast to intermittent stream.
r постоянный водоток
Водоток, текущий непрерывно в течение всего года, в противоположность временному водотоку.
f cours *m* d'eau pérenne; cours *m* d'eau permanente
s curso *m* de agua permanente; corriente *f* fluvial permanente
g perennierender Fluß *m*; perennierender Wasserlauf *m*

4491 pericline
A small anticline in which the strata are arched up into a dome so that the beds dip away on all sides from a central point.
r периклиналь
Небольшая антиклиналь, в которой пласты образуют купол и опускаются во все стороны от центральной точки.
f périclinal *m*
s periclinal *m*
g Periklinale *f*

4492 perigee
The point in the orbit of a planet (especially of the Moon) or satellites when it is nearest to the Earth.
r перигей
Ближайшая к Земле точка орбиты Луны или искусственного спутника Земли.
f périgée *m*
s perigeo *m*
g Erdnähe *f*; Perigäum *n*

4493 periglacial *adj*
Applied to areas adjacent to an ice sheet or glacier; applied to processes and phenomena characteristic of glacial epochs and developing mainly in cold and dry conditions under the impact of both general climatic changes and climatic influence of the glacier.
r перигляциальный

Относящийся к областям, прилегающим к ледникам или ледниковым покровам; характеризующий процессы и явления, свойственные ледниковым эпохам и развивающиеся преимущественно в холодных и сухих условиях, в том числе под влиянием самих ледников.

f périglaciaire *adj*
s periglaciar *adj*
g periglazial *adj*

4494 periglacial deposits; extraglacial deposits
Vast deposits formed near the Pleistocene glaciers or the deposits formed near a glacier edge mainly by its melt waters.

r перигляциальные отложения; экстрагляциальные отложения
Отложения, характерные для обширных приледниковых областей в плейстоцене, или отложения, образуемые вблизи края ледника, преимущественно его талыми водами.

f sédiments *mpl* périglaciaires; dépôts *mpl* extraglaciaires
s depósitos *mpl* periglaciares; depósitos *mpl* extraglaciares
g periglaziale Ablagerungen *fpl*; extraglaziale Ablagerungen *fpl*

* **periglacial drainage channel → 3678**

4495 periglacial lake
A lake near the glacier margin caused by streams of glacial melt water, more often during the degradation of a glacier.

r приледниковое озеро
Озеро у края ледника, образуемое потоками талых вод, вытекающих с ледника, особенно в фазу его деградации.

f lac *m* périglaciaire
s lago *m* periglacial
g periglazialer See *m*

4496 periglacial zone
A fringe of land located in the nearest proximity to ice sheets and glaciers (the Pleistocene or modern) characterized by specific climatic conditions, geological phenomena and landforms.

r перигляциальная зона
Полоса суши, непосредственно примыкающая к ледниковым покровам (плейстоценовым или современным), для которой характерен специфический климат, геологические процессы и образования.

f zone *f* périglaciaire
s zona *f* periglaciar
g periglaziale Zone *f*

4497 perihelion
The point nearest to the Sun in the orbit of a planet, comet or other celestial body rotating around the Sun.

r перигелий
Ближайшая к Солнцу точка орбиты планеты, кометы или другого небесного тела, обращающегося вокруг Солнца.

f périhélie *f*
s perihelio *m*
g Perihel *n*; Sonnenhöhe *f*

4498 period of a mudflow risk
The period when the formation of mudflows is most feasible in a region.

r селеопасный период
Время возможного схода селей.

f période *f* risquée des coulées de boue
s período *m* de riesgo de coladas fangosas
g schlammstromgefährdete Zeit *f*

4499 periodic migration; cyclic migration
Regular movements of people who always return to their permanent residences, when a worker, usually male, moves to industrial, urbanized area for a year or two and sends money back to his wife and children.

r периодическая миграция; циклическая миграция
Регулярные отъезды людей (чаще мужчин) с возвращением к месту проживания, когда рабочий едет на заработки из деревни в урбанизированные районы (иногда на год и больше) и посылает заработанные деньги жене и детям.

f migration *f* périodique; migration *f* cyclique
s migración *f* periódica; migración *f* cíclica
g periodische Migration *f*; zyklische Migration *f*

4500 peripheral settlement
A settlement within a zone of influence of a central settlement, often dominated by the latter.

r периферийное поселение
Поселение, расположенное в сфере влияния центрального поселения и нередко ему подчиненное.

f peuplement *m* périphèrique
s población *f* periférica; asentamiento *m* periférico
g Randsiedlung *f*

4501 periphery
A marginal, relatively weakly developed regions having functional ties with the centre, as in the core-periphery model.

r периферия
Окраинная относительно слабо развитая территория, имеющая функциональные связи со своим центром.

f périphérie *f*
s periferia *f*
g Peripherie *f*; Randgebiet *n*

4502 periphyton
Water organisms which cover any objects submerged into water objects.

r перифитон

Водные организмы, которые покрывают любые предметы, погруженные в водные объекты.
f periphyton *m*
s perifiton *m*
g Periphyton *n*; Aufwuchs *m*

4503 periplus
The ancient coastal description giving navigation details and sailing directions.
r перипл
Античная лоция берегов с описанием путешествия и направлений плавания.
f periplus *m*
s periplo *m*
g Peripl *n*

4504 permafrost; permanently frozen ground
The rocks and grounds, frozen (for at least two consecutive years) due to negative temperature and cemented by frozen moisture. → Fig. 246
r многолетнемерзлый грунт; многолетнемерзлая горная порода; вечная мерзлота
Горная порода (грунт), охлажденная до отрицательной температуры и сцементированная замерзшей в ней влагой. → Рис. 246
f pergélisol *m*
s permafrost *m*
g Permafrost *m*

4505 permafrost aggradation
A naturally caused increase in the thickness and areal extent of permafrost.
r аградация вечной мерзлоты

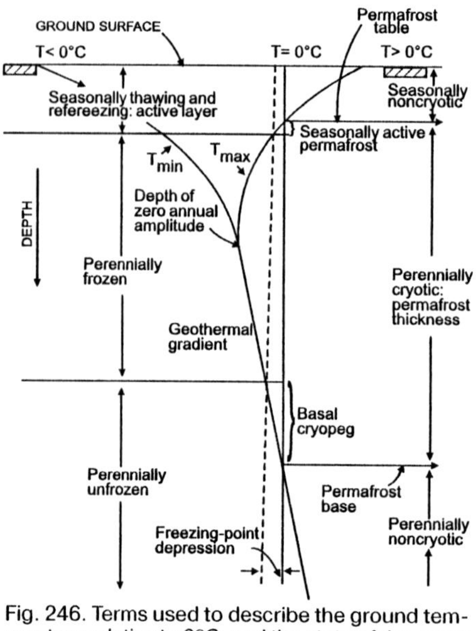

Fig. 246. Terms used to describe the ground temperature relative to 0°C, and the state of the water vs depth, in a **permafrost** environment

Расширение площади и увеличение мощности толщ многолетнемерзлых пород, вызванное природными причинами.
f aggradation *f* du pergélisol; expansion *f* de pergélisol
s agradación *f* del permafrost
g Permafrost-Aggradation *m*

4506 permafrost base
The lower boundary surface of permafrost above which the temperature of ground is below 0°C (cryotic) and below which the temperature is perennially above 0°C (noncryotic).
r подошва вечной мерзлоты
Основание толщи многолетнемерзлых пород, над которым температура грунта отрицательна, а ниже всегда положительна.
f plancher *m* du pergélisol
s base *f* de permafrost
g Permafrostuntergrenze *f*

4507 permafrost degradation
A naturally or artificially caused decrease in the thickness and area extent of permafrost.
r деградация вечной мерзлоты
Сокращение площади и уменьшение мощности толщ многолетнемерзлых пород, вызванное природными причинами или искусственно.
f dégradation *f* du pergélisol; régression *f* du pergélisol
s degradación *f* del permafrost
g Degradierung *f* der Permafrost; Permafrost-Degradation *f*

4508 permafrost limit
The maximum latitudinal (the closest to the equator) or the lowest altitudinal limit of the permafrost extent.
r граница вечной мерзлоты
Граница самого низкого (в горах) и самого дальнего к экватору географического распространения вечной мерзлоты.
f frontière *f* du pergélisol
s frontera *f* del permafrost
g Permafrostgrenze *f*

4509 permafrost studies; cryopedology; geocryology
The study of frost action and permanently frozen ground, including the study of the processes of freezing and thawing, and the engineering devices and techniques which may be employed to overcome the physical problems in such conditions.
r геокриология; мерзлотоведение
Наука о мерзлых грунтах и горных породах, природных явлениях, связанных с процессами промерзания и оттаивания многолетнемерзлых пород, а также инженерных проблемах, свойственных этим условиям.
f géocryologie *f*
s geocriología *f*
g Geokryologie *f*; Frostbodenkunde *f*

4510 permafrost table; permafrost top

The upper boundary surface of permafrost above which there is a layer of seasonally frozen ground or thawed ground.

r кровля вечной мерзлоты

Верхняя поверхность многолетнемерзлых пород, отделенная от дневной поверхности слоем сезонномерзлого грунта (сливающаяся мерзлота) или талого грунта, полностью не промерзающего (несливающаяся мерзлота).

f plafond *m* du pergélisol

s tope *m* de permafrost

g Permafrostobergrenze *f*

4511 permafrost thickness

The vertical distance between the permafrost table and the permafrost base, it includes short-term, seasonally and permanently frozen layers and may range from 1 m to more than 1000 m.

r толща промерзания; мощность вечной мерзлоты

Поверхностная толща земной коры, где гравитационные воды зимой превращаются в лед, различают толщи кратковременного, сезонного и многолетнего промерзания.

f épaisseur *f* du pergélisol

s espesor *m* de permafrost

g Permafrostmächtigkeit *f*

* **permafrost top → 4510**

4512 permafrost zone

The upper layer of the earthrs crust with negative temperature of soils and rocks and the presence or possible existence of ground ice. → Fig. 247

r криолитозона; зона вечной мерзлоты

Верхний слой земной коры с отрицательной температурой грунтов и горных пород и наличием или возможностью существования подземного льда. → Рис. 247

f zone *f* du pergélisol

s zona *f* de permafrost

g Permafrostzone *f*; Frostbodenzone *f*

4513 permanent snow; eternal snow

The accumulation over many years of snow and ice in polar regions and in mountains above the snow line.

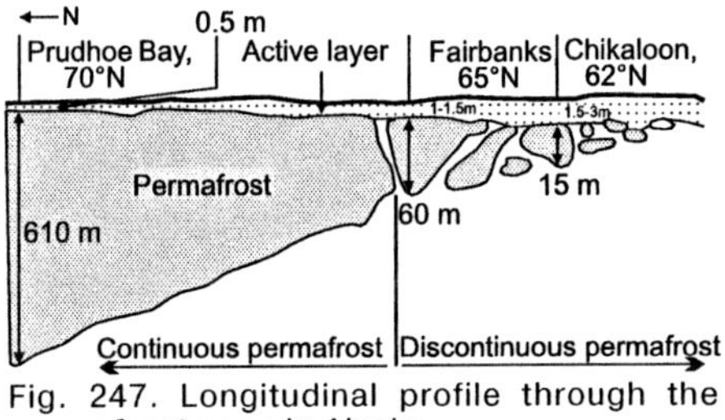

Fig. 247. Longitudinal profile through the **permafrost zone** in Alaska

r вечные снега

Многолетние скопления снега и льда в полярных странах и в горах выше снеговой линии.

f neige *f* permanente; neige *f* éternelle; neige *f* persistante

s nieve *f* permanente; nieves *fpl* perpetuas; nieve *f* eterna

g ewiger Schnee *m*

* **permanent snow stretch → 4489**

* **permanent stream → 4490**

* **permanently frozen ground → 4504**

4514 permeability

The ease with which liquids or gases can pass through rocks or a layer of soil.

r проницаемость

Способность горных пород или почв пропускать воду или воздух.

f perméabilité *f*

s permeabilidad *f*

g Durchlässigkeit *f*

4515 permeable rock

A rock that allows the free passage of groundwater owing to its porosity and fissures.

r водопроницаемая порода

Пористые или трещиноватые горные породы, по которым возможно движение подземных вод.

f roche *f* perméable

s roca *f* permeable

g wasserführendes Gestein *n*; wasserdurchlässiges Gestein *n*

4516 Permian

The last geological period of the Palaeozoic era, it began 280 million years ago and lasted for 55 million years. → Table.1, p. 284

r пермь

Последний период палеозойской эры геологической истории Земли, начался 280 млн лет назад и продолжался 55 млн лет. → Табл. 1, с. 284

f Permien *m*; Dyas *m*

s Pérmico *m*

g Perm *n*; Dyas *n*

4517 permissible environmental impact; threshold environmental pressure

The human effect which does not change the quality of the environment or changes it only slightly providing the preservation or increase of productivity of the environment.

r допустимая экологическая нагрузка

Влияние человека, которое не изменяет качество окружающей среды или меняет его в допустимых пределах, обеспечивая сохранение или повышение продуктивности окружающей среды.

f impact *m* écologique admissible

s impacto *m* ecológico admisible

g zulässige Umweltbelastung *f*

4518 perpetual frost climate; frost climate
Climate of the glaciers and ice sheet regions where temperatures are sufficiently low for ablation not to exceed annual accumulation of snow and ice.
r климат вечного мороза; нивальный климат
Климат ледников и ледниковых покровов, где температуры достаточно низкие для того, чтобы абляция не превышала годовое накопление снега и льда.
f climat *m* de pergélisol
s clima *m* de hielos perpetuos; clima *m* de glaciares
g Klima *n* der eviger Frost

4519 perspective projection
A type of map projection got by perspective projecting of the points of the spheric surface of the Earth on a plane, which includes azimuthal, conical and cylindrical perspective projections are distinguished.
r перспективная проекция
Картографическая проекция, получаемая перспективным проектированием точек поверхности шара на плоскость, различают азимутальные, конические и цилиндрические перспективные проекции.
f projection *f* perspective
s proyección *f* perspectiva
g perspektivischer Kartennetzentwurf *m*

4520 pesticides
Chemical substances used to destroy insects and other pests which are harmful to crops, domestic animals etc.
r пестициды
Вещества, используемые для уничтожения вредителей, паразитов, сорняков и пр. в целях защиты культурных растений.
f pesticides *mpl*
s pesticidas *mpl*
g Pestizide *mpl*

4521 petrifaction; petrification
The process by which any organic object is turned to stone, achieved by the gradual replacement of the fissured by mineral matter.
r окаменение
Процесс, при котором любой органический объект превращается в камень, что достигается путем постепенного заполнения трещин минеральным веществом.
f pétrifaction *f*
s petrificación *f*
g Petrifikation *f*; Versteinerung *f*

* **petrification → 4521**

4522 petrochemical ***adj***
Derived from petroleum or natural gas or from derivative of either, extensively used industrially in the producton of plastics, detergents, fertilizers, insecticides, etc.
r нефтехимический
Изготовленный из нефти или натурального газа или из производных этих субстанций и широко используемый в производстве пластиков, детергентов, удобрений, инсектицидов и пр.
f pétrochimique *adj*
s petroquímico *adj*
g petrolchemisch *adj*; erdölchemisch *adj*

4523 petrography; petrology
The description and systematic classification of rocks, the study of their chemical and mineralogical composition, structure, texture, forms of bedding, distribution, genesis and changes in the Earth's crust and on the Earth's surface.
r петрография; петрология
Наука о горных породах, их химическом и минералогическом составе, структурах и текстурах, формах залегания, распространении, происхождении и изменениях в земной коре и на поверхности Земли.
f pétrographie *f*
s petrografía *f*
g Petrographie *f*; Gesteinskunde *f*

4524 petroleum; crude oil
A complex mixture of hydrocarbons in a solid (bitumen), liquid or gaseous (natural gas) state, occurring in porous sedimentary rocks interbedded with shales and other rocks.
r сырая нефть
Сложная смесь углерода и водорода в твердой (битум), жидкой или газообразной (природный газ) форме, находящаяся в пористых осадочных породах, переслоенных со сланцами и другими породами.
f pétrole *m*; pétrole *m* brut
s petróleo *m*; petróleo *m* crudo
g Erdöl *n*

* **petrology → 4523**

4525 pH index; potential hydrogen
A number used to express the hydrogen-ion concentration of a solution and hence its degree of acidity, widely used to denote the acidity or alkalinity of soils.
r водородный показатель; индекс pH
Величина, характеризующая концентрацию иона водорода в растворах, а отсюда степень кислотности, широко используется для определения кислотности или щелочности почв.
f pH-index i
s pH-índice *m*
g pH-Wert *m*

4526 phaeozem
A soil that has thick, rich in humus, dark surface horizon and indications of deep leaching of calcium carbonate from profile, it is widespread in the central part of the USA, in

Argentina and Uruguay, as well as in north-eastern China.

r файозем
Почва, имеющая мощный, богатый гумусовым веществом, темный поверхностный горизонт и признаки глубокого вымывания из профиля карбоната кальция, распространена в центральной части США, в Аргентине и Уругвае, а также в северо-восточном Китае.

f phaeosem *m*
s faeozem *m*
g Phaeosem *m*

4527 **Phanerozoic**
The longest time of geological history of the Earth embracing the Palaeozoic, Mesozoic and Cainozoic eras with the total duration of 570 million years. → Table. 1, p. 284

r фанерозой
Крупнейший этап геологической истории Земли, охватывающий палеозойскую, мезозойскую и кайнозойскую эры, общей продолжительностью 570 млн лет. → Табл. 1, с. 284

f Phanérozoïque *m*
s era *f* phanerozoica; Phanerozoica *f*
g Phanerozoikum *n*

4528 **phases of the Moon**
Different forms of the aspect of the Moon as seen from the Earth are due to changes in the relative position of the Earth, the Moon and the Sun (new moon, first quarter, full moon, and gibbous Moon or last quarter, when tree-quarters of the Moon are illuminated by the Sun).

r фазы Луны
Различные формы видимой части Луны, обусловленные изменением взаимного расположения Земли и Луны относительно Солнца, различают новолуние, первую четверть, полнолуние и последнюю четверть.

f phases *fpl* de la Lune
s fases *fpl* de la luna
g Mondphasen *fpl*

4529 **phenological aspect**
The view of a phytocenosis that changes during a year according to plant development phases and especially to mass flowering of different plants.

r фенологический аспект; сезонный аспект растительности
Внешний вид фитоценоза, изменяющийся на протяжении года в соответствии с чередованием фаз развития растений и особенно массовым цветением того или иного растения.

f aspect *m* phenologique
s aspecto *m* fenológico
g phänologischer Aspekt *m*

4530 **phenological phases**
The stages of the seasonal development of nature, including the seasonal development of species of plants (as florescence, fruitification, etc.) and typical seasonal phenomena in animal life such as bird migration, nesting, etc.

r фенологические фазы
Стадии сезонного развития природы, сроки наступления которых определяются как особенностями данного вида, так и метеорологическими условиями года.

f phases *fpl* phénologiques
s fases *fpl* fenológicas
g phänologische Phasen *fpl*

4531 **phenological spectrum**
A graph showing the seasonal chainge of phenological phases of plant species or their communities.

r фенологический спектр
Графическое изображение сезонного чередования фенологических фаз отдельных видов растений или их сообществ в целом.

f spectre *m* phénologiques
s espectro *m* fenológico
g phänologisches Spektrum *n*

4532 **phenology**
The study of the seasonal (periodic) phenomena in nature, such as bird migration or plant flowering in various locality, temporal patterns of their occurrence and the causes determining them.

r фенология
Изучение сезонных явлений природы, сроков их наступления и причин, определяющих эти сроки, например, миграций перелетных птиц или времени цветения растений.

f phénologie *f*
s fenología *f*
g Phänologie *f*

4533 **phosphorescence**
Production of light with no heat, either caused by oxidation of phosphorous (as in sea water) or generated by some animals such as glow-worms.

r фосфоресценция
Свечение, не дающее тепла, вызываемое окислением фосфора (обычно в морской воде) или генерируемое некоторыми животными, например, светляками.

f phosphorescence *f*
s fosforescencia *f*
g Phosphoreszenz *f*; Nachleuchten *n*

4534 **photic zone; photo layer; euphotic zone**
The shallow water layer of the ocean lor any ather water body lying above the level of 200 m depth, in which light can easily penetrate and make photosynthesis possible. → Fig. 244, p. 534

r эвфотическая зона; фотическая зона; фотический слой
Поверхностная зона океана или любого иного водоема, в которую проникает доста-

точное количество солнечного света для поддержания фотосинтеза. → Рис. 244, с. 534
f couche *f* photique; photocouche *f*; zone *f* euphotique; étage *m* photique; étage *m* phytal
s capa *f* fótica; piso *m* fital; piso *m* fótico; fotocapa *f*; piso *m* eufótico
g euphotische Zone *f*; photishe Schicht *f*

* **photo layer → 4534**

4535 photochemical fog
A state of poor visibility caused by the chemical reaction of sunlight on hydrocarbons in the atmosphere.
r фотохимический туман
Состояние слабой видимости в результате химической реакции углеводородных соединений под действием солнечного света.
f brouillard *m* photochimique
s niebla *f* fotoquímica
g photochemischer Nebel *m*

4536 photogrammetry
The technique of using photographs to obtain measurements of any natural feature; a way of making a topographical map from air photograph.
r фотограмметрия
Метод использования фотографии для измерений природных объектов; способ изготовления топографических карт на базе фотографического материала.
f photogrammétrie *f*
s fotogrametría *f*; metrofotografía *f*
g Fotogrammetrie *f*; Photogrammetrie *f*; Lichtbildmessung *f*

4537 photographic hill shading
The hillshading used on the geographical maps and produced by photographing an obliquely-illuminated relief model.
r фоторельеф
Полутоновое изображение рельефа на географических картах, полученное фотографированием его объемной модели при косом освещении.
f estompage *m* photographique
s esfumado *m* fotomecánico
g photomechanische Schummerung *f*

4538 photomap
Map on which cartographic signs are placed on the background of a photograph.
r фотокарта
Карта, на которой условно-знаковое изображение размещается на фоне фотоснимка.
f photocarte *f*
s fotomapa *m*
g Luftbildkarte *f*

4539 photomosaic
A group of overlapping aerial photographs joined together according to common terrain features and oriented according to a map.
r фотосхема
Группа аэроснимков, смонтированных по общим контурам, ориентированных по карте и наклеенных на какую-либо основу.
f photomosaïque *f*
s fotomosaico *m*
g Luftbildmosaik *n*

4540 photoplan
A set of rectified photographs reproduced as a controlled photomosaic.
r фотоплан
Группа трансформированных аэроснимков, смонтированных на геодезической основе как контрольная фотосхема.
f photoplan *m*; photomosaïque *f* contrôlée
s fotoplano *m*; fotomosaico *m* controlado
g Luftbildplan *m*

4541 photosynthesis
The process by which green plants convert carbon dioxide and water into sugar and oxygen using sunlight energy.
r фотосинтез
Превращение зелеными растениями и некоторыми микроорганизмами лучистой энергии Солнца в энергию роста и развития растений.
f photosynthèse *f*
s fotosíntesis *f*
g Photosynthese *f*

4542 phototheodolite survey
Making topographic maps by stereopair of ground photographs obtained from the points of the Earth surface.
r фототеодолитная съемка
Создание топографических карт по стереопарам фотоснимков, полученных с точек земной поверхности.
f levé *m* au photothéodolite
s levantamiento *m* por fototeodolito
g Phototheodolitaufnahme *f*

4543 phototriangulation
The determination of fixed points and further making of a geodetic network on the basis of aerial photographs in laboratory with the help of photogrammetric instruments or graphic constructions.
r фототриангуляция
Определение опорных точек и построения геодезической сети на аэрофотоснимках в камеральных условиях при помощи фотограмметрических приборов или графических построений.
f phototriangulation *f*
s fototriangulación *f*
g Phototriangulation *f*

4544 phototrophic *adj*
Applied to an organism that obtains energy from sunlight.
r фототрофический
Применяется к организму, который получает энергию от солнечного света.

f phototrophique *adj*
s fototrófico *adj*
g phototroph *adj*

4545 **phototropism**
The response of a plant or animal to a source of bright light, e.g. the orientation of a flowering plant towards the Sun, or the flight of an insect towards a light at night.
r фототропизм
Реакция растения или животного на источник яркого света, например, ориентация цветущего растения в сторону Солнца, или полет насекомого ночью в сторону света.
f phototropisme *m*
s fototropismo *m*
g Phototropismus *m*

4546 **phreatic *adj***
Referring to groundwater lying below the water table.
r фреатический
Относящийся к водам, залегающим ниже зеркала грунтовых вод.
f phréatique *adj*
s freático *adj*
g phreatisch *adj*

4547 **phreatic divide; underground watershed**
A conventional line dividing groundwater flows moving in different directions.
r водораздел подземных вод
Условная линия, разделяющая потоки подземных вод, движущихся в разных направлениях.
f ligne *f* de partage des eaux souterraines
s línea *f* de separación de aguas subterráneas
g Grundwasserscheide *f*

4548 **phreatic eruption**
A type of volcanic eruption poduced when underground lava comes into contact with groundwater, thereby generating a sudden, violent emission of steam and mud at the ground surface.
r фреатическое извержение
Тип вулканического извержения, когда лава входит в контакт с грунтовыми водами, что приводит к внезапному появлению большого количества водяного пара и грязи на поверхности грунта.
f éruption *f* phréatique
s erupción *f* freática
g phreatische Eruption *f*

4549 **phreatic water**
Ground water in the zone of saturation.
r фреатические воды
Грунтовые воды, находящиеся в зоне насыщения.
f eau *f* phreatique
s agua *f* freática
g phreatisches Wasser *n*

4550 **phrigana**
Shrub communitities on dry southern mountain slopes in the Mediterranean region.
r фригана
Кустарниковые сообщества на сухих южных склонах гор в Средиземноморье.
f phrygana *f*
s frigana *f*
g Phrygana *n*

* **physical conditions → 4049**

* **physical environment → 4070**

4551 **physical frontier; natural boundary**
A natural limit that divides areas with different natural conditions.
r природная граница
Естественный рубеж, разделяющий местности, характеризующиеся неодинаковыми природными условиями.
f frontière *f* naturelle
s frontera *f* natural
g Naturgrenze *f*

4552 **physical geography**
One of the two main divisions of gheography which studies the components of the Earth's physical environment (lithosphere, atmosphere, hydrosphere, biosphere), their relationships, distribution over the Earth's surface and changes over time caused by narural processes or by human impact, it includes geomorphology, oceanography, climatology, land hydrology, glaciology, geocryology, biogeography, soil geography, landscape study, paleogeography.
r физическая география
Наука, изучающая географическую оболочку и ее компоненты (литосферу, атмосферу, гидросферу, биосферу), их взаимосвязи, распределение по земной поверхности и изменения во времени под влиянием природных процессов и антропогенных воздействий, включает геоморфологию, океанологию, климатологию, гидрологию суши, гляциологию, геокриологию, биогеографию, географиию почв, землеведение, ландшафтоведение, палеогеографию.
f géographie *f* physique
s geografía *f* física
g physische Geographie *f*; Physiogeographie *f*

4553 **physical map**
A small scale map, representing relief and drainage network.
r физическая карта
Мелкомасштабная карта, главным содержанием которой служит изображение рельефа и гидрографии.
f carte *f* physique
s mapa *m* físico
g physikalische Karte *f*; physische Karte *f*

* **physical meteorology → 4555**

4554 physical oceanography; marine hydrophysics
A branch of *oceanography* studying the structure and physical properties of sea water together with physical processes determining their features and changes in time and space.
r физическая океанология; физика моря
Раздел *океанологии*, исследующий строение и физические свойства водных толщ океанов и морей, а также физические процессы, определяющие их особенности и изменения во времени и пространстве.
f océanographie *f* physique
s oceanografía *f* física
g physische Ozeanographie *f*; Physik *f* des Meeres

* **physical weathering → 3754**

* **physical-geographical region → 4061**

4555 physics of atmosphere; physical meteorology; atmospheric physics
The physical properties of the atmosphere and the branch of meteorology concerned with this study.
r физика атмосферы
Физические процессы в атмосфере и физические свойства атмосферного воздуха, а также изучающая их наука.
f physique *f* de l'atmosphère
s física *f* de la atmósfera
g Physik *f* der Atmosphäre; Atmosphärenphysik *f*

4556 physiological drought
A condition of soil water bting sufficient, but as when the water is frozen, or when the rate of evapotranspiration exceeds the rate of uptake of water by a plant, which affects plants during daytime, causing them to wilt, although the soil may have an adequate supply of water.
r физиологическая засуха; физиологическая сухость почв
Недостаток влаги у растений, проявляющийся в дневное время и приводящий их к завяданию из-за того, что потеря воды путем транспирации происходит быстрее, чем растения впитывают ее своими корнями, хотя почва может содержать достаточное количество влаги, либо состояние почвы бореальных областей, когда имеющаяся в почве влага не может быть использована растениями из-за низких температур почвы, недостатка кислорода и пр.
f sécheresse *f* physiologique; sécheresse *f* physiologique du sol
s sequía *f* fisiológica; sequedad *f* fisiológica del suelo
g physiologische Trockenheit *f*; physiologische Bodentrockenheit *f*

4557 phytoclimate; plant climate
Microclimate conditions of the atmospheric layers in which plants live.
r фитоклимат
Микроклиматические условия в среде обитания растений.
f phytoclimat *m*
s fitoclima *m*
g Phytoklima *n*

* **phytocoenosis → 4600**

4558 phytogeography
A branch of *biogeography* studying the distribution of plants and vegetation communities over the Earth's surface.
r фитогеография; география растений; ботаническая география
Раздел *биогеографии*, изучающий распространение растений и растительных сообществ по земной поверхности.
f phytogéography *f*; géographie *f* botanique
s fitogeografía *f*; geografía *f* botánica
g Pflanzengeographie *f*; Vegetationsgeographie *f*

4559 phytomass
A part of biomass, the total mass of all plants, or of a particular species of plants referred to a unit area or volume of the habitat.
r фитомасса
Часть биомассы, общая масса всех растительных организмов, какой-либо их группы или отдельных растений на единицу площади или объема местообитания.
f phytomasse *f*
s fitomasa *f*
g Phytomasse *f*

4560 phytophage
An animal feeding on herbaceous vegetation.
r фитофаг
Животное, питающееся травянистой растительностью.
f phytophage *m*
s fitofago *m*
g Phytophag *m*

4561 phytoplankton
Microscopic plant organisms living and drifting in water (diatomic, green, and blue-green algae).
r фитопланктон
Совокупность растительных организмов, входящих в состав планктона (диатомовые, зеленые и синезеленые водоросли).
f phytoplancton *m*
s fitoplanctón *m*
g Phytoplankton *n*

* **phytoplankton bloom → 6642**

* **phytosociology → 4601**

* **pie diagram → 1660**

* **pie graph → 1660**

* **piedmont → 4565**

4562 **piedmont benches**
Step-like foothills found on the periphery of a mountain system or mountain range.
r предгорная лестница
Ступенчатые предгорья, расположенные по периферии горной системы или хребта.
f escalier *m* de piémont; piedmonttreppe *f*
s piedmonttreppe *f*; escalera *f* de piamonte
g Piedmonttreppe *f*; Rumpftreppe *f*

* **piedmont flat → 4565**

4563 **piedmont glacier**
A rather large glacier at the foot of a mountain formed by the coalescence of several valley glaciers, each with its own area of accumulation.
r предгорный ледник
Обширный ледник, образованный слиянием нескольких долинных ледников при выходе их к подножью склона.
f glacier *m* de piémont
s glaciar *m* de piedemonte; glaciar *m* de piamonte
g Piedmontgletscher *m*; Vorlandgletscher *m*

4564 **piedmont lake**
A lake located near the lower extremity of alpine glaciers, which advanced during the Pleistocene glaciation to the adjoining plain.
r окраинное озеро
Концевые водоемы у альпийских ледников, выползавших во время плейстоценового оледенения на прилежащую равнину.
f lac *m* de bordure; lac *m* de piémont
s lago *m* de piedemonte
g Randsee *m*

4565 **piedmont plain; piedmont flat; piedmont**
An inclined plain at the foot of a mountain with which it was closely connected in its formation and development.
r подгорная равнина; предгорная равнина; пьедмонт
Наклонная равнина, прилегающая к горам и тесно связанная с ними в процессе своего образования и развития.
f plaine *f* de piédmont; piédmont *m*
s llanura *f* piamonte; piamonte *m*; piedemonte *m*
g Pediment *n*

4566 **pier; jetty**
A narrow metal or concrete structure which extends into navigable water to be used as a landing place for getting on and off boats, promenade, etc.
r пирс
Причальное сооружение, выступающее в портовую акваторию и служащее для швартовки судов с двух сторон.
f quai *m* d'escale
s muelle *m* de atraque
g Pier *m*; Anlegekai *m*

* **piezometric head → 2853**

4567 **piezometric level; piezometric surface**
The level to which water from a confined aquifer will rise under its own pressure in a borehole.
r пьезометрический уровень
Уровень подземных вод, устанавливающийся в скважинах при вскрытии напорных вод.
f surface *f* piézométrique; niveau *m* hydroctatique; niveau *m* piézométrique
s nivel *m* piezométrico; nivel *m* hidrostático
g piezometrische Oberfläche *f*

* **piezometric surface → 4567**

4568 **pig-breeding**
The breeding of pigs used in foodstuffs and raw materials for light and food industries.
r свиноводство
Отрасль животноводства, занимающаяся разведением свиней с целью получения продуктов питания и сырья для легкой и пищевкусовой промышленности.
f élevage *m* des porcs; élevage *m* porcin
s cría *f* de cerdos
g Schweinezucht *f*; Schweinehaltung *f*

* **pile-dwelling → 3362**

* **pilgrimage → 5013**

4569 **pillow lava**
Lava effused into water and consolidated so as to resemble a jumbled mass of pillows.
r пиллоу-лава
Поток лавы, излившийся в воду и застывший в виде стекловидных шаров.
f pillow-lava *f*
s pillow-lava *f*
g Pillow-Lava *f*; Kissenlava *f*

* **pilot book → 5164**

4570 **pine forest**
A light coniferous forest with pure pine trees stand; growing on poor dry sandy soils of plains. → Fig. 248, p. 546
r сосновый лес; бор[1]; сосняк
Чистый сосновый лес на бедных сухих песчаных почвах равнин. → Рис. 248, с. 546
f pinède *f*
s bosque *m* de pinos; pinar *m*
g Kiefernwald *m*; Föhrenwald *m*

4571 **pingo; bulgunyakh**
A domed perennial ice-cored mound of cryogenic heaving occurring due to uneven ice

Fig. 248. **Pine forest**: the drawing by Russian artist Ivan Shishkin

formation when taliks freeze within the permafrost zone. → Fig. 249

r пинго; булгуннях

Многолетний бугор криогенного пучения, возникающий в результате неравномерного льдообразования при промерзании таликов и формировании многолетнемерзлых пород. → Рис. 249

f pingo *m*

s pingo *m*

g Pingo *n*

* **pinnacled iceberg → 4836**

4572 pinnate drainage pattern

A drainage pattern that resembles a feather in that the tributaries to the main stream are closely spaced and meet the main stream of acute angles. → Fig. 95, p. 202

r перистая речная сеть

Речная сеть, имеющая в плане перистую форму, когда притоки главной реки расположены близко друг к другу и впадают в главную реку под острым углом. → Рис. 95, с. 202

f réseau *m* hydrographique penné

s red *f* hidrográfica pinada

g federförmiges Flußnetz *n*

* **pioneer community → 4574**

4573 pioneer region

A newly-developed economic region.

r пионерный район

Экономический район нового освоения.

f région *f* pionnière

s región *f* pionera

g Pionierregion *f*; Erschließungsregion *f*

4574 pioneer vegetation; primitive plant aggregation; pioneer community

The earliest vegetation that develops on bare lands, i.e. occupies newly available site such as a cooled lava-glow, a newly emerged shoreline, a freshly deposited flood sediment or a new sand-dune, that is characterized by a random composition of plant species, the lack of closed cover and interaction between the plants.

Fig. 249. **Pingo** (bulgunyakh) in the North Siberia

r пионерная растительность; растительная группировка

Ранняя растительность, заселяющая оголенную территорию, например, охлаждающиеся после извержения лавовые поля, недавно поднятое побережье, вновь возникшие песчаные отложения и т.п., отличается случайным составом растений, отсутствием их сомкнутости и взаимовлияния.

f végétation *f* pionniére

s vegetación *f* pionera

g Pioniervegetation *f*; primitive Pflanzengruppe

4575 pipe

1. A volcanic channel or conduit filled with solidified magma, the vent opening into crater. 2. In hydrology, a subsurface channel through which water passes.

r трубообразная полость

1. Вулканический канал, заполненный затвердевшей магмой, жерло, открывающееся в кратер вулкана. 2. В гидрологии, подземный канал, по которому стекает вода.

f tube *m*

s tubo *m*

g Rohr *n*

4576 pipeline; conduit

A tube which is used to transport liquid, gaseous or solid goods under pressure produced by pumping stations.

r трубопровод; трубопроводный транспорт

Передача на расстояние жидких, газообразных или твердых грузов по трубам под давлением, которое обеспечивается насосными станциями.

f conduite *f*; pipe-line *m*

s tubería *f*; conducción *f*; conducto *m*

g Rohrleitung *f*

4577 piping

Underground erosion caused by the percolation of the atmospheric waters and which washes out solid particles from detrital rocks resulting in pceudokarst.

r пайпинг

Подземная эрозия, начало которой дают просачивающиеся атмосферные воды, вы-

носящие твердые частицы из обломочных отложений, что приводит к псевдокарсту.
f piping *m*
s piping *m*
g Piping *n*; Fugenerosion *f*

* **pipkrake → 4087**

4578 **pisciculture; fish breeding; fish farming**
The rearing of fish in natural and man-made pools for commercial purposes (especially developed in Japan, China, the USA, Canada, Norway, Russia, and Great Britain).
r рыбоводство
Отрасль хозяйства по разведению рыбы, увеличению и улучшению рыбных запасов в естественных и искусственных водоемах, особенно развито в Японии, Китае, США Канаде, Норвегии, России, Великобритании.
f pisciculture *f*
s piscicultura *f*; cría *f* de peces
g Fischzucht *f*

4579 **place**
A particular area of land or water characterized by its geographical situation, it is something integral, having characteristics distinct from those of surrounding places.
r место
Участок территории (акватории), характеризуемый географическим положением и представляющий нечто целостное, отличающееся от окружающей местности.
f lieu *m*
s paraje *m*
g Lage *f*; Stelle *f*; Platz *m*; Ort *n*

* **place-name → 6252**

4580 **placer**
An accumulation of small detrital rock material (sand, gravel, etc.) with particles of valuable minerals, especially of gold, tin, diamond, weathered from rocks and washed down by a stream.
r россыпь
Скопление мелких обломков горных пород (песка, гравия и т.п.) с частицами минералов, особенно с частицами золота, олова, алмазов, выветренных из коренных пород и вынесенных текучими водами.
f placer *m*
s placer *m*
g Feld *n*; Seife *f*; Lager *n*

4581 **plaggen**
A man-made pitch-dark soil horizon that has been developing under a periodical excessive application of organic matter for centuries.
r плагген
Антропогенный очень темный почвенный горизонт, формирующийся при повышенном периодическом внесении в почву органического вещества в течение столетий.
f plaggen *m*
s plaggen *m*
g Plaggen *m*

* **plagioclimax → 1622**

4582 **plain; flat**
A vast, relatively flat or gently rolling terrain with a slight range of heights and small inclinations, without prominent elevations and depressions.
r равнина
Относительно ровные поверхности, иногда весьма обширные, с небольшими колебаниями высот и малыми уклонами.
f plaine *f*
s llanura *f*
g Ebene *f*; Flachland *n*

4583 **plain built of horizontal strata**
A plain built up by horizontal or slightly inclined beds of the sedimentary cover.
r пластовая равнина
Равнина, сложенная напластованиями платформенного чехла, залегающими почти горизонтально или слегка наклонно.
f plaine *f* stratifée
s llanura *f* de capas
g Flözebene *f*

* **plain of aggradation → 127**

4584 **plan**
1. A large-scale map of a small area or a large scale having constant scale at any point and in all directions. 2. A formulated scheme of action, or the way in which it is proposed or intended to carry out some proceeding or course of action to achieve some goal, e.g. in connexion with economic and social development.
r план
1. Крупномасштабная карта ограниченного участка земной поверхности, сохраняющая постоянный масштаб в любой точке и по всем направлениям. 2. Сформулированная схема действий, преследующая цель достижения некоторой цели, например, в связи с экономическим и социальным развитием.
f plan *m*
s plano *m*
g Plan *m*

4585 **planation**
The process by which an area of the Earth's surface is denuded to produce a flat surface (planation surface) under the impact of denudation in the elevated area and as a result of accumulation in the area of low relief.
r планация; выравнивание рельефа
Уменьшение контрастности рельефа под воздействием процессов денудации в области поднятия и аккумуляции в области опускания.
f planation *f*; aplanissement *m*

s planación *f*; aplanamiento *m*
g Planieren *n*; Einebuung *f*; Planation *f*

4586 planation surface; erosion surface
A relatively flat surface produced by exogenic processes of denudation and accumulation both in mountainous regions and on platform plains.
r поверхность выравнивания; эрозионная поверхность
Относительно выровненная экзогенными процессами денудационная и аккумулятивная территории в горах и на платформенных равнинах.
f surface *f* d'aplanissement; surface *f* d'érosion
s superficie *f* de aplanamiento; superficie *f* de erosión
g Einebnungsfläche *f*

4587 planet
One of the celestial bodies, including the Earth, which revolve around the Sun along an elliptic orbit and shine only by reflected light.
r планета
Одно из небесных тел, включая Землю, обращающееся вокруг Солнца по эллиптической орбите и светящееся отраженным светом.
f planète *f*
s planeta *f*
g Planet *m*

4588 plane-table
Sheet of paper or photoplan sticked on a hard base and intended for graphical constructions when topographic survey is carried out.
r мензула; планшет
Лист бумаги или фотоплан, наклеенный на жесткую основу и предназначенный для графических построений при топографической съемке.
f planchette *f*
s plancheta *f*
g Mensel *f*; Messtisch *m*; Mensul *f*

4589 plane-table survey
A topographic survey caried out with the help of a plane-table and alidade (instrument for the measurement of angles).
r мензульная съемка
Топографическая съемка, выполняемая при помощи простейших инструментов: мензулы и кипрегеля (прибора для измерения углов).
f levé *m* à la planchette
s levantamiento *m* de plancheta
g Messtischaufnahme *f*

4590 planetary *adj*
Related to the planets, specifically to the Earth as a planet.
r планетарный
Относящийся к планете или связанный с ней в целом, особенно с планетой Земля.
f planétaire *adj*

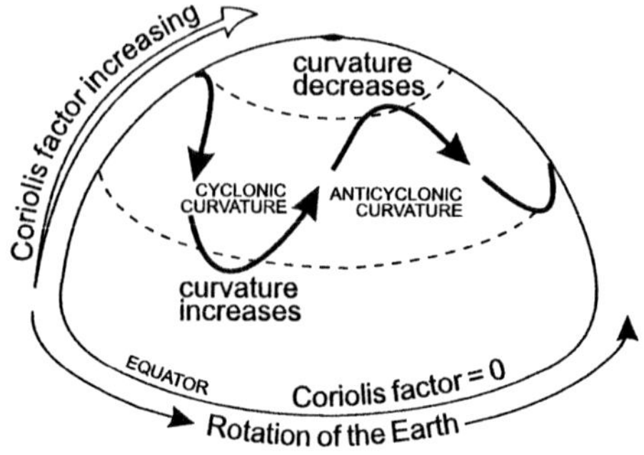

Fig. 250. **Planetary wave**

s planetario *adj*
g Planeten-; planetarisch *adj*

* **planetary boundary layer → 772**

4591 planetary geomorphology
A branch of *geomorphology* studying the main characteristics of the surface forms of the Earth as a planet and surface forms of the planets of the solar system.
r планетарная геоморфология
Раздел *геоморфологии*, изучающий основные особенности рельефа Земли как планеты и рельеф планет Солнечной системы.
f géomorphologie *f* planétaire
s geomorfología *f* planetaria
g Planetengeomorphologie *f*

4592 planetary wave; Rossby wave
The wave in the atmospheric circulation, in one of the principal zones of westerly winds, which is characterised by a great length and significant amplitude. → Fig. 250
r планетарная волна; волна Россби
Длинная атмосферная волна в основном западно-восточном переносе средних широт, характеризующаяся большой длиной и значительной амплитудой. → Рис. 250
f grande onde *f*; onde *f* planetaire; onde *f* de Rossby
s onda *f* larga; onda *f* planetaria; onda *f* de Rossby
g Rossby-Welle *f*

4593 planetary winds
The major latitudinal winds (trade winds, westerlies, etc.) in the atmospheric circulation in contrast to local winds.
r планетарные ветры
Основные широтные ветры (пассаты, западный перенос и т.д.) в системе атмосферной циркуляции, в отличие от местных ветров.
f vents *mpl* planétaires; circulation *f* planétaire
s vientos *mpl* planetarios; circulación *f* planetaria
g planetarische Winde *mpl*

4594 planimeter
A mechanical or electronic instrument used in measuring distances and areas on a map.
r планиметр
Механическое или электронное устройство для измерения площадей объектов по планам и картам.
f planimètre *m*
s planímetro *m*
g Planimeter *n*

4595 planimetric feature
Two-dimensional spatial object described by two co-ordinates, it includes three main types: point, linear, and polygonal.
r плоский объект
Двухмерный пространственный объект, описываемый двумя координатами, включает три основных типа: точечные, линейные и площадные (полигональные) объекты.
f symbole *m* plane
s símbolo *m* plano
g Flachobjekt *n*

4596 planina
Each of the flat-topped, lenghty limestone ridges which follow the structural grain of the Dalmatian karst country.
r планина
Плосковершинные, вытянутые в соответствии со структурным ядром, известняковые цепи в карстовом массиве Далмации.
f planina *f*
s planina *f*
g Planina *f*

4597 planisphere; map of the world
Cartographic depiction of the whole Earthrs surface in a form of an ellipsis, circle, star, and other figures with a graticule of meridians and parallels representing the Earthrs sphericity but without the division into hemispheres.
r планисфера; карта мира
Картографическое изображение всей земной поверхности (без разбиения на полушария) в виде эллипса, круга, звезды и других фигур с сетью меридианов и параллелей, передающее шарообразность Земли.
f planisphère *f*
s planisferio *m*
g Planisphäre *f*

4598 plankton
All forms of minute plants and animals freely floating in the ocean or in fresh water and drifting with water movements, they are essential food for fish, marine mammals, sea birds, etc.
r планктон
Совокупность растительных и животных организмов, населяющих толщу воды пресных и соленых водоемов и пассивно переносимых водой, служат пищей для рыб, водных млекопитающих, морских птиц и др.
f plancton *m*
s planctón *m*
g Plankton *n*

* **plankton bloom → 5255**

4599 planosol
Any of intrazonal group of soils found in humid climates that have a strongly leached upper layer over a compacted clay or silt horizon and occur on smooth flat uplands of Argentina, South Brazil, Africa and Australia.
r планосоль
Почва с резкой сменой механического состава с легкого на тяжелый, что связано с периодическим переувлажнением, имеет отбеленный, регулярно насыщаемый влагой горизонт, встречается в Аргентине, Южной Бразилии, Африке и Австралии.
f planosol *m*
s planosol *m*; planosuelo *m*
g Planosol *m*

* **plant → 2088**

* **plant association → 4600**

* **plant climate → 4557**

4600 plant community; plant association; phytocoenosis; vegetation community
Naturally occurring group of different plants or plant species occurying a common environment, interacting with each other, open, closed and stable communities are distinguished.
r растительное сообщество; растительная ассоциация; фитоценоз
Исторически сложившееся на некоторой территории сочетание растений с определенным видовым составом, структурой и местообитанием, различают открытые, закрытые и устойчивые сообщества.
f communauté *f* végétale; association *f* végétale; groupement *m* végétal; phytocénose *f*
s comunidad *f* vegetal; asociación *f* vegetal; agrupamiento *m* vegetal; fitocenosis *f*
g Pflanzengemeinschaft *f*; Pflanzengesellschaft *f*; Phytozönose *f*

4601 plant ecology; phytosociology
The science dealing with plant communities and vegetation cover formed by them, their composition, structure, development, productivity, use, and transformations.
r геоботаника; фитоценология
Наука, изучающая растительные сообщества и создаваемый ими растительный покров, их состав, строение, развитие, а также продуктивность, использование и преобразования.
f géobotanique *f*; phytocénologie *f*; sociologie *f* végétale; phytosociologie *f*

s geobotánica *f*; fitosociología *f*; sociología *f* vegetal
g Geobotanik *f*; Pflanzensoziologie *f*; Phytozönologie *f*

4602 plant formation
A taxonomic unit comprising plant communities having the same species-edificator (as pine forests of Scotch pine), or formed by several species without the predominance of one of them (as herb steppes).
r растительная формация
Таксономическая единица, объединяющая растительные ассоциации с одним видом-эдификатором, или ассоциации, формирующиеся несколькими видами без преобладания какого-либо одного из них, к первым относятся, например, все сосновые леса из сосны обыкновенной, ко вторым – разнотравные степи.
f formation *f* végétale
s formación *f* vegetal
g Pflanzenformation *f*

* **plant indicator → 3058**

* **plant litter → 3525**

* **plant residue → 3525**

* **plant succession → 5954**

4603 plantation
A farm or estate, especially in the tropics, with large scale planting of cash crops, especially on a monocultural basis, often with a large labour force, e.g. as on cotton, tea or rubber plantations.
r плантация
Большой участок земли, преимущественно в тропиках и субтропиках, используемый для выращивания в коммерческих целях урожая монокультуры, например, хлопка, чая или каучуконосов, требующего большого приложения труда, в прошлом нередко рабского.
f plantation *f*
s plantación *f*
g Plantage *f*

4604 plantation agriculture; plantation system
A system of agricultural production with the use of a large labour force (formerly of slave labour) to produce one or two commercial crops on a large scale.
r плантационное хозяйство
Крупнотоварное земледельческое хозяйство, специализирующееся на выращивании продовольственных и технических культур, преимущественно тропических и субтропических.
f agriculture *f* de plantation
s agricultura *f* de plantación
g Plantagewirtschaft *f*; Pflanzung *f*

* **plantation system → 4604**

* **plant-eating animal → 2754**

* **plantgrowing of waterbody → 2143**

4605 planting of greenery; verdurization
Growing wild or cultivated plants in the spaces free from buildings and roads in settlements to improve the environment.
r озеленение
Посадки на свободных от застройки и дорог пространствах населенных мест дикорастущих или окультуренных растений для улучшения качества среды.
f aménagement *m* d'espaces verts; verdissement *m*
s creación *f* de zonas verdes
g Begrünung *f*

4606 plastic soil
A soil that is capable of being deformed or moulded by moderate pressure into a different shape.
r пластичная почва
Почва, способная деформироваться или формоваться под давлением.
f sol *m* plastique
s suelo *m* plástico
g plastischer Boden *m*

4607 plate tectonics; continental drift theory
The theory that the lithosphere consists of several large and some small plates which move very slowly, the movement causes seismic and volcanic activity, continental drift, sea floor spreading, etc. → Fig. 251, p. 551
r тектоника плит;
глобальная тектоника плит
Теория, предполагающая, что литосфера разбита на крупные плиты, которые перемещаются по астеносфере в горизонтальном направлении, что вызывает сейсмическую и вулканическую активность, дрейф материков, спрединг и др. → Рис. 251, с. 551
f tectonique *f* des plaques
s tectónica *f* de placas
g Plattentektonik *f*

4608 plateau; highland
A vast area of high mountain relief that is composed of plateaus, mountain ridges and flat massifs, and wide valleys.
r плоскогорье; нагорье; плато
Обширные участки горного рельефа с преобладанием плоских или слабоволнистых междуречий, иногда разделенных узкими глубоко врезанными долинами.
f hautes terres *fpl*
s meseta *f*; altiplano *m*
g Hochebene *f*; Bergland *n*

4609 plateau glacier
A mountain glacier complex lying on a highland or a high plateau being also an area of accumulation of valley glaciers flowing down.

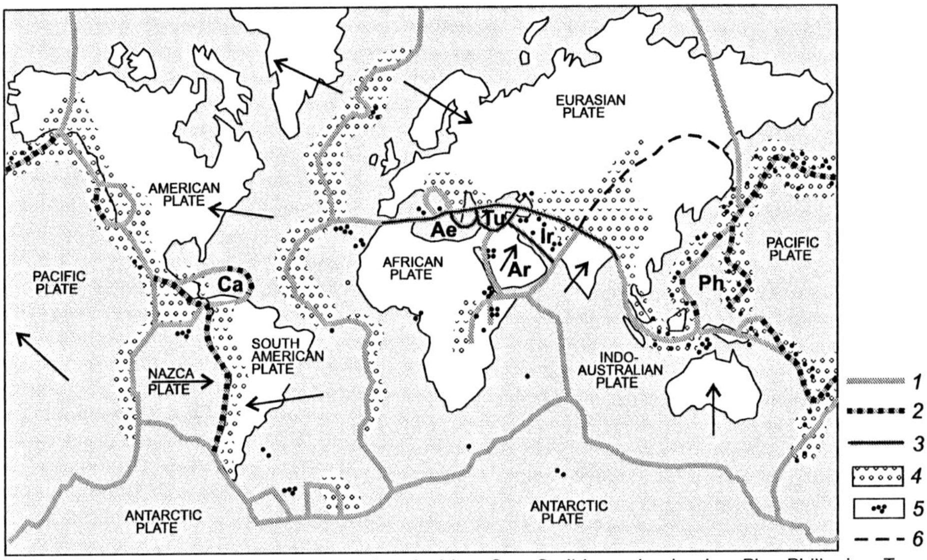

Fig. 251. **Plate tectonics**: Ae – Aegean, Ar – Arabian, Ca – Caribbean, Ir – Iranian, Ph – Philippine, Tu – Turkish. 1 – spreading ridge, 2 – collision zone, 3 – subduction zone, 4 – earthquake zone, 5 – volcanoes, 6 – uncertain plate boundary

r ледник-плато
Горный ледниковый комплекс на плоскогорье, представляющий собой область питания стекающих вниз долинных ледников.
f glacier *m* de plateau
s glaciar *m* de meseta
g Plateaugletscher *m*

4610 platform; craton
A large, relatively stable section of the Earth's crust characterized by plateau or plain relief, it is composed of a fold basement beneath a sedimentary cover and has remained relatively unaffected by orogenic activity for a very long period of time.
r платформа; кратон
Обширный участок земной коры, обладающий сравнительно малой подвижностью, равнинным или платообразным рельефом, а также двухъярусным строением, состоящим из складчатого основания и осадочного чехла.
f plate-forme *f*; craton *m*
s plataforma *f* estructural; cratón *m*
g Plattform *f*; Kraton *m*

4611 platform mantle; sedimentary cover
The thickness of slightly dislocated sedimentary rocks which overlie older folded and metamorphosed basement.
r платформенный чехол;
осадочный чехол
Толща слабо дислоцированных осадочных пород платформенных областей, залегающая на более древнем складчатом и метаморфизованном фундаменте.
f converture *f* sédimentaire
s manto *m* detrítico; manto *m* de derrubios
g Plattformdecke *f*

4612 plavni
Swamped or flooded for a considerable part of a year and densely overgrown flood-plains of lower reaches of large rivers in the south of European Russia and the Ukraine. → Fig. 252
r плавни
Заболоченные или затопленные значительную часть года и густо заросшие поймы низовий крупных рек на юге Европейской России и Украины. → Рис. 252
f plavni *m*
s estero *m*
g Plaven *n*

4613 playa
A level area occupying the centre of an enclosed basin in a desert in which a lake forms periodically.
r плайя

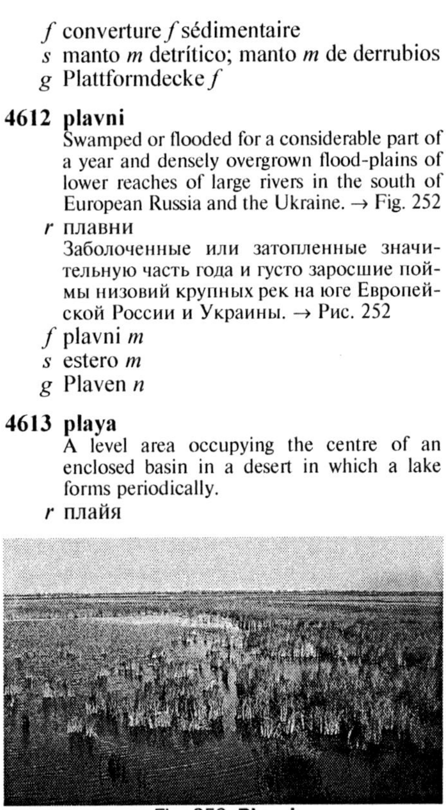

Fig. 252. **Plavni**

Плоское днище небольшой бессточной впадины в пустыне, временами превращающейся в озеро.
f playa *f*
s playa *f*
g Playa *f*

4614 Pleistocene
The longest epoch of the Quaternary characterised by the cooling of the Earth's climate and periodical appearance of vast continental ice sheets in the middle latitudes. → Table. 1, p. 284
r плейстоцен
Наиболее длительная эпоха четвертичного периода, характеризуется общим похолоданием климата Земли и периодическим возникновением в средних широтах обширных материковых ледниковых покровов. → Табл. 1, с. 284
f Pléistocène *m*
s Pleistoceno *m*
g Pleistozän *n*

* **Plinian eruption → 4428**

4615 plinthisol
A hydromorphic soil containing irreversibly hardening horizon (*plinthite*), it occurs mainly in tropics, sometimes on ancient surfaces of subtropical and temperate areas.
r плинтисоль
Переувлажненная почва, содержащая необратимо затвердевающий горизонт (*плинтит*), встречается преимущественно в тропиках, но иногда на древних поверхностях в субтропических и умеренных областях.
f plinthisol *m*
s plintisol *m*
g Plinthisol *m*

4616 plinthite
A soil horizon composed of a mixture of clay, quartz and other components, and enriched with iron, but depleted of humus which becomes transformed intoa hard crust formed under the alternation of wet and dry conditions.
r плинтит
Почвенный горизонт, представляющий собой смесь глины, кварца и других компонентов, обогащенный железом и обедненный гумусом и трансформирующийся в необратимый железистый панцирь при чередовании увлажнения и высыхания.
f plinthite *m*
s plintita *f*
g Plinthit *m*

4617 Pliocene
The upper epoch of the Neogene period which began about 7 million years ago and lasted more than 5 million years. → Table. 1, p. 284
r плиоцен
Верхняя эпоха неогенового периода (неогена), началась около 7 млн лет назад и продолжалась более 5 млн лет. → Табл. 1, с. 284
f Pliocène *m*
s Plioceno *m*
g Pliozän *n*

4618 plough pan; plough sole; tillage pan
The soil layer packed and solidified as a result of repeated ploughing at one and the same depth that hampers the penetration of roots and downward infiltration of water.
r плужная подошва
Подпахотный слой почвы, уплотненный и затвердевший в результате повторяющейся вспашки плугом всегда на одну и ту же глубину, что препятствует проникновению корней и инфильтрации воды вниз.
f semelle *f* de labour
s pie *m* de capa arable de suelo
g Pflugsohle *f*

* **plough sole → 4618**

4619 ploughing block
A boulder that moves downslope by frost creep and/or solifluction in a periglacial environment.
r блуждающий валун
Валун, движущийся вниз по склону под действием десерпции или солифлюкции в перигляциальных условиях.
f bloc *m* de labourage
s bloque *m* de labranza
g Wanderblock *m*

* **ploughland → 354**

4620 plucking
Glacial erosion or quarrying consisting in splitting off and breaking out of rocks from a glacier bed, carried out by glacier ice whereby the blocks are mechanically pulled from the rock floor when the basal ice freezes to them (it is one of the several types of exaration).
r плакинг
Ледниковая эрозия, заключающаяся в отщеплении и выламывании пород ложа ледника, производимая ледниковым льдом, когда его блоки механически движутся по коренному ложу, а базальный лед время от времени примерзают к нему (это один из нескольких видов экзарации).
f plucking *m*
s plucking *m*
g Plucking *n*

4621 plume
1. In plate tectonics, the upward movement of magma in a convection current caused by a locally very hot area lying between the core and the mantle of the Earth. 2. An underwater cloud-like effluent of liquid or gaseous fluid emitted from hydrothermal springs in the ocean floor.
r плюм
1. В тектонике плит, направленное вверх движение магмы в конвективном токе, формирующееся в очень горячей области

между ядром и мантией Земли. 2. Подводное облако жидкого или газообразного флюида, извергаемого из гидротермальных источников на океанском дне.

f plume *f*
s pluma *f*
g Fahne *f*

* **plumes → 3063**

4622 **plunge pool**
A hollow at the base of a waterfall, thought to have been formed largely by cavitation.
r кавитационная ванна
Впадина у основания водопада, выбитая падающей водой, преимущественно в результате кавитации.
f creu *m* de cavitation; mouille *f* de cavitation
s hueco *m* de cavitación; mouille *f* de cavitación
g wasserfallende Wanne *f*

4623 **plural society**
A society whithin which there are two or more elements, social orders or cultural groups which, in many areas of social behaviour, do not mix because each group wishes to retain its identity, its beliefs, traditions, etc.
r многонациональное общество
Общество, внутри которого существуют две или больше социальных или культурных групп, не смешивающихся друг с другом и сохраняющих их собственные культуру, верования, традиции и другие черты самобытности.
f société *f* plurielle
s sociedad *f* plural
g vielfältige Gesellschaft *f*

4624 **plutonic *adj***
Pertaining to any process associated with great heat deep in the Earth's crust.
r плутонический
Относящийся к любому процессу, связанному с тепловой энергией внутри земной коры.
f plutonique *adj*; plutonien *adj*
s plutónico *adj*
g plutonisch *adj*

* **plutonic rock → 3181**

4625 **plutonism**
The geological idea that relied upon the power of endogenic process, especially volcanism to form and transform rocks and to raise continents.
r плутонизм
Геологическая концепция, в которой основное значение в формировании и преобразовании пород земной коры придается действию эндогенных процессов, в особенности вулканизму.
f plutonisme *m*
s plutonismo *m*
g Plutonismus *m*

* **pluvial → 4627**

4626 **pluvial *adj*; pluviose *adj***
Referring to rain or atmospheric precipitation in general.
r плювиальный; дождевой
Относящийся к дождю или вообще к атмосферным осадкам.
f pluvial *adj*
s pluvial *adj*
g Pluvial-; Regen-; pluvial *adj*

4627 **pluvial period; pluvial**
In the Quaternary the stage of intensive growth of humidity of climate due to the increase of liquid precipitation in non-glacial areas, as opposed to relatively dry epochs (interpluvials).
r плювиальная эпоха; плювиал
В четвертичное время этап интенсивного увлажнения климата вследствие увеличения жидких осадков во внеледниковых областях, противопоставляются относительно сухим (аридным, или ксеротермическим) эпохам.
f période *f* pluviale; pluvial *m*
s período *m* pluvial; pluvial *m*
g Pluvialzeit *f*

4628 **pluviometric coefficient**
The ratio between the mean rainfall total of a particular month and the hypothetical amount equivalent to each monthrs rainfall were the total rainfall to be equally distributed throughout the year.
r плювиометрический коэффициент
Отношение действительной суммы осадков за некоторый месяц к той, которую этот месяц имел бы при вполне равномерном распределении годового количества осадков.
f coefficient *m* pluviométrique
s coeficiente *m* pluviométrico
g pluviometrischer Koeffizient *m*

* **pluviose *adj* → 4626**

4629 **poaching**
Catching or killing of wild animals llegally, violating the rules for hunting and fishery, and laws of wildlife protection.
r браконьерство
Добыча или уничтожение диких животных с нарушением правил охоты, рыболовства и других требований законодательства об охране животного мира.
f braconnage *m*
s caza *f* furtiva
g Wilddieberei *f*; Wildern *n*; Jagdfrevel *m*

4630 **podbur**
A soil of tundra and taiga areas with coarse-humus and illuvial humus-iron horizons, and without a distinguished bleached horizon.

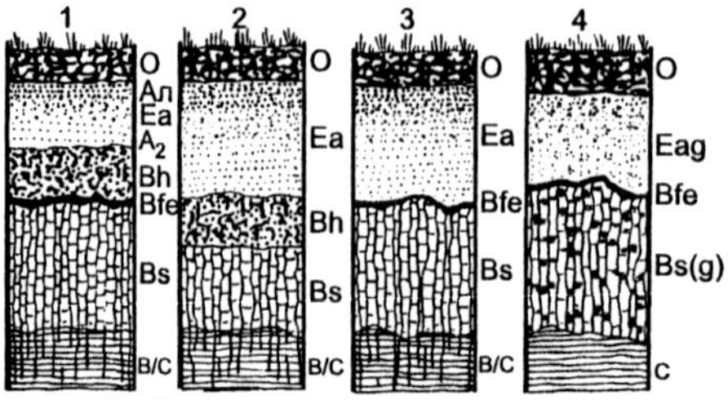

Fig. 253. **Podzols**: 1 – iron-humus podzol, 2 – humus podzol, 3 – iron podzol, 4 – peaty gleyed podzol; O – surface litter and decomposed organic matter; A-horizon: Ea – dark mineral soil (A_1) underlain by light-coloured leached layer (A_2), Eag – gleyed horizon; B-horizon: Bfe – iron pan, Bh – horizon enriched by humus, Bs – horizon enriched by clay minerals and sesquioxides (g=gleying), B/C – transitional horizon; C-horizon: C – weathered parent material

r подбур
Почва тундровых и таежных областей с грубогумусовым и иллювиальным гумусово-железистым горизонтами без самостоятельного осветленного горизонта.
f podbur *m*
s podbur *m*
g Podbur *m*

* **podsol → 4631**

4631 podzol; podsol; podzolic soil
A soil with an illuvial-humus-ferrous horizon and often a bleached horizon, the soil forms on light and gravel substrates in cool, seasonally humid climates where leaching is a dominant process under the vegetation of coniferous forests. → Fig. 253
r подзол
Почва с иллювиально-гумусово-железистым и, как правило, с осветленным горизонтами на легких и щебнистых субстратах, формирующаяся в холодном, сезонно влажном климате, где в хвойных лесах доминирует процесс выщелачивания. → Рис. 253
f podzol *m*; sol *m* podzolique
s podzol *m*; suelo *m* podzólico
g Podsol *m*; Bleicherde *f*; Podsolboden *m*

* **podzolic soil → 4631**

4632 podzolization
The formation of a podzol, the destruction of primary and secondary soil minerals through the agency of microorganisms and organic acids formed in the process of decomposing of organic substances, and the accumulation of minerals in the lower part of soil profile or beyond.
r подзолистый процесс; подзолообразование; оподзоливание почв
Разрушение первичных и вторичных минералов почвы под действием микроорганизмов, органических кислот, образующихся при разложении органических остатков, и вынос продуктов разрушения в нижнюю часть почвенного профиля или за его переделы.
f podzolisation *f*
s podzolización *f*
g Podsolierung *f*; Podsolbildung *f*

4633 podzolized soil
A soil showing signs of podzolization (as light colour, washed quartz grains, etc.) in the upper mineral horizons.
r оподзоленная почва
Почва с признаками оподзоливания (светлая окраска, отмытые кварцевые зерна) в верхних минеральных горизонтах.
f sol *m* podzolisé
s suelo *m* podzolizado
g podsolierter Boden *m*

4634 poikilotherm; exotherm; cold-blooded animal; poikilothermic animal
An animal whose body temperature varies according to the temperature of their environment, among these are all invertebrates, fish, amphibians and reptiles.
r пойкилотермное животное; холоднокровное животное
Животное с непостоянной температурой тела, меняющейся в зависимости от температуры внешней среды: беспозвоночные, рыбы, земноводные и пресмыкающиеся.
f animal *m* poïkilotherme; animal *m* à sang froid
s animal *m* poiquilotermo
g poikilothermes Tier *n*

* **poikilothermic animal → 4634**

* **point → 3706**

4635 point bar; side bar; meander bar; meander scroll
The deposit of sand or gravel developing on the inside of meander bend. → Fig. 254
r прирусловая отмель; побочень
Накопление аллювия у выпуклых (внутренних) участков берега меандрирующей реки. → Рис. 254, с. 554
f croissant *m*
s barra *f* lateral; banco *m* lateral

Fig. 254. **Point bar**

g Uferbank *f*

4636 **point feature**
A point symbol which represents a particular feature which is too small to be shown in outline at the scale of the map.
r точечный объект
Объект (изображаемый на географической карте точкой), размеры которого не выражаются в масштабе карты.
f symbole *m* ponctuel
s fenómeno *m* puntual; símbolo *m* puntual
g Punktobjekt *n*

4637 **point symbol**
Cartographic symbols used to show on a map the objects which are not expressed on a map scale.
r внемасштабный знак
Картографические условные знаки, применяемые для изображения объектов, площади которых не выражаются в масштабе карты.
f signe *m* symbolique
s signo *m* símbólico; símbolo *m* simbólico
g Signatur *f*

4638 **points of zero distortion**
Points on a geographical map at which there are no some or all kinds of distortions.
r точки нулевых искажений
Точки на географической карте, в которых отсутствуют отдельные или все виды искажений.
f isocentres *mpl*; centres *mpl* d'homologie
s centros *mpl* de homología
g Berührungspunkte *mpl*; Fokalpunkte *mpl*

4639 **polar *adj***
Applied to the natural or social phenomena found in the regions close to the poles.
r полярный
Относящийся к природному или социальному явлению, характерному для околополюсных районов.
f polaire *adj*
s polar *adj*
g polar *adj*

* **polar air → 4640**

4640 **polar air mass; polar air**
An airmass characterized by cool temperature, which originates between latitudes 40-60°, either over the ocean or over a continental interior.
r полярная воздушная масса; полярный воздух
Воздушная масса, характеризующаяся низкими температурами, формирующаяся между широтами 40-60° и над океаном, и внутри континента.
f masse *f* d'air polaire; air *m* polaire
s masa *f* de aire polar; aire *m* polar
g Polarluftmasse *f*; Polarluft *f*; polare Luftmasse *f*

4641 **polar anticyclone; polar high**
A persistent area of high atmospheric pressure, an anticyclone which forms over polar regions due to the cold ice and snow surface.
r полярный антициклон
Область повышеного давления, возникающая в полярных районах из-за сильного выхолаживания поверхности снега и льда.
f anticyclone *m* polaire
s anticiclón *m* polar
g Polarhoch *n*

4642 **polar circle**
Parallels situated at the latitudes 66°33′ which form the limits of the polar day and night areas (northward from the equator passes the Arctic Circle, southward – the Antarctic Circle).
r полярный круг
Параллели, отстоящие от экватора на 66°33′ и определяющие границы зон полярного дня и ночи (к северу от экватора расположен Северный полярный круг, к югу – Южный полярный круг).
f cercle *m* polaire
s círculo *m* polar
g Polarkreis *m*

4643 **polar climate**
Climate with long and cold winter, short and cool summer, it is typical for Arctic and Antarctic regions.
r полярный климат
Климат с длительной и холодной зимой, коротким и прохладным летом, характерен для арктических и антарктических районов.
f climat *m* polaire
s clima *m* polar
g Polarklima *n*

4644 **polar day**
The period of a year in polar regions during which the Sun does not set below the horizon.
r полярный день
Часть года в полярных областях, когда солнце не заходит за горизонт.
f jour *m* polaire
s día *m* polar
g Polartag *m*

4645 **polar distance**
One of coordinates in the equatorial system of celestial coordinates, an arc of a circle of declination from the North celestial pole to celestial body.
r полярное расстояние
Одна из координат в экваториальной системе небесных координат, дуга круга склонения от Северного полюса мира до небесного светила.
f distance *f* polaire
s distancia *f* polar
g Poldistanz *f*

4646 **polar easterlies**

A belt of eastrely winds (generally sporadic and of low velocity) blowing in a zone between the westerly depression track and the polar anticyclones.

r полярный восточный перенос
Пояс восточных ветров (случающихся время от времени и не достигающих больших скоростей), дующих в полосе между западным переносом умеренных широт и областью полярных антициклонов.

f zone *f* des vents polaires d'est
s zona *f* de vientos polares del este
g Polarostwinddrift *f*

4647 polar fauna
Animals inhabiting polar regions both in the Arctic and Antarctica.

r полярная фауна
Совокупность видов животных, обитающих в приполярных областях земного шара, включает и арктическую и антарктическую фауну.

f faune *f* polaire
s fauna *f* polar
g Polarfauna *f*; polare Fauna *f*

4648 polar front
The front or frontal zone which separates tropical and polar air masses in the North Atlantic and the North Pacific oceans, along which a series of depressions is generated which determine north European weather.

r полярный фронт
Фронт или фронтальная зона, разделяющая тропические и полярные воздушные массы в северных частях Атлантического и Тихого океанов, вдоль нее возникает серия барических депрессий, которые определяют погоду на севере Европы.

f front *m* polaire
s frente *m* polar
g Polarfront *f*

* **polar glacier → 1113**

* **polar high → 4641**

4649 polar lake; cold lake
A lake having the surfece temperature lower or about 4°C all the year round.

r полярное озеро; холодное озеро
Озеро, где в течение всего года температура поверхности воды ниже или около 4°C.

f lac *m* du type polaire
s lago *m* polar; lago *m* frío
g Polarsee *m*; kalter See *m*

4650 polar low; cold low
A persistent area of low atmospheric pressure located in the upper atmosphere over high latitudes.

r полярная область пониженного давления
Устойчивая область пониженного давления в верхней атмосфере над высокими широтами.

f dépression *f* polaire
s baja *f* polar
g Polartief *n*

* **polar motion → 4656**

4651 polar night
The period of a year in polar regions when the Sun does not rise above the horizon and there is no direct solar radiation.

r полярная ночь
Период года в полярных областях, когда Солнце не поднимается над горизонтом и прямое солнечное излучение отсутствует.

f nuit *f* polaire
s noche *f* polar
g Polarnacht *f*

4652 polar orbital satellite
A satellite used for remote sensing, which orbits the Earth along the meridians of longitude.

r спутник с полярной орбитой
Спутник, используемый для дистанционного зондирования Земли, орбита которого проходит вдоль одного из земных меридианов.

f satellite *m* orbital polaire
s satélite *m* orbital polar
g Polarorbitsatellit *m*

4653 polar outbreak
The occasional incursion of a cold mass, bringing strong winds and cool weather, from middle latitudes into lower latitudes.

r холодное вторжение
Случайное вторжение холодной воздушной массы, приносящее сильные ветры и холодную погоду из средних широт в более низкие широты.

f irruption *f* polaire
s irrupción *f* polar
g Kälteeinbruch *m*

4654 polar projection
Any circular map projection which is centred on one of the geographic poles.

r полярная проекция
Любая круговая картографическая проекция, центрированная в одном из географических полюсов.

f projection *f* polaire
s proyección *f* polar
g Polarprojektion *f*

4655 polar station
A place of observation and research established on an island or continental coast of the Arctic ocean or in Antarctic.

r полярная станция
Место постоянных налюдений на побережье континента или острове Северного Ледовитого океана, а также в Антарктике.

f station *f* polaire
s estación *f* polar

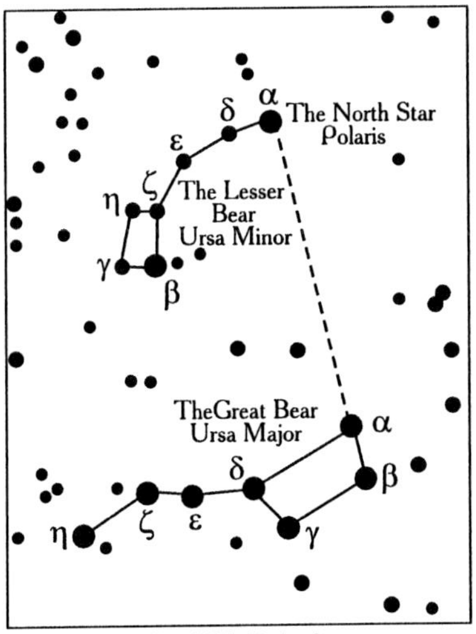

Fig. 255. **Polaris**

g Polarstation *f*

* **polar vortex → 1003**

4656 **polar wandering; polar motion**
A slow movement of geographical poles over the Earth's surface because the Earth axis does not keep the invariable position.
r перемещение полюсов;
движение полюсов Земли
Медленное перемещение географических полюсов по земной поверхности из-за того, что земная ось не сохраняет относительно тела планеты неизменного положения.
f mouvement *m* du pôle;
déplacement *m* du pôle
s movimiento *m* polar;
movimiento *m* de polos
g Polbewegung *f*

4657 **polar zone**
Geographical regions to the north of 40-65° N and to the south of 42-58° S with severe conditions, long polar day in summer and long polar night in winter.
r полярный пояс
Географические пояса, расположенные севернее 40-65° с.ш. и южнее 42-58° ю.ш., с суровыми природными условиями, длинным полярным днем в летнее время и долгой полярной ночью зимою.
f zone *f* polaire
s zona *f* polar
g Polarzone *f*

4658 **Polaris; Pole Star**
The brightest star in the constellation Ursa Minor, seen in the zenith at the North Pole, helps to find true North from any point in the Northern Hemisphere, the hight of the Pole Star above the horizon seen from that given point being equal to the latitude (like the *Southern Cross* in the Southern Hemisphere). → Fig. 255
r Полярная звезда
Яркая звезда в созвездии Малой Медведицы, расположена вблизи Северного полюса мира и сохраняет почти неизменное положение на небе при видимом суточном вращении небесной сферы, помогает найти направление на север из любой точки Северного полушария, а высота звезды над горизонтом указывает примерную широту данной точки (как и созвездие *Южный крест* в Южном полушарии). → Рис. 255
f étoile *f* polaire
s Estrella *f* Polar
g Polarstern *m*; Nordstern *m*

4659 **polder**
A low-lying land, usually below sea-level, reclaimed from the sea, lake, or floodplain of a river by embanking and draining, often supported by continued pumping.
r польдер
Осушенные и возделанные участки маршей, защищенные дамбами от затопления морскими и речными водами.
f polder *m*
s pólder *m*
g Polder *m*

4660 **pole; geographical pole**
The point of crossing of the Earth's rotation axis with the Earth's surface, i.e. the North Pole and South Pole where geographical meridians meet.
r полюс; географический полюс
Точка пересечения оси вращения Земли с земной поверхностью, различают Северный и Южный полюсы, здесь сходятся географические меридианы.
f pôle *m*; pôle *m* géographique
s polo *m*; polo *m* geográfico
g Erdpol *m*

4661 **pole of inaccessibility**
Regions which are the most difficult or impossible to reach in the Arctic and Antarctica situated in the central part of the Arctic Ocean and on Sovetskoye Plateau in Central Antarctica.
r полюс относительной недоступности
Наиболее труднодоступные области в Арктике или Антарктике, расположенные в центральной части Северного Ледовитого океана и на плато Советском в Центральной Антарктиде.
f pôle *m* de inaccessibilité
s polo *m* de inaccesibilidad
g Pol *m* der Unzugänglichkeit

* **Pole Star → 4658**

* **political border → 4035**

4662 **political geography**

A branch of human geography that studies how the world political map is being formed how political forces interact with economic, social, political and cultural life of people and how geographical factors influence political activities and problems.

r политическая география
Изучение формирования политической карты мира, взаимодействия политических сил с пространственной организацией экономической, социальной, политической и духовной жизни общества.

f géographie *f* politique
s geografía *f* política
g politische Geographie *f*

4663 political landscape
Spatial distribution of any political feature (e.g. electoral preferences) over the territory of a region.

r политический ландшафт
Пространственное распределение политических сил или какой-либо политической характеристики (например, электоральных предпочтений) по территории региона.

f paysage *m* politique
s paisaje *m* político
g politische Landschaft *f*

4664 political map
Geographic map of the globe, continents or large regions with state boundaries shown in it.

r политическая карта
Географическая карта земного шара, материка или крупного региона, на которой показаны границы между государствами.

f carte *f* politique
s mapa *m* político
g politische Karte *f*

4665 polje
A closed depression in karst terrain with steep sides and a flat floor up to 400 km^2 in area, is thought to have been formed by the coalescense of collapsed cave systems. → Fig. 177, p. 393

r полье
Замкнутая впадина карстового происхождения с крутыми бортами и плоским днищем площадью от нескольких до 400 км^2, образуется путем обрушения подземной полости. → Рис. 177, с. 393

f poljé *m*; dépression *f* fermée
s polje *m*; depresión *f* cerrada
g Polje *f*

* **pollen analysis → 5729**

4666 pollen diagram
A diagram made on the basis of the results of spores and pollen analysis of some samples taken from deposits, it shows the successive development of vegetation on a certain territory over a geological period under investigation.

r спорово-пыльцевая диаграмма
Диаграмма, построенная по результатам спорово-пыльцевых анализов отдельных образцов из толщи отложений, показывающая последовательное развитие растительности в данной местности за исследуемое геологическое время.

f diagramme *m* pollinique
s diagrama *f* de polen
g Pollendiagramm *n*

4667 pollutant
Natural or man-made harmful substances which get into the environment or appear inside it in abnormally high concentrations as compared to their natural level.

r загрязнитель;
загрязняющее вещество
Природные или антропогенные вещества, попадающие в окружающую среду или возникающие в ней в количествах, аномально высоких по сравнению с их естественным уровнем.

f polluant *m*
s contaminante *m*; producto *m* contaminante
g Schadstoff *m*

4668 polluted area; polluted zone
An area of land or sea in which pollutants are found.

r зона загрязнения
Территория или акватория, в пределах которой распространены или куда привнесены загрязнители.

f aire *f* de contamination; zone *f* de contamination; zone *f* de pollution; aire *f* de pollution; aire *f* contaminée
s área *f* contaminada
g verschmutztes Gebiet *n*; Belastungsgebiet *n*

* **polluted zone → 4668**

4669 pollution; environmental contamination; environmental pollution; contamination
The introduction of substances, organisms or energy undesirable for the environment, the formation or increase of the concentration of harmful elements over the average level, the exsessive occurrence of a process, action in an inappropriate place (e.g. acid rain, oil spillage, sewage outfall, industrial effluent, noise etc.) caused by human activity or natural hazards.

r загрязнение;
загрязнение окружающей среды
Привнесение в окружающую среду или образование в ней новых, обычно несвойственных ей веществ и соединений, либо превышение уровня их естественной концентрации, вызванное естественными или антропогенными причинами, например, нефтяные разливы, сбросы сточных вод и промышленных отходов, шум и пр.

f contamination *f* de l'environnement; pollution *f* de l'environnement
s contaminación *f* de medio ambiente; polución *f* de medio ambiente
g Verschmutzung *f*; Belastung *f*; Verunreinigung *f*

4670 pollution dome
A mass of polluted air in and above a city or industrial complex which is prevented from rising by the presence of an inversion above it (winds may alongate the dome into a pollution plume).
r купол загрязнения
Масса загрязненного воздуха внутри и над городом или промышленным комплексом, она остается на месте благодаря инверсии над этим воздухом, но под действием ветра может превратиться в шлейф загрязнений.
f coupole *f* de pollution
s cúpula *f* de polución
g Verschmutzungkuppel *f*

4671 pollution plume
The narrow streamer of polluted air carried downwind from a city or other major source of pollution during periods of moderate to strong air movement.
r шлейф загрязнений; факел выбросов
Узкий поток загрязненного воздуха, выносимого от города или иного источника загрязнений в периоды перехода от умеренного к сильному ветру.
f plume *f* de pollution
s pluma *f* de polución
g Verschmutzungfahne *f*

4672 polycentralisation
A type of urban development in which the functions of the centre of a state or territory are distributed among several places as it happens on the countrywide scale (the Netherlands) or on the city level (Tokyo) or the functions are evenly distributed over the territory.
r полицентризм
Тип урбанистического развития, когда функции центра государства или данной территории распределяются между несколькими местами, как это происходит в масштабах страны в Нидерландах или в масштабах крупного города в Токио, либо сводятся к равновесному распределению функций на всей территории.
f polycentrisme *m*
s policentrismo *m*
g Polyzentrismus *m*

* **polycentric world → 4685**

4673 polyconic projection
A type of map projection in which the parallels are represented by a system of non-concentric circular arcs with their centres lying on the straight line representing the central meridian, the other meridians are curves. → Fig. 256

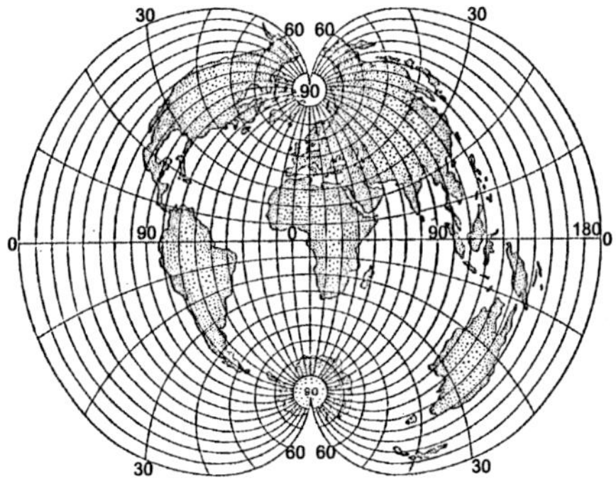

Fig. 256. **Polyconic projection**

r поликоническая проекция
Картографическая проекция, в которой параллели нормальной сетки изображаются эксцентрическими окружностями, средний меридиан – прямой, на которой расположены центры параллелей, остальные меридианы – кривыми. → Рис. 256
f projection *f* homolographique
s proyección *f* policónica
g polykonischer Entwurf *m*; Vielkegelentwurf *m*

4674 polygenetic landscape
Any landscape or tract of land the geomorphological features of which have been developed under different types of climates and different types of processes at various periods of its formation.
r полигенетический ландшафт
Любой ландшафт или участок территории, геоморфологические черты которого были сформированы в условиях разного климата и разных типов процессов в период его становления.
f paysage *m* polygénétique; modelé *m* polygénique
s paisaje *m* poligenético; modelado *m* poligénico
g polygenetische Landschaft *f*

* **polygenetic soil → 4675**

4675 polygenic soil; polygenetic soil
A soil formed by two or more different processes so that none of the horizons are genetically related.
r полигенетическая почва
Почва, сфомированная под влиянием двух или более различных процессов, так что ни один из ее горизонтов не может быть четко генетически определен.
f sol *m* polygénétique; sol *m* complexe
s suelo *m* poligénico
g polygenetischer Boden *m*

4676 polyglacialism

The concept of repeated glaciations.
r полигляциализм
Концепция множественности оледенений.
f polyglaciation *f*
s poliglacialismo *m*
g Polyglazialismus *m*

* **polygon → 1360**

4677 polygonal feature
An object of rather big size that can be expressed at the scale of the map.
r площадной объект; полигональный объект
Объект значительной площади, размеры которого выражаются в масштабе карты.
f symbole *m* polygonal
s símbolo *m* poligonal
g polygonal Objekt *n*

4678 polygonal point
Geodetic point the location of which on the Earth's surface in the unified system of coordinates is determined by a system of interconnected polygons, in which all angles and sides are measured.
r полигонометрический пункт
Геодезический пункт, положение которого на земной поверхности в принятой системе координат определено путем проложения системы взаимосвязанных полигонов, в которых измеряются все углы и стороны.
f point *m* polygonométrique
s punto *m* de polígonación
g polygonometrischer Punkt *m*

4679 polygonal relief
A kind of cryogenic relief resulting from frost fracturing of rocks and filling of the cracks with the secondary wedge ice.
r полигонально-жильный рельеф
Тип криогенного рельефа, формирующийся под воздействием морозного трещинообразования с заполнением трещин повторножильным льдом.
f relief *m* polygonal
s relieve *m* polygonal
g polygonales Gelände *n*

4680 polygonal tundra
Tundra with typical polygonal fissuring on the surface where the vegetation grows along the cracks, and the surface of polygons is almost bare.
r полигональная тундра
Тундра с характерной полигональной трещиноватостью на поверхности, где растительность селится вдоль трещин, а поверхность полигонов почти обнажена.
f toundra *f* polygonale
s tundra *f* poligonal
g polygonale Tundra *f*

* **polygonometry → 6332**

4681 polymictic lake
A lake in which water repeatedly mixes up to the bottom due to convection and dynamic processes.
r полимиктическое озеро
Озеро, в котором вода многократно перемешивается до дна благодаря конвекции и динамическим процессам.
f lac *m* polymictique
s lago *m* polimíctico
g polymiktischer See *m*

4682 polymineral rock
A rock consisting of several minerals which are present in approximately equal or comparable proportion.
r полиминеральная горная порода
Порода, состоящая из нескольких минералов, которые присутствуют примерно в равных или сопоставимых соотношениях.
f roche *f* polyminérale
s roca *f* polimineral
g polymineralisches Gestein *n*

4683 polynya; opening in sea-ice; pool of water
Any non-linear opening in sea-ice, i.e. a water area which is surrounded by sea ice, or between fast ice and pack-ice.
r полынья
Устойчивое пространство чистой воды среди льдов, на их границе или у берега.
f polynia *f*; polynie *f*; clairière *f* de glace
s polynie *f*; claro *m* de hielo; polinia *f*
g Polynja *n*; Eisloch *n*; Wake *f*; Eisblanke *f*

4684 polypedon
An area of soil (1-10 m^2) big enough to be classified on the basis of all the soil hotizons present on their proportions with respect to each other.
r элементарный почвенный ареал
Элементарный объем почвы площадью около 1-10 $м^2$, достаточный для выявления всех почвенных горизонтов и их соотношения.
f polypedon *m*
s polipedón *m*
g Polypedon *n*

4685 polypolar world; polycentric world; multipolar world
Geopolitical arrangement of the world with several (more than thee) centres of economic and political influence of the largest states.
r многополярный мир
Геополитическое устройство мира, в котором существует несколько (больше трех) центров экономического и политического влияния крупнейших государств.
f monde *m* multipolaire
s mundo *m* multipolar
g multipolare Welt *f*; mehrpolige Welt *f*

4686 polysaprobe

Fig. 257. **Pond**. The drawing by Russian artist Vasily Polenov

Anaerobic organisms (some species of bacteria and protozoa) living in oxygen-poor waters that are heavily polluted with easily decomposing organic matter, they are used for wastewater purification.

r полисапроб

Организмы, обитающие в водах, сильно загрязненных легко разлагающимися органическими веществами и очень бедных кислородом, преобладают бактерии и их потребители, с помощью которых производят очистку сточных вод.

f polysaprobe *m*

s polisaprobe *m*

g Polysaprob *m*

4687 pond

A water body lying in a natural hollow or in a depression formed by digging or by embanking a natural hollow. → Fig. 257

r пруд

Искусственный водоем, выкопанный или созданный путем постройки плотины в долине небольшой реки. → Рис. 257

f étang *m*

s estanque *m*

g Teich *m*

4688 ponor; water sink (*amer*); swallet; swallow; swallow-hole

A natural hole in the surface of a karstic area which absorbs rainwater, meltwater and river water and takes it to the depth.

r понор

Естественное отверстие на поверхности закарстованного массива, поглощающее и отводящее в глубину дождевую, талую снеговую и речную воду.

f ponor *m*; puits *m* absorbent; embut *m*

s ponor *m*; pozo *m* absorvente

g Ponor *m*

4689 pool$_1$

A still, usually deep, area in a stream or river. → Fig. 203, p. 447

r омут

Тихий и относительно глубокий участок на реке. → Рис. 203, с. 447

f profond *m*

s hondón *m*; hondura *f*

g Untiefe *f*

4690 pool$_2$

A relatively deep, still part of a river channel alternating with riffles, forming pool-and-riffle pattern of river bedforms.

r плёс

Относительно глубокий участок русла реки между перекатами.

f mouille *f*

s mouille *f*; hueco *m* en el lecho del río

g Flußstrecke *f*; Flußabschnitt *m*

4691 pool and riffle

A pattern developed on the bed of a stream by a sequence of alternating scoured pools and shallow gravel bars termed riffles, even if the channel is straight and the bed uniform.

r плёс и перекат

Особенность ложа речного потока, в котором чередуются глубокие участки (плёсы) и сложенные гравием отмели (перекаты), даже в случае прямолинейного русла и ровного днища.

f mouille *f* et banc *m*

s mouille *f* y banco *m*

g Flußstrecke *f* und Furt *f*

4692 pool in bog

The lowering in a raised bog between mounds and ridges usually overgrown by rare low grasses.

r мочажина

Понижение в верховых болотах между буграми и грядами, обычно заросшее редкими низкими травами.

f flaque *f* d'eau dans les tourbières

s charca *f* en la turbera

g Schlenke *f*

* **pool of water → 4683**

* **poor soil → 1521**

* **poorga → 818**

* **popular ecology → 2240**

4693 populated area; residence

Any place of permanent and temporal human habitation from an individual house to a megalopolis.

r населенное место; резиденция

Любое место временного и постоянного проживания людей от одиночно стоящего дома до мегалополиса.

f localité *f* habitée; lieu *m* habité; résidence *f*

s localidad *f* habitada; residencia *f*

g besiedelter Ort *m*; Wohnsitz *m*; Residenz *f*

4694 population

1. The total number of people living in an area at a particular time. 2. The total number or a group of people, animals or plants of the

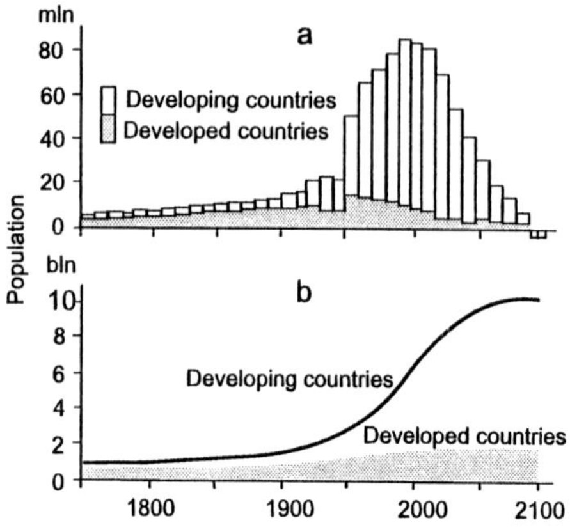

Fig. 258. Changes of **population₁** in developed and developing countries: a – annual increase in population, b – cumulative increase in population

same species living in a certain area at a particular time. → Fig. 258

r население; народонаселение[1]; популяция$_2$

1. Совокупность людей, живущих в пределах конкретной территории. 2. Обособленная совокупность разновозрастных особей одного вида, занимающая определенное пространство. → Рис. 258

f peuplement$_1$ *m*; population$_2$ *f*

s población$_1$ *f*; populación$_2$ *f*

g Bevölkerung$_1$ *f*; Population$_2$ *f*

4695 population density

The number of individuals occupying a particular unit of area (people, plants, animals).

r плотность населения; плотность популяции

Численность особей (людей, растений, животных) на конкретной территории, приходящаяся на единицу площади.

f densité *f* de population

s densidad *f* de población

g Bevölkerungsdichte *f*; Populationsdichte *f*

4696 population explosion

A great, rapid increase in size of the population of species due to almost a complete lack of limiting factors.

r популяционный взрыв

Резкое увеличение численности и плотности популяции в условиях почти полного отсутствия лимитирующих факторов, для многих видов грызунов (например, леммингов) и насекомых (например, саранча) носит волнообразный характер.

f explosion *f* de population

s explosión *f* de población

g Populationsexplosion *f*

4697 population geography; geography of population

The study of composition, spatial distribution and development of human population and human settlements in different natural and social conditions including the study of migration flows between territories and between different settlement types.

r география населения

Изучение закономерностей и пространственных особенностей формирования и развития современного состава населения и населенных мест в разных природных и социальных условиях, включая изучение миграционных потоков между отдельными территориями и разными типами поселений.

f géographie *f* de population

s geografía *f* de población

g Bevölkerungsgeographie *f*

4698 population growth; natural movement of population

The changes of the amount of population in a given region over a definite period of time.

r прирост населения; естественное движение населения

Изменение количества населения на какой-либо территории в целом за определенный промежуток времени.

f accroissement *m* de population; croissance *f* démographique

s crecimiento *m* de población; movimiento *m* natural de población; crecimiento *m* demográfico

g Bevölkerungszuwachs *m*; natürliche Bevölkerungsbewegung *f*

4699 population growth by migration

The increase of population due to migration of people from other regions.

r механический прирост населения

Прирост населения в данном регионе за счет миграции.

f accroissement *m* migratoire de population

s crecimiento *m* demográfico migratorio

g Migrationszuwachs *m*; Migrationsbevölkerungszuwachs *m*

* **population mobility → 3895**

4700 population pyramid; age-sex pyramid

A diagrammatic representation in the form of a pyramid showing the distribution of human population according to the age and sex. → Fig. 259, p. 563

r демографическая пирамида; возрастная пирамида

Графическое изображение распределения людей по возрасту и полу, на котором число людей каждого возраста и пола или доля их в населении изображены горизонтальными полосами определенного масштаба. → Рис. 259, с. 563

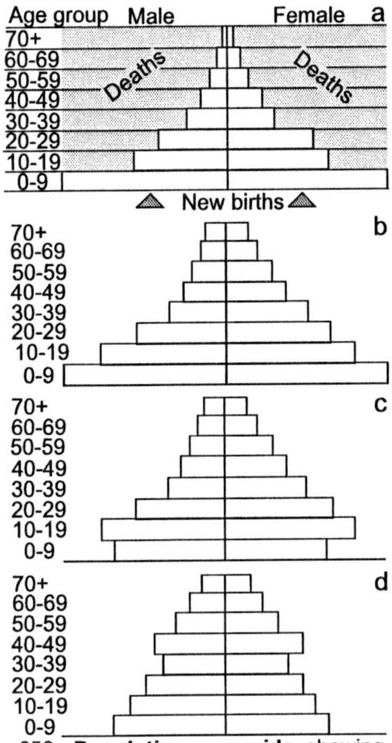

Fig. 259. **Population pyramids** showing the effects of reductions in birth and death rates: a – high death rate, b – effect of reduction in death rate; effect of reduction in birth rate after 10 years (c) and 40 years (d)

f pyramide *f* des âges
s pirámide *f* de edades
g Bevölkerungspyramide *f*; demographische Pyramide *f*

4701 pore; void
A cavity or space in a soil, snow or rock not occupied by solid mineral and through which fluids may pass through or be stored within it.
r пора
Пустота или пространство в почве, снеге или горной породе, не занятое твердым минеральным веществом, что позволяет жидкости проникать через поры и накапливаться в них.
f pore *m*
s poro *m*
g Pore *f*

4702 pore ice; ice-cement
Ice contained in the pores of rocks in the form of a thin film and separate crystals surrounding the soil particles.
r лед-цемент
Лед в дисперсных горных породах в виде тонких пленок и отдельных кристаллов, обволакивающих частицы скелета породы.
f glace *f* interstitielle
s hielo *m* intersticial
g Poreneis *n*

4703 pore water pressure
The pressure on rock and soil particles produced by water contained in the interstices between the particles, it can cause failure of rocks.
r поровое давление воды
Давление на горную породу или почву, вызываемое водой, содержащейся в промежутках между частицами, может приводить к оседанию и обвалам.
f pression *f* d′eau de pore
s presión *f* de agua de poro
g Porewasserdruck *m*

4704 pororoca
The tidal bore at the mouth of the Amazon river.
r поророка
Приливная волна, распространяющаяся вверх по течению реки Амазонки.
f pororoca *f*
s pororoca *f*
g Pororoka *f*

4705 porosity
The volume of the pore spaces occupied by air in relation to the total volume of the rock expressed as a percentage.
r пористость
Объем занятых воздухом полостей в грунте или горной породе в процентах от общего объема образца.
f porosité *f*
s porosidad *f*
g Porosität *f*

4706 porous structure
The texture of rocks with the pore spaces not filled with secondary minerals, e.g. volcanic tuff.
r пористая текстура
Текстура горных пород с пустотами, не заполненными вторичными минералами, например, вулканический туф.
f texture *f* poreuse
s textura *f* porosa
g poröse Struktur *f*

4707 port
A place where ships may anchor to load and unload cargo, together with the complex of constructions in a coastal zone of a sea, lake, reservoir, river and adjacent water surface aimed at handling freights, regarding and supplying vessels with fuel and other services. → Fig. 260, p. 564
r порт
Комплекс сооружений с прилегающими участками берега моря, озера, водохранилища или реки и акваторий, предназначенный для погрузки и разгрузки судов, снабжения их топливом, ремонта и других услуг. → Рис. 260

Fig. 260. **Port** of Havre, France

f port *m*
s puerto *m*
g Hafen *m*

4708 portage
The overland transport of a boat and its goods from one navigable waterway to another as well as a place of the minimum distance between two navigable bodies of water where in the past boats and goods were carried overland from one river to the other.
r волок
Транспортировка по суше речного судна и товаров из одной судоходной реки в другую, а также место наибольшего сближения двух судоходных рек, где в далеком прошлом кратчайшим путем перетаскивали по суше суда и грузы из одной реки в другую.
f portage *m*
s porte *m*
g Tragstelle *f*

4709 portolan
Written description of ports and routes which connect them, known in the ancient classical world and the Early Middle Ages.
r портолан
Текстовое описание портов и путей между ними в период античности и раннего средневековья.
f portulan *m*
s portulano *m*
g Portulan *n*

4710 portolan chart
A marine nautical chart with detailed image of a shoreline used from XIII to XVI century for trade seafaring.
r карта портолан; компасная карта
Морская навигационная карта с детальным изображением береговой линии и компасными координатными сетками, употреблявшаяся с XIII по XVI в. для торгового мореплавания.
f carte *f* portulan
s carta *f* portulana
g Portulankarte *f*

4711 positioning; GPS measurement
The determination of co-ordinates of an object in three-dimensional Earth's space by satellite positioning systems (GPS).
r позиционирование
Определение координат объекта в трехмерном земном пространстве с помощью глобальных спутниковых систем.
f positionnement *m*
s posicionamiento *m*
g Positionierung *f*; Positionieren *n*

4712 positive difference of glacierization
The difference in altitude marks between the highest point of the glacier basin and the snow line.
r положительная разность оледенения
Разница в отметках наивысшей точки в бассейне ледника и снеговой линии.
f différence *f* positive de glaciation
s diferencia *f* de glaciación positiva
g positive Vergletscherungsdifferenz *f*

4713 positive feedback
The situation when the result of natural process stimulates the process causing it. → Fig. 261, p. 565
r положительная обратная связь
Ситуация, когда результат природного процесса стимулирует процесс, ее вызывающий. → Рис. 261, с. 565
f rapport *m* de causalité inverse positive
s realimentación *f* positiva; retrocomunicación *f* positiva
g positive Rückkopplung *f*

* **possibilism → 2423**

* **possible insolation duration → 4714**

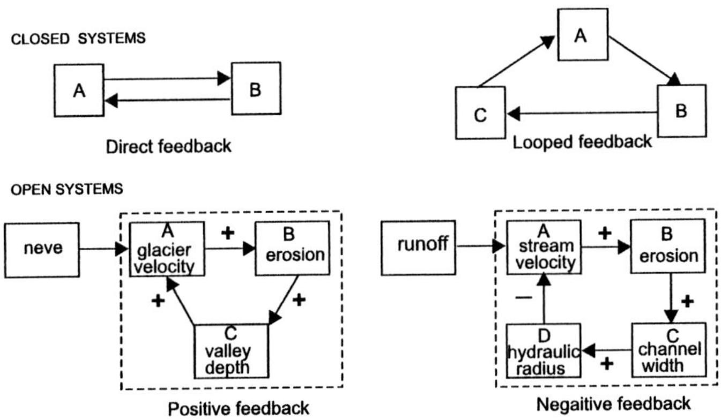

Fig. 261. **Positive feedback**

4714 possible sunshine duration; possible insolation duration
The maximum time interval during which the solar radiation can reach a given surface.
r возможная продолжительность солнечного сияния
Максимальная продолжительность времени, в течение которого солнечная радиация может падать на данную поверхность.
f durée *f* d'insolation éventuelle
s duración *f* de la insolación eventual
g mögliche Sonnenscheindauer *f*

4715 postclimax
A climax community still existing in a place where the environmental conditions are no longer stable.
r посткликмакс
Климаксная растительность, сохранившаяся лишь в данном местообитании благодаря некоторым местным условиям.
f postclimax *m*
s postclímax *m*
g Postklimax *m*

4716 post-Fordism
A new system of industrial organization characterized by flexibility both of labour and machinery that replaced Fordism, a form of production of small lots of a wide range of goods constantly modified to consumer demands.
r постфордизм
Новая система организации промышленного производства, пришедшая на смену фордизму, заключающаяся в мелкосерийном выпуске широкого ассортимента товаров, постоянно модифицируемых с учетом запросов потребителей.
f post-Fordisme *m*
s postfordismo *m*
g Postfordismus *m*

4717 postglacial *adj*
Referring directly to the period after the end of glaciation in the Pleistocene.
r послеледниковый
В плейстоцене относящийся непосредственно ко времени после завершения ледниковой эпохи.
f postglaciaire *adj*
s postglaciar *adj*
g postglazial *adj*

4718 post-industrial city
A city which it is thought will become typical of the post-industrial society, it is characterized with rather low population density, abundant recreational facilities, etc.
r постиндустриальный город
Город с небольшой плотностью населения, обширными рекреационными площадями и пр., который, как полагают, будет типичен для постиндустриального общества.
f ville *f* postindustrielle
s ciudad *f* postindustrial
g postindustrielle Stadt *f*

4719 post-industrial society
The stage of socio-economic development of a society following the industrial one in which the service industries, finances, management, and information rather than manufacturing production dominate in the economy.
r постиндустриальное общество
Новая стадия социально-экономического развития, сменяющая индустриальное общество, когда главной составной частью экономики становится уже не материальное производство, а сфера услуг, финансы, управление и информационная среда.
f société *f* post-industrielle
s sociedad *f* postindustrial
g postindustrielle Gesellschaft *f*

4720 potamology
The scientific study of rivers including their geomorphological, hydrological and hydrobiological processes.
r потамология
Учение о реках, включающее геоморфологические, гидрологические и гидробиологические аспекты.
f potamologie *f*

s potamología *f*
g Potamologie *f*; Flußkunde *f*

* **potential evaporation → 2029**

4721 potential evaporation; evaporativity
The maximum amount of water vapour which could be evaporated by the surface of pure water per unit surface and unit time, under existing atmospheric conditions.
r испаряемость;
потенциально возможное испарение
Количество водяного пара, которое могло бы испариться с единицы поверхности чистой воды за единицу времени при существующих условиях.
f évaporation *f* potentielle; evaporativité *f*
s evaporación *f* potencial
g Verdunstungsstärke *f*; potentielle Verdunstung *f*; Verdunstungsfähigkeit *f*

4722 potential evapotranspiration
The highest amount of water that could be evaporated or transpired from plants if the plants had an unlimited water supply.
r потенциальная эвапотранспирация
Наибольшее количество воды, которое могло быть испариться с растения с участием процесса транспирации в случае, если растение получает неограниченное количество влаги.
f évapotranspiration *f* potentielle
s evapotranspiración *f* potencial
g potentielle Evapotranspiration *f*

* **potential hydrogen → 4525**

4723 potential instability
The state of air layer when its rise at a front or over a relief barrier causes the increase of instability of the layer stratification.
r потенциальная неустойчивость;
конвективная неустойчивость
Состояние воздушного слоя, когда его подъем вызывает возрастание неустойчивости стратификации слоя.
f instabilité *f* potentielle;
instabilité *f* convective
s inestabilidad *f* potencial;
inestabilidad *f* convectiva
g potentielle Instabilität *f*;
konvektive Instabilität *f*

4724 potential model
A model of spatial organization used for the mathematical analysis of settlement patterns in which demographic indices are used as masses, distances are expressed as time and cost of transportation, while the exponent is determined by a *gravitational model.*
r модель потенциалов
Модель пространственного взаимодействия, используемая для математического анализа систем расселения, когда в качестве масс применяются демографические показатели, для расстояния используются показатели времени передвижения или транспортных издержек, а показатель степени определяется по *гравитационной модели.*
f modèle *m* du potential
s modelo *m* de potencial
g potential Modell *n*

4725 potential of territory;
capacity of territory
The possibility to use a certain territory for economic purposes.
r потенциал территории
Возможность использования данной территории в хозяйственных целях (с учетом всевозможных затрат, открытия новых ресурсов, технического прогресса и пр.).
f potentiel *m* du territoire
s potencia *f* del territorio;
capacidad *f* del territorio
g Potential *n* des Territoriums

4726 pothole; aven
A system of vertical karst cavities alternating with almost horizontal and inclined galleries which begins with a natural shaft.
r карстовая пропасть
Система вертикальных карстовых полостей, чередующихся с почти горизонтальными и наклонными галереями (начинается естественной шахтой).
f abîme *m* karstique
s abismo *m* kárstico
g Karstschlucht *f*; Karstloch *n*

4727 pothole lake; kettle lake
A lake occupying a depression created by a whirlpool by foot of a former waterfall.
r эворзионное озеро
Озеро, занимающее впадину, созданную водоворотом у подножия бывшего водопада.
f lac *m* de évorsion
s lago *m* de evorsión
g Strudelloch *n*; Kolkloch *n*

4728 poultry farming
Breeding birds for eggs and meat.
r птицеводство
Отрасль животноводства, занимающаяся разведением птиц для получения яиц и мяса.
f aviculture *f*
s avicultura *f*; cría *f* de aves
g Geflügelzucht *f*

4729 poverty
The absence or inadequacy in the lives of people of the diets, amenities, standards, services and activities which are common or customary in the society in which they are living.
r бедность; нищета
Осутствие или недостаток в повседневной жизни людей питания, удобств, жизненных

стандартов, услуг и деятельности, обычных в обществе, где живут эти люди.

f pauvreté *f*
s pobreza *f*
g Armut *f*

4730 **poverty cycle**
The recurrent transmission of poverty and deprivation from one generation to the next.
r цикл бедности
Постоянный переход бедности и лишений от одного поколения к следующему.
f cycle *m* de pauvreté
s ciclo *m* de pobreza
g Armutszyklus *m*

4731 **powder avalanche; powder snow avalanche; loose snow avalanche**
An avalanche of dry snow or firn accompanied by a cloud of powdery snow.
r пылевая лавина; снежно-пылевая лавина
Лавина из сухого снега или фирна, сопровождающаяся облаком снежной пыли.
f avalanche *f* de neige poudreuse; avalanche *f* sèche; avalanche *f* poudreuse
s avalancha *f* en polvo; avalancha *f* seca
g Staublawine *f*; Pulverlawine *f*; Pulverschneelawine *f*

4732 **powder cloud**
The mixture of air and snow that commonly accompanies a powder avalanche release.
r пылевое облако
Смесь воздуха и снега, обычно сопровождающая сход пылевой лавины.
f panache *m* de neige; souffle *m* d'avalanche
s nube *f* de nieve
g Schneestaubwolke *f*

4733 **powder snow**
A very loose snow precipitating under the conditions of a windless weather and light frost and composed of undisturbed snow crystals, most common in polar latitudes and in high mountain areas.
r пушистый снег
Очень рыхлый снег, отлагающийся при полном безветрии и небольшом морозе, состоящий из неразрушенных снежинок, обычен в полярных районах и высоко в горах.
f neige *f* poudreuse
s nieve *f* en polvo; nieve *f* vaporosa
g Pulverschnee *m*; Flockschnee *m*

* **powder snow avalanche → 4731**

* **powedery snow → 1585**

* **power consumption per unit → 1904**

4734 **power engineering**
A branch of national economy that deals with energy resources, production, conversion, transmission and use of all types of energy.
r энергетика
Область народного хозяйства, охватывающая энергетические ресурсы, выработку, преобразование, передачу и использование различных видов энергии.
f énergie *f*
s energía *f*; generación *f* de energía eléctrica
g Energetik *f*; Energiewirtschaft *f*

* **power plant → 4735**

4735 **power station; power plant**
An enterprise where electrical or heat energy is produced.
r электростанция
Предприятие, производящее электрическую, а в отдельных случаях и тепловую энергию.
f centrale *f* électrique
s central *f* eléctrica
g Kraftwerk *n*

4736 **power system; energy system**
A system of electric power stations connected by power lines that jointly supply consumers with electrical power.
r энергосистема; электроэнергетическая система
Объединение электростанций, связанных линиями электропередач и совместно питающих потребителей электроэнергии.
f système *m* énergétique
s sistema *m* energético
g Energiesystem *n*

4737 **practicability; passability**
Possibility to move through a country without special roads, sometimes using country roads.
r проходимость
Возможность передвигаться по данной местности без специальных дорог, иногда используя проселочные дороги.
f viabilité *f*
s viabilidad *f*; practicabilidad *f*
g Gangbarkeit *f*; Passierbarkeit *f*

4738 **prairie**
A large area with grassland and occasional trees as natural vegetation on the plains of North America where rainfall is low and summer temperatures are high (similar conditions are found in the South American *pampas*, the Russian *steppes*, and South African *veld*).
r прерия
Обширные пространства, покрытые травой и изредка деревьями на равнинах Северной Америки, где осадков выпадает немного и лето весьма жаркое (аналогичные условия характерны для *пампы* Южной Америки, *степей* России и *велда* Южной Африки).
f prairie *f*
s pradera *f*; pradera *f* norteamericana
g Prärie *f*

4739 prairie grey soil
A soil that forms in the semi-desert zone among grey desert soils under the conditions of ground water moistening, it differs from grey desert soils by greater humus content and sometimes by gleization of lower horizons.
r лугово-сероземная почва
Почва, формирующаяся в зоне полупустынь среди сероземов в условиях грунтового увлажнения, от сероземов отличается большей гумусностью и иногда оглеением нижних горизонтов.
f siérozem *m* de prairie
s suelo *m* gris de pradera
g Wiesen-Serosemboden *m*

4740 prairie soil
A soil of the wetter prairies, resembling chernozem in its high humus content and its development under grassland, however increased rainfall leads to the leaching of calcium and slight acidity of the soil.
r почва прерий
Почва влажных прерий, напоминающая чернозем высоким содержанием гумуса и развитием под травяной растительностью, однако более высокое количество осадков ведет к выщелачиванию кальция и повышению кислотности почвы.
f sol *m* de prairie
s suelo *m* de pradera
g Prärieboden *m*

* **prairie soil → 810**

* **Pre-Boreal → 4741**

4741 Preboreal; Pre-Boreal
The initial stage of the Holocene characterised by the warming of climate after the melting of ice caps of the last glaciation and which caused the growth of birch-pine forest vegetation on the plains of middle latitudes of Europe appeared.
r пребореальное время
Начальный этап голоцена, характеризующийся потеплением климата после стаивания льдов последнего оледенения и появлением лесной растительности на равнинах умеренных широт Европы.
f Préboréal *m*
s Preboreal *m*
g Präboreal *n*

4742 Precambrian
All geological time which preceeded the Cambrian period (Palaeozoic, Phanerozoic) and the corresponding rock layers without the remains of skeleton fauna, it lasted for 4 billon years and finished 570 million years ago, subdivided into the Archaean and the Proterozoic. → Table 1, p. 284
r докембрий; криптозой
Все геологическое время, предшествовавшее кембрию (палеозою, фанерозою), и соответствующие ему толщи горных пород, лишенные явных остатков скелетной фауны, продолжавшееся до 4 млрд лет и завершившееся 570 млн лет назад, подразделяется на архей и протерозой. → Табл. 1, с. 284
f Précambrien *m*
s Precámbrico *m*
g Präkambrium *n*

4743 precession; precession of the equinoxes
The slow change of the relative positions of the Earth's equator and the ecliptic as the axis of the Earth's rotation describes a cone-shaped rotation causing the north pole to move in a circle with the period of 23-26 thousand years. → Fig. 262
r прецессия; предварение равноденствия
Медленное изменение относительного положения земного экватора и эклиптики, которое приводит к тому, что земная ось производит конусообразное вращение, благодаря чему Северный полюс описывает круг за 23-26 тыс. лет. → Рис. 262
f précession *f*
s precesión *f*; precesión *f* de los equinoccios
g Präzession *f*

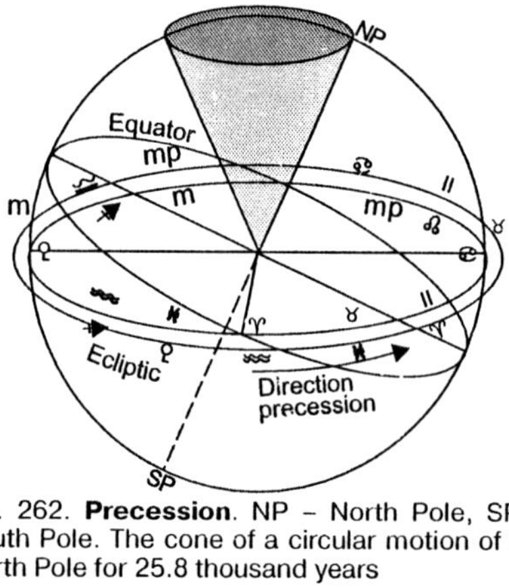

Fig. 262. **Precession**. NP – North Pole, SP – South Pole. The cone of a circular motion of the North Pole for 25.8 thousand years

* **precession of the equinoxes → 4743**

4744 precipitable water; precipitable water vapour
The theoretical index of the moisture content of air above a specific point, i.e. the total content of water vapour in a vertical column of unit cross-sectional area extending between any two specified levels, commonly expressed as the thickness of water layer, which could form after a complete condensation.
r осажденная вода
Теоретический индекс влагосодержания воздуха над данной точкой, т.е. общее содержание водяного пара в единичном вертикальном столбе атмосферы, выраженное толщиной слоя воды, который образовался бы при конденсации всей этой воды.

f eau *f* précipitable
s agua *f* precipitable
g wiederschlagbares Wasser *n*

* **precipitable water vapour → 4744**

4745 precipitation
The deposition of moisture from the atmosphere onto the Earth's surface, including dew, hail, rain, snow, etc.
r атмосферные осадки; осадки$_1$
Отложение влаги из атмосферы на земной поверхности, включая дождь, снег, град, росу и т.п.
f précipitation *f*
s precipitación *f*
g Niederschläge *mpl*

* **precipitation shadow → 4887**

4746 precipitation-day; rain day; rainy day
A period of 24 hours during which at least 0.25 mm of rainfall or 0.25 cm of snowfall are measured.
r день с осадками
Сутки, в течение которых выпало по крайней мере 0,25 мм дождя или 0,25 см снега.
f jour *m* de précipitation
s día *m* con precipitación
g Niederschlagstag *m*

4747 precipitator; setting tank; setting basin
A construction for gravitational extraction of suspended or fatty matter from sewage that accumulate at the bottom or float to the surface.
r отстойник
Сооружение для гравитационного выделения из сточных вод взвешенных или жировых веществ, оседающих на дно или всплывающих на поверхность.
f bassin *m* de décantation; réservoir *m* décanteur
s decantador *m*; tanque *m* de decantación; sedimentador *m*; cámara *f* de sedimento
g Klärbecken *n*; Absatzbecken *n*

* **pre-climax → 5887**

* **predator → 884**

4748 preglacial *adj*
Referring directly to the period before the ice age in the Pleistocene.
r доледниковый
В плейстоцене относящийся непосредственно к предледниковому времени.
f préglaciaire *adj*
s preglaciar *adj*
g präglazial *adj*

4749 preglacial relief
The landforms created mainly by erosion and denudation right before the glaciation.
r доледниковый рельеф
Формы рельефа, созданные преимущественно эрозионно-денудационной деятельностью в период, непосредственно предшествовавший оледенению.
f relief *m* préglaciaire
s relieve *m* preglaciar
g präglaziales Relief *n*

4750 pre-industrial city
A city of the past or present serving a population mostly engaged in agriculture, it has no central busieness district.
r доиндустриальный город
Прошлый или современный город, работающее население которого занято в сельском хозяйстве, в таком городе нет центрального делового района.
f ville *f* préindustrielle
s ciudad *f* preindustrial
g vorindustrielle Stadt *f*

4751 pre-industrial society
The stage of the development of a society and social relationships preceeding the industrial revolution characterized by the predominance of raw-materials production (primary industry).
r доиндустриальное общество
Этап развития общества и общественных отношений, предшествовавших промышленной революции, когда в промышленном производстве господствовали сырьевые отрасли хозяйства (первичный сектор экономики).
f société *f* pré-industrielle
s sociedad *f* preindustrial
g preindustrielle Gesellschaft *f*

4752 preliminary layout of a map
A graphic document illustrating the proposed arrangement of the main elements of a geographical map, including marginal information.
r макет компоновки карты
Чертеж, на котором показаны рамки проектируемой географической карты, расположение картографируемой области и основных элементов карты.
f ébauche *f*
s anteproyecto *m*
g Vorentwurf *m*

* **pre-soil → 4768**

* **pressure gradient → 574**

4753 pressure head; head
The pressure exerted by the water upon the height of the water surface above the point where the pressure is determined.
r напор
Давление воды, зависящее от высоты водного столба над точкой, где производится измерение давления.
f pression *f* d'eau

s altura *f* de presión de agua; carga *f* de presión de agua
g Wasserdruck *m*

4754 **pressure ice; hummocked ice; pressure ridge; hummocky ice**
Sea, river or lake ice which has been deformed or altered by the lateral stresses, tides, waves, and surf, this may include the ice pressed against the shore, or one piece of ice lying upon another.
r сжатый лед; всторошенный лед; торосистый лед
Морской, речной или озерный лед, деформированный или видоизмененный боковыми сжатиями (подвижками), вызванными ветром, течениями, приливными волнами.
f glace *f* comprimée; glace *f* entassante
s hielo *m* comprimido; hielo *m* amontonado; hielo *m* apretado
g Kompressionseis *n*

4755 **pressure melting-point**
The temperature at which ice and other minerals may be induced to melt by the introduction of pressure.
r точка таяния под давлением
Температура, при которой лед или иные минералы могут начать таять (плавиться) под влиянием давления окружающей среды.
f point *m* de fonte á pression
s punto *m* de fusión bajo presión
g Drucktaupunkt *m*

4756 **pressure rampart**
An upfold that forms in front of a landslide as a result of the deformation of surface layers.
r напорный вал
Вал, образующийся перед движущимся оползнем в результате деформации поверхностных слоев горных пород.
f rempart *m* de pression
s caballón *m* de presión
g Schwelle *f* eines Bergrutsches

* **pressure ridge → 2840**

* **pressure step → 576**

4757 **pressure system**
Individual cyclonic-scale feature of atmospheric circulation, used to denote either a high (anticyclon) or a low (cyclon) pressure.
r барическая система
Отдельная часть атмосферной циркуляции циклонического масштаба, обычно используемая для обозначения областей высокого (антициклон) или низкого (циклон) давления.
f système *m* de pression
s sistema *m* de presión
g Drucksystem *n*

4758 **pressure tendency**
The rate of surface pressure change at a selected observation station.
r барическая тенденция
Скорость изменения атмосферного давления у земной поверхности на данной метеорологической станции,
f tendance *f* de la pression; tendance *f* barométrique
s tendencia *f* de la presión; tendencia *f* barométrica
g barometrische Tendenz *f*

4759 **prevailing wind**
A wind blowing most frequently from a specific direction in a particular area.
r преобладающий ветер; господствующий ветер
Направление ветра, наблюдаемое чаще всего в данной местности.
f vent *m* dominant; flux *m* permanent
s viento *m* dominante; viento *m* general; flujo *m* permanente
g vorherrschender Wind *m*

4760 **price**
The money for which a commodity or service is bought or sold.
r цена
Деньги, истраченные на покупку или продажу товаров или услуг.
f prix *m*
s precio *m*
g Preis *m*

* **primary activity → 4764**

4761 **primary *adj***
Related to any pimarily existed element of the environment formed in the previous time and subjected by modern natural processes.
r коренной
Относящийся к изначально существующему элементу окружающей среды, возникшему в предшествующее время и подверженному действию современных природных процессов, например, коренная порода, коренное ложе, коренная растительность.
f primaire *adj*
s primario *adj*
g primär *adj*

4762 **primary area; primary range**
A comparatively small territory where the development of a species begins, and from where its spreading starts.
r первичный ареал
Сравнительно небольшая территория, где шло развитие вида и откуда начинается его распространение.
f aire *f* primaire
s área *f* primaria
g primäres Areal *n*

4763 **primary biological production**
The amount of organic matter produced by

autotrophs (*producers*) over a certain time period by photosynthesis and chemosynthesis.

r первичная биологическая продукция
Количество органического вещества, произведенное автотрофами (*продуцентами*) за определенный отрезок времени в результате фотосинтеза и хемосинтеза.

f production *f* biologique primaire; productivité *f* biologique primaire

s productividad *f* biológica primaria

g primäre biologische Produktion *f*

* **primary forest → 4293**

4764 primary industry; primary activity; primary sector; primary production
Economic activity concerned with obtaining raw materials, extraction of natural resources, or making them available, e.g. mining, agriculture, fishing industry, and forestry.

r первичный сектор экономики
Сырьевые отрасли хозяйства, включая добычу полезных ископаемых, сельское, лесное и рыбное хозяйство.

f secteur *m* primaire; production *f* primaire

s sector *m* primario

g primärer Wirtschaftssektor *m*

* **primary production → 4764**

* **primary range → 4762**

* **primary sector → 4764**

4765 primary succession
The succession of a plant community which develops as a «pioneer» in a place where nothing has lived before as on bare rocks, lava fields, newly formed sand dunes, etc.

r первичная сукцессия
Сукцессия растительного сообщества, возникшего на субстратах, еще не затронутых почвоообразованием: скальных породах, лавовых полях и т.п.

f succession *f* primaire

s sucesión *f* primaria

g primäre Sukzession *f*; Primärsukzession *f*

* **primary vegetation → 4295**

4766 primate city
A city which is first in the rank of city sizes in a country, being larger than any other, so that the gap between its population total and that of the city or cities of the second rank is exceptionally wide, e.g. London in the Great Britain, or Moscow in Russia.

r главный город
Город, первый в ряду городов данной страны, значительно больший, чем любые другие, так что разрыв в общем количестве жителей этого города и любого города второго ранга исключительно велик, например, Лондон в Великобритании или Москва в России.

f prime ville *f*

s ciudad *f* primaria

g Hauptstadt *f*

4767 prime meridian; meridian of origin; meridian of Greenwich; initial meridian
The meridian on the Earth's surface passing through the Greenwich observatory in England from which longitudes are measured.

r начальный меридиан; Гринвичский меридиан; нулевой меридиан
Меридиан, проходящий через прежнее место нахождения Гринвичской обсерватории в Англии, от которого ведется счет долгот на земном шаре.

f méridien *m* d'origine; méridien *m* de Greenwich

s meridiano *m* de origen; meridiano *m* de Greenwich

g Anfangsmeridian *m*; Meridian *m* von Greenwich

* **primitive plant aggregation → 4574**

4768 primitive soil; pre-soil
A soil at an early stage of soil formation in which soil horizons are not yet formed.

r примитивная почва
Почва на ранней стадии почвообразования, в которой еще не сформировались почвенные горизонты.

f sol *m* jeune; sol *m* initial

s suelo *m* jóven; suelo *m* inicial

g einfacher Boden *m*

4769 principal directions
Two orthogonal directions at any point on a sphere or spheroid which remain as orthogonal directions when transferred to a plane map of that point.

r главные направления
Два ортогональных направления в любой точке на сфере или сфероиде, которые остаются ортогональными направлениями той же точки при проекции на плановую основу.

f directions *fpl* principales

s direcciones *fpl* principales

g Hauptrichtungen *fpl*

4770 principal front
The atmospheric front between two air masses of different geographical origin (arctic front, polar front, intertropical front, etc.).

r главный фронт
Атмосферный фронт, разделяющий воздушные массы основных (зональных) географических типов: арктическую, полярную, внутритропическую и др.

f front *m* principal

s frente *m* principal

g Hauptfront *f*

4771 principal meridian
The central meridian on which a rectangular grid system on a map is based.

r главный меридиан

Центральный меридиан, на котором базируется прямоугольная картографическая сетка.
f méridien *m* principal
s meridiano *m* principal
g Hauptmeridian *m*

4772 **printing industry; graphic industry**
A branch of industry engaged in publishing books, newspapers, and magazines.
r полиграфическая промышленность
Отрасль промышленности, занимающаяся производством книг, газет и журналов.
f industrie *f* polygraphique; industrie *f* graphique
s industria *f* de artes gráficas
g polygraphische Industrie *f*

4773 **private sector**
One of the two major divisions of a national economy (the other being the public sector) which is not owned by the state but includes the economic activities of private persons, industrial and commercial companies, institutions, trade unions, charities and churches, etc.
r частный сектор
Один из двух основных секторов народного хозяйства (другой – общественный сектор), включающий экономическую деятельность частных лиц, промышленных и коммерческих компаний, профсоюзов, благотворительных и церковных организаций и т.п.
f secteur *m* privé
s sector *m* privado
g Privatsektor *m*

* **privatization → 1506**

4774 **processing industry; manufacturing industry**
Branches of industry that put raw materials and semi-finished products through a manufacturing process.
r обрабатывающая промышленность
Отрасли промышленности, занимающиеся обработкой или переработкой сырья и полуфабрикатов.
f industrie *f* de transformation; industrie *f* manufacturère
s industria *f* transformadora; industria *f* de transformación
g verarbeitende Industrie *f*; verarbeitendes Gewerbe *n*

4775 **proclimax**
A stable stage of the development of a plant succession similar to the climax, but reached due to the influence of factors other than the existing climate, e.g. fire or overgrazing may lead to a proclimax as animals leave unpalatable plants which are then overpresented.
r проклимакс
Устойчивое состояние растительности, достигнутое в результате сукцессии, аналогичное климаксу, но обязанное не климатическим, а иным факторам, например, пожары или перевыпас могут приводить к проклимаксу, так как животные оставляют некоторые несъедобные растения, которые затем сильно разрастаются.
f proclimax *m*
s proclímax *m*
g Proklimax *f*

* **produce → 4329**

4776 **producer; autotroph**
An organism that uses solar energy (photosynthesis) or energy of some chemical reactions (chemosynthesis) to synthesize organic matter from inorganic compounds.
r продуцент; автотроф
Организм, способный синтезировать необходимое для жизни органическое вещество из неорганического благодаря энергии солнца (фотосинтез) или энергии некоторых химических реакций (хемосинтез).
f producteur *m*; autotrophe *m*
s productor *m*; autotrofo *m*
g Produzent *m*; Autotroph *m*

* **producer goods → 3751**

* **production → 4329**

4777 **production assets; producer goods; capital goods**
The tools and objects of labour, and industrial buildings used in material production.
r производственные фонды
Орудия и предметы труда, здания и сооружения предприятий, необходимые для материального производства.
f fonds *mpl* de production
s fondos *mpl* de producción
g Produktionsfonds fonds *mpl*

* **production co-operation → 1288**

4778 **productive economy**
The way of obtaining food by ancient people by the primitive methods of agriculture and animal rearing which 12-11 thousand years ago replaced the appropriative economy.
r производящий способ хозяйства
Пришедший 12-11 тыс. лет назад на смену присваивающему типу хозяйства способ добычи продуктов питания древними людьми с помощью примитивных методов земледелия и животноводства.
f économie *f* de production
s economía *f* de producción
g erzeugende Wirtschaft *f*

4779 **productivity**
The output of an economic activity seen in terms of the economic inputs such as capital, labour, and raw materials (social and environvental costs must also be considered).
r продуктивность
Результат экономической деятельности,

выраженный в полученном продукте (капитале, труде, сырье и пр.), при этом должны быть учтены также социальные и экологические затраты.
f productivité *f*
s productividad *f*
g Produktivität *f*

4780 productivity of land; land potential
The yield from any given plot of land under the most favourable conditions.
r продуктивность земли
Величина урожая с данного земельного участка при наиболее благоприятных условиях.
f productivité *f* de terre; rendement *m* de la terre
s productividad *f* de la tierra; rendimiento *m* de la tierra
g Landesproduktivität *f*

4781 productivity of the environment
The amount of matter and energy produced in the environment over a certain time period.
r продуктивность окружающей среды
Количество вещества и энергии, производимой в окружающей среде за определенный интервал времени.
f productivité *f* d'environnement
s productividad *f* de medio ambiente
g Leistungsfähigkeit *f* der Umwelt

4782 profile; section
The line of crossing of the Earth's surface or other surface by a vertical plane.
r профиль
Линия сечения земной или другой поверхности вертикальной плоскостью.
f profil *m*
s perfil *m*
g Schnitt *m*; Profil *n*

4783 profile of equilibrium
A section of a slope, a depositional shoreline or a long profile of a river, depicting the stage at which they are in a state just capable of maintaining a balance between erosion, transport and deposition.
r профиль равновесия
Часть склона, формирующаяся береговая линия или продольный профиль реки, находящиеся в состоянии, при котором указанные элементы рельефа сохраняют баланс между эрозией, переносом материала и его отложением.
f profil *m* de équilibre
s perfil *m* de equilibrio
g Profil *m* des Gleichgewichtes

* **profitable capacity → 886**

4784 profundal
The deep part of a water body into which wind wave and sun light do not penetrate.
r профундаль
Глубинная часть водоема, в которую не проникают ветровое волнение и солнечный свет.
f profundal *m*
s profundal *m*
g Profundal *n*

4785 proglacial *adj*
Located in front of a glacier, applied to time and position.
r предледниковый
Расположенный или располагавшийся перед фронтом ледника.
f proglaciaire *adj*
s proglaciar *adj*
g proglazial *adj*

4786 proglacial lake
A lake formed between terminus of the ice and the higher ground which is often in the form of a terminal moraine; a vast lake arising along the margin of the Quaternary ice sheets in Europe and North America by damming by the glacier of the rivers, running northwards.
r предледниковое озеро
Озеро, формирующее между концом ледника и возвышенностями прилегающего рельефа, чаще всего это моренные гряды; обширные озера, образовывавшиеся вдоль края четвертичных ледниковых покровов в Европе и Северной Америке вследствие подпруды ледником рек, стекающих на север.
f lac *m* proglaciaire
s lago *m* proglaciar
g Gletscherrandstausee *m*

4787 proglacial runoff
The runoff of melt waters of the Pleistocene ice sheets through marginal canals along their edges.
r предледниковый сток
Сток талых вод плейстоценовых ледниковых покровов по маргинальным каналам вдоль их краев.
f écoulement *m* proglaciaire
s escorrentía *f* proglaciar
g proglazialer Abfluss *m*

4788 progradation; progradation of shore
A process in which the shore is extended seaward by the accumulation of beach material by the action of waves on a seashore, continuing so long as the current builds up the sea bottom offshore.
r выдвижение берега
Процесс нарастания берега в сторону моря под действием волн, которые приносят материал и откладывают его на морском дне у береговой линии.
f progradation *f* de rivage
s progradación *f* de costa
g Progradation *f* der Küste; Verlandung *f*

* **progradation of shore → 4788**

4789 programme of the atlas
The document determining the purpose of an atlas, its structure, themes, volume, format, and contents by section.
r программа атласа
Документ, устанавливающий назначение атласа, его структуру, тематику, объем, формат и содержание по разделам.
f programme *m* d'atlas
s programa *m* de atlas
g Programm *n* eines Atlases

4790 programme of the map
The ocument determining type and purpose of a map, its mathematical basis, contents, principles of generalisation, conventional signs, recommended sources and technology of map compilation.
r программа карты
Документ, устанавливающий тип и назначение карты, ее математическую основу, содержание, принципы генерализации, условные знаки, рекомендуемые источники и технологию изготовления карты.
f programme *m* de carte
s programa *m* de mapa
g Programm *n* einer Karte

* **projection → 3670**

4791 projective cover
A horizontal projection of the aboveground parts of plants on the Earth's surface expressed in the percentage of the area under consideration.
r проективное покрытие
Горизонтальная проекция надземных частей растений на поверхность почвы, выраженная в процентах от рассматриваемой площади.
f recouvrement *m* projectif; couverture *f* projective
s recubrimiento *m* proyectivo
g projektive Deckung *f*

* **promontory → 860**

* **propulsive branche → 4792**

4792 propulsive industry; propulsive branche
A vigorous, fast-growing industry characterized by a high level of technology, management, and an extensive market, which drives the economy forward at a given stage of economic growth, e.g. coal and ferrous metal industries during industrialization, microelectronics, biotechnology and information technology today.
r пропульсивные отрасли
Энергично развивающиеся отрасли, характеризующиеся высоким уровнем технологий и обширным рынком, которые являются ведущими на данном цикле экономического развития, в период индустриализации это были каменноугольная промышленность и черная металлургия, на нынешнем этапе – микроэлектроника, биотехнология, информатика во всех сферах ее приложения.
f branches *fpl* propulsives
s sectores *mpl* propulsivos
g propulzive Zweige *mpl*

4793 prostrate plant
Plants with horizontal sprouts trailing on the ground capable of rooting in the process of growth, they occur in sub-polar and alpine zones.
r стелющееся растение
Растение с горизонтальными приземными побегами, которые полегают и могут укореняться в процессе роста, встречаются главным образом в субполярных и высокогорных областях.
f plante *f* traçante; plante *f* couchée
s planta *fpl* rastrera
g Rankenpflanze *f*

* **prostrate shrubs → 3329**

* **prostrate wood → 3329**

4794 protalus rampart
A linear or curvilinear accumulation of coarse angular rock debris, resembling a moraine, consisting of material that has slipped down from perennial snow patch, and lying parallel to the slope that produced it.
r осыпной вал
Аккумуляция угловатых скальных обломков, напоминающих морену, но состоящих из материала, скользящего вниз с перелетовывающих снежников и отлагающегося параллельно склону, к которому они раньше принадлежали.
f rempart *m* d'éboulement
s caballón *f* de derrumbamiento
g Geröllwall *m*

4795 protected natural area
An area of land or sea set aside for nature conservation with legal protection aimed at the preservation of species inhabiting there.
r охраняемая природная территория
Территория или акватория, где действует правовой или административный режим, направленный на сохранение и восстановление природных ресурсов, включая обитающие здесь виды растений и животных.
f zone *f* naturelle protégée
s território *m* natural protegido; área *f* protegida; área *f* natural protegida
g Schutzgebiet *n*; Schutzbereich *n*

4796 protected species
Animal and plant species legally protected from human impact.
r охраняемый вид
Вид животных или растительных организмов, воздействие на который со стороны людей находится под законодательным запретом.

f espèce *f* protégée
s especie *f* protegida
g geschützte Art *f*

4797 protection of nature; nature conservation; environmental conservation; conservation
Active management of the Earth's natural resources and environment to ensure their quality is maintained and that they are wisely used.
r охрана природы;
охрана окружающей среды
Комплекс мер по сохранению, рациональному использованию и восстановления природных ресурсов, в том числе видового разнообразие флоры и фауны, богатства недр, чистоты вод и атмосферы.
f protection *f* de la nature;
conservation *f* de la nature
s protección *f* de la naturaleza;
conservación *f* de la naturaleza
g Naturschutz *m*

4798 protection regime
The regime of protection of natural territories and water bodies which guarantees their partial or temporary withdrawal from economic use.
r заказной режим
Условия охраны природных территорий и акваторий, обеспечивающие их частичное или временное изъятие из хозяйственного использования.
f régime *m* de protection
s régimen *m* de protección
g Schutzregime *n*; Umweltschutzregime *n*

4799 protectionism
The economic policy of estate aimed at supporting the national economy by limiting the imports with customs tariffs, quotas and other means.
r протекционизм
Экономическая политика государства, направленная на поддержку национальной экономики, в частности, путем ограничения импорта иностранных товаров с помощью таможенных пошлин, квот и других средств.
f protectionnisme *m*
s proteccionismo *m*
g Protektionismus *m*; Schutzzollsystem *n*

4800 protective afforestation
The planting of forest shelter belts and woodlands to protect agricultural lands, water bodies, canals and settlements from the impact of unfavourable environmental factors.
r защитное лесонасаждение
Искусственные насаждения деревьев и кустарников в виде лесных полос и массивов, служащих для защиты сельскохозяйственных угодий, водоемов, дорог, каналов и населенных пунктов от неблагоприятных природных воздействий.
f protection *f* par des rideaux d'arbres
s repoblación *f* forestal para protección
g Schutzpflanzungen *fpl*

4801 protective forest
Natural or planted trees or bushes intendeed for the improvement of the environment.
r защитный лес
Естественная или посаженная древесно-кустарниковая растительность, предназначенная для улучшения природных условий окружающей территории.
f forêt *f* de protection
s bosque *m* de protección
g Schutzwald *m*

4802 protectorate
A form of colonial dependence when one state partly controls, but has not annexed another dependent state, e.g. is responsible for foreign affairs.
r протекторат
Форма колониальной зависимости государств, при которой государство протектор осуществляет внешние сношения и решает другие важные вопросы политики подчиненного государства.
f protectorat *m*
s protectorado *m*
g Protektorat *n*; Schutzherrschaft *f*

4803 Proterozoic
The upper of two largest subdivisions of the Precambrian (Cryptozoic) extend up to the Archaean, continued for over 2 billion years and finished about 57 million years ago. → Table 1, p. 284
r протерозой
Верхнее из двух крупнейших подразделений докембрия (криптозоя), внизу граничащее с археем, продолжался более 2 млрд лет и завершился около 570 млн лет назад. → Табл. 1, с. 284
f Protérozoïque *m*
s Proterozoico *m*
g Proterozoikum *m*

4804 protruding delta
A fan-shaped river delta advancing over the general shore line into the sea. → Fig. 83, p. 178
r выдвинутая дельта
Речная дельта, выступающая за общую черту берега в открытый водоем. → Рис. 83, с. 178
f delta *m* avancé
s delta *m* avanzado
g hinausragendes Delta *n*

4805 province
A historically established unit, e.g. in Italy or France; a modern administrative district of such countries as China, Argentina, Italy, Ireland, etc; the parts of a country outside the capital city or the large cities.

r провинция
Исторически сложившаяся территориальная единица, например, в Италии или Франции; основная единица современного административно-территориального деления в ряде стран (Китае, Аргентине, Италии, Ирландии и др.); части страны, расположенные вдалеке от столицы и крупных городов.
f province *f*
s provincia *f*; comunidad *f*
g Provinz *f*

4806 provinciality
The regular change of natural conditions away from the ocean within one geographical zone.
r провинциальность
Закономерное изменение природных условий по мере удаления от океана в пределах одной географической зоны.
f provincialité *f*
s provincialidad *f*
g Provinzialität *f*

4807 proximal ***adj***
Internal or close to the place of origin.
r проксимальный
Внутренний или ближний к месту происхождения.
f proximal *adj*
s próximo *adj*
g proximal *adj*

4808 psammon
Organisms inhabiting moist sand near water edge.
r псаммон
Организмы, обитающие во влажном песке вблизи уреза воды.
f psammon *m*
s psammon *m*
g Psammon *n*

4809 psammophile
Animals that inhabit sands (plants adapted to growth on friable sands, are called psammophytes).
r псаммофил
Организмы, обитающие в песках (растения, имеющие специальные приспособления для жизни на подвижных песках, называют псаммофиты).
f psammophile *m*
s psamófilo *m*
g Psammophil *m*

4810 pseudo-adiabatic process
Saturation adiabatic process in which the condenced water particles (precipitation) are removed from the rising air.
r псевдоадиабатический процесс
Влажноадиабатический процесс, при котором частицы сконденсировавшейся воды (осадки) удаляются из поднимающегося воздуха.
f processus *m* pseudo-adiabatique
s proceso *m* (p)seudoadiabático
g pseudoadiabatischer Prozess *m*

4811 pseudo-azimuthal projection
A map projection that cosists of concentric circular parallels and curved meridians which converge at the pole at their true angular value, a variant of the azimuthal projection.
r псевдоазимутальная проекция
Картографическая проекция, состоящая из концентрических круглых параллелей и изогнутых меридианов, которые сходятся у полюсов, сохраняя истинные величины углов, вариант азимутальной проекции.
f projection *f* mériazimutale
s proyección *f* pseudoacimutal
g unechtazimutaler Kartennetzentwurf *m*

4812 pseudo-bedding
Structures in igneous rocks which resemble the bedding of sedimentary rocks.
r ложное напластование
Структура изверженных пород, напоминающая залегание осадочных пород.
f disposition *f* fausse des couches
s disposición *f* falsa de capas; estratificación *f* falsa
g falsche Lagerung *f*

4813 pseudo-conical projection
A map projection in which all parallels are represented by arcs of concentric circles, the central meridian is a straight line, on which the center of all parallels is situated, the rest meridians are curve lines. → Fig. 10, p. 33
r псевдоконическая проекция
Картографическая проекция, в которой все параллели изображаются дугами концентрических окружностей, средний меридиан - прямой линией, на которой лежит центр всех параллелей, остальные меридианы – кривыми линиями. → Рис. 10, с. 33
f projection *f* méриconique
s proyección *f* pseudocónica
g unechtkegeliger Kartennetzentwurf *m*

4814 pseudo-cylindrical projection
A map projection in which all parallels are represented by parallel straight lines, the central meridian is a straight line perpendicular to them, the rest meridians are curve lines. → Fig. 263, p. 577
r псевдоцилиндрическая проекция
Картографическая проекция, в которой все параллели изображаются параллельными прямыми, средний меридиан – перпендикулярной им прямой, остальные меридианы – кривыми линиями. → Рис. 263, с. 577
f projection *f* méricylindrique
s proyección *f* pseudocilíndrica
g unechter Zylinderentwurf *m*

4815 pseudo-front

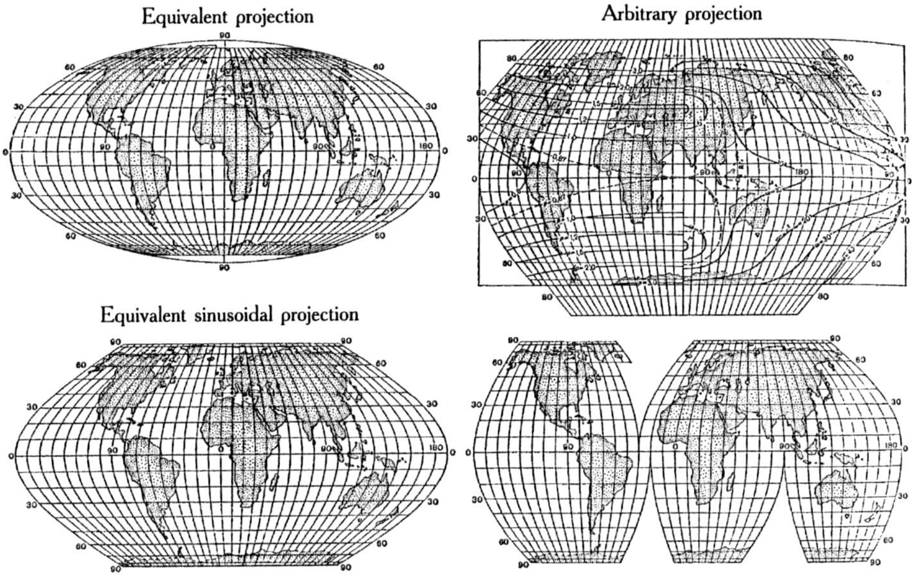

Fig. 263. **Pseudo-cylindrical projection**

A line or surface of discontinuity of temperature or other meteorological element in the lower atmosphere caused by the underlying surface and not related to any atmospheric front.

r мнимый фронт
Линия или поверхность разрыва в распределении температуры или иного метеорологического элемента в нижних слоях воздуха, обусловленная влиянием подстилающей поверхности и не связанная с атмосферным фронтом.
f pseudo-front *m*
s pseudofrente *m*
g Pseudofront *f*

4816 pseudokarst
Landforms which resemble true karst but which have developed on non-limestone rocks and are the result of other processes.
r псевдокарст; ложный карст
Формы рельефа, внешне напоминающие карст, но развивающиеся не в известняках и обусловленные другими процессами.
f pseudo-karst *m*
s pseudo-karst *m*
g Pseudokarst *m*

4817 pseudomorph
Mineral formations having an outward form which does not correspond to their composition and structure, e.g. a wedge-shaped form in the ground formed due to the thawing of ice vein and filling the space with sediments.
r псевдоморфоза
Минеральные образования, внешняя форма которых не соответствует их составу и внутреннему строению, например, клиновидная форма в грунте, возникшая в результате вытаивания ледяной жилы и заполнения образовавшейся полости грунтом.
f pseudomorphose *f*
s pseudomorfo *m*
g Pseudomorphose *f*

4818 pseudourbanization
The growth of the number of urban dwellers in the developing countries accompanied by urban sprawl over large area lacking adequate urban environment and way of life.
r ложная урбанизация; псевдоурбанизация
Характерный для развивающихся стран процесс роста численности городского населения и территориальное расширение городов, не сопровождающиеся формированием адекватной им городской среды и городского образа жизни.
f pseudourbanisation *f*
s pseudourbanización *f*
g Pseudourbanisation *f*

4819 psychrometer
An instrument used to measure the relative humidity of the air.
r психрометр
Инструмент, прдназначенный для измерения относительной влажности воздуха.
f psychromètre *m*
s psicrómetro *m*
g Psychrometer *m*

4820 psychrophyte; hekistothermal plant
Plants growing in the areas with the average annual temperature below 0°C, mainly plants

of polar countries and alpine zone of mountains.

r психрофит;
растение холодного климата
Растения, обитающие в областях со средней годовой температурой ниже 0°C, главным образом растения полярных стран и альпийского пояса гор.

f psychrophyte *m*; plante *f* hékistotherme

s psicrofita *m*; planta *f* hekistoterma

g Psychropyt *m*

4821 **public goods**
Goods freely available, either naturally, like air, or from state as social goods that cannot be divided and sold to individuals, e.g. national defence, fire services, hospital services, street lighting, etc.

r всенародное достояние
Доступные всем блага, как природные, например, воздух, так и социальные, предоставляемые государством, они не могут быть разделены и проданы отдельным людям, они принадлежат всем гражданам, например, национальная оборона, пожарная охрана, медицина, уличное освещение и т.п.

f biens *mpl* publics; biens *mpl* libres; biens *mpl* collectives

s bienes *mpl* públicos; bienes *mpl* libres

g öffentliches Vermögen *n*; allgemeines Wohl *n*

4822 **public resources**
Natural resources in public possession which use is limited by law.

r общественные ресурсы
Природные ресурсы, находящиеся в общественной собственности, доступ к которым законодательно ограничен, например, леса и водохранилища.

f ressource *f* publique

s recursos *mpl* públicos

g gesellschaftliche Resourcen *fpl*

4823 **public sector**
Economic activities partially or completely controlled and owned by the government, it includes nationalized industries, national and local government services and public corporations (as opposed to private sector).

r общественный сектор; государственный сектор экономики
Отрасли хозяйства, находящиеся полностью или в значительной степени в руках государства, они включают национализированные предприятия, правительственные и муниципальные службы, общественные объединения (в противоположность частному сектору).

f secteur *m* public; secteur *m* d'Etat

s sector *m* público; sector *m* económico de estado

g gesellschaftlicher Sektor *m*; Staatssektor *m* der Wirtschaft

* **pudding rock → 1207**

4824 **pull factor**
A positive factor exerted by the locality towards which people move, e.g. the granting of new land for farmers, assisted passages and other government inducements, freedom of speech and religions, etc.

r фактор привлечения
Положительный фактор миграции на новые земли, причинами чего могут быть: раздача новых земель крестьянам, помощь в переезде и государственные гарантии, свобода слова и религий и мн. др.

f facteur *m* de traction

s factor *m* de tracción

g Zugfaktor *m*

4825 **pulsation hypothesis**
A concept suggesting the existence of global epochs of compression and expansion of the Earth and on this basis explaining the Earthrs magnetism, transgressions and regressions of seas, folding epochs, etc.

r пульсационная гипотеза
Концепция, утверждающая существование глобальных эпох сжатия и расширения Земли и на этой основе объясняющая явления магматизма, трансгрессий и регрессий Мирового океана, эпох складчатости и пр.

f hypothèse *f* de pulsation de la Terre

s hipótesis *f* de pulsación de la Tierra

g Pulsationshypothese *f*

* **pulsation period → 5995**

4826 **pumice**
A very light, fine-grained, and cellular rock produced when the froth on the surface of lava solidifies.

r пемза
Очень легкая, мелкозернистая, пористая горная порода, образующаяся при застывании пены на поверхности твердеющей лавы.

f pierre *f* ponce

s pómez *f*; piedra *f* pómez

g Bimsstein *n*

4827 **pumped storage power station; pumped storage scheme**
A hydroelectric power station that pumps water from a lower to a higher reservoir when the energy demand is low, while at peak demand, the water is allowed to fall back to the lower level and produce electricity.

r гидроаккумулирующая электростанция; ГАЭС
Гидроэлектростанция, которая перекачиванием воды из нижнего бассейна в верхний накапливает избыточную энергию, вырабатываемую другими электростанциями, когда спрос на электроэнергию мал, и преобразует потенциальную энергию запасенной воды в электрическую в часы пиковых нагрузок в энергосистеме.

f centrale *f* hydroélectrique d'accumulation

s central *f* hidroeléctrica de acumulación
g Speicherkraftwerk *n*

* **pumped storage scheme → 4827**

4828 puna
An altitudinal belt of semi-desert and desert intermontane plateaux of the Central Andes lying between 3,500 and 4,600 m, and characterized by extremely large seasonal and diurnal ranges of temperature, resulting in high-mountain tropical semi-desert vegetation with evergreen xerophyllous shrubs and grasses.
r пуна
Высотный пояс полупустынных и пустынных внутренних плоскогорий Центральных Анд между 14-28° ю.ш. на высотах 3500-4600 м, характеризующийся исключительно большими сезонными и суточными контрастами температуры, в результате здесь формируется высокогорная тропическая полупустыня с вечнозелеными ксерофитными кустарниками и древовидными злаками.
f puna *f*
s puna *f*
g Puna *f*

* **purga → 818**

4829 purification
The removing of impurities from the environment and its components such as water, air, soil.
r очистка окружающей среды
Устранение посторонних и нежелательных веществ из окружающей среды и ее компонентов: воды, воздуха, почвы.
f purification *f* d'environnement; épuration *f* d'environnement
s purificación *f* de medio ambiente; depuración *f* de medio ambiente
g Reinigung *f* der Natur

4830 push factor
In migration, any adverse factor which causes movement away from the place of residence, e.g. changes in land tenure, political persecution, economic crisis, etc.
r фактор выдавливания
Любой неблагоприятный фактор, вызывающий миграцию населения из мест его проживания вследствие перемен в земельной собственности, политического преследования, экономического кризиса и мн. др.
f facteur *m* de pression
s factor *m* de presión
g Andruckfaktor *m*

4831 push-moraine; thrust moraine
The landforms of bow-shaped ridges produced by the bulldozing effect of an ice sheet advancing across the glacial drift from an earlier glaciation.
r морена напора
Крупные дугообразные гряды, образованные гляциодислокациями вблизи ледников.
f moraine *f* de poussée
s morrena *f* de empuje
g Stauchmoräne *f*; Vorstossmoräne *f*

4832 pusza
A dense, virtually impenetrable forest, often a forest reserve, in European Russia, Poland and Byelorussia, e.g. Belovezhskaya Pusza.
r пуща
Густой, труднопроходимый, часто заповедный лес в Европейской части России, Польше и Белоруссии, например, Беловежская Пуща.
f pusza *f*
s pusza *f*
g Pusza *f*

4833 puszta
Lowland grass steppes of Middle Europe that develop on sand chernozems.
r пушта
Злаковые степи равнин Средней Европы, развивающиеся на песчаных черноземах.
f puszta *f*
s puszta *f*
g Puszta *f*

4834 putrefaction
Anaerobic decomposition of organic matters by microorganisms with the formation of gases, first of all methane; it is characteristic of marshy regions.
r гниение
Характерное для болотистых районов анаэробное разложение органических веществ под действием микроорганизмов, сопровождающееся выделением газов, прежде всего метана.
f putréfaction *f*
s putrefacción *f*; pudrición *f*
g Putrefaktion *f*; Putreszenz *f*; Fäulnis *f*

4835 pycnocline
The layer in a lake having the maximum vertical gradient of water density.
r пикноклин
Горизонт в водоеме с максимальным вертикальным градиентом плотности воды.
f pycnocline *f*
s picnoclina *f*
g Pyknokline *f*

4836 pyramidal iceberg; pinnacled iceberg
An iceberg resembling a pyramid, or having several peaks. → Fig. 264, p. 580
r пирамидальный айсберг; островершинный айсберг
Айсберг пирамидальной формы или с несколькими острыми вершинами. → Рис. 264, с. 580
f iceberg *m* pyramidal
s iceberg *m* piramidal
g Pyramidaleisberg *m*

4837 pyramidal peak

Fig. 264. **Pyramidal iceberg** in the Southern Ocean

Fig. 265. **Pyramidal peak** Matterhorn in Swiss Alps

An isolated mountain summit or a rock with a number of sides that meet in a point at the top. → Fig. 265

r пирамида

Горная вершина или скала, имеющая многогранную форму. → Рис. 265

f pyramide *f*

s pirámide *f*

g Pyramide *f*

* **pyroclast → 4838**

* **pyroclastic flow → 4839**

4838 pyroclastic material; pyroclast

Loose volcanic rocks formed from the volcanic debris ejected by an explosive volcanic eruption.

r пирокластический материал

Рыхлая вулканическая порода, состоящая из вулканических обломков разной величины, выброшенных во время извержения.

f matérial *m* pyroclastique

s piroclasto *m*; material *m* piroclástica

g pyroklastisches Material *n*

4839 pyroclastic surge; pyroclastic flow

A sudden blast of suffocating superheated gases and ash ejected in a volcanic eruption.

r пирокластический выброс

Внезапный выброс струи удушливых перегретых газов и пепла во время вулканического извержения.

f jet *m* pyroclastique

s chorro *m* piroclástico

g pyroklastischer Ausbruch *m*

4840 pyrogenic community

A community of plants which appeares under the impact of fire and consists of plants well adapted to withstand fires.

r пирогенное сообщество; пирогенный фитоценоз

Растительное сообщество, возникшее под воздействием пожара, состоит из вторичной растительности, вырастающей после лесных пожаров.

f groupement *m* pyrogène

s comunidad *f* pirógena

g pyrogene Pflanzengesellschaft *f*

4841 pyrogenic landscape

A geographical landscape that forms on the place of a large forest or steppe fire.

r пирогенный ландшафт

Географический ландшафт, возникающий на месте крупного лесного или степного пожара.

f paysage *m* pyrogène; paysage *m* de brûlis; paysage *m* d'incendie

s paisaje *m* pirógeno; paisaje *m* de quema; paisaje *m* de rozas

g pyrogene Landschaft *f*

4842 pyrogenic succesion

The sequential change of vegetation caused by the fires of grass, peat, forest, etc.

r пирогенная сукцессия

Восстановительная смена растительности после травяного, торфяного, лесного или какого-либо другого пожара.

f succession *f* pyrogène

s sucesión *f* pirogénica

g pyrogene Sukzession *f*

4843 pyrophyte

A plant adapted to withstand or to achieve a competitive advantage from fire, e.g. giant redwood trees have developed very thick bark.

r пирофит

Растение, адаптировавшееся к лесным пожарам, например, гигантские красные деревья имеют очень толстую кору.

f pyrophyte *m*

s pirofita *m*; pirófito *m*

g Pyrophyt *m*

Q

4844 quadrangle; degree square
A four-side figure which is bounded on the right-and-left margins by meridians, and on the top and bottom by parallels which are a specified numbers of minutes or degrees apart.
r картографическая трапеция
Четырехугольник (трапеция), образованный по правому и левому краям меридианами, а сверху и снизу параллелями, которые имеют подписи в градусах или минутах.
f degré *m* carré
s trapecio *m* cartográfico
g Gradfeld *n*

4845 quadrature
The times at which the Moon and the Sun are so situated that rays drawn from each to the Earth make an angle of 90°, such a situation occurs twice every synodic month causing relatively small tidal ranges (neap tides).
r квадратура
Время, когда Луна и Солнце расположены таким образом, что их лучи падают на Землю под прямым углом, такая ситуация случается дважды за синодический месяц и совпадает с квадратурным приливом.
f quadrature *f*
s cuadratura *f*
g Quadratur *f*

* **quag → 4846**

* **quagmire → 4846**

4846 quaking bog; mire; quagmire; quag
An area of especially soft and wet bog in which under the layer of moss, grass and peat there is water or liquid mud which quakes or trembles when trodden on.
r топь; трясина
Зыбкое топкое болото, в котором под толщей мхов, травы и торфа находится слой воды или жидкого ила.
f bourbier *m*
s tremedal *m*; cenagal *m*
g Morast *m*; Sumpf *m*

4847 quality of life
The comformity of the conditions of life with human requirements expressed by three characteristics: the mean life expectancy, people´s health index, and the proportion of diseases in different groups of population.
r качество жизни
Мера соответствия условий жизни потребностям человека, выражаемая тремя показателями: средней продолжительностью жизни, показателем здоровья людей и уровнем заболеваемости разных групп населения.
f qualité *f* de vie
s calidad *f* de vida
g Lebensqualität *f*

4848 quality of the environment; environmental quality
The degree of correspondence of environmental conditions to biological requirements of humans and other living organisms.
r качество окружающей среды
Степень соответствия природных условий биологическим потребностям человека и других организмов.
f qualité *f* d´environnement
s calidad *f* de medio ambiente
g Umweltqualität *f*

4849 quantitative background
A cartographic method of colouring or hatching used to show the quantitative differences of those phenomena which have a continuous distribution within a given region.
r количественный фон
Картографический способ изображения количественных различий явлений сплошного распространения в пределах выделенных районов с помощью окраски или штриховки.
f fond *m* quantitatif
s fondo *m* cuantitativo
g quantitativer Hintergrund *m*

4850 quantitative revolution
The approach to geographical studies in the 1950s and 1960s concerned with the use of statistical and mathematical methods and techniques to analyse data in order to make objective systems of classification and reliable theories of spatial organization of nature and human society.
r количественная революция
Новый подход в географических исследованиях в 1950-60-х годах, связанный с использованием статистических и математических методов в попытке получить объективные системы классификации и теории пространственной организации природы и общества.
f révolution *f* quantitative
s revolución *f* cuantitativa
g quantitative Revolution *f*

4851 quarry
An open excavation on the surface of the Earth worked by open-cast mining. → Fig. 266, p. 582
r карьер
Выемка в земной коре, образованная при добыче полезных ископаемых открытым способом. → Рис. 266, с. 582
f carrière *f*
s cantera *f*
g Steinbruch *m*; Tagebau *m*; Grube *f*

4852 quartz

Fig. 266. Space image of a **quarry** at the North-East of the South-African Republic

The crystalline form of the mineral silicon dioxide (SiO_2), one of the commonest minerals in nature, which is transparent and uncoloured but has several varieties including amethyst, yellow, rose, smoky quartz, etc.

r кварц
Кристаллическая форма минерала диоксида кремния (SiO_2), один из самых распространенных минералов в природе, прозрачный и бесцветный, имеющий много разновидностей: аметист, желтый, розовый, дымчатый кварц и др.

f quartz *m*
s cuarzo *m*
g Quarz *m*

* **quasi-biennial oscilation → 651**

* **quasi-stationary front → 5787**

* **quasy-permanent anticyclone → 5342**

4853 **Quaternary; Anthropogene**
The last geological period in the Earth's history lasting from 700 thousand to 2.5 million years till the present time, it is subdivided into the Pleistocene and Holocene. → Table 1, p. 284

r четвертичный период; антропоген
Последний геологический период в истории Земли, продолжающийся от 700 тыс. до 2,5 млн лет вплоть до настоящего времени, подразделяется на плейстоцен и голоцен. → Табл. 1, с. 284

f période *f* quaternaire; Anthropogène *m*
s período *m* cuaternario; Cuaternario *m*
g Quartär *n*

* **quaternary industry → 4854**

4854 **quaternary industry; quaternary sector; quaternary industry**
Economic activity dealing with information, computer and other new technologies that have been introduced into industry since the last quarter of the XX century.

r четвертичный сектор экономики
Отрасли хозяйства, связанные с информационными, компьютерными и иными новыми технологиями, внедряемыми в промышленность с последней четверти XX в.

f secteur *m* quaternaire
s sector *m* cuaternario
g quartiärer Wirtschaftssektor *m*

* **quaternary sector → 4854**

4855 **Quaternary studies**
The study of the Quaternary rocks and the corresponding period of the Earth's history.

r четвертичные исследования
Изучение четвертичного периода истории Земли.

f étude *f* quaternaire
s estudio *m* cuaternario
g Quartäruntersuchung *f*

4856 **quickclays; floating layer**
Clays analogous to a quicksand because of their lack of cohesion under certain conditions of stress, they have been deposited in a body of standing water to form stratified or varved sediments.

r плывун
Глины, аналогичные зыбучим пескам из-за отсутствия сцепления между отдельными частицами, их отложение происходило в стоячей воде, где формировались стратифицированные или ленточные отложения.

f argile *f* coulante
s arcilla *f* fluente
g Treibton *m*

4857 **quicksand**
A thick mass of unstable, fine loose sand, sometimes mixed with mud, supersaturated with water, occurring on some coasts and near river mouths, liable to suck down any heavy object that comes to rest on it.

r зыбучие пески
Распространенные на морских и речных берегах незакрепленные тонкозернистые пески, перенасыщенные водой, способные всасывать предметы, находящиеся на их поверхности.

f sables *mpl* mouvants
s arenas *fpl* movedizas; arenas *fpl* fluentes
g Schwimmsand *m*; Triebsand *m*

4858 **quota**
A fixed level indicating the maximum amount of imported goods which the state will allow in (import quota) or out (export quota) in a given period of time.

r квота
Фиксированная норма, показывающая максимальное количество товаров, которое государство разрешает ввозить (импортная квота) или вывозить (экспортная квота) за конкретный период времени.

f quota *m*
s cuota *f*
g Quote *f*

R

4859 race[1]
A group of people having common inherited physical features: the colour of skin, hair and eyes, shape of nose and lips, height and body proportions, etc., sufficient to characterize it as a distinct human type; a group of animals or plants, connected by common descent of origin.
r раса
Исторически сложившаяся группа людей, имеющих сходные, передаваемые по наследству внешние признаки – цвет кожи, волос и глаз, форма носа и губ, рост и пропорции тела; группа животных или растений, связанных общим происхождением.
f race *f*
s raza *f*
g Rasse *f*

4860 race[2]
A rapid flow of water through a restricted channel, either through a water-mill (mill-race) or in the sea (tidal-race).
r быстроток
Быстрое течение воды в канале, проложенном к водяной мельнице или в море.
f chenal *m*
s canal *m* de molino
g Mühlrinne *f*

* **racial segregation → 328**

4861 racism
The assumption that the abilities and characteristics of a person are determined by race, and that biologically one race is inherently superior to another, as well as a political programme or social system based on such assumption.
r расизм
Теория о том, что способности и характеристики людей определяются их принадлежностью к той или иной расе, что биологически одна раса может превосходить другую, а отсюда политическая программа или социальная система, базирующиеся на подобной концепции.
f racisme *m*
s racismo *m*
g Rassismus *m*

* **radial dissection → 4862**

4862 radial drainage pattern; radial dissection
A pattern of drainage produced when streams flow down and outward from a central dome or cone-shaped upland, analogous with the spokes of a wheel occurring especially in mountainous areas or on volcanic cones. → Fig. 95, p. 202
r радиальная речная сеть; радиальное расчленение гор
Речная сеть, образующаяся в том случае, когда потоки стекают с одиночной вершины, чаще всего с вулканического конуса или конусообразного плато, а речные долины, начинающиеся в наиболее высокой центральной части, расходятся по радиусам. → Рис. 95, с. 202
f réseau *m* hydrographique radial; dissection *f* radiale
s red *f* hidrográfica radial; disección *f* radial
g radiales Flußnetz *n*; radiale Gliederung *f*

4863 radiant *adj*
Emitting energy in the form of electromagnetic waves.
r излучающий
Испускающий энергию в форме электромагнитных волн.
f rayonnant *adj*
s radiante *adj*
g Strahlungs-

4864 radiant energy
The energy of electromagnetic radiation of the Sun mainly related to the visible part of the Sun radiation spectrum.
r лучистая энергия
Энергия электромагнитного излучения Солнца, относящаяся в основном к видимой части спектра.
f énergie *f* rayonnante
s energía *f* radiante
g Strahlungsenergie *f*

4865 radiation balance of the atmosphere
The difference of radiation flows absorbed and emitted by the atmosphere. → Fig. 267, p. 584
r радиационный баланс атмосферы
Разница между потоками радиации, поглощаемыми и излучаемыми атмосферой. → Рис. 267, с. 584
f bilan *m* radiatif de l'atmosphère
s balance *m* de radiación de la atmósfera
g Strahlungsbilanz *f* der Atmosphäre

4866 radiation balance of the Earth's surface; net radiation
A measure of the difference between incoming and outcoming radiation near the Earth's surface.
r радиационный баланс земной поверхности
Разница между приходящей и уходящей солнечной радиацией у поверхности Земли.
f bilan *m* radiatif de la surface terrestre; rayonnement *m* total resultant; radiation *f* nette
s radiación *f* neta
g Strahlungsbilanz *f* der Erdoberfläche; reine Strahlung *f*

* **radiation cooling → 4869**

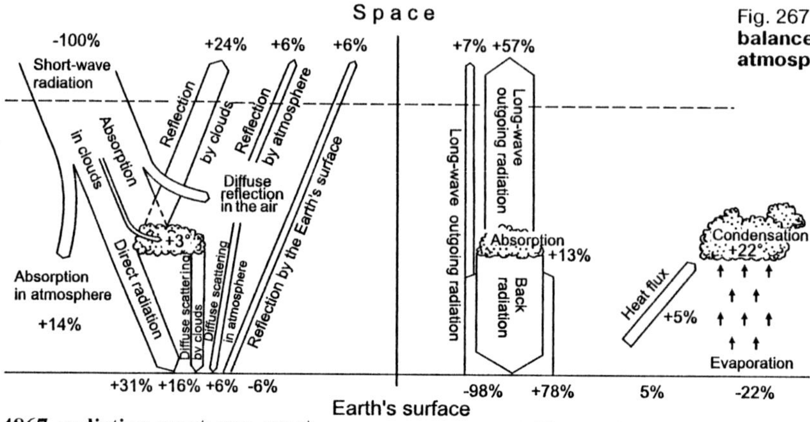

Fig. 267. **Radiation balance of the atmosphere**

4867 radiation crust; sun crust
A snow crust formed by the refreezing of the surface snow crystals which have already been thawed out by the sun.
r радиационная корка
Снежная корка, образованная путем повторного замерзания снежных кристаллов, оттаявших на солнце.
f croûte *f* de radiation
s costra *f* de radiación
g Sonnenstrahlungskruste *f*

4868 radiation fog; ground fog; ground mist
A layer of fog formed near to the ground surface due to nocturnal cooling and loss of heat by terrestrial radiation when the sky is clear and there is little wind.
r радиационный туман; приземный туман
Туман, образующийся вблизи подстилающей поверхности благодаря ночному выхолаживанию и потери тепла земным излучением при ясном небе и небольшом ветре.
f brouillard *m* de rayonnement
s niebla *f* por radiación
g Strahlungsnebel *m*

4869 radiative cooling; radiation cooling
The decrease of temperature at the Earth's surface or within the atmosphere due to negative radiation balance.
r радиационное охлаждение
Понижение температуры земной поверхности или атмосферы вследствие отрицательного радиационного баланса.
f refroidissement *m* radiatif; refroidissement *m* par rayonnement
s enfriamiento *m* radiativo; enfriamiento *m* por radiación
g Strahlungsabkühlung *f*

4870 radiative inversion
The temperature inversion in the atmosphere caused by radiative cooling of the underlying surface.
r радиационная инверсия
Инверсия температуры в атмосфере, обусловленная радиационным выхолаживанием подстилающей поверхности.
f inversion *f* de rayonnement
s inversión *f* de radiación
g Strahlungsinversion *f*

4871 radical geography
A trend in human geography stating that the spatial organization of a territory is determined by the class structure of the society.
r радикальная география
Направление в социально-экономической географии, утверждающее, что организация территории находится в прямой зависимости от социально-классовой структуры общества.
f géographie *f* radicale
s geografía *f* radical
g Radikalgeographie *f*; radikale Geographie *f*

4872 radioactive contamination
The pollution caused by exceeding background radioactivity in the environment as a result of human activity.
r радиоактивное загрязнение
Загрязнение окружающей среды в результате превышения естественного уровня содержания радиоактивных веществ.
f pollution *f* radioactive; contamination *f* radioactive
s polución *f* radioactiva; contaminación *f* radioactiva
g radioaktive Verschmutzung *f*; radioaktive Verunreinigung *f*

* **radioactive fallout → 2095**

4873 radioactive waste; nuclear waste
Radioactive materials produced as waste by nuclear power stations, industrial plants, hospitals etc.
r радиоактивные отходы; ядерные отходы
Отработанные радиоактивные материалы из ядерных реакторов и иных произ-

водств, использующих радиоактивные вещества.
f déchets *m* radioactives
s desechos *mpl* radioactivos; residuos *mpl* radioactivos; residuos *mpl* nucleares
g radioaktive Abfälle *mpl*

4874 radioactive water
Natural waters with the increased content of radioactive substances.
r радиоактивные воды
Природные воды, содержащие повышенное количество радиоактивных веществ.
f eaux *fpl* radioactives
s aguas *f* radioactivas
g radioactives Wasser *n*

4875 radiocarbon dating
A method used to determine the age of organic remains, such as wood, bones, sells, etc., based on the relative content of carbon ^{14}C isotope which makes it possible to calculate the absolute age of a sample (up to 20-30 thousand years in age).
r радиоуглеродное датирование
Способ определения относительного содержания изотопа углерода ^{14}C в органических остатках: деревьях, костях, раковинах и пр., используемый для расчета их абсолютного возраста (не более 20-30 тыс. лет).
f datation *f* au radiocarbone
s datación *f* radiocarbónica
g Radiokarbonmethode *f*

* **radiometric age → 14**

* **radiometric dating → 16**

4876 radiosonde
A free-flying balloon carrying meteorological instruments used to measure meteorological elements and provided with a radio transmitter which sends information to ground stations.
r радиозонд
Прибор, поднимаемый на шаре в атмосферу, снабженный устройствами для измерения метеорологических элементов и передачи информации о них на наблюдательную станцию с помощью радиопередатчика.
f radiosonde *f*
s radiosonda *f*
g Radiosonde *f*; Funksonde *f*

4877 rafted ice; shuffled ice; telescoped ice
Pressure ice in which one floe has been pushed on top another.
r наслоенный лед; натасованный лед
Двухслойные и многослойные участки плавучего льда, образующиеся при надвигании льдин друг на друга под действием сжатий и давлений.
f glace *f* empilée
s hielo *m* estratificado
g aufgeschichtetes Eis *n*

4878 rafting
1. The transportation of material (seeds, fauna, soil, rock) by floating ice and icebergs or by means of attachment to floating vegetable matter. 2. A type of the tourist entertainment, floating with a raft along rivers full of rapids.
r рафтинг
1. Разнос обломков горных пород и иных частиц плавучими льдами, айсбергами и плавающими частями кустов и деревьев. 2. Вид туристского развлечения, сплав на плотах по порожистым рекам.
f rafting *m*
s rafting *m*
g Rafting *n*

* **$rafting_3$ → 2181**

* **railroad (*amer*) → 4879**

4879 railway; railroad (*amer*)
The metal track built for fast travel between towns and cities.
r железная дорога
Рельсовый путь движения железнодорожного транспорта.
f chemin *m* de fer; voie *f* ferrée
s vía *f* férrea; ferrocarril *m*
g Eisenbahn *f*

4880 railway gauge
The width in the clear between the top flanges of the rails, the parallel metal lines on which trains run, the standard gauge is 1435 mm in majority of European countries and in North America and 1650 mm in Russia and surrounding countries.
r ширина колеи
Расстояние между внутренними краями верхней части металлических рельс, по которым передвигаются поезда, в большинстве европейских стран и в Северной Америке оно равно 1435 мм, а в России и сопределеных странах – 1650 мм.
f écartement *m* des rails
s ancho *m* de vía férrea
g Spurweite *f*

4881 railway junction
A point where several railways meet and a considerable volume of goods is transported.
r железнодорожный узел
Пункт, в котором сходятся или пересекаются несколько железнодорожных линий со значительным объемом перевозок.
f noeud *m* ferroviaire
s nudo *m* ferroviario
g Eisenbahnknotenpunkt *m*

4882 railway transport
A type of transport that moves passengers and goods along railways.
r железнодорожный транспорт
Вид транспорта, занятый перевозкой пассажиров и грузов по железным дорогам.

f transport *m* ferroviaire
s transporte *m* ferroviario
g Eisenbahnverkehr *m*

4883 rain
A form of precipitation consisting of liquid water particles, either in the form of drops more than 0.5 mm in diameter or of smaller widely scattered droplets, large enough to fall under the influence of gravity from clouds to the Earth's surface.
r дождь
Осадки жидких частиц воды либо в виде капель диаметром более 0,5 мм, либо в виде более мелких сильно рассеянных капель, выпадающих из облаков и опускающихся к земной поверхности под действием силы тяжести.
f pluie *f*
s lluvia *f*
g Regen *m*

4884 rain alimentation
The income of surface and ground waters into a water body coming from the rain precipitation.
r дождевое питание
Поступление в водный объект поверхностных и грунтовых вод, обусловленное выпадением атмосферных осадков в виде дождей.
f alimentation *f* pluviale
s alimentación *f* pluvial
g Speisung *f* durch Regenwasser

* **rain and snow mixed → 5473**

* **rain day → 4746**

4885 rain forest; hylea
A dense moist evergreen forest typical for rainy regions (annual rainfall is more than 1,800 mm) whithin the tropics having high floristic diversity: up to 70-200 tree species per 0,25 ha. → Fig. 268
r влажный тропический лес; тропический дождевой лес; гилея
Лес, типичный для дождливых районов тропической зоны, состоящий преимущественно из вечнозеленых влаголюбивых древесных пород, флористически очень богат: до 70-200 видов деревьев на 0, 25 га. → Рис. 268
f forêt *f* pluviale; hylaea *f*; pluviselve *f*; pluviselva *f*; forêt *f* tropicale humide
s bosque *m* lluvioso; bosque *m* pluvial tropical; bosque *m* tropical humedo; hylea *f*
g tropischer Regenwald *m*; Hyläa *f*

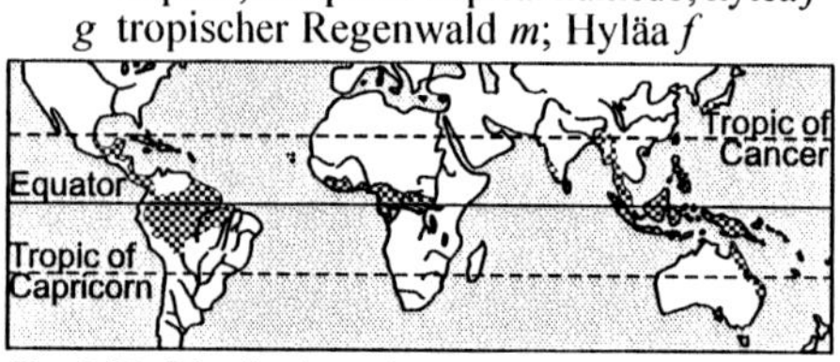

Fig. 268. Distribution of the **rain forests** over the globe

4886 rain gauge
An instrument used to measure rainfall at a meteorological station.
r осадкомер; плювиометр
Прибор для измерения дождевых осадков на метеорологической станции.
f pluviomètre *m*
s pluviómetro *m*
g Regenmesser *m*; Pluviometer *m*

4887 rain shadow; precipitation shadow
The area on the lee side of a mountain or mountain range where the precipitation is relatively low than on the windward side.
r дождевая тень
Область с подветренной стороны горы или горного хребта, где количество осадков заметно меньше, чем на подветренной стороне.
f ombre *f* pluviométrique
s sombra *f* pluvimétrica
g Regenschatten *m*

4888 rainbow
An arc consisting of concentric spectral coloured bands arranged from red inside to violet outside, usually seen on the sky wshen the sun is behind and rain is in front of an observer, caused by the refraction and difraction of light from the Sun or Moon.
r радуга
Группа концентрических дуг, имеющих окраску от красного внутри до фиолетового снаружи, наблюдаемая на фоне капель дождя, образующаяся в результате рефракции и дифракции в лучах Солнца или Луны.
f arc-en-ciel *m*
s arco *m* iris
g Regenbogen *m*

4889 raindrop
A drop of water of more than 0.5 mm in diameter falling from clouds to the Earth's surface.
r дождевая капля
Капля воды диаметром более 0,5 мм, выпадающая из облаков на земную поверхность.
f goutte *f* de pluie
s gota *f* de lluvia
g Regentropfen *m*

4890 raindrop erosion
Soil erosion caused by large raindrops falling on bare earth, when soil particles are mechanically detached and transported by surface runoff, occurring most intensively in tropical and semiarid areas.
r дождевая эрозия
Почвенная эрозия, вызываемая выпадением крупных дождевых капель на голую землю, когда частицы почвы вымываются и уносятся дождевыми водами, особенно часто случается в тропических и семиаридных областях.

f érosion *f* de goutte de pluie; érosion *f* fluviale; érosion *f* fluviatile
s erosión *f* de gota de lluvia; erosión *f* pluvial
g Regenerosion *f*; Tropfenerosion *f*

4891 rainfall
The water equivalent of all forms of atmospheric precipitation received in a raingauge as falling in a certain time within a given area, usually expressed in millimetres.
r дождевые осадки
Водный эквивалент всех форм осадков, измеряемых с помощью осадкомера, за определенный преиод времени на данной территории, обычно выражается в миллиметрах.
f pluviosité *m*; hauter *m* d'eau recueillie
s pluviosidad *f*; cantidad *f* de agua recogida
g Regenfall *m*

4892 rainfall flood
Flood appearing when precipitation fall evenly over the whole catchment as one insulated rainfall per time unit.
r дождевой паводок; единичный паводок; элементарный паводок
Паводок, возникающий в том случае, когда осадки выпадают равномерно по всему водосбору в виде одного изолированного дождя за единицу времени.
f crue *f* pluviale; crue *f* d'averse
s crecida *f* de chubasco; crecida *f* pluvial
g Regenhochwasser *n*; Regenflut *f*

* **rainfall intensity → 3130**

4893 rainfall runoff
The overland and downslope flow of rainwater into channeled flow.
r дождевой сток
Сток дождевой воды вниз по склону, где начинается канализированнычй сток.
f écoulement *m* pluvial
s escorrentía *f* pluvial
g Regenabfluß *m*

4894 rain-making
A process which attempts to stimulate rainfall by artificial means.
r искусственное вызывание дождя
Процесс стимулирования выпадения дождевых осадков посредством засева облаков.
f encouragement *m* de pluie
s incentivación *f* de lluvia
g Stimulierung *f* des Regens

* **rainy day → 4746**

4895 rainy season
A time of the year characterized by the rainfalls which is preceded and followed by relatively dry periods.
r дождливый сезон
Ежегодно повторяющийся период с большим количеством дождей, который чередуется с периодами относительно сухой погоды.
f saison *m* des pluies
s estación *f* de lluvias
g Regenzeit *f*; Regenperiode *f*

4896 raised beach
A shoreline and its littoral deposits elevated above present sea-level either by positive isostatic movements or by fall in sea-level.
r приподнятый берег
Береговая линия и литоральные отложения, оказавшиеся выше современного уровня моря вследствие изостатического поднятия суши или падения уровня моря.
f plage *f* soulevée
s playa *f* elevada
g Hebungsküste *f*

4897 raised bog; oligotrophic bog; moor
A deep bog having poor mineral nourishment formed in the conditions of poorly drained or undrained flat depressions on watershed divides (the surface of bog takes a convex form due to thick vegetation of mosses).
r верховое болото; моховое болото; олиготрофное болото
Болото с бедным минеральным питанием, формирующееся в условиях застаивания поверхностных вод на плоских понижениях водоразделов (поверхность болота принимает выпуклую форму из-за толстого слоя мхов в его центральной части).
f tourbière *f* haute; tourbière *f* oligotrophe
s turbera *f* oligótrofa; turbera *f* alta
g Hochmoor *n*; oligotrophes Moor *n*

4898 raised tundra
A type of tundra with perennial peat hillocks 2-3 m high and up to 20 m in diameter which develops as a result of the degradation of polygonal ice veines.
r бугристая тундра
Тундра, отличающаяся бугристой поверхностью, образованной многолетними торфяными буграми высотой 2-3 м и диаметром до 20 м с мерзлым ядром внутри.
f toundra *f* des tertres
s tundra *f* con montículos
g Torfhügeltundra *f*

4899 ramen
Dense dark highly productive spruce forests growing on dry, well drained soils of East European Plain.
r рамень
Густые темные высокопродуктивные еловые леса на сухих, хорошо дренируемых почвах Восточно-Европейской равнины.
f ramen *m*
s ramen *m*
g Ramen *m*

4900 rampart; stone border
A bank of stones bordering stone polygons and formed as a result of joint action of fracturing, heaving and frost sorting of the ground.
r каменный бордюр
Валик из угловатых камней, обрамляющий каменные полигоны и формирующийся в результате совокупного действия растрескивания, пучения и мерзлотной сортировки грунтов.
f bordure *f* de pierres
s orladura *f* de piedras
g Steinbordüre *f*

4901 ranch
An estate in Latin America, in the U.S. a large farm where cattle, horses or sheep are bred and reared.
r ранчо
Усадьба в странах Латинской Америки, в США – крупная ферма, как правило, скотоводческая.
f rancho *m*
s rancho *m*
g Ranch *f*

4902 randkluft
A gap between the moving glacier and the rock wall of a valley formed due to the warming effect of the wall and the action of melt water streams.
r рандклюфт
Зазор между телом ледника и склоном долины, образующийся в результате теплового воздействия склона и стока талых вод.
f rimaye *f*
s rimaya *f*
g Randkluft *f*

4903 range$_1$
The difference between the least and the greatest of a series of numerical values, e.g. of pressure, elevation, temperature.
r диапазон; размах; амплитуда
Разница между наименьшим и наибольшим значениями количественной величины, например, давления воздуха, высоты, температуры.
f amplitude *f*
s amplitud *f*
g Reihe *f*

* **range$_2$ → 374**

* **range$_3$ → 3967**

* **range action → 2708**

4904 range of dissection; relative relief
The difference between the maximum and minimum heights of a terrain.
r глубина расчленения; размах рельефа
Разность абсолютных высот местности, относительное превышение максимальных высот над минимальными.

Fig. 269. **Rapids** in Iceland

f amplitude *f* de dissection
s amplitud *f* de disección
g Reliefamplitude *f*

* **range of visibility → 6571**

4905 ranker
A thin soil that develops on a mountain slope on dense non-calcareous rocks and has a well developed humus horizon.
r ранкер
Маломощная почва на горном склоне, развивающаяся на плотных бескарбонатных породах и имеющая хорошо выраженный гумусовый горизонт.
f ranker *m*
s ranker *m*
g Ranker *m*

4906 rapids; chute
A rocky obstruction in a river bad formed by outcrops of hard rocks where the water is relatively shallow and the rate of flow is accelerated by increased slope of the stream bed. → Fig. 269
r порог$_1$; быстрина; стремнина
Мелководный каменистый или скалистый участок в русле реки, образованный выходами плотных горных пород и обладающий большим падением и быстрым течением воды. → Рис. 269
f rapide *m*
s rápido *m*; rabión *m*
g Stromschnelle *f*

4907 rare species
A plant or animal species which is not under the direct threat of extinction but has a small population and is found on a limited territory.
r редкий вид
Вид растений или животных, не находящийся под прямой угрозой вымирания, но имеющий низкую численность и распространенный на ограниченной территории.
f espèce *f* rare
s especie *f* rara
g seltene Art *f*

* **rational environmental management → 1935**

* **ravine → 2671**

4908 **raw materials**
Natural or partly processed basic commodities, i.e. the substances and materials used for manufacturing finished products (such as ore for metal making, wood for making furniture, etc.).
r сырье
Вещества и материалы, используемые в производстве для получения готовой продукции.
f matières *fpl* premières
s materias *fpl* primas
g Rohstoff *m*

4909 **raw-materials crisis**
A structural economic crisis that reflects disproportions between the growth of consumption of resources and the volume of their production in the world economy.
r сырьевой кризис
Структурный экономический кризис, отражающий диспропорции между ростом потребления сырья и объемом его производства в мировом хозяйстве.
f crise *f* en matières primières; crise *f* de ressources
s crisis *f* de materias primas; crisis *f* de recursos
g Rohstoffekrise *f*

* **Rayleigh scattering → 5241**

4910 **reach**
An uninterrupted stretch of a straight section of a river, especially if navigable, between two bends, or between locks on a canal.
r прямой участок реки
Непрерывный, как правило, судоходный, участок реки или канала между двумя изгибами русла или между двумя шлюзами.
f section *f* droite
s sección *f* recta del río
g gerader Abschnitt *m*

4911 **reafforestation; arboriculture**
Natural regeneration or restoration of the forest cover by planting trees on clearings, fire-damages, and other previously forested areas.
r лесовозобновление
Естественное восстановление или искусственное выращивание леса на вырубленных, выжженных и других лесных площадях.
f régénération *f* de forêt; reboisement *m*; reforestation *f*
s forestación *f*; arborización *f*; repoblación *f* forestal
g Waldverjüngung *f*; Wiederaufforstung *f*

4912 **real evaporation**
The difference between the evaporation from a given water surface and the atmospheric precipitation falling on it.
r видимое испарение
Разница между испарением с водной поверхности и осадками, выпавшими на эту поверхность.
f évaporation *f* actuelle
s evaporación *f* actual
g aktuelle Verdunstung *f*

4913 **reanalysis of meteorological data**
Obtaining of model global fields of meteorological quantities based on using of full complete sets of the data available by the moment of the model calculations, tha data used are received from different platforms including those coming with a certain delay.
r реанализ метеорологических полей
Получение модельных глобальных полей метеорологических величин с использованием полных комплектов доступных к моменту расчетов данных наблюдений с различных платформ, в том числе данных, поступающих в задержанном режиме.
f réanalyse *f* des données météorologiques
s reanálisis *m* de datos meteorológicos
g Reanalyse *f* der meteorologischen Daten

4914 **rear of a depression**
The rear with regard to the direction of movement sector of a middle or high latitude depression in its mature or declining stage.
r тыл циклона
Тыловая по отношению к направлению движения часть циклона умеренных или высоких широт, представленная надвигающимся холодным воздухом, характерная для циклона на стадии зрелости или затухания.
f traîne *f* de cyclone
s parte *f* posterior de ciclón
g Rückseite *f* eines Tiefs; Tiefdruckrücken *m*

4915 **receiving waters**
A water body into which wastewater or treated sewage is discharged.
r сбросный водоем
Водный объект, куда спускают сточные воды.
f dépôt *m* de réception d'eau; bassin *m* à décharge
s depósito *m* de recepción de agua
g Vorfluter *m*

4916 **recession; economic depression; economic recession; depression**
A slow-down or sudden decline of economic activity accompanied by reduced production that leads to economic stagnation.
r экономическая депрессия; спад производства
Состояние экономического объекта или района, характеризующееся уменьшением объема производства, завершающимся его экономической стагнацией.
f depréssion *f* économique; récession *f* économique

s depresión *f* económica; recesión *f* económica
g ökonomische Depression *f*

* **recession of glaciers → 1478**

* **recessional moraine → 5761**

4917 **recharge**
A process in which water in an aquifer, e.g. in an artesian basin, is replenished by the sinking of precipitation into the land surface, or artificially by pumping water upwards to the flanks of the upper slopes of an artesian basin.
r заполнение водой; водозарядка
Поступление воды в водоносный горизонт артезианского бассейна путем ее постепенного просачивания после выпадения осадков на земную поверхность или искусственно, путем закачивания воды в верхней части склона артезианского бассейна.
f charge *f* d'eau
s carga *f* de agua
g Nachfüllen *n*

* **recharge area → 4918**

4918 **recharge zone; recharge area**
The area of the most intensive absorption of atmospheric and surface waters on the Earth's surface for the replenishment of ground water.
r очаг питания подземных вод
Место наиболее интенсивного поглощения атмосферных и поверхностных вод на земной поверхности, питающих подземные воды.
f zone *f* de charge d'aquifère
s zona *f* de recarga de acuífero
g Grundwasseranreicherungsgebiet *n*; Infiltrationsgebiet *n*

* **reclamation → 3376**

4919 **reconnaissance**
A preliminary survey and examination of some area in order to choose the routes and key areas for the further expedition field investigations.
r рекогносцировка
Предварительный осмотр и обследование местности с целью выбора маршрутов и ключевых участков дальнейших более детальных экспедиционных исследований.
f reconnaissance *f*
s reconocimiento *m* del terreno
g Erkundung *f*; Rekognoszierung *f*

* **recovery → 4934**

4920 **recreation**
The recovery of people's health and capacity for work by way of resting in nature or health resorts, as well as by tourisme.
r рекреация
Восстановление и поддержание здоровья и трудоспособности людей посредством туризма, отдыха на природе и в оздоровительных центрах.
f récréation *f*
s recreación *f*
g Rekreation *f*

* **recreation area → 6272**

* **recreation geography → 2430**

4921 **recreational capacity; recreational carrying capacity; tourism potential**
The ability of an area to provide a certain number of holiday-makers with necessary comfort without any negative effect on nature, i.e. the amount of recreation which a site can take without any deterioration of its quality.
r рекреационная емкость; туристический потенциал
Способность территории или акватории обеспечивать определенному числу отдыхающих необходимый комфорт без отрицательных последствий для природной среды.
f capacité *f* récréative; potentiel *m* touristique
s capacidad *f* recreativa; potencial *m* turístico
g Erholungskapazität *f*; Touristenpotential *n*

* **recreational carrying capacity → 4921**

4922 **recreational impact; impact of recreation on the landscape; recreational pressure**
The degree of impact of tourists upon the environment and recreational objects.
r рекреационная нагрузка
Степень влияния отдыхающих людей на окружающую среду и рекреационные объекты.
f impact *m* touristique
s impacto *m* turístico; presión *f* turística
g Erholungsbelastung *f*

* **recreational pressure → 4922**

4923 **recreational resources**
A combination of natural objects and resources which can be used for recreation and leisure, and especially for tourism.
r рекреационные ресурсы; туристические ресурсы
Совокупность природных объектов и условий, используемых в рекреационных, в том числе туристических целях.
f ressource *f* de récréation; ressource *f* touristique
s recursos *mpl* de recreación; recursos *mpl* turísticos
g Erholungsressourcen *fpl*; Touristenressourcen *fpl*

4924 recreational water body
A natural or artificially made water object used for mass people recreation and equipped specially for this purpose.
r рекреационный водоем
Естественный или искусственно созданный водный объект, используемый для массового отдыха населения и специально для этого обустроенный.
f étendue *f* d'eau de récréation
s extensión *f* de agua de recreación
g Erholungsgewässer *n*

4925 recrystallization ice
The ice formed on glaciers due to the processes of recrystallization without a liquid phase accompanying the settling of ice and metamorphism during the ice flow.
r рекристаллизационный лед
Лед, формирующийся на ледниках в результате процессов рекристаллизации, сопровождающих оседание толщи и метаморфизм во время течения льда.
f glace *f* de recristallisation
s hielo *m* de recristalización
g Rekristallisationseis *n*; umkristallisiertes Eis *n*

4926 recrystallization of ice
The changes in the original structure of ice or snow caused by the processes of metamorphism without a liquid phase.
r перекристаллизация льда; рекристаллизация льда
Изменение структуры льда или снега в результате процессов метаморфизма, без участия жидкой фазы.
f recrystallisation *f* de la glace
s recristalización *f* de hielo
g Umkristallisation *f* von Eis

* **recrystallization zone → 1736**

4927 rectangular drainage pattern
A drainage pattern formed under the influence of a rectilinear joint pattern and characterized by the tributaries meeting larger streams mainly at right angles, and all streams having sections of approximately the same length between junctions. → Fig. 95, p. 202
r прямоугольная речная сеть
Речная сеть, обычно формирующаяся под влиянием прямоугольной трещиноватости коренных пород, когда притоки впадают в главную реку под прямым углом и имеют приблизительно одинаковую длину до своего устья. → Рис. 95, с. 202
f réseau *m* hydrographique rectangle
s red *f* hidrográfica rectangular
g rechtwinkliges Flußnetz *n*

4928 rectangular grid
A network of squares printed on a map to assist in the location of a particular point.
r прямоугольная сетка
Сетка плоских прямоугольных координат, нанесенная на карту с целью более простого определения положения любой точки.
f quadrillage *m*
s cuadriculado *m*
g Kartengitter *n*

4929 rectifying of aerial photographs
Trhe tansformation of plan or perspective aerial photographs into horizontal ones including their reduction to the prescribed scale and decrease of distortions determined by the topography.
r трансформирование аэрофотоснимков
Преобразование плановых или перспективных аэрофотоснимков в горизонтальные, включающее приведение их к заданному масштабу и уменьшение искажений, обусловленных рельефом местности.
f transformation *f* photographique
s transformación *f* de fotografía aérea
g Luftbildentzerrung *f*

4930 rectilinear slope
A slope, or part of a slope, which has a constant angle, i.e. the slope is straight and has uniform conditions on all its parts.
r прямолинейный склон
Склон или часть склона с одинаковым углом наклона, т.е. прямой склон, имеющий схожие физические свойства во всех своих частях.
f versant *m* rectiligne; versant *m* réglé
s vertiente *f* rectilínea; vertiente *f* regulada
g geradliniger Abhang *m*

4931 recultivation of land
The restoration of land disturbed by economic activity.
r рекультивация земель
Восстановление земель, нарушенных хозяйственной деятельностью.
f recultivation *f*
s recultivación *f*; rehabilitación *f* de tierras erosionadas
g Rekultivierung *f*

4932 recultivation of soil
The restoration of soil cover disturbed under exploitation of mineral resources or deprived of natural soil cover as a result of some other activities.
r рекультивация почв
Восстановление почвенного покрова на землях, нарушенных при разработке полезных ископаемых или лишенных естественного покрова почв иным способом.
f recultivation *f* des terrains
s recultivación *f* del suelo
g Bodenrekultivierung *f*

4933 recumbent fold
A type of overfold, resulting from very intense Earth movements, in which the limbs of the fold are nearly horizontal. → Fig. 270, p. 592

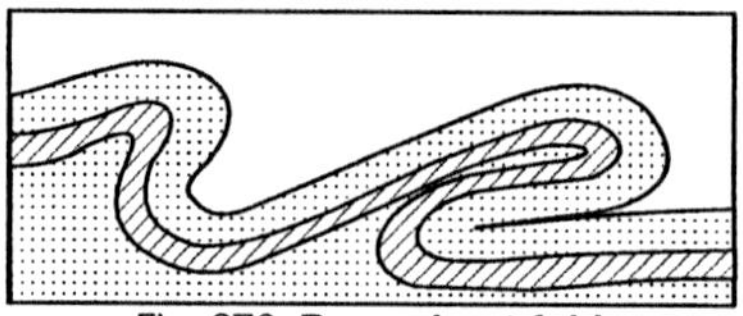

Fig. 270. **Recumbent fold**

r лежачая складка
Тип опрокинутой складки, возникающей в результате интенсивных движений земной поверхности, когда крылья складки почти горизонтальны. → Рис. 270
f pli *m* couché
s pliegue *m* acostado
g liegende Falte *f*

4934 recuperation; recovery
The extraction and reuse of valuable substnces and energy in the same technological process.
r рекуперация
Извлечение и повторное использование ценных веществ и энергии в том же технологическом процессе.
f récupération *f*
s recuperación *f*
g Rekuperation *f*

* **recurrence interval → 5057**

4935 recycling
The processing of waste material to convert it back into a useful product so that it can be used again in order to reduce waste and environmental disturbance.
r утилизация отходов; рециклизация
Переработка отходов, образующихся в процессе производства и потребления, таким образом, чтобы сделать возможным их повторное использование и тем самым уменьшить общую массу отходов и нарушения в окружающей среде.
f recyclage *m*
s reciclaje *m*
g Wiederaufbereitung *f*

4936 red algae
Multicellular or one-celled algae, typical of deep-water benthos, the colour of which varies from violet to almost scarlet.
r красные водоросли; багрянки
Многоклеточные, реже одноклеточные водоросли, типичные для глубоководного бентоса, у которых разнообразие окраски – от фиолетовой до почти алой – объясняется различным соотношением пигментов.
f algues *fpl* rouges
s algas *fpl* rojas
g Rotalgen *fpl*

4937 red clay; red oceanic clay
Ocean-floor sediments of deep ocean (below 4000 m) consisting of terrigenous material and clayish products derived from volcanic ash and pumice with ferrous-manganese nodule inclusions.
r красная глина
Донные отложения глубоких частей океана (свыше 4000 м), состоящие из терригенного материала и глинистых продуктов разложения вулканического пепла и пемзы с включением железо-марганцевых конкреций.
f argile *f* rouge
s arcilla *f* roja
g roter Tiefseeton *m*

4938 Red Data Book
The list of names of rare and threatened plant and animal species which need special protection.
r красная книга
Списки редких и находящихся под угрозой исчезновения видов растений и животных, для охраны которых принимаются особые меры.
f Livre *m* rouge
s libro *m* rojo
g Rotes Buch *n*; Rote Liste *f*

4939 red earth
A soil coloured red by iron compounds which form under broad-leaved forests of humid subtropics or under tropical savannas where leaching regime is typical.
r краснозем
Почва, формирующаяся под широколиственными лесами влажных субтропиков и частично под тропическими саваннами в условиях промывного водного режима.
f sol *m* rouge
s suelo *m* rojo
g Roterde *f*

* **red oceanic clay → 4937**

4940 red rain
A rain of reddish colour caused by red sand or dust particles containing iron oxide, carried by high-level winds from arid to more humid areas, e.g. from the Sahara to southern Europe.
r красный дождь; кровавый» дождь
Дождь, окрашенной в красноватый цвет частицами пыли, содержащими оксиды железа, которые переносятся высоко в тропосфере ветрами, дующими из аридных в более влажные районы, например, из Сахары в южную Европу.
f pluie *f* rouge
s lluvia *f* roja
g roter Regen *m*

4941 red snow
Snow coloured reddish due to the presence of red algae in its upper layer, occurs in high mountains or in polar regions.
r красный снег
Встречающийся высоко в горах и в полярных странах снег красноватого оттенка

из-за развития в его верхнем слое особых водорослей.
f neige *f* rouge
s nieve *f* roja
g roter Schnee *m*

4942 **red tide; harmful algae blossoming**
A phenomenon where vast areas of water in littoral zones of oceans become red, caused by the blossoming of a type of very small algae or phytoplankton (this caused the Red Sea name).
r красный прилив
Процесс, при котором морская вода в прибрежных частях океанов на обширной площади меняет цвет вследствие сильного увеличения численности фитопланктона (отсюда название Красное море).
f marée *f* rouge
s marea *f* roja
g rote Flut *f*

4943 **red-brown savanna soil**
A soil of tropical grass savannas with humus horizon containing only 1.5-3% of humus, often with carbonate horizon coloured reddish-brown, it is widespread in Africa, North Australia and some regions of South America.
r красно-бурая почва
Почва тропических остепненных саванн с гумусовым (содержание гумуса всего 1,5-3%) и часто с карбонатным горизонтами, имеющая красновато-бурую окраску, широко распространена в Африке, Северной Австралии и некоторых районах Южной Америки.
f sol *m* rouge-brun des savanes
s suelo *m* rojo-pardo de sabanas
g rotbrauner Boden *m*

4944 **red-brown soil**
A soil found under dry tropical deciduous forests which are widespread on the Plateau of Brazil, in the basins of the Congo and the Zambezi, in the central part of Indochina.
r красно-коричневая почва
Почва сухих тропических листопадных лесов и кустарников, распространена на Бразильском плоскогорье, в бассейнах рек Конго и Замбези, в центральной части полуострова Индокитай.
f sol *m* rouge-marron
s suelo *m* rojo-marrón
g rotbrauner Boden *m*

4945 **reddish-black soil**
A fertile soil that develops under dense herbaceous vegetation in the regions of semi-humid subtropical climate, has humus horizon 20-40 cm thick (2-5% of humus) and reddish clayish horizon.
r красновато-черная почва
Плодородная почва, формирующаяся под мощной травянистой растительностью в районах с полувлажным субтропическим климатом, имеет гумусовый горизонт мощностью 20-40 см (2-5% гумуса) и красноватый глинистый горизонт.
f sol *m* noir-rougeâtre
s suelo *m* negro-rojizo
g rötlich-schwarzer Boden *m*

4946 **reddish-brown soil**
A soil that forms in the transitional zone from savannas to tropical deserts, under non-leaching water regime, widespread mainly in Africa where it is used as pastures when irrigated it may be used in agriculture.
r красновато-бурая почва
Почва, образующаяся в зоне перехода от саванн к тропическим пустыням, развивается в условиях непромывного водного режима, распространена преимущественно в Африке, используется как пастбища, в земледелии лишь при орошении.
f sol *m* brun-rougeâtre
s suelo *m* pardo rojizo
g rötlich-brauner Boden *m*

* **redevelopment → 6504**

4947 **reduced air temperature**
The calculated temperature taken from observed temperature at a given height and reduced to the temperature at the mean sea level on the basis of a mean seasonal lapse rate of temperature.
r приведенная температура воздуха
Вычисленная температура воздуха, полученная из наблюденного на высоте станции значения и приведенная к значению, соответствующему среднему уровню моря на основе принятого стандартного вертикального градиента температуры.
f température *f* de réduction
s temperatura *f* de reducción
g reduzierte Temperatur *f*

4948 **reduced pressure**
The pressure calculated at some height by the barometric equation from the value measured at a given level by taking into account the theoretical effect of the air column between the two levels under consideration.
r приведенное давление
Давление воздуха, вычисленное с помощью барометрической формулы на некотором уровне, исходя из измеренного давления на рассматриваемом (исходном) уровне.
f pression *f* de réduction
s presión *f* de reducción
g reduzierter Luftdruck *m*

4949 **reduction**
1. A chemical process in which oxygen is removed from a compound. 2. In cartography, the scale of a map is reduced by means of a pantograph or by photography.
r восстановление
1. Химический процесс, при котором кислород удаляется из природного химического соединения. 2. В картографии,

уменьшение масштаба карты на пантографе или путем фотографического процесса.
f réduction *f*
s reducción *f*
g Reduktion *f*

4950 red-yellow soil
An acid soil of rainy tropical and equatorial forests that develops on ferralitic weathering crusts and has a low content of silica and high content of iron and aluminium due to great weathering of substrate, widespread in South America, Africa, and South-East Asia.
r красно-желтая почва
Кислая почва дождевых тропических и экваториальных лесов на ферралитовых корах выветривания, с низким содержанием кремнезема и высоким – железа и алюминия из-за высокой выветрелости субстрата, распространена в Южной Америке, Африке, Юго-Восточной Азии.
f sol *m* rouge jaune
s suelo *m* rojo-amarillo
g rotgelber Boden *m*

4951 reed swamp
A lowland swamp with the predominance of reed vegetation.
r тростниковое болото
Низинное болото с господством тростника.
f tourbière *f* roselière
s turbera *f* juncal
g Schilfmoor *n*

4952 reef
A line of rocks in the tidal zone of a coast, submerged at high water but partly uncovered at low water.
r риф
Череда скал в приливной зоне побережья, оказывающихся под водой при приливе и лишь частично остающихся под водой при отливе.
f écueil *m*
s escollo *m*
g Riff *n*; Klippe *f*

* **reef-knoll → 676**

4953 re-export
The resale of commodities, usually raw materials, which have been imported from other countries.
r реэкспорт
Перепродажа товаров (обычно сырья), ранее импортированного из других стран.
f réexportation *f*
s reexportación *f*
g Reexport *m*; Wiederausfuhr *f*

* **reference atmosphere → 5769**

* **reference grid → 2635**

* **reference plane → 963**

4954 reference point; landmark
A well seen motionless object or relief feature.
r ориентир
Хорошо видимый на местности неподвижный предмет или элемент рельефа.
f point *m* terrestre de repère
s punto *m* terrestre de referencia
g Orientierungspunkt *m*; Bezugspunkt *m*

* **reflected global radiation → 4955**

4955 reflected solar radiation; reflected global radiation
A part of total radiation reflected by the Earth's surface.
r отраженная солнечная радиация
Часть суммарной солнечной радиации, отраженная земной поверхностью.
f rayonnement *m* solaire réfléchi
s radiación *f* solar reflejada
g reflektierte Sonnenstrahlung *f*

4956 refraction diagram; wave-refraction diagram
A drawing which shows the changing positions of the waves over a given area for a given period and direction.
r диаграмма рефракции волн
Рисунок, показывающий изменение положения волн в данном районе за данный промежуток времени при заданном направлении волнения.
f plan *m* des vagues
s plano *m* de olas
g Wellenrefraktiondiagramm *n*

* **refraction of waves → 6723**

4957 refuge; refugium
An isolated area with favourable conditions where the species, which used to be widespread in the region, but now are nearly extinct, can escape environmental change, such as glaciation, cooling or warming of climate, etc.
r рефугиум
Местность, в которой благодаря наличию благоприятных условий смогли сохраниться виды растений и животных, ранее широко распространенные, но затем вымершие на большей площади ареала вследствие изменения экологических условий, например, во время оледенений, похолоданий или резких потеплений.
f refuge *m*
s refugio *m*
g Refugialgebiet *n*; Rückzugsgebiet *n*; Refugium *n*

4958 refugee; asylum seeker
A person who has left the place of permanent residence as a result of hostilities or natural calamities or fear of persecution for reasons of race, religion, nationality or political opinion.
r беженец

Человек, покинувший место своего постоянного проживания в результате военных действий или стихийных бедствий, либо из-за опасения преследований по причине расовой, религиозной, националистической или политической нетерпимости.

f réfugié *m*
s refugiado *m*
g Flüchtling *m*; Flüchtige *f*

* **refugium → 4957**

4959 **reg**
A bare, stony desert or a completely denuded desert plain covered with rock debris and tightly packed gravel overlying sand and clayed sand.
r рег
Пустынная полностью денудированная равнина или твердая почва, покрытая щебнем и гравием, с небольшим количеством песка (первоначально – равнинная каменистая пустыня в Алжирской Сахаре).
f reg *m*
s reg *m*
g Reg *m*

4960 **regelation**
The refreezing of ice which has melted under pressure at the contacts of ice crystals under local stress or heat influx.
r режеляция льда
Смерзание льда при повторной кристаллизации воды, появляющейся на контактах ледяных кристаллов в результате приложения местной нагрузки или притока тепла.
f regel *m* de la glace
s regelación *f* de hielo
g Regelation *f* des Eises

4961 **regelation ice**
Ice the structure of which is transformed by thermal and dynamic metamorphism.

Fig. 271. **Regenerated glacier** (down)

r режеляционный лед
Лед, структура которого преобразована в процессе термо- и динамометаморфизма.
f glace *f* de regel
s hielo *m* de regelación
g Regelationseis *n*

4962 **regenerated glacier**
A valley glacier deprived of firn basin and fed by the ice falls from the glacier hanging above, or from the valley glacier located at a higher altitude. → Fig. 271
r возрожденный ледник
Долинный ледник, лишенный фирнового бассейна и питающийся обвалами льда с висячего или более высоко расположенного долинного ледника. → Рис. 271
f glacier *m* régénéré
s glaciar *m* regenerado
g regenerierter Gletscher *m*

* **regeneration of a depression → 4963**

4963 **regeneration of cyclone; regeneration of a depression**
The deepening of a cyclon which is in the state of filling due to the income of new masses of cold air and further increase of thermal contrasts.
r регенерация циклона
Углубление заполняющегося циклона вследствие притока свежих масс холодного воздуха и обострения вследствие этого термических контрастов.
f régénération *f* de depression
s regeneración *f* de una depresión; regeneración *f* de ciclón
g Neubildung *f* eines Tiefdruckgebietes; Tiefdruckregeneration *f*

4964 **regime**
The seasonal fluctuation in respect of precipitation, or of the volume of a glacier, or of the volume of water in a river, as well as the recurring seasonal pattern of climatic change.
r режим
Сезонные колебания атмосферных осадков, ледников, речного стока, а также текущие колебания климата и погоды.
f régime *m*
s régimen *m*
g Regime *n*

4965 **regime of flow; regime of runoff**
The changes of water discharge in time determined by the climatic conditions and peculiarities of the basin.
r режим стока
Изменения стока во времени, обусловленные климатическими условиями и особенностями бассейна.
f régime *m* d´écoulement
s régimen *m* escorrentía
g Abflußverhältnisse *pl*

4966 regime of reserve
The way of protection of natural territories and water bodies which guaranties their withdrawal from traditional economic use.
r заповедный режим
Условия охраны природных территорий и акваторий, обеспечивающие их изъятие из всякого традиционного хозяйственного использования.
f régime *m* de réservation
s régimen *m* de reservación
g Schutzgebietsregime *n*

* **regime of runoff → 4965**

4967 region
A land or water area, considerable in size, having some unity of one or more features (natural or the result of human activity); an area which is a unit of administration in certain countries.
r регион; область; район
Значительная по размерам территория (акватория) с присущими ей природными и антропогенными особенностями и ресурсами, обладающая некоторой общностью природных условий и характеризуемая протяженностью или площадью; одна из основных единиц в иерархии административного деления ряда стран.
f région *f*; pays *f*
s región *f*
g Gebiet *n*; Land *n*; Landesteil *m*; Region *f*

4968 region of development
An economic region where in a certain period the rates of economic growth are higher than elsewhere across the country.
r район развития
Экономический район, где темпы роста хозяйства в определенный период выше, чем в среднем по стране.
f région *f* de développement
s región *f* de fomento; región *f* de crecimiento; región *f* de desarrollo
g Eniwicklungsgebiet *n*

4969 region of economic gravitation
The region with the most intensive territorial ties with the neighbouring regions that frequently develops into a centre of attraction for productive forces.
r район тяготения
Район, имеющий наиболее интенсивные территориальные связи с окружающими районами, нередко становящийся районом притяжения производительных сил.
f région *f* de gravitation économique
s región *f* de atracción económica
g Attraktionsgebiet *n*

4970 region of stagnation
An economic region where the progress is very slow.
r стагнирующий район
Экономический район, где очень слабы или совсем отсутствуют признаки поступательного развития.
f région *f* stagnante
s región *f* de estancamiento
g stagnierendes Gebiet *n*

4971 regional analysis
An integrated analysis of nature, population, economy, culture, social organization and possible ways of the development of a particular region.
r региональный анализ
Комплексный анализ природы, населения, хозяйства, культуры, социальной организации конкретного региона и его дальнейшего развития.
f analyse *f* régionale
s análisis *m* regional
g Regionalanalyse *f*

4972 regional atlas
An atlas the maps of which are devoted to different aspects of a particular region of a country or continent.
r региональный атлас
Атлас какой-либо крупной территориальной единицы.
f atlas *m* régional
s atlas *m* regional
g Regionalatlas *m*; Landesatlas *m*

4973 regional development; local development
Positive changes in social and economic spheres of a region based on a rational use of local nature, resources and workforce.
r региональное развитие; местное развитие
Положительные изменения в социальной и экономической жизни отдельных регионов страны, основанные, в частности, на рациональном использовании местных природных, материальных и трудовых ресурсов.
f développement *m* regional; développement *m* local
s desarrollo *m* regional; crecimiento *m* regional
g Regionalentwicklung *f*; lokale Entwicklung *f*

4974 regional disparity; regional inequality
Considerable differences in the standard of living, industrial production and services between large economic regions of a country or macroregions of the global economic system.
r региональные диспропорции
Заметные различия в уровне жизни, производстве, сфере услуг между крупными экономическими районами одной страны или макрорегионами глобальной экономической системы.
f disparités *fpl* régionales

s desproporciones *fpl* regionales
g regionale Disproportionen *fpl*; regionale Disparitäten *fpl*

4975 **regional economics**
Distribution of enterprises, markets, organization of urbanized areas, etc. in a particular region.
r региональная экономика
Размещение предприятий, ареалов сбыта продукции, организации урбанизированных районов и пр. на конкретной территории.
f economie *f* régionale
s economía *f* regional
g Regionalökonomie *f*

4976 **regional geography; country study**
The integrated study of nature, population, economy, culture and social organization of a particular country or region.
r страноведение; региональная география
Комплексное изучение природы, населения, хозяйства, культуры и социальной организации отдельных стран и крупных территорий.
f géographie *f* régionale; études *fpl* des pays; étude *f* de la contrée
s geografía *f* regional; estudio *m* de país
g Länderkunde *f*

4977 **regional identity; territorial identity**
The awareness and appreciation of specific features of the region by its population and self-understanding of each inhabitant as the region's part which result in the desire to reach the regional prosperity and development.
r региональное самосознание; территориальное самосознание
Осознание и оценка населением особенностей своего региона и самого себя как части этой территории, забота о ее благоденствии и развитии.
f conscience *f* d'identité régionale; identité *f* territoriale; identité *f* régionale
s identidad *f* regional; identidad *f* territorial
g Regionalbewußtsein *n*; regionale Identität *f*

* **regional inequality → 4974**

4978 **regional planning; territorial planning**
Making regional plans of land use for various purposes (urban development, rural development, development of infrastructure) that consider natural and socio-economic conditions.
r организация территории; территориальное планирование; региональное планирование; районная планировка
Составление в региональных масштабах (с учетом природных и социально-экономических условий) планов использования земель для разных целей: развития городов, благоустройства сельских районов, развития инфраструктуры.
f aménagement *m* régional
s planeamiento *m* regional; planificación *f* regional; planificación *f* territorial
g Raumordnung *f*; territorial Planung *f*; Landesplanung *f*; Regionalplanung *f*

4979 **regional policy**
A system of measures and legal acts designed and adopted by the government for economic and social management of individual regions and administrative units.
r региональная политика
Система мер и законодательных актов государства по управлению экономическим и социальным развитием отдельных регионов и административных единиц страны.
f politique *f* régionale
s política *f* regional
g Regionalpolitik *f*

4980 **regional science; area studies**
An interdisciplinary study which concentrates on the integrated analysis of economic and social phenomena in a region, to understand regional change and to plan future regional development.
r регионалистика; регионоведение
Междисциплинарный анализ социально-экономических явлений в конкретном регионе с целью изучения региональных изменений и планирования будущего регионального развития.
f études *fpl* régionales
s estudios *mpl* regionales
g Regionalkunde *f*

4981 **regionalism**
Loyalty to a region, concentration on local or regional interests rather than on the interests of the state; a move to foster or protect an indigenous culture in a particular region.
r регионализм; местничество
Выдвижение на первый план узкоместных, региональных интересов в ущерб общенациональным или федеральным; деятельность, преследующая цель способствовать или защитить культуру местных народов в конкретном регионе.
f régionalisme *m*
s regionalismo *m*
g Regionalismus *m*

4982 **regionalization**
The demarcation of regions in such a way that there is little variation within each region while each region is sharply distinct from the others (the regions may be delimited according to different criteria: natural or man-made).
r районирование
Выделение по принятым критериям (как естественным, так и антропогенным) на исследуемой территории (акватории) районов таким образом, чтобы в их пределах рассматриваемое явление или процесс было достаточно однородным, но заметно отличалось от таковых в соседних районах.

f régionalisation *f*
s regionalización *f*
g Regionalisierung *f*; Rayonierung *f*

4983 **register of population**
A record of the major events in the lives of a population: births, deaths, marriages, divorces, and adoptions.
r запись гражданского состояния
Регистрация главных событий в жизни людей: рождений, смерти, заключений брака, разводов, усыновлений.
f registre *m* de population
s registro *m* de población
g Volksregister *n*

4984 **regolith**
The layer of loose broken, rocky material (soil, sediments, broken rock, volcanic ash, wind-blown material etc.) overlying the solid bedrock.
r реголит
Чехол рыхлого материала (почва, рыхлые отложения, разрушенные скальные породы, вулканический пепел, эоловый материал и т.п.), перекрывающий твердые коренные породы.
f régolite *m*
s regolita *f*
g Regolith *m*

* **regoor → 4991**

4985 **regosol; abrazem**
A poorly developed soil without diagnostic horizons that forms as a result of erosion of loose rocks, for example loess or dune sands, it is found all over the world, but does not form large tracts.
r регосоль; абразем
Почва, лишенная верхних горизонтов в результате естественной и антропогенной эрозии и механического срезания, формируется при эрозии рыхлых пород, например, лёссов или дюнных песков, распространена по всему миру, но не образует крупных массивов.
f régosol *m*; abrazem *m*
s regosol *m*; regosuelo *m*; abrazem *m*
g Regosol *m*

* **regosolic forest soil → 4415**

4986 **regression**
The retreat of a shallow sea from a land mass owing to a negative movement of sea-level due to decreasing of the water volume in the ocean, e.g. during the glacial epochs, or a positive isostatic movement of the land (upward tectonic rise).
r регрессия
Медленное отступание моря от первоначальных берегов вследствие поднятия суши, опускания океанического дна или уменьшение объема воды в океане, например, во время ледниковых эпох.
f régression *f*
s regresión *f*
g Regression *f*; Rückgang *m*

4987 **regular grid; tessellation**
Regular geometrical grid used for division of the Earth's surface or georepresentation into cells: squares, equilateral triangles, hexagons, etc.
r регулярная сеть; грид
Геометрически правильная сеть, используемая для разбиения земной поверхности или геоизображения на ячейки: квадраты, равносторонние треугольники, шестиугольники и т.п.
f quadrillage *m* rectangulaire
s red *f* ordenada
g Gitter *n*; Gitternetz *n*

4988 **regular park; formal symmetrical park**
A park having the strict, often symmetric, design.
r регулярный парк
Парк со строгой, нередко симметричной, планировкой.
f parc *m* régulier
s parque *m* regular
g Regulärpark *m*

4989 **regulated flow**
A river flow with relatively levelled distribution over the year, smoothed floods and comparatively high debit in the period of low-water.
r зарегулированный сток
Сток реки с относительно равномерным распределением в течение года, сглаженными паводками и сравнительно высокими расходами в период межени.
f débit *m* régulier
s caudal *m* regulado
g gleichmässiger Abfluß *m*; geregelter Abfluß *m*

4990 **regulated river**
A river the regime of which acquired new features due to taken engineering measures aimed at its economic use.
r зарегулированная река
Река, режим которой приобрел существенно новые черты в результате инженерных мероприятий, направленных на ее хозяйственное использование.
f rivière *f* régulière
s río *m* regulado; río *m* regularizado
g regulierter Fluß *m*

4991 **regur; black cotton soil; regoor**
A dark-coloured fertile clayey, fissured tropical soil which has low-humus and swels greatly when moistened occurring on basalt covers of the Deccan Plateau in India.
r регур
Распространенная на базальтовых покровах Деканского плоскогорья в Индии темноцветная тропическая почва, малогумус-

ная, глинистая и трещиноватая, сильно набухающая при увлажнении.
f régur *m*
s regur *m*
g Regur *m*; Baumwollboden *m*

4992 **reinforced ice**
An artificial ice formed by freezing of water with special admixtures increasing the ice strength.
r армированный лед
Замороженная вода с добавками различных веществ, повышающих механическую прочность льда.
f glace *f* renforcée
s hielo *m* armado
g bewehrtes Eis *n*; armiertes Eis *n*

* **rejuvenation head → 3323**

4993 **rejuvenation of relief**
The development of younger surface forms, appropriate to the earlier stages of the cycle of erosion and the increase of geomorphic contrasts of relief manifested in deepening of valleys, increase of inclination of slopes and relative heights which is caused either by the intensive tectonic elevation or lowering of base level of erosion.
r омоложение рельефа
Увеличение контрастности рельефа, проявляющееся в углублении долин, возрастании крутизны склонов и относительных высот, что вызывается интенсивным тектоническим поднятием либо понижением базиса эрозии.
f rajeunissement *m* du relief
s rejuvenecimiento *m* de relieve
g Verjüngung *f* des Reliefs

4994 **rejuvenationof a river**
The state of a river when its energy has been revitalized, thereby leading to an increase in its erosional capability.
r омоложение реки
Состояние реки, когда по той или иной причине произошло возрождение энергии ее потока, что приводит к вступлению реки в новый цикл эрозии.
f rajeunissement *m* de rivière
s rejuvenecimiento *m* del río
g Flußverjüngung *f*

4995 **relations of production**
Material economic relationships between people in the productive process and the movement of goods from production to consumption.
r производственные отношения
Материальные экономические отношения между людьми в процессе производства и движения общественного продукта от производства до потребления.
f relations *fpl* de production; rapports *mpl* de production
s relaciones *fpl* de producción
g Produktionsverhältnisse *npl*

4996 **relative height; elevation**
The vertical distance between two points on the Earth's surface.
r относительная высота; превышение
Разность абсолютных высот двух точек.
f dénivelée *f*; élévation *f*; altitude *f* relative
s desnivel *m*; diferencia *f* de niveles; altitud *f* relativa
g relative Höhe *f*

4997 **relative humidity**
The ratio of the actual vapour density which indicates the amount of water vapour present in the air unit to the theoretical maximum (saturation) vapour density at the same temperature, expressed as a percentage.
r относительная влажность
Процентное отношение упругости водяного пара, содержащегося в единице объема воздуха, к упругости насыщающего пара при той же температуре.
f humidité *f* relative
s humedad *f* relativa
g relative Luftfeuchtigkeit *f*

4998 **relative overpopulation; latent overpopulation**
A relative excess of the number of active population living in a given territory over the local demand for labour resources.
r относительная перенаселенность; скрытая перенаселенность
Относительный избыток трудоспособного населения, проживающего на данной территории, по сравнению со спросом на рабочую силу.
f surpeuplement *m* relatif; surpopulation *f* relative; surpeuplement *m* latent
s superpoblación *f* relativa; superpoblación *f* latente
g relative Übervölkerung *f*; verborgene Übervölkerung *f*

* **relative relief → 4904**

4999 **relative transparency of water**
The value of water transparency determined by water layer thickness through which a white disk 30 cm in diameter is still distinguished by eye.
r относительная прозрачность воды
Величина прозрачности воды, определяемая по толщине водного слоя, сквозь который еще различим глазом белый диск диаметром 30 см.
f transparence *f* relative de l'eau
s transparencia *f* relativa de agua
g relative Wasserdurchsichtigkeit *f*; relative Wassertransparenz *f*

5000 **release of avalanche**
Avalanche occurrence induced by a natural triggering mechanism or by artificial way. → Fig. 272, p. 600

Fig. 272. **Release of avalanche**

r сход лавины
Процесс падения снежной лавины, вызванный естественными причинами или искусственно. → Рис. 272
f déclenchement *m* d'avalanche
s desencadenamiento *m* de avalancha; salida *f* espontánea de alud
g Lawinensturz *m*

5001 release of reservoir
The process of alowing water needed for economic purposes to drain away from the reservoir.
r сработка водохранилища
Спуск воды из водохранилища, необходимый в хозяйственных целях.
f assèchement *m* de réservoir
s desagüe *m* de embalse
g Ablaß *m* des Staubeckens

5002 reliability diagram
An indication in the margin of a map of the dates and the quality of the source material from which the map has been compiled.
r зарамочные сведения
Данные, приводимые за рамкой географической карты, включающие дату составления и оценку качества исходных материалов.
f diagramme *f* d'authenticité
s diagrama *f* de autenticidad
g Zuverlässigkeitsdiagramm *n*

* **relic → 5003**

5003 relict; relic
An animal or plant species which are the remnants of the flora or fauna of the past geological epochs and was at an earlier time common in, or chracteristic of, this area.
r реликт
Виды животных или растений, входящие в состав современной биоты как пережитки флор или фаун минувших геологических эпох.
f relique *f*
s relicto *m*
g Relikt *n*; Überbleibsel *n*

* **relict biocoenosis → 5006**

5004 relict fauna
Animal species which survived from past climatic epochs in areas with the favourable environmental conditions.
r реликтовая фауна
Древняя фауна, сохранившаяся в местах с благоприятными условиями существования.
f faune *f* relique; faune *f* résiduelle
s fauna *f* relicta
g Reliktfauna *f*

5005 relict flora
Plants which survived from past climatic epochs in favourable conditions.
r реликтовая флора
Растения былых климатических эпох, сохранившиеся в особенно благоприятных условиях.
f flore *f* résiduelle
s flora *f* relicta
g Reliktflora *f*

5006 relict habitat; relict biocoenosis
A small isolated area of former wider area where the original relic flora and fauna remain in new environmental conditions.
r реликтовое местообитание; реликтовый биоценоз
Небольшой островной участок прежнего, более широкого ареала, в котором сохранились реликтовые виды в новых условиях среды.
f habitat *m* relique; biocénose *f* relique
s hábitat *m* relicto; biocenosis *f* relicta
g Reliktstandort *m*; Reliktbiozönose *f*

5007 relict ice; fossil ice
Big masses of ground ice formed under past climatic conditions more severe than those of today.
r ископаемый лед; реликтовый лед
Крупные массивы подземного льда, сформировавшегося в более суровых условиях климата по сравнению с современным.
f glace *f* relique; glace *f* fossile
s hielo *m* relicto; hielo *m* fosil
g reliktisches Eis *n*; fossiles Eis *n*

5008 relict lake
A water body continuing to exist on the place of retreated sea, either completely isolated from it or connected with it by an outlet channel.
r реликтовое озеро
Водоем, продолжающий существовать на месте отступившего моря, изолированный от него или сообщающийся с ним протокой.
f lac *m* relique
s lago *m* relicto
g Restsee *m*

5009 relict landform
A landform that was created by past processes which are no longer operative in the region.

r реликтовый рельеф
Формы земной поверхности, сформировавшиеся в прошлые геологические эпохи процессами, прекратившими свою деятельность в данном регионе.
f formes *fpl* reliques
s relieve *m* relicto
g Reliktrelief *n*

5010 relict permafrost
Places of permanently frozen ground existing today at lower levels than the zone of constant temperatures, which preserved from the previous geological epoch which have had favourable climatic conditions for the formation of permafrost.
r реликтовая вечная мерзлота
Участки многолетнемерзлых пород, сохранившихся ниже зоны постоянных температур от прежней геологической эпохи, когда в данном районе существовали благоприятные для нее климатические условия.
f pergélisol *m* relique
s permafrost *m* relicto
g reliktischer Permafrost *m*

5011 relief; landform
The character of the surface of the Earth which comprises a variety of positive and negative forms of different shape, size, origin, age, and the history of development, that results from the interaction of physical processes (e.g. fluvial action, glacial action, weathering) and crustal movements.
r рельеф
Характер земной поверхности, представляющей собой совокупность положительных и отрицательных форм, неодинаковых по очертаниям, размерам, происхождению, возрасту и истории развития, что служит результатом физических процессов (например, речных и ледниковых процессов, выветривания), а также движений земной коры.
f relief *m*; modelé *m*
s relieve *m*; relieve *m* accidentado; modelado *m*
g Relief *n*; Geländegestalt *f*

5012 relief model
A geographical map moulded into a three-dinensional form with a certain ratio of vertical and horizontal scales.
r рельефная карта
Географическая карта, изображающая рельеф в трехмерной форме с соблюдением определенного соотношения вертикального и горизонтального масштабов.
f relief *m* habillé; carte *f* en relief
s mapa *m* en relieve
g Kartenrelief *n*

* **relief rainfall → 4307**

5013 religious tourism; pilgrimage
Travelling of believers to sacred places, monasteries and temples as an act of devotion and worship.
r религиозный туризм; паломничество
Странствие верующих для поклонения святым местам и посещения монастырей и храмов.
f tourisme *m* religieux; pélerinage *m*
s turismo *m* religioso; peregrinación *f*
g Religionstourismus *n*; Wallfahrtstourismus *n*; Pilgerschaft *f*

5014 remote sensing
The process of getting information on the Earth's surface and other celestial bodies and objects situated on them from the distance, by non-contact methods: by artificial satellites, planes, probes, etc.
r дистанционное зондирование
Процесс получения информации о поверхности Земли и других небесных тел и расположенных на них объектах неконтактными методами – с искусственных спутников, самолетов, зондов и пр.
f télédétection *f*
s sondeo *m* remoto; percepción *f* remota
g Fernerkundung *f*

5015 remote sensing from satellites; satellite imagery
Getting the images of the Earth's surface, celestial bodies, and space phenomena by instruments from outside the Earth atmosphere.
r космическая съемка
Получение изображений земной поверхности, небесных тел и космических явлений приборами, находящимися за пределами земной атмосферы.
f levé *m* d'espace
s levantamiento *m* espacial
g Satellitenaufnahme *f*

* **rendsina → 5016**

5016 rendzina; rendsina
An intrazonal soil forming on carbonate rocks under coniferous and mixed forests of the temperate belt by the action of leaching, it has well developed humus horizon with humus content of 5-9%.
r дерново-карбонатная почва; рендзина
Интразональная почва, формирующаяся на карбонатных породах под хвойными и смешанными лесами умеренного пояса в условиях промывного режима, имеет хорошо развитый гумусовый горизонт с содержанием гумуса 5-9%.
f rendzine *m*
s rendzina *f*
g Rendzina *f*; Humuskarbonatboden *m*

* **renewable energy sources → 221**

5017 renewable natural resources; renewable resource

Natural resources that can be replenished through natural environmental processes, e.g. forests.

r возобновлямые природные ресурсы
Природные ресурсы, способные к возобновлению естественным путем, например, леса.

f ressource *f* naturelle renouvelable; ressource *f* renouvelable

s recursos *mpl* naturales renovables; recursos *mpl* renovables

g erneuerbare Resourcen *fpl*

* **renewable resource → 5017**

5018 **rent**
The price paid for the use of any durable good (usually land or buildings) at certain specified or customary times, or for a certain period of time.

r рента
Плата за любой товар долгосрочного пользования (обычно земля или строения) в расчете на конкретный период времени.

f rente *f*; loyer *m*

s renta *f*

g Rente *f*; Mietpreis *m*

* **replacement of population → 5020**

* **replacement rate → 5020**

* **repose slope → 5481**

5019 **repository; burial site; graveyard**
A place where domestic or industrial toxic and radioactive waste is disposed of.

r могильник
Специально подготовленное хранилище для захоронения бытовых или промышленных токсичных и радиоактивных отходов.

f cimetière *m*; stockage *m* de déchets radioactifs

s cementerio *m* para desechos radioactivos

g Endlager *m*

5020 **reproduction rate; replacement rate; replacement of population**
Any of the calculations used to indicate trends in numbers of human population, when birth rate exceeds mortality rate, the population reproduction is described as extended, when mortality exceeds birth rate the population reproduction is described as narrowed.

r воспроизводство населения
Любые вычисления, используемые для расчетов тенденции роста или убыли населения, при превышении рождаемости над смертностью – расширенное воспроизводство населения, при превышении смертности – суженное.

f reproduction *f* de population; replacement *m*

s reproducción *f* de población; reemplazamiento *m*

g Bevölkerungsreproduktion *f*

5021 **republic**
A state in which supreme political power lies in the hands of an elected body, the head of the state is elected directly by the population or through elected representatives governing according to law.

r республика
Форма государства, когда высшая власть принадлежит выборному органу, а глава государства избирается населением или специальной избирательной коллегией.

f république *f*

s república *f*

g Republik *f*

5022 **research and development city**
A city or its suburbs where high-tech productions is located, e.g. Silicon Valley in the USA, Tsukuba in Japan.

r технополис
Город или его окраинная часть, где сосредоточены высокотехнологичные производства, или специализированный главным образом на научно-технических разработках (наукоград).

f technopole *f*; technopôle *f*

s tecnopoli *f*

g Technopole *f*

5023 **research and science park; science park**
A tract of land, in some cases park-like, planned and officially designated for the accomodation of individuals and organizations engaged in research, development and scientific production.

r научно-исследовательский парк; научный парк
Участок земли, нередко имеющий облик парка, спланированный и построенный для размещения людей и организаций, занимающихся научными исследованиями и инновациями.

f parc *m* des recherches scientifiques; parc *m* scientifique

s parque *m* de investigación científica; parque *m* científico

g Forschungspark *m*

* **research ship → 5024**

5024 **research vessel; research ship**
A ship specially equipped for different kinds of observations in oceans, seas, and lakes.

r научно-исследовательское судно; гидрографическое судно
Судно, предназначенное для научно-исследовательских работ на водных объектах.

f navire *m* scientifique

s barco *m* científico

g Forschungsschiff *n*

5025 **resequent stream; secondary consequent stream; resequent valley**

A tributary of a subsequent stream flowing in the valley running in the same direction as the initial consequent stream, i.e. in accordance with the dip of the rock strata. → Fig. 68, p. 145

r ресеквентная река;
ресеквентная долина
Приток субсеквентной реки, текущий в долине в том же направлении, что и первоначальная консеквентная река, т.е. в согласии с падением слоев горных пород. → Рис. 68, с. 145

f fleuve *m* reséquent; vallée *f* reséquente
s río *m* resecuente; valle *m* resecuente
g resequenter Fluß *m*; resequentes Tal *n*

* **resequent valley → 5025**

5026 **reservation$_1$**
A tract of land set aside for the use of ethnic minorities (native population) on which they continue living in their traditional ways and engaged in natural economy.

r резервация
Специальная территория, выделенная для жизни и ведения натурального хозяйства малым народностям – коренным жителям данного региона.

f réserve$_1$ *f*
s reserva$_1$ *f*
g Reservation *f*

* **reservation$_2$ → 5027**

5027 **reserve; nature preserve; reservation$_2$**
An area of land set aside for nature conservation and associated scientific research usually with strong legal protection from other uses and with restricted public access.

r заповедник; резерват;
особо охраняемая территория
Охраняемая природная территория, полностью исключенная из любой хозяйственной деятельности ради сохранения здесь природы в нетронутом виде.

f réserve$_2$ *f*; réserve *f* naturelle
s reserva$_2$ *f*; reserva *f* natural
g Naturschutzgebiet *n*; Naturreservat *n*

* **reserves → 3866**

5028 **reservoir**
An artificial water body created in order to regulate the discharge and the use of the water in irrigation, in producing hydroelectricity, for industrial and domestic use and recreational purposes. → Fig. 273

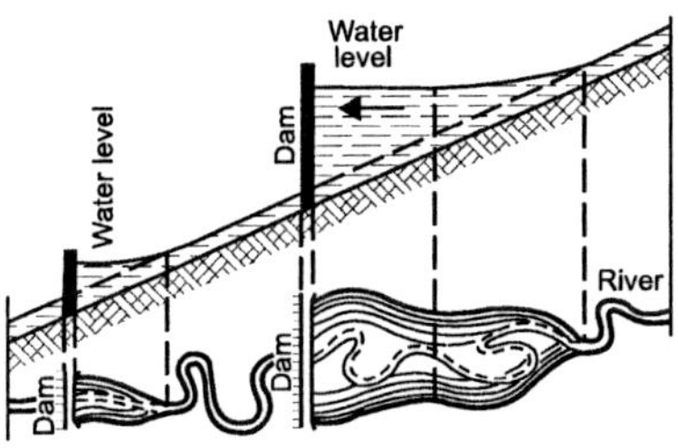

Fig. 273. **Reservoir**

r водохранилище
Искусственный водоем, созданный для регулирования стока воды в целях ирригации, гидроэнергетики, промышленного и местного использования, для рекреационных целей. → Рис. 273

f réservoir *m*; retenue *f*; lac *m* artificiel
s embalse *m*
g Staubecken *n*; Stausee *m*

5029 **resettlement**
The movement of people from one place to another organized by authorities in order to change the residence for a relatively long period of time or permanently.

r переселение
Организованное властями передвижение людей из одного места в другое с переменой места жительства на более или менее длительное время или навсегда.

f déplacement *m* de population; migration *f* organisée; transferts *mpl* de population
s desplazamiento *m* de población; traslados *mpl* de población
g Umsiedlung *f*

* **residence → 4693**

5030 **resident birds**
The birds which do not do seasonal migrations.

r оседлые птицы
Птицы, не совершающие сезонных перелетов или кочевок.

f oiseaux *mpl* sédentaires
s aves *fpl* sedentarias
g seßhafte Vögel *mpl*; ansässige Vögel *mpl*

* **residential area → 1678**

* **residential district → 1678**

* **residual debris → 5031**

5031 **residual deposit; residual debris**
An accumulation of rock waste, from clays to boulders in grain-size, caused by the weathering of rocks in situ, it forms part of the *regolith*.

r остаточные отложения
Материал, остающийся на поверхности скал после длительного периода химического выветривания, представляет собой часть *реголита*.

f débris *mpl* résiduel
s escombro *m* residual
g Verwitterungsgesteine *mpl*

5032 **residual soil; sedentary soil**
A soil formed in situ by the decomposition of the underlying solid rocks by weathering.

r первичная почва
Почва, сформировавшаяся из разрушенных остатков твердых горных пород, лежащих непосредственно под ней.

f sol *m* résiduel
s suelo *m* residual
g Verwitterungsboden *m*

5033 resistant rock
A rock with physical and chemical properties which make it able to withstand the processes of erosion and weathering.
r стойкая горная порода
Горная порода, физические и химические свойства которой позволяют ей противостоять процессам эрозии и выветривания.
f roche *f* résistante
s roca *f* resistente
g resistentes Gestein *n*

* **resolution → 5706**

5034 resort
A place with attractive natural conditions for leasure, recreation and medical treatment.
r курорт
Местность, обладающая привлекательными природными условиями для отдыха или лечения.
f station *f* balnéaire
s bernario *m*
g Kurort *m*

5035 resource conservation
The careful management and maintenance of natural resources.
r сохранение ресурсов
Осторожное управление природными ресурсами, сдерживающее их быстрое потребление.
f conservation *f* des ressources
s conservación *f* de recursos
g Verwahrung *f* der Naturschätze

5036 resource cycle
Transformations and territorial movements of a natural substance or group of substances at the stages of their use.
r ресурсный цикл;
природно-ресурсный цикл
Совокупность превращений и территориальных перемещений природного вещества или группы веществ, происходящих на всех этапах использования его человеком.
f cycle *m* de ressources naturelles
s ciclo *m* de recursos naturales
g Ressourcenzyklus *m*

5037 resource intensity
The amount of resources used to produce a unit of production or services.
r ресурсоемкость
Количество ресурсов (энергии и материалов), используемых для получения единицы производимого продукта или услуг.
f consommation *f* des ressources
s consumo *m* de recursos
g Ressourceskapazität *f*

5038 resource management
The skillful control of resources by those who ensure that it is used economically and with forethought, and determine the present need for and the value of such a use, balance benefits and costs, taking into account environmental constrains, social, economic and political implications, national policy, etc.
r управление ресурсами
Разработка и контроль ресурсов с учетом их современной необходимости и будущей потребности в них, принимая во внимание ограниченные возможности использования, природные условия, социальные, экономические и политические обстоятельства, национальную политику и пр.
f management *m* des ressources
s gestión *f* de recursos
g Verwaltung *f* der Naturschätze

5039 resource map
A geographic map showing the distribution of natural and other resources.
r ресурсная карта
Географическая карта, отображающая размещение и запасы природных и иных ресурсов.
f carte *f* de ressources naturelles
s mapa *f* de recursos naturales
g Ressourcenkarten *fpl*

5040 resource orientation
The tendency of an enterprise or a particular industry to be located close to the source of its raw materials.
r ресурсная ориентация
Тенденция конкретного предприятия или промышленной отрасли располагаться недалеко от источника сырья.
f orientation *f* à ressources
s orientación *f* a recursos
g Rohstofforientierung *f*

5041 resource saving
The production with the minimum consumption of materials and energy at all stages of production.
r ресурсосбережение
Производство и использование конечных продуктов с минимальным расходом материалов и энергии на всех этапах производственного цикла.
f épargne *f* de ressources
s ahorro *m* de recursos
g Ressourceneinsparung *f*

5042 resources
A source of supply or support, the means (or collective means) of meeting a need or deficiency, especially an economic or social need or deficiency.
r ресурсы
Источник ценностей, запасов, возможностей, средств и доходов для жизни и деятельности людей, в числе которых важнейшее место занимают природные ресурсы.

f ressources *fpl*
s recursos *mpl*
g Ressourcen *fpl*

5043 resources of the ocean
Mineral and biological resources which are beeing extracted, or can be potentially extracted, from the ocean bed or sea water.
r ресурсы Мирового океана
Совокупность минеральных и биологических ресурсов, добываемых или перспективных для добычи со дна или из толщи Мирового океана
f ressources *fpl* de l'océan
s recursos *mpl* de océano
g Ozeansressourcen *fpl*

5044 respiration of soil
The liberation of carbon dioxide and other gases from soil as a result of the functioning of soil biota.
r дыхание почвы
Выделение углекислоты и других газов из почвы, связанное с функционированием почвенной биоты.
f respiration *f* du sol
s respiración *f* del suelo
g Bodenatmung *f*

5045 rest area of reserve
The internal part of a reserve where any interference of man in the natural processes, including his presence, is prohibited.
r зона покоя заповедника
Внутренняя часть заповедника, на территории которой запрещено любое вмешательство человека в природные процессы, включая его присутствие.
f zone *f* de repos de réserve naturelle
s zona *f* de reposo de reserva natural
g Ruhezone *f* des Naturschutzgebiets

5046 restoration
The establishment of the number and initial area of an extinct plant or animal species on a certain territory.
r реакклиматизация; восстановление вида
Искусственное возвращение в какую-либо местность ранее исчезнувшего там вида животных или растений.
f réacclimatation *f*
s reaclimatación *f*; reintroducción *f* de especie
g Wiederherstellung *f* der Arten

5047 restored species
A biological species, the number of individuals or diversity of populations of which has reached the level required to avoid the risk of extinction.
r восстановленный вид
Биологический вид, число особей или разнообразие популяций которого достигло уровня, безопасного в отношении угрозы вымирания.
f espèce *f* rétablie
s especie *f* reforzada; especie *f* reintroducida
g wiederhergestellte Art *f*

5048 resurgence
The reappearance at the Earth's surface of a stream which originated on impermeable rock but which disappears underground on reaching calcareous strata.
r возрождение потока
Появление вновь на поверхности потока, который протекал по водонепроницаемым породам, но затем исчез под землей, попав на известняковые отложения.
f régénération *f* de torrent
s regeneración *f* de torrente
g Stromerneuerung *f*

5049 retail trade; retailing
The sale of goods and services to the general public.
r розничная торговля
Часть внутренней торговли, реализующая товары и услуги населению.
f commerce *m* de détail
s comercio *m* al por menor
g Einzelhandel *m*; Kleinhandel *m*

* **retailing → 5049**

5050 retarding basin
An artificially maintained flood-control area into which floodwaters are deflected by means of a dam placed across a river channel.
r регулирующий бассейн
Участок речного бассейна с искусственным регулированием паводков путем создания дамбы, перегораживающей основное речное русло.
f bassin *m* retardateur
s cuenca *f* retardataria; cuenca *f* de regulación
g Rückhaltebecken *n*

5051 retention of snow; snow retention
Methods and techniques for retaining snow on fields so that to increase the water storage in soils and to protect wintering plants.
r снегозадержание
Приемы и методы удержания снега от сноса ветром на полях и его накопление для увеличения запасов воды в почве и утепления зимующих растений.
f retenue *f* de neige
s retención *f* de nieve en los campos; retención *f* nival
g Schneeaufhalten *n*

5052 retention of water; water retention
A system of measures (forest plantations, snow retention, terracing of slopes, etc.) taken at catchments in order to decrease surface runoff and hold the moisture in soil.
r водозадержание

Система мер на водосборах (лесопосадки, снегозадержание, террасирование склонов и пр.), имеющая цель уменьшить поверхностный сток и сохранить влагу в почве.
f rétention *f* d'eau
s retención *f* de agua
g Wasserrückhaltung *f*

*** reticular glaciation → 5053**

5053 reticular glaciers; reticular glaciation
A system of ice-filled valleys with ice caps which occurs on watersheds and alternate with ice-free summits and crests.
r сетчатое оледенение; горно-покровное оледенение
Система заполненных льдом сквозных долин с ледниковыми куполами на водораздельных участках, чередующихся с неоледенелыми вершинами и гребнями.
f glaciation *f* réticulée
s glaciación *f* reticular
g netzförmige Vergletscherung *f*; Eisstromnetz *n*

*** retreat of a glacier → 2529**

5054 retreat of slope; backwasting; slope retreat; parallel retreat of slope; slope retreat
A progressive backward movement of a slope with very little change of gradient, the results of rock weathering and removal of debris.
r отступание склона; параллельное отступание склона
Попятное перемещение склона в результате выветривания горных пород и последующего удаления продуктов выветривания от подножия, обычно при этом угол склона почти не меняется.
f recul *m* de versant; retrait *m* de versant
s retroceso *m* del vertiente
g Hangrückzug *m*

5055 retrievable water consumption; revocable water consumption
The use of water in such a way that it is returned into water bodies after being used for domestic or economic purposes.
r возвратное водопотребление
Тип водопотребления, когда использованные в хозяйстве или быту воды вновь попадают в водные объекты.
f consommation *f* d'eau avec restitution; utilisation *f* d'eau avec restitution
s consumo *m* de agua de retorno; consumo *m* de agua de restitución
g zurückgehender Wasserverbrauch *m*

5056 return of cold
The intrusion of cold air in spring causing drastic drop of temperature, after a long period of warm weather,
r возврат холодов

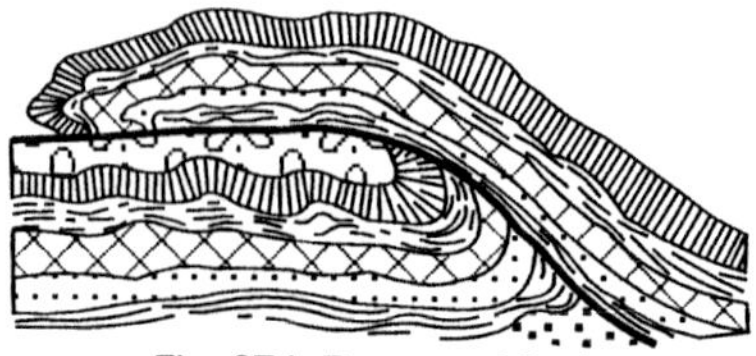
Fig. 274. **Reversed fault**

Вторжение холодного воздуха весной, после длительного периода теплой погоды, вызывающее резкое понижение температуры.
f retour *m* de froid
s retorno *m* de frío
g Kälterückfall *m*

5057 return period; recurrence interval
The average lengh of time separating events of a similar magnitude such as droughts, floods, earthquakes.
r периодичность
Средняя продолжительность между событиями одинаковой интенсивности, такими как засухи, паводки, землетрясения и пр.
f période *f* de retour
s perodicidad *f*; intervalo *m* de retorno
g Rückkehrperiode *f*

5058 return waters
Surface or ground waters which get back into water bodies after being used for industrial or domestic purposes.
r возвратные воды
Поверхностные и подземные воды, попадающие в водные объекты после их хозяйственного или бытового использования.
f écoulement *m* restitu
s aguas *fpl* de restitución
g Rücklaufwasser *n*

5059 reverse fault; thrust fault
The displacement of rocks up the plane of the fault. → Fig. 274
r взброс
Смещение горных пород по разлому, связанное с поднятием одного блока земной коры относительно другого. → Рис. 274
f faille *f* inverse; faille *f* chevauchante
s falla *f* inversa; falla *f* cabalgante
g Aufschiebung *f*

5060 reversible change of the environment
The changes in the environment of a fluctuating character with the restoration of the previous life conditions.
r обратимое изменение окружающей среды
Перемены в окружающей среде, носящие колебательный характер с восстановлением прежних условий жизни.
f changement *m* réversible d'environnement
s cambio *m* reversible de medio ambiente
g umkehrbare Umweltveränderung *f*

*** reversing current → 6208**

5061 **reversing dune**
A dune that grows upwards but stays in virtually the same position due to seasonal changes in wind direction causing it to move backwards and forwards over time.
r реверсивная дюна
Растущая песчаная дюна, сохраняющаяся примерно на том же месте из-за сезонных изменений направления ветра, вызывающего ее движение то в одном, то в противоположном направлениях.
f dune *f* réversible
s duna *f* reversible
g reversible Düne *f*

5062 **reversion**
The action of returning to a former condition or ancestral type, e.g. natural vegetation tends to regain its former position when human interference ceases; the changing of essential plant-nutrient elements from soluble into less-soluble forms by reaction with the soil.
r обратимость
Возвращение к прошлым условиям или наследственному типу, например, естественная растительность стремится вернуться к своему прежнему положению, как только прекращается антропогенное вмешательство; переход питательных элементов растения из растворимой в нерастворимые формы благодаря химическим реакциям, происходящим в почве.
f réversion *f*
s reversión *f*
g Reversion *f*

* **revestment → 1098**

* **revocable water consumption → 5055**

5063 **revolution of the Earth**
The movement of the Earth along elliptical orbit around the Sun under its gravitation.
r обращение Земли
Движение Земли по эллиптической орбите вокруг Солнца под действием его притяжения.
f révolution *f* de la Terre
s revolución *f* de la Tierra; translación *f* de la Tierra alrededor del Sol
g Erdumlauf *m*

5064 **rheophil**
Organisms living in running waters, which resisthe flow by way of intensive movement (fishes), accretion with (sponges, pearlworts) or attachment to substrate (molluscs, leeches, lampreys).
r реофил
Организмы, обитающие в текучих водах, они противостоят течению, активно двигаясь (рыбы), прирастая (губки, мшанки) или прикрепляясь к субстрату (моллюски, пиявки, миноги).
f rhéophile *m*; rhéobie *m*

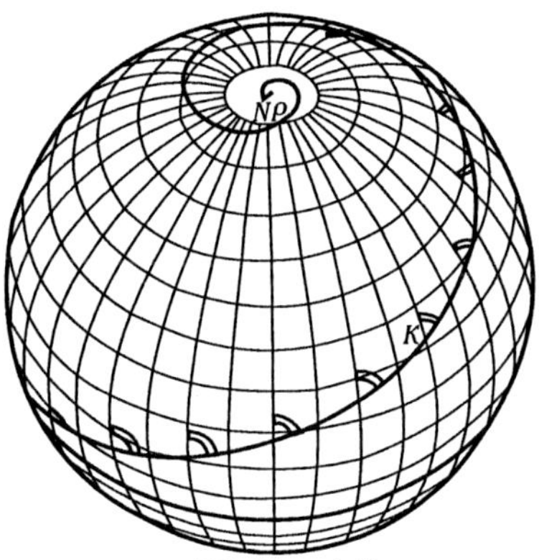

Fig. 275. **Rhumb line**

s reofilo *m*
g Rheophil *m*

5065 **rhizosphere**
The soil immediately surrounding plant roots which is altered by their growth, respiration excange of nutriens, etc., it is characterized by considerable biological activity because of the abundance of soil bacteria and other microflora.
r ризосфера
Слой почвы толщиной 2-5 мм, окружающий корни растений и отличающийся значительной биологической активностью, обусловленной большим количеством почвенных бактерий и иной микрофлоры.
f rhizosphère *f*
s rizosfera *f*
g Rhizosphäre *f*

5066 **rhumb line; loxodrome; line of constant bearing**
A line on the Earth's surface which crosses successive meridians at a constant angle (it looks like a straight line on navigation charts). → Fig. 275
r локсодромия
Линия на поверхности земного шара, пересекающая все меридианы под одним и тем же углом, на морских навигационных картах изображается прямой. → Рис. 275
f loxodromie *f*; ligne *f* loxodromique
s línea *f* loxodrómica
g Loxodromie *f*

5067 **rhythms**
Recurrence of certain natural processes or fluctuations of their intensity of not strictly periodical character.
r ритмы; ритмичность
Повторения определенных природных процессов или колебания их интенсивности, не имеющие строго периодического характера.
f rythmes *mpl*
s ritmos *mpl*

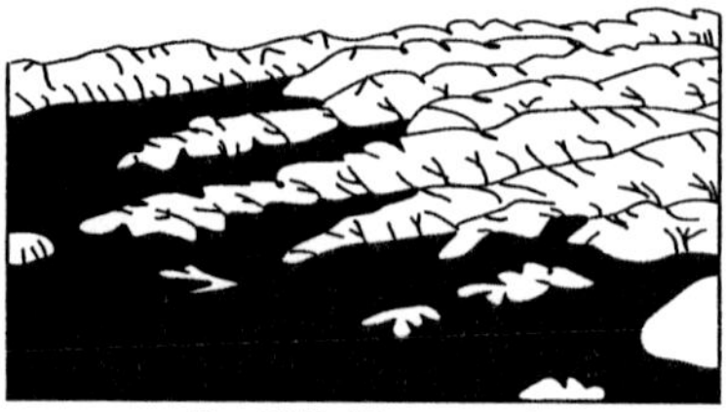

Fig. 276. **Rias coast**

g Rhythmen *mpl*; Rhythmik *f*

5068 rhytmic lamination

The stratification characterised by the regularly repeated layers of sedimentary rocks in the profile.

r ритмичная слоистость

Слоистость, характеризующаяся правильной повторяемостью слоев осадочных пород в разрезе.

f litage *m* rythmique

s litaje *m* rítmico

g Repetitionsschichtung *f*; rhythmische Schichtung *f*

5069 ria

A funnel-shaped indentation on a coast, narrowing inland, increasing in depth seaward, a submerged river valley formed as a result of sea ingression on a mountainous sea coast.

r риас

Воронкообразный залив, постепенно сужающийся и мелеющий к своему верховью, образовавшийся в результате ингрессии моря в низовьях речной долины на гористом морском побережье.

f ria *f*

s ría *f*

g Ria *f*

5070 rias coast

A type of shore characterised by the alternation of capes and long narrow meandering bays. → Fig. 276

r риасовый берег

Тип берега, характеризующийся чередованием мысов и длинных узких извилистых заливов. → Рис. 276

f côte *f* à rias

s costa *f* con rias

g Riasküste *f*

5071 ribbon development

A building-up area along a main road running outwards from the city centre.

r ленточная застройка

Застроенная полоса вдоль главной дороги, идущей из центра города к его периферии.

f développement *m* en bande

s desarrollo *m* en banda

g Streifenentwicklung *f*

5072 ribbon lake

A temporary narrow body of water impounded between the front of an ice sheet and its stadial moraines.

r ленточное озеро

Временный мелкий узкий водоем между фронтом ледникового покрова и его стадиальными моренами, заполняемый талыми водами.

f lac *m* de bande

s lago *m* de banda

g Streifensee *m*

5073 Richter scale

A scale, ranging from 0 to 10, used in measuring the magnitude of earthquakes, based on instrumental recordings of a standard seismograph 100 km from the epicentre.

r шкала Рихтера

Шкала, градуированная в арабских цифрах от 0 до 10, используемая для измерения силы землетрясения стандартным сейсмографом в 100 км от его эпицентра.

f échelle *f* de Richter

s escala *f* de Richter

g Richter-Skala *f*

5074 ride; cleared path; cutting

A narrow cleared strip cut in a forest as a firebreak or to mark the boundaries between patches of the forest, to make paths for power lines and pipelines.

r просека

Узкая полоса, прорубаемая в лесном массиве для обозначения границ лесных кварталов, линий электропередач, трубопроводов, а также для борьбы с лесными пожарами.

f trouée *f*; percée *f*

s entresaca *f*

g Waldweg *m*; Schneise *f*

5075 ridge; mountain ridge

A linear steep-sided upland of considerable height and length.

r горный хребет

Линейно вытянутое горное сооружение значительной высоты и протяженности.

f cordillère *f*

s cordillera *f*

g Gebirgszug *m*; Kamm *m*; Rücken *m*

5076 ridge of high pressure; wedge of high pressure

An elongated area of the atmosphere in which the pressure is high relative to the surrounding area at the same level, it is represented on a synoptic chart by a system of nearly parallel isobars or contours, of approximately U-shape, which are concave towards the anticyclone.

r гребень высокого давления; отрог антициклона

Область атмосферы, давление в которой выше, чем в окружающих областях на том же высотном уровне, на синоптической карте представлена системой почти параллельных изобар приблизительно U-образной формы, выпуклой по отношению к антициклону.

f crête *f* de haute pression; crête *f* barométrique

s cresta *f* de alta presión

g Hochdruckrücken *m*

5077 ridge relief

Erosion, accumulative or aeolian relief notable for alternating ridges and depressions.

r грядовый рельеф

Эрозионный, аккумулятивный или эоловый рельеф, отличающийся чередованием гряд и понижений между ними.

f relief *m* des barrages; relief *m* différencié

s relieve *m* diferencial

g Kettenrelief *n*

5078 ridge-and-pool bog

A bog with narrow parallel peat ridges, alternating with shallow depressions, many of which are filled with water.

r грядово-мочажинное болото

Болото с параллельными узкими грядами торфа, чередующимися с неглубокими понижениями, многие из которых заполнены водой.

f tourbière *f* cordée

s turbera *f* acordonada

g Schlenkemoor *n*

* **ridge-and-ravine terrain → 2673**

5079 riegel; rock bar

A transverse rock bar or rock step across the bottom of a glacial valley.

r ригель

Поперечный скалистый уступ на дне ледниковой долины.

f verrou *m*

s umbral *m*; viga *f* de refuerzo

g Riegel *m*

5080 riffle; cross-over

A shallow gravel bar in the river channel over which water flows rapidly, it has a gentle slope turned upstream, and steep slope turned downstream. → Fig. 277

r перекат

Мелководный участок речного русла, имеющий вид вала с пологим скатом, обращенным против течения, и крутым по течению. → Рис. 277

f banc *m*

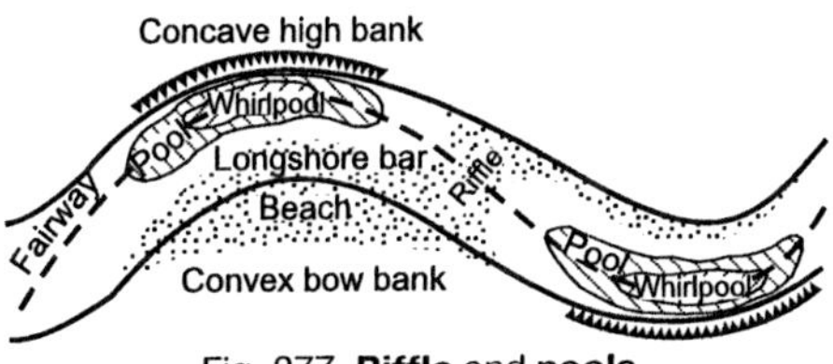

Fig. 277. **Riffle** and **pools**

s banco *m*

g Furt *f*

5081 rift

A large linear tectonic structure hundreds and thousands kilometres long formed by a system of parallel faults with throws in opposite directions thus appearing as a long, flat looked valley with steep sides.

r рифт

Крупная линейная тектоническая структура протяженностью в сотни и тысячи километров, образованная системой параллельных сбросов при горизонтальном растяжении земной коры.

f fossé *m* tectonique; rift *m*

s fosa *f* tectónica; rift *m*

g Rift *m*

5082 rift valley

A linear depression created by the sinking of the intermediate crustal rocks between parallel fault zones.

r рифтовая долина

Депрессия, возникшая в результате опускания линейно вытянутого участка земной коры (рифта), ограниченного параллельными зонами разломов.

f vallée *f* de rift

s valle *m* de rift; valle *m* axial

g Rifttal *n*; Senkungsgraben *m*

5083 rift zone; axial trough; axial rift

An axial part of a mid-oceanic ridge having the form of a deep ditch stretched out far along the ridge.

r рифтовая зона

Осевая часть срединно-океанического хребта, имеющая характер глубокого грабена, вытянутого вдоль хребта на большое расстояние.

f zone *f* de rift médio-océánique

s zona *f* de rift medio-oceánica; zona *f* de rift de dorsales oceánicas

g Riftzone *f* (der mittelozeanischen Schwelle)

* **rights of common → 1150**

* **rill → 802**

5084 rill; delve; ditch; furrow

A narrow channel produced by a rupture or washout of the earthrs surface.

r ров; рытвина; водороина; промоина$_2$

Небольшая узкая ложбина, образованная разрывом или размывом земной поверхности.

f rigole *f*

s zanja *f*; trinchera *f*; bache *m*

g Rösche *f*; Rille *f*; Furche *f*

5085 rill erosion

The removal of surface material, usually soil, by the action of running water in numerous tiny channels (rills), a few centimetres in depth, most of which carry water only during storm.

r струйчатая эрозия

Снос поверхностного материала, главным образом почвы, под действием стекающей воды в многочисленных маленьких канальцах глубиной в несколько сантиметров, возникающих во время сильных ливней.

f ruissellement *m*

s erosión *f* de hilo

g Rinnsalerosion *f*

5086 **rill marks**

Small furrows or channels running down the slope of the beach, formed by the seaward flow of sea water that has percolated into the upper part of the beach zone during a high tide after a storm.

r пляжные бороздки

Небольшие бороздки или канальцы на мористой части пляжа, возникающие в его тыловой части при струйчатом стекании воды в море после высокого штормового нагона.

f rigoles *fpl* de plage

s surcos *mpl* de playa

g Strandrille *f*

5087 **rime**

A white or milky and opaque granular deposit of ice formed by the rapid freezing of supercooled water drops as they impinge upon an exposed object which is colder than the air (it is denser and harder than hoarfrost, but softer and less transparent than glaze).

r иней

Тонкий неравномерный слой льда, образующийся путем сублимации водяного пара из воздуха на поверхности почвы, снежного покрова и разных предметов в результате их радиационного охлаждения до отрицательных температур, более низких, чем температура воздуха.

f givre *m* mou; givre *m*; gelée *f* blanche

s escarcha *f*

g Reif *m*; Rauhreif *m*

5088 **rime formation**

An accumulation of ice crystals due to the process of direct sublimation of water vapour from the ambient air on the surface of sufficiently cooled objects (e.g. trees, fences, rocks, telegraph poles, etc.).

r заиневение

Образование на поверхности охлажденного предмета мелких кристаллов льда (инея) путем сублимации, например, на деревьях, телеграфных столбах, заборах, обнаженных скалах и т.п.

f formation *f* du givre mou

s formación *f* de escarcha

g Rauhreifbildung *f*

5089 **ring**

A mesoscale whirl (200-300 km in diameter) with a cold core surrounded by a narrow circle of warm water formed by the separation of jets from high-velocity currents.

r кольцо; ринг

Мезомасштабный вихрь (диаметром 200-300 км) с холодным ядром, окруженный узким кольцом теплых вод, образующийся при отделении струй быстрых течений.

f ring *m*

s ring *m*

g Ring *m*

5090 **ring road**

A road that goes around a large town used as a bypass for keeping traffic away the town centre.

r кольцевая автодорога; окружная автодорога

Дорога вокруг города, разгружающая его центр от транспортных перегрузок.

f périphérique *m*; voie *f* d'évitement

s carretera *f* circular; vía *f* de circunvalación

g Kreisbahn *f*; Ringstraße *f*

5091 **rip current**

A strong, narrow current moving seawards in the near-shore zone which compensates the water piled up on the shore by incoming waves and onshore winds.

r разрывное течение

Направленный от берега в сторону моря интенсивный поток воды в виде отдельных струй, компенсирующий нагон, создаваемый волновым переносом.

f courant *m* d'arrachement

s corriente *f* de arranque

g diskontinnierliche Strömung *f*

* **riparian *adj* → 5112**

5092 **ripeness of soil**

The availability of soil for cultivation, sawing and planting of crops, it is determined by the optimum moistening and sufficiently high temperature.

r спелость почвы

Готовность почвы к ее обработке или посеву и посадке культурных растений, определяется оптимальным увлажнением и достаточно высокими температурами.

f maturité *f* du sol

s madurez *f* del suelo

g Bodenreife *f*

5093 **ripple**

The initial form of waves under the impact of wind.

r рябь

Начальная форма развития волнения под действием ветра.

f rides *fpl*

s rizado *m*

g Kräuseln *n*; Rippeln *npl*

5094 ripple mark; current ripple; sand ripple
A series of small more or less parallel ridges of sand or other unconsolidated sediments produced by wind, currents or waves (sometimes they can be found in the fossilized state).
r знаки ряби; рифели; песчаная рябь
Прямые или изогнутые параллельные ряды валиков, образованные ветром, водными течениями или волнами на поверхности рыхлых осадков (встречаются в ископаемом состоянии).
f ride *f* de sable; rides *fpl* sableuses; rides *fpl* de plage
s ripple-marks *mpl*; arrugas *fpl* de arena; arrugas *fpl* de playa
g Rippelmarken *fpl*; Wellenfurchen *fpl*; Sandrippeln *npl*

* **rising tide → 2768**

5095 river; stream
A natural stream of water flowing in a definite channel and nourished by the runoff from its catchment.
r река
Естественный водный поток, текущий в выработанном им русле, питающийся за счет стока с его водосбора.
f fleuve *m*; rivière *f*
s río *m*
g Fluß *m*; Strom *m*

* **river bank → 545**

5096 river basin; river drainage basin
An area of land from which water is drained both over the surface and underground into a given river. → Fig. 278
r речной бассейн; речной водосбор
Территория, с которой вода по поверхности и подземным путем стекает в данную реку. → Рис. 278
f bassin *m* fluvial; bassin *m* versant
s cuenca *f* fluvial
g Flußbecken *n*; Stromgebiet *n*; Flußeinzugsgebiet *n*

* **river capture → 871**

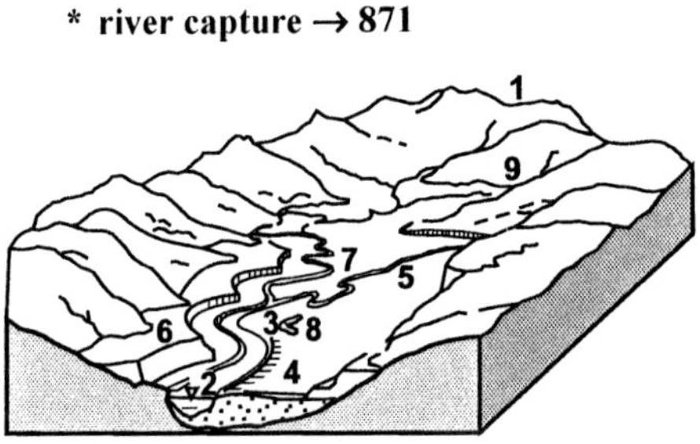

Fig. 278. **River basin**: 1 – water divide, 2 – river, 3 – river channel, 4 – flood-plane, 5 – tributary, 6 – terrace, 7 – meander, 8 – ox-bow, 9 – secondary catchment

5097 river channel
A natural course for running water; the deepest part of a river valley along which water runs in the periods of low water.
r русло
Самая низкая часть речной долины, по которой происходит сток воды в межпаводочные периоды.
f chenal *m*; lit *m* fluvial
s lecho *m* del río
g Strombett *n*; Flußbett *n*

* **river diversion flow → 5098**

* **river drainage basin → 5096**

5098 river flow diversion; river diversion flow; river flow transfer; water transfer
The change of the natural direction of a river with its diversion into some other drainage basin with the help of engineering constructions.
r переброска стока; территориальное перераспределение стока
Изменение природного направления стока рек с выводом его в другой водосборный бассейн при помощи гидротехнических сооружений.
f dérivation *f* des cours d´eau; transfert *m* des cours d´eau
s derivación *f* del curso de río; transferencia *f* del curso de río
g Umsetzung *f* der Abflusses

* **river flow transfer → 5098**

5099 river hydrology
Natural processes and phenomena in rivers and their scientific study.
r гидрология рек
Природные процессы и явления на реках и их изучение.
f hydrologie *f* fluviale
s hidrología *f* fluvial
g Flußhydrologie *f*

5100 river ice
The ice formed on rivers by the freezing of water.
r речной лед
Лед, образующийся на реках при замерзании воды.
f glace *f* de rivière
s hielo *m* de río
g Flußeis *n*

5101 river junction
A meeting point of two streams.
r слияние рек
Соединение двух речных потоков в одно русло.
f confluent *m*
s confluencia *f* de ríos
g Zusammenfluß *m*; Zusammenfließen *n*

5102 river load; load
Solid particles formed as a result of erosion of catchments and channels as well as abrasion of water bodies shores transported by watercourses and currents and forming their bed.
r наносы
Твердые частицы, образованные в результате эрозии водосборов и русел, а также абразии берегов водоемов, переносимые водотоками и течениями и формирующие их ложе.
f charge *f*
s carga *m*
g Feststoffe *mpl*

5103 river load balance
The relation between the mass of suspended material and bed load transported into a river and deposited on its bed over a certain period of time.
r баланс наносов
Соотношение массы взвешенных и влекомых наносов, поступивших в реку и отложившихся на ее дне за определенный отрезок времени.
f balance *f* de charge fluviale
s balance *m* de carga fluvial
g Geschiebebilanz *f*

5104 river mouth; mouth
A place where a stream enters a sea, lake or a larger stream.
r устье
Место впадения водотока в море, озеро или более крупную реку.
f embouchure *f*
s desembocadura *f*
g Einmündung *f*; Mündung *f*

5105 river mouth cone
Stream deposits in the bottom of a water body dropped by the inflowing river in its delta.
r устьевой конус выноса
Область речных отложений водоема в дельтовой части впадающей в него реки.
f cône *m* de delta
s cono *m* de delta;
sedimento *m* de embouchure
g Mündungsschuttkegel *m*

*** river piracy → 871**

5106 river port
A port situated on a river, usually at the point farthest from the mouth where the water is deep enough for navigation by trading vessels.
r речной порт
Порт на реке, на небольших реках обычно в самой удаленной точке от устья, где речные глубины еще позволяют проходить торговым судам.
f port *m* de rivière; port *m* fluvial
s puerto *m* fluvial
g Binnenhafen *m*

5107 river profile
A section of a river valley, either longitudinal (along the course of the river) or transverse (across the valley at right angles to the stream).
r профиль реки
Разрез через речную долину, как продольный (вдоль течения реки), так и поперечный (пересекающий долину).
f profil *m* de rivière
s perfil *m* de río
g Flußprofil *n*

5108 river regulation
Redistribution in time of a river runoff in the outlet according the requirements of different branches of economy.
r регулирование стока
Перераспределение во времени объема речного стока в замыкающем створе в соответствии с потребностями разных отраслей хозяйства.
f régulation *f* de fleuve; distribution *f* d'eau
s regulación *f* del río; distribución *f* de agua
g Abflußregulierung *f*

5109 river system
The main river flowing into a sea or lake and all its tributaries and inflowing streams.
r речная система
Главная река, впадающая в море или озеро, и все водотоки, собирающие в нее воду.
f système *m* fluvial
s sistema *m* fluvial
g Flußsystem *n*

5110 river terrace
A terrace, formed by the action of a river and composed of alluvium, which is a part of former floodplain of the river, left on the side of the valley as the stream cut down its bed at a level higher than that of the present channel. → Fig. 279
r речная терраса
Терраса, сформированная деятельностью реки и, как правило, сложенная аллювием, представляющая собой часть прежней речной поймы, обычно на левой стороне долины, где речной поток размывал русло на более высоком уровне по сравнению с современным. → Рис. 279

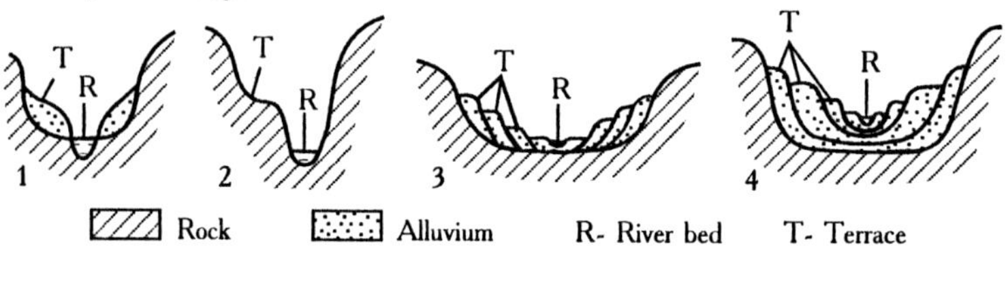

Fig. 279. **River terrace**:
1 – accumulative,
2 – erosional,
3 – superimposed,
4 – enclosure

f terrasse *f* fluviale
s terraza *f* fluvial
g Flussterrasse *f*

5111 river transport
A type of transport which moves passengers and goods along inland waterways such as navigable rivers and canals.
r речной транспорт
Вид транспорта, занятый перевозкой пассажиров и грузов по внутренним водным путям – судоходным рекам и каналам.
f transport *m* fluvial
s transporte *m* fluvial
g Binnenschiffahrt *f*

* **river valley → 6520**

* **riverain *adj* → 5112**

* **riverbed flow → 954**

* **riverine *adj* → 5112**

5112 riverside; riverain *adj*; riverine *adj*; riparian *adj*
Related to the land alongside a river, a river bank.
r приречная полоса; приречный
Относящийся к территории, расположеннной вдоль реки.
f riverain *adj*; ripicole *adj*
s ribereño *adj*; ripícolo *adj*
g Flußufer *n*; Fluß-

5113 riviera
A generic term for a coastal balneological resort with a mild climate limited by mountains or any coastline of outstanding natural beauty, the term is derived from the resort of Cфte d'Azur of France, in Russia the Caucasian Riviere is very popular.
r ривьера
Обобщающее название приморского бальнеологического курорта, в тыловой части ограниченного горами, происходит от курортной местности на Лазурном берегу Франции, в России широко известна Кавказская Ривьера.
f riviera *f*
s riviera *f*
g Riviera *f*

* **rivulet → 802**

5114 road
A land communication way usually paved used for the passage of people, vehicles, and animals.
r дорога
Наземный путь сообщения, предназначенный для передвижения людей, транспорта и грузов.
f chemin *m*; voie *f*; route *f*
s camino *m*; vía *f*; ruta *f*
g Weg *m*; Straße *f*

5115 road transport
A type of transport moving passengers and goods along non-rail roads.
r автомобильный транспорт
Вид транспорта, занятый перевозкой пассажиров и грузов по безрельсовым путям.
f transport *m* routier
s transporte *m* por carretera; transporte *m* rodado
g Kraftverkehr *m*; Autoverkehr *m*

5116 roadstead
An offshore anchorage in a bay or estuary where some degree of protection from storms is provided for any shipping unable to get into a harbour.
r рейд; якорная стоянка
Прибрежная якорная стоянка в заливе или эстуарии, где существует некоторая степень защиты от штормов для любых кораблей, которые не могут отшвартоваться в гавани.
f rade *f*
s rada *f*
g Reede *f*; Ankerplatz *m*

5117 roar of the surf; thundering of the breakers; rote (*amer*)
Noise caused by the coastal breaking of waves.
r рев прибоя; шум прибоя; грохот прибоя
Шум, вызываемый прибрежным накатом волн.
f bruit *m* de mer
s ruido *m* de oleaje
g Brandungsrauchen *n*

5118 Roaring Forties
The region of the Southern Hemisphere between latitudes 40° and 50° where few land masses occur as barriers to an uninterrupted flow of the westerlies which blow with great regularity and strength across the open oceans bringing storminess, cloudiness and rough seas, a notoriously hazardous shipping zone.
r ревущие сороковые
Пояс с преобладанием западных штормовых ветров в широтах 40-50° Южного полушария, где возникает барьер на пути океанических водушных масс, вызывающий штормовую погоду, облачность и волнение на море, так что эти широты оказываются сложными и опасными для судоходства.
f quarantièmes *mpl* rugissants
s cuadragésimos *mpl* mugidos
g brüllende Vierziger *fpl*

5119 roche moutonnée
Oblong asymmetric hillocks of hard rocks smoothed and rounded owing to the effect of abrasion, and marked with striae by the glacier. → Fig. 280, p. 614
r бараньи лбы; курчавые скалы

Продолговатые асимметричные холмы из твердых пород со следами интенсивной ледниковой обработки. → Рис. 280

f roche *f* moutonnée

s roca *f* aborregada

g Rundhöcker *m*; glazialer Felsenbuckel *m*

5120 rock

1. Any naturally formed aggregate of mineral particles, whether it is hard, relatively soft, unconsolidated or in compact, which constitutes an integral part of the lithosphere. 2. A large, hard, consolidated, compact part of the Earth's crust, also a large piece of material protruding from the land or sea.

r горная порода

1. Природный агрегат минералов в земной коре более или менее постоянного минералогического и химического состава. 2. Крупная, твердая, консолидированная и компактная часть земной коры либо обширная часть каменного материала, выступающая из земли или из моря.

f roche *f*

s roca *f*

g Gestein *n*

* **rock bar → 5079**

5121 rock city

Peculiar bedrock formations in deserts which look like ruins due to intensive physical weathering and mechanical impact of wind.

r каменный город; эоловый город

Причудливые формы коренных пород в пустынях, приобретающие вид городских развалин благодаря интенсивному физическому выветриванию и механическому воздействию ветра.

f relief *m* ruiniforme

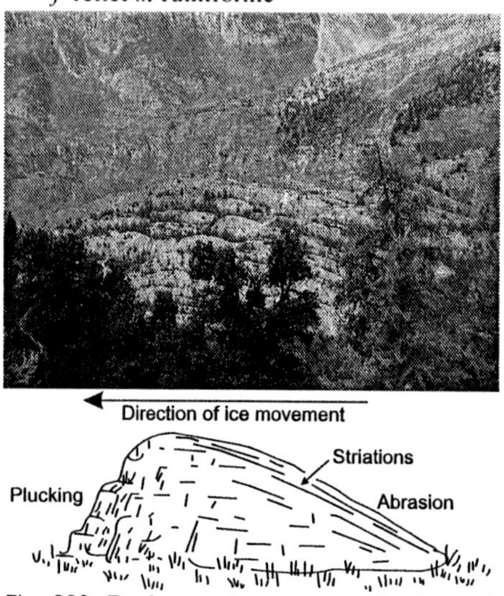

Fig. 280. **Roche moutonnee** near the front of Grossaletschgletscher (to the right) in Swiss Alps

Fig. 281. **Rock glacier**

s relieve *m* ruiniforme

g Felsenstadt *f*

5122 rock creep; écoulement

The gravitational sliding of large masses of rock as a result of tectonic deformation, as opposed of a *rock fall*, rockslide and *landslide*, because all of these are caused by surface denudation.

r стекание горных пород

Гравитационное скольжение больших масс горных пород в результате тектонических деформаций, в отличие от *камнепадов* и *оползней*, которые вызываются процессами поверхностной денудации.

f écoulement *m* par gravité

s escorrentía *f* por gravedad

g Gesteinrutschen *n*

* **rock exposure → 4319**

5123 rock fall

The detachment and descent of a large rock mass from a high and steep slope.

r горный обвал; обвал

Падение большой массы горной породы с крутых высоких склонов.

f éboulement *m*

s desmoronamiento *m*; derrumbamiento *m*; derrumbe *m*

g Bergsturz *m*; Einsturz *m*; Einbruch *m*

5124 rock glacier; rock stream (*amer*)

A glacier-like mass of large rock fragments and other finer material that contains either interstitial ice or an ice core and has evidence of former or present movement. → Fig. 281

r каменный глетчер

Напоминающая ледник масса крупных каменных обломков и более тонкого материала, включающая лед и нередко ледяное ядро, имеющая признаки прошлого или настоящего движения. → Рис. 281

f glacier *m* rocheux
s glaciar *m* rocoso
g Blockgletscher *m*; Steinstrom *m*

5125 **rock milk**
A white paste or crumble white powder occurring on floors and walls of limestone caves due to the rock disintegration by microorganisms or the freezing of the limestone with ice in which CO_2 is expelled.
r пещерное молоко
Белесая паста или зернистая белая пудра, появляющаяся на полу и стенах известняковых пещер вследствие разрушения горных пород микрорганизмами или замерзания известняка со льдом, из которого при этом выделяется CO_2.
f lait *m* de roche
s leche *f* de roca
g Höhlenmilch *f*

* **rock pillar → 981**

* **rock stream (*amer*) → 5124**

5126 **rock-building mineral**
The most widespread minerals of the Earth's crust which compose rocks, e.g. quartz or calcite.
r породообразующий минерал
Наиболее распространенные минералы земной коры, входящие в состав горных пород, например, кварц или кальцит.
f mineral *m* essentiel de roche
s mineral *m* esencial de roca
g gesteinsbildendes Mineral *n*

5127 **rock-fall slope**
A slope with the morphological traces of rock falls: niches of detachment and heaps of large fragments at the foot.
r обвальный склон
Склон с морфологическими следами горных обвалов: нишами отрыва и грудами крупных обломков у подножия.
f versant *m* de éboulement
s pendiente *f* de derrumbamiento
g Sturzhang *m*

5128 **rocky desert**
A desert developed on poorly weathered bedrocks on low-mountain plateaus and hummocky relief, the surface is covered with rock debris rubbles and pebbles, with almost no soil and vegetation cover.
r каменистая пустыня
Пустыня, развитая на слабо выветрелых коренных породах на низкогорных плато и мелкосопочниках, с поверхностью, покрытой щебнем и галькой и почти лишенной почвенно-растительного покрова.
f désert *m* pierreux
s desierto *m* pedregoso
g Steinwüste *f*

5129 **roller**
A large ocean wave that breaks on exposed ocean coastlines where the length of the fetch is considerable.
r вал зыби
Большая океанская волна, обрушивающаяся на открытый морской берег со значительной областью разгона волн.
f lame *f* solitaire
s lámina *f* solitaria
g Woge *f*

* **rolling plain → 2386**

5130 **rookery; seashore colony of birds**
A breeding ground of gregarious birds that are usually sit on bluffs, mainly in polar regions.
r птичий базар
Колониальное гнездовье морских птиц на карнизах крутых скал, скалистых береговых обрывах, преимущественно в полярных странах.
f rassemblement *m* d'oiseaux; roquerie *f*; rooquerie *f*
s bazar *m* de aves; posadero *m* de aves
g Vogelkolonie *f*

5131 **root action**
The process by which the roots of vegetation penetrate the soil and help to break up the underlying bedrock by pressure exerted in the fissures.
r корневое воздействие
Процесс, при котором корни растений внедряются в почву и помогают разрушать верхний слой горных пород в результате дополнительного напряжения в трещинах.
f action *f* des racines
s acción *f* de raíces
g Wurzeleinwirkung *f*

* **ropak → 5776**

5132 **ropy lava; corded lava; pahoehoe**
Lava which has solidified so that the surface of the flow is glassy and smooth with surface shapes resembling ropes or cords.
r волнистая лава
Застывшая базальтовая маловязкая лава с волнообразной стекловатой поверхностью.
f lave *f* cordée; pahoehoe *f*
s lava *f* acorazonada; pahoehoe *f*
g wellenförmige Lava *f*

* **Rossby wave → 4592**

* **rotation of crops → 3379**

5133 **rotation of the Earth**
The revolving movement of the Earth around the axis from west to east.
r вращение Земли;
суточное вращение Земли
Вращательное движение Земли вокруг оси с запада на восток.

f rotation *f* de la Terre
s rotación *f* de la Tierra
g Erdrotation *f*; Erdumdrehung *f*

5134 rotational slip
The downward movement of a mass of rock or ice on a slip-plane, in which the solid mass appears to rotate on a pivot as it descends.
r вращательное скольжение
Направленное вниз по склону движение масс горных пород или льда по плоскости скольжения, в процессе которого твердая масса приобретает составляющую вращения.
f glissade *f* rotatoire
s deslizamiento *m* rotativo
g Drehgleiten *n*

5135 rotatory stream
Tidal current whose direction rotates (clockwise in the Northern Hemisphere, and counter-clockwise in the Southern Hemisphere) with the period equal to one lunar day, or a half of it.
r вращательное течение
Приливное течение в открытом море, направление которого вращается (в Северном полушарии по часовой стрелке, в Южном – против) с периодом, равным половине или одним лунным суткам.
f rotation *f* amphidromique
s rotación *f* anfidrómica
g Drehstrom *m*

* **rote (*amer*) → 5117**

5136 rotten ice
Loose sea ice at the last stage of its degradation.
r гнилой лед
Рыхлый морской лед, находящийся на последней стадии разрушения.
f glace *f* pourrie; glace *f* en chandelle
s hielo *m* pútrido
g verrottetes Eis *n*; morsches Eis *n*

5137 rotten stone; gruss
Loose mass of not rounded rock debris and minerals from 2 to 10 mm in size.
r дресва
Рыхлая масса неокатанных обломков горных пород и минералов размером от 2 до 10 мм.
f gravier *m*
s arena *f* gruesa
g Grus *m*

5138 roughness coefficient; friction coefficient of river bed
The value which numerically characterises the channel resistance to the stream, which is generated by roughness element, such as bed and bank features.
r коэффициент шероховатости русла
Величина, численно характеризующая сопротивление, оказываемое руслом протекающему в нем потоку.
f coefficient *m* de rugosité du lit
s coeficiente *m* de rugosidad del lecho
g Rauhigkeitswerf *m* des Flußbetts

5139 roughness of channel; bed roughness; channel resistance
The irregularities of a river bed, such as bars, ripples, antidunes, channel sinuosity, which oppose downstream water motion and result in energy loss.
r шероховатость русла
Неровности речного русла (отмели, рябь, антидюны, искривление русла), оказывающие сопротивление движению воды.
f rugosité *m* du lit
s rugosidad *f* del lecho
g Bettrauhigkeit *f*

5140 roughness of sea surface
Sea surface irregularity caused by the wind.
r шероховатость морской поверхности
Неровности поверхности моря, вызываемые воздействием ветра.
f rugosité *f* de la surface océanique
s rugosidad *f* de superficie oceánica
g Rauhigkeit *f* der Meeresoberfläche

* **rounded iceberg → 1669**

5141 roundness; roundness index
The degree of smoothness of the initial edges of debris fragments due to the abrasion while transported by water, glacier or wind.
r окатанность
Степень сглаженности первоначальных ребер обломков вследствие их истирания при переносе водой, ледником или ветром.
f arrondissement *m*; indice *m* d'émoussé; degré *m* arrondissement
s redondeamiento *m*; grado *m* de redondeamiento; índice *m* de desgaste
g Rundung *f*

* **roundness index → 5141**

5142 route sketching; strip survey
A survey carried out along a definite route while compiling and updating topographic and thematic maps.
r маршрутная съемка
Съемка местности по отдельному маршруту при создании и обновлении топографических и тематических карт.
f levé *m* d'itinéraire
s levantamiento *m* de itinerario
g Routenaufnahme *f*

5143 rubbish; litter; garbage; trash
Things thrown away which are no longer useful, such as packaging, containers, glass, plastics, they are harmful pollutants of the environment.
r мусор
Негодные остатки бумаги, стекла, пластмассовые и металлические изделия, загрязняющие окружающую среду.

f ordures *fpl*
s basura *f*
g Müll *m*

5144 rubble
Loose coarse angular rock fragments 10-100 mm in size.
r щебень
Рыхлые остроугольные обломки пород размером 10-100 мм.
f blocaille *f*
s cascajo *m*; cascote *m*
g Schotter *m*

5145 rubefaction; rubification
A soil-forming process involving the transformation of ferruginous minerals, it occurs in ferralitic soils and gives them bright-red colour.
r рубефикация
Почвообразовательный процесс, характеризующийся трансформацией железосодержащих минералов, преобладает в ферралитных почвах и придает почве ярко-красный цвет.
f rubéfaction *f*
s rubefacción *f*
g Rubefizierung *f*

* **rubification → 5145**

5146 ruderal plant; ruderal vegetation
Plants growing on waste or disturbed places, e.g. in the streets, along fences and road verges, on littered areas, dumps, wastes, etc.
r рудеральное растение; мусорник
Сорные растения, произрастающие на улицах, вдоль заборов и дорог, на замусоренных участках, отвалах, пустырях и т.п.
f plante *f* rudérale
s planta *f* ruderal
g Ruderalpflanze *f*

* **ruderal vegetation → 5146**

* **runnel → 802**

5147 running water
Water flows capable to carry away solid particles and cause linear or sheet erosion.
r текучая вода

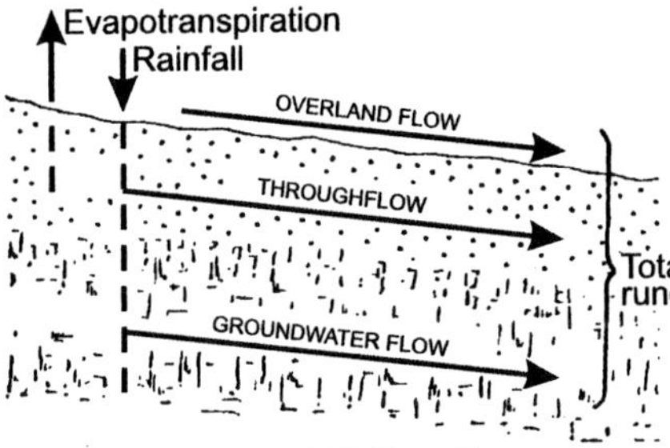

Fig. 282. **Runoff**

Вода потоков, способная увлекать с собой твердые частицы и вызывать линейную или плоскостную эрозию.
f eau *f* courante
s agua *f* corriente
g Fließgewässer *n*

5148 runoff; flow$_2$
Flowing of rain and melt waters into water bodies and depressions on the earthr's surface (surface runoff, overland flow) and underground (subsurface flow). → Fig. 282
r сток; водный сток
Стекание дождевых и талых вод в водоемы и понижения рельефа по земной поверхности (поверхностный сток) и под землей (подземный сток). → Рис. 282
f écoulement *m*
s escorrentía *f*
g Abfluß *m*

5149 runoff channel
The middle part of a mountain brook channel having the slope of a deep rut along which water runs in consolidated flow.
r канал стока
Средняя часть русла горного ручья, имеющая вид глубокой рытвины, по которой во время дождя вода стекает единым потоком.
f canal *m* d'écoulement
s canal *m* de escorrentía
g Abflußkanal *m*

5150 runoff coefficient; impermeability factor
The ratio of the amount of runoff to the sum of precipitation which induced the runoff.
r коэффициент стока
Отношение величины стока к сумме осадков, обусловивших возникновение стока.
f coefficient *m* d'écoulement
s coeficiente *m* de escorrentía; coeficiente *m* de desagüe; coeficiente *m* de escorrentía
g Abflußbeiwert *m*; Abflußkoeffizient *m*

5151 runoff depletion
A part of hydrograph showing the decrease of runoff value due to the depletion of water reserves in the river basin.
r кривая истощения стока
Участок гидрографа, показывающий уменьшение величины стока вследствие истощения запасов воды в речном бассейне.
f courbe *f* de tarissement d'écoulement
s curva *f* de agotamiento
g Trockenwetterganglinie *m*

5152 runoff plot
A part of a slope edged by borders and equipped with devices to determine and monitore the runoff and soil loss.
r стоковая площадка
Участок склона, ограниченный бортиками и оборудованный устройствами для

перехвата и учета поверхностного стока и почвенной эрозии.
f parcelle *f* experimentale d'études de l'écoulement
s área *f* de escorrentía
g Abflußmessstelle *f*

* **rupicolous plant → 5231**

* **rupture → 2283**

5153 **rural *adj***
Relating to the country or country life, in contrast to the town or urban life.
r сельский; деревенский
Относящийся к сельской местности и сельскому образу жизни, в отличие от города и городского образа жизни.
f rural *adj*
s rural *adj*
g Land-; ländlich *adj*

* **rural area → 1323**

5154 **rural depopulation**
The decrease of population of rural areas, whether by migration or falling birth rates as young people move away to towns.
r сельская депопуляция
Сокращение населения в сельской местности, как в результате миграции, так и благодаря падению рождамости из-за того, что часть молодежи уезжает из деревни в город.
f dépeuplement *m* rurale
s despoblamiento *m* rural; despoblación *f* rural
g Landdepopulation *f*

* **rural planning → 5159**

5155 **rural population**
The population living permanently in the countryside, not in a town.
r сельское население
Население, постоянно проживающее в сельской местности.
f population *f* rurale
s población *f* rural
g ländliche Bevölkerung *f*

5156 **rural settlement**
A settlement in the rural area where agricultural production is the main sphere of human activity.
r сельское поселение
Населенный пункт в сельской местности, где главная сфера приложения труда – сельскохозяйственное производство.
f localité *f* rurale; habitat *m* rural
s asentamiento *m* rural; hábitat *m* rural
g ländliche Siedlung *f*

5157 **rural tourism**
Tourism which takes place in the countryside, the main attractions are the beauty of the natural landscape and farm life, tourists participate in the agricultural production experience and recreational activities.
r сельский туризм
Туризм в сельские районы для знакомства с технологиями хозяйствования и участия в сельских работах.
f tourisme *m* rural
s turismo *m* rural
g Bauernhoftourismus *n*

5158 **rural urbanization; rurbanization**
The spread of the urban living conditions and forms of life to the countryside, which forms part of the general process of urbanization.
r рурбанизация
Процесс распространения городских форм и условий жизни на сельскую местность, часть общего процесса урбанизации.
f rurbanisation *f*
s rurbanización *f*
g Rurbanisation *f*

5159 **ruralisation; rural planning**
A policy aimed at keeping spontaneous urbanization in check by developing rural communities, modernizing rural settlement pattern and developing agricultural production.
r руралиэация
Политика, направленная на сдерживание неуправляемой урбанизации путем развития сельских общин, модернизации сельского расселения и подъема сельскохозяйственного призводства.
f ruralisation *f*
s ruralización *f*
g Ruralisation *f*

5160 **rural-urban continuum; urban-rural continuum**
The merging of town and country, in general there is rarely, either physically or socially, a sharp division, a clearly marked boundary, between the two, with one part of the population wholly urban, the other wholly rural.
r континуум «город-деревня»
Соединение города с деревней, хотя изредка и существует резкое физическое или социальное разделение с хорошо видимой границей, по разные стороны которой население либо целиком городское, либо целиком сельское.
f continuum *m* rural-urbain
s continuo *m* rural-urbano
g riuraler-urbaner Kontinuum *m*

* **rurbanization → 5158**

S

5161 **saddle**

1. A lowering on a mountain ridge between two summits. 2. The area of slack low pressure and low windspeeds which has a saddle-like form on a weather map between two diametrically opposed anticyclones and two areas of even lower pressure.

r седловина

1. Понижение в гребне горного хребта, разделяющее две соседние вершины. 2. Область низкого давления и слабых ветров, имеющая седлообразную форму на погодных картах между двумя диаметрально расположенными антициклонами и двумя областями пониженного давления.

f ensellement *m*

s ensillamiento *m*; depresión *f* entre cumbres

g Einsattelung *f*; Gebirgssattel *m*; Joch *n*

5162 **Sahel**

A natural semi-desert zone, transitional from desert to savannah landscapes streetching along the fringes of the deserts of Northern Africa from the Red Sea to the Atlantic coast.

r Сахель

Природная полупустынная область вдоль окраины пустынь Северной Африки, переходная от ландшафтов пустынь к ландшафтам саванн.

f Sahel *m*; zone *f* de Sahel

s Sahel *m*; zona *f* de Sahel

g Sahelzone *f*

5163 **sai**

A small valley with a permanent or temporary watercourse in Central Asia and Kazakhstan, it is called wadi in Africa, and creek in Australia.

r сай

Небольшая долина с постоянным или временым водотоком в Средней Азии и Казахстане, в Африке называется вади, в Австралии – крик.

f sai *m*

s sai *m*

g Sai *m*

5164 **sailing directions; pilot book**

A guide for navigation with a detailed description of shores, sea floor topography, hydrological and meteorological conditions and dangeres in certain water bodies, ports and anchorages.

r лоция

Руководство для плавания, содержащее подробное описание берегов, рельефа дна, навигационных опасностей, портов и якорных стоянок, гидрологических и метеорологических условий на определенной акватории.

f routier *m*

s derrotero *m*; libro *m* de piloto

g Lotsenhandbuch *n*

* **Saint Elmo´s fire → 5749**

5165 **saline aufeis**

An aufeis formed when brine pours through to the surface due to the multistratified freezing of the solutions of hydrohalite, mirabilite or mixed salts.

r соляная наледь

Наледь, формирующаяся вблизи выходов на поверхность рассолов при послойном намораживании растворов гидрогалита, мирабилита или смешанных солей.

f glaçage *m* salin

s naledj *m* salina

g Salzaufeis *n*

5166 **saline ice**

Ice with the mineralization from 1 to 50 g/l.

r соленый лед

Лед с минерализацией от 1 до 50 г/л.

f glace *f* salée

s hielo *m* salino

g Salzeis *n*

5167 **saline lake; brackish lake**

A lake with the salt content up to 25 g/l.

r солоноватое озеро

Озеро с минерализацией воды до 25 г/л.

f lac *m* de salinité basse

s lago *m* salobre

g Brackwassersee *m*

5168 **saline regime of soil**

The variations of the content and composition of salts in soils as a result of stable periodical descending or ascending migration of salts.

r солевой режим почвы

Совокупность изменений содержания и количественного состава солей в почвах, связанных с устойчивыми периодическими нисходящими или восходящими перемещениями солей.

f régime *m* salin du sol

s régimen *m* salino del suelo

g Salzbhaushalt *m* des Bodens

5169 **saline soil**

A soil having an increased (over 0.25%) content of easily soluble salts, and forming in arid regions as a result of the accumulation of salts in soil and in ground waters.

r засоленная почва

Почва, характеризующаяся повышенным (свыше 0,25%) содержанием легкорастворимых солей, формирующаяся в засушливых районах в результате накопления солей в почве и грунтовых водах.

f sol *m* halomorphe; sol *m* salin

s suelo *m* halomorfo; suelo *m* salino

g Salzboden *m*

5170 saline water; brackish water
Natural waters having mineralization from 1 to 10 g/l.
r солоноватые воды
Природные воды с минерализацией от 1 до 10 г/л.
f eau *f* saumâtre
s agua *f* salobre
g Brackwasser *n*

5171 salinity
The degree of concentration of salts dissolved in water, usually expressed in parts per thousand by mass.
r соленость
Мера концентрации в воде растворенных солей, обычно в промилле.
f salinité *f*
s salinidad *f*
g Salzgehalt *m*

5172 salinization
The accumulation of soluble salts on land surface, in soil or in water occuring naturally in arid and semi-arid areas due to evaporation, or as a result of excessive irrigation which leads to disturbance of water-salt balance of soils (secondary salinization). → Fig. 283
r засоление
Накопление растворимых солей на земной поверхности или в почвенном слое, происходящее в аридных или семиаридных областях вследствие сильного испарения, либо в результате неправильного применения орошения, приводящего к нарушению водно-солевого баланса почв (вторичное засоление). → Рис. 283
f salinisation *f*
s salinización *f*
g Versalzung *f*

5173 salinization of lake
The process of salts accumulation in a lake resulting in the increase of concentration of minerals up to 1000 mg/l and over.
r засоление озера
Процесс накопления солей в озере, приводящий к росту минерализации воды до 1000 мг/л и более.
f salinisation *f* de lac
s salinización *f* del lago
g Versalzung *f* eines Sees

Fig. 283. **Salinization of soil** at the bottom of the former Aral Sea

5174 salinization of rivers
The intrusion of saline waters into deep river beds and delta arms during low-water, high tides and wind-induced surges.
r осолонение рек
Проникновение в глубокие русла рек и рукава дельт осолоненных вод во время межени, приливов и нагонов.
f salinisation *f* de fleuves
s salinización *f* de ríos
g Versalzung *f* der Flüsse

5175 salt
A white crystalline compound (common salt, sodium hloride NaCl), widespread in nature as a solid rock salt around margins of salt lakes, in salt domes, or in solution in seawater.
r соль
Белое кристаллическое соединение, широко распространенное в природе в виде твердой соленой горной породы у берегов соляных озер, в ледяных куполах или в виде раствора в морской воде.
f sel *m*
s sal *f*
g Salz *n*

5176 salt crust
A hard surface layer of a dried salt lake or solonchak which forms in a desert due to the evaporation of the mineralized groundwater; the accumulation of salt on the ice surface while the ice crystals grow and make the salt concentration increase.
r соляная корка
Корка на поверхности высохших соляных озер и солончаков, возникающая в условиях пустыни при испарении минерализованных грунтовых вод; отложение соли на поверхности морского ледяного покрова при росте кристаллов молодого льда и выжимании соли вверх.
f croûte *f* de sel; croûte *f* saline
s costra *f* salina
g Salzkruste *f*

5177 salt desert
A desert area which develops under an intensive evaporation of shallow, highly mineralized groundwater in relief depressions or due to the crystallization of salts out of marine sediments on coasts, where solonchak soils often with salt crust are typical.
r солончаковая пустыня
Участки пустыни, формирующиеся при интенсивном испарении неглубоко залегающих сильно минерализованных грунтовых вод в понижениях рельефа или за счет солей из морских отложений на побережьях,

где характерны солончаковые почвы, нередко с соляными корками.
f désert *m* salé
s desierto *m* salino
g Salzwüste *f*

5178 salt dome; salt plug
A rounded hill produced by the upward doming of rock strata as a result of the almost circular mass of rocksalt forced upward from a great depth in the Earth's crust and piercing the overlying sedimentary rocks.
r соляной купол
Округлый холм в осадочных породах с ядром из соли, возникший в результате поднятия масс соли из глубин земной коры в вышележащие горные породы.
f dôme *m* de sel
s domo *m* de sal
g Salzdom *m*

5179 salt dust
Crystals of soluble salts windblown from a solonchak and suspended in the air.
r соляная пыль
Взвешенные в воздухе кристаллики растворимых солей, сдуваемые ветром с поверхности солончаков.
f poussière *f* de sel
s polvo *m* de sal; polvo *m* salino
g Salzstaub *m*

5180 salt efflorescence
Thin films and crusts of salt occurring on soil surface and rocks at seacoasts of arid climate which are formed as a result of upward migration of soil solutions and their evaporation.
r выцветы солей
Встречающиеся в аридном климате и вблизи морских берегов налёты и корочки на поверхности почвы и горных пород, образующиеся при капиллярном поднятии и испарении почвенных растворов.
f efflorescences *fpl* salines
s eflorescencias *fpl* salinas
g Salzausblühungen *fpl*

* **salt flat → 5184**

5181 salt flowers
Dendritic growth of brittle ice crystals having been formed on the sea ice surface around a salt-crystal nucleus when the brine freezes.
r солевые цветы
Хрупкие веточки из кристаллов льда и солей, образующиеся на поверхности морского льда при замерзании выступившего рассола.
f fleurs *fpl* de saumure
s flores *fpl* de salmuera
g Salzblumen *fpl*

5182 salt lake; saltwater lake
A lake having high concentration of salts dissolved in water, which ossurs in arid or semiarid environments.
r соленое озеро; минеральное озеро
Озеро с большой концентрацией в воде растворенных солей, встречающееся в пустынных и полупустынных условиях.
f lac *m* salé
s lago *m* salino
g Salzsee *m*

5183 salt mine
A mine where a natural deposit of rock salt is worked.
r соляная шахта
Шахта, в которой добывается природная каменная соль.
f mine *f* de sel gemme
s mina *f* de sal
g Salzgrube *f*

5184 salt pan; salt flat
A stretch of unbroken, saltencrusted horizontal land, the bed of a former salt lake, now dried out, usually permanently but sometimes only temporarily.
r соляная котловина
Небольшой не дренируемый естественный бассейн, в прошлом соленое озеро, в котором испаряется вода, оставляя отложения соли.
f salar *m*
s depresión *f* salina
g Salzpfanne *f*

* **salt plug → 5178**

5185 salt runoff; ion runoff; chemical runoff
The amount of dissolved chemical compounds removed by a river from its catchment over a certain time interval.
r солевой сток; ионный сток; химический сток
Количество растворенных химических соединений, выносимых рекой с водосбора за данный интервал времени.
f écoulement *m* chimique
s escorrentía *m* química
g chemischer Abfluß *m*

5186 salt settling
The deposition of crystallized salts on the botom of a salt lake.
r садка солей; самосадка
Осаждение кристаллизующихся солей на дно соленого озера.
f précipitation *f* de sels
s asentamiento *m* de sales
g Salzfällung *f*

5187 salt tectonics
The deformations in sedimentary rocks induced by the intrusion of salt-bearing layers inside the rocks.
r соляная тектоника
Деформации осадочных пород, обусловленные перемещениями залегающих в них соленосных толщ.

f tectonique *f* de sel; halocinèse *f*
s tectónica *f* de sal
g Salztektonik *f*

5188 salt tidal marsh; salting
A natural coastal marsh, supporting halophytes, regularly covered by the sea water at high tide.
r соленый марш
Естественный прибрежный марш с галофитной растительностью, регулярно затапливаемый морской водой во время приливов.
f marais *m* salant
s marisma *f* salina
g Salzmarsch *m*; Salzwiese *f*

5189 salt water
Natural waters having the concentration of salt from 10 to 50 g/l.
r соленые воды
Природные воды с концентрацией солей от 10 до 50 г/л.
f eau *f* saline
s agua *f* salina
g Salzwasser *n*

5190 salt weathering; haloclasty
The disintegration of rocks produced by the crystallisation or thermal expansion of salts in the fissures and pores.
r солевое выветривание
Разрушение почв и горных пород при кристаллизации в их трещинах и порах солей из почвенных растворов.
f haloclastie *f*
s haloclastia *f*
g Salzverwitterung *f*

5191 saltation
The mode of transportation of sediments bouncing along a surface under the impact of running water or wind, such as of material carried by a water stream, grains of sand propelled by the wind, and snow particles transported in a wind-blown snow stream. → Fig. 297, p. 659
r сальтация
Скачкообразное движение частиц над поверхностью земли под действием текущей воды или ветра, в частности, перемещение наносов по дну реки, песчаных зерен в ветровом потоке, частиц снега в снеговетровом потоке. → Стр. 297, с. 659
f saltation *f*
s saltación *f*
g Saltation *f*

*** salting → 5188**

*** saltwater lake → 5182**

*** samoon → 5448**

5192 sample
A small part taken from the whole object (rock, soil, stream water etc.) by which the characteristics of that object can be deduced.
r образец; проба
Небольшая часть, взятая из всего объекта (горной породы, почвы, речной воды и пр.), анализ которой позволяет дать характеристику всего этого объекта.
f échantillon *m*
s muestra *f*
g Probe *f*

*** sample plot → 3315**

*** samun → 5448**

5193 sanctuary; wildlife refuge; animal sanctuary; wildlife reserve
A special area with limitated economic activity in order to preserve animal and plant species including valuable game animals.
r заказник
Участок природной территории с ограничением хозяйственной деятельности с целью сохранения отдельных видов флоры и фауны, а также воспроизводства особо ценных животных для охоты.
f sanctuaire *m*; refuge *m* d'animaux
s coto *m* forestal de caza, de pesca; vedado *m* forestal de caza, de pesca; reserva *f* biológica
g Naturschutzgebiet *n*; Zuflucht *f*; Zufluchtsort *m*

5194 sand
Sedimentary loose rock with particles ranging from 0.1 to 2 mm in diameter, usually formed of mineral quartz.
r песок
Осадочная рыхлая горная порода с частицами диаметром от 0,1 до 2 мм, обычно образованная из кварца.
f sable *m*
s arena *f*
g Sand *m*

*** sand cone → 1443**

5195 sand desert
A desert with a sandy surface that forms as a result of the degradation of local bedrock or

Fig. 284. **Sand desert** Karakum

re-dispersing the deposits of ancient alluvial plains. → Fig. 284, p. 622

r песчаная пустыня

Пустыня с песчаной поверхностью, образующейся в результате разрушения местных коренных пород или перевевания отложений древнеаллювиальных равнин. → Рис. 284, с. 622

f désert *m* sableux

s desierto *m* arenoso

g Sandwüste *f*

5196 sand drift

An accumulation of windblown sand behind an obstacle.

r песчаный нанос

Отложение переносимого ветром песка за естественными или искусственными препятствиями.

f alluvion *f* sablonneuse; dunes *fpl* d'obstacle

s aluvión *m* arenoso; dunas *fpl* de obstáculo

g Sandaufschüttung *f*

*** sand dune → 1745**

5197 sand lens

A discontinuous sand layer in a sedimentary sequence formed as a former runoff channel deposit.

r песчаная линза

Прерывистый слой песка в осадочных отложениях, представляющий собой осадок прежнего канала стока.

f lentille *f* de sable

s lenteja *f* de arena

g Sandlinse *f*

*** sand ribbon → 5200**

*** sand riching of soil → 5204**

5198 sand ridge

Overgrown or exposed chain of sand up to several dozens kilometres long stretched in the direction of prevailing winds.

r песчаная гряда; грядовые пески

Полузаросшая или оголенная цепь песчаных наносов длиною до нескольких десятков километров, вытянутая в направлении господствующих ветров.

f flèche *f* dunaire

s flecha *f* dunar

g Sandrücken *m*

*** sand ripple → 5094**

*** sand storm → 1756**

5199 sand stream

A broad tract (tens of kilometres wide) of sand moving parallel to the coast below low tide, driven by strong tidal streams.

r песчаный поток

Широкая полоса (десятки километров в ширину) песка, движущегося параллельно берегу ниже уровня малой воды, что вызывается сильными приливными течениями.

f flux *m* sablonneux

s flujo *m* arenoso

g Sandstrom *m*

5200 sand wave; sand ribbon

Rows of symmetric and smoothed sand bars up to 1 m high and 20 m wide formed on shallow continental shelf, predominantly in tidal seas or in coastal areas subjected to wind-induced surge.

r подводная песчаная гряда

Ряды симметричных и сглаженных песчаных валов высотой менее 1 м и шириной до 20 м, формирующихся на подводном склоне у отмелых берегов преимущественно приливных морей или на участках, подверженных нагонам.

f onde *f* de sable

s onda *f* de arena

g Sandwelle *f*

5201 sand wedge

The infilling by sand of a fissure in the ground surface formerly occupied by an ice wedge.

r песчаный клин

Заполненная песком трещина в грунте, прежде занятая ледяной жилой.

f coin *m* de sable

s cuña *f* de arena

g Sandkeil *m*

5202 sand whirlwind; dust devil; dust whirlwind

A local swirl of wind with vertical axis carrying dust or sand picked up from the earth's surface in sandy deserts.

r песчаный вихрь; пыльный вихрь

Воздушный вихрь с вертикальной осью, захватывающий песок с земной поверхности в песчаных пустынях.

f tourbillon *m* de sable; tourbillon *m* de poussière

s torbellino *m* de arena; remolino *m* de polvo; tolvanera *f*

g Sandwirbel *m*; Staubwirbel *m*

*** sandblasting → 721**

5203 sand-dune stabilization

A series of techniques (the use of vegetation, spreading gravel over the surface, erecting fences, walls to divert wind, etc.) applied in order to stop the advancing of dunes which threaten farm land, settlements, airports.

r укрепление дюн

Группа методов (использование растительности, разбрасывание гравия на дюнной поверхности, возведение защитных заборов и стен для отведения ветра и пр.), применяемые для того, чтобы остановить продвижение дюн, засыпающих фермы, поселки, аэропорты.

f stabilisation *f* des dunes de sable

s estabilización *f* de dunas de arena
g Stabilisierung *f* der Sanddünen

5204 sanding of soil; sand riching of soil
Treating of upper layer of hard clayey soils with sand to improve their water-physical properties, it is applied in vegetable-growing, gardening and floriculture.
r пескование почвы
Улучшение водно-физических свойств тяжелой глинистой почвы путем обогащения верхнего слоя почвы песком, применяется в овощеводстве, садоводстве и цветоводстве.
f sablonnement *m* du sol
s arenación *f* del suelo;
aporte *m* de arena al suelo
g Bodenbesandung *f*

* **sandr** ***m*** **→ 4330**

5205 sandstone
A porous, arenaceous sedimentary rock composed of grains of chiefly quartz and feldspar consolidated and cemented by such substances as clay or silica.
r песчаник
Осадочная горная порода, состоящая преимущественно из сцементированных зерен кварца и полевого шпата.
f grès *m*
s arenisca *f*; gres *m*
g Sandstein *m*

* **sandur → 4330**

5206 sandy desert soil
A non-saline soil of the desert zone with a weakly pronounced humus horizon and sand layers of various degrees of cohesion.
r песчаная пустынная почва
Незасоленная почва пустынной зоны со слабо выраженным гумусовым горизонтом и слоями песка разной степени связности.
f sol *m* sablé désertique
s suelo *m* arenoso de desierto
g Wüstensandboden *m*

5207 sandy ice
A type of reinforced ice produced by the multistratified freezing of water with the admixture of differently grained sand.
r песколед
Разновидность армированного льда, изготовленного послойным намораживанием воды с добавлением разнозернистого песка.
f glace *f* renforcée par le sable
s hielo *m* arenoso
g Sandeis *n*

5208 sandy loam
A loam which contains between 50 and 70% sand and between 10 and 20% clay, this mixture produces a free-draining, friable soil.
r легкий суглинок
Суглинок, содержащий 50-70% песчаных и 27-40% глинистых частиц, что формирует рыхлую, хорошо дренируемую почву.
f limon *m* sableux
s limo *m* arenoso
g sandiger Lehmboden *m*

5209 saprobiciness
Pollution of water bodies with decaying organic substances.
r сапробность вод
Загрязненность водных объектов разлагающимися органическими веществами.
f saprobition *f*
s saprobición *f*
g Saprobität *f*

5210 saprobiont; saprophyte; saprophagous animal; saprotroph
Plant and animal organisms feeding on dead wood, decaying remains manure, etc., which live in waters and soils polluted with organic matter, and consuming organic residues and excrements.
r сапробионт; сапротроф; сапрофит; сапрофаг
Растительные и животные организмы, обитающие в водоемах и в почве, загрязненных органическими веществами, использующие в пищу органические остатки и экскременты.
f saprobionte *m*; saprotrophe *m*; saprophyte *m*
s saprobionto *m*; saprofita *m*; saprótrofo *m*
g Saprobiont *m*; Saprophag *m*; Saprotroph *m*; Saprophyt *m*

5211 sapropel; decay ooze
Sludge or mud which collects in swamps or shallow marine basins, rich in organic matter, formed by slow anaerobic decomposition of remains of small organisms, such as diatoms and plankton.
r сапропель
Ил или грязь, откладывающаяся в болотах или в мелких морских бассейнах, богатых органическим веществом, образующимся в процессе анаэробного разложения остатков мелких организмов: диатомовых водорослей и планктона.
f sapropel *m*; sapropèle *m*
s sapropel *m*; sapropelium *m*
g Faulschlamm *m*; Sapropel *m*

5212 sapropelit
Fossil coals formed as a result of the decay of organic matter in bottom sediments of water bodies.
r сапропелит
Ископаемые угли, образовавшиеся в результате выгнивания органических веществ в донных отложениях водоемов.
f sapropélite *m*
s sapropelita *m*
g Sapropelite *m*

* **saprophagous animal → 5210**

* **saprophyte → 5210**

* **saprotroph → 5210**

5213 **Sargassaceae; sargasso**
The sea brown algae inhabiting littoral zones of warm seas (huge floating aggregations of Sargassaceae are known in the western part of the Atlantic Ocean, once being torn off from the substrate, they multiply autonomically).
r саргассы
Морские бурые водоросли, обитающие на дне у берегов теплых морей (в западной части Атлантики известны огромные плавающие скопления саргассов, некогда оторвавшиеся от субстрата и размножающиеся вегетативно).
f sargassaceae *fpl*
s sargazo *m*
g Sargassaceae *fpl*

* **sargasso → 5213**

5214 **sarysu**
Rivers of Central Asia with typical yellow or whitish color of water caused by intensive washing out of loess and clay.
r сарысу
Реки в Средней Азии с характерным желтым или белесым цветом, обусловленным интенсивным размывом лёссовых и глинистых толщ.
f sarysu *m*
s sarysu *m*
g Sarysu *n*

5215 **sastrugi; wind ridges; zastrugi**
A series of long, frequently sharp, ridges of hard snow transported by wind and oriented perpendicular to the prevailing wind, and usually having a gentle slope to the windward and a steep slope to the leeward.
r заструги
Серии длинных, часто острых, жестких снежных гребней, ориентированных перпендикулярно к направлению ветра и обычно имеющих пологие склоны с наветренной стороны и крутые – с подветренной.
f sastrugi *mpl*; zastrouguis *mpl*; rides *mpl* sculptées par le vent
s sastrugi *mpl*; zastrouguis *mpl*; arrugas *fpl* talladas por el viento
g Sastruga *f*; Windrippe *f*

5216 **satellite; sputnik**
A natural or artificial celestial body constrained by gravitation and moving in orbit around another more massive heavenly body (artificial satellites in orbit around the Earth are used in remote sensing).
r спутник
Естественное или искусственное небесное тело, удерживаемое гравитацией и движущееся по орбите вокруг более массивного небесного тела (искусственные спутники на околоземной орбите используются для дистанционнного зондирования).
f satellite *m*; spoutnik *m*
s satélite *m*; spútnik *m*
g Satellit *m*; Sputnik *m*

5217 **satellite altimetry**
The identification of heights on the earth and water surface with regard to the Earth's ellipsoid by using satellite data.
r спутниковая альтиметрия
Определение высоты земной поверхности и акватории Мирового океана относительно земного эллипсоида с помощью измерений с искусственного спутника Земли.
f altimétrie *f* par satellite
s altimetría *f* por satélite
g Satellitenaltimetrie *f*

5218 **satellite geodesy**
Measurements of parameters of the Earth and position of the Earth's surface features on the basis of satellite data.
r спутниковая геодезия
Измерение параметров Земли и пространственного положения объектов земной поверхности на основе данных искусственных спутников Земли.
f géodésie *f* par satellite
s geodesia *f* por satélite
g Satellitengeodäsie *f*

* **satellite image → 5694**

* **satellite imagery → 5015**

5219 **satellite meteorology**
The study of the atmosphere by means of meteorological data obtaining from artificial Earth's satellites.
r спутниковая метеорология
Изучение атмосферы с помощью метеорологических данных, получаемых с искусственных спутников Земли.
f météorologie *f* par satellite
s meteorología *f* por satélite
g Satellitenmeteorologie *f*

5220 **satellite oceanography**
The study of seas and oceans by means of oceanographic data obtained from satellites.
r спутниковая океанология
Изучение морей и океанов с помощью океанографических данных, получаемых со спутников Земли.
f océanographie *f* par satellite
s oceanografía *f* por satélite
g Satelitenozeanographie *f*

5221 **satellite town; new town**
An urban settlement close to a major city to which it is related economically and culturally and has good communications, but located well beyond the limits of that city and operating as a discrete, self-contained entity.

r город-спутник
Городское поселение близ большого города, тяготеющее к нему в хозяйственном и культурно-бытовом отношениях, но расположенное достаточно далеко за его пределами и живущее как самодостаточная образование.
f ville-satellite *f*; ville *f* nouvelle
s ciudad *f* satélite; ciudad *f* nueva
g Satellitenstadt *f*

5222 **satellite triangulation**
The development of three-dimensional geodetic network on the basis of synchronous observations from the geodetic Earth's satellite at the determined and initial points, marked on a satellite image.
r спутниковая триангуляция
Развитие трехмерной геодезической сети на основе синхронных наблюдений геодезического искусственного спутника Земли на определяемом и исходном пунктах, отмеченных на космическом снимке.
f triangulation *f* par satellite
s triangulación *f* por satélite
g Satellitentriangulation *f*

5223 **saturated adiabatic lapse rate; wet adiabatic lapse rate; moist adiabatic lapse rate**
The rate of loss of temperature with increasing height occurring in a moist or saturated body of air as it ascends adiabatically.
r влажноадиабатический градиент
Скорость понижения температуры с высотой во влажном или насыщенном объеме воздуха, когда он испытывает адиабатическое восхождение, вследствие высвобождения скрытой теплоты конденсации он меньше градиента в сухом воздухе.
f gradient *m* adiabatique saturé
s gradiente *m* de saturación adiabático
g feuchtadiabatischer Gradient *m*

5224 **saturated adiabatic process**
Adiabatic process in humid saturated air.
r влажноадиабатический процесс
Адиабатический процесс во влажном насыщенном воздухе.
f processus *m* adiabatique saturé
s proceso *m* adiabático del aire saturado
g feuchtadiabatischer Prozeß *m*

5225 **saturated air**
A wet air having saturation vapour pressure, and relative humidity equal to 100%.
r насыщенный воздух
Влажный воздух, в котором упругость водяного пара соответствует стадии насыщения, его относительная влажность равна 100%.
f air *m* saturé
s aire *m* saturado
g gesättigte Luft *f*

5226 **saturated zone; zone of saturation**
A part of the Earth's crust lying lower than the ground water table where the pores are completely filled with water.
r зона насыщения
Часть земной коры, расположенная ниже зеркала грунтовых вод, где поры целиком заполнены водой.
f zone *f* de saturation
s zona *f* de saturación
g Sättigungszone *f*

5227 **saturation**
1. The state of the atmosphere in which a parcel of moist air is in equilibrium with an open water surface at the same pressure and temperature. 2. The condition in which the majority of pores, cracks and joints of a rock or soil are filled with water.
r насыщение
1. Состояние атмосферы, когда некоторый объем влажного воздуха находится в равновесии с открытой водной поверхностью при одинаковом давлении и температуре. 2. Условия, при которых большинство пор и трещин в горной породе или почве заполнены водой.
f saturation *f*
s saturación *f*
g Sättigung *f*

5228 **saturation deficit**
The amount of water vapour that is required to convert a parcel of non-saturated air at a given pressure and temperature to the state of saturation.
r дефицит насыщения
Количество водяного пара, необходимое для полного насыщения некоторого объема ненасыщенного воздуха при данном давлении и температуре.
f déficit *m* de saturation
s déficit *m* de saturación
g Sättigungsdefizit *n*

5229 **savanna; savanna zone; tropical grassland; savannah**
A zonal type of landscape of tropical grasslands flanking each side of the equatorial forest, a combination of tropical and subtropical xerophylous grasses with thinly scattered trees or groups of trees and thickets of shrubs, it is widespread in Africa, South and Central America, Asia and Australia.
r саванна; саванновая зона
Зональный тип ландшафта, образованный сочетанием тропической и субтропической ксерофитной травянистой растительности с отдельными деревьями или группами деревьев и кустарниковыми зарослями, широко распространен в Африке, Южной и Центральной Америке, Азии и Австралии.
f savane *f*; zone *f* des savanes
s sabana *f*; zona *f* de sabanas
g Savanne *f*; Savannenzone *f*; savannah *f*

5230 savanna forest
Grassy woofland with sparse clusters of trees occurring in herbaceous vegetation of savanna.
r саванновый лес
Редкие группы деревьев, разбросанные в виде островков среди травянистой растительности саванны.
f savane *f* arborée
s sabana *f* arbolada
g Savannenwald *m*

* **savanna grassland → 2606**

* **savanna zone → 5229**

* **savannah → 5229**

5231 saxicolous plant; rupicolous plant
The plants of rocky habitats growing in cracks where silty products of weathering accumulate.
r скальное растение; хасмофит
Растения каменистых местообитаний, растущие в расселинах скал, укореняются в трещинах, где накапливаются мелкоземистые продукты выветривания.
f plante *f* saxicole; plante *f* rupicole
s planta *f* de roca
g Felspflanze *f*; Chasmophyt *m*

5232 saz
A bog or humid meadow fed by groundwater in the mountains and foothills of Central Asia.
r саз
Болото или влажный, подпитываемый грунтовыми водами луг в горах и предгорьях Средней Азии.
f saz *m*
s saz *m*
g Saz *m*

5233 scabland
A landscape with flat-topped hills or plateaus, bare or covered with a thin soil consisting of angular debris formed in situ and supporting a sparse vegetation, formed by the glacial erosion of a hard basalt surface.
r скэбленд
Ландшафт плосковершинных холмов или плато, покрытых тонким слоем почвы, содержащим грубообломочный материал, и скудной растительностью, возникший в результате ледниковой эрозии твердой базальтовой поверхности.
f scabland *m*
s scabland *m*
g Scabland *n*

5234 scald; scalded flat
A wide flat depression in Australia covered with scanty vegetation.
r скальд
Широкая плоская депрессия в Австралии, покрытая скудной растительностью.
f scald *m*
s scald *m*
g Scald *m*

* **scalded flat → 5234**

5235 scale
The ratio of the distance on a map, globe, or vertical section to the actual distance on the ground.
r масштаб
Отношение расстояния на карте и глобусе или на вертикальном разрезе к реальному расстоянию на местности.
f échelle *f*
s escala *f*
g Maßstab *m*

5236 scale accuracy
The smallest distance on the ground which can be measured on a map of a given scale.
r точность масштаба карты
Наименьшее расстояние на местности, которое можно измерить на карте данного масштаба.
f précision *f* d'échelle
s precisión *f* de escala
g Maßstabsgenauigkeit *f*

5237 scale of horizontal equivalent; scale of slope
A graph making it possible to determine the angle of a slope in accordance with the selected direction by measuring a horizontal equivalent on topographic map. → Fig. 285
r шкала заложений
График, позволяющий по измеренному на топографической карте заложению горизонталей определить крутизну склона. → Рис. 285
f échelle *f* de pente
s escala *f* de pendiente; escala *f* de elevaciones
g Böschungsdiagramm *n*; Böschungsmaßstab *m*

5238 scale of seismic intensity
A scale used in measuring the magnitude or intensity of an earthquake.
r сейсмическая шкала
Шкала, используемая для оценки интенсивности или магнитуды землетрясения.

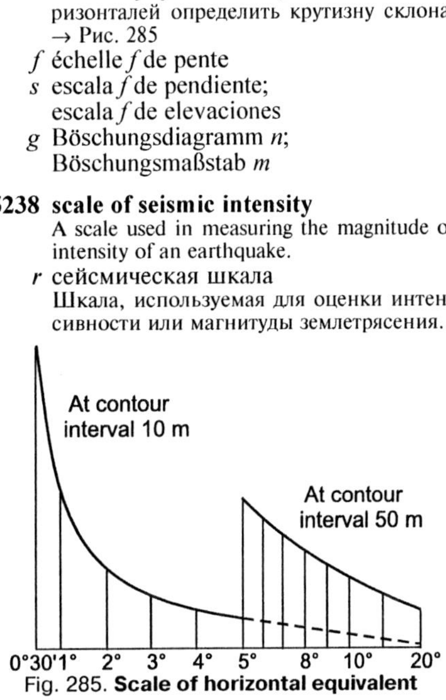

Fig. 285. **Scale of horizontal equivalent**

f échelle *f* d'intensité seismique
s escala *f* sísmica
g Erdbebenskala *f*

* **scale of slope → 5237**

5239 Scandinavian ice sheet
Continental ice sheet which existed in the Pleistocene in the north of Europe having the centre located in the Scandinavian shield.
r Скандинавский ледниковый покров
Материковый ледниковый покров, существовавший в плейстоцене на севере Европейского континента с центром на Скандинавском кристаллическом щите.
f bouclier *m* glaciaire scandinave
s casquete *m* glaciar escandinava
g skandinavischer Eisschild *m*

* **scarp → 2112**

5240 scarp; escarpment
A steep slope or cliff that separates surfaces with smaller gradients located at different heights.
r уступ; эскарп
Крутой склон или отвесный обрыв, разделяющий более пологие поверхности, расположенные на разной высоте.
f escarpement *m*
s escarpe *m*
g Abstufung *f*; Landstufe *f*; Steilwand *f*; Absturz *m*

* **scattered radiation → 5241**

5241 scattering in the atmosphere; scattered radiation; Rayleigh scattering
The interaction between sunlight in the atmosphere and air molecules or suspended particles, resulting in the diffusion of the radiation in all directions but involving no loss of energy, i.e. the dispersal of incoming sunlight by air molecules, which scatter shorter (blue) wavelengths better than longer (red) ones, hence the blue colour of cloudless sky.
r рассеяние в атмосфере
Отклонения распространяющегося в атмосфере светового пучка в результате взаимодействия с молекулами воздуха или взвешенными в нем частицами (без потерь энергии луча), т.е. рассеяние приходящей солнечной радиации молекулами воздуха, причем коротковолновая радиация (голубые лучи) рассеивается лучше длинноволновой (красные лучи), поэтому безоблачное небо имеет голубой цвет.
f diffusion *f* dans l'atmosphère
s difusión *f* en la atmósfera; difusión *f* de Rayleigh
g atmospherische Streuung *f*

* **scavenger → 4084**

5242 scenic *adj*
Providing the beautiful views of nature, or displaying fine natural scenery.
r живописный
Определяющий особую привлекательность ландшафта или окружающей природы.
f panoramique *adj*
s pintoresco *adj*; panorámico *adj*
g malerisch *adj*

5243 schematic map; sketch map
A map with free scale and projection as well as simplified and generalised representation of content elements.
r картосхема
Карта с произвольным масштабом и проекцией, упрощенным и обобщенным изображением элементов содержания.
f croquis *m* cartographique; carte *f* schématique
s mapa *m* esquemático; croquis *m* cartográfico
g Kartenskizze *f*; schematische Karte *f*

* **scherm coast → 5244**

5244 scherms; scherm coast
Short bays of angular outlines separated from each other by linear parts of coast, often incised into coral reefs adjacent to the bedrock coast.
r шермы; шермовый берег
Короткие бухты угловатых очертаний, отделенные одна от другой прямолинейными участками берега, как правило, врезанные в линию коралловых рифов, прилегающих к коренному берегу.
f scherme *m*; côte *f* à schermes
s scherma *m*; costa *f* arrecifal; costa *f* con schermas
g Scherme *mpl*; Schermküste *f*

* **schistosity → 1027**

* **science park → 5023**

5245 scientific and technical revolution
Radical changes that occurred in the industrial production in the XX century as the science became the leading factor of production development and industrial growth.
r научно-техническая революция
Происходившие в XX в. коренные преобразования производительных сил на основе превращения науки в ведущий фактор развития общественного производства.
f révolution *f* technologique
s revolución *f* científica y técnica
g wissenschaftlich-technische Revolution *f*

5246 scientific expedition
Field studies by travel over the territory under investigation on foot or with the use of transportation facilities (horses, cars, water and air transport, etc.).
r научная экспедиция
Полевые исследования, связанные с передвижениями по исследуемой территории

и применением транспортных средств (лошадей, автомашин, водного и воздушного транспорта и пр.).

f expédition *f* scientifique
s expedición *f* científica
g wissenschaftliche Expedition *f*; Forschungsreise *f*

5247 **sciophyte; heliophobic plant**
A shade-loving plant which avoids direct sunlight, e.g. the plants of the lower etages of the forest.
r теневое растение; сциофит; тенелюбивое растение
Растения, которые не переносят большого освещения и могут нормально развиваться лишь при рассеянном свете в условиях затенения, например, растения нижних ярусов леса.
f plante *f* sciaphile; sciaphyte *m*; plante *f* sciadophyte
s esciáfilo *m*; esciáfita *m*
g Schattenpflanze *f*

* **scirocco → 5456**

* **sclerophyllous forest → 2710**

* **sclerophyllous plant → 5248**

5248 **sclerophyte; sclerophyllous plant**
Drought-resistant plants with tough leathery leaves and stalks which loose very little moisture, e.g. cork oak, olive tree, oleander.
r склерофит
Засухоустойчивые растения с жесткими, кожистыми листьями и стеблями, эффективно задерживающими испарение воды, например, пробковый дуб, оливковое дерево, олеандр.
f sclérophyte *m*
s esclerófita *m*
g Sklerophyt *m*

5249 **scour$_1$**
A small area of open water in the ice cover formed by a rapid current or flowing warm waters.
r промоина$_1$
Небольшой открытый участок воды, образующийся в ледяном покрове под влиянием сильного течения, а также в местах поступления теплых вод.
f trou *m* d'eau
s claro *m* entre los hielos
g Eisrinne *f*

5250 **scour$_2$**
Clearing away deposits such as mud, sand and gravel by the action of current or wind flow, in river channels mainly during floods, high water and mudflows; erosive action of boulders frozen into the base of an ice sheet or glacier.
r размыв; русловая эрозия; выпахивание
Захват и вынос потоком или ветровым волнением частиц грунта, образующего речные или морские отложения, на реках в основном во время половодий, паводков, селей; эрозионное действие валунов, вмороженных в основание ледника или ледникового покрова.
f érosion *f* de lit
s derrubio *m*; erosión *f* del lecho
g Abbruch *m*; Flußbetterosion *f*

5251 **scouring basin**
Overdeepened parts of trough valleys separated one from another by riegels.
r бассейн выпахивания
Переуглубленные участки троговых долин, разделенных ригелями.
f bassin *m* d'érosion glaciaire
s cuenca *f* de erosión glaciar; cuenca *f* de socavamiento
g glazial ubertieftes Tal *n*; glaziales Übertiefungsgebiet *n*

* **scree → 6051**

* **scroll → 5252**

5252 **scroll bar; scroll**
Low sand ridges which stretch as a belt along the river channel on a floodplain.
r прирусловые дюны
Бугристые всхолмления, сложенные песком и тянущиеся полосой вдоль прирусловой части поймы.
f dunes *fpl* en bordure de lit
s dunas *fpl* en bordo de lecho
g Uferdünen *fpl*

5253 **scrub; scrubland**
An area of land covered with small trees and busches (as eucalyptus, acacia) associated with poor soils, exposed locations, in which the species are stunted, gnarled or specially adapted to seasonal drought.
r скрэб
Заросли низкорослых ксерофильных кустарников (эвкалиптов, акаций). связанные с бедными почвами, открытыми местообитаниями, где растения растут медленно, искривляются и тем самым приспосабливаются к сезонным засухам.
f formation *f* arbustive; brousse *f*; scrub *m*
s scrub *m*
g Skrub *m*

* **scrubland → 5253**

5254 **sea**
A great body of salt water on the Earthrs surface, as opposed to land; any large body of salt water smaller than an ocean, and partly or entirely enclosed by land.
r море
Масса соленой воды на земной поверхности, противопоставляется суше; часть океана, более или менее обособленная от него сушей или возвышениями подводного рельефа.

f mer *f*
s mar *m*
g Meer *n*; See *f*

5255 **sea bloom; plankton bloom**
Mass growth and accumulation of plankton algae causing different hues of the water colour in seas and oceans.
r цветение моря
Массовое скопление планктонных водорослей, обусловливающее тот или иной оттенок цвета воды в морях и океанах.
f floraison *f* de mer
s floración *f* de mar
g Meeresblühen *n*

5256 **sea breeze; lake breeze**
A local breeze that blows usually during the afternoon from the sea or from the large lake to the land, owing to the differential heating and cooling of land and water. → Fig. 46, p. 95
r морской бриз
Местный ветер, дующий после полудня с моря или с большого озера на сушу вследствие разного прогревания земли и воды. → Рис. 46, с. 95
f brise *f* de mer
s brisa *f* marina
g Meeresbrise *f*

* **sea current → 4208**

5257 **sea farm; oceanic farm**
A part of sea where fish, shellfish or marine plants are grown for the market.
r морская ферма
Ферма по выращиванию рыбы, моллюсков, морских животных и растений в прибрежных частях моря, например, выращивание лососевых во фьордах у берегов Норвегии.
f ferme *f* aquacole; ferme *f* maritime
s granja *f* marina
g Seefarm *f*

5258 **sea ferry**
A big boat carrying vehicles, people and cargoes from one point to another on a regular route.
r морской паром
Специальное судно, совершающее регулярные рейсы между морскими портами, предназначенное для перевозки транспортных средств, людей и грузов.
f ferry-boat *m*; ferry *m*
s buque *m* transbordador; ferry boat *m*
g Fährschiff *n*

5259 **sea foam**
The whitish mixture of water with air bubbles being formed on the wave crests or in the sea surf.
r морская пена
Белесая смесь воды с воздушными пузырьками, образующаяся на гребнях волн или в морском прибое.
f écume *f* de mer
s espuma *f* del mar
g Gischt *m*

* **sea fog → 90**

5260 **sea ice**
Ice formed by the in situ freezing of sea water and later added to by snow accumulation above and further sea water freezing below.
r морской лед
Лед, образующийся при замерзании морской воды, в дальнейшем с участием снега, выпадающего на его поверхность, и намерзания снизу морской воды.
f glace *f* de mer
s hielo *m* marino
g Meereis *n*; Seeeis *n*

* **sea icing → 5734**

5261 **sea level; mean sea-level**
The average level of the surface of the sea, for all stages of the tide over a long period, being determined from hourly readings of tidal oscillation, from which all heights are calculated.
r уровень моря; средний уровень моря
Многолетняя средняя высота уровня моря (с учетом приливных характеристик за длительный период), вычисленная по его измерениям за длительный интервал времени.
f niveau *m* de la mer; niveau *m* marin
s nivel *m* del mar
g Meeresspiegel *m*; mittlere Meeereshöhe *f*; Mittelwasser *n*

5262 **sea level rise**
The increase of sea level, that occurs everywhere throughout the world in a different extent, induced by the increase of water mass and its thermal expansion, which results in the ingression of sea on land.
r поднятие уровня моря
Повышение уровня Мирового океана, неодинаковое в разных его частях, вызываемое увеличением водной массы и тепловым расширением ее верхнего слоя, ведущий к ингрессии моря на береговую сушу.
f montée *f* de niveau de la mer
s elevación *f* de nivel del mar
g Erhöhung *f* des Meeresspiegels

5263 **sea mill**
A cavity in a rocky sea bottom or between coastal karsted rocks into which a surf water is strongly engulfed forming a whirlpool.
r морская мельница
Отверстие скалистого морского дна или прибрежных закарстованных скал, в которое во время прибоя с большой силой уходит вода, образуя водоворот.
f moulin *m* de mer
s molino *m* de mar
g Seeschwinde *n*

5264 **sea mount; submarine mount; seamount**
An isolated mountain peak of different genesis, usually a volcano with a pointed summit, towering above oceanic bottom but without extending above sea level or this is an individual top of an underwater range.
r подводная гора
Изолированная гора разного генезиса, поднимающаяся со дна океана, чаще всего вершина вулкана, или отдельная вершина подводного хребта.
f piton *m* sous-marin; mont *m* sous-marin
s pico *m* submarino
g submariner Berg *m*; untermeerischer Berg *m*; Unterwasserberg *m*

5265 **sea necton**
Actively floating organisms inhabiting pelagic zones of seas and oceans.
r морской нектон
Активно плавающие организмы, обитающие в водной толще пелагической области морей и океанов.
f necton *m* de mer
s necton *m* de mar
g Nekton *m*

* **sea pastureland → 2123**

5266 **sea resort; seaside resort**
A resort the main characteristics of which are favourable marine climatic conditions.
r морской курорт; приморский курорт
Курорт, основным достоинством которого служат благоприятные для лечения приморские климатические условия: морские купания, солнечная радиация и пр.
f station *f* marine; station *f* balnéaire marine
s estación *f* marina
g Seebad *n*

5267 **sea smoke**
A fog produced due to the contact of cold air with a relatively warm water, it occurs over newly-formed leads, or leeward of the ice edge.
r морозное парение
Туман над полыньями, разводьями, каналами во льду и у кромки льда, проявляющийся при соприкосновении холодного воздуха с более теплой водой.
f brume *f* d'évaporation
s humo *m* de helada; humo *m* de evaporación
g Frostrauch *m*; Frostdunst *m*

* **sea stack → 5759**

5268 **sea state**
The state of sea (ocean) surface, i.e. the stage of sea surface waving.
r состояние моря
Состояние поверхности моря (океана), т.е. интенсивность волнения морской поверхности.
f état *m* de mer
s estado *m* de mar
g Seeverhältnisse *npl*; Meeresstand *m*

5269 **sea stew**
A "marine pasture" specially arranged in coastal areas to grow fish and marine animals.
r морской садок
«Морское пастбище», специальные устройства в прибрежной части моря для выращивания рыбы и морских животных.
f vivier *m*
s vivero *m*
g Fishbehälter *m*

5270 **sea transport**
A type of transport moving passengers and goods in vessels along oceans, seas and sea channels.
r морской транспорт
Вид транспорта, занятый перевозкой пассажиров и грузов с помощью судов по океанам, морям и морским каналам.
f transport *m* maritime
s transporte *m* marítimo
g Seefahrtsverkehr *m*

5271 **sea valley; submarine canyon; submarine channel**
Linearly extended sea-floor form morphologically similar to a river valley on land (within a shelf zone they are relics of river valleys of a former land).
r подводная долина; подводный каньон
Линейно вытянутая форма рельефа дна морей и океанов, морфологически сходная с речными долинами на суше (на шельфе это реликт затопленной морем речной долины бывшей суши).
f vallée *f* sous-marine; canyon *m* sous-marin
s valle *m* submarino; cañon *m* submarino
g submarines Tal *n*; Unterwassertal *n*

5272 **sea water**
The water of seas and oceans having constant salinity, i.e. constant concentration of dissolved salts (normal salinity is 34.33‰).
r морская вода
Вода, сосредоточенная в морях и океанах, имеющая постоянный солевой состав (обычно 34,33‰).
f eau *f* de mer; eau *f* marine
s agua *f* de mar; agua *f* marina; agua *f* salada
g Seewasser *n*

5273 **sea water colour**
Water colouring of the upper layer as seen from above which is determined by the amount and type of particles suspended in water and other causes.
r цвет моря
Окраска воды верхнего слоя моря при взгляде сверху, определяемая количеством и типом взвешенных в воде частиц и другими причинами.

f couleur *f* de la mer
s color *m* de mar
g Meerwasserfarbe *f*

* **sea-charts → 4071**

* **sea-floor spreading → 5274**

5274 sea-floor spreading; spreading zone; sea-floor spreading
The sideways movement of lithospheric plates in the area of the rifts of mid-oceanic ridges with constant formation of new oceanic crust due to the outpouring of molten rock from the mantle.
r спрединг; зона раздвижения
Раздвигание литосферных плит в области рифтов срединно-океанических хребтов с постоянным воспроизводством земной коры океанического типа вследствие подъема мантийного вещества.
f expansion *f* des fondes océaniques
s expansión *f* de fondos oceánicos
g Ausdehnung *f* des Ozeanbodens; Spreizungszone *f*

5275 sea-gauge; sounding rod; gauging rod
A device installed at gauging stations of rivers, lakes, seas in order to carry out observations over water level.
r футшток
Рейка, устанавливаемая на водомерных постах рек, озер и морей для наблюдения за уровнем воды.
f maréographe *m*
s mareógrafo *m*
g Flutmesser *m*; Lattenpegel *m*

* **seamount → 5264**

5276 seaquake
Shakings and strokes on the surface and within water mass of oceans and seas caused by submarine earthquakes.
r моретрясение
Сотрясения и удары, возникающие на поверхности и внутри толщи воды океанов и морей при подводных землетрясениях.
f tremblement *m* de mer
s maremoto *m*
g Seebeben *n*

* **seashore → 5406**

* **seashore colony of birds → 5130**

* **seaside resort → 5266**

5277 seaside tourism; sun-and-beach tourism; beach tourism
Tourism organised in coastal regions in order to provide recreation in the beach.
r купально-пляжный туризм
Туризм, организуемый в прибрежных районах с целью отдыха на море.
f tourisme *m* balnéaire
s turismo *m* costero
g Sonnen- und Strandtourismus *m*

5278 season
A time period of the year lasting for several months having some characteristic conditions related in the intensity of solar radiation or differencies of humidity (in midlatitudes there are four seasons – winter, spring, summer, autumn, in tropical regions there are two seasons – rainy and dry seasons, and in equatorial regions there is little differentiation of season).
r сезон; время года
Периоды в течение года продолжительностью в несколько месяцев, имеющие характерные климатические особенности, обусловленные изменениями интенсивности солнечной радиации или различиями в увлажнении (в умеренном поясе различают четыре сезона – зиму, весну, лето и осень, в тропиках два - дождливый и сухой сезоны, а в постоянно влажных экваториальных широтах – всего один сезон в году).
f saison *f* de l'année
s estación *f* del año; estacionalidad *f*
g Jahreszeit *f*

5279 season of bad roads; slush spring thaws
A period of spring characterised by floods and sludge due to snow melt.
r распутица
Состояние подстилающей поверхности, характеризующееся разливами рек и грязью на дорогах в связи с таянием снега.
f routes *fpl* defoncées; chemin *m* impracticable en automne et printemps
s temporada *f* de malos caminos
g Schlammzeit *f*; Schlammwetter *n*

5280 seasonal *adj*
Related to natural phenomena that are changed according to different seasons.
r сезонный
Относящийся к явлениям природы, изменяющимся от сезона к сезону.
f saisonnier *adj*
s estacional *adj*
g Saison-; jahreszeitlich *adj*

5281 seasonal baric depression
Semi-permanent depression strongly pronounced in a certain season and visible on mean pressure charts from the frequent appearance of cyclones in this region (the following seasonal depressions are widely known: Australian, Asiatic, Aleutian, Icelandic, Subantarctic, Subpolar, South-American, and South-African).
r сезонная барическая депрессия
Область пониженного давления, проявляющаяся в конкретный сезон и обнаруживаемая на картах среднего давления благодаря частому образованию циклонов в

данном регионе (широко известны австралийская, азиатская, алеутская, исландская, субантарктическая, субполярная, южно-американская и южноафриканская барические депрессии).
f dépression *f* barométrique saisonnier
s depresión *f* barométrica estacional
g jahreszeitliches Tiefdruckgebiet *n*

5282 seasonal freezing
The occurrence in the ground and rock of temperatures of 0°C or lower during the cold part of the year leading to ice formation if the water is present.
r сезонное промерзание
Распространение в грунтах и горных породах отрицательной и нулевой температуры в холодный период года, приводящее к их цементации льдом.
f gel *m* saisonnier
s congelamiento *m* estacional
g saisonaler Gefrieren *n*

5283 seasonal migration
Movements of people from cities to villages to take part in summer and autumn rural works; seasonal moving of animals fron one place to another one.
r сезонная миграция
Перемещения людей из города в деревню в теплое время года для участия в сельскохозяйственных работах; сезонные перемещения хивотных с одного места их обитания на другое.
f migration *f* saisonnière; migration *f* temporaire
s migración *f* estacional
g jahrzeitliche Migration *f*

5284 seasonal snow line
The lower limit of snow-covered highland at a given moment.
r граница сезонного снега; сезонная снеговая линия
Самый низкий уровень залегания снежного покрова в данный момент.
f limite *f* saisonnière de la neige
s límite *m* de nieve estacional
g temporäre Schneegrenze *f*

5285 seasonally frozen ground; seasonally frozen layer
Ground outside the permafrost zone that freezes during the cold season of the year.
r сезонномерзлый грунт; сезонномерзлый слой; сезонномерзлая горная порода; сезонная мерзлота
Грунт на немерзлых горных породах, промерзающий в холодный период года.
f gélisol *m* saisonnier
s suelo *m* congelado estacional
g saisonaler Frostboden *m*; Winterfrosthorizont *m*

*** seasonally frozen layer → 5285**

*** seasonally thawed layer → 63**

5286 seaway
A way across the open ocean used regularly by shipping.
r морской путь
Направление (маршрут) в открытом океане, регулярно используемое для судоходства.
f voie *f* maritime; ligne *f* de navigation maritime
s vía *f* marítima; línea *f* de navegación marítima
g Seeweg *m*

5287 sebkha; alkali flat
A closed flat-bottomed depression with solonchak and no outflow which is found in the deserts of north Africa (it is synonymous to *takyr* in Central Asia).
r себха
Замкнутое бессточное понижение рельефа с плоским днищем, занятым солончаками, в пустынях Северной Африки, то же, что *такыр* в Средней Азии.
f sebkha *m*; alkali flat *m*
s sebja *m*; sebkha *m*
g Sebcha *f*

5288 seclusion
A rare stage of the evolution of a cyclone which occurs before the occlusion, it consists in the junction of the warm front and cold front at some distance from the centre and leaves an isolated mass of warm air completely surrounded by colder air.
r секклюзия
Редкая стадия в развитии циклона, предшествующая окклюзию и заключающаяся в смыкании теплого и холодного фронтов на некотором расстоянии от центра, в результате чего образуется изолированная масса теплого воздуха, полностью окруженная холодным воздухом.
f séclusion *f*
s seclusión *f*
g Seklusion *f*

5289 second home
A property, usually found in rural areas, occasionally used by a household whose normal place of residence is elsewhere, often used for recreation and pleasure.
r второе жилье
Недвижимость, как правило, сельская, используемая хозяином, у которого есть другой постоянный дом, от случая к случаю, чаще всего в целях рекреации и отдыха.
f résidence *f* secondaire
s residencia *f* segundaria
g die zweite Wohnung *f*

5290 **secondary biological production**
The amount of organic matter produced by heterotrophic organisms (*consumers*) per unit time.
r вторичная биологическая продукция
Количество органического вещества, созданного за единицу времени гетеротрофными организмами – *консументами*.
f production *f* biologique secondaire; productivité *f* biologique secondaire
s productividad *f* biológica segundaria
g sekundäre biologische Produktion *f*

* **secondary consequent stream → 5025**

5291 **secondary consumer**
An animal, such as carnivore, which eats other consumers in the food chain.
r вторичный консумент; вторичный потребитель
Животное, поедающее других потребителей в трофической цепи.
f consommateur *m* secondaire
s consumidor *m* secundario
g Sekundärkonsument *m*

* **secondary contamination → 5296**

5292 **secondary depression**
An area of low pressure, smaller than the primary depression on the margin of which it forms prior to travelling around its perimeter (anticlockwise in the Northern Hemisphere, clockwise in the Northern Hemisphere).
r вторичный циклон
Область низкого давления, меньшая, чем главный циклон, по ее периметру формируются воздушные потоки (против часовой стрелки в Северном полушарии, по часовой стрелке – в Южном).
f dépression *f* secondaire
s depresión *f* secundaria
g Sekundärtief *n*; Randtief *n*

5293 **secondary divide**
A second-order watershed which separates the drainage systems of the tributaries of a river.
r боковой водораздел
Водораздел второго порядка, разделяющий бассейны притоков одной и той же реки.
f ligne *f* secondaire de partage des eaux
s línea *f* secundaria de división de aguas
g Talwasserscheide *f*

5294 **secondary front**
The atmospheric front between two air masses which, though being of the same origin, have different thermal characteristics, it is typical for rear part of a cyclone.
r вторичный фронт
Фронт между двумя воздушными массами, имеющими одинаковое происхождение, но разные термические характеристики, часто возникает в тылу циклона при затоках более холодного воздуха.
f front *m* secondaire
s frente *m* secundario
g Sekundärefront *f*

5295 **secondary industry; manufacturing industry; secondary sector**
The mechanized and usually large scale processing of raw materials into partly finished or finished products.
r вторичный сектор экономики
Механизированная и крупномасштабная переработка сырьевых материалов в почти готовую или полностью готовую продукцию.
f secteur *m* secondaire
s sector *m* secundario
g sekundärer Wirtschaftssektor *m*; verarbeitende Industrie *f*

5296 **secondary pollution; secondary contamination**
The formation of dangerous pollutants due to physical and chemical processes in the polluted environment.
r вторичное загрязнение
Образование опасных загрязнителей в ходе физико-химических процессов в загрязненной среде.
f pollution *f* secondaire; contamination *f* secondaire
s polución *f* secundaria; contaminación *f* secundaria
g Sekundärverschmutzung *f*

5297 **secondary resources**
The materials produced by recycling and used again in economy.
r вторичные ресурсы
Материалы, полученные в результате рециклизации и повторно используемые в хозяйственной деятельности.
f ressource *f* réitérative
s recursos *mpl* reterados
g Sekundärresourcen *fpl*

5298 **secondary salinization**
The salinization of soil as a result of excessive irrigation.
r вторичное засоление
Засоление земель, вызванное чрезмерным поливом.
f salinisation *f* secondaire
s salinización *f* secundaria
g sekundäre Versalzung *f*

* **secondary sector → 5295**

5299 **secondary soil**
A transported soil formed on transported material, as opposed to a residual soil.
r вторичная почва
Почва, сформировавшаяся из принесенного материала, в противоположность первичной почве местного происхождения.

f sol *m* secondaire
s suelo *m* secundario
g Sekundärboden *m*

5300 secondary succession
The succession of a plant community which develops in a place where a previous phytocenosis has been disturbed (as a result of fire, forest slash, drought, erosion, volcanic eruption, etc).
r вторичная сукцессия
Сукцессия, происходящая на месте сформировавшегося фитоценоза после его нарушения (в результате пожара, вырубки леса, засухи, эрозии, вулканического извержения и др.).
f succession *f* secondaire
s sucesión *f* segundaria
g sekundäre Sukzession *f*; Sekundärsukzession *f*

5301 secondary vegetation
The natural vegetation, which appeared on land that was once cleared of primary natural vegetation as a result of human activity, e.g. after cutting down the forests, fire, or ploughing-up.
r вторичная растительность; производная растительность
Естественная растительность, возникшая на месте первичной растительности в результате хозяйственной деятельности, например, после рубки леса, пожара, распашки.
f végétation *f* secondaire
s vegetación *f* segundaria
g sekundäre Vegetation *f*; Sekundärvegetation *f*

5302 secondary water pollution
The pollution of a water body caused by the decay of dead organisms or decomposition of deposited pollutants.
r вторичное загрязнение вод
Загрязнение, вызванное распадом отмерших организмов или загрязняющих веществ, отложившихся в водоеме.
f contamination *f* secondaire d'eau
s contaminación *f* secundaria de agua
g sekundäre Wasserverschmutzung *f*

5303 second-generation migrant
A descendant in the emigrant family who came to this country earlier.
r мигрант второго поколения
Потомок в семье мигрантов, приехавших в данную страну ранее.
f migrant *m* de seconde génération
s migrante *m* de segunda generación
g Migrant *m* der zweite Generation

5304 second-year ice; two-year ice
Drifting polar ice lasted for more than one year and remained for the second year of its existence. → Fig. 286

Fig. 286. Aerial photograph of the **second-year ice** on the surface of a sea

r двухлетний лед
Плавучий полярный лед, просуществовавший более года и остающийся на поверхности моря на второй год. → Рис. 286
f glace *f* de deux ans
s hielo *m* de dos años
g zweijähriges Eis *n*

*** section → 4782**

*** sector of economy → 1831**

5305 secular *adj*
Occurring very rarely, e.g. once in a century or other very long period of time, as well as lasting over such a long period.
r вековой
Случающийся очень редко, например, один раз в сто лет или еще реже, а также продолжающийся в течение этого долгого времени.
f séculaire *adj*
s secular *adj*
g hundertjährig *adj*

5306 secular trend in climate
A slow change in the values of climatic elements observed over a very long period of time.
r вековой ход
Изменения в величинах климатических элементов, наблюдаемые в течение очень долгого периода времени.
f tendance *f* séculaire du climat
s tendencia *f* secular del clima
g Säkularänderung *f* des Klimas

5307 secular variation
Changes over a lengthy time period of the properties of the Earth's magnetic field.
r вековое колебание
Изменения за длительный период времени свойств магнитного поля Земли.
f variation *f* séculaire
s variación *f* secular
g säkulare Schwankung *f*

5308 secular variations of climate
Regular changes of climate going on during

many dozens or hundreds years and having no well-manifested trend.
r вековые колебания климата
Периодические изменения климата, происходящие на протяжении многих десятков или сотен лет и не имеющие ярко выраженного тренда.
f variations *fpl* séculaires du climat
s variaciónes *fpl* seculares de clima
g säkulare Klimaschwankungen *fpl*

5309 secular variations of sea-level
Slow changes of sea level occurring over a long period of time and related to vertical movement of the Earth's crust or change of water amount in the ocean.
r вековые колебания уровня моря
Медленные изменения уровня моря, происходящие в течение длительного времени под влиянием колебательных движений земной коры и изменения количества воды в Мировом океане.
f variations *fpl* séculaires du niveau de la mer
s variaciónes *fpl* seculares de nivel del mar
g säkulare Meeresspiegelschwankungen *fpl*

5310 secularization
Putting the property of church, mainly land)under the control of the state.
r секуляризация
Обращение государством церковной собственности, преимущественно земли, в светскую.
f sécularisation *f*
s secularización *f*
g Sekularisierung *f*

5311 sedentarization; settled way of life; sedentary agriculture
In contrast to nomadism the way of life and agriculture where a peasant works in the same place.
r оседлость
В противоположность кочевничеству тип сельского хозяйства, когда крестьянин трудится на одном и том же месте.
f sédentarisation *f*; vie *f* sédentaire
s sedentarización *f*; vida *f* sedentaria
g Seßhaftigkeit *f*; Ansässigkeit *f*

5312 sedentary *adj*
Staying in one place, i.e. not migratory, not transported.
r оседлый
Остающийся на одном месте, т.е. не мигрирующий, не переносимый.
f sédentaire *adj*
s sedentario *adj*
g sedentär *adj*

*** sedentary agriculture → 5311**

*** sedentary soil → 5032**

*** sediment → 1527**

*** sediment discharge → 5314**

5313 sediment sorting
The process by which sediments are graded by the flow of water according to their shape, size, density, etc.
r сортировка наносов; сепарация наносов
Разделение наносов по плотности, форме и крупности в процессе их движения под воздействием гидродинамических факторов.
f triage *m* de sédiments
s selección *f* de sedimentos; selección *f* de depósitos
g Sedimenttrennung *f*

5314 sediment yield; sediment discharge
The amount of suspended material and bed load transported through a cross section of a stream in a given unit of time.
r расход наносов; сток наносов
Количество взвешенных и влекомых наносов, проносимое через живое сечение потока в единицу времени.
f débit *m* solide
s caudal *m* sólido; caudal *m* de sedimentos
g Geschiebeführung *f*; Feststofführung *f*

5315 sedimentary *adj*
Related to the processes of deposition of substances in the environment as well as to the subsequent deposits of rocks.
r осадочный
Относящийся к процессам осаждения вещества в разных природных средах и соответствующим отложениям горных пород.
f sédimentaire *adj*
s sedimentario *adj*
g Sediment-

*** sedimentary cover → 4611**

5316 sedimentary layer of the Earth's crust
The upper part of the Earth's crust 15 km thick consisting of the sedimentary and volcanic rocks.
r осадочный слой земной коры
Верхняя часть земной коры толщиной до 15 км, состоящая из осадочных и вулканических горных пород.
f couche *m* sédimentaire de l'écorce terrestre
s estrato *m* sedimentario de la corteza terrestre
g Sedimentschicht *f* der Erdkruste

5317 sedimentary rock
A rock composed of the fragments and particles of pre-existing rocks which have been eroded, transported and deposited by wind or water as well as due to glaciers activity.
r осадочная горная порода
Горные породы, формирующиеся путем осаждения вещества в водной среде, реже из воздуха, а также в результате деятельности ледников.

f roche *f* sédimentaire
s roca *f* sedimentaria
g Ablagerungsgestein *n*; Sedimentgestein *n*

* **sedimentation → 1530**

5318 sedimentogenesis; lithogenesis
Natural processes of formation and subsequent changes of sedimentary rocks.
r седиментогенез; литогенез
Совокупность природных процессов образования и последующих изменений осадочных горных пород.
f sédimentogenése *f*; lithogenése *f*
s sedimentogénesis *f*; litogénesis *f*
g Sedimentogenese *f*; Lithogenese *f*

* **sedimentology → 3520**

* **seeding of clouds → 1074**

5319 seepage
The very slow percolation or oozing out of a fluid through a porous body, or along a fault or joint-plane, e.g. slight oozing out of petroleum on the ground surface or the percolation of surface water into the soil.
r вытекание
Очень медленное просачивание или истечение жидкости через пористое тело или вдоль разломов, например, высачивание нефти на земную поверхность, или просачивание поверхностной воды в почву.
f suintement *m*
s rezumadero *m*
g Durchsickern *n*

5320 segetal vegetation
A type of semi-natural vegetation growing where there has been human interference with the natural growth.
r сегетальная растительность
Тип полуестественной растительности, растущей повсюду, где было вмешательство человека в естественный рост растений.
f végétation *f* segetale
s vegetación *f* segetal
g Segetalvegetation *f*

5321 segregated ice
A kind of ground ice formed by the freezing of wet coarse-grained sediments accompanied by the migration of water towards the areas of ice formation.
r сегрегационный лед
Вид подземного льда, формирующегося при промерзании увлажненных тонкодисперсных пород, сопровождающемся подтягиванием воды к участкам льдовыделения.
f glace *f* de ségrégation
s hielo *m* de segregación
g Segregationseis *n*

5322 segregation
1. The separation of the subgroups of a large population, particularly into distinct residential areas, it may be based on ground of income, race, religion, or language. 2. The separation of minerals into discrete masses during crystallization of igneous rocks or the concentration of minerals into bands in metamorphic rocks. 3. The formation of ice lenses in the ground.
r сегрегация
1. Выделение отдельных групп населения, особенно в специальных регионах, основанное на различиях в доходах, расе, религии, языке. 2. Выделение минералов в отдельные массы во время кристаллизации изверженных пород или концентрация минералов в отдельных поясах в метаморфических горных породах. 3. Формирование ледяных линз в грунте.
f ségrégation *f*
s segregación *f*
g Segregation *f*

5323 seiche
A stationary or standing wave in an enclosed body of water, such as lake, which usually occurs as a result of intense storm activity. → Fig. 287
r сейша
Стационарная или стоячая волна в закрытом водоеме, например в озере, возникающая в результате штормового волнения. → Рис. 287
f seiche *f*
s seiche *f*

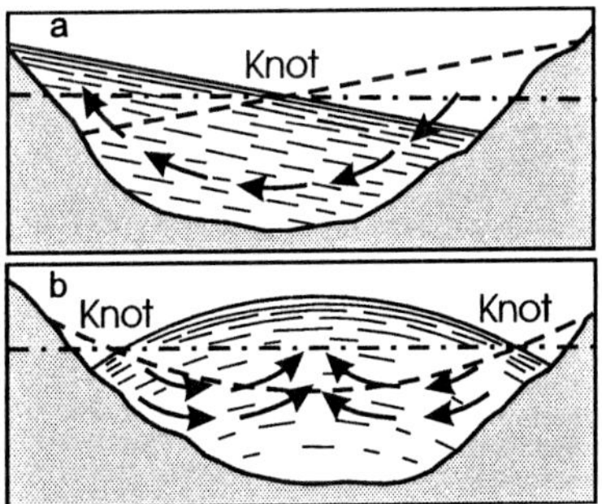

Fig. 287. **Seiche**: 1 – one-modal, 2 – bimodal

g Seiche *f*; selbständige Gezeit *f*

* **seif-dune → 3578**

* **seismic activity → 5328**

5324 seismic *adj*
Related to the movement within the Earth, particularly to earthquakes.
r сейсмический
Относящийся к движениям внутри Земли, особенно к землетрясениям.
f séismique *adj*; sismique *adj*
s sísmico *adj*
g seismisch *adj*

5325 **seismic focus; seismic origin; hypocentre**
The place in the Earth's crust, under the surface, from which an earthquake shock originates.
r очаг землетрясения; гипоцентр
Область в недрах Земли, где происходит быстрое перемещение масс вдоль тектонического разрыва и высвобождается накопленная энергия.
f foyer *m* sismique; hypocentre *m*
s hipocentro *m*
g Erdbebenherd *m*; Hypozentum *n*

5326 **seismic hazard**
The likelihood that an earthquake will be felt in a particular area (e.g. Japan and California).
r сейсмическая опасность
Подверженность территории регулярным сильным землетрясениям.
f risque *f* sismique
s riesgo *m* sísmico
g Erdbebengefahr *f*

* **seismic origin → 5325**

5327 **seismic waves; shake waves; earthquake waves**
Longitudinal and transverse elastic vibrations spreading in the Earth from seismic focuses, explosions and other sources.
r сейсмические волны
Продольные и поперечные упругие колебания, распространяющиеся в Земле от очагов землетрясений, взрывов и других источников.
f ondes *fpl* séismiques; ondes *fpl* sismiques
s ondas *fpl* sísmicas
g Erdbebenwellen *fpl*

5328 **seismicity; seismic activity**
The susceptibility of a given region to earthquakes expressed in the intensity and frequency of earthquakes in the area.
r сейсмичность;
сейсмическая активность
Подверженность Земли или отдельных территорий землетрясениям, определяемая их интенсивностью и частотой в данном регионе.
f séismicité *f*; activité *f* sismique
s sismisidad *f*
g Seismizität *f*

5329 **seismograph**
An instrument used in recording the duration, magnitude and direction (horizontal and vertical) of earth tremors, natural (earthquake) or artificial.
r сейсмограф
Прибор для регистрации продолжительности, магнитуды и направления (горизонтального и вертикального) сотрясения земли, как естественного (землетрясения), так и вызванного искусственно.
f séismographe *m*
s sismógrafo *m*
g Seismograph *m*; Erdbebenmesser *m*

5330 **seismology**
The scientific study of earthquakes, their after-effects, relationship between earthquakes and tectonic processes and the possibility of their forecast as well as earth tremors produced artificially.
r сейсмология
Наука, изучающая причины землетрясений, их последствия, связь землетрясений с тектоническими процессами и возможность их предсказания, а также сейсмические явления, вызывемые искусственно.
f séismologie *f*; sismologie *f*
s sismología *f*
g Seismologie *f*

5331 **selective logging**
The felling, at intervals, of the mature trees in a mixed age forest.
r выборочная рубка
Заготовка время от времени спелых деревьев в разновозрастном лесу.
f abattage *m* sélectif
s corta *f* selectiva
g Ausforsten *n*

* **self-induced oscillations → 473**

5332 **self-purification of the environment; natural purification of the environment**
Natural processes of utilization and destruction of substances polluting the environment.
r самоочищение окружающей среды
Происходящий естественным путем процесс утилизации и обезвреживания веществ, загрязняющих окружающую среду.
f autopurification *f* de l'environnement
s autodepuración *f* de medio ambiente
g Selbstreinigung *f* der Natur

5333 **self-purification of water**
Natural processes getting rid of undesirable elements and to restore the initial properties and composition of water.
r самоочищение воды
Совокупность природных процессов, приводящих к восстановлению первоначальных свойств и состава воды.
f autopurification *f* de l'eau
s autopurificición *f* del agua
g Selbstreinigung *f* des Wassers

* **selfregulation → 2796**

5334 **selfsettling salt lake**
A water body in which the crystallization and settling of salts take place.
r самосадочное озеро; соляное озеро
Водоем, в котором происходит кристаллизация и осаждение солей.
f lac *m* autosédimentaire

s lago *m* de asentamiento espontáneo de sales
g Salzsee *m* mit Salzenscheidung

5335 selka
A collective term for ridge-like landforms in Karelia, Finland and Estonia.
r сельги
Общее название грядоподобных форм рельефа в Карелии, Финляндии и Эстонии.
f selka *f*
s selca *f*
g Selka *f*

* **selva → 1968**

5336 semi-arid *adj*
Related to the transional zone (climate, vegetation and others) lying between savanna and the true hot desert, or between the hot desert and a Mediterranean climatic region, characteristically supporting patchy xerophilous vegetation.
r полупустынный
Относящийся к транзитной зоне (а также соответствующим климату, растительности и пр.), лежащей между саванной и настоящей жаркой пустыней, или между жаркой пустыней и средиземноморским районом, отличающийся своеобразной ксерофитной растительностью.
f semi-aride *adj*
s semiárido *adj*
g semiarid *adj*

* **semi-arid area → 5338**

* **semi-closed sea → 171**

5337 semi-deciduous vegetation
Plant communities in which trees composing the upper etage shed their leaves for a short dry period while those of lower etage remain foliage, such forests occur in those parts of tropics where the dry period does not exceed 1-1.5 months.
r полулистопадная растительность
Растительные сообщества, верхний ярус которых сбрасывает листву на короткий сухой период, а находящаяся под его пологом растительность остается вечнозеленой, произрастают в тропических широтах, где есть засушливый период продолжительностью 1-1,5 месяца.
f vegetation *f* semi-caducifoliée
s vegetación *f* semicaducifolia
g halblaubabwerfende Vegetation *f*

5338 semi-desert; semi-desert zone; semi-arid area
A type of landscape that develops in arid climate, it is characterized by combination of steppe and desert landscapes, form the transition zone between true desert and more thickly vegetated zone such as savanna, etc.
r полупустыня
Тип ландшафта, формирующийся в условиях аридного климата, отличается сочетанием степных и пустынных ландшафтов, образует переходную зону между настоящей пустыней и более плотной растительностью, как, например, в саванне.
f semi-désert *m*; zone *f* des semi-déserts; zone *f* semi-aride
s semi-desierto *m*; zona *f* de semi-desiertos; zona *f* semiárida
g Halbwüste *f*; Halbwüstenzone *f*

* **semi-desert zone → 5338**

5339 semi-diurnal tide
A tide having two high waters and two low waters each lunar day, each occurs at an interval of approximately 12 h 25 min.
r полусуточный прилив; правильный прилив
Прилив, в цикле которого в течение лунных суток имеются две фазы полной воды и две фазы малой воды, равные или слегка различающиеся по длительности.
f marée *f* semi-diurne
s marea *f* semidiurna
g Halbtagstide *f*; halbtägige Gezeiten *fpl*

5340 semi-enclave
An enclave with a marine outlet such as Monaco or the Kaliningrad Region of the Russian Federation.
r полуанклав
Анклав, имеющий выход к морю, например, Монако или Калининградская область Российской Федерации.
f demienclave *f*
s semienclave *m*
g Semi-Enklave *f*

5341 semi-natural vegetation
Vegetation not actually planted by human hand, but resulting directly or indirectly from the activities of human beings or their livestock.
r полуестественная растительность
Растительность, не выращенная непосредственно руками людей, но ставшая результатом прямой или косвенной деятельности людей или их скота.
f végétation *f* semi-naturelle
s vegetación *f* semi-natural
g halbnatürliche Vegetation *f*

5342 semi-permanent anticyclone; quasy-permanent anticyclone
The region where high pressures largely predominate during a better part of the year and where the anticyclone appears on the mean monthly pressure charts, they are known as Azores, Bermuda, North-Atlantic, South-Atlantic, Siberian highs.
r квазипостоянный антициклон
Район, в котором в течение большей части года преобладает высокое давление и

антициклоническая циркуляция сохраняется на средних месячных картах атмосферного давления, к подобным антициклонам относятся азорский, бермудский, северо-атлантический, южно-атлантический, сибирский.
f anticyclone *m* semi-permanent; anticyclone *m* pseudopermanent
s anticiclón *m* semiestacionario
g semipermanentes Hoch *n*

5343 semi-permanent centre of action of the atmosphere
Centre of the atmosphere action existing for only a certain season.
r сезонный центр действия атмосферы
Центр действия атмосферы, существующий в течение одного определенного сезона.
f centre *m* d'action atmosphérique semi-permanent
s centro *m* de acción atmosférico semipermanente
g jahreszeitliches Aktionszentrum *m* der Atmosphäre

5344 semi-permanent depression
Region where low pressures largely predominate during a major portion of the year and for which a depression appears on the mean monthly pressure charts, they are known Aleutian and Icelandic lows.
r квазипостояный циклон
Район, в котором в течение большей части года преобладает низкое давление и циклоническая циркуляция сохраняется на средних месячных картах атмосферного давления, сюда относятся алеутский и исландский минимумы.
f dépression *f* pseudopermanente; dépression *f* semi-permanente
s depresión *f* semiestacionaria; depresión *f* semipermanente
g semipermanentes Tief *n*

5345 sensible heat
Atmospheric heat energy which can be transferred within the Earth/atmosphere heat-balance system either by connection, in which warm air rises and mixes, or by direct conduction.
r явная теплота
Атмосферная тепловая энергия, которая передается в системе теплового баланса Земля–атмосфера как путем конвекции, когда теплый воздух поднимается и смешивается с холодным, так и путем прямой теплопроводности.
f chaleur *m* sensible
s calor *m* sensible
g sensible Wärme *f*; fühlbare Wärme *f*

5346 sensible temperature
Cold or heat as felt by the human body, depending not only on actual temperature but also on relative humidity and wind intensity.
r ощутимая температура
Ощущение холода или тепла человеческим телом, зависящее не только от реальной температуры, но также от относительной влажности и ветра.
f température *f* sensible
s temperatura *f* sensible
g fühlbare Temperatur *f*

5347 sensitivity of a landscape
The likelihood that a change in the controls of a landscape will produce a recognizable response.
r чувствительность ландшафта
Степень возможной реакции ландшафта на изменение воздействия на него извне.
f sensibilité *f* de paysage
s sensibilidad *f* de paisaje
g Empfindlichkeit *f* der Landschaft

5348 separatism; national separatism
The demand of a particular group of people that a particular area should become separate in territory and political sovereignity from the state within which it lies, e.g. the Kurds in Turkey and Basques in Spain.
r сепаратизм
Стремление отдельных групп населения и местной элиты к обособлению данной территории вплоть до образования отдельного государства, например, курды в Турции или баски в Испании.
f séparatisme *m*
s separatismo *m*
g Separatismus *m*

5349 sequential landforms
The landforms created by modifications of the initial landform, no matter what the degree of change that has been wrought by denudation.
r неунаследованный рельеф
Рельеф, созданный путем развития первоначальной топографии, сильно измененной последующей денудацией.
f relief *m* conséquent
s relieve *m* consecuente
g konsequentes Relief *n*

5350 serac; ice pedestal
A pinnacle of ice formwd in the part of a glacier where crevasses intersect, it is resulted from the uneven melting and breakdown of broken ice blocks on icefalls. → Fig. 288, p. 641
r серак
Ледяные пики и зубцы на поверхности ледников, образующиеся в результате неравномерного таяния и обрушения глыб на ледопадах. → Рис. 288, с. 641
f sérac *m*
s serac *m*
g Serac *m*; Eisturm *m*

5351 serclimax
The terminal phase in the development of a plant community in which all further progress

towards a true climax is inhibited by the constant repetition of a natural intervention, e.g. regular flooding or wetting by salt spray.

r серклимакс
Конечная фаза развития растительного сообщества, в котором дальнейшему прогрессу к стадии климакса препятствует регулярно повторяющиеся природные явления, например, частые паводки или выпадения солей.

f serclimax *m*
s serclímax *m*
g Serklimax *f*

5352 **sere**
A particular and easily recognized stage of a succession of a plant community.

r сере
Легко различимая стадия сукцессии растительного сообщества.

f sere *f*
s sere *f*
g Sere *f*

* **serfage → 5353**

5353 **serfdom; serfage**
The system when a member of a class of agricultural labourers in a feudal society, bound in service to a lord, and transferred with the land they worked if its ownership changed hands.

r крепостное право
Прикрепление крестьян к земле и подчинение административной и судебной власти феодала.

f servage *m*
s servidumbre *f*
g Leibeigenschaft *f*; Hörigkeit *f*

5354 **sericulture; silk culture**
The breeding of silkworms to produce cocoons used as raw material in silk production.

r шелководство
Разведение шелкопрядов с целью получения шелковичных коконов – сырья для изготовления натурального шелка.

Fig. 288. **Serac**

f magnanerie *f*; séricículture *f*
s sericicultura *f*
g Seidenbau *m*; Seidenzucht *f*

5355 **series**
Geological formations characterised by some common character of composing rocks.

r толща
Совокупность отложений горных пород, происходивших в сходных природных условиях и поэтому характеризующихся некоторыми общими признаками и свойствами.

f série *f* sédimentaire; assise *f*
s serie *f* sedimentaria; serie *f* de capas
g Serie *f*

5356 **serir**
The vast, flat stony and gravel-covered deserts in eastern Sahara.

r серир
Обширные каменисто-щебнистые плоские равнинные пустыни в Восточной Сахаре.

f sérir *m*
s serir *m*
g Serir *m*

* **serozem → 5434**

5357 **service industry**
Any economic activity intended to satisfy human needs but not constituting of item of goods, including wholesailing, transport, retailing, etc.

r сфера услуг; обслуживающие отрасли
Сфера хозяйства, занимающаяся удовлетворением потребностей населения в услугах разного рода.

f services *mpl*
s servicios *mpl*
g Dienstleistungsbereich *m*; Dienstindustrie *f*

5358 **services**
The product of human activity intended to satisfy a human need or needs but not constituting an item of goods.

r услуги
Продукт деятельности людей, имеющий в виду удовлетворение человеческих потребностей, но не предполагающий конкретного товара.

f services *mpl*
s servicios *mpl*
g Bedienung *f*

5359 **seston**
Small plankton organisms as well as inorganic and organic particles suspended in water.

r сестон
Мелкие планктонные организмы и взвешенные в воде неорганические и органические частицы.

f seston *m*
s sestón *m*
g Seston *n*

5360 set of scales
A succession of scales used for a group of map series.
r масштабный ряд
Перечень масштабов, принятых для какой-либо группы карт.
f série *f* des échelles
s serie *f* de escalas
g Maßstabsreihe *f*

5361 setting basin
A lake or reservoir used to purify turbid waters from suspended particles incoming with the river runoff.
r отстойный бассейн
Озеро или водохранилище, очищающие мутные воды от взвешенных частиц, поступающих сюда с речным стоком.
f bassin *m* de décantation
s estanque *m* decantador
g Ablagerungbecken *n*

*** setting tank → 4747**

5362 setting-up of snow cover
The beginning of the period of stable snow cover in a region.
r установление снежного покрова
Начало периода залегания устойчивого снежного покрова.
f stabilisation *f* du manteau neigeux
s formación *f* de manto nivoso
g Beginn *m* der Schneebedeckung

5363 settled ice
A portion of ice cover near the coast or in shallow water, grounded at the bottom due to the drop of the water level.
r осевший лед
Участок ледяного покрова вблизи берега или на мелководье, осевший на дно при снижении уровня воды.
f glace *f* échouée
s hielo *m* asentado
g abgelagerte Eisdecke *f*

*** settled way of life → 5311**

5364 settlement
Any form of human habitation either seasonal or permanent, normally with a geographical name; the act of peopling a formerly uninhabited land.
r поселение; населенный пункт
Постоянное или сезонно обитаемое место проживания населения с географическим наименованием; заселение людьми ранее необитаемого места.
f peuplement[2] *m*; établissement *m* humain
s poblamiento *m*; pueblo[2] *m*; asentamiento *m*; población[2] *f*
g Siedlung *f*

*** settlement network → 5365**

5365 settlement pattern; settlement network
The nature of the distribution of population settlements of varying sizes.
r расселение; система расселения
Процесс распределения населения по территории и его результат – сеть поселений и ее формы.
f système *m* de peuplement
s sistema *m* de asentamientos; distribución *f* espacial de asentamientos
g Ansiedelung *f*; Siedlungssystem *n*; Besiedlung *f*

5366 settling
The descent of particles through water or air, e.g. a dust particle through the atmosphere or a particle of clay through water.
r осаждение
Опускание частиц в воде или воздухе, например, частиц пыли через атмосферу или глинистых частиц через воду.
f abaissement *m*
s descendimiento *m*
g Fällung *f*

5367 settling of snow
The decrease in the thickness of a snow mass as a result of its compaction.
r оседание снега
Уменьшение толщины снежного покрова в результате уплотнения снега.
f tassement *m* de la neige
s asentamiento *m* de nieve
g Schneesetzung *f*

5368 settling velocity
The uniform speed attained by a particle as it descends through aerial or water environment, e.g. a dust particle through the atmosphere or a particle of clay through water.
r скорость осаждения
Скорость, с которой частица опускается в воздушной или водной среде, например, частица пыли в воздухе, или грязевая частица в воде.
f vélocité *f* de précipité
s velocidad *f* de precipitación
g Sinkgeschwindigkeit *f*

*** setup phenomena → 6793**

5369 sewage; waste water; effluent$_2$; sewage effluent; sewage water
Liquid waste matter generated in the process of human economic activity.
r сточные воды; стоки
Жидкие отходы, образующиеся в результате хозяйственной деятельности.
f effluents *mpl* d'égout; effluent *m*
s aguas *fpl* residuales; efluente *m*
g Abwässer *npl*

*** sewage aeration → 99**

*** sewage effluent → 5369**

5370 **sewage outfall**
Planned or accidental release of sewage into rivers, lakes or other natural or man-made bodies of water.
r сброс сточных вод
Планируемое или аварийное поступление сточных вод в естественные или искусственные водные объекты.
f rejet *m* d'eaux usées; effluent *m* d'eaux usées
s vertido *m* de aguas residuales; efluente *m* de aguas residuales
g Abwässerableitung *f*

* **sewage water → 5369**

5371 **sewer**
A pipe or canal along which water is removed outside the drained area.
r коллектор
Дренажная труба или канал, по которым вода отводится за пределы осушаемой территории.
f tuyaux *m* de drenage
s collector *m*; tubería *f* de drenaje
g Kollektor *m*; Abzugskanal *m*; Abwasserkanal *m*

5372 **sewer water**
Surface rain and melt waters carried away through sewers from the streets of cities.
r канализационные воды
Поверхностные, дождевые и талые воды, отводимые с территории города с помощью канализации.
f eaux *fpl* d'égout
s aguas *fpl* de acantarillado
g Kanalisationsabwässer *npl*

5373 **shade temperature**
The temperature normally taken in meteorological observations, away from direct insolation or radiation from nearby objects.
r температура в тени
Температура, принятая в метеорологических наблюдениях, ее измеряют вне прямой инсоляции и радиации от окружающих объектов.
f température *f* à l'ombre
s temperatura *f* de sombra
g Schattentemperatur *f*

5374 **shade-tolerant plant; shading species**
Plants that grow in sunlight, yet can endure shade, e.g. the forest-forming species of temperate zone.
r теневыносливое растение
Растения, растущие на свету, но способные выносить и затенение, например, основные лесообразующие породы умеренных широт.
f plante *f* d'ombre; plante *f* résistant à l'ombre
s planta *f* tolerante a la sombra; especie *f* de sombra
g schattenertragende Pflanze *f*

* **shading species → 5374**

* **shadow economy → 716**

5375 **shaft**
A vertical mine working of a small section.
r шурф
Вертикальная горная выработка небольшого сечения.
f fouille *f*
s pozo *m* de sección pequeña
g Schurf *m*

* **shake waves → 5327**

5376 **shallow karst**
Karst in which karstic rocks and the zone of vertical circulation of water do not exceed 50 m, and underground waters are not deep.
r мелкий карст
Карст, в котором карстующаяся толща и зона вертикальной циркуляции вод не превышает 50 м, а подземные воды лежат неглубоко.
f karst *m* peu profond
s karst *m* poco profundo
g seichter Karst *m*

5377 **shallows**
A part of a river, sea, lake much shallow than the surrounding, in many cases it is dangerous for navigation.
r мель
Участок водоема или водотока с глубинами значительно меньше окружающего водного пространства, служащий серьезным препятствием для судоходства.
f bas-fond *m*
s bajío *m*; bajo *m*
g Untiefe *f*

5378 **shallow-water deposit; shallow-water sediment**
Marine sediments laid down at the depths less than 200 m.
r мелководные отложения
Морские осадки, отлагающиеся на глубинах до 200 м.
f dépôt *m* thalassique
s depósito *m* de aguas poco profundas
g Flachseeablagerungen *fpl*; Flachseesedimente *npl*

* **shallow-water sediment → 5378**

* **shallow-water wave → 3575**

* **shanty town → 5486**

* **shanty-town → 5486**

5379 **shape distortion**
The difference in an outline on a geographical map and that on the surface of the Earth's ellipsoid.

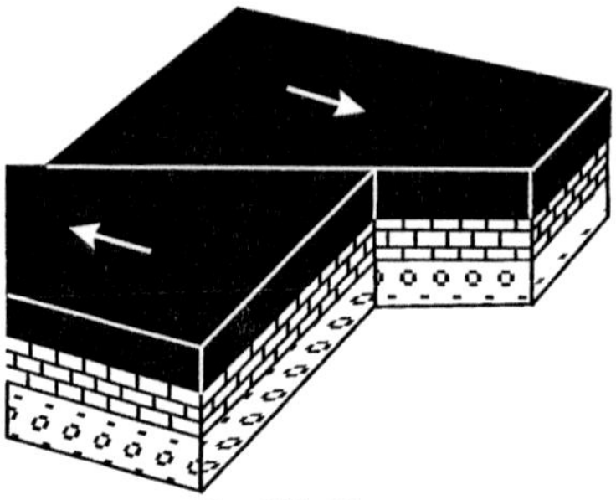

Fig. 289. **Shear**

r искажение форм
Различие очертаний фигур на географической карте и соответствующих им фигур на поверхности земного эллипсида.
f déformation *f* de configuration
s deformación *f* de contornos
g Gestaltenverzerrung *f*

5380 sharecropping
An agricultural tenancy system in which the tenant renders rent to the landlord in the form of produce rather than cash.
r плата урожаем
Система имущественных отношений в сельском хозяйстве, когда арендатор возвращает ренту помещику в виде части полученного им урожая.
f métayage *m*
s división *f* de cosecha;
pagamiento *m* con cosecha
g Ernteverteilung *f*

5381 shear; shear-fault
A fault produced by tangential stress which shears one part of a rock past its adjoining part, along a *shear plane*. → Fig. 289
r сдвиг
Смещение блоков горных пород в горизонтальном направлении по разлому (*плоскости сдвига*). → Рис. 289
f faille *f* de déplacement horizontal
s falla *f* de desplazamiento horizontal
g Verschiebung *f*

5382 shear plain
The surface along which *shearing* occurs during the application of tangential stress to a material.
r плоскость сдвига
Поверхность, вдоль которой происходит сдвиг (*шарьяж*) при воздействии тангенциальных сил на материал.
f plan *m* de déplacement
s plano *m* de desplazamiento
g Verschiebungfläche *f*

*** shear-fault → 5381**

5383 shearing; charriage
The deformation of a solid material caused by the application of tengential stresses so that its layers move laterally over each other.
r шарьяж
Деформация твердого материала, вызванная приложением к нему тангенциальных сил, в результате чего происходит сдвиг вбок одного слоя относительно другого.
f charriage *m*
s corrimiento *m*
g Überschiebung *f*

*** sheep tracks» → 6099**

*** sheep-breeding → 5384**

5384 sheep-husbandry; sheep-breeding
Breeding sheep for foodstuffs and raw materials for light and food industries.
r овцеводство
Отрасль животноводства, занимающаяся разведением овец с целью получения продуктов питания и сырья для легкой и пищевкусовой промышленности.
f élevage *m* ovin
s cría *f* de ovejas; ganadería *f* ovina
g Schafzucht *f*

5385 sheet erosion; sheet wash; surface wash
Very slow erosion, i.e. the detachment and downslope movement on low-gradient slopes of ground particles and aggregates as well as a surface soil by a thin film of water.
r поверхностный смыв; плоскостной смыв
Перенос вниз по склону мелких минеральных частиц (реголита) и поверхнострных горизонтов почвы струйками талых и дождевых вод, образующими сплошную водную пленку.
f érosion *f* pelliculaire; érosion *f* superficielle
s erosión *f* pelicular; derrubio *m*;
desplazamiento *m* superficial;
erosión *f* laminar
g Flächenspülung *f*

5386 sheet line system
The system by which a geographical map series is divided into individual map sheets.
r нарезка карты; разграфка карты
Система разделения многолистной географической карты на отдельные листы.
f découpage *m*
s división *f* en hojas; recorte *m*
g Blatteinteilung *f*

*** sheet of water → 6688**

*** sheet wash → 5385**

*** sheetflow → 4346**

5387 sheeting
The splitting away of rock from the upper surface of a massive rock resulting from the expansion of the rock by the release of pressure.
r расщепление

Откалывание горной породы от верхней поверхности массивных скал в результате ее расширения при воздействии внешнего давления.

f scindement *m*
s hendimiento *m*
g Zerspalten *n*

* **shelf → 1253**

* **shelf break → 5388**

5388 shelf edge; depth dump; shelf break
An external boundary of the continental shelf behind which the continental slope begins and the sea floor sharply deepens.
r граница материковой отмели;
свал глубин;
уступ континентального шельфа
Относительно резкий уступ морского дна, за которым начинается материковый склон и глубины резко возрастают.
f rupture *f* de la pente continentale
s ruptura *f* de la pendiente continental
g Kontinentalrand *m*; Tiefensprung *m*

5389 shelf sea
A marginal sea with the depths usually less than 200 m located within the continental shelf.
r шельфовое море
Окраинное море с глубинами обычно менее 200 м, расположенное в пределах шельфа.
f mer *f* de plate-forme continentale
s mar *m* de plataforma continental
g Schelfmeer *n*

5390 shell limestone
Limestone consisting mainly of comminuted shell fragments and shells.
r ракушечник
Известняк, состоящий преимущественно из мало раздробленных раковин моллюсков.
f lumachelle *f*; calcaire *m* coquillier
s lumaquela *f*; caliza *f* conchífera
g Muschelkalk *m*

5391 shelter belt; forest strip; windbreak
A row of trees and bush vegetation intended for the protection of roads and fields from the wind snow, sand drifts and other dangerous natural phenomena.
r лесополоса; лесозащитная полоса
Широкая полоса лесной и кустарниковой растительности, предназначенная для защиты дорог и полей от снежных и песчаных заносов и других опасных стихийных явлений.
f bande *f* forestière; rideau-abri *m*;
rideau *m* boisé; écran *m* forestier
s franja *f* forestal
g Schutzstreifen *m*; Waldschutzpflanzung *f*

5392 sheltered desert
A desert induced by foehn occurring near mountain ridges, it is typical for Patagonia, Iran, Central Asia, Rocky Mountains in North America.
r закрытая пустыня
Пустыня, лежащая у горных хребтов, связанная с эффектом фёна, распространена в Патагонии, Иране, Центральной Азии, Скалистых горах Северной Америки.
f désert *m* d'abri
s desierto *m* de abrigo
g Deckungswüste *f*

5393 shiblyak
A community of deciduous shrubs and low trees dominated by oaks, hawthorn, brier and etc. in the Mediterranean region.
r шибляк
Сообщество листопадных кустарников и низкорослых деревьев в Средиземноморье, образовано дубами, боярышником, шиповником и др.
f chibliak *m*
s shibliak *m*
g Schibljak *m*

5394 shield
A large outcrop of the old crystalline basement, such as the Fennoscandian Shield of Northern Europe, or the Laurentian Shield of Canada.
r щит
Крупный выход на поверхность древнего кристаллического фундамента, как, например, Фенноскандинавский щит в Северной Европе или Лаврентийский щит в Канаде.
f bouclier *m*
s escudo *m*
g Schild *m*

5395 shield volcano
A volcano shaped like a shield, formed of successive eruptions of free-flowing lava which creates a gently sloping, broad dome tens of kilometres across and more than one kilometre high, e.g. Mauna Loa in Hawaii.
r щитовой вулкан
Вулкан, возникающий при излиянии маловязкой базальтовой лавы, растекающейся на большие расстояния, что приводит к формированию плоского широкого купола поперечником в десятки километров и высотой до 1 км, например вулкан Мауна Лоа на Гавайских островах.
f volcan *m* effusif; volcan-bouclier *m*
s volcán *m* efusivo; volcán-escudo *m*
g Schildvulkan *m*

* **shifting agriculture → 5396**

* **shifting cultivation → 3479**

5396 shifting cultivation; shifting agriculture
Periodical changing of plots of land under cultivation for different reasons (soil deple-

tion, waterlogging, salinization, etc.) and returning to them when the soil recovers its fertility.
r переложное земледелие
Периодическая смена участков, используемых под пашню, практикуемая по разным причинам (истощение почв, переувлажнение, засоление и пр.), и возвращение на них после естественного восстановления плодородия.
f agriculture *f* itinérante; culture *f* itinérante
s agricultura *f* itinerante
g Wanderfeldbau *m*

5397 shifting sands
Large sand massives in desert, without vegetation cover, which have big sand ridges, barchans and dunes, moving in the direction of prevailing winds.
r движущиеся пески
Большие песчаные массивы в пустыне, лишенные растительности, с крупными грядами, барханами и дюнами, передвигающимися в направлении господствующих ветров.
f sables *mpl* déplaçables
s arenas *fpl* movibles
g beweglicher Sand *m*

5398 shingle; pebble
Rock fragments rounded and polished by running water, wind, or waves the size of which is from 1 to 10 cm.
r галька
Окатанные и отшлифованные текучей водой или волнами обломки горных пород размером от 1 до 10 см.
f galet *m*
s guija *f*; canto *m*
g Geröll *n*; Kiesel *m*

5399 shingle bed
An accumulation of loose sedimentary rock consisting mainly of pebbles.
r галечник
Рыхлая крупнообломочная осадочная порода, состоящая в основном из галек.
f couche *f* de galets; cailloutis *m*
s guijarro *m*
g Schottergestein *n*

* **ship canal → 2366**

5400 ship displacement
A measure of capacity of ships, calculated in tonnage.
r водоизмещение судна
Мера емкости судна, измеряемая в тоннах.
f déplacement *m* de vaisseau
s desplazamiento *m* de barco
g Wasserverdrängung *f*; Schiffsmasse *f*

5401 ship turnover
A characteristic of the work of ports calculated by the number of calling ships and their combined gross or net tonnage.
r судооборот
Показатель работы портов, исчисляется числом заходящих и уходящих судов и их суммарной валовой или чистой вместимостью.
f trafic *m* portuaire
s tráfico *m* portuario
g Hafenumlauf *m*

5402 ship waves
Waves generated by a moving ship.
r корабельные волны
Волны, образующиеся при движении судна.
f vagues *fpl* de navire
s olas *fpl* de buque
g Schiffwellen *fpl*

* **shipping → 4076**

5403 shoal
A shallow part of a river, sea, lake; an accumulation of sand, mud, pebbles creating such shallow water and in many cases dangerous to navigation.
r мелководье
Мелкая часть реки, озера, моря; отложения песка, ила, гальки, приводящие к обмелению и во многих случаях очень опасные для судоходства.
f basses eaux *fpl*
s bajos fondos *mpl*
g niedriger Wasserstand *m*; seichte Stelle *f*

5404 shoalbank
A part of a sea floor where the depths are smaller than of neighbouring area.
r банка
Участок морского дна с глубинами значительно меньше окружающих.
f banc *m*
s banco *m* de arena; banco *m*
g Bank *f*

5405 shopping centre
A self-service store consisting of many separate shops and attendant enterprises.
r торговый центр
Торговый комплекс самообслуживания, включающий много отдельных магазинов и сопутствующих предприятий.
f centre *m* commercial
s centro *m* comercial
g Handelszentrum *m*

* **shore → 1087**

5406 shore; seashore
The land adjoining a large body of water or next to the sea.
r побережье
Часть суши, непосредственно прилегающая к крупному водоему или морю.
f rivage *m*; côte *m*
s costa *f*; ribera *f*

g Küste *f*; Uferland *n*

5407 shore ice bar
A ridge of hummocked ice piled up along a gently sloping coast, where the ice has been pushed from the sea.
r береговой вал льда
Торосистое образование, возникшее на пологом берегу в результате напора льда со стороны моря.
f cordon *m* glaciaire de plage
s cordón *m* de hielo costero
g gestrandete Eisbarre *f*; gestrandeter Eiswall *m*

5408 shore lead; flaw lead
A linear opening in sea-ice between the pack ice and the fast ice, or between the floating ice and the shore which forms before the ice in the river breaks up or the fast ice in the sea fractures.
r водяной заберег; закраина; прогалина
Полоса открытой воды между ледяным покровом и берегом, образующаяся перед вскрытием реки или взломом припая в море.
f chenal *m*
s claro *m*; franja *f* de agua
g Durchgang *m* im Festeis; Wasserrand *m*

* **shore platform → 11**

5409 shore protection dike
A concrete or bulk dam aimed at protecting against the sea waves of the shores or the entrance to a canal.
r берегозащитная дамба
Бетонная или насыпная дамба для защиты берега или входа в канал от морского волнения.
f digue *f* protecteur de côte
s dique *f* protector de costa
g Uferschutzdamm *m*

5410 shore slope
A high and usually steep land slope towards the seacoast.
r береговой склон
Высокий и обычно крутой склон суши к берегу моря.
f versant *m* littoral
s vertiente *f* litoral
g Küsten(ab)hang *m*

5411 shoreline; coastline
The line where the land meets the water surface.
r береговая линия
Линия, разграничивающая сушу и водное пространство.
f ligne *f* de rivage; ligne *f* riveraine
s línea *f* de costa
g Küstenlinie *f*; Strandlinie *f*; Uferlinie *f*

5412 shoreline smoothing
The simplification of coastline curvature as a result of erosion and sedimentation.
r выравнивание берега
Уменьшение кривизны береговой линии в результате эрозии и аккумуляции наносов.
f égalisation *f* de côte
s alineamiento *m* de costa
g Küstenausgleich *m*

* **short grass vegetation → 5413**

5413 short herbaceous vegetation; short grass vegetation
A layer or community of short (up to 10-15 cm) mainly herbaceous plants growing under the conditions of a deficient warmth or moisture.
r низкотравье
Ярус или сообщество невысоких (до 10-15 см), преимущественно травянистых растений, произрастающих в условиях недостатка тепла или влаги.
f végétation *f* à herbes courtes
s hierba *f* baja
g Kurzstaudenflur *f*

5414 short wave; deep-water wave
An ocean wave the length of which is less than twice the depth of the water.
r короткая волна
Морская волна, длина которой меньше половины глубины моря, и скорость ее распространения не зависит от этой глубины.
f vague *f* courte
s ola *f* corta
g Kurzwelle *f*; kurzperiodische Welle *f*

5415 short-grass prairie
Steppe-type grass vegetation of the Great Plains of North America.
r низкотравная прерия
Травянистые растительные сообщества степного типа, занимающие Великие Равнины Северной Америки.
f prairie *f* à herbes courtes
s pradera *f* de hierba corta
g Kurzgrasprärie *f*

5416 short-range weather forecast
Weather prediction for a period from 1 to 3 days.
r краткосрочный прогноз погоды
Прогноз погоды на срок от 1 до 3 суток.
f prévision *f* à courte échéance
s predicción *f* a corto plazo; pronóstico *m* a corto plazo
g kurzfristige Vorhersage *f*

5417 short-wave outgoing radiation
Solar radiation reflected, back-scattered and re-emitted by the Earth's surface, atmosphere and cloud tops into space.
r уходящая коротковолновая радиация
Солнечная радиация, отраженная, рассеянная в обратном направлении и переиз-

лученная земной поверхностью, атмосферой и облаками в космос.
f rayonnement *m* sortant à ondes courtes
s radiación *f* saliente de onda corta
g ausgehende Kurzwellenstrahlung *f*; ausgehende kurzwellige Strahlung *f*

5418 **short-wave radiation**
Radiation with wavelengths from 0.3 to about 4 micron.
r коротковолновая радиация
Радиация в интервале длин волн от 0,3 приблизительно до 4 микрон.
f rayonnement *m* de courtes longueurs d'onde
s radiación *f* de onda corta
g Kurzwellenstrahlung *f*; kurzwellige Strahlung *f*

5419 **shott**
A fluctuating shallow brackish or saltwater lake in north Africa, especially in Tunisia and Algeria, dry for much of the year, water-filled in winter and fed from underground sources, in contrast to *sebka*.
r шотт
Мелкое солоноватое или соленое озеро в Северной Африке, особенно в Тунисе и Алжире, сухое бульшую часть года, но зимой заполненное водой из подземных источников, противопоставляется *себхе*.
f chott *m*
s chott *m*
g Chott *m*

* **shoulder → 6385**

5420 **shower; torrential rain**
Precipitation, often short-lived and heavy, falling from convective (Cumulonimbus) clouds, such as rain, hail or snow.
r ливень; проливной дождь
Осадки, часто не продолжительные и обильные, выпадающие из конвективных (кучево-дождевых) облаков в жидком или твердом виде.
f averse *f*; pluie *f* torrentielle; pluie *f* battante
s chaparrón *m*; aguacero *m*; lluvia *f* torrencial
g Schauerregen *m*; Regenguß *m*; Platzregen *m*; Wolkenbruch *m*

* **shrinkage crack → 1555**

* **shrinking of glaciers → 1478**

5421 **shrinking of soil**
The decrease of the soil volume as a result of moisture reduction in dry season which leads to the compaction and fissuring of soil.
r усыхание почвы
Уменьшение объема почвы по мере ее высыхания при отсутствии дождей, что влечет за сбой уплотнение и растрескивание почвы.
f séchage *m* du sol
s secamiento *m* del suelo; encogimiento *m* del suelo
g Bodenvertrocknung *f*

* **shrinkler irrigation → 5735**

5422 **shrub; bush**
A perennial woody plant up to 5-6 m high that branches below or near ground level into several main stems, so has no clear trunk.
r кустарник
Многолетнее деревянистое растение высотой до 5-6 м с главным стволом, выраженным лишь в начале жизни, и последовательно сменяющимися многолетними побегами.
f arbrisseau *m*; frutescente *m*; plante *f* frutescente
s arbusto *m*; planta *f* frutescente; mata *m*; frutescente *m*
g Strauch *m*; Busch *m*

5423 **shrub layer**
In the stratified structure of a forest, the layer of woody plants comprising shrubs and saplings of trees.
r кустарниковый ярус
Ярус кустарников в лесном сообществе.
f strate *f* arbustive
s estrato *m* arbustivo
g Strauchschicht *f*

5424 **shrubbery**
A plant community formed mainly of shrub and subshrub, e.g. osiery, low birch thicket, maquis, garrigue, shibliak, etc.
r кустарниковое сообщество
Растительное сообщество, состоящее в основном из кустарника и полукустарника, например, ивняк, ерник, маквис, гаррига, шибляк и пр.
f communauté *f* arbustive
s arbustal *m*
g Strauchgesellschaft *f*

* **shuffled ice → 4877**

5425 **shuga; frazil**
An accumulation of spongy lumps of white ice on the surface of water.
r шуга
Скопление на поверхности воды пористых белых комков внутриводного льда.
f shuga *f*; sorbet *m*
s hielo *m* friable
g Eisbrei *m*; Matscheis *n*

5426 **shuga-drift**
The movement of shuga on the surface of water or inside the water flow.
r шугоход
Движение шуги в поверхностном слое или внутри водного потока.
f passage *m* de sorbet

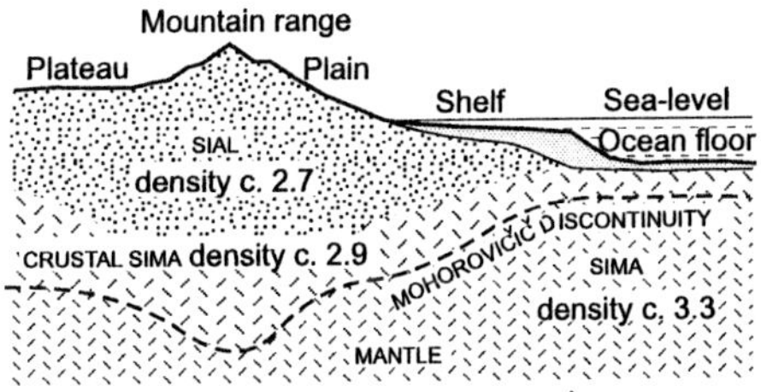

Fig. 290. Two main parts of the Earth's crust: **sial** and **sima**

s movimiento *m* de hielo friable
g Eisbreidrift *f*

5427 sial
That part of the Earth's continental crust which is composed of granitic material rich in silica and alumina, light in colour and having lower density (between 2.65 and 2.70 kg/m^3) than the *sima*. → Fig. 290
r сиаль
Часть континентальной земной коры, состоящая из гранитного материала, богатого кремнием и алюминием, с плотностью меньшей (от 2,65 до 2,70 кг/м^3), чем плотность *симы*. → Рис. 290
f sial *m*
s sial *m*
g Sial *n*

5428 siallitic soil
A soil containing a medium to high ratio of silica to alluminium, it is uncommon in zones of tropical weathering but is typical of the temperate zone.
r сиаллитная почва
Почва, содержащая значительное количество алюмосиликатов, редко встречается в тропических областях, но обычна в умеренных широтах.
f sol *m* siallitique
s suelo *m* sialítico
g sialitischer Boden *m*

5429 siallitization
Partial hydrolysis of aluminosilicate rocks with resulting formation of new stratified clay silicates (kaolinite, etc.), it occurs in soils of temperate climate under medium amount of precipitation.
r сиаллитизация
Почвенный процесс частичного гидролиза алюмосиликатных пород, при котором образуются новые глинистые слоистые силикаты (каолинит и др.), происходит в условиях умеренного климата со средним количеством осадков.
f siallitisation *f*
s sialitización *f*
g Siallitisation *f*

* **Siberian anticyclone → 5430**

5430 Siberian high; Siberian anticyclone
A persistent area of high pressure during winter in the north of Central Asia.
r сибирский антициклон
Устойчивый антициклон, расположенный зимой над северной частью Азии.
f anticyclone *m* sibérien
s anticiclón *m* de Siberia
g sibirisches Hoch *n*

* **side bar → 4635**

* **side moraine → 2172**

5431 sidereal day
The time interval equivalent to one complete rotation of the Earth in relation to the stars, its length of 23 hours 56 min 4.09 sec is almost 4 min shorter than that of a mean solar day.
r сидерические сутки
Промежуток времени, эквивалентный одному полному обороту Земли по отношению к звездам, продолжаются 23 часа 56 мин 4,09 сек, что почти на 4 мин короче, чем средние солнечные сутки.
f jour *m* sidéral
s día *m* sideral
g siderischer Tag *m*

5432 sidereal month
The time interval during which the Moon makes a complete revolution around the Earth and takes the initial position with regard to stars, it lasts for 27.3217 average solar days.
r сидерический месяц; звездный месяц
Промежуток времени, в течение которого Луна совершает полный оборот вокруг Земли и занимает исходное положение относительно звезд, продолжается 27,3217 средних солнечных суток.
f mois *m* sidéral
s mes *m* sideral
g siderischer Monat *m*

5433 sidereal year
The average time taken by the Earth to complete one revolution in its orbit round the sun measured with respect to the fixed stars, it is equal to 365.2564 average solar days, i.e. 365 days, 6 hours, 9 min and 9,54 sec of solar time.
r сидерический год
Промежуток времени, в течение которого Солнце совершает видимый годичный путь по небесной сфере относительно звезд, его продолжительность 365,2564 средних солнечных суток.
f année *f* sidérale
s año *m* sideral
g siderisches Jahr *n*

* **sierozem → 5434**

5434 sierozem; serozem; grey earth; grey desert soil
A soil that forms on piedmont inclined plains under herbaceous vegetation of desert-steppe

(semi-arid) zone, it has light-grey humus horizon and clayed illuvial-carbonate horizon as well as a layer of accumulation of gypsum and easily soluble salts.
r серозем
Почва, формирующаяся на подгорных наклонных равнинах под травянистой растительностью в пустынно-степной зоне, имеет светло-серый гумусовый и оглиненный иллювиально-карбонатный горизонты, а также слой с накоплением гипса и легко растворимых солей.
f siérozem *m*; sol *m* gris désertique
s sierozem *m*; serozem *m*; suelo *m* gris de desierto
g Grauboden *m*; Serosem *m*; grauer Wüstenboden *m*

5435 sierra
Mountains with jagged crests in the Spanish-speaking countries of Europe and Latin America.
r сьерра
Остроконечные горные хребты в испаноязычных странах Европы и Латинской Америки.
f sierra *f*
s sierra *f*
g Sierra *f*

5436 sieve map
A series of maps drawn on transparent material, each showing the distribution of a selected factor, used by superimposing which allows to «sieve out» the factors needed.
r факторные карты
Серия карт, составленных на прозрачной основе, каждая из которых содержит информацию об определенном факторе, они используются путем наложения друг на друга и выбора необходимых для анализа факторов.
f carte *f* de criblage
s mapa *m* de cribado
g Siebkarte *f*

5437 silage
Green crops, such as grass and clover, which are compressed, fermented, and stored for use as animal foddes.
r силос
Зеленый корм для скота (включающий траву, клевер, люцерну, зеленую кукурузу), сохраняемый в силосном хранилище или упакованный в блоки для скармливания скоту по мере надобности.
f silo *m*
s silo *m*; forraje *m* ensilado
g Silofutter *n*

5438 silica
The silicate mineral (silicon dioxide) content of a rock, related to the largest group of compounds in the Earth's crust, among the several varieties comprising the silica group of quartz, chalcedony and opal.
r кремний
Силикатный минерал (диоксид кремния), широко распространенный в земной коре, среди нескольких его разновидностей кварцевая группа, халцедон и опал.
f silicium *m*
s silíceo *m*
g Silizium *m*

5439 silicate
A group of minerals based around the highly stable SiO_4, constituing some 95 per cent of the Earth's crust, and including clay minerals, feldspar, quartz etc. as members.
r силикат
Группа минералов, состоящих в основном из SiO_4, составляющих до 95% земной коры и включающих глинистые минералы, полевой шпат, кварц и др.
f silicate *m*
s silicato *m*
g Silikat *n*

5440 silicification
The process by which silica replaces existing structures or minerals within an organic and inorganic bodies being generated either by groundwater solutions or from igneous sources.
r окремнение
Процесс, при котором кремний замещает существующие структуры и минералы внутри органических и неорганических тел, поступая с грунтовыми водами и продуктами вулканических извержений.
f silicification *f*
s silicificación *f*
g Silifizierung *f*; Verkieselung *f*

* **silk culture → 5354**

5441 sill
An intrusion of igneous rock of tabular form which spreads along the bedding planes of sedimentary or volcanic formation, in a nearby horizontal sheet concordant with the strata.
r пластовая интрузия
Интрузия изверженных пород столовой формы, когда очень жидкая магма внедряется между пластами осадочной или вулканической формации, т.е. она залегает согласно с существующими напластованиями.
f sill *m*; filon-couche *m*
s sill *m*; intrusión *f* en forma de capa; filón-capa *m*
g Schichtintrusion *f*

* **silt → 4001**

5442 silt delta
A river delta formed at the mouths half-dosed from the sea by the bar and not advancing seawards over the bar.
r дельта выполнения

Речная дельта, формирующаяся на полузакрытых устьевых взморьях и не выступающая за пределы устьевого бара.

f delta *m* de comblement

s delta *m* de aluvión; delta *f* de limo

g Schlammdelta *n*

5443 siltation; silting

The accumulation of fine-grained material of silt in a body of water and on river flood-plain when the flood water is quiet; method of improving sandy soil by adding silt.

r заиление

Отложение приносимых водой наносов на дне водоемов, водотоков и на речной пойме во время паводков; метод мелиорации песчаных земель путем внесения ила.

f envasement *m*; comblement *m*

s enfangamiento *m*;
cubrimiento *m* de fango; elodamiento *m*

g Verschlammung *f*

* **silting → 5443**

5444 silting area

An area of shallowing and rising the bottom of water bodies as a result of silt accumulation.

r зона заиления

Участок обмеления и повышения дна водоемов в результате процесса заиления.

f zone *f* de envasement

s zona *f* de enfangamiento

g Verlandungszone *f*

5445 silty loam

A loam which contains between 20 and 50% sand, less than 30% clay and a number of silty particles, this mixture produces a friable soil with some permeability and with particles of different sizes.

r бурый суглинок

Суглинок, содержащий 20-50% песка, менее 30% глины и большое количество мелкоземистых частиц, что формирует более или менее рыхлую почву с ограниченной проницаемостью и частицами разного размера.

f limon *m* de sable fin

s limo *m* de arena fina

g feiner Lehmboden *m*

5446 Silurian

The third geological period of the Palaeozoic era which began 440 million years ago and lasted for 45 million years. → Table. 1, p. 284

r силур

Третий снизу период палеозойской эры геологической истории Земли, начался 440 млн лет назад и продолжался 45 млн лет. → Табл. 1, с. 284

f Silurien *m*

s Silúrico *m*

g Silur *n*

5447 sima

That part of the Earthrs oceanic crust which is composed mostly of silica and magnesium, relatively heavier in density than the sial. → Fig. 290, p. 649

r сима

Часть океанической земной коры, состоящая в основном из кремния и магния, имеющая бóльшую плотность, чем сиаль. → Рис. 290, с. 649

f sima *m*

s sima *m*

g Sima *n*

5448 simoon; samoon; samun

A dry, hot, and dust-laden wind occurring in the northern Sahara and the deserts of Arabia.

r самум

Сухой горячий и пыльный ветер в Сахаре, странах Ближнего Востока и пустынях Аравии.

f simoun *m*

s simún *m*

g Samum *m*

5449 simulation model

Either a physical construction (e.g. scale model of relief) or a mathematical model constructed for the purposes of experimentation into workings of a system.

r имитационная модель

Физическая конструкция (например, уменьшенная модель рельефа) или математическая модель, созданная в целях системного моделирования.

f modèle *m* de simulation

s modelo *m* de simulación

g Simulationmodell *n*

5450 sink-hole; swallow-hole

A funnel-shaped depression, a hollow in which surface water collects and escapes through a deep vertical hole or opening in th earth, especially one produced by the solution of rocks in limestone country, down which a surface stream or rainfall disappears. → Fig. 291

r провал

Воронкообразная депрессия, глубокая открытая вертикальная полость, особенно часто встречающаяся в карстовых областях, в которой нередко исчезает река и потоки дождевых вод. → Рис. 291

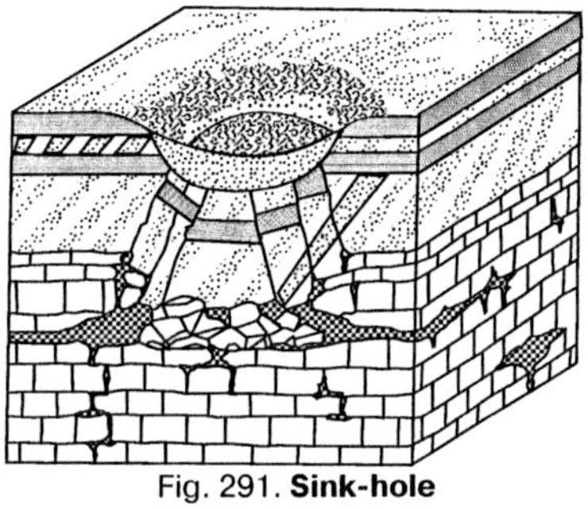

Fig. 291. **Sink-hole**

f dépression *f* fermée de éffondrement
s depresión *f* cerrada de desmoronamiento; hundimiento *m* natural del terreno
g Bruchstelle *f*; Einsturzstelle *f*

* **sinking → 1687**

* **sinking velocity → 2092**

5451 **sinter; flowstone**
A deposit of silica formed around a geyser or hot spring, the material having been previously held in solution in the water.
r натёк
Кремнистые отложения около гейзера или горячего источника, выпадающие в осадок при излиянии воды.
f sinter *m*
s incrustación *f* sobre rocas; sinter *m*
g Tropfstein *m*; Sinter *m*

5452 **sinuosity of river**
The twisting character of a river channel caused by meandering, i.e. the ratio of the distance between two points along the river flow including the lengh of all its meanders to the shortest distance between the same points.
r извилистость реки; гидрографическая извилистость
Извилистый характер речного русла, возникающий вследствие меандрирования, т.е. отношение длины реки между двумя точками течения со всеми ее излучинами к кратчайшему расстоянию между теми же точками.
f sinuosité *f* de fleuve
s sinuosidad *f* del río
g Windungsgrad *m* eines Flusses; Sinuasität *f* eines Flusses

5453 **sinuosity of shoreline**
The relation of the real length of a shoreline with all its curves to that of its direct line.
r извилистость береговой линии
Отношение реальной длины береговой линии со всеми ее изгибами и извилинами к спрямленной линии побережья.
f sinuosité *f* de ligne de rivage
s sinuosidad *f* de línea de costa
g Ausbuchtung *f* der Uferlinie

5454 **siphon**
A knee-shaped bend in the vertical plane of the channel of an underground karst river. → Fig. 292

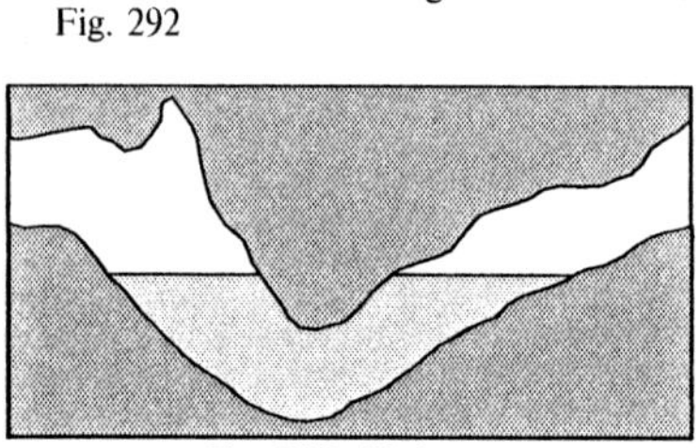

Fig. 292. **Siphon**

r сифон
Коленообразный изгиб в вертикальной плоскости канала подземной карстовой реки. → Рис. 292
f siphon *m*
s sifón *m*
g Siphon *m*

5455 **siphon spring**
A types of intermittent spring in a karstic terrain which operates from time to time when karst cavity and siphon canal connecting the cavity with the Earthrs surface get filled with water.
r сифонный источник
Источник подземных вод, действующий периодически после наполнения водой карстовой полости и сифонного канала, соединяющего полость с земной поверхностью.
f source *f* à siphon
s fuente *m* de sifón
g Siphonquelle *f*

5456 **sirocco; scirocco**
A very hot, humid southerly wind which blows in the Mediterranean Sea basin.
r сирокко
Теплый влажный южный ветер в бассейне Средиземного моря.
f sirocco *m*
s siroco *m*
g Scirokko *n*

5457 **site**
Fixed position where an object, structure or tissue is placed or where something occurs, e.g. the position on the ground of a place, town, building, etc. in the past, present or future.
r местоположение
Фиксированное положение объекта, структуры или сети, например, положение на земной поверхности какого-либо места, города, сооружения и пр. в настоящее время, в прошлом и возможном будущем.
f site *m*
s sitio *m*
g Lage *f*

5458 **site of ancient town**
The rests of ancient fortified town of primitive man.
r городище
Остатки древнего укрепленного поселения.
f vestiges *mpl* d'une ville
s vestigios *mpl* de una ciudad
g prähistorische Siedlung *f*; vorgeschichtliche Siedlung *f*

5459 **skeletal soil**
A soil with a high content of rock debris mixed with fine earth, increased aeration and water permeability are its distinct features.
r скелетная почва; щебнистая почва

Почва с высоким содержанием щебня, смешанного с мелкоземом, отличается повышенной аэрацией и водопроницаемостью.

f sol *m* squelettique
s suelo *m* esquelético
g Skelettboden *m*; Schotterboden *m*

5460 skerries
lnes of low, small rocky islands near a sea coast dissected in a complicated manner, found in the areas of Pleistocene glaciation. → Fig. 60, p. 130
r шхеры
Небольшие скалистые острова и подводные скалы у сложно расчлененных морских берегов в областях плейстоценового оледенения. → Рис. 60, с. 130
f skajars *fpl*
s escollos *mpl*
g Schären *fpl*

5461 skerry-guard
The area of calm water between the line of skerries and the mainland.
r тишь за шхерами
Область тихой воды между линией шхер и коренным берегом.
f skargaard *m*
s skargaard *m*
g Schärenschutz *m*

*** sketch map → 5243**

5462 sketching; outline$_2$
A sketch map representing the terrain in a simplified form with all its features.
r абрис; кроки
Схематический чертеж участка местности со всеми находящимися здесь предметами.
f esquisse *f*; schéma *m* topographique; croquis *m*; minute *f* topographique
s esquema *m* topográfico; matriz *f*; minuta *f* topográfica; croquis *m*
g Geländeskizze *f*; Kroki *n*; Umriss *m*

5463 skin-layer
A thin surface layer of sea water which actively accumulates pollution and impeding exchange of the ocean with the atmosphere.
r скин-слой
Тонкий поверхностный слой морской воды, активно аккумулирующий загрязнения и препятствующий обмену океана с атмосферой.
f peau-couche *f*
s piel-capa *f*
g Skinschicht *f*

5464 sky
The atmosphere enveloping the Earth, with or without clouds.
r небо
Атмосфера, окружающая Землю, с облаками или без облаков.
f ciel *m*
s cielo *m*
g Himmel *m*

*** sky radiation → 1600**

*** skylab → 5698**

5465 skylight
The duffused and reflected light of the sky.
r свечение неба
Рассеянный и отраженный свет неба.
f lumière *f* de ciel
s luz *f* de cielo
g Himmelblau *n*

5466 skyline; local horizon; apparent horizon
A line of a visible contact of sky with surrounding natural or man-made objects, e.g. skyline of New York.
r местный горизонт
Линия пересечения небесного свода с окружающими природными или созданными человеком объектами.
f horizon *m* local
s horizonte *m* local
g lokaler scheinbarer Horizont *m*

5467 slab avalanche
A snow avalanche resulted from a simultaneous rupture of a coherent mass of snow slab over extended area.
r лавина из снежной доски
Снежная лавина, возникшая в результате одновременного разрушения снежной доски на значительной площади.
f avalanche *f* en plaque
s avalancha *f* en placa
g Schneebrettlawine *f*

*** slack → 5468**

*** slack tide → 5468**

5468 slack water; slack; slack tide
The state of tidal current when it is almost motionless, usually reached in the intermediate period between ebb and flood currents.
r кроткая вода; смена приливных течений
Состояние моря, когда приливо-отливные течения почти прекратились, обычно близко к моменту полной или малой воды, когда нет ни прилива, ни отлива.
f eau *f* immobilisée
s agua *f* inmobilizada
g Stauwasser *n*

5469 slag
Solid waste in the form of caked or fused particles of ash left after using coal in furnaces.
r шлак
Твердые отходы в виде частичек золы, спекшейся или сплавленной в куски, при использовании в топках каменных углей.

f scorie *f*
s escoria *f*
g Schlacke *f*

* **slag heap → 6630**

* **slag-heap → 3875**

* **slash and burn cultivation → 5470**

5470 slash-and-burn; slash and burn cultivation; swidden farming
A primitive system of agriculture when crops are grown on a land cleared from the forest and tge rest of the vegetation is fired, when the fertility of the soil is exhausted in 2-3 years, the land is abandoned and another clearing is made.
r подсечно-огневое земледелие
Примитивная система земледелия, когда на освобожденных от леса вырубкой и поджогом землях 2-3 года выращивают сельскохозяйственные растения, а затем, после утраты плодородия, участок забрасывают и осваивают новый.
f culture *f* sur brûlis
s cultivo *m* de rozas
g Brandwirtschaft *f*

* **slate → 5471**

5471 slatestone; slate
A fine-grained metamorphic rock with almost parallel location of elongated rock-forming minerals, it has a property of being split into thin layers.
r сланец
Мелкозернистая метаморфическая горная порода, обладающая способностью раскалываться на тонкие пластинки.
f schiste *m*
s esquisto *m*
g Schiefer *m*

5472 slavery
A form of exploitation, the system by which people are owned by other people as slaves.
r рабство
Форма эксплуатации, при которой раб наряду с орудиями производства является собственностью своего хозяина.
f esclavage *m*
s esclavitud *f*
g Sklaverei *f*; Knechtschaft *f*

5473 sleet; rain and snow mixed
Precipitation consisting of wet snow and rain mixed, or of partially thawed snow, usually developing when temperatures are just above freezing point.
r дождь со снегом; снег с дождем
Осадки, состоящие из смеси дождя и мокрого снега, обычно выпадают при температуре, близкой к точке замерзания.
f giboulée *f*
s aguanieve *f*
g Schneeregen *m*

5474 slick
A light mirror-like smooth spot or band on the surface of a sea or lake wavy surface which are probably caused by spiral-like whirls with horizontal axes.
r штилевое пятно; слик
Светлые зеркальные пятна или полосы на волнистой поверхности моря или озера, вызываемые, вероятно, спиралеобразными вихрями с горизонтальными осями.
f slic *m*
s slic *m*
g Lichtstelle *f*; Lichtblick *m*

* **slide → 3406**

5475 sliding horizon
A layer of loose recrystallized snow of small cohesion along which avalanches move down.
r горизонт скольжения
Слой рыхлого перекристаллизованного снега малой связности, по которому сходят лавины.
f couche *f* de glissement
s capa *f* de deslizamiento
g Gleitschicht *f* im Schnee

5476 slip
1. The actual relative movement along the fault plane, either in the direction of the strike, termed strike-slip, or in the direction of the dip of the fault plane, termed dip-slip. 2. The landslide in which a mass of rock or surface debris moves as a whole down a slip-plane.
r сползание
1. Сдвиг горных пород вдоль плоскости разлома как по простиранию слоев, так и по их падению. 2. Оползень, в котором масса горных пород или скальных обломков движется по поверхности сдвига как единое целое.
f glissement *m*
s deslizamiento *m*
g Rutschen *n*

5477 slip-off slope
The gentle slope on the convex inner curve of a meander opposite to the steep bank of a river cliff on the concave curve (outer side).
r пологий склон излучины
Мягкий склон на выпуклой внутренней стороне речной излучины напротив крутого берега реки на ее вогнутой внешней стороне.
f pente *f* douce de méandre
s pendiente *f* suave de meandro
g sanfter Abhang *m* der Flußwindung

5478 slope; hillslope
An inclined surface, the gradient of which is determined by the amount of its inclination from the horizontal, and the length of which is determined by the inclined distance between its crest and its foot.

r склон; косогор; откос
Наклонный участок земной поверхности, угол наклона которого определяется отклонением от горизонтали, а длина – наклонным расстоянием между гребнем склона и его подножьем.
f versant *m*; pente *m*
s vertiente *f*; pendiente *f*
g Abhang *n*; Hang *m*; Böschung *f*; Leite *f*

* **slope angle → 3054**

5479 **slope decline; downwearing**
A type of slope evolution in which the upper part of a slope decreases in angle as a result of denudation and accumulation of material at the foot of the slope.
r выполаживание склона
Снижение и уменьшение угла наклона склона под воздействием денудации его верхней части и аккумуляции материала у подножья.
f aplanissement *m* de versant
s aplanamiento *m* de vertiente
g Hangverflachtung *f*

* **slope exposure → 2066**

5480 **slope glacier**
A glacier covering a large area on a weakly dissected gentle mountain slope.
r склоновый ледник
Ледник, занимающий обширное пространство слабо расчлененного горного склона.
f glacier *m* de pente
s glaciar *m* de pendiente
g Hanggletscher *m*

5481 **slope of repose; repose slope**
The maximum permissible angle of incidence of a slope when loose deposits composing it are in equilibrium (do not crumble).
r угол естественного откоса; естественный откос
Предельная крутизна склона, при которой слагающие его рыхлые отложения находятся в равновесии (не осыпаются).
f angle *m* de repos
s ángulo *m* de talud natural
g natürliche Böschung *f*; natürlicher Böschungswinkel *m*

5482 **slope processes**
The processes effecting the profile and evolution of slopes: rock falls, landslides, etc.
r склоновые процессы
Совокупность процессов, воздействующих на форму и эволюцию склонов: горные обвалы, оползни и т.п.
f processus *mpl* de versant; modelé *m* des versants
s modelado *m* de vertientes
g Hangprozesse *mpl*

5483 **slope profile**
The two-dimensional form of a hillslope down the steepest gradient.
r профиль склона
Форма склона в поперечном разрезе, т.е по линии падения.
f profil *m* de versant
s perfil *m* de vertiente
g Hangprofil *n*

* **slope retreat → 5054**

5484 **«slope-direction tick»**
A tick at a contour line on map indicating the direction of the exposition of slope.
r бергштрих
Указатель (черточка) направления склона на карте с изображением рельефа горизонталями.
f flèche *f* de cuvette
s flecha *f* de depresión; señal *f* de sentido de pendiente
g Fallstrich *m*

* **slope-wash → 1127**

5485 **sludge; spent slurry; slurry**
Insoluble material obtained in the process of sewage treatment which contains up to 60-70% of mineral and organic matter.
r шлам
Извлеченный из сточных вод в процессе их очистки илистый осадок, содержащий до 60-70% минеральных частиц и органического материала.
f boue *f*; schlamm *m*
s barro *m*; fango *m*; lodo *m*
g Schlamm *m*; Klärschlamm *m*

5486 **slum; squatter settlement; shanty town**
An area of unauthorized housing generally at the edge of a city with houses made from available cheap material such as packing cases, metal cans etc.
r трущобы; бидонвиль
Бедные кварталы, как правило, на городской окраине, построенные из подручных средств и не всегда официально разрешенные.
f bidonville *m*; favelle *f*; taudis *m*
s favela *f*; chabola *f*; chabolismo *m*
g Elendsviertel *n*; Bidonville *n*

5487 **slum clearance**
The demolition of substandard housing accompanied by rehabilitation and redevelopment.
r расчистка трущоб
Снос бедных нестандартных построек, сопровождаемый реабилитацией территории и городским обновлением.
f déblaiement *m* de bidonville
s limpieza *f* de favela
g Aufräumen *n* des Elendsviertels

* **slump → 5488**

5488 **slumping; slump**
The downward, usually distinctly rotational, slipping under gravity of a mass of rock, torn away from its base, over a curved slip fault, leaving a scar on the slope surface.
r оползание; осов
Скольжение масс горной породы под действием силы тяжести вниз по вогнутому склону на участке разлома, приобретающее заметный вращательный характер и оставляющее глубокий след на поверхности склона.
f glissance *f*
s resbalón *m*
g Abrutschen *n*; kleiner Erdrutsch *m*

* **slurry → 5485**

5489 **slush**
A water-saturated snow, the mixture of snow with water caused by rain or thaw, when the water fills all pores in snow and destroys the bonds between snow grains.
r водоснежная смесь; снежура; слякоть; снежная каша
Смесь снега с водой, когда все поры снежного покрова заполнены ею, что приводит к потере связей между снежными зернами.
f neige *f* fondante
s nieve *f* parcialmente fundida
g Schneebrei *m*; Schneeschlamm *m*

* **slush avalanche → 5491**

* **slush spring thaws → 5279**

5490 **slush zone; infiltration zone**
The zone on a glacier where amount of melt water exceeds half of the annual accumulation but the firn is retained due to accumulation of previous more cold and more snowy years.
r зона промачивания; инфильтрационная зона
Зона на леднике, где стаивает более половины годового снегонакопления, но фирн сохраняется за счет накопления предыдущих более холодных и снежных годов.
f zone *f* d'infiltration
s zona *f* de infiltración
g Infiltrationszone *f*

5491 **slush-flow; snow mudflow; slush avalanche**
A mudflow carrying a water-snow mixture with rock fragments.
r водоснежный поток; снежный сель
Селевой поток, представленный смесью снега с водой, а также обломками горных пород.
f cours *m* d'eau et de neige
s alud *m* de agua y nieve
g Wasser-Schnee-Strom *m*

5492 **small circle**

Fig. 293. **Small floes** on the surface of polar sea

A circle on the Earth's surface whose plain does not pass through the centre of the Earth (in contrast to the plane of a great circle), all the lines of latitude but the equator are small circles.
r малый круг
Круг на земной поверхности, плоскость которого не проходит через центр Земли (в противоположность плоскости большого круга), все параллели кроме экватора – это малые круги.
f cercle *m* petit
s círculo *m* pequeño
g kleiner Kreis *m*; Kleinkreis *m*

5493 **small floe; broken ice**
A fragment of ice smaller than 100 m across. → Fig. 293
r битый лед
Льдины до 100 м в поперечнике. → Рис. 293
f glaçon *m* flottant; banc *m* de glace; floe *m* petit
s bloque *m* de hielo flotante; floe *m* pequeño
g kleine Eisschollen *fpl*

* **small hail → 2978**

5494 **small river**
A river with the basin area less than 2,000 km^2 usually where the whole river basin is situated in the same geographical zone.
r малая река
Река, бассейн которой не превышает 2000 $км^2$ и обычно располагается в одной географической зоне.
f rivière *f* petite
s río *m* pequeño
g kleiner Fluß *m*; Kleinfluß *m*

5495 **small town**
A settlement officially regarded as a town with a relatively small number of dwellers dealing with non-agricultural activities.
r поселок
Сравнительно небольшое поселение с преобладанием производства несельскохозяйственного профиля, иногда официально отнесенное к категории городов.

f bourg *m*
s pequeña ciudad *f*; pueblo[2] *m*
g Städtchen *n*; Ortschaft *f*

5496 small wood; copse; coppice
A small area of woodland separated by glades from other forest areas, or a tract of thin forest in a denser woodland.
r перелесок
Небольшой массив леса, отделенный полянами от других лесных участков, либо редкий лес между густыми лесными массивами.
f boqueteau *m*; petit bois *m*
s bosquecillo *m*; bosquete *m*; bosque *m* pequeño
g Waldstück *n*; Wäldchen *n*

* **small-grained** ***adj*** **→ 2150**

5497 small-leaved forest; parvifoliate forest
The deciduous forest where different species of birch and aspen are dominant, as a rule, it is a secondary forest growing in place of broad-leaved and coniferous forests.
r мелколиственный лес
Летнезеленый лес с господством различных видов березы и осины, как правило, вторичный – на месте широколиственных и хвойных лесов.
f forêt *f* parvifoliée
s bosque *m* parvifoliado
g kleinblätteriger Wald *m*

5498 small-scale map
A geographical map on the scale less than 1:1,000,000.
r мелкомасштабная карта
Географическая карта масштаба мельче 1:1 000 000.
f carte *f* à petite échelle
s mapa *m* a pequeña escala
g kleinmaßstäbliche Karte *f*

5499 smog
The fog having a high pollution content (a combination of smoke and fog), occurring in large cities and industrial areas. → Fig. 294
r смог
Туман с высокой степенью загрязнения (смесь дыма и тумана), характерный для больших городов и промышленных центров. → Рис. 294
f smog *m*; brouillard *m* épais
s smog *m*
g Smog *m*

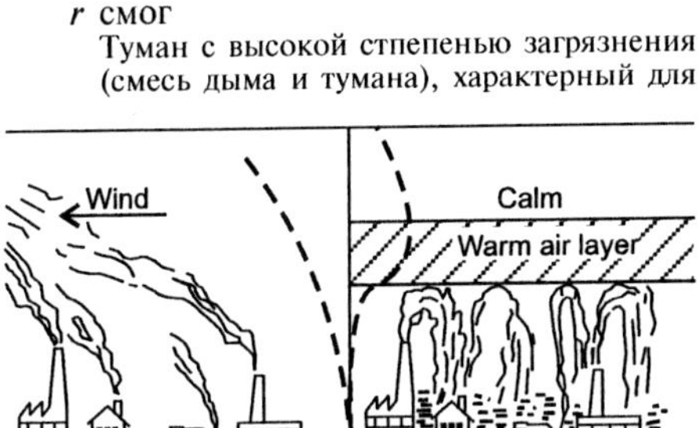

Fig. 294. Formation of **smog**: left – in the normal conditions, right – in the condition of inversion above settlement

5500 smoke
The gaseous products of burning carbon-containing materials made visible by the presence of small particles of carbon, a suspension of the particles in the air, formed from incomplete combustion.
r дым
Газовые продукты горения углеродосодержащих материалов, видимые из-за присутствия мельчайщих частичек углерода, взвесь частиц в воздухе, образованных от неполного сгорания.
f fumée *f*
s humo *m*
g Rauch *m*

5501 smoke haze
The cloudiness of air in the lower atmosphere caused by the particles of soot and smoke from forest, peat, and steppe fires as well as human economic activity.
r дымная мгла
Помутнение воздуха в нижних слоях атмосферы, вызванное повышением содержания частичек гари и дыма от лесных, торфяных, степных пожаров, а также хозяйственной деятельности.
f brume *f* de pollution; brouillard *m* de pollution
s calina *f* fumosa; velo *m*
g Rauchschleier *m*

5502 snag
A tree or big branch embedded in a lake or stream bed and constituting a danger to navigation.
r топляк
Сплавная древесина, препятствующая плаванию по реке при образовании скоплений и задержки другого плывущего материала, либо вследствие частичного заякоривания на дне реки, когда стволы торчат вверх и почти достигают поверхности воды.
f souche *f*
s tronco *m* hundido
g Baumknorren *m*; Senkholz *n*; Stumpf *m*

5503 snow
Precipitation in the form of delicate, featherlight, various patterned ice crystals, often aggregated into snowflakes, formed when water vapour in the atmosphere condenses quickly at the temperature below freezinf point.
r снег
Осадки в форме ледяных кристаллов самой разной формы или их скоплений в виде снежных хлопьев, формирующихся путем быстрой конденсации атмосферных водяных паров при отрицательных температурах.

f neige *f*
s nieve *f*
g Schnee *m*

*** snow depth → 1537**

*** snow thickness → 1537**

5504 snow accumulation; cumulation of snow
The accumulation of snow on the ground, soil or ice.
r снегонакопление; накопление снега
Накопление снега на поверхности грунта, почвы или льда.
f accumulation *f* de la neige
s acumulación *f* de nieve
g Schneeakkumulation *f*; Schneezuwachs *m*

5505 snow adhesion
A bond between the particles of snow and the surface of a slope, preventing the snow motion.
r адгезия снега
Связь между частицами снежного покрова и поверхностью склона, препятствующая смещению снега.
f adhésion *f* de la neige
s adhesión *f* de nieve
g Schneeadhasion *f*

5506 snow alimentation; feeding by snowmelt
The incoming of surface and ground waters into a water body determined by snowmelt.
r снеговое питание
Поступление в водный объект поверхностных и грунтовых вод, обусловленное снеготаянием.
f alimentation *f* neigeuse; alimentation *f* nivale
s alimentación *f* de nieve
g Speisung *f* durch Schneeschmelzwasser

*** snow and ice penitent → 4474**

5507 snow and ice resources
The storage of moisture in all forms of natural ice and snow on the Earth's surface, in the lithosphere and hydrosphere.
r снежно-ледовые ресурсы
Запасы влаги во всех видах природных льдов и снега на поверхности земли, в литосфере и гидросфере.
f ressources *fpl* en neige et glace
s recursos *mpl* de nieve y hielo
g Schnee- und Eisressourcen *fpl*

*** snow avalanche → 478**

5508 snow barchan
An isolated crescentic dune-form deposition of snow resembling a sand barchan with the horns directed with wind, created by wind-driven snow in polar regions, where wind direction is fairly constant. → Fig. 295
r снежный бархан
Скопление отложенного снега серповидной формы с рогами по направлению ветра, напоминающее песчаный бархан, встречающееся чаще всего в полярных районах, где ветер отличается устойчивой направленностью. → Рис. 295
f barkhane *f* de neige
s barkhana *f* de nieve; barjana *f* de nieve
g Schneebarchan *m*

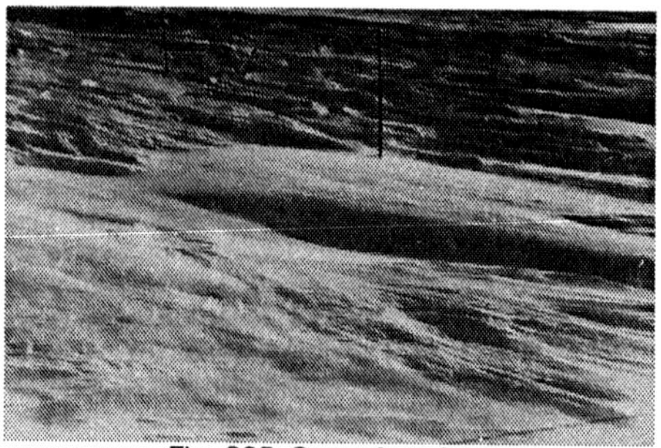

Fig. 295. **Snow barchan**

5509 snow blindness
Impared vision or temporary blindness caused by sunlight reflected from fresh snow usually in Arctic regions or in high mountain areas.
r снежная слепота
Временная (полная или неполная) потеря зрения, вызванная действием отраженного солнечного света от чистой и сухой поверхности снежного покрова.
f éblouissement *m* nival
s ceguera *f* de nieve
g Schneeblindheit *f*

5510 snow bridge
A snow cover over a crevasse formed in winter.
r снежный мост
Перекрытие зимой ледниковой трещины снегом, нанесенным метелью.
f pont *m* de neige
s puente *m* de nieve
g Schneebrücke *f*

5511 snow cannon
A device in the form of a cannon which produces artificial snow by spraying frozen water droplets, it is commonly used to prepare the runs for skiing.
r снежная пушка
Установка в виде пушки, распыляющая искусственный снег, используемый для подготовки лыжных трасс.
f canon *m* de neige
s cañon *m* de nieve
g Schneekanone *f*

5512 snow concrete
A mixture of snow and water, rammed and frozen.
r снегобетон
Утрамбованная смесь снега с водой.
f béton *m* de neige
s hormigón *m* de nieve

Table 3. Main shapes of **snow crystals**

English	Russian	French	German	Spanish
plate	пластинка	plaquette *f*	Plättchen *n*	planchuela *f*
star	звезда	étoile *f*	Stern *m*	estrella *f*
column crystal	столбик	colonne *f*	Säulenkristall *m*	pillar *m*
needle	игла	aiguille *f*	Schneenadel *f*	aguja *f*
snowflake	пушинка	plume *f* de duvet	Flocke *f*	copo *m* de nieve
hedgehog	ёж	cristal *m* en rose des	Igelkristall *m*	cristal *m* con
stud	запонка	bouton *m* de col	Kragenknopfkristall	pasador *m*

g Schneebeton *m*

5513 snow cornice
Great accumulation of packed snow formed by strong winds on the upper edges of steep leeward slopes or cliff faces. → Fig. 296
r снежный карниз
Мощное скопление плотного снега, образующееся на верхнем крае крутых подветренных склонов или обрывов гор при сильных метелевых ветрах. → Рис. 296
f corniche *f* de neige
s cornisa *f* de nieve
g Schneekarnies *n*

5514 snow cover; snow pack
A layer of snow covering the ground, either completely or partially.
r снежный покров
Слой снега на земной поверхности, покрывающий ее полностью или частично.
f manteau *m* neigeux; couche *f* de neige; tapis *m* nival; manteau *m* nival
s manto *m* nivoso; tapiz *m* nival; capa *f* de nieve; manto *m* nevado; manto *m* nival
g Schneedecke *f*

5515 snow creep
A property of snow to have continuous plastic deformations under the permanent stress.
r ползучесть снега
Свойство снега подвергаться непрерывным пластическим деформациям под воздействием постоянной нагрузки.
f fluage *m* de neige
s fluencia *f* de nieve
g Schneekriechen *n*

5516 snow crystal
Crystals of atmospheric ice precipitated in the form of snow (principal types of snow crystals are shown in the Table 3).

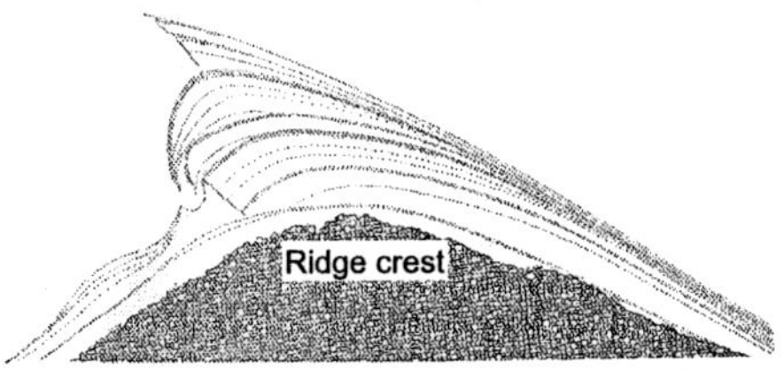

Fig. 296. Basic structure of a **snow cornice**

r снежный кристалл; снежинка
Кристаллы атмосферного льда, выпадающие в виде снега (основные виды снежных кристаллов приведены в табл. 3).
f cristal *m* de neige
s cristal *m* de nieve
g Schneekristall *m*

5517 snow drifting; snow-storm
A strong wind which can carry the falling snow and remove the deposited snow. → Fig. 297
r метель
Сильный ветер, переносящий массы выпадающего и ранее отложенного снега. → Рис. 297
f chasse-neige *f*; tempête *f* de neige; poudrerie *f*
s ventisca *f*; tormenta *f* de nieve
g Schneegestöber *n*; Schneesturm *m*

5518 snow drifting rose
A diagramm of distributon and frequency of snow-drifting winds from the eight chief points of the compass.
r роза метелей
Распределение по румбам повторяемости и силы метелевых ветров.
f rose *f* des chasse-neiges
s rosa *f* de ventiscas
g Windrose *f* der Schneegestöber

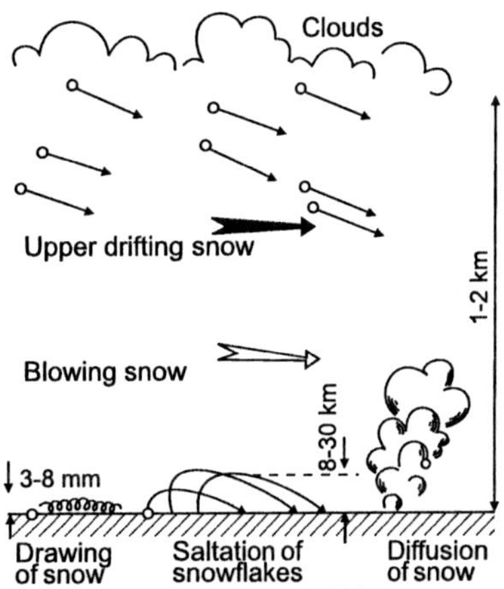

Fig. 297. **Snow drifting**

5519 snow dune
An accumulation of snow up to 1 m high and 5-10 m long formed on top of an ice and snow crust after the deposition of loose snow often falling as snow pellets.
r снежная дюна
Гряда снега высотой до 1 м и длиной 5-10 м, образующаяся на ледяной корке или насте в результате отложения рыхлого снега, часто выпадающего в виде крупы.
f dune *f* de neige
s duna *f* de nieve
g Schneedüne *f*

5520 snow fence
Any fence made of stone, wood or reinforced concrete to protect against snow drifts.
r снеговой забор; снегозадерживающий щит
Стенка, сооружаемая из камня, дерева, железобетона и пр. для защиты от снежных заносов.
f barrière *f* contre la neige
s valla *f* paranieves
g Schneeschutzzaum *m*

5521 snow fontain
The rise of snow up to 20 m above wide crevasses of a large snow-covered glacier when a strong wind blowing along the crevasse produces a blowing snow.
r снежный фонтан
Подъем снега на высоту до 20 м над зияющими отверстиями крупных занесенных снегом ледниковых трещин, когда вдоль них дует сильный ветер, вызывающий низовую метель.
f fontaine *f* de neige
s fuente *m* de nieve
g Schneefontäne *f*

5522 snow garlands
Chains of frozen snow flockson the branches of trees, cornices and other exposed surfaces formed of snow and hoar.
r снежные гирлянды
Гирлянды из смерзшегося снега на ветвях деревьев, карнизах и других выступающих предметах, образующиеся из снега, налепи и изморози.
f guirlandes *fpl* de neige
s guirnaldas *fpl* de nieve
g Schneegirlanden *fpl*

* **snow grains → 5528**

* **snow limit → 5523**

5523 snow line; snow limit
The altitude which delimits the actual lower level of permanent snow, i.e. the zone where the winter snowfall exceeds the amount removed by summer melting and evaporation. → Fig. 298
r снеговая линия; снеговая граница
Уровень в горах, выше которого годовое накопление твердых атмосферных осадков на земной поверхности преобладает над их таянием и испарением. → Рис. 298
f limite *f* des neiges permanentes; limite *f* des neiges; limite *f* d'enneigement
s límite *m* de nieves perpetuas
g Schneegrenze *f*

5524 snow line depression
The descending of the snow line in mountains caused by the climate change favourable for the existence of glaciers.
r депрессия снеговой линии
Снижение снеговой линии вследствие изменений климата, благоприятных для существования ледников.
f abaissement *m* de la limite de neiges
s depresión *f* de límite de nieves perpetuas
g Schneegrenzedepression *f*

5525 snow load
The weight of snow deposited on a structure.
r снеговая нагрузка
Нагрузка, испытываемая сооружениями от массы снега.
f charge *f* de neige

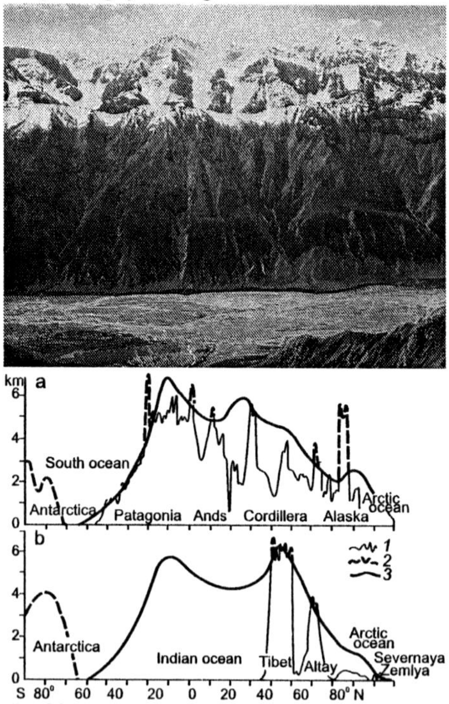

Fig. 298. **Snow line** in the Tien Shan Mountains. Down → Snow line over latitudes: a – Northern Hemisphere, b – Southern Hemisphere; 1 – height of the Earth's surface, 2 – glaciers, 3 – snow line

Fig. 299. **Snow patches** in the Pamir Mountains

s carga *f* de nieve
g Schneelast *f*

* **snow measurement → 5541**

* **snow melting → 3774**

5526 **snow micro-relief**
Surface irregularities produced on the snow surface by the action of wind and snow melting.
r снежный микрорельеф
Рельеф снежной поверхности, формирующийся под действием ветра и снеготаяния.
f microrelief *m* de la neige
s microrelieve *m* de nieve
g Mikrorelief *n* der Schneefläche

* **snow mudflow → 5491**

* **snow pack → 5514**

5527 **snow patch**
An isolated motionless mass of snow and ice which retains after the seasonal melting of snow cover. → Fig. 299
r снежник
Неподвижное скопление снега и льда, сохраняющееся после схода снежного покрова. → Рис. 299
f plaque *f* de neige; névé *m*
s lugar *m* de nieve; nevero *m*
g Schneebahn *f*; Schneefleck *m*; Firnfeld *n*

5528 **snow pellet; snow grains**
White, opaque snow balls of irregular form with the diameter up to 5 mm.
r снежная крупа
Матово-белые снегоподобные ядра неправильной округлой формы диаметром до 5 мм.
f grésil *m*
s grano *m* de nieve
g Graupeln *fpl*; Reifgraupeln *fpl*

5529 **snow pit**
A vertical hole made in a snow cover for the investigation and taking the snow samples. → Fig. 300
r снежный шурф
Вертикальная выработка в снежном покрове для изучения его строения и отбора образцов снега. → Рис. 300
f puits *m* dans la neige
s pozo *m* de nieve
g Schneeschurf *m*

5530 **snow protection**
Measures taken to protect roads, buildings and industrial areas against snowdrifts including various ways of snow removal.
r снегозащита; снегоборьба
Мероприятия по защите дорог, населенных пунктов и промышленных объектов от снежных заносов, включают разнообразные средства и способы борьбы с заносами, выдуванием и уборки снега.
f protection *f* contre la neige
s protección *f* contra la nieve
g Schneeschutz *m*

5531 **snow puddle**
An accumulation of fresh water on an ice cover surface.
r снежница
Скопление пресной воды на поверхности ледяного покрова.
f mare *f* superficielle
s poza *f* superficial
g Eispfütze *f*

* **snow retention → 5051**

5532 **snow ridge**
Snow bank, up to 30 m high, forming on slopes of ice sheets due to the interaction of permanent winds with the mesorelief of the glacier surface.
r снежная гряда
Снежный вал высотой до 30 м, возникающий на склоне ледниковых щитов при взаимодействии постоянных ветров с мезорельефом ледниковой поверхности.
f crête *f* de neige; banc *m* de neige

Fig. 300. **Snow pit**

s cresta *f* de nieve; banco *m* de nieve
g Schneewall *m*

5533 **snow ripple**
Elongated small ridges of new snow, 2-3 cm high and 10-15 cm long, facing the wind and moved by it along a flat surface of old snow.
r снежная рябь
Вытянутые поперек ветра валики свежего снега высотой 2-3 см и длиной 10-15 см, медленно передвигающиеся под действием ветра по ровной поверхности старого снега.
f rides *fpl* de neige
s ondulaciones *fpl* de nieve
g Schneerippen *fpl*

5534 **snow runoff**
The flow of snow melt water either over the surface of the ground or within it.
r снеговой сток
Стекание талой снеговой воды по земной поверхности и в толще почвогрунтов.
f écoulement *m* de l'eau de fusion
s escorrentía *f* del agua de fusión
g Schneeschmelzwasserabfluß *m*

5535 **snow shower**
A very intensive snow precipitation falling from a cumuliform clouds.
r ливневой снег
Ливневые осадки в форме снега или мокрого снега, выпадающие из кучевообразного облака.
f averse *f* de neige
s chaparrón *m* de nieve
g Schneeschauer *m*

* **snow slab → 6786**

5536 **snow slip**
A reduced friction at moving along the surface of a snow cover caused by the melting of snow.
r скользкость снега
Пониженное трение при движении по снежному покрову, вызываемое подтаиванием снега.
f glissement *m* de la neige
s resbaladizo *m* de nieve
g Gleitfähigkeit *f* von Schnee

5537 **snow slump; surface avalanche**
A small downhill movement of snow.
r снежный осов
Снежный оползень на поверхности слаборасчлененного склона.
f coulée *f* de neige superficiele
s avalancha *f* de la capa superficial
g Schneerutsch *m*; stumpfe Lawine *f*

5538 **snow stability on a slope**
The ability of snow cover on a slope to keep its equilibrium despite the action of external forces.
r устойчивость снега на склоне
Способность снежного покрова, лежащего на склоне, сохранять равновесие при действии внешних сил.
f stabilité *f* de la neige sur pente
s estabilidad *f* de nieve en pendiente
g Schneestabilität *f* auf einem Hang

5539 **snow storage**
A mass of water in solid and liquid state contained at a given moment in the snow cover, usually expressed in g/cm^2 or cm of water equivalent.
r снегозапас
Масса воды в твердом и жидком виде, содержащаяся в данный момент в снежном покрове, обычно выражается в г/см^2 или в см водного эквивалента.
f réserve *f* de neige
s reserva *f* de nieve
g Schneevorrat *m*

5540 **snow studies**
The branch of *glaciology* devoted to the study of snow and snow cover.
r снеговедение
Отрасль *гляциологии*, изучающая снег и снежный покров.
f étude *f* de la neige; nivologie *f*
s estudio *m* de nieve; nivología *f*
g Schneekunde *f*; Nivologie *f*

5541 **snow survey; snow measurement**
The process of determining the depth and water content of snow cover at representative points.
r снегомерная съемка; снегосъемка
Измерение толщины и плотности снежного покрова в репрезентативных точках с целью определения снегозапасов.
f nivométrie *f*; surveillance *f* du manteau neigeux
s nivometría *f*; medición *f* de manto de nieve
g Schneedeckaufnahme *f*

5542 **snow swamp**
An area of water-saturated snow cover having a soupy consistency.
r снежное болото
Плоский участок пропитанного водой снежного покрова.
f marécage *m* de neige
s ciénaga *f* de nieve
g Schneesumpf *m*; Schneematsch *m*

5543 **snow temperature inversion; spring inversion**
A temperature inversion occurring in the layers of air adjacent to the surface of snow cover.
r снежная инверсия; весенняя инверсия
Выхолаживание нижних слоев водуха, прилегающих к поверхности снега.
f inversion *f* par la neige
s inversión *f* de nieve

g Frühjahrsinversion *f*

5544 snow-and-ice road
A temporary road laid in winter across snowbound expanses, frozen rivers and lakes.
r зимник; снежно-ледяная дорога
Временная дорога, прокладываемая зимой по снежной целине, речному или озерному льду.
f piste *f* en glace et neige
s pista *f* de nieve y hielo
g Schnee- und Eisrollgang *m*

*** snow-capped mountain → 5553**

5545 snowdrift
An accumulations of snow piled in heaps by the wind during snow storms and snowfalls on the roads and over built-up territories, created behind the obstacles or irregularities of the surface.
r снежный занос; сугроб
Отложение снега во время метелей и снегопадов на дороге или застроенной территории, возникающее на подветренной стороне препятствия в результате местного снижения скорости ветра.
f congère *f*; amas *m* de neige
s montón *m* de nieve
g Schneeverwehung *f*; Schneewehe *f*

5546 snowdrift glacier
A small glacier, which is actually a semi-permanent mass of firn, formed by drifted snow in the depressions of the ground or behind obstructions.
r навеянный ледник
Ледник, питание которого в основном происходит за счет навевания снега.
f glacier *m* alimenté par la chasse-neige
s glaciar *m* de montañas de nieve
g Schneewehengletscher *m*

5547 snowfall
The precipitation of snow from clouds.
r снегопад
Выпадение снега из облаков.
f chute *f* de neige
s nevada *f*; nevasca *f*
g Schneefall *m*

5548 snowfield
An area of permanent snow with a relatively level, smooth surface, occurring in shallow depressions in mountainous areas or on high plateaus.
r снежное поле
Снежник с относительно ровной поверхностью, занимающий плоские котловины в горах или на высоких плато.
f champ *m* de neige
s campo *m* de nieve
g Schneefeld *n*

5549 snowflakes
Aggregations of several single crystals that become attached together while falling.
r снежные хлопья
Группы снежинок, механически соединенные между собой в процессе выпадения.
f flocon *m* de neige
s copo *m* de nieve
g Schneeflocken *fpl*

5550 snowiness
A characteristic related to the presence of snow cover on an area, usually expressed as a percentage.
r снежность
Характеристика природных условий территории, связанных с наличием снежного покрова, нередко выражается в процентах покрытия ее снегом.
f enneigement *m*; nivosité *f*
s nivosidad *f*
g Schneevorrat *m*

*** snowmelt → 3774**

5551 snow-retaining structures
Anti-avalanche structures aimed at maintaining the snow stability on an avalanche-hazardous slope.
r снегоудерживающие сооружения
Противолавинные сооружения, предназначенные для обеспечения устойчивости снега на лавиноопасном склоне.
f ouvrages *mpl* de fixation de la neige
s construcciones *fpl* de fijación de nieve
g Schneeanrissverbauungen *fpl*

*** snow-storm → 5517**

5552 snowy *adj*
Related to snow or composed of snow.
r снежный; снеговой
Состоящий из снега, сложенный снегом (например, снежный покров), а также связанный со снегом, но из него не состоящий (снеговое половодье).
f neigeux *adj*
s nevoso *adj*; nevado *adj*; nivoso *adj*
g Schnee- *adj*; verschneit *adj*; beschneit *adj*

5553 snowy mountain; snow-capped mountain
Mountain top covered with snow all the year round.
r снежная гора; снежная вершина
Горы, покрытые в вершинных частях снегом в течение круглого года.
f montagne *f* neigeuse; montagne *f* couronnée de neige
s nevado *m*
g Schneegebirge *n*

5554 soaking up of water
The initial stage of water infiltration into the ground under the conditions of its incomplete saturation with moisture.

r впитывание воды
Начальная стадия просачивания воды в грунт в условиях его неполного насыщения влагой.
f absorption *f* d'eau à terrain
s absorción *f* del agua en el suelo
g Wasseraufsaugen *n*

5555 social *adj*
Related to the phenomena and processes directly connected with various human societies and their relationships.
r социальный; общественный
Относящийся к явлениям и процессам, непосредственно связанным с разными общностями людей и их взаимоотношениями.
f social *adj*
s social *adj*
g sozial *adj*

5556 social desert
Early developed part of a country where, due to different reasons, a depopulation took place (a process, similar to that happened in XX century in Russia over the territory between its two capitals Moscow and Saint-Petersburg.
r социальная пустыня
Пространство в некогда населенной части страны, где в силу разных причин произошла депопуляция, как это случилось в XX столетии на территории России между двумя ее столицами – Москвой и Санкт-Петербургом.
f désert *m* social
s desierto *m* social
g soziale Wüste *f*

5557 social ecology
The study of the influence of the physical environment on the social processes of human society.
r социальная экология
Изучение влияния природных условий на социальные процессы человеческого общества.
f écologie *f* sociale
s ecología *f* social
g Sozialökologie *f*

5558 social geography
The study of spatial processes and forms of the social organization of people and their culture.
r социальная география
Изучение территориальных процессов и территориальных форм социальной организации жизни людей и их культуры.
f géographie *f* sociale
s geografía *f* social
g Sozialgeographie *f*

5559 social group
A group of people united by common features or relationships, such as age, education, social status, etc.
r социальная группа
Совокупность людей, объединяемых общими признаками или отношениями: по возрасту, образованию, социальному положению и пр.
f groupe *m* social
s grupo *m* social
g soziale Gruppe *f*

5560 social infrastructure
A combination on a given territory the institutions of health care, education, culture centres and administration.
r социальная инфраструктура
Сочетание на данной территории предприятий здравоохранения, просвещения, культуры, бытового обслуживания.
f infrastructure *f* sociale
s infraestructura *f* social
g soziale Infrastruktur *f*

5561 social integration
The process whereby a minority group, particularly an ethnic minority, adapts to the host society and where it is accorded equal rights with the rest of the community.
r социальная интеграция
Общественный процесс, когда национальное меньшинство адаптируется к основному сообществу и получает права, равные со всеми слоями общества.
f intégration *f* sociale
s integración *f* social
g Sozialintegration *f*

5562 social methods of management
The management of national economy and people's life through society's extra-administrative pressure on the structures of power.
r социальные методы управления
Руководство народным хозяйством и жизнью населения с использованием форм внеадминистративного воздействия общества на властные структуры.
f méthodes *fpl* sociales de géstion
s métodos *mpl* sociales de gestión
g gesellschaftliche Methoden *fpl* der Leitung

5563 social mobility
The ability of individuals to change their socio-economic status, i.e. to move between social classes and income groups.
r социальная мобильность
Способность людей изменять свое социально-экономическое положение, т.е. перемещаться из одного социального слоя в другой.
f mobilité *f* sociale
s movilidad *f* social
g Sozialmobilität *f*; soziale Mobilität *f*

5564 social polarization
The consequence of such divergence of segregated groups in a society that those large discrete groups (representing widely different

socio-economic status, cultures etc.) form at the extremities of the social spectrum and there is little or no social contact between them.

r социальная поляризация
Состояние общества, в котором крупные социальные группы находятся на противоположных полюсах социально-экономического и культурного спектра.
f polarisation *f* sociale
s polarización *f* social
g soziale Polarisation *f*; soziale Polarisierung *f*

5565 social segregation
The spatial separation of people whithin an area, on the basis of social distinctions.
r социальное расслоение
Пространственное разъединение людей на данной территории на основе социальных различий.
f ségrégation *f* sociale
s segregación *f* social
g soziale Segregation *f*

5566 socialism
A political, social and economic system that aims to create a society in which everyone has equal opportunities and in which the major part of the means of production is owned and controlled by the state.
r социализм
Политическая, социальная и экономическая система, в которой преобладает государственная собственность и плановая экономика.
f socialisme *m*
s socialismo *m*
g Sozialismus *m*

5567 society
A large group of people associated together geographically, culturally and otherwise, with common interest, shared laws, customs, etc. and with a particular organisation.
r общество
Совокупность исторически сложившихся форм совместной деятельности людей на данной территории.
f societé *f*
s sociedad *f*
g Gesellschaft *f*

5568 socio-ecological system
Natural system which is rather vulnerable to the anthropogenic and social drivers.
r социально-экологическая система
Природная система, весьма чувствительная к антропогенным воздействиям и социальным изменениям.
f système *m* socio-écologique
s sistema *m* social y ecológico
g sozialökologisches System *n*

5569 socioeconomic *adj*
Related to social and economic conditions or invoving a combination of social and economic factors.
r социально-экономический
Относящийся к социальным и экономическим условиям или подразумевающий комбинацию социальных и экономических факторов.
f socio-économique *adj*
s socioeconómico *adj*
g sozialekonomisch *adj*

*** socioeconomic geography → 2830**

5570 socioeconomic system
The system of social and economic relations in a society, including the relations between social groups, property right, legal system, etc.
r социально-экономическая система
Структура социальных и экономических отношений в обществе, включающая взаимоотношения между социальными группами, право собственности, судебную систему и пр.
f système *m* socio-économique
s sistema *m* socioeconómico
g sozialökonomisches System *n*

5571 sociology
The study of all sorts of communities and groups of people and their behaviour within the human society.
r социология
Изучение различных общностей и групп людей и их поведения в рамках человеческого общества.
f sociologie *f*
s sociología *f*
g Soziologie *f*

5572 sociosphere
The Earth's sphere in which mankind and social laws exist side by side with the environment.
r социосфера
Земная сфера, в которой наряду с окружающей природой существует человечество и действуют социальные законы.
f sociosphère *f*
s sociosfera *f*
g Soziosphäre *f*

5573 socle plain
A denudation plain formed on the dislocated rocks of the crystalline basement.
r цокольная равнина
Денудационная равнина, образованная на дислоцированных породах кристаллического фундамента платформы.
f plaine *f* à socle rocheux
s llanura *f* de zócalo
g Sockelebene *f*

5574 sod; turf

The top layer of the soil with thick network of roots and spouts, rich in organic nutrients, it is characteristic of meadows and virgin steppe.

r дёрн$_1$; дернина; дерновина
Верхний горизонт почвы, густо пронизанный переплетенными живыми и отмершими корнями и побегами растений, богатый питательными органическими веществами, нередко с тесным скоплением на почве листьев и стеблей травянистых растений.

f gazon *m*; motte *f*; touffe *f*
s césped *m*
g Plagge *f*; Rasen *m*

5575 soda lake
A lake having a high proportion of sodium in water and in which sodium carbonate (natural soda) can be deposited.

r содовое озеро
Озеро, в котором происходит кристаллизация и осаждение карбоната натрия (природной соды).

f lac *m* de soude
s lago *m* sódico
g Sodasee *m*; Natronsee *m*

5576 sodded karst
A type of karst developing in karstic rocks covered by a layer of soil and sod.

r задернованный карст
Тип карста, развивающийся в карстующихся породах, покрытых слоем почвы и дерна.

f karst *m* couvert avec gazon
s karst *m* cubierto con césped
g bodenbedeckter Karst *m*

5577 sodding
The natural overgrowing of barren land, clearings, tramped former pastures and other lands with grass vegetation or sowing grass in order to increase the productivity of forage land and to prevent soil and wind erosion.

r залужение
Длительный процесс самозарастания травянистой растительностью оголенных участков, вырубок, выбитых пастбищ и других угодий (естественное залужение), а также посев трав с целью повышения продуктивности кормовых угодий и борьбы с почвенной и ветровой эрозией (искусственное залужение).

f enherbement *m*; création *f* des prairies
s creación *f* de pradera; enhierbamiento *m*
g Berasung *f*; Beweidung *f*

5578 soddy forest soils
Soils with dark humus horizon that develop by the action of sod-forming process under forest vegetation on rich parent rocks.

r дерновые лесные почвы
Почвы, формирующиеся при воздействии дернового процесса под лесной растительностью на богатых почвообразующих породах, имеют темный гумусовый горизонт.

f sols *mpl* gazonnés forestiers
s suelos *mpl* de bosque de césped; suelos *mpl* forestales de césped
g Rasenböden *mpl* (des Waldes)

5579 soddy podzolic soil
A podzolic soil having well pronounced leaf-litter 3-5 cm thick, humus horizon (3-7% of humus), bleached podzolic horizon underlain by illuvial horizon, it is widespread in southern taiga of Eurasia and North America.

r дерново-подзолистая почва
Подзолистая почва, где четко выделяются лесная подстилка мощностью 3-5 см, гумусовый горизонт (3-7% гумуса), осветленный подзолистый горизонт, подстилаемый иллювиальным, распространена в южной тайге Евразии и Северной Америки.

f sol *m* gazonné-podzolique
s suelo *m* podzólico de césped
g podsoliger Rasenboden *m*

5580 soddy-gley soil
Soils of boreal plains of Eurasia and North America that form in depressions under herbaceous forests and meadows under the conditions of plenty fulmoistening, they have dark humus horizon underlain by gley horizon.

r дерново-глеевая почва
Почвы бореальных равнин Евразии и Северной Америки, формирующиеся в понижениях под травянистыми лесами и лугами в условиях большого увлажнения, имеют темный гумусовый горизонт, подстилаемый глеевым горизонтом.

f sol *m* à gley gazonné
s suelo *m* de gley y césped
g gleyiger Rasenboden *m*

5581 sodication; sodicity
A combination of processes responsible for the enrichment of soil with exchangeable sodium when contacting with the solutions of neutral sodium salts without any noticeable destruction of soil structure and high increase of pH values.

r осолонцевание
Совокупность процессов обогащения почвы обменным натрием при контакте ее с растворами нейтральных натриевых солей без заметного разрушения структуры и сильного повышения значений pH.

f sodisation *f*
s sodisación *f*
g Sodisierung *f*

* **sodicity → 5581**

5582 soft water
Water containing less than 60 mg/l of dissolved calcium and magnesium carbonates.

r мягкая вода
Вода с малым (менее 60 мг/л) содержанием карбонатов кальция и магния.

f eau *f* douce

s agua *f* blanda
g weiches Wasser *n*

5583 soil
The naturally occurring thin layer of unconsolidated material on the Earth's surface that originated as a result of transformation of the surface layers of the lithosphere under joint impact of water, air and living organisms to produce a medium suitable for the growth of land plants.
r почва
Природное тело, возникшее в результате преобразования поверхностных слоев литосферы под совместным воздействием воды, воздуха, живых организмов и обладающее плодородием.
f sol *m*
s suelo *m*
g Boden *m*

5584 soil acidity
The increased content of hydrogen ions in soil solution which makes it possible for the soil to neutralize alkali solutions and acidify water and solutions of neutral salts.
r кислотность почвы
Повышенное содержание в почвенном растворе водородных ионов, что позволяет почве нейтрализовать растворы со щелочной реакцией и подкислять воду и растворы нейтральных солей.
f acidité *f* du sol
s acidez *f* del suelo; basicidad *f* del suelo
g Bodenazidität *f*; Bodensäuregrad *m*

5585 soil aeration
Gas exchange of soil air with the atmospheric air that favours the development of plants and increases soil fertility.
r аэрация почвы
Газообмен почвенного воздуха с атмосферным, способствующий развитию растений и плодородию почвы.
f aération *f* du sol; aérage *m* du sol
s aeración *f* del suelo
g Bodenbelüftung *f*; Bodendurchlüftung *f*

5586 soil age
The time passed since the beginning of soil development (incipient, young, mature and ancient soils are distinguished).
r возраст почвы
Время, прошедшее с начала развития почвы, различают зарождающиеся, молодые, зрелые и древние почвы.
f âge *m* du sol
s edad *f* del suelo
g Bodenalter *n*

5587 soil aggregate; aggregate of soil
A single mass or cluster of soil particles which adhere together in such a way that they behave mechanically as a unit forming the crumb structure of soil.
r почвенный агрегат
Отдельные сростки почвенных частиц, которые, механически соединяясь, формируют крупитчатую структуру почвы.
f agrégat *m* du sol
s agregado *m* del suelo
g Bodenaggregat *n*

5588 soil air
The air filling in the soil pores and cavities not occupied with water (as compared with the atmospheric air it is depleted in oxygen and enriched with carbon dioxide).
r почвенный воздух
Воздух, заполняющий поры и пустоты почвы, не занятые водой (по сравнению с атмосферным обеднен кислородом и обогащен углекислотой).
f air *m* du sol
s aire *m* del suelo
g Bodenluft *f*

5589 soil argillization
Clay formation in situ in any part of a soil profile as a result of soil-forming processes.
r оглинивание почвы
Процесс образования глины in situ в той или иной части почвенного профиля как следствие процессов почвообразования.
f argilisation *f* du sol
s arcillización *f* del suelo
g Verlehmung *f*

5590 soil association
A group of soil series developed on a similar parent material derived from similar rocks or on a combination of rocks, it is usually used for the purpose of their depiction on a map.
r почвенная ассоциация
Группа почвенных серий, развитых на одних и тех же материнских породах, или их комбинации, обычно используется в целях почвенного картографирования.
f association *f* de sols
s asociación *f* de suelos
g Bodenassoziation *f*

5591 soil biology
The study of living organisms inhabiting soils.
r биология почвы
Изучение живых организмов, обитающих в почве.
f biologie *f* du sol
s biología *f* del suelo
g Bodenbiologie *f*

5592 soil biota
All living organisms inhabiting soil and having direct or indirect impact upon it, includes microorganisms, fungi, invertebrates, small vertebrate animals, and roots of green vegetation.
r почвенная биота; почвенные организмы
Совокупность живых организмов, населяющих почву и оказывающих на нее прямое или косвенное воздействие, включает

микроорганизмы, грибы, беспозвоночные и мелкие позвоночные животные, корни зеленой растительности.
f biote *f* du sol
s biota *f* del suelo
g Bodenbiota *f*; Bodenorganismen *mpl*

5593 **soil chemistry**
Chemical composition and properties of organic and mineral constituents of soils.
r химия почв
Химический состав и свойства органической и минеральной частей почв.
f chimie *f* de sol
s química *f* del suelo
g Bodenchemie *f*

5594 **soil chronosequence**
Soils the properties of which differ depending only on the time of soil formation.
r хроноряд почв
Ряд почв, свойства которых различаются только в зависимости от времени почвообразования.
f chronoséquence *f* des sols
s cronosucesión *f* de suelos
g chronologische Bodenabfolge *f*

5595 **soil class**
One of the categories in taxonomic soil classifications uniting a number of soil types distinguished by certain characteristics, such as the character of moistening, soil-forming process, etc.
r класс почв
Одна из категорий в таксономических классификациях почв, объединяющая ряд типов почв, выделяемых по определенным признакам: характеру увлажнения, особенностям почвообразовательного процесса и т.п.
f classe *f* des sols
s clase *f* de suelos
g Bodenklasse *f*

* **soil classification → 5632**

5596 **soil climate**
Long-term regime of temperature, humidity and air content in soil that depends on soil properties, atmospheric climate, vegetation, relief, properties of soil-forming rocks and human activity.
r почвенный климат
Многолетний режим температуры, влажности и воздухосодержания почвы, зависящий от ее свойств, климата атмосферы, растительности, рельефа, свойств почвообразующих пород и деятельности человека.
f climat *m* du sol; pédoclimat *m*
s clima *m* del suelo; pedoclima *m*
g Bodenklima *n*

5597 **soil colloids**
Fine organic, mineral and organic-mineral soil particles less than 0.2 micron in size.
r почвенные коллоиды
Тонкодисперсные органические, минеральные и органо-минеральные частицы почвы размером менее 0,2 микрон.
f colloides *mpl* du sol
s coloides *mpl* del suelo
g Bodenkolloide *npl*

5598 **soil complex**
Combination of more or less regularly alternating small and genetically related areas (from one to dozens meters) of soils on different elements of microrelief.
r почвенный комплекс; пятнистость почв
Комбинация более или менее регулярно чередующихся мелких генетически связанных ареалов (от одного до десятков метров) почвы на разных элементах микрорельефа.
f complexe *m* du sol; complexe *m* edaphique
s complejo *m* del suelo; complejo *m* edáfico
g Bodenkomplex *m*

5599 **soil conservation; soil protection**
Methods to protect soil from harmful effects of water and wind erosion, pollution, salinization, etc, which destroy soils and decrease their fertility.
r сохранение почв
Средства охраны почв от водной и ветровой эрозии, загрязнения и других неблагоприятных воздействий, разрушающих почвы или снижающих их плодородие.
f conservation *f* des sols
s conservación *f* de suelos
g Bodenverwahrung *f*

* **soil contamination → 5624**

5600 **soil cover**
Non-continuous cover of the Earth in the interface between the lithosphere and atmosphere created as a result of impact of climate and living organisms upon rocks.
r почвенный покров; педосфера
Несплошная оболочка Земли на границе литосферы и атмосферы, возникающая в результате воздействия климата и живых организмов на горные породы.
f couche *m* du sol
s manto *m* del suelo
g Bodendecke *f*

* **soil degradation → 1481**

* **soil depletion → 1524**

* **soil development → 5604**

5601 **soil drainage**
The removal of excessive moisture from soil and from its surface in order to improve water and air regime of soils.
r осушение земель
Удаление избыточной влаги из почвы и с ее поверхности с целью улучшения водного и воздушного режима почв.

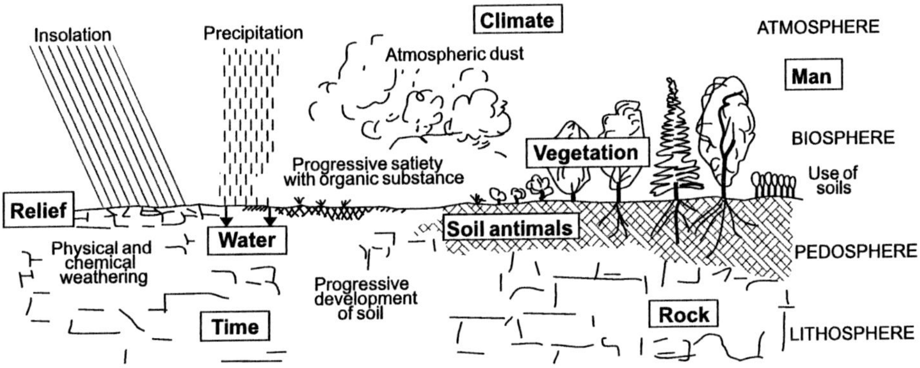

Fig. 301. **Soil formation**

f assèchement *m* du sol
s desecación *f* del suelo; drenaje *m* del suelo
g Bodenentwässerung *f*

5602 soil engineering
The study of grounds as a basis for the construction of different structures and roads.
r грунтоведение
Изучение грунтов с точки зрения возможностей строительства на них различных сооружений и дорог.
f génie *m* du sol
s ingeniería *f* del suelo
g Bodenbauwesen *n*

5603 soil erosion
The removal of soil by wind, water and by the movement of soil downslope, the erosion getting more intensive when the vegetation acting as a soil protection is destroyed by human activities and grazing animals.
r почвенная эрозия
Перемещение и удаление почвы вследствие эрозии, нередко сопровождаемое деятельностью людей и пастбищных животных, особенно при уничтожении растительности, служащей почвенной защитой.
f érosion *f* du sol
s erosión *f* del suelo
g Bodenerosion *f*; Bodenabtrag *m*

5604 soil evolution; soil development
Progressive or regressive development of soil under the impact of climatic, biological or other natural factors.
r эволюция почвы; развитие почвы
Поступательное развитие почвы под влиянием климатических, биологических и иных природных факторов, может быть прогрессивным и регрессивным.
f évolution *f* du sol
s evolución *f* de suelo; desarrollo *m* de suelo
g Bodenevolution *f*; Bodenentwicklung *f*

5605 soil extract
A solution remained after the separation of solid particles from soil suspension by centrifugation or filtration, it is used for laboratory analysis of soil.
r вытяжка из почвы
Раствор, остающийся после отделения твердых частиц из почвенной суспензии центрифугированием или фильтрованием, используется для лабораторных анализов почвы.
f extrait *m* du sol
s extracto *m* del suelo
g Bodenextrakt *m*

5606 soil fauna
A set of animals inhabiting soil (numerous Protozoan, worms, larvae of some insects, small vertebrates).
r почвенная фауна; эдафон
Совокупность животных, обитающих в почве, представлена многочисленными простейшими, червями, личинками некоторых насекомых, а также мелкими позвоночными.
f faune *f* du sol
s fauna *f* de suelo; edafofauna *f*
g Bodenfauna *f*; Edaphon *n*

5607 soil ferruginization
The accumulation of non-silicate iron in a soil profile or its horizons as a result of the eluviation (removal) of other components or due to the influx of iron from outside.
r ожелезнение почвы
Накопление в почвенном профиле или отдельных его горизонтах несиликатных форм железа, связанное как с выносом других компонентов, так и привносом железа извне.
f ferrugination *f* du sol
s ferruginación *f* del suelo
g Bodenvereisung *f*

5608 soil fertility; fertility of soil
The capacity of soil to provide nutrients, moisture, etc. essential for plant growth and to produce crops.

r плодородие почвы
Способность почвы обеспечивать растения питательными веществами, влагой и пр. и давать урожай.
f fertilité *f* du sol
s fertilidad *f* del suelo
g Bodenfruchtbarkeit *f*

5609 **soil formation; pedogenic processes; soil-forming process; pedogenesis**
Soil inception and evolution affected by natural and man-induced processes. → Fig. 301, p. 669
r почвообразование; педогенез; генезис почвы
Зарождение, становление и эволюция почвы под влиянием природных и антропогенных процессов. → Рис. 301, с. 669
f formation *f* du sol; pédogenèse *f*
s formación *f* del suelo; edafogénesis *f*; génesis *f* del suelo; pedogénesis *f*
g Bodenbildung *f*

5610 **soil geography**
A branch of physical geography and pedology, studying the distribution of soils over the Earth's surface and factors determining the distribution and peculiar features of soils.
r география почв
Отрасль физической географии и почвоведения, изучающая распространение почв на поверхности Земли и факторы, определяющие размещение и особенности почв.
f géographie *f* des sols
s geografía *f* de suelos
g Bodengeographie *f*

5611 **soil horizon; horizon$_2$**
A distinct soil layer relatively homogenious with well-defined characteristics that became differentiated in the course of soil formation, genetically related soil horizons constitute the soil profile.
r почвенный горизонт
Относительно однородный слой почвы, обособившийся в процессе почвообразования, генетически связанные горизонты составляют почвенный профиль.
f horizon *m* du sol
s horizonte *m* del suelo
g Bodenhorizont *m*

5612 **soil hygroscopicity**
The ability of soil to absorb water vapour from air and to retain it on the surface of its particles due to molecular forces (such moisture is not available for plants).
r гигроскопичность почвы
Способность почвенных частиц поглощать из воздуха влагу, которая прочно удерживается на поверхности частиц молекулярными силами и не может быть использована растениями.
f hygroscopie *f* du sol
s higroscopicidad *f* del suelo
g Hygroskopizität *f* des Bodens

* **soil improvement → 3048**

5613 **soil management**
Preparation of lands for further purposeful use in economy that includes grading, applying of soils upon lands under re-cultivation, soil improvement, construction of roads and water-development works.
r культивация почв
Подготовка земель для последующего целевого использования в хозяйстве, включающая планировку, нанесение почв на рекультивируемые земли, мелиорацию, строительство дорог и гидротехнических сооружений.
f culture *f* du sol; aménagement *m* agricole
s cultivo *m* del suelo; ordenación *f* agrícola
g Bodenkultivierung *f*; Bodenbehandlung *f*

5614 **soil map**
The map which represents the distribution of soils on the Earth's surface (according to a certain soil classification).
r почвенная карта
Карта, на которой показано распространение разновидностей почв в соответствии с принятой классификацией.
f carte *f* pédologique
s mapa *m* pedológico
g Bodenkarte *f*

5615 **soil mechanics**
The mechanical properties of soils and mechanical processes going on in soils.
r механика почв
Механические свойства почв и протекающие в почвах механические процессы.
f mécanique *f* du sol
s mecánica *f* del suelo
g Bodenmechanik *f*

5616 **soil microorganisms**
The tiny organisms which help to incorporate plant residues into the soil in the form of humus.
r почвенные микроорганизмы
Мельчайшие организмы, перерабатывающие остатки растений в почвенный гумус.
f microorganismes *mpl* de sol
s microorganismos *mpl* de suelo
g Bodenorganismen *mpl*

5617 **soil migration; translocation**
Transportation of organic-metal complexes (cheluviation) or suspended particles (lessivage) under the impact of water moving in soil.
r почвенная миграция
Перемещение органо-металлических комплексов (хелювирование) или взвешенных частиц (лессиваж) под влиянием движущейся в почве воды.
f migration *f* du sol; translocation *f*

s migración *f* del suelo; traslación *f*
g Bodenmigration *f*; Verlegung *f*

* **soil moisture → 5635**

* **soil moisture storage → 6681**

5618 soil monolith
A vertical soil sample taken without disturbing its structure from a side of the soil section.
r почвенный монолит
Вертикальный образец почвы, отобранный (без нарушения ее строения) из стенки почвенного разреза.
f monolithe *m* du sol
s monolito *m* del suelo
g Bodenmonolith *m*

5619 soil morphology
A set of external characteristics of soils such as the structure of soil profile, thickness of horizons, their colors, compositions, structures, presence of inclusions, pores, roots, etc.
r морфология почвы
Совокупность внешних признаков почв: строение почвенного профиля, мощность горизонтов, их окраска, сложение, структура, присутствие включений, пор, корней и пр.
f morphologie *f* du sol
s morfología *f* del suelo
g Bodenmorphologie *f*

5620 soil mosaic
A combination of different soils not related genetically within a limited territory that is determined usually by the differences between soil-forming rocks.
r мозаика почв
Комбинация на ограниченной территории разных почв, не обладающих генетической общностью и связанных, как правило, с различиями почвообразующих пород.
f mosaïque *f* du sol
s mosáico *m* de suelos
g Bodenmosaik *n*

5621 soil nomenclature
The system of classes, types and kinds of soils used when mapping soil cover and making soil classifications.
r номенклатура почв
Система обозначений отдельных классов, типов и видов почв, применяемая при картировании почвенного покрова и создании почвенных классификаций.
f nomenclature *f* des sols; nomenclature *f* pédologique
s nomenclatura *f* de suelos; nomenclatura *f* edafológica
g Bodennomenklatur *f*

5622 soil order
The highest category of soil classification that unites soils on the basis of similar properties associated with the main diagnostic horizons.
r порядок почв
Высшая категория классификации, объединяющая почвы на основе общности их свойств, связанных с основными диагностическими горизонтами.
f ordre *m* des sols
s orden *m* de suelos
g Bodenordnung *f*

5623 soil plasticity
The property of soil to be deformed without cracking by moderate the pressure.
r пластичность почвы
Способность почвы под влиянием нагрузки необратимо менять форму без образования трещин.
f plasticité *f* du sol
s plasticidad *f* del suelo
g Bodenplastizität *f*

5624 soil pollution; soil contamination
Accumulation of toxic and harmful substances and organisms in soil leading to soil degradation.
r загрязнение почвы
Накопление в почве токсических и вредных веществ и организмов, вызывающих деградацию почвенного покрова.
f pollution *f* du sol; contamination *f* du sol
s polución *f* del suelo; contaminación *f* del suelo
g Bodenverschmutzung *f*; Bodenverunreinigung *f*

5625 soil profile
A vertical section through all the constituent horizons of a soil from the surface to the relatively unaltered parent rock. → Fig. 302
r почвенный профиль
Совокупность почвенных генетических горизонтов, последовательно сменяющих

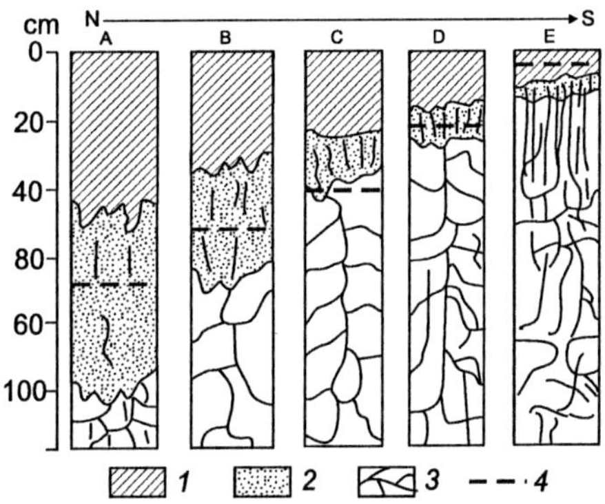

Fig. 302. **Soil profile**. Soil changes with latitudes (from north to south): A – typical chernozem, B – usual chernozem, C – southern chernozem, D – dark chestnut soil, E – light chestnut soil; 1 – humus-accumulating layers, 2 – illuvial horizons, 3 – parent rock, 4 – level of the boiling up of carbonates under the action of hydrochloric acid

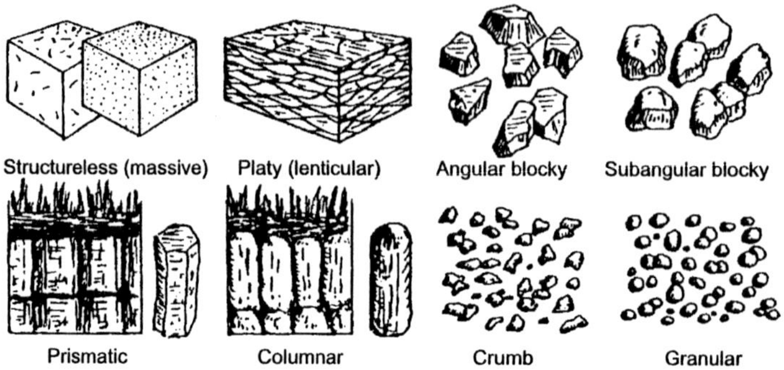

Fig. 303.
Soil structure

друг друга от поверхности до материнской породы. → Рис. 302, с.671
f profil *m* du sol
s perfil *m* del suelo
g Bodenprofil *n*

* **soil protection → 5599**

5626 **soil science; pedology**
The science that studies soils as phenomena which occur naturally, their properties, origin and formation, geographical distribution, the ways of efficient use and increase of fertility.
r почвоведение
Наука о почвах, их свойствах, формировании и развитии, географическом распространении, способах рационального использования и повышения плодородия.
f pédologie *f*
s edafología *f*; pedología *f*
g Bodenkunde *f*

5627 **soil section**
A vertical side of a pit that demonstrates soil profile.
r почвенный разрез
Вертикальная стенка шурфа, вскрывающая почвенный профиль.
f coupe *f* du sol
s sección *f* del suelo
g Bodenschnitt *m*

5628 **soil series**
A group of soils with similar profiles developed from similar parent materials (similar lithology and similar granulometric composition) under comparable climatic and vegetation conditions, the main taxonomic unit in American and French soil classifications.
r почвенная серия
Группировка, объединяющая почвы с единым типом профиля, развивающиеся на однотипном по литологии и гранулометрическому составу исходном материале, основная таксономическая категория в американской и французской почвенных классификациях.
f série *f* des sols
s serie *f* de suelos
g Bodenserie *f*

5629 **soil skeleton**
Solid soil particles presented by primary minerals and rock debris over 2 mm in size (in Russia over 1 mm).
r скелет почвы
Твердые частицы почвы, представленные первичными минералами и оболомками горных пород размером более 2 мм (в России более 1 мм).
f squelette *m* du sol
s esqueleto *m* del suelo
g Bodenskelett *n*

* **soil structural decline → 1481**

5630 **soil structure**
The character of a soil expressed in terms of its mode of aggregation or the binding properties, i.e. shape, dimensions, relative position and orientation of individual particles or secondary units, the soil is composed of. → Fig. 303
r структура почвы
Форма, размеры, относительное положение и ориентация агрегатов, на которые распадается почвенное тело. → Рис. 303
f structure *f* du sol
s estructura *f* del suelo
g Bodenstruktur *f*

5631 **soil taxon**
The smallest taxonomic unit of soil classifications reflecting such features of soil-forming process as degree of podzolization, salinization, amount of humus, thickness of humus horizon, etc.
r вид почв
Низшая таксономическая категория почвенных классификаций, отражающая особенности процесса почвообразования: степень оподзоленности, засоленности, количество гумуса, мощность гумусового горизонта и т.п.
f taxon *m* des sols
s taxón *m* de suelos
g Bodenart *f*

5632 **soil taxonomy; soil classification**
A hierarchial classification of soils ranging from the most general grouping (soil order) to the most detailed (soil series and taxons).

r таксономия почв; почвенная классификация
Иерархическая классификация почв, ранжирующая их от наиболее общих групп (порядки почв) до самых детальных (почвенных серий и таксонов).
f taxonomie *f* des sols; systématique *f* des sols; classification *f* des sols
s taxonomía *f* de suelos; sistemática *f* de suelos; clasificación *f* de suelos
g Bodentaxinomie *f*

* **soil texture → 6131**

5633 **soil thixotropy**
The capacity of some soils and grounds of becoming fluid when over-moistened and disturbed mechanically (as by shaking, mixing), and of restoring the solid state when left at rest.
r тиксотропность почвы
Способность некоторых почв и грунтов в переувлажненном состоянии приобретать текучесть под влиянием механических воздействий (при встряхивании, перемешивании) и снова переходить в твердообразное состояние в покое.
f thixotropie *f* du sol
s tixotropía *f* del suelo
g Thixotropie *f* des Bodens

5634 **soil type**
A group of soils that develop under similar biological, climatic and hydrological conditions characterized by vivid manifestation of the main soil-forming process, the main taxonomic unit of genetic soil classifications.
r тип почв
Группа почв, развивающихся в однотипных биологических, климатических и гидрологических условиях с ярким проявлением основного процесса почвообразования, главная таксономическая категория генетических почвенных классификаций.
f type *m* des sols
s tipo *m* de suelos
g Bodentyp *m*

5635 **soil water; soil moisture**
Water that is contained in soil in various forms and characterized by its mobility and availability for plants.
r почвенная влага
Вода, содержащаяся в почве в разных формах, отличающаяся подвижностью и доступностью для растений.
f eau *f* du sol; humidité *f* du sol
s agua *f* del suelo; humedad *f* del suelo
g Bodenwasser *n*

* **soil water storage → 6681**

5636 **soil wedge**
A wedge-shaped body formed as a result of filling in of the surface cracks which occur due to the freezing with fine sediments, carried-in by melt waters.
r грунтовая жила
Клиновидное тело, образующееся в результате заполнения формирующихся при промерзании поверхностных трещин мелкоземом, заносимым в них талыми водами.
f coin *m* de sol
s cuña *f* de suelo
g Mollisolfrostkeit *f*

5637 **soil zone**
Widely stretched areas of soil types the development of which is determined mainly by zonal climatic factors while local features have almost no impact upon it.
r почвенная зона
Широко протянувшиеся ареалы почвенных типов, развитие которых обусловлено в основном зональными климатическими факторами и мало зависит от локальных условий.
f zone *f* du sol
s zona *f* del suelo
g Bodenzone *f*

* **soil-forming process → 5609**

5638 **soil-ground**
The combination of soil and underlying rocks from 2 to 10 m thick.
r почвогрунт
Совокупность почвы и подстилающией ее толщи горных пород мощностью от 2 до 10 м.
f sol-terrain *m*
s suelo-terreno *m*
g Bodengrund *m*

* **soil-individual → 4462**

5639 **solar activity**
A regular (with the period of about 11 years) appearance of characteristic formations and phenomena in the Sun's atmosphere such as sunspots, prominences, etc. causing magnetic storms and the ionization of gases in the Earth's atmosphere.
r солнечная активность
Регулярное (с периодом около 11 лет) возникновение в атмосфере Солнца характерных образований и явлений: солнечных пятен, протуберанцев и пр., вызывающих магнитные бури и ионизацию газов в атмосфере Земли.
f activité *f* solaire
s actividad *f* solar
g Sonnenaktivität *f*

5640 **solar architecture**
The design of buildings with solar energy as the main or additional source of energy.
r солнечная архитектура
Проектирование зданий, в которые закладывается использование солнечной радиации в качестве основного или вспомогательного источника энергии.

f architecture *f* solaire
s arquitectura *f* solar
g Sonnenarchitektur *f*

* **solar atmospheric tide → 441**

5641 **solar battery; solar cell; solar panel**
A device which absorbs the sunrs radiation and converts it into electricity.
r солнечная батарея;
солнечный коллектор
Устройство для выработки электроэнергии в результате поглощения и преобразования солнечной радиации.
f batterie *f* solaire; cellule *f* solaire
s batería *f* solar; pila *f* solar
g Solarkollektor *m*

* **solar cell → 5641**

5642 **solar climate**
Theoretical climate calculated under the assumption that no atmosphere exists and the Earth is exposed only to solar radiation.
r солярный климат
Теоретический климат, рассчитанный в предположении, что атмосфера отсутствует и на Землю действует только солнечная радиация.
f climat *m* solaire
s clima *m* solar
g Solarklima *n*

5643 **solar constant**
The amount of solar radiation incident at the outer limit of the atmosphere at the mean distance of the Earth from the Sun.
r солнечная постоянная
Интенсивность потока солнечной радиации на верхней границе атмосферы при среднем расстоянии Земли от Солнца.
f constante *f* solaire
s constanta *f* solar
g Sonnenkonstante *f*

5644 **solar day**
The time interval between two successive upper or lower culminations of the Sun varying slightly at different times of the year because the orbit of the Earth round the Sun is elliptical and inclined to the equator.
r солнечные сутки
Продолжительность времени между двумя последовательными верхними или нижними кульминациями Солнца, не остающаяся постоянной из-за эллиптической формы земной орбиты, наклоненной к земному экватору.
f jour *m* solaire
s día *m* solar
g Sonnentag *m*

5645 **solar energy; solar power**
Any form of energy produced from the radiation of the Sun.
r солнечная энергия
Альтернативный источник энергии, получаемой от солнечной радиации.
f énergie *f* solaire
s energía *f* solar
g Sonnenenergie *f*

5646 **solar farm; solar park**
A tract of land where solar bateries are located.
r солнечная электростанция$_1$
Участок земли, на котором стоят солнечные батареи.
f centrale *f* électrique solaire; centrale *f* solaire
s central *f* de energía solar
g Sonnenkraftwerk *n*

* **solar panel → 5641**

* **solar park → 5646**

* **solar power → 5645**

* **solar power plant → 5647**

5647 **solar power station; solar power plant**
A power station that converts solar energy into electrical energy.
r солнечная электростанция$_2$
Установка, преобразующая солнечную энергию в электрическую.
f centrale *f* électrique solaire
s central *f* de energía solar
g Sonnekraftwerk *n*

5648 **solar radiation**
The total electromagnetic radiation emitted by the Sun which reaches the Earth.
r солнечная радиация
Полный поток электромагнитной радиации, излучаемой Солнцем и попадающий на Землю.
f radiation *f* solaire
s radiación *f* solar
g Sonnenstrahlung *f*

5649 **solar system**
The system of gravitationally connected celestial bodies consisting of a central massive body, the star and 9 large planets moving around the Sun with their satellites, many small planets, comets and meteor bodies.
r солнечная система
Система гравитационно связанных небесных тел, состоящая из центрального массивного тела – Солнца и движущихся вокруг него 9 больших планет с их спутниками, множеством малых планет, комет и метеорных тел.
f système *m* solaire
s sistema *m* solar
g Sonnensystem *n*

5650 **solar tide**
That portion of a tide which is caused by the tide-producting force of the Sun.
r солнечный прилив

Составляющая прилива, вызываемая приливообразующей силой Солнца (она в 2,2 раза меньше лунной составляющей – *лунного прилива*).
f marée *f* solaire
s marea *f* solar
g Sonnenflut *f*; Sonnentide *f*

5651 solar time; apparent time
The time as shown by the apparent movement of the Sun.
r солнечное время; истинное время
Система счета времени, в которой в качестве основной единицы принят интервал (солнечные сутки) между двумя последовательными верхними (или нижними) кульминациями Солнца, которые называют соответственно истинным полднем и истинной полночью.
f temps *m* solaire; heure *f* vraie
s tiempo *m* solar; hora *f* solar; hora *f* verdadera
g Sonnenzeit *f*; wahre Sonnenzeit *f*

*** solar year → 5652**

5652 solar year; tropical year
The time interval between two successive passings of the Sun's centre through the point of the vernal equinox, it is equal to 365.2422 mean solar days, i.e. 365 days 5 hours 48 min 45.51 seconds.
r солнечный год; тропический год
Промежуток времени между двумя последовательными прохождениями центра Солнца при его видимом годичном движении по небу через точку весеннего равноденствия, его продолжительность равна 365,2422 средних солнечных суток.
f année *f* tropique
s año *m* tropical; año *m* solar
g tropisches Jahr *n*

5653 solfatare
The phenomenon when acid gases are emitted through canals and fissures in a crater and on the slopes of a volcano usually in areas where violent volcanism has ceased.
r сольфатара
Природное явление и местность, где струи сернистого газа и сероводорода с примесью паров воды и других газов выделяются из мелких каналов и трещин в кратере и на склонах вулкана.
f solfatare *f*
s solfatara *f*
g Solfatare *f*

5654 solid waste
Waste matter, which is hard and not liquid, generated in the process of human economic activity.
r твердые отходы
Различные твердые вещества или предметы, возникающие в результате хозяйственной деятельности.

Fig. 304. **Solifluction**

f déchets *mpl* solides
s desechos *mpl* sólidos; residuos *mpl* sólidos
g feste Abfälle *mpl*

5655 solid precipitation
The deposition of water in a solid form (snow and ice crystals, snow pellet, hail, etc.) on the Earth's surface from atmospheric sources.
r твердые осадки
Осадки, состоящие из любых форм замерзшей воды, т.е. снежные и ледяные кристаллы, снежная крупа, град и т.п.
f précipitation *f* solide
s precipitacion *f* sólida
g feste Niederschläge *mpl*

5656 solifluction; solifluxion
The slow viscous-plastic downslope flowing of the loosely-bonded rock debris material and soil saturated with water, which move under the force of gravity and as a result of freezing and thawing. → Fig. 304
r солифлюкция
Вязко-пластическое течение увлажненных тонкодисперсных грунтов и почвы на склонах под действием силы тяжести и в результате их промерзания и протаивания. → Рис. 304
f solifluxion *f*
s solifluxión *f*
g Solifluktion *f*

5657 solifluction apron
An accumulated cone in the lower part of a slope formed as a result of the movement of debris over a surface of frozen material.
r солифлюкционный шлейф
Аккумулятивный конус в нижней части склона, формирующийся в результате смещения грязекаменных масс по мерзлому основанию.
f cône *m* de solifluxion
s cono *m* de solifluxión
g Solifluktionszunge *f*

5658 solifluction protalus

A rampart or ridge of fine material that has been moved down by solifluction over frozen ground.

r солифлюкционный вал
Вал или гряда мелкоземистого материала, перемещенного солифлюкцией по еще не оттаявшему слою мерзлоты.
f bourrelet *m* de solifluxion
s bulón *m* de solifluxión
g Fließfaltung *f*; Schutzwulst *m*

* **solifluction terrace → 225**

* **solifluxion → 5656**

5659 solitary wave; soliton
A finite-amplitude gravity wave consisting of a single crest; under certain conditions, a permanent wave.
r одиночная волна; солитон
Непериодическая волна, иногда наблюдающаяся на мелководье и имеющая только гребень, где заключена основная часть энергии и объема воды.
f vague *f* solitaire
s ola *f* solitaria
g Einzelwelle *f*; Soliton-Welle *f*

* **soliton → 5659**

* **solod → 5660**

5660 solod soil; solod; soloth
A soil forming in forest-steppes, steppes and semi-deserts of the temperate climate in flat-bottom depressions under long-term surface and ground excessive moistening, it includes a solodic whitish poorly humificated horizon, an illuvial-clay horizon as well as a carbonate one.
r солодь
Почва, формирующаяся в лесостепях, степях и полупустынях умеренного пояса в западинах в условиях длительного поверхностного и грунтового переувлажнения, имеющая осолоделый белесый слабо гумусированный, иллювиально-глинистый и карбонатный горизонты.
f solod *m*; soloth *m*
s solod *m*; soloth *m*
g Solod *f*

5661 solodization
The last stage of solonets degradation that gives rise to the development of solod soil as a result of clay deflocculation and eluviation in illuvial horizons, and of surface acidification.
r осолодение
Последняя стадия деградации солонцов, заключающаяся в диспергировании и выносе глины в иллювиальные горизонты и поверхностном подкислении, что приводит к формированию солоди.
f solodisation *f*
s solodización *f*
g Solodisierung *f*

Fig. 305. **Solonchak**

* **solonchak → 5663**

* **solonets → 5662**

5662 solonetz; solonets
A soil developing on formerly saline rocks and occurring from steppes to deserts, it has bleached eluvial and illuvial horizons with a columnar structure and high content of exchange sodium and/or magnesium.
r солонец
Почва на исходно засоленных породах, встречающаяся от степей до пустынь, имеющая осветленные элювиальный и иллювиальный горизонты со столбчатой структурой и высокое содержание обменного натрия и/или магния.
f solonetz *m*
s solonetz *m*
g Solonez *m*

5663 solontchak; solonchak
A soil forming under salt-resistant vegetation where saline ground water is within 2 m of the surface, the mineralized ground waters, provide the supply of soluble salts, carbonates and gypsum accumulated in the soil profile. → Fig. 305
r солончак
Почва, формирующаяся под солеустойчивой растительностью в условиях близкого залегания минерализованных грунтовых вод, из-за испарения которых в почвенном профиле аккумулируются легкорастворимые соли, карбонаты и гипс. → Рис. 305
f solontchak *m*; salenque *f*
s solonchak *m*; solontchak *m*; salitral *m*
g Solontschak *m*; Salar *m*; Salzboden *m*

* **soloth → 5660**

5664 solstice
The time when the Sun in its visible annual movement along the ecliptic passes through the point farthest from the equator either the most northern point (summer solstice June 21-22), or the most southern point (winter solstice December 21-22).
r солнцестояние
Момент времени, в который Солнце в своем видимом годовом движении по эклиптике проходит либо через самую се-

верную ее точку (летнее солнцестояние 21-22 июня), либо через самую южную (зимнее солнцестояние 21-22 декабря).

f solstice *m*
s solsticio *m*
g Solstitium *n*; Sonnenwende *f*

5665 soluble *adj*
Related to salts or rocks which are quickly dissolved under the water action.

r растворимый
Относящийся к солям или горным породам, легко растворимым под действием воды.

f soluble *adj*
s soluble *adj*
g löslich *adj*

5666 solution
The process whereby a fluid, usually water, as a solvent, picks up and dissolves particles of a solid, e.g. in the chemical weathering of rocks.

r растворение
Процесс, при котором жидкость, обычно вода, растворяет твердые частицы горных пород, как, например, это имеет место при химическом выветривании.

f dissolution *f*
s disolución *f*
g Auflösung *f*; Lösung *f*

* **solution basin → 4287**

5667 solution cavity
An underground hollow which forms in soluble rock when circulating water carries away the matter.

r полость растворения
Подземная полость, образованная в растворимой горной породе вследствие выноса вещества в растворе циркулирующими здесь водами.

f poche *f* de dissolution
s bolsa *f* de disolución; cavidad *f* de disolución
g Auswaschungshöhle *f*

5668 somma
A circular crest of an old volcano inside which there is a new volcanic cone. → Fig. 306

r сомма
Кольцевой гребень старого кратера, внутри которого находится новый вулканический конус. → Рис. 306

f somma *f*
s somma *f*
g Somma *f*

* **sonar → 1788**

5669 sonde
An apparatus designed to measure and record conditions in the atmosphere at certain altitudes, variously equipped with a lifting device and sensing and recording instruments.

Fig. 306. **Somma** of Vesuvio

r зонд
Прибор для измерений и записи физических условий на разных высотных уровнях в атмосфере, снабженный приспособлениями для подъема в воздухе, измерительной и записывающей аппаратурой.

f sonde *f*
s sonda *f*
g Sonde *f*

* **sonic depth finger → 1788**

5670 sor
Solonchak, shallow salt lake with clayey bottom in Central Asia, Kazakhstan, Azerbaijan.

r шор; сор
Солончак, пересыхающее мелководное соленое озеро с глинистым дном в Средней Азии, Казахстане, Азербайджане.

f sor *m*
s sor *m*
g Schor *m*

5671 sorted polygon; stone polygon
Cryogenic polygon that is equidimensional in several directions, with a dominantly polygonal outline, and a sorted appearance commonly due to a border of stones surrounding a central area of finer material.

r сортированный полигон; каменный полигон; каменный многоугольник
Криогенный полигон, ячейки которого имеют в основном форму многоугольников и выглядят сортированными благодаря скоплениям камней по их границам.

f polygone *m* trié
s polígono *m* de piedras; polígono *m* seleccionado
g Steinringboden *m*; Steinpolygon *n*

5672 sorting
The deposition of sediments graded according to size, shape, density, etc., by the natural processes of flowing water and of wind, when the larger particles usually settle first.

r сортировка
Отложение частиц в водной или воздушной среде в соответствии с их размерами, формой, плотностью и пр., причем первыми оседают более крупные частицы.

f triage *m*
s selección *f*
g Sortierung *f*

* **sound → 5836**

5673 **sound channel**
A zone in the oceanic water column at a depth of about 1500 m where the velocity of sound is at a minimum value which serves as a waveguide for sound waves as they propagate over long distances almost without distortion.
r звуковой канал; звуковой волновод
Слой в океане на глубине около 1500 м, вдоль которого звуковые волны без искажений распространяются на большие расстояния.
f canal *m* sonore
s canal *m* sonoro
g Schallkanal *m*

5674 **sound scattering layer; deep scattering**
A water layer in the ocean with very high ability to scatter acoustic signals usually due to the aggregation of small fishes and other marine organizms.
r звукорассеивающий слой; ложное дно
Слой в океане с исключительно высоким рассеиванием акустических сигналов, вызванным присутствием мелких рыб и других морских организмов.
f couche *f* diffusante profonde
s capa *f* difusora profunda
g Echostreuschicht *f*

5675 **sounding**
A method by which the depth of a sea or lake can be determined, formerly by a weighted line dropped overboard, now usually by an echo-sounder (sonar).
r зондирование; эхолокация
Метод измерения глубины моря или озера, в прошлом с помощью лота, опускаемого на тросе, ныне с помощью эхолота.
f sondage *m*
s sondaje *m*
g Sondierung *f*

* **sounding lead → 3454**

5676 **sounding of the atmosphere**
The determination of one or several upper-air meteorological elements by means of instruments lifted by a balloon, aircraft, kite, glider, rocket, etc.
r зондирование атмосферы
Определение метеорологических элементов на высотах с помощью приборов, поднимаемых воздушным шаром, самолетом, змеем, ракетой и т.п.
f sondage *m* aérologique
s sondeo *m* de atmósfera
g Atmosphärensondierung *f*

* **sounding rod → 5275**

* **source area → 2722**

5677 **source of a river**
The place at which a river begins its flow, the source of a river at the head of its catchment area.
r исток реки
Место, с которого появляется постоянное течение воды в русле.
f source *f* de rivière
s nacimiento *m* del río
g Flußquelle *f*

5678 **source of water supply; water supply source**
A water body which is used or planned to be used for water supply.
r источник водоснабжения
Водный объект, используемый или намечаемый к использованию для водоснабжения.
f source *f* d'alimentation en eau; adduction *f* d'eau
s fuente *m* de abastecimiento de agua
g Wasserversordnungsquelle *f*

5679 **south**
One of the four cardinal points, directly opposite the north, lying on the right side of a person facing due east; the southern part, especially of a country.
r юг
Одна из четырех главных сторон горизонта, прямо противоположная северу, расположенная справа от наблюдателя, стоящего лицом к востоку.
f sud *m*
s sur *m*
g Süden *m*; Süd *m*

5680 **south *adj*; southern *adj***
Relating to, belonging to, situated towards, coming from the south, e.g. winds blowing from the south.
r южный
Относящийся, расположенный или приходящий с юга, например, южный ветер.
f du sud; austral *adj*
s del sur; austral *adj*; meridional *adj*
g Süd-; südlich *adj*

5681 **South Pole**
The point of crossing of the Earth's rotation axis with the Earth's surface in the Southern Hemisphere.
r Южный полюс
Точка пересечения оси вращения Земли с земной поверхностью в Южном полушарии.
f pôle *m* sud; pôle *m* austral
s polo *m* sur; polo *m* austral
g Südpol *m*

5682 **southeast trades**
Persistent south-east trade wind of tropics in the Southern Hemisphere.
r юго-восточный пассат
Постоянный юго-восточный ветер тропиков Южного поолушария.
f alizé *m* du sud-est
s alisio *mpl* del sudeste
g Südostpassat *m*

* **southern** ***adj*** **→ 5680**

5683 Southern Cross
A five-star constellation forming a Christian cross, with a bright star at each extremity, the longer axis pointing towards the South Pole, it is used to find true south and latitude from any point in the Southern Hemisphere (comparable with the *Polaris* in the Northern Hemisphere).
r Южный крест
Созвездие Южного полушария, состоящее из пяти звезд, образующих форму христианского креста с длинной осью, направленной на Южный полюс мира, помогает найти направление на юг из любой точки Южного полушария (как и *Полярная звезда* в Северном полушарии), а высота созвездия над горизонтом указывает примерную широту данной точки.
f Croix *f* du Sud
s Cruz *f* del Sur
g Südkreuz *n*

5684 Southern Hemisphere
The half of the Earth south of the equator.
r Южное полушарие
Половина земного шара к югу от экватора.
f hémisphère *m* austral
s hemisferio *m* sur; hemisferio *m* austral
g südliche Halbkugel *f*

5685 southern oscillation
A fluctuation of the intertropical atmospheric circulation in which the air moves between the subtropical high and equatorial low when pressure is high over the Pacific Ocean and low in the Indian Ocean (the phenomenon is strongly linked to El Niño).
r южное колебание
Нерегулярные межгодовые колебания атмосферного давления над морем между восточной и западной частями на юге Тихого океана (это явление строго связано с эль-ниньо).
f oscillation *f* du sud
s oscilación *f* del sur
g südliche Oszillation *f*

5686 south-west monsoon
Summer oceanic monsoon in the northern part of the Indian Ocean, including the Indostan peninsula.
r юго-западный муссон
Летний океанический муссон в северной части Индийского океана, включая полуостров Индостан.
f mousson *m* sud-ouest
s monzón *f* del sudoeste
g Südwestmonsun *m*

5687 sovereign
A supreme ruler in the monarchy which is independent of the authority of other states.
r суверен
Носитель верховной власти в монархическом государстве, независимый от власти другого государства.
f souverain *m*
s soberano *m*
g Souverän *m*

5688 sovereign state
A state with the supreme authority held within the state itself, a state which is therefore independent and fully selfgoverning.
r суверенное гоосударство
Государство с сильной верховной властью и поэтому независимое во всех своих действиях.
f état *m* souverain
s estado *m* soberano
g souveräner Staat *m*

5689 sovereignty
The independence of a state in the foreign policy and the leadership in the domestic affairs, i.e. the authority of the state is autonomous and not subject to legal control by other states.
r суверенитет
Независимость государства во внешних и верховенство во внутренних делах, т.е. государственная власть автономна и неподконтрольна какому-либо другому государству.
f souveraineté *f*
s soberanía *f*
g Souveränität *f*; Staatshoheit *f*

* **sovkhoz → 5780**

5690 spa resort
A type of resort where mineral spring water and mud are used as the main medical treatment.
r бальнеологический курорт
Курорт, где основным видом лечения служит использование минеральных вод и грязей.
f station *f* balnéaire; station *f* thermale
s estación *f* balnearia;
baños *mpl* de aguas medicinales
g balneologischer Kurort *m*

5691 space
1. The extent of an area, usually expressed in terms of the Earth's surface. 2. The region which is beyond the limit of the Earth atmosphere.
r пространство;
космическое пространство; космос
1. Территория, площадь на земной поверхности. 2. Область в мировом пространстве, находящаяся за пределами земной атмосферы.
f espace *m*
s espacio *m*
g Raum *m*

* **space$_2$ → 6466**

5692 space Earth science
The study of the Earth from space using remote sensing and visual observation.
r космическое землеведение
Исследование Земли из космоса с помощью аэрокосмических методов и визуальных наблюдений.
f étude *f* cosmique de la Terre
s estudio *m* cósmico de la tierra
g Weltraumerdkunde *f*

5693 space geodesy
The determination of shapes and dimensions of cosmic bodies by astronomic and radiometric observations.
r космическая геодезия
Определение формы и размеров космических тел с помощью астрономических и радиометрических наблюдений.
f géodésie *f* spatiale
s geodesia *f* espacial; geodesia *f* cósmica
g Raumgeodäsie *f*

5694 space image; satellite imagery; satellite image; image
A solid or optical copy which is the product of remote sensing, a photograph obtained by the instrument for electronic imagery, operating in various wave bands making it possible to stady the Earthrs surface or other planets.
r космический снимок
Бумажная или цифровая копия изображения поверхности Земли или иной планеты, полученная путем дистанционного зондирования с космического аппарата в разных диапазонах волн.
f image *f* de satellite
s imagen *f* de satélite
g Satellitenbild *n*; Weltraumbild *n*

5695 space industry
A branch of machine-building industry which produces spaceships and rockets.
r ракетно-космическая промышленность
Отрасль машиностроения, производящая летательные космические аппараты и ракеты различного назначения.
f industrie *f* aérospatiale
s industria *f* aeroespacial
g Raumfahrtindustrie *f*

5696 space map
Topographic or thematic maps compiled by the data of remote sensing the basis of which is plane photograph.
r космокарта
Топографическая или тематическая карта, создаваемая по данным дистанционного зондирования, в основе которой лежит плановое фотоизображение.
f carte *f* spatiale
s mapa *m* espacial; minuta *f* fotogramétrica
g Raumkarte *f*

5697 space mapping
Compilation and renovation of topographic and thematic maps directly on the basis of space images.
r космическое картографирование
Составление и обновление топографических и тематических карт непосредственно на основе изображений, полученных с искусственных спутников Земли.
f cartographie *f* spatiale
s cartografía *f* espacial
g Satellitenkartierung *f*

5698 space station; skylab
An artificial satellite, serving as a work station, with a crew abord, orbiting the Earth in space, equipped with a laboratory and remote sensing instruments.
r орбитальная станция
Искусственный спутник на орбите вокруг Земли, служащий рабочей станцией с космонавтами на борту, лабораториями и инструментами для дистанционного зондирования.
f station *f* spatiale
s estación *f* espacial; laboratorio *m* espacial
g Weltraumstation *f*; Orbitalstation *f*

* **sparging → 555**

5699 spate
A sudden increase in the velocity and volume of a river owing to an influx of water, generally caused by a spell of intensive rainfall or a sudden melting of snow in the upper reaches.
r разлив реки
Внезапное увеличение скорости и объема воды в реке, вызванное сильным дождем или интенсивным таянием снега в верховьях.
f grandes crues *fpl*
s crecida *f*; desbordamiento *m*
g Überschwemmung *f*; Hochwasser *n*; Wasserflut *f*

5700 spatial adj
1. Related to space (as distinct to temporal) or controlled by the conditions of space on the Earth´s surface. 2. Related to the space which is beyond the limit of the atmosphere.
r пространственный$_1$; космический$_2$
1. Относящийся к пространству (в отличие от временно́го) или контролируемый условиями данной территории. 2. Относящийся к мировому пространству, находящемуся за пределами земной атмосферы.
f spatial *adj*
s espacial *adj*
g räumlich *adj*; Raum-

* **spatial$_2$ *adj* → 1312**

5701 spatial analysis
A type of geographical analysis which explains the spatial patterns of natural and

human phenomena in terms of mathematics and geometry.

r пространственный анализ
Совокупность аналитических методов географического исследования, связанных с изучением пространственных структур: методы географических полей, картометрический анализ и мн. др.

f analyse *f* spatiale
s análisis *m* espacial
g Raumanalyse *f*

5702 spatial data
Data on geographical objects including data on their situation and properties, spatial and non-spatial attributes.

r пространственные данные; географические данные
Данные о географических объектах, включающие сведения об их местоположении и свойствах, пространственных и непространственных атрибутах.

f données *fpl* spatiales
s datos *mpl* espaciales
g Geodaten *pl*

5703 spatial interaction
The interaction between objects and phenomena in geographic space including the exange of matter, energy, information, people, etc.

r пространственное взаимодействие
Взаимодействие между объектами и явлениями в географическом пространстве, включающее обмен веществом, энергией, информацией, людьми и пр.

f interaction *f* spatiale
s interacción *f* espacial
g räumliche Wechselwirkung *f*

5704 spatial modelling
Design and making of cartographic images on the basis of visualisation of the initial, derived or summarised materials.

r геомоделирование; пространственное моделирование
Проектирование и создание картографических изображений на основе визуализации исходных, производных или итоговых материалов.

f simulation *f* cartographique spatiale
s simulación *f* cartográfica espacial
g Raummodellierung *f*

5705 spatial object
In geoinformatics, a digital model of a geographical object containing the indication of its location and set of properties, characteristics, attributes.

r пространственный объект
В геоинформатике цифровая модель географического объекта, содержащая его местоуказание и набор свойств, характеристик, атрибутов.

f symbole *m* spatial
s símbolo *m* espacial; fenómeno *m* espacial
g Raumobjekt *n*

5706 spatial resolution; ground resolution; resolution
The minimum linear or angular distance between two objects at which they continue to be seen as distinct and separate on an image or photograph.

r пространственное разрешение; разрешение на местности
Наименьшее линейное или угловое расстояние между двумя объектами, на котором они продолжают различаться раздельно на космическом или аэроснимке.

f résolution *f* spatiale
s resolución *f* espacial
g geometrische Auflösung *f*; Erdauflösung *f*

5707 spatter cone; driplet cone
A small but steep-sided mound of lava, up to 15 m in height, built by small eruptions around a volcanic vent or along a fissure.

r брызговой конус
Небольшой крутосклонный холм из лавы высотой до 15 м, возникший в результате быстрого затвердения лавы после небольших извержений вокруг жерла вулкана или вдоль трещины.

f cône *m* de jaillissement
s cono *m* de salpicadura
g Spritzkegel *m*

* **spawning ground → 792**

5708 specialization of production
Concentration of production of similar products at specialized enterprises which is leading to the growth of labour productivity and production efficiency through the increase of mass production and flow production.

r специализация производства
Концентрация производства однородной продукции на специализированных предприятиях, ведущая к росту производительности труда и эффективности производства путем повышения его массовости и поточности.

f spécialisation *f* de production
s especialización *f* de producción; especialización *f* funcional
g Spezialisierung *f* der Produktion

5709 specialization of the region
The orientation of the economy of a region at producing certain commodities and services to be sold outside the region within the system of the territorial division of labour.

r специализация района
Ориентация хозяйства района на производство товаров и услуг для внерайонного потребления в общей системе территориального разделения труда.

f spécialisation *f* de la région
s especialización *f* de la región

g Spezialisation *f* der Region

5710 **species**
A group of living things having common gene pool, inhabiting a particular environment and capable of interbreeding with each other, but not with other species.
r биологический вид
Группа живых организмов, имеющих общий генофонд, населяющих определенный ареал и способных к скрещиванию.
f espèce *f*
s especie *f*
g Art *f*

* **species at risk of extinction → 1895**

5711 **specific discharge**
The runoff volume per unit of time from a unit of a watershed.
r модуль стока
Объем стока в единицу времени с единицы площади водосбора.
f débit *m* spécifique
s caudal *m* específico
g Abflußspende *f*; Wasserspende *f*

5712 **specific humidity**
The ratio of the weight of water vapour in a parcel of the atmosphere to the total weight of moist air (i.e. including water vapour).
r удельная влажность
Отношение веса водяного пара в единице объема атмосферы к общему весу влажного воздуха, включая и водяной пар.
f humidité *f* spécifique
s humedad *f* específica
g spezifische Feuchtigkeit *f*

5713 **specific retention**
The volume of water that is retained in a soil or rock (against the influence of gravity) following drainage after saturation.
r удельное удержание воды
Объем воды, который способна удержать почва или горная порода после ее насыщения водой и последующего ее стекания под влиянием гравитации.
f rétention *f* spécifique
s retención *f* específica
g spezifische Retention *f*

5714 **spectral *adj***
Related to a part of the total range of electromagnetic radiation and usually given in wavelength intervals.
r спектральный
Относящийся к части общего потока электромагнитного излучения, обычно выражаемый в определенной полосе длины волн.
f spectral *adj*
s espectral *adj*
g Spektral-

5715 **spectral albedo**
Albedo within a certain wave band of spectrum.
r спектральное альбедо
Альбедо в пределах той или иной полосы спектра.
f albédo *m* spectral
s albedo *m* espectral
g spektrale Albedo *f*

5716 **spectrazonal aerial survey; multispectral survey**
The photography of the Earth's surface from aerial and space flying vehicles simultaneously in several wave-bands.
r спектрозональная аэрофотосъемка; многоспектральная съемка
Фотографирование местности с воздушных и космических летательных аппаратов одновременно в нескольких спектральных диапазонах.
f photographie *f* à séparation de couleurs; levé *m* multispectral
s fotografía *f* aérea espectrozonal
g Spektrozonalluftaufnahme *f*; Multispektralaufnahme *f*

5717 **speed-up distance of snow drifting**
The distance along which a wind-blown snow stream is completely saturated with snow.
r длина разгона метели
Расстояние, на котором происходит полное насыщение снегом снеговетрового потока.
f distance *f* d'accélération du chasse-neige
s distancia *f* de aceleración de ventisca
g Sättigungslänge *f* des Schneestroms

* **spelaeology → 5718**

5718 **speleology; spelaeology**
The scientific study of caves, their formation and processes.
r спелеология
Составная часть карстоведения, изучающая пещеры.
f spéléologie *f*
s espeleología *f*
g Speläologie *f*; Höhlenforschung *f*

5719 **spent fuel**
Radioactive fuel which has been used in a nuclear reactor and can not support now a chain reaction.
r отработанное топливо
Топливо, которое было использовано в ядерном реакторе до такой степени, что больше не может поддерживать цепную реакцию.
f combustible *m* usé
s combustible *m* consumido
g aufgebrannter Brennstoff *m*

* **spent slurry → 5485**

5720 **sphagnum bog**
A raised bog with the predominance of peat-mosses.

r сфагновое болото
Моховое болото с преобладанием сфагновых мхов.
f tourbière *f* à sphaignes
s turbera *f* de sphaignes
g Torfmoosmoor *n*; Sphagnummoor *n*

5721 sphere of influence
1. A territory or part of a territory in which a foreign power has political and economic interests, special rights and privileges. 2. An area which depends on an urban centre for various services, or with which it has special relations.
r сфера влияния
1. Территория другой страны, на которую распространяется политическое, экономическое и культурное влияние данного государства. 2. Некоторая область, сфера обслуживания которой зависит от ближайшего урбанизированного центра, или она имеет с ним особые связи.
f sphère *f* d'influence
s esfera *f* de influencia
g Einflussbereich *m*

5722 spillway; weir; escape
The area below a dam or a natural obstruction over which excess water from a reservoir or lake above is allowed to drain away; a hydraulic construction in the form of a channel intended to remove the excess water from a reservoir and a passage for the overflow of water from a reservoir.
r водослив; водосброс
Место перелива воды при переполнении озера или водохранилища, а также гидротехническое устройство, предназначенное для сброса излишков воды из водохранилища или пруда.
f déversoir *m*
s lugar *m* de desagüe; aliviadero *m*
g Überlauf *m*

5723 spit
A low and narrow accumulation of sand and pebbles projecting away from the shore into a large body of water as a result of longshore drift.
r коса; стрелка; пересыпь
Низкая и узкая намывная полоса суши из песка и гальки на берегу моря или озера, причлененная к берегу и вдающаяся клином в пределы акватории.
f flèche *f* littorale; flèche *f* aiguillée
s flecha *f* litoral; flecha *f* agujada; barra *f* costera; barra *f* litoral
g Nehrung *f*; Landzunge *f*

5724 splash
Sudden appearance of spray when the water meets an obstacle.
r всплеск
Внезапное появление брызг, когда вода обходит препятствие.
f houle *f* gaufrée
s oleaje *m* encañonado
g Plätschern *n*

5725 spodosol
A soil that has a horizon of accumulation of compounds of aluminum and iron, and is similar to podzols.
r сподосоль
Почва с горизонтом накопления соединений алюминия и железа, сходная с подзолами.
f spodosol *m*
s spodosol *m*; spodosuelo *m*
g Spodosol *m*

5726 spoil
Waste material from mining or quarrying operations, piled up in mining dumps, embankments, spoil dums or waste piles.
r пустая порода
Отходы из шахт и карьеров, скапливающиеся в виде отвалов, насыпей, терриконов, предназначенных для хранения отходов.
f barle *f*; décombres *mpl*; gaine *f*
s ganga *f*; roca *f* estéril; escombrera *f*
g taubes Gestein *n*; Gangart *f*

* **spoil bank → 3875**

* **spoil heap → 3875**

5727 sporadic avalanche
A snow avalanche which occurs in a given place only once or twice for 50-100 years period.
r спорадическая лавина
Снежная лавина, сходящая в данном месте не чаще одного раза в 50-100 лет.
f avalanche *f* sporadique
s avalancha *f* esporádica
g sporadische Lawine *f*

5728 sporadic permafrost; island permafrost
A zone where individual areas of permafrost are surrounded by unfrozen ground.
r островная вечная мерзлота
Область, в которой встречаются отдельные участки многолетнемерзлых пород.
f pergélisol *m* sporadique
s permafrost *m* esporádico
g sporadischer Permafrost *m*

5729 spores and pollen analysis; pollen analysis
The study of the distribution of spores and pollen grains preserved in the deposits in order to infer the vegetation and climatic conditions in the past geological epochs.
r спорово-пыльцевой анализ
Изучение закономерностей распределения спор и пыльцы в отложениях с целью восстановления растительности и климатических условий прошлых геологических эпох.
f analyse *m* pollinique
s análisis *m* de esporas y polen

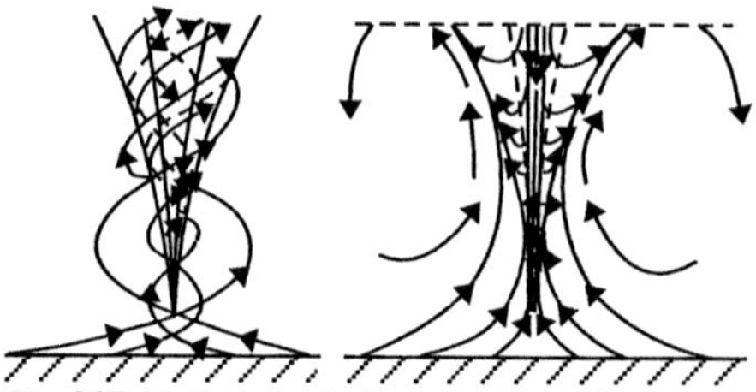

Fig. 307. Vertical distribution of air fluxes in a **spout**

g Sporen- und Pollenanalyse *f*

5730 spot height
An exact point on a map with an accompanying elevation printed alongside.
r высотная отметка
Цифра, указывающая абсолютную высоту характерной точки местности на карте.
f cote *f* d'altitude
s cota *f* de altitud
g Höhenkote *f*

5731 spotted tundra
A type of tundra where the bare multi-sided sites or round bare spots are surrounded by the terrain covered with tundra vegetation.
r пятнистая тундра
Тип тундры, в которой участки небольших многоугольников или округлых пятен, лишенных растительности, вкраплены в покрытую растительностью тундру.
f toundra *f* maculée
s tundra *f* esporádica
g sporadische Tundra *f*

5732 spout; whirlwind
A rapidly rotating column of air under a funnel-shaped cloud which advances over tha land along a narrow path. → Fig. 307
r смерч
Атмосферный вихрь в виде столба с воронкообразными расширениями сверху и снизу, возникает под грозовым кучево-дождевым облаком и распространяется до земной поверхности. → Рис. 307
f trombe *f*
s tromba *f*; vórtice *m*; manga *f*
g Windhose *f*; Trombe *f*

5733 spray
Fine water drops spurting off the waves and squirting in the air.
r брызги
Мелкие частицы воды, срывающиеся с волны в воздух.
f embruns *mpl*
s gotas *fpl*; agua *f* en suspensión en el aire
g Spitzer *mpl*

5734 spray freezing; sea icing
The formation of ice coating on exposed objects produced by the freezing of water droplets carried by strong wind from water bodies in cold weather.
r брызговое обледенение; морская наледь
Образование льда на предметах от замерзания водяной пыли, переносимой штормовым ветром с водоемов в морозную погоду.
f givre *m* par gouttelettes
s escarcha *f* por gotitas
g Vereisung *f* durch Ansprühen; Meeraufeis *n*

5735 spray irrigation; shrinkler irrigation
Artificial watering of agricultural fields by rain produced with special devicies.
r дождевание
Искусственное орошение сельскохозяйственных полей в виде дождя, создаваемого специальными установками.
f aspersion *f*
s aspersión *f*
g Beregnung *f*

* **spreading zone → 5274**

5736 spring crops
Cultivated annual plants such as cereals, melons, etc., which develop well when sowed in spring to secure harvest in autumn, in contrast to *winter crops.*
r яровые культуры
Однолетние растения (зерновые, технические, бахчевые, овощные и др.), нормально развивающиеся, в отличие от *озимых культур*, при посеве весной и дающие урожай в год посева.
f cultures *fpl* de printemps
s cultivos *mpl* de primavera
g Sommergetreibe *n*

* **spring equinox → 6548**

* **spring flood → 5737**

5737 spring high water; spring flood
High and long rise of water in plain rivers caused by spring snowmelt.
r весеннее половодье; снеговое половодье
Высокий и длительный подъем воды на равнинных реках, вызванный весенним таянием снега.
f crue *f* de printemps
s crecida *f* de primavera
g Frühjahrshochwasser *n*

* **spring inversion → 5543**

5738 spring tide
The tidal condition during new moon and full moon (*syzygies*), characteristic of a period when the Moon is at its perigee, so that its gravitational pull is great, resulting in a high tide that is higher, a low tide that is lower, and a tidal range greater than usual.
r сизигийный прилив; перигейный прилив
Прилив во время новолуний и полнолу-

ний (*сизигий*), когда Луна находится в перигее и ее приливная сила максимальна, благодаря чему высокий прилив повышается, а низкий понижается, и величина прилива становится больше обычной.
f grande marée *f*; marée *f* de vives eaux
s marea *f* viva
g Springflut *f*; Springtide *f*

5739 spring[1]
The season of growth and development occurring between winter and summer, the period astronomically defined from the vernal equinox to the summer solstice (from March 20-21 to June 21-22 in Northern Hemisphere, from September 23 to December 22 in the Southern Hemisphere).
r весна
Время года, переходный сезон между зимой и летом, в астрономическом понимании – промежуток времени от момента весеннего равноденствия до летнего солнцестояния (с 20-21 марта до 21-22 июня в Северном полушарии, с 23 сентября до 22 декабря в Южном).
f printemps *m*
s primavera *f*
g Frühling *m*; Frühjahr *n*

5740 spring[2]
The natural issue of ground water to the earth's surface at the point where the water-table intersects the surface. → Fig. 308
r источник; ключ; родник
Естественный выход подземных вод на земную поверхность в том месте, где водное зеркало пересекается дневной поверхностью. → Рис. 308
f source *f*
s manantial *m*; fuente *f*
g Quelle *f*

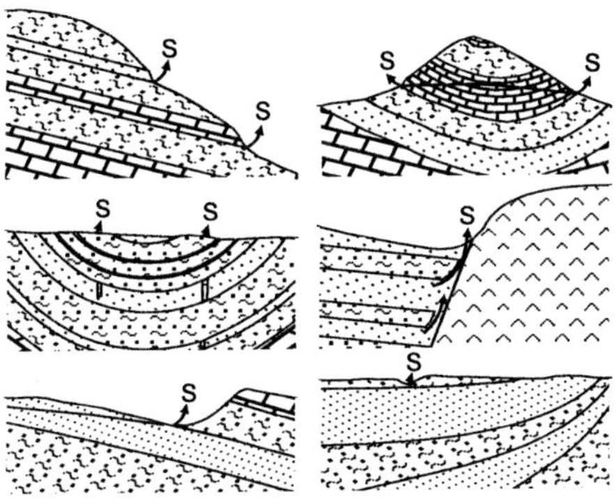

Fig. 308. Different locations of natural **springs[2]** (S)

Fig. 309. **Spruce forest**

5741 spring-line
A line of springs occurring roughly at the level where, by reason of deposition of the strata, the water table reaches the surface, as at the foot of an escarpment, such springs attracted early settlements.
r выход источников
Череда источников, появляющаяся в том месте, где водное зеркало в силу геологического напластования достигает поверхности, как это бывает у подножья склона, в таких местах возникали ранние поселения.
f ligne *f* des sources
s línea *f* de fuentes
g Quellenlinie *f*

5742 spruce forest
Dark coniferous forest growing on rich substrate and dominated by common spruce (Picea abies) in Europe, Siberian spruce (Picea obovata) in Siberia. → Fig. 309
r еловый лес; ельник
Темнохвойный лес на богатом субстрате, с господством ели обыкновенной (Picea abies) в Европе, ели сибирской (Picea obovata) в Сибири. → Рис. 309
f sapinière *f*; épicéa *m*; sapinaie *f*; abetal *m*; sapineraire *f*
s bosque *m* abetal
g Fichtenwald *m*; Fichtenforst *m*; Tannenwald *m*

5743 spur; offshoot
The ridge which extends from the main mountain range.
r отрог
Хребет второго порядка, отходящий от главной цепи и имеющий иное простирание.
f chaînon *f*
s estribación *f*
g Gebirgsausläufer *m*

* **sputnik → 5216**

5744 squall
An atmospheric phenomena characterized by sudden and violent gusts of wind which last for several minutes, often accompanied with a shower or thunderstorm.
r шквал

Атмосферное явление, характеризующееся резким усилением скорости ветра продолжительностью в несколько минут, часто сопровождается ливнем или грозой.
f grain *m*
s turbonada *f*
g Windbö *f*; Stoßwind *m*

5745 **squall line; squall-line thunderstorm; line squall**
A cluster of storm-bearing covection cells, measuring in total some 100 km, located along a non-frontal line or belt; a very stormy phase of weather of short duration, extending along a sharply defined line a head of a cold front.
r линия шквала
Узкая грозовая полоса, не всегда связанная с фронтом и не всегда сопровождающаяся шквалами; кратковременная фаза очень штормовой погоды на резко выраженной линии вдоль холодного фронта.
f ligne *f* de grains;
orages *mpl* de la ligne de grains
s línea *f* de turbonadas;
tormentas *fpl* de la línea de turbonadas
g Böenlinie *f*

* **squall-line thunderstorm → 5745**

* **squatter settlement → 5486**

5746 **squatting$_1$**
Haphazard building in cities in the past frequently reflected in their modern arrangements; the illegal commandeering of housing or land.
r хаотическая застройка; захват
Стихийное строительство городов в прошлом, нередко находящее отражение и в их современном облике; незаконный захват земли и зданий.
f habitat *m* spontané
s hábitat *m* espantáneo
g ungeordnete Bebauung *f*

5747 **squatting$_2$**
The occupation of lands that have no definite ownership for agricultural purposes as well as a landed plot with houses and outbuildings built on newly developed lands.
r заимка
Занятие под сельскохозяйственную обработку никому не принадлежащих земель, а также земельный участок с жилой и хозяйственной постройками на нем, возникшие на вновь осваиваемой земле.
f squatter *m*
s casería *f*
g Anwesen *n*

5748 **squeezing**
The pressing out of water-saturated debris material into glacial fissures and tunnels under the action of glacier load.
r сквизинг
Выдавливающее нагнетание водонасыщенного обломочного материала в ледниковые трещины и туннели под действием ледниковой нагрузки.
f squeezing *m*
s squeezing *m*
g Squeezing *n*

5749 **St Elmo's fire; corona discharge; Saint Elmo's fire**
A more or less continuous luminous electrical discharge in the atmosphere, emanating from elevated objects at the Earth's surface or from aircraft in flight, sometimes acoompanied by a crackling noise.
r огни святого Эльма; коронный разряд
Более или менее постоянный светящийся электрический разряд в атмосфере, исходящий от возвышающихся над землей предметов или от летящих воздушных судов, иногда сопровождается треском.
f feu *m* Saint-Elme;
décharge *f* lumineuse silencieuse
s fuego *m* de San Telmo;
descarga *f* en corona
g St. Elms Feuer *n*; Koronaentladung *f*

* **St. Martin summer → 3057**

* **stability limit → 5750**

5750 **stability threshold; stability limit**
The maximum impact which an organism, community or environment could bear keeping in action their structure and functional duties.
r порог устойчивости;
предел устойчивости
Максимально возможное воздействие извне, переносимое организмом, сообществом, окружающей средой при сохранении их структуры и функциональных свойств.
f seuil *m* de stabilité
s umbral *m* de estabilidad
g Stabilitätsgrenze *f*

5751 **stabilization lagoon**
A pond used to purify sewage by allowing sunlight to fall on the mixture of sewage and water.
r очистной пруд
Пруд, служащий для очистки сточных вод путем использования солнечного света для разложения их загрязненной составляющей.
f étang *m* stabilisateur; bassin *m* de stabilisation; bassin *m* d'épuration
s estanque *m* estabilizador
g Stabilisierungsbecken *n*

5752 **stabilized dune**
A dune that has become fixed oe anchored by vegetation and is therefore protected from further aelian action.
r закрепленная дюна

Дюна, ставшая неподвижной вследствие развития густого растительного покрова.
f dune *f* fixée; dune *f* morte; dune *f* stabilisée
s duna *f* fijada; duna *f* estabilizada; duna *f* muerta
g unbewegliche Düne *f*

* **stable atmospheric equilibrium → 440**

5753 **stable channel**
A channel in which load deposition is balanced by the washing out or is not subject to washing at all due to the resistance of rocks.
r устойчивое русло
Русло, в котором отложение наносов уравновешивается их размывом, либо вообще не подвергается размыву из-за стойкости горных пород.
f lit *m* stable
s lecho *m* estable
g stabiles Flußbett *n*

5754 **stable community**
A long-existing community the changes in which go on slowly, during centuries and thousand years.
r устойчивое сообщество; климаксное сообщество; устойчивый биоценоз
Длительно существующее, сложившееся сообщество, изменения в котором протекают очень медленно – в течение столетий и тысячелетий.
f groupement *m* durable
s comunidad *f* estable
g Dauergesellschaft *f*

* **stable current → 5785**

5755 **stable ecosystem**
An ecosystem which will maintain or return to its original condition following any disturbance.
r устойчивая экосистема
Экосистема, которая остается или возвращается к своему первоначальному состоянию, после того, как она из него выведена.
f écosystème *m* stable
s ecosistema *f* estable
g stabile Ökosystem *n*

* **stable population → 5789**

5756 **stable population; zero population growth; stationary population**
The ending population growth when the number of depths equals the number of births.
r стабильное население
Неизменяемая общая численность населения, когда количество смертей в целом равно количеству рождений.
f population *f* stable
s población *f* estable
g Stabilbewölkerung *f*

5757 **stable slope**
A slope on which rapid mass-movement does not occur, because the strength of the rock and regolith is greater than the applied stresses which tend to cause slope failure.
r устойчивый склон
Склон, на котором не происходит быстрых движений масс грунта из-за значительной твердости слагающих пород, которую не могут преодолеть любые воздействия, приводящие к обвалам и оползням.
f versant *m* stable
s vertiente *f* estable
g stabiler Abhang *m*

5758 **stable snow cover**
A snow cover which remains unthawed over the whole winter.
r устойчивый снежный покров; установившийся снежный покров
Снежный покров, лежащий в течение всей зимы или с небольшими перерывами.
f manteau *m* neigeux stabilisé
s manto *m* nivoso estabilizado
g beständige Schneedecke *f*

5759 **stack; sea stack**
An isolated mass of rock near a coastline, detached from the main mass usually by marine erosion (especially by wave action), rising steeply from the surrounding sea.
r прибрежная скала
Изолированный останец горных пород вблизи береговой линии, отделенный от берега волновой эрозией.
f roche *f* littorale
s peña *f*; cabeza *f* de roca; forallón *m*
g Küstefelsen *m*; Schornstein *m*

5760 **stadial**
A time of glaciation when glaciers advanced and periglacial coditions extended, but not as significantly as in ice age (glacial).
r стадиал
Стадия оледенения, когда ледники продвигались вперед и разрастались перигляциальные условия, но не так существенно, как в ледниковую эпоху.
f stade *m*
s estadio *m*
g Stadial *n*

5761 **stadial moraine; recessional moraine**
A moraine marking a recessional phase in the overall decline of an ice sheet or glacier.
r стадиальная морена
Конечно-моренные гряды, сформированные при наступаниях ледника, вызванных вековыми и внутривековыми колебаниями климата.
f moraine *f* stadiaire
s morrena *f* estadiaria
g Stadialmoräne *f*; Rückzugsmorane *f*

* **stagnant ice → 1435**

5762 stagnant water
Water accumulated in a closed depression on the surface of impermeable rock or underground and deprived of inflow and outflow and having no vertical circulation of water.
r застойная вода
Вода в замкнутом понижении поверхностного или подземного водоупорного слоя, лишенная притока и оттока и соответственно вертикальной циркуляции.
f eau *f* stagnante
s agua *f* estancada
g stehendes Gewässer *n*

5763 stagnic cambisol
A soil of the taiga zone that has a bleached horizon (in which iron is absent) and is devoid of illuvial-clay horizon.
r светлозем
Почва тайги с осветленным горизонтом, в котором отсутствует железо, и без иллювиально-глинистого горизонта.
f svetlozem *m*
s svetlozem *m*
g Swetlosem *m*

5764 stalactite
A dripstone deposit in the form of an icicle or fringe hanging down from the roof of a karst cave, created by the re-precipitation of carbonate in calcite form from percolating groundwater in a karst environment.
r сталактит
Натечно-капельное образование в виде сосульки или бахромы, свешивающихся с потолка карстовой пещеры, возникающее при постоянном поступлении карбоната в форме кальцита из просачивающихся грунтовых вод.
f stalactite *m*
s estalactita *f*
g Stalaktit *m*

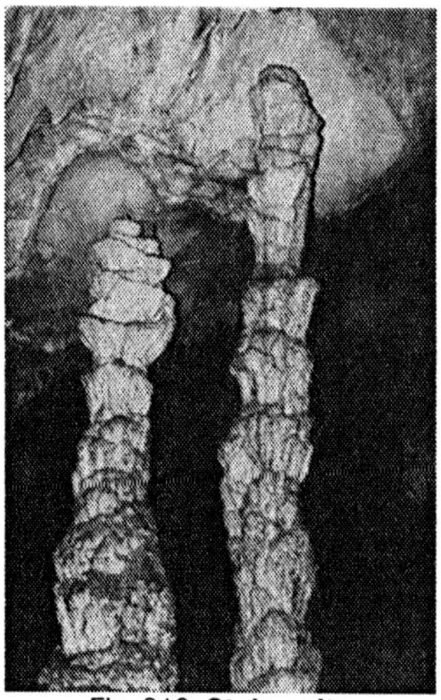
Fig. 310. **Stalagmite**

5765 stalagmite
A columnar or conical dripstone deposit rising from the floor of a cave, created from the re-precipitation of carbonate in calcite form perpendicularly beneath a constant source of groundwater that drips off the lower tip of a stalactite or percolates through the roof of a cave in a karst environment. → Fig. 310
r сталагмит
Натечно-капельное образование в виде конуса или столба, поднимающееся со дна карстовой пещеры, возникающее при постоянном поступлении карбоната в форме кальцита из просачивающихся грунтовых вод, падающих вниз со сталактита. → Рис. 310
f stalagmite *m*
s estalagmita *f*
g Stalagmit *m*

5766 stalagnate
A dripstone formation in a karst cave which forms when a stalactite and a stalagmite meet and form a continuous column.
r сталагнат
Натечно-капельное образование в виде колонны, возникающее в карстовой пещере при срастании сталактита и сталагмита.
f stalagnate *m*
s estalagnata *f*
g Stalagnat *m*

5767 stall-feeding of livestock; zero grazing
Keeping cattle in stalls, i.e. indoor maintenance of cattle.
r стойловое содержание скота
Содержание крупного рогатого скота вне пастбищ.
f elevage *f* à l´étable
s ganadería *f* en establo
g Stallhaltung *f* des Viehs

5768 stand; forest stand
The trees which form forest, an area occupied by a collection of trees which are structurally or floristically homogeneous.
r древостой
Совокупность деревьев, образующих лес, где собрание деревьев структурно и флористически весьма однородно.
f peuplement *m* forestier
s conjunto *m* de arboles; poblamiento *m* arbóreo
g Bestand *m*

5769 standard atmosphere; reference atmosphere
Conventional reference atmosphere representing an approximate average distribution of pressure, temperature and density at succesive altitudes, used as a unit of measure in calibrating instruments, such as altimeter.

r стандартная атмосфера
Условная атмосфера, в которой распределение по высоте температуры, давления и плотности воздуха характеризует ее среднее годовое состояние, используется для калибровки приборов, например, альтиметра.
f atmosphère *f* standard; atmosphère *f* typique
s atmósfera *f* de referencia
g Standardatmosphäre *f*; Normalatmosphäre *f*

5770 standard depth of measurements
Fixed depths of oceanographic observations accepted by the International Association of Physical Oceanography.
r стандартный горизонт (гидрологических наблюдений)
Фиксированные глубины океанографических наблюдений, принятые Международной ассоциацией физической океанографии.
f profondeur *m* de référence
s profundidad *f* de referencia
g Standardtiefe *f*

*** standard level → 4163**

5771 standard line; standard parallel
A line on a map along which the principal scale is retained.
r линия нулевых искажений
Линия на карте, во всех точках которой сохраняется главный масштаб.
f isométre *m*
s línea *f* de sección; paralelo *m* base
g Berührungslinie *f*

*** standard mean ocean water → 5773**

5772 standard ocean
A convenitional notion accepted in oceanology corresponding to the ocean with temperature of 0°C, salinity equal to 35‰ and pressure proportional to the depth.
r стандартный океан
Условное понятие, принятое в океанологии, соответствующее состоянию океана при температуре 0°C, солености 35‰ и давлении, пропорциональном глубине.
f océan *m* standard
s océano *m* (e)standartizado
g Standardozean *m*

*** standard of living → 3538**

*** standard parallel → 5771**

5773 standard sea water; standard mean ocean water
Purified sea water with precisely fixed chlorine content (about 19.38% that corresponds to salinity of 35‰) used as the international standard for identification of a sea water salinity.
r стандартная морская вода; стандарт средней морской воды
Очищенная морская вода с точно установленным содержанием хлора (около 19,38%, что соответствует солености 35‰), используемая в качестве международного эталона для определения солености морской воды.
f eau *f* marine normalle
s agua *f* marina normal
g Normalseewasser *n*

5774 standard time
The system of counting time by meridional time zones with longitudes divisible by 15° (the time in contiguous zones differ by one hour), the time zones extending on either side of the standard meridian, and the places within that zone are governed by the same standard time.
r поясное время
Система счета времени по часовым поясам, простирающимся вдоль меридианов с долготой, кратной 15° (в смежных поясах время различается на один час), часовой пояс протягивается по обе стороны от стандартного меридиана и повсюду в этом поясе время одинаково.
f heure *f* légale
s hora *f* legal
g Zonenzeit *f*

5775 standards of environmental quality; environmental quality standards
The threshold amounts of pollutants which are permitted to be present in the environment without altering its quality.
r нормативы качества окружающей среды
Пороговые величины допустимой концентрации загрязнителей в окружающей среде, не нарушающие ее природных свойств и качества.
f standards *mpl* de qualité d'environnement
s standardes *mpl* de calidad de medio ambiente
g Umweltqualitätsstandards *mpl*; Umweltqualitätsnormen *fpl*

5776 standing floe; ropak
An isolated floe standing vertically or slanting above a relatively smooth surface of an ice field.
r ропак
Отдельная льдина, стоящая вертикально или наклонно среди относительно ровной поверхности плавучего льда.
f plaque *m* de glace «debout»
s bloque *m* de hielo vertical
g Ropak *m*

5777 standing wave; stationary wave
A type of wave in which the surface oscillates vertically between fixed points called «nodes», without any movement forward (the crest at one moment becomes the trough at

the next and so on), it is formed in the ocean when a wave reflected from a sea cliff matches the incoming wave.

r стоячая волна
Тип волны, когда поверхность колеблется вертикально между фиксированными точками, называемыми «узлами», без продвижения вперед (в какой-то момент гребень становится ложбиной волны и наоборот), подобная волна формируется в океане при отражении волн от берегового уступа.

f vague *f* stationnaire; onde *f* stationnaire
s ola *f* estacionaria; onda *f* estacionaria
g stehende Schwingung *f*

* **starting zone of avalanche → 492**

5778 **state; country; nation$_2$**
A territorial unit with clearly defined and internationally accepted boundaries, having the sovereignty or being under the power of another state, such as colony or protectorate.

r государство; страна
Территория, имеющая границы, признанные международным сообществом, пользующаяся государственным суверенитетом или находящаяся под властью другого государства (колонии, подопечные территории).

f état *m*; pays *m*
s estado *m*; país *m*; nación$_2$ *f*
g Staat *m*; Land *n*

5779 **state capitalism**
An economic system in which the state owns or controls a major part of the economy, will consistently supporting the interests of private capital.

r государственный капитализм
Экономическая система, в которой государство владеет или контролирует основную часть экономики, вместе с тем поддерживая интересы частного капитала.

f capitalisme *m* d´etat
s capitalismo *m* de estado
g Staatskapitalismus *m*

5780 **state farm; sovkhoz; state farming; sovkhoz**
A state-owned agricultural enterprise the property of which is transferred to the farm for economic activities.

r совхоз
Сельскохозяйственное государственное предприятие, собственность которого принадлежит государству и закрепляется за совхозом на правах хозяйственного ведения.

f sovkhoze *m*; ferme *f* d´Etat; exploitation *f* d´Etat
s sovjós *m*; explotación *f* agraria estatal; sovjoz *m*
g Sowhos *m*

* **state farming → 5780**

5781 **state-archipelago**
A state which consists of one or more archipelagos and the water between them.

r государство-архипелаг
Государство, территория которого полностью состоит из одного или более архипелагов вместе с водами, заключенными между ними.

f état-archipel *m*
s estado-archipiélago *m*
g Archipelstaat *m*

5782 **station**
A place with its associated buildings on a route where buses, trains, etc. habitually stop to take on and discharge passengers, goods, mail.

r станция
Пункт на дороге со всеми сопутствующими строениями, где обычно останавливаются автобусы, поезда и пр., и происходит обслуживание пассажирских и грузовых перевозок.

f gare *f*; station *f*
s estación[1] *f*
g Bahnhof *m*; Station *f*

5783 **stationary anticyclone**
A slowly moving anticyclone remaining in the same place for a long time.

r стационарный антициклон
Малоподвижный антициклон, остающийся в течение длительного времени на одном месте.

f anticyclone *m* stationnaire
s anticiclón *m* estacionario
g stationäres Hoch *n*

5784 **stationary climatic series**
A series of climatic or meteorological variables, the statistical characteristics of which do not change with time.

r стационарный климатический ряд
Ряд климатических или метеорологических величин, статистические характеристики которого не меняются со временем.

f serie *f* climatique stationnaire
s serie *f* climática estable
g stationäre Klimareihe *f*

5785 **stationary current; stable current**
A water flow the average speed of which remains constant over a long time.

r стационарное течение; установившееся течение; устойчивое течение
Водный поток, средняя скорость которого остается постоянной в течение длительного времени.

f courant *m* stationnaire
s corriente *f* estable
g stationäre Strömung *f*; stetige Strömung *f*

5786 **stationary cyclone; stationary depression**
A slow moving cyclone remaining at the same region for a long time.

r стационарный циклон
Малоподвижный циклон, длительное время сохраняющийся в одном и том же районе.
f dépression *f* stationnaire
s depresión *f* estacionaria
g stationäres Tiefdruckgebiet *n*

* **stationary depression → 5786**

5787 stationary front; quasi-stationary front
A front whose position on succesive synoptic charts is unchanged or almost unchanged.
r стационарный фронт
Атмосферный фронт, почти не меняющий своего положения на последовательных синоптических картах.
f front *m* stationnaire; front *m* quasi-stationnaire
s frente *m* estacionario; frente *m* casi estacionario
g stationäre Front *f*; ortsfeste Front *f*

5788 stationary method of research
The study of natural processes and geographical objects, e.g. landscapes, in the same location for a long time.
r стационарный метод исследований
Изучение природных процессов и географических объектов, например, ландшафтов, в одном и том же месте на протяжении длительного времени.
f méthode *f* stationnaire de exploration
s método *m* de exploración estable
g stationäre Forschungsmethode *f*

* **stationary population → 5756**

5789 stationary population; zero population growth; stable population
A population where fertility and mortality are constant hence the population grows at a constant rate.
r демографическое равновесие
Состояние населения на данной территории, когда его численность не изменяется (в том числе по возрастам), так как смертность примерно соответствует рождаемости.
f équilibre *m* démographique; population *f* stable
s crecimiento *m* demográfico cero; estabilización *f* demográfica; populación *f* estable
g demographischer Equilibrium *m*; demographisches Gleichgewicht *n*

* **stationary wave → 5777**

* **statute mile → 3856**

* **steady precipitation → 1261**

5790 steam fog; evaporation fog
A type of fog formed when cold air passes over the surface of warmer fresh water, the moisture from the water condenses into tiny visible droplets in the air so that its surface seems to steam (steam fog is common in winter over rivers where the air is more than 10°C colder than the water).
r надводный туман; туман парения
Туман, возникающий при прохождении холодного воздуха над поверхностью более теплой пресной воды, когда поднимающаяся с воды влага конденсируется в воздухе в мельчайшие видимые капли, так что водная поверхность кажется подернутой дымкой (обычен зимой над рекой, когда воздух над водой на 10°C холоднее).
f brouillard *m* d'evaporation
s niebla *f* de evaporación
g Dampfnebel *m*

5791 steel industry; ferrous metallurgy; iron-and-steel industry
A branch of heavy industry engaged in producing ferrous metals, pipes, rolled stock, etc. and is the basis for machine-building.
r черная металлургия
Отрасль тяжелой промышленности, занимающаяся производством черных металлов, труб, проката и пр., основа машиностроения.
f sidérurgie *f*
s siderurgia *f*; metalurgia *f* de hierro y acero
g Eisen- und Stahlindustrie *f*

5792 steep slope
A mountain slope with sharp features widespread near summits and crests of ridges as well as at sides of deeply cut valleys.
r крутой склон
Горный склон с резкими очертаниями, типичен для предвершинных участков гор и бортов глубоко врезанных долин.
f pente *f* escarpée; pente *f* raide
s pendiente *f* escarpada; pendiente *f* propícia
g Steilhang *m*

5793 steepness of slope
A measure of the inclination of the earth's surface.
r крутизна склона
Степень наклона земной поверхности.
f raideur *f* de la pente
s inclinación *f* de pendiente; pendiente *f*
g Hangneigung *f*

5794 stenobathic animal
An aquatic animal whose natural habitat is limited by certain depths, e.g. they live only in the abyssal or littoral.
r стенобатное животное
Водное животное, обитающее только на определенной глубине водоема, например, в абиссали или литорали.
f animal *m* stenobathique
s animal *m* estenobático
g stenobathisches Tier *n*

5795 stenohaline organism
An organism which tolerates only slight fluctuations of salinity.
r стеногалинный организм
Организм, переносящий лишь небольшие колебания солености.
f organisme *m* sténohalin
s organismo *m* estenohalino
g stenohaliner Organismus *m*

5796 stenohydrobiont
An organism adapted to strictly specified conditions of moistening.
r стеногигробионт
Организм, способный существовать лишь в узкой зоне увлажненности.
f sténohydrobionte *m*
s estenohidrobionte *m*
g Stenohydrobiont *m*

5797 stenooxybiont
An organism adapted to live only under slight fluctuations of oxygen content in the air or in the water.
r стенооксибионт
Организм, способный существовать лишь при незначительных колебаниях содержания кислорода в воздухе или кислорода, растворенного в воде.
f sténooxybionte *m*
s estenooxibionte *m*
g Stenooxybiont *m*

5798 stenophagous animal
An animal that feeds on a limited range of food, e.g. crossbills feeding only on the seeds of coniferous species.
r стенофаг
Животное узкой пищевой специализации, питание которого ограничено немногими видами пищи, например, клест, питающийся семенами исключительно хвойных пород.
f animal *m* sténophage
s animal *mpl* estenófago
g Stenophag *m*

5799 stenotherm
An organism adapted to a narrow range of temperatures, e.g. the inhabitants of thermal springs (blue-green algae and bacteria).
r стенотермный организм
Организм, приспособленный к жизни в узком температурном интервале, например, обитатели термальных источников (синезеленые водоросли и бактерии).
f sténotherm *m*
s estenotermo *m*; estenotermófilo *m*
g stenothermer Organismus *m*

5800 stenotopic organism
An organism adapted to a narrow range of environmental conditions (e.g. sandy deserts, solonchaks, sphagnum bogs, etc.).
r стенобионт
Организм, приспособленный к существованию в строго определенных условиях (песчаных пустынях, солончаках, сфагновых болотах и т.п.).
f organisme *m* sténotopique; sténobionte *m*
s organismo *m* estenotopo; estenobionte *m*
g stenotoper Organismus *m*; Stenobionte *m*

5801 step fault
A series of parallel faults each with a throw that projects in the same directions, but to a greater distance than the one above, thus producing step-like changes of level of strata.
r ступенчатый сброс
Серия параллельных разломов с вертикальным смещением того же направления, но с увеличивающимся расстоянием по сравнению с предыдущим разломом, так что возникает ступенчатое залегание пластов.
f escalier *m* de failles
s escalera *f* de fallas
g Staffelbruch *m*

* **steppe → 2608**

5802 steppe expansion
The expansion of the areas of steppe landscapes at the expense of adjacent areas due to natural and anthropogenic factors.
r остепнение
Расширение площади степных ландшафтов за счет смежных территорий под воздействием естественных и антропогенных факторов.
f steppisation *f*; formation *f* de steppe
s sabanización *f*; estepización *f*; formación *f* de estepa;
g Versteppung *f*

5803 steppe fauna
The animals inhabiting steppes, e.g. ungulates, rodents, reptiles.
r степная фауна
Совокупность видов животных, обитающих в степях, обычны копытные, грызуны, пресмыкающиеся.
f faune *f* de steppe
s fauna *f* de estepa
g Steppenfauna *f*

* **steppe flora → 5805**

5804 steppe meadow
A meadow with a large participation of steppe grasses and low shrubs, it is often the result of overgrazing.
r остепненный луг
Луг, в котором существенную роль играют степные злаки и мелкие степные кустарнички, часто результат чрезмерного выпаса скота.
f prairie *f* steppisée
s pradera *f* estepizada
g Steppenwiese *f*; Steppenrasen *m*

5805 steppe vegetation; steppe flora
The vegetation characterized by the communities of xerophilous grasses – gramineous plants and forbs, as well as steppe bushes in depressions.
r степная растительность
Растительность степи, представленная сообществами ксерофильных трав (злаками) и в меньшей степени разнотравьем, а также степными кустарниками в западинах.
f végétation *f* de steppe
s vegetación *f* de estepa
g Steppenvegetation *f*

*** steppe zone → 2608**

5806 stepped relief
The terrain characterised by stepped slopes which connect several planation surfaces or terraces.
r ступенчатый рельеф
Рельеф, отличающийся ступенчатыми склонами, соединяющими несколько уровней поверхностей выравнивания или террас.
f relief *m* des gradins; relief *m* des replats
s relieve *m* escalonado; relieve *m* de gradas; relieve *m* de rellanos
g Stufenrelief *f*

5807 stepped slope
A slope with alternating subhorizontal and inclined parts.
r ступенчатый склон
Склон с чередующимися субгоризонтальными и наклонными участками.
f versant *m* á gradins
s vertiente *f* escalonada
g Stufenhang *m*

5808 stepwise migration
A type of migration which occurs in a series of movements, e.g. from a hamlet to a village, from a village to a town, and from a town to a city, that is, up the hierarchy.
r ступенчатая миграция
Тип миграции, когда перемещение людей следует иерархической лестнице: из маленькой деревушки в село, затем из села в небольшой город, и, наконец, из небольшого в большой город.
f migration *f* à marches
s migración *f* escalonada
g Stufenmigration *f*

5809 stereographic projection
A kind of azimuthal perspective map projection got by projecting from a point situated on a sphere's surface.
r стереографическая проекция
Азимутальная перспективаная картографическая проекция, получаемая проектированием из точки, расположенной на поверхности шара.
f projection *f* stéréographique
s proyección *f* estereográfica
g stereographische Projektion *f*

5810 stereomodel
Three-dimensional model of an object of a remote sensing got on a stereopair and serving as the basis for making digital models of relief by photogrammetry methods.
r стереоскопическая модель; стереомодель
Объемная модель объекта дистанционной съемки, полученная по стереопаре и служащая основой для фотограмметрического построения цифровой модели рельефа.
f modèle *m* stéréoscopique; stéréominute *f*
s modelo *m* estereoscópico; estereominuta *f*
g Stereomodell *n*; stereoskopisches Modell *n*

5811 stereopair
Two aerial photographs of the Earth's surface overlapping not less than 55%.
r стереопара
Два аэрофотоснимка земной поверхности, снятых с продольным перекрытием не менее 55%.
f stéréogramme *f*
s estereograma *f*
g Stereopaar *n*

5812 stereophotogrammetric survey
The main method of topographic survey when the representation of terrain on maps is got as a result of the processing of stereopairs.
r стереотопографическая съемка
Основной метод топографической съемки, когда изображение местности получают в результате обработки стереопар.
f topographie *f* par restitution stéréophotogrammétrique
s levantamiento *m* estereofotogramétrico
g stereophotogrammetrische Aufnahme *f*

5813 stereophotogrammetry
The determination of shapes, sizes and spatial location of the Earth's surface features by the measurements taken on the stereopair of their images.
r стереофотограмметрия
Определение формы, размеров и положения объектов земной поверхности путем измерения их изображения на стереопарах аэрофотоснимков.
f stéréophotogrammétrie *f*
s estereofotogrametría *f*
g Stereophotogrammetrie *f*; Raumbildmessung *f*

5814 steric variations of sea level
Sea level fluctuations caused by the change of the sea water density.
r стерические колебания уровня моря
Изменения уровня моря, обусловленные изменением плотности морской воды.
f variations *fpl* steriques du niveau de la mer
s variaciones *fpl* estéricas de nivel del mar

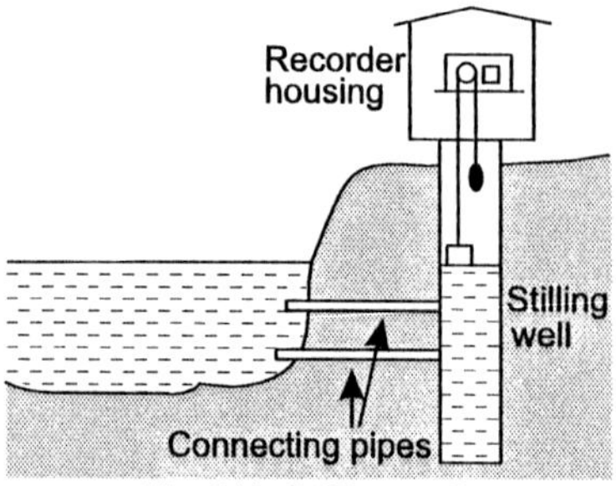

Fig. 311. **Stilling well**

g sterische Schwankungen *fpl* des Meeresspiegels

5815 stilling well
A large-diameter vertical tube on the bank of a river connected to the water in the channel aimed to measure the still-water surface level in the river channel. → Fig. 311
r успокоительный колодец
Вертикальная труба большого диаметра на берегу реки, соединенная с речным руслом с целью измерения спокойного уровня воды в реке. → Рис. 311
f puits *m* rassurant
s pozo *m* calmante
g Tosbecken *n*

5816 stock exchange
An organized market where commodities, securities, foreign currencies and labour force are traded, there are, accordingly, commodity, currency (foreign exchange) and labour markets.
r биржа
Регулярно действующий оптовый рынок товаров, ценных бумаг, иностранной валюты и рабочей силы (соответственно различаются товарная, фондовая, валютная биржи и биржа труда).
f bourse *f*
s bolsa *f*
g Börse *f*

5817 stock of maps
A collection of maps held by any particular agency.
r картографический фонд
Собрание картографических произведений в каком-либо учреждении.
f archivage *m* cartographique
s archivo *m* cartográfico
g Kartensammlung *f*

5818 stock¹
A relatively small irregularly shaped intrusion of igneous rock, smaller than a batholith cuts across the strata of the rock.
r шток
Интрузивное тело относительно небольшого размера (меньше батолита) неправильной формы, близкой к цилиндрической.
f stock¹ *m*
s stock¹ *m*
g Stock *m*

5819 stock²
An accumulation of goods held for future use or maintained as a constant source of supply.
r запасы
Фонд товаров, предназначенных для будущего использования или в качестве постоянного источника снабжения.
f stock² *m*
s stock² *m*
g Vorrat² *m*

* **stock-raising → 916**

5820 stone
Hard mineral matter, a piece of rock of imprecise size, but larger than a particle of coarse sand, smaller than a boulder.
r камень₁
Твердый обломок горной породы произвольного размера.
f pierre *f*
s piedra *f*
g Stein *m*

5821 Stone Age
The period in human development extending from the beginning of the Pleistocene to the beginning of the Bronze Age when tools and weapons were made of stone, wood and bones, it is subdivided into three culture periods:old (Paleolithic), middle (Mesolithic) and new (Neolithic).
r каменный век
Древнейший период в развитии человечества, когда орудия и оружие изготавливались из камня, дерева и кости, подразделяется на древний (палеолит), средний (мезолит) и новый (неолит), сменился бронзовым веком.
f âge *m* de Pierre
s edad *f* de piedra
g Steinzeit *f*

5822 stone block
A fragment of rock more than 1 m in diameter.
r глыба
Обломок горной породы поперечником более 1 м.
f bloc *m*
s bloque *m*
g Scholle *f*; Block *m*

* **stone border → 4900**

5823 stone garland
A curved stripe of coarse, stony material lying across a slope in high latitudes and high mountains.
r каменная гирлянда
Полоса грубообломочного материала, фестончато изогнутая поперек склона в высоких широтах или в высокогорьях.

f guirlande *f* des pierres; sol *m* en guirlandes
s guirnalda *f* de piedra; suelo *m* en guirnaldas
g Steingirlande *f*

5824 stone pavement
Rock debris accumulation in which the stones located on top are influenced by frost action, particularly freezing-out, forming a mosaic.
r каменная мостовая
Скопление обломков горных пород, в котором поверхностные камни под влиянием мерзлотных процессов и, в частности, вымораживания, уложены наподобие мозаики.
f dallage *m* nival
s pavimento *m* de piedras
g Steinpflaster *n*

* **stone polygon → 5671**

5825 stone ring; stony earth circle
A type of cryogenic relief consisting of a system of small rings or polygons of coarse-grained matter surrounding a central part of fine debris.
r каменное кольцо; каменный венок
Тип криогенного рельефа в виде системы небольших колец или полигонов, сложенных в центральной части мелкоземом, а по периферии грубообломочным материалом.
f cercle *m* de pierres
s anillo *m* de piedras; círculo *m* de piedras
g Steinkreis *m*

5826 stone stripes; blockstripes
Alternating stripes of coarse-grained and fine-grained debris occurring due to freezing-out on steep slopes.
r каменные полосы
Чередующиеся полосы крупно- и мелкообломочного материала, возникающие в результате вымораживания на крутых склонах.
f sols *mpl* striés
s franjas *fpl* de piedras
g Steinstreifen *mpl*

5827 stoniness
The relative proportion of stones in a soil or on its surface.
r каменистость
Относительное количество камней в почве или на ее поверхности.
f pierrelité *f*
s pedregosidad *f*; pedregal *m*
g Steinigität *f*

* **stony earth circle → 5825**

5828 stony soils
Soils of different genesis that contain rock debris with the particles of different size and various degree of roundness, classified as mountain soils.
r каменистые почвы
Различные по генезису почвы, включающие множество разнообразных по величине и степени окатанности обломков горных пород, обычно их относят к горным почвам.
f sols *mpl* caillouteux
s suelos *mpl* pedregosos
g Steinböden *mpl*

* **storage site → 5829**

5829 store area; storage site; entrepot
A place to which goods, raw materials, equipment, finished products are brought to be stored temporarily while awaiting transfer (transhipment) and transportation.
r складская зона
Концентрация в одном месте хранилищ материалов, сырья, оборудования, готовой продукции и пр. с целью более рациональной перегрузки и перевозки их транспортными средствами.
f zone *f* d'entrepôt; stockage *m*; entrepôt *m*
s zona *f* de almacén; almacén *m*
g Lagerungszone *f*

5830 store of cold in ice
The heat necessary to warm a given mass of ice or frozen rock up to the melting point.
r запас холода во льду
Количество тепла, необходимое для нагревания данного объема льда или льдистой породы до температуры плавления.
f stock *m* de froid en la glace
s stock *m* del frío en el hielo
g Kälteinhalt *m* im Eis; Frostinhalt *m* im Eis

5831 storm
A very violent wind (greater 20 m/s) which causes large waves in the sea and destruction on land.
r буря; шторм
Очень сильный ветер (более 20 м/с), сопровождающийся большим волнением на море и разрушениями на суше.
f tempête *f*
s tempestad *f*; tormenta *f*
g Sturm *m*; Unwetter *n*

* **storm beach → 5834**

* **storm eye → 2084**

5832 storm runoff
The runoff appearing due to storm rains; the total water volume after a rain passing through a certain control section.
r ливневой сток
Сток, возникающий в результате выпадения ливней; суммарный объем воды от дождя, проходящей через данный створ.
f écoulement *m* direct
s escorrentía *f* de chubasco
g Abfluß *m* wegen eines Platzregens

5833 storm surge; storm tide
An abnormal rapid rise of the sea level along a shore as a result of storm winds, pushing sea water on to a windward coast.

r штормовой нагон
Экстремальное повышение уровня воды в береговой зоне, вызванное штормовым ветром.
f onde *f* de tempête; raz *m* de marée; onde *f* de cyclone
s onda *f* de borrasca; marejada *f*; onda *f* de ciclón
g Sturmflut *f*

* **storm tide → 5833**

5834 storm-beach; storm beach
A deposit of coarse and rounded material, including cobbles and boulders, accumulated high up on the shore by unusually strong waves in a storm.
r пляж после шторма
Отложение грубого прибрежного материала, включая крупную гальку, булыжники и валуны, заброшенные вдаль от берега редкими сильными волнами во время шторма.
f plage *f* de tempête
s playa *f* de tormenta
g Sturmstrand *m*

5835 stow; landscape unit
A part of geographical landscape not always with marked limits (e.g. a moraine hill, solonchak depression), it consists of genetically and territorially connected facies.
r урочище
Часть географического ландшафта, не всегда имеющая четко выраженные границы (например, моренный холм, солончаковая впадина), состоит из генетически и территориально связанных фаций.
f unité *f* paysagique
s unidad *f* de paisaje
g Landschaftseinheit *f*; Urotschistsche *n*

5836 strait; sound; channel
A narrow passage of water or rather wide waterway (sound) connecting two bodies of water. → Fig. 312
r пролив
Относительно узкое водное пространство (или более широкое), разделяющее участки суши и соединяющее смежные водоемы или их части. → Рис. 312
f détroit *m*
s estrecho *m*
g Sund *m*; Meerenge *f*; Meerstraße *f*

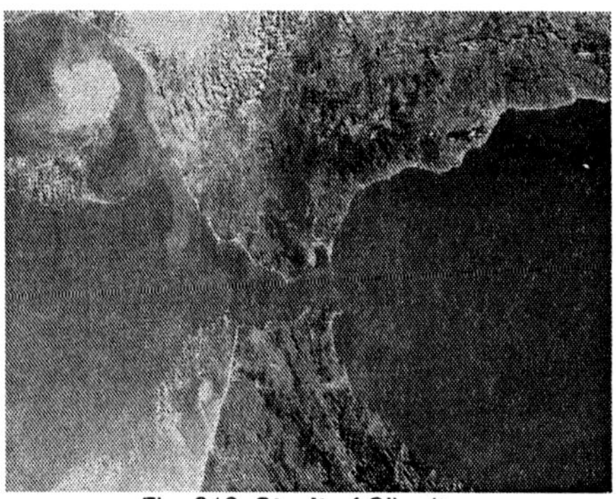
Fig. 312. **Strait** of Gibraltar

* **stranded hummock → 2659**

* **strandflat → 11**

5837 strath; glen
A broad flat-floored river valley deeply filled with alluvial deposits and often with glacigenic deposits, very common in Scotland.
r страт
Типичная для Шотландии широкая горная долина, заполненная мощными аллювиальными отложениями, нередко ледникового происхождения.
f strath *m*
s strath *m*
g Strath *m*

5838 stratification; bedding
1. The structure of rocks in the form of layers superimposing each other, differing in composition, colour and other characteristics. 2. The arrangement of vegetation in layers.
r стратификация; напластование
1. Строение горных пород в виде налегающих один на другой слоев, различающихся по составу, цвету и другим признакам. 2. Ярусное расположение растительного покрова.
f stratification *f*
s estratificación *f*
g Stratifikation *f*; Bänderung *f*

5839 stratification of the atmosphere
The distribution of temperature in the atmosphere with height characterised by vertical temperature gradient.
r стратификация атмосферы
Распределение температуры в атмосфере с высотой, характеризующееся вертикальным градиентом температуры.
f stratification *f* atmosphérique
s estratificación *f* atmosférica
g Schichtung *f* der Atmosphäre; Atmosphärenschichtung *f*

5840 stratification of water
The division of the whole layer of water in a water body into smaller layers of different density, temperature, salinity etc.
r стратификация вод
Разделение водной толщи на слои разной плотности, температуры, солености и т.п.
f stratification *f* des eaux
s estratificación *f* de agua
g Wasserschichtung *f*

* **stratification plane → 630**

5841 stratified *adj*
Arranged or formed in strata, in layers.

r стратифицированный
Отложенный в форме напластования, т.е. слоями.
f stratifié *adj*
s estratificado *adj*
g geschichtet *adj*

5842 **stratiform cloud**
Clouds spread out at extensive horizontal sheet in the form of regular layer (symbol: str), presenting the appearance of Stratus.
r слоистообразные облака
Облака в форме равномерных слоев большого протяжения, похожие на слоистые (символ: str).
f stratiformis *m*
s stratoforme *m*; estratoforme *m*
g stratiforme Wolken *fpl*

5843 **stratigraphic correlation**
The process by which two stratigraphic units or formations may be equated, usually in a time relationship, even though they are spatially separated.
r стратиграфическая корреляция
Сопоставление двух стратиграфических подразделений, нередко пространственно разобщенных, в отношении времени их формирования.
f corrélation *f* stratigraphique
s correlación *f* estratigráfica
g stratigraphische Korrelation *f*

5844 **stratigraphic discordance; stratigraphic hiatus**
The disturbance of normal chronological succession of rock layers bedding due to an interval in sedimentation or washing out of early deposited rocks.
r стратиграфическое несогласие; стратиграфический перерыв
Нарушение нормальной возрастной последовательности залегания слоев горных пород из-за выпадения из разреза какой-то их части, свидетельствующего о перерыве в накоплении осадков.
f discordance *f* de stratification; lacune *f* stratigraphique
s discordancia *f* estratigráfica; laguna *f* sedimentaria; laguna *f* estratigráfica; interrupción *f* estratigráfica
g stratigraphische Diskordanz *m*; stratigraphische Lücke *f*

* **stratigraphic hiatus → 5844**

5845 **stratigraphic section**
A vertical section showing the succession of bedding and correlation of different rock formations.
r стратиграфический разрез
Вертикальный разрез или графическое изображение последовательности залегания и соотношения разных толщ горных пород.
f coupe *f* stratigraphique
s sección *m* estratigráfica
g stratigraphischer Schnitt *m*

5846 **stratigraphy**
A branch of *geology* concerned with the study of the occurrence, composition, sequence, and correlation of rock strata and the chronological order of succession of rock formations.
r стратиграфия
Раздел *геологии*, изучающий последовательность формирования горных пород и их первичные пространственные взаимоотношения.
f stratigraphie *f*
s estratigrafía *f*
g Stratigraphie *f*

5847 **stratocumulus**
A type of low-level cloud (symbol: Sc), predominantly stratiform, in the form of relatively low grey and/or whitish layer, sheet or patch, its elements are often arranged in bands or rolls that lie across the wind (light rain, snow or drizzle may fall from stratocumulus).
r слоисто-кучевые облака
Облака нижнего яруса (символ: Sc), преимущественно слоистообразные в виде серых или белых слоев и гряд, лежащих против ветра, иногда из них выпадают слабые осадки в виде дождя, снега или мороси.
f stratocumulus *m*
s stratocúmulo *m*; estratocúmulo *m*
g Stratocumuluswolken *fpl*; Stratokumulus *m*; Stratokumulus-Wolken *fpl*; Stratocumulus *m*

5848 **stratopause**
The boundary of the upper stratosphere at about 50-55 km. → Fig. 21, p. 51
r стратопауза
Верхняя граница стратосферы на высотах 50-55 км. → Рис. 21, с. 51
f stratopause *f*
s estratopausa *f*
g Stratopause *f*

5849 **stratosphere**
The layer of the atmosphere between the tropopause and the stratopause in which the temperature generally increases with height. → Fig. 21, p. 51
r стратосфера
Слой атмосферы между тропопаузой и стратопаузой, в котором температура обычно растет с высотой. → Рис. 21, с. 51
f stratosphère *f*
s estratosfera *f*
g Stratosphäre *f*

* **strato-volcano → 5850**

5850 **stratovolcano; strato-volcano; composite volcano**

Fig. 313. **Stratovolcano**

A volcanic cone of bedded structure composed of alternating layers of solidified lava and accumulations of loose material. → Fig. 313

r стратовулкан
Вулканический конус слоистого строения, сложенный чередующимися потоками затвердевшей лавы и накоплениями рыхлого материала. → Рис. 313

f volcan *m* stratifié; strato-volcan *m*
s estratovolcano *m*; estrato-volcán *m*
g Stratovulkan *m*

5851 stratum; layer; bed$_2$
A layer of soil or rock distinguished from adjacent layers by its composition, structure or texture and separated from the overlying and underlying layers by well marked bedding planes.

r слой; пласт
Часть стратифицированного разреза почв или горных пород с общими литологическими признаками, ограниченная кровлей пласта сверху и подошвой пласта снизу.

f couche *f*; lit *m*; strate *f*; assise *f*
s capa *f*; estrato$_1$ *m*; asiento *m*; lecho *m*
g Schicht *f*; Lage *f*; Flöz *n*

5852 stratus
A type of low-level cloud (symbol: St) in the form of low-altitude, light to dark grey cloud layer with rather uniform base sometimes bringing drizzle, ice crystals or fine snow grains, but never heavier rainfall.

r слоистые облака
Облака нижнего яруса (символ: St) в форме светлого серого или однородного облачного слоя с довольно однородным основанием, обычно не дают осадков, но иногда из них выпадают мелкие частицы мороси, ледяные кристаллы или мельчайшие снежные зерна.

f stratus *m*
s estrato$_2$ *m*
g Stratuswolken *fpl*; Stratus-Wolken *fpl*; Stratus *m*

* **stream → 5095**

5853 stream; watercourse
A river natural and artificial body of flowing water having more or less formed channel.

r водоток; русловой поток; река
Общее название для всех естественных и искусственных водных потоков, протекающих в более или менее сформированном русле.

f cours *m* d'eau
s curso *m* de agua; corriente *f* fluvial
g Wasserlauf *m*; Fließgewässer *n*

* **stream bifurcation → 652**

5854 stream discharge
The total amount of water transported by a river over some period of time.

r водоносность реки; водность реки
Общее количество воды, проносимое рекой за какой-либо период времени.

f abondance *f* de rivière
s abundancia *f* del río
g Wasserführung *f*

5855 stream gradient
The degree of slope of a stream.

r уклон реки
Отношение разности высот уровня реки в двух точках, расположенных по течению (падение реки), к расстоянию между ними.

f pente *f* d'une rivière
s pendiente *f* del río
g Flußgefälle *n*

5856 stream order
A hierarchical classification of streams based on the magnitude of their channels and position in a drainage area. → Fig. 314

r порядок реки
Положение реки в иерархии притоков. → Рис. 314

f ordre *m* de rivière
s orden *m* del río
g Flußordnung *f*

5857 streambed alluvium
Mineral and organic alluvial sediments laid down in stream channels.

r русловые отложения
Минеральные и органические осадки, отлагающиеся в русле водотоков.

f dépôts *mpl* alluviaux de lit

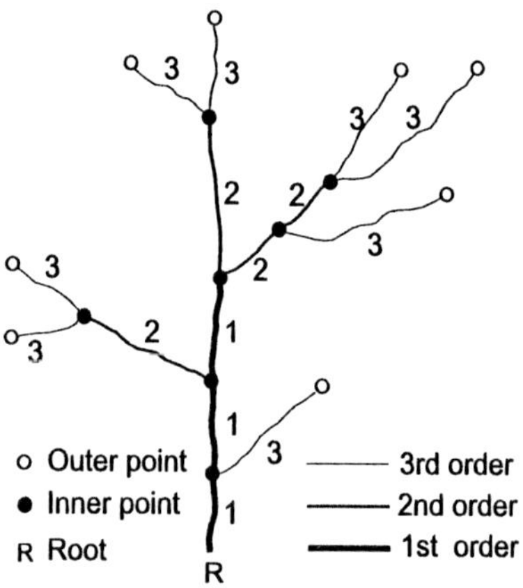

Fig. 314. **Stream order**

s depósitos *mpl* aluviales de lecho
g Flußbettablagerungen *fpl*

5858 streamflow
The runoff of water, loads, dissolved substances and heat in the channel of a certain river.
r речной сток
Сток воды, наносов, растворенных веществ и тепла в русле данной реки.
f écoulement *m* fluvial
s escorrentía *f* fluvial
g fluvialer Abfluß *m*

5859 streamline; flow line
The direction of movement of all parcels of air, water or ice in an area, providing an immediate general picture of motion.
r линия тока
Направление потока воздуха, воды или льда, дающее представление об общей картине движения.
f ligne *f* de flux
s línea *f* de flujo; línea *f* de corriente
g Stromlinie *f*

5860 strength of rock
The ultimate force that rock can withstand without destruction.
r прочность породы
Предельная нагрузка, при которой горная порода разрушается.
f résistance *f* de roche
s resistencia *f* de roca
g Gesteinsfestigkeit *f*

5861 stretch of water; water area
The area of water limited by natural, artificial or conventional boundaries.
r акватория
Участок водного пространства, ограниченный естественными, искусственными или условными границами.
f étendue *f* d'eau
s espacio *m* de agua; espacio *m* acuoso
g Wasserfläche *f*

5862 striated boulder
A boulder having scratches, striation and polishing made during its transportation by a glacier.
r штрихованный валун
Валун с царапинами, ледниковой штриховкой и полировкой, полученными при его переносе ледником.
f cailloux *m* striés
s canto *m* rodado estriado
g gestrichelte Geschiebe *n*; geschrammte Geschiebe *n*

5863 striation; glacial striation; glacial scoring
Long parallel scratches cut on hard rock surfaces of the bed by the fragments of rock frozen into the ice mass, which demonstrate the direction of former ice movement. → Fig. 315

Fig. 315. **Striation**

r ледниковая штриховка
Системы тонких параллельных царапин, которые нанесены на твердые породы ложа обломками, включенными в нижнюю поверхность ледника, и тем самым показывают направление прошлого движения ледника. → Рис. 315
f stries *f* glaciaires
s estría *f* glaciar; estracción *f* glaciar
g Gletscherkritzen *fpl*; Gletscherschrammen *fpl*

5864 stries de Forel
Minute linear grooves on the facets of ice crystals related to outlets of their basal planes.
r штриховка Фореля
Тонкая прямолинейная ребристость на поверхности ледяных кристаллов, связанная с выходами их базисных плоскостей.
f stries *f* de Forel
s estría *f* de Forel
g Forel-Striche *mpl*

5865 strike; strike line
The direction of a horizontal line on an inclined rock stratum at right angles to the direction of the true dip of the rocks. → Fig. 90, p. 192
r простирание; линия простирания
Направление горизонтальной линии на поверхности пласта к направлению его действительного падения. → Рис. 90, с. 192
f direction *f* des strates
s dirección *f* de estratos
g Streichen *n*; Streichlinie *f*

* **strike line → 5865**

* **strike valley → 3581**

* **string bog → 5866**

5866 string fen; string bog
A marshy area which consists of roughly parallel narrow ridges of peat and vegetation iseparated by shallow depressions, occupied by ponds and lakes.
r грядово-мочажинное болото
Болото с узкими параллельными грядами торфяника, перемежающимися с неглубокими депрессиями, многие из которых содержат мелкие участки открытой воды.

f tourbière *f* réticulée
s turbera *f* cordada
g Strangmoor *n*

5867 strip cultivation; open field$_2$
1. The distribution of fields within one land possession that consists of several plots (parcels), i.e. the system of dividing a large field into strips, each worked by a separate owner. 2. The growing of different crops in contiguous strip along the contours of a hillside, to counteract soil erosion.
r чересполосица
1. Разбросанность полей в одном землевладении, состоящем из нескольких участков (парцелл), а также система разделения большого поля на полосы, каждая из которых обрабатывается отдельным собственником. 2. Выращивание урожая на смежных полосах склона холма, закладываемых вдоль горизонталей, чтобы препятствовать почвенной эрозии.
f lanière *f* parcellaire; open field *m*
s parcelación *f* alargada; openfield *m*
g Gemengelage *f*; Streifenbearbeitung *f*

5868 strip map
A map showing only a narrow band of land or sea in which the user is interested, e.g. on each side of a road.
r маршрутная карта
Карта, показывающая лишь узкую полосу территории, интересующую потребителя, например, по обе стороны от дороги.
f carte *f* de bande
s mapa *m* de ruta; mapa *m* itinerario
g Marsch(routen)karte *f*; Wegkarte *f*

*** strip survey → 5142**

5869 Strombolian eruption
A volcanic eruption characterized by the effusion of relatively liquid basic lava and the regular ejection of volcanic dust and bombs from the open vent of the volcano. → Fig. 187, p. 411
r стромболианский тип извержения
Излияние относительно жидкой лавы основного состава с ритмично повторяющимися взрывами в открытом жерле вулкана. → Рис. 187, с. 411
f éruption *f* strombolienne
s erupción *f* estromboliana
g strombolianischer Eruptionstyp *m*

5870 structural crisis; morphogenic crisis
The disproportion in economy caused by the growth of contradictions between the developed and developing countries and by the aggravation of ecological problems and involves economic, energy, raw material, and monetary crises.
r структурный кризис
Нарушение пропорций общественного производства, возникающее в результате обострения противоречий между развитыми и развивающимися странами, а также из-за роста экологических проблем, включает экономический, энергетический, сырьевой, валютный кризисы.
f crise *f* structurelle; crise *f* morphogénique
s crisis *f* estructural; crisis *f* morfogénica
g Strukturkrise *f*

5871 structural geomorphology
A branch of *geomorphology* studying the interrelations between large landforms, geological structure and the movements of the Earth's crust.
r структурная геоморфология
Раздел геоморфологии, исследующий взаимоотношения между крупными чертами рельефа, геологическим строением и движениями земной коры.
f géomorphologie *f* structurale
s geomorfología *f* estructural
g Strukturgeomorphologie *f*

5872 structural glaciology
A branch of *glaciology* studying crystalline, petrographic and tectonic structure and the texture of natural ice.
r структурная гляциология
Раздел *гляциологии*, рассматривающий кристаллическую, петрографическую и тектоническую структуру и текстуру льдов.
f glaciologie *f* structurale
s glaciología *f* estructural
g strukturelle Glaziologie *f*

5873 structural terrace
A terrace on a slope formed by an outcrop of a horizontal layer of resistant rock smoothed by denudation.
r денудационная терраса; структурная терраса
Терраса на склоне, обусловленная выходом на поверхность горизонтального пласта твердых пород, отпрепарированного денудацией.
f terrasse *f* structurale; replat *m* structural; terrasse *f* de dénudation
s terraza *f* estructural; rellano *m* estructural; terraza *f* de denudación
g Denudationsterrase *f*

5874 structure of rock
The features of rock (including ice) related to the shape, dimensions and arrangement of morhological elements, e.g. columnar structure.
r структура горной породы
Особенности строения горной породы (включая лед и почву), обусловленные формой, размерами и ориентировкой морфологических элементов, например, столбчатая структура.
f structure *f* de roche
s estructura *f* de roca
g Gesteinsstruktur *f*

5875 subaerial *adj*; superficial *adj*

Related to any feature, substance or process which occurs or operates on the Earth's surface, as distinct from subterranean or submarine phenomena.

r поверхностный
Относящийся к природным явлениям и процессам, происходящим на земной поверхности, в отличие от подземных и подводных явлений.

f superficiel *adj*
s superficial *adj*
g Oberflächen-; oberflächlich *adj*

5876 subalpine belt
An altitudinal belt of subalpine meadows in humid mountains in temperate and subtropical latitudes, it is found below alpine and above mountain-forest belts in humid regions and above steppe mountain belt – in drier ones.

r субальпийский пояс
Высотный пояс субальпийских лугов в хорошо и равномерно увлажняемых горах умеренных и субтропических широт, расположен ниже альпийского и выше горнолесного поясов во влажных районах и выше степного горного пояса – в более сухих.

f étage *m* subalpin
s piso *m* subalpino
g subalpine Stufe *f*

5877 subalpine vegetation
The vegetation of the subalpine belt of the mountain regions, mainly in the temperate zone, represented by the communities of low bushes, crooked forests and open woodlands, meadows, and subalpine high grasses.

r субальпийская растительность
Растительность горных стран, главным образом в умеренном поясе, представленная сообществами низкорослых кустарников, криволесьем и редколесьем, а также разнотравно-злаковыми лугами и субальпийским высокотравьем.

f végétation *f* subalpine
s vegetación *f* subalpina
g subalpine Vegetation *f*

5878 subantarctic *adj*
Related to an area of ocean and islands located in the Southern Hemisphere between latitudes 58-60° and 65-67° S.

r субантарктический
Относящийся к региону, расположенному в Южном полушарии между 58-60° и 65-67° ю.ш.

f subantarctique *adj*
s subantártico *adj*
g subantarktisch *adj*

5879 subantarctic zone
A latitudinal geographical region in the Southern Hemisphere covering the ocean and a few islands in the South Ocean between latitudes 58-60° and 65-67° S, it is characterized by cold climate with strong winds and fogs.

r субантарктический пояс
Широтный географический пояс Южного полушария, охватывающий акваторию и немногочисленные острова Южного океана между 58-60° и 65-67° ю.ш., характерен холодный климат с сильными ветрами и туманами.

f zone *f* subantarctique
s zona *f* subantártica
g subantarktische Zone *f*

5880 sub-aquatic landscape
A landscape developing on a river flood-plain periodically flooded by freshwater.

r субаквальный ландшафт
Ландшафт, развивающийся в пойме реки, периодически затапливаемой пресными водами.

f paysage *m* subaquatique
s paisaje *m* subacuático
g subaquatische Landschaft *f*

* **subaqueous relief → 5913**

5882 subarctic *adj*
Related to an area located in the Northern Hemisphere between 60-65° and 67-73° N.

r субарктический
Относящийся к региону, расположенному в Северном полушарим между 60-65° и 67-73° с.ш.

f subarctique *adj*
s subártico *adj*
g subarktisch *adj*

5883 subarctic sod soil
An acid soil with the horizon of sod and dry peat, and that of a humus; this soil is found under light forests and meadows of cold subarctic climate.

r субарктическая дерновая почва
Кислая почва разреженных лесов и лугов в условиях холодного субарктического климата с горизонтом дернины и сухого торфа, а также гумусовым горизонтом.

f sol *m* gazonné subarctique
s suelo *m* de césped subártico
g subarktischer Rasenboden *m*

5884 subarctic zone
A latitudinal geographical region in the Northern Hemisphere between 60-65° and 67-73° N, it is characterized by cold climate and permafrost.

r субарктический пояс
Широтный географический пояс Северного полушария, расположенный от 60-65° до 67-73° с.ш., характерен холодный климат и вечная мерзлота.

f zone *f* subarctique
s zona *f* subártica
g subarktische Zone *f*

5885 Subatlantic; Sub-Atlantic period
The late stage of the Holocene, which began

about 2300 years ago in Europe which is still continuing and characterised by a relatively humid and temperate climate.

r субатлантическое время
Поздний этап голоцена, начавшийся около 2300 лет назад, отличающийся в Европе относительно влажным и умеренным климатом, продолжается поныне.

f Sub-atlantique *m*
s Subatlántico *m*
g Subatlantikum *n*

* **Sub-Atlantic period → 5885**

5886 Subboreal; Sub-Boreal period
The stage of the Holocene (about 4-2 thousand years ago) following the Atlantic and characterised by cooler but drier conditions.

r суббореальное время
Этап голоцена (около 4-2 тыс. лет назад), следовавший за атлантическим временем и характеризовавшийся по сравнению с ним более сухим климатом.

f Sub-boréal *m*
s Subboreal *m*
g Subboreal *n*

* **Sub-Boreal period → 5886**

5887 subclimax; pre-climax
The stage in the development of a plant community where the development stops before a climax is reached due to the effects of different catastrophic (as fires), biotic or edaphic factors.

r субклимакс
Одна из стадий сукцессии, которая не смогла дойти до климаксного состояния вследствие действия различных стихийных (например, пожары), биотических или эдафических факторов.

f subclimax *m*; antéclimax *m*
s subclímax *m*; autoclímax *m*
g Subklimax *m*

5888 subcontinent
A large land mass forming part of a continent, and having a certain geographical entity, e.g. the Indian subcontinent.

r субконтинент
Значительная часть континента, отличающаяся некоторой общностью природных условий и отделенная от других частей континента естественными границами, например, Индостанский субконтинент.

f subcontinent *m*
s subcontinente *m*
g Subkontinent *m*

5889 subdominant species
A plant species prevailing in a particular layer of a community.

r субдоминантный вид; субдоминант
Вид растений, доминирующий в отдельных ярусах сообщества.

f espèce *f* subdominante
s especie *f* subdominante
g subdominante Art *f*

5890 subduction
Underthrust of lithosphere plates of the oceanic crust and mantle rocks under edges of other plates.

r субдукция
Поддвигание литосферных плит океанической коры и пород мантии под края других плит.

f subduction *f*
s subducción *f*
g Subduktion *f*

5891 subduction zone; destructive margin; destructive plate margin
An area where the rocks making up the seafloor are forced beneath continental rocks at a plate margin and are reincorporated in the magma beneath the Earth's crust, it is marked by volcanoes and earthquakes.

r зона субдукции
Область, где породы океанической коры поддвигаются под край литосферной плиты и вторгаются в магматический слой земной коры, для этой зоны характерны вулканы и землетрясения.

f zone *f* de subduction
s zona *f* de subducción
g Subduktionzone *f*

5892 subequatorial *adj*
Related to an area in the Northern and Southern hemispheres between the tropical and equatorial zones.

r субэкваториальный
Относящийся к региону, расположенному в Северном и Южном полушариях между тропическим и экваториальным поясами.

f subéquatorial *adj*
s subecuatorial *adj*
g subäquatorial *adj*

5893 subequatorial zone
Latitudinal geographical regions in the Northern and Southern hemispheres between the tropical and equatorial zones, characterized by constant high temperature and equatorial monsoons with dry winters and humid summers.

r субэкваториальный пояс
Широтные географические пояса, расположенные в Северном и Южном полушариях между тропическим и экваториальным поясами и характеризующиеся постоянно высокой температурой и господством экваториальных муссонов с сухой зимой и влажным летом.

f zone *f* subéquatoriale
s zona *f* subecuatorial
g subäquatoriale Zone *f*

5894 **subglacial ablation**
The loss of ice mass due to melting at the bed of a glacier.
r подледниковая абляция
Убыль массы льда в результате таяния на границе ледника с ложем.
f ablation *m* sous-glaciaire
s ablación *f* subglaciar
g subglaziale Ablation *m*

5895 **subglacial *adj***
Related to the environment at the base of a glacier.
r подледниковый
Относящийся к природным явлениям, происходящим под ледником.
f sous-glaciaire *adj*
s subglaciar *adj*
g unterglazial *adj*

5896 **subglacial channel**
A melt-water channel cut in the rocks and deposits of a glacier bed, made by a stream of melt water flowing beneath a glacier or ice sheet.
r подледниковый канал
Врез в коренных породах и отложениях ледникового ложа, выработанный потоком воды, стекающей под ледником или ледниковым покровом.
f canal *m* sous-glaciaire; conduit *m* sous-glaciaire
s canal *m* subglaciar
g subglaziales Gerinne *n*

5897 **subglacial drainage; subglacial stream**
Water streams flowing from a tunnel at the bottom of a glacier or ice sheet.
r подледниковый сток
Водные потоки, вытекающие из-под концов горных ледников или края ледникового покрова.
f écoulement *m* sous-glaciaire
s desagüe *m* subglaciar
g subglazialer Abfluß *m*

5898 **subglacial lake**
A water reservoir in the bedrock depression beneath a glacier.
r подледное озеро
Скопление воды во впадине коренного ложа под ледником.
f lac *m* sous-glaciaire
s lago *m* subglaciar
g subglazialer See *m*

* **subglacial moraine → 2652**

* **subglacial stream → 5897**

5899 **subglacial water**
Water mass or streams at the bedrock beneath a glacier.
r подледниковые воды
Жидкая вода и ее потоки на ложе ледника.
f eau *f* sous-glaciaire
s agua *f* subglaciar
g subglaziales Wasser *n*

5900 **subhumid *adj***
Applied to a climate in relation to its precipitation which is too dry for tree growth but humid enough for grass growth, it is characteristic for savanna and steppe climate.
r субвлажный
Применяется к степени влажности климата, которая недостаточна для роста деревьев, но достаточна для произрастания трав, что характерно для климата саванны или степи.
f subhumide *adj*
s subhúmedo *adj*
g subhumid *adj*

5901 **subject of federation**
A constituent entity of a federal state (republic, state, canton, land) the independence of which is limited within the federation.
r субъект федерации
Государственное образование (республики, штаты, кантоны, земли), имеющие ограниченную самостоятельность в рамках федеративного государства,
f sujet *m* de fédération; état *m* de fédération
s sujeto *m* de federación; república *f* federada; estado *m* federado; unidad *f* federada
g Föderationssubjekt *n*; Bundessubjekt *n*

5902 **sublimation**
A direct change of state from a solid to a gas omitting the liquid stage, or of a gas directly to a solid, as in the formation of ice from water vapour.
r сублимация (водяного пара)
Переход воды из твердого состояния в газообразное, минуя жидкую фазу, или непосредственно из газобразного состояния в твердое, как это бывает при образовании льда из водяного пара.
f sublimation *f*
s sublimación *f*
g Sublimation *f*

5903 **sublimation ice**
An ice formed directly from water vapour, without passing through a liquid phase.
r сублимационный лед
Лед, возникающий из парообразной влаги, минуя жидкую фазу.
f glace *f* de sublimation
s hielo *m* de sublimación
g Sublimationseis *n*

* **sublittoral zone → 4100**

5905 **submarine bar**
Cumulose relief form on a shelf, usually composed of sand, such ramparts extend in parallel to the shoreline over a distance up to several kilometers, rising above the bottom by 1-4 m.

r подводный вал
Кумулятивная форма рельефа на шельфе, обычно сложенная песком, такие валы простираются параллельно береговой линии на расстояние до нескольких километров, возвышаясь над дном на 1-4 м.
f barre *f* sous-marine
s barra *f* submarina
g untermeerische Barre *f*; Unterwasserbarre *f*

* **submarine canyon → 5271**

* **submarine channel → 5271**

* **submarine contour → 3223**

5906 submarine earthquake
An earthquake on the ocean floor occuring often in narrow mobile zones of underwater continental margins or in mid-oceanic ridges.
r подводное землетрясение
Землетрясение на дне или под дном океанов или морей, чаще всего в узких подвижных зонах подводных окраин материков или в срединно-океанических хребтах.
f séisme *m* sous-marin
s seísmo *m* submarino
g submarines Erdbeben *n*; unterseeisches Beben *n*

5907 submarine elevation
A wide elevation of oceanic bottom not always clearly separated from the neighbouring depressions.
r подводная возвышенность
Обширное поднятие океанического дна, не всегда резко обособленное от соседних котловин.
f élévation *f* sous-marine
s elevación *f* submarina
g submarine Erhebung *f*; untermeerische Erhebung *f*

5908 submarine fan
A fan- or cone-shaped mass of sediment deposited by turbidity currents on to the lower part of the continental slope at the exit of a submarine canyon.
r подводный конус выноса
Аккумуляция в устье подводного каньона материала, выносимого мутьевыми потоками в нижнюю часть материкового склона.
f cône *m* de déjection subaquatique
s cono *m* de deyección submarina
g untermeerischer Schuttkegel *m*

* **submarine geomorphology → 3705**

5909 submarine landslide
A slump of bottom sediments from the upper part of a continental slope or from the slopes of submarine mountains and submarine canyons, as well as the form of submarine relief created by this process, sometimes it causes the turbidity currents.
r подводный оползень; подводный обвал
Оползание масс донных осадков с верхней части материкового склона или со склонов подводных гор и подводных каньонов, а также форма подводного рельефа, созданная этим процессом, иногда служащая причиной образования мутьевых потоков.
f éboulement *m* sous-marin
s derrumbe *m* submarino
g submarine Rutschung *f*; untermeerische Rutschung *f*

5910 submarine moraine
Unsorted and unstratified loose rock debris formed under a floating glacier not far from its grounding line.
r подводная морена
Несортированная и неслоистая рыхлая грубообломочная порода, образованная под плавучим ледником недалеко от его линии налегания.
f moraine *f* sous-marine
s morrena *f* submarina
g submarine Moräne *f*

* **submarine mount → 5264**

5911 submarine plateau; oceanic plateau
An extensive, topographically high area of an ocean floor that rises to within 2-3 km of the sea surface above the abyssal floor and often serves as a submarine platform of islands, such as Iceland, Galbpagos, Azores, etc.
r подводное плато; океаническое плато
Подводная возвышенность с четкими морфологическими уступами и относительно выровненной вершинной поверхностью, нередко протягивающаяся вдоль внешних частей осевой зоны срединно-океанических хребтов и служит коренным основанием архипелагов.
f plateau *m* sous-marin; support *m* d'archipels
s meseta *f* submarina; soporte *m* de archipiélagos
g untermeerisches Plateau *n*; Unterwasserplateau *n*

5912 submarine quarry
A part of sea floor where deposits are systematically extracted as a material.
r подводный карьер
Участок дна, откуда наносы систематически добываются в качестве строительного материала.
f carrière *f* sous-marine
s cantera *f* submarina
g Unterwassergrube *f*

* **submarine range → 5914**

5913 submarine relief; subaqueous relief
The bottom relief in any water body, its features are exerted by tectonics, waves and currents.
r подводный рельеф

Рельеф дна в любом водоеме, формируется под действием тектонических процессов, волн и течений.
f relief *m* subaquatique
s relieve *m* subacuático
g submarines Relief *n*; Unterwasserrelief *n*; untermeerisches Relief *n*

5914 **submarine ridge; submarine range**
A linear mountain elevation in the ocean floor or on a marginal sea floor occurring in the zone of transition from continents to oceans forming island arcs.
r подводный хребет
Горные поднятия на ложе океанов и в котловинах окраинных морей, встречающиеся в зоне перехода от материков к океанам и образующие островные дуги.
f créte *f* sous-marine
s cordillera *f* submarina
g submariner Gebirgszug *m*; untermeerischer Rücken *m*; submariner Rücken *m*

5915 **submarine rift**
Zone of deep disturbances of the crust in oceans, several thousand kilometres long and 100-200 km wide, with large (up to 1000 km) horizontal displocations of linear magnetic anomalies.
r подводный разлом
Зона глубоких нарушений земной коры в океанах длиной в несколько тысяч и шириной 100-200 км с крупными (до 1000 км) горизонтальными смещениями линейных магнитных аномалий.
f faille *f* sous-marine
s falla *f* submarina
g submarine Störungszone *f*

5916 **submarine spring**
A spring of ground water on the bottom and walls of water bodies including seas or watercourses.
r подводный источник; субаквальный источник
Источник подземных вод на дне и бортах водоема или водотока, включая моря.
f source *f* sous-marine
s fuente *m* submarino
g subaquatische Quelle *f*; Unterwasserquelle *f*

5917 **submarine volcano; underwater volcano**
A volcano situated on an oceanic floor (its eruptions take place in water and sometimes form a new island).
r подводный вулкан
Вулкан, расположенный на дне океана, извержения которого происходят под водой (иногда в результате подводных извержений образуются острова).
f volcan *m* sous-marin
s volcán *m* submarino
g untermeerischer Vulkan *m*; Unterwasservulkan *m*; submariner Vulkan *m*

5918 **submerged coast; submerged shoreline**
A coastline formed when a rise of sea level in relation to the land or a negative movement of the land surface lead to the flooding of the former land surface.
r погружаемый берег
Береговая линия, формирующаяся в условиях, когда подъем уровня моря или опускание береговой линии вызывает затопление прежней поверхности суши.
f côte *f* de submersion
s costa *f* de sumersión
g versunkenes Ufer *n*

5919 **submerged delta**
An area of delta depressions covered with water as a result of flooding and storm surges of seawater.
r затопленная дельта
Пониженные участки дельты, заполняемые водой в результате половодья или паводков (заливание дельты), а также штормовых нагонов морской воды.
f delta *m* sous-marin; delta *m* de submersion
s delta *m* sumergido; delta *m* submarino
g überschwemmtes Delta *n*

5920 **submerged forest**
The remains of a former forest, now completely covered by the sea except occasionally at very low tide, resulting from the submergence of the coast.
r затопленный лес
Остатки прежнего леса, ныне полностью покрытого морем (за исключением небольших участков, видимых во время отлива) в результате погружения побережья.
f forêt *f* submergée
s bosque *m* inundado; bosque *m* sumergido
g versunkener Wald *m*

5921 **submerged ice**
A part of floating ice covered with a layer of melt water.
r затопленный лед
Участок плавучего льда, покрытый сплошным слоем талой воды.
f glace *f* inondée; glace *f* submergée
s hielo *m* sumergido; hielo *m* inundado
g untergetauchtes Eis *n*; übergeschwemmtes Eis *n*

5922 **submerged reef; bank reef**
A coral construction sunk due to vertical motions of the Earth's crust deeper 50 m where reef-forming corals cannot survive.
r погруженный коралловый риф
Коралловое сооружение, опустившееся вследствие вертикальных движений земной коры глубже 50 м, где рифообразующие кораллы обитать не могут.

f récif *f* submergé; plature *f* corallienne
s arrecife *m* sumergido; plataforma *f* coralina
g versunkenes Riff *n*

* **submerged shoreline → 5918**

5923 **submergence**
The lowering of the shoreline caused by the sea level rise and the flooding of the former land surface.
r погружение берега
Опускание береговой линии вследствие подъема уровня моря и затопления прежней поверхности суши.
f ennoiment *m*
s sumersión *f*
g Untertauchen *n*

* **submergible → 5924**

5924 **submersible; submergible; underwater vehicle; undersea vehicle**
A diving apparatus, manned or unmanned, autonomous or driven, used for oceanographic investigation at great depths, since it is able to withstand great pressure.
r подводный аппарат
Плавучее средство для погружения в глубины моря или озера в научных или спортивных целях, может быть необитаемым или обитаемым, автономным или управляемым извне.
f appareil *m* sous-marin
s aparato *m* submarino
g Unterwassergerät *n*

* **subordinate volcano → 4417**

5925 **subpermafrost water**
Ground water normally under pressure occurring below the permafrost.
r подмерзлотные воды
Подземные воды, обычно напорные, расположенные под многолетнемерзлыми породами.
f eau *f* sous-pergélisol
s agua *m* de subpermafrost
g Subpermafrostwasser *n*

5926 **subpolar glacier**
A glacier with permanent subfreezing temperatures inside it but some melting is possible in summer.
r субполярный ледник
Ледник, в толще которого господствует отрицательная температура, но летом возможно таяние.
f glacier *m* sub-polaire
s glaciar *m* subpolar
g subpolarer Gletscher *m*

5927 **subregion**
Any of the divisions of a large region.
r субрегион
Часть крупного региона, объединяющая несколько соседних стран с общими особенностями развития.
f subrégion *f*
s subregión *f*
g Subregion *f*

5928 **subsea permafrost**
The permafrost occurring beneath the sea floor.
r субмаринная мерзлота
Многолетнемерзлые породы под морским дном.
f pergélisol *m* sous-marin
s permafrost *m* submarino
g submariner Permafrost *m*

5929 **subsequent stream; subsequent valley**
A tributary of a primary, consequent river, usually running at right angle to consequent stream along the line of a fault. → Fig.68, p. 145
r субсеквентная река; субсеквентная долина
Приток первичной, консеквентной реки, текущий в долине, обычно совпадающей с простиранием моноклинально залегающих пластов. → Рис. 68, с. 145
f fleuve *m* subséquent; vallée *f* subséquente
s río *m* subsecuente; valle *m* subsecuente
g subsequenter Fluß *m*; subsequentes Tal *n*

* **subsequent valley → 5929**

5930 **subshrub; dwarfshrub; undershrub**
A perennial plant smaller than a shrub (up to 50-60 cm high) has a short woody arborizing stem and abundant growth branching upwards from the base, the upper stems dying back at the end of each growing season, e.g. heather, cowberrys, cranberrys.
r кустарничек
Низкорослое многолетнее растение (высотой до 50-60 см) с древеснеющим ветвящимся стволом и меньшей по сравнению с кустарником длительностью жизни наземных побегов, например, вереск, брусника, клюква.
f sous-arbuste *m*; sous-arbrisseau *m*
s subarbusto *m*
g Zwergstrauch *m*

5931 **subshrub layer**
In the stratified structure of a forest, the layer of subshrubs.
r кустарничковый ярус
Ярус кустарничков в лесном сообществе.
f strate *f* sous-arbustive; strate *f* frutescente
s estrato *m* subarbustivo;
estrato *m* frutescente
g Zwergstrauchschicht *f*

5932 **subsidence**
1. A small-scale lowering of the surface following cave-roof collapse in a karst area; a large-scale structural readjustment of the Earth's surface, as in basin downwarping or rift valley formation; the gradual settling of

the ground resulting from the extraction of mineral deposits, e.g. salt or coal subsidence. 2. The slow descent of a large air mass, as in an anticyclone.

r опускание
1. Понижение поверхности вслед за обвалом карстовой полости; снижение поверхности в результате крупных структурных изменений, как это имеет место при формировании рифтовой долины; постепенное оседание грунта в результате добычи полезных ископаемых, например, при добыче соли или каменного угля. 2. Медленное опускание большой воздушной массы, как это бывает при антициклоне.

f subsidence *f*
s subsidencia *f*
g Senkung *f*

* **subsidiary volcano → 4417**

5933 **subsidy**
Allocations from the state budget to cover the losses by enterprises and to support certain branches of the economy.
r дотация
Ассигнования из государственного бюджета на покрытие убытков предприятий и на финансовую поддержку отдельных отраслей хозяйства.
f dotation *f*
s dotación *f*; subsidio *m*
g Stützung *f*; Staatszuwendung *f*

5934 **subsistence agriculture; subsistence farming**
Farming in which the products are grown or raised primarily for the support of the farmer and the farmer′s dependants, not primarily for sale or trading.
r натуральное хозяйство; потребительское сельское хозяйство
Тип хозяйства, при котором продукты труда производятся для удовлетворения потребностей самих производителей, а не для продажи на рынке.
f agriculture *f* de subsistance
s agricultura *f* de subsistencia
g Konsumlandwirtschaft *f*

5935 **subsistence crop**
A crop grown for the use of the farmer and for the farmerr's family.
r потребительский урожай
Урожай, выращиваемый для использования семьей самого производителя.
f récolte *f* d′autoconsommation
s cosecha *f* de autoconsumo
g Konsumernte *f*

5936 **subsistence economy**
An economic system in which there is little if any buying and selling, although there may be bartering, in contrast to market economy.
r потребительская экономика
Экономическая система, при которой купля–продажа занимает очень незначительное место, хотя возможен бартер, в отличие от рыночной экономики.
f économie *f* de subsistance
s economía *f* de subsistencia
g Konsumökonomik *f*; Konsumwirtschaft *f*

* **subsistence farming → 5934**

5937 **subsoil; bottom soil**
The layer og soil resting directly above the parent rock and containing little organic matter due to its leaching from this layer.
r подпочва
Сло почвы, лежащий непосредственно над материнской породой и содержащий мало органики, так как органические вещества отсюда вымыты.
f sous-sol *m*
s subsuelo *m*
g Untergrund *m*

5938 **subsoil irrigation**
The increase or renewal of moisture reserves in soil by way of capillary nourishment from the regular unconfined groundwater level.
r подпочвенное орошение
Увеличение или восполнение запаса влаги в почве путем капиллярного подпитывания от регулируемого уровня грунтовых вод.
f irrigation *f* par élévation des eaux souterraines
s irrigación *f* por elevación de aguas subterráneas
g Untergrundbewässerung *f*

5939 **subsurface brine; brackish groundwater**
Ground waters containing dissolved minerals in high concentration.
r подземный рассол
Подземные воды, содержащие растворенные минеральные вещества в повышенных концентрациях.
f saumure *f* souterraine; eau *f* salée souterraine
s salmuera *f* subterránea
g unterirdische Sole *f*

5940 **subsurface current; deepwater current; depth current**
A current in a water body below the thermocline and direct wind impact (usually at depths from 10 to 500 m) induced by the difference of water density or as a compensation of a surface water outflow caused by wind.
r глубинное течение; подповерхностное течение
Поток в водоеме вне сферы непосредственного воздействия ветра (обычно на глубинах от 10 до 500 м), вызываемый разностью плотности воды, а также компенсацией оттока поверхностных вод, сгоняемых ветром.

f courant *m* de profondeur
s corriente *f* de profundidad;
corriente *f* de fondo
g Tiefenströmung *f*; Tiefenstrom *f*

* **subsurface flow → 2662**

5941 subsurface wash
The processes relating to sediment transport within the regolith, with the actual flow of water referred to surface flow.
r подповерхностное вымывание
Перенос твердых частиц в пределах реголита с участием водного потока, текущего по земной поверхности.
f écoulement *m* hypodermique
s derrubio *m* hipodérmico
g Unterhautauswaschung *f*

* **subtidal zone → 4100**

5942 subtropical *adj*
Related to areas in the Northern and Southern hemispheres lying between latitudes 30° and 40° and to the natural phenomena (climate, forest, etc.) characteristic for these latitudes.
r субтропический
Относящийся к региону, расположенному в Северном и Южном полушариях приблизительно между 30° и 40° широты, и к распространенным здесь природным явлениям (климат, лес и т.п.).
f subtropical *adj*
s subtropical *adj*
g subtropisch *adj*

* **subtropical anticyclone → 5945**

5943 subtropical forest
A dense broad-leaved forest with evergreen trees and shrubs that is widespread in the subtropics (mainly in the Mediterranean region) and in the mountains of tropics and subtropics, e.g. in the Himalayas.
r субтропический лес
Густой широколиственный лес с вечнозелеными древесными и кустарниковыми породами, распространенный в субтропических поясах (главным образом в Средиземноморье) и в горах тропиков и субтропиков, например, в Гималаях.
f forêt *f* subtropicale
s bosque *m* subtropical
g subtropischer Wald *m*

5944 subtropical forest zone
Forest zones of subtropics of the Northern and Southern hemispheres, they include hemihylaea, monsoon mixed forests and Mediterranean forests.
r лесная субтропическая зона
Совокупность лесных зон субтропиков Северного и Южного полушария, включает гемигилею, зоны муссонных смешанных лесов и средиземноморских лесов.
f zone *f* des forêts subtropicales
s zona *f* de bosques subtropicales
g Zone *f* der subtropischen Wälder

5945 subtropical high;
subtropical anticyclone
A series of high pressure cells occurring along meridians in both hemispheres, the axis is normally along about 35° latitude and has slight annual meridional fluctuations.
r субтропическая зона высокого давления;
субтропический антициклон
Серии небольших ячеек высокого давления, расположенных в обоих полушариях вдоль меридианов на широтах около 35° и испытывающих небольшие годовые колебания.
f ceinture *f* subtropicale de haute pression;
anticyclone *m* subtropical
s cinturón *m* subtropical de altas presiones;
anticiclón *m* subtropical
g subtropische Hochdruckzone *f*;
subtropische Antizyklone *f*

5946 subtropical jet stream
A western high velocity air flow in the troposphere of subtropical latitudes, the most stable and strong among tropospheric jet streams.
r субтропическое струйное течение
Западное струйное течение в тропосфере субтропических широт, наиболее устойчивое и сильное из тропосферных струйных течений.
f courant-jet *m* subtropical
s corriente *f* de chorro subtropical
g subtropischer Strahlstrom *m*

5947 subtropical vegetation
The plant life of subtropical zone with evergreen broad-leaved forests.
r субтропическая растительность
Совокупность растительных сообществ субтропического пояса с вечнозелеными широколиственными лесами.
f végétation *f* subtropicale
s vegetación *f* subtropical
g subtropische Vegetation *f*

5948 subtropical zone
The latitudinal zones in the Northern and Southern hemispheres approximately between latitudes 30° and 40° characterized by the alternation of temperate (winter) and tropical (summer) thermal regimes.
r субтропический пояс; субтропики
Широтные географические пояса в Северном и Южном полушариях приблизительно между 30° и 40° широты, отличаются чередованием по полугодиям умеренного (зимой) и тропического (летом) термического режимов.
f zone *f* subtropicale
s zona *f* subtropical
g subtropische Zone *f*

*** suburb → 5952**

5949 suburban ***adj***
Pertaining to the suburbs or having qualities considered to be characteristic of the suburbs or of the people who inhabit them.
r пригородный
Относящийся к пригороду или имеющий специфические черты пригорода и характеризующий проживающее здесь население.
f suburbain *adj*
s suburbano *adj*
g stadtnah *adj*; Vorort-

5950 suburban agriculture
Agricultural production around big cities which specializes in perishable products: milk, vegetables, fruits, berries, etc.
r пригородное хозяйство
Сельскохозяйственное производство вокруг крупных городов, специализирующееся на получении скоропортящихся и малотранспортабельных видов продукции: молока, овощей, плодов, ягод и т.п.
f agriculture *f* suburbaine
s agricultura *f* suburbana
g suburbane Landwirtschaft *f*

5951 suburbanization
The growth and development of built-up areas around large cities which leads to the formation of urban agglomerations.
r субурбанизация
Процесс роста и развития пригородной зоны крупных городов, ведущий к формированию городских агломераций.
f pértiurbanisation *f*; suburbanisation *f*
s suburbanización *f*
g Suburbanisierung *f*; Subverstädterung *f*

*** suburbia → 5952**

5952 suburbs; suburb; suburbia
Mainly residential part of a built-up urban area located at the edge of the city, usually characterized by a lower housing density and the provision of open space.
r пригород; предместье
В основном жилая часть застроенной урбанизированной территории, расположенная на окраине города, с меньшей плотностью застройки и некоторым запасом свободного пространства.
f banlieue *f*; faubourg *m*
s suburbio *m*; banlieue *f*
g Vorort *m*; Ergnzungsgebiet *n*

5953 subvention
An amount of money given to local authorities by the state for specific aims.
r субвенция
Вид денежного пособия местным органам власти со стороны государства (в отличие от дотации предоставляется на финансирование с конкретными целями).

Fig. 316. **Succession** in a forest

f subvention *f*
s subvención *f*
g Subvention *f*

*** subway (*amer*) → 6448**

5954 succession; plant succession
The progressive natural development of vegetation on a certain area from the initial pioneer community to the final stable stage (climax), one community being gradually replaced by another under the influence of different factors. → Fig. 316
r сукцессия
Последовательная смена одних растительных сообществ другими на данной территории, например, образование болот при заболачивании лесов или зарастании водоемов. → Рис. 316
f succession *f*
s sucesión *f*
g Sukzession *f*

5955 succulent
A plant with thick fleshy leaves and stems that conserve a lot of water which allows it to withstand drought.
r суккулент
Растение с толстыми, мясистыми надземными органами, содержащими большое количество воды в своих тканях.
f succulente *m*; plante *f* succulente
s planta *f* suculenta; planta *f* grasa; suculenta *f*
g Sukkulente *m*

5956 sudd
A floating mass of vegetation, comprising plants that have broken away from neighbouring swamps, which regularly blocks the main channel of the White Nile in Sudan and causes navigation hazards.
r садд
Плотная плавающая масса растений, принесенных из окрестных болот, которая регулярно перекрывает главное русло Бе-

лого Нила в Судане и служит причиной многих сложностей навигации.
f sudd *m*
s sudd *m*
g Sudd *m*

5957 suffocation
Mass dying of aquatic animals in a river or lake caused by the lack of oxygen in water because of winter freezing or rapid development of blue-green algae in warm period (water bloom).
r замор
Массовая гибель водных животных в реке или озере из-за недостатка кислорода в воде, скованной в зимнее время льдом, или бурного развития синезеленых водорослей (цветение водоема).
f suffocation *f*
s sofocación *f*
g Erstickung *f*

5958 suffosion
An erosional process occurring in areas where limestone bedrock is overlain by unconsolidated superficial materials, the sediments slump down into widened joints and cavities in the bedrock surface, producing an irregular land-surface.
r суффозия
Вынос мелких минеральных частиц потоками грунтовых вод, фильтрующихся в толще горных пород, этот процесс характерен для областей, где известковые породы перекрываются рыхлым поверхностным материалом, который в процессе растворения известняка проваливается в трещины и полости.
f suffosion *f*
s sufusión *f*
g Suffosion *f*

*** sugar hoar → 1533**

*** sukhovey → 1738**

5959 sulphur
A non-metallic yellow-coloured element (symbol: S) occcurring in the vicinity of volcanic vents, hot springs and fumaroles.
r сера
Неметаллический элемент (символ: S) желтого цвета, встречающийся в чистом виде в окрестностях жерла вулкана, у горячих источников и фумарол.
f soufre *f*
s azufre *f*
g Schwefel *m*

5960 sum of active temperatures
Sum of average daily air or soil temperatures exceeding the biological temperature minimum established for a certain stage of plant development.
r сумма активных температур
Сумма средних суточных температур воздуха или почвы, превышающих биологический минимум температуры, установленный для определенного периода развития растения.
f somme *f* de température active
s suma *f* de temperatura activa
g Summe *f* der aktiven Temperaturen

5961 sum of effective temperatures
A characteristic of heat regime for the period of vegetation which is the sum of mean daily temperatutres for the same period, or the sum of the mean daily temperature excesses over a conditional value of the vegetation low limit.
r сумма эффективных температур
Характеристика теплового режима за вегетационный период, представляющая собой сумму средних суточных температур за этот период или превышений среднесуточной температуры воздуха над условной величиной нижнего предела вегетации растений.
f somme *f* de température effective
s suma *f* de temperatura efectiva
g Summe *f* der effektiven Temperaturen

5962 summer
The season of the year with the highest air temperatures for a certain territory, the period astronomically defined from the summer solstice to the autumn equinox (from June 21-22 to September 23 in the Northern Hemisphere, from December 21-22 to March 20-21 in the Southern Hemisphere).
r лето
Время года, климатический сезон с наиболее высокими для данной территории температурами воздуха, в астрономическом понимании – промежуток времени от летнего солнцестояния до осеннего равноденствия (с 21-22 июня до 23 сентября в Северном полушарии, с 21-22 декабря до 20-21 марта в Южном полушарии).
f été *m*
s verano *m*
g Sommer *m*

5963 summer flood
A short-term rise of water level in rivers in summer as a result of high mountain snow and glaciers melting as well as of rainfalls.
r летний паводок
Краткосрочный подъем уровня воды в реке летом в результате таяния высокогорных снегов и ледников, выпадения дождей.
f crue *f* d'été
s crecida *f* de verano
g Sommerhochwasser *m*

5964 summer solstice
21-22 June (the longest day in the Northern Hemisphere) when the Sun passes through the most northern point of ecliptic shining vertically at noon over latitude 23°30′ N, the Tropic of Cancer. → Fig. 25, p. 58
r летнее солнцестояние

Момент времени, когда Солнце проходит через самую северную точку эклиптики (21или 22 июня) – это самый длинный день в Северном полушарии, а на тропике Рака в полдень сонечные лучи падают вертикально. → Рис. 25, с. 58
f solstice *m* d'été
s solsticio *m* de verano; solsticio *m* de Cancer
g Sommersonnenwende *f*; Sommersolstitium *n*

5965 summer surface
The surface of a glacier at the end of ablation period.
r летняя поверхность
Поверхность ледника в момент окончания периода абляции.
f surface *f* estivale
s superficie *f* estival
g Sommeroberfläche *f*

5966 summit; top
The highest part of a mountain, massif or ridge crest.
r горная вершина
Наиболее высокая часть горы, массива или гребня хребта.
f sommet *m*
s cumbre *f*
g Gipfel *m*; Spitze *f*

5967 Sun
The central body of the Solar system round which the Earth and other planets revolve, serving as the source of energy for all processes on the Earth.
r Солнце
Центральное тело Солнечной системы, источник энергии для всех процессов, происходящих на Земле.
f Soleil *m*
s sol *m*
g Sonne *f*

* **sun crust → 4867**

* **sun dog → 4422**

* **sun-and-beach tourism → 5277**

5968 sunrise
The time at which the Sun apparently rises above the horizon, due to the Earth's rotation on its axis.
r восход
Момент появления Солнца из-под горизонта при его видимом суточном движении.
f soleil *m* levant
s salida *f* del sol
g Sonnenaufgang *m*

5969 sunset
The time at which the Sun apparently sets in the evening below the horizon, due to the Earth's rotation on its axis.
r заход
Момент исчезновения Солнца за горизонтом при его видимом суточном движении.
f coucher *m* du soleil
s puesta *f* del sol
g Sonnenuntergang *m*

5970 sunshine
An element comprising the direct radiation (insolation) received from the Sun, the amount varying according to latitude and to degrees of cloudiness.
r солнечный свет
Элемент прямой радиации (инсоляции), получаемой Землей от Солнца, ее количество изменяется в соответствии с широтой и наличием облачности.
f lumière *f* du soleil
s luz *f* solar; luz *f* del sol
g Sonnenschein *m*

* **sunshine duration → 3117**

5971 sunspot
A dark area on the visible surface of the Sun causing an increase in the solar radiation received on the Earth, and particularly affecting the Earth's magnetic field and ionosphere, the number of sun spots usually reaches a maximum every eleven years.
r солнечное пятно
Темное пространство (пятно) на видимой стороне Солнца, развитие которого вызывает увеличение солнечной радиации, получаемой Землей, и воздействует на магнитное поле и ионосферу Земли (количество пятен достигает максимума каждые 11 лет).
f taches *fpl* solaires
s mancha *f* solar
g Sonnenfleck *m*

5972 superadiabatic lapse rate
Temperature lapse rate exceeding the dry adiabatic one, i.e. greater 1°/100 m.
r сверхадиабатический градиент
Вертикальный градиент температуры в атмосфере, превышающий сухоадиабатический, т.е. больший, чем 1°/100 м.
f gradient *m* superadiabatique
s gradiente *m* superadiabático
g überadiabatischer Gradient *m*

5973 superaquatic landscape
A landscape developing under the constant influence of groundwater where aerobic soil conditions and development of hydrophilous vegetation are typical.
r супераквальный ландшафт
Ландшафт, развивающийся в условиях постоянного влияния близко залегающих грунтовых вод, для которого характерны аэробные условия в почвах, развитие гидрофильных растительных сообществ.
f paysage *m* superaquatique

s paisaje *m* superacuático
g superaquatische Landschaft *f*

5974 supercooled rain
Rain composed of water droplets falling at a temperature below freezing point.
r переохлажденный дождь
Дождь из капель, имеющих температуру ниже точки замерзания.
f pluie *f* surfondue
s lluvia *f* subfundida
g unterkühlter Regen *m*

5975 supercooled water
Water which has been cooled to a temperature below its normal freezing-point (0°C), but which remains in liquid form if undisturbed.
r переохлажденная вода
Вода, охлажденная до температуры ниже точки ее замерзания (0°C), но остающаяся в жидком состоянии, пока это состояние не нарушено.
f eau *f* sous-refroidie
s agua *f* sobreenfriada
g unterkühltes Wasser *n*

5976 supercooling
Cooling of liquid water below its normal freezing point without causing it to freeze.
r переохлаждение
Охлаждение жидкой воды до температуры ниже ее обычной точки замерзания без замораживания ее.
f surfusion *f*
s subfusión *f*; sobreenfriamiento *m*
g Unterkühlung *f*

* **superficial *adj* → 5875**

5977 superficial deposit
Unconsolidated Pleistocene and Holocene material lying on the surface of the Earth's crust, not formed in situ from underlying bedrock, but carried out and set down in position by the agencies of wind, water, ice, or the gravity.
r поверхностные отложения
Рыхлый плейстоценовый и голоценовый материал, лежащий на поверхности земной коры, не сформированный in situ из нижележащих коренных пород, а принесенный и отложенный под действием ветра, воды, льда или силы тяжести.
f dépôt *m* superficiel
s depósito *m* superficial
g Oberflächenablagerung *f*

5978 superglacial *adj*; supraglacial *adj*
Relating to the surface or to the environment at the surface of a glacier.
r наледниковый
Относящийся к поверхности ледника или природным явлениям на поверхности ледника.
f supra-glaciaire *adj*
s superglaciar *adj*; supraglaciar *adj*
g supraglazial *adj*

* **superglacial sediments → 5992**

5979 superglacial stream
A rivulet of meltwater flowing in summer in a deep runnel on the surface of a glacier and descending into a crevasse.
r наледниковый поток
Ручей с талой водой, текущий летом в глубокой ложбине на поверхности ледника, пока он не проваливается в трещину.
f écoulement *m* supra-glaciaire; cours *m* d'eau supra-glaciaire
s corriente *m* superglaciar
g Strom *m* auf der Gletscheroberfläche

5980 superimposed drainage
A drainage pattern appearing to be independent of the structure of the underlying rocks because it originally developed on a former rock surface, that has now disappeared owing of denudation.
r наложенная речная сеть
Речная сеть, кажущаяся независимой от подстилающих ее структур, так как она возникла на прошлой поверхности горных пород, с тех пор сильно измененной процессами денудации.
f réseau *m* hydrographique surimposé
s red *f* hidrográfica sobreimpuesta
g überlagertes Flußnetz *n*

5981 superimposed ice
The ice formed on the surface of a glacier due to the freezing of melt water mixed with snow, when water from melting snow comes into contact with the cold surface of a glacier.
r наложенный лед
Лед, формирующийся на поверхности ледника от замерзания талых вод в смеси со снегом, когда талая вода приходит в контакт с холодной поверхностью ледника.
f glace *f* surimposée
s hielo *m* sobreimpuesto
g aufgelagertes Eis *n*

5982 supermarket
A self-service store stocked with food and usually some small household goods displayed on open shelves with the total trade area of up to 2500 m^2.
r супермаркет
Торговый комплекс самообслуживания с реализацией в основном пищевых товаров и некоторых товаров, необходимых для дома, с размещением всего на открытых полках и суммарной торговой площадью до 2500 $м^2$.
f supermarché *m*
s supermercado *m*
g Supermarket *m*

5983 supersaturation
The state of air when its relative humidity exceeds 100%, i.e. when it contains more water vapour than is required to saturate it.
r перенасыщение
Состояние воздуха, при котором содержание в нем водяного пара превышает количество, необходимое для его насыщения.
f sursaturation *f*
s supersaturación *f*; sobresaturación *f*
g Übersättigung *f*

* **superstore → 2912**

5984 supplementary contour
An extra contour included on certain relief maps where the terrain is very irregalar and where the normal contour interval would be insufficient to illustrate the character of the relief.
r дополнительная горизонталь
Добавочная горизонталь на некоторых географических картах со сложным рельефом, где обычных горизонталей недостаточно, чтобы показать характерные черты рельефа.
f courbe *f* de niveau supplémentaire
s curva *f* de nivel suplementaria
g zusätzliche Höhenlinie *f*

5985 supply
One of two (supply and demand) market forces that govern prices, operating freely in a market economy, apparent through the price mechanism as responces in quantities offered for sale. → Fig. 317
r предложение
Одна из двух (предложение и спрос) категорий товарного производства, соотношение которых – главный ценообразующий фактор, предложение представляет собой совокупность товаров и услуг, которые могут быть реализованы на рынке. → Рис. 317
f offre *f*
s oferta *f*
g Angebot *n*

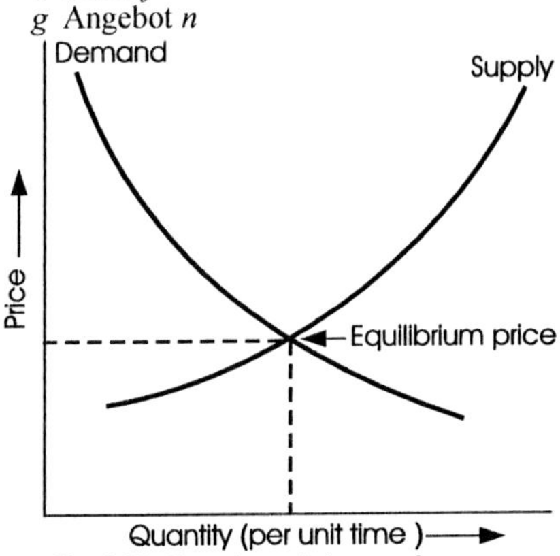

Fig. 317. **Supply** and **demand** curves

* **supraglacial** *adj* **→ 5978**

5986 supralittoral zone
The seashore zone immediately above the littoral fringe and beyond the reach of tidal submergence, through affected by sea spray.
r супралиторальная зона; супралитораль
Полоса суши, расположенная выше литорали, отличающаяся высокой влажностью из-за частого попадания брызг прибоя и заплеска волн.
f étage *m* supralittoral
s piso *m* supralitoral
g Supralitoral *n*

5987 suprapermafrost water
Ground gravitational waters occuring above frozen ground which serves as an impermeable table.
r надмерзлотные воды
Подземные гравитационные воды, залегающие на мерзлых породах как на водоупорном ложе.
f eau *f* de suprapergélisol
s agua *m* de suprapermafrost
g Suprapermafrostwasser *n*

* **surf → 6013**

5988 surface ablation
The removal of snow and ice from the surface of a glacier.
r поверхностная абляция
Убыль снега и льда на поверхности ледника.
f ablation *f* superficielle
s ablación *f* superficial
g Oberflächenablation *f*

* **surface avalanche → 5537**

5989 surface current
The motion of a surface water layer of an ocean generated by different forces, but mainly by winds.
r поверхностное течение
Поступательное или вращательное движение поверхностного слоя воды в океане, обусловленное различными силами, но главным образом силой ветра.
f courant *m* de surface
s corriente *f* de superficie
g Oberflächenströmung *f*

5990 surface detention
The process by which an amount of water is retained in temporary storage both during and immediately after a rainstorm before moving downslope as overland flow.
r задержка стока
Процесс, при котором некоторое количество воды остается в грунте как во время, так и непосредственно после сильного ливня, прежде чем начнется поверхностный сток.

Fig. 318. **Surging glacier** (left) in the Pamirs

f détention *f* superficielle
s detención *f* superficial
g Oberflächenretention *f*

* **surface inversion → 2651**

5991 **surface irrigation**
The form of irrigation when water is distributed over the soil surface as a unbroken layer of water or surface flows.
r поверхностное орошение
Способ орошения, при котором вода распространяется по поверхности почвы в виде сплошного слоя или поверхностных струй.
f irrigation *f* superficielle
s irrigación *f* superficial
g Oberflächenbewässerung *f*

* **surface layer of air → 771**

5992 **surface moraine; superglacial sediments**
Rock fragments lying on the surface of a glacier and carried by it.
r поверхностная морена
Обломочный материал, переносимый на поверхности ледника.
f moraine *f* superficielle
s morrena *f* superficial; morrena *f* externa
g Oberflächenschutt *m*; Aussenmoräne *f*

5993 **surface roughness**
The Earth's surface irregularities which cause the air flow in the adjacent layer to be more or less turbulent.
r шероховатость поверхности
Неровности земной поверхности, создающие турбулентность в прилежащем к поверхности воздушном потоке.
f rugosité *f* de la surface
s rugosidad *f* de la superficie
g Oberflächenrauigkeit *f*

* **surface runoff → 4346**

* **surface wash → 5385**

5994 **surface water**
Waters constantly or temporarily present on the Earth's surface in the form of rivers, lakes, bogs, etc.
r поверхностные воды
Воды, постоянно или временно находящиеся на земной поверхности в виде водных объектов: рек, озер, болот и др.
f eaux *fpl* de surface; eaux *fpl* superficielles
s agua *f* superficial; agua *f* de superficie
g Oberflächengewässer *pl*

* **surge → 2533**

5995 **surge cycle; pulsation period**
A time interval between the same phases of two glacier surges following one after another (it divides into two stages: surging stage and restoration stage).
r период пульсации
Время от завершения одной из подвижек пульсирующего ледника до завершения последующей, подразделяется на две части – стадию подвижки и стадию восстановления.
f période *f* de pulsation
s período *m* de pulsación
g Pulsationsperiode *f*

* **surge-type glacier → 5996**

5996 **surging glacier; surge-type glacier**
A glacier that is characterized by periodic rapid advances which redistribute the glacier material causing no change in its total mass. → Fig. 318
r пульсирующий ледник
Ледник, которому свойственны периодические быстрые продвижения, перераспределяющие вещество без изменения общей массы. → Рис. 318
f glacier *m* de pulsation
s glaciar *m* de pulsación
g Pulsierungsgletscher *m*

* **surplus area → 43**

* **survey mark → 640**

5997 **survey signal; tower beacon**
A construction marking the location of a geodetic point on the ground.
r геодезический знак; геодезический сигнал
Устройство или сооружение, обозначающее положение геодезического пункта на местности.
f signal *m* géodésique
s señal *f* geodésica
g Pfeiler *m*; Signal *n*

5998 **surveying peg**
Point of geodetic survey the position of which is determined relatively to the position of the instrument.

r пикет
Закрепленная на местности точка, положение которой определяют относительно позиции инструмента при геодезической съемке данного участка местности.
f point *m* jeté; piquet *m*
s poste *m*; piquete *m*
g Vermessungspunkt *m*

5999 survival rate
The percentage of newly born individuals which survived in a population over a certain period of time (a season, year, etc.), is calculated for various age and sex groups.
r выживаемость
Доля новорожденных особей в популяции, доживших до определенного возраста.
f survivance *f*
s supervivencia *f*
g Überlebensrate *f*

6000 suspended load
The fine solid particles transported by a stream in suspension without the aid of saltation.
r взвешенные наносы; взвеси
Твердые частицы, переносимые водным потоком во взвешенном состоянии, без помощи сальтации.
f charge *f* en suspension
s carga *f* en suspensión
g Schwebstofffracht *f*; Trübe *f*

6001 suspension
The state in which small particles of an insoluble material are evenly distributed within a fluid such as water or air.
r суспензия
Состояние, при котором мелкие нерастворимые частицы парят в воде или воздухе.
f suspension *f*
s suspensión *f*
g Suspension *f*

6002 suspension roiling
The transition of load to a suspended state.
r взмучивание наносов
Переход наносов во взвешенное состояние.
f suspension *f* de aluvion
s suspensión *f* de aluvión
g Trübden *n* durch Geschiebe

6003 sustainable agriculture
Environmentally friendly methods of farming that allow to produce crops or livestock when land resources are used at maximum rate but without damaging the ecosystem.
r устойчивое сельское хозяйство
Ведение сельского хозяйства, при котором максимально используется ресурсный потенциал земель, но при этом обеспечивается экологическая безопасность и постоянное возобновление плодородия экосистемы.
f agriculture *f* durable
s agricultura *f* sustentable; agricultura *f* sostenible
g umweltverträgliche Landwirtschaft *f*

6004 sustainable development
The development which balances the satisfaction of people's immediate needs and the protection of future generationsr' interests to meet their own needs.
r устойчивое развитие
Развитие окружающей среды и общества, при котором достигается удовлетворение жизненных потребностей ныне живущих людей и такая возможность сохраняется для будущих поколений.
f développement *m* durable
s desarrollo *m* sustentable; desarrollo *m* sostenible; desarrollo *m* durable
g nachhaltige Entwicklung *f*; dauerhafte Entwicklung *f*

6005 sustainable development indicators
Ecological, economic, and social characteristics linking the quality of the environment with the socio-economic development, e.g. the state of biodiversity, consumption of energy and materials per product unit, quality of life, etc.
r показатели устойчивого развития
Экологические, экономические и социальные характеристики, увязывающие качество окружающй среды с социально-экономическим развитием, например, состояние биоразнообразия, расход энергии и ресурсов на единицу конечного продукта, качество жизни и пр.
f indices *mpl* de développement durable
s índices *mpl* de desarrollo sustentable; índices *mpl* de desarrollo sostenible
g Parameter *mpl* der nachhaltigen Entwicklung

6006 sustained yield forestry
Any practice or technique (e.g. patch cutting) used in forestry which aims to conserve the forest.
r устойчивое лесное хозяйство
Любая практика или технология (например, пятнистая вырубка), используемая в лесном хозяйстве, имеющая цель сохранить лес.
f exploitation *f* forestière durable
s silvicultura *f* durable; silvicultura *f* sustentable
g umweltverträgliche Forstwirtschaft *f*

6007 swale
A shallow, closed, flat-bottomed depression, often created by suffosion.
r западина
Мелкая замкнутая плоскодонная котловина, часто суффозионного происхождения (в степях Евразии называется *степное блюдце*).
f cuvette *f*
s cubeta *f*; hundimiento *m*
g Talkessel *m*

Fig. 319. **Swamp**

* **swallet → 4688**

* **swallow → 4688**

6008 swallow hole
A vertical or near-vertical shaft down which a stream disappears in areas of limestone topography.
r поглощающая впадина
Вертикальная или почти вертикальная шахта, в которой исчезает речной поток, текущий в карстовой области.
f puits *m* d'engloutissement
s cavidad *f* de absorbencia
g Aufsaugenhöhle *f*

* **swallow-hole → 5450**

6009 swamp; wetland; bog; fen; marsh
A wet spongy land saturated with water and having stagnant drainage all over the year. → Fig. 319
r болото
Участок суши с обильным застойным или слабопроточным увлажнением грунта в течение большей части года. → Рис. 319
f marais *m*; marécage *m*; tourbière *f*
s ciénaga *f*; pantano *m*; aguazal *m*; turbera *f*
g Sumpf *m*; Moor *n*

6010 swamp percentage
The ratio of the area occupied by bogs to the whole area under consideration.
r болотистость; заболоченность
Отношение площади, занятой болотами, ко всей площади рассматриваемой территории.
f pourcentage *m* des marais
s porcentaje *m* de ciénagas
g Moordichte *f*

6011 swamp plain
A flat surface which is permanently waterlogged and covered with hydrophilous vegetation.
r заболоченная равнина
Плоская поверхность, на значительной части насыщенная водой и покрытая водолюбивой растительностью.
f plaine *f* de marécage
s llanura *f* inundable; llanura *f* pantanosa
g Sumpfflachland *n*

6012 swarding
Fixation of natural and artificial slopes by creating a dense grass cover.
r задернение
Придание устойчивости естественным и искусственным откосам путем создания на них сплошного травяного покрова.
f plaçage *m*; enherbement *m*; gazonnement *m*
s encespedamiento *m*
g Verunkrautung *f*; Berasung *f*

6013 swash; surf
A mass of broken foaming water in the breaker zone of a shoreline, produced by the deformation of a powerful wave as it breaks on rocks: the steepness of wave slope sharply increases, and the wave crests capsize on the shore and spume. → Fig. 320
r прибой; накат
Масса воды в бурунистой прибрежной зоне, возникающая в результате деформации волн, разбивающихся о береговую черту, когда резко увеличивается крутизна склона волн, гребни которых опрокидываются на берег и пенятся. → Рис. 320
f déferlement *m*
s rompiente *f*
g Brandung *f*; Strandwelle *f*

6014 swash flow
Flat water flow running-up onto a coastal zone after a wave destruction.
r накат волны; прямой поток заплеска
Плоский поток воды, распространяющийся вверх по береговой зоне после разрушения волны.
f cours *m* direct d'eau
s flujo *m* directo de agua
g Schwall *m*; landwärtiger Brandungsstrom *m*

6015 swell$_1$
A regular movement of waves which do not break created by wind stress in a lazrge body

Fig. 320. **Swash**

of water, especially in the open ocean which travel considerable distances away from the generating field (the waves have relatively smooth and generally unbroken crests and regular wavelength).

r зыбь; плоская волна
Волны на поверхности океанов и морей, образующиеся из ветровых волн после полного прекращения воздействия на них ветра (обычно зыбь длиннее ветровых волн и более правильна).

f houle *f*; vague *f* libre; système *m* des vagues à courtes crêtes

s oleaje *m*; ola *f* libre; sistema *m* de olas de crestas cortas

g Wanderwellen *fpl*; Seegang *m*

* **swell$_2$ → 546**

6016 swell scale
A scale for the visual estimation of the waving degree in the water bodies.

r шкала волнения
Шкала для визуальной оценки степени волнения в водоемах.

f échelle *f* d'agitation d'eau

s escala *f* de agitación de agua

g Seegangskala *f*

6017 swelling of soil
The expansion of soil or ground in volume when wetted and moisturized.

r набухание почвы
Увеличение объема почвы или грунта в процессе смачивания и увлажнения.

f gonflement *m* du sol

s henchimiento *m* del suelo

g Bodenanschwellung *f*

* **swidden farming → 5470**

6018 symbiosis; symbiotic relationship
1. The very close mutually beneficial association of organisms of two or more species. 2. The very close, mutually advantageous relations among dissimilar members of a human group or among dissimilar groups within a larger group.

r симбиоз
1. Тесное взаимовыгодное сожительство организмов двух или более видов. 2. Очень близкие обоюдовыгодные отношения между различными членами одной группы людей или между различными группами людей в их более обширном сообществе.

f symbiose *f*

s simbiosis *f*

g Symbiose *f*

* **symbiotic relationship → 6018**

6019 symmetrical fold
A fold of rock layers with the limbs, which dip away at equal angles from the axis of fold.

r симметричная складка; прямая складка
Складка в горных породах с крыльями, падающими под одинаковыми углами к оси складки.

f pli *m* droit

s pliegue *m* recto

g gerade Falte *f*

6020 synanthropic organism
Animal or plant the life cycle of which is related with humans, their homes or landscape formed or modified by people.

r синантропный организм
Животное или растение, жизненный цикл которого связан с человеком, его жильем или антропогенным ландшафтом.

f organisme *m* synanthropique

s organismo *m* sinántropo

g synanthroper Organismus *m*

* **synclinal fold → 6022**

6021 synclinal valley
A valley which follows the syncline axis, and the rock strata of the slope dip towards the talweg. → Fig. 13, p. 38

r синклинальная долина
Речная долина, заложившаяся вдоль оси синклинали, причем пласты пород на склонах долины имеют падение к ее тальвегу. → Рис. 13, с. 38

f vallée *f* synclinale

s valle *m* sinclinal

g Synklinaltal *n*; Muldental *n*

6022 syncline; synclinal fold
A downfold in the strata of the Earth's crust formed by compression, the rocks dipping inwards to a central axis.

r синклиналь
Складка с изгибом горных слоев, обращенных выпуклостью вниз, с более молодыми отложениями в центральной части.

f pli *m* synclinal; synclinal *m*

s sinclinal *m*

g Synklinale *f*; Mulde *f*

6023 synclinorium
A complex syncline, a broad downfold of the rocks on which numerous minor upfolds and downfolds occur, formed within the limits of geosynclines as a result of deformation of sedimentary strata.

r синклинорий
Сложная складчатая структура синклинального строения, возникающая обычно в пределах геосинклиналей в результате деформации осадочных толщ, заполняющих крупные прогибы.

f synclinorium *m*

s sinclinorio *m*

g Synklinorium *n*

6024 syndicate

A form of business alliance in which a group of enterprises co-operates in some activities (pricing, buying raw materials, marketing, etc.) but which remain economically and legally independent.
r синдикат
Одна из форм объединения предприятий, ведущих общую коммерческую деятельность (определение цен, закупка сырья, сбыт продукции и т.п.) при сохранении производственной и юридической самостоятельности.
f syndicat *m*
s sindicato *m*
g Syndicat *n*

6025 syneclise
A gently concave platform structure, hundreds of kilometres across, overlain by thick strata of sedimentary rocks.
r синеклиза
Плоская вогнутая платформенная структура поперечником в сотни километров с мощной толщей осадочных пород.
f synéclise *f*
s sineclisa *f*
g Syneklise *f*

6026 syngenetic ice wedge
An ice wedge developed simultaneously with the deposition of surrounding sediments.
r сингенетическая ледяная жила
Ледяная жила, формирующаяся одновременно с накоплением вмещающих осадков.
f coin *m* de glace syngénétique
s cuña *f* de hielo singenética
g syngenetischer Eiskeil *m*

* **synodic month → 3617**

* **synoptic chart → 6738**

6027 synoptic eddy
A non-stationary whirl-like disturbance in oceanic currents, 100-200 km in horizontal size, extending into the depth of the ocean for hundreds and thousands of meters (and sometimes even to the bottom) and moving with translational velocity of several cm/s.
r синоптический вихрь
Нестационарное вихреобразное возмущение в системе океанских течений горизонтальным размером 100-200 км, распространяющееся в глубину океана на сотни и тысячи метров (а иногда и до дна) и перемещающееся с поступательной скоростью в несколько см/с.
f tourbillon *m* synoptique
s torbellino *m* sinóptico
g synoptischer Wirbel *m*

6028 synoptic meteorology
The branch of *meteorology* studying atmospheric conditions over a large area up to 1000 km in size and lasting about one day by using charts on which synoptic observations are plotted for the purpose of weather analysis and forecasting.
r синоптическая метеорология
Раздел *метеорологии*, изучающий атмосферные явления в крупном регионе размером до 1000 км, продолжающиеся не менее суток, на основе карт с данными синоптических наблюдений с целью анализа и прогноза погоды.
f météorologie *f* synoptique
s meteorología *f* sinóptica
g Synoptik *f*

6029 synoptic processes
Large scale atmospheric processes determining weather conditions over vast territories and studied by means of synoptic charts.
r синоптические процессы
Макромасштабные атмосферные процессы, определяющие режим погоды на больших территориях и изучаемые с помощью синоптических карт.
f processus *mpl* synoptiques
s procesos *mpl* sinópticos
g synoptische Prozesse *mpl*

6030 synsedimentary folding
The folding of rocks formed simultaneously with the sedimentation.
r конседиментационная складчатость
Складчатость горных пород, формирующаяся одновременно с осадконакоплением.
f plissement *m* synsédimentaire
s plegamiento *m* sin sedimentos
g synsedimentäre Faltung *f*

6031 synthetic map
A thematic map representing a phenomenon as a single whole on the basis of a number of integrated parameters.
r синтетическая карта
Тематическая карта, отображающая явление как единое целое на основе совмещения ряда показателей.
f carte *f* synthétique
s mapa *m* sintético
g synthetische Karte *f*

6032 synusia
A structural part of a phytocenosis that consists of ecologically close species belonging to

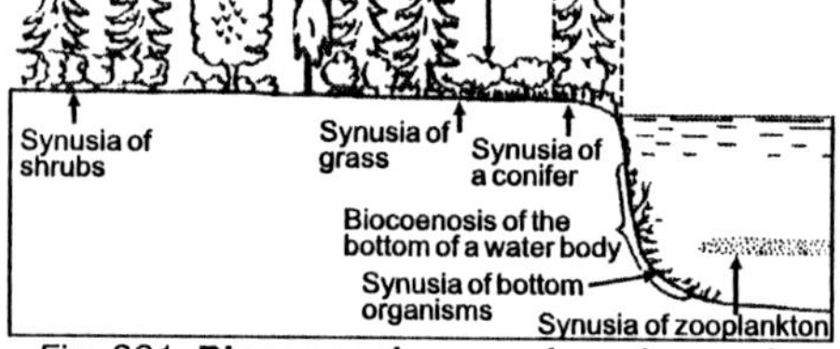

Fig. 321. **Biocoenosis**, **parcels** and **synusia**

the same life form and to the same element of vertical structure, e.g. synusia of bushes, dwarf shrubs, mosses, etc. → Fig. 321, p. 718

r синузия

Структурная часть фитоценоза, характеризующаяся определенным составом экологически близких видов, относящихся к одной жизненной форме и элементу вертикальной структуры, например, синузии кустарников, кустарничков, мхов и т.п. → Рис. 321, с. 718

f synusie *f*

s sinusía *f*

g Synusie *f*

6033 **system of farming; farming system**

A system of interrelated agrotechnical, organizational and other land use measures formed in specific social, historical and geographical conditions.

r система земледелия

Система взаимосвязанных агротехнических, мелиоративных, организационных и других мероприятий по использованию земли, сложившаяся в конкретных общественно-исторических и географических условиях.

f système *m* agraire; système *m* de culture

s sistema *m* agrário; sistema *m* de cultivo

g Landwirtschaftssystem *n*; Ackerbausystem *n*

6034 **systems analysis; systems approach**

The applied study of the functional and structural relationships of a natural phenomenon as a system which is identified by defining its boundaries, its purpose, its position and role in a larger system.

r системный анализ; системный подход

Исследование функциональных и структурных взаимосвязей природных явлений, рассматриваемых в качестве системы, в которой определяются границы, возможности использования, а также положение и роль в следующей по рангу природной системе.

f analyse *f* systématique

s análisis *m* sistémico

g Systemanalyse *f*

* **systems approach → 6034**

6035 **syzygy**

The new Moon or full Moon when the Sun, the Earth and the Moon are in the same straight line.

r сизигия

Моменты новолуния и полнолуния, когда Солнце, Земля и Луна оказываются на одной прямой.

f syzygie *f*

s sicigia *f*

g Syzygie *f*

T

* **table iceberg** → 6038

* **table mountain → 6037**

6036 **tableland**
A flat-topped or undulating area of high relief, usually bounded by steeply descending slopes.
r столовая страна; столовый рельеф
Обширные территории с преобладанием в рельефе плато и высоких равнин, разделенных глубокими долинами, часто со ступенчатыми склонами.
f relief *m* tabulaire
s relieve *m* tabular
g Tableland *n*

6037 **tablemount; table mountain**
An isolated flat-topped mountain with steep slopes, e.g. the famous mountain near Capetown.
r столовая гора
Изолированная плоская вершина с крутыми склонами, например, известная гора над Кейптауном.
f mont *m* tabulaire; montagne *f* tabulaire
s monte *m* tabular; montaña *f* tabular
g Tafelberg *m*

6038 **tabular iceberg; barrier iceberg; table iceberg**
An immense horizontal floating mass of ice, often exceeding 20 km in lengh, an iceberg with vertical sides and flat tops which has broken off from an ice shelf. → Fig. 322
r столовый айсберг
Обширный плоский айсберг с вертикальными стенками, нередко в поперечнике больше 20 км, отколовшийся от шельфового ледника. → Рис. 322
f iceberg *m* tabulaire
s iceberg *m* tabular
g Tafeleisberg *m*

6039 **tacheometry; tachymetry**
A surveying technique developed to produce rapid determination of direction, elevation and distance by the use of specially adapted theodolite, known as a tacheometer.
r тахеометрическая съемка
Техника быстрого определения направления, высоты и расстояния с помощью специального теодолита, называемого тахеометром.
f levé *m* tachéométrique
s levantamiento *m* taquimétrico
g tachymetrische Aufnahme *f*

Fig. 322. **Tabular iceberg** in the Southern Ocean

* **tachymetry → 6039**

6040 **tafone; tafoni *pl***
A deep hollow on a rock surface formed by differential chemical cavernous weathering and deflation.
r котел выдувания; ниша выдувания
Глубоко вдающиеся в скалы углубления, возникшие в результате избирательного химического выветривания и дефляции.
f taffoni *m*; cavité *f* de déflation
s tafoni *m*; cavidad *f* de deflación
g Ausblasungskessel *m*; Tafoni *pl*

* **tafoni *pl* → 6040**

6041 **taiga; boreal forest**
The predominantly coniferous forests stretching in a broad zone in the Northern Hemisphere, adjacent to the tundra.
r тайга; бореальный лес
Хвойные леса умеренного пояса, расположенные к югу от тундры, распространены в Евразии и Северной Америке.
f taïga *f*; forêt *f* résineuse boréale
s taiga *f*; bosque *m* boreal; aciculisilva *f*
g Taiga *f*; borealer Nadelwald *m*

6042 **taiga pergelisol**
The group of soils having permafrost horizon that develop under taiga vegetation.
r мерзлотно-таежные почвы
Собирательный термин для почв с горизонтом многолетней мерзлоты под таежной растительностью.
f pergélisol *m* de taïga
s pergelisol *m* de taiga
g Frost- und Taigaböden *mpl*

6043 **taiga soils**
Soils forming under taiga vegetation in Eurasia and North America, may belong to various types, but mainly to podzols, taiga pergelisol, etc.
r таежные почвы
Почвы, формирующиеся под таежной растительностью в Евразии и Северной Америке, могут принадлежать к разным типам, но прежде всего к подзолам.
f sols *mpl* de taïga
s suelos *mpl* de taiga
g Taigaböden *mpl*

6044 **tail dune**

A sand dune of varying length, formed leeward of an obstacle and tapering away from it.

r дюнный шлейф

Песчаная дюна разной длины, образующаяся на подветренной стороне препятствия и постепенно сходящая на конус.

f flèche *f* dunaire

s flecha *f* dunar

g Dünenschleife *f*

6045 tailing storage; tip-heap

A variety of water intake facilities aimed at purification from suspensions of liquid *tailings* usually supplied to it as pulp.

r хвостохранилище

Разновидность водоприемника, созданного для очистки от взвесей жидких *хвостов*, подаваемых к нему обычно в виде пульпы.

f stockage *m* des queues; bassin *m* à déchets de lavage

s depósito *m* de colas

g Abpängebecken *n*

6046 tailings

The waste left after mining operations and processing of minerals which is regarded as uneconomic to work.

r хвосты

Отходы, возникающие при добыче и обогащении полезных ископаемых или других технологических процессах.

f queues *fpl*; déchets *mpl* de lavage; schistes *mpl*

s colas *fpl*

g Abgänge *mpl*; Abfallerz *n*

* **taking a bearing → 624**

6047 takyr

A flat depression in a clay or stony desert of Central Asia with a cracked surface devoid of vegetation occurring in dry season after the drying up of seasonal lakes.

r такыр

Плоская, лишенная растительности растресканная на мелкие многоугольники поверхность в понижениях рельефа песчаных, глинистых и каменистых пустынь Средней Азии, возникающая в сухой сезон после высыхания сезонных озер.

f takyr *m*

s takyr *m*

g Takyr *m*

6048 takyr-soil

A soil formed in flat clay periodically flooded depressions in deserts and semi-deserts under ephemeral vegetation.

r такыровидная почва

Почва, образующаяся в плоских глинистых периодически затопляемых водой понижениях в пустынях и полупустынях под эфемерной растительностью.

f sol *m* de takyr

s suelo *m* de takyr

g Takyrboden *m*

6049 talik

A layer or a body of unfrozen ground occurring between the permafrost and the seasonally frozen layer.

r талик

Участок немерзлого грунта среди многолетнемерзлых пород.

f talik *m*

s talik *m*

g Talik *n*

* **tall grass vegetation → 6050**

* **tall grassland → 3574**

6050 tall herbaceous vegetation; tall grass vegetation; tall-grassland

A layer or community of long (2-4 m) herbaceous plants growing under the conditions of higher humidity of air and soil.

r высокотравье

Ярус или сообщество высоких (2-4 м) травянистых растений, развивающихся при повышенной влажности воздуха и почвы.

f végétation *f* altherbeuse

s hierba *f* alta

g Hochstaudenflur *f*; Hochstaudenwiese *f*

* **tall-grassland → 6050**

6051 talus; scree

An accumulation of angular rock debris on a steep slope as a result of the falling of fragments destroyed by weathering.

r осыпь

Скопление угловатых скальных обломков на крутом склоне в результате осыпания разрушенного выветриванием обломочного материала.

f éboulis *m*

s acumulación *f* de escombros

g Schutthalde *f*; Felsschutt *m*; Hangschutt *m*

6052 talus cone

A cone-shaped mass of rock debris with the top pointing upward at the foot of a steep slope.

r конус осыпи

Скопление обломочных продуктов выветривания у подножья крутого скалистого склона в виде полуконуса, обращенного вершиной вверх.

f talus *m* d'éboulis

s talud *m* de escombros

g Schuttkegel *m*; Geröllkegel *m*

6053 talus slope

The slope subject to downhill sliding, rolling and falling of small rock fragments formed due to physical weathering.

r осыпной склон

Склон, на котором сильно развито осыпание и падение небольших обломков, об-

разующихся вследствие физического выветривания.
f versant *m* d'éboulement continu
s vertiente *f* de derrumbamiento continuo
g Geröllhang *m*; Schutthang *m*

* **talweg → 6134**

6054 **tanker**
A vessel used for the carrying of liquids in bulk.
r танкер
Судно, предназначенное для перевозки жидких грузов наливом.
f pétrolier *m*; tanker *m*
s petrolero *m*; buque *m* tanque
g Tankschiff *n*; Tanker *m*

6055 **taoism; daoism**
A philosophical concept originated in China in the VI century B.C. which declares that the human life should not disturb the harmony of the natural environment.
r даосизм
Философская концепция, возникшая в Китае в VI в. до н.э., которая провозглашает жизнь людей, не нарушающую гармонию окружающего мира.
f daoisme *m*; taoisme *m*
s taoísmo *m*
g Daoismus *m*; Taoismus *m*

6056 **taphrogenesis**
Vertical movements of the Earth's crust leading to the formation of major faults and rift valleys.
r тафрогенез
Вертикальные движения земной коры, приводящие к формированию крупных разломов и рифтовых долин.
f taphrogenèse *f*
s tafrogénesis *f*
g Taphrogenese *f*

6057 **tariff**
A scale of duties imposed by a government on goods exported from or imported into the territory under its jurisdiction.
r тариф
Размер государственной пошлины, накладываемой на товары, экспортируемые или импортируемые на территорию, находящуюся под юрисдикцией данного государства.
f tarif *m*
s tarifa *f*
g Tarif *m*

6058 **tariff bappier; customs barrier; tariff wall**
An additional export duty for some commodities to increase the cost of imported goods in order to protect domestic industry from foreign competition.
r тарифный барьер; таможенный барьер
Дополнительная экспортная пошлина на некоторые товары с целью защитить национальное производство данного продукта.
f obstacle *f* tarifaire; barrière *f* douanière
s barrera *f* tarifaria
g Zollmauer *f*; Zollschranke *f*

* **tariff wall → 6058**

6059 **taryn**
A big aufeis occurring in Siberian rivers.
r тарын
Крупная многолетняя наледь на сибирских реках.
f taryn *m*
s taryn *m*
g Taryn *m*

6060 **tax**
A compulsory payment the government (central and local authorities) collects from natural or legal persons to be used for public spending.
r налог
Обязательные платежи, взимаемые государством (центральными и местными органами власти) с физических и юридических лиц в государственный и местные бюджеты.
f taxe *f*
s impuesto *m*
g Steuer *f*

6061 **tear fault; transcurrent fault**
A strike-slip fault that trends transverse to the strike of the folded rocks.
r раздвиг
Разрыв в земной коре, сопровождающийся растяжением блоков горных пород без заметного их смещения по вертикали.
f faille *f* béante
s falla *f* abierta
g Bruchstelle *f*

6062 **technical ice**
Ice produced by a layered or spray freezing of water, or by cutting off an ice block from the natural ice massif.
r технический лед
Лед, изготавливаемый послойным или брызговым намораживанием, либо выпиливанием из естественных ледяных массивов.
f glace *f* technique
s hielo *m* técnico
g technisches Eis *n*

6063 **technogenic landscape; industrial landscape**
A geographical landscape in which industrial and other constructive elements play essential role.
r техногенный ландшафт; индустриальный ландшафт
Географический ландшафт, в функционировании которого существенную роль

играют промышленные и иные подобные элементы.

f paysage *m* technogénique; paysage *m* industriel
s paisaje *m* tecnogeno; paisaje *m* industrial
g technogene Landschaft *f*; industrielle Landschaft *f*

* **technosphere → 300**

6064 tectogenesis
The combination of tectonic movements and processes forming the structure of the Earth's crust.
r тектогенез
Совокупность тектонических движений и процессов, формирующих структуры земной коры.
f tectogenèse *f*
s tectógénesis *f*
g Tektogenese *f*

6065 tectonic *adj*
Related to the processes which build up or form features of the Earth's crust.
r тектонический
Относящийся к процессам, формирующим основные особенности земной коры.
f tectonique *adj*
s tectónico *adj*
g tektonisch *adj*

6066 tectonic epochs
Relatively short (millions of years) periods of increased seismic activity in the Earth's history, manifested in the intensive formation of elevations, troughs, fractures and folds.
r тектонические эпохи
Относительно кратковременные (миллионы лет) фазы повышенной сейсмической активности в истории Земли, проявляются в интенсивном образовании поднятий, прогибов, разломов и складок.
f époques *fpl* tectoniques
s épocas *fpl* tectónicas
g tektonische Phasen *fpl*

6067 tectonic fracture
A break in the Earth's crust formed under the tectonic movements and deformations of rocks. → Fig. 323
r тектонический разрыв
Разлом в земной коре, образовавшийся при тектонических движениях и деформациях горных пород. → Рис. 323
f rupture *f* tectonique; fracture *f*
s ruptura *f* tectónica; fractura *f*
g tektonischer Bruch *m*

6068 tectonic lake
A lake formed by tectonic processes.
r тектоническое озеро
Водоем, котловина которого создана тектоническими процессами.
f lac *m* tectonique
s lago *m* tectónico
g tektonischer See *m*

6069 tectonic mountains
The mountains formed as a result of tectonic movements of the Earth's crust which deform the rocks and create folds, mountain ranges and faults.
r тектонические горы
Горы, возникшие в результате тектонических движений земной коры, которые приводят к деформациям горных пород, возникновению складок, горных хребтов и впадин.
f montagnes *fpl* tectoniques
s montañas *fpl* tectónicas
g tektonisches Gebirge *n*

6070 tectonic movements
The movements of the Earth's crust produced by the endogenetic processes (secular and recent tectonic movements are distinguished).
r движения земной коры; тектонические движения
Движения внешней твердой оболочки Земли, происходящие под воздействием эндогенных сил (различают вековые и современные движения земной коры).
f mouvements *mpl* de l'écorce terrestre
s movimientos *mpl* de la corteza terrestre
g Erdkrustenbewegungen *fpl*

6071 tectonic plate; lithospheric plate
A rigid segment of the Earth's crust which can «float» across the heavier, semi-molten rock below. → Fig. 251, p. 551
r тектоническая плита; литосферная плита
Крупный жесткий блок литосферы Земли, ограниченный сейсмически и тектонически активными зонами разломов, согласно тектонике плит, такие блоки перемещаются по астеносфере. → Рис. 251, с. 551
f plaque *f* lithosphérique
s placa *f* litosférica

Fig. 323. **Tectonic fracture** in Iceland on the boundary between European and American tectonic plates

g lithosphärische Platte *f*;
Lithosphänplatte *f*

6072 tectonic valley
A river valley formed in a relief depression of tectonic origin.
r тектоническая долина
Речная долина, заложившаяся в понижении рельефа тектонического происхождения.
f vallée *f* tectonique
s valle *m* tectónico;
valle *m* de fosa tectónica
g tektonisches Tal *n*

6073 tectonics
A branch of *geology* studying the development of structural features of the Earth's crust and their changes.
r тектоника
Отрасль *геологии*, изучающая развитие глубинных структур земной коры и их изменения.
f tectonique *f*
s tectónica *f*
g Tektonik *f*

6074 tectonosphere
The external layer of the Earth embracing the Earth's crust and the upper mantle in which tectonic as well as magma and metamorphic processes occur.
r тектоносфера
Внешняя оболочка Земли, охватывающая земную кору и верхнюю мантию, в которой в основном и происходят тектонические, а также магматические и метаморфические процессы.
f tectonosphère *f*
s tectonosfera *f*
g Tektonosphäre *f*

6075 teleconnections
Atmospheric events that occur simultaneously but in areas geographically distant from each other, e.g. the southern oscillation.
r дальние связи
Атмосферные события, случающиеся одновременно в районах, расположенных далеко друг от друга, например, южное колебание.
f téléconnexion *f*
s teleconexiónes *fpl*
g Fernverbin

* **telescoped ice → 4877**

* **temblor (*amer*) → 1775**

6076 temperate *adj*
Having mild climate, without extremes of temperature or related to middle latitudes and to the natural phenomena (climate, forest, etc.) characteristic for this region.
r умеренный
Обладающий мягким климатом, относящийся к средним широтам и распространенным здесь природным явлениям (климат, лес и т.п.).
f tempéré *adj*
s templado *adj*
g gemäßigt *adj*

6077 temperate climate
Climate of middle latitudes of both hemispheres, which is under the impact of westerlies prevailing during a year, having no great extremes of temperature, with marked seasonality of natural processes.
r умеренный климат
Климат умеренных широт обоих полушарий, находящийся под воздействием господствующего в течение года западного переноса воздушных масс, без резких колебаний температуры и с заметной сезонностью природных процессов.
f climat *m* tempéré
s clima *m* templado
g gemässigtes Klima *n*

6078 temperate forest zone
Forest zones of temperate belts of the Northern and Southern hemispheres, they include mixed and broad-leaved forests, and taiga in the Northern Hemisphere.
r лесная зона умеренного пояса
Совокупность лесных зон умеренных поясов Северного и Южного полушарий, включает зоны смешанных и широколиственных лесов, а в Северном полушарии также таежную зону.
f zones *f* tempérées des forêts
s zona *f* templada de bosques
g Zone *f* der gemässigten Wälder;
Zone *f* der temperierten Wälder

6079 temperate glacier; warm glacier; isothermal glacier
A glacier whose ice is at the melting point temperature due to pressure across the whole glacier body below its active layer.
r теплый ледник;
изотермический ледник
Ледник, у которого лед находится на точке плавления по всей толще ниже активного слоя.
f glacier *m* tempéré
s glaciar *m* templado
g temperierter Gletscher *m*

6080 temperate zone; mid-latitude zone
Geographical regions between 40-65° N and 42-58° S with marked seasonality of natural processes, in the Northern Hemisphere are between subtropical and subarctic belts, in the Southern Hemisphere between subtropical and subantarctic belts.
r умеренный пояс
Географические пояса, расположенные между 40-65° с.ш. и 42-58° ю.ш., ограни-

чены в Северном полушарии субтропическим и субарктическим поясами, а в Южном – субтропическим и субантарктическим поясами, характерна четкая сезонная ритмичность природных процессов.
f zone *f* tempérée
s zona *f* templada
g temperierte Zone *f*; gemäßigte Zone *f*

6081 temperature
The degree of sensible heat or cold in the environment, measured on various scales.
r температура
Степень тепла или холода в окружающей среде, измеряемая в разных шкалах.
f température *f*
s temperatura *f*
g Temperatur *f*

6082 temperature anomaly; thermal anomaly
The difference between the mean temperature of a place and the mean temperature along its parallel of latuitude.
r аномалия температуры
Отклонение средней температуры в данном месте от средней температуры, характерной для данной широты.
f anomalie *f* de température
s anomalía *f* de temperatura
g termische Anomalie *f*

6083 temperature coefficient of freezing
The relation between the mass of ice that can be obtained by freezing of water in successive layers, and the sum of negative mean daily air temperature.
r температурный коэффициент намораживания
Отношение массы льда, которая может быть получена послойным намораживанием воды, к сумме отрицательных средних суточных температур воздуха.
f coefficient *m* thermique de congélation
s coeficiente *m* térmico de congelación
g Temperaturkoeffizient *m* des Auffrierens

6084 temperature coefficient of melting
The thickness of melt water layer corresponding to 1°C of positive daily air temperature.
r температурный коэффициент таяния
Толщина слоя талой воды, приходящаяся на 1°C положительной температуры воздуха в сутки.
f coefficient *m* thermique de fusion
s coeficiente *m* térmico de deshielo
g Temperaturkoeffizient *m* der Schmelze; Schmelzpunkt *m*

* **temperature gradient → 6558**

* **temperature inversion → 3189**

6085 temporal ***adj***
Related or pertaining to time (as distinct from *spatial*), e.g. temporal change that occurs over a period of time.
r временной
Относящийся ко времени (в отличие от *пространственного*).
f temporaire *adj*
s temporal *adj*
g temporär *adj*; Zeit-; zeitlich *adj*

6086 tenth wave; wreck wave; freak wave
The strongest and most disastrous wave regularly occurring during a storm in seas and oceans.
r девятый вал
Самая сильная и опасная волна, регулярно возникающая во время шторма в морях и океанах.
f neuvième vague *f*
s novena ola *f*
g stärkste Woge *f*

6087 tephra
A collective term for all the solid material ejected into the air from a volcano during an eruption, it may range in size from volcanic bombs to fine dust and ash.
r тефра
Общее название обломков разных размеров и формы, образующихся при извержениях вулканов (они могут разниться от крупных вулканических бомб до мелкой пыли и пепла).
f tephra *f*
s tefra *f*
g Tephra *f*

6088 tephrochronology
The dating of past events by means of their association with layers of volcanic ash from former volcanic eruptions.
r тефрохронология
Датирование прошлых природных событий на основе их связи со слоями вулканического пепла от вулканических извержений.
f téphrochronologie *f*
s tefrocronología *f*
g Tephrochronologie *f*

6089 terai
A swampy jungle at the southern foothills of the Himalayas between the mountains and low piedmont plain.
r тераи
Заболоченные джунгли у южных подножий Гималаев, находящиеся на границе между горами и низменной подгорной равниной.
f terai *f*
s terai *f*
g Terai *mpl*

6090 terminal basin
A plane surface on the place of the wide front of a Pleistocene glacier that extended from the mountains to the adjoining plain.

r концевой бассейн
Ровная поверхность на месте расширенных концов плейстоценовых ледников, выходивших из гор на прилегающую равнину.
f bassin *m* terminal
s cuenca *f* terminal
g Zungenbecken *n*

6091 terminal moraine; end moraine
A linear ridge of glacial debris (moraine) deposited at the terminus of a glacier or at the end of an ice sheet marking the furthest extent of the ice. → Fig. 324
r конечная морена
Линейная моренная гряда, отложенная у конца ледника или ледникового покрова, отмечающая максимальное продвижение льда. → Рис. 324
f moraine *f* terminale
s morrena *f* terminal
g Endmoräne *f*

6092 terminal port
The final destination of cargo-carrying ships (some of them specialize on handling predominantly one particular type of traffic and include container, ferry, fishing and naval ports).
r порт назначения
Конечный порт для грузовых морских перевозок, нередко специализируемый на определенном виде операций (различают контейнерные, паромные, рыбные и военные порты).
f port *m* de destination
s puerto *m* de destino
g Bestimmungshafen *m*

6093 termination
The final stage of cold epochs in the Quaternary accompanied with the retreating of terrestrial glaciation (deglaciation).
r терминация
Завершающая стадия холодных эпох в четвертичном периоде, сопровождавшаяся сокращением наземного оледенения (дегляциацией).
f terminaison *f*
s terminación *f*
g Terminieren *n*

* **termitarium → 6094**

6094 termite mound; termitarium
A nest made by termites to live in, formed as the shape of a dome-like or conical hill of hardened clay, characteristic of tropical landscapes with a dry season (in Australia and Africa).
r термитник
Построенное из глины подземное или надземное гнездо термитов в виде узкого крутосклонного конического сооружения иногда высотой до 15 м, характерны для ландшафтов тропиков с сухим периодом года (в Австралии и Африке).

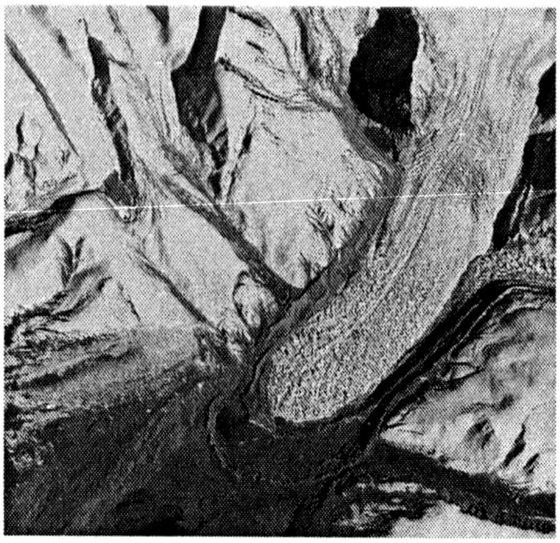

Fig. 324. **Terminal moraine** (down) near the glacier tongue

f termitière *f*
s termitero *m*
g Termitennest *n*; Termitenbau *m*; Termitenhügel *m*

6095 terra fusca
A brown clay-loam soil developed under a warm, seasonally dry climate, on limestone, it is commonly found beneath a maquis type of vegetation in some part of the Mediterranean and may represent a degraded terra rossa.
r терра фуска
Коричневая глинистая почва, развивающаяся на известняках в теплом климате с сухим сезоном, встречается под маквисом в некоторых районах Средиземноморья и может представлять собой деградировавшую терра росса.
f terra fusca *f*
s terra fusca *f*
g Terra fusca *f*

6096 terra rossa
A red-coloured soil that is the product of limestone weathering under Mediterranean climates (humid subtropical with a dry season), it occurs in the Mediterranean region, Eastn Africa, etc.
r терра росса
Почва красного цвета, формирующаяся в условиях субтропического влажного с сухим сезоном средиземноморского климата на известняках, встречается в Средиземноморье, Восточной Африке и др.
f terra *f* rossa
s terra *f* rossa
g Terra Rossa *f*

6097 terrace
1. A bench-like form of relief of valley slopes, lake and sea shores. 2. One of a series of horizontal steps cut into a hillside to provide cul-

tivable land in an area of steep relief, and to reduce soil erosion.

r терраса
1. Ступенеобразная форма рельефа склонов долин, берегов озер и морей. 2. Одна из серии горизонтальных ступеней, сделанных на крутом склоне для посевов и уменьшения почвенной эрозии.

f terrasse *f*
s terraza *f*
g Terrasse *f*

6098 terrace cultivation
A method of farming practiced on man-made steps cut into a hillside where soil is retained by artificial banks to provide cultivable land and to reduce soil erosion, sometimes irrigation is used.

r террасное земледелие
Система земледелия на террасах, созданных человеком на склонах холмов и гор, где почва удерживается искусственно воздвигнутыми стенками, а дополнительно к осадкам иногда используется орошение.

f agriculture *f* en terrasses de culture; culture *f* en terrasse
s agricultura *f* de terrazas de cultivo
g Terrassenlandwirtschaft *f*

6099 terracette; «sheep tracks»
A type of micro-relief in the form of paths or narrow terraces which rise above each other in steps on turfed slopes.

r тропинчатость склонов; «овечьи тропы»
Формы микрорельефа в виде тропинок или узких террасок, расположенных ступенчато на задернованных склонах холмов и гор.

f sentiers *mpl* de vaches»; piste *f*
s senderos *mpl* de obejas
g Schafspfade *mpl*

6100 terracing
The work of making of artificial horizontal steps for terrace cultivation in areas of steep relief in order to enlarge cultivated lands, make them more suitable for irrigation or diminish the effect of soil erosion caused by running water. → Fig. 325

r террасирование склона
Создание на склоне искусственных террас для увеличения площади обрабатываемых земель, удобства орошения и борьбы с водной эрозией почв. → Рис. 325

f aménagement *m* des versantes en terrasses de culture
s construcción *f* de terrazas de cultivo
g Terrasierung *f*; Terrasieren *n*; Stufenformierung *f*

Fig. 325. **Terracing** in Hymalaya Mountains

6101 terrain; locality
A particular area, place with all its natural components (relief, soils, water, etc.) as well as transportation lines, sociol-economic and cultural features.

r местность
Часть земной поверхности со всеми ее природными компонентами (рельефом, почвами, водами и пр.), а также путями сообщения, социально-экономическими и культурными объектами.

f localité *f*; lieu *m*
s lugar *m*; localidad *f*
g Örtlichkeit *f*; Gelände *n*

* **terrain analysis → 3373**

6102 terrestrial *adj*
1. Relating to land as opposed to water. 2. Relating to the Earth as opposed to celestial.

r наземный; земной
1. Относящийся к суше, а не к водным пространствам. 2. Относящийся к Земле, но не к внеземной сфере.

f terrestre *adj*
s terrestre *adj*
g Erd-; terrestrisch *adj*

6103 terrestrial fauna; land animals
All animal species inhabiting land in contrast to aquatic animals.

r сухопутная фауна
Совокупность видов животных, населяющих сушу земного шара.

f faune *f* terrestre
s fauna *f* terrestre
g Landfauna *f*

6104 terrestrial magnetism; Earth´s magnetic field; geomagnetism
The magnetic field of the Earth as a whole, weak and varying in intensity and direction, which causes the needle of a magnetic compass, indicating the north and south magnetic poles.

r земной магнетизм
Магнитное поле Земли, рассматриваемое в целом, разнящееся по своей интенсивности и направлению, воздействующее на стрелку магнитного компаса, который указывает на северный геомагнитный полюс.

f magnétisme *m* terrestre
s magnetismo *m* terrestre; geomagnetismo *m*
g Erdmagnetismus *m*

* **terrestrial radiation → 3587**

6105 **terrestrial radiation;**
outgoing long-wave radiation
Radiation emitted out to space from Earth's surface together with the atmosphere radiation mainly as the infrared radiation.
r земное излучение;
уходящая длинноволновая радиация
Излучение Земли как планеты (вместе с излучением атмосферы) в космическое пространство, главным образом в виде инфракрасного излучения.
f rayonnement *m* terrestre
s radiación *f* terrestre
g Erdstrahlung *f*;
ausgehende Langwellenstrahlung *f*

* **terricone → 6630**

6106 **terrigenous *adj***
Derived from the land.
r терригенный
Образованный на суше.
f terrigène *adj*
s terrigene *adj*
g terrigen *adj*

6107 **terrigenous deposits;**
terrigenous sediments
Inorganic deposits (sand, gravel, pebbles, etc.) derived from the denudation of the land and laid down in the littoral zone of the sea floor, as distinct from pelagic deposits.
r терригенные отложения
Осадки, состоящие в основном из обломков горных пород, возникших в результате денудации суши, снесенные и отложенные в литоральной зоне моря (отличаются от пелагических отложений, сформировавшихся здесь же).
f dépôts *mpl* terrigénes
s depósitos *mpl* terrígenos
g terrigene Ablagerungen *fpl*;
terrigene Sedimente *npl*

6108 **terrigenous runoff**
The income of dissolved, colloid and suspended material carried from the land to ocean mainly with river waters.
r терригенный сток
Поступление с суши растворенных, коллоидных и взвешенных веществ в океан, преимущественно с речными водами.
f écoulement *m* terrigène
s escorrentía *f* terrígena
g terrigener Abfluß *m*

* **terrigenous sediments → 6107**

6109 **territorial *adj***
Related to a certain area or national territory of a state.
r территориальный
Относящийся к конкретной территории или к национальной территории государства.
f territorial *adj*
s territorial *adj*
g territorial *adj*

6110 **territorial capacity**
The possibility to expand economic activity on any given territory without additional investments or with additional capital investments.
r потенциал территории
Возможность расширения хозяйственной деятельности на данной территории без крупных дополнительных затрат на ее благоустройство, либо с дополнительными капитальными затратами.
f potentiel *m* du territoire
s capacidad *f* del território
g Territorialkapazität *f*

* **territorial claims → 6114**

6111 **territorial concentration of production;**
agglomeration of enterprises;
localization of enterprises
The concentration of industrial and agricultural enterprises and the infrastructure in certain districts, areas, regions, and countries.
r территориальная концентрация производства
Сосредоточение промышленных, сельскохозяйственных предприятий и связанных с ними сооружений производственной инфрастуктуры в отдельных пунктах, ареалах, районах и странах.
f concentration *f* territoriale de production;
localisation *f* industrielle;
agglomération *f* industrielle
s concentración *f* territorial de producción;
aglomeración *f* industrial
g territoriale Produktionskonzentration *f*

6112 **territorial connections**
The relationships between geographical objects manifested in the regular exchange of matter, energy, information always realized through a certain distance.
r территориальные связи
Отношения между географическими объектами, выражающиеся в регулярном обмене веществом, энергией, информацией, их реализация обязательно проявляется через преодоление расстояния.
f connexions *fpl* territoriales
s conexiones *fpl* territoriales
g territoriale Verbindungen *fpl*

6113 **territorial differentiation**
The spatial diversity of geographical objects and phenomena manifested in their alternation, combination and various links between them.
r территориальная дифференциация
Пространственное разнообразие географических объектов и явлений, проявляющееся в их чередовании, сочетаемости и наличии разнообразных отношений между ними.

f différenciation *f* territorial
s diferenciación *f* territorial
g territoriale Differenzierung *f*

6114 territorial disputes; territorial claims
Contradictions between two neighbouring states over a contested territory which is currently under the jurisdiction of one of these states.
r территориальные споры
Разногласия двух соседних государств по поводу спорной территории, находящейся в данный момент под юрисдикцией одной из этих стран.
f litiges *mpl* territoriaux; conflits *mpl* territoriaux
s litigios *mpl* territoriales
g territoriale Anschprüche *mpl*; territoriale Stritte *mpl*

6115 territorial division of labour; geographical division of labour
Production specialization of individual economic regions combined with the interregional cooperation, exchange of products and services.
r территориальное разделение труда; межрайонное разделение труда
Производственная специализация отдельных экономических регионов в сочетании с межрайонной кооперацией, обменом продукцией и услугами.
f division *f* territoriale du travail
s división *f* territorial del trabajo
g territoriale Arbeiteilung *f*

* **territorial identity → 4977**

6116 territorial interest
A historically rooted interest a country has in a neighbouring territory outside its jurisdiction.
r территориальные интересы
Исторически сложившиеся интересы какой-либо страны к соседней территории, не находящейся под ее юрисдикцией.
f intérêts *mpl* territoriaux
s intereses *mpl* territoriales
g territoriale Interessen *npl*

6117 territorial organization of society
A combination of territorial patterns of settlements, industries, use of nature contributing to the administration of society on any given territory.
r территориальная организация общества
Сочетание территориальных систем расселения людей, производства, природопользования, способствующее управлению обществом на данной территории.
f organisation *f* territoriale de la société
s organización *f* territorial de la sociedad
g Territorialorganisation *f* der Gesellschaft

6118 territorial pattern
The spatial location of interconnected elements of the environment and human society on a definite territory, e.g. settlement pattern.
r территориальная система
Пространственная близость взаимосвязанных элементов окружающей среды и общества, взаимодействующих на определенной территории, например, система расселения.
f système *m* territorial
s sistema *m* territorial
g Territorialsystem *n*

* **territorial planning → 4978**

6119 territorial production complex; industrial agglomeration; industrial complex
The association of industrial enterprises which share geographic location, transportation system, and industrial infrastructure (the term refers to planned, rather than spontaneously arising, industrial complexes).
r территориально-производственный комплекс; ТПК; промышленная группировка
Совокупность промышленных предприятий на определенной территории (символ: ТПК), объединенных общностью транспортно-географического положения и совместно использующих производственную инфраструктуру (термин скорее относится к плановой экономике, чем к спонтанно развивающимся производственным комплексам).
f agglomération *f* industrielle; complexe *m* industriel
s aglomeración *f* industrial; complejo *m* industrial
g industrielle Agglomeration *f*; Industriekomplex *m*

6120 territorial production links
Spatial relations between productive enterprises, associations and their branches, as well as among the groups of enterprises and associations working in various branches within certain areas.
r территориально-производственные связи
Пространственные отношения между производственными предприятиями, объединениями и их филиалами, а также между группами предприятий и объединений разных отраслей материального производства в границах территориальных единиц разного ранга.
f liaisons *fpl* territoriales de production
s vinculaciones *fpl* territoriales de producción; relaciónes *fpl* interindustriales
g Produktionsverbindungen *fpl*

* **territorial seas → 6122**

6121 territorial structure of economy

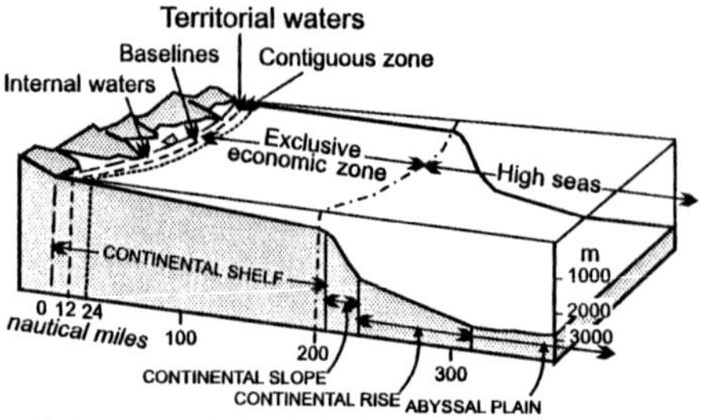

Fig. 326. **Territorial waters**: maritime zones

The spatial distribution of production forces over economic regions united into a single system of national economy.

r территориальная структура хозяйства
Размещение производительных сил по экономическим районам, связанным в единой системе национальной экономики.

f structure *f* territoriale de l'économie
s estructura *f* territorial de la economía
g Territorialstruktur *f* der Wirtschaft

6122 territorial waters; territorial seas
The coastal waters with the sea bed below and the air space above with all that lies or lives therein forming part of the state territory over which the coastal state has sovereignty. → Fig. 326

r территориальные воды
Морские воды, а также морское дно под ними и прилегающее воздушное пространство, примыкающие к побережью или внутренним водам государства и составляющие часть его территории, которая находится под его суверинитетом. → Рис. 326

f eaux *fpl* territoriales
s aguas *fpl* territoriales; aguas *fpl* jurisdiccionales
g Hoheitsgewässer *n*; Territorialgewässer *n*

6123 territory
1. The area of land and adjacent sea, a region, a district, with undefind boundaries. 2. An area dependent on a sovereign state, but having some autonomy, e.g. an area supervised by a sovereign state on behalf jf the United Nations.

r территория
1. Площадь суши с прилегающими водными пространствами, регион с неопределенными границами. 2. Область, зависимая от суверенного государства, но имеющая некоторую автономию, например, область, находящаяся под надзором суверенной страны по поручению Организации Объединенных Наций.

f territoire *m*
s territorio *m*
g Territorium *m*

6124 tertiarization
The decline of machinery and manufacturing industry and growth of tertiary sector in economy, which provides services.

r развитие третичного сектора
Сокращение машинного производства и рост сервисных отраслей, характерный для третичного сектора экономики.

f tertiarisation *f*
s terciarisación *f*
g Terziarisation *f*

6125 Tertiary
The earlier period of the Cainozoic era which began 65 million years ago and lasted about 63 million years, subdivided into the Palaeogene and Neogene. → Table 1, p. 284

r третичный период
Ранний период кайнозойской эры геологической истории Земли, начался 65 млн лет назад и продолжался около 63 млн лет, подразделяется на палеоген и неоген. → Табл. 1, с. 284

f Tertiaire *m*
s era *f* terciaria
g Tertiär *n*

6126 tertiary industry; service industry; tertiary sector
Economic activity that is not concerned with manufactured goods but which provides services such as banking, trade, tourism, etc.

r третичный сектор экономики
Отрасли хозяйства, выполняющие разнообразные услуги, такие как банковское дело, торговля, туризм и пр.

f secteur *m* tertiaire
s sector *m* terciario
g terziarer Wirtschaftssektor *m*; Dienstleistungssektor *m*

* **tertiary sector → 6126**

* **tessellation → 4987**

6127 testing area
Part of a territory serving as an object of special investigations.

r полигон
Участок территории, служащий объектом специального исследования.

f polygone *m*
s polígono *m* de investigación
g Polygon *n*

6128 Tethys; Tethys Ocean
The system of vast sea basins formed at the end of the Precambrian and the beginning of the Palaeozoic on the place of the contemporary continent of Eurasia which divided Gondwanaland from Laurasia. → Fig. 327, p. 731

r тетис
Система обширных морских бассейнов, образовавшаяся в конце докембрия – начале палеозоя на месте современного материка Евразии, которая отделяла Гондвану от Лавразии. → Рис. 327, с. 731

f Téthys *m*

s Tethys *m*
g Tethys *f*

* **Tethys Ocean → 6128**

6129 **textile industry**
A branch of light industry that produces yarn, threads, cloths from natural and artificial fibres.
r текстильная промышленность
Отрасль легкой промышленности, занятая переработкой растительных, животных, искусственных и синтетических волокон в пряжу, нитки, ткани.
f industrie *f* textile
s industria *f* textilera; industria *f* textil
g Textilindustrie *f*

6130 **texture of rock**
The geometrical aspects of the particles constituing a rock, surface deposit or ice, such as the size, shape, arrangement and distribution of its components.
r текстура горной породы
Особенности строения толщи горных пород, включая лед, обусловленные пространственным расположением ее составных частей.
f texture *f* de roche
s textura *f* de roca
g Gesteinstextur *f*

6131 **texture of soil; soil texture**
A measure of different particle size characteristics of soils achieved by passing the samples through a succession of sieves in order to recognize the proportions of sand, silt and clay present expressed as a percentage to soil dry weight.
r текстура почвы; механический состав почвы; гранулометрический состав почвы
Мера соотношения почвенных частиц разного размера, которую выясняют путем просеивания почвенного образца через соответствующие сита и получают соотношение в почве частиц песка, ила и глины, выраженное в процентах к весу сухой почвы.

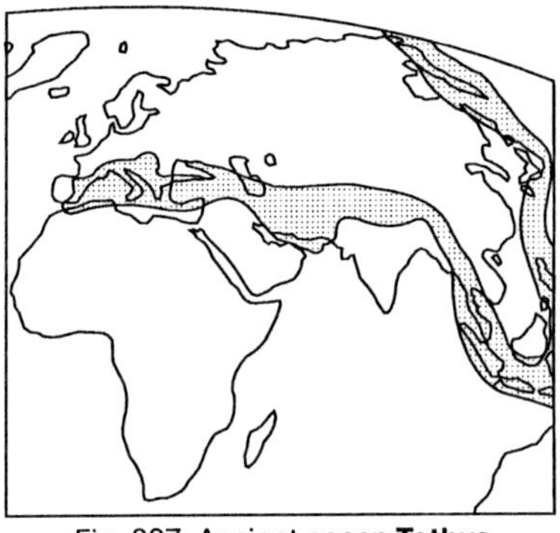

Fig. 327. Ancient ocean **Tethys**

f texture *f* du sol
s textura *f* del suelo
g Bodentextur *f*

6132 **thalassocratic period**
A period in the geological history of the Earth when there was a considerable extension of seas and oceans, as opposed to a *geocratic period.*
r талассократический период
Период в геологической истории Земли, когда происходило значительное расширение площади морей, противопоставляется *геократическому периоду.*
f période *f* talassocratique
s período *m* talasocrático
g thalassokratische Phase *f*

6133 **thalassocraton**
Tectonically relatively stable, not mobile area of an ocean bed on which abyssal plains occur.
r талассократон
Тектонически относительно устойчивая, малоподвижная область ложа океанов, в пределах которой развиты абиссальные равнины.
f thalassocraton *m*
s talasocratón *m*
g Thalassokraton *n*

6134 **thalweg; talweg**
A line joining the lowest points along a river valley or gully.
r тальвег
Осевая, наиболее глубокая часть дна речной долины, балки или оврага.
f talweg *m*
s talweg *m*; thalweg *m*
g Talweg *m*

6135 **thaw**
The rise in temperature above the freezing point, usually due to the advection of a warm air in the period of steady negative temperature, which leads to the melting of snow and ice.
r оттепель
Повышение температуры воздуха до 0°C и выше в результате адвекции теплого воздуха на фоне установившихся отрицательных температур, что приводит к таянию снега и льда.
f dégel *m*
s deshielo *m*
g Tauwetter *n*; Tauen *n*

* **thaw lake → 6181**

6136 **thaw slumping**
A rapid movement of thawed ground downslope when frost no longer cements it.
r криогенное оплывание
Быстрое стекание грунта по склону при оттаивании сковывавшей его мерзлоты.
f glissement *m* au dégel
s deslizamiemto *m* por descongelamiento

g auftaubedingtes Rutschen *n*

6137 thawed patch
A snow-free part of ground formed by thawing surrounded by the snow cover.
r проталина
Свободный от снега участок поверхности, появившийся в результате протаивания.
f endroit *m* où la neige à fondu; endroit *m* décongéle
s lugar *m* deshelado
g ausgetaute Stelle *f* in der Schneedecke

6138 thawing; defreezing of ground
The conversion of frozen ground into the melted state when the temperature rises and ice changes into water.
r протаивание; оттаивание; размораживание грунта
Переход мерзлого грунта в талое состояние при повышении температуры.
f dégel *m*; décongélation *f*
s descongelamiento *m*
g Auftauen *n*

6139 thawing front
The advancing boundary between frozen and thawed ground.
r фронт протаивания
Поверхность раздела между мерзлым и талым грунтом.
f front *m* de dégel
s frente *m* de descongelamiento
g Auftau-Front *f*

6140 thematic atlas
The geographical atlas consisting mainly of thematic maps which show the spatial distribution of a certain phenomenon.
r тематический атлас
Географический атлас, состоящий преимущественно из карт одной тематики, показывающих пространственное распределение определенных процессов или явлений.
f atlas *m* thématique
s atlas *m* temático
g thematischer Atlas *m*

6141 thematic cartography; thematic mapping
Theory and practice of compilation and use of thematic maps.
r тематическая картография; тематическое картографирование
Теория и практика создания и использования тематических карт.
f cartographie *f* thématique
s cartografía *f* temática
g thematische Kartographie *f*

6142 thematic map
A map of any scale representing a spatial distribution of a natural or social phenomenon in detail on the geographical basis, e.g. soil map, map of land use.
r тематическая карта
Географическая карта, содержание которой посвящено какой-либо конкретной теме, например, почвенная карта, карта землепользования.
f carte *f* thématique; carte *f* spéciale
s mapa *m* temático; mapa *m* especial
g thematische Karte *f*

* **thematic mapping → 6141**

6143 theocratic state
A state in which political and spiritual power belong to the church.
r теократическое государство
Государство, в котором как политическая, так и духовная власть сосредоточены в руках церкви.
f état *m* théocratique
s estado *m* teocrático
g theokratischer Staat *m*

6144 theodolite
An optical surveying instrument used to measure vertical and horizontal angles by means of a small telescope, a spirit level, and graduated arcs, mounted on a tripod.
r теодолит
Оптический съемочный инструмент, предназначенный для измерения вертикальных и горизонтальных углов посредством небольшой подзорной трубы, ватерпаса и градуированных кругов, помещенных на треногу.
f théodolithe *m*
s teodolito *m*
g Theodolit *m*

6145 theodolite survey; theodolite surveying
Making the plan of a land with theodolite.
r теодолитная съемка
Составление плана местности с помощью теодолита.
f levé *m* au théodolite
s levantamiento *m* por teodolito
g Theodolitaufnahme *f*

* **theodolite surveying → 6145**

6146 thermal$_1$
1. Drop-wise volume of water appearing under the rupture and coagulation of water in the surface layer of water bodies in conditions of microconvection. 2. Steady and strong upward current of air in the atmosphere.
r термик
1. Каплевидный объем воды, возникающий при разрыве и свертывании водной пленки в поверхностном слое водоемов при микроконвекции. 2. Устойчивые и сильные восходящие токи конвекции в атмосфере.
f thermique$_1$ *f*
s térmico$_1$ *m*
g Thermik *f*

6147 thermal abrasion
The process of wearing away of sea coasts, lake and reservoir banks which are composed of ice or permafrost.
r термоабразия; термическая абразия
Процесс разрушения берегов морей, озер и водохранилищ, сложенных льдом или многолетнемерзлыми породами.
f abrasion *f* thermique
s abrasión *f* térmica
g Thermoabrasion *f*

6148 thermal$_2$ *adj*
Applied to temperature and heat.
r термальный; тепловой; термический
Относящийся к температуре или потоку тепла.
f thermique$_2$ *adj*
s térmico$_2$ *adj*
g Wärme-; termisch *adj*

6149 thermal advection
The change of air temperature due to the invasion (advection) of a new air mass. → Fig. 328
r термическая адвекция
Изменение температуры воздуха благодаря вторжению новой воздушной массы. → Рис. 328
f advection *f* thermique
s advección *f* térmica
g Wärmeadvektion *f*

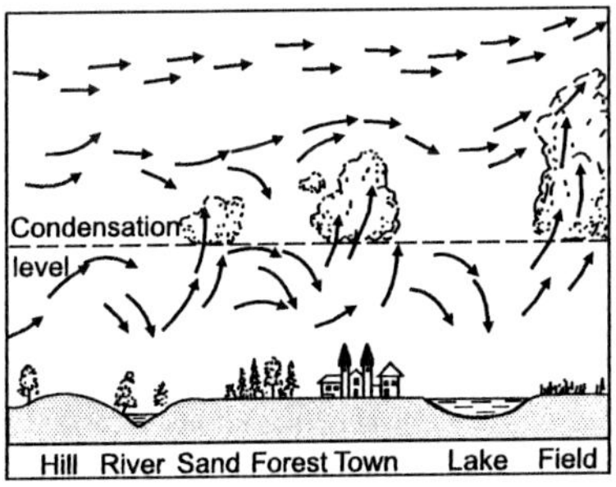

Fig. 328. **Thermal advection**

* **thermal anomaly → 6082**

6150 thermal balance of water body
The relation of income and outcome of heat in a water body taking into account the change of its thermal content over a given period of time.
r тепловой баланс водоема
Соотношение прихода и расхода тепла в водном объекте с учетом изменения его теплозапаса за рассматриваемый интервал времени.
f bilan *m* thermique de bassin d'eau
s balance *m* térmico de depósito de agua
g Wärmebilanz *f* des Gewässers

6151 thermal bar
A narrow zone in a deep lake of temperate latitudes in which occurs the submergence of water having the greatest density.
r термический бар; термобар
Узкая зона в глубоком озере умеренных широт, в которой происходит погружение воды, имеющей наибольшую плотность,
f barre *f* thermique
s barra *f* térmica
g thermische Barre *f*

6152 thermal belts
Latitudinal belts of the Earth with certain characteristics of total solar radiation and air temperature: hot zone, temperate and polar zones of the both hemispheres.
r тепловые пояса; радиационно-термические пояса
Широтные пояса Земли с определенными градациями суммарной солнечной радиации и температуры воздуха, обычно выделяют жаркий пояс, умеренные и полярные пояса обоих полушарий.
f zones *fpl* thermiques
s zonas *fpl* térmicas
g thermische Zonen *fpl*

6153 thermal content in water body
The heat accumulated in a water body exceeding the content of heat in the whole water mass taken under 0°C.
r теплозапас водоема
Количество тепла, аккумулированное в водоеме, превышающее запас тепла во всей массе воды, которая была бы в нем при температуре 0°C.
f réserve *f* thermique de réservoir d'eau
s reserva *f* térmica de depósito de agua
g Wärmeinhalt *m* des Gewässers

6154 thermal denudation
The combination of cryogenic wearing away of rocks on the land surface and the removal of debris to the lower levels.
r термоденудация
Совокупность криогенного разрушения горных пород на поверхности земли и переноса продуктов разрушения в пониженные участки.
f dénudation *f* thermique; planation *f* thermique
s denudación *f* térmica
g Thermodenudation *f*

* **thermal depression → 6159**

6155 thermal equator; heat equator
A line or narrow zone encircling the low latitudes of the globe joining up the point on each meridian which exhibits the highest mean annual air temperature.
r термический экватор; тепловой экватор
Линия или узкая зона с самыми высокими средними годовыми температурами, опоясывающая земной шар.

f équateur *m* thermique
s ecuador *m* térmico
g Wärmeäquator *m*; termische Äquator *m*

6156 thermal erosion; thermo-erosion
A type of erosion occurring in areas of permafrost where, if the organic layer protecting the land surface is removed, the ground ice melts and the land surface breaks up.
r термоэрозия
Тип эрозии, случающийся в криолитозоне, когда подземный лед, не покрытый слоем с богатой органикой, протаивает и земная поверхность разрушается.
f érosion *f* thermique
s erosión *f* térmica
g Thermoerosion *f*

* **thermal expansion → 3118**

6157 thermal expansion of the ocean; ocean thermal expansion
The increase of the volume of water in the ocean due to the warming up of its upper layer, which leads to the rise of sea level.
r тепловое расширение океана
Увеличение объема воды в Мировом океане вследствие потепления его верхнего слоя, что ведет к повышению уровня моря.
f expansion *f* thermique de l'océan
s expansión *f* térmica de océano
g Wärmeexpansion *f* des Ozeans

6158 thermal front
A zone in an ocean or sea across which the temperature gradient is great, it separates water masses of different origin.
r термический фронт
Линия или полоса резкого перепада температуры, разделяющая водные массы разного происхождения.
f front *m* thermique
s frente *m* térmico
g thermische Front *f*

6159 thermal low; thermal depression
An area of low pressure which results from high temperatures caused by very intensive solar heating of the Earth's surface.
r термическая депрессия
Область пониженного атмосферного давления, температура воздуха в которой выше, чем в окружающей тропосфере, что вызвано интенсивным прогреванием земной поверхности.
f dépression *f* thermique
s depresión *f* térmica
g thermische Tief(druckgebiet) *n*

6160 thermal metamorphism of ice
The transformation of petrographic structure and texture of natural ice due to the effect of a heat influx from outside.
r термометаморфизм льда
Преобразование петрографической структуры и текстуры природных льдов под влиянием притока тепла извне.
f thermométamorphisme *m* de la glace
s termometamorfismo *m* de hielo
g Thermometamorphose *f* des Eises

6161 thermal pollution
The heating of part of the environment by the discharge of substances with high temperatures and the emission of radiant heat, such as the heat from the cooling towers of a power station or from discharge of condenser cooling water into rivers and lakes which leads to the increase of temperature of water and air.
r тепловое загрязнение
Нагревание участков окружающей среды в результате выброса веществ с высокой температурой и выделения тепла в результате хозяйственной деятельности, например, от охлаждающих башен тепловых электростанций или сброса горячих вод в реки и озера, что ведет к росту температуры воды и воздуха.
f pollution *f* thermique; contamination *f* thermique
s polución *f* térmica; contaminación *f* térmica
g thermische Verschmutzung *f*; thermische Belastung *f*

6162 thermal regime of soil
Heat state of soil and its changes in time, having secular, long-term, seasonal and daily cycles.
r тепловой режим почвы
Характер теплового состояния почвы и его изменения во времени, обладающего вековой, многолетней, сезонной и суточной цикличностью.
f régime *m* thermique du sol
s régimen *m* térmico del suelo
g Bodenwärmehaushalt *m*

* **thermal sensing → 3092**

6163 thermal spring; geothermal spring
A continuous flow from the ground of hot water with the temperature above 20°C, usually associated with present of a former volcanic activity.
r термальный источник
Постоянное истечение из грунта горячей воды с температурой выше 20°C, обычно связанное с прошлой вулканической активностью.
f source *f* thermale
s fuente *f* termal
g Thermalquelle *f*

6164 thermal stratification
The succession of well-defined layers of water of different temperatures lying at various depths in the ocean or deep lake, the top layer being the warmest, except under ice, where there may be an iversion. → Fig. 329, p. 735

r термическая стратификация
Ясно выраженная последовательность слоев воды с разной температурой на разных глубинах в океане или глубоком озере, как правило, самый теплый слой – верхний, за исключением водоема, покрытого льдом, в толще которого может иметь место инверсия температуры. → Рис. 329
f stratification *f* thermique
s estratificación *f* térmica
g Wärmeschichtung *f*; Wasserstratifikation *f*

6165 thermal vegetation
The plants growing in thermal springs and around them, among which the relicts of warmer epochs occur.
r термальная растительность
Растения, произрастающие в источниках с повышенной температурой и вокруг них, встречаются реликты прошлых более теплых эпох.
f végétation *f* thermique
s vegetación *f* térmica
g thermische Vegetation *f*

6166 thermal waste
The heat released into the environment due to human economic activity.
r тепловые отходы
Тепло, образующееся в результате хозяйственной деятельности человека и поступающее в окружающую среду.
f déchets *mpl* thermiques
s desechos *mpl* térmicos; residuos *mpl* térmicos
g thermische Abfälle *mpl*

6167 thermal water
Ground waters having the temperature more than 20°C.
r термальные воды
Подземные воды с температурой выше 20°С.
f eau *f* thermale
s agua *f* termal
g Thermalwasser *n*

6168 thermal wind
The shear vector difference between the geostrophic winds at the base and top of an atmospheric layer.

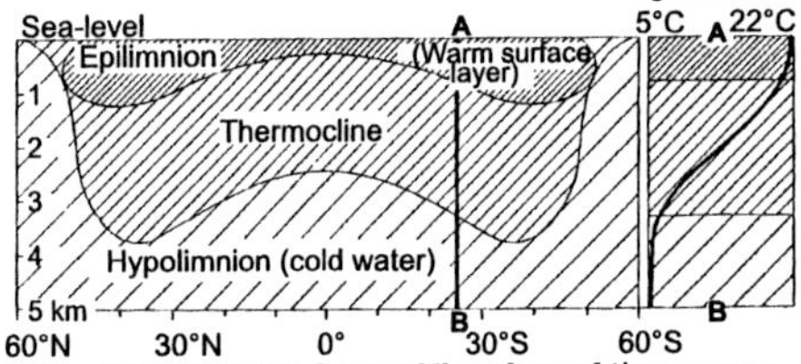

Fig. 329. **Thermal stratification** of the ocean

r термический ветер
Векторная разность между геострофическим ветром на верхней и нижней границах слоя атмосферы.
f vent *m* thermique
s viento *m* térmico
g Thermalwind *m*

6169 thermic power plant
An electricity-generating power station that produces electrical energy by converting heat energy from burning fossil fuel.
r тепловая электростанция
Электростанция, вырабатывающая электрическую энергию в результате преобразования тепловой энергии, выделяющейся при сжигании органического топлива.
f centrale *f* thermique
s central *f* térmica
g Wärmekraftwerk *n*

6170 thermoabrasional niche
A niche formed at the base of a river bank or cliff which is composed of the locally thawed frozen sediments.
r термоабразионная ниша
Ниша, образующаяся в нижней части берегового уступа, сложенного мерзлыми вытаивающими отложениями.
f niche *f* d'abrasion thermique
s nicho *m* de abrasión térmica
g Thermo-Abrasionnische *f*

6171 thermobiont
Some algae and bacteria which are permanent inhabitants of hot (up to 95-100°C) springs.
r термобионт
Некоторые водоросли и бактерии, постоянно обитающие в горячих (до 95-100°С) источниках.
f thermobiont *m*
s termobionte *m*
g Thermobiont *m*

* **thermoclasty → 3118**

6172 thermocline
An intermediate layer of water in an ocean or a deep lake, lying between the disturbed upper layer (*epilimnion*) and the relatively stagnant lower layer (*hypolimnion*).
r термоклин; слой скачка
Промежуточный слой воды в океане или глубоком озере, лежащий между взволнованным верхним слоем (*эпилимнионом*) и относительно спокойным нижним слоем (*гиполимнионом*).
f thermocline *f*; couche *f* de discontinuité thermique
s termoclina *f*; termoclino *m*
g Thermokline *f*; Sprungschicht *f*

6173 thermo-electricity
Power produced from a range of fuels such as coal, oil, gas, peat, lignite, nuclear fuels, or geothermal heat which drive steam turbines.

r теплоэнергетика
Отрасль энергетики, занимающаяся преобразованием теплоты при сгорании угля, нефти, природного газа, торфа, бурого угля, ядерного топлива, а также геотермического тепла в энергию вращающихся турбин, производящих электроэнергию.
f électricité *f* thermique
s electricidad *f* térmica
g Thermalenergiewirtschaft *f*; Wärmeenergetik *f*

* **thermo-erosion → 6156**

6174 thermohaline analysis; T-S analysis
Method of isolation and investigation of sea water masses, their interaction and transformation on the basis of characteristic state diagrams (T-S-diagrams).
r температурно-соленостный анализ; термохалинный анализ; T-S-анализ
Метод выделения и изучения водных масс Мирового океана, их взаимодействия и трансформации на основе характеристической диаграммы состояния (T-S-диаграммы).
f analyse *f* de température et de salinité; analyse *f* T-S
s análisis *f* de temperatura y de salinidad
g TS-Analyse *f*

* **thermohaline circulation → 6175**

6175 thermohaline convection; thermohaline circulation
Ocean-water circulation resulting from differences of temperature and salinity, which is one of the chief mechanisms of ocean currents.
r термохалинная конвекция; термогалинная конвекция; конвективное перемешивание
Циркуляция воды в океане, вызванная различиями температуры и солености, один из главных механизмов морских течений.
f convection *f* thermohaline
s convección *f* termohalina
g thermohaline Konvektion *f*

6176 thermoisopleth
A line of equal values of temperature in rock, water or air mass. → Fig. 330
r термоизоплета
Линия равных значений температуры в толще горной породы, воды или воздуха. → Рис. 330
f thermoisoplethe *f*
s termoisopleta *f*
g Thermoisoplethe *f*

6177 thermokarst
The formation of depressions and underground cavities due to the melting of ground ice or thawing of frozen ground, the features include alas, thaw lakes, collapsed pingos, etc.
r термокарст
Образование просадочных, провальных форм рельефа и подземных пустот в результате вытаивания подземного льда или оттаивания мерзлого грунта, таких как аласы, озера талой воды, пинго и пр.
f thermokarst *m*
s termocarst *m*
g Thermokarst *m*

6178 thermokarst cave-in
A cylindrical depression up to several metres across formed due to ground subsidence in places where the ground ice thawes out.
r термокарстовый провал
Цилиндрическое понижение рельефа до нескольких метров в диаметре, возникшее в результате просадок грунта при протаивании подземного льда.
f caverne *f* de thermokarst
s caverna *f* de termocarst
g Thermokarsthöhle *f*

6179 thermokarst depression
A small low-sloped depression formed as a result of thermokarst ground subsidence.
r термокарстовая западина
Мелкое пологое понижение рельефа, возникающее в результате термокарстовых просадок грунта.
f dépression *f* de thermokarst
s depresión *f* de termocarst
g Thermokarstdepression *f*

6180 thermokarst funnel
A small, deep depression, round in plan, formed as a result of the local development of thermokarst.
r термокарстовая воронка
Небольшое округлое в плане глубокое понижение, возникающее при локальном проявлении термокарста.
f entonnoir *f* thermokarstique
s embudo *m* de termocarst
g Thermokarsttrichter *m*

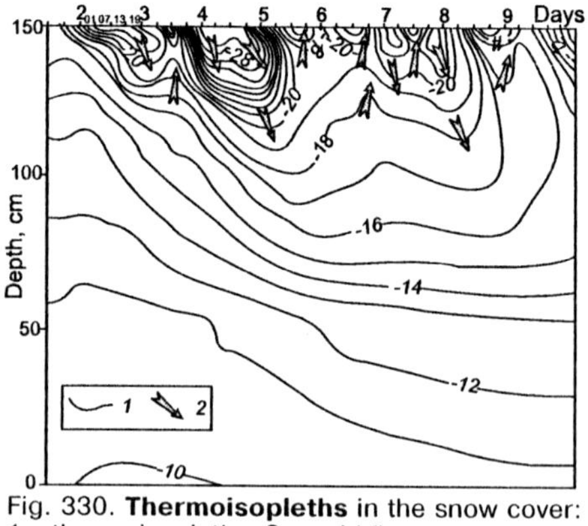

Fig. 330. **Thermoisopleths** in the snow cover: 1 – thermoisopleths, 2 – cold fluxes

6181 thermokarst lake; thaw lake
A lake occupying a closed depression which have been formed or enlarged by thawing of ice-rich permafrost or the melting of massive ice veins, it also may occur in collapsed pingos and on the flat floors of alases.
r термокарстовое озеро; талое озеро
Озеро, занимающее изолированную депрессию, сформировавшуюся в результате таяния богатых льдом многолетнемерзлых грунтов или мощных повторных ледяных жил, нередко в результате разрушения пинго или на плоской поверхности аласа.
f lac *m* de thermokarst; lac *m* dégelé
s lago *m* de termocarst; lago *m* deshelado
g Thermokarstsee *n*; Tausee *m*

6182 thermokarst mound
A hill 0.5-1 m high formed in wet swampy areas when the ground is freezing in the zone of discontinuous permafrost with a thick active layer.
r бугор-могильник
Бугор высотой 0,5-1 м, образующийся на заболоченных местах при замерзании грунтов в зоне прерывистых многолетнемерзлых пород с мощным деятельным слоем.
f butte *m* de thermokarst
s domo *m* de termocarst
g Thermokarsthügel *m*; Grabhügel *m*

6183 thermokarst terrain
A landscape of irregular depressions caused by the irregular heaving and melting of ice and numerous lakes formed by the processes of thermokarst.
r термокарстовый ландшафт
Сочетание пониженных и повышенных участков рельефа и обилия озер, вызванных термокарстовыми процессами.
f modelé *m* thermokarstique
s paisaje *m* de termocarst
g Thermokarstlandschaft *f*

6184 thermophile
An organism that is adapted to permanently high temperatures, usually above 45°C.
r термофил; теплолюбивый организм
Организм, предпочитающий жить в условиях постоянно высоких температур, как правило, выше 45°C.
f thermophile *m*
s termofilo *m*
g Thermophil *m*

* **thermophobe → 1370**

6185 thermosphere
The layer of the upper atmosphere above the mesopause in which temperature increases with height. → Fig. 21, p. 51
r термосфера
Слой верхней атмосферы, расположенный над мезопаузой, в котором температура обычно повышается с высотой. → Рис. 21, с. 51
f thermosphère *f*
s termosfera *f*
g Thermosphäre *f*

6186 thermostad
A deep water layer in ocean which is considerably thick and is characterized by quasi-stable temperature along its vertical.
r термостад
Глубинный слой воды в океане значительной толщины с квазипостоянной температурой по вертикали.
f thermostade *m*
s termostad *m*
g Thermostad *m*

* **thicket → 811**

6187 thickness of aquifer
The distance from the water table to the surface of impermeable layer, in case of confined aquifer it is the distance between two impermeable layers.
r мощность водоносного горизонта
Расстояние от зеркала воды до поверхности водоупорного ложа, для напорных вод – между двумя водоупорами.
f épaisseure *f* de nappe aquifère
s espesor *m* de capa aquífera
g Mächtigkeit *f* der Grundwasserschicht *m*

* **thin forest → 4271**

6188 thin section
A thin slice of rock or mineral intended for the study under a microscope.
r шлиф
Тончайшая пластинка горной породы или минерала, предназначенная для изучения под микроскопом.
f coupe *f* mince
s placa *f* fina de roca; schliff *m*
g Dünnschliff *m*

6189 thinning
The extraction of some of the young trees in a forest to enable those that to remain grow and develop fully and give much timber.
r прореживание
Удаление некоторых молодых деревьев в лесу с тем, чтобы остальные развивались лучше и давали больше древесины.
f dépressage *m*
s despejo *m*; entresaca *f*
g Lichten *n*

6190 Third World
In the «cold war» period of 1950s the term was applied to countries committed to neither of the two major power blocks, now is generally applied to countries considered to be not yet fully developed primarily in economic but also in social terms.
r страны третьего мира
В 1950-е годы холодной войны термин от-

носился к странам, не входившим ни в один из двух военно-политических блоков, ныне он относится к странам, еще недостаточно развитым в экономическом и социальном отношениях.
f troisième monde *m*
s tercer mundo *m*
g dritte Welt *f*

6191 thorn forest; thorn woodland
Tropical or subtropical forest or woodland of small, xerophilous thorny bushes, the thorns giving partial protection form browsing animals.
r колючий кустарник
Тропические или субтропические заросли небольших сухолюбивых деревьев с колючками, защищающими растения от поедания животными молодых побегов.
f buisson *f* épineuse
s espino *m*; matorral *m* espinoso
g Dorngebüsch *n*

* **thorn woodland → 6191**

* **threatement species → 1895**

* **threecentric world → 6355**

6192 three-field system of farming
A farming system used in Medieval Europe whereby two of the three fields were cultivated while the third was left fallow to recover its fertility.
r трехпольная система земледелия
Применявшийся в крестьянских хозяйствах некоторых стран Европы севооборот с чередованием культур: пар, озимые, яровые.
f assolement *m* triennal
s rotación *f* trienal
g dreifeldisches Landwirtschaftssystem *n*

6193 threshold
The point at which a stimulus of increasing strength is first perceived or provokes a special respond, the point of transition from one state to another.
r порог$_2$; критический уровень
Точка, на которой процесс развития достигает некоторого предела или вызывает обратную реакцию, момент перехода из одного состояния процесса в другое его состояние.
f seuil *m*; limite *f*
s umbral *m*
g Schwelle *f*

* **threshold concentration → 3731**

* **threshold emission → 3732**

* **threshold environmental pressure → 4517**

* **threshold impact → 3733**

6194 threshold population
The minimum number of people needed in an area to support a function, service or provision of goods.
r пороговое население
Минимальное количество людей, необходимое на данной территории для выполнения важнейших функций, обслуживания и снабжения товарами.
f population *f* seule
s población *f* umbral
g Schwellenbewölkerung *f*

6195 throughflow; inter-flow
The flow of water downslope through the regolith, as distinct from overland flow. → Fig. 282, p. 617
r внутригрунтовый сток; внутрипочвенный сток
Стекание воды вниз по склону через реголит, т.е. в почвенной толще непосредственно под поверхностью, в отличие от поверхностного стока. → Рис. 282, с. 617
f écoulement *m* subsuperficiel; écoulement *m* du sous-sol
s escorrentía *f* subsuperficial; escorrentía *f* hipodérmica
g Untergrundabfluß *m*; Zwischenabfluß *m*

* **throw → 4162**

* **thrust fault → 5059**

6196 thrust fault; overfault
A type of reverse fault that is characterized by a low angle of inclination with reference to the horizontal plane.
r надвиг
Смещение по разлому одних масс горных пород по другим.
f chevauchement *m*; faille *f* chevauchante conforme; faille *f* de charriage
s falla *f* cabalgante conforme; cabalgamiento *m*
g Überschiebung *f*

* **thrust moraine → 4831**

6197 thrust plane
The low angle fault face of rocks over which the movement of upper layers occurs.
r плоскость надвига
Поверхность горных пород, обычно наклоненная под малым углом, по которой происходит сдвиг верхних напластований.
f plan *m* de chevauchement
s plano *m* de cabalgamiento
g Stoßfläche *f*

6198 thufur
Tundra microrelief of elongated hummocks up to 0.5 m high, formed due to cryogenic heaving of ground.
r туфур
Мелкобугристый тундровый микрорель-

еф, представленный буграми продолговатой формы высотой до 0,5 м, формирующийся в результате криогенного пучения суглинистых грунтов.

f butte *f* terreuse; butte *f* gazonnée
s tufur *m*
g Thufur *n*

6199 thunder
A sharp or rumbling sound accompanying a lightning flash, it results from the rapidly expending gases along the channel of the lightning stroke.
r гром
Резкий или раскатистый звук, сопровождающий молнию, образуется быстрым расширением газов вдоль канала разряда молнии.
f tonnerre *m*
s trueno *m*
g Donner *m*

* **thundering of the breakers → 5117**

6200 thunderstorm
A storm produced by a cumulonimbus cloud and always accompanied by lightning and thunder, strong winds gusts, heavy rain, and sometimes hail.
r гроза
Буря, порождаемая кучеводождевыми облаками, всегда сопровождающаяся разрядами молний и громом, сильными порывами ветра, обильными осадками и градом.
f orage *m*
s tormenta *f*
g Gewitter *n*

6201 thunderstorm electricity
Electrical discharges formed in the atmosphere in connection with covective clouds and producing lightning.
r грозовое электричество
Электрические заряды, образующиеся в атмосфере в связи с облаками и дающие молнии.
f électricité *f* d'orage
s electricidad *f* de tormenta
g Gewitterelektrizität *f*

6202 thunderstorm squall
A sharp intensification of wnd accompanied by a thunderstorm.
r грозовой шквал
Резкое усиление ветра, сопровождаемое грозой.
f grain *m* orageux
s turbonada *f* tormentosa
g Gewitterbö *f*

6203 thundery cloud system
Cloud system, usually not associated with a cyclone or atmospheric front, the formation of which is caused by strong convection and instability of the air mass. → Fig. 331
r грозовая облачная система

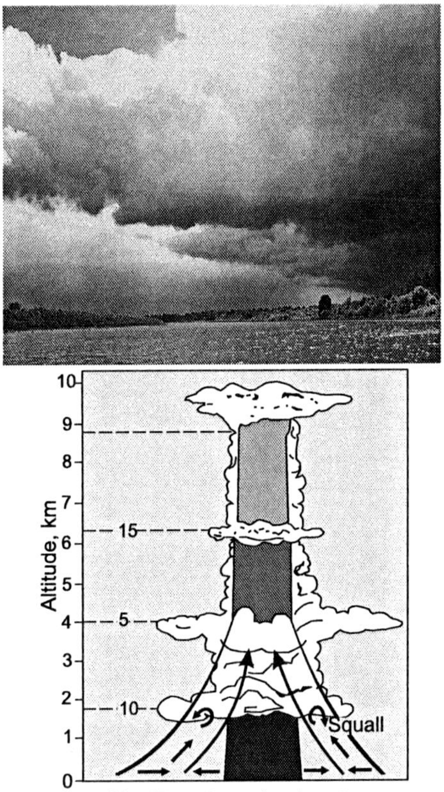

Fig. 331. **Thundery cloud system**

Облачная система, не связанная с циклоном и фронтом, формирование которой вызвано сильной адвекцией и неустойчивостью воздушной массы. → Рис. 331
f système *m* de nuages orageux
s sistema *m* nuboso de tormenta
g gewitteriges Wolkensystem *n*

6204 tidal ***adj***
Related to, due to, affected by the tide.
r приливный
Относящийся к природным явлениям, связанным с приливно-отливными процессами в морях и океанах.
f de marée
s de marea ascedente
g Flut-

6205 tidal basin
A basin filled with water at high tide that can be held and released at low tide, the force of the outflownig water scouring the neighbouring harbour.
r приливный бассейн;
приливная акватория
Прибрежный участок суши, заполняемый водой во время прилива и освобождаемый

ею при отливе, при этом сила уходящей воды промывает прилегающую гавань.
f bassin *m* de marée
s cuenca *f* de marea
g Flutbecken *n*

6206 tidal bore; bore; eagre
A rapidly advancing tidal wave which moves upstream in any shallow estuary experiencing a large tidal range, on the Atlantic coast of France it is called mascaret, on the Amazon river – pororoca.
r бор²
Одиночная длинная волна, быстро распространяющаяся вверх по течению при приливах в некоторых устьях рек и эстуариях, на атлантическом побережье Франции называется маскарэ, на реке Амазонка – поророка.
f mascaret *m*
s barra *f* de agua; macareo *m*
g Sprungwelle *f*; Flutwelle *f*

6207 tidal creek
A narrow passage of water between headlands, islands, and reefs available for through navigation.
r проход
Узкий участок водного пространства между мысами, островами, рифами, доступный для сквозного плавания.
f passage *m* marin; chenal *m* de marée
s paso *m* marino
g Priel *n*

6208 tidal current; reversing current
The normal movement of seawater in response to the rise and fall of the tide: the inward flow with the flood tide, the outward with the ebb tide.
r приливное течение;
реверсивное течение
Возвратно-поступательное течение в океане и особенно в небольших бухтах, изменяющее свое направление вследствие приливов и отливов.
f courant *m* de marée
s corriente *f* de marea
g Gezeitenstrom *m*

6209 tidal delta
Underwater geomorphologic structure formed near a lagoon-type shore subjected to tide action, it is a result of the tidal current entering the lagoon and bringing there loads depositing in form of the delta.
r приливная дельта
Дельта, формирующаяся у лагунных берегов, подверженных приливам, образуется в результате приливного течения, которое проникает в лагуну и приносит сюда наносы, откладывающиеся в спокойных водах лагуны в виде дельты.
f delta *m* de marée
s delta *f* de marea
g Gezeitendelta *n*

6210 tidal energy; tidal power
Energy obtained from the movement of the tides, i.e. produced by the turbines driven by the force of the tides.
r энергия приливов
Альтернативный источник энергии на берегах морей и океанов, вызванной деятельностью приливно-отливных течений.
f énergie *f* de marée; énergie *f* marémotrice
s energía *f* de mareas; energía *f* maremotriz
g Gezeitenenergie *f*

6211 tidal flat; mudflat
An area of sand or mud uncovered at low tide.
r приливная зона;
осыхающая отливная отмель
Песчаная или илистая часть берега, обнажающаяся при отливе.
f zone *f* de marée
s zona *f* de marea
g Gezeitenzone *f*; Wattfläche *f*; Flutsenke *f*

6212 tidal glacier; tidewater glacier
A valley glacier that reaches the sea, and from which icebergs calve, or ice-floes form.
r приливный ледник
Наземный долинный ледник, конец которого достигает моря и продуцирует айсберги.
f glacier *m* soumis á la marée
s glaciar *m* de marea
g Gezeitengletscher *m*

* **tidal marsh → 6708**

* **tidal power → 6210**

6213 tidal power station
A power station that converts tidal energy into electrical energy, as the tides rush out and into an enclosed reservoir (tidal barrage) to drive hydraulic turbines which generate electricity.
r приливная электростанция
Электростанция, преобразующая энергию морских приливов в электрическую, для чего сооружается специальный приливно-отливный резервуар с гидравлическими турбинами, вырабатывающими электроэнергию.
f centrale *f* marémotrice; usine *f* marémotrice
s central *f* maremotriz
g Gezeitenkraftwerk *n*

6214 tidal pump
A natural mechanism rocking a glacier at its grounding line due to the effect of periodic (24-hour and 12-hour) tidal sea level change.
r приливный насос
Природный механизм, раскачивающий ледник у линии налегания вследствие периодических (суточных и полусуточных) приливных колебаний уровня моря.

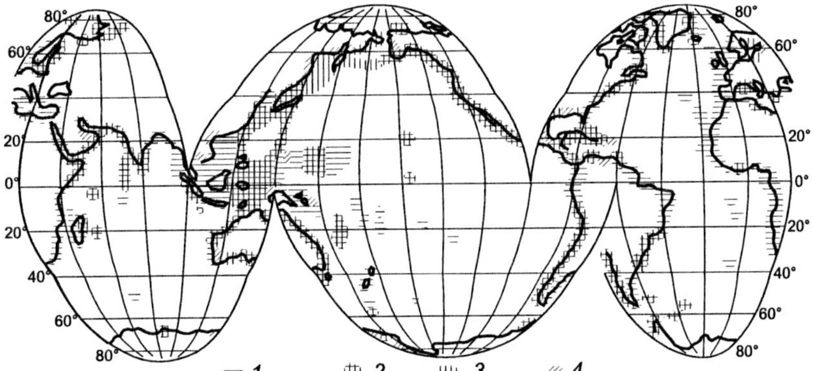

Fig. 332. **Tide** distribution over the Earth: 1 – semidiurnal tide, 2 – mixed tide with predominance of a semidiurnal tide, 3 – mixed tide with predominance of a diurnal tide, 4 – diurnal tide

f pompe *f* de marée
s bomba *f* de marea
g Gezeitenpumpe *f*

6215 tidal range; tide range
The difference between the level of sea water at high tide and low tide at a place, varying from day to day.
r высота прилива
Разница в уровне воды при приливе и отливе, изменяющаяся день ото дня.
f amplitude *f* marée; marnage *m* de marée
s amplitud *f* de la marea; desnivel *m* de la marea
g Fluthöhe *f*

6216 tidal sea
A sea where the tide effects are well pronounced.
r приливное море
Море, в котором наблюдаются существенные приливы.
f mer *f* de marée
s mar *f* de marea
g Gezeitenmeer *n*

6217 tidal wave
An unusually large wave at high water resulting from tidal movement.
r приливная волна
Необычно крупная волна высокой воды, возникающая в процессе приливного цикла.
f vague *f* de marée
s ola *f* de marea
g Flutwelle *f*

6218 tide
The regular rise and fall of water level in the worldrs oceans, resulting from the gravitational attraction of the Sun and the Moon, the latter being the more powerful (the rising of the water is termed the flood tide, or high tide, the falling is the ebb tide, or low tide). → Fig. 332
r прилив; морской прилив
Периодические колебания уровня моря (подъемы и опускания), вызываемые гравитационными силами Луны и Солнца, причем лунный прилив значительнее солнечного (подъем воды носит название прилив, а понижение воды – отлив). → Рис. 332
f marée *f*
s marea *f*
g Flut *f*; Tide *f*

6219 tide crack in ice
A crack between the fixed ice foot and the vertically moving fast ice, it may widen to form a shore lead.
r приливная трещина
Сквозная трещина между неподвижной подошвой припая и самим припаем, подверженным вертикальным колебаниям.
f crevasse *f* de marée dans glace
s grieta *f* de marea en el hielo
g Gezeitenspalt *m*

6220 tide cycle
A time period during which all phases of tides occur, there are semidiurnal (about 12.42 hours) and diurnal (about 24.84 hours) cycles.
r приливный цикл
Период времени, в течение которого происходят все фазы приливов, выделяют полусуточные (около 12,42 часа) и суточные (около 24,84 часа) циклы.
f cycle *m* de la marée montante
s ritmo *m* de la marea; período *m* de la marea
g Gezeitenzyklus *m*

*** tide range → 6215**

6221 tideless sea
Any sea where no pronounced tides occur.
r бесприливное море
Море со слабо выраженными приливами.
f mer *f* sans marée

s mar *m* sin marea
g gezeitenloses Meer *n*

* **tidewater glacier → 6212**

6222 tierra caliente
The lowest altitudinal belt in the mountains of Central America (up to 800 m) and low latitudes of South America (up to 1,000-1,500 m), it has hot humid climate, with the natural vegetation of tropical forests and crops of bananas, sugar, cocoa, maize and coffee. → Fig. 7, p. 28
r тьерра кальенте
Нижний высотный пояс в горах Центральной Америки и низких широтах Южной Америки, располагающийся соответственно на высотах до 800 м и 1000-1500 м, имеет жаркий и влажный климат, растительность тропических лесов и плантации бананов, сахарного тростника, какао, кукурузы и кофе. → Рис. 7, с. 28
f tierra *f* caliente
s tierra *f* caliente
g Tierra *f* Caliente

6223 tierra fria
The upper altitudinal belt with moderately cold climate in the mountains of Central America (between 1,700 and 3,000 m), and low latitudes of South America (from 2,200-2,800 to 3,500 m), with the natural vegetation of coniferous forest, giving way to grassland and crops of wheat and vegetables. → Fig. 7, p. 28
r тьерра фриа
Верхний высотный пояс в горах Центральной Америки и низких широт Южной Америки, располагающийся соответственно на высотах от 1700 до 3000 м и от 2200-2800 до 3500 м, имеет умеренно холодный климат и естественную растительность хвойных лесов, а также участки степей, овощных культур и полей пшеницы. → Рис. 7, с. 28
f tierra *f* fria
s tierra *f* fría
g Tierra *f* Fria

6224 tierra helada
Subnival and nival altitudinal belt at 3,000 m in the mountains of Central America, and from 3,200-3,500 m in South America, with cold high mountain climate and snow covered peaks. → Fig. 7, p. 28
r тьерра элада
Субнивальный и нивальный высотные пояса, располагающиеся на высотах от 3000 м в Центральной Америке и от 3200-3500 м в Южной Америке, имеют холодный высокогорный климат и покрытые снегом вершины. → Рис. 7, с. 28
f tierra *f* helada
s tierra *f* helada
g Tierra *f* Helada

6225 tierra templada
The intermediate altitudinal belt with moderately warm climate in the mountains of Central America (between 800 and 1,700 m), and low latitudes of South America (from 1,000-1,500 up to 2,200-2,800 m), known as he zone of perpetual spring, with the natural vegetation of savanna and open forest and maize, coffee and tobacco as crops. → Fig. 7, p. 28
r тьерра темплада
Средний высотный пояс в горах Центральной Америки и низких широт Южной Америки, располагающийся соответственно на высотах 800-1700 м и от 1000-1500 до 2200-2800 м, имеет умеренно теплый климат (известен как пояс постоянной весны), саванновую растительность и плантации кукурузы, кофе и табака. → Рис. 7, с. 28
f tierra *f* templada
s tierra *f* templada
g Tierra *f* Templada

6226 tight-closed fracture
A crack in rocks with closely pressed opposite walls.
r закрытая трещина
Трещина в горных породах с плотно прижатыми стенками разрыва.
f fissure *f* fermée
s fisura *f* cerrada
g geschlossene Spalte *f*; geschlossener Spalt *m*

6227 till$_1$
An unstratified, unconsolidated material consisting of a heterogeneous mixture of angular and/or subangular clay, sand, gravel and boulders carried by ice and deposited, with little or no subsequent sorting or transportation by water.
r тилл$_1$
Нестратифицированный, рыхлый осадок, состоящий из неоднородной смеси несортированных глинистых частиц, песка, гравия и валунов, перенесенных льдом и отложенных, иногда с участием воды.
f till *m*
s till *m*
g Till *m*

* **till^2 → 3938**

6228 till plain; till plateau
An area of low relief in the region of the Pleistocene glaciation which is created by the ground moraine covering the irregularities in the bedrock relief.
r первичная моренная равнина
Равнинная поверхность среди холмисто-западинного рельефа областей плейстоценового покровного оледенения, выстланная слоем основной морены.
f plaine *f* morainique initiale
s llanura *f* morrénica inicial
g Grundmoränenebene *f*

* **till plateau → 6228**

6229 tillage$_1$
The process of cultivating land so as to make in fit for raising crops; arable land ploughed or hoed in the current year.
r пахота
Процесс подготовки земли к выращиванию урожая; земля, вспаханная или разрыхленная в текущем году.
f labour *m*; labourage *m*
s labranza *f*; labor *f*
g Pflügen *n*; Ackern *n*; Beackern *n*

* **tillage$_2$ → 354**

* **tillage pan → 4618**

6230 tillite
A moraine formed of ancient unsorted coarse-grained till and glacio-marine deposits.
r тиллит
Древние несортированные грубообломочные морены и ледниково-морские отложения.
f tillite *m*
s tilita *m*
g Tillit *m*; fossile Moräne *f*

6231 tilloid
A chaotic often partly indurated sedimentary mixture of large sub-angular rocks in a clay-rich matrix resulting from the action of landslides, mudslides and glacial outbursts.
r тиллоид
Хаотическая, нередко затвердевшая масса крупных округлых обломков горной породы в глинистой цементирующей среде, возникшая в результате оползней или ледниковых паводков
f tilloide *m*
s tiloida *m*
g Tilloid *n*

6232 timber; lumber (*amer*)
A wood suitable for or processed for use of construction as well as a tree yielding such wood.
r древесина; строевой лес
Лесоматериал, пригодный или используемый в хозяйстве, а также дерево, служащее сырьем для такого материала.
f bois *m* de construction
s madera *f* de construcción
g Holz *n*

* **timber-industry complex → 2275**

* **timber-line (*amer*) → 6339**

6233 time
Time intervals measured by comparing them with certain periodical processes: rotation of the Earth with regard to stars (sidereal time), movement of the Sun (solar time), time for certain longitude (local time), the mean time of a meridian (standard time) and for the meridian 0° (Greenwich Mean Time), daylidht saving, etc.
r время
Интервалы времени, отсчитываемые путем сравнения их с принятыми периодическими процессами: вращением Земли относительно звезд (звездное время), движением Солнца (солнечное время), различают время для заданной долготы (местное время) и для Гринвичского меридиана (всемирное время), для удобства во многих странах принято поясное время и летнее время.
f temps[1] *m*
s tiempo[1] *m*; hora *f*
g Zeit *f*

6234 time equation
The difference in value between the mean time and apparent solar time, it changes gradually during a year from −16.4 up to +14.3 min.
r уравнение времени
Разность между средним и истинным солнечным временем, плавно изменяющаяся в течение года от -16,4 до +14,3 мин.
f équation *f* de temps
s ecuación *f* de tiempo
g Zeitgleichung *f*

6235 time scale
Relationship of the demonstration period and the real time applied in dynamic geoimages (animation, virtual maps), e.g. scale 1:86 000 means that 1 second of demonstration is approximately equals to a day.
r временной масштаб
Соотношение времени демонстрации и реального времени, применяемое в динамических геоизображениях (анимационных, виртуальных картах), например, масштаб 1:86 000 означает, что 1 секунда демонстрации приближенно равна суткам.
f échelle *f* de temps
s escala *f* de tiempo; escala *f* temporal
g Zeitmaßstab *m*

6236 time series
Succesive observations of the same phenomenon over a period of time.
r серия наблюдений
Последовательные наблюдения одного и того же природного явления за определенный период времени.
f série *f* temporelle
s serie *f* temporal
g temporale Serie *f*

6237 time zone
A time division represented by 15° longitude (less in small countries) whithin which the mean time of the central (or near central) meridian is selected to represent the whole division, actual boundaries of time zones are determined with account of state and administrative frontiers. → Fig. 333, p. 744
r часовой пояс

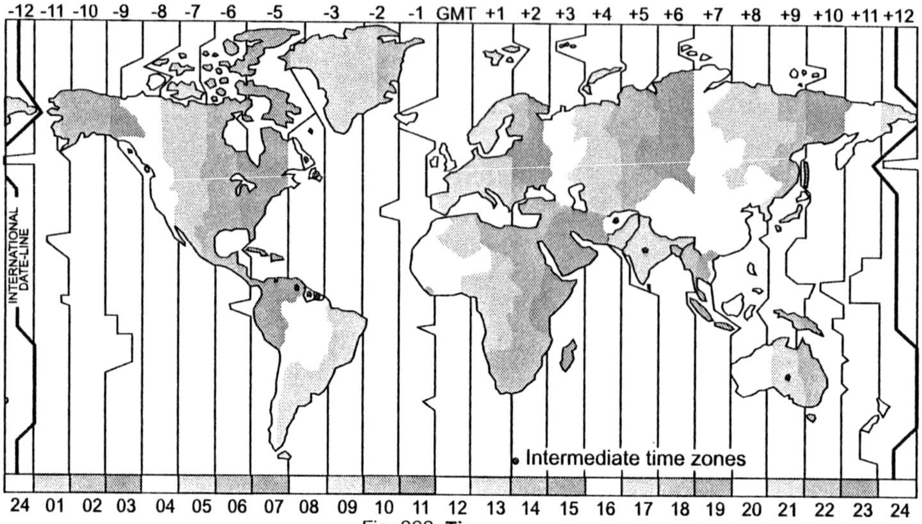

Fig. 333. **Time zones**

Долготный сектор шириной в 15° (может быть меньше в небольших странах), в котором среднее время центрального меридиана (или близкого к центральному) принимается для всего этого сектора, реальные границы часовых поясов проведены с учетом государственных и административных границ. → Рис. 333

f fuseau *m* horaire
s huso *m* horario
g Zeitzone *f*

* **time-lag → 3346**

* **tip-heap → 6045**

6238 tirs
A compact soil (vertisol) that occurs in poorly drained depressions of North Africa, it is dark-coloured and often contains carbonate crust in deep parts of profile.
r тирс
Слитая почва (вертисоль) в Северной Африке, встречается в слабодренированных депрессиях, имеет темную окраску и в глубоких частях профиля часто содержит карбонатную корку.
f tirs *m*
s tirs *m*
g Tirs *m*

* **Tissot's indicatrix → 1874**

6239 tolar
A high-mountain tropical desert in the Southern Hemisphere with sparse thorny shrubs and cushion-like Opuntia.
r толар
Высокогорная тропическая пустыня в Южном полушарии с редкими колючими кустарниками и подушковидными опунциями.
f tolar *m*
s tolar *m*
g Tolar *m*

6240 tolerance
The degree of adaptation of an organism or population to the life and development under changing environmental conditions.
r толерантность; выносливость
Степень приспособленности организмов к изменениям окружающей среды и способность популяций сохранять свое существование в меняющихся условиях среды.
f tolérance *f*
s tolerancia *f*
g Toleranz *f*; Lebensfähigkeit *f*

* **toll road → 6421**

6241 toltre
Deeply dissected and karstificated, rocky limestone hills stretched as parallel ridges in Romania and Moldova.
r толтры
Скалистые сильно расчлененные и закарстованные известняковые холмы, протягивающиеся параллельными грядами в Румынии и Молдавии.
f toltre *m*
s toltre *m*
g Toltre *f*

* **tombolo → 1215**

6242 tomillar
Communities of low sclerophyllous shrubs and subshrubs typical of the Mediterranean region, dominated by such evergreen aromatic plants as thyme, lavender, etc.
r томилляр

Распространенные в Средиземноморье сообщества жестколистных кустарничков и полукустарников с преобладанием вечнозеленых ароматических растений: тимьяна, лаванды и др.
f tomillar *m*; garrigue *f* claire
s tomillar *m*; garriga *f* clara
g Tomillar *m*

6243 tonnage
The carrying capacity of a vessel measured in tons.
r тоннаж
Грузоподъемность судна, выраженная в тоннах.
f tonnage *m*
s tonelaje *m*
g Tonnage *f*

*** top → 5966**

6244 topographic barrier
A mountain ridge or chain of heights which influeces atmospheric processes and induces orographic precipitation.
r орографическое препятствие
Неровности рельефа (горный хребет или цепь возвышенностей), влияющие на атмосферные процессы, что приводит, например, к формированию орографических осадков.
f barrière *f* topographique
s barrera *f* topográfica
g topographische Barriere *f*

6245 topographic base
A geographical map which presents just sufficient geographic information to permit the correlation of additional data placed upon it.
r бланковая карта; географическая основа
Географическая карта, предназначенная для нанесения на нее тех или иных данных, обычно отпечатанная одним бледным цветом.
f fond *m* topographique; fond *m* de carte; carte *f* en blanc
s fondo *m* cartográfico; fondo *m* topográfico; fondo *m* de mapa
g topographischer Unterdruck *m*; topographische Grundlage *f*

6246 topographic map
A map, usually on a fairly scale (e.g. 1:50 000), representing surface features, e.g. landforms and other natural phenomena as well as features produced by human activities.
r топографическая карта
Крупно- и среднемасштабная географическая карта, показывающая черты земной поверхности, например, рельеф, речную сеть и пр., а также некоторые объекты человеческой деятельности.
f carte *f* topographique
s mapa *m* topográfico
g topographische Karte *f*

6247 topographic survey
Field measurements for making topographic maps on the base of ground surveys.
r топографическая съемка
Измерения на местности, проводимые для создания топографических карт путем наземных съемок.
f levé *m* topographique
s levantamiento *m* topográfico
g topographische Aufnahme *f*

6248 topographical manuscript
Any plan surveyed hurriedly by field sketching, or with the use of simple instruments, such as compass.
r брульон
Всякий план, снятый наскоро, глазомерно или с помощью простых приборов, например, буссоли.
f brouillon *m*
s borrón *m*
g topographischer Entwurf *m*

6249 topography
The study of techniques of the topographic survey and making topographic maps and plans on its basis.
r топография
Изучение и разработка методов съемки местности с целью изображения ее на топографической карте.
f topographie *f*
s topografía *f*
g Topographie *f*; Vermessungswesen *n*

6250 topological *adj*
Applied to relative position of geographical object rather than to actual distance and orientation of it, as it is used in the process of compilation of topological maps (anamorphoids).
r топологический
Используемый при изображении относительного положения географического объекта на основе каких-либо его особенностей, пренебрегая реальными расстояниями и ориентацией на местности, как это применяется при составлении топологических карт (картоидов).
f topologique *adj*
s topologico *adj*
g topologisch *adj*

6251 topological map
A map designed to show only a selected feature, such as the routes of public transportation in a city, using dots, lines, etc. without real orientation and scale.
r топологическая схема
Карта, показывающая лишь отдельные черты, как, например, схему городского транспорта, с использованием точек, линий и без соблюдения реальной ориентации и масштаба.
f carte *f* topologique

s mapa *m* topológico
g topologische Karte *f*

6252 **toponym; place-name; geographic name**
A proper name of a geographical object or geographical locality.
r топоним; географическое название
Собственное имя, название географического объекта.
f toponyme *m*; nom *m* géographique
s topónimo *m*; nombre *m* geográfico
g Ortsname *m*

6253 **toponymy**
All geographic names (toponyms) of some locality as well as the study of their meaning, structure, origin and distribution.
r топонимика
Совокупность географических названий (топонимов) какой-либо местности, а также изучение их значения, структуры, происхождения и ареалов распространения.
f toponymie *f*
s toponimia *f*
g Ortsnamenkunde *f*

6254 **topsoil**
The surface soil layer usually containing organic matter which is turned over under cultivation and from which many substances have been eluviated into the lower horizons (as opposed to subsoil).
r надпочва
Верхний слой почвы, как правило, содержащий органические остатки и разрушаемый при сельскохозяйственной обработке, из которого многие вещества вымываются в нижележащие горизонты.
f couche *f* superficielle du sol
s capa *f* superficial del suelo
g Oberboden *m*

6255 **tor**
An upstanding mass of rocks or boulders which rises above the gentler slopes and is formed by frost shattering in periglacial conditions.
r тор
Выступающая среди пологого склона масса обломков горной породы или валунов, образованная морозным разрушением породы в перигляциальных условиях.
f tor *m*
s tor *m*
g Tor *m*

6256 **tornado**
A violent, anticlockwise, very destructive, short live revolving tropical storm, usually accompanied by rain and thunder, associated with an intensive low pressure system, small in diameter and with a dark, funnel-shaped cloud.
r торнадо
Яростный и разрушительный, вращающийся против часовой стрелки, короткоживущий тропический шторм, сопровождающийся дождем и громом, связанный с глубоким циклоном, как правило, небольшого диаметра, с темным воронкообразным облаком, аналог *смерча* в умеренных широтах.
f tornade *f*
s tornado *m*
g Tornado *m*

* **tornado cloud → 2349**

6257 **torrent; flashy flow**
A high-velocity flow of water down a stream channel; a fast-moving flow of lava.
r бурный поток; горный поток
Высокоскоростной поток воды, стекающей вниз по руслу; быстродвижущийся лавовый поток.
f torrent *m*; courant *m* torrentiel
s torrente *m*; corriente *f* torrencial
g Sturzbach *m*; reißende Strömung *f*

* **torrential rain → 5420**

* **torrid belt → 6258**

6258 **torrid zone; torrid belt**
The geographical region with the highest temperatures which includes equatorial, subequatorial and tropical belts; its boundary is usually drown along a mean annual isotherm of 20°C.
r жаркий пояс
Совокупность географических поясов с наиболее высокими температурами воздуха; охватывает экваториальный, субэкваториальный и тропический пояса, его границу обычно проводят по среднегодовой изотерме 20°C.
f zone *f* torride
s zona *f* tórrida
g heiße Zone *f*

6259 **total capacity of reservoir**
The sum of effective and dead capacities of a reservoir.
r полный объем водохранилища
Сумма полезного и мертвого объемов водохранилища.
f capacité *f* total de réservoir
s capacidad *f* total de embalse
g Gesamtkapazität *f* des Staubeckens

* **total cloud cover → 6260**

6260 **total cloudiness; total cloud cover**
Combination of all visible clouds hiding the celestial dome.
r общая облачность
Общее количество облаков, покрывающих небесный свод, без разделения по ярусам.
f nébulosité *f* totale
s nubosidad *f* total

g Gesamtbewölkung *f*; Wolkendecke *f*

6261 total radiation
Sum of solar radiation and terrestrial radiation.
r полная радиация
Сумма солнечной и земной радиации.
f rayonnement *m* total
s radiación *f* total
g Gesamtstrahlung *f*

6262 total runoff
General term which includes all types of runoff: overland flow, throughflow, and groundwater flow.
r суммарный сток
Понятие, объединяющее все виды стока: поверхностный, внутригрунтовый и подземный.
f écoulement *m* total
s escorrentía *f* total
g Gesamtabfluß *m*

6263 tour; travel; trip; touristic trip
A journey made for pleasure or in order to visit places of interest, and which ends in the place of origin.
r туристическая поездка; путешествие; тур
Путешествие по определенному маршруту с целью отдыха и спорта или знакомства с новыми интересными регионами и объектами.
f tour *m*; circuit *m* touristique; voyage *m* touristique
s viaje *m* turístico
g Reise *f*; Reisetour *f*

6264 tourism
Leisure-time activity away from home for pleasure or for reasons of doing sport, travelling and visiting places of interest or medical treatment.
r туризм
Активный отдых вдалеке от дома для удовольствия, а также со спортивными или оздоровительными целями, объект широко развитой индустрии.
f tourisme *m*
s turismo *m*
g Tourismus *m*

*** tourism potential → 4921**

6265 tourist arrival
The act of a person of coming to a place for touristic activities.
r туристское прибытие
Приезд человека в данную страну или территорию с туристическими целями.
f arrivée *f* touristique; entrée *f* touristique
s llegada *f* turística
g Touristenankunft *f*

6266 tourist departure
The act of a person of leaving the place of residence for touristic activities.
r туристское отбытие
Отъезд человека из места его постоянного проживания с туристическими целями.
f départ *m* touristique; sortie *f* touristique
s partida *f* turística; salida *f* turística
g Touristenabreise *f*; Touristenabfahrt *f*

6267 tourist flow
The total amount of tourists moving in a definite direction or coming to a particular area per time unit (year, season), it is expressed in tourist arrivals and tourist departures. → Fig. 334
r туристский поток
Суммарное количество туристов, перемещающихся в определенном направлении или прибывающих в данную местность, выражается в количестве туристских отбытий и прибытий за конкретный отрезок времени (год, сезон). → Рис. 334
f flux *m* touristique
s flujo *m* turístico; corriente *f* turística
g Touristenstrom *m*

6268 tourist industry
A branch of the industry including the complex of touristic services such as accomodation, travel arrangements, sightseeing trips,

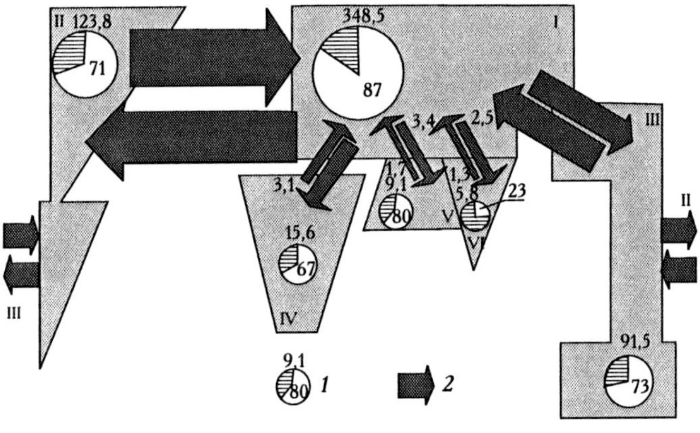

Fig. 334. International **tourist flows**: I – Eupope and whole Russia, II – America, III – Asia and Pacifica, IV – Africa, V – Middle East, VI – South Asia. 1 – number of tourist arrivals, mln; regional tourism, per cent of total number of tourist arrivals in the region. 2 – intercontinental tourism, millions of tourist arrivals (thickness of arrow is proportional to the total number of tourists)

etc. for those travelling or staying in a place for a relatively limited period of time.

r индустрия туризма
Отрасль хозяйства, включающая весь комплекс туристических услуг: жилье, транспорт, организацию туров и др.

f industrie *f* du tourisme
s industria *f* de turismo; industria *f* turística
g Touristenindustrie *f*

6269 tourist map
A geographical map representing objects of interest for tourism and usually accompanied by text, photographs and figures.

r туристская карта
Географическая карта, отображающая объекты, представляющие интерес для туризма и обычно сопровождаемая текстом, фотографиями и рисунками.

f carte *f* touristique
s mapa *m* turístico
g touristische Karte *f*

6270 tourist season
A part of the year with maximal amount of tourists in a given region.

r туристский сезон
Часть года, когда в том или ином туристском центре наблюдается максимальный поток туристов.

f saison *f* touristique
s temporada *f* turística
g Touristenzeit *f*

*** touristic area → 6272**

6271 touristic attraction
A natural, historical or modern man-made site used for recreational activities.

r рекреационный объект; туристический объект
Природный, исторический или современный рукотворный объект, используемый в целях туризма и отдыха.

f attraction *f* touristique
s atracción *f* turística
g Erholungsort *n*

*** touristic region → 6272**

6272 touristic site; touristic zone; recreation area; touristic area; touristic region
A place or whole region intended for leisure and recreation.

r туристический центр; рекреационный район
Место или целый регион, предназначенный для многоцелевого отдыха или туризма.

f site *m* touristique; région *f* de tourisme; région *f* touristique; centre *m* touristique; zone *f* touristique
s lugar *m* turístico; área *f* turística; espacio *m* de ocio
g Touristenzentrum *n*

*** touristic trip → 6263**

*** touristic zone → 6272**

*** tower beacon → 5997**

6273 tower karst; turmkarst
Form of tropical karst presented by columns with steep nearly vertical walls and gently domed or flattened summits. → Fig. 335

r башенный карст
Форма тропического карста, представленная крутостенными башнями с уплощенными вершинами. → Рис. 335

f karst *m* à tour; karst *m* de mogotes
s karst *m* de torretos; karst *m* de mogotes
g Turmkarst *m*

6274 town
A place larger than a village consisting of a compact agglomeration of dwellings, shops, offices, public buildings, and a community pursuing a distinctive, urban way of life.

r малый город
Поселение размером больше села с компактным размещением жилых и общественных зданий, магазинами и оффисами, где население ведет городской образ жизни.

f petite ville *f*
s ciudad *f* pequeña
g Kleinstadt *f*

*** town planning → 6501**

6275 towpath
A narrow coastal belt at the foot of a high river bank, limited by low and high water marks, formerly a path alongside a navigable river or canal, used by draught animals or people towing boats.

r бечевник

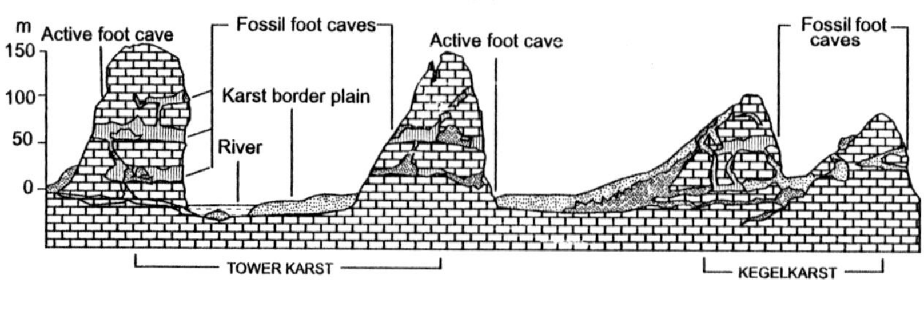

Fig. 335. **Tower karst**

Узкая береговая полоса у подошвы высокого берега реки, ограниченная по высоте уровнями межени и половодья, в прошлом полоса вдоль судоходной реки или канала, использовавшаяся бурлаками, которые тащили суда вверх по течению.
f chemin *m* de halage
s banda *f* de sirga
g Strandpfad *m*

6276 toxic waste
Waste products containing radioactive materials or harmful chemical substances which can cause death or different diseases of people or other living things.
r токсичные отходы
Отходы, содержащие радиоактивные вещества или вредные химические примеси, способные вызвать различные поражения человека или других живых существ.
f déchets *mpl* toxiques
s desechos *mpl* tóxicos; residuos *mpl* tóxicos
g giftige Abfälle *mpl*

6277 trace element
A chemical element which occurs in relatively small quantities in the Earth's crust, and essential in very small amount for the normal health of plants, animals and people.
r редкоземельный элемент
Химический элемент, который присутствует в земной коре в относительно малых количествах, но играет важную роль в нормальном развитии и здоровье растений, животных и людей.
f elément-trace *f*
s elemento *m* raro
g seltenes Element *n*

6278 tracer
A chemical compound marked or labelled with a paint or isotope and used to determine the course of a water current or air flow.
r трассер
Химическое вещество, помеченное красителем или изотопом, используемое для слежения за траекторией потока (или течения) воды или воздуха.
f traceur *m*
s trazador *m*
g Tracer *m*

6279 traction
A form of bed load transportation in which particles are rolling and sliding along parallel with and close to the bed.
r волочение наносов
Форма перемещения донных наносов, при которой частицы перекатываются по дну, лишь иногда не надолго отрываясь от него.
f roulage *m*; traínage *m*
s arrastre *m*; rodamiento *m* del materiales; rodamiento *m*
g Geschiebetransport *m*

* **traction load → 627**

6280 trade; commerce
The movement of goods from producer to consumer, a branch of national economy that realizes commodities by purchase and sale process, it includes internal trade (within one country) and foreign trade (among countries).
r торговля
Движение товара от продавца к покупателю, отрасль хозяйства, реализующая товары путем купли-продажи, включает внутреннюю (в пределах данной страны) и внешнюю (между странами) торговлю.
f commerce *m*
s comercio *m*
g Handel *m*

6281 trade winds; trade-wind circulation; northeast trades
Persistent winds which blow over vast regions from the tropical high pressure belts toward the equatorial region (more strongly over the ocean than over the continents), from the northeast in the Northern Hemisphere and from southeast in the Southern Hemisphere.
r пассат; пассатная циркуляция
Устойчивые ветры, дующие над обширными пространствами из субтропической области высокого давления к экватору (более сильные над океаном, чем над континентами), с северо-востока в Северном полушарии и с юго-востока в Южном.
f alizés *mpl*; alizé *m*; circulation *f* d'alizé
s alisios *mpl*; circulación *f* de alisio
g Passat *m*; Passatzirkulation *f*

* **trade-wind circulation → 6281**

* **trade-wind current → 1967**

6282 trade-wind desert
A desert that forms under a strong impact of trade winds, e.g. the Sahara or the deserts of Arabian Peninsula.
r пассатная пустыня
Пустыня, формирующаяся под большим воздействием пассатов, например, Сахара или пустыни Аравийского полуострова.
f désert *m* d'alizé
s desierto *m* de alisios
g Passatwüste *f*

6283 trade-wind front
The front occurring in the warm season between the fresh outburst of oceanic trade winds and the warm air of the neighbouring continent.
r пассатный фронт
Фронт, возникающий в течение теплого сезона при свежем вторжении пассатов из океана в теплую континентальную воздушную массу.
f front *m* des alizés
s frente *m* de los alisios

g Passatwindfront *f*

6284 trade-wind inversion
A temperature inversion occurring in the area of predominant trade winds, at heights of 1-2 km above the Earth's surface and with temperature difference of 3-5°C (this inversion acts as a ceiling for pollution, an effect which is marked in cities with heavy traffic).
r пассатная инверсия
Инверсия температуры в области господства пассатов, наблюдающаяся в среднем на высотах 1-2 км над земной поверхностью с перепадом температуры в 3-5°C (эта инверсия служит потолком для загрязнения, особенно заметного в городах с интенсивным автомобильным движением).
f inversion *f* des alizés
s inversión *f* de los alisios
g Passatinversion *f*

6285 trade-wind zone
Latitudinal zones in tropical and partly subtropical latitudes of the Northern and Southern hemispheres, which are better marked over the oceans where easterly flow of tropical air towards the equator takes place.
r пассатная зона
Широтные зоны в тропических и отчасти субтропических широтах Северного и Южного полушарий, лучше всего выраженные над океанами, здесь происходит пассатный перенос тропического воздуха по направлению к экватору.
f zone *f* des alizés
s zona *f* de alisio
g Passatregion *f*

6286 traffic
The movement of people and vehicles along a routeway.
r дорожное движение
Движение людей и транспортных средств по дороге.
f trafic *m*
s tráfico *m*
g Verkehr *m*

6287 traffic capacity
The maximum number of vehicles which can pass over a route in a given time.
r пропускная способность
Максимальное количество транспортных средств, которое может проехать на данном участке дороги за конкретное время.
f capacité *f* de trafic
s capacidad *f* de tráfico
g Durchlaßfähigkeit *f*

6288 traffic congestion; traffic jam
A difficulty of the road transport movement induced by the overload of a highway.
r транспортный затор; пробка
Затруднение движения автомобильного транспорта из-за перегруженности автострады.
f encombrement *m* de trafic; embouteillage *m*
s congestion *f* de tráfico; embotellamiento *m*
g Verkehrsstockung *f*

6289 traffic intensity
The existing number of vehicles which pass over a route in a given time.
r интенсивность движения;
густота движения
Количество транспортных средств, которые проходят по дороге за определенное время.
f intensité *f* de trafic
s intensidad *f* de tráfico
g Verkehrsdichte *f*

* **traffic jam → 6288**

* **trail → 4447**

6290 transboundary impact
Harmful impact upon the environment in a country from the source which is under the jurisdiction of another state.
r трансграничное воздействие
Вредное воздействие на окружающую среду той или иной страны от источника, находящегося под юрисдикцией другого государства.
f impact *m* transfrontière;
action *f* transfrontière
s impacto *m* transfronterizo
g grenzüberschreitende Umweltbelastung *f*

6291 transboundary pollutants transfer
The movement of pollutants with air or water flows outside the frontiers of the states where the sources of pollution are situated.
r трансграничный перенос загрязнителей
Распространение загрязнителей с воздушными или водными потоками за пределы границ государств, где находятся источники загрязнения.
f transfert *m* transfrontalier des polluants
s transferencia *f* transfronteriza de polución
g grenzenüberschreitender Schadstofftransfer *m*

6292 transcription system of place-names
A set of rules for the rendering of geographical names from one language script into another one taking into consideration phonetical and orphographical peculiarities of both languages.
r транскрипция географических наименований
Передача географических названий какого-либо языка средствами другого языка с учетом фонетических и орфографических особенностей обоих языков.
f transcription *f* des toponymes
s transcripción *f* de topónimos
g Transkriptionssystem *n* des Ortsnahmes

* **transcurrent fault → 6061**

6293 **transect**
A section taken across a tract of country for the purpose of studying the vegetation in relation to soil and relief.
r трансект
Поперечный разрез через участок местности для изучения связей растительности с рельефом и почвами.
f transect *m*
s transecto *m*
g Transekt *m*; Probestreifen *m*; Querschnitt *m*

6294 **transection glacier; transfluent glacier; twinned glacier**
Two or more glaciers located on the opposite slopes of a mountain range and having a common accumulation area on its saddle.
r переметный ледник
Два или несколько ледников, расположенных на противоположных склонах хребта и имеющих общую область аккумуляции на его седловине.
f glacier *m* siamois
s glaciare *m* anastomosado
g Jochgletscher *m*

6295 **transfer costs**
Total transport costs in moving goods and people combined with the extra costs of overcoming other obstacles, such as tariffs, insurance, etc.
r трансфертные затраты
В международной торговле комбинация транспортных затрат и стоимости преодоления возникающих препятствий при перевозке товаров, например, тарифных барьеров, с траховки и пр.
f frais *mpl* de transfert
s gastos *mpl* de transferencia
g Transferkosten *pl*

6296 **transfer of funds**
Moving foreign currency or gold from one country to another.
r трансферт
Передача иностранной валюты или золота из одной страны в другую.
f transfert *m* des fonds
s transferencia *f* de fondos
g Transfer *m*; Üdertragung *f*

6297 **transfluence of glacier; glacial transfluence**
The flowing of a glacial tongue or its branch into a nedighbouring valley through a depression in a dividing crest of the watershed.
r трансфлюэнция ледника
Перетекание ледникового языка или его ветви в соседнюю долину через понижение в водораздельном гребне.
f transfluence *f* de glacier
s transfluencia *f* de glaciar
g Transfluenz *f* eines Gletschers

* **transfluent glacier → 6294**

6298 **transformation of an air mass**
The modification of the properties of an air mass by physical or dynamic processes in such a way that in changes into the air mass of a different type.
r трансформация воздушной массы
Изменение характеристик воздушной массы под действием физических или динамических процессов таким образом, что она приобретает свойства воздушной массы другого типа.
f transformation *f* d'une masse d'air
s transformación *f* de una masa de aire
g Transformation *f* der Luftmasse; Luftmassentransformation *f*

6299 **transformation of nature**
A serious man-made change of nature in order to increase biological productivity or economic value of the natural environment.
r преобразование природы
Искусственное изменение природной среды с целью увеличения биологической продуктивности или хозяйственной ценности природных комплексов.
f transformation *f* du milieu naturel; transformation *f* de l'environnement
s transformación *f* de naturaleza; transformación *f* de medio ambiente
g Erneuerung *f* der Natur

6300 **transgression**
An advance of the sea to cover former land areas, due to the rise in the sea level relative to the land (a positive eustatic movement) or downward tectonic movements (a negative isostatic movement).
r трансгрессия
Наступление моря на сушу в результате опускания земной коры под влиянием нисходящих тектонических движений или повышения уровня океана.
f transgression *f*
s transgresión *f*
g Transgression *f*; Überschwennung *f*

6301 **transgressive bedding**
The deposition of marine sediments on a washed out surface of more ancient rocks formed in the conditions of the transgression of sea to the land.
r трансгрессивное залегание
Залегание морских отложений на размытой поверхности более древних пород, возникающее в условиях наступления моря на сушу.
f discordance *f* transgressive
s estratificación *f* de transgresión
g transgredierende Lagerung *f*

6302 **transhumance; cattle driving**
A periodic or seasonal movement of pastoral farmers and livestock between different graz-

ing grounds in the areas of different climatic conditions.

r перегон скота

Перемещение скота, главным образом овец, в теплое время года на высокогорные пастбища, иногда далеко от мест их зимнего содержания.

f transhumance *f*

s traslado *m*; trashumancia *f*

g Hinübertreiben *n*; Transhumance *f*

6303 **transit**

The movement of passengers and goods through intermediate stations (countries) situated between the departure and destination points.

r транзит

Перевозка пассажиров и грузов через промежуточные станции (страны), расположенные между пунктами отправления и назначения.

f transit *m*

s tránsito *m*

g Transit *m*; Durchfuhr *f*

6304 **transition economy**

The economy of the transitional period which combines planned and market economy.

r переходная экономика

Экономика переходного периода, когда в стране одновременно действуют элементы и плановой и рыночной экономики.

f économie *f* de transition

s economía *f* de transición

g Transitionswirtschaft *f*; Übergangwirtschaft *f*

6305 **transition zone; transitory zone**

1. Structural zone between continent and ocean composed of basins of marginal seas, mountainous archipelagos of island arcs and deepwater tranches from the ocean′s side. 2. The zone surrounding the central business district, with residential areas occupied by business and light manufacturing.

r переходная зона

1. Структурная зона между материком и океаном, слагающаяся из котловин окраинных морей, гористых архипелагов островных дуг и глубоководных желобов со стороны океана. 2. Зона, окружающая центральный деловой район, включая жилые массивы, занятая в основном предприятиями легкой промышленности.

f zone *f* de transition

s zona *f* de transición

g Übergangszone *f*

* **transitory zone → 6305**

6306 **translocation**

The migration of material in solution or suspension from one soil horizon to another.

r перемещение

Миграция материала в виде раствора или взвеси из одного почвенного горизонта в другой.

f transfèrement *m*

s transferencia *f*

g Verlegung *f*

6307 **transmission capacity of soil**

Tha capacity of soil to allow water to pass through it.

r пропускная способность почвы

Способность почвы пропускать воду через свою толщу.

f capacité *f* de transmission du sol

s capacidad *f* de transmisión del suelo

g Bodendurchflusskapazität *f*

6308 **transmission line**

A system of wires or cables and auxiliary devices designed to transmit and distribute electric power from power stations to the users.

r линия электропередачи; ЛЭП

Система проводов или кабелей и вспомогательных устройств, предназначенная для передачи или распределения электроэнергии от электростанции к потребителям.

f ligne *f* de transmission électrique

s línea *f* eléctrica; línea *f* de transporte de energía

g Kraftstromleitung *f*

6309 **transnational corporation; multinational corporation**

Very large firms like IBM or General Motors which own or control business (production facilities) in more than one country.

r транснациональная корпорация

Очень крупные фирмы (такие как IBM или General Motors), которые владеют или контролируют выпуск продукции более чем в одной стране.

f entreprise *f* transnationale; entreprise *f* multinationale

s empresa *f* multinacional; empresa *f* transnacional

g transnationale Korporation *f*

6310 **transparency of water; water transparency**

The property of water to let the light beams pass through the water mass.

r прозрачность воды

Способность воды пропускать вглубь световые лучи.

f transparence *f* de l′eau

s transparencia *f* de agua

g Wasserdurchsichtigkeit *f*; Wassertransparenz *f*

6311 **transparent ice**

Pure ice containing no air bubbles, brine cells, particles of salt or ground, or other mineral or organic inclusions.

r прозрачный лед

Лед, не имеющий пузырьков воздуха, ячеек рассола, частичек соли, грунта, других минеральных или органических веществ.
f glace *f* transparente
s hielo *m* transparente
g transparentes Eis *n*; durchsichtiges Eis *n*

6312 **transpiration**
The loss of water vapour from a plant through the tiny pores on their leaves and other parts of plants.
r транспирация
Потеря водяного пара растениями через мельчайшие поры на листьях или других частях растений.
f transpiration *f*
s transpiración *f*
g Transpiration *f*

6313 **transpiration ratio**
The ratio of water mass transpired by a plant during its growing season to the mass of dry matter produced.
r коэффициент транспирации
Отношение массы воды, расходуемой растениями на транспирацию за вегетационный период, к массе сухого вещества.
f coefficient *m* de transpiration
s coeficiente *m* de transpiración
g Transpirationskoeffizient *m*

6314 **transport; transportation (*amer*)**
1. A branch of economy dealing with the carrying of material or people from one place to another. 2. A phase of the process of denudation concerned with the conveying of loose material of the Earth's crust by a natural agents (other than mass movement by gravity, i.e. by running water, ice, wind, the ocean) to the site of deposition.
r транспорт
1. Отрасль народного хозяйства, выполняющая перевозку пассажиров и грузов. 2. Стадия процесса денудации, связанная с переносом рыхлого материала земной коры природными агентами (не гравитационным движением масс, а текущей водой, льдом, процессами в океане) к месту его отложения.
f transport *m*
s transporte *m*
g Transport *m*; Verkehr *m*

6315 **transport corridor**
A relatively narrow strip of land having a considerable number of highways and/or railways connecting densely populated regions.
r транспортный коридор
Сравнительно узкая полоса территории, на которой сосредоточено значительное число авто- и/или железных дорог, связывающих густо населенные регионы.
f axe *m* de transport
s eje *m* de transporte
g Transportkorridor *m*

* **transport costs → 1317**

6316 **transport geography; transportation geography; geography of transport**
A branch of human geography studying the territorial structure of transport, its spatial distribution, the volume of people and goods carried, transport flows, transport facilities and the role of transport in economic, political and social development.
r география транспорта
Отрасль социально-экономической географии, изучающая территориальную структуру транспорта, особенности его размещения, количество перевозимых людей и грузов, транспортные потоки, степень транспортной обеспеченности территорий и роль транспорта в экономическом, политическом и социальном развитии.
f géographie *f* des transports
s geografía *f* de transporte
g Verkehrsgeographie *f*

* **transport hub → 6317**

6317 **transport junction; transport node; transport hub**
A point where several lines of interacting means of transport meet.
r транспортный узел
Пункт, в котором сходятся не менее 2-3 линий взаимодействующих видов транспорта.
f noeud *m* des voies de communication
s nudo *m* de transporte
g Verkehrsknotenpunkt *m*

6318 **transport network; communication network**
A set of interconnecting routes: roads, waterways, pipelines and aviation lines.
r транспортная сеть
Общая сеть путей сообщения всех видов транспорта: наземных дорог, водных путей, трубопроводов, авиалиний.
f réseau *m* des transports
s red *f* de transporte
g Verkehrsnetz *n*

* **transport node → 6317**

6319 **transport providing**
A geographic characteristic of a territory that reflects the ratio between the needs in communication routes and means of transport and the available transport network and means of transport.
r транспортная обеспеченность
Географическая характеристика территории, отражающая соотношение между потребностями в путях сообщения и транспортных средствах и наличием транспортной сети и парков транспортных средств.

f densité *f* des voies de communication
s abastecimiento *m* por vías de communicación
g Verkehrsversorgung *f*

6320 **transport system**
The network of one means of communication that operates regularly and uses definite routes coordinated in volume, place and time.
r транспортная система
Транспортная сеть одного средства связи, операции по которой проводятся регулярно и по фиксированным маршрутам, согласованным по объему, месту и времени.
f système *m* des transports
s sistema *m* de transporte
g Verkehrssystem *n*

* **transportation (*amer*) → 6314**

* **transportation geography → 6316**

6321 **transportation slope**
A slope on which at each point the amount of material received from points upslope is balanced by the loss of material passing downslope.
r транзитный склон
Склон, в каждой точке которого количество материала, принесенного с верхней части склона, сбалансировано потерей материала, уносимого в нижнюю часть склона.
f versant *m* transporteur
s pendiente *f* transportadora
g Transportabhang *m*

6322 **transportation way; ways of communication**
A particular road, path or route used for carrying people and goods between any territories.
r путь сообщения
Разные виды дорог, служащих для перевозок людей и грузов между отдельными территориями.
f voies *f* de communication
s vía *f* de transporte; ruta *f* de comunicación
g Verkehrswege *mpl*; Verkehrsstraßen *fpl*

* **transported debris → 6323**

6323 **transported moraine; transported debris**
Rock fragments carried by a glacier. → Fig. 214, p. 470

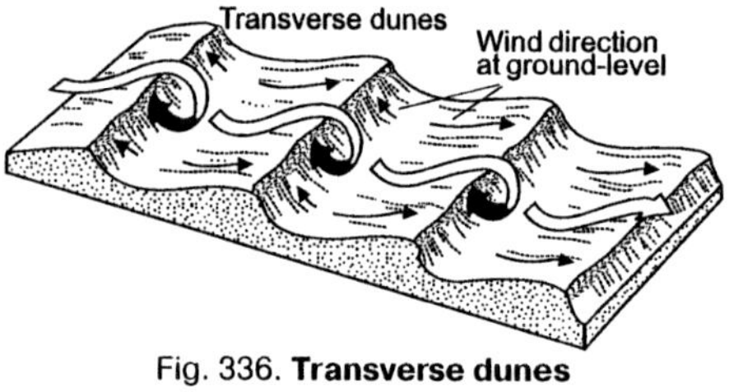

Fig. 336. **Transverse dunes**

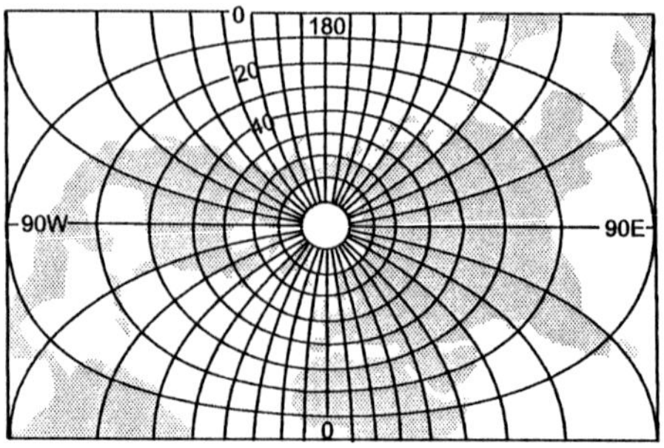

Fig. 337. **Transverse Mercator projection**

r влекомая морена
Переносимые ледником обломки горных пород. → Рис. 214, с. 470
f moraine *f* transportée
s morrena *f* transportada
g geschobene Moräne *f*; Wandermoräne *f*

6324 **transsedimentation**
The redeposition of soft material roiled in shallows by waves during storms and transferred by currents to other places.
r трансседиментация
Переотложение в водоемах рыхлых отложений, взмучиваемых на мелководьях волнением во время штормов и переносимых течениями в другие районы.
f transsédimentation *f*
s transsedimentación *f*
g Transsedimentation *f*

6325 **transverse *adj***
Oriented at right angle to the strike of strata or other elements of the environment.
r поперечный
Ориентированный перпендикулярно к простиранию пластов или других элементов окружающей среды.
f transversal *adj*
s transversal *adj*
g querlaufend *adj*; Quer-

* **transverse coast → 1628**

6326 **transverse dune**
A cross-wind dune having an asymmetric cross profile, with its crest running at right angles to the direction of wind blow. → Fig. 336
r поперечная дюна
Несимметричная в поперечном профиле внепустынная дюнная гряда, вытянутая перпендикулярно господствующим ветрам. → Рис. 336
f dune *f* transversale
s duna *f* transversal
g Querdüne *f*

6327 **transverse Mercator projection**
The transverse aspect of the Mercator pojection in which the line of zero distorsion is coincident with the meridian. → Fig. 337

r поперечная проекция Меркатора
Поперечный вариант проекции Меркатора, в котором линии нулевых искажений совпадают с меридианами. → Рис. 337, с. 754
f projection *f* de Mercator transverse
s proyección *f* transversal de Mercator
g querachsige Mercatorprojektion *f*

6328 transverse projection
A map projection where a transverse system of polar spherical coordinates is applied, the pole of which lies at the equator.
r поперечная проекция
Картографическая проекция, получаемая в случае применения поперечной системы полярных сферических координат, полюс которой расположен на экваторе.
f projection *f* transverse
s proyección *f* transversal
g querachsiger Kartennetzentwurf *m*

* **transverse section → 1345**

6329 transverse valley
Any valley which cuts at a right or oblique angle across the direction of folds.
r поперечная долина
Речная долина, протягивающаяся под прямым или косым углом к линии простирания складок горных пород.
f vallée *f* transversale
s valle *m* transversal
g Quertal *n*

6330 trap; trap rock
Igneous rocks of basic composition forming vast covers with characteristic steps.
r трапп
Магматические породы основного состава, образующие обширные покровы с характерными ступенчатыми отдельностями.
f trapp *m*
s trapp *m*
g Trapps *m*

6331 trap landscape
A terrain characterized, by a step appearance, created by a vertical succession of horizontal basaltic lava-flows which have been fashioned into step-like benches by denudation.
r трапповый ландшафт
Характерный ландшафт с вертикальными отдельностями, сформированными потоками базальтовой лавы и преобразованными денудацией в ступенчатый рельеф.
f paysage *m* de trapp
s paisaje *m* de trapp
g Trappslandschaft *f*

* **trap rock → 6330**

* **trash → 5143**

* **travel → 6263**

6332 traversing; polygonometry
The method to determine the position of geodetic points by way of plotting on the ground of an open polygon or system of interconnected open polygons in which all angles and sides are measured.
r полигонометрия
Метод определения положения геодезических пунктов путем проложения на местности ломаной линии или системы связанных между собой ломаных линий, в которых измеряются все углы и стороны.
f polygonation *f*;
cheminement *m* géodésique
s poligonación *f*
g Polygonzug *m*;
polygonometrische Messungen *fpl*

6333 travertine; tufa; calcareous tuff
A light-coloured concretionary deposit of calcium carbonate deposited by precipitation from highly impregnated groundwater around a hot spring.
r травертиновые отложения; травертин; известковый туф
Слегка окрашенные отложения карбоната кальция, выпадающие из концентрированного раствора грунтовых вод около горячего источника.
f travertin *m*; tuf *m* calcaire
s travertino *m*; toba *f* calcárea
g Travertinablagerungen *fpl*; Kalktuff *m*

6334 travertine mound
A hilly sediment of calcium carbonate (travertine) deposited by precipitation around a hot spring.
r травертиновый холм
Холмистые отложения известкового туфа (травертина) в местах выхода на поверхность углекислых источников.
f colline *f* de travertin
s colina *f* de travertino
g Sinterhügel *m*; Travertinhügel *m*;
Kalktuffhügel *m*

6335 travertine terrace
A terrace composed of calcium carbonate (travertine) which is deposited around a hot spring.
r травертиновая терраса
Терраса, сложенная известковым туфом (травертином) и часто возникающая у выходов на поверхность горячих источников.
f terrasse *f* de travertin
s terraza *f* de travertino
g Kalksinterterrasse *f*; Travertinterrasse *f*

6336 treaty port
A sea or river port opened by treaty for the purpose of foreign trade (applied especially in the XIX century to certain ports in China, Japan, Korea).
r открытый порт

Речной или морской порт, открытый по договору для внешней торговли, как это было в XIX в. для нескольких портов в Китае, Японии, Корее.

f port *m* de traité

s puerto *m* de tratado

g Vertragshafen *m*; Freihafen *m*

6337 tree

A perennial plant with a woody stem, the trunk, it is the major component of forests.

r дерево

Многолетнее растение с одревесневающим стеблем-стволом, главная составляющая лесов.

f arbre *m*

s árbol *m*

g Baum *m*

6338 tree layer

In the stratified structure of a forest community the layer formed by trees.

r древесный ярус

Ярус, образованный деревьями в лесном растительном сообществе.

f strate *f* arborée; strate *f* arborescente

s estrato *m* arbolado; estrato *m* arborescente; estrato *m* arbóreo

g Baumschicht *f*

* **tree ring → 275**

* **tree-limit → 6339**

6339 tree-line; timber-line (*amer*); tree-limit

The upper limit of tree growth in mountains. → Fig. 338

r граница леса

Верхняя граница распространения древесной растительности в горах. → Рис. 338

f limite *f* des arbres; limite *f* de la végétation arborescente; limite *f* supérieure de la forêt

s límite *m* de árboles

g Baumgrenze *f*; Waldgrenze *f*

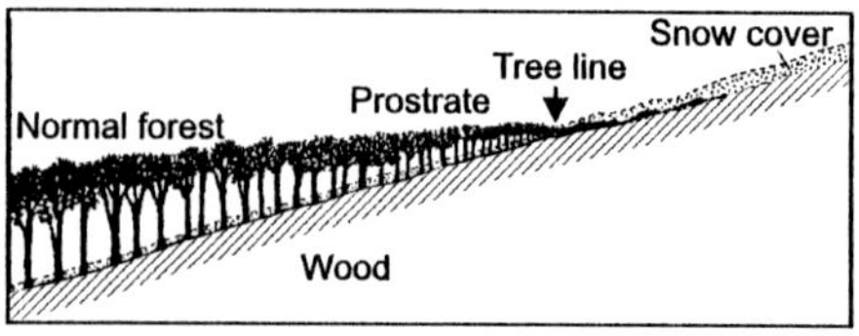

Fig. 338. **Tree-line**

6340 trellis drainage pattern

A rectilinear drainage occurring particularly in areas of folded sedimentary rocks, where consequent, obsequent and secondary consequent streams cut channels through the less resistant rocks at right angles to the initial slope. → Fig. 95, p. 202

r решетчатая речная сеть; шпалерная речная сеть

Прямолинейная речная сеть, особенно развитая в областях складчатых осадочных пород с крутыми откосами, где консеквентные, обсеквентные и вторичные консеквентные реки прорезали долины в менее прочных породах под прямыми углами к первоначальному склону. → Рис. 95, с. 202

f réseau *m* hydrographique rectangulaire

s red *f* hidrográfica rectangular

g gitterartiges Flußnetz *n*

6341 trench

A long narrow depression, a valley between two mountain ranges with relatively steep slopes, especially rift valley or U-shaped valley.

r желоб

Длинное узкое понижение, долина между двумя горными хребтами с относительно крутыми склонами, как правило, рифтовая или троговая долина.

f fossé *f*

s fosa *f*

g Rinne *f*

* **trend azimuth → 507**

6342 triangular delta; cuspate delta

A triangular-shaped delta with the apex advancing seawards formed on a straight coast by a river transporting small amount of load where strong wave action distributes the sediments evenly on each side of the river mouth. → Fig. 83, p. 178

r клювовидная дельта

Речная дельта с длинными приустьевыми косами, возникающая при относительно малом количестве приносимых рекой наносов. → Рис. 83, с. 178

f delta *m* en forme de bec

s delta *m* triangular

g Schaufeldelta *n*

* **triangular survey → 6343**

6343 triangulation; triangular survey

The method of construction of a geodetic framework for topographic survey in the form of triangles in which their angles and sides are measured.

r триангуляция

Метод построения геодезической сети для топографической съемки в виде треугольников, в которых измерены их углы и некоторые из сторон.

f triangulation *f*

s triangulación *f*

g Triangulation *f*; Dreiecksaufnahme *f*

6344 Trias; Triassic

The earliest (first) geological period of the Mesozoic era, it began 225 million years ago and lasted for about 30 million years. → Table. 1, p. 284

r триас

Нижний (первый) период мезозойской эры геологической истории Земли, начался 225 млн лет и продолжался около 30 млн лет. → Табл. 1, с. 284

f Trias *m*
s Triásico *m*
g Trias *f*

* **Triassic → 6344**

6345 tribalism
Strong loyalty to culturally and socio-politically isolated group that takes the form of tribal enmity existing mainly in Africa.
r трайбализм
Приверженность к культурно-бытовой, культовой и общественно-политической племенной обособленности, проявляющаяся в межплеменной вражде, главным образом в Африке.
f hostilité *f* de tribu
s hostilidad *f* tribual
g Stammfeinseligkeit *f*

6346 tribe
A social group of related families of pre-class society living in the same territory and sharing a common language, religion, economy and self-identity.
r племя
Социальная группа близких семей в доклассовом обществе, имеющая общую территорию, единый язык, религию, некоторую экономическую общность и самосознание.
f tribu *f*
s tribu *f*
g Stamm *m*

6347 tributary; influent; affluent
A stream flowing into a larger stream or lake.
r приток
Водоток, впадающий в более крупный водоток или озеро.
f affluent *m*; tributaire *m*
s afluente *m*; tributario *m*
g Nebenfluß *m*; Zufluß *m*

6348 trichohygrophyte
The plant of arid areas that obtains water from soil capillaries, through which ground water rises upwards.
r трихогигрофит
Растение аридных областей, использующее воду капилляров почвы, которая поднимается от сравнительно неглубоко залегающих грунтовых вод.
f trichohygrophyte *m*
s tricohidrófita *m*
g Trichohygrophyt *m*

6349 trigonometric levelling
Determination of altitude differences of the Earth surface points by measured angle of inclination of axis of sight from one point to another one and the distance between the points.
r геодезическое нивелирование; тригонометрическое нивелирование
Определение разностей высот точек земной поверхности по измеренному углу наклона линии визирования с одной точки на другую и расстоянию между ними.
f nivellement *m* trigonométrique
s nivelación *f* trigonométrica; medición *f* trigonométrica
g trigonometrische Höhenmessung *f*

6350 trigonometric point
Geodetic point the co-ordinates of which are obtained by method of construction of geodetic network in the form of triangles where their angles and some sides are measured.
r тригонометрический пункт
Геодезический пункт, координаты которого получены методом построения геодезической сети в виде треугольников, в которых измерены их углы и некоторые стороны.
f point *m* trigonométrique
s punto *m* trigonométrico
g trigonometrischer Punkt *m*

6351 trigonometric survey
A survey carried out by triangulation and trigonometrical calculation.
r тригонометрическая съемка
Съемка на местности, выполненная путем триангуляции и последующих тригонометрических вычислений.
f levé *m* trigonométrique
s levantamiento *m* trigonométrico
g trigonometrische Aufnahme *f*

6352 trilateration
Method of construction of a geodetic network in the form of triangles in which all sides are measured, while the angles and co-ordinates of the tops are calculated.
r трилатерация
Метод построения геодезической сети в виде треугольников, в которых измерены все их стороны, а углы и координаты вершин вычисляются.
f trilatération *f*
s trilateración *f*
g Trilateration *f*

6353 trimline

Fig. 339. **Trimline** in the glacier valley in Alps

A line separating vegetated lateral moraine on a valley side from the bare moraine lower down the slope marking the level of the recent glacier advance. → Fig. 339, p. 757

r тримлайн

Резкая граница, отделяющая покрытый растительностью склон ледниковой долины от лежащей ниже обнаженной боковой морены, маркирующая уровень, на котором располагалась поверхность ледника во время недавнего его наступания. → Рис. 339, с. 757

f trimligne *f*

s trimlínea *f*

g Trimlinie *f*

* **trip → 6263**

6354 **triple point; triple state**

The thermodynamic state at which three phases of a substance exist in equilibrium (0°C for water).

r тройная точка

Термодинамическое состояние, при котором три фазы вещества находятся в равновесии (для воды 0°C).

f point *m* triple

s punto *m* triple

g Tripelpunkt *m*; Tripelzustand *m*

* **triple state → 6354**

6355 **tripolar world; threecentric world**

The geopolitical arrangement of the world with three main centres of economic and political influence: the U.S., Western Europe and Japan.

r трехполярный мир

Геополитическое устройство мира, в котором существует три главных центра экономического и политического влияния: США, Западная Европа и Япония.

f monde *m* tripolaire

s mundo *m* tripolar

g dreipolare Welt *f*

6356 **trophic *adj***

Related to the nourishment of organisms in the environment.

r трофический

Относящийся к питанию организмов в окружающей среде.

f trophique *adj*

s trófico *adj*

g trophisch *adj*

* **trophic chain → 2241**

6357 **trophic level**

1. The level at which energy in the form of food is transferred from one organism to another as part of the food chain. 2. The nutrient level of a body of water, especially in relation to nitrate and phosphate content.

r трофический уровень

1. Уровень, на котором энергия в форме пищи передается от одного организма к другому как часть трофической цепи. 2. Уровень распространения питательных веществ в водоеме, особенно по отношению к содержанию в воде нитратов и фосфатов.

f niveau *m* trophique

s nivel *m* trófico

g Ernährungsstand *m*

6358 **trophic zone**

Surface layer in a water body where the biomass of plants which serve as food for water bacteria and animals grows.

r трофогенный слой

Поверхностный слой водоема, в котором происходит увеличение массы водорослей, служащих пищей для водных бактерий и животных.

f zone *f* trophogéne

s zona *f* trofogénica

g trophogene Schicht *f*

6359 **tropholytic zone**

Deep layer of a water body in which the decay of organic matter exceeds its formation in the process of photosynthesis.

r трофолитический слой

Глубинный слой водоема, в котором разложение органического вещества преобладает над его образованием в процессе фотосинтеза.

f zone *f* tropholytique

s zona *f* trofolítica

g tropholytische Schicht *f*

6360 **tropic**

One of the two parallels of latitude of 23°27′, i.e. the angle equal to the angle between the plane of the Earth's equator and that of its orbit (tropic of Cancer in the Northern Hemisphere and tropic of Capricorn in the Southern Hemisphere).

r тропик

Параллель, отстоящая от экватора на 23°27′, т.е. на угол, равный углу между плоскостями экватора Земли и ее орбиты (различают тропик Рака в Северном полушарии и тропик Козерога в Южном).

f tropique *m*

s trópico *m*

g Wendekreis *m*

6361 **tropic of Cancer**

The tropic of Northern Hemisphere which lies approximately along 23°27′ N (at points situated on this latitude the Sun passes through zenith on the day of summer solstice on 21-22 June).

r тропик Рака; Северный тропик

Тропик Северного полушария, проходящий по 23°27′ с.ш. (в пунктах, лежащих на этой широте, в день летнего солнцестояния Солнце проходит через зенит).

f tropique *m* du Cancer

s trópico *m* de Cáncer

g Wendekreis *m* des Krebses

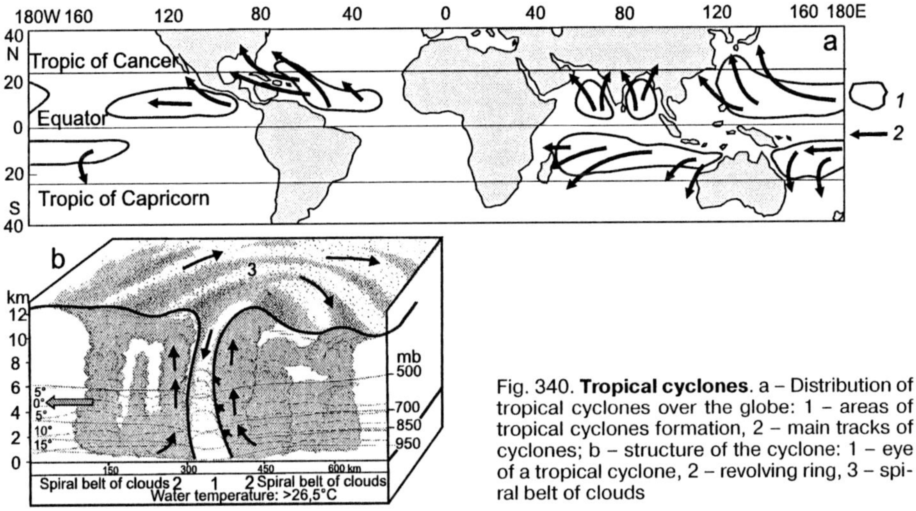

Fig. 340. **Tropical cyclones**. a – Distribution of tropical cyclones over the globe: 1 – areas of tropical cyclones formation, 2 – main tracks of cyclones; b – structure of the cyclone: 1 – eye of a tropical cyclone, 2 – revolving ring, 3 – spiral belt of clouds

6362 tropic of Capricorn

The tropic of the Southern Hemisphere which lies approximately along 23°27′ S (at points situated on this latitude the Sun passes through zenith on the day of winter solstice on 22-23 December).

r тропик Козерога; Южный тропик

Тропик Южного полушария, проходящий по 23°27′ ю.ш. (в пунктах, лежащих на этой широте, в день зимнего солнцестояния Солнце проходит через зенит).

f tropique *m* du Capricorne

s trópico *m* de Capricornio

g Wendekreis *m* des Steinbocks

6363 tropical *adj*

Related to the tropics, to the zone lying between 23°30′ N and 23°30′ S and adjacent areas in the Northern and Southern hemispheres and to the natural phenomena (climate, forest, etc.) characteristic for these latitudes.

r тропический

Относящийся к региону, расположенному в Северном и Южном полушариях между 20 и 30° широты и к природным явлениям (климат, лес и т.п.), распространенным в этих широтах.

f tropical *adj*

s tropical *adj*

g tropisch *adj*; Tropen-

* **tropical air → 6364**

6364 tropical air mass; tropical air

An air mass originating from the belt of subtropical anticyclones: tropical continental over the continental interiors, tropical maritime over the oceans.

r тропическая воздушная масса; тропический воздух

Воздушная масса, сформировавшаяся над поясом субтропических антициклонов: тропическая континентальная – над внутренними частями континентов, тропическая морская – над океанами.

f masse *f* d'air tropical; air *m* tropical

s masa *f* de aire tropical; aire *m* tropical

g tropische Luftmasse *f*; Tropikluft *f*; Tropenluft *f*

* **tropical belt → 6376**

6365 tropical black soil

A dark coloured, heavy textured soil, characterized by the presence of montmorillonite, which causes swelling during the rainy seasons but drying out and cracking during dry periods, it is similar to ***regur*** but has less organic material.

r тропическая черная почва

Темноцветная почва с присутствием монтмориллонита, что вызывает ее набухание в сезоны дождей, но высыхание и растрескивание в сухие периоды, она близка к ***регуру***, но включает меньше органического материала.

f sol *m* noir tropical

s suelo *m* negro tropical

g tropischer schwarzer Boden *m*

6366 tropical climate

The climate occufrring in the tropics, characterized by the lack of cool season, with no month experiencing a mean temperature below 20°C (can be subdivided into a number of distinct climatic regimes such as equatorial, subequatorial, monsoon and trade-wind climates).

r тропический климат

Климат тропических широт, характеризующийся отсутствием холодного сезона и температурами, не опускающимися ниже 20°C (различают несколько разновидностей этого климата: экваториальный, субэкваториальный, муссонный и пассатный климаты).

f climat *m* tropical

s clima *m* tropical
g tropisches Klima *n*

6367 **tropical cyclone; tropical storm; tropical revolving storm**
A cyclone occurring in tropical latitudes, several hundreds of kilometres in diameter, with very low atmospheric pressure in its centre (the eye), very violent winds, torrential rain and often thunder and lightning (in the areas affected the storms are given special names: hurricane, typhoon, willy-willy). → Fig. 340, p. 759
r тропический циклон
Мощный циклонический вихрь, зарождающийся в тропических широтах, диаметром в несколько сот километров, с очень низким атмосферным давлением в центре (глаз бури), сильнейшими ветрами и проливными дождями (в районах, подверженных тропическим циклонам, они носят местные названия: ураган, тайфун, вилли-вилли). → Рис. 340, с. 759
f cyclone *m* tropical
s ciclón *m* tropical
g tropische Zyklone *f*

6368 **tropical diseases; tropical endemic disease**
The diseases found in tropical climate, e.g. malaria or sleeping sickness caused by the bite of tsetse fly (these deseases determine life expectancy and health of the population of developing countries).
r тропические болезни
Болезни, распространенные в странах с жарким климатом, определяющие продолжительность жизни и уровень здоровья населения развивающихся стран, например малярия или сонная болезнь, вызываемая укусом мухи це-це.
f maladies *fpl* tropicales; endémie *f* tropicale
s enfermedades *fpl* tropicales
g Tropenkrankheiten *fpl*

6369 **tropical dry forest**
Deciduous and semi-deciduous forests characteristic of the tropical and subequatorial belts with a dry season up to 2,5 months long (in less dry regions monsoon forests grow, in drier regions savannah forests are typical).
r сухой тропический лес
Листопадные и полулистопадные леса, свойственные тропическому и субэкваториальному поясам с сухим периодом продолжительностью до 2,5 месяцев (в менее сухих районах развиты муссонные леса, в более сухих – саванновые леса).
f forêt *f* tropicale sèche
s bosque *m* tropical seco
g tropischer Trockenwald *m*

* **tropical endemic disease → 6368**

6370 **tropical fauna**
The animals inhabiting tropics, e.g. apes, elephants, rhinoceroses, giraffes, iguanas, crocodiles.
r тропическая фауна
Совокупность видов животных, обитающих в тропиках, в их числе человекообразные обезьяны, слоны, носороги, жирафы, игуаны, крокодилы.
f faune *f* tropicale
s fauna *f* tropical
g tropische Fauna *f*; Tropenfauna *f*

6371 **tropical forest**
The forests that grow in the tropical, equatorial and subequatorial zones, they consist of tropical rain forest, dry tropical forest, monsoon forest, and mangroves.
r тропический лес
Леса в тропическом, экваториальном и субэкваториальном поясах, включающие многие типы леса: влажный тропический лес, сухой тропический лес, муссонный лес, мангры.
f forêt *f* tropicale
s bosque *m* tropical
g Tropenwald *m*; tropischer Wald *m*

* **tropical grassland → 5229**

6372 **tropical karst**
A type of karst terrain in which steep-sided hills rise from the basal surface.
r тропический карст
Тип карста тропиков с преобладанием башнеобразных возвышений, расположенных на базальной поверхности.
f karst *m* tropical
s karst *m* trópico
g tropischer Karst *m*

6373 **tropical lake**
A lake having the surface temperature higher than 4°C all the year round and direct thermal stratification.
r тропическое озеро
Озеро, имеющее в течение всего года поверхностную температуру воды выше 4°C и прямую термическую стратификацию.
f lac *m* de type tropical
s lago *m* tropical
g tropischer See *m*

6374 **tropical monsoon**
Seasonal change of flows of the tropical and equatorial air: in summer equatorial air flows to tropical latitudes, and in winter tropical air moves towards the equator.
r тропический муссон
Сезонная смена потоков тропического и экваторального воздуха: летом поток экваториального воздуха направлен в тропические широты, а зимой тропический воздух переносится в сторону экватора.
f mousson *f* tropicale
s monzón *m* tropical
g tropischer Monsun *m*

* **tropical revolving storm → 6367**

* **tropical storm → 6367**

6375 **tropical vegetation**
Plant communities typical for tropics which include tropical forests, open woodland, savanna, brushwood of evergreen shrubs and deserts.
r тропическая растительность
Совокупность растительных сообществ, характерных для тропиков, представленная тропическими лесами и редколесьем, саваннами, зарослями вечнозеленых кустарников и пустынями.
f végétation *f* tropicale
s vegetación *f* tropical
g tropische Vegetation *f*

* **tropical year → 5652**

6376 **tropical zone; tropical belt**
Geographical regions in the Northern and Southern hemispheres between 20° and 30° of the latitude where trade-wind circulation determines and deserts and semi-deserts occupy large areas.
r тропический пояс
Географические пояса, расположенные в Северном и Южном полушариях между 20° и 30° широты, где господствует пассатная циркуляция и большие площади занимают пустыни и полупустыни.
f zone *f* tropicale; zone *f* intertropicale
s zona *f* tropical; zona *f* intertropical
g tropische Zone *f*; Tropenzone *f*

6377 **tropics**
The zone of the Earthr's surface lying between the Tropic of Cancer and the Tropic of Capricorn (i.e. between 23°30′ N and 23°30′ S), where the Sun′s rays strike vertically at noon at least two days in the year.
r тропики
Пояс на земной поверхности, лежащий между тропиками Рака и Козерога (т.е. между 23°30′ с.ш. и 23°30′ ю.ш.), где солнечные лучи падают вертикально по крайней мере два дня в году.
f tropiques *mpl*; pays *mpl* tropicaux
s trópicos *mpl*; paises *mpl* tropicales
g Tropen *pl*

6378 **tropopause**
The boundary between the troposphere and stratosphere at the elevation about 18 km usually characterized by an abrupt change of the lapse rate, i.e. temperatures cease to decrease with increasing height. → Fig. 341
r тропопауза
Граница между тропосферой и стратосферой на высоте около 18 км, где обычно происходит резкое изменение вертикального градиента температуры, т.е. температура перестает понижаться с высотой. → Рис. 341
f tropopause *f*
s tropopausa *f*
g Tropopause *f*

6379 **tropopause funnel**
A funnel-shaped or bowl-like deformation of the tropopause above a deep slow-moving cyclone.
r воронка тропопаузы
Воронкообразная или чашеобразная деформация (понижение) тропопаузы над глубоким малоподвижным циклоном.
f entonnoir *m* de la tropopause
s embudo *m* de la tropopausa
g Tropopausentrichter *m*

6380 **tropophyte**
Plants that are adapted to the seasonal change of wet and dry periods by shedding their leaves or changing the size of their leaves.
r тропофит
Растения, приспособившиеся к обитанию в областях с регулярной ежегодной сменой влажного и сухого периодов путем регулирования своего водного обмена: одни растения сбрасывают на сухой период листья, другие меняют большие листья на маленькие.
f tropophyte *m*
s tropofita *m*
g Tropophyt *m*

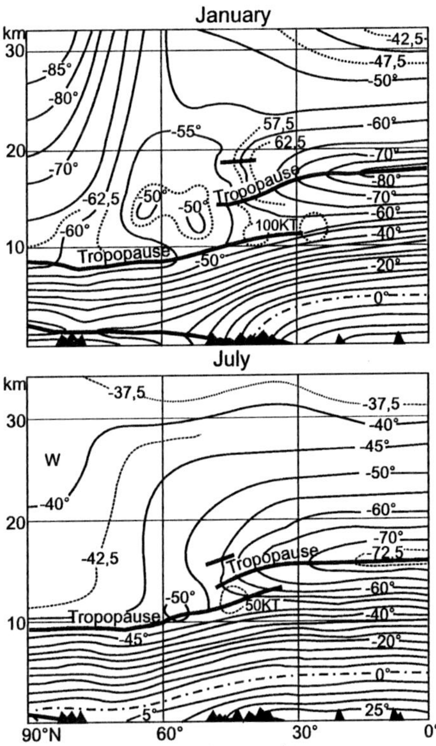

Fig. 341. **Troposphere** and **tropopause**

6381 **troposphere**
The lowest layer of the atmosphere extending from the Earthrs surface up to a height varying from about 9 km at the poles to about 17 km over the equator, in which temperature decreases with height and most of the weather activity take place. → Fig. 341, p. 761
r тропосфера
Нижняя основная часть атмосферы, простирающаяся от поверхности Земли до высоты приблизительно 9 км на полюсах и 17 км на экваторе, характеризуется падением температуры с высотой. → Рис. 341, с. 761
f troposphère *f*
s troposfera *f*
g Troposphäre *f*

* **tropospheric aerosol → 111**

6382 **troug; glaciated valley; vallée *f* en U; glacial valley; U-shaped valley; hglacial trough**
An valley that has been overdeepened and straightened by a glacier that once occupied it.
r трог; троговая долина; ледниковая долина; U-образная долина
Горная долина U-образной или корытообразной формы, углубленная и спрямленная занимавшим ее некогда ледником.
f auge *f* glaciaire; vallée *f* glaciaire
s artesa *f* glaciar; valle *m* en U; valle *m* glaciar
g Gletschertrog *m*; Trogtal *n*; Gletschertal *n*; U-artiges Tal *n*

6383 **trough end**
The precipitous rock wall which closes off the upper end of an overdeepend glacial valley and around which a group of cirques frequently occurs. → Fig. 342
r тыловая часть трога
Очень крутая скалистая стена, возвышающаяся в верхней части переуглубленной ледниковой долины, в которой нередко формируется несколько каров. → Рис. 342
f bord *m* d'auge glaciaire
s borde *m* de artesa glaciar
g Ende *n* des Gletschertrogs

* **trough lake → 2151**

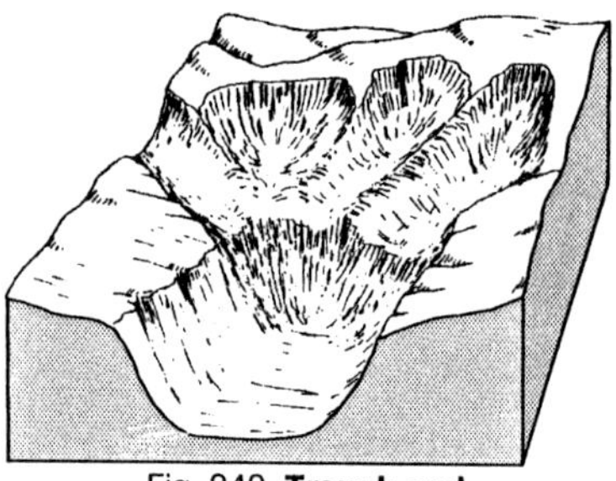
Fig. 342. **Trough end**

6384 **trough of wave**
The depression between any two crests of a regular wave motion.
r подошва волны; ложбина волны
Понижение между двумя последовательными гребнями волн при обычном волнении моря.
f creu *m* d'onde
s seno *m* de ola; hueco *m* de onda
g Wellenfuß *m*

6385 **trough shoulder; shoulder**
Escarp on a slope of the glacial valley where the steeper slope becomes more gently sloping one situated above the glaciation level, a valleyside bench above the level of glacial overdeepening.
r плечо трога
Уступ на склоне ледниковой долины, где более крутой склон переходит в более пологий, расположенный выше уровня оледенения, формируется в результате переуглубления долины ледником.
f épaulement *m* de l'auge; bord *m* d'auge glaciaire
s borde *m* de artesa; hombrera *f* de artesa glaciar
g Trogschulter *f*

6386 **trough valley of glaciated shelf**
Underwater trench or channel formed on the sea bottom by former glaciers.
r желоб-трог гляциального шельфа
Похожий на троговую долину подводный желоб континентального шельфа, испытавшего покровное оледенение.
f vallée *f* submergée d'érosion glaciaire
s valle *m* submarino de erosión glacial
g unterseeische Backen *fpl* von vergletscherten Schelfen

* **truck farming (*amer*) → 2824**

6387 **true dip**
The maximum inclination of a rock stratum, as distinct from *apparent dip.*
r действительное падение
Максимальное падение пласта горных пород, в отличие от *видимого падения.*
f pendage *m* vrai
s buzamiento *m* verdadero
g regelrechtes Einfallen *n*

6388 **true horizon**
The big circle of the celestial sphere formed when it is crossed by the plane passing through the eye of an observer perpendicular to the point of observation.
r истинный горизонт; астрономический горизонт; математический горизонт
Большой круг небесной сферы, образующийся при пересечении ее с плоскостью, проходящей через глаз наблюдателя перпендикулярно к отвесной линии в точке наблюдений.

f horizon *m* rationnel; horizon *m* astronomique
s horizonte *m* racional; horizonte *m* astronómico; horizonte *m* verdadero
g wahrer Horizont *m*; astronomischer Horizont *m*

6389 true north
The direction of geographic north from the observer, i.e. the direction along a meridian towards the North Pole.
r истинный север
Направление от наблюдателя на географический север, т.е. вдоль меридиана к Северному полюсу.
f nord *m* rationnel
s norte *m* racional
g wahrer Norden *m*

6390 true south
The direction of geographic south from the observer, i.e. the direction along a meridian towards the South Pole.
r истинный юг
Направление от наблюдателя на географический юг, т.е. вдоль меридиана к Южному полюсу.
f sud *m* rationnel
s sur *m* racional
g wahrer Süden *m*

6391 truncated cyclone
A cyclonic circulation existing in the upper troposphere which does not exist near the Earth's surface.
r высотный циклон
Холодный циклон, выраженный в средней и верхней тропосфере, но отсутствующий у земной поверхности.
f dépression *f* isolée
s depresión *f* aislada
g Höentief *n*

6392 truncated high
An anticyclonic circulation existing in the upper troposphere which does not exist near the Earth's surface.
r высотный антициклон
Антициклон, выраженный в средней и верхней тропосфере, но отсутствующий у земной поверхности.
f anticyclone *m* isolé
s anticiclón *m* aislado
g Höhenhoch *n*

6393 truncated soil
A soil subjected to erosion and has lost all or part of its soil horizons.
r смытая почва; усеченная почва; эродированная почва
Почва, подверженная эрозии, в результате которой полностью или частично уничтожена (смыта) часть почвенных горизонтов.
f sol *m* tronqué; sol *m* décapé
s suelo *m* truncado
g erodierter Boden *m*; ausgewaschener Boden *m*; abgekürzter Boden *m*

6394 truncated spur
A steep bluff on the side of a trough, protruding between tributary, possibly hanging, valleys, resulting of mainly vertical glacial erosion.
r усеченная шпора
Крутой обрыв на склоне ледниковой долины, торчащий между долинами-притоками, нередко висячими, выработанный в основном вертикальной ледниковой эрозией.
f éperon *m* tronqué
s espuela *f* truncada
g abgestumpfter Sporn *m*

6395 trunk glacier
The main stream of a large compound valley (dendritic) glacier.
r главный ледник
Основной, магистральный поток крупного горно-долинного (дендритового) ледника.
f glacier *m* de tronc
s glaciar *m* troncal
g Hauptgletscher *m*

*** trunk stream → 3643**

6396 trunk valley
The main valley occupied by the dominant stream of a river system (trunk stream) or the largest glacier of a valley glacier complex (trunk glacier).
r магистральная долина
Главная долина, занятая доминирующим потоком в речной системе или самым большим ледником, принимающим боковые притоки.
f vallée *f* principale
s valle *m* principal
g Haupttal *n*

*** trust territory → 6397**

6397 trusteeship territory; trust territory
A non-self-governing territory administered in accordance with a system of International Trusteeship set up after World War II on an instruction and in the name of the UN, e.g. the Marschall, Mariana and Caroline islands (U.S. trusteeship).
r подопечная территория
Территория, управляемая в соответствии с созданной после второй мировой войны системой международной опеки по поручению и от имени ООН, например, Маршалловы, Марианские и Каролинские острова (опека США).
f territoire *m* sous administration
s territorio *m* en fideicomiso
g Treuhandgebiet *n*

*** T-S analysis → 6174**

6398 tsunami

An ocean wave produced by a submarine earthquake, landslide or volcanic eruption, which may reach enormous dimensions and have sufficient energy to travel across entire ocean, it is capable of considerable destruction in certain coastal regions.

r цунами; морская сейсмическая волна
Океанская волна, вызванная подводным землетрясением или вулканическим извержением, способная достигать гигантских размеров, обладающая достаточной энергией для пересечения океана и вызывающая многочисленные разрушения в прибрежной зоне.

f tsunami *m*
s tsunami *m*
g Tsunami *n*; Erdbebenflutwelle *f*

* **tube → 6448**

* **tufa → 6333**

6399 tuff; volcanic tuff
A rock formed from compacted or cemented pyroclastic material (fine volcanic dust, ash, etc.) thrown out of a volcano in eruption.

r вулканический туф
Плотная горная порода, образованная из спресованного пирокластического материала, выброшенного при вулканическом извержении.

f tuf *m* volcanique
s tuf *m* volcánico; toba *f* volcánica; tuf *m*
g vulkanischer Tuff *m*; Vulkantuff *m*

6400 tuff-breccia
Volcanic rock formed by way of cementation of lava debris by volcanic sand and dust.

r туфобрекчия
Вулканическая горная порода, образованная путем цементирования обломков лавы и шлака среди вулканического песка и пыли.

f brèche *f* volcanique
s brecha *f* volcánica; brecha *f* ígnea
g Tuffbreccie *f*

6401 tufflava
Volcanic rock consisting of lava, ash, and glass fragments.

r туфолава
Вулканогенная горная порода, состоящая из лавы, пепла и обломков стекла.

f tufolave *f*
s tufalava *f*
g Tufflava *f*

6402 tugai
Leaf-bearing forests and bushes occurring on flood plains of the river valleys of Central Asia with insufficient moisture.

r тугаи
Пойменные лиственные леса и кустарники в долинах рек Центральной Азии, протекающих по территориям с недостаточным увлажнением.

Fig. 343. **Tundra** near the Polar Urals Mountains

f tougai *f*
s tugai *f*
g Tugai *mpl*

6403 tumulus; burial mound; barrow
A man-made watch or burial hill in the steppe zone of Eurasia 0.5-20 m high and 4-100 m in diameter.

r курган
Насыпной сторожевой или могильный холм в степной зоне Евразии высотой от 0,5 до 20 м, поперечником от 4 до 100 м.

f tumulus *m*; tertre *m* funéraire; kourgane *m*
s túmulo *m*; altosano *m*; montículo *m*
g Hügelgrab *m*; Hünengrab *m*

6404 tundra; tundra zone
A zonal type of treeless landscape, characterized by mosses, lichens, dwarf shrubs, and no trees, it is found in Arctic and sub-Arctic belts, mainly in the Northern Hemisphere between the polar region of perpetual snow and ice and the northern limit of taiga. → Fig. 343

r тундра; тундровая зона
Зональный тип ландшафта, в котором отсутствует лес и широко развиты моховолишайниковые и кустарниковые сообщества, распространен в арктическом и субарктическом поясах, главным образом в Северном полушарии между областью постоянных снегов и льдов и северной границей тайги. → Рис. 343

f toundra *f*; zone *f* de toundra
s tundra *f*; zona *f* de tundra
g Tundra *f*; Tundrazone *f*

6405 tundra fauna
The animals inhabiting tundra, e.g.reindeer, polar foxes, lemmings, white and rock ptarmigans, waterfowl (in summer).

r тундровая фауна
Совокупность видов животных, обитающих в тундре, характерны северный олень, песец, лемминги, белая и тундряная куропатки, летом – водоплавающие птицы.

f faune *f* de toundra
s fauna *f* de tundra
g Tundrafauna *f*

* **tundra flora → 6408**

6406 tundra gley soil
A soil of tundra zone that develops under tundra vegetation, the soil profile is constituted

by litter, thin humus horizon and mineral horizon gleyed to a different degree.

r тундровая глеевая почва
Почва тундровой зоны, формирующаяся под тундровой растительностью, с подстилкой, маломощным гумусовым горизонтом и минеральным горизонтом, оглеенным в разной степени.

f sol *m* à gley de toundra
s suelo *m* de gley de tundra
g Tundragleyboden *m*

6407 tundra soils
Soils of different types that develop under tundra vegetation, they usually have little development of horizons with a thick, peat layer of poorly decomposed vegetation, which is usually underlain by a frozen layer of soil.

r тундровые почвы
Почвы разных типов, формирующиеся под тундровой растительностью, обычно маломощны, с малым содержанием гумуса, часто с признаками мерзлотных явлений.

f sols *mpl* de toundra
s suelos *mpl* de tundra
g Tundraböden *mpl*

6408 tundra vegetation; tundra flora
The vegetation which is restricted by the intense winter cold, insufficient summer heat, and waterlogged soil, characteristically herbaceous perennials with scattered trees, mosses, lichens, and sedges.

r тундровая растительность; тундровая флора
Растительность, формирующаяся в условиях большого зимнего холода, недостатка летнего тепла и перенасыщенных влагой почв, характерны многолетние травянистые растения, отдельные деревья, мхи, лишайники и осоки.

f végétation *f* de toundra
s vegetación *f* de tundra
g Tundravegetation *f*

* **tundra zone → 6404**

6409 tunnel
A man-made passageway through an obstruction or underground used for transportation, water flow, underground communication lines, etc. (tunnels can be mountainous, underwater and urban ones).

r туннель
Искусственное подземное сооружение, иногда проложенное через препятствие, служащее для транспортных целей, перемещения воды, прокладки подземных коммуникаций и т.п. (различают горные, подводные и городские туннели).

f tunnel *m*
s túnel *m*
g Tunnel *m*; Stollen *m*

6410 tunnel valley
A valley which is formed by subglacial streams during the later stages of the Pleistocene glaciation, it has narrow flat floor and steep sides which rise up to 100 m from the valley bottom.

r туннельная долина
Долина, сформированная подледниковыми потоками в одну из последних стадий плейстценового оледенения, имеет узкое плоское дно и крутые склоны, поднимающиеся на 100 м над дном долины.

f vallée *f* de tunnel
s valle *m* de túnel
g Tunneltal *n*

6411 turbidite
A sediment deposited in water by a turbidity current, it contains the material of different size and roundness.

r турбидит
Отложения мутьевых потоков на дне морей и океанов, представленные осадками разной размерности и степени окатанности.

f turbidit *f*
s turbidita *f*
g Turbidit *m*

6412 turbidity
Cloudiness in the water or air caused by the disturbance which results in the holding in suspension of finely-divided particles.

r мутность
Помутнение в воде или воздухе, вызванное присутствием в них взвеси мельчайших частиц.

f turbidité *f*
s turbidez *f*
g Trübheit *f*

6413 turbidity current
A submarine flow of a dense mixture of water and sediment capable of moving at very high speeds, it can be trigged off by earthquakes or submarine slides on the continental slope and is capable of eroding channels on the ocean floor.

r мутьевой поток; суспензионный поток
Придонное течение плотной смеси морской воды и донных отложений, нередко достигающее большой скорости, оно вызывается землетрясениями или подводными оползнями на континентальном склоне и приводит к выработке каналов и ложбин на морском дне.

f courant *m* de turbidité
s corriente *f* de turbidez
g Dichtestrom *m*

6414 turbidity of water
Cloudiness of water caused by the holding in suspension of particles of different origin or the stirring of sediment by the movement of water.

r мутность воды

Непрозрачность воды из-за примесей взвешенных веществ разного происхождения или взмучивания осадков при движении воды.
f turbidité *f* de l'eau
s turbidez *f* de agua
g Wassertrübung *f*

6415 **turbosphere**
The region of the atmosphere in which turbulence frequently exists and the mixing of the atmosphere occurs.
r турбосфера
Нижняя часть атмосферы, в которой часто возникает турбулентность и перемешивание воздуха.
f turbosphère *f*
s turbosfera *f*
g Turbosphäre *f*

6416 **turbulent flow**
Eddy motion of water or air the particles of which follow independent, irregular paths so that mixing occurs, in contrast to laminar flow.
r турбулентное течение; турбулентность
Форма течения воды или воздуха, при которой их частицы совершают неупорядоченные движения по сложным траекториям, что приводит к интенсивному перемешиванию.
f courant *m* turbulent; coulement *m* turbulent; turbulence *f*
s corriente *m* turbulente; flujo *m* turbulento; turbulencia *f*
g Wirbelströmung *f*; Verwirbelung *f*; turbulenzströmung *f*; Turbulenz *f*; Böigkeit *f*

* **turf → 5574**

6417 **turf; peat**
Unconsolidated black or dark-brown soil material consisting largely of slightly decomposed or undecomposed fibrous vegetable matter that has accumulated in a waterlogged environment, it may be dried and used for fuel.
r торф; дёрн$_2$; торфяник
Неплотная черная и темнокоричневая масса, состоящая в основном из слегка разложившихся или вовсе неразложившихся волокнистых растительных остатков, накапливавшихся в насыщенной водой среде, высушенные блоки торфяника используются в качестве топлива.
f tourbe *f*
s turba *f*
g Torf *m*

6418 **turf cover**
The degree of density of a layer of grass and herbaceous plants which create a mat of vegetable material on the soil surface.
r задернованность
Степень сомкнутости травянистого покрова, образующего сплошной слой растительного материала на повехности почвы.
f gazonnement *m*
s cespedera *f*
g Verfilzung *f*; Vergrasung *f*

6419 **turf process**
The soil-forming process going on under herbaceous vegetation, but sometimes under forests, and resulting in the accumulation of humus in the upper soil horizons and formation of sod soils.
r дерновый процесс; задернение почвы
Почвообразовательный процесс под травянистой растительностью, но иногда и под лесами, ведущий к накоплению в верхних горизонтах почвы гумуса и формированию дерновых почв.
f processus *m* gazonné; gazonnement *m* du sol
s proceso *m* de césped
g Rasenprozeß *m*

* **turkestan-type glacier → 3943**

* **turmkarst → 6273**

* **turnover of goods → 883**

6420 **turnover water supply; water recycling**
The system of water reuse in technological cycles and domestic usage.
r оборотное водоснабжение
Система многократного использования воды в технологических циклах и для бытовых целей.
f alimentation *f* en eau de circulation
s abastecimiento de agua circulante; reutilización *f* del agua; reciclaje *m* del agua
g Wasserwiederverwertung *f*

6421 **turnpike; toll road**
A road that people have to pay to drive on.
r платная дорога
Дорога, за проезд по которой надо платить.
f route *f* à péage; barrière *f* de péage
s camino *m* de portazgo; barrera *f* de portazgo
g bezahlte Straße *f*

6422 **tussock**
An area of raised solid ground of height up to 1 m and diameter 20-50 cm in a bog that is formed by roots of grasses and sedges. → Fig. 344, p. 767
r кочка
Возвышение на болоте или на лугу высотой до 1 м с диаметром 20-50 см, формируемое плотнодерновинными злаками и осоками. → Рис. 344, с. 767
f motte *f*

Fig. 344. **Tussock**

s terrón *m*;
montecillo *m* de hierbas crecientes
g Grasbüschel *n*; Hügelchen *n*

6423 tussocky bog
Bog in the permafrost area with perennial peat mounds 2-3 m high and 10-20 m in diameter.
r кочковатое болото
Болото в области многолетнемерзлых пород с многолетними торфяными буграми высотой 2-3 м и диаметром 10-20 м.
f tourbière *f* mamelonnée;
tourbière *f* cordée
s turbera *f* acordonada
g Moor *n* mit Torfkugeln

6424 tuz
Salt lakes and solonchaks in Turkey, Middle and Central Asia.
r туз
Соленые озера и солончаки в Турции, Средней и Центральной Азии.
f tuz *m*
s tuz *m*
g Tus *n*

6425 twig-fall
The shedding of annual sprouts of bushes and trees, e.g. of saxaul, in hot summer or autumn period.
r веткопад
Опадение в жаркий период лета или осенью однолетних побегов у кустарников и деревьев, например у саксаула.
f chute *f* des branches
s caída *f* de las ramas
g Astfall *m*

6426 twilight
An intermediate period of illumination of the sky before sunrise and after sunset, there are three types of twilight: civil, nautical and astronomical.
r сумерки
Период освещенности неба непосредственно до восхода или после заката Солнца, существуют три разновидности сумерек – гражданские, морские и астрономические.
f crépuscule *m*
s crepúsculo *m*
g Dämmerung *f*

6427 twilight colours
The colourful glow which illuminates the sky before sunrise or after sunset, owing to the reflection of sunlight on to the Earth from the upper layers of the atmosphere where the Sun is below the horizon.
r заря
Разноцветное свечение неба перед восходом Солнца или после его захода, вызываемое отражением солнечных лучей от верхних слоев атмосферы, когда Солнце находится ниже горизонта.
f teintes *fpl* crépusculaires
s colores *mpl* crepusculares
g Dämmerungsfarben *fpl*

*** twinned glacier → 6294**

6428 two-field system of farming
The system of cultivation used in the past in which a half of the land was cultivated while the other one was left fallow for a year.
r двухпольная система земледелия
Применявшаяся в прошлом система, когда в один и тот же год половина земельного клина обрабатывалась, а другая половина оставалась под паром.
f assolement *m* biennal
s rotación *f* bienal
g zweifeldisches Landwirtschaftssystem *n*

*** two-year ice → 5304**

6429 Tyndall figures
Small water-filled cavities, often of hexagonal shape, which appear in the interior of ice masses upon which light is falling, their formation results from the melting of ice by radiative absorption at specifis points.
r цветы Тиндаля
Фигуры, как правило, гексагональные, возникающие в ледяных кристаллах при внутреннем плавлении за счет поглощаемой лучистой энергии.
f fleurs *fpl* de Tyndall
s flores *fpl* de Tyndall
g Tyndall-Figuren *fpl*

6430 type of landscape
A taxonomic unit in landscape classification which unites landscapes with similar zonal features.
r тип ландшафта
Одна из высших таксономических единиц в систематике ландшафтов, объединяющая группу однородных ландшафтов по зональным признакам.
f type *m* de paysage
s tipo *m* de paisaje
g Landschaftstyp *m*

6431 type of terrain
A general typological cocept not limited by any taxonomic order, e.g. mountain terrain.
r тип местности
Общее типологическое понятие, не ограниченное каким-либо таксономическим рангом, например, горная местность.
f type *m* de terrain
s tipo *m* de terreno
g Ortstyp *m*

6432 type of vegetation
A taxonomic category of vegetation cover that unites plant formations with the common requirements to environmental conditions and typical dominant species, e.g. steppe, meadow, forest, bog, etc.
r тип растительности
Таксономическая категория растительного покрова, объединяющая формации со сходными требованиями к условиям среды и характерными доминантными видами, например, степь, луг, лес, болото и т.п.
f type *m* de végétation
s tipo *m* de vegetación
g Vegetationstyp *m*

6433 typhoon
Tropical cyclone, a violent tropical revlving system of the China Sea and, more generally, of the north-west Pacific.
r тайфун
Тропический циклон Китайского моря и всей северо-восточной части Тихого океана.
f typhon *m*
s tifón *m*
g Taifun *n*

U

6434 ubac
The shaded side in a mountain terrain, in contrast to *adret.*
r теневой склон
Находящийся в тени, холодный склон холма или речной долины, в противоположность *солнечному склону.*
f ubac *m*
s umbría *f*
g Schattenseite *f*

6435 ubiquist; eurytopic organism; cosmopolite
Plant and animal species that live in different ecological conditions (e.g. bur reed grows from tropics to Arctic, in water bodies and on land), as opposed to *endemic species.*
r убиквист; эвритопобионт; космополит
Виды растений и животных, обитающие в разных условиях и имеющие обширные ареалы (например, тростник обыкновенный растет от тропиков до Арктики, в водоемах и на суше), противопоставляются *эндемичным видам.*
f ubiquiste *m*; cosmopolite *m*; organisme *m* eurytopique
s ubicuista *m*; organismo *m* euritópico
g Ubiquist *m*; eurytoper Organismus *m*; Kosmopolit *m*

* **ubiquitous materials → 6436**

6436 ubiquity; ubiquitous materials
Resources which are available everywhere such as air or some building materials (sand, gravel, crushed stone, etc.), and therefore does not influence the selection of location of the industry.
r убиквитет
Повсеместно распространенный ресурс, например, воздух или некоторые стройматериалы: песок, гравий, щебень и пр.
f ubiquité *f*
s ubiquidad *f*; recursos *mpl* ubicuos; ubicuidad *f*
g Ubiquität *f*

6437 ultisol
A deeply weathered and relatively infertile soil, lacking base minerals, having red and yellow clay constituents, associated with humid temperate to tropical climates.
r ултисоль
Глубоко выветрелая и относительно бесплодная почва, потерявшая основные минералы и сохраняющая красные и желтые глинистые элементы, встречается во влажных умеренных и тропических климатах.
f ultisol *m*
s ultisol *m*; ultisuelo *m*
g Ultisol *m*

6438 ultra-abyssal zone; hadal zone
The deepest part of the ocean (below 6,000-7,000 m), the area of deep oceanic trenches. → Fig. 244, p. 533
r ультраабиссаль; хадальная зона; ультраабиссальная зона
Наиболее глубокая часть океана (более 6000-7000 м), область океанических желобов. → Рис. 244, с. 533
f étage *m* ultra-abyssal; étage *m* hadal
s zona *f* ultra-abisal; piso *m* hadal; piso *m* ultra-abisal
g Ultraabyssal *n*

6439 ultraviolet radiation
Invisible electromagnetic radiation with very short wavelengths, plays an important role in some life processes, but its excess can cause diseases or be lethal to organic life.
r ультрафиолетовая радиация
Невидимое электромагнитное излучение с очень короткики волнами, играет важную роль в некоторых жизненных процессах, но его излишек может вызвать болезни и даже привести к смерти.
f rayonnement *m* ultraviolet
s radiación *f* ultravioleta
g ultraviolette Strahlung *f*; Ultraviolettstrahlung *f*

6440 umbrisol
An acid soil with thick, dark surface horizon, rich in organic matter, it is widespread in cold, humid, usually mountain regions.
r умбрисоль
Кислая почва, имеющая мощный, богатый органическим веществом, темный поверхностный горизонт, распространена в холодных, влажных, обычно горных районах.
f umbrisol *m*
s umbrisol *m*
g Ubrisol *m*

* **umland → 6503**

6441 unavailable water
Water that is present in the soil but which can not be taken up by plant roots because it is strongly adsorbed on to the surface of soil particles.
r недоступная влага
Влага, присутствующая в почве, которую не могут использовать корни растений, так как она сильно абсорбирована поверхностью почвенных частиц.
f eau *f* inaccessible
s agua *f* inaccesible
g unzugängliches Wasser *n*

6442 unconfind aquifer; free groundwaters

The waters on the land surface as well as underground gravity waters having a water table the pressure on which is equal to the atmospheric one.

r безнапорные воды
Воды в наземных водоемах и водотоках, а также подземные гравитационные воды, имеющие зеркало вод, давление на которое равно атмосферному.

f nappe *f* aquifère libre; nappe *f* phréatique
s capa *f* acuífera libre
g freies Grundwasser *n*

6443 unconfined avalanche; non-channeled avalanche
A snow avalanche which moves along the open even slope having no clearly defined channel.

r неканализованная лавина
Снежная лавина, которая движется по ровному склону, лишенному четко выраженного в рельефе русла.

f avalanche *f* de versant
s avalancha *f* sin confinar
g Flachenlawine *f*

* **unconformity → 1625**

6444 unconsolidated rock; loose rock
Easily scattering rock the particles or fragments of which are not tightly connected.

r рыхлая горная порода; несцементированная горная порода
Легко рассыпающаяся горная порода, частицы или обломки которой не связаны прочно между собой.

f roche *f* incoherente
s roca *f* incoherente; roca *f* no consolidada
g Lockergestein *n*

* **underchannel water → 6447**

6445 undercutting
The erosive action of a river or an ocean wave at the base of a cliff, i.e. the destruction of sea cliffs by waves and the concave banks of meandering stream by the current.

r подмыв
Разрушение морских обрывов и склонов, а также берегов реки, особенно на вогнутых берегах излучин, водными потоками и волнами.

f affouillement *m*
s socavación *f*
g Unterhöhlung *f*; Unterspülung *f*

6446 under-developed country
A country in which the existing resources has not been exploited, the industrial sector is small, GDP per capita is extremely low, etc.

r слаборазвитая страна
Страна, в которой слабо используются собственные ресурсы, хозяйство находится на низкой ступени развития, и поэтому население пребывает в бедности.

f pays *m* sous-développé
s país *m* subdesarrollo
g unterentwickeltes Land *n*

6447 underflow; underchannel water
Underground stream flowing in alluvial deposits of river channel.

r подрусловой поток; подрусловые воды
Сток воды в толще аллювиальных отложений, слагающих русло реки.

f infèro-flux *m*; sous-écoulement *m*
s sub-escorrentía *f*; infero-escorrentía *f*
g Unterströmung *f*; Tiefenstrom *m*

6448 underground; metro; tube; subway (*amer*)
An urban railway system in a city running underground and outside the streets.

r метро; метрополитен
Городская внеуличная железная дорога для массовых скоростных перевозок пассажиров.

f métro *m*; métropolitain *m*
s metro *m*; metropolitano *m*
g Untergrundbahn *f*; U-bahn *f*

* **underground lake → 921**

6449 underground stream
A water course flowing in large fissures, caves and other underground cavities, mainly in karst ones.

r подземная река
Река, протекающая в крупных трещинах, пещерах и других подземных пустотах, чаще всего карстовых.

f rivière *f* souterraine; courant *m* souterrain
s río *m* subterráneo; coriente *f* subterránea
g unterirdischer Fluß *m*

* **underground watershed → 4547**

6450 undergrowth; young growth; underwood
The low plants growing under the trees in a forest including herbs, saplings, shrubs and small trees.

r подлесок; подрост
Кустарники и невысокие деревья, растущие в лесу под пологом более высоких деревьев.

f sous-bois *m*
s sotobosque *m*
g Unterwuchs *m*; Unterholz *n*; Nachwuchs *m*; Jungwuchs *m*

6451 underlying surface
The earth surface interacting with the atmosphere in the process of heat- and water exchange.

r подстилающая поверхность; деятельная поверхность
Поверхность земли, взаимодействующая с атмосферой в процессе тепло- и влагообмена.

f superficie *f* inférieure
s superficie *f* inferior
g Unterlage *f*

6452 underpopulation
The population in an area where the available resources are not used so fully as they might be and the economic potential is not developed fully because there arfe too few people living there.
r недостаточное население
Население на территории, где доступные ресурсы и экономический потенциал не используются полностью из-за нехватки проживающих здесь людей.
f population *f* insuffisante; sous-peuplement *m*
s población *f* insuficiente; infrapoblación *f*
g Bevölkerungsmangel *m*

* **undersea vehicle → 5924**

6453 undershrub
A perennial plant with woody lower parts of sprouts and herbaceous top parts which die off during the unfavourable period.
r полукустарник
Многолетнее растение, у которого только нижние части побегов древеснеют и перезимовывают, а верхние остаются травянистыми и ежегодно отмирают в неблагоприятный для роста период.
f buisson *m*; buisson *m* nain; buisson *m* bas
s sub-arbusto *m*; planta *f* subfrutescente
g Halbstrauch *m*

6454 underthrust
The shift ot the blocks of a rockalong a fracture which is morphologically similar to an overthrust, but which resulted from the active movement of the downthrough block.
r поддвиг
Смещение блоков гоорной породы по разлому, морфологически подобное надвигу, но образовавшееся при активном движении не висячего, а нижележащего крыла складки.
f faille *f* inverse contraire
s falla *f* inversa contraria
g Unterschiebung *f*

* **underwater vehicle → 5924**

* **underwater volcano → 5917**

* **underwood → 6450**

6455 unemployment; unemployment level; unemployment rate
The proportion of the total population of working age who are not engaged in a paid work in the national economy.
r безработица; уровень безработицы
Доля населения в трудоспособном возрасте, не занятого в народном хозяйстве.
f chômage *m*; taux *m* de chômage
s desocupación *f*; paro *m*; desempleo *m*; tasa *f* de paro
g Arbeitslosigkeit *f*

* **unemployment level → 6455**

* **unemployment rate → 6455**

6456 unequilibrium models
The reconstruction of any natural conditions the parameters of which at a certain moment are not in complete correspondence with the conditions determining the development of these phenomena in the past due to the inertia of the system.
r неравновесные модели
Воспроизведение тех или иных природных условий, значение параметров которых в данный момент, в силу инерционности системы, не достигло полного соответствия с условиями, определявшими развитие этих явлений в прошлом.
f modèles *mpl* inéquilibrés
s modelos *mpl* inequiponderantes
g Ungleichgewichtsmodelle *npl*

6457 uneven development
The condition of the development of economic and social systems of the regions, when some of their parts progress and prosper while others are static or in relative decline.
r неравномерное развитие
Особенности экономического и социального развития отдельных регионов, когда некоторые их части прогрессируют, а другие стагнируют или даже откатываются назад.
f développement *m* inégal
s desarrollo *m* irregular; desarrollo *m* desigual
g ungleichmäßige Entwicklung *f*

6458 unfrozen ground; non-frozen ground
The ground in areas where temperature can potentially be negative but at the moment contains water in the liquid state.
r немерзлый грунт
Грунт в областях, где температура опускается ниже 0°C, в данный момент содержащий влагу в жидком состоянии.
f sol *m* non gelé
s suelo *m* no congelado
g ungefrorener Boden *m*

6459 uniclinal shifting
The process of asymmetrical development of a valley due to a stream which, following the line of the geological strike, tends to cut sideways in the direction of the dip, i.e. the valley is gradually eroded and moved laterally down the dip.
r моноклинальное смещение
Процесс асимметричного развития долины благодаря речному потоку, который, следуя простиранию геологических плас-

тов, размывает их в направлении падения, т.е. долина в результате эрозии постепенно сдвигается в направлении падения пластов.

f déplacement *m* uniclinal
s desplazamiento *m* uniclinal
g uniclinale Verschiebung *f*

6460 uniform steady flow
The state of stream-flow when its discharge, width, cross-section area and velocity are constant at every section in a channel reach.
r равномерное течение
Состояние речного потока, когда его расход, ширина, поперечное сечение и скорость течения постоянны в каждом сечении русла.
f flux *m* uniforme
s flujo *m* uniforme
g gleichmäßiger Strom *m*

6461 uniformitarianism
The view that processes currently acting upon the Earth are virtually the same as those which have been acting throughout geological time, hence the interpretations of Earth's history can be based on the present-day evidence of natural processes, which is wrong.
r униформизм
Концепция, согласно которой в геологическом прошлом действовали те же силы и с такой же интенсивностью, что и в настоящее время, откуда делался вывод, что характеристики современных геологических явлений можно распространять в прошлое без изменений, что ошибочно.
f uniformitarianisme *f*
s uniformitarianismo *m*
g Prinzip *n* des Uniformismus

6462 uninverted relief
A landscape where the surface relief reflects the underlying geological structure, where hill ridges coincide with anticlines and valleys with synclines, the opposite of *inverted relief.*
r необращенный рельеф
Ландшафт, в котором рельеф поверхности отражает нижележащие геологические структуры: гряды холмов совпадают с антиклиналями, а речные долины с синклиналями, противоположен *обращенному рельефу.*
f relief *m* non inverté
s relieve *m* no invertido
g nicht umgekehrtes Relief *n*

6463 union
A form of unification of several monarchies under one sovereign, e.g. a union between Sweden and Norway in 1814-1905.
r уния
Форма объединения монархических государств под короной одного суверена, например, уния Швеции и Норвегии в 1814-1905 гг.
f union *f*
s unión *f*
g Union *f*

* **unipolar world → 3924**

6464 unitary state
A state with no federal units (states, lands, etc) in which authoritative control is vested in a central or national government which delegates power to local government authorities, such as departments, regions, districts (France or Poland can serve as examples).
r унитарное государство
Государство, на территории которого нет федеративных единиц (штатов, земель и пр.), но часть властных полномочий передана административно-территориальным единицам: департаментам, областям, районам, например Франция или Польша.
f état *m* unitaire; état *m* centralisé
s estado *m* unitario; estado *m* centralizado
g Einheitsstaat *m*

6465 universal time
The time scale (symbol: UT) used, regardless of location, to define time on the Earth, it is essentially the same as Greenwich mean time and measured now by the atomic clock.
r всемирное время
Шкала времени, используемая для Земли в целом, мало отличающаяся от среднего Гринвичского времени и в настоящее время базирующаяся на атомных часах.
f temps *m* universel
s tiempo *m* universal
g Universalzeit *f*

6466 universe; cosmos; space$_2$
The whole existing material world infinite in time and space and infinitely diverse in forms which substance takes in the process of its development, that which is beyond the limit of the Earth's atmosphere.
r Вселенная; космос
Весь существующий материальный мир, безграничный во времени и пространстве и бесконечно разнообразный по формам, которые принимает материя в процессе своего развития, пространство за пределами земной атмосферы.
f univers *m*; cosmos *m*
s universo *m*; cosmos *m*; espacio *m* cósmico
g Universum *m*; All *n*; Weltraum *m*

* **unpaved road → 1769**

6467 unsaturated air
Humid air whose vapour pressure is smaller than the saturation vapour pressure at the same temperature.
r ненасыщенный воздух
Влажный воздух, в котором упругость водяного пара меньше упругости насыщения при данной температуре.

f air *m* insaturé
s aire *m* insaturado
g ungesättigte Luft *f*

6468 unsorted polygon
Cryogenic soil pattern with mainly polygonal structures but without any stone borders.
r несортированный полигон
Криогенный полигон с преимущественно полигональными ячейками, без каменных бордюров.
f sols *mpl* polygonaux non structurés
s polígono *m* no estructurado
g unsortierter Polygonboden *m*

6469 unstable air mass
An air mass with a high water vapour content, liable to spontaneous convectional activity which results in heavy showers and thunderstorms.
r неустойчивая воздушная масса
Воздушная масса с большим содержанием водяного пара, подверженная стихийной конвекции, что вызывает сильные ливни и грозы.
f masse *f* d'air instable
s masa *f* del aire inestable
g labile Luftmasse *f*

6470 unsteady flow
The state of flow of a river or canal when it is affected by a flood wave or surge.
r неравномерное течение
Состояние течения реки, когда на него воздействует волна паводка.
f courant *m* irrégulier
s corriente *f* irregular
g ungleichmäßige Strömung *f*

6471 unstratified *adj*
Arranged or formed not in layers.
r нестратифицированный
Отложенный не в форме напластования, т.е. с отсутствием ясно выраженных слоев.
f non-stratifié *adj*
s noestratificado *adj*
g ungeschichtet *adj*

6472 updating of map
The process of bringing the information on a map up to date.
r обновление карты
Приведение содержания карты в соответствие с современным состоянием картографируемой территории.
f tenue *f* à jour d'une carte; révision *f* d'une carte
s actualización *f* de mapa; revisión *f* del mapa
g Laufendhaltung *f*

6473 updraught
A forced air ascending when the air mass approaches mountains, resulting in the increase of precipitation at windward side.
r предвосхождение
Вынужденный подъем воздуха при приближении воздушных масс к горам, приводящий к увеличению осадков с наветренной стороны.
f ascendance *f*; courant *m* ascendant
s corriente *f* ascedente
g Luftaufstieg *m*; Aufwind *m*

6474 up-floating of glacier
The detachment of a glacier from its bed while it advances into the sea or lake forced by the difference in the specific weights of water and ice.
r всплывание ледника
Отрыв ледника от ложа при его продвижении в море или озеро вследствие разности удельного веса воды и льда.
f remontée *f* de glacier
s remonte *m* de glaciar
g Aufschwimmen *n* von Gletscher

* **upfreezing → 2332**

6475 upland
A part of a plain that is raised above the surrounding terrain.
r возвышенность
Участок равнины, приподнятый относительно окружающей территории.
f haute plaine *f*
s llanura *f* alta; país *m* elevado
g Hochland *n*

6476 upper atmosphere
The atmosphere below the exosphere and above the troposphere.
r верхняя атмосфера; высокие слои атмосферы
Слой атмосферы, расположенный между тропосферой и экзосферой.
f atmosphère *f* d'en haut
s alta atmósfera *f*
g hohe Atmosphäre *f*

6477 upper bedding plane of stratum
Stratigraphically the upper surface limiting a stratum as well as a rock overlying the stratum.
r кровля пласта
Стратиграфически верхняя поверхность, ограничивающая пласт, а также горная порода, лежащая непосредственно над пластом.
f toit *m* de couche
s techo *m* de capa
g Flözdach *n*

* **upper course → 2722**

6478 upper front
The atmospheric front which exists at higher levels but does not reach the Earth's surface.
r верхний фронт
Атмосферный фронт, существующий на верхних уровнях, но не достигающий по-

верхности Земли.
f front *m* supérieur
s frente *m* superior
g Höhenfront *f*

6479 upper frontal zone
Frontal zone in middle or upper troposphere separating a high warm anticyclone and high cold cyclone.
r высотная фронтальная зона
Фронтальная зона в средней или верхней тропосфере, разделяющая высокий теплый антициклон и высокий холодный циклон.
f zone *f* frontale supérieure
s zona *f* frontal superior
g Höhenfrontalzone *f*

6480 upper pool
Part of a river or canal up-stream of a backing construction (dam, lock).
r верхний бьеф
Участок реки или канала выше подпорного сооружения (плотины, шлюза).
f bief *m* d'amont
s aguas *fpl* arriba
g Oberhaltung *f*

6481 upper snow drifting
A snow storm carrying snow before it touches the earthrs surface.
r верховая метель
Перенос ветром снежинок, еще не касавшихся земной поверхности.
f chasse-neige *f* élevée
s ventisca *f* alta
g hohe Schneedrift *f*

6482 upper water mass
Quasi-homogeneous layer in an ocean or sea formed by the destruction of surface waves, wind induced mixing and vertical density convection.
r поверхностный слой океана
Слой морской воды, порождаемый перемешиванием воды при разрушении поверхностных волн, ветровым перемешиванием и вертикальной плотностной конвекцией.
f couche *f* océanique supérieur
s capa *f* oceánica superior
g Oberflächenschicht *f* des Ozeans

* **upper-air anticyclone → 6483**

* **upper-air cyclone → 6484**

6483 upper-level anticyclone; upper-air anticyclone
Anticyclone keeping the system of closed isobars in high layers of the atmosphere, at least in the middle part of the troposphere.
r высокий антициклон
Антициклон, сохраняющий систему замкнутых изобар в достаточно высоких слоях атмосферы, по крайней мере в средней части тропосферы.
f anticyclone *m* en altitude
s anticiclón *m* en altos niveles; anticiclón *m* en altitud; alto anticiclón *m*
g hochreichende Antizyklone *f*; Höhenantizyklone *f*

6484 upper-level cyclone; upper-air cyclone
A cyclone spreading up to high-altitude levels in the troposphere, sometimes it penetrates through the whole tropospere.
r высокий циклон
Циклон, сохраняющий систему замкнутых изобар во всей толще тропосферы, а иногда и в нижней стратосфере.
f cyclone *m* en altitude
s ciclón *m* en altos niveles; ciclón *m* en altitud; alto ciclón *m*
g hochreichendes Tief *n*; Höhenzyklone *f*

* **uprush → 6488**

6485 upscaling
The way of generalization of meteorological parameters on a gross grid on the basis of their values shown on a detailed grid by using certain rules and taking into account the specific features of the region.
r апскейлинг
Процедура обобщения метеорологических параметров на грубой сетке по данным на мелкой сетке в соответствии с определенными правилами и учетом природных условий данного региона.
f upscaling *m*
s upscaling *m*
g Upscaling *m*

6486 upset of environmental balance; ecological imbalance
An imbalanced development of natural processes induced naturally or by human activity that destroys the environmental balance of an ecosystem.
r нарушение экологического равновесия; экологический дисбаланс
Естественный или вызванный хозяйственной деятельностью несбалансированный ход природных процессов, ведущий к нарушениям устойчивого состояния окружающей среды (геосистем).
f déséquilibre *m* écologique
s desequilibrio *m* ecológico
g ökologisches Ungleichgewicht *n*

6487 upstream *adv adj*
In the direction opposite to the flow of a stream and towards the riverrs source from the point further down the river.
r вверх по течению
Указывает направление к истоку реки или против течения реки.
f amont *adv*
s aguasarriba *adv*; contra la corriente

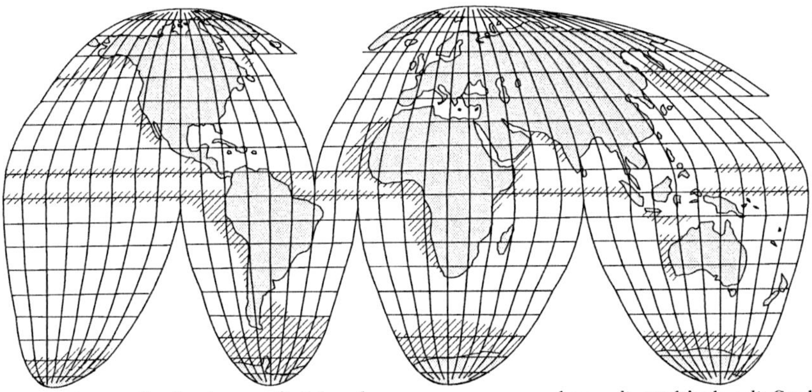

Fig. 345. Zones of **upwelling** und high productivity

g stromauf *adv*; stromaufwärts *adv*

6488 upwash; uprush
Water flow formed as swash and backwash as a result of the complete destruction of waves.
r прибойный поток
Поток воды, образующийся в результате полного разрушения волны, состоящий из наката и отката.
f jet *m* de rive
s flujo *m* de oleaje
g Brandungsstrom *m*

6489 upwelling
The rise of colder water from the lower layers of the ocean upwards to the higher layers, which is typical for the zones of divergence and giving rise to cold currents, rich in nutrients. → Fig. 345
r апвеллинг
Подъем глубинных холодных морских вод на поверхность, вызываемый в прибрежных районах сгонными ветрами, а в открытом океане происходящий в зонах дивергенциии и порождающий холодные течения, богатые питательными веществами. → Рис. 345
f upwelling *m*; remontée *f* d'eau
s upwelling *m*; subida *f* de agua
g Upwelling *n*

*** ural glacier → 4117**

6490 urban *adj*
Related to a city or town, to a place consisting of a compact agglomeration of dwellings, shops, offices, public buildings etc., usually with paved roads, street light, public services, and a population engaged in non-agricultural occupations, the opposite of rural.
r городской; урбанизированный
Относящийся к территории, застроенной жилыми кварталами с магазинами, офисами, общественными зданиями, мощеными дорогами, уличным освещением, системой обслуживания населения, которое не занято в сельском хозяйстве.
f urbain *adj*
s urbano *adj*
g urban *adj*; städtisch *adj*; Stadt-

6491 urban agglomeration
The concentration or merging of urban settlements, united by close economic, labour, cultural and everyday ties.
r городская агломерация; агломерация населенных пунктов
Скопление, а местами срастание населенных пунктов, объединенных тесными хозяйственными, трудовыми и культурно-бытовыми связями.
f agglomération *f* urbaine
s aglomeración *f* urbana
g Stadtagglomeration *f*

6492 urban area; urbanized territory
A densely built-up area in a city or in the urban agglomeration.
r урбанизированная территория; урбанизированный ареал
Зона сплошной застройки в городе или городской агломерации.
f territoire *m* urbanisé; espace *m* urbain
s territorio *m* urbanizado; espacio *m* urbano; área *f* urbanizada
g urbanisierte Gebiet *n*; Verstädterungsgebiet *n*

6493 urban climate
The local climate of a build-up area forming under the impact of urban constructions, industry, transport, etc.
r городской климат
Местный климат, формирующийся под влиянием городской застройки, промышленности, транспорта и т.п.
f climat *m* urbain
s clima *m* urbano
g urbanes Klima *n*

6494 urban containment
The policy of limiting urban sprawl by restricting out-of-town development.
r сдерживание роста города
Политика ограничения расползания города путем сдерживания развития на пери-

фории городской агломерации.
f rétention *f* urbaine
s detención *f* urbana
g urbane Hemmung *f*

* **urban district → 1648**

6495 **urban ecology**
The study of urban areas as geosystems with the purpose of rational land-use planning of the city.
r экология города
Изучение урбанизированных территорий как геосистем с целью рационального планирования землепользования в городе.
f écologie *f* urbaine
s ecología *f* urbana
g Stadtökologie *f*

* **urban field → 6503**

6496 **urban fringe**
The area of social change around a town or metropolitan area, where urban development impinges on agricultural land, population density increases and land values rise.
r городское окаймление
Территория вокруг города, на которую распространяются социальные изменения, где урбанизация захватывает сельскохозяйственные земли, плотность населения растет и ценность земли увеличивается.
f frange *f* urbaine
s franja *f* urbana
g Stadtrand *m*

6497 **urban geography**
The study of morphology, evolution, spatial patterns, classification of towns and other urban settlements, their relations with the natural environment and human society.
r география городов; геоурбанистика
Изучение морфологии, эволюции, пространственных черт, классификации городских поселений и их связей с окружающей средой и обществом.
f géographie *f* urbaine
s geografía *f* urbana
g Stadtgeographie *f*

* **urban growth → 6506**

6498 **urban landscape**
An urbanized geographical landscape, the structure and functioning of which are connected with city development. → Fig. 346
r городской ландшафт
Урбанизированный географический ландшафт, в структуре и функционировании которого ведущее место занимают процессы, связанные с развитием города. → Рис. 346
f paysage *m* urbain
s paisaje *m* urbano
g Stadtlandschaft *f*

6499 **urban macrocephaly**
An exessive development of the main city of a country or a region, e.g. Paris or Moscow.
r макроцефалия
Гипертрофированное развитие главного города страны или данного региона, например, Париж или Москва.
f macrocéphalie *f* urbaine
s macrocefalia *f* urbana
g Makrozephalie *f*

6500 **urban morphology**
The origin, growth, form, plan, structure, functions and development of a city and the

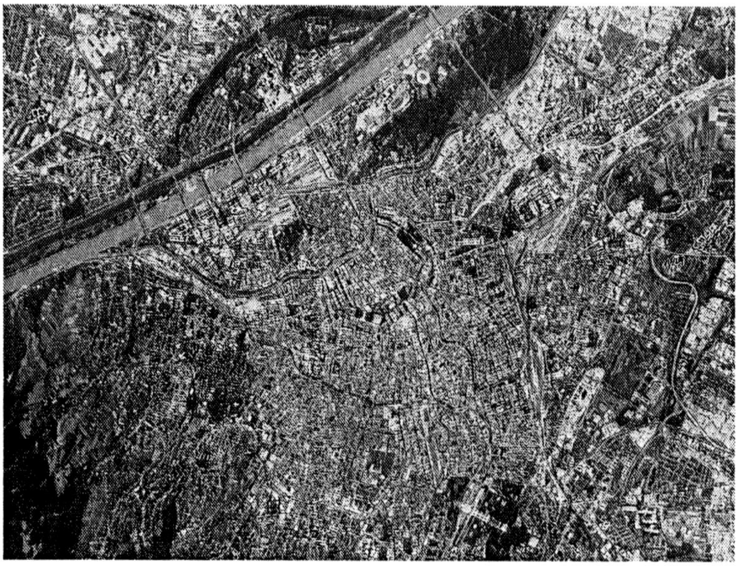

Fig. 346. **Urban landscape**. Central part of Vienna

urban habitat, and the study of these features, including their development over time.

r морфология города
Структура, планировка, функции, развитие города и городских поселений, а также их изучение, включая изменения во времени.

f morphologie *f* urbaine
s morfología *f* urbana
g Stadtmorphologie *f*

6501 urban planning; town planning
An attempt to manage the city in a rational way in order to avoid or alleviate common urban problems, such as inner city decay, overcrowding, traffic congestion, etc.

r городская планировка
Упорядочение пространства в городах с тем, чтобы предотвратить или смягчить проблемы урбанизации, такие как избыток шума, скученность, транспортные проблемы и пр.

f planification *m* urbaine; aménagement *m* urbain
s planificación *f* urbanística; planeamiento *m* urbanístico
g Stadtplanung *f*

6502 urban population
People living in cities and towns.

r городское население
Население, проживающее в городах.

f population *m* urbaine
s población *f* urbana
g Stadtbevölkerung *f*

6503 urban region; urban field; umland
The sphere of influence of a town, the territory around a town with which it is functionally linked.

r урбанизированный регион
Сфера влияния города, территория вокруг города, с которой он функционально связан.

f région *f* urbaine
s región *f* urbana
g urbane Region *f*

6504 urban renewal; redevelopment
A process in which the obsolete fabric of an urban area is restored, renovated and improved in order to meet contemporary needs or standards.

r городское обновление
Процесс, при котором устаревшее устройство урбанизированной территории реставрируется, обновляется и улучшается в целях достижения современных потребностей и стандартов.

f rénovation *f* urbaine
s renovación *f* urbana
g Stadterneuerung *f*

6505 urban settlement
A relatively densely built-up area with its associated open spaces where the population is engaged mainly in non-agricultural occupations.

r городское поселение
Сравнительно населенное место, выполняющее преимущественно несельскохозяйственные функции.

f habitat *m* urbain; peuplement *m* urbain
s asentamiento *m* urbano; hábitat *m* urbano; poblamiento *m* urbano; pueblo *m* urbano
g Stadtsiedlung *f*

6506 urban sprawl; urban growth; city expansion
The process of fast irregular and unplanned spread of large cities as more housing and shopping centres are built around them and city transportation means develop, resulting in the extension of the city into the countryside.

r расползание городов
Процесс быстрого разрастания территории крупных городов в ходе строительства жилья и торговых центров и развития сети транспортных магистралей.

f étalement *m* urbain; croissance *f* urbaine
s derramamiento *m* urbano; derrame *m* urbano; expansión *f* urbana; crecimiento *m* urbano
g Erweiterung *f* der Städte; Zersiedelung *f*

6507 urban system
Any network of towns and cities, and their hinterlands, which can be seen as a system, since it depends on the movements of labour, goods, services and capital through the network.

r урбанизированная система
Любая сеть городов и городских поселений, представляющая собой единую систему, в которой происходит постоянное передвижение рабочей силы, товаров, услуг и капитала, что сохраняет ее единство.

f système *m* urbain
s sistema *m* urbano
g urbanes System *n*

6508 urban transport
All types of transport moving passengers and goods in a city and its suburbs and within the whole urban agglomeration.

r городской транспорт
Разные виды транспорта, перевозящие пассажиров и грузы на территории города и ближайших пригородов в пределах городской агломерации.

f transport *m* urbain
s transporte *m* urbano
g Stadttransport *m*

6509 urban village
A residential area of the city containing a cluster of individuals of similar culture and interests.

r городская деревня
Жилой район города, где проживает груп-

па горожан с близкими культурными потребностями и интересами.
f village *m* urbain
s aldea *f* urbana
g urbanes Dorf *n*

6510 urban-industrial agglomeration
An urban agglomeration in which industrial enterprises play the main town-forming function.
r промышленная агломерация
Городская агломерация, в которой промышленное производство выполняет основную градообразующую функцию.
f agglomération *f* industrielle et urbaine
s aglomeración *f* industrial y urbana
g industrieller Verdichtungsraum *m*

6511 urbanism
A set of features typical of a city or a city dweller, and a way of life associated with urban dwelling.
r урбанизм
Совокупность признаков, свойственных городу или его жителю, и образ жизни, связанный с проживанием в гороode.
f urbanisme *m*
s urbanismo *m*
g Urbanität *f*; Urbanismus *m*

6512 urbanization
A historical process of the growth of the role of towns and cities in social development which affect the population migration patterns, way of life, culture, etc., this process includes the rise of a share of urban population and the increase of amount of cities and towns.
r урбанизация
Исторический процесс повышения роли городов в развитии общества, охватывающий изменения в расселении людей, их жизни, культуре и пр., этот процесс включает увеличение доли городского населения и рост общего количества городов.
f urbanisation *f*
s urbanización *f*
g Verstädterung *f*; Urbanisation *f*

*** urbanized territory → 6492**

*** urban-rural continuum → 5160**

6513 urman
The local name of dense taiga forests dominated by fir, spruce and Siberian cedar that is characteristic of Western and Central Siberia.
r урман
Густые таежные леса с преобладанием пихты, ели и кедра сибирского в Западной и Средней Сибири.
f ourmane *f*
s urman *m*
g Urman *m*

6514 uryoma
Moisture-loving forest vegetation characteristic of river flood-plains in the steppe and forest-steppe zones consisting of poplars, willow thickets, and scrubs.
r урёма

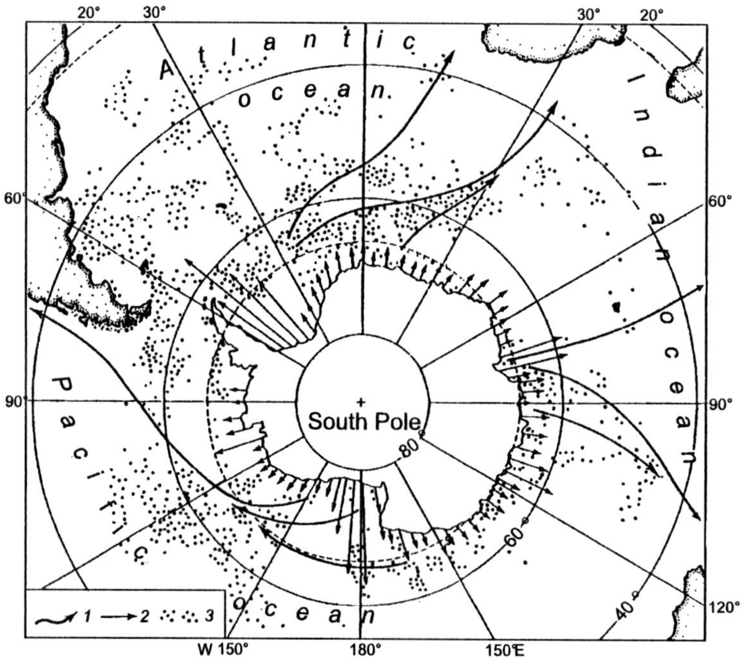

Fig. 347. **Utilization of icebergs**: 1 – routs of the iceberg transportation, 2 – the mass of icebergs, annually calving from the coast line of 200-km length, 3 – places where icebergs have been observed

Влаголюбивая, преимущественно лесная растительность, развитая на речных поймах в степной и лесостепной зонах, характерны тополя, заросли ив, кустарников.
f ouréma *f*
s urema *f*
g Urema *f*

6515 usable ground water resources
The discharge of underground waters of a given quality which it is possible to provide for water consumption during a long time.
r эксплуатационные ресурсы подземных вод
Расход подземных вод, который можно обеспечить для водопотребления в течение длительного времени при соблюдении заданного качества воды.
f ressource *f* de explotation d'eau souterraine
s recursos *mpl* de explotación de agua subterránea
g nutzbare Grundwasserressourcen *mpl*

* **U-shaped valley → 6382**

6516 utilization of icebergs
The use of icebergs as a source of fresh water. → Fig. 347, p. 778
r утилизация айсбергов
Использование айсбергов для получения пресной воды. → Рис. 347, с. 778
f utilisation *f* d'icebergs
s utilización *f* de icebergs
g Ausnutzung *f* der Eisberge

6517 uvala; karst basin
A large trough-like depression formed by the merging of dolines, usually stretched along the strike of the beds or along a crack.
r увала; карстовая котловина
Возникающее при слиянии карстовых воронок мульдообразное понижение, замкнутое на обоих концах и обычно вытянутое по простиранию пластов или вдоль трещин.
f sotch *m*; ouvala *f*; dépression *f* karstique
s uvala *f*; sotch *m*; depresión *f* cársica
g Karstmulde *f*; Uvala *f*

6518 uzboi
A dry channel with episodic storm runoff or without runoff, or a relict river valley in the desert areas of Central Asia.
r узбой
Сухое русло с эпизодическим ливневым стоком или вообще не имеющее стока, а также реликтовая речная долина в пустынных областях Средней Азии.
f uzboi *f*

V

6519 vadose water

Underground waters of atmospheric origin which move freely under the influence of gravity above the water table in permeable rocks, in contrast with *phreatic water.*

r водозные воды

Подземные воды атмосферного происхождения, свободно движущиеся под влиянием силы тяжести над водным зеркалом в водопроницаемых горных породах, в противоположность *фреатическим водам.*

f eau *f* vadeuse; eau *f* vadose

s agua *f* vadosa

g vadoses Wasser *n*

* **vallée *f* en U → 6382**

6520 valley; river valley

A linear depression sloping down from the headwaters to the lower reaches created by flivial erosion, usually (but not always) occupied by a river. → Fig. 348

r долина; речная долина

Линейно вытянутое углубление рельефа с уклоном от верховья к низовью, сформированное в результате эрозионной деятельности реки, в настоящее время, как правило (но не всегда), занятая текущей рекой. → Рис. 348

f vallée *f*

s valle *m*

g Tal *n*; Flußtal *n*

* **valley bog → 514**

* **valley breeze → 3966**

6521 valley floor; floor of a valley

The relatively flat bottom part of a valley occupied by the floodplain and river channel.

r дно долины

Относительно ровная нижняя часть долины, где находятся пойма и русло реки.

f fond *m* de vallée

s fondo *m* de valle

g Talsohle *f*

6522 valley glacier

A glacier filling partly or completely a mountain valley, by which the form and flow direction of the glacier are conditioned.

r долинный ледник

Ледник, частично или полностью занимающий горную долину, которая определяет его форму и направление движения.

f glacier *m* de vallée

s glaciar *m* de valle

g Talgletscher *m*

6523 valley train

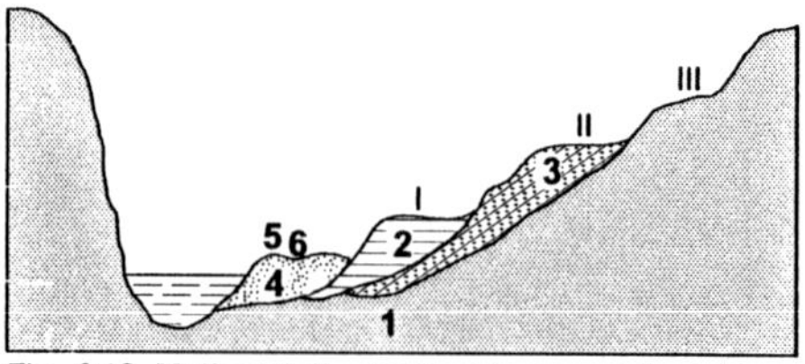

Fig. 348. **Valley**: I, II – first and second fluvial terraces above flood-plane, III – third erosion terrace; 1 – bedrock, 2, 3 – alluvium of the first and second fluvial terraces, 4 – alluvium of a flood-plane, 5 – natural levee, 6 – flood-plane deposits

A plain within a valley sloping down and away from the site of a glacier snout, composed of sands and gravels and containing pebbles and boulders.

r долинный шлейф

Плоские расширения в долине, наклоненные вниз по ее течению и расположенные на значительном расстоянии от конца ледника, сложенные флювиогляциальным материалом: песком, гравием с включениями валунов.

f traîne *f* de vallée

s cola *f* de valle

g Talschleppe *f*

* **valley wind → 3966**

6524 valley-and-gully terrain

The erosion relief formed by river valleys with wide gentle slopes and rounded interfluves.

r долинно-балочный рельеф

Эрозионный рельеф, образованный сочетанием речных долин с широкими пологими задернованными склонами и нерезкими междуречиями.

f relief *m* des vallées

s relieve *m* de valles

g Tälerrelief *n*

6525 value

The monetary equivalent of a product or a factor of production which satisfies three criteria, i.e. it is capable of being owned, it has utility (satisfies needs or desires) and is in limited supply.

r стоимость; ценность

Денежный эквивалент произведенной продукции, удовлетворающей трем критериям: принадлежать кому-то, быть полезной, т.е. удовлетворять чьи-то потребности, и иметь ограниченные запасы.

f valeur *f*

s valor *m*

g Wert *m*

6526 value added

The value of the gross product minus the amount by which resources, machinery, etc. lose their value.

r добавленная стоимость;
чистая продукция

Стоимость валовой продукции за вычетом потребленных материалов и износа основных фондов.
f valeur *f* ajoutée
s valor *m* añadido
g Nettoproduktion *f*

* **vapour concentration → 17**

6527 vapour pressure
The pressure exerted by the water vapour in the atmosphere.
r давление водяного пара
Давление, вызываемое наличием водяного пара в атмосфере.
f pression *f* de vapeur
s presión *f* de vapor
g Dampfdruck *m*

6528 variability
The property of a physical characteristic or phenomenon to deviate from its average value.
r изменчивость
Свойство физической величины или явления (например, климата) отклоняться от среднего состояния.
f variabilité *f*
s variabilidad *f*
g Variabilität *f*; Veränderlichkeit *f*

6529 variable cloudness
The clouds which change the sky coverage during a certain period of time.
r переменная облачность
Облачность со значительными изменениями количества баллов в течение некоторого промежутка времени.
f nébulosité *f* variable
s nubosidad *f* variable
g wechselnde Bewölkung *f*

* **Variscian folding → 2755**

* **varve → 6530**

6530 varved clay; varve
A distinct two-layered sediment of clay deposited annually in a lake, it consists of the thin, very fine-grained layer of winter and thicker, coarse-grained layer of summer.
r ленточные глины; варв
Годичные ленты глинистых отложений озер, состоящие из двух слоев – тонкого и мелкозернистого зимнего и более толстого крупнозернистого летнего.
f argile *f* varvée; varve *f*
s arcilla *f* varvada; varva *f*
g Bändertone *mpl*; Warve *f*

6531 varzea
The landscapes of high flood-plains in the Amazon Basin inundated only during the highest floods, and usually covered with high many-layered tropical rain forests.
r варзеа
Ландшафты высоких речных пойм в бас-

Fig. 349. **Vauclusian spring**

сейне р. Амазонки, затапливаемые лишь в самые высокие паводки, обычно на них произрастают высокие многоярусные влажные тропические леса.
f várzea *f*
s varzea *f*
g Varzea *f*

6532 vauclusian spring
Spring in karst areas with a big discharge and continuous flow in low water periods. → Fig. 349
r воклюз
Источник в карстовой области с большим дебитом и непрекращающимся стоком в периоды маловодья. → Рис. 349
f source *f* vauclusienne
s fuente *f* vauclusiana
g Vauclusequelle *f*; Vaucluse-Duelle *f*

6533 veering of wind
A change of direction of wind, in a clockwise direction, i.e. from north through east to south in the Northern Hemisphere, equivalent to a backing wind in the Southern Hemisphere.
r правое вращение ветра
Изменение направления ветра по часовой стрелке, т.е. с севера через восток на юг в Северном полушарии, и в обратном направлении – в Южном полушарии.
f vent *m* dextrogyre
s viento *m* con giro horario
g Winddrehung *f*

6534 vegetable-growing
The intensive cultivation of vegetables.
r овощеводство
Выращивание овощных культур.
f culture *f pl* maraîchères
s horticultura *f*
g Gemüseanbau *m*

6535 vegetated sand dune
A small (up to 1 m) accumulation of wind-blown sand around a low shrub on a flat desert surface.
r бугристые пески; кучугуры; кучевые пески
Мелкие (до 1 м) скопления приносимого ветром песка вокруг низких кустарников на плоской пустынной поверхности.
f sables *m* en tas
s dunas *fpl* vegetadas
g Sandhügel *mpl*; Sandhäufen *mpl*

6536 vegetation; vegetation cover
The total plant cover of the Earth or a particular region, comprising one or more floral communities which is controlled by several factors: climate, soil, terrain, and human impact.
r растительность; растительный покров
Растительный покров Земли или отдельных ее регионов, состоящий из одного или нескольких растительных сообществ, формирующихся под влиянием ряда природных факторов: климата, рельефа, почв, а также деятельности человека.
f végétation *f*; manteau *m* végétal; couverture *f* végétale; couverture *f* vivante
s vegetación *f*
g Vegetation *f*; Pflanzendecke *f*

* **vegetation community → 4600**

* **vegetation cover → 6536**

6537 vegetation indices
Ratios between coefficients of reflection by the Earth's surface in different spectral bands used to recognize the leaf area density and state of vegetation from satellites.
r вегетационные индексы
Соотношения коэффициентов отражения земной поверхностью в разных спектральных интервалах, используемые для распознавания со спутников плотности листовой поверхности и состояния растительного покрова.
f indices *mpl* de végétation
s índices *mpl* de vegetación
g Vegetationindizes *mpl*

* **vegetation infill of a lake → 2143**

6538 vegetation of solonchaks
The plant cover consisting mainly of halophytes, growing on saline soils in the depressions of arid areas, and on sea shores.
r солончаковая растительность
Растения, преимущественно галофиты, развитые на засоленных почвах, в основном в понижениях рельефа аридных областей, а также по морским побережьям.
f végétation *f* halophite
s vegetación *f* halófita
g Solontschakvegetation *f*; Salzbodenvegetation *f*

* **vegetation season → 2666**

6539 vegetation zone; zone of vegetation
Horizontal alternation of plant life from the equator towards the poles, as well as from margins towards inland parts of continents.
r растительная зона
Полосы растительности, сменяющие одна другую в широтном направлении от экватора к полюсам, а также от окраин в глубь континентов.
f zone *f* de végétation
s zona *f* de vegetación
g Vegetationszone *f*

6540 vein; lode
A plate-like geological body differing in composition and genesis from the rock.
r жила
Пластинообразное геологическое тело, отличающееся по составу и генезису от вмещающих пород.
f filon *m*; veine *f*
s filón *m*
g Gang *m*; Ader *f*

6541 vein ice
The ice filling the fissures of rocks and grounds in permafrost regions.
r жильный лед
Лед в трещинах горных пород и грунтов в областях распространения вечной мерзлоты.
f glace *f* de fissure
s hielo *m* de vena
g Veneneis *n*

6542 veld
Flat or hilly uplands in South Africa covered with xerophylous grasses and bushes, suitable for pasture.
r велд
Выровненные или холмистые возвышенности в Южной Африке, покрытые ксерофитными злаками и кустарниками, обычно используются как пастбища.
f veld *m*
s veld *m*
g Veld *n*

6543 velocity head; dynamic head
Head caused by the flow of water and depending on the kinetic energy of the flow.
r скоростной напор
Напор, возникающий при движении воды и зависящий от кинетической энергии потока.
f énergie *f* cinétique
s altura *f* de presión; potencial *m* de caída
g Staudruck *m*

6544 velocity profile
The graph which depicts the variation of velocity in relation to the depth or width of a river or glacier.

r профиль скорости
График, показывающий различия скорости движения по глубине и ширине в реке или леднике.
f profil *m* de vélocité
s perfil *m* de velocidad
g Geschwindigkeitprofil *n*

* **vent → 6601**

* **verdurization → 4605**

6545 verglas; glazed frost; glaze
A coating of ice on the earth's surface produced by a frost after a partial thaw or following rain.
r гололед; гололедица
Слой льда на земной поверхности, образующийся после оттепели, дождя или мокрого снега в результате похолодания.
f verglas *m*
s verglas *m*
g Glatteis *n*; Reifglätte *f*; Blachfrost *m*

6546 verification of forecast
The degree of correspondence of a forecast to the real weather conditions.
r оправдываемость прогноза
Степень соответствия прогноза фактическим условиям погоды.
f vérification *f* de prévisions météorologiques
s verificación *f* de previsión del tiempo
g Verifizierung *f* der Wettervorhersage

6547 vernal *adj*
Referring to the season of spring.
r весенний
Относящийся к природным явлениям и процессам, связанным с весной.
f printanier *adj*; vernal *adj*
s primaveral *adj*; vernal *adj*
g Frühlings-; Frühjahrs-

6548 vernal equinox; spring equinox
The moment when the Sun crosses the celestial equator during its apparent annual movement from south to north on March 20 or 21 which is the beginning of the astronomic spring in the Northern Hemisphere. → Fig. 25, p. 58
r весеннее равноденствие
Момент времени, когда Солнце пересекает небесный экватор при движении из Южного полушария в Северное 20 или 21 марта, что является началом астрономической весны в Северном полушарии. → Рис. 25, с. 58
f équinoxe *m* vernal; équinoxe *m* de printemps
s equinoccio *m* vernal; equinoccio *m* de primavera
g Frühlingsäquinoktikum *m*; Frühlingstagundnachtgleiche *f*

6549 vernalization
The renewal of plant growth due to the increasing temperatures during the spring months, following the dormant winter period.
r яровизация
Возобновление роста растений после зимних месяцев, когда растения находятся в покое, благодаря увеличению температур весной.
f vernalisation *f*
s vernalización *f*
g Jarowisation *f*; Vernalisation *f*

6550 vertical
The large circle of the celestial sphere passing through the zenith and a given point.
r вертикал
Большой круг небесной сферы, проходящий через зенит и заданную точку.
f vertical *m*
s vertical *f*
g Höhenkreis *m*

6551 vertical aerial photograph
An aerial photogpaph or space image taken when the optical axis of the camera is approximately perpendicular to the Earth's surface (with the deviation not more than 3°). → Fig. 3, p. 14
r плановый аэрофотоснимок
Аэрофото- или космический снимок, сделанный при отвесном (или с отклонением не более 3°) положении оси фотографирования. → Рис. 3, с. 14
f photographie *f* aérienne verticale
s fotografía *f* aérea vertical
g Steilbild *n*

6552 vertical circulation of water
The movement of water in a water mass in the vertical direction caused by density convection, wind-driven near-coastal currents, vortex phenomena, etc.
r вертикальная циркуляция воды
Движение морских вод по вертикали вследствие плотностной конвекции, сгонно-нагонных явлений вблизи берегов, вихревых образований и т.п.
f circulation *f* verticale d'eau
s circulación *f* vertical de agua
g Vertikalzirkulation *f* des Wassers

6553 vertical columns in snow
The texture of recrystallized snow consisting of vertical elongated aggregations of crystals.
r вертикальная столбчатость снега
Текстура перекристаллизованного снега, представляющая собой удлинение сростков кристаллов в вертикальном направлении.
f colonnettes *fpl* verticales dans la neige
s columnas *fpl* verticales de la nieve
g Schneepfeiligkeit *f*; Schneesäuligkeit *f*

6554 vertical dissection of mountains
A relative elevation of peaks and crests above

valleys that results from the dissection of mountains by exogenous processes.

r вертикальное расчленение гор
Относительное превышение вершин и гребней над долинами, обусловленное расчленением гор экзогенными процессами.

f dissection *f* verticale
s disección *f* vertical
g vertikale Gliederung *f*

6555 vertical exaggeration
The increase necessary in the vertical scale in comparison with the horizontal scale in order fairly to represent the height of an elevation in a relief model.

r вертикальное преувеличение
Необходимое увеличение вертикального масштаба по сравнению с горизонтальным, чтобы зрительно представить высоту в модели рельефа.

f exagération *f* verticale
s exageración *f* vertical
g vertikale Übertreibung *f*

6556 vertical profile
The distribution of a meteorological element with height.

r вертикальный профиль
Распределение метеорологического элемента по высоте.

f profil *m* vertical
s perfil *m* vertical
g vertikales Profil *n*

6557 vertical scale
The ratio of vertical distances on three-dimensional representations to the corresponding distances in nature.

r вертикальный масштаб
Отношение длины вертикального отрезка на трехмерном изображении к длине соответствующего вертикального отрезка в натуре.

f échelle *f* des hauteurs
s escala *f* vertical
g Höhenmaßstab *m*

6558 vertical temperature gradient; temperature gradient; lapse rate
Rate of decrease of air temperature with increasing height, averaging some 0.6°C per 100 m.

r вертикальный градиент температуры; температурный градиент
Величина, характеризующая понижение температуры воздуха с ростом высоты, в среднем равная 0,6°C на 100 м высоты.

f gradient *m* thermique vertical; gradient *m* thermique
s gradiente *m* vertical de temperatura
g vertikales Temperaturgefälle *n*; vertikaler Temperaturgradient *m*

* **vertical zonation → 230**

* **vertical zone → 228**

* **vertisols → 1157**

6559 vesicle
A small, generally rounded or oval cavity in a mineral or rock, particularly in glassy volcanic rock, formed by the trapping of steam or gas bubbles in the molten material as it solified.

r везикула
Небольшая округлая или овальная полость в минерале или горной породе, особенно в вулканическом стекле, образованная при замыкании пузырьков пара или газа в расплавленной массе по мере ее застывания.

f vésicule *f*
s vesicula *f*
g Vesikula *f*

6560 viaduct
A bridge built at a crossing of two or more transportation lines for unhampered movement on all levels.

r путепровод; виадук
Мост, сооружаемый на пересечении двух или более транспортных магистралей для беспрепятственного движения в разных уровнях.

f viaduc *m*
s viaducto *m*
g Bahnüberfürung *f*; Straßenüberfürung *f*

6561 vicarious species
Closely related species of plants or animals, which geographically or ecologically replace each other, originated due to the isolation of some parts of the primary area, e.g. larch in Northern Eurasia.

r викарирующий вид
Близкородственные виды растений и животных, географически или экологически замещающие друг друга, возникшие в результате изоляции отдельных районов первичного ареала, например, лиственница в Северной Евразии.

f espèce *f* vicariente
s especie *f* vicaria
g vikariierende Art *f*

6562 viewshed analysis
The processing of digital landform models in order to estimate the visibility of the relief separate parts from some point of view.

r анализ видимости/невидимости
Обработка цифровых моделей рельефа, обеспечивающая оценку рельефа местности с точки зрения видимости отдельных его частей с некоторой точки обзора.

f analyse *f* de visibilité/invisibilité
s análisis *m* de visibilidad/invisibilidad
g Sicht/Unsichtbarkeitanalyse *f*

6563 village; hamlet

A settlement consisting of a collection of dwelling houses and other buildings in rural surroundings usually having a church and village shop.
r деревня; село
Крупное сельское поселение, обычно имеющее церковь и магазины (небольшое село, обычно без церкви, в России называют деревней).
f village *m*
s aldea *f*; pueblo² *m*
g Dorf *n*

* **vine → 3480**

6564 vine-growing; viticulture
A branch of agriculture dealing with the growing of grapes, usually for wine-making.
r виноградарство
Отрасль растениеводста, занимающаяся выращиванием винограда.
f viticulture *f*
s viticultura *f*
g Weinbau *m*; Winzerei *f*

6565 vineyard
A plot of land which usually occupies a non-steep slope, where grape vines are cultivated.
r виноградник
Участок земли, как правило на некрутом склоне, где выращивают виноград.
f vigne *f*; vignoble *m*
s viña *f*; viñal *m*; viñedo *m*
g Weinberg *m*; Weingarten

* **violent → 6566**

6566 violent species; violent
The biological species which is the strongest in terms of the ability to dominate in its own community or to penetrate into other communities, this species competes successfully due to more efficient use of environmental resources and gradually invades the area, e.g. oak.
r виолент
Биологический вид, способный доминировать в сообществе или внедряться в другие сообщества, энергично развиваясь, он подавляет соперников полнотой использования ресурсов окружающей среды, например, дуб.
f violent *m*
s violento *m*
g Violent *m*

6567 virgation
The spreading out of several mountain ranges from a central mountain knot.
r виргация
Расхождение нескольких горных цепей от центрального горного узла.
f virgation *f*
s virgación *f*
g Virgation *f*

* **virgin forest → 4293**

6568 virgin land; virgin soil
Tha land, suitable for growing crops, that either has never been cultivated or was cultivated long ago and presently covered with natural (mainly steppe) vegetation.
r целинные земли; целина
Пригодные для земледелия земли, покрытые естественной, главным образом степной, растительностью, которые либо никогда не распахивались, либо использовались много лет назад.
f terres *fpl* vierges
s tierra *f* virgen
g Neuland *n*

* **virgin soil → 6568**

6569 virtual map
Map made by means of computer graphics.
r виртуальная карта
Карта, создаваемая средствами компьютерной графики.
f carte *f* virtuelle
s mapa *m* virtual
g virtuelle Karte *f*

6570 viscous *adj*
Pertaining to a substance which is sticky, adhesive, and flows slowly down, e.g. asphalt, wax and certain types of lava.
r вязкий
Относящийся к липкому веществу, которое медленно течет вниз по склону, например, асфальт, воск и некоторые типы лавы.
f visqueux *adj*
s viscoso *adj*
g klebrig *adj*

6571 visibility; range of visibility
The clarity of the atmosphere characterized by the distance at which an object under observation disappears from the sight in the day time and becomes invisible for the observer.
r видимость; дальность видимости
Прозрачность атмосферы, характеризуемая расстоянием, на котором днем наблюдаемый объект становится невидимым для наблюдателя.
f visibilité *f*; portée *f* de visibilité
s visibilidad *f*
g Sichtbarkeit *f*; Sichtweite *f*

6572 visible ground ice
The ice which is seen in rocks, when thawed it converts into free water.
r эвидентный лед
Видимый подземный лед, при протаивании переходит в свободную воду.
f glace *f* souterraine évidente
s hielo *m* de suelo evidente
g sichtbares Bodeneis *n*

6573 visible light; light radiation; visible radiation
Electromagnetic radiation in wave length interval of 0.39-0.76 micron perceived by human eyes.
r видимая радиация; световая радиация
Электромагнитная радиация в интервале длин волн 0,39-0,76 микрон, воспринимаемая человеческим глазом.
f rayonnement *m* visible
s radiación *f* visible; luz *m* visible
g sichtbare Strahlung *f*

*** visible radiation → 6573**

6574 visual pollution
Presence in the environment of buildings and objects the view of which causes unpleasant sensations in humans.
r визуальное загрязнение
Присутствие в окружающей среде сооружений и объектов, вид которых вызывает у людей неприятные ощущения.
f pollution *f* visuelle
s polución *f* visual; contaminación *f* visual
g visuelle Verschmutzung *f*

6575 vital rates
Rates of those components, such as birth, marriage, fertility, and death which indicate the nature and possible changes in a population.
r жизненные показатели
Значения таких компонентов как рождаемость, замужество, смертность, показывающих природу и возможные изменения населения.
f normes *fpl* de la vie
s normas *fpl* vitales
g lebenswichtige Rate *f*

6576 vital space; living space
Neighbouring territory over which a strong state expands its economic and political interests.
r жизненное пространство
Соседняя территория, на которую распространяются экономические и политические интересы сильного государства.
f espace *m* vital
s espacio *m* vital
g Lebensraum *m*

6577 vitalism
The doctrine, developed in the XVII-XVIII centures, stressing the qualitative difference between living and non-living nature, as well as the fact that it is impossible to ascribe processes in living nature to physical and chemical laws of non-living nature.
r витализм
Сложившееся в XVII-XVIII вв. учение о качественном отличии живой природы от неживой, о принципиальной несводимости жизненных процессов к физико-химическим законам неживой природы.
f vitalisme *m*
s vitalismo *m*
g Vitalismus *m*

*** viticulture → 6564**

6578 vlei
In South Africa a circle or oval depression which forms in the loose surface deposits and has a swamp or shallow lake which often connected with the neighbouring river system.
r влей
В Южной Африке впадина округлой или овальной формы в поверхностных рыхлых отложениях, в которой обычно существует болото или мелкое озеро, нередко связанное с соседней речной системой.
f vlei *m*
s vlei *m*
g Vlei *n*

6579 voice of sea
Sound waves appearing above the sea surface under strong winds as a result of whirl formation in the air above wave crests.
r голос моря
Звуковые волны, возникающие над поверхностью моря при сильном ветре в результате вихреобразования в воздухе над гребнями волн.
f voix *f* de mer
s voz *f* de mar
g Stimme *f* des Meeres

*** void → 4701**

6580 void ratio
The ratio of the volume of void space to the volume of solid particles in a given mass of soil.
r коэффициент пористости почвы
Отношение объема пор к объему твердых частиц в данной почвенной массе.
f facteur *m* de porosité du sol
s factor *m* de porosidad de suelo
g Bodenporenkoeffizient *m*

*** volcanic activity → 6602**

6581 volcanic *adj*
Related to natural phenomena induced by the eruptions of volcanoes.
r вулканический
Относящийся к природным явлениям, связанным с извержениями вулканов.
f volcanique *adj*
s volcánico *adj*
g vulkanisch *adj*

*** volcanic ash → 400**

6582 volcanic belt
A linearly stretched zone several thousand kilometers long with a great number of volcanoes.
r вулканический пояс

Линейно вытянутая на земном шаре полоса длиной до нескольких тысяч километров с большим количеством вулканов.
f zone *f* volcanique
s zona *f* volcánica
g Vulkangürtel *m*

6583 volcanic bomb
A block of lava ejected from a volcano explosively during the eruption in a liquid or plastic form and solidified in the air.
r вулканическая бомба
Комок лавы, выброшенный во время извержения вулкана в жидком или пластичном состоянии и застывший во время полета в воздухе.
f bombe *f* volcanique
s bomba *f* volcánica
g Vulkanbombe *f*; vulkanische Bombe *f*

6584 volcanic bottom material
Erupted rocks participated in the formation of chemical composition of pelagic deposits on limited areas of the mid-ocean ridges.
r вулканогенный донный материал
Изверженные породы, участвующие в формировании химического состава пелагических осадков на ограниченных участках срединно-океанических хребтов.
f dépôts *mpl* de fond volcanogènes
s depósitos *mpl* del fondo volcanogéneos
g vulkanogene Bodenablagerungen *fpl*

6585 volcanic cloud
An atmospheric cloud, usually cumuliform, often formed above a volcano during its eruption.
r вулканическое облако
Кучевообразное облако, нередко возникающее над вулканом во время извержения.
f nuage *m* d'éruption volcanique
s nube *f* devida a erupción volcánica
g vulkanische Wolke *f*

6586 volcanic conduit
A vertical pipe through which magma moves upwards in a volcano towards the vent.
r вулканический канал
Внутривулканический вертикальный канал, по которому магма течет вверх к жерлу вулкана.
f conduit *m* volcanique
s conducto *m* vulcánico
g vulkanisches Rohr *n*

6587 volcanic cone
A volcanic peak with a roughly circular base tapering to a point (often with a cut top), formed as a result of accumulation of erupted material around the volcanic vent. → Fig. 350
r вулканический конус
Положительная форма рельефа в виде конуса, часто со срезанной вершиной, формирующаяся в результате накопления продуктов извержения вокруг жерла вулкана. → Рис. 350

Fig. 350. **Volcanic cone**: Avacha and Koryaksky volcanos at Kamchtka

f cône *m* volcanique
s cono *m* volcánico
g Vulkankegel *m*

6588 volcanic dome
A mass of erupted rocks in the form of a bell-like body formed as a result of pressing out of viscous lava from a volcanic vent.
r вулканический купол
Залегание изверженных пород в виде колоколообразного тела, возникшего в результате выдавливания из жерла вулкана вязкой лавы.
f dôme *m* volcanique; coupole *f* volcanique
s cúpula *f* volcánica; domo *m* volcánico
g Staukuppe *f*

6589 volcanic dust
The finest particles of pyroclastic material thrown out of a volcano in eruption high in the air and carried great distances by winds.
r вулканическая пыль
Мельчайшие частицы пирокластического материала, попавшие в атмосферу в результате вулканического извержения и отнесенные ветром на значительное расстояние от вулкана.
f poussiére *f* volcanique
s polvo *m* volcánico
g vulkanischer Staub *m*; Vulkanstaub *m*

6590 volcanic earthquake
The light shaking and trembling of the Earth's surface caused by magma motion before and during the volcanic eruption.
r вулканическое землетрясение
Слабые колебания земной поверхности, вызываемые движением магмы при подготовке или в момент вулканического извержения.

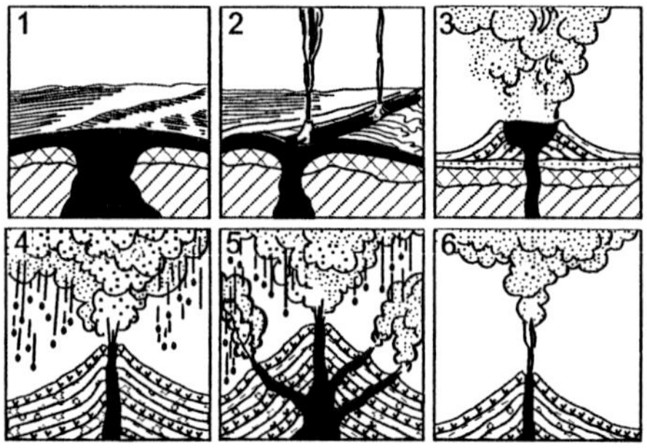

Fig. 351. **Volcanic eruption**: 1 – eruption at the large area, 2 – fissure eruption, 3 – Hawaiian eruptian, 4 – Strombolian eruption, 5 – explosive eruption, 6 – Pelean eruption

f séisme *m* volcanique
s seísmo *m* volcánico
g vulkanisches Erdbeben *n*

* **volcanic effusion → 3440**

6591 volcanic eruption; eruption
The process by which solid, liquid or gaseous material is ejected or emitted gently or explosively from the vent of a volcano or from fissures in the Earth's crust as a result of volcanic activity. → Fig. 351
r вулканическое извержение; извержение
Выброс из вулкана раскаленных или горячих твердых, жидких и газообразных вулканических продуктов и излияние лавы. → Рис. 351
f éruption *f* volcanique; éruption *f*
s erupción *f* volcánica; erupción *f*
g Eruption *f*; Vulkanausbruch *m*

* **volcanic explosion → 2060**

6592 volcanic gases
The gases emitted by volcanoes during eruption (eruptive) and in the period of their quiet activity (fumarols and solfatares).
r вулканические газы
Газы, выделяемые вулканами во время извержения (эруптивные) и в период их спокойной деятельности (фумарольные и сольфатарные).
f gazes *mpl* volcaniques
s gases *mpl* volcánicos
g vulkanische Gase *npl*; Vulkangase *npl*

6593 volcanic island
An island formed as a result of volcanic eruption on sea bottom.
r вулканический остров
Остров, возникший в результате вулканической деятельности на дне моря.
f île *f* volcanique
s isla *f* volcánica
g vulkanische Insel *f*; Vulkaninsel *f*

6594 volcanic lake; crater lake
A lake occupying a crater of an extinct volcano. → Fig. 352
r вулканическое озеро; кратерное озеро
Озеро, располагающееся в кратере потухшего вулкана. → Рис. 352
f lac *m* volcanique; lac *m* de cratère
s lago *m* volcánico; lago *m* de cráter
g vulkanischer See *m*; Kratersee *m*; Vulkansee *m*

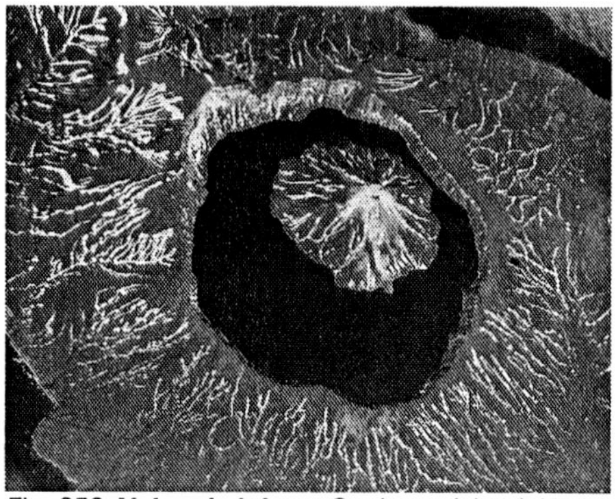

Fig. 352. **Volcanic lake** at Onekotan island, Kurily islands

6595 volcanic mountains
The highlands and ranges created by the coalescence of volcanic cones.
r вулканические горы
Нагорья и хребты, возникшие в результате слияния отдельных вулканических конусов
f montagnes *fpl* volcaniques
s montañas *fpl* volcánicas
g vulkanisches Gebirge *n*; Vulkangebirge *n*

* **volcanic neck → 4083**

* **volcanic pipe → 6601**

6596 volcanic plateau
A vast elevated plain formed as a result of effusion of large lava masses, which filled in the roughness of the former relief.
r вулканическое плато; лавовое плато
Обширная возвышенная равнина, образовавшаяся в результате излияния больших масс лавы, заполнивших неровности прежнего рельефа.
f plateau *m* volcanique
s meseta *f* volcánica
g vulkanisches Plateau *n*; Vulkanplateau *n*

6597 volcanic rock
Rocks formed as a result of volcanic eruptions, consisting of solified material which has issued in the molten state from the depths of the Earth: effusive, pyroclastic or exstrusive rocks.
r вулканическая горная порода
Горные породы, образующиеся в результате вулканических извержений: эффу-

зивные (изверженные), пирокластические, либо экструзивные.
f roche *f* volcanique
s roca *f* volcánica
g vulkanisches Gestein *n*

6598 volcanic slag
Porous lava pieces erupted from a volcanic crater during explosions or solidified in lava flows with intensive emission of gases.
r вулканический шлак
Пористые куски лавы, выброшенные из кратера вулкана при взрывах или застывшие в лавовых потоках с бурным выделением газов.
f scorie *f* volcanique
s escoria *f* volcánica
g Vulkanschlacke *f*; vulkanische Schlacke *f*

* **volcanic soil → 264**

6599 volcanic spine
A pointed rock formed as a result of pressing out of a viscous lava from a volcanic vent.
r вулканический обелиск
Остроконечная скала, возникшая в результате выжимания из жерла вулкана вязкой лавы.
f aiguille *f* volcanique
s aguja *f* volcánica
g vulkanische Felsnadel *f*

6600 volcanic thunderstorm
A thunderstorm induced by a volcanic eruption.
r вулканическая гроза
Гроза, возникающая при извержении вулкана.
f orage *m* volcanique
s tormenta *f* volcánica
g vulkanisches Gewitter *n*

* **volcanic tuff → 6399**

6601 volcanic vent; volcanic pipe; vent
A vertical or almost vertical cylindrical canal or the opening in the crust through which magma erupts during volcanic activity.
r жерло вулкана
Вертикальный или почти вертикальный цилиндрический канал, по которому поднимается магма из очага вулкана.
f cheminée *f* volcanique
s chimenea *f* volcánica
g Vulkanschlot *m*

6602 volcanism; volcanic activity
Processes and phenomena connected with the extrusion of material onto the Earth's surface.
r вулканизм; вулканическая деятельность
Совокупность процессов и явлений, связанных с излиянием магмы на поверхности Земли.

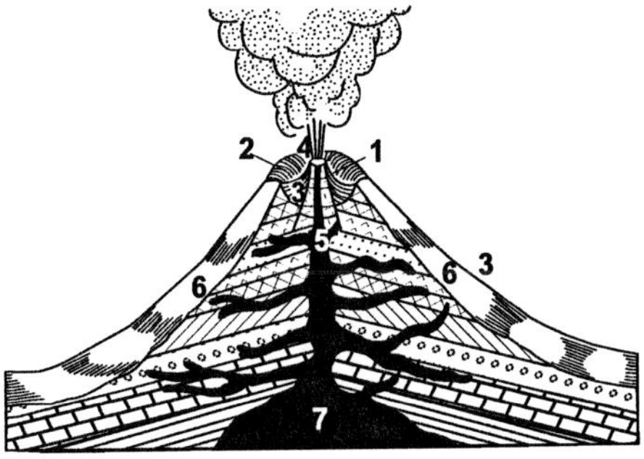

Fig. 353. **Volcano**: 1 – caldera, 2 – somma, 3 – cone, 4 – crater, 5 – volcanic vent funnel, 6 – lava flow, 7 – lava site

f volcanisme *m*; activité *f* volcanique
s volcanismo *m*; vulcanismo *m*
g Vulkanismus *m*

6603 volcano
An opening or vent in the Earth's crust through which molten lava, ash, volcanic gases, steam, etc. are ejected and solidify on the Earth's surface, or the landform produced by the erupted material. → Fig. 353
r вулкан
Поднятие, возникающее над каналами и трещинами в земной коре, по которым извергаются на земную поверхность из глубинных магматических очагов лава, горячие газы, пары воды и обломки горных пород. → Рис. 353
f volcan *m*
s volcán *m*
g Vulkan *m*

* **volcanogenic mudflow → 3350**

6604 volcanology; vulcanology
The study of volcanic activity, the structure and distribution of volcanoes, and composition of erupted material.
r вулканология
Наука, изучающая вулканическую деятельность, строение и размещение вулканов, структуру и состав продуктов извержений.
f volcanologie *f*; vulcanologie *f*
s volcanología *f*; vulcanología *f*
g Vulkanologie *f*

6605 volume of runoff
The amount of water flowing from a watershed over some period of time.
r объем стока
Количество воды, стекающей с водосбора за какой-либо интервал времени.
f débit *m* total; volume *m* débité
s volumen *m* de escorrentía
g Abflußsumme *f*; Abflußwasserfracht *f*

6606 vortex; eddy; whirl

The manifestation of a rapid spiralling movement of fluid or vapour in a circular direction around a central axis; a swirling movement of fluid within a larger mass of fluid, in a direction contrary to that of the main flow, e.g. as in depressions or highs in the atmosphere, or in the water as it encounters an obstruction in its flow.

r вихрь
Масса воздуха или воды, охваченная вращательным движением около некоторой оси; вращательное движение жидкости или воздуха внутри значительного их объема в направлении, противоположном общему движению, как это имеет место в циклонах или антициклонах, или в воде, когда она обтекает препятствие на пути своего движения.

f tourbillon *m*; vortex *m*
s torbellino *m*; remolino *m*; vórtice *m*
g Wirbel *m*

6607 vortex chain
A system of individual whirl-like water patterns appearing in the ocean while the currents flow around islands extend linearly.
r вихревая цепочка
Система отдельных вихрей, возникающих в океане при обтекании островов течениями и расположенных в определенном линейно-вытянутом порядке.
f chaîne *f* de tourbillons
s cadena *f* de torbellinos
g Wirbelkette *f*

6608 vortex entrainment
The process of capturing of water masses by large oceanic whirls (rings) occurring on their peripheries or in the zones of convergence of ocean currents.
r вихревое вовлечение
Захват крупными океанскими вихрями (кольцами) окружающих их вод на периферии мезомасштабных вихрей и в зонах конвергенции течений.
f entraînement *m* pour tourbillon
s atracción *f* por torbellino
g Wirbeleinbeziehung *f*

6609 vorticity
The rotational movement of a fluid about an axis, either in the form of whirlpools in the turbulent flow of a river or as vortices in the atmosphere.
r вихревое движение
Вращательное движение жидкости вокруг оси, либо в форме водоворотов в турбулентном течении реки, либо в форме вихрей в атмосфере.
f vorticité *f*
s vorticidad *f*
g Verwirbelung *f*

6610 voyage round the World
A journey by ship when all the meridians of the Earth are crossed.
r кругосветное плавание
Плавание, при котором пересекаются все меридианы Земли.
f voyage *m* de circumnavigation; voyage *m* autour du monde
s viaje *m* de circunnavegación; viaje *m* alrededor de la tierra; viaje *m* alrededor del mundo
g Weltumschiffung *f*

6611 V-shaped valley
A valley eroded by a river, having V-shaped cross-section in contrast to *U-shaped valley* modified by the glacial erosion.
r V-образная долина
Долина, эродированная рекой, имеющая V-образное поперечное сечение, отличное от *U-образной долины*, модифицированной гляциальной эрозией.
f vallée *f* en V
s valle *m* en V; valle *m* en uve
g V-artiges Tal *n*

6612 Vulcanian eruption
A volcanic eruption in which the lava surface rapidly solidifies because of its high viscosity.
r вулканианский тип извержения
Вулканическое извержение, при котором лава очень быстро затвердевает из-за ее большой вязкости.
f éruption *f* vulcanienne
s erupción *f* vulcaniana
g vulkanischer Eruptionstyp *m*

* **vulcanology → 6604**

6613 vulnerability
The ability of the environment (living things and species) to be damaged or destroyed by even a slight impact of external forces.
r уязвимость окружающей среды
Неспособность окружающей среды (прежде всего живых организмов) сохранить свое прежнее состояние даже при небольшом внешнем воздействии.
f vulnérabilité *f*
s vulnerabilidad *f*
g Verwundbarkeit *f*

6614 vulnerable area; environmentally sensitive area
The area in which even a slight human impact causes serious changes of the environment, i.e. the natural environment can easily be harmed.
r экологически чувствительный регион; уязвимая территория
Территория, на которой даже незначительные антропогенные воздействия вызывают крупные изменения окружающей среды.
f région *f* vulnérable; région *f* écologiquement fragile; région *f* d'écologie fragile
s área *m* vulnerable; medio *m* ambiente sensible
g umweltempfindliche Gebiete *pl*

W

* **wadden** → 6708

6615 **wadi; oued**
A steep-sided, flat-floored valley very occasionally occupied by an intermittent stream in hot desert or semi-arid areas, especially of Arabian Peninsula and North Africa. → Fig. 354
r вади; уэд
Сухие большую часть года крутосклонные и плоскодонные долины временных водотоков в пустынях Аравийского полуострова и Северной Африки. → Рис. 354
f ennedi *m*; oued *m*
s uadi *m*; ennedy *m*; oued *m*; ued *m*
g Wadi *m*

6616 **walking tourisme; hiking**
A travel to countryside in order to do hiking along the organized paths to enjoy natural beauties and visit touristic attractions.
r пешеходный туризм
Распространенный вид туризма, включающий пешие походы по проложенным тропам с посещением памятных или аттрактивных мест.
f tourisme *m* pédestre
s turismo *m* pedestre
g Fußtourismus *m*

* **wall-sided glacier → 959**

* **waning slope → 6735**

* **war contamination → 6617**

6617 **war pollution; war contamination**
The pollution of the environment caused by war industry and military operations, as a rule, being out of control.
r военное загрязнение
Загрязнение окружающей среды в результате работы военно-промышленного комплекса и при ведении военных действий, как правило, не контролируемое.
f pollution *f* militaire; contamination *f* militaire
s polución *f* militar; contaminación *f* militar
g Militärverschmutzung *f*

6618 **warm air mass**
An air mass the temperature of which is higher than surrounding air masses, or an air mass moving on a colder underlying surface.
r теплая воздушная масса
Масса воздуха с более высокой температурой, чем соседние воздушные массы, или масса воздуха, надвигающаяся на более холодную подстилающую поверхность.

Fig. 354. **Wadi** at Sinai Peninsula

f masse *f* d'air chaud
s masa *f* del aire caliente
g warme Luftmasse *f*; Warmluftmasse *f*

6619 **warm current**
A sea or ocean stream carring water which is warmer than the surrounding water masses.
r теплое морское течение
Течение, средняя температура воды которого выше средней температуры окружающих вод.
f courant *m* chaud
s corriente *f* caliente; corriente *f* cálida
g warme Meeresströmung *f*

6620 **warm front**
Any non-occluded front which moves in such a way that an advancing mass of warmer air is overriding and rising above a mass of colder air. → Fig. 355
r теплый фронт
Любой неокклюдированный фронт, движущийся таким образом, что теплый воздух натекает на холодный. → Рис. 355
f front *m* chaud
s frente *m* caliente; frente *m* cálido
g Warmfront *f*

* **warm glacier → 6079**

6621 **warm occlusion**
An occlusion where the cold, overtaking air is not so cold as the air mass ahead of it.
r теплая окклюзия

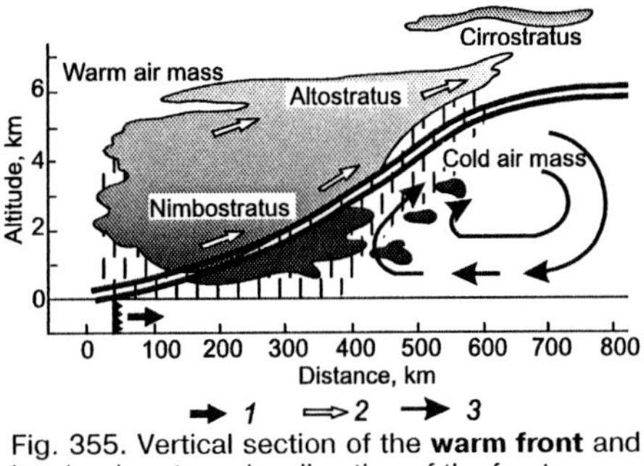

Fig. 355. Vertical section of the **warm front** and its cloud system: 1 – direction of the front movement, 2, 3 – circulation of warm and cold air

Окклюзия, при которой поступающий воздух не такой холодный, как воздушная масса перед ним.

f occlusion *f* chaude;
occlusion *f* à caractère de front chaud

s oclusión *f* cálida;
oclusión *f* de carácter de frente cálido

g warme Okklusion *f*

6622 warm pole
The area with the highest registered air temperature near the earth surface in one of the hemispheres or on the whole globe.

r полюс тепла
Область с самыми высокими температурами воздуха у земной поверхности в данном полушарии или на всем земном шаре.

f pôle *m* de chaud

s polo *m* de calor

g Wärmepol *m*

6623 wash
The process involved in the downslope movement of fine material by water; an area of sand and mud washed by the tide or the surge of water up a beach following the breaking of a wave.

r размывание
Движение мелкозема вниз по склону, вызванное потоками воды; участки песка и ила на берегу моря и крупных водоемов, смываемые приливо-отливными течениями или прибрежным волнением.

f affouillement *m*

s derrubio *m*; deslave *m*

g Auswaschung *f*; Unterspülung *f*

6624 washing out
The process of extraction of substances from soil, rocks and bottom sediments by water, because water is able to dissolve them and carry to other places or layers.

r вымывание
Процесс извлечения веществ из почв, пород и донных отложений водой, обусловленный их растворением в воде.

f lavage *m*

s lavado *m*

g Abspülung *f*

6625 wash-out
The destruction caused by a torrential river in spate following a storm, it refers to bank erosion, levee breaching, dam bursting and bridge destruction.

r смыв
Разрушение, вызванное быстро текущей рекой во время наводнения, сопровождаемого сильным ветром, термин обычно используют в отношении эрозии берегов, разрушений насыпи, дамбы или моста через реку.

f affouillement *m*

s derrubio *m*; erosión *f* por la corriente torrencial

g Auswaschung *f*; Abwaschung *f*

* **washplain → 204**

6626 waste; waste product
Substances or energy not used up in a technological process and returned to the environment.

r отходы
Часть вещества или энергии, не реализованная в технологическом цикле и возвращаемая в окружающую среду.

f déchets *mpl*; déchet *m*

s desechos *mpl*; residuos *mpl*

g Abfälle *mpl*; Abfallstoff *m*

6627 waste burial
Placing of highly dangerous and toxic wastes into specially equipped storages without intention to take them out.

r захоронение отходов
Удаление особо опасных и токсичных отходов в специально подготовленные хранилища, без намерения их извлечения.

f enfouissement *m* des déchets

s enterramiento *m* de desechos

g Endlagerung *f*

* **waste disposal → 6632**

6628 waste land; barren lands
An uncultivated, open area of land with a natural vegetation of low shrubs that tend to develop on poor soils, it is often the result of human impact – fires, felling, grazing, or previously used in other way but later abandoned.

r пустошь; заброшенная земля
Невозделанные пространства, образовавшиеся преимущественно на месте лесов, главным образом после вырубок и пожаров, характерны для регионов с холодным и влажным климатом.

f friche *f*; kande *f*; lande *f*;
terrain *m* incultivé; gâtine *f*

s erial *m*; baldío *m*

g Ödland *n*

6629 waste neutralization
The removal of harmful and poisonous toxic substances or radioactive material from waste and from the environment, if they have already got into it.

r обезвреживание отходов
Очистка природных объектов и отходов от вредных материалов, в частности, отравляющих и радиоактивных, попавших в окружающую среду.

f neutralisation *f* des déchets;
décontamination *f* des déchets

s neutralización *f* de desechos;
decontaminación *f* de residuos

g Schadstoffentsorgung *f*;
Abfallneutralisation *f*

6630 **waste pile; slag heap; terricone**
An accumulation of spoil deposited in the form of a cone-shaped mound.
r террикон
Отвалы пустой породы из шахт на земной поверхности, как правило, конусовидной формы.
f terril *m*
s terrero *m*; colina *f* de tierra; montón *m*
g Haldenberg *m*

6631 **waste processing; waste treatment**
Mechanical, physical, chemical and biological transformation of industrial and domestic waste to make it reusable or so it may be disposed of safely.
r переработка отходов
Механическое, физико-химическое и биологическое преобразование промышленных и бытовых отходов с целью извлечения из них полезных компонентов, тепла и пр.
f traitement *m* des déchets
s reutilización *f* de desechos; tratamiento *m* de residuos; recuperación *f* de desechos
g Abfallverwertung *f*

* **waste product → 6626**

* **waste site → 1742**

6632 **waste storage; waste disposal**
The storage of waste in special tanks or waste storage sites.
r хранение отходов
Содержание отходов в специальных емкостях или хранилищах.
f disposition *f* des déchets
s desposición *f* de desechos
g Abfallablagerung *f*

6633 **waste storage site**
A specially equipped place for keeping waste to reduce harm to the environment.
r хранилище отходов
Специально оборудованное место складирования отходов.
f dépôt *m* de déchets; entreposage *m* de déchets
s depósito *m* de desechos
g Abfalllager *n*

* **waste treatment → 6631**

* **waste water → 5369**

6634 **wasteless technology; non-waste technology; low-waste technology**
The methods of production giving a minimum amount of solid, liquid, gaseous, and thermal waste and emissions.
r безотходная технология; малоотходная технология
Технология, дающая минимальный объем твердых, жидких, газообразных и тепловых отходов и выбросов.
f technologie *f* sans déchets
s tecnología *f* sin desperdicios; tecnología *f* sin desechos
g abfallfreie Technologie *f*

6635 **water; natural water**
Water (H_2O) existing in the environment as one of its essential components and naturally containing solid, liquid and gaseous admixtures.
r вода; природная вода
Одна из основных субстанций окружающей среды (H_2O) с содержащимися в ней твердыми, жидкими и газообразными веществами.
f eau *f*
s agua *f*
g Wasser *n*

6636 **water abrasion; marine abrasion; fluvial abrasion**
The destruction and washout of seashores, banks of lakes and reservoirs by waves and surf.
r водная абразия
Размывание берегов морей, озер и водохранилищ действием волн и прибоя.
f abrasion *f*
s abrasión *f*
g Abrasion *f*

* **water aeration → 100**

6637 **water ana salt balance**
The relation of water mass and weighted mean salt content in rhe flows incoming to a water body and those discharging from it.
r водно-солевой баланс
Соотношение объема водной массы и средневзвешенного содержания солей в потоках, поступающих в водный объект и выходящих из него.
f bilan *m* d'eau et sels
s balance *m* de agua y sales
g Wasser und Salzbilanz *f*

* **water area → 5861**

6638 **water balance**
The correlation of income and discharge components of water cycle in a certain area.
r водный баланс
Соотношение приходной и расходной частей круговорота воды на каком-либо пространстве.
f bilan *m* d'eau; bilan *m* hydrologique
s balance *m* hidrológico; balance *m* hídrico; balance *m* de agua
g Wasserhaushalt *m*; Wasserbilanz *f*

6639 **water balance of a glacier**
The sum of water influx and runoff in a given glacier with due regard for the changes of the water content over a certain time interval.
r водный баланс ледника

Сумма поступлений и убыли воды для данного ледника с учетом изменения ее запасов за выбранный интервал времени.
f bilan *m* des eaux du glacier
s balance *m* de agua de glaciar
g Wasserbilanz *f* des Gletschers

6640 water balance of the atmosphere
The input and loss of water within a certain area of the atmosphere or in the atmosphere as a whole.
r водный баланс атмосферы
Приход и расход влаги внутри определенной области атмосферы или во всей атмосфере в целом.
f bilan *f* hidrologique de l'atmosphère
s balance *m* hidrológico de atmósfera
g Wasserhaushalt *m* der Atmosphäre; Wasserbilanz *f* der Atmosphäre

6641 water bearing capacity
The amount of free water carried by an aquifer.
r водоносность горизонта
Объем или масса гравитационной воды в водоносном горизонте.
f capacité *f* d'aquaifére
s acuosidad *f* de aquífero
g Wasserführung *f* des Grundwasserschichts

6642 water bloom; algae blooming; phytoplankton bloom
A sudden increase of the amount of algae in a body of water induced by natural nutrient enrichment of waters or due to pollution.
r цветение водоема
Массовое развитие фитопланктона и водорослей, приводящее к изменению окраски воды, вызываемое ростом питательных веществ в водоеме, а также загрязнением воды.
f floraison *f* d'eau; floraison *f* algale
s floración *f* de agua
g Wasserblüte *f*

* **water body → 747**

6643 water body; body of water
A permanent or intermittent mass of water in a hollow of land surface characterized by a rather slow water exchange.
r водоем
Постоянное или временное скопление воды в углублении земной поверхности, характеризующееся замедленным водообменом.
f étendue *f* d'eau; bassin *m* d'eau; réservoir *m* d'eau
s depósito *m* de agua; masa *f* de agua; extensión *f* de agua
g Wasserbecken *n*; Gewässer *n*; Wasserkörper *n*

* **water capacity → 3904**

6644 water cloud
A cloud composed only of water droplets, supercooled or not, and containing no ice crystals.
r водяное облако; жидко-капельное облако
Облако, состоящее исключительно из водяных капель (включая переохлажденные) и не содержащее кристаллов льда.
f nuage *m* aqueux
s nube *f* de agua
g Wasserwolke *f*

* **water conditioning → 6691**

6645 water consumer
A branch of economy or an enterprise which takes up water from water bodies or water courses, uses it and returns it back in some other place in a reduced amount and often having worse quality.
r водопотребитель
Отрасли хозяйства и отдельные предприятия, которые изымают воду из водных объектов, используют ее и возвращают в другом месте, в меньшем количестве и часто худшего качества.
f consommateur *m* d'eau
s consumidor *m* de agua
g Wasserverbraucher *m*

6646 water consumption; water use
The use of natural water for economic and domestic purposes.
r водопотребление; водопользование
Использование природных вод в хозяйственной деятельности и в быту.
f consommation *f* d'eau; utilisation *f* d'eau
s consumo *m* de agua; utilización *f* del agua
g Wasserverbrauch *m*

6647 water consumption norm
A set standard of water amount consumed per capita or per a production unit.
r норма водопотребления
Установленное количество воды, потребляемой одним жителем, либо потребляемой на единицу произведенной продукции.
f norme *f* de consommation d'eau
s norma *f* de consumo de agua
g Wasserbedarf *m*

* **water contamination → 6671**

6648 water content in a glacier
The amount of water contained on a glacier surface and within its body.
r водность ледника
Количество воды, содержащейся на поверхности ледника и в его толще.
f stock *m* d'eau dans un glacier
s stock *m* de agua en el glaciar
g Wasserspeicherung *f* im Gletscher

6649 water content of cloud

The amount of water in the liquid or solid state contained in a unit volume of air in the cloud.

r водность облака
Количетво воды в жидком и твердом состоянии, которое находится в единичном объеме воздуха в облаке.

f teneur *f* d'eau en nuage
s contenido *m* de agua en las nubes
g Wassergehalt *m* der Wolken

* **water content of soil → 3906**

6650 **water content of the air**
The amount of water content in the liquid, solid and gaseous states contained in a unit volume of the atmospheric air.

r влагосодержание воздуха
Содержание воды в единичном объеме воздуха во всех ее агрегатных состояниях.

f teneur *f* d'eau en air
s contenido *m* de agua en el aire
g Wassergehalt *m* der Luft

6651 **water conveying capacity**
The amount of water which can flow through an effective cross section of a watercourse per time unit.

r водопропускная способность
Количество воды, которое может протекать через живое сечение водотока в единицу времени.

f capacité *f* de transport de l'eau
s capacidad *f* de transporte del agua
g Wasserdurchlässigkeit *f*

* **water cycle → 2882**

6652 **water desinfection**
The process of the removal of pathogenic microorganisms from water which is used for economic and domestic purposes.

r обеззараживание воды
Процесс удаления из воды, используемой для хозяйственно-бытовых целей, болезнетворных микроорганизмов.

f désinfection *f* d'eau
s desinfección *f* de agua
g Wasserdesinfektion *f*

6653 **water disposal; water removal**
The removal of melt, storm and waste waters from settlements and industrial objects.

r водоотведение
Удаление талых, ливневых и сточных вод из населенных мест и промышленных объектов.

f évacuation *f* d'eau
s evacuación *f* del agua
g Wasserableitung *f*

6654 **water economy**
A branch of economy cosisting of the interconnected water bodies and hydraulic structures which provide each region with water for industrial and domestic use, and protect the waters from any harmful impact.

r водное хозяйство;
водохозяйственная система
Отрасль хозяйства, состоящая из связанных между собой водных объектов и гидротехнических сооружений, обеспечивающая каждый регион водой для промышленного и бытового использования, а также охрану вод и борьбу с их вредным воздействием.

f économie *f* d'eau
s economía *f* hidráulica; economía *f* de agua
g Wasserwirtschaft *f*

* **water edge → 1850**

6655 **water energy; hydroelectric energy; hydroelectric power**
Energy of water flowing in streams or accumulated in lakes or reservoirs used for producing of electicity.

r водная энергия; гидроэнергия
Энергия воды, движущейся в потоках или аккумулированной в озерах и водохранилищах, используемая для выработки электроэнергии.

f énergie *f* hydroélectrique;
énergie *f* hydraulique
s energía *f* hidroeléctrica;
energía *f* hidráulica
g Wasserenergie *f*; hydraulische Energie *f*

6656 **water engineering; hydraulic engineering**
The industrial use of water resources and means of protecting water from any destructive water impact control.

r гидротехника
Промышленное использование водных ресурсов и борьба с разрушительным действием вод.

f hydrotechnique *f*
s hidrotécnica *f*
g Wasserbau *m*; Wassertechnik *f*;
Hydrotechnik *f*

* **water equivalent → 1536**

6657 **water equivalent of snow cover**
The amount of water that would result from a complete melting of a sample of deposited snow, expressed in mm of the water layer thickness.

r водный эквивалент снежного покрова
Масса снега, выраженная в эквивалентном слое воды, полученной от ее таяния.

f équivalent *m* en eau du couche de neige
s equivalente *m* en agua de capa de nieve
g Wasseräquivalent *n* der Schneedecke

6658 **water exchange**
The replacement of water mass in a water body by other water masses incoming from contiguous water bodies (external water

excange) or from other parts and layers of the same water body (internal water excange).

r водообмен

Замещение водных масс, находящихся в водном объекте, другими водными массами, поступающими из сопредельных объектов гидросферы (внешний водообмен) или из других слоев данного водного объекта (внутренний водообмен).

f échange *m* d'eau; échange *m* hydrique

s cambio *m* de agua

g Wasseraustausch *m*

6659 water exchange index

The ratio of the volume yearly inflow to the volume of water in a given water body.

r коэффициент водооообмена

Отношение годового притока воды в водоем к объему воды в нем.

f coefficient *m* d'échange d'ea

s coeficiente *m* de camb

g Koeffizient *m* des Wassera

6660 water exchange time

The period of a complete change of water in a water body.

r время водообмена

Период полной смены воды в водном объекте.

f temps *m* de l'échange d'eau

s período *m* de cambio de agua

g Wasseraustauschzeit *f*

6661 water fnd ice balance

Moisture balance in a glacier basin including water and ice balances.

r водно-ледовый баланс

Баланс влаги в ледниковом бассейне, включающий баланс воды и баланс льда.

f bilan *m* des eaux et glace

s balance *m* de agua y hielo

g Wasser-Eis-Bilanz *f*

6662 water hole

A hollow where water gathers, especially in savanna and hot desert lands, in some cases fed by a spring; a hole on the surface of ice.

r лужа

Небольшое понижение, в котором собирается вода, особенно в саванне и полупустыне, как правило, сохранившаяся с весны; озерцо на поверхности льда.

f mare *f*

s balsa *f*

g Pfütze *f*

6663 water intake

An engineering a structure aimed at taking water from a water object for water supply, power generation, irrigation; the process of taken water for economic purposes.

r водозабор

Гидротехническое сооружение для изъятия воды из водного объекта с целью ее использования для водоснабжения, энергетики, ирригации; само изъятие воды для хозяйственных целей.

f prélèvement *m*; prise *f* d'eau

s toma *f* de agua

g Wasserentnahme *f*

6664 water lag

The interval of time during which the water moves within a drainage area: along the slopes (slope lag), and along talwegs of channels (channel lag).

r добегание воды

Время перемещения водных масс в пределах водосбора: по его склонам (склоновое добегание) и по тальвегам русловых потоков (русловое добегание).

f décalage *m* d'eau

s retención *f* de agua; respuesta *f* de agua

g Wasserablauf *m*

6665 water lag time

The time difference in the hydrograph between the peak of a storm rainfall or snowmelt and the consequent later occurring peak in the discharge of the stream.

r время добегания

Разница во времени на гидрографе между пиком сильного дождя или таяния снега и вызванным ими позднее ростом расхода воды в потоке.

f temps *m* de réponse;
temps *m* de décalage

s tiempo *m* de retención

g Ablaufzeit *f*

6666 water level

The height of free water surface in water bodies with regard to a conventional horizontal surface or sea level.

r уровень воды

Высота свободной поверхности воды в водоемах относительно условной горизональной поверхности или уровня моря.

f niveau *m* d'eau; plan *m* d'eau;
hauteur *f* d'eau

s nivel *m* de agua

g Wasserstand *m*; Wasserniveau *n*

6667 water mass

A great amount of water characterized by the homogeneity of its main characteristics and located within the object of water.

r водная масса

Большой объем воды внутри водного объекта, характеризующийся однородностью основных своих характеристик.

f masse *f* d'eau

s masa *f* de agua

g Wassermasse *f*

* **water meadow → 2199**

* **water parting → 1662**

6668 water permeability

The ability of soil or rock to absorb or pass water through.
r водопроницаемость
Способность почвы или горной породы впитывать и пропускать воду.
f perméabilité *f* par eau
s permeabilidad *f* al agua
g Wasserdurchlässigkeit *f*

6669 **water permeability of soil**
The ability of soil, as a body which contains pores, to absorb water and to pass it through (it is expressed in the amount of water incoming to soil through its surface per time unit).
r водопропускная способность почвы
Свойство почвы как пористого тела впитывать и пропускать воду, выражается величиной слоя воды, поступающей в почву через ее поверхность в единицу времени.
f perméabilité *f* du sol par eau
s permeabilidad *f* del suelo al agua
g Wasserdurchlässigkeit *f* des Bodens

6670 **water pocket**
A film of water enveloping air bubbles in temperate glaciers.
r водяная сумка
Пленка воды, обволакивающая воздушные пузырьки в теплых ледниках.
f pellicule *f* d'eau
s película *f* de agua; bolsa *f* de agua
g Wassertasche *f*

6671 **water pollution; water contamination**
Pollution of rivers, lakes, seas, underground waters, etc. with harmful substances which makes water dangerous to use.
r загрязнение воды;
загрязнение водных объектов
Загрязнение рек, озер, морей, подземных вод веществами, обычно не присутствующими в них, которые делают воду мало пригодной для использования.
f pollution *f* des eaux;
contamination *f* des eaux
s polución *f* de aguas;
contaminación *f* de aguas
g Wasserverschmutzung *f*;
Wasserverunreinigung *f*

6672 **water protection area**
An area around the sources of ground and surface drinking water set aside for protecting water from being contaminated and where any economic activity is prohibited.
r водоохранная зона;
водоохранная полоса
Территория вокруг подземных или поверхностных источников питьевого водоснабжения, где запрещена всякая хозяйственная деятельность.
f aire *f* de protection d'eau
s zona *f* de protección de agua;
faja *f* de protección de agua
g Wasserschutzgebiet *n*

6673 **water protection forest**
Forest growing around a water body which is set aside to protect its natural regime and control erosion.
r водоохранный лес
Лес вокруг водных объектов, предназначенный для сохранения их природного режима, предотвращения эрозии.
f forêt *f* de protection d'eau
s bosque *m* de protección de agua
g Wasserschutzwald *m*

6674 **water purification**
A system of technological processes which allows to clean the natural water used for water supply from undesirable elements and to bring it to the set norms.
r водоочистка
Система технологических процессов, с помощью которых природная вода, используемая для водоснабжения, доводится до установленных норм.
f purification *f* d'eau; épuration *f* des eaux
s purificación *f* del agua;
depuración *f* del agua
g Wasserreinigung *f*

6675 **water quality**
The conformity of water to health and technological requirements which is determined by a number of indices.
r качество воды
Степень соответствия воды потребностям людей и технологическим требованиям, которая определяется по ряду показателей.
f qualité *f* d'eau
s calidad *f* de agua
g Wasserqualität *f*

6676 **water quality index**
A generalised numerical value of water quality estimated by the combination of the main paraneters and types of water use.
r индекс качества воды
Обобщенная числовая оценка качества воды по совокупности основных показателей и видам водопользования.
f indice *m* de qualité d'eau
s índice *m* de calidad de agua
g Wasserqualitätsindex *m*

6677 **water quality parameters**
Biological, physical and chemical characteristics of water indicating its safety for use: salinity, hardness, concentration of harmful substances, etc.
r показатели качества воды
Биологические и физико-химические характеристики (соленость, жесткость, концентрация вредных веществ и т.п.), показывающие пригодность воды для использования.
f indices *mpl* de qualité de l'eau

s caracteristicas *mpl* de calidad de agua
g Wasserqualitätsparameter *mpl*

6678 water receiver
A depression, watercourse or water body receiving water of a drainage system from the adjacent drained territory.
r водоприемник
Лощина, водоток или водоем, принимающие воду осушительной системы с прилегающей мелиорируемой территории.
f bassin *m* de reception d'eau
s receptor *m* de agua
g Einlass *m*

*** water recycling → 6420**

*** water regime → 2883**

6679 water regime of soil
The change of soil moisture in time determined by its income and movement, as well as by the variation of its physical properties and consumption.
r водный режим почвы
Совокупность изменений влажности почв во времени, определяемых поступлением, передвижением, изменением состояния и расхода почвенной влаги.
f régime *m* hydrique du sol; comportement *m* hydrodynamique du sol; régime *m* hydrodynamique du sol
s régimen *m* hidrodinámico del suelo; comportamiento *m* hidrodinámico del suelo
g Bodenwasserhaushalt *m*

6680 water register
Classified data on water reserves.
r водный кадастр
Систематизированный свод сведений о водных ресурсах.
f cadastre *m* d'eau
s inventário *m* de agua
g Wasserkataster *m*

*** water removal → 6653**

6681 water reserve in soil; soil water storage; soil moisture storage
Absolute amount of water contained in a certain soil layer.
r запас влаги в почве; влагозапас
Абсолютное количество влаги, содержащееся в определенном слое почвы.
f réserve *f* d'eau à sol
s reserva *f* de agua en suelo
g Bodenwasservorrat *m*

*** water reservoir-cooler → 1286**

6682 water resources
Ground and surface water reserves fit for the use in economic activity.
r водные ресурсы
Запасы подземных и поверхностных вод, пригодные для использования в хозяйственной деятельности.
f ressources *mpl* en eau
s recursos *mpl* de agua
g Wasserressourcen *mpl*

*** water retention → 5052**

*** water sink (*amer*) → 4688**

6683 water sky
The dark layer on the underside of a cloud which indicates the presence of open water in or behind sea ice.
r водяное небо
Темные полосы на нижней стороне низких облаков, указывающие на чистую воду среди морского льда или за льдом.
f ciel *m* de clairière; ciel *m* d'eau
s cielo *m* de agua
g Wasserhimmel *m*

6684 water spill
The release of water from a reservoir in order to regulate the discharge on a lower part of the watercourse or water level in the reservoir itself. → Fig. 356
r попуск
Сброс воды из водохранилища для регулирования расхода воды на нижележащем участке водотока или уровня воды в самом водохраниище. → Рис. 356
f sortie *f* d'eau
s vertedero *m* de agua; salida *f* de agua; descarga *f* de agua
g Wasserabsenkung *f*

Fig. 356. **Water spill** (right down) at the Nurek hydroelectric power station, Tadjikistan

6685 water spout
The marine equivalent of a tornado, of shorter duration and lower intensity, the product of an intense, localized small-scale cyclonic storm occurring over an ocean or lake, usually in tropical and subtropical regions.
r водяной смерч
Морской эквивалент торнадо, но меньшей интенсивности и продолжительности, смерч, возникающий в небольшом, но

интенсивном тропическом циклоне над океаном или большим озером, обычно в тропических или субтропических широтах.
f trombe *f* marine
s tromba *f* marina
g Wasserhose *f*

6686 water stratification
The succession of water layers of different temperatures, the top layer being the warmest, the lowest layer being the coldest (under certain conditions may be disturbed up to the inverse thermal stratification).
r стратификация воды
Понижение температуры воды от поверхности ко дну водоема (при некоторых условиях может нарушаться, вплоть до обратной термической стратификации).
f stratification *f* des eaux
s estratificación *f* de agua
g Wasserschichtung *f*

6687 water supply; water-supply system
The process of the intake, purification and delivery of water to the place of consumption, and the complex of constructions which provides this.
r водоснабжение;
система водоснабжения
Процесс забора воды из источника, ее очистки и подачи к месту потребления и комплекс сооружений, обеспечивающих это.
f approvisionnement *m* en eau
s abastecimiento *m* de agua
g Wasserversordnung *f*

* **water supply source → 5678**

6688 water surface; sheet of water
The surface of water bodies and watercources.
r водная поверхность
Поверхность воды в водоемах и водотоках.
f surface *f* d'eau
s superficie *f* del agua
g Wasseroberfläche *f*

6689 water table
The surface below which permeable rocks are saturated with water.
r зеркало грунтовых вод
Поверхность, ниже которой водопроницаемые породы насыщены водой.
f niveau *m* de la nappe aquifère;
surface *f* piézometrique
s techo *m* de capa acuífera;
techo *m* de aguas freáticas
g Grundwasserspiegel *m*

* **water transfer → 5098**

* **water transparency → 6310**

6690 water transport
A type of transport moving passengers and goods along natural (rivers, lakes, seas, and oceans) and artificial (canals, reservoirs) waterways.
r водный транспорт
Вид транспорта, занятый перевозкой пассажиров и грузов по естественным (реки, озера, моря и океаны) и искусственным (каналы, водохранилища) водным путям.
f transport *m* fluvial et maritime
s transporte *m* fluvial y marítimo
g Wassertransport *m*; Wasserverkehr *m*

6691 water treatment; water conditioning
Processing of natural waters in order to make its quality to meet consumers' demands.
r водоподготовка
Совокупность способов обработки природной воды, применяемых для приведения ее качества в соответствие с требованиями потребителей.
f traitement *m* d'eau
s tratamiento *m* de agua
g Wasseraufbereitung *f*

6692 water use
The use of water as a source of energy, in industry, agriculture and for domestic purposes.
r использование воды
Применение воды в качестве важного ресурса и источника энергии в промышленности, сельском хозяйстве и в быту.
f utilisation *f* de l'eau
s utilización *f* de agua
g Wasserausnutzung *f*

6693 water user
A branch of economy which uses hydroelectric energy, or uses water as a landscape element, rather than consumes water.
r водопользователь
Отрасли хозяйства, которые используют не саму воду, а ее энергию, а также воду как среду или элемент ландшафта.
f usager *m* d'eau
s usuario *m* de agua
g Wassernutzer *m*

6694 water vapour; aqueous vapour
Water in the Earth's atmosphere in the vapour state and below the critical temperature for water.
r водяной пар
Вода, находящаяся в земной атмосфере в парообразной фазе и ниже критической температуры для воды.
f vapeur *f* d'eau
s vapor *m* de agua; vapor *m* acuoso
g Wasserdampf *m*

6695 water yield
The ability of the ground to release a part of water when the level of ground waters lows down, or the flow of excessive melt or rain water from snow.
r водоотдача

Fig. 357. **Waterfall** Skogafoss in Iceland

Способность грунта отдавать часть воды при понижении уровня подземных вод или стекание из снега избыточной талой или дождевой воды.

f rendement *m* de l'eau
s rendimento *m* de agua
g Wasserabgabe *f*

6696 water-and-ice pulp
A mixture of water and finely crushed ice, used for the cooling of water which has the excess of heat.
r льдоводяная пульпа
Смесь воды и мелкодисперсного льда, применяемая для охлаждения воды, содержащей излишки тепла.
f mélange *m* de glace et eau
s pasta *f* de hielo y agua
g Eis-Wasser-Pulpe *f*

* **water-bearing layer → 352**

* **water-built terrace → 13**

* **watercourse → 5853**

* **water-cut bench → 13**

6697 waterfall
Vertical or steep descent of water in the bed of a river occurring where the flow of the river is broken by the sharp edge of a cliff. → Fig. 357
r водопад
Вертикальное или круто наклонное падение воды с уступа, пересекающего речное русло. → Рис. 357
f chute *f* d'eau
s salto *m* de agua
g Wasserfall *m*

6698 waterfront
Land or a section of a town fronting or bordering on a body of water: a river, lake or sea.
r водный фасад
Часть города, обращенная или граничащая с водоемом: рекой, озером или морем.
f front *m* d'eau
s frente *m* de agua
g Wasserfront *f*

6699 water-gap
A narrow, deep part of a valley where the river crosses an upland or mountain ridge.
r сквозная долина; долина прорыва; клюз
Узкий и глубокий участок долины, в котором река пересекает возвышенность или горный хребет на всю их ширину.
f cluse *f*
s cluse *f*
g Durchbruchstal *n*; Klamm *f*

6700 watering place
1. A place to which animals were taken or come for drinking. 2. A spa or locality with a mineral spring to which people went to drink the waters or to bathe.
r водопой$_1$; водный курорт$_2$; воды$_2$
1. Место, куда животные обычно приходят, чтобы попить воды. 2. Курорт или местность с минеральным источником, используемым для лечения.
f abreuvoir *m*
s aguada *f*
g Tränke *f*; Tränkstelle *f*

6701 water-level weathering
The development and enlargement of tidal pools by weathering and by the action of rock-grinding animals.
r береговое выветривание
Разрастание приливной полосы вследствие выветривания и разрушения прибрежных скал живущими здесь животными.
f météorisation *f* littorale
s meteorización *f* litoral
g Strandverwitterung *f*

6702 waterlogged *adj*
Applied to the state of a soil when it is saturated with water.
r затопленный; подтопленный; заболоченный
Применяется к состоянию почвы, когда ее поры заполнены водой.
f inondé *adj*; uligineux *adj*; uliginaire *adj*; vaseux *adj*
s inundado *adj*; cenagoso *adj*; anegado *adj*
g gesättigt *adj*

6703 waterlogging
The state of impeded drainage of soil and excessive saturation of land induced by the rise of groundwater level due to building dams, reservoirs, canals and other hydraulic constructions.
r подтопление земель

Избыточное увлажнение земель вследствие подъема уровня грунтовых вод в результате сооружения плотин, водохраниилищ, каналов и других гидротехнических сооружений.
f sous-submersion *f* du terrain
s empantanamiento *m* del terreno; anegamiento *m*; anegación *f*
g Unterschwemmung *f* des Landes

* **water-logging → 4396**

6704 waterriching of soil
The water that incomes to soil at a time, it may be natural watering (rains, melt waters) or man-induced water riching (irrigation) usually used before the beginning of vegetative season aimed at the restoration of soil water storage.
r влагозарядка почвы
Единовременное поступление в почву влаги, может быть естественной (дожди, поступление талых вод) или искусственной (орошение), проводимой, как правило, до начала вегетационного периода с целью восстановления запасов влаги в почве.
f humidification *f* du sol; humectation *f* du sol
s humidificación *f* del suelo; humectación *f* del suelo
g Bodenbewässerung *f*

* **watershed (*amer*) → 1690**

6705 watershed; divide
The boundary between the adjacent drainage basins of rivers or water bodies which usually runs along the highest points of the interfluves.
r водораздел
Линия, разделяющая смежные водосборные бассейны рек или водоемов, обычно проходящая по самым высоким точкам междуречья.
f ligne *f* de partage des eaux
s divisoria *f* de aguas; línea *f* de división de aguas
g Wasserscheide *f*

6706 watershed area
A plain area with a vaguely defined watershed.
r водораздельное пространство
Равнинная область с неясно выраженной в рельефе линией водораздела.
f région *f* de partage des eaux
s área *f* divisoria de aguas; área *f* de división de aguas
g Wasserscheidegebiet *n*

* **water-supply system → 6687**

6707 waterway
A stretch of water, i.e. of a lake, river, canal, which is expanse used for navigation and rafting.
r водный путь

Fig. 358. **Wave**. The drawing by Russian artist Ivan Aivazovsky

Водное пространство озера, реки или канала, используемое для судоходства и сплава леса.
f voie *f* navigable; voie *f* fluviale
s vía *f* navegable; vía *f* fluvial
g Wasserstraße *f*; Wasserweg *m*

6708 watten; tidal marsh; mud flat; wadden
A tract of lower-lying coast of tidal seas flooded and dried in the course of tidal cycle and composed of silt and clay to a considerable depth, common at the North Sea coast of Germany, in the Netherlands, etc.
r ватты; илистая равнина
Полоса низменных побережий приливных морей, затопляемая при приливе и осушающаяся при отливе, сложена илом и глинистым материалом на значительную глубину, характерны для побережья Северного моря в Германии, а также в Голландии и др.
f vasière *f*; marais *m* littoral
s cenagal *m*; marisma *f*
g Watten *npl*

6709 wave
A deformation of a water surface in the form of an oscillatory movement which manifests itself by an alternating rise and fall of that surface generated by the wind's pressure on the water surface, by submarine earthquakes or landslides, by the gravitational pull of the Sun and Moon. → Fig. 358
r волна
Деформация водной поверхности в форме колебательного движения, или последовательного подъема и падения этой поверхности, вызываемая давлением ветра на поверхность воды, подводными землетрясениями или оползнями, а также гравитационным взаимодействием Солнца и Луны.→ Рис. 358
f vague *f*
s ola *f*
g Welle *f*

6710 wave amplitude
The vertical distance between the level of still water and the crest of a wave which is half of the wave height.
r амплитуда волны

Вертикальное расстояние от уровня спокойной воды до гребня волны (т.е. половина высоты волны).
f amplitude *f* de vague
s amplitud *f* de ola
g Wellenamplitude *f*

6711 wave approach angle
The angle between a wave path and coastal line.
r угол подхода волн
Угол между направлением волны и береговой линией.
f angle *m* de l'approche de vague
s ángulo *m* de acceso de ola
g Wellenwinkel *m*

6712 wave base
The lowest limit of orbital motion beneath waves and the depth at which wave action ceases to disturb the sea-floor sediments.
r волновая база
Самая нижняя точка орбитального движения волн и соответствующая глубина, на которой прекращается волновое разрушение донных осадков.
f base *f* des vagues
s base *f* de olas
g Wellenbasis *f*

6713 wave cyclone
The frontal cyclone at the initial stage of its development and having a wide warm sector.
r волновой циклон
Фронтальный циклон в начальной стадии развития с широким теплым сектором.
f cyclone *m* d'onde
s ciclón *m* de ondas
g Wellentief *n*

6714 wave deformation
The changes of the parameters of a wave with the decrease of the sea depth and the approaching of the wave to the shore.
r разрушение волны; деформация волны
Изменения параметров волны по мере уменьшения глубины и приближения волны к берегу.
f déformation *f* de vague; modification *f* de vague
s deformación *f* de ola; modificación *f* de ola
g Wellendeformation *f*; Wellenbruch *m*

* **wave diffraction → 1598**

6715 wave direction
The direction from which the wave comes.
r направление волны
Направление, откуда идет волна.
f direction *f* de vague
s dirección *f* de ola
g Wellenausbreitrichtung *f*

6716 wave drag layer
A layer of loads involved into motion by waves or flows determined by size of particles and the intensity of water movement.
r активный слой наносов
Вовлекаемый в движение под действием волн или течения слой наносов, параметры которого определяются размерами частиц и интенсивностью движения воды.
f dépôt *m* actif
s depósito *m* activo
g aktive Schicht *f* der Anschwemmungen

6717 wave erosion; marine abrasion
The destructive action of waves on a seashore.
r волновая эрозия
Разрушительное действие волн у побережья.
f érosion *f* marine
s erosión *f* marina
g Brandungserosion *f*; Wellenerosion *f*

6718 wave extinction; wave weakening
Decrease of wave strength and surge in a near-coastal area.
r затухание волнения; гашение волны; затухание волн
Уменьшение волнения и волнового нагона в прибрежной зоне.
f extinction *f* des vagues; amortissement *m* de vague
s extinción *f* de olas; amortiguación *f* de olas
g Abklingung *f* des Wellenganges; Wellendämpfung *f*; Wellenlöschung *f*

6719 wave height
The height difference between the crest of wave and the preceding trough of wave, it is twice the wave amplitude.
r высота волны
Вертикальное расстояние от гребня волны до нижней точки подошвы волны.
f hauteur *f* de vague; hauteur *f* d'onde
s altura *f* de ola; altura *f* de onda
g Wellenhöhe *f*

6720 wave overturning
Partial destruction of a wave in the shallow water or at its meeting with a head-on sea current when its crest overturns into the trough of wave.
r забурунивание волны; опрокидывание волны
Частичное разрушение волны на мелководье или на встречном течении путем опрокидывания верхней части ее гребня.
f basculement *m* de vague
s basculamiento *m* de ola
g Wellenumsturz *m*

6721 wave period
The time during which two successive crests of wave pass a fixed point.
r период волны
Интервал времени между прохождением двух последовательных гребней волн через фиксированную вертикальную линию.

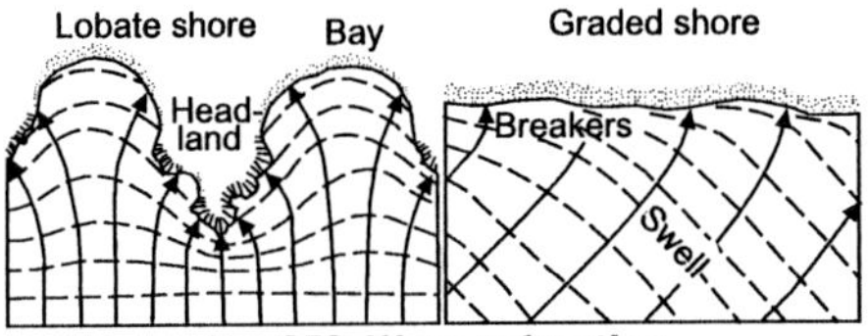

Fig. 359. **Wave refraction**

f période *f* d'onde; période *f* de vague
s período *m* de ola; período *m* de onda
g Wellenperiode *f*

6722 wave reflection
The backward motion of a sea wave from a cliff or coastal constructions, such as a pier or breakwater.
r отражение волны
Обратное движение волны от обрывистого берега или береговых сооружений, например, от мола.
f réflexion *f* de vague
s reflexión *f* de ola
g Wellenreflexion *f*

6723 wave refraction; refraction of waves
The process by which wave crests change direction as they approach a shoreline, owing to the shallowing of water. → Fig. 359
r рефракция волн
Процесс, посредством которого гребни волн изменяют направление по мере приближения к берегу из-за уменьшения глубины моря. → Рис. 359
f réfraction *f* de vague
s refracción *f* de ola
g Wellenrefraktion *f*

6724 wave repulsing wall
A massive concrete construction usually with uneven front surface reflecting waves and, thus, protecting the shore being washed out.
r волноотбойная стенка
Конструкция из массивных бетонных элементов, обычно с криволинейным очертанием передней грани, отражающая набегающую волну и тем самым предохраняющая берег от размыва.
f mur *m* brise-lames
s pared *f* de guía de ondas; pared *f* rompeolas
g Wellenschutzwand *f*

6725 wave shadow
A natural obstacle in the path of waves (island, cape, sea bed elevation) behind which the energy of the waves is considerably weakened.
r волновая тень
Естественное препятствие на пути волн (остров, мыс, подводное поднятие), за которым воздействие волн ослаблено.
f ombre *f* de vague
s sombra *f* de ola
g Wellenschatten *m*

* **wave splash → 785**

6726 wave stabilisation
A relatively stable regime of a sea with the fixed position of wave destruction zone and respective weakening of exchange of waters and loads between the beach and underwater slope.
r стабилизация волнения
Относительно устойчивый режим волнения с постоянным положением зоны разрушения волн и, соответственно, ослабление обмена вод и наносов между пляжем и подводным склоном.
f stabilisation *f* des vagues
s estabilización *f* de olas
g Stabilisierung *f* des Wellenganges

6727 wave strengthening
The increase of the parameters of waves and surge towards the shore and, accordingly, the increase of the compensating water outflow towards the sea and parallel to the coastal line.
r усиление волнения; нарастание волнения
Возрастание параметров волн и нагона воды к берегу, а также, соответственно, компенсационного оттока в сторону моря и параллельно береговой линии.
f accroissement *m* des vagues
s acrecentamiento *m* de olas
g Verstärkung *f* des Wellenganges

6728 wave train
A series of waves with the same direction.
r волновая система; цуги волн
Серия волн одного направления.
f train *m* des vagues
s tren *m* de olas
g Wellenzug *m*

* **wave weakening → 6718**

6729 wave-built island; alluvian island
A low island formed by the accumulation of sea or river loads.
r намывной остров
Низменный остров, образующийся вследствие аккумуляции морских или речных наносов.
f île *f* construite
s isla *f* de aluvión
g Schwemmlandinsel *f*

6730 wave-built shore; alluvian coast
A low flat shore of a sea or lake formed by the accumulation of loose material.
r намывной берег
Низкий плоский берег моря или озера, образованный рыхлыми наносами.
f côte *f* d'alluvionnement
s costa *f* de aluvión
g Anschwemmungsküste *f*; Schwemmlandküste *f*

6731 **wave-cut niche**
A hollow made by the upwash at the basis of a cliff.
r волноприбойная ниша
Углубление, выработанное прибойным потоком в основании берегового уступа.
f encoche *f*
s entalladura *f*
g Brandungshohlkehle *f*

* **wave-cut platform → 11**

6732 **wavelength**
The distance between the highest poins of two succesive wave crests, or other equivalent points in a wave.
r длина волны
Расстояние по горизонтали между двумя последовательными гребнями или подошвами волны.
f longueur *f* de vague
s longitud *f* de ola
g Wellenlänge *f*

* **wave-refraction diagram → 4956**

* **waves → 6734**

6733 **waves of life**
Periodical or spontaneous increases and decreases of natural populations characteristic of all species.
r волны жизни
Присущие всем видам периодические и непериодические подъемы и спады численности природных популяций.
f vagues *fpl* de vie
s olas *fpl* de vida
g Lebenswellen *fpl*

6734 **waving; waves**
Regular oscillations in the water surface of large water bodies caused by the pushing effect of wind.
r волнение; ветровое волнение
Сложное колебательное движение поверхностного слоя водоема, вызванное ветром.
f houle *f*
s oleaje *m*
g Wellenschlag *m*; Wellengang *m*

6735 **waxing slope; waning slope**
The convex element in the upper part of a slope profile, situated above the constant slope.
r выступающий склон
Выпуклый элемент верхней части профиля склона, расположенный выше стабильного склона.
f pente *f* bombée
s pendiente *f* abombada
g erhabener Abhang *m*

6736 **way of life**
The way in which human societies organize their common life according to the mode of production and national traditions.
r жизненный уклад
Система социальных отношений в обществе, связанная со способом производства и национальными традициями.
f mode *m* de vie; genre *m* de vie
s modo *m* de vida
g Lebenshaltung *f*

* **ways of communication → 6322**

6737 **weather**
The state of the atmosphere at a particular time characterized by several meteorological elements, such as humidity, temperature, sunshine hours, cloud cover, visibility, and precipitation.
r погода
Состояние атмосферы в данный момент времени, характеризуемое различными метеорологическими элементами, к которым относятся температура, влажность, продолжительность солнечного сияния, облачность, видимость, атмосферные осадки.
f temps^2 *m*; météo *f*
s tiempo^2 *m*; temporal *m*
g Wetter *m*; Wetterlage *f*

6738 **weather chart; weather map; synoptic chart**
A chart giving a summary, a general view of the meteorological conditions (isobars, temperatures, winds, etc.) over a large area at a given time, an essential tool in weather forecasting.
r синоптическая карта; метеорологическая карта; карта погоды
Географическая карта, на которой представлен общий обзор метеорологических условий (изобары, температуры, ветры и т.п.) на большой территории в определенное время, важный инструмент прогноза погоды.
f carte *f* synoptique; carte *f* météorologique; carte *f* de temps
s mapa *m* sinóptico; mapa *m* del tiempo
g Wetterkarte *f*

6739 **weather discomfort**
A characteristic of weather from the viewpoint of its unfavourable influence upon living conditions.
r дискомфортность погоды
Характеристика погоды с точки зрения ее неблагоприятного воздействия на условия жизни.
f inconfort *m* de temps
s incomodidad *f* de tiempo
g Wetterbeschwerden *fpl*

6740 **weather forecast; weather forecasting; meteorological forecast**
A statement of expected meteorological conditions for a specific period and area.

r прогноз погоды
Составление научно обоснованных предположений о будущем состоянии погоды для определенного периода времени и места.
f prévision *f* météorologique; prévision *f* du temps
s predicción *f* meteorológica; pronóstico *m* meteorológico; previsión *f* del tiempo; previsión *f* meteorológica
g Wettervorhersage *f*

* **weather forecasting → 6740**

* **weather map → 6738**

* **weather satellite → 3817**

6741 **weather type**
A series of generalized synoptic situations with typical general characteristics of weather.
r тип погоды
Серия обобщенных синоптических ситуаций с типичным комплексом характеристик погоды.
f type *m* de temps
s tipo *m* de tiempo
g Wettertyp *m*

6742 **weathering**
The mechanical destruction and chemical change of rocks at or near the Earth's surface by temperature fluctuations and exposure to water, the atmosphere, organic matter. → Fig. 360
r выветривание
Механическое разрушение и химическое изменение горных пород на земной поверхности под влиянием колебаний температуры, химического и механического воздействия атмосферы, воды и организмов. → Рис. 360
f météorisation *f*

Fig.360. **Weathering**

s meteorización *f*
g Verwitterung *f*

6743 **weathering crust**
A sheet of continental deposits formed on the Earth's surface as a result of decomposition and disintegration of rocks by weathering.
r кора выветривания
Плащ континентальных отложений, возникающий на земной поверхности в результате выветривания горных пород.
f pellicule *f* détririque meuble; manteau *m* de débris
s manto *m* de derrubios; manto *m* detrítico
g Verwitterungsdecke *f*; Verwitterungskruste *f*

6744 **weathering front**
The zone of contact of the regolith with the underlying rock, i.e. between weathered and unweathered rock.
r фронт выветривания
Зона контакта реголита и нижележащей породы, т.е. полоса между выветрелой и невыветрелой горной породой.
f front *m* de météorisation
s frente *m* de meteorización
g Verwitterungsfront *f*

* **wedge of high pressure → 5076**

6745 **wedging out of a stratum**
The decrease of a rock layer thickness till its disappearance.
r выклинивание слоя
Уменьшение мощности пласта горных пород до полного его исчезновения.
f amincissement *m* de couche; amenuisement *m* de couche
s acuñamiento *m* de estrato
g Verdünnung *f* der Schicht

6746 **weeds**
Plants growing on the areas where they are not desirable, e.g. on agricultural lands among the cultivated crops (segetal plants) or on the areas transformed by human activity, e.g. on dumps and wastes (ruderal plants).
r сорная растительность; сорняк
Растения, произрастающие на сельскохозяйственных землях и засоряющие их (сегетальные растения), а также на свалках и пустырях (рудеральные растения).
f mauvaises herbes *fpl*
s hierbas *fpl* malas; maleza *f*
g Unkräuter *npl*

* **weir → 5722**

6747 **welfare; well-being**
The state or condition of being well, thriving, prospering, and the bringing about of this state for the members of a community by providing medical care, education, public housing, pay for the aged, etc.

r благосостояние
Состояние или условия процветания и достатка в обществе за счет обеспечения всех членов общества образованием, медицинским обслуживанием, жильем, пенсией и пр.
f bien-être *m*
s bienestar *m*; prosperidad *f*
g Wohlstand *m*; Wohlergehen *n*

6748 welfare geography
An approach in human geography which features inequalities in social well-being and social justice, looking at the areal differentiations and spatial organization of human activity from the point of view of the welfare of the people involved.
r география благосостояния
Направление в общественной географии, рассматривающее неравенство качества жизни и социальной справедливости, их территориальную дифференциацию, а также пространственную организацию деятельности людей с точки зрения их благосостояния.
f géographie *f* de bien-être
s geografía *f* de bienestar
g Wohlstandgeographie *f*

6749 well$_1$
A hole in the ground from which a supply of water is extracted.
r колодец
Узкая и глубокая вертикальная яма, вырытая до водоносного слоя и предназначенная для добывания воды.
f puits *m*
s pozo *m*
g Brunnen *m*

* **well$_2$ → 1713**

* **well-being → 6747**

6750 west
One of the four cardinal points, directly opposite the east on the side of someone facing due north, i.e. the direction of the setting sun at the equinox.
r запад
Одна из четырех главных сторон горизонта, расположенная слева от наблюдателя, стоящего лицом к северу, точка захода Солнца в период равноденствия.
f ouest *m*
s oeste *m*
g West(en) *m*

* **west *adj* → 6752**

* **West Wind Drift → 286**

6751 westerlies; Brave West Winds
The dominant west-to-east motion of the atmosphere centered over the middle latitudes of both hemispheres. → Fig. 361
r западный перенос
Преобладающий перенос воздуха с запада на восток в тропосфере и стратосфере средних широт, а также в верхней тропосфере и стратосфере тропических и полярных широт. → Рис. 361
f zone *f* des westerlies
s zona *f* de westerlies
g Westwinddrift *f*

6752 western *adj*; occidental *adj*; west *adj*
Relating to, belonging to, situated towards, coming from the west, e.g. winds blowing from the west.
r западный
Относящийся, расположенный или приходящий с запада, например, ветер, дующий с этого направления.
f d'ouest; occidental *adj*
s del oeste; occidental *adj*
g West-; westlich *adj*

* **wet adiabatic lapse rate → 5223**

6753 wet snow
A type of snow which occurs when temperatures are high enough to bond crystals into large flakes dy regelation; deposited snow that contains a great amount of water drops.
r мокрый снег
Снег, выпадающий при достаточно высоких температурах, когда отдельные кристаллы превращаются в большие снежинки благодаря режеляции; отложенный снег, содержащий капли воды.
f neige *f* trempée
s nieve *f* mojada
g Naßschnee *f*; Feuchtschnee *f*

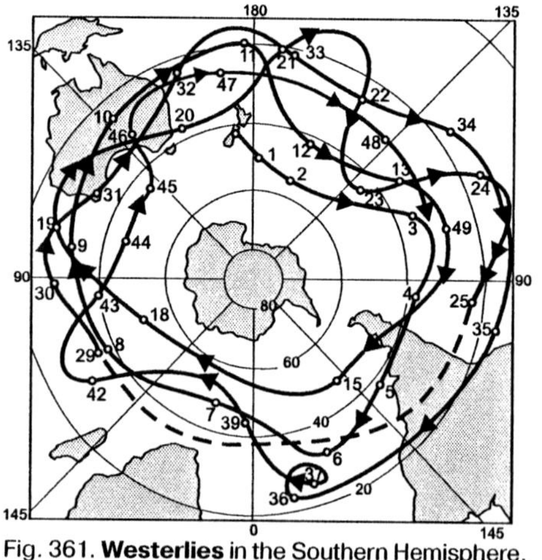

Fig. 361. **Westerlies** in the Southern Hemisphere. Typical trajectories of meteorological sonds; figures near points are number of days after the beginning of flights, the distance between points equals one day

6754 wet snow avalanche; headwater avalanche
An avalanche forming by a wet or very wet snow caused by the thawing, spring snow melting and rainfall; a gravitational gliding of the snow saturated with water.
r мокрая лавина; гидронапорная лавина
Лавины из влажного, увлажненного или мокрого снега, образующиеся во время оттепелей, весеннего снеготаяния, выпадения дождя; гравитационное сползание снежных масс, пропитанных водой.
f avalanche *f* de neige humide
s avalancha *f* de nieve húmeda
g Naßschneelawine *f*

6755 wetland
Any land partly or completely covered with a thin layer of water fresh or salt and has developed a specially adapted vegetation where the soil is waterlogged: lagoons and marshes, flood-plains, bogs and lakes overfilling in with plants.
r ветланд; переувлажненные земли; водно-болотные угодья
Территория, частично или полностью покрытая тонким слоем воды, пресной или соленой, и особой растительностью, адаптированной к таким условиям, сюда относятся заболоченные понижения рельефа, лагуны и марши, поймы, болота и зарастающие озера.
f wetland *m*
s wetland *m*; humedal *m*
g Feuchtgebiet *n*

6756 wetting front
The lowest limit to which a soil becomes saturated by infiltration from above, especially in arid regions.
r влажный фронт
Нижняя граница, выше которой почва насыщена просачивающейся дождевой водой, особенно в аридных районах.
f front *m* mouillé
s frente *m* mojado
g Befeuchtungsfront *f*

6757 whaleback
A large barchan ridge parallel to the wind direction with a flat top crossed by chains of smaller barchans, they are often occur in Sahara.
r китовая спина
Характерные для Сахары крупные барханные продольные ветру гряды со сглаженным гребнем и пересекающими их поперечными или диагональными «ребрами» барханных цепей.
f dos *m* de baleine
s lomo *m* de ballena
g Walfischrücken *m*

6758 wharf
A landing stage to which barges and ships may be moored while loading and unloading.
r пристань; причал
Место на берегу, куда могут причалить суда и баржи для разгрузки и загрузки грузов и пассажиров.
f quai *m*; débarcadère *m*
s desembarcadero *m*; atracadero *m*
g Anlegestelle *f*; Landungsplatz *m*

* **wheelgraph → 1660**

* **whirl → 6606**

6759 whirlpool
A circular eddy movement of water appearing in some parts of a water body, produces in a river by the configuration of the channel, in the sea by the effect of winds or tides, or by the meeting of currents.
r водоворот
Круговое движение воды, возникающее на отдельных участках водоемов и водотоков, вызванное на реке конфигурацией русла, а на море воздействием ветра или приливоотливных течений, а также встречей течений разного направления.
f tourbillon *m*
s torbellino *m*; remolino *m*
g Kolkbecken *n*; Strudel *m*

6760 whirlpool notch
A hollow formed on the rocky walls of river gorges when kettle holes merge together.
r водоворотная ниша
Ниша, образовавшаяся на скалистых бортах ущелий бурных горных рек в результате слияния исполиновых котлов.
f trous *f* de remous; niche *f* de remous
s nicho *m* de remolino
g Strudelnische *f*

6761 whirlwind
A rapidly rapidly rotating column of air revolving around a local centre of low pressure, caused by local surface heating and exceptinally strong convection.
r воздушный вихрь
Небольшая масса воздуха, охваченная вращательным движением по вертикальной оси вокруг центра низкого давления, образованного местным прогревание поверхности и исключительно сильной конвекцией.
f tourbillon *m* de vent
s torbellino *m* de viento; torbellino *m* de aire
g Wirbelwind *m*

6762 white caps; flock of sheep; horses
White foam clusters which occur by surfing of wave chests.
r барашки
Сгустки белой пены, образующейся на гребнях волн при их забурунивании.
f moutons *mpl*; taches *fpl* d'écume; lames *fpl*
s cabrillas *fpl* manchas *fpl* de espuma; borregos *mpl*

Fig. 362. **White nights** at Kola Peninsula

g Schaumköpfe *mpl*; Schaumwellen *fpl*; Schwäne *mpl*

* **white frost → 2786**

6763 **white nights**
Summer nights in subpolar and polar latitudes during which twilight is continuous. → Fig. 362
r белые ночи
Летние ночи в субполярных и полярных широтах, в течение которых сумерки не прекращаются. → Рис. 362
f nuits *fpl* blanches
s noches *fpl* blancas
g weiße Nächte *fpl*

* **white rainbow → 2231**

* **white sea ice → 2158**

6764 **white smoker**
A plume of hydrothermal fluid containing finely dispersed particles of white sulphides issuing from a submarine vent on a mid-oceanic ridge.
r белый курильщик
Вертикальная струя или «султан» гидротермального раствора, содержащего мелкие частицы белых сульфидов, извергающаяся из подводной трещины или жерла на срединно-океаническом хребте.
f fumeur *m* blanc
s fumador *m* blanco
g weißer Raucher *m*

6765 **white squall**
A squall without precipitation, sometimes even without cloud.
r белый шквал
Шквал без осадков, иногда даже без облаков.
f grain *m* blanc
s turbonada *f* blanca
g Bö *f* ohne Niederschläge

6766 **white-out$_1$**
A condition in a snow storm when the falling snow is so great that visibility is reduced to the minimum owing to multiple reflection from cloud and snow crystals, and the snow cover merges into the sky without a perceptible horizon.
r снежная мгла
Помутнение воздуха во время метели и непосредственно после ее окончания, вызванное снежными частицами в воздухе, переносимыми ветром.
f brume *f* de neige
s bruma *f* de nieve
g Schneeblendung *f*

6767 **white-out$_2$**
An atmospheric optical phenomenon of the polar regions in which the observer appears to be engulfed in a uniformly white glow, owing to lack of contrast between the sky and the surface.
r белая мгла; белая «тьма»
Атмосферное оптическое явление в полярных районах, при котором наблюдатель кажется окутанным однородно белым светом из-за отсутствия контраста между небом и землей.
f blanc dehors *m*; temps *m* laiteux
s paisaje *m* blanco
g Weißblendung *f*

6768 **wholesale trade**
Buying and selling goods in large quantities.
r оптовая торговля
Часть внутренней торговли, охватывающая продажу крупных партий товаров.
f commerce *m* en gros
s comercio *m* al por mayor
g Großhandel *m*

* **wholesome production → 1816**

6769 **wilderness**
An area which has generally been affected more by natural forces than by human agency, a region little affected by people.
r дикая местность; девственная природа
Территория, подверженная естественным процессам и мало тронутая человеком.
f terrain *m* vierge
s lugar *m* salvaje
g Wildnis *f*; unkultiviertes Gebiet *n*

6770 **wildfires; brushfires; bushfires; forest fire**
Conflagrations occurring in desiccated vegetation encouraged by hot, dry weather conditions and spread by strong winds.
r лесной пожар
Пожар, вызванный жаркой и сухой погодой и распространяющийся по лесной площади благодаря сильным ветрам.
f feu *m* de forêt
s fuego *m* de bosque
g Waldbrand *m*

* **wildlife refuge → 5193**

* **wildlife reserve → 5193**

6771 wilting point
Minimal possible soil water content at which the liquid water is able to move and the soil moisture is absorbed by plant roots.
r влажность завядания
Минимально возможное влагосодержание почвы, при котором вода в жидком виде может перемещаться и всасываться корнями растений.
f point *m* de flétrissure
s punto *m* de marchitamiento
g Welkepunkt *m*

6772 wind
A horizontal movement of air relative to the Earth's surface induced by irregular distribution of the atmospheric pressure resulting from differential heating of the global surface and dynamic factors of the atmosphere itself.
r ветер
Движение воздуха над земной поверхностью, вызванное неравномерным распределением атмосферного давления, возникающим из-за неодикового нагревания поверхности Земли и динамических факторов самой атмосферы.
f vent *m*
s viento *m*
g Wind *m*

6773 wind chill; chill factor
A parameter used to determine a combined influence of air temperature, wind speed and air humidity on human sensation of temperature discomfort, it is measured in the loss of a body heat in watts per square meter of dry human skin.
r фактор охлаждения;
индекс охлаждения ветром
Показатель, демонстрирующий комплексное влияние температуры, скорости ветра и влажности воздуха на теплоощущение человека, выражается потерей тепла телом в ваттах на квадратный метр поверхности кожи.
f indice *m* de refroidissement du vent; refroidissement *m* éolien
s factor *m* del enfriamiento por viento; factor *m* del frío sentido
g Kühlfaktor *m*; Windkühlfaktor *m*

6774 wind corrasion
The mechanical wearing of rocks by wind-blown sand.
r ветровая корразия; эоловая шлифовка
Механическое воздействие на горные породы переносимого ветром песка.
f polissage *m* éolien; corrasion *f* éolienne
s pulimento *m* eólico; corrasión *f* eólica
g Windschliff *m*; Sandschliff *m*

6775 wind direction
The direction from which the wind is blowing.
r направление ветра
Направление, откуда дует ветер.
f direction *f* de vent
s dirección *f* de viento
g Windrichtung *f*

6776 wind divide
A semipermanent feature of the atmospheric circulation (usually a ridge of high pressure) on opposite sides of which the prevailing wind directions differ greatly.
r ветрораздел
Более или менее устойчивая ситуация в циркуляции атмосферы (обычно гребень высокого давления), по обе стороны которой преобладающие направления ветра сильно различаются.
f ligne *f* de partage des vents
s línea *f* divisoria del viento
g Windscheide *f*

* **wind drift → 1707**

6777 wind energy; wind power
Mechanical or electrical power generated by the rotor of a windmill driven by wind.
r ветровая энергия
Альтернативный источник энергии, обусловленной действием ветра.
f énergie *f* de vent; énergie *f* éolienne
s energía *f* de viento; energía *f* eólica
g Windenergie *f*

* **wind farm → 6780**

6778 wind farm; eolic park
A group of generators which produce the electric energy.
r ветряная электостанция
Группа ветряков, производящих электроэнергию.
f centrale *f* électrique éolienne; centrale *f* éolienne
s parque *m* eólico; central *f* eólica
g äolisches Kraftwerk *n*

6779 wind induced mixing
Mixing of water in the seas and lakes as a result of its disturbance produced by wind.
r ветровое перемешивание
Перемешивание воды в морях и озерах в результате волнения, вызываемого ветром.
f mélange *m* venteux
s mezcla *f* por viento
g Windvermischung *f*

* **wind load → 6781**

* **wind power → 6777**

6780 wind power station; wind farm
A place where wind turbines convert wind energy into electrical energy.
r ветровая электростанция
Установка, преобразующая энергию ветра в электрическую.

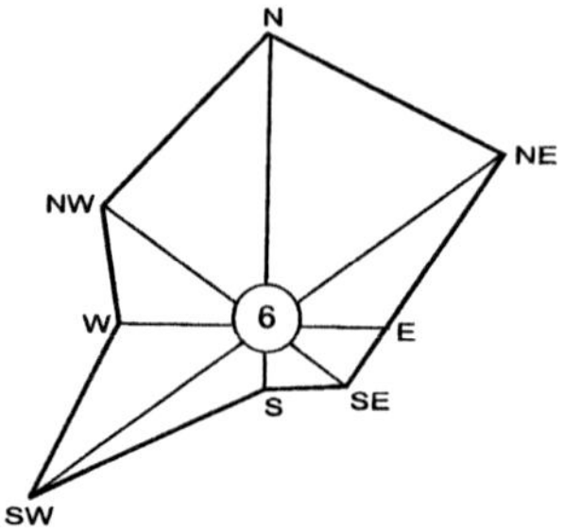

Fig. 363. **Wind rose**

f centrale *f* électrique éolienne
s central *f* eólica;
generador *m* de energía eólica
g Windkraftwerk *n*

6781 wind pressure; wind load
The total force exerted by moving air upon a structure situated on its way.
r давление ветра; ветровая нагрузка
Суммарное давление, производимое движущимся воздухом на предметы или поверхность, находящиеся на его пути.
f pression *f* de vent; charge *f* due au vent
s presión *f* del viento; carga *f* del viento
g Winddruck *m*; Windbelastung *f*

6782 wind resistance
The ability of tree species with deep and branching roots to withstand storms and hurricanes.
r ветроустойчивость
Способность древесных пород с глубокой и разветвленной корневой системой противостоять бурям и ураганам.
f résistance *f* aux vents
s resistencia *f* al viento
g Windbeständigkeit *f*

*** wind ridges → 5215**

6783 wind rose
A diagram, a star-shaped graphic representation of to show the frequency (and usually the speed) with which wind blows from the various points of the compass in a certain place. → Fig. 363
r роза ветров
Графическое изображение, диаграмма в виде звезды, показывающая преобладающую частоту и нередко скорость, с которой ветер обычно дует в данном месте. → Рис. 363
f rose *f* des vents; diagramme *f* de vents
s rosa *f* de vientos
g Windrose *f*; Windstern *m*

6784 wind setup; wind-induced surge
The rise of water level under the impact of wind.
r нагон
Повышение уровня воды, вызванное воздействием ветра.
f courant *m* d'afflux
s corriente *m* de aflujo
g Windstau *m*

6785 wind shadow
A zone of weak winds in the leeside of a mountain, mountain range, or a large construction sheltered from direct force of the wind.
r ветровая тень
Область слабых ветров на подветренной стороне горы, горного хребта или крупного сооружения.
f ombre *f* du vent
s sombra *f* del viento
g Windschatten *m*

6786 wind slab; snow slab
A dry, hard layer of fine-grained snow hardened by the wind which is prone to sliding and hence often causes snow avalanches.
r ветровая доска; снежная доска
Пласт мелкозернистого снега на поверхности снежного покрова, состоящий из плотно уложенных ветром кристаллов и нередко служащий причиной схода снежных лавин.
f plaque *f* de neige
s placa *f* de nieve
g Packschnee *m*; windgepresster Schnee *m*; Schneebrett *n*

*** wind slab snow → 2342**

6787 wind stress
The drag or tangential force per unit area exerted on the Earth's surface by the adjacent layer of moving air.
r напряжение ветра
Сопротивление или касательная сила на единицу площади, действующие на поверхность Земли прилегающим слоем движущегося воздуха.
f tension *f* du vent
s tensión *f* del viento
g Windspannung *f*

*** wind turbine → 6795**

6788 wind wane
A device which indicates wind direction by means of pivoted horizontal arm rotating freely on a vertical spindle.
r флюгер
Прибор, состоящий из свободно вращающегося крыла, закрепленного на вертикальной оси, показывающего направление дующего ветра.
f girouette *f*
s giralda *m*; veleta *f*; giraldilla *f*
g Wettwefahne *f*; Windfahne *f*

6789 wind wave
A wave resulting from the action of wind on a water surface. → Fig. 364, p. 811

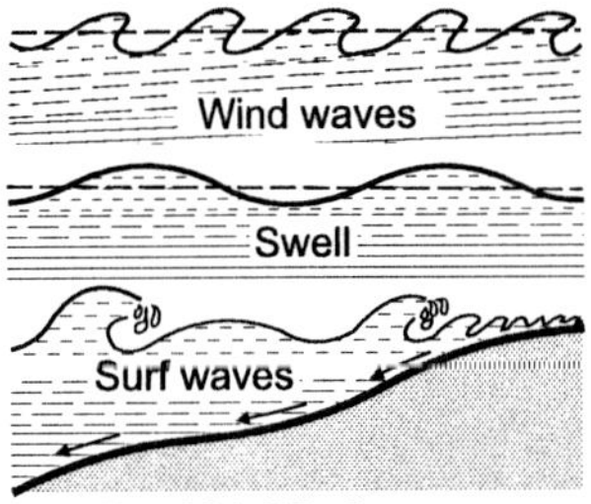

Fig. 364. **Wind waves**

r ветровая волна
Колебательные движения в слое воды глубиной до 50-60 м, вызываемые энергией ветра при его непосредственном воздействии на водную поверхность. → Рис. 364
f vague *f* forcée
s ola *f* forzada
g Windwelle *f*

6790 **wind-blown furrows**
Deep grooves between the sharp ridges of hard snow sastrugi formed by the erosive action of windborne ice particles.
r ветровые борозды
Глубокие рытвины между застругами, возникающие в результате эрозии плотного снега снежными частицами, переносимыми ветром.
f rigoles *fpl* de vent; cannelures *mpl* sculptées par le vent
s canalículos *mpl* tallados por el viento
g Windfurchen *fpl*; Schuppenschnee *m*

6791 **wind-blown snow stream**
An air stream with suspended snow particles transported by the wind.
r снеговетровой поток; метелевый поток
Воздушный поток со взвесью в виде снежинок, переносимых ветром.
f courant *m* du chasse-neige
s corriente *m* de ventisca
g Driftschneestrom *m*

* **windbreak → 5391**

6792 **wind-drift current; wind-driven current**
An ocean current produced by wind stress on the surface of the water.
r ветровое течение; волновое течение
Морское течение, вызванное давлением ветра на поверхностный слой воды.
f courant *m* de houle
s corriente *f* de oleaje; corriente *f* de viento
g Wellenströmung *f*; Windströmung *f*

* **wind-driven current → 1707**

6793 **wind-driven effects; setup phenomena**
Water level fluctuations in a water body produced by strong permanent wind blowing over the water surface, which often results in disastrous floods on flat coasts.
r сгонно-нагонные явления
Колебания уровня воды озер, морей и океанов, вызываемые изменениями циркуляции вод под действием ветра, нередко служащие причиной катастрофических наводнений на пологих берегах.
f afflux-reflux phénomenès *mpl*
s aflujo-reflujo fenómenos *mpl*
g Ablauf-Auflauferscheinungen *f*; Windstauerscheinungen *fpl*

6794 **wind-gap**
A valley or saddle in a ridge through which no stream passes, i.e. it contrasts with a *water-gap*.
r ветровой проход
Долина или седловина в горном хребте без текущей здесь реки, в отличие от *сквозной долины*.
f passage *m* de vent
s pasaje *m* de viento
g Windpassge *f*

* **wind-induced surge → 6784**

6795 **windmill; wind turbine**
Formerly a mill operated by rotating sails which are turned by the wind, at present time it is used to produce electricity.
r ветряк
В прошлом ветряная мельница с лопастями, которые вращались под действием ветра, ныне используется для получения электроэнергии.
f éolienne *f*; moteur *m* à vent; moulin *m* à vent
s motor *m* eólico; aeromotor *m*; molino *m* de viento
g Windmotor *m*; Windmühle *f*

* **window → 445**

* **window frost → 2330**

6796 **window ice**
A thin deposit of ice forming by the freezing of many tiny drops of water which have condensed on the indoor side of a cold window surface. → Fig. 365, p. 812
r оконный лед
Слои льда на внутренней стороне окон, формирующиеся в результате намерзания влаги, конденсирующейся на поверхности охлажденных стекол. → Рис. 365, с. 812
f glace *f* de fenêtre
s hielo *m* de ventana
g Fenstereis *n*

6797 **windthrow**
The uprooting of trees by strong winds.
r бурелом
Вырывание с корнем деревьев под действием сильных ветров.
f chablis *m*; abattis *m*
s árboles *mpl* derribados por el viento
g Windbruch *m*

Fig. 365. **Window ice**

6798 **windward** ***adj***
Turned to the side from which wind blows, facing into the wind, as opposed to *leeward*.
r наветренный
Обращенный в ту сторону, откуда дует ветер.
f au-vent *adj*
s de barlovento *adj*
g luvseitig *adj*; windseitig *adj*; Luf-

6799 **windward slope**
The slope exposed to the direction from which the wind is blowing, facing into the wind, as opposed to *leeward slope*.
r наветренный склон
Склон, обращенный в сторону, откуда дует ветер, в противоположность *подветренному склону*.
f versant *m* au-vent; côté *f* du vent
s vertiente *f* al viento; lado *m* del viento; barlovento *m*
g Luvhang *m*; windzugewandter Hang *m*

6800 **winged headland**
A coastal cliff with two bays or spits on both sides.
r крылатый мыс
Выступ берега с заливами или косами с обеих сторон.
f promontoire *m* de deux fléches
s cabo *m* de dos flechas
g Kap *n* mit zweiseitig Nehrungen

6801 **winter**
The season of the year with the lowest air temperatures for a certain territory, the period astronomically defined from the winter solstice to the vernal equinox (from December 21-22 to March 20-21 in the Northern Hemisphere, from June 21-22 to September 23 in the Southern Hemisphere).
r зима
Время года, климатический сезон с наиболее низкими для данной территории температурами воздуха, продолжается от момента зимнего солнцестояния до весеннего равноденствия (с 21-22 декабря до 20-21 марта в Северном полушарии, с 21-22 июня до 23 сентября в Южном).
f hiver *m*
s invierno *m*
g Winter *m*

6802 **winter crops**
Cultivated annual plants such as winter wheat, rye, barbey, which develop well when sowed in autumn to secure harvest next year.
r озимые культуры
Однолетние растения (озимые пшеница, рожь, ячмень и др.), нормально развивающиеся при осеннем посеве и дающие урожай на следующий год.
f cultures *fpl* d'hiver; blés *mpl* d'automne
s cultivos *mpl* de invierno; cereales *mpl* de inverno; sementera *f* de otoño
g Wintergetreibe *n*; Winterkulturen *fpl*

6803 **winter flood**
Rapid and short rise of the water level in a river as a result of snowmelt during winter thaws.
r зимний паводок
Быстрый и кратковременный подъем уровня воды в реке в результате снеготаяния во время зимних оттепелей.
f crue *f* d'hiver
s crecida *f* nival; crecida *f* de invierno
g Winterhochwasser *n*

6804 **winter hut; winter quarters**
A house or small village in a polar region where people live during a severe winter season.
r зимовье
Отдельная хижина или небольшой поселок в полярной области, где люди живут в суровое зимнее время.
f lieu *m* d'hivernage
s invernadero *m*
g Winterhütte *f*; Winterlager *n*

* **winter quarters → 6804**

6805 **winter resistance**
The ability of plants to withstand unfavourable conditions of winter and early spring (alternation of frosts and thaws, ice crusts, etc.) without much damage.
r зимостойкость
Способность растений переносить неблагоприятные условия зимы и ранней весны (чередование морозов и оттепелей, ледяные корки, выпревание и пр.) без существенных повреждений.
f résistance *f* au l'hiver
s resistencia *f* al invierno
g Winterfestigkeit *f*

6806 winter resort; winter sports resort
A picturesque place in the mountains on the slopes of which the runs for skiing have been built as well as hotels and other facilities for doing winter sports.
r зимний курорт; горнолыжный курорт
Живописная и удобная местность в горах, на склонах которых организованы горнолыжные трассы, а в долине построены пансионаты и гостиницы.
f station *f* d'hiver;
station *f* de sports d'hiver
s estación *f* de inverno;
estación *f* de deporte de inverno
g Winterkurort *m*; Wintersportkurort *m*

6807 winter solstice
21-22 December (the shortest day in the Northern Hemisphere) when the Sun passes through the most southern point of ecliptic shining vertically at noon over latitude 23°30′ S, the Tropic of Capricorn. → Fig. 25, p. 58
r зимнее солнцестояние
Момент времени, когда Солнце проходит через самую южную точку эклиптики (21 или 22 декабря) – это самый короткий день в Северном полушарии, а на тропике Козерога в полдень сонечные лучи падают вертикально. → Рис. 25, с. 58
f solstice *m* d'hiver
s solsticio *m* de invierno;
solsticio *m* de Capricornio
g Wintersonnenwende *f*; Wintersolstitium *n*

* **winter sports resort → 6806**

6808 wintergreen plant
Herbaceous plants remaining green in winter under snow cover and showing little growth of sprouts, they shed leaves in spring after the new ones appear.
r зимнезеленое растение
Травянистые растения, сохраняющие зимой под снегом зеленые листья и обнаруживающие слабый рост побегов, однако весной они теряют листья после появления новых.
f plante *f* hivernale; plante *f* d'hiver verte
s planta *f* invernal verde
g wintergrüne Pflanze *f*

6809 wisps of mist
Individual flocks of fog formed predominantly over a water surface.
r клочья тумана
Разорванный туман, отдельные пятна которого чаще приурочены к водной поверхности.
f lambeaux *mpl* de brouillard
s pedazos *mpl* de niebla
g Nebelfetzen *mpl*

* **wood → 2263**

* **wood debris → 1439**

6810 wood industry
Wood processing and paper making (paper industry), furniture production (furniture-making industry).
r деревообрабатывающая промышленность
Отрасли промышленности, занимающиеся обработкой древесины и изготовлением из нее бумаги (бумажная промышленность), мебели (мебельная промышленность) и других изделий.
f industrie *f* du bois
s industria *f* maderera
g holzbearbeitende Industrie *f*

6811 wooded *adj*
Covered with trees.
r лесистый
Покрытый лесом.
f boisé *adj*
s boscoso *adj*; selvoso *adj*
g waldig *adj*; waldreich *adj*

6812 work sheet of a map
Graphic design of a geographic map prepared usually disregarding accuracy of the compilation and demands of design.
r авторский макет карты
Графический проект географической карты, подготовленный обычно без соблюдения точности составления и требований оформления.
f modèle *m* cartographique d'auteur
s modelo *m* cartográfico de autor
g Autorenvorlage *f*

* **workforce → 3338**

6813 working age
The age of people when, according to the labour laws of a given country, they can work for hire.
r трудоспособный возраст
Возраст людей, когда, согласно законодательству данной страны, они могут трудиться в какой-либо производственной отрасли.
f âge *m* apte au travail
s edad *f* apto para el trabajo
g arbeitsfähigalter *n*

* **working population → 64**

6814 world atlas
Geographical atlas containing maps whichc cover all the areas of the globe.
r мировой атлас
Географический атлас, содержащий карты всего земного шара.
f atlas *m* mondial
s atlas *m* mundial
g Weltatlas *m*

6815 world city; global city
A city characterized by the range of its economic, financial, cultural and political power

and influence on a global scale, e.g. New-York, London, Moscow.

r мировой город
Город, обладающий экономическим, финансовым, культурным и политическим влиянием мирового значения, например, Нью-Йорк, Лондон, Москва.

f cité *f* mondial
s ciudad *f* mundial
g Weltstadt *f*

6816 world development models
The estimation of different parameters of the future development of civilization on the Earth on the basis of computer calculations.

r модели мирового развития
Оценка на основе компьютерных расчетов различных показателей будущего развития цивилизации на Земле.

f modèles *mpl* de développement mondial
s modelos *mpl* de desarrollo mundial
g Modelle *npl* der Weltentwicklung

6817 world economy
National economies of all countries, their production and economic interrelationships in the form of foreign trade, export of capital, migration of workforce, exchange of scientific and technical information, etc.

r мировое хозяйство; мировая экономика
Совокупность национальных хозяйств всех стран, их производственных и экономических взаимосвязей в форме внешней торговли, вывоза капитала, миграции рабочей силы, обмена научно-технической информацией и пр.

f économie *f* mondiale
s economía *f* mundial
g Weltwirtschaft *f*

6818 World Heritage Area;
World Heritage Site
Site that has natural or cultural properties of outstanding value and included by a special procedure into the UNESCO World Heritage List (symbol: WHA) to be protected against the threat of damage.

r объект всемирного наследия
Памятник культуры или природы, включенный на основе специальной процедуры в международный список всемирного наследия ЮНЕСКО, предполагающий сохранение и защиту этой территории.

f site *f* du patrimoine mondial;
site *f* du patrimoine de l'Humanité
s lugar *m* de patrimonio mundial;
lugar *m* de patrimonio de la humanidad
g Welterbegebiet *n*

* **World Heritage Site → 6818**

6819 world market
A wide and sustainable international exchange of commodities and services that includes national markets of individual countries interconnected by trade and economic agreements.

r мировой рынок
Широкий и устойчивый международный обмен товарами и услугами, включающий национальные рынки отдельных стран, связанные друг с другом торгово-экономическими соглашениями.

f marché *m* mondial
s mercado *m* mundial
g Weltmarkt *m*

6820 World Ocean
The body of salt water which covers 70.78 per cent of the Earth's surface.

r Мировой океан
Непрерывная водная оболочка Земли, окружающая материки и острова, занимающая около 70,78% поверхности земного шара.

f Océan *m* Mondial
s Océano *m* Mundial
g Weltmeer *n*

6821 World political map
A map showing all the countries of the world; in a figurative meaning a historically formed system of state borders and relations between the countries of the world;

r политическая карта мира
Карта, на которой показаны все страны, существующие на земном шаре; в переносном смысле исторически сложившаяся система государственных границ и взаимоотношений между странами.

f carte *f* politique du monde
s mapa *m* político del mundo
g politische Karte *f* der Welt

* **wreck wave → 6086**

X

6822 xenophobia
A strong fear and dislike of people from other countries and cultures.
r ксенофобия
Сильная боязнь и нелюбовь к людям из другой страны или культуры.
f xénophobie *f*
s xenofobia *f*
g Xenophobie *f*

6823 xeric *adj*; xerophitic *adj*; xerophilous *adj*
Having a low or inadequate supply of moisture to sustain plant life; adjusted to arid conditions, applies particularly to a plant or animal having such a quality.
r засушливый; ксерофитный; засухоустойчивый
Имеющий низкое или недостаточное количество влаги, необходимой для жизни растений; приспособившийся к аридным условиям, относящийся к растениям и животным, обитающим в подобных условиях.
f xérique *adj*
s xerofitic *adj*
g dürr *adj*; dürrefest *adj*

6824 xeromorphism
Morphological adaptation of plants, e.g. cacti and feather grasses, to arid conditions by means of preventing water losses by the leaf area reduction, waxy leaf coating, formation of thorns, etc., or by and storing great amount of water in fleshy leaves.
r ксероморфизм
Морфологическая адаптация растений, например, у ковылей и кактусов, к сухим условиям, выражающая в редукции листвы, накоплении больших запасов воды, образовании колючек, шипов и пр.
f xeromorphisme *m*
s xeromorfismo *m*
g Xeromorphismus *m*

6825 xerophile
Organisms of arid regions.
r ксерофил
Организмы, обитающие в аридных районах.
f xérophile *m*
s xerófilo *m*
g Xerophil *m*

*** xerophilous *adj* → 6823**

*** xerophitic *adj* → 6823**

6826 xerophyte
Plants of steppes and semi-deserts, that are adapted for insufficient moisture and able to withstand prolonged drought.
r ксерофит
Растения степей, полупустынь и пустынь, живущие в условиях недостаточного увлажнения, способные переносить длительную засуху.
f xérophyte *m*
s xerófita *m*
g Xerophyt *m*; Trockenpflanze *f*

6827 xerosere
The sequential development of a plant community in a very dry habitat.
r аридная сукцессия
Последовательное развитие растительных сообществ в очень сухих условиях местообитания.
f xerosere *f*
s xerosere *f*
g Xerosere *f*

6828 xerothermic *adj*
Related to a season or place which is dryer and hotter than the surrounding area or period of time, e.g. xerothermic period.
r ксеротермический
Относящийся к сезону или месту, более сухому и жаркому, чем окружающая местность или другое время, например, ксеротермический период.
f xérothermique *adj*
s xerotérmico *adj*
g xerothermisch *adj*

6829 xerothermic period
A part of the Holocene with dry and warm climate corresponding to Subboreal (about 4.5-2.5 thousand years ago).
r ксеротермический период
Часть голоцена с сухим и теплым климатом, соответствующая суббореальному времени (около 4,5-2,5 тыс. лет назад).
f période *f* xérothermique
s período *m* xerotérmico
g xerotherme Periode *f*

6830 xylophage
An organism feeding on wood.
r ксилофаг
Организм, питающийся древесиной.
f xylophage *m*
s xilófago *m*
g Xylophag *m*; Holzfresser *m*

Y

6831 **yachting**
Navegation for pleasure, the activity of racing or sailing a yacht.
r плавание на яхте
Водная прогулка для удовольствия с использованием парусного судна.
f navigation *f* de plaisance; yachting *m*
s navegación *f* de recreo; navegación *f* en yate; yachting *m*
g Segeln *n*

6832 **yardang**
A sharp-crested linear ridge of rock with steep asymmetrical slopes stretched in the direction of the prevailing winds, which has been isolated by the removal of rocks on either side, in a clay desert of Central Asia.
r ярданга
Удлиненная узкая гряда с острыми гребнями и крутыми асимметричными склонами, вытянутая в направлении господствующих ветров в глинистых пустынях Центральной Азии.
f yardang *m*
s yardang *m*
g Jardang *m*

6833 **yazoo**
A tributary stream which does not join the main stream directly but runs parallel to it for some distance, usually because it cannot breach the levees which flank the main stream.
r параллельный приток
Приток, который не сразу впадает в главную реку, а течет параллельно ей, как правило, из-за того, что не может преодолеть прирусловой вал, тянущийся вдоль главной реки.
f yazoo *m*
s yazoo *m*
g Yazoo *n*

6834 **year**
The average time taken by the Earth to complete one orbit round the Sun, calendar, sidereal, tropical years, etc. are distinguished.
r год
Промежуток времени, приблизительно равный периоду обращения Земли вокруг Солнца, различают календарный, сидерический, тропический годы и др.
f année *f*
s año *m*
g Jahr *n*

6835 **yellow brown forest soil**
A soil that is of transitional type from brown forest soil to yellow soil.
r желто-бурая лесная почва
Почва, переходная по свойствам от бурой лесной почвы (бурозема) к желтозему.
f sol *m* jaune-brun forestier
s suelo *m* amarillo-pardo de bosque
g gelbbrauner Waldboden *m*

6836 **yellow soil**
A sois coloured yellowish and forming on the products of weathering of sedimentary, acid effusive and metamorphic rocks under the forests and the vegetation of evergreen species growing in the areas of humid subtropical climate.
r желтозем
Почва желтоватой окраски, формирующаяся на продуктах выветривания осадочных, кислых изверженных и метаморфических пород под лесами со значительным участием вечнозеленых пород в условиях влажного субтропического климата.
f sol *m* jaune
s suelo *m* amarillo
g Gelbboden *m*; Gelberde *f*

6837 **yernik**
The community of low shrubs dominated by dwarf birch trees, occuring in polar tundra, high mountains, on bogs of Northern Eurasia and North America.
r ёрник
Сообщество низкорослых или стелющихся кустарников с господством карликовых березок, распространено в приполярных тундрах, высокогорьях, на болотах Северной Евразии и Северной Америки.
f yernik *m*
s ernik *m*
g Jernik *m*

6838 **yield**
The amount of output production expressed in relation to units of land or livestock or to units of capital or labour applied.
r выработка; продуктивность
Количество выпущенной готовой продукции, в сельском хозяйстве по отношению к единице земли или поголовью домашнего скота, в промышленности по отношению к затраченному труду или капиталу.
f rendement *m*
s rendimiento *m*
g Produktionsmenge *f*

6839 **Yoldia Sea**
A sea basin, which appeared in the place of the contemporary Baltic Sea about 10 thousand years ago after the emptying of Baltic Ice Lake.
r Иольдиевое море
Морской бассейн, возникший на месте современного Балтийского моря около 10 тыс. лет назад после спуска Балтийского ледникового озера.

Fig. 366. Aerial photoghaph of the **young ice** on the surface of a sea

f mer *f* à Yoldia
s mar *m* de Yoldia
g Yoldia Meer *n*; Yoldiameer *n*

* **young fold mountains → 6841**

* **young growth → 6450**

6840 young ice
A newly formed, flat, sea or lake ice usually between 8 and 20 cm thick. → Fig. 366
r молодой лед
Вновь образованный морской лед в следующей после ниласа возрастной стадии толщиной 8-20 см. → Fig. 366
f glace *f* jeune
s hielo *m* joven; young-ice *m*; hielo *m* nuevo
g Jungeis *n*; dünnes Eis *n*

6841 young mountains; young fold mountains
The mountains with jagged peaks which has not yet been smoothed by the agents of erosion, formed by the Alpine orogeny.
r молодые горы
Горы с резкими пиками, еще слабо затронутыми эрозией, складчатая структура которых сформировалась в последнюю (альпийскую) эпоху горообразования.
f montagnes *fpl* jeunes
s montañas *fpl* jovenes
g junges Gebirge *n*

6842 young shore ice
A strip of ice frozen into the shore of river, sea, lake, or reservoir while the better part of the water surface is unfrozen.
r заберег; ледяной заберег
Полоса льда, смерзшаяся с берегом реки, моря, озера и водохранилища при незамерзшей основной части водного пространства.
f glace *f* jeune de berge
s hielo *m* joven costero
g Randeis *n*; Küsteneis *n*

Z

* **zastrugi → 5215**

6843 **zenith**
The point on the celestial sphere, that is vertically above the observer, the opposite of nadir. → Fig. 51, p. 110
r зенит
Наивысшая воображаемая точка небесной сферы, находящаяся над головой наблюдателя, противоположна надиру. → Рис. 51, с. 110
f zénith *m*
s cenit *m*
g Scheitelpunkt *m*; Zenit *m*

6844 **zenith distance**
The angle distance of a celestial body or any object on the Earth from zenith read along heights circle from 0 up to 180°, one of the ccoordinates in the horizontal system of celestial coordinates.
r зенитное расстояние
Угловое расстояние небесного светила или земного предмета от зенита, отсчитываемое вдоль круга высот от 0 до 180°, одна из координат в горизонтальной системе небесных координат.
f distance *f* zénithale
s distancia *f* cenital
g Zenitdistanz *f*; Zenitabstand *m*

* **zenithal projection → 508**

6845 **zenithal rains**
In tropics and subtropics, rainy seasons which occur annually or semi-annually (equinoctial rains) when the Sun is at zenith.
r зенитальные дожди
Дожди, регулярно повторяющиеся в тропиках и субтропиках раз в год или дважды в год (равноденственные дожди) около времени, когда Солнце проходит через зенит.
f pluies *fpl* zénithales
s lluvias *fpl* cenitales
g Zenitalregen *m*

6846 **zephyr**
Any soft, gentle breeze.
r зефир
Слабый теплый и влажный бриз.
f zéphir *m*
s céfiro *m*
g Zephir *m*

6847 **zero curtain**
The zone immediately above the permafrost table, in which temperature 0°C lasts a considerable period of time (as long as 115 days a year) during freezing and thawing of overlyiong ground.
r нулевая завеса
Область непосредственно выше кровли вечной мерзлоты с температурой 0°C, сохраняющейся во время ежегодного промерзания (а иногда и во время таяния) активного слоя.
f zone *f* zéro
s zona *f* cero
g Nullschleier *m*

* **zero grazing → 5767**

* **zero population growth → 5789**

6848 **zero-balance line**
The line on a glacier surface, along which, at a particular moment, ablation is equal to accumulation.
r линия нулевого баланса
Линия на поверхности ледника, на которой в данный момент абляция равна аккумуляции.
f ligne *f* d'équilibre instantanée
s línea *f* de balance equilibre
g Gleichgewichtslinie *f*

* **zeugen → 2102**

6849 **zodiacal light**
A cone of light extending upward from the horizon in the direction of the ecliptic (or zodiac), it is seen in tropical latitudes for a few hours after sunset or before sunrise.
r зодиакальный свет
Световой конус, простирающийся вверх от горизонта в направлении эклиптики (или зеркального круга), наблюдается в тропиках в течение нескольких часов после заката или до восхода Солнца.
f lumière *f* zodiacale
s luz *f* zodiacal
g Zodiakallicht *n*; Tierkreislicht *n*

6850 **zonal** ***adj***
Related to the processes and phenomena whose characteristics are dominated by the influence of climate and differ according to the latitudes, hence occurring in bands parallel with lines of latitude.
r зональный
Относящийся к явлениям и процессам, которые определяются в основном климатом и имеют широтные различия.
f zonal *adj*
s zonal *adj*
g zonal *adj*; Zonen-

* **zonal circulation → 6852**

6851 **zonal current**
An ocean current moving mainly in latitudinal direction, e.g. the Antarctic Circumpolar Current (West Wind Drift).
r зональное течение

Движение воды преимущественно в широтном направлении, например, Антарктическое циркумполярное течение (Южное океаническое кольцо).
f courant *m* zonal
s corriente *f* zonal
g zonale Strömung *f*

6852 zonal flow; zonal winds; zonal circulation
Atmospheric circulation in which the dominant airflow follows the lines of latitude, e.g. trade winds, westerlies, in contrast to *meridional flow*.
r зональная циркуляция
Циркуляция атмосферы, при которой потоки воздуха преимущественно следуют вдоль широты, например, пассаты или западный перенос, в противоположность *меридиональной циркуляции*.
f circulation *f* zonale
s circulación *f* zonal; vientos *mpl* zonales
g Zonalzirkulation *f*

6853 zonal index
A measure of the strength of the zonal circulation over a given length of time in a particular region.
r зональный индекс
Показатель интенсивности зональной (преимущественно западной) циркуляции атмосферы за данный период времени в конкретном регионе.
f index *m* zonal
s índice *m* zonal
g Zonalindex *m*

6854 zonal soils
Soils whose characteristics are dominated by the influence of climate of a certain latitudinal zone and which develop on well-drained plains under typical zonal vegetation.
r зональные почвы
Почвы, в которых генетические свойства и почвообразовательный процесс более всего соответствуют физико-географическим условиям данной зоны, развиваются на хорошо дренированных равнинах под типичной зональной растительностью.
f sols *mpl* zonaux
s suelos *mpl* zonales
g zonale Böden *mpl*

6855 zonal vegetation
Plant communities which have the typical features of adaptation to the environment of a given geographic zone, and occupy most typical habitats such as plain elevated well drained interfluves.
r зональная растительность
Растительные сообщества, лучше всего адаптированные к природным условиям данной географической зоны и не встречающиеся в других зонах, произрастают чаще всего на выровненных, приподнятых и хорошо дренированных междуречьях.
f végétation *f* zonale
s vegetación *f* zonal
g zonale Vegetation *f*

* **zonal winds → 6852**

6856 zonality; latitudinal zonation
Latitudinal differentiation of the Earth's surface in the form of the succesive change of geographical belts, zones, and subzones, caused by the latitudinal differentiation of the radiant energy and uneven moisture. → Fig. 367, p. 820
r географическая зональность
Широтная дифференциация географической оболочки Земли, проявляющаяся в последовательной смене географических поясов, зон и подзон, обусловленной изменением прихода лучистой энергии Солнца по широтам и неравномерностью увлажнения. → Рис. 367, с. 820
f zonation *f*; zonalité *f* latitudinale
s zonación$_1$ *f*; zonalidad *f*
g geographische Zonierung *f*

6857 zonda
A hot dry and dusty descending western wind in the west of Argentina, most usual in spring.
r зонда
Жаркий сухой и пыльный нисходящий западный ветер на западе Аргентины, чаще всего дующий весной.
f zonda *m*
s zonda *m*
g Zonda *n*

6858 zone of aeration; aerated zone
The upper layers between the Earth's surface and the surface of ground waters where there are temporary aquifers and perched groundwaters.
r зона аэрации
Верхние слои между дневной поверхностью и поверхностью грунтовых вод, где находятся временные водоносные пласты и проходит верховодка.
f zone *f* d'aération
s zona *f* de aeración
g Aerationszone *f*

6859 zone of ecological disaster
An officially declared territory where deep irreversible changes of the environment took place, which caused a considerable deterioration of human life.
r зона экологического бедствия
Официально объявляемая территория, на которой произошли глубокие необратимые изменения окружающей среды, повлекшие за собой существенное ухудшение жизни населения.
f zone *f* de catastrophe environnementale

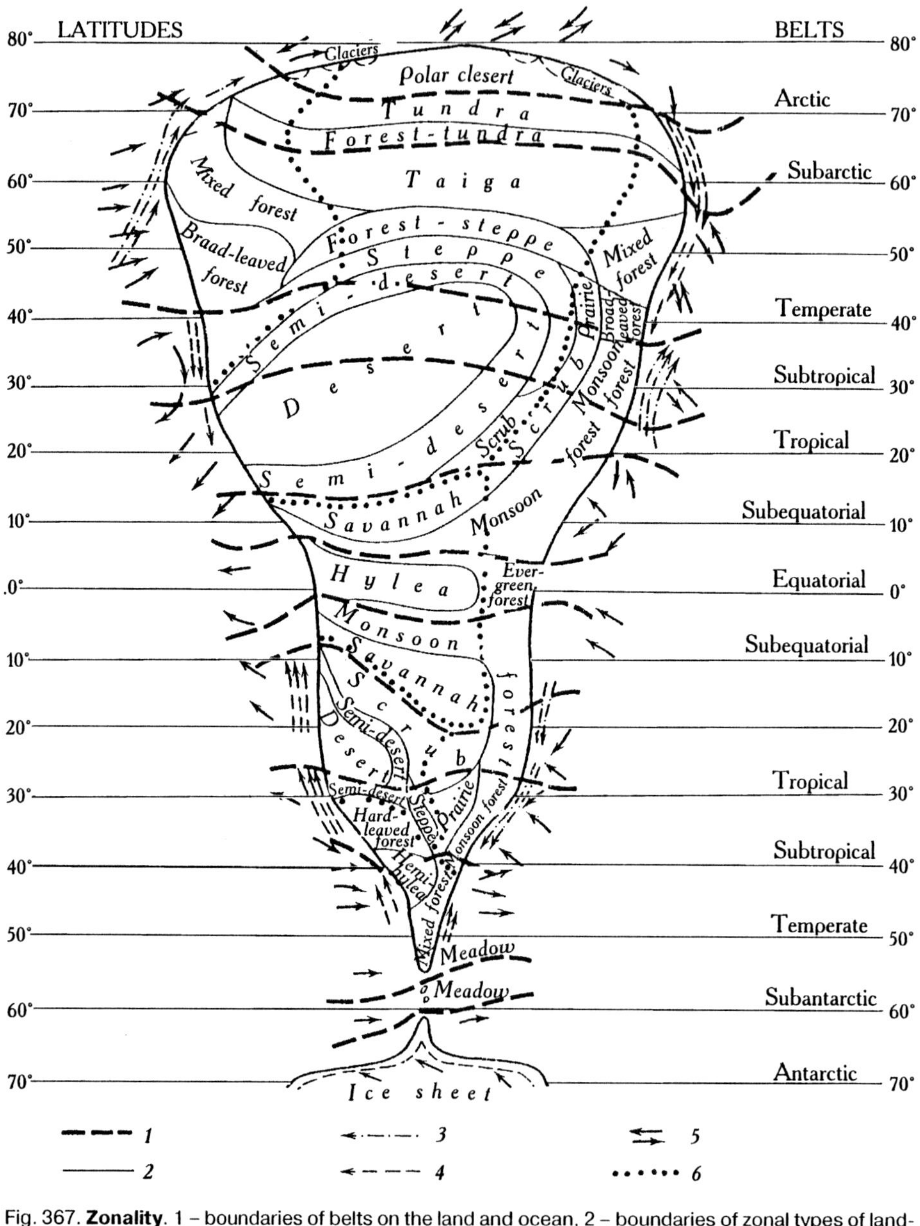

Fig. 367. **Zonality**. 1 – boundaries of belts on the land and ocean, 2 – boundaries of zonal types of landscapes, 3 – warm oceanic currents, 4 – cold oceanic currents, 5 – prevailing winds, 6 – boundaries of provinciality

s zona *f* de catástrofe ecológica;
zona *f* de catástrofe medioambiental
g Gebiet *n* der Umweltkatastrophe

6860 zone of environmental risk
Places on land and ocean where human activity can create dangerous ecological situation.
r зона экологического риска
Места на поверхности суши и океана, где велика вероятность того, что человеческая деятельность нарушит устойчивость окружающей среды.

f zone *f* de risque environnemental
s zona *f* de riesgo medioambiental; zona *f* de riesgo ecológico
g Gebiet *n* mit hohen Umweltrisiko

* **zone of saturation → 5226**

6861 zone of superimposed ice
The zone of a glacier between the firn line and the equilibrium line where the ice formation is completed over just one year.
r зона ледяного питания
Зона на леднике, располагающаяся между фирновой линией и границей питания, где льдообразование завершается в течение одного года.
f zone *f* de glace de regel
s zona *f* de infiltración y congelación
g Superimposed Ice Zone *f*

* **zone of vegetation → 6539**

6862 zoning
The singling out of areas with different intensity of the same phenomenon.
r зонирование
Выделение участков территории с различной интенсивностью какого-либо явления.
f zonage *m*
s delimitación *f* zonal; zonación$_2$ *f*
g Zonierung *f*

6863 zoo; zoological garten
A collection of living animals displayed to the public where animals are kept for the study and restoration of the gene pool of endangered species.
r зоологический сад; зоопарк
Коллекция диких животных, размещенных с целью их демонстрации, изучения, сохранения генофонда и воспроизводства редких и исчезающих видов животных.
f zoo *m*
s parque *m* zoológico
g Tiergarten *m*; Zoo *m*

6864 zoochore
Plants, the seeds and spores of which are dispersed by animals.
r зоохор
Растения, семена и споры которых распространяются животными.
f zoochore *m*
s zoócoro *m*
g Zoochor *m*; Tierwanderer *m*

6865 zoocoenosis
The community of animals that inhabit a certain area (biotope), e.g. zoocenosis of sand desert.
r зооценоз
Совокупность взаимосвязанных популяций животных, населяющих определенную территорию – биотоп, например, зооценоз песчаной пустыни.
f zoocénose *f*
s zoocenosis *f*
g Zoozönose *f*

6866 zoogenic microrelief
Landforms created as a result of the activity of animals such as rodents (molehills), insects (anthills), mammals (boars′ hollows).
r зоогенный микрорельеф
Рельеф, образованный в результате деятельности животных, например, грызунов (кротовины), насекомых (муравьиные колонии), млекопитающих (копани кабанов).
f microrelief *m* zoogène
s microrelieve *m* zoógeno
g zoogenes Kleinrelief *n*

6867 zoogenic succession
The sequential change of vegetation caused by the unusually strong impact of animals, e.g. their mass reproduction, overgrazing, and income of great amount of excretion products.
r зоогенная сукцессия
Смена растительности, вызванная необычно сильным воздействием животных, например, их массовым размножением, перевыпасом, привносом большого объема продуктов жизнедеятельности.
f succession *f* zoogène
s sucesión *f* zoogénica
g zoogene Sukzession *f*

6868 zoogeography
A branch of *biogeography* dealing with the distribution of animals over the Earth′s surface and their interaction with the environment.
r зоогеография; география животных
Раздел *биогеографии*, изучающий распространение животных и их взаимоотношения с окружающей средой.
f zoogéographie *f*; geographie *f* des animaux
s zoogeografía *f*
g Zoogeographie *f*; Tiergeographie *f*

* **zoological garten → 6863**

6869 zoophyte
An animal resembling a plant in appearance or growth, e.g. coral.
r зоофит
Животное, имеющее внешнее сходство с растением, например, коралл.
f zoophyte *m*
s zoofita *m*
g Zoophyt *m*

6870 zooplankton
The animals which spend their life freely floating or drifting in water.
r зоопланктон
Совокупность животных, обитающих и дрейфующих в воде.

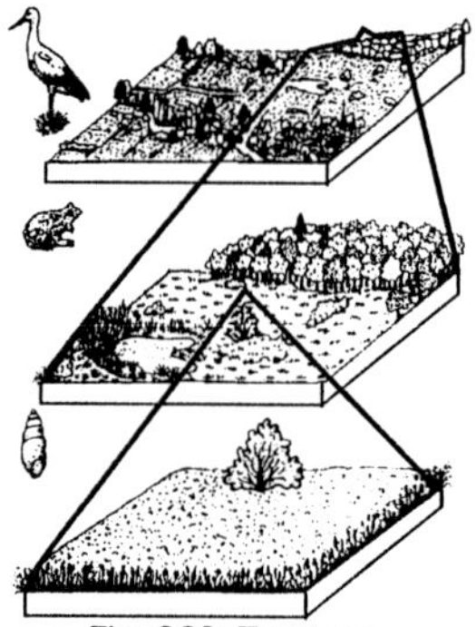
Fig. 368. **Zootope**

f zooplancton *m*
s zooplancton *m*
g Zooplankton *n*

6871 zootope
A piece of land or water with characteristic uniformity of environmental conditions that is occupied by one zoocenosis. → Fig. 368
r зоотоп
Однородный в экологическом отношении участок земной поверхности (территории или акватории), занятый одним зооценозом. → Рис. 368
f zootope *f*
s zootopo *m*
g Zootop *n*

Русский

Français

Español

Deutsch

Edwards Brothers Malloy
Ann Arbor MI. USA
May 21, 2012